PRECALCULUS

Second Edition

Teacher Edition

Part 1

bju press®

Greenville, South Carolina

Note:

The fact that materials produced by other publishers may be referred to in this volume does not constitute an endorsement of the content or theological position of materials produced by such publishers. Any references and ancillary materials are listed as an aid to the student or the teacher and in an attempt to maintain the accepted academic standards of the publishing industry.

PRECALCULUS Teacher Edition

Second Edition

Coordinating Writer
Mark Wetzel, MEd

Writers
Ben Adams, MEd
Gene Bucholtz, MEd
Jeffrey A. Jaeger, MEd
Timothy King
Tamera Knisely, MEd
Steve McKisic, MEd
Kathy Pilger, EdD
Sarah Ream, MPA

Academic Oversight
Jeff Heath, EdD

Editors
Heather Lonaberger, MA
Abigail Sivyer

Biblical Worldview
Vern Poythress, PhD
Bryan Smith, PhD
Tyler Trometer, MDiv

Previous Edition Writers
Larry Hall, MS
Kathy Pilger, EdD
Ron Tagliapietra, EdD

Consultants
Steve McKisic, MEd
Kathy Pilger, EdD

Permissions
Sylvia Gass
Carrie Hanna
Ashleigh Schieber
Carrie Walker

Project Coordinator
Kyla J. Smith

Page Layout
Dzign Associates,
 Patricia Tirado

Designers and Illustrators
Chris Barnhart
Dzign Associates,
 Patricia Tirado
Drew Fields
Josh Frederick
Emily Heinz
Pixel Mouse House,
 David Casas

Cover Designers
Chris Barnhart
Drew Fields
Josh Frederick

p. 471 "Triangulo Pascal" by Blaise Pascal/Wikimedia Commons/Public Domain
Photograph credits appear on pages 696–97.
Text acknowledgements appear on pages 698–99.

© 2020 BJU Press
Greenville, South Carolina 29609
First Edition © 2002, 2009 BJU Press

Printed in the United States of America
All rights reserved
ISBN 978-1-62856-332-0
15   14   13   12   11   10   9   8   7   6   5   4   3   2   1

Contents

To the Teacher

PRECALCULUS serves as a capstone course that provides a solid foundation for students preparing for higher education in science or math. It also prepares students pursuing other majors, such as business or education, in which they will encounter mathematics of finance, probability, and statistics. Throughout the text, a biblical worldview is emphasized when interpreting data and making related decisions. Christian service is integrated within real-world applications to prepare students for ministry in their future vocations.

Careful research and evaluation of state and national standards and other texts have ensured this text aligns with the highest standards. Each lesson is crafted to be accessible to students through carefully sequenced instruction, examples, and exercises.

We desire that students will grow in their knowledge of Christ, understand the applications of mathematics from a biblical worldview, and develop a greater desire to serve God and others. We also desire that God will work in the hearts of unsaved students, bringing them to recognize their need of Jesus Christ as Savior.

What's New in Precalculus?

- Significant restructuring provides improved flow, scope, and sequencing.
  1. New Chapters 1–6 introduce foundational precalculus concepts and provide essential review of polynomial, rational, exponential, logarithmic, and trigonometric functions and equations.
  2. Sections on linear programming and partial fraction decomposition have been added to Chapter 7, Systems and Matrices.
  3. Sections on rotated conics, parametric representations, and polar equations of conics have been added to Chapter 8, Analytic Geometry.
  4. The new Chapter 11 introduces inferential statistics.
  5. Sections introducing indefinite and definite integrals are now included in Chapter 12, Limits, Derivatives, and Integrals.
- Enriched explanations and additional examples have been incorporated into each lesson.
- Expanded exercise sets provide greater flexibility for customizing student assignments while emphasizing real-life applications and critical thinking.
- Expanded cumulative reviews have been strategically designed to review foundational concepts for upcoming lessons, encourage retention of key concepts, and prepare students for standardized testing.
- New interactive Technology Corner features provide instruction for hands-on exploration of concepts using the TI-84 Plus family of graphing calculators, the powerful Desmos® Internet graphing calculator, and Excel® spreadsheets.
- Three new features in each chapter emphasize a biblical view of mathematics: Historical Connection, Biblical Perspective of Mathematics, and Data Analysis.

Historical Connection features consider why mathematical concepts were developed at a particular time and place.

Biblical Perspective of Mathematics features explore the philosophical basis of mathematics from a biblical worldview.

Data Analysis features use engaging hands-on exercises to illustrate how concepts from the chapter are used in modern applications in keeping with a biblical worldview.

- Internet keyword searches in the *Student Edition* provide quick access to tools, remedial help, and enrichment, while those in the *Teacher Edition* provide tools for the teacher and interactive activities.
- Complete solutions are included with answers in the *Assessments Answer Key*.

Supplements

- The *Teacher Edition* provides guidance for each lesson, including presentation suggestions, common student errors, one-on-one strategies, motivational ideas, Internet keyword searches, interactive activities, additional exercises, solutions to exercises, and tips for selected exercises.
- *Assessments* contains quizzes, chapter tests, and quarterly exams. Tests are designed for the standard track but can be modified for the minimum or extended tracks. The *Assessments Answer Key* provides complete solutions.
- TeacherToolsOnline.com provides a wealth of valuable extras, including PowerPoint presentations, lesson visuals, projection-ready answers (PDF format), a curriculum map, and an eTextbook. An *ExamView* database of approximately 1200 test bank items, referenced by section and classified by difficulty level, is also included. Customized tests and quizzes can be quickly created. Add and edit items to personalize your test bank.
- AfterSchoolHelp.com provides video lessons and practice for major concepts, complementing the BJU Press textbooks.

Goals of BJU Press Secondary Math

1. To glorify God by recognizing value and purpose for His created order in the study of mathematics
2. To develop a biblical worldview of mathematics and to identify elements of nonbiblical worldviews in mathematical applications
3. To prepare students for Christian service in a wide variety of vocations
4. To develop analytical thinking and reasoning skills
5. To develop knowledge of mathematical concepts including number systems and their operations, functions, algebra, geometry, probability, and statistics
6. To develop problem-solving abilities
7. To develop speed and accuracy in computation
8. To promote an interest in and an appreciation of mathematics
9. To encourage the use of technology to enhance learning, incorporate multiple representations, and remove computational constraints

The most important element of these goals is the development of a biblical worldview, which provides the purpose and guiding principles for the attainment and application of mathematical knowledge. The development of reasoning and problem-solving skills in *Precalculus* is crucial for a Christian's vocation and service. The attainment of these goals enables success in the standardized tests used for acceptance and placement in higher education.

These goals involve both content and process. Certain topics must be covered to prepare students for standardized tests, but this alone is not sufficient. The teacher should exemplify and promote mathematical processes. Students must understand the reasoning behind the procedures in order to adapt acquired mathematical knowledge to new situations.

Course Objectives

1. To determine key characteristics of algebraic and transcendental functions and to draw graphs illustrating those characteristics
2. To solve algebraic, exponential, logarithmic, and trigonometric equations and inequalities
3. To apply trigonometric functions to solve problems involving measures in triangles and vectors, graph polar equations, and complete operations on complex numbers
4. To use matrices to solve systems of equations
5. To determine key characteristics of conic sections and to draw graphs illustrating those characteristics

6. To derive and apply functions representing sequences and series
7. To use statistics to represent data and make inferences from that data
8. To calculate limits and simple derivatives and integrals
9. To model real-world data and situations to solve problems within the context of a biblical worldview
10. To evaluate statements regarding the development and nature of mathematics
11. To articulate a biblical view of mathematics

Teaching the Course

Precalculus and its supplements are designed to provide a wealth of materials accommodating a variety of academic needs. Suggested assignments for minimum, standard, and extended tracks are provided along with many activities, resources, and special features. Exercise sets are presented at A, B, and C levels. The C-level exercises provide critical thinking challenges. Tests and quizzes focus on A- and B-level exercises but usually include one or two C-level items. Customized assessments can be generated from the *ExamView* test bank.

Precalculus begins by introducing foundational concepts of calculus with a review of algebraic, exponential, logarithmic, and trigonometric functions and equations. Restructured Chapters 1–6 reflect the standard scope and sequences of most precalculus courses. Chapter 1, Analyzing Functions, reviews essential algebraic concepts while informally introducing limits, continuity, and parametric equations. Chapter 2, Radical, Polynomial, and Rational Functions, reviews descriptions of these families of functions and methods of solving related equations and inequalities. Chapter 3, Exponential and Logarithmic Functions, describes these families of functions, their inverse relationship, and methods of solving related equations.

Chapters 4–6 focus on trigonometry. Chapter 4, Trigonometric Functions, introduces degree and radian measures of angles and defines the six trigonometric functions in a right triangle before expanding their definitions to other angles and real numbers. The graphs of sinusoids and other trigonometric functions are examined and the inverse trigonometric functions are defined. In Chapter 5, Trigonometric Identities and Equations, identities are derived, verified, and applied to solve equations. The Law of Sines and Law of Cosines are derived and used to solve real-life applications. Chapter 6, Vectors, Polar Graphs, and Complex Numbers, combines various applications of trigonometry.

Chapters 7–9 provide a firm foundation in several essential areas for higher-level mathematics and sciences. Chapter 7, Systems and Matrices, includes new sections on linear programming and finding partial fraction decompositions. Chapter 8, Analytic Geometry, includes new sections on rotated conics, parametric representations, and polar equations of conics. Chapter 9, Sequences and Series, provides several financial applications typically used in businesses.

Chapters 10–12 equip the students with a solid framework in statistics and calculus. Chapter 10, Descriptive Statistics, reviews basic counting principles and probability concepts before progressing into data distributions. Chapter 11, Inferential Statistics, describes how statistics from a sample are used to draw conclusions about similar values for the entire population. Chapter 12, Limits, Derivatives, and Integrals, adds the calculation of indefinite and definite integrals to introductory calculus concepts.

Implementing supplemental resources and ideas from the *Student Edition* and the *Teacher Edition* can help you customize the content for your students. Utilizing the Technology Corner features can be invaluable in solidifying student understanding.

While the teacher is the key to the students' learning, a textbook also represents significant authority to the students. Therefore, it is vitally important that the textbook is in harmony with the teacher's worldview. PRECALCULUS provides many biblical worldview-shaping opportunities that supplement the content being taught.

The Historical Connection and Data Analysis series allow students to connect the mathematical concepts to their historical development and modern real-world applications. The Biblical Perspective of Mathematics series culminates the product line's objective of developing a biblical philosophy of mathematics.

Suggested Presentation Schedules

The standard track is recommended, but the teacher is encouraged to adjust the course to meet individual class needs. Selected review lessons on fundamental concepts necessary for success in the course can be found in Appendix Lessons A.1–A.5. AfterSchoolHelp.com provides video lessons and practice for a wider range of topics.

❯ Standard

The following table summarizes the Standard Lesson Plan Overview, which allocates 180 class periods (including quarterly reviews and tests) over two semesters. Consider omitting lessons labeled as optional in the minimum track table.

Chapter	Class Periods	Chapter	Class Periods
Intro	1	7	16
1	18	8	14
2	13	9	12
3	11	10	14
4	13	11	13
5	14	12	15
6	14		
Quarterly/Semester Reviews and Exams			12
Total			180

The text contains 77 sections, and two days are allocated to cover each of 36 of these sections. This schedule allows time for students to master concepts and for the inclusion of the special features. One or more days of review are allocated before each quarterly or semester exam.

Suggested modifications for the extended and minimum tracks follow.

❯ Extended

The extended track is intended for honors or advanced classes and should cover the entire *Student Edition*. The suggested assignments include more C-level exercises, and the teacher should include all additional textbook features.

❯ Minimum

The minimum track omits several optional sections, Chapter 11, and some Data Analysis features while retaining minimum requirements for precalculus credit. Minimum track assignments place greater emphasis on A- and B-level exercises. If all optional sections in the table are omitted, 67 sections will be covered. The spare 25 days can be used to focus on foundational concepts and review.

Chapter	Optional Sections	Class Periods
Intro–1	none	19
2	2.7	11
3–5	none	38
6	6.3	13
7	7.7	14
8	8.4; DA	12
9–10	DA	24
11	all	0
12	12.5; DA	12
Quarterly/Semester Reviews and Exams		12
Total		155

Features of the *Student Edition*

The sections of each chapter are organized topically and build sequentially, with clear development of mathematical terminology and theory. Each section includes the following.

- **Thorough explanations** encourage understanding of each concept.
- **Examples** in a two-column format show each step and the supporting reasoning.
- **Exercise sets** are arranged according to level of difficulty.
  - A. **Exercises** assess basic concepts from the lesson.
  - B. **Exercises** are more challenging and may require steps or concepts from previous sections.
  - C. **Exercises** are designed to challenge higher-level thinking skills.
- **Cumulative Review** exercises at the end of each exercise set are strategically designed to encourage retention of key concepts and aid in preparation for the next lesson. Multiple-choice items help prepare students for standardized tests.
- **Chapter Review** exercises are designed to review the concepts covered in the chapter and can be used to assess student understanding, encourage retention, and prepare for the chapter test.
- **Selected Odd Answers** are on reduced pages in the back of the text. A PDF of these selected odd answers can be found on AfterSchoolHelp.com. Click on "Math" from the "Subjects" menu and then select "View Overview" for *Precalculus*.

- **Historical Connection** features provide fascinating insights on the important role a biblical worldview has played throughout the history of mathematics.
- **Biblical Perspective of Mathematics** features proclaim a biblical perspective of mathematics and its origin.
- **Data Analysis** features use real mathematical data to evaluate real-world issues from a biblical perspective.
- **Technology Corner** features provide detailed instruction for the use of technology. (Instruction is integrated into the lessons in Chapters 10–12.)
- **Keyword Searches** guide students to related reinforcement or enrichment materials.

Additional Resources

A wealth of precalculus activities can be found using the Internet keyword searches given throughout the Presentation sections of the Teacher Edition. The Internet keyword searches *Mathematics Teacher NCTM*, *TI-84 precalculus activities*, or similar searches will help locate many other precalculus activities. Use the Internet keyword search *free precalculus worksheets* to find multiple sites that facilitate the creation of customized practice worksheets.

Using the Teacher Edition

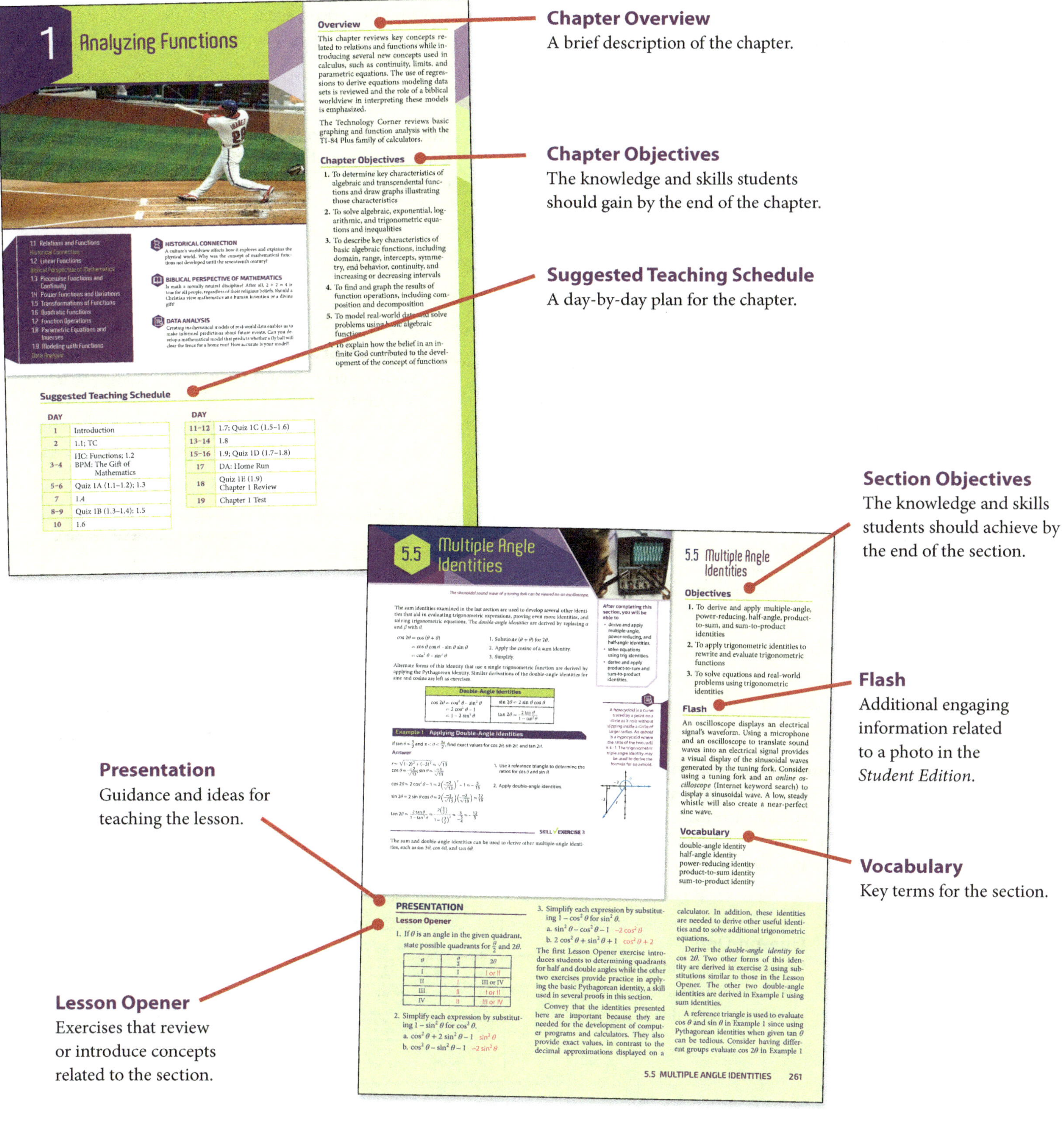

Chapter Overview
A brief description of the chapter.

Chapter Objectives
The knowledge and skills students should gain by the end of the chapter.

Suggested Teaching Schedule
A day-by-day plan for the chapter.

Section Objectives
The knowledge and skills students should achieve by the end of the section.

Flash
Additional engaging information related to a photo in the *Student Edition*.

Vocabulary
Key terms for the section.

Presentation
Guidance and ideas for teaching the lesson.

Lesson Opener
Exercises that review or introduce concepts related to the section.

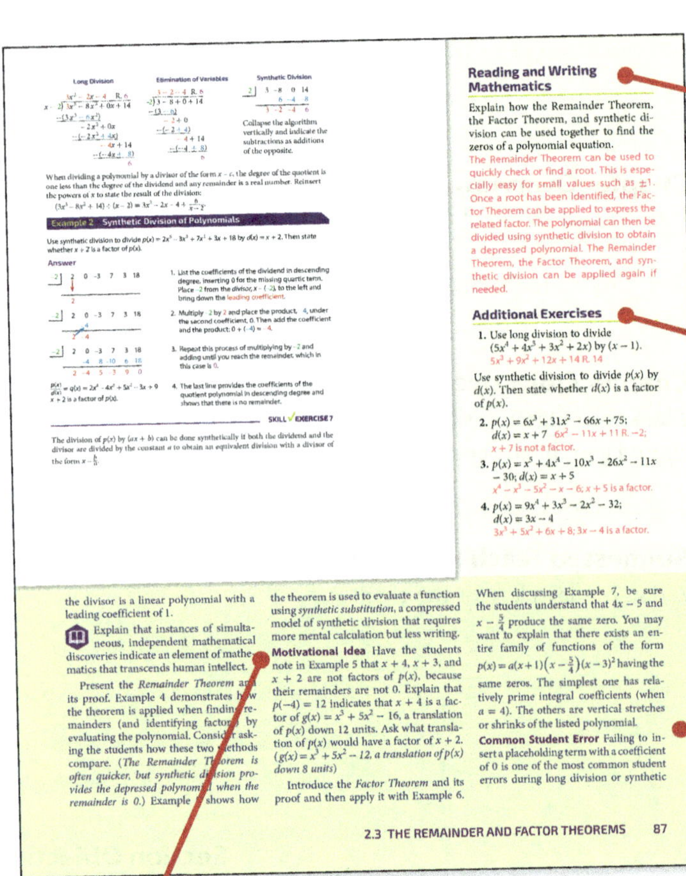

Reading and Writing Mathematics

Exercises that encourage students to convey mathematical principles in their own words.

Additional Exercises

Exercises that can be used for reinforcement, review, quizzes, or discussion.

Common Student Error

Insight on errors that students are likely to make.

Motivational Idea

Engaging activities, demonstrations, and teaching techniques that build student interest.

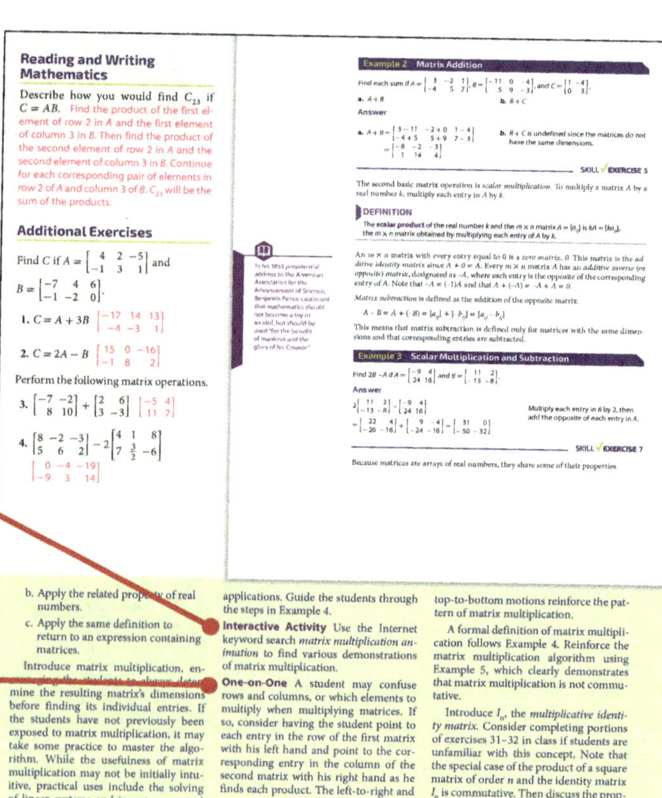

Interactive Activity

Teaching ideas that involve Internet applets that can be used with a data projector or an interactive board.

One-on-One

Suggestions for remediation, enrichment, and differentiating instruction for individual students.

Answers

Short answers printed in magenta on the reduced student pages. Longer answers are given with the solutions in the margins. Projection-ready answers are provided on TeacherToolsOnline.com.

Assignments

Suggested assignments for each track.

Solutions

Step-by-step solutions to exercises. Final answers are often expressed as rounded values, but exact values are used throughout multi-step solutions.

Assessment

Related printed assessments.

Internet Keyword Search

Suggested keywords that provide quick access to teacher tools, interactive activities, and more.

Tips

Insights for selected exercises that can be shared with students or used when grading or explaining solutions.

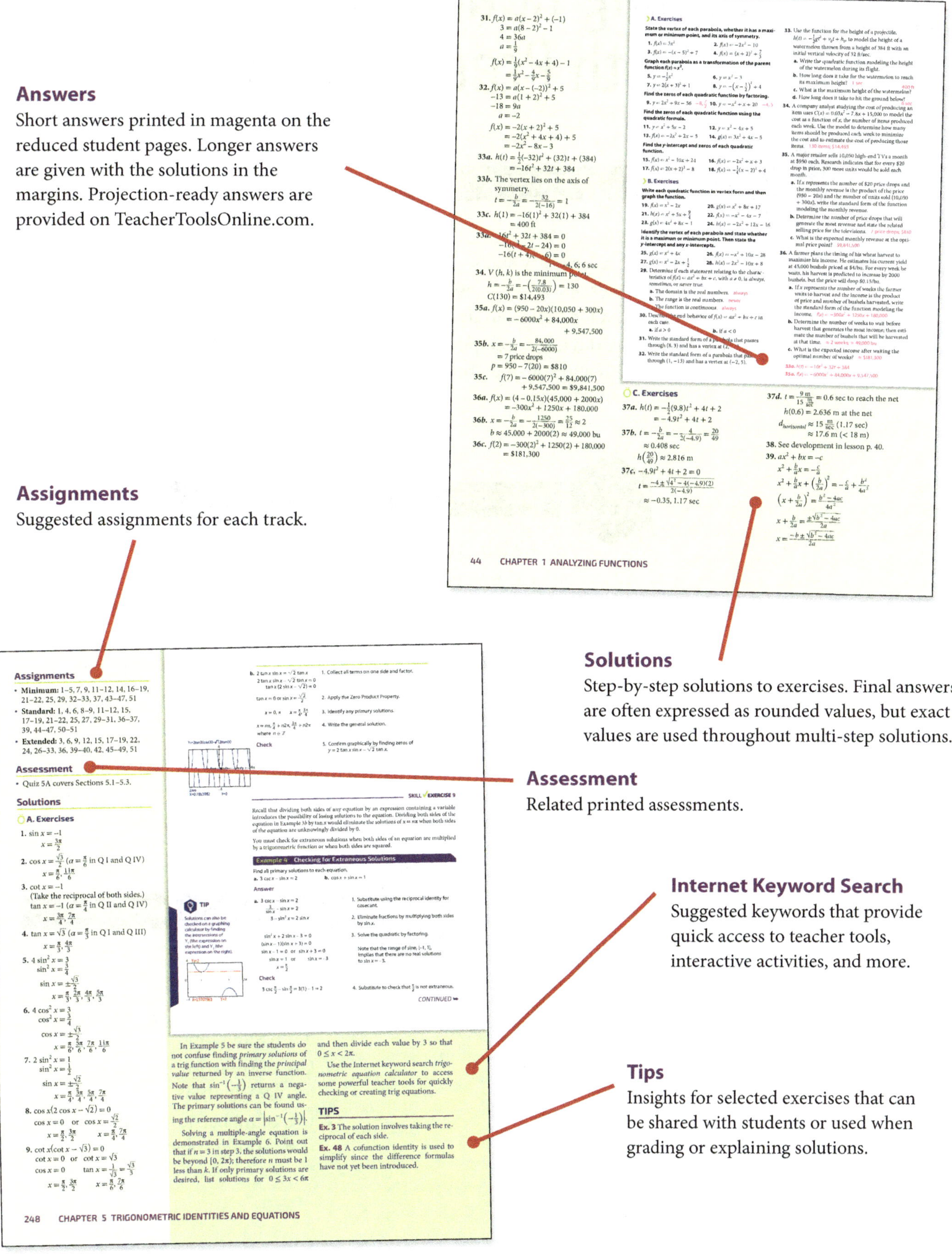

Building Academic Rigor with Precalculus

1 DESIRED LEARNING OUTCOMES

- Mathematical literacy
- Twenty-first-century skills—collaboration, creativity, critical thinking, problem solving, technological literacy
- Graph creation and analysis
- Biblical worldview development regarding the history, philosophy, and use of mathematics

2 TEACHING AND LEARNING SUPPORTS

- Models of mastery
- Corrections for common student errors
- Strategic Internet keyword searches to further learning
- Cross-disciplinary studies
- Mathematical literacy practice through Reading and Writing Mathematics features
- Case studies that emphasize data analysis and creative problem solving
- Guidance for course tracking

3 EVIDENCE OF UNDERSTANDING

- Scaffolded exercise sets for formative assessment
- High-level thinking exercises, case studies, discussions, and reviews
- Use of spreadsheets and graphing calculators for data analysis
- Use of higher-level mathematics to solve real-world problems

A Panorama of Academic Rigor

THE LEARNING ENVIRONMENT

Learning happens in a context. Physical and social environments create this context. For students to be stimulated and engaged and to perform at high levels, they need an environment that connects with and shapes their interests and values. Students need a flexible environment that adapts to remedy their deficits and capitalize on their proficiencies. They need an interactive environment that intrinsically motivates them by giving them structure, freedom, and choices in learning. They need a safe environment that invites questioning, risk-taking, and honest discussion. Most importantly, students need positive relationships with mentors and peers and one-on-one time with caring, responsible instructors and parents. This kind of environment does not happen by accident; an educational climate like this must be crafted.

THE TEACHER'S ROLE

The teacher is the person who crafts this learning environment. The teacher creates a learning community with high educational expectations, ignites interest and passion to help students persevere in challenging educational tasks, and inspires them by providing models of learning. The teacher instructs, supports, critiques, and praises the efforts of the growing learners. Then he helps his students transfer their knowledge, understanding, and skills to life with the purpose of shaping his students' worldview.

THE LEARNING EXPERIENCE

The textbook opens doors to the world of the learning experience. It captures student interest and significantly contributes to the educational environment. Educational and technological resources enable the teacher to develop knowledge, understanding, and skills in a scaffolded sequence to build a foundation that students can transfer to life. Teachers use educational tools to craft relevant, authentic learning experiences that develop creativity and problem-solving skills in a variety of contexts. These kinds of learning experiences allow students to take responsibility for their learning and discover the joy of serving God with their vocational knowledge and skills.

Math from a Biblical Worldview

Basics of a Biblical Worldview

A biblical worldview can be summarized by the story of Creation, Fall, and Redemption. Every smaller story fits into this big story presented by the Bible and everything that plays a part in this story, math included, must be understood through this lens.

CREATION

God made everything fundamentally good. Math is part of that good creation because the Creator built it into the way our world works.

FALL

Adam's fall into sin twisted the goodness of God's creation. As a result of the Fall, math can be confusing to us and people may use and interpret mathematics in ways that are wrong and dangerous.

REDEMPTION

God is at work to restore His fallen creation to its original purpose. All of creation longs to be restored to the way that God made it (Rom. 8:22–23). Math allows people to serve others and take care of this world in unprecedented ways. A biblical worldview allows us to recognize wrong views of mathematics and interpret and use math for God's glory instead.

Applying a Biblical Worldview to Math

This book employs three series of features to apply a biblical worldview to precalculus: Historical Connection, Biblical Perspective of Mathematics, and Data Analysis. These features, found in each of the 12 chapters, present foundational truths about the nature of mathematics and enable students to examine and apply math from a biblical worldview.

HISTORICAL CONNECTION

The Historical Connection features survey the history of key mathematical concepts. These features provide rich opportunities for seeing how math has developed. Historical Connection features often point to the creational nature of math by showing how math concepts like calculus and geometry were discovered independently in separate parts of the world. These features give examples of how math is used to create models and how different models have worked effectively throughout history. Students will also learn how math has been misused or misunderstood throughout history, such as when René Descartes declared that math was the only source of truth. The Historical Connection features also show the redemptive effect of a biblical worldview on key mathematical concepts like infinity and artificial numbers.

BIBLICAL PERSPECTIVE OF MATHEMATICS

The Biblical Perspective of Mathematics features introduce students to a biblical philosophy of mathematics. Since the Creator made this world good, we expect to see math work very well in this world. However, people can also misinterpret mathematics when attempting to support unbiblical conclusions. Students will evaluate different philosophies of mathematics, including Platonism, logicism, intuitionism, and formalism. They will learn that math can be discovered because God built math into the fabric of His plan for the world. The students will learn that people can often create mathematical models that accurately describe the world and its underlying structure. Students will also be guided through paradoxes like the unity and diversity of math and will tackle the use of indirect and direct proofs from a biblical worldview.

DATA ANALYSIS

The Data Analysis features illustrate how to apply a biblical philosophy of mathematics to real-world problems. Students will learn that math is a powerful tool for wise stewardship of God's world and for serving others. They will grapple with the power and possible abuses of math in real-world situations such as cryptology and medical screenings. Students will also apply a biblical worldview and mathematical skills to recommend proper responses to topics such as the volatile housing market, the world population crisis, and fiber optics.

Precalculus Standard Lesson Plan Overview

The Standard Lesson Plan allocates 90 days for each semester.

Key:

- Assessments
- *Student Edition* features:
  - **TC:** Technology Corner
  - **HC:** Historical Connection
  - **BPM:** Biblical Perspective of Mathematics
  - **DA:** Data Analysis
- Biblical worldview topics in the *Teacher Edition*

DAY	TOPIC	PAGES	ASSESSMENT	BIBLICAL WORLDVIEW
	Chapter 1 Analyzing Functions			
1	Introduction	v–1		
2	**1.1** Relations and Functions **TC:** Multiple Representations	2–8 6		• gaining a better understanding of infinity through knowing God • declaring God's infinite nature and recognizing His creation (Ps. 90:2)
3–4	**HC:** Functions **1.2** Linear Functions **BPM:** The Gift of Mathematics	9 10–16 17		• explaining the laws of the physical world from a biblical worldview (Gen. 1:1, 27) • recognizing the flaws in Descartes's reasoning • viewing mathematics as a divine gift, an intrinsic part of His creation (Ps. 104:24)
5–6	**1.3** Piecewise Functions and Continuity	18–24	Quiz 1A (1.1–1.2)	
7	**1.4** Power Functions and Variations	25–31		
8–9	**1.5** Transformations of Functions	32–38	Quiz 1B (1.3–1.4)	
10	**1.6** Quadratic Functions	39–45		
11–12	**1.7** Function Operations	46–51	Quiz 1C (1.5–1.6)	• identifying underlying fundamental laws evident in God's creation • recognizing how technology is making the fulfillment of prophecy more imminent (Rev. 13:17)
13–14	**1.8** Parametric Equations and Inverses	52–58		
15–16	**1.9** Modeling with Functions	59–66	Quiz 1D (1.7–1.8)	
17	**DA:** Home Run	67		• recognizing the human limitation of mathematics and mathematical models
18	Chapter 1 Review	68–70	Quiz 1E (1.9)	
19	Chapter 1 Test			

DAY	TOPIC	PAGES	ASSESSMENT	BIBLICAL WORLDVIEW
Chapter 2 Radical, Polynomial, and Rational Functions				
20	**2.1** Radical Functions and Equations	72–77		
21	**2.2** Polynomial Functions	78–85		
22	**TC:** Discontinuities	85	Quiz 2A (2.1–2.2)	
23	**2.3** The Remainder and Factor Theorems **HC:** Symbols	86–92 93		• understanding that elements of mathematics transcend human intellect • creating math symbols to describe what God has created
24–25	**2.4** Zeros of Polynomial Functions **BPM:** Math's Divine Nature	94–101 102–3	Quiz 2B (2.3–2.4)	• understanding how Christianity's fundamental doctrines apply to math • learning about key common attributes of God and math • understanding infallible transcendent mathematics
26	**2.5** Rational Functions	104–11		• recognizing how our limited understanding inspires awe and worship of God (Job 38–41)
27	**2.6** Solving Rational Equations	112–17		• understanding universal mathematical truth and the goodness in its applications
28–29	**2.7** Nonlinear Inequalities	118–23	Quiz 2C (2.5–2.6)	
30	**DA:** The Housing Market	124	Quiz 2D (2.7)	• using biblical wisdom and discernment in home buying (Prov. 22:3, 7)
31	Chapter 2 Review	125–26		
32	Chapter 2 Test			

DAY	TOPIC	PAGES	ASSESSMENT	BIBLICAL WORLDVIEW
Chapter 4 **Trigonometric Functions**				
46	**4.1** Angle Measure and Arc Length	172–78		
47	**4.2** Right Triangle Trigonometry	179–86		
48	**HC:** Trigonometry	187	Quiz 4A (4.1–4.2)	
49	**4.3** Extending Trigonometric Functions **TC:** Parametric Graphs of Trigonometric Functions	188–95 196		• identifying periodic functions in creation (Gen. 8:22)
50–51	**4.4** Sinusoidal Functions	197–205		• recognizing that life begins before birth (Jer. 1:4–5) • evaluating cyclic patterns in nature from a biblical perspective (Deut. 17:2–3; 2 Kings 23:5)
52	**4.5** Graphing Other Trigonometric Functions	206–13		
53	**BPM:** The Utility and Value of Mathematics	214–15	Quiz 4B (4.3–4.5)	• exploring how mathematical truth applies to the real world through recognizing God as Creator of both the universe and the human ability to describe it
54	**4.6** Inverse Trigonometric Functions	216–22		• acknowledging our inability to count the stars and grains of sand (Heb. 11:12) • recognizing the involvement of math in a God-given sign of answered prayer (2 Kings 20:8–11)
55	**4.7** Analyzing Combinations of Sinusoidal Functions	223–27		• seeing how mathematical patterns appeal to our God-given sense of beauty
56	**DA:** Sunspots and Solar Flares	228–29	Quiz 4C (4.6–4.7)	• responding biblically to climate change (Gen. 1:27–28; 8:20–22; Mark 12:31; Ps. 46:2–3)
57	Chapter 4 Review	230–31		
58	Chapter 4 Test			

DAY	TOPIC	PAGES	ASSESSMENT	BIBLICAL WORLDVIEW
Chapter 5	**Trigonometric Identities and Equations**			
59	**5.1** Fundamental Identities	233–39		• exploring unity and beauty in mathematics established by God in His creation
60	**5.2** Verifying Trigonometric Identities	240–45		• using logical reasoning developed through math to live biblically
61	**5.3** Solving Trigonometric Equations	246–51		
62	**BPM:** The Unity and Beauty of Mathematics	252–53	Quiz 5A (5.1–5.3)	• acknowledging evidence of God's unity and beauty in His universe
63	**5.4** Sum and Difference Identities	254–60		
64–65	**5.5** Multiple Angle Identities	261–67		
66	**HC:** Harmonic Analysis	268	Quiz 5B (5.4–5.5)	• recognizing math as more than a product of the human mind • learning about a mathematician's testimony of faith in Christ
67–68	**5.6** Law of Sines **TC:** Exploring the Law of Sines	269–75 273		
69	**5.7** Law of Cosines	276–81		
70	**DA:** Refraction, Reflection, and Fiber Optics	282–83	Quiz 5C (5.6–5.7)	• understanding the nature and source of light (Gen. 1:3–4; James 1:17)
71	Chapter 5 Review	284–85		
72	Chapter 5 Test			

DAY	TOPIC	PAGES	ASSESSMENT	BIBLICAL WORLDVIEW
Chapter 6	**Vectors, Polar Graphs, and Complex Numbers**			
73	**6.1** Vectors in the Plane	287–93		
74–75	**6.2** Dot Products	294–300		
76	**BPM:** Truth and Mathematical Proof	301–2	Quiz 6A (6.1–6.2)	• acknowledging mathematical proof as an instrument of discovering truth (Phil. 4:8) • seeking truth through God's Word v. worldly thinking (John 17:17; Phil. 4:8)
77	**6.3** Vectors in Space	303–8		• recognizing manmade mathematical models fail to completely describe God's creation
78–79	**6.4** Polar Coordinates **TC:** Graphing Polar Functions	309–14 313	Quiz 6B (6.3)	
80	**6.5** Graphs of Polar Equations	315–22		
81	**HC:** Complex Numbers	323	Quiz 6C (6.4–6.5)	• discovering how mathematical models reflect structure and design in the world
82–83	**6.6** Polar Forms of Complex Numbers	324–30		
84	**DA:** Fractals	331–32		• understanding how similar design points to a common Creator • discussing chaos theory and the importance of a person's worldview
85	Chapter 6 Review	333–34	Quiz 6D (6.6)	
86	Chapter 6 Test			
87–89	Review for Second Quarter Exam (Chapters 4–6) or First Semester Exam (Chapters 1–6)			
90	Second Quarter Exam or First Semester Exam			

Reaching the Pinnacle

In 1958 the first recorded ascent of the 3000 foot rock face of El Capitan in Yosemite National Park took the team of three climbers 47 days. The entire project took eighteen months as the climbers established multiple camps along the route—all connected by ropes. In 2017 Alex Honnold, using only a small bag of chalk strapped to his waist, climbed El Capitan free solo in under four hours.

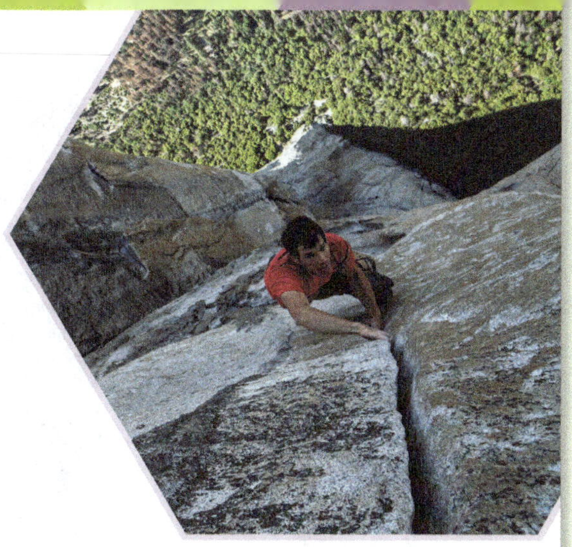

Honnold did not accomplish this feat on a whim. He practiced climbing from a young age, noting that others had more natural ability, but he just kept on working. He trained hard (hanging by just his fingertips up to an hour every day), carefully plotted his route, and practiced particularly challenging parts of the climb by repeatedly rappelling down and climbing up.

In Precalculus you will hone and combine skills you have developed in algebra and geometry. For many of you, this course will be the final preparation you need before tackling a college-level calculus course. For others, it may be your final math class before you turn your focus to other areas. Either way, if you invest the necessary effort, you will expand both your skill in solving problems and your understanding of mathematics—its history, current use, and relevance from a biblical perspective. In the current Information Age with its data-driven algorithms, these skills are essential to understanding complex issues in health care, law enforcement, social media, space flight, and myriad other areas.

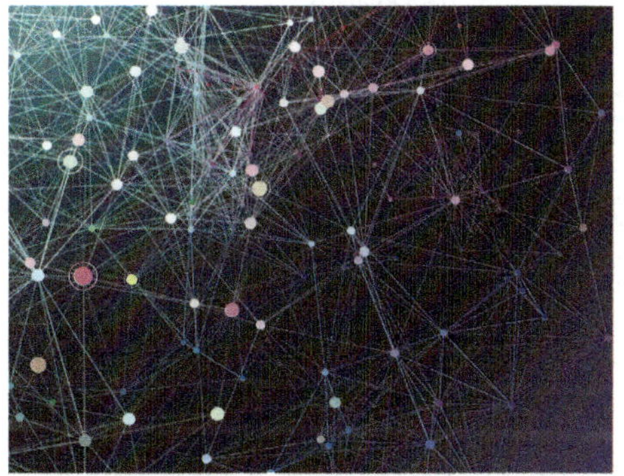

Over the next few pages you will be introduced to some of the special features contained in this book. The text integrates answers to critical questions regarding the genesis of mathematical ideas, modern applications of these ideas, and the importance of these matters in today's world. Frequent technology instruction, tips, and step-by-step examples will assist you in your journey. In addition, margin notes in each section provide interesting bits of information that directly relate to the development or application of ideas in the accompanying text.

The Historical Connection, Data Analysis, and Biblical Perspective of Mathematics features found in each chapter provide context, application, and meaning to the mathematical ideas as they are introduced and will improve comprehension and retention. Woven together, these three features present a strong biblical worldview of mathematics. The Historical Connection features provide more than a timeline of events; they explore the factors leading to the development of particular mathematical ideas and the reasons why they were discovered at a particular time and place. The Data Analysis features offer an opportunity to work on real-world problems using the tools learned in that chapter. While the entire text is written from a biblical worldview that rejects the moral neutrality of mathematics, particular worldview topics are addressed in detail in the Biblical Perspective of Mathematics features.

We hope you are excited about this course! The material may be challenging, but the accomplishments are well worth the effort. Take advantage of the many features that are included in the text to help you along the way. Our goal is that you find this book useful, rigorous, and consistent with a biblical worldview.

Presentation

The introduction to *Precalculus* is intentionally concise. Consider having students read the facing page in class or including it as part of the first homework assignment. The following discussion questions may be helpful in emphasizing the key concepts from the introduction.

1. What similarities do you see between a feat such as climbing El Capitan and completing a course in precalculus?
Both require a lot of preparation and are challenging, but both are also very rewarding.

2. List and describe the text's three series of special features that integrate a biblical worldview of mathematical ideas in their historical context or in current applications.
Historical Connection; Data Analysis; Biblical Perspective of Mathematics; (See the student text for descriptions.)

3. Discuss: Why is precalculus worth the effort?   Answers will vary.

Using This Book

Chapter Introduction
Preview the flow of lessons in the chapter and read a brief introduction to each of the three special features.

Biblical Perspective of Mathematics
Explore the philosophical basis of mathematics from a biblical worldview.

Historical Connection
Consider why mathematical concepts were developed at a particular time and place.

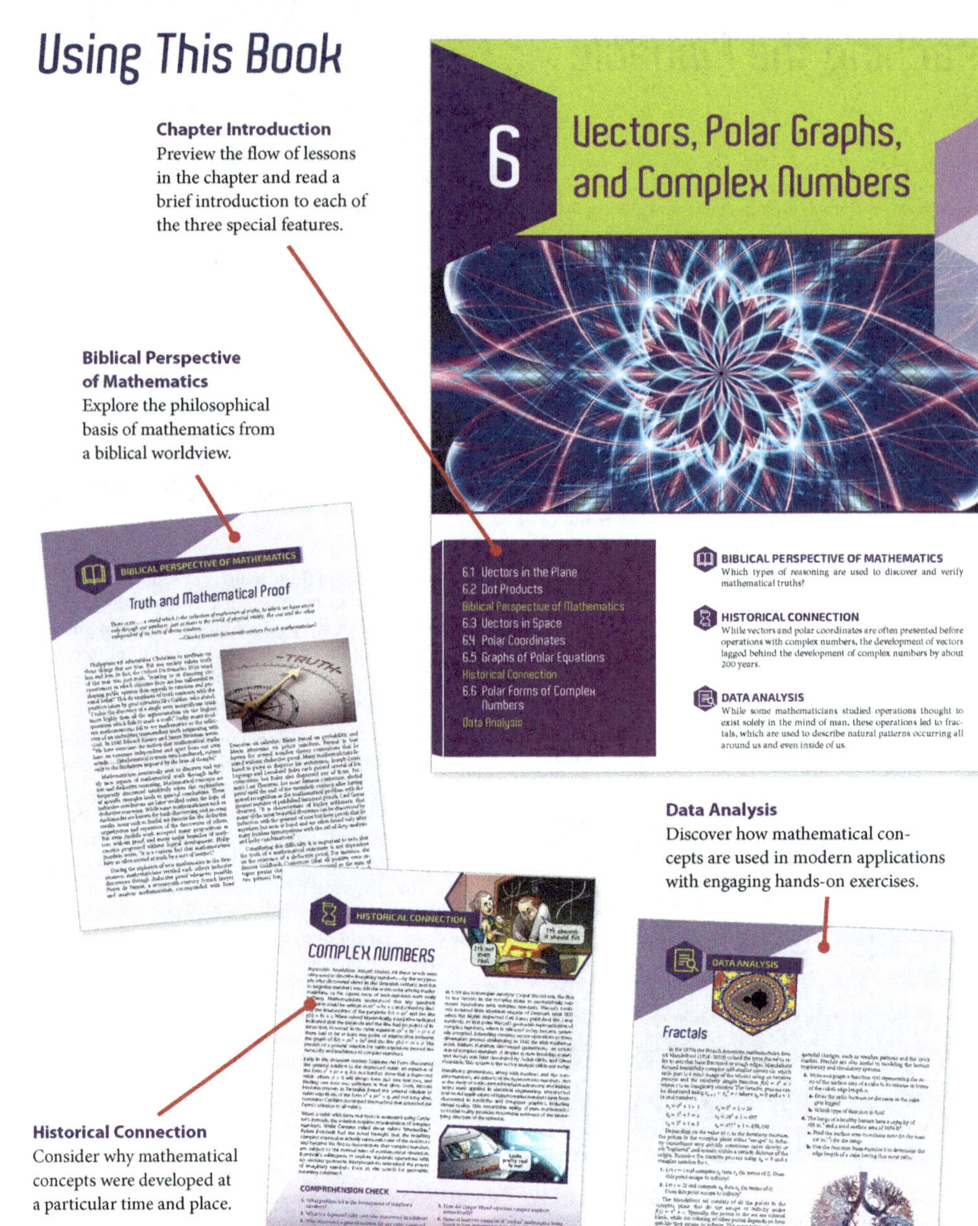

Data Analysis
Discover how mathematical concepts are used in modern applications with engaging hands-on exercises.

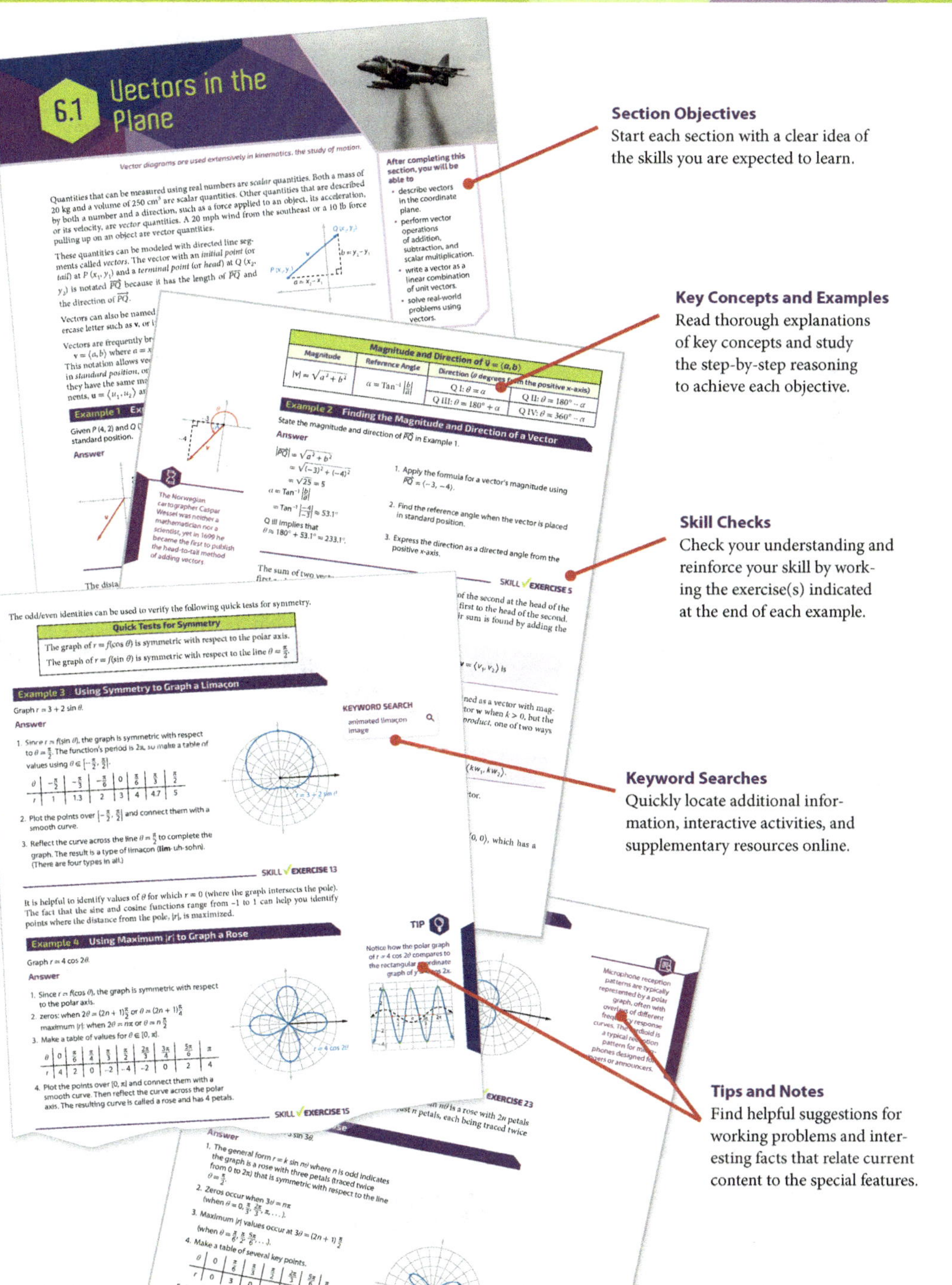

Section Objectives
Start each section with a clear idea of
the skills you are expected to learn.

Key Concepts and Examples
Read thorough explanations
of key concepts and study
the step-by-step reasoning
to achieve each objective.

Skill Checks
Check your understanding and
reinforce your skill by working the exercise(s) indicated
at the end of each example.

Keyword Searches
Quickly locate additional information, interactive activities, and
supplementary resources online.

Tips and Notes
Find helpful suggestions for
working problems and interesting facts that relate current
content to the special features.

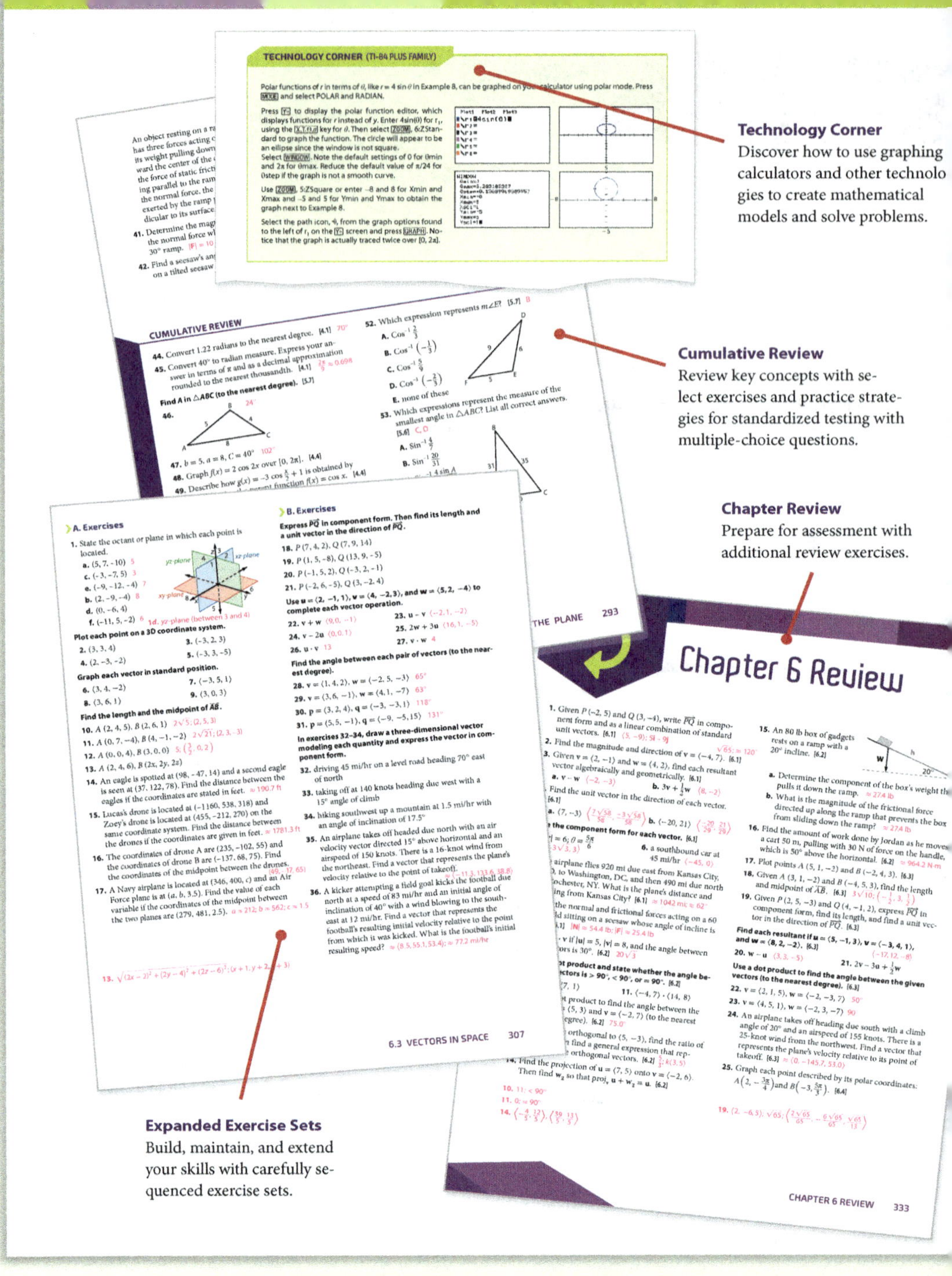

Technology Corner

Discover how to use graphing calculators and other technologies to create mathematical models and solve problems.

Cumulative Review

Review key concepts with select exercises and practice strategies for standardized testing with multiple-choice questions.

Chapter Review

Prepare for assessment with additional review exercises.

Expanded Exercise Sets

Build, maintain, and extend your skills with carefully sequenced exercise sets.

1 Analyzing Functions

HISTORICAL CONNECTION
A culture's worldview affects how it explores and explains the physical world. Why was the concept of mathematical functions not developed until the seventeenth century?

BIBLICAL PERSPECTIVE OF MATHEMATICS
Is math a morally neutral discipline? After all, $2 + 2 = 4$ is true for all people, regardless of their religious beliefs. Should a Christian view mathematics as a human invention or a divine gift?

DATA ANALYSIS
Creating mathematical models of real-world data enables us to make informed predictions about future events. Can you develop a mathematical model that predicts whether a fly ball will clear the fence for a home run? How accurate is your model?

Overview

This chapter reviews key concepts related to relations and functions while introducing several new concepts used in calculus, such as continuity, limits, and parametric equations. The use of regressions to derive equations modeling data sets is reviewed and the role of a biblical worldview in interpreting these models is emphasized.

The Technology Corner reviews basic graphing and function analysis with the TI-84 Plus family of calculators.

Chapter Objectives

1. To determine key characteristics of algebraic and transcendental functions and draw graphs illustrating those characteristics

2. To solve algebraic, exponential, logarithmic, and trigonometric equations and inequalities

3. To describe key characteristics of basic algebraic functions, including domain, range, intercepts, symmetry, end behavior, continuity, and increasing or decreasing intervals

4. To find and graph the results of function operations, including composition and decomposition

5. To model real-world data and solve problems using basic algebraic functions

6. To explain how the belief in an infinite God contributed to the development of the concept of functions

Suggested Teaching Schedule

DAY		DAY	
1	Introduction	10	1.6
2	1.1 TC: Multiple Representations	11–12	Quiz 1C (1.5–1.6) 1.7
3–4	HC: Functions 1.2 BPM: The Gift of Mathematics	13–14	1.8
		15–16	Quiz 1D (1.7–1.8) 1.9
5–6	Quiz 1A (1.1–1.2) 1.3	17	DA: Home Run
7	1.4	18	Quiz 1E (1.9) Chapter 1 Review
8–9	Quiz 1B (1.3–1.4) 1.5	19	Chapter 1 Test

Objectives

1. To utilize multiple representations of relations and functions
2. To identify relations that are functions
3. To use interval notation to state the domain and range for relations and functions

Flash

Knowledge of tsunamis dates back to the Peloponnesian War. During the summer of 426 BC a series of earthquakes shook Greece, followed by a tsunami and massive flooding that halted Sparta's invasion of Attica. Thucydides determined that the flooding was a direct result of the earthquakes. Mathematical formulas to measure the intensity and magnitude of a tsunami were not developed until the mid-twentieth century. Modern scientists can predict the speed of a tsunami based on the depth of the water at the location of the earthquake.

Vocabulary

abscissa
closed interval
dependent variable
domain
function
function notation
half-open interval
half-closed interval
independent variable
interval notation
open interval
ordinate
range
relation
Vertical Line Test

1.1 Relations and Functions

The speed of a tsunami is modeled with a mathematical function in exercise 35.

After completing this section, you will be able to

- use multiple representations of relations.
- identify relations that are functions.
- use interval notation.

While René Descartes (1596–1650) developed the idea of graphing relations, he likely studied the writings of Nicole Oresme, a fourteenth-century scholastic who plotted the results of heat transfer experiments using horizontal and vertical axes.

The idea of a quantity changing as a function of another quantity is a relatively recent mathematical concept. Its roots lie in the scientific exploration of the physical world that flourished under a Christian worldview in the seventeenth century. The study of relations and functions led to the development of calculus and is the basis of most of modern mathematics. The pairing of numbers in a relation or function is frequently represented by a set of ordered pairs.

> **DEFINITION**
>
> A **relation** is any set of ordered pairs.

The set $A = \{(4, -2), (1, -1), (0, 0), (1, 1), (4, 2)\}$ is a relation since it is a set of ordered pairs. A relation can also be represented by a table, a mapping diagram, or a graph.

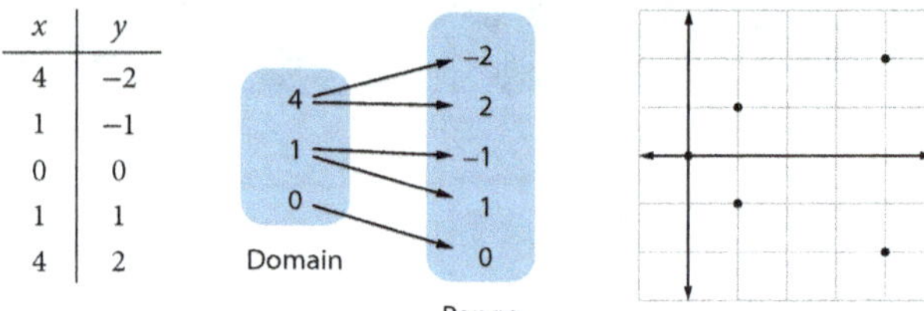

The mapping diagram for relation A illustrates its *domain*, the set of first elements $D = \{4, 1, 0\}$, and its *range*, the set of second elements $R = \{-2, -1, 0, 1, 2\}$. In mathematics, we are typically concerned with relations in which each first coordinate is associated with exactly one second coordinate.

> **DEFINITION**
>
> A **function** is a relation in which every first coordinate has one and only one second coordinate associated with it.

Relation A is not a function because the element 4 in the domain is associated with both -2 and 2 in the range. Relation $B = \{(-2, -3), (0, 1), (1, 3)\}$ is a function because each first coordinate is associated with exactly one second coordinate.

PRESENTATION

Lesson Opener

Solve.

1. $a^2 - 8a = 0$ $\quad a = 0, 8$
2. $r^2 - 36 = 0$ $\quad r = \pm 6$
3. $\sqrt{3x - 6} = 0$ $\quad x = 2$
4. $7 - z > 0$ $\quad z < 7$
5. For what values of x is $\sqrt{12 - 4x}$ a real number? $\quad x \leq 3$

Lesson Openers can be used as a quick review or assessment, and as a way to engage students at the beginning of class.

Consider beginning by drawing a curve on a coordinate plane and explaining that it represents a relation. A mathematical relation describes the relationship between two sets. A graph illustrates the relationship between the set of x-values and the set of y-values. Define a *relation* and use set A to illustrate multiple representations of a relation. Use the mapping diagram to illustrate its *domain* and *range*. Define a *function* as a special case of a relation and illustrate how set A fails the definition while set B satisfies the definition.

You may want to ask students if relation B would be a function if the last ordered pair were $(1, -3)$. (*yes*) Discuss how each of the representations shown

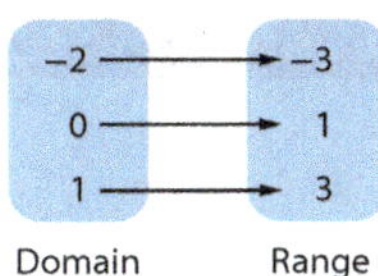 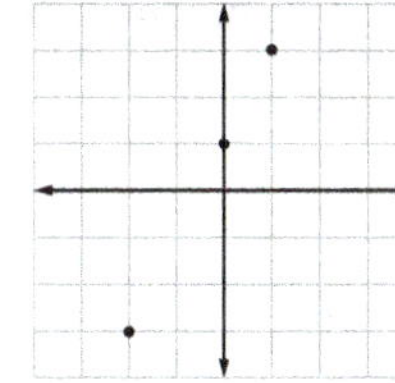

Domain   Range

The *Vertical Line Test* can be used to determine whether a graphed relation is a function.

VERTICAL LINE TEST

If two or more points of a graphed relation lie on the same vertical line, the relation is not a function.

Example 1 Using the Vertical Line Test

Determine whether each relation is a function.

a. 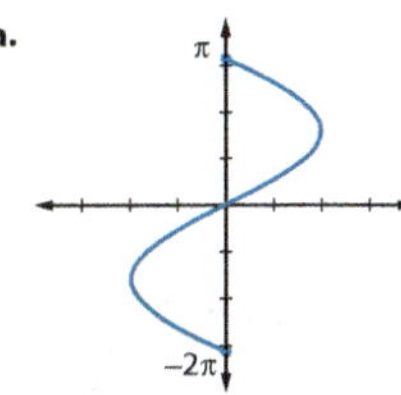

b. 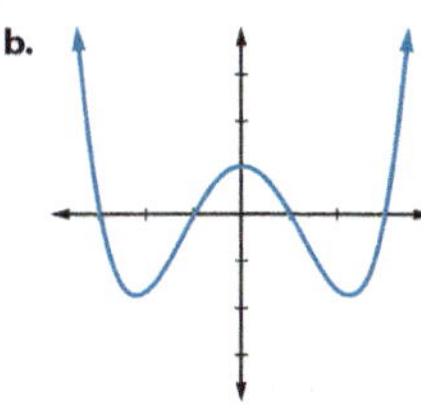

Answer

a. No; a vertical line can intersect the graph at more than one point.

b. Yes; no vertical line intersects the graph at more than one point.

SKILL ✔ EXERCISE 5

The relationship between the first coordinate (the *abscissa*) and the second coordinate (the *ordinate*) in a relation is usually described by an equation. For example, $y = 2x + 1$ describes the relationship between the ordered pairs in relation B. Functions are named with lowercase letters (particularly f and g) and their equations are frequently written in *function notation*, where y is replaced by $f(x)$, read "f of x": $f(x) = 2x + 1$.

A function can be defined by its function rule and a statement of its domain:
$g(x) = x^2 - 1; D = \{x \mid -2 < x < 2\}$.

You can determine as many of the function's ordered pairs as needed by choosing a value for x from the domain and using the function rule to calculate the corresponding y-value. Because the value of y depends on the value chosen for x, y is called the *dependent variable* and x is called the *independent variable*. Plotting several ordered pairs of a function and recognizing a pattern often allows us to sketch the function's graph.

Reading and Writing Mathematics

Explain the difference between a relation and a function. Then explain why the characteristics of a function make it an important mathematical tool.

A relation associates every element of a set with one or more elements in a second set. A function associates every element of a set with only one element of a second set.

Functions ensure a unique output value for a given input value. If the square root symbol did not imply a function (the principal square root), we could not have a unique answer for the square root of a number.

Additional Exercises

Figures for Additional Exercises throughout the chapter can be found at TeacherToolsOnline.com.

1. Write the domain and range of
   $f(x) = \{(7, 2), (-5, 3), (1, 4), (3, 3)\}$
   using set-builder notation.

   $D = \{x \mid x = -5, 1, 3, 7\}$
   $R = \{y \mid y = 2, 3, 4\}$

Determine whether each relation is a function.

2. $x = -8$ not a function

3. $y = x^5$ function

4. a relation in which the input value x is the hour of the day and the output value y is the number of occupants in each of the Good Cook'n restaurants   not a function

---

would change. The fact that two different values in the domain would be mapped to the same value in the range would not disqualify the relation from being a function.

Explain how the *Vertical Line Test* graphically demonstrates the alternative definition of a function as a set of ordered pairs in which no two points have the same first coordinate.

One-on-One Consider describing a function using the illustration of a change machine. The machine "functions" properly if, when a dollar bill is inserted (input), it dispenses four quarters (output). The machine "malfunctions" if, when a dollar bill is inserted (input), the machine dispenses either four or five quarters (two possible outputs for a specific input).

Present Example 1 and then, using $y = 2x + 1$, review *function notation* and the terms related to *independent* and *dependent variables*.

Use Example 2 to illustrate how any function can be plotted by evaluating enough points to detect a pattern and connecting the points with a curve. Explain that *interval notation* is often more convenient than using an inequality to describe the domain or range of a function.

Interactive Activity Use the Internet keyword search *domain range demo* to find visual demonstrations of domain and range.

Discuss the use of parentheses and brackets to indicate open and closed endpoints of an interval and review the chart summarizing bounded and unbounded intervals. Then check student comprehension of interval notation with Example 3.

Describe each graph using interval notation.

5. 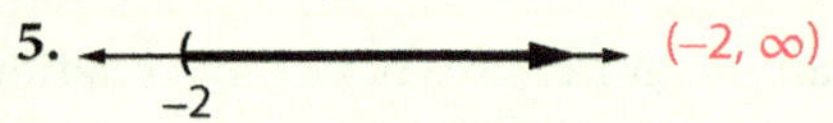 $(-2, \infty)$

-2

6. $-2 \quad 4$ $[-2, 4) \cup (4, \infty)$

Graph each function within the specified domain.

7. $f(x) = x + 3; D = [-1, 3]$

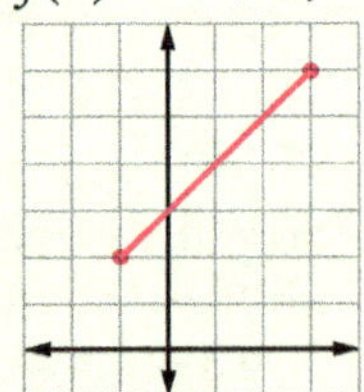

8. $g(x) = -x - 2; D = (0, 5]$

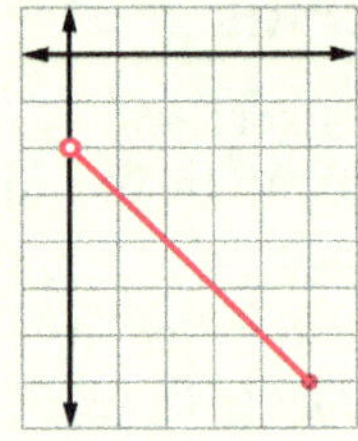

Use the graph of $y = \dfrac{1}{x - 5}$ for exercises 9–10.

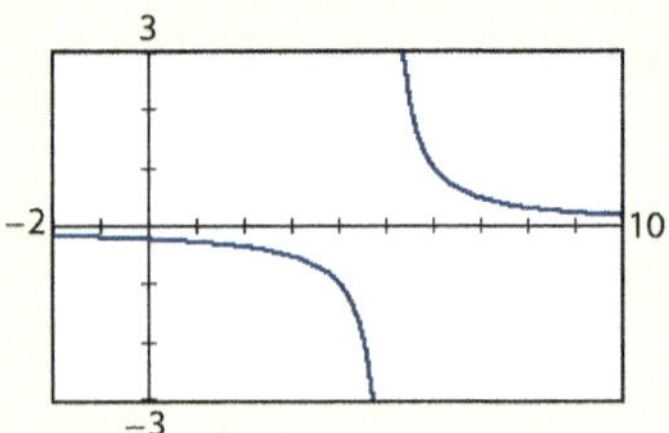

9. Is the relation a function? yes

10. State the domain using interval and set-builder notation.

$D = (-\infty, 5) \cup (5, \infty); D = \{x \mid x \neq 5\}$

Assignments

- **Minimum:** 1, 3–6, 9–13, 15–16, 18–19, 27–28, 35, 42–46, 48, 50
- **Standard:** 2, 4, 7–9, 11–13, 17, 20, 24–26, 28, 31, 34–35, 41–42, 45–48, 50
- **Extended:** 2, 4, 7–8, 11–12, 14, 17–18, 22, 25, 28–32, 35, 36(a, d), 37–38, 40, 47–50

Graph $g(x) = x^2 - 1; D = \{x \mid -2 \leq x \leq 2\}$.

Answer

$g(-2) = (-2)^2 - 1 = 3$
$g(-1) = (-1)^2 - 1 = 0$
$g(0) = (0)^2 - 1 = -1$
$g(1) = (1)^2 - 1 = 0$
$g(2) = (2)^2 - 1 = 3$

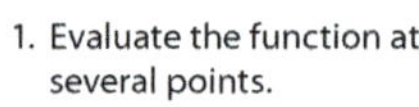

1. Evaluate the function at several points.
2. Plot the ordered pairs.
3. Since the domain contains all real numbers from −2 to 2, connect the points with a smooth curve.

SKILL ✔ **EXERCISE 19**

TIP

The context determines whether the notation represents an ordered pair or an interval.

The range of $g(x)$ in Example 2 can be described with an inequality: $\{y \mid -1 \leq y \leq 3\}$. A shorter *interval notation* can also be used. *Closed* intervals include both endpoints and are indicated by brackets: $[-1, 3]$. *Open* intervals, such as $0 < x < 5$, exclude both endpoints and are indicated with parentheses: $(0, 5)$. Intervals may also be *half-open* (or *half-closed*). The symbols used in interval notation are frequently used instead of open circles and solid points when the interval is graphed on a number line. Unbounded intervals extend infinitely and are indicated using ∞ for positive infinity or $-\infty$ for negative infinity.

Bounded Intervals			Unbounded Intervals		
Graph	Inequality	Interval	Graph	Inequality	Interval
$a \quad b$	$a \leq x \leq b$	$[a, b]$	a	$x \geq a$	$[a, \infty)$
$a \quad b$	$a < x < b$	(a, b)	a	$x > a$	(a, ∞)
$a \quad b$	$a \leq x < b$	$[a, b)$	b	$x < b$	$(-\infty, b)$
$a \quad b$	$a < x \leq b$	$(a, b]$	b	$x \leq b$	$(-\infty, b]$

Example 3   Representing Intervals

Write each inequality in interval notation, then graph the interval on a number line.

a. $-1 < x \leq 3$ **b.** $x \geq 2$

Answer

a. $(-1, 3]$ $-1 \quad 0 \quad 3$

b. $[2, \infty)$ $0 \quad 2$

SKILL ✔ **EXERCISE 11**

When the domain of a function is not stated, its domain is assumed to be the real numbers except those values that would cause the dependent variable to be undefined. Determining the range of a function algebraically is more difficult. It can be helpful to examine the function's graph.

Have the students use technology to graph the functions in Example 4. The Technology Corner reviews basic graphing techniques. It also illustrates how a calculator can be used to verify whether a function is defined at a particular value of x. Some calculators may connect the last points plotted just to the left and right of $x = 2$, giving the appearance of a nearly vertical line at $x = 2$. While $x = 2$ is a vertical asymptote (discussed in later sections), this nearly vertical line is actually an error caused by the calculator's "plot and connect" method of graphing.

Motivational Idea Use the Internet keyword search *domain range calculator* to find applications that determine the domain and range of a function.

TIPS

Ex. 17–18 You may want to explain that $D = \{x \mid x = 3\}$ and $R = \{y \mid y = 3\}$ are typically not expressed in interval notation.

Ex. 41–50 See Appendix 1 if the students need more review of sets, set operations, and common subsets of real numbers.

State the domain and range of each function.

a. $g(x) = \sqrt{2x - 3}$　　　　**b.** $f(x) = \dfrac{1}{x + 2}$

Answer

a. $g(x) = \sqrt{2x - 3}$

$2x - 3 \geq 0$

$x \geq \dfrac{3}{2}$

$\therefore D = [1.5, \infty)$

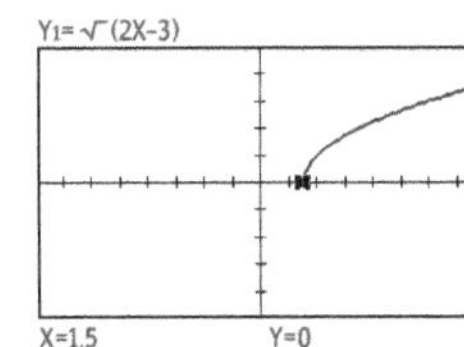

$\therefore R = [0, \infty)$

1. The radicand of an even root must be greater than or equal to 0.
2. Solve.
3. Write the domain in interval notation.
4. Use technology to graph $g(x)$ and confirm the domain.
5. Use the graph to determine that there are no points with negative y-values and that the y-values increase without bound.

b. $f(x) = \dfrac{1}{x + 2}$

$x + 2 \neq 0$

$x \neq -2$

$\therefore D = \{x \mid x \neq -2\}$ or
$D = (-\infty, -2) \cup (-2, \infty)$

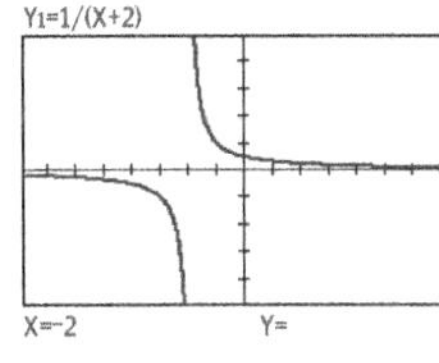

$0 = \dfrac{1}{x + 2}$

$0(x + 2) = 1$

$0 = 1$

$\therefore R = \{x \mid x \neq 0\}$ or
$R = (-\infty, 0) \cup (0, \infty)$

1. Since division by 0 is undefined, the denominator cannot be equal to 0.
2. Solve.
3. Write the domain in set-builder notation or in interval notation.
4. Use technology to graph $g(x)$ and confirm that y is undefined when $x = -2$.
5. It appears that the y-values of the function extend forever in each direction but may not include 0. This can be checked algebraically.

   a. Set the function value, $f(x)$ or y, equal to 0.
   b. Multiply both sides by $x + 2$ and simplify.

   c. Since $0 \neq 1$, $f(x) \neq 0$.

——————————— SKILL ✔ **EXERCISE 29**

Solutions

❯ A. Exercises

1.

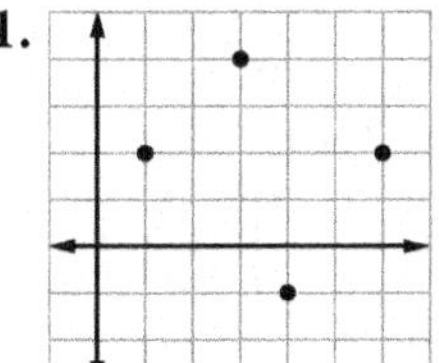

2.

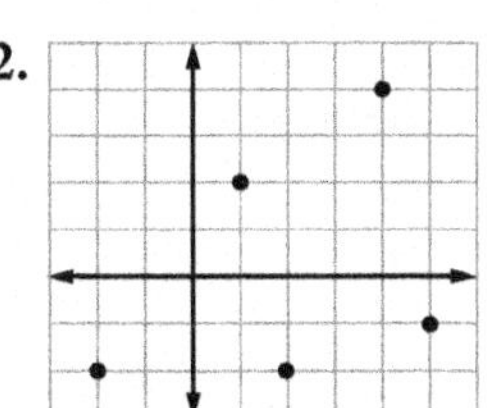

5. The relation passes the Vertical Line Test.

6. The relation fails the Vertical Line Test.

7–8. The relation passes the Vertical Line Test.

9. $(-\infty, 12)$

10. $[-7, \infty)$

11. $(-4, 10]$

12. $(-\infty, 5) \cup (18, \infty)$

B. Exercises

19. $f(-1) = 2(-1) - 5 = -2 - 5 = -7$
$f(1) = 2(1) - 5 = 2 - 5 = -3$
$f(4) = 2(4) - 5 = 8 - 5 = 3$
$\{(-1, -7), (1, -3), (4, 3)\}$

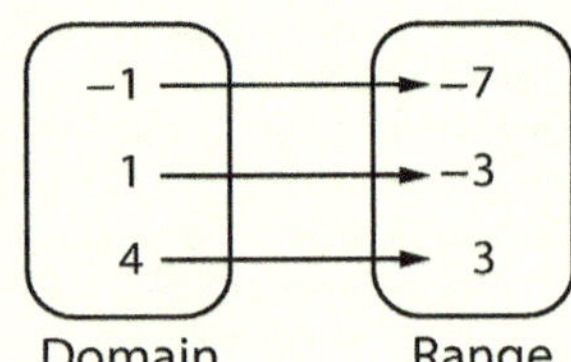

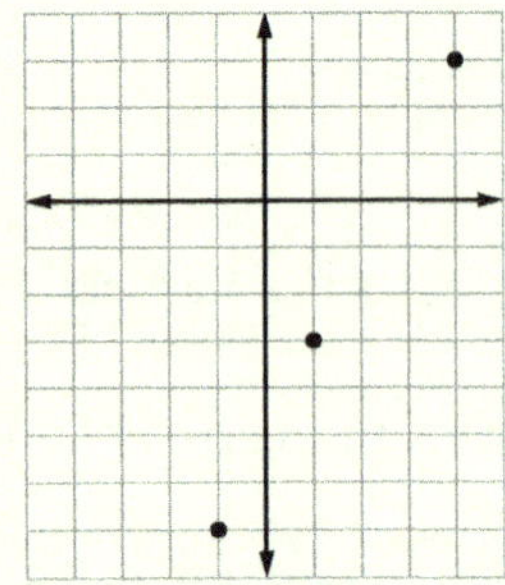

20. $g(-2) = 2(-2)^2 - 3 = 2(4) - 3 = 5$
$g(-1) = 2(-1)^2 - 3 = 2(1) - 3 = -1$
$g(0) = 2(0)^2 - 3 = 2(0) - 3 = -3$
$g(1) = 2(1)^2 - 3 = 2(1) - 3 = -1$
$g(2) = 2(2)^2 - 3 = 2(4) - 3 = 5$
$\{(-2, 5), (-1, -1), (0, -3), (1, -1), (2, 5)\}$

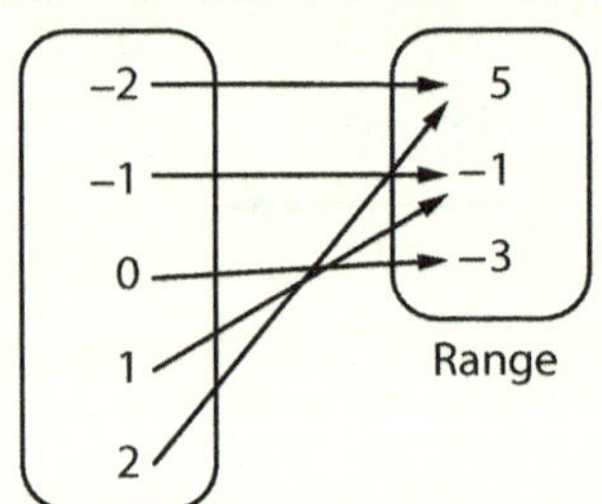

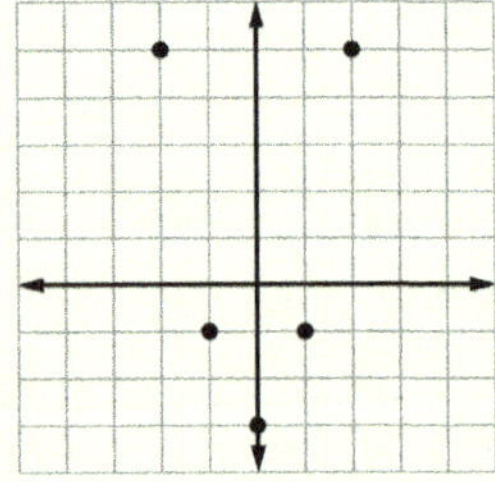

21. $h(-2) = -(-2) + 4 = 2 + 4 = 6$
$h(-1) = -(-1) + 4 = 1 + 4 = 5$
$h(0) = -(0) + 4 = 0 + 4 = 4$
$h(3) = -(3) + 4 = -3 + 4 = 1$
$\{(-2, 6), (-1, 5), (0, 4), (3, 1)\}$

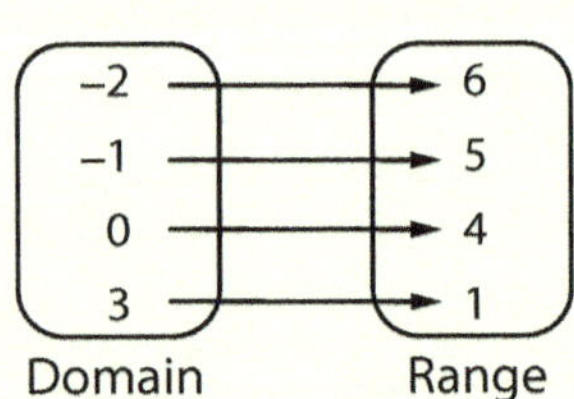

TECHNOLOGY CORNER (TI-84 PLUS FAMILY)

Your graphing calculator can be used to display several representations of a function. Be sure that the calculator is in function mode by selecting FUNCTION on the MODE menu. Then press Y= to enter the function editor and enter the function rule $f(x) = \dfrac{1}{x + 2}$ as Y1.

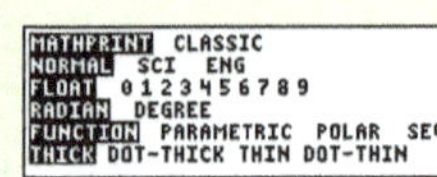
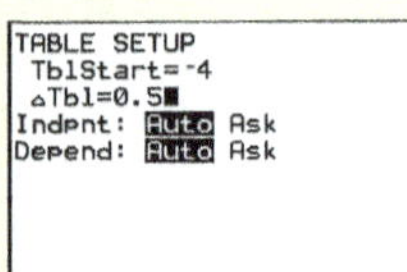
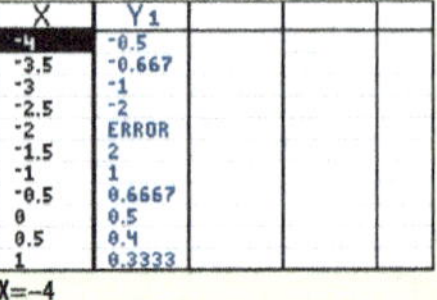

Use the [TABLE] command ([2nd], [GRAPH]) to view the ordered pairs for the defined function. The [TBLSET] command ([2nd], [WINDOW]) displays the TABLE SETUP menu. TblStart and ΔTbl define the initial value and the increment for the independent variable. Setting Indpnt to Ask enables you to search for individual x-values in the table.

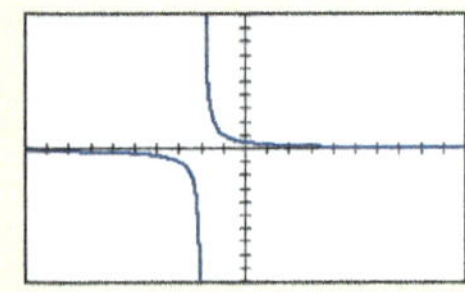
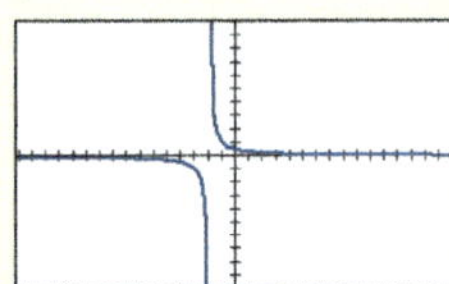

Use the [GRAPH] command to display a graph of the function. The [ZOOM] command can be used to adjust the portion of the graph that is viewed. Selecting ZStandard from the [ZOOM] menu causes the function to be graphed in a window with $-10 \leq x \leq 10$ and $-10 \leq y \leq 10$. Use ZSquare to produce a graph where the scales of the axes are the same.

Selecting ZBox from the [ZOOM] menu allows you to zoom in and further examine a rectangular region that is defined using the cursor keys and [ENTER] to position the opposite vertices. The [TRACE] command allows you to see the function's coordinates as you move across the graph or enter a particular value for x.

The [WINDOW] menu allows you to enter an exact minimum, maximum, and scale for each axis. With experience, you will be able to quickly determine the best window to display the graph's key characteristics.

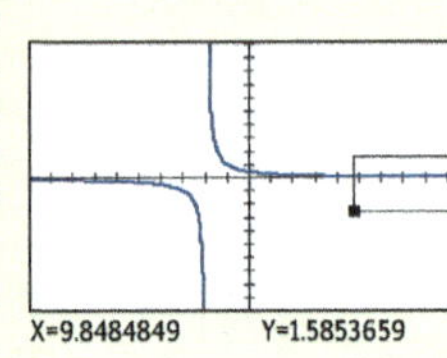
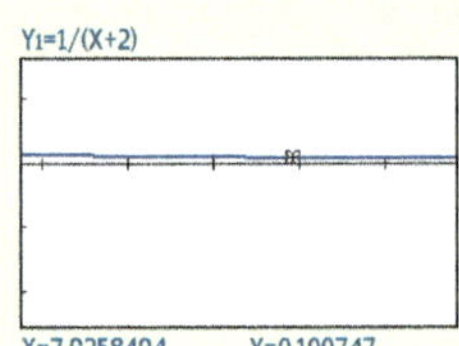

A. Exercises

Graph each relation and state its domain and range using set-builder notation.

1. $\{(1, 2), (3, 4),$
$(6, 2), (4, -1)\}$
$D = \{x \mid x = 1, 3, 4, 6\}$
$R = \{y \mid y = -1, 2, 4\}$

2.

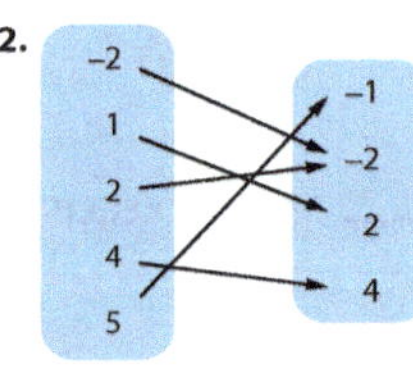

$D = \{x \mid x = -2, 1, 2, 4, 5\}$
$R = \{y \mid y = -2, -1, 2, 4\}$

3.

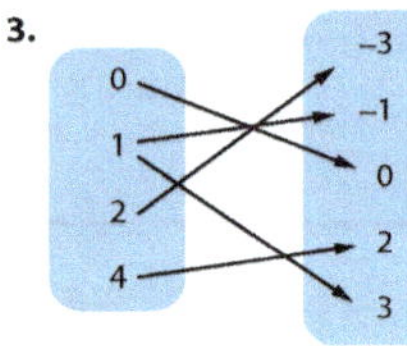

$\{(0, 0), (1, -1), (1, 3),$
$(2, -3), (4, 2)\}$
not a function

4.

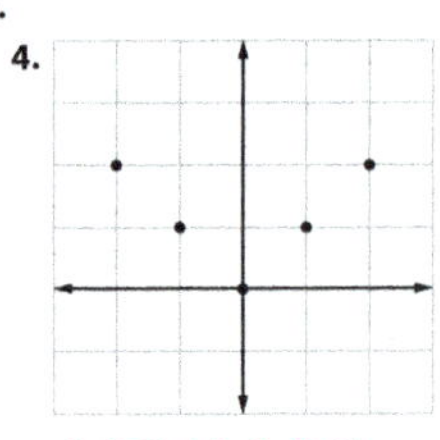

$\{(-2, 2), (-1, 1), (0, 0),$
$(1, 1), (2, 2)\}$
function

Write each relation as a set of ordered pairs and determine whether it is a function.

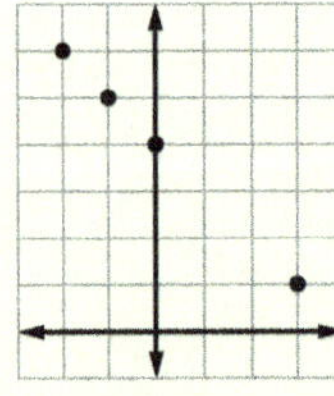

22. $l(-1) = (-1)^3 - 2(-1)^2 - 4 = -7$
$l(0) = (0)^3 - 2(0)^2 - 4 = -4$
$l(1) = (1)^3 - 2(1)^2 - 4 = -5$
$l(2) = (2)^3 - 2(2)^2 - 4 = -4$
$l(3) = (3)^3 - 2(3)^2 - 4 = 5$
$\{(-1, -7), (0, -4), (1, -5),$
$(2, -4), (3, 5)\}$

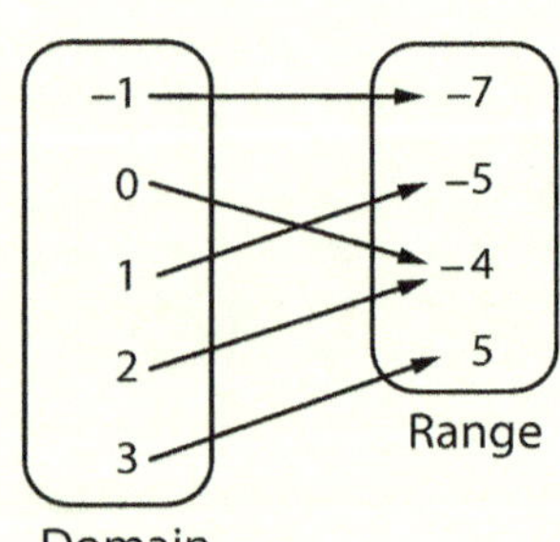

Determine whether each relation is a function.

5. 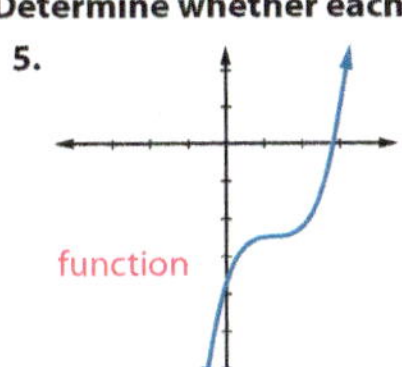
function

6. 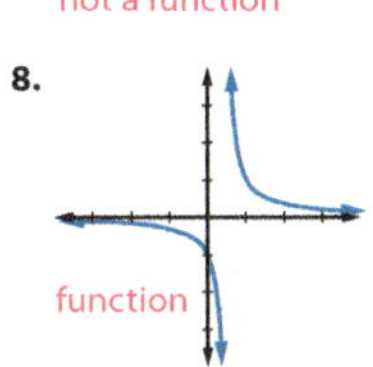
not a function

7. 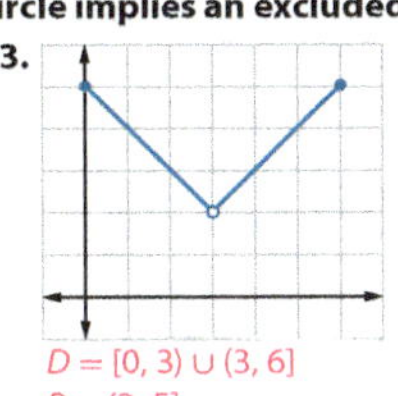
function

8. 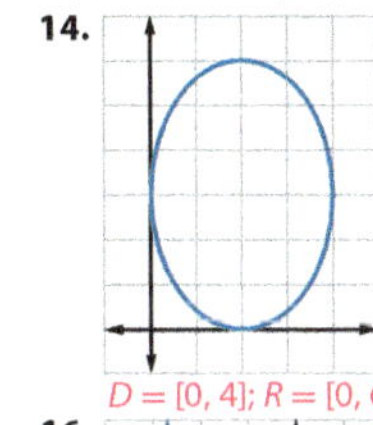
function

Write each inequality in interval notation, then graph the interval on a number line.

9. $x < 12$

10. $x \geq -7$

11. $-4 < x \leq 10$

12. $x < 5$ or $x > 18$

State the domain and range of each function, using interval notation where possible. Remember that an open circle implies an excluded endpoint in the graph.

13. 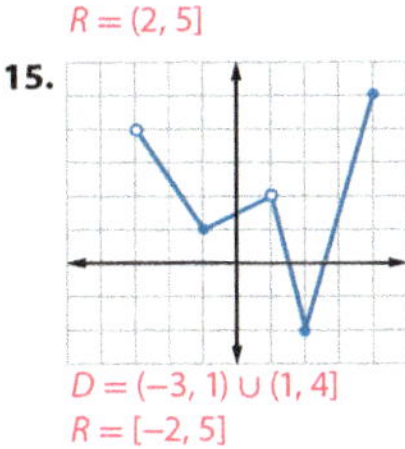
$D = [0, 3) \cup (3, 6]$
$R = (2, 5]$

14.
$D = [0, 4]; R = [0, 6]$

15. 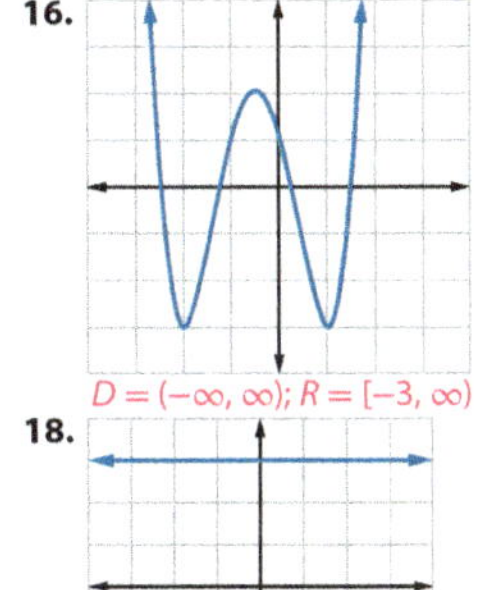
$D = (-3, 1) \cup (1, 4]$
$R = [-2, 5]$

16.
$D = (-\infty, \infty); R = [-3, \infty)$

17. 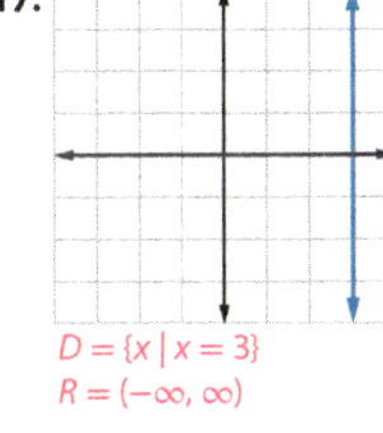
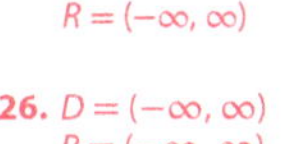
$D = \{x \mid x = 3\}$
$R = (-\infty, \infty)$

18.
$D = (-\infty, \infty)$
$R = \{y \mid y = 3\}$

26. $D = (-\infty, \infty)$
$R = (-\infty, \infty)$

27. $D = (-\infty, \infty)$
$R = \{y \mid y = -3\}$

Represent each function as a set of ordered pairs, a mapping diagram, and a graph.

19. $f(x) = 2x - 5$ where $D = \{x \mid x = -1, 1, 4\}$

20. $g(x) = 2x^2 - 3$ where $D = \{-2, -1, 0, 1, 2\}$

21. $h(x) = -x + 4$ where $D = \{-2, -1, 0, 3\}$

22. $l(x) = x^3 - 2x^2 - 4$ where $D = \{-1, 0, 1, 2, 3\}$

23. Give an example of a relation consisting of 4 ordered pairs.

24. Give an example of a function consisting of 4 ordered pairs.

25. Give an example of a relation that is not a function.

Use technology to graph each function. Then state the domain and range of each function, using interval notation where possible.

26. $f(x) = 3x - 4$

27. $g(x) = -3$

28. $h(x) = \sqrt{x + 4}$

29. $f(x) = \dfrac{3}{x - 1}$

Determine whether each relation is a function.

30. $y = x$ function

31. $x = \dfrac{1}{y^2}$ not a function

32. a relation where the input value x is the day of the month and the output value y is the dollar amount for individual debit transactions for a hardware store
not a function

33. a pairing of a golf hole number and the score recorded by a golfer during a round   function

34. Jenni started a home business making backpacks and spent $300 to purchase enough supplies to make 15 backpacks. Her profit can be modeled by $f(n) = 50n - 300$ where n represents the number of bags sold.

  a. How much is she charging for each bag? $50

  b. How much did she spend on supplies for each bag? $20

  c. How many bags must she sell to break even? 6 bags

  d. State the function's domain and range.

35. The speed at which a tsunami moves across the ocean can be calculated using the formula $s = \sqrt{9.8d}$ where s is the speed in meters per second and d is the ocean's depth in meters.

  a. Describe the domain and range using interval notation. $D = [0, \infty); R = [0, \infty)$

  b. Find the speed of a tsunami in kilometers per hour at an ocean depth of 3600 m. ≈ 676.2 km/hr

28. $D = [-4, \infty)$
$R = (0, \infty)$

29. $D = (-\infty, 1) \cup (1, \infty)$
$R = (-\infty, 0) \cup (0, \infty)$

34d. $D = \{0, 1, 2, 3, \ldots 15\}$
$R = \{-300, -250, -200, \ldots 450\}$

---

28. 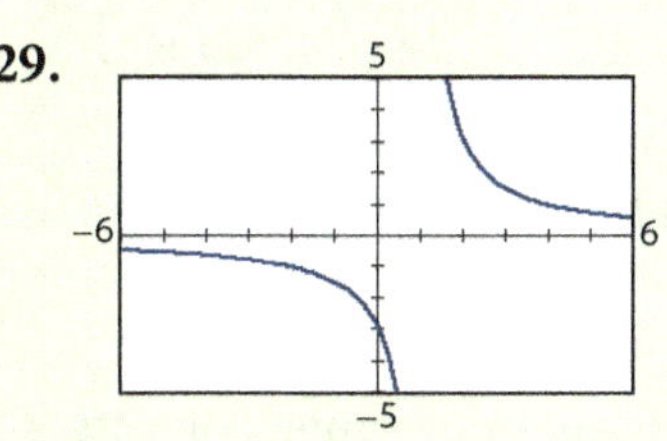

29.

30. $\{\ldots, (-1, -1), (0, 0), (2, 2), (3, 3), \ldots\}$
Each x-coordinate is paired with exactly one y-coordinate.

31. $\{\ldots, (0.25, -2), (1, -1),$ (undefined, 0), $(1, 1), (0.25, 2), \ldots\}$
At least one x-coordinate is paired with more than one y-coordinate.

32. More than one hardware store transaction can take place on any day.

33. There is only one score recorded per hole in a round of golf.

34b. $\dfrac{\$300}{15 \text{ bags}} = \$20$ per bag

34c. $\quad 0 = 50n - 300$
$\quad 50n = 300$
$\quad\quad n = 6$ bags

35b. $s = \sqrt{9.8(3600)}$
$\quad \approx 187.83 \; \dfrac{\text{m}}{\text{sec}} \left(\dfrac{3600 \text{ sec}}{1 \text{ hr}} \right) \left(\dfrac{1 \text{ km}}{1000 \text{ m}} \right)$
$\quad \approx 676.2$ km/hr

36a. $g\left(\dfrac{1}{4}\right) = \dfrac{\left(\frac{1}{4}\right)}{2} + 1 = \dfrac{1}{8} + 1 = \dfrac{9}{8}$

36b. $g(-b) = \dfrac{-b}{2} + 1 = -\dfrac{b}{2} + 1$

36c. $g(4x) = \dfrac{4x}{2} + 1 = 2x + 1$

36d. $g(6x - 6) = \dfrac{6x - 6}{2} + 1$
$\quad = 3x - 3 + 1 = 3x - 2$

37a. $\dfrac{(x + h - 2) - (x - 2)}{h}$
$\quad = \dfrac{x + h - 2 - x + 2}{h} = \dfrac{h}{h} = 1$

37b. $\dfrac{(x + h)^2 + 2(x + h) - 4 - (x^2 + 2x - 4)}{h}$
$\quad = \dfrac{x^2 + 2xh + h^2 + 2x + 2h - 4 - x^2 - 2x + 4}{h}$
$\quad = \dfrac{2xh + h^2 + 2h}{h} = 2x + h + 2$

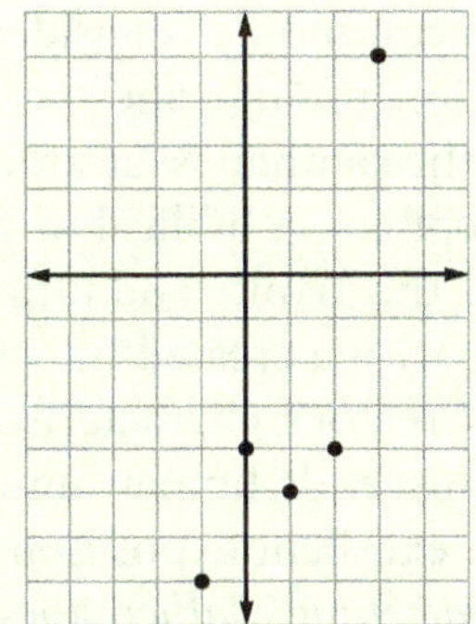

23. any set of ordered pairs, such as $\{(2, 4), (1, 3), (0, 5), (1, 9)\}$

24. any set of ordered pairs in which all the x-coordinates are unique, such as $\{(1, 3), (2, 4), (3, 3), (7, 5)\}$

25. any set of ordered pairs in which at least two of the x-coordinates are shared, such as $\{(1, 3), (2, 4), (3, 3), (1, 5)\}$

26. 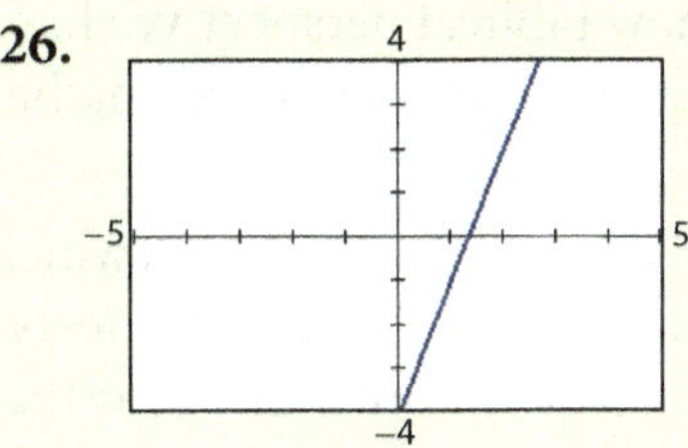

27. 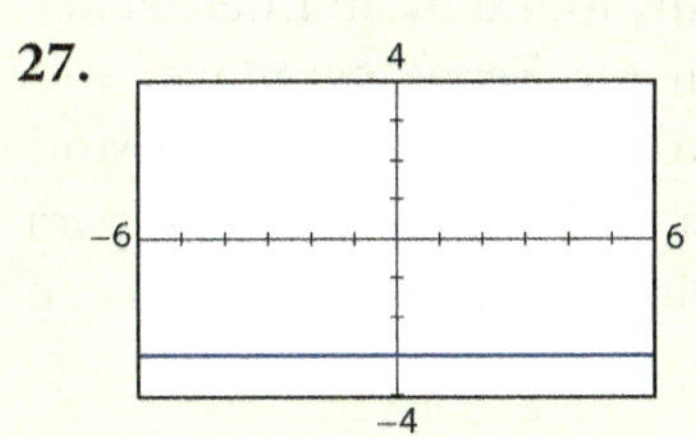

38. No; the radicand of an even root may be equal to 0. Including $x = \frac{5}{3}$ in the domain makes the denominator 0, so the domain is $\left(\frac{5}{3}, \infty\right)$.

39. $D_f = \left\{ x \mid x \neq -\frac{11}{2} \right\}$

$D_f = \left(-\infty, -\frac{11}{2}\right) \cup \left(-\frac{11}{2}, \infty\right)$

$D_g = \{x \mid x < 6\}; \ D_g = (-\infty, 6)$

Set-builder notation is more concise for $f(x)$, but interval notation is more concise for $g(x)$.

40. when a domain or range consists of separate (non-continuous or discrete) values, such as when the domain is the set of integers, $D = \{x \mid x \in \mathbb{Z}\}$

> **Cumulative Review**

43–46. $\mathbb{Z}$ = integers
$\mathbb{N}$ = natural (counting) numbers
$\mathbb{W}$ = whole numbers
$\mathbb{Q}$ = rational numbers
$\mathbb{Q}'$ = irrational numbers
$\mathbb{R}$ = real numbers

48. $A \cap B = \{2, 4\}$
$(A \cap B)' = \{0, 1, 3, 5, 6, 7, 8, 9\}$

49. $U \subset \mathbb{W}$

50. $A \cup B = \{1, 2, 3, 4, 5, 6, 7, 8\} \neq U$

> **C. Exercises**

36. If $g(x) = \frac{x}{2} + 1$, find the following.
  a. $g\left(\frac{1}{4}\right)$ b. $g(-b)$
  c. $g(4x)$ d. $g(6x - 6)$

37. Find $\dfrac{f(x + h) - f(x)}{h}$ for each function.
  a. $f(x) = x - 2$ b. $f(x) = x^2 + 2x - 4$

38. Explain: Is Stacy correct in claiming that the domain of $f(x) = \dfrac{1}{\sqrt{3x - 5}}$ is $\left[\frac{5}{3}, \infty\right)$? Explain why or why not.

39. Analyze: State the domain of $f(x) = \dfrac{1}{2x + 11}$ and $g(x) = \sqrt{6 - x}$ in both set-builder notation and interval notation. Describe which notation you prefer for each function and explain why.

40. Discuss: Describe a case in which interval notation would not be appropriate for stating a domain or range.

CUMULATIVE REVIEW

List any of the following subsets of the real numbers to which each number belongs: $\mathbb{Z}$, $\mathbb{N}$, $\mathbb{W}$, $\mathbb{Q}$, $\mathbb{Q}'$, $\mathbb{R}$. [Appendix 1]

41. $1.191191119\ldots$ $\mathbb{Q}', \mathbb{R}$ **42.** π $\mathbb{Q}', \mathbb{R}$

Choose which of the following subsets is the most specific subset of the real numbers to which each number belongs: $\mathbb{Z}$, $\mathbb{N}$, $\mathbb{W}$, $\mathbb{Q}$, $\mathbb{Q}'$, $\mathbb{R}$. [Appendix 1]

43. -8 $\mathbb{Z}$ **44.** 0 $\mathbb{W}$ **45.** $\sqrt{6}$ $\mathbb{Q}'$ **46.** $\frac{2}{9}$ $\mathbb{Q}$

Use the Venn diagram to complete exercises 47–50. [Appendix 1]

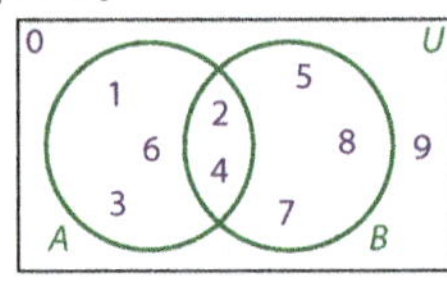

47. Which set represents the universal set? C
  A. $\{0, 9\}$ D. $\{x \mid x \leq 9, x \in \mathbb{Z}\}$
  B. $\{x \mid 0 \leq x \leq 9\}$ E. none of these
  C. $\{x \mid x \leq 9, x \in \mathbb{W}\}$

48. Find $(A \cap B)'$. E
  A. U D. $\{1, 3, 5, 6, 7, 8\}$
  B. $\{0, 9\}$ E. $\{0, 1, 3, 5, 6, 7, 8, 9\}$
  C. $\{2, 4\}$

49. Which of the following statements is true? D
  I. $A \subset U$ II. $U \in \mathbb{W}$ III. $7 \in A'$
  A. I only   D. I and III only
  B. II only   E. II and III only
  C. I and II only

50. Which of the following statements is true? E
  I. $A \cup B = U$ II. $A \cup B \subset U$ III. $A' \subset U$
  A. I only   D. I and III only
  B. II only   E. II and III only
  C. III only

PRESENTATION

The Historical Connection features provide interesting insights on the history of math and how biblical perspectives have influenced the development of mathematics.

None of Zeno's paradoxes (of which there are thought to be around forty) survived in his own words. The four that are commonly found by an Internet keyword search for *Zeno's paradoxes* were attributed to Zeno by Aristotle. Aristotle seems never to have resolved the paradoxes, but simply dismissed them as unimportant.

Though not commonly considered today, the question of "the one" versus "the many" was the seminal issue argued by Greek philosophers. A biblical worldview that accepts the Trinity and infinity has no problems with a created universe that reflects its Creator's characteristics, even when it transcends human understanding. For an excellent exposition on this topic, see *Redeeming Mathematics* by Vern Poythress.

Scholastics such as Oresme were the academics of medieval times. Scholasticism emphasized dialectical reasoning and inference.

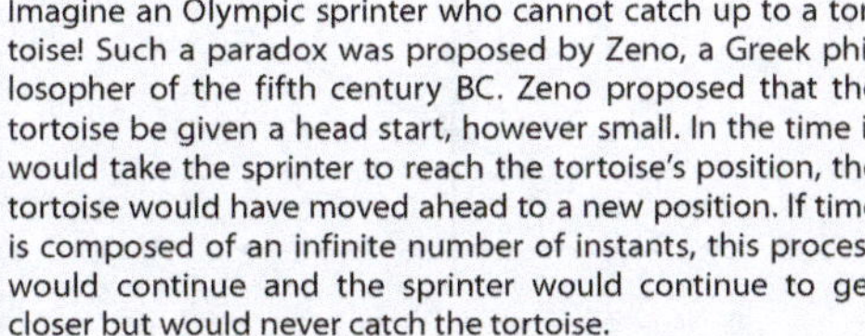

FUNCTIONS

Imagine an Olympic sprinter who cannot catch up to a tortoise! Such a paradox was proposed by Zeno, a Greek philosopher of the fifth century BC. Zeno proposed that the tortoise be given a head start, however small. In the time it would take the sprinter to reach the tortoise's position, the tortoise would have moved ahead to a new position. If time is composed of an infinite number of instants, this process would continue and the sprinter would continue to get closer but would never catch the tortoise.

Zeno was attempting to show that a belief in infinity leads to absurdity. His famous paradoxes were written in support of Parmenides's view that reality consists of "the one" as opposed to "the many," a perspective that rejected the ideas of infinity and motion. Ancient Greek philosophers deified man's reasoning ability and rejected anything they could not fully understand since it would transcend their god. To Parmenides, any motion was simply an illusion. However, a

biblical worldview accepts the idea that God is three persons in one (triune) and recognizes that His creation exhibits characteristics of both "the one" and "the many."

The ancient Greeks demonstrated magnificent reasoning in their development of geometry, but because of their worldview, they did not develop the idea of a function. Their geometry was static, exploring unchanging shapes. If motion was considered at all, it was circular motion that had a beginning and an end, never linear motion that was infinite. Since functions were developed primarily through the study of motion, this necessary prerequisite of calculus awaited a worldview that was not afraid of infinity.

A biblical worldview recognizes an infinite Creator, a finite creation, and a triune God. The Western world's acceptance of these ideas explains, at least in part, the development of the idea of a function in the Middle Ages, when according to math historian Howard Eves, "the meditations of scholastic philosophers led to subtle theorizing on motion, infinity, and the continuum, all of which are fundamental concepts in modern mathematics . . . [and] may, to some extent, account for the remarkable transformation from ancient to modern mathematical thinking."

Nicole Oresme, a fourteenth-century French scholastic, studied motion and heat transfer by plotting functions on a rectangular coordinate system centuries before Descartes. Scholastics like Oresme were the academics of the Middle Ages. They generally held a biblical worldview and expected natural processes to be orderly and to follow observable laws laid down by a sovereign Creator.

COMPREHENSION CHECK

1. Name the Greek philosopher whose paradoxes questioned the idea of infinity.

2. Describe in your own words the paradox of the sprinter and the tortoise. Why would this paradox be viewed as an argument against the idea of infinity?

3. In what century did Oresme plot functions?

4. What was the predominant worldview of Western cultures during the Middle Ages?

5. Why were scholastics more willing than Greek philosophers to explore changing natural processes involving concepts of motion and infinity?

Historical Connection

Objectives

1. To explain the reasoning used in the paradox of the sprinter and the tortoise

2. To describe the predominant worldview of Western cultures in the Middle Ages

3. To explain why the scholastics of the Middle Ages were more willing than Greek philosophers to explore infinity

Solutions

1. Zeno

2. sample answer: In the time it takes the sprinter to reach the place where the tortoise was, the tortoise would have moved farther ahead. If time is composed of an unlimited number of instants (which may be infinitely small), then the sprinter can never reach the tortoise. Since we know a sprinter can catch a tortoise, there must not be an infinite number of instants.

3. the fourteenth century

4. biblical

5. Scholastics were more willing to explore ideas they could not fully explain (e.g., infinity, the one and the many) because of their acceptance of biblical teachings regarding an infinite, triune, sovereign Creator.

Many of Oresme's mathematical contributions resulted from his interest in music. He proved that the infinite harmonic series, $1 + \frac{1}{2} + \frac{1}{3} + \frac{1}{4} + \cdots + \frac{1}{n}$, is divergent by starting with the third term and grouping 2, 4, 8, 16 . . . terms, each having a sum greater than $\frac{1}{2}$.

$$1 + \frac{1}{2} + \left(\frac{1}{3} + \frac{1}{4}\right) + \left(\frac{1}{5} + \frac{1}{6} + \frac{1}{7} + \frac{1}{8}\right) + \cdots$$

This indicates that the harmonic series is greater than $1 + \frac{1}{2} + \frac{1}{2} + \frac{1}{2} + \cdots$, and since an infinite sum of terms equal to or greater than $\frac{1}{2}$ would diverge, the harmonic series diverges.

Alternatively, note that the sum of the third and fourth terms must be greater than $2\left(\frac{1}{4}\right)$, the sum of the fifth through eighth terms must be greater than $4\left(\frac{1}{8}\right)$, the sum of the ninth through sixteenth terms must be greater than $8\left(\frac{1}{16}\right)$, etc.

Some students may wrestle with the concept of infinity, and specifically the concept of God's infinite existence. How can anything be infinite? It is a scientific fact that you cannot form something from nothing. Therefore, either matter has always existed (is infinite), or it was created. If matter was created, then God existed before matter and is indeed infinite. Either matter is infinite, or God

is infinite. The Bible declares that God is infinite (Ps. 90:2) and that matter was created by Him (Gen. 1:1). Humans have always struggled with the concept of infinity. When Job wrestled with these kinds of truths, God responded by asking where Job was when He laid the foundations of the earth, reminding Job that created beings cannot understand everything about God's creation (Job 38).

1.2 Linear Functions

Objectives

1. To graph linear equations
2. To find intercepts graphically and algebraically
3. To write an equation of a given line
4. To apply the distance formula and the midpoint formula

Flash

Bungee jumping is an extreme sport originally inspired by land divers on Pentecost Island, Vanuatu. These men would tie tree vines around their ankles and jump from towers as high as 30 m tall. The first documented bungee jump took place on April 1, 1979, on the Clifton Suspension Bridge in Bristol, England. Since then, jump sites have included mobile cranes, hot air balloons, and dams.

Vocabulary

acceleration due to gravity
constant function
general linear form
point-slope form
slope
slope-intercept form

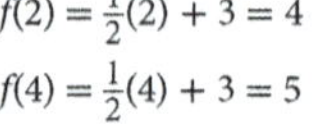
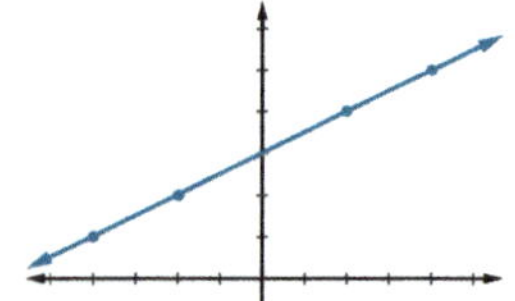

1.2 Linear Functions

The vertical velocity of an object in free fall is modeled by a linear equation in Example 7.

After completing this section, you will be able to

- graph linear equations.
- find intercepts.
- write an equation for a line.
- apply the distance and midpoint formulas.

The graph of a linear function, such as $f(x) = \frac{1}{2}x + 3$, can be constructed by plotting several ordered pairs from the function, noticing that the points lie on a line, and then drawing the line through the points to illustrate all the ordered pairs that define the function.

$$f(-4) = \tfrac{1}{2}(-4) + 3 = 1$$
$$f(-2) = \tfrac{1}{2}(-2) + 3 = 2$$
$$f(2) = \tfrac{1}{2}(2) + 3 = 4$$
$$f(4) = \tfrac{1}{2}(4) + 3 = 5$$

Since any real number could have been chosen for x, the domain of $f(x)$ is the real numbers. The range of the function is also the real numbers since the line continues infinitely.

Remembering key characteristics of linear functions will help you quickly graph and describe these functions. Recall that the slope of a line describes its incline or steepness.

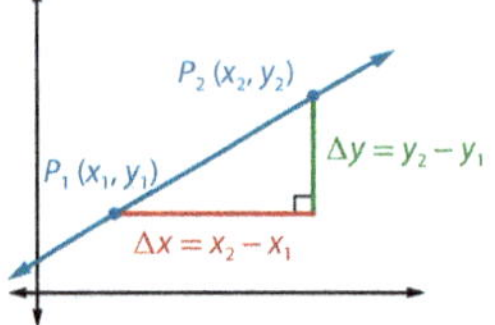

DEFINITION

The **slope** of a line, m, is the ratio of its *rise* (vertical change) to the corresponding *run* (horizontal change).

$$m = \frac{rise}{run} = \frac{\Delta y}{\Delta x} = \frac{y_2 - y_1}{x_2 - x_1}$$

When Nicole Oresme plotted data from heat transfer experiments in the Middle Ages, he recognized that a constant slope indicated uniform variance in the tested material.

Example 1   Finding the Slope of a Line

Find the slope of the line passing through $(-2, 2)$ and $(4, 5)$.

Answer

$$m = \frac{y_2 - y_1}{x_2 - x_1} = \frac{5 - 2}{4 - (-2)} = \frac{3}{6} = \frac{1}{2}$$

1. Substitute and simplify.

2. Confirm graphically.

$$m = \frac{rise}{run} = \frac{\Delta y}{\Delta x} = \frac{3}{6} = \frac{1}{2}$$

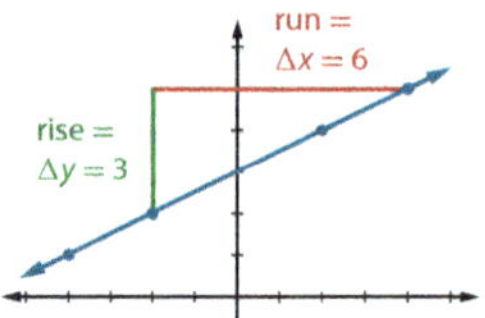

SKILL ✔ **EXERCISE 3**

PRESENTATION

Lesson Opener

Write each expression in interval notation.

1. $-2 < x \le 7$ $(-2, 7]$
2. $0 \le x \le 20$ $[0, 20]$

Solve for y.

3. $3x + 2y = 12$ $y = -\frac{3}{2}x + 6$
4. $x - 4y = -10$ $y = \frac{1}{4}x + \frac{5}{2}$

While most students should be familiar with linear functions, many will benefit from the review in this section. Remind them that the graph of any function can be constructed by plotting and connecting points as illustrated.

Present the definition and formula for the *slope* of a line. Use Example 1 to find the slope of a line algebraically and graphically. Then guide the students through Example 2, graphing a linear function using the y-intercept and slope.

Familiarity with the *general linear form* is valuable. Emphasize that the slope and both intercepts can be determined directly without transforming the equation. Derive these formulas and use Example 3 to show how being able to quickly state these values directly from the general form saves time.

Consider discussing Descartes's secular perspective presented in the margin note on page 11. When we derive formulas such as those for intercepts and slopes of lines, it is easy to see how Descartes would suggest that mathematical truths come from the human mind. This flawed reasoning is discussed in the Biblical Perspective of Mathematics feature following this section. Interestingly, Descartes's second conclusion after determining that someone exists because he thinks (*cogito ergo sum*) is that God exists. Unfortunately, Descartes concluded that the revealed truths that lead to heaven are beyond our understanding.

Notice that the line in Example 1 is the linear function $f(x) = \frac{1}{2}x + 3$ where the coefficient of x is the line's slope, $\frac{1}{2}$, and the constant, 3, is the y-value of its y-intercept, the point where the line intersects the y-axis. The function notation for a linear function with slope m and y-intercept $(0, b)$ is called the *slope-intercept form* of a line.

$$f(x) = mx + b \text{ or } y = mx + b$$

Example 2   Graphing a Linear Function

Graph $f(x) = -3x + 2$. Then state its domain and range.

Answer

$m = -3$ and $b = 2$

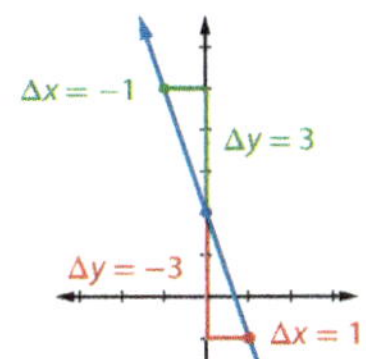

$D = \mathbb{R}; R = \mathbb{R}$

1. Compare $y = -3x + 2$ to $y = mx + b$.
2. Plot the y-intercept $(0, b) = (0, 2)$.
3. Use the y-intercept and $m = \frac{\Delta y}{\Delta x} = \frac{-3}{1} = \frac{3}{-1}$ to graph at least one more point, $(1, -1)$ or $(-1, 5)$, and draw the line through these points.
4. The graph shows that all real numbers are included in both the domain and range.

SKILL ✔ EXERCISE 7

Every line in a plane, including vertical lines, can be written in *general linear form*: $Ax + By = C$ where A is not negative and the coefficients A and B and the constant C are expressed as integers if possible.

A line's x-intercept, the point at which the line crosses the x-axis, can be determined by letting $y = 0$ and solving for x.

$$Ax + B(0) = C$$
$$x = \frac{C}{A}$$

$\therefore$ the x-intercept is $\left(\frac{C}{A}, 0\right)$.

The slope and y-intercept of a line in general form can be determined by converting the equation to slope-intercept form.

$$Ax + By = C$$
$$By = -Ax + C$$
$$y = -\frac{A}{B}x + \frac{C}{B}$$

$\therefore$ the y-intercept is $\left(0, \frac{C}{B}\right)$ and the slope $m = -\frac{A}{B}$.

Note that the y-intercept could also be found by letting $x = 0$ and solving for y.

In *Discourse on the Method of Rightly Conducting the Reason, and Seeking Truth in the Sciences*, the French philosopher and mathematician René Descartes proposed that mathematical truths are established by human reasoning.

Additional Exercises

1. Find the slope of the line passing through $(-1, 6)$ and $(5, -8)$. $m = -\frac{7}{3}$

2. Graph $y = -\frac{3}{2}x + 4$ and state its domain and range.

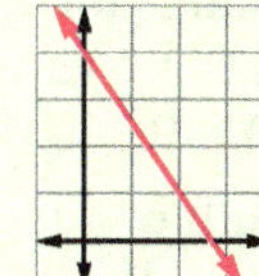

$D = \mathbb{R}; R = \mathbb{R}$

3. Graph $2x - y = 4$ using the x- and y-intercepts and then find the slope of the line.

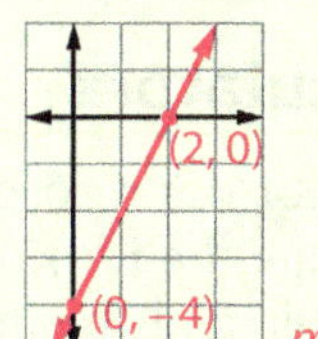

$m = 2$

Write the equation of each described line.

4. passing through $(1, 6)$ and $(-5, -3)$
$y = \frac{3}{2}x + \frac{9}{2}$ or $3x - 2y = -9$

5. passing through $(2, 3)$ and perpendicular to $2x - y = 5$
$y = -\frac{1}{2}x + 4$ or $x + 2y = 8$

6. passing through $(-4, -3)$ and parallel to $3x + 4y = 2$
$y = -\frac{3}{4}x - 6$ or $3x + 4y = -24$

Introduce the *point-slope form* of a linear equation, showing how it is simply a transformation of the slope formula. Explain that it can be used to find the equation of a line when a point and its slope are known. In Example 4, first the slope is determined from two points. Then the slope can be used with either of the points to determine the equation.

Discuss the characteristics of *constant functions* (horizontal lines) and vertical lines, which are not functions.

One-on-One Equations for horizontal and vertical lines may confuse students due to the absence of a variable. Explaining that the coefficient of the missing variable is 0 can help students see the characteristic values related to these equations. For example, a slope of 0 can be easily seen by expressing $y = 3$ as $y = 0x + 3$.

Review the relationships between slopes of parallel and perpendicular lines and illustrate their application with Example 5. Then present the distance and midpoint formulas and apply them using Example 6.

Real-life applications related to projectile motion will be encountered throughout the chapter. Example 7 presents the linear model of the vertical velocity of a projectile.

The various forms of linear equations and related formulas can be reinforced with the summary table.

Find the distance between each pair of points. Then find the midpoint of the segment connecting them.

7. $(5, -4)$ and $(-1, -6)$ $2\sqrt{10}; (2, -5)$

8. $(-2, 6)$ and $(5, -1)$ $7\sqrt{2}; \left(\frac{3}{2}, \frac{5}{2}\right)$

9. Find the length of the radius for a circle that is centered at $(-3, 7)$ and contains the point $(5, 1)$.   10

10. Find the center and length of a radius for the circle whose diameter has endpoints $(2, 1)$ and $(-8, 8)$.

$C\left(-3, \frac{9}{2}\right); r = \frac{\sqrt{149}}{2} \approx 6.1$

For Graphing Calculators

If your students have not done so in a previous course, challenge them to program the distance formula and the midpoint formula into their calculators. Use the Internet keyword search *distance formula program TI-84* to view sample programs if needed.

Assignments

- **Minimum:** 1–6, 8–9, 12–13, 15–16; 19–25 odd; 28–29, 33–34, 39, 47–49, 53

- **Standard:** 2–4, 7–9, 12, 15, 18–19, 22–23, 27, 30, 33–34, 36, 38–40, 43, 49–50, 54

- **Extended:** 4, 7–8, 11, 15, 18–19, 21, 24, 28, 32, 34–46, 50–51, 54, 56

Assessment

- Quiz 1A covers Sections 1.1–1.2.

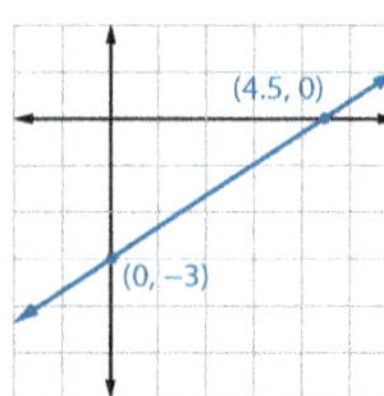

Many real-world functions can be modeled using linear equations. Even the rate of change for a non-linear function at a specific point is calculated using the slope of the tangent to the curve at that point.

Find the intercepts and slope of $2x - 3y = 9$. Then graph the equation.

Answer

x-intercept: $\left(\frac{9}{2}, 0\right) = (4.5, 0)$ 1. Use $\left(\frac{C}{A}, 0\right)$ to find the x-intercept.

y-intercept: $\left(0, \frac{9}{-3}\right) = (0, -3)$ 2. Use $\left(0, \frac{C}{B}\right)$ to find the y-intercept.

$m = -\frac{2}{(-3)} = \frac{2}{3}$ 3. Use $m = -\frac{A}{B}$ to find the slope.

4. Plot the intercepts and graph the line. Start at the y-intercept and use $m = \frac{2}{3}$ to confirm that the point 2 units up and 3 units to the right, $(3, -1)$, is also on the line.

SKILL ✔ EXERCISE 29

The *point-slope form* of a line through a particular point (x_1, y_1) with slope m can be derived by letting (x, y) represent any other point on the line and applying the slope formula.

$$m = \frac{y - y_1}{x - x_1}; \therefore y - y_1 = m(x - x_1)$$

Write the slope-intercept form equation of the line passing through $(1, 6)$ and $(3, -1)$.

Answer

$m = \frac{-1 - 6}{3 - 1} = -\frac{7}{2}$ 1. Find the slope.

$y - y_1 = m(x - x_1)$ 2. Apply the point-slope form.

$y - 6 = -\frac{7}{2}(x - 1)$

$y - 6 = -\frac{7}{2}x + \frac{7}{2}$

$y = -\frac{7}{2}x + \frac{19}{2}$

SKILL ✔ EXERCISE 19

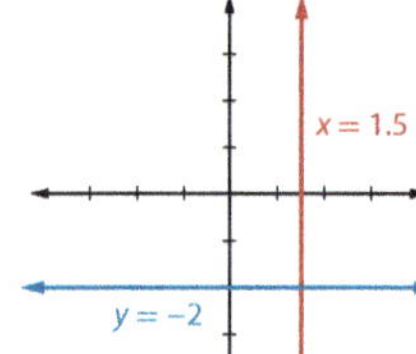

Since every point of the graph of $y = -2$ has a y-coordinate of -2, the graph is a horizontal line with a slope of $m = \frac{0}{run} = 0$. Similar linear functions of the form $f(x) = 0x + k$ or $y = k$ are called *constant functions* and their graphs are horizontal lines with a y-intercept of $(0, k)$.

Likewise, every point of the graph of $x = 1.5$ has an x-coordinate of 1.5, so the graph is a vertical line with a slope of $m = \frac{rise}{0}$, which is undefined. Similar equations of the form $x = k$ are vertical lines with an x-intercept of $(k, 0)$. These equations do not represent functions and cannot be written in function notation or in slope-intercept form.

In previous classes you have seen the relationship between the slopes of parallel and perpendicular lines.

1. Two non-vertical lines are parallel if and only if they have equal slopes: $m_1 = m_2$.

2. Two non-vertical lines are perpendicular if and only if the product of their slopes is -1: $m_1 \cdot m_2 = -1$. This implies the slopes are negative reciprocals of each other: $m_1 = -\frac{1}{m_2}$.

Do not confuse the zero slope of horizontal lines with the undefined slope of vertical lines.

Example 5 Finding Equations of Parallel and Perpendicular Lines

Write the equation of each line in slope-intercept form.

a. the line parallel to $2x + 3y = 6$ and through $(-2, -1)$

b. the line perpendicular to $y = \frac{3}{4}x - 5$ and through $(3, -2)$

Answer

a. $m_G = -\frac{A}{B} = -\frac{2}{3}$ — 1. Find the slope of the given line.

$m_\parallel = m_G = -\frac{2}{3}$ — 2. Parallel lines have the same slope.

$y - (-1) = -\frac{2}{3}[x - (-2)]$ — 3. Substitute the slope and the known coordinates into $y - y_1 = m(x - x_1)$ and convert to slope-intercept form.

$y + 1 = -\frac{2}{3}x - \frac{4}{3}$

$y = -\frac{2}{3}x - \frac{7}{3}$

b. $m_\perp = -\frac{1}{m_G} = -\frac{1}{\left(\frac{3}{4}\right)} = -\frac{4}{3}$ — 1. Find the negative reciprocal of the slope of the given line.

$-2 = -\frac{4}{3}(3) + b$ — 2. Find the y-intercept by substituting the slope and the known coordinates into $y = mx + b$ and solving for b.

$-2 = -4 + b$

$2 = b$

$y = -\frac{4}{3}x + 2$ — 3. Use the slope and y-intercept to write the equation.

SKILL ✓ **EXERCISES 21, 23**

Review the following formulas for the distance between two points and the midpoint of a segment in the coordinate plane. The distance formula is derived using the Pythagorean Theorem and the midpoint's coordinates are averages of the endpoints' coordinates.

Distance Formula	Midpoint Formula
The distance, d, between $P_1\,(x_1, y_1)$ and $P_2\,(x_2, y_2)$ is $d = \sqrt{(x_2 - x_1)^2 + (y_2 - y_1)^2}.$	The midpoint of the segment connecting $P_1\,(x_1, y_1)$ and $P_2\,(x_2, y_2)$ is $M\left(\dfrac{x_1 + x_2}{2},\ \dfrac{y_1 + y_2}{2}\right).$

Solutions

▶ A. Exercises

5. 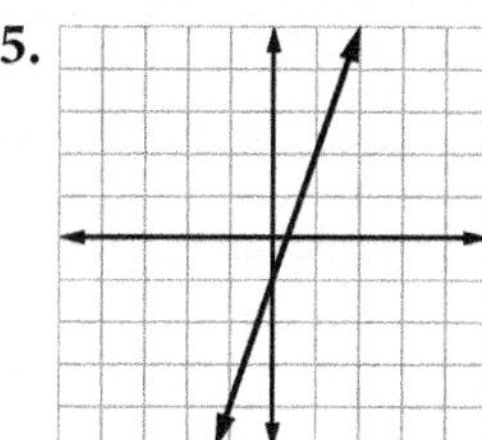

6. 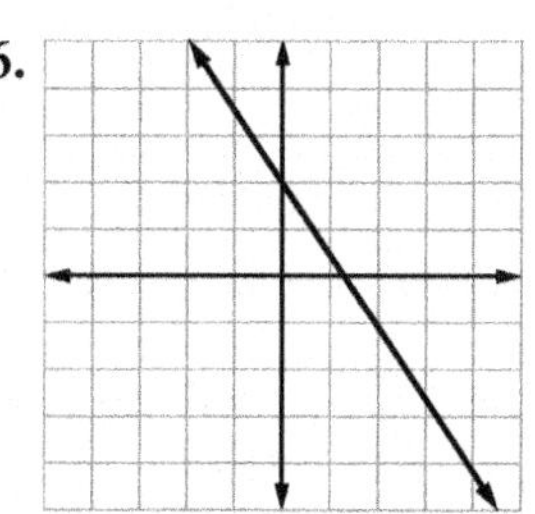

7. 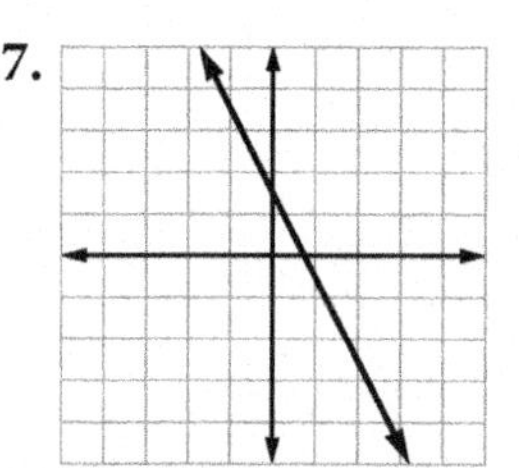

8. $-4y = -x + 12$

$y = \frac{1}{4}x - 3$

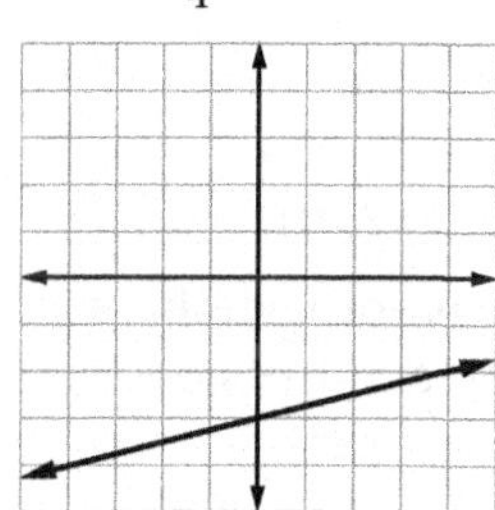

9. 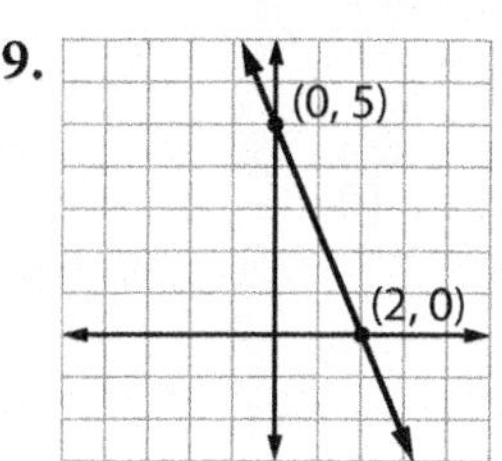

10. 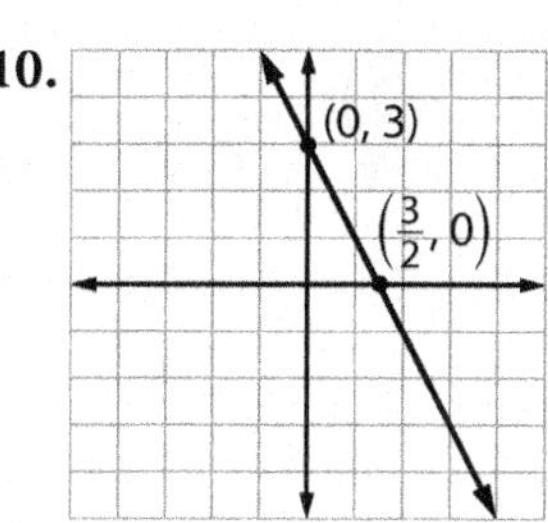

11. 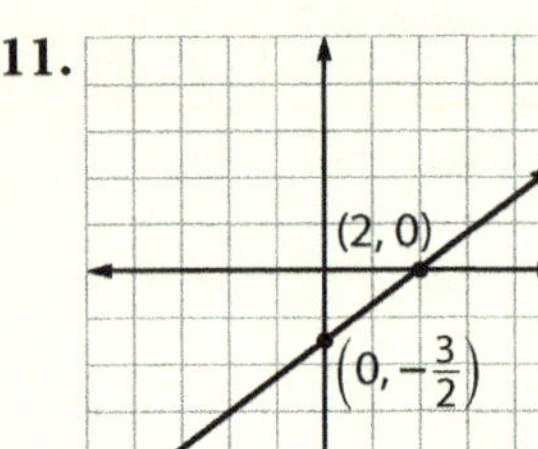

12. 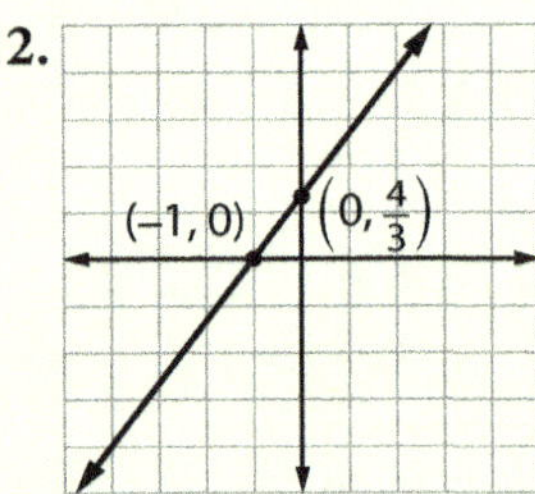

13. $\left(\frac{C}{A}, 0\right) = \left(\frac{2}{1}, 0\right) = (2, 0)$

$\left(0, \frac{C}{B}\right) = \left(0, \frac{2}{2}\right) = (0, 1)$

$m = -\frac{A}{B} = -\frac{1}{2}$

14. $5x + 2(0) = 8$

$5x = 8$

$x = \frac{8}{5}$; x-int.: $\left(\frac{8}{5}, 0\right)$

$5(0) + 2y = 8$

$2y = 8$

$y = 4$; y-int.: $(0, 4)$

$m = -\frac{A}{B} = -\frac{5}{2}$

15. $2x = -10$

$x = -5$; x-int.: $(-5, 0)$

Since the line is vertical, there is no y-intercept, and the slope is undefined.

16. Since the line is horizontal, there is no x-intercept.

$3y = 15$

$y = \frac{15}{3} = 5$; y-int.: $(0, 5)$

$m = 0$

17. $(0) = -\frac{2}{3}x + 1$

$-1 = -\frac{2}{3}x$

$x = \frac{3}{2}$; x-int.: $\left(\frac{3}{2}, 0\right)$

y-int.: $(0, 1)$

$m = -\frac{2}{3}$

18. $(0) = \frac{2}{5}x - \frac{1}{3}$

$\frac{1}{3} = \frac{2}{5}x$

$x = \frac{5}{2}\left(\frac{1}{3}\right) = \frac{5}{6}$; x-int.: $\left(\frac{5}{6}, 0\right)$

y-int.: $\left(0, -\frac{1}{3}\right)$

$m = \frac{2}{5}$

Example 6 Applying the Distance and Midpoint Formulas

Find the length and midpoint of $\overline{AB}$ with A (−3, 7) and B (2, −4).

Answer

$AB = \sqrt{(2 - (-3))^2 + (-4 - 7)^2} = \sqrt{5^2 + (-11)^2} = \sqrt{146}$

$M\left(\frac{-3 + 2}{2}, \frac{7 + (-4)}{2}\right) = \left(-\frac{1}{2}, \frac{3}{2}\right)$

SKILL ✔ EXERCISE 23

The vertical velocity of a projectile (with negligible air resistance) can be modeled with the following linear function of the time it has been in the air: $v(t) = -gt + v_0$. Vertical velocities and the initial vertical velocity, v_0, are positive when the projectile is traveling up from the earth and negative when the projectile is traveling down toward the earth. The magnitude of g, the *acceleration due to gravity*, is about 32 ft/sec^2 or 9.8 m/sec^2.

Example 7 Modeling Vertical Velocity

Use the linear function to model the vertical velocity of a firework shot upward with an initial vertical velocity of 144 ft/sec and determine how long it takes the firework to reach its highest point.

Answer

$v(t) = -32t + 144$ 1. Substitute $g = 32$ ft/sec^2 and $v_0 = 144$ ft/sec into $v(t) = -gt + v_0$.

$0 = -32t + 144$ 2. At its peak, the firework has a vertical velocity of 0.

$32t = 144$ 3. Solve for t.

$t = 4.5$ sec

SKILL ✔ EXERCISE 33

Review the following summary of equations before completing the exercises.

Forms of Linear Equations		Slope Formulas	
$Ax + By = C$	general linear	$m = \frac{y_2 - y_1}{x_2 - x_1}$	definition
$y - y_1 = m(x - x_1)$	point-slope	$m_2 = m_1$	parallel lines
$y = mx + b$	slope-intercept	$m_2 = -\frac{1}{m_1}$	perpendicular lines

> **B. Exercises**

19. $m = \frac{-4 - 4}{-9 - 3} = \frac{-8}{-12} = \frac{2}{3}$

$y - 4 = \frac{2}{3}(x - 3)$

$y - 4 = \frac{2}{3}x - 2$

$y = \frac{2}{3}x + 2$

20. $m = \frac{-9 - 2}{6 - (-5)} = \frac{-11}{11} = -1$

$y - 2 = -1(x - (-5))$

$y - 2 = -x - 5$

$y = -x - 3$

21. $m_{\parallel} = m_G = \frac{4}{3}$

$-3 = \frac{4}{3}(2) + b$

$-3 = \frac{8}{3} + b$

$b = -\frac{17}{3}$; $y = \frac{4}{3}x - \frac{17}{3}$

22. $m_{\parallel} = m_G = -\frac{A}{B} = -\frac{3}{-1} = 3$

$y - 2 = 3(x - (-10))$

$y - 2 = 3x + 30$

$y = 3x + 32$

23. $m_G = \frac{-4 - 3}{-7 - 9} = \frac{-7}{-16} = \frac{7}{16}$

$m_{\perp} = -\frac{16}{7}$; y-int.: $(0, 5)$

$y = -\frac{16}{7}x + 5$

A. Exercises

1. The graph of a function in the form $f(x) = ax + b$ ($a, b \in \mathbb{R}$) is always a ___line___.

2. State the domain and range of any non-constant linear function. $D = \mathbb{R}; R = \mathbb{R}$

State the slope of each line.

3. 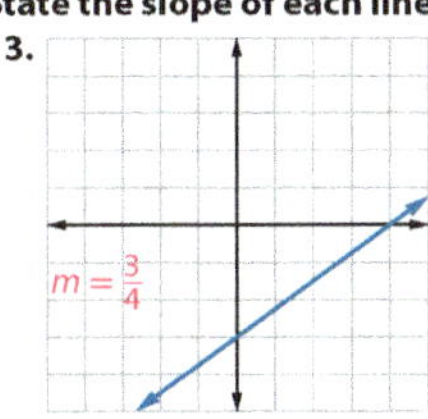
   $m = \frac{3}{4}$

4. 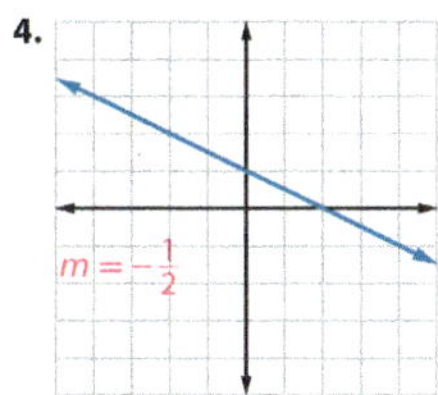
   $m = -\frac{1}{2}$

Graph each linear equation using its slope and y-intercept.

5. $y = 3x - 1$
6. $y = -\frac{3}{2}x + 2$
7. $y = -2x + \frac{3}{2}$
8. $x - 4y = 12$

Graph each linear equation using its x- and y-intercepts.

9. $5x + 2y = 10$
10. $6x + 3y = 9$
11. $3x - 4y = 6$
12. $4x - 3y = -4$

State the x-intercept, y-intercept, and slope of each line.

13. $x + 2y = 2$
14. $5x + 2y = 8$
15. $2x + 3 = -7$
16. $3y - 4 = 11$
17. $y = -\frac{2}{3}x + 1$
18. $y = \frac{2}{5}x - \frac{1}{3}$

B. Exercises

Write the slope-intercept form equation of each described line.   21. $y = \frac{4}{3}x - \frac{17}{3}$

19. passing through $(3, 4)$ and $(-9, -4)$ $y = \frac{2}{3}x + 2$
20. passing through $(-5, 2)$ and $(6, -9)$ $y = -x - 3$
21. passing through $(2, -3)$ and parallel to $y = \frac{4}{3}x - 1$
22. passing through $(-10, 2)$ and parallel to $3x - y = 12$ $y = 3x + 32$
23. passing through $(0, 5)$ and perpendicular to the line containing $(9, 3)$ and $(-7, -4)$ $y = -\frac{16}{7}x + 5$
24. passing through $(6, -2)$ and perpendicular to $x + 2y = 7$ $y = 2x - 14$

Find the distance between each pair of points. Then find the midpoint of the connecting segment.

25. $(5, 6)$ and $(-4, -6)$
26. $(5, 8)$ and $(2, 10)$
27. $(-11, 3)$ and $(2, 19)$
28. $(-3, 7)$ and $(2, -4)$

29. Find the length of a radius of the circle that is centered at $(2, -3)$ and contains the point $(-10, 2)$.

25. $15; \left(\frac{1}{2}, 0\right)$ 26. $\sqrt{13}; \left(\frac{7}{2}, 9\right)$
27. $5\sqrt{17}; \left(-\frac{9}{2}, 11\right)$ 28. $\sqrt{146}; \left(-\frac{1}{2}, \frac{3}{2}\right)$

30. Find the length of a diameter of the circle that is centered at $(-3, 1)$ and contains the point $(4, -6)$.

31. Find the center and length of a radius of the circle whose diameter has endpoints $(-3, 2)$ and $(1, 5)$.

32. Find the center and length of a radius of the circle whose diameter has endpoints $(1, 4)$ and $(6, -1)$.

In exercises 33–34 model the vertical velocity of a projectile using the linear function $v(t) = -gt + v_0$ where g, the magnitude of the acceleration due to gravity, is 32 ft/sec² and v_0 is the initial vertical velocity.

33. Determine the velocity at which a watermelon strikes the ground if it takes 2.5 sec to fall when John drops it from a cliff. Why is the final velocity negative? −80 ft/sec; The watermelon is traveling down.

34. John launches a watermelon from a cliff with an initial vertical velocity of 44 ft/sec.

    a. How long will it take the watermelon to reach its highest point? 1.375 sec

    b. If the time it takes to fall from its highest point back to the height of the cliff is the same as the time it took to reach its highest point, at what velocity does the watermelon pass the cliff on its way down to the ground? −44 ft/sec

    c. Determine the final vertical velocity if it takes the watermelon 4.2 sec from its initial launch to strike the ground. −90.4 ft/sec

35. The cost of a ride-sharing service is a function of the distance traveled and the type of car. Write a linear function modeling the cost and find the cost for a 5 mi ride.

    a. ShareCarX has a $2.00 base fare with a $3.00/mi charge. $f(x) = 3x + 2$; $17.00

    b. ShareCarXL has a $3.00 base fare with a $3.25/mi charge. $f(x) = 3.25x + 3$; $19.25

    c. ShareCarSelect has a $5.00 base fare with a charge of $3.50/mi. $f(x) = 3.5x + 5$; $22.50

    d. ShareCarPremium has an $8.00 base fare with a $4.00/mi. $f(x) = 4x + 8$; $28.00

36. **Write:** State the equation of a line that is not a function and describe this type of line.

37. **Write:** State the equation of a line whose range is not the set of real numbers and describe this type of line.

38. **Explain:** Why is a linear function in the form $y = k$ called a *constant function*?

29. $r = 13$ 30. $d = 14\sqrt{2}$
31. $C\left(-1, \frac{7}{2}\right); r = 2.5$ 32. $C\left(\frac{7}{2}, \frac{3}{2}\right); r = \frac{5\sqrt{2}}{2}$

---

31. $C\left(\frac{-3+1}{2}, \frac{2+5}{2}\right) = \left(-1, \frac{7}{2}\right)$
$$d = \sqrt{(1+3)^2 + (5-2)^2}$$
$$= \sqrt{25} = 5; r = \frac{d}{2} = \frac{5}{2}$$

32. $C\left(\frac{1+6}{2}, \frac{4+(-1)}{2}\right) = \left(\frac{7}{2}, \frac{3}{2}\right)$
$$d = \sqrt{(6-1)^2 + (-1-4)^2}$$
$$= \sqrt{50} = 5\sqrt{2}; r = \frac{d}{2} = \frac{5\sqrt{2}}{2}$$

33. $v(2.5) = -32(2.5) + 0 = -80$ ft/sec

34a. $v(t) = 0 = -32t + 44$
$$t = \frac{44}{32} = 1.375 \text{ sec}$$

34b. $v(2(1.375)) = -32(2(1.375)) + 44$
$$= -44 \text{ ft/sec}$$

34c. $v(4.2) = -32(4.2) + 44$
$$= -90.4 \text{ ft/sec}$$

35a. $f(5) = 3(5) + 2 = \$17.00$

35b. $f(5) = 3.25(5) + 3 = \$19.25$

35c. $f(5) = 3.50(5) + 5 = \$22.50$

35d. $f(5) = 4(5) + 8 = \$28.00$

36. sample answer: $x = 3$
This equation is a vertical line.

37. sample answer: $y = -2$
This equation is a horizontal line where $R = \{y \mid y = -2\}$.

38. For any value in the domain, the range will always be the same constant value, k.

39a. $(x - 0)^2 + (y - 0)^2 = 7^2$
$$x^2 + y^2 = 49$$

39b. $(x - 1)^2 + (y - (-4))^2 = 2^2$
$$(x - 1)^2 + (y + 4)^2 = 4$$

40a.

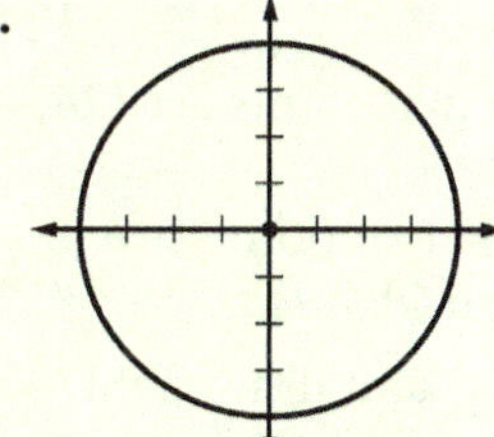

40b.

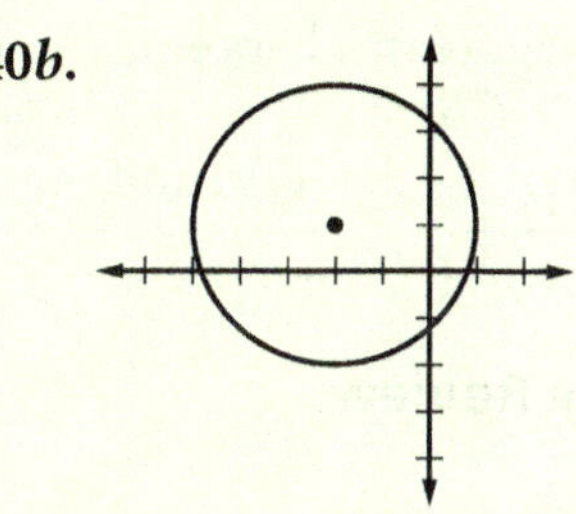

---

24. $m_G = -\frac{A}{B} = -\frac{1}{2}; m_\perp = 2$
$$-2 = 2(6) + b$$
$$-2 = 12 + b$$
$$b = -14$$
$$y = 2x - 14$$

25. $d = \sqrt{(-4 - 5)^2 + (-6 - 6)^2}$
$$= \sqrt{(-9)^2 + (-12)^2}$$
$$= \sqrt{225} = 15$$
$$M\left(\frac{5 + (-4)}{2}, \frac{6 + (-6)}{2}\right) = \left(\frac{1}{2}, 0\right)$$

26. $d = \sqrt{(2 - 5)^2 + (10 - 8)^2}$
$$= \sqrt{(-3)^2 + 2^2} = \sqrt{13}$$
$$M\left(\frac{5 + 2}{2}, \frac{8 + 10}{2}\right) = \left(\frac{7}{2}, 9\right)$$

27. $d = \sqrt{(2 - (-11))^2 + (19 - 3)^2}$
$$= \sqrt{13^2 + 16^2}$$
$$= \sqrt{425} = 5\sqrt{17}$$
$$M\left(\frac{-11 + 2}{2}, \frac{3 + 19}{2}\right) = \left(-\frac{9}{2}, 11\right)$$

28. $d = \sqrt{(2 - (-3))^2 + ((-4) - 7)^2}$
$$= \sqrt{5^2 + (-11)^2} = \sqrt{146}$$
$$M = \left(\frac{-3 + 2}{2}, \frac{7 + (-4)}{2}\right) = \left(-\frac{1}{2}, \frac{3}{2}\right)$$

29. $r = \sqrt{(-10 - 2)^2 + (2 - (-3))^2}$
$$= \sqrt{(-12)^2 + 5^2} = \sqrt{169} = 13$$

30. $r = \sqrt{(4 - (-3))^2 + (-6 - 1)^2}$
$$= \sqrt{7^2 + (-7)^2} = \sqrt{98} = 7\sqrt{2}$$
$$d = 2r = 14\sqrt{2}$$

41.

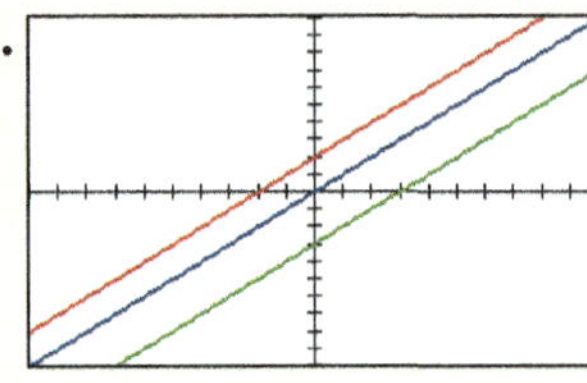

Adding or subtracting a constant moves the y-intercept up or down the y-axis while maintaining the same slope.

42.

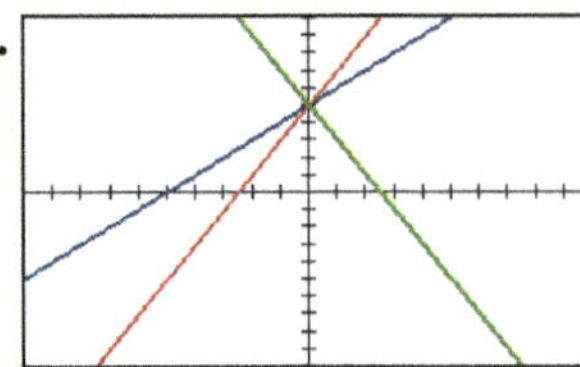

Changing the value of the coefficient of x changes the slope without changing the y-intercept.

43. $M\left(\dfrac{-10+2}{2}, \dfrac{-7+9}{2}\right) = (-4, 1)$

$m_G = \dfrac{-7-9}{-10-2} = \dfrac{-16}{-12} = \dfrac{4}{3}$

$m_\perp = -\dfrac{3}{4}$

$y - 1 = -\dfrac{3}{4}(x - (-4))$

$y - 1 = -\dfrac{3}{4}x - 3$

$y = -\dfrac{3}{4}x - 2$

44. $d = \sqrt{(8-3x)^2 + (x-(-4))^2} = 20$

$(8-3x)^2 + (x+4)^2 = 20^2$

$64 - 48x + 9x^2 + x^2 + 8x + 16 = 400$

$10x^2 - 40x + 80 = 400$

$x^2 - 4x - 32 = 0$

$(x-8)(x+4) = 0$

$x = 8$ or $x = -4$

If $x = 8$, the two points are $(24, -4)$ and $(8, 8)$.

If $x = -4$, the two points are $(-12, -4)$ and $(8, -4)$.

45. Plot $C(x_2, y_1)$ and draw right $\triangle ABC$.

By the Pythagorean Theorem,

$AB^2 = AC^2 + CB^2$,

$d^2 = (x_2 - x_1)^2 + (y_2 - y_1)^2$, and

$d = \sqrt{(x_2 - x_1)^2 + (y_2 - y_1)^2}$.

48. $8^{-2} = \dfrac{1}{8^2} = \dfrac{1}{64}$

49. $8^{\frac{1}{3}} = \sqrt[3]{8} = 2$

50. $(2^5)^{\frac{2}{3}} = 2^{\frac{10}{3}} = \sqrt[3]{2^{10}}$

$= 2^3 \cdot \sqrt[3]{2} = 8\sqrt[3]{2}$

Applying the distance formula to the definition of a circle and squaring both sides produces the following equation for a circle centered at (h, k) with radius r:
$(x - h)^2 + (y - k)^2 = r^2$.

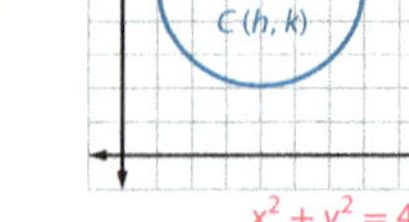

39. Write an equation for each circle. $x^2 + y^2 = 49$

  a. a circle centered at the origin with a radius of 7

  b. a circle centered at $(1, -4)$ with a radius of 2 $(x-1)^2 + (y+4)^2 = 4$

40. Graph each circle.

  a. $x^2 + y^2 = 16$ **b.** $(x+2)^2 + (y-1)^2 = 9$

> **C. Exercises**

Graph each system of equations. Then describe how the change in the function rule from $f(x)$ to $g(x)$ and $h(x)$ affects the slope and y-intercept of the graph of $f(x)$.

41. $f(x) = x$

  $g(x) = x + 2$

  $h(x) = x - 3$

42. $f(x) = x + 5$

  $g(x) = 2x + 5$

  $h(x) = -2x + 5$

CUMULATIVE REVIEW

Simplify each exponential expression. [Appendix 4]

47. 8^0 1   **48.** 8^{-2} $\dfrac{1}{64}$ **49.** $8^{\frac{1}{3}}$ 2

50. Write $32^{\frac{2}{5}}$ as a simplified radical. [Appendix 4] $8\sqrt[5]{2}$

51. Which type of line is not a function? [1.1]   vertical

52. Simplify i^{15}. [Appendix 5] $-i$

53. Simplify $\sqrt[3]{4320}$. [Appendix 4]   D

  A. $12\sqrt{30}$ **C.** $2\sqrt[3]{90}$ **E.** none of

  B. $12\sqrt{5}$ **D.** $6\sqrt[3]{20}$ these

54. What is the area of the shaded part of the square? [Geometry]   C

  A. 12π u^2

  B. $(9\pi + 27)$ u^2

  C. $(36 - 9\pi)$ u^2

  D. $(36 - 6\pi)$ u^2

  E. none of these

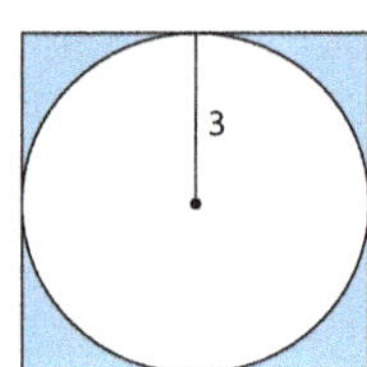

---

43. Write the slope-intercept form equation of the perpendicular bisector of a segment with endpoints $(2, 9)$ and $(-10, -7)$. $y = -\frac{3}{4}x - 2$

44. Find two pairs of points in the form of $(3x, -4)$ and $(8, x)$ where the distance between the points is 20 units. $(24, -4)$ and $(8, 8)$; $(-12, -4)$ and $(8, -4)$

Use the figure for exercises 45–46.

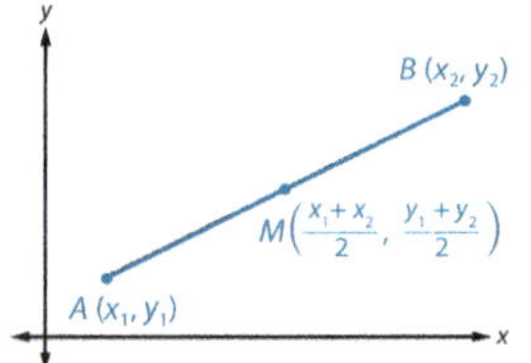

45. Prove the distance formula.

46. Verify the midpoint formula by showing that $AM = MB = \frac{1}{2}AB$.

Write and solve an equation or inequality to complete the following exercises. [Appendix 2]

55. What is the product of two consecutive odd numbers whose sum is 48?   D

  A. 551   **C.** 567   **E.** none of

  B. 572   **D.** 575   these

56. What is the largest of three consecutive even numbers whose sum is -144?   A

  A. -46 **C.** -50 **E.** none of

  B. -42 **D.** -47 these

46.
$$AM \overset{?}{=} MB$$

$$\sqrt{\left(\frac{x_1+x_2}{2} - x_1\right)^2 + \left(\frac{y_1+y_2}{2} - y_1\right)^2} \overset{?}{=} \sqrt{\left(x_2 - \frac{x_1+x_2}{2}\right)^2 + \left(y_2 - \frac{y_1+y_2}{2}\right)^2}$$

$$\sqrt{\left(\frac{x_2-x_1}{2}\right)^2 + \left(\frac{y_2-y_1}{2}\right)^2} = \sqrt{\left(\frac{x_2-x_1}{2}\right)^2 + \left(\frac{y_2-y_1}{2}\right)^2}$$

$$\sqrt{\left(\frac{x_2-x_1}{2}\right)^2 + \left(\frac{y_2-y_1}{2}\right)^2} = \sqrt{\frac{1}{4}\left[(x_2-x_1)^2 + (y_2-y_1)^2\right]}$$

$$= \frac{1}{2}\sqrt{(x_2-x_1)^2 + (y_2-y_1)^2} = \frac{1}{2}AB$$

52. $i^{15} = (i^4)^3 \cdot (i^3) = 1^3(-i) = -i$

53. $\sqrt[3]{4320} = \sqrt[3]{2^3 \cdot 2^2 \cdot 3^3 \cdot 5}$

$= 2 \cdot 3\sqrt[3]{2^2 \cdot 5} = 6\sqrt[3]{20}$

54. $A_{sq} = s^2 = (6)^2 = 36$ u^2

$A_\odot = \pi r^2 = \pi(3)^2 = 9\pi$ u^2

$A_{shaded} = A_{sq} - A_\odot = (36 - 9\pi)$ u^2

55. $n + (n+2) = 48$

$2n + 2 = 48$

$n = 23; n + 2 = 25$

$23(25) = 575$

56. $n + (n+2) + (n+4) = -144$

$3n + 6 = -144$

$n = -50; n + 2 = -48$

$n + 4 = -46$

The Gift of Mathematics

The miracle of appropriateness of the language of mathematics for the formulation of the laws of physics is a wonderful gift which we neither understand nor deserve.

—Eugene Wigner (recipient of the 1963 Nobel Prize in Physics)

Many modern secular mathematicians maintain that mathematics is simply a creation of human minds. But God created the world (Gen. 1:1) with a system of orderly, transcendent principles that govern creation. Through the centuries, human minds have developed the symbols and terms of mathematics as they quantitatively described the created world. God is the origin of the underlying principles that provide the foundation and consistency for their work.

The practical arithmetic operations of addition and multiplication work the same from culture to culture, even though the abstract notation used may vary. The same is true for the independent discoveries of calculus by Newton in England using "fluxions" and by Leibniz in Europe using "differentials," terms referring to the same calculus concept. Important formulas such as $F = ma$ and $E = mc^2$ are man's descriptions of universal laws put in place by God. Eugene Wigner could not understand how the human-created language of mathematics could be so successful in modeling our physical universe. His reference to the appropriateness of mathematics as a "wonderful gift" points to the underlying connections established by God.

The mathematics created by human minds owes its origin to God, who created man in His image (Gen. 1:27) as a thinking, reasoning being. It is God who makes it possible for sinful humans to develop logical mathematical systems. Famous Protestant theologian John Calvin concluded, ". . . since it is manifest that men whom the Scriptures term carnal, are so acute and clear-sighted in the investigation of inferior things, their example should teach us how many gifts the Lord has left in possession of human nature . . . . But if the Lord has been pleased to assist us by the work and ministry of the ungodly in physics, dialectics, mathematics, and other similar sciences, let us avail ourselves of it, lest, by neglecting the gifts of God spontaneously offered to us, we be justly punished for our sloth." Calvin understood that human mathematical and scientific abilities are inherent gifts from our Creator.

Many prominent men have recognized that their important discoveries are not their exclusive creations. Heinrich Hertz, the German physicist who first proved the existence of Maxwell's electromagnetic waves, stated that "one cannot escape the feeling that these mathematical formulas have an independent existence and an intelligence of their own, that they are wiser than we are, wiser even than their discoverers, that we get more out of them than was originally put into them." James Nickel, a prominent Christian mathematician, contrasted man's knowledge with God's, stating that "man's mathematical knowledge will never exhaust the infinite panorama of God's knowledge (I Sam. 2:3; Ps. 147:5). At best, man is gifted with an infinitesimal subset of God's exhaustive wisdom and knowledge (Ps. 104:24)."

▸ Exercises

1. What is the original source of mathematical laws and truth?

2. Describe the view of the origin of mathematics held by many modern secular mathematicians.

3. What two terms did Newton and Leibniz use to describe the same calculus concept?

4. How did Eugene Wigner describe "the miracle of the appropriateness of the language of mathematics" in developing the laws of physics?

5. Why did John Calvin say that even the ungodly can make great discoveries in mathematics and the sciences?

6. **Discuss:** Why do mathematical systems designed by man owe their origin to God?

7. **Discuss:** Explain the statement "Mathematics is a gift from God."

8. **Discuss:** Should man's role in the development of mathematics be described as creating, discovering, or both? Explain why.

Biblical Perspective of Mathematics

Objectives

1. To identify God as the source of mathematical truth

2. To explain how mathematics is a gift from God

3. To summarize our dual role in the development of mathematics as creator and discoverer

Assignments

- **Minimum:** 1–5, 8
- **Standard:** 1–8
- **Extended:** 1–8

Answers

1. God

2. Mathematics is simply a creation of human minds.

3. "fluxions" and "differentials"

4. as a wonderful gift

5. He understood that these abilities were gifts from God when He created mankind.

6. When God created humans in His image, He gave us the reasoning ability to create mathematical systems. We develop descriptions of universal abstract concepts already in existence. For example, the concept of the number 3 has absolute meaning throughout the universe.

7. God designed the universe with consistent, orderly, transcendent principles that govern it. When God created humans in His image, He gave us the creative, intellectual ability necessary to develop mathematical systems that describe these transcendent principles. This gift helps us complete God's commands to have dominion over the earth and to serve Him by demonstrating love for others.

8. both; People create mathematical vocabulary and notation and use them to discover the underlying principles and mathematical laws governing our universe, as well as other universal abstract concepts.

PRESENTATION

This first Biblical Perspective of Mathematics feature presents mathematics as it relates to Creation in three ways.

1. God designed the universe and the mathematical principles that govern it.

2. God created humans in His image, with intelligence and reasoning abilities, making it possible for us to create mathematical language to describe absolute concepts.

3. God has given us the ability to apply mathematics to solve real-world problems and fulfill our responsibilities in exercising dominion over the earth, serving Him, and serving others.

Instruct the students to read the feature and complete the exercises. Review their answers and lead a discussion related to the three discussion questions to emphasize these points.

1.3 Piecewise Functions and Continuity

Objectives

1. To define and graph the basic greatest integer and absolute value functions

2. To graph piecewise functions with and without technology

3. To express the rule, domain, and range of graphed piecewise functions

4. To evaluate limits of functions at points of continuity and discontinuity

Vocabulary

absolute value function
continuous function
continuous on an interval
greatest integer function
infinite discontinuity
jump discontinuity
left-hand limit
limit
one-sided limit
piecewise function
point (removable) discontinuity
right-hand limit

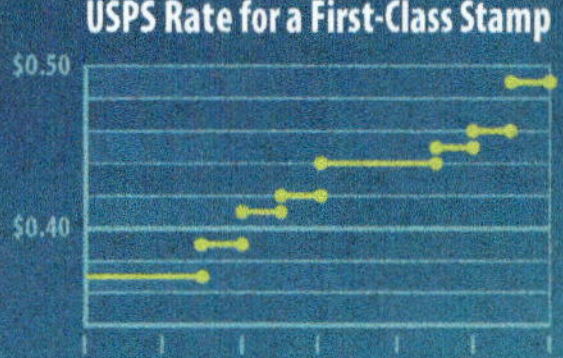

Many graphs cannot be defined by a single rule.

After completing this section, you will be able to

- define and graph the basic absolute value and greatest integer functions.
- graph and analyze piecewise functions.
- use limits to identify points of discontinuity.

A *piecewise function* is defined by two or more function rules for various intervals of its domain. The correct rule is selected based on the value of x. The *absolute value function* is likely the first piecewise function you have encountered.

$$f(x) = |x| = \begin{cases} x \text{ if } x \geq 0 \\ -x \text{ if } x < 0 \end{cases}$$

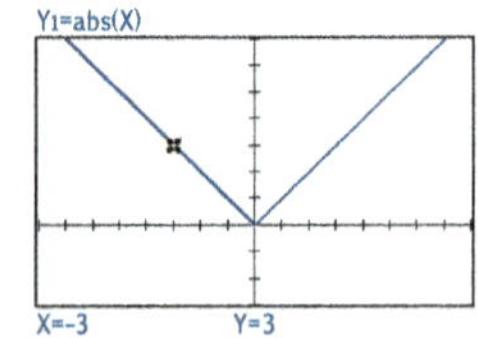

If $x \geq 0$, then $|x|$ is x. If $x < 0$ (a negative number), then $|x|$ is the opposite of x (a positive number). This implies that an absolute value is always nonnegative.

The graph of the absolute value function consists of two linear pieces: $y = -x$ when $x < 0$ and $y = x$ when $x \geq 0$. Notice that the domain is all real numbers and the range is the nonnegative real numbers.

Example 1   Evaluating an Absolute Value Function

Find $g(x) = |2x - 3|$ with $D = \{-4, -2, 0, 1, 2, 4\}$.

Answer

$g(-4) = |2(-4) - 3| = |-11| = 11$	1. Evaluate the function at each member of the domain.
$g(-2) = |2(-2) - 3| = |-7| = 7$	
$g(0) = |2(0) - 3| = |-3| = 3$	
$g(1) = |2(1) - 3| = |-1| = 1$	
$g(2) = |2(2) - 3| = |1| = 1$	
$g(4) = |2(4) - 3| = |5| = 5$	
$g = \{(-4, 11), (-2, 7), (0, 3), (1, 1), (2, 1), (4, 5)\}$	2. List the set of ordered pairs.

SKILL ✓ EXERCISE 1

In Excel spreadsheets the greatest integer function, INT, is a special case of a FLOOR function, while rounding up to the next integer is an example of a CEILING function.

The graph of the *greatest integer function* consists of an infinite number of pieces. Each range value is the greatest integer less than or equal to the given domain value. The function is represented using square brackets: $f(x) = [x]$. Its domain is the set of real numbers, but its range consists of the set of integers. The rule for the greatest integer function can be written as a piecewise function.

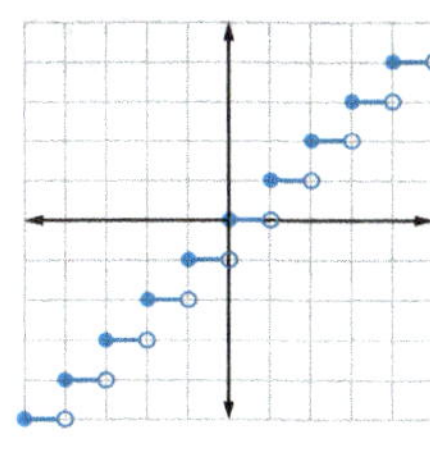

$$f(x) = [x] = \begin{cases} \vdots \\ -2 \text{ if } -2 \leq x < -1 \\ -1 \text{ if } -1 \leq x < 0 \\ 0 \text{ if } 0 \leq x < 1 \\ 1 \text{ if } 1 \leq x < 2 \\ \vdots \end{cases}$$

Since the y-value of each piece is constant, each of the infinitely many pieces is a portion of a horizontal line. Notice how the open circles and solid dots are used to indicate whether the endpoint of each portion is included or excluded in the graph.

Example 2 Evaluating the Greatest Integer Function

Find $f(x) = [x]$ where $D = \left\{ -5, -\frac{3}{2}, -\frac{3}{4}, 0, \frac{1}{4}, \frac{5}{2} \right\}$.

Answer

$f(-5) = [-5] = -5$ 1. The greatest integer less than or equal to -5 is -5.

$f\left(-\frac{3}{2}\right) = \left[-\frac{3}{2}\right] = -2$ 2. The greatest integer less than or equal to -1.5 is -2.

$f\left(-\frac{3}{4}\right) = \left[-\frac{3}{4}\right] = -1$ 3. The greatest integer less than or equal to -0.75 is -1.

$f(0) = [0] = 0$ 4. The greatest integer less than or equal to 0 is 0.

$f\left(\frac{1}{4}\right) = \left[\frac{1}{4}\right] = 0$ 5. The greatest integer less than or equal to 0.25 is 0.

$f\left(\frac{5}{2}\right) = \left[\frac{5}{2}\right] = 2$ 6. The greatest integer less than or equal to 2.5 is 2.

$f = \left\{ (-5, -5), \left(-\frac{3}{2}, -2\right), \left(-\frac{3}{4}, -1\right), (0, 0), \left(\frac{1}{4}, 0\right), \left(\frac{5}{2}, 2\right) \right\}$

SKILL ✔ EXERCISE 3

A function that has no gaps, jumps, or holes is a *continuous function*. You can graph a continuous function without lifting your pencil from the paper. The absolute value function is continuous, while the greatest integer function is discontinuous because of the jumps between pieces. Many piecewise functions are discontinuous.

The definition of a mathematical function has broadened considerably since the development of calculus in the eighteenth century. While the term initially referred only to continuous functions, these are a small subset of the vast number of relations that are considered to be functions today.

Example 3 illustrates the graphing of the three distinct rules in each interval of a piecewise function. Have the students identify each interval and apply the associated rule for each. Ask the students whether the function appears to be continuous or discontinuous. (*discontinuous*) Graph or have the students graph the piecewise function using technology, emphasizing the use of the interval in the denominator.

Use Example 3 to present the definitions of *one-sided limits* and "two-sided" limits. Discuss the types of discontinuities and how they relate to the limit of the function. Infinite and point discontinuities will be discussed further in Section 2.5.

The students should be able to identify discontinuity types by examining limits in Example 4. The $\boxed{\text{TRACE}}$ or [TABLE] functions of a graphing calculator could be used to quickly evaluate the function. Note that although ∞ is not a real number and therefore not a limit, the notation $\lim f(x) = \infty$ or $\lim f(x) = -\infty$ is commonly used to express the idea that the limit is approaching infinity.

- **Minimum:** 2–5, 7, 9, 11, 14, 16, 18, 21–23, 25–27, 29, 31–32, 35, 43–46, 50
- **Standard:** 1, 3–4, 8–9, 11, 14–19, 21–23, 25–27, 29, 31–33, 35, 37, 39–40, 43–47, 49–50
- **Extended:** 3–4, 6, 8, 10, 15–23, 25–33, 35–37, 39–40, 43; 44–52 even

KEYWORD SEARCH

piecewise grapher

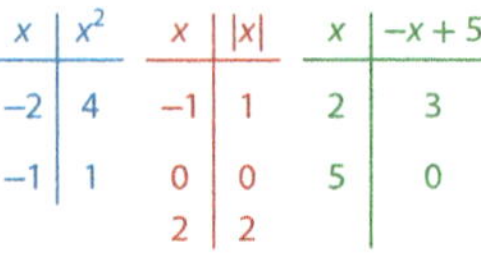

TIP

Use TEST ([2nd], [MATH]) to enter inequalities or logical operators. To [TRACE] along the graph of a piecewise function, use ▲ or ▼ to select the desired function rule.

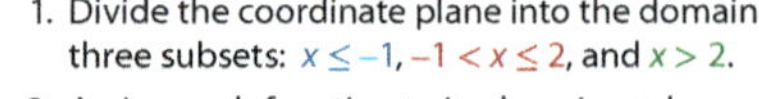

Graph $f(x) = \begin{cases} x^2 & \text{if } x \le -1 \\ |x| & \text{if } -1 < x \le 2 \\ -x + 5 & \text{if } x > 2 \end{cases}$.

Then list the function's domain and range and state whether the function appears to be continuous.

Answer

x	x^2
-2	4
-1	1

x	$|x|$
-1	1
0	0
2	2

x	$-x + 5$
2	3
5	0

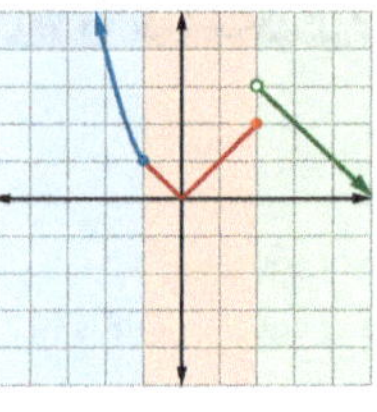

$D = \mathbb{R}; R = \mathbb{R}$
The function appears to be discontinuous when $x = 2$.

Check

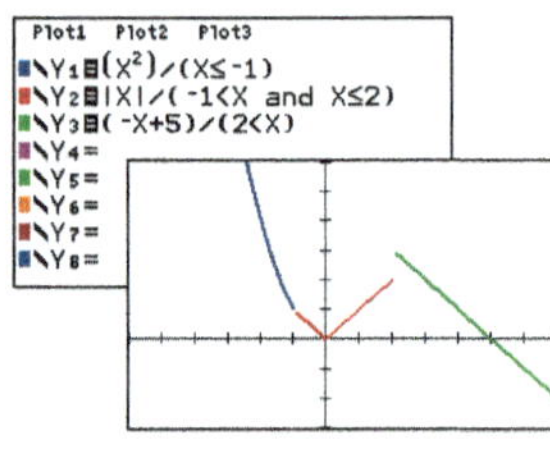

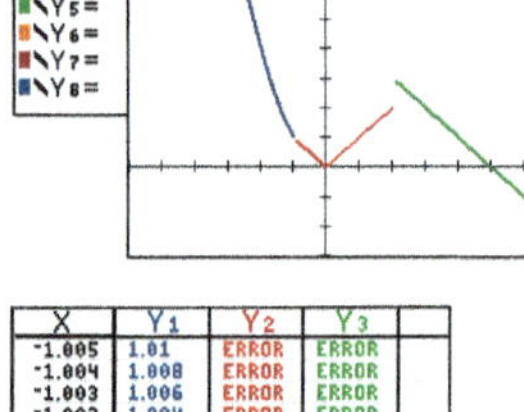

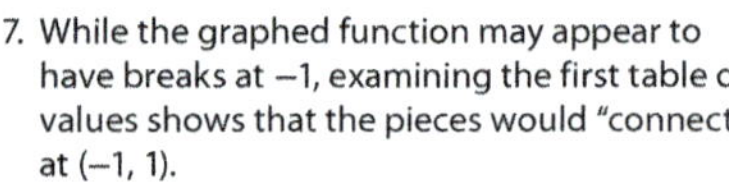

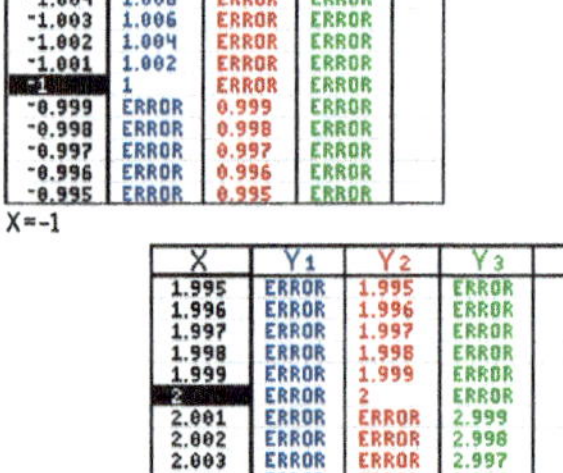

1. Divide the coordinate plane into the domain's three subsets: $x \le -1$, $-1 < x \le 2$, and $x > 2$.

2. Assign each function to its domain and evaluate it at selected values within the domain or at an excluded endpoint.

3. Graph each piece of the function.

4. Verify whether the endpoint of each piece is open or closed.

5. The function is defined for all real numbers. The graph implies that the function ranges from $-\infty$ to ∞.

6. To graph each piece of the function with a graphing calculator, divide its function rule by the inequality representing its domain as shown. The expressions in the denominator evaluate to 1 when the inequality is true and to 0 when the inequality is false.

7. While the graphed function may appear to have breaks at -1, examining the first table of values shows that the pieces would "connect" at $(-1, 1)$.

8. Examining the second table around $x = 2$ shows that the function's value jumps from 2 to almost 3.

SKILL ✔ **EXERCISE 11**

TIPS

Ex. 8 The x-value for the turning point of this graph can be found by setting the expression within the absolute value function equal to 0 and solving.

If a function $f(x)$ is continuous at $x = a$, then the value of the function must approach $y = f(a)$ from both the left and the right. The number that a function approaches, even if that number is never reached, is called a *limit*.

DEFINITIONS

The **left-hand limit** of $f(x)$ as x approaches a, denoted $\lim\limits_{x \to a^-} f(x)$, is the number to which $f(x)$ gets closer and closer as x gets closer to a from the left.

The **right-hand limit** of $f(x)$ as x approaches a, denoted $\lim\limits_{x \to a^+} f(x)$, is the number to which $f(x)$ gets closer and closer as x gets closer to a from the right.

Left- and right-hand limits are often referred to as *one-sided limits*.

In Example 3, the left-hand limit is $\lim\limits_{x \to 2^-} f(x) = 2$ since $f(x) = |x|$ gets closer and closer to 2 as x approaches 2 from the left. The right-hand limit is $\lim\limits_{x \to 2^+} f(x) = 3$ since $f(x) = -x + 5$ gets closer and closer to 3 as x approaches 2 from the right.

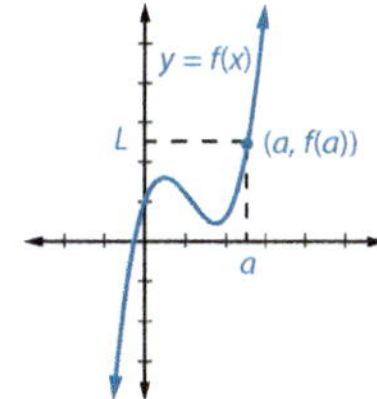

DEFINITION

The **limit** of $f(x)$ as x approaches a, denoted $\lim\limits_{x \to a} f(x)$, is L if and only if
$$\lim_{x \to a^-} f(x) = \lim_{x \to a^+} f(x) = L.$$

This "two-sided" limit of the piecewise function in Example 3 at $x = -1$ exists and $\lim\limits_{x \to -1} f(x) = 1$ since $\lim\limits_{x \to -1^-} f(x) = \lim\limits_{x \to -1^+} f(x) = 1$. Because $\lim\limits_{x \to -1} f(x) = f(-1)$, the function is continuous at $x = -1$.

Notice that $\lim\limits_{x \to 2} f(x)$ does not exist since $\lim\limits_{x \to 2^-} f(x) = 2$ and $\lim\limits_{x \to 2^+} f(x) = 3$. This difference between left- and right-hand limits indicates a jump in the value of the function.

The following table illustrates different possible types of discontinuity at $x = 2$.

TIP

An infinite discontinuity indicates the presence of a vertical asymptote. A point discontinuity is also called a removable discontinuity.

Jump Discontinuity	Infinite Discontinuity	Point Discontinuity	
$\lim\limits_{x \to a^-} f(x) \neq \lim\limits_{x \to a^+} f(x)$	$f(a)$ increases or decreases infinitely.	$\lim\limits_{x \to a} f(x) = L$ but $f(a)$ is undefined.	$\lim\limits_{x \to a} f(x) \neq f(a)$

DEFINITION

A function is **continuous** at $x = a$ if $f(a)$ and $\lim\limits_{x \to a} f(x)$ exist and $\lim\limits_{x \to a} f(x) = f(a)$.
A function is *continuous on an interval* (a, b) if and only if it is continuous at every point in (a, b).

Solutions

A. Exercises

1. $g(-4) = |4(-4)| = |-16| = 16$

$g\left(-\frac{1}{2}\right) = \left|4\left(-\frac{1}{2}\right)\right| = \left|-\frac{4}{2}\right|$
$= |-2| = 2$

$g(0) = |4(0)| = |0| = 0$

$g\left(\frac{3}{4}\right) = \left|4\left(\frac{3}{4}\right)\right| = |3| = 3$

$g(2) = |4(2)| = |8| = 8$

$\{(-4, 16), \left(-\frac{1}{2}, 2\right), (0, 0),$

$\left(\frac{3}{4}, 3\right), (2, 8)\}$

2. $f(-4) = |(-4) - 4| = |-8| = 8$

$f\left(-\frac{1}{2}\right) = \left|\left(-\frac{1}{2}\right) - 4\right| = \left|-\frac{9}{2}\right| = \frac{9}{2}$

$f(0) = |(0) - 4| = |-4| = 4$

$f\left(\frac{3}{4}\right) = \left|\left(\frac{3}{4}\right) - 4\right| = \left|-\frac{13}{4}\right| = \frac{13}{4}$

$f(2) = |(2) - 4| = |-2| = 2$

$\{(-4, 8), \left(-\frac{1}{2}, \frac{9}{2}\right), (0, 4),$

$\left(\frac{3}{4}, \frac{13}{4}\right), (2, 2)\}$

3. $h(-4) = [-4] = -4$

$h\left(-\frac{1}{2}\right) = \left[-\frac{1}{2}\right] = -1$

$h(0) = [0] = 0$

$h\left(\frac{3}{4}\right) = \left[\frac{3}{4}\right] = 0$

$h(2) = [2] = 2$

$\{(-4, -4), \left(-\frac{1}{2}, -1\right), (0, 0),$

$\left(\frac{3}{4}, 0\right), (2, 2)\}$

4. $g(-4) = [3(-4) + 1] = [-11] = -11$

$g\left(-\frac{1}{2}\right) = \left[3\left(-\frac{1}{2}\right) + 1\right] = \left[-\frac{1}{2}\right]$
$= -1$

$g(0) = [3(0) + 1] = [1] = 1$

$g\left(\frac{3}{4}\right) = \left[3\left(\frac{3}{4}\right) + 1\right] = \left[\frac{9}{4} + 1\right]$
$= \left[\frac{13}{4}\right] = 3$

$g(2) = [3(2) + 1] = 7$

$\{(-4, -11), \left(-\frac{1}{2}, -1\right), (0, 1),$

$\left(\frac{3}{4}, 3\right), (2, 7)\}$

5. $h(-4) = -6(-4) = 24$

$h\left(-\frac{1}{2}\right) = -6\left(-\frac{1}{2}\right) = 3$

$h(0) = (0) + 2 = 2$

$h\left(\frac{3}{4}\right) = \left(\frac{3}{4}\right) + 2 = \frac{11}{4}$

$h(2) = (2) + 2 = 4$

$\{(-4, 24), \left(-\frac{1}{2}, 3\right), (0, 2),$

$\left(\frac{3}{4}, \frac{11}{4}\right), (2, 4)\}$

6. $f(-4) = 2(-4) + 5 = -3$

$f\left(-\frac{1}{2}\right) = -\frac{1}{2}$

$f(0) = 0$

$f\left(\frac{3}{4}\right) = \frac{3}{4}$

$f(2) = (2) - 1 = 1$

$\{(-4, -3), \left(-\frac{1}{2}, -\frac{1}{2}\right), (0, 0),$

$\left(\frac{3}{4}, \frac{3}{4}\right), (2, 1)\}$

7.

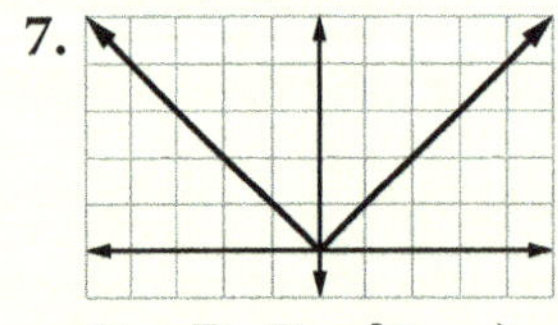

$D = \mathbb{R};\ R = [0, \infty)$; cont.

8.

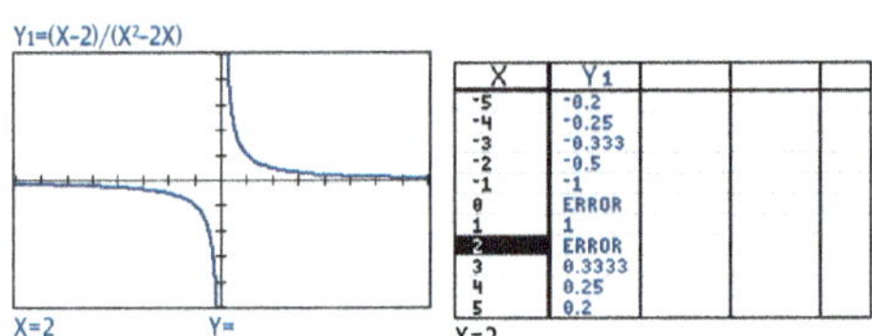

$D = \mathbb{R};\ R = [0, \infty)$; cont.

9.

$D = \mathbb{R};\ R = [1, \infty)$; cont.

Determine if $f(x) = \frac{x - 2}{x^2 - 2x}$ is continuous at each value. If it is discontinuous, identify the type of discontinuity.

a. $x = -2$ **b.** $x = 0$ **c.** $x = 2$

Answer

a. At $x = -2$, $f(-2) = \frac{-4}{8} = -\frac{1}{2}$.

		$x \to -2^-$				$x \to -2^+$	
x	-2.1	-2.01	-2.001	-2	-1.999	-1.99	-1.9
$f(x)$	-0.4762	-0.4975	-0.4998	-0.5	-0.5003	-0.5025	-0.5263

$f(x) \longrightarrow -0.5 \longleftarrow f(x)$

$\lim\limits_{x \to -2} f(x) = -0.5$ since $\lim\limits_{x \to -2^-} f(x) = -0.5$ and $\lim\limits_{x \to -2^+} f(x) = -0.5$.

The function is continuous at $x = -2$ since $\lim\limits_{x \to -2} f(x) = f(-2)$.

b. At $x = 0$, $f(0) = \frac{-2}{0}$, which is undefined.

		$x \to 0^-$				$x \to 0^+$	
x	-0.1	-0.01	-0.001	0	0.001	0.01	0.1
$f(x)$	-10	-100	-1000		1000	100	10

$f(x) \longrightarrow -\infty \qquad \infty \longleftarrow f(x)$

$\lim\limits_{x \to 0} f(x)$ does not exist since $\lim\limits_{x \to 0^-} f(x) = -\infty$ and $\lim\limits_{x \to 0^+} f(x) = \infty$.

The function has an infinite discontinuity at $x = 0$.

c. At $x = 2$, $f(2) = \frac{0}{0}$, which is undefined.

		$x \to 2^-$				$x \to 2^+$	
x	1.9	1.99	1.999	2	2.001	2.01	2.1
$f(x)$	0.5263	0.5025	0.5003		0.4998	0.4975	0.4762

$f(x) \longrightarrow 0.5 \longleftarrow f(x)$

$\lim\limits_{x \to 2} f(x) = 0.5$ since $\lim\limits_{x \to 2^-} f(x) = 0.5$ and $\lim\limits_{x \to 2^+} f(x) = 0.5$.

The function has a point discontinuity at $x = 2$.

Check

Graphing the function clearly illustrates the infinite discontinuity at $x = 0$. Use the trace function or view a table of values to confirm the point discontinuity at $x = 2$.

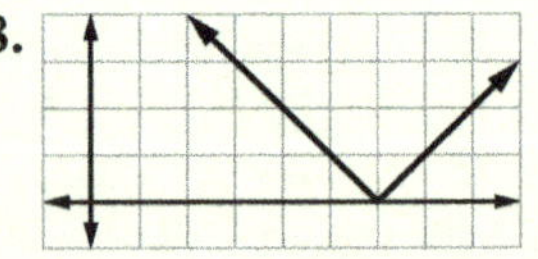

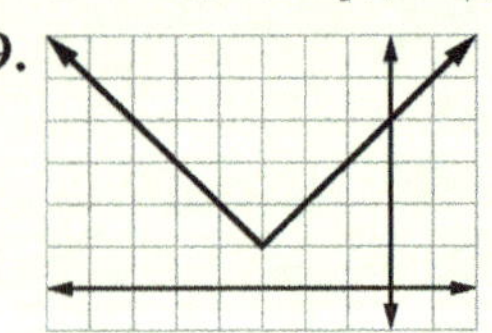

The intervals of continuity for the function in Example 4 are $(-\infty, 0)$, $(0, 2)$, and $(2, \infty)$.

SKILL ✔ **EXERCISE 36**

Find the set of ordered pairs for each function rule when $D = \{-4, -\frac{1}{2}, 0, \frac{3}{4}, 2\}$.

1. $g(x) = |4x|$
2. $f(x) = |x - 4|$
3. $h(x) = [x]$
4. $g(x) = [3x + 1]$
5. $h(x) = \begin{cases} -6x & \text{if } x < 0 \\ x + 2 & \text{if } x \geq 0 \end{cases}$
6. $f(x) = \begin{cases} 2x + 5 & \text{if } x < -1 \\ x & \text{if } -1 \leq x \leq 1 \\ x - 1 & \text{if } x > 1 \end{cases}$

Graph each function. Then state its domain, range, and whether it is continuous or discontinuous.

7. $f(x) = |x|$
8. $h(x) = |x - 6|$
9. $f(x) = |x + 3| + 1$
10. $p(x) = \begin{cases} 0 & \text{if } x = 3 \\ 5 & \text{if } x \neq 3 \end{cases}$
11. $g(x) = \begin{cases} -x & \text{if } x < 1 \\ 2x - 3 & \text{if } x \geq 1 \end{cases}$
12. $f(x) = \begin{cases} x + 3 & \text{if } x < -2 \\ 3^x & \text{if } x > -1 \end{cases}$

Classify each function as continuous or discontinuous. Further classify any discontinuities as jump, infinite, or point.

13. 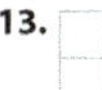
discontinuous; jump

14.
continuous

15. 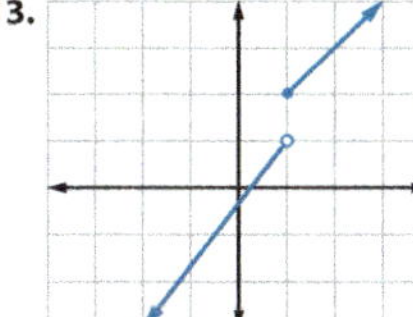
discontinuous; point

16. 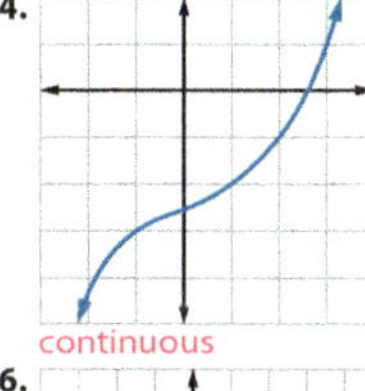
discontinuous; jump

17. 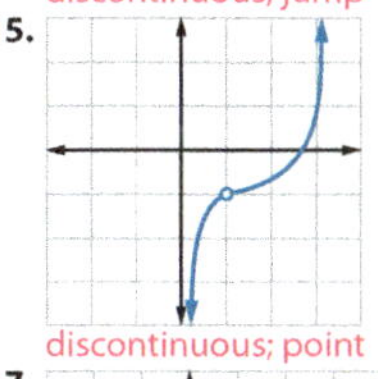
discontinuous; infinite

18.
discontinuous; point

19. Which is a continuous function?   B
   A. the number of people attending church as a function of the week of the year
   B. the area of a triangle as a function of its height
   C. the cost of gasoline as a function of time

20. Which is a discontinuous function?   C
   A. the cost of filling your tank with gasoline as a function of the number of gallons
   B. the distance traveled as a function of time
   C. the number of minutes charged to your cell phone account as a function of the length of the call

Graph each function. Then state its domain, range, and whether it is continuous or discontinuous.

21. $g(x) = [x]$
22. $y = [x - 4]$
23. $y = \begin{cases} x & \text{if } x < 1 \\ 2x - 4 & \text{if } 1 \leq x \leq 4 \\ \frac{1}{3}x & \text{if } x > 4 \end{cases}$ $D = \mathbb{R}; R = \mathbb{R};$ discontinuous
24. $f(x) = \begin{cases} x^2 & \text{if } -2 < x < 2 \\ 4 & \text{otherwise} \end{cases}$ $D = \mathbb{R}; R = [0, 4];$ continuous

Use the graph of $f(x)$ to complete exercises 25–33.

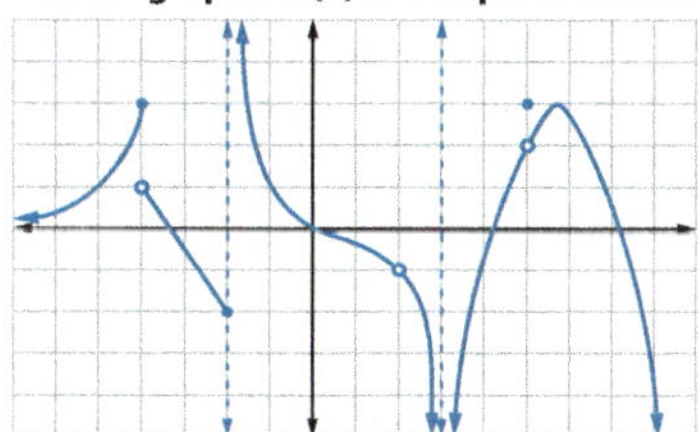

Find the one-sided limits.

25. $\lim_{x \to -4^-} f(x)$ and $\lim_{x \to -4^+} f(x)$ 3; 1
26. $\lim_{x \to -2^-} f(x)$ and $\lim_{x \to -2^+} f(x)$ $-2; \infty$
27. $\lim_{x \to 2^-} f(x)$ and $\lim_{x \to 2^+} f(x)$ $-1; -1$
28. $\lim_{x \to 5^-} f(x)$ and $\lim_{x \to 5^+} f(x)$ 2; 2

Use the graph of $f(x)$ above to find each limit. Justify your answer using the definition of a limit.

29. $\lim_{x \to 2} f(x)$ -1
30. $\lim_{x \to 3} f(x)$ $-\infty$
31. $\lim_{x \to -2} f(x)$ does not exist
32. List the values for which a limit of $f(x)$ does not exist. $-4; -2$
33. Name the value of x for which the limit of $f(x)$ is not equal to the defined value of the function.   5

21. 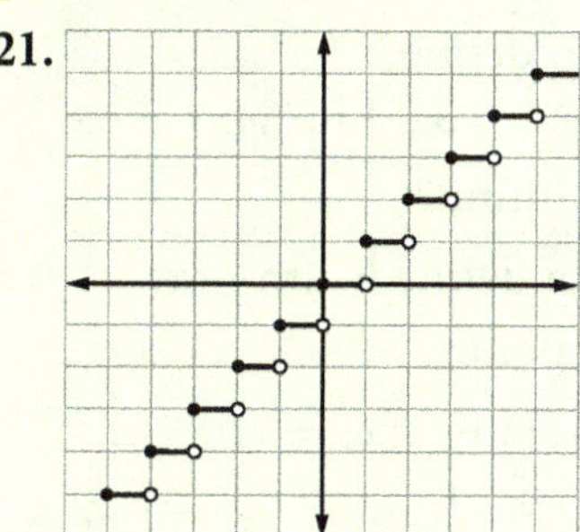
$D = \mathbb{R}; R = \mathbb{Z};$ discont.

22. 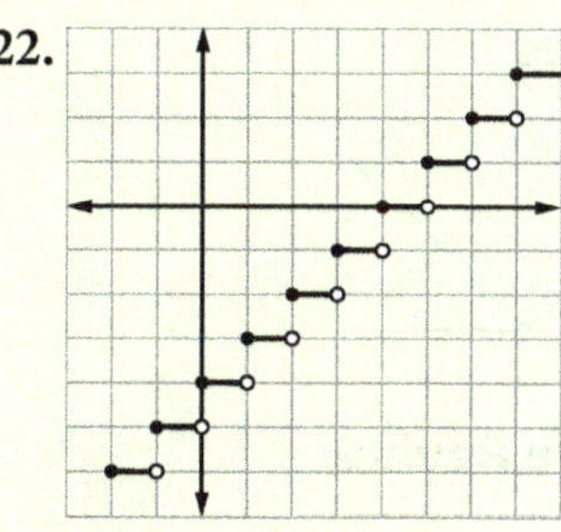
$D = \mathbb{R}; R = \mathbb{Z};$ discont.

23. 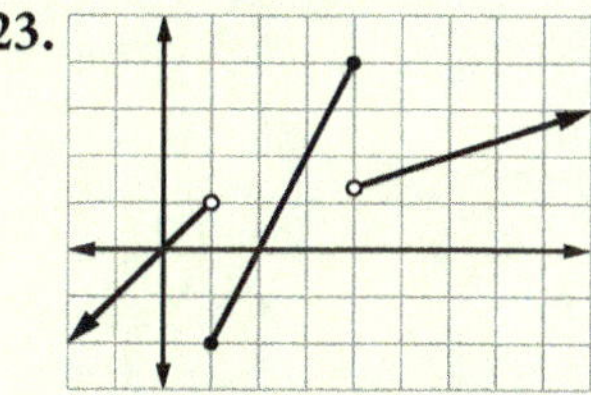

24. 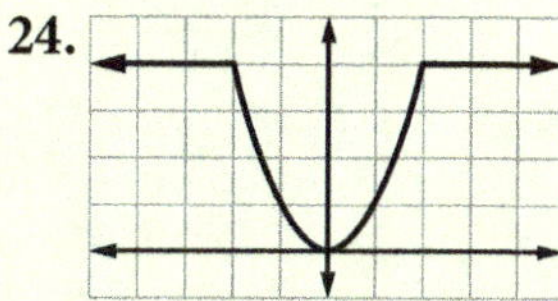

29. $\lim_{x \to 2^-} f(x) = \lim_{x \to 2^+} f(x) = -1$
30. $\lim_{x \to 3^-} f(x) = \lim_{x \to 3^+} f(x) = -\infty$
31. $\lim_{x \to -2^-} f(x) = -2$, but $\lim_{x \to -2^+} f(x) = \infty$; does not exist
32. See solutions for exercises 25–26 and note that the two-sided limits are different. This implies that the limits do not exist as x approaches -4 and -2.
33. $f(5) = 3; \lim_{x \to 5} f(x) = 2$

10. 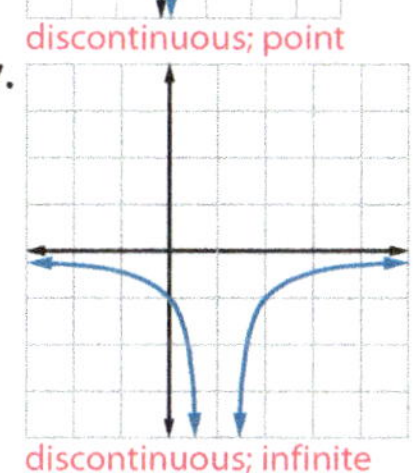
$D = (-\infty, \infty); R = \{0, 5\};$ discont.

11. 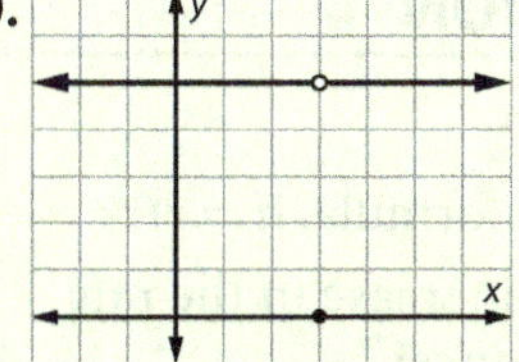
$D = \mathbb{R}; R = [-1, \infty);$ cont.

12. 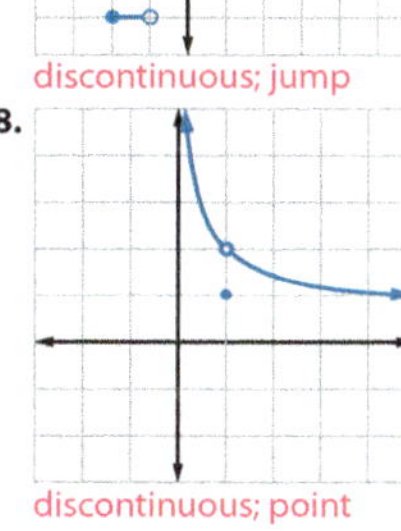
$D = (-\infty, -2) \cup (-1, \infty); R = \mathbb{R};$ discont.

34. $f(-2) = 2; \lim\limits_{x \to -2} f(x) = 2$; cont.

35. $g(-2)$ is undefined;
$\lim\limits_{x \to -2^-} g(x) = -\infty; \lim\limits_{x \to -2^+} g(x) = \infty$;
discont.; infinite

36. $h(-2)$ is undefined; $\lim\limits_{x \to -2} h(x) = -0.3$;
discont.; point

37. $r(-2) = -1; \lim\limits_{x \to -2} r(x) = -1$; cont.

38. $\lim\limits_{x \to -2^-} t(x) = -3; \lim\limits_{x \to -2^+} t(x) = -2$;
discont.; jump

❭ C. Exercises

41.

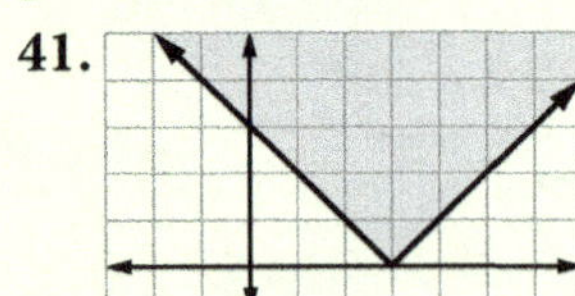

❭ Cumulative Review

45. $\dfrac{x}{8} = \dfrac{3}{4}$
$4x = 24$
$x = 6$

46. $y = 0x - 2; m = 0$

47. $m = \dfrac{-6 - 3}{-2 - 1} = 3$
$3 = 3(1) + b$
$b = 0$
$y = 3x$

48. $m = \dfrac{8 - 5}{3 - 0} = 1$
$b = 5$
$y = x + 5$

49. $20n > 12n + 3200$
$8n > 3200$
$n > 400$

50. $\dfrac{x + 5}{2} = -1$
$x + 5 = -2$
$x = -7$
$\dfrac{y + 1}{2} = 2$
$y + 1 = 4$
$y = 3$

51. $\dfrac{2}{3}(n + 9) = 2n$
$2n + 18 = 3(2n)$
$4n = 18$
$n = \dfrac{9}{2}$

52. $3n + 5 \le -16$
$3n \le -21$
$n \le -7$

State whether each function is continuous or discontinuous at $x = -2$ and classify any discontinuities as jump, infinite, or point. Justify your answer using the definition of continuity.

34. $f(x) = \dfrac{x}{x + 1}$ **35.** $g(x) = \dfrac{x}{x^2 - 4}$

36. $h(x) = \dfrac{x^2 + 6x + 8}{x^2 - 2x - 8}$ **37.** $r(x) = \dfrac{4x}{x^2 + 4}$

38. $t(x) = [x]$

❭ C. Exercises

39. Classify each type of function as *always*, *sometimes*, or *never* continuous.
  a. linear   **b.** greatest integer   **c.** piecewise
    always   never   sometimes

CUMULATIVE REVIEW

43. Given $g(x) = -3x^3$, find $g(2), g(-1), g(0),$ and $g(5)$. [1.1]   −24, 3, 0, −375
44. Given $f(x) = 3x^{-1}$, find $f(2), f(-1), f(0),$ and $f(5)$. [1.1] $\frac{3}{2}$, −3, undefined, $\frac{3}{5}$
45. Solve $\dfrac{x}{8} = \dfrac{21}{28}$. [Appendix 2] $x = 6$
46. State the slope of the line $y = -2$. [1.2] $m = 0$

Write a function rule describing the line through each pair of points. [1.2]
47. (1, 3) and (−2, −6) $f(x) = 3x$ **48.** (0, 5) and (3, 8) $f(x) = x + 5$

49. Producing a Wi-Fi device costs \$12 per device plus a one-time setup expense of \$3200. How many Wi-Fi devices must a company sell at \$20 each to make a profit? [1.2]   E
  A. > 160   **C.** > 200   **E.** > 400
  B. < 200   **D.** > 300

50. If $M(-1, 2)$ is the midpoint of $\overline{CD}$ with D at (5, 1), what are the coordinates of C? [1.2]   A
  A. (−7, 3)   **C.** (−7, 1)   **E.** none of these
  B. (4, 1)   **D.** (4, 3)

---

40. Determine whether $f(x) = \sqrt{x + 1}$ is continuous in each case.
  a. at $x = -2$ no
  b. over its domain   yes
  c. over the real numbers   no

41. Graph $y \ge |x - 3|$.

42. Refer to the graph of $f(x)$ in exercises 25–33.
  a. Find $\lim\limits_{x \to -\infty} f(x)$.   0
  b. What characteristic of the graph is indicated by this limit?   a horizontal asymptote

Write and solve an equation or inequality to complete the following exercises. [Appendix 2]

51. Find a number such that two-thirds of the sum of the number and 9 is twice the number.   C
  A. $\dfrac{9}{4}$ **C.** $\dfrac{9}{2}$ **E.** none of these
  B. $\dfrac{27}{4}$ **D.** $\dfrac{27}{2}$

52. Which interval describes the set of real numbers such that the sum of 3 times a number and 5 is no more than −16?   D
  A. $\left(-\infty, -\dfrac{11}{3}\right)$ **D.** $(-\infty, -7]$
  B. $\left(-\infty, -\dfrac{11}{3}\right]$ **E.** none of these
  C. $(-\infty, -7)$

PRESENTATION

Lesson Opener

Use the distance formula, $d = rt$.

1. How does an increase in the rate affect the distance?
   The distance increases.

2. Solve the formula for time. $t = \dfrac{d}{r}$

3. How does a decrease in the rate affect the time?   The time increases.

Present the definition of a *power function*. This section focuses on power functions where the exponent is a nonzero integer. Power functions that have other rational exponents will be covered in Section 2.1.

1.4 Power Functions and Variations

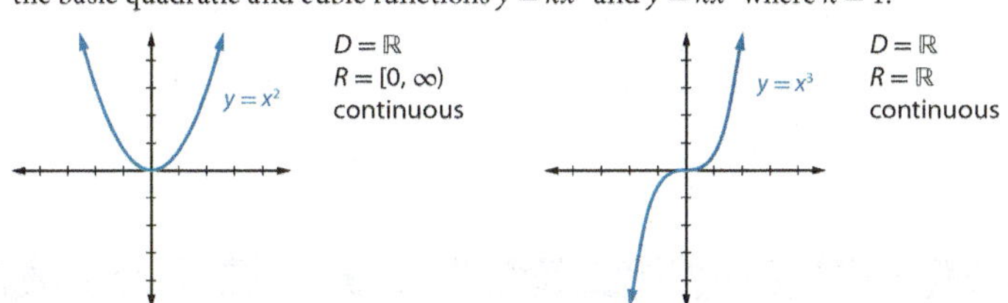

The electrostatic force between two charged objects varies inversely with the square of the distance between them.

Power functions include several basic functions that serve as building blocks for other functions.

> **DEFINITION**
>
> A **power function** is a function in the form $f(x) = kx^n$ where k and the power n are non-zero constants.

Functions such as $y = -3x^2$, $y = 2x^{\frac{1}{3}}$, and $y = \frac{1}{5x^4}$ are power functions, but $y = 2^x$ and $y = 6$ are not. While the exponent n can be any rational or irrational value, we will limit our discussion in this section to power functions with integral exponents. The simplest power function, $y = kx$, is a linear function through the origin with a slope of k. Examine the basic quadratic and cubic functions $y = kx^2$ and $y = kx^3$ where $k = 1$.

$y = x^2$
$D = \mathbb{R}$
$R = [0, \infty)$
continuous

$y = x^3$
$D = \mathbb{R}$
$R = \mathbb{R}$
continuous

The *end behavior* of a function can be described as limits of the values of $f(x)$ as x increases or decreases without bound. Notice how end behaviors of the quadratic and cubic functions are expressed using the notation of limits.

Examining End Behavior			
$f(x) = x^2$		$f(x) = x^3$	
Left-End	Right-End	Left-End	Right-End
$\lim\limits_{x \to -\infty} x^2 = \infty$	$\lim\limits_{x \to \infty} x^2 = \infty$	$\lim\limits_{x \to -\infty} x^3 = -\infty$	$\lim\limits_{x \to \infty} x^3 = \infty$

Examining the graph of a function will help you find intervals where the function is *increasing, decreasing,* or *constant* as you move from left to right. Notice that $f(x) = x^2$ decreases over $(-\infty, 0]$ and increases over $[0, \infty)$, and that $f(x) = x^3$ increases over its entire domain.

Objectives

1. Use key characteristics to identify power functions
2. Use limits to describe the end behaviors of functions
3. Describe intervals in which a function is increasing, decreasing, or constant
4. Classify functions as even, odd, or neither
5. Use power functions to model and solve problems involving direct and inverse variations

Flash

Electrostatic refers to a resting electric charge. The electrostatic force between two objects is measured in newtons. A buildup of static electricity can discharge when the charged objects come close to each other or when the buildup becomes so great that it overcomes the resistance between the objects, as is seen in lightning during a storm.

Common Student Error Students may confuse power functions with exponential functions. While their forms are similar, a power function has a variable base raised to a fixed exponent, and an exponential function has a fixed base raised to a variable exponent.

Have the students examine the characteristics of the graphs of the basic quadratic and cubic functions, drawing attention to their domain, range, and *end behavior.*

Use the graphs of $y = x^2$ and $y = x^3$ to illustrate the definitions of *increasing* and *decreasing* intervals within a function. Note that $x = 0$ is an endpoint of such an interval in either of these functions and is part of the increasing or decreasing interval. In calculus students will learn that the slope of the tangent to the curve at $x = 0$ is 0 for both $y = x^2$ and $y = x^3$, and that these functions are neither increasing nor decreasing at that specific point.

Define *even* and *odd* functions, emphasizing that even functions are symmetric with respect to (SWRT) the y-axis and odd functions are symmetric with respect to the origin (which is equivalent to 180° rotational symmetry).

Have the students describe characteristics of the power functions with positive exponents in Example 1.

Consider discussing *vertical asymptotes* when introducing power functions with negative exponents; these will be formally introduced in Chapter 2. After discussing Example 2, you may want to generalize the characteristics of $y = x^n$ when odd $n < 0$ and even $n < 0$, as well as the effect of the constant k.

Use the Data Analysis margin note to introduce Example 3. Most students should be familiar with direct and inverse variations from previous courses. You may want to ask students to state the more familiar version of the formula ($d = rt$) and the meaning of k, the *constant of proportionality*, in this equation.

Vocabulary

constant of variation (or proportion)
directly proportional
decreasing function
even function
increasing function
inversely proportional
end behavior
odd function
power
power function

Additional Exercises

Graph the function. Then describe its characteristics, including: domain; range; end behavior; intervals of continuity; whether the function is even, odd, or neither; and any symmetry.

1. $g(x) = -\frac{1}{2}x^4$

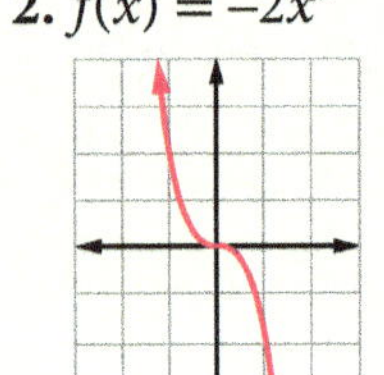

$D = \mathbb{R}; R = (-\infty, 0];$
$\lim\limits_{x \to -\infty} g(x) = -\infty,$
$\lim\limits_{x \to \infty} g(x) = -\infty;$ cont.: $\mathbb{R};$
even; SWRT y-axis

2. $f(x) = -2x^3$

$D = \mathbb{R}; R = \mathbb{R};$
$\lim\limits_{x \to -\infty} f(x) = \infty,$
$\lim\limits_{x \to \infty} f(x) = -\infty;$
cont.: $\mathbb{R};$ odd; SWRT origin

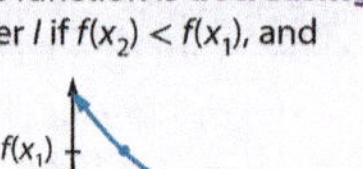

Given a function $f(x)$ and an interval I in which $x_2 > x_1$ for all x_1 and $x_2 \in I$,

the function is **increasing** over I if $f(x_2) > f(x_1)$,

the function is **decreasing** over I if $f(x_2) < f(x_1)$, and

the function is **constant** over I if $f(x_2) = f(x_1)$.

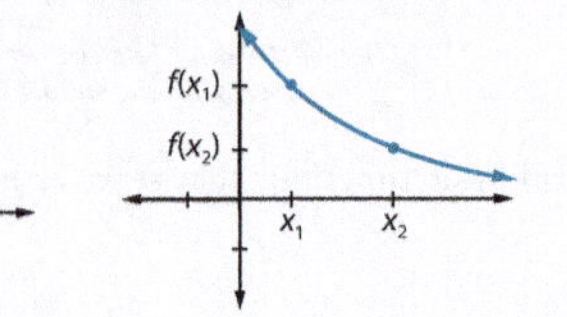
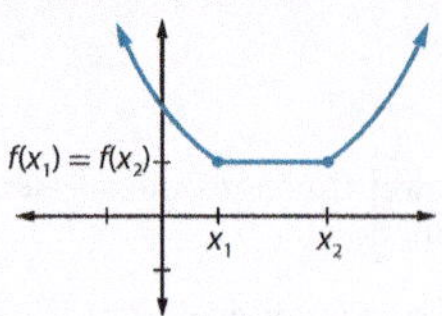

Many power functions can be classified as either even or odd.

DEFINITIONS

A function is **even** when $f(-x) = f(x)$ for all $x \in D$.
A function is **odd** when $f(-x) = -f(x)$ for all $x \in D$.

A function in which $f(-x) \neq f(x)$ and $f(-x) \neq -f(x)$ is neither even nor odd. The basic quadratic function $f(x) = x^2$ is an even function since $f(-x) = (-x)^2 = x^2 = f(x)$. Notice that its graph exhibits line symmetry with respect to the y-axis, a characteristic of any even function. The basic cubic function $f(x) = x^3$ is an odd function since $f(-x) = (-x)^3 = -x^3 = -f(x)$. Notice that its graph exhibits point symmetry with respect to the origin, a characteristic of any odd function. In exercises 40–41 you will show that all power functions of even degree are even functions and that all power functions of odd degree are odd functions.

Example 1   Describing Power Functions with Positive Exponents

Graph each function. Then describe its domain; range; end behavior; intervals of continuity; intervals for which the function is increasing, decreasing, or constant; whether the function is even, odd, or neither; and any symmetry.

a. $f(x) = \frac{1}{4}x^5$ **b.** $g(x) = -2x^4$

Answer

a.

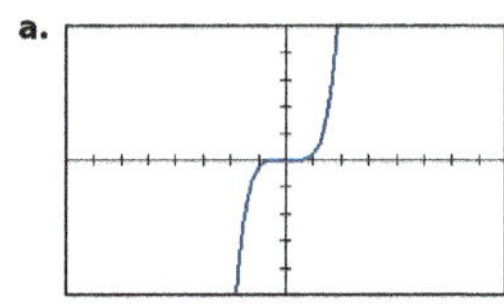

$D = \mathbb{R}; R = \mathbb{R}$
$\lim\limits_{x \to -\infty} \frac{1}{4}x^5 = -\infty, \lim\limits_{x \to \infty} \frac{1}{4}x^5 = \infty$
The function is continuous and increases over its entire domain.
The function is odd and exhibits point symmetry with respect to the origin.

b.

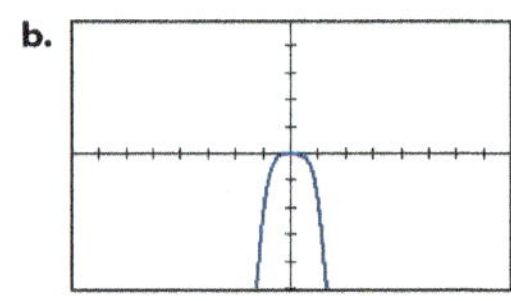

$D = \mathbb{R}; R = (-\infty, 0]$
$\lim\limits_{x \to -\infty} -2x^4 = -\infty, \lim\limits_{x \to \infty} -2x^4 = -\infty$
The function is continuous over its domain.
It increases over $(-\infty, 0]$ and decreases over $[0, \infty)$.
The function is even and exhibits line symmetry with respect to the y-axis.

SKILL ✓ **EXERCISE 19**

(the constant speed of the object)

In Examples 3–5 the strategy is to find the *constant of variation* and then use it to find the missing value. You may want to explain that *directly proportional* quantities maintain a constant ratio. This fact allows direct variation problems like those demonstrated in Examples 3–4 to be solved using a proportion.

$$\frac{V_1}{r_1^{\,3}} = \frac{V_2}{r_2^{\,3}}; \frac{288\pi \text{ cm}^3}{(6 \text{ cm})^3} = \frac{V_2}{(12 \text{ cm})^3}$$

$$V_2 = \frac{288\pi (12)^3}{(6)^3} \text{ cm}^3 = 2304\pi \text{ cm}^3$$

Students have studied joint variations in previous courses. They have not been included in this section because they are neither a power nor a reciprocal function. You may want to use Additional Exercises 8–9 to review joint variations.

Even inaccurate models can increase our understanding of natural phenomena through efforts to explain discrepancies and improve predictions.

Consider reminding the students that indirect variation problems like Example 5 can be solved using the fact that the product remains constant.

$$I_1 d_1^{\,2} = I_2 d_2^{\,2}$$
$$(150 \text{ lumen})(2 \text{ m})^2 = I_2(6 \text{ m})^2$$
$$I_2 = \frac{150(4)}{36} \text{ lumen} \approx 16.7 \text{ lumen}$$

The following chart can be used to summarize variations.

Variation	Function	Exponent
direct	$y = kx$	1
	$y = kx^n$	n
indirect	$y = \frac{k}{x}$	-1
	$y = \frac{k}{x^n}$	$-n$

TIPS

Ex. 3–11 Communicate your expectations for the method of graphing and the degree of accuracy for these sketched graphs.

The graph of a power function $y = kx^2$ with $k < 0$ is a reflection across the x-axis of the graph of the corresponding power function with $k > 0$. For example, $f(x) = 2x^4$ approaches ∞ as x approaches both $-\infty$ and ∞, but $f(x) = -2x^4$ approaches $-\infty$ as x approaches both $-\infty$ and ∞.

While the graphs of power functions with positive exponents are continuous functions, the graphs of power functions with negative exponents contain a discontinuity. Notice that the graph of $f(x) = x^{-1} = \frac{1}{x}$ has an infinite discontinuity at $x = 0$, where the function is undefined. The end behaviors of power functions with negative exponents also differ from those of power functions with positive exponents.

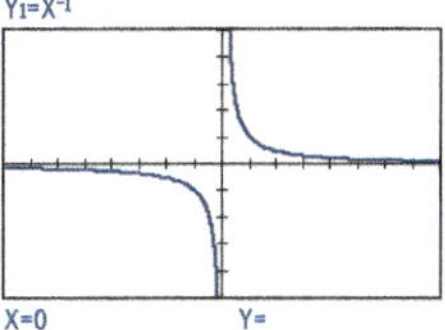

$f(x) = x^{-1}$	
Left-End	Right-End
$\lim\limits_{x \to -\infty} x^{-1} = 0$	$\lim\limits_{x \to \infty} x^{-1} = 0$

Example 2 Describing Power Functions with Negative Exponents

Graph each function. Then describe the domain; range; end behavior; intervals of continuity; intervals for which the function is increasing, decreasing, or constant; whether the function is even, odd, or neither; and any symmetry.

a. $f(x) = 3x^{-2}$ **b.** $g(x) = -\frac{1}{4}x^{-3}$

Answer

a.
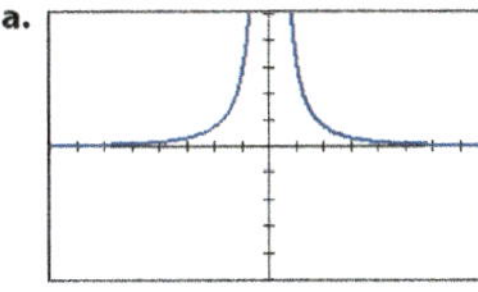
$D = \{x \mid x \neq 0\}; R = \{y \mid y > 0\}$
$\lim\limits_{x \to -\infty} 3x^{-2} = 0, \lim\limits_{x \to \infty} 3x^{-2} = 0$
The function has an infinite discontinuity at $x = 0$.
It increases over $(-\infty, 0)$ and decreases over $(0, \infty)$.
The function is even and exhibits line symmetry with respect to the y-axis.

b.
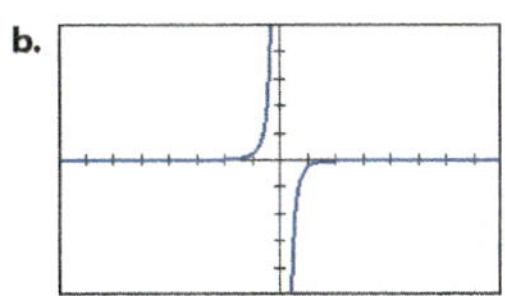
$D = \{x \mid x \neq 0\}; R = \{y \mid y \neq 0\}$
$\lim\limits_{x \to -\infty} -\frac{1}{4}x^{-3} = 0, \lim\limits_{x \to \infty} -\frac{1}{4}x^{-3} = 0$
The function has an infinite discontinuity at $x = 0$.
It increases over $(-\infty, 0)$ and over $(0, \infty)$.
The function is odd and exhibits point symmetry with respect to the origin.

— SKILL ✔ **EXERCISE 21**

You have studied direct and inverse variations in previous courses. Any variation of two variables can be viewed as a power function of the form $y = kx^n$ where k, the *constant of variation* (or *constant of proportion*), is positive.

If y varies *directly* with (or is *directly proportional* to) x, then $y = kx$. The fact that the circumference of a circle varies directly with the length of its radius is described by the function $c = 2\pi r$ where the constant of variation is 2π.

3. $h(x) = -2x^{-3}$
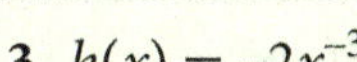
$D = \{x \mid x \neq 0\};$
$R = \{y \mid y \neq 0\};$
$\lim\limits_{x \to -\infty} h(x) = 0,$
$\lim\limits_{x \to \infty} h(x) = 0;$
cont.: $(-\infty, 0)$ and $(0, \infty)$;
odd; SWRT origin

4. $g(x) = \dfrac{1}{|x|}$
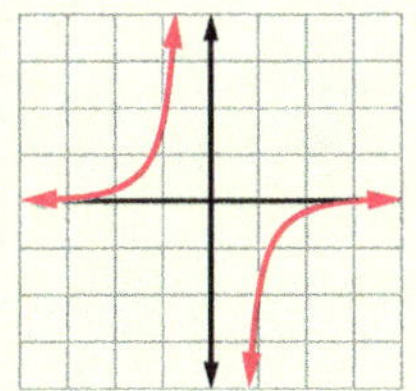
$D = \{x \mid x \neq 0\};$
$R = \{y \mid y > 0\};$
$\lim\limits_{x \to -\infty} g(x) = 0,$
$\lim\limits_{x \to \infty} g(x) = 0;$
cont.: $(-\infty, 0)$ and $(0, \infty)$;
even; SWRT y-axis

5. $f(x) = \begin{cases} x - 2 & \text{if } x \leq 3 \\ 1 & \text{if } 3 < x < 6 \\ \frac{1}{2}x - 2 & \text{if } x \geq 6 \end{cases}$
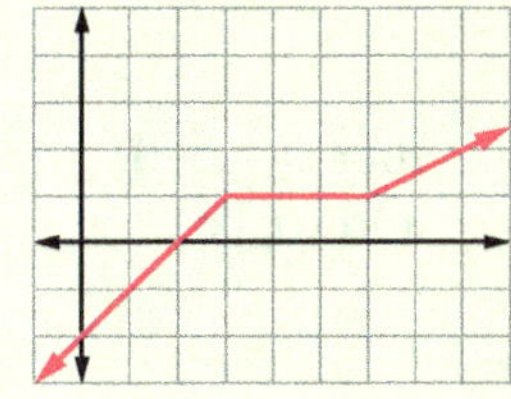
$D = \mathbb{R}; R = \mathbb{R};$
$\lim\limits_{x \to -\infty} f(x) = -\infty,$
$\lim\limits_{x \to \infty} f(x) = \infty;$ cont.: $\mathbb{R};$
neither; no symmetry

6. If s varies directly as the square of t, find the constant of variation if $s = 64$ when $t = 4$. $k = 4$

7. State and graph the variation function if y varies inversely with x and the constant of variation is 3.
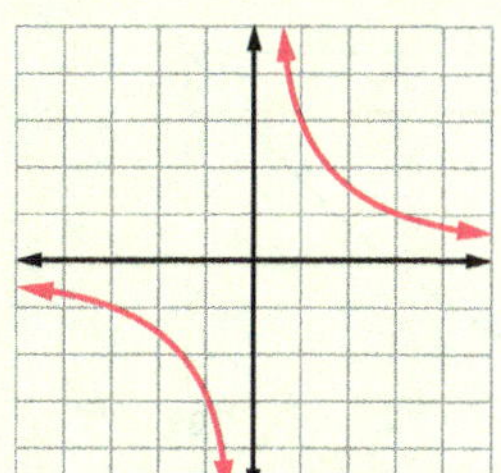

8. Suppose d varies jointly with r and t. If $d = 420$ when $r = 15$ and $t = 4$, find the values of k and d when $r = 75$ and $t = 20$. $k = 7; d = 10{,}500$

9. The volume of a right circular cylinder varies jointly with the height and the square of the radius. If a can containing 12 fl oz has a radius of 3 cm and a height of 12.5 cm, find the constant of variation and then the volume of a bucket with a radius of 15 cm and a height of 45 cm.

$k = \frac{8}{75} \approx 0.107; V = 1080$ oz or ≈ 8.4 gal

10. The value of the US dollar d varies inversely with the consumer price index (CPI). If the dollar is worth \$0.74 when the CPI is 135.14, find k and the worth of the dollar when the CPI is 220. $k = 100.0036; d \approx \$0.45$

For Graphing Calculators

The kinetic energy (KE) of a moving body varies directly with the square of the velocity (v). Assume the kinetic energy of a motorcyclist moving at 20 m/sec is 10,000 joules.

1. Write the direct variation.
$KE = 25v^2$ joules

2. Graph the variation in a $[-1, 25]$ by $[-1, 15{,}000]$ window.

3. Use CALC, 1:value to find the motorcyclist's kinetic energy when traveling 10 m/sec.   2500 joules
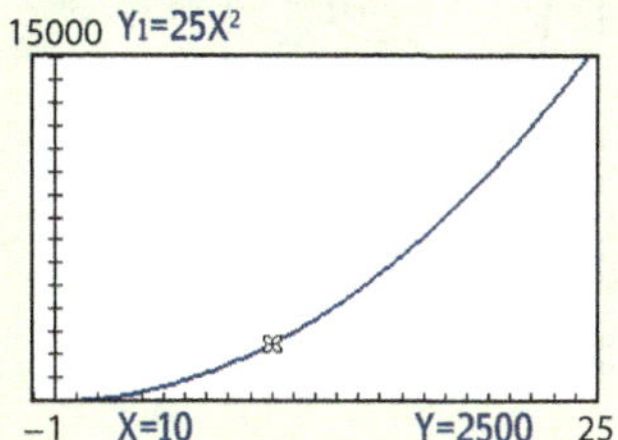

4. Find the velocity needed to obtain 12,500 joules of kinetic energy.

≈ 22.4 m/sec

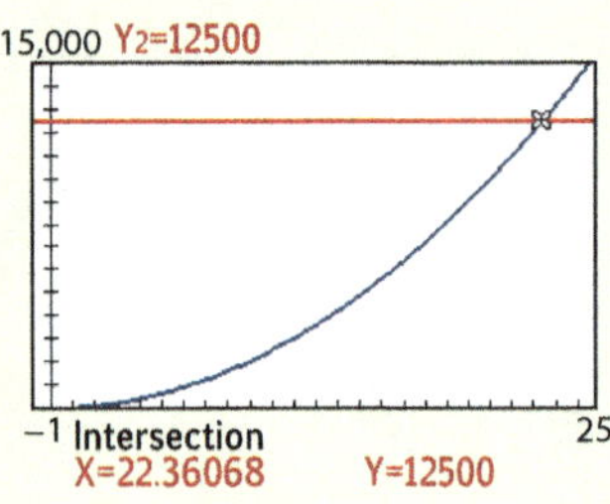

Assignments

- **Minimum:** 1, 3–4, 6–10, 15–18, 27–30, 32, 35–36, 44–45, 48–49, 52–53
- **Standard:** 1–4, 6–7, 10–14, 16, 18, 20–26, 29, 33, 35, 38–39, 41; 45–53 odd
- **Extended:** 1–2, 5, 8, 10–12, 15, 18, 20–25, 28, 30–32, 34, 36–40, 43, 46, 48–50, 53

Assessment

- Quiz 1B covers Sections 1.3–1.4.

Solutions

❯ A. Exercises

2a. There is no independent variable.

2b. The independent variable is the exponent, not the base.

3.

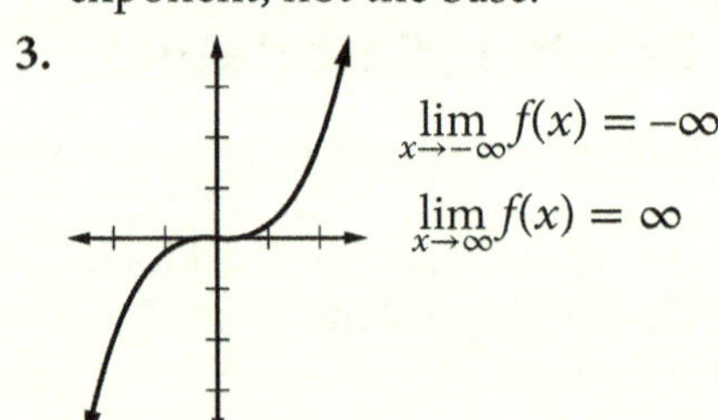

$$\lim_{x \to -\infty} f(x) = -\infty$$
$$\lim_{x \to \infty} f(x) = \infty$$

4.

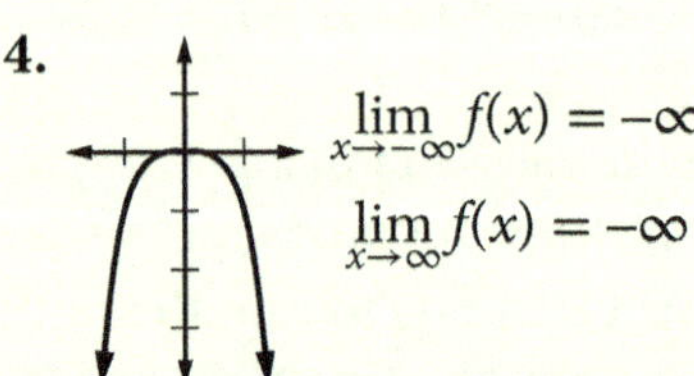

$$\lim_{x \to -\infty} f(x) = -\infty$$
$$\lim_{x \to \infty} f(x) = -\infty$$

5.

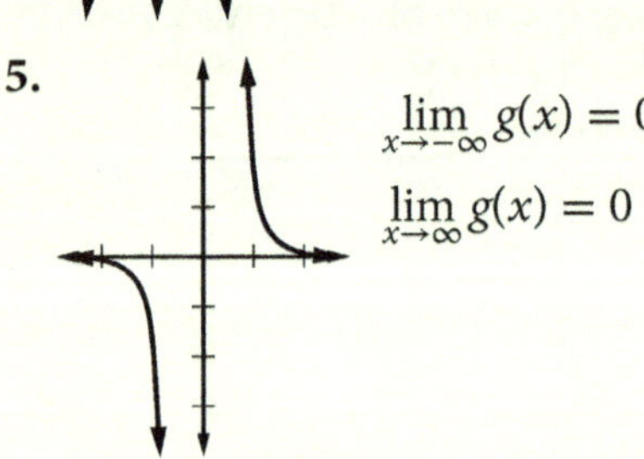

$$\lim_{x \to -\infty} g(x) = 0$$
$$\lim_{x \to \infty} g(x) = 0$$

When an object is traveling at a constant speed, the distance traveled varies directly with the elapsed time. If it took 2 hr to travel 50 mi, find the distance traveled in 7 hr.

Answer

$d = kt$ — 1. Distance varies directly with time.

$k = \dfrac{d}{t} = \dfrac{50 \text{ mi}}{2 \text{ hr}} = 25$ mi/hr — 2. Solve for the constant of variation, k.

$d = \left(25 \dfrac{\text{mi}}{\text{hr}}\right)t$ — 3. Use the variation from step 1 to write a function modeling the variation.

$= \left(25 \dfrac{\text{mi}}{\text{hr}}\right)(7 \text{ hr}) = 175$ mi — 4. Evaluate the function for $t = 7$ hr.

SKILL ✔ EXERCISE 27

A variable may also vary directly with a power of another variable. For example, the fact that the area of a circle varies directly with the square of its radius can be modeled with the function $A = \pi r^2$, where the constant of variation is π.

The volume of a sphere is directly proportional to the cube of its radius. If a sphere with a radius of 6 cm has a volume of 288π cm^3, find the volume of a sphere with a radius of 12 cm.

Answer

$V = kr^3$ — 1. The volume of a sphere varies directly with the cube of its radius.

$k = \dfrac{V}{r^3} = \dfrac{288\pi \text{ cm}^3}{(6 \text{ cm})^3} = \dfrac{4}{3}\pi$ — 2. Solve for k.

$V = \dfrac{4}{3}\pi r^3$ — 3. Write a function modeling the variation.

$= \dfrac{4}{3}\pi(12 \text{ cm})^3 = 2304\pi$ cm^3 — 4. Evaluate the function when $r = 12$ cm.

SKILL ✔ EXERCISE 29

In a direct variation the exponent in the power function is positive, but in an inverse variation, the exponent of the power function is negative. If *y varies inversely* with (or is *inversely proportional* to) *x*, then $y = kx^{-1}$ or $y = \dfrac{k}{x}$. The fact that the gravitational force between two objects varies inversely with the square of the distance between their centers can be modeled by the equation $F = kd^{-2}$ or $F = \dfrac{k}{d^2}$. Rearranging this equation shows that the product of inversely proportional quantities remains constant: $Fd^2 = k$.

6.

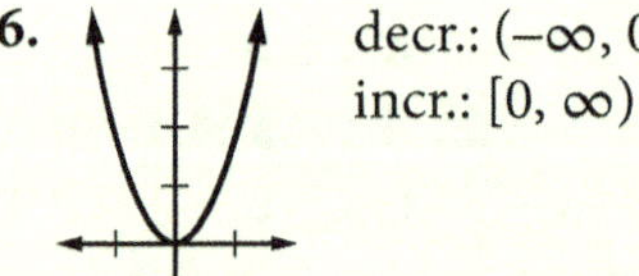

decr.: $(-\infty, 0]$
incr.: $[0, \infty)$

7.

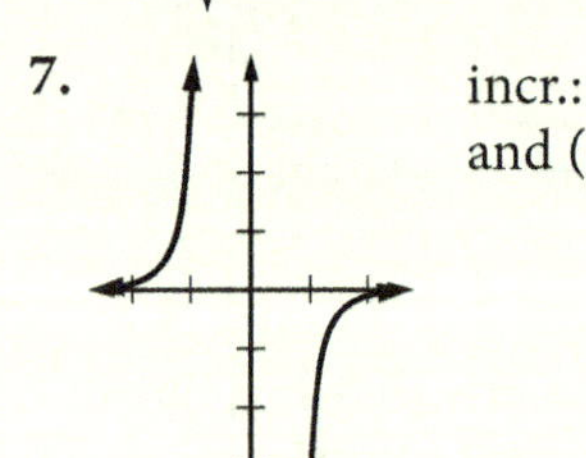

incr.: $(-\infty, 0)$ and $(0, \infty)$

8.

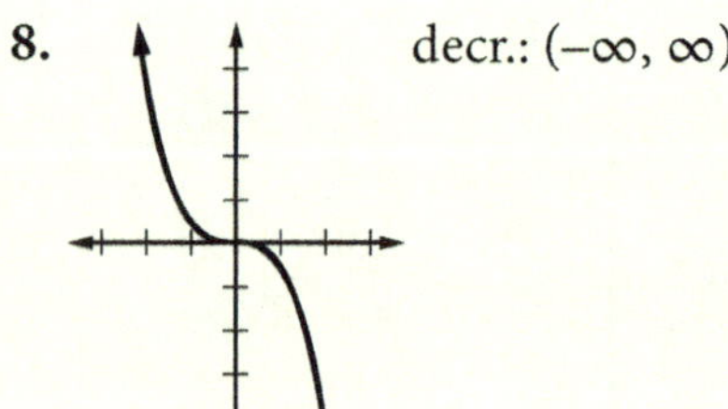

decr.: $(-\infty, \infty)$

9.

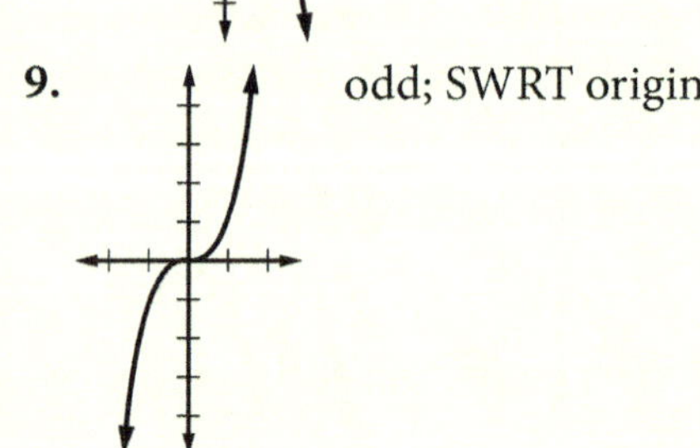

odd; SWRT origin

SKILL ✓ EXERCISE 31

Example 5   Modeling an Inverse Variation

The illumination, I, of an object is inversely proportional to the square of its distance, d, from the light source. What is the illumination of an object 6 m from a lamp if the illumination 2 m from the lamp is 150 lumen/m²?

Answer

$I = kd^{-2} = \dfrac{k}{d^2}$ 1. Illumination varies inversely with the square of the distance from the light source.

$k = Id^2 = \left(150\,\dfrac{\text{lumen}}{\text{m}^2}\right)(2\text{ m})^2$ 2. Solve for k.
$\quad = 600\text{ lumen}$

$I = \dfrac{600\text{ lumen}}{d^2}$ 3. Write a function modeling the variation.

$I = \dfrac{600\text{ lumen}}{(6\text{ m})^2} \approx 16.7\text{ lumen/m}^2$ 4. Evaluate the function when $d = 6$ m.

Notice that tripling the distance caused the illumination to be one-ninth of the original illumination.

❯ A. Exercises

1. State the power and constant of variation for each power function.
  a. $f(x) = -\frac{1}{4}x^3$ $3; -\frac{1}{4}$ **b.** $g(x) = -\frac{2}{3x^7}$ $-7; -\frac{2}{3}$
  c. $P = \frac{22.4}{V}$ $-1; 22.4$ **d.** $A = \frac{\sqrt{3}}{4}x^2$ $2; \frac{\sqrt{3}}{4}$

2. Explain why each function is not a power function.
  a. $y = 10$ **b.** $h(x) = 5^x$

Sketch a graph of each function. Then use limit notation to state its end behavior.

3. $f(x) = \frac{1}{4}x^3$ **4.** $f(x) = -x^4$
5. $g(x) = 2x^{-5}$

Sketch a graph of each function. Then describe intervals for which the function is increasing, decreasing, or constant.

6. $f(x) = 2x^2$ **7.** $f(x) = -3x^{-5}$
8. $g(x) = -\frac{x^3}{2}$

Sketch a graph of each function. Then state whether the function is even, odd, or neither and describe any symmetry.

9. $g(x) = 2x^3$ **10.** $f(x) = x^{-2}$
11. $f(x) = 4x^{-3}$

Write a power function modeling each variation.

12. q varies directly with r; $q = 3$ when $r = 12$ $q = \frac{1}{4}r$
13. m varies directly with the cube of n; $m = 96$ when $n = 2$ $m = 12n^3$

14. r varies inversely with p; $r = 15$ when $p = 3$ $r = 45p^{-1}$
15. d varies inversely with the square of a;
  $d = 8$ when $a = \frac{1}{2}$ $d = 2a^{-2}$

❯ B. Exercises

Use technology to graph each function. Then describe the domain; range; end behavior; intervals of continuity; intervals for which the function is increasing, decreasing, or constant; whether the function is even, odd, or neither; and any symmetry.

16. $f(x) = \frac{1}{2}x^{-1}$ **17.** $f(x) = -\frac{1}{8}x^2$
18. $f(x) = -4x^{-2}$ **19.** $f(x) = -0.25x^3$
20. $f(x) = \frac{1}{4}x^4$ **21.** $f(x) = 3x^{-3}$

Match the values given for $f(x) = kx^n$ to each described graph.

A. $k > 0$, n is even   **C.** $k < 0$, n is even
B. $k > 0$, n is odd   **D.** $k < 0$, n is odd

22. symmetric with respect to the origin and $\lim\limits_{x \to -\infty} f(x) = \infty$ D
23. symmetric with respect to the y-axis and $\lim\limits_{x \to \infty} f(x) = -\infty$ C
24. symmetric with respect to the y-axis and $\lim\limits_{x \to \infty} f(x) = \infty$ A
25. symmetric with respect to the origin and $\lim\limits_{x \to -\infty} f(x) = -\infty$ B

17.

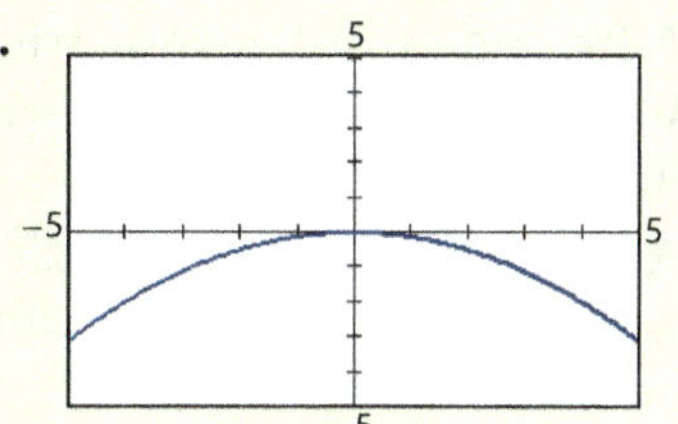

$D = \mathbb{R}$; $R = (-\infty, 0]$;
$\lim\limits_{x \to -\infty} f(x) = -\infty$; $\lim\limits_{x \to \infty} f(x) = -\infty$;
cont.: $\mathbb{R}$; incr.: $(-\infty, 0]$; decr.: $[0, \infty)$;
even; SWRT y-axis

18.

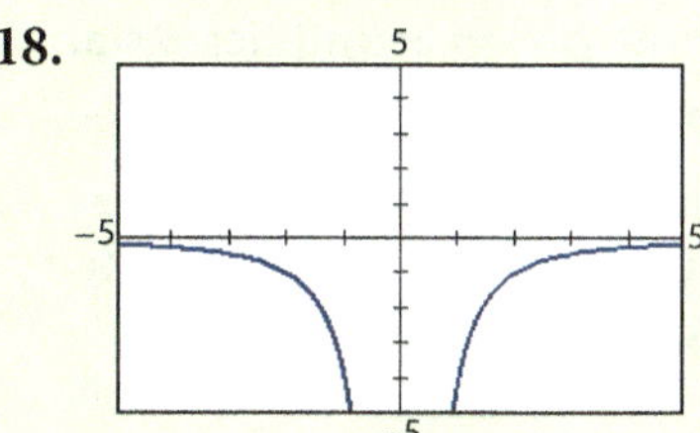

$D = \{x \mid x \neq 0\}$; $R = (-\infty, 0)$;
$\lim\limits_{x \to -\infty} f(x) = 0$; $\lim\limits_{x \to \infty} f(x) = 0$;
cont.: $(-\infty, 0)$ and $(0, \infty)$;
decr.: $(-\infty, 0)$; incr.: $(0, \infty)$;
even; SWRT y-axis

19.

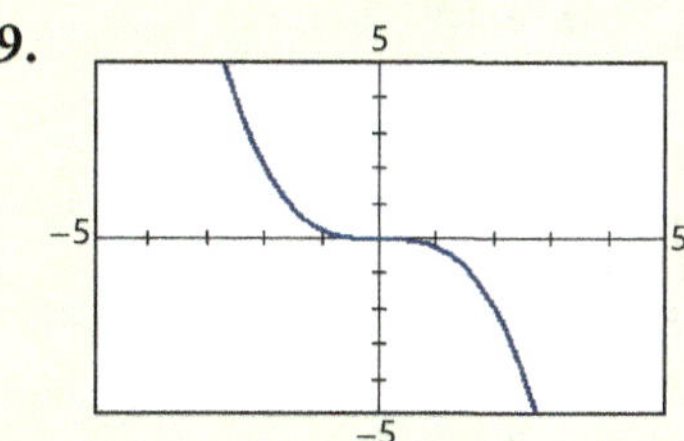

$D = \mathbb{R}$; $R = \mathbb{R}$; $\lim\limits_{x \to -\infty} f(x) = \infty$;
$\lim\limits_{x \to \infty} f(x) = -\infty$; cont.: $\mathbb{R}$; decr.: $\mathbb{R}$;
odd; SWRT origin

20.

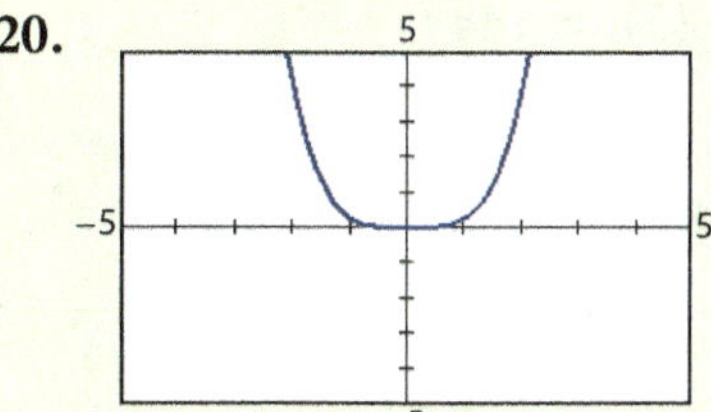

$D = \mathbb{R}$; $R = [0, \infty)$; $\lim\limits_{x \to -\infty} f(x) = \infty$;
$\lim\limits_{x \to \infty} f(x) = \infty$; cont.: $\mathbb{R}$;
decr.: $(-\infty, 0]$; incr.: $[0, \infty)$;
even; SWRT y-axis

21.

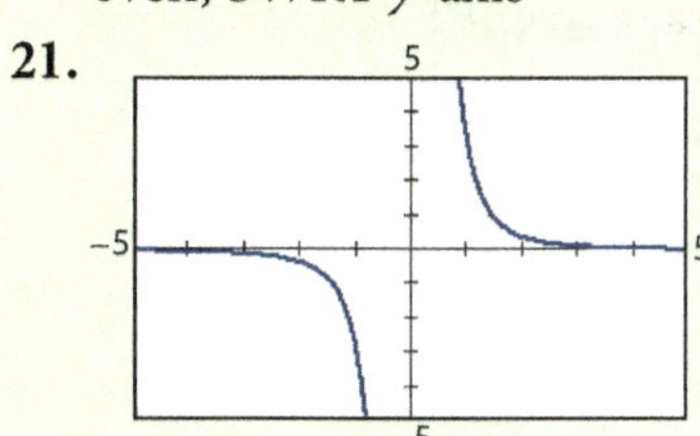

$D = \{x \mid x \neq 0\}$; $R = \{y \mid y \neq 0\}$;
$\lim\limits_{x \to -\infty} f(x) = 0$; $\lim\limits_{x \to \infty} f(x) = 0$;
cont.: $(-\infty, 0)$ and $(0, \infty)$;
decr.: $(-\infty, 0)$ and $(0, \infty)$; odd;
SWRT origin

10.

even; SWRT y-axis

11.

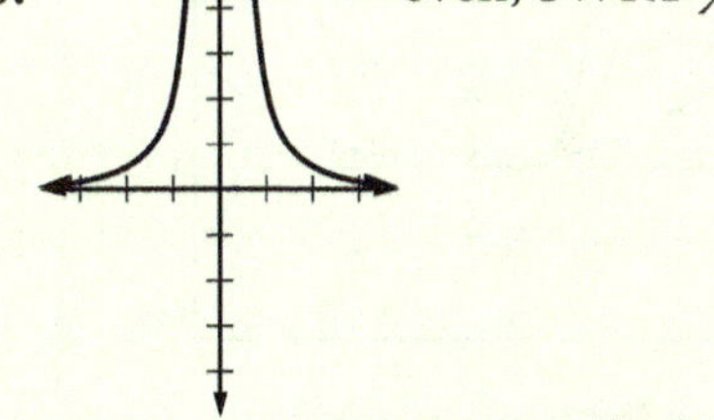

odd; SWRT origin

12. $q = kr$; $3 = k(12)$; $k = \frac{1}{4}$

13. $m = kn^3$; $96 = k(2)^3$; $k = 12$

14. $r = kp^{-1}$; $15 = \frac{k}{3}$; $k = 45$

15. $d = ka^{-2}$; $8 = \dfrac{k}{\left(\frac{1}{2}\right)^2}$; $k = 2$

❯ B. Exercises

16.

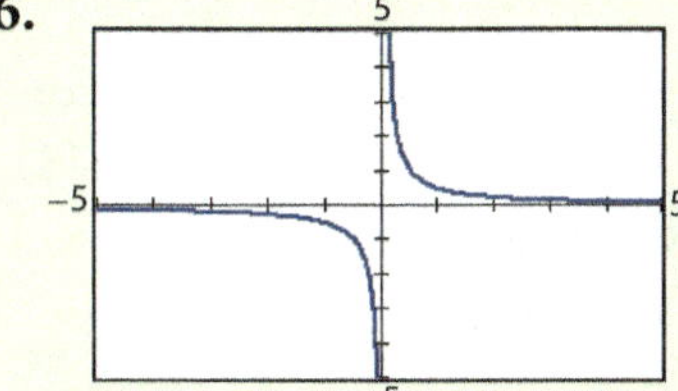

$D = \{x \mid x \neq 0\}$; $R = \{y \mid y \neq 0\}$;
$\lim\limits_{x \to -\infty} f(x) = 0$; $\lim\limits_{x \to \infty} f(x) = 0$;
cont.: $(-\infty, 0)$ and $(0, \infty)$;
decr.: $(-\infty, 0)$ and $(0, \infty)$;
odd; SWRT origin

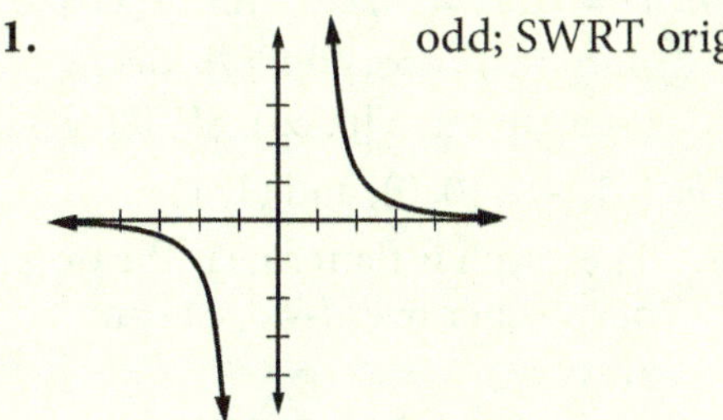

22, 25. A function that is symmetric with respect to the origin will have an odd power n.

23–24. A function that is symmetric with respect to the y-axis will have an even power n.

22. The $\lim\limits_{x \to -\infty} f(x) = \infty$ implies a reflection across the x-axis. So $k < 0$.

23. The $\lim\limits_{x \to -\infty} f(x) = -\infty$ implies a reflection across the x-axis. So $k < 0$.

24. The $\lim\limits_{x \to -\infty} f(x) = \infty$ implies a standard position. So $k > 0$.

25. An odd power function with the $\lim\limits_{x \to -\infty} f(x) = -\infty$ implies a standard position. So $k > 0$.

26. $c = kV$
$3 = k(10); k = \frac{3}{10}$
$5 = \frac{3}{10}V; V = 5\left(\frac{10}{3}\right) \approx 16.7$ volts

27. $d = kF$
$1.4 = k(12); k = \frac{1.4}{12}$
$d = \frac{1.4}{12}(42) = 4.9$ in.

28. $d = \frac{k}{w}$
$54 = \frac{k}{4}; k = 216$
$d = \frac{216}{18} = 12$ days

29. $d = ks^2$
$96 = k(40)^2; k = 0.06$
$d = 0.06(60)^2 = 216$ ft

30. $d = kt^2$
$4 = k(0.5)^2; k = 16$
$d = 16(3)^2 = 144$ ft

31. $f = \frac{k}{d^2}$
$8 \times 10^{-20} = \frac{k}{(0.02)^2}; k = 3.2 \times 10^{-23}$
$d = \frac{3.2 \times 10^{-23}}{(0.08)^2} = 5 \times 10^{-21}$ N

32. $a = \frac{k}{t}$
$2.8 = \frac{k}{18}; k = 50.4$
$3.5 = \frac{50.4}{t}; t = \frac{50.4}{3.5} = 14.4$ hr/week

33. $S_1 = \frac{k}{d_1^{\,3}}; k = S_1 d_1^{\,3}$
$S_2 = \frac{S_1 d_1^{\,3}}{\left(\frac{d_1}{2}\right)^3} = 8S_1$

34. $I = \frac{k}{d^2}; 1 = \frac{k}{1^2}; k = 1$
$I_2 = \frac{1}{2^2} = \frac{1}{4} = 25\%$
$I_5 = \frac{1}{5^2} = \frac{1}{25} = 4\%$
$I_{10} = \frac{1}{10^2} = \frac{1}{100} = 1\%$

---

26. Ohm's law states that current in an electric circuit with a given resistance varies directly with the voltage. If the current is 3 amps at 10 volts, what is the voltage if the current is 5 amps? ≈ 16.7 volts

27. Hooke's law states that the distance a spring stretches is directly proportional to the force applied to the spring. Find the distance a spring stretches under 42 lb of force if a force of 12 lb stretches the spring 1.4 in. 4.9 in.

28. If the number of days it takes to complete a project is inversely proportional to the number of workers hired for the project, and if it takes 54 days for 4 workers to complete the project, how many days are needed to complete a project with 18 workers? 12 days

29. If the stopping distance of a motor vehicle is directly proportional to the square of the vehicle's speed, and if it takes 96 ft for a vehicle to stop when traveling 40 mi/hr, find the stopping distance for the vehicle when traveling 60 mi/hr. 216 ft

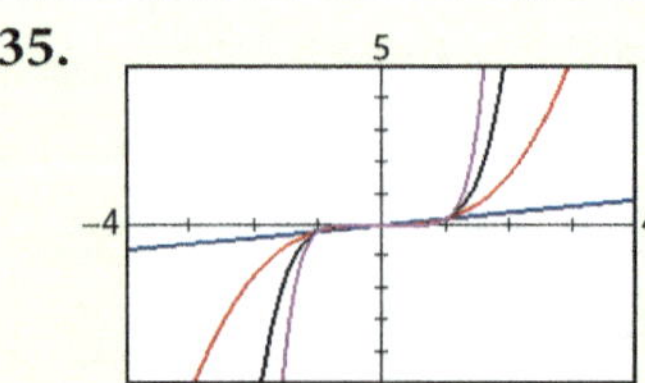

30. If the distance a stone falls varies directly with the square of the time it has been falling, and if a stone falls 4 ft in the first 0.5 sec, how far does the stone fall in 3 sec? 144 ft

31. The electrostatic force between two charged particles is inversely proportional to the square of the distance between them. If the force between two particles 0.02 m apart is 8×10^{-20} N, find the force when the particles are 0.08 m apart. 5×10^{-21} N

32. Bill's grade point average (GPA) varies inversely with the time he spends playing video games. Bill has a GPA of 2.8 when playing video games for 18 hr/week. How many hours should Bill limit his video game playing to if he wants to achieve a 3.5 GPA? 14.4 hr/week

33. The strength of a magnetic field is inversely proportional to the cube of the distance from the magnet. By what factor does the strength of the field increase when the distance to the magnet is halved? 8

34. The relative illumination of a photograph's subject is inversely proportional to the square of its distance from the light source. If the subject's relative illumination at 1 m is 100%, what is its relative illumination at 2 m? 5 m? 10 m? 25%; 4%; 1%

35. Graph the following power functions in the same window: $y = x; y = x^3; y = x^5; y = x^7$.
  a. State the domain; range; end behavior; intervals of continuity; intervals for which the function is increasing, decreasing, or constant; and the symmetry for any odd power function with $k = 1$.
  b. What points are included in all the graphs?
  c. Describe how the graph changes as the exponent increases.

36. Graph the following power functions in the same window: $y = x^2; y = x^4; y = x^6; y = x^8$.
  a. State the domain; range; end behavior; intervals of continuity; intervals for which the function is increasing, decreasing, or constant; and the symmetry for any even power function for which $k = 1$.
  b. What points are included in all the graphs?
  c. Describe how the graph changes as the exponent increases.

37. In what ways would the description of each function differ from the description of the even power functions in exercise 36a?
  a. $y = x^0$ **b.** $y = kx^0$

35.

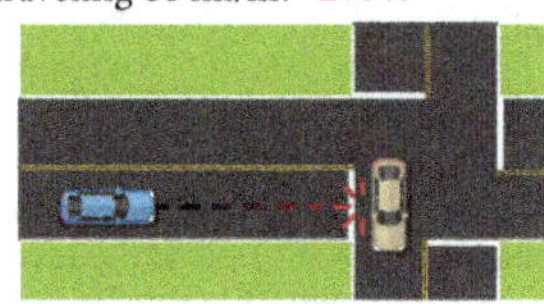

35a. $D = \mathbb{R}; R = \mathbb{R}; \lim\limits_{x \to -\infty} f(x) = -\infty;$
$\lim\limits_{x \to \infty} f(x) = \infty;$ cont.: $\mathbb{R}$; incr.: $\mathbb{R}$; SWRT origin

35b. $(-1, -1)$, $(0, 0)$, and $(1, 1)$

35c. The graph is flatter near the origin but steeper over $(-\infty, -1]$ and $[1, \infty)$.

36.

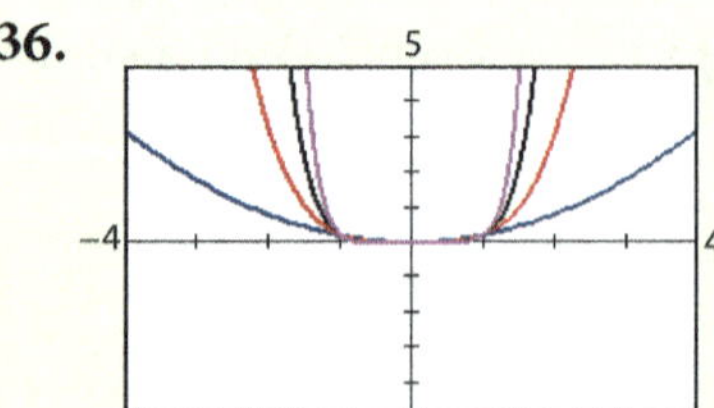

36a. $D = \mathbb{R}; R = [0, \infty); \lim\limits_{x \to -\infty} f(x) = \infty;$
$\lim\limits_{x \to \infty} f(x) = \infty;$ cont.: $\mathbb{R}$; decr.: $(-\infty, 0]$; incr.: $[0, \infty)$; SWRT y-axis

36b. $(-1, -1)$, $(0, 0)$, and $(1, 1)$

36c. The graph is flatter near the origin but steeper over $(-\infty, -1]$ and $[1, \infty)$.

37a. $R = \{y \mid y = 1\}; \lim\limits_{x \to -\infty} f(x) = 1;$
$\lim\limits_{x \to \infty} f(x) = 1;$ constant: $\mathbb{R}$

38. State whether each function is a power function. If not, give the reason. Then state whether the function is even, odd, or neither.

   a. $f(x) = 5\sqrt[3]{x}$
   b. $f(x) = 2x^0$
   c. $f(t) = 5^x$
   d. $f(x) = x^{\sqrt{3}}$

39. Use the graph of each function to state whether the function is even, odd, or neither. Then verify your answer algebraically.

   a. $y = 6x$ odd
   b. $y = 3x + 5$ neither
   c. $y = x^4 - 3x^2 + 5$ even
   d. $y = x^3 - 4x$ odd
   e. $y = x^2 - 5x - 2$ neither

40. **Prove:** Show that power functions in which the exponent is even are even functions.

41. **Prove:** Show that power functions in which the exponent is odd are odd functions.

42. **Prove:** Prove that if functions f and g are increasing functions, then $f + g$ is also an increasing function.

43. Kepler's third law of planetary motion states that the square of a planet's orbital period is directly proportional to the cube of the planet's average distance from the sun.

   a. Using T for the planet's orbital period and R for its average distance from the sun, express the relation as a variation. $T^2 = kR^3$

   b. If Earth has an orbital period of 365.2 days and an average distance of 93 million miles from the sun, find k for our sun's solar system.

   c. Use the data about Earth's orbit to estimate Neptune's orbital period (in Earth years) if Neptune's average distance from the sun is 2.8 billion miles.
   ≈ 165 yr

   43b. $k \approx 1.66 \times 10^{-19} \, \dfrac{\text{days}^2}{\text{mi}^3}$

CUMULATIVE REVIEW

Write each quadratic expression as a binomial squared.
[Algebra]

44. $x^2 - 8x + 16$ $(x - 4)^2$
45. $9x^2 + 12x + 4$ $(3x + 2)^2$
46. Simplify $(x - 1)^2 - 3(x - 1) + 5$. [Algebra] $x^2 - 5x + 9$
47. Write the linear function rule for $f(x) = 2x + 1$
   $f = \{(x, y) \mid (-4, -7), (-1, -1), (3, 7)\}$. [1.2]
48. Graph $g(x) = \begin{cases} -2x & \text{if } x < 0 \\ x - 2 & \text{if } x \geq 0 \end{cases}$. [1.3]
49. Find each limit of $g(x)$ from exercise 48. [1.3]
   a. $\lim\limits_{x \to 0^+}$ -2
   b. $\lim\limits_{x \to 0^-}$ 0
   c. $\lim\limits_{x \to 0}$ does not exist
50. Describe the continuity of $g(x)$ from exercise 48 at $x = 0$. [1.3]   C
   A. continuous
   B. infinite discontinuity
   C. jump discontinuity
   D. point discontinuity
   E. none of these
51. Which point is on the line passing through (4, 7) with a slope of $-\frac{5}{2}$? [1.2]   B
   A. (2, 2)
   B. (2, 12)
   C. (6, 12)
   D. (-1, 5)
   E. none of these

Write and solve an equation or inequality to complete the following exercises.   [Appendix 2]

52. Which interval describes the numbers such that the difference of 5 and twice the number is at least 37?   D
   A. $(-\infty, 16)$
   B. $[-16, \infty)$
   C. $(16, \infty)$
   D. $(-\infty, -16]$
   E. none of these

53. A commercial flight allows 5 hr for the plane to travel 2380 nautical miles while cruising at an average air speed of 500 knots. What is the average headwind speed that is expected during the flight?   C
   A. 120 knots
   B. 34 knots
   C. 24 knots
   D. 220 knots
   E. none of these

39d. $f(-x) = (-x)^3 - 4(-x)$
$= -x^3 + 4x = -(x^3 - 4x)$
$= -f(x)$

39e. $f(-x) = (-x)^2 - 5(-x) - 2$
$= x^2 + 5x - 2$
$-f(x) = -(x^2 - 5x - 2)$
$= -x^2 + 5x + 2$
$f(-x) \neq f(x)$ and $f(-x) \neq -f(x)$

40. Since $(-1)^n = 1$ when n is even, $f(-x)$
$= k(-x)^n = k(-1)^n x^n = kx^n = f(x)$.

41. Since $(-1)^n = -1$ when n is odd,
$f(-x) = k(-x)^n = k(-1)^n x^n = -kx^n$
$= -f(x)$.

42. Assume $x_1 < x_2$. Since f and g are increasing, $f(x_1) < f(x_2)$ and $g(x_1) < g(x_2)$ for all $x_1, x_2 \in \mathbb{R}$. Therefore $f(x_1) + g(x_1) < f(x_2) + g(x_2)$ and $(f + g)(x_1) < (f + g)(x_2)$.

43b. $k = \dfrac{(365.2 \text{ days})^2}{(9.3 \times 10^7 \text{ mi})^3} \approx 1.66 \times 10^{-19}$

43c. $T = \sqrt{\dfrac{(1 \text{ yr})^2}{(9.3 \times 10^7 \text{ mi})^3} (2.8 \times 10^9 \text{ mi})^3}$
≈ 165.2 yr

or $= \sqrt{\left(1.66 \times 10^{-19} \dfrac{\text{days}^2}{\text{mi}^3}\right)^2 (2.8 \times 10^9 \text{ mi})^3}$
$\approx 60{,}329$ days ≈ 165.2 yr

> **Cumulative Review**

46. $(x^2 - 2x + 1) - 3x + 3 + 5$
$= x^2 - 5x + 9$

47. $m = \dfrac{7 - (-1)}{3 - (-1)} = 2$
$y - 7 = 2(x - 3)$
$y - 7 = 2x - 6$
$y = 2x + 1$

48.

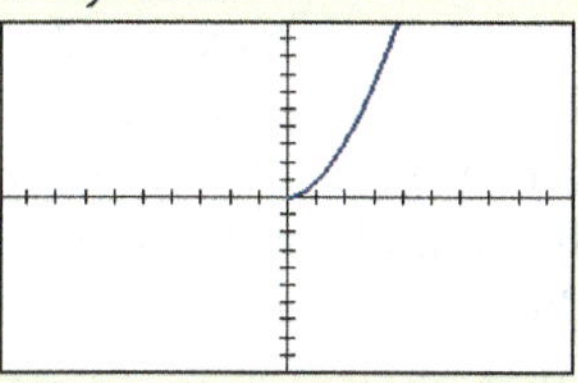

49c. Since $\lim\limits_{x \to 0^+} g(x) \neq \lim\limits_{x \to 0^-} g(x)$, $\lim\limits_{x \to 0} g(x)$ does not exist.

51. $m = \dfrac{7 - 12}{4 - 2} = -\dfrac{5}{2}$

52. $5 - 2n \geq 37$
$-2n \geq 32$
$n \leq -16$

53. $5(500 - w) = 2380$
$2500 - 5w = 2380$
$w = 24$

37b. $R = \{y \mid y = k\};$ $\lim\limits_{x \to -\infty} f(x) = k;$
$\lim\limits_{x \to \infty} f(x) = k;$ constant: $\mathbb{R}$

> **C. Exercises**

38a. yes; $5\sqrt[3]{x} = 5x^{\frac{1}{3}};$ Both 5 and $\frac{1}{3}$ are nonzero constants.; odd, since
$f(-x) = 5\sqrt[3]{-x} = 5\sqrt[3]{(-1)^3 x} = -5\sqrt[3]{x}$
$= -f(x)$

38b. no; The exponent cannot be zero.; even, since $f(x) = f(-x) = 2$

38c. no; The independent variable is the exponent.; neither, since $f(-x) = 5^{-x} \neq f(x)$ and $-f(x) = -5^x \neq f(x)$

38d. yes; $k = 1$ and $n = \sqrt{3};$ Both are nonzero constants.

The graph of $f(x) = x^{\sqrt{3}}$ is not symmetric with respect to the origin or the y-axis.

39a. $f(-x) = 6(-x) = -6x = -(6x)$
$= -f(x)$

39b. $f(-x) = 3(-x) + 5 = -3x + 5$
$-f(x) = -(3x + 5) = -3x - 5$
$f(-x) \neq f(x)$ and $f(-x) \neq -f(x)$

39c. $f(-x) = (-x)^4 - 3(-x)^2 + 5$
$= x^4 - 3x^2 + 5 = f(x)$

1.5 Transformations of Functions

Objectives

1. To graph rigid transformations (vertical and horizontal translations, reflections) of parent functions
2. To graph nonrigid transformations (stretching or shrinking) of parent functions
3. To graph compositions of rigid and nonrigid transformations
4. To write function rules for transformations of parent functions

Flash

The Sydney Harbour Bridge, the world's largest (though not longest) steel bridge, was built between 1924 and 1932. It is made up of approximately 58,500 tons of steel and is held together with about 6 million steel rivets. The bridge cost 4.2 million Australian dollars. Engineers allowed for 18 cm of expansion for the arches to accommodate extreme temperatures.

Vocabulary

family of functions
nonrigid transformation
parent function
reflection
rigid transformation
translation

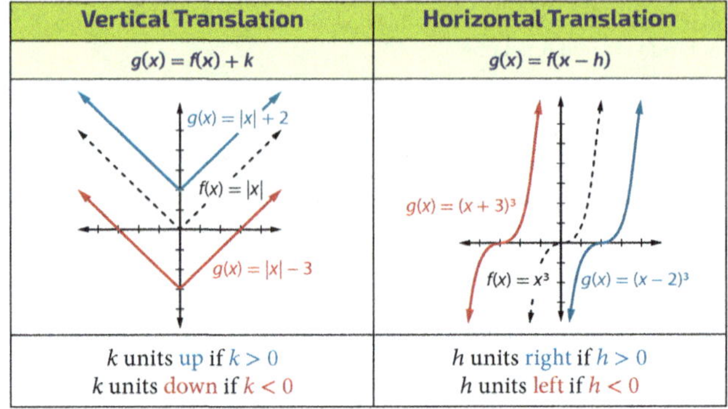

The arches of the Sydney Harbour Bridge can be modeled by transformations of the basic quadratic function.

After completing this section, you will be able to

- graph rigid and nonrigid transformations of parent functions.
- write function rules for transformations of parent functions.

Imagine texting without abbreviations or acronyms! The study of mathematics would be especially difficult without generalized algebraic notation. Much of this notation was not developed until the writings of François Viète were published in the sixteenth century.

A *family of functions* is a group of functions whose graphs share similar characteristics. The basic function with the simplest graph within the family is called the *parent function*. This section explores several ways that the graph of a parent function can be transformed to acquire the graph of a related function.

A *translation* of the parent function occurs when the graph is shifted horizontally or vertically.

Vertical Translation	Horizontal Translation
$g(x) = f(x) + k$	$g(x) = f(x - h)$
$g(x) = |x| + 2$ $f(x) = |x|$ $g(x) = |x| - 3$	$g(x) = (x + 3)^3$ $f(x) = x^3$ $g(x) = (x - 2)^3$
k units up if $k > 0$ k units down if $k < 0$	h units right if $h > 0$ h units left if $h < 0$

If $g(x) = f(x - h) + k$ (with h and $k \neq 0$), then $f(x)$ has been translated both horizontally and vertically.

Example 1 Translating a Parent Function

Graph $g(x) = (x + 1)^2 - 2$, then state the function's domain and range.

Answer

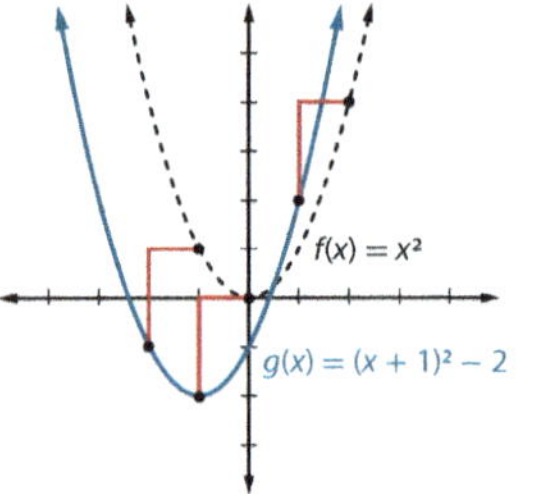

1. Sketch a graph of the parent function $f(x) = x^2$.
2. Since $g(x) = f(x - (-1))^2 + (-2)$ with $h = -1$ and $k = -2$, slide the graph of the parent function 1 unit left and 2 units down.
3. $D = \mathbb{R}; R = \{y \mid y \geq -2\}$

SKILL ✔ **EXERCISE 11**

A *reflection* across a coordinate axis produces a mirror image. Notice that a "vertical" reflection occurs when the "output" of the function is negated and a "horizontal" reflection occurs when the "input" of the function is negated.

Reflection in the *x*-axis	Reflection in the *y*-axis
$g(x) = -f(x)$	$g(x) = f(-x)$
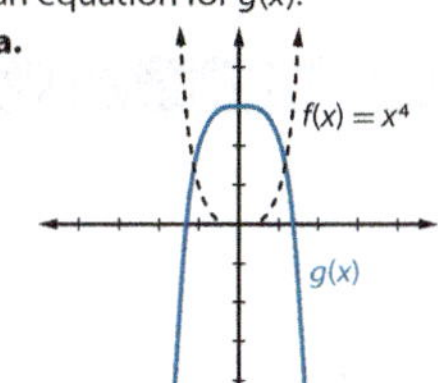	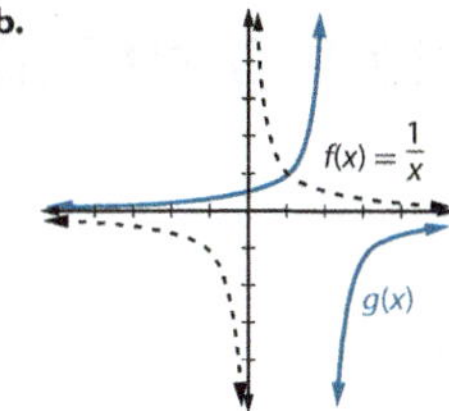

Example 2 Reflecting and Translating Parent Functions

Describe the transformation of the graph of $f(x)$ that results in the graph of $g(x)$. Then write an equation for $g(x)$.

a.

b.

Answer

a. The graph of $g(x)$ is a reflection of $f(x) = x^4$ in the *x*-axis followed by a translation 3 units up.

$$\therefore g(x) = -x^4 + 3$$

b. The graph of $g(x)$ is a translation of $f(x) = \frac{1}{x}$ two units to the right followed by a reflection in the *x*-axis.

$$\therefore g(x) = -\frac{1}{x-2}$$

SKILL ✔ **EXERCISE 13**

Translations and reflections are called *rigid transformations* since they maintain the shape of the graph, changing only the graph's position in the plane. The graph of a function can also be stretched or shrunk vertically or horizontally. These *nonrigid transformations* distort the shape of the graph.

graph of $g(x)$ is also a translation of $f(x)$ two units to the left, followed by a reflection in the *y*-axis. The end of the lesson provides more instruction on combining transformations.

Refer to the illustrations of the *nonrigid* (stretching and shrinking) *transformations*. Point out that multiplying the output of a function results in a vertical stretching or shrinking, while multiplying the input of a function results in a horizontal stretching or shrinking. Examples 3a and 3b illustrate a vertical and a horizontal stretch, respectively.

Combining several transformations can be confusing, especially when com-bining horizontal transformations, which are performed in an order that is opposite the typical order of operations. Carefully illustrate each transformation in Example 4.

One-on-One Some students may benefit from translating the vertex (for even functions) or the point of inflection (for odd functions) first. Once the translation is completed, the student can then complete any reflections and dilations.

Finally, use Example 5 to discuss how a function rule can be written by interpreting the graph as a transformation of a parent function.

Without graphing, describe each transformation of the parent function $f(x) = |x|$.

1. $g(x) = |x + 3| + 2$

$f(x)$ shifted 3 units left and 2 units up

2. $h(x) = -2|x| - 5$

$f(x)$ reflected across the *x*-axis, stretched vertically by a factor of 2, and then shifted 5 units down

3. $m(x) = 3|2 - x| + 1$

$f(x)$ shifted 2 units left, reflected across the *y*-axis, stretched vertically by a factor of 3, and then shifted 1 unit up

Graph each function as a transformation of a parent function. Then state the function's domain and range.

4. $f(x) = \dfrac{1}{x - 3} + 2$

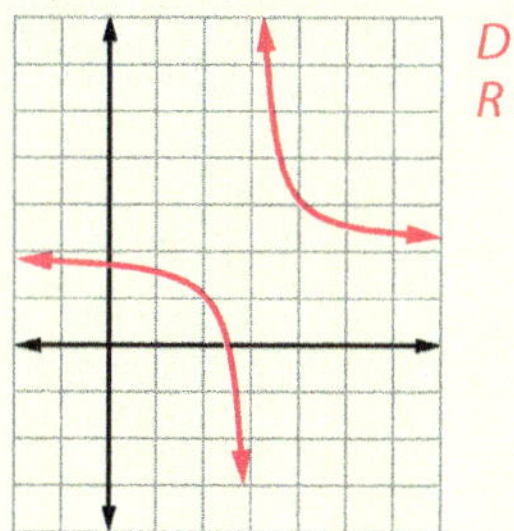

$D = \{x \mid x \neq 3\}$
$R = \{y \mid y \neq 2\}$

5. $f(x) = 3[x] + 1$

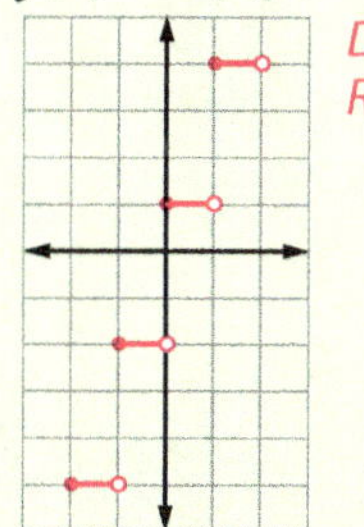

$D = \mathbb{R}$
$R = \{3x + 1 \mid x \in \mathbb{Z}\}$

6. $f(x) = -[2x]$

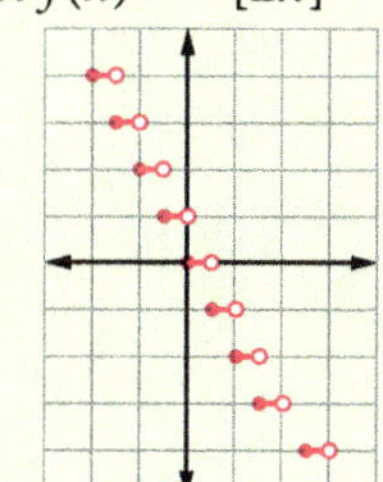

$D = \mathbb{R}$
$R = \mathbb{Z}$

7. $f(x) = 2(3 - x)^3 + 2$

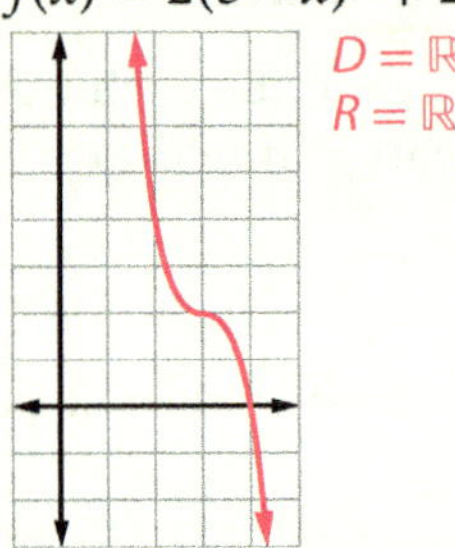

$D = \mathbb{R}$
$R = \mathbb{R}$

8. Write an equation for $g(x)$ as a transformation of $f(x) = |x|$.

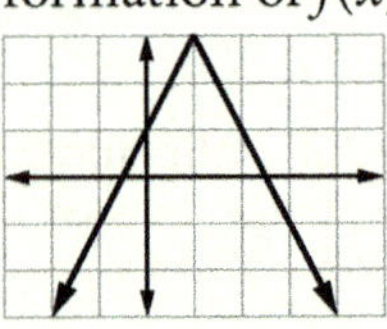

$g(x) = -2|x - 1| + 3$

 **TIP**

To stretch a function vertically by a factor of 3, triple the y-coordinate (output) of each ordered pair. To stretch a function horizontally by a factor of 3, triple the x-value (input) of each ordered pair.

Vertical Stretch or Shrink	Horizontal Stretch or Shrink
$g(x) = af(x)$	$g(x) = f(bx)$
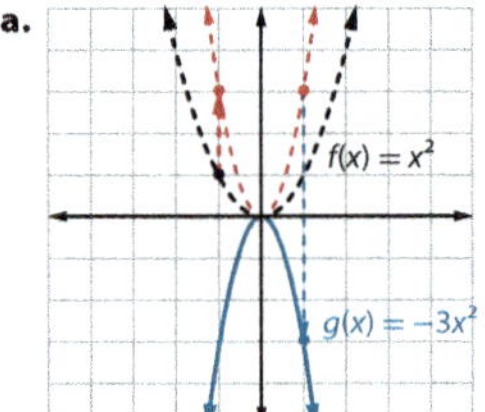	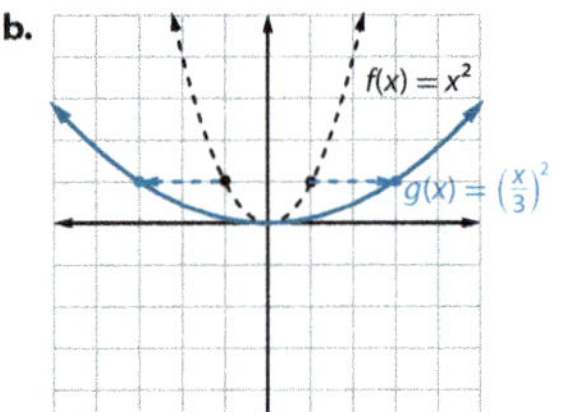
stretch by a factor of a if $|a| > 1$ or shrink by a factor of a if $0 \le |a| < 1$	shrink by a factor of $\frac{1}{|b|}$ if $|b| > 1$ or stretch by a factor of $\frac{1}{|b|}$ if $0 \le |b| < 1$

The horizontal shrink of $h(x) = \sqrt{2x}$ could also be viewed as a vertical stretch when the function is written as $g(x) = 1.414\sqrt{x}$. Similarly, the horizontal stretch of $h(x) = \sqrt{0.5x}$ can be viewed as a vertical shrink when the function is written as $g(x) = 0.707\sqrt{x}$.

Example 3 Stretching and Shrinking Parent Functions

Sketch the graph of the parent function $f(x) = x^2$ and use it to graph each function.

a. $g(x) = -3x^2$

b. $h(x) = \left(\frac{x}{3}\right)^2$

Answer

a.

b.

The graph of $f(x)$ is stretched vertically by a factor of 3 and then reflected in the x-axis.

The graph of $f(x)$ is stretched horizontally by a factor of $\frac{1}{b} = \frac{1}{\left(\frac{1}{3}\right)} = 3$.

SKILL ✔ EXERCISE 15

What happens when several transformations are combined? Consider the following illustrations of transformations of $f(x) = x^2$. The vertical transformations in $g(x)$ are the result of operations on the output of the parent function $f(x) = x^2$ and the horizontal transformations in $h(x)$ are the result of operations on the input of $f(x) = x^2$.

TIPS

Ex. 17 The set of even integers can be described as $\{x \mid x = 2n, n \in \mathbb{Z}\}$, and the set of odd integers can be described as $\{x \mid x = 2n + 1, n \in \mathbb{Z}\}$.

Ex. 39 The absolute value function ABS is found using MATH, NUM. The graph of an equation can be turned off by moving the cursor to the graph's equal sign in the Y= menu and pressing ENTER to deselect the function. Selecting different graph styles such as Dot, ˙·, or Path, ⫯, from the options found to the left of the functions can help differentiate graphs that overlap.

Ex. 40 Students should complete exercise 39 to discover general principles that help with this exercise.

Ex. 49 This question assumes Euclidean geometry, in which a line cannot be parallel to itself.

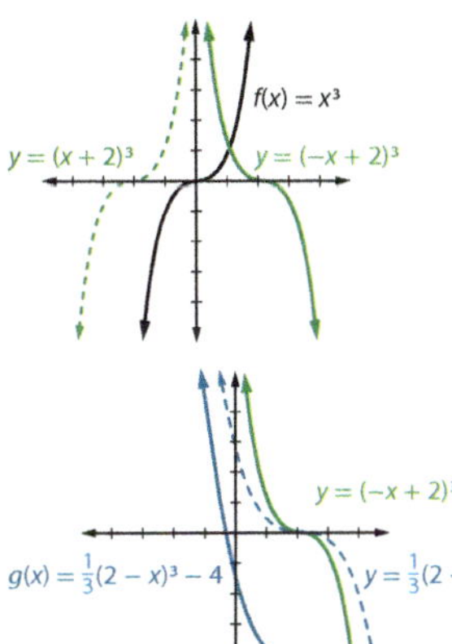

$g(x) = -2x^2 + 3$

1. a reflection in the x-axis and a vertical stretch by a factor of 2
2. then a slide up 3 units

$h(x) = (-2x + 3)^2$

1. a slide left 3 units
2. then a reflection in the y-axis and a horizontal shrink by a factor of $\frac{1}{2}$

The vertical transformations related to the output of the parent function follow the order of operations. You may have noticed that the horizontal shift and shrink seem to be opposites of the vertical shift and stretch. These horizontal transformations are related to the input of the parent function (within the parentheses). They are performed in an order that is opposite of the order of operations. While horizontal and vertical transformations do not affect each another, horizontal transformations are typically completed first.

Example 4   Combining Transformations

Use transformations of a basic parent function to sketch the graph of $g(x) = \frac{1}{3}(2 - x)^3 - 4$.

Answer

1. Sketch a graph of the parent function $f(x) = x^3$.
2. Complete the horizontal transformations implied by operations on the input of the parent function: $(2 - x)$ or $(-x + 2)$.
   a. a shift 2 units left
   b. then a reflection in the y-axis

3. Complete the vertical transformations implied by operations on the output of the parent function: $\frac{1}{3}f(x) - 4$.
   a. a vertical shrink by a factor of 3
   b. then a shift 4 units down

—————————————————— SKILL ✔ EXERCISE 21

Describing the graph of a function as a transformation of a known parent function can help you write a function rule for the graph.

KEYWORD SEARCH

parent function chart 🔍

Solutions

❯ A. Exercises

1. $f(x)$ translated 3 units down

2. $f(x)$ translated 2 units left

3. $f(x)$ reflected in the x-axis and shrunk vertically by a factor of $\frac{1}{4}$

4. $f(x)$ stretched horizontally by 3 (or shrunk vertically by a factor of $\frac{1}{9}$)

7. $f(x)$ stretched horizontally by 2 (or shrunk vertically by a factor of $\frac{1}{8}$) and translated 5 units down

8. $f(x)$ translated 1 unit right and 2 units up

9. $f(x)$ stretched vertically by 2 and translated 3 units down

10. $f(x)$ translated 4 units left, reflected in the x-axis, and shrunk vertically by a factor of $\frac{3}{4}$

11. 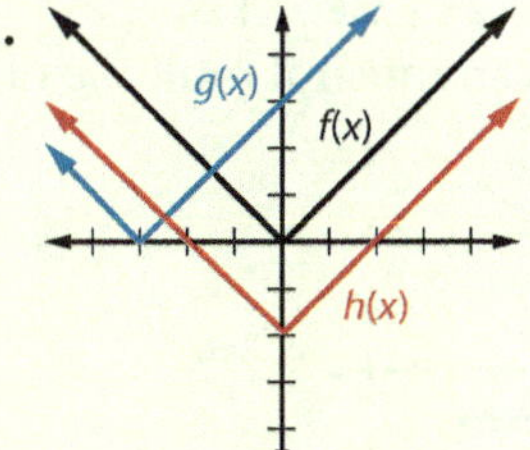

12. 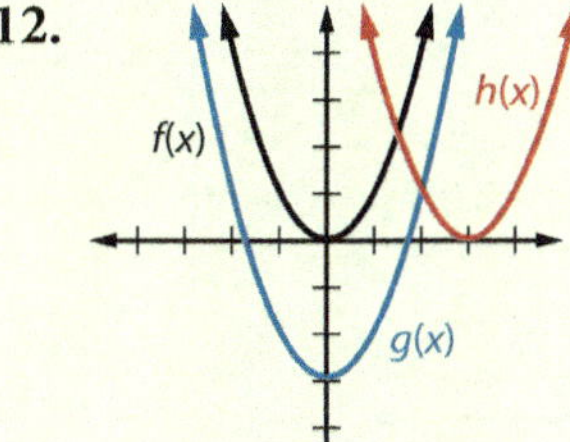

13. 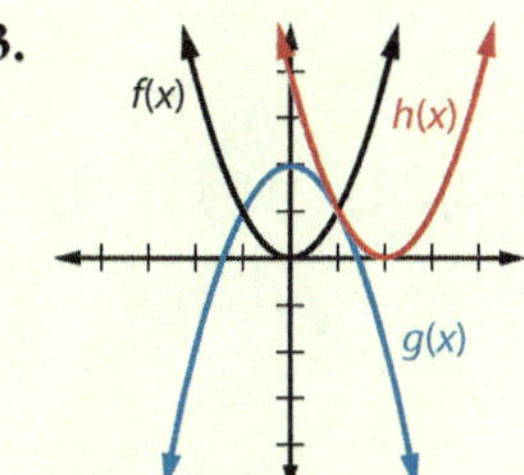

14. 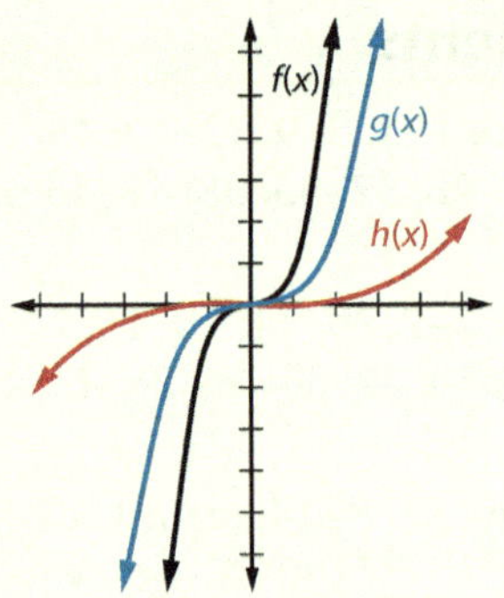

15. 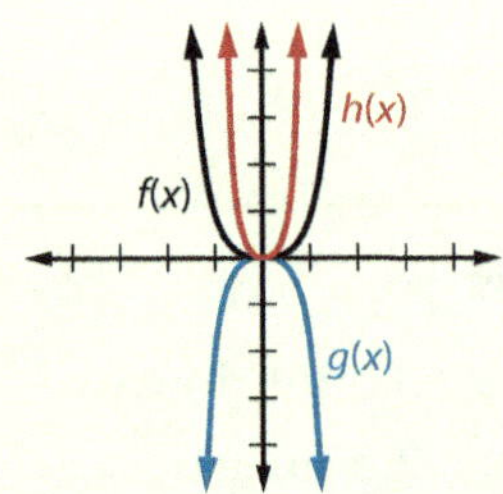

16. 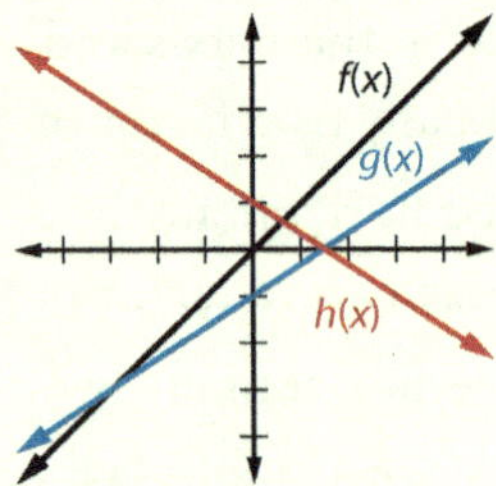

> ## B. Exercises

17. 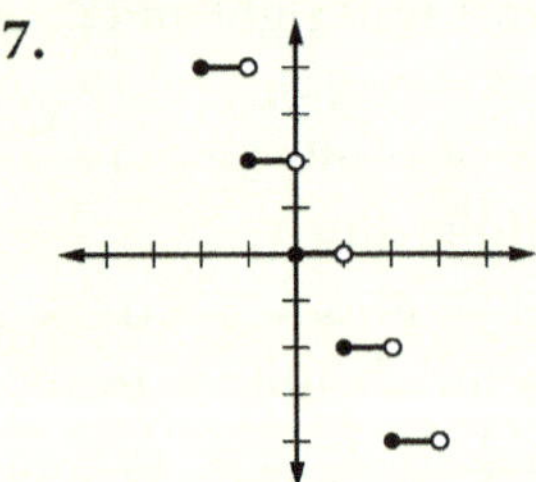

$D = \mathbb{R}; R = \{2x \mid x \in \mathbb{Z}\}$ or
$R = \{x \mid x \text{ is an integral multiple of } 2.\}$

18. 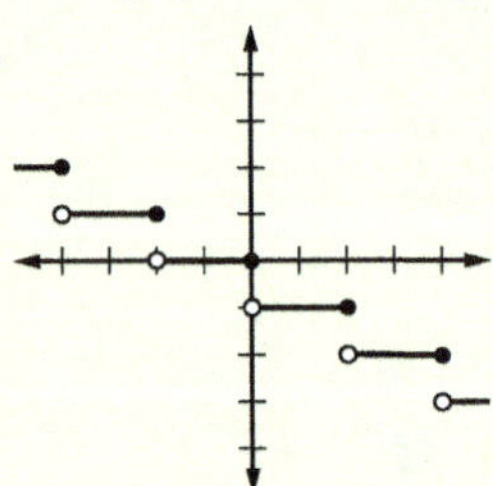

$D = \mathbb{R}; R = \mathbb{Z}$

19. 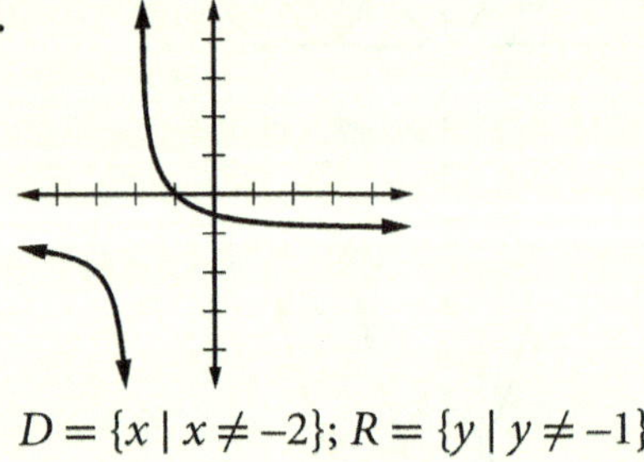

$D = \{x \mid x \neq -2\}; R = \{y \mid y \neq -1\}$

Example 5   Writing Function Rules

Describe each graph as a transformation of $f(x) = x^2$.
Then write its function rule.

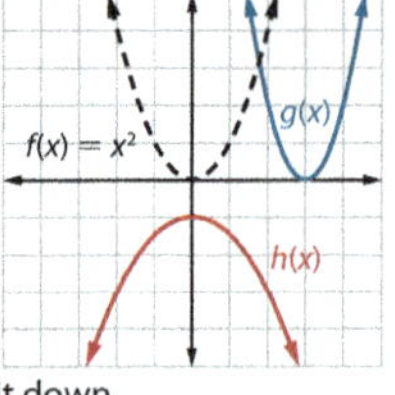

Answer

a. The graph for $g(x)$ was translated 3 units right and
vertically stretched by a factor of 2.
$\therefore g(x) = 2(x - 3)^2$

b. The graph for $h(x)$ was reflected in the x-axis and
vertically shrunk by a factor of $\frac{1}{2}$ before being translated 1 unit down.
$\therefore h(x) = -\frac{1}{2}x^2 - 1$

SKILL ✔ EXERCISE 25

Knowing the general shape of parent functions helps to determine the equation that best models collected data.

> ## A. Exercises

Describe the graph of $g(x)$ as a transformation of $f(x) = x^2$. Confirm your answer by graphing $f(x)$ and $g(x)$ in the same window.

1. $g(x) = x^2 - 3$
2. $g(x) = (x + 2)^2$
3. $g(x) = -\frac{1}{4}x^2$
4. $g(x) = \left(\frac{x}{3}\right)^2$

5. Match each function to the described transformation of $f(x)$.

a. $f(x) + c$ III   **I.** reflection in the x-axis
b. $f(x + c)$ IV   **II.** reflection in the y-axis
c. $-f(x)$ I   **III.** vertical translation
d. $f(-x)$ II   **IV.** horizontal translation

6. Match each function to the described transformation of $g(x)$.

a. $g(cx); |c| > 1$ IV   **I.** vertical stretch
b. $g(cx); 0 < |c| < 1$ III   **II.** vertical shrink
c. $cg(x); 0 < |c| < 1$ II   **III.** horizontal stretch
d. $cg(x); |c| > 1$ I   **IV.** horizontal shrink

Describe the graph of each function as a transformation of $f(x) = x^3$. Confirm your answer by graphing $f(x)$ and $g(x)$ in the same window.

7. $g(x) = \left(\frac{x}{2}\right)^3 - 5$
8. $g(x) = (x - 1)^3 + 2$
9. $g(x) = 2x^3 - 3$
10. $g(x) = -\frac{3}{4}(x + 4)^3$

Draw a graph of the parent function $f(x)$. Then graph $g(x)$ and $h(x)$ as transformations of $f(x)$ on the same coordinate plane.

11. $f(x) = |x|; g(x) = |x + 3|; h(x) = |x| - 2$
12. $f(x) = x^2; g(x) = x^2 - 3; h(x) = (x - 3)^2$
13. $f(x) = x^2; g(x) = -x^2 + 2; h(x) = (-x + 2)^2$
14. $f(x) = x^3; g(x) = \frac{x^3}{4}; h(x) = \left(\frac{x}{4}\right)^3$

15. $f(x) = x^4; g(x) = -2x^4; h(x) = (-2x)^4$
16. $f(x) = x; g(x) = \frac{2}{3}x - 1; h(x) = -\frac{2}{3}x + 1$

> ## B. Exercises

Graph each function as a transformation of a parent function. Then state the function's domain and range.

17. $g(x) = -2[x]$
18. $h(x) = \left[-\frac{x}{2}\right]$
19. $m(x) = \frac{1}{x + 2} - 1$
20. $n(x) = \frac{2}{(x - 3)^2}$
21. $p(x) = -2(x - 3)^2 + 4$
22. $q(x) = \frac{1}{2}|x + 1| - 3$

Write a function rule for each illustrated transformation of $f(x) = \sqrt{x}$.

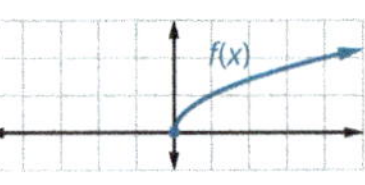

23. 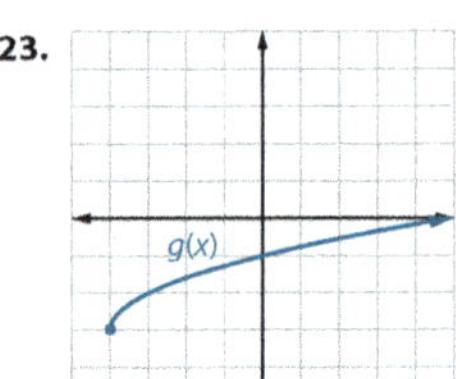

$g(x) = \sqrt{x + 4} - 3$

24. 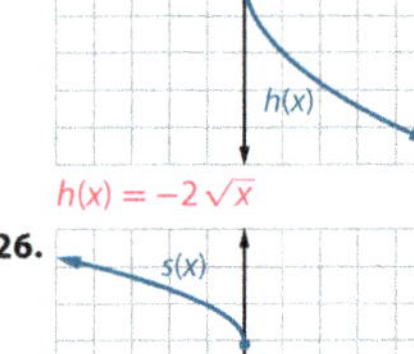

$h(x) = -2\sqrt{x}$

25. 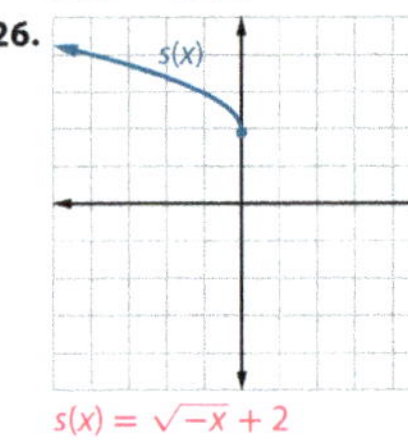

$r(x) = \frac{1}{2}\sqrt{x + 3}$

26.

$s(x) = \sqrt{-x} + 2$

---

20. 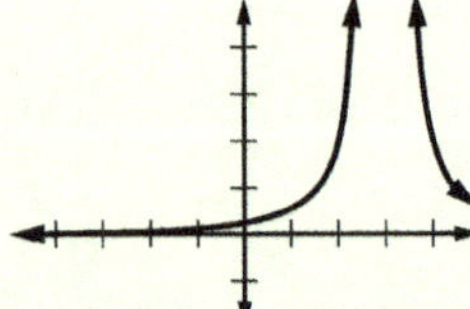

$D = \{x \mid x \neq 3\}; R = \{y \mid y > 0\}$

21. 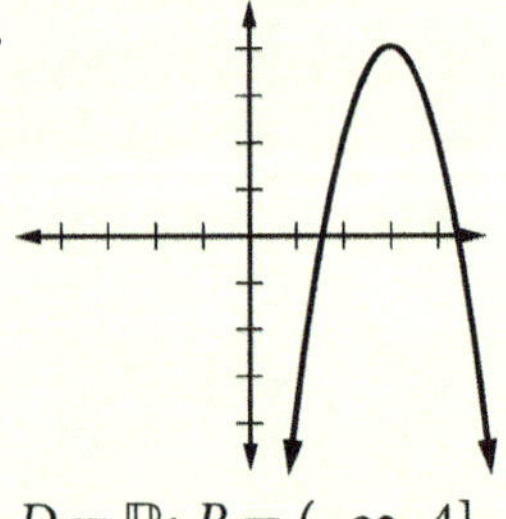

$D = \mathbb{R}; R = (-\infty, 4]$

22. 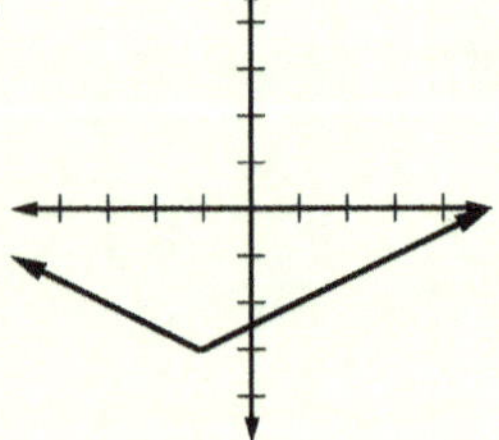

$D = \mathbb{R}; R = [-3, \infty)$

33. The vertex is moved from $(-2, -1)$ to $(5, 3)$.

34. The vertex is moved from $(-3, 0)$ to $(4, 0)$ and the graph is reflected in the x-axis.

Write a function rule for $g(x)$, the described transformation of $f(x)$. Then confirm your answer by graphing $f(x)$ and $g(x)$ in the same window.

27. $f(x) = |x|$ translated 2 units left and 3 units down $g(x) = |x + 2| - 3$

28. $f(x) = |x|$ vertically stretched by a factor of 4 and reflected in the x-axis $g(x) = -4|x|$

29. $f(x) = x^2$ translated 5 units right and vertically shrunk by a factor of $\frac{1}{3}$ $g(x) = \frac{1}{3}(x - 5)^2$

30. $f(x) = x^3$ vertically shrunk by a factor of $\frac{1}{2}$ and then translated 5 units up $g(x) = \frac{x^3}{2} + 5$

31. $f(x) = x^4$ reflected in the x-axis and then translated 3 units up $g(x) = -x^4 + 3$

32. $f(x) = x^2$ reflected in the x-axis and stretched vertically by a factor of 2, then translated 3 units left and 1 unit up $g(x) = -2(x + 3)^2 + 1$

Describe how the graph of $f(x)$ is transformed into the graph of $g(x)$. You may want to graph each function to verify your answer.   translated 7 units right and 4 units up

33. $f(x) = |x + 2|^2 - 1$ and $g(x) = |x - 5|^2 + 3$

34. $f(x) = (x + 3)^2$ and $g(x) = -(x - 4)^2$
translated 7 units right and reflected in the x-axis

> **C. Exercises**

35. Write a function rule for each described transformation of $f(x) = x^2$. Then confirm each answer by using technology to graph the new function in the same window with $f(x)$.

  a. $g(x)$: reflected in the x-axis and then translated 2 units up $g(x) = -x^2 + 2$

  b. $h(x)$: translated 2 units up and then reflected in the x-axis $h(x) = -x^2 - 2$

36. Analyze: Describe the graph of $y = mx + b$ as a transformation of the parent function $y = x$.

37. Prove: Show algebraically that the result of reflecting $f(x) = (x - 2)^2 + 1$ in both the x- and y-axes is $h(x) = -(x + 2)^2 - 1$.

38. Suppose a parabolic arch is used to form a building that is 25 ft tall and 30 ft wide at ground level.

  a. Write the equation for the transformation of $f(x) = x^2$ that has been translated 15 units right, reflected across the x-axis, and then translated 25 units up. $f(x) = -(x - 15)^2 + 25$

  b. Graph your function and explain why it is not a good model for the building's arch.

  c. Adjust the coefficient of the squared term to vertically stretch or shrink the function until it passes through the origin and (30, 0).

39. Use technology to graph $f(x) = (x - 2)^2 - 3$ and each composition of the absolute value function and $f(x)$ in the same window. Then match the composed function to the described transformation of $f(x)$.

  a. $|f(x)|$ I   **b.** $f(|x|)$ IV

  I. Any portion of $f(x)$ below the x-axis is reflected in the x-axis.

  II. Any portion of $f(x)$ to the left of the y-axis is reflected in the y-axis.

  III. The portion of $f(x)$ below the x-axis is replaced by a reflection in the x-axis of the portion of $f(x)$ that is above the x-axis.

  IV. The portion of $f(x)$ to the left of the y-axis is replaced by a reflection in the y-axis of the portion of $f(x)$ that is to the right of the y-axis.

39.

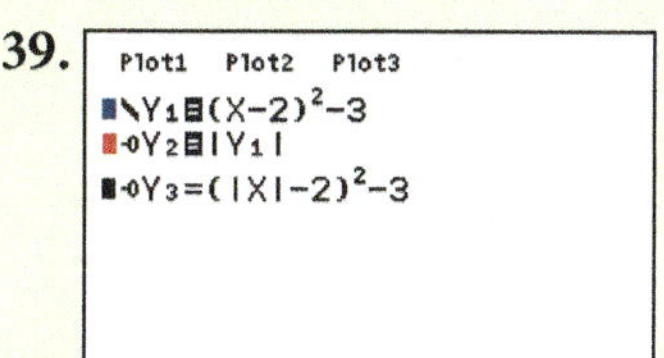

39a.

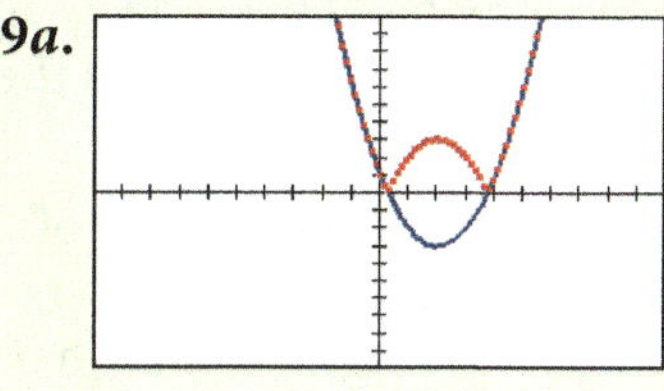

39b.

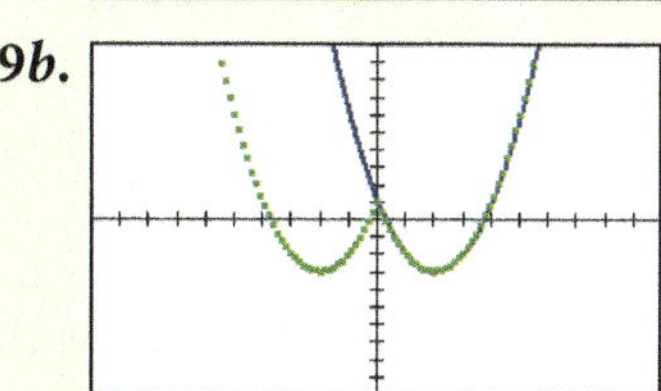

38b. It is not wide enough at ground level.

38c. $f(x) = -\frac{1}{9}(x - 15)^2 + 25$

> **C. Exercises**

35b. $h(x) = -(x^2 + 2) = -x^2 - 2$

36. If $m < 0$, the line is reflected in the x-axis. The line is stretched vertically by a factor of $|m|$ if $|m| > 1$ or shrunk vertically by a factor of $|m|$ if $0 < |m| < 1$. The graph is then translated b units up if $b > 0$ or down if $b < 0$.

37. $g(x) = f(-x)$ is the reflection in the y-axis, and $h(x) = -g(x) = -f(-x)$ is the reflection in both axes.
$$-f(-x) = -((-x - 2)^2 + 1)$$
$$= -(-x - 2)^2 - 1$$
$$= -((-1)(x + 2))^2 - 1$$
$$= -(-1)^2(x + 2)^2 - 1$$
$$= -(x + 2)^2 - 1$$

38b.

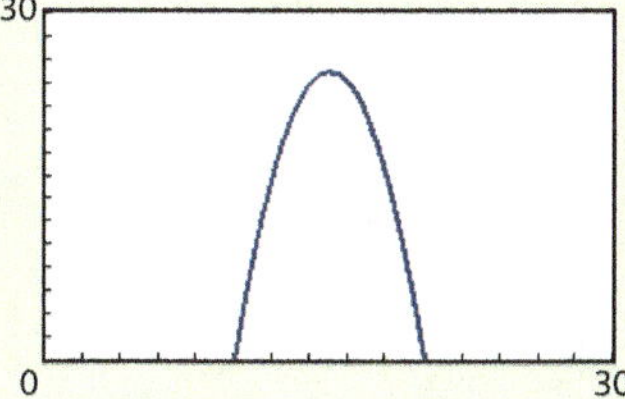

38c. Substituting $(0, 0)$, solve for a.
$$0 = a((0) - 15)^2 + 25$$
$$225a = -25$$
$$a = -\frac{1}{9}$$
$$f(x) = -\frac{1}{9}(x - 15)^2 + 25$$

40a. $-f(x)$ reflects the graph in the x-axis.

40b. $f(-x)$ reflects the graph in the y-axis.

40c. $|f(x)|$ takes the absolute value of all the y-coordinates. This reflects all points below the x-axis across the x-axis.

40d. $f(|x|)$ takes the absolute value of all the input values. This reflects all values to the right of the y-axis across the y-axis.

40e. Begin with the graph for $f(-x)$ found in choice II. Taking the absolute value of each output value $|f(-x)|$ reflects all points below the x-axis across the x-axis.

40f. Begin with the graph for $f(|x|)$ found in choice IV. Negating the output values reflects the entire graph across the x-axis.

> **Cumulative Review**

41. $(x + 12)(x - 2) = 0$
$x + 12 = 0$ or $x - 2 = 0$
$x = -12, 2$

42. $2(3y + 4)(y - 8) = 0$
$3y + 4 = 0$ or $y - 8 = 0$
$y = -\frac{4}{3}, 8$

43. $2a^2 = 64$
$a^2 = 32$
$a = \pm\sqrt{32} = \pm 4\sqrt{2}$

44. $(x + 5)^2 = 16$
$x + 5 = \pm 4$
$x = -5 \pm 4$
$x = -9, -1$

45. $x^2 - 4x + 4 = 45 + 4$
$(x - 2)^2 = 49$
$x - 2 = \pm 7$
$x = 2 \pm 7$
$x = -5, 9$

46. $y^2 + 6y + 9 = -1 + 9$
$(y + 3)^2 = 8$
$y + 3 = \pm\sqrt{8}$
$y = -3 \pm 2\sqrt{2}$

47. $4x = 5y + 16$
$x = \frac{5}{4}y + 4$

49. $m_{AB} = \dfrac{12 - 4}{3 - (-1)} = 2$
$12 = 2(3) + b$
$b = 6$
$y = 2x + 6$
So $y = 2x - 4$ is parallel to $\overleftrightarrow{AB}$.

---

40. Match each function to its graph given the graph of $f(x)$.

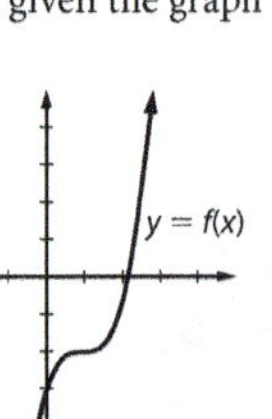

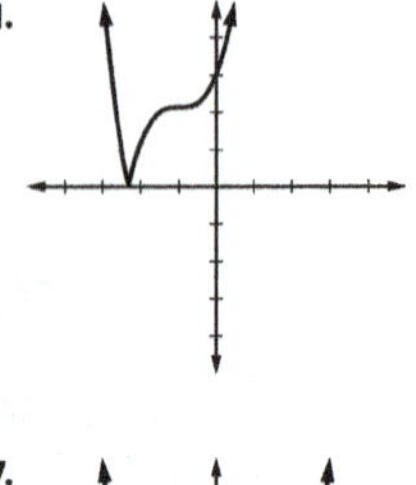

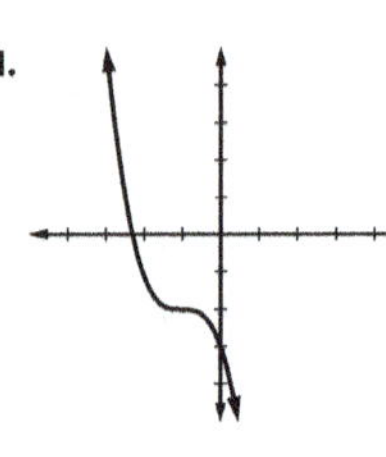

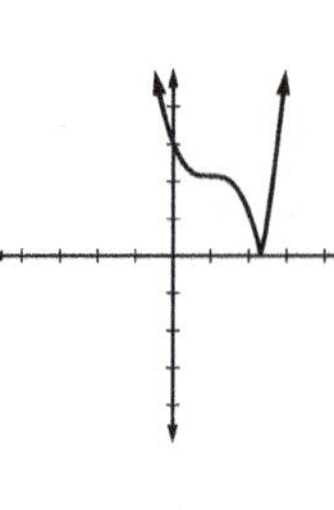

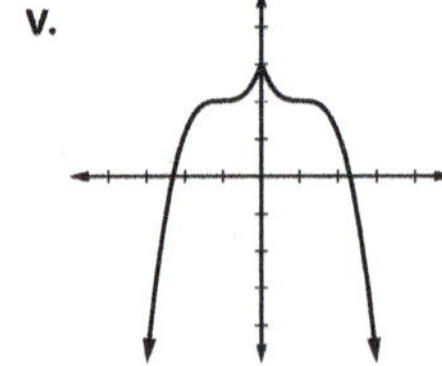

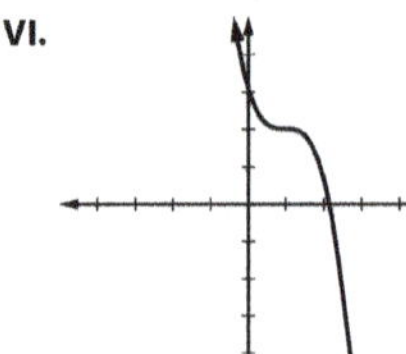

VI.

a. $g(x) = -f(x)$ VI
b. $g(x) = f(-x)$ II
c. $g(x) = |f(x)|$ III
d. $g(x) = f(|x|)$ IV
e. $g(x) = |f(-x)|$ I
f. $g(x) = -f(|x|)$ V

CUMULATIVE REVIEW

Solve by factoring. [Algebra]

41. $x^2 + 10x - 24 = 0$ $x = -12, 2$

42. $6y^2 - 40y - 64 = 0$ $y = -\frac{4}{3}, 8$

Solve by taking roots. [Algebra]

43. $2a^2 - 64 = 0$ $a = \pm 4\sqrt{2}$

44. $3(x + 5)^2 = 48$ $x = -9, -1$

Solve by completing the square. [Algebra]

45. $x^2 - 4x = 45$ $x = -5, 9$

46. $y^2 + 6y = -1$ $y = -3 \pm 2\sqrt{2}$

47. Express x in terms of y if $5y = 4x - 16$.   [Appendix 4]   B

A. $x = \frac{5}{4}y + 16$

B. $x = \frac{5}{4}y + 4$

C. $x = \frac{4}{5}y + 4$

D. $x = \frac{4}{5}y - 4$

E. $x = \frac{4}{5}y + 16$

48. The graph of $3x - 4y = 12$ lies in which quadrants of the coordinate plane?   [1.2]   B
A. I, II, and III
B. I, III, and IV
C. II, III, and IV
D. II and IV only
E. I and III only

49. Which equation represents a line parallel to $\overleftrightarrow{AB}$ with $A\ (-1, 4)$ and $B\ (3, 12)$?   [1.2]   B
A. $y = -2x + 2$
B. $y = 2x - 4$
C. $y = 2x + 6$
D. $y = -\frac{1}{2}x + 6$
E. $y = \frac{1}{2}x + 3$

50. Which graph represents an odd function?   [1.4]   D

A.

C.

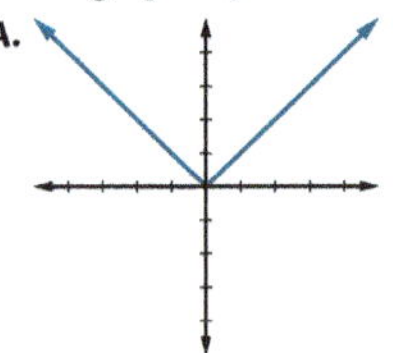

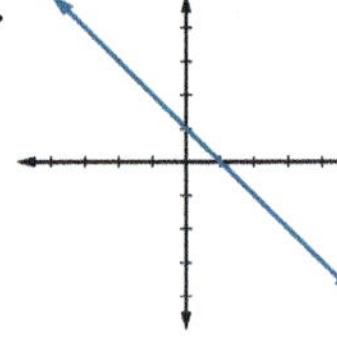

B.

D.

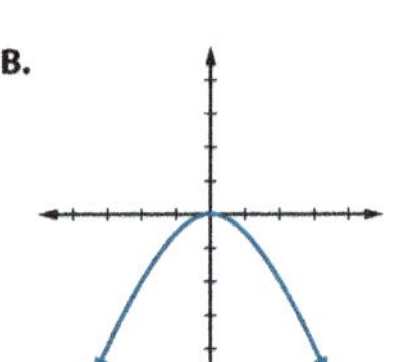

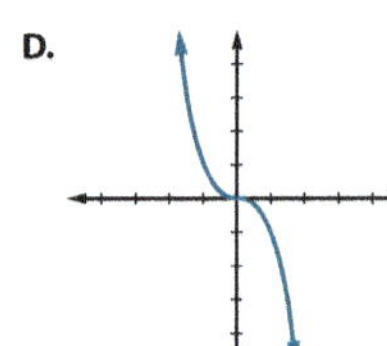

E. none of these

1.6 Quadratic Functions

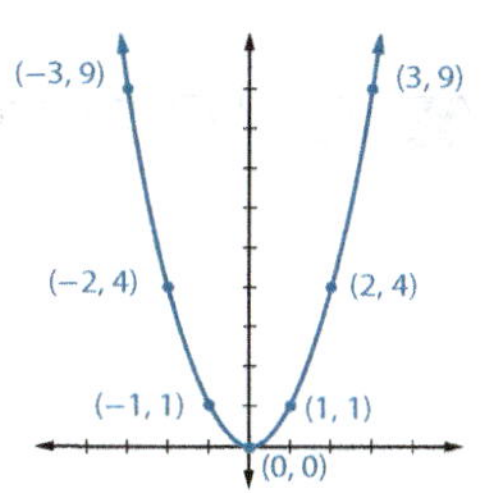

The height of a served volleyball can be modeled by a quadratic function.

The graph of the basic quadratic function $f(x) = x^2$ was examined in Section 1.4. Horizontal and vertical shifts, reflections, stretches, and shrinks of this parent function produce a family of *quadratic functions* whose graphs are called *parabolas*.

The *axis of symmetry* divides a parabola into mirror-image halves and intersects the parabola at its *vertex*, which is either a minimum or maximum point. The axis of symmetry for the parent function $f(x) = x^2$ is the line $x = 0$ (the y-axis) and its vertex is the origin, which is a minimum point.

After completing this section, you will be able to

- graph quadratic functions.
- identify the vertex and classify it as a maximum or minimum.
- find the zeros of a quadratic function.

Example 1 Transforming the Parent Quadratic Function

Transform the graph of $f(x) = x^2$ to obtain the graph of $g(x) = -2(x - 3)^2 + 4$. Then identify the new parabola's domain, range, axis of symmetry, vertex, and whether the vertex is a maximum or minimum point.

Answer

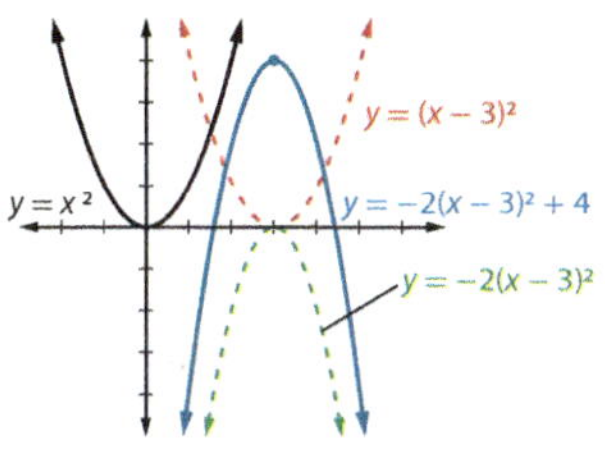

1. Transform the graph.
   a. Shift the graph 3 units right.
   b. Then reflect it in the x-axis and stretch it vertically by 2.
   c. Finally, translate it 4 units up.
2. Describe the graph.
   a. $D = \mathbb{R}; R = (-\infty, 4]$
   b. The axis of symmetry is $x = 3$.
   c. The vertex (3, 4) is a maximum point.

SKILL ✔ EXERCISE 7

When written in *vertex form*, $f(x) = a(x - h)^2 + k$, a parabola's axis of symmetry $x = h$ and vertex (h, k) are easily identified. If $a > 0$ the parabola opens upward and the vertex is a minimum point. If $a < 0$ the parabola opens downward and the vertex is a maximum point.

Quadratic functions are also frequently expressed in *standard form* as a sum of a quadratic term, a linear term, and a constant.

When supply and demand graphs are linear, the total revenue can be modeled by a quadratic function. The parabola's vertex indicates the number of units that provides the greatest revenue.

PRESENTATION

Lesson Opener

1. Sketch the graph of $f(x) = x^2$ by plotting points.

Complete the square to express each equation in *vertex form*: $y = a(x - h)^2 + k$.

2. $y = x^2 + 4x - 10$ $y = (x + 2)^2 - 14$
3. $y = 2x^2 + 12x + 21$ $y = 2(x + 3)^2 + 3$

This section provides an important review of quadratic functions and emphasizes viewing quadratic functions as transformations of the parent function $f(x) = x^2$, as illustrated in Example 1. Be sure students can quickly identify the vertex and *axis of symmetry* when the function is written in this form. Some prefer writing the vertex form as $y - k = a(x - h)^2$.

Interactive Activity Use dynamic geometry software or the Internet keyword search *interactive parabola* for interactive parabolic graphs, some with sliders for dynamic value changes.

Introduce the *standard form* of a quadratic function and use Example 2 to review *completing the square*, which provides a means for converting to an equivalent vertex form as an aid to graphing the function. Have the students differentiate completing the square in an expression

1.6 Quadratic Functions

Objectives

1. To graph quadratic functions
2. To identify the vertex of the graph of a quadratic function and classify it as a maximum or minimum
3. To identify the zeros (roots, x-intercepts) of a quadratic function using factoring or the quadratic formula

Flash

The more top spin is placed on a served volleyball, the more the path of the ball will deviate from the expected parabolic path, making the reception of the ball more challenging for the opponent.

Vocabulary

axis of symmetry
completing the square
height of a projectile
parabola
quadratic function
standard form
vertex
vertex form
zero (root) of a function
Zero Product Property

Reading and Writing Mathematics

Explain why a projectile does not follow the exact path of a quadratic function. Then explain why a quadratic model is useful even though it is not perfect. There are several varying factors affecting the flight path of a projectile, including wind gusts, changes in air density, and the aerodynamics of the object. Even though a quadratic mathematical model cannot incorporate every factor, especially changing factors, it can provide the best predictions that we can obtain, many of which may be more than sufficient for our needs.

1. For $f(x) = -2(x-4)^2$, identify the vertex, whether the vertex is a maximum or minimum point, and the axis of symmetry.

   $V(4, 0)$; max; $x = 4$

2. Graph $f(x) = \frac{1}{2}x^2 + 2$.

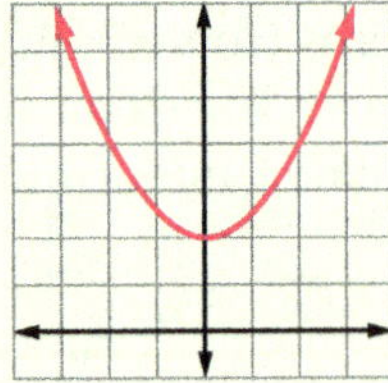

3. Find the vertex, any zeros, and the y-intercept for each function.

   a. $y = 2(x + 2)^2$ $V(-2, 0)$; zeros: -2; y-int.: $(0, 8)$

   b. $y = (x - 1)^2 - 4$ $V(1, -4)$; zeros: $-1, 3$; y-int.: $(0, -3)$

   c. $y = \frac{1}{2}x^2 - 4x + 9$ $V(4, 1)$; no zeros; y-int.: $(0, 9)$

Write each quadratic function in vertex form and then graph the function.

4. $y = x^2 - 6x + 8$ $y = (x - 3)^2 - 1$

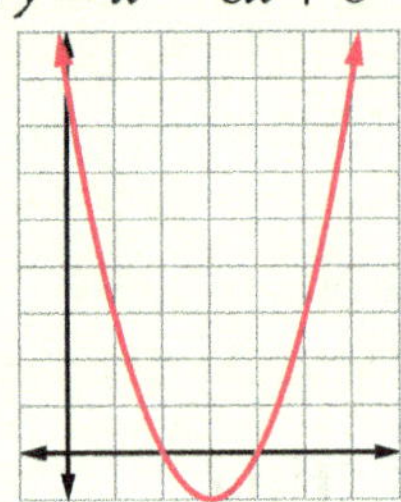

5. $y = \frac{1}{2}x^2 - 5x + 8$ $y = \frac{1}{2}(x - 5)^2 - \frac{9}{2}$

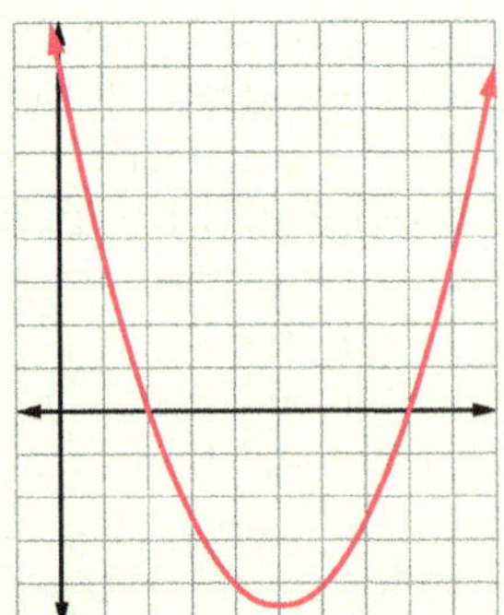

For Graphing Calculators

Use technology to graph the quadratic function $y = 0.4x^2 + 3.2x + 3.4$ and then use the [CALC] menu to find the vertex, y-intercept, and zeros. $V(-4, -3)$; y-int.: $(0, 3.4)$; zeros: $\approx -6.74, -1.26$

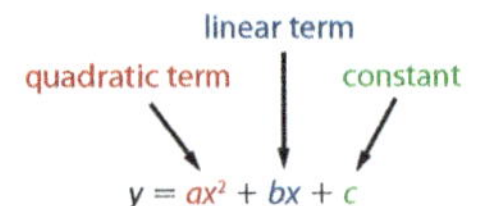

KEYWORD SEARCH

interactive completing square proof

DEFINITION

A **quadratic function** is a function that can be written in the form $f(x) = ax^2 + bx + c$ where a, b, and $c \in \mathbb{R}$ and $a \neq 0$.

Converting a standard form quadratic function to its vertex form will help when analyzing and graphing the function. This conversion usually involves a process called *completing the square*. Recall that completing the square of a binomial in the form $x^2 + dx$ is done by adding $\left(\frac{1}{2}d\right)^2$.

Example 2   Graphing a Quadratic Function in Standard Form

Graph $f(x) = 3x^2 + 12x + 8$. Then identify the domain, range, axis of symmetry, vertex, and whether the vertex is a maximum or minimum point.

Answer

$f(x) = 3x^2 + 12x + 8$

$f(x) = 3(x^2 + 4x + \underline{\quad}) + 8$

$f(x) = 3(x^2 + 4x + 4) + 8 - 12$

$f(x) = 3(x + 2)^2 - 4$

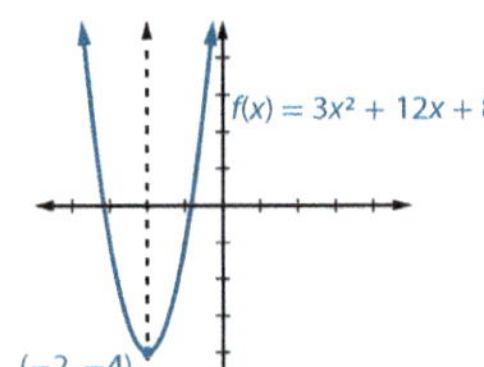

1. Convert to vertex form by completing the square.

   a. Factor the leading coefficient out of the quadratic and linear terms, leaving an expression in the form of $x^2 + dx$.

   b. Add $\left(\frac{1}{2}d\right)^2 = \left(\frac{4}{2}\right)^2 = 4$ within the parentheses. Since any amount added within the parentheses is multiplied by the leading coefficient, compensate by subtracting $3(4) = 12$.

   c. Rewrite the expression within the parentheses as a binomial squared and simplify the remaining terms.

2. Use the function's vertex form, $f(x) = 3(x - (-2))^2 + (-4)$, to analyze the function and graph the parabola.

   a. Since $a = 3$, the parabola opens upward and is obtained by stretching $y = x^2$ vertically by a factor of 3.

   b. The axis of symmetry is $x = -2$ and the vertex $(-2, -4)$ is a minimum point.

   c. $D = \mathbb{R}$; $R = [-4, \infty)$

— SKILL ✔ EXERCISE 23

By completing the square for $f(x) = ax^2 + bx + c$, we can develop formulas for h and k.

$f(x) = a\left(x^2 + \frac{b}{a}x\right) + c$

$= a\left(x^2 + \frac{b}{a}x + \frac{b^2}{4a^2}\right) + c - \frac{b^2}{4a}$

$= a\left(x + \frac{b}{2a}\right)^2 + \left(c - \frac{b^2}{4a}\right)$

$h = -\frac{b}{2a}$ and $k = c - \frac{b^2}{4a}$

1. Factor the leading coefficient from the quadratic and linear terms.

2. Complete the square by adding $\left(\frac{1}{2} \cdot \frac{b}{a}\right)^2 = \frac{b^2}{4a^2}$ within the parentheses. Since $a\left(\frac{b^2}{4a^2}\right) = \frac{b^2}{4a}$ is added to the function, subtract $\frac{b^2}{4a}$ from the constant term.

3. Factor the trinomial.

4. Comparing the resulting function to $f(x) = a(x - h)^2 + k$ provides general expressions for h and k.

---

versus an equation. (*In an expression, the amount added to complete the square must be subtracted to maintain the same value. In an equation, that amount can be added to the other side to maintain the equality.*)

Complete the same steps on the general form to derive the formulas for h and k. Explain that the formula for h is often used to identify the axis of symmetry, but k is usually found as $f(h)$.

One-on-One Given a quadratic equation in vertex form such as $y = 3(x - 2)^2 + 5$, some students will mistakenly think $h = -2$ instead of 2. Remind them that in the form of $x - h$, h is 2.

Explain that the y-intercept can always be found by evaluating $f(0)$ and that the x-values of the x-intercepts (the *zeros* of the function) are found by solving (finding the *roots* of) $f(x) = 0$. Example 3 illustrates solving by factoring and applying the *Zero Product Property*.

One-on-One Since the *zero* of a function $f(x)$ is a root of the equation $f(x) = 0$, the terms *zero* and *root* are often used interchangeably. *Zero* is a more graphical description (where $f(x) = 0$ on the graph), while *root* is a more algebraic description.

The quadratic formula is used to find the zeros in Example 4. Consider deriv-

The formula for h is easy to remember and use, but it is often easier to find the value of k by evaluating $f(h)$.

Another key characteristic describing a parabola is the location of the x- and y-intercepts. The y-intercept of any function has an x-coordinate of 0, and its y-coordinate is found by evaluating $f(0)$. Therefore the y-intercept of $f(x) = ax^2 + bx + c$ is $(0, c)$. Since the x-intercepts of a graph have a y-coordinate of 0, the x-intercepts can be found by letting $f(x) = 0$ and solving the resulting equation.

> **DEFINITION**
>
> A **zero** (or **root**) of a function is any value of x for which $f(x) = 0$.

Quadratic functions can have real or imaginary zeros. The real zeros of a function are the x-coordinates of the parabola's x-intercepts. When the function has only one zero, the x-intercept is the parabola's vertex. The graph of a function with imaginary zeros does not intersect the x-axis.

Some quadratic equations can be solved by factoring and applying the Zero Product Property.

> **ZERO PRODUCT PROPERTY**
>
> A product of real numbers is 0 if and only if one or more of its factors is 0.
> Symbolically, $pq = 0$ iff $p = 0$ or $q = 0$.

Example 3 Finding Zeros by Factoring

Find the y-intercept and the zeros of $y = -x^2 + 6x$. Then graph the function.

Answer

y-intercept: $(0, 0)$

$-x^2 + 6x = 0$

$-x(x - 6) = 0$

$-x = 0$ or $x - 6 = 0$

$x = 0; x = 6$

The zeros are 0 and 6.

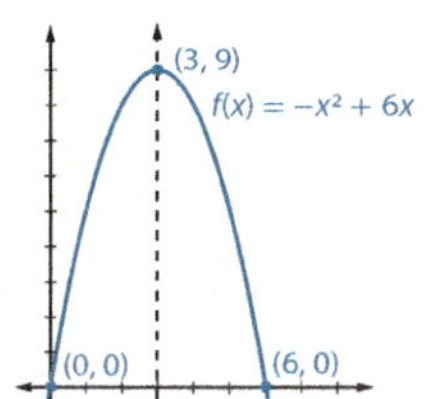

1. The y-intercept of $f(x) = ax^2 + bx + c$ is $(0, c)$.
2. Set the function equal to 0.
   a. Factor.
   b. Apply the Zero Product Property.
   c. Solve.
3. The roots of $f(x) = 0$ are the zeros of the function.
4. Use the standard form $y = ax^2 + bx + c$ to analyze the function and graph the parabola.
   a. Since $a = -1$, the graph of $y = x^2$ is reflected in the x-axis and opens downward.
   b. The axis of symmetry is $x = h = -\frac{b}{2a} = -\frac{6}{2(-1)} = 3$.
   c. The vertex (h, k) lies on the axis of symmetry. $k = f(h) = f(3) = -(3)^2 + 6(3) = 9; V(3, 9)$
   d. The x-intercepts are $(0, 0)$ and $(6, 0)$.

SKILL ✔ EXERCISE 9

You can also use the quadratic formula (derived in exercise 39) to find the zeros of a quadratic function.

If $ax^2 + bx + c = 0$ and $a \neq 0$, then $x = \dfrac{-b \pm \sqrt{b^2 - 4ac}}{2a}$.

Notice that any real zeros are equidistant from the axis of symmetry, $x = -\frac{b}{2a}$.

ing the quadratic formula by completing the square on the standard form of a quadratic equation, assigning exercise 39, or both. Remind the students that complex zeros imply that there are no x-intercepts.

Example 5 illustrates writing the equation of a *parabola* when the vertex and another point are known.

One-on-One Some students may find exercises similar to Example 5 difficult. Encourage students to use the basic strategy of writing an equation or formula—plugging in all the known values and then solving for the unknown, as demonstrated in the example.

Example 6 is a real-world application involving a quadratic function of a projectile. Students should become familiar with the formula for the *height of a projectile*. Consider explaining that time is the independent variable, and is therefore graphed on the horizontal axis.

Motivational Idea Use the Internet keyword search *projectile motion video* to demonstrate projectile paths. Point out that the path can be modeled by a quadratic equation, but the path is not a perfect parabola.

Interactive Activity Exercises 36–37 provide good opportunities for working in groups of the same or differing ability.

8. $a = -1 \Rightarrow$ a reflection in the x-axis and a translation $\frac{1}{2}$ unit right and 4 units up

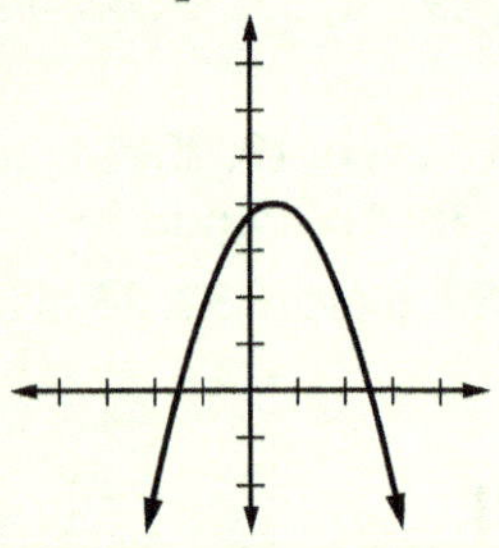

9. $(x + 8)(2x - 7) = 0$
$x + 8 = 0$ or $2x - 7 = 0$
$x = -8, \frac{7}{2}$

10. $-(x^2 - x - 20) = 0$
$-(x - 5)(x + 4) = 0$
$x - 5 = 0$ or $x + 4 = 0$
$x = 5, -4$

11. $x^2 + 5x - 2 = 0$
$x = \dfrac{-(5) \pm \sqrt{(5)^2 - 4(1)(-2)}}{2(1)}$
$= \dfrac{-5 \pm \sqrt{33}}{2}$
$\approx -5.37, 0.37$

12. $x^2 - 4x + 5 = 0$
$x = \dfrac{-(-4) \pm \sqrt{(-4)^2 - 4(1)(5)}}{2(1)}$
$= \dfrac{4 \pm \sqrt{-4}}{2} = \dfrac{4 \pm 2i}{2} = 2 \pm i$

13. $-2x^2 + 2x - 5 = 0$
$x = \dfrac{-(2) \pm \sqrt{(2)^2 - 4(-2)(-5)}}{2(-2)}$
$= \dfrac{-2 \pm \sqrt{-36}}{-4} = \dfrac{2 \pm 6i}{4} = \dfrac{1}{2} \pm \dfrac{3}{2}i$

14. $3x^2 + 4x - 5 = 0$
$x = \dfrac{-(4) \pm \sqrt{(4)^2 - 4(3)(-5)}}{2(3)} = \dfrac{-4 \pm \sqrt{76}}{6}$
$= \dfrac{-4 \pm 2\sqrt{19}}{6} = \dfrac{-2 \pm \sqrt{19}}{3}$
$\approx -2.12, 0.79$

15. $f(0) = (0)^2 - 10(0) + 24 = 24$
$(x - 4)(x - 6) = 0$
$x - 4 = 0$ or $x - 6 = 0$
$x = 4, 6$

16. $f(0) = -2(0)^2 + (0) + 3 = 3$
$-(2x^2 - x - 3) = 0$
$-(2x - 3)(x + 1) = 0$
$2x - 3 = 0$ or $x + 1 = 0$
$x = \dfrac{3}{2}, -1$

17. $f(x) = 2(x^2 + 4x + 4) - 8 = 2x^2 + 8x$
$f(0) = 2(0)^2 + 8(0) = 0$
$2x^2 + 8x = 0$
$2x(x + 4) = 0$
$2x = 0$ or $x + 4 = 0$
$x = 0, -4$

18. $f(x) = -\dfrac{1}{4}(x^2 - 4x + 4) + 4$
$= -\dfrac{1}{4}x^2 + x + 3$
$f(0) = -\dfrac{1}{4}(0)^2 + (0) + 3 = 3$
$-\dfrac{1}{4}x^2 + x + 3 = 0$
$x^2 - 4x - 12 = 0$
$(x - 6)(x + 2) = 0$
$x - 6 = 0$ or $x + 2 = 0$
$x = 6, -2$

Example 4 Finding Zeros Using the Quadratic Formula

Find the y-intercept and zeros of $f(x) = \frac{1}{2}x^2 + 2x + 3$. Then graph the function.

Answer

y-intercept: $(0, 3)$

$\frac{1}{2}x^2 + 2x + 3 = 0$

$x = \dfrac{-b \pm \sqrt{b^2 - 4ac}}{2a}$

$= \dfrac{-(2) \pm \sqrt{(2)^2 - 4\left(\frac{1}{2}\right)(3)}}{2\left(\frac{1}{2}\right)}$

$= -2 \pm \sqrt{-2}$
$= -2 \pm \sqrt{2}\,i$

1. The y-intercept of $f(x) = ax^2 + bx + c$ is $(0, c)$.

2. Set $f(x) = 0$.

   a. Solve the equation using the quadratic formula since $f(x)$ does not factor easily.

   b. Express the imaginary number $\sqrt{-2}$ as $\sqrt{2}\,i$ and state the zeros in the standard form for a complex number, $a + bi$.

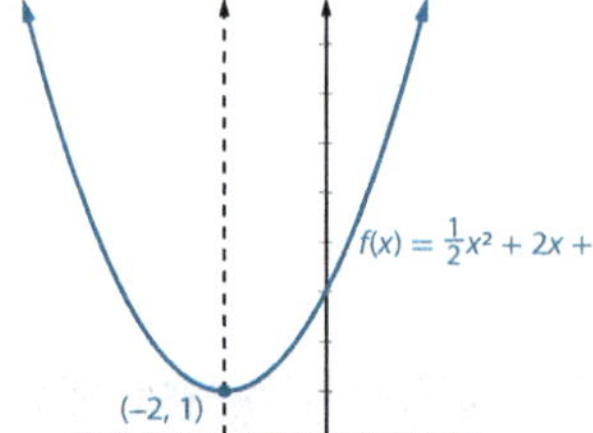

3. Use the standard form $y = ax^2 + bx + c$ to analyze the function and graph the parabola.

   a. Since $a = \frac{1}{2}$, the graph of $y = x^2$ shrinks vertically by $\frac{1}{2}$ and opens upward.

   b. The axis of symmetry is
$x = h = -\dfrac{b}{2a} = -\dfrac{(2)}{2\left(\frac{1}{2}\right)} = -2.$

   c. The vertex (h, k) lies on the axis of symmetry.
$k = f(h) = f(-2) = \frac{1}{2}(-2)^2 + 2(-2) + 3 = 1$
$V\,(-2, 1)$

   d. The imaginary zeros confirm that the parabola does not cross the x-axis.

SKILL ✔ **EXERCISE 13**

If the vertex and one other point on the parabola are known, the vertex form can be used to derive the quadratic function.

Example 5 Writing a Quadratic Function Rule

Write the standard form of a parabola passing through $(2, 7)$ with a vertex at $(-4, 1)$.

Answer

$y = a(x - h)^2 + k$
$7 = a(2 - (-4))^2 + 1$

$7 = a(2 + 4)^2 + 1$

$6 = 36a$

$a = \dfrac{1}{6}$
$f(x) = \dfrac{1}{6}(x + 4)^2 + 1$

$= \dfrac{1}{6}(x^2 + 8x + 16) + 1$

$f(x) = \dfrac{1}{6}x^2 + \dfrac{4}{3}x + \dfrac{11}{3}$

1. Substitute the coordinates of the vertex (h, k) and any other point on the parabola (x, y) into the vertex form.

2. Solve to find a.

3. Write the equation of the parabola in vertex form.

4. Expand and simplify to write the function in standard form.

CONTINUED ➡

B. Exercises

19. $f(x) = (x^2 - 2x + 1^2) - 1$
$= (x - 1)^2 - 1$

20. $g(x) = (x^2 + 8x + 4^2) + 17 - 16$
$= (x + 4)^2 + 1$

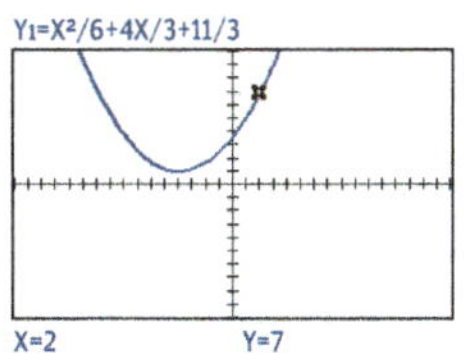

5. Graph the function to confirm that it passes through $(2, 7)$ and has a vertex at $(-4, 1)$.

_______________ SKILL ✔ **EXERCISE 31**

In Section 1.2 the vertical velocity of a projectile was modeled by the linear function $v(t) = -gt + v_0$. The *height of a projectile* can be modeled by the quadratic function $h(t) = -\frac{1}{2}gt^2 + v_0 t + h_0$ where g, the acceleration due to gravity, is approximately 32 ft/sec^2 or 9.8 m/sec^2, v_0 is the initial vertical velocity, and h_0 is the initial height.

Example 6 — Modeling the Height of a Projectile

A ball is thrown upward from a height that is 10 m above the ocean with an initial vertical velocity of 25 m/sec. Find the time it takes for the ball to hit the ocean and the maximum height of the ball (above the ocean).

Answer

$v_0 = 25$ m/sec; $h_0 = 10$ m, $g = 9.8$ m/sec^2

$h(t) = -\frac{1}{2}(9.8)t^2 + 25t + 10$

$\quad = -4.9t^2 + 25t + 10$

$-4.9t^2 + 25t + 10 = 0$

$t = \dfrac{-(25) \pm \sqrt{(25)^2 - 4(-4.9)(10)}}{2(-4.9)}$

$\quad \approx -0.373$ or 5.475

It takes the ball about 5.5 sec to hit the ocean.

$t = -\dfrac{25}{2(-4.9)} \approx 2.55$ sec

$h(2.55) = -4.9(2.55)^2 + 25(2.55) + 10$

$\quad \approx 41.9$ m

1. Use $h(t) = -\frac{1}{2}gt^2 + v_0 t + h_0$ to model the height of the ball.

2. To determine when the ball's height is 0 m, apply the quadratic formula.

3. The maximum height lies on the axis of symmetry: $t = -\frac{b}{2a}$.

4. Evaluate the function at the time the ball is at its maximum height.

5. Graph the function and use the Max and Zero functions to confirm your results.

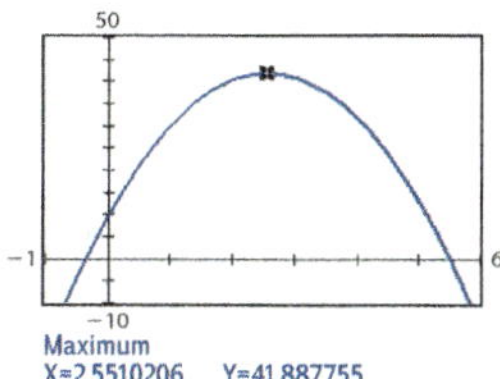

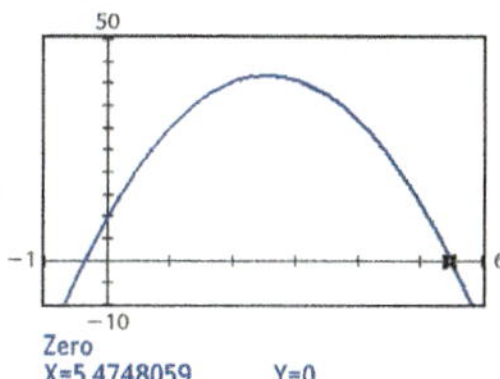

_______________ SKILL ✔ **EXERCISE 33**

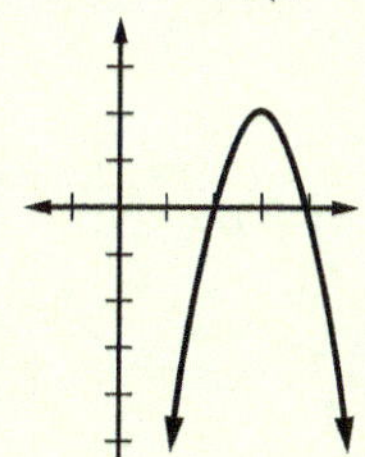

21. $h(x) = \left(x^2 + 5x + \left(\frac{5}{2}\right)^2\right) + \frac{9}{4} - \frac{25}{4}$

$\quad = \left(x + \frac{5}{2}\right)^2 - 4$

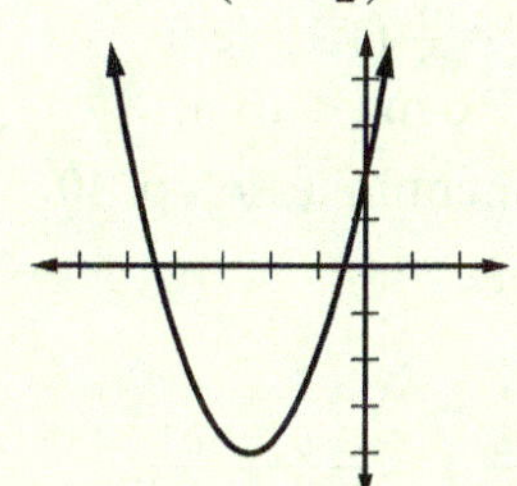

22. $f(x) = -(x^2 + 4x + 2^2) - 7 - 4(-1)$

$\quad = -(x + 2)^2 - 3$

23. $g(x) = 4(x^2 + 2x + 1^2) - 1 - (1)4$

$\quad = 4(x + 1)^2 - 5$

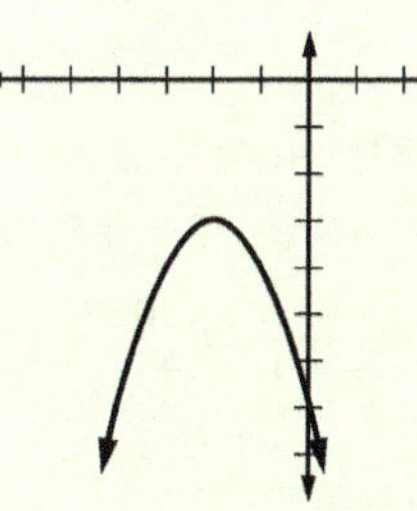

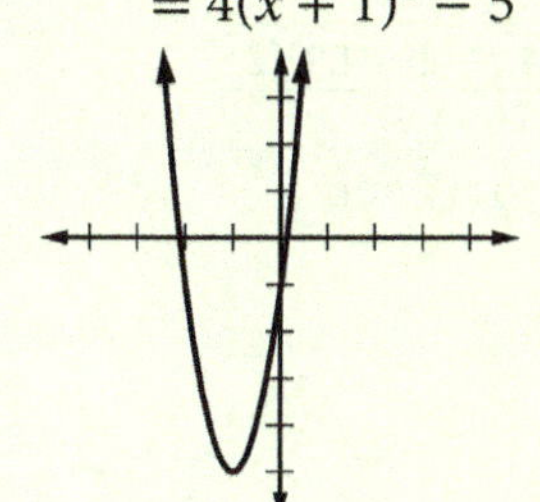

24. $h(x) = -2(x^2 - 6x + 3^2) - 16 - 9(-2)$

$\quad = -2(x - 3)^2 + 2$

25. $g(x) = (x^2 + 4x + 2^2) - 4$

$\quad = (x + 2)^2 - 4$

$V(-2, -4)$; min

$g(0) = 0$; y-int.: $(0, 0)$

$g(x) = x^2 + 4x = 0$

$\quad x(x + 4) = 0$

$\quad\quad x = 0, -4$

x-int.: $(0, 0)$ and $(-4, 0)$

26. $f(x) = -1(x^2 - 10x + 5^2) - 28 - 25(-1)$

$\quad = -(x - 5)^2 - 3$

$V(5, -3)$; max

no x-int. (complex zeros)

$f(0) = -28$; y-int.: $(0, -28)$

27. $g(x) = (x^2 - 2x + 1^2) + \frac{1}{2} - 1$

$\quad = (x - 1)^2 - \frac{1}{2}$

$V\left(1, -\frac{1}{2}\right)$; min

$g(0) = \frac{1}{2}$; y-int.: $\left(0, \frac{1}{2}\right)$

$g(x) = x^2 - 2x + 0.5 = 0$

$x = \dfrac{-(-2) \pm \sqrt{(-2)^2 - 4(1)(0.5)}}{2(1)}$

$\quad = \dfrac{2 \pm \sqrt{2}}{2} = 1 \pm \dfrac{\sqrt{2}}{2}$

x-int.: $\left(1 \pm \frac{\sqrt{2}}{2}, 0\right)$ or $\approx (1.71, 0)$ and $(0.29, 0)$

28. $h(x) = 2\left(x^2 - 5x + \left(\frac{5}{2}\right)^2\right) + 8 - 2\left(\frac{25}{4}\right)$

$\quad = 2\left(x - \frac{5}{2}\right)^2 - \frac{9}{2}$

$V\left(\frac{5}{2}, -\frac{9}{2}\right)$; min

$h(0) = 8$; y-int.: $(0, 8)$

$h(x) = 2(x^2 - 5x + 4) = 0$

$\quad 2(x - 1)(x - 4) = 0$

$\quad\quad x = 1, 4$

x-int.: $(1, 0)$ and $(4, 0)$

30a. $\lim\limits_{x \to -\infty} f(x) = \infty$, $\lim\limits_{x \to \infty} f(x) = \infty$

30b. $\lim\limits_{x \to -\infty} f(x) = -\infty$, $\lim\limits_{x \to \infty} f(x) = -\infty$

31. $f(x) = a(x-2)^2 + (-1)$
$$3 = a(8-2)^2 - 1$$
$$4 = 36a$$
$$a = \frac{1}{9}$$

$$f(x) = \frac{1}{9}(x^2 - 4x + 4) - 1$$
$$= \frac{1}{9}x^2 - \frac{4}{9}x - \frac{5}{9}$$

32. $f(x) = a(x - (-2))^2 + 5$
$$-13 = a(1+2)^2 + 5$$
$$-18 = 9a$$
$$a = -2$$

$$f(x) = -2(x+2)^2 + 5$$
$$= -2(x^2 + 4x + 4) + 5$$
$$= -2x^2 - 8x - 3$$

33a. $h(t) = \frac{1}{2}(-32)t^2 + (32)t + (384)$
$$= -16t^2 + 32t + 384$$

33b. The vertex lies on the axis of symmetry.
$$t = -\frac{b}{2a} = -\frac{32}{2(-16)} = 1$$

33c. $h(1) = -16(1)^2 + 32(1) + 384$
$$= 400 \text{ ft}$$

33d. $-16t^2 + 32t + 384 = 0$
$$-16(t^2 - 2t - 24) = 0$$
$$-16(t+4)(t-6) = 0$$
$$t = -4, 6 \text{ sec}$$

34. $V(h, k)$ is the minimum point.
$$h = -\frac{b}{2a} = -\left(\frac{7.8}{2(0.03)}\right) = 130$$
$$C(130) = \$14{,}493$$

35a. $f(x) = (950 - 20x)(10{,}050 + 300x)$
$$= -6000x^2 + 84{,}000x$$
$$+ 9{,}547{,}500$$

35b. $x = -\frac{b}{2a} = -\frac{84{,}000}{2(-6000)}$
$$= 7 \text{ price drops}$$
$$p = 950 - 7(20) = \$810$$

35c. $f(7) = -6000(7)^2 + 84{,}000(7)$
$$+ 9{,}547{,}500 = \$9{,}841{,}500$$

36a. $f(x) = (4 - 0.15x)(45{,}000 + 2000x)$
$$= -300x^2 + 1250x + 180{,}000$$

36b. $x = -\frac{b}{2a} = -\frac{1250}{2(-300)} = \frac{25}{12} \approx 2$
$$b \approx 45{,}000 + 2000(2) \approx 49{,}000 \text{ bu}$$

36c. $f(2) = -300(2)^2 + 1250(2) + 180{,}000$
$$= \$181{,}300$$

> **C. Exercises**

37a. $h(t) = -\frac{1}{2}(9.8)t^2 + 4t + 2$
$$= -4.9t^2 + 4t + 2$$

37b. $t = -\frac{b}{2a} = -\frac{4}{2(-4.9)} = \frac{20}{49}$
$$\approx 0.408 \text{ sec}$$
$$h\left(\frac{20}{49}\right) \approx 2.816 \text{ m}$$

37c. $-4.9t^2 + 4t + 2 = 0$
$$t = \frac{-4 \pm \sqrt{4^2 - 4(-4.9)(2)}}{2(-4.9)}$$
$$\approx -0.35, 1.17 \text{ sec}$$

37d. $t = \frac{9 \text{ m}}{15 \frac{\text{m}}{\text{sec}}} = 0.6$ sec to reach the net
$$h(0.6) = 2.636 \text{ m at the net}$$
$$d_{horizontal} \approx 15 \frac{\text{m}}{\text{sec}} (1.17 \text{ sec})$$
$$\approx 17.6 \text{ m } (< 18 \text{ m})$$

38. See development in lesson p. 40.

39. $ax^2 + bx = -c$
$$x^2 + \frac{b}{a}x = -\frac{c}{a}$$
$$x^2 + \frac{b}{a}x + \left(\frac{b}{2a}\right)^2 = -\frac{c}{a} + \frac{b^2}{4a^2}$$
$$\left(x + \frac{b}{2a}\right)^2 = \frac{b^2 - 4ac}{4a^2}$$
$$x + \frac{b}{2a} = \frac{\pm\sqrt{b^2 - 4ac}}{2a}$$
$$x = \frac{-b \pm \sqrt{b^2 - 4ac}}{2a}$$

37. Use the function for the height of a projectile, $h(t) = -\frac{1}{2}gt^2 + v_0t + h_0$, to model the height of a volleyball served from a height of 2 m with an initial vertical velocity of 4 m/sec.

   a. Write the quadratic function modeling the height of the volleyball during its flight.

   b. Determine the maximum height of the ball and how long it takes to reach that height.

   c. How long does it take the volleyball to hit the floor? ≈ 1.17 sec

   d. If the ball's horizontal velocity during its flight is 15 m/sec, does the ball clear the 2.24 m tall net located 9 m from the serve? If so, does the ball land within the out-of-bounds line 18 m from where it was served? Explain your reasoning.

38. Convert $y = ax^2 + bx + c$ to vertex form. Then state the coordinates of the vertex and the axis of symmetry.

39. Derive the quadratic formula by completing the square to solve the standard form of a quadratic equation, $ax^2 + bx + c = 0$.

Complete the square to identify the vertex of each parabola and state whether it is a maximum or minimum point. Then state the y-intercept and any x-intercepts.

40. $f(x) = \frac{1}{2}x^2 - 12x + 15$ **41.** $f(x) = -\frac{3}{2}x^2 + 12x - 18$

CUMULATIVE REVIEW

Express the domain of each function using interval notation. [1.1] $\left(-\infty, \frac{1}{2}\right) \cup \left(\frac{1}{2}, \infty\right)$

42. $f(x) = \sqrt{4 - x^2}$ $[-2, 2]$ **43.** $g(x) = \frac{3}{2x - 1}$

44. Find the equation of the line through $(2, 3)$ and perpendicular to $x = 2$. [1.2] $y = 3$

Write a function rule for each variation. [1.4]

45. The value of y varies directly with the cube of x, and $y = 128$ when $x = 4$. $y = 2x^3$

46. The value of y varies inversely with x, and $y = 20$ when $x = 4$. $y = \frac{80}{x}$

47. If the annual interest earned varies directly with the amount invested and \$1500 invested earns \$108 annually, how much must be invested in order to earn \$180? [1.4]   \$2500

48. Simplify i^{54}. [Appendix 5]   B

   A. 1   **C.** $-i$ **E.** none of

   B. -1 **D.** i these

49. Which function passes through $(0, a)$ and $(a, 0)$? [1.2]   C

   A. $y = ax$ **C.** $y = -x + a$ **E.** none of

   B. $y = x - a$ **D.** $y = x + a$ these

50. If $g(x)$ is a translation of $f(x) = x^2$, which equation represents $g(x)$? [1.5]   C

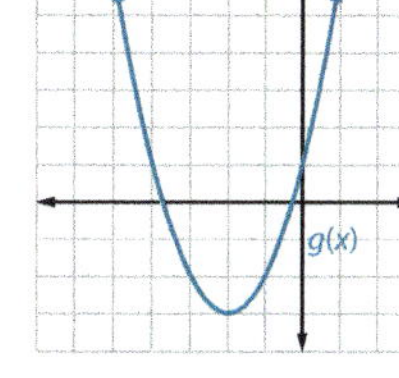

   A. $g(x) = (x - 2)^2 - 3$

   B. $g(x) = (x - 3)^2 - 2$

   C. $g(x) = (x + 2)^2 - 3$

   D. $g(x) = (x - 2)^2 + 3$

   E. $g(x) = (x + 2)^2 + 3$

51. Which function represents $f(x) = (x + 4)^2 - 9$ after it has been shifted 3 units left and 5 units down? [1.5]   A

   A. $g(x) = (x + 7)^2 - 14$ **D.** $g(x) = (x + 9)^2 - 12$

   B. $g(x) = (x + 1)^2 - 14$ **E.** none of these

   C. $g(x) = (x - 1)^2 - 12$

---

42. $4 - x^2 \geq 0$
$$4 \geq x^2$$
$$-2 \leq x \leq 2$$

43. $2x - 1 \neq 0; x \neq \frac{1}{2}$

44. $x = 2$ is a vertical line. The horizontal line through $(2, 3)$ is $y = 3$.

45. $y = kx^3; 128 = k(4^3)$
$$k = \frac{128}{4^3} = 2$$
$$y = 2x^3$$

46. $y = \frac{k}{x}; 20 = \frac{k}{4}; k = 80$
$$y = \frac{80}{x}$$

47. $y = kx; 108 = 1500k$
$$k = \frac{108}{1500} = 0.072$$
$$180 = 0.072x$$
$$x = \$2500$$

48. $i^{54} = i^{4(13)} \cdot i^2 = 1^{13} \cdot -1 = -1$

49. $m = \frac{0 - a}{a - 0} = -1$ and $b = a$
$$\therefore y = -x + a$$

50. $V(-2, -3)$
$$y = (x - (-2))^2 - 3 = (x + 2)^2 - 3$$

51. $f(x) = (x + 4 + 3)^2 - 9 - 5$
$$= (x + 7)^2 - 14$$

37a. $h(t) = -4.9t^2 + 4t + 2$

37b. ≈ 2.82 m; ≈ 0.41 sec

37d. yes; The ball is ≈ 2.64 m high at the time it travels over the net, and it travels ≈ 7.4 m during the ≈ 1.17 sec it takes to hit the floor.

38. $y = \left(x + \frac{b}{2a}\right)^2 + \left(c - \frac{b^2}{4a}\right); V\left(-\frac{b}{2a}, c - \frac{b^2}{4a}\right); x = -\frac{b}{2a}$

---

40. $f(x) = \frac{1}{2}(x^2 - 24x + 12^2)$
$$+ 15 - 144\left(\frac{1}{2}\right)$$
$$f(x) = \frac{1}{2}(x - 12)^2 - 57$$
$$V(12, -57); \text{min}$$
$$f(0) = 15; y\text{-int.}: (0, 15)$$
$$f(x) = \frac{1}{2}x^2 - 12x + 15 = 0$$
$$x = \frac{-(-12) \pm \sqrt{(-12)^2 - 4(0.5)(15)}}{2(0.5)}$$
$$= 12 \pm \sqrt{114} \approx 1.3, 22.7$$
$$x\text{-int.}: (12 \pm \sqrt{114}, 0)$$

41. $f(x) = -\frac{3}{2}(x^2 - 8x + 4^2)$
$$- 18 - 16\left(-\frac{3}{2}\right)$$
$$f(x) = -\frac{3}{2}(x - 4)^2 + 6$$
$$V(4, 6); \text{max}$$
$$f(0) = -18; y\text{-int.}: (0, -18)$$
$$f(x) = -\frac{3}{2}x^2 + 12x - 18 = 0$$
$$-\frac{3}{2}(x^2 - 8x + 12) = 0$$
$$(x - 2)(x - 6) = 0$$
$$x = 2, 6$$
$$x\text{-int.}: (2, 0) \text{ and } (6, 0)$$

1.7 Function Operations

Objectives

1. To perform function operations including composition

2. To determine restrictions on the domain of the result of a function operation

3. To decompose a composite function

4. To model and solve real-world problems using combinations and compositions of functions

Flash

Mobile POS (point of sale) systems enable credit and debit transactions on a mobile device. Many believe that these systems will soon replace credit cards and even cash. Revelation 13:17 seems to describe a similar transaction system, in which simply freezing an individual's account would prevent him from buying or selling.

Vocabulary

composition
decomposition

1.7 Function Operations

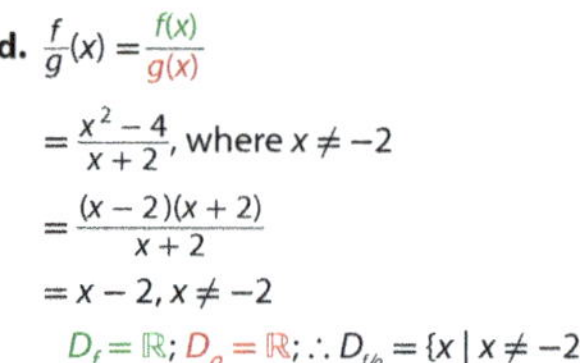

Consumer costs can often be represented with a composition of functions.

After completing this section, you will be able to

- perform function operations including composition.
- determine the domain for combinations of functions.
- decompose a composite function.
- model and solve real-world problems using combinations and compositions of functions.

Many functions are algebraic combinations of simpler functions. The four basic operations of addition, subtraction, multiplication, and division can be performed on functions as well as real numbers.

Function Operations	
addition	$(f + g)(x) = f(x) + g(x)$
subtraction	$(f - g)(x) = f(x) - g(x)$
multiplication	$(fg)(x) = f(x) \cdot g(x)$
division	$\dfrac{f}{g}(x) = \dfrac{f(x)}{g(x)}$, where $g(x) \neq 0$

The domain of the resulting function is the intersection of the domains of the two functions, except where the domain is further limited by the fact that division by 0 is undefined.

Example 1 Arithmetic Operations with Functions

If $f(x) = x^2 - 4$ and $g(x) = x + 2$, write a function rule representing the result of each function operation and state the resulting function's domain.

a. $(f + g)(x)$ **b.** $(f - g)(x)$ **c.** $(fg)(x)$ **d.** $\dfrac{f}{g}(x)$

Answer

a.
$$(f + g)(x) = f(x) + g(x)$$
$$= (x^2 - 4) + (x + 2)$$
$$= x^2 + x - 2$$
$$D_f = \mathbb{R}; D_g = \mathbb{R}; \therefore D_{f+g} = \mathbb{R}$$

b.
$$(f - g)(x) = f(x) - g(x)$$
$$= (x^2 - 4) - (x + 2)$$
$$= x^2 - x - 6$$
$$D_f = \mathbb{R}; D_g = \mathbb{R}; \therefore D_{f-g} = \mathbb{R}$$

c.
$$(fg)(x) = f(x) \cdot g(x)$$
$$= (x^2 - 4)(x + 2)$$
$$= x^3 + 2x^2 - 4x - 8$$
$$D_f = \mathbb{R}; D_g = \mathbb{R}; \therefore D_{fg} = \mathbb{R}$$

d.
$$\frac{f}{g}(x) = \frac{f(x)}{g(x)}$$
$$= \frac{x^2 - 4}{x + 2}, \text{ where } x \neq -2$$
$$= \frac{(x - 2)(x + 2)}{x + 2}$$
$$= x - 2, x \neq -2$$
$$D_f = \mathbb{R}; D_g = \mathbb{R}; \therefore D_{f/g} = \{x \mid x \neq -2\}$$

_____ SKILL ✔ EXERCISES 3, 7

The value for $(f + g)(-1)$ in Example 1 is $f(-1) + g(-1) = -3 + 1 = -2$. Alternately, the resulting function rule can be used: $(f + g)(-1) = (-1)^2 + (-1) - 2 = -2$.

PRESENTATION

Lesson Opener

Substitute 3 for x and simplify.

1. $x^2 + x + 4$ 16
2. $(x + 1)^2 + 2x$ 22

Substitute $a + 1$ for x and simplify.

3. $x^2 + x - 1$ $a^2 + 3a + 1$
4. $(x - 1)^2 + 2x + 3$ $a^2 + 2a + 5$

Preview the four basic function operations in the table, then apply these in Example 1. Emphasize the necessity of including the restriction on the domain in Example 1*d* so that the final form of the function operation has the removable discontinuity that it would not have otherwise.

Motivational Idea A visual activity may help students better understand function operations. Have students graph $f(x) = x - 3$ and then $g(x) = x + 1$ on the same coordinate plane. At any specific value of x, have the students add the y-coordinates of the two functions and then plot $(x, y_1 + y_2)$. Repeat this with two more points and then draw a line through the points. Use the y-intercept and slope to state the equation of the new line, $y = 2x - 2$. Show that this is the graph of the algebraic sum $f(x) + g(x) = 2x - 2$. A similar procedure can be used to illustrate subtraction, multiplication, and division of functions.

Example 2 introduces function composition by simply substituting an expression in a function. Define *composition* of functions, and then illustrate function composition with Example 3. Note that composition is not commutative.

One-on-One Function composition can be illustrated by a sequence of function "machines" where the output of the first machine is fed into the second machine. Note that if the illustrated machines were switched, the resulting output would be 45 instead of 118, confirming that the composition of functions is not commutative.

Another function operation involves substituting an algebraic expression into a function.

Example 2 Substituting Expressions

If $f(x) = 5x - 7$ and $g(x) = x^2 + 3x - 2$, evaluate each function.

a. $f(a + b)$ **b.** $f(x^2 - 9)$ **c.** $g(4a)$ **d.** $g(3x + 1)$

Answer

a. $f(a + b) = 5(a + b) - 7$
$= 5a + 5b - 7$

b. $f(x^2 - 9) = 5(x^2 - 9) - 7$
$= 5x^2 - 45 - 7$
$= 5x^2 - 52$

c. $g(4a) = (4a)^2 + 3(4a) - 2$
$= 16a^2 + 12a - 2$

d. $g(3x + 1) = (3x + 1)^2 + 3(3x + 1) - 2$
$= (9x^2 + 6x + 1) + (9x + 3) - 2$
$= 9x^2 + 15x + 2$

SKILL ✔ EXERCISE 11

This fifth function operation is *composition*. To compose two functions, the function rule for the second function is substituted into the first function. The range (output) of the second function becomes the domain (input) of the first function.

DEFINITION

The **composition** of two functions f and g, denoted $f \circ g$, is defined as $(f \circ g)(x) = f(g(x))$, read "f of g of x."

Example 3 Composing Functions

If $f(x) = x^2 - 9$ and $g(x) = x + 3$, write a function rule representing each composition.

a. $(f \circ g)(x)$ **b.** $(g \circ f)(x)$

Answer

a. $(f \circ g)(x) = f(g(x))$ 1. Apply the definition of composition.
$= (x + 3)^2 - 9$ 2. Substitute $g(x) = x + 3$ for x in $f(x)$.
$= x^2 + 6x + 9 - 9$ 3. Simplify.
$= x^2 + 6x$

b. $(g \circ f)(x) = g(f(x))$ 1. Apply the definition of composition.
$= (x^2 - 9) + 3$ 2. Substitute $f(x) = x^2 - 9$ for x in $g(x)$.
$= x^2 - 6$ 3. Simplify.

SKILL ✔ EXERCISE 17

The value for the composition $(f \circ g)(2)$ in Example 3 is $f(g(2))$, where $g(2) = 5$ and $f(5) = 16$. Alternately, the resulting function rule can be used: $(f \circ g)(2) = (2)^2 + 6(2) = 16$.

The domain of $(f \circ g)(x)$ is not the intersection of the domains of $f(x)$ and $g(x)$. Instead, it is the domain of g, the inner function, less any values that are undefined in the rule for $f(g(x))$, the composite function.

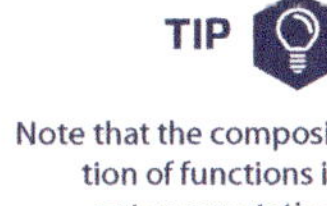

TIP

Note that the composition of functions is not commutative.

Additional Exercises

Let $f(x) = x^2 + 2x$, $g(x) = \frac{1}{x}$, and $h(x) = 5x - 3$.

Write the function rule for each operation and state the resulting function's domain.

1. $(fg)(x)$ $x + 2; D = \mathbb{R}$
2. $(f + h)(x)$ $x^2 + 7x - 3; D = \mathbb{R}$
3. $(h \circ f)(x)$ $5x^2 + 10x - 3; D = \mathbb{R}$
4. $(f \circ g)(x)$ $\frac{1}{x^2} + \frac{2}{x}; D = \{x \mid x \neq 0\}$
5. $(g \circ (f \circ h))(x)$ $\frac{1}{25x^2 - 20x + 3}$; $D = \{x \mid x \neq 0.2 \text{ or } 0.6\}$
6. Use functions f and h from above to evaluate $(f \circ h)(-2)$.   143

For Graphing Calculators

The composition of two or more functions can be illustrated graphically on a calculator.

Let $f(x) = 0.5x^2 - 2.4$ and $g(x) = 2x - 3.5$.

a. Enter the composition $(f \circ g)(x)$ in the Y= editor as shown.

```
Plot1   Plot2   Plot3
\Y1■0.5X^2-2.4
\Y2■2X-3.5
\Y3■Y1(Y2)
\Y4=■
\Y5=
\Y6=
\Y7=
\Y8=
```

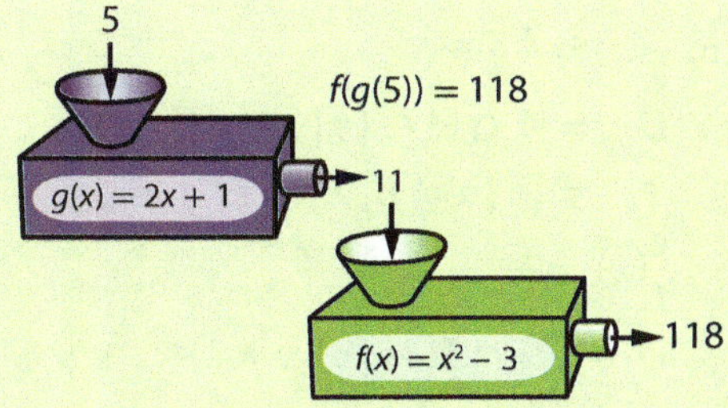

In Step 5 of Example 4*a*, be sure the students understand why the absolute value symbols can be omitted. (*because the restricted domain eliminates any negative radicand*) Have the students perform the composition on their graphing calculators to verify the answer.

One-on-One If a student confuses which function gets substituted into the other, encourage the use of the definition of function composition: $(f \circ g)(x) = f(g(x))$ as in the first step in Example 4*a* and 4*b*.

Explain that *decomposing* a function is a skill used when applying the Chain Rule, an important theorem in calculus. Present the composite functions in Example 5, encouraging the students to look for more than one possible decomposition. When decomposing functions, it is assumed that neither function is the identity function, $f(x) = x$ or $g(x) = x$.

Note that more than one decomposition may be possible, just as a composite number may have more than one set of possible factors.

Consider using the quote from Fourier to reinforce that worldview makes a difference, even in mathematics. Fourier searched for underlying principles in natural phenomena because he believed the world was created by a wise and orderly God. Consider using the margin note to emphasize that there is an underlying order to God's creation that we attempt to describe mathematically.

Use Example 6 to illustrate a real-world application of composite functions.

b. Graph the functions and then use the $\boxed{\text{TRACE}}$ function on Y_3 to evaluate $(f \circ g)(2.6)$. -0.955

Y3=Y1(Y2)

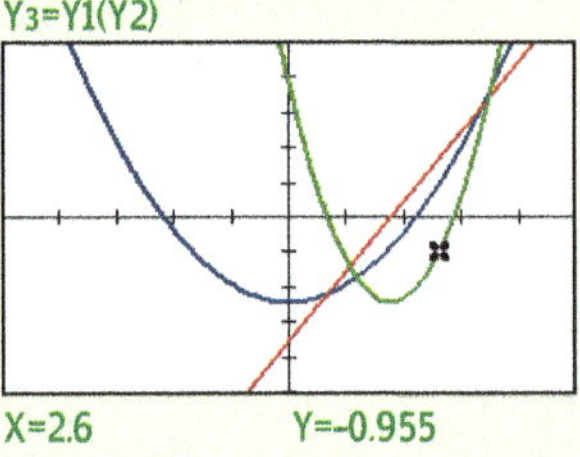

c. Find $(f \circ g)(x)$ algebraically and enter it as Y_4. Verify your composition by checking that the graphs of Y_4 and Y_3 coincide.

$f(g(x)) = 2x^2 - 7x + 3.725$

d. Evaluate $(f \circ g)(2.6)$ by entering both $Y_1(Y_2(2.6))$ and $Y_4(2.6)$ on the home screen.

```
Y1(Y2)(2.6)
                    -2.483
Y4(2.6)
                    -0.955
```

Assignments

- **Minimum:** 1, 3, 5–6, 9–11, 15, 17, 20–21, 23, 27, 29, 31–32, 37, 43, 45–52
- **Standard:** 1, 3, 5, 8, 10–12, 15–17; 19–25 odd; 27–32, 34, 37; 45–51 odd
- **Extended:** 1, 4, 7–8, 11, 13, 16, 18; 19–25 odd; 28–30, 32–33, 35–40, 42, 46, 49, 51

Solutions

❯ A. Exercises

1. $(f + g)(x) = (-2x + 7) + (x - 9)$
$\qquad = -x - 2$
$(g - f)(x) = (x - 9) - (-2x + 7)$
$\qquad = 3x - 16$
$(fg)(x) = (-2x + 7)(x - 9)$
$\qquad = -2x^2 + 25x - 63$
$D_f \cap D_g = \mathbb{R} \cap \mathbb{R} = \mathbb{R}$

2. $(f + g)(x) = (-2x + 7) + (5x^2)$
$\qquad = 5x^2 - 2x + 7$
$(g - f)(x) = 5x^2 - (-2x + 7)$
$\qquad = 5x^2 + 2x - 7$
$(fg)(x) = (-2x + 7)(5x^2)$
$\qquad = -10x^3 + 35x^2$
$D_f \cap D_g = \mathbb{R} \cap \mathbb{R} = \mathbb{R}$

3. $(f + g)(x) = 5x^2 + \frac{1}{x}$

$(g - f)(x) = \frac{1}{x} - 5x^2$

$(fg)(x) = (5x^2)\left(\frac{1}{x}\right) = 5x$

$D_f \cap D_g = \mathbb{R} \cap \{x \mid x \neq 0\}$
$\qquad = \{x \mid x \neq 0\}$

4. $(f + g)(x) = \sqrt{x} + (x - 4)^2$
$\qquad = x^2 - 8x + \sqrt{x} + 16$
$(g - f)(x) = (x - 4)^2 - \sqrt{x}$
$\qquad = x^2 - 8x - \sqrt{x} + 16$
$(fg)(x) = \sqrt{x}(x - 4)^2$
$\qquad = x^{\frac{1}{2}}(x^2 - 8x + 16)$
$\qquad = x^{\frac{5}{2}} - 8x^{\frac{3}{2}} + 16x^{\frac{1}{2}}$
$D_f \cap D_g = \{x \mid x \geq 0\} \cap \mathbb{R}$
$\qquad = \{x \mid x \geq 0\}$

5. $\dfrac{f}{g}(x) = \dfrac{x - 2}{x}$

$D_{f/g} = \mathbb{R} \cap \mathbb{R} \cap \{x \mid x \neq 0\}$
$\qquad = \{x \mid x \neq 0\}$

$\dfrac{g}{f}(x) = \dfrac{x}{x - 2}$

$D_{g/f} = \mathbb{R} \cap \mathbb{R} \cap \{x \mid x \neq 2\}$
$\qquad = \{x \mid x \neq 2\}$

6. $\dfrac{f}{g}(x) = \dfrac{x + 7}{2x - 3}$

$D_{f/g} = \mathbb{R} \cap \mathbb{R} \cap \left\{x \mid x \neq \frac{3}{2}\right\}$
$\qquad = \left\{x \mid x \neq \frac{3}{2}\right\}$

$\dfrac{g}{f}(x) = \dfrac{2x - 3}{x + 7}$

$D_{g/f} = \mathbb{R} \cap \mathbb{R} \cap \{x \mid x \neq -7\}$
$\qquad = \{x \mid x \neq -7\}$

Example 6 Applying a Composite Function

A spherical water balloon is filled at a faucet whose flow rate is 45 cm^3/sec. Use the formula for the volume of a sphere, $V = \frac{4}{3}\pi r^3$, to write a rule for the diameter of the balloon as a function of time. Then use the function to determine the diameter of a balloon after being filled for 10 sec and the time required to fill a balloon to a width of 12 cm.

Answer

$V = \frac{4}{3}\pi r^3$

$\frac{3}{4\pi}V = r^3$

$r = \sqrt[3]{\frac{3V}{4\pi}}$

1. Solve the volume formula for the radius, r.

Since $d = 2r$,

$d(V) = 2\sqrt[3]{\frac{3V}{4\pi}}$.

2. Write a function rule for diameter in terms of volume.

$V(t) = 45t$

3. Write a function rule for volume in terms of time.

$d(V(t)) = 2\sqrt[3]{\frac{3(45t)}{4\pi}}$

$d(V(t)) = 6\sqrt[3]{\frac{5t}{4\pi}}$

4. Find $(d \circ V)(t)$ to express diameter in terms of time.

$d(V(10)) = 6\sqrt[3]{\frac{5(10)}{4\pi}}$

5. Evaluate the composite function when $t = 10$.

$= 6\sqrt[3]{\frac{25}{2\pi}} \approx 9.5$ cm

$12 = 6\sqrt[3]{\frac{5t}{4\pi}}$

$2^3 = \frac{5t}{4\pi}$

6. Substitute a diameter of 12 cm into the composite function and solve for t.

$t = \frac{32\pi}{5} \approx 20.1$ sec

SKILL ✔ EXERCISE 33

The French mathematician Joseph Fourier, when considering the work of Archimedes, Galileo, and Newton, concluded that "they have taught us that the most diverse phenomena are subject to a small number of fundamental laws which are reproduced in all the acts of nature."

❯ A. Exercises

Write the function rule for $(f + g)(x)$, $(g - f)(x)$, and $(fg)(x)$ and state each domain.

1. $f(x) = -2x + 7$
   $g(x) = x - 9$

2. $f(x) = -2x + 7$
   $g(x) = 5x^2$

3. $f(x) = 5x^2$
   $g(x) = \frac{1}{x}$

4. $f(x) = \sqrt{x}$
   $g(x) = (x - 4)^2$

Write the function rule for $\frac{f}{g}(x)$ and $\frac{g}{f}(x)$ and state each domain.

5. $f(x) = x - 2$
   $g(x) = x$

6. $f(x) = x + 7$
   $g(x) = 2x - 3$

7. $f(x) = x^2 - 4$
   $g(x) = \frac{2}{x}$

8. $f(x) = x^2 - 9$
   $g(x) = \sqrt{x - 2}$

Evaluate each expression when $f(x) = -2x + 7$, $g(x) = 5x^2$, and $h(x) = x - 9$.

9. $f(x^2)$ $-2x^2 + 7$

10. $h(x - 4)$ $x - 13$

11. $g(3a + b)$ $45a^2 + 30ab + 5b^2$

12. $f(x^2 + 4)$ $-2x^2 - 1$

13. $f(a^2 + 4a - 9)$ $-2a^2 - 8a + 25$

14. $g(2a)$ $20a^2$

Use the graphs of $f(x) = x^2 + 2x + 1$ and $g(x) = x - 2$ for exercises 15–16.

a. Sketch the graph of each function by adding or multiplying the y-coordinates of $f(x)$ and $g(x)$.

b. State the function rule for the resulting function.

c. Confirm your answer by comparing your sketch to a graph of the function rule on your calculator.

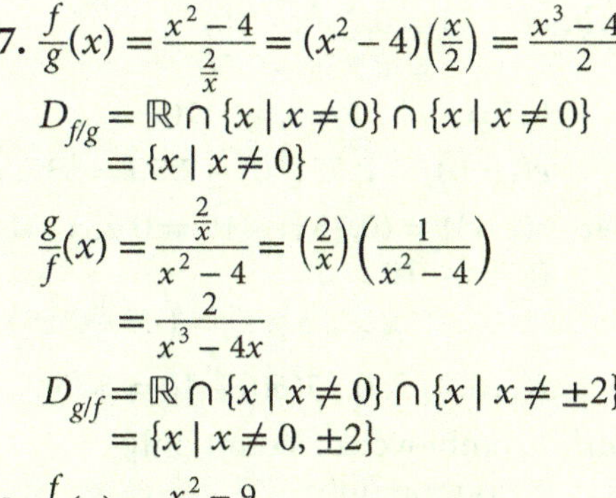

15. $(f + g)(x)$ $(f+g)(x) = x^2 + 3x - 1$

16. $(fg)(x)$ $(fg)(x) = x^3 - 3x - 2$

Write the function rule for $(f \circ g)(x)$ and $(g \circ f)(x)$. Then find $(f \circ g)(3)$ and $(g \circ f)(3)$.

17. $f(x) = x - 7$
    $g(x) = x^2 + 8$

18. $f(x) = 5x - 4$
    $g(x) = x^2 + 8$

15a. 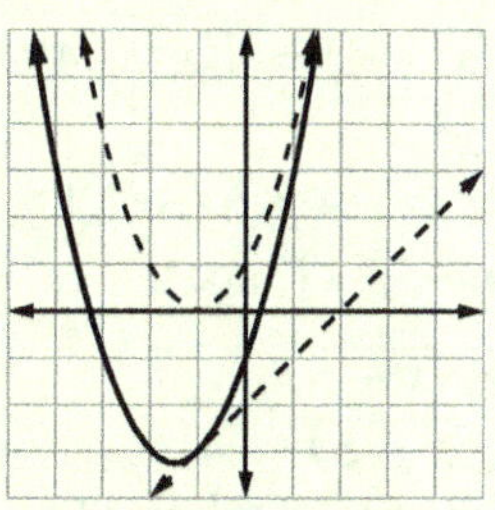

16a. 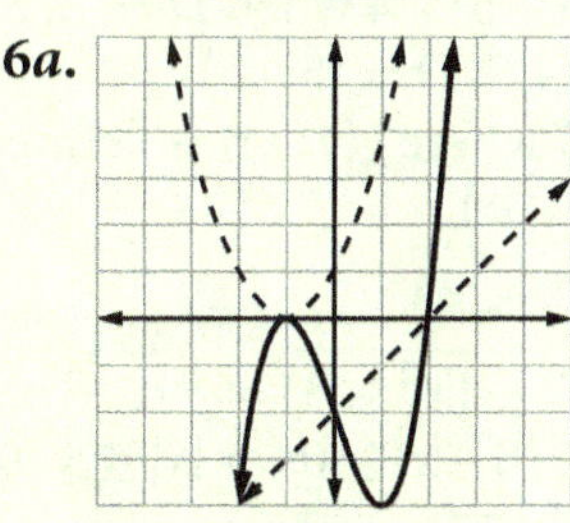

17. $f(g(x)) = f(x^2 + 8) = (x^2 + 8) - 7$
    $= x^2 + 1$
    $g(f(x)) = g(x - 7) = (x - 7)^2 + 8$
    $= x^2 - 14x + 57$
    $f(g(3)) = (3)^2 + 1 = 10$
    $g(f(3)) = (3)^2 - 14(3) + 57 = 24$

18. $f(g(x)) = f(x^2 + 8) = 5(x^2 + 8) - 4$
    $= 5x^2 + 36$
    $g(f(x)) = (5x - 4)^2 + 8$
    $= 25x^2 - 40x + 24$
    $f(g(3)) = 5(3)^2 + 36 = 81$
    $g(f(3)) = 25(3)^2 - 40(3) + 24 = 129$

19. Commutative Property; $f \circ g \neq g \circ f$ (Changing the order in the composition produced different results.)

20a. $(p + w)(d) = 5.7d + 22.5$
     $(p - w)(d) = 4.9d - 5.5$

20b. The family's combined power and water bills for a 31-day billing cycle average \$199.20. The difference between the family's power bill and water bill for a 29-day billing cycle averages \$136.60.

❯ B. Exercises

21. $f(g(x)) = (x + 7)^2 - 4(x + 7) - 24$
    $= x^2 + 10x - 3$
    $D_g = \mathbb{R}; D_{f \circ g} = \mathbb{R}$
    $g(f(x)) = (x^2 - 4x - 24) + 7$
    $= x^2 - 4x - 17$
    $D_f = \mathbb{R}; D_{g \circ f} = \mathbb{R}$

22. $f(g(x)) = f(x - 7) = \frac{9}{(x - 7)} = \frac{9}{x - 7}$
    $D_g = \mathbb{R}$, but $x - 7 \neq 0$, so
    $D_{f \circ g} = \{x \mid x \neq 7\}$.

    $g(f(x)) = g\left(\frac{9}{x}\right) = \left(\frac{9}{x}\right) - 7 = \frac{9}{x} - 7$
    $D_f = \{x \mid x \neq 0\}$, and $x \neq 0$, so
    $D_{g \circ f} = \{x \mid x \neq 0\}$.

7. $\frac{f}{g}(x) = \frac{x^2 - 4}{\frac{2}{x}} = (x^2 - 4)\left(\frac{x}{2}\right) = \frac{x^3 - 4x}{2}$

$D_{f/g} = \mathbb{R} \cap \{x \mid x \neq 0\} \cap \{x \mid x \neq 0\}$
$= \{x \mid x \neq 0\}$

$\frac{g}{f}(x) = \frac{\frac{2}{x}}{x^2 - 4} = \left(\frac{2}{x}\right)\left(\frac{1}{x^2 - 4}\right)$
$= \frac{2}{x^3 - 4x}$

$D_{g/f} = \mathbb{R} \cap \{x \mid x \neq 0\} \cap \{x \mid x \neq \pm 2\}$
$= \{x \mid x \neq 0, \pm 2\}$

8. $\frac{f}{g}(x) = \frac{x^2 - 9}{\sqrt{x - 2}}$

$D_{f/g} = \mathbb{R} \cap \{x \mid x \geq 2\} \cap \{x \mid x > 2\}$
$= \{x \mid x > 2\}$

$\frac{g}{f}(x) = \frac{\sqrt{x - 2}}{x^2 - 9}$

$D_{g/f} = \{x \mid x \geq 2\} \cap \mathbb{R} \cap \{x \mid x \neq \pm 3\}$
$= [2, 3) \cup (3, \infty)$

9. $f(x^2) = -2(x^2) + 7 = -2x^2 + 7$

10. $h(x - 4) = (x - 4) - 9 = x - 13$

11. $g(3a + b) = 5(3a + b)^2$
    $= 5(9a^2 + 6ab + b^2)$
    $= 45a^2 + 30ab + 5b^2$

12. $f(x^2 + 4) = -2(x^2 + 4) + 7$
    $= -2x^2 - 1$

13. $f(a^2 + 4a - 9) = -2(a^2 + 4a - 9) + 7$
    $= -2a^2 - 8a + 25$

14. $g(2a) = 5(2a)^2 = 5(4a^2) = 20a^2$

23. $f(g(x)) = f(x^2 - 4) = \sqrt{(x^2 - 4) - 5}$
$$= \sqrt{x^2 - 9}$$
$D_g = \mathbb{R}$, but $x^2 - 9 \geq 0$ (or $|x| \geq 3$), so $D_{f \circ g} = (-\infty, -3] \cup [3, \infty)$.
$$g(f(x)) = g(\sqrt{x - 5})$$
$$= (\sqrt{x - 5})^2 - 4$$
$$= (x - 5) - 4 \text{ when } D = [5, \infty)$$
$$= x - 9; D = [5, \infty)$$
$D_f = [5, \infty)$ where $g(f(x))$ is defined, so $D_{g \circ f} = [5, \infty)$.

24. $f(g(x)) = f(x^2 - 3)$
$$= \sqrt{7 - (x^2 - 3)} = \sqrt{10 - x^2}$$
$D_g = \mathbb{R}$, but $10 - x^2 \geq 0$ (or $|x| \leq \sqrt{10}$), so $D_{f \circ g} = [-\sqrt{10}, \sqrt{10}]$.
$$g(f(x)) = g(\sqrt{7 - x}) = (\sqrt{7 - x})^2 - 3$$
$$= |7 - x| - 3$$
$$= (7 - x) - 3 \text{ when } 7 - x \geq 0$$
$$= -x + 4; D = (-\infty, 7]$$
$D_f = (-\infty, 7]$ where $g(f(x))$ is defined, so $D_{g \circ f} = (-\infty, 7]$.

25. $f(g(x)) = f\left(\frac{1}{x - 2}\right) = \left(\frac{1}{x - 2}\right)^2 - 10$
$$= \frac{1}{(x - 2)^2} - 10$$
$$\text{or } \frac{-10x^2 + 40x - 39}{(x - 2)^2}$$
$D_g = \{x \mid x \neq 2\}$; $D_{f \circ g} = \{x \mid x \neq 2\}$
$$g(f(x)) = g(x^2 - 10) = \frac{1}{(x^2 - 10) - 2}$$
$$= \frac{1}{x^2 - 12}$$
$D_f = \mathbb{R}$, but $x^2 - 12 \neq 0$, so $D_{g \circ f} = \{x \mid x \neq \pm 2\sqrt{3}\}$.

26. $f(g(x)) = f\left(\frac{1}{x}\right) = \sqrt{\frac{1}{x} - 2}$
$D_g = \{x \mid x \neq 0\}$,
but $\frac{1}{x} - 2 \geq 0$ and $\frac{1}{x} \geq 2$.
If $x > 0$, then $1 \geq 2x$;
or if $x < 0$, then $1 \leq 2x$.
$\therefore \left(0, \frac{1}{2}\right] \cup \varnothing$; $D_{f \circ g} = \left(0, \frac{1}{2}\right]$
$$g(f(x)) = g(\sqrt{x - 2}) = \frac{1}{\sqrt{x - 2}}$$
$D_f = [2, \infty)$, but $\sqrt{x - 2} \neq 0$, so $D_{g \circ f} = (2, \infty)$.

29. $h(x) = (x + 4x + 2^2) - 3 - 4$
$$= (x + 2)^2 - 7$$

30. $h(x) = 3(x - 12x + 6^2) + 100 - 3(36)$
$$= 3(x - 6)^2 - 8$$

31b. $(m \circ c)(x) = 1000(100x)$
$$= 100{,}000x$$

31c. $(m \circ c)(5) = 100{,}000(5)$
$$= 500{,}000 \text{ cm}$$

32b. $V = \pi\left(\frac{5}{2}\right)^2 H; H = \frac{4}{25\pi}V$

---

19. Write: In exercises 17–18, which property is shown not to apply in compositions? Explain your reasoning.

20. The Smith family's average power bill is modeled by $p(d) = 5.3d + 8.5$ and their average water bill is modeled by $w(d) = 0.4d + 14$ where d is the number of days in the billing cycle.
   a. Find $(p + w)(d)$ and $(p - w)(d)$.
   b. Find $(p + w)(31)$ and $(p - w)(29)$ and explain what each value represents.

Write the function rule for $(f \circ g)(x)$ and $(g \circ f)(x)$ and state each domain.

21. $f(x) = x^2 - 4x - 24$
   $g(x) = x + 7$

22. $f(x) = \frac{9}{x}$
   $g(x) = x - 7$

23. $f(x) = \sqrt{x - 5}$
   $g(x) = x^2 - 4$

24. $f(x) = \sqrt{7 - x}$
   $g(x) = x^2 - 3$

25. $f(x) = x^2 - 10$
   $g(x) = \frac{1}{x - 2}$

26. $f(x) = \sqrt{x - 2}$
   $g(x) = \frac{1}{x}$

Find $f(x)$ and $g(x)$ such that $h(x) = (f \circ g)(x)$.

27. $h(x) = \sqrt{x^2 - 3}$

28. $h(x) = \frac{1}{\sqrt{x - 4}}$

29. $h(x) = x^2 + 4x - 3$

30. $h(x) = 3x^2 - 36x + 100$

31. There are 1000 m in 1 km and 100 cm in 1 m.
   a. Write a function rule $m(x)$ to change kilometers to meters and a function rule $c(x)$ to convert meters to centimeters.
   b. Find the composition of the function rules that converts kilometers directly to centimeters.
   c. Use the composite function to find the number of centimeters in 5 km.

32. A cylindrical dunk tank with a diameter of 5 ft is being filled at 3 ft³/min.
   a. Write a function, $V(t)$, for the volume of water after t minutes.
   b. Use $V = \pi r^2 H$ to write a function, $H(V)$, for the water's height in terms of its volume.
   c. Find a composite function for the water's height in terms of the time the tank is being filled.
   d. Use the composite function to determine the depth of the water in the tank after being filled for 20 min.
   e. How long will it take for the tank to be filled to a depth of 5.5 ft?

---

33. A programmer animating travel through a tunnel increases the radius of a circle from an initial length of 3 mm at a rate of 2 mm/sec.
   a. Write function $A(r)$ describing the area of the circle in terms of its radius, and a second function $r(t)$ describing the radius in terms of the time.
   b. Find $A \circ r$. What does this composition represent?
   c. How long will it take for the area of the circle to be 169π mm²?

34. A store offers an employee discount of 25% as well as a coupon for $10 off any purchase over $10.
   a. Write function $e(x)$ that calculates the price with the employee discount, and function $c(x)$ that calculates the price of any purchase over $10 with the coupon.
   b. If the coupon is applied before the employee discount, find the composite function that calculates the final price and state its domain. Then use the function to calculate the final price of a $60 sweater.
   c. If the employee discount is applied before the coupon, find the composite function that calculates the final price and state its domain. Then use the function to calculate the final price of a $60 sweater.
   d. If line a represents the regular price, state the type of discount represented by each of the other lines.

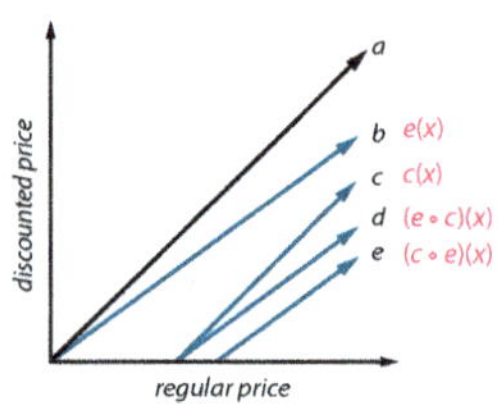

35. Given $f(x) = x^2 + 5x - 6$ and $g(x) = x - 1$, find $\frac{f}{g}(x)$ and state its domain and range. Then graph the function and describe its end behavior and any discontinuities.

21. $f(g(x)) = x^2 + 10x - 3; D = \mathbb{R}$
   $g(f(x)) = x^2 - 4x - 17; D = \mathbb{R}$

22. $f(g(x)) = \frac{9}{x - 7}; D = \{x \mid x \neq 7\}$
   $g(f(x)) = \frac{9}{x} - 7; D = \{x \mid x \neq 0\}$

23. $f(g(x)) = \sqrt{x^2 - 9}; D = (-\infty, -3] \cup [3, \infty)$
   $g(f(x)) = (\sqrt{x - 5})^2 - 4 = x - 9; D = [5, \infty)$

27. $f(x) = \sqrt{x}, g(x) = x^2 - 3$
   or $f(x) = \sqrt{x - 3}, g(x) = x^2$

28. $f(x) = \frac{1}{x}, g(x) = \sqrt{x - 4}$
   or $f(x) = \frac{1}{\sqrt{x}}, g(x) = x - 4$

29. $f(x) = x^2 - 7, g(x) = x + 2$
   or $f(x) = x - 7, g(x) = (x + 2)^2$

30. $f(x) = 3x^2 - 8, g(x) = x - 6$
   or $f(x) = 3x - 8, g(x) = (x - 6)^2$

32c. $H(V(t)) = \frac{4}{25\pi}(3t) = \frac{12}{25\pi}t$

32d. $H(V(t)) = \frac{12}{25\pi}(20) \approx 3.1$ ft

32e. $5.5 = \frac{12}{25\pi}t; t = \frac{5.5(25)\pi}{12} \approx 36.0$ min

33a. $A(r) = \pi r^2; r(t) = 2t + 3$

33b. $A(r(t)) = \pi(2t + 3)^2$
$$= \pi(4t^2 + 12t + 9)$$
$$= 4\pi t^2 + 12\pi t + 9\pi$$
the area of the circle in terms of time (in seconds)

33c. $169\pi = \pi(2t + 3)^2$
$$169 = (2t + 3)^2$$
$$13 = 2t + 3; t = 5 \text{ sec}$$

34b. $e(c(x)) = 0.75(x - 10)$
$$= 0.75x - 7.50$$
$$0.75x - 7.50 > 0; x > 10$$
$$e(c(60)) = 0.75(60) - 7.50 = \$37.50$$

34c. $c(e(x)) = (0.75x) - 10 = 0.75x - 10$
$$0.75x - 10 > 0$$
$$x > 13\tfrac{1}{3}; D = \left(13\tfrac{1}{3}, \infty\right)$$
$$c(e(60)) = 0.75(60) - 10 = \$35$$

34d. b: employee discount only
c: coupon only
d: coupon before employee discount
e: employee discount before coupon

Determine whether each statement is *always*, *sometimes*, or *never* true. Explain your reasoning.

36. The domain of $(f + g)(x)$ is the union of the domains of $f(x)$ and $g(x)$.   sometimes

37. The domain of $(fg)(x)$ is the intersection of the domains of $f(x)$ and $g(x)$.   always

38. The domain of $\frac{f}{g}(x)$ is the domain of $f(x)$.   sometimes

39. The domain of $(f \circ g)(x)$ is the domain of $g(x)$.   sometimes

Use $f(x) = 2x + 3$, $g(x) = \frac{1}{x}$, and $h(x) = \sqrt{x}$ to state the function rule and domain for each composition.

40. Examine $(f \circ g \circ h)(x)$ by finding the following.
  a. $(f \circ (g \circ h))(x)$ **b.** $((f \circ g) \circ h)(x)$

41. Examine $(g \circ h \circ f)(x)$ by finding the following.
  a. $(g \circ (h \circ f))(x)$ **b.** $((g \circ h) \circ f)(x)$

42. Write: Which property of composition is illustrated in exercises 40–41? Explain your reasoning.

CUMULATIVE REVIEW

Solve each equation for the indicated variable.
[Algebra]

43. $p = 2l + 2w$ for l **44.** $b = \frac{a - c}{a + c}$ for a

45. $x = y^2 - z$ for y

46. Find the slope of a line that is perpendicular to $3x + 5y = 6$.   [1.2]

47. Write a function rule for the line through $(2, 5)$ that is parallel to $f(x) = -3x + 7$.   [1.2] $f(x) = -3x + 11$

48. Graph $f(x) = \begin{cases} -1 & \text{if } x < -1 \\ x + 2 & \text{if } -1 \le x \le 1 \\ -\frac{1}{2}x + 2 & \text{if } x > 1 \end{cases}$ [1.3]

49. Describe the continuity of $y = \frac{x + 2}{x^2 - x - 6}$ at $x = 3$.   [1.3]   B

  A. continuous   **D.** point discontinuity
  B. infinite discontinuity   **E.** none of these
  C. jump discontinuity

50. The US Postal Service increased the per-ounce rate of a standard-sized domestic letter 17 times in the twentieth century. What type of continuity would be present in a line graph of twentieth-century postal rates?   [1.3]   C

  A. continuous   **D.** point discontinuity
  B. infinite discontinuity   **E.** none of these
  C. jump discontinuity

51. Which could be the graph of $f(x) = -x^2 - 4x - 2$? [1.6]   A

A. 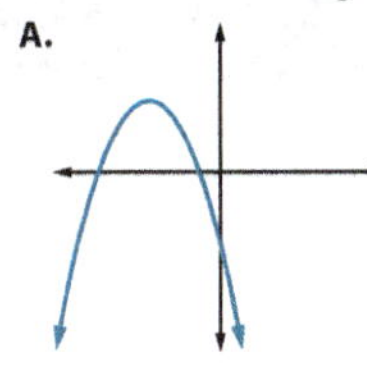   **C.** 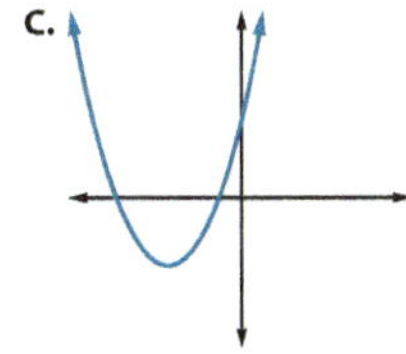

B. 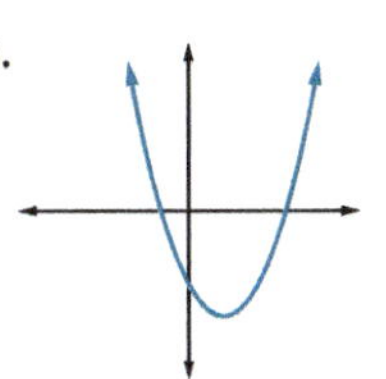   **D.** 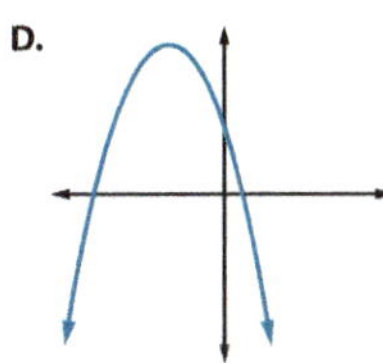

E. none of these

52. Describe the solution(s) of $f(x) = 0$.   [1.6]   E

  A. 2 positive
  B. 2 negative
  C. 1 negative
  D. 1 positive
  E. 1 negative and 1 positive

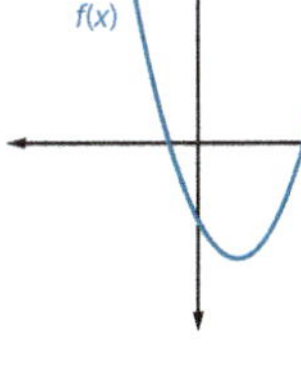

40a. $\frac{2}{\sqrt{x}} + 3; D = (0, \infty)$ **41a.** $\frac{1}{\sqrt{2x + 3}}; D = \left(-\frac{3}{2}, \infty\right)$ **43.** $l = \frac{1}{2}p - w$ **45.** $y = \pm\sqrt{x + z}$

40b. $\frac{2}{\sqrt{x}} + 3; D = (0, \infty)$ **41b.** $\frac{1}{\sqrt{2x + 3}}; D = \left(-\frac{3}{2}, \infty\right)$ **44.** $a = \frac{bc + c}{1 - b}$ **46.** $\frac{5}{3}$

40b. $f(g(x)) = 2\left(\frac{1}{x}\right) + 3 = \frac{2}{x} + 3$
$$f(g(h(x))) = 2\frac{1}{(\sqrt{x})} + 3$$
$$= \frac{2}{\sqrt{x}} + 3 \text{ or } \frac{3x + 2\sqrt{x}}{x}$$
$D_g = (-\infty, 0) \cup (0, \infty) = D_{f(g)}$, but $x \ge 0$ in $f(g(h(x)))$, so $D_{f(g(h))} = (0, \infty)$.

41a. $h(f(x)) = \sqrt{2x + 3}$
$D_f = \mathbb{R}$ and $D_{h(f)} = [-\frac{3}{2}, \infty)$
$g(h(f(x))) = \frac{1}{\sqrt{2x + 3}}$ or $\frac{\sqrt{2x + 3}}{2x + 3}$;
but $2x + 3 > 0$ in $f(g(h(x)))$,
so $D_{f(g(h))} = (-\frac{3}{2}, \infty)$.

41b. $g(h(x)) = \frac{1}{\sqrt{x}}$
$D_h = [0, \infty)$, $D_{g(h)} = (0, \infty)$
$g(h(f(x))) = \frac{1}{\sqrt{2x + 3}}$ or $\frac{\sqrt{2x + 3}}{2x + 3}$;
$D_f = \mathbb{R}$, but $2x + 3 > 0$ in $f(g(h(x)))$,
so $D_{f(g(h))} = (-\frac{3}{2}, \infty)$.

Note that restricting $D_f = (-\frac{3}{2}, \infty)$ produces $R_f = (0, \infty)$, the restricted $D_{g(h)}$.

42. Associative Property; The grouping of the functions does not change the composition (when the ordering of the functions remains constant).

⟩ Cumulative Review

43. $2l = p - 2w$
$$l = \frac{p - 2w}{2} = \frac{1}{2}p - w$$

44. $ab + bc = a - c$
$$ab - a = -bc - c$$
$$a(b - 1) = -bc - c$$
$$a = \frac{-bc - c}{b - 1} = \frac{-(bc + c)}{b - 1} = \frac{bc + c}{1 - b}$$

45. $y^2 = x + z; y = \pm\sqrt{x + z}$

46. $5y = -3x + 6$
$$y = -\frac{3}{5}x + \frac{6}{5}$$
$$m_\perp = \frac{5}{3}$$

47. $m_G = -3 = m_{||}$
$$5 = -3(2) + b$$
$$b = 11; y = -3x + 11$$

48. 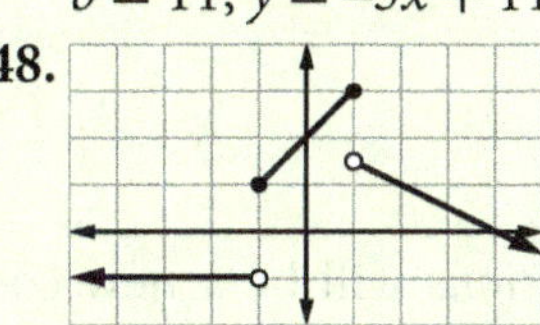

⟩ C. Exercises

35. $\frac{f}{g}(x) = \frac{x^2 + 5x - 6}{x - 1}$ where $x \ne 1$
$$= \frac{(x + 6)(x - 1)}{x - 1}$$
$$= x + 6, x \ne 1$$
$$D = \{x \mid x \ne 1\}; R = \{y \mid y \ne 7\}$$

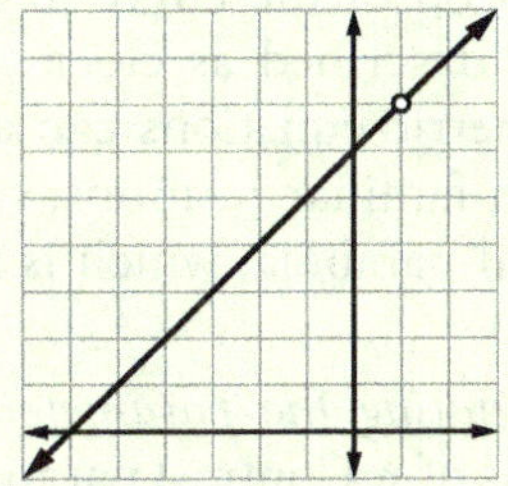

$$\lim_{x \to -\infty} f(x) = -\infty; \lim_{x \to \infty} f(x) = \infty$$
PD at $x = 1$

36. sometimes; only when $D_f = D_g$

37. always; by definition

38. sometimes; only when $D_f = D_g$ and no zeros of $g(x)$ are elements of D_f or D_g

39. sometimes; only when the composition does not further limit D_g

40a. $g(h(x)) = \frac{1}{\sqrt{x}}$
$$f(g(h(x))) = 2\left(\frac{1}{\sqrt{x}}\right) + 3$$
$$= \frac{2}{\sqrt{x}} + 3 \text{ or } \frac{3x + 2\sqrt{x}}{x}$$
$D_h = [0, \infty)$, $D_{g(h)} = (0, \infty)$, and $D_{f(g(h))} = (0, \infty)$

continued in Answers and Solutions Overflow

1.8 Parametric Equations and Inverses

Objectives

1. To define relations and functions parametrically

2. To find a rule for the inverse of a function when given a rule for the function

3. To graph the inverse of a function by reflecting the function's graph in the line $y = x$

4. To determine whether the inverse of a graphed function is a function by using the Horizontal Line Test

5. To determine whether two functions are inverses by applying composition

Vocabulary

eliminating the parameter
Horizontal Line Test
inverse functions
inverse relation
one-to-one function
parameter
parametric equations

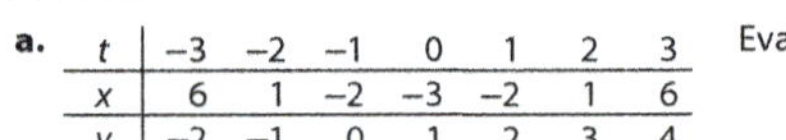

1.8 Parametric Equations and Inverses

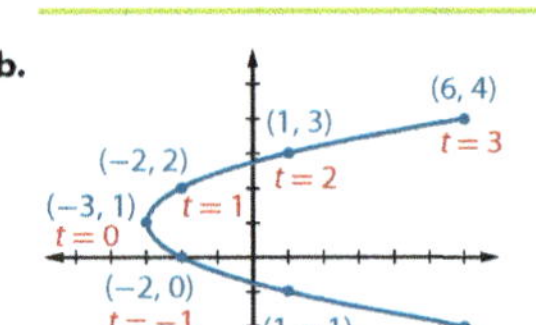

Projectile motion is often modeled by parametric equations.

After completing this section, you will be able to

- define relations and functions parametrically.
- find a rule for the inverse of a function.
- graph the inverse of a function.
- determine whether the inverse of a graphed function is a function.
- determine whether two functions are inverses.

The correlation between two variables may be caused by their relationship to a third variable. For example, a convenience store may see that ice-cream sales are directly proportional to sales of sunscreen, but this correlation is probably caused by their common relationship to daily high temperatures. In these cases it may be desirable to define each coordinate of the relation's ordered pairs with a pair of *parametric equations*, both in terms of the third variable, called a *parameter*.

Example 1 Graphing a Relation Represented by Parametric Equations

Given the relation P defined by $x = t^2 - 3$ and $y = t + 1$ with $-3 \le t \le 3$;

a. find the ordered pairs for the relation when $t = -3, -2, -1, 0, 1, 2, 3$;

b. plot the points and draw the graph of the relation; and

c. state whether the relation is a function.

Answer

a.

t	-3	-2	-1	0	1	2	3
x	6	1	-2	-3	-2	1	6
y	-2	-1	0	1	2	3	4

Evaluate x and y for each value of t.

b.

Plot the resulting points and connect them with a smooth curve.

c. The relation is not a function.

The graph fails the Vertical Line Test.

SKILL ✔ EXERCISE 3

The process of expressing y in terms of x when given a set of parametric equations is called *eliminating the parameter*. Solve for t in the equation for x and substitute the solution into the equation for y. This process is illustrated using $x = t^2 - 3$ and $y = t + 1$ from Example 1.

$$x = t^2 - 3 \qquad\qquad y = t + 1$$
$$t^2 = x + 3 \qquad\qquad y = \pm\sqrt{x + 3} + 1$$
$$t = \pm\sqrt{x + 3}$$

PRESENTATION

Lesson Opener

Write the values of x and y as an ordered pair for each value of t if $x = 3t + 1$ and $y = t^2$.

1. $t = 2$ (7, 4)

2. $t = -5$ (−14, 25)

3. $t = 0$ (1, 0)

Parametric equations will be a new concept for most students. The Lesson Opener introduces students to this concept, with t as the *parameter* (or *parametric variable*) that determines the values of both x and y.

Motivational Idea Parametric equations are useful in the study of physics. Use the Internet keyword search *parametric projectile video* for some interesting videos, or use the keyword search *parametric equation grapher* to create graphic demonstrations.

One-on-One Some students may struggle with understanding the purpose of parametric equations. Explain that a parameter is just another variable that affects the data. A data model for the amount of corn planted and profit from corn sales may be useful, but including other variables (or parameters), such as rainfall or the cost of seed, provides a more useful, realistic model.

Guide the students through the steps of Example 1. Consider discussing the curve's orientation (the direction that points move along the curve as t increases). The orientation of the curve in Example 1 can be described as clockwise. Note that parametric equations can also be used to graph familiar relations, such as this horizontal parabola, which is not a function.

Explain *eliminating the parameter* to express y in terms of x. Contrast parametric equations, which express height, horizontal distance, and time, with standard

The path of a ball thrown from a height of 6 ft at an approximate angle of 37° above the horizontal with a speed of 80 ft/sec can be modeled by $y = -\frac{1}{256}x^2 + \frac{3}{4}x + 6$. A limitation of this representation is that it does not indicate the time when the object is at a particular point on its parabolic path.

When the horizontal displacement x and vertical displacement (or height) y are modeled as functions of time t, the parametric equations allow the x- and y-coordinates to be determined at any time during the projectile's flight.

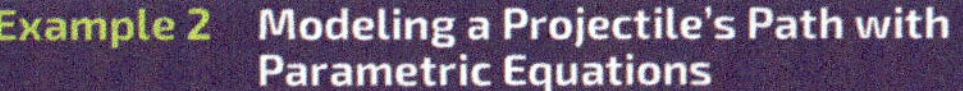

Example 2 — Modeling a Projectile's Path with Parametric Equations

Use your calculator and the parametric equations below to graph the trajectory of a ball thrown with an initial speed of 80 ft/sec at an approximate angle of 37° above horizontal. Then eliminate the parameter to show that the parametric equations are equivalent to the original functional representation above.

$$x = 64t \text{ and } y = -16t^2 + 48t + 6$$

Answer

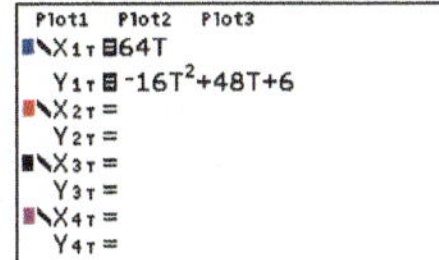

1. Change your calculator's mode from function to parametric, and enter the equations for x and y as separate functions in the [Y=] screen. (Use the [X,T,Θ,n] key to enter the parameter t.)

2. [GRAPH] the function and resize the [WINDOW] to show the ball's trajectory.

   a. Set Tmin = 0 and Tmax = 5 to ensure that sufficient points are plotted. Use Tstep = 0.05 to adjust the incrementation of T values.

   b. The [TRACE] function can be used to display the coordinates at a given time t.

   c. The [CALC] functions zero, min, max, and intersect are not available in parametric mode.

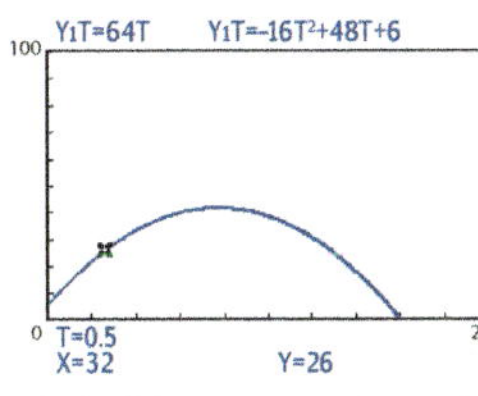

3. Solve $x = 64t$ for t, and substitute the resulting expression for t into the equation for y.

$$x = 64t; \, t = \frac{x}{64}$$

$$y = -16\left(\frac{x}{64}\right)^2 + 48\left(\frac{x}{64}\right) + 6$$

$$= -\frac{1}{256}x^2 + \frac{3}{4}x + 6$$

SKILL ✔ EXERCISE 37

When a relation is expressed parametrically, it is easy to find parametric equations for its *inverse*.

DEFINITION

If a relation S contains the ordered pair (a, b), its **inverse relation** S^{-1} contains the ordered pair (b, a).

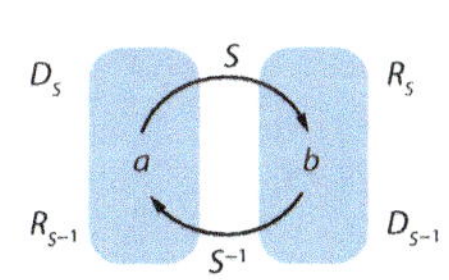

Reading and Writing Mathematics

Use the Internet keyword search *parametric equations projectile motion* to write a paragraph explaining how parametric equations are related to projectile motion.

The position of a projectile can be described by two variables, (x, y), but the motion of the projectile is also related to a third variable, t, or time. The x-position and the y-position are each related to time. Both x and y can be written in terms of t.

Additional Exercises

1. Find the ordered pairs defined by the parametric equations $x = 3t + 1$ and $y = -t^2$ where $t = \{0, 2, 5\}$.

   $(1, 0), (7, -4), (16, -25)$

2. Graph the relation represented by the parametric equations $x = t^2 - 3$ and $y = -2t + 3$ where $-2 \le t \le 2$.

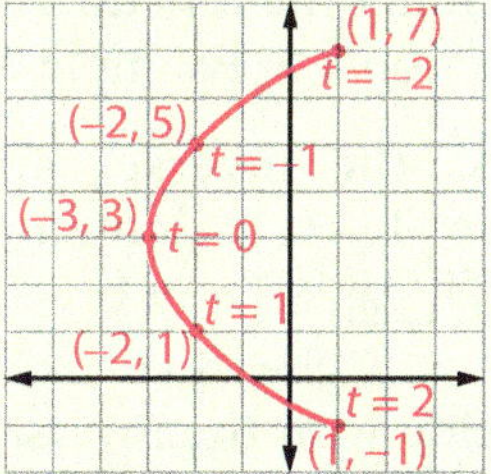

3. Eliminate the parameter in exercise 2 and express y in terms of x. Use technology to graph the function and confirm your answer.

   $x = t^2 - 3; \, t = \pm\sqrt{x + 3}$
   $y = -2(\pm\sqrt{x + 3}) + 3$
   $y = \pm 2\sqrt{x + 3} + 3$

The impact that modeling dynamic versus static processes had on the development of calculus is seen in Isaac Newton's use of the term *fluxion* (Latin for "flow"). He thought of curves as being formed by a point as it moved along a path.

---

equations, which express only height and horizontal distance.

Direct attention to the equation for the path of a thrown ball. Example 2 illustrates graphing the equivalent parametric equations. Note that the [TABLE] function can be used while in PARAMETRIC mode, but the [CALC] functions of zero, min, max, and intersect are available only in FUNCTION mode.

Motivational Idea Consider sharing that the same graph can be represented by different parametric equations. In such a case, the value of the parametric variable, say a time of $t = 5$ seconds, may be at a different point on the graph than it would be for another set of parametric equations. Therefore, different sets of parametric equations could be used to represent the same object traveling along a given path at different times.

Present the definition for an *inverse relation* and explain how to find the inverse of a relation. Introduce the *Horizontal Line Test* and explain that the graph of the inverse is a reflection in the line $y = x$; therefore, applying the Horizontal Line Test to the original relation provides the same information as applying the Vertical Line Test to its inverse.

Emphasize that all relations have an inverse relation, and therefore all functions have an inverse, but not all functions have an inverse function. Example 1 and the discussion following the definition of an *inverse function* show that some relations that are not functions have inverses that are functions.

In Example 3, stress the first step of interchanging the variables. Point out that by swapping x and y we are swapping the independent and dependent variables, thus swapping the input (domain) and output (range). Use Example 4 to demonstrate how the function rule of the inverse may require a restricted domain.

One-on-One If a student confuses the inverse function notation, $f^{-1}(x)$, with a

Find $(f \circ g)(x)$ and $(g \circ f)(x)$ and state whether $f(x)$ and $g(x)$ are inverse functions.

4. $f(x) = \frac{1}{3}x - 3$ and $g(x) = 3x + 3$

$f(g(x)) = x - 2; g(f(x)) = x - 6;$ not inverses

5. $f(x) = \frac{2x}{x + 3}$ and $g(x) = -\frac{3x}{x - 2}$

$f(g(x)) = g(f(x)) = x;$ inverses

Use the Horizontal Line Test to determine whether the inverse relation is a function. Write the rule for the inverse.

6. $f(x) = -1 + \frac{1}{x + 3}$

yes; $f^{-1}(x) = \frac{1}{x + 1} - 3$

7. $f(x) = x^2 - 2$ no; $y = \pm\sqrt{x + 2}$

For Graphing Calculators

Graph the function, its inverse, and $y = x$ in the same standard window. Using ZOOM, 5:ZSquare will display the graph so it looks symmetric about $y = x$. When necessary, a domain can be restricted by dividing the function by its domain. (See Section 1.3, Example 3.)

To check work, the inverse can be graphed on the TI-84 calculator by pressing 2nd, PRGM, 8:DrawInv, then entering Y_1. If the inverse is correct, the graph will coincide with the inverse entered as Y_2.

1. $f(x) = \frac{2}{x - 4}$ $f^{-1}(x) = \frac{2}{x} + 4$

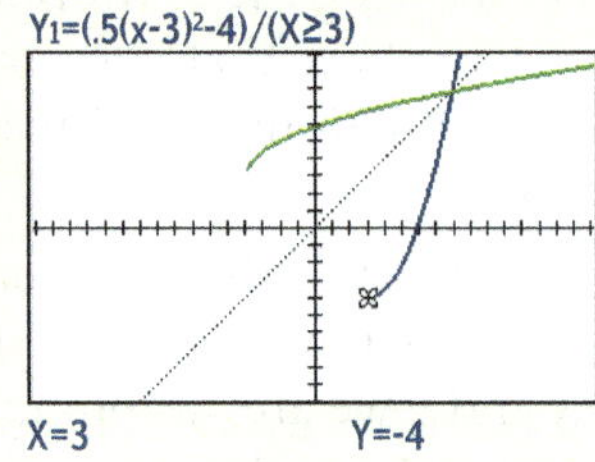

2. $f(x) = \frac{1}{2}(x - 3)^2 - 4$

Restrict the range of $f^{-1}(x)$ for $D = \{x \mid x \geq 3\}$. $f^{-1}(x) = \sqrt{2x + 8} + 3$

The range of the relation is the domain of its inverse, and the domain of the relation is the range of its inverse. The inverse of the relation in Example 1 can be found by interchanging the equations for the first and second coordinates.

$$P: x = t^2 - 3 \text{ and } y = t + 1 \text{ with } -3 \leq t \leq 3$$

$$P^{-1}: x = t + 1 \text{ and } y = t^2 - 3 \text{ with } -3 \leq t \leq 3$$

When the relation is expressed as a single equation in terms of x and y, the inverse can be found by interchanging the variables and solving for y.

$$P: y = \pm\sqrt{x + 3} + 1$$
$$P^{-1}: x = \pm\sqrt{y + 3} + 1$$
$$x - 1 = \pm\sqrt{y + 3}$$
$$(x - 1)^2 = y + 3$$
$$y = (x - 1)^2 - 3$$

Notice that the graphs of P and P^{-1} are reflections of each other in the line $y = x$, and that the inverse is a function since it passes the Vertical Line Test. The fact that the inverse passes the Vertical Line Test can be predicted by a Horizontal Line Test of the original relation.

HORIZONTAL LINE TEST

If a horizontal line can be drawn through more than one point of a relation's graph, the relation's inverse is not a function.

Every relation has an inverse, which may or may not be a function. The Horizontal Line Test confirms that the inverse of the relation in Example 1 is a function. Likewise, every function has an inverse, which may or may not be a function. Notice how the Horizontal Line Test indicates that the inverse of the function in Example 2 is not a function.

Example 3   Finding the Inverse

Find the inverse of $f(x) = \frac{1}{3}x + 1$. Then graph $f(x)$ and its inverse and determine whether the inverse is a function.

Answer

$$y = \frac{1}{3}x + 1$$
$$x = \frac{1}{3}y + 1$$

1. Express the function in terms of x and y and interchange the independent and dependent variables.

$$x - 1 = \frac{1}{3}y$$
$$y = 3x - 3$$

2. Solve for y.

CONTINUED ➡

---

negative exponent, point out that the superscript -1 is part of the function notation, not a power of the variable.

Discuss *one-to-one functions* using the illustrations of functions g and h. State the definition of inverse functions and present Example 5, showing how the definition is used to test whether two functions are inverses of each other.

TIPS

Ex. 20 Students should recognize this function as the traditional temperature conversion formula with different variables.

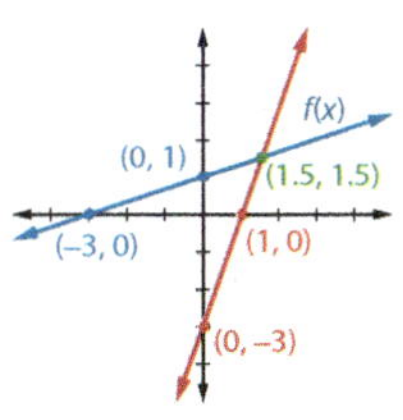

3. Graph the function $y = \frac{1}{3}x + 1$ and its inverse $y = 3x - 3$, noting that they are reflections of each other in the line $y = x$.
Note that the y-intercept becomes the x-intercept and the x-intercept becomes the y-intercept.

The inverse $y = 3x - 3$ is a function.

4. The graph of the original function passes the Horizontal Line Test.

SKILL ✓ **EXERCISES 9, 15**

Restrictions on the domain of a function can cause difficulty when finding the inverse.

Example 4 Finding the Inverse with a Restricted Domain

Find an equation for the inverse of $y = \sqrt{x + 2}$. Then determine whether the inverse is a function, state the inverse's domain and range, and sketch its graph.

Answer

$x = \sqrt{y + 2}$
$x^2 = y + 2$
$y = x^2 - 2$
The inverse is a function.

$D_{inv} = [0, \infty); R_{inv} = [-2, \infty)$

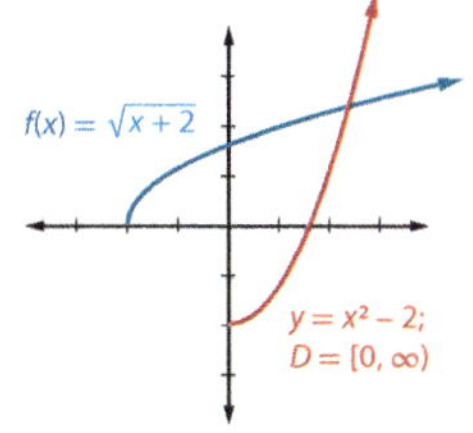

1. Switch the variables and solve for y.

2. The graph of the original function passes the Horizontal Line Test.

3. Switch the original function's domain, $D_f = [-2, \infty)$, and range, $R_f = [0, \infty)$, to obtain the domain and range of the inverse function.

4. Graph the inverse.
Notice that without its restricted domain, $y = x^2 - 2$ is not the inverse of the original function (a reflection across $y = x$).

SKILL ✓ **EXERCISE 33**

TIP 💡

The notation $f^{-1}(x)$ does not indicate the reciprocal of the function, which must be indicated as $\frac{1}{f(x)}$.

The inverse of $f(x) = \sqrt{x + 2}$ in Example 4 is a function since each member of its domain is paired with exactly one element of its range. When the inverse of a function f is also a function, the inverse is notated as f^{-1} and read "f inverse": $f^{-1}(x) = x^2 - 2$, $D = [0, \infty)$. Functions that pass the Horizontal Line Test are further classified as *one-to-one functions* since each member of the range is paired with exactly one member of the domain.

Assignments

- **Minimum:** 1, 3–4, 7–11, 13, 15–17, 20–21, 24–25, 27–28, 33, 37, 45, 48–53
- **Standard:** 2, 5–7, 10, 13–15, 17, 19; 20–28 even; 29, 32, 34, 37–38, 40, 42, 45, 50–53
- **Extended:** 2, 5, 8, 11, 14, 17–21, 25–26, 28–29, 32, 35–42, 46, 50–54

Assessment

- Quiz 1D covers Sections 1.7–1.8.

Solutions

❯ A. Exercises

1. $x = -2(3) - 3 = -9$
   $y = 3(3) - 4 = 5$

2. $x = (-4) + 3 = -1$
   $y = (-4)^2 - 5 = 11$

3. 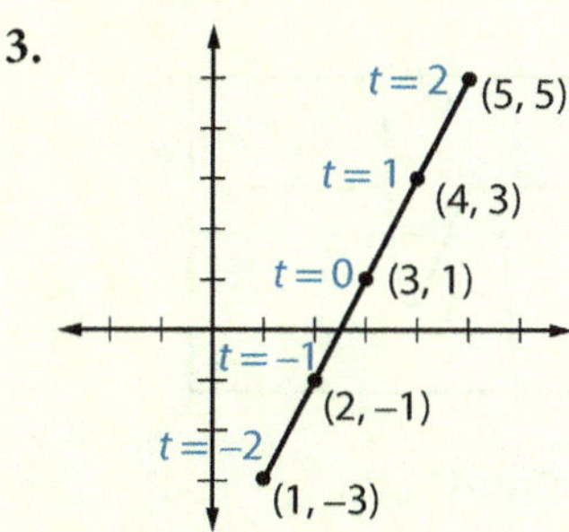

4. 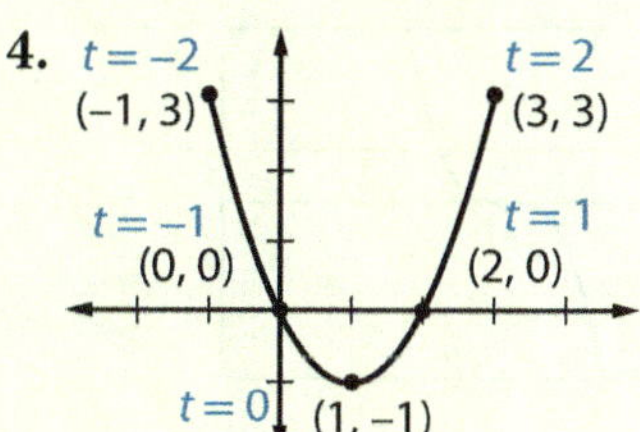

5. 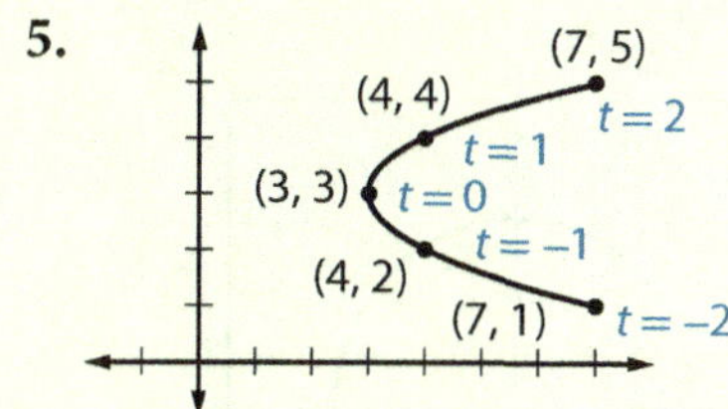

6. 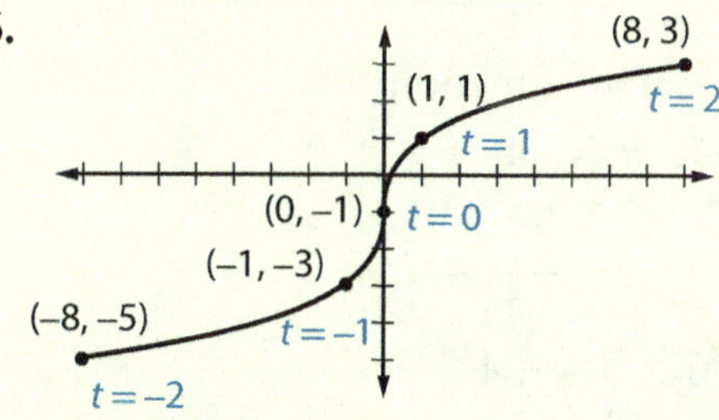

7. P is a function. $P^{-1} = \{(7, 4), (-3, 2), (7, 5), (8, 1)\}$ is not a function because 7 is mapped to both 4 and 5.

8. Q is a one-to-one function.
   $\therefore Q^{-1} = \{(7, -1), (2, -6), (4, 3)\}$ is also a function.

9. The relation is a function since it passes the Vertical Line Test. Since the relation fails the Horizontal Line Test, its inverse is not a function.

10. The relation is not a function since it fails the Vertical Line Test. Since the relation passes the Horizontal Line Test, its inverse is a function.

11.

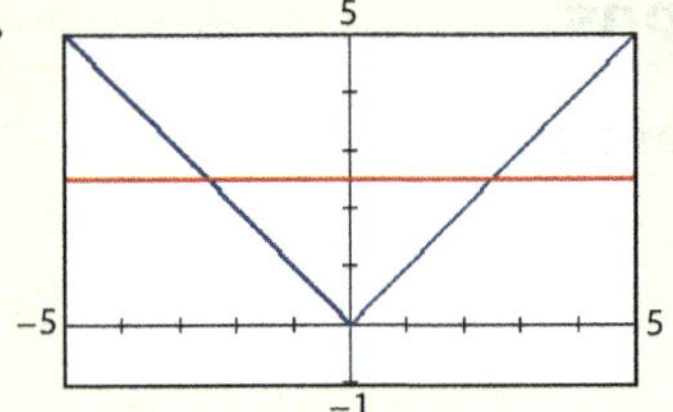

not a function

12.

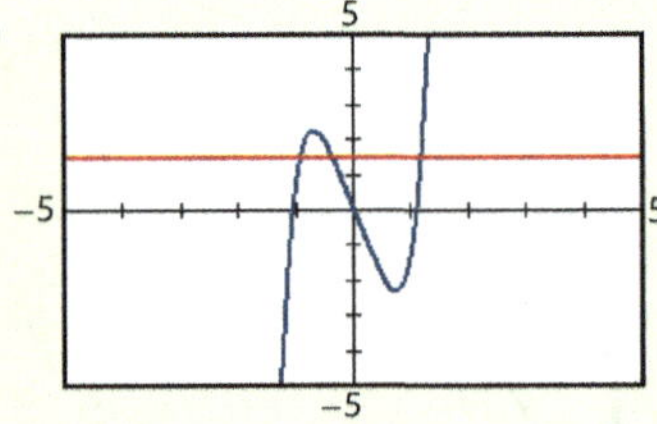

not a function

13.

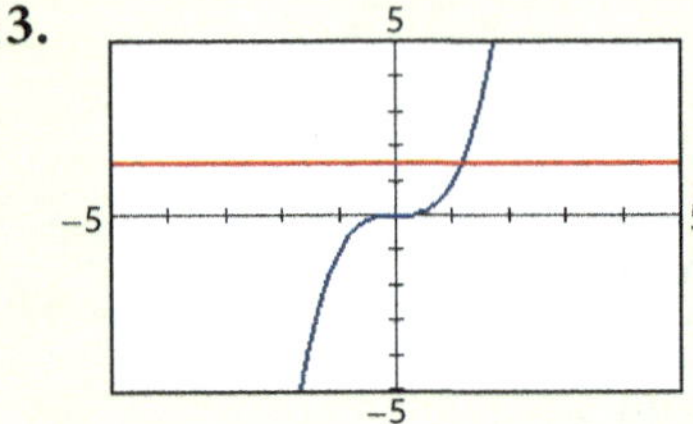

function

14.

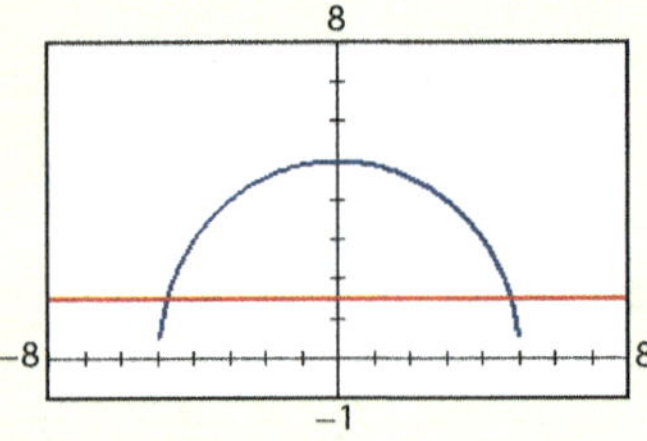

not a function

15. $x = -4y + 6$
$4y = -x + 6$
$y = -\frac{1}{4}x + \frac{3}{2}$

16. $x = \frac{y}{2} + 3$
$-\frac{y}{2} = -x + 3$
$y = 2x - 6$

17. $x = y^3 + 3$
$y^3 = x - 3$
$y = \sqrt[3]{x - 3}$

18. $x = \frac{y + 1}{2y}$
$2xy = y + 1$
$2xy - y = 1$
$y(2x - 1) = 1$
$y = \frac{1}{2x - 1}$

19a. $f^{-1}(x): x = 0.2642y$
$y = \frac{x}{0.2642} \approx 3.785x$

19b. $f^{-1}(4.5) = \frac{4.5}{0.2642} \approx 17.03$ L

---

11. not a function
12. not a function
13. function
14. not a function
15. $f^{-1}(x) = -\frac{1}{4}x + \frac{3}{2}$
16. $g^{-1}(x) = 2x - 6$
17. $h^{-1}(x) = \sqrt[3]{x - 3}$
18. $f^{-1}(x) = \frac{1}{2x - 1}$
19a. $f^{-1}(x) = \frac{x}{0.2642} \approx 3.785x$
20a. $f^{-1}(x) = \frac{5}{9}(x - 32) = \frac{5}{9}x - \frac{160}{9}$

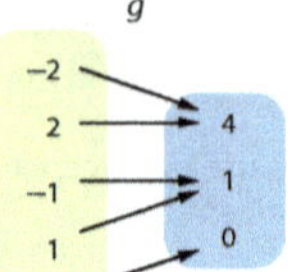
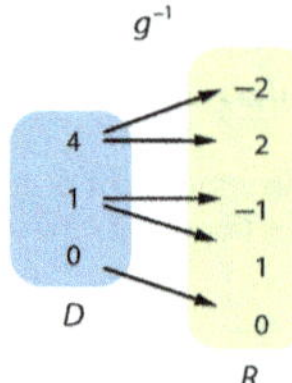
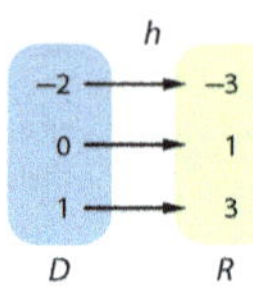
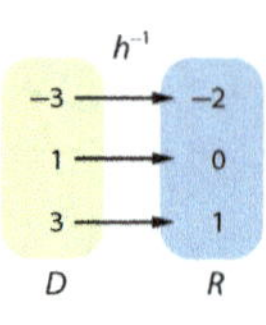

Notice that function g is not one-to-one. Since the domain of its inverse has a value that is paired with two range values, its inverse is not a function. Since h is a one-to-one function, $h(1) = 3$ and $h^{-1}(3) = 1$. Therefore, $(h^{-1} \circ h)(1) = 1$ and $(h \circ h^{-1})(3) = 3$. In fact, $(h^{-1} \circ h)(x) = x$ and $(h \circ h^{-1})(x) = x$ for all x in their domains. When composed, a function and its inverse function "reverse" each other.

> **DEFINITION**
>
> The functions $f(x)$ and $g(x)$ are **inverse functions** if $(f \circ g)(x) = (g \circ f)(x) = x$ for all x in their domains.

Example 5 Verifying Inverse Functions

Show that $f(x) = \frac{1}{3}x + 1$ and $g(x) = 3x - 3$ are inverse functions.

Answer

$(f \circ g)(x) = f(g(x))$
$= \frac{1}{3}(3x - 3) + 1$
$= x - 1 + 1$
$= x$

1. Find $(f \circ g)(x)$ by substituting $g(x)$ into $f(x)$.

$(g \circ f)(x) = g(f(x))$
$= 3\left(\frac{1}{3}x + 1\right) - 3$
$= x + 3 - 3$
$= x$

2. Find $(g \circ f)(x)$ by substituting $f(x)$ into $g(x)$.

$f(x)$ and $g(x)$ are inverse functions.

3. $(f \circ g)(x) = (g \circ f)(x) = x$

SKILL ✔ EXERCISE 21

> **A. Exercises**

Find the ordered pair defined by the parametric equations.

1. $x = -2t - 3$ and $y = 3t - 4$, $t = 3$ $(-9, 5)$

2. $x = t + 3$ and $y = t^2 - 5$, $t = -4$ $(-1, 11)$

Graph the relation represented by each pair of parametric equations and state whether each relation is a function.

3. $x = t + 3$ and $y = 2t + 1$, $-2 \le t \le 2$ function
4. $x = t + 1$ and $y = t^2 - 1$, $-2 \le t \le 2$ function
5. $x = t^2 + 3$ and $y = t + 3$, $-2 \le t \le 2$ not a function
6. $x = t^3$ and $y = 2t - 1$, $-2 \le t \le 2$ function

State whether each relation is a function. Then determine whether the inverse is a function, and if not, explain why.

7. $P = \{(4, 7), (2, -3), (5, 7), (1, 8)\}$
8. $Q = \{(-1, 7), (-6, 2), (3, 4)\}$

9. 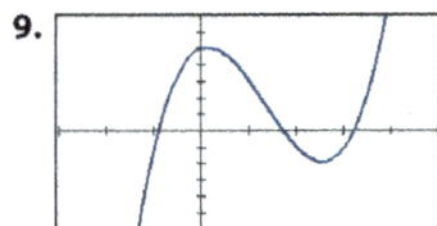

10. 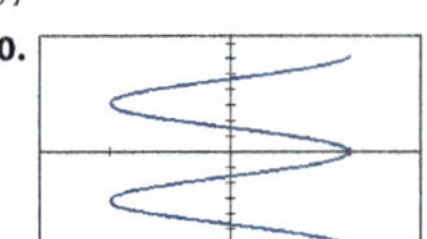

20a. $f^{-1}(x):\ x = \frac{9}{5}y + 32$
$\frac{9}{5}y = x - 32$
$y = \frac{5}{9}(x - 32) = \frac{5}{9}x - \frac{160}{9}$

20b. $f^{-1}(98.6) = \frac{5}{9}(98.6 - 32) = 37°\text{C}$

> **B. Exercises**

21. $f(g(x)) = 3\left(\frac{x - 9}{3}\right) + 9$
$= x - 9 + 9 = x$

$g(f(x)) = \frac{(3x + 9) - 9}{3} = x$

22. $f(g(x)) = \frac{\left(\sqrt[5]{2x + 1}\right)^5 - 1}{2}$
$= \frac{(2x + 1) - 1}{2} = x$

$g(f(x)) = \sqrt[5]{2\left(\frac{x^5 - 1}{2}\right) + 1}$
$= \sqrt[5]{(x^5 - 1) + 1} = x$

Use technology to graph each function and then use the Horizontal Line Test to determine if the inverse is a function.

11. $f(x) = |x|$ **12.** $y = 3x^5 - 4x$

13. $y = x^3$ **14.** $y = \sqrt{25 - x^2}$

Find the inverse of each function.

15. $f(x) = -4x + 6$ **16.** $g(x) = \frac{x}{2} + 3$

17. $h(x) = x^3 + 3$ **18.** $f(x) = \frac{x+1}{2x}$

19. The function $f(x) = 0.2642x$ converts a number of liters, x, to an equivalent number of gallons.
  a. Find $f^{-1}(x)$, which converts a number of gallons to an equivalent number of liters.
  b. Express a volume of 4.5 gal in liters. ≈ 17.03 L

20. The function $f(x) = \frac{9}{5}x + 32$ converts a temperature in degrees Celsius, x, to degrees Fahrenheit.
  a. Find $f^{-1}(x)$, which converts a temperature in degrees Fahrenheit to degrees Celsius.
  b. Express a temperature of 98.6°F in degrees Celsius. 37°C

❯ B. Exercises

Confirm that $f(x)$ and $g(x)$ are inverse functions by showing that $(f \circ g)(x) = x = (g \circ f)(x)$.

21. $f(x) = 3x + 9$ **22.** $f(x) = \frac{x^5 - 1}{2}$
  $g(x) = \frac{x-9}{3}$ $g(x) = \sqrt[5]{2x + 1}$

23. $f(x) = \frac{x-1}{x+7}$
  $g(x) = \frac{-7x-1}{x-1}$

Find $(f \circ g)(x)$ and $(g \circ f)(x)$ and state whether $f(x)$ and $g(x)$ are inverse functions.

24. $f(x) = x + 6$
  $g(x) = x - 6$ inverses

25. $f(x) = 2x + 1$
  $g(x) = \frac{x}{2} - 1$ not inverses

26. $f(x) = \frac{x}{x+1}$
  $g(x) = \frac{x}{1-x}$ inverses

Sketch the graph of the inverse of each function.

27.

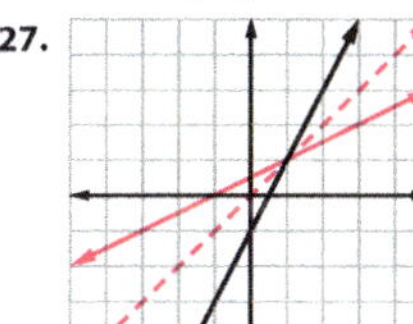

28.

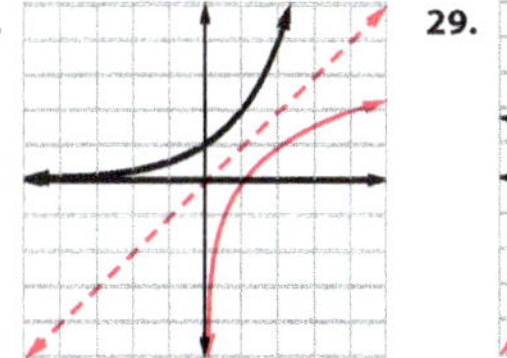

29.

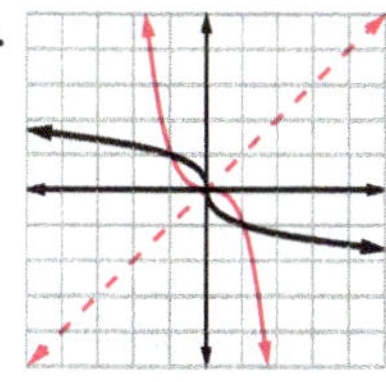

Find $f^{-1}(x)$ for each function. Confirm your results by using technology to graph the function and its inverse.

30. $f(x) = 2x - 3$ **31.** $f(x) = 3x^5 - 4$

32. $f(x) = \frac{1}{x+2}$

Find the inverse of each function and specify any restricted domains. Confirm your results by using technology to graph the function and its inverse.

33. $h(x) = \sqrt{x - 2}$ **34.** $g(x) = \frac{3}{\sqrt{x-4}}$

35. $f(x) = \frac{3}{2\sqrt{x+1}}$ **36.** $h(x) = \sqrt{\frac{3(x+4)}{2}}$

37. The horizontal distance traveled and height of the served volleyball described in Section 1.6 exercise 37 can be modeled (in meters) by the following parametric equations where the service line acts as the origin.
$$x = 15t$$
$$y = -4.9t^2 + 4t + 2$$
  a. Use technology to graph the parametric function. Then determine the ordered pair describing the position of the volleyball at each quarter-second until the ball hits the floor.
  b. Eliminate the parameter to find a function, $f(x)$, for the ball's height in terms of the horizontal distance traveled. Use technology to graph the function and use the ordered pairs from part a to confirm your answer.
  c. Find the maximum height of the serve and the horizontal distance it traveled to that point.
  d. How far from the service line does the ball land?

38. The men's PE instructor wants to write a linear formula to grade the results of the 12-minute run. A 1.75 mi run would earn a 95% (1.75, 95) and a 1 mi run would earn a 65% (1, 65).
  a. Find the linear function $f(x)$ that assigns a percentage grade for various recorded distances.
  b. Find $f^{-1}(x)$ and describe what its independent and dependent variables represent.
  c. Find the percentage grade for a student who runs 1.125 mi.
  d. Find the distance needed to earn an 85%.

30. $f^{-1}(x)$: $x = 2y - 3$
$$2y = x + 3$$
$$y = \frac{1}{2}x + \frac{3}{2}$$

31. $f^{-1}(x)$: $x = 3y^5 - 4$
$$y^5 = \frac{x+4}{3}$$
$$y = \sqrt[5]{\frac{x+4}{3}}$$

32. $f^{-1}(x)$: $x = \frac{1}{y+2}$
$$x(y + 2) = 1$$
$$xy + 2x = 1$$
$$y = \frac{1 - 2x}{x}$$

33. $D_h = (2, \infty), R_h = (0, \infty)$
$h^{-1}(x)$: $x = \sqrt{y - 2}$
$$x^2 = y - 2$$
$$y = x^2 + 2; D = [0, \infty)$$

34. $D_g = (4, \infty), R_g = (0, \infty)$
$g^{-1}(x)$: $x = \frac{3}{\sqrt{y-4}}$
$$\sqrt{y - 4} = \frac{3}{x}$$
$$y - 4 = \frac{9}{x^2}$$
$$y = \frac{9}{x^2} + 4; D = (0, \infty)$$

35. $D_f = (-1, \infty), R_f = (0, \infty)$
$f^{-1}(x)$: $x = \frac{3}{2\sqrt{y+1}}$
$$\sqrt{y + 1} = \frac{3}{2x}$$
$$y = \frac{9}{4x^2} - 1; D = (0, \infty)$$

36. $D_h = (-4, \infty), R_h = [0, \infty)$
$h^{-1}(x)$: $x = \sqrt{\frac{3(y+4)}{2}}$
$$2x^2 = 3y + 12$$
$$y = \frac{2x^2 - 12}{3}; D = [0, \infty)$$

37a.

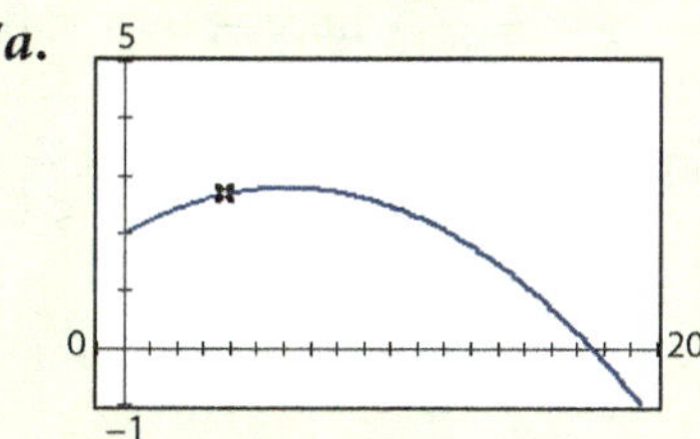

using $\boxed{\text{TRACE}}$ and entering $t = 0, 0.25, 0.5, 0.75, 1$:
$(0, 2), \approx (3.75, 2.69), \approx (7.5, 2.78), \approx (11.25, 2.24), (15, 1.1)$

37b. $t = \frac{x}{15}$
$$y = -4.9\left(\frac{x}{15}\right)^2 + 4\left(\frac{x}{15}\right) + 2$$
$$f(x) = -\frac{49}{2250}x^2 + \frac{4}{15}x + 2$$

27.

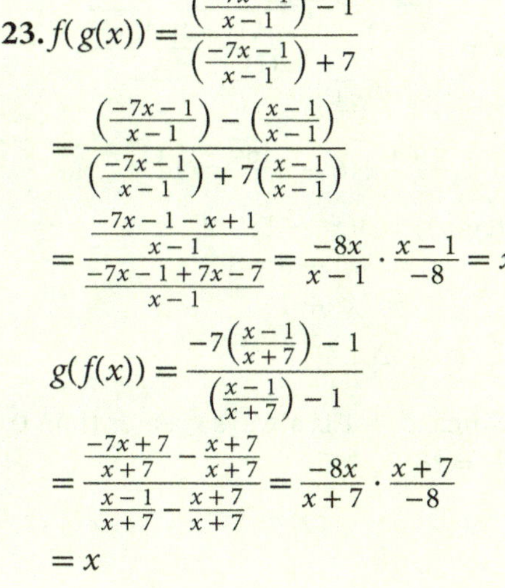

37a. (0, 2), (3.75, 2.69), (7.5, 2.78), (11.25, 2.24), (15, 1.1)

37b. $f(x) = -\frac{49}{2250}x^2 + \frac{4}{15}x + 2$

37c. ≈ 2.82 m high; ≈ 6.12 m

37d. ≈ 17.49 m

38a. $f(x) = 40x + 25$

38b. $f^{-1}(x) = 0.025x - 0.625$

38c. 70%

38d. 1.5 mi

23. $f(g(x)) = \dfrac{\left(\frac{-7x-1}{x-1}\right) - 1}{\left(\frac{-7x-1}{x-1}\right) + 7}$

$= \dfrac{\left(\frac{-7x-1}{x-1}\right) - \left(\frac{x-1}{x-1}\right)}{\left(\frac{-7x-1}{x-1}\right) + 7\left(\frac{x-1}{x-1}\right)}$

$= \dfrac{\frac{-7x-1-x+1}{x-1}}{\frac{-7x-1+7x-7}{x-1}} = \frac{-8x}{x-1} \cdot \frac{x-1}{-8} = x$

$g(f(x)) = \dfrac{-7\left(\frac{x-1}{x+7}\right) - 1}{\left(\frac{x-1}{x+7}\right) - 1}$

$= \dfrac{\frac{-7x+7}{x+7} - \frac{x+7}{x+7}}{\frac{x-1}{x+7} - \frac{x+7}{x+7}} = \frac{-8x}{x+7} \cdot \frac{x+7}{-8} = x$

24. $f(g(x)) = (x - 6) + 6 = x$
$g(f(x)) = (x + 6) - 6 = x$

25. $f(g(x)) = 2\left(\frac{x}{2} - 1\right) + 1$
$= x - 2 + 1 = x - 1 \neq x$

$g(f(x)) = \dfrac{(2x+1)}{2} - 1 = \dfrac{2x+1-2}{2}$
$= \dfrac{2x-1}{2} \neq x$

26. $f(g(x)) = \dfrac{\left(\frac{x}{1-x}\right)}{\left(\frac{x}{1-x}\right) + 1} = \dfrac{\frac{x}{1-x}}{\frac{x}{1-x} + \frac{1-x}{1-x}}$

$= \dfrac{x}{1-x} \cdot \dfrac{1-x}{x+1-x} = x$

$g(f(x)) = \dfrac{\left(\frac{x}{x+1}\right)}{1 - \left(\frac{x}{x+1}\right)} = \dfrac{\frac{x}{x+1}}{\frac{x+1}{x+1} - \frac{x}{x+1}}$

$= \dfrac{x}{x+1} \cdot \dfrac{x+1}{1} = x$

37c. $V(h, f(h))$:

$$h = -\frac{\frac{4}{15}}{2\left(-\frac{49}{2250}\right)} = \frac{300}{49} \approx 6.12 \text{ m}$$

$$f\left(\frac{300}{49}\right) \approx 2.82 \text{ m high}$$

37d. $-\frac{49}{2250}x^2 + \frac{4}{15}x + 2 = 0$

$$x = \frac{-\left(\frac{4}{15}\right) \pm \sqrt{\left(\frac{4}{15}\right)^2 - 4\left(-\frac{49}{2250}\right)(2)}}{2\left(-\frac{49}{2250}\right)}$$

$$\approx -5.25,\ 17.49$$

38a. $m = \dfrac{95 - 65}{1.75 - 1} = 40$

$$y - 65 = 40(x - 1)$$
$$y = 40x + 25$$

38b. $f^{-1}(x):\quad x = 40y + 25$
$$40y = x - 25$$
$$y = \frac{1}{40}x - \frac{5}{8}$$

The independent variable represents the percentage grade and the dependent represents the corresponding distance.

38c. $f(1.125) = 40(1.125) + 25 = 70\%$

38d. $f^{-1}(85) = 0.025(85) - 0.625$
$$= 1.5 \text{ mi}$$

❯ C. Exercises

44. $D_1 = (-\infty, 2];\ R_1 = (-\infty, -3]$
$D_2 = (2, \infty);\ R_2 = (-3, 0)$

$f_1^{-1}(x):\ x = \dfrac{y - 14}{4}$
$$4x = y - 14$$
$$y = 4x + 14 \text{ where } x \le -3$$

$f_2^{-1}(x):\ x = \dfrac{9}{1 - 2y}$
$$x - 2xy = 9$$
$$-2xy = 9 - x$$
$$y = \frac{x - 9}{2x} \text{ where } -3 < x < 0$$

$$f^{-1}(x) = \begin{cases} 4x + 14 & \text{if } x \le -3 \\ \dfrac{x - 9}{2x} & \text{if } -3 < x < 0 \end{cases}$$

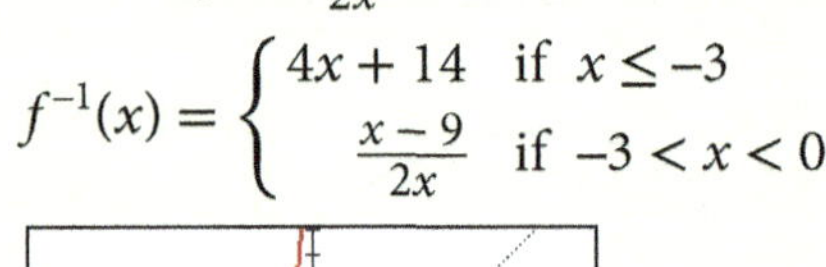

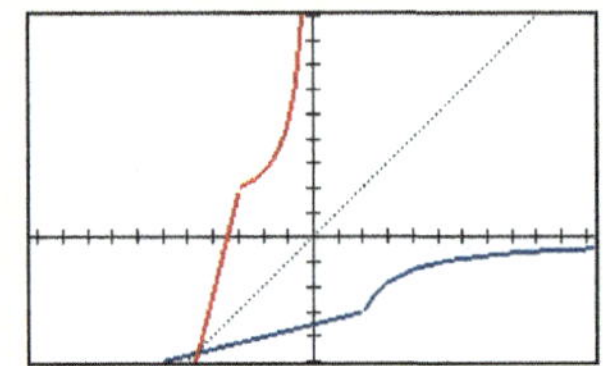

❯ Cumulative Review

45.

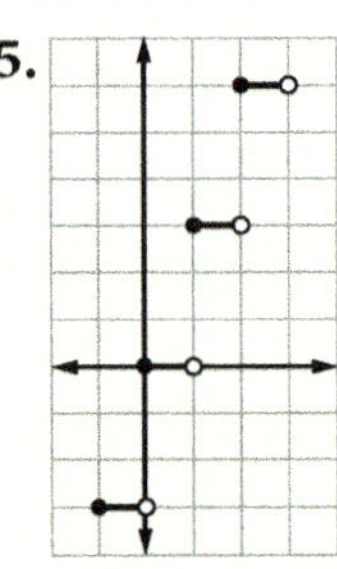

❯ C. Exercises

39. Write: Find expressions for the slope of a linear function given in standard form, $ax + by = c$, and the slope of its inverse. Describe the relationship between the slopes.

State whether each statement is *always*, *sometimes*, or *never* true. Explain your reasoning.

40. The inverse of a linear function is also a function.

41. The inverse of an odd function is an even function.

42. The inverse of an increasing function is an increasing function.

43. Explain: Why is the inverse of $g(x)$ not a function? State a least restrictive domain of $g(x)$ for which the inverse is a function.

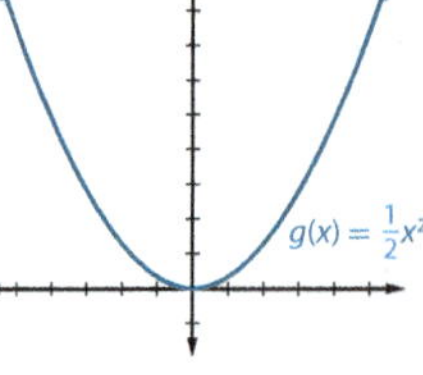

44. Find the inverse of $f(x) = \begin{cases} \dfrac{x - 14}{4} & \text{if } x \le 2 \\ \dfrac{9}{1 - 2x} & \text{if } x > 2 \end{cases}$.

Confirm your result by using technology to graph the function and its inverse.

CUMULATIVE REVIEW

Graph each function. [1.3–1.4]

45. $f(x) = 3[x]$

46. $g(x) = \left[\frac{1}{2}x\right]$

47. Determine whether $f(x) = \begin{cases} |x| & \text{if } x \le 2 \\ x^2 & \text{if } x > 2 \end{cases}$ is continuous or discontinuous. Explain your reasoning. [1.3]

Classify each function as even, odd, or neither. [1.4]

48. $f(x) = 3x^2$ even

49. $f(x) = 2x^5$ odd

50. $f(x) = x^2 + 2$ even

51. Which equation describes $g(x)$, a transformation of $f(x)$? [1.5]   E

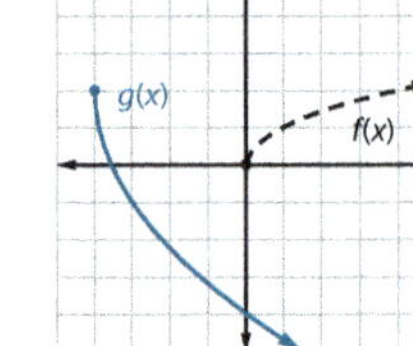

  A. $g(x) = 3f(x - 4) + 2$
  B. $g(x) = -3f(x - 4) + 2$
  C. $g(x) = -3f(x + 4) - 2$
  D. $g(x) = 3f(x + 4) + 2$
  E. $g(x) = -3f(x + 4) + 2$

52. What is the sum of the roots of the equation $4x^2 - 12x - 25 = 3x^2 - 20$? [1.6]   A

  A. 12   **C.** -18 **E.** -15
  B. -12 **D.** 18

53. Find $f(g(x))$ if $f(x) = 3x - 2$ and $g(x) = x^2 + 1$. [1.7]   B

  A. $f(g(x)) = 9x^2 + 5$
  B. $f(g(x)) = 3x^2 + 1$
  C. $f(g(x)) = 9x^2 - 6x + 5$
  D. $f(g(x)) = 3x^3 - 2x^2 + 3x - 2$
  E. none of these

54. If $f(x) = 3x - 2$ and $g(x) = x^2 + 1$, what is the domain of $\dfrac{f}{g}(x)$? [1.7]   E

  A. $\{x \mid x \ne -1\}$ **D.** $\{x \mid x > -1\}$
  B. $\left\{x \mid x > \frac{2}{3}\right\}$ **E.** $\mathbb{R}$
  C. $\left\{x \mid x \ne \frac{2}{3}\right\}$

39. The slope of $ax + by = c$ is $-\frac{a}{b}$, and the slope of its inverse, $bx + ay = c$, is $-\frac{b}{a}$. The slopes are reciprocals.

40. sometimes; The inverse of a non-constant linear function is always a function, but the inverse of a constant function is a vertical line, which is not a function.

41. never; It is always an odd function.

42. always; For instance, the slope of a line and its inverse have the same sign.

43. The graph does not pass the Horizontal Line Test. Either $D_g = (-\infty, 0]$ or $D_g = [0, \infty)$ allows $g(x)$ to pass the Horizontal Line Test.

47. discontinuous; $\lim\limits_{x \to 2^-} f(x) = 2$ but $\lim\limits_{x \to 2^+} f(x) = 4$

46.

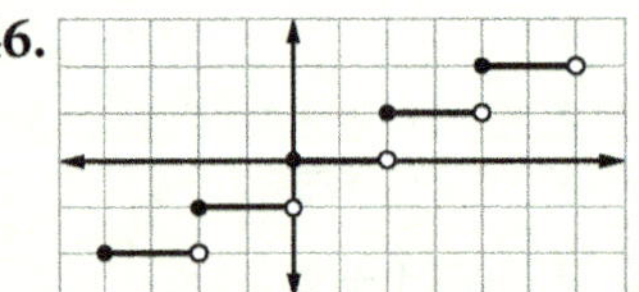

48. even; It is symmetric with respect to the y-axis.

49. odd; It is symmetric with respect to the origin.

50. even; It is translated up 2 units but is still symmetric with respect to the y-axis.

51. The function f has been moved 4 units left, reflected across the y-axis, and stretched vertically by a factor of 3 before being moved 2 units up.

52. $x^2 - 12x - 5 = 0$
$$x = \frac{-(-12) \pm \sqrt{(-12)^2 - 4(1)(-5)}}{2(1)}$$
$$x = 6 \pm \frac{\sqrt{164}}{2}$$
$$6 + \frac{\sqrt{164}}{2} + 6 - \frac{\sqrt{164}}{2} = 12$$

53. $f(g(x)) = 3(x^2 + 1) - 2$
$$= 3x^2 + 3 - 2$$
$$= 3x^2 + 1$$

54. $\dfrac{f}{g}(x) = \dfrac{3x - 2}{x^2 + 1}$

Since $x^2 + 1$ is always greater than 0, $D = \mathbb{R}$.

1.9 Modeling with Functions

The American alligator can grow up to a foot a year in the wild and up to three feet a year on an alligator farm. A biologist studying alligator growth patterns recorded the lengths and weights of fourteen alligators.

Data containing two related variables, such as length and weight, is called *bivariate* data. A *scatterplot* of the ordered pairs can be drawn to illustrate the relationship between the two variables. A *trend line* can be drawn to model the relationship between the data and make predictions related to other values of either variable. This trend line should minimize the distances from the data points. While trend lines drawn by visual inspection can provide imprecise estimates, a variety of technologies can be used to draw a scatterplot and complete a *linear regression* using a statistical algorithm to find the *line of best fit*.

Length (in.)	Weight (lb)
58	28
63	33
86	90
69	36
147	640
78	57
114	197
128	366
85	84
88	70
72	61
74	54
90	106
94	130

After completing this section, you will be able to

- find linear and quadratic models for data.
- evaluate how well a function models data.
- use a model to make predictions.

Joseph Fourier, in the preliminary discourse to his 1822 *The Analytical Theory of Heat*, wrote that "profound study of nature is the most fertile source of mathematical discoveries."

Graphing Calculator	Spreadsheet	Internet
See the Technology Corner in Section 6.6 of BJU Press's *Algebra 1*, 3rd ed., or use the Internet keyword search *TI-84 regression*.	See the Technology Corner in Section 2.4 of BJU Press's *Algebra 2*, 3rd ed., or use the Internet keyword search *spreadsheet trend line*.	Use the Internet keyword search *best fit line applet* or *regression app*.

Example 1 Finding the Line of Best Fit

Use technology to create a scatterplot and determine the equation of the line of best fit for the data collected by the biologist. Then use the equation to predict the following.

a. the weight of a 100 in. long alligator

b. the weight of a 3 ft long alligator

c. the length of an alligator that weighs 100 lb

Answer

Let x = length (in inches) and $y = f(x)$ = weight (in pounds).

1. Weight is more likely to be a function of length.

CONTINUED ➡

1.9 Modeling with Functions

Objectives

1. To use technology to find linear and quadratic functions that model data
2. To evaluate how well a function models data
3. To use a model to make predictions

Flash

There are about 5 million American alligators in the southeastern US, about 25% of which are found in Florida. The American alligator is the largest reptile in North America. An alligator can be distinguished from the crocodile by its broader head.

Vocabulary

bivariate data
correlation coefficient
coefficient of determination
extrapolation
interpolation
linear correlation
linear regression
outliers
quadratic regression
residuals
scatterplot
trend line

PRESENTATION

Lesson Opener

The height in feet of a thrown shot put ball is modeled by $f(d) = -0.0225d^2 + d + 5.2$, where d is the distance traveled in feet. Find the height for each distance that the shot put ball travels.

1. 10 ft   12.95 ft
2. 30 ft   14.95 ft
3. 48 ft   1.36 ft

Use the alligator data to introduce the italicized terms in the section's introduction. These terms should be review for most students.

The graphing calculators in the TI-84 family are powerful tools for creating functions that model data sets. A spreadsheet is also an excellent tool for quickly entering and verifying data. While the trace function may not be available in spreadsheets, the regression formula can be used in place of tracing. Detailed instructions on modeling data with either of these tools are readily available on the Internet. An Internet app or widget may also be a useful tool for data modeling.

Later in this same discourse, Fourier states that mathematical analysis (including calculus) interprets all natural phenomena in the same language and attests to "the unity and simplicity of the plan of the universe."

Have the students complete Example 1, creating a *scatterplot*, finding the line of best fit, and determining the requested values for the function.

Consider having students find $f(114)$. Note that this estimate (≈ 295.7) is different from the actual y-value for (114, 197). Have the students calculate how much the value deviates from the model's prediction.

Motivational Idea If your students are already proficient at using a graphing calculator to complete regressions, encourage them to use a spreadsheet or an Internet

Additional Exercises

x	0	2	4	6	8
y	−10	−6	−1	3	10

1. Use technology to create a scatter-plot for the data set and determine the line of best fit. $y = 2.45x - 10.6$

2. Describe the function as increasing, decreasing, or both. increasing

3. State the coefficient of determination. $r^2 \approx 0.989$

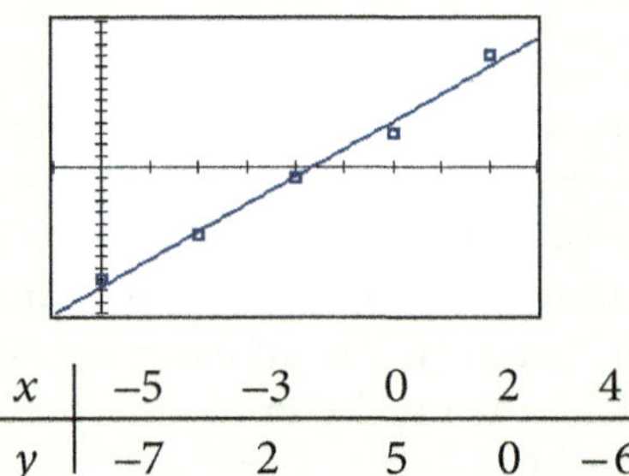

x	−5	−3	0	2	4
y	−7	2	5	0	−6

4. Use technology to create a scatter-plot for the data set and determine the function that best models the data. $y = -0.56x^2 - 0.566x + 4.6$

5. Describe the function as increasing, decreasing, or both. both

6. State the coefficient of determination. $r^2 \approx 0.974$

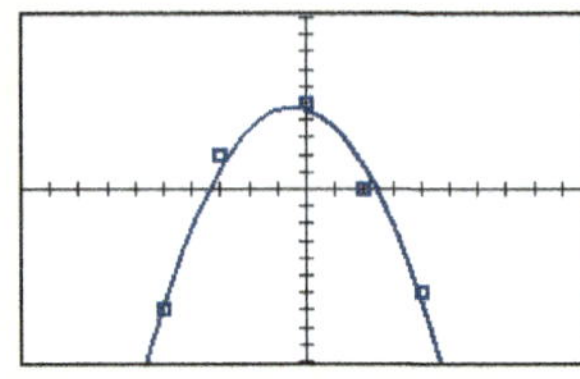

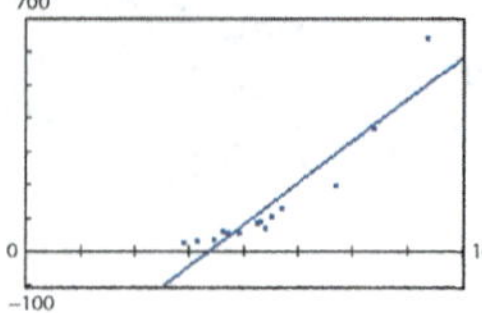

$f(x) \approx 6.252x - 417.000$

a. $f(100) \approx 6.252(100) - 417.00 = 208.2$ lb

b. $f(36) \approx 6.252(36) - 417.00 = -191.9$ lb

c. $f(x) \approx 6.252x - 417.00 = 100$

$x \approx \dfrac{517.00}{6.252} \approx 82.7$ in.

2. Input the data into a graphing calculator, spreadsheet, or Internet applet and complete the linear regression to determine the equation of the line of best fit.

3. Graph the function on the scatterplot.

4. Find $f(100)$.

5. Find $f(36)$, using 3 ft = 36 in.

6. Set $f(x) = 100$ and solve for x.

In this model, the prediction within the range of data points (*interpolation*) is fairly reasonable while the prediction outside the range of data points (*extrapolation*) is not reasonable. In general, interpolation tends to produce better estimates than extrapolation.

The *linear correlation* of the data describes how closely the data points are clustered along the trend line. The *correlation coefficient*, r, where $-1 \le r \le 1$, numerically describes the degree to which the two variables are related. A positive correlation indicates that the two quantities increase or decrease together. A negative value of r represents an inverse correlation, where one variable increases as the other decreases.

The closer $|r|$ is to 1, the stronger the correlation is between the variables. If $r = 1$ or -1, all the data points will be on a line with a positive or negative slope, respectively. If r is near 0, the points appear to be randomly scattered. The correlation is usually considered to be strong when $|r| > 0.8$ and weak when $|r| < 0.5$, but these guidelines vary with the context of the data.

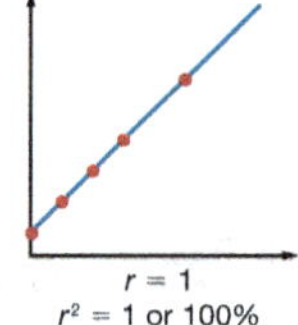
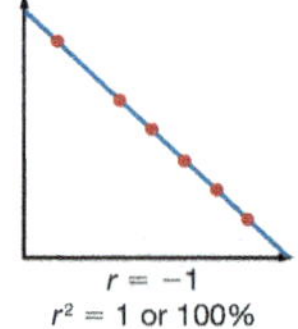
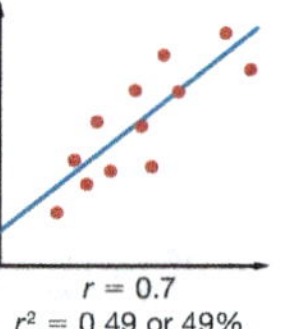
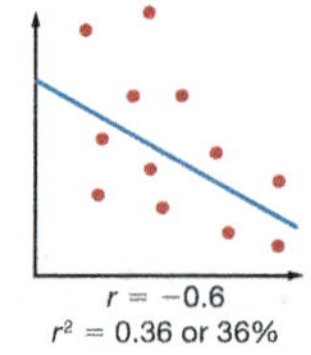

Many statistical applications also state the value of r^2, the *coefficient of determination*. If r is -0.8, then r^2 is 0.64. This indicates that 64% of the variation in the dependent variable can be explained by variation in the independent variable.

Outliers are points that do not seem to fit the trend of the scatterplot. These points typically represent extreme variations or unusual circumstances and are often ignored when mathematical models are constructed. We should ask if the last one or two data points in Example 1 are outliers or if a different function would serve as a better model for the data.

The *residuals*, the differences between the actual y-value of each data point and its value predicted by the modeling function, also indicate how well the model fits the data. If there is a pattern within the residuals, continue to search for a better model.

---

app to experience the advantages and disadvantages of various technologies.

Introduce the terms *interpolation* and *extrapolation*. These terms will be very important in the Data Analysis exercises.

Explain the terms relating to correlation and use the figures to identify various classifications of correlation. Stress that the significance of a specific *coefficient of determination* may differ with the context of the data. For instance, a 97% coefficient of determination may be acceptable in one study but not in another.

When defining *outliers*, explain that an outlier is sometimes an obvious extreme deviation of a value in the data, while at other times the value may border between normal and extreme, in which case its classification as an outlier becomes a judgment call. Consider having the students identify what two points might be considered outliers in the linear model in Example 1. *(74, 54) and (94, 130)*

Define *residuals*. A *linear regression* finds the line that minimizes the total error represented by the squares of the residuals at the data points. Explain that an analysis of the residuals can help determine whether a model is acceptable.

Example 2 examines both the *correlation coefficient* and the residuals to determine whether the linear model is a good fit. If DiagnosticsOn has been selected on your graphing calculator, it will display the r^2 value when a regression is completed. In an Excel spreadsheet, simply check "Display R-squared value on chart" when inserting the *trend line*.

An Excel spreadsheet will not display the residuals unless a statistics add-in has been installed. The trend line formula can be used to create a column of predicted values, which are then subtracted from the observed values to generate a list of residuals.

Interactive Activity Consider having the students delete the last two points in the data set and observe the effects on the model.

Example 2   Evaluating a Linear Model

Evaluate the linear model of the line of best fit for the alligator data.

a. State and interpret the values of the coefficient of determination and the linear correlation coefficient.

b. Plot the residuals and interpret the graph.

Answer

a. $r^2 \approx 0.866$
86.6% of the variation in weight is explained by variation in length.
$r \approx 0.931$
There is a strong positive correlation between length and weight.

Use technology to display the correlation coefficient, r, or the coefficient of determination, r^2, or both. You may have to calculate r from r^2, using the slope of the line to determine whether r is positive or negative.

b.

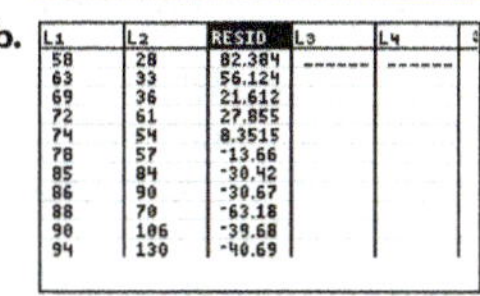

L1	L2	RESID L3	L4	
58	28	82.384	------	------
63	33	56.124		
69	36	21.612		
72	61	27.855		
74	54	8.3515		
78	57	-13.66		
85	84	-30.42		
86	90	-30.67		
88	70	-63.18		
90	106	-39.68		
94	130	-40.69		

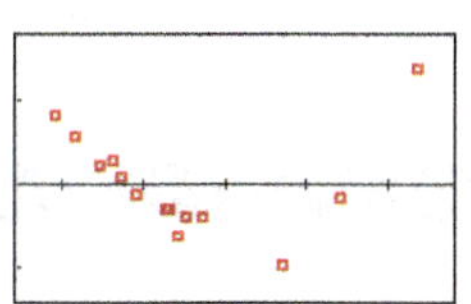

The curved pattern of the residuals suggests that a quadratic function might be a better model.

1. To view the residuals calculated by your graphing calculator while completing the regression, display the list editor ([STAT], [1]), use [▶] and [▲] to select L3, and select [INS] to place an empty column in the list editor. Then select [LIST] ([2nd], [STAT]), choose RESID from the list name menu, and press [ENTER] to name the list and display the residual for each data point.
2. Use [STAT PLOT] to turn off Plot1 and turn on Plot2, using L_1 for the Xlist and defining the Ylist by selecting LIST and choosing RESID from the NAMES submenu. Press [Y=], [◀], [ENTER] to turn off the graph of Y1, and then select ZoomStat ([ZOOM], [9]) to display a graph of the residuals.

TIP 

When using other technologies, you may have to calculate the list of residuals before creating a similar scatterplot.

SKILL ✔ EXERCISE 21

The process of completing a *quadratic regression* with technology is similar to the process used to complete a linear regression. The coefficient of determination, r^2, is calculated, but a correlation coefficient, r, is not since this measure applies only to linear models.

Example 3   Creating and Evaluating a Quadratic Model

Complete a quadratic regression to find a quadratic function that models the alligator data. Then examine and interpret the coefficient of determination and a plot of the residuals.

Answer

$f(x) \approx 0.088x^2 - 11.656x + 425.841$

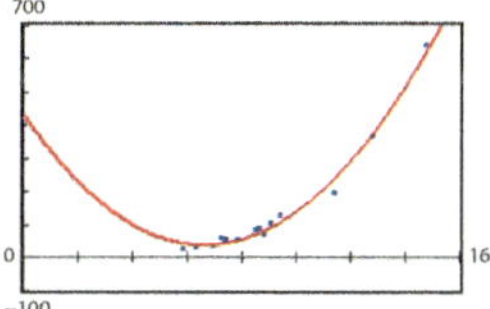

1. Complete a quadratic regression and plot the parabola on the data's scatterplot.

CONTINUED ➡

Match each scatterplot in excercises 7–10 to its best algebraic model.

A. $y = -\frac{1}{2}x^2 + 3x + 5.5$

B. $y = 3x + 6$

C. $y = \frac{1}{2}(x - 3)^2 - 9$

D. $y = -3x + 6$

E. $y = \frac{1}{2}x^2 + 3x + 5.5$

F. $y = x + 6$

7.

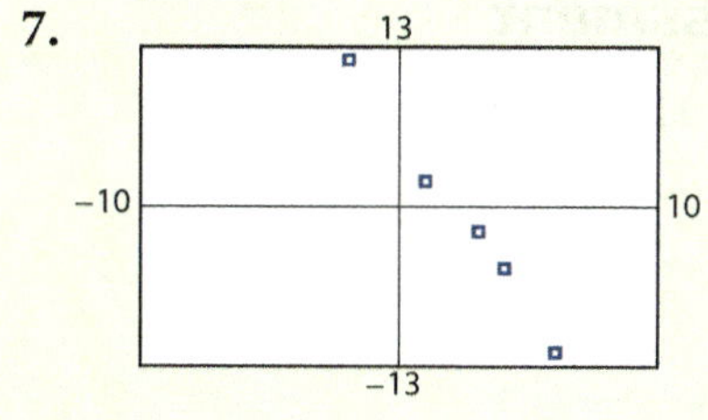

D

8.

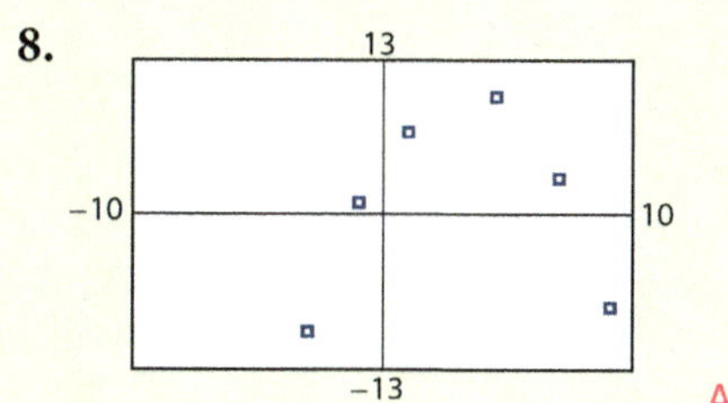

A

9.

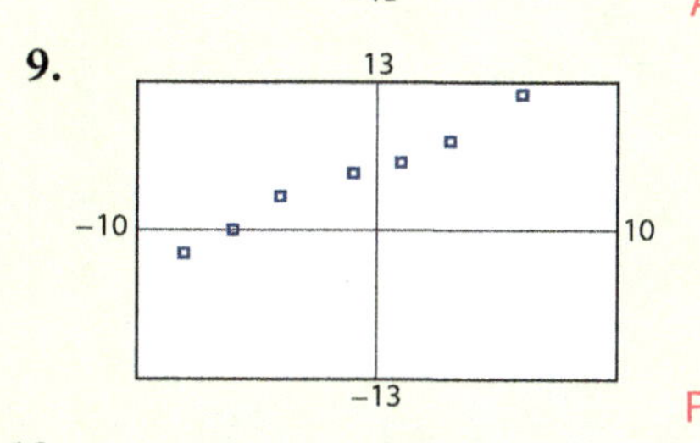

F

10.

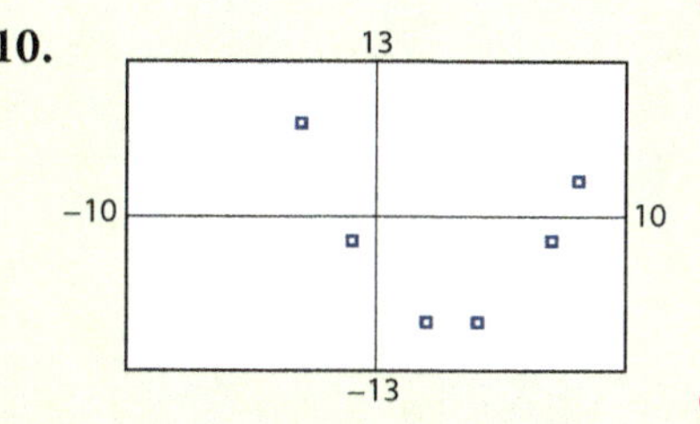

C

Demonstrate plotting and analyzing the residuals of a *quadratic regression* with Example 3. Be sure students turn off Plot2 (the residuals) and turn on Plot1 (the original data) in the STAT PLOT menu before graphing the quadratic regression.

Data transformations like that in Example 4 are frequently used when analyzing data that includes large numbers. These transformations can help avoid errors in predictions due to the rounding of coefficients in the regression equation. Be sure the students use a window that includes all the values that will be examined. Exercises include specific instructions for transforming data when appropriate.

Consider discussing some factors that may affect population growth, such as the economy and employment opportunities, which can sometimes make predicting populations difficult. Students will learn later that population growth is frequently modeled with exponential functions.

TIPS

Ex. 8 If the students use 9:ZoomStat, they will need to adjust Xmax in [WINDOW] to avoid an error message in response to typing a value of 20 in [TRACE].

Assignments

- **Minimum:** 1–8, 13–17, 20–27, 45–49, 51, 53
- **Standard:** 1–2, 9–16, 19, 21–27, 30–35, 46–47, 49–53
- **Extended:** 2, 4, 13–17, 26–27, 36–43, 50–53

Assessment

- Quiz 1E covers Section 1.9.

$r^2 \approx 0.987$

With this model, 98.7% of the variation in weight is explained by variation in length.

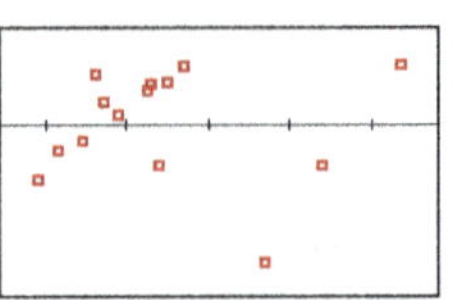

A random distribution suggests that a quadratic function might be the best model.

SKILL ✔ **EXERCISE 22**

2. Use technology to display the coefficient of determination, r^2.

3. Use Plot2 to display a scatterplot of the updated residuals.

While r^2 values and plots of residuals are indicators of the value of a model, the usefulness of the model still needs to be examined. Our quadratic model appears to be a good fit within the data set and may continue to work for longer alligators. However, the fact that the parabola's minimum value occurs when $x \approx 66$ in. makes it clear that it is not a good model for alligators much shorter than this. In later chapters, we will study other functions that may provide an even better model for this type of data.

Raw data is frequently transformed when creating a model. Years are frequently expressed as the number of years after a convenient reference point, and large numbers can be expressed in thousands, millions, or billions.

Example 4   Transforming Data

Find a function modeling the population of New York City during the nineteenth century. Then use the model to predict its population in 1900.

New York City					
Year	Population	Year	Population	Year	Population
1800	60,515	1840	312,710	1880	1,206,299
1810	96,373	1850	515,547	1890	1,515,301
1820	123,706	1860	813,669		
1830	202,589	1870	942,292		

Answer

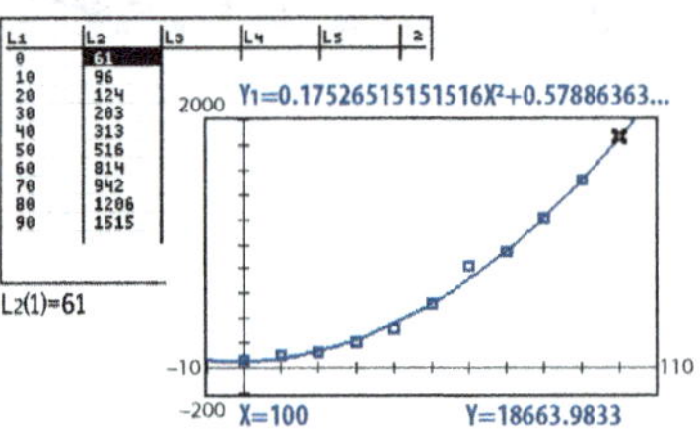

$f(x) \approx 0.175x^2 + 0.579x + 53.445$

The model predicts a population of approximately 1,864,000 in 1900.

1. Let x be the number of years since 1800 and let y be the population (in thousands).

2. The scatterplot suggests a quadratic model, and an r^2 value of 0.99 implies that the resulting quadratic regression produces a good fit for the data.

3. Use TRACE on Y1 and enter 100 to find the approximate population in 1900.

SKILL ✔ **EXERCISES 26–29**

> **TIP**
>
> Using the exact regression equation eliminates rounding errors, which can be significant when the x-values are relatively large.

Ex. 20 Some students may choose a linear model, but the r^2 values (0.7882 for linear and 0.8999 for quadratic) indicate the quadratic model would likely be better. Residuals from both regressions exhibit patterns that indicate a better model (cubic) may exist.

Ex. 21 Even though there is an obvious outlier, the residual pattern is nearly parabolic.

Ex. 27 The function $q(x)$ can be expressed in the more familiar form $f(n) = \dfrac{n(n-1)}{2}$. This function also models the number of handshakes that take place in a room full of people if each person shakes hands once with each other person.

Ex. 34–35 The instruction to "let y represent the population (in thousands)" does not mean that the population should be entered as a rounded value. For 1860, enter $(-40, 0.678)$.

Ex. 37–43 Students should be cautious when working with transformed data values. For example, they need to remember that 3 may represent 1930 (3 decades after 1900).

We must be careful not to assume that a high coefficient of determination implies a cause-and-effect relationship between the independent variable and the dependent variable. The cause-and-effect relationship may be reversed, both variables might be related to a third variable or a combination of other variables, or the relationship may be coincidental.

❯ A. Exercises

Match each scatterplot to the best description of its linear correlation.

A. strong positive **C. weak positive**
B. strong negative **D. weak negative**

1.   A   2.   D

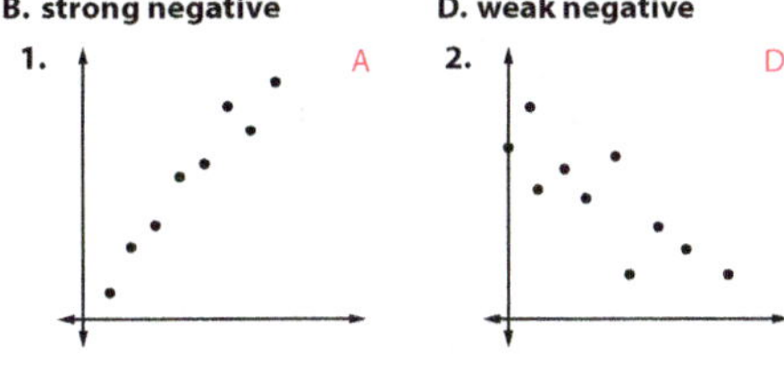

3.   C   4.   B

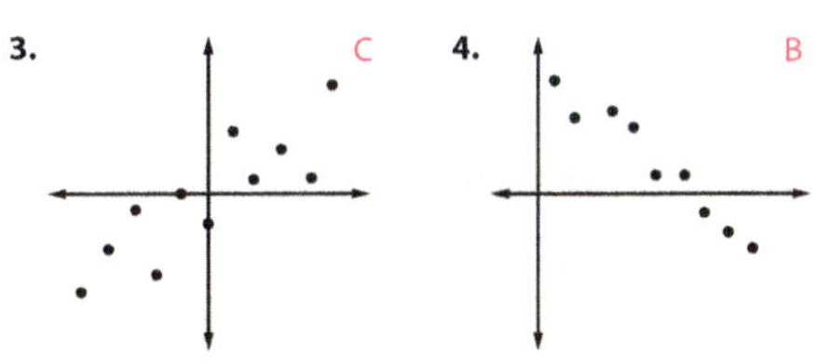

Refer to the data set for exercises 5–8.

x	0	2	4	6	8	10	12	14
y	21	17	14	11	7	2	1	−3

5. Is the function increasing, decreasing, or both? What does this imply about the linear correlation?

6. Use technology to create a scatterplot and determine the linear function that best models the data. Round coefficients to the nearest thousandth.

7. State the coefficient of determination. What does this value indicate? ≈ 0.992

8. Use your model to predict the value of y (to the nearest tenth) when x is 20. Is this an example of interpolation or extrapolation? $f(20) \approx -13.5$; extrapolation

5. decreasing; It is negative.
6. $f(x) \approx -1.714x + 20.75$

Refer to the data set for exercises 9–12.

x	−5	−3	−1	1	3	5
y	−6	−3	0	2	5	11

$f(x) \approx 1.586x + 1.5$

9. Use technology to create a scatterplot and determine the linear function that best models the data. Round coefficients to the nearest thousandth.

10. State the coefficient of determination. What does this value indicate? ≈ 0.970

11. How does the regression equation algebraically confirm the type of linear correlation?

12. Use your model to predict the value of y (to the nearest tenth) when x is 4. Is this an example of interpolation or extrapolation? $f(4) \approx 7.8$; interpolation

Refer to the data set for exercises 13–16.

x	2	5	7	10	13	15	18	20
y	2	8	16	20	23	21	13	3

13. Is the function increasing, decreasing, or both? What does this imply about the linear correlation?

14. Use technology to create a scatterplot and determine the quadratic function that best models the data. Round coefficients to the nearest thousandth.

15. State the coefficient of determination. What does this value indicate? ≈ 0.937

16. Use your model to predict the value of y (to the nearest tenth) when $x = 13$; then find the difference between the actual and predicted values. What is this difference called? $f(13) \approx 21.4$; ≈ 1.6; a residual

11. The positive slope confirms a positive correlation.
13. both; It is weak.
14. $f(x) \approx -0.241x^2 + 5.599x - 10.761$

Solutions

❯ A. Exercises

6.

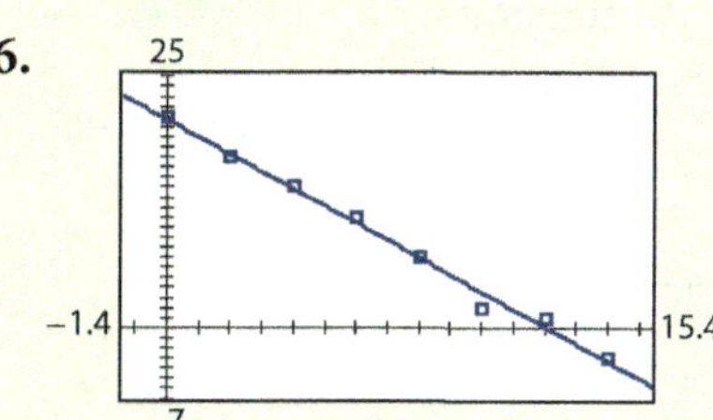

7. 99.2% of the variation in y is explained by variation in x.

8. $f(20) \approx -1.714(20) + 20.75 \approx -13.5$

9.

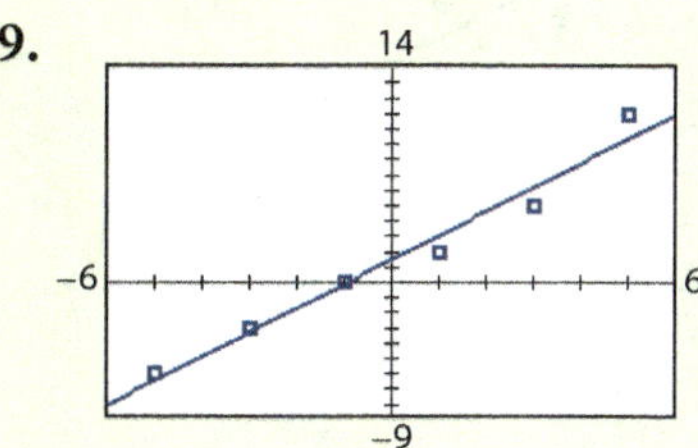

10. 97.0% of the variation in y is explained by variation in x.

12. $f(4) \approx 1.586(4) + 1.5 \approx 7.8$

14.

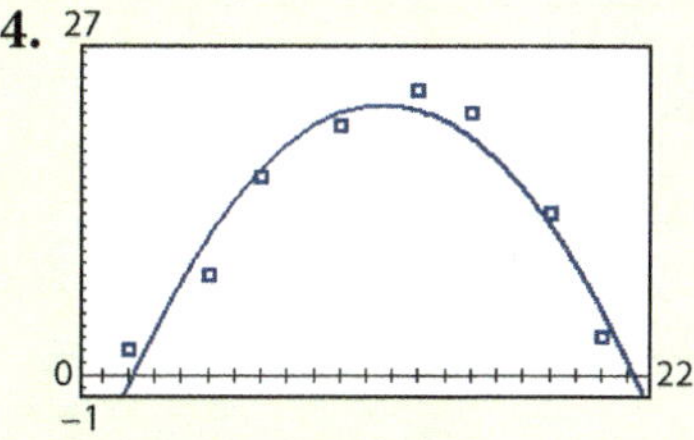

15. 93.7% of the variation in y is explained by variation in x.

16. $f(13) \approx -0.241(13)^2 + 5.599(13)$
$$- 10.761 \approx 21.4$$
$$23 - 21.4 = 1.6$$

⟩ **B. Exercises**

17–20. Use ZoomStat to set window dimensions.

17.

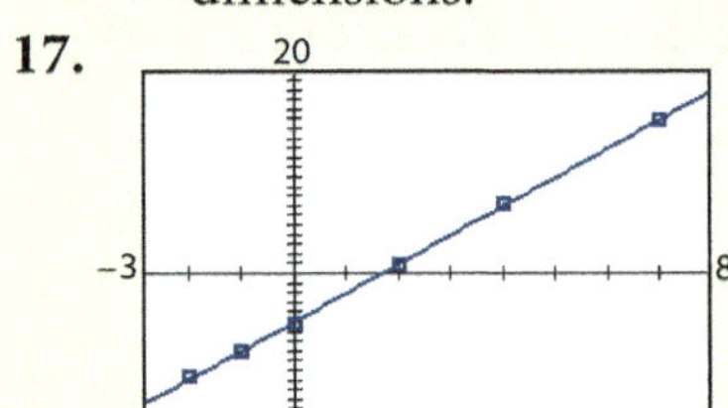

18.

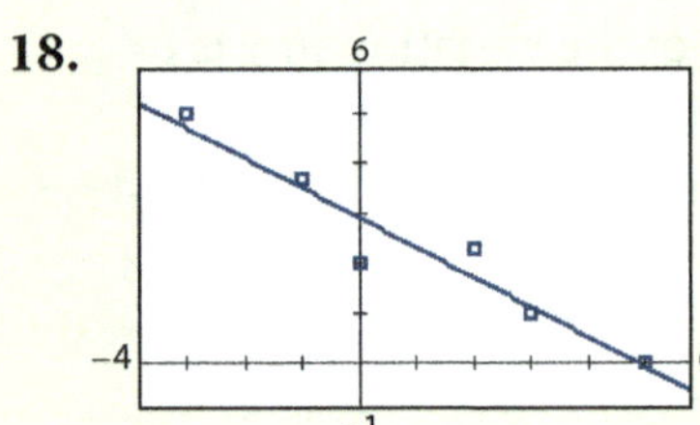

19.

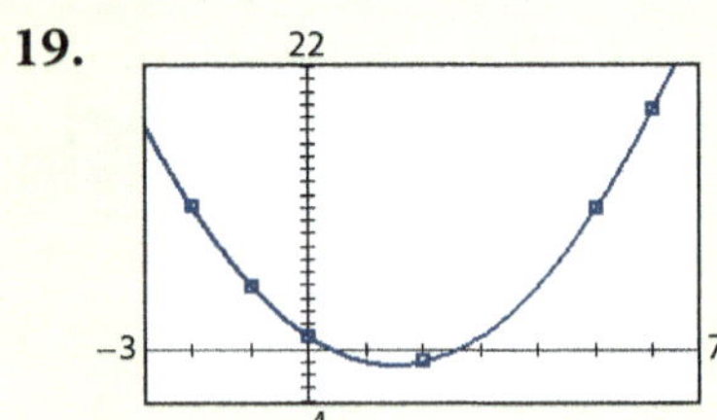

20.

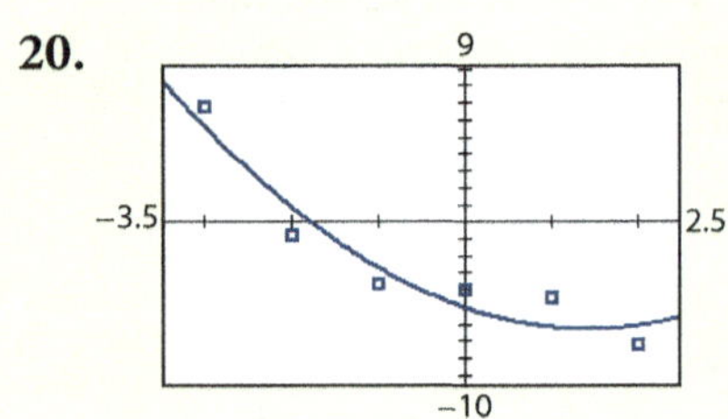

21–24.

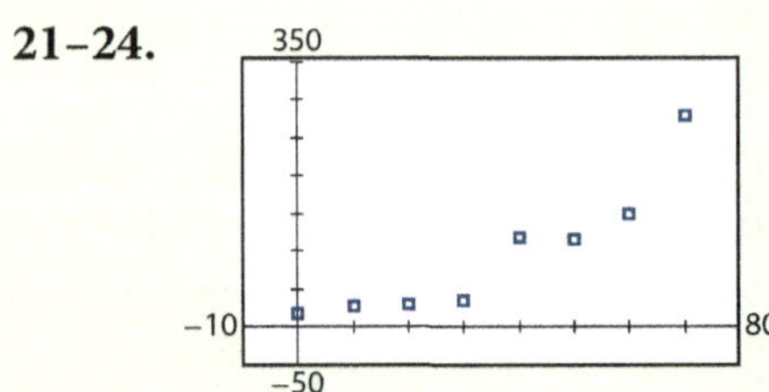

21. Xlist: L1; Ylist: RESID

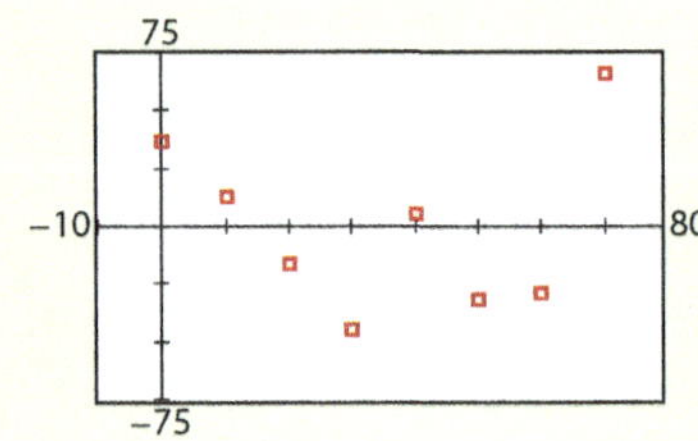

22.

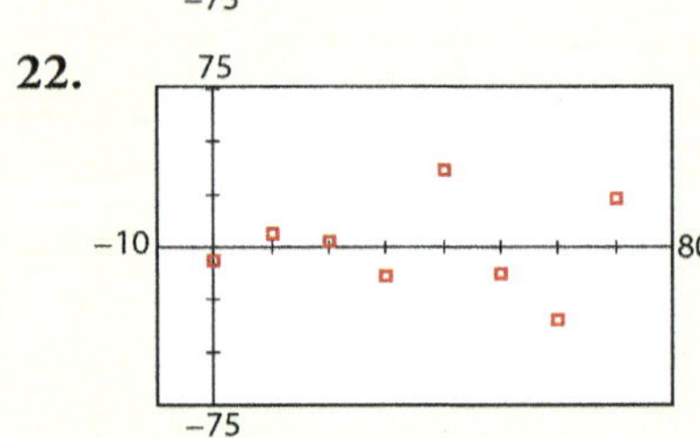

23a. $q(55) \approx 0.0621(55)^2 - 1.0381(55) + 24.5833 \approx 1.55$

⟩ **B. Exercises**

Use technology to create a scatterplot of each set of data. Then determine the linear or quadratic function that best models the data. Round coefficients to the nearest thousandth.

17.

x	y
-2	-10.7
-1	-8.0
0	-5.2
2	0.9
4	7.1
7	15.9

18.

x	y
-3	5.0
-1	3.7
0	2.0
2	2.3
3	1.0
5	0.0

19.

x	y
-2	11.1
-1	4.9
0	1.0
2	-0.9
5	11.0
6	18.7

20.

x	y
-3	6.8
-2	-0.8
-1	-3.6
0	-4.0
1	-4.4
2	-7.2

For exercises 21–25, use technology to create a scatterplot of gasoline prices. Let x represent the number of years since 1940 and let y represent the price per gallon (in cents).

Average US Gas Prices	
Year	$/gal
1940	0.18
1950	0.27
1960	0.31
1970	0.36
1980	1.19
1990	1.15
2000	1.51
2010	2.79

21. Find the best linear function modeling the data; then find the corresponding coefficient of determination and create a graph of the residuals. Do the residuals appear to form a pattern?

17. $f(x) = 2.981x - 4.968$
18. $f(x) = -0.598x + 2.931$
19. $f(x) = 0.990x^2 - 2.983x + 1.057$
20. $f(x) = 0.6x^2 - 1.72x - 4.96$

---

22. Find the best quadratic function modeling the data; then find the corresponding coefficient of determination and create a graph of the residuals. Do the residuals appear to form a pattern?

23. Using the better model from exercises 21 and 22, estimate the average price of gasoline for each year.
   a. 1995 $\approx$ \$1.55   **b.** 2015 $\approx$ \$2.96   **c.** this year

24. The actual average price for gasoline was \$1.15/gal in 1995 and \$2.45/gal in 2015. Was the interpolation or extrapolation more accurate in exercise 23? Is this unusual?

25. Discuss: State several reasons why gas price predictions may not be accurate. Then state why mathematical models are valuable even if their predictions are not as accurate as we would like.

Communications networks use various *topologies*, or configurations of connectivity. A full mesh topology increases network reliability by connecting every *node* (device such as computer or printer) to every other node so that communications can be redirected when a connection fails. The number of connections often makes such configurations expensive.

Nodes	Connections
2	1
3	3
4	6
5	10
6	15

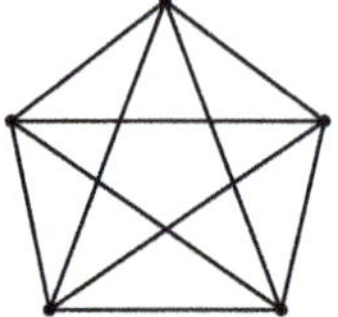

26. Complete the table by listing the number of connections in full-mesh networks with each number of nodes.

27. Create a scatterplot of the data and complete linear and quadratic regressions to find best-fit linear and quadratic models for the data. State each function and corresponding r^2 value.

28. Analyze: What does the coefficient of determination for the quadratic model indicate?

29. Determine the number of connections needed for a full-mesh network of 100 nodes. 4950 connections

21. $l(x) \approx 3.312x - 18.917; r^2 \approx 0.822;$ yes
22. $q(x) = 0.0621x^2 - 1.0381x + 24.5833; r^2 \approx 0.9374;$ no
27. $l(x) = 3.5x - 7; r^2 \approx 0.9722; q(x) = 0.5x^2 - 0.5x; r^2 \approx 1$
28. When $r^2 = 1$ the function fits the data exactly.

23b. $q(75) \approx 0.0621(75)^2 - 1.0381(75) + 24.5833 \approx 2.96$

24. 1995: $1.55 - 1.15 = 0.40;$ \$0.40 higher
2015: $2.96 - 2.45 = 0.51;$ \$0.51 higher
The interpolated value was closer, which is not unusual.

25. sample answer: When $r^2 \neq 1$, we do not expect the model to be a perfect fit. Gas prices are affected by many factors other than time, such as supply and demand, the strength of national and global economies, foreign relations, and the growing prevalence of alternative-fuel vehicles.

A mathematical model provides a reasonable and explainable basis for making predictions if one realizes its inherent limitations. It can also serve as a starting point for a more refined model as additional factors are considered.

27.

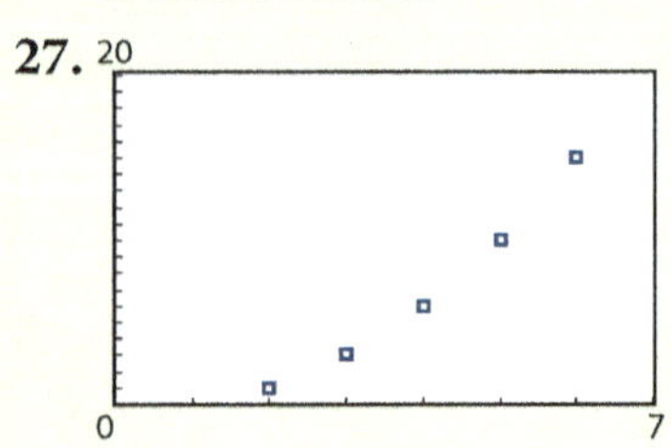

29. $f(100) = 0.5(100)^2 - 0.5(100) = 4950$

Dallas, Texas, had the ninth largest population in a 2015 survey of US cities. Use technology to create a scatterplot of the following population data. Let x represent the number of years since 1900 and let y represent the population (in thousands).

Dallas, Texas			
Year	Population	Year	Population
1920	158,976	1970	844,401
1930	260,475	1980	904,075
1940	294,734	1990	1,006,877
1950	434,462	2000	1,188,580
1960	679,684		

30. In which decade did Dallas experience its greatest population increase?   the 1950s

31. Find the best linear function modeling the data and the corresponding coefficient of determination. Then create a graph of the residuals. Do they appear to form a pattern? $f(x) \approx 13.315x - 157.567$; $r^2 \approx 0.9793$; no

32. Use the linear model found in exercise 31 to estimate the population of Dallas for each year.
 a. 2010 $\approx 1,307,000$ **b.** 1900 $\approx -158,000$ **c.** 1860 $\approx -690,000$

33. Analyze: Compare the predicted populations in exercise 32 to the actual populations in the table below and the actual 2010 population of 1,197,816. Does the linear model appear to be a good fit for the data? Explain your reasoning.

Use technology to create a scatterplot of a second set of population data for Dallas. Let x represent the number of years since 1900 and let y represent the population (in thousands).

Dallas, Texas			
Year	Population	Year	Population
1860	678	1940	294,734
1880	10,358	1960	679,684
1900	42,639	1980	904,075
1920	158,976	2000	1,188,580

34. Find the best quadratic function modeling the data and the corresponding coefficient of determination. Then create a graph of the residuals. Do they appear to form a pattern?

35. Use the quadratic model found in exercise 34 to estimate the population of Dallas for each year.
 a. 2010 $\approx 1,424,000$ **b.** 1990 $\approx 1,047,000$ **c.** 1930 $\approx 260,000$

34. $q(x) = 0.0715x^2 + 4.5388x + 59.3492$; $r^2 \approx 0.9898$; no

36. Analyze: Calculate the percent error of each predicted population from exercise 35 using the table for exercises 30–33 and the actual 2010 population of 1,197,816. Which values were more accurate: the interpolated values or the extrapolated values? Is this unusual? $\approx 18.8\%$, $\approx 4.0\%$, $\approx -0.2\%$; interpolated; no

❯ C. Exercises

Use the life expectancy data to create scatterplots for both males and females. Let x represent the number of decades since 1900.

US Life Expectancy		
Year Born	Males	Females
1930	58.1	61.6
1940	60.8	65.2
1950	65.6	71.1
1960	66.6	73.1
1970	67.1	74.7
1980	70.0	77.4
1990	71.8	78.8
2000	74.3	79.7

37. Referring to the numeric data in the table, in which decade did life expectancy increase the most for males? for females?   1940s; 1940s

38. Complete linear regressions to determine the best linear functions modeling the male and the female life expectancies and state the coefficient of determination for each model.

39. Analyze: What does comparing the slopes of the lines allow us to conclude about the life expectancy of the genders?

40. Use the linear models found in exercise 38 to estimate the US life expectancies for both males and females born in each year.
 a. 1985 ≈ 71.1; ≈ 77.8 **b.** 2010 ≈ 76.5; ≈ 84.2

41. Complete quadratic regressions to determine the best quadratic function modeling the male and the female life expectancies and state the coefficient of determination for each model.

42. Determine the vertex for each regression equation and interpret its significance.

38. $m(x) \approx 2.168x + 52.696$; $r^2 \approx 0.964$
 $f(x) = 2.562x + 56.048$; $r^2 \approx 0.936$

41. $m(x) \approx -0.0946x^2 + 3.3982x + 49.1946$; $r^2 \approx 0.9714$
 $f(x) = -0.3060x^2 + 6.5393x + 44.7274$; $r^2 \approx 0.9895$

34. 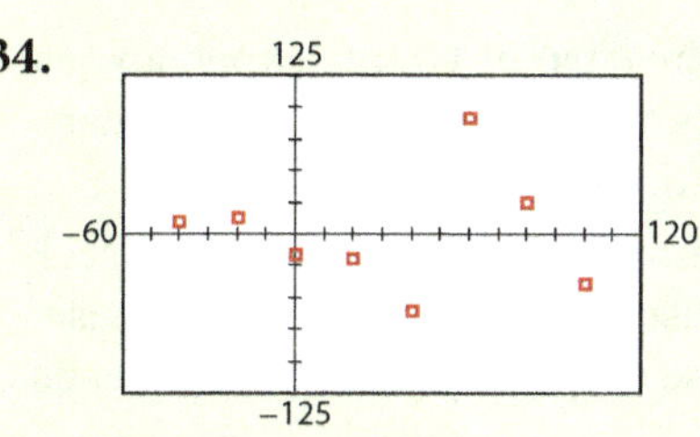

35a. $f(110) \approx 1423.575$

35b. $f(90) \approx 1046.863$

35c. $f(30) \approx 259.849$

36. 2010 % error $\approx \dfrac{1,423,575 - 1,197,816}{1,197,816}$
 $\approx 18.8\%$
 1990 % error $\approx \dfrac{1,046,863 - 1,006,877}{1,006,877}$
 $\approx 4.0\%$
 1930 % error $\approx \dfrac{259,849 - 260,475}{260,475}$
 $\approx -0.2\%$

❯ C. Exercises

37–40. 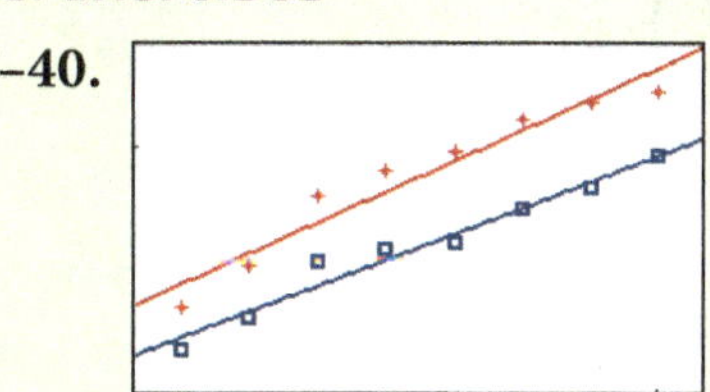

37. M: 1950 (4.8 yr); F: 1950 (5.9 yr)

39. Female life expectancy increased at a faster rate than male life expectancy.

40. Evaluate $m(x)$ and $f(x)$ at 8.5 and 11.

41–43. 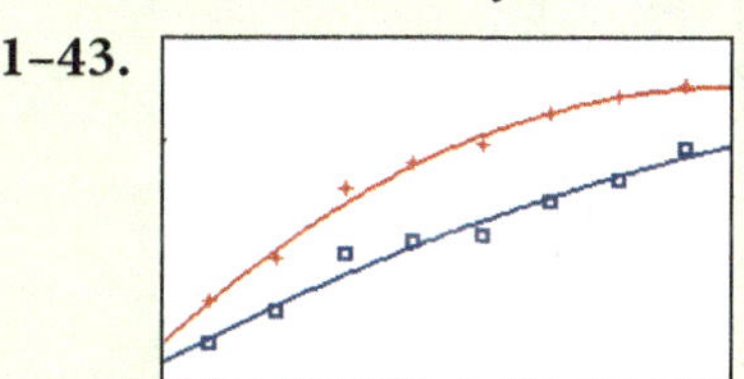

41. $m(x) \approx -0.0946x^2 + 3.3982x + 49.1946$; $r^2 \approx 0.9714$
 $f(x) = -0.3060x^2 + 6.5393x + 44.7274$; $r^2 \approx 0.9895$

42. M: $h = -\dfrac{3.3982}{2(-0.0946)} \approx 17.96$
 $m(17.96) \approx 79.7$
 The model predicts male life expectancy will maximize at ≈ 79.7 for males born in ≈ 2080.
 F: $h = -\dfrac{6.5393}{2(-0.3060)} \approx 10.69$
 $f(10.69) \approx 79.7$
 The model predicts female life expectancy will maximize at ≈ 79.7 for females born in 2007.

---

30. $434,000 - 295,000 = 246,000$

31–33. 
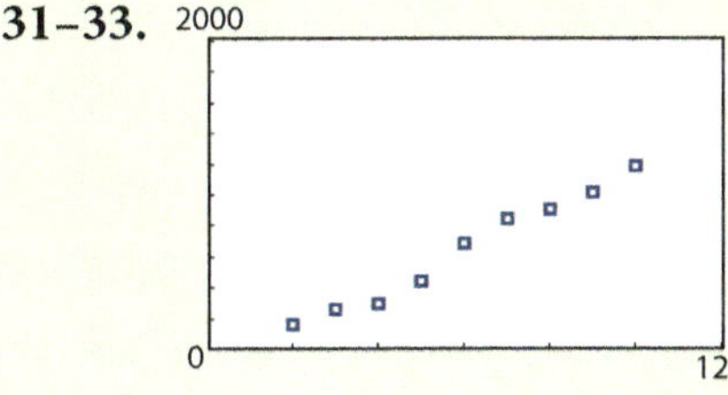

31. 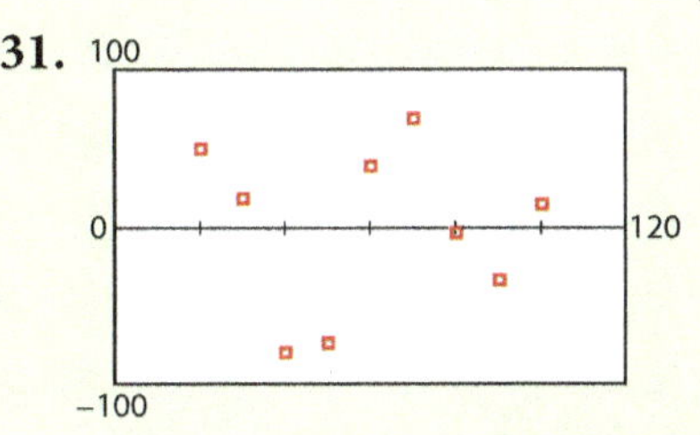

32a. $f(110) \approx 13.315(110) - 157.567$
 ≈ 1307

32b. $f(0) \approx 13.315(0) - 157.567 \approx -158$

32c. $f(-40) \approx 13.315(-40) - 157.567$
 ≈ -690

33. The linear model appears to be a good fit within the data set, but extrapolated values are likely not accurate.

34–36. 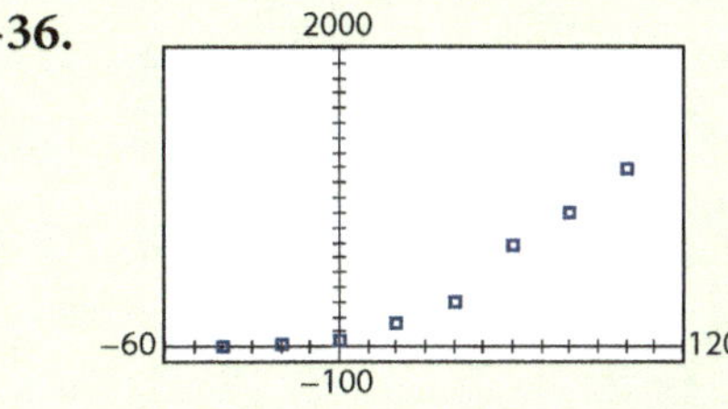

43a. The current US life expectancies are very close to the 70–80 year estimate.

43b. sample answer: that we may seek wisdom and strive to make wise use of the days that God gives us

❯ Cumulative Review

44. 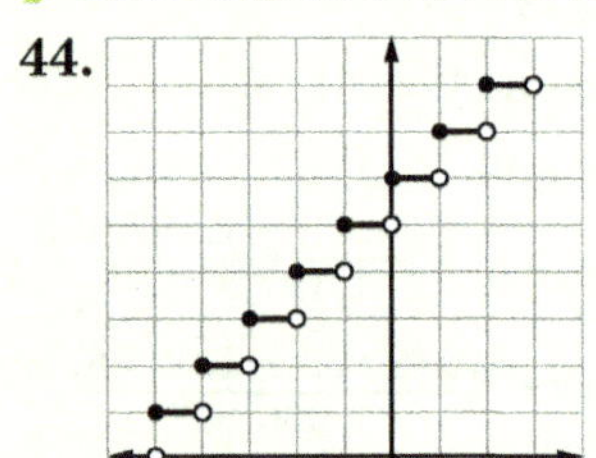

45. 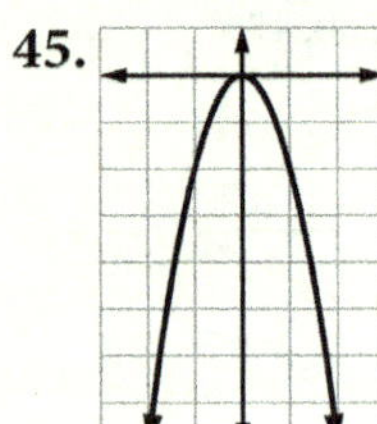

46. 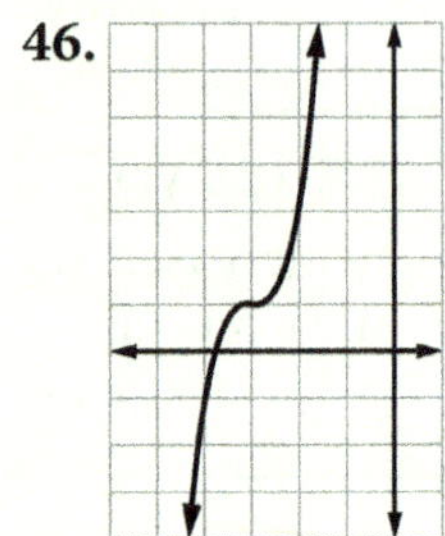

47. 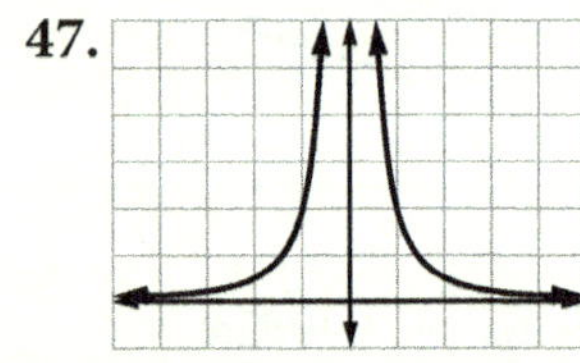

50. The vertex is at $(-4, -2)$, so the maximum value is -2 when $x = -4$.

51. If $f(x)$ and $g(x)$ are inverse functions, $g(f(x)) = x$.

52. Functions f, g, and r fail the Horizontal Line Test, while the linear function h passes the Horizontal Line Test.

53. $(0) = t - 7; t = 7$
$x = 2(7) + 1$
$x = 15; (15, 0)$

---

43. Many factors affect life expectancy. The psalmist considers life expectancy in Psalm 90:10–12.
  a. How do the models from exercises 38 and 40 compare with these estimates made thousands of years ago?
  b. According to this passage and Ephesians 5:16, what is the purpose of recognizing our life expectancy?

CUMULATIVE REVIEW

Graph each function and then list its domain and range. Use limit notation to describe its end behavior. [1.3–1.6]

44. $y = [x] + 6$ **45.** $y = -2x^2$

46. $y = 2(x + 3)^3 + 1$ **47.** $y = 2x^{-2}$

Write a function rule for $g(x)$, the described transformation of $f(x)$. [1.5]

48. $f(x) = x^2$ translated 2 units right and reflected in the x-axis $g(x) = -(x - 2)^2$

49. $f(x) = x^3$ stretched vertically by a factor of 2 and then translated 3 units down and 1 unit left

50. What is the maximum value of $f(x) = -3(x + 4)^2 - 2$? [1.6]   C
  A. -4 **C.** -2 **E.** 4
  B. -3 **D.** 2

51. If $f(x)$ and the $g(x)$ are inverse functions, what is $g(f(5))$? [1.7]   A
  A. 5   **C.** $\frac{1}{5}$ **E.** 25
  B. -5 **D.** $-\frac{1}{5}$

52. Which function has an inverse function? [1.8]   D
  A. $f(x) = 3x^2$ **D.** $h(x) = 2x - 4$
  B. $g(x) = |x - 2|$ **E.** all of these
  C. $r(x) = 3x^4$

53. If the equations $x = 2t + 1$ and $y = t - 7$ represent a parametric function, what is the x-intercept? [1.8]   C
  A. $(7, 0)$ **C.** $(15, 0)$ **E.** $\left(-\frac{1}{2}, 7\right)$
  B. $(0, 7)$ **D.** $\left(0, -\frac{15}{2}\right)$

---

44. $D = \mathbb{R}; R = \mathbb{Z}; \lim\limits_{x \to -\infty} f(x) = -\infty, \lim\limits_{x \to \infty} f(x) = \infty$

45. $D = \mathbb{R}; R = \{y \mid y \leq 0\}; \lim\limits_{x \to -\infty} f(x) = -\infty, \lim\limits_{x \to \infty} f(x) = -\infty$

46. $D = \mathbb{R}; R = \mathbb{R}; \lim\limits_{x \to -\infty} f(x) = -\infty, \lim\limits_{x \to \infty} f(x) = \infty$

47. $D = \{x \mid x \neq 0\}; R = \{y \mid y > 0\}; \lim\limits_{x \to -\infty} f(x) = 0, \lim\limits_{x \to \infty} f(x) = 0$

49. $g(x) = 2(x + 1)^3 - 3$

Data Analysis:

5a. The predicted height is less than the height of two data points. The prediction is short by at least $86.2 - 83.2 = 3.0$ ft.

5b. $464 - 397 = 67$ ft

5c. The effect of air resistance causes the ball's trajectory to be non-parabolic.

6. Limitations in the recording of data limit the accuracy of the model. The data set is not representative of the entire trajectory. A quadratic model does not include the effects of wind resistance (or the spin of the ball).

7. Obtaining more data from the latter portion of the trajectory would significantly increase the accuracy of the model. Another mathematical function besides a quadratic function might provide a better model.

8. While recognizing the limitations of mathematical models, we can plan based on fairly accurate predictions.

9. The fact that mathematical models can be used to make predictions points to the consistent underlying principles of the universe established by the Creator. The fact that the mathematics created by humans does not perfectly model God's creation points to our limited, fallen nature.

Home Run

Mathematical models are powerful tools for making predictions and informing decisions. They can be used to estimate unknown values within a data set (interpolation) and to predict values beyond a data set (extrapolation). However, you need to be aware of the limitations of mathematical models so that the resulting estimates are properly interpreted and applied.

Aristotle's assertion that all motion on earth is linear, circular, or a combination of the two was believed and taught for about 2000 years. Galileo studied Aristotle's laws of motion but later showed that projectile motion can be modeled by a parabolic path. You will investigate both linear and parabolic models of the flight of a batted ball, predict whether the ball clears the outfield wall for a home run, and evaluate the usefulness of your model.

1. Use technology to complete a scatterplot and linear regression of the data. Record the equation of the line of best fit, the correlation coefficient, and the coefficient of determination.
   a. Describe the linear correlation of the data.
   b. Interpret the coefficient of determination for the linear model.
   c. Explain why the line of best fit would not be a good model for the ball's flight.
2. Use technology to graph the residuals. What characteristic of this graph indicates that you should attempt to find a different model?
3. Complete a quadratic regression of the data. Record the resulting quadratic function and its coefficient of determination and graph your quadratic model on the scatterplot.
   a. Interpret the coefficient of determination for the quadratic model.
   b. Does the quadratic model appear to be a good model for the data? yes
4. Use the quadratic model to make predictions. ≈ 83.2 ft

Horizontal Distance (ft)	Height (ft)
0.0	3.4
50.3	28.7
100.6	52.1
149.6	72.8
198.7	85.6
250.2	86.2
300.5	72.4

   a. Predict the maximum height of the ball.
   b. Predict whether the batter hit a home run if the ball is headed to the 10 ft tall left-center field wall (381 ft from home plate). Justify your prediction.
   c. Predict how far from home plate the ball will land (assuming level ground and no obstructions). ≈ 464 ft
5. Evaluate your predictions.
   a. How can you know that the model's prediction for the maximum height is not exact? Determine the smallest possible error in the model's prediction of the maximum height.
   b. The batted ball was actually a home run landing 397 ft from home plate. Determine the amount of error in the model's prediction. 67 ft
   c. When evaluating a model, consider factors that are not included by the model. What is likely the most significant factor excluded by the quadratic model?

While our quadratic model is not perfect, quadratic equations are frequently used to model the trajectory of a projectile. There are many other factors that can be included to make the model more accurate. Video game developers utilize mathematical models like these to simulate the path of projectiles. While some games strive to be as realistic as possible, others play with physics by intentionally altering the gravitational force or eliminating air resistance or friction.

6. What factors affect the accuracy of the mathematical model of the baseball's trajectory?
7. What could help provide a more accurate mathematical model?
8. Why is mathematical modeling an important tool even if some predictions are not perfectly accurate?
9. How do the strengths and weaknesses of mathematical models relate to the roles of God and humans in the development of mathematics?

Data Analysis

Objectives

1. To determine a mathematical function modeling a data set
2. To determine if the mathematical model can be used to predict the outcome of a batted ball
3. To evaluate the usefulness and limitations of a mathematical model

Assignments

- **Minimum:** 1–5; 6–8 (in class)
- **Standard:** 1–6; 7–8 (in class)
- **Extended:** 1–8

Solutions

1. $y \approx 0.2535x + 19.2861$
   $r \approx 0.8764; r^2 \approx 0.7680$

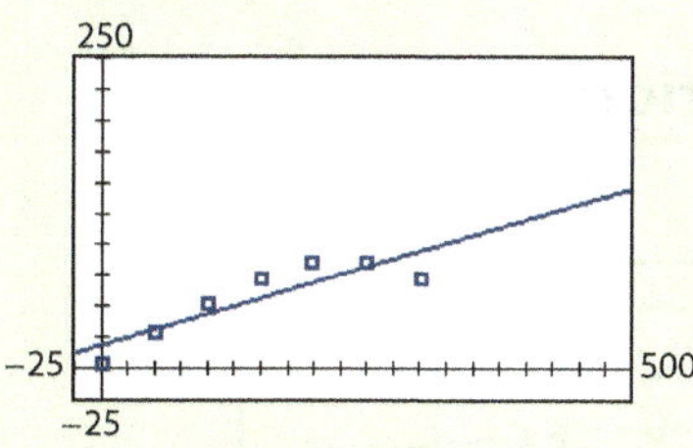

1a. The linear correlation coefficient, $r \approx 0.8764$, implies a fairly strong positive correlation.

1b. Approximately 77% of the variation in height is explained by a variation in horizontal distance.

1c. The ball must return to earth.

2. 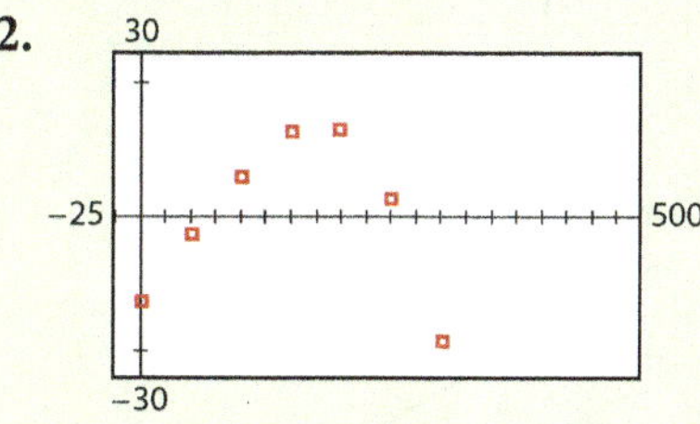

There is a pattern in the scatterplot of residuals.

continued in Answers and Solutions Overflow

PRESENTATION

The Data Analysis features present unique experiences in mathematical modeling, integrating data analysis and biblical principles in the interpretation and application of engaging, real-world scenarios. This Data Analysis feature provides a solid foundation for the mathematical modeling of real-world data and evaluating the usefulness and limitations of a mathematical model.

The conic sections, including the parabola and the ellipse, had been studied since the time of Euclid as an exercise in geometric reasoning with no practical application. By the end of the seventeenth century, however, Johannes Kepler had shown that planets follow an elliptical path, and Galileo had demonstrated that projectiles follow a parabolic path. Both men were able to recognize these special curves because of their knowledge of the conic sections. See the Biblical Perspective of Mathematics feature in Chapter 4 for more discussion of how the concepts developed in pure mathematics often find practical uses in real-world applications.

TIPS

Ex. 6–8 Usefulness, limitations, proper interpretation, and application will be key elements in modeling throughout *Precalculus*. These discussion topics set a foundation for later development.

Objective

To prepare for evaluation

Vocabulary

See Appendix A.

Assignments

- **Minimum:** 2–3, 5–10, 13–24, 28–29, 31–33, 37, 39–40, 43, 45–50
- **Standard:** 1–49 odd; 2–50 even (in class)
- **Extended:** 3–5, 7–10, 12–15, 17–18, 20–25, 28–31, 33–35, 37, 39–42, 45–50

Assessment

- Chapter 1 Test

Solutions

5. $\{(-4, 6), (-2, 4), (0, 2), (2, 0), (4, 2)\}$

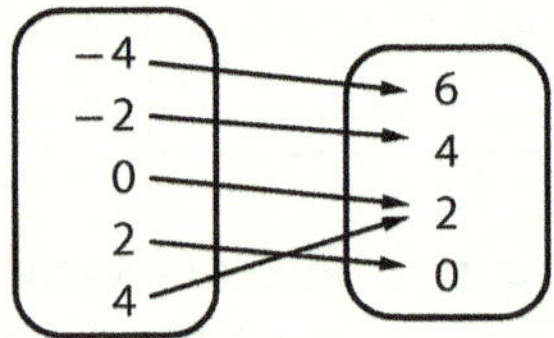

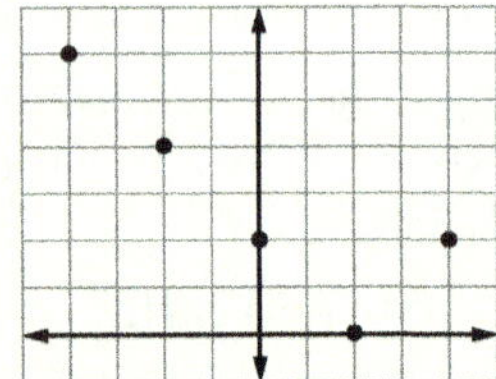

6. $0 = -\frac{3}{5}x + 6; \ \frac{3}{5}x = 6; \ x = 10$

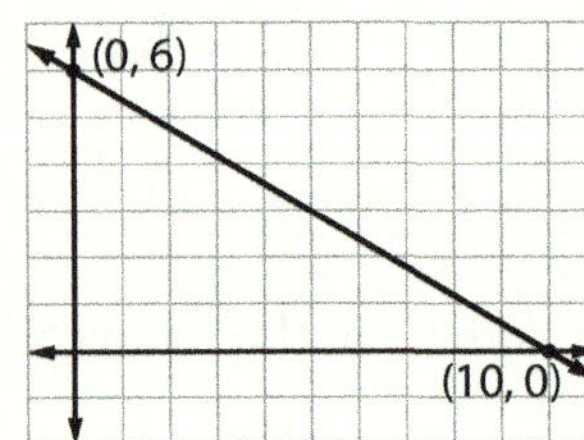

State the domain and range, using interval notation when possible. Determine whether the relation is a function. [1.1]

1.

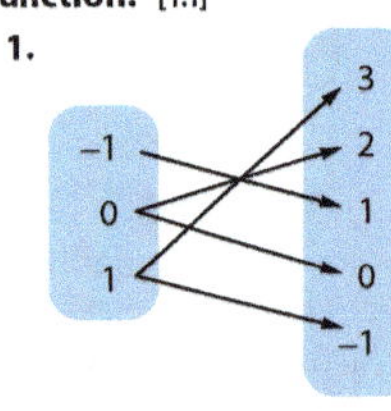

2. $\{(1, 3), (4, 6), (-3, -1), (0, 3)\}$

3. $y = \dfrac{1}{x + 5}$

4. $y^2 = x - 3$

5. Represent $f(x) = |x - 2|$ where $D = \{-4, -2, 0, 2, 4\}$ as a set of ordered pairs, with a mapping diagram, and with its graph. [1.1]

State the x-intercept, y-intercept, and slope of each line. Then graph the function. [1.2]

6. $y = -\frac{3}{5}x + 6$

7. $2x - 3y = 12$

8. Given $A\,(9, -5)$ and $B\,(-5, -3)$, find AB and the midpoint of $\overline{AB}$. [1.2] $10\sqrt{2}; (2, -4)$

9. Write the slope-intercept form equation of the line passing through $(-3, 7)$ and $(6, 1)$. [1.2] $y = -\frac{2}{3}x + 5$

10. Write the general form equation of the line passing through the point $(-4, -5)$ and perpendicular to $x - 2y = 6$. [1.2] $2x + y = -13$

Graph each function and state its domain and range. Then identify the function as continuous or discontinuous. [1.3, 1.5]

11. $f(x) = |x - 1|$

12. $g(x) = [x] + 2$

13. $h(x) = \begin{cases} -1 & \text{if } x < 0 \\ x^2 & \text{if } 0 \le x \le 2 \\ -x + 6 & \text{if } x > 2 \end{cases}$

14. Find each limit of the function in exercise 13. Justify your answer. [1.3]

    a. $\lim\limits_{x \to 0} h(x)$ does not exist
    b. $\lim\limits_{x \to 2} h(x)$ 4

1. $D = \{-1, 0, 1\}; R = \{-1, 0, 1, 2, 3\}$; not a function

2. $D = \{-3, 0, 1, 4\}; R = \{-1, 3, 6\}$; function

3. $D = (-\infty, -5) \cup (-5, \infty); R = (-\infty, 0) \cup (0, \infty)$; function

4. $D = [3, \infty); R = (-\infty, \infty)$; not a function

6. $(10, 0); (0, 6); m = -\frac{3}{5}$

7. $(6, 0); (0, -4); m = \frac{2}{3}$

15. Classify each discontinuity below as jump, infinite, or point. [1.3]

a. 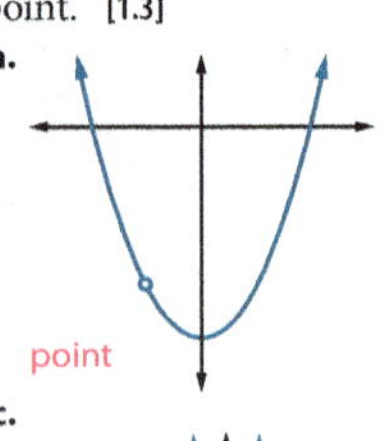

b. 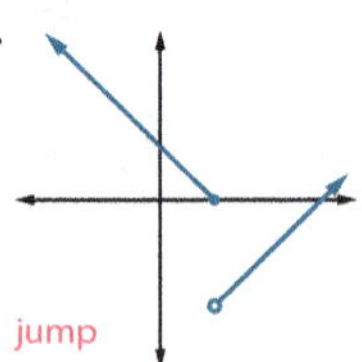

c. 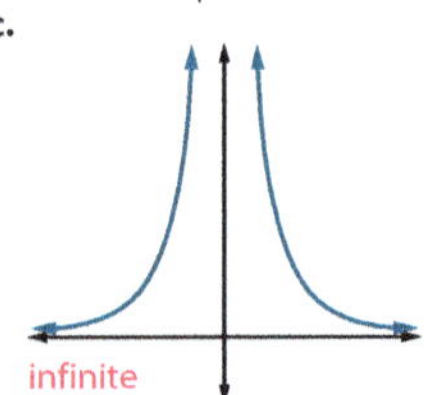

Use technology to graph each function. Then describe its domain; range; end behavior; intervals of continuity; intervals for which the function is increasing, decreasing, or constant; whether the function is even, odd, or neither; and any symmetry. [1.4]

16. $y = -2x^4$

17. $y = \dfrac{x^3}{2}$

18. $y = -3x^{-2}$

Use a power function to model each variation in exercises 19–21. [1.4]

19. If Jennie pays \$15 dollars for supplies to clean 3 homes and the cost varies directly with the number of homes, what is the cost for the supplies needed to clean 10 homes?   \$50

20. The number of days needed to complete a project is inversely proportional to the number of workers on the project. Determine the number of days needed to complete a project with 2 workers if it takes 3 days for 7 workers to complete the project.   10.5 days

21. The volume of a hemisphere is directly proportional to the cube of its radius. Determine the volume of a hemisphere with a radius of 14 ft if a hemisphere with a 2 ft radius has a volume of $\frac{16\pi}{3}$ ft^3. ≈ 5747 ft^3

PRESENTATION

Consider asking the students if there are any concepts in the chapter that they would like to review briefly. This first chapter is a good time to discuss and share different methods that students can use in preparing for the chapter tests, beginning with assessing their understanding and skill with the Chapter Review. Point out that each exercise is labeled with the related section number so they can find related examples if needed.

Skimming through the chapter to recall key concepts, doing select odd-numbered exercises, and reviewing notes are just a few other methods of preparing for the test.

22. Write a function rule for $g(x)$, the described transformation of $f(x)$. Confirm your answer by graphing $f(x)$ and $g(x)$ in the same window. [1.5]

  a. $f(x) = |x|$ translated 4 units right and 2 units up

  b. $f(x) = x^3$ vertically stretched by a factor of 2 and then translated 1 unit down

  c. $f(x) = x^2$ reflected across the x-axis, vertically shrunk by a factor of $\frac{1}{3}$, and then translated 5 units up

Graph each function as a transformation of a parent function. Then state the function's domain and range. [1.5–1.6]

23. $g(x) = -2|x - 3| + 1$ **24.** $g(x) = \left(\frac{x}{2}\right)^3 - 2$

25. $g(x) = 2[x]$ **26.** $f(x) = -2(x - 1)^2 + 3$

27. $h(x) = \frac{1}{2}(x + 2)^2$ **28.** $g(x) = 3x^2 - 18x + 25$

Identify the vertex of each parabola and state whether it is a maximum or minimum point. Then state the y-intercept and any zeros. [1.6]

29. $f(x) = -(x - 1)^2 + 4$ **30.** $f(x) = 2x^2 + 5x - 12$

Use $f(x) = x^2 - 4x + 4$ and $g(x) = x - 2$ to write function rules for each function operation in exercises 31–33. [1.7]

31. Find $(f + g)(x)$, $(g - f)(x)$, and $(fg)(x)$.

32. Find $(f \circ g)(x)$ and $(g \circ f)(x)$.

33. Find $\frac{f}{g}(x)$, graph the function, state its domain and range, and describe any discontinuities.

34. If $f(x) = \sqrt{x - 3}$ and $g(x) = x^2 - 1$, find $(f \circ g)(x)$ and $(g \circ f)(x)$ and state the domain of each. [1.7]

35. Find $f(x)$ and $g(x)$ such that $h(x) = (f \circ g)(x) = -2x^2 + 12x - 13$. [1.7]

Use composition to determine whether $f(x)$ and $g(x)$ are inverse functions. [1.8]

36. $f(x) = \frac{x}{3} - 2$ **37.** $f(x) = \frac{x^5 + 1}{4}$

$g(x) = 3x + 2$ $g(x) = \sqrt[5]{4x - 1}$

Find the inverse of each function. Confirm your results by using technology to graph the function and its inverse in the same square window. [1.8]

38. $f(x) = \frac{x - 3}{4}$ **39.** $f(x) = \frac{1}{x - 3}$

40. Use technology to graph $y = -x^4 + 5x^2 + x + 1$. Determine whether the inverse is a function and justify your answer. [1.8]

41. A relation is defined by the parametric equations $x = t + 4$ and $y = t^2 - 1$ with $-3 \leq t \leq 3$. [1.8]

  a. Draw a graph of the relation.

  b. Eliminate the parameter to express y in terms of x.
  $y = x^2 - 8x + 15$

42. The horizontal distance traveled and height of a punted football are modeled (in feet) by the following parametric equations. [1.8]
$$x = 45t$$
$$y = -16t^2 + 60t + 3$$

  a. Use technology to graph the parametric functions and determine the ordered pair describing the position of the football at each second before it hits the ground.

  b. Eliminate the parameter to find a function expressing the football's height in terms of the horizontal distance traveled. Use technology to graph the function and use the ordered pairs from part a to confirm your answer.

  c. Find the maximum height of the football. 59.25 ft

  d. Find the horizontal distance the football travels before hitting the ground (in yards). $\approx$ 57 yd

43. A swimming pool with a diameter of 20 ft and a depth of 4 ft is being filled by a garden hose at 4 ft^3/min. [1.7]

  a. Write a function $V(t)$ for the volume of water in the pool after t minutes. $V(t) = 4t$

  b. Use $V = \pi r^2 H$ to write a function $H(V)$ that expresses the water's height in terms of its volume.

  c. Compose the functions to find a function that expresses the water's height in terms of the time.

  d. Find the height of the water after 3 hr. $\approx$ 2.3 ft

  e. How long will it take for the pool to be filled to a depth of 4 ft? $\approx$ 314 min or 5 hr 14 min

44. A quarterback throws a pass from a height of 6 ft at the 20 yd line. The pass is caught at the same height at the 50 yd line after the football reached a maximum height of 15 ft. Write a quadratic function modeling the path of the ball, using the team's goal line as the origin. [1.6]

45. The height of a projectile can be modeled by the quadratic function $h(t) = -\frac{1}{2}gt^2 + v_0 t + h_0$ where g, the acceleration due to gravity, is approximately 32 ft/sec^2. A firework is launched with an initial vertical velocity of 270 ft/sec from a barge at a point 15 ft above the river. Ignoring the effect of air resistance, predict the maximum height the firework reaches (above the river) and the time it takes to reach that height. [1.6] $\approx$ 1154 ft; $\approx$ 8.4 sec

11.
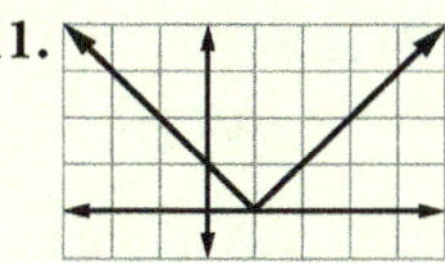

$D = \mathbb{R}$; $R = [0, \infty)$; cont.

12.
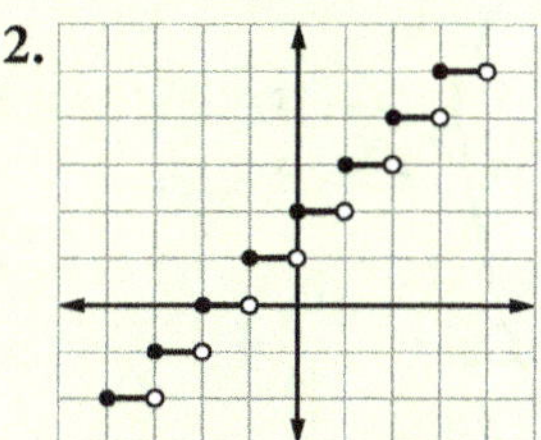

$D = \mathbb{R}$; $R = \mathbb{Z}$; discont.

13.
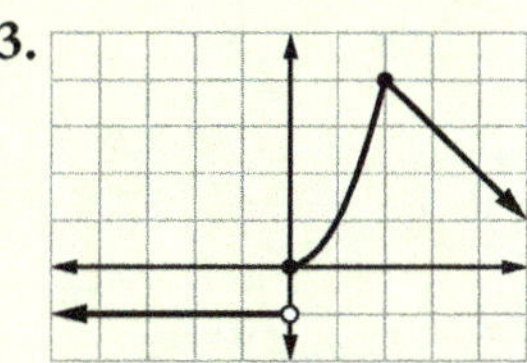

$D = \mathbb{R}$; $R = (-\infty, 4]$; discont.

14a. $\lim\limits_{x \to 0^-} h(x) = -1$ and $\lim\limits_{x \to 0^+} h(x) = 0$;

$\therefore \lim\limits_{x \to 0^-} h(x) \neq \lim\limits_{x \to 0^+} h(x)$

14b. $\lim\limits_{x \to 2^-} h(x) = \lim\limits_{x \to 2^+} h(x) = 4$

16. $D = \mathbb{R}$; $R = [0, -\infty)$;

$\lim\limits_{x \to -\infty} f(x) = -\infty$; $\lim\limits_{x \to \infty} f(x) = -\infty$;

cont.: $\mathbb{R}$; incr.: $(-\infty, 0]$; decr. $[0, \infty)$;

even; SWRT y-axis

17. $D = \mathbb{R}$; $R = \mathbb{R}$;

$\lim\limits_{x \to -\infty} f(x) = -\infty$; $\lim\limits_{x \to \infty} f(x) = \infty$;

cont.: $\mathbb{R}$; incr.: $(-\infty, \infty)$; odd;

SWRT origin

18. $D = (-\infty, 0) \cup (0, \infty)$; $R = (-\infty, 0)$;

$\lim\limits_{x \to -\infty} f(x) = 0$; $\lim\limits_{x \to \infty} f(x) = 0$;

cont.: $(-\infty, 0)$ and $(0, \infty)$;

decr.: $(-\infty, 0)$; incr.: $(0, \infty)$; even;

SWRT y-axis

19. $c = kh$

$15 = k(3)$; $k = 5$

$c = 5(10) = \$50$

20. $d = \frac{k}{w}$

$3 = \frac{k}{7}$; $k = 21$

$d = \frac{21}{2} = 10.5$ days

21. $V = kr^3$

$\frac{16\pi}{3} = k(2)^3$; $k = \frac{2}{3}\pi$

$V = \frac{2}{3}\pi(14^3) = \frac{5488}{3}\pi \approx 5747$ ft^3

22a. $g(x) = |x - 4| + 2$

22b. $g(x) = 2x^3 - 1$

22c. $g(x) = -\frac{1}{3}(x - 0)^2 + 5 = -\frac{x^2}{3} + 5$

7. x-int.: $\frac{C}{A} = 6$; y-int.: $\frac{C}{B} = -4$;

$m = -\frac{a}{b} = \frac{2}{3}$

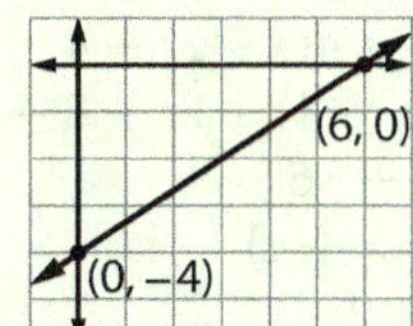

8. $AB = \sqrt{(-5 - 9)^2 + [-3 - (-5)]^2}$
$= \sqrt{(-14)^2 + 2^2} = 10\sqrt{2} \approx 14.1$

$M\left(\frac{9 + (-5)}{2}, \frac{-5 + (-3)}{2}\right) = (2, -4)$

9. $m = \frac{1 - 7}{6 - (-3)} = -\frac{2}{3}$

$y - 1 = -\frac{2}{3}(x - 6)$

$y - 1 = -\frac{2}{3}x + 4$

$y = -\frac{2}{3}x + 5$

10. $m_G = -\frac{A}{B} = -\frac{1}{-2} = \frac{1}{2}$; $m_\perp = -2$

$y - (-5) = -2(x - (-4))$

$y + 5 = -2x - 8$

$y = -2x - 13$

$2x + y = -13$

23. 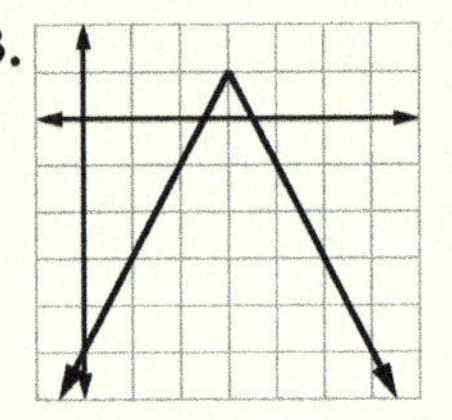 $D = \mathbb{R}$
$R = (-\infty, 1]$

24. 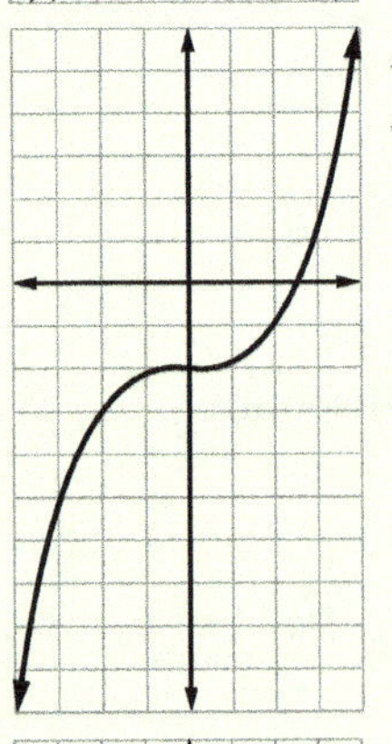 $D = \mathbb{R}$
$R = \mathbb{R}$

25. 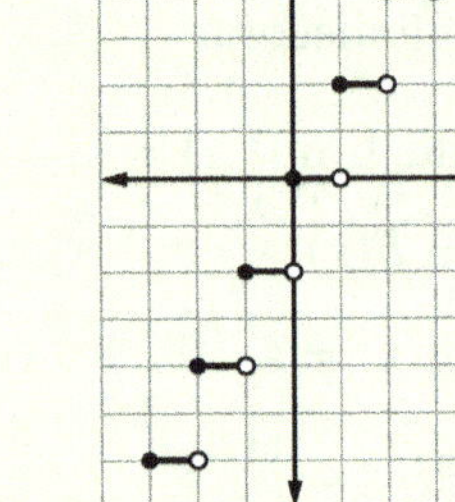 $D = \mathbb{R}$
$R = \{2x \mid x \in \mathbb{Z}\}$

26. 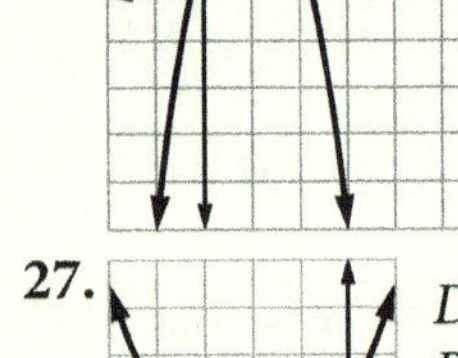 $D = \mathbb{R}$
$R = (-\infty, 3]$

27. 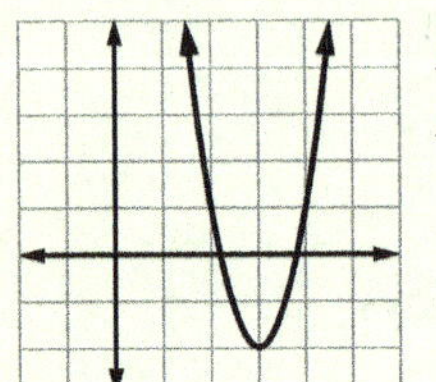 $D = \mathbb{R}$
$R = [0, \infty)$

28. $f(x) = 3x^2 - 18x + 25$
$= 3(x^2 - 6x + 3^2) + 25 - 9(3)$
$= 3(x - 3)^2 - 2$

$D = \mathbb{R}$
$R = [2, \infty)$

29. $V(1, 4)$; max
$f(x) = -(x^2 - 2x + 1) + 4$
$= -x^2 + 2x + 3$
$f(0) = 3$; y-int.: $(0, 3)$
$0 = -1(x^2 - 2x - 3)$
$0 = (x - 3)(x + 1)$
$x = 3, -1$
x-int.: $(3, 0), (-1, 0)$

Use technology to create a scatterplot of median new home prices in the US. Let x represent the number of years since 1970, and let y represent the price in thousands of dollars. [1.9]

Median New Home Prices in the US	
Year	Price
1970	\$23,400
1975	\$39,300
1980	\$64,600
1985	\$84,300
1990	\$122,900
1995	\$133,900
2000	\$169,000

30. $f(0) = -12$; y-int.: $(0, -12)$
$0 = (2x - 3)(x + 4)$
$x = \frac{3}{2}, -4$
x-int.: $(-4, 0)$ and $\left(\frac{3}{2}, 0\right)$
$f(x) = 2x^2 + 5x - 12$
$= 2\left(x^2 + \frac{5}{2}x + \left(\frac{5}{4}\right)^2\right) - 12 - \frac{25}{16}(2)$
$y = 2\left(x + \frac{5}{4}\right)^2 - \frac{121}{8}$
$V\left(-\frac{5}{4}, -\frac{121}{8}\right)$; min

31. $(f + g)(x) = (x^2 - 4x + 4) + (x - 2)$
$= x^2 - 3x + 2$
$(g - f)(x) = (x - 2) - (x^2 - 4x + 4)$
$= -x^2 + 5x - 6$
$(fg)(x) = (x^2 - 4x + 4)(x - 2)$

46. Find the best linear function modeling the data and the corresponding coefficient of determination.

47. Create a graph of the residuals. Do the residuals appear to form a pattern? no

48. Find the best quadratic function modeling the data and the corresponding coefficient of determination.

49. Use the linear model to estimate the median price of a new home in 1983. \$81,300

50. Use both the linear model and the quadratic model to estimate the median price of a new home in 2010. Then determine the difference between each estimate and the actual 2010 median price of \$221,300.
$\approx$ \$213,300 ($\approx$ \$8,000 low); $\approx$ \$228,800 ($\approx$ \$7500 high)

46. $p(x) \approx 4.8879x + 17.7393$; $r^2 \approx 0.9880$
48. $p(x) \approx 0.0297x^2 + 3.9979x + 21.4476$; $r^2 \approx 0.9908$

$= x^3 - 4x^2 + 4x - 2x^2 + 8x - 8$
$= x^3 - 6x^2 + 12x - 8$

32. $f(g(x)) = f(x - 2)$
$= (x - 2)^2 - 4(x - 2) + 4$
$= (x^2 - 4x + 4) - 4x + 8 + 4$
$= x^2 - 8x + 16$
$g(f(x)) = g(x^2 - 4x + 4)$
$= (x^2 - 4x + 4) - 2$
$= x^2 - 4x + 2$

continued in Answers and Solutions Overflow

Radical, Polynomial, and Rational Functions

Overview

Radical, polynomial, and rational functions play an important role in modeling, engineering, predicting, and problem solving. In this chapter the students will analyze the characteristics of these functions, including their end behaviors, zeros, extrema, and discontinuities.

The Technology Corner examines some of the limitations and interpretations of discontinuities and asymptotes on graphing calculators.

Chapter Objectives

1. To graph basic radical, polynomial, and rational functions and their reflections, translations, and dilations

2. To describe key characteristics of more advanced algebraic functions including domain, range, intercepts, symmetry, end behavior, continuity, and increasing or decreasing intervals

3. To solve radical, polynomial, and rational equations

4. To use advanced algebraic functions to model real-world data and solve problems

5. To solve nonlinear inequalities in one variable

6. To explain how the belief in a triune God contributed to the development of the concept of functions

Flash

The average home price in the United States is more than \$200,000. Data analysis can provide insight into one of the biggest financial investments in a person's life.

HISTORICAL CONNECTION
During the Middle Ages, interest in using algebraic techniques to solve problems in science, commerce, and finance flourished. Gutenberg's printing press (ca. 1440) led to the proliferation of mathematical texts. This encouraged the adoption and standardization of symbolic notation, a key development in the study of functions.

BIBLICAL PERSPECTIVE OF MATHEMATICS
Do mathematical laws evidence the divine nature of God? Are attributes of God found in simple arithmetical truths?

DATA ANALYSIS
Mathematicians analyze the key characteristics of functions that model raw data, enabling informed predictions and decisions to be made. You will learn how to find zeros, asymptotes, point discontinuities, end behavior, and relative minimum and maximum points of several types of nonlinear functions.

Suggested Teaching Schedule

DAY		DAY	
1	2.1	7	2.5
2	2.2	8	2.6
3	TC: Discontinuities Quiz 2A (2.1–2.2)	9–10	Quiz 2C (2.5–2.6) 2.7
4	2.3 HC: Symbols	11	DA: The Housing Market Quiz 2D (2.7)
5	2.4	12	Chapter 2 Review
6	Quiz 2B (2.3–2.4) BPM: Math's Divine Nature	13	Chapter 2 Test

2.1 Radical Functions and Equations

Objectives

1. To graph and describe the key characteristics of a radical function
2. To interpret a radical function as a power function with a rational exponent and as the inverse of a power function
3. To solve radical equations

Vocabulary

extraneous root
radical equation
radical function

Reading and Writing Mathematics

Explain why $x^{\frac{2}{3}} = 16$ has two solutions but $x^{\frac{3}{2}} = 8$ has one solution.

Fractional powers do not indicate the principal root. So $\left(x^{\frac{2}{3}}\right)^{\frac{3}{2}} = 16^{\frac{3}{2}} = x = (\pm 4)^3 = \pm 64.$

However, $\left(x^{\frac{3}{2}}\right)^{\frac{2}{3}} = 8^{\frac{2}{3}} = 2^2 = 4.$

Additional Exercises

Figures for Additional Exercises throughout the chapter can be found at Teacher ToolsOnline.com.

An athlete's hang time is a function of the height of the jump.

After completing this section, you will be able to
- graph a radical function and describe its key characteristics.
- relate radical functions and their inverse power functions.
- solve radical equations.

Fractional exponents were first used by Nicole Oresme in the fourteenth century. He expressed $2^{\frac{1}{2}}$ as $\frac{1}{2}\,2^p$ and $4^{\frac{3}{2}}$ as $\left|1\,p\,\frac{1}{2}\right|\,4$.

Power functions with integral exponents were examined in Section 1.4. Recall that the radical expression $\sqrt[r]{x}$ can also be written as a power with a rational exponent, $x^{\frac{1}{r}}$. Therefore, a function of the form $f(x) = k\sqrt[r]{x}$ is a power function with a fractional exponent.

> **DEFINITION**
>
> A **radical function** contains a radical expression with the independent variable in the radicand.

One of the simplest radical functions is the square root function, $f(x) = \sqrt{x}$ or $f(x) = x^{\frac{1}{2}}$. The domain is restricted to $[0, \infty)$ since negative values for x result in imaginary values for y. The range is also $[0, \infty)$ since the notation indicates the principal (or positive) square root. The function is continuous and increasing over its domain.

Example 1 Analyzing Radical Functions

Use technology to graph each radical function. State its domain and range and then determine the intervals in which the function is increasing or decreasing. Then state whether the function is continuous or discontinuous and whether it is odd, even, or neither.

a. $g(x) = \sqrt[3]{x}$ **b.** $h(x) = \sqrt[4]{x}$

Answer

a.

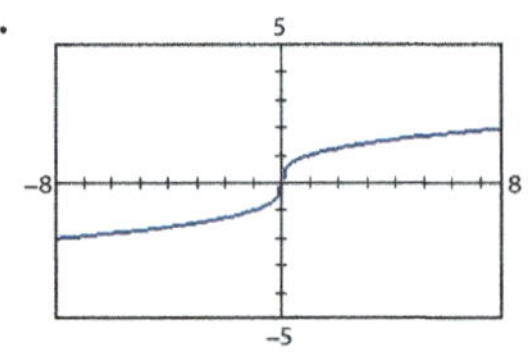

$D = \mathbb{R};\ R = \mathbb{R}$
increasing: $\mathbb{R}$
continuous; odd

b.

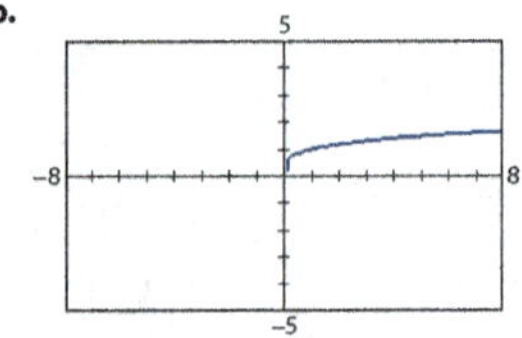

$D = [0, \infty);\ R = [0, \infty)$
increasing: $[0, \infty)$
continuous; neither

SKILL ✔ EXERCISE 3

The graph of $g(x) = \sqrt[3]{x} = x^{\frac{1}{3}}$ in Example *1a* suggests that $g(x)$ is the inverse function of $f(x) = x^3$. This can be verified by showing that $f(g(x)) = \left(x^{\frac{1}{3}}\right)^3 = x$ and $g(f(x)) = (x^3)^{\frac{1}{3}} = x.$

PRESENTATION

Lesson Opener

Solve for x.

1. $\sqrt{2x} = 16$ 128
2. $\sqrt{x+5} = 20$ 395

Solve for y.

3. $x = \sqrt{y-8}$ $y = x^2 + 8$
4. $x = y^2 + 3$ $y = \pm\sqrt{x-3}$
5. $x = \dfrac{y^2 - 4}{4} - 1$ $y = \pm 2\sqrt{x+2}$

Present the definition of a *radical function*, stressing that the independent variable must be in the radicand. Consider explaining why an equation such as $y = \sqrt{5}x$ is not a radical function. Point out that radical expressions can also be written as exponential expressions, and that radical functions are therefore a subset of power functions.

Use the radical functions in Example 1 to analyze the characteristics of basic radical functions involving even and odd roots. These can be entered on a graphing calculator using the $\sqrt[x]{}$ option under $\boxed{\text{MATH}}$, or as a rational exponent. Be sure fractional exponents are in parentheses.

Note that $g(x) = x^{\frac{1}{3}}$ is the inverse function of $f(x) = x^3$ and that the functions will "undo" each other. This can be illustrated using $f(3) = 27$ and $g(27) = 3$.

Emphasize that the rule for the inverse of $f(x) = \sqrt[r]{x}$ must include a restricted domain.

Use the graph preceding Example 2 to discuss the first quadrant characteristics of basic power functions with various exponents. Stress that a radical function with an even root (or a rational exponent with an even denominator) will have a limited domain since the even root of a negative number is a complex number and cannot be graphed in the coordinate plane.

Interactive Activity Consider exploring exercise 9 together by projecting the graphs and discussing characteristics of specific radical functions. Dynamic

Since $f(x) = x^4$ is an even function and does not pass the Horizontal Line Test, its inverse is not a function unless the function's domain is restricted. When the domain of $f(x) = x^4$ is restricted to $D_f = [0, \infty)$, its inverse function is $h(x) = \sqrt[4]{x} = x^{\frac{1}{4}}$.

Notice the first-quadrant shapes of the illustrated parent power functions of the form $f(x) = x^n$. The functions all contain the point $(1, 1)$. When $n > 0$, $\lim_{x \to 0} x^n = 0$ and $\lim_{x \to \infty} x^n = \infty$; but when $n < 0$, $\lim_{x \to 0^+} x^n = \infty$ and $\lim_{x \to \infty} x^n = 0$. When n is a rational exponent, such as $x^{\frac{p}{r}}$ where $\frac{p}{r}$ is simplified, $f(x) = \sqrt[r]{x^p}$. The radical function $f(x) = x^{\frac{p}{r}}$ is restricted to $D = [0, \infty)$ when r is even. The shape of the remaining portion of each graph will be explored in exercise 9.

Many radical functions can be graphed by transforming a parent function of the form $f(x) = \sqrt[r]{x^p}$.

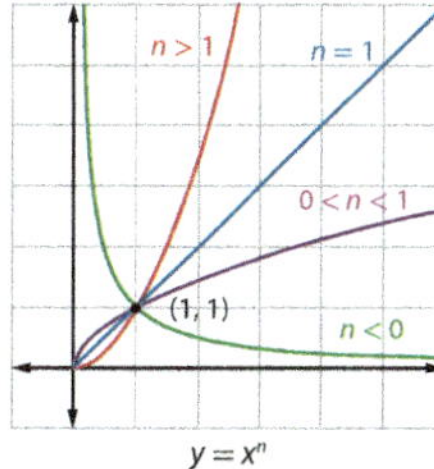

Example 2   Graphing the Transformation of a Radical Function

Use transformations of a parent radical function to graph $g(x) = 4 - \sqrt{x + 3}$. Then state its domain and range.

Answer

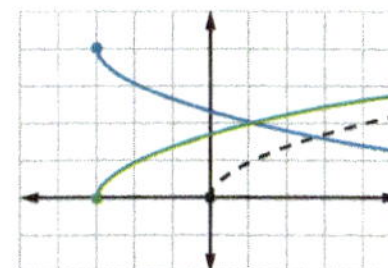

1. Rewrite the function $g(x) = -\sqrt{x + 3} + 4$.
2. Sketch a graph of the parent function $f(x) = \sqrt{x}$.
3. Translate the graph 3 units left.
4. Reflect the graph in the x-axis and translate the graph 4 units up.
5. The radicand of an even root cannot be negative.

$x + 3 \geq 0; x \geq -3;$
$D = [-3, \infty); R = (-\infty, 4]$

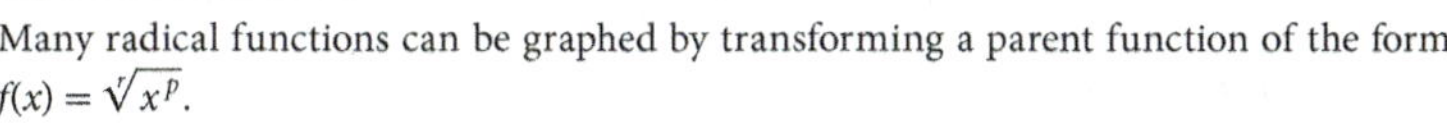

SKILL ✓ EXERCISE 19

Note also that the inverse of $f(x) = x^{\frac{p}{r}}$ is $g(x) = x^{\frac{r}{p}}$ with any necessary restrictions on the domains.

Example 3   Finding and Graphing the Inverse of a Radical Function

Find the inverse function of $f(x) = -\sqrt{x - 1}$. Then graph $f(x)$ and $f^{-1}(x)$.

Answer

$y = -\sqrt{x - 1}$
$D = [1, \infty); R = (-\infty, 0]$

1. State the function's domain and range.

$x = -\sqrt{y - 1}$
$x^2 = \left(-\sqrt{y - 1}\right)^2$
$x^2 = y - 1$
$y = x^2 + 1$
$f^{-1}(x) = x^2 + 1; D = (-\infty, 0]$

2. Find the inverse by switching the variables and solving for y. Recall that squaring both sides eliminates the radical.

3. Use the range of $f(x)$ to restrict the domain of the parabola, $y = x^2 + 1$, so that the function is the inverse.

CONTINUED ➡

In exercises 1–2, use transformations of a parent function to graph each function and state its domain and range.

1. $g(x) = -\sqrt[3]{x - 2} + 4$

$D = \mathbb{R}; R = \mathbb{R}$

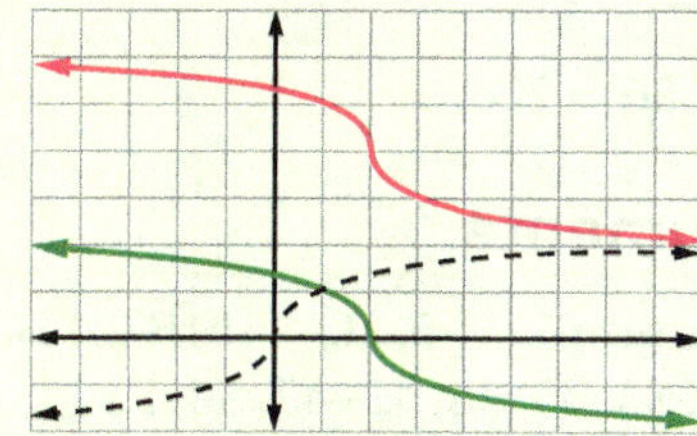

2. $g(x) = -\sqrt{x + 4} - 3$

$D = [-4, \infty); R = (-\infty, -3]$

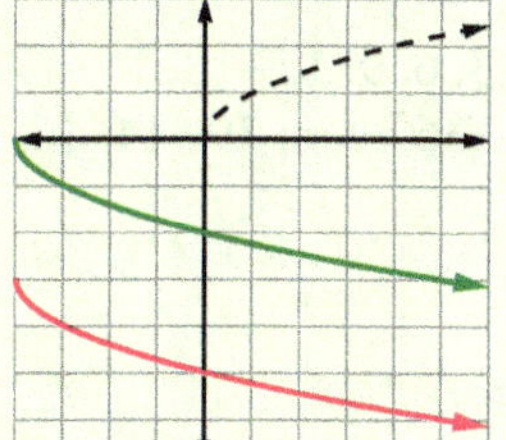

3. Find the inverse of $f(x) = \sqrt{x + 2}$, then graph $f(x)$ and $f^{-1}(x)$.

$f^{-1}(x) = x^2 - 2, D = [0, \infty)$

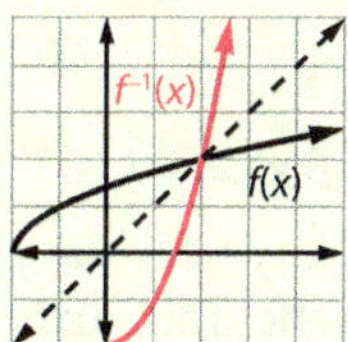

Solve.

4. $\sqrt{5x^2 + 19} + x = 7$ $x = \frac{3}{2}, -5$

5. $\sqrt{x^2 + 4x + 6} = x + 2$ no solution

6. $\sqrt{4x + 17} = 4 - \sqrt{x + 1}$
$x = -\frac{8}{9}, 8$ (extraneous)

graphs using sliders can be found using the Internet keyword search *graphing calculator sliders*.

Illustrate the graphing of a transformed radical function with Example 2. In Example 3 show that identifying the function's range is critical to identifying the inverse's limited domain. If the restricted range is not listed as part of the function rule, $y = x^2 + 1$ is not the inverse of the radical function since the parabola is not a reflection of the graph of $f(x)$ in the line $y = x$. A limited domain provides a one-to-one correspondence.

The second part of the lesson reviews solving *radical equations*, a topic studied in previous courses. Consider having the students solve the equation in Example 4 before presenting it.

Common Student Error A student may attempt to square both sides of the equation before isolating the radical term. Consider illustrating how this fails to eliminate the radical, leaving it in the middle term.

Encourage the students to check their answers by using technology to graph radical equations. The visualization will increase familiarity with radical functions while confirming solutions.

Use Example 5 to demonstrate solving a radical equation with multiple radicals.

In step 3 you may want to demonstrate that both sides could be squared once the term containing the radical has been isolated. This would produce an equation with larger coefficients in this case but could help avoid fractions in other cases.

Common Student Error Students may incorrectly square each term of a binomial, such as $\sqrt{x - 5} + 9$. Remind them to use the FOIL method, in which the first term is the entire radical.

In Example 6 be sure the students understand why the absolute value is used when p is even. You may want to use a simple example of $4^{\frac{1}{2}} = \sqrt{4} = \pm 2$, whereas $8^{\frac{1}{3}} = \sqrt[3]{8} = 2$.

7. $\sqrt{x^2 - 5x + 10} - \sqrt{x^2 - 3x} = 2$

$x = -1, 3$

8. $\sqrt{x^2 - 11} - 2 = \dfrac{15}{\sqrt{x^2 - 11}}$

$x = \pm 6, \pm 2\sqrt{5}$ (extraneous)

9. $\sqrt{x^5} - 6 = 26$ $x = 4$

10. $x^{\frac{4}{3}} - 56 = 200$ $x = \pm 64$

Assignments

- **Minimum:** 1, 4–5, 10–11; 15–21 odd; 25, 29, 33, 35, 38, 45–47, 50, 52
- **Standard:** 2, 5, 8–10, 13, 15–16, 18–19, 22–23, 27–29, 31, 33, 36, 40–41, 46–47, 53–54
- **Extended:** 2–3, 6, 9, 12–13, 16–17, 20, 23–24; 26–36 even; 39–44, 46–47, 50, 53

Solutions

❯ A. Exercises

5. $4 - x \geq 0; \; x \leq 4$

6. $2x + 4 \geq 0$

$2x \geq -4$

$x \geq -2$

10–13. The inverse of $f(x) = x^{\frac{p}{r}}$ is $g(x) = x^{\frac{r}{p}}$ with any necessary restrictions on the domain.

10. $R_f = \mathbb{R} = D_g$

$x = y^3$

$\sqrt[3]{x} = y$

$y = g(x) = \sqrt[3]{x}$

11. $R_f = [0, \infty) = D_g$

$x = y^{\frac{3}{4}}$

$x^{\frac{4}{3}} = \left(y^{\frac{3}{4}}\right)^{\frac{4}{3}}$

$y = x^{\frac{4}{3}}$ or $\sqrt[3]{x^4}$

12. $R_f = [0, \infty) = D_g$

$x = \sqrt{y^3}$

$x^{\frac{2}{3}} = \left(y^{\frac{3}{2}}\right)^{\frac{2}{3}}$

$y = x^{\frac{2}{3}}$ or $\sqrt[3]{x^2}$

13. $R_f = \mathbb{R} = D_g$

$x = \sqrt[3]{y^5}$

$x^{\frac{3}{5}} = \left(y^{\frac{5}{3}}\right)^{\frac{3}{5}}$

$y = x^{\frac{3}{5}}$ or $\sqrt[5]{x^3}$

14. $3x + 4 = 25$

$3x = 21$

$x = 7$

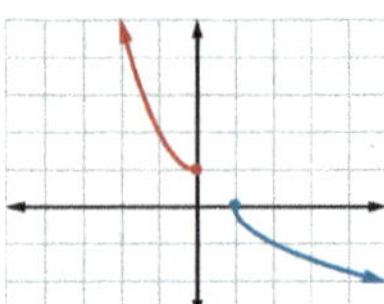

4. Graph $f(x)$ and $f^{-1}(x)$.

A *radical equation* contains a radical expression with a variable in the radicand. To solve a radical equation, isolate the radical expression and raise both sides to the appropriate power. You must check each solution of the resulting equation, since *extraneous roots* can appear when you raise both sides to a power.

Example 4 Solving a Radical Equation

Solve $\sqrt{x + 5} + 1 = x$.

Answer

$\sqrt{x + 5} = x - 1$ 1. Subtract 1 from each side to isolate the radical.

$x + 5 = x^2 - 2x + 1$ 2. Square both sides.

$0 = x^2 - 3x - 4$ 3. Solve for x.

$0 = (x - 4)(x + 1)$

$x = 4$ or $x = -1$

$\sqrt{(4) + 5} + 1 \; ? \; (4)$ $\sqrt{(-1) + 5} + 1 \; ? \; (-1)$ 4. Substitute into the original equation to check for extraneous roots.

$3 + 1 = 4$ $2 + 1 \neq -1$

solution: $x = 4$ 5. Exclude -1 since it is extraneous.

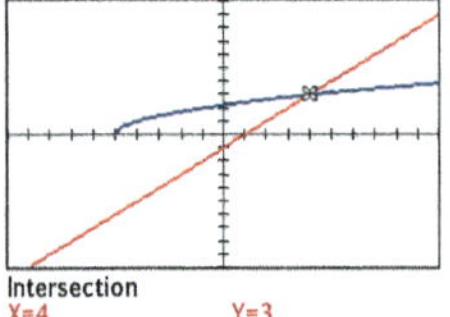

The solutions from Example 4 can also be checked by graphing the left side of the equation as Y_1 and the right side as Y_2. The x-coordinate of any intersection is a solution to the equation (use 5:intersect from the [CALC] menu). If the graphs do not intersect, there are no real solutions.

If an equation has several radicals containing the variable, isolate one of them, raise both sides to the appropriate power, isolate a remaining radical, and repeat the process. The solutions can be checked graphically.

One-on-One Some students may be mystified by the presence of *extraneous roots*. Demonstrate how the false statement $2 = -2$ can be made a true statement by squaring both sides. The squaring of both sides of an equation assumes that they are equal and may introduce solutions that are not true in the original equation.

TIPS

Ex. 9 These functions can also be expressed as

$f(x) = x^{2n}, n \in \mathbb{Z}$,

$f(x) = x^{2n + 1}, n \in \mathbb{Z}$,

$f(x) = x^n, n \in \mathbb{Q}'$,

$f(x) = x^{\frac{2p}{2r + 1}}, r, p \in \mathbb{Z}$, and

$f(x) = x^{\frac{2p + 1}{2r + 1}}, r, p \in \mathbb{Z}$.

Solve $\sqrt{x^2 + 5x + 3} - \sqrt{x^2 + 3x} = 1$.

Answer

$\sqrt{x^2 + 5x + 3} = \sqrt{x^2 + 3x} + 1$	1. Isolate one of the radicals.
$x^2 + 5x + 3 = x^2 + 3x + 2\sqrt{x^2 + 3x} + 1$	2. Square both sides.
$2x + 2 = 2\sqrt{x^2 + 3x}$	3. Simplify and isolate the remaining radical.
$x + 1 = \sqrt{x^2 + 3x}$	
$x^2 + 2x + 1 = x^2 + 3x$	4. Square both sides again.
$1 = x$	5. Solve for x.
$\sqrt{(1)^2 + 5(1) + 3} - \sqrt{(1)^2 + 3(1)} \overset{?}{=} 1$	6. Check the solution.
$3 - 2 = 1$	

———————————————————— SKILL ✔ **EXERCISE 29**

Equations of the form $x^{\frac{p}{r}} = k$ are also radical equations and can be solved by raising each side to the power of $\frac{r}{p}$. Notice that $\left(x^{\frac{p}{r}}\right)^{\frac{r}{p}} = \sqrt[p]{\left(x^{\frac{p}{r}}\right)^r} = \sqrt[p]{x^p}$. Therefore when p is even, $\left(x^{\frac{p}{r}}\right)^{\frac{r}{p}} = \sqrt[p]{x^p} = |x| = \pm x$, and when p is odd, $\left(x^{\frac{p}{r}}\right)^{\frac{r}{p}} = \sqrt[p]{x^p} = x$.

Solve $x^{\frac{2}{3}} - 1 = 15$.

Answer

$x^{\frac{2}{3}} = 16$	1. Isolate the variable.
$\left(x^{\frac{2}{3}}\right)^{\frac{3}{2}} = 16^{\frac{3}{2}}$	2. Raise both sides to the power of $\frac{3}{2}$.
$\|x\| = \left(4^2\right)^{\frac{3}{2}} = 4^3$	3. Solve the resulting equation.
$x = \pm 64$	

$(64)^{\frac{2}{3}} - 1 \overset{?}{=} 15 \qquad (-64)^{\frac{2}{3}} - 1 \overset{?}{=} 15$	4. Check.
$4^2 - 1 \overset{?}{=} 15 \qquad (-4)^2 - 1 \overset{?}{=} 15$	
$16 - 1 = 15 \qquad 16 - 1 = 15$	

———————————————————— SKILL ✔ **EXERCISE 33**

15. $\sqrt[3]{x} = -4$
$\left(\sqrt[3]{x}\right)^3 = (-4)^3$
$x = -64$

16. $\sqrt{x + 4} = x - 2$
$x + 4 = x^2 - 4x + 4$
$x^2 - 5x = 0$
$x(x - 5) = 0$
$x = 5, 0 \text{ (extraneous)}$

17. $\quad 2x - 3 = \sqrt{7x - 3}$
$4x^2 - 12x + 9 = 7x - 3$
$4x^2 - 19x + 12 = 0$
$(4x - 3)(x - 4) = 0$
$x = 4, \frac{3}{4} \text{ (extraneous)}$

18. $\sqrt[3]{n^2 + 2} = 3$
$n^2 + 2 = 27$
$n^2 = 25$
$n = \pm 5$

> **B. Exercises**

19. Translate $f(x) = \sqrt[3]{x}$ up 2 units.
$D = \mathbb{R}; R = \mathbb{R}$

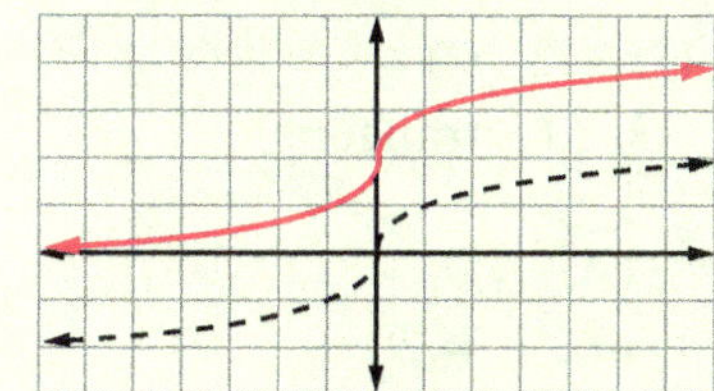

20. Translate $f(x) = \sqrt{x}$ right 2 units and reflect it in the x-axis.
$D = [2, \infty); R = (-\infty, 0]$

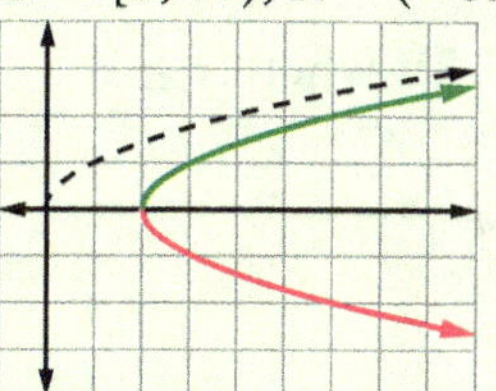

21. Translate $f(x) = \sqrt{x}$ left 3 units and stretch it vertically by a factor of 2.
$D = [-3, \infty); R = [0, \infty)$

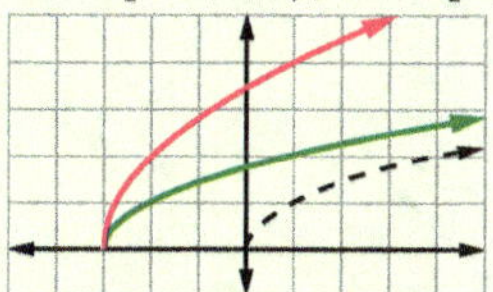

22. Translate $f(x) = \sqrt[3]{x}$ left 1 unit and 3 units up. $D = \mathbb{R}; R = \mathbb{R}$

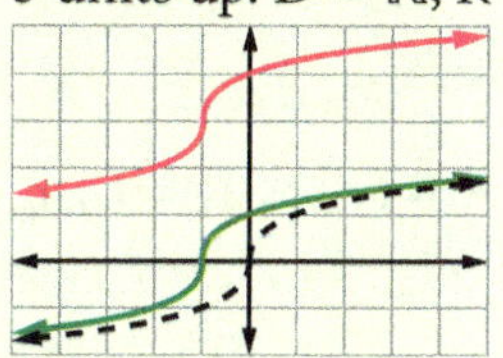

23. Translate $f(x) = \sqrt{x}$ left 2 units and reflect it in the y-axis.
$D = (-\infty, 2]; R = [0, \infty)$

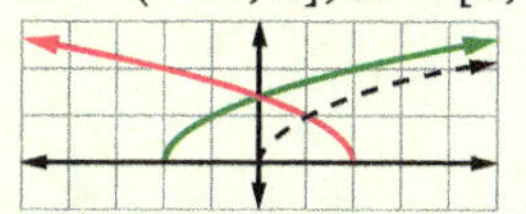

24. Reflect $f(x) = \sqrt[3]{x}$ in the x-axis and stretch it vertically by a factor of 2; then translate it 3 units up.
$D = \mathbb{R}; R = \mathbb{R}$

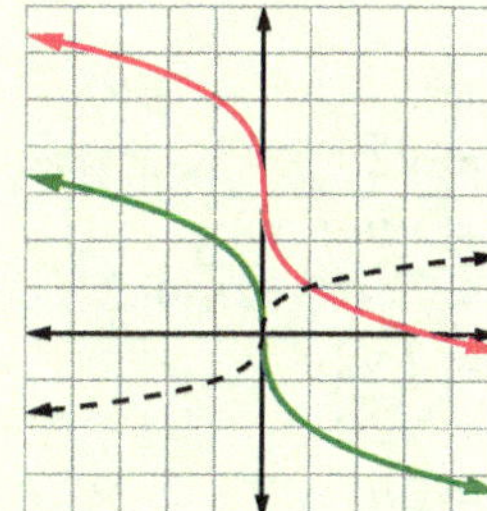

25. $D_f = [1, \infty); R_f = [0, \infty) = D_{f^{-1}(x)}$
$x = \sqrt{y - 1}$
$y - 1 = x^2$
$y = x^2 + 1 \text{ where } x \geq 0$

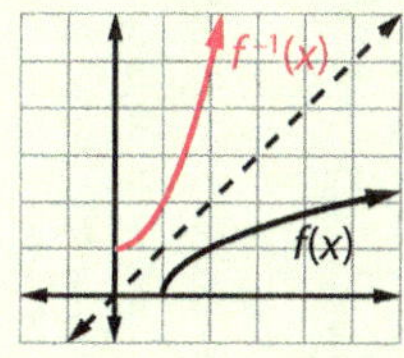

26. $D_f = [0, \infty); R_f = [2, \infty) = D_{f^{-1}(x)}$

$$x = \sqrt{y} + 2$$
$$x - 2 = \sqrt{y}$$
$$y = (x - 2)^2 \text{ where } x \geq 2$$

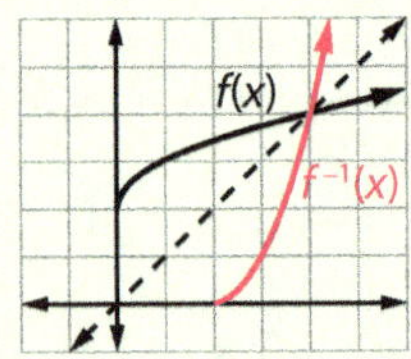

27. $D_f = \mathbb{R}; R_f = \mathbb{R} = D_{f^{-1}(x)}$

$$x = \sqrt[3]{y - 2}$$
$$y - 2 = x^3$$
$$y = x^3 + 2$$

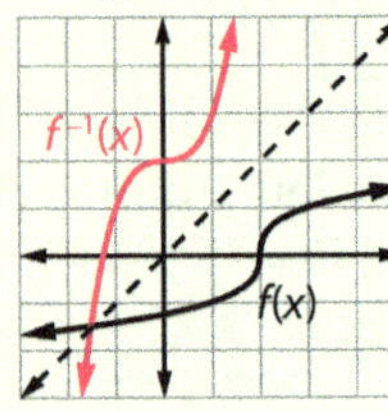

28. $D = [0, \infty); R = (-\infty, 2] = D_{f^{-1}(x)}$

$$x = 2 - \sqrt{y}$$
$$\sqrt{y} = 2 - x$$
$$y = (2 - x)^2 = [(-1)(x - 2)]^2$$
$$= (x - 2)^2 \text{ where } x \leq 2$$

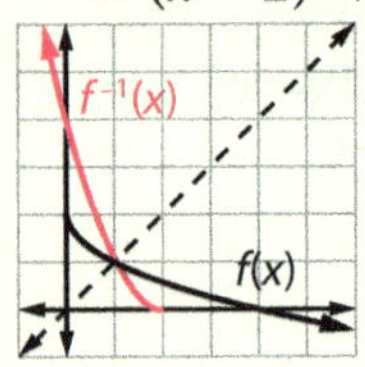

29. $\sqrt{x + 5} = 5 - \sqrt{x}$

$$x + 5 = 25 - 10\sqrt{x} + x$$
$$-20 = -10\sqrt{x}$$
$$2 = \sqrt{x}$$
$$x = 4$$

30. $\sqrt{n + 6} - \sqrt{n} = \sqrt{2}$

$$\sqrt{n + 6} = \sqrt{n} + \sqrt{2}$$
$$n + 6 = n + 2\sqrt{2n} + 2$$
$$4 = 2\sqrt{2n}$$
$$2 = \sqrt{2n}$$
$$4 = 2n$$
$$n = 2$$

31. $\sqrt{4x + 5} = 17 - 2\sqrt{x - 3}$

$$4x + 5 = 289 - 68\sqrt{x - 3} + 4x - 12$$
$$-272 = -68\sqrt{x - 3}$$
$$4 = \sqrt{x - 3}$$
$$16 = x - 3$$
$$x = 19$$

A. Exercises

Use technology to graph each radical function. State its domain, range, and the intervals for which the function is increasing or decreasing. State whether the function is continuous or discontinuous and whether it is odd, even, or neither.

1. $f(x) = \sqrt[3]{x}$ **2.** $f(x) = \sqrt[5]{x}$

3. $f(x) = x^{\frac{5}{2}}$ **4.** $f(x) = x^{\frac{3}{7}}$

5. $g(x) = \sqrt{4 - x}$ **6.** $g(x) = \sqrt[3]{2x + 4}$

7. $h(x) = \sqrt[3]{x - 5}$ **8.** $h(x) = \sqrt[3]{11 - 2x}$

9. Explore: Use technology to graph several power functions to illustrate each case of $f(x) = x^n$ or $f(x) = x^{\frac{p}{r}}$ where $\frac{p}{r}$ is simplified. Then state whether each type of function is even, odd, or neither.

a. $f(x) = x^n, n \in$ even $\mathbb{Z}$ even

b. $f(x) = x^n, n \in$ odd $\mathbb{Z}$ odd

c. $f(x) = x^n, n \in \mathbb{Q}'$ neither

d. $f(x) = x^{\frac{p}{r}}, r \in$ even $\mathbb{Z}$ neither

e. $f(x) = x^{\frac{p}{r}}, r \in$ odd $\mathbb{Z}, p \in$ even $\mathbb{Z}$ even

f. $f(x) = x^{\frac{p}{r}}, r \in$ odd $\mathbb{Z}, p \in$ odd $\mathbb{Z}$ odd

State a rule for $g(x)$, the inverse of $f(x)$, and use technology to graph both $f(x)$ and $g(x)$. State the restricted domain for either $f(x)$ or $g(x)$ that is required for $f(x)$ and $g(x)$ to be inverses of each other.

10. $f(x) = x^3$ **11.** $f(x) = x^{\frac{3}{4}}$

12. $f(x) = \sqrt{x^3}$ **13.** $f(x) = \sqrt[3]{x^5}$

Solve. Be sure to check for extraneous solutions.

14. $\sqrt{3x + 4} = 5$ $x = 7$ **15.** $\sqrt[3]{x} + 6 = 2$ $x = -64$

16. $\sqrt{x + 4} = x - 2$ $x = 5$ **17.** $2x = 3 + \sqrt{7x - 3}$ $x = 4$

18. $\sqrt[3]{n^2 + 2} + 1 = 4$ $n = \pm 5$

B. Exercises

Use transformations of a parent radical function to graph each function. Then state its domain and range.

19. $g(x) = \sqrt[3]{x} + 2$ **20.** $h(x) = -\sqrt{x - 2}$

21. $m(x) = 2\sqrt{x + 3}$ **22.** $n(x) = 3 + \sqrt[3]{x + 1}$

23. $r(x) = \sqrt{2 - x}$ **24.** $t(x) = 3 - 2\sqrt[3]{x}$

Find the inverse of each function, then graph $f(x)$ and $f^{-1}(x)$.

25. $f(x) = \sqrt{x - 1}$ **26.** $f(x) = \sqrt{x} + 2$

27. $f(x) = \sqrt[3]{x - 2}$ **28.** $f(x) = 2 - \sqrt{x}$

Solve.

29. $\sqrt{x} + \sqrt{x + 5} = 5$ $x = 4$

30. $\sqrt{n + 6} - \sqrt{n} = \sqrt{2}$ $n = 2$

31. $\sqrt{4x + 5} + 2\sqrt{x - 3} = 17$ $x = 19$

32. $\sqrt{r^2 + 3r + 3} + \sqrt{r^2 + r - 1} = 2$ $r = -2$

Solve by raising powers.

33. $x^{\frac{2}{3}} + 6 = 10$ $x = \pm 8$ **34.** $x^{\frac{3}{4}} + 1 = 9$ $x = 16$

35. $\sqrt{x^3} + 10 = 74$ $x = 16$ **36.** $100 - \sqrt[3]{x^4} = 19$ $x = \pm 27$

37. Solve $d = 2\sqrt{\dfrac{V}{h\pi}}$ for V and use the resulting equation to find the volume of a 5 ft tall cylinder with a diameter of 20 ft. 500π ft$^3 \approx 1570.8$ ft^3

38. Solve $r = \sqrt[3]{\dfrac{3V}{4\pi}}$ for V and use the resulting equation to find the volume of a soccerball whose diameter is 22 cm. $\dfrac{5324\pi}{3}$ cm$^3 \approx 5575.3$ cm^3

39. The equation $I = \sqrt{\dfrac{P}{R}}$ expresses electrical current I (in amps) as a function of power P (in watts) and resistance R (in ohms). Solve the equation for R and use the result to find the resistance of a motor drawing 10 amps of current and using 1100 watts of power.   11 ohms

40. Kadour Ziani, a professional basketball dunker, set a world record with a 5 ft vertical leap. His hang time, the total time he was airborne, can be modeled by $t = \dfrac{\sqrt{h}}{2}$ where t is in seconds and h is in feet.

a. What is the hang time for a vertical leap of 2 ft? ≈ 0.71 sec

b. What was the hang time for Ziani's world record jump? ≈ 1.12 sec

c. Find the vertical leap of an athlete with a hang time of 1 sec.   4 ft

d. Approximate the height of a punted football with a 4.40 sec hang time.   77.44 ft

C. Exercises

Solve.

41. $\sqrt[3]{7x + 1} = x + 1$ $x = 0, 1, -4$

42. $\sqrt{n} + \sqrt{n - 3} = \dfrac{3}{\sqrt{n - 3}}$ $n = 4$

State a rule for the inverse of $f(x)$ and graph both $f(x)$ and its inverse. If the inverse is not a function, restrict the domain of $f(x)$ so that its inverse is a function $f^{-1}(x)$ and state the rule for $f^{-1}(x)$.

43. $f(x) = x^2 - 4$ **44.** $f(x) = x^2 + 2x + 1$

$\quad y = \pm\sqrt{x + 4}$ $y = \pm\sqrt{x} - 1$

---

32. $\sqrt{r^2 + 3r + 3} = 2 - \sqrt{r^2 + r - 1}$

$$r^2 + 3r + 3$$
$$= 4 - 4\sqrt{r^2 + r - 1} + r^2 + r - 1$$
$$2r = -4\sqrt{r^2 + r - 1}$$
$$r = -2\sqrt{r^2 + r - 1}$$
$$r^2 = 4(r^2 + r - 1)$$
$$0 = 3r^2 + 4r - 4$$
$$0 = (3r - 2)(r + 2)$$
$$r = -2, \tfrac{2}{3} \text{ (extraneous)}$$

33. $\left(x^{\frac{2}{3}}\right)^{\frac{3}{2}} = 4^{\frac{3}{2}}$

$$|x| = (2^2)^{\frac{3}{2}} = 2^3$$
$$x = \pm 8$$

34. $\quad x^{\frac{3}{4}} = 8$

$$\left(x^{\frac{3}{4}}\right)^{\frac{4}{3}} = (2^3)^{\frac{4}{3}} = 2^4$$
$$x = 16$$

35. $\quad x^{\frac{3}{2}} = 64$

$$\left(x^{\frac{3}{2}}\right)^{\frac{2}{3}} = (4^3)^{\frac{2}{3}} = 4^2$$
$$x = 16$$

36. $\quad x^{\frac{4}{3}} = 81$

$$\left(x^{\frac{4}{3}}\right)^{\frac{3}{4}} = (3^4)^{\frac{3}{4}} = 3^3$$
$$|x| = 27$$
$$x = \pm 27$$

37. $d^2 = 4\left(\dfrac{V}{h\pi}\right)$

$$V = \dfrac{d^2 h\pi}{4}$$
$$= \dfrac{(20)^2 (5)\pi}{4} = 500\pi \text{ ft}^3 \approx 1570.8 \text{ ft}^3$$

45. Multiply $(-5a^3 + 7b)(-5a^3 - 7b)$. [Algebra] $25a^6 - 49b^2$

Factor. [Algebra] $4(2x + 7)(x - 1)$ $(x + 2)(x - 2)(x - 7)$

46. $8x^2 + 20x - 28$ **47.** $x^3 - 7x^2 - 4x + 28$

48. Draw a mapping diagram of $R = \{(0, -3), (2, 2), (-\tfrac{1}{2}, -3), (1, 5)\}$ and state whether the relation is a function or not. [1.1]

49. A city's population is found to vary inversely with the square root of its unemployment rate. If the population is 137,800 with 5% unemployment, predict the population if unemployment rises to 6%. [1.4] 125,794

50. Write the equation of the transformation of $f(x) = x^2$ that has been translated 5 units left, stretched vertically by a factor of 3, and translated 1 unit up. [1.5]

51. Which of the following relations represents a function? [1.1] C

A. $x = 2$
B. $x = \dfrac{1}{y^2}$
C. $x = y$
D. $\{(-1, 0), (2, 0.5), (1, -3), (-1, 1)\}$
E. none of these

52. Which function results when $f(x) = \sqrt{x}$ is reflected across the x-axis and translated 6 units up? [1.5] B

A. $g(x) = \sqrt{-x} + 6$
B. $g(x) = -\sqrt{x} + 6$
C. $g(x) = -\sqrt{x} - 6$
D. $g(x) = \sqrt{x} - 6$
E. none of these

53. Write the standard form of a parabola that passes through $(2, 4)$ and has a vertex at $(1, 1)$. [1.6] D

A. $f(x) = 3x^2 - 2x + 1$
B. $f(x) = 3x^2 - 6x + 3$
C. $f(x) = 3x^2 + 6x + 4$
D. $f(x) = 3x^2 - 6x + 4$
E. none of these

54. Which of the following statements is true when $f(x) = x^2 - 3$ and $g(x) = \sqrt{x + 3}$? [1.7] C

A. The domain of $f(g(x))$ is $(0, \infty)$.
B. $f(g(x)) = x$ for all real numbers
C. $g(f(x)) = |x|$
D. $f(g(x)) = g(f(x))$
E. none of these

---

1. $D = \mathbb{R}; R = \mathbb{R};$ increasing $(-\infty, \infty);$ continuous; odd

2. $D = [0, \infty); R = [0, \infty);$ increasing $(0, \infty);$ continuous; neither

3. $D = [0, \infty); R = [0, \infty);$ increasing $(0, \infty);$ continuous; neither

4. $D = \mathbb{R}; R = \mathbb{R};$ increasing $(-\infty, \infty);$ continuous; odd

5. $D = (-\infty, 4]; R = [0, \infty);$ decreasing $(-\infty, 4];$ continuous; neither

6. $D = [-2, \infty); R = [0, \infty);$ increasing $[-2, \infty);$ continuous; neither

7. $D = \mathbb{R}; R = \mathbb{R};$ increasing $(-\infty, \infty);$ continuous; neither

8. $D = \mathbb{R}; R = \mathbb{R};$ decreasing $(-\infty, \infty);$ continuous; neither

10. $g(x) = \sqrt[3]{x};$ none

11. $g(x) = x^{\frac{4}{3}}$ or $\sqrt[3]{x^4}; D_g = [0, \infty)$

12. $g(x) = \sqrt[3]{x^2}; D_g = [0, \infty)$

13. $g(x) = \sqrt[5]{x^3};$ none

19. $D = \mathbb{R}; R = \mathbb{R}$

20. $D = [2, \infty); R = (-\infty, 0]$

21. $D = [-3, \infty); R = [0, \infty)$

22. $D = \mathbb{R}; R = \mathbb{R}$

23. $D = (-\infty, 2]; R = [0, \infty)$

24. $D = \mathbb{R}; R = \mathbb{R}$

25. $f^{-1}(x) = x^2 + 1, D = [0, \infty)$

26. $f^{-1}(x) = (x - 2)^2, D = [2, \infty)$

27. $f^{-1}(x) = x^3 + 2, D = \mathbb{R}$

28. $f^{-1}(x) = (x - 2)^2, D = (-\infty, 2]$

48. It is a function.

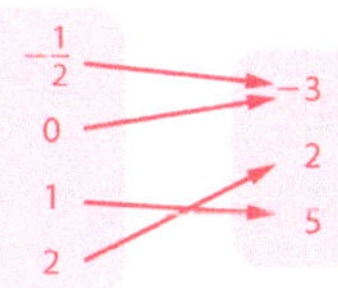

50. $g(x) = 3(x + 5)^2 + 1$

---

43. $x = y^2 - 4$
$y^2 = x + 4$
$y = \pm\sqrt{x + 4}$

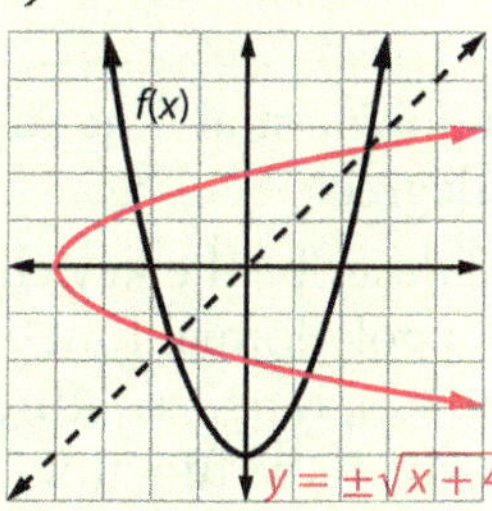

If $D_f = [0, \infty)$, then $f^{-1}(x) = \sqrt{x + 4}$.
If $D_f = (-\infty, 0]$, then $f^{-1}(x) = -\sqrt{x + 4}$.

44. $f(x) = (x + 1)^2$
$x = (y + 1)^2$
$y + 1 = \pm\sqrt{x}$
$y = \pm\sqrt{x} - 1$

If $D_f = [-1, \infty)$, then $f^{-1}(x) = \sqrt{x} - 1$. If $D_f = (-\infty, -1]$, then $f^{-1}(x) = -\sqrt{x} - 1$.

❭ Cumulative Review

46. $4(2x^2 + 5x - 7)$
$= 4(2x + 7)(x - 1)$

47. $x^2(x - 7) - 4(x - 7)$
$= (x^2 - 4)(x - 7)$
$= (x + 2)(x - 2)(x - 7)$

49. $p = \dfrac{k}{\sqrt{u}}$
$137,800 = \dfrac{k}{\sqrt{0.05}}$
$k = 137,800\sqrt{0.05}$
$p_2 = \dfrac{137,800\sqrt{0.05}}{\sqrt{0.06}} \approx 125,794$

53. $y = a(x - h)^2 + k$
$4 = a(2 - 1)^2 + 1$
$4 = a + 1$
$a = 3$
$y = 3(x - 1)^2 + 1$
$y = 3(x^2 - 2x + 1) + 1$
$y = 3x^2 - 6x + 4$

54. $f(g(x)) = (\sqrt{x + 3})^2 - 3 = x$
$D = [-3, \infty)$
$g(f(x)) = \sqrt{(x^2 - 3) + 3} = \sqrt{x^2} = |x|$
$D = \mathbb{R}$

---

38. $r^3 = \dfrac{3V}{4\pi}$
$V = \dfrac{4}{3}\pi r^3$
$= \dfrac{4}{3}\pi(11)^3 = \dfrac{5324\pi}{3}$ cm^3
≈ 5575.3 cm^3

39. $I = \sqrt{\dfrac{P}{R}}; I^2 = \dfrac{P}{R}; R = \dfrac{P}{I^2}$
$R = \dfrac{1100}{10^2} = 11$ ohms

40a. $f(2) = \dfrac{\sqrt{2}}{2} \approx 0.71$ sec

40b. $f(5) = \dfrac{\sqrt{5}}{2} \approx 1.12$ sec

40c. $1 = \dfrac{\sqrt{h}}{2}$
$\sqrt{h} = 2$
$h = 4$ ft

40d. $4.40 = \dfrac{\sqrt{h}}{2}$
$\sqrt{h} = 8.8$
$h = 77.44$ ft

❭ C. Exercises

41. $7x + 1 = x^3 + 3x^2 + 3x + 1$
$x^3 + 3x^2 - 4x = 0$
$x(x + 4)(x - 1) = 0$
$x = 0, 1, -4$

42. $\sqrt{n - 3}(\sqrt{n} + \sqrt{n - 3}) = 3$
$\sqrt{n^2 - 3n} + n - 3 = 3$
$(\sqrt{n^2 - 3n})^2 = (6 - n)^2$
$n^2 - 3n = 36 - 12n + n^2$
$9n = 36$
$n = 4$

2.2 Polynomial Functions

Objectives

1. To identify a polynomial function and state its degree

2. To graph and describe the key characteristics of a polynomial function, including domain, range, continuity, intervals for which the function is increasing or decreasing, and whether it is odd, even, or neither

3. To find the zeros of a polynomial function by factoring

4. To apply the Intermediate Value Theorem to approximate zeros of continuous functions

Flash

While video games are often associated with young people, approximately 25% of gamers are at least 50 yr old. Americans spend more than \$25 billion a year on video games (new and used). The United States, China, and Japan top the video game spending list, with Japan spending the most per person (about \$120 per year). By the age of 21, the average gamer in the United States will have spent over 10,000 hr playing video games.

Vocabulary

Intermediate Value Theorem
leading coefficient
Leading Term Test
monomial functions
multiplicity
polynomial function of degree n
relative extrema
relative maximum
relative minimum
standard form
test intervals

Video game sales can be modeled with a polynomial function.

After completing this section, you will be able to

- graph and describe the key characteristics of a polynomial function.
- find zeros of a polynomial function by factoring.
- apply the Intermediate Value Theorem.

Power functions of the form $f(x) = kx^n$ where n is a natural number and constant functions of the form $f(x) = c$ where c is a real number are called *monomial functions*. You learned how to graph transformations of parent power functions of the form $g(x) = a(x - h)^3 + k$ in Section 1.5. Constant functions such as $f(x) = c$, linear functions such as $f(x) = mx + b$, and quadratic functions such as $f(x) = ax^2 + bx + c$ were also studied in Chapter 1. All of these are examples of a larger class of functions called *polynomial functions*.

DEFINITION

A **polynomial function** of degree n is a function that can be written in the form
$p(x) = a_n x^n + a_{n-1} x^{n-1} + a_{n-2} x^{n-2} + \cdots + a_1 x + a_0$ where n is a whole number, $a_n, a_{n-1}, \ldots, a_0 \in \mathbb{R}$, and the *leading coefficient* $a_n \neq 0$. When its terms are written in order of descending degree, a polynomial function is said to be in *standard form*.

Quadratic functions have a degree of 2, linear functions have a degree of 1, and constant functions have a degree of 0 since $a_0 x^0 = a_0(1) = a_0$. The zero function has no degree since $f(x) = 0 = 0x^n$ and a specific power of x cannot be determined.

Example 1 — Identifying Polynomial Functions

State the degree and leading coefficient of each polynomial function. If the function is not a polynomial function, explain why.

a. $f(x) = 2x^3 - 24x^2 + 96x - 127$ **c.** $h(x) = \sqrt{2}x - x^4$

b. $g(x) = 3x^{-1} + x - 5$ **d.** $j(x) = \sqrt{x^2 + 4}$

Answer

a. degree: 3; leading coefficient: 2

b. $g(x)$ is not a polynomial function since the exponent of the first term is not a whole number.

c. $h(x) = -x^4 + \sqrt{2}x$;
degree: 4; leading coefficient: -1

d. $j(x)$ cannot be written as a polynomial. $\left(\sqrt{x^2 + 4} \neq x + 2\right)$

SKILL ✓ EXERCISE 1

TIP

Polynomial functions of degrees 3, 4, and 5 are called cubic, quartic, and quintic functions, respectively.

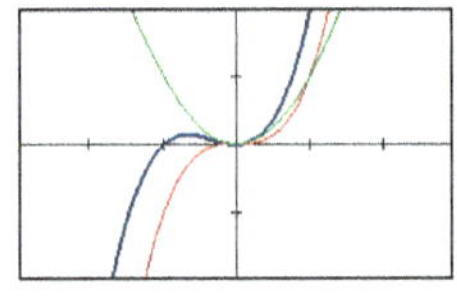

Polynomial functions can be viewed as the sum or difference of one or more monomial functions. The graph of $f(x) = x^3 + x^2$ illustrates that it is not a simple transformation of either $y = x^3$ or $y = x^2$. Like monomial functions, polynomial functions are defined and continuous over the real numbers. Unlike the absolute value function, the graph of a polynomial function is smooth, with no sharp corners.

Another key characteristic of polynomial functions is illustrated by the graph of $f(x) = x^3 + x^2$. The function's end behavior can be predicted using the end behavior of the leading term.

PRESENTATION

Lesson Opener

1. Factor $6x^2 - 13x - 5$. $(2x - 5)(3x + 1)$

Sketch a graph of each function.

2. $y = -2(x + 1)^2$

3. $y = \frac{1}{4}x^3 - 2$

Explain that the transformations of power functions graphed in the Lesson Opener are part of the larger classification of functions called *polynomial functions*. Explain that the definition of a polynomial requires a sum of terms with real number coefficients (either rational or irrational) and nonnegative integral exponents. Power functions with whole number exponents and constant functions (including the zero function) are *monomial functions*. Show how quadratic and linear functions (including the constant function) fulfill the definition of a polynomial function. Note the degree of cubic, quartic, and quintic polynomials and complete Example 1.

While some polynomial functions are transformations of power functions, many are not. All can be viewed as sums of monomial functions and are continuous over the real numbers with smooth turning points. You may want to have the students enter $Y_1 = x^2$, $Y_2 = x^3$, and

The end behavior of a polynomial function
$$p(x) = a_n x^n + a_{n-1} x^{n-1} + a_{n-2} x^{n-2} + \cdots + a_1 x + a_0$$
is determined by its degree, n, and its leading coefficient, a_n.

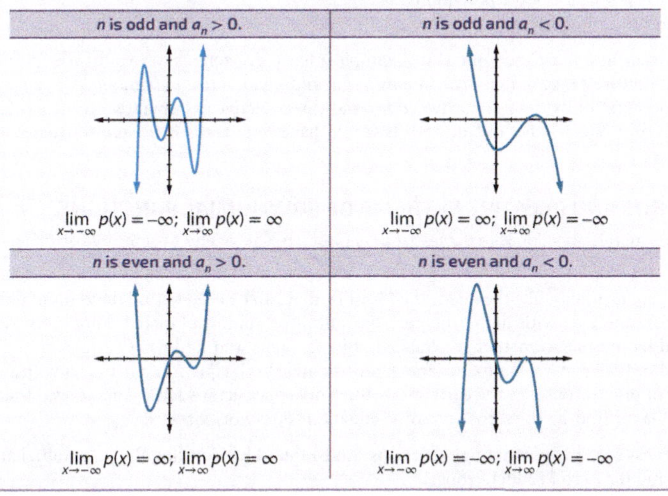

Example 2 Applying the Leading Term Test

Describe the end behavior of each polynomial function.

a. $f(x) = -2x^3 + 5x$

b. $g(x) = -3x^2 + x^4 - 2$

Answer

a. Since $n = 3$ is odd and $a_n = -2 < 0$,
$$\lim_{x \to -\infty} f(x) = \infty \text{ and } \lim_{x \to \infty} f(x) = -\infty.$$

b. Writing the function in standard form, $g(x) = x^4 - 3x^2 - 2$, makes it easy to see that $n = 4$ is even and $a_n = 1 > 0$.
$$\lim_{x \to -\infty} g(x) = \infty \text{ and } \lim_{x \to \infty} g(x) = \infty$$

Check

The graphs of the functions confirm the predicted end behavior.

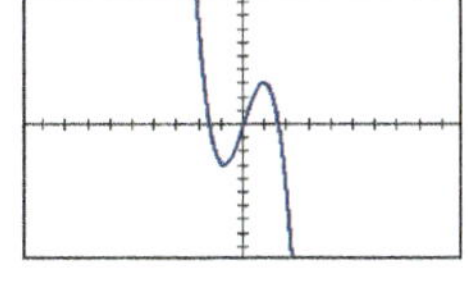 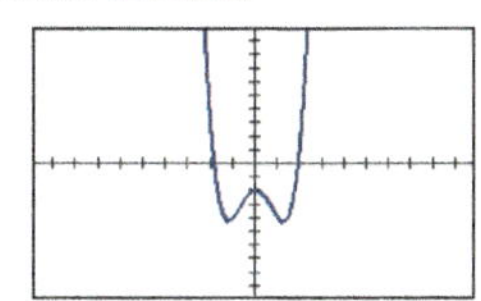

SKILL ✓ EXERCISE 7

Additional Exercises

Use the Leading Term Test and limits to describe each function's end behavior.

1. $p(x) = -3x^7 - 5x^2 + 6$
$\lim_{x \to -\infty} p(x) = \infty,\ \lim_{x \to \infty} p(x) = -\infty$

2. $p(x) = 6x^8 - x^3 + 6$ $\lim_{x \to \pm\infty} p(x) = \infty$

3. Find the zeros of
$p(x) = (x^2 - 25)(x + 2)^2$.
$\pm 5, -2$ (multiplicity 2)

State the maximum number of real zeros and relative extrema for each function. Then find the real zeros.

4. $p(x) = x^5 + 2x^4 + x^3$ $5; 4; x = 0$ (multiplicity 3), -1 (multiplicity 2)

5. $p(x) = 2x^3 - 32x$ $3; 2; x = 0, \pm 4$

6. Use technology to find the zeros and relative extrema of
$p(x) = 8x^3 + 5x^2 - 2$.
$x \approx 0.5$; rel. min.: $(0, -2)$
rel. max.: $\approx (-0.4, -1.7)$

Identify the function's zeros, y-intercepts, and end behavior. Then sketch its graph.

7. $p(x) = 2x^3 - 12x^2 + 18x$
zeros: $0, 3$ (multiplicity 2)
y-int.: $(0, 0)$
$\lim_{x \to -\infty} p(x) = -\infty,\ \lim_{x \to \infty} p(x) = \infty$

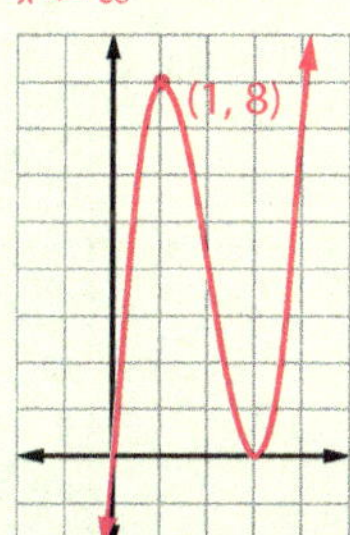

$Y_3 = x^3 + x^2$ and then examine the table of values.

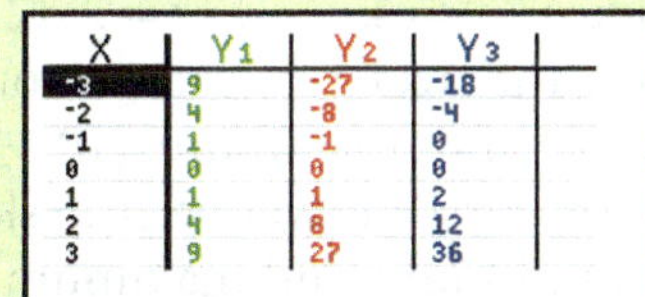

Introduce the *Leading Term Test* by explaining that the term of highest degree eventually grows so much faster than the other terms that it tends to dominate the behavior of the function. This can be illustrated by zooming out of a graphed function until the graph resembles the graph of the leading term. Encourage the students to explore the end behavior in each of the four cases using actual values of the first term as $x \to \pm\infty$ before describing the characteristics of the graph.

Some students may benefit from the following shorthand notation used in BJU Press's *Algebra 2* to describe end behavior.

n is odd.	$a_n > 0$	(down, up)
	$a_n < 0$	(up, down)
n is even.	$a_n > 0$	(up, up)
	$a_n < 0$	(down, down)

Apply the Leading Term Test using the two functions in Example 2. Encourage students to investigate end behaviors before graphing a function with technology.

One-on-One Some students may recall the Leading Term Test more easily by visualizing the characteristics of the parent functions $y = x^2$ and $y = x^3$ and their reflections.

Discuss the equivalent implications of $(c, 0)$ being an x-intercept of $p(x)$. Note that we cannot assume the converse of this conditional since polynomials can have a complex zero (discussed in Section 2.4). If c is a real zero, the four statements relating x-intercepts, roots, factors, and zeros are equivalent (all true if one is true, and all false if one is false).

8. $p(x) = -3x^4 - 2x^3 + 5x^2$

zeros: $-\frac{5}{3}$, 0 (multiplicity 2), 1
y-int.: $(0, 0)$
$\lim_{x \to -\infty} p(x) = -\infty$, $\lim_{x \to \infty} p(x) = -\infty$

Use the Intermediate Value Theorem to complete exercises 9–10.

9. Show that there is a zero of $p(x) = x^4 - 4x^2 + 2x + 1$ within the interval $[-1, 0]$. The sign change between $p(-1) = -4$ and $p(0) = 1$ implies there must be a zero within $[-1, 0]$.

10. Determine between which consecutive integers the real zeros occur in $p(x) = -4x^3 + 3x^2 + 5x - 3$.
Real zeros occur between -2 and -1, 0 and 1, and 1 and 2.

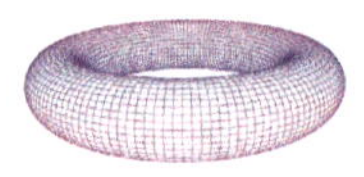

In addition to modeling data sets, quartic equations are used by robotic milling machines where the intersection of a line and a torus must be computed.

KEYWORD SEARCH

Niels Abel

The zeros and number of turning points are key characteristics of the graph of a polynomial. If $(c, 0)$ is an x-intercept of the function's graph, the following statements are also true:

$x = c$ is a solution to the equation $p(x) = 0$,
$(x - c)$ is a factor of the polynomial $p(x)$,
and c is a zero of $p(x)$.

Note that the cubic function $f(x)$ in Example 2 has three zeros and two turning points, called *relative extrema* (one *relative minimum* and one *relative maximum*). The quartic function $g(x)$ in Example 2 has two zeros and three relative extrema (two relative minimums and one relative maximum). It would have four zeros if it were translated up 3 units.

ZEROS AND RELATIVE EXTREMA OF POLYNOMIAL FUNCTIONS
A polynomial function with degree n has at most n real zeros and $n - 1$ relative extrema.

Are there formulas, like the quadratic formula, that can be used to find the zeros of polynomial functions with higher degree? Formulas for cubic and quartic functions were found by the Italian mathematicians Niccolò Tartaglia and Lodovico Ferrari and published in 1545 by Gerolamo Cardano. Unfortunately both equations are too complicated to be of practical use. In the early 1800s the Norwegian genius Niels Abel proved that a formula for finding the zeros of quintic equations does not exist.

The zeros of many polynomial functions can be found by factoring the polynomial and applying the Zero Product Property.

Example 3 Finding Zeros by Factoring

State the maximum number of real zeros and relative extrema for $p(x) = -2x^3 + 4x^2$. Then find the real zeros.

Answer

A cubic function has at most 3 real zeros and 2 relative extrema.

$-2x^3 + 4x^2 = 0$
$-2x^2(x - 2) = 0$
$-2x^2 = 0$ or $x - 2 = 0$
$x = 0$ $x = 2$

Check

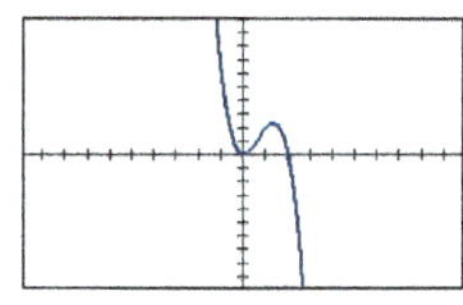

1. A polynomial function with degree n has at most n real zeros and $n - 1$ relative extrema.
2. Set $p(x) = 0$.
3. Factor.
4. Apply the Zero Product Property.
5. Solve the resulting equations.

6. Graphing the function confirms the zeros.

SKILL ✔ EXERCISE 19

Discuss the number of real zeros and *relative extrema* in a polynomial function. In Section 2.4 students will learn that an nth degree polynomial function always has n roots, with any other roots being complex.

Explain that finding the zeros of polynomial functions of a degree higher than 2 can be difficult since there are no simple, generalized formulas as there are for quadratic functions.

After mentioning Abel's proof, explain that proving that something does not exist is very difficult in all fields, not just in mathematics. Consider discussing atheists' attempts to prove that God does not exist. Atheists exercise blind faith (lacking proof) when they believe that God does not exist.

Example 3 illustrates how some polynomial functions can be factored and solved by applying the Zero Product Property.

Example 4 demonstrates how technology can be used to find decimal approximations of zeros and relative extrema. The calculator displays an approximation of 0 in scientific notation. For example, the y-value of $1\text{E}-12$ should be interpreted as 0. Note that exact zeros can be found by factoring the quartic polynomial in quadratic form.

Motivational Idea You may want to explain that in calculus the zeros of the first derivative $p'(x) = 4x^3 - 14x = 2x(2x^2 - 7)$ produces the exact x-values for local minimums and maximums $(0, \pm\sqrt{3.5})$ because the first derivative (the slope) is 0 at the minimums and maximums.

Discuss the effects of odd and even multiplicities of repeated zeros on the function's graph. Consider $x^3 - 5x^2 - 8x + 48 = (x + 3)(x - 4)^2$. Any factor whose exponent is even, such as $(x - 4)^2$, will touch the x-axis at the zero but will not cross it. This can be seen algebraically since there must be a sign change for the function to cross the x-axis, but a factor

If the completely factored form of $p(x)$ contains a repeated factor $(x - c)^m$, c is a repeated zero with a *multiplicity* of m. When m is odd, the graph crosses the x-axis at $(c, 0)$, but when m is even, the graph touches the x-axis at $(c, 0)$ without crossing the axis. The graph of $p(x) = (x - 2)^3(x + 1)^2$ has a repeated zero of 2 with a multiplicity of 3 and a repeated zero of -1 with a multiplicity of 2.

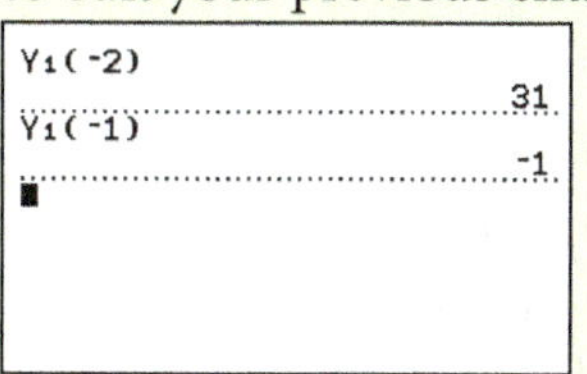

Use technology to find the real zeros and relative extrema for $p(x) = x^4 - 7x^2 + 6$.

Answer

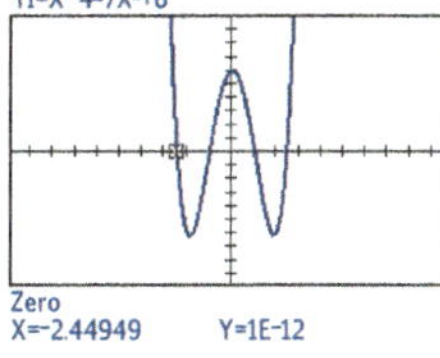

zeros at $x \approx \pm 2.45$ and $x = \pm 1$

rel. max.: $(0, 6)$
rel. min.: $\approx (-1.87, -6.25)$
$\approx (1.87, -6.25)$

Check

$(x^2)^2 - 7(x^2) + 6 = 0$
$(x^2 - 6)(x^2 - 1) = 0$
$x^2 - 6 = 0 \quad$ or $\quad x^2 - 1 = 0$
$x^2 = 6 \qquad\qquad x^2 = 1$
$x = \pm\sqrt{6} \approx \pm 2.45 \qquad x = \pm 1$

1. Graph the function in a window that illustrates the function's x-intercepts and relative extrema.

2. Select the zero function from the [CALC] menu. Then respond to the prompts to enter a search interval and an estimated zero. Repeat the process to find all the zeros.

3. Select the maximum or minimum function from the [CALC] menu and enter a search interval and an estimate for the x-values of the local extrema.

4. Factoring and applying the Zero Product Property produces exact values for the zeros and verifies the approximations found using technology.

SKILL ✔ EXERCISE 25

Polynomials that can be written as $au^2 + bu + c$, such as $p(x)$ in Example 4 where $u = x^2$, are said to be in quadratic form.

Since a polynomial function is continuous, the only place the value of the function can change signs from positive to negative or from negative to positive is at a zero of the function.

Placing the distinct real zeros on the x-axis divides the function's domain into *test intervals* in which the function is always positive or always negative. You can sketch the basic shape of a polynomial function by determining its end behavior, finding the real zeros and their multiplicity, and plotting a few additional points.

with an even power is always positive. Consider guiding the students to discover this truth.

Use Example 5 to illustrate how the end behavior, zeros, *test intervals*, and *multiplicity* of roots (including a multiplicity of 1) can be used to sketch the graph.

The last example shows how a special case of the *Intermediate Value Theorem* can be used to approximate zeros. The proof of this theorem is presented in calculus courses. Explain that this technique can be used to estimate the location of the zeros as precisely as desired.

TIPS

Ex. 42 The solution to part *b* shows that there are no other points of intersection in the locality of $x = 2$ and that the line is also tangent to the curve at $x = 2$.

For Graphing Calculators

There are several ways to evaluate functions with your graphing calculator. Begin by entering the function rule for Additional Exercise 10 as Y_1.

1. On the home screen, enter Y_1 using [VARS], Y-VARS, 1:Function, [1], and then enter the value to be evaluated within parentheses. Use [2nd], [ENTER] to edit your previous entry.

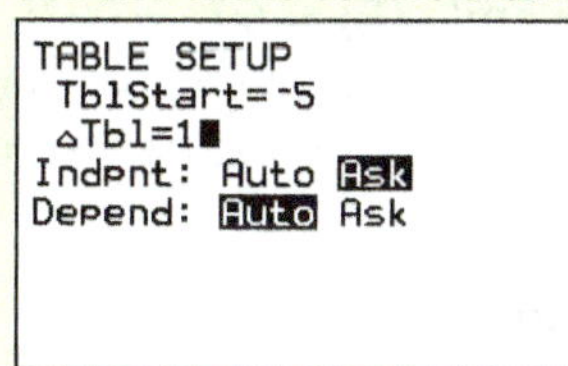

2. The starting value and increment between x-values displayed by the TABLE function can be adjusted using TBLSET ([2nd], π). You can also set the table to the Ask mode.

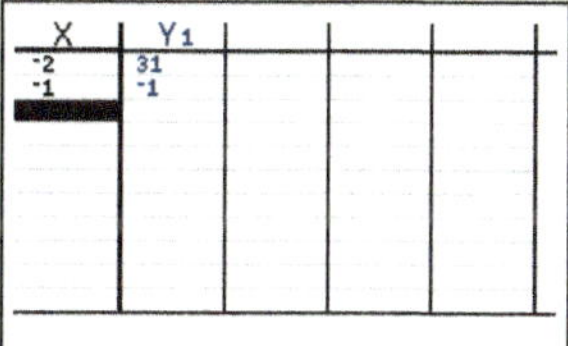

Then access the table ([2nd], [GRAPH]) and enter each value to be evaluated.

Assignments

- **Minimum:** 1–2, 5, 7–8, 10–11, 14–17, 20, 22, 25, 28, 35, 37, 42; 43–49 odd; 50
- **Standard:** 1, 3, 6, 9, 12–13, 15–20, 24, 26, 28, 31, 34–39, 44, 47, 49–50
- **Extended:** 5, 8, 11, 14–16, 18, 21–23, 27, 30–40, 42, 49–52

Assessment

- Quiz 2A covers Sections 2.1–2.2.

> ### A. Exercises

1a. degree 5; $a_n = 7$

1b. It is not a polynomial function since the exponent $-1 \notin \mathbb{W}$.

1c. $f(x) = -6x^4 + 2x$
degree 4; $a_n = -6$

1d. $f(x) = 28x^0 = 28(1) = 28$
degree 0; $a_n = 28$

10. $\pm 4, \pm 2$

11. 6 (multiplicity 4)

12. 0 (multiplicity 2)

13. 4 (multiplicity 2)
-1 (multiplicity 3)

14. $p(-1) = 1$ and $p(0) = -2$
The sign change between $x = -1$ and $x = 0$ implies there must be a zero within $[-1, 0]$.

15. $p(1) = -3$ and $p(2) = 69$
The sign change between $x = 1$ and $x = 2$ implies there must be a zero within $[1, 2]$.

> ### B. Exercises

16. 2; 1
$$(x - 4)(x - 7) = 0$$
$$x = 4, 7$$

17. 2; 1
$$4(x^2 - 2x + 1) = 0$$
$$4(x - 1)^2 = 0$$
$$x = 1$$

18. 4; 3
$$3x^2(x^2 - 4) = 0$$
$$x = 0, \pm 2$$

19. 3; 2
$$x(16x^2 + 8x + 1) = 0$$
$$x(4x + 1)^2 = 0$$
$$x = 0, -\frac{1}{4}$$

20. 2; 1
$$(2x - 3)(x + 2) = 0$$
$$x = \frac{3}{2}, -2$$

21. 5; 4
$$x(x^4 - 2x^2 + 1) = 0$$
$$x(x^2 - 1)(x^2 - 1) = 0$$
$$x = 0 \quad \text{or} \quad x^2 - 1 = 0$$
$$x = 0, \pm 1$$

22. 4; 3
$$(3x^2 - 4)(x^2 - 2) = 0$$
$$3x^2 - 4 = 0 \quad \text{or} \quad x^2 - 2 = 0$$
$$x^2 = \frac{4}{3} \qquad\qquad x^2 = 2$$
$$x = \pm\sqrt{\frac{4}{3}} \qquad x = \pm\sqrt{2}$$
$$x = \pm\frac{2\sqrt{3}}{3}$$

23. 5; 4
$$x(4x^2 - 9)(x^2 - 1) = 0$$
$$x = 0 \text{ or } 4x^2 - 9 = 0 \text{ or } x^2 - 1 = 0$$
$$x^2 = \frac{9}{4} \qquad\qquad x^2 = 1$$
$$x = \pm\frac{3}{2} \qquad\qquad x = \pm 1$$

24. zeros: $-3, 0, 2$
rel. min.: $\approx (-1.79, -8.21)$
rel. max.: $\approx (1.12, 4.06)$

25. zeros: $-2, \frac{1}{2}, 5$
rel. min.: $\approx (3.21, -50.55)$
rel. max.: $\approx (-0.88, 18.18)$

26. zeros: $-1.75, 3.44$
rel. min.: $\approx (0.09, 2.95)$
rel. max.: $\approx (-1.07, 6.03)$,
$\approx (2.47, 23.95)$

27. zeros: -0.5
rel. min.: $(-0.5, 0), \approx (1.35, 7.46)$
rel. max.: $\approx (0.65, 10.29)$

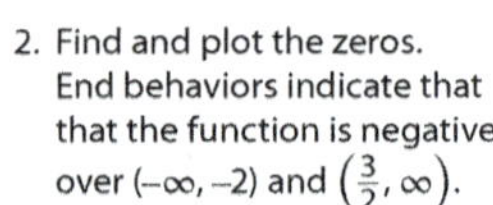

Sketch the graph of $p(x) = -2x^4 - x^3 + 6x^2$.

Answer

Since n is even and $a_n < 1$,
$$\lim_{x \to -\infty} p(x) = -\infty \text{ and}$$
$$\lim_{x \to \infty} p(x) = -\infty.$$

$$-x^2(2x^2 + x - 6) = 0$$
$$-x^2(2x - 3)(x + 2) = 0$$
$$x = 0, \frac{3}{2}, -2$$

x	-1	1
y	5	3

1. Use the Leading Term Test to determine the function's end behavior.

2. Find and plot the zeros. End behaviors indicate that that the function is negative over $(-\infty, -2)$ and $\left(\frac{3}{2}, \infty\right)$.

3. The multiplicity of roots implies that the function is positive over $(-2, 0)$ and $\left(0, \frac{3}{2}\right)$.

Plot at least one point between each zero and sketch the graph.

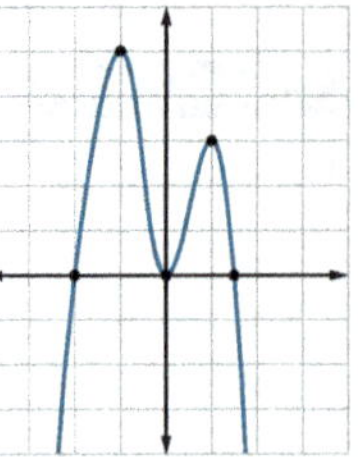

— SKILL ✔ **EXERCISE 29**

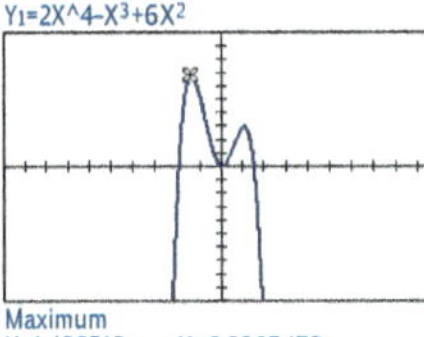

The even multiplicity of the repeated root 0 implies that $(0, 0)$ is a relative minimum of the function. We cannot assume the additional points $(-1, 5)$ and $(1, 3)$ are relative maximums, but the function must contain these points. Technology can be used to obtain a more accurate graph and to find the relative maximums of approximately $(-1.43, 6.83)$ and $(1.05, 3.03)$. Calculus provides a direct method of locating local extrema.

We have seen that a sign change in the value of the function indicates the presence of a zero. This statement is a specific application of the more general Intermediate Value Theorem.

▌**INTERMEDIATE VALUE THEOREM**

If a function is continuous over $[a, b]$ and there exists a value n that is between $f(a)$ and $f(b)$, then c exists in $[a, b]$ such that $f(c) = n$.

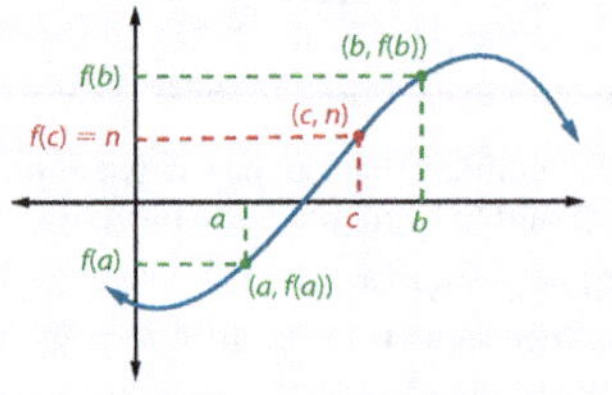

This theorem guarantees that the function includes every value between $f(a)$ and $f(b)$ in the interval $[a, b]$. When either $f(a)$ or $f(b)$ is negative and the other is positive, applying this theorem with $n = 0$ guarantees the presence of c such that $f(c) = 0$, a real root.

Use the Intermediate Value Theorem to determine which consecutive integers the real zeros of $p(x) = 3x^3 - 2x^2 - 6x + 4$ occur between.

Answer

x	-2	-1	0	1	2
$p(x)$	-16	5	4	-1	8

Real zeros occur between
a. -2 and -1,
b. 0 and 1, and
c. 1 and 2.

There are no other zeros.

1. Evaluate the polynomial at several points.

2. Apply the Intermediate Value Theorem where
   a. $p(-2)$ is negative and $p(-1)$ is positive,
   b. $p(0)$ is positive and $p(1)$ is negative, and
   c. $p(1)$ is negative and $p(2)$ is positive.

3. We have located 3 zeros, the maximum number for a cubic polynomial function.

Simon Stevin, a Dutch scientist born in 1548, was the first proponent of a decimal system. He developed a decimal expansion algorithm to find the roots of any polynomial. This same algorithm (in binary form) was used by Augustin-Louis Cauchy in his proof of the Intermediate Value Theorem.

—— SKILL ✓ **EXERCISES 15, 33**

Note that the lack of a sign change between two values does not guarantee the absence of real zeros. The function could cross the x-axis twice within the defined interval.

❯ A. Exercises

1. State the degree and leading coefficient of each polynomial function. If the function is not a polynomial function, explain why.
   a. $f(x) = 7x^5 + 3x^3 - 4x$ b. $f(x) = 2x^{-1} + 1$
   c. $f(x) = 2x - 6x^4$ d. $f(x) = 28$

State whether the degree, n, of the graphed polynomial is odd or even and whether its lead coefficient, a_n, is positive or negative.

2. 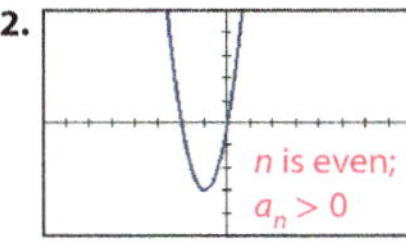

3. 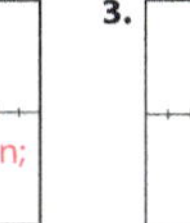

4. 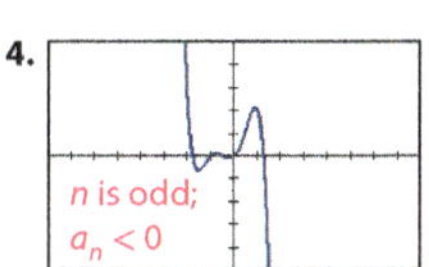

5. 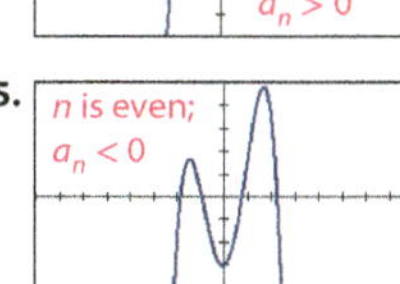

Use the Leading Term Test to describe each function's end behavior.

6. $p(x) = -x^2 + 3x + 1$ 7. $p(x) = -2x^5 - 3x$
8. $p(x) = x^3 - x^2 + 4x + 2$ 9. $p(x) = 3x^4 + 6x^2 - 4$

6. $\lim\limits_{x \to -\infty} p(x) = -\infty; \ \lim\limits_{x \to \infty} p(x) = -\infty$
7. $\lim\limits_{x \to -\infty} p(x) = \infty; \ \lim\limits_{x \to \infty} p(x) = -\infty$
8. $\lim\limits_{x \to -\infty} p(x) = -\infty; \ \lim\limits_{x \to \infty} p(x) = \infty$
9. $\lim\limits_{x \to -\infty} p(x) = \infty; \ \lim\limits_{x \to \infty} p(x) = \infty$

Find the zeros for each polynomial function, including the multiplicity of any repeated zeros.

10. $p(x) = (x^2 - 16)(x^2 - 4)$ 11. $p(x) = (x - 6)^4$
12. $p(x) = x^2$ 13. $p(x) = (x - 4)^2(x + 1)^3$

Use the Intermediate Value Theorem to show that there is a zero within the given interval.

14. $p(x) = -x^4 + 5x^2 + x - 2; \ [-1, 0]$
15. $p(x) = x^5 + 3x^4 - x^3 + x^2 - 7; \ [1, 2]$

❯ B. Exercises

State the maximum number of real zeros and relative extrema for each function. Then find the real zeros.

16. $p(x) = x^2 - 11x + 28$ 17. $p(x) = 4x^2 - 8x + 4$
18. $p(x) = 3x^4 - 12x^2$ 19. $p(x) = 16x^3 + 8x^2 + x$
20. $p(x) = 2x^2 + x - 6$ 21. $p(x) = x^5 - 2x^3 + x$
22. $p(x) = 3x^4 - 10x^2 + 8$ 23. $p(x) = 4x^5 - 13x^3 + 9x$

Use technology to graph each function in a window that shows all the zeros and relative extrema. Then state each zero and any relative extrema (to the nearest hundredth).

24. $p(x) = -x^3 - x^2 + 6x$
25. $p(x) = 2x^3 - 7x^2 - 17x + 10$
26. $p(x) = -x^4 + 2x^3 + 5x^2 - x + 3$
27. $p(x) = 8x^4 - 16x^3 - 2x^2 + 14x + 5$

30. $p(x) = \frac{1}{2}x(x^2 - 4)$
$\frac{1}{2}x = 0$ or $x^2 = 4$
zeros: $0, \pm 2$
y-int.: $(0, 0)$
$\lim\limits_{x \to -\infty} p(x) = -\infty, \ \lim\limits_{x \to \infty} p(x) = \infty$

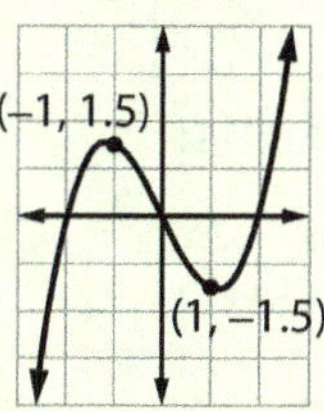

31. $p(x) = (x^2 - 2)(x^2 - 4)$
$x^2 = 2$ or $x^2 = 4$
zeros: $\pm\sqrt{2}, \pm 2$
y-int.: $(0, 8)$
$\lim\limits_{x \to \pm\infty} p(x) = \infty$

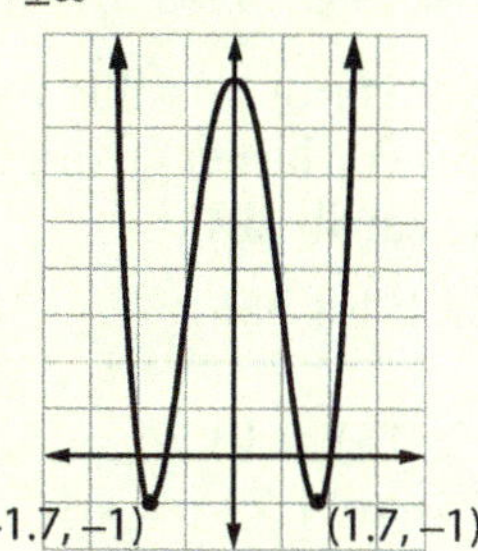

32.

x	-5	-4	-3	-2	-1	0	1
y	-1	15	17	11	3	-1	5

Real zeros occur between -5 and -4, -1 and 0, and 0 and 1.

33.

x	-1	0	1	2	3	4
y	5	-2	-1	2	1	-10

Real zeros occur between -1 and 0, 1 and 2, and 3 and 4.

28. $p(x) = x^2(x^2 - 4)$
$x^2 = 0$ or $x^2 = 4$
zeros: $0, \pm 2$
y-int.: $(0, 0)$
$\lim\limits_{x \to \pm\infty} p(x) = \infty$

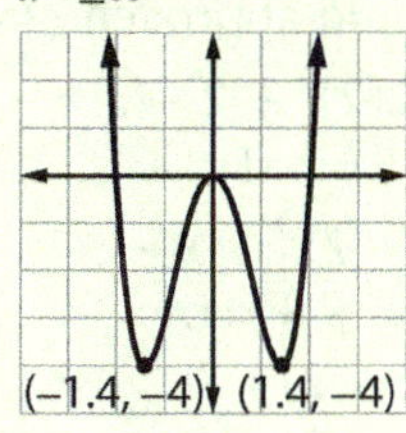

29. $p(x) = x(2x + 3)(x - 2)$
zeros: $-\frac{3}{2}, 0, 2$
y-int.: $(0, 0)$
$\lim\limits_{x \to -\infty} p(x) = -\infty, \ \lim\limits_{x \to \infty} p(x) = \infty$

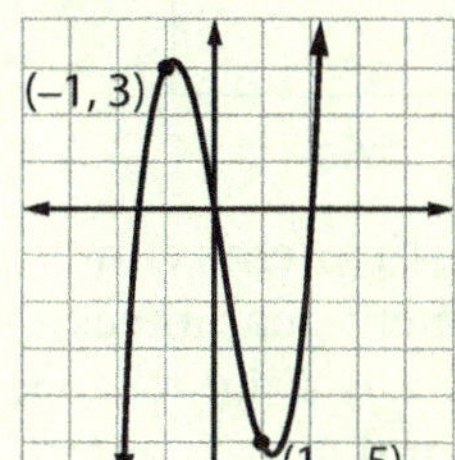

34. $Y_1 = -37.5X^4 + 452X^3 - 1846X^2 + 3150X - 1...$

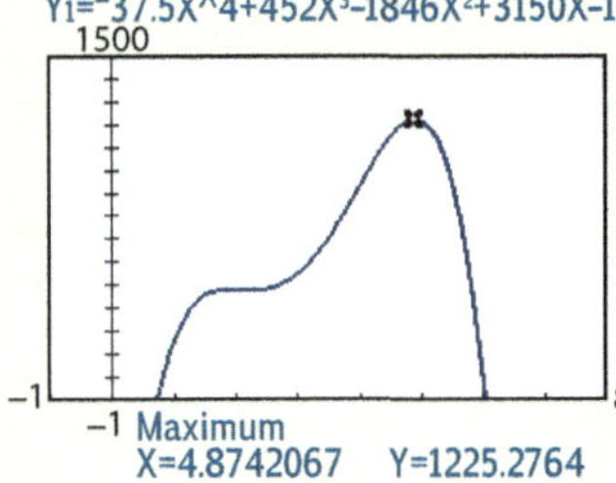

35a. $V = x(18 - 2x)(20 - 2x)$
$V = 360x - 76x^2 + 4x^3$

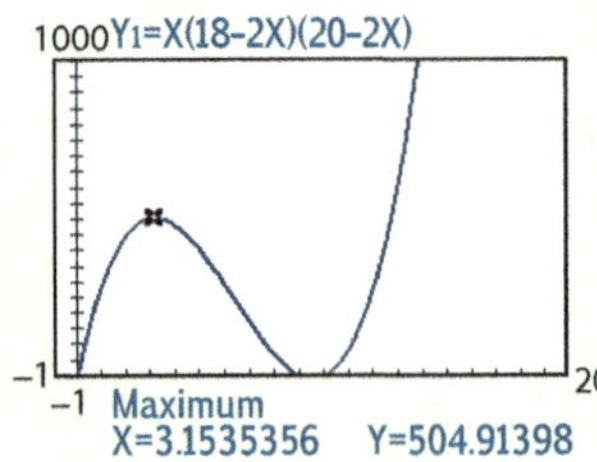

35b. $D = (0, 9)$; It is not possible to cut 9 in. squares (or larger) from both sides of the cardboard.

35c. See the graph for 35a.
$x \approx 3.15$
$l \approx 20 - 2(3.15) \approx 13.7$
$18 - 2(3.15) \approx 11.7$

36a. The graph is symmetric with respect to the y-axis.

36b. The graph is symmetric with respect to the origin.

36c. The graph is not symmetric with respect to the origin or the y-axis.

37a. $p(-x) = 3(-x)^5 - 4(-x)^2 - 5$
$= -3x^5 - 4x^2 - 5 \neq p(x)$
$\qquad\qquad\qquad$ or $-p(x)$

37b. $p(-x) = 3(-x)^6 - 4(-x)^4 - 5(-x)^2$
$= 3x^6 - 4x^4 - 5x^2 = p(x)$

37c. $p(-x) = 3(-x)^7 - 4(-x)^5 - 5(-x)^3$
$= -3x^7 + 4x^5 + 5x^3 = -p(x)$

38a. Every term (including the constant term) has an even exponent.

38b. Every term has an odd exponent.

❯ C. Exercises

39. $x^{n-1}(x - 1) = 0$
$x^{n-1} = 0$ and $x - 1 = 0$
$x = 0, 1$

40a. $p(0) = 1, p(1) = 1$; No; both values are positive.

Identify the function's zeros, y-intercepts, and end behavior. Then sketch its graph.

28. $p(x) = x^4 - 4x^2$
29. $p(x) = 2x^3 - x^2 - 6x$
30. $p(x) = \frac{1}{2}x^3 - 2x$
31. $p(x) = x^4 - 6x^2 + 8$

Use the Intermediate Value Theorem to determine which consecutive integers the real zeros occur between.

32. $p(x) = x^3 + 5x^2 - 1$

33. $p(x) = -x^3 + 4x^2 - 2x - 2$

34. Sales in the video game market from August 2015 through January 2016 can be modeled by $p(x) = -37.5x^4 + 452x^3 - 1846x^2 + 3150x - 1447$ where x is the month, beginning with August, and y is the sales in millions of dollars.
   a. According to the model, which month had the highest sales? November (month 4.87)
   b. What was the dollar amount for these sales? $1.225 billion

35. A box without a lid is formed by cutting squares of length x in. from each corner of an 18 in. by 20 in. sheet of cardboard.

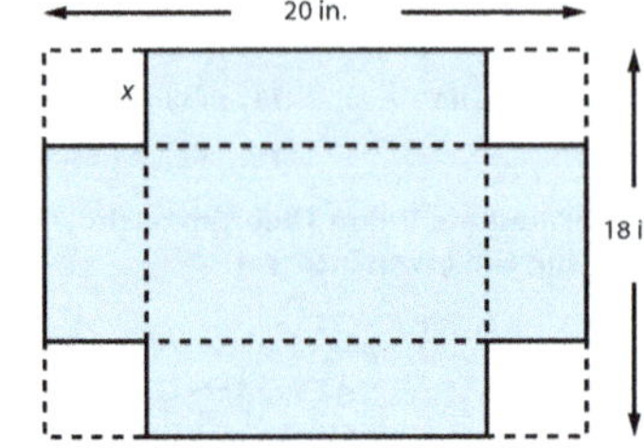

   a. Find the polynomial function that models the volume of the box and graph the function.
   b. Considering the context, state a reasonable domain for the function. Explain your reasoning.
   c. Find the maximum volume for the box and the corresponding dimensions (to the nearest tenth of an inch). 504.9 in³; 13.7 in. × 11.7 in. × 3.2 in.

36. Explore: Graph each function using technology and examine its symmetry to classify it as even, odd, or neither.
   a. $p(x) = 3x^4 - 4x^2 - 5$ even
   b. $p(x) = 3x^5 - 4x^3 - 5x$ odd
   c. $p(x) = 3x^4 - 4x - 5$ neither

40b.

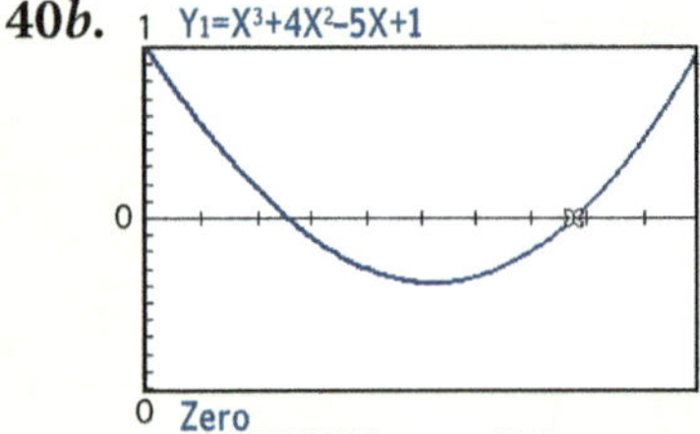

40c. No; While the theorem's conditional statement guarantees a zero when there is a sign change between $p(a)$ and $p(b)$, it does not address the presence or absence of zeros if both $p(a)$ and $p(b)$ have the same sign.

---

37. Explore: Examine $p(-x)$ for each function to classify the function as even, odd, or neither.
   a. $p(x) = 3x^5 - 4x^2 - 5$ neither
   b. $p(x) = 3x^6 - 4x^4 - 5x^2$ even
   c. $p(x) = 3x^7 - 4x^5 - 5x^3$ odd

38. Analyze: Use the results of exercises 36–37 to make a conjecture describing the terms in each type of polynomial function.
   a. an even polynomial function
   b. an odd polynomial function

❯ C. Exercises

39. Determine the zeros of the binomial function $p(x) = x^n - x^{n-1}$. Then state the multiplicity of each zero. 0 (multiplicity $n - 1$), 1 (multiplicity 1)

40. Explore: Consider $p(x) = x^3 + 4x^2 - 5x + 1$.
   a. Evaluate $p(0)$ and $p(1)$. Does the Intermediate Value Theorem predict a zero within $[0, 1]$?
   b. Use technology to graph the function and find any zeros within $[0, 1]$. $x \approx 0.26, 0.78$
   c. Do the results of steps a and b demonstrate an inconsistency in the Intermediate Value Theorem? Explain why or why not.

41. Explain: Is it possible for a quartic function to have just two relative extrema? Explain your answer.

42. A method for finding the tangent to a curve at a point is learned in calculus. Consider $p(x) = -x^3 + 2x^2 + x - 3$ and the line $f(x) = -3x + 5$.

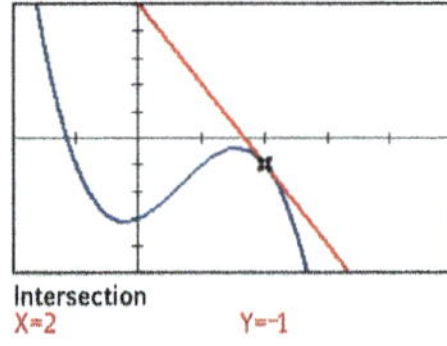

   a. Algebraically verify that $(2, -1)$ is a solution to both equations.
   b. Algebraically find a second point of intersection of $f(x)$ and $p(x)$.
   c. Use technology to verify the second point of intersection.
   d. Use the two points of intersection to verify the slope of the secant line that represents the curve's average rate of change between the two points.

41. No; Since the end behaviors of a quartic must match, it can have only 1 or 3 relative extrema (turning points). If it had just two relative extrema, one end would approach ∞ and the other would approach $-\infty$.

42a. $f(2) = -(2)^3 + 2(2)^2 + 2 - 3 = -1$
$g(2) = -3(2) + 5 = -1$

42b. $-3x + 5 = -x^3 + 2x^2 + x - 3$
$x^3 - 2x^2 - 4x + 8 = 0$
$x^2(x - 2) - 4(x - 2) = 0$
$(x^2 - 4)(x - 2) = 0$
$x = \pm 2$
$f(-2) = -(-2)^3 + 2(-2)^2$
$\qquad\qquad + (-2) - 3 = 11$

Multiply. [Algebra]

43. $(x - 3)(x^2 + 3x + 9)$
$x^3 - 27$

44. $(x^2 - 9)(x^2 - 3x + 9)$
$x^4 - 3x^3 + 27x - 81$

Factor. [Algebra]

45. $2x^2 + 9x - 18$
$(2x - 3)(x + 6)$

46. $x^3 + 3x^2 - 9x - 27$
$(x - 3)(x + 3)^2$

Divide. [Algebra]

47. $(x^3 + 9x^2 + 29x + 66) \div (x + 6)$ $x^2 + 3x + 11$

48. $(-x^3 - 3x + 18) \div (x - 3)$ $-x^2 - 3x - 12$ R. -18

49. If $f(x) = \dfrac{5}{x - 4}$, find $\lim\limits_{x \to 4^-} f(x)$ and $\lim\limits_{x \to 4^+} f(x)$. [1.3] C

A. ∞, ∞ **D.** $-\infty, -\infty$

B. $\infty, -\infty$ **E.** undefined

C. $-\infty, \infty$

50. Which of the following functions is *not* a power function? [1.4] D

A. $y = \sqrt{x}$ **D.** $y = 5^x$

B. $y = -\dfrac{3}{5x^4}$ **E.** $y = -6x^{-3}$

C. $y = x$

51. Given $h(x) = \dfrac{1}{\sqrt{x^2 - 4}}$, find $f(x)$ and $g(x)$ such that $h(x) = (f \circ g)(x)$. [1.7] D

A. $f(x) = \dfrac{1}{\sqrt{x - 4}}$; $g(x) = x^2$

B. $f(x) = \dfrac{1}{\sqrt{x}}$; $g(x) = x^2 - 4$

C. $f(x) = \dfrac{1}{x}$; $g(x) = \sqrt{x^2 - 4}$

D. all of these

E. none of these

52. Which equation represents the inverse of $f(x) = \sqrt[5]{x^2}$ where $x \geq 0$? [2.1] E

A. $y = \dfrac{5}{2}\sqrt{x}$

B. $y = \sqrt{x^{\frac{2}{5}}}$

C. $y = x^{\frac{2}{5}}$

D. $y = x^{-\frac{2}{5}}$

E. $y = x^{\frac{5}{2}}$

Your graphing calculator may have several limitations when graphing functions. Graphing $f(x) = \dfrac{x^4 - 2x^3 - x^2 + 2x}{2x^2 + 2x - 4} = \dfrac{x(x - 1)(x + 1)(x - 2)}{2(x - 1)(x + 2)}$ in the standard window is not sufficient to illustrate the entire graph. Increasing the range of the window to [−10, 30] reveals the true shape of the graph.

When graphing rational functions containing discontinuities, the calculator will not draw vertical asymptotes. In Connected mode, some calculators may connect plotted points on either side of a vertical asymptote in an attempt to produce a smooth curve. For example, notice the nearly vertical line drawn near $x = -2$. This line is often misinterpreted as an asymptote.

The point discontinuity at $x = 1$ may not be accurately represented in the graph. Examining the table of functional values, an ERROR occurs at each attempted division by 0 and indicates either a vertical asymptote (at $x = -2$) or a point discontinuity (at $x = 1$). The exact number of zeros may be difficult to determine from a graph unless you zoom in. The magnified view helps define the search interval when using the zero function found on the CALC menu. The table also confirms zeros at $x = -1$, 0, and 2.

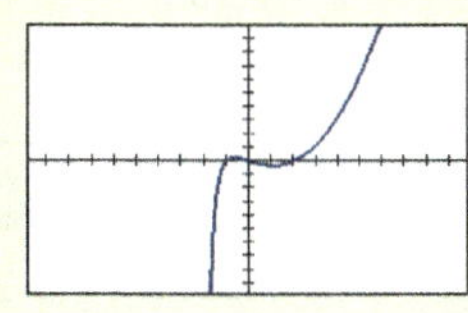

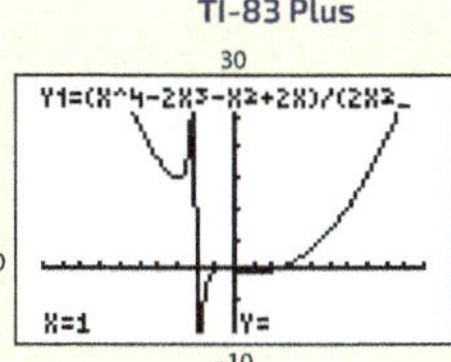

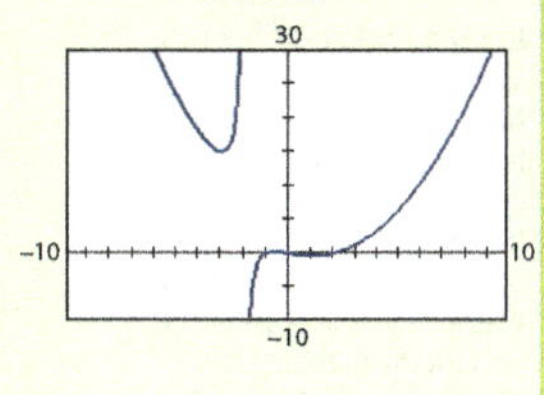
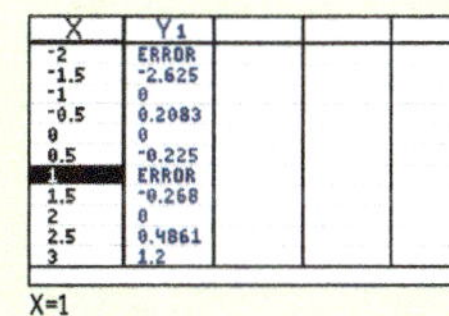
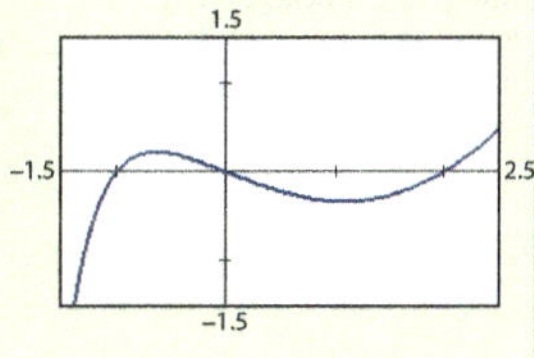

PRESENTATION

There are many graphing technologies, including calculators, online applications, and mobile device apps. Each tool has unique limitations, and many have difficulty displaying point discontinuities. This Technology Corner increases student awareness of these limitations and provides guidance in navigating these challenges.

42c.

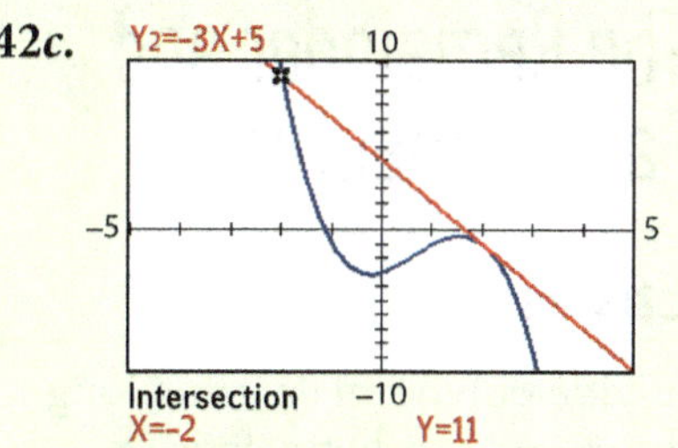

42d. $m = \dfrac{\Delta y}{\Delta x} = \dfrac{11 + 1}{-2 - 2} = -\dfrac{3}{1}$

Cumulative Review

43. $x^3 + 3x^2 + 9x - 3x^2 - 9x - 27$
$= x^3 - 27$

44. $x^4 - 3x^3 + 9x^2 - 9x^2 + 27x - 81$
$= x^4 - 3x^3 + 27x - 81$

45. $ac = -36$ and $b = -3 + 12 = 9$
$2x^2 - 3x + 12x - 18$
$= x(2x - 3) + 6(2x - 3)$
$= (2x - 3)(x + 6)$

46. $x^2(x + 3) - 9(x + 3)$
$= (x^2 - 9)(x + 3)$
$= (x - 3)(x + 3)^2$

47.
$$
\begin{array}{r}
x^2 + 3x + 11 \\
x + 6 \overline{)\ x^3 + 9x^2 + 29x + 66} \\
-(x^3 + 6x^2) \\
\hline
3x^2 + 29x \\
-(3x^2 + 18x) \\
\hline
11x + 66 \\
-(11x + 66) \\
\hline
0
\end{array}
$$

48.
$$
\begin{array}{r}
-x^2 - 3x - 12 \\
x - 3 \overline{)\ -x^3 + 0x^2 - 3x + 18} \\
-(-x^3 + 3x^2) \\
\hline
-3x^2 - 3x \\
-(-3x^2 + 9x) \\
\hline
-12x + 18 \\
-(-12x + 36) \\
\hline
-18
\end{array}
$$

49.

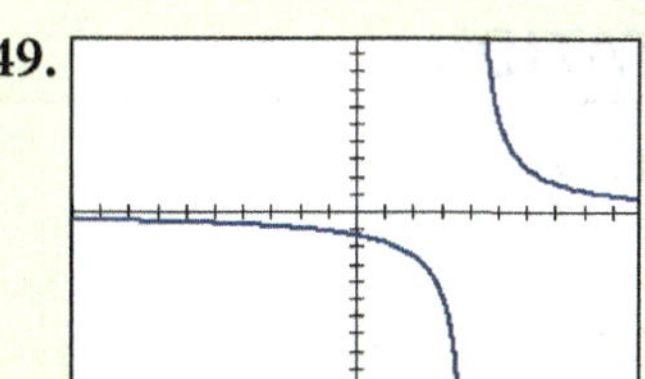

50. The exponent is a variable.

52. $R_f = [0, \infty) = D_g$
$x = \sqrt[5]{y^2}$
$x^{\frac{5}{2}} = \left(y^{\frac{2}{5}}\right)^{\frac{5}{2}}$
$|y| = x^{\frac{5}{2}}$
$y = \pm x^{\frac{5}{2}}$

Objectives

1. To divide polynomials using long division and synthetic division
2. To evaluate polynomial functions using the Remainder Theorem
3. To determine linear binomial factors of a polynomial using the Factor Theorem
4. To write a least-degree polynomial function with given rational zeros

Flash

There are more microorganisms in one teaspoon of soil than there are people on Earth.

Vocabulary

depressed polynomial
Factor Theorem
monic polynomial
Remainder Theorem
synthetic division
synthetic substitution

2.3 The Remainder and Factor Theorems

Cubic polynomials are frequently used to model a volume as in exercise 34.

After completing this section, you will be able to

- divide polynomials using long division and synthetic division.
- evaluate polynomial functions using the Remainder Theorem.
- determine the linear factors of a polynomial using the Factor Theorem.
- write a polynomial function with given zeros.

In 1804 Paolo Ruffini's new synthetic division won him the gold medal in an Italian Scientific Society contest to present the best method of finding roots of equations. Ruffini was also a philosopher and medical doctor, publishing articles on the immortality of the soul and contagious typhoid.

The factoring of a polynomial of higher degree often involves dividing the polynomial by another polynomial. The long-division algorithm can be used when the dividend has a degree higher than that of the divisor.

Example 1   Long Division of Polynomials

Divide $p(x) = 2x^3 + 7x^2 - 15$ by $d(x) = 3 + 2x$ and state whether $d(x)$ is a factor of $p(x)$.

Answer

$$\begin{array}{r} x^2 + 2x - 3 \ \ \text{R.} \ -6 \\ 2x + 3 \overline{\smash{\big)}\, 2x^3 + 7x^2 + 0x - 15} \\ \underline{-(2x^3 + 3x^2)} \\ 4x^2 + 0x \\ \underline{-(4x^2 + 6x)} \\ -6x - 15 \\ \underline{-(-6x - \ \ 9)} \\ -6 \end{array}$$

$d(x)$ is not a factor of $p(x)$.

1. Express the dividend $p(x)$ and divisor in standard form and insert a place-holding term of $0x$ in $p(x)$.
2. Use the steps of the long-division algorithm to divide, multiply, and subtract until the remaining expression has a lower degree than the divisor.
3. The division produces a remainder, so the divisor is not a factor of the dividend.

SKILL ✓ EXERCISE 1

The result of the division in Example 1 can be expressed as

$$(2x^3 + 7x^2 - 15) \div (2x + 3) = x^2 + 2x - 3 - \frac{6}{2x + 3} \text{ or}$$

$$\frac{p(x)}{d(x)} = q(x) + \frac{r(x)}{d(x)} \text{ where } q(x) = x^2 + 2x - 3 \text{ and } r(x) = -6.$$

The fact that $p(x) = d(x)\left[q(x) + \frac{r(x)}{d(x)}\right] = d(x) \cdot q(x) + r(x)$ can be used to check the division.

$$(2x + 3)(x^2 + 2x - 3) - 6 = 2x^3 + 7x^2 - 15$$

Synthetic division, a shortened version of the long-division algorithm, can be used when dividing a polynomial by a linear polynomial of the form $x - c$. You will understand the shortcut better if you compare the colored coefficients in each shortened version of the following long division. Notice that c, the zero of $x - c$, is used in the shortened versions instead of $-c$.

PRESENTATION

Lesson Opener

Divide.
1. $(8x^3 - 12x^2 + 5x - 9) \div (x - 2)$
   $8x^2 + 4x + 13 + \frac{17}{x - 2}$
2. $(8x^3 - 12x^2 + 5x - 9) \div (x^2 - 3)$
   $8x - 12 + \frac{29x - 45}{x^2 - 3}$
3. Identify each of the following in Lesson Opener exercise 1.
   a. dividend $8x^3 - 12x^2 + 5x - 9$
   b. divisor $x - 2$
   c. quotient $8x^2 + 4x + 13$
   d. remainder 17

PRESENTATION

Lesson Opener

Divide.
1. $(8x^3 - 12x^2 + 5x - 9) \div (x - 2)$
   $8x^2 + 4x + 13 + \frac{17}{x - 2}$
2. $(8x^3 - 12x^2 + 5x - 9) \div (x^2 - 3)$
   $8x - 12 + \frac{29x - 45}{x^2 - 3}$
3. Identify each of the following in Lesson Opener exercise 1.
   a. dividend $8x^3 - 12x^2 + 5x - 9$
   b. divisor $x - 2$
   c. quotient $8x^2 + 4x + 13$
   d. remainder 17

Most students will probably recall using long division of polynomials and the abbreviated form, *synthetic division.* Stress that long division of polynomials is often necessary when factoring a polynomial and finding its zeros.

Example 1 reviews the polynomial long-division process. Explain that a remainder of 0 indicates that the divisor is a factor of the dividend, just as with real numbers. When searching for factors, it is not necessary to express the quotient with the remainder over the divisor. Point out that division can be checked by multiplying the divisor times the quotient. Discuss the generalizations leading to $p(x) = d(x) \cdot q(x) + r(x)$.

Present synthetic division using the illustrated elimination of the variables. Point out that the use of the zero of $x - c$ for the divisor permits the use of addition instead of subtraction throughout the process.

Consider sharing that Ruffini used calculus to prove his method and that his paper was published by the society in a 175-page book. Later more concise revisions used a strictly algebraic proof.

Be sure the students follow each step of the process in Example 2 and in Example 3, in which a factor is indicated by a remainder of 0. Remind the students that synthetic division can be used only when

Long Division	Elimination of Variables	Synthetic Division

$$\begin{array}{r} 3x^2 - 2x - 4 \quad \text{R. } 6 \\ x-2\overline{)\,3x^3 - 8x^2 + 0x + 14} \\ \underline{-(3x^3 - 6x^2)} \\ -2x^2 + 0x \\ \underline{-(-2x^2 + 4x)} \\ -4x + 14 \\ \underline{-(-4x + 8)} \\ 6 \end{array}$$

$$\begin{array}{r} 3 - 2 - 4 \quad \text{R. } 6 \\ -2\overline{)\,3 - 8 + 0 + 14} \\ \underline{-(3 - 6)} \\ -2 + 0 \\ \underline{-(-2 + 4)} \\ -4 + 14 \\ \underline{-(-4 + 8)} \\ 6 \end{array}$$

$$\begin{array}{r|rrrr} 2 & 3 & -8 & 0 & 14 \\ & & 6 & -4 & -8 \\ \hline & 3 & -2 & -4 & 6 \end{array}$$

Collapse the algorithm vertically and indicate the subtractions as additions of the opposite.

When dividing a polynomial by a divisor of the form $x - c$, the degree of the quotient is one less than the degree of the dividend and any remainder is a real number. Reinsert the powers of x to state the result of the division:

$$(3x^3 - 8x^2 + 14) \div (x - 2) = 3x^2 - 2x - 4 + \frac{6}{x - 2}.$$

Example 2 Synthetic Division of Polynomials

Use synthetic division to divide $p(x) = 2x^5 - 3x^3 + 7x^2 + 3x + 18$ by $d(x) = x + 2$. Then state whether $x + 2$ is a factor of $p(x)$.

Answer

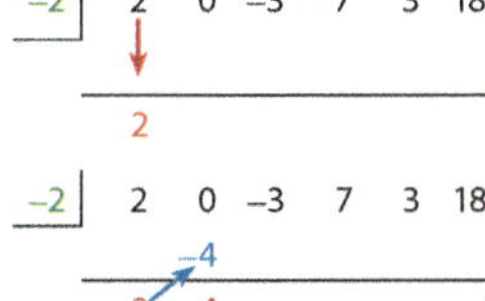

$$\begin{array}{r|rrrrrr} -2 & 2 & 0 & -3 & 7 & 3 & 18 \\ & & & & & & \\ \hline & 2 & & & & & \end{array}$$

1. List the coefficients of the dividend in descending degree, inserting 0 for the missing quartic term. Place -2 from the *divisor*, $x - (-2)$, to the left and bring down the leading coefficient.

$$\begin{array}{r|rrrrrr} -2 & 2 & 0 & -3 & 7 & 3 & 18 \\ & & -4 & & & & \\ \hline & 2 & -4 & & & & \end{array}$$

2. Multiply -2 by 2 and place the product, -4, under the second coefficient, 0. Then add the coefficient and the product: $0 + (-4) = -4$.

$$\begin{array}{r|rrrrrr} -2 & 2 & 0 & -3 & 7 & 3 & 18 \\ & & -4 & 8 & -10 & 6 & -18 \\ \hline & 2 & -4 & 5 & -3 & 9 & 0 \end{array}$$

3. Repeat this process of multiplying by -2 and adding until you reach the remainder, which in this case is 0.

$\frac{p(x)}{d(x)} = q(x) = 2x^4 - 4x^3 + 5x^2 - 3x + 9$

$x + 2$ is a factor of $p(x)$.

4. The last line provides the coefficients of the quotient polynomial in descending degree and shows that there is no remainder.

SKILL ✔ EXERCISE 7

The division of $p(x)$ by $(ax + b)$ can be done synthetically if both the dividend and the divisor are divided by the constant a to obtain an equivalent division with a divisor of the form $x - \frac{b}{a}$.

Reading and Writing Mathematics

Explain how the Remainder Theorem, the Factor Theorem, and synthetic division can be used together to find the zeros of a polynomial equation.

The Remainder Theorem can be used to quickly check or find a root. This is especially easy for small values such as ± 1. Once a root has been identified, the Factor Theorem can be applied to express the related factor. The polynomial can then be divided using synthetic division to obtain a depressed polynomial. The Remainder Theorem, the Factor Theorem, and synthetic division can be applied again if needed.

Additional Exercises

1. Use long division to divide $(5x^4 + 4x^3 + 3x^2 + 2x)$ by $(x - 1)$.
   $5x^3 + 9x^2 + 12x + 14$ R. 14

Use synthetic division to divide $p(x)$ by $d(x)$. Then state whether $d(x)$ is a factor of $p(x)$.

2. $p(x) = 6x^3 + 31x^2 - 66x + 75$; $d(x) = x + 7$ $6x^2 - 11x + 11$ R. -2; $x + 7$ is not a factor.

3. $p(x) = x^5 + 4x^4 - 10x^3 - 26x^2 - 11x - 30$; $d(x) = x + 5$
   $x^4 - x^3 - 5x^2 - x - 6$; $x + 5$ is a factor.

4. $p(x) = 9x^4 + 3x^3 - 2x^2 - 32$; $d(x) = 3x - 4$
   $3x^3 + 5x^2 + 6x + 8$; $3x - 4$ is a factor.

---

the divisor is a linear polynomial with a leading coefficient of 1.

Explain that instances of simultaneous, independent mathematical discoveries indicate an element of mathematics that transcends human intellect.

Present the *Remainder Theorem* and its proof. Example 4 demonstrates how the theorem is applied when finding remainders (and identifying factors) by evaluating the polynomial. Consider asking the students how these two methods compare. (*The Remainder Theorem is often quicker, but synthetic division provides the depressed polynomial when the remainder is 0.*) Example 5 shows how

the theorem is used to evaluate a function using *synthetic substitution*, a compressed model of synthetic division that requires more mental calculation but less writing.

Motivational Idea Have the students note in Example 5 that $x + 4$, $x + 3$, and $x + 2$ are not factors of $p(x)$, because their remainders are not 0. Explain that $p(-4) = 12$ indicates that $x + 4$ is a factor of $g(x) = x^3 + 5x^2 - 16$, a translation of $p(x)$ down 12 units. Ask what translation of $p(x)$ would have a factor of $x + 2$. ($g(x) = x^3 + 5x^2 - 12$, a translation of $p(x)$ down 8 units)

Introduce the *Factor Theorem* and its proof and then apply it with Example 6.

When discussing Example 7, be sure the students understand that $4x - 5$ and $x - \frac{5}{4}$ produce the same zero. You may want to explain that there exists an entire family of functions of the form $p(x) = a(x + 1)\left(x - \frac{5}{4}\right)(x - 3)^2$ having the same zeros. The simplest one has relatively prime integral coefficients (when $a = 4$). The others are vertical stretches or shrinks of the listed polynomial.

Common Student Error Failing to insert a placeholding term with a coefficient of 0 is one of the most common student errors during long division or synthetic

Use the Remainder Theorem to find the remainder when $p(x)$ is divided by the binomial.

5. $p(x) = 3x^2 + 6x^4 - x^5;\ x + 1$
$p(-1) = 10$

6. $p(x) = x^3 - 25x^2 + 50x + 40;\ x - 3$
$p(3) = -8$

Use synthetic substitution and the Remainder Theorem to evaluate $p(-2)$, $p(-1)$, and $p(2)$ for each polynomial.

7. $p(x) = x^3 - 2x^2 - 5x + 6$
$p(-2) = 0,\ p(-1) = 8,\ p(2) = -4$

8. $p(x) = x^4 - 13x^2 - 12x$
$p(-2) = -12,\ p(-1) = 0,\ p(2) = -60$

9. Completely factor $p(x) = 27x^4 + 117x^3 - 21x^2 - 637x - 686$ given that -2 and $\frac{7}{3}$ are zeros. Then list all the real zeros of $p(x)$.
$p(x) = (x + 2)(3x - 7)(3x + 7)^2$
$x = -2, \pm\frac{7}{3}$

10. Write a polynomial function of least degree with integral coefficients that has zeros of $\frac{3}{2}$ and -1 (multiplicity 3).
$2x^4 + 3x^3 - 3x^2 - 7x - 3$

Assignments

- **Minimum:** 1–2, 5, 7–8, 11, 14, 19–21, 23–24, 27, 29, 31, 34, 41–44, 46, 48

- **Standard:** 3, 6, 8–9, 13, 15, 18, 20–25, 27, 29–30, 35, 37, 40, 43–44, 46–48, 50

- **Extended:** 4–5, 8, 10, 12, 17, 22–23, 26–28, 30–31, 37–38, 40, 43, 47–50

Solutions

A. Exercises

1.
$$x + 4 \overline{\smash{\big)}\ 2x^2 + 5x + 1}$$
quotient $2x - 3$
$$-(2x^2 + 8x)$$
$$-3x + 1$$
$$-(-3x - 12)$$
$$13$$

2.
$$3x - 1 \overline{\smash{\big)}\ 3x^3 + 8x^2 - 15x + 4}$$
quotient $x^2 + 3x - 4$
$$-(3x^3 - x^2)$$
$$9x^2 - 15x$$
$$-(9x^2 - 3x)$$
$$-12x + 4$$
$$-(-12x + 4)$$
$$0$$

Example 3 Factoring with Synthetic Division

Use synthetic division to divide $p(x) = 2x^3 - 3x^2 - 4x + 6$ by $d(x) = 2x - 3$. Then state whether $2x - 3$ is a factor of $p(x)$.

Answer

$$\frac{(2x^3 - 3x^2 - 4x + 6) \div 2}{(2x - 3) \div 2} = \frac{x^3 - \frac{3}{2}x^2 - 2x + 3}{x - \frac{3}{2}}$$

$$\begin{array}{r|rrrr} \frac{3}{2} & 1 & -\frac{3}{2} & -2 & 3 \\ & & \frac{3}{2} & 0 & -3 \\ \hline & 1 & 0 & -2 & 0 \end{array}$$

$$\frac{p(x)}{d(x)} = q(x) = x^2 - 2$$

$2x - 3$ is a factor of $p(x)$.

1. Divide both $p(x)$ and $d(x)$ by 2 to derive an equivalent division with a divisor of the form $x - \frac{b}{a}$.

2. Complete the equivalent division by $x - \frac{3}{2}$.

3. The last line provides the coefficients of the quotient polynomial in descending degree and shows that there is no remainder.

SKILL ✔ EXERCISES 9, 23

When the remainder is 0, the quotient $q(x)$ is called the *depressed polynomial* and is also a factor of $p(x)$ since $p(x) = d(x) \cdot q(x)$.

Given the polynomial function $p(x) = 3x^3 - 6x^2 + 2x - 4$, compare the value of $p(3)$ and the remainder of $p(x) \div (x - 3)$.

$$p(3) = 3(3)^3 - 6(3)^2 + 2(3) - 4$$
$$= 81 - 54 + 6 - 4$$
$$= 29$$

and using synthetic division:

$$\begin{array}{r|rrrr} 3 & 3 & -6 & 2 & -4 \\ & & 9 & 9 & 33 \\ \hline & 3 & 3 & 11 & 29 \end{array}$$
$$3x^2 + 3x + 11 \text{ R. } 29$$

The fact that $p(3)$ equals the remainder when $p(x)$ is divided by $x - 3$ is guaranteed by the Remainder Theorem.

REMAINDER THEOREM

If a polynomial $P(x)$ is divided by a binomial $x - c$, the remainder $R = P(c)$.

Proof:

Given $p(x) = d(x) \cdot q(x) + r(x)$ where $d(x) = x - c$, $r(x)$ must be a constant, R, with a degree of 0, since $r(x)$ has a degree less than the divisor $x - c$, which has a degree of 1.
$\therefore p(x) = (x - c) \cdot q(x) + R$.
Substituting c for x:
$p(c) = (c - c) \cdot q(c) + R$,
$p(c) = 0 \cdot q(c) + R$, and
$p(c) = R$.

Synthetic division and the Remainder Theorem provide alternate methods of evaluating a polynomial and finding a remainder.

division. Emphasize the need to identify any missing terms as a first step.

Interactive Activity The Internet keyword searches *polynomial long-division calculator* and *synthetic division calculator* can be used to locate calculators for checking, illustrating, or exploring synthetic division.

TIPS

Ex. 28–33 Multiples of the provided answers would also have the stated zeros.

Ex. 40 Assigning exercise 22 will help prepare the students for this exercise.

Example 4 Finding a Remainder with the Remainder Theorem

Use the Remainder Theorem to find the remainder when $p(x) = x^4 - 6x^3 - x^2 + 2x + 7$ is divided by $x - 1$.

Answer

$$R = p(1) = 1 - 6 - 1 + 2 + 7 \qquad \text{Find } R \text{ by evaluating } p(1).$$
$$= 3$$

SKILL ✓ **EXERCISE 13**

A compressed form of synthetic division omits the intermediate step of recording each product before the product is added mentally, leaving just the coefficients of $q(x)$ and the remainder. This can be especially helpful when using synthetic division to evaluate several values of a function, a process sometimes called *synthetic substitution*.

Example 5 Synthetic Substitution

Given $p(x) = x^3 + 5x^2 - 4$, apply the Remainder Theorem to evaluate $p(-4)$, $p(-3)$, $p(-2)$, and $p(-1)$.

Answer

c	1	5	0	-4	
-4	1	1	-4	12	$p(-4) = 12$
-3	1	2	-6	14	$p(-3) = 14$
-2	1	3	-6	8	$p(-2) = 8$
-1	1	4	-4	0	$p(-1) = 0$

Use compressed synthetic division to find $p(c) = R$ when $p(x)$ is divided by $(x - c)$.

SKILL ✓ **EXERCISE 15**

The Remainder Theorem and synthetic substitution can be used to find zeros of a function. The Factor Theorem formally states the relationship between the zeros of a polynomial function and its factors. Recall that the proof of a biconditional $p \leftrightarrow q$ requires the proof of both conditionals, $p \rightarrow q$ and $q \rightarrow p$.

▌FACTOR THEOREM

$p(c) = 0$ if and only if $(x - c)$ is a factor of $p(x)$.

Proof:

Using the Remainder Theorem, if $p(c) = 0$,
then $p(x) \div (x - c) = q(x) + 0$ and $(x - c)$ divides $p(x)$ exactly.
Therefore, $p(x) = (x - c) \cdot q(x)$ and $(x - c)$ is a factor of $p(x)$.

Conversely, if $(x - c)$ is a factor of $p(x)$,
then $p(x) = (x - c) \cdot q(x) + 0$ and
the Remainder Theorem guarantees that $p(c) = 0$ if $R = 0$.

Using synthetic division, the Remainder Theorem, and the Factor Theorem together allows you to check whether $(x - c)$ is a factor of a polynomial function and provides the coefficients of the depressed polynomial when it is a factor.

3.
$$
\begin{array}{r}
4x^2 + 22x + 121 \\
2x - 11 \overline{\smash{\big)}\ 8x^3 + 0x^2 + 0x - 1331} \\
\underline{-(8x^3 - 44x^2)} \\
44x^2 + 0x \\
\underline{-(44x^2 - 242x)} \\
242x - 1331 \\
\underline{-(242x - 1331)} \\
0
\end{array}
$$

4.
$$
\begin{array}{r}
x^2 - 3x + 5 \\
x^2 + 4 \overline{\smash{\big)}\ x^4 - 3x^3 + 9x^2 - 12x + 17} \\
\underline{-(x^4 + 4x^2)} \\
-3x^3 + 5x^2 - 12x \\
\underline{-(-3x^3 - 12x)} \\
5x^2 + 17 \\
\underline{-(5x^2 + 20)} \\
-3
\end{array}
$$

5.
$$
\begin{array}{r|rrrrr}
4 & 1 & -7 & 10 & 9 & -4 \\
& \downarrow & 4 & -12 & -8 & 4 \\
\hline
& 1 & -3 & -2 & 1 & 0
\end{array}
$$
$$x^3 - 3x^2 - 2x + 1$$

6.
$$
\begin{array}{r|rrrrr}
3 & 1 & 0 & -5 & -17 & 15 \\
& \downarrow & 3 & 9 & 12 & -15 \\
\hline
& 1 & 3 & 4 & -5 & 0
\end{array}
$$
$$x^3 + 3x^2 + 4x - 5$$

7.
$$
\begin{array}{r|rrrrrr}
-1 & -1 & 0 & -1 & 3 & 4 & 0 \\
& \downarrow & 1 & -1 & 2 & -5 & 1 \\
\hline
& -1 & 1 & -2 & 5 & -1 & 1
\end{array}
$$
$$-x^4 + x^3 - 2x^2 + 5x - 1 \text{ R. } 1$$

8.
$$
\begin{array}{r|rrrrr}
-2 & -6 & 0 & 4 & 0 & 0 \\
& \downarrow & 12 & -24 & 40 & -80 \\
\hline
& -6 & 12 & -20 & 40 & -80
\end{array}
$$
$$-6x^3 + 12x^2 - 20x + 40 \text{ R. } -80$$

9. $\left(3x^3 - \frac{13}{2}x^2 + \frac{9}{2}x - 1\right) \div \left(x - \frac{1}{2}\right)$
$$
\begin{array}{r|rrrr}
\frac{1}{2} & 3 & -\frac{13}{2} & \frac{9}{2} & -1 \\
& \downarrow & \frac{3}{2} & -\frac{5}{2} & 1 \\
\hline
& 3 & -5 & 2 & 0
\end{array}
$$
$$3x^2 - 5x + 2$$

10. $\left(x^3 - \frac{7}{3}x^2 - 3x - \frac{2}{3}\right) \div \left(x + \frac{2}{3}\right)$
$$
\begin{array}{r|rrrr}
-\frac{2}{3} & 1 & -\frac{7}{3} & -3 & -\frac{2}{3} \\
& \downarrow & -\frac{2}{3} & 2 & \frac{2}{3} \\
\hline
& 1 & -3 & -1 & 0
\end{array}
$$
$$x^2 - 3x - 1$$

11a. $p(1) = (1)^4 - 8(1)^2 + 16 = 9$

11b. $p(-2) = (-2)^4 - 8(-2)^2 + 16 = 0$

11c. $p(2) = (2)^4 - 8(2)^2 + 16 = 0$

12a. $p(1) = (1)^3 - 8 = -7$

12b. $p(-2) = (-2)^3 - 8 = -16$

12c. $p(2) = (2)^3 - 8 = 0$

13a. $p(-1) = (-1)^3 - 4(-1)^2 - 20(-1) - 7 = 8$

13b. $p(7) = (7)^3 - 4(7)^2 - 20(7) - 7 = 0$

13c. $p(1) = (1)^3 - 4(1)^2 - 20(1) - 7 = -30$

14a. $p(-1) = (-1)^5 - 1 = -2$

14b. $p(-2) = (-2)^5 - 1 = -33$

14c. $p(1) = (1)^5 - 1 = 0$

15.
$$
\begin{array}{r|rrr l}
c & 3 & 1 & -2 & \\
1 & 3 & 4 & 2 & p(1) = 2 \\
2 & 3 & 7 & 12 & p(2) = 12 \\
-1 & 3 & -2 & 0 & p(-1) = 0
\end{array}
$$

16.
$$
\begin{array}{r|rrrr l}
c & 1 & 2 & -7 & -8 & 12 & \\
1 & 1 & 3 & -4 & -12 & 0 & p(1) = 0 \\
2 & 1 & 4 & 1 & -6 & 0 & p(2) = 0 \\
-1 & 1 & 1 & -8 & 0 & 12 & p(-1) = 12
\end{array}
$$

17.
$$
\begin{array}{r|rrr l}
c & -3 & 7 & -2 & \\
1 & -3 & 4 & 2 & p(1) = 2 \\
2 & -3 & 1 & 0 & p(2) = 0 \\
-1 & -3 & 10 & -12 & p(-1) = -12
\end{array}
$$

18.
$$
\begin{array}{r|rrrr l}
c & 1 & 3 & -1 & -3 & \\
1 & 1 & 4 & 3 & 0 & p(1) = 0 \\
2 & 1 & 5 & 9 & 15 & p(2) = 15 \\
-1 & 1 & 2 & -3 & 0 & p(-1) = 0
\end{array}
$$

B. Exercises

20. No; Since $p(3) = 48$, the remainder is 48, not 0. Therefore $(x - 3)$ is not a factor.

21a.
$$-4 \,\big|\; 1 \quad 0 \quad 0 \quad 64$$
$$\underline{\downarrow \;-4 \quad 16 \quad -64}$$
$$1 \quad -4 \quad 16 \quad 0$$

$$x^2 - 4x + 16$$

22a.
$$1 \,\big|\; 1 \quad 0 \quad 0 \quad 0 \quad 0 \quad -1$$
$$\underline{\downarrow \;1 \quad 1 \quad 1 \quad 1 \quad 1}$$
$$1 \quad 1 \quad 1 \quad 1 \quad 1 \quad 0$$

$$p(x) = (x - 1) \cdot (x^4 + x^3 + x^2 + x + 1)$$

22b. $p(1) = (1)^5 - 1 = 0$, so $p(x) \div (x - 1)$ has $R = 0$ and $(x - 1)$ is a factor.

23.
$$1 \,\big|\; 1 \quad -3 \quad -6 \quad 8$$
$$1 \quad -2 \quad -8 \quad 0$$

$$p(x) = (x - 1)(x^2 - 2x - 8)$$
$$= (x - 1)(x - 4)(x + 2)$$
$$x = 1, 4, -2$$

24.
$$-3 \,\big|\; 2 \quad 7 \quad 2 \quad -3$$
$$2 \quad 1 \quad -1 \quad 0$$

$$p(x) = (x + 3)(2x^2 + x - 1)$$
$$= (x + 3)(2x - 1)(x + 1)$$
$$x = -3, \tfrac{1}{2}, -1$$

25.
$$1 \,\big|\; 2 \quad -5 \quad 1 \quad 5 \quad -3$$
$$\;2 \quad -3 \quad -2 \quad 3 \quad 0$$
$$1 \,\big|\; 2 \quad -3 \quad -2 \quad 3 \quad 0$$
$$\;2 \quad -1 \quad -3 \quad 0$$

$$p(x) = (x - 1)^2(2x^2 - x - 3)$$
$$= (x - 1)^2(2x - 3)(x + 1)$$
$$x = 1 \text{ (multiplicity 2)}, \tfrac{3}{2}, -1$$

26.
$$1 \,\big|\; 1 \quad 4 \quad -13 \quad -40 \quad 48$$
$$\;1 \quad 5 \quad -8 \quad -48 \quad 0$$
$$-4 \,\big|\; 1 \quad 1 \quad -12 \quad 0$$

$$p(x) = (x - 1)(x + 4)(x^2 + x - 12)$$
$$= (x - 1)(x + 4)(x + 4)(x - 3)$$
$$= (x - 1)(x - 3)(x + 4)^2$$
$$x = 1, 3, -4 \text{ (multiplicity 2)}$$

27.
$$\tfrac{7}{3} \,\big|\; 27 \quad 0 \quad 0 \quad -343$$
$$\phantom{\tfrac{7}{3}\,|\,}27 \quad 63 \quad 147 \quad 0$$

$$p(x) = \left(x - \tfrac{7}{3}\right)(27x^2 + 63x + 147)$$
$$= 3\left(x - \tfrac{7}{3}\right)(9x^2 + 21x + 49)$$
$$= (3x - 7)(9x^2 + 21x + 49)$$
$$x = \tfrac{7}{3} \text{ (Other zeros are not real.)}$$

Example 6   Factoring Polynomials

If 2 is a zero of $p(x) = x^3 + 6x^2 - x - 30$, factor $p(x)$ completely and list all of its zeros.

Answer

$$2 \,\big|\; 1 \quad 6 \quad -1 \quad -30$$
$$1 \quad 8 \quad 15 \quad 0$$

$$\therefore p(x) = (x - 2)(x^2 + 8x + 15)$$
$$= (x - 2)(x + 3)(x + 5)$$

$$(x - 2)(x + 3)(x + 5) = 0$$
$$x = 2, -3, -5$$

1. Since 2 is a zero, $x - 2$ is a factor. Use compressed synthetic division to find the depressed polynomial.

2. Continue to factor the polynomial.

3. Apply the Zero Product Property and list all the zeros.

SKILL ✔ **EXERCISE 25**

The Factor Theorem enables you to construct polynomials with known zeros. While there are many polynomials that could have the given roots, we will find a polynomial with the least possible degree. While the leading coefficient of a polynomial is often 1 (a *monic polynomial*), we usually find a polynomial in which all the coefficients are integers.

Example 7   Writing a Polynomial Function with Given Zeros

Write a polynomial function of least degree with integral coefficients that has zeros of $-1, \tfrac{5}{4},$ and 3 (with a multiplicity of 2).

Answer

$$p_1(x) = (x + 1)\left(x - \tfrac{5}{4}\right)(x - 3)^2$$

1. Use the Factor Theorem to list all the factors of a least degree polynomial with the stated roots. The multiplicity of a zero is assumed to be 1 unless otherwise stated.

$$p_2(x) = (x + 1)(4x - 5)(x - 3)^2$$

2. Multiplying the second factor by 4 produces another polynomial (without fractions) with the same zeros.

$$= (4x^2 - x - 5)(x^2 - 6x + 9)$$
$$= 4x^4 - 25x^3 + 37x^2 + 21x - 45$$

3. Multiply the factors, writing the polynomial in standard form.

SKILL ✔ **EXERCISE 31**

A. Exercises

Use long division to find each quotient.

1. $(2x^2 + 5x + 1) \div (x + 4)$ $2x - 3$ R. 13

2. $(3x^3 + 8x^2 - 15x + 4) \div (3x - 1)$ $x^2 + 3x - 4$

3. $(8x^3 - 1331) \div (2x - 11)$ $4x^2 + 22x + 121$

4. $(x^4 - 3x^3 + 9x^2 - 12x + 17) \div (x^2 + 4)$ $x^2 - 3x + 5$ R. -3

Use synthetic division to find each quotient.

5. $(x^4 - 7x^3 + 10x^2 + 9x - 4) \div (x - 4)$ $x^3 - 3x^2 - 2x + 1$

6. $(x^4 - 5x^2 - 17x + 15) \div (x - 3)$ $x^3 + 3x^2 + 4x - 5$

7. $(3x^2 - x^5 + 4x - x^3) \div (x + 1)$ $-x^4 + x^3 - 2x^2 + 5x - 1$ R. 1

8. $(4x^2 - 6x^4) \div (x + 2)$ $-6x^3 + 12x^2 - 20x + 40$ R. -80

9. $(6x^3 - 13x^2 + 9x - 2) \div (2x - 1)$ $3x^2 - 5x + 2$

10. $(3x^3 - 7x^2 - 9x - 2) \div (3x + 2)$ $x^2 - 3x - 1$

Use the Remainder Theorem to find the remainder when $p(x)$ is divided by each binomial.

11. $p(x) = x^4 - 8x^2 + 16$
  a. $x - 1$ 9
  b. $x + 2$ 0
  c. $x - 2$ 0

12. $p(x) = x^3 - 8$
  a. $x - 1$ -7
  b. $x + 2$ -16
  c. $x - 2$ 0

28. $p(x) = (x - 1)(x - 4)$
$$= x^2 - 5x + 4$$

29. $p(x) = (x - 2)^3$
$$= (x^2 - 4x + 4)(x - 2)$$
$$= x^3 - 6x^2 + 12x - 8$$

30. $p(x) = 6\left(x - \tfrac{1}{6}\right) \cdot 7\left(x - \tfrac{5}{7}\right)$
$$= (6x - 1)(7x - 5)$$
$$= 42x^2 - 37x + 5$$

31. $p(x) = \left[2\left(x - \tfrac{1}{2}\right)\right]^2(x - 5)$
$$= (2x - 1)^2(x - 5)$$
$$= (4x^2 - 4x + 1)(x - 5)$$
$$= 4x^3 - 24x^2 + 21x - 5$$

32. $p(x) = 3\left(x - \tfrac{2}{3}\right)(x - 3)(x + 2)^2$
$$= (3x - 2)(x - 3)(x + 2)^2$$
$$= (3x^2 - 11x + 6)(x^2 + 4x + 4)$$
$$= 3x^4 + x^3 - 26x^2 - 20x + 24$$

33. $p(x) = (x - 1)(x + 1) \cdot 3\left(x - \tfrac{1}{3}\right) \cdot 3\left(x + \tfrac{1}{3}\right)$
$$= (x - 1)(x + 1)(3x - 1)(3x + 1)$$
$$= (x^2 - 1)(9x^2 - 1)$$
$$= 9x^4 - 10x^2 + 1$$

34a. Let $3x$ represent the increase in both the width and length.
$$V(x) = (1 + x)(2 + 3x)(5 + 3x)$$
$$= (3x^2 + 5x + 2)(3x + 5)$$
$$= 9x^3 + 30x^2 + 31x + 10$$

13. $p(x) = x^3 - 4x^2 - 20x - 7$

   a. $x + 1$ 8
   b. $x - 7$ 0
   c. $x - 1$ −30

14. $p(x) = x^5 - 1$

   a. $x + 1$ −2
   b. $x + 2$ −33
   c. $x - 1$ 0

Use synthetic substitution to evaluate $p(1)$, $p(2)$, and $p(-1)$ for each function.

15. $p(x) = 3x^2 + x - 2$ 2, 12, 0

16. $p(x) = x^4 + 2x^3 - 7x^2 - 8x + 12$ 0, 0, 12

17. $p(x) = -2 + 7x - 3x^2$ 2, 0, −12

18. $p(x) = x^3 + 3x^2 - x - 3$ 0, 15, 0

❯ B. Exercises

19. Which of the following must be true if $x - 4$ is a factor of $p(x)$? List all correct answers.   B, D

   A. $x + 4$ is a factor of $p(x)$.   **D.** $p(4) = 0$
   B. 4 is a zero of $p(x)$.   **E.** $p(-4) = 0$
   C. −4 is a zero of $p(x)$.

20. Explain: A classmate claims that $p(x) = x^3 + 3x^2 - x - 3$ has a factor of $x - 3$. Is this correct? Use the Remainder Theorem to explain why or why not.

21. Consider $(x^3 + 64) \div (x + 4)$.
   a. Use synthetic division to find the quotient. $x^2 - 4x + 16$
   b. Write the factorization of $x^3 + 64$. $(x + 4)(x^2 - 4x + 16)$

22. Use each of the following to show that $x - 1$ is a factor of $p(x) = x^5 - 1$.
   a. synthetic division
   b. the Remainder Theorem

Completely factor each polynomial using the given information. Then list all the real zeros of the function.

23. $p(x) = x^3 - 3x^2 - 6x + 8$ if $p(1) = 0$

24. $p(x) = 2x^3 + 7x^2 + 2x - 3$ if $x + 3$ is a factor

25. $p(x) = 2x^4 - 5x^3 + x^2 + 5x - 3$ if 1 is a zero (multiplicity 2)

26. $p(x) = x^4 + 4x^3 - 13x^2 - 40x + 48$ if $x - 1$ and $x + 4$ are factors

27. $p(x) = 27x^3 - 343$; $p\left(\frac{7}{3}\right) = 0$

Write a standard form polynomial function with integral coefficients having the following zeros.

28. 1 and 4

29. 2 (multiplicity 3)

30. $\frac{1}{6}$ and $\frac{5}{7}$

31. $\frac{1}{2}$ (multiplicity 2) and 5

32. 3, −2 (multiplicity 2), $\frac{2}{3}$

33. ± 1, $\pm \frac{1}{3}$

34. Joe wants to enlarge his raised garden box, which is 2 ft wide by 5 ft long and 1 ft high. He plans to increase the length and width by three times the amount that he increases the height.
   a. Write a polynomial function, $V(x)$, for the volume of the enlarged box in terms of its increase in height. $V(x) = 9x^3 + 30x^2 + 31x + 10$
   b. Use the function to find the volume (in ft^3) if the enlarged box is 27 in. tall. $\approx 113.2 \text{ ft}^3$
   c. What is the length and width of the new box?

35. The table lists the median values of homes in Massachusetts.

Year	x	Median Value ($/\text{ft}^2$)
1996	0	101
1999	3	122
2002	6	181
2005	9	322
2008	12	268
2011	15	204
2014	18	271

   a. Letting x represent the number of years since 1996, complete a regression to find a quartic model for the data.
   b. Use the model to approximate the cost in 2000 and in 2007. $\approx \$137/\text{ft}^2$; $\$289/\text{ft}^2$
   c. Use the model to approximate the year that housing prices began to decrease and the year they began to increase again.   2005, 2011

❯ C. Exercises

Find k so that $d(x)$ is a factor of $p(x)$.

36. $p(x) = 2x^2 + 9x + k$; $d(x) = x + 5$ $k = -5$

37. $p(x) = x^4 - x^3 - kx^2 + 11x + 6$; $d(x) = (x + 3)$ $k = 9$

38. Find k if $g(x) = x^2 + 6x - 1$ and $h(x) = kx^3 - x^2 - 2x - 1$ have the same remainder when divided by $x - 2$. $k = 3$

39. Find the values of j and k if the function $p(x) = -2x^2 - jx + k$ has a remainder of 7 when divided by $x + 1$ and a remainder of 5 when divided by $x + 2$. $j = 4$; $k = 5$

40. Prove: Show that $x - 1$ is always a factor of $x^n - 1$.

❯ C. Exercises

36. $p(-5) = 2(-5)^2 + 9(-5) + k = 0$
$$50 - 45 + k = 0$$
$$k = -5$$

37. $p(-3) = (-3)^4 - (-3)^3 - k(-3)^2 + 11(-3) + 6 = 0$
$$81 + 27 - 9k - 33 + 6 = 0$$
$$-9k = -81$$
$$k = 9$$

38. $g(2) = (2)^2 + 6(2) - 1 = 15$
$h(2) = k(2)^3 - (2)^2 - 2(2) - 1 = 15$
$$8k - 9 = 15$$
$$8k = 24$$
$$k = 3$$

39. $p(-1) = 7;\ 7 = -2 + j + k$
$$k = 9 - j$$
$p(-2) = 5;\ 5 = -8 + 2j + k$
$$13 = 2j + k$$
using substitution:
$$13 = 2j + (9 - j)$$
$$j = 4$$
$$k = 9 - 4 = 5$$

40. using the Remainder Theorem:
$p(1) = (1)^n - 1$
$$= 1 - 1 = 0$$

34b. 27 in. = 2.25 ft = $1 + x$
$$x = 1.25 \text{ ft}$$
$$V(1.25) \approx 113.2 \text{ ft}^3$$

34c. length: $5 + 3(1.25) = 8.75$ ft or 105 in.
width: $2 + 3(1.25) = 5.75$ ft or 69 in.

35a. $p(x) = 0.0568x^4 - 2.0444x^3 + 21.6662x^2 - 49.8841x + 105.6926$

35b. $p(4) \approx \$136.52$; $p(11) \approx \$289.33$

35c. 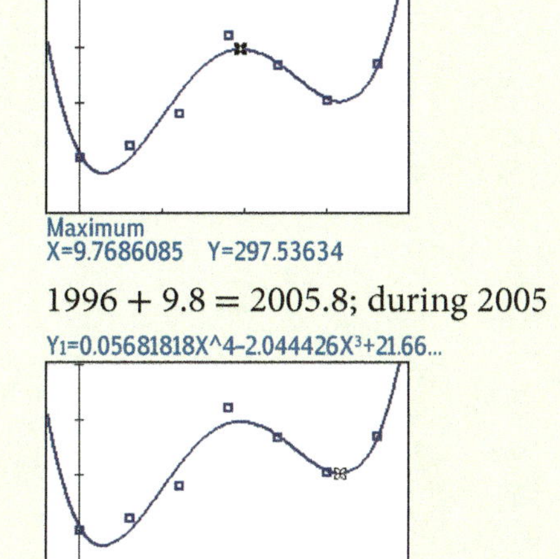

$1996 + 9.8 = 2005.8$; during 2005

$1996 + 15.8 = 2011.8$; during 2011

❯ Cumulative Review

42. $\sqrt{-104} = \sqrt{-4 \cdot 26} = 2i\sqrt{26}$

43. $4x^2 = -9$

$x^2 = -\dfrac{9}{4}$

$x = \pm\sqrt{-\dfrac{9}{4}} = \pm\dfrac{3}{2}i$

44. $(x+9)(x^2 - 9x + 81) = 0$

$x = -9$ or

$x = \dfrac{-(-9) \pm \sqrt{(-9)^2 - 4(1)(81)}}{2(1)}$

$= \dfrac{9 \pm \sqrt{-243}}{2} = \dfrac{9 \pm 9i\sqrt{3}}{2}$

45.

$(-\infty, -2) \cup \left[\dfrac{1}{4}, \infty\right)$

46. $\dfrac{f(x)}{g(x)} = \dfrac{(5x-1)(x+7)}{(x+7)}$

$= 5x - 1,\ x \neq -7$

47. $\dfrac{x(5x^2+3) + y(5x^2+3)}{(x^2+y^2)(x+y)(x-y)}$

$= \dfrac{(x+y)(5x^2+3)}{(x+y)(x^2+y^2)(x-y)}$

$= \dfrac{5x^2+3}{(x^2+y^2)(x-y)}$

48. $y = \dfrac{2}{3}x - \dfrac{5}{6}$

$6y = 4x - 5$

$4x - 6y = 5$

49. $g(x) = 2(x-2)^2 - 3$

$= 2(x^2 - 4x + 4) - 3$

$= 2x^2 - 8x + 5$

50. $x = \dfrac{-3 \pm \sqrt{3^2 - 4(4)(10)}}{2(4)} = \dfrac{-3 \pm \sqrt{-151}}{8}$

$= \dfrac{-3 \pm i\sqrt{151}}{8}$

CUMULATIVE REVIEW

41. Simplify $(3x^5 - 2x^4 + x - 4) - (4x^4 - x^3 + x - 7)$.
[Algebra] $3x^5 - 6x^4 + x^3 + 3$

42. Simplify $\sqrt{-104}$. [Appendix 5] $2i\sqrt{26}$

Solve. [1.6]

43. $x^2 + 2x + 5 = -3x^2 + 2x - 4$ [1.6] $x = \pm\frac{3}{2}i$

44. $x^3 + 729 = 0$ [1.6] $x = -9, \frac{9 \pm 9\sqrt{3}i}{2}$

45. Graph the compound inequality $x \geq \frac{1}{4}$ or $x < -2$ on a number line and express it using interval notation. [1.1]

46. If $f(x) = 5x^2 + 34x - 7$ and $g(x) = x + 7$, find $\frac{f}{g}(x)$. [1.7]

47. Simplify $\dfrac{5x^3 + 3x + 5x^2 y + 3y}{x^4 - y^4}$. [Algebra] A

A. $\dfrac{5x^2 + 3}{(x^2 + y^2)(x - y)}$

B. $\dfrac{5x^2 + 3}{(x^2 + y^2)(x + y)}$

C. $\dfrac{5x^2 + 3y}{(x^2 + y^2)(x + y)}$

D. $\dfrac{5x^2 + 3}{(x^2 - y^2)(x - y)}$

E. none of these

48. Which equation represents the line $y = \frac{2}{3}x - \frac{5}{6}$? [1.2] D

A. $4x - 5y = 6$

B. $4x + 6y = 5$

C. $4x + 6y = -5$

D. $4x - 6y = 5$

E. none of these

49. What is the result when $f(x) = 2x^2$ is translated 2 units right and 3 units down? [1.5] C

A. $g(x) = 2x^2 + 8x + 5$

B. $g(x) = 2x^2 - 2x - 3$

C. $g(x) = 2x^2 - 8x + 5$

D. $g(x) = 2x^2 + 8x - 3$

E. none of these

50. Find the zeros of $f(x) = 4x^2 + 3x + 10$. [1.6] A

A. $x = \dfrac{-3 \pm i\sqrt{151}}{8}$

B. $x = \dfrac{-3 \pm \sqrt{151}}{8}$

C. $x = \dfrac{-3 \pm 13i}{8}$

D. $x = -2, \frac{5}{4}$

E. none of these

45. $(-\infty, -2) \cup \left[\frac{1}{4}, \infty\right)$

46. $\frac{f}{g}(x) = 5x - 1,\ x \neq -7$

SYMBOLS

Can you calculate the zenzizenzizenzic of 2?

Modern mathematicians express this intimidating notation as the much more manageable 2^8. The introduction and acceptance of symbols such as 2^8 in mathematical publications was driven by the study of algebra and enabled by the invention of movable-type printing.

The first published use (in English) of the now familiar addition and subtraction signs occurred in *The Whetstone of Wit*, an algebra text published by Robert Recorde in 1557. Recorde also introduced the sign for equality in this text, commenting that nothing could be more equal than two parallel lines.

The full title of Recorde's book is *The whetstone of wit, which is the second part of Arithmetic: containing the extraction of roots: The Cossic practice, with the rule of Equation: and the works of Surd Numbers*.

The archaic German word for the unknown value in an equation is *coss*. So prolific were German mathematical authors in the sixteenth century that for a while *coss* was used instead of

algebra throughout Europe when referring to the mathematics of unknown quantities.

The first German algebra text, commonly known as *The Coss*, was published in 1525. In 1553 Michael Stifel published a new edition with extensive notes. Stifel was an Augustinian monk who, partly because of his distaste for the sale of indulgences, became a follower of Martin Luther. Nearly 200 years later, Leonhard Euler would be taught mathematics by his father from Stifel's edition.

The Coss was written by Christoff Rudolff and contained the first use of the modern radical symbol to indicate roots. The vinculum (horizontal bar) was added by René Descartes in his 1637 volume on geometry. Indices to indicate other roots besides the square root were not adopted until the end of the century.

Nicole Oresme introduced the idea of a mathematical function in the fourteenth century, but it would be another 200 years before the scientific explorations of Galileo and Kepler led to the fuller development of the power of functions in modeling natural phenomena. In the interim, Gutenberg published the first Bible printed on a movable-type press, Columbus discovered the New World, Magellan circumnavigated the globe, and Martin Luther nailed his Ninety-Five Theses to the church door in Wittenberg.

As the concept of a function developed and symbols became more standardized and useful, mathematicians began to study functions as abstract ideas in their own right, even when they saw no practical application. This development of "pure mathematics" led some mathematicians to conclude that all math is created by man.

COMPREHENSION CHECK

1. When did the first use of addition, subtraction, and equality signs occur in print in English?
2. Why would a sixteenth-century algebra book be called *The Coss*?
3. What famous eighteenth-century mathematician learned mathematics from *The Coss*?
4. What key invention of the fifteenth century helped standardize mathematical notation?
5. Why did the use and development of functions flourish during a time of scientific exploration?
6. State a factor contributing to the belief by some that mathematics is solely a human creation.

Historical Connection 

Objectives

1. To list several driving forces behind the standardization of mathematical notation
2. To describe the role of algebra in the development of functions
3. To explain why the development of functions accelerated during an age of scientific exploration
4. To state a factor contributing to the belief that mathematics is solely a human creation

Answers

1. 1557 in *The Whetstone of Wit*
2. *Coss* is an archaic word meaning "the thing." It was used to indicate unknown values in an equation.
3. Leonard Euler
4. the movable-type printing press
5. Functions proved useful in modeling natural phenomena.
6. the development of "pure mathematics" with no apparent practical application

PRESENTATION

The zenzizenzizenzic of 2 is 256. Zenzizenzizenzic was a mathematical notation that was used before superscript notation was introduced for powers. *Zenzic* is the German spelling of an Italian root word meaning "squared." Thus, *zenzi-* (square) *-zenzi-* (square again) *-zenzic* (squared) represents a value "squared" three times, or $\left(\left(x^2\right)^2\right)^2 = x^8$.

Discuss the Historical Connection and why the standardization of symbols would be difficult prior to the use of Gutenberg's printing press. Point out that the hand copying of manuscripts led to several different scribal interpretations of intended notation.

Emphasize that while symbols, systems, and notations are created by man to express mathematical truth, mathematics is a direct result of God's creation, and therefore a reflection of God's mind and design.

Objectives

1. To use the Rational Zero Theorem to identify the possible rational zeros of a polynomial function

2. To apply Descartes's Rule of Signs and the Upper and Lower Bound Tests to limit the search for real zeros

3. To find complex zeros of polynomial functions using the Fundamental Theorem of Algebra and the Conjugate Root Theorem

4. To find a least-degree polynomial function for a given set of complex zeros

Flash

Germany also honored Carl Gauss (1777–1855) on three postage stamps, including one issued on the 100th anniversary of his death and another issued on the 200th anniversary of his birth.

Vocabulary

Complete Linear Factorization
  Theorem
Conjugate Root Theorems
Corollary of the Fundamental
  Theorem of Algebra
Descartes's Rule of Signs
existence theorems
Fundamental Theorem of Algebra
irreducible quadratic factor
linear factors
lower bound
Lower Bound Test
Rational Zero Theorem
upper bound
Upper Bound Test

2.4 Zeros of Polynomial Functions

Germany honored Carl Gauss, "the Prince of Mathematics," on the 10 deutsche mark note in 1991.

After completing this section, you will be able to

- identify the possible rational zeros of a polynomial.
- apply Descartes's Rule of Signs and the Upper and Lower Bound Tests.
- find complex zeros of a polynomial function.
- find a polynomial function of least degree for a given set of zeros.

Christianity is built on fundamental doctrines, such as the infallibility of God's Word, the deity of Christ, Christ's resurrection, and salvation through Christ alone. These fundamentals are foundational to other truths. Similarly, there are several fundamental theorems in mathematics that form the foundation for other theorems. These include the Fundamental Theorem of Algebra, first proved by Carl Friedrich Gauss.

FUNDAMENTAL THEOREM OF ALGEBRA

Every polynomial function of degree greater than 0 has at least one complex zero.

The complex root(s) may be real or imaginary. This theorem is used to prove a second theorem describing the factorization of polynomial functions and leads to a corollary describing the exact number of zeros.

COMPLETE LINEAR FACTORIZATION THEOREM

Every polynomial function of positive degree n can be factored into n *linear factors* so that $p(x) = a(x - z_1)(x - z_2) \ldots (x - z_n)$ where $z_1, z_2, \ldots, z_n$ are the complex zeros of $p(x)$.

Proof:

The Fundamental Theorem of Algebra guarantees that $p(x)$ has at least one complex zero, z_1.

Thus $p(z_1) = 0$, and by the Factor Theorem $p(x) = (x - z_1)[q_1(x)]$, where $q_1(x)$ is a depressed polynomial.

The Fundamental Theorem also guarantees that $q_1(x)$ has a complex zero; let it be z_2. Therefore, $p(x) = (x - z_1)(x - z_2)[q_2(x)]$.
Repeating this process decreases the degree of $q(x)$ by 1 until $q(x)$ is of degree 0 (a constant) and $p(x) = a(x - z_1)(x - z_2) \ldots (x - z_n)$.

COROLLARY OF THE FUNDAMENTAL THEOREM OF ALGEBRA

A polynomial function of positive degree n has exactly n zeros, which may be repeated. Therefore, there are at most n roots to the equation $p(x) = 0$.

These *existence theorems* guarantee that the zeros and factors exist, but do not explain how to find them. The rest of this section discusses several ways to find the zeros of certain polynomial functions, whether they are rational, irrational, or imaginary.

The linear function $f(x) = 3x - 5$ has one zero: $x = \frac{5}{3}$. Factoring the cubic $g(x) = x^3 - 4x = x(x^2 - 4)$ allows you to quickly find its three zeros: $0, \pm 2$. Factoring the quadratic $h(x) = 2x^2 - x - 6 = (x - 2)(2x + 3)$ produces its two rational zeros: 2 and $-\frac{3}{2}$. Notice that the numerators of these zeros, 2 and -3, are factors of the polynomial's constant, -6, and

---

PRESENTATION

Lesson Opener

Find each value for $f(x) = 3x^2 - 2x + 17$.

1. $f(0)$ 17

2. $f(-2)$ 33

3. What does the value $f(0)$ represent on the graph of the function?
the *y*-coordinate of the *y*-intercept

4. What does a root of $f(x) = 0$ represent on the graph of the function?
a zero of the function

Engage the students by explaining that the zeros of a function are one of its most important characteristics. They reveal the function's nature and the number of *x*-intercepts, which can represent possible solutions to real-world problems or reveal that there are no possible solutions to the scenario.

Share that the term *fundamental* implies "foundational," not "simple" or "basic." In fact, the proofs for many fundamental theorems are very complex. You may want to mention other fundamental theorems of mathematics, including the Fundamental Theorem of Arithmetic and the Fundamental Theorem of Calculus, before presenting the *Fundamental Theorem of Algebra*.

RATIONAL ZERO THEOREM

If $p(x) = a_nx^n + a_{n-1}x^{n-1} + \cdots + a_2x^2 + a_1x + a_0$ with integral coefficients and $\frac{c}{d}$ is a reduced rational zero of $p(x)$, then c is a factor of a_0 and d is a factor of a_n.

Example 1 Using the Rational Zero Theorem

Find the zeros of $p(x) = 6x^3 + 7x^2 - 1$.

Answer

factors of -1: ± 1

factors of 6: $\pm 1, \pm 2, \pm 3, \pm 6$

possible zeros: $\frac{c}{d} = \pm 1, \pm\frac{1}{2}, \pm\frac{1}{3}, \pm\frac{1}{6}$

1. Make a list of possible rational zeros by listing possible factors of the constant term and the leading coefficient.

$p(1) = 6 + 7 - 1 = 12$
$p(-1) = -6 + 7 - 1 = 0$

2. The Remainder Theorem can be used to quickly check some of the simpler possible zeros.

$$\begin{array}{r|rrrr} -1 & 6 & 7 & 0 & -1 \\ & & -6 & -1 & 1 \\ \hline & 6 & 1 & -1 & 0 \end{array}$$

3. Use synthetic division to divide $p(x)$ by $(x + 1)$ and find the coefficients of the depressed polynomial.

$p(x) = (x + 1)(6x^2 + x - 1)$
$(x + 1)(3x - 1)(2x + 1) = 0$

4. Set $p(x) = 0$ and factor the depressed polynomial.

$\therefore x = -1, \frac{1}{3}, -\frac{1}{2}$

5. Solve for the zeros using the Zero Product Property.

SKILL ✔ EXERCISE 5

Example 2 Solving a Polynomial Equation

Solve $x^4 + 9x^3 + 15x^2 + 9x = -14$.

Answer

$x^4 + 9x^3 + 15x^2 + 9x + 14 = 0$

1. Set the equation equal to 0.

factors of 14: $\pm 1, \pm 2, \pm 7, \pm 14$

factors of 1: ± 1

possible zeros: $\frac{c}{d} = \pm 1, \pm 2, \pm 7, \pm 14$

2. Make a list of possible rational zeros by listing possible factors of the constant term and the leading coefficient.

Test only $-1, -2, -7$, and -14.

3. Note that all the coefficients are positive and no positive number will make $p(x) = 0$.

$$\begin{array}{r|rrrrr} & 1 & 9 & 15 & 9 & 14 \\ -1 & 1 & 8 & 7 & 2 & 12 \\ -2 & 1 & 7 & 1 & 7 & 0 \end{array}$$

4. Use compressed synthetic division to test these possible roots.
Since -2 is a zero, $(x + 2)$ is a factor of $p(x)$.

$(x + 2)(x^3 + 7x^2 + x + 7) = 0$
$(x + 2)[x^2(x + 7) + (x + 7)] = 0$
$(x + 2)(x + 7)(x^2 + 1) = 0$

5. Factor $p(x)$ completely into linear factors. You can factor the cubic by grouping or by further synthetic division, trying only the possible negative roots of the depressed polynomial, -1 and -7.

$\therefore x = -2, -7$, and $\pm i$

6. Solve using the Zero Product Property.

SKILL ✔ EXERCISE 19

Additional Exercises

Find all the zeros of each polynomial function.

1. $f(x) = 2x^3 - 3x^2 - 5x + 6$ $1, 2, -\frac{3}{2}$

2. $f(x) = x^4 + x^3 - 11x^2 - 5x + 30$
$2, -3, \pm\sqrt{5}$

3. $f(x) = x^3 - \frac{11}{12}x^2 - \frac{5}{2}x + \frac{2}{3}$ $2, -\frac{4}{3}, \frac{1}{4}$

4. $f(x) = x^4 - \frac{11}{6}x^3 - \frac{7}{2}x^2 + \frac{22}{3}x - 2$
$\pm 2, \frac{3}{2}, \frac{1}{3}$

Analyze the real roots of the function $p(x) = 4x^4 - 5x^3 + x^2 + 6x - 5$.

5. Apply Descartes's Rule of Signs to describe the possible number of positive and negative zeros for $p(x)$.
3 or 1 positive zero; 1 negative zero

6. Use the Upper and Lower Bound Tests to show that any real zeros lie within the interval $[-2, 2]$.

$$\begin{array}{ccrrrrr} & & 4 & -5 & 1 & 6 & -5 \\ \text{U.B.} & 2 & 4 & 3 & 7 & 20 & 35 \text{ pass} \\ \text{L.B.} & -2 & 4 & -13 & 27 & -48 & 91 \text{ pass} \end{array}$$

Use the given zero to find the remaining zeros for each polynomial function.

7. $p(x) = x^3 - 6x^2 + 9x - 2$; $x = 2 - \sqrt{3}$
$2, 2 + \sqrt{3}$

8. $p(x) = 2x^4 - x^3 - 15x^2 + 23x + 15$; $x = 2 + i$ $-\frac{1}{2}, -3, 2 - i$

Consider using the prime factoring of a composite number to show how dividing by a factor produces another factor. Relate this process to the factoring of a polynomial when presenting the *Complete Linear Factorization Theorem* and the *Corollary of the Fundamental Theorem of Algebra*, which essentially states that an *n*th degree polynomial will have *n linear factors* and therefore *n* complex (real or imaginary) zeros.

While technology may be used to identify integral zeros and some rational zeros, exact values of irrational zeros and other rational zeros usually cannot be found from the calculator's decimal approximations. Explain that the *Rational Zero Theorem* defines a limited number of possible rational zeros. Use Example 1 and Example 2 to apply and review the preceding theorems.

Some texts present an Integral Root Theorem, stating that when the leading coefficient is 1, any rational zero will be an integer that is a factor of the constant term. We treat this corollary of the Rational Zero Theorem as a special case in which the only possible denominators are ± 1.

Example 3 demonstrates how the theorem can be applied when the coefficients are rational fractions. Have the students note that the zeros of $p(x)$ are identical to those of $24p(x)$, which is a vertical stretching of $p(x)$.

Interactive Activity Using technology, project or have the students graph $f(x) = x^2 - 8x + 14$. Then graph $4f(x)$ to demonstrate that the zeros remain the same through the vertical stretch. This can also be done dynamically with sliders.

Be sure the students recognize that the *n* roots of an *n*th-power polynomial can be rational, irrational, or imaginary. Point out that *Descartes's Rules of Signs* and the *Upper and Lower Bound Tests* help the students to limit their search for

Write a standard form polynomial function of least degree with integral coefficients that has the given zeros.

9. $x = 3 + 2i, \sqrt{2}$

$\quad p(x) = x^4 - 6x^3 + 11x^2 + 12x - 26$

10. $x = 3 - i, 2 - \sqrt{3}, 1$ (multiplicity 2)

$\quad p(x) = x^6 - 12x^5 + 56x^4 - 126x^3$
$\qquad\qquad + 137x^2 - 66x + 10$

Assignments

- **Minimum:** 1–3, 6–7, 9, 12–14, 19–20, 22–23, 25, 29, 31, 37–38, 45, 48, 50–51
- **Standard:** 2, 5, 8–11, 14, 16, 19, 22–23, 27, 29; 30–40 even; 41, 45–48, 52–54
- **Extended:** 4, 7, 9; 11, 14, 17, . . . , 32; 35–37, 40, 42, 44–45, 47, 51–54

Assessment

- Quiz 2B covers Sections 2.3–2.4.

While the Rational Zero Theorem requires integral coefficients, Example 3 illustrates how it can be used when the polynomial's coefficients are rational.

Example 3 Using the Rational Zero Theorem with Rational Coefficients

Find the zeros of $p(x) = x^3 - \frac{13}{12}x^2 + \frac{3}{8}x - \frac{1}{24}$.

Answer

$24p(x) = 24x^3 - 26x^2 + 9x - 1$

1. Multiply by the least common denominator to obtain integral coefficients.

factors of -1: ± 1

factors of 24: $\pm 1, \pm 2, \pm 3, \pm 4, \pm 6, \pm 8, \pm 12, \pm 24$

possible zeros: $\pm 1, \pm \frac{1}{2}, \pm \frac{1}{3}, \pm \frac{1}{4}, \pm \frac{1}{6}, \pm \frac{1}{8}, \pm \frac{1}{12}, \pm \frac{1}{24}$

2. Make a list of the possible rational zeros for $24p(x)$.

$$
\begin{array}{r|rrrr}
 & 24 & -26 & 9 & -1 \\
\frac{1}{2} & 24 & -14 & 2 & 0 \\
\end{array}
$$

3. Compressed synthetic division reveals that $\left(x - \frac{1}{2}\right)$ is a factor of $24p(x)$ and provides the coefficients of the depressed polynomial.

$24p(x) = \left(x - \frac{1}{2}\right)(24x^2 - 14x + 2)$

4. Factor $24p(x)$.

$24p(x) = 2\left(x - \frac{1}{2}\right)(12x^2 - 7x + 1)$

a. Factor 2 out of the depressed polynomial.

$24p(x) = 2\left(x - \frac{1}{2}\right)(3x - 1)(4x - 1)$

b. Factor the remaining quadratic.

$p(x) = \frac{1}{12}\left(x - \frac{1}{2}\right)(3x - 1)(4x - 1)$

5. Divide both sides by 24 to solve for $p(x)$.

The zeros of $p(x)$ are $x = \frac{1}{2}, \frac{1}{3}$, and $\frac{1}{4}$.

6. Apply the Zero Product Property.

SKILL ✔ **EXERCISE 25**

When all possible roots suggested by the Rational Zero Theorem fail, you know that the zeros are irrational or imaginary. For example, it can be shown (either by synthetic division or substitution) that ± 1 and ± 3 are not zeros of $p(x) = x^2 + 3x - 3$. Fortunately, the quadratic formula can be used to find the function's two irrational roots: $\frac{-3 \pm \sqrt{21}}{2}$.

The following theorems can help limit your search for real zeros of a polynomial function whether they are rational or irrational.

DESCARTES'S RULE OF SIGNS

If $p(x) = a_n x^n + a_{n-1}x^{n-1} + \cdots + a_2 x^2 + a_1 x + a_0$ with real coefficients, then

- the number of positive real zeros of $p(x)$ is the number of variations in the sign of $p(x)$ or less than that number by an even number and
- the number of negative real zeros of $p(x)$ is the number of variations in the sign of $p(-x)$ or less than that number by an even number.

This rule allows you to describe the possible number of positive and negative zeros.

real roots. These tests are used in Examples 4 and 5. When applying either test, the synthetic division can be terminated as soon as the test fails. Consider explaining that these tests are conditionals, and that a number may fail the test but still be a bound for the zeros. (See exercise 44.)

Introduce the *Conjugate Root Theorems*, emphasizing the coefficient restrictions for each theorem. In Example 6 you may want to use evaluation or synthetic substitution to show that none of the possibilities suggested by the Rational Zero Theorem (± 1 and ± 2) are zeros. This indicates that all zeros would have to be irrational or complex. In step 4, remind the students that synthetic division is restricted to divisors of the form $x - c$. An alternative solution would be to use consecutive synthetic division by $x - i$ and $x + i$ before using the quadratic formula.

When writing the factorization of $p(x)$ in Example 6, $x^2 + 2x - 2$ is said to be irreducible (or prime) over the rational numbers, and $x^2 + 1$ is said to be irreducible (or prime) over the real numbers (as well as the rational numbers).

Example 7 illustrates finding a polynomial function with given rational, irrational, and imaginary zeros.

TIPS

Ex. 15 Descartes's Rule of Signs implies that there are no positive roots and 3 or 1 negative root. But there cannot be 3 negative roots since a quartic can have only 4 complex roots and at least 2 are imaginary. Therefore there is 1 negative root, and the fourth root must be 0 since any other irrational or imaginary roots would occur in conjugate pairs.

Ex. 39 The Upper Bound Test will fail for $x = 2$, which is a zero of the function. This is investigated in exercise 44.

Example 4 — Applying Descartes's Rule of Signs

Describe the possible number of positive and negative zeros of
$p(x) = 3x^4 - 5x^3 + 4x^2 - 10x - 4$.

Answer

$p(x)$ has 3 or 1 positive zero.

1. There are 3 sign changes in the coefficients of $p(x)$.
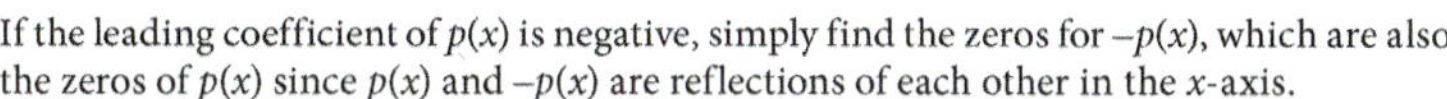
$p(x) = 3x^4 - 5x^3 + 4x^2 - 10x - 4$

$p(x)$ has 1 negative zero.

2. The coefficients of $p(-x)$ have only 1 sign change.

$p(-x) = 3x^4 + 5x^3 + 4x^2 + 10x - 4$

— SKILL ✓ **EXERCISE 9a**

The Upper and Lower Bound Tests can establish an interval containing all of a function's real zeros.

UPPER AND LOWER BOUND TESTS

If $p(x)$ is a polynomial function of degree $n \geq 1$ with a positive leading coefficient and $p(x)$ is divided by $x - c$ using synthetic division,

(1) $c \geq 0$ is an **upper bound** of the zeros if every coefficient of the depressed polynomial is nonnegative and

(2) $c \leq 0$ is a **lower bound** of the zeros if the coefficients of the depressed polynomial and its remainder are alternately nonnegative and nonpositive.

The Upper and Lower Bound Tests can ensure that a graphing window illustrates all the real zeros of a function.

If the leading coefficient of $p(x)$ is negative, simply find the zeros for $-p(x)$, which are also the zeros of $p(x)$ since $p(x)$ and $-p(x)$ are reflections of each other in the x-axis.

Example 5 — Applying the Upper and Lower Bound Tests

Show that any real zeros of $p(x) = 3x^4 - 5x^3 + 4x^2 - 10x - 4$ lie within the interval $[-1, 2]$.

Answer

$$2 \;\Big|\; \begin{array}{ccccc} 3 & -5 & 4 & -10 & -4 \\ & 3 & 1 & 6 & 2 \quad 0 \end{array}$$

1. Since all the coefficients of the depressed polynomial when $p(x) \div (x - 2)$ are nonnegative, 2 is an upper bound of the zeros for $p(x)$.

$$-1 \;\Big|\; \begin{array}{ccccc} 3 & -5 & 4 & -10 & -4 \\ & 3 & -8 & 12 & -22 \quad 18 \end{array}$$

2. Since the coefficients of the depressed polynomial when $p(x) \div [x - (-1)]$ are alternately nonnegative and nonpositive, -1 is a lower bound of the zeros for $p(x)$.

— SKILL ✓ **EXERCISE 9b**

The synthetic division testing the upper bound of 2 in Example 5 reveals that 2 is also a zero and that $p(x)$ can be factored into

$$p(x) = (x - 2)(3x^3 + x^2 + 6x + 2).$$

The graph of $p(x)$ suggests that $-\frac{1}{3}$ is a zero and can be used to factor the depressed polynomial.

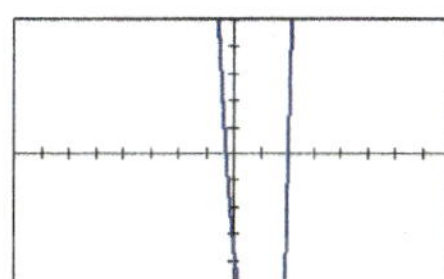

$$-\frac{1}{3} \;\Big|\; \begin{array}{cccc} 3 & 1 & 6 & 2 \\ & 3 & 0 & 6 \quad 0 \end{array}$$

$$p(x) = (x - 2)\left(x + \tfrac{1}{3}\right)(3x^2 + 6)$$

$$p(x) = 3(x - 2)\left(x + \tfrac{1}{3}\right)(x^2 + 2)$$

Solutions

▶ A. Exercises

1. $\dfrac{\text{factors of } 2}{\text{factors of } 2} = \dfrac{\pm 1, \pm 2}{\pm 1, \pm 2} = \pm 1, \pm 2, \pm \tfrac{1}{2}$

2. $\dfrac{\text{factors of } 4}{\text{factors of } 6} = \dfrac{\pm 1, \pm 2, \pm 4}{\pm 1, \pm 2, \pm 3, \pm 6}$

   $= \pm 1, \pm 2, \pm 4, \pm \tfrac{1}{2}, \pm \tfrac{1}{3}, \pm \tfrac{2}{3}, \pm \tfrac{4}{3}, \pm \tfrac{1}{6}$

3. $(x - 2)(x + 2)(x - 4)(x + 4) = 0$

   $x = \pm 2, \pm 4$

4. possible zeros: $\pm 1, \pm 2, \pm 3, \pm 6$

$$1 \;\Big|\; \begin{array}{cccc} 1 & -2 & -5 & 6 \\ & 1 & -1 & -6 \quad 0 \end{array}$$

   $(x - 1)(x^2 - x - 6) = 0$
   $(x - 1)(x - 3)(x + 2) = 0$
   $x = -2, 1, 3$

5. possible zeros: $\pm 1, \pm 2, \pm 3, \pm 4, \pm 6, \pm 12$

$$-1 \;\Big|\; \begin{array}{cccc} 1 & 0 & -13 & -12 \\ & 1 & -1 & -12 \quad 0 \end{array}$$

   $(x + 1)(x^2 - x - 12) = 0$
   $(x + 1)(x - 4)(x + 3) = 0$
   $x = -3, -1, 4$

6. possible zeros: $\pm 1, \pm 2, \pm 4, \pm 8$

$$2 \;\Big|\; \begin{array}{cccc} 1 & -6 & 12 & -8 \\ & 1 & -4 & 4 \quad 0 \end{array}$$

   $(x - 2)(x^2 - 4x + 4) = 0$
   $(x - 2)(x - 2)^2 = 0$
   $(x - 2)^3 = 0$
   $x = 2$ (multiplicity 3)

7. possible zeros: $\pm 1, \pm 2, \pm 4, \pm 8, \pm 16$

$$\begin{array}{l} 1 \;\Big|\; \begin{array}{ccccc} 1 & -10 & 33 & -40 & 16 \\ & 1 & -9 & 24 & -16 \quad 0 \end{array} \\ 1 \;\Big|\; \begin{array}{cccc} 1 & -8 & 16 & 0 \end{array} \end{array}$$

   $(x - 1)(x - 1)(x^2 - 8x + 16) = 0$
   $(x - 1)^2(x - 4)^2 = 0$
   $x = 1, 4$ (both multiplicity 2)

8. possible zeros: $\pm 1, \pm 2, \pm 3, \pm 4, \pm 6, \pm 12$

$$\begin{array}{l} -1 \;\Big|\; \begin{array}{ccccc} 1 & 0 & -9 & 4 & 12 \\ & 1 & -1 & -8 & 12 \quad 0 \end{array} \\ 2 \;\Big|\; \begin{array}{cccc} 1 & 1 & -6 & 0 \end{array} \end{array}$$

   $(x + 1)(x - 2)(x^2 + x - 6) = 0$
   $(x + 1)(x - 2)(x - 2)(x + 3) = 0$
   $x = -3, -1, 2$ (multiplicity 2)

9a. $p(x)$ has 4 sign changes, so 4, 2, or 0 positive zeros; 0 negative zeros
$p(-x) = x^4 + 4x^3 + 4x^2 + 4x + 3$
and has no sign changes, so 0 negative zeros

9b.

$$\begin{array}{ll} & \begin{array}{ccccc} 1 & -4 & 4 & -4 & 3 \end{array} \\ \text{U.B.} \; 4 \;\Big|\; & \begin{array}{ccccc} 1 & 0 & 4 & 12 & 51 \end{array} \\ \text{L.B.} \; 0 \;\Big|\; & \begin{array}{ccccc} 1 & -4 & 4 & -4 & 3 \end{array} \end{array}$$

4 passes the Upper Bound Test and 0 passes the Lower Bound Test.

10a. $p(x)$ has 3 sign changes, so 3 or 1 positive zeros
$p(-x) = x^4 - x^3 - x^2 - x - 2$
and has 1 sign change, so 1 negative zero

10b.

$$\begin{array}{ll} & \begin{array}{ccccc} 1 & 1 & -1 & 1 & -2 \end{array} \\ \text{U.B.} \; 2 \;\Big|\; & \begin{array}{ccccc} 1 & 3 & 5 & 11 & 20 \end{array} \\ \text{L.B.} \; -3 \;\Big|\; & \begin{array}{ccccc} 1 & -2 & 5 & -14 & 40 \end{array} \end{array}$$

2 passes the Upper Bound Test and -3 passes the Lower Bound Test.

11. $p(x) = (x - 3)(x - \sqrt{2})(x + \sqrt{2})$
$= (x - 3)(x^2 - 2)$
$= x^3 - 3x^2 - 2x + 6$

12. $p(x) = (x - 1)(x - 4i)(x + 4i)$
$= (x - 1)(x^2 + 16)$
$= x^3 - x^2 + 16x - 16$

13–17. Recall that irrational and complex zeros appear in conjugate pairs.

13. $p(x) = 2(x^3 - 6x^2 - 3x + 18)$
$(x - \sqrt{3})(x + \sqrt{3}) = x^2 - 3$

$$
\begin{array}{r}
x - 6 \\
x^2 - 3 \overline{) x^3 - 6x^2 - 3x + 18} \\
-\underline{(x^3 \qquad - 3x)} \\
-6x^2 \qquad + 18 \\
-\underline{(-6x^2 \qquad + 18)} \\
0
\end{array}
$$

$p(x) = 2(x - \sqrt{3})(x + \sqrt{3})(x - 6)$
$x = \pm\sqrt{3}, 6$

14. $(x - i)(x + i) = x^2 + 1$

$$
\begin{array}{r}
x + 3 \\
x^2 + 1 \overline{) x^3 + 3x^2 + x + 3} \\
-\underline{(x^3 \qquad + x)} \\
3x^2 \qquad + 3 \\
-\underline{(3x^2 \qquad + 3)} \\
0
\end{array}
$$

$p(x) = (x - i)(x + i)(x + 3)$
$x = \pm i, -3$

15. $p(x) = x(x^3 + x^2 + 4x + 4)$
$(x + 2i)(x - 2i) = x^2 + 4$

$$
\begin{array}{r}
x + 1 \\
x^2 + 4 \overline{) x^3 + x^2 + 4x + 4} \\
-\underline{(x^3 \qquad + 4x)} \\
x^2 \qquad + 4 \\
-\underline{(x^2 \qquad + 4)} \\
0
\end{array}
$$

$p(x) = x(x + 2i)(x - 2i)(x + 1)$
$x = 0, \pm 2i, -1$

16. $\left[x - (2 - \sqrt{2}) \right]\left[x - (2 + \sqrt{2}) \right]$
$= \left[(x - 2) + \sqrt{2} \right]\left[(x - 2) - \sqrt{2} \right]$
$= (x - 2)^2 - 2 = x^2 - 4x + 2$

$$
\begin{array}{r}
x - 5 \\
x^2 - 4x + 2 \overline{) x^3 - 9x^2 + 22x - 10} \\
-\underline{(x^3 - 4x^2 + 2x)} \\
-5x^2 + 20x - 10 \\
-\underline{(-5x^2 + 20x - 10)} \\
0
\end{array}
$$

$p(x) = (x - 2 + \sqrt{2})(x - 2 - \sqrt{2})$
$\qquad\qquad\qquad\qquad \cdot (x - 5)$

$x = 2 \pm \sqrt{2}, 5$

17. $[x - (-1 - 2i)][x - (-1 + 2i)]$
$[(x + 1) + 2i][(x + 1) - 2i]$
$(x + 1)^2 - 4i^2 = x^2 + 2x + 5$

$$
\begin{array}{r}
x - 1 \\
x^2 + 2x + 5 \overline{) x^3 + x^2 + 3x - 5} \\
-\underline{(x^3 + 2x^2 + 5x)} \\
-x^2 - 2x - 5 \\
-\underline{(-x^2 - 2x - 5)} \\
0
\end{array}
$$

$p(x) = (x + 1 + 2i)(x + 1 - 2i)$
$\qquad\qquad\qquad\qquad \cdot (x - 1)$

$x = -1 \pm 2i, 1$

18. possible zeros: $\pm 1, \pm 3$

$$
\begin{array}{r}
1 \;-1 \;\;\; 3 \;-3 \\
1 \overline{) 1 \quad 0 \quad 3 \quad 0}
\end{array}
$$

$(x - 1)(x^2 + 3) = 0$
$x = 1 \quad \text{or} \quad x^2 = -3$
$x = 1, \pm\sqrt{3}i$

19. $x^4 - 3x^3 - 4x^2 + 12x = 0$
$x(x^3 - 3x^2 - 4x + 12) = 0$

$$
\begin{array}{r}
1 \;-3 \;-4 \;\; 12 \\
2 \overline{) 1 \;-1 \;-6 \quad 0}
\end{array}
$$

$x(x - 2)(x^2 - x - 6) = 0$
$x(x - 2)(x - 3)(x + 2) = 0$
$x = 0, \pm 2, 3$

The remaining two zeros, $x = \pm\sqrt{2}i$, are complex numbers, which are neither positive nor negative.

The following theorems describe the occurrence of irrational and complex roots for polynomials with appropriate coefficients.

▌CONJUGATE ROOT THEOREMS

1. If $p(x)$ is a polynomial with rational coefficients and $a + c\sqrt{b}$ is a zero of $p(x)$, then its conjugate, $a - c\sqrt{b}$, is also a zero of $p(x)$.

2. If $p(x)$ is a polynomial with real coefficients and $a + bi$ is a zero of $p(x)$, then its conjugate, $a - bi$, is also a zero of $p(x)$.

Example 6 Applying a Conjugate Root Theorem

List all the zeros of $p(x) = x^4 + 2x^3 - x^2 + 2x - 2$ if $-i$ is a zero.

Answer

Note that i is also a zero of $p(x)$.

1. Apply the second Conjugate Root Theorem since all coefficients are real.

$(x - i)$ and $(x + i)$ are linear factors of $p(x)$.

2. Apply the Factor Theorem using the zeros $\pm i$.

$p(x) = (x + i)(x - i) \cdot q(x)$
$p(x) = (x^2 + 1) \cdot q(x)$

3. Express $p(x)$ as the product of known factors and an unknown quotient polynomial, $q(x)$.

$$
\begin{array}{r}
x^2 + 2x - 2 \\
x^2 + 1 \overline{) x^4 + 2x^3 - x^2 + 2x - 2} \\
-\underline{(x^4 \qquad + x^2)} \\
2x^3 - 2x^2 + 2x \\
-\underline{(2x^3 \qquad + 2x)} \\
-2x^2 \qquad - 2 \\
-\underline{(-2x^2 \qquad - 2)} \\
0
\end{array}
$$

4. Divide to find the factor $q(x)$. Synthetic division cannot be used since the divisor has a degree of 2.

$q(x) = x^2 + 2x - 2$
$x = \dfrac{-2 \pm \sqrt{2^2 - 4(1)(-2)}}{2(1)}$
$= -1 \pm \sqrt{3} \approx -2.73, 0.73$

5. Since $q(x)$ cannot be factored with integral coefficients, use the quadratic formula to find its irrational zeros.

$\therefore p(x) = 0$ when $x = \pm i, -1 \pm \sqrt{3}$

SKILL ✔ **EXERCISE 17**

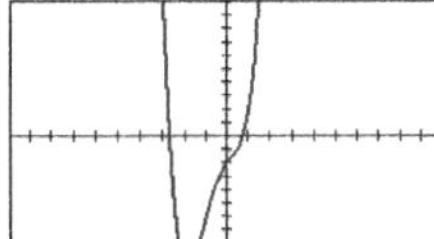

Note that Descartes's Rule of Signs predicts that the function in Example 6 has either 3 or 1 positive real root and 1 negative real root, and that the Upper and Lower Bound Tests can confirm that there are no real zeros outside the interval $[-3, 1]$.

Bound Tests		1	2	–1	2	–2	
Upper	1	1	3	2	4	2	(all positive coefficients)
Lower	–3	1	–1	2	–4	10	(alternating nonnegative and nonpositive coefficients)

The graph of the function confirms the two irrational zeros. The Complete Linear Factorization Theorem guarantees that the four zeros can be used to write the complete linear factorization of the polynomial.

$$p(x) = (x - i)[x - (-i)]\left[x - \left(-1 + \sqrt{3}\right)\right]\left[x - \left(-1 - \sqrt{3}\right)\right]$$

If the zeros occur in conjugate pairs, those factors are frequently combined and written as an *irreducible quadratic factor.*

$$p(x) = (x^2 + 1)(x^2 + 2x - 2)$$

The Conjugate Root Theorems and the Factor Theorem enable you to construct a polynomial of least degree with integral coefficients from the known zeros.

Write a standard form polynomial function of least degree with integral coefficients that has zeros of 0 (multiplicity 3), $\frac{2}{3}$, $1 + \sqrt{2}$, and $3 - 2i$.

Answer

$(x - 0)^3 = x^3$

1. The multiplicity of 3 implies that the factor $(x - 0)$ appears 3 times.

$3\left(x - \frac{2}{3}\right) = 3x - 2$

2. Multiply the factor from the zero of $\frac{2}{3}$ by a constant to eliminate fractions.

$\left[x - \left(1 + \sqrt{2}\right)\right]\left[x - \left(1 - \sqrt{2}\right)\right]$
$= \left[(x - 1) - \sqrt{2}\right]\left[(x - 1) + \sqrt{2}\right]$
$= \left[(x - 1)^2 - 2\right] = x^2 - 2x - 1$

3. A zero of $1 + \sqrt{2}$ implies that its conjugate $1 - \sqrt{2}$ is also a zero. Note how the Associative Property is used to multiply the related factors using the difference of squares pattern.

$[x - (3 - 2i)][x - (3 + 2i)]$
$= [(x - 3) + 2i][(x - 3) - 2i]$
$= (x - 3)^2 - 4i^2 = x^2 - 6x + 13$

4. The complex zero $3 - 2i$ implies that its conjugate $3 + 2i$ is also a zero. The Associative Property is used to multiply the related factors using the difference of squares pattern.

$p(x) = x^3(3x - 2)(x^2 - 2x - 1)(x^2 - 6x + 13)$
$= (3x^4 - 2x^3)(x^4 - 8x^3 + 24x^2 - 20x - 13)$
$= 3x^8 - 26x^7 + 88x^6 - 108x^5 + x^4 + 26x^3$

5. Multiply the factors to find a least degree polynomial with integral coefficients.

SKILL ✔ EXERCISE 31

A. Exercises

List all the possible rational zeros for each polynomial function.

1. $p(x) = 2x^3 + 3x^2 - x + 2$ $\pm 1, \pm 2, \pm \frac{1}{2}$

2. $p(x) = 6x^3 + 5x^2 - 12x + 4$ $\pm 1, \pm 2, \pm 4, \pm \frac{1}{2}, \pm \frac{1}{3}, \pm \frac{2}{3}, \pm \frac{4}{3}, \pm \frac{1}{6}$

Find all the zeros for each polynomial function. Indicate any multiplicities greater than 1.

3. $p(x) = (x^2 - 4)(x^2 - 16)$ $x = \pm 2, \pm 4$

4. $p(x) = x^3 - 2x^2 - 5x + 6$ $x = -2, 1, 3$

5. $p(x) = x^3 - 13x - 12$ $x = -3, -1, 4$

6. $p(x) = x^3 - 6x^2 + 12x - 8$ $x = 2$ (multiplicity 3)

7. $p(x) = x^4 - 10x^3 + 33x^2 - 40x + 16$ $x = 4, 1$ (both multiplicity 2)

8. $p(x) = x^4 - 9x^2 + 4x + 12$ $x = -3, -1, 2$ (multiplicity 2)

Given the polynomial function $p(x)$,
a. use Descartes's Rule of Signs to describe the possible number of positive and negative zeros and
b. use Upper and Lower Bound Tests to show that any real zeros lie within the given interval.

9. $p(x) = x^4 - 4x^3 + 4x^2 - 4x + 3$, $[0, 4]$

10. $p(x) = x^4 + x^3 - x^2 + x - 2$, $[-3, 2]$

Write a polynomial function of least degree with integral coefficients that has the given zeros.

11. $x = 3, \pm \sqrt{2}$

12. $x = 1, \pm 4i$

20. $x(x^2 - x - 1) = 0$
$x = 0$ or $x^2 - x - 1 = 0$
$$x = \frac{-(-1) \pm \sqrt{(-1)^2 - 4(1)(-1)}}{2(1)} = \frac{1 \pm \sqrt{5}}{2}$$

21. $2\left(x^3 + \frac{7}{2}x^2 - 14x + 6\right) = 0$
$2x^3 + 7x^2 - 28x + 12 = 0$
possible zeros: $\pm 1, \pm \frac{1}{2}, \pm 2, \pm 3, \pm \frac{3}{2}, \pm 4, \pm 6, \pm 12$

$$\begin{array}{r|rrrr} & 2 & 7 & -28 & 12 \\ 2 & & 2 & 11 & -6 \\ \hline & 2 & 11 & -6 & 0 \end{array}$$

$(x - 2)(2x^2 + 11x - 6) = 0$
$(x - 2)(2x - 1)(x + 6) = 0$
$x = 2, \frac{1}{2}, -6$

22. possible zeros: $\pm 1, \pm 2, \pm 3, \pm 4, \pm 6, \pm 12$

$$\begin{array}{r|rrrr} & 1 & -1 & -8 & 12 \\ 2 & & 2 & 1 & -6 \\ \hline & 1 & 1 & -6 & 0 \end{array}$$

$(x - 2)(x^2 + x - 6) = 0$
$(x - 2)(x + 3)(x - 2) = 0$
$x = -3, 2$ (multiplicity 2)

23. possible zeros: $\pm 1, \pm 3, \pm 5, \pm 9, \pm 15, \pm 45$

$$\begin{array}{r|rrrr} & 1 & 3 & -1 & 45 \\ -5 & & 1 & -2 & 9 \\ \hline & 1 & -2 & 9 & 0 \end{array}$$

$(x + 5)(x^2 - 2x + 9) = 0$
$x = -5$ or
$$x = \frac{-(-2) \pm \sqrt{(-2)^2 - 4(1)(9)}}{2(1)}$$
$$= \frac{2 \pm \sqrt{-32}}{2} = \frac{2 \pm 4\sqrt{2}\,i}{2}$$
$$= 1 \pm 2\sqrt{2}\,i$$
$x = -5, 1 \pm 2\sqrt{2}\,i$

24. possible zeros: $\pm 1, \pm 3, \pm 9, \pm \frac{1}{5}, \pm \frac{3}{5}, \pm \frac{9}{5}$

$$\begin{array}{r|rrrrr} & 5 & -16 & 18 & -48 & 9 \\ 3 & & 5 & -1 & 15 & -3 \\ \hline \frac{1}{5} & 5 & -1 & 15 & -3 & 0 \\ & 5 & 0 & 15 & 0 & \end{array}$$

$(x - 3)\left(x - \frac{1}{5}\right)(5x^2 + 15) = 0$
$5(x - 3)\left(x - \frac{1}{5}\right)(x^2 + 3) = 0$
$x = 3, \frac{1}{5}, \pm \sqrt{3}\,i$

25. $4p(x) = 4x^4 - 11x^3 + 9x^2 - x - 1$
possible zeros: $\pm 1, \pm \frac{1}{2}, \pm \frac{1}{4}$

$$\begin{array}{r|rrrrr} & 4 & -11 & 9 & -1 & -1 \\ 1 & & 4 & -7 & 2 & 1 \\ \hline 1 & 4 & -7 & 2 & 1 & 0 \\ & 4 & -3 & -1 & 0 & \end{array}$$

$(x - 1)^2(4x^2 - 3x - 1) = 0$
$(x - 1)^2(4x + 1)(x - 1) = 0$
$(x - 1)^3(4x + 1) = 0$
$x = -\frac{1}{4}, 1$ (multiplicity 3)

26. $30p(x) = 30x^3 - 17x^2 - 3x + 2$
possible zeros: $\pm 1, \pm 2, \pm \frac{1}{2}, \pm \frac{1}{3}, \pm \frac{1}{5}, \pm \frac{1}{6}, \pm \frac{1}{10}, \pm \frac{1}{15}, \pm \frac{1}{30}, \pm \frac{2}{3}, \pm \frac{2}{5}, \pm \frac{2}{15}$

$$\begin{array}{r|rrrr} & 30 & -17 & -3 & 2 \\ \frac{1}{2} & & 30 & -2 & -4 \\ \hline & 30 & -2 & -4 & 0 \end{array}$$

$30p(x) = 2\left(x - \frac{1}{2}\right)(15x^2 - x - 2)$
$0 = (2x - 1)(5x - 2)(3x + 1)$
$x = \frac{1}{2}, \frac{2}{5}, -\frac{1}{3}$

27. $24p(x) = 24x^4 + 44x^3 + 6x^2 - 11x - 3$

possible zeros: $\pm1, \pm\frac{1}{2}, \pm\frac{1}{3}, \pm\frac{1}{4}, \pm\frac{1}{6},$

$\pm\frac{1}{8}, \pm\frac{1}{12}, \pm\frac{1}{24}, \pm3, \pm\frac{3}{2}, \pm\frac{3}{4}, \pm\frac{3}{8}$

$$\begin{array}{c|ccccc} & 24 & 44 & 6 & -11 & -3 \\ \hline \frac{1}{2} & 24 & 56 & 34 & 6 & 0 \\ \hline -\frac{1}{2} & 24 & 44 & 12 & 0 & \end{array}$$

$\left(x - \frac{1}{2}\right)\left(x + \frac{1}{2}\right)(24x^2 + 44x + 12) = 0$

$4\left(x - \frac{1}{2}\right)\left(x + \frac{1}{2}\right)(6x^2 + 11x + 3) = 0$

$(2x - 1)(2x + 1)(2x + 3)(3x + 1) = 0$

$x = \pm\frac{1}{2}, -\frac{3}{2}, -\frac{1}{3}$

28. $16p(x) = 16x^4 - 44x^3 + 8x^2 + 29x + 6$

possible zeros: $\pm1, \pm2, \pm3, \pm6, \pm\frac{1}{2},$

$\pm\frac{3}{2}, \pm\frac{1}{4}, \pm\frac{3}{4}, \pm\frac{1}{8}, \pm\frac{3}{8}, \pm\frac{1}{16}, \pm\frac{3}{16}$

$$\begin{array}{c|ccccc} & 16 & -44 & 8 & 29 & 6 \\ \hline 2 & 16 & -12 & -16 & -3 & 0 \\ \hline -\frac{1}{2} & 16 & -20 & -6 & 0 & \end{array}$$

$2(x - 2)\left(x + \frac{1}{2}\right)(8x^2 - 10x - 3) = 0$

$(x - 2)(2x + 1)(4x + 1)(2x - 3) = 0$

$x = 2, -\frac{1}{2}, -\frac{1}{4}, \frac{3}{2}$

29. $p(x) = (x - 2)\left[x - (1 + \sqrt{3})\right]$

$\qquad\qquad \cdot \left[x - (1 - \sqrt{3})\right]$

$= (x - 2)[(x - 1) - \sqrt{3}]$

$\qquad\qquad \cdot [(x - 1) + \sqrt{3}]$

$= (x - 2)[(x - 1)^2 - 3]$

$= (x - 2)(x^2 - 2x - 2)$

$= x^3 - 4x^2 + 2x + 4$

30. $p(x) = (x + 2)[x - (3 + 2i)]$

$\qquad\qquad \cdot [x - (3 - 2i)]$

$= (x + 2)[(x - 3) - 2i][(x - 3) + 2i]$

$= (x + 2)[(x - 3)^2 + 4]$

$= (x + 2)(x^2 - 6x + 13)$

$= x^3 - 4x^2 + x + 26$

31. $p(x) = 4\left(x + \frac{3}{4}\right)[x - (4 - 3i)]$

$\qquad\qquad \cdot [x - (4 + 3i)]$

$= (4x + 3)[(x - 4) + 3i]$

$\qquad\qquad \cdot [(x - 4) - 3i]$

$= (4x + 3)[(x - 4)^2 + 9]$

$= (4x + 3)(x^2 - 8x + 25)$

$= 4x^3 - 29x^2 + 76x + 75$

Use the given zero to find the remaining zeros for each polynomial function.

13. $p(x) = 2x^3 - 12x^2 - 6x + 36; x = \sqrt{3}$ $x = -\sqrt{3}, 6$

14. $p(x) = x^3 + 3x^2 + x + 3; x = i$ $x = -i, -3$

15. $p(x) = x^4 + x^3 + 4x^2 + 4x; x = 2i$ $x = 0, -1, -2i$

16. $p(x) = x^3 - 9x^2 + 22x - 10; x = 2 - \sqrt{2}$ $x = 2 + \sqrt{2}, 5$

17. $p(x) = x^3 + x^2 + 3x - 5; x = -1 - 2i$ $x = -1 + 2i, 1$

Solve.

18. $x^3 - x^2 + 3x - 3 = 0$ $x = 1, \pm\sqrt{3}i$

19. $x^4 - 3x^3 - 4x^2 + 12x = 0$ $x = 0, \pm2, 3$

20. $x^3 - x^2 = x$ $x = 0, \frac{1 \pm \sqrt{5}}{2}$

21. $x^3 + \frac{7}{2}x^2 = 14x - 6$

Find all the zeros of each polynomial function. $x = -6, \frac{1}{2}, 2$

22. $p(x) = x^3 - x^2 - 8x + 12$ $x = -3, 2$ (multiplicity 2)

23. $p(x) = x^3 + 3x^2 - x + 45$ $x = -5, 1 \pm 2\sqrt{2}i$

24. $p(x) = 5x^4 - 16x^3 + 18x^2 - 48x + 9$ $x = 3, \frac{1}{5}, \pm\sqrt{3}i$

25. $p(x) = x^4 - \frac{11}{4}x^3 + \frac{9}{4}x^2 - \frac{1}{4}x - \frac{1}{4}$ $x = -\frac{1}{4}, 1$ (multiplicity 3)

26. $p(x) = x^3 - \frac{17}{30}x^2 - \frac{1}{10}x + \frac{1}{15}$ $x = \frac{1}{2}, \frac{2}{5}, -\frac{1}{3}$

27. $p(x) = x^4 + \frac{11}{6}x^3 + \frac{1}{4}x^2 - \frac{11}{24}x - \frac{1}{8}$ $x = \pm\frac{1}{2}, -\frac{3}{2}, -\frac{1}{3}$

28. $p(x) = x^4 - \frac{11}{4}x^3 + \frac{1}{2}x^2 + \frac{29}{16}x + \frac{3}{8}$ $x = 2, -\frac{1}{2}, -\frac{1}{4}, \frac{3}{2}$

Write a standard form polynomial function of least degree with integral coefficients that has the given zeros.

29. $x = 2, 1 \pm \sqrt{3}$

30. $x = -2, 3 \pm 2i$

31. $x = -\frac{3}{4}, 4 - 3i$

32. $x = -\frac{1}{2}, \frac{2}{3}, 2 + i$

33. $x = -2$ (multiplicity 2), $1 + \sqrt{5}, 2 - 3i$

Use technology to find a zero. Then use synthetic division to factor the polynomial.

34. $p(x) = 21x^3 + 29x^2 - 24x + 4$ $(x + 2)(3x - 1)(7x - 2)$

35. $p(x) = 5x^3 + 2x^2 - 10x - 4$ $(5x + 2)(x^2 - 2)$

Use the Upper and Lower Bound Tests to determine if the window illustrates all the real zeros of each function. If not, state and verify another integral interval that contains all the real zeros of the function.

36. $p(x) = x^5 - 6x^4 - 10x^3 + 5x^2 + 8x + 1$

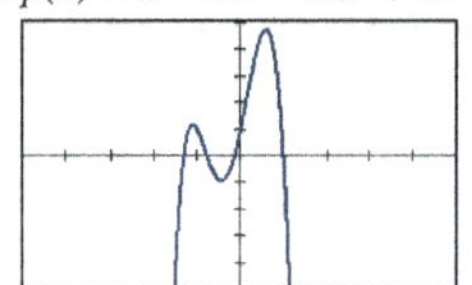

no; Answers will vary.

---

37. $p(x) = x^4 + x^3 - x^2 + x - 2$ yes

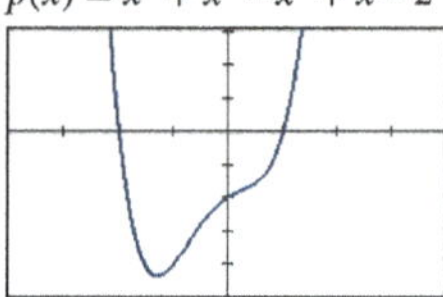

38. $p(x) = x^3 + 8x^2 - 6x - 3$ no; Answers will vary.

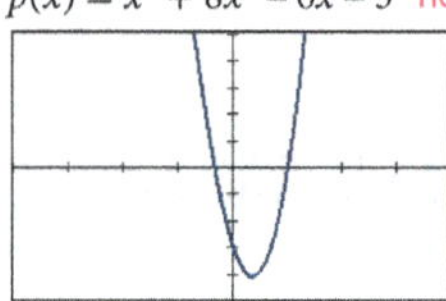

39. $p(x) = -x^4 - 4x^3 + 13x^2 + 4x - 12$

no; Answers will vary.

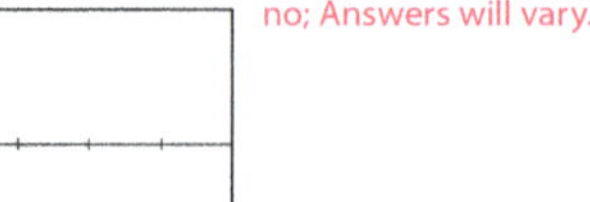

40. Explain: If a fifth-degree polynomial function with rational coefficients has zeros at $x = \pm1$ and $x = \sqrt{3}$, what is the maximum multiplicity for each given zero? Explain your reasoning.

41. Discover: Consider a polynomial function whose only zeros are $x = \sqrt{3}$ and $x = -2$.

  a. Find a quadratic function with these zeros.

  b. Show that $x = -\sqrt{3}$ is not a zero of the function found in step *a*.

  c. Explain why your findings are not a contradiction of the first Conjugate Root Theorem.

42. Discover: Consider a polynomial function whose only zeros are $x = -2i$ and $x = 1$.

  a. Find a quadratic function with these zeros.

  b. Show that $x = 2i$ is not a zero of the function found in step *a*.

  c. Explain why your findings are not a contradiction of the second Conjugate Root Theorem.

29. $p(x) = x^3 - 4x^2 + 2x + 4$

30. $p(x) = x^3 - 4x^2 + x + 26$

31. $p(x) = 4x^3 - 29x^2 + 76x + 75$

32. $p(x) = 6x^4 - 25x^3 + 32x^2 + 3x - 10$

33. $p(x) = x^6 - 2x^5 - 3x^4 + 34x^3 - 24x^2 - 248x - 208$

---

32. $p(x) = 6\left(x + \frac{1}{2}\right)\left(x - \frac{2}{3}\right)$

$\qquad\qquad \cdot [x - (2 + i)][x - (2 - i)]$

$= (2x + 1)(3x - 2)[(x - 2) - i]$

$\qquad\qquad \cdot [(x - 2) + i]$

$= (2x + 1)(3x - 2)[(x - 2)^2 + 1]$

$= (6x^2 - x - 2)(x^2 - 4x + 5)$

$= 6x^4 - 25x^3 + 32x^2 + 3x - 10$

33. $p(x) = (x + 2)^2[x - (1 + \sqrt{5})]$

$\qquad \cdot [x - (1 - \sqrt{5})][x - (2 - 3i)]$

$\qquad\qquad \cdot [x - (2 + 3i)]$

$= (x^2 + 4x + 4)[(x - 1) - \sqrt{5}]$

$\qquad \cdot [(x - 1) + \sqrt{5}][(x - 2) + 3i]$

$\qquad\qquad \cdot [(x - 2) - 3i]$

$= (x^2 + 4x + 4)[(x - 1)^2 - 5]$

$\qquad\qquad \cdot [(x - 2)^2 + 9]$

$= (x^2 + 4x + 4)(x^2 - 2x - 4)$

$\qquad\qquad \cdot (x^2 - 4x + 13)$

$= x^6 - 2x^5 - 3x^4 + 34x^3 - 24x^2$

$\qquad\qquad - 248x - 208$

34.

$$\begin{array}{c|cccc} & 21 & 29 & -24 & 4 \\ -2 & 21 & -13 & 2 & 0 \end{array}$$

$p(x) = (x + 2)(21x^2 - 13x + 2)$

$p(x) = (x + 2)(3x - 1)(7x - 2)$

35.

$$\begin{array}{c|cccc} & 5 & 2 & -10 & -4 \\ -\frac{2}{5} & 5 & 0 & -10 & 0 \end{array}$$

$p(x) = \left(x + \frac{2}{5}\right)(5x^2 - 10)$

$= 5\left(x + \frac{2}{5}\right)(x^2 - 2)$

$= (5x + 2)(x^2 - 2)$

43. Generalize: Find the general form of a quadratic function having the given zero.
$$x^2 - 2ax + a^2 - bc^2$$
a. $a + bi$ $x^2 - 2ax + a^2 + b^2$ **b.** $a - c\sqrt{b}$

44. Use technology to graph $p(x) = x^3 + 2x^2 - 5x - 6$.
  a. State the function's zeros. $-3, -1, 2$
  b. Does the greatest positive zero pass the Upper Bound Test? Does the least negative zero pass the Lower Bound Test? yes; no
  c. Some might think that the results of part *b* show that one of the tests is invalid. Explain why this is not true. The fact that -3 fails the hypothesis of the test does not prevent it from being a lower bound. The test simply does not prove that it is a lower bound.

CUMULATIVE REVIEW

45. Given $A\ (-2, 5)$ and $B\ (4, -1)$, find AB and the midpoint of $\overline{AB}$. [1.2] $6\sqrt{2}, (1, 2)$

46. Classify $f(x) = \frac{2}{x^2}$ as a continuous or discontinuous function. Further classify any discontinuities as jump, infinite, or point. [1.3] discontinuous; infinite discontinuity at $x = 0$

47. Describe how the graph of $f(x) = \frac{1}{x^2}$ is transformed to get the graph of $g(x) = \frac{1}{x^2 - 4} + 3$. [1.5]

48. Write the function rule for $\frac{g}{f}(x)$ if $f(x) = x + 1$ and $g(x) = 7x^2 + 15x + 8$. [1.7] $\frac{g}{f}(x) = 7x + 8, x \neq -1$

Use $p(x) = 2x^4 - 3x^3 + x^2 - 73x + 5$ **for exercises 49–50.**

49. Use the Leading Term Test to describe the function's end behavior. [2.2] $\lim\limits_{x \to \pm\infty} p(x) = \infty$

50. Use the Remainder Theorem to find the remainder when $p(x)$ is divided by $x - 4$. [2.3] 49

51. Factor the polynomial $f(x) = x^3 - 27$. [Algebra] D
  A. $(x + 3)(x^2 - 3x + 9)$ **D.** $(x - 3)(x^2 + 3x - 9)$
  B. $(x - 3)(x^2 - 3x + 9)$ **E.** none of these
  C. $(x + 3)(x^2 + 3x + 9)$

52. State the domain of $f(x) = \sqrt{x - 5}$. [1.1] B
  A. $(-5, \infty)$ **D.** $(0, \infty)$
  B. $[5, \infty)$ **E.** $[0, \infty)$
  C. $(-\infty, \infty)$

53. Describe the continuity of $g(x) = \frac{x^2 - 9}{x + 3}$ at $x = -3$. [1.3] D
  A. continuous   **D.** point discontinuity
  B. jump discontinuity   **E.** none of these
  C. infinite discontinuity

54. Find the zeros of $f(x) = -3x^2 - x + 4$. [1.6] B
  A. $x = -\frac{3}{4}, 4$ **D.** $x = -\frac{3}{4}, 1$
  B. $x = -\frac{4}{3}, 1$ **E.** none of these
  C. $x = \frac{4}{3}, 1$

47. The graph of $f(x)$ is translated 4 units right and 3 units up.

40. Since there are at most 5 zeros and $x = -\sqrt{3}$ is a zero (per the Conjugate Zero Theorem), $x = 1$ or $x = -1$ (but not both) could have a multiplicity of 2. The zero of $x = \sqrt{3}$ cannot be repeated (a maximum multiplicity of 1), as it would also imply a repeated zero at $x = -\sqrt{3}$, for a total of at least 6 zeros.

C. Exercises

41a. $p(x) = (x - \sqrt{3})(x + 2)$
$$= x^2 + 2x - \sqrt{3}x - 2\sqrt{3}$$
$$= x^2 + (2 - \sqrt{3})x - 2\sqrt{3}$$

41b. $p(-\sqrt{3}) = (-\sqrt{3})^2 + (2 - \sqrt{3})$
$$\cdot (-\sqrt{3}) - 2\sqrt{3}$$
$$= 3 - 2\sqrt{3} + 3 - 2\sqrt{3}$$
$$= 6 - 4\sqrt{3} \neq 0$$

41c. The first Conjugate Root Theorem applies only when the polynomial's coefficients are rational.

42a. $p(x) = (x + 2i)(x - 1)$
$$= x^2 - x + 2ix - 2i$$
$$= x^2 + (-1 + 2i)x - 2i$$

42b. $p(2i) = (2i)^2 + (-1 + 2i)(2i) - 2i$
$$= -4 - 2i - 4 - 2i$$
$$= -8 - 4i \neq 0$$

42c. The second Conjugate Root Theorem applies only when the polynomial's coefficients are real.

43a. $[x - (a - bi)][x - (a + bi)]$
$$= [(x - a) - bi][(x - a) + bi]$$
$$= (x - a)^2 - (bi)^2$$
$$= x^2 - 2ax + a^2 - b^2i^2$$
$$= x^2 - 2ax + a^2 + b^2$$

43b. $[x - (a - c\sqrt{b})][x - (a + c\sqrt{b})]$
$$= [(x - a) + c\sqrt{b}][(x - a) - c\sqrt{b}]$$
$$= (x - a)^2 - (c\sqrt{b})^2$$
$$= x^2 - 2ax + a^2 - bc^2$$

36.

		1	−6	−10	5	8	1
U.B.	5	1	−1				fail
	8	1	2	6	53	432	3457 pass
L.B.	−5	1	−11	45	−220	1108	−5539 pass

$[-5, 8]$ or any interval beyond $[-2, 8]$

37.

		1	1	−1	1	−2
U.B.	4	1	5	19	77	306 pass
L.B.	−4	1	−3	11	−43	170 pass

38.

		1	8	−6	−3
U.B.	4	1	12	42	165 pass
L.B.	−4	1	4		fail
	−9	1	−1	3	−30 pass

any interval beyond $[-9, 4]$

39. Examine $-p(x) = x^4 + 4x^3 - 13x^2 - 4x + 12$.

		1	4	−13	−4	12
U.B.	6	1	10	47	278	1680 pass
L.B.	0	1	4			fail
	−7	1	−3	8	−60	432 pass

$[-7, 6]$ or any interval beyond $[-7, 3]$

44.

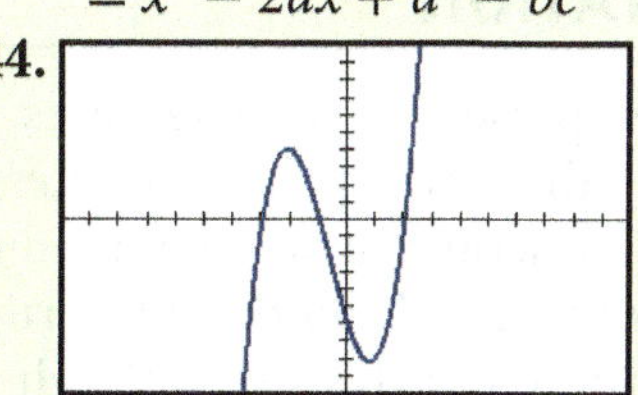

44b.

		1	2	−5	−6
U.B.	2	1	4	3	0 pass
L.B.	−3	1	−1	−2	fail

continued in Answers and Solutions Overflow

Biblical Perspective of Mathematics

Objectives

1. To describe key attributes of God that are also characteristics of the laws of mathematics
2. To distinguish between fallible human mathematics and infallible transcendent mathematics
3. To defend the position that the nature of mathematical truth is divine

Assignments

- **Minimum:** 1–8; 9 (in class)
- **Standard:** 1–8; 9 (in class)
- **Extended:** 1–9

Math's Divine Nature

2 + 2 = 4 is true at all times and at all places. We have classic terms to describe this situation: the truth is omnipresent (present at all places) and eternal (there at all times) . . . . The attributes of omnipresence and eternity are only the beginning. On close examination, other divine attributes seem to belong to arithmetical truths.

—*Vern S. Poythress (American theologian, mathematician, and author)*

Mathematics can be defined as the science of number, quantity, and space. The inadequacy of the human intellect to fully understand these concepts should be immediately obvious as we contemplate the infinite set of natural numbers, the infinite quantity of real numbers between 0 and 1, or the infinite space suggested by our universe. Our mathematics contains transcendent truth originating with our infinite God, who has been gracious in making mathematical knowledge accessible to mankind. We should not be surprised that our mathematics reflects several characteristics of God.

Vern S. Poythress characterizes mathematical truth as both omnipresent and eternal. In *Redeeming Mathematics*, he explains that mathematical facts have other divine attributes because they come from God. For example, the fact 2 + 2 = 4 does not change with the passage of time; it is immutable. Furthermore, "two apples and two apples always make four apples. No event escapes the 'hold' or dominion of arithmetical laws. The power of these laws is absolute, in fact, infinite. In classical language, the law is omnipotent ('all powerful')."

The divine nature of mathematical truth is clarified by a comparison to God's Word. Poythress explains: "The Bible indicates that God rules the world through his speech. He speaks, and it is done [Ps. 33:6] . . . . We may then conclude that the same principle applies in particular to numerical truths about the world. God governs *everything*, including numerical truth. His word specifies what is true. The apples in a group of four apples are created things. What God says about them is divine."

With that in mind we can say that "2 + 2 = 4 is both transcendent and immanent. It transcends the creatures of the world by exercising power over them, conforming them to its dictates. It is immanent in that it touches and

holds in its dominion even the smallest bits of this world. 2 + 2 = 4 transcends the galactic clusters and is immanently present in the behavior of the electrons surrounding a beryllium nucleus. Transcendence and immanence are characteristics of God."

A Christian view of mathematics realizes that our transcendent God controls all things, including mathematics, and that He has authority over all mathematical truth. Because God created finite man in His image, "as human beings we are capable of thinking God's thoughts after him. In particular, we can know that 2 + 2 = 4, a truth that is in God's mind before it is in ours." Our immanent God gives this mathematical truth to us; it is observed in our created universe and comprehended in our finite minds.

The mathematics crafted and applied by mankind is derived from God's perfect body of mathematical truth. Our understanding of God's truths has grown over time. For example, the mathematical notation developed by

Excerpts taken from *Redeeming Mathematics: A God-Centered Approach* by Vern S. Poythress, © 2015, pp. 1, 16, 17, 20, 57, 58. Used by permission of Crossway, a publishing ministry of Good News Publishers, Wheaton, IL 60187, www.crossway.org.

PRESENTATION

This feature presents characteristics of mathematics that reflect the characteristics of God. Encourage further reading from *Redeeming Mathematics* by Vern S. Poythress (quoted in this feature), which can be found online as a PDF. Emphasize the quotations that describe how mathematics is eternal, omnipresent, immutable, omnipotent, transcendent, and immanent.

Spend some time discussing the fact that the mathematics created and applied by human beings is less reliable than the mathematics that is part of God's transcendent truth. Point out that some of the mathematics produced by the world's greatest mathematicians contained logic errors or gaps that had to be corrected or filled later. For example, Leonhard Euler, the most prolific of all mathematicians, gave a proof of the Fundamental Theorem of Algebra that Carl Gauss showed to be incorrect in his 1799 doctoral dissertation, which contained a proof frequently cited as the first correct proof of that theorem. But even Gauss's 1799 proof had gaps and is not considered rigorous by today's standards. (Gauss later provided three other proofs of the theorem that are considered correct and complete.) As a second prominent example, the original development of calculus by Isaac Newton and Gottfried Leibniz lacked a foundation for nearly 150 years until Augustin-Louis Cauchy and Karl Weierstrass used the definitions of continuity and limits in their work.

Encourage students with the fact that mathematical and logical errors occur regularly, even amongst excellent mathematicians. Consider using the Internet keyword search *Marilyn game show problem* and discussing the problem's counterintuitive solution. The Internet keyword search *Monty Hall simulation* yields possible interactive activities.

Conclude the lesson with the discussion of exercise 9 and the fallibility of

past civilizations may be ineffective. Performing multiplication and division is certainly more difficult with Roman numerals than it is with decimal numbers. Each construction of a superior notation or an illuminating concept is a blessing from God.

While human knowledge increases as we interact with God's revelation in His creation, our understanding of His knowledge is still limited. Our work in mathematics provides undeniable evidence of these limitations. We often make logical or computational errors as we create proofs or apply theory. The work of a professional mathematician is thoroughly examined for errors by peers before it is accepted by the mathematics community. Mathematical modeling uses data from the past to make predictions about future events, sometimes with unreliable results. Inferential statistics presents conclusions with margins of error. These limitations contrast with the omniscient character of God. God does not operate in a world of uncertainty; His sovereignty is complete and unfailing.

Despite the limitations of our understanding, humanly crafted mathematics can express transcendent truths that originate with God. When our mathematics agrees with and expresses transcendent truth, it reflects God's own faithfulness. 2 + 2 = 4 is certainly and always true. God's knowledge of 2 + 2 = 4 exceeds ours, but God has allowed us to know this truth, reflecting His mind and the certainty of who He is.

› Exercises

1. What example does Poythress use to illustrate the immanence of mathematics in the "the smallest bits of the world"?

2. Which prominent past civilization utilized an inferior number system? Roman

3. Describe two types of errors made in geometric proofs and algebraic solutions.

4. Why is the work of a professional mathematician reviewed by peers before gaining acceptance in the mathematics community?

5. Which mathematical discipline describes conclusions in terms of a margin of error? inferential statistics

6. When is humanly crafted mathematics true?

7. Which divine characteristic provides the sense of certainty for mathematical truths like 2 + 2 = 4?

8. Match each divine characteristic of mathematical truth to its description.

a. all powerful IV
b. present at all places V
c. existing forever I
d. not subject to the limitations of the universe VI
e. involved in the universe II
f. unchanging over time III

I. eternal
II. immanent
III. immutable
IV. omnipotent
V. omnipresent
VI. transcendent

9. Discuss: Contrast human mathematics with God's transcendent mathematical truths.

1. the behavior of the electrons surrounding a beryllium nucleus

3. logical and computational

4. Professional mathematicians can make errors, and review by other reputable mathematicians verifies the accuracy and reliability of their work.

6. when it agrees with God's transcendent truth

7. God's faithfulness

9. Human mathematics is fallible; transcendent mathematics is not. Math students and professional mathematicians can make computational errors in solutions and logical errors in proofs. Mathematical modeling often produces unreliable answers. Human mathematics exhibits divine characteristics to the extent that it accurately reflects transcendent mathematical truth.

mathematical models. Examples may include failed stock market predictions and the failed predictions of the US presidential elections in 2000 (by Florida TV networks on election night) and in 2016 (by professional pollsters in several Rust Belt states in the days leading up to the election).

2.5 Rational Functions

Objectives

1. To define a rational function as a quotient of polynomials

2. To graph rational functions expressed as transformations of basic reciprocal functions

3. To graph proper and improper rational functions and identify their key characteristics

Flash

In digital cameras, solid-state sensors take the place of film. These sensors convert light striking their surface into electrical signals, much as our eyes do. The signals are then converted into digital form.

Vocabulary

asymptote
horizontal asymptote
hyperbola
improper rational function
infinite discontinuity
nonlinear asymptote
proper rational function
rational function
reduced rational function
slant asymptote
vertical asymptote

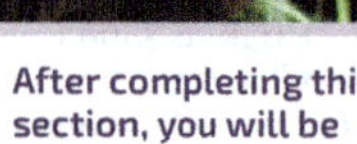

A photographer changes the distance between the camera's lens and the image sensor to bring the image into focus.

After completing this section, you will be able to

- define a rational function.
- graph rational functions expressed as transformations of basic reciprocal functions.
- graph proper and improper rational functions and identify their key characteristics.

 **TIP**

Each continuous portion of these graphs is called a *branch*. The graph of $f(x) = \frac{1}{x}$ is a *hyperbola*.

In the seventeenth century Evangelista Torricelli showed that the solid created by rotating the graph of the rational function $f(x) = \frac{1}{x}$ where $x \geq 1$ about the *x*-axis has a finite volume but an infinite surface area. Such mysteries emphasize our limited understanding of creation (Job 38–41) and should inspire awe and worship of the Creator.

In mathematics, the word *rational* conveys the idea of a ratio. Just as the rational numbers, $\mathbb{Q} = \{\frac{a}{b} \mid a, b \in \mathbb{Z}, b \neq 0\}$, are ratios of integers, rational functions are ratios of polynomials.

> **DEFINITION**
>
> A **rational function** is a function that can be expressed as $f(x) = \frac{a(x)}{b(x)}$ where $a(x)$ and $b(x)$ are polynomials and $b(x) \neq 0$.

The domain of any rational function excludes the zeros of $b(x)$. Examining the end behaviors and the value of the function near the excluded values provides a general idea of the shape of the function's graph. Reciprocal power functions such as $f(x) = \frac{1}{x}$ and $g(x) = \frac{1}{x^2}$ are some of the simplest rational functions. Notice that the graphs of these functions approach specific *x*- and *y*- values called *asymptotes*.

$$f(x) = \frac{1}{x}$$

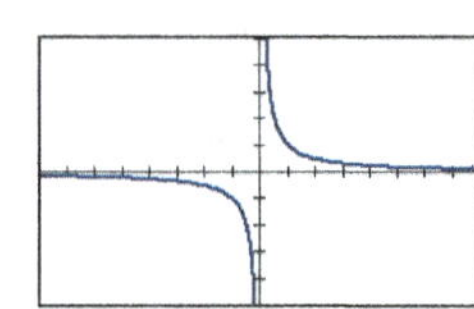

$$g(x) = \frac{1}{x^2}$$

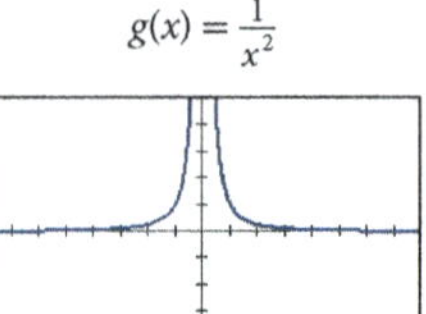

$D = \{x \mid x \neq 0\}; R = \{y \mid y \neq 0\}$
end behavior:
$\lim_{x \to -\infty} f(x) = 0; \lim_{x \to \infty} f(x) = 0$
one-sided limits for $x = 0$:
$\lim_{x \to 0^-} f(x) = -\infty; \lim_{x \to 0^+} f(x) = \infty$

$D = \{x \mid x \neq 0\}; R = \{y \mid y > 0\}$
end behavior:
$\lim_{x \to -\infty} g(x) = 0; \lim_{x \to \infty} g(x) = 0$
one-sided limits for $x = 0$:
$\lim_{x \to 0^-} g(x) = \infty; \lim_{x \to 0^+} g(x) = \infty$

> **VERTICAL AND HORIZONTAL ASYMPTOTES**
>
> The line $x = c$ is a **vertical asymptote** if $\lim_{x \to c^-} f(x) = \pm\infty$ or $\lim_{x \to c^+} f(x) = \pm\infty$.
>
> The line $y = c$ is a **horizontal asymptote** if $\lim_{x \to -\infty} f(x) = c$ or $\lim_{x \to \infty} f(x) = c$.

A function of the form $f(x) = \frac{a}{(x - h)^n} + k$ where $n \in \mathbb{N}$ is a rational function that can be graphed as a transformation of a power function with a negative exponent.

Example 1 Graphing a Transformed Reciprocal Power Function

Identify the asymptotes of $g(x) = \dfrac{2}{(x-1)^2} + 3$. Then graph the function.

Answer

$g(x) = 2\left(\dfrac{1}{(x-1)^2}\right) + 3$

$= 2f(x-1) + 3$

vertical asymptote: $x = 1$
horizontal asymptote: $y = 3$

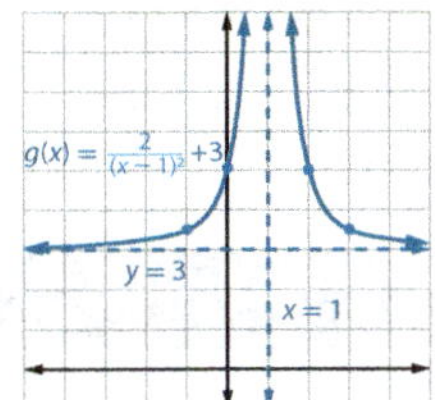

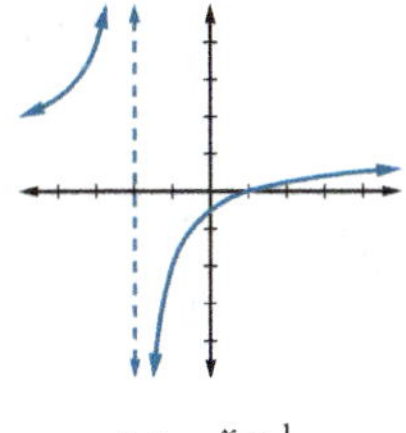
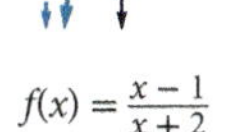

1. Express $g(x)$ as a transformation of $f(x) = \dfrac{1}{x^2}$. Note that $g(x)$ is $f(x)$ translated 1 unit right, stretched vertically by a factor of 2, and then translated 3 units up.

2. Use the translations to identify the vertical asymptote and horizontal asymptote. $D_g = \{x \mid x \neq 1\}$ confirms the vertical asymptote.

3. The graph of $f(x)$ has been stretched vertically from its horizontal asymptote by a factor of 2. Evaluating several points ensures an accurate graph.

Check

4. Confirm your results by examining the end behavior, $\lim\limits_{x \to -\infty} g(x) = 3$ and $\lim\limits_{x \to \infty} g(x) = 3$, and the one-sided limits at the domain's excluded value, $\lim\limits_{x \to 1^-} g(x) = \infty$ and $\lim\limits_{x \to 1^+} g(x) = \infty$.

SKILL ✔ EXERCISE 5

Rational functions that are not transformations of basic power functions can be graphed by identifying vertical asymptotes and point discontinuities, describing the end behavior, and identifying several strategic points before sketching the graph. Analyzing the discontinuities of a rational function $f(x) = \dfrac{a(x)}{b(x)}$ caused by the zeros of $b(x)$ is essential to understanding the shape of the graph.

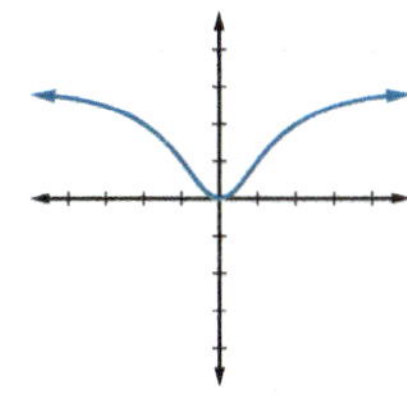

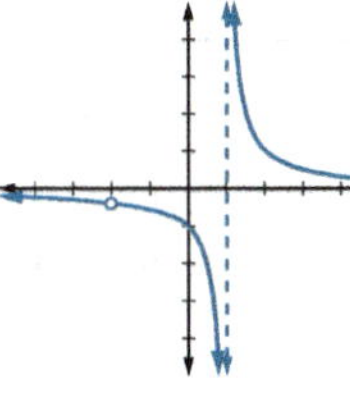

$f(x) = \dfrac{x-1}{x+2}$ $g(x) = \dfrac{3x^2}{x^2+2}$ $h(x) = \dfrac{x+2}{x^2+x-2}$

The function $f(x) = \dfrac{x-1}{x+2}$, which could be rewritten as a transformed reciprocal function $f(x) = \dfrac{-3}{x+2} + 1$, has an infinite discontinuity at its vertical asymptote, $x = -2$. Since the denominator in $g(x) = \dfrac{3x^2}{x^2+2}$ has no real zeros, it is a continuous function.

Additional Exercises

1. Identify the asymptotes of $f(x) = \dfrac{1}{2(x-4)^2} + 3$ and graph the function as a transformation of a reciprocal power function.

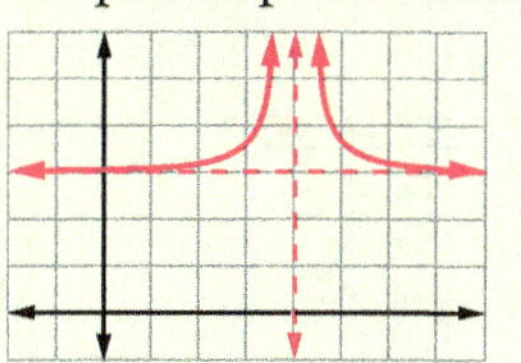

Classify each function as proper or improper and then identify any asymptotes and point discontinuities.

2. $f(x) = \dfrac{(x+2)(x-1)}{(x+2)(x^2-1)}$ proper;
HA: $y = 0$; VA: $x = -1$; PD at $x = -2, 1$

3. $f(x) = \dfrac{(x+3)(3-x)^2}{(x^2-9)(x+1)}$ improper;
HA: $y = 1$; VA: $x = -1$; PD at $x = \pm 3$

Use the characteristics of each function to sketch its graph.

4. $g(x) = \dfrac{8}{x^2-4}$
HA: $y = 0$; VA: $x = \pm 2$; y-int.: $(0, -2)$

---

branch to ensure an accurate graph. The limits describing the function's end behavior can be found by examining a table of values or determined logically with the following reasoning.

as $x \to -\infty$:
$x - 1 \to -\infty$ (huge negative numbers),
$(x-1)^2 \to \infty$ (huge positive numbers),
$\dfrac{2}{(x-1)^2} \to 0^+$ (tiny positive numbers),
and $\dfrac{2}{(x-1)^2} + 3 \to 3^+$ (numbers slightly more than 3)

Similarly, $\lim\limits_{x \to 1} g(x)$ can be reasoned as follows. as $x \to 1^-$:
$x - 1 \to 0^-$ (tiny negative numbers),
$(x-1)^2 \to 0^+$ (tiny positive numbers),

$\dfrac{2}{(x-1)^2} \to \infty$ (huge positive numbers),
and $\dfrac{2}{(x-1)^2} + 3 \to \infty$ (huge positive numbers)

Interactive Activity Use the Internet keyword search *limit calculator* for exploring limits of various functions.

Be sure the students understand that they need to state any domain restriction when writing the *reduced rational function*. The domain restrictions correspond with discontinuities of the function. Explain how to identify *vertical asymptotes* and *point discontinuities* by examining zeros of $a(x)$ and $b(x)$. Use Example 2 to illustrate the identification of these key

characteristics. Note that the restriction $x \neq 2$ is still implied in the reduced expression and does not need to be stated explicitly. Refer to the Technology Corner on page 85 to discuss possible inaccuracies in the graphs of some calculators.

Explain that point discontinuities can occur where the unreduced rational function evaluates to $\dfrac{0}{0}$, such as $\dfrac{x-2}{x-2}$ when $x = 2$. Vertical asymptotes occur whenever the reduced rational function has a 0 in the denominator.

You may want to review why $\dfrac{0}{0}$ is undefined. The fact that $0 \cdot a = 0 \; \forall \, a$ can be rewritten as a division, $\dfrac{0}{0} = a \; \forall \, a$. Since dividing 0 by itself can result in any real

5. $h(x) = \dfrac{3x^3 - 9x^2 - 30x}{5x^3 - 10x^2 - 40x}$

HA: $y = \frac{3}{5}$; VA: $x = 4$; PD at $x = -2, 0$;

y-int.: $\left(0, \frac{3}{4}\right)$; x-int.: $(5, 0)$

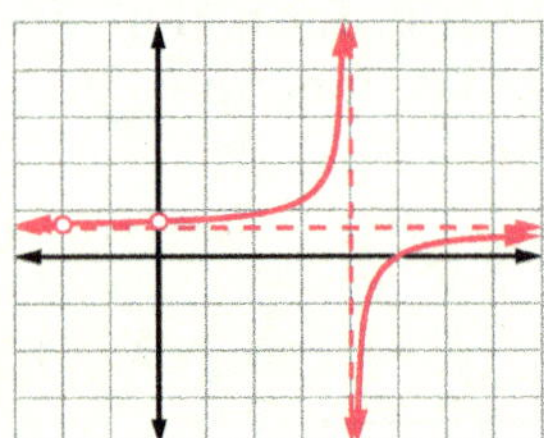

6. $g(x) = \dfrac{x^2 - 2x + 1}{x}$ VA: $x = 0$;

SA: $y = x - 2$; x-int.: $(1, 0)$; y-int.: none

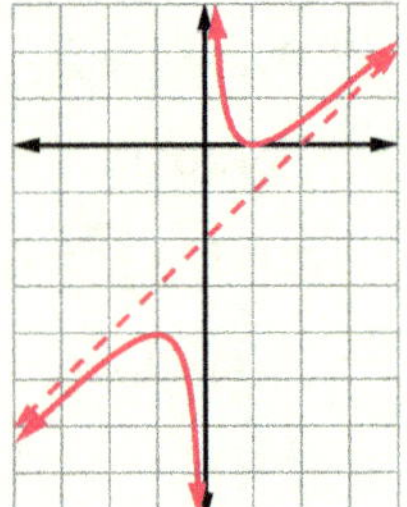

7. $g(x) = \dfrac{x^3 + x^2 - 8x - 12}{x + 1}$

VA: $x = -1$; NLA: $y = x^2 - 8$;

x-int.: $(-2, 0)$ and $(3, 0)$; y-int.: $(0, -12)$

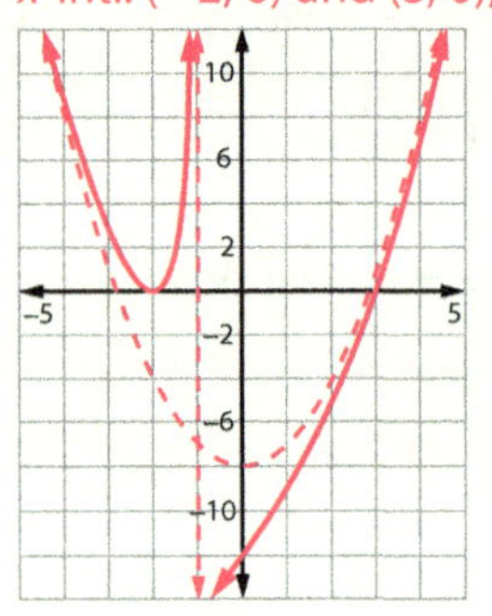

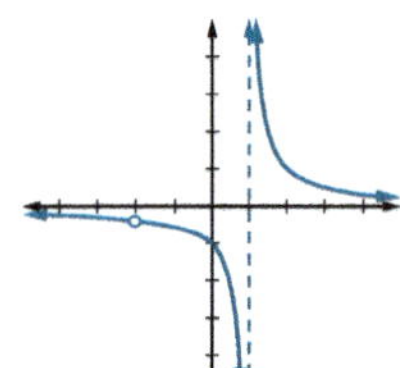

The numerator and denominator of $h(x) = \dfrac{x + 2}{x^2 + x - 2}$ contain a common factor; therefore the function can be rewritten as a *reduced rational function* with a stated restriction on the domain:

$$h(x) = \frac{x + 2}{(x + 2)(x - 1)} = \frac{1}{x - 1}, \; x \neq -2.$$

This equivalent form of $h(x)$ indicates that the graph is a translated hyperbola that is undefined when $x = -2$. Therefore, the zeros of the denominator in $h(x) = \dfrac{x + 2}{(x + 2)(x - 1)}$ indicate different types of discontinuity. The function has an infinite discontinuity at its vertical asymptote, $x = 1$, and a point discontinuity, or *hole*, at $x = -2$.

Discontinuities of $f(x) = \dfrac{a(x)}{b(x)}$	
vertical asymptote	at any zero of $b(x)$ without a matching zero in $a(x)$
point discontinuity	where zeros of $b(x)$ have matching zeros in $a(x)$

Example 2 Finding Vertical Asymptotes and Point Discontinuities

Identify the discontinuities in the graph of $f(x) = \dfrac{(x + 3)(x - 2)}{(x + 3)(x - 2)(x - 2)}$.

Answer

$f(x) = \dfrac{(x + 3)(x - 2)}{(x + 3)(x - 2)(x - 2)} = \dfrac{1}{x - 2}, \; x \neq -3$

$b(x): x = -3, 2$ (multiplicity 2)
$a(x): x = -3, 2$

1. Identify the zeros of $b(x)$ and $a(x)$.

point discontinuity at $x = -3$
vertical asymptote: $x = 2$

2. The zero of -3 in $b(x)$ is matched by a zero of $a(x)$, but one of the repeated zeros of 2 in $b(x)$ has no match in $a(x)$.

SKILL ✔ EXERCISE 9

The end behavior of a rational function can be described by a horizontal, slant, or non-linear asymptote, depending on the degrees of $a(x)$ and $b(x)$. In *proper rational functions*, such as $f(x)$ in the following table, the numerator's degree is less than the denominator's degree. In *improper rational functions*, such as $g(x)$, $h(x)$, and $j(x)$ in the following table, the numerator's degree is the same as or greater than the denominator's degree.

number, it is impossible to uniquely define this division. Therefore, $\frac{x - 2}{x - 2} = 1$ only if $x \neq 2$, and the reduced expression must include this restriction if it is to be equivalent to the unreduced expression.

Consider reminding the students of the difference between proper and improper fractions. Rational functions are also classified as proper or improper, depending on the relative degrees of the numerator and denominator, and are usually written in a "reduced" form.

Explain that the end behavior of every *proper rational function* is described by a *horizontal asymptote* of the x-axis. This is because the denominator has a higher degree than the numerator, so the denominator increases faster than the numerator, causing the fraction to approach 0 as x approaches $\pm\infty$.

Go over the table of end behaviors with the students. Discuss the key characteristics of each function's graph, pointing out comparisons of proper and *improper rational functions*. The graphs also provide a good opportunity to reinforce identifying vertical asymptotes.

Use Examples 3–5 to demonstrate the various asymptotes that describe the end behavior of rational functions, as shown in the end behavior table. You may want to have the students practice using lim-

its as well. In Example 3, $\lim\limits_{x \to -\infty} f(x) = 0^-$ and $\lim\limits_{x \to +\infty} f(x) = 0^+$. The students should not have to find many ordered pairs for each graph. Have them refer back to the table to become more familiar with comparing rational functions with various degrees of $a(x)$ and $b(x)$.

In Example 5, the following reasoning can help students find $\lim\limits_{x \to 1^-} h(x)$.

as $x \to 1^-$:
$x^3 \to 1^-$ (numbers just short of 1) and
$1 - x^2 \to 0^+$ (tiny positive numbers),

so $\dfrac{x^3}{1 - x^2} \to \infty$

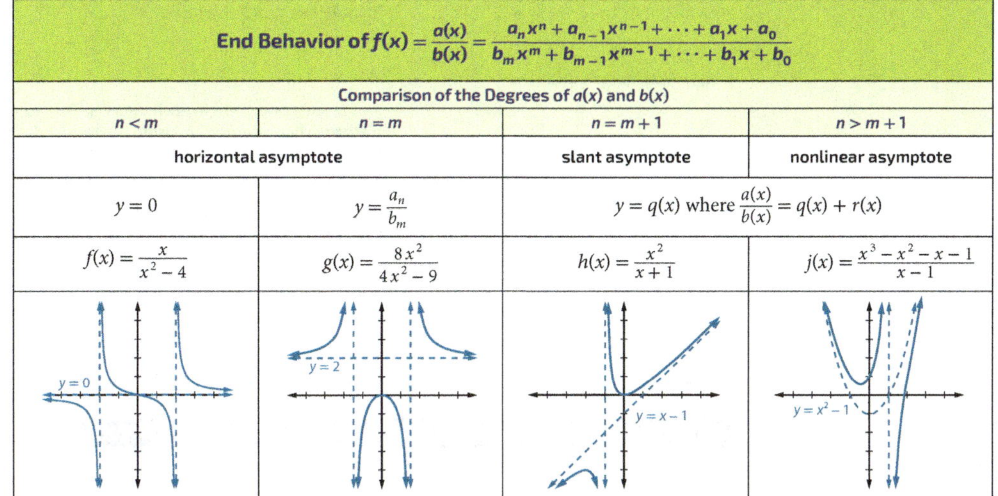

		End Behavior of $f(x) = \dfrac{a(x)}{b(x)} = \dfrac{a_n x^n + a_{n-1} x^{n-1} + \cdots + a_1 x + a_0}{b_m x^m + b_{m-1} x^{m-1} + \cdots + b_1 x + b_0}$	
		Comparison of the Degrees of $a(x)$ **and** $b(x)$	
$n < m$	$n = m$	$n = m + 1$	$n > m + 1$
horizontal asymptote	horizontal asymptote	slant asymptote	nonlinear asymptote
$y = 0$	$y = \dfrac{a_n}{b_m}$	$y = q(x)$ where $\dfrac{a(x)}{b(x)} = q(x) + r(x)$	
$f(x) = \dfrac{x}{x^2 - 4}$	$g(x) = \dfrac{8x^2}{4x^2 - 9}$	$h(x) = \dfrac{x^2}{x + 1}$	$j(x) = \dfrac{x^3 - x^2 - x - 1}{x - 1}$

While a function's graph may cross a non-vertical asymptote (such as the graph of $f(x)$ above), these asymptotes indicate the function's end behavior. The horizontal asymptote of a proper rational function is the x-axis, $y = 0$. Improper rational functions having a numerator and denominator with the same degree have a horizontal asymptote $y = k$, where k is the ratio of the leading coefficients.

When the degree of the numerator is greater than the degree of the denominator, the asymptote is the quotient polynomial from the stated division (without any remainder). Completing the division in $h(x) = \dfrac{x^2}{x+1}$ shows that $h(x) = x - 1 + \dfrac{1}{x+1}$. Since the value of the fractional term gets closer to 0 as $x \to \pm\infty$, the function approaches the linear slant asymptote, $q(x) = x - 1$. Completing the division in $j(x)$ shows that $j(x) = x^2 - 1 - \dfrac{2}{x-1}$ and reveals that the function approaches the parabola $y = x^2 - 1$ asymptotically.

The sketch of the graph of a rational function should indicate any asymptotes or point discontinuities and the x- and y-intercepts. The x-intercepts can be found by examining the zeros of the numerator, $a(x)$, and the y-intercept can be found by evaluating $f(0)$ when it exists.

After being taught from *The Coss*, Leonhard Euler wrote a complete algebra text intended for self-study in 1765. Since Euler had gone blind by this time, he dictated the book to a young man he had hired as a servant.

Example 3   Graphing a Proper Rational Function ($n < m$)

Graph $f(x) = \dfrac{x - 2}{x^2 - 5x + 6}$.

Answer

$f(x) = \dfrac{x - 2}{(x - 2)(x - 3)} = \dfrac{1}{x - 3}, x \neq 2$

vertical asymptote: $x = 3$
point discontinuity at $x = 2$

horizontal asymptote: $y = 0$

1. Find the excluded values of the domain and identify the location of any vertical asymptotes or point discontinuities.

2. Identify the horizontal asymptote of a proper rational function ($n < m$).

CONTINUED ➡

One-on-One Graphing rational functions can be summarized with these five basic steps.

1. Identify any vertical asymptotes or discontinuities.

2. Describe the end behavior by identifying a *horizontal*, *slant*, or *nonlinear* asymptote.

3. Identify any x- and y-intercepts.

4. Plot at least one point in each interval defined by vertical asymptotes.

5. Sketch the graph.

Additional exercise 7 can be used to illustrate graphing a rational function with a nonlinear asymptote ($n > m + 1$).

Interactive Activity Use the Internet keyword search *interactive graph rational function* to find dynamic rational function graphing activities.

TIPS

Ex. 18–25 Encourage the students to use technology to check their answers with a graph of the function.

Ex. 37 You may want to remind students of the difference-of-cubes factoring pattern: $a^3 - b^3 = (a - b)(a^2 + ab + b^2)$.

For Graphing Calculators

Graphical solutions are helpful when finding zeros and vertical asymptotes that have nonintegral values.

a. Graph each rational function and its slant asymptote and identify any x- and y-intercepts.

b. Find the vertical asymptotes by graphing the denominator.

1. $y = \dfrac{2x^3 + 14x^2 - 3x + 25}{x^2 + 5x - 7}$

SA: $y = 2x + 4$

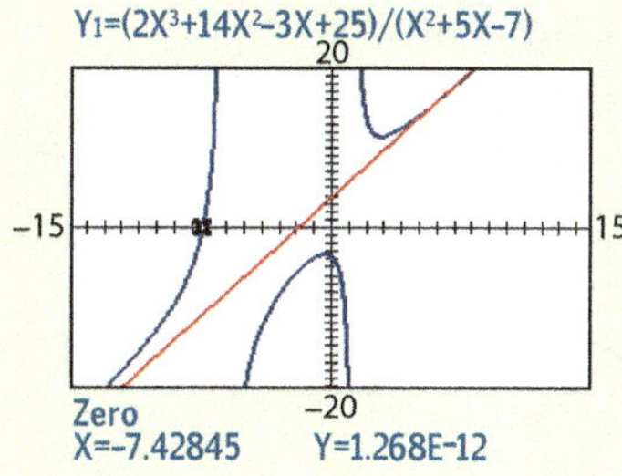

x-int.: $\approx (-7.4, 0)$; y-int.: $\approx (0, -3.6)$

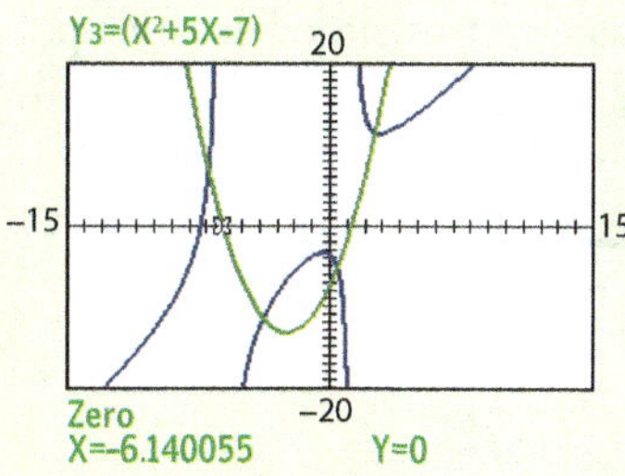

VA: $x \approx -6.1, x \approx 1.1$

2. $y = \dfrac{0.3x^3 + 0.8x^2 - 1.5x + 2}{x^2 + 5x - 9}$

SA: $y = 0.3x - 0.7$

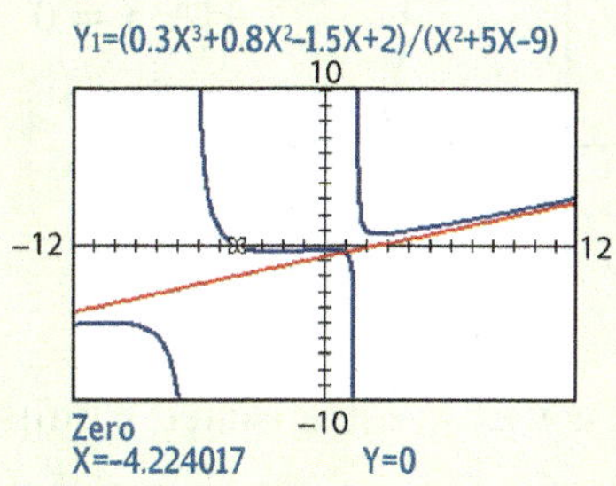

x-int.: $\approx (-4.2, 0)$; y-int.: $\approx (0, -0.2)$

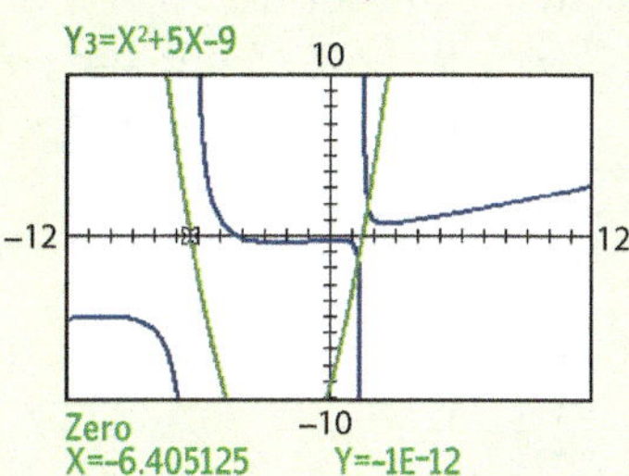

VA: $x \approx -6.4, x \approx 1.4$

Assignments

- **Minimum:** 1–4, 7, 10–17, 19, 22, 25–27, 29, 31, 35, 38–39, 47, 49–53
- **Standard:** 1–3, 5–6, 9–18, 23, 25, 27–28, 30–31, 35, 37–41, 43, 48–56
- **Extended:** 5–7, 10–17, 20, 23–24, 27, 29–30, 36–37, 40, 42–45; 48–56 even

Solutions

❯ A. Exercises

3. $g(x)$ is $f(x) = \dfrac{1}{x^2}$ translated 1 unit left and 2 units down.

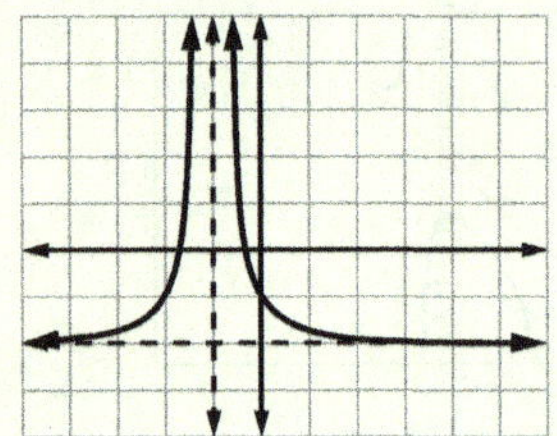

VA: $x = -1$
HA: $y = -2$

4. $g(x)$ is $f(x) = \dfrac{1}{x^3}$ reflected in the x-axis and translated 1 unit up.

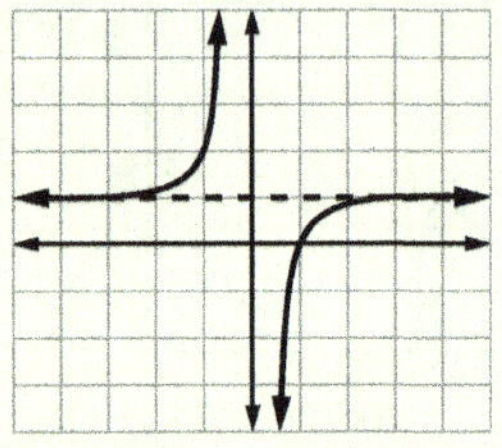

VA: $x = 0$
HA: $y = 1$

5. $g(x)$ is $f(x) = \dfrac{1}{x^3}$ translated 3 units right and stretched vertically by a factor of 2.

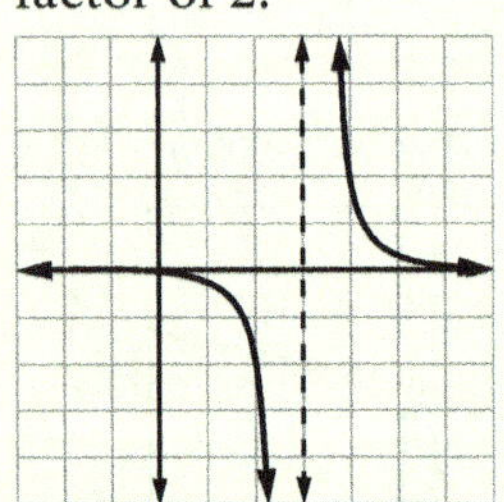

VA: $x = 3$
HA: $y = 0$

6. $g(x)$ is $f(x) = \dfrac{1}{x^2}$ translated 2 units left, reflected in the x-axis, and then translated 1 unit down.

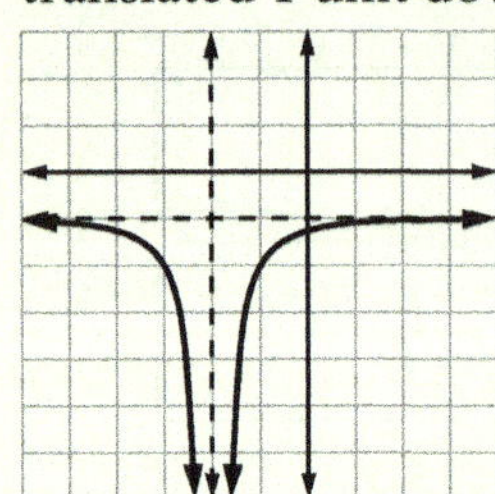

VA: $x = -2$
HA: $y = -1$

---

y-intercept: $f(0) = \dfrac{1}{(0) - 3} = -\dfrac{1}{3}$;

$\therefore \left(0, -\dfrac{1}{3}\right)$

x-intercept: $a(x) = 1 \neq 0; \therefore$ none

3. Identify any x- or y-intercepts.

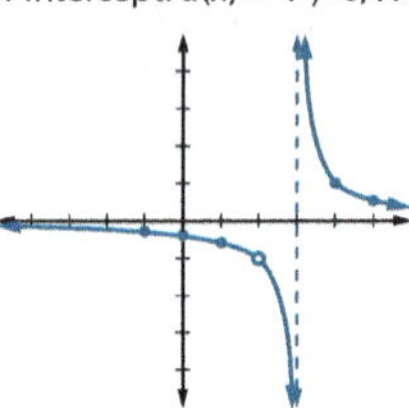

4. Plot at least one point in each interval defined by the vertical asymptote. Using the reduced function allows you to find the exact location of the **point discontinuity**.

x	-1	1	2	4	5
$y = \dfrac{1}{x-3}$	$-\dfrac{1}{4}$	$-\dfrac{1}{2}$	-1	1	$\dfrac{1}{2}$

Note that $\lim\limits_{x \to 3^-} f(x) = -\infty$, $\lim\limits_{x \to 3^+} f(x) = \infty$, and the absence of any x-intercepts indicates that neither branch crosses the x-axis.

——————————— SKILL ✔ **EXERCISE 29**

Example 4   Graphing an Improper Rational Function ($n = m$)

Graph $g(x) = \dfrac{1 - 4x^2}{x^2}$.

Answer

$g(x) = \dfrac{-4x^2 + 1}{x^2}$

vertical asymptote: $x = 0$

1. The zero of $b(x)$ is not a zero of $a(x)$, indicating a vertical asymptote.

horizontal asymptote: $y = -4$

2. When $n = m$, there is a horizontal asymptote at $y = \dfrac{a_n}{b_m}$.

y-intercept: $g(0) = \dfrac{-4(0) + 1}{(0)^2}$; undefined.

$\therefore$ none

x-intercept: $a(x) = (1 + 2x)(1 - 2x) = 0$;

$\therefore \left(\pm\dfrac{1}{2}, 0\right)$

3. Identify any x- or y-intercepts.

💡 TIP

Dividing to express
$g(x) = \dfrac{1}{x^2} - 4$ illustrates
that $g(x)$ is $f(x) = \dfrac{1}{x^2}$
translated down 4 units.

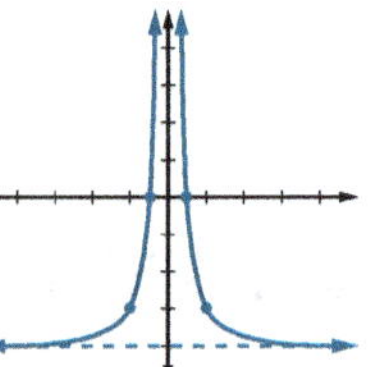

4. Plot at least one point in each interval defined by the vertical asymptote.

x	-1	1
y	-3	-3

Note that $\lim\limits_{x \to 0^-} g(x) = \infty$ and $\lim\limits_{x \to 0^+} g(x) = \infty$ and the x-intercepts indicate that each branch crosses the x-axis just once. Since $g(x)$ is an even function, its graph is symmetric with respect to the y-axis.

——————————— SKILL ✔ **EXERCISE 33**

---

7. $f(x) = \dfrac{x - 2}{(x - 2)(x + 2)}$
zeros of $a(x)$: 2
zeros of $b(x)$: ± 2
VA: $x = -2$; PD at $x = 2$

8. $f(x) = \dfrac{(x + 1)(x - 5)}{(x + 1)(x - 5)(x - 5)}$
zeros of $a(x)$: $-1, 5$
zeros of $b(x)$: $-1, 5$ (multiplicity 2)
VA: $x = 5$; PD at $x = -1$

9. $f(x) = \dfrac{(x + 2)(x - 4)}{(x + 2)(x - 4)(x + 4)}$
zeros of $a(x)$: $-2, 4$
zeros of $b(x)$: $\pm 2, \pm 4$
VA: $x = -4$; PD at $x = -2, 4$

10. $f(x) = \dfrac{(x - 1)(x + 3)}{(x - 3)(x + 3)(x + 3)}$
zeros of $a(x)$: $1, -3$
zeros of $b(x)$: $3, -3$ (multiplicity 2)
VA: $x = \pm 3$

Graph $h(x) = \dfrac{x^3}{1 - x^2}$.

Answer

$b(x) = 1 - x^2 = 0; x = \pm 1$
vertical asymptote: $x = \pm 1$

$$\begin{array}{r} -x \qquad\quad \text{R. } x \\ -x^2 + 0x + 1\ \overline{)\ x^3 + 0x^2 + 0x + 0} \\ \underline{-(x^3 + 0x^2 - \ x)} \\ x + 0 \end{array}$$

$h(x) = -x + \dfrac{x}{1 - x^2}$

oblique asymptote: $y = -x$

y-intercept: $h(0) = \dfrac{(0)}{1 - (0)^2} = 0; \therefore (0, 0)$

x-intercept: $a(x) = x^3 = 0; \therefore (0, 0)$

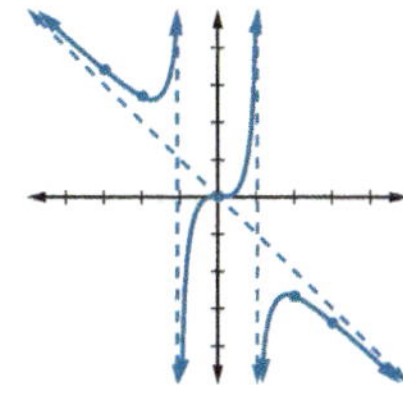

1. The zeros of $b(x)$ are not zeros of $a(x)$, indicating vertical asymptotes.

2. When $n > m$, complete the stated division to find $q(x)$, the equation of the oblique asymptote. Notice that as $x \to \pm\infty$, $\dfrac{x}{1 - x^2} \to 0$ since x^2 increases much faster than x. Therefore $h(x) \to -x$ as $x \to \pm\infty$.

3. Identify the x- and y-intercepts. Note that the origin is the only x-intercept and its multiplicity of 3 indicates that the graph crosses the axis at this point.

4. Plot at least one point in each interval defined by the asymptotes.

x	-3	-2	2	3
y	$\dfrac{27}{8}$	$\dfrac{8}{3}$	$-\dfrac{8}{3}$	$-\dfrac{27}{8}$

Note that $\lim\limits_{x \to 1^-} h(x) = \infty$ and $\lim\limits_{x \to 1^+} h(x) = -\infty$.

Since $h(-x) = \dfrac{(-x)^3}{1 - (-x)^2} = \dfrac{-x^3}{1 - x^2} = -h(x)$, it is an odd function whose graph is symmetric with respect to the origin.

SKILL ✓ **EXERCISE 35**

❯ A. Exercises

Identify any asymptotes and state each function's domain and range.

1. $f(x) = -\dfrac{1}{x^2}$ **2.** $f(x) = \dfrac{1}{x^3}$

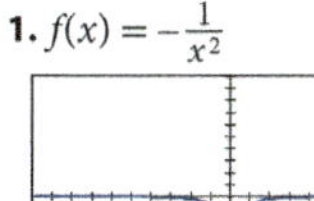

Identify the asymptotes and graph each function as a transformation of a reciprocal power function.

3. $g(x) = \dfrac{1}{(x+1)^2} - 2$ **4.** $g(x) = -\dfrac{1}{x^3} + 1$

5. $g(x) = \dfrac{2}{(x-3)^3}$ **6.** $g(x) = -\dfrac{1}{(x+2)^2} - 1$

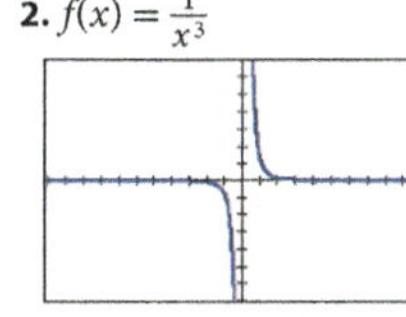

1. VA: $x = 0$; HA: $y = 0$;
$D = \{x \mid x \neq 0\}$;
$R = \{y \mid y < 0\}$

2. VA: $x = 0$; HA: $y = 0$;
$D = \{x \mid x \neq 0\}$;
$R = \{y \mid y \neq 0\}$

Identify any vertical asymptotes or point discontinuities of each function.

7. $f(x) = \dfrac{x - 2}{x^2 - 4}$ VA: $x = -2$; PD at $x = 2$

8. $f(x) = \dfrac{(x+1)(x-5)}{(x+1)(x^2 - 10x + 25)}$ VA: $x = 5$; PD at $x = -1$

9. $f(x) = \dfrac{(x+2)(x-4)}{(x+2)(x^2 - 16)}$ VA: $x = -4$; PD at $x = -2, 4$

10. $f(x) = \dfrac{(x-1)(x+3)}{(x^2 - 9)(x+3)}$ VA: $x = \pm 3$

Match each description of a rational function $f(x) = \dfrac{a(x)}{b(x)}$ to the characteristic exhibited by its graph.
A. a zero at $x = k$
B. vertical asymptote at $x = k$
C. point discontinuity at $x = k$

11. A factor of $(x - k)$ in $b(x)$ is not matched in $a(x)$. B

12. The factor of $(x - k)$ in $a(x)$ is not in $b(x)$. A

13. All factors of $(x - k)$ are matched in $a(x)$ and $b(x)$. C

24. $f(x) = \dfrac{x^3}{x + 2}$
$n > m + 1$ and
$\dfrac{x^3}{x + 2} = x^2 - 2x + 4 - \dfrac{8}{x + 2}$
NLA: $y = x^2 - 2x + 4$
zeros of $a(x)$: 0; zeros of $b(x)$: -2
VA: $x = 2$

25. $f(x) = \dfrac{x(x-1)(x+2)}{x - 2}$
$n > m + 1$ and
$\dfrac{x^3 + x^2 - 2x}{x - 2} = x^2 + 3x + 4 + \dfrac{8}{x - 2}$
NLA: $y = x^2 + 3x + 4$
zeros of $a(x)$: 0, 1, -2; zeros of $b(x)$: 2
VA: $x = 2$

26. $f(x) = \dfrac{x + 1}{(x - 1)(x + 1)} = \dfrac{1}{x - 1}, x \neq -1$
y-int.: $f(0) = -1; (0, -1)$
x-int.: $1 \neq 0; \therefore$ none

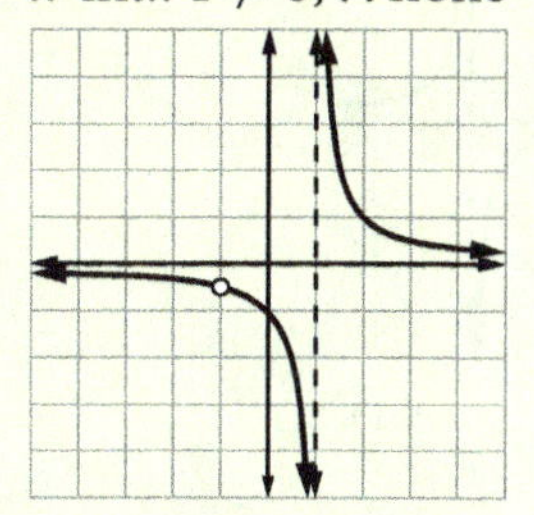

HA: $y = 0$
VA: $x = 1$
PD at $x = -1$

27. $f(x) = \dfrac{x + 2}{x - 1}$
y-int.: $f(0) = -2; (0, -2)$
x-int.: $x + 2 = 0$ when $x = -2$;
$(-2, 0)$

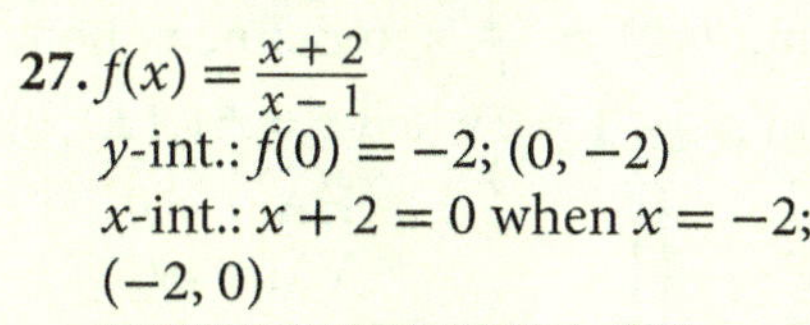

HA: $y = \dfrac{1}{1} = 1$
VA: $x = 1$

28. $f(x) = \dfrac{x - 3}{-x + 1}$
y-int.: $f(0) = -3; (0, -3)$
x-int: $3 - x = 0$ when $x = 3; (3, 0)$

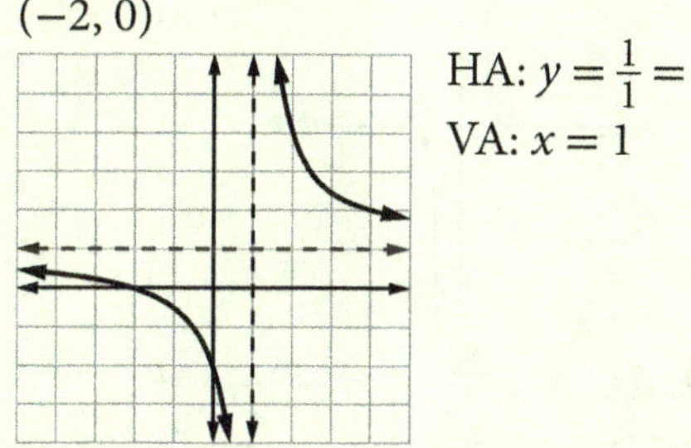

HA: $x = \dfrac{1}{-1} = -1$
VA: $x = 1$

❯ B. Exercises

18. $f(x) = \dfrac{x}{(x + 3)(x - 3)}$
$n < m$; HA: $y = 0$
zeros of $a(x)$: 0; zeros of $b(x)$: ± 3
VA: $x = \pm 3$

19. $f(x) = \dfrac{x + 4}{(x + 4)(x - 4)} = \dfrac{1}{x - 4}; x \neq -4$
$n < m$; HA: $y = 0$
zeros of $a(x)$: -4; zeros of $b(x)$: ± 4
VA: $x = 4$; PD at $x = -4$

20. $g(x) = \dfrac{(x + 3)(x + 1)}{x(x + 3)}$
$n = m$; HA: $y = \dfrac{a_n}{b_m} = \dfrac{1}{1} = 1$
zeros of $a(x)$: $-3, -1$
zeros of $b(x)$: $0, -3$
VA: $x = 0$; PD at $x = -3$

21. $g(x) = \dfrac{6x^2 - 5}{3(x + 2)(x - 2)}$
$n = m$; HA: $y = \dfrac{a_n}{b_m} = \dfrac{6}{3} = 2$
zeros of $a(x)$: $\pm\sqrt{\dfrac{5}{6}}$; zeros of $b(x)$: ± 2
VA: $x = \pm 2$

22. $n = m + 1$ and $\dfrac{x^2}{x - 2} = x + 2 + \dfrac{4}{x - 2}$
SA: $y = x + 2$
zeros of $a(x)$: 0; zeros of $b(x)$: 2
VA: $x = 2$

23. $h(x) = \dfrac{x^3}{(x + 1)(x - 1)}$
$n = m + 1$ and $\dfrac{x^3}{x^2 - 1} = x + \dfrac{x}{x^2 - 1}$
SA: $y = x$
zeros of $a(x)$: 0; zeros of $b(x)$: $-1, 1$
VA: $x = \pm 1$

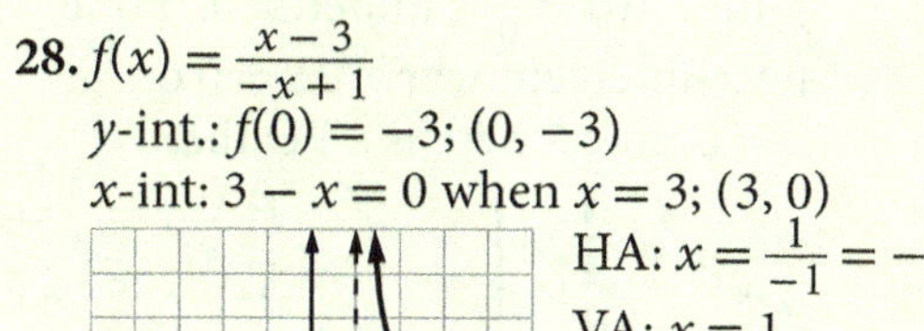

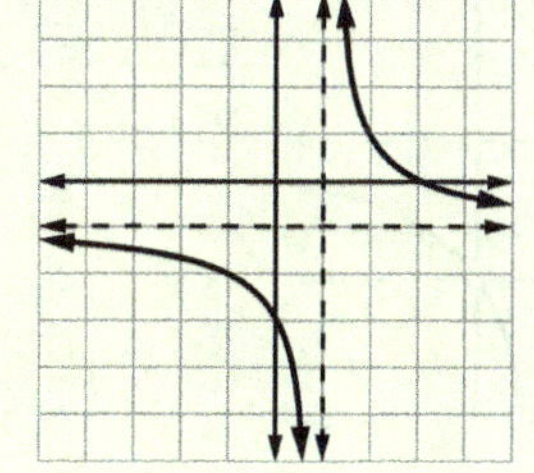

29. $f(x) = \dfrac{1}{x(x-4)}$

y-int.: $f(0) = \dfrac{1}{0}$ is undefined.; none

x-int.: $1 \neq 0$; none

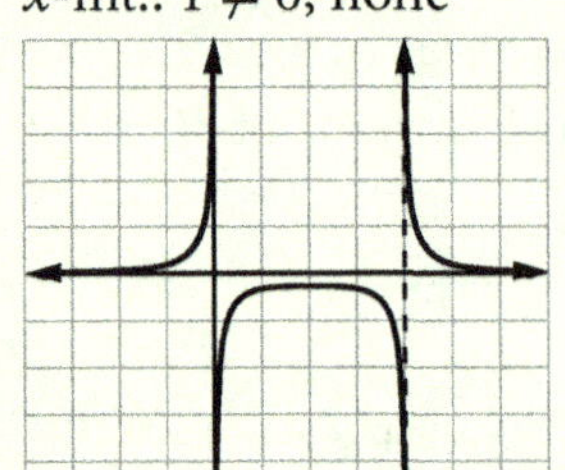

HA: $y = 0$
VA: $x = 0, 4$

30. $f(x) = \dfrac{x+1}{(x-2)(x+2)}$

y-int.: $f(0) = -\dfrac{1}{4}$; $\left(0, -\dfrac{1}{4}\right)$

x-int.: $x + 1 = 0$ when $x = -1$; $(-1, 0)$

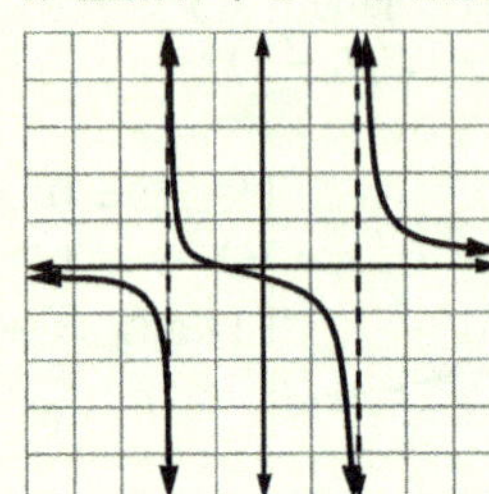

HA: $y = 0$
VA: $x = \pm 2$

31. $h(x) = \dfrac{(x-1)(x+1)}{x^3}$

y-int.: $h(0) = \dfrac{-1}{0}$ is undefined.; none

x-int.: $x^2 - 1 = 0$ when $x = \pm 1$; $(\pm 1, 0)$

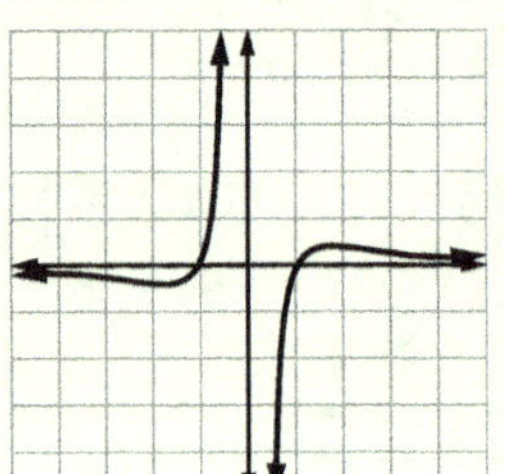

HA: $y = 0$
VA: $x = 0$

32. $f(x) = \dfrac{3x^2}{x(x+2)} = \dfrac{3x}{x+2}$, $x \neq 0$

y-int.: $f(0) = \dfrac{0}{0}$ is undefined.; none;

no x-intercept where $3x = 0$ or
$x = 0$, since there is a PD there

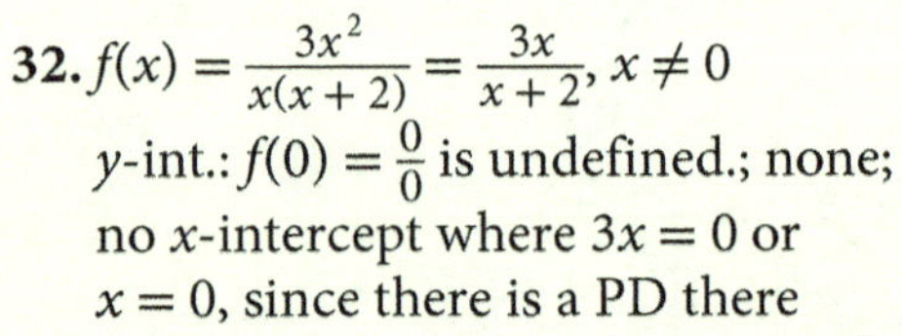

HA: $y = \dfrac{3}{1} = 3$
VA: $x = -2$
PD at $x = 0$

Match each description of a rational function

$$f(x) = \frac{a_n x^n + a_{n-1}x^{n-1} + \cdots + a_1 x + a_0}{b_m x^m + b_{m-1}x^{m-1} + \cdots + b_1 x + b_0}$$ **to the characteristic exhibited by its graph.**

14. $n < m$ B   **A.** horizontal asymptote at $y = \dfrac{a_n}{b_m}$

15. $n = m$ A   **B.** horizontal asymptote at $y = 0$

16. $n = m + 1$ D   **C.** nonlinear asymptote

17. $n > m + 1$ C   **D.** slant asymptote

> **B. Exercises**

Classify each rational function as proper or improper. Then identify all asymptotes and any point discontinuities.

18. $f(x) = \dfrac{x}{x^2 - 9}$ **19.** $f(x) = \dfrac{x+4}{x^2 - 16}$

20. $g(x) = \dfrac{x^2 + 4x + 3}{x^2 + 3x}$ **21.** $g(x) = \dfrac{6x^2 - 5}{3x^2 - 12}$

22. $h(x) = \dfrac{x^2}{x - 2}$ **23.** $h(x) = \dfrac{x^3}{x^2 - 1}$

24. $f(x) = \dfrac{x^3}{x + 2}$ **25.** $f(x) = \dfrac{x^3 + x^2 - 2x}{x - 2}$

Identify any asymptotes, point discontinuities, and intercepts for each function. Then use these characteristics to sketch its graph.

26. $f(x) = \dfrac{x+1}{x^2 - 1}$ **27.** $f(x) = \dfrac{x+2}{x-1}$

28. $f(x) = \dfrac{3-x}{x-1}$ **29.** $f(x) = \dfrac{1}{x^2 - 4x}$

30. $f(x) = \dfrac{x+1}{x^2 - 4}$ **31.** $h(x) = \dfrac{x^2 - 1}{x^3}$

32. $f(x) = \dfrac{3x^2}{x^2 + 2x}$ **33.** $f(x) = \dfrac{9 - x^2}{x^2 - 3x}$

34. $f(x) = \dfrac{x^2 + 2}{2 - 2x}$ **35.** $f(x) = \dfrac{x^3 - x^2 - 2x}{x^2 - 3x + 2}$

36. $f(x) = \dfrac{x^4 + x^2 - 2}{2x^3 + 4x^2 + 4x + 2}$ **37.** $f(x) = \dfrac{x^4 - 3x^2 - 4}{x^3 - 8}$

38. How does the graph of $f(x) = \dfrac{(x-3)(x+1)}{x+1}$ compare to the graph of $g(x) = x - 3$?

39. Algebraically determine if the graph of each function is linear or parabolic. Explain your reasoning.

   a. $f(x) = \dfrac{2x^2 + x - 15}{x + 3}$ **b.** $f(x) = \dfrac{2x^3 - x^2 - x}{2x + 1}$

18. proper; HA: $y = 0$; VA: $x = \pm 3$

19. proper; HA: $y = 0$; VA: $x = 4$; PD at $x = -4$

20. improper; HA: $y = 1$; VA: $x = 0$; PD at $x = -3$

21. improper; HA: $y = 2$; VA: $x = \pm 2$

---

40. Write a rational function with the given characteristics.

   a. a horizontal asymptote at $y = 2$, a vertical asymptote at $x = -5$, and a point discontinuity at $x = 3$

   b. a nonlinear asymptote, a vertical asymptote at $x = 1$, and a point discontinuity at $x = -2$

41. Model: The function $f(x) = \dfrac{5x}{x - 5}$ models the distance from a particular camera lens to the focused image in terms of the distance x (in centimeters) from the lens to the photographed object.

   a. Use technology to graph the function. Then state a reasonable domain for the function. $\ D = (5, \infty)$

   b. State the equation of any asymptotes and explain their significance.

   c. How far from the lens is the focused image when the photographed object is 2 m in front of the lens? ≈ 5.13 cm

42. Model: Complete the following steps to derive and apply a function modeling the percent concentration of the resulting solution when x mL of pure acid is added to 150 mL of a 20% acid solution.

   a. Write expressions for the amount of acid and the total volume of the final solution.

   b. Write a function for the percent concentration of the solution. State a reasonable domain for the function.

   c. Use technology to graph the function. Then state the equation of any asymptotes and explain their significance.

   d. How many milliliters of pure acid must be added to reach a final concentration of 50%? 90 mL

> **C. Exercises**

Identify any asymptotes, point discontinuities, and intercepts for each function. Then use these characteristics to sketch its graph.

43. $f(x) = \dfrac{x^3 - x^2 + x}{1 - x}$ **44.** $f(x) = \dfrac{2x^4 - x^3 - x^2}{2x^2 - 3x + 1}$

45. Explain: Why can every polynomial function also be classified as a rational function?

46. Reduce the rational function $f(x) = \dfrac{x^{n+2} + 2x^{n+1} + 4x^n}{x^{n+2} - 8x^{n-1}}$. Then state all asymptotes, point discontinuities, and intercepts of the function.

22. improper; SA: $y = x + 2$; VA: $x = 2$

23. improper; SA: $y = x$; VA: $x = \pm 1$

24. improper; NLA: $y = x^2 - 2x + 4$; VA: $x = 2$

25. improper; NLA: $y = x^2 + 3x + 4$; VA: $x = 2$

---

33. $f(x) = \dfrac{-(x-3)(x+3)}{x(x-3)} = \dfrac{-(x+3)}{x}$, $x \neq 3$

no y-intercept since $f(0)$ is undefined

x-int.: $-(x + 3) = 0$ when $x = -3$; $(-3, 0)$

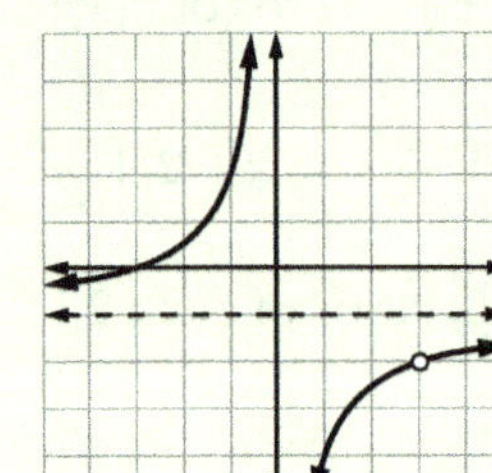

HA: $y = \dfrac{-1}{1} = -1$
VA: $x = 0$
PD at $x = 3$

34. $f(x) = \dfrac{x^2 + 2}{2(1 - x)} = -\dfrac{1}{2}x - \dfrac{1}{2} + \dfrac{3}{2 - 2x}$

y-int.: $f(0) = 1$; $(0, 1)$

x-int.: $x^2 + 2 \neq 0$; none

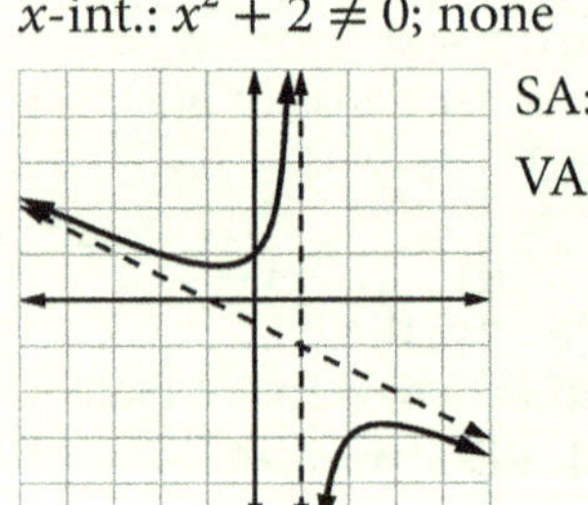

SA: $y = -\dfrac{1}{2}x - \dfrac{1}{2}$
VA: $x = 1$

Add or subtract. [Algebra]

47. $\dfrac{a+3}{a^2-1} + \dfrac{2a}{a+1}$

48. $\dfrac{x+3}{x^2+3x-10} - \dfrac{5x}{x^2-4}$

Solve. [1.6]

49. $x^2 - 10x + 9 = -x^2 - 5x + 6$ $x = 1.5, 1$

50. $5x^2 + x - 3 = 0$ $x = \dfrac{-1 \pm \sqrt{61}}{10}$

Let $f(x) = 3x^5 - 23x^4 + 61x^3 - 61x^2 + 8x + 12$. [2.4]

51. List all possible integral zeros. $\pm1, \pm2, \pm3, \pm4, \pm6, \pm12$

52. Factor $f(x)$. $f(x) = (x-1)(x-2)^2(x-3)(3x+1)$

53. If $f(x) = -x^3 + 7$, what is $f(-3)$? [1.1] E

A. -2 **D.** 20

B. 16   **E.** 34

C. -20

54. Which of the following functions is both odd and continuous? [1.4] B

A. $y = 3x^2$ **D.** $y = 5$

B. $y = \dfrac{6}{x^{-3}}$ **E.** none of these

C. $y = \dfrac{2}{7}(x-3)^{-5}$

55. Which function has an inverse function? [1.8] C

A. $f(x) = 4|x|$ **D.** $y = \sqrt{16 - x^2}$

B. $y = 6x^5 - 4x$ **E.** none of these

C. $y = x^3$

56. Use technology to find the linear or quadratic function that best models the data. Round coefficients to the nearest thousandth. [1.9] B

x	y
-5	-1
-2	3
0	7
5	9
8	7
12	2
15	0

A. $y = -0.010x + 3.904$

B. $y = -0.092x^2 + 0.909x + 5.970$

C. $y = -0.092x + 3.904$

D. $y = 0.036x^2 + 0.410x + 3.904$

E. none of these

38. They are both the same linear function, except $f(x)$ has a point discontinuity at -1.

39a. linear; Factors of $x+3$ cancel, leaving $f(x) = 2x - 5$, $x \neq -3$, a line with a point discontinuity.

39b. parabolic; Factors of $2x+1$ cancel, leaving $f(x) = x^2 + x$, $x \neq -\dfrac{1}{2}$, a parabola with a point discontinuity.

42a. acid: $x + 30$; total volume: $x + 150$

42b. $f(x) = \dfrac{x+30}{x+150}$; $D = [0, \infty)$

45. The polynomial function $p(x)$ can be viewed as $f(x) = \dfrac{p(x)}{b(x)}$ where $b(x) = 1 = 1x^0$, a polynomial function of degree 0.

47. $\dfrac{2a^2 - a + 3}{a^2 - 1}$

48. $\dfrac{-4x^2 - 20x + 6}{(x+2)(x-2)(x+5)}$

35. $f(x) = \dfrac{x(x-2)(x+1)}{(x-1)(x-2)} = \dfrac{x^2+x}{x-1}$, $x \neq 2$

$= x + 2 + \dfrac{2}{x-1}$, $x \neq 2$

y-int.: $f(0) = 0$; $(0, 0)$

x-int.: $x^2 + x = 0$ when $x = 0, -1$; $(0, 0)$ and $(-1, 0)$

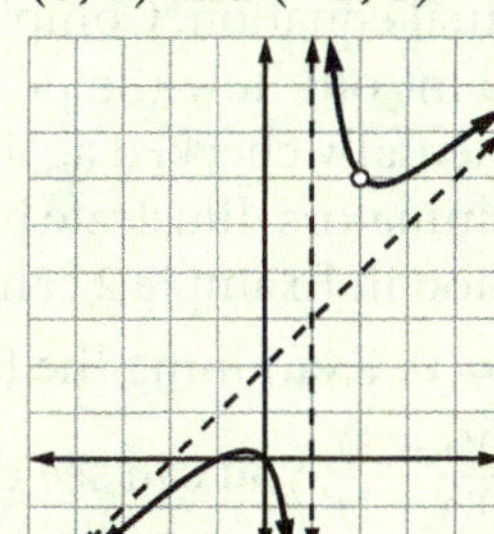

SA: $y = x + 2$
VA: $x = 1$
PD at $x = 2$

36. $f(x) = \dfrac{(x-1)(x+1)(x^2+2)}{2(x+1)(x^2+x+1)}$

$= \dfrac{(x-1)(x^2+2)}{2(x^2+x+1)}$, $x \neq -1$

$= \dfrac{x}{2} - 1 + \dfrac{3x}{2(x^2+x+1)}$

SA: $y = \dfrac{1}{2}x - 1$

VA: $x^2 + x + 1 \neq 0 \Rightarrow$ none

PD at $x = -1$

y-int.: $f(0) = -1$; $(0, -1)$

x-int.: $(x-1)(x^2+2) = 0$ when $x = 1$; $(1, 0)$

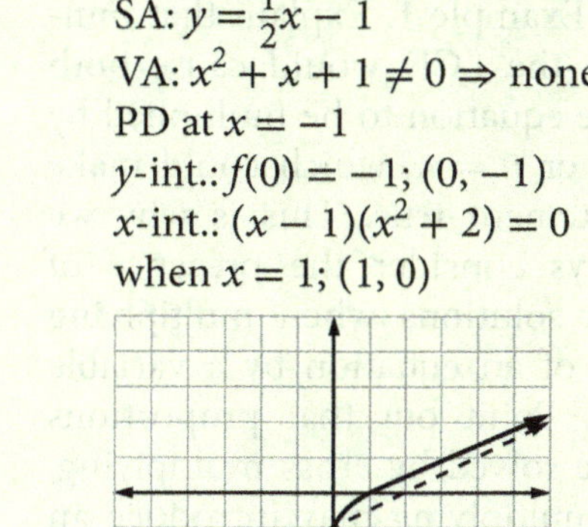

37. $f(x) = \dfrac{(x+2)(x-2)(x^2+1)}{(x-2)(x^2+2x+4)}$

$= \dfrac{(x+2)(x^2+1)}{(x^2+2x+4)}$, $x \neq 2$

$= x + \dfrac{-3x+2}{x^2+2x+4}$

SA: $y = x$

VA: $x^2 + 2x + 4 \neq 0 \Rightarrow$ none

PD at $x = 2$

y-int.: $f(0) = \dfrac{1}{2}$; $\left(0, \dfrac{1}{2}\right)$

x-int.: $(x+2)(x^2+1) = 0$ when $x = -2$; $(-2, 0)$

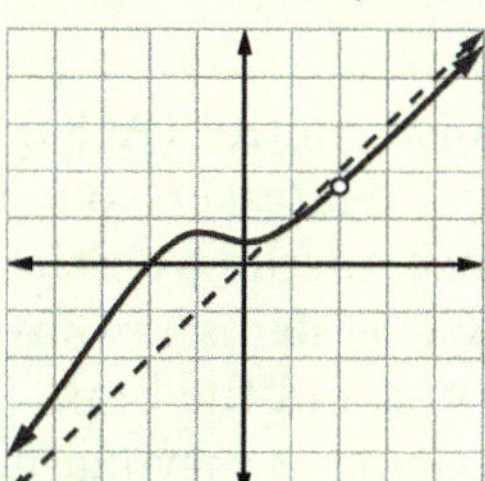

40a. sample answer: $f(x) = \dfrac{2x(x-3)}{(x+5)(x-3)}$

Note that $\dfrac{a_n}{b_m} = 2$, $n = m$, $a(x)$ and $b(x)$ have the same number of factors of $x - 3$, and $b(x)$ has at least one more factor of $x + 5$ than $a(x)$.

40b. sample answer: $f(x) = \dfrac{x^3(x+2)}{(x-1)(x+2)}$

Note that $n > m + 1$, $a(x)$ and $b(x)$ have the same number of factors of $x + 2$, and $b(x)$ has at least one more factor of $x - 1$.

41a. 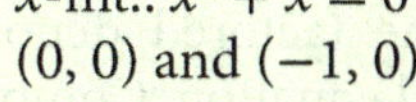

Negative distances are not meaningful; $D = (5, \infty)$

41b. VA: $x = 5$; As the object's distance from the lens approaches 5 cm, the distance to the focused image approaches infinity. HA: $y = 5$; As the object's distance from the lens approaches infinity, the focused image approaches 5 cm from the lens.

continued in Answers and Solutions Overflow

2.6 Solving Rational Equations

Objectives

1. To solve rational equations, identifying any extraneous roots
2. To use rational functions to model and solve real-life problems

Flash

The adaptive cruise control (ACC) system increases and decreases your car's speed in response to the vehicle you are following. ACC systems can involve both the accelerator and the brake. Adjusting vehicle spacing at varying speeds involves rational equations.

Vocabulary

rational equation

Reading and Writing Mathematics

Use the Internet to find definitions for *extraneous*. Then explain *extraneous solution* in your own words.

An extraneous solution is an irrelevant, superfluous, or invalid solution that was introduced through the solving process.

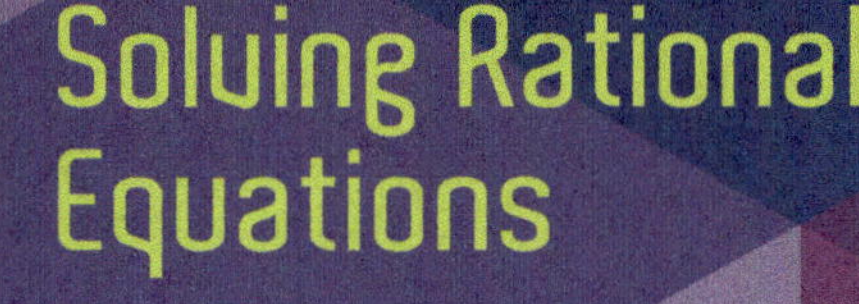

2.6 Solving Rational Equations

After completing this section, you will be able to
- solve a rational equation, identifying any extraneous solutions.
- use rational functions to model and solve real-life problems.

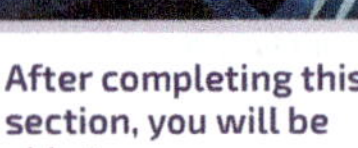

Rational equations such as $P(s) = \frac{1}{ms + b}$ are used to express the dynamics of an adaptive cruise control (ACC) system.

A *rational equation* contains at least one ratio of polynomials, $\frac{a(x)}{b(x)}$, where $b(x) \neq 0$. When solving a rational equation, multiplying both sides of the equation by the least common multiple of the denominators of the rational expressions (the LCD) eliminates the fractions. When an equation is multiplied by an expression containing variables, there is a possibility that you will introduce some answers that are not solutions of the rational equation. You *must* check for extraneous solutions resulting from such a multiplication.

Example 1 Solving a Rational Equation

Solve $\frac{3n}{n-5} = \frac{3n+1}{n-4}$.

Answer

$$(n-5)(n-4)\left[\frac{3n}{n-5} = \frac{3n+1}{n-4}\right]$$

1. Multiply both sides by the LCD, $(n-5)(n-4)$.

$$3n(n-4) = (n-5)(3n+1)$$

$$3n^2 - 12n = 3n^2 - 14n - 5$$

2. Simplify and solve.

$$2n = -5$$

$$n = -\frac{5}{2}$$

Check

$$\frac{3\left(-\frac{5}{2}\right)}{-\frac{5}{2}-5} = \frac{3\left(-\frac{5}{2}\right)+1}{-\frac{5}{2}-4}$$

3. Substitute the answer into the original equation.

$$\frac{-\frac{15}{2}}{-\frac{15}{2}} = \frac{-\frac{13}{2}}{-\frac{13}{2}}$$

$$1 = 1$$

SKILL ✔ EXERCISE 1

A proportion, such as the equation in Example 1, is a special type of rational equation. In this case, setting the product of the extremes equal to the product of the means (cross-multiplying) quickly produces the same result as multiplying both sides by the least common denominator (LCD).

$$\text{If } \frac{18}{x+3} = \frac{12}{x}, \text{ then } 18x = 12(x+3).$$

Example 2 illustrates another way of checking for extraneous roots. Any result that is an excluded value of a rational expression in the original equation is an extraneous solution and must be omitted.

Example 2   Using Domains to Identify Extraneous Solutions

Solve $\dfrac{x+2}{x-4} - \dfrac{30}{x^2-3x-4} - \dfrac{4}{x+1} = 0$.

Answer

$\dfrac{x+2}{x-4} - \dfrac{30}{(x-4)(x+1)} - \dfrac{4}{x+1} = 0$

1. Factor the denominators to identify the LCD, $(x-4)(x+1)$.

$(x-4)(x+1)\left[\dfrac{x+2}{x-4} - \dfrac{30}{(x-4)(x+1)} - \dfrac{4}{x+1} = 0\right]$

$(x+1)(x+2) - 30 - 4(x-4) = 0$

2. Multiply both sides by the LCD to eliminate the rational expressions.

$(x^2+3x+2) - 30 - 4x + 16 = 0$

$x^2 - x - 12 = 0$

$(x+3)(x-4) = 0$

$x = -3, 4$

3. Simplify the resulting equation and solve.

Check

$\dfrac{-3+2}{-3-4} - \dfrac{30}{(-3)^2-3(-3)-4} - \dfrac{4}{-3+1} = 0$

$\dfrac{1}{7} - \dfrac{15}{7} + 2 = 0$

4. Since -3 is in the domain of each expression, it is not an extraneous solution.

If $x = 4$, two of the original expressions are undefined, so this solution is extraneous.

5. The extraneous solution is introduced in step 2, where both sides are multiplied by 0 when $x = 4$.

The only solution is $x = -3$.

SKILL ✔ EXERCISE 17

Any rational equation can be expressed in the form $\dfrac{c(x)}{d(x)} = 0$. The solutions are then the zeros of $f(x) = \dfrac{c(x)}{d(x)}$. Adding the rational expressions in Example 2,

$\dfrac{(x+2)(x+1)}{(x-4)(x+1)} - \dfrac{30}{(x-4)(x+1)} - \dfrac{4(x-4)}{(x+1)(x-4)} = 0$

$\dfrac{(x^2+3x+2) - 30 - 4x + 16}{(x-4)(x+1)} = 0$

$\dfrac{x^2 - x - 12}{x^2 - 3x - 4} = 0.$

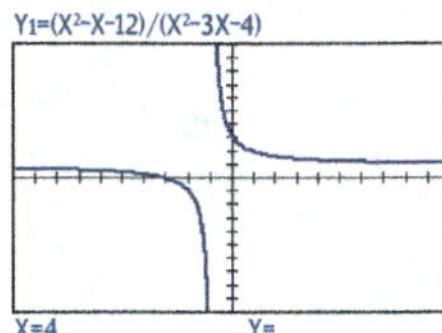

The graph of $f(x) = \dfrac{x^2-x-12}{x^2-3x-4} = \dfrac{(x-4)(x+3)}{(x-4)(x+1)}$ indicates a zero at $x = -3$ and contains a point discontinuity at $x = 4$.

Example 3   Applying a Rational Equation

Find every number whose opposite is equal to the reciprocal of the sum of the number and 3.

Answer

$-x = \dfrac{1}{x+3}$

1. Let x represent the number and translate the statement into an equation.

$(x+3)\left[-x = \dfrac{1}{x+3}\right]$

$-x^2 - 3x = 1$

2. Multiply both sides by the LCD to clear the fraction.

$0 = x^2 + 3x + 1$

$x = \dfrac{-(3) \pm \sqrt{(3)^2 - 4(1)(1)}}{2(1)}$

$= \dfrac{-3 \pm \sqrt{5}}{2} \approx -2.62, -0.38$

3. Solve the resulting quadratic equation.

CONTINUED ➡

---

want to ask students to use this form to identify another key characteristic of the function. (*VA: $x = -1$*) Also consider discussing how Example 1 could be checked graphically using this procedure.

Example 3 illustrates the use of a rational equation in solving a word problem.

Motivational Idea Consider showing an alternative graphical solution to this example. The x-values for the intersections

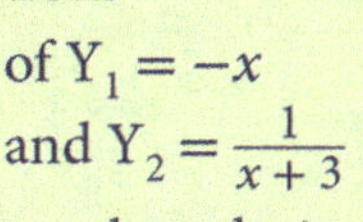

of $Y_1 = -x$ and $Y_2 = \dfrac{1}{x+3}$ are the solutions for the equation.

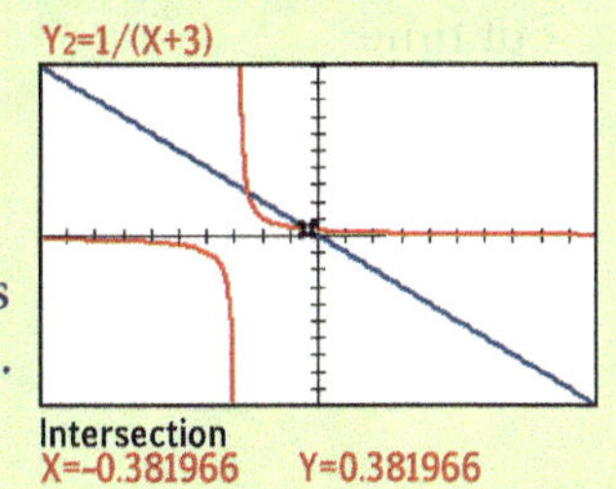

Example 4 reviews solving rate-of-work problems from previous courses. Emphasize that the combined rate is assumed to be the sum of the individual rates. Consider discussing factors that may cause two or more people working together to accomplish the task in more or less time. Encourage students to evaluate the reasonableness of their answers. If two people are working together, the job should be completed in less time than it would take for either to do the job alone. In Example 4, the time for the small pipe alone must be more than 2.4 hr, the time required for both pipes to fill the tank.

Point out that the unknown distance in Example 5 is not needed. The fact that

Additional Exercises

Solve.

1. $\dfrac{5}{x} + \dfrac{10x}{24} = \dfrac{20}{3x}$ $x = \pm 2$

2. $\dfrac{x-4}{x+3} - \dfrac{3x-7}{5x^2+6x-27} - \dfrac{x+3}{5x-9} = 0$
   $x = 1, 8.5$

3. $\dfrac{x+6}{x-3} + \dfrac{x}{5} = \dfrac{x(x+3)}{5(x-3)} - \dfrac{5}{3-x}$ $x = 5$

4. $\dfrac{x+2}{x-1} + \dfrac{x}{x-3} = 5$ $x = 3 \pm \sqrt{2}$

5. $\dfrac{4}{x+1} + \dfrac{x}{1-x} = \dfrac{1}{x-1} + \dfrac{x^2-3}{1-x^2}$ $x = 4$

6. $\dfrac{x+1}{x-2} - \dfrac{25}{x^2+x-6} = \dfrac{5}{x+3}$ $x = 4$

7. $\dfrac{3x}{x-5} + 3 = \dfrac{7x-20}{x-5}$ $\varnothing$

8. $\dfrac{4x}{x+1} + 8 = \dfrac{12x^2+20x+8}{x^2+2x+1}$
   $(-\infty, -1) \cup (-1, \infty)$ or $\{x \in \mathbb{R} \mid x \neq -1\}$

9. Jared takes 2 hr longer than his father to mow his family's acreage. Together they can mow the acreage in 4 hr. How long does it take each working alone?
   father ≈ 7.1 hr; Jared ≈ 9.1 hr

10. Find two numbers whose sum is 9 and the sum of whose reciprocals is $\dfrac{1}{2}$.   3 and 6

Assignments

- **Minimum:** 1, 4, 7–8, 11, 14, 17–19, 21, 23, 26, 29, 32, 39–41, 43, 45, 47
- **Standard:** 2, 5, 8, 11, 14–15, 17, 20–22, 25, 28, 30, 32–34, 36; 39–47 odd
- **Extended:** 3, 6, 9, 12, 14–15, 18, 21, 24, 27, 30; 31–37 odd; 38, 41, 44–48

Assessment

- Quiz 2C covers Sections 2.5–2.6.

Solutions

❯ A. Exercises

1. $(x-2)(x-5)\left[\dfrac{x}{x-2} = \dfrac{x-4}{x-5}\right]$
   $x(x-5) = (x-4)(x-2)$
   $x^2 - 5x = x^2 - 6x + 8$
   $x = 8$

2. $10x\left[\dfrac{3}{x} + \dfrac{x}{10} = \dfrac{11}{2x}\right]$
   $30 + x^2 = 55$
   $x^2 = 25$
   $x = \pm 5$

3. $(x-8)(x+8)\left[\dfrac{1}{x-8} = \dfrac{2x}{(x-8)(x+8)}\right]$

$$x + 8 = 2x$$
$$x = 8 \ (\text{extraneous})$$

4. $x\left[x - \dfrac{10}{x} = -3\right]$

$$x^2 - 10 = -3x$$
$$x^2 + 3x - 10 = 0$$
$$(x + 5)(x - 2) = 0$$
$$x = -5, 2$$

5. $(x^2 - 2x + 1)\left[\dfrac{2x^2 + x + 2}{x^2 - 2x + 1} = 2\right]$

$$2x^2 + x + 2 = 2x^2 - 4x + 2$$
$$5x = 0$$
$$x = 0$$

6. $7(4n + 1)(n - 2)$

$\cdot \left[\dfrac{3}{(4n+1)(n-2)} - \dfrac{2}{n-2} = \dfrac{4}{7(4n+1)}\right]$

$$21 - 14(4n + 1) = 4(n - 2)$$
$$21 - 56n - 14 = 4n - 8$$
$$15 = 60n$$
$$n = \dfrac{1}{4}$$

7. $3(r + 4)(r - 1)$

$\cdot \left[\dfrac{2}{(r+4)(r-1)} = \dfrac{2}{3(r+4)} - \dfrac{r}{r-1}\right]$

$$6 = 2(r - 1) - 3r(r + 4)$$
$$6 = 2r - 2 - 3r^2 - 12r$$
$$3r^2 + 10r + 8 = 0$$
$$(r + 2)(3r + 4) = 0$$
$$r = -2, -\dfrac{4}{3}$$

8. $(x - 1)(x + 2)$

$\cdot \left[\dfrac{6}{x-1} = \dfrac{x}{x+2} - \dfrac{18}{(x-1)(x+2)}\right]$

$$6(x + 2) = x(x - 1) - 18$$
$$6x + 12 = x^2 - x - 18$$
$$x^2 - 7x - 30 = 0$$
$$(x - 10)(x + 3) = 0$$
$$x = 10, -3$$

9. $(3n - 2)(n - 1)$

$\cdot \left[\dfrac{11 - n^2}{(3n-2)(n-1)} = \dfrac{2n+3}{3n-2} - \dfrac{n-3}{n-1}\right]$

$$11 - n^2 = (2n + 3)(n - 1)$$
$$ - (n - 3)(3n - 2)$$
$$11 - n^2 = 2n^2 + n - 3 - 3n^2$$
$$ + 11n - 6$$
$$20 = 12n$$
$$n = \dfrac{5}{3}$$

10. $(t + 1)(t + 5)(t - 5)$

$\cdot \left[\dfrac{1}{(t+1)(t+5)(t-5)} + \dfrac{2}{(t+5)(t-5)}\right.$

$\left. - \dfrac{3}{(t+5)(t+1)} = 0\right]$

$$1 + 2(t + 1) - 3(t - 5) = 0$$
$$1 + 2t + 2 - 3t + 15 = 0$$
$$t = 18$$

Check

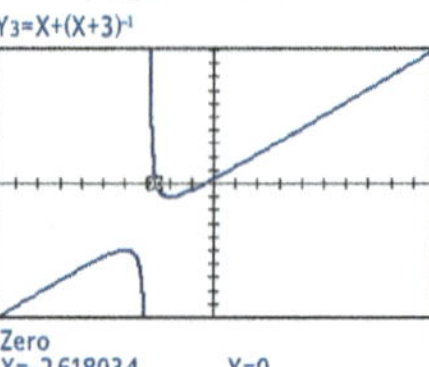

4. The zeros of $f(x) = x + \dfrac{1}{x+3}$ confirm that these solutions are not excluded values for the rational expression.

<hr>

SKILL ✔ EXERCISE 23

Rational equations are often used to solve problems involving the rate at which work is accomplished. If a job can be completed in 8 hr, the rate is $\dfrac{1 \text{ job}}{8 \text{ hr}} = \dfrac{1}{8}$ of the job per hour. The rate of working together is assumed to be the sum of the individual rates.

Example 4   Solving a Rate-of-Work Problem

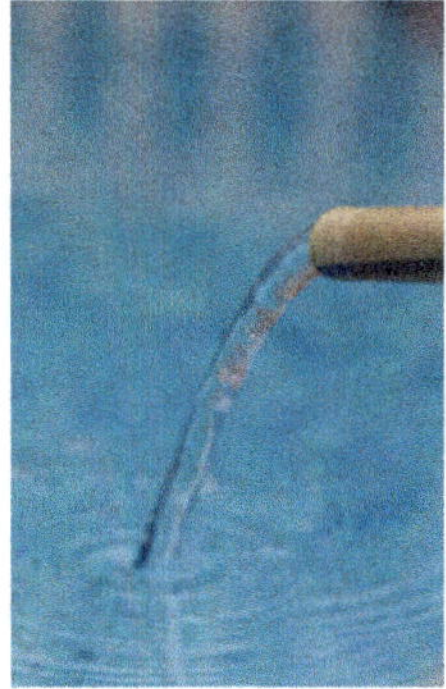

A tank can be filled by two pipes in 2.4 hr. If the large pipe alone takes two fewer hours to fill the tank than the small pipe, how long would it take the small pipe to fill the tank?

Answer

$$rate_{large} + rate_{small} = rate_{together}$$
$$\dfrac{1}{t-2} + \dfrac{1}{t} = \dfrac{1}{2.4}$$

1. Letting t represent the time it takes the small pipe to fill the tank, write an equation relating the rates of work.

$$2.4t(t-2)\left[\dfrac{1}{t-2} + \dfrac{1}{t} = \dfrac{1}{2.4}\right]$$
$$2.4t + 2.4(t - 2) = t(t - 2)$$

2. Multiply both sides by the LCD.

$$4.8t - 4.8 = t^2 - 2t$$
$$0 = t^2 - 6.8t + 4.8$$

3. Solve the resulting quadratic equation and interpret the answer.

$$t = \dfrac{-(-6.8) \pm \sqrt{(-6.8)^2 - 4(1)(4.8)}}{2(1)}$$
$$= 6, 0.8$$

The small pipe will fill the tank in 6 hr.

4. A time of 0.8 hr for the small pipe would cause the time for the large pipe to be negative.

Check

$$\dfrac{1}{4} + \dfrac{1}{6} = \dfrac{3}{12} + \dfrac{2}{12} = \dfrac{5}{12} = \dfrac{10}{24} = \dfrac{1}{2.4}$$

5. Substitute into the original equation.

<hr>

SKILL ✔ EXERCISE 27

Some distance problems are solved by using $r = \dfrac{d}{t}$ or $t = \dfrac{d}{r}$ and then writing and solving a rational equation relating the rates or times of each part of a trip.

the distances can be canceled in step *3a* indicates that the equation is true for any distance. Step *3b* illustrates that any common denominator can be used when clearing fractions, although a less complicated equation will result if the LCD is used.

Discuss the margin note, which reinforces the Biblical Perspective of Mathematics feature in this chapter.

TIPS

Ex. 12 If students solve by graphing $Y_1 = \dfrac{-2x}{x+4}$ and $Y_2 = \dfrac{5.5x - 7}{x + 5}$ in the standard window without considering the asymptotes and overall shape of the graphs, they can easily miss the negative solution.

Ex. 24 The two numbers can be represented as x and $x + 4$ or as x and $x - 4$.

Ex. 35 Some students may be surprised that the total resistance (≈ 7.06 ohms) is less than the smallest resistor. This is comparable to a store having multiple checkout lanes, which reduces the checkout time.

Due to construction, Sarah averages only 25 mi/hr during the first fourth of her drive from Tryon to Anderson. What average speed does she maintain for the rest of the trip if she averages 40 mi/hr over the entire trip?

Answer

Let d = the entire distance driven and r = Sarah's average speed for the rest of the trip.

	Rate	Time	Distance
First Part	25	$\dfrac{\frac{1}{4}d}{25} = \dfrac{d}{4(25)}$	$\frac{1}{4}d$
Last Part	r	$\dfrac{\frac{3}{4}d}{r} = \dfrac{3d}{4r}$	$\frac{3}{4}d$
Total Trip	40	$\dfrac{d}{40}$	d

$$\frac{d}{100} + \frac{3d}{4r} = \frac{d}{40}$$

$$\frac{1}{100} + \frac{3}{4r} = \frac{1}{40}$$

$$400r\left[\frac{1}{100} + \frac{3}{4r} = \frac{1}{40}\right]$$
$$4r + 300 = 10r$$

$$300 = 6r$$
$$r = 50 \text{ mi/hr}$$

Check

$$\frac{d}{\frac{d}{4(25)} + \frac{3d}{4(50)}} \cdot \frac{200}{200} = \frac{200d}{2d + 3d} = \frac{200d}{5d} = 40 \text{ mi/hr}$$

1. Assign variables to represent the unknowns.

2. Using a table to organize the information, write simplified expressions for each time using $t = \frac{d}{r}$.

3. Write and solve an equation relating the times for the trip.

   a. Divide each term by d. Note that we do not need to know the total distance.

   b. Multiply both sides by a common denominator to clear the fractions.

   c. Solve the resulting equation.

4. The average speed $= \dfrac{\text{total distance}}{\text{total time}}$.

———————————————— SKILL ✔ **EXERCISE 31**

▶ A. Exercises

Solve.

1. $\dfrac{x}{x-2} = \dfrac{x-4}{x-5}$ $x = 8$

2. $\dfrac{3}{x} + \dfrac{x}{10} = \dfrac{11}{2x}$ $x = \pm 5$

3. $\dfrac{1}{x-8} = \dfrac{2x}{x^2 - 64}$ no solution

4. $x - \dfrac{10}{x} = -3$ $x = -5, 2$

5. $\dfrac{2x^2 + x + 2}{x^2 - 2x + 1} = 2$ $x = 0$

6. $\dfrac{3}{4n^2 - 7n - 2} + \dfrac{2}{2-n} = \dfrac{4}{28n + 7}$ $n = \frac{1}{4}$

7. $\dfrac{2}{r^2 + 3r - 4} = \dfrac{2}{3r + 12} - \dfrac{r}{r-1}$ $r = -2, -\frac{4}{3}$

8. $\dfrac{6}{x-1} = \dfrac{x}{x+2} - \dfrac{18}{x^2 + x - 2}$ $x = -3, 10$

9. $\dfrac{11 - n^2}{3n^2 - 5n + 2} = \dfrac{2n+3}{3n-2} - \dfrac{n-3}{n-1}$ $n = \frac{5}{3}$

10. $\dfrac{1}{t^3 + t^2 - 25t - 25} + \dfrac{2}{t^2 - 25} = \dfrac{3}{t^2 + 6t + 5}$ $t = 18$

Use technology to graph each rational equation and identify the real roots of the function. Round answers to the nearest thousandth.

11. $\dfrac{8x^2 + 2x - 5}{3x + 4} = 0$ $x = -0.925, 0.675$

12. $\dfrac{-2x}{x+4} = \dfrac{5.5x - 7}{x+5}$ $x = -4.218, 0.885$

The universal consistency of arithmetical truths allows for diverse applications. From creating safer automobiles with adaptive cruise control to modeling motion, this consistency is an indicator of the "goodness" of mathematics.

▶ B. Exercises

13. $(x-1)(x+1)\left[\dfrac{x+4}{(x-1)(x+1)} = \dfrac{3x}{x+1}\right]$
$$x + 4 = 3x^2 - 3x$$
$$0 = 3x^2 - 4x - 4$$
$$0 = (3x + 2)(x - 2)$$
$$x = 2, -\frac{2}{3}$$

14. $x\left[x - \dfrac{1}{x} = -2\right]$
$$x^2 - 1 = -2x$$
$$x^2 + 2x - 1 = 0$$
$$x = \frac{-(2) \pm \sqrt{(2)^2 - 4(1)(-1)}}{2(1)}$$
$$x = \frac{-2 \pm 2\sqrt{2}}{2} = -1 \pm \sqrt{2}$$

15. $(x-3)(x+5)$
$$\cdot \left[\dfrac{x}{x-3} - \dfrac{4}{x+5} = \dfrac{11}{(x+5)(x-3)}\right]$$
$$x(x+5) - 4(x-3) = 11$$
$$x^2 + 5x - 4x + 12 = 11$$
$$x^2 + x + 1 = 0$$
$$x = \frac{-(1) \pm \sqrt{(1)^2 - 4(1)(1)}}{2(1)}$$
$$= \frac{-1 \pm \sqrt{-3}}{2} = -\frac{1}{2} \pm \frac{\sqrt{3}}{2}i$$

16. $y(y-1)\left[\dfrac{y+2}{y} - \dfrac{y+1}{y-1} = 4\right]$
$$(y+2)(y-1) - y(y+1) = 4y^2 - 4y$$
$$y^2 + y - 2 - y^2 - y = 4y^2 - 4y$$
$$4y^2 - 4y + 2 = 0$$
$$y = \frac{-(-4) \pm \sqrt{(-4)^2 - 4(4)(2)}}{2(4)}$$
$$= \frac{4 \pm \sqrt{-16}}{8} = \frac{4 \pm 4i}{8} = \frac{1}{2} \pm \frac{1}{2}i$$

17. $(r-5)(r+2)$
$$\cdot \left[\dfrac{r}{r-5} = \dfrac{7r}{(r-5)(r+2)} - \dfrac{3}{r+2}\right]$$
$$r(r+2) = 7r - 3(r-5)$$
$$r^2 + 2r = 7r - 3r + 15$$
$$r^2 - 2r - 15 = 0$$
$$(r+3)(r-5) = 0$$
$$r = -3, 5 \text{ (extraneous)}$$

18. $x(x+3)\left[\dfrac{2}{x+3} + \dfrac{6}{x(x+3)} = \dfrac{2-x}{x}\right]$
$$2x + 6 = (2-x)(x+3)$$
$$2x + 6 = -x^2 - x + 6$$
$$x^2 + 3x = 0$$
$$x(x+3) = 0$$
$$x = 0, -3 \text{ (both extraneous)}$$

19. $(2x+1)(x+4)$
$$\cdot \left[\dfrac{x^2 + 3x - 60}{(2x+1)(x+4)} = \dfrac{8x}{2x+1} - \dfrac{2x}{x+4}\right]$$
$$x^2 + 3x - 60 = 8x(x+4) - 2x(2x+1)$$
$$x^2 + 3x - 60 = 8x^2 + 32x - 4x^2 - 2x$$
$$3x^2 + 27x + 60 = 0$$
$$x^2 + 9x + 20 = 0$$
$$(x+5)(x+4) = 0$$
$$x = -5, -4 \text{ (extraneous)}$$

11. 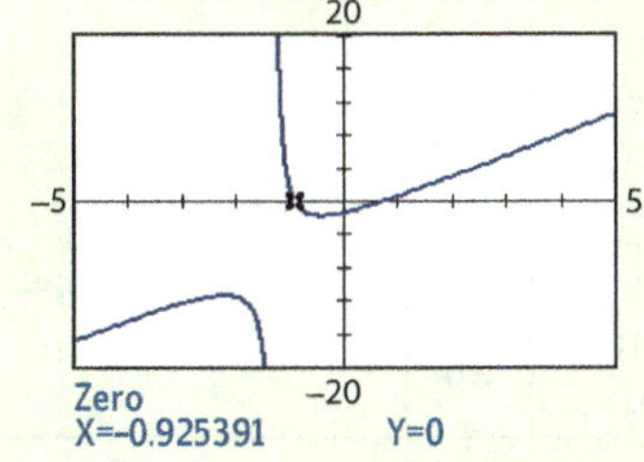

12. $Y_1 = \dfrac{-2x}{x+4}$, $Y_2 = \dfrac{5.5x - 7}{x+5}$
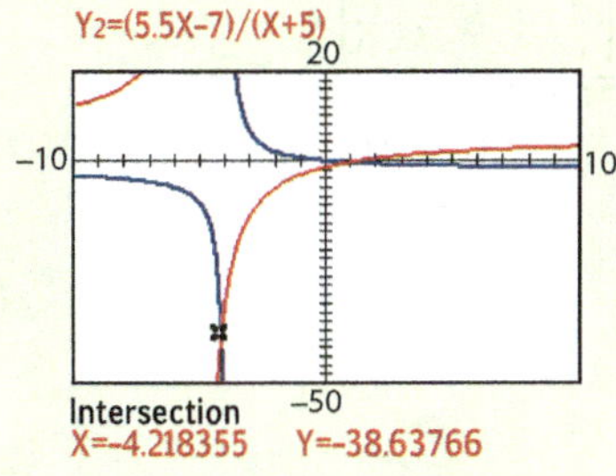

or finding the zeros of
$$Y_1 = \frac{5.5x - 7}{x + 5} + \frac{2x}{x + 4}$$

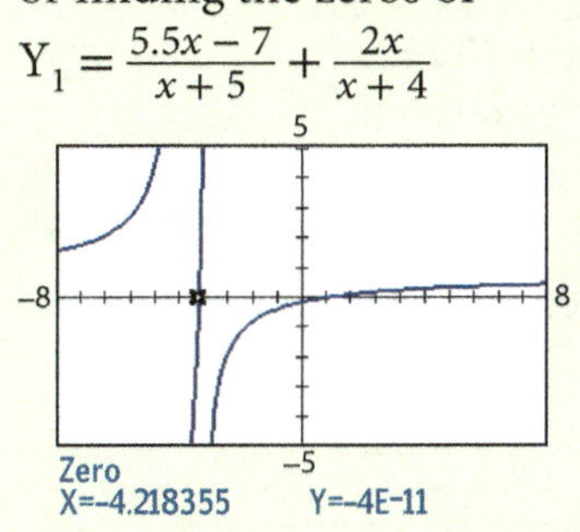

20. $b(b-1)(b-2)$
$$\cdot\left[\frac{3}{b}+\frac{5}{b-1}-\frac{6}{b-2}=\frac{-5}{(b-1)(b-2)}\right]$$
$$3(b^2-3b+2)+5b(b-2)$$
$$-6b(b-1)=-5b$$
$$3b^2-9b+6+5b^2-10b$$
$$-6b^2+6b=-5b$$
$$2b^2-8b+6=0$$
$$b^2-4b+3=0$$
$$(b-3)(b-1)=0$$
$$b=3,\,1\text{ (extraneous)}$$

21a. The graph intercepts the x-axis at each x-value solution.

21b. The graph is undefined at each x-value along the x-axis, indicating either an asymptote or a point discontinuity.

21c. The graph has no x-intercepts.

22. $6x\left[\frac{1}{x}+\frac{1}{3x}=\frac{1}{6}\right]$
$$6+2=x$$
$$x=8;\ 3x=24$$

23.
$$\frac{2+x}{17+x}=\frac{2}{5}$$
$$5(17+x)\left[\frac{2+x}{17+x}=\frac{2}{5}\right]$$
$$10+5x=34+2x$$
$$3x=24$$
$$x=8$$

24. Let x and $x+4$ be the numbers.
$$3x(x+4)\left[\frac{1}{x}+\frac{1}{x+4}=\frac{2}{3}\right]$$
$$3(x+4)+3x=2x(x+4)$$
$$3x+12+3x=2x^2+8x$$
$$2x^2+2x-12=0$$
$$x^2+x-6=0$$
$$(x+3)(x-2)=0$$
$$x=-3,2$$
If $x=-3$, then $x+4=1$.
If $x=2$, then $x+4=6$.

25. Let the numbers be n and $2n+4$.
$$3n(2n+4)\left[\frac{1}{n}+\frac{1}{2n+4}=\frac{1}{3}\right]$$
$$3(2n+4)+3n=n(2n+4)$$
$$6n+12+3n=2n^2+4n$$
$$2n^2-5n-12=0$$
$$(2n+3)(n-4)=0$$
$$n=-\frac{3}{2},4$$
If $n=4$, then $2(4)+4=12$.
If $n=-\frac{3}{2}$, then $2\left(-\frac{3}{2}\right)+4=1$.

26. $24x\left[\frac{1}{12}+\frac{1}{x}=\frac{1}{8}\right]$
$$2x+24=3x$$
$$x=24$$

27. $40x^2\left[\frac{1}{x}+\frac{1}{2x}=\frac{1}{40}\right]$
$$40x+20x=x^2$$
$$x^2-60x=0$$
$$x(x-60)=0$$
$$x=60,0\text{ (extraneous)}$$

28. $12x\left[\frac{1}{4}-\frac{1}{x}=\frac{1}{6}\right]$
$$3x-12=2x$$
$$x=12$$

29. $18x(x+3)\left[\frac{1}{x}-\frac{1}{x+3}=\frac{1}{18}\right]$
$$18(x+3)-18x=x(x+3)$$
$$18x+54-18x=x^2+3x$$
$$x^2+3x-54=0$$
$$(x+9)(x-6)=0$$
$$x=-9,6$$

30.

	R	T	D
First Part	18	$\frac{\left(\frac{d}{2}\right)}{18}=\frac{d}{2(18)}$	$\frac{1}{2}d$
Last Part	r	$\frac{\left(\frac{d}{2}\right)}{r}=\frac{d}{2r}$	$\frac{1}{2}d$
Total Trip	20	$\frac{d}{20}$	d

$$\frac{d}{36}+\frac{d}{2r}=\frac{d}{20}$$
$$90r\left[\frac{1}{18}+\frac{1}{r}=\frac{1}{10}\right]$$
$$5r+90=9r$$
$$90=4r$$
$$r=22.5$$

Solve.

13. $\frac{x+4}{x^2-1}=\frac{3x}{x+1}$ $x=2,-\frac{2}{3}$

14. $x-\frac{1}{x}=-2$ $x=-1\pm\sqrt{2}$

15. $\frac{x}{x-3}-\frac{4}{x+5}=\frac{11}{x^2+2x-15}$ $x=-\frac{1}{2}\pm\frac{\sqrt{3}}{2}i$

16. $\frac{y+2}{y}-\frac{y+1}{y-1}=4$ $y=\frac{1}{2}\pm\frac{1}{2}i$

17. $\frac{r}{r-5}=\frac{7r}{r^2-3r-10}-\frac{3}{r+2}$ $r=-3$

18. $\frac{2}{x+3}+\frac{6}{x^2+3x}=\frac{2-x}{x}$ no solution

19. $\frac{x^2+3x-60}{2x^2+9x+4}=\frac{8x}{2x+1}-\frac{2x}{x+4}$ $x=-5$

20. $\frac{3}{b}+\frac{5}{b-1}-\frac{6}{b-2}=\frac{-5}{b^2-3b+2}$ $b=3$

21. Explain what each type of solution of $f(x)=0$ implies about the graph of $f(x)$.
   a. real solutions
   b. extraneous solutions
   c. imaginary solutions only   8 and 24

22. One number is three times another number. Find the two numbers if the sum of their reciprocals is $\frac{1}{6}$.

23. What number must be added to both the numerator and the denominator of $\frac{2}{17}$ to produce a fraction equivalent to $\frac{2}{5}$?   8

24. Find all numbers whose difference is 4 and the sum of whose reciprocals is $\frac{2}{3}$. -3 and 1; 2 and 6

25. A number is 4 more than twice another number. Find all such numbers if the sum of their reciprocals is $\frac{1}{3}$.

26. Josiah and his younger brother Lucas volunteered to paint a community center. They did the job in 8 hr. Josiah could do the job alone in 12 hr. How long would it have taken Lucas to do the job alone?   24 hr

---

27. Karleigh can detail a car in half the time that it would take her trainee, Mallory. Working together they can do the job in 40 min. How long would it take Mallory to do the job alone?   1 hr

28. A swimming pool that usually takes 4 hr to fill took 6 hr to fill because a small drain was left open. How long would it take the drain to empty a full pool?   12 hr

29. It took 18 hr to fill a tank when the drain was accidentally left open. Determine how long it normally takes to fill and to empty the tank if it normally takes 3 hr longer to empty the tank than it takes to fill it.   6 hr; 9 hr

30. An avid cyclist sets out to break his personal record by averaging 20 mi/hr on a ride but realizes that he has averaged only 18 mi/hr at the halfway point. What will his average speed need to be for the second half of the ride if he is to meet his goal?   22.5 mi/hr

31. Two planes that fly at 600 mi/hr in still air leave the Kansas City airport in opposite directions. If the plane flying with the wind flies 990 mi in 20 min less time than the plane flying into the wind, find the wind speed.   60 mi/hr

32. Determine the average speed of a boat in still water if it makes a 280 mi trip downriver but takes an extra 8 hr to return upriver against a 2 mi/hr current.   12 mi/hr

33. Analyze: Explain the advantage of multiplying both sides of the equation in exercise 13 by the LCD instead of using cross multiplication to solve the proportion.

34. If a flask in a medical lab contains a mixture of 5 cm^3 of medicine and 11 cm^3 of water, how much medicine must be added for the mixture to be $\frac{2}{3}$ medicine?   17 cm^3

35. A circuit contains three parallel resistors of 10 ohms, 40 ohms, and 60 ohms. The total resistance for n parallel resistors is $\frac{1}{R_T}=\frac{1}{R_1}+\frac{1}{R_2}+\frac{1}{R_3}+\cdots+\frac{1}{R_n}$.
   a. Calculate the total resistance of the parallel resistors in the circuit. ≈ 7.06 ohms
   b. If a total resistance of 15 ohms is desired, what size resistor should replace the 10 ohm resistor?   40 ohms

25. 4 and 12; $-\frac{3}{2}$ and 1

---

36. Write a rational function $p(w)$ for the perimeter of a 100 ft^2 rectangle in terms of its width. Then use technology to graph the function and determine the minimum perimeter for such a rectangle.

37. Write a rational function $p(w)$ representing the perimeter of a 37 ft^2 rectangle in terms of its width.

  a. Find the exact dimensions of such a rectangle whose perimeter is 44 ft.

  b. Verify that the area of the rectangle is 37 ft^2.

36. $p(w) = 2w + \frac{200}{w}$; 40 ft

37. $p(w) = 2w + \frac{74}{w}$

37a. $(11 + 2\sqrt{21})$ ft $\times$ $(11 - 2\sqrt{21})$ ft

38. A cyclist desiring to average 20 mi/hr during a training ride wants a function that expresses the required average speed during the second half of a ride based on the average speed to the halfway point.

  a. Write a function for the second half's average speed (r_2) in terms of the first half's average speed (r_1).

  b. Use technology to graph the function and to find the second half's averages for first-half averages of 15 mi/hr, 13 mi/hr, and 10 mi/hr.

  c. State a reasonable domain for the function. Explain your reasoning. $D = (10, \infty)$

  d. State the equation and explain the meaning of any asymptotes. VA: $x = 10$; HA: $y = 10$

38a. $r_2 = \frac{10r_1}{r_1 - 10}$

38b. 30 mi/hr; $43\frac{1}{3}$ mi/hr; not possible

State whether the value of $p(x) = (x+1)^4(x-2)^5$ changes its sign at the given zero. [2.2]

39. $x = -1$ no   **40.** $x = 2$ yes

Use limit notation to describe each function's end behavior. [2.2, 2.5]

41. $f(x) = -3x^3 + 2x - 1$ **42.** $f(x) = \frac{2}{(x-3)^5}$

Add or subtract. [Algebra]

43. $\frac{1}{x^2 - 9} + \frac{x - 3}{x^2 - 6x + 9}$ **44.** $\frac{x}{x^2 + 5x + 6} - \frac{6}{x^2 + x - 2}$

45. Find the slope-intercept form of the line passing through $(1, 2)$ and perpendicular to $x + y = 3$. [1.2] D

  A. $y = x - 1$ **D.** $y = x + 1$

  B. $y = -x - 1$ **E.** none of these

  C. $y = -x + 1$

46. Find a function rule for the relation containing $\{(-1, -1), (0, 0), (1, 1), (2, 8), (3, 27)\}$. [1.3] C

  A. $f(x) = x$ **D.** $f(x) = [x]$

  B. $f(x) = -x^2$ **E.** none of these

  C. $f(x) = x^3$

47. Which is the inverse of $f(x) = \frac{1}{x + 5}$? [1.8] D

  A. $f^{-1}(x) = x - 5$ **D.** $f^{-1}(x) = \frac{1 - 5x}{x}$

  B. $f^{-1}(x) = \frac{1}{x - 5}$ **E.** none of these

  C. $f^{-1}(x) = \frac{1}{5x}$

48. Which of the following statements about modeling with functions is true? [1.9] D

  A. If $|r| > 0.5$, there is a strong linear correlation.

  B. The correlation coefficient is represented by r^2.

  C. The value of r^2 indicates the percentage of the variation in the independent variable that is explained by variation in the dependent variable.

  D. Outliers are points that do not seem to fit the trend of the scatterplot.

  E. all of these

41. $\lim\limits_{x \to -\infty} f(x) = \infty$; $\lim\limits_{x \to \infty} f(x) = -\infty$

42. $\lim\limits_{x \to \pm\infty} f(x) = 0$

43. $\frac{x + 4}{x^2 - 9}$

44. $\frac{x - 9}{(x + 3)(x - 1)}$

33. Cross multiplication results in a cubic equation, which requires long division to solve and produces an extraneous solution. Multiplying by the LCM results in a quadratic equation, which can be solved directly and produces no extraneous solutions.

❯ C. Exercises

34. initial mixture: $\frac{5}{5 + 11}$

$$\frac{5 + m}{16 + m} = \frac{2}{3}$$
$$3(16 + m)\left[\frac{5 + m}{16 + m} = \frac{2}{3}\right]$$
$$3(5 + m) = 2(16 + m)$$
$$15 + 3m = 32 + 2m$$
$$m = 17 \text{ cm}^3$$

35a.
$$\frac{1}{R_T} = \frac{1}{10} + \frac{1}{40} + \frac{1}{60}$$
$$120R_T\left[\frac{1}{R_T} = \frac{1}{10} + \frac{1}{40} + \frac{1}{60}\right]$$
$$120 = 12R_T + 3R_T + 2R_T$$
$$120 = 17R_T$$
$$R_T \approx 7.06$$

35b.
$$\frac{1}{15} = \frac{1}{x} + \frac{1}{40} + \frac{1}{60}$$
$$120x\left[\frac{1}{15} = \frac{1}{x} + \frac{1}{40} + \frac{1}{60}\right]$$
$$8x = 120 + 3x + 2x$$
$$3x = 120$$
$$x = 40$$

36. $lw = 100$; $l = \frac{100}{w}$
$$p = 2w + 2l$$
$$p(w) = 2w + \frac{200}{w}$$

Minimum
X=10   Y=40

37. $lw = 37$; $l = \frac{37}{w}$
$$p = 2w + 2l$$
$$p(w) = 2w + 2\left(\frac{37}{w}\right)$$

37a.
$$44 = 2w + \frac{74}{w}$$
$$2w^2 - 44w + 74 = 0$$
$$w = 11 \pm 2\sqrt{21}$$

using $w = 11 - 2\sqrt{21}$:
$$l = \frac{37}{11 - 2\sqrt{21}} \cdot \frac{(11 + 2\sqrt{21})}{(11 + 2\sqrt{21})}$$
$$= \frac{37(11 + 2\sqrt{21})}{121 - 4(21)}$$
$$= \frac{37(11 + 2\sqrt{21})}{37} = 11 + 2\sqrt{21}$$

continued in Answers and Solutions Overflow

31.

	R	T	D
With Wind	$600 + x$	$\frac{990}{600 + x}$	990
Against Wind	$600 - x$	$\frac{990}{600 - x}$	990

$$\frac{990}{600 + x} = \frac{990}{600 - x} - \frac{1}{3}$$
$$\left(20 \text{ min} = \frac{1}{3} \text{ hr}\right)$$
$$3(600 + x)(600 - x)$$
$$\cdot \left[\frac{990}{600 + x} = \frac{990}{600 - x} - \frac{1}{3}\right]$$
$$2970(600 - x) =$$
$$\quad 2970(600 + x) - (360{,}000 - x^2)$$
$$1{,}782{,}000 - 2970x =$$
$$\quad 1{,}782{,}000 + 2970x - 360{,}000 + x^2$$
$$x^2 + 5940x - 360{,}000 = 0$$
$$(x - 60)(x + 6000) = 0$$
$$x = 60, -6000$$

32.

	R	T	D
Downstream	$r + 2$	$\frac{280}{r + 2}$	280
Upstream	$r - 2$	$\frac{280}{r - 2}$	280

$$\frac{280}{r - 2} = \frac{280}{r + 2} + 8$$
$$(r - 2)(r + 2)\left[\frac{280}{r - 2} = \frac{280}{r + 2} + 8\right]$$
$$280(r + 2) = 280(r - 2)$$
$$\qquad\qquad\qquad + 8(r - 2)(r + 2)$$
$$280r + 560 = 280r - 560 + 8(r^2 - 4)$$
$$1120 = 8r^2 - 32$$
$$8r^2 = 1152$$
$$r^2 = 144$$
$$r = \pm 12$$

2.7 Nonlinear Inequalities (Extended)

Objectives

1. To solve polynomial inequalities algebraically and graphically
2. To solve rational inequalities algebraically and graphically

Flash

Techniques from the ancient art of origami are now used in micro-robots that can move through the body performing medical tasks. Use the Internet keyword search *micro-robot origami* for more details.

Vocabulary

polynomial inequality
rational inequality
sign chart

Additional Exercises

Find the zeros of the polynomial; then determine the intervals where the function's values are (a) positive and (b) negative.

1. $p(x) = (x + 5)(x + 3)(x - 5)$
   zeros: $\pm 5, -3$; a. $(-5, -3) \cup (5, \infty)$;
   b. $(-\infty, -5) \cup (-3, 5)$

2. $p(x) = (x - 3)(x + 2)(x + 4)^2$
   zeros: $-4, -2, 3$; a. $(-\infty, -4) \cup (-4, -2) \cup (3, \infty)$; b. $(-2, 3)$

Solve.

3. $4x^4 - 28x^3 + 53x^2 + 10x - 75 \geq 0$
   $(-\infty, -1] \cup \left\{\frac{5}{2}\right\} \cup [3, \infty)$

4. $-2x^2 + x + 10 < 0$
   $(-\infty, -2) \cup (2.5, \infty)$

5. $-2x^3 - 3x^2 + 5x + 6 \geq 0$
   $(-\infty, -2] \cup [-1, 1.5]$

6. $3x^4 - 4x^2 \geq -8x^2 - 1$ $\mathbb{R}$

7. $\dfrac{x^2 - 3x - 18}{3x^2 + 13x + 12} \geq 0$
   $(-\infty, -3) \cup \left(-3, -\frac{4}{3}\right) \cup [6, \infty)$
   (PD at $x = -3$)

8. $\dfrac{x^2 + 3x - 10}{x - 1} < 10$ $(-\infty, 0) \cup (1, 7)$

9. $\dfrac{x + 1}{|x - 5|} > 0$ $(-1, 5) \cup (5, \infty)$

10. $\dfrac{\sqrt{x + 1}}{x + 6} \geq 0$ $[-1, \infty)$

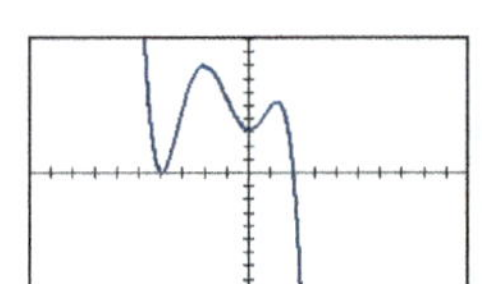

A polynomial inequality can be solved to find an acceptable range of values for the height of an origami box.

After completing this section, you will be able to

- solve polynomial inequalities.
- solve rational inequalities.

A *polynomial inequality* can be written in the form $p(x) > 0$, $p(x) \geq 0$, $p(x) < 0$, $p(x) \leq 0$, or $p(x) \neq 0$ where $p(x)$ is a polynomial function.

Since a polynomial function is continuous, a change in the sign of the function's value can only occur at a zero. Expressing the inequality as a polynomial on one side with 0 on the other allows us to use the zeros to define intervals for which the function takes on exclusively positive or exclusively negative values. A *sign chart* showing the sign values of the function over each interval can be used to state solutions to the following inequalities.

$p(x) > 0$ over $(-\infty, -4) \cup (-4, 2)$ $p(x) \geq 0$ over $(-\infty, 2]$
$p(x) < 0$ over $(2, \infty)$ $p(x) \leq 0$ over $[2, \infty)$

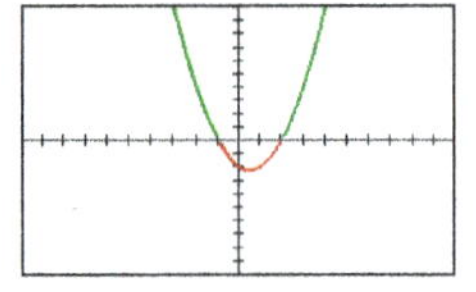

By examining the sign of each factor for a single x-value in each interval, you can quickly determine the sign of the function's values over that entire interval.

Example 1 Solving a Polynomial Inequality

Solve $x^2 > x + 2$.

Answer

$x^2 - x - 2 > 0$
1. Express as an equivalent inequality in the form $p(x) > 0$.

$(x + 1)(x - 2) = 0$
The zeros are at $x = -1, 2$.
2. Solve $p(x) = 0$ to find the zeros of the function.

$p(x) > 0$ $p(x) < 0$ $p(x) > 0$
$(-)(-)$ -1 $(+)(-)$ 2 $(+)(+)$

3. Make a sign chart by examining the signs of the factors within each interval.
   a. When $x = -3$, $(x + 1)$ and $(x - 2)$ are negative,
      $\therefore p(x)$ is positive over $(-\infty, -1)$.
   b. When $x = 0$, $(x + 1)$ is positive and $(x - 2)$ is negative,
      $\therefore p(x)$ is negative over $(-1, 2)$.
   c. When $x = 3$, both $(x + 1)$ and $(x - 2)$ are positive,
      $\therefore p(x)$ is positive over $(2, \infty)$.

$x \in (-\infty, -1) \cup (2, \infty)$
4. State the solution to $x^2 > x + 2$.

Check

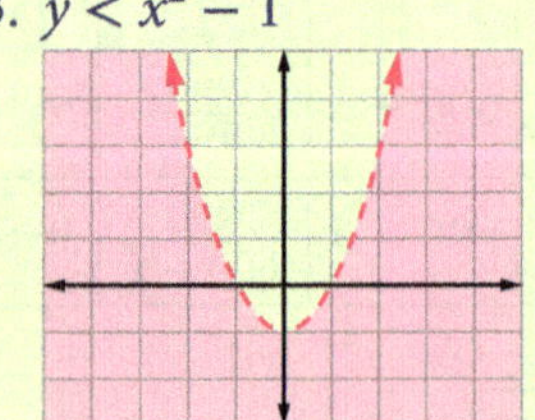

5. Note how the graph of $p(x) = x^2 - x - 2$ confirms this result.

SKILL ✔ EXERCISE 5

PRESENTATION

Lesson Opener

Sketch a graph of each inequality.

1. $-2 > x \geq 1$

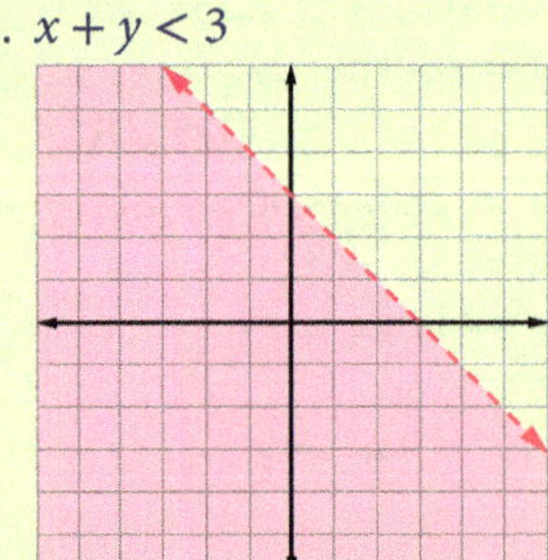

2. $x + y < 3$

3. $y < x^2 - 1$

Section 2.7 can be used for the extended track. It explores *polynomial inequalities* in Examples 1–3 and *rational inequalities* in Examples 4–5. Define a polynomial inequality, emphasizing that we are focusing on the sign (positive or negative) of the dependent variable. Use the illustrated polynomial function preceding Example 1 to demonstrate how a *sign chart* can be used to summarize the sign

Recall that the graph of a polynomial function crosses the x-axis at zeros with odd multiplicity but touches the x-axis without crossing it at zeros with even multiplicity. There are no x-intercepts or sign changes associated with complex zeros. Combining these facts with the function's end behavior provides a second method of predicting the sign of the function's values over each interval.

Example 2 Solving a Polynomial Inequality

Solve $(2 - x)(x^2 + 1)(x + 4)^2 \le 0$.

Answer

$x = 2, -4$ (multiplicity 2)
$x^2 + 1$ has only complex zeros.

$\lim\limits_{x \to -\infty} p(x) = \infty;\ \lim\limits_{x \to \infty} p(x) = -\infty$

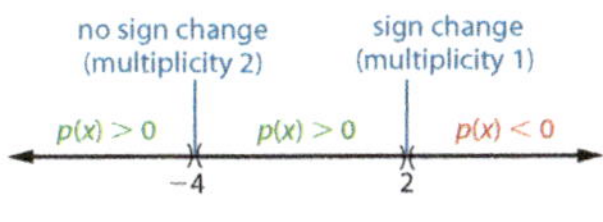

$x \in \{-4\} \cup [2, \infty)$

1. Find the zeros, noting the multiplicity of any repeated zeros.
2. Note the end behavior of a fifth-degree polynomial with a negative leading coefficient.
3. Make a sign chart using the end behavior and the sign changes indicated by the multiplicity of each real zero.
4. State the solution to the inequality.

Check

5. Evaluating when each factor is positive produces the same sign chart for $p(x)$.
$(2 - x)$ is positive when $x < 2$.
$(x^2 + 1)$ is always positive.
$(x + 4)^2$ is never negative.

SKILL ✔ EXERCISE 7

The solution for a polynomial inequality where $p(x)$ has no real zeros is either the set of all real numbers or the empty set, $\varnothing$.

Example 3 Solving a Polynomial Inequality with No Real Zeros

Solve $4x^4 + 9 \le -37x^2$.

Answer

$4x^4 + 37x^2 + 9 \le 0$
$(4x^2 + 1)(x^2 + 9) = 0$
$x = \pm\frac{i}{2}, \pm 3i$

$\lim\limits_{x \to \pm\infty} p(x) = \infty$

$\therefore$ There are no solutions.

1. Find the zeros of the equivalent inequality in the form $p(x) \le 0$.
2. Note the end behavior of a fourth-degree polynomial with a positive leading coefficient.
3. The end behavior and the fact that there are no real zeros implies that $p(x) > 0$ for all x.

Check

$4x^2 + 1 > 0;$ $x^2 + 9 > 0;$
always   always

4. Examining each factor confirms that $p(x) > 0$ for all x.

SKILL ✔ EXERCISE 29

Solutions

❯ A. Exercises

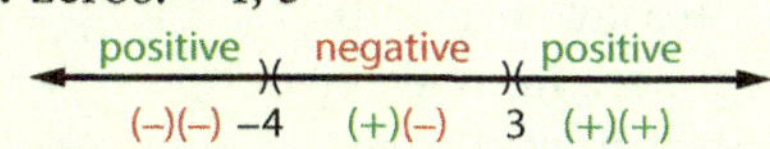

1. zeros: $-4, 3$

2. zeros: $-2, 1, 3$

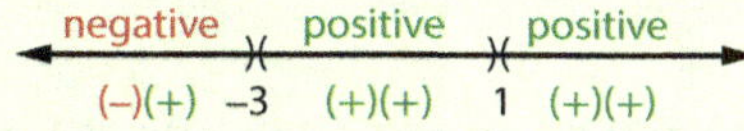

3. zeros: $-3, 1$

4. zeros: $1.5, 3$

5. $(2x + 1)(x - 6) \ge 0$
zeros: $-\frac{1}{2}, 6$

---

of the function within each interval defined by the function's zeros.

After finding the zeros of the function in Example 1, demonstrate how to make a sign chart by identifying the sign of each factor within each interval. Use the sign chart to state the solution. Students can check their work by using technology to graph the function.

An alternative graphical representation is found in Example 4. Graph $Y_1 = x^2$ and $Y_2 = x + 2$, use the x-values of their intersections to define the intervals, and note that the solution consists of those intervals where $Y_1 > Y_2$.

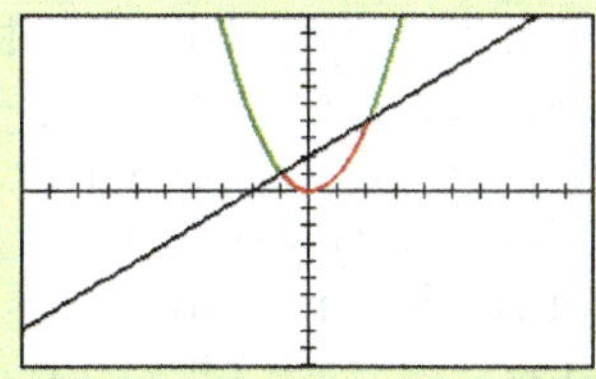

Example 2 describes another method for identifying the sign of the function's values within each interval. Consider applying this method in Example 1, either before or after presenting Example 2. Note that the solution using this method can be checked using the first method or by observing the graph of the function.

Explain that when the graph of a function has no real zeros (x-intercepts), it must be entirely above the x-axis (positive values) or below it (negative values). Therefore, the polynomial inequality will always be true or always be false with a solution of all real numbers or the empty set, respectively. Consider demonstrating how the graph of $p(x) = 4x^4 + 37x^2 + 9$ in Example 3 is entirely above the x-axis.

The key to solving rational inequalities, as in Example 4, is to express the function as one term on one side with 0 on the other side. Be sure the students understand why -2 is included in the interval while -3 is not.

Common Student Error Students may attempt to multiply both sides of the in-

6.

$$\begin{array}{r} 1 \quad -2 \quad -1 \quad 2 \\ 2\overline{)\,1 \quad 0 \quad -1 \quad 0\,} \end{array}$$

$$(x-2)(x^2-1) \le 0$$
$$(x-2)(x-1)(x+1) \le 0$$
zeros: $\pm 1, 2$

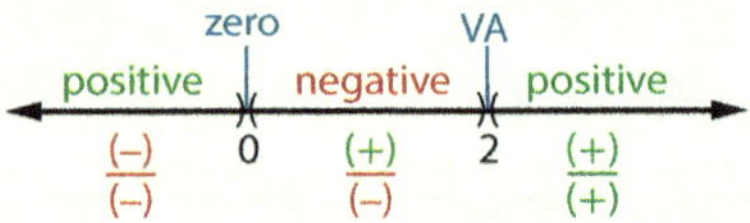

7.

$$\begin{array}{r} 1 \quad -3 \quad 0 \quad 4 \\ 2\overline{)\,1 \quad -1 \quad -2 \quad 0\,} \end{array}$$

$$(x-2)(x^2-x-2) < 0$$
$$(x-2)(x-2)(x+1) < 0$$
$$(x+1)(x-2)^2 < 0$$
zeros: $-1, 2$ (multiplicity 2)

8. $(x^2+3)(x^2+1) < 0$
There are no real zeros
and $\displaystyle\lim_{x\to\pm\infty} p(x) = \infty$.

9. zeros of $a(x)$: $x = 0$, the zero of $f(x)$
zeros of $b(x)$: $x = 2$, a VA of $f(x)$

$f(x) > 0$ when $x \in (-\infty, 0) \cup (2, \infty)$
$f(x) < 0$ when $x \in (0, 2)$

10. $f(x) = \dfrac{x-2}{(x-1)(x+1)}$
zeros of $a(x)$: $x = 2$, the zero of $f(x)$
zeros of $b(x)$: $x = \pm 1$, VAs of $f(x)$

$f(x) > 0$ when $x \in (-1, 1) \cup (2, \infty)$
$f(x) < 0$ when $x \in (-\infty, -1) \cup (1, 2)$

11. $f(x) = \dfrac{x+1}{(5x-4)(x+1)} = \dfrac{1}{5x-4}$
where $x \ne -1$
zeros of $a(x)$: $x = -1$, a PD of $f(x)$
zeros of $b(x)$: $x = -1$, a PD of $f(x)$
and $x = \dfrac{4}{5}$, a VA of $f(x)$

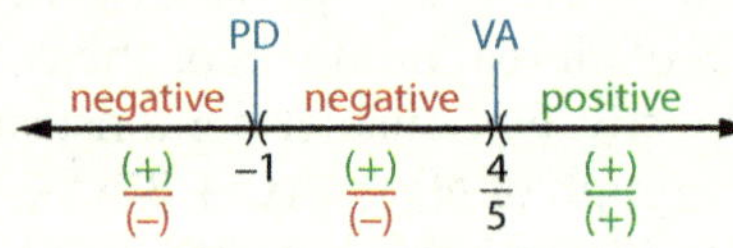

$f(x) > 0$ when $x \in \left(\dfrac{4}{5}, \infty\right)$
$f(x) < 0$ when $x \in (-\infty, -1) \cup \left(-1, \dfrac{4}{5}\right)$

When solving a *rational inequality* (an inequality containing rational expressions), write the inequality with a single rational function on one side and 0 on the other. Since the value of a rational function can change its sign at either a zero or at a discontinuity, the zeros of both the numerator and the denominator are included in the sign chart.

Example 4   Solving a Rational Inequality

Solve $\dfrac{x}{x+3} \le -2$.

Answer

$$\frac{x}{x+3} + 2 \le 0$$
1. Express the inequality with 0 on one side.

$$\frac{x}{x+3} + 2\left(\frac{x+3}{x+3}\right) \le 0$$
2. Combine the terms to express the nonzero side as a single rational function, $f(x) = \dfrac{a(x)}{b(x)}$.

$$\frac{3x+6}{x+3} \le 0$$

zeros of $a(x)$: -2, a zero of $f(x)$
zeros of $b(x)$: -3, a discontinuity
3. Find the zeros of the numerator and the denominator.

4. Make a sign chart by examining values within each interval.

a. When $x = -4$, both $(3x+6)$ and $(x+3)$ are negative.
b. When $x = -2.5$, $(3x+6)$ is negative and $(x+3)$ is positive.
c. When $x = 0$, both $(3x+6)$ and $(x+3)$ are positive.

$x \in (-3, -2]$
5. State the solution to $\dfrac{x}{x+3} \le -2$.

Check

6. Note how the graphs of $Y_1 = \dfrac{x}{x+3}$ and $Y_2 = -2$ confirm the result.

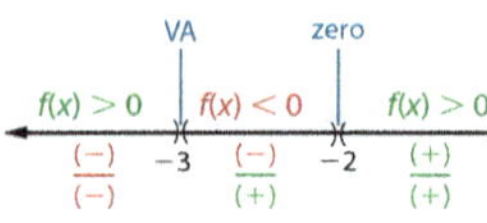

SKILL ✔ EXERCISE 21

Graphical solutions are used in many real-life applications of polynomial and rational inequalities.

Example 5   Applying a Rational Inequality

The Cantin Corporation is designing a 500 cm^3 soup can. Find the range of possible radii lengths (to the nearest hundredth of a centimeter) for the can if its surface area cannot exceed 380 cm^2.

Answer

$$V = \pi r^2 H$$
$$H = \frac{V}{\pi r^2} = \frac{500}{\pi r^2}$$
$$SA = 2\pi r^2 + 2\pi r H$$
$$= 2\pi r^2 + 2\pi r\left(\frac{500}{\pi r^2}\right)$$
$$= 2\pi r^2 + \frac{1000}{r}$$
1. Letting r represent the radius of the can, find an expression for the height and surface area of the can in terms of the radius.

$$2\pi r^2 + \frac{1000}{r} \le 380$$
$$2\pi r^2 - 380 + \frac{1000}{r} \le 0$$
$$\frac{2\pi r^3 - 380r + 1000}{r} \le 0$$
2. Write a rational inequality modeling the restriction on the can's surface area and combine the terms on the left side.

CONTINUED ➡

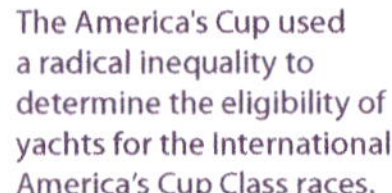

The America's Cup used a radical inequality to determine the eligibility of yachts for the International America's Cup Class races.

---

equality in Example 4 by the LCD and solve $x \le -2x - 6$ to get $x \le -2$. This method ignores instances where $x + 3 \le 0$ and the inequality would need to be reversed or rendered meaningless.

Graphing $Y_1 = \dfrac{3x+6}{x+3}$ and finding the intervals where $Y_1 \le 0$, as illustrated in Example 1, also confirms the result for Example 4. Notice that the vertical asymptote and the x-intercept define the intervals.

Complete Example 5 with the class, emphasizing that graphical analysis of polynomial and rational inequalities is a critical element in solving real-world problems.

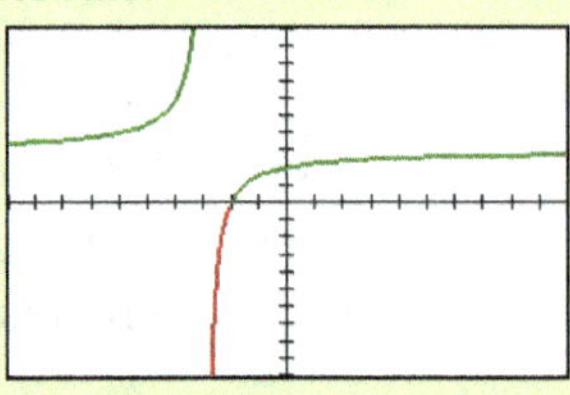

3. Solve graphically using the zero option from the [CALC] menu.

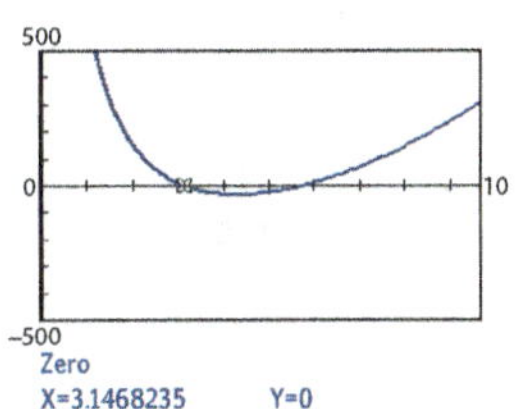

$r \in [3.15 \text{ cm}, 5.71 \text{ cm}]$

A. Exercises

Find the zeros for each polynomial function and then use a sign chart to determine intervals where the function's values are (a) positive and (b) negative.

1. $p(x) = (x + 4)(x - 3)$ a. $(-\infty, -4) \cup (3, \infty)$; b. $(-4, 2)$
2. $p(x) = (x + 2)(x - 1)(x - 3)$ a. $(-2, 1) \cup (3, \infty)$; b. $(-\infty, -2) \cup (1, 3)$
3. $p(x) = (x + 3)^3(x - 1)^2$ a. $(-3, 1) \cup (1, \infty)$; b. $(-\infty, -3)$
4. $p(x) = (2x - 3)(x^2 + 1)(x - 3)$ a. $(-\infty, 1.5) \cup (3, \infty)$; b. $(1.5, 3)$

Solve each polynomial inequality by factoring and using a sign chart.

5. $2x^2 - 11x - 6 \geq 0$ $\left(-\infty, -\frac{1}{2}\right] \cup [6, \infty)$
6. $x^3 - 2x^2 - x + 2 \leq 0$ $(-\infty, -1] \cup [1, 2]$
7. $x^3 - 3x^2 + 4 < 0$ $(-\infty, -1)$
8. $x^4 + 4x^2 + 3 < 0$ $\varnothing$

Find the zeros and points of discontinuity for each rational function. Then use a sign chart to determine the intervals where the function's values are (a) positive and (b) negative.

9. $f(x) = \dfrac{x}{x - 2}$
10. $f(x) = \dfrac{x - 2}{x^2 - 1}$
11. $f(x) = \dfrac{x + 1}{5x^2 + x - 4}$
12. $f(x) = \dfrac{x + 5}{x - 1} - 2$

13. Which statements are true of sign charts?   A, C
   A. Unique linear factors change signs only once.
   B. The sign of the function must change at a point of discontinuity.
   C. Odd multiplicity of a linear factor implies a sign change.

14. At which value(s) is $f(x) = \dfrac{x + 1}{x^2 - 1}$ discontinuous?   C
   A. $x = 1$ C. $x = \pm 1$
   B. $x = -1$ D. none of these

B. Exercises

Solve each inequality using a sign chart.

15. $\dfrac{x - 3}{2x^2 - 3x - 9} \leq 0$ $\left(\infty, -\frac{3}{2}\right)$
16. $\dfrac{x - 2}{x^2 - 9} > 0$ $(-3, 2) \cup (3, \infty)$
17. $\dfrac{x^2 + 5x - 14}{2x^2 - 5x + 2} \geq 0$
18. $\dfrac{x^2 + 2x - 15}{x^2 + 7x + 10} < 0$ $(-2, 3)$
19. $\dfrac{2x + 11}{x^2 + 3} \geq 1$ $[-2, 4]$
20. $\dfrac{x^2 - 10}{2x - 5} \leq 3$ $[1, 2.5) \cup [5, \infty)$
21. $\dfrac{(2x + 1)(x + 2)}{(x + 5)(x - 1)} \geq 2$ $(-\infty, -5) \cup (1, 4]$
22. $\dfrac{2x^2 - x - 27}{x^2 - x - 12} \leq 1$
23. $x^3 + 2x^2 > 8x$
24. $x^3 - 3x^2 + 3x - 1 < 0$ $(-\infty, 1)$
25. $2x^3 + 6 \leq 3x^2 + 11x$ $(-\infty, -2] \cup \left[\frac{1}{2}, 3\right]$
26. $x^4 + 20 \leq 2x^3 + 15x^2 + 4x$ $[1, 5]$
27. $x^4 + 2x^3 - 11x^2 - 12x + 36 \geq 0$ $\mathbb{R}$
28. $2x^4 + 2x^3 + 4x \geq x^6 + 2x^5 + x^2 + 4$ $x = -2, 1$
29. $x^5 + x^3 - 4x < 4x^4 - 10x^2 + 8$ $(-\infty, -1) \cup (-1, 2]$
30. A church grounds crew is designing a playground whose area must be no more than 4000 ft^2. Find the largest possible dimensions (to the nearest foot) if its length is to be 10 ft longer than its width.   58 ft × 68 ft
31. The Cantin Corporation is redesigning its large fruit cans to hold 3.6 L (3600 cm^3). Find the range of possible radii lengths (to the nearest hundredth of a centimeter) for the can if its surface area cannot exceed 1325 cm^2.   [7.21 cm, 9.50 cm]
32. The cost per square foot for a three-bedroom home in Denver, Colorado, between 2006 and 2016 can be modeled by the polynomial function $f(x) = -0.0058x^4 + 0.58x^3 - 12.26x^2 + 92x - 41$ where x represents the number of years since 2000. Use technology to determine when the cost per square foot was less than \$185.   mid-2007 through mid-2012

---

17. $\dfrac{(x + 7)(x - 2)}{(2x - 1)(x - 2)} = \dfrac{x + 7}{2x - 1}$ where $x \neq 2$

  zero — VA — PD

  positive | negative | positive | positive

  $\frac{(-)}{(-)}$ −7 $\frac{(+)}{(-)}$ $\frac{1}{2}$ $\frac{(+)}{(+)}$ 2 $\frac{(+)}{(+)}$

18. $\dfrac{(x + 5)(x - 3)}{(x + 5)(x + 2)} = \dfrac{x - 3}{x + 2}$ where $x \neq -5$

  PD — VA — zero

  positive | positive | negative | positive

  $\frac{(-)}{(-)}$ −5 $\frac{(-)}{(-)}$ −2 $\frac{(-)}{(+)}$ 3 $\frac{(+)}{(+)}$

19. $\dfrac{2x + 11}{x^2 + 3} - 1\left(\dfrac{x^2 + 3}{x^2 + 3}\right) \geq 0$

$$\dfrac{-x^2 + 2x + 8}{x^2 + 3} \geq 0$$

$$\dfrac{-1(x - 4)(x + 2)}{x^2 + 3} \geq 0$$

  zero — zero

  negative | positive | negative

  $\frac{(-)(-)(-)}{(+)}$ −2 $\frac{(-)(-)(+)}{(+)}$ 4 $\frac{(-)(+)(+)}{(+)}$

20. $\dfrac{x^2 - 10}{2x - 5} - 3\left(\dfrac{2x - 5}{2x - 5}\right) \geq 0$

$$\dfrac{x^2 - 6x + 5}{2x - 5} \geq 0$$

$$\dfrac{(x - 5)(x - 1)}{2x - 5} \geq 0$$

  zero — VA — zero

  negative | positive | negative | positive

  $\frac{(-)(-)}{(-)}$ 1 $\frac{(-)(+)}{(-)}$ $\frac{5}{2}$ $\frac{(-)(+)}{(+)}$ 5 $\frac{(+)(+)}{(+)}$

21. $\dfrac{2x^2 + 5x + 2}{x^2 + 4x - 5} - 2\left(\dfrac{x^2 + 4x - 5}{x^2 + 4x - 5}\right) \geq 0$

$$\dfrac{-3x + 12}{x^2 + 4x - 5} \geq 0$$

$$\dfrac{-3(x - 4)}{(x + 5)(x - 1)} \geq 0$$

  VA — VA — zero

  positive | negative | positive | negative

  $\frac{(-)(-)}{(-)(-)}$ −5 $\frac{(-)(-)}{(+)(-)}$ 1 $\frac{(-)(-)}{(+)(-)}$ 4 $\frac{(-)(+)}{(+)(+)}$

22. $\dfrac{2x^2 - x - 27}{x^2 - x - 12} - 1\left(\dfrac{x^2 - x - 12}{x^2 - x - 12}\right) \leq 0$

$$\dfrac{x^2 - 15}{(x - 4)(x + 3)} \leq 0$$

  zero — VA — zero — VA

  positive | negative | positive | negative | positive

  $\frac{(+)}{(-)(-)}$ −3.87 $\frac{(-)}{(-)(-)}$ −3 $\frac{(-)}{(-)(+)}$ 3.87 $\frac{(+)}{(-)(+)}$ 4 $\frac{(+)}{(+)(+)}$

23. $x^3 + 2x^2 - 8x > 0$

  $x(x + 4)(x - 2) > 0$

  zero — zero — zero

  negative | positive | negative | positive

  $(-)(-)(-)$ −4 $(-)(+)(-)$ 0 $(+)(+)(-)$ 2 $(+)(+)(+)$

---

12. $f(x) = \dfrac{x + 5}{x - 1} - 2\left(\dfrac{x - 1}{x - 1}\right) = \dfrac{7 - x}{x - 1}$

  zeros of $a(x)$: $x = 7$, a zero of $f(x)$
  zeros of $b(x)$: $x = 1$, a VA of $f(x)$

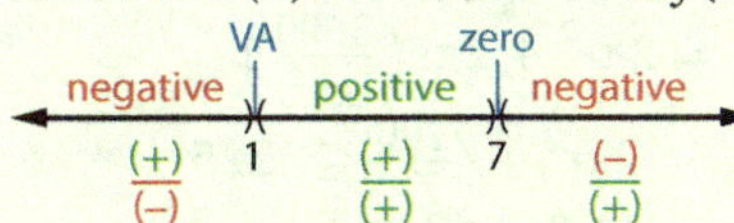

  $f(x) > 0$ when $x \in (1, 7)$
  $f(x) < 0$ when $x \in (-\infty, 1) \cup (7, \infty)$

14. $f(x) = \dfrac{x + 1}{(x - 1)(x + 1)} = \dfrac{1}{x - 1}$

  where $x \neq -1$
  PD at $x = -1$; VA: $x = 1$

B. Exercises

15. $\dfrac{x - 3}{(2x + 3)(x - 3)} = \dfrac{1}{2x + 3}$ where $x \neq 3$

  VA — PD

  negative | positive | positive

  $\frac{(+)}{(-)}$ $-\frac{3}{2}$ $\frac{(+)}{(+)}$ 3 $\frac{(+)}{(+)}$

16. $\dfrac{x - 2}{(x - 3)(x + 3)} > 0$

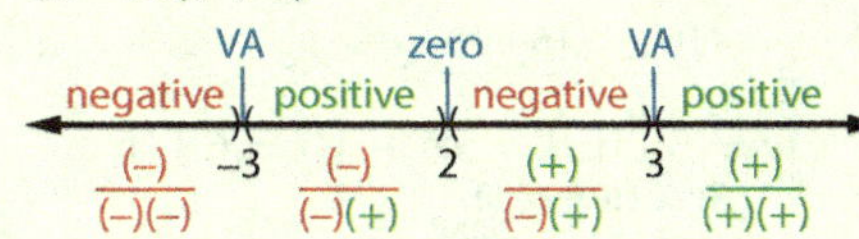

24.

$$\begin{array}{c|rrrr} & 1 & -3 & 3 & -1 \\ 1 & 1 & -2 & 1 & 0 \end{array}$$

$$(x-1)(x^2 - 2x + 1) < 0$$
$$(x-1)^3 < 0$$

zero (mult: 3)

negative | positive

(−) 1 (+)

25. $2x^3 - 3x^2 - 11x + 6 \le 0$

$$\begin{array}{c|rrrr} & 2 & -3 & -11 & 6 \\ -2 & 2 & -7 & 3 & 0 \end{array}$$

$$(x+2)(2x^2 - 7x + 3) \le 0$$
$$(x+2)(2x-1)(x-3) \le 0$$

zero   zero   zero

negative | positive | negative | positive

(−)(−)(−) −2 (+)(−)(−) $\frac{1}{2}$ (+)(+)(−) 3 (+)(+)(+)

26. $x^4 - 2x^3 - 15x^2 - 4x + 20 \le 0$

$$\begin{array}{c|rrrrr} & 1 & -2 & -15 & -4 & 20 \\ 1 & 1 & -1 & -16 & -20 & 0 \\ -2 & 1 & -3 & -10 & 0 \end{array}$$

$$(x-1)(x+2)(x^2 - 3x - 10) \le 0$$
$$(x+2)^2(x-1)(x-5) \le 0$$

zero (mult: 2) zero   zero

positive | positive | negative | positive

(+)(−)(−) −2 (+)(−)(−) 1 (+)(+)(−) 5 (+)(+)(+)

27.

$$\begin{array}{c|rrrrr} & 1 & 2 & -11 & -12 & 36 \\ 2 & 1 & 4 & -3 & -18 & 0 \\ -3 & 1 & 1 & -6 & 0 \end{array}$$

$$(x-2)(x+3)(x^2 + x - 6) \ge 0$$
$$(x+3)^2(x-2)^2 \ge 0$$

zero   zero

positive | positive | positive

(+)(+) −3 (+)(+) 2 (+)(+)

28. $x^6 + 2x^5 - 2x^4 - 2x^3 + x^2 - 4x$
$$+ 4 \le 0$$

$$\begin{array}{c|rrrrrrr} & 1 & 2 & -2 & -2 & 1 & -4 & 4 \\ 1 & 1 & 3 & 1 & -1 & 0 & -4 & 0 \\ -2 & 1 & 1 & -1 & 1 & -2 & 0 \\ -2 & 1 & -1 & 1 & -1 & 0 \\ 1 & 1 & 0 & 1 & 0 \end{array}$$

$$(x+2)^2(x-1)^2(x^2+1) \le 0$$

zero (mult: 2)   zero (mult: 2)

positive | positive | positive

(+)(+)(+) −2 (+)(+)(+) 1 (+)(+)(+)

29. $x^5 - 4x^4 + x^3 + 10x^2 - 4x - 8 < 0$

$$\begin{array}{c|rrrrrr} & 1 & -4 & 1 & 10 & -4 & -8 \\ -1 & 1 & -5 & 6 & 4 & -8 & 0 \\ -1 & 1 & -6 & 12 & -8 & 0 \\ 2 & 1 & -4 & 4 & 0 \end{array}$$

$$(x+1)^2(x-2)(x^2 - 4x + 4) < 0$$
$$(x+1)^2(x-2)^3 < 0$$

33. An origami gift box can be made by folding an 8″ × 11″ piece of cardstock. Find the range of possible heights for the box if its volume needs to be at least 55 in.[3] ≈ [1.04 in., 2.06 in.]

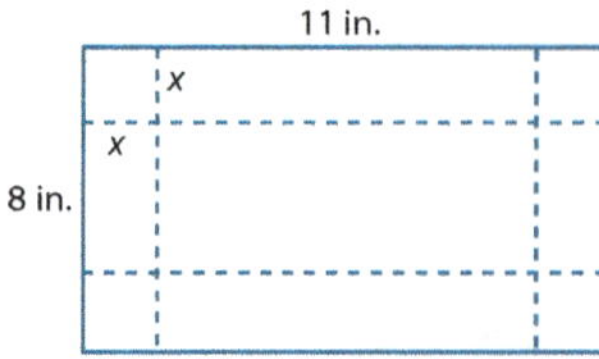

> **C. Exercises**

Solve each inequality using a sign chart.

34. $\dfrac{x+3}{2x+5} \ge \dfrac{x+2}{x-3}$ **35.** $\dfrac{x^2 - 49}{x^2 - x - 42} \le \dfrac{6 - x}{x^2 + 6x}$

Sign charts can be adapted for other types of inequalities. Use a modified sign chart to solve each inequality.

36. $\dfrac{x}{\sqrt{x+3}} \ge 0$ $[0, \infty)$ **37.** $\dfrac{2-x}{\sqrt{x-5}} < 0$ $(5, \infty)$

38. $\dfrac{x}{|x+3|} < 0$ **39.** $\dfrac{\sqrt{5-2x}}{|x-1|} > 0$

$(-\infty, -3) \cup (-3, 0)$ $(-\infty, 1) \cup \left(1, \frac{5}{2}\right)$

34. $\left[\dfrac{-9 - \sqrt{5}}{2}, \dfrac{-9 + \sqrt{5}}{2}\right] \cup \left(-\dfrac{5}{2}, 3\right)$

35. $[-4 - \sqrt{22}, -6) \cup (0, -4 + \sqrt{22}]$

40. The formula $\dfrac{L + 1.25 \sqrt{S} - 9.8 \sqrt[3]{DSP}}{0.686} \le 24.00$ m was used to determine the eligibility of a yacht for the International America's Cup Class races. In the formula, L is the rated length in meters, S is the rated sail area in square meters, and DSP is the volume of water displaced in cubic meters. Determine the minimum displacement for a yacht that is 25 m long with a rated sail area of 325 m². $DSP \approx 31.87$ m³

41. An Internet company is planning a new cloud storage facility of at least 12 acres on their 800-acre campus. The engineers recommend that the structure have a length 150 ft longer than twice its width. Use the fact that 1 acre = 43,560 ft² to write a polynomial inequality modeling the area of the planned facility in square feet. Use technology to find the smallest dimensions for the storage facility (to the nearest foot). 475 ft × 1100 ft

42. Adam has decided that his rectangular cornfield will have a length twice as long as its width, with an option to add 50 yd to the length in later years. He eventually wants to plant 25 acres of corn. Write an inequality that would represent the total area of the enlarged field. Use technology to find the smallest dimensions that would meet the requirements. (1 acre = 4840 yd²) ≈ 233.8 yd × 517.6 yd

zero (mult: 2)   zero (mult: 3)

negative | negative | positive

(+)(−) −1 (+)(−) 2 (+)(+)

30.

$$A(w) = (w + 10)w$$
$$w^2 + 10w \le 4000$$
$$w^2 + 10w - 4000 \le 0$$

$$\text{zeros: } w = \frac{-10 \pm \sqrt{10^2 - 4(1)(-4000)}}{2(1)}$$

$$= \frac{-10 \pm \sqrt{16{,}100}}{2} \approx -68.4,\ 58.4$$

$$w = 58 \text{ ft}; \ l = 58 + 10 = 68 \text{ ft}$$

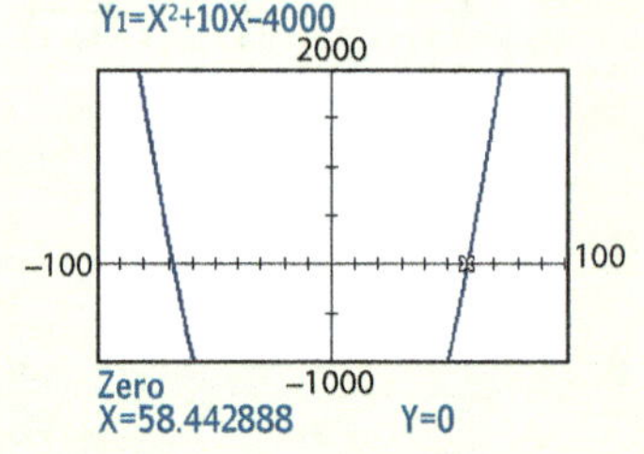

31. $V = \pi r^2 H$
$$H = \frac{V}{\pi r^2} = \frac{3600}{\pi r^2}$$
$$SA = 2\pi r^2 + 2\pi r H$$
$$= 2\pi r^2 + 2\pi r \left(\frac{3600}{\pi r^2}\right)$$
$$2\pi r^2 + \frac{7200}{r} \le 1325$$
$$\frac{2\pi r^3 - 1325r + 7200}{r} \le 0$$

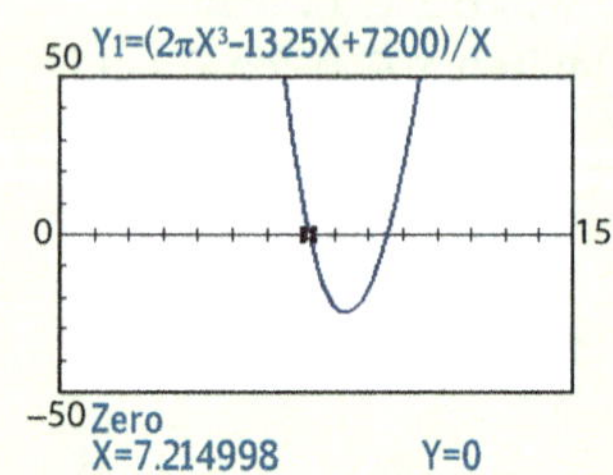

43. Graph the function $f(x) = \begin{cases} 2x^2 - 4x & \text{if } x \le 2 \\ 2 & \text{if } x > 2 \end{cases}$. Then state its domain, range, and whether it is continuous or discontinuous. [1.3] $D = (-\infty, \infty);$ $R = [-2, \infty);$ discontinuous

44. Write $f(x) = -3x^2 + 12x - 9$ in vertex form and graph the function. [1.6] $f(x) = -3(x-2)^2 + 3$

45. Write the function rule and domain for $\frac{f}{g}(x)$ if $f(x) = x^3 + 3x^2 + 2x + 6$ and $g(x) = x^2 + 2$. [1.7]

46. Classify each statement as *always*, *sometimes*, or *never* true. [1.8]
  a. A relation has an inverse. always
  b. A relation has an inverse that is a function. sometimes
  c. A function has an inverse. always
  d. A function has an inverse function. sometimes
  e. A one-to-one function has an inverse function. always

47. Complete a linear regression to find a model for the data. [1.9] $y = 1.118x + 1.5273$

x	y
-3	-1
0	2
4	4
6	7
8	10
9	14

48. Use limits to describe the end behavior of $f(x) = -x^6 + 7x - 1$. [2.2] $\lim\limits_{x \to \pm\infty} f(x) = -\infty$

49. Charles's law states that the volume of gas is directly proportional to the temperature of the gas (in Kelvin). A helium-filled balloon has a volume of 2.86 L at 293 K. At what temperature (to the nearest degree) will its volume be 2.72 L? [1.4] D
  A. 292 K   **E.** none of these
  B. 287 K
  C. 268 K
  D. 279 K

45. $\frac{f}{g}(x) = x + 3; D = \mathbb{R}$

50. Which equation describes $g(x)$ as a transformation of $f(x)$? [1.5] D

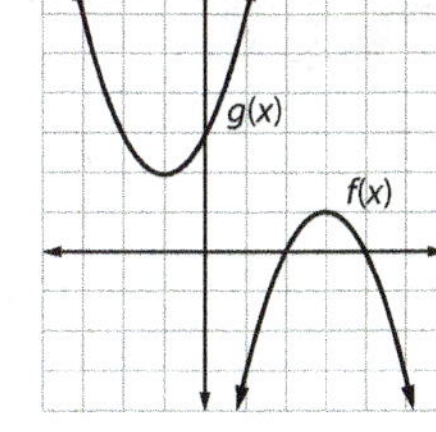

  A. $g(x) = f(x - 4) + 1$
  B. $g(x) = -f(x + 4) + 1$
  C. $g(x) = f(x + 4) + 3$
  D. $g(x) = -f(x + 4) + 3$
  E. none of these

51. Find $h(x) = (fg)(x)$ if $f(x) = \sqrt{x}$ and $g(x) = \sqrt{x^3 + 2x^2 + x}$. [1.7] C
  A. $h(x) = x\sqrt{x^2 + 2x + 1}; D = \mathbb{R}$
  B. $h(x) = x\sqrt{x(x + 1)}; D = [0, \infty)$
  C. $h(x) = x(x + 1); D = [0, \infty)$
  D. $h(x) = x^2(x + 1); D = \mathbb{R}$
  E. none of these

52. Solve $\sqrt{x + 9} = x - 3$. [2.1] C
  A. no solution   **D.** $x = 0, 7$
  B. $x = 0$ **E.** none of these
  C. $x = 7$

32. $-0.0058x^4 + 0.58x^3 - 12.26x^2 + 92x - 41 < 185$

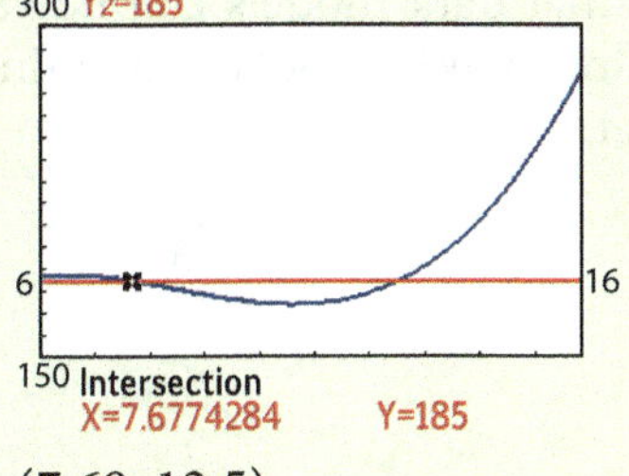

$(7.68, 12.5)$

33. $l = 11 - 2x; w = 8 - 2x$
$V = lwh = (11 - 2x)(8 - 2x)x \ge 55$

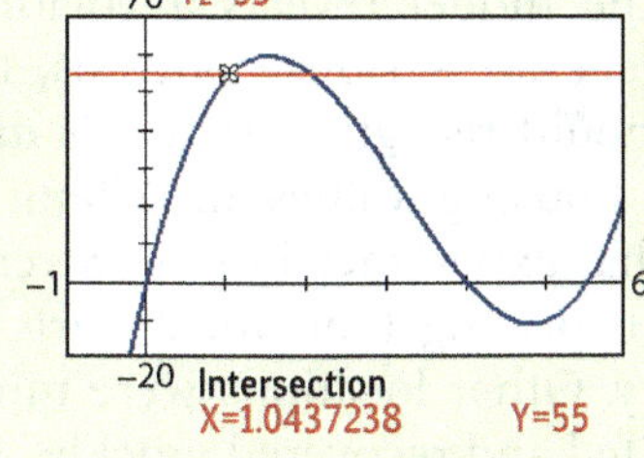

▷ C. Exercises

34. $\dfrac{(x + 3)(x - 3) - (x + 2)(2x + 5)}{(2x + 5)(x - 3)} \ge 0$

$\dfrac{(x^2 - 9) - (2x^2 + 9x + 10)}{(2x + 5)(x - 3)} \ge 0$

$\dfrac{-x^2 - 9x - 19}{(2x + 5)(x - 3)} \ge 0$

zeros of $a(x)$: $x = \dfrac{-9 \pm \sqrt{5}}{2}$
$\approx -3.4, -5.6$

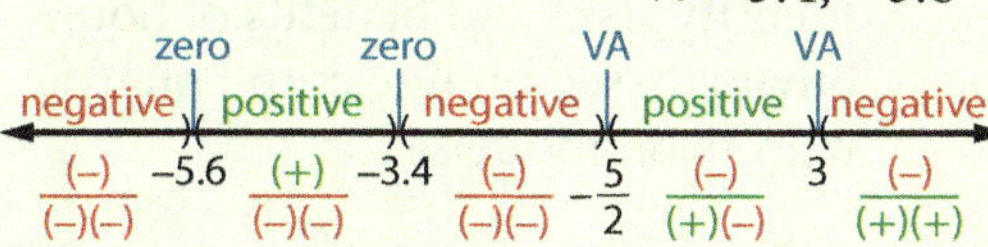

$\left[\dfrac{-9 - \sqrt{5}}{2}, \dfrac{-9 + \sqrt{5}}{2}\right] \cup \left(-\dfrac{5}{2}, 3\right)$

35. $\dfrac{(x + 7)(x - 7)}{(x - 7)(x + 6)} - \dfrac{6 - x}{x(x + 6)} \le 0$

$\dfrac{x(x + 7)}{x(x + 6)} + \dfrac{x - 6}{x(x + 6)} \le 0; x \ne 7$

$\dfrac{x^2 + 8x - 6}{x(x + 6)} \le 0; x \ne 7$

zeros of $a(x)$: $x = -4 \pm \sqrt{22}$
$\approx -8.7, 0.7$

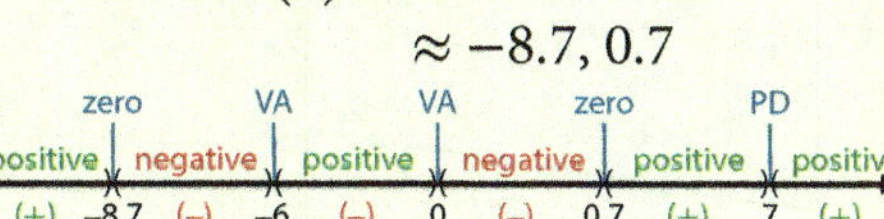

$[-4 - \sqrt{22}, -6) \cup (0, -4 + \sqrt{22}]$

36.

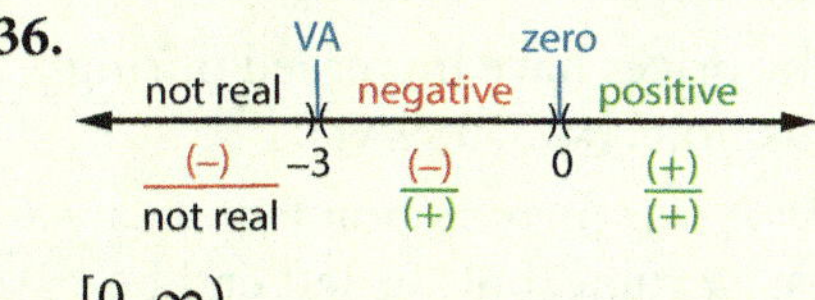

$[0, \infty)$

37.

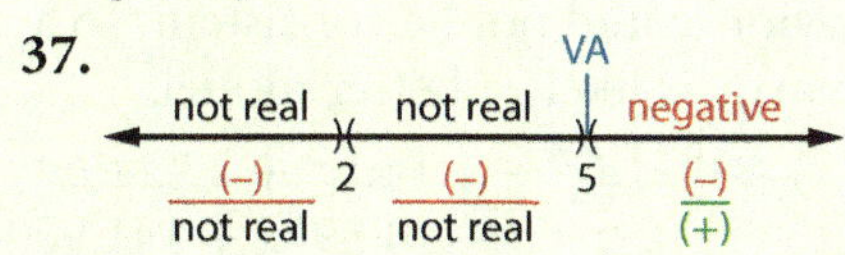

$(5, \infty)$

38.

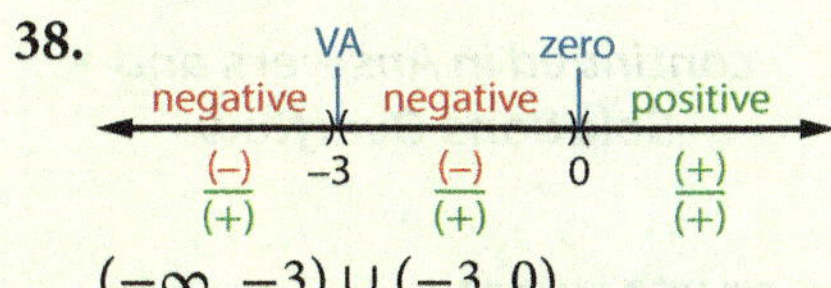

$(-\infty, -3) \cup (-3, 0)$

39.

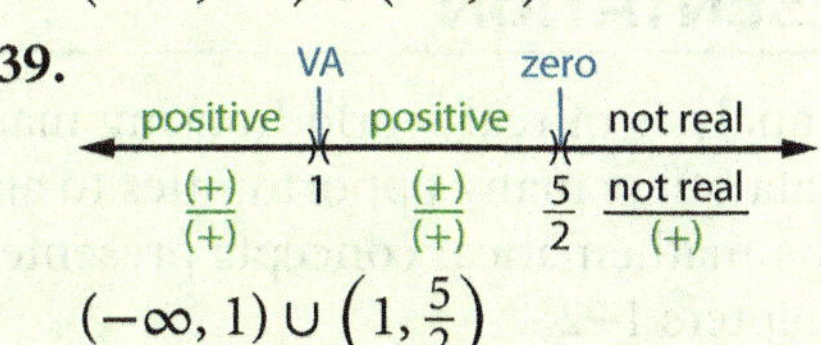

$(-\infty, 1) \cup \left(1, \dfrac{5}{2}\right)$

40. $\dfrac{25 + 1.25\sqrt{325} - 9.8\sqrt[3]{DSP}}{0.686} \le 24$

$-9.8\sqrt[3]{DSP} \le -31.07$

$\sqrt[3]{DSP} \ge 3.17$

$DSP \ge 31.87 \text{ m}^3$

continued in Answers and Solutions Overflow

Data Analysis

Objectives

1. To use linear and quartic models to analyze real-world data relating to housing prices
2. Identify risks and benefits of home buying based on linear and quartic regression models

Assignments

- **Minimum:** 1–3; 4–9 (in class)
- **Standard:** 1–3; 4–9 (in class)
- **Extended:** 1–9

Solutions

1. 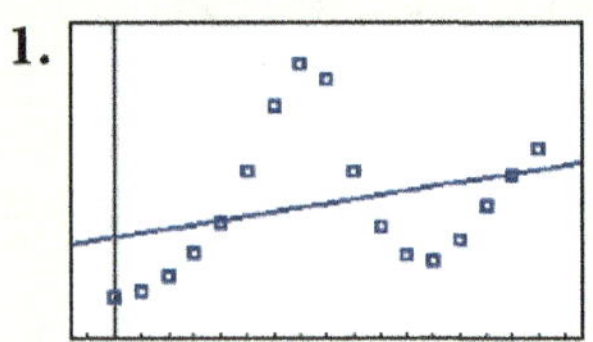

2a. $f(x) \approx 5.389x + 226.565$

2b. The prices have increased during this time. ($\approx$ \$5389/yr)

3a. Two clear turning points indicate at least a cubic, but the left end behavior would not be consistent, so a quartic is likely a better model.

3b. $f(x) \approx 0.113x^4 - 3.140x^3 + 23.225x^2 - 14.897x + 141.900$

continued in Answers and Solutions Overflow

Year	Median Home Price ($1000s)	Year	Median Home Price ($1000s)
2000	149.8	2009	309.5
2001	157.4	2010	240.2
2002	176.7	2011	203.3
2003	205.5	2012	195.4
2004	243.4	2013	222.1
2005	310.2	2014	266.6
2006	392.1	2015	304.2
2007	445.1	2016	337.4
2008	425.6		

The Housing Market

To buy or not to buy? The prospect of purchasing your own home can be exciting, but it can also involve some risk. The housing market was booming during the years leading up to 2008. Then suddenly the market crashed, putting millions of individuals and businesses in financial jeopardy.

Some attribute the 2008 housing market crash to greed. As housing values increased in the early 2000s, lenders made riskier loans, some home buyers overextended their means, and speculators purchased homes hoping to make quick profits.

This feature examines several mathematical models of three-bedroom home prices in Miami, Florida, during the years before and after the market crash of 2008. Mathematical models of the housing market can provide insight into what will likely become your most significant financial investment.

1. Use technology to complete a scatterplot where x represents the number of years since 2000.
2. Although the plot clearly is not linear, a linear trend line indicates the general trend of prices over time.
   a. Find the linear function that best models the data.
   b. What does the slope of the trend line indicate?
3. Complete the following steps to find a better model for the pricing data.
   a. Would a quadratic, cubic, or quartic polynomial appear to be the best model for the graphed data?
   b. Complete a regression and state the polynomial function that best models the data.
   c. Recall that an outlier is an extreme or unusual value. The classification of a data point as an outlier is often subjective and must be evaluated in context. Examining the scatterplot and the graph of the polynomial model, list several years that could be considered outliers.

As data analysis tools, mathematical models are used for much more than determining unknown data and making estimates and predictions. Models can help us evaluate when, how, and why certain events happened. Combined with wisdom and discernment, especially from God's Word, models can provide insights on the past and prepare us for the future.

4. Describe the end behavior of the polynomial model. What does its end behavior tell us about the reliability of the model outside the known data set?
5. Note the sharp increase before the relative maximum and the sharp decrease before the relative minimum. What events are indicated by this portion of the model? How might this have affected homeowners who had made recent home purchases?
6. Describe the characteristics of a model that would be realistic and useful for estimating home values outside the known data set.
7. What conclusions can be drawn about investing in the housing market by looking at the linear model? the polynomial model?
8. From your analysis of the polynomial model, when is the best time to buy a home? What other factors would a buyer need to consider?
9. Buying a home can be a good investment, but caution is important. Explain how Proverbs 22:3, 7 should influence your perspective when purchasing a home.

Chapter 2 Review

1. Which of the following are true for $f(x) = \sqrt[r]{x}$? [2.1]
   A. The domain of $f(x)$ is $[0, \infty)$ when r is even. A, C, D
   B. The function $f(x)$ is an even function when r is even.
   C. The domain of $f(x)$ is $(-\infty, \infty)$ when r is odd.
   D. The function $f(x)$ is odd when r is odd.

Use technology to graph each function. State its domain, range, and the intervals for which the function is increasing or decreasing. State whether the function is continuous or discontinuous and whether it is odd, even, or neither. [2.1]

2. $g(x) = \sqrt[4]{x-3} - 1$ 3. $h(x) = x^{\frac{2}{3}} + 2$

4. Graph $g(x) = \sqrt[3]{x+3} - 2$ as a transformation of a parent radical function. Then state its domain and range. [2.1]

5. Find the inverse of $f(x) = \sqrt{x} - 1$. Then graph $f(x)$ and $f^{-1}(x)$. [2.1] $f^{-1}(x) = (x+1)^2, D = [-1, \infty)$

Solve. [2.1]

6. $\sqrt{x+4} + 1 = \sqrt{2x+1}$ $x = 12$

7. $\sqrt[3]{x^2} + 15 = 24$ $x = \pm 27$

8. Solve $m^2 = \sqrt[3]{\dfrac{rp}{g^2}}$ for p. [2.1] $p = \dfrac{m^6 g^2}{r}$

Use the Leading Term Test to describe each function's end behavior. [2.2]

9. $p(x) = -x^3 + 2x^2 - x - 7$ $\lim\limits_{x \to -\infty} p(x) = \infty;\ \lim\limits_{x \to \infty} p(x) = -\infty$

10. $p(x) = x^6 - x^3 - 4$ $\lim\limits_{x \to \pm\infty} p(x) = \infty$

11. Use technology to graph $p(x) = x^4 + x^3 - 3x^2$ and find the relative extrema. [2.2]
    max.: (0, 0); min.: $\approx$ (1.66, −5.25), $\approx$ (−0.91, −1.05)

Factor to find all the real zeros of each polynomial function. Indicate any multiplicities greater than 1. [2.2]

12. $p(x) = 4x^2 + 12x + 9$ 13. $p(x) = x^4 - 5x^2 - 24$

Identify each function's zeros, y-intercepts, and end behavior. Then sketch its graph. [2.2]

14. $p(x) = 2x^4 - x^3 - 6x^2$ 15. $p(x) = -x^4 + 5x^2 - 4$

16. Use long division to find $(3x^2 + 7x - 5) \div (x + 3)$. [2.2]
    $3x - 2$ R. 1

Use synthetic division to find each quotient. [2.3]

17. $(2x^3 - x^2 - 5x - 1) \div (x - 2)$ $2x^2 + 3x + 1$ R. 1

18. $(2x^3 - 9x^2 + x + 12) \div (2x - 3)$ $x^2 - 3x - 4$

19. Use the Remainder Theorem to find the remainder when $p(x) = x^4 - 3x^3 + x^2 - 3x$ is divided by each binomial. [2.3]
    a. $x - 2$ b. $x - 3$ c. $x + 1$
    $p(2) = -10$ $p(3) = 0$ $p(-1) = 8$

Completely factor each polynomial function using the given information. [2.3]

20. $p(x) = 2x^4 - x^3 - 33x^2 - 56x - 20$ if −2 is a zero (multiplicity 2)

21. $p(x) = 27x^3 - 54x^2 + 36x - 8$ if $p\left(\frac{2}{3}\right) = 0$

Write a polynomial function of least degree with integral coefficients that has the given zeros. [2.3]

22. −2 (multiplicity 4) [2.3]   23. $\frac{1}{2}, -\frac{3}{4}$ [2.3]

24. Analyze $p(x) = 2x^3 - x^2 - 6x + 5$. [2.4]
    a. List all possible rational zeros. $\pm 1, \pm 5, \pm\frac{1}{2}, \pm\frac{5}{2}$
    b. Use Descartes's Rule of Signs to describe the possible number of positive and negative real roots.
    c. Do the Upper and Lower Bound Tests show that all the real zeros are within $[-2, 2]$? If not, state and verify another integral interval that contains all the real zeros of the function. yes

Find all the zeros of each polynomial function. [2.4]

25. $p(x) = x^3 - 2x^2 - 5x + 6$ −2, 1, 3

26. $p(x) = 2x^4 - 9x^3 + 17x^2 - 19x - 15$ $3, -\frac{1}{2}, 1 \pm 2i$

Write a polynomial function of least degree with integral coefficients that has the given zeros. [2.4]

27. $2 + \sqrt{3}, -5$ 28. $1, 3 + 4i$

Identify each function's zeros, y-intercepts, and end behavior. Then sketch its graph. [2.4]

29. $p(x) = x^3 - 2x^2 - 5x + 6$

30. $p(x) = -4x^3 + 4x^2 + 5x - 3$

27. $x^3 + x^2 - 19x + 5$ 28. $x^3 - 7x^2 + 31x - 25$

Chapter 2 Review

Objective

To prepare for evaluation

Vocabulary

See Appendix A.

Assignments

- **Minimum:** 1–7, 9, 11, 14–20, 23, 26–27, 30–32, 34–37, 39, 42–45; 47–53 odd
- **Standard:** 1–53 odd; 2–54 even (in class)
- **Extended:** 1, 3, 5–9, 11, 15, 17–18, 21, 23–26, 28, 30–34, 36; 37–47 odd; 51, 53

Assessment

- Chapter 2 Test

Solutions

2. $D = [3, \infty);\ R = [-1, \infty);$ incr.: $[3, \infty)$; cont.; neither

3. $D = (-\infty, \infty);\ R = [2, \infty);$ decr.: $(-\infty, 0]$; incr.: $[0, \infty)$; cont.; even

4. Translate $f(x) = \sqrt[3]{x}$ 3 units left and 1 unit down. $D = \mathbb{R};\ R = \mathbb{R}$

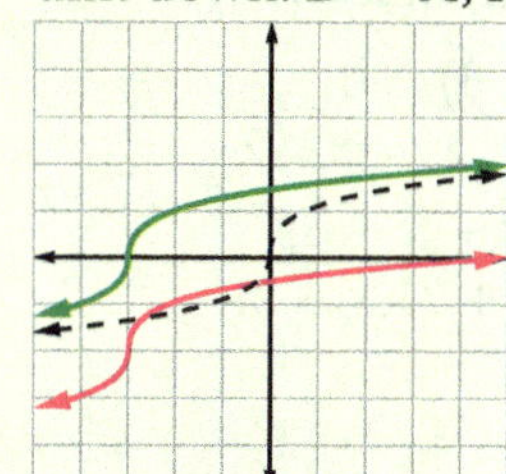

5. $D_f = [0, \infty);\ R_f = [-1, \infty) = D_{f^{-1}(x)}$
$$x = \sqrt{y} - 1$$
$$x + 1 = \sqrt{y}$$
$$y = (x+1)^2$$

PRESENTATION

Be sure to communicate which technologies the students will be allowed to use and which they are expected to have for the chapter test.

Consider giving an overview of the chapter by reviewing section objectives and taking related questions. This may help you identify areas where students need more focused review.

6. squaring both sides:
$$(x + 4) + 2\sqrt{x + 4} + 1 = 2x + 1$$
$$2\sqrt{x + 4} = x - 4$$
$$\sqrt{x + 4} = \frac{x - 4}{2}$$
$$x + 4 = \frac{x^2 - 8x + 16}{4}$$
$$4x + 16 = x^2 - 8x + 16$$
$$x^2 - 12x = 0$$
$$x(x - 12) = 0$$
$$x = 12,\ 0\ \text{(extraneous)}$$

7. $x^{\frac{2}{3}} = 9$
$$\left(x^{\frac{2}{3}}\right)^{\frac{3}{2}} = (3^2)^{\frac{3}{2}} = 3^3$$
$$|x| = 27$$
$$x = \pm 27$$

8. $m^6 = \dfrac{rp}{g^2}$
$$p = \frac{m^6 g^2}{r}$$

9. The degree is odd and $a_n < 0$.

10. The degree is odd and $a_n > 0$.

11. Y₁=X^4+X³-3X²
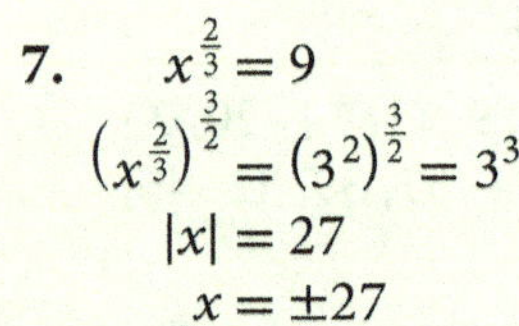

12. $(2x + 3)(2x + 3) = 0$
$$x = -\frac{3}{2}\ \text{(multiplicity 2)}$$

13. $(x^2 - 8)(x^2 + 3) = 0$
$$x = \pm 2\sqrt{2}$$

14. $p(x) = x^2(2x^2 - x - 6)$
$$= x^2(2x + 3)(x - 2)$$
zeros: $0,\ -\dfrac{3}{2},\ 2$
y-int.: $(0, 0)$
Since n is even and $a_n > 0$,
$$\lim_{x \to \pm\infty} p(x) = \infty.$$

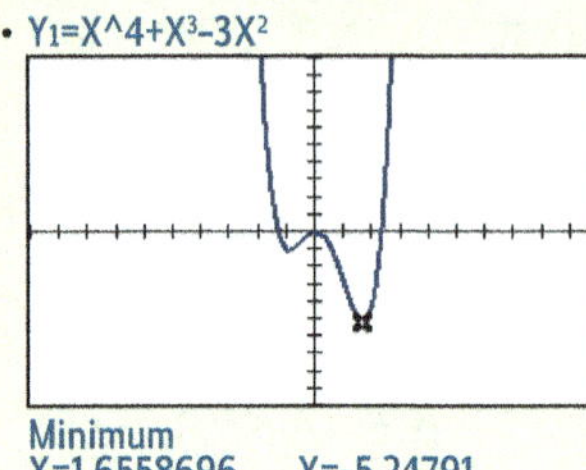

15. $p(x) = -1(x^4 - 5x^2 + 4)$
$$= -1(x^2 - 1)(x^2 - 4)$$
zeros: $\pm 1,\ \pm 2$
y-int.: $(0, -4)$
Since n is odd and $a_n < 0$,
$$\lim_{x \to \pm\infty} p(x) = -\infty.$$

Describe the rational function
$$f(x) = \frac{a(x)}{b(x)} = \frac{a_n x^n + a_{n-1}x^{n-1} + \cdots + a_1 x + a_0}{b_m x^m + b_{m-1}x^{m-1} + \cdots + b_1 x + b_0}$$
if its graph has the stated characteristic. [2.5]

31. a horizontal asymptote at $y = \dfrac{a_n}{b_m}$ **B**

    A. $n < m$ **C.** $n = m + 1$

    B. $n = m$ **D.** $n > m + 1$

32. a point discontinuity at $x = k$ **C**

    A. A factor of $(x - k)$ in $b(x)$ is not matched in $a(x)$.

    B. The factor of $(x - k)$ in $a(x)$ is not in $b(x)$.

    C. All factors of $(x - k)$ are matched in $a(x)$ and $b(x)$.

Identify the asymptotes and graph each function as a transformation of a reciprocal power function. [2.5]

33. $g(x) = 1 - \dfrac{1}{x^3}$ **34.** $g(x) = \dfrac{2}{(x - 3)^2} - 4$

Classify each rational function as proper or improper. Then identify all asymptotes and any point discontinuities. [2.5]

35. $f(x) = \dfrac{x + 6}{x^2 - 36}$ **36.** $f(x) = \dfrac{x^4 + x^3}{x^2 - 4x - 5}$

Identify any asymptotes, point discontinuities, and intercepts for each function. Then use these characteristics to sketch its graph. [2.5]

37. $f(x) = \dfrac{x - 2}{x^2 - 4}$ **38.** $f(x) = \dfrac{x^2}{2x^2 - 2x}$

39. $f(x) = \dfrac{x^3 - x}{x^2 + 2x + 1}$

Solve. [2.6]

40. $\dfrac{x + 1}{x - 5} = \dfrac{x}{x + 3}$ $x = -\dfrac{1}{3}$

41. $\dfrac{4}{n^2 - n - 6} + \dfrac{3}{n^2 - 2n - 3} = \dfrac{8}{n^2 + 3n + 2}$ $n = 34$

42. $\dfrac{4}{2r^2 + r - 1} = \dfrac{1}{2r - 2} - \dfrac{r}{r^2 - 1}$ $r = -\dfrac{7}{2}$

43. Find two numbers whose sum is 18 and the sum of whose reciprocals is $\dfrac{1}{4}$.   6 and 12

44. Olivia can process 300 applications twice as fast as Ava can. If they work together, they can finish the job in 8 hr. How much time would it take each of them to do the job alone?   Olivia: 12 hr; Ava: 24 hr

35. proper; HA: $y = 0$; VA: $x = 6$; PD at $x = -6$

36. improper; NLA: $y = x^2 + 5x + 25$; VA: $x = 5$; PD at $x = -1$

48. a. $\left(-\infty, -\dfrac{4}{3}\right) \cup (5, \infty)$; b. $\left(-\dfrac{4}{3}, 5\right)$

49. a. $\left(-3, -\dfrac{1}{2}\right) \cup (1, \infty)$; b. $(-\infty, -3) \cup \left(-\dfrac{1}{2}, 1\right)$

---

45. A small outlet can drain a water tank in 8 hr. If a large outlet is also opened, the tank can be drained in 3 hr. How long would it take to drain the tank with just the large outlet?   4.8 hr

46. A boat that travels 15 mi/hr in still water takes 40% more time to make a trip upriver than it takes to make the trip back downriver. Determine the average speed of the river's current.   2.5 mi/hr

47. Jen ran the first third of her race at 7 mi/hr (a pace of 8:34 min/mi). If her goal is to average 8 mi/hr for the race (a pace of 7:30 min/mi), at what speed should she run the rest of the course? What is the equivalent pace (in min/mi)? ≈ 8.6 mi/hr; $\approx 6:58$ per mi.

Use a sign chart to determine the intervals where the function's values are (a) positive and (b) negative. [2.7]

48. $p(x) = (3x + 4)(x^2 + 2)(x - 5)$

49. $f(x) = \dfrac{(x + 3)}{(x - 1)(2x + 1)}$

Solve each polynomial inequality by factoring and then using a sign chart. [2.7]

50. $2x^3 + x^2 - 6x < 0$ $(-\infty, -2) \cup \left(0, \dfrac{3}{2}\right)$

51. $2x^4 + 68x \le 3x^3 + 24x^2 + 48$ $\left[-4, \dfrac{3}{2}\right] \cup \{2\}$

52. $\dfrac{x^2 + x - 6}{x^2 - 9x + 14} \ge 0$ $(-\infty, -3] \cup (7, \infty)$

53. $\dfrac{x^2 + 16}{3x + 4} \le 2$ $\left(-\infty, -\dfrac{4}{3}\right) \cup [2, 4]$

54. A large family-size box for Alge-Crunch cereal is being designed. The 12.5 oz box is 10" × 8" × 2" and the company wants to increase the dimensions as illustrated. Find the range of possible values of x if the volume of the new box is to be no more than 425 in.³ [2.7]   (0, 1.114)

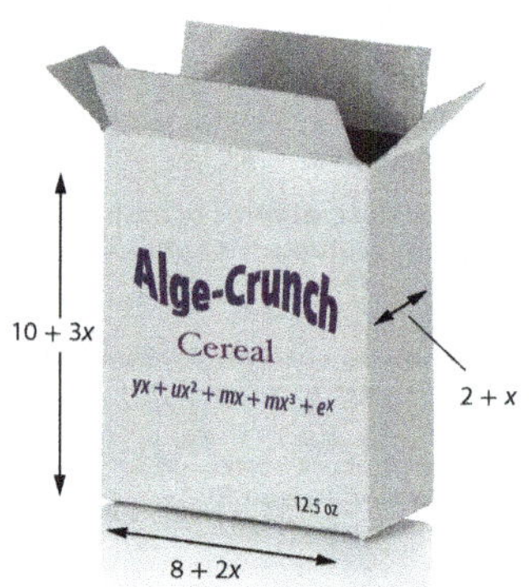

16.
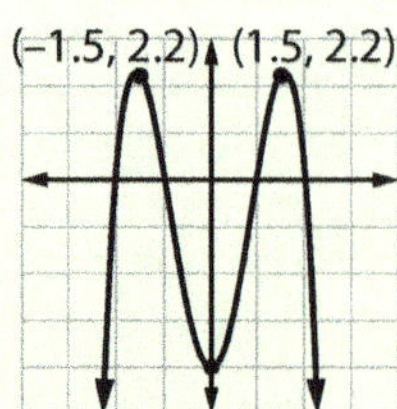
$$x + 3 \overline{)\,3x^2 + 7x - 5}$$
quotient: $3x - 2$
$$-(3x^2 + 9x)$$
$$-2x - 5$$
$$-(-2x - 6)$$
$$1$$

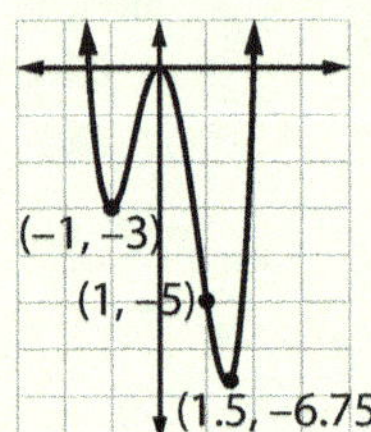

17.

$$\begin{array}{r|rrrr} 2 & 2 & -1 & -5 & -1 \\ & & 4 & 6 & 2 \\ \hline & 2 & 3 & 1 & 1 \end{array}$$

18. $\left(x^3 - \dfrac{9}{2}x^2 + \dfrac{1}{2}x + 6\right) \div \left(x - \dfrac{3}{2}\right)$

$$\begin{array}{r|rrrr} \frac{3}{2} & 1 & -\frac{9}{2} & \frac{1}{2} & 6 \\ & & \frac{3}{2} & -\frac{9}{2} & -6 \\ \hline & 1 & -3 & -4 & 0 \end{array}$$

20.

$$\begin{array}{r|rrrrr} & 2 & -1 & -33 & -56 & -20 \\ -2 & & -2 & -5 & -23 & -10 \\ \hline & 2 & & \\ -2 & 2 & -9 & -5 & 0 \\ \hline & 2 & -5 & -23 & -10 & 0 \end{array}$$

$$p(x) = (x+2)^2(2x^2 - 9x - 5)$$
$$= (x+2)^2(2x+1)(x-5)$$

21.

$$\begin{array}{r|rrrr} \frac{2}{3} & 27 & -54 & 36 & -8 \\ & & 18 & -24 & 8 \\ \hline & 27 & -36 & 12 & 0 \end{array}$$

$$p(x) = \left(x - \tfrac{2}{3}\right) \cdot 3(9x^2 - 12x + 4)$$
$$= (3x - 2)(3x - 2)^2 = (3x - 2)^3$$

22. $(x+2)^4 = x^4 + 8x^3 + 24x^2 + 32x + 16$

23. $2\left(x - \tfrac{1}{2}\right) \cdot 4\left(x + \tfrac{3}{4}\right) = 0 \cdot 8$

$$p(x) = (2x - 1)(4x + 3)$$
$$= 8x^2 + 2x - 3$$

24a. $\dfrac{\text{factors of } 5}{\text{factors of } 2} = \dfrac{\pm 1, \pm 5}{\pm 1, \pm 2}$

$$= \pm 1, \pm 5, \pm\tfrac{1}{2}, \pm\tfrac{5}{2}$$

24b. $p(x)$ has 2 sign changes, which implies 2 or 0 positive zeros, and
$p(-x) = -2x^3 - x^2 + 6x + 5$
has 1 sign change, which implies 1 negative zero.

24c.

$$\begin{array}{rr|rrrr} & & 2 & -1 & -6 & 5 \\ \text{U.B.} & 2 & 2 & 3 & 0 & 5 \quad \text{pass} \\ \text{L.B.} & -2 & 2 & -5 & 4 & -3 \quad \text{pass} \end{array}$$

25. possible zeros: $\pm 1, \pm 2, \pm 3, \pm 6$

$$\begin{array}{r|rrrr} 1 & 1 & -2 & -5 & 6 \\ \hline & 1 & -1 & -6 & 0 \end{array}$$

$$p(x) = (x - 1)(x^2 - x - 6)$$
$$= (x - 1)(x - 3)(x + 2)$$
$$x = 1, 3, -2$$

26. possible zeros: $\pm 1, \pm 3, \pm 5, \pm 15, \pm\tfrac{1}{2},$
$\pm\tfrac{3}{2}, \pm\tfrac{5}{2}, \pm\tfrac{15}{2}$

$$\begin{array}{r|rrrrr} & & 2 & -9 & 17 & -19 & -15 \\ 3 & & 2 & -3 & 8 & 5 & 0 \\ -\tfrac{1}{2} & & 2 & -4 & 10 & 0 \end{array}$$

$$p(x) = (x - 3)\left(x + \tfrac{1}{2}\right) \cdot 2(x^2 - 2x + 5)$$

$x^2 - 2x + 5 = 0$ when

$$x = \frac{-(-2) \pm \sqrt{(-2)^2 - 4(1)(5)}}{2(1)} = \frac{2 \pm 4i}{2}$$
$$= 1 \pm 2i$$

$$p(x) = (x - 3)(2x - 1)(1 + 2i) \cdot (1 - 2i)$$

$$x = 3, -\tfrac{1}{2}, 1 \pm 2i$$

27. If $2 + \sqrt{3}$ is a zero, then $2 - \sqrt{3}$ is a zero.

$$p(x) = \left[x - (2 + \sqrt{3})\right]\left[x - (2 - \sqrt{3})\right] \cdot (x + 5)$$
$$= \left[(x - 2) - \sqrt{3}\right]\left[(x - 2) + \sqrt{3}\right] \cdot (x + 5)$$
$$= \left[(x - 2)^2 - 3\right](x + 5)$$
$$= (x^2 - 4x + 1)(x + 5)$$
$$= x^3 + x^2 - 19x + 5$$

28. If $3 + 4i$ is a zero, then $3 - 4i$ is a zero.

$$p(x) = \left[x - (3 + 4i)\right]\left[x - (3 - 4i)\right] \cdot (x - 1)$$
$$= \left[(x - 3) + 4i\right]\left[(x + 3) - 4i\right] \cdot (x - 1)$$
$$= \left[(x - 3)^2 + 16\right](x - 1)$$
$$= (x^2 - 6x + 25)(x - 1)$$
$$= x^3 - 7x^2 + 31x - 25$$

29. possible zeros: $\pm 1, \pm 2, \pm 3, \pm 6$

$$\begin{array}{r|rrrr} 1 & 1 & -2 & -5 & 6 \\ \hline & 1 & -1 & -6 & 0 \end{array}$$

$$p(x) = (x - 1)(x^2 - x - 6)$$
$$= (x - 1)(x - 3)(x + 2)$$

zeros: $-2, 1, 3$

y-int.: $(0, 6)$

$$\lim_{x \to -\infty} p(x) = -\infty \text{ and } \lim_{x \to \infty} p(x) = \infty$$

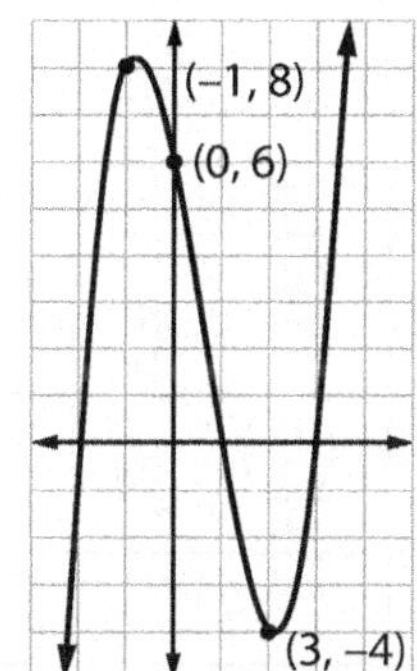

30. possible zeros: $\pm 1, \pm 3, \pm\tfrac{1}{2}, \pm\tfrac{3}{2}, \pm\tfrac{1}{4}, \pm\tfrac{3}{4}$

$$\begin{array}{r|rrrr} & -4 & 4 & 5 & -3 \\ -1 & -4 & 8 & -3 & 0 \end{array}$$

$$p(x) = (x + 1)(-1)(4x^2 - 8x + 3)$$
$$= -(x + 1)(2x - 1)(2x - 3)$$

zeros: $-1, \tfrac{1}{2}, \tfrac{3}{2}$

y-int.: $(0, -3)$

$$\lim_{x \to -\infty} p(x) = \infty \text{ and } \lim_{x \to \infty} p(x) = -\infty$$

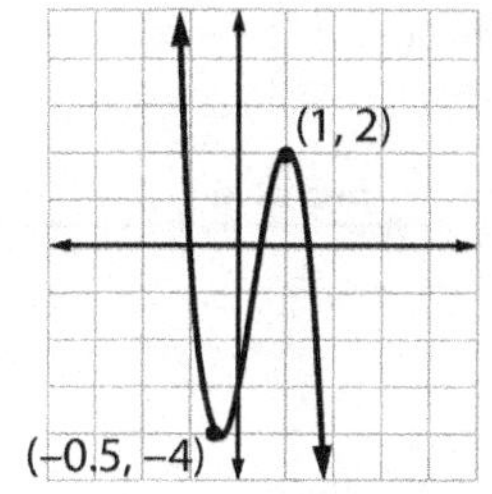

33. $g(x) = -\dfrac{1}{x^3} + 1$ is $f(x) = \dfrac{1}{x^3}$ reflected in the x-axis and translated 1 unit up.
VA: $x = 0$; HA: $y = 1$

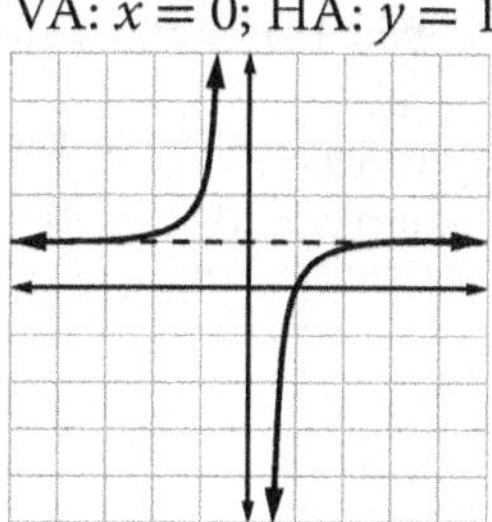

34. $g(x)$ is $f(x) = \dfrac{1}{x^2}$ translated 3 units right, stretched vertically by a factor of 2, and then translated 4 units down.
VA: $x = 3$; HA: $y = -4$

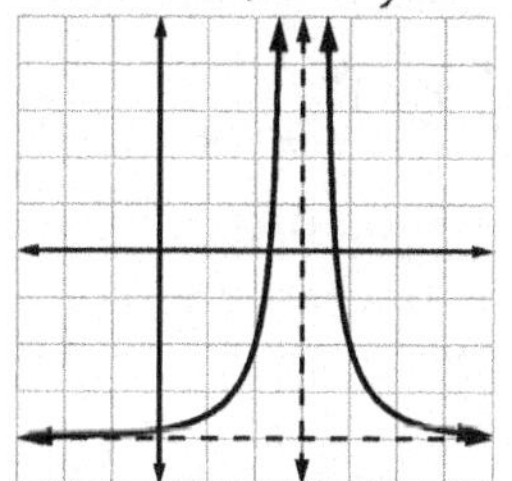

35. $f(x) = \dfrac{x + 6}{(x + 6)(x - 6)} = \dfrac{1}{x - 6}, x \neq -6$
$n < m$; HA: $y = 0$
zeros of $a(x)$: -6
zeros of $b(x)$: ± 6
VA: $x = 6$; PD at $x = -6$

36. $f(x) = \dfrac{x^3(x + 1)}{(x - 5)(x + 1)} = \dfrac{x^3}{x - 5}, x \neq -1$
$n > m + 1$ and
$$\dfrac{x^3}{x - 5} = x^2 + 5x + 25 - \dfrac{125}{x - 5}$$
NLA: $y = x^2 + 5x + 25$
zeros of $a(x)$: $-1, 0$
zeros of $b(x)$: $-1, 5$
VA: $x = 5$; PD at $x = -1$

37. $f(x) = \dfrac{x - 2}{(x - 2)(x + 2)} = \dfrac{1}{x + 2}, x \neq 2$
HA: $y = 0$
VA: $x = -2$; PD at $x = 2$
y-int.: $f(0) = \tfrac{1}{2}$; $\left(0, \tfrac{1}{2}\right)$
x-int.: $1 \neq 0$; $\therefore$ none

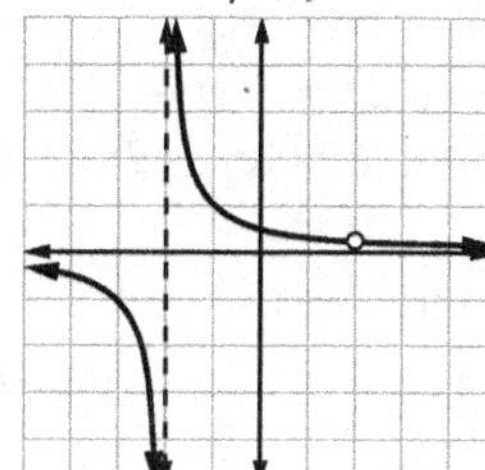

continued on next page

38. $f(x) = \dfrac{x^2}{2x(x-1)} = \dfrac{x}{2(x-1)}$

when $x \neq 0$

HA: $y = \frac{1}{2}$; VA: $x = 1$; PD at $x = 0$

y-int.: $f(0) = \dfrac{(0)^2}{2(0)(0-1)} = \dfrac{0}{0}$; none

x-int.: $x^2 = 0$ when $x = 0$, but PD at $x = 0$

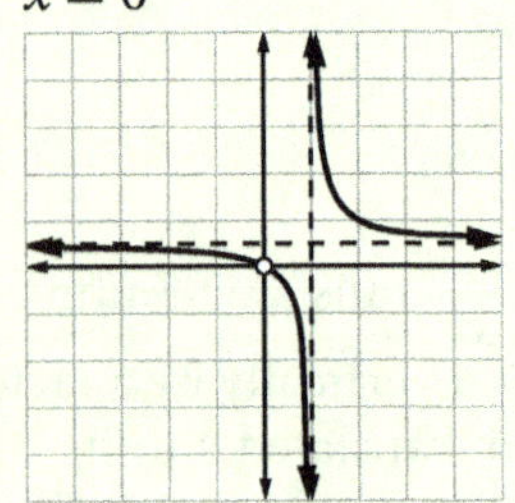

39. $f(x) = \dfrac{x(x+1)(x-1)}{(x+1)(x+1)} = \dfrac{x(x-1)}{x+1}$

$= x - 2 + \dfrac{2}{x+1}$

SA: $y = x - 2$; VA: $x = -1$

y-int.: $f(0) = \dfrac{0}{1} = 0$; $(0, 0)$

x-int.: $x^3 - x = 0$ when $x = 0$ or 1

$(0, 0)$ and $(1, 0)$

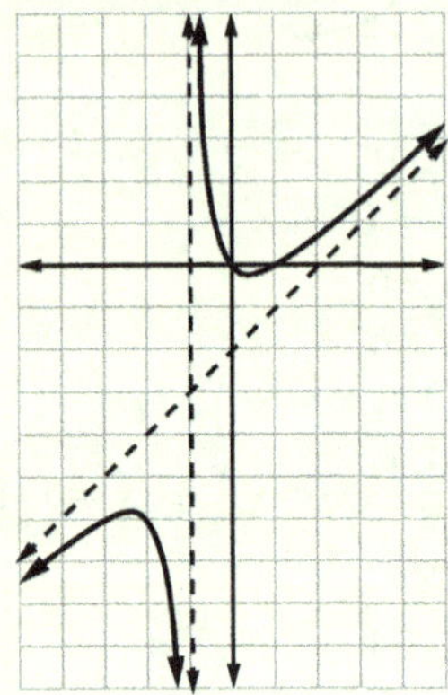

40. $(x-5)(x+3)\left[\dfrac{x+1}{x-5} = \dfrac{x}{x+3}\right]$

$(x+3)(x+1) = x(x-5)$

$x^2 + 4x + 3 = x^2 - 5x$

$9x = -3$

$x = -\dfrac{1}{3}$

41. $(n-3)(n+2)(n+1)$

$\cdot \left[\dfrac{4}{(n-3)(n+2)} + \dfrac{3}{(n-3)(n+1)}\right.$

$\left. = \dfrac{8}{(n+2)(n+1)}\right]$

$4(n+1) + 3(n+2) = 8(n-3)$

$4n + 4 + 3n + 6 = 8n - 24$

$n = 34$

42. $2(2r-1)(r+1)(r-1)$

$\cdot \left[\dfrac{4}{(2r-1)(r+1)} = \dfrac{1}{2(r-1)}\right.$

$\left. - \dfrac{r}{(r+1)(r-1)}\right]$

$8(r-1) = (2r-1)(r+1)$

$\qquad\qquad\qquad - 2r(2r-1)$

$8r - 8 = 2r^2 + r - 1 - 4r^2 + 2r$

$2r^2 + 5r - 7 = 0$

$(2r+7)(r-1) = 0$

$r = -\dfrac{7}{2}, 1 \text{ (extraneous)}$

43. Let x and $18 - x$ be the numbers.

$4x(18-x)\left[\dfrac{1}{x} + \dfrac{1}{18-x} = \dfrac{1}{4}\right]$

$4(18-x) + 4x = x(18-x)$

$72 - 4x + 4x = 18x - x^2$

$x^2 - 18x + 72 = 0$

$(x-6)(x-12) = 0$

$x = 6, 12$

44. Let $x =$ Olivia's time alone

and $2x =$ Ava's time alone.

$16x\left[\dfrac{1}{2x} + \dfrac{1}{x} = \dfrac{1}{8}\right]$

$8 + 16 = 2x$

$2x = 24$

$x = 12$

45. $24x\left[\dfrac{1}{8} + \dfrac{1}{x} = \dfrac{1}{3}\right]$

$3x + 24 = 8x$

$5x = 24$

$x = 4.8 \text{ hr or 4 hr 48 min}$

46.

	R	T	D
Upstream	$15 - c$	$\dfrac{d}{15-c}$	d
Down-stream	$15 + c$	$\dfrac{d}{15+c}$	d

$\dfrac{d}{15-c} = 1.4\left(\dfrac{d}{15+c}\right)$

$(15+c)(15-c)\left[\dfrac{1}{15-c} = \dfrac{1.4}{15+c}\right]$

$15 + c = 21 - 1.4c$

$2.4c = 6$

$c = 2.5 \text{ mi/hr}$

47.

	R	T	D
First Part	7	$\dfrac{\left(\frac{d}{3}\right)}{7} = \dfrac{d}{21}$	$\dfrac{1}{3}d$
Second Part	r	$\dfrac{\left(\frac{2d}{3}\right)}{r} = \dfrac{2d}{3r}$	$\dfrac{2}{3}d$
Total Race	8	$\dfrac{d}{8}$	d

$\dfrac{d}{21} + \dfrac{2d}{3r} = \dfrac{d}{8}$

$21(8)r\left[\dfrac{1}{21} + \dfrac{2}{3r} = \dfrac{1}{8}\right]$

$8r + 112 = 21r$

$112 = 13r$

$r = \dfrac{112}{13} \approx 8.6 \text{ mi/hr}$

$\text{pace} = \dfrac{13 \text{ hr}}{112 \text{ mi}}\left(\dfrac{60 \text{ min}}{1 \text{ hr}}\right)$

$\approx 6.96 \text{ min/mi or 6:58 per mile}$

48.

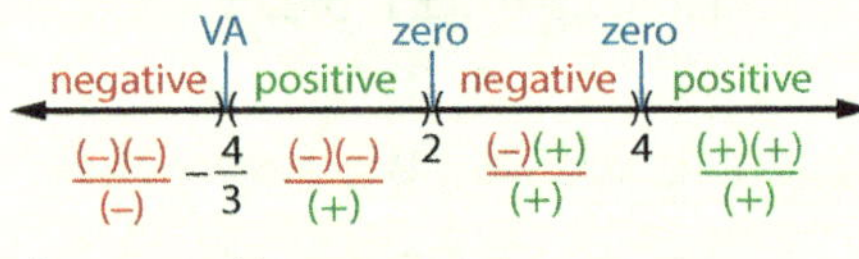

49.

50. $x(2x^2 + x - 6) < 0$

$x(2x-3)(x+2) < 0$

$(-\infty, -2) \cup \left(0, \dfrac{3}{2}\right)$

51. $2x^4 - 3x^3 - 24x^2 + 68x - 48 \leq 0$

$$
\begin{array}{r|rrrrr}
 & 2 & -3 & -24 & 68 & -48 \\
2 & 2 & 1 & -22 & 24 & 0 \\
\hline
-4 & 2 & -7 & 6 & 0 &
\end{array}
$$

$(x-2)(x+4)(2x^2 - 7x + 6) \leq 0$

$(x-2)^2(x+4)(2x-3) \leq 0$

$\left[-4, \dfrac{3}{2}\right] \cup \{2\}$

52. $\dfrac{(x-2)(x+3)}{(x-2)(x-7)} \geq 0$

$\dfrac{x+3}{x-7}$ where $x \neq 2$

$(-\infty, -3] \cup (7, \infty)$

53. $\dfrac{x^2 + 16}{3x + 4} - 2\left(\dfrac{3x + 4}{3x + 4}\right) \leq 0$

$\dfrac{x^2 - 6x + 8}{3x + 4} \leq 0$

$\dfrac{(x-4)(x-2)}{3x + 4} \leq 0$

$\left(-\infty, -\dfrac{4}{3}\right) \cup [2, 4]$

54. $V = (2+x)(8+2x)(10+3x)$

$6x^3 + 56x^2 + 168x + 160 \leq 425$

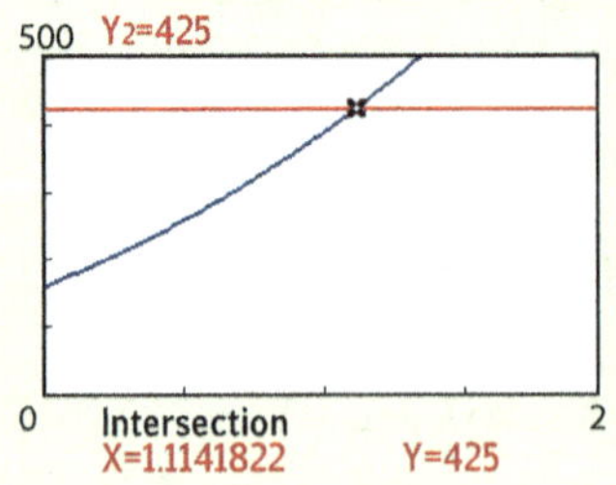

3 Exponential and Logarithmic Functions

3.1 Exponential Functions
3.2 Logarithmic Functions
Historical Connection
3.3 Properties of Logarithms
3.4 Exponential and Logarithmic Equations
Biblical Perspective of Mathematics
3.5 Exponential, Logistic, and Logarithmic Models
Data Analysis

HISTORICAL CONNECTION
In 1919 the mathematical historian Florian Cajori stated that "the miraculous powers of modern calculation are due to three inventions: the Arabic Notation, Decimal Fractions, and Logarithms."

BIBLICAL PERSPECTIVE OF MATHEMATICS
John Napier, who developed logarithms, is a prime example of a man who recognized that his gifts were from God and used them in a conscious effort to serve others, thus bringing glory to God.

DATA ANALYSIS
While populations of living organisms may initially experience exponential growth, limited resources within their environments tend to restrict their growth. Is it reasonable to apply logistic functions and their carrying capacities to human populations?

Overview

Exponential and logarithmic functions have been a vital part of mathematics for centuries. Exponential growth and decay models are used to make predictions that can affect millions of people. While the students will learn how estimations can help us take actions to prevent avoidable adversities, they will also see how a person's worldview can affect his responses.

The Technology Corner provides detailed steps for transforming data and plotting the transformed data.

Flash

Mathematical models are commonly used to estimate population. The world population at the time of Christ was estimated to be 300 million. The current world population estimate can be found using the Internet keyword search *world population clock*.

Chapter Objectives

1. To apply the definition and properties of exponents and logarithms
2. To graph basic exponential and logarithmic functions and their reflections, translations, and dilations
3. To describe key characteristics of exponential and logarithmic functions
4. To solve exponential and logarithmic equations
5. To model real-world data and solve problems using exponential and logarithmic functions
6. To describe how the belief in the orderliness of a created universe contributed to the development of logarithms

Suggested Teaching Schedule

DAY	
1	3.1
2	3.2
3	Quiz 3A (3.1–3.2) HC: Napier's Artificial Numbers
4	3.3
5–6	3.4 BPM: Glorifying God with Math

DAY	
7–8	Quiz 3B (3.3–3.4) 3.5; TC: Transformed Data
9	Quiz 3C (3.5) DA: A World Population Crisis?
10	Chapter 3 Review
11	Chapter 3 Test

3.1 Exponential Functions

Objectives

1. To identify exponential functions

2. To graph exponential functions

3. To model and solve real-world problems involving exponential growth and decay

Vocabulary

algebraic function
annual percentage rate (APR)
annual percentage yield (APY)
compounded interest
decay factor
exponential decay
exponential function
exponential growth
growth factor
growth rate
half-life
natural base
natural exponential function
transcendental function

The resurgent population of bald eagles can be modeled using an exponential function.

After completing this section, you will be able to

- identify exponential functions.
- graph exponential functions.
- model and solve real-world problems involving exponential growth and decay.

In Chapters 1–2 we studied polynomial, rational, radical, and power functions with rational exponents. These *algebraic functions* involve only the algebraic operations of addition, subtraction, multiplication, division, and raising a variable to a rational power. In Chapters 3–5 we will study exponential, logarithmic, and trigonometric functions. These functions are called *transcendental functions* because they go beyond, or transcend, the basic algebraic operations.

While exponential functions may appear to be similar to power functions, they are distinguished by the location of the variable. In a power function the variable is in the base and the exponent is a constant. In an *exponential function* the base is a constant and the variable is in the exponent.

$$f(x) = x^4 \text{ is a power function with a degree of 4, and}$$
$$g(x) = 4^x \text{ is an exponential function with a base of 4.}$$

> **DEFINITION**
>
> A function of the form $f(x) = ab^x$ where $a \neq 0$, $b > 0$, and $b \neq 1$ is an **exponential function** with base b.

The coefficient $a = 0$ is excluded since $f(x) = 0 \cdot b^x = 0$ is the zero function. The base $b = 1$ is excluded since $f(x) = a \cdot 1^x = a$ is a constant function.

Example 1 Graphing an Exponential Function

Graph each exponential function. Then state each function's domain and range, any asymptotes and intercepts, its end behavior, and the intervals in which the function is increasing or decreasing.

a. $f(x) = 2^x$ **b.** $g(x) = \left(\frac{1}{2}\right)^x$

Answer

1. Make a table of ordered pairs for each function.

x	$f(x)$	$g(x)$
-3	$\frac{1}{8}$	8
-2	$\frac{1}{4}$	4
-1	$\frac{1}{2}$	2
0	1	1
1	2	$\frac{1}{2}$
2	4	$\frac{1}{4}$
3	8	$\frac{1}{8}$

2. Plot the ordered pairs and draw the graphs.

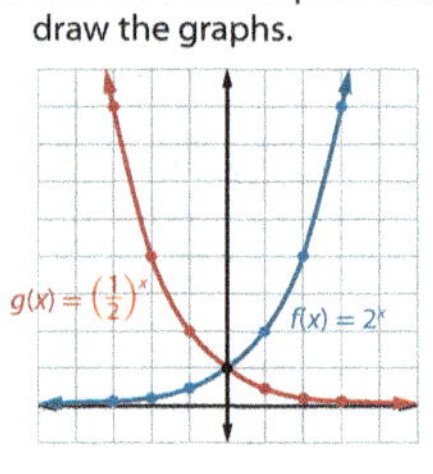

3. Analyze each function.

$f(x) = 2^x$	$g(x) = \left(\frac{1}{2}\right)^x$
$D = \mathbb{R}$; $R = (0, \infty)$	$D = \mathbb{R}$; $R = (0, \infty)$
y-intercept: $(0, 1)$	y-intercept: $(0, 1)$
asymptote: $y = 0$	asymptote: $y = 0$
$\lim\limits_{x \to -\infty} g(x) = 0$	$\lim\limits_{x \to -\infty} g(x) = \infty$
$\lim\limits_{x \to \infty} g(x) = \infty$	$\lim\limits_{x \to \infty} g(x) = 0$
increasing $(-\infty, \infty)$	decreasing $(-\infty, \infty)$

SKILL ✔ EXERCISE 7

PRESENTATION

Lesson Opener

Evaluate.

1. 3^{-1} $\frac{1}{3}$

2. 4^{-2} $\frac{1}{16}$

3. -2^{-3} $-\frac{1}{8}$

4. $\left(\frac{1}{3}\right)^{-2}$ 9

Introduce the lesson by discussing the difference between *algebraic* and *transcendental functions*. You may want to share that transcendental functions go beyond algebraic functions in that a transcendental function cannot be expressed as a finite sequence of terms using the algebraic operations of addition, multiplication, and roots.

Point out that an *exponential function* has a variable in the exponent and that the base is a constant. Use the examples given to compare a power function and an exponential function; then present the definition of an exponential function.

The graphs in Example 1 will allow the students to visually observe the parent functions $y = 2^x$ and $y = \left(\frac{1}{2}\right)^x$ and the relationship between these functions. (*They are reflections across the x-axis.*) Consider asking the students how $y = -3^x$ and $y = -\left(\frac{1}{3}\right)^x$ would compare to these graphs. (*They would be reflected across the x-axis and stretched vertically.*)

Point out that irrational exponents are defined for exponential functions. This can be confusing to some students because they are used to defining 5^3 as three factors of 5. Although $5^{\sqrt{2}}$ cannot be defined as $\sqrt{2}$ factors of 5, exponential functions are continuous and are therefore defined for irrational values of x. Demonstrate the estimated value of $f(\pi) = 2^\pi$ by examining the progressive values as x approaches π. Understanding the continuous nature of exponential functions is important in realizing that their inverse functions, the logarithmic functions cov-

The arithmetic operations of exponentiation and finding roots allow us to evaluate exponential functions, such as $f(x) = 2^x$, for rational values of x. For example, $f(3) = 2^3 = 8$ and $f(1.5) = 2^{\frac{3}{2}} = \sqrt{8} \approx 2.83$.

The fact that exponential functions are defined and continuous over the real numbers enables us to extend our evaluations to irrational values of x, such as $f(\pi) = 2^\pi$, using successively closer rational approximations of x.

x	3	3.1	3.14	3.141	3.1415	3.14159
$f(x)$	8	≈ 8.5742	≈ 8.8152	≈ 8.8214	≈ 8.8244	≈ 8.8250

You can see that $f(\pi) = 2^\pi \approx 8.825$.

Example 1 illustrates several general characteristics of exponential functions of the form $y = b^x$. Since $b^0 = 1$, the y-intercept of the graph is $(0, 1)$. There is no x-intercept since there is a horizontal asymptote at $y = 0$. If $b > 1$, the function is increasing and can model *exponential growth* with a *growth factor* of b. If $0 < b < 1$, the function is decreasing and can model *exponential decay* with a *decay factor* of b.

The fact that $g(x) = \left(\frac{1}{2}\right)^x$ is a reflection of $f(x) = 2^x$ in the y-axis can be verified algebraically by rearranging $g(x) = (2^{-1})^x = 2^{(-1)x} = 2^{-x}$. Exponential functions of the form $g(x) = ab^x$ are vertical stretches or compressions of the parent function $f(x) = b^x$ and have a y-intercept of $(0, a)$. Horizontal and vertical transformations are easily identified in functions of the form $g(x) = ab^{x-h} + k$.

Example 2 Transforming a Parent Exponential Function

Describe each transformation of $f(x) = 3^x$. Then sketch $g(x)$.

a. $g(x) = -2 \cdot 3^x$ **b.** $g(x) = 3^{x-2}$

c. $g(x) = 3^x - 2$ **d.** $g(x) = 3^{-x}$

Answer

a. Reflect $f(x)$ in the x-axis and stretch it vertically by a factor of 2.

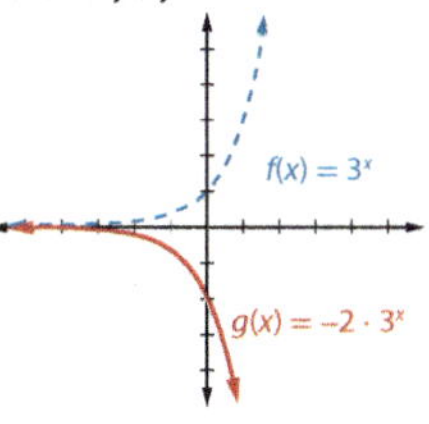

b. Translate $f(x)$ right 2 units.

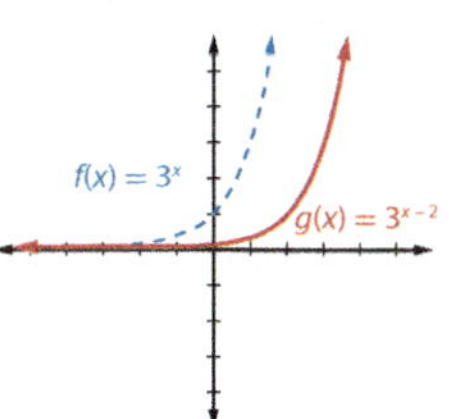

c. Translate $f(x)$ down 2 units.

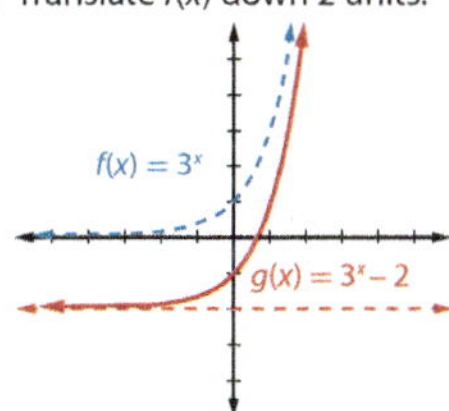

d. Reflect $f(x)$ in the y-axis.

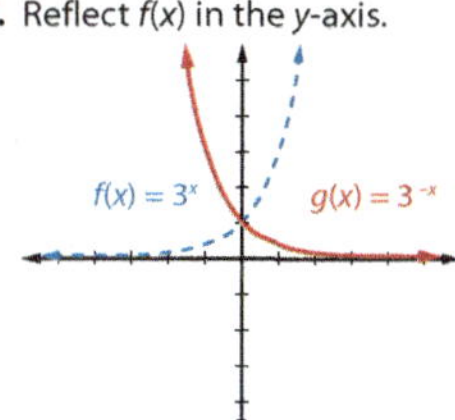

SKILL ✓ EXERCISE 11

1. Classify each function as exponential growth or decay.

  a. $f(x) = \left(\frac{4}{3}\right)^x$ growth

  b. $f(x) = \left(\frac{3}{4}\right)^x$ decay

  c. $f(x) = \frac{1}{2}\left(\frac{3}{4}\right)^{-x}$ growth

  d. $f(x) = 4\left(\frac{3}{4}\right)^x$ decay

2. Describe each transformation of $f(x) = 3\left(\frac{1}{2}\right)^x$. Then sketch $g(x)$.

  a. $g(x) = -3\left(\frac{1}{2}\right)^x$

  a reflection in the x-axis

  b. $g(x) = \frac{3}{2}\left(\frac{1}{2}\right)^x$

  a vertical shrink by a factor of $\frac{1}{2}$

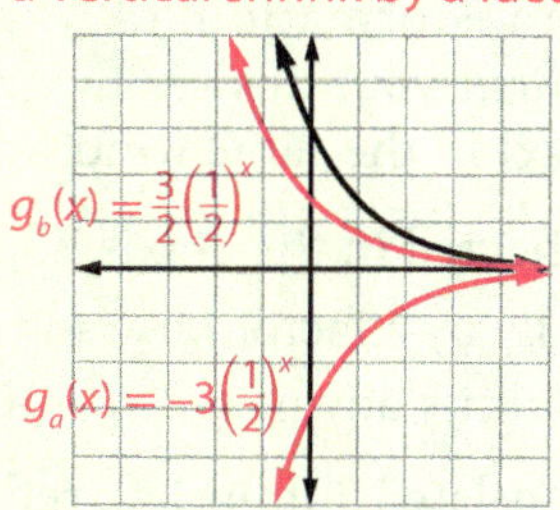

3. Describe each transformation of $f(x) = 3\left(\frac{1}{2}\right)^x$. Then sketch $g(x)$.

  a. $g(x) = 3\left(\frac{1}{2}\right)^{x-1} + 2$

  a shift 1 unit right and 2 units up

  b. $g(x) = 3\left(\frac{1}{2}\right)^{x+2} - 1$

  a shift 2 units left and 1 unit down

ered in the next section, are also continuous and include irrational values.

Discuss *exponential growth* and *exponential decay*, including their related forms and terms. Discuss the general form for translations of exponential functions and each transformation in Example 2. Consider asking students to identify the function's initial value (*the y-coordinate of the y-intercept*) in part a (-2) and in part c (-1). Point out how the horizontal asymptote in part c is translated down 2 units and that the function in part d is an example of exponential decay.

Introduce the number e and the *natural exponential function*, $f(x) = e^x$.

Motivational Idea It is easy to memorize an approximation of e if you group the digits as 2.7 1828 1828. You can then append the familiar digits 45 90 45 to quickly extend the approximation to 15 decimal places.

Use Example 3 to illustrate transformations of the natural exponential function.

Leonhard Euler is widely recognized as the most prolific mathematician of all time. His more than 800 published papers covered every area of mathematics known in his day. The symbols e, i, and π are all attributed to Euler. You may want to encourage the students to research the work of Leonhard Euler and share their findings.

Continuously *compounded interest* can be modeled with an exponential function with the *natural base e*. Guide the students through the derivation of the formula for interest compounded annually. Students may need to be reminded that $I = Prt$ and

$$A = P + I$$
$$= P + Prt$$
$$= P(1 + rt)$$
$$= P(1 + r) \text{ since } t = 1.$$

Explain the adjustments to the formula for other compounding periods, $A(t) = P\left(1 + \frac{r}{n}\right)^{nt}$, and show how this formula transforms to $A(t) = Pe^{rt}$ when interest is compounded continuously. The

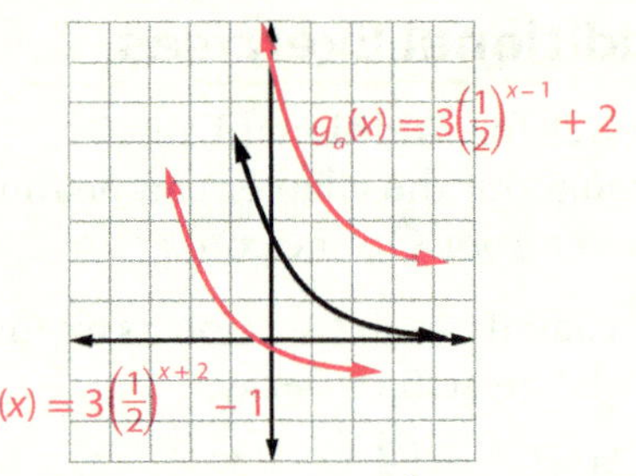

4. Describe $g(x) = 3\left(\frac{1}{2}\right)^{2-x}$ as a transformation of $f(x) = 3\left(\frac{1}{2}\right)^{x}$. Then sketch $g(x)$. a shift 2 units left and a reflection in the y-axis (or a reflection in the y-axis and a shift 2 units right)

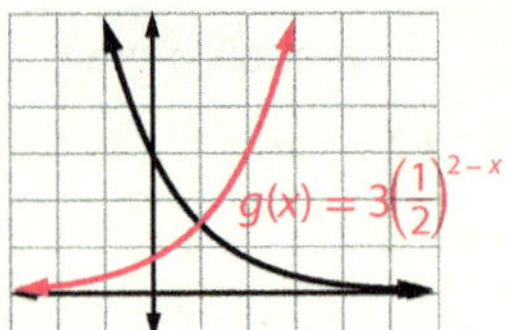

Write a function rule for $g(x)$, the described transformation of $f(x) = 2^x$. Then confirm your answer by graphing $f(x)$ and $g(x)$ in the same window.

5. $f(x)$ reflected in the x-axis, shrunk vertically by a factor of $\frac{1}{3}$, and translated 4 units down $g(x) = -\frac{1}{3}(2^x) - 4$

6. $f(x)$ translated 3 units left, reflected in the y-axis, and translated 1 unit down $g(x) = 2^{3-x} - 1$

Approximating e

x	$\left(1+\frac{1}{x}\right)^x$
1	2
10	2.59374
100	2.70481
1000	2.71692
10,000	2.71814

$$e \approx 2.71828$$

TIP

A more precise estimate of e can be found using e^x ([2nd], [LN]) on your calculator with $x = 1$.

The symbol e owes its acceptance primarily to Leonhard Euler. His first use of e occurred in a paper that he wrote at the age of twenty exploring the mathematics of cannon fire.

Many real-life instances of continuous exponential growth can be modeled using the irrational number e as the base. The *natural base* is defined as $e = \lim\limits_{x \to \infty}\left(1+\frac{1}{x}\right)^x$ and $e \approx 2.71828$. The function $f(x) = e^x$ is called the *natural exponential function* and has several properties that simplify calculations done in calculus.

Example 3 Transforming the Natural Exponential Function

Describe each transformation of $f(x) = e^x$. Then sketch the transformed function.

a. $g(x) = \frac{1}{4}e^x$ **b.** $h(x) = e^{-x} + 2$

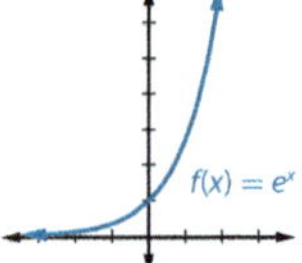

Answer

a. Shrink $f(x)$ vertically by a factor of $\frac{1}{4}$.

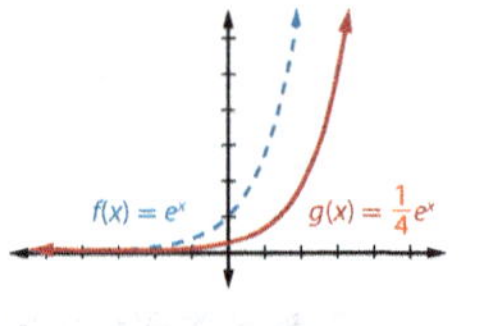

b. Reflect $f(x)$ in the y-axis and translate the result up 2 units.

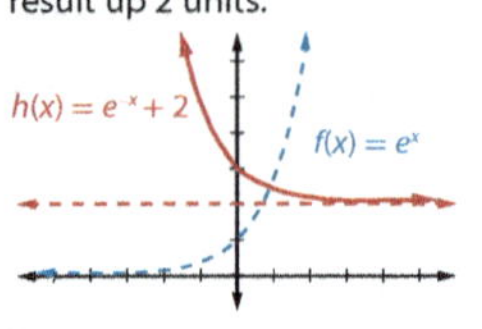

SKILL ✓ EXERCISE 17

Compound interest is modeled by an exponential function. While simple interest is paid only on the original investment, interest is frequently reinvested, or compounded, at the end of a given period so that in subsequent periods it earns interest along with the original investment.

Examine the account balance after t years when a principal P is invested at an annual interest rate r (frequently called the *annual percentage rate*, or APR) and is compounded annually.

Year	Balance
0	$A_0 = P$
1	$A_1 = P + rP = P(1 + r)$
2	$A_2 = A_1(1 + r) = P(1 + r)(1 + r) = P(1 + r)^2$
3	$A_3 = A_2(1 + r) = P(1 + r)^2(1 + r) = P(1 + r)^3$
$\vdots$	$\vdots$
t	$A_t = A_{t-1}(1 + r) = P(1 + r)^{t-1}(1 + r) = P(1 + r)^t$

If the interest is compounded n times a year (quarterly, monthly, or daily),
the rate for each compounding period is $\frac{r}{n}$,
the number of compounding periods is nt, and
the function rule for the account balance after t years is $A(t) = P\left(1 + \frac{r}{n}\right)^{nt}$.

Power Property of Exponents is reviewed in Appendix 4.

When presenting Example 4, explain that leap years are usually ignored in this type of calculation. Direct attention to the minimal difference of $5.45 between daily compounded interest and continuously compounded interest over 15 yr. Consider discussing the assumption that there were no additional deposits and no withdrawals from the account. This assumption has important implications when applying the concept to radiometric dating methods.

Example 5 uses the annual growth model, since eagles breed once a year. You may want to explain that P_0 (read "P sub naught") is a shortened notation for the initial value, $P(0)$. Remind the students that using the exact value for the *growth rate, r*, produces the best prediction. The actual FWS count in 2006 was 9789 nesting pairs. The bald eagle was reclassified as a threatened species in 1995 and removed from the list of threatened species in 2007.

Interactive Activity Use the Internet keyword search *exponential growth applet* or *exponential decay applet* for interactive graphs and related visuals.

Share that exponential growth or decay is often modeled with a continuous compounding function. Present the *half-life* exponential function and apply it in Example 6.

Modeling growth or decay with an exponential function assumes that conditions are ideal for consistent growth or decay over an interval of time. In reality, environments are complex and conditions can vary significantly. Consider discussing factors that affect consistent population growth or decline.

Motivational Idea Use the Internet keyword search *inaccuracies in radiocarbon dating* to research this topic. Share some of your findings and discuss the assumptions involved in exponential growth and decay models.

A formula for continuously compounded interest can be derived by letting $\frac{n}{r} = x$ and letting the number of compoundings per year approach infinity.

$$A(t) = P\left(1 + \frac{r}{n}\right)^{nt} \qquad \text{Note that } n = xr \text{ and that } \frac{r}{n} = \frac{1}{x}.$$
$$= P\left(1 + \left(\frac{1}{x}\right)\right)^{(xr)t} \qquad \text{Substitute.}$$
$$= P\left(\left(1 + \frac{1}{x}\right)^{x}\right)^{rt} \qquad \text{Power Property of Exponents}$$
$$= Pe^{rt} \qquad \text{definition of the natural base, } e$$

Compound Interest Formulas	
compounded n times annually: $A(t) = P\left(1 + \frac{r}{n}\right)^{nt}$	compounded continuously: $A(t) = Pe^{rt}$

Example 4 Compounding Interest

If $30,000 is invested at 6% APR, calculate the account balance after 15 yr with each compounding.

a. annually   **b.** quarterly   **c.** monthly

d. daily   **e.** continuously

Answer

using $A(t) = P\left(1 + \frac{r}{n}\right)^{nt}$:

a. $n = 1; A(15) = \$30{,}000\left(1 + \frac{0.06}{1}\right)^{1(15)} \approx \$71{,}896.75$

b. $n = 4; A(15) = \$30{,}000\left(1 + \frac{0.06}{4}\right)^{4(15)} \approx \$73{,}296.59$

c. $n = 12; A(15) = \$30{,}000\left(1 + \frac{0.06}{12}\right)^{12(15)} \approx \$73{,}622.81$

d. $n = 365; A(15) = \$30{,}000\left(1 + \frac{0.06}{365}\right)^{365(15)} \approx \$73{,}782.64$

e. using $A(t) = Pe^{rt}: A(15) = \$30{,}000e^{0.06(15)} \approx \$73{,}788.09$

 SKILL ✓ EXERCISE 27

Populations and other quantities frequently experience exponential growth or decay, which can both be modeled as compounding annually or continuously where r is the *growth rate*.

Exponential Growth ($r > 0$) and Decay ($r < 0$) Formulas	
compounded annually: $P(t) = P_0(1 + r)^t$	compounded continuously: $P(t) = P_0e^{rt}$

TIP

Since $f(x) = ab^x$ or $P(t) = P_0(1 + r)t$, $P_0 = a$ and $1 + r = b$.

TIPS

Ex. 8 This exercise provides an opportunity to illustrate the equivalence of the vertical stretch in $g(x) = 2 \cdot 2^x$ and the translation 1 unit left in $h(x) = 2^{x+1}$.

7. Find the value of a $5000 bank deposit after 9 yr and 9 mo at each interest rate.

  a. 5% compounded monthly
    $8132.96

  b. 8% compounded quarterly
    $10,823.72

8. A bacteria culture grows from 75 bacteria at 3% per minute.

  a. Write a continuous-growth function to model the growth phase after m minutes. $P(m) = 75e^{0.03m}$

  b. Find the bacteria count after 30 min.
    $P(30) = 75e^{0.03(30)} \approx 184$ bacteria

9. 1.5 g of nitrogen-13 (N-13) has a half-life of 9.965 min.

  a. Find the amount of N-13 left after 30 min. ≈ 0.186 g

  b. Find the percentage of N-13 left after 30 min. $\approx 12.4\%$

Assignments

- **Minimum:** 1–9, 12–15, 18, 21–27, 29, 31, 33; 45–53 odd
- **Standard:** 1–8, 10, 12–15, 18–19, 21–26, 28, 30, 33–36, 38, 43, 45, 47–48, 51–54
- **Extended:** 2, 6–7, 10–11, 13–14, 16–17, 19–21, 23–26, 30, 32–36, 39, 43–45, 49–52, 54

Solutions

❯ A. Exercises

2a. $b > 1$

2b. $0 < b < 1$

2c. $p(x) = 3\left(\frac{1}{5}\right)^x; 0 < b < 1$

2d. $b > 1$

5. $D = \mathbb{R}; R = (0, \infty); b > 1 \Rightarrow$ growth

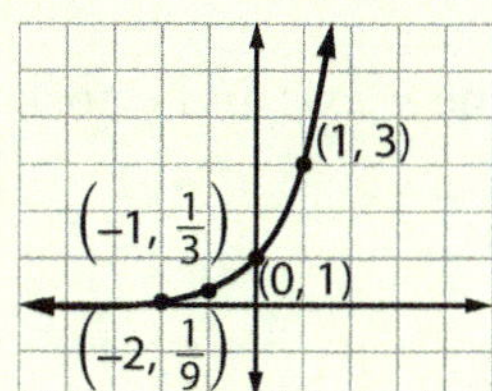

6. $D = \mathbb{R}$; $R = (0, \infty)$
$0 < b < 1 \Rightarrow$ decay

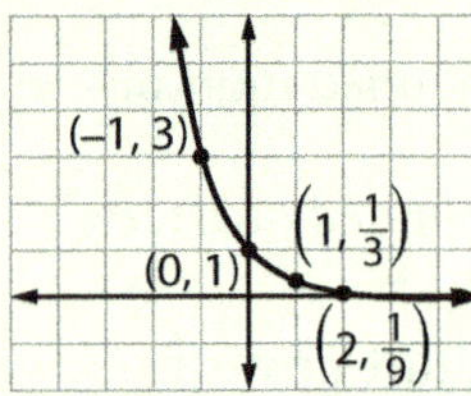

7. $D = \mathbb{R}$; $R = (0, \infty)$
$0 < b < 1 \Rightarrow$ decay

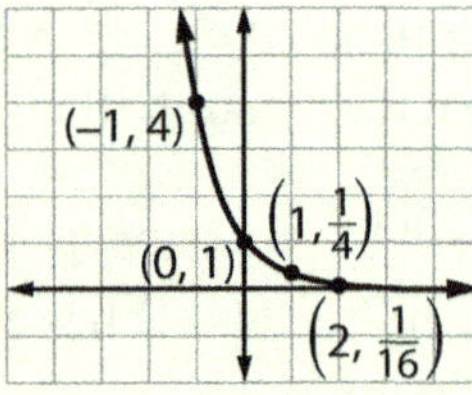

8. $D = \mathbb{R}$; $R = (0, \infty)$; $b > 1 \Rightarrow$ growth

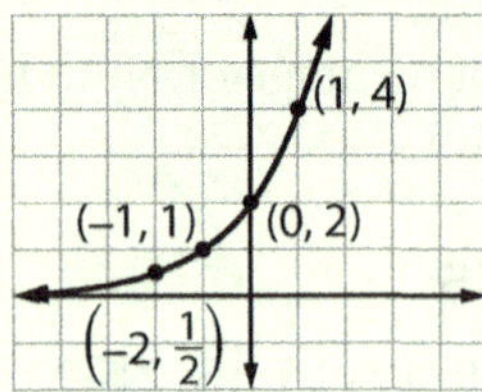

9. translated 3 units left and 2 units down

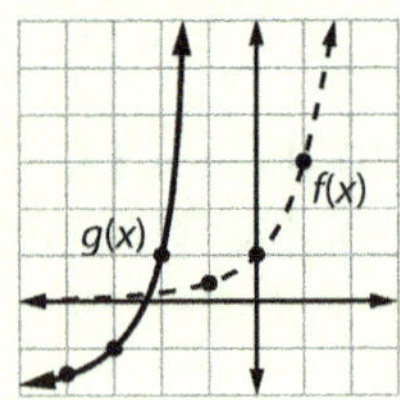

10. translated 4 units right and stretched vertically by a factor of 2

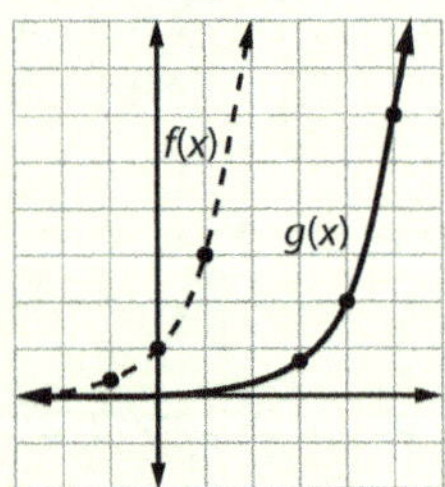

11. reflected in the x-axis and translated 3 units up

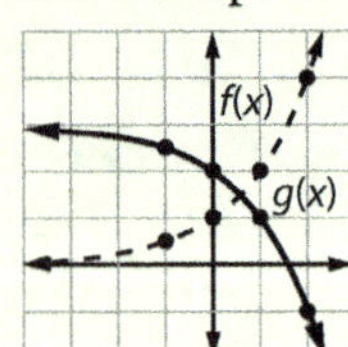

12. translated 1 unit right, reflected in the x-axis, and shrunk vertically by a factor of $\frac{1}{4}$

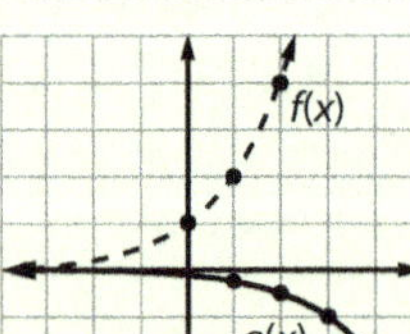

13. or $g(x) = \left(\frac{1}{3}\right)^x + 2$

15. $D = \mathbb{R}$; $R = (0, \infty)$; x-int.: none
y-int.: $(0, 1)$; HA: $y = 0$
$\lim_{x \to -\infty} f(x) = 0$, $\lim_{x \to \infty} f(x) = \infty$

16. $D = \mathbb{R}$; $R = (-\infty, 1)$; x-int.: $(3.39, 0)$
y-int.: $(0, 0.98)$; HA: $y = 1$
$\lim_{x \to -\infty} f(x) = 1$, $\lim_{x \to \infty} f(x) = -\infty$

In 1963 the US Fish and Wildlife Service counted 487 pairs of nesting bald eagles in the 48 contiguous United States. After being placed on the US endangered species list in 1967, their number increased to 3035 nesting pairs in 1990. Assuming exponential growth, find the average annual percent increase in nesting pairs from 1963 to 1990. Then use an exponential growth model to predict the number of nesting pairs in 2006.

Answer

$$P(t) = P_0(1 + r)^t$$
$$3035 = 487(1 + r)^{27}$$
$$\frac{3035}{487} = (1 + r)^{27}$$
$$\sqrt[27]{\frac{3035}{487}} = 1 + r$$
$$r = \sqrt[27]{\frac{3035}{487}} - 1$$
$$\approx 0.0701$$
or 7.01% annually

1. Let $t = 0$ in 1963 so that $t = 27$ in 1990, $P_0 = 487$, and $P(27) = 3035$.

2. Solve for r.

$$P(t) = P_0(1 + r)^t$$
$$P(43) \approx 487(1 + 0.0701\ldots)^{43}$$
$$\approx 8975 \text{ nesting pairs}$$

3. Substitute $t = 43$ and the unrounded annual rate of increase into the exponential growth function and evaluate. Using the rounded rate of increase produces a slightly different estimate.

SKILL ✓ **EXERCISE 31**

The speed at which a radioactive element decays is described by its *half-life*, H, the time it takes for half the radioactive isotope to decay into a stable isotope of another element. The number of half-lives in a given time period t is found using $\frac{t}{H}$, and the amount of the radioactive element remaining is modeled by the following exponential decay equation:

$$P(t) = P_0\left(1 - \frac{1}{2}\right)^{\frac{t}{H}} = P_0\left(\frac{1}{2}\right)^{\frac{t}{H}}.$$

The mathematical model makes several assumptions, including a constant rate of decay.

The half-life of carbon-14 (C-14) is estimated to be 5730 yr. Organic remains near Ale's Stones in Sweden have been used to date the monument near AD 600. Suppose a sample of the remains contains 68 mg of C-14.

a. Estimate the original amount of C-14 in the sample 1400 yr ago.

b. Predict the amount of C-14 remaining in the sample 50 yr from now.

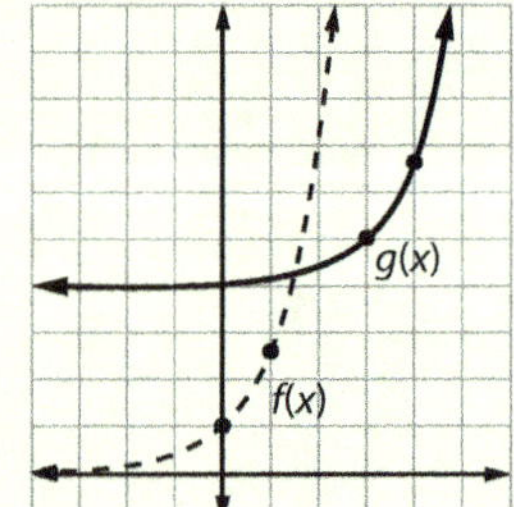

These 59 large boulders, weighing up to 4000 lb each, form the oval outline of a ship that is 220 ft long.

TIP

Part *b* could also be solved by evaluating $P(50)$, using 68 mg as the initial amount of C-14.
$$P(50) = 68\left(\frac{1}{2}\right)^{\frac{50}{5730}}$$
$$\approx 67.6 \text{ mg}$$

Answer

a. $P(t) = P_0\left(\frac{1}{2}\right)^{\frac{t}{H}}$

$$68 = P_0\left(\frac{1}{2}\right)^{\frac{1400}{5730}}$$

$$P_0 = \frac{68}{\left(\frac{1}{2}\right)^{\frac{1400}{5730}}} \approx 80.5 \text{ mg}$$

Substitute the current amount, the number of elapsed years, and the isotope's half-life into the function modeling radioactive decay; then solve for P_0.

b. $P(1450) \approx 80.5\left(\frac{1}{2}\right)^{\frac{1450}{5730}}$

$$\approx 67.6 \text{ mg}$$

Evaluate $P(1450)$ using P_0 from part *a*.

SKILL ✓ **EXERCISE 33**

> **B. Exercises**

17. translated 3 units right and 4 units up; $D = \mathbb{R}$; $R = (4, \infty)$
HA: $y = 4$; x-int.: none
y-int.: $g(0) = e^{-3} + 4 \approx 4.05$, $(0, 4.05)$
$\lim_{x \to -\infty} f(x) = 4$, $\lim_{x \to \infty} f(x) = \infty$

1. Classify each function as a power function or an exponential function.
   a. $f(x) = 7^x$ exponential   b. $g(x) = x^7$ power
   c. $h(x) = \left(\frac{2}{3}\right)^x$ exponential   d. $j(x) = x^e$ power

2. Classify each function as modeling exponential growth or decay.
   a. $f(x) = 2.4^x$ growth   b. $k(x) = 35\left(\frac{1}{2}\right)^x$ decay
   c. $p(x) = 3 \cdot 5^{-x}$ decay   d. $g(x) = 500(1.25)^x$ growth

3. Evaluate.
   a. $f(-1)$ when $f(x) = 3^x$ $\frac{1}{3}$
   b. $p(3)$ when $p(x) = 4\left(\frac{1}{2}\right)^x$ $\frac{1}{2}$
   c. $q(-2)$ when $q(x) = \left(\frac{3}{5}\right)^x$ $\frac{25}{9}$

4. Use a calculator to evaluate, rounding your answer to the nearest ten thousandth.
   a. $g(0.001)$ when $g(x) = 4^x$ 1.0014
   b. $h(\sqrt{3})$ when $h(x) = 2\left(\frac{1}{3}\right)^x$ 0.2983
   c. $j(-e)$ when $j(x) = \left(\frac{5}{2}\right)^x$ 0.0828

Graph the following exponential functions by finding several ordered pairs. State each function's domain and range, then classify the function as exponential growth or decay.

5. $y = 3^x$
6. $y = \left(\frac{1}{3}\right)^x$
7. $y = 4^{-x}$
8. $g(x) = 2 \cdot 2^x$

Describe each transformation of $f(x) = 3^x$ and then graph $f(x)$ and $g(x)$.

9. $g(x) = 3^{x+3} - 2$
10. $g(x) = 2 \cdot 3^{x-4}$

Describe each transformation of $f(x) = 2^x$ and then graph $f(x)$ and $g(x)$.

11. $g(x) = -2^x + 3$
12. $g(x) = -\frac{1}{4} \cdot 2^{x-1}$

Write a function rule for $g(x)$, the described transformation of $f(x) = 3^x$. Then confirm your answer by graphing $f(x)$ and $g(x)$ in the same window.

13. $f(x)$ reflected in the y-axis and then translated 2 units up $g(x) = 3^{-x} + 2$

14. $f(x)$ reflected in the x-axis, stretched vertically by a factor of 4, and translated 5 units to the left $g(x) = -4 \cdot 3^{x+5}$

Use technology to graph each function and then state its domain, range, intercepts, asymptotes, and end behavior.

15. $f(x) = 1.23^x$
16. $f(x) = -2\pi^{x-4} + 1$

Describe each transformation of $f(x) = e^x$. Then sketch the graph and state the translated function's domain and range, any asymptotes and intercepts, and the function's end behavior.

17. $g(x) = e^{x-3} + 4$
18. $g(x) = -\frac{1}{3}e^{x+2}$
19. $g(x) = e^{4-x}$
20. $g(x) = 2e^{-x} + 1$

Describe the transformation of $f(x) = 2^x$ and then graph $f(x)$ and $g(x)$.

21. $g(x) = 3 \cdot 2^{x-3} + 1$
22. $g(x) = \frac{1}{3} \cdot 2^{-x} - 1$

23. Use technology to graph $f(x) = 2^x$, $g(x) = 3^x$, $h(x) = 5^x$, and $j(x) = 10^x$ in the same window. Describe the rate of growth in $f(x) = b^x$ in the first quadrant as b gets larger.

24. State the domain, range, and intercepts of the exponential function $f(x) = ab^x$.

25. Describe the end behavior of $f(x) = ab^x + k$.
    a. when $a > 0$ and $b > 1$
    b. when $a > 0$ and $0 < b < 1$

26. State whether each function represents exponential growth, exponential decay, or neither.
    a. $y = ab^x$, where $a > 0$ and $b > 1$ growth
    b. $y = ab^x$, where $a > 0$ and $0 < b < 1$ decay
    c. $y = ab^{-x}$, where $a > 0$ and $b > 1$ decay
    d. $y = ab^x$, where $a > 0$ and $b < 0$ neither
    e. $y = ab^x$, where $a > 0$ and $b = 1$ neither

Determine the account balance of each investment when compounded (a) annually, (b) quarterly, (c) monthly, (d) daily, and (e) continuously.

27. $10,000 at 1.25% annual interest for 5 yr

28. $4000 deposited at 6% annual interest in 2000 and withdrawn on the same date in 2042

29. From which bank will you earn more money in 8 yr if Bank A compounds monthly and Bank B compounds semiannually? Explain your reasoning.
    a. if both banks offer the same interest rate Bank A
    b. if Bank A offers 4% and Bank B offers 5% Bank B

30. In 2015 the average student loan debt was $30,000. How much will a student with this debt compounded daily at 6.3% APR owe if payment is deferred for 5 yr? for 10 yr? $41,106.66; $56,325.26

20. $f(x)$ is reflected in the y-axis, stretched vertically by a factor of 2, then translated 1 unit up.; $D = \mathbb{R}$
$R = (1, \infty)$; HA: $y = 1$; x-int.: none
y-int.: $g(0) = 2e^0 + 1 = 3$, $(0, 3)$
$\lim_{x \to -\infty} f(x) = \infty$; $\lim_{x \to \infty} f(x) = 1$

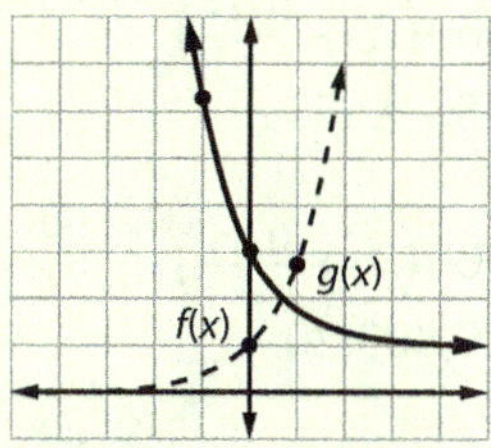

21. translated 3 units right, stretched vertically by a factor of 3, then translated 1 unit up

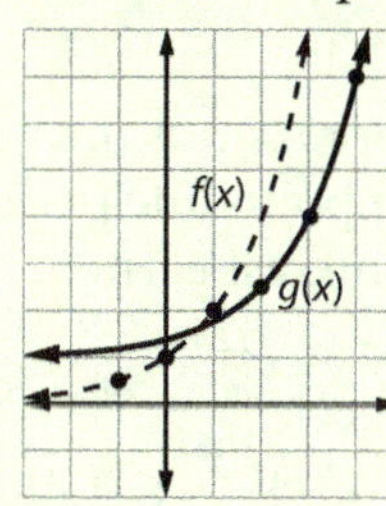

22. reflected in the y-axis, shrunk vertically by a factor of $\frac{1}{3}$, then translated 1 unit down

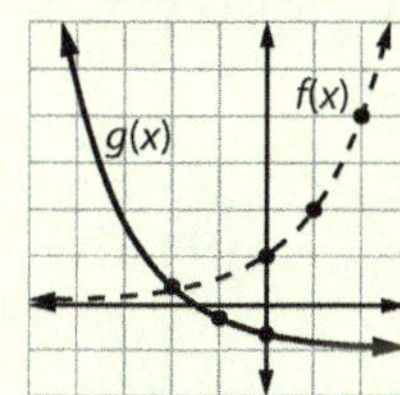

23. The growth is faster for larger values of b.

24. $D = \mathbb{R}$; $R = (0, \infty)$ if $a > 0$, but $R = (-\infty, 0)$ if $a < 0$; y-int.: $(0, a)$
x-int.: none

25a. $\lim_{x \to -\infty} f(x) = k$, $\lim_{x \to \infty} f(x) = \infty$
25b. $\lim_{x \to -\infty} f(x) = \infty$, $\lim_{x \to \infty} f(x) = k$

27a. $n = 1$
$A(5) = 10{,}000(1 + 0.0125)^5$
$\approx \$10{,}640.82$

27b. $n = 4$
$A(5) = 10{,}000\left(1 + \frac{0.0125}{4}\right)^{(5)(4)}$
$\approx \$10{,}643.91$

27c. $n = 12$
$A(5) = 10{,}000\left(1 + \frac{0.0125}{12}\right)^{(5)(12)}$
$\approx \$10{,}644.60$

27d. $n = 365$
$A(5) = 10{,}000\left(1 + \frac{0.0125}{365}\right)^{(5)(365)}$
$\approx \$10{,}644.93$

18. translated 2 units left, shrunk vertically by a factor of $\frac{1}{3}$, and reflected in the x-axis; $D = \mathbb{R}$; $R = (-\infty, 0)$
HA: $y = 0$; x-int.: none
y-int.: $g(0) = -\frac{1}{3}e^2 \approx -2.46$, $(0, -2.46)$
$\lim_{x \to -\infty} f(x) = 0$, $\lim_{x \to \infty} f(x) = -\infty$

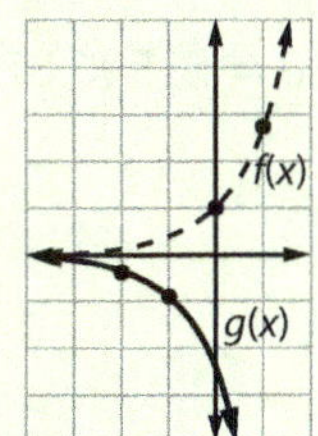

19. $g(x) = e^{(-x+4)}$ implies that $f(x)$ is translated 4 units left and then reflected in the y-axis,
or $g(x) = e^{-(x-4)}$ implies that $f(x)$ is reflected in the y-axis and then translated 4 units right.; $D = \mathbb{R}$
$R = (-\infty, 0)$; HA: $y = 0$; x-int.: none
y-int.: $g(0) = e^4 \approx 54.60$, $(0, 54.60)$
$\lim_{x \to -\infty} f(x) = \infty$, $\lim_{x \to \infty} f(x) = 0$

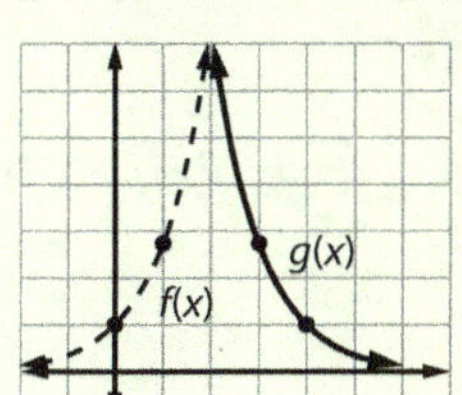

27e. $A(5) = 10{,}000e^{(5)(0.0125)}$
$\approx \$10{,}644.94$

28a. $n = 1$
$A(42) = 4000(1 + 0.06)^{42}$
$\approx \$46{,}228.13$

28b. $n = 4$
$A(42) = 4000\left(1 + \dfrac{0.06}{4}\right)^{(42)(4)}$
$\approx \$48{,}792.73$

28c. $n = 12$
$A(42) = 4000\left(1 + \dfrac{0.06}{12}\right)^{(42)(12)}$
$\approx \$49{,}403.20$

28d. $n = 365$
$A(42) = 4000\left(1 + \dfrac{0.06}{365}\right)^{(42)(365)}$
$\approx \$49{,}704.10$

28e. $A(42) = 4000e^{0.06(42)}$
$\approx \$49{,}714.39$

29a. Bank A earns more, since the more frequent the compounding periods, the greater the interest earned.

29b. $A_A(8) = P\left(1 + \dfrac{0.04}{12}\right)^{(12)(8)} \approx 1.38P$
$A_B(8) = P\left(1 + \dfrac{0.05}{2}\right)^{(2)(8)} \approx 1.48P$

30. $A(5) = 30{,}000\left(1 + \dfrac{0.063}{365}\right)^{(5)(365)}$
$\approx \$41{,}106.66$
$A(10) = 30{,}000\left(1 + \dfrac{0.063}{365}\right)^{(10)(365)}$
$\approx \$56{,}325.26$

31. $P(t) = P_0(1 + r)^t$
$70{,}174 = 18{,}313(1 + r)^{31}$
$\sqrt[31]{\dfrac{70{,}174}{18{,}313}} = 1 + r$
$r = \sqrt[31]{\dfrac{70{,}174}{18{,}313}} - 1$
$r \approx 4.43\%$
$P(38) = 18{,}313(1 + 0.0443)^{38}$
$\approx 95{,}087$ sea lions

32. Note that the World Health Organization does not consider this level of emission harmful to human health.

using $P(t) = 800\left(\dfrac{1}{2}\right)^{\frac{t}{30}}$:

$P(50) = 800\left(\dfrac{1}{2}\right)^{\frac{50}{30}} \approx 251.98$ TBq

$P(150) = 800\left(\dfrac{1}{2}\right)^{\frac{150}{30}} = 25$ TBq

33. $t = 3.25$ hr
$P(t) = 750\left(\dfrac{1}{2}\right)^{\frac{t}{4}}$
$P(3.25) = 750\left(\dfrac{1}{2}\right)^{\frac{3.25}{4}} \approx 427.05$ mg

34. $P(m) = 250(1 + 1)^{\frac{m}{20}} = 250(2)^{\frac{m}{20}}$

34a. $P(60) = 250(2)^{\frac{60}{20}} = 2000$ bacteria

31. In 1979 the Steller sea lion was placed on the endangered species list with an estimated population of 18,313 sea lions. It was removed from the list after the population increased to an estimated 70,174 in 2010. Determine the population's average annual rate of increase from 1979 to 2010. Then use an exponential growth model to predict the population in 2017. $\approx 4.43\%$; $\approx 95{,}087$ sea lions

32. A radiation source with an activity of 1 becquerel (Bq) transforms one nucleus per second. Scientists estimate that 800 terabecquerels (TBq) of cesium-137 (Cs-137) reached California in 2016 as a result of the 2011 Fukushima nuclear disaster. The half-life of Cs-137 is approximately 30 yr. Find the radiation emission of this amount of Cs-137 after 50 yr and after 150 yr. ≈ 252 TBq; 25 TBq

33. The antibiotic ciprofloxacin has a half-life of 4 hr. If a patient is administered a dose of 750 mg at 10:30 AM, estimate the amount present in the patient's body at 1:45 PM. ≈ 427 mg

34. If a culture of 250 initial bacteria doubles every 20 min, write a function modeling its growth. Then use the model to estimate the number of bacteria in the culture after each amount of time.

   a. 1 hr   **b.** 3 hr   **c.** 9 hr

2000 bacteria   128,000 bacteria $\approx 3.36 \times 10^{10}$ bacteria

Interest rates with different compounding periods can be compared using the *annual percentage yield* (APY). The APY is the equivalent annual rate for an investment at a given annual percentage rate (APR) compounded n times after 1 yr.

$$P(1 + APY) = P\left(1 + \frac{APR}{n}\right)^n$$

35. Solve the equation above to derive a formula for the APY. Explain why the APY does not depend on the amount invested.

36. Find the APY and account balance of a $3600 investment compounded daily at 6% APR after 5 yr.

37. Bobby is considering investing $50,000 in either a certificate of deposit at 2% APR compounded daily or a money market account with 2.25% APR compounded quarterly. Find the APY of both accounts and determine the difference between the investments after 10 yr.
$APY_{CD} \approx 2.020\%$; $APY_{MM} \approx 2.269\%$; $\$1506.87$

Use the properties of exponents to write each function in the form $y = a \cdot b^x$. Then state which functions are equivalent.

38. $f(x) = 4 \cdot 2^{2x}$, $g(x) = 2^{4x+2}$, and $h(x) = 4^{2x+1}$
$g(x)$ and $h(x)$

39. $f(x) = \dfrac{1}{27} \cdot 3^{2x}$, $g(x) = \left(\dfrac{1}{9}\right)^{2x}$, and $h(x) = 3^{2x-3}$
$f(x)$ and $h(x)$

40. Use technology to solve each inequality.
   a. $2^x < 3^x$ $x \in (0, \infty)$ **b.** $2^x > 3^x$ $x \in (-\infty, 0)$

41. Use technology to solve each inequality.
   a. $\left(\dfrac{1}{2}\right)^x < \left(\dfrac{1}{5}\right)^x$ $(-\infty, 0)$ **b.** $\left(\dfrac{1}{2}\right)^x > \left(\dfrac{1}{5}\right)^x$ $(0, \infty)$

42. Gail purchased a used car in 2012 for $12,500. After 5 yr Gail used an appraisal website to determine that the value of her car was $4555. Determine the car's annual rate of depreciation and estimate the car's value in 2020. $\approx 18\%$; $\$2485.62$

43. Ron is comparing two 3-yr-old used cars. Determine the depreciation rate for each car and each car's value 3 yr from now. Car A: $\approx 19.6\%$, $\approx \$5180$; Car B: $\approx 20.3\%$, $\approx \$4467$

Car	Original Price	Price After 3 Yr
A	$19,190	$9970
B	$17,400	$8816

44. A metal bar at 875°C is placed in 20°C water to cool. Its temperature can be modeled using Newton's law of cooling: $T(t) = T_S + (T_0 - T_S)e^{-kt}$ where T_0 is the object's initial temperature, T_S is the surrounding medium's temperature (assumed to be constant), k is the cooling constant, and t is the time.

   a. State the equation modeling the object's temperature if the cooling constant $k = 0.1$ sec^{-1}.

   b. What is the temperature of the bar after 45 sec? $\approx 29.5°C$

   c. Find the initial temperature of a second bar placed in the water if its temperature after 1 min is 22.3°C. $\approx 947.9°C$

   d. State one factor that may cause this model to be inaccurate. The model does not account for an increase in the surrounding medium's temperature.

35. $APY = \left(1 + \dfrac{APR}{n}\right)^n - 1$; The principal cancels from each side of the equation.

36. $\approx 6.18\%$; $\$4859.37$

44a. $T(t) = 20 + 855e^{-0.1t}$

34b. $P[3(60)] = 250(2)^{\frac{180}{20}} = 128{,}000$ bacteria

34c. $P[9(60)] = 250(2)^{\frac{540}{20}} \approx 3.355 \times 10^{10}$ bacteria

36. using $APY = \left(1 + \dfrac{APR}{n}\right)^n - 1$:
$APY = \left(1 + \dfrac{0.06}{365}\right)^{365} - 1 \approx 6.18\%$
$A(5) \approx 3600(1.0618)^5 \approx \$4859.37$

37. $APY_{CD} = \left(1 + \dfrac{0.02}{365}\right)^{365} - 1$
$\approx 2.020\%$
$A_{CD}(10) \approx 50{,}000(1.02020)^{10}$
$\approx 61{,}069.80$
$APY_{MM} = \left(1 + \dfrac{0.0225}{4}\right)^4 - 1$
$\approx 2.269\%$
$A_{MM}(10) \approx 50{,}000(1.02269)^{10}$
$\approx 62{,}579.67$
$62{,}579.67 - 61{,}069.80 = \$1506.87$

45. Write the equation of a line passing through $(-3, 1)$ and parallel to $5x - 8y = 10$. [1.2] $5x - 8y = -23$

46. State another classification for any power function with degree 1. [1.4] linear function

47. When is a quadratic function also a power function? [1.4]

Evaluate each function. [2.1]

48. $f(x) = x^{\frac{2}{3}}$ when $x = -27$ 9

49. $g(x) = \sqrt[3]{x}$ when $x = \frac{8}{343}$ $\frac{2}{7}$

50. $h(x) = x^{\frac{4}{5}}$ when $x = 32$ 16

51. Which of the following is equivalent to $\dfrac{5x\sqrt{54y}}{3\sqrt{6x^2}}$? [Algebra] B

A. $15\sqrt{y}$ **D.** $5y$

B. $5\sqrt{y}$ **E.** none of these

C. $\dfrac{5\sqrt{y}}{3}$

52. Which of the following functions is odd and increasing over the real numbers? [2.1] C

A. $f(x) = 3x^2$ **D.** $h(x) = x^{\frac{2}{3}}$

B. $g(x) = -2x^5$ **E.** none of these

C. $j(x) = \sqrt[3]{x}$

53. Which describes the end behavior of $f(x) = -4x^5 + 3x$? [2.2] C

A. $\lim\limits_{x \to \pm\infty} f(x) = -\infty$

B. $\lim\limits_{x \to \pm\infty} f(x) = \infty$

C. $\lim\limits_{x \to \infty} f(x) = \infty,\ \lim\limits_{x \to -\infty} f(x) = -\infty$

D. $\lim\limits_{x \to \infty} f(x) = -\infty,\ \lim\limits_{x \to -\infty} f(x) = \infty$

E. none of these

54. Which of the following is not a factor of $f(x) = x^3 - 3x^2 - 10x + 24$? [2.3] A

A. $x + 1$ **D.** $x + 3$

B. $x - 2$ **E.** none of these

C. $x - 4$

47. When $f(x) = ax^2 + bx + c$ has $b = 0$ and $c = 0$, it is a second-degree power function.

❯ C. Exercises

38. $f(x) = 2^2 \cdot 2^{2x} = 2^{2x + 2}$
$g(x) = (2^4)^x \cdot 2^2 = 2^{4x + 2}$
$h(x) = (2^2)^{2x + 1} = 2^{4x + 2}$
$g(x)$ and $h(x)$

39. $f(x) = (x) = 3^{-3} \cdot 3^{2x} = 3^{2x - 3}$
$g(x) = (3^{-2})^{2x} = 3^{-4x}$
$h(x) = 3^{2x - 3}$
$f(x)$ and $h(x)$

40. 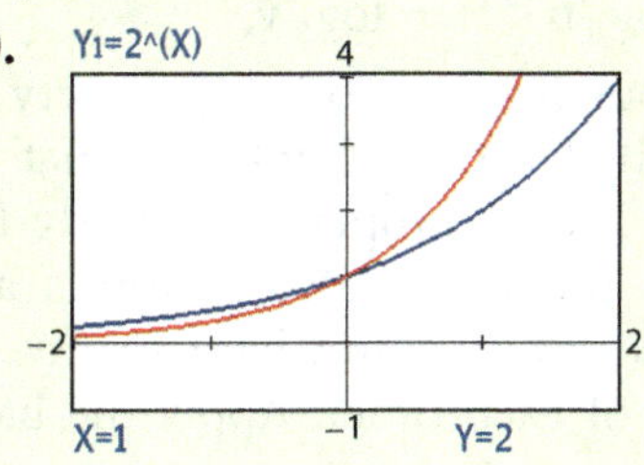

41. 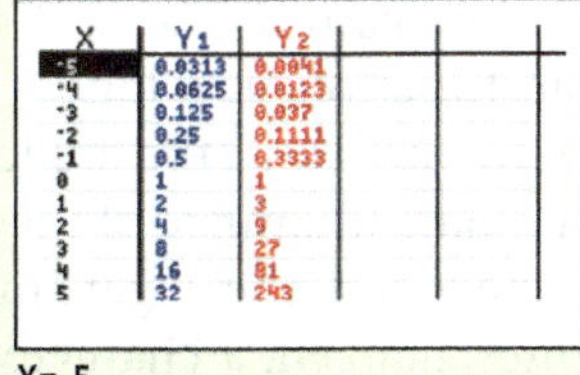

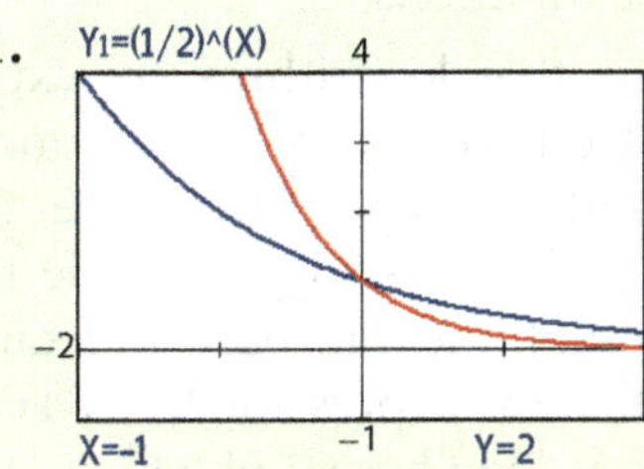

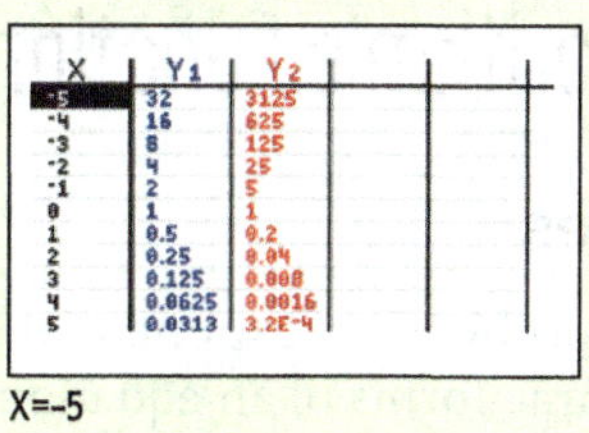

42.
$$4555 = 12{,}500(1 + r)^5$$
$$\frac{4555}{12{,}500} = (1 + r)^5$$
$$\sqrt[5]{\frac{4555}{12{,}500}} = 1 + r$$
$$r = \sqrt[5]{\frac{4555}{12{,}500}} - 1 \approx -0.183$$
$$A(8) = 12{,}500(0.817)^8 \approx \$2485.62$$

43. Car A:
$$9970 = 19{,}190(1 + r)^3$$
$$r_A = \sqrt[3]{\frac{9970}{19{,}190}} - 1 \approx -0.196 \text{ or } -19.6\%$$
$$A_A \approx 19{,}190(1 - 0.196)^6 \approx \$5179.83$$
Car B:
$$8816 = 17{,}400(1 + r)^3$$
$$r_B = \sqrt[3]{\frac{8816}{17{,}400}} - 1 \approx -0.203 \text{ or } -20.3\%$$
$$A_B \approx 17{,}400(1 - 0.203)^6 \approx \$4466.77$$

44a. $T(t) = 20 + (875 - 20)e^{-0.1t}$
$\qquad = 20 + 855e^{-0.1t}$

44b. $T(45) = 20 + 855e^{-0.1(45)} \approx 29.5°\text{C}$

44c.
$$T(60) = 20 + (T_0 - 20)e^{-0.1(60)}$$
$$22.3 = 20 + (T_0 - 20)e^{-6}$$
$$2.3 + 20e^{-6} = T_0 e^{-6}$$
$$T_0 = \frac{2.3 + 20e^{-6}}{e^{-6}}$$
$$\approx 947.9°\text{C}$$

❯ Cumulative Review

45. $m_{\parallel} = m_G = \frac{5}{8}$
$$y - 1 = \frac{5}{8}(x + 3)$$
$$8y - 8 = 5x + 15$$
$$5x - 8y = -23$$

48. $f(-27) = \left(-27^{\frac{1}{3}}\right)^2 = (-3)^2 = 9$

49. $g\left(\frac{8}{343}\right) = \sqrt[3]{\frac{(2)^3}{(7)^3}} = \frac{2}{7}$

50. $h(32) = \left((2^5)^{\frac{1}{5}}\right)^4 = 2^4 = 16$

51. $\dfrac{5x\sqrt{54y}}{3\sqrt{6x^2}} = \dfrac{15x\sqrt{6y}}{3x\sqrt{6}} = 5\sqrt{y}$

52. $f(x)$ is even,
$g(x)$ is decreasing over $(-\infty, \infty)$,
and $h(x)$ is decreasing over $(-\infty, 0]$.

53. $f(x)$ is odd with $a < 0$.

54. $f(1) = 30; \therefore f(x) \div (x + 1)$ has $R = 30$.

3.2 Logarithmic Functions

Objectives

1. To convert between exponential and logarithmic forms of an equation

2. To evaluate common and natural logarithms

3. To graph logarithmic functions

4. To model and solve real-world problems using logarithmic functions

Flash

Road damage is common during an earthquake. Some streets dropped 11 ft or more during the 1964 Alaska earthquake. At 9.2 on the Richter scale, it was the largest recorded earthquake in North America and the second-largest in the world.

Vocabulary

common logarithmic function
logarithmic function
natural logarithmic function

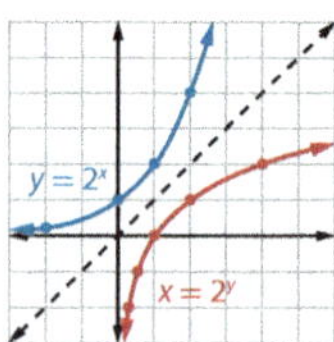

The magnitude of an earthquake is measured by the logarithmic Richter scale.

After completing this section, you will be able to

- convert between exponential and logarithmic forms of an equation.
- evaluate common and natural logarithms.
- graph logarithmic functions.
- model and solve real-world problems using logarithmic functions.

Consider the graph of the exponential function $y = 2^x$. Since its graph passes the horizontal line test, the function is one-to-one and its inverse is also a function. Recall that the graph of a function's inverse is the reflection of the function across the line $y = x$, and that the function rule for the inverse is found by interchanging x and y and solving for y. Doing this with an exponential function, such as $f(x) = 2^x$, presents a challenge. How do you solve $x = 2^y$ for y? Logarithmic functions provide the means for solving such an equation.

> **DEFINITION**
>
> A **logarithmic function** with base b is the inverse of an exponential function of the form $y = b^x$ where x and b are positive numbers and $b \neq 1$: $y = \log_b x$ if and only if $b^y = x$.

The notation $y = \log_b x$ is read "y equals the logarithm of x with base b." The base b logarithm of a number is the exponent when the number is expressed as a power of b. The logarithmic function $f(x) = \log_2 x$ is the inverse of the exponential function $f(x) = 2^x$.

the exponent

$$y = \log_b x \text{ iff } b^y = x$$

the base   the number

Example 1   Evaluating Logarithms

Evaluate each logarithm.

a. $\log_2 64$ **b.** $\log_4 \frac{1}{16}$ **c.** $\log_5 \sqrt[3]{5}$ **d.** $\log_3 1$

Answer

Set each expression equal to y, convert it to exponential form, and then solve.

a. $\log_2 64 = y$
$2^y = 64$
$2^y = 2^6$
$y = 6$
$\therefore \log_2 64 = 6$

b. $\log_4 \frac{1}{16} = y$
$4^y = \frac{1}{16}$
$4^y = 4^{-2}$
$y = -2$
$\therefore \log_4 \frac{1}{16} = -2$

c. $\log_5 \sqrt[3]{5} = y$
$5^y = 5^{\frac{1}{3}}$
$y = \frac{1}{3}$
$\therefore \log_5 \sqrt[3]{5} = \frac{1}{3}$

d. $\log_3 1 = y$
$3^y = 1$
$3^y = 3^0$
$y = 0$
$\therefore \log_3 1 = 0$

SKILL ✔ **EXERCISES 5, 9**

The following properties of logarithms can be demonstrated by converting from logarithmic to exponential form, or from exponential form to logarithmic form.

PRESENTATION

Lesson Opener

Express each of the following in exponential form.

1. 100 as a power of 10 10^2

2. 16 as a power of 2 2^4

3. $\frac{1}{64}$ as a power of 4 4^{-3}

4. 1 as a power of 10 10^0

5. $\sqrt[3]{9}$ as a power of 3 $3^{\frac{2}{3}}$

Refer to the graph of the exponential function $y = 2^x$ and its inverse $x = 2^y$ illustrated in the text. Discuss the need to define new notation enabling us to solve for y. Present the definition of a *logarithmic function* and explain that $x = 2^y$ and the logarithmic function $y = \log_2 x$ are equivalent statements. Have the students note that the domain and range of the exponential function and the logarithmic function are interchanged, a characteristic of inverse functions.

Emphasize that logarithms are exponents. Point out that y is the exponent in $b^y = x$, and $y = \log_b x$; therefore, the exponent in $b^y = x$ is $\log_b x$. Guide the students through the four parts of Example 1. Note that the expression $\log_2 64$ can also be read as "log base 2 of 64" or "the base 2 log of 64."

Discuss the basic properties of logarithms. Ask the students what characteristic of the graph of $y = \log_b x$ is implied by the $\log_b 1 = 0$ property. (*The graph's x-intercept is always (1, 0).*)

Students may need help justifying the fourth basic property. It may help to view the property as taking the base b logarithm of both sides of the equation,

$$\log_b b^{\log_b y} = \log_b y,$$

before applying the third property to simplify the left side. Emphasize that the third and fourth properties directly follow from the fact that logarithms and exponents with the same base are inverse functions of each other. Apply the basic

> **Basic Properties of Logarithms**
>
> If x and b are positive numbers and $b \neq 1$,
>
> 1. $\log_b 1 = 0$ since $b^0 = 1$;
> 2. $\log_b b = 1$ since $b^1 = b$;
> 3. $\log_b b^x = x$ since $b^x = b^x$; and
> 4. $b^{\log_b y} = y$ since $\log_b y = \log_b y$.

The last two properties follow directly from the fact that logarithms and exponents with the same base are inverse functions of each other. These properties allow the evaluation of certain logarithmic and exponential expressions.

Example 2 Applying Properties of Logarithms

Evaluate each expression.

a. $\log_2 1$ **b.** $\log_7 7$ **c.** $\log_2 64$ **d.** $6^{\log_6 3.5}$

Answer

a. $\log_2 1 = 0$ **b.** $\log_7 7 = 1$ **c.** $\log_2 2^6 = 6$ **d.** $6^{\log_6 3.5} = 3.5$
since $2^0 = 1$ since $7^1 = 7$

SKILL ✔ EXERCISES 11, 13

While the base of a logarithmic function can be any positive number except 1, base 10 is commonly used. The base 10 logarithmic function is called the *common logarithmic function* and is usually written without the base. Therefore, $y = \log x$ and $y = 10^x$ are inverse functions. The basic properties of logarithms apply to common logs.

1. $\log 1 = 0$ 2. $\log 10 = 1$ 3. $\log 10^x = x$ 4. $10^{\log y} = y$

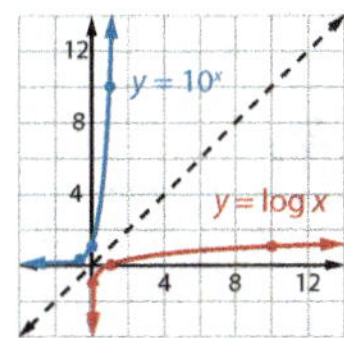

Common logs of powers of ten can be evaluated using these properties. The $\boxed{\text{LOG}}$ key on your calculator is used to find decimal approximations of other common logs.

Example 3 Common Logarithms

Evaluate each expression.

a. $\log 1000$ **b.** $\log 0.01$ **c.** $\log 4.39$

d. $\log \frac{1}{4}$ **e.** $10^{\log 63}$ **f.** $\log(-3)$

Answer

a. $\log 10^3 = 3$ **b.** $\log 10^{-2} = -2$ **c.** using your calculator:
$\log 4.39 \approx 0.6425$

d. using your calculator: **e.** using inverse functions: **f.** Not a real number.
$\log \frac{1}{4} \approx -0.6021$ $10^{\log 63} = 63$ Logarithmic functions have $D = (0, \infty)$.

SKILL ✔ EXERCISES 19, 21

Common logarithms are sometimes called Briggsian logarithms after Henry Briggs, the Oxford mathematician who published the first table of common logs in 1619.

KEYWORD SEARCH

common log table 🔍

Reading and Writing Mathematics

Use an example to explain each restriction on b in $f(x) = \log_b x$.

Answers will vary.

1. $b \neq 0$ Written in exponential form, $y = \log_0 5$ is $0^y = 5$. But any nonzero power of 0 is 0, so this is not possible.

2. $b \neq 1$ Written in exponential form, $y = \log_1 3$ is $1^y = 3$. But any power of 1 is 1, so this is not possible.

3. $b \not< 0$ Written in exponential form, $y = \log_{-2} 8$ is $(-2)^y = 8$. But y must be even to produce a positive value, and there is no even power of -2 that produces 8.

Additional Exercises

1. Convert to logarithmic form.
a. $2^6 = 64$ $\log_2 64 = 6$
b. $3^{-3} = \frac{1}{27}$ $\log_3 \frac{1}{27} = -3$
c. $\left(\frac{1}{2}\right)^{-4} = 16$ $\log_{\frac{1}{2}} 16 = -4$

2. Convert to exponential form.
a. $\log_2 1024 = 10$ $2^{10} = 1024$
b. $\ln 65 \approx 4.1744$ $e^{4.1744} \approx 65$

3. Evaluate each logarithm.
a. $\log_4 4096$ 6
b. $\log_5 \frac{1}{125}$ -3
c. $\log -10$ not real
d. $\log 0.00001$ -5

4. Evaluate each natural logarithm.
a. $\ln \sqrt[7]{e^4}$ $\frac{4}{7}$
b. $\ln 0.001$ -6.9078

properties of logarithms in Example 2. Students should work toward being able to mentally evaluate logs similar to those in Examples 1 and 2.

Introduce common logarithms (base 10). Emphasize that while the common log of a power of 10 is easily determined, the $\boxed{\text{LOG}}$ key on a calculator or other technology is used to find the common logs of other values.

Direct attention to the graph of the basic *common logarithmic function* and ask the students over what interval common logarithms are negative. (*between 0 and 1*)

Interactive Activity Having the students mentally estimate common logs can help them become more comfortable with the concept of logarithms. Ask them to estimate log 550. Guide the students to recognize that if log 100 = 2 and log 1000 = 3, then log 550 is between 2 and 3. If the function were linear, log 550 would equal 2.5. However, logarithmic functions initially increase rapidly before their growth slows; so log 550 ≈ 2.7. After estimating, have the students find the decimal approximation on their calculators. You may then want to have them estimate log 20,000. (≈ 4.3)

Discuss Example 3. Consider showing the students how they can verify the answer to part *c* by evaluating $10^{\wedge}$(ANS)

($\boxed{\text{2nd}}$, $\boxed{\text{LOG}}$, $\boxed{\text{2nd}}$, $\boxed{(\text{-})}$). Since $10^0 = 1$ and $10^1 = 10$, we would expect $1 < 10^{0.6425} < 10$. For the expression in part *f*, most calculators will display an error message, while some may return the complex number $\log 3 + \pi i$. The logarithms of negative values are investigated in higher levels of math.

Introduce the *natural logarithmic function*. Inform the students that although any positive number except 1 can be a logarithmic base, real-life problems are almost exclusively limited to common and natural logarithms. For this reason, calculators typically have keys for these two functions.

5. Verify that $y = \log x$ and $y = 10^x$ are inverses by graphing the functions on the same graph. What indicates that they are inverses?   The graphs are reflections across the line $y = x$.

6. Write a function rule for $g(x)$, the transformation of $f(x) = \ln x$ when $f(x)$ is stretched vertically by a factor of 2, reflected in the x-axis, and then translated 1 unit up. $g(x) = 1 - 2\ln x$

7. Describe $g(x) = \log(x + 1)$ as a transformation of $f(x) = \log x$. Then state the domain and range of $g(x)$, any asymptotes and intercepts, and the end behavior of $g(x)$.
$g(x)$ is $f(x)$ translated 1 unit left.
$D = (-1, \infty); R = \mathbb{R}; \text{VA: } x = -1$
x-int. and y-int.: $(0, 0)$
$\lim\limits_{x \to \infty} f(x) = \infty, \ \lim\limits_{x \to -1^+} f(x) = -\infty$

8. Describe $g(x) = -\log_2(x - 2)$ as a transformation of $f(x) = \log_2 x$. Then graph $f(x)$ and $g(x)$ on the same graph. $f(x)$ is reflected in the x-axis and translated 2 units right.

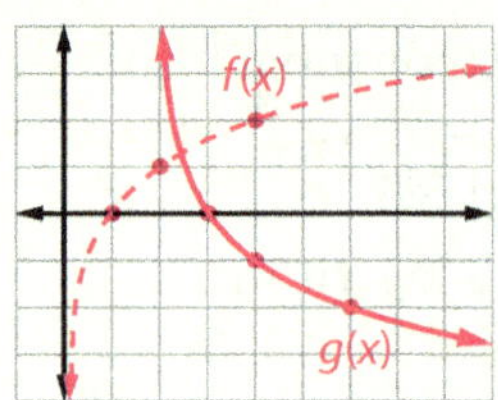

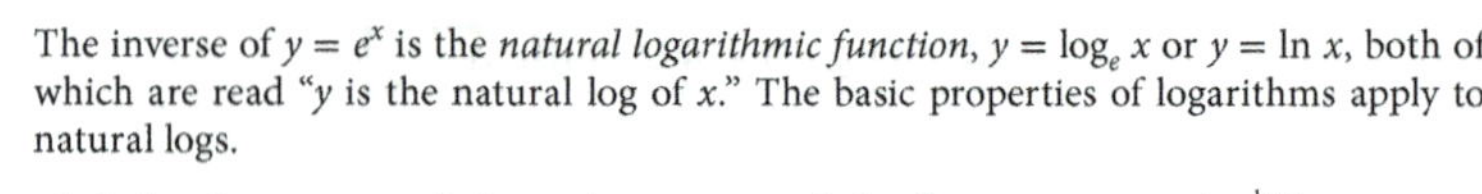

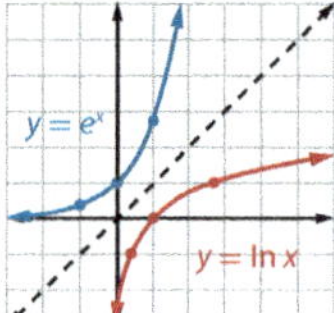

The inverse of $y = e^x$ is the *natural logarithmic function*, $y = \log_e x$ or $y = \ln x$, both of which are read "y is the natural log of x." The basic properties of logarithms apply to natural logs.

1. $\ln 1 = 0$ 2. $\ln e = 1$ 3. $\ln e^x = x$ 4. $e^{\ln y} = y$

The $\boxed{\text{LN}}$ key of your calculator is used to find decimal approximations of natural logs.

Example 4   Natural Logarithms

Evaluate each expression.

a. $\ln e^2$ **b.** $\ln e^{-1}$ **c.** $\ln 4.39$
d. $\ln \frac{1}{4}$ **e.** $e^{\ln 3}$ **f.** $\ln 0$

Answer

a. $\ln e^2 = 2$

b. $\ln e^{-1} = -1$

c. using a calculator:
$\ln 4.39 \approx 1.479$

d. using a calculator:
$\ln \frac{1}{4} \approx -1.3863$

e. using inverse functions:
$e^{\ln 3} = 3$

f. Not a real number. Logarithmic functions have $D = (0, \infty)$.

SKILL ✔ **EXERCISES 23, 25**

KEYWORD SEARCH

natural log table 🔍

The fact that $y = b^x$ and $y = \log_b x$ are inverse functions can help when graphing basic logarithmic functions.

Example 5   Graphing a Logarithmic Function

Graph $f(x) = \log_3 x$.

Answer

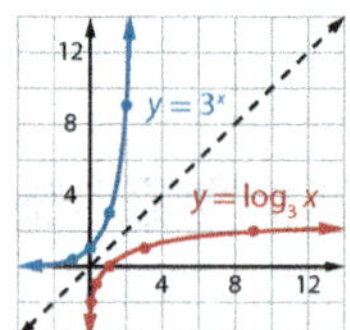

a. Make a table of values for the inverse function $g(x) = 3^x$, plot the points, and connect them with a smooth curve to graph the exponential function.

x	-2	-1	0	1	2
$g(x) = 3^x$	$\frac{1}{9}$	$\frac{1}{3}$	1	3	9

b. Since $f(x)$ is the inverse of $g(x)$, plot the ordered pairs $(g(x), x)$ and connect them with a smooth curve to graph the logarithmic function $f(x) = \log_3 x$.

c. Notice that the graphs are reflections of each other in the line $y = x$.

SKILL ✔ **EXERCISE 27**

$y = b^x$	$y = \log_b x$
$D = \mathbb{R}; R = (0, \infty)$	$D = (0, \infty); R = \mathbb{R}$
HA: $y = 0$	VA: $x = 0$
no x-intercept; y-intercept: $(0, 1)$	no y-intercept; x-intercept: $(1, 0)$
contains $(1, b)$ and $\left(-1, \frac{1}{b}\right)$	contains $(b, 1)$ and $\left(\frac{1}{b}, -1\right)$

Instruct the students to practice evaluating the natural logarithmic expressions in Example 4. Be sure they understand how to interpret parts *a*, *b*, and *f* without the use of a calculator. Share that natural logarithms are often used in scientific and financial applications.

One-on-One A student may question the use of *ln* for natural logs, wondering why *nl* is not used instead. Many believe *ln* originates from Euler's use of the French phrase *le logarithme naturel*. An Internet search presents several other theories.

Example 5 shows how a logarithmic function can be sketched by graphing the inverse of the function and reflecting it across $y = x$. Explain that logarithmic and exponential functions can readily be converted between equivalent forms because they are inverse functions. An alternate approach to Example 5 is to convert $f(x)$ to exponential form, $x = 3^y$, and create a table using y as the independent variable.

Discuss the transformations in Example 6. Example 7 illustrates the advantage of using logarithmic functions when representing large quantities or comparing vastly different quantities.

Motivational Idea Explain that the slide rule (invented in 1622) was a common calculating tool used by mathematicians and scientists before computers. The slide rule is based on logarithmic scales. Slide rules were used in high school math and science classes until the 1970s, when calculators became common. Prior to this, students often had to look up logs in tables in the back of their math textbooks. Use the Internet keyword search *virtual slide rule* to find examples.

The graphs of other basic logarithmic functions with $b > 1$ have similar characteristics, while logarithmic functions with $0 < b < 1$ are rare. The graphs of logarithmic functions of the form $g(x) = a \log_b (x - h) + k$ can be drawn as transformations of the parent function $f(x) = \log_b x$.

Example 6　Graphing Transformed Logarithmic Functions

Describe the transformation of $f(x) = \log_3 x$ that produces each function. Then sketch both the parent function and the transformed function on the same coordinate plane.

a. $g(x) = \log_3 (x + 2)$　　**b.** $h(x) = \log_3 x + 2$　　**c.** $j(x) = -2 \log_3 x$

Answer

a. $f(x)$ translated 2 units left　**b.** $f(x)$ translated 2 units up　**c.** $f(x)$ reflected in the x-axis and stretched vertically by a factor of 2

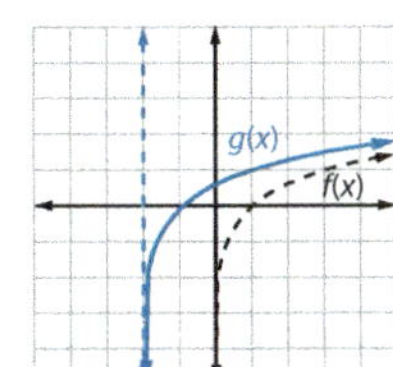 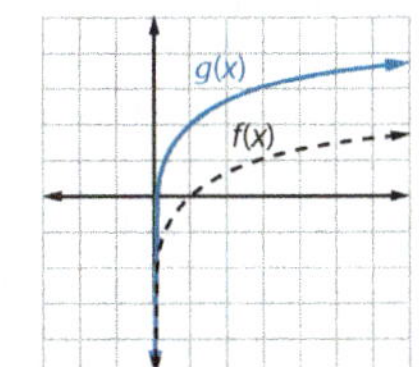 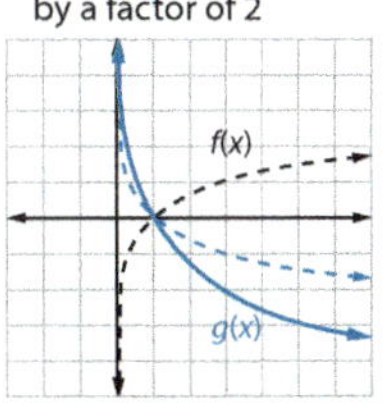

—— SKILL ✔ **EXERCISE 31**

It can be difficult to measure quantities that vary by extremely large amounts, such as the intensity of sound, the amplitude of seismic waves, or the concentration of ions in acids and bases. Logarithmic measures such as decibels, the Richter scale, pH, or pOH are used to describe the vast differences between measured quantities in terms of powers of ten or some other base.

Example 7　Applying a Logarithmic Measure

The Richter magnitude scale uses a logarithmic function, $R = \log \frac{A}{A_0}$, to compare the amplitude of an earthquake, A, to a small reference amplitude, A_0.

a. What does an earthquake measure on the Richter scale if its amplitude is 10,000 times the small reference amplitude?

b. By what factor was A_0 magnified in the 1964 Alaskan earthquake, which measured 9.2 on the Richter scale?

Answer

a. $R = \log \dfrac{(10{,}000\,A_0)}{A_0}$　　　　1. Find R when $A = 10{,}000 A_0$.

$ = \log 10{,}000 = \log 10^4 = 4.0$

b. $R = \log \dfrac{A}{A_0} = 9.2$　　　　1. Substitute for R.

$\dfrac{A}{A_0} = 10^{9.2}$　　　　2. Convert to exponential form.

$A = 10^{9.2} A_0$　　　　3. Solve for A in terms of A_0.

The earthquake's amplitude was $10^{9.2}$ (about 1.6 billion) times the reference amplitude.

—— SKILL ✔ **EXERCISE 39**

TIPS

Ex. 20–21, 25–26 These expressions are evaluated using a calculator.

Ex. 30 Consider using online graphing technology such as the Desmos® graphing calculator, where $\log_a$ is found using the misc tab under the functions menu.

Assignments

- **Minimum:** 1–3(a, c), 4–7, 9, 11–13, 16–17, 19–20, 22–25, 27, 32–33, 35–36, 38–39, 43, 47, 49, 52, 55–57
- **Standard:** 1–4(b, c); 6–30 even; 31–37 odd; 38, 40–43, 47–48; 49–57 odd
- **Extended:** 1–4(b, d); 5, 8, 10–11, 14, 16–17, 20, 23, 26, 29–30, 32–33, 35, 37–38, 41, 43, 46–49; 52–58 even

Assessment

- Quiz 3A covers Sections 3.1–3.2.

Solutions

❭ A. Exercises

5. $\log_2 2^8 = y$　　　　**6.** $\log_4 4^3 = y$
$\ 2^y = 2^8$　　　　$\ 4^y = 4^3$
$\ \ \ y = 8$　　　　$\ \ \ y = 3$

7. $\log_2 2^{-3} = y$　　　**8.** $\log_6 6^{-4} = y$
$\ 2^y = 2^{-3}$　　　$\ 6^y = 6^{-4}$
$\ \ \ y = -3$　　　$\ \ \ y = -4$

9. $\log_7 7^{\frac{1}{2}} = y$　　**10.** $\log_2 \left(2^2\right)^{\frac{1}{3}} = y$
$\ 7^y = 7^{\frac{1}{2}}$　　$\ 2^y = 2^{\frac{2}{3}}$
$\ \ \ y = \dfrac{1}{2}$　　$\ \ \ y = \dfrac{2}{3}$

18. $\log 10^5 = 5$　　　**19.** $\log 10^{-4} = -4$

23. $\ln e^{-7} = -7$　　　**24.** $\ln e^{\frac{2}{5}} = \dfrac{2}{5}$

❭ B. Exercises

27.

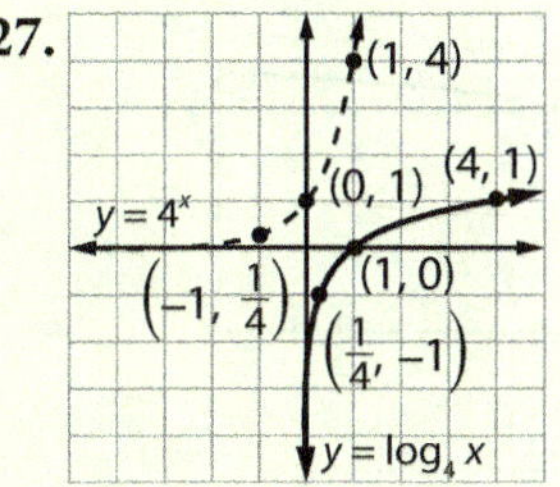

28.

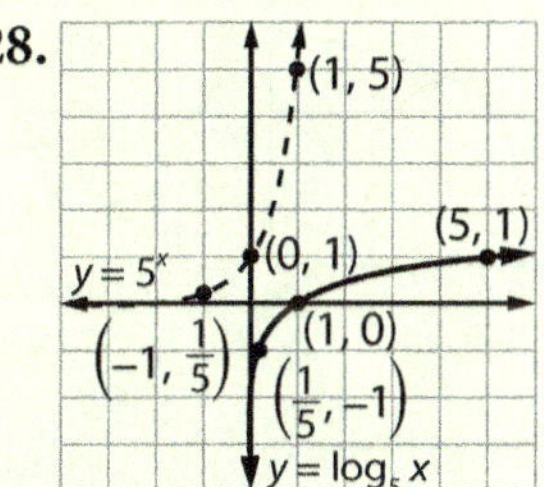

29.

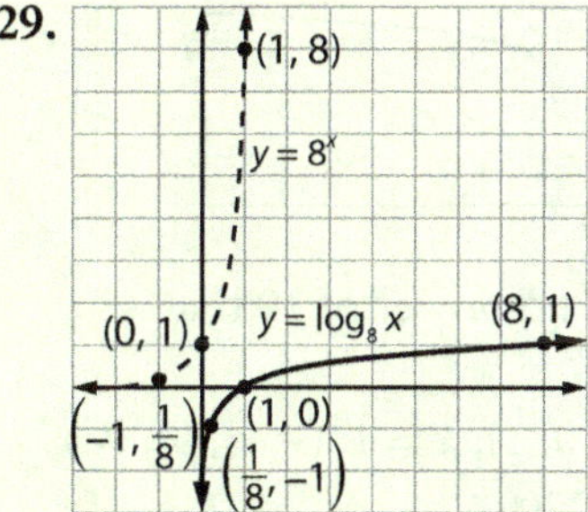

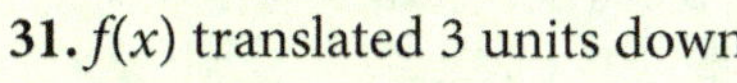

31. $f(x)$ translated 3 units down

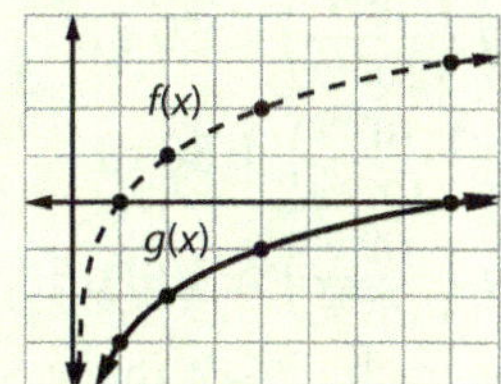

32. $f(x)$ translated 2 units left and 4 units up

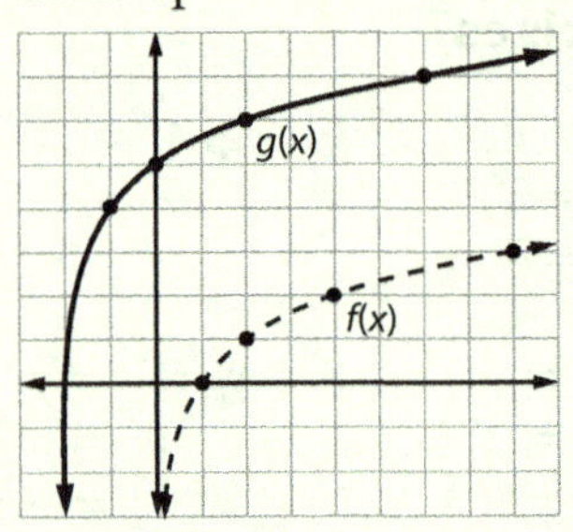

33. $f(x)$ reflected in the x-axis and vertically shrunk by a factor of $\frac{1}{2}$

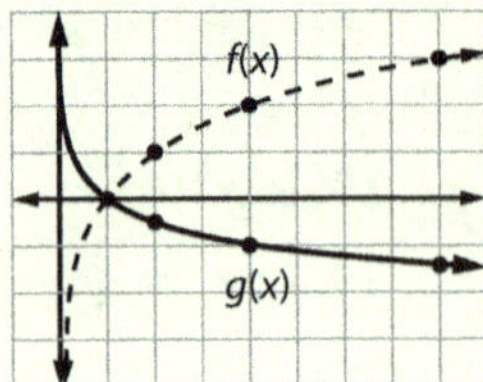

34.

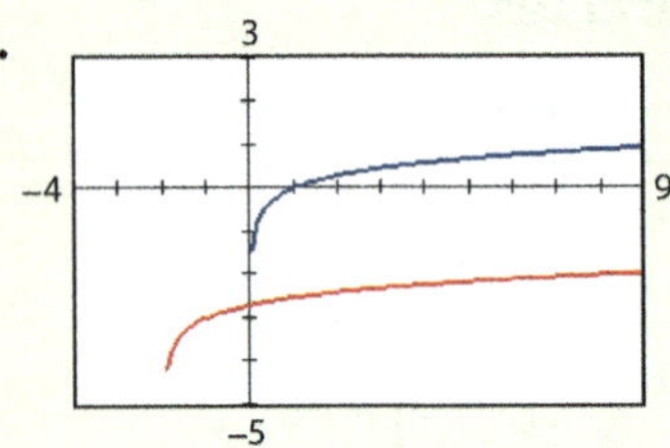

35.

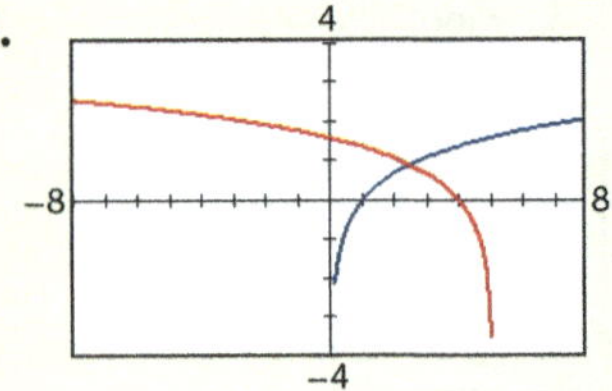

36.

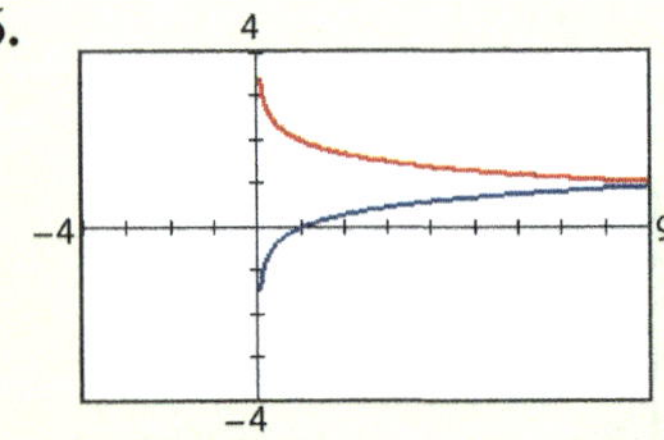

37. There is no power of 10 that equals 0.

38. $(1, \infty)$; $(0, 1)$; $x = 1$; $(-\infty, 0)$; yes

39. $R = \log \dfrac{1200\,A_0}{A_0} = \log 1200 \approx 3.1$

40a. $L_{nc} = 10 \log \left(\dfrac{10^{-7}}{10^{-12}} \right) \approx 50$ dB

$L_{vc} = 10 \log \left(\dfrac{10^{-4}}{10^{-12}} \right) \approx 80$ dB

$L_{so} = 10 \log \left(\dfrac{10^{-2}}{10^{-12}} \right) \approx 100$ dB

$L_{ch} = 10 \log \left(\dfrac{1}{10^{-12}} \right) \approx 120$ dB

41a. $f(0) = 78 - 17 \log (0 + 1) \approx 78$

41b. $f(2) = 78 - 17 \log (2 + 1) \approx 70$

A. Exercises

1. Convert to logarithmic form.
   a. $5^3 = 125$ $\log_5 125 = 3$ b. $8^2 = 64$ $\log_8 64 = 2$
   c. $2^{-4} = \frac{1}{16}$ $\log_2 \frac{1}{16} = -4$ d. $\left(\frac{1}{3}\right)^2 = \frac{1}{9}$ $\log_{\frac{1}{3}} \frac{1}{9} = 2$

2. Convert to exponential form.
   a. $\log_2 8 = 3$ $2^3 = 8$ b. $\log_5 25 = 2$ $5^2 = 25$
   c. $\log_{12} 144 = 2$ $12^2 = 144$ d. $\log_{\frac{1}{4}} \frac{1}{16} = 2$ $\left(\frac{1}{4}\right)^2 = \frac{1}{16}$

3. Convert to exponential form.
   a. $\log 1 = 0$ $10^0 = 1$ b. $\log 1000 = 3$ $10^3 = 1000$
   c. $\log 7 \approx 0.8451$ $10^{0.8451} \approx 7$ d. $\log 0.01 = -2$ $10^{-2} = 0.01$

4. Convert to exponential form.
   a. $\ln 1 = 0$ $e^0 = 1$ b. $\ln 5 \approx 1.6094$ $e^{1.6094} \approx 5$
   c. $\ln e = 1$ $e^1 = e$ d. $\ln \frac{1}{e} = -1$ $e^{-1} = \frac{1}{e}$

Evaluate each logarithm.

5. $\log_2 256$ 8
6. $\log_4 64$ 3
7. $\log_2 \frac{1}{8}$ -3
8. $\log_6 \frac{1}{1296}$ -4
9. $\log_7 \sqrt{7}$ $\frac{1}{2}$
10. $\log_2 \sqrt[3]{4}$ $\frac{2}{3}$

Use the properties of logarithms to evaluate each expression.

11. $\log_4 4$ 1
12. $\log_5 1$ 0
13. $2^{\log_2 \frac{1}{4}}$ $\frac{1}{4}$
14. $5^{\log_5 2}$ 2
15. $10^{\log 0.001}$ 0.001
16. $e^{\ln 1}$ 1

Evaluate each common log. Round to the nearest ten thousandth if necessary.

17. $\log 10$ 1
18. $\log 100{,}000$ 5
19. $\log 0.0001$ -4
20. $\log 25$ 1.3979
21. $\log e$ 0.4343

Evaluate each natural log. Round to the nearest ten thousandth if necessary.

22. $\ln 1$ 0
23. $\ln \frac{1}{e^7}$ -7
24. $\ln \sqrt[5]{e^2}$ $\frac{2}{5}$
25. $\ln 100$ 4.6052
26. $\ln \frac{2}{3}$ -0.4055

B. Exercises

Graph each logarithmic function. Then state its domain and range.

27. $y = \log_4 x$ $D = (0, \infty);$ $R = \mathbb{R}$
28. $y = \log_5 x$ $D = (0, \infty);$ $R = \mathbb{R}$
29. $y = \log_8 x$ $D = (0, \infty); R = \mathbb{R}$

30. The growth is slower for higher values of b.

41c. $f(6) = 78 - 17 \log (6 + 1) \approx 64$

C. Exercises

42. reflected in the x-axis, vertically shrunk by a factor of $\frac{1}{2}$, and translated 4 units left
$D = (-4, \infty);$ $R = \mathbb{R};$ VA: $x = -4$
x-int.: $(-3, 0);$ y-int.: $(0, -1)$
$\lim\limits_{x \to \infty} g(x) = -\infty,$ $\lim\limits_{x \to -4^+} g(x) = \infty$

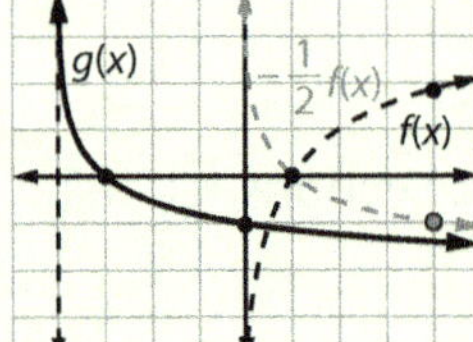

---

30. **Analyze:** Compare the graphs of $y = \log_2 x$, $y = \ln x$, $y = \log x$, and your graphs from exercises 21–23. Describe the rate of growth of $f(x) = \log_b x$ beyond its x-intercept as the value of b increases.

Describe each transformation of $f(x) = \log_2 x$. Then use a graph of $f(x)$ to sketch the graph of $g(x)$.

31. $g(x) = \log_2 x - 3$ 32. $g(x) = \log_2 (x + 2) + 4$
33. $g(x) = -\frac{1}{2} \log_2 x$

Write a function rule for $g(x)$, the described transformation of $f(x)$. Then confirm your answer by graphing $f(x)$ and $g(x)$ in the same window. $g(x) = \log (x + 2) - 3$

34. $f(x) = \log x$ translated 2 units left and 3 units down
35. $f(x) = \ln x$ translated 5 units left and then reflected in the y-axis $g(x) = \ln (5 - x)$
36. $f(x) = \log x$ reflected in the x-axis and translated 2 units up $g(x) = -\log x + 2$

37. **Explain:** Why is 0 not in the domain of $f(x) = \log x$?

38. Specify the intervals of x for which $f(x) = \ln x$ is (a) positive, (b) negative, (c) zero, and (d) undefined. Are these intervals the same for $f(x) = \log x$?

39. Example 7 illustrates how the Richter scale is used to compare earthquakes. Earthquakes with a magnitude less than 3.0 are often not felt, while the most severe earthquakes may measure 7.0 or higher.
   Use $R = \log \dfrac{A}{A_0}$ to find the magnitude of the 2017 earthquake near Brenas, Puerto Rico, if its amplitude was 1200 times as great as A_0.   3.1

40. **Model:** Prolonged exposure to sounds greater than 85 decibels (dB) is known to contribute to noise-induced hearing loss. The loudness of a sound (in decibels) is calculated using $L = 10 \log \dfrac{I}{I_0}$, where I is the sound's intensity (in watts per square meter) and I_0 is the intensity of sound at the threshold of hearing.
   a. Find the missing measures to the nearest decibel.

Sound Source	Intensity (W/m²)	Loudness (dB)
threshold of hearing	10^{-12}	0
normal conversation	10^{-7}	50
vacuum cleaner	10^{-4}	80
symphonic orchestra	10^{-2}	100
car horn	10^{0}	120

   b. Prolonged exposure to which of the listed sound sources can contribute to hearing loss?   orchestra, car horn

43. $g(x) = \log_4 (-x + 4) \Rightarrow$ translated 4 units left, then reflected in the y-axis, or $g(x) = \log_4 -(x - 4) \Rightarrow$ reflected in the y-axis, then translated 4 units right
$D = (-\infty, 4);$ $R = \mathbb{R};$ VA: $x = 4$
x-int.: $(3, 0);$ y-int.: $(0, \log 4)$
$\approx (0, 0.60);$ $\lim\limits_{x \to -\infty} g(x) = \infty,$
$\lim\limits_{x \to 4^-} g(x) = -\infty$

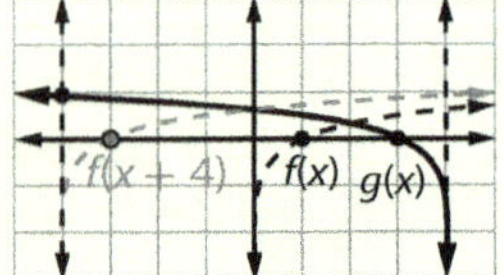

41. Students at Smithville High School took their statistics exam on January 5. Researchers found that when students took an equivalent exam t months later without studying, the average score could be modeled by the function $f(t) = 78 - 17 \log (t + 1)$ where $0 \le t \le 12$. Use the model to estimate the following average scores.

  a. the score on the original exam 78

  b. the score after 2 months 70

  c. the score after 6 months 64

❯ C. Exercises

Describe $g(x)$ as a transformation of $f(x)$, then use the graph of $f(x)$ to sketch the graph of $g(x)$. State the transformed function's domain and range, any asymptotes and intercepts, and the end behavior.

42. $f(x) = \log_2 x$ and
$g(x) = -\frac{1}{2} \log_2 (x + 4)$

43. $f(x) = \log x$ and
$g(x) = \log (4 - x)$

44. $f(x) = \ln x$ and
$g(x) = 2 \ln (-x)$

45. Explain: Determine whether the following statements are true or false. Justify your answers algebraically.

  a. $e^{\log 10^e} = e^e$ true

  b. $\log_2 \frac{1}{2\sqrt{2}} = \frac{4}{3}$ false

  c. $\log \frac{1}{\sqrt{0.00001}} = \frac{5}{2}$ true

46. Compare: Describe the relationship between the graphs of each pair of functions. Verify your answer algebraically.

  a. $f(x) = \log_2 x$ and $g(x) = \log_{\frac{1}{2}} x$

  b. $g(x) = \log_{\frac{1}{2}} x$ and $h(x) = -\log_2 x$

47. The number of months M required to repay a loan balance B at an annual percentage rate r with monthly payments of P is modeled by the equation

$$M = -\frac{\log \left(1 - \frac{rB}{12P}\right)}{\log \left(1 + \frac{r}{12}\right)}.$$ Determine the number of months required to repay a \$20,000 car loan at 6% APR for each payment amount.

P	M
\$300	82
\$350	68
\$400	58
\$450	51
\$500	45

48. The number of years T that it takes to double an investment at a given annual percentage yield is given by the formula $T = \frac{\ln 2}{\ln (1 + APY)}$. How many years will it take Cathy to double \$100,000 invested in a certificate of deposit with each annual percentage yield?

  a. 5% APY ≈ 14.2 yr

  b. 3% APY ≈ 23.4 yr

  c. 1.5% APY ≈ 46.6 yr

46a. The graphs are reflections of each other across the x-axis.

46b. The graphs are the same.

CUMULATIVE REVIEW

Simplify each expression. [Algebra]

49. $3^x \cdot 81^2$ 3^{x+8}

50. $2^x \cdot 16^{x+1}$ 2^{5x+4}

51. $\frac{128^{x+2}}{4}$ 2^{7x+12}

Solve. [2.1]

52. $\sqrt{2x + 5} = \sqrt{3}$ $x = -1$

53. $\sqrt[4]{\frac{x}{2}} - 5 = -2$ $x = 162$

54. $\sqrt{3x + 4} = x$ $x = 4$ (–1 is extraneous.)

55. Which of the following is true if $(x + 2)$ is a factor of $p(x)$? [2.3] D

  A. $x - 2$ is a factor of $p(x)$.

  B. 2 is a zero of $p(x)$.

  C. The degree of $p(x)$ is at least 2.

  D. $p(-2) = 0$

  E. none of these

56. Describe the possible number of positive and negative zeros of $p(x) = 2x^4 + 5x^3 - 2x^2 + x - 8$. [2.4] B

  A. 1 positive zero; 1 or 3 negative zeros

  B. 1 or 3 positive zeros; 1 negative zero

  C. 0, 2, or 4 positive zeros; 0 negative zeros

  D. 3 positive zeros; 1 or 3 negative zeros

  E. none of these

57. Which of the following functions represents exponential growth? [3.1] C

  A. $f(x) = 3 \cdot \left(\frac{1}{5}\right)^x$

  B. $g(x) = 3 \cdot (-5)^x$

  C. $h(x) = 3 \cdot 5^x$

  D. $j(x) = 3 \cdot 5^{-x}$

  E. none of these

58. If \$20,000 is invested at 3% APR, which of the following represents the account balance after 10 yr compounding monthly? [3.1] C

  A. \$26,878.33

  B. \$26,970.46

  C. \$26,987.07

  D. \$26,997.18

  E. none of these

46b. The functions can be represented by the same exponential equations since $g(x) = \log_2 x$ is equivalent to $\left(\frac{1}{2}\right)^y = x$ or $2^{-y} = x$ and $h(x) = -\log_2 x$ or $-y = \log_2 x$ is equivalent to $2^{-y} = x$.

47. Enter $Y_1 = -\dfrac{\log \left(1 - \frac{(0.06)(20,000)}{12P}\right)}{\log \left(1 + \frac{0.06}{12}\right)}$ or

$Y_1 = -\dfrac{\log \left(1 - \frac{100}{P}\right)}{\log 1.005}$ and use [TBLSET] to set TblStart to 300 and DTbl = 50.

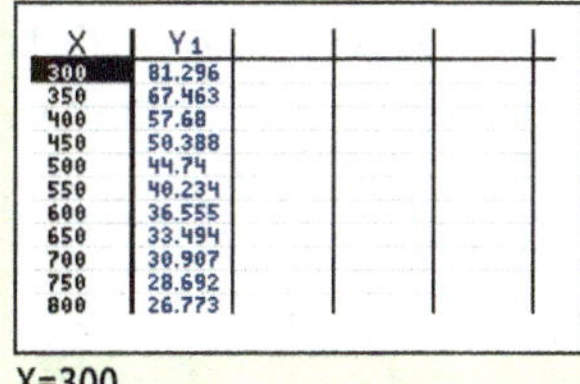

X=300

Decimals are rounded up due to the need for a final payment.

48a. $T = \dfrac{\ln 2}{\ln (1 + 0.05)} \approx 14.2$ yr or 14 yr 2 mo

48b. $T = \dfrac{\ln 2}{\ln (1 + 0.03)} \approx 23.4$ yr or 23 yr 5 mo

48c. $T = \dfrac{\ln 2}{\ln (1 + 0.015)} \approx 46.6$ yr or 46 yr 7 mo

❯ Cumulative Review

49. $3^x \cdot \left(3^4\right)^2 = 3^x \cdot 3^8 = 3^{x+8}$

50. $2^x \cdot \left(2^4\right)^{x+1} = 2^x \cdot 2^{4x+4} = 2^{5x+4}$

51. $\dfrac{\left(2^7\right)^{x+2}}{2^2} = 2^{7x+14} \cdot 2^{-2} = 2^{7x+12}$

52. $2x + 5 = 3$
$2x = -2$
$x = -1$

53. $\left(\sqrt[4]{\frac{x}{2}}\right)^4 = 3^4$
$\frac{x}{2} = 81$
$x = 162$

54. $3x + 4 = x^2$
$x^2 - 3x - 4 = 0$
$(x - 4)(x + 1) = 0$
$x = 4, -1$ (extraneous)

56. $p(x)$ has 3 sign changes, and $p(-x) = 2x^4 - 5x^3 - 2x^2 - x - 8$ has 1 sign change.

57. $f(x)$ and $j(x)$ represent exponential decay; $g(x)$ is not an exponential function.

58. $A(10) = \$20,000 \left(1 + \frac{0.03}{12}\right)^{12(10)}$
$\approx \$26,987.07$

---

44. reflected in the y-axis and vertically stretched by a factor of 2
$D = (-\infty, 0)$; $R = \mathbb{R}$; VA: $x = 0$
x-int.: $(-1, 0)$; y-int.: none
$\lim_{x \to -\infty} g(x) = -\infty$, $\lim_{x \to 0^-} g(x) = \infty$

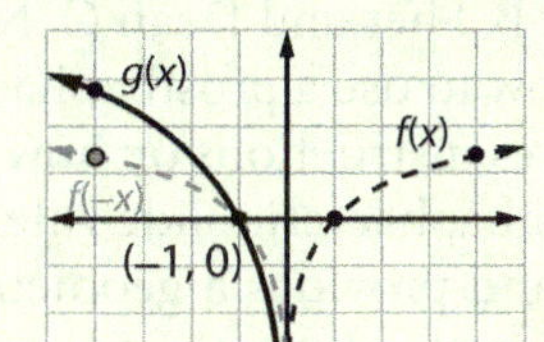

45a. Since $\log 10^e = e$ (Basic Property #3), $e^{\log 10^e} = e^e$.

45b. $\log_2 \frac{1}{2\sqrt[3]{2}} = \log_2 \frac{1}{\sqrt[3]{2^4}} = \log_2 \frac{1}{2^{\frac{4}{3}}}$
$= \log_2 2^{-\frac{4}{3}} = -\frac{4}{3}$

45c. $\log \dfrac{1}{\left(10^{-5}\right)^{\frac{1}{2}}} = \log \dfrac{1}{(10)^{-\frac{5}{2}}} = \log 10^{\frac{5}{2}} = \frac{5}{2}$

46a. $f(x) = \log_2 x$ is equivalent to $2^y = x$, and $g(x) = \log_2 x$ is equivalent to $\left(\frac{1}{2}\right)^y = x$ or $2^{-y} = x$. The y-values associated with any x-value are opposites. Therefore, the graphs are reflections of each other across the x-axis.

Historical Connection

Objectives

1. To summarize the development of Napier's logarithms
2. To explain the usefulness of logarithms as a tool
3. To evaluate Napier's use of math in *A Plaine Discovery*

Answers

1. John Napier and Henry Briggs
2. a method of converting multiplication to addition by means of a trigonometric equation
3. the book of Revelation
4. They greatly reduced the time and errors associated with calculating products, quotients, powers, and roots of large numbers.
5. No; Napier expected mathematics and reasoning to help him discover truth that Scripture clearly indicates is not able to be determined by man. There are many truths that cannot be determined mathematically. Truths revealed in Scripture should be accepted on its authority as God's Word.

NAPIER'S ARTIFICIAL NUMBERS

Imagine an invention that would allow you to work twice as fast as you do now. That would mean two-and-a-half day school weeks, a two-year college degree, and twenty-hour work weeks!

The logarithms proposed by John Napier seemed almost this revolutionary to the mathematicians and astronomers who lauded their introduction. Johannes Kepler, who discovered laws describing the elliptical orbits of planets, even dedicated one of his published works to Napier in praise of his logarithms. Such enthusiasm is understandable, considering the tedious and error-prone nature of multiplying, dividing, and extracting roots of very large numbers by hand.

Napier did not "stumble upon" his discovery, but rather set out purposefully to ease the burden of calculating products, quotients, powers, and roots of large numbers. While developing his ideas and tables over the course of twenty years, Napier initially called his logarithms "artificial numbers" but coined the term *logarithms* from the Greek words *logos* (ratio) and *arithmos* (number) before publishing his results in 1614.

Rather than thinking in terms of the inverses of exponential functions, Napier developed logarithms by noting the relationship between arithmetic and geometric sequences. In fact, the historian Florian Cajori says, "It is one of the greatest curiosities of the history of science that Napier constructed logarithms before exponents were used."

Before Napier invented his logarithms, the method of prosthaphaeresis was used to convert multiplication to addition by means of a trigonometric equation. The Danish astronomer Tycho Brahe and his assistant Paul Wittich simplified calculations with the formula $2 \sin A \sin B = \cos (A - B) - \cos (A + B)$.

While history remembers Napier for his logarithms, his first book, published in 1593, was a commentary titled *A Plaine Discovery of the whole Revelation of Saint John*. He believed this was his most important work and the one for which he would be remembered. Using mathematics and reasoning, he constructed a chronology of the events in Revelation and predicted that the Second Coming of Christ and the destruction of the world would occur in 1688. Napier dedicated his book to the Scottish ruler King James VI (who would later commission the King James Version of the Bible), but the dedication was dropped after the king ignored Napier's encouragement to purge the court of ungodly influences.

The English mathematician Henry Briggs was so fascinated by Napier's logarithms that he made a difficult four-day journey to meet with him personally in 1615. Briggs and Napier eventually agreed that the most useful logarithmic system would be based on the powers of ten, where log 10 = 1 and log 1 = 0. Due to Napier's failing health, the completion of the resulting logarithmic tables was left to Briggs, who published them in 1617. These tables endured as the primary means of calculation for more than three and a half centuries until they and calculating devices that used them (such as slide rules) were replaced by computers and handheld calculators.

COMPREHENSION CHECK

1. Name the two men most responsible for developing logarithms.
2. What is prosthaphaeresis?
3. What was the subject of Napier's first published book?
4. Why were logarithms so readily accepted and frequently utilized for hundreds of years?
5. **Discuss:** Referencing Matthew 24:36, evaluate Napier's attempt to decode a mathematical system in Revelation to predict the date of Christ's return. Did he approach his study of Revelation appropriately? Why or why not?

PRESENTATION

The history surrounding John Napier and his invention of logarithms provides a treasure trove of material to supplement lessons in this chapter.

Nicknamed "Marvelous Merchiston" (he was the 8th Laird of Merchiston in Scotland), Napier was not a mathematician per se, but a well-educated man who pursued many interests, including theology, agriculture, and alchemy. He lived in a time of religious wars between Catholics and Protestants in Europe and believed that these wars would culminate in the return of the Lord. He considered it his duty to share the special understanding of Revelation that he believed he had received from God.

Use the Internet keyword search *Alexander Corrigan thesis on Napier PDF* to find a thesis containing information on John Napier and his book *A Plaine Discovery*. The Internet keyword search *European religious wars* provides information about Napier's times, including information about the Spanish Armada, the St. Bartholomew's Day Massacre, and the Spanish Blanks.

Another of Napier's inventions was "Napier's Bones," a set of rods with powers of numbers printed on them that simplified multiplication problems.

The astronomer Tycho Brahe used prosthaphaeresis extensively to simplify calculations. While this method was a big step forward, it was still tedious.

"The Prosthaphaeretic Slide Rule," an article by David B. Sher and Dean C. Nataro, explains how to use a prosthaphaeretic formula, gives instructions on how to construct a prosthaphaeretic slide rule (a modern idea), and provides a geometric proof of the prosthaphaeretic formula (*Mathematics and Computer Education* 38 (Winter 2004): 37–43).

3.3 Properties of Logarithms

A logarithmic function relates the frequency of played notes (what we hear as pitch) to their relative position on the keyboard.

The properties of logarithms have historically been used to simplify complicated products, quotients, powers, and roots. While calculators now assist us with these tasks, these properties still have other practical applications. Because logarithms are the exponents when numbers are expressed as a power of the given base, their properties follow directly from the properties of exponents.

Property	of Logarithms*	of Exponents
Equality	$\log_b m = \log_b n$ iff $m = n$	$b^m = b^n$ iff $m = n$
Product	$\log_b mn = \log_b m + \log_b n$	$b^m \cdot b^n = b^{m+n}$
Quotient	$\log_b \frac{m}{n} = \log_b m - \log_b n$	$\frac{b^m}{b^n} = b^{m-n}$
Power	$\log_b m^p = p \log_b m$	$(b^m)^p = b^{mp}$

*where $m > 0$ and $n > 0$

The Product Property of Logarithms is derived from the Product Property of Exponents.

Let $\log_b m = x$ and $\log_b n = y$.

$b^x = m$ and $b^y = n$	Convert to exponential form.
$mn = b^x \cdot b^y$	Substitute.
$mn = b^{x+y}$	Product Property of Exponents
$\log_b mn = x + y$	Convert to logarithmic form.
$\log_b mn = \log_b m + \log_b n$	Substitute.

These properties can be used to expand logarithmic expressions, converting multiplication and division into addition and subtraction and converting the evaluation of powers and roots into multiplication and division.

Example 1 Expanding a Logarithmic Expression

Expand each logarithmic expression.

a. $\log \frac{a^2 b}{c^4}$ **b.** $\ln \frac{\sqrt[3]{2c-1}}{b^2}$

Answer

a.
$$\log \frac{a^2 b}{c^4} = \log a^2 b - \log c^4$$
$$= \log a^2 + \log b - \log c^4$$
$$= 2 \log a + \log b - 4 \log c$$

1. Quotient Property
2. Product Property
3. Power Property

b.
$$\ln \frac{\sqrt[3]{2c-1}}{b^2} = \ln (2c-1)^{\frac{1}{3}} - \ln b^2$$
$$= \frac{1}{3} \ln (2c-1) - 2 \ln b$$

1. Quotient Property
2. Product Property; Note that $\ln (2c-1)$ cannot be further expanded.

SKILL ✔ EXERCISE 5

After completing this section, you will be able to

- evaluate, expand, and condense logarithmic expressions.
- use the change of base formula.
- model and solve real-world problems using logarithmic functions.

Before devising his logarithms, John Napier designed a system of rods imprinted with numbers that could be arranged in sequence to aid in multiplication. These popular seventeenth-century computing devices were called "Napier's Bones" because of their appearance.

KEYWORD SEARCH

Napier's Bones 🔍

Objectives

1. To use the properties of logarithms to evaluate, simplify, or expand logarithmic expressions

2. To use the change of base formula to rewrite and evaluate logarithmic expressions

3. To use logarithmic functions to model and solve real-world problems

Flash

A piano keyboard is divided into 12-note octaves. A_4 (the A above middle C) has a frequency of 440 Hertz (Hz). To find the frequency of consecutive notes, one can multiply or divide the frequency by the 12th root of 2. For example, the frequency of $A\sharp_4$ is $440 \cdot \sqrt[12]{2} \approx 466.16$ Hz. Using this information, a function can be written to determine the frequency based upon the key number.

Vocabulary

change of base formula

PRESENTATION

Lesson Opener

Simplify.

1. $b^x \cdot b^y$ b^{x+y}

2. $\frac{b^x}{b^y}$ b^{x-y}

3. $(b^x)^y$ b^{xy}

Convert to exponential form.

4. $\log 100 = 2$ $10^2 = 100$

5. $y = \log N$ $10^y = N$

Consider explaining that before calculators, logarithms were used to perform difficult operations such as multiplication or long division with large quantities. Remind the students that the logarithm of a number is the exponent when a quantity is expressed as a power of the given base. Instead of multiplying or dividing, mathematicians would essentially add or subtract exponents of powers with like bases.

Motivational Idea While logarithms have historically been used to multiply and divide very large or very small numbers, they still have a variety of essential applications, such as modeling radioactive decay, bacterial growth, population growth, financial information, and pH measurements.

Introduce the properties of logarithms, relating each property to the corresponding property of exponents. The Equality Property of Logarithms is sometimes called the One-to-One Property. The derivation of the Product Property can help the students complete exercises 41–42. Examples 1–3 illustrate applications of the properties of logarithms. In Example 1, note that $\ln (2c - 1)$ cannot be expanded.

Common Student Error A student may incorrectly convert $\log (a - b)$ to $\frac{\log a}{\log b}$, confusing it with $\log a - \log b = \frac{\log a}{\log b}$.

Students may be surprised if they try to verify Example 3a by evaluating $7^{1.2950} \approx 12.4280$. This approximation

Reading and Writing Mathematics

Answer each question in your own words. A sample answer is given for each.

1. What is a logarithm?

A logarithm is the power to which a specified base must be raised in order to obtain the stated value.

2. What is the base of a logarithm?

The base of a logarithm is the number which is raised to a power (the logarithm) in order to obtain the stated value.

3. What is the difference between a common log and a natural log?

A common log is the power to which 10 is raised to obtain a desired value. A natural log is the power to which the number e is raised to obtain the stated value.

Additional Exercises

Expand each logarithmic expression. Assume all variables are positive values.

1. $\ln 5y^3$ $\ln 5 + 3 \ln y$

2. $\log x^3 y^2 z^5$ $3 \log x + 2 \log y + 5 \log z$

3. $\ln \dfrac{x^2 y}{z^3}$ $2 \ln x + \ln y - 3 \ln z$

4. $\log \sqrt[3]{\dfrac{10x^2}{y^2}}$ $\dfrac{1}{3}(1 + 2 \log x - 2 \log y)$

or $\dfrac{1}{3} + \dfrac{2}{3} \log x - \dfrac{2}{3} \log y$

An expression involving logarithms may need to be expressed as a single logarithm.

Example 2   Condensing a Logarithmic Expression

Write each expression as a single logarithm.

a. $3 \log ab - \dfrac{1}{2} \log c$ **b.** $\ln (b + 1) + 5 \ln b$

Answer

a. $3 \log ab - \dfrac{1}{2} \log c$

$= \log (ab)^3 - \log c^{\frac{1}{2}}$ 1. Power Property

$= \log \dfrac{(ab)^3}{\sqrt{c}}$ or $\log \dfrac{a^3 b^3}{\sqrt{c}}$ 2. Quotient Property

b. $\ln (b + 1) + 5 \ln b$

$= \ln (b + 1) + \ln b^5$ 1. Power Property

$= \ln b^5(b + 1)$ or $\ln (b^6 + b^5)$ 2. Product Property

SKILL ✓ EXERCISE 9

These properties allow logarithms to be expressed in terms of other logarithms. This skill is critical when solving logarithmic equations.

Example 3   Applying Properties of Logarithms

Using $\log_7 2 \approx 0.3562$ and $\log_7 3 \approx 0.5646$, find a decimal approximation of each logarithm.

a. $\log_7 12$ **b.** $\log_7 \dfrac{49\sqrt{3}}{2}$

Answer

a. $\log_7 12 = \log_7 (2^2 \cdot 3)$
$= \log_7 2^2 + \log_7 3$
$= 2 \log_7 2 + \log_7 3$
$\approx 2(0.3652) + 0.5646$
$= 1.2950$

b. $\log_7 \dfrac{49\sqrt{3}}{2} = \log_7 7^2 + \log_7 3^{\frac{1}{2}} - \log_7 2$
$= 2 \log_7 7 + \dfrac{1}{2} \log_7 3 - \log_7 2$
$\approx 2(1) + \dfrac{1}{2}(0.5646) - (0.3562)$
$= 1.9261$

SKILL ✓ EXERCISE 23

Working with logarithms with bases other then 10 or e can be difficult since most calculators have only LOG and LN keys. Converting $\log_3 8 = y$ to exponential form and rewriting it in terms of common logs enables us to find a decimal approximation of the logarithm.

Let $\log_3 8 = y$.

$3^y = 8$ Convert to exponential form.

$\log 3^y = \log 8$ Equality Property of Logarithms

$y \log 3 = \log 8$ Power Property of Logarithms

$y = \dfrac{\log 8}{\log 3}$ Multiplication Property of Equality

≈ 1.8928 Evaluate using technology.

Generalizing this process produces a formula that converts a logarithmic expression into one with a more convenient base.

CHANGE OF BASE FORMULA

For positive a, b, and x with $a \neq 1$ and $b \neq 1$, $\log_b x = \dfrac{\log_a x}{\log_a b}$.

Logs of other bases are usually converted to common or natural logs.

$$\log_b x = \frac{\log x}{\log b} \quad \text{or} \quad \log_b x = \frac{\ln x}{\ln b}$$

Example 4 Applying the Change of Base Formula

Find $\log_2 5.89$.

a. using common logs

b. using natural logs

Answer

a. $\log_2 5.89 = \dfrac{\log 5.89}{\log 2}$
≈ 2.5583

b. $\log_2 5.89 = \dfrac{\ln 5.89}{\ln 2}$
≈ 2.5583

Check

Use technology to verify that $2^{2.5583} \approx 5.89$.

SKILL ✔ **EXERCISE 15**

The change of base formula can be used to show that the graph of any logarithmic function $g(x) = \log_b x$ is a vertical stretch or shrink of $f(x) = \ln x$ by a factor of $\dfrac{1}{\ln b}$.

$$g(x) = \log_b x = \frac{\ln x}{\ln b} = \left(\frac{1}{\ln b}\right)\ln x$$

Example 5 Graphing Base b Logarithmic Functions

Describe how the graph of $f(x) = \ln x$ is transformed to obtain the graph of each function.

a. $g(x) = \log_7 x$

b. $h(x) = \log_{\frac{1}{7}} x$

Answer

a. Using the change of base formula, $g(x) = \log_7 x = \dfrac{1}{\ln 7}\ln x$.

Therefore, $g(x)$ is a vertical shrink of $f(x)$ by a factor of $\dfrac{1}{\ln 7} \approx 0.5$.

b. $h(x) = \log_{\frac{1}{7}} x$

$= \dfrac{\ln x}{\ln \frac{1}{7}} = \dfrac{1}{\ln 7^{-1}}\ln x$ change of base formula

$= \dfrac{1}{-\ln 7}\ln x$ Power Property of Logarithms

Therefore, $h(x)$ is a vertical shrink of $f(x)$ by a factor of $\dfrac{1}{\ln 7} \approx 0.5$ that is reflected in the x-axis.

Check

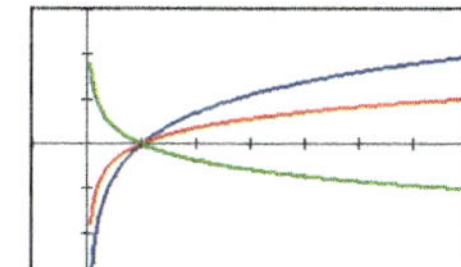

Use technology to graph and compare the functions.
$Y_1 = \ln(X)$
$Y_2 = \ln(X)/\ln(7)$
$Y_3 = \ln(X)/\ln(1/7)$

SKILL ✔ **EXERCISE 33**

Write each expression as a single logarithm.

5. $\log a + 4 \log b$ $\log ab^4$

6. $4 \log M - 2 \log N$ $\log \dfrac{M^4}{N^2}$

7. $\dfrac{1}{2}\log a + \dfrac{1}{4}\log b - \dfrac{1}{2}\log c$ $\log \sqrt[4]{\dfrac{a^2 b}{c^2}}$

Evaluate each logarithm.

8. $\log_3 2.725$ $\dfrac{\log 2.725}{\log 3} \approx 0.9125$

9. $\log_5 28.9$ $\dfrac{\log 28.9}{\log 5} \approx 2.0901$

10. $\log_4 389$ $\dfrac{\log 389}{\log 4} \approx 4.3018$

For Graphing Calculators

Be aware that many calculators have the ability to calculate logarithms of bases other than 10 or e without using the change of base formula.

To evaluate $\log_8 1000$ using a calculator in the TI-84 Plus family, enter MATH, A:logBASE, 8, ▸, 1000, ENTER.

```
log₈(1000)
                   3.321928095
■
```

Some older calculators use the same menu options but reverse the order of entering the function's argument and base: MATH, A:logBASE, (, 1000, ,, 8,), ENTER.

```
logBASE(1000,8)
                   3.321928095
```

Assignments

- **Minimum:** 1, 4, 7, 10, 12–13, 15, 18, 21–23, 25, 28–29, 33, 35–37*a*, 38, 45, 47–49, 52

- **Standard:** 2, 5, 8–9, 12–13, 16, 19, 22–23, 26, 28, 31, 34, 36–38, 40, 46–48, 50, 52, 54

- **Extended:** 3, 6–7, 11–12, 14, 17, 20, 23–25, 28, 32–33, 37–44, 46–48, 52–54

You may want students to research the logarithmic spiral mentioned in the Data Analysis margin note. Jakob Bernoulli called logarithmic spirals miraculous and requested that one be engraved on his tombstone with a Latin inscription that translates, "Though changed, I shall arise the same." Unfortunately, the spiral engraved on his tombstone is an Archimedean spiral.

One-on-One Some students may struggle with logarithms because they are an unfamiliar concept, but many will become more comfortable with them after applying logarithmic properties to expressions. Encourage struggling students by reminding them that they will become more comfortable with logarithms through continued practice.

TIPS

Ex. 21–24 These exercises review the properties of logarithms and also show how logarithms can be evaluated using the logarithms of only prime numbers.

Ex. 37*b* This exercise previews Section 3.4.

Ex. 41–43 These exercises could be done in class in small groups, or the proofs could be written individually and modified after discussion with a classmate.

Solutions

❯ A. Exercises

1. $\log x + \log y^3 = \log x + 3 \log y$

2. $\ln a + \ln b^2 + \ln c = \ln a + 2 \ln b$
$$+ \ln c$$

3. $\ln 7 + \ln x^4 = \ln 7 + 4 \ln x$

4. $\log 3 - \log x^2 = \log 3 - 2 \log x$

5. $\ln a^4 - \ln b^2 = 4 \ln a - 2 \ln b$

6. $\log x^a + \log y^b - \log z$
$$= a \log x + b \log y - \log z$$

7. $\log \dfrac{x^{\frac{3}{5}}}{y} = \log x^{\frac{3}{5}} - \log y$
$$= \frac{3}{5} \log x - \log y$$

8. $\log \left(a^2 b^5\right)^{\frac{1}{3}} = \frac{1}{3} \log a^2 b^5$
$$= \frac{1}{3}(\log a^2 + \log b^5)$$
$$= \frac{1}{3}(2 \log a + 5 \log b)$$
$$\text{or } \frac{2}{3} \log a + \frac{5}{3} \log b$$

9. $\log x + \log y^2 = \log xy^2$

10. $\log x^4 + \log 5^2$
$$= \log x^4 + \log 25 = \log 25x^4$$

11. $\ln y^3 - \ln 3^2 = \ln \dfrac{y^3}{9}$

12. $\ln x^2 + \ln 2^3 - \ln y = \ln \dfrac{8x^2}{y}$

13. $\frac{1}{3}(\log x^2 + \log 1000) = \frac{1}{3} \log (1000x^2)$
$$= \log (1000 x^2)^{\frac{1}{3}} = \log 10 x^{\frac{2}{3}}$$

14. $\frac{1}{5}(\log xy^3) - \log 10^2$
$$= \log \sqrt[5]{xy^3} - \log 100 = \log \dfrac{\sqrt[5]{xy^3}}{100}$$

15. $\log_2 50 = \dfrac{\log 50}{\log 2} \approx 5.6439$

16. $\log_5 175 = \dfrac{\log 175}{\log 5} \approx 3.2091$

17. $\log_8 95.7 = \dfrac{\log 95.7}{\log 8} \approx 2.1935$

18. $\log_2 50 = \dfrac{\ln 50}{\ln 2} \approx 5.6439$

19. $\log_4 212 = \dfrac{\ln 212}{\ln 4} \approx 3.8640$

20. $\log_7 34.75 = \dfrac{\ln 34.75}{\ln 7} \approx 1.8234$

❯ B. Exercises

21. $\log_7 2^{-1} = -\log_7 2 \approx -0.3562$

22. $\log_7 (2 \cdot 5)^2 = 2 (\log_7 2 + \log_7 5)$
$$= 2(0.3562 + 0.8271) \approx 2.3666$$

23. $\log_7 (7 \cdot 5 \cdot 2)^2$
$$= 2(\log_7 7 + \log_7 5 + \log_7 2)$$
$$\approx 2(1 + 0.8271 + 0.3562) \approx 4.3666$$

Note that $\log_{\frac{1}{7}} x = -\log_7 x$ since $Y_3 = \log_{\frac{1}{7}} x$ is a reflection of $Y_2 = \log_7 x$ across the x-axis. The properties of logarithms and the change of base formula can be used to show that when $b > 1$ (and therefore $0 < \frac{1}{b} < 1$), $\log_{\frac{1}{b}} x = -\log_b x$.

Understanding the properties of logs and the change of base formula enables mathematicians to easily work with mathematical models of natural phenomena.

Example 6 Modeling Frequencies of the Piano

The frequency of a note played by a key on the piano is related to the key's position on the keyboard. The number of notes, n, that one key is above another is modeled by the function $n = 12 \log_2 \left(\dfrac{f_2}{f_1}\right)$ where f_2 is the higher frequency and f_1 is the lower frequency.

a. The note C_5 is one octave higher than middle C and has twice its frequency. How many notes above middle C is C_5?

b. If middle C has a frequency of 261.63 Hertz (Hz), how many notes above middle C is A_4, which has a frequency of 440 Hz?

Answer

a. $n = 12 \log_2 \left(\dfrac{2f_1}{f_1}\right) = 12 \log_2 2$ 1. Substitute $2f_1$ for f_2.

$\quad = 12(1) = 12$ notes   2. Apply the fact that $\log_n n = 1$.

b. $n = 12 \log_2 \left(\dfrac{440}{261.63}\right)$ 1. Substitute the frequencies into the function rule.

$\quad = \dfrac{12}{\ln 2} \ln \left(\dfrac{440}{261.63}\right) = 9$ notes   2. Evaluate by applying the change of base formula.

— SKILL ✔ **EXERCISE 37**

❯ A. Exercises

Expand each logarithmic expression. Assume all variables are positive values.

1. $\log xy^3$ **2.** $\ln ab^2c$ **3.** $\ln 7x^4$

4. $\log \dfrac{3}{x^2}$ **5.** $\ln \dfrac{a^4}{b^2}$ **6.** $\log \dfrac{x^a y^b}{z}$

7. $\log \dfrac{\sqrt[5]{x^3}}{y}$ **8.** $\log \sqrt[4]{a^2 b^5}$

Write each expression as a single logarithm.

9. $\log x + 2 \log y$ $\log xy^2$

10. $4 \log x + 2 \log 5$ $\log 25x^4$

11. $3 \ln y - 2 \ln 3$ $\ln \dfrac{y^3}{9}$

12. $2 \ln x + 3 \ln 2 - \ln y$ $\ln \dfrac{8x^2}{y}$

13. $\frac{1}{3}(2 \log x + \log 1000)$ $\log 10x^{\frac{2}{3}}$

14. $\frac{1}{5}(\log x + 3 \log y) - 2 \log 10$ $\log \dfrac{\sqrt[5]{xy^3}}{100}$

Use common logs to evaluate each logarithm. Round answers to the nearest ten thousandth.

15. $\log_2 50$ **16.** $\log_5 175$ **17.** $\log_8 95.7$

Use natural logs to evaluate each logarithm. Round answers to the nearest ten thousandth.

18. $\log_2 50$ **19.** $\log_4 212$ **20.** $\log_7 34.75$

❯ B. Exercises

Use $\log_7 2 \approx 0.3562$, $\log_7 3 \approx 0.5646$, and $\log_7 5 \approx 0.8271$ to find a decimal approximation of each logarithm.

21. $\log_7 \frac{1}{2} \approx -0.3562$ **22.** $\log_7 100 \approx 2.3666$

23. $\log_7 4900 \approx 4.3666$ **24.** $\log_7 \dfrac{5\sqrt{6}}{7} \approx 0.2875$

Expand each logarithmic expression. Assume all variables are positive values.

25. $\log 100 \sqrt[3]{x^2}$ **26.** $\log_3 \sqrt{27xy^2}$

27. $\log_7 \dfrac{49 \sqrt{y}}{x}$ **28.** $\log_2 \dfrac{\sqrt{2x}}{16y}$

Write each expression as a single logarithm.

29. $x \log 3 - 2 \log y + \frac{1}{3} \log z$ $\log \dfrac{3^x \sqrt[3]{z}}{y^2}$

30. $x \ln 3 + y \ln 2 - z \ln 3$ $\ln (2^y \cdot 3^{x-z})$

31. $\ln 2x + 3 \ln 4x^2 - \ln 8x^4$ $\ln 16x^3$

32. $3 \log \frac{3}{2} + 2 \log 6x - 2 \log 3$ $\log \dfrac{27}{2} x^2$

---

24. $\log_7 5 + \log_7 (3 \cdot 2)^{\frac{1}{2}} - \log_7 7$
$$= \log_7 5 + \frac{1}{2}(\log_7 3 + \log_7 2) - \log_7 7$$
$$\approx 0.8271 + \frac{1}{2}(0.5646 + 0.3562) - 1$$
$$= 0.2875$$

25. $\log 10^2 + \log x^{\frac{2}{3}} = 2 + \frac{2}{3} \log x$

26. $\frac{1}{2}(\log_3 3^3 + \log_3 x + \log_3 y^2)$
$$= \frac{1}{2}(3 + \log_3 x + 2 \log_3 y)$$
$$= \frac{3}{2} + \frac{1}{2} \log_3 x + \log_3 y$$

27. $\log_7 7^2 + \log_7 y^{\frac{1}{2}} - \log_7 x$
$$= 2 + \frac{1}{2} \log_7 y - \log_7 x$$

28. $\log_2 \dfrac{(2x)^{\frac{1}{2}}}{(2)^4 y} = \frac{1}{2}(\log_2 2 + \log_2 x)$
$$- (4 \log_2 2 + \log_2 y)$$
$$= -\frac{7}{2} \log_2 2 + \frac{1}{2} \log_2 x - \log_2 y$$
$$= -\frac{7}{2} + \frac{1}{2} \log_2 x - \log_2 y$$

29. $\log 3^x - \log y^2 + \log z^{\frac{1}{3}} = \log \dfrac{3^x \sqrt[3]{z}}{y^2}$

30. $\ln 3^x + \ln 2^y + \ln 3^{-z}$
$$= \ln (3^x \cdot 2^y \cdot 3^{-z}) = \ln (2^y \cdot 3^{x-z})$$

31. $\ln 2x + \ln (4x^2)^3 - \ln 8x^4$
$$= \ln \dfrac{2x(64x^6)}{8x^4} = \ln 16x^3$$

Describe the graph of $g(x)$ as a transformation of $f(x) = \ln x$.

33. $g(x) = \log_{20} x$ **34.** $g(x) = \log_{\frac{1}{3}} x$

35. Rewrite the Richter scale formula, $R = \log \frac{A}{A_0}$, in terms of natural logs. $R = \dfrac{\ln A - \ln A_0}{\ln 10}$

36. The loudness of sound is measured in decibels using the equation $L = 10 \log \frac{I}{I_0}$ where I is the intensity of the sound and I_0 is the intensity of sound at the threshold of hearing, 10^{-12} W/m^2. Write the expanded form of the equation. Then use the equation to find the loudness of a mosquito's buzz, which has an intensity of 10^{-8} W/m^2. $L = 10 \log I + 120;\ L = 40$ dB

37. A composer must understand which combinations of frequencies create a desired sound.

Use $n = 12 \log_2 \left(\frac{f_2}{f_1}\right)$ (introduced in Example 6) to answer the following questions.

   a. A clarinet is playing the F♯ below middle C at a frequency of 370 Hz while a flute is playing a note at a frequency of 1976 Hz. How many notes above the clarinet's note is the flute's note?   29 notes

   b. A bassoon is playing a G_2 with a frequency of 98 Hz. What is the frequency of a clarinet playing two octaves (24 notes) above the bassoon?   392 Hz

Use the following table to answer questions 38–39.

Substance	pH	pOH
vinegar	2.2	11.8
tomato	4.5	9.5
milk	6.6	7.4
distilled water	7.0	7.0
baking soda	8.3	5.7
ammonia	11.0	3.0
lye	13.0	1.0

38. The acidity of a solution is measured on the pH scale by pH $= -\log [\mathrm{H_3O^+}]$ where $[\mathrm{H_3O^+}]$ represents the concentration of hydronium ions (in moles/L).

   a. Which substance listed in the table has $[\mathrm{H_3O^+}] = 6.3 \times 10^{-12}$ moles/L?   ammonia

   b. What is the concentration of hydronium ions in distilled water? $[\mathrm{H_3O^+}] = 10^{-7.0}$ moles/L

   c. What is the concentration of hydronium ions in vinegar? Is this more or less than the concentration of $\mathrm{H_3O^+}$ in distilled water?

   d. Approximately how many times greater is $[\mathrm{H_3O^+}]$ in baking soda than in lye?   50,000 times greater

39. The pOH scale quantifies the basicity of a solution using pOH $= -\log [\mathrm{OH^-}]$ where $[\mathrm{OH^-}]$ represents the concentration of hydroxide ions (in moles/L). The sum of a solution's pH and pOH is 14.

   a. What is the concentration of hydroxide ions in distilled water? $[\mathrm{OH^-}] = 10^{-7.0}$ moles/L

   b. What is the concentration of hydroxide ions in vinegar? Is this more or less than $[\mathrm{OH^-}]$ in distilled water?

   c. Given that the pH of an unknown substance is 3.7, find $[\mathrm{OH^-}]$. 5.01×10^{-11}

   d. Derive an equation relating $[\mathrm{H_3O^+}]$ and $[\mathrm{OH^-}]$.

40. Prove each property of logarithmic functions.

   a. $\log_b \frac{1}{x} = -\log_b x$ **b.** $\log_{\frac{1}{b}} x = -\log_b x$

❯ C. Exercises

41. Prove the Quotient Property of Logarithms: $\log_b \frac{m}{n} = \log_b m - \log_b n$.

42. Prove the Power Property of Logarithms: $\log_b m^p = p \log_b m$.

43. Prove the change of base formula: $\log_b x = \frac{\log_a x}{\log_a b}$.

44. Explain: Use technology to graph $f(x) = \log (x - 2) - \log x$ and $g(x) = \log \left(\frac{x-2}{x}\right)$. Why are the functions not equivalent?

1. $\log x + 3 \log y$

2. $\ln a + 2 \ln b + \ln c$

3. $\ln 7 + 4 \ln x$

4. $\log 3 - 2 \log x$

5. $4 \ln a - 2 \ln b$

6. $a \log x + b \log y - \log z$

7. $\frac{3}{5} \log x - \log y$

8. $\frac{1}{3}(2 \log a + 5 \log b)$

15. 5.6439

16. 3.2091

17. 2.1935

18. 5.6439

19. 3.8640

20. 1.8234

25. $2 + \frac{2}{3} \log x$

26. $\frac{3}{2} + \frac{1}{2} \log_3 x + \log_3 y$

27. $\frac{1}{2} \log_7 y - \log_7 x + 2$

28. $\frac{1}{2} \log_2 x - \log_2 y - \frac{7}{2}$

38c. $[\mathrm{H_3O^+}] = 10^{-2.2}$ moles/L $\approx 6.3 \times 10^{-3}$; more

39b. $[\mathrm{OH^-}] = 10^{-11.8}$ moles/L $\approx 1.6 \times 10^{-12}$ moles/L; less

39d. $[\mathrm{H_3O^+}][\mathrm{OH^-}] = 10^{-14}$

37b. $24 = \dfrac{12}{\ln 2} \ln \left(\dfrac{f_2}{98}\right)$

$2 \ln 2 = \ln \left(\dfrac{f_2}{98}\right)$

$2 \ln 2 = \ln f_2 - \ln 98$

$2 \ln 2 + \ln 98 = \ln f_2$

$\ln (2^2 \cdot 98) = \ln f_2$

$f_2 = 392$ Hz

38a. pH $= -\log (6.3 \times 10^{-12})$
≈ 11.2; ammonia

38b. $-\log [\mathrm{H_3O^+}] = 7.0$
$\log [\mathrm{H_3O^+}] = -7.0$
$[\mathrm{H_3O^+}] = 10^{-7.0}$ moles/L

38c. $-\log [\mathrm{H_3O^+}] = 2.2$
$\log [\mathrm{H_3O^+}] = -2.2$
$[\mathrm{H_3O^+}] = 10^{-2.2}$
$\approx 6.3 \times 10^{-3}$ moles/L
$10^{-2.2} > 10^{-7.0}$; more

38d. baking soda: $-\log [\mathrm{H_3O^+}] = 8.3$
$[\mathrm{H_3O^+}] = 10^{-8.3}$

lye: $-\log [\mathrm{H_3O^+}] = 13.0$
$[\mathrm{H_3O^+}] = 10^{-13.0}$

$\dfrac{[\mathrm{H_3O^+}]}{[\mathrm{OH^-}]} = \dfrac{10^{-8.3}}{10^{-13.0}}$

$\approx 50{,}119$ or $50{,}000$

39a. $-\log [\mathrm{OH^-}] = 7.0$
$\log [\mathrm{OH^-}] = -7.0$
$[\mathrm{OH^-}] = 10^{-7.0}$

39b. $-\log [\mathrm{OH^-}] = 11.8$
$\log [\mathrm{OH^-}] = -11.8$
$[\mathrm{OH^-}] = 10^{-11.8}$
$\approx 1.6 \times 10^{-12}$
$10^{-11.8} < 10^{-7.0}$

39c. $3.7 + \mathrm{pOH} = 14$
$\mathrm{pOH} = 10.3$
$10.3 = -\log [\mathrm{OH^-}]$
$[\mathrm{OH^-}] = 10^{-10.3} \approx 5.01 \times 10^{-11}$

39d. $\mathrm{pH} + \mathrm{pOH} = 14$
$-\log [\mathrm{H_3O^+}] - \log [\mathrm{OH^-}] = 14$
$-(\log [\mathrm{H_3O^+}] + \log [\mathrm{OH^-}]) = 14$
$\log ([\mathrm{H_3O^+}][\mathrm{OH^-}]) = -14$
$[\mathrm{H_3O^+}][\mathrm{OH^-}] = 10^{-14}$

40a. $y = \log_b \frac{1}{x}$ (given)
$= \log_b x^{-1}$ (def. neg. exp.)
$= -\log_b x$ (Power Prop. of Logs)

40b. $y = \log_{\frac{1}{b}} x$ (given)
$= \dfrac{\log_a x}{\log_a b^{-1}}$ (change of base)
$= \dfrac{\log_a x}{-\log_a b}$ (Power Prop. of Logs)
$= -\dfrac{\log_a x}{\log_a b}$
$= -\log_b x$ (change of base)

---

32. $\log \left(\frac{3}{2}\right)^3 + \log (6x)^2 - \log 3^2$

$= \log \left(\frac{27}{8}\right)(36x^2) - \log 9$

$= \log \dfrac{\left(\frac{27}{8}\right)(36x^2)}{9} = \log \frac{27}{2}x^2$

33. $\log_{20} x = \dfrac{\ln x}{\ln 20} = \dfrac{1}{\ln 20} \ln x$
a vertical shrink of $f(x)$ by a factor of $\frac{1}{\ln 20} \approx 0.33$

34. $\log_{\frac{1}{3}} x = \dfrac{\ln x}{\ln \frac{1}{3}} = \dfrac{1}{\ln 3^{-1}} \ln x = -\dfrac{1}{\ln 3} \ln x$
a vertical shrink of $f(x)$ by a factor of $\frac{1}{\ln 3} \approx 0.91$ reflected across the x-axis

35. $R = \dfrac{\ln \left(\frac{A}{A_0}\right)}{\ln 10}$ or $\dfrac{\ln A - \ln A_0}{\ln 10}$

36. $L = 10 \log \left(\dfrac{I}{10^{-12}}\right)$
$= 10(\log I - \log 10^{-12})$
$= 10(\log I + 12 \log 10)$
$= 10(\log I + 12)$
$= 10 \log I + 120$
$L = 10 \log 10^{-8} + 120$
$= -80 + 120 = 40$ dB

37a. $n = 12 \log_2 \left(\dfrac{1976}{370}\right) = \dfrac{12}{\ln 2} \ln \left(\dfrac{1976}{370}\right)$
≈ 29 notes

> ## C. Exercises

41. Let $\log_b m = x$ and $\log_b n = y$.
$b^x = m$ and $b^y = n$ (exp. form)
$\dfrac{m}{n} = \dfrac{b^x}{b^y}$ (subst.)
$\dfrac{m}{n} = b^{x-y}$ ($\div$ Prop. of Exp.)
$\log_b \dfrac{m}{n} = x - y$ (log form)
$\log_b \dfrac{m}{n} = \log_b m - \log_b n$ (subst.)

42. Let $\log_b m = x$.
$b^x = m$ (exp. form)
$b^{xp} = m^p$ (Power Prop. of Logs)
$\log_b m^p = xp$ (log form)
$\log_b m^p = p \log_b m$ (subst. $\log_b m$ for x)

43. Let $\log_b x = y$.
$b^y = x$ (exp. form)
$\log_a b^y = \log_a x$ (Prop. of $=$ for Logs)
$y \log_a b = \log_a x$ (Power Prop. of Logs)
$y = \dfrac{\log_a x}{\log_a b}$ ($\times$ Prop. of $=$)
$\therefore \log_b x = \dfrac{\log_a x}{\log_a b}$ (subst.)

44.

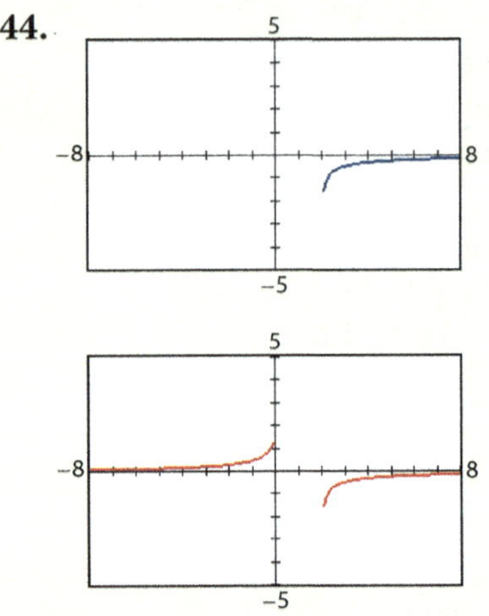

The properties of logarithms require that the input of a logarithmic function be positive. Therefore $f(x)$ has a domain of $(2, \infty)$. Unless the domain of $g(x)$, $D = (-\infty, 0) \cup (2, \infty)$, is restricted so that negative values of x are not included, $g(x)$ is not an equivalent function.

> ## Cumulative Review

45. $f(g(x)) = 2\left(\dfrac{x}{2} + 3\right) - 6$
$= x + 6 - 6 = x$
$g(f(x)) = \dfrac{(2x - 6)}{2} + 3$
$= x - 3 + 3 = x$

Determine whether $f(x)$ and $g(x)$ are inverses of each other. [1.8]

45. $f(x) = 2x - 6$
$g(x) = \dfrac{x}{2} + 3$ inverses

46. $f(x) = \dfrac{7}{x+2}$
$g(x) = \dfrac{x+2}{7}$ not inverses

47. List the possible rational zeros for the function $f(x) = x^3 + 2x^2 - 5x - 6$ and then factor $f(x)$. [2.4]

48. Identify the asymptotes of $g(x) = -\dfrac{1}{(x+1)^2} + 3$. Then graph the function as a transformation of a reciprocal power function. [2.5]

49. Solve $\dfrac{5}{3x^2 + 16x - 12} + \dfrac{3x-2}{x+6} = \dfrac{x}{3x-2}$. [2.6] $x = \dfrac{3}{2}, \dfrac{3}{4}$

50. What is best description of the linear correlation of the following graph? [1.9] D

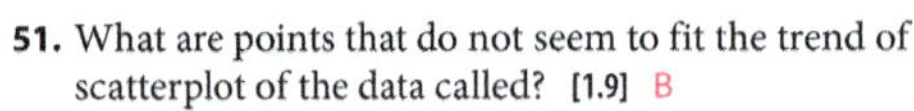

A. strong positive
B. strong negative
C. weak positive
D. weak negative
E. none of these

51. What are points that do not seem to fit the trend of a scatterplot of the data called? [1.9] B
A. correlators   **D.** regressions
B. outliers   **E.** determinates
C. residuals

47. $\pm 1, \pm 2, \pm 3, \pm 6; f(x) = (x + 1)(x + 3)(x - 2)$

52. Which of the following cannot be true in the exponential function $f(x) = ab^x$? [3.1] C
A. $a < 0$ **D.** $0 < b < 1$
B. $0 < a < 1$ **E.** all of these
C. $b < 0$

53. Between what two integers does $\log_2 3$ lie? [3.2] D
A. 0, 1   **D.** 1, 2
B. 3, 4   **E.** none of these
C. 2, 3

54. A biology class took an exam on June 1. When students took an equivalent exam without preparation in the subsequent months, their average score could be modeled by the function $f(t) = 74 - 16 \log(t + 1)$ where $0 \le t \le 12$. Use the model to estimate the average score 6 months after the original exam. [3.2] B
A. 61%   **D.** 42%
B. 60%   **E.** none of these
C. 45%

---

46. $f(g(x)) = \dfrac{7}{\left(\frac{x+2}{7}\right) + 2} = \dfrac{7}{\frac{x+2+14}{7}}$
$= \dfrac{7}{1} \cdot \dfrac{7}{x + 16} \neq x$ or
$g(f(x)) = \dfrac{\left(\frac{7}{x+2}\right) + 2}{7} = \dfrac{\frac{7 + 2x + 4}{x+2}}{7}$
$= \dfrac{7 + 2x + 4}{7x + 14} \neq x$

47. $\dfrac{c}{d}: \pm 1, \pm 2, \pm 3, \pm 6$

$$\begin{array}{r|rrrr}
-1 & 1 & 2 & -5 & -6 \\
   &   & -1 & -1 & 6 \\
\hline
   & 1 & 1 & -6 & 0
\end{array}$$

$f(x) = (x + 1)(x^2 + x - 6)$
$\quad\; = (x + 1)(x + 3)(x - 2)$

48. $g(x)$ is $f(x) = \dfrac{1}{x^2}$ translated 1 unit left, reflected in the x-axis, and translated 3 units up.
VA: $x = -1$; HA: $y = 3$

continued in Answers and Solutions Overflow

Newton's law of cooling can be used to determine how long ice cream should be out of a commercial freezer before it is served.

In an *exponential equation*, such as $e^x = 9$, the variable occurs in an exponent. In a *logarithmic equation*, such as $\log x = -2$, the variable occurs in the argument (input) of a logarithmic expression. Many simple exponential and logarithmic equations can be solved by simply converting to the alternative form.

If $\log x = -2$ then $x = 10^{-2} = 0.01$, and if $e^x = 9$ then $x = \ln 9 \approx 2.1972$.

A variety of strategies are used when solving more complicated exponential equations. When both sides of the equation can be expressed as powers of the same base, apply the *Equality Property of Exponents* ($b^x = b^y$ if and only if $x = y$) and solve the resulting equation.

Example 1 Solving by Using Like Bases

Solve $27^x = \left(\frac{1}{9}\right)^{x+5}$.

Answer

$(3^3)^x = (3^{-2})^{x+5}$ 1. Express each base as a power of 3.

$3^{3x} = 3^{(-2x-10)}$ 2. Power Property of Exponents

$3x = -2x - 10$ 3. Equality Property of Exponents

$5x = -10$ 4. Solve for x.

$x = -2$

SKILL ✓ **EXERCISE 3**

When the two sides of an exponential equation cannot be expressed using the same base, solve by finding a logarithm of each side. This is an application of the *Equality Property of Logarithms*:

$\log_b x = \log_b y$ if and only if $x = y$. Use the base of the exponential expression when it has a base of 10 or e.

Example 2 Solving by Finding Logarithms

Solve each exponential equation.

a. $10^{x+1} = 4$ **b.** $e^{-5x} = 17$

Answer

a. $\log 10^{x+1} = \log 4$ **b.** $\ln e^{-5x} = \ln 17$ 1. Take the common or natural log of both sides of the equation.

$x + 1 = \log 4$ $-5x = \ln 17$ 2. Simplify using the inverse property of logarithms.

$x = -1 + \log 4$ $x = -\frac{1}{5}\ln 17$ 3. Solve for x.

≈ -0.3979 ≈ -0.5666 4. Use a calculator to find a decimal approximation.

SKILL ✓ **EXERCISE 5**

3.4 Exponential and Logarithmic Equations

Objectives

1. To solve exponential equations by applying the Exponential Property of Equality or by taking the logarithm of both sides

2. To solve logarithmic equations using properties of logarithms

3. To model and solve real-world problems using exponential and logarithmic equations

Flash

Isaac Newton conducted many experiments related to the cooling of an object such as a red-hot iron. He desired to obtain a specific law for the cooling process. Newton's law of cooling essentially says that an object's heat loss is proportional to the difference between its temperature and the temperature of its surroundings.

Vocabulary

Equality Property of Exponents
Equality Property of Logarithms
exponential equation
logarithmic equation

After completing this section, you will be able to

- solve exponential and logarithmic equations.
- model and solve real-world problems using exponential and logarithmic equations.

PRESENTATION

Lesson Opener

Convert to exponential form.

1. $\log_3 81 = 4$ $3^4 = 81$

2. $\log_b N = x$ $b^x = N$

Convert to logarithmic form.

3. $10^2 = 100$ $\log 100 = 2$

4. $x^{a+2} = y$ $\log_x y = a + 2$

Evaluate.

5. $2\log_3 3 + \log_3 9$ 4

6. $2\log_2 6 - \log_2 9$ 2

Discuss the position of the variables in *exponential* and *logarithmic* equations such as $e^x = 9$ and $\log x = -2$. This section presents strategies used to solve these two types of equations. Convey to the students that they have already solved many simple exponential and logarithmic equations by converting to the alternate form.

Explain that some exponential equations, like those in Example 1, can be solved by obtaining like bases and applying the *Equality Property of Exponents*.

Other exponential equations are generally solved by taking the log of both sides. Present Example 2, emphasizing that the first step is justified by the second conditional in the *Equality Property of Logarithms*. Discuss how converting to logarithmic form can shorten the solution.

Emphasize that either common or natural logs can be used to solve the equations in Example 3. Clarify that any base can be used, but $\log x$ and $\ln x$ are usually the easiest bases to evaluate with a calculator. Students can check the decimal approximation in Example 3a by entering 5, $\boxed{\wedge}$, [ANS] ($\boxed{2nd}$, $\boxed{(-)}$). Consider explaining that simplified exact answers are typically preferred when the solution is an intermediate step in a larger problem. Approximations are typically used to make a final answer easier to comprehend.

Additional Exercises

Solve. Round decimal answers to the nearest thousandth.

1. $4^{x^2} = 1024$ $x = \pm\sqrt{5}$

2. $24\left(\frac{1}{2}\right)^{\frac{x}{3}} = 384$ $x = -12$

3. $e^{-\frac{x}{4}} = 27$ $x \approx -13.183$

4. $6^{x-1} = 3^{2x}$ $x \approx -4.419$

5. $2e^{2x} - 11e^x + 5 = 0$
   $x \approx 1.609, \approx -0.693$

6. $\log_8 x = 3$ $x = 512$

7. $\log_5 (x + 2) = 3$ $x = 123$

8. $\log_5 12 = \log_5 (x - 4) + \log_5 x$
   $x = 6, -2$ (extraneous)

9. $10^{2x} = e^{3x+1}$ $x = \dfrac{1}{2\ln 10 - 3} \approx 0.623$

10. $x^{\log_5 9} = 81$ $x = 25$

For Graphing Calculators

Checking solutions graphically can help students understand why certain solutions are extraneous. Enter each side of the equation as a function in the $\boxed{Y=}$ equation editor. The x-value of any intersection of the functions is a solution. See Additional Exercise 8.

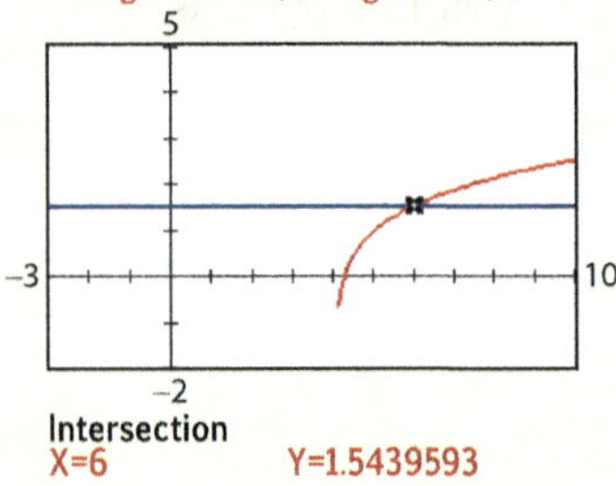

This process can also be viewed as converting to logarithmic form and then solving for x.

$$10^{x+1} = 4 \qquad\qquad e^{-5x} = 17$$
$$x + 1 = \log 4 \qquad -5x = \ln 17 \qquad \text{Write the equation in logarithmic form.}$$
$$x = -1 + \log 4 \qquad x = -\tfrac{1}{5}\ln 17 \qquad \text{Solve for } x.$$

When the exponential equation has a base other than 10 or e, the strategy of finding either the natural or common log of each side will produce the same result.

Example 3   Solving by Finding Logarithms

Solve each exponential equation.

a. $5^x = 7$ b. $3^{2x-1} = 5^x$

Answer

a.
$$\log 5^x = \log 7$$
$$x \log 5 = \log 7$$
$$x = \frac{\log 7}{\log 5}$$
$$x \approx 1.209$$

b.
$$\ln 3^{2x-1} = \ln 5^x$$
$$(2x - 1)\ln 3 = x \ln 5$$
$$2x \ln 3 - \ln 3 = x \ln 5$$
$$2x \ln 3 - x \ln 5 = \ln 3$$
$$x(2 \ln 3 - \ln 5) = \ln 3$$
$$x = \frac{\ln 3}{\ln 9 - \ln 5}$$
$$x \approx 1.869$$

1. Take the logarithm of both sides.
2. Power Property of Logarithms
3. Solve for x.

(The power property can be used to express $2 \ln 3$ as $\ln 9$.)

4. Use a calculator to find a decimal approximation.

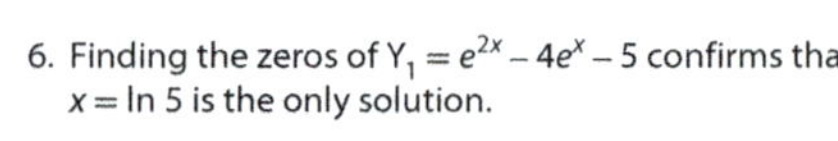

In Example 3a you could convert directly to logarithmic form, obtaining $x = \log_5 7$. While this is an equivalent solution, you would need to use the change of base formula to express $\log_5 7 = \frac{\log 7}{\log 5}$ or $\frac{\ln 7}{\ln 5}$ before using a calculator to evaluate the expression.

Exponential equations in quadratic form can be solved by factoring or using the quadratic formula.

Example 4   Solving by Factoring

Solve $e^{2x} - 4e^x - 5 = 0$.

Answer

$$(e^x)^2 - 4(e^x) - 5 = 0$$
$$(e^x - 5)(e^x + 1) = 0$$
$$e^x - 5 = 0 \quad \text{or} \quad e^x + 1 = 0$$
$$e^x = 5 \qquad\qquad e^x = -1$$
$$x = \ln 5 \qquad\qquad x = \ln (-1)$$
$$x \approx 1.6094$$

1. Write the equation in the quadratic form.
2. Factor.
3. Zero Product Property
4. Solve the resulting exponential equations.
5. The natural log of a negative number is not real.

Check

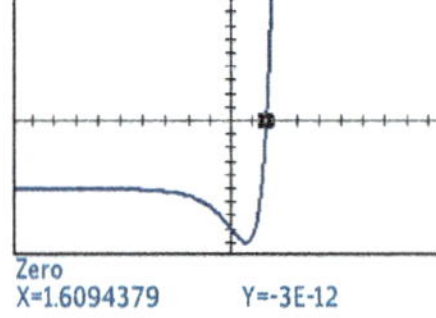

6. Finding the zeros of $Y_1 = e^{2x} - 4e^x - 5$ confirms that $x = \ln 5$ is the only solution.

SKILL ✔ EXERCISE 27

Consider having the students list advantages and disadvantages of the alternate method of converting to logarithmic form (with base 5).

Interactive Activity Have the students complete Example 3 using a base other than 10, e, or 5 and a calculator that evaluates logarithms of other bases. See the Reading and Writing Mathematics feature on page 144.

Carefully illustrate how the exponential equation in Example 4 is expressed in quadratic form and factored. Reinforce the need to convert the exponential equation to logarithmic form in order to find the value of the variable with the calculator.

Examples 5 and 6 illustrate strategies used to solve logarithmic equations. The second step of Example 5 can also be viewed as exponentiating both sides, $e^{\ln x^2} = e^{\ln 9}$, and then applying the inverse property. A similar approach could be applied in Example 6.

$$5^{\log_5 x} = 5^{1 - \log_5 (2x-3)}$$
$$x = \frac{5^1}{5^{\log_5 (2x-3)}}$$
$$x = \frac{5}{2x - 3}$$
$$2x^2 - 3x - 5 = 0$$

When presenting Example 7, point out that the exponential equation can be solved using natural logs instead of common logs. Remind the students that

parentheses must be placed around the product in the denominator when it is entered in the calculator.

Motivational Idea Share the importance of logarithms in today's world. Example 7 illustrates the use of logarithms in finances. Computer programmers use logarithms in their code to develop financial apps that are used by people every day.

There are several strategies used to solve more complicated logarithmic equations. When both sides of the equation can be expressed using a logarithmic expression, apply the Equality Property of Logarithms and solve the resulting equation.

Example 5 Solving with the Equality Property of Logarithm

Solve $2 \ln x = \ln 9$.

Answer

$\ln x^2 = \ln 9$	1. Power Property of Logarithms
$x^2 = 9$	2. Equality Property of Logarithms
$x = \pm 3$	3. Take the square root of each side.
The solution is $x = 3$.	4. Since the domain of $\ln x$ is $(0, \infty)$, -3 is an extraneous solution.

SKILL ✔ **EXERCISE 15**

Other logarithmic equations can be solved by converting to exponential form after writing the equation as a single logarithmic expression isolated on one side of the equation.

Example 6 Solving by Converting to Exponential Form

Solve $\log_5 x = 1 - \log_5 (2x - 3)$.

Answer

$\log_5 x + \log_5 (2x - 3) = 1$	1. Isolate the logarithmic functions on one side.
$\log_5 x(2x - 3) = 1$	2. Write that side as a single logarithmic expression using the Product Property of Logarithms.
$x(2x - 3) = 5^1$	3. Convert to exponential form.
$2x^2 - 3x = 5$	
$2x^2 - 3x - 5 = 0$	4. Solve for x.
$(2x - 5)(x + 1) = 0$	
$x = \frac{5}{2}$ or $x = -1$	
The solution is $x = \frac{5}{2}$.	5. Since -1 is not in the domain of either original logarithmic expression, it is extraneous.

Check

6. Confirm by finding the intersections of
$Y1 = \dfrac{\log x}{\log 5}$ and $Y2 = 1 - \dfrac{\log (2x - 3)}{\log 5}$.

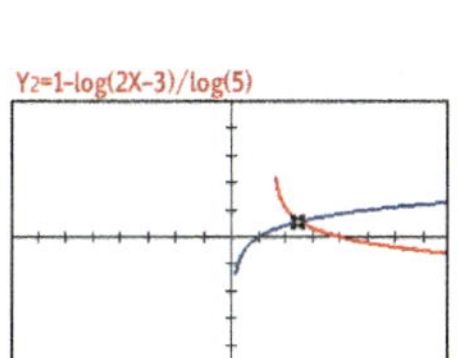

SKILL ✔ **EXERCISE 37**

The ability to solve exponential and logarithmic equations enables the solving of many real-life problems.

One-on-One It can be difficult to determine which strategy should be used to solve a specific equation. Encourage the students to analyze the characteristics of an equation at each step. The time required to find a good strategy decreases with experience.

TIPS

Ex. 21, 25 Simplifying the logarithmic expression makes the final calculation easier.

Ex. 30 This alternate solution for the second factor avoids the undefined result.

$$e^{2x} - 1 = 0$$
$$e^{2x} = 1$$
$$2x = \ln 1$$
$$x = \tfrac{1}{2} \ln 1 = 0$$

Assignments

- **Minimum:** 1–2, 4–5, 8, 10–11, 13–15, 17, 21, 33, 35, 43, 49–50; 51–57 odd
- **Standard:** 1–5, 7, 9–11, 13–15, 18–19; 21–41 odd; 40, 42–43, 47–49, 52–57
- **Extended:** 2–3, 5–6, 9–10, 13, 15–16, 19; 20–30 even; 31, 33–39, 42–46; 48–58 even

Assessment

- Quiz 3B covers Sections 3.3–3.4.

Solutions

❯ A. Exercises

1. $2^x = 2^6$; $x = 6$

2. $3^x = 3^{-5}$; $x = -5$

3. $3^{3(x - 3)} = 3^{2(x - 1)}$
   $3x - 9 = 2x - 2$; $x = 7$

4. $\log 10^x = \log 77$
   $x \approx 1.886$

5. $\ln e^{x + 1} = \ln 10$
   $x + 1 = \ln 10$
   $x = \ln 10 - 1 \approx 1.303$

6. $e^x = 5$
   $\ln e^x = \ln 5$
   $x = \ln 5 \approx 1.609$

Exercises 7–10 can also be solved using natural logs.

7. $\log 6^x = \log 12$
   $x \log 6 = \log 12$
   $x = \dfrac{\log 12}{\log 6} \approx 1.387$

8. $\log 5^x = \log \frac{1}{12}$
   $x \log 5 = \log \frac{1}{12}$
   $x = \dfrac{\log \frac{1}{12}}{\log 5} \approx -1.544$

9. $\log 3^{5x} = \log 2$
   $5x \log 3 = \log 2$
   $x = \dfrac{\log 2}{5 \log 3} \approx 0.126$

10. $\log 5^{x + 2} = \log 15$
    $(x + 2) \log 5 = \log 15$
    $x = \dfrac{\log 15}{\log 5} - 2 \approx -0.317$

11. $3^2 = 4x - 7$
    $16 = 4x$; $x = 4$

12. $3^2 = x + 1$; $x = 8$

13. $27^{\frac{1}{3}} = x$
    $x = \sqrt[3]{27} = 3$

14. $2x + 1 = 4x - 5$; $x = 3$

15. $\log_3 x^2 = \log_3 8$
$x^2 = 8; \ x = \pm 2\sqrt{2}$

x must be positive in the original equation; $\therefore x = 2\sqrt{2}$.

16. $\log_3 x = 4$
$3^4 = x; \ x = 81$

17. $R = \log \dfrac{A}{A_0} = 6.2$

$\dfrac{A}{A_0} = 10^{6.2}$

$A = A_0 10^{6.2} \approx A_0(1.6 \text{ million})$

18. $10 \log \dfrac{I}{10^{-12}} = 64$

$\log \dfrac{I}{10^{-12}} = 6.4$

$10^{6.4} = \dfrac{I}{10^{-12}}$

$I = 10^{6.4}(10^{-12}) = 10^{-5.6}$
$\approx 2.512 \times 10^{-6} \text{ W/m}^2$

B. Exercises

19. $\left(\dfrac{2}{3}\right)^x = \left(\dfrac{3}{2}\right)^4$

$\left(\dfrac{2}{3}\right)^x = \left(\dfrac{2}{3}\right)^{-4}$

$x = -4$

20. $5^{-4(2x - 10)} = 5^{2(10x)}$

$-8x + 40 = 20x$
$40 = 28x$
$x = \dfrac{10}{7} \approx 1.429$

Exercises 21, 22, 25, and 26 can also be solved using common logs.

21. $\ln 4^{1-x} = \ln 5^x$

$(1 - x)\ln 4 = x \ln 5$
$\ln 4 - x \ln 4 = x \ln 5$
$\ln 4 = x(\ln 5 + \ln 4)$
$\ln 4 = x \ln 20$
$x = \dfrac{\ln 4}{\ln 20} \approx 0.463$

22. $\ln 3^{x+1} = \ln 17^{2x}$

$(x + 1)\ln 3 = 2x \ln 17$
$x \ln 3 + \ln 3 = 2x \ln 17$
$x \ln 3 - 2x \ln 17 = -\ln 3$
$x(\ln 3 - 2\ln 17) = -\ln 3$
$x = \dfrac{-\ln 3}{\ln 3 - 2\ln 17} \approx 0.241$

23. $\ln e^x = \ln 10^{x+1}$

$x = (x + 1)\ln 10$
$x = x \ln 10 + \ln 10$
$x - x \ln 10 = \ln 10$
$x(1 - \ln 10) = \ln 10$
$x = \dfrac{\ln 10}{1 - \ln 10} \approx -1.768$

24. $\ln e^{2x-1} = \ln 10^x$

$2x - 1 = x \ln 10$
$2x - x \ln 10 = 1$
$x(2 - \ln 10) = 1$
$x = \dfrac{1}{2 - \ln 10} \approx -3.305$

25. $\ln 12^{x-4} = \ln 3^{x-2}$

$(x - 4)\ln 12 = (x - 2)\ln 3$
$x \ln 12 - 4 \ln 12 = x \ln 3 - 2 \ln 3$
$x \ln 12 - x \ln 3 = 4 \ln 12 - 2 \ln 3$
$x(\ln 12 - \ln 3) = 4 \ln 12 - 2 \ln 3$
$x = \dfrac{\ln 12^4 - \ln 9}{\ln 4} \approx 5.585$

26. $\log 3^{-x-1} = \log 6^{-x+2}$

$(-x - 1)\log 3 = (-x + 2)\log 6$
$-x \log 3 - \log 3 = -x \log 6 + 2 \log 6$
$x \log 6 - x \log 3 = 2 \log 6 + \log 3$
$x (\log 6 - \log 3) = \log 62 + \log 3$
$x = \dfrac{\log 108}{\log 2} \approx 6.755$

27. $(e^x)^2 - 27e^x + 72 = 0$
$(e^x - 24)(e^x - 3) = 0$
$e^x = 24 \quad$ or $\quad e^x = 3$
$x = \ln 24 \qquad x = \ln 3$
$\qquad \approx 3.178 \qquad\quad \approx 1.099$

How long does it take for \$4000 invested at 6% annual interest to grow to \$8000 if interest is compounded monthly?

Answer

$A(t) = P\left(1 + \dfrac{r}{n}\right)^{nt}$

$8000 = 4000\left(1 + \dfrac{0.06}{12}\right)^{12t}$

$2 = 1.005^{12t}$

$\log 2 = \log 1.005^{12t}$

$\log 2 = 12t \log 1.005$

$t = \dfrac{\log 2}{12 \log 1.005} \approx 11.6 \text{ years}$

1. Substitute the given information into the compound interest formula.

2. Simplify the exponential equation.

3. Take the common log of both sides.

4. Solve for t.

Check

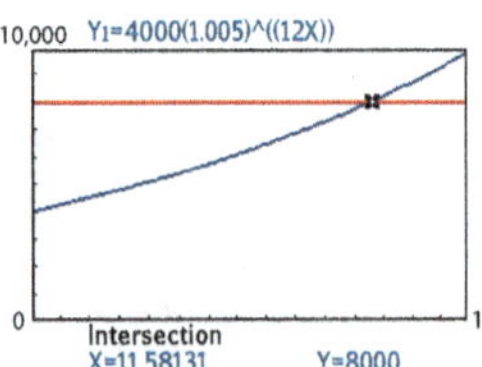

5. Confirm by finding the intersection of $Y_1 = 4000(1.005)^{12t}$ and $Y_2 = 8000$.

SKILL ✓ **EXERCISE 41**

While the charging of interest on loans to a fellow Israelite in need is forbidden (Deut. 15:1–7, 23:19), the parable of the talents teaches that investing what has been entrusted to us is expected if we are to be faithful servants (Matt. 25:14–30).

A. Exercises

Solve each exponential equation. Round answers to the nearest thousandth if necessary.

1. $2^x = 64$ 6
2. $3^x = \dfrac{1}{243}$ −5
3. $27^{x-3} = 9^{x-1}$ 7
4. $10^x = 77$ 1.886
5. $e^{x+1} = 10$ 1.303
6. $4e^x = 20$ 1.609
7. $6^x = 12$ 1.387
8. $5^x = \dfrac{1}{12}$ −1.544
9. $3^{5x} = 2$ 0.126
10. $5^{x+2} = 15$ −0.317

Solve each logarithmic equation.

11. $\log_3 (4x - 7) = 2$ 4
12. $\log_3 (x + 1) = 2$ 8
13. $\log_{27} x = \dfrac{1}{3}$ 3
14. $\ln (2x + 1) = \ln (4x - 5)$ 3
15. $2 \log_3 x = \log_3 8$ $2\sqrt{2}$
16. $\dfrac{1}{2} \log_3 x = 2$ 81

17. The formula for the Richter magnitude scale, $R = \log \dfrac{A}{A_0}$, compares an earthquake's amplitude A to a small reference amplitude, A_0. If the April 2017 earthquake in Atacama, Chile, had a magnitude than 6.2, approximately how many times larger than A_0 was its amplitude? $10^{6.2}$ or ≈ 1.6 million times larger

18. The loudness of a sound is measured in decibels using the equation $L = 10 \log \dfrac{I}{I_0}$, where I is the intensity of the sound and I_0 is the intensity of sound at the threshold of hearing, 10^{-12} W/m². If a phone app warns users when the volume of the device is at or above 64 dB, find the lowest sound intensity that triggers the warning. $\approx 2.512 \times 10^{-6}$ W/m²

B. Exercises

Find all real solutions for x. Round answers to the nearest thousandth if necessary.

19. $\left(\dfrac{2}{3}\right)^x = \dfrac{81}{16}$ −4
20. $\left(\dfrac{1}{625}\right)^{2x-10} = 25^{10x}$ $\dfrac{10}{7}$ or 1.429
21. $4^{1-x} = 5^x$ 0.463
22. $3^{x+1} = 17^{2x}$ 0.241
23. $e^x = 10^{x+1}$ −1.768
24. $e^{2x-1} = 10^x$ −3.305
25. $12^{x-4} = 3^{x-2}$ 5.585
26. $3^{-x-1} = 6^{-x+2}$ 6.755
27. $e^{2x} - 27e^x + 72 = 0$ 1.099, 3.178
28. $e^{2x} + 5e^x = 6$ 0
29. $e^{2x} + 11e^x = -24$ no solution
30. $3e^{4x} - 11e^{2x} + 8 = 0$ 0, 0.490

Find all real solutions for x.

31. $\dfrac{1}{4} \log_4 x^2 = \log_4 8$ ± 64
32. $\log_2 3x = \log_2 (x + 4) - \log_2 3$ $\dfrac{1}{2}$
33. $\log_3 x = 2 \log_3 5 + \log_3 16$ 400
34. $\ln x + \ln(x - 3) = \ln 54$ 9
35. $\log_2 8^x = -3$ −1
36. $\log_{16} x^2 = -\dfrac{1}{2}$ $\pm \dfrac{1}{2}$
37. $\log_2 (x - 2) = 3 - \log_2 x$ 4
38. $\log_3 \dfrac{1}{4} - \log_3 \dfrac{3}{4} = x$ −1

39. Explore: Complete the following to explore the properties of equality for powers and exponents.

  a. If $x^2 = 3^2$, is $x = 3$ the only solution? Explain why.

  b. If $x^3 = 3^3$, is $x = 3$ the only solution? Explain why.

  c. Compare the graphs of $f(x) = x^2$ and $g(x) = x^3$. Which properties of the graphs explain the difference between the answers in part a and part b?

  d. According to the Equality Property for Exponents, if $b^x = b^n$, then $x = n$. What characteristic of the graph of $h(x) = b^x$ supports this conclusion?

40. Contrast: Solve the following equations. Then explain the general process used to solve each equation.

  a. $\log_5 x = 7$ **b.** $\log_7 5 = x$ **c.** $\log_x 5 = 7$

41. Model: In a Sierpinski fractal, the number of black triangles T surrounding the central white triangle is modeled by the equation $T = 3^n$ where n is number of iterations. How many iterations are needed to produce 1,594,323 black triangles?   13

Iterations

  0   1   2   3

42. How many years will it take for a \$60,000 investment with 3% APR compounded monthly to match an \$80,000 investment with 2% APR compounded quarterly? Verify your answer graphically. ≈ 28.7 yr

39a. no; $x = -3$ is also possible.

39b. yes; There is only one cube root of 27, $x = 3$.

39c. $f(x)$ is not one-to-one, but $g(x)$ is one-to-one.

39d. It is a one-to-one function (so its inverse is also a function).

Find all real solutions for x.

43. $\ln (2x + 5) + \ln (3x + 1) = \ln 8$ $\frac{1}{6}$

44. $\log_2 \sqrt{\frac{3x + 4}{x}} = 0$ -2

45. $\log_4 x^8 = \frac{4}{3} \log_4 64$ ± 2

46. $x^{\log_3 5} = 125$ 27

47. An object's temperature may be modeled using Newton's law of cooling (and heating):
$T(t) = T_S + (T_0 - T_S)e^{-kt}$ where T_0 is the object's initial temperature, T_S is the surrounding medium's temperature (assumed to be constant), k is the heating constant, and t is the time. A tub of ice cream stored in a freezer at $-20°$C is placed in a room at $25°$C.

  a. Find k if the temperature of the ice cream increased to $-15°$C after 5 min. ≈ 0.0236

  b. Write a function modeling the temperature of the ice cream after t min in the room.

  c. Estimate the temperature of the ice cream after 20 min in the room. $\approx -3.09°$C

48. The periodic payment P for a car loan of B dollars can be calculated using the equation $P = \frac{iB}{1 - (1 + i)^{-n}}$ where i is the periodic interest rate and n is the number of payments. Use the formula and an APR of 3% to complete the following.

  a. Find the monthly payment on a 3 yr loan of \$12,500.   \$363.52

  b. Solve the equation for B. Then use the result to find the maximum 3 yr loan with a monthly payment of \$300.

  c. Solve the equation for n. Then determine the number of monthly \$300 payments needed to repay a \$9000 car loan.

47b. $T(t) = 25 - 45e^{\left(\frac{1}{5}\ln\frac{8}{9}\right)t}$

48b. $B = P\left(\frac{1 - (1 + i)^{-n}}{i}\right)$; \$10,315.94

48c. $n = -\frac{\log\left(1 - \frac{iB}{P}\right)}{\log (1 + i)}$; 32 payments

32. $\log_2 3x = \log_2 \frac{x + 4}{3}$
$$3x = \frac{x + 4}{3}$$
$$9x = x + 4; \; x = \frac{1}{2}$$

33. $\log_3 x = \log_3 5^2 + \log_3 4^2$
$$\log_3 x = \log_3 (25)(16)$$
$$x = 400$$

34. $\ln x(x - 3) = \ln 54$
$$x^2 - 3x = 54$$
$$x^2 - 3x - 54 = 0$$
$$(x - 9)(x + 6) = 0$$
$$x = 9, -6 \text{ (extraneous)}$$

35. $2^{-3} = 8^x$
$$2^{-3} = 2^{3x}$$
$$-3 = 3x; \; x = -1$$

36. $16^{-\frac{1}{2}} = x^2$
$$\frac{1}{4} = x^2$$
$$x = \pm\frac{1}{2}$$

37. $\log_2 (x - 2) + \log_2 x = 3$
$$\log_2 (x - 2)x = 3$$
$$x^2 - 2x = 2^3$$
$$x^2 - 2x - 8 = 0$$
$$(x - 4)(x + 2) = 0$$
$$x = 4, -2 \text{ (extraneous)}$$

38. $\log_3 \frac{\frac{1}{4}}{\frac{3}{4}} = x$
$$\log_3 \frac{1}{3} = x$$
$$3^x = \frac{1}{3}; \; x = -1$$

40a. $x = 5^7 = 78,125$; solved by simplifying an exponential expression

40b. $x = \frac{\log 5}{\log 7}$ or $\frac{\ln 5}{\ln 7} \approx 0.827$; solved by applying the change of base formula

40c. $x^7 = 5$; $x = \sqrt[7]{5} \approx 1.2585$; solved by finding the seventh root

41. $3^n = 1,594,323$
$$\log 3^n = \log 1,594,323$$
$$n = \frac{\log 1,594,323}{\log 3} = 13$$

42. $60,000\left(1 + \frac{0.03}{12}\right)^{12t}$
$$= 80,000\left(1 + \frac{0.02}{4}\right)^{4t}$$
$$3(1.0025)^{12t} = 4(1.005)^{4t}$$
$$0.75 = \frac{1.005^{4t}}{1.0025^{12t}}$$
$$\ln 0.75 = \ln \frac{1.005^{4t}}{1.0025^{12t}}$$
$$\ln 0.75 = 4t \cdot \ln 1.005 - 12t \cdot \ln 1.0025$$
$$t = \frac{\ln 0.75}{4 \cdot \ln 1.005 - 12 \cdot \ln 1.0025}$$
$$\approx 28.7 \text{ yr}$$

28. $(e^x)^2 + 5e^x - 6 = 0$
$$(e^x + 6)(e^x - 1) = 0$$
$$e^x = -6 \text{ or } e^x = 1$$
$$x = \ln (-6) \quad x = \ln 1 = 0$$
$$\text{(extraneous)}$$

29. $(e^x)^2 + 11e^x + 24 = 0$
$$(e^x + 8)(e^x + 3) = 0$$
$$e^x = -8 \text{ or } e^x = -3$$
$$x = \ln (-8) \quad x = \ln (-3)$$
$$\text{(Neither solution is real.)}$$

30. $3(e^{2x})^2 - 11e^{2x} + 8 = 0$
$$(3e^{2x} - 8)(e^{2x} - 1) = 0$$
$$(3e^{2x} - 8)(e^x - 1)(e^x + 1) = 0$$
$$3e^{2x} = 8 \text{ or } e^x = 1 \text{ or } e^x = -1$$
$$e^{2x} = \frac{8}{3} \quad x = \ln 1 \quad x = \ln (-1)$$
$$2x = \ln \frac{8}{3} \quad x = 0 \quad \text{(undefined)}$$
$$x = \frac{1}{2} \ln \frac{8}{3}$$
$$\approx 0.490$$

31. $\log_4 x^2 = 4 \log_4 8$
$$\log_4 x^2 = \log_4 8^4$$
$$x^2 = 8^4$$
$$x = \pm 8^2 = \pm 64$$

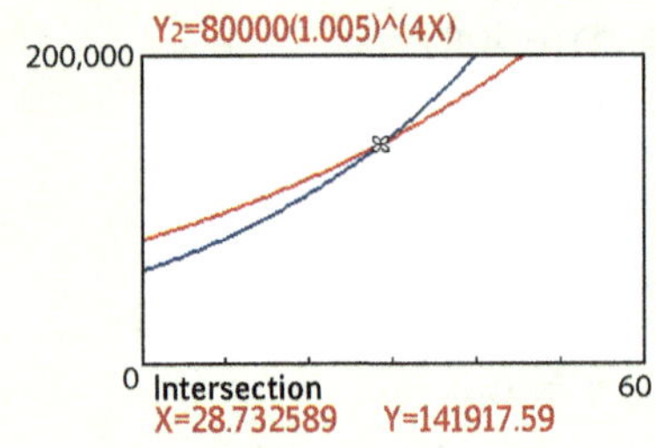

▶ C. Exercises

43. $\ln(2x+5)(3x+1) = \ln 8$
$\ln(6x^2 + 17x + 5) = \ln 8$
$6x^2 + 17x + 5 = 8$
$6x^2 + 17x - 3 = 0$
$(x+3)(6x-1) = 0$
$x = \frac{1}{6}, -3 \text{ (extraneous)}$

44. $2^0 = \sqrt{\dfrac{3x+4}{x}}$
$(1)^2 = \left(\sqrt{\dfrac{3x+4}{x}}\right)^2$
$1 = \dfrac{3x+4}{x}$
$x = 3x + 4$
$x = -2$

45. $\log_4 x^8 = \log_4 (2^6)^{\frac{4}{3}}$
$\log_4 x^8 = \log_4 2^8$
$x^8 = 2^8$
$x = \pm 2$

46. $x^{\frac{\log 5}{\log 3}} = 125$
$\dfrac{\log 5}{\log 3}\log x = \log 125$
$\log x = \dfrac{\log 3}{\log 5} \cdot \log 125$
$x = 10^{\frac{\log 3 \,\cdot\, \log 125}{\log 5}} = 27$

47a. $T(5) = 25 + (-20 - 25)e^{-k(5)}$
$= -15$
$-15 = 25 + (-45e^{-5k})$
$\dfrac{-40}{-45} = e^{-5k}$
$\ln \dfrac{8}{9} = -5k$
$k = -\dfrac{1}{5}\ln\dfrac{8}{9} \approx 0.0236$

47b. $T(t) = 25 - 45e^{-\left(-\frac{1}{5}\ln\frac{8}{9}\right)t}$
$= 25 - 45e^{\left(\frac{1}{5}\ln\frac{8}{9}\right)t}$

47c. $T(20) = 25 - 45e^{\left(\frac{1}{5}\ln\frac{8}{9}\right)(20)}$
$\approx -3.09°C$

48a. $n = 3(12) = 36; \ i = \dfrac{0.03}{12} = 0.0025$
$P = \dfrac{(0.0025)(12,500)}{1 - (1.0025)^{-36}} \approx \$363.52$

48b. $B = P\left(\dfrac{1 - (1+i)^{-n}}{i}\right)$
$= 300\left[\dfrac{1 - (1.0025)^{-36}}{0.0025}\right]$
$\approx \$10,315.94$

48c. $B = P\left(\dfrac{1 - (1+i)^{-n}}{i}\right)$
$\dfrac{Bi}{P} = 1 - (1+i)^{-n}$
$(1+i)^{-n} = 1 - \dfrac{Bi}{P}$
$\log(1+i)^{-n} = \log\left(1 - \dfrac{Bi}{P}\right)$
$-n\log(1+i) = \log\left(1 - \dfrac{Bi}{P}\right)$
$n = -\dfrac{\log\left(1 - \dfrac{Bi}{P}\right)}{\log(1+i)}$
$= -\dfrac{\log\left(1 - \dfrac{(9000)(0.0025)}{300}\right)}{\log 1.0025}$
≈ 31.2

Solve. [2.6]

49. $\dfrac{x+1}{x-1} = \dfrac{6x+1}{x^2-1}$ $x = 0, 4$

50. $\dfrac{4}{x-6} + \dfrac{8x}{x+3} = \dfrac{3}{x^2-3x-18}$ $x = \dfrac{11 \pm \sqrt{103}}{4}$

Describe each transformation of $f(x) = \log_5 x$. Then use a graph of $f(x)$ to sketch the graph of $g(x)$. [3.2]

51. $g(x) = \log_5 x + 3$ **52.** $g(x) = \log_5(x-4) + 2$

Expand each logarithmic expression. Assume all variables are positive values. [3.3]

53. $\log 10\sqrt[3]{x^2 y}$ **54.** $\log_5 \dfrac{25\sqrt{x}}{yz}$

55. Identify any asymptotes of $f(x) = \dfrac{x^3 - x^2 - 2x}{x^2 - 3x + 2}$. [2.5] B
   A. HA: $y = 1$; VA: $x = 2$ and $x = 1$
   B. VA: $x = 1$; SA: $y = x + 2$
   C. VA: $x = 2$ and $x = 1$; SA: $y = x + 2$
   D. VA: $x = 2$ and $x = 1$; SA: $y = x + 2$
   E. HA: $y = 1$; VA: $x = 2$ and $x = 1$; SA: $y = x$

56. Find the solution of $\dfrac{x^2 - 2x - 15}{2x^2 + 9x + 9} \geq 0$. [2.7] A
   A. $(-\infty, -3) \cup \left(-3, -\dfrac{3}{2}\right) \cup [5, \infty)$
   B. $\left(-3, -\dfrac{3}{2}\right)$
   C. $(-\infty, -3) \cup (-\dfrac{3}{2}, 5]$
   D. $\left(-\infty, -\dfrac{3}{2}\right) \cup [5, \infty)$
   E. $(-\infty, -3) \cup \left(-\dfrac{3}{2}, 5\right)$

57. Find the account balance in 12 yr if \$4500 is invested at 6.5% compounded semiannually. [3.1] C
   A. \$4801.37 **C.** \$9695.58 **E.** \$9816.63
   B. \$9580.93 **D.** \$9795.98

58. Which describes $f(x) = \log_5 x$ translated 2 units left, reflected across the x-axis, and then translated 4 units up? [3.2] D
   A. $g(x) = 2\log_5 x + 2 + 4$
   B. $g(x) = \log_5(-x + 2) + 4$
   C. $g(x) = -\log_5(x - 2) + 4$
   D. $g(x) = -\log_5(x + 2) + 4$
   E. $g(x) = -\log_5(x - 4) + 2$

51. $f(x)$ is translated 3 units up.

52. $f(x)$ is translated 4 units right and 2 units up.

53. $1 + \dfrac{2}{3}\log x + \dfrac{1}{3}\log y$

54. $2 + \dfrac{1}{2}\log_5 x - \log_5 y - \log_5 z$

▶ Cumulative Review

49. $\left[\dfrac{x+1}{x-1} = \dfrac{6x+1}{x^2-1}\right](x+1)(x-1)$
$x^2 + 2x + 1 = 6x + 1$
$x^2 - 4x = 0$
$x(x-4) = 0$
$x = 0, -4$

50. $\left[\dfrac{4}{x-6} + \dfrac{8x}{x+3} = \dfrac{3}{x^2-3x-18}\right](x-6)$
$\cdot (x+3)$
$4x + 12 + 8x^2 - 48x = 3$
$8x^2 - 44x + 9 = 0$
$x = \dfrac{-(-44) \pm \sqrt{(-44)^2 - 4(8)(9)}}{2(8)}$
$= \dfrac{44 \pm 4\sqrt{103}}{16} = \dfrac{11 \pm \sqrt{103}}{4}$

continued in Answers and Solutions Overflow

Glorifying God with Math

The work of the sixteenth, seventeenth, and even some eighteenth-century mathematicians was a religious quest. . . . The search for the mathematical laws of nature was an act of devotion. It was the study of the ways and nature of God which would reveal the glory and grandeur of his handiwork. . . . Man could not hope to perceive the divine plan as clearly as God himself understood it, but man could with humility and modesty seek to at least approach the mind of God. . . . Each discovery of a law of nature was hailed as evidence testifying more to God's brilliance than to the ingenuity of the investigator.

—Morris Kline (American mathematics historian and philosopher)

The German mathematician and astronomer Johannes Kepler (1571–1630) stated that astronomers "ought to keep in their minds not the glory of their own intellect, but the glory of God above everything else." Everything in a Christian's life should bring glory to God (1 Cor. 10:31). Morris Kline observed that mathematicians before the nineteenth century had, if not a strictly biblical worldview, at least a religious worldview. Their goal was to glorify God as they worked. Colossians 1:18 provides additional focus for Christians, stating that our Lord Jesus Christ should have the preeminence in all things. But how can we glorify God with mathematics?

Acknowledging God as the origin of mathematical truth allows us to praise Him as the reason for the divine characteristics apparent in the subject. Christian mathematician Larry Zimmerman observed that the knowledge of mathematics "unveils not only vistas of beauty and power unsuspected before but also an order, symmetry and infinitude which stuns and awes the beholder." As we study mathematics and marvel at its beauty and order, we should praise God, the giver of that beauty and order. Everything we do should be done in thanksgiving to God (Col. 3:17). We can glorify God by recognizing the power of algebra, the symmetry of geometry, and the infinitude of calculus as reflections of our Creator.

We can also glorify God with mathematics by using it to fulfill the Creation Mandate (Gen. 1:26–28, Ps. 8). Gottfried Leibniz, one of the founders of calculus, stated that "the principal goal of the whole of mankind must be the knowledge and development of the wonders of God, and that this is the reason that God gave him the empire of the globe." The knowledge, skills, and principles used in proper stewardship of the earth and its resources are due in large part to mathematics. American educator H. F. Fehr asserted that "mathematics serves as a handmaiden for the explanation of the quantitative situations in other subjects, such as economics, physics, navigation, finance, biology and even the arts." Our mathematical knowledge has enabled advances in science, medicine, and technology, resulting in opportunities to better manage our God-given domain.

Many careers depend on mathematical proficiencies acquired through diligent study of mathematics. Galileo declared, "I do not feel obliged to believe that the same God who has endowed us with senses, reason, and intellect has intended us to forgo their use and by some other means to give us knowledge which we can attain by them." Your senses, reason, and intellect come from God. You should use them for His glory.

❯ Exercises

1. According to Morris Kline, in which centuries did mathematicians view their work as a religious quest?

2. What is the primary purpose of math for a Christian?

3. What did Kepler admonish scientists and mathematicians not to glory in? their own intellect

4. According to Leibniz, why did God give people dominion over the earth?

5. Which mathematician believed that God would not use other means to reveal to us those things that we could determine using the intellect that He gave us? Galileo

6. Cite two Scripture passages that state that everything we do should glorify God.

7. Cite two Scripture passages describing the mandate that people are to exercise dominion over the earth.

8. According to Colossians 1:18, who specifically should have the preeminence in all that we do?

9. According to Zimmerman, what five characteristics of mathematics should stun and awe the beholder?

10. **Discuss:** Describe several specific ways that God can be glorified through mathematics.

Biblical Perspectiue of Mathematics

Objectives

1. To state a Christian's primary purpose for mathematics

2. To analyze quotes of Renaissance mathematicians who glorified God

3. To describe ways to glorify God with mathematics

Assignments

- **Minimum:** 1–9; 10 (in class)
- **Standard:** 1–9; 10 (in class)
- **Extended:** 1–10

Answers

1. the sixteenth, seventeenth, and eighteenth centuries

2. to glorify God

4. so we would pursue the knowledge and development of the wonders of God

6. 1 Corinthians 10:31, Colossians 3:17

7. Genesis 1:26–28, Psalm 8

8. the Lord Jesus Christ

9. its beauty, power, order, symmetry, and infinitude

10. acknowledging God as the source of mathematics; praising God as we see His characteristics reflected in mathematics; and using mathematics to obey God in fulfilling the Creation Mandate

---

PRESENTATION

Discuss the introductory quote by secular mathematician Morris Kline. Point out that three of the greatest mathematicians of all time (Kepler, Galileo, and Newton) had religious worldviews. Newton stated, "When I wrote my treatise [*Mathematical Principles of Natural Philosophy*] about our system, I had an eye on such principles as might work with considering men for the belief in a Deity; and nothing can rejoice me more than to find it useful for that purpose." And Kepler stated, "The chief aim of all investigations of the external world should be to discover the rational order and harmony which has been imposed on it by God and which He revealed to us in the language of mathematics."

Discuss Bible verses that emphasize glorifying God and discuss how we can use mathematics to glorify God. Use Larry Zimmerman's quote to review how characteristics of math often reflect attributes of God. Encourage the students to look for order, consistency, power, and beauty as they study math. Consider using the Internet keyword search *mathematical beauty* to further the discussion. Zimmerman's *Truth & the Transcendent: The Origin, Nature, & Purpose of Mathematics* and James Nickel's *Mathematics: Is God Silent?* are excellent resources on this topic.

Discuss using mathematics to fulfill the Creation Mandate. Ask students to name careers that are dependent on different areas of mathematics from basic arithmetic, geometry, and algebra to college calculus, probability, and statistics.

Mention the need to be a good steward of what God entrusts to us by developing math skills. Ask students to consider whether God may be leading them into careers involving mathematics, especially if God has given them special ability and interest in math.

3.5 Exponential, Logistic, and Logarithmic Models (Extended)

Objectives

1. To use exponential, logistic, and logarithmic functions to model data and solve problems

2. To use transformations to linearize data

Flash

Denver, Colorado, is known as the Mile-High City because its elevation is 1 mi above sea level. Home to one of four US mints, it is located on the High Plains along the South Platte River and is the most populous city in Colorado. Denver boasts a 14,000-acre mountain park system, which includes trails and conservation areas.

Vocabulary

carrying capacity
logistic function

3.5 Exponential, Logistic, and Logarithmic Models

Exponential, logistic, and logarithmic models help further our understanding of population growth.

After completing this section, you will be able to
- use exponential, logistic, and logarithmic functions to model data and solve problems.
- use transformations to linearize data.

Compare the growth of $f(x) = \frac{1}{2}x + 1$ and $g(x) = 2\left(\frac{3}{2}\right)^x$.

x	-2	-1	0	1	2
$f(x)$	0	$\frac{1}{2}$	1	$\frac{3}{2}$	2

$+\frac{1}{2}\quad +\frac{1}{2}\quad +\frac{1}{2}\quad +\frac{1}{2}$

x	-2	-1	0	1	2
$g(x)$	$\frac{8}{9}$	$\frac{4}{3}$	2	3	$\frac{9}{2}$

$\times\frac{3}{2}\quad \times\frac{3}{2}\quad \times\frac{3}{2}\quad \times\frac{3}{2}$

The linear function $f(x)$ grows by the consistent amount of $\frac{1}{2}$, but the exponential function grows by the consistent factor of $\frac{3}{2}$, or 1.5, which can also be expressed as a growth rate of 50%. If a population is experiencing exponential growth or decay, a function modeling its growth can be determined given two data points.

Example 1 Modeling Exponential Growth

The Denver metro area grew from 2.4 million residents in 2000 to 2.8 million in 2010.

a. Determine the average annual percent increase between 2000 and 2010 and write an exponential function modeling the population.

b. Use your exponential model to estimate the population in 2015.

c. Use your exponential model to predict the year in which the population will reach 4 million.

Answer

a. $P(t) = ab^t$ where $a = 2.4$

1. Let $P(t)$ represent the population t years after 2000 so that $P(0) = ab^0 = 2.4$.

$2.8 = 2.4b^{10}$

$b^{10} = \frac{2.8}{2.4}$

2. Use $P(10) = 2.8$ to find b, the growth factor.

$b = \sqrt[10]{\frac{2.8}{2.4}} \approx 1.0155$

$r = b - 1 \approx 0.0155$ or 1.55%

3. Use $b = 1 + r$ to find r, the growth rate.

$P(t) \approx 2.4(1.0155)^t$

4. State the exponential model.

b. $P(15) \approx 2.4(1.0155)^{15}$
≈ 3.0 million people

Evaluate $P(15)$.

c.
$4 \approx 2.4(1.0155)^t$

$\frac{4}{2.4} \approx 1.0155^t$

$\ln\frac{4}{2.4} \approx \ln 1.0155^t$

$\ln 4 - \ln 2.4 \approx t \ln 1.0155$

$t \approx \frac{\ln 4 - \ln 2.4}{\ln 1.0155} \approx 33$

Substitute 4 into $P(t) \approx 2.4(1.0155)^t$ and solve for t.

The model predicts that the population of the Denver metro area will reach 4 million in 2033.

SKILL ✔ EXERCISE 17

PRESENTATION

Lesson Opener

1. Classify each function as a power, exponential, or logarithmic function.

   a. $y = ab^x$ exponential

   b. $y = a + b \ln x$ logarithmic

   c. $y = ax^b$ power

Identify each function as exponential growth or exponential decay.

2. $f(x) = 5(0.4)^x$ decay

3. $g(x) = 0.6(1.2)^x$ growth

Use the tables of values to compare the consistent amount of linear growth and the consistent factor in exponential growth.

Example 1 reviews how to use two points to determine an exponential growth model—a skill first taught in Section 3.1. In part c students solve an exponential equation using logarithms and their properties—a skill introduced in Section 3.4. Consider explaining that using the exact function resulting from the regression and the rounded modeling function may produce different estimates, especially when rounding the base of an exponential function.

In Example 2 an exponential regression is performed using a population data set. Encourage the students to store the resulting function as Y1.

Consider asking the students if populations can grow indefinitely without limits. (*no*) Introduce *logistic functions* as a more realistic function for modeling long-term population growth. Present the general form of the logistic function. You may want to explain that when $b < 0$, the function models logistic decay, which is not addressed in this text. The logistic function can also be expressed as $f(x) = \frac{c}{1 + ad^x}$ where $d = e^{-b} < 1$.

Consider having the students reason through the limits describing the end behaviors of $f(x) = \frac{c}{1 + e^{-x}}$.

When there are multiple data points from a population experiencing exponential growth or decay, an exponential regression can be performed using technology to determine a function modeling the population over time.

Example 2 Using an Exponential Regression

Find a function $P(t)$ modeling the population of the Denver metro area t years after 1900 by completing an exponential regression for the data in the table. Then use the model to estimate the population in 2015 and the year in which the population will reach 4 million.

Year	Population (millions)	Year	Population (millions)
1910	0.28	1970	1.24
1920	0.33	1980	1.62
1930	0.39	1990	1.85
1940	0.45	2000	2.40
1950	0.62	2010	2.78
1960	0.93		

Answer

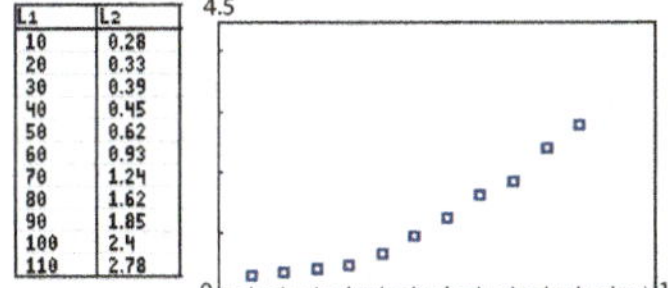

1. Enter the data, using $L_1 =$ years since 1900 and $L_2 =$ the population. The scatterplot indicates that an exponential model may be a good fit.

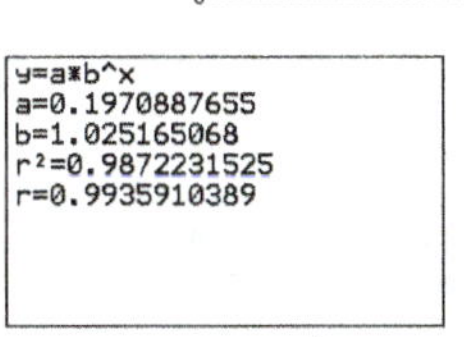

$P(t) \approx 0.1971(1.0252)^t$

2. Complete an exponential regression ([STAT], [CALC], 0:ExpReg) with Diagnostics On and store the regression equation as Y_1.

   The high coefficient of determination implies that the exponential model is a good fit.

   Note the average annual growth rate: $b - 1 \approx 1.025 - 1 = 0.025$ or 2.52%.

$P(115) \approx 0.1971(1.0252)^{115}$
≈ 3.44 million people

3. Evaluate $Y_1(115)$ to estimate the 2015 population.

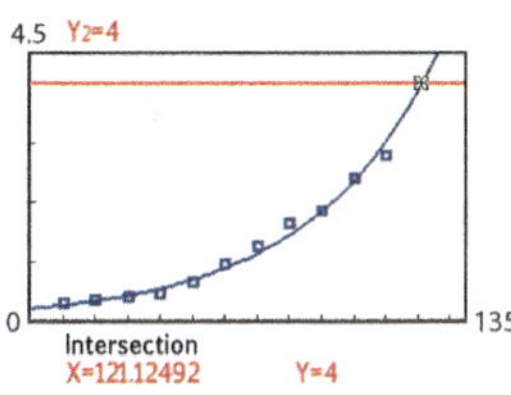

4. To solve $4 \approx 0.1971(1.0252)^t$ graphically, find the intersection of the exponential regression equation (stored as Y_1) and $Y_2 = 4$. Then interpret its x-value of ≈ 121.1.

The model predicts that the population will reach 4 million in 2021.

The most recent regression equation can be entered in the [Y=] equation editor using [VARS], 5:Statistics, EQ, and selecting 1:RegEQ. Other related constants, including a, b, r^2, and r, can be entered in other contexts using this menu.

SKILL ✓ EXERCISES 19–21

Reading and Writing Mathematics

1. Explain why logistic models are well suited for modeling population growth. Every population has a carrying capacity. Logistic models interpret a decreasing growth pattern as a leveling off at the carrying capacity.

2. State several factors that can determine the carrying capacity of bacteria growth in a test tube. Answers may include nutrients, water (moisture), oxygen, or pH.

3. Explain the conditions of a population for which an exponential growth model would provide estimates of nearly the same accuracy as a logistic model. Population estimates would need to be done when factors affecting the population's carrying capacity had not yet been reached.

Additional Exercises

1. An account grows exponentially from \$20,000 to \$30,000 in 10 yr.
   a. Determine the average annual growth rate and use it to write an exponential function to model the growth. $\approx 4.138\%$; $g(t) \approx 20{,}000(1.04138)^t$
   b. Use the exponential model to find what the investment will be worth after 12 yr and then when the investment will reach \$50,000. \$32,534; ≈ 22.6 yr

As $x \to -\infty$, $e^{-x} \to +\infty$, so $\dfrac{c}{1 + e^{-x}} \to 0^+$.
As $x \to \infty$, $e^{-x} \to 0^+$, so $1 + e^{-x} \to 1^+$, and $\dfrac{c}{1 + e^{-x}} \to c^-$.

Have the students perform a logistic regression on the data set as illustrated in Example 3. Have them identify the model's limit of growth and discuss why a logistic model is often the best fit for population growth. The estimate for the population in a given year can also be found graphically using the trace function. Consider asking how the year that the population reaches 4 million could be found algebraically. (*by solving the exponential equation*

$$34.6385e^{-0.0333t} = \frac{5.2649}{4} - 1)$$

Discuss the margin note regarding the ability of humans to alter the *carrying capacity* of their environment. Humans can do this by making changes in the availability of food and water resources, constructing shelter from the environment, providing medical care, providing protection from predators, etc.

Explain that rapid initial growth that decreases over time can be modeled by a logarithmic function. Example 4 illustrates this pattern by modeling weight loss. Challenge the students to share some other real-world examples where a logarithmic regression would be a suitable model. (*Examples may include learning*

to read, time studying versus test scores, athletic improvement, musical skills, rate of cooling, or sales.)

While the linearization of nonlinear data is a new concept that can be challenging, advances in computing technology have made this powerful process possible for your students. Discuss the linearization data for power, exponential, and logarithmic functions as given in the table. Consider sharing that quadratic, cubic, and quartic data can also be linearized using $(x, \sqrt{y})$, $(x, \sqrt[3]{y})$, and $(x, \sqrt[4]{y})$, respectively.

Complete a power regression and an exponential regression for each data set. State the better model and its coefficient of determination.

2.

x	2	3	4	9	12
y	2.4	3.9	5.7	15.6	22.3

power: $y \approx 1.004x^{1.248}$; $r^2 \approx 0.99992$

3.

x	1	5	8	10	13
y	1.5	7.6	25.7	57.7	194.6

exponential: $y \approx 1.0008 \cdot 1.5^x$
$r^2 \approx 0.9999996$

Complete a power regression and a logarithmic regression for each data set. State the better model and its coefficient of determination.

4.

x	2	5	8	10	13
y	5.2	8.2	9.7	10.4	11.2

logarithmic: $y \approx 3.0 + 3.2 \ln x$
$r^2 \approx 0.9998$

5.

x	1	3	5	6	8
y	5	6.6	7.5	7.8	8.4

power: $y \approx 5x^{0.25}$; $r^2 \approx 0.9998$

Complete the following steps for Additional Exercises 6–7:

a. Create a scatterplot of the data and determine the likely type of modeling function.

b. Linearize the data to find a linear model for the transformed data.

c. Substitute for X and Y to derive a function modeling the original data.

6.

x	1	2	3	4	5
y	1.75	3.06	5.36	9.38	16.41

exponential; $Y \approx 0.5597X - 0.0003$
$y \approx 1.75^x$

7.

x	2	3	4	5	6
y	0.841	1.140	1.414	1.672	1.917

power; $Y \approx 0.7499X - 0.6930$
$y \approx 0.5x^{0.75}$

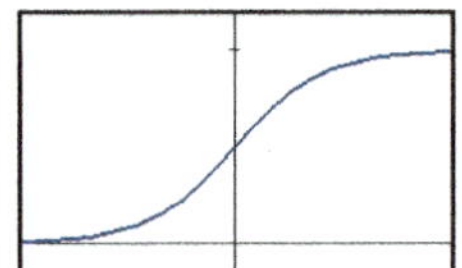

Exponential growth requires a consistent rate of growth. It is difficult for a population to sustain unrestricted exponential growth as supporting resources are stressed. In reality, its growth eventually slows and the graph levels out, approaching a maximum sustainable population. A *logistic function* provides a good model for the restricted growth of a population.

DEFINITION

A **logistic growth function** can be written in the form $f(x) = \dfrac{c}{1 + ae^{-bx}}$ where a, b, and c are positive constants.

Notice that the function has two horizontal asymptotes: $y = 0$ describes the left end behavior; the right end behavior is described by $y = c$, the limit of growth when the *carrying capacity* of the supporting environment is reached.

Example 3   Using a Logistic Regression

Using the data from Example 2, find a logistic function modeling the restricted population growth of the Denver metro area t years after 1900. Then use the model to estimate the population in 2015 and the year in which the population will reach 4 million.

Answer

$$P(t) \approx \frac{5.2649}{1 + 34.6385\,e^{-0.0333t}}$$

$Y_3(115) \approx 3.01$ million people

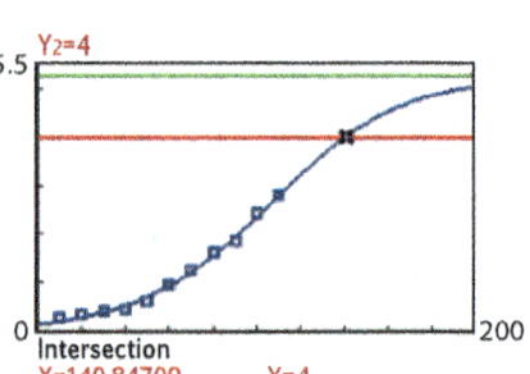

1. Complete a logistic regression of the data ([STAT], [CALC], B:Logistic) and store the regression equation as Y_3.

   There is no coefficient of determination associated with a logistic regression.

   Note that the model estimates the limit of growth for the Denver metro area population to be about 5.26 million.

2. Evaluate $Y_3(115)$ to estimate the 2015 population.

3. Find the intersection of the logistic regression equation (stored as Y_3) and $Y_2 = 4$. Then interpret the x-value of ≈ 140.85.

The model predicts that the population will reach 4 million in 2040.

SKILL ✔ EXERCISE 23

Notice we have developed three different models to make predictions regarding the Denver metro area population. The actual 2015 population of about 2.8 million may indicate that the logistic function is the best model to predict when the population would reach 4 million.

While carrying capacity can be determined experimentally for bacteria or animal populations, no consensus exists on how to determine the carrying capacity for a human population, primarily because people can exercise stewardship over their environment.

Once a linearized data set has been found and modeled with a linear function, substitute for X and Y and solve for the appropriate function representing the original data. This process provides excellent practice in applying the properties of exponents and logarithms. The results can be verified by performing the appropriate regression on the original data set.

Example 5 illustrates how Kepler's third law of planetary motion can be derived using orbital data from the six planets known to Kepler. While the data may at first appear logarithmic, the $(\ln x, y)$ data set (not illustrated) is not linear. Have students note the extremely high coefficient of determination for the linear regression of the $(\ln x, \ln y)$ data set. Some students may benefit from a demonstration of this alternative solution of $\ln y \approx 0.6677 \ln x + 1.0686$.

$$\ln y \approx \ln x^{0.6677} + 1.0686$$
$$\ln y - \ln x^{0.6677} \approx 1.0686$$
$$\ln \frac{y}{x^{0.6677}} \approx 1.0686$$
$$\frac{y}{x^{0.6677}} \approx e^{1.0686}$$
$$y \approx 2.9113x^{0.6677}$$

Common Student Error When performing linearization of data it is easy to use the incorrect data transformation for (X, Y), such as $(x, \ln y)$ instead of $(\ln x, \ln y)$.

Logarithmic functions are used to model rapid initial growth followed by slower growth over time. Measures of productivity, language or skill acquisition, and weight loss frequently exhibit logarithmic growth.

Example 4 Using a Logarithmic Regression

Mark has instituted a 13-week eating and exercise plan in order to improve his overall fitness. Use the data to create a function $W(t)$ modeling his overall weight loss. Then use the model to predict how many additional weeks are required to reach his goal of losing 10 lb.

Answer

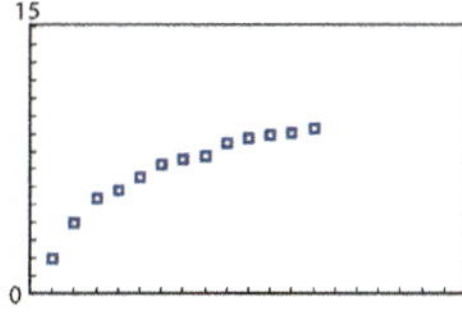

1. Complete a scatterplot of the data. The shape indicates that a logarithmic model may be a good fit.

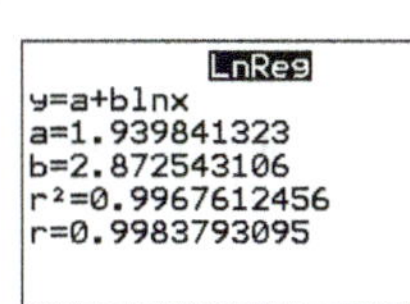

$$W(t) \approx 1.9398 + 2.8725 \ln t$$

2. Complete a logarithmic regression of the data (STAT, CALC, 9:LnReg) and store the regression equation as Y_1.

   Note the high coefficient of determination.

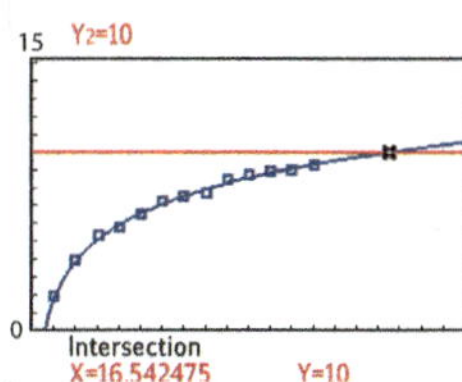

3. Find the intersection of the logarithmic regression equation (stored as Y_1) and $Y_2 = 10$ and interpret the x-value.

Week	Cumulative Weight Loss (lb)
1	1.9
2	3.9
3	5.3
4	5.8
5	6.5
6	7.2
7	7.5
8	7.7
9	8.4
10	8.7
11	8.9
12	9.0
13	9.2

The model predicts that a weight loss of 10 lb will be achieved around week 16.5, so another 3.5 weeks is needed for Mark to reach his goal.

_______________________ SKILL ✓ **EXERCISE 25**

The preceding example illustrates how the natural tendency to assume linear growth when the growth is actually logarithmic can lead to frustration as weight loss slows over time.

A visual inspection of a scatterplot may not be sufficient to determine whether the data is best modeled by a power function, an exponential function, or a logarithmic function. In this case, exploring transformed data sets using $X = \ln x$, $Y = \ln y$, or both and determining whether they are linear can help establish the appropriate model for the original data.

Derive the formula for each set of linearized data where $Y = aX + b$.

8. logarithmic: $(X, Y) = (\ln x, y)$
   $$y = a \ln x + b$$

9. exponential: $(X, Y) = (x, \ln y)$
   $$\ln y = ax + b$$
   $$y = e^{(ax + b)}$$
   $$y = e^b \cdot (e^a)^x$$

10. power: $(X, Y) = (\ln x, \ln y)$
    $$\ln y = a \ln x + b$$
    $$y = e^{(a \ln x + b)}$$
    $$y = (e^{\ln x})^a \cdot e^b$$
    $$y = e^b \cdot x^a$$

Assignments

- **Minimum:** 1–9, 11, 13, 15–17, 22–24, 28, 39, 45, 47, 49–54
- **Standard:** 1–9, 11, 13–14, 16–21, 25–27, 29–35, 38–39, 46–47, 49, 51–54
- **Extended:** 1–10, 13–14, 16–21, 25–27, 29–31, 35–44, 46, 51, 53–54

Assessment

- Quiz 3C covers Section 3.5.

It is also easy to use the incorrect list on the calculator when drawing scatterplots and calculating regressions. Caution students to carefully check their work.

TIPS

Ex. 17–18 Students can use MATH, $5: \sqrt[x]{}$ or raise the quotient to the power of $(1/10)$ or 10^{-1}.

Ex. 27 using regression constants a and b from VARS, 5:Statistics, EQ:
$$68 = a + b \ln x$$
$$68 - a = b \ln x$$
$$\ln x = \frac{68 - a}{b}$$
$$x = e^{\frac{68 - a}{b}} \approx 1318.51$$

Linearization of a Nonlinear Data Set		
Original Data Model		**Linear Transformed Data**
power function	$y = ax^b$	$(X, Y) = (\ln x, \ln y)$
exponential function	$y = ab^x$	$(X, Y) = (x, \ln y)$
logarithmic function	$y = a + b \ln x$	$(X, Y) = (\ln x, y)$

A scatterplot of the following data indicates the data may best be modeled by a power function (with $b > 1$) or an exponential function.

x	y
2	11.25
4	25.30
6	56.95
8	128.15
10	288.35

Use technology to create the transformed data sets $X = \ln x = L_3$ and $Y = \ln y = L_4$. The non-linear scatterplot of $(X, Y) = (\ln x, \ln y)$ indicates that a power function may not be the best model. The linear scatterplot of $(X, Y) = (x, \ln y)$ indicates that an exponential model will be a good fit for the original data.

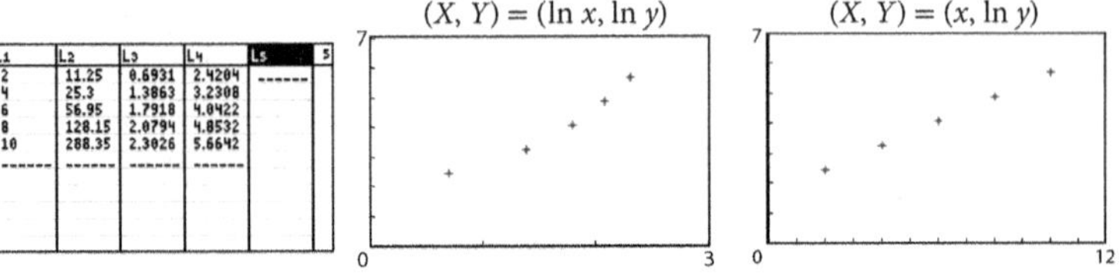

Completing a linear regression using L_1 and L_4 for (X, Y) produces the modeling function $Y \approx 0.4055X + 1.6091$. Substituting into this equation using $X = x$ and $Y = \ln y$ allows us to derive an equation modeling the original data.

$$Y \approx 0.4055X + 1.6091$$

$\ln y \approx 0.4055x + 1.6091$ Substitute.

$y \approx e^{(0.4055x + 1.6091)}$ Convert to exponential form.

$y \approx (e^{0.4055})^x e^{1.6091}$ Apply properties of exponents to obtain the form $y = ab^x$.

$y \approx (1.5)^x (4.9983)$ Simplify.

or $y \approx 5(1.5)^x$

You may have wondered why correlation coefficients, which apply only to linear regressions, are listed with the results of power, exponential, and logarithmic regressions. This is because technology uses the process of linearizing the data in order to complete these other regressions.

Use the data to determine the relationship between a planet's orbital period (the time it takes to orbit the sun) and its average orbital radius (its distance from the sun).

	Orbital Period (days)	Mean Radius (10^6 km)
Mercury	88.0	57.9
Venus	224.7	108.2
Earth	365.2	149.6
Mars	687.0	227.9
Jupiter	4331	778.6

Answer

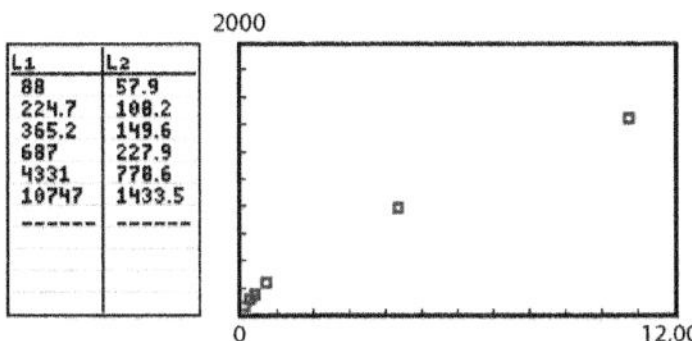

1. A visual inspection of a scatterplot of the data indicates that either a power function ($b < 1$) or a logarithmic function may be the best model for the relationship between the variables.

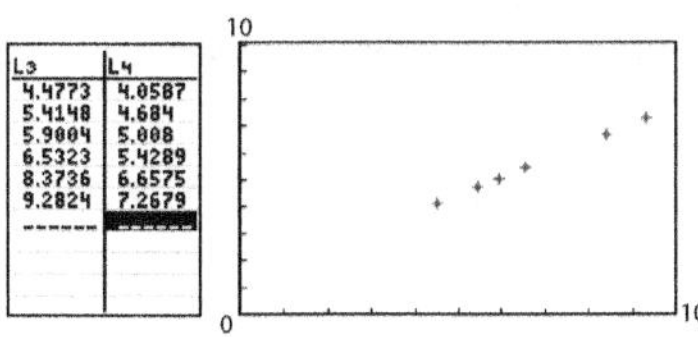

2. A scatterplot of $(X, Y) = (\ln x, \ln y)$ illustrates the linear relationship of the transformed data and indicates that a power function will be a good model for the original data.

$Y \approx 0.6677X + 1.0686$

3. Find the linear regression equation for the transformed data.

$\ln y \approx 0.6677 \ln x + 1.0686$

4. Use $X = \ln x$ and $Y = \ln y$ to substitute into the linear equation.

$$y \approx e^{0.6677 \ln x + 1.0686}$$
$$y \approx (e^{\ln x})^{0.6677} e^{1.0686}$$
$$y \approx x^{0.6677} e^{1.0686}$$
$$y \approx 2.9113\, x^{0.6677}$$
$$\text{or } y \approx 2.9113 x^{\frac{2}{3}}$$

5. Convert to exponential form and simplify to derive a power function modeling the original data.

SKILL ✓ **EXERCISES 29–31**

Cubing both sides of our result produces the equation $y^3 \approx 24.6752x^2$ where x represents the orbital period and y represents the average radius. This is a mathematical expression of Kepler's third law of planetary motion, which states that the square of the orbital period of a planet is directly proportional to the cube of the orbit's average radius.

Solutions

B. Exercises

17a. $P(0) = 8{,}008{,}278 = ab^0 = a$

$P(10) = 8{,}175{,}133 = 8{,}008{,}278b^{10}$

$$b^{10} = \frac{8{,}175{,}133}{8{,}008{,}278}$$

$$b = \sqrt[10]{\frac{8{,}175{,}133}{8{,}008{,}278}}$$

$$\approx 1.0021$$

$r = b - 1 \approx 0.0021 \text{ or } 0.21\%$

17b. $P(17) \approx 8{,}008{,}278(1.0021)^{17}$

$\approx 8{,}299{,}027$

17c. $8{,}500{,}000 = 8{,}008{,}278(1.0021)^t$

$$\frac{8{,}500{,}000}{8{,}008{,}278} = 1.0021^t$$

$$\ln \frac{8{,}500{,}000}{8{,}008{,}278} = \ln 1.0021^t$$

$\ln 8{,}500{,}000 - \ln 8{,}008{,}278$
$= t \ln 1.0021$

$$t = \frac{\ln 8{,}500{,}000 - \ln 8{,}008{,}278}{\ln 1.0021}$$

$\approx 28.4 \text{ yr}$

18a. $P(0) = 3{,}620{,}962 = ab^0 = a$

$P(10) = 3{,}550{,}404 = 3{,}620{,}962b^{10}$

$$b^{10} = \frac{3{,}550{,}404}{3{,}620{,}962}$$

$$b = \sqrt[10]{\frac{3{,}550{,}404}{3{,}620{,}962}}$$

$$\approx 0.9980$$

$r = b - 1 \approx 0.9980 - 1$

$\approx -0.0020 \text{ or a } 0.20\% \text{ decrease}$

18b. $P(5) = 3{,}620{,}962(0.9980)^5$

$\approx 3{,}585{,}000$

18c. $3{,}200{,}000 = 3{,}620{,}962(0.9980)^t$

$$\frac{3{,}200{,}000}{3{,}620{,}962} = 0.9980^t$$

$$\ln \frac{3{,}200{,}000}{3{,}620{,}962} = \ln 0.9980^t$$

$\ln 3{,}200{,}000 - \ln 3{,}620{,}962$
$= t \ln 0.9980$

$$t = \frac{\ln 3{,}200{,}000 - \ln 3{,}620{,}962}{\ln 0.9980}$$

$\approx 61.7 \text{ yr}$

19.

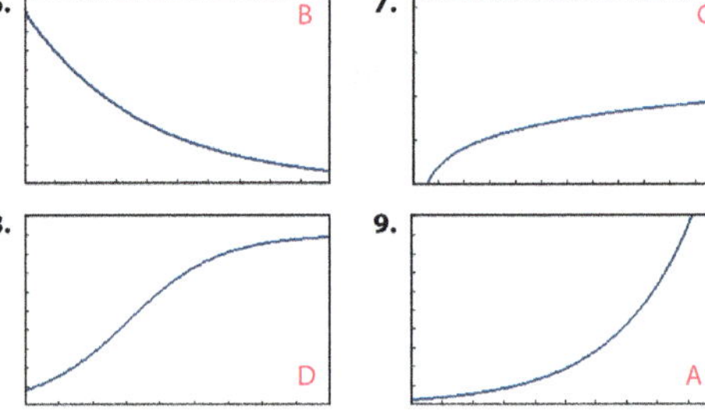

20. $f(36) = 14.9773(0.9604)^{36} \approx 3.50$

When transforming the data on page 160, lists L_3 and L_4 can be quickly generated by selecting the column header, pressing CLEAR, and then entering $\ln(L_1)$ and $\ln(L_2)$, respectively.

To graph the transformed data $(X, Y) = (\ln x, \ln y)$, be sure all stat plots are turned off except Plot 2. Enter L_3 for the Xlist and L_4 for the Ylist, using 2nd, 3 and 2nd, 4. Then use ZOOM, 9:ZoomStat to display the graph. The test for a power function does not appear to be linear.

To graph the transformed data $(X, Y) = (x, \ln y)$, return to Plot 2 and change the Xlist to L_1. Use 9:ZoomStat to display the new scatterplot. The test for an exponential model appears to be linear. Use L_1 and L_4 when completing the linear regression of the transformed data.

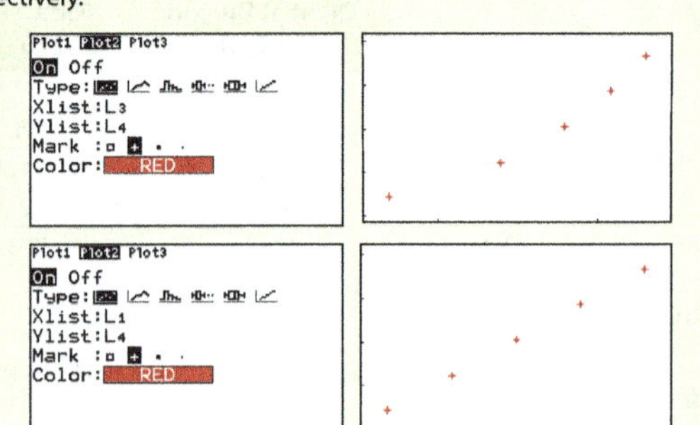

A. Exercises

1. Identify the model that best fits each type of growth.

  I. exponential   **III.** logarithmic

  II. linear   **IV.** logistic

  a. Values regularly change by the same amount. II

  b. Values regularly change by the same factor. I

  c. Growth eventually slows and the graph approaches a maximum value. IV

  d. There is rapid initial growth with slower growth over time. III

Classify each function.
A. exponential growth   **C. logarithmic growth**
B. exponential decay   **D. logistic growth**

2. $y = \dfrac{c}{1 + ae^{-bx}}$ D   **3.** $y = a + b \ln x$ C

4. $y = ab^x$ with $0 < b < 1$ B   **5.** $y = ab^x$ with $b > 1$ A

6. B   **7.** C

8. D   **9.** A

Make a scatterplot of each data set. Then complete a regression to find an exponential function that models the data, rounding each constant to the nearest ten thousandth. Finally, use the model to estimate $f(14)$.

10.

x	y
1	9
2	14
3	20
4	30
5	45
6	68

$y \approx 6.0978(1.4927)^x$
$f(14) \approx 1662$

11.

x	y
0	76.8
2	49.4
4	31.6
6	20.1
8	12.9
10	8.3

$y \approx 76.9191(0.8002)^x$; $f(14) \approx 3.4$

Make a scatterplot of each data set. Then complete a regression to find a logistic function that models the data, rounding each constant to the nearest ten thousandth. Finally, use the model to estimate $f(14)$.

12.

x	y
1	2.9
3	6.4
5	11.6
7	16.0
9	18.9
11	20.1

$y \approx \dfrac{20.8881}{1 + 10.2380e^{-0.5059t}}$
$f(14) \approx 20.7$

13.

x	y
0	2.3
2	4.7
4	8.7
6	14.4
8	20.1
10	24.6
12	27.2

$y \approx \dfrac{29.8494}{1 + 12.0516e^{-0.4023t}}$
$f(14) \approx 28.6$

21.

$$3 = 14.9773(0.9604)^x$$

$$\frac{3}{14.9773} = 0.9604^x$$

$$\ln \frac{3}{14.9773} = \ln 0.9604^x$$

$$\ln 3 - \ln 14.9773 = x \ln 0.9604$$

$$x = \frac{\ln 3 - \ln 14.9773}{\ln 0.9604} \approx 39.8$$

or

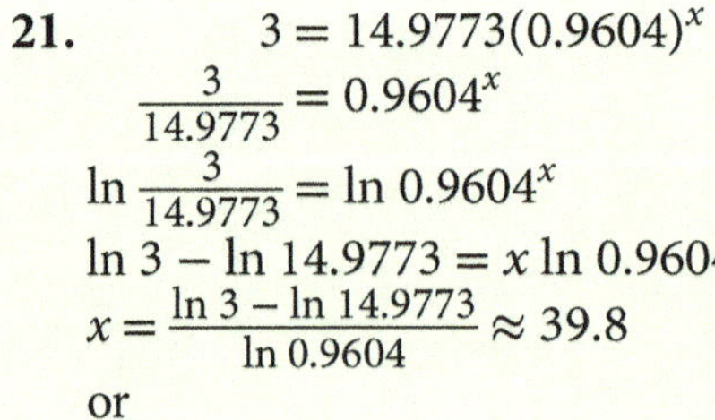

22. $f(106) \approx 9500$; $f(110) \approx 12{,}800$

or

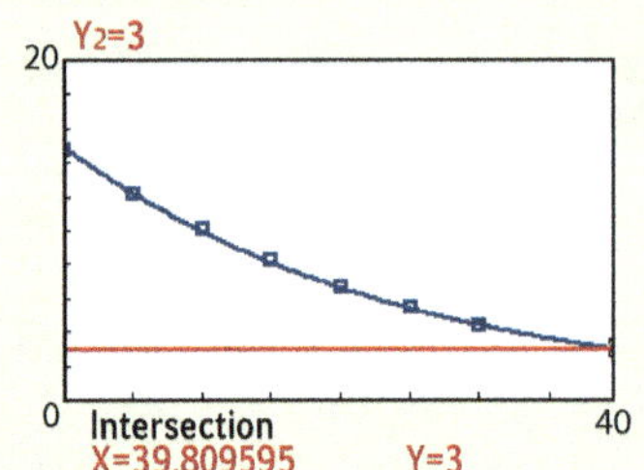

Make a scatterplot of each data set. Then complete a regression to find a logarithmic function that models the data, rounding each constant to the nearest ten thousandth. Finally, use the model to estimate $f(19)$.

14.

x	y
1	8
2	15
3	19
4	22
5	24
6	26

$y \approx 8.0241 + 10.0096 \ln x$; $f(19) \approx 37$

15.

x	y
1	2.0
3	10.7
5	14.9
7	17.5
9	19.7
11	21.0

$y \approx 1.9927 + 7.9839 \ln x$
$f(19) \approx 25.5$

16. Match each type of growth to the best modeling function.

   I. exponential   **II.** logistic   **III.** logarithmic

   a. rapid initial growth that slows over time   III
   b. slow initial growth that increases over time   I
   c. slow initial growth that increases before the growth slows as it approaches a limit   II

> **B. Exercises**

17. In 2000, New York City had a population of 8,008,278, making it the first US city to exceed 8 million people. The city's population in 2010 was 8,175,133.

   a. Using t to represent the number of years since 2000, find the city's average annual rate of growth from 2000 to 2010 and write an exponential function modeling the population. $\approx 0.21\%$; $P(t) \approx 8,008,278(1.0021)^t$

   b. Use your exponential model to estimate the city's population in 2017 (to the nearest hundred thousand). ≈ 8.3 million

   c. Use your exponential model to predict the year in which the city's population will reach 8.5 million.   2028

18. Chicago, Illinois, experienced its first ever population drop when its population declined from 3,620,962 in 1950 to 3,550,404 in 1960.

   a. Using t to represent the number of years since 1950, find the population's average annual rate of decline from 1950 to 1960 and write an exponential function modeling the population.

   b. Use your exponential model to estimate the city's population (to the nearest thousand) in 1955. $\approx 3,585,000$

   c. If the decline in population continued at the same average annual rate, in which year would the population have fallen to 3.2 million?   2011

18a. $\approx 0.20\%$ decrease; $P(t) = 3,620,962(0.9980)^t$

The table lists typical atmospheric pressures at various altitudes.

Altitude (1000 ft)	Atmospheric Pressure (psi)
0	14.70
5	12.23
10	10.11
15	8.30
20	6.76
25	5.46
30	4.37

19. Use technology to create a scatterplot and complete an exponential regression to find a function that models atmospheric pressure in terms of thousands of feet in altitude. $f(x) = 14.9773(0.9604)^x$

20. Use the function to estimate the atmospheric pressure at 36,000 ft, the altitude of a cruising airliner. ≈ 3.50 psi

21. Estimate the altitude (to the nearest hundred feet) with an atmospheric pressure of 3 psi. $\approx 39,800$ ft

In 1782 when the United States adopted it as the national bird, the American bald eagle population consisted of an estimated 100,000 nesting pairs. By 1963, the bald eagle was placed on the endangered species list with an estimated 487 nesting pairs. Letting x represent the number of years since 1900, create a scatterplot of the data.

Year	Nesting Pairs
1963	487
1974	791
1981	1188
1984	1757
1990	3035
1996	5094
2000	6471

22. Complete a regression to find an exponential function that models the data. Then use the model to predict (to the nearest hundred) the number of nesting pairs in 2006 and 2010. $f(x) \approx 3.7226(1.0769)^x$; 9500; 12,800

23. Complete a regression to find a logistic function that models the data. Then use the model to predict (to the nearest hundred) the number of nesting pairs in 2006 and 2010. $f(x) \approx \dfrac{17,514.32}{1 + 50,258.15 \cdot e^{-0.1031x}}$; 9200; 11,000

24. Which model produces lower estimates? Explain why.

23. $f(106) \approx 9200$; $f(110) \approx 11,000$

or

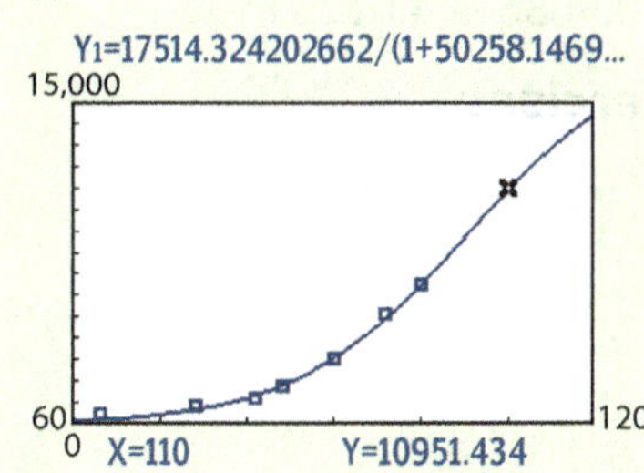

24. The logistic model produces lower estimates. The exponential model assumes that the population will continue to grow at the same rate, while the logistic model assumes that limiting factors will restrict the population's growth.

25.

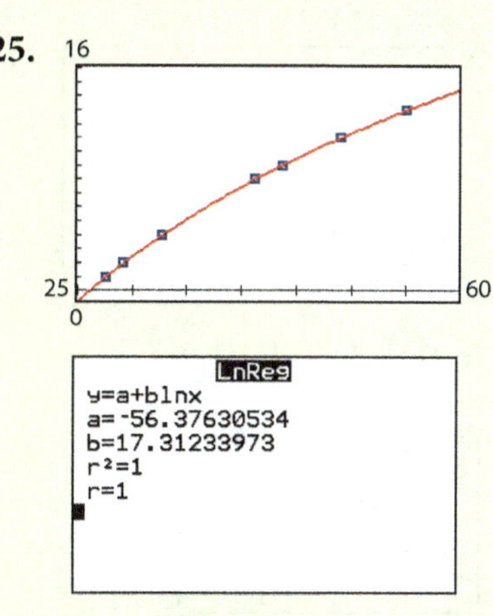

26. $f(3951.07) = 87$

Since there are 12 notes within the octave, divide by 12. The remainder will tell you the note (note 3, or B) within its octave.

27.
$$68 = -56.3763 + 17.3123 \ln x$$
$$124.3763 = 17.3123 \ln x$$
$$\ln x = \frac{124.3763}{17.3123}$$
$$x = e^{\frac{124.3763}{17.3123}} \approx 1318.53$$

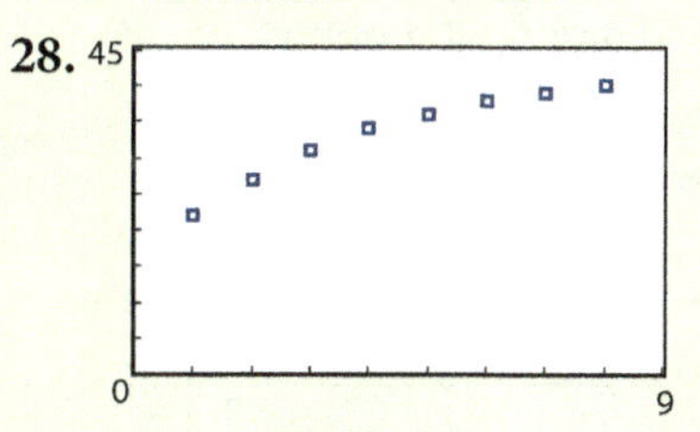

28.

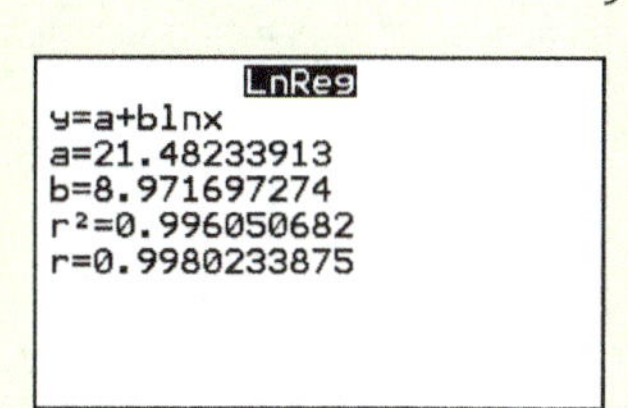

$$f(2.5) \approx 21.4823 + 8.9717 \ln (2.5)$$
$$\approx 29.7$$

29–31. 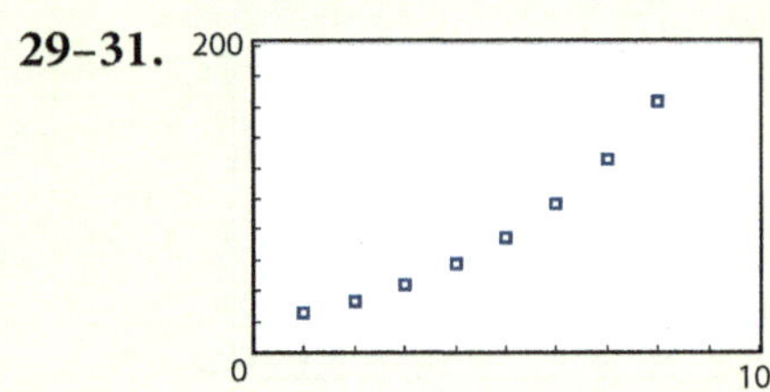

30. $L_3 = \ln (L_1)$; $L_4 = \ln (L_2)$

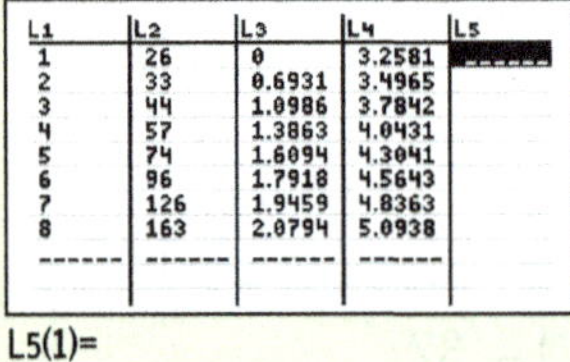

L1	L2	L3	L4	L5
1	26	0	3.2581	
2	33	0.6931	3.4965	
3	44	1.0986	3.7842	
4	57	1.3863	4.0431	
5	74	1.6094	4.3041	
6	96	1.7918	4.5643	
7	126	1.9459	4.8363	
8	163	2.0794	5.0938	
------	------	------	------	

L5(1)=

exponential: $(X, Y) = (x, \ln y)$
$= (L_1, L_4)$

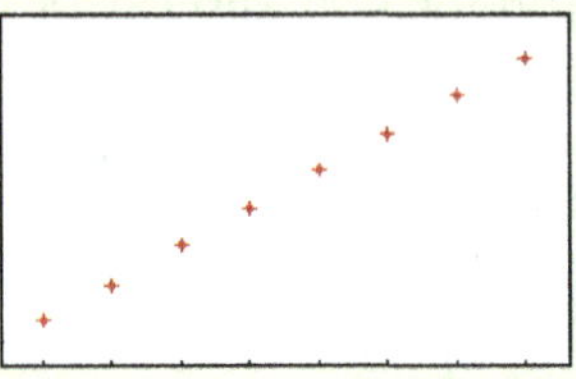

power: $(X, Y) = (\ln x, \ln y) = (L_3, L_4)$

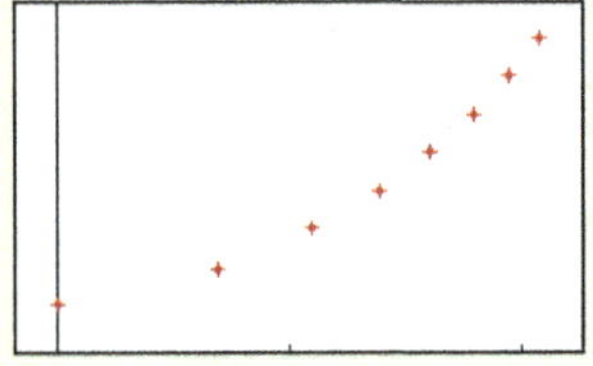

31.
$$Y \approx 0.2637X + 2.9859$$
$$\ln y \approx 0.2637x + 2.9859$$
$$e^{\ln y} \approx e^{(0.2637x + 2.9859)}$$
$$y \approx e^{0.2637x}e^{2.9859}$$
$$y \approx (1.3017)^x(19.8050)$$
$$y \approx 19.8050(1.3017)^x$$

32–34. 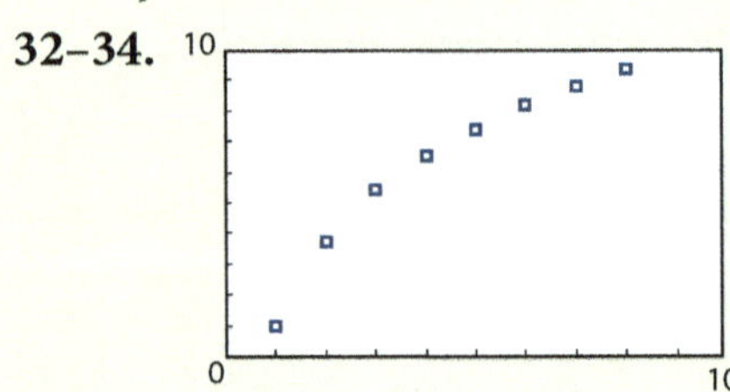

33. $L_3 = \ln (L_1)$; $L_4 = \ln (L_2)$

L1	L2	L3	L4	L5
1	1	0	0	
2	3.7	0.6931	1.3083	
3	5.4	1.0986	1.6864	
4	6.5	1.3863	1.8718	
5	7.4	1.6094	2.0015	
6	8.2	1.7918	2.1041	
7	8.8	1.9459	2.1748	
8	9.3	2.0794	2.23	
------	------	------	------	

L5(1)=

The table lists the frequencies for several keys on the piano.

Note	Frequency (Hz)	Key Number
A_0	27.5000	1
$A\sharp/B\flat_0$	29.1352	2
C_1	32.7032	4
E_1	41.2034	8
F_1	43.6536	9
G_1	48.9994	11
A_1	55.0000	13

25. Use technology to create a scatterplot and complete a regression to find a logarithmic function that models each note's key number on the piano in terms of its frequency. $f(x) \approx 17.3123 \ln x - 56.3763$

26. Use the function to find the key on the piano with a frequency of 3951.07 Hz.   note 87 (B_7)

27. Find the frequency for piano key number 68. ≈ 1318.5 Hz

28. Concrete gets stronger as it cures over time. The table lists the maximum compressive pressures (in megapascals, MPa) that columns of fabric-formed concrete were able to withstand while curing. Create a scatterplot and complete a regression to find a logarithmic function that models the data. Then use the model to estimate the maximum pressure the columns can withstand 2.5 weeks after being formed.

Curing Time (weeks)	Max. Pressure (MPa)
1	22
2	27
3	31
4	34
5	36
6	38
7	39
8	40

28. $y \approx 21.4823 + 8.9717 \ln x$
≈ 30 MPa

29. exponential or power (with $n > 1$)

30. exponential $(x, \ln y)$
$Y = 0.2637X + 2.9859$

Create a scatterplot of the data for exercises 29–31.

x	y
1	26
2	33
3	44
4	57
5	74
6	96
7	126
8	163

29. Which two types of modeling functions does the scatterplot suggest?

30. Use technology to create linearization data and scatterplots to test which type of function provides the best model. Use your results to complete a linear regression and state the linear function modeling the transformed data. Round constants to the nearest ten thousandth.

31. Substitute for X and Y in the linear model of the transformed data and derive a function modeling the original data. $y \approx 19.805(1.3017)^x$

Create a scatterplot of the data for exercises 32–34.

x	y
1	1
2	3.7
3	5.4
4	6.5
5	7.4
6	8.2
7	8.8
8	9.3

32. Which two types of modeling functions does the scatterplot suggest?   logarithmic or power (with $0 < n < 1$)

33. Use technology to create linearization data and scatterplots to test which type of function provides the best model. Use your results to complete a linear regression and state the linear function modeling the transformed data. Round constants to the nearest ten thousandth.   logarithmic $(\ln x, y)$; $Y \approx 4.0125X + 0.9686$

34. Substitute for X and Y in the linear model of the transformed data and derive a function modeling the original data. $y \approx 0.9686 + 4.0125 \ln x$

35. **Explain:** Why are logistic functions typically used to model population growth instead of exponential functions?

36. **Analyze:** Identify the carrying capacity of the logistic function $f(x) = \dfrac{5}{1 + 3e^{-2x}}$.   5

37. **Research:** List several primary factors that affect an environment's carrying capacity for a population.

> **C. Exercises**

38. Derive a formula for average annual rate of growth from an initial amount P_i to a final amount P_f in n years.

39. **Explain:** What is the result of completing a logarithmic regression on the following data? Which value causes this result? Explain why.

x	15	7	3	1	0
y	6	5	4	3	2

---

logarithmic: $(X, Y) = (\ln x, y)$
$= (L_3, L_2)$

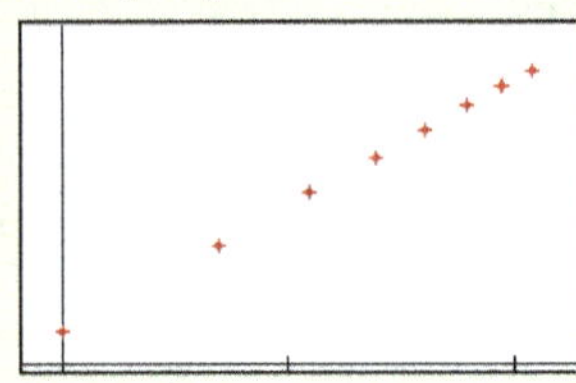

power: $(X, Y) = (\ln x, \ln y) = (L_3, L_4)$

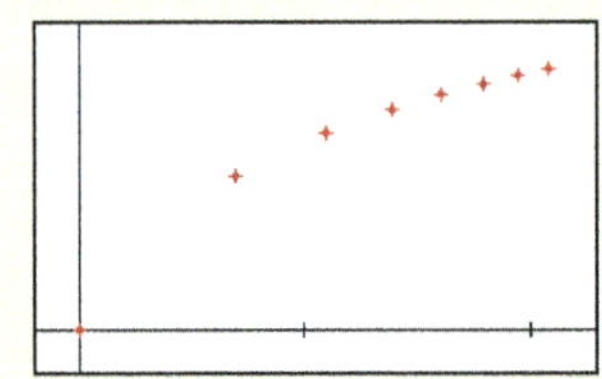

34. $Y \approx 4.0125X + 0.9686$
$y \approx 4.0125 \ln x + 0.9686$
$y \approx 0.9686 + 4.0125 \ln x$

> **C. Exercises**

38. $P(0) = P_i = ab^0 = a$
$P(n) = P_f = P_i b^n$
$\dfrac{P_f}{P_i} = b^n$
$b = \sqrt[n]{\dfrac{P_f}{P_i}}$
$r = b - 1 = \sqrt[n]{\dfrac{P_f}{P_i}} - 1$

Create a scatterplot of the data for exercises 40–44.

40. Which two possible types of modeling functions does the scatterplot suggest?

41. Use technology to create linearization data and scatterplots to test which type of function provides the best model. Use your results to complete a linear regression and state the linear function modeling the transformed data and the coefficient of determination. Round constants to the nearest ten thousandth.

x	y
2	0.3010
3	0.4771
4	0.6021
5	0.6990
6	0.7782
7	0.8451
8	0.9031
9	0.9542
10	1.0000

42. Substitute for X and Y in the linear model of the transformed data and derive a function modeling the original data. $y = 0.4343 \ln x$

43. Use the modeling function to evaluate $f(100)$, $f(1000)$, and $f(10,000)$. $\approx 2; \approx 3; \approx 4$

44. Use the change of base formula to express $f(x) = \log x$ in terms of natural logs. What does your solution reveal about the data set in the table?
$\approx 0.4343 \ln x$; The data represents the common log function $f(x) = \log x$.

CUMULATIVE REVIEW

Identify all asymptotes and point discontinuities in each function. [2.5]

45. $f(x) = \dfrac{5 - x}{x^2 - 25}$

46. $f(x) = \dfrac{x^2 - x - 6}{x^3 + x^2 - x - 1}$

Solve each inequality by factoring and using a sign chart. [2.7]

47. $2x^2 + 3x - 20 \le 0$

48. $x^3 - 3x^2 + 2x - 6 > 0$

49. Determine the account balance of a \$12,000 investment compounded quarterly at 1.5% interest for 10 yr. [3.1] \$13,938.10

50. Write a function rule for $g(x)$, the described transformation of $f(x) = 5^x$, if $f(x)$ is translated 2 units left, reflected in the y-axis, stretched vertically by a factor of 3, and translated 1 unit down. [3.1]

51. Which of the following are not inverse functions? [1.8, 3.2] B

  A. $y = 2x$ and $y = \frac{1}{2}x$
  B. $y = x^2$ and $y = \sqrt{x}$
  C. $y = x^3$ and $y = \sqrt[3]{x}$
  D. $y = 3^x$ and $y = \log_3 x$
  E. These are all inverse functions.

52. Which expression is equivalent to $3 \ln x + 2 \ln 3 - 4 \ln y$? [3.3] C

  A. $\ln \dfrac{x^3 + 9}{y^4}$
  B. $\ln \dfrac{6x^3}{y^4}$
  C. $\ln \dfrac{9x^3}{y^4}$
  D. $\ln 9x^3 y^4$
  E. none of these

53. Solve $\left(\frac{1}{3}\right)^{x-2} = 81$. [3.4] A

  A. $x = -2$
  B. $x = 2$
  C. $x = 6$
  D. $x = 4$
  E. none of these

54. Solve $\log_3 (x + 1) + \log_3 (x + 3) = 1$. [3.4] D

  A. $x = 4$
  B. $x = -4$
  C. $x = 0, -4$
  D. $x = 0$
  E. none of these

35. In many cases logistic functions will be more accurate since population growth is eventually restricted.

37. food and water supplies, living space (Answers will vary.)

38. $r = \sqrt[n]{\dfrac{P_f}{P_i}} - 1$

39. an error message; 0; It is not in the domain of $f(x) = a + b \ln x$.

40. logarithmic or power (with $0 < n < 1$)

41. logarithmic $(\ln x, y)$; $Y \approx 0.4343X$; $r^2 \approx 1.0000$

45. HA: $y = 0$; VA: $x = -5$ PD: $x = 5$

46. HA: $y = 0$; VA: $x = \pm 1$ PD: none

47. $\left[-4, \frac{5}{2}\right]$

48. $(3, \infty)$

50. $g(x) = 3 \cdot 5^{2-x} - 1$

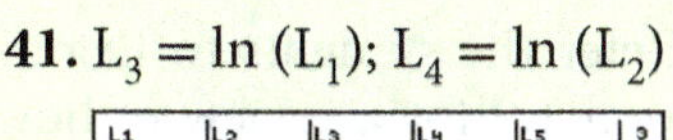

40–44.

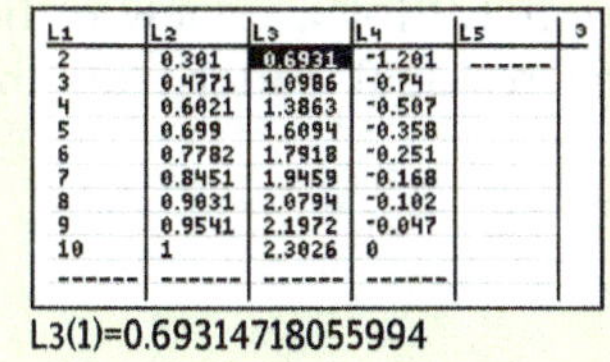

41. $L_3 = \ln (L_1)$; $L_4 = \ln (L_2)$

L1	L2	L3	L4	L5
2	0.301	0.6931	-1.201	------
3	0.4771	1.0986	-0.74	
4	0.6021	1.3863	-0.507	
5	0.699	1.6094	-0.358	
6	0.7782	1.7918	-0.251	
7	0.8451	1.9459	-0.168	
8	0.9031	2.0794	-0.102	
9	0.9541	2.1972	-0.047	
10	1	2.3026	0	
------	------	------	------	

L3(1)=0.69314718055994

logarithmic: $(X, Y) = (\ln x, y) = (L_3, L_2)$

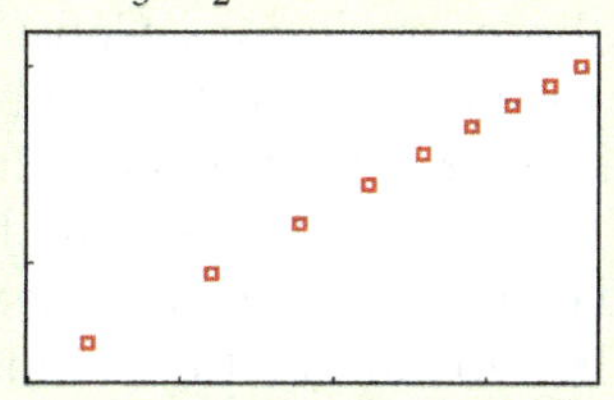

power: $(X, Y) = (\ln x, \ln y) = (L_3, L_4)$

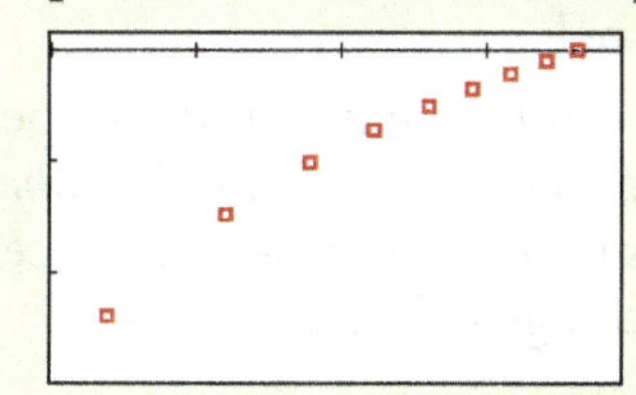

42. $Y \approx 0.4343X$
$y \approx 0.4343 \ln x$

43. $f(100) = 0.4343 \ln 100 \approx 2$
$f(1000) = 0.4343 \ln 1000 \approx 3$
$f(10,000) = 0.4343 \ln 10,000 \approx 4$

44. $f(x) = \log x = \dfrac{\ln x}{\ln 10} = \dfrac{1}{\ln 10} \ln x$
$\approx 0.4343 \ln x$

Cumulative Review

45. $f(x) = \dfrac{-(x-5)}{(x-5)(x+5)}$
$n < m, \therefore$ HA: $y = 0$
zeros of $a(x)$: 5
zeros of $b(x)$: $-5, 5$
VA: $x = -5$; PD at $x = 5$

46. $f(x) = \dfrac{(x+2)(x-3)}{(x+1)(x+1)(x-1)}$
$n < m, \therefore$ HA: $y = 0$
zeros of $a(x)$: $-2, 3$
zeros of $b(x)$: $-1, 1$
VA: $x = \pm 1$; PD: none

47. $(2x - 5)(x + 4) \le 0$
zeros: $\frac{5}{2}, -4$; $\left[-4, \frac{5}{2}\right]$

positive	negative	positive
$(-)(-)$ -4	$(-)(+)$ $\frac{5}{2}$	$(+)(+)$

48. $(x - 3)(x^2 + 2) > 0$
zeros: 3; $(3, \infty)$

negative	positive
$(-)(+)$ 3	$(+)(+)$

49. $n = 4$
$A(10) = 12{,}000\left(1 + \dfrac{0.015}{4}\right)^{(10)(4)}$
$\approx \$13{,}938.10$

51. The inverse of $y = x^2$ is $y = \pm\sqrt{x}$, which is not a function.

52. $\ln x^3 + \ln 3^2 - \ln y^4 = \ln \dfrac{9x^3}{y^4}$

53. $(3^{-1})^{x-2} = 3^4$
$3^{2-x} = 3^4$
$2 - x = 4$
$x = -2$

54. $\log_3 (x^2 + 4x + 3) = 1$
$x^2 + 4x + 3 = 3$
$x^2 + 4x = 0$
$x(x + 4) = 0$
$x = 0, -4$ (extraneous)

Data Analysis

Objectives

1. To find, apply, and analyze a mathematical model for world population
2. To describe inappropriate and appropriate responses to the growth in world population

Assignments

- **Minimum:** 1–4, 7–9; 10–16 (in class)
- **Standard:** 1–9; 10–16 (in class)
- **Extended:** 1–9; 10–16 (in class)

Solutions

1. 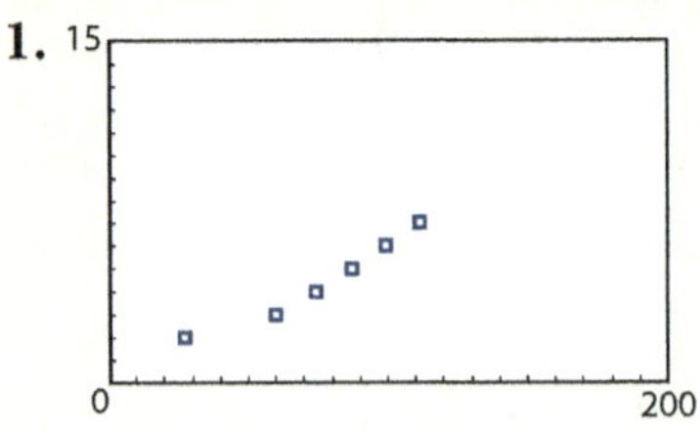

3a. $m = 0.0606 \frac{\text{billion}}{\text{year}} \approx 60.6$ million

3b. $P(130) = 0.0606(130) + 0.1285$
$= 7.75$

3c. $9 = 0.0606x - 0.1285$
$x = \frac{9.1285}{0.0606} \approx 150.6$

4b. 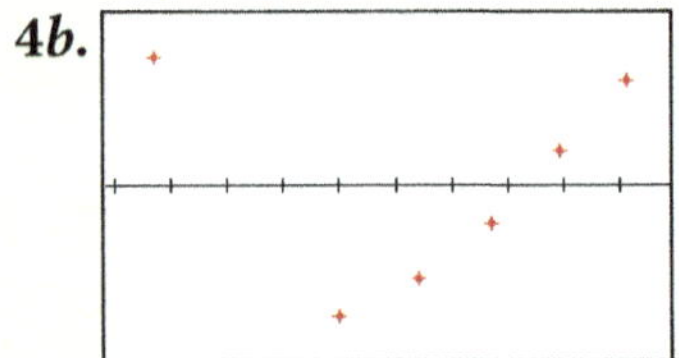

Year	Population (billions)
1927	2
1960	3
1974	4
1987	5
1999	6
2011	7

A World Population Crisis?

How long can the earth sustain its current population growth? Will we run out of space for people to live? Will we experience widespread food and water shortages? Many people are looking for answers to these questions. We can use mathematical analysis of world population data to see patterns in population growth and to predict future populations. At the same time, God's Word provides insight into how we should respond to these concerns.

1. Use technology to complete a scatterplot of the world population data where x represents the number of years since 1900. Which type of function(s) might be used to model world population?

2. Complete a linear regression and state the resulting linear function and its coefficient of determination. Round constants to the nearest ten thousandth.

3. Apply the linear model.
   a. What is the average annual population increase (in millions) during this time? 60.6 million
   b. Estimate the world's population in 2030. ≈ 7.75 billion
   c. Predict the year in which the world's population will reach 9 billion. 2050

4. Analyze the linear model.
   a. Does the coefficient of determination indicate a strong correlation? yes
   b. Use technology to plot the residuals. What conclusion can be drawn from this graph?

5. Using technology, create linearization data and scatterplots to test whether a power, exponential, or logarithmic function provides the best model. Which model is suggested by these graphs? Use your results to complete a linear regression and state the linear function modeling the transformed data. Round constants to the nearest ten thousandth.

6. Substitute for X and Y in the linear model of the transformed data and derive a function modeling the original data. $P(x) \approx 1.2776(1.0155)^x$

7. Complete a regression for the original data to find the equation for the best-fit model and state its coefficient of determination. Compare this function to the one derived in exercise 6.

8. Apply the nonlinear model.
   a. What is the average annual population growth rate during this time? 1.55%
   b. Estimate the world's population in 2030. ≈ 9.44 billion
   c. Predict the year in which the world's population will reach 9 billion. 2026

9. Analyze the nonlinear model.
   a. Does the coefficient of determination indicate a stronger correlation? yes
   b. Use technology to plot the residuals. What conclusion can be drawn from this graph?
   c. The US Census Bureau estimates that the world population will reach 8 billion in 2024 and 9 billion in 2042. Do they expect the population to grow at about the same rate, at a significantly faster rate, or at a significantly slower rate than your exponential model predicts?

10. State reasons for and against the use of a logistic function to model the world population.

11. List several factors that might restrict human population growth.

12. Since bacteria, plants, and animals have very little control over their environment, they cannot voluntarily affect the carrying capacity for their populations. How does the fact that we are created in God's image allow us to affect the carrying capacity of human populations? State several examples.

13. How can population growth models help us fulfill our biblical responsibilities?

While data can help us make informed decisions, our worldview influences our response to world population growth. Responses shaped by biblical and secular worldviews, such as making clean water more available, can be similar. However, there are times when these worldviews generate very different responses.

14. List several unethical practices that have affected human population growth.

15. What are some ways in which a biblical worldview will affect our response to a growing world population?

16. Consider this quote from former president George H. W. Bush: "Every human being represents hands to work, and not just another mouth to feed." How should we relate the biblical commands to exercise dominion and fill the earth (Gen. 1:28) to concerns about the world's population growth?

1. linear, power, or exponential (Answers may vary.)

2. $P(x) = 0.0606x - 0.1285$; $r^2 = 0.9528$

4b. There may be a better function to model the data.

5. exponential $(x, \ln y)$; $Y \approx 0.0154X + 0.2450$

7. $P(x) = 1.2776(1.0155)^x$; $r^2 = 0.9934$; They are the same.

9b. The lack of a pattern in the residuals supports the use of the exponential model.

9c. at a significantly slower rate

10. A logistic model is a good fit since population growth eventually slows and approaches a limit. However, a logistic model will likely be inaccurate if the data does not include the interval where growth begins to be restricted, or if the decrease is temporary. The world's population does not yet appear to be restricted and does not appear to be near its carrying capacity.

11. depletion of natural resources (including food and fresh water), disease, crime, war, economic strains (Answers will vary.)

12. Because God created us as rational and creative beings, we can change our environment and enhance living conditions. We have engineered new forms of habitation and advanced methods in agriculture, food processing, and harvesting other natural resources. Humans can engineer drugs to combat diseases that plague over-populated areas. We can also build new infrastructures to overcome previously reached capacities. This human intervention is a fulfillment of the Creation Mandate of Genesis 1:28.

13. Population growth models can help us identify patterns, make predictions, and identify causes in population changes. Solutions to these constraints can then be devised in order to meet future needs. By providing needed food, water, and shelter, we can better love our neighbors and more effectively share the gospel.

14. government policies limiting the number of children, including higher taxation for parents with "excess children"; forced sterilization; abortion; euthanasia; eugenics

15. We will trust God for our physical needs (Matt. 6:31–32), acknowledge the sanctity of human life (Gen. 1:27, Jer. 1:5), serve others (Gal. 5:13–14), live righteously (Phil. 4:9), spread the gospel (Matt. 28:19), and look forward to Christ's return (1 Thess. 4:16).

16. Our ability to fulfill these two aspects of the Creation Mandate are interdependent. As we manage the earth's resources, the potential for a larger population grows, and with it, a great ability and responsibility to wisely exercise dominion to sustain the growing population.

5.

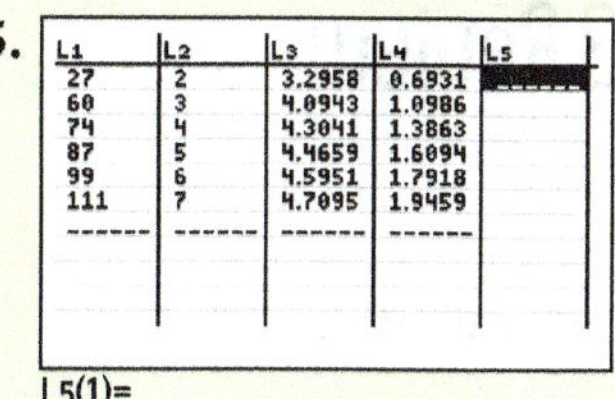

L5(1)=

power: $(X, Y) = (\ln x, \ln y)$

exponential: $(X, Y) = (x, \ln y)$

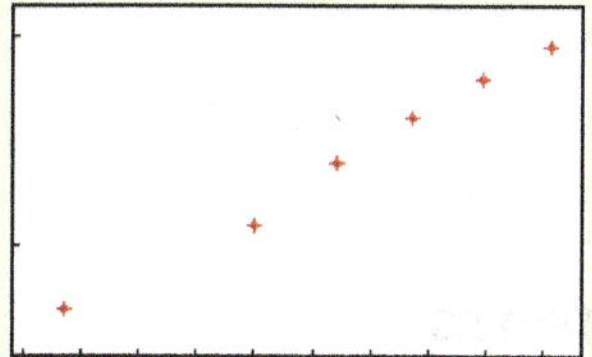

logarithmic: $(X, Y) = (\ln x, y)$

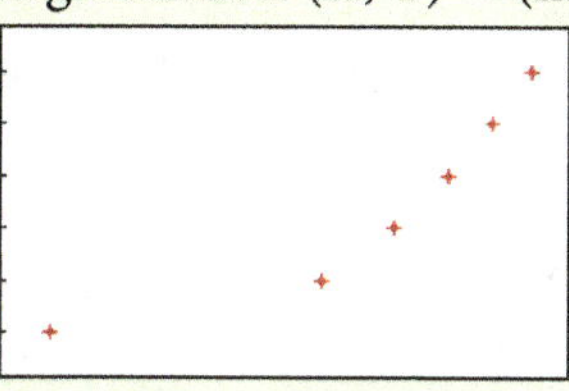

exponential; $Y \approx 0.0154X + 0.2450$

6.
$$Y \approx 0.0154X + 0.2450$$
$$\ln y \approx 0.0154x + 0.2450$$
$$y \approx e^{0.0154x + 0.2450}$$
$$\approx e^{0.0154x} \cdot e^{0.2450}$$
$$\approx (e^{0.0154})^x \cdot e^{0.2450}$$
$$\approx (1.0155)^x \cdot 1.2776$$
$$\approx 1.2776(1.0155)^x$$

8a. $r = b - 1 \approx 1.0155 = 0.0155$ or 1.55%

8b. $P(130) \approx 1.2776(1.0155)^{130} \approx 9.4359$

8c.
$$9 = 1.2776(1.0155)^x$$
$$1.0155^x = \frac{9}{1.2776}$$
$$\ln 1.0155^x = \ln \frac{9}{1.2776}$$
$$x \ln 1.0155 = \ln 9 - \ln 1.2776$$
$$x = \frac{\ln 9 - \ln 1.2776}{\ln 1.0155} \approx 126.9$$

9b.

Chapter 3 Review

Objective

To prepare for evaluation

Vocabulary

See Appendix A.

Assignments

- **Minimum:** 1–3, 5–13, 18, 20–22, 24, 26, 27a–30a, 31–35, 38–44
- **Standard:** 1–47 odd; 2–46 even (in class)
- **Extended:** 1–3, 5, 7–17, 19–21, 23, 25–27, 28b–30b, 31–47

Assessments

- Chapter 3 Test
- First Quarter Exam covers Chapters 1–3.

Solutions

1a. exponential

1b. exponential

1c. power

2.

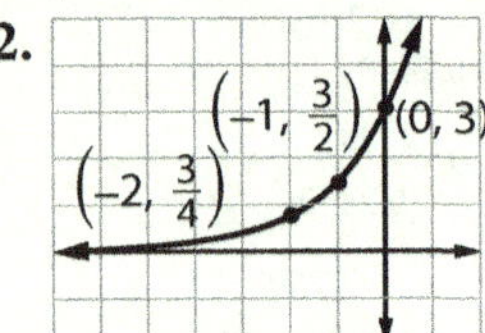

$D = \mathbb{R}$; $R = (0, \infty)$; growth

3.

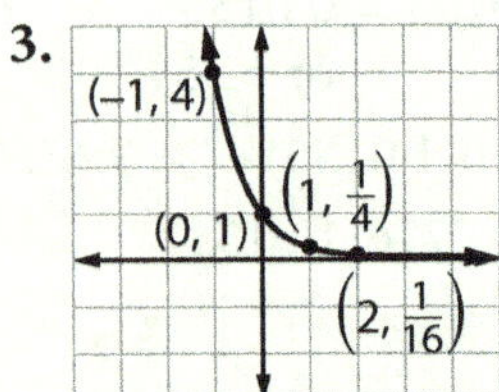

$D = \mathbb{R}$; $R = (0, \infty)$; decay

4. $f(x)$ is translated 3 units right and 1 unit up.

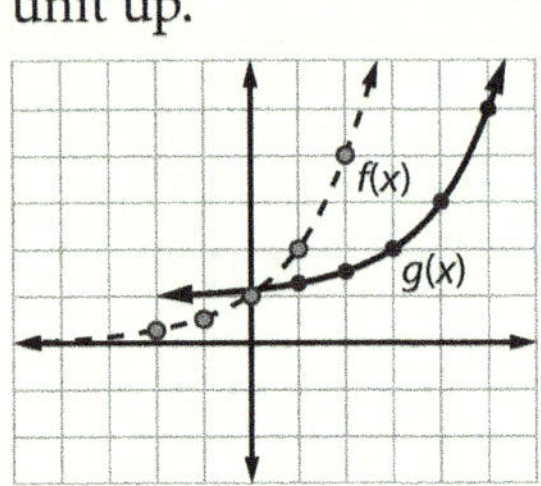

$D = \mathbb{R}$; $R = (1, \infty)$

1. Classify each function as a power function, an exponential function, or neither. [3.1]

   a. $f(x) = 4^x$ **b.** $h(x) = \left(\frac{1}{6}\right)^x$ **c.** $q(x) = \frac{1}{2}x^3$

Graph each exponential function by finding several ordered pairs. State the function's domain and range, then classify it as exponential growth or decay. [3.1]

2. $g(x) = 3 \cdot 2^x$ 3. $y = \left(\frac{1}{4}\right)^x$

4. Describe how the graph of $f(x) = 2^x$ is transformed to obtain the graph of $g(x) = 2^{x-3} + 1$. Use the graph of $f(x)$ to sketch the graph of $g(x)$ and then state the domain and range of $g(x)$. [3.1]

5. Describe how the graph of $f(x) = 2^x$ is transformed to obtain the graph of $g(x) = -4 \cdot 2^{x-3}$. Use the graph of $f(x)$ to sketch the graph of $g(x)$ and then state the domain, range, intercepts, asymptotes, and end behavior of $g(x)$. [3.1]

Write a function rule for $g(x)$, the described transformation of $f(x)$. Then confirm your answer by graphing $f(x)$ and $g(x)$ in the same window. [3.1]

6. $f(x) = 5^x$ is translated 4 units to the left and 1 unit up.

7. $f(x) = e^x$ is reflected in the x-axis, vertically stretched by a factor of 3, then translated 2 units to the right and 1 unit down.

8. Determine the account balance of an investment of $25,000 at 1.5% APR after 10 yr when compounded (a) annually, (b) monthly, and (c) continuously.

9. In 1989 the population of the Florida panther was estimated to be 46, and the species was placed on the endangered list. By 2017 the panther's population had increased to approximately 100. Determine the population's average annual rate of increase during this time. Then use an exponential growth model to predict the population in 2025. [3.1]

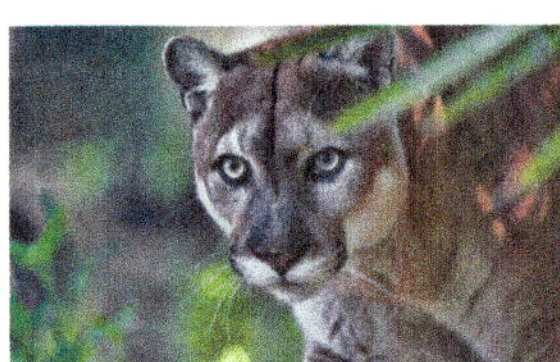

10. A sample is determined to contain 27 g of barium-139, an isotope with a half-life of 83 min. How much barium-139 remains 2.5 hr later? [3.1] ≈ 7.71 g

11. Convert to logarithmic form. [3.2]

   a. $7^2 = 49$ $\log_7 49 = 2$ **b.** $4^3 = 64$ $\log_6 \frac{1}{36} = -2$

   c. $6^{-2} = \frac{1}{36}$ $\log_4 64 = 3$ **d.** $\left(\frac{1}{5}\right)^3 = \frac{1}{125}$ $\log_{\frac{1}{5}} \frac{1}{125} = 3$

12. Convert to exponential form. [3.2]

   a. $\log_4 16 = 2$ $4^2 = 16$ **b.** $\log_{\frac{1}{3}} \frac{1}{27} = 3$ $10^{-5} = 0.0000$

   c. $\log 0.00001 = -5$ $\left(\frac{1}{3}\right)^3 = \frac{1}{27}$ **d.** $\ln \frac{1}{e^2} = -2$ $e^{-2} = \frac{1}{e^2}$

Evaluate each expression. Round answers to the nearest ten thousandth if necessary. [3.2]

13. $\log_3 81$ 4   14. $\ln \sqrt{e}$ $\frac{1}{2}$

15. $e^{\ln 5}$ 5   16. $\log 50$ 1.6990

17. Graph $f(x) = \log_6 x$. Then state its domain and range. [3.2] $D = (0, \infty)$; $R = \mathbb{R}$

18. Describe $g(x) = \log_2 (x + 3)$ as a transformation of $f(x) = \log_2 x$, then graph both functions. [3.2]

19. Describe $g(x) = -2 \log_3 x + 2$ as a transformation of $f(x) = \log_3 x$, then use a graph of $f(x)$ to sketch the graph of $g(x)$. State the transformed function's domain and range, any asymptotes and intercepts, and the end behavior. [3.2]

20. Use the investment doubling formula, $T = \dfrac{\ln 2}{\ln (1 + APY)}$, to determine how many years, T, it will take Robert to double his $10,000 investment in a certificate of deposit (CD) with an APY of 2.5%. [3.2] ≈ 28.1 yr

Expand each logarithmic expression. Assume all variables are positive values. [3.3]

21. $\log x^2 y$ $2 \log x + \log y$ 22. $\log \frac{x^3}{15}$ $3 \log x - \log 15$

23. $\ln \sqrt[3]{\frac{a^2}{b^4}}$ $\frac{1}{3}(2 \ln a - 4 \ln b)$

Write each expression as a single logarithm. [3.3]

24. $\log x + 2 \log 3$ $\log 9x$

25. $\frac{1}{2}(4 \log x + \log y) - 3 \log z$ $\log \frac{x^2 \sqrt{y}}{z^3}$

26. $2 \ln 5x + \ln 10y - \ln 5$ $\ln 50x^2 y$

5. $f(x)$ is translated 3 units right, reflected in the x-axis, and vertically stretched by a factor of 4.

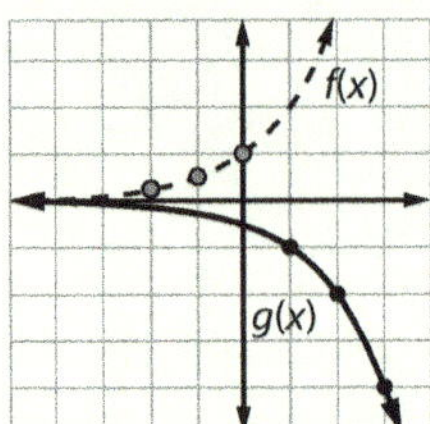

$D = \mathbb{R}$; $R = (-\infty, 0)$; x-int.: none; y-int.: $(0, -0.5)$; HA: $y = 0$; $\lim_{x \to -\infty} g(x) = 0$, $\lim_{x \to \infty} g(x) = -\infty$

6. $g(x) = 5^{x+4} + 1$

7. $g(x) = -3e^{x-2} - 1$

8a. $A(10) = 25{,}000(1 + 0.015)^{10}$
    $\approx \$29{,}013.52$

PRESENTATION

Consider reviewing any concepts the students may need to have reinforced. Communicate whether they will be expected to memorize the three linearized data transformations for the chapter test.

TIPS

Ex. 19–20, 39–40 These exercises review C-level exercises from Sections 3.2 and 3.4. You may not want to assign these exercises if similar exercises were not completed in earlier lessons.

27. Use $\log_7 2 \approx 0.3562$ and $\log_7 3 \approx 0.5646$ to find a decimal approximation of each logarithm. [3.3]

  a. $\log_7 24$ ≈ 1.6332   **b.** $\log_7 14\sqrt{2}$ ≈ 1.5343

28. Use common logs to evaluate each logarithm. [3.3]

  a. $\log_2 40$ ≈ 5.3219   **b.** $\log_6 108$ ≈ 2.6131

29. Use natural logs to evaluate each logarithm. [3.3]

  a. $\log_4 24$ ≈ 2.2925   **b.** $\log_3 21$ ≈ 2.7712

30. Describe the graph of $g(x)$ as a transformation of $f(x) = \ln x$. [3.3]

  a. $g(x) = \log_5 x$ **b.** $g(x) = \log_{\frac{1}{2}} x$

Solve each exponential equation. [3.4]

31. $5^x = \frac{1}{625}$ $x = -4$ **32.** $8^x = \left(\frac{1}{32}\right)^{-3x+12}$ $x = 5$

Solve each logarithmic equation. [3.4]

33. $\log_2 (3x + 11) = 5$ **34.** $\ln (x - 5) = \ln x - \ln 6$
           $x = 7$ $x = 6$

Find all real solutions for x. Round answers to the nearest thousandth. [3.4]

35. $5^{x+3} = e^{4x}$ $x = 2.020$

36. $e^{2x} + 2e^x - 48 = 0$ $x = 1.792$

37. $\log_5 2 - \log_5 \frac{2}{5} = 12x$ $x = \frac{1}{12}$

38. $\ln x + \ln (x + 5) = \ln 14$ $x = 2$

39. A customer was served a cup of 186°F coffee in a café whose temperature was 71°F. The coffee cooled to 171°F in 3 min. Its temperature can be modeled using Newton's law of cooling: $T(t) = T_S + (T_0 - T_S)e^{-kt}$, where T_0 is the object's initial temperature, T_S is the surrounding medium's temperature, k is the cooling constant, and t is the time. [3.4] $k = -\frac{1}{3} \ln \frac{20}{23} \approx 0.0466$

  a. Find the value of k and state the function modeling the coffee's temperature. $T(t) = 71 + 115e^{\left(\frac{1}{3} \ln \frac{20}{23}\right)t}$

  b. Estimate the temperature of the coffee 10 min after being served. ≈ 143°F

40. The periodic payment P for a home mortgage of B dollars can be calculated using the equation $P = \frac{iB}{1 - (1 + i)^{-n}}$ where i is the periodic interest rate and n is the number of payments. [3.4]

  a. Calculate the monthly payment for a 30 yr mortgage of \$300,000 with a 4.2% APR. \$1467.05

  b. Solve the equation for n and determine the number of payments required to pay off the loan with monthly payments of \$1700. $n = -\dfrac{\log\left(1 - \frac{iB}{P}\right)}{\log (1 + i)}$; 276 payments

---

≈ 3.50%; $P(t) = 465{,}622(1.0350)^t$

41. The population around Austin, Texas, grew from 465,622 people in 1990 to 656,562 people in 2000. Determine the average annual rate of growth during this time and use it to write an exponential function modeling the population. [3.5]

  a. Use the exponential function to estimate the population (to the nearest thousand) in 2015. 1,100,000

  b. Use the exponential function to predict the year in which the population will reach 1.25 million. 2018

Use the population data for Austin, Texas, (rounded to the nearest thousand) to complete exercises 42–43. [3.5]

Year	Population	Year	Population
1920	35,000	1970	254,000
1930	53,000	1980	346,000
1940	88,000	1990	466,000
1950	132,000	2000	657,000
1960	187,000	2010	790,000

42. Letting t represent the number of years since 1900, complete a regression to find an exponential function $P(t)$ that models the population of Austin, Texas (in thousands). $P(t) \approx 20.6351(1.0353)^t$

  a. Use your model to estimate the 2015 population (to the nearest thousand). 1,112,000

  b. Use your model to estimate the year in which the population will reach 1.25 million. 2018

43. Letting t represent the number of years since 1900, complete a regression to find a logistic function $P(t)$ that models the population of Austin, Texas (in thousands). $P(t) \approx \dfrac{1706.53}{1 + 95.3535e^{-0.0403t}}$

  a. State the carrying capacity predicted by your model (to the nearest thousand). 1,707,000

  b. Use your model to graphically estimate the 2015 population (to the nearest thousand). 884,000

  c. Use your model to graphically estimate the year in which the population will reach 1.25 million. 2038

44. Create a scatterplot and complete a regression to find a logarithmic function that models the data. Then estimate $f(12)$. [3.5]

x	y
1	5.00
2	13.30
3	18.18
4	21.64
5	24.31
6	26.50

45. Identify the type of function associated with each linearized data transformation. [3.5]

  a. $(X, Y) = (\ln x, y)$ logarithmic

  b. $(X, Y) = (\ln x, \ln y)$ power

  c. $(X, Y) = (x, \ln y)$ exponential

44. $y \approx 4.9935 + 12.0027 \ln x$; $f(12) \approx 34.82$

18.

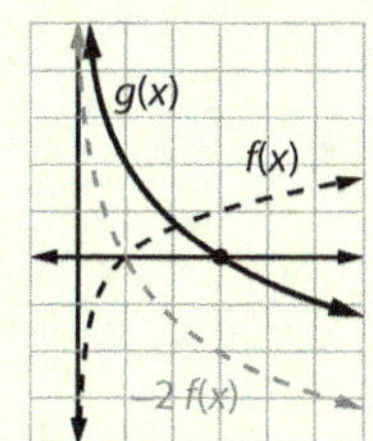

The graph of $f(x)$ is translated 3 units left.

19. $f(x)$ reflected in the x-axis, vertically stretched by a factor of 2, and translated 2 units up

$D = (0, \infty)$; $R = \mathbb{R}$; VA: $x = 0$

x-int.: $(3, 0)$; y-int.: none

$\lim\limits_{x \to 0^+} f(x) = \infty$, $\lim\limits_{x \to \infty} f(x) = -\infty$

20. $T = \dfrac{\ln 2}{\ln (1 + 0.025)} \approx 28.1$ yr

21. $\log x^2 + \log y = 2 \log x + \log y$

22. $\log x^3 - \log 15 = 3 \log x - \log 15$

23. $\frac{1}{3}(\ln a^2 - \ln b^4)$

  $= \frac{1}{3}(2 \ln a - 4 \ln b)$ or

  $\frac{2}{3} \ln a - \frac{4}{3} \ln b$

24. $\log x + \log 3^2 = \log 9x$

25. $\frac{1}{2}(\log x^4 y) - \log z^3$

  $= \log \sqrt{x^4 y} - \log z^3$

  $= \log \dfrac{x^2 \sqrt{y}}{z^3}$

26. $\ln (5x)^2 + \ln 10y - \ln 5$

  $= \ln \left(\dfrac{25x^2 \cdot 10y}{5}\right)$

  $= \ln 50x^2 y$

27a. $\log_7 (2^3 \cdot 3)$

  $= \log_7 2^3 + \log_7 3$

  $= 3 \log_7 2 + \log_7 3$

  $\approx 3(0.3562) + 0.5646 \approx 1.6332$

27b. $\log_7 (7 \cdot 2) + \log_7 \sqrt{2}$

  $= \log_7 7 + \log_7 2 + \frac{1}{2} \log_7 2$

  $\approx 1 + 0.3562 + \frac{1}{2}(0.3562) \approx 1.5343$

28a. $\log_2 40 = \dfrac{\log 40}{\log 2} \approx 5.3219$

28b. $\log_6 108 = \dfrac{\log 108}{\log 6} \approx 2.6131$

29a. $\log_4 24 = \dfrac{\ln 24}{\ln 4} \approx 2.2925$

29b. $\log_3 21 = \dfrac{\ln 21}{\ln 3} \approx 2.7712$

8b. $A(10) = 25{,}000\left(1 + \dfrac{0.015}{12}\right)^{(10)(12)}$

  $\approx \$29{,}043.14$

8c. $A(10) = 25{,}000e^{0.015(10)} \approx \$29{,}045.86$

9. $P(t) = P_0(1 + r)^t$

  $100 = 46(1 + r)^{28}$

  $\sqrt[28]{\dfrac{100}{46}} = 1 + r$

  $r = \sqrt[28]{\dfrac{100}{46}} - 1$

  ≈ 0.0281 or 2.81%

  $P(36) \approx 46(1 + 0.0281)^{36}$

  $P(36) \approx 124.75$; 125 panthers

10. $P(t) = 27\left(\dfrac{1}{2}\right)^{\frac{t}{83}}$

  $P(150) = 27\left(\dfrac{1}{2}\right)^{\frac{150}{83}} \approx 7.71$ g

13. $\log_3 3^4 = y$

  $3^y = 3^4$

  $y = 4$

14. $\ln e^{\frac{1}{2}} = y$

  $e^y = e^{\frac{1}{2}}$

  $y = \frac{1}{2}$

17.

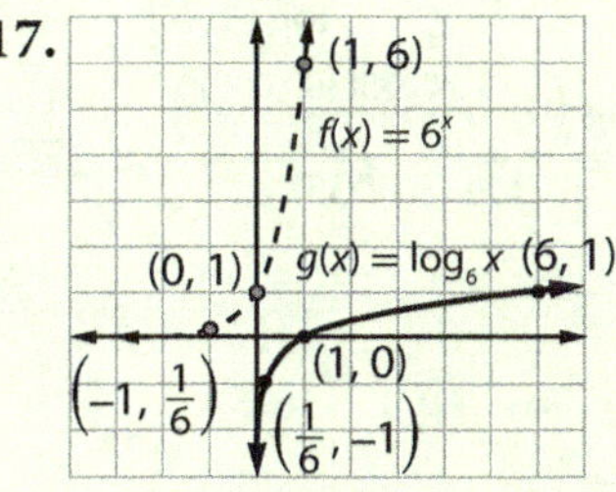

30a. $\log_5 x = \dfrac{\ln x}{\ln 5} = \dfrac{1}{\ln 5}\ln x$

a vertical shrink of $f(x)$ by a factor of $\dfrac{1}{\ln 5} \approx 0.62$

30b. $\log_{\frac{1}{2}} x = \dfrac{\ln x}{\ln \frac{1}{2}} = \dfrac{1}{\ln 2^{-1}}\ln x$

$= -\dfrac{1}{\ln 2}\ln x$

a vertical shrink of $f(x)$ by a factor of $\dfrac{1}{\ln 2} \approx 1.44$ reflected across the x-axis

31. $5^x = 5^{-4};\ x = -4$

32. $2^{3x} = 2^{-5(-3x + 12)}$

$3x = 15x - 60;\ x = 5$

33. $3x + 11 = 2^5$

$3x = 21;\ x = 7$

34. $\ln (x - 5) = \ln \dfrac{x}{6}$

$x - 5 = \dfrac{x}{6}$

$6x - 30 = x$

$5x = 30;\ x = 6$

35. $\ln 5^{x+3} = \ln e^{4x}$

$(x + 3) \ln 5 = 4x$

$x \ln 5 + 3 \ln 5 = 4x$

$x \ln 5 - 4x = -3 \ln 5$

$x (\ln 5 - 4) = -3 \ln 5$

$x = \dfrac{-3 \cdot \ln 5}{\ln 5 - 4} \approx 2.020$

36. $(e^x)^2 + 2e^x - 48 = 0$

$(e^x - 6)(e^x + 8) = 0$

$e^x = 6 \quad$ or $\quad e^x = -8$

$x = \ln 6 \qquad x = \ln -8$ (extraneous)

$x \approx 1.792$

37. $\log_5 \dfrac{2}{\frac{2}{5}} = 12x$

$\log_5 5 = 12x$

$1 = 12x;\ x = \dfrac{1}{12}$

38. $\ln (x^2 + 5x) = \ln 14$

$x^2 + 5x - 14 = 0$

$(x - 2)(x + 7) = 0$

$x = 2,\ -7$ (extraneous)

39a. $171 = 71 + (186 - 71)e^{-k(3)}$

$100 = 115e^{-3k}$

$\dfrac{100}{115} = e^{-3k}$

$\ln \dfrac{100}{115} = -3k$

$k = -\dfrac{1}{3}\ln \dfrac{20}{23} \approx 0.0466$

39b. $T(10) = 71 + 115e^{\left(\frac{1}{3}\ln \frac{20}{23}\right)10} \approx 143.2$

40a. $n = 30(12) = 360;\ i = \dfrac{0.042}{12} = 0.0035$

$P = \dfrac{(0.0035)(300,000)}{1 - 1.0035^{-360}} = \$1467.05$

40b. $\dfrac{iB}{P} = 1 - (1 + i)^{-n}$

$(1 + i)^{-n} = 1 - \dfrac{iB}{P}$

$\log (1 + i)^{-n} = \log \left(1 - \dfrac{iB}{P}\right)$

$-n \log (1 + i) = \log \left(1 - \dfrac{iB}{P}\right)$

$n = -\dfrac{\log \left(1 - \dfrac{iB}{P}\right)}{\log (1 + i)}$

$n = -\dfrac{\log \left(1 - \dfrac{0.0035(300,000)}{1700}\right)}{\log (1.0035)} \approx 275.2$

41. $656,562 = 465,622b^{10}$

$b^{10} = \dfrac{656,562}{465,622}$

$b \approx \sqrt[10]{\dfrac{656,562}{465,622}} \approx 1.0350$

$r = b - 1 \approx 0.0350$ or 3.50%

41a. $P(15) = 465,622(1.0350)^{25}$

$\approx 1,100,379$

41b. $1,250,000 = 465,622(1.0350)^t$

$\dfrac{1,250,000}{465,622} = 1.0350^t$

$\ln \dfrac{1,250,000}{465,622} = t \ln 1.0350$

$t = \dfrac{\ln 1,250,000 - \ln 465,622}{\ln 1.0350}$

≈ 28.7 yr

continued in Answers and Solutions Overflow

Use the following data to complete exercises 46–47.
[3.5]

46. Use technology to create a scatterplot for the data. Then create linearization data and scatterplots to test whether a power, exponential, or logarithmic function provides the best model. Which model is suggested by your graphs? power function

47. Complete a regression for the linear transformed data set and state the linear function modeling the transformed data. (Round constants to 4 decimal places.) Then substitute into the linear equation and derive a function modeling the original data.

$Y \approx 4.0223X - 2.1096;\ y \approx 0.1213x^{4.0223}$

x	y
1	0.12
2	2.00
3	10.13
4	32.00
5	78.12
6	163.00

Trigonometric Functions

4.1 Angle Measure and Arc Length
4.2 Right Triangle Trigonometry
Historical Connection
4.3 Extending Trigonometric Functions
4.4 Sinusoidal Functions
4.5 Graphing Other Trigonometric Functions
Biblical Perspective of Mathematics
4.6 Inverse Trigonometric Functions
4.7 Analyzing Combinations of Sinusoidal Functions
Data Analysis

HISTORICAL CONNECTION

Trigonometry, born out of an effort to understand movement in the "celestial sphere," initially had as much to do with circles and spheres as it did with triangles.

BIBLICAL PERSPECTIVE OF MATHEMATICS

The development of abstract mathematics has allowed us to explore the vastness of God's creation. Stanley L. Jaki notes in *The Relevance of Physics* that "the science of trigonometry was in a sense a precursor of the telescope. It brought faraway objects within the compass of measurement and first made it possible for man to penetrate in a quantitative manner the far reaches of space."

DATA ANALYSIS

Regularly recurring (cyclic) patterns are found in abundance in natural and manmade systems. Since trigonometric functions are cyclical, they are used to model many practical applications, including the occurrence of sunspots.

Overview

Key concepts in this chapter include degree and radian measures of angles, especially for key points on the unit circle, the six trigonometric functions and their graphs, and the graphs of combinations of polynomial and trigonometric functions. The level of previous exposure to trigonometric concepts may dictate the level of review needed, although many new concepts are also presented. Students can explore various combinations of sinusoidal functions in this chapter's final (extended) section.

The Technology Corner uses parametric equations to present animated graphs of trigonometric functions.

Chapter Objectives

1. To solve problems related to the measures of angles, arcs, and sectors
2. To solve right triangles using trigonometric ratios
3. To evaluate trigonometric and inverse trigonometric functions
4. To graph basic trigonometric functions, their transformations, and their inverses
5. To determine key characteristics of trigonometric functions
6. To solve equations containing trigonometric expressions
7. To model real-world data and solve problems using trigonometric functions
8. To explain biblical views of and biblical responses to climate change

Flash

A solar eclipse occurs when the moon blocks the sun, casting a shadow across the earth. The sun is about 400 times larger than the moon, but because it is also about 400 times farther away, the moon blocks the entire sun from Earth's view during a total eclipse. A total solar eclipse will occur at a certain location on Earth about once every 400 yr.

Suggested Teaching Schedule

DAY	
1	4.1
2	4.2
3	Quiz 4A (4.1–4.2) HC: Trigonometry
4	4.3 TC: Parametric Graphs of Trig Functions
5–6	4.4
7	4.5

DAY	
8	Quiz 4B (4.3–4.5) BPM: The Utility and Value of Mathematics
9	4.6
10	4.7
11	Quiz 4C (4.6–4.7) DA: Sunspots and Solar Flares
12	Chapter 4 Review
13	Chapter 4 Test

4.1 Angle Measure and Arc Length

Objectives

1. To draw an angle in standard position
2. To convert between degrees-minutes-seconds (DMS), degree-decimal (DD), and radian measures of angles
3. To identify coterminal angles
4. To calculate arc length and sector area using angle measure and radius length
5. To calculate angular and linear speeds

Flash

This memorial to the French explorer Samuel de Champlain depicts him using an astrolabe. Invented by the Greeks and improved by the Arabs, the astrolabe uses the angle of elevation of the sun or stars above the horizon to determine position. It was widely used on ships during the Middle Ages. Using the astrolabe, Champlain was able to complete more than 25 round-trip voyages across the Atlantic without losing a single ship.

Vocabulary

acute
angular speed
arc length
arc length formula

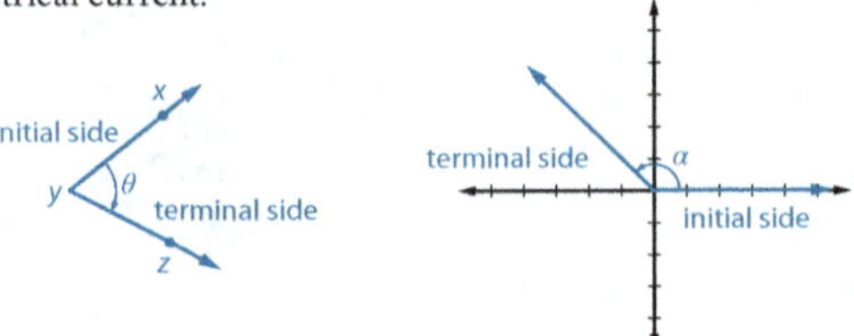

4.1 Angle Measure and Arc Length

This statue on Nepean Point in Ottawa depicts the French explorer Samuel de Champlain holding an astrolabe, which would have helped him navigate by using the sun's angle of elevation above the horizon.

After completing this section, you will be able to

- draw an angle in standard position.
- convert between degrees, minutes, and seconds; decimal degrees; and radian measures of angles.
- identify coterminal angles.
- calculate arc length and sector area using angle measure and radius length.
- calculate angular and linear speeds.

The term *trigonometry* is derived from Greek words meaning "triangle measure." Ancient civilizations used the relationships between the sides and angles of right triangles in surveying, navigation, engineering, and astronomy. Trigonometry was then expanded to explain circular and harmonic phenomena, including orbits, vibrating strings, pendulums, and electrical current.

In geometry, an angle is often defined as the union of two rays with a common endpoint, or *vertex*. An angle can also be viewed as the sweep of a ray moving from an initial position to a final position, similar to the sweep of a clock's hand in a given period of time. An angle is in *standard position* when it is placed on the coordinate plane with its vertex at the origin and the initial position of the ray aligned with the positive x-axis. The final position of the sweeping ray is the *terminal side* of the angle.

An angle is often named by its vertex, such as $\angle Y$. When more than one angle has its vertex at Y, three letters are used to avoid confusion, as in $\angle XYZ$. In this text, $m\angle Y$ is frequently notated simply as Y. Lowercase Greek letters such as α (alpha), β (beta), or θ (theta) are also used to name an angle or indicate its measure.

When an angle is viewed as a rotation, its measure includes both the direction and amount of rotation. In mathematics, a counterclockwise rotation is described as a *positive angle* and a clockwise rotation is described as a *negative angle*. The amount of rotation can be measured in *degrees*, where $1°$ represents $\frac{1}{360}$ of a circle. In the illustration, $\alpha = 120°$ and $\theta = -68°$. Degrees can be further divided using decimal degrees (DD) or degrees, minutes, and seconds (DMS), in which a *minute* ($1'$) is $\frac{1}{60}$ of a degree and a *second* ($1''$) is $\frac{1}{60}$ of a minute (or $\frac{1}{3600}$ of a degree). In navigation, angles representing bearings are measured from due north where clockwise rotations are positive.

PRESENTATION

Lesson Opener

Identify the angle described by the amount of each rotation about the origin.

1. one quadrant 90°
2. two quadrants 180°
3. three quadrants 270°
4. $\frac{1}{2}$ a quadrant 45°
5. $\frac{4}{3}$ a quadrant 120°

Challenge the students to recall the definition of an angle from *GEOMETRY*. Introduce the *standard position* of an angle. Emphasizing angle measure as the amount of rotation will help students grasp *radian measure* and *angular speed*.

Use Example 1 to review algebraic conversion from DMS form to DD form. If students have mastered these conversions, demonstrate the functions on the calculator or use apps, which can be found using the Internet keyword search *dms converter*.

Present the definitions of *radian* and radian measure. Ask the students to describe how many "radius lengths" comprise the circumference of a circle. (2π or ≈ 6.28) Consider displaying the following figure or use the Internet keyword search *circle radian gif* to find a helpful animation.

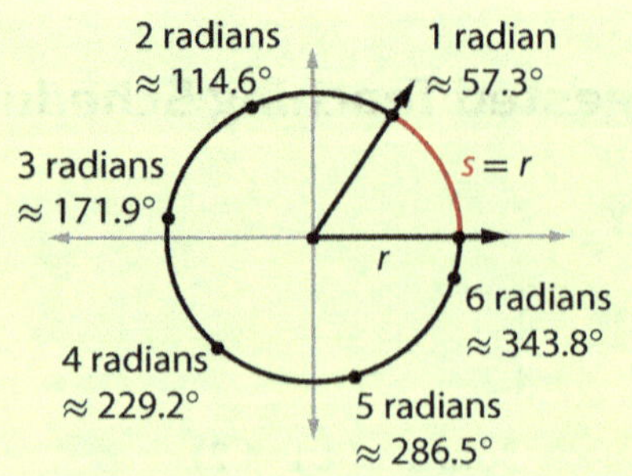

Derive the unit multipliers used to convert between *degrees* and radians and demonstrate the conversion with Example 2. Explain that the label "radians" is usually omitted, as the radian measure is a real-number ratio with no units. You may want to emphasize the decimal equiva-

Represent a ship's bearing, given in DMS form as 127°30′9″, as a standard position angle in DD form.

Answer

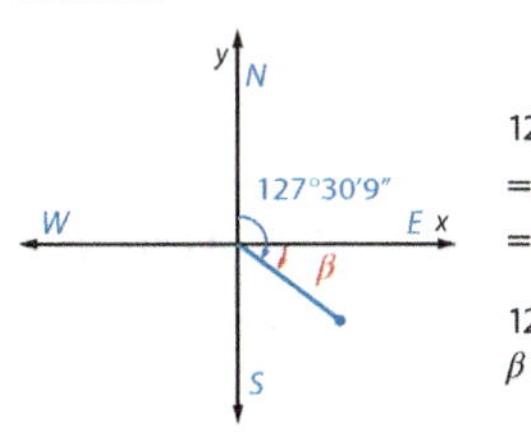

1. Sketch the ship's bearing.

$127° + 30'\left(\frac{1°}{60'}\right) + 9''\left(\frac{1°}{3600''}\right)$
$= 127° + 0.5° + 0.0025°$
$= 127.5025°$

2. Convert DMS form to DD form.

$127.5025° - 90° = 37.5025°$
$β = -37.5025°$

3. Express the terminal ray as a negative rotation from the positive x-axis.

Check

$0.5025°\left(\frac{60'}{1°}\right) = 30.15'$

4. Convert the decimal portion of the degrees to minutes.

$30' + 0.15'\left(\frac{60''}{1'}\right) = 30'\,9''$

5. Convert the decimal portion of the minutes to seconds.

SKILL ✔ **EXERCISE 9**

While degrees work well for navigation and construction, many applications involving trigonometric functions require the domain to be real numbers. Using *radian measure* allows both the domain and the range to be expressed with real numbers having similar units.

> **DEFINITIONS**
>
> The **radian measure** of an angle $θ$ is the ratio of its intercepted arc's length s to the circle's radius r: $θ = \frac{s}{r}$. A central angle that intercepts an arc whose length is the same as the circle's radius has a measure of 1 **radian**.

Note that a radian measure is a real number with no units since it is a ratio of two lengths. The relationship between radian and degree measures can be derived by considering the arc length of one complete rotation of 360°: $s = 2πr$, the circumference of a circle.

$θ = \frac{s}{r} = \frac{2πr}{r} = 2π$ radians $= 360°$
$π$ radians $= 180°$

This relationship produces two unit multipliers that can be used to convert between radians and degrees.

$\frac{π \text{ radians}}{180°} = 1$ and $\frac{180°}{π \text{ radians}} = 1$

The DMS divisions of a circle are rooted in the Babylonian sexagesimal (base 60) number system and the ease of dividing a circle's circumference into equal arcs with 6 equilateral triangles. The fact that a year is approximately 360 days may also have been an influencing factor.

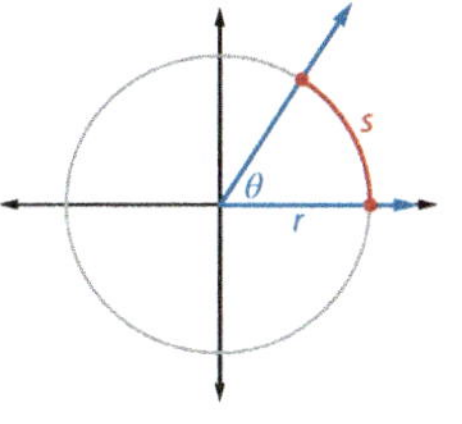

TIP 

This text occasionally includes the word *radians* with a radian measure for clarity.

coterminal angle
degree
linear speed
minute (of a degree)
negative angle
obtuse
positive angle
quadrantal angle
radian
radian measure
second (of a degree)
sector
standard position
terminal side
uniform circular motion
vertex

Reading and Writing Mathematics

Navigation is greatly enhanced by the Global Positioning System (GPS). Originally developed for military purposes, GPS uses 24 earth-orbiting satellites (and additional spare satellites). It became fully operational in April 1995. Research how GPS works and explain why a GPS device must have contact with at least 3 satellites. Signals from 1 satellite provide the distance from the satellite to the device and determine a circle of possible locations for the device. Signals from 2 satellites generate 2 possible locations, the 2 intersections of their circles of possible locations. Using 3 (nonlinear) satellites generates 3 circles whose single point of intersection indicates the exact location.

---

lents for radian measures of $\frac{π}{2}$, $π$, $\frac{3π}{2}$, and $2π$ (1.57, 3.14, 4.71, and 6.28).

You may want to demonstrate how to convert between radians and degrees using the proportion $\frac{n \text{ radians}}{m \text{ degrees}} = \frac{π}{180}$. When a radian measure is expressed in terms of $π$, students could also replace $π$ with 180° and simplify.

Angle measures entered into your calculator are assumed to be entered in the current mode (degree or radian). If the calculator is in degree mode and $π$ radians ($π^r$) is entered using [ANGLE], ③, the calculator will display 180 as the degree measure. When in radian mode, entering 180° using [ANGLE], ①, causes

the calculator to display 3.14159… as the radian measure.

Describe *acute*, *obtuse*, and *quadrantal* angles in terms of their radian measures. Then discuss *coterminal angles* and Example 3.

Show how measuring angles in radians provides a simplified arc length formula and demonstrate its application by finding the "length" of a nautical mile in Example 4.

Common Student Error Students often forget to convert angle measure from degrees to radians before applying $s = rθ$.

Introduce *uniform circular motion*, making sure students understand the

difference between angular speed (the rate of change in the central angle as an object rotates) and *linear speed* (the distance traveled per unit of time). Be sure students can convert angular speed given in revolutions per minute (rpm) to radians per minute or radians per second. Including the label *radians* helps to clarify the otherwise confusing unit of $\min^{-1}$ or $\sec^{-1}$.

Emphasize this difference when discussing Example 5 by asking the students to interpret 3.49 $\frac{\text{radians}}{\text{sec}}$ (*just over half a revolution in 1 sec*) and asking them to describe how the linear speed varies with the radius. (*v is directly proportional to r*

Additional Exercises

Figures for Additional Exercises throughout the chapter can be found at TeacherToolsOnline.com.

1. Represent a ship's bearing of 210.42° as a standard position angle in DMS form. $-120°25'12''$

2. Convert 103°32'56'' to DD form to the nearest tenth. $103.5°$

3. Convert 100° to a radian measure in terms of π and then as a decimal approximation. $\frac{5\pi}{9}; \approx 1.745$

4. Convert 1.2 to a degree measure. $\approx 68.8°$

5. Which angle is not coterminal with $\alpha = \frac{7\pi}{6}$? D

   A. $-150°$ C. $-\frac{5\pi}{6}$ E. $570°$

   B. $\frac{19\pi}{6}$ D. $150°$

6. Find the missing arc length s, radius measure r, or angle measure θ (in radians).

	s	r	θ
a.	36 in.	8 in.	4.5
b.	≈ 90.8 cm	16 cm	325°
c.	14.75 m	1.25 m	11.8
d.	60 ft	40 ft	1.5

7. A fan spins at 3300 rpm. Find the angular speed in radians per second. Then find the linear speed (in feet per second) at the tip of the blade if the length of the fan blade is 6 in.

   $\approx 345.575 \frac{radians}{sec}; \approx 172.788 \frac{ft}{sec}$

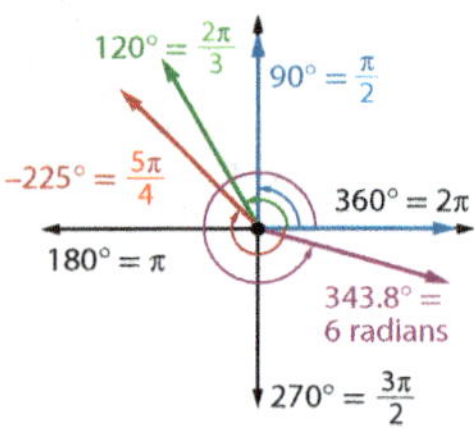

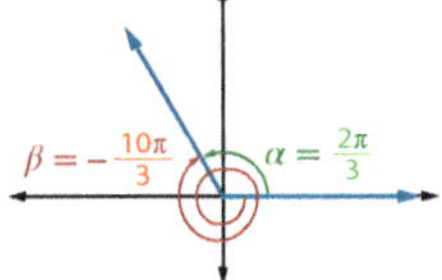

Example 2 Converting between Degree and Radian Measures

Convert each degree measure to radians and each radian measure to degrees.

a. 90°   b. −225°   c. $\frac{2\pi}{3}$ d. 6

Answer

a. $90°\left(\frac{\pi\ radians}{180°}\right) = \frac{\pi}{2} \approx 1.57$

b. $-225°\left(\frac{\pi\ radians}{180°}\right) = -\frac{5\pi}{4} \approx -3.93$

c. $\frac{2\pi}{3}\left(\frac{180°}{\pi\ radians}\right) = 120°$

d. $6\left(\frac{180°}{\pi\ radians}\right) = \frac{1080}{\pi} \approx 343.8°$

SKILL ✓ EXERCISE 11

The terminal side of an *acute* angle (radian measure $0 < \theta < \frac{\pi}{2}$) lies in Quadrant I, and the terminal side of an *obtuse* angle ($\frac{\pi}{2} < \theta < \pi$) lies in Quadrant II. The terminal side of a *quadrantal angle* ($\theta = n\frac{\pi}{2}$, $n \in \mathbb{Z}$) lies on one of the axes.

Two angles in standard position with the same terminal ray are called *coterminal angles*. Since the dynamic view of angles allows a ray to sweep in either direction or to sweep through one or more complete circles before taking its terminal position, any angle has an infinite number of coterminal angles. Coterminal angles can be found by adding or subtracting multiples of 360° or 2π radians. For example, a 300° angle is coterminal with angles of −60°, 660°, and −420°.

Example 3 Finding Coterminal Angles

Write an expression for all angles coterminal with $\theta = -\frac{4\pi}{3}$. Then specify one positive and one negative coterminal angle.

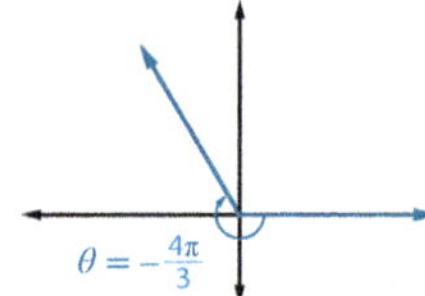

Answer

$-\frac{4\pi}{3} + n2\pi, n \in \mathbb{N}$

1. Coterminal angles can be found by adding and subtracting multiples of 2π.

$\alpha = -\frac{4\pi}{3} + (1)2\pi = \frac{2\pi}{3}$

2. Let $n = 1$ and add 2π to find a positive coterminal angle.

$\beta = -\frac{4\pi}{3} - (1)2\pi = -\frac{10\pi}{3}$

3. Let $n = -1$ and subtract 2π to find a negative coterminal angle.

SKILL ✓ EXERCISE 21

Solving the definition of radian measure, $\theta = \frac{s}{r}$, for r leads directly to a formula for *arc length*.

▶ ARC LENGTH FORMULA

A central angle of radian measure θ in a circle with radius length r intercepts an arc with length $s = r\theta$.

Note that when $r = 1$, the arc length is equal to the radian measure of the central angle. When the central angle is given in degrees, convert to radian measure before applying the arc length formula.

---

with ω as the constant of proportionality.)
One-on-One The movement of a clock's minute hand can also be used to help differentiate angular speed from linear speed. Explain that every point along the big hand of the clock moves at an angular speed of $2\pi \frac{radians}{hr}$, but points closer to the tip of the hand move at a greater linear speed, since they move a greater linear distance during each hour.

Students may be interested in researching the Jewish calendar or the general development of calendars throughout history. Students will find that the Jewish calendar was a significant topic in portions of the Dead Sea Scrolls. Our current calendar, the Gregorian calendar, was implemented by and named after Pope Gregory XIII in 1582. It is also known as the Western calendar and as the Christian calendar. The Gregorian calendar was not adopted by the American colonies until 1752, resulting in the "loss" of eleven days in September. George Washington was born in Virginia on February 11, 1731, according to the then-used Julian calendar, but once the Gregorian calendar was adopted, Washington's birth date was moved a year and 11 days to February 22, 1732.

Motivational Idea Compare circular media devices that utilize constant angular velocity (CAV) and those that utilize constant linear velocity (CLV). Record players usually spin at CAVs of $33\frac{1}{3}$, 45, or 78 rpm, and a typical hard disk drive spins at a CAV of 7200 rpm. Optical disc drives for CDs and DVDs read and write data at a CLV by spinning at varying angular speeds as data is written at different distances from the center. Encourage students to investigate CLV using the Internet keyword search *constant linear velocity*.

Explain that the formula for the area of a *sector* can be recalled as a fractional part of the circle $A_{sector} = \left(\frac{\theta}{2\pi}\right)\pi r^2$ before simplifying to $A_{sector} = \frac{1}{2}r^2\theta$. Demonstrate this application with Example 6.

Example 4　Finding Arc Length

A nautical mile has historically been defined as 1 minute of latitude at the earth's equator. Determine the relationship between a (statute) mile and a nautical mile (nm), given that the earth's average radius at the equator is 3959 mi.

Answer

$\theta = \frac{1°}{60}\left(\frac{\pi \text{ radians}}{180°}\right) = \frac{\pi}{10{,}800}$

$s = r\theta = (3959 \text{ mi})\left(\frac{\pi}{10{,}800}\right)$
$\quad \approx 1.15 \text{ mi}$

$\therefore 1 \text{ nm} \approx 1.15 \text{ mi}$

1. Convert 1 minute to radian measure.

2. Apply the arc length formula.

——————— SKILL ✔ **EXERCISE 37**

The study of *uniform circular motion* analyzes the linear and angular speeds of an object moving in a circular path at a constant speed. The *angular speed*, usually denoted with the lowercase Greek letter omega (ω), is the ratio of the angle of rotation (in radians) per unit of time: $\omega = \frac{\theta}{t}$. Angular speed stated in terms of revolutions can be converted to radians using the unit multiplier $\frac{2\pi \text{ radians}}{1 \text{ revolution}}$. The *linear speed* is the ratio of arc length, or distance traveled, per unit of time: $v = \frac{s}{t}$. Since $s = r\theta$, $v = \frac{r\theta}{t} = r\left(\frac{\theta}{t}\right) = r\omega$.

📖 UNIFORM CIRCULAR MOTION FORMULAS

If θ is the radian measure of the angle of rotation,
then the angular speed is $\omega = \frac{\theta}{t}$ and the linear speed is $v = \frac{s}{t} = r\omega$.

Example 5　Analyzing Uniform Circular Motion

Karyn is playing an LP vinyl record at $33\frac{1}{3}$ rpm (revolutions per minute) on her vintage record player. Determine the angular speed of the record (in radians per second). Then find the linear speed (in inches per second) of a point at the edge of the record (6 in. from the center) and the linear speed of a point at the edge of its label (2 in. from the center).

Answer

$\omega = \frac{100 \text{ revolutions}}{3 \text{ min}}\left(\frac{2\pi \text{ radians}}{1 \text{ revolution}}\right)\left(\frac{1 \text{ min}}{60 \text{ sec}}\right)$
$\quad = \frac{10\pi \text{ radians}}{9 \text{ sec}} \approx 3.49 \text{ radians/sec}$

$v_6 = r\omega = (6 \text{ in.})\left(\frac{10\pi}{9 \text{ sec}}\right)$
$\quad = \frac{20\pi \text{ in.}}{3 \text{ sec}} \approx 20.94 \text{ in./sec}$

$v_2 = (2 \text{ in.})\left(\frac{10\pi}{9 \text{ sec}}\right) = \frac{20\pi \text{ in.}}{9 \text{ sec}} \approx 6.98 \text{ in./sec}$

1. Convert the angular speed from revolutions per minute to radians per second.

2. Substitute into $v = r\omega$ to find each linear speed.
   Since the radian measure is a ratio of two lengths, it is a real number and the label is not needed.

——————— SKILL ✔ **EXERCISE 33**

Radian measure provides a convenient formula for the area of a *sector* of a circle, the region bounded by two radii and the intercepted arc. Recall from *GEOMETRY* that the ratio of the area of a sector and the circle's area is equal to ratio of the central angle and a complete rotation: $\frac{A_{sector}}{\pi r^2} = \frac{\theta}{2\pi}$. Solving for the area of the sector, $A_{sector} = \left(\frac{\theta}{2\pi}\right)\pi r^2 = \frac{1}{2}r^2\theta$.

TIPS

Ex. 40 Use the Internet keyword search *circular field images* or zoom in on Champion, Nebraska, using Google Earth® to view circular fields.

Ex. 41 First find the angular velocity of the rear wheel. This will equal the angular velocity of the rear sprocket.

---

8. The radii of similar sectors are 3 cm and 2 cm. Find the area defined by the larger sector if the smaller sector has an area of 2.6 cm². **5.85 cm²**

Assignments

- **Minimum:** 1–2(*a*, *c*), 4, 7, 10–13, 16–19, 22–26, 30, 32–33, 40, 44–45, 48–49, 51

- **Standard:** 1–2(*c*, *d*), 5, 8–9, 11, 14, 17–18, 21–23, 26–28, 30, 33–34, 37–40; 43–51 odd

- **Extended:** 1*d*, 2*c*, 4, 7, 10, 13, 16–19, 23, 26–28; 29–35 odd; 36–37, 39, 41; 42–52 even

Creating an accurate calendar during Old Testament times was a difficult task requiring a thorough understanding of both mathematics and astronomy. Months based on cycles of the moon were about 10 days short of a solar year, so an extra month was periodically added to keep the months aligned with the seasons. Even with leap years, our current model must be adjusted every 400 years.

Solutions

❯ **A. Exercises**

1*a*.

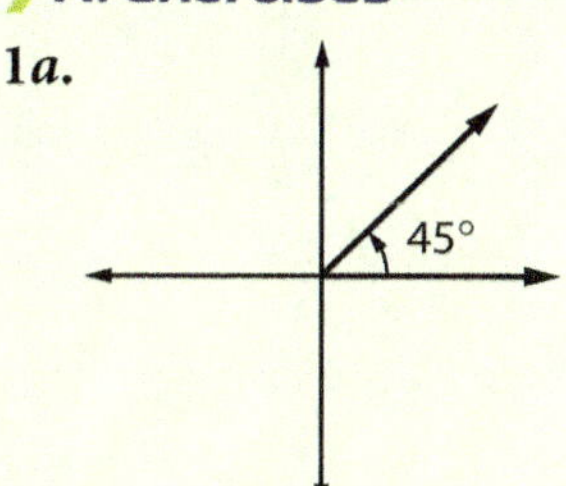

1*b*.

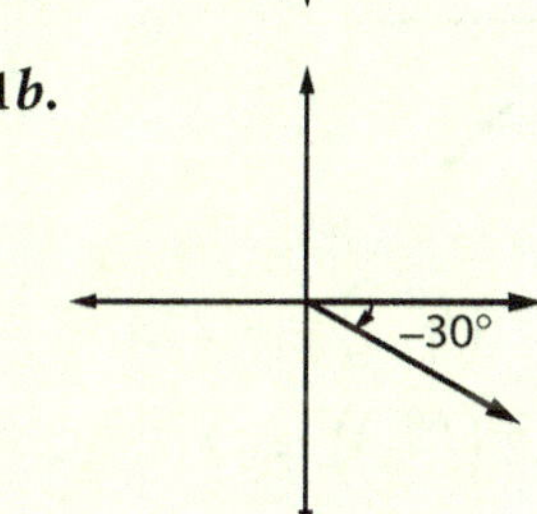

1*c*.

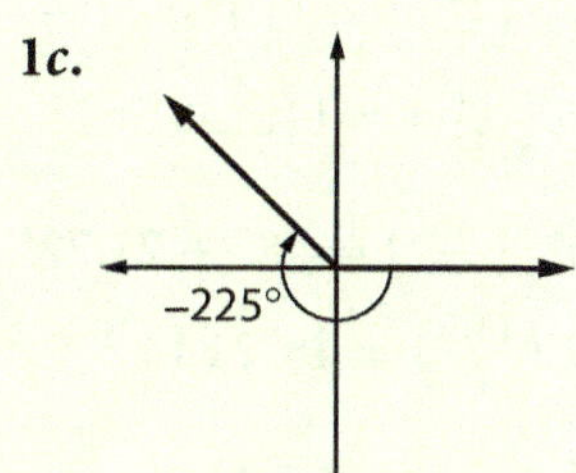

1*d*.

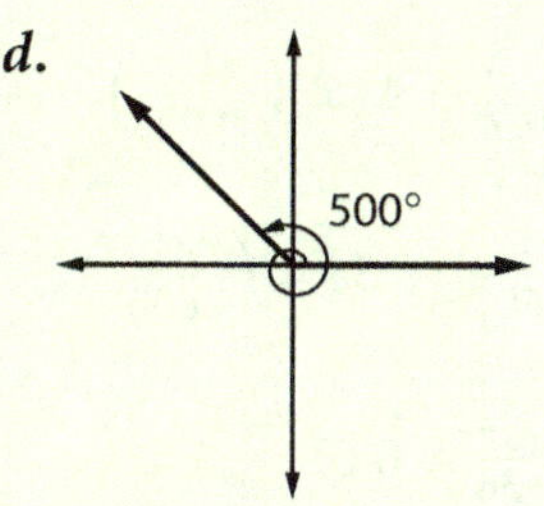

2a.

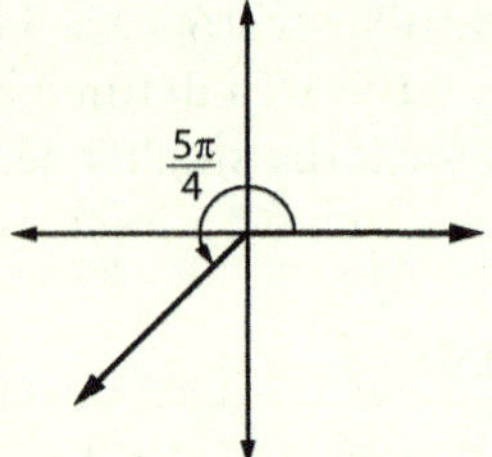

2b.

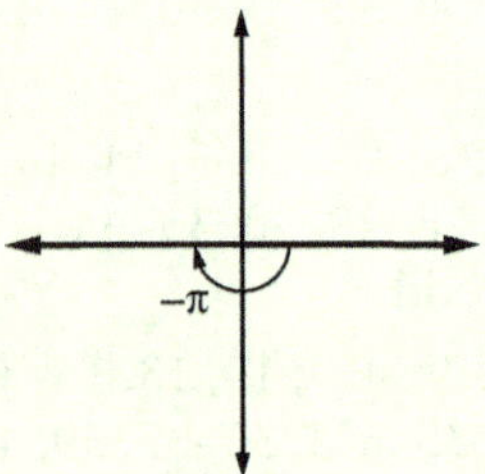

2c.

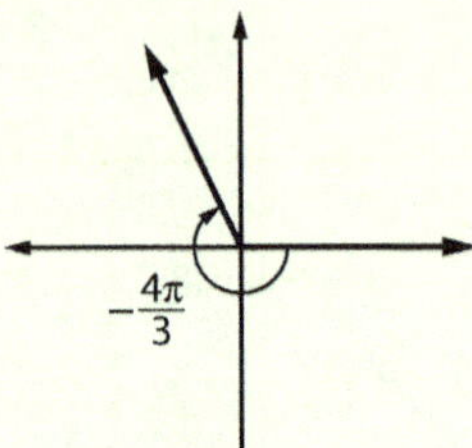

2d.

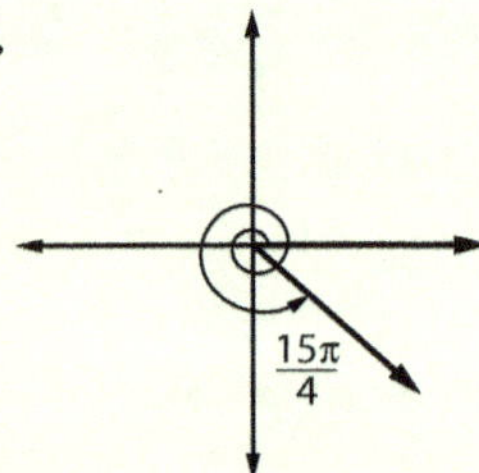

3. $-\left(2° + 0.87°\left(\frac{60'}{1°}\right)\right) = -(2° + 52.2')$

$-\left(2°52' + 0.2'\left(\frac{60''}{1'}\right)\right) = -2°52'12''$

4. $110° + 0.51°\left(\frac{60'}{1°}\right) = 110° + 30.6'$

$110°30' + 0.6'\left(\frac{60''}{1'}\right) = 110°30'36''$

5. $48° + 0.362°\left(\frac{60'}{1°}\right) = 48° + 21.72'$

$48°21' + 0.72'\left(\frac{60''}{1'}\right) = 48°21'43.2''$

6. $-\left(58° + 54'\left(\frac{1°}{60'}\right)\right) = -58.9°$

7. $98° + 45'\left(\frac{1°}{60'}\right) + 43.2''\left(\frac{1°}{3600''}\right)$
$= 98.762°$

8. $135° + 35'\left(\frac{1°}{60'}\right) + 38.4''\left(\frac{1°}{3600''}\right)$
$= 135.594°$

9. $35°\left(\frac{\pi}{180°}\right) = \frac{7\pi}{36} \approx 0.61$

10. $-40°\left(\frac{\pi}{180°}\right) = -\frac{2\pi}{9} \approx -0.70$

11. $1080°\left(\frac{\pi}{180°}\right) = 6\pi \approx 18.85$

12. $154.5°\left(\frac{\pi}{180°}\right) = \frac{1545\pi}{1800} = \frac{103\pi}{120} \approx 2.70$

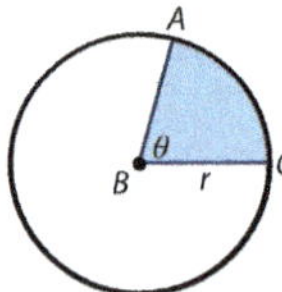

If the shaded sector has an area of 24π m² and $BA = 12$ m, find the radian measure and the degree measure of $\angle ABC$.

Answer

$A_{sector} = \frac{1}{2}r^2\theta$

$\theta = \frac{2\,A_{sector}}{r^2}$

1. Solve the formula for the area of a sector for θ, the radian measure of the angle.

$\theta = \frac{2(24\pi\ \text{m}^2)}{(12\ \text{m})^2} = \frac{48\pi\ \text{m}^2}{144\ \text{m}^2} = \frac{\pi}{3}$

2. Substitute and simplify to express θ in radians.

$\frac{\pi}{3}$ radians $\left(\frac{180°}{\pi\ \text{radians}}\right) = 60°$

3. Convert radians to degrees.

SKILL ✓ **EXERCISE 31**

❯ A. Exercises

1. Sketch a standard position angle with each degree measure.

  a. 45°   **b.** −30°

  c. −225°   **d.** 500°

2. Sketch a standard position angle with each radian measure.

  a. $\frac{5\pi}{4}$ **b.** $-\pi$

  c. $\frac{-4\pi}{3}$ **d.** $\frac{15\pi}{4}$

Convert to degrees, minutes, and seconds (DMS).

3. −2.87° _−2°52′12″_ **4.** 110.51° _110°30′36″_ **5.** 48.362° _48°21′43.2″_

Convert to decimal degrees (DD).

6. −58°54′ _−58.9°_ **7.** 98°45′43.2″ _98.762°_ **8.** 135°35′38.4″ _135.594°_

Convert each degree measure to radian measure. Express the answer in terms of π and as a decimal approximation rounded to the nearest hundredth.

9. 35° _$\frac{7\pi}{36} \approx 0.61$_   **10.** −40° _$-\frac{2\pi}{9} \approx -0.70$_

11. 1080° _$6\pi \approx 18.85$_   **12.** 154.5° _$\frac{103\pi}{120} \approx 2.70$_

Convert each radian measure to degrees. Round answers to the nearest tenth of a degree if necessary.

13. $\frac{\pi}{5}$ _36°_   **14.** $-\frac{7\pi}{2}$ _−630°_

15. $\frac{\pi}{12}$ _15°_   **16.** 5 _286.5°_

17. Identify the missing degree or radian measure of the standard position angle whose terminal side passes through each point (A–K).

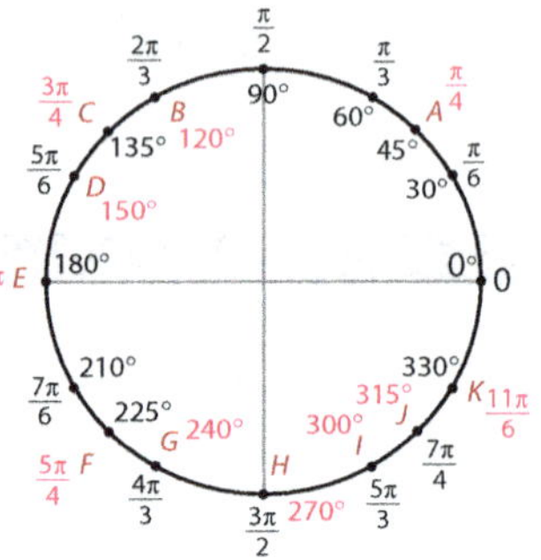

❯ B. Exercises

18. How many degrees (to the nearest tenth) are in one radian? How many radians (to the nearest ten thousandth) are in one degree? _57.3°; 0.0175_

For each angle, write an expression for all coterminal angles. Then state one positive and one negative coterminal angle. _Answers may vary._

19. $\alpha = 148°$ **20.** $\alpha = -200°$

21. $\alpha = \frac{\pi}{4}$ **22.** $-\frac{5\pi}{3}$

23. Which of the following angles are coterminal to $\alpha = \frac{\pi}{5}$? Select all that apply. _B, C, D_

  A. $\frac{4\pi}{5}$ **B.** $\frac{11\pi}{5}$ **C.** 756°   **D.** −324°

---

13. $\frac{\pi}{5}\left(\frac{180°}{\pi}\right) = 36°$

14. $-\frac{7\pi}{2}\left(\frac{180°}{\pi}\right) = -630°$

15. $\frac{\pi}{12}\left(\frac{180°}{\pi}\right) = 15°$

16. $5\left(\frac{180°}{\pi}\right) \approx 286.48°$

❯ B. Exercises

18. 1 radian $= 1\left(\frac{180°}{\pi}\right) \approx 57.3°$

$1° = 1°\left(\frac{\pi}{180°}\right) \approx 0.0175$

19. $148° \pm n360°$

$\beta = 148 + (1)360 = 508°$
$\theta = 148 - (1)360 = -212°$

20. $-200° \pm n360°$

$\beta = -200 + (1)360 = 160°$
$\theta = -200 - (1)360 = -560°$

21. $\frac{\pi}{4} + n2\pi$

$\beta = \frac{\pi}{4} + (1)2\pi = \frac{9\pi}{4}$
$\theta = \frac{\pi}{4} - (1)2\pi = -\frac{7\pi}{4}$

22. $-\frac{5\pi}{3} + n2\pi$

$\beta = -\frac{5\pi}{3} + (1)2\pi = \frac{\pi}{3}$
$\theta = -\frac{5\pi}{3} - (1)2\pi = -\frac{11\pi}{3}$

23a. $\frac{4\pi}{5} - \frac{\pi}{5} = \frac{3\pi}{5}$; not a multiple of 2π

23b. $\frac{11\pi}{5} - \frac{\pi}{5} = 2\pi$; a multiple of 2π

Note: $\frac{\pi}{5}\left(\frac{180°}{\pi\ \text{radians}}\right) = 36°$

Find the missing arc length, radius measure, or angle measure (in radians).

	Arc (s)	Radius (r)	Angle (θ)
24.	2π	8	$\frac{\pi}{4}$
25.	6π	4	$\frac{3\pi}{2}$
26.	$\frac{\pi}{2}$	3	$\frac{\pi}{6}$
27.	3π	6	$\frac{\pi}{2}$

28. An air traffic controller notes that an airplane flying from Kansas City to Los Angeles at a bearing of 259°38′6″ will be landing shortly. Represent the plane's bearing as a standard position angle in decimal degrees (DD). −169.635° or 190.365°

29. If the pendulum of a grandfather clock is 94 cm long, how far does the end of the pendulum travel as it swings through an angle of 6°? If the pendulum is lengthened by 2 cm to prevent the clock from running too fast, how much farther does the end swing through its 6° arc? Round answers to the nearest hundredth of a centimeter. 9.84 cm; 0.21 cm

30. If a sprinkler rotates 60° and the nozzle streams water 5 ft away, find the area watered (to the nearest tenth of a square foot). 13.1 ft²

31. A 24 in. diameter prize wheel is divided into 16 equal sectors. Find the arc length and the area of one sector. Round answers to the nearest tenth. 4.7 in.; 28.3 in.²

32. A compact circular saw operates at a maximum speed of 3500 rpm and has a blade with a diameter of 4.5 in. Find the angular speed of the blade (in radians per second) and the linear speed (in inches per second) for a tip of the sawblade. Round answers to the nearest tenth. 366.5 radians/sec; 824.7 in./sec

33. A floor buffer with a 20 in. radius operates at 1500 rpm. Find the angular speed of the buffer (in radians per second) and the difference of the linear speeds (in miles per hour) for a point on the outer edge of the buffer pad and a point on the edge of the 3 in. diameter hole in the center of the pad. Round answers to the nearest tenth. 157.1 radians/sec; 165.1 mi/hr

34. Show algebraically that every angle coterminal with 2π is an even multiple of π.

35. Show algebraically that every angle coterminal with π is an odd multiple of π.

36. Write an equation expressing the difference of any two coterminal angles α and θ.

❭ **C. Exercises**

37. Minneapolis, Minnesota, is located at 44°59′ N and Springfield, Missouri, is located at 37°13′ N on the same longitudinal line. Estimate the distance between the cities (to the nearest mile), using 3959 mi as the radius of the earth. 537 mi or 536 mi

38. Samuel is making a 3D pie chart with a 3 ft diameter for a presentation and plans to cover each slice with fabric. If the pie chart is 6 in. thick, how much fabric (to the nearest tenth of a square foot) will be used to cover the top and the sides of the slice that represents 40% of the spending? 6.2 ft²

39. An amusement park is planning the *Centrihuge*, a new ride that presses passengers against the wall of a circular cylinder as it spins at 24 rpm.
   a. Determine the angular speed of the ride (in radians per second). ≈ 2.51 $\frac{radians}{sec}$
   b. Find the linear speed (in miles per hour) of a person on the wall for rides with diameters of 45 ft and 48 ft. ≈ 38.6 $\frac{mi}{hr}$; ≈ 41.4 $\frac{mi}{hr}$

40. A 1300 ft long center-pivot irrigation system with 10 equally spaced trusses takes 72 hr for each rotation over a circular field.

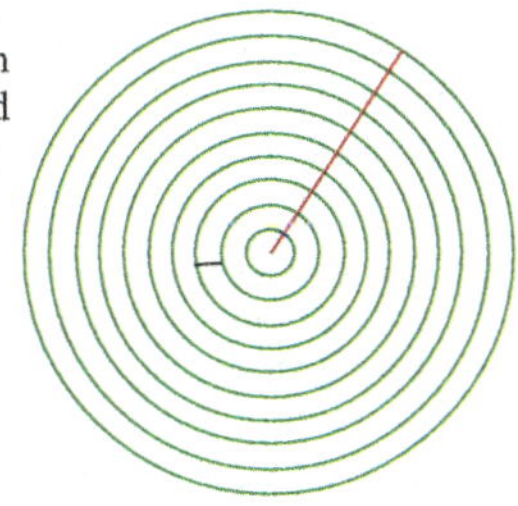

   a. Find the distance that the third irrigation truss from the center travels in 9 hr. 306.31 ft
   b. How many acres are irrigated between the third and outermost trusses in 9 hr? (1 acre = 43,560 ft²) 13.86 acres
   c. Find the angular speed (in radians per hour) of the irrigation system. $\frac{\pi}{36}$ radians/hr
   d. Find the linear speed (in feet per hour) of the first truss and the ninth truss. 11.34 ft/hr; 102.10 ft/hr

30. $\theta = 60°\left(\frac{\pi}{180°}\right) = \frac{\pi}{3}$
$A = \frac{1}{2}(5)^2\left(\frac{\pi}{3}\right) \approx 13.1 \text{ ft}^2$

31. $\theta = \left(\frac{360°}{16}\right)\left(\frac{\pi}{180°}\right) = \frac{\pi}{8}$
$s = 12\left(\frac{\pi}{8}\right) \approx 4.7 \text{ in.}$
$A = \frac{1}{2}(12)^2\left(\frac{\pi}{8}\right) \approx 28.3 \text{ in.}^2$

32. $\omega = 3500 \, \frac{rev}{min}\left(\frac{1 \text{ min}}{60 \text{ sec}}\right)\left(\frac{2\pi \text{ radians}}{1 \text{ rev}}\right)$
$= \frac{350\pi}{3} \frac{radians}{sec} \approx 366.5 \frac{radians}{sec}$
$v = 2.25 \text{ in.}\left(\frac{350\pi}{3} \frac{radians}{sec}\right)$
$\approx 824.7 \frac{in.}{sec}$

33. $\omega = 1500 \, \frac{rev}{min}\left(\frac{1 \text{ min}}{60 \text{ sec}}\right)\left(\frac{2\pi \text{ radians}}{1 \text{ rev}}\right)$
$= 50\pi \frac{radians}{sec} \approx 157.080 \frac{radians}{sec}$
$v_2 - v_1 = (r_2 - r_1)\omega$
$= (20 - 1.5) \text{ in.}\left(50\pi \frac{radians}{sec}\right)$
$= 925\pi \frac{in.}{sec}$
$925\pi \frac{in.}{sec}\left(\frac{1 \text{ ft}}{12 \text{ in.}}\right)\left(\frac{1 \text{ mi}}{5280 \text{ ft}}\right)\left(\frac{60^2 \text{ sec}}{1 \text{ hr}}\right)$
$\approx 165.1 \text{ mi/hr}$

34. $\alpha = 2\pi + n2\pi, \, n \in \mathbb{Z}$
$\alpha = 2(1 + n)\pi$, and $2(1 + n)$ is even.

35. $\alpha = \pi + n2\pi, \, n \in \mathbb{Z}$
$\alpha = (1 + 2n)\pi$, and $2n + 1$ is odd.

36. $\alpha = \theta + n2\pi, \, n \in \mathbb{Z}$
$\therefore \alpha - \theta = n2\pi, \, n \in \mathbb{Z}$
or $\alpha - \theta = n360°, \, n \in \mathbb{Z}$

❭ **C. Exercises**

37. $\theta = 44°59′ - 37°13′ = 7°46′$
$s = r\theta$
$= 3959 \text{ mi}\left(7° + 46′\left(\frac{1°}{60′}\right)\right)\left(\frac{\pi}{180°}\right)$
$\approx 536.7 \text{ mi}$
or $7°\left(\frac{60′}{1°}\right) + 46′ = 466′$ or 466 nm
$466 \text{ nm}\left(\frac{1.15 \text{ mi}}{1 \text{ nm}}\right) = 535.9 \text{ mi}$

38. $\theta = 0.40(2\pi) = 0.8\pi$ or $\frac{4\pi}{5}$ radians
$A_{sector} = \frac{1}{2}(1.5)^2(0.8\pi) \approx 2.83 \text{ ft}^2$
$L = ph = (r\theta + 2r)h$
$= [0.8\pi(1.5) + 2(1.5)](0.5)$
$\approx 3.38 \text{ ft}^2$
$2.83 + 3.38 = 6.21 \text{ ft}^2$

39a. $\theta = \frac{24 \text{ rev}}{1 \text{ min}}\left(\frac{2\pi \text{ radians}}{1 \text{ rev}}\right)\left(\frac{1 \text{ min}}{60 \text{ sec}}\right)$
$= \frac{4\pi}{5} \frac{radians}{sec} \approx 2.51 \frac{radians}{sec}$

23c. $756 - 36 = 720°$; a multiple of 360°

23d. $-324 - 36 = 360°$; a multiple of 360°

24. $s = 8\left(\frac{\pi}{4}\right)$
$s = 2\pi$

25. $6\pi = 4\theta$
$\theta = \frac{6\pi}{4} = \frac{3\pi}{2}$

26. $\frac{\pi}{2} = r\left(\frac{\pi}{6}\right)$
$r = \frac{\pi}{2} \cdot \frac{6}{\pi} = 3$

27. $3\pi = 6\theta$
$\theta = \frac{3\pi}{6} = \frac{\pi}{2}$

28.

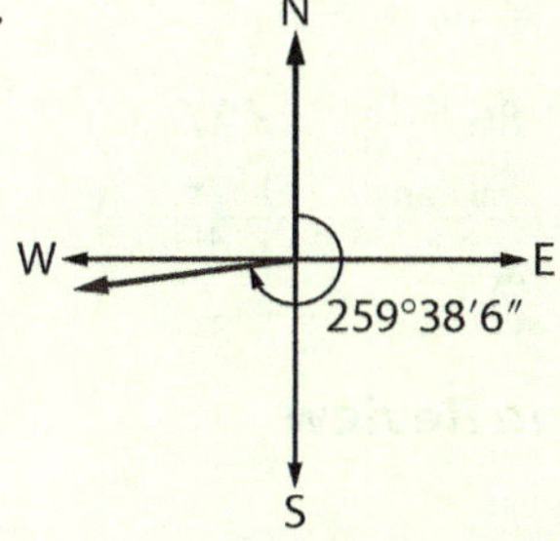

$259° + 38′\left(\frac{1°}{60′}\right) + 6″\left(\frac{1°}{3600″}\right)$
$= 259.635°$
$259.635 - 90 = 169.635°$
$\beta = -169.635°$ or 190.365°

29. $s_1 = 94\left(6°\left(\frac{\pi}{180°}\right)\right) \approx 9.84 \text{ cm}$
$s_2 - s_1 = r_2\theta - r_1\theta = (r_2 - r_1)\theta$
$= (96 - 94)\left(6°\left(\frac{\pi}{180°}\right)\right) \approx 0.21 \text{ cm}$

39b. $v_{22.5} = (22.5 \text{ ft})\left(\frac{4\pi}{5 \text{ sec}}\right) = 18\pi \frac{\text{ft}}{\text{sec}}$

$\approx 56.55 \frac{\text{ft}}{\text{sec}}$

$18\pi \frac{\text{ft}}{\text{sec}}\left(\frac{1 \text{ mi}}{5280 \text{ ft}}\right)\left(\frac{60^2 \text{ sec}}{1 \text{ hr}}\right)$

$\approx 38.6 \text{ mi/hr}$

$v_{24} = (24 \text{ ft})\left(\frac{4\pi}{5 \text{ sec}}\right) = \frac{96\pi}{5} \frac{\text{ft}}{\text{sec}}$

$\frac{96\pi}{5} \frac{\text{ft}}{\text{sec}}\left(\frac{1 \text{ mi}}{5280 \text{ ft}}\right)\left(\frac{60^2 \text{ sec}}{1 \text{ hr}}\right)$

$\approx 41.1 \text{ mi/hr}$

40a. $r = 3\left(\frac{1300}{10}\right) = 390 \text{ ft}$

$\theta = \frac{9}{72}(2\pi) = \frac{\pi}{4}$

$s = 390\left(\frac{\pi}{4}\right) \approx 306.31 \text{ ft}$

40b. $A = \frac{1}{2}r_L^2\theta - \frac{1}{2}r_S^2\theta = \frac{1}{2}\theta(r_L^2 - r_S^2)$

$= \frac{1}{2}\left(\frac{\pi}{4}\right)(1300^2 - 390^2)$

$\approx 603{,}932 \text{ ft}^2\left(\frac{1 \text{ acre}}{43{,}560 \text{ ft}^2}\right)$

$\approx 13.86 \text{ acres}$

40c. $\omega = \frac{1 \text{ rev}}{72 \text{ hr}}\left(\frac{2\pi}{1 \text{ rev}}\right) = \frac{\pi}{36} \frac{\text{radians}}{\text{hr}}$

40d. $v_1 = (130 \text{ ft})\left(\frac{\pi}{36 \text{ hr}}\right) \approx 11.34 \frac{\text{ft}}{\text{hr}}$

$v_9 = ((9)130 \text{ ft})\left(\frac{\pi}{36 \text{ hr}}\right) \approx 102.10 \frac{\text{ft}}{\text{hr}}$

41. $v_{20} = r_{20}\omega_{20}$

$\omega_{20} = \frac{v_{20}}{r_{20}} = \frac{24 \text{ mi}}{\text{hr}}\left(\frac{5280 \text{ ft}}{1 \text{ mi}}\right)\left(\frac{12 \text{ in.}}{1 \text{ ft}}\right)$

$\left(\frac{1 \text{ hr}}{3600 \text{ sec}}\right)\left(\frac{1 \text{ radian}}{10 \text{ in.}}\right) \approx 42.24 \frac{\text{radians}}{\text{sec}}$

$\omega_3 = \omega_{20} \approx 42.24 \frac{\text{radians}}{\text{sec}}$

$v_3 = \omega_3 r_3$

$= \frac{42.24 \text{ radians}}{\text{sec}}\left(\frac{1.5 \text{ in.}}{1}\right) = 63.36 \frac{\text{in.}}{\text{sec}}$

$= v_6$

$\omega_6 = \frac{v_6}{r_6} \approx \frac{63.36 \frac{\text{in.}}{\text{sec}}}{3 \text{ in.}} = 21.12 \frac{\text{radians}}{\text{sec}}$

42a. $\omega = \frac{v}{r}$

$= \frac{60 \text{ mi}}{\text{hr}}\left(\frac{1 \text{ radian}}{12.9 \text{ in.}}\right)\left(\frac{12 \text{ in.}}{1 \text{ ft}}\right)$

$\cdot \left(\frac{5280 \text{ ft}}{1 \text{ mi}}\right)\left(\frac{1 \text{ hr}}{60^2 \text{ sec}}\right)$

$\approx 81.86 \frac{\text{radians}}{\text{sec}}$

$\approx 81.86 \frac{\text{radians}}{\text{sec}}\left(\frac{1 \text{ rev}}{2\pi \text{ radians}}\right)$

$\approx 13.03 \frac{\text{rev}}{\text{sec}}$

42b. $\omega = \frac{v}{r}$

$= \frac{202 \text{ mi}}{\text{hr}}\left(\frac{1 \text{ radian}}{12.9 \text{ in.}}\right)\left(\frac{12 \text{ in.}}{1 \text{ ft}}\right)$

$\cdot \left(\frac{5280 \text{ ft}}{1 \text{ mi}}\right)\left(\frac{1 \text{ hr}}{60^2 \text{ sec}}\right)$

$\approx 275.60 \frac{\text{radians}}{\text{sec}}$

$\approx 275.60 \frac{\text{radians}}{\text{sec}}\left(\frac{1 \text{ rev}}{2\pi \text{ radians}}\right)$

$\approx 43.86 \frac{\text{rev}}{\text{sec}}$

or

---

41. A freestyle bike competitor pedals at his maximum speed of 24 mi/hr as he approaches a ramp on an obstacle course. If the diameters of his rear wheel, wheel sprocket, and pedal sprocket are 20 in., 3 in., and 6 in., respectively, find the angular speed (in radians per second) of each component. *Hint:* Every point along the chain moves at the same linear speed.

$\approx 42.2 \text{ radians/sec};$
$\approx 42.2 \text{ radians/sec};$
$\approx 21.1 \text{ radians/sec}$

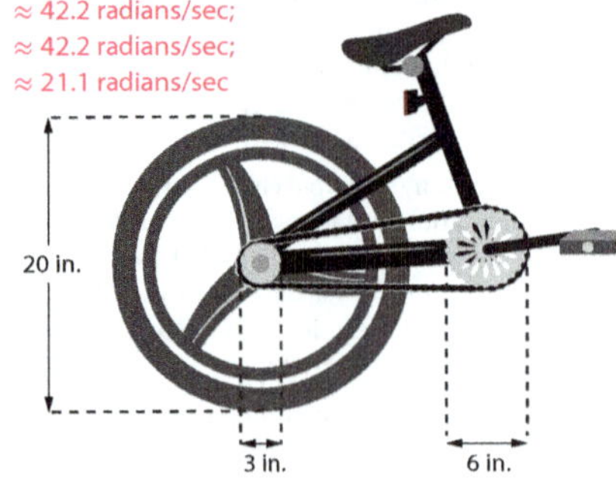

42. Use the table listing the maximum speed for each gear of a racecar with 25.8 in. diameter tires to find the maximum angular speed of its tires in each gear (in radians per second and revolutions per second).

Gear	1st	2nd	3rd	4th	5th
Max Speed (mi/hr)	60	92	124	158	202

a. 1st
$\approx 82 \frac{\text{radians}}{\text{sec}}; \approx 13 \frac{\text{rev}}{\text{sec}}$

b. 5th
$\approx 276 \frac{\text{radians}}{\text{sec}}; \approx 44 \frac{\text{rev}}{\text{sec}}$

CUMULATIVE REVIEW

State the domain of each function. [2.1]

43. $f(x) = \sqrt{x + 7}$ $D = [-7, \infty)$

44. $f(x) = \sqrt{x - 12}$ $D = [12, \infty)$

List all the possible rational zeros for each polynomial function. [2.4]

45. $p(x) = x^4 + 16$ $\pm 1, \pm 2, \pm 4, \pm 8, \pm 16$

46. $p(x) = 4x^3 - 5x^2 - 7x + 2$ $\pm 1, \pm 2, \pm\frac{1}{2}, \pm\frac{1}{4}$

47. A fishpond has an expected annual growth rate of 8%. Write an exponential function that models the fish population, given a current population of 2000 fish. [3.1] $P(t) = 2000(1.08)^t$

48. Solve $3^{x-7} = 26.2$. [3.4] $x \approx 9.973$

49. Solve for x. [Algebra]   D

A. $2\sqrt{13}$ **D.** $4\sqrt{13}$

B. $2\sqrt{17}$ **E.** none of these

C. $4\sqrt{17}$

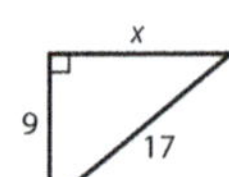

50. Which of the following is true if $x + 3$ is a factor of $p(x)$? [2.3]   B

A. $x - 3$ is a factor of $p(x)$.

B. -3 is a zero of $p(x)$.

C. 3 is a zero of $p(x)$.

D. $p(3) = 0$

E. none of these

51. Determine the account balance of a $25,000 investment compounded quarterly at 3.75% interest for 15 yr. [3.1]   A

A. $43,761.55   **D.** $43,875.10

B. $43,648.43   **E.** none of these

C. $43,837.90

52. Use the properties of logarithms to evaluate $3^{\log_3 \frac{1}{9}}$. [3.2]   B

A. 3   **D.** $\frac{1}{3}$

B. $\frac{1}{9}$ **E.** none of these

C. 9

$\frac{\omega_5}{\omega_1} = \frac{v_5}{v_1}$; so $\omega_5 = \frac{202}{60}\omega_5$

$\left(\frac{202}{60}\right)81.86 \frac{\text{radians}}{\text{sec}} \approx 275.60 \frac{\text{radians}}{\text{sec}}$

$\approx 275.60 \frac{\text{radians}}{\text{sec}}\left(\frac{1 \text{ rev}}{2\pi \text{ radians}}\right)$

$\approx 43.86 \frac{\text{rev}}{\text{sec}}$

❯ Cumulative Review

43. $x + 7 \geq 0$

$x \geq -7$

44. $x - 12 \geq 0$

$x \geq 12$

45. $\frac{\text{factors of } 16}{\text{factors of } 1} = \frac{\pm 1, \pm 2, \pm 4, \pm 8, \pm 16}{\pm 1}$

$= \pm 1, \pm 2, \pm 4, \pm 8, \pm 16$

46. $\frac{\text{factors of } 2}{\text{factors of } 4} = \frac{\pm 1, \pm 2}{\pm 1, \pm 2, \pm 4}$

$= \pm 1, \pm 2, \pm\frac{1}{2}, \pm\frac{1}{4}$

47. $a = 2000, r = 0.08, b = 1 + r = 1.08$

$P(t) = ab^t = 2000(1.08)^t$

48. $\log 3^{x-7} = \log 26.2$

$(x - 7)\log 3 = \log 26.2$

$x = \frac{\log 26.2}{\log 3} + 7 \approx 9.973$

49. $x = \sqrt{17^2 - 9^2} = \sqrt{208} = 4\sqrt{13}$

51. $A(15) = \$25{,}000\left(1 + \frac{0.0375}{4}\right)^{4(15)}$

$\approx \$43{,}761.55$

52. $3^{\log_3 3^{-2}} = 3^{-2\log_3 3} = 3^{-2(1)} = \frac{1}{9}$

Right Triangle Trigonometry

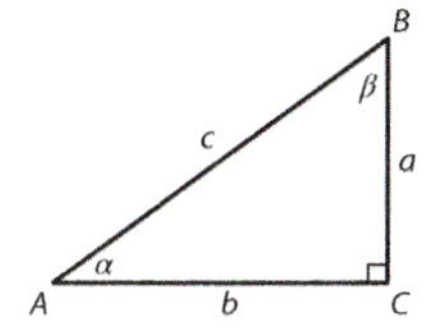

The change in altitude when climbing from Lake Phelps to the peak of the Grand Teton can be estimated using trigonometry.

We will initially limit our study of *trigonometric functions* (frequently abbreviated as *trig functions*) to the acute angles found in right triangles. The hypotenuse of the illustrated right triangle is the side opposite right $\angle C$. The angles are commonly designated by uppercase letters such as A and B, or by Greek letters such as α and β. Lowercase letters a, b, and c represent the lengths of sides opposite the corresponding vertices A, B, and C. Recall that the lengths of the three sides in the illustrated right triangle are related by the Pythagorean Theorem: $a^2 + b^2 = c^2$.

After completing this section, you will be able to

- evaluate the six trigonometric ratios of an acute angle.
- solve right triangles.
- solve real-world problems involving angles of elevation and depression.

The trigonometric functions describe the six possible ratios of side lengths in terms of their relationship to an acute angle of a right triangle.

Basic Trigonometric Functions	Reciprocal Trigonometric Functions
sine of $\angle A = \sin A = \dfrac{\text{leg opposite } \angle A}{\text{hypotenuse}}$ $= \dfrac{opp.}{hyp.} = \dfrac{a}{c}$	**cosecant** of $\angle A = \csc A = \dfrac{1}{\sin A}$ $= \dfrac{hyp.}{opp.} = \dfrac{c}{a}$
cosine of $\angle A = \cos A = \dfrac{\text{leg adjacent } \angle A}{\text{hypotenuse}}$ $= \dfrac{adj.}{hyp.} = \dfrac{b}{c}$	**secant** of $\angle A = \sec A = \dfrac{1}{\cos A}$ $= \dfrac{hyp.}{adj.} = \dfrac{c}{b}$
tangent of $\angle A = \tan A = \dfrac{\text{leg opposite } \angle A}{\text{leg adjacent } \angle A}$ $= \dfrac{opp.}{adj.} = \dfrac{a}{b}$	**cotangent** of $\angle A = \cot A = \dfrac{1}{\tan A}$ $= \dfrac{adj.}{opp.} = \dfrac{b}{a}$

The tangent and cotangent functions can be expressed as quotients of the sine and cosine functions.

$$\tan A = \frac{\sin A}{\cos A} \text{ since } \frac{\sin A}{\cos A} = \frac{\frac{a}{c}}{\frac{b}{c}} = \frac{a}{b} = \tan A, \text{ and}$$

$$\cot A = \frac{\cos A}{\sin A} \text{ since } \frac{\cos A}{\sin A} = \frac{\frac{b}{c}}{\frac{a}{c}} = \frac{b}{a} = \cot A.$$

4.2 Right Triangle Trigonometry

Objectives

1. To apply the Pythagorean Theorem to find a missing side length in a right triangle
2. To evaluate the six trigonometric ratios of an acute angle in a right triangle
3. To solve right triangles using trigonometric ratios and their inverses
4. To solve real-world problems involving angles of elevation and depression

Flash

The Teton Range is a mountain range in Wyoming just south of Yellowstone National Park. Most of the east side of the range is located in Grand Teton National Park. The sharp 7000 ft rise from the valley floor provides a majestic view as well as a great challenge for climbing enthusiasts.

Vocabulary

angle of depression
angle of elevation
cofunction
cosecant function
cosine function
cotangent function
inverse trigonometric function

PRESENTATION

Lesson Opener

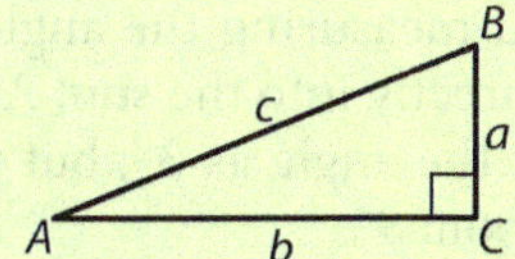

1. If $a = 6$ and $b = 8$, find c.   10
2. If $a = 8$ and $c = 17$, find b.   15
3. If $a = 4$ and $b = 6$, find c. $2\sqrt{13}$
4. If $b = 12$ and $c = 16$, find a. $4\sqrt{7}$

Most students have had some exposure to right triangle trigonometry, but many will benefit from a review. Display the general right $\triangle ABC$ and ask whether the students can define the six *trigonometric functions* in terms of ratios of the sides of the right triangle.

Then display the triangle from Example 1 and have the students find each trigonometric ratio without looking at the example. Students should be able to state exact answers in radical form or as rounded decimals.

Review the relationship between side length ratios for a 45-45-90 (right isosceles) triangle and a 30-60-90 triangle and have the students identify the related trigonometric ratios. The students should be able to sketch these figures and know the three basic trigonometric ratios for these common angles.

The paragraph following the table introduces the relationship of a *cofunction* to its complement. This statement will be proved formally in Chapter 5.

Direct attention to the margin note regarding Plimpton 322. It is possible that the prophet Daniel was instructed in what we know as the Pythagorean Theorem before Pythagoras was taken to Babylon in 525 BC.

Example 2 demonstrates finding the other trigonometric ratios of an acute angle when one trigonometric ratio is known. Some students may recognize that 0.8 is equal to $\frac{4}{5}$ and use the 3, 4, 5 Pythagorean triple to quickly state the ratios in fractional form.

secant function
sine function
solving a triangle
tangent function
trigonometric functions

Reading and Writing Mathematics

The 70-yr-long survey of the entire Indian subcontinent used only 10 direct measurements—the rest being calculated using triangulation. Since the accuracy of the direct measurements would affect the entire survey, it was important to get them right. The metal rods used to measure were subject to contraction and expansion due to temperature changes. To compensate, George Everest used the Colby apparatus, designed by the Irish surveyor Major General Thomas Colby. Research the Colby apparatus and use the diagram to briefly describe how it worked.

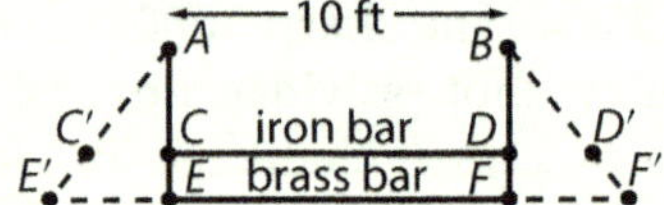

Answers will vary. Since the expansion of brass and iron are in a known ratio of 5 : 3, the apparatus is constructed so that AC : AE and BD : BF are in a ratio of 3 : 5. If the iron bar expands to $C'D'$ then the brass bar will expand to $E'F'$. Since BD : DD' = BF : FF', triangles BDD' and BFF' are similar, and B does not move. Since the same is true on the other end of the apparatus, AB remains constant at 10 ft regardless of the expansion or contraction of the bars.

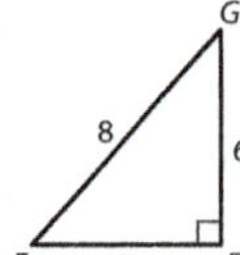

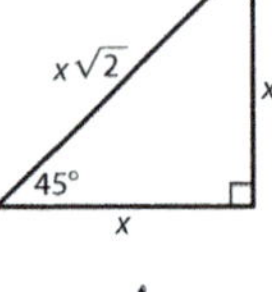

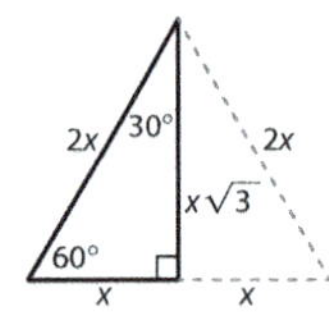

A ca. 1700 BC clay tablet known as Plimpton 322 shows that ancient Babylonians were aware of the relationship expressed by the Pythagorean Theorem and that they could derive many Pythagorean triples, including 4601, 4800, and 6649, well over 1000 yr before the Pythagoreans (ca. 500 BC).

Example 1   Evaluating Trigonometric Functions

Evaluate the six trigonometric functions for $\angle G$.

Answer

$g^2 + 6^2 = 8^2$

$g = \sqrt{8^2 - 6^2} = \sqrt{28} = 2\sqrt{7}$

1. Use the Pythagorean Theorem to find g.

$\sin G = \dfrac{opp.}{hyp.} = \dfrac{2\sqrt{7}}{8}$

$\csc G = \dfrac{1}{\sin G} = \dfrac{1}{\frac{\sqrt{7}}{4}} = \dfrac{4}{\sqrt{7}}$

$= \dfrac{\sqrt{7}}{4} \approx 0.6614$

$= \dfrac{4\sqrt{7}}{7} \approx 1.5119$

2. Use the definitions to find the six trigonometric ratios.

$\cos G = \dfrac{adj.}{hyp.} = \dfrac{6}{8}$

$\sec G = \dfrac{1}{\cos G} = \dfrac{1}{\frac{3}{4}}$

$= \dfrac{3}{4} = 0.75$

$= \dfrac{4}{3} \approx 1.3333$

$\tan G = \dfrac{opp.}{adj.} = \dfrac{2\sqrt{7}}{6}$

$\cot G = \dfrac{1}{\tan G} = \dfrac{1}{\frac{\sqrt{7}}{3}} = \dfrac{3}{\sqrt{7}}$

$= \dfrac{\sqrt{7}}{3} \approx 0.8819$

$= \dfrac{3\sqrt{7}}{7} \approx 1.1339$

SKILL ✔ EXERCISE 3

The Angle-Angle Similarity Postulate guarantees that any two right triangles are similar if they have a corresponding pair of congruent acute angles. Since ratios of corresponding sides of similar figures are equal, these trigonometric ratios of a given angle are the same for similar right triangles of any size.

Recall that the leg : leg : hypotenuse ratio in an isosceles right triangle is $1 : 1 : \sqrt{2}$ and that the short leg : long leg : hypotenuse ratio in a 30-60-90 triangle is $1 : \sqrt{3} : 2$. These ratios allow you to state exact trigonometric ratios for 45°, 30°, and 60° angles without the use of a calculator.

$m\angle A$	$\sin A$	$\cos A$	$\tan A$	$\csc A$	$\sec A$	$\cot A$
30° or $\frac{\pi}{6}$	$\frac{1}{2}$	$\frac{\sqrt{3}}{2}$	$\frac{\sqrt{3}}{3}$	2	$\frac{2\sqrt{3}}{3}$	$\sqrt{3}$
45° or $\frac{\pi}{4}$	$\frac{\sqrt{2}}{2}$	$\frac{\sqrt{2}}{2}$	1	$\sqrt{2}$	$\sqrt{2}$	1
60° or $\frac{\pi}{3}$	$\frac{\sqrt{3}}{2}$	$\frac{1}{2}$	$\sqrt{3}$	$\frac{2\sqrt{3}}{3}$	2	$\frac{\sqrt{3}}{3}$

Notice that $\sin 60° = \cos 30° = \frac{\sqrt{3}}{2}$; $\tan 60° = \cot 30° = \sqrt{3}$, and that $\sec 60° = \csc 30° = 2$. In fact, the trigonometric function of any acute angle is equal to the *cofunction* of its complement. This generalized statement will be proved in later sections using the definitions of sine, tangent, secant, and their respective cofunctions cosine, cotangent, and cosecant.

If one trigonometric ratio of an acute angle is known, the other five trigonometric ratios of that angle can be found.

Example 3 illustrates how to evaluate trigonometric functions on the TI-84 Plus family of graphing calculators. Forgetting to switch to the correct mode or to use the [ANGLE] menu to label the angle measure as needed causes the calculator to evaluate the function for an unintended angle measure. When evaluating csc 30°15′, the conversion and the evaluation can be combined as 1/sin ((30 + 15/60)°) but care must be taken in the placement of parentheses. Students may prefer to switch calculator modes instead of using ° or ʳ and the required parentheses when the calculator is in the other mode. Caution students to be sure they know whether they are working with ra-

dians or degrees if they use other technologies, such as Internet apps.

Introduce *inverse trigonometric functions*. Consider having the students find sin 40° on the calculator and then find $\sin^{-1}$ (ANS). Explain that [SIN] returns the sine ratio for an angle, while [SIN^{-1}] returns the angle measure that has the given ratio of $\dfrac{opp.}{hyp.}$. Inverse trigonometric functions are needed to solve a right triangle when the value of neither acute angle is known, as is demonstrated in Example 4.

Consider sharing that the reasoning behind Aristarchus's calculations was sound but that the calculations were dif-

ficult to execute since the exact timing for a half-moon is hard to determine. The angle from the sun to the moon is very small, and measuring the angle requires looking directly into the sun. Aristarchus measured the angle as 3°, but the actual angle is about 9′.

Present Example 5, stressing that the known values and the values to be found dictate which trigonometric ratio should be used. There may be more than one valid trigonometric ratio option.

Review *angles of elevation* and *depression* before demonstrating how to solve the system of equations derived from the overlapping right triangles in Example 6.

Example 2 Using One Ratio to Find the Other Trig Ratios

Given acute θ where $\sin \theta = 0.8$, find the other five trigonometric ratios for θ.

Answer

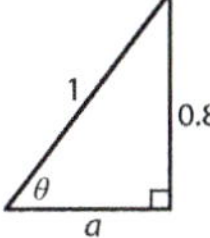

$$a^2 + 0.8^2 = 1^2$$
$$a = \sqrt{1^2 - 0.8^2}$$
$$= \sqrt{0.36} = 0.6$$

$$\cos \theta = \frac{0.6}{1} = 0.6$$
$$\tan \theta = \frac{0.8}{0.6} = \frac{4}{3} \approx 1.3333$$

$$\csc \theta = \frac{1}{\sin \theta} = \frac{1}{0.8} = 1.25$$
$$\sec \theta = \frac{1}{\cos \theta} = \frac{1}{0.6} \approx 1.6667$$
$$\cot \theta = \frac{1}{\tan \theta} = \frac{1}{\frac{4}{3}} = \frac{3}{4} = 0.75$$

1. Sketch a right triangle with θ such that $\sin \theta = \frac{opp.}{hyp.} = \frac{0.8}{1}$.

2. Use the Pythagorean Theorem to find the length of the adjacent leg.

3. Apply the definitions to find the values of the other two basic trigonometric functions.

4. Find the values of the reciprocal trigonometric functions.

— SKILL ✔ EXERCISE 15

It is important to know whether your calculator is set to radian or degree mode when evaluating trigonometric functions. Most calculators require cotangents, secants, and cosecants to be evaluated using their reciprocal functions.

Example 3 Using a Calculator to Evaluate Trig Functions

Evaluate $\cos 37°$, $\tan 61°40'$, and $\sin \frac{\pi}{6}$ with the calculator in degree mode. Then evaluate $\tan \frac{\pi}{4}$, $\csc 30°15'$, and $\sec \frac{\pi}{6}$ with the calculator in radian mode.

Answer

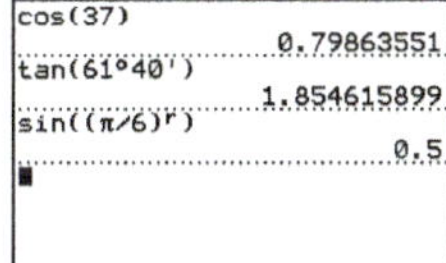
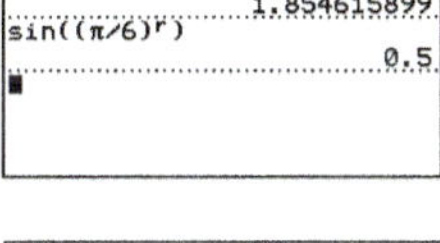

1. Press [MODE] and check that Degree is selected. Then select [QUIT] ([2nd], [MODE]).

2. Enter [COS], [3], [7], [ )] to find $\cos 37°$.

3. Evaluate $\tan 61°40'$, using the [ANGLE] menu ([2nd], [APPS]) to enter the function in DMS.

4. Use the Angle menu to identify ($\pi/6$) as a radian measure.

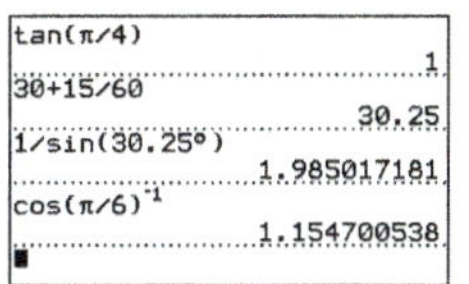

5. Press [MODE], highlight Radian, and press [ENTER], to set the calculator to Radian mode.

6. Enter [TAN], [2nd], [^], [÷], [4], [ )] to evaluate $\tan \frac{\pi}{4}$.

7. Express $30°15'$ in decimal degrees. Use the Angle menu to indicate degree measure when evaluating $\csc 30.25° = \dfrac{1}{\sin 30.25°}$ in radian mode.

8. Evaluate $\sec \frac{\pi}{6}$ using the reciprocal key, [x^{-1}], to represent the function as the reciprocal of $\cos \frac{\pi}{6}$.

TIP 

Be careful not to use one of the inverse functions ($\sin^{-1}$, $\cos^{-1}$, or $\tan^{-1}$) when attempting to evaluate reciprocal functions.

— SKILL ✔ EXERCISE 7

Note that this relatively simplistic method produces a reasonably accurate estimate of the actual 7144 ft change in elevation.

Interactive Activity Use the Internet keyword search *animated trigonometry* to find interactive demonstrations of basic trigonometric functions.

Ex. 42 The actual height of Lady Liberty is 151 ft 1 in.

Ex. 43–44 Exercise 43 could work as a collaboration problem for students using the standard-level assignment. Likewise, exercise 44 could be used for the extended level.

TIPS

Ex. 7–8 While students may choose to simply switch modes in other cases, it can be beneficial for them to understand the full functionality of their graphing calculators.

Find the six trigonometric ratios in exact form for $\angle B$ in each triangle.

1. 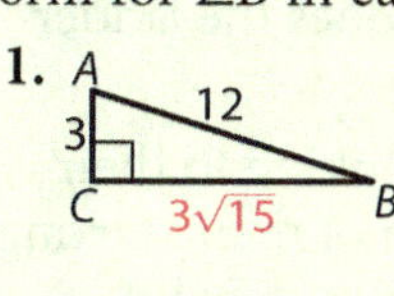

$$\sin B = \frac{1}{4}; \ \csc B = 4$$
$$\cos B = \frac{\sqrt{15}}{4}; \ \sec B = \frac{4\sqrt{15}}{15}$$
$$\tan B = \frac{\sqrt{15}}{15}; \ \cot B = \sqrt{15}$$

2. 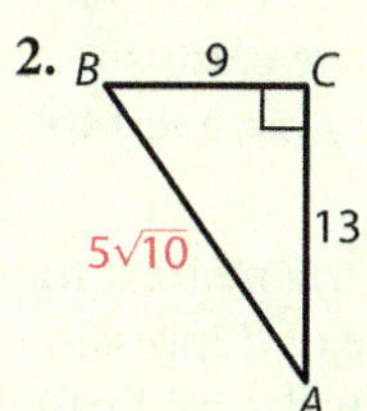

$$\sin B = \frac{13\sqrt{10}}{50}; \ \csc B = \frac{5\sqrt{10}}{13}$$
$$\cos B = \frac{9\sqrt{10}}{50}; \ \sec B = \frac{5\sqrt{10}}{9}$$
$$\tan B = \frac{13}{9}; \ \cot B = \frac{9}{13}$$

Use the given trigonometric ratio for an acute angle θ to find the other two basic trigonometric ratios.

3. $\sin \theta = \frac{3}{4}$ $\cos \theta = \frac{\sqrt{7}}{4}; \ \tan \theta = \frac{3\sqrt{7}}{7}$

4. $\tan \theta = 1.875$
   $\sin \theta \approx 0.8824; \ \cos \theta \approx 0.4706$

5. Evaluate $\sin 37°30'$, $\csc 60°$, and $\cot \frac{\pi}{6}$ with the calculator in degree mode. Repeat the calculations with the calculator in radian mode to verify your answers.

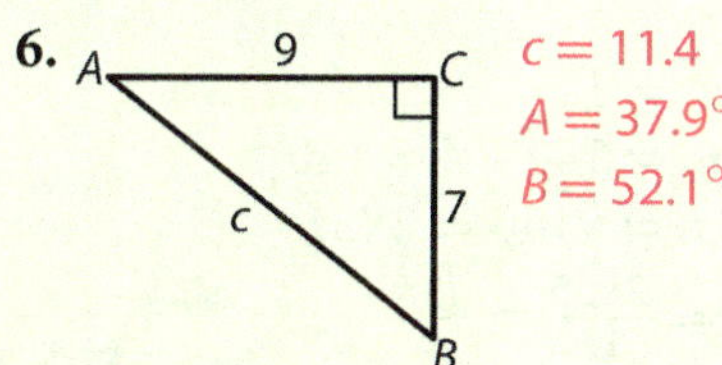
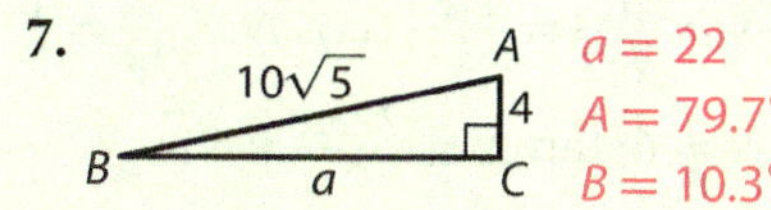

Solve $\triangle ABC$, stating measures to the nearest tenth.

6.

$$c = 11.4$$
$$A = 37.9°$$
$$B = 52.1°$$

7.

$$a = 22$$
$$A = 79.7°$$
$$B = 10.3°$$

8. A 28 ft ladder leaning against the wall of a building reaches a point 24 ft above the ground. At approximately what angle does the ladder meet the building? 31°

9. Mike and Tim are hiking to their campsite due south of their starting point. To avoid a lake, they hike 2.5 mi at a bearing of 221°40′ until they are due west of the campsite. Find the distance they have left to hike. ≈ 1.7 mi

10. Find the altitude (to the nearest foot) of a plane if the angle of depression to a landmark is 6°, and 2 mi farther into the flight the angle of depression is 7°. ≈ 7708 ft

Assignments

- **Minimum:** 1–2, 5–7a, 8–9, 11, 14–16; 19–33 odd; 34, 36–38; 45–51 odd; 52, 54
- **Standard:** 3, 5, 7–13; 15–25 odd; 26, 29–32, 34–40, 42–43, 46–49, 51, 53–54
- **Extended:** 2, 6–9, 12, 16; 17–25 odd; 29–31, 34–39, 41–44; 46–54 even

Assessment

- Quiz 4A covers Sections 4.1–4.2.

Solutions

❯ A. Exercises

1. $\sin A = \dfrac{\sqrt{21}}{5}$; $\csc A = \dfrac{5}{\sqrt{21}} = \dfrac{5\sqrt{21}}{21}$

$\cos A = \dfrac{2}{5}$; $\sec A = \dfrac{5}{2}$

$\tan A = \dfrac{\sqrt{21}}{2}$; $\cot A = \dfrac{2}{\sqrt{21}} = \dfrac{2\sqrt{21}}{21}$

2. $2^2 + n^2 = 12^2$

$n^2 = 144 - 4 = 140$

$n = \sqrt{140} = 2\sqrt{35}$

$\sin N = \dfrac{2\sqrt{35}}{12} = \dfrac{\sqrt{35}}{6}$

$\csc N = \dfrac{6}{\sqrt{35}} = \dfrac{6\sqrt{35}}{35}$; $\cos N = \dfrac{2}{12} = \dfrac{1}{6}$

$\sec N = 6$; $\tan N = \dfrac{2\sqrt{35}}{2} = \sqrt{35}$

$\cot N = \dfrac{1}{\sqrt{35}} = \dfrac{\sqrt{35}}{35}$

3. $5^2 + q^2 = 15^2$

$q^2 = 225 - 25 = 200$

$q = \sqrt{200} = 10\sqrt{2}$

$\sin Q = \dfrac{10\sqrt{2}}{15} = \dfrac{2\sqrt{2}}{3}$

$\csc Q = \dfrac{3}{2\sqrt{2}} = \dfrac{3\sqrt{2}}{4}$

$\cos Q = \dfrac{5}{15} = \dfrac{1}{3}$; $\sec Q = 3$

$\tan Q = \dfrac{10\sqrt{2}}{5} = 2\sqrt{2}$

$\cot Q = \dfrac{1}{2\sqrt{2}} = \dfrac{\sqrt{2}}{4}$

4. $\sin C = \dfrac{2}{5} = 0.4$; $\csc C = \dfrac{5}{2} = 2.5$

$\cos C = \dfrac{\sqrt{21}}{5} \approx 0.9165$

$\sec C = \dfrac{5}{\sqrt{21}} \approx 1.0911$

$\tan C = \dfrac{2}{\sqrt{21}} \approx 0.4364$

$\cot C = \dfrac{\sqrt{21}}{2} \approx 2.2913$

5. $n = \sqrt{140}$

$\sin L = \dfrac{2}{12} \approx 0.1667$; $\csc L = \dfrac{12}{2} = 6$

$\cos L = \dfrac{\sqrt{140}}{12} \approx 0.9860$

$\sec L = \dfrac{12}{\sqrt{140}} \approx 1.0142$

$\tan L = \dfrac{2}{\sqrt{140}} \approx 0.1690$

$\cot L = \dfrac{\sqrt{140}}{2} \approx 5.9161$

6. $q = \sqrt{200}$

$\sin P = \dfrac{5}{15} \approx 0.3333$; $\csc P = \dfrac{15}{5} = 3$

$\cos P = \dfrac{\sqrt{200}}{15} \approx 0.9428$

$\sec P = \dfrac{15}{\sqrt{200}} \approx 1.0607$

$\tan P = \dfrac{5}{\sqrt{200}} \approx 0.3536$

$\cot P = \dfrac{\sqrt{200}}{5} \approx 2.8284$

When the lengths of two sides in a right triangle are known, the measures of the acute angles can be determined. *Inverse trigonometric functions* are defined by reversing the functional relationship between angle measure and the ratio of the sides in a right triangle.

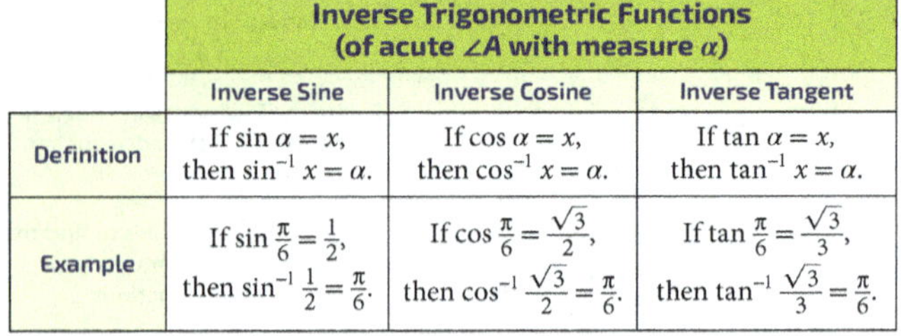

Inverse Trigonometric Functions (of acute ∠A with measure α)			
	Inverse Sine	**Inverse Cosine**	**Inverse Tangent**
Definition	If $\sin \alpha = x$, then $\sin^{-1} x = \alpha$.	If $\cos \alpha = x$, then $\cos^{-1} x = \alpha$.	If $\tan \alpha = x$, then $\tan^{-1} x = \alpha$.
Example	If $\sin \dfrac{\pi}{6} = \dfrac{1}{2}$, then $\sin^{-1} \dfrac{1}{2} = \dfrac{\pi}{6}$.	If $\cos \dfrac{\pi}{6} = \dfrac{\sqrt{3}}{2}$, then $\cos^{-1} \dfrac{\sqrt{3}}{2} = \dfrac{\pi}{6}$.	If $\tan \dfrac{\pi}{6} = \dfrac{\sqrt{3}}{3}$, then $\tan^{-1} \dfrac{\sqrt{3}}{3} = \dfrac{\pi}{6}$.

Inverse trigonometric functions are frequently used when *solving a triangle* (finding any unknown angle measures or side lengths). A right triangle can be solved if one side length and either of the acute angles or a second side length is known.

Example 4 Solving a Right Triangle

Solve $\triangle ABC$, stating angle measures to the nearest tenth of a degree.

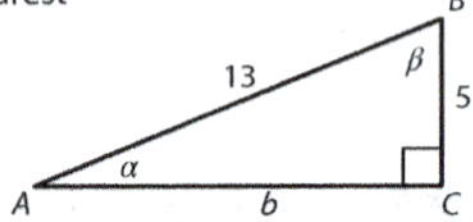

Answer

$5^2 + b^2 = 13^2$

$b = \sqrt{13^2 - 5^2} = \sqrt{144} = 12$

1. Use the Pythagorean Theorem to find b.

$\sin \alpha = \dfrac{5}{13}$

$\sin^{-1} \dfrac{5}{13} = \alpha \approx 22.6°$

2. Use the inverse sine function to find α. (Be sure the calculator is in degree mode.)

$\cos \beta = \dfrac{5}{13}$

$\cos^{-1} \dfrac{5}{13} = \beta \approx 67.4°$

3. Use the inverse cosine function to find β.

Check

$22.6° + 67.4° = 90°$

The acute angles of a right triangle are complementary.

SKILL ✔ EXERCISE 27

Many practical problems can be solved by finding the measure of an angle or a side in a right triangle modeling the scenario.

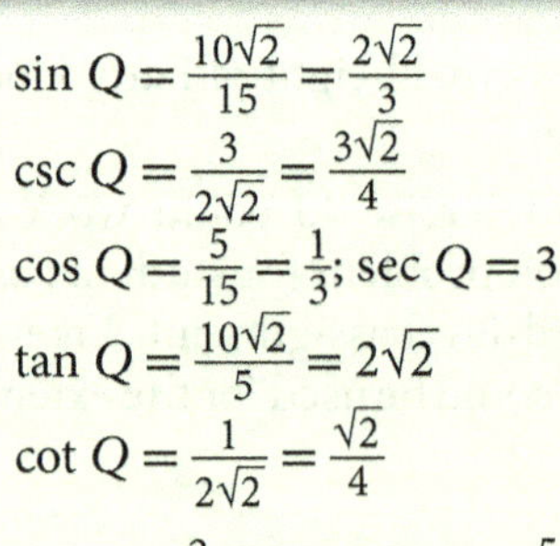

The first recorded estimate of the relative distances of the moon and sun from the earth was made by Aristarchus (ca. 300 BC) using the right triangle formed by the sun, the earth, and a half-moon.

A ship is spotted at a bearing of 90° from a lighthouse and at a bearing of 137°20′ from a Coast Guard station that is 10 km due north of the lighthouse. Determine the distance to the ship from the lighthouse and from the Coast Guard station.

Answer

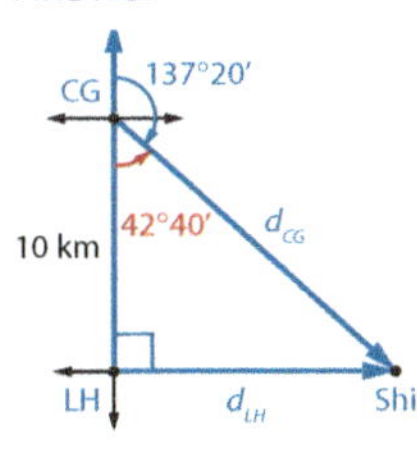

1. Sketch a right triangle modeling the scenario.

$$180° - 137°20′ = 42°40′$$

2. Find the measure of an acute angle in the triangle.

$$\tan 42°40′ = \frac{d_{LH}}{10}$$
$$10 \tan 42°40′ = d_{LH}$$
$$d_{LH} \approx 9.2 \text{ km}$$

3. Use the tangent ratio to find the length of the leg opposite the given angle when you know the length of the adjacent leg.

$$10^2 + 9.2^2 = d_{CG}^2$$
$$d_{CG} = \sqrt{184.95} \approx 13.6 \text{ km}$$

4. Use the Pythagorean Theorem to solve for the distance between the ship and the Coast Guard station.

SKILL ✓ **EXERCISE 33**

Solving practical problems often requires the use of angles of elevation and depression. The *angle of elevation* is the angle formed by a horizontal line and the line of sight as the observer looks at an object above the horizontal. The *angle of depression* is the angle formed by a horizontal line and the line of sight as the observer looks at an object below the horizontal.

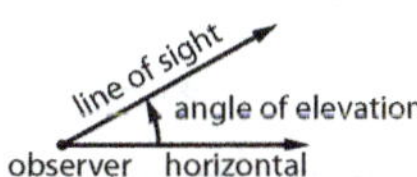

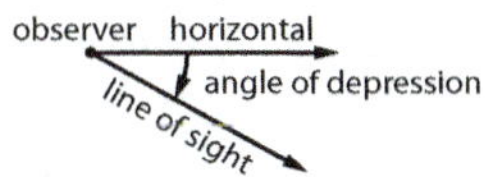

Tyler views the peak of Grand Teton from Phelps Lake at an 11° angle of elevation. After kayaking 1.5 mi toward Grand Teton, the angle of elevation is 14°. Calculate the peak's height above the lake (to the nearest foot).

Answer

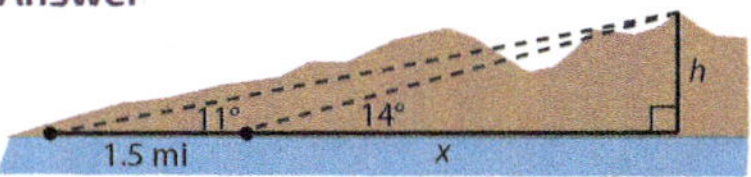

1. Sketch a diagram of the scenario.

$$\tan 11° = \frac{h}{x + 1.5}$$
$$h = (x + 1.5) \tan 11°$$
$$h = x \tan 11° + 1.5 \tan 11°$$
$$x = \frac{h - 1.5 \tan 11°}{\tan 11°}$$

$$\tan 14° = \frac{h}{x}$$
$$x = \frac{h}{\tan 14°}$$

2. Use the overlapping right triangles to write equations relating each angle of elevation to the peak's height and solve each equation for x.

CONTINUED ➡

7.

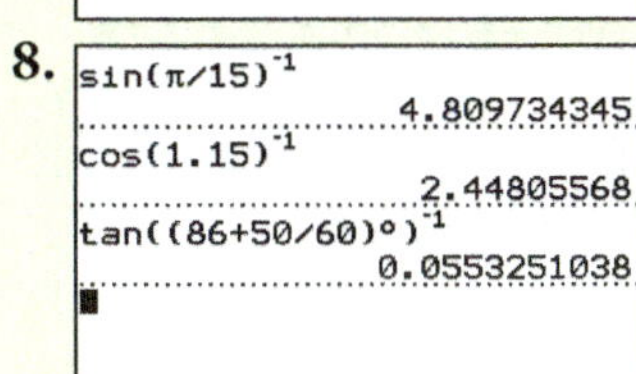

```
cos(26°20')
              0.8962285103
tan(17)
              0.3057306815
sin((2π/5)ʳ)
              0.9510565163
```

8.

```
sin(π/15)⁻¹
              4.809734345
cos(1.15)⁻¹
              2.44805568
tan((86+50/60)°)⁻¹
              0.0553251038
▪
```

9. $\tan 23° = \frac{x}{8}$

$$x = 8 \tan 23° \approx 3.4$$

10. $\cos 47° = \frac{x}{10}$

$$x = 10 \cos 47° \approx 6.8$$

11. $\sin A = \frac{5}{11}$

$$A = \sin^{-1} \frac{5}{11} \approx 27°$$

12. $\tan A = \frac{6}{10.39}$

$$A = \tan^{-1} \frac{6}{10.39} \approx 30°$$

13.

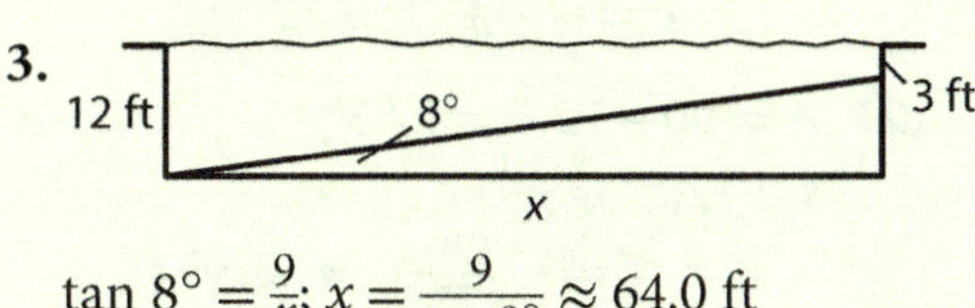

$$\tan 8° = \frac{9}{x}; \ x = \frac{9}{\tan 8°} \approx 64.0 \text{ ft}$$

14.

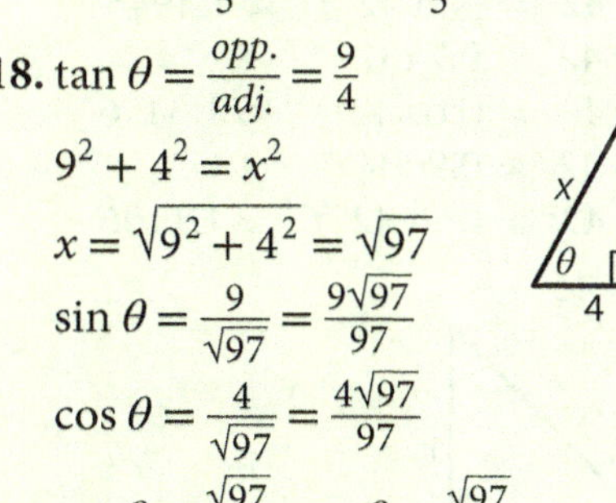

$$\tan x = \frac{150}{32}$$
$$x = \tan^{-1} \frac{150}{32} \approx 78.0°$$

❯ B. Exercises

15. $\cos \theta = \frac{adj.}{hyp.} = \frac{7}{25}$

$$7^2 + x^2 = 25^2$$
$$x = \sqrt{25^2 - 7^2} = 24$$
$$\sin \theta = \frac{24}{25}; \ \tan \theta = \frac{24}{7}$$
$$\csc \theta = \frac{25}{24}; \ \sec \theta = \frac{25}{7}; \ \cot \theta = \frac{7}{24}$$

16. $\tan \theta = \frac{opp.}{adj.} = \frac{8}{15}$

$$8^2 + 15^2 = x^2$$
$$x = \sqrt{15^2 + 8^2} = 17$$
$$\sin \theta = \frac{8}{17}; \ \cos \theta = \frac{15}{17}$$
$$\csc \theta = \frac{17}{8}; \ \sec \theta = \frac{17}{15}; \ \cot \theta = \frac{15}{8}$$

17. $\cos \theta = \frac{1}{\sec \theta} = \frac{12}{13} = \frac{adj.}{hyp.}$

$$12^2 + x^2 = 13^2$$
$$x = \sqrt{13^2 - 12^2} = 5$$
$$\sin \theta = \frac{5}{13}; \ \tan \theta = \frac{5}{12}$$
$$\csc \theta = \frac{13}{5}; \ \cot \theta = \frac{12}{5}$$

18. $\tan \theta = \frac{opp.}{adj.} = \frac{9}{4}$

$$9^2 + 4^2 = x^2$$
$$x = \sqrt{9^2 + 4^2} = \sqrt{97}$$
$$\sin \theta = \frac{9}{\sqrt{97}} = \frac{9\sqrt{97}}{97}$$
$$\cos \theta = \frac{4}{\sqrt{97}} = \frac{4\sqrt{97}}{97}$$
$$\csc \theta = \frac{\sqrt{97}}{9}; \ \sec \theta = \frac{\sqrt{97}}{4}$$

19. $\sin \theta = \frac{opp.}{hyp.} = \frac{0.5}{1}$

$$0.5^2 + x^2 = 1^2$$
$$x = \sqrt{1^2 - 0.5^2} \approx 0.8660$$
$$\cos \theta \approx \frac{0.8660}{1} = 0.8660$$
$$\tan \theta \approx \frac{0.5}{0.8660} \approx 0.5774$$
$$\sec \theta \approx \frac{1}{0.8660} \approx 1.1547$$
$$\csc \theta = \frac{1}{0.5} = 2$$
$$\cot \theta \approx \frac{0.8660}{0.5} = 1.732$$

20. $\cos\theta = \dfrac{adj.}{hyp.} = \dfrac{1}{\sec\theta} = \dfrac{1}{1.25} = 0.8$

$1^2 + x^2 = 1.25^2$

$\quad x = \sqrt{1.25^2 - 1^2}$

$\quad\quad = 0.75$

$\sin\theta = \dfrac{0.75}{1.25} = 0.6$

$\tan\theta = \dfrac{0.75}{1} = 0.75$

$\csc\theta = \dfrac{1.25}{0.75} \approx 1.6667$

$\cot\theta = \dfrac{1}{0.75} \approx 1.3333$

21. $\dfrac{\pi}{3} = 60°$

$\sin 60° = \dfrac{\sqrt{3}}{2}; \ \csc 60° = \dfrac{2}{\sqrt{3}} = \dfrac{2\sqrt{3}}{3}$

$\cos 60° = \dfrac{1}{2}; \ \sec 60° = \dfrac{2}{1} = 2$

$\tan 60° = \dfrac{\sqrt{3}}{1} = \sqrt{3}$

$\cot 60° = \dfrac{1}{\sqrt{3}} = \dfrac{\sqrt{3}}{3}$

22. $\sin 30° = \dfrac{1}{2}; \ \csc 30° = 2$

$\cos 30° = \dfrac{\sqrt{3}}{2}$

$\sec 30° = \dfrac{2}{\sqrt{3}} = \dfrac{2\sqrt{3}}{3}$

$\tan 30° = \dfrac{1}{\sqrt{3}} = \dfrac{\sqrt{3}}{3}$

$\cot 30° = \dfrac{\sqrt{3}}{1} = \sqrt{3}$

23. Use technology.

$\sin 42° \approx 0.6691$

$\csc 42° = (\sin 42°)^{-1} \approx 1.4945$

$\cos 42° \approx 0.7431$

$\sec 42° = (\cos 42°)^{-1} \approx 1.3456$

$\tan 42° \approx 0.9004$

$\cot 42° = (\tan 42°)^{-1} \approx 1.1106$

24. $\dfrac{\pi}{4} = 45°$

$\sin 45° = \dfrac{1}{\sqrt{2}} = \dfrac{\sqrt{2}}{2} = \cos 45°$

$\csc 45° = \dfrac{\sqrt{2}}{1} = \sqrt{2} = \sec 45°$

$\tan 45° = \dfrac{1}{1} = 1 = \cot 45°$

25. $5^2 + b^2 = 9^2$

$\quad b = \sqrt{9^2 - 5^2} = \sqrt{56}$

$\quad\quad = 2\sqrt{14} \approx 7.48$

$\sin A = \dfrac{5}{9}$

$\quad A = \sin^{-1}\dfrac{5}{9} \approx 33.7°$

$\cos B = \dfrac{5}{9}$

$\quad B = \cos^{-1}\dfrac{5}{9} \approx 56.3°$

$\dfrac{h - 1.5\tan 11°}{\tan 11°} = \dfrac{h}{\tan 14°}$

$\tan 14°(h - 1.5\tan 11°) = h\tan 11°$

$h\tan 14° - 1.5\tan 11°(\tan 14°) = h\tan 11°$

$h(\tan 14° - \tan 11°) = 1.5\tan 11°(\tan 14°)$

$h = \dfrac{1.5\tan 11°(\tan 14°)}{\tan 14° - \tan 11°} \approx 1.323 \text{ mi}$

$1.323 \text{ mi}\left(\dfrac{5280\text{ ft}}{1\text{ mi}}\right) \approx 6986 \text{ ft}$

3. Solve the system of equations using substitution.

4. Convert miles to the nearest foot.

7a. 0.8962   **7b.** 0.3057   **7c.** 0.9511   **8a.** 4.8097   **8b.** 2.4481   **8c.** 0.0553

❯ A. Exercises

Find the six trigonometric ratios in exact fractional form for the given acute angles in each triangle.

1. $\angle A$ **2.** $\angle N$ **3.** $\angle Q$

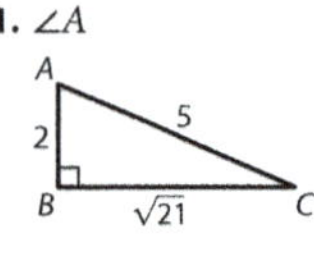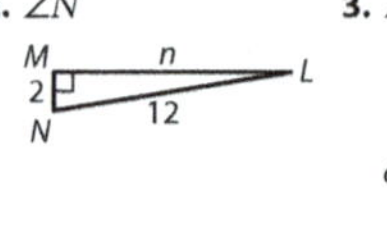

Use the triangles from exercises 1–3 to find the six trigonometric ratios for each angle in decimal form. Round answers to the nearest ten thousandth when necessary.

4. $\angle C$ **5.** $\angle L$ **6.** $\angle P$

7. Use a calculator in degree mode to evaluate the following trigonometric ratios (to the nearest ten thousandth).

   a. $\cos 26°20'$ **b.** $\tan 17°$ **c.** $\sin\dfrac{2\pi}{5}$

8. Use a calculator in radian mode to evaluate the following trigonometric ratios (to the nearest ten thousandth).

   a. $\csc\dfrac{\pi}{15}$ **b.** $\sec 1.15$ **c.** $\cot 86°50'$

Find each indicated length (to the nearest tenth).

9.   **10.**

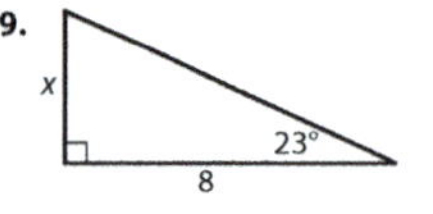

Find the measure of $\angle A$ (to the nearest degree).

11.   **12.**

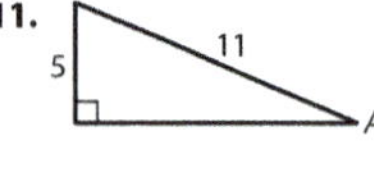

9. $x = 3.4$ **10.** $x = 6.8$

11. $\angle A = 27°$ **12.** $\angle A = 30°$

Solve. Round answers to the nearest tenth.

13. The bottom of a rectangular pool rises at a constant 8° grade from 12 ft at the deep end to 3 ft at the shallow end. How long is the pool? 64.0 ft

14. A wire supporting a 150 ft radio antenna stretches from the top of the antenna to the ground 32 ft away from the base of the antenna. Find the angle formed by the wire and the ground. 78.0°

❯ B. Exercises

Find the remaining trigonometric ratios for each acute angle θ in a right triangle.

15. $\cos\theta = \dfrac{7}{25}$ **16.** $\tan\theta = \dfrac{8}{15}$ **17.** $\sec\theta = \dfrac{13}{12}$

18. $\cot\theta = \dfrac{4}{9}$ **19.** $\sin\theta = 0.5$ **20.** $\sec\theta = 1.25$

Find the six trigonometric ratios for each angle. Use exact values when possible.

21. $\dfrac{\pi}{3}$ **22.** 30°   **23.** 42°   **24.** $\dfrac{\pi}{4}$

Solve right $\triangle ABC$. State side lengths to the nearest hundredth and angle measures to the nearest tenth of a degree.

25. $a = 5, c = 9$

26. $\angle A = 49°, a = 6$

27. $\angle B = 12.7°, a = 20$

28. $c = 14, b = 8$

29. $\angle A = 64.2°, c = 16$

30. Use $\triangle ABC$ to prove each statement.

   a. $\sin A = \cos B$

   b. $\sin^2 A + \cos^2 A = 1$

   (Note that $\sin^2\alpha$ is a simplified notation for $(\sin\alpha)^2$.)

31. **Explain:** Why are the trigonometric ratios the same for all 30-60-90 triangles?

---

26. $B = 90 - 49 = 41°$

$\sin 49° = \dfrac{6}{c}$

$\quad c = \dfrac{6}{\sin 49°} \approx 7.95$

$\tan 49° = \dfrac{6}{b}$

$\quad b = \dfrac{6}{\tan 49°} \approx 5.22$

27. $A = 90 - 12.7 = 77.3°$

$\sin 77.3° = \dfrac{20}{c}$

$\quad c = \dfrac{20}{\sin 77.3°} \approx 20.50$

$\tan 77.3° = \dfrac{20}{b}$

$\quad b = \dfrac{20}{\tan 77.3°} \approx 4.51$

28. $8^2 + a^2 = 14^2$

$\quad a^2 = 196 - 64 = 132$

$\quad a = \sqrt{14^2 - 8^2} = \sqrt{132} \approx 11.49$

$\cos A = \dfrac{8}{14}$

$\quad A = \cos^{-1}\dfrac{8}{14} \approx 55.2°$

$\sin B = \dfrac{8}{14}$

$\quad B = \sin^{-1}\dfrac{8}{14} \approx 34.8°$

29. $B = 90 - 64.2 = 25.8°$

$\sin 64.2° = \dfrac{a}{16}$

$\quad a = 16\sin 64.2° \approx 14.41$

$\sin 25.8° = \dfrac{b}{16}$

$\quad b = 16\sin 25.8° \approx 6.96$

Solve. Round answers to the nearest tenth unless specified otherwise.

32. Safety guidelines recommend that the angle formed by an extension ladder and level ground not exceed 76°. How high against the side of a building can a 20 ft long ladder be safely placed? How far from the building should the ladder be placed to reach this height?   19.4 ft; 4.8 ft

33. A plane leaves airport A traveling at 210 mi/hr. After 1.5 hr, the plane is due east of airport B. If airport A is 230.4 mi due south of airport B, what was the bearing of the plane as it took off from airport A?   43.0°

34. A ship's captain sights an anchored ship at a bearing of 128°. After the ship travels south for 63 nautical miles, the anchored ship is due east of the moving ship. How far apart are the two ships at the second sighting?   80.6 nm

35. A pilot flying at 3000 ft approaches a runway with an angle of depression of 16° to the start of the runway and 9°20′ to the end of the runway. Find the length of the runway (to the nearest foot).   7791 ft

36. A missile travels at a speed of 900 ft/sec at a constant angle of elevation of 73°. After 5 sec what is its altitude (to the nearest foot)?   4303 ft

37. The angle of depression from the top of a 75 ft lighthouse to a ship at sea is 0°40′. How far is the ship from the lighthouse?   6445.5 ft or 1.2 mi

> **C. Exercises**

Solve for x, y, and z (to the nearest tenth).

38.
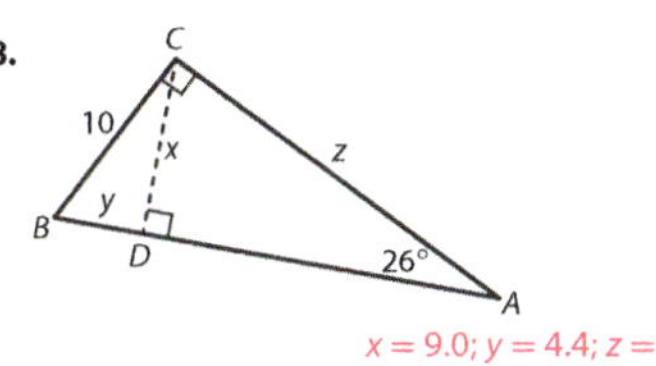

$x = 9.0; y = 4.4; z = 20.5$

39.
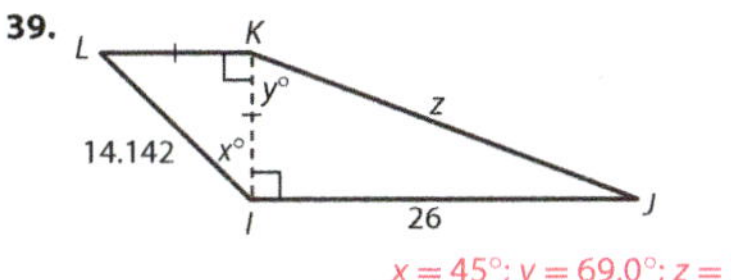

$x = 45°; y = 69.0°; z = 27.9$

40. Samantha is checking the height of a tethered blimp. She measures a 30° angle of elevation, then walks 60 ft toward the blimp and measures a 35° angle of elevation. If her line of sight is 5 ft above the ground, estimate the height of the blimp.   ≈ 202 ft

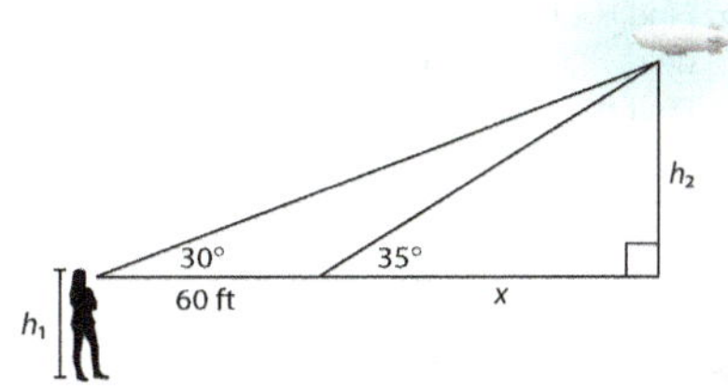

41. When a hot-air balloon is directly over the highway, the pilot measures a 4° angle of depression to the base of the town's tallest building. Estimate the height of the balloon (to the nearest foot) if the angle of depression is 5° after flying 1 mi toward the building without changing altitude.   = 1839 ft

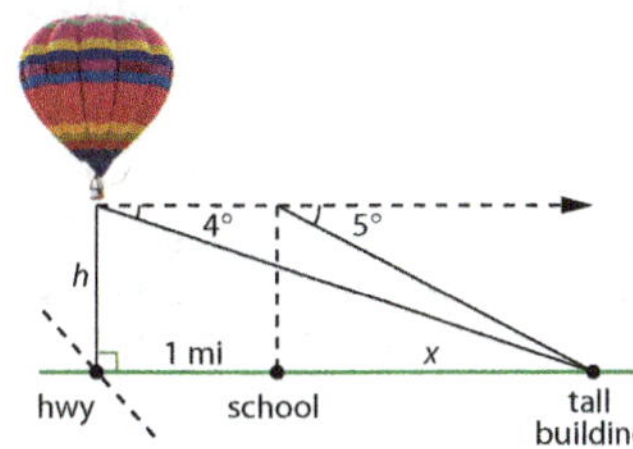

42. From your location on a ferry, the angle of elevation to the top of the Statue of Liberty's torch is 7°, and the angle of elevation to the base of her feet is 3.5°. Determine the height of the statue (y) and your ferry's distance from the monument if the height of the entire monument is 305.5 ft. Round answers to the nearest foot.   153 ft; 2488 ft

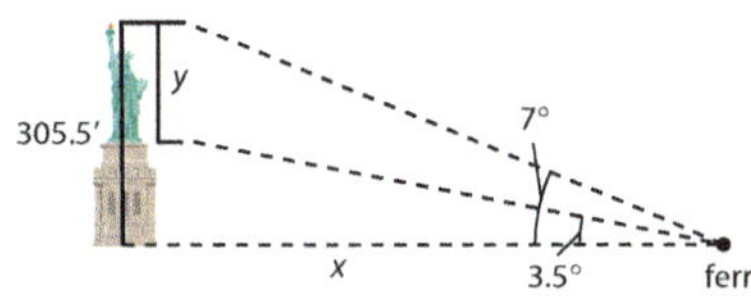

---

36. $d = 900 \dfrac{\text{ft}}{\text{sec}}(5 \text{ sec})$
$= 4500 \text{ ft}$
$\sin 73° = \dfrac{x}{4500}$
$x = 4500 \sin 73°$
$\approx 4303.4 \text{ ft}$

37.
$\tan 40' = \dfrac{75}{x}$
$x = \dfrac{75}{\tan 0°40'} \approx 6445.5 \text{ ft}$
$6445.5 \text{ ft} \left(\dfrac{1 \text{ mi}}{5280 \text{ ft}} \right) \approx 1.2 \text{ mi}$

> **C. Exercises**

38. $m\angle B = 90 - 26 = 64°$
$\sin 64° = \dfrac{x}{10}$
$x = 10 \sin 64° \approx 9.0$
$\cos 64° = \dfrac{y}{10}$
$y = 10 \cos 64° \approx 4.4$
$\tan 26° = \dfrac{10}{z}$
$z = \dfrac{10}{\tan 26°} \approx 20.5$

39. $\overline{LK} \cong \overline{KI} \Rightarrow \angle L \cong \angle LIK$
$m\angle L = m\angle LIK = 45°; x = 45°$
$LI^2 = 2KI^2$
$\dfrac{(14.142)^2}{2} = KI^2$
$KI = \sqrt{\dfrac{(14.142)^2}{2}} \approx 10.0$
$\tan y = \dfrac{26}{10}$
$y = \tan^{-1} \dfrac{26}{10} \approx 69.0°$
$z = \sqrt{10^2 + 26^2} \approx 27.9$

40. $\tan 35° = \dfrac{h_2}{x}$
$x = \dfrac{h_2}{\tan 35°}$
$\tan 30° = \dfrac{h_2}{x + 60}$
$h_2 = (x + 60) \tan 30°$
$h_2 = x \tan 30° + 60 \tan 30°$
$x = \dfrac{h_2 - 60 \tan 30°}{\tan 30°}$
$\dfrac{h_2 - 60 \tan 30°}{\tan 30°} = \dfrac{h_2}{\tan 35°}$
$h_2 \tan 35° - 60 \tan 30° \tan 35° = h_2 \tan 30°$
$h_2 \tan 35° - h_2 \tan 30° = 60 \tan 35° \tan 30°$
$h_2 = \dfrac{60 \tan 30° \tan 35°}{\tan 35° - \tan 30°} \approx 197 \text{ ft}$
$h_1 + h_2 \approx 5 + 197 = 202 \text{ ft}$

30a. $\sin A = \dfrac{a}{c}$ (def. of sine)
$\cos B = \dfrac{a}{c}$ (def. of cosine)
$\sin A = \cos B$ (Transitive Property)

30b. $\sin^2 A + \cos^2 A$
$= \left(\dfrac{a}{c} \right)^2 + \left(\dfrac{b}{c} \right)^2$
$= \dfrac{a^2 + b^2}{c^2} = \dfrac{c^2}{c^2} = 1$

31. Since all 30-60-90 triangles are similar (AA ~ Postulate), their sides are proportional.

32. $\sin 76° = \dfrac{h}{20}$
$h = 20 \sin 76°$
$\approx 19.4 \text{ ft}$
$\cos 76° = \dfrac{d}{20}$
$d = 20 \cos 76°$
$\approx 4.8 \text{ ft}$

33. $d_a = rt = 210(1.5)$
$= 315 \text{ mi}$
$\cos x = \dfrac{230.4}{315}$
$x = \cos^{-1} \dfrac{230.4}{315}$
$\approx 43.0°$

34. $180 - 128 = 52°$
$\tan 52° = \dfrac{x}{63}$
$x = 63 \tan 52°$
$\approx 80.6 \text{ nm}$

35. $\tan 16° = \dfrac{3000}{y}$
$y = \dfrac{3000}{\tan 16°} \approx 10,462.2 \text{ ft}$
$\tan 9°20' = \dfrac{3000}{x + y}$
$x + y = \dfrac{3000}{\tan 9°20'} \approx 18,253.3 \text{ ft}$
$x \approx 18,253.3 - 10,462.2 \approx 7791 \text{ ft}$

41. $\tan 5° = \dfrac{h}{x}$

$$x = \dfrac{h}{\tan 5°}$$

$$\tan 4° = \dfrac{h}{x+1}$$
$$h = (x+1)\tan 4°$$
$$h = x\tan 4° + \tan 4°$$
$$x = \dfrac{h - \tan 4°}{\tan 4°}$$
$$\dfrac{h - \tan 4°}{\tan 4°} = \dfrac{h}{\tan 5°}$$
$$h\tan 5° - \tan 4° \tan 5° = h\tan 4°$$
$$h\tan 5° - h\tan 4° = \tan 4° \tan 5°$$
$$h = \dfrac{\tan 4° \tan 5°}{\tan 5° - \tan 4°} \approx 0.348 \text{ mi}$$
$$\approx 1839 \text{ ft}$$

42. $\tan 7° = \dfrac{305.5}{x}$

$$x = \dfrac{305.5}{\tan 7°} \approx 2488 \text{ ft}$$
$$\tan 3.5° = \dfrac{305.5 - y}{2488}$$
$$2488 \tan 3.5° = 305.5 - y$$
$$y = 305.5 - 2488\tan 3.5°$$
$$y \approx 153 \text{ ft}$$

43a.

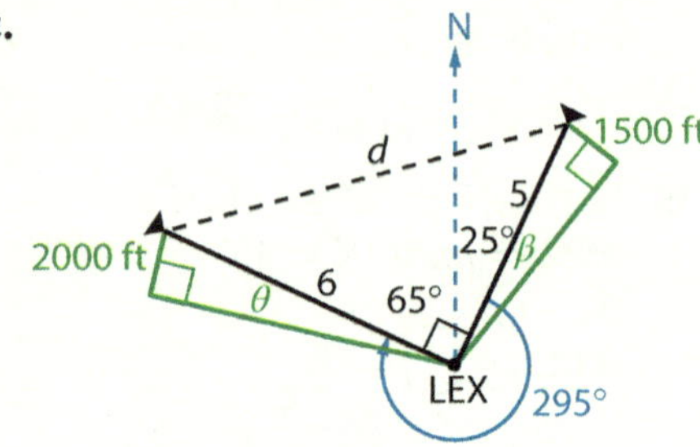

$$d_{horizontal} = \sqrt{6^2 + 5^2} \approx 7.8 \text{ nm}$$

The difference in altitude is 500 ft or 0.08 nm.

$$d_{actual} = \sqrt{(d_{horizontal})^2 + 0.08^2}$$
$$\approx \sqrt{61 + 0.08^2} \approx 7.81$$

43b. $\sin \alpha = \dfrac{2000}{6(6076)}$

$$\alpha = \sin^{-1}\dfrac{250}{4557} \approx 3.1°$$
$$\sin \beta = \dfrac{1500}{5(6076)}$$
$$\beta = \sin^{-1}\dfrac{75}{1519} \approx 2.8°$$

44.

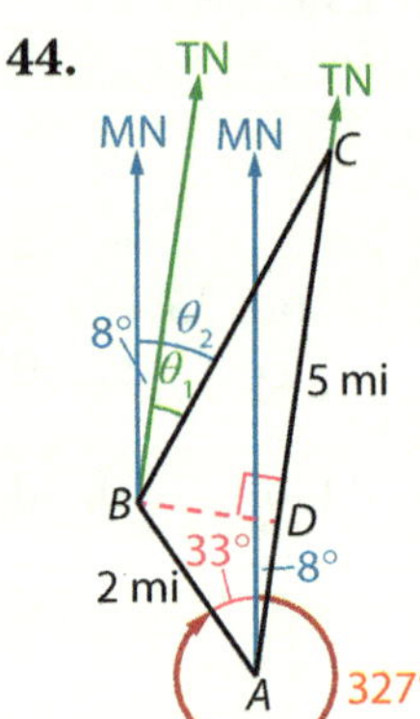

$$m\angle DAB = 360 - 327 + 8 = 41°$$
$$m\angle ABD = 90 - 41 = 49°$$
$$\cos 49° = \dfrac{BD}{2}$$

---

43. A plane 6 nm from the Lexington airport approaches the airport from a bearing of 295° at an altitude of 2000 ft. A second plane 5 nm from the airport approaches at an altitude of 1500 ft from a bearing of 25°. (1 nm $\approx$ 6076 ft)

a. Find the horizontal distance between the planes (to the nearest tenth of a nautical mile). 7.8 nm

b. Find the angle of descent (depression) for each plane (to the nearest tenth of a degree). 3.1°; 2.8°

44. A compass points toward the magnetic north pole instead of true north, the direction of the geographic North Pole. At Trent's location, true north is 8° east of magnetic north. He follows a compass bearing of 327° for 2 mi to avoid a lake, but his final destination is a point 5 mi true north of his starting point. After the detour, what compass bearing should he follow to reach his final destination? How far will he have traveled?

Algebraically find the *y*-intercept, the domain, and any vertical asymptotes for each function. Then use technology to verify your answers. [2.5]

45. $f(x) = \dfrac{6}{x+2}$ y-int.: $(0, 3)$; $D = \{x \mid x \neq -2\};$ VA: $x = -2$

46. $f(x) = \dfrac{x+3}{x^2 - 4}$ y-int.: $\left(0, -\dfrac{3}{4}\right)$; $D = \{x \mid x \neq \pm 2\};$ VA: $x = \pm 2$

Use the properties of logarithms to find the value of *x*. [3.4]

47. $\log_2 (x^2 + 3x + 4) = 1$ $x = -1, -2$

48. $\log (x^2 - 144) = 1 + \log (x + 12)$ $x = 22$

Convert each degree measure to radian measure. Express the answer in terms of π and then as a decimal approximation rounded to the nearest hundredth. [4.1]

49. 36° $\dfrac{\pi}{5} \approx 0.63$

50. −148° $-\dfrac{37\pi}{45} \approx -2.58$

51. Solve $\sqrt{x} - \sqrt{2x - 2} + 4 = 5$. [2.1] A

A. $x = 1$
B. $x = 1, 9$
C. $x = -1, -9$
D. $x = 9$
E. none of these

52. Solve $e^{6x} = 36$. [3.4] C

A. $\ln 2$
B. $\dfrac{1}{2}\ln 6$
C. $\dfrac{1}{3}\ln 6$
D. $\dfrac{1}{2}\ln 3$
E. none of these

53. Solve $3e^{2x+1} - 6 = 0$. [3.4] E

A. $\ln 2 - \dfrac{1}{2}$
B. $\dfrac{1}{2}(\ln 2 - 2)$
C. $-\ln 2 + 1$
D. $\ln 2$
E. none of these

54. Which of the following angles is not coterminal with $\dfrac{5\pi}{4}$? [4.1] C

A. $\dfrac{13\pi}{4}$
B. $-\dfrac{3\pi}{4}$
C. $\dfrac{9\pi}{4}$
D. $-\dfrac{11\pi}{4}$
E. $\dfrac{21\pi}{4}$

$$BD = 2\cos 49° \approx 1.3 \text{ mi}$$
$$\sin 49° = \dfrac{DA}{2}$$
$$DA = 2\sin 49° \approx 1.5 \text{ mi}$$
$$CD = CA - DA = 5 - 1.5 = 3.5 \text{ mi}$$
$$m\angle CBD = \tan^{-1}\dfrac{3.5}{1.3} \approx 69.6°$$
$$\theta_1 \approx 90 - 69.6 = 20.4°$$
$$\theta_2 = 20.4 + 8 = 28.4°$$
$$BC = \sqrt{1.3^2 + 3.5^2} \approx 3.7 \text{ mi}$$
total distance: $2 + 3.7 = 5.7$ mi

▶ Cumulative Review

45. $f(0) = \dfrac{6}{0+2} = 3$; y-int.: $(0, 3)$

$x + 2 \neq 0$; $x \neq -2$

$\therefore D = \{x \mid x \neq -2\}$; VA: $x = -2$

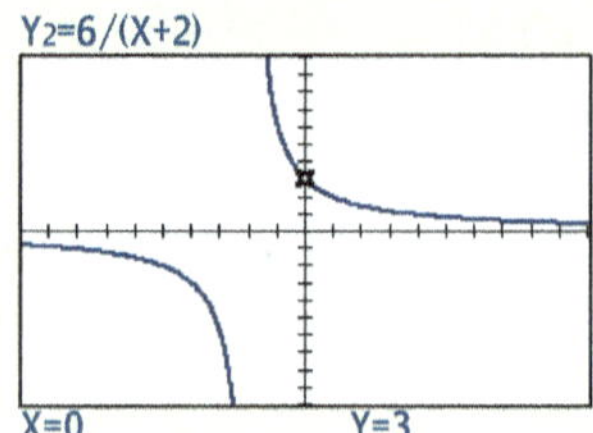

continued in Answers and Solutions Overflow

TRIGONOMETRY

If you have ever camped far from the city and artificial lights, you have likely experienced the awe of a brilliant night sky (Ps. 19:1). When you consider that ancient civilizations regularly enjoyed such views, it is no wonder that they realized the constellations maintained their relative positions as they moved across the sky. This consistency proved useful in navigation and developing calendars. However, a few of the heavenly lights were different—they moved among the fixed stars, and even seemed to move backwards at times! The Babylonians (like the Greeks, Arabs, and Europeans after them) assumed that the earth was stationary at the center of the universe and that the stars were on a sphere (or spheres) that revolved around the earth. Attempting to explain the retrograde motion of the planets occupied astronomers for millennia. Trigonometry developed as an aid in their study of celestial motion and the mapping of the night sky.

In his second-century treatise *Almagest*, Ptolemy references the oldest known Greek trigonometric table, produced by the astronomer Hipparchus ca. 140 BC. Ptolemy's description of the table's construction includes a theorem that now bears his name: "In an inscribed quadrilateral, the sum of the products of the opposite sides is equal to the product of the diagonals." When using Ptolemy's table, it was often necessary to find half the chord of twice the angle. For convenience, Hindu astronomers added a half-chord column to their trigonometric tables. The Sanskrit word for *half* was mistranslated into an

Arabic word meaning a "bay" or "fold." Latin translations of these texts replaced the word with the Latin term for "bay" or "fold," *sinus*, which eventually gave us the English term *sine*.

Ptolemy also authored works on astrology, optics, and geography. Maps in his text *Geography* included latitude and longitude and influenced cartography for over a thousand years. Ptolemy's maps were based on an 18,000 mi estimate of the earth's circumference attributed to the Stoic philosopher Posidonius, instead of Eratosthenes's more accurate calculation of roughly 25,000 mi. Since the maps used by Christopher Columbus in the fifteenth century were still based on Ptolemy's projections, they greatly underestimated the distance to India. Many sources suggest that Columbus may not have attempted his epic journey if the maps had been more accurate.

Trigonometry began with the astronomy of the Babylonians and Greeks, was further developed by Hindus and Arabs, and eventually came to Europe in Latin translations of the ancient texts. In the fifteenth century Johann Müller summarized this heritage in his five-volume work *On Triangles of Every Kind*. This text was studied by Nicolaus Copernicus, who developed a heliocentric model of the universe. Copernicus placed the sun at the center of the universe but preserved much of Ptolemy's model, which explained retrograde motion using circular orbits and epicycles (small circles traversed by a planet as it makes its larger orbit around the sun). Nevertheless, Copernicus's model could not completely solve the mystery of retrograde motion. The accurate plotting of planetary positions provided by trigonometry and the development of logarithms as an aid in calculations eventually enabled Johannes Kepler to discover the key to planetary motion—elliptical orbits.

COMPREHENSION CHECK

1. Why was trigonometry developed?
2. Who published the oldest known Greek trigonometric table?
3. State Ptolemy's Theorem.
4. What English word for a common trigonometric function has its origin in the Hindu word for "half"?
5. Explain how Ptolemy's *Geography* may have influenced Columbus's discovery of America.
6. **Discuss:** Explain the role of trigonometry in the development of the current model of the solar system.

PRESENTATION

This Historical Connection emphasizes the real-world motivations behind the development of trigonometry from its beginnings in ancient astronomy to its utilization in the wave theory of electromagnetic energy. For further exploration, consider the development of the modern calendar, sundials, astrolabes, chronometers, harmonographs, and the Great Trigonometrical Survey.

Historical Connection

Objectives

1. To summarize the history of the development of trigonometry
2. To evaluate the impact of Ptolemy's maps
3. To explain the role of trigonometry in the development of heliocentric theory

Answers

1. to aid astronomers in the study of celestial objects
2. Hipparchus
3. In an inscribed quadrilateral, the sum of the products of the opposite sides is equal to the product of the diagonals.
4. sine
5. Ptolemy's maps, which underestimated the circumference of the earth, still influenced maps in Columbus's day, leading Columbus to believe that India was much closer to Europe than it actually was. If he had known the true distance, Columbus may not have attempted the voyage.
6. Ancient civilizations laid the foundations of trigonometry as they attempted to understand the movement of celestial bodies in the night sky. Müller's compilation of trigonometric knowledge may have influenced Copernicus's heliocentric theory. The mathematical tools of trigonometry and logarithms were essential in Kepler's discovery of elliptical orbits.

4.3 Extending Trigonometric Functions

Objectives

1. To evaluate trigonometric functions for any standard position angle or Cartesian point
2. To use reference angles to find trigonometric values
3. To use the unit circle to evaluate trigonometric functions for multiples of $\frac{\pi}{6}$ and $\frac{\pi}{4}$
4. To use the unit circle to explain and apply symmetry and the periodic nature of trigonometric functions

Flash

The first exhibition of a Foucault pendulum was in Paris in 1851. The movement of the pendulum reveals that the earth is rotating. Foucault pendulums are located all over the world. Use the Internet keyword search *list of Foucault pendulums* to find a nearby location.

Vocabulary

circular function
periodic function
reference angle
reference triangle
unit circle

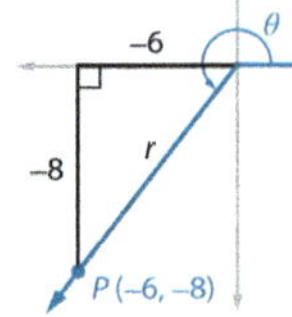

4.3 Extending Trigonometric Functions

The change in height of this Foucault pendulum in Paris is modeled by $\Delta h = L(1 - \cos\theta)$.

After completing this section, you will be able to

- evaluate trig functions using coordinates of a point on its terminal side, its reference angle, or the unit circle.
- use the unit circle to explain symmetry (odd and even) and the periodic nature of trig functions.

In this section we will expand the definitions of trigonometric functions beyond angles found in right triangles to angles of any measure. We will then use radian measure to extend the domain of trigonometric functions to real numbers.

The trigonometric functions can be defined for any angle drawn in standard position using a point $P(x, y)$ on its terminal side.

DEFINITIONS

Given point $P(x, y)$ on the terminal side of angle θ in standard position and $r = \sqrt{x^2 + y^2}$, the six trigonometric ratios are defined as follows.

$$\sin\theta = \frac{y}{r} \qquad \cos\theta = \frac{x}{r} \qquad \tan\theta = \frac{y}{x}$$

$$\csc\theta = \frac{r}{y} \qquad \sec\theta = \frac{r}{x} \qquad \cot\theta = \frac{x}{y}$$

Example 1 — Evaluating Trig Functions Using a Point on the Terminal Side

State the six trigonometric functions of an angle θ in standard position if its terminal side passes through $(-6, -8)$.

Answer

The German clergyman Bartholomaeus Pitiscus was the first to use the word *trigonometry* in the title of a book. His *Trigonometriae sive de dimensione triangulorum libri quinque* (Trigonometry, the properties of triangles in five books) was published in Frankfurt in 1595.

1. Sketch the angle in standard position.

$$r = \sqrt{(-6)^2 + (-8)^2}$$
$$= \sqrt{100} = 10$$

2. Determine r, the distance from the origin to P.

$$\sin\theta = \frac{y}{r} = \frac{-8}{10} = -\frac{4}{5} \qquad \csc\theta = \frac{r}{y} = \frac{10}{-8} = -\frac{5}{4}$$

$$\cos\theta = \frac{x}{r} = \frac{-6}{10} = -\frac{3}{5} \qquad \sec\theta = \frac{r}{x} = \frac{10}{-6} = -\frac{5}{3}$$

$$\tan\theta = \frac{y}{x} = \frac{-8}{-6} = \frac{4}{3} \qquad \cot\theta = \frac{x}{y} = \frac{-6}{-8} = \frac{3}{4}$$

3. Apply the expanded definitions of the trigonometric functions.

SKILL ✔ EXERCISE 3

PRESENTATION

Lesson Opener

Identify each missing length.

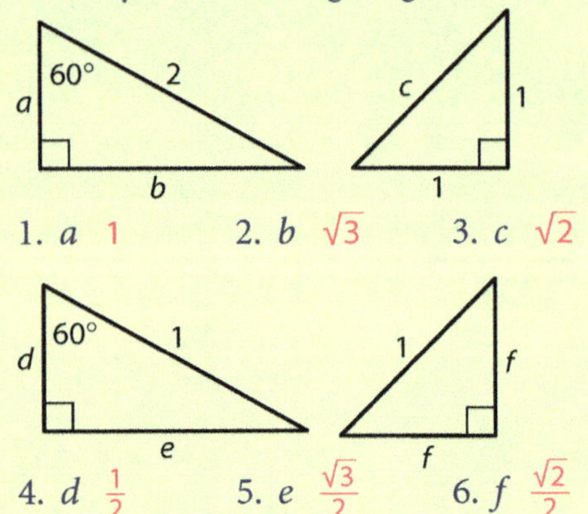

1. a **1**   2. b **$\sqrt{3}$**   3. c **$\sqrt{2}$**

4. d **$\frac{1}{2}$**   5. e **$\frac{\sqrt{3}}{2}$**   6. f **$\frac{\sqrt{2}}{2}$**

Explain the need to extend the definitions of the trigonometric functions beyond acute angles and use the illustration to define the six trigonometric functions in terms of angle θ in standard position on the Cartesian coordinate plane. Then have the students evaluate the functions in Examples 1–2. Remind the students that the calculator will give an error message for $\tan\frac{3\pi}{2}$ and $\sec\frac{3\pi}{2}$ because these values are undefined.

After completing Example 2, ask the students which trigonometric functions can be undefined at a quadrantal angle and have them explain why. (*all but sine and cosine; They contain a denominator in which either $x = 0$ or $y = 0$. Note that sine and cosine have a denominator of r, which is defined as a nonzero distance.*)

Discuss *reference angles* and *reference triangles*. Consider completing exercise 32 together in class; in this exercise students generalize finding the reference angle in each quadrant. The students should be able to quickly determine the sign of the trigonometric functions in each quadrant as shown in the figure. Review the ratios of side lengths in 45-45-90 and 30-60-90 triangles found in the Lesson Opener and then present Example 3, reinforcing reference angles, reference triangles, special triangle ratios, and the sign of trigonometric functions in a given quadrant.

The terminal side of a quadrantal angle lies on one of the coordinate axes. The expanded definitions allow us to quickly evaluate the trigonometric functions of these angles whose measures are multiples of $\frac{\pi}{2}$ or 90°.

Example 2 Evaluating Trig Functions for a Quadrantal Angle

Evaluate the six trigonometric functions for an angle measuring $\frac{3\pi}{2}$.

Answer

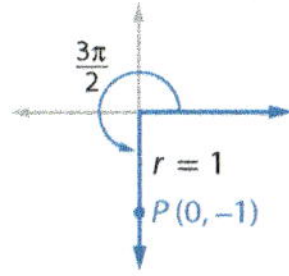

$P = (0, -1)$

$r = 1$

1. Sketch the angle and choose a convenient point on the terminal ray.
2. The point P is 1 unit from the origin.

$\sin \theta = \frac{y}{r} = \frac{-1}{1} = -1$ $\csc \theta = \frac{r}{y} = \frac{1}{-1} = -1$

$\cos \theta = \frac{x}{r} = \frac{0}{1} = 0$ $\sec \theta = \frac{r}{x} = \frac{1}{0};$ undefined

$\tan \theta = \frac{y}{x} = \frac{-1}{0};$ undefined $\cot \theta = \frac{x}{y} = \frac{0}{-1} = 0$

3. Apply the definitions. Note that the secant and tangent functions are not defined for $\frac{3\pi}{2}$.

SKILL ✔ EXERCISE 5

Any nonquadrantal angle θ has an associated *reference angle* α, the acute angle formed by the terminal side and the x-axis when θ is in standard position. The triangle formed by a perpendicular drawn to the x-axis from $P(x, y)$ on the terminal side of a nonquadrantal angle forms a *reference triangle*.

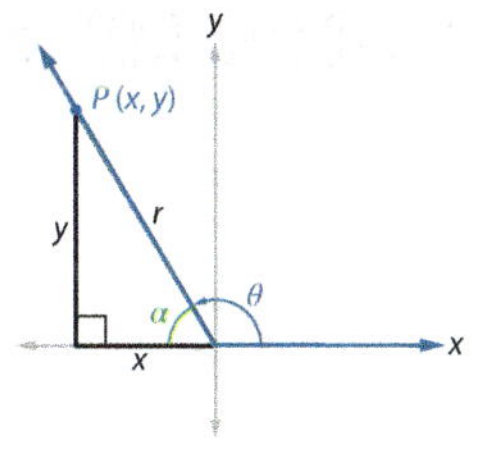

$\theta = 120°; \alpha = 60°$

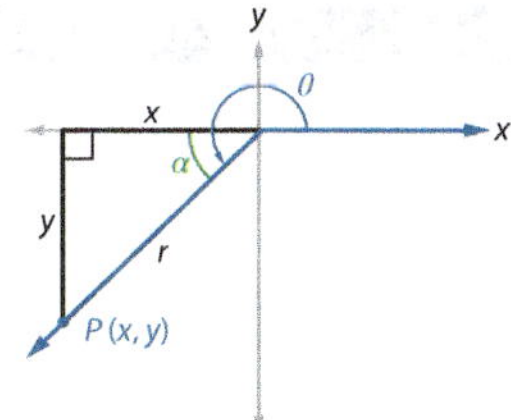

$\theta = 225°; \alpha = 45°$

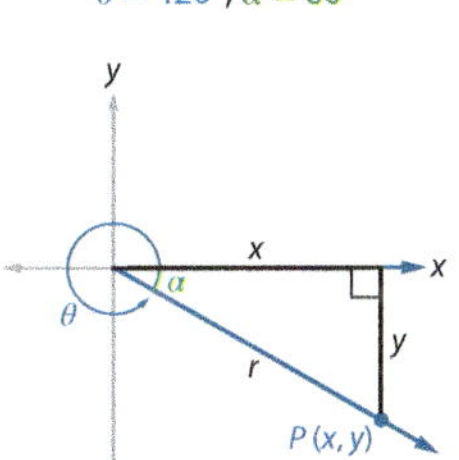

$\theta = 330°; \alpha = 30°$

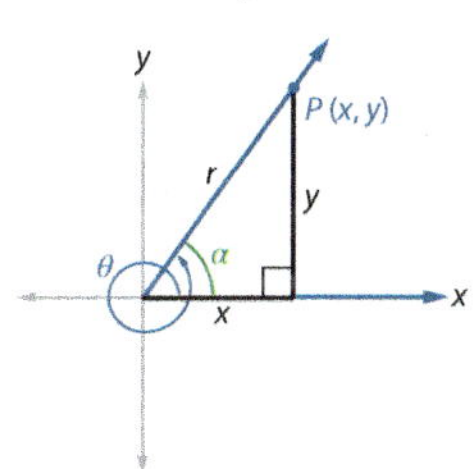

$\theta = 420°; \alpha = 60°$

Reading and Writing Mathematics

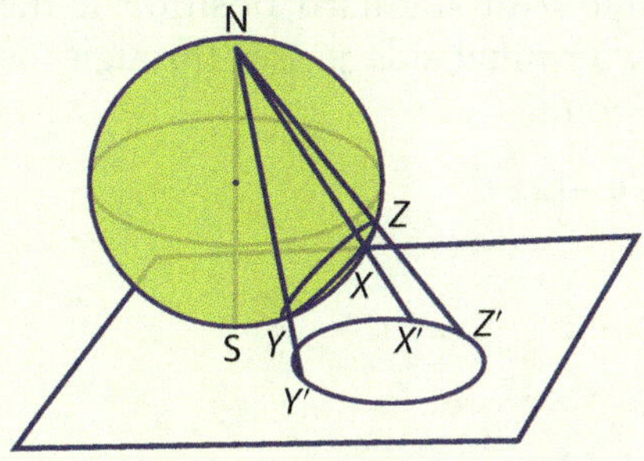

Research a stereographic projection map. The diagram below shows a stereographic mapping of point B from the sphere to the plane at point B'. Assuming the sphere has a diameter of 1 unit, explain why $AB' = \tan\left(45° + \frac{\theta}{2}\right)$.

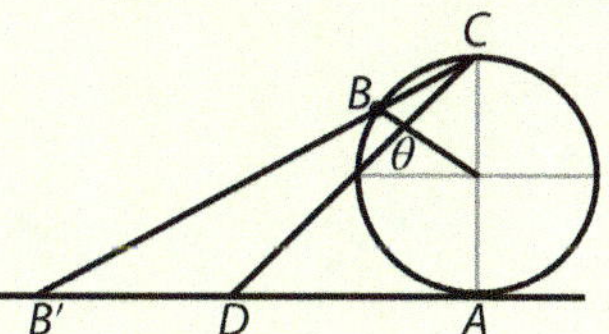

Since the measure of an inscribed angle is $\frac{1}{2}$ the measure of the intercepted arc, $m\angle ACD = \frac{1}{2}(90°) = 45°$ and $m\angle DCB' = \frac{1}{2}\theta$. Since $CA = 1$, $AB' = \tan \angle ACB'$ and $m\angle ACB' = m\angle ACD + \angle DCB' = 45° + \frac{1}{2}\theta$. Therefore $AB' = \tan\left(45° + \frac{\theta}{2}\right)$.

Have the students evaluate the trigonometric functions in Example 4 without referencing the provided answers. Then encourage the students to investigate any incorrect answers.

Example 5 demonstrates how our expanded definitions require us to know more than the value of one trigonometric function to determine the value of other trigonometric functions for the angle. Emphasize the need to determine which quadrant contains the angle's terminal side.

Explain that wrapping a number line around the *unit circle* allows us to extend the definitions of trigonometric functions to include a domain of real numbers, t. Emphasize that $P(x, y)$, the endpoint of the wrapped arc length, has coordinates $x = \cos t$ and $y = \sin t$. The other four trigonometric functions can be evaluated using these coordinates. Direct attention to the side lengths of the two special reference triangles that can be used within the unit circle since their hypotenuse is 1 unit. These ratios are used in Example 6.

Consider beginning exercise 37 together in class. Explain that being able to quickly determine the sine and cosine of these special angles is critical to success in trigonometry.

Interactive Activity Use the Internet keyword search *unit circle image* for detailed images, *interactive unit circle* for demonstrations, or *blank unit circle* for handouts.

Show why cosine is an even function and sine is an odd function. You may want to guide the students in determining the symmetry of the other trigonometric functions (exercise 40).

Discuss the definition of *periodic* and the specific periods of the six trigonometric functions. In Example 7, a reference angle is determined using the fact that the basic sine function has a period of 2π.

Additional Exercises

State the six trigonometric functions of an angle θ in standard position if the angle's terminal side passes through the given point.

1. $(-5, -12)$

$\sin \theta = -\frac{12}{13}; \csc \theta = -\frac{13}{12}$

$\cos \theta = -\frac{5}{13}; \sec \theta = -\frac{13}{5}$

$\tan \theta = \frac{12}{5}; \cot \theta = \frac{5}{12}$

2. $(-1, 0)$

$\sin \theta = 0; \csc \theta = \frac{1}{0}$ (undefined)

$\cos \theta = -1; \sec \theta = -1$

$\tan \theta = 0; \cot \theta = \frac{-1}{0}$ (undefined)

Use reference angles to find the exact value of the three basic trigonometric functions.

3. $\theta = \frac{7\pi}{6}$ $\sin \theta = -\frac{1}{2}$

$\cos \theta = -\frac{\sqrt{3}}{2}; \tan \theta = \frac{\sqrt{3}}{3}$

4. $\theta = -\frac{3\pi}{4}$ $\sin \theta = -\frac{\sqrt{2}}{2}$

$\cos \theta = -\frac{\sqrt{2}}{2}; \tan \theta = 1$

5. Use your calculator (in radian mode) to evaluate $\sin 121°$, $\cos -\frac{3\pi}{5}$, and $\cot 3$.

```
sin(121°)
              0.8571673007
cos(-3π/5)
             -0.3090169944
tan(3)⁻¹
             -7.015252551
1/tan(3)
             -7.015252551
```

6. If $\sin \theta = \frac{1}{3}$ and $\cos \theta < 0$, find $\tan \theta$ and $\sec \theta$. $\tan \theta = -\frac{\sqrt{2}}{4}; \sec \theta = -\frac{3\sqrt{2}}{4}$

Use the unit circle to state the exact value of each expression.

7. $\csc 60°$ $\frac{2\sqrt{3}}{3}$

8. $\cot \frac{\pi}{6}$ $\sqrt{3}$

9. $\cos \left(-\frac{23\pi}{6}\right)$ $\frac{\sqrt{3}}{2}$

10. $\sin \left(-\frac{19\pi}{4}\right)$ $-\frac{\sqrt{2}}{2}$

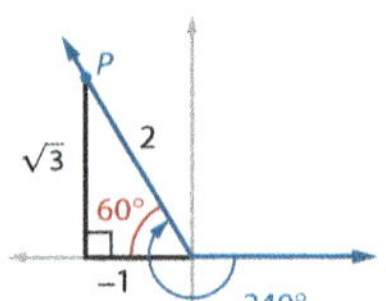

Because r is always positive, the sign of a trigonometric function is determined by the signs of x and y in the quadrant containing the terminal side. Reference angles and the ratios of side lengths in a 30-60-90 or 45-45-90 triangle allow us to quickly evaluate the exact values of trigonometric functions for angles whose measures are multiples of 30° or 45° $\left(\frac{\pi}{6} \text{ or } \frac{\pi}{4}\right)$.

Example 3   Using Reference Angles to Evaluate Trig Functions

State the exact values of the three basic trigonometric functions for each angle.

a. $-240°$ **b.** $\frac{7\pi}{4}$

Answer

a.

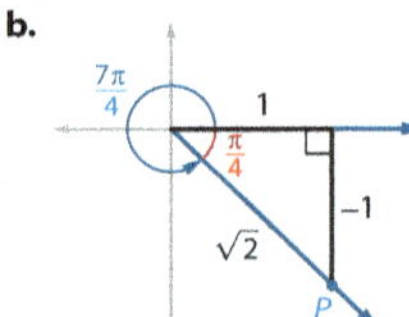

b.

1. Sketch each angle, its reference angle, and an associated reference triangle.

$\sin(-240°) = \frac{\sqrt{3}}{2}$

$\cos(-240°) = \frac{-1}{2} = -\frac{1}{2}$

$\tan(-240°) = \frac{\sqrt{3}}{-1} = -\sqrt{3}$

$\sin \frac{7\pi}{4} = \frac{-1}{\sqrt{2}} = -\frac{\sqrt{2}}{2}$

$\cos \frac{7\pi}{4} = \frac{1}{\sqrt{2}} = \frac{\sqrt{2}}{2}$

$\tan \frac{7\pi}{4} = \frac{-1}{1} = -1$

2. Apply the expanded definitions:
$\sin \theta = \frac{y}{r}$, $\cos \theta = \frac{x}{r}$, and $\tan \theta = \frac{y}{x}$.

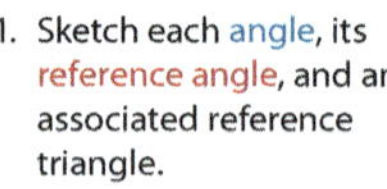

——— SKILL ✓ EXERCISE 11

Your calculator will give decimal approximations for trigonometric functions of other nonacute angles.

Example 4   Using a Calculator to Evaluate Trig Functions

Use your calculator (in radian mode) to evaluate each function.

a. $\cos 337°$ **b.** $\sin \left(-\frac{4\pi}{5}\right)$ **c.** $\cot 2$

Answer

```
cos(337°)
              0.9205048535
sin(-4π/5)
             -0.5877852523
tan(2)⁻¹
             -0.4576575544
1/tan(2)
             -0.4576575544
```

Be sure your calculator is in Radian mode.

a. Enter cos(337°), using the [ANGLE] menu to insert the degree symbol.

b. Enter sin(−4π/5).

c. Find cot 2 by entering tan(2)⟦x⁻¹⟧, or 1/tan(2).

Check

Use the quadrant of the terminal side to verify that the sign of each value is correct.

a. $270° < 337° < 360°$; The cosine is positive in Q IV.

b. $-\pi < -\frac{4\pi}{5} < -\frac{\pi}{2}$; The sine is negative in Q III.

c. $\frac{\pi}{2} < 2 < \pi$ (or $1.57 < 2 < 3.14$); The cotangent is negative in Q II.

——— SKILL ✓ EXERCISE 15

Consider sharing Genesis 8:22 and discussing some of the cyclic phenomena of nature mentioned in the text.

Motivational Idea Share that many periodic functions are present in nature. Some are the result of God's design of our solar system and the universe. Others, such as the heartbeat, are evident in the human body. Many manmade periodic functions are the result of circular engineering. Examples include wheels, engines (the spinning of the crankshaft, flywheel, drivetrain, and pulleys), power generators, electric motors, windmills, and carnival rides. The existence of many God-created and manmade func-tions supports the importance of studying periodic functions in math, science, and engineering.

TIPS

Ex. 16 This exercise extends Example 4. Some students may prefer to complete part *c* by converting to degrees before evaluating the function in order to avoid confusion about the placement of the parenthesis and the ^{r} symbol.

If the value of one of the trigonometric functions of an angle and the quadrant of the terminal side is known, you can use the reference triangle to determine the values of the other trigonometric functions of the angle.

If $\cos \theta = -\frac{2}{3}$ and $\sin \theta > 0$, find $\tan \theta$ and $\csc \theta$.

Answer

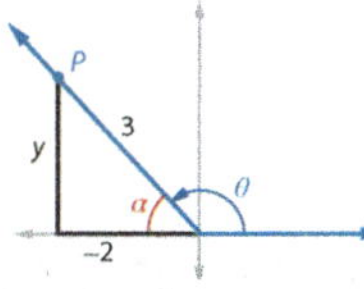

$2^2 + y^2 = 3^2$
$y = \pm\sqrt{9 - 4} = \sqrt{5}$

$\tan \theta = \frac{y}{x} = \frac{\sqrt{5}}{-2} = -\frac{\sqrt{5}}{2}$ $\csc \theta = \frac{r}{y} = \frac{3}{\sqrt{5}} = \frac{3\sqrt{5}}{5}$

1. Sketch θ and the reference triangle in Q II since it is the only quadrant where $\cos \theta < 0$ and $\sin \theta > 0$.

2. Use $\cos \theta = \frac{x}{r}$ and the Pythagorean Theorem to find y, which is positive in Q II.

3. Use x, y, and r to find the other trigonometric ratios.

SKILL ✔ EXERCISE 17

While we have extended the definitions of trigonometric functions to angles of any measure, calculus and many applications require that the trigonometric functions use domains containing real numbers instead of angle measures. This is accomplished using the *unit circle*.

DEFINITION

The **unit circle** is the circle centered at the origin and having a radius of 1 unit. Its equation is $x^2 + y^2 = 1$.

Placing a vertical real number line so that its origin is at $A\,(1, 0)$ and wrapping it around the unit circle allows a real number t to be mapped to a point $P\,(x, y)$ on the unit circle where $t =$ the length of $\overset{\frown}{AP}$. Point P defines the standard position angle θ with radian measure of $\theta = \frac{s}{r} = \frac{t}{1} = t$. This allows the independent variable of a trigonometric function to be viewed as either an angle measure of θ radians or a real number t.

DEFINITIONS

If t is a real number on the number line wrapped around the unit circle so that $t =$ the length of $\overset{\frown}{AP}$, the trigonometric functions of t are defined in terms of $P\,(x, y)$ as follows.

$\sin t = y$ $\cos t = x$ $\tan t = \frac{y}{x}, x \neq 0$

$\csc t = \frac{1}{y}, y \neq 0$ $\sec t = \frac{1}{x}, x \neq 0$ $\cot t = \frac{x}{y}, y \neq 0$

Assignments

- **Minimum:** 1–2; 5–9 odd; 11–13, 15, 16*b*, 17, 20–24, 29–30, 32; 33–37 odd; 42–45, 48

- **Standard:** 1, 4–15, 16*b*, 18; 19–31 odd; 32–39, 44–46, 48–50

- **Extended:** 2–10 even; 11, 14–18; 20–30 even; 33–41, 44–47, 49, 51

Solutions

❯ A. Exercises

1. $r = \sqrt{1^2 + 4^2} = \sqrt{17}$

$\sin \theta = \frac{4}{\sqrt{17}} = \frac{4\sqrt{17}}{17}$

$\csc \theta = \frac{\sqrt{17}}{4}$

$\cos \theta = \frac{1}{\sqrt{17}} = \frac{\sqrt{17}}{17}$

$\sec \theta = \frac{\sqrt{17}}{1} = \sqrt{17}$

$\tan \theta = \frac{4}{1} = 4$; $\cot \theta = \frac{1}{4}$

2. 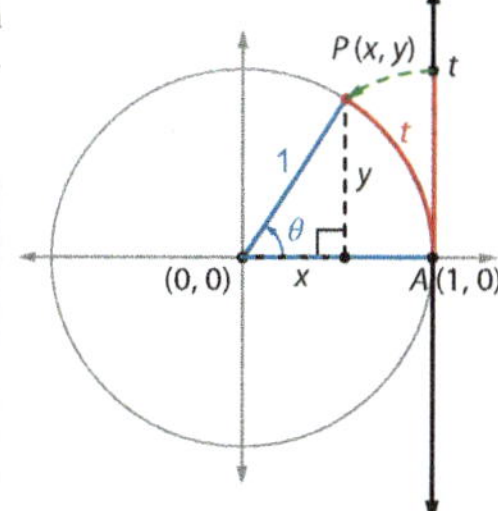

$r = \sqrt{(-6)^2 + (-1)^2} = \sqrt{37}$

$\sin \theta = \frac{-1}{\sqrt{37}} = -\frac{\sqrt{37}}{37}$

$\csc \theta = \frac{\sqrt{37}}{-1} = -\sqrt{37}$

$\cos \theta = \frac{-6}{\sqrt{37}} = -\frac{6\sqrt{37}}{37}$

$\sec \theta = \frac{\sqrt{37}}{-6} = -\frac{\sqrt{37}}{6}$

$\tan \theta = \frac{-1}{-6} = \frac{1}{6}$; $\cot \theta = \frac{6}{1} = 6$

3. 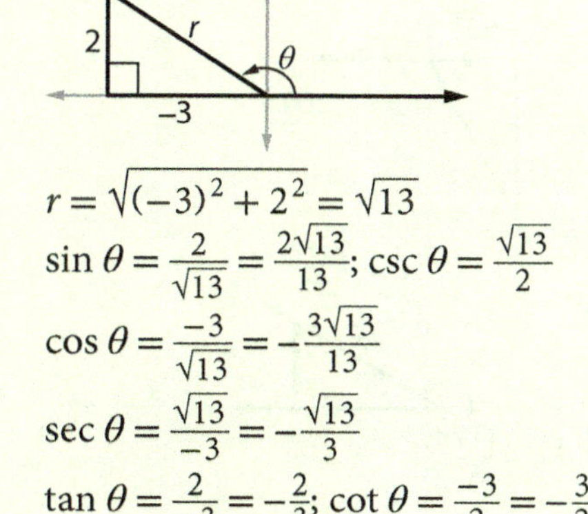

$r = \sqrt{(-3)^2 + 2^2} = \sqrt{13}$

$\sin \theta = \frac{2}{\sqrt{13}} = \frac{2\sqrt{13}}{13}$; $\csc \theta = \frac{\sqrt{13}}{2}$

$\cos \theta = \frac{-3}{\sqrt{13}} = -\frac{3\sqrt{13}}{13}$

$\sec \theta = \frac{\sqrt{13}}{-3} = -\frac{\sqrt{13}}{3}$

$\tan \theta = \frac{2}{-3} = -\frac{2}{3}$; $\cot \theta = \frac{-3}{2} = -\frac{3}{2}$

4. 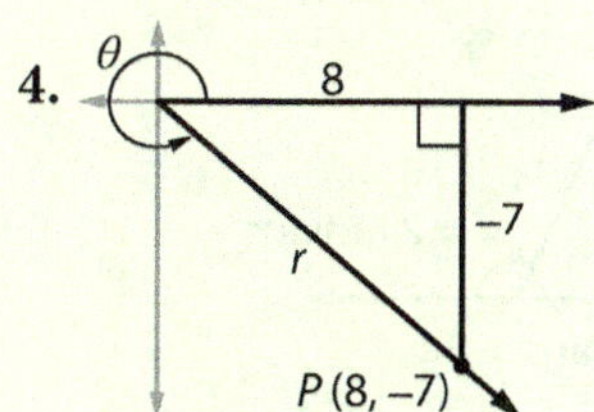

$r = \sqrt{8^2 + (-7)^2} = \sqrt{113}$

$\sin \theta = \frac{-7}{\sqrt{113}} = -\frac{7\sqrt{113}}{113}$

$\csc \theta = \frac{\sqrt{113}}{-7} = -\frac{\sqrt{113}}{7}$

$\cos \theta = \frac{8}{\sqrt{113}} = \frac{8\sqrt{113}}{113}$; $\sec \theta = \frac{\sqrt{113}}{8}$

$\tan \theta = \frac{-7}{8} = -\frac{7}{8}$; $\cot \theta = \frac{8}{-7} = -\frac{8}{7}$

5.

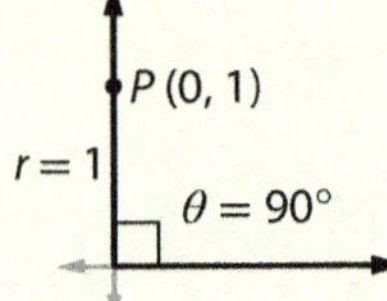

$$\sin\theta = \frac{1}{1} = 1; \quad \csc\theta = \frac{1}{1} = 1$$

$$\cos\theta = \frac{0}{1} = 0$$

$$\sec\theta = \frac{1}{0} \ (\text{undefined})$$

$$\tan\theta = \frac{1}{0} \ (\text{undefined}); \quad \cot\theta = \frac{0}{1} = 0$$

6.

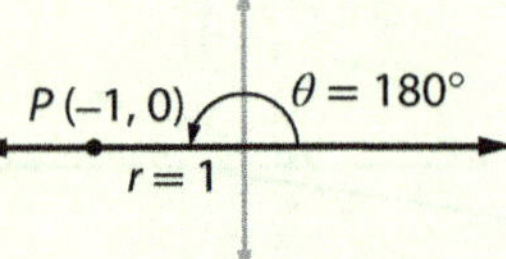

$$\sin\theta = \frac{0}{1} = 0; \quad \csc\theta = \frac{1}{0} \ (\text{undefined})$$

$$\cos\theta = \frac{-1}{1} = -1; \quad \sec\theta = \frac{1}{-1} = -1$$

$$\tan\theta = \frac{0}{-1} = 0$$

$$\cot\theta = \frac{-1}{0} \ (\text{undefined})$$

7.

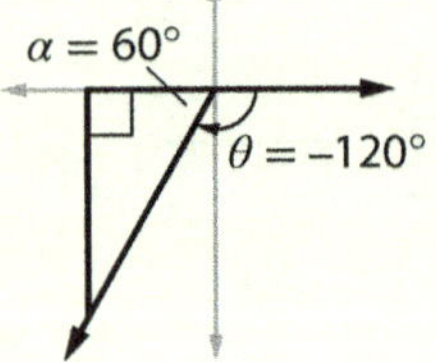

8.

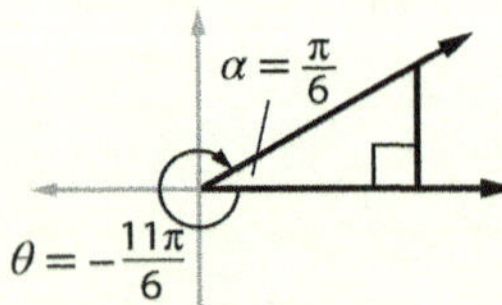

9.

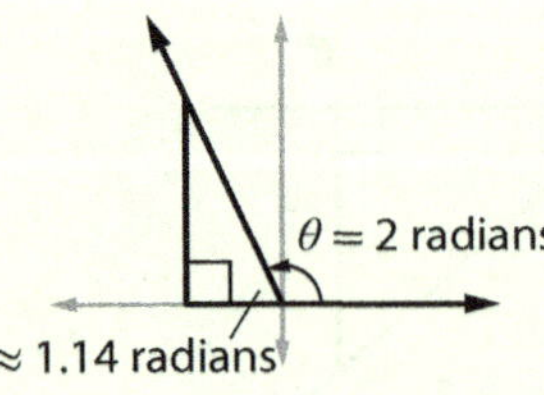

10.

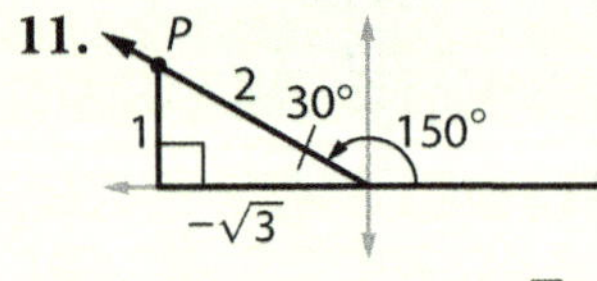

$$\alpha \approx 1.14 \text{ radians}$$

$$\alpha = \pi - 2 \approx 1.14$$

11.

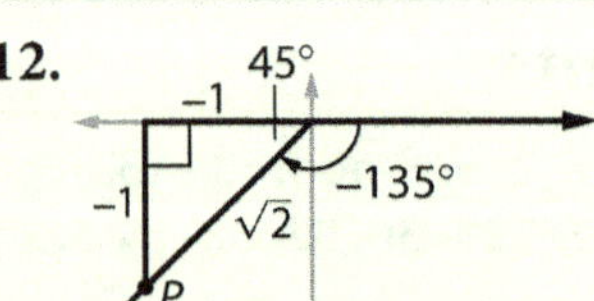

$$\cos 150° = \frac{x}{r} = \frac{-\sqrt{3}}{2} = -\frac{\sqrt{3}}{2}$$

12.

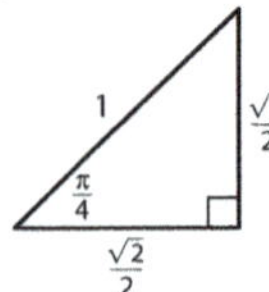

$$\sin(-135°) = \frac{y}{r} = \frac{-1}{\sqrt{2}} = -\frac{\sqrt{2}}{2}$$

13.

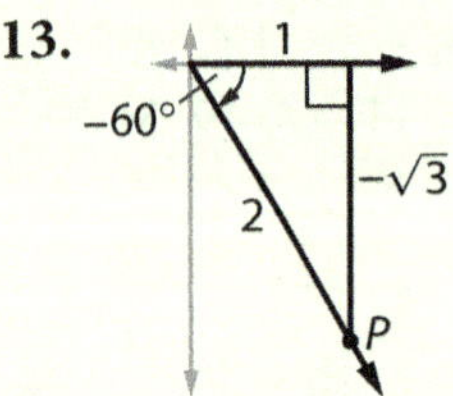

$$\tan(-60°) = \frac{y}{x} = \frac{-\sqrt{3}}{1} = -\sqrt{3}$$

14.

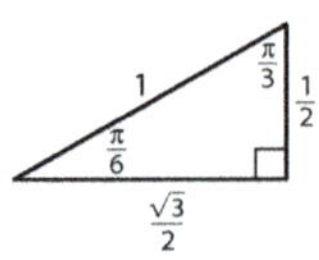

$$\csc 315° = \frac{r}{y} = \frac{\sqrt{2}}{-1} = -\sqrt{2}$$

15.

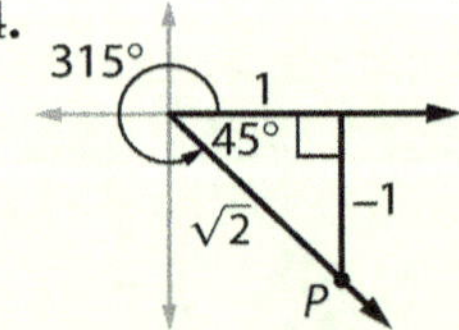

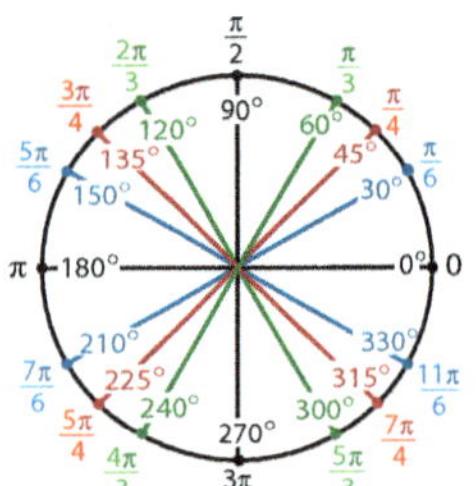

Since a unit circle is used in these definitions, these trigonometric functions are also called *circular functions*.

Notice that $P(x, y)$ has coordinates $(\cos t, \sin t)$. The following reference triangles and quadrantal angles can be used to quickly identify these coordinates on the unit circle and the exact values of each trigonometric function for the illustrated angle measures and arc lengths.

Example 6   Using the Unit Circle to Evaluate Trig Functions

Use the unit circle to state the exact value of each expression.

a. $\cos\dfrac{2\pi}{3}$ **b.** $\csc 225°$ **c.** $\tan\pi$

Answer

a.

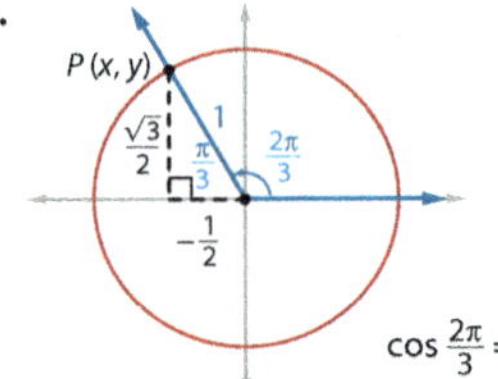

$$\cos\frac{2\pi}{3} = -\frac{1}{2}$$

1. Sketch the angle and the related reference triangle within the unit circle.
2. Identify $P(x, y) = \left(-\dfrac{1}{2}, \dfrac{\sqrt{3}}{2}\right)$.
3. Since $P(x, y) = (\cos t, \sin t)$, $\cos\theta = x$.

b.

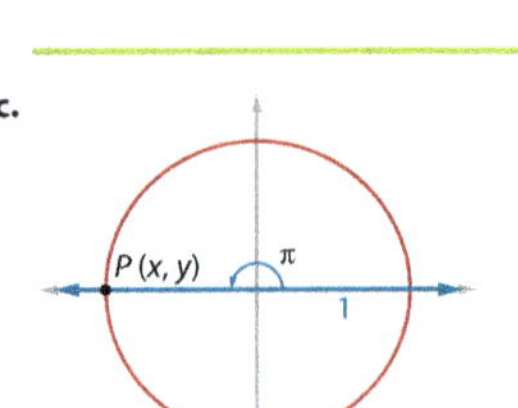

$$\csc 225° = \frac{1}{-\frac{\sqrt{2}}{2}} = -\sqrt{2}$$

1. Sketch the angle and the related reference triangle within the unit circle.
2. Identify $P(x, y) = \left(-\dfrac{\sqrt{2}}{2}, -\dfrac{\sqrt{2}}{2}\right)$.
3. Use $P(x, y) = (\cos t, \sin t)$ and $\csc t = \dfrac{1}{\sin t} = \dfrac{1}{y}$.

c.

$$\tan\pi = \frac{0}{-1} = 0$$

1. Sketch the quadrantal angle within the unit circle.
2. Identify $P(x, y) = (-1, 0)$.
3. Use $P(x, y) = (\cos t, \sin t)$ and $\tan t = \dfrac{\sin t}{\cos t} = \dfrac{y}{x}$.

SKILL ✔ EXERCISE 23

> **TIP**
>
> The following patterns can be used to remember the values of sine and cosine functions for special values.

θ	$\cos\theta$	$\sin\theta$
0	$\dfrac{\sqrt{4}}{2}$	$\dfrac{\sqrt{0}}{2}$
$\dfrac{\pi}{6}$	$\dfrac{\sqrt{3}}{2}$	$\dfrac{\sqrt{1}}{2}$
$\dfrac{\pi}{4}$	$\dfrac{\sqrt{2}}{2}$	$\dfrac{\sqrt{2}}{2}$
$\dfrac{\pi}{3}$	$\dfrac{\sqrt{1}}{2}$	$\dfrac{\sqrt{3}}{2}$
$\dfrac{\pi}{2}$	$\dfrac{\sqrt{0}}{2}$	$\dfrac{\sqrt{4}}{2}$

Viewing $f(t) = \cos t$ and $g(t) = \sin t$ as the x- and y-coordinates of the endpoint of an arc on the unit circle helps to explain several characteristics of trigonometric functions. Both functions have a domain of the real numbers and a range of $[-1, 1]$. The domains and ranges of other trigonometric functions will be investigated in future sections.

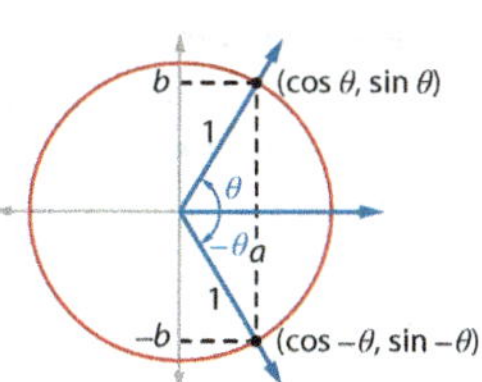

The figure illustrates that $\cos(-\theta) = \cos\theta$ and $\sin(-\theta) = -\sin\theta$. Therefore, cosine is an even function and sine is an odd function. It can be shown that secant is an even function and that tangent, cotangent, and cosecant are odd functions by expressing these functions in terms of sine and cosine or in terms of x, y, and r.

The definitions related to the unit circle make it clear that evaluating the same trigonometric function for coterminal angles will produce the same result. If $n \in \mathbb{Z}$, $\sin(\theta + n2\pi) = \sin\theta$ and $\cos(\theta + n2\pi) = \cos\theta$. A function is *periodic* if, for some given constant c, $f(x + c) = f(x)$ for all x. The smallest such positive value of c is called the *period* of the function. All six trigonometric functions are periodic, with sine, cosine, secant, and cosecant having periods of 2π. In Section 4.5 you will learn why the period of the tangent and cotangent functions are both π.

Example 7 Using the Period of a Trig Function

Find the exact value of $\sin\left(-\frac{29\pi}{3}\right)$.

Answer

$\sin\left(-\frac{29\pi}{3}\right) = -\sin\frac{29\pi}{3}$
    1. Apply the fact that sine is an odd function.

$= -\sin\left(\frac{5}{3}\pi + \frac{24}{3}\pi\right)$
    2. Use the fact that sine has a period of 2π to rewrite the expression in terms of a coterminal angle within $[0, 2\pi)$.

$= -\sin\left(\frac{5}{3}\pi + (4)2\pi\right)$

$= -\sin\frac{5\pi}{3}$

$-\sin\frac{29\pi}{3} = -\sin\frac{5\pi}{3} = -\left(-\frac{\sqrt{3}}{2}\right) = \frac{\sqrt{3}}{2}$
    3. Evaluate $\sin\frac{5\pi}{3} = -\frac{\sqrt{3}}{2}$ using a Q IV reference angle of $\frac{\pi}{3}$.

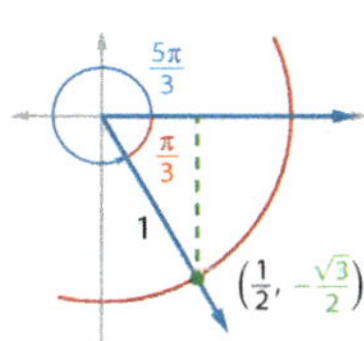

SKILL ✔ EXERCISE 29

God's creation has many periodic phenomena that can be modeled with trigonometric functions. He promised in Genesis 8:22 that the yearly cycle of the seasons and the daily cycle of sunlight and darkness would continue throughout the earth's existence. Other cycles in nature include lunar phases, tides, sunspots, comets, prevailing winds, and animal migrations.

Seventeenth-century scientists used pendulums and spring bobs (springs with weights attached) to measure time since these devices, like the sun and the moon, have cyclical properties. Hooke's law, which describes the linear relationship between the force applied to a spring and the spring's displacement, is the basis for the balance wheel found in wristwatches.

19. $\sec\theta = \frac{r}{x} = \frac{7}{3}$; so $\cos\theta = \frac{x}{r} = \frac{3}{7}$

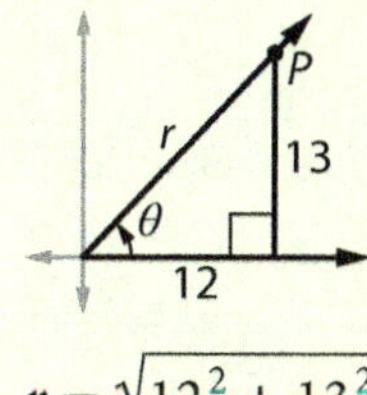

$y = -\sqrt{7^2 - 3^2} = -\sqrt{40} = -2\sqrt{10}$

$\sin\theta = \frac{y}{r} = \frac{-2\sqrt{10}}{7} = -\frac{2\sqrt{10}}{7}$

$\cos\theta = \frac{x}{r} = \frac{3}{7}$

$\tan\theta = \frac{y}{x} = \frac{-2\sqrt{10}}{3} = -\frac{2\sqrt{10}}{3}$

20. $\cot\theta = \frac{x}{y} = \frac{12}{13}$; so $\tan\theta = \frac{y}{x} = \frac{13}{12}$

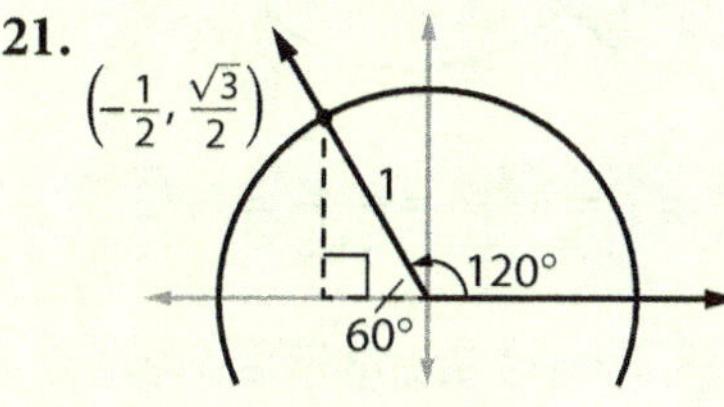

$r = \sqrt{12^2 + 13^2} = \sqrt{313}$

$\sin\theta = \frac{y}{r} = \frac{13}{\sqrt{313}} = \frac{13\sqrt{313}}{313}$

$\cos\theta = \frac{x}{r} = \frac{12}{\sqrt{313}} = \frac{12\sqrt{313}}{313}$

21.

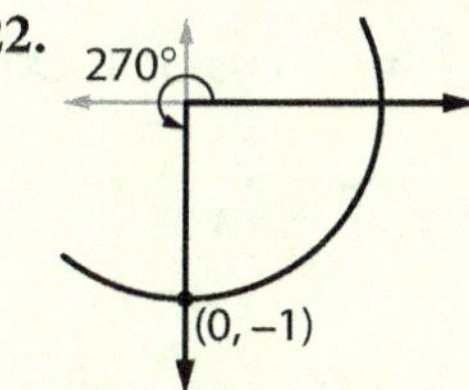

$\sin 120° = y = \frac{\sqrt{3}}{2}$

22.

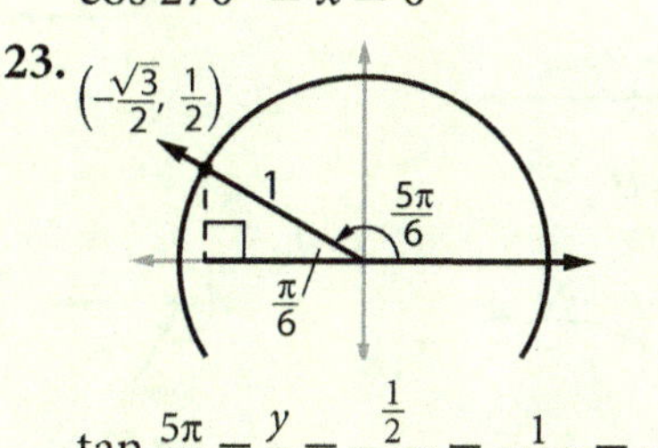

$\cos 270° = x = 0$

23.

$\tan\frac{5\pi}{6} = \frac{y}{x} = \frac{\frac{1}{2}}{-\frac{\sqrt{3}}{2}} = \frac{1}{-\sqrt{3}} = -\frac{\sqrt{3}}{3}$

16.

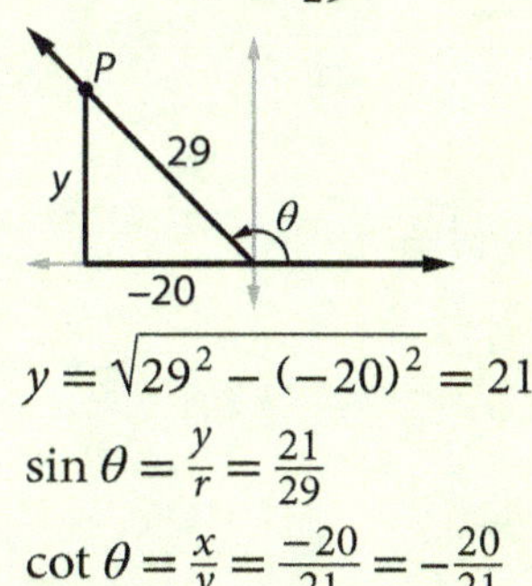

› B. Exercises

17. $\cos\theta = \frac{x}{r} = \frac{-20}{29}$

$y = \sqrt{29^2 - (-20)^2} = 21$

$\sin\theta = \frac{y}{r} = \frac{21}{29}$

$\cot\theta = \frac{x}{y} = \frac{-20}{21} = -\frac{20}{21}$

18. $\sin\theta = \frac{y}{r} = \frac{-8}{17}$

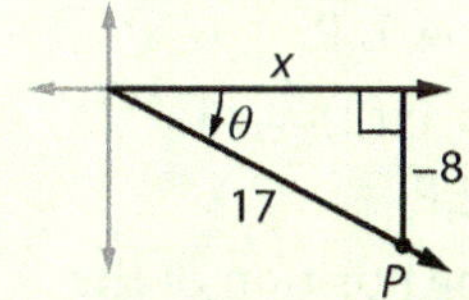

$x = \sqrt{17^2 - (-8)^2} = 15$

$\tan\theta = \frac{y}{x} = \frac{-8}{15} = -\frac{8}{15}$

$\sec\theta = \frac{r}{x} = \frac{17}{15}$

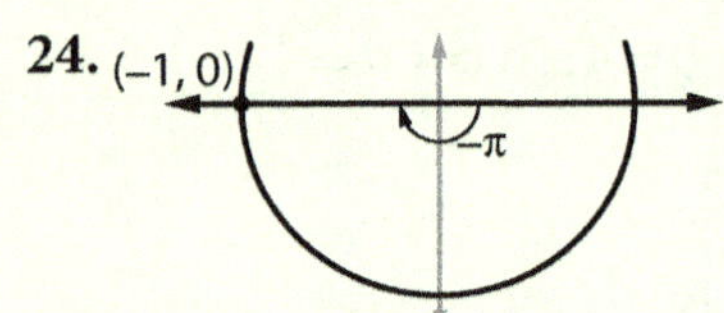

24. $(-1, 0)$, $-\pi$

$$\cot(-\pi) = \frac{x}{y} = \frac{-1}{0} \text{ (undefined)}$$

25. 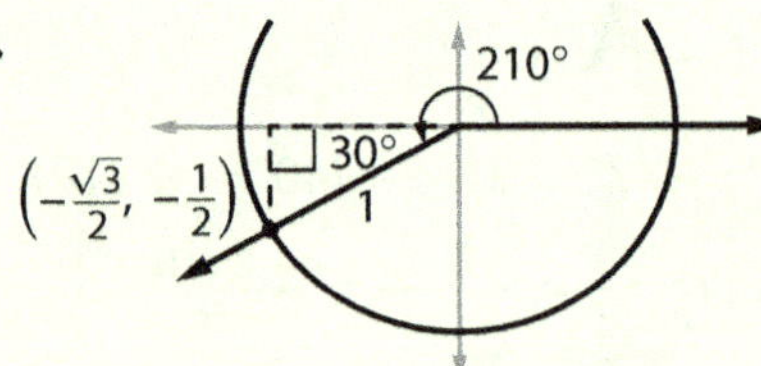

$210°$, $30°$, $\left(-\frac{\sqrt{3}}{2}, -\frac{1}{2}\right)$, 1

$$\sec 210° = \frac{1}{x} = -\frac{2}{\sqrt{3}} = -\frac{2\sqrt{3}}{3}$$

26. 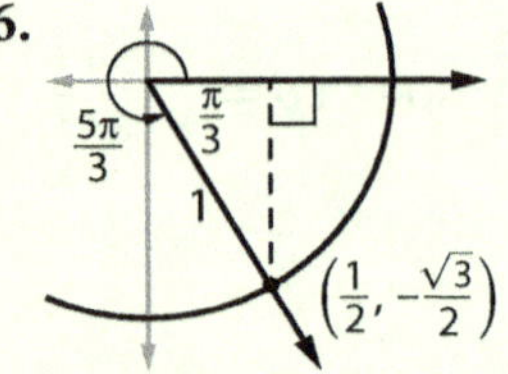

$\frac{5\pi}{3}$, $\frac{\pi}{3}$, 1, $\left(\frac{1}{2}, -\frac{\sqrt{3}}{2}\right)$

$$\cot \frac{5\pi}{3} = \frac{x}{y} = \frac{\frac{1}{2}}{-\frac{\sqrt{3}}{2}} = -\frac{1}{\sqrt{3}} = -\frac{\sqrt{3}}{3}$$

27. 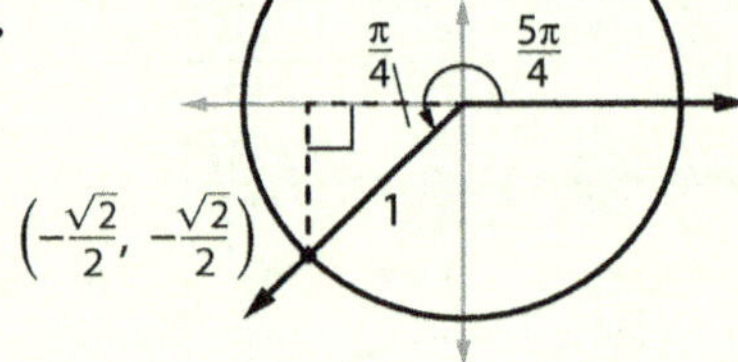

$\frac{\pi}{4}$, $\frac{5\pi}{4}$, $\left(-\frac{\sqrt{2}}{2}, -\frac{\sqrt{2}}{2}\right)$, 1

$$\csc \frac{5\pi}{4} = \frac{1}{y} = \frac{1}{-\frac{\sqrt{2}}{2}} = -\frac{2}{\sqrt{2}} = -\sqrt{2}$$

28. $\sin\left(\frac{4\pi}{3} + \frac{12}{3}\pi\right) = \sin \frac{4\pi}{3} = -\frac{\sqrt{3}}{2}$

29. $\tan\left(\frac{\pi}{2} + \frac{16}{2}\pi\right) = \tan \frac{\pi}{2}$ (undefined)

30. $\cos(-3\pi) = \cos(3\pi) = \cos(\pi + 2\pi)$
$= \cos \pi = -1$

31. $-\sin \frac{11\pi}{2} = -\sin\left(\frac{3\pi}{2} + \frac{8}{2}\pi\right)$
$= -\sin \frac{3\pi}{2} = 1$

32a. 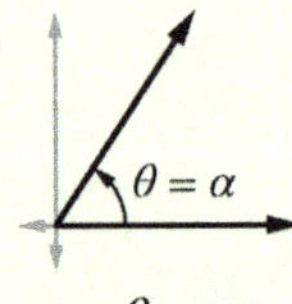

$\theta = \alpha$

$\alpha = \theta$

32b. 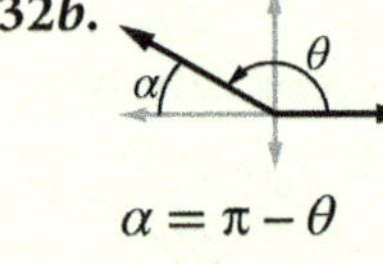

$\alpha = \pi - \theta$

32c. 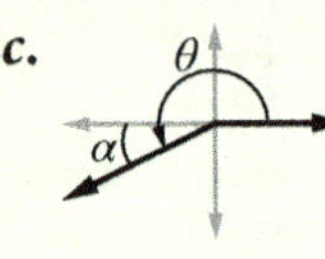

$\alpha = \theta - \pi$

32d. 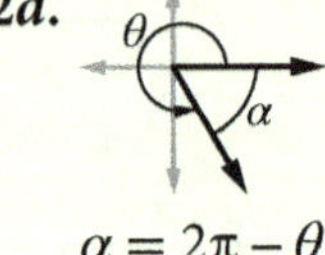

$\alpha = 2\pi - \theta$

35. $f(t) = -3 \cos \frac{2\pi}{0.5}t = -3 \cos 4\pi t$

35a. $f(0.125) = -3 \cos 4\pi(0.125) = 0$
at E.P.

35b. $f(0.25) = -3 \cos 4\pi(0.25) = 3$
3 in. above E.P.

Evaluate the six trigonometric functions for an angle in standard position whose terminal ray passes through each point.

1. $(1, 4)$

2. $(-6, -1)$

3. $(-3, 2)$

4. $(8, -7)$

Evaluate the six trigonometric functions for each quadrantal angle.

5. $\theta = \frac{\pi}{2}$

6. $\theta = 180°$

Sketch each angle θ and a reference triangle. Then state the measure of the reference angle α.

7. $\theta = 160°$ $\alpha = 20°$

8. $\theta = -120°$ $\alpha = 60°$

9. $\theta = -\frac{11\pi}{6}$ $\alpha = \frac{\pi}{6}$

10. $\theta = 2$ $\alpha = \pi - 2 \approx 1.14$

Use reference angles to find the exact value of each expression.

11. $\cos 150°$ $-\frac{\sqrt{3}}{2}$

12. $\sin(-135°)$ $-\frac{\sqrt{2}}{2}$

13. $\tan(-60°)$ $-\sqrt{3}$

14. $\csc 315°$ $-\sqrt{2}$

15. Use a calculator in radian mode to evaluate each expression (to four decimal places).

  a. $\sin(-100°)$ -0.9848
  b. $\sec 2.75$ -1.0819
  c. $\cot \frac{3\pi}{8}$ 0.4142

16. Use a calculator in degree mode to evaluate each expression (to four decimal places).

  a. $\tan(-50°)$ -1.1918
  b. $\cot 117°$ -0.5095
  c. $\sin \frac{9\pi}{5}$ -0.5878

17. If $\cos \theta = -\frac{20}{29}$ and $\tan \theta < 0$, find $\sin \theta$ and $\cot \theta$.

18. If $\sin \theta = -\frac{8}{17}$ and $\cos \theta > 0$, find $\tan \theta$ and $\sec \theta$.

19. If $\sec \theta = \frac{7}{3}$ and $\sin \theta < 0$, find $\sin \theta$, $\cos \theta$, and $\tan \theta$.

20. If $\cot \theta = \frac{12}{13}$ and $\sin \theta > 0$, find $\sin \theta$, $\cos \theta$, and $\tan \theta$.

Use the unit circle to find the exact value of each expression.

21. $\sin 120°$ $\frac{\sqrt{3}}{2}$

22. $\cos 270°$ 0

23. $\tan \frac{5\pi}{6}$ $-\frac{\sqrt{3}}{3}$

24. $\cot(-\pi)$ undefined

25. $\sec 210°$ $-\frac{2\sqrt{3}}{3}$

26. $\cot \frac{5\pi}{3}$ $-\frac{\sqrt{3}}{3}$

27. $\csc \frac{5\pi}{4}$ $-\sqrt{2}$

Evaluate each expression.

28. $\sin \frac{16\pi}{3}$ $-\frac{\sqrt{3}}{2}$

29. $\tan \frac{17\pi}{2}$ undefined

30. $\cos(-3\pi)$ -1

31. $\sin\left(-\frac{11\pi}{2}\right)$ 1

32. Sketch each angle θ in standard position and label its reference angle α. Then express α in terms of θ.

  a. $0 < \theta < \frac{\pi}{2}$ $\alpha = \theta$
  b. $\frac{\pi}{2} < \theta < \pi$ $\alpha = \pi - \theta$
  c. $\pi < \theta < \frac{3\pi}{2}$ $\alpha = \theta - \pi$
  d. $\frac{3\pi}{2} < \theta < 2\pi$ $\alpha = 2\pi - \theta$

33. Find $\sin(-\theta)$ if $\sin \theta = \frac{7}{8}$. $-\frac{7}{8}$

34. Find $\cos(-\theta)$ if $\cos \theta = \frac{5}{13}$. $\frac{5}{13}$

An object hanging from a spring is pulled down k inches below its equilibrium position and released. Its height (relative to its equilibrium position) t seconds after its release is modeled by $f(t) = -a \cos \frac{2\pi}{p}t$ where p is the time it takes to return to the released position.

35. Find and interpret the height of the object at each time if $a = 3$ and $p = 0.5$.

  a. 0.125 sec
  b. 0.25 sec
  c. 0.6 sec
  d. 2.0 sec

36. Use $t = 0$ to explain why $g(t) = -k \sin \frac{2\pi}{c}t$ does not correctly model the object's height.

37. Make a copy of the unit circle illustrating the degree and radian measures of all angles $\theta \in [0, 2\pi)$ whose measures are multiples of $\frac{\pi}{6}$ or $\frac{\pi}{4}$. Then label each endpoint of the related arc lengths with the ordered triplet $(\cos \theta, \sin \theta, \tan \theta)$.

For a pendulum with length L swinging through relatively small angles of θ, the formula relating the change in height Δh and the angle θ it makes with its stationary position is $\Delta h = L(1 - \cos \theta)$.

38. Foucault's pendulum suspended from the dome of the Pantheon in Paris was 67 m long.

  a. Find Δh when $\theta = 19°$. ≈ 3.65 m
  b. Find θ when $\Delta h = 2$ m. $\approx 14.0°$

39. Use the figure to derive the formula for Δh.

40. Using the fact that cosine is an even function and sine is an odd function, algebraically prove each statement.

  a. Tangent is an odd function.
  b. Secant is an even function.

35c. $f(0.6) = -3 \cos 4\pi(0.6) = -0.927$
   0.927 in. below E.P.

35d. $f(2) = -3 \cos 4\pi(2) = -3$
   3 in. below E.P.

36. At $t = 0$ sec, the position of the spring is k in. below the equilibrium position, but $g(0) = -k \sin \frac{2\pi}{c}(0)$
$= -k \sin(0) = -k(0) = 0$, the E.P.

41. Use the drawing and unit circle definitions to prove each relationship.

a. $\cos(\pi - \theta) = -\cos\theta$

b. $\sin(\pi - \theta) = \sin\theta$

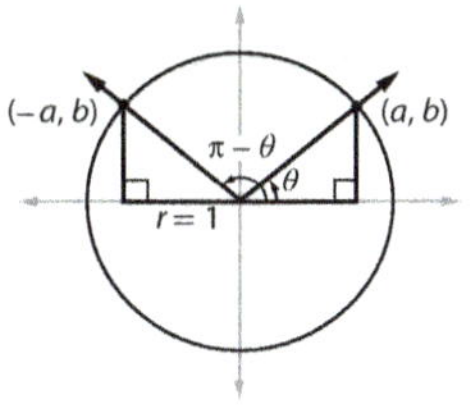

42. $f(x)$ is translated 8 units to the right.

43. $f(x)$ is reflected in the x-axis and stretched vertically by a factor of 4.

CUMULATIVE REVIEW

Describe the graph of $g(x)$ as a transformation of $f(x) = x^2$. [1.5]

42. $g(x) = (x - 8)^2$ **43.** $g(x) = -4x^2$

44. Write a function rule for $g(x)$, a reflection of $f(x) = x^3$ in the x-axis that has been translated 7 units left and 9 units up. [1.5] $g(x) = -(x + 7)^3 + 9$

45. Determine the type of discontinuity for each function at $x = 3$. [1.3, 2.5]

I. jump **II.** infinite **III.** point

a. $f(x) = \dfrac{x+3}{x^2 - 6x + 9}$ II

b. $g(x) = \dfrac{x-3}{x^2 - x - 6}$ III

c. $h(x) = [3x]$ I

Determine whether each function is even, odd, or neither. [2.2]

46. $p(x) = 8x^6 + 2x^3 - 5$ neither **47.** $p(x) = 7x^8 - 6x^2$ even

48. Find the radian measure for 42°. [4.1] B

A. $\frac{7\pi}{6}$ **D.** $\frac{17\pi}{30}$

B. $\frac{7\pi}{30}$ **E.** none of these

C. $\frac{5\pi}{6}$

49. The radii of the two similar shaded sectors have a ratio of 4 : 3. If the larger sector has an area of 10 cm², find the arc length defined by the smaller sector. [4.1] C

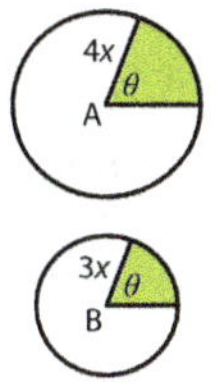

A. 15 cm **D.** 2.5 cm

B. 7.5 cm **E.** none of these

C. 3.75 cm

50. If $\csc\theta = \frac{17}{15}$, find $\sec\theta$. [4.2] A

A. $\frac{17}{8}$ **D.** $\frac{15}{17}$

B. $\frac{8}{17}$ **E.** none of these

C. $\frac{16}{17}$

51. Find m if a line perpendicular to $y = mx + b$ has a slope of $-\sec\theta$. [4.2] D

A. $\csc\theta$ **D.** $\cos\theta$

B. $-\cos\theta$ **E.** none of these

C. $-\sin\theta$

1. $\sin\theta = \frac{4\sqrt{17}}{17}$; $\csc\theta = \frac{\sqrt{17}}{4}$

$\cos\theta = \frac{\sqrt{17}}{17}$; $\sec\theta = \sqrt{17}$

$\tan\theta = 4$; $\cot\theta = \frac{1}{4}$

2. $\sin\theta = -\frac{\sqrt{37}}{37}$; $\csc\theta = -\sqrt{37}$

$\cos\theta = -\frac{6\sqrt{37}}{37}$; $\sec\theta = -\frac{\sqrt{37}}{6}$

$\tan\theta = \frac{1}{6}$; $\cot\theta = 6$

3. $\sin\theta = \frac{2\sqrt{13}}{13}$; $\csc\theta = \frac{\sqrt{13}}{2}$

$\cos\theta = -\frac{3\sqrt{13}}{13}$; $\sec\theta = -\frac{\sqrt{13}}{3}$

$\tan\theta = -\frac{2}{3}$; $\cot\theta = -\frac{3}{2}$

4. $\sin\theta = -\frac{7\sqrt{113}}{113}$; $\csc\theta = -\frac{\sqrt{113}}{7}$

$\cos\theta = \frac{8\sqrt{113}}{113}$; $\sec\theta = \frac{\sqrt{113}}{8}$

$\tan\theta = -\frac{7}{8}$; $\cot\theta = -\frac{8}{7}$

5. $\sin\theta = 1$; $\csc\theta = 1$;
$\cos\theta = 0$; $\sec\theta$ is undefined;
$\tan\theta$ is undefined; $\cot\theta = 0$

6. $\sin\theta = 0$; $\csc\theta$ is undefined;
$\cos\theta = -1$; $\sec\theta = -1$;
$\tan\theta = 0$; $\cot\theta$ is undefined

17. $\sin\theta = \frac{21}{29}$; $\cot\theta = -\frac{20}{21}$

18. $\tan\theta = -\frac{8}{15}$; $\sec\theta = \frac{17}{15}$

19. $\sin\theta = -\frac{2\sqrt{10}}{7}$; $\cos\theta = \frac{3}{7}$;

$\tan\theta = -\frac{2\sqrt{10}}{3}$

20. $\sin\theta = \frac{13\sqrt{313}}{313}$; $\cos\theta = \frac{12\sqrt{313}}{313}$

35a. $f(0.125) = 0$; at E.P.

35b. $f(0.25) = 3$; 3 in. above E.P.

35c. $f(0.6) = -0.927$; 0.927 in. below E.P.

35d. $f(2) = -3$; 3 in. below E.P.

37.

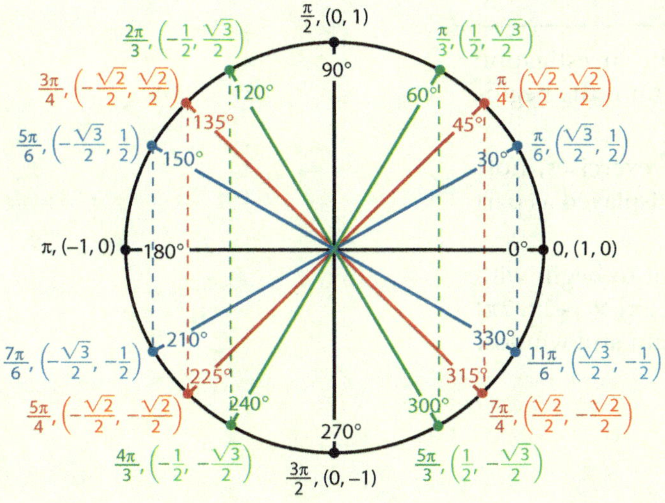

C. Exercises

38a. $\Delta h = 67(1 - \cos 19°) \approx 3.65$

38b.
$$2 = 67(1 - \cos\theta)$$
$$2 = 67 - 67\cos\theta$$
$$-65 = -67\cos\theta$$
$$\frac{65}{67} = \cos\theta$$
$$\theta = \cos^{-1}\frac{65}{67} \approx 14.0°$$

39.

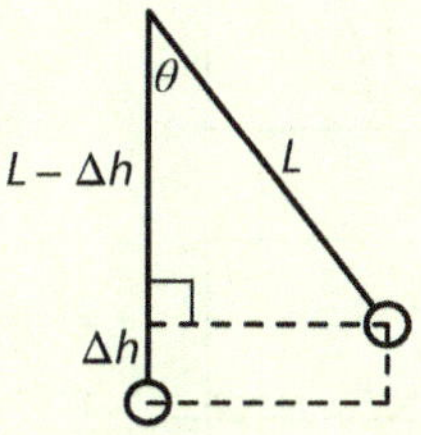

$$\cos\theta = \frac{adj.}{hyp.} = \frac{L - \Delta h}{L}$$
$$L\cos\theta = L - \Delta h$$
$$\Delta h = L - L\cos\theta$$
$$= L(1 - \cos\theta)$$

40a. $\tan(-\theta) = \dfrac{\sin(-\theta)}{\cos(-\theta)} = \dfrac{-\sin\theta}{\cos\theta}$
$= -\tan\theta$

40b. $\sec(-\theta) = \dfrac{1}{\cos(-\theta)} = \dfrac{1}{\cos\theta} = \sec\theta$

41a. $\cos\theta = a$ and
$\cos(\pi - \theta) = -a = -\cos\theta$

41b. $\sin\theta = b$ and
$\sin(\pi - \theta) = b = \sin\theta$

Cumulative Review

45a. $f(x) = \dfrac{x+3}{(x-3)^2}$; The factor $(x - 3)$ is not matched in the numerator.

45b. $g(x) = \dfrac{x-3}{(x-3)(x+2)}$; The factor $(x - 3)$ is matched in the numerator.

45c. $h(x) = [3x]$ jumps from 8 to 9.

46. $p(-x) = 8(-x)^6 + 2(-x)^3 - 5$
$= 8x^6 - 2x^3 - 5 \neq -p(x)$ or $p(x)$

47. $p(-x) = 7(-x)^8 - 6(-x)^2$
$= 7x^8 - 6x^2 = p(x)$

48. $42°\left(\frac{\pi}{180°}\right) = \frac{7\pi}{30}$

49.
$$A_L = \frac{1}{2}r^2\theta_L$$
$$10 = \frac{1}{2}(4)^2\theta_L$$
$$\theta_L = 1.25 = \theta_S$$
$$s_S = r\theta_S = 3(1.25) = 3.75 \text{ cm}$$

50.
$$\csc\theta = \frac{17}{15} = \frac{r}{y}$$
$$x^2 + 15^2 = 17^2$$
$$x = \sqrt{17^2 - 15^2} = 8$$
$$\sec\theta = \frac{r}{x} = \frac{17}{8}$$

51.
$$m_g = -\sec\theta = -\frac{r}{x}$$
$$m_\perp = -\frac{1}{m_g} = -\left(-\frac{x}{r}\right) = \frac{x}{r} = \cos\theta$$

Technology Corner

Solutions

1. $y = \tan x$

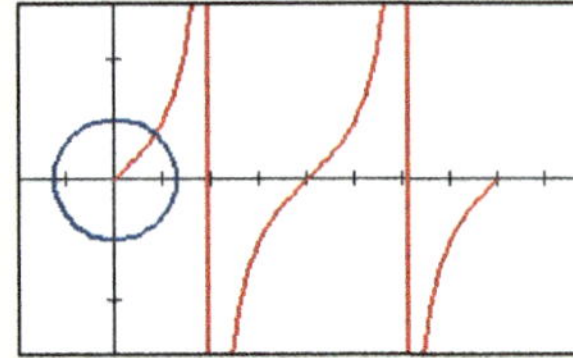

$y = \cot x$

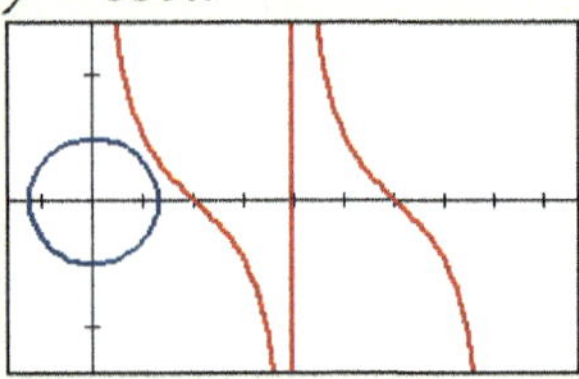

$y = \sec x$

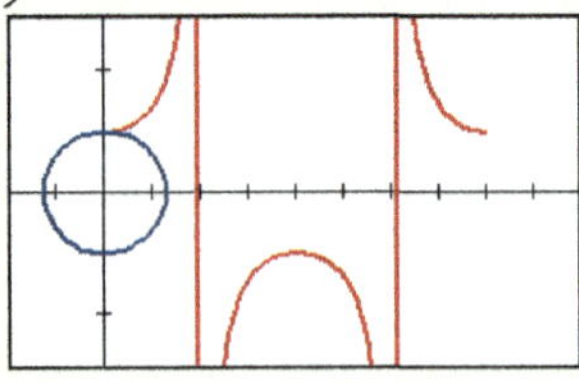

$y = \csc x$

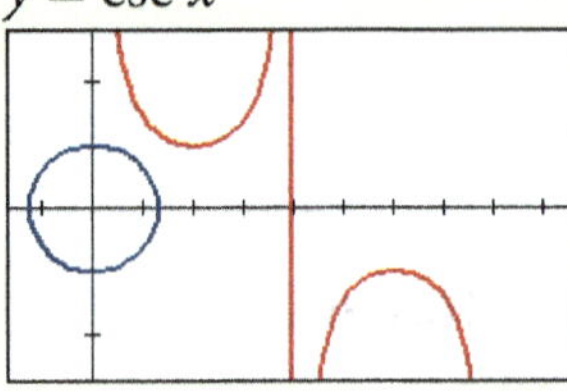

2.

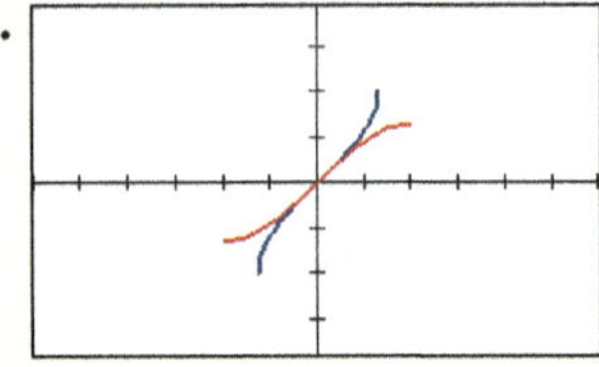

3. $\text{Tmin} = 0$, $\text{Tmax} = \pi$

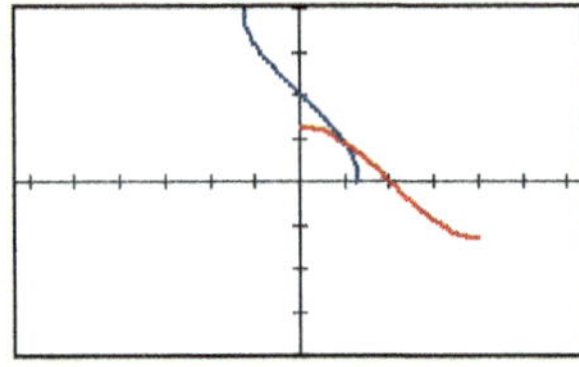

$\text{Tmin} = -\pi/2$, $\text{Tmax} = \pi/2$

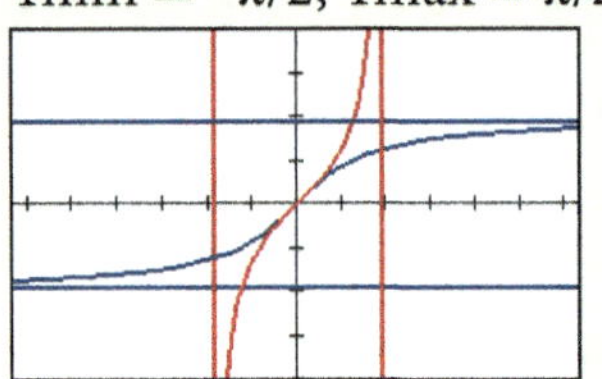

Drawing graphs of the parent trigonometric functions and the unit circle can help you understand how the shapes of the graphs are generated. Press MODE and select RADIAN, PAR, and SIMUL to graph parametric equations simultaneously.

Press Y= and enter $X_{1T} = \cos T$ and $Y_{1T} = \sin T$ for the unit circle. Enter $X_{2T} = T$ and $Y_{2T} = \sin T$ for the sine function. Cursor to the left of each set of parametric equations and select the path icon. Then press WINDOW and enter the following window values before graphing the functions.

$\text{Tmin} = 0$, $\text{Tmax} = 2\pi$, $\text{Tstep} = \pi/36$
$\text{Xmin} = -\pi/2$, $\text{Xmax} = 5\pi/2$, $\text{Xscl} = \pi/4$
$\text{Ymin} = -2.9$, $\text{Ymax} = 2.9$, $\text{Yscl} = 1$

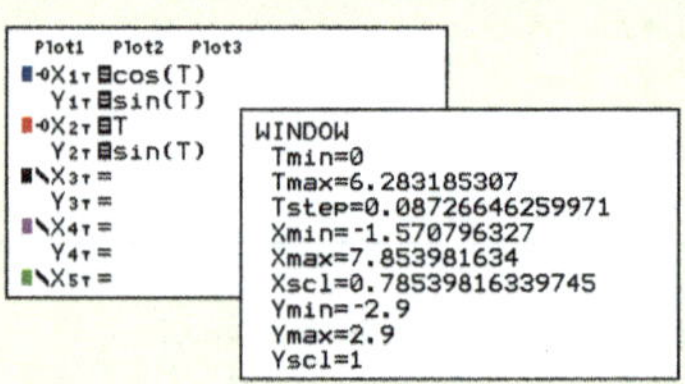

Notice how the value of the sine function is the height of the point moving along the unit circle.

1. Draw the graphs of the other trigonometric functions by entering each function in Y_{2T}.

2. Explore the graph of the inverse of $y = \sin x$ by entering $X_{2T} = T$ and $Y_{2T} = \sin T$ and switching these expressions in X_{1T} and Y_{1T}. Adjust the window as needed. Can you change the domain of the trigonometric function using Tmin and Tmax so that the inverse is also a function?

3. Explore the graphs of the inverse of $y = \cos x$ and the inverse of $y = \tan x$.

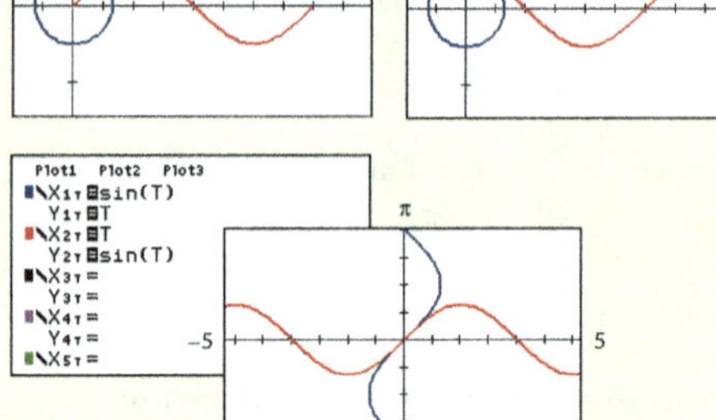

PRESENTATION

You may want to save the investigation of inverse trig functions until the beginning of Section 4.6.

In the first and third exercises, note the error of asymptotes displayed as part of the graphs.

The students may want to begin with window settings of $[-2\pi, 2\pi] \times [-2\pi, 2\pi]$ and then adjust the domain and window as needed.

4.4 Sinusoidal Functions

A bore tide (a large wave or series of waves coming in with the tide) can be modeled by a sinusoidal function of time.

Just as we used the unit circle to extend the definitions of the trigonometric functions to real numbers, it can be used to draw graphs of the basic sine and cosine functions. Recall from Section 4.3 that wrapping a real number around the unit circle allows the value of sine to be viewed as the y-coordinate (the height) of the endpoint of an arc on the unit circle. Similarly, the cosine value may be viewed as the x-coordinate (the directed horizontal distance) of the endpoint. In Section 4.3 we saw that the period (the length of one complete cycle of values) of both functions is 2π.

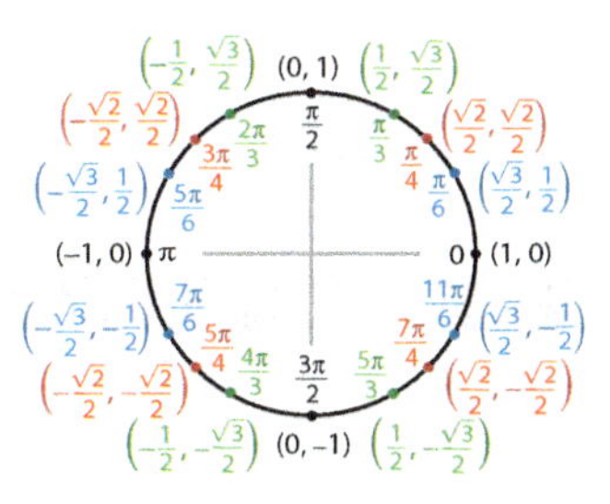

After completing this section, you will be able to

- analyze and graph sinusoidal functions.
- write function rules for sinusoidal functions.
- model real-world data and solve problems using sinusoidal functions.

Plotting the coordinates from the 16 special values on the unit circle illustrates the pattern of values during the first period of each function. This cycle of values can be repeated to extend the graphs. The Technology Corner at the end of the previous section shows how to produce parametric representations of these functions.

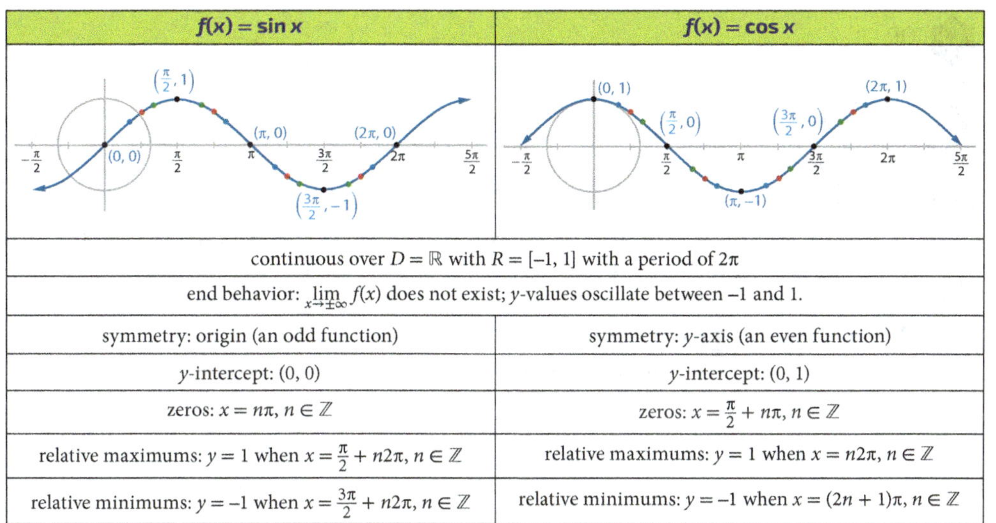

$f(x) = \sin x$	$f(x) = \cos x$
continuous over $D = \mathbb{R}$ with $R = [-1, 1]$ with a period of 2π	
end behavior: $\lim\limits_{x \to \pm\infty} f(x)$ does not exist; y-values oscillate between -1 and 1.	
symmetry: origin (an odd function)	symmetry: y-axis (an even function)
y-intercept: $(0, 0)$	y-intercept: $(0, 1)$
zeros: $x = n\pi, n \in \mathbb{Z}$	zeros: $x = \frac{\pi}{2} + n\pi, n \in \mathbb{Z}$
relative maximums: $y = 1$ when $x = \frac{\pi}{2} + n2\pi, n \in \mathbb{Z}$	relative maximums: $y = 1$ when $x = n2\pi, n \in \mathbb{Z}$
relative minimums: $y = -1$ when $x = \frac{3\pi}{2} + n2\pi, n \in \mathbb{Z}$	relative minimums: $y = -1$ when $x = (2n + 1)\pi, n \in \mathbb{Z}$

Objectives

1. To graph basic and transformed sinusoidal functions
2. To identify the amplitude, period, phase shift, and vertical shift of sinusoidal functions
3. To write function rules for transformations of basic sinusoidal functions
4. To model real-world data and solve problems using sinusoidal functions

Flash

A tidal bore is a rare occurrence. The bore waves appear only when a significant incoming tide flows into a shallow river or bay and encounters a river's current flowing in the opposite direction.

Vocabulary

amplitude
frequency
hertz
midline
period
phase shift
sinusoid
sinusoidal function
wavelength

PRESENTATION

Lesson Opener

State the equivalent radian measure for each degree measure.

1. $90°$ $\frac{\pi}{2}$
2. $180°$ π
3. $270°$ $\frac{3\pi}{2}$
4. $360°$ 2π
5. $45°$ $\frac{\pi}{4}$
6. $30°$ $\frac{\pi}{6}$
7. $60°$ $\frac{\pi}{3}$

Consider using the Technology Corner preceding the lesson to introduce the graphs of the basic sine and cosine functions. Alternately, use the unit circle to plot key points and sketch the graphs by hand. Stress that x-values are the radian measures of the points around the unit circle and that each sine value is the y-coordinate of the point. Compare this to the cosine function, which has the same x-values but returns the x-coordinate of each point.

In this text, the graphs of trigonometric functions are usually done in radians (expressed either in terms of π or as integers). Students should be sure their graphing technology is set for radians instead of degrees.

Motivational Idea Use the Internet keyword search *animated unit circle* for an animated demonstration of the creation of the sine and cosine waves.

Examine the characteristics of the graphs of parent sine and cosine functions in the table. Note that the zeros could also be thought of as even or odd multiples of $\frac{\pi}{2}$: $x = 2n\left(\frac{\pi}{2}\right)$ for sin x and $x = (2n + 1)\left(\frac{\pi}{2}\right)$ for cos x. Some students may better visualize odd multiples of $\frac{\pi}{2}$ as $\frac{\pi}{2} + n\pi$.

Direct attention to the fact that the cosine function is the sine function translated $\frac{\pi}{2}$ to the left. In physics, this is

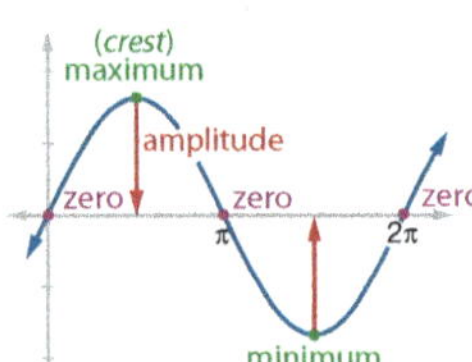

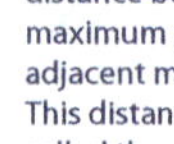

The period of a sinusoid can be measured as the distance between adjacent maximum points or adjacent minimum points. This distance is often called the *wavelength*.

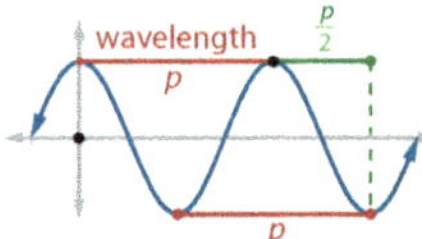

Notice that the graph of the cosine function can be obtained by shifting the graph of the sine function $\frac{\pi}{2}$ units to the left. Any function that is a transformation of a sine function is called a *sinusoid*. A *sinusoidal function* can be written as $y = a\sin b(x - h) + k$ or $y = a\cos b(x - h) + k$. We will examine how the values of each constant a, b, h, and k transform the graph of the parent function.

In Chapter 1 we saw that the graph of $g(x) = a \cdot f(x)$ is a vertical stretch (if $|a| > 1$) or shrink (if $0 < |a| < 1$) of the graph of $f(x)$. If $a < 0$, the graph is also reflected across the x-axis. When the output of a sinusoidal function is multiplied by a constant, the vertical stretch or shrink affects the function's *amplitude*.

DEFINITION

The **amplitude** of a sinusoid is one-half the difference of the function's maximum and minimum values. The amplitude of $y = a\sin x$ and $y = a\cos x$ is $|a|$.

Example 1   Graphing a Vertical Stretch or Shrink

State the amplitude of $g(x) = -\frac{1}{2}\sin x$. Then describe how $g(x)$ is obtained by transforming the parent function $f(x) = \sin x$ and graph the function over $[-2\pi, 2\pi]$.

Answer

amplitude $= |a| = \left|-\frac{1}{2}\right| = \frac{1}{2}$

The graph of $f(x)$ is vertically shrunk by a factor of $\frac{1}{2}$ and reflected across the x-axis.

$f(x) = \sin x$	$(0, 0)$	$\left(\frac{\pi}{2}, 1\right)$	$(\pi, 0)$	$\left(\frac{3\pi}{2}, -1\right)$	$(2\pi, 0)$
$g(x) = -\frac{1}{2}\sin x$	$(0, 0)$	$\left(\frac{\pi}{2}, -\frac{1}{2}\right)$	$(\pi, 0)$	$\left(\frac{3\pi}{2}, \frac{1}{2}\right)$	$(2\pi, 0)$

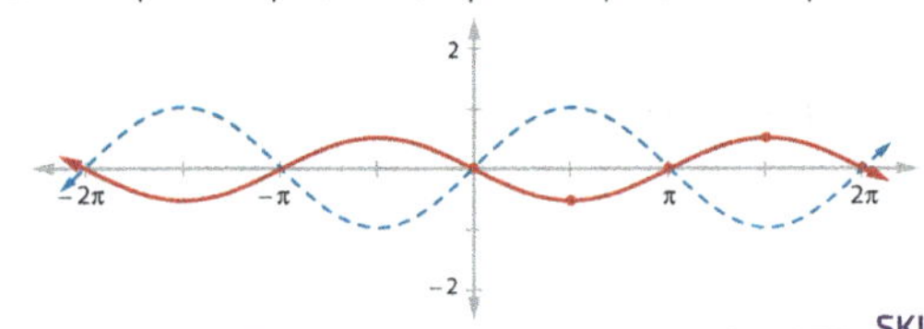

SKILL ✔ EXERCISE 11

Recall that the graph of $g(x) = f(bx)$ is a horizontal stretch (if $0 < |b| < 1$) or shrink (if $|b| > 1$) of the graph of $f(x)$ by a factor of $\frac{1}{|b|}$. Similarly, the constant b in $y = a\sin bx$ and $y = a\cos bx$ changes the period of these functions to $\frac{2\pi}{|b|}$. When b is negative, either $\sin(-x) = -\sin x$ or $\cos(-x) = \cos x$ can be applied to produce an equivalent function where b is positive.

DEFINITION

The **period** p of a periodic function is the horizontal length of one cycle. The period of $y = \sin bx$ and $y = \cos bx$ is $\frac{2\pi}{|b|}$.

State the amplitude and period of $h(x) = 3 \cos(-2x)$. Then describe how $h(x)$ is obtained by transforming the parent function $f(x) = \sin x$ and graph $h(x)$ over $[-2\pi, 2\pi]$.

Answer

Since cosine is an even function, $g(x) = 3 \cos(-2x) = 3 \cos 2x$.

amplitude: $|a| = |3| = 3$ period: $p = \frac{2\pi}{|b|} = \frac{2\pi}{|2|} = \pi$

The graph of $f(x)$ is horizontally shrunk by a factor of $\frac{1}{2}$ and vertically stretched by a factor of 3.

$y_1 = \cos x$	$(0, 1)$	$\left(\frac{\pi}{2}, 0\right)$	$(\pi, -1)$	$\left(\frac{3\pi}{2}, 0\right)$	$(2\pi, 1)$
$y_2 = \cos 2x$	$(0, 1)$	$\left(\frac{\pi}{4}, 0\right)$	$\left(\frac{\pi}{2}, -1\right)$	$\left(\frac{3\pi}{4}, 0\right)$	$(\pi, 1)$
$y_3 = 3 \cos 2x$	$(0, 3)$	$\left(\frac{\pi}{4}, 0\right)$	$\left(\frac{\pi}{2}, -3\right)$	$\left(\frac{3\pi}{4}, 0\right)$	$(\pi, 3)$

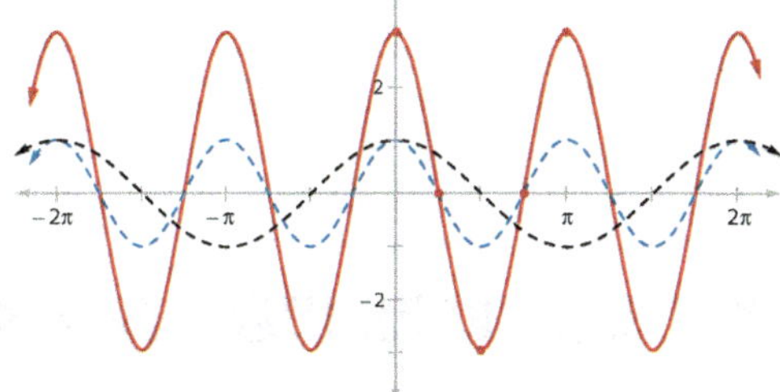

SKILL ✔ **EXERCISE 15**

Wave phenomena are modeled by sinusoids. The *frequency* (f) of a sinusoid is the reciprocal of the period: $f = \frac{1}{p} = \frac{|b|}{2\pi}$. The frequency for the cosine function in Example 2 is $\frac{1}{p} = \frac{1}{\pi} \approx 0.32$. This implies the graph completes 1 cycle over any x-interval of length π, or about 0.32 of a cycle per every unit of x. In many applications, frequencies are measured in *hertz* (Hz), or cycles per second. The loudness of a sound, which is measured in decibels, is related to the amplitude of the sound wave. The frequency of a sound wave is heard as its pitch.

The term *hertz* formally replaced cycles per second in 1960 in honor of the German physicist Heinrich Hertz, who in 1888 became the first person to transmit and receive radio waves, thus confirming their existence.

4. State the amplitude, period, and frequency of $f(x) = -3 \sin \frac{x}{2}$. Then graph $f(x)$ over $[-2\pi, 2\pi]$.

$|a| = 3; p = 4\pi; f = \frac{1}{4\pi}$

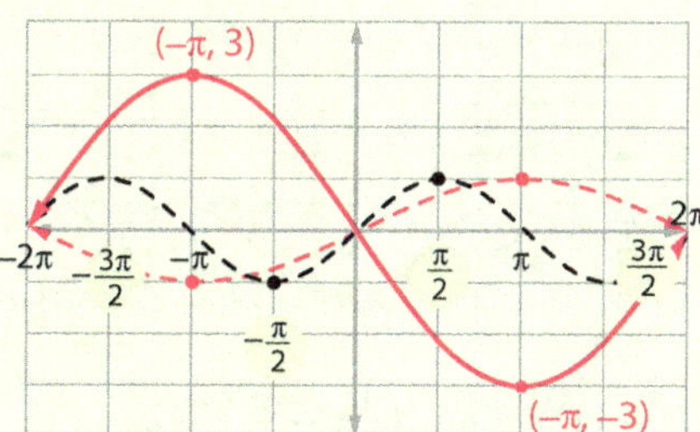

5. G_3 (G below middle C) has a frequency of 196 Hz. Find the period for G_3 in cycles per millisecond. Write a sine function modeling the sound of G_3 played with an amplitude of 0.9. Use technology to graph the function.

$p = \frac{1}{196} \cdot 1000 \approx 5.1$ ms

$p \approx 5.1$ ms $= \frac{2\pi}{|b|}; |b| = 0.392\pi$

$f(t) = 0.9 \sin 0.392\pi t$ or

$f(t) \approx 0.9 \sin 1.232t$

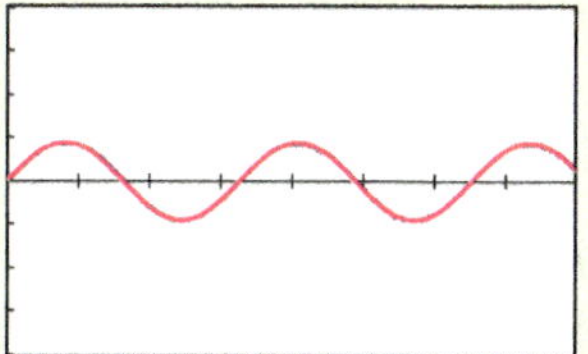

Example 2 illustrates a change in both amplitude and period. Guide the students through algebraic results for the horizontal shrink by a factor of $\frac{1}{2}$ (dividing x-values by 2) and the vertical stretch by a factor of 3 (multiplying y-values by 3).

Introduce the *frequency* of a sinusoid as the reciprocal of the period. In Example 3, either the negative or positive value of both a and b can be used. The function $f(x) = -60 \sin(-220\pi x)$ produces the same wave, while $f(x) = -60 \sin 220\pi x$ and $f(x) = 60 \sin(-220\pi x)$ are reflections across the x-axis. Any of these would be correct.

One-on-One Encourage students to keep the reciprocal relationship between period and frequency in mind when graphing trigonometric functions. The shorter the period, the more cycles are completed within a given time (interval on the x-axis). Therefore, the shorter the period, the higher the frequency, and vice versa.

Discuss the vertical and horizontal translations and use Example 4 to illustrate these translations after a horizontal stretch. Direct attention to the *midline* of the basic sine function and the midline of the translated sine function.

Students may struggle when combining horizontal dilations and translations, especially when the input is written as $(bx + c)$ instead of $b(x - h)$. They may benefit from focusing on finding the interval for one cycle of the function as described in the paragraph after Example 4. Consider demonstrating the solving of the compound inequality $0 \leq bx + c \leq 2\pi$. A similar approach can be used for the factored form $g(x) = a \sin b(x - h) + k$, where one cycle is completed over $0 \leq b(x - h) \leq 2\pi$ or $h \leq x \leq \frac{2\pi}{b} + h$.

Summarize the characteristics of sinusoidal functions and present Example 5. Consider asking the students how many correct possible answers exist and discuss

State the amplitude, period, phase shift, and vertical shift of each function. Then graph the function over $[-2\pi, 2\pi]$.

6. $f(x) = -\cos\left(x + \frac{\pi}{6}\right) - 1$

$|a| = 1; p = 2\pi; h = -\frac{\pi}{6}; k = -1$

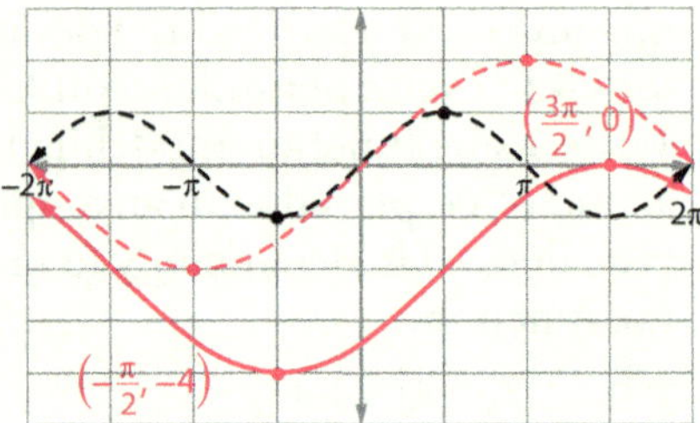

7. $f(x) = 2\sin\left(\frac{1}{2}x - \frac{\pi}{4}\right) - 2$

$|a| = 2; p = 4\pi; h = \frac{\pi}{2}; k = -2$

Write a cosine function that fits the given characteristics.

8. $|a| = 1; p = \frac{\pi}{2}; h = 2$

$y = \pm\cos 4\left(x - \frac{1}{2}\right)$ or

$y = \pm\cos(4x - 2)$

9. $|a| = 4; p = 3; h = -3$; translated 1 unit up

$y = \pm 4\cos\frac{2\pi}{3}(x + 3) + 1$ or

$y = \pm 4\cos\left(\frac{2\pi}{3}x + 2\pi\right) + 1$

Humans can generally hear sound frequencies from 20 Hz to 20 kHz. Write a sine function modeling a 110 Hz note (A_2) played with an amplitude of 60 dB.

Answer

$a = \pm 60$

1. amplitude: $|a| = 60$

$|b| = 2\pi f = 2\pi(110) = 220\pi$

$b = \pm 220\pi$

2. Use $f = \frac{1}{p} = \frac{|b|}{2\pi}$ to find the value of b.

$f(x) = 60\sin 220\pi x$

3. Write an equation in the form $f(x) = a\sin bx$, choosing positive values for a and b.

_______________________ SKILL ✔ **EXERCISE 17**

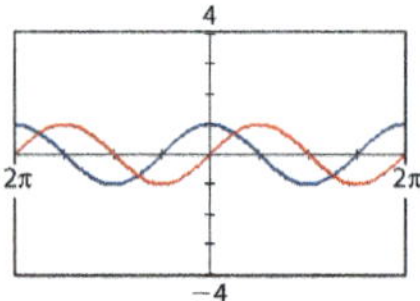

In previous chapters you have seen how $g(x) = f(x - h) + k$ is a translation of $f(x)$ by $|h|$ units right (if $h > 0$) or left (if $h < 0$) and $|k|$ units up (if $k > 0$) or down (if $k < 0$). The horizontal translation of a sinusoid is called its *phase shift*. Notice that translating the graph of $f(x) = \cos x$ right $\frac{\pi}{2}$ units produces $g(x) = \cos\left(x - \frac{\pi}{2}\right) = \sin x$. Translating the graph of $g(x) = \sin x$ left $\frac{\pi}{2}$ units produces $f(x) = \sin\left(x + \frac{\pi}{2}\right) = \cos x$.

When the sinusoid is written in the form $y = a\sin(bx + c) + k$ or $y = a\cos(bx + c) + k$, factoring out b yields $y = a\sin b\left(x + \frac{c}{b}\right) + k$ or $y = a\cos b\left(x + \frac{c}{b}\right) + k$, where $h = -\frac{c}{b}$ is the horizontal phase shift and k is the vertical shift.

State the amplitude, period, and phase shift of $g(x) = \sin\left(\frac{x}{2} + \frac{\pi}{4}\right) + 2$. Then describe how $g(x)$ is obtained by transforming the parent function $f(x) = \sin x$ and graph $g(x)$ over $[-2\pi, 4\pi]$.

Answer

Factor out the coefficient of x: $g(x) = \sin\left(\frac{x}{2} + \frac{\pi}{4}\right) + 2 = \sin\frac{1}{2}\left(x + \frac{\pi}{2}\right) + 2$.

amplitude: $|a| = |1| = 1$ period: $p = \frac{2\pi}{|b|} = \frac{2\pi}{\left|\frac{1}{2}\right|} = 4\pi$ phase shift: $h = -\frac{c}{b} = -\frac{\frac{\pi}{4}}{\frac{1}{2}} = -\frac{\pi}{2}$

The graph of $f(x)$ is horizontally stretched by a factor of 2, and then translated $\frac{\pi}{2}$ units left and 2 units up.

$y_1 = \sin x$	$(0, 0)$	$\left(\frac{\pi}{2}, 1\right)$	$(\pi, 0)$	$\left(\frac{3\pi}{2}, -1\right)$	$(2\pi, 0)$
$y_2 = \sin\frac{1}{2}x$	$(0, 0)$	$(\pi, 1)$	$(2\pi, 0)$	$(3\pi, -1)$	$(4\pi, 0)$
$y_3 = \sin\frac{1}{2}\left(x - \left(-\frac{\pi}{2}\right)\right) + 2$	$\left(-\frac{\pi}{2}, 2\right)$	$\left(\frac{\pi}{2}, 3\right)$	$\left(\frac{3\pi}{2}, 2\right)$	$\left(\frac{5\pi}{2}, 1\right)$	$\left(\frac{\pi}{2}, 2\right)$

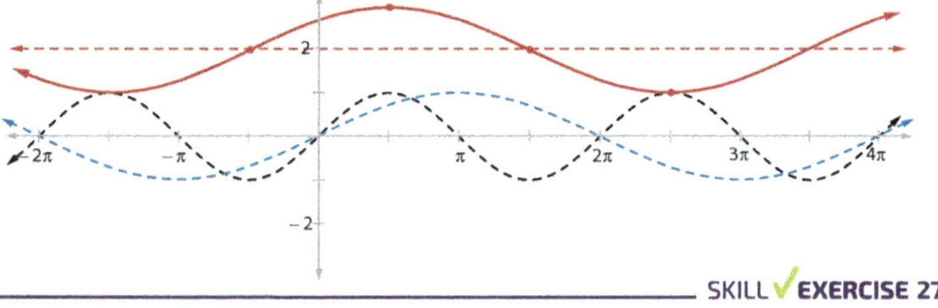

_______________________ SKILL ✔ **EXERCISE 27**

several possibilities. (*infinite; negative values for amplitude or period, any phase shift of* $2\sin\frac{2\pi}{5}(x + 3)$ *that is a multiple of 2.5, which is half the period*)

Example 6 provides a real-world application using two data points. Point out in step 4 that the horizontal distance between two consecutive maximum points is one cycle, but the horizontal distance between two consecutive extrema (a maximum and a minimum) is half a cycle.

Consider mentioning that many people throughout history have been fascinated by the cyclic (or periodic) nature of the phases of the moon. Remind students that these things should cause us to praise God for His creation and provision (Gen. 1:16, Deut. 33:13–14). Warn that some people promote unbiblical views of the sun and the moon, such as those of astrology. Point out that God clearly condemns such a worldview (Deut. 17:2–3; 2 Kings 23:5).

Motivational Idea Share that communication engineers frequently encounter the amplitude, period, and *phase shift* of sinusoidal waves. AM (amplitude modulation) radio uses waves in which the audio is encoded as changes in amplitude. Interference causes distortions in the amplitude, resulting in static. FM (frequency modulation) radio uses waves in which the audio is encoded as changes in frequency. Circuits can limit distortions in amplitude while maintaining the changes in frequency. This is why FM radio is clearer than AM radio. Digital signals, such as digital TV signals, encode the signals as only two different frequencies, representing 0 and 1, and provide high fidelity.

TIPS

Ex. 43–44 The factorial function is entered using MATH, PROB, 4: !.

The *midline* of a sinusoid is the horizontal line halfway between the maximum and minimum values of the function, $y = \frac{max - min}{2}$. The midline of the basic sinusoidal functions is the x-axis. The midline of a translated sinusoid is $y = k$. Since one cycle of $f(x) = \sin x$ is completed over $0 \leq x \leq 2\pi$, the corresponding cycle of $g(x) = a \sin(bx + c) + k$ is completed over $0 \leq bx + c \leq 2\pi$ or $-\frac{c}{b} \leq x \leq \frac{2\pi}{b} - \frac{c}{b}$. In Example 4, the midline is $y = 2$ and the function completes one cycle over $0 \leq \frac{x}{2} + \frac{\pi}{4} \leq 2\pi$ or $-\frac{\pi}{2} \leq x \leq \frac{7\pi}{2}$.

<table>
<tr><td colspan="3">Characteristics of Sinusoidal Functions</td></tr>
<tr><td colspan="3">functions of the form $f(x) = a \sin b(x - h) + k$ or
$f(x) = a \cos b(x - h) + k$ (where $a \neq 0$ and $b \neq 0$)</td></tr>
<tr><td>amplitude: $|a|$</td><td>period: $p = \frac{2\pi}{|b|}$</td><td>frequency: $f = \frac{1}{p} = \frac{|b|}{2\pi}$</td></tr>
<tr><td>phase shift: h</td><td>vertical shift: k</td><td>midline: $y = k$</td></tr>
<tr><td colspan="3">When written as $f(x) = a \sin(bx + c) + k$ or
$f(x) = a \cos(bx + c) + k$, the phase shift $h = -\frac{c}{b}$.</td></tr>
</table>

Example 5 Writing the Equation of a Sinusoid

Write an equation of a sine function passing through (–3, 0) with a period of 5 and an amplitude of 2.

Answer

Let $a = 2$.

Let $b = \frac{2\pi}{5}$.

Let $h = -3$.

$\therefore f(x) = 2 \sin \frac{2\pi}{5}(x + 3)$ or

$f(x) = 2 \sin\left(\frac{2\pi}{5}x + \frac{6\pi}{5}\right)$

Check

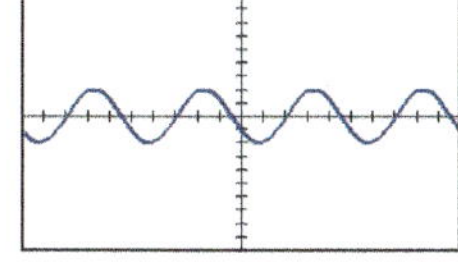

1. amplitude: $|a| = 2$, so $a = \pm 2$

2. period: $p = \frac{2\pi}{|b|} = 5$, so $|b| = \frac{2\pi}{5}$

3. The graph of $f(x) = 2 \sin \frac{2\pi}{5}x$ passes through (0, 0), so the graph needs to be shifted 3 units left.

4. Write the equation.

5. Graph the function and check that it fulfills the requirements. Can you see why $g(x) = -2 \sin \frac{2\pi}{5}(x + 3)$ and $h(x) = 2 \sin \frac{2\pi}{5}(x - 2)$ are two of many other sine functions that also fulfill the requirements?

SKILL ✔ EXERCISE 29

Carl Hertz, the son of a nephew of Heinrich Hertz, was instrumental in developing medical ultrasound equipment and procedures. This technology is used by doctors in diagnosing and treating patients, and is now a standard part of prenatal care. The resulting sonograms vividly illustrate that life begins before birth (Jer. 1:4–5).

10.

sample answers: $y = 2 \cos 3\left(x - \frac{\pi}{2}\right)$

or $y = 2 \cos 3\left(x + \frac{\pi}{6}\right)$

Assignments

- **Minimum:** 1–5, 7, 8a, 9–15, 17–18, 21–23, 25, 29, 33, 36–37, 45–48, 50–53
- **Standard:** 2–3, 5, 8–10; 13–19 odd; 20–32 even; 33, 36–39, 42; 46–54 even
- **Extended:** 2, 5, 8, 11, 14–16, 18, 21, 24, 28–29, 31, 34, 38–44, 46, 48–49, 52–54

Solutions

❯ A. Exercises

1. $(0, 0)$, $\left(\frac{\pi}{6}, \frac{1}{2}\right)$, $\left(\frac{\pi}{4}, \frac{\sqrt{2}}{2}\right)$, $\left(\frac{\pi}{3}, \frac{\sqrt{3}}{2}\right)$, $\left(\frac{\pi}{2}, 1\right)$, $\left(\frac{2\pi}{3}, \frac{\sqrt{3}}{2}\right)$, $\left(\frac{3\pi}{4}, \frac{\sqrt{2}}{2}\right)$, $\left(\frac{5\pi}{6}, \frac{1}{2}\right)$, $(\pi, 0)$, $\left(\frac{7\pi}{6}, -\frac{1}{2}\right)$, $\left(\frac{5\pi}{4}, -\frac{\sqrt{2}}{2}\right)$, $\left(\frac{4\pi}{3}, -\frac{\sqrt{3}}{2}\right)$, $\left(\frac{3\pi}{2}, -1\right)$, $\left(\frac{5\pi}{3}, -\frac{\sqrt{3}}{2}\right)$, $\left(\frac{7\pi}{4}, -\frac{\sqrt{2}}{2}\right)$, $\left(\frac{11\pi}{6}, -\frac{1}{2}\right)$, $(2\pi, 0)$

2. $(0, 1)$, $\left(\frac{\pi}{6}, \frac{\sqrt{3}}{2}\right)$, $\left(\frac{\pi}{4}, \frac{\sqrt{2}}{2}\right)$, $\left(\frac{\pi}{3}, \frac{1}{2}\right)$, $\left(\frac{\pi}{2}, 0\right)$, $\left(\frac{2\pi}{3}, -\frac{1}{2}\right)$, $\left(\frac{3\pi}{4}, -\frac{\sqrt{2}}{2}\right)$, $\left(\frac{5\pi}{6}, -\frac{\sqrt{3}}{2}\right)$, $(\pi, -1)$, $\left(\frac{7\pi}{6}, -\frac{\sqrt{3}}{2}\right)$, $\left(\frac{5\pi}{4}, -\frac{\sqrt{2}}{2}\right)$, $\left(\frac{4\pi}{3}, -\frac{1}{2}\right)$, $\left(\frac{3\pi}{2}, 0\right)$, $\left(\frac{5\pi}{3}, \frac{1}{2}\right)$, $\left(\frac{7\pi}{4}, \frac{\sqrt{2}}{2}\right)$, $\left(\frac{11\pi}{6}, \frac{\sqrt{3}}{2}\right)$, $(2\pi, 1)$

3. $|a| = \left|\frac{1}{4}\right| = \frac{1}{4}$; $p = \frac{2\pi}{|b|} = 2\pi$

4. $|a| = |-3| = 3$; $p = \frac{2\pi}{|b|} = \frac{2\pi}{|2|} = \pi$

5. $|a| = |-2| = 2$; $p = \frac{2\pi}{|b|} = \frac{2\pi}{|4\pi|} = \frac{1}{2}$

6. $|a| = \left|\frac{7}{3}\right| = \frac{7}{3}$; $p = \frac{2\pi}{|b|} = \frac{2\pi}{\left|\frac{2}{3}\right|} = 3\pi$

7a. The graph of $f(x)$ is vertically stretched by a factor of 3.

7b. The graph of $f(x)$ is reflected in the x-axis, vertically stretched by a factor of 4, and horizontally stretched by a factor of 2.

7c. Since the sine function is odd, $g(x) = -\sin 2x$. The graph of $f(x)$ is reflected in the x-axis and horizontally shrunk by a factor of $\frac{1}{2}$.

8a. The graph of $f(x)$ is horizontally stretched by a factor of 4.

8b. The graph of $f(x)$ is reflected in the x-axis, vertically shrunk by a factor of $\frac{1}{3}$, and horizontally shrunk by a factor of $\frac{1}{2}$.

8c. Since the cosine function is even, $g(x) = 4 \cos x$. The graph of $f(x)$ is vertically stretched by a factor of 4.

9a. $|a| = 1$; $p = \frac{2\pi}{|b|} = \frac{2\pi}{4} = \frac{\pi}{2}$; $f = \frac{1}{p} = \frac{2}{\pi}$

$g(x)$ completes 2 cycles over any x-interval of length π (or ≈ 0.64 of a cycle per unit of x).

9b. $|a| = 5$; $p = \frac{2\pi}{|b|} = \frac{2\pi}{6\pi} = \frac{1}{3}$; $f = \frac{1}{p} = 3$

$r(x)$ completes 3 cycles per unit of x.

9c. $|a| = 1; p = \dfrac{2\pi}{|b|} = \dfrac{2\pi}{\frac{3\pi}{2}} = \dfrac{4}{3}; f = \dfrac{1}{p} = \dfrac{3}{4}$

$v(x)$ completes 3 cycles over any x-interval of length 4 (or 0.75 of a cycle per unit of x).

10. $|a| = 4; p = \dfrac{2\pi}{1} = 2\pi$

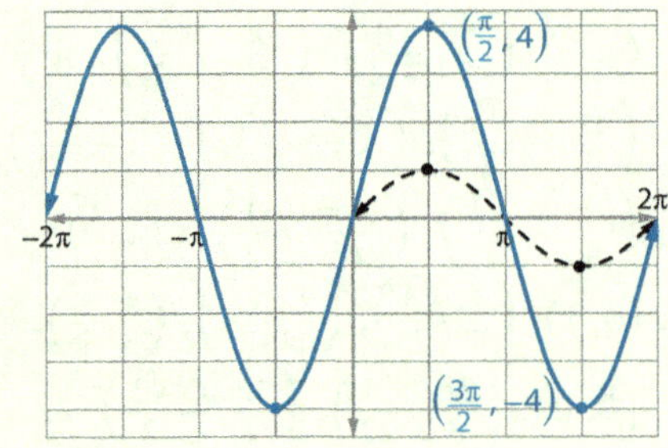

11. $|a| = \dfrac{1}{2}; p = \dfrac{2\pi}{1} = 2\pi$

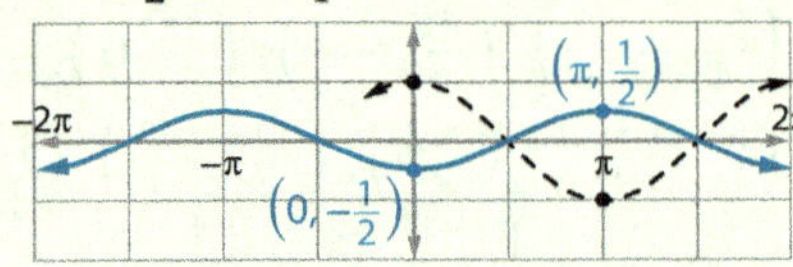

12. $|a| = 1; p = \dfrac{2\pi}{1} = 2\pi$

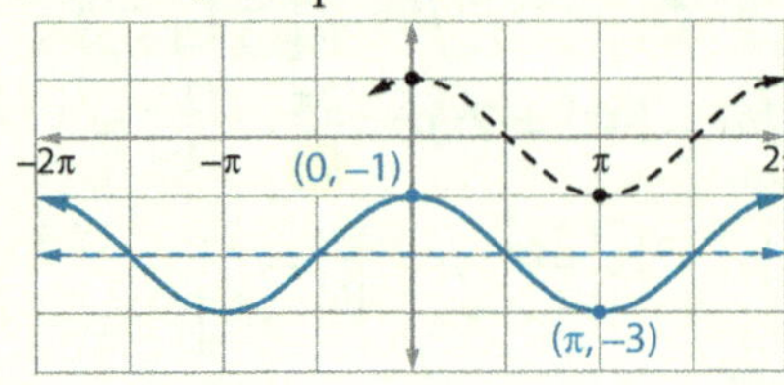

13. $|a| = 1; p = \dfrac{2\pi}{1} = 2\pi; h = \dfrac{\pi}{2}$

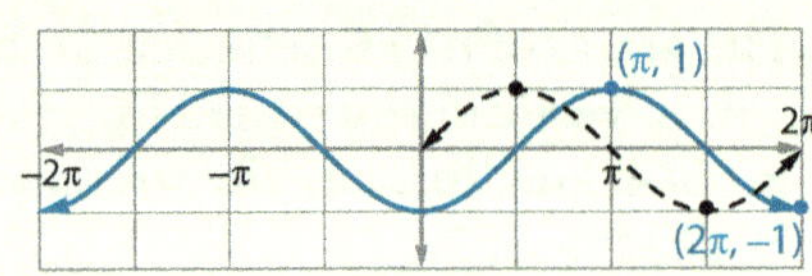

14. $|a| = 1; p = \dfrac{2\pi}{1} = 2\pi$
$h = -\dfrac{\pi}{4}; k = 3$

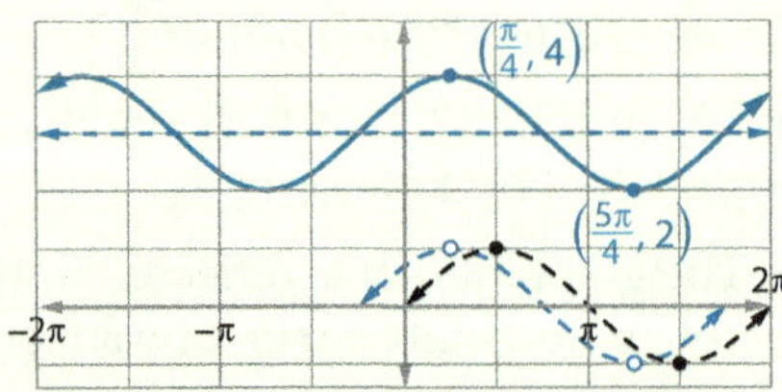

15. $|a| = 2; p = \dfrac{2\pi}{3}$

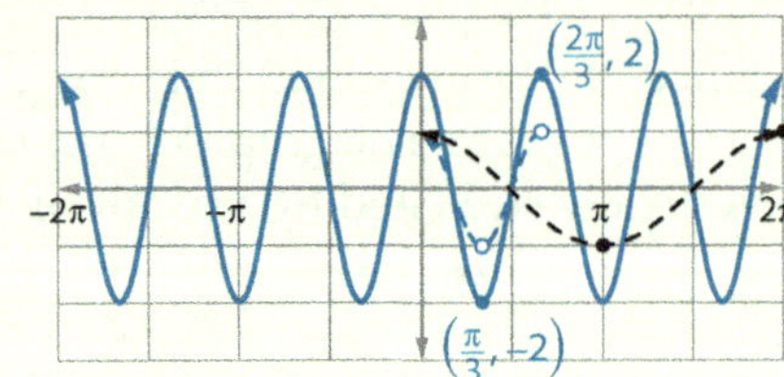

16a. $V_i = 186.6 \sin(377t); p = \dfrac{2\pi}{377}$

16b. $f = \dfrac{377}{2\pi} \approx 60$ Hz

Sinusoidal functions frequently model periodic phenomena in God's creation.

Example 6 Modeling Data with a Periodic Function

Write a function modeling the illumination of the moon during 2018 if a full moon (100% illumination) occurred on January 2, 2018, and the next new moon (0% illumination) occurred on January 17, 2018. Then use the model to predict the percentage of the moon illuminated on July 9, the 190th day of that year.

Answer

maximum at (2, 1.00) and minimum at (17, 0.00)

1. Letting x represent the day of the year and y represent the illumination of the moon, express the data as ordered pairs.

$a = \dfrac{1}{2}(1.00 - 0.00) = \dfrac{1}{2}$

2. Use $a = \dfrac{1}{2}(y_{max} - y_{min})$ to determine the value of a.

$k = \dfrac{1}{2}(1.00 + 0.00) = \dfrac{1}{2}$

3. Use $k = \dfrac{1}{2}(y_{max} + y_{min})$ to determine the value of k.

$p = 2|x_{max} - x_{min}|$
$ = 2|2 - 17| = 30$

4. Determine the period of the function using the fact that the horizontal distance between consecutive maximum and minimum points is half a cycle.

$|b| = \dfrac{2\pi}{p} = \dfrac{2\pi}{30};$ Let $b = \dfrac{\pi}{15}.$

5. Use $p = \dfrac{2\pi}{|b|}$ to determine the value of b.

$h = 2$

6. Since the first maximum point of $f(x) = \dfrac{1}{2} \cos \dfrac{\pi}{15}x + \dfrac{1}{2}$ is at (0, 1), the graph needs to be shifted right by 2 units.

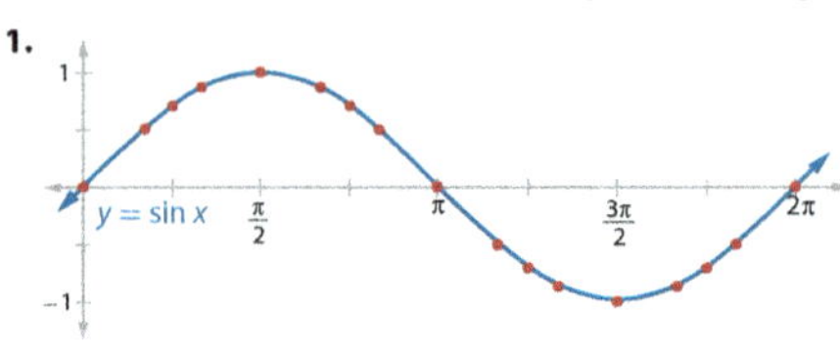

$f(x) = \dfrac{1}{2} \cos \dfrac{\pi}{15}(x - 2) + \dfrac{1}{2}$

7. Write the equation of the modeling function.

$f(x) \approx 0.45$ or $\approx 45\%$

8. Evaluate $f(190)$ to predict the illumination percentage of the moon on July 9, 2018.

SKILL ✓ EXERCISE 37

❯ A. Exercises

Use the 16 special values from the unit circle to state the exact coordinates of each illustrated point on the graph.

1.

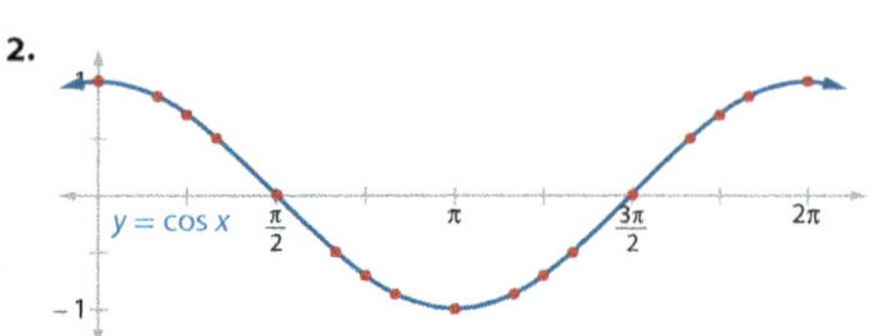

2.

17. $|a| = 105$

$p = \dfrac{1}{f} = \dfrac{1}{470}; \dfrac{1}{470} = \dfrac{2\pi}{|b|}$
$|b| = 2\pi(470) = 940\pi$
$y = 105 \sin(940\pi t)$

❯ B. Exercises

18. $|a| = 2; p = \dfrac{2\pi}{\frac{1}{2}} = 4\pi$

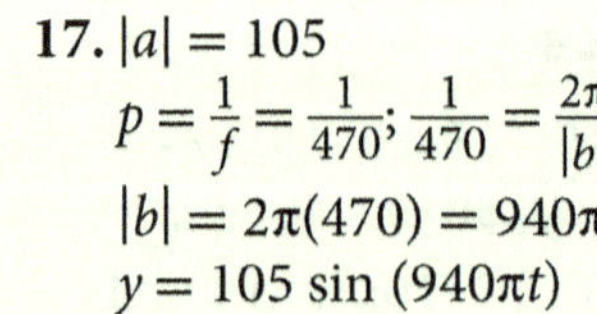

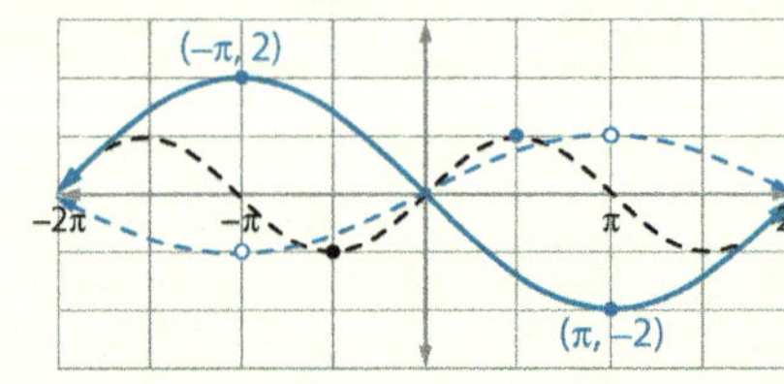

19. $|a| = 1.5; p = \dfrac{2\pi}{\frac{2}{3}} = 3\pi$

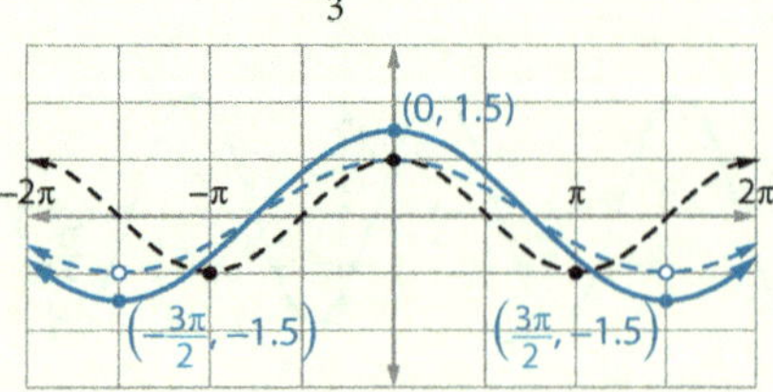

20. $|a| = 1; p = \dfrac{2\pi}{3}; k = 2$

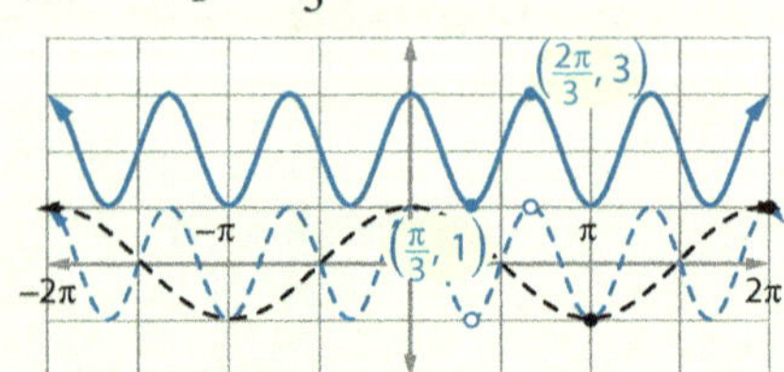

State the amplitude and period of each function.

3. $y = \frac{1}{4}\cos x$ $\frac{1}{4}$; 2π
4. $y = -3\sin 2x$ 3; π
5. $y = -2\cos 4\pi x$ 2; $\frac{1}{2}$
6. $y = \frac{7}{3}\cos \frac{2x}{3}$ $\frac{7}{3}$; 3π

7. Describe the graph of $g(x)$ as a transformation of $f(x) = \sin x$.
   a. $g(x) = 3\sin x$
   b. $g(x) = -4\sin \frac{x}{2}$
   c. $g(x) = \sin(-2x)$

8. Describe the graph of $g(x)$ as a transformation of $f(x) = \cos x$.
   a. $g(x) = \cos 0.25x$
   b. $g(x) = -\frac{1}{3}\cos 2x$
   c. $g(x) = 4\cos(-x)$

9. Find the amplitude, period, and frequency of each function and explain what the frequency means.
   a. $g(x) = \sin 4x$
   b. $r(x) = -5\cos 6\pi x$
   c. $v(x) = -\sin \frac{3\pi}{2}x$

State the amplitude, period, and any phase shift or vertical shift for each function. Then graph the function over $[-2\pi, 2\pi]$.

10. $f(x) = 4\sin x$
11. $g(x) = -\frac{1}{2}\cos x$
12. $h(x) = \cos x - 2$
13. $k(x) = \sin\left(x - \frac{\pi}{2}\right)$
14. $f(x) = \sin\left(x + \frac{\pi}{4}\right) + 3$
15. $g(x) = 2\cos 3x$

16. The instantaneous voltage (V_i) of an alternating current generator can be modeled by $V_i = V_m \sin(\omega t)$ where V_m is the maximum voltage and ω is the angular speed of the generator's rotor in radians per second.
   a. What is the period of the function if the maximum voltage is 186.6 volts and the angular speed is 377 $\frac{\text{radians}}{\text{sec}}$? $p = \frac{2\pi}{377}$
   b. What is the frequency of the function in Hz?   60 Hz

17. An elephant call can start with a low rumble at 25 Hz and progress to a roar pitched at 470 Hz with an amplitude of 105 dB. Write a sine function that models the final sound wave. $y = 105\sin(940\pi t)$

❯ **B. Exercises**

State the amplitude, period, and any phase shift or vertical shift for each function. Then graph the function over $[-2\pi, 2\pi]$.

18. $h(x) = -2\sin \frac{1}{2}x$
19. $h(x) = 1.5\cos \frac{2x}{3}$
20. $f(x) = \cos 3x + 2$
21. $f(x) = 2\sin(x + \pi)$

State the amplitude, period, phase shift, and vertical shift for each function. Then describe the function's transformation from the parent function $f(x) = \sin x$ or $f(x) = \cos x$. Confirm your answer by using technology to graph the function and its parent function.

22. $g(x) = 2\cos 5x - 1$ $|a| = 2,\ p = \frac{2\pi}{5},\ h = 0,\ k = -1$
23. $h(x) = \sin\left(3x - \frac{\pi}{2}\right) + 3$ $|a| = 1,\ p = \frac{2\pi}{3},\ h = \frac{\pi}{6},\ k = 3$
24. $r(x) = 2\cos(\pi - 2x)$ $|a| = 2,\ p = \pi,\ h = \frac{\pi}{2},\ k = 0$

Graph two periods of each function.

25. $f(x) = -2\cos\left(x - \frac{\pi}{2}\right)$
26. $g(x) = 3\sin\left(2x + \frac{\pi}{2}\right)$
27. $h(x) = -3\sin(2x - \pi) + 2$
28. $k(x) = 2\cos(4x - \pi) - 3$

Write a sine function with the given characteristics.

29. amplitude: 1, period: $\frac{\pi}{2}$, phase shift: $\frac{\pi}{8}$, translated 3 units up   Answers may vary.
30. amplitude: 4, period: 5π, passing through $(3, 0)$ Answers may vary.

Write a cosine function with the given characteristics.

31. amplitude: 3, period: π, phase shift: $-\frac{\pi}{6}$, translated 4 units up
32. passing through a relative maximum at $\left(\frac{1}{4}, \frac{5}{2}\right)$ and a relative minimum at $\left(\frac{5}{4}, \frac{3}{2}\right)$

Write a sine function rule for each graph.   Answers may vary.

33. 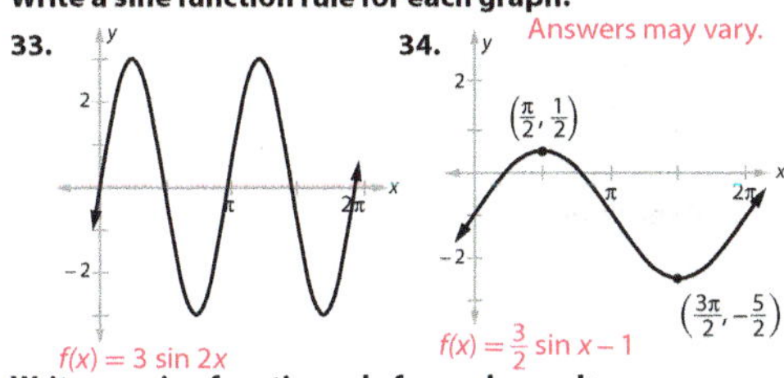

34.

$f(x) = 3\sin 2x$

$f(x) = \frac{3}{2}\sin x - 1$

Write a cosine function rule for each graph.   Answers may vary.

35. 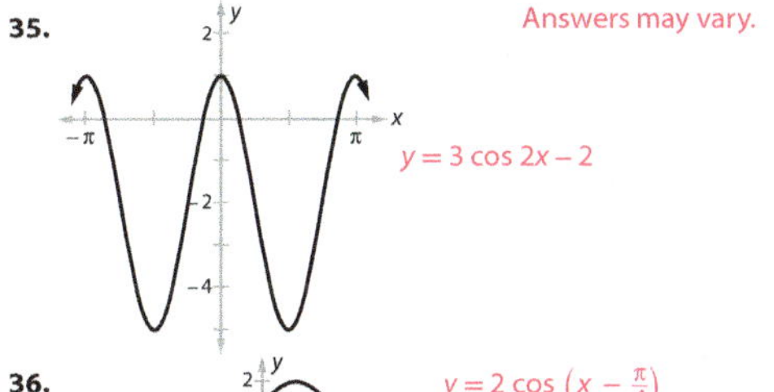

$y = 3\cos 2x - 2$

36. 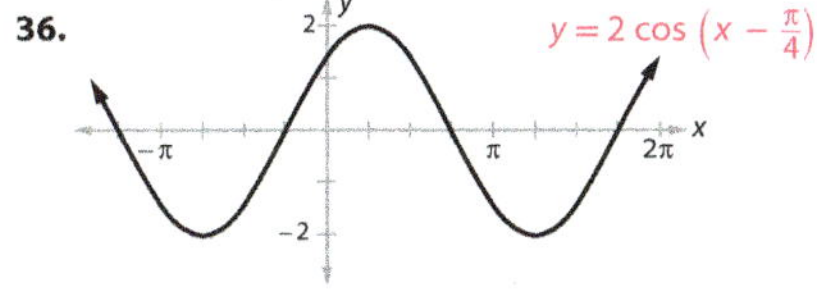

$y = 2\cos\left(x - \frac{\pi}{4}\right)$

25. $|a| = 2$; reflected in the x-axis; translated $\frac{\pi}{2}$ units right

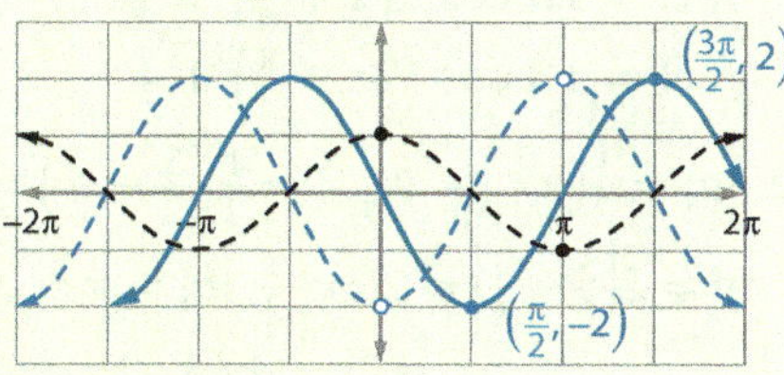

26. $g(x) = 3\sin 2\left(x + \frac{\pi}{4}\right)$
    $p = \pi$; $|a| = 3$; translated $\frac{\pi}{4}$ units left

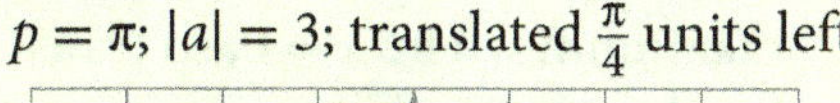
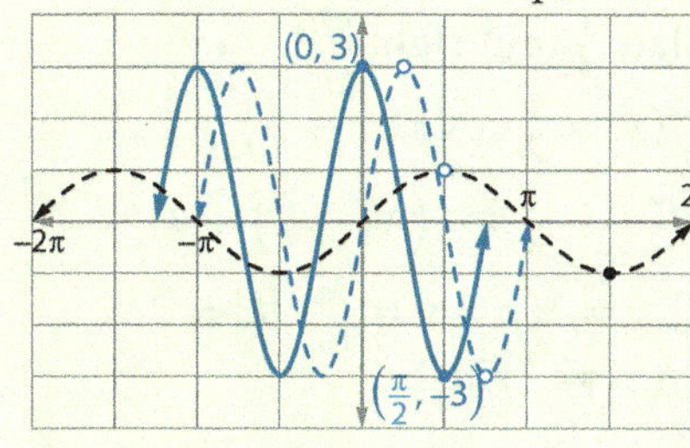

27. $h(x) = -3\sin 2\left(x - \frac{\pi}{2}\right) + 2$
    $p = \pi$; $|a| = 3$; reflected in the x-axis; translated $\frac{\pi}{2}$ units right and 2 units up

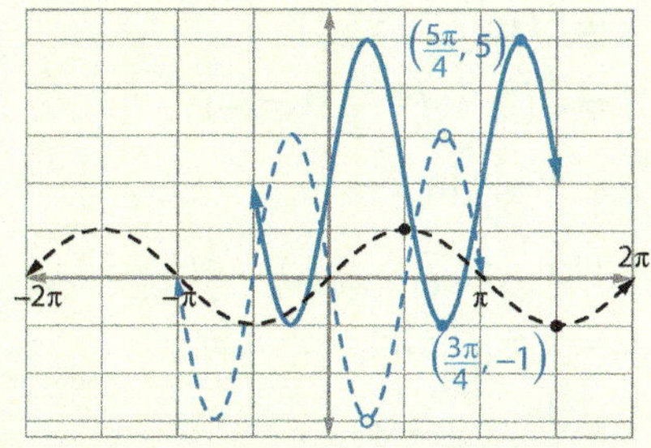

28. $k(x) = 2\cos 4\left(x - \frac{\pi}{4}\right) - 3$
    $p = \frac{\pi}{2}$; $|a| = 2$; translated $\frac{\pi}{4}$ units right and 3 units down

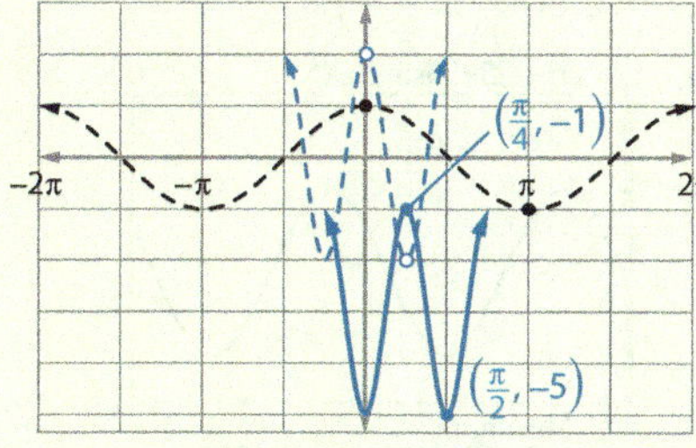

21. $|a| = 2$; $p = 2\pi$; $h = -\pi$

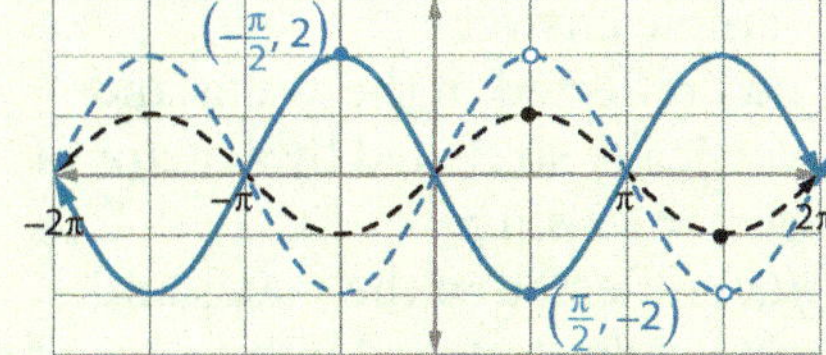

22. $|a| = 2$; $p = \frac{2\pi}{5}$; $h = 0$; $k = -1$; The graph of $f(x)$ is horizontally shrunk by a factor of $\frac{1}{5}$, vertically stretched by a factor of 2, and then translated 1 unit down.

23. $h(x) = \sin 3\left(x - \frac{\pi}{6}\right) + 3$
    $|a| = 1$; $p = \frac{2\pi}{3}$; $h = \frac{\pi}{6}$; $k = 3$; The graph of $f(x)$ is horizontally shrunk by a factor of $\frac{1}{3}$ and translated $\frac{\pi}{6}$ units right and 3 units up.

24. $r(x) = 2\cos\left(-2\left(x - \frac{\pi}{2}\right)\right)$
    $= 2\cos 2\left(x - \frac{\pi}{2}\right)$
    (Cosine is an even function.)
    $|a| = 2$; $p = \frac{2\pi}{2} = \pi$; $h = \frac{\pi}{2}$; $k = 0$; The graph of $f(x)$ is horizontally shrunk by a factor of $\frac{1}{2}$, translated $\frac{\pi}{2}$ units right, and vertically stretched by a factor of 2.

29. $p = \frac{\pi}{2} = \frac{2\pi}{|b|}$; $|b| = 4$
    $f(x) = \pm\sin 4\left(x - \frac{\pi}{8}\right) + 3$ or
    $f(x) = \pm\sin\left(4x - \frac{\pi}{2}\right) + 3$

30. $p = \frac{2\pi}{|b|} = 5\pi$; $|b| = \frac{2}{5}$
    $f(x) = \pm 4\sin \frac{2x}{5}$ passes through $(0, 0)$, so translate 3 units right.
    $f(x) = \pm 4\sin \frac{2}{5}(x - 3)$ or
    $f(x) = \pm 4\sin\left(\frac{2}{5}x - \frac{6}{5}\right)$

31. $p = \pi = \frac{2\pi}{|b|}$; $|b| = 2$

$f(x) = \pm 3 \cos 2\left(x + \frac{\pi}{6}\right) + 4$ or

$f(x) = \pm 3 \cos \left(2x + \frac{\pi}{3}\right) + 4$

32. $|a| = \frac{1}{2}(y_{max} - y_{min}) = \frac{1}{2}\left(\frac{5}{2} - \frac{3}{2}\right) = \frac{1}{2}$

$p = 2\left|\frac{5}{4} - \frac{1}{4}\right| = 2 = \frac{2\pi}{|b|}$; $|b| = \pi$

$k = \frac{1}{2}(y_{max} + y_{min}) = \frac{1}{2}\left(\frac{5}{2} + \frac{3}{2}\right) = 2$

$f(x) = \frac{1}{2}\cos \pi x + 2$ has a relative maximum where $x = 0$, so translate $\frac{1}{4}$ unit right.

$f(x) = \frac{1}{2}\cos \pi\left(x - \frac{1}{4}\right) + 2$ or

$f(x) = \frac{1}{2}\cos \left(\pi x - \frac{\pi}{4}\right) + 2$

33. $|a| = 3$; $p = \pi = \frac{2\pi}{|b|}$; $|b| = 2$

$f(x) = 3 \sin 2x$

34. $|a| = \frac{1}{2}(y_{max} - y_{min})$

$= \frac{1}{2}\left(\frac{1}{2} - \left(-\frac{5}{2}\right)\right) = \frac{3}{2}$

$p = 2\pi = \frac{2\pi}{|b|}$; $|b| = 1$

$k = \frac{1}{2}(y_{max} + y_{min})$

$= \frac{1}{2}\left(\frac{1}{2} + \left(-\frac{5}{2}\right)\right) = -1$

$f(x) = \frac{3}{2}\sin x - 1$

35. $|a| = \frac{1}{2}(y_{max} - y_{min}) = \frac{1}{2}(1 - (-5)) = 3$

$p = \pi = \frac{2\pi}{|b|}$; $|b| = 2$; $h = 0$

$k = \frac{1}{2}(y_{max} + y_{min})$

$= \frac{1}{2}(1 + (-5)) = -2$

$y = 3 \cos 2x - 2$

36. $|a| = 2$; $p = 2\pi$; $|b| = 1$; $h = \frac{\pi}{4}$; $k = 0$

$y = 2 \cos \left(x - \frac{\pi}{4}\right)$

37a. 170 Y₁=125+35sin(2πX)

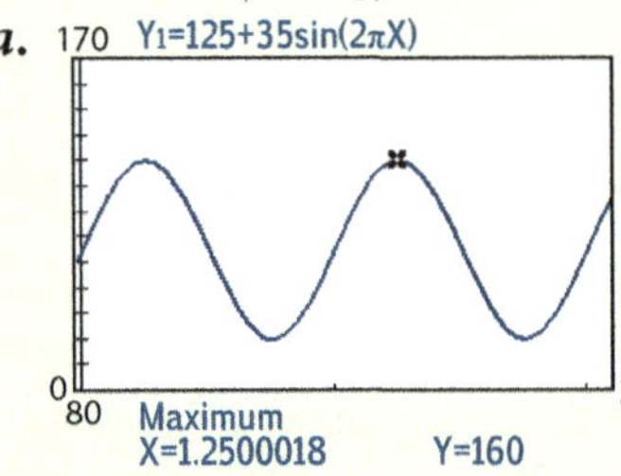

37b. $|a| = \frac{1}{2}(160 - 90) = 35$

$k = \frac{1}{2}(160 + 90) = 125$

37c. $p = \frac{2\pi}{2\pi} = 1$ sec

38a. rel. min.: height − diameter = 30 ft

$|a| = \frac{1}{2}(550 - 30) = 260$

$k = \frac{1}{2}(550 + 30) = 290$

$p = 30 = \frac{2\pi}{|b|}$; $|b| = \frac{\pi}{15}$

$h(t) = -260 \cos \frac{\pi}{15}t + 290$

---

37. The blood pressure of a person with hypertension is modeled in millimeters of mercury (mm Hg) by a function of time (in minutes):
$p(t) = 125 + 35 \sin 2\pi t$.

  a. Use technology to graph two periods of the function. State the person's highest and lowest blood pressure.   160 mm Hg; 90 mm Hg

  b. State the function's amplitude and its vertical shift. $|a| = 35$; $k = 125$

  c. Find the function's period. How many times does the person's heart beat in one minute?   1 sec; 60 times

38. The world's tallest observation wheel (as of 2016) has a height of 550 ft and a diameter of 520 ft. Riders board and depart its glass observation cabins while the wheel takes approximately 30 min to complete one revolution.

  a. Derive a function $h(t)$ that models the height of an observation cabin as a function of the time (in minutes) after a rider enters at the bottom of the wheel.

  b. How high would a passenger be after 5 min?   160 ft

  c. How much time would elapse before the passenger returned to that height?   25 min   **38a.** $h(t) = -260 \cos \frac{\pi}{15}t + 290$

❭ C. Exercises

39. Discuss: Write three sine function rules for $g(x)$.

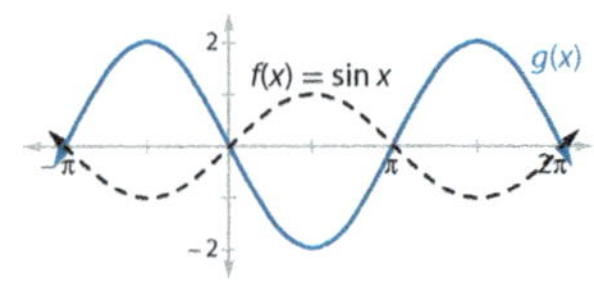

40. Discuss: Write two cosine function rules and two sine function rules for $g(x)$.

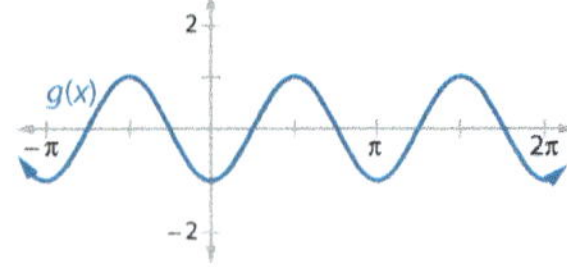

38b. $h(5) = 160$ ft

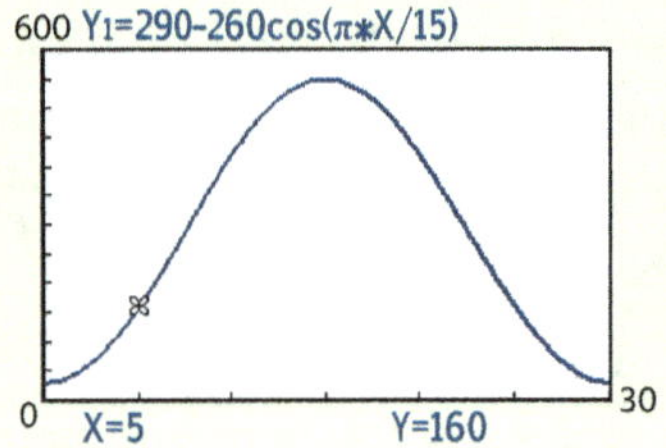

38c. $h(25) = 160$ ft

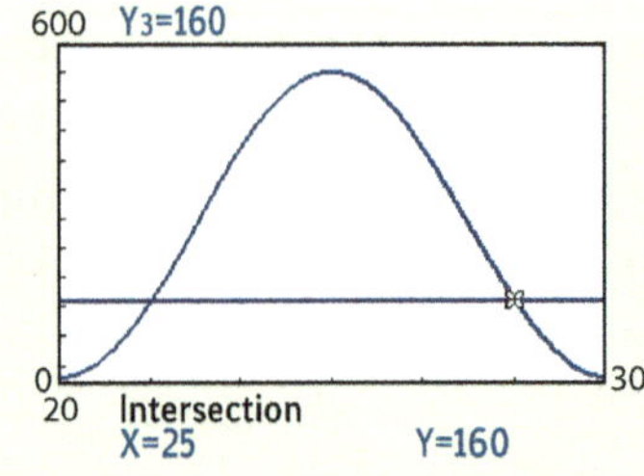

---

41. Turnagain Arm, a waterway in northwestern Alaska, boasts one of the world's largest bore tides. The water depth can be modeled by a sinusoidal function of time with a period of 12 hr 24 min. On a certain day a high tide of 25.87 ft occurs at 4:00 AM after the previous evening's low tide of 5.47 ft.   9:48 PM; 10:12 AM

  a. At what time did the previous evening's low tide occur, and when will the next low tide occur?

  b. Write a sinusoidal function $d(t)$ that models the depth of the water that day.

  c. Estimate the depth of the water at 3:00 PM that day.   23.41 ft

  d. When was the depth of the water 15.67 ft that day?   12:54 AM, 7:06 AM, 1:18 PM, 7:30 PM

42. The table lists the average high temperatures on the 15th of each month in Minneapolis, Minnesota.

Month	1	2	3	4	5	6
High (°F)	21	28	40	57	70	79
Month	7	8	9	10	11	12
High (°F)	84	81	71	59	40	26

  a. Write a sinusoidal function $h(t)$ that models the high temperature as a function of time.

  b. Use technology to graph the data and your model. Which month has the largest residual?   October

  c. Use the function to estimate the high temperature on April 30.   61°F

Taylor polynomial approximations are used in calculus to estimate trigonometric functions. For exercises 43–44, complete the following steps to explore each approximation.
Note that $n! = n$ factorial $= n(n-1)(n-2) \dots 1$.

  a. Use technology to graph the trigonometric function as Y_1 and the given Taylor polynomial approximation as Y_2 over $[-2\pi, 2\pi]$. In what interval does the approximation appear to be a good model for the function?

  b. Define the residual function $Y_3 = Y_1 - Y_2$, then examine a table of values with Tstep = 0.1 to find the interval for which $|Y_3| \leq 0.01$.

43. $f(x) = \sin x \approx x - \frac{x^3}{3!} + \frac{x^5}{5!}$

44. $f(x) = \cos x \approx 1 - \frac{x^2}{2!} + \frac{x^4}{4!}$

41b. $d(t) = 10.2 \cos \frac{\pi}{6.2}(t - 4) + 15.67$

42a. $h(t) = 31.5 \cos \frac{\pi}{6}(t - 7) + 52.5$

❭ C. Exercises

39. sample answers:

$f(x)$ reflected in the x-axis and vertically stretched by a factor of 2:
$g(x) = -2 \sin x$

$f(x)$ reflected in the y-axis and vertically stretched by a factor of 2:
$g(x) = 2 \sin (-x)$

$f(x)$ vertically stretched by a factor of 2 and translated π units right or left:
$g(x) = 2 \sin (x \pm \pi)$

40. sample answers:

$f(x) = \cos x$ reflected in the x-axis and horizontally shrunk by a factor of $\frac{1}{2}$: $g(x) = -\cos 2x$

$f(x) = \cos x$ horizontally shrunk by a

45. Write a polynomial function with roots of multiplicity 2 at -3 and at 4. [3.1]
$f(x) = x^4 - 2x^3 - 23x^2 + 24x + 144$

Find the degree of each function if $P(x)$ is a quadratic function and $Q(x)$ is a cubic function. [1.7, 2.4]

46. $P(x) + Q(x)$ 3 **47.** $P(x) \cdot Q(x)$ 5 **48.** $\dfrac{Q(x)}{P(x)}$ 1

49. When is the sum of two quadratic functions not quadratic? when the leading coefficients are additive inverses

50. Identify any vertical asymptote(s) for each reciprocal function. [2.5]

a. $f(x) = \dfrac{1}{x+3}$ $x = -3$ **b.** $h(x) = \dfrac{1}{x^2 + 5x - 14}$ $x = -7, x = 2$

c. $g(x) = \dfrac{1}{x^2 + 1}$ no vertical asymptote

51. Which of the following is irrational? [4.3] B

A. $\tan \frac{\pi}{4}$ **D.** $\sin \frac{7\pi}{6}$

B. $\cos \frac{\pi}{6}$ **E.** none of these

C. $\tan \frac{3\pi}{2}$

52. Which of the following is not true if $\sin \theta = \frac{9}{41}$ and $\tan \theta > 0$? [4.3] E

A. $\csc \theta = \frac{41}{9}$ **D.** $\cos \theta > \frac{1}{2}$

B. $\tan \theta = \frac{9}{40}$ **E.** These are all true.

C. $\csc \theta > \sec \theta$

53. Find $\cos(-\theta) + \sec \theta$ if $\cos \theta = 0.8$. [4.3] D

A. 0   **D.** 2.05

B. 0.45   **E.** none of these

C. 1.6

54. For which angle θ is $\sin \theta > 0$ and $\tan \theta < 0$? B

A. 88°   **D.** 345°

B. 146°   **E.** none of these

C. 225°

---

factor of $\frac{1}{2}$ and translated $\frac{\pi}{2}$ units left or right: $g(x) = \cos 2\left(x \pm \frac{\pi}{2}\right)$

$f(x) = \sin x$ horizontally shrunk by a factor of $\frac{1}{2}$ and translated $\frac{\pi}{4}$ units right: $g(x) = \sin 2\left(x - \frac{\pi}{4}\right)$

$f(x) = \sin x$ horizontally shrunk by a factor of $\frac{1}{2}$ and translated $\frac{3\pi}{4}$ units left: $g(x) = \sin 2\left(x + \frac{3\pi}{4}\right)$

41a. $\frac{p}{2} = 6$ hr 12 min

4:00 AM $-$ 6 hr 12 min $=$ 9:48 PM

4:00 AM $+$ 6 hr 12 min $=$ 10:12 AM

41b. $|a| = \frac{1}{2}(25.87 - 5.47) = 10.2$

$p = 12$ hr 24 min or 12.4 hr $= \dfrac{2\pi}{|b|}$

$|b| = \dfrac{\pi}{6.2}$

$h = 4$, the x-value of the first relative maximum

$k = \frac{1}{2}(25.87 + 5.47) = 15.67$

$d(t) = 10.2 \cos \dfrac{\pi}{6.2}(t - 4) + 15.67$

41c. $d(15) \approx 23.41$ ft

41d.

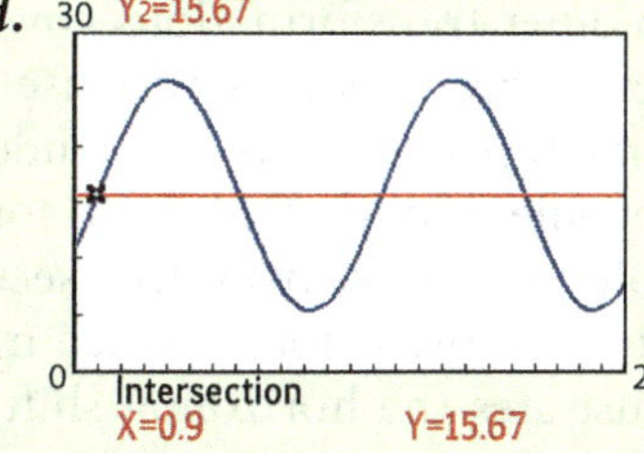

$t = 0.9$ hr $=$ 00:54, or 12:54 AM

$t = 0.9 + 6.2 = 7.1$ hr $=$ 07:06, or 7:06 AM

$t = 7.1 + 6.2 = 13.3$ hr $= 13{:}18$, or 1:18 PM

$t = 13.3 + 6.2 = 19.5$ hr $= 19{:}30$, or 7:30 PM

42a. $|a| = \frac{1}{2}(84 - 21) = 31.5$

$k = \frac{1}{2}(84 + 21) = 52.5$

$p = 12 = \dfrac{2\pi}{|b|}$; $|b| = \dfrac{\pi}{6}$

using cos, $h = 7$, and the x-value of the maximum:

$h(t) = 31.5 \cos \dfrac{\pi}{6}(t - 7) + 52.5$

42b.

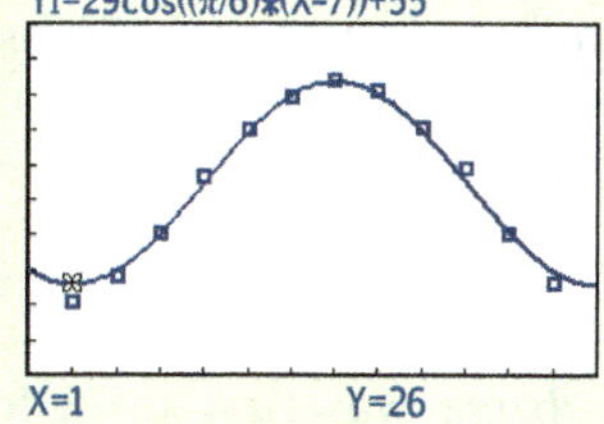

Apr. residual $= |57 - 52.5| = 4.5°$F
Oct. residual $\approx |59 - 52.5| = 6.5°$F

42c. $h(4.5) \approx 61°$F

43a.

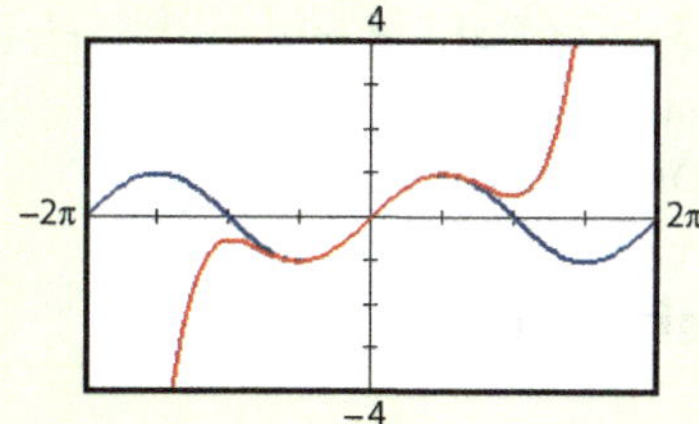

Answers may vary. A good maximum interval is $\left[-\dfrac{5\pi}{6}, \dfrac{5\pi}{6}\right]$.

43b.

X	Y1	Y2	Y3
-1.8	-0.974	-0.985	0.0116
-1.7	-0.992	-0.999	0.0078
-1.6	-1	-1.005	0.0051
-1.5	-0.997	-1.001	0.0033
-1.4	-0.985	-0.987	0.002
-1.3	-0.964	-0.965	0.0012
-1.2	-0.932	-0.933	7E-4
-1.1	-0.891	-0.892	3.8E-4
-1	-0.841	-0.842	2E-4
-0.9	-0.783	-0.783	9.4E-5
-0.8	-0.717	-0.717	4.1E-5

X=-1.7

Since both functions are odd and symmetric with respect to the origin, the upper limit is 1.7.

$[-1.7, 1.7]$

44a.

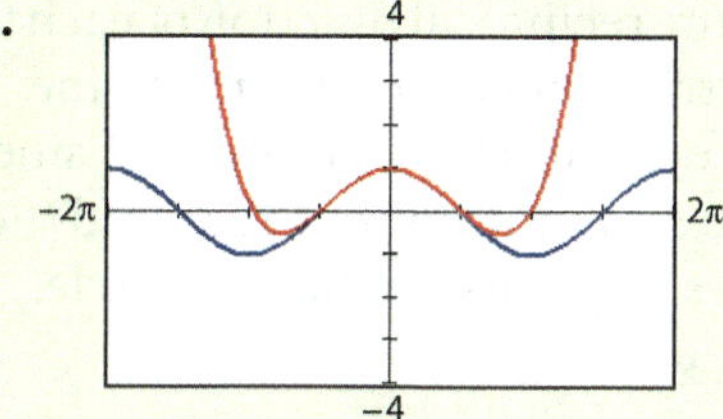

Answers may vary. A good maximum interval is $\left[-\dfrac{3\pi}{4}, \dfrac{3\pi}{4}\right]$.

continued in Answers and Solutions Overflow

4.5 Graphing Other Trigonometric Functions

Objectives

1. To graph the tangent function and its transformations

2. To graph reciprocal trig functions and their transformations

3. To write function rules for transformations of tangent and reciprocal functions

Flash

The term *drone* was first used for an unmanned, radio-controlled aircraft in 1946. These unmanned aircraft were used in WWII for target practice. Today, military drone usage includes target practice for fighter jets and surface-to-air systems, surveillance, supply drops, and bombing missions.

Vocabulary

none

The distance between an observer and a drone flying overhead can be modeled with a reciprocal trigonometric function.

After completing this section, you will be able to

- graph the tangent function and its transformations.
- graph the reciprocal trigonometric functions and their transformations.
- write function rules for transformations of tangent and reciprocal functions.

The tangent, cotangent, cosecant, and secant functions can be expressed as quotients using the sine and cosine functions. Whenever the expression in the denominator is 0, the function is undefined and the graph contains a vertical asymptote. After examining the graphs of tangent functions, the graphs of cotangent, cosecant, and secant functions will be considered as reciprocals of the basic sine, cosine, and tangent functions.

$$\tan x = \frac{\sin x}{\cos x} \qquad \cot x = \frac{\cos x}{\sin x} \qquad \sec x = \frac{1}{\cos x} \qquad \csc x = \frac{1}{\sin x}$$

Since these are periodic functions with either $\cos x$ or $\sin x$ in the denominator, they will have an infinite number of discontinuities. Consider completing the Technology Corner at the end of Section 4.3 to produce animated graphs of these functions.

Using key values from the unit circle, we can sketch a graph of the tangent function. Since $f(x) = \tan x = \frac{\sin x}{\cos x}$, its graph has vertical asymptotes where $\cos x = 0$ (at every odd multiple of $\frac{\pi}{2}$) and x-intercepts where $\sin x = 0$ (at every multiple of π). Note also that the function evaluates to 1 or −1 at odd multiples of $\frac{\pi}{4}$, and that the function increases from $-\infty$ to ∞ within each interval defined by consecutive asymptotes.

x	0	$\frac{\pi}{4}$	$\frac{\pi}{2}$	$\frac{3\pi}{4}$	π	$\frac{5\pi}{4}$	$\frac{3\pi}{2}$	$\frac{7\pi}{4}$	2π
$\tan x$	0	1	*	−1	0	1	*	−1	0

* undefined

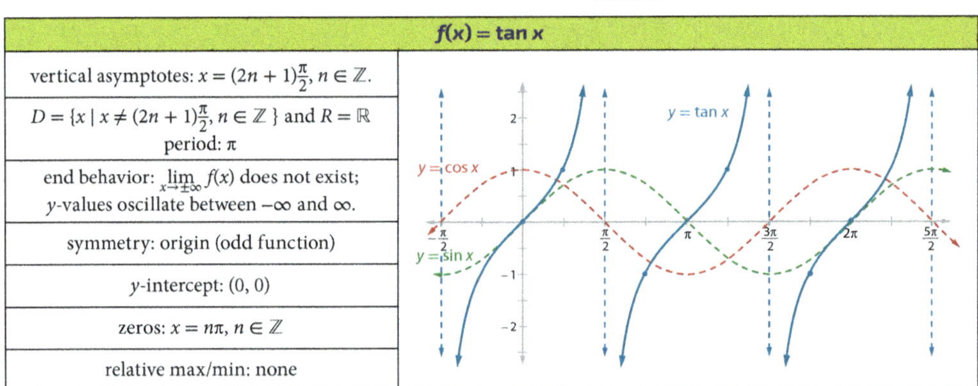

$f(x) = \tan x$	
vertical asymptotes: $x = (2n + 1)\frac{\pi}{2}, n \in \mathbb{Z}$.	
$D = \{x \mid x \neq (2n + 1)\frac{\pi}{2}, n \in \mathbb{Z}\}$ and $R = \mathbb{R}$ period: π	
end behavior: $\lim\limits_{x \to \pm\infty} f(x)$ does not exist; y-values oscillate between $-\infty$ and ∞.	
symmetry: origin (odd function)	
y-intercept: $(0, 0)$	
zeros: $x = n\pi, n \in \mathbb{Z}$	
relative max/min: none	

The graphs of functions of the form $g(x) = a \tan b(x - h) + k$ are transformations of the parent function $f(x) = \tan x$. While amplitude is not defined for tangent functions, their graphs are stretched or shrunk vertically by a factor of $|a|$ and reflected in the x-axis if $a < 0$. Since b causes a horizontal stretch ($|b| < 0$) or shrink ($|b| > 0$), the period is $\frac{\pi}{|b|}$. The constants h and k cause horizontal and vertical shifts as in sinusoids.

PRESENTATION

Lesson Opener

State the reciprocal function of each trigonometric function. Then define each reciprocal function in terms of sine and cosine and in terms of the x- and y-coordinates of points on the unit circle.

1. $\sin x \quad \csc x = \frac{1}{\sin x} = \frac{1}{y}$

2. $\cos x \quad \sec x = \frac{1}{\cos x} = \frac{1}{x}$

3. $\tan x \quad \cot x = \frac{\cos x}{\sin x} = \frac{x}{y}$

Explain that key values from the unit circle will be used to determine the graphs of the parent tangent, cotangent, cosecant, and secant functions. The unit circle on page 704, a handout, or a unit circle from the Internet can be used. Students should focus on the location of zeros and vertical asymptotes.

Introduce the general form for the transformations of the parent tangent function, $y = a \tan b(x - h) + k$, which is similar to other transformations covered in this text. Some sources will use the form $y = a \tan (bx + c) + d$. The students should become familiar with both forms and be able to convert from this second form to the "factored" form, which more clearly illustrates the horizontal shift for the functions.

Using Example 1, encourage the students to identify the vertical asymptotes and zeros, along with a few key points of one period, and to sketch the graph of that period. The other two periods can then be sketched by copying the graph of that period.

Identify key characteristics of the parent cotangent function and demonstrate graphing the transformation of a cotangent function in Example 2.

A close study of the graphs of the polynomial function $f(x) = x^2 - 4$ and its reciprocal function, $g(x) = \frac{1}{x^2 - 4}$, can help the students understand the graphs of the parent reciprocal trig functions, cotangent, secant, and cosecant. Carefully explain each bulleted relationship. Then

Example 1 Graphing a Tangent Function

State the period of $g(x) = \tan \frac{x}{2}$. Then describe how $g(x)$ is obtained by transforming the parent function $f(x) = \tan x$ and sketch three periods of $g(x)$.

Answer

$$\text{period} = \frac{\pi}{|b|} = \frac{\pi}{\left|\frac{1}{2}\right|} = 2\pi$$

The graph is horizontally stretched by a factor of 2.

$f(x) = \tan x$	VA: $x = -\frac{\pi}{2}$	$\left(-\frac{\pi}{4}, -1\right)$	$(0, 0)$	$\left(\frac{\pi}{4}, 1\right)$	VA: $x = \frac{\pi}{2}$
$g(x) = \tan \frac{x}{2}$	VA: $x = -\pi$	$\left(-\frac{\pi}{2}, -1\right)$	$(0, 0)$	$\left(\frac{\pi}{2}, 1\right)$	VA: $x = \pi$

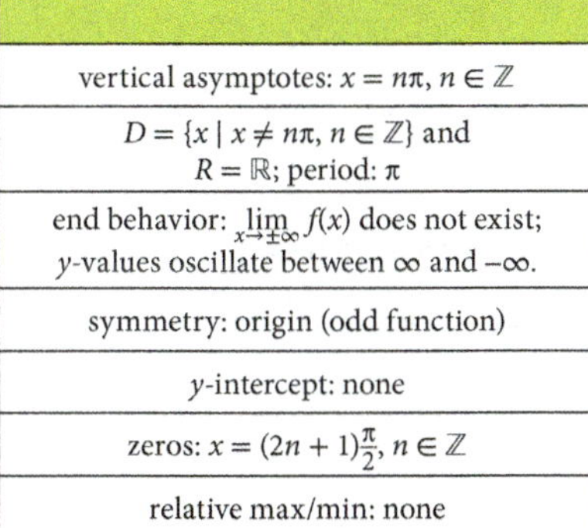

Note:
VA: $x = (2n + 1)\pi$
zeros: $x = 2n\pi$

SKILL ✔ EXERCISE 7

While one cycle of $f(x) = \tan x$ is completed over $-\frac{\pi}{2} \leq x \leq \frac{\pi}{2}$, the corresponding cycle of $g(x) = \tan \frac{x}{2}$ in Example 1 is completed over $-\frac{\pi}{2} \leq \frac{x}{2} \leq \frac{\pi}{2}$ or $-\pi \leq x \leq \pi$.

Using key values from the unit circle, we can also sketch a graph of the cotangent function $f(x) = \cot x = \frac{\cos x}{\sin x}$. The graph has vertical asymptotes where $\sin x = 0$ (at every multiple of π) and x-intercepts where $\cos x = 0$ (at every odd multiple of $\frac{\pi}{2}$). Note that $\cot x = \tan x = \pm 1$ at odd multiples of $\frac{\pi}{4}$. Note that the function decreases from $-\infty$ to ∞ within each interval defined by consecutive asymptotes.

x	0	$\frac{\pi}{4}$	$\frac{\pi}{2}$	$\frac{3\pi}{4}$	π	$\frac{5\pi}{4}$	$\frac{3\pi}{2}$	$\frac{7\pi}{4}$	2π
$\cot x$	*	1	0	−1	*	1	0	−1	*

* undefined

$f(x) = \cot x$	
vertical asymptotes: $x = n\pi, n \in \mathbb{Z}$	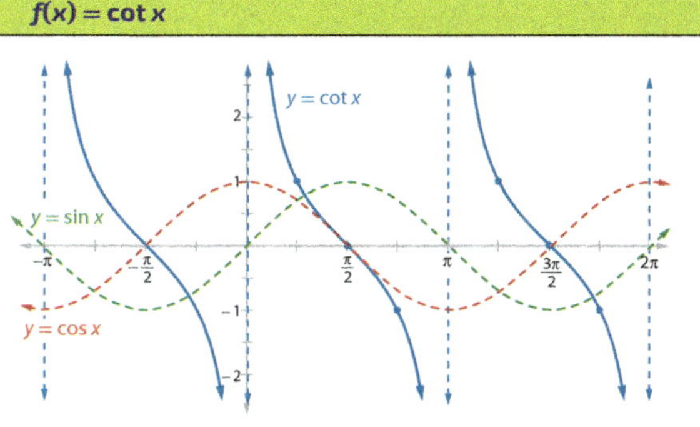
$D = \{x \mid x \neq n\pi, n \in \mathbb{Z}\}$ and $R = \mathbb{R}$; period: π	
end behavior: $\lim\limits_{x \to \pm\infty} f(x)$ does not exist; y-values oscillate between ∞ and $-\infty$.	
symmetry: origin (odd function)	
y-intercept: none	
zeros: $x = (2n + 1)\frac{\pi}{2}, n \in \mathbb{Z}$	
relative max/min: none	

Additional Exercises

Sketch a graph of each function over $[-2\pi, 2\pi]$.

1. $y = \tan \dfrac{3x}{2}$

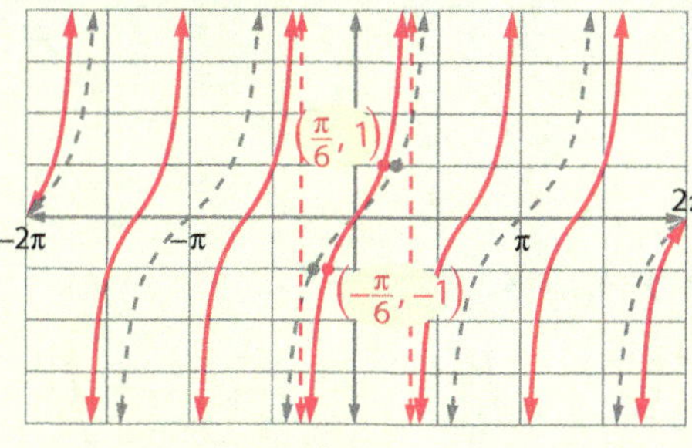

2. $y = \cot \left(x + \dfrac{\pi}{6}\right)$

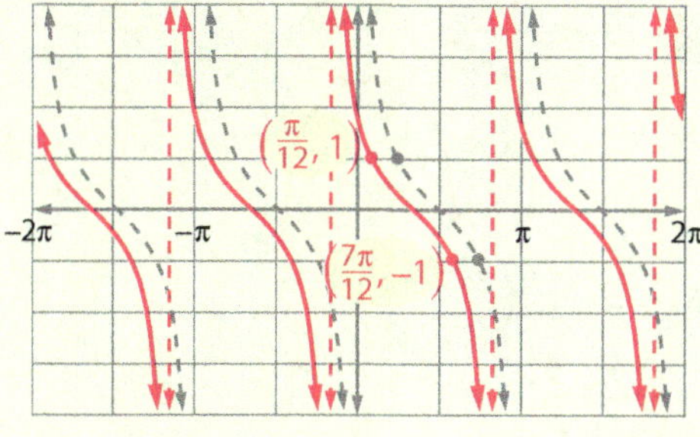

3. $y = \dfrac{1}{2} \csc x$

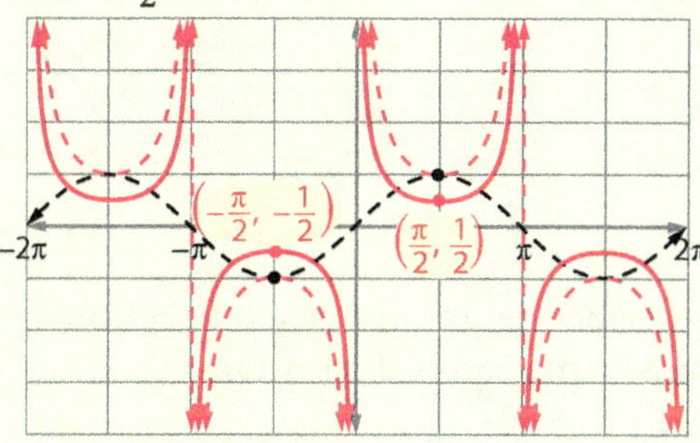

show how these relationships apply in the graphs of $f(x) = \tan x$ and $g(x) = \cot x$. Present the characteristics of the parent cosecant and secant functions. Reinforce these relationships while completing Example 3.

The presence of both a horizontal stretch and a horizontal translation can make graphing the function in Example 4 a challenge. Using the graph of the related sine or cosine function is demonstrated. Consider demonstrating the following alternative approaches for determining the graph of the transformed function.

- Determine the graph for one period, the transformation of $f(x) = \sec x$ over $\left(-\frac{\pi}{2}, \frac{3\pi}{2}\right)$, and then repeat that pattern over the desired interval.

- After describing the transformations, have the students sequentially graph the parent function, the result of the horizontal stretch, and then the final result with the phase shift.

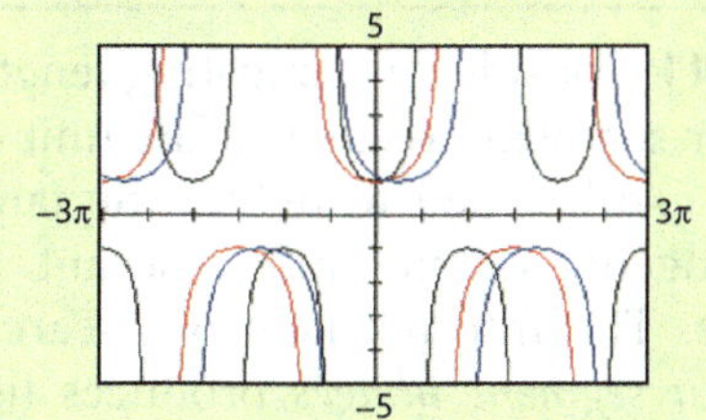

Interactive Activity After the students have sketched the graph of the basic tangent function, consider using the Technology Corner in Section 4.4 to guide them in creating an animated tangent graph. Other animated graphs generated from the unit circle can be found using the Internet keyword search *unit circle tangent graph*.

Motivational Idea Challenge the students to experiment by creating some interesting trigonometric equations that will produce various designs and patterns, using the graphing calculator or Internet graphing technology. Consider demonstrating some or all of the following functions to encourage exploration.

4. $y = 2 \tan \dfrac{x}{3}$

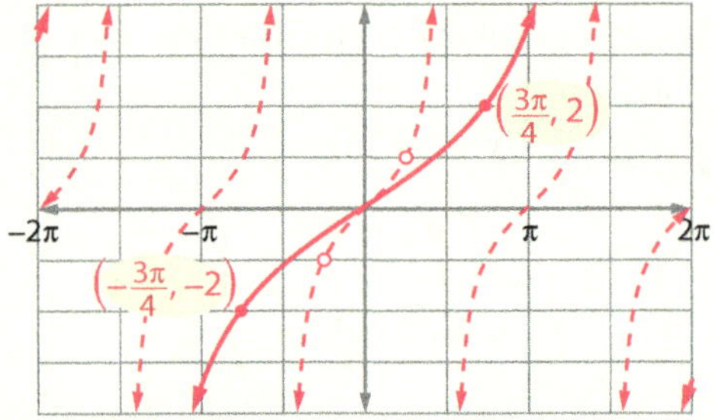

5. $y = -\sec \dfrac{\pi}{4}x$

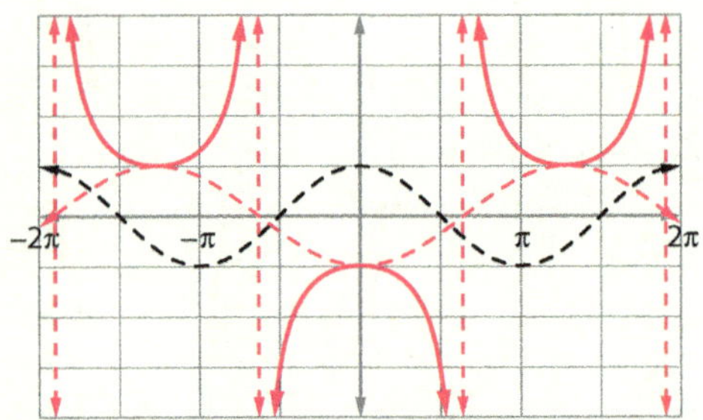

6. $y = \csc\left(\dfrac{3x}{4} - \dfrac{\pi}{4}\right)$

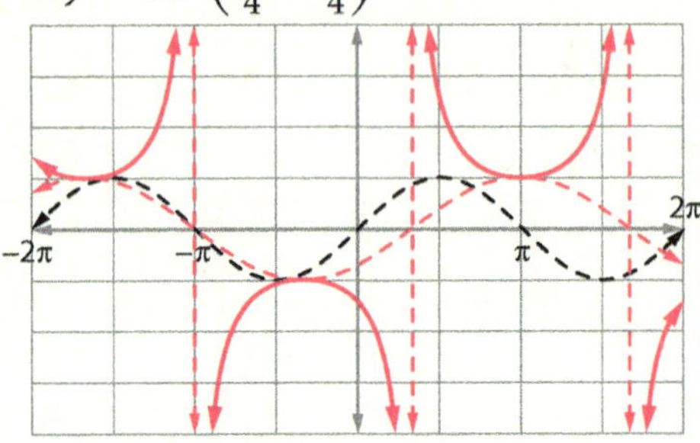

Describe the graph of each function as a transformation of $f(x) = \tan x$.

7. $g(x) = -2\tan(x - 3) + 5$

The graph of $f(x)$ is shifted 3 units right, stretched vertically by a factor of 2, reflected across the x-axis, and shifted 5 units up.

8. $g(x) = 3\tan\left(3x + \dfrac{\pi}{2}\right)$

The graph of $f(x)$ is horizontally shrunk by a factor of $\frac{1}{3}$, shifted $\frac{\pi}{6}$ units left, and vertically stretched by a factor of 3.

Write a rule for $g(x)$, the described transformation of $f(x) = \sec x$.

9. The graph of $f(x)$ is vertically shrunk by $\frac{1}{3}$, reflected in the x-axis, and horizontally stretched so its period is 8. $g(x) = -\frac{1}{3}\sec \frac{\pi}{4}x$

10. The graph of $f(x)$ is vertically stretched by 4, horizontally shrunk so its period is 6, and shifted 3 units left and 2 units up.

$g(x) = 4\sec \frac{\pi}{3}(x + 3) + 2$ or

$g(x) = 4\sec\left(\frac{\pi}{3}x + \pi\right) + 2$

State the period and phase shift of $g(x) = -2\cot\left(x + \frac{\pi}{4}\right)$. Then describe how $g(x)$ is obtained by transforming the parent function $f(x) = \cot x$ and sketch the graph of $g(x)$ over $[-2\pi, 2\pi]$.

Answer

period $= \dfrac{\pi}{|b|} = \dfrac{\pi}{|1|} = \pi$; phase shift $= h = -\dfrac{\pi}{4}$

The graph is shifted $\frac{\pi}{4}$ units left, reflected in the y-axis and stretched vertically by a factor of 2.

$y_1 = \cot x$	VA: $x = 0$	$\left(\frac{\pi}{4}, 1\right)$	$\left(\frac{\pi}{2}, 0\right)$	$\left(\frac{3\pi}{4}, -1\right)$	VA: $x = \pi$
$y_2 = \cot\left(x + \frac{\pi}{4}\right)$	VA: $x = -\frac{\pi}{4}$	$(0, 1)$	$\left(\frac{\pi}{4}, 0\right)$	$\left(\frac{\pi}{2}, -1\right)$	VA: $x = \frac{3\pi}{4}$
$y_3 = -2\cot\left(x + \frac{\pi}{4}\right)$	VA: $x = -\frac{\pi}{4}$	$(0, -2)$	$\left(\frac{\pi}{4}, 0\right)$	$\left(\frac{\pi}{2}, 2\right)$	VA: $x = \frac{3\pi}{4}$

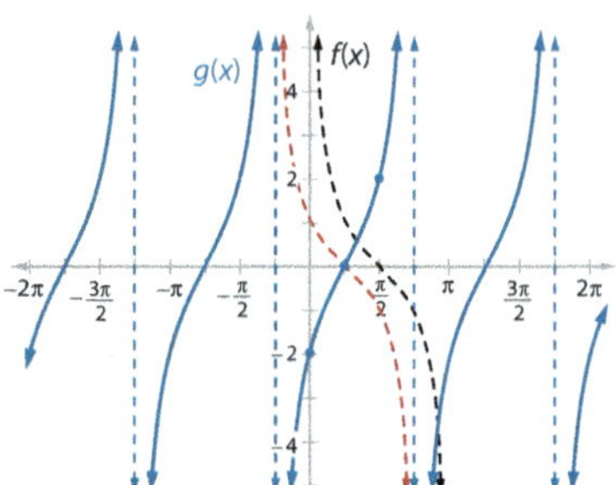
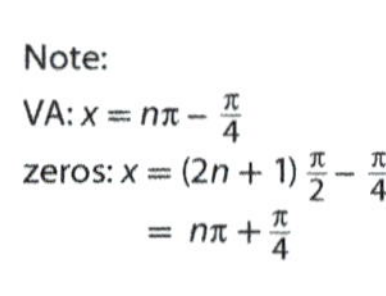

Note:
VA: $x = n\pi - \frac{\pi}{4}$
zeros: $x = (2n + 1)\frac{\pi}{2} - \frac{\pi}{4}$
$= n\pi + \frac{\pi}{4}$

SKILL ✔ EXERCISE 23

Since one cycle of $f(x) = \cot x$ is completed over $0 \le x \le \pi$, the corresponding cycle of $g(x) = -2\cot\left(x + \frac{\pi}{4}\right)$ in Example 2 is completed over $0 \le x + \frac{\pi}{4} \le \pi$ or $-\frac{\pi}{4} \le x \le \frac{3\pi}{4}$.

Understanding general principles relating the graphs of $f(x)$ and its reciprocal function $g(x) = \frac{1}{f(x)}$ can help you graph secant, cosecant, and cotangent functions. Analyzing the graphs of $f(x) = x^2 - 4$ and its reciprocal $g(x) = \frac{1}{f(x)} = \frac{1}{x^2 - 4}$ reveals several key relationships.

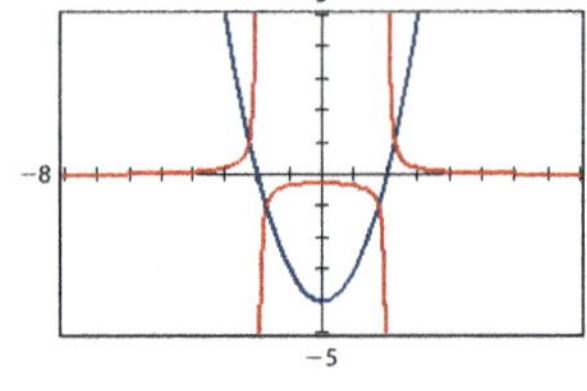

- If $f(x) = 0$, then $\frac{1}{f(x)}$ is undefined and $g(x)$ has a vertical asymptote.
- As $f(x) \to 0^+$, $\frac{1}{f(x)} \to \infty$, and as $f(x) \to 0^-$, $\frac{1}{f(x)} \to -\infty$.
- As $f(x) \to \infty$, $\frac{1}{f(x)} \to 0^+$, and as $f(x) \to -\infty$, $\frac{1}{f(x)} \to 0^-$.
- If $f(x) = 1$, then $\frac{1}{f(x)} = 1$, and if $f(x) = -1$, then $\frac{1}{f(x)} = -1$.
- If $f(x) > 1$, then $0 < \frac{1}{f(x)} < 1$, and if $0 < f(x) < 1$, then $\frac{1}{f(x)} > 1$.
- If $f(x) < -1$, then $-1 < \frac{1}{f(x)} < 0$, and if $-1 < f(x) < 0$, then $\frac{1}{f(x)} < -1$.

Promote the sharing of ideas and findings.

$y = \sin x + \sin 5\pi x$

$y = \tan x + \sin 7\pi x$

$y = \sin 2x + \csc 30\pi x$

$y = 2\sin\left(\dfrac{1}{\sin x + 1}\right)$

TIPS

Ex. 41 Consider investigating lengths of other segments related to the unit circle that can be used to define the tangent, cotangent, secant, and cosecant functions. The Internet keyword search *cosecant segment images* produces figures that provide a variety of segments that could be used.

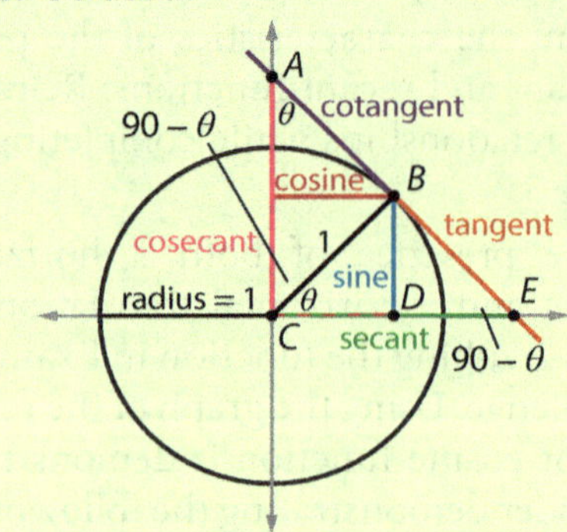

You can see how these principles apply to $f(x) = \tan x$ and $g(x) = \cot x = \frac{1}{\tan x}$. Since $y = \csc x = \frac{1}{\sin x}$ and $y = \sec x = \frac{1}{\cos x}$, their graphs can be drawn as reciprocals of the basic sinusoids. The relative maximums of the related sinusoid (its crests) become relative minimums of the reciprocal function and the relative minimums of the sinusoid (its troughs) become the relative maximums of the reciprocal function. Note also that if $f(x) = \pm\frac{1}{2}$, then $\frac{1}{f(x)} = \pm 2$.

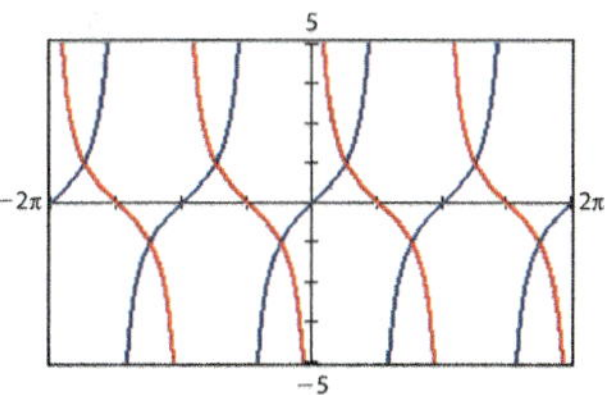

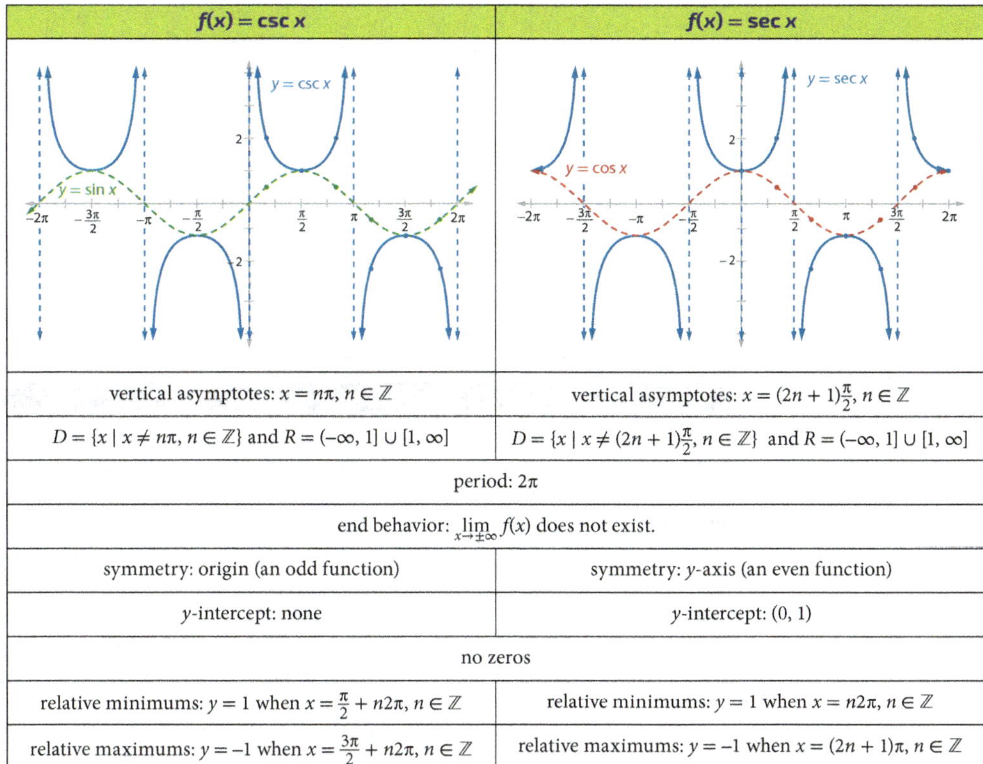

$f(x) = \csc x$	$f(x) = \sec x$
vertical asymptotes: $x = n\pi,\ n \in \mathbb{Z}$	vertical asymptotes: $x = (2n + 1)\frac{\pi}{2},\ n \in \mathbb{Z}$
$D = \{x \mid x \neq n\pi,\ n \in \mathbb{Z}\}$ and $R = (-\infty, 1] \cup [1, \infty)$	$D = \{x \mid x \neq (2n + 1)\frac{\pi}{2},\ n \in \mathbb{Z}\}$ and $R = (-\infty, 1] \cup [1, \infty)$
period: 2π	
end behavior: $\lim\limits_{x \to \pm\infty} f(x)$ does not exist.	
symmetry: origin (an odd function)	symmetry: y-axis (an even function)
y-intercept: none	y-intercept: $(0, 1)$
no zeros	
relative minimums: $y = 1$ when $x = \frac{\pi}{2} + n2\pi,\ n \in \mathbb{Z}$	relative minimums: $y = 1$ when $x = n2\pi,\ n \in \mathbb{Z}$
relative maximums: $y = -1$ when $x = \frac{3\pi}{2} + n2\pi,\ n \in \mathbb{Z}$	relative maximums: $y = -1$ when $x = (2n + 1)\pi,\ n \in \mathbb{Z}$

Graphing a related basic trig function can help you draw the graphs of tranformed reciprocal functions of the form $g(x) = a \csc b(x - h) + k$ or $g(x) = a \sec b(x - h) + k$.

Example 3 Graphing a Cosecant Function

Sketch a graph of $g(x) = 2 \csc x$ over $[-2\pi, 2\pi]$. Then state the period, domain, and range of the function.

Answer

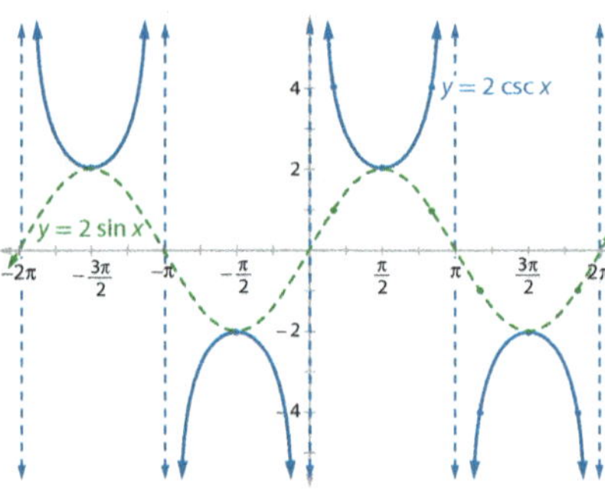

1. Graph $y = 2 \sin x$ over $[-2\pi, 2\pi]$.

2. Draw vertical asymptotes where $y = 2 \sin x = 0$ (at $x = n\pi$).

3. Plot relative minimums and maximums of $g(x) = 2\left(\dfrac{1}{\sin x}\right)$ at crests and troughs of $y = 2 \sin x$.

4. When $\sin x = \pm\dfrac{1}{2}$ (at $x = \dfrac{\pi}{6}, \dfrac{5\pi}{6}, \dfrac{7\pi}{6}, \dfrac{11\pi}{6}, \ldots$), $g(x) = 2\left(\dfrac{1}{\sin x}\right) = \pm 4$. Plotting these points shows how quickly $g(x) \to \pm\infty$ as $\sin x \to 0$.

5. Sketch the graph of $g(x) = 2 \csc x$. Note that the graph is a vertical stretch of $f(x) = \csc x$ by a factor of 2.

The period is 2π.
$D = \{x \mid x \neq n\pi, n \in \mathbb{Z}\}; R = (-\infty, -2] \cup [2, \infty)$

6. State the period, domain, and range.

SKILL ✔ EXERCISE 9

Example 4 Graphing a Secant Function

State the period and phase shift of $g(x) = \sec\left(\dfrac{2}{3}x - \dfrac{\pi}{6}\right)$ and describe how $g(x)$ is obtained by transforming the parent function $f(x) = \sec x$. Sketch a graph illustrating two periods of $g(x)$ and verify your graph using technology.

Answer

$g(x) = \sec\left(\dfrac{2}{3}x - \dfrac{\pi}{6}\right) = \sec\dfrac{2}{3}\left(x - \dfrac{\pi}{4}\right)$

1. Factor out the coefficient of x.

$\text{period} = \dfrac{2\pi}{\left|\frac{2}{3}\right|} = 3\pi$; phase shift: $h = \dfrac{\frac{\pi}{6}}{\frac{2}{3}} = \dfrac{\pi}{4}$

2. Find the period and phase shift using $p = \dfrac{2\pi}{|b|}$ and $h = \dfrac{c}{b}$.

The graph of $f(x)$ is stretched horizontally by a factor of $\dfrac{3}{2}$ and shifted $\dfrac{\pi}{4}$ units right.

3. Describe the transformations.

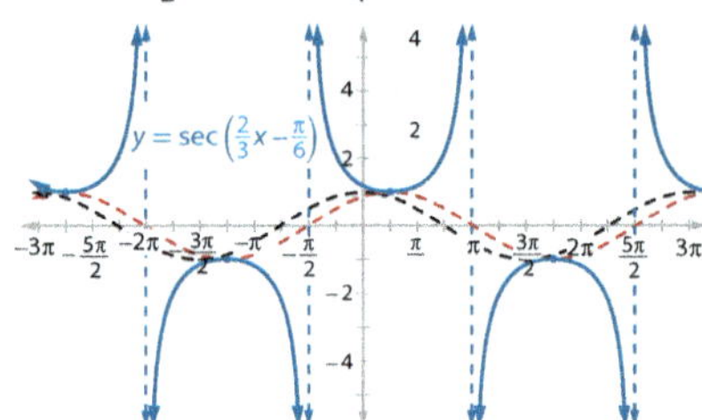

4. Graph $y = \cos\left(\dfrac{2}{3}x - \dfrac{\pi}{6}\right)$ over $[-3\pi, 3\pi]$ by shifting the graph of $y = \cos\dfrac{2}{3}x$ to the right by $\dfrac{\pi}{4}$ units.

5. Graph $g(x) = \sec\left(\dfrac{2}{3}x - \dfrac{\pi}{6}\right)$.

  a. Vertical asymptotes of $g(x)$ occur at zeros of the related sinusoid.

  b. Relative minimums and maximums of $g(x)$ occur at the crests and troughs of the related sinusoid, respectively.

CONTINUED ➡

Solutions

❯ A. Exercises

6. $p = \dfrac{2\pi}{\frac{1}{2}} = 4\pi$

horizontally stretched by a factor of 2; VA: $x = \pi \pm n2\pi, n \in \mathbb{Z}$

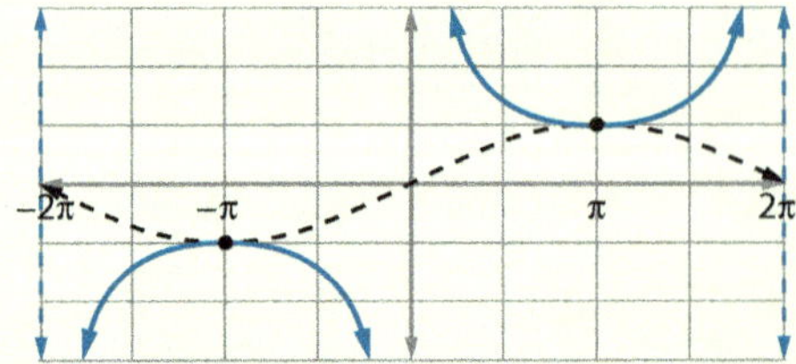

7. $p = \dfrac{\pi}{1} = \pi$

vertically shrunk by a factor of $\dfrac{1}{2}$
VA: $x = (2n + 1)\dfrac{\pi}{2}, n \in \mathbb{Z}$

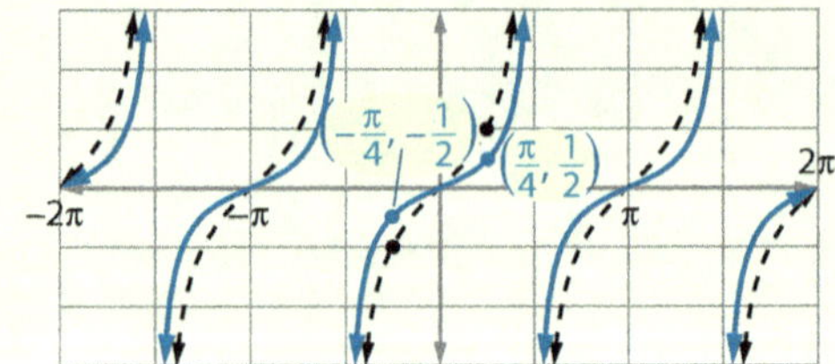

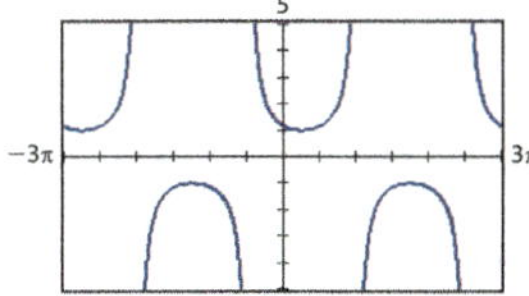

6. Graph $Y_1 = 1/\cos(2x/3 - \pi/6)$.

SKILL ✔ EXERCISE 27

Since one cycle of $f(x) = \sec x$ is defined by the vertical asymptotes at $-\frac{\pi}{2}$ and $\frac{3\pi}{2}$, the vertical asymptotes defining the corresponding cycle of $g(x) = \sec\left(\frac{2}{3}x - \frac{\pi}{6}\right)$ in Example 4 are found by solving $\frac{2}{3}x - \frac{\pi}{6} = -\frac{\pi}{2}$ and $\frac{2}{3}x - \frac{\pi}{6} = \frac{3\pi}{2}$ to get $x = -\frac{\pi}{2}$ and $x = \frac{5\pi}{2}$.

Ernst Chladni, the Father of Acoustics, found that sound could be visualized by sprinkling sand on a metal plate and then bowing the plate with a violin bow. The sand moves to the areas on the plate that vibrate the least (the nodes), forming different patterns at different frequencies. Some violin makers still utilize Chladni patterns when designing their instruments.

❯ A. Exercises

Use the special values from the unit circle to state the exact coordinates of each point on the graph.

1. $y = \tan x$

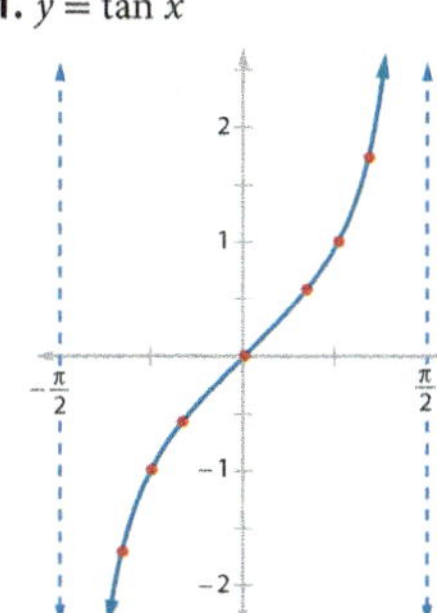

1. $\left(-\frac{\pi}{3}, -\sqrt{3}\right), \left(-\frac{\pi}{4}, -1\right),$ $\left(-\frac{\pi}{6}, -\frac{\sqrt{3}}{3}\right), (0, 0),$ $\left(\frac{\pi}{6}, \frac{\sqrt{3}}{3}\right), \left(\frac{\pi}{4}, 1\right), \left(\frac{\pi}{3}, \sqrt{3}\right)$

2. $y = \cot x$

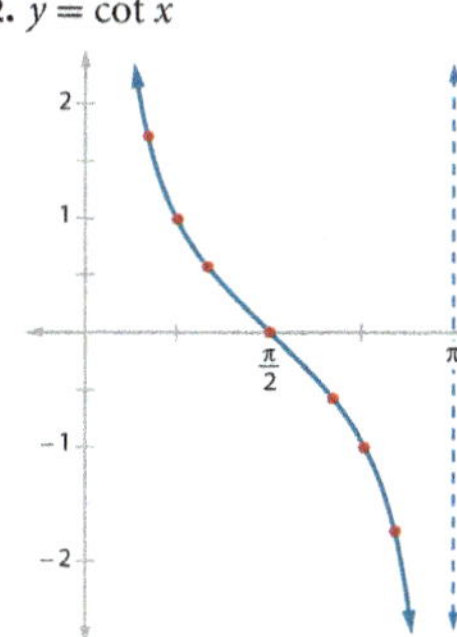

2. $\left(\frac{\pi}{6}, \sqrt{3}\right), \left(\frac{\pi}{4}, 1\right), \left(\frac{\pi}{3}, \frac{\sqrt{3}}{3}\right),$ $\left(\frac{\pi}{2}, 0\right), \left(\frac{2\pi}{3}, -\frac{\sqrt{3}}{3}\right),$ $\left(\frac{3\pi}{4}, -1\right), \left(\frac{5\pi}{6}, -\sqrt{3}\right)$

3. $\left(\frac{\pi}{6}, 2\right), \left(\frac{\pi}{4}, \sqrt{2}\right), \left(\frac{\pi}{3}, \frac{2\sqrt{3}}{3}\right),$ $\left(\frac{\pi}{2}, 1\right), \left(\frac{2\pi}{3}, \frac{2\sqrt{3}}{3}\right), \left(\frac{3\pi}{4}, \sqrt{2}\right),$ $\left(\frac{5\pi}{6}, 2\right), \left(\frac{7\pi}{6}, -2\right),$ $\left(\frac{5\pi}{4}, -\sqrt{2}\right), \left(\frac{4\pi}{3}, -\frac{2\sqrt{3}}{3}\right),$ $\left(\frac{3\pi}{2}, -1\right), \left(\frac{5\pi}{3}, -\frac{2\sqrt{3}}{3}\right),$ $\left(\frac{7\pi}{4}, -\sqrt{2}\right), \left(\frac{11\pi}{6}, -2\right)$

4. $\left(-\frac{\pi}{3}, 2\right), \left(-\frac{\pi}{4}, \sqrt{2}\right),$ $\left(-\frac{\pi}{6}, \frac{2\sqrt{3}}{3}\right), (0, 1),$ $\left(\frac{\pi}{6}, \frac{2\sqrt{3}}{3}\right), \left(\frac{\pi}{4}, \sqrt{2}\right),$ $\left(\frac{\pi}{3}, 2\right), \left(\frac{2\pi}{3}, -2\right),$ $\left(\frac{3\pi}{4}, -\sqrt{2}\right), \left(\frac{5\pi}{6}, -\frac{2\sqrt{3}}{3}\right),$ $(\pi, -1), \left(\frac{7\pi}{6}, -\frac{2\sqrt{3}}{3}\right),$ $\left(\frac{5\pi}{4}, -\sqrt{2}\right), \left(\frac{4\pi}{3}, -2\right)$

3. $y = \csc x$

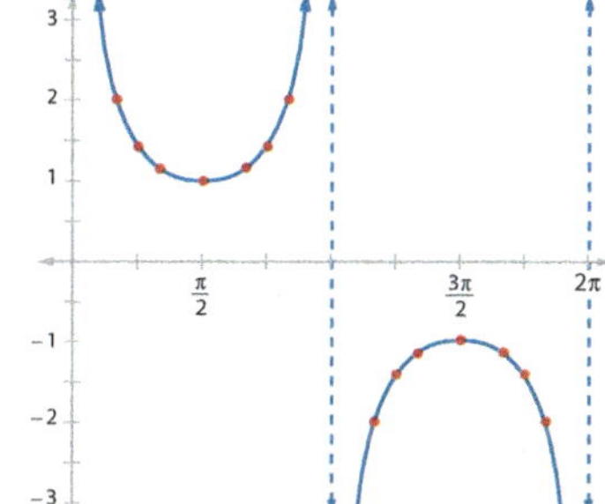

4. $y = \sec x$

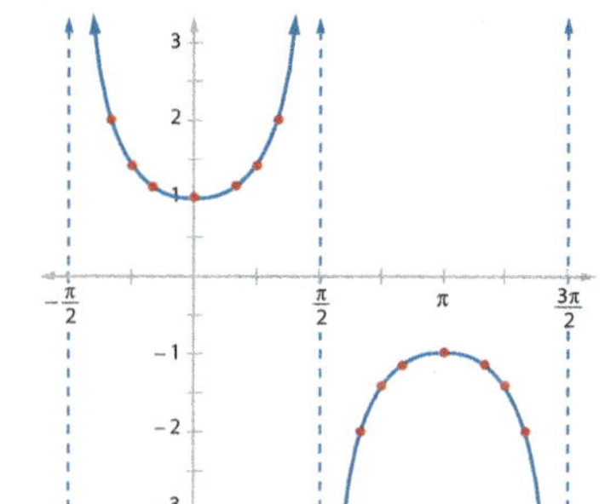

8. $p = \frac{\pi}{1} = \pi; h = \frac{\pi}{4}$
VA: $x = \frac{\pi}{4} + n\pi, n \in \mathbb{Z}$

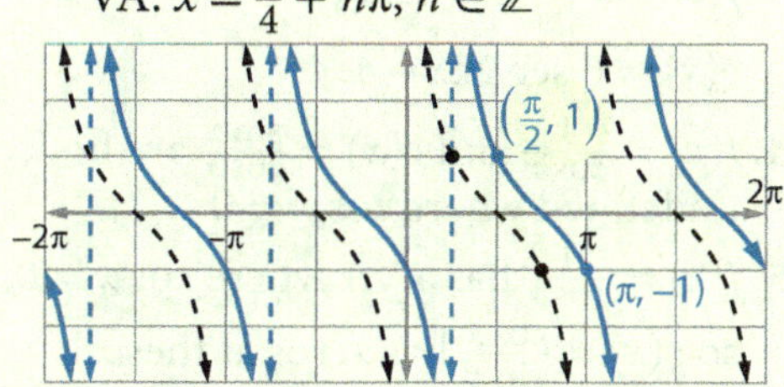

9. $p = \frac{2\pi}{\pi} = 2$
horizontally shrunk by a factor of π
VA: $x = n, n \in \mathbb{Z}$

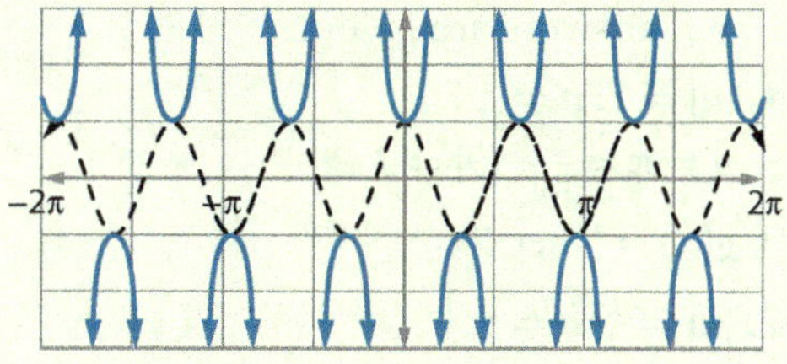

❯ B. Exercises

21. $p = \frac{\pi}{1} = \pi$; vertically shrunk by a factor of $\frac{1}{2}$ and reflected across the x-axis

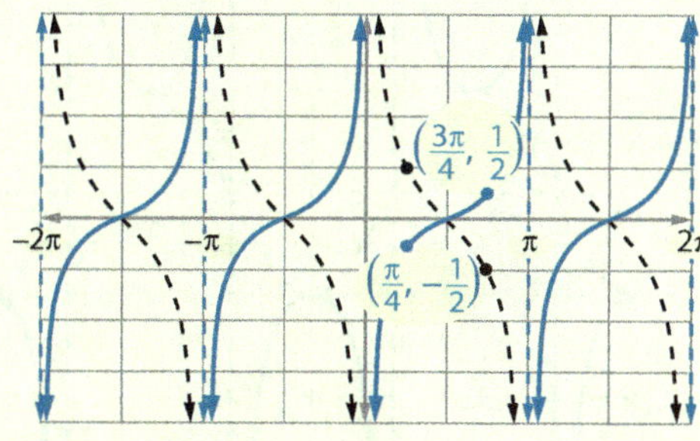

22. $p = \frac{\pi}{\frac{1}{2}} = 2\pi$; stretched vertically and horizontally by a factor of 2

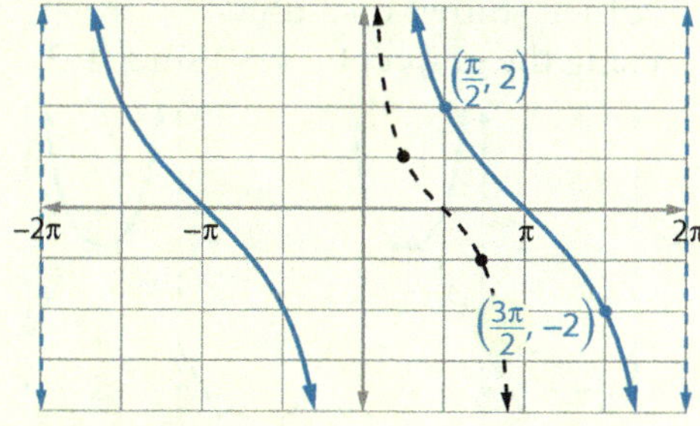

23. $p = \frac{\pi}{\frac{1}{2}} = 2\pi$; stretched vertically and horizontally by a factor of 2

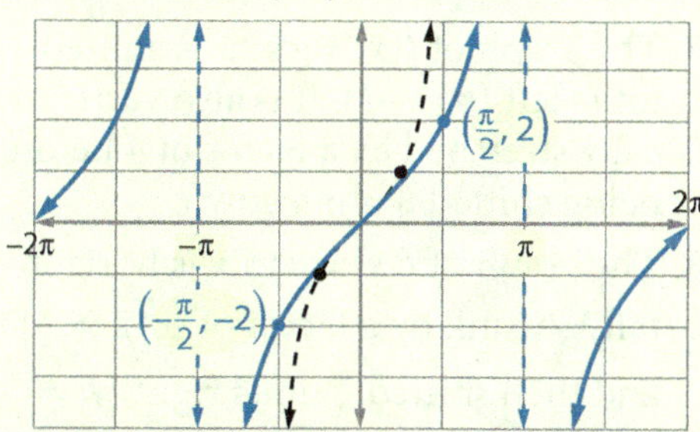

24. $p = \frac{2\pi}{1} = 2\pi; h = \frac{\pi}{2}$
using the graph of $y = \sin\left(x - \frac{\pi}{2}\right)$:

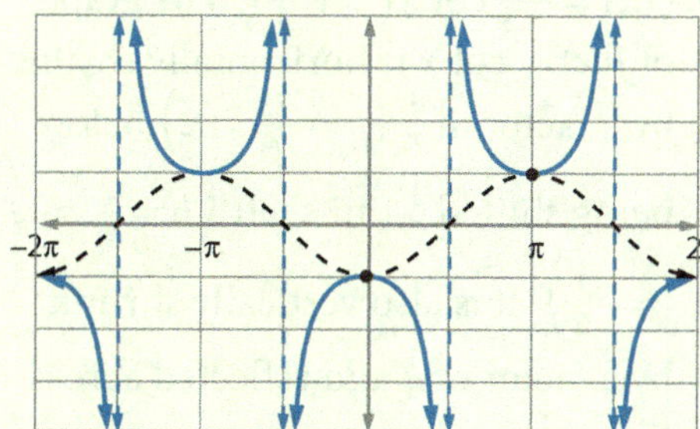

25. $p = \frac{2\pi}{1} = 2\pi$; $h = \pi$; vertically stretched by a factor of 2 using the graph of $y = 2\cos(x + \pi)$:

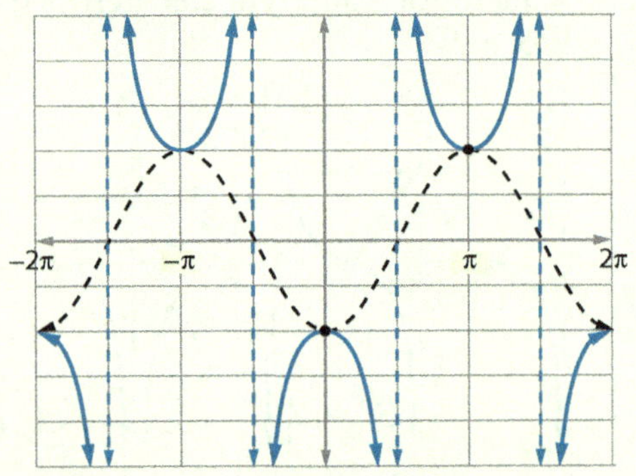

26. $p = \frac{2\pi}{1} = 2\pi$; reflected in the x-axis and vertically stretched by a factor of 2; then shifted up 3 units using the graph of $y = -2\sin x + 3$:

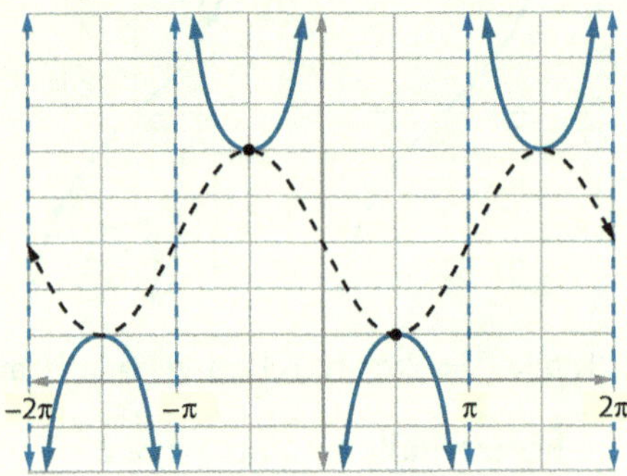

27. The graph of $f(x) = \csc x$ is shifted 2 units left ($h = -2$). It is also vertically stretched by a factor of 4 before being shifted 8 units down.

28. The graph of $f(x) = \tan x$ is horizontally shrunk by a factor of $\frac{1}{3}$ $\left(p = \frac{\pi}{3}\right)$ and then shifted $\frac{\pi}{3}$ units right $\left(h = \frac{\pi}{3}\right)$. It is also vertically stretched by a factor of 5.

29. $g(x) = -\frac{1}{3}\cot\pi\left(x + \frac{1}{3}\right)$; The graph of $f(x) = \cot x$ is horizontally shrunk by a factor of $\frac{1}{\pi}$ $\left(p = \frac{2\pi}{\pi} = 2\right)$ before being shifted $\frac{1}{3}$ units left $\left(h = \frac{c}{b} = \frac{-\frac{\pi}{3}}{\pi}\right)$ $= -\frac{1}{3}$. It is also vertically shrunk by a factor of $\frac{1}{3}$ and reflected across the x-axis.

30. $g(x) = -0.2\sec\frac{1}{5}\left(x - \frac{5\pi}{2}\right) + 0.6$; The graph of $f(x) = \sec x$ is horizontally stretched by a factor of 5 $\left(p = \frac{2\pi}{\frac{1}{5}} = 10\pi\right)$ before being shifted $\frac{5\pi}{2}$ units right. It is also vertically shrunk by a factor of 0.2 and re-

---

5. State the period, vertical asymptotes, and zeros of each function's graph.

   a. sine   **b.** cosine   **c.** tangent

   d. cotangent   **e.** cosecant   **f.** secant

Sketch a graph of each function over $[-2\pi, 2\pi]$.

6. $y = \csc\frac{x}{2}$ **7.** $y = \frac{1}{2}\tan x$

8. $y = \cot\left(x - \frac{\pi}{4}\right)$ **9.** $y = \sec\pi x$

Select all the functions that match each description.

  A. $y = \sin x$ **C.** $y = \tan x$ **E.** $y = \csc x$

  B. $y = \cos x$ **D.** $y = \cot x$ **F.** $y = \sec x$

10. has no zeros   E, F

11. has a period of π C, D

12. has vertical asymptotes where $\sin\theta = 0$ D, E

13. has zeros at odd multiples of $\frac{\pi}{2}$ B, D

14. has zeros where $y = \csc x$ has vertical asymptotes   A

❭ B. Exercises

Without using technology, match each function with its graph.

15. $y = \tan 2x$ C   **16.** $y = 2\tan x$ E

17. $y = \csc 2x$ A   **18.** $y = 2\sec\frac{x}{2}$ B

19. $y = \csc(x - \pi)$ D   **20.** $y = \tan\left(x - \frac{\pi}{2}\right)$ F

A. 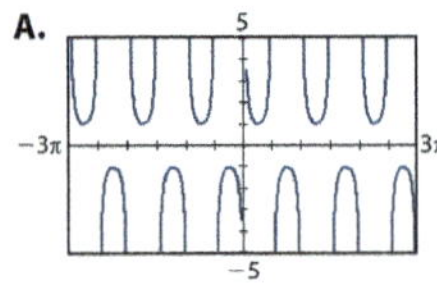   **D.** 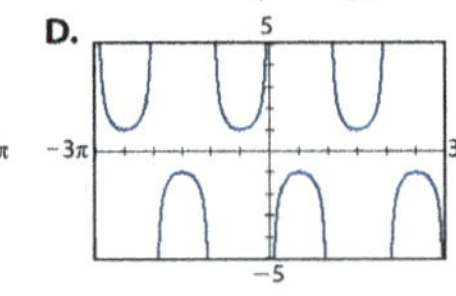

B. 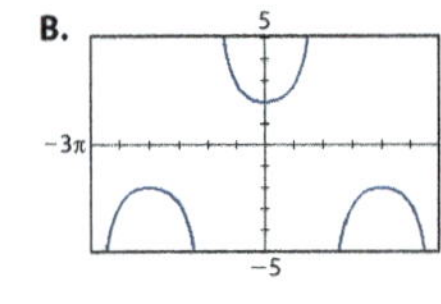   **E.** 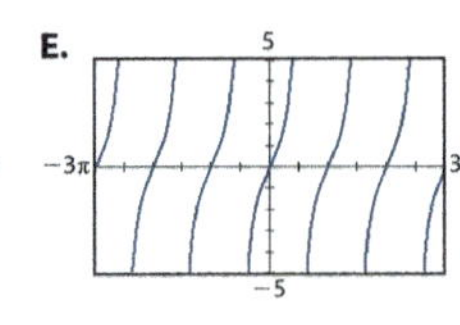

C. 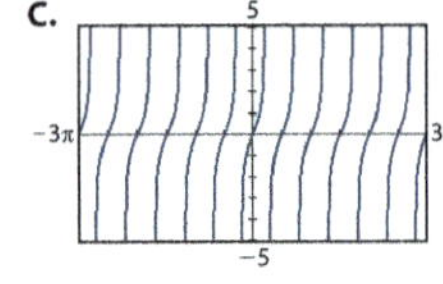   **F.** 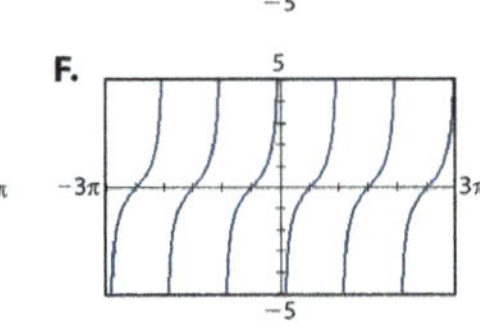

5a. 2π, none, $n\pi$ **5d.** π, $x = n\pi$, $(2n+1)\frac{\pi}{2}$

5b. 2π, none, $(2n+1)\frac{\pi}{2}$ **5e.** 2π, $x = n\pi$, none

5c. π, $x = (2n+1)\frac{\pi}{2}$, $n\pi$ **5f.** 2π, $x = (2n+1)\frac{\pi}{2}$, none

---

Sketch the graph of each function from -2π to 2π.

21. $y = -\frac{1}{2}\cot x$ **22.** $y = 2\cot\frac{x}{2}$

23. $y = 2\tan\frac{x}{2}$ **24.** $y = \csc\left(x - \frac{\pi}{2}\right)$

25. $y = 2\sec(x + \pi)$ **26.** $y = -2\csc x + 3$

Describe the graph of each function $g(x)$ as a transformation of its parent function.

27. $g(x) = 4\csc(x + 2) - 8$

28. $g(x) = 5\tan 3\left(x - \frac{\pi}{3}\right)$

29. $g(x) = -\frac{1}{3}\cot\left(\pi x + \frac{\pi}{3}\right)$

30. $g(x) = -0.2\sec\left(\frac{x}{5} - \frac{\pi}{2}\right) + 0.6$

Write a rule for $g(x)$, the described transformation of $f(x)$. Answers may vary.

31. The function $f(x) = \sec x$ is vertically stretched by 7 and horizontally shrunk so its period is π. $g(x) = 7\sec 2x$

32. The function $f(x) = \tan x$ is horizontally shrunk so its period is $\frac{\pi}{3}$, shifted 2 units left, and vertically shrunk by a factor of $\frac{1}{4}$. $g(x) = \frac{1}{4}\tan(3x + 6)$

33. The function $f(x) = \sec x$ is horizontally shrunk so its period is 1, vertically stretched by a factor of 6, and shifted 4 units down. $g(x) = 6\sec 2\pi x - 4$

34. Analyze: Which parent trigonometric function has asymptotes wherever $f(x) = \sec x$ has an asymptote? Explain your reasoning. $y = \tan x$

35. Analyze: Which parent trigonometric function has zeros where $f(x) = \sec x$ has an asymptote? Explain your reasoning. $y = \cos x$ or $y = \cot x$

36. Analyze: State the value of $g(x) = \tan x$ for any value of x where the graph of $f(x) = \csc x$ has a vertical asymptote. Explain your reasoning. $g(x) = 0$

37. When Rebecca boards a glass elevator on the second floor of a building, the angle of depression as she views her niece standing on the first floor 20 ft from the base of the elevator is 35°.

  a. Write a function that models the elevator's distance above the first floor in terms of Rebecca's angle of depression. $d(\theta) = \frac{20}{\tan(90 - \theta)}$

  b. Find the elevator's height above ground level (to the nearest foot).   14 ft

  c. From what height (to the nearest foot) and what floor will Rebecca be viewing her niece when the angle of depression is 71°? Assume each story in the building has the same height.   58 ft; the fifth floor

---

flected across the x-axis before being shifted up 0.6 units.

31–33. Answers may vary. Sample answers are given.

31. $|a| = 7$; $a = \pm 7$

$p = \pi = \frac{2\pi}{|b|}$; $|b| = 2$

$g(x) = 7\sec 2x$

32. $|a| = \frac{1}{4}$; $a = \pm\frac{1}{4}$

$p = \frac{\pi}{3} = \frac{\pi}{|b|}$; $b = \pm 3$; $h = -2$

$g(x) = \frac{1}{4}\tan 3(x + 2)$ or

$g(x) = \frac{1}{4}\tan(3x + 6)$

33. $|a| = 6$; $a = \pm 6$

$p = 1 = \frac{2\pi}{|b|}$; $b = \pm 2\pi$; $k = -4$

$g(x) = 6\sec 2\pi x - 4$

34. $f(x) = \frac{1}{\cos x}$ and $g(x) = \frac{\sin x}{\cos x}$ are both undefined where $\cos x = 0$.

35. $f(x) = \frac{1}{\cos x}$ has a VA where $\cos x = 0$, so $g(x) = \frac{\cos x}{\sin x}$ has zeros at these values of x.

36. Vertical asymptotes of $f(x) = \frac{1}{\sin x}$ occur where $\sin x = 0$, so $g(x) = \frac{\sin x}{\cos x} = 0$ for those values of x.

37a. $\cot\theta = \frac{d}{20}$; $d(\theta) = 20\cot\theta$ or $\tan(90 - \theta) = \frac{20}{d}$; $d(\theta) = \frac{20}{\tan(90 - \theta)}$

---

38. A drone performs a low-altitude airdrop of medical supplies at 80 ft and passes directly over a marker (M) in the drop zone.

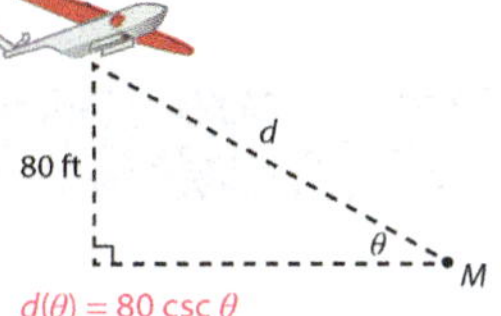

a. Write a function that $d(\theta) = 80 \csc \theta$ models the distance from the marker to the drone in terms of the angle of elevation from the marker.

b. Graph the function from 0° to 90° and describe how the distance changes over this interval.

c. Find the distance when the drone is sighted at a 45° angle of elevation. ≈ 113 ft

39. Analyze: Write function rules describing two different transformations of $f(x) = \sin x$ that result in $g(x) = \cos x$.

40. Analyze: Write function rules describing two different transformations of $f(x) = \tan x$ that result in $g(x) = \cot x$.

41. Prove: Use the figure to show that $\tan \theta$ is also the y-coordinate of the intersection of the angle's terminal side and the tangent drawn to the unit circle at $(1, 0)$.

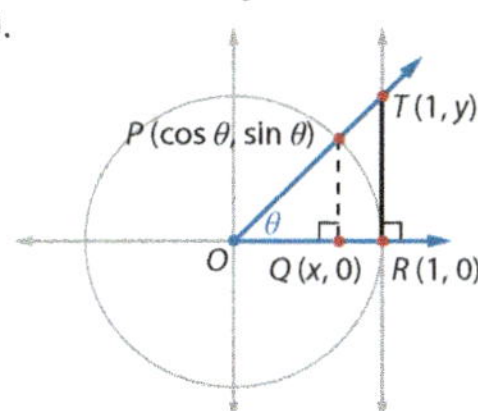

CUMULATIVE REVIEW

Identify any asymptotes and state each function's domain and range. [2.5]

42. $f(x) = \dfrac{1}{x^4}$

43. $f(x) = \dfrac{1}{(x+1)^3}$

44. Find the radian measure of θ or the related arc length in a circle with the given radius. [4.1]

a. $r = 6$ cm, $s = 8\pi$ cm, $\theta = ?$ $\dfrac{4\pi}{3}$

b. $r = 4$ ft, $\theta = \dfrac{\pi}{6}$, $s = ?$ $\dfrac{2\pi}{3}$ ft

45. Write an expression for all angles coterminal with $\theta = \dfrac{\pi}{5}$. Then state one positive and one negative coterminal angle. [4.1] $\dfrac{\pi}{5} \pm n2\pi; \dfrac{11\pi}{5}; -\dfrac{9\pi}{5}$

46. Evaluate the six trigonometric functions for $\angle B$. [4.2]

47. Find x to the nearest tenth. [4.2] 1.9

48. Find the angular speed (to the nearest radian per second) of the brush head of a sonic toothbrush that rotates at 30,000 rpm. [4.1] D

A. 1000 radians/sec

B. 314 radians/sec

C. 500 radians/sec

D. 3142 radians/sec

E. none of these

49. A cosine function has a maximum value of –2 and an amplitude of 5. What is its minimum value? [4.4] D

A. –7

B. –9

C. –10

D. –12

E. none of these

50. What is the frequency of $y = -2 \sin\left(\dfrac{x}{3}\right)$? [4.4] A

A. $\dfrac{1}{6\pi}$

B. 3π

C. 6π

D. $\dfrac{1}{3}$

E. none of these

51. Which equation is represented by the graph? [4.4] C

A. $y = 3 \sin x$

B. $y = 3 \sin\dfrac{x}{2}$

C. $y = 3 \sin 2x$

D. $y = \sin 2x$

E. none of these

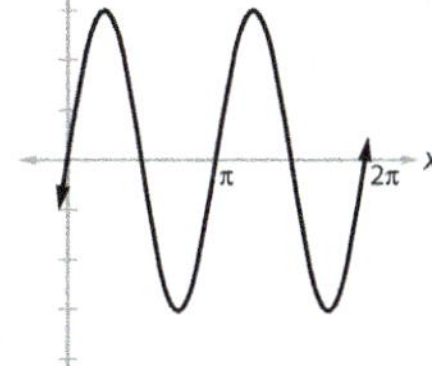

42. VA: $x = 0$; HA: $y = 0$; $D = \{x \mid x \neq 0\}$; $R = \{y \mid y > 0\}$

43. VA: $x = -1$; HA: $y = 0$; $D = \{x \mid x \neq -1\}$; $R = \{y \mid y \neq 0\}$

46. $\sin B = \dfrac{\sqrt{10}}{10}$; $\csc B = \sqrt{10}$; $\cos B = \dfrac{3\sqrt{10}}{10}$; $\sec B = \dfrac{\sqrt{10}}{3}$; $\tan B = \dfrac{1}{3}$; $\cot B = 3$

---

39. any phase shift of $\dfrac{\pi}{2} + n2\pi$, $n \in \mathbb{Z}$
$$g(x) = \sin\left(x + \dfrac{\pi}{2} + n2\pi\right) = \cos x \text{ or}$$
any reflection in the x-axis with a phase shift of $-\dfrac{\pi}{2} + n2\pi$, $n \in \mathbb{Z}$
$$g(x) = -\sin\left(x - \dfrac{\pi}{2} + n2\pi\right) = \cos x$$

40. a refection in the x-axis with any phase shift of $(2n+1)\dfrac{\pi}{2}$, $n \in \mathbb{Z}$
$$g(x) = -\tan\left(x + (2n+1)\dfrac{\pi}{2}\right) = \cot x$$

41. since $\triangle POQ \sim \triangle TOR$ by the AA $\sim$ Post., $\dfrac{TR}{PQ} = \dfrac{OR}{OQ}$; $\therefore$
$$\dfrac{TR}{\sin \theta} = \dfrac{1}{\cos \theta} \text{ and } TR = \dfrac{\sin \theta}{\cos \theta} = \tan \theta$$

Cumulative Review

44a. $s = r\theta$
$$8\pi \text{ cm} = (6 \text{ cm})\theta$$
$$\theta = \dfrac{8\pi \text{ cm}}{6 \text{ cm}} = \dfrac{4\pi}{3}$$

44b. $s = (4 \text{ ft})\left(\dfrac{\pi}{6}\right) = \dfrac{2\pi}{3}$ ft

45. $\dfrac{\pi}{5} \pm n2\pi$
$$\beta = \dfrac{\pi}{5} + (1)2\pi = \dfrac{11\pi}{5}$$
$$\theta = \dfrac{\pi}{5} - (1)2\pi = -\dfrac{9\pi}{5}$$

46. $6^2 + b^2 = \left(2\sqrt{10}\right)^2$
$$b = \sqrt{4(10) - 36} = 2$$
$$\sin B = \dfrac{2}{2\sqrt{10}} = \dfrac{\sqrt{10}}{10}$$
$$\csc B = \dfrac{2\sqrt{10}}{2} = \sqrt{10}$$
$$\cos B = \dfrac{6}{2\sqrt{10}} = \dfrac{3\sqrt{10}}{10}$$
$$\sec B = \dfrac{2\sqrt{10}}{6} = \dfrac{\sqrt{10}}{3}$$
$$\tan B = \dfrac{2}{6} = \dfrac{1}{3}; \cot B = 3$$

47. $\cos 72° = \dfrac{x}{6}$
$$x = 6 \cos 72° \approx 1.9$$

48. $\omega = 30{,}000 \dfrac{\text{rev}}{\text{min}}\left(\dfrac{1 \text{ min}}{60 \text{ sec}}\right)\left(\dfrac{2\pi \text{ radians}}{1 \text{ rev}}\right)$
$$= 1000\pi \dfrac{\text{radians}}{\text{sec}} \approx 3142 \dfrac{\text{radians}}{\text{sec}}$$

49. $|a| = \dfrac{1}{2}(y_{max} - y_{min})$
$$5 = \dfrac{1}{2}(-2 - y_{min})$$
$$10 = -2 - y_{min}$$
$$y_{min} = -12$$

50. $p = \dfrac{2\pi}{|b|} = \dfrac{2\pi}{\frac{1}{3}} = 6\pi$
$$f = \dfrac{1}{p} = \dfrac{1}{6\pi}$$

37b. $d = \dfrac{20}{\tan 55°} \approx 14.0$

37c. $d = \dfrac{20}{\tan 19°} \approx 58.1$; $\dfrac{58.1}{14} = 4.15$ floors above the first floor, therefore the fifth floor

38a. $\sin \theta = \dfrac{80}{d}$; $d = \dfrac{80}{\sin \theta}$
$$d(\theta) = 80 \csc \theta$$

38b.

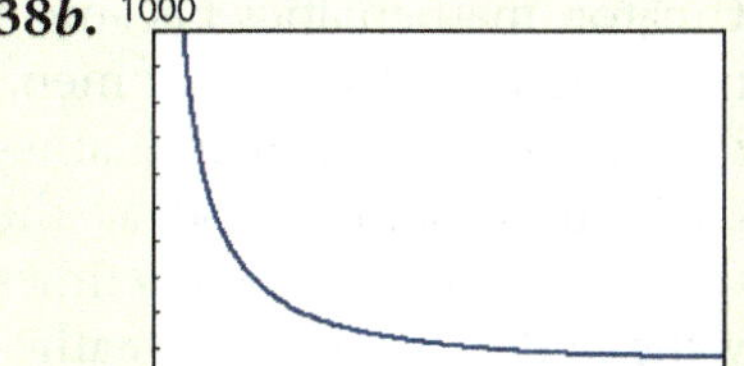

The distance decreases as the angle of elevation to the drone increases from 0° to $\dfrac{\pi}{2}$, or 90°, where it reaches a minimum of 80 ft when it is directly above the marker.

38c. $d(45°) = 80 \csc 45° \approx 113$ ft

Biblical Perspective of Mathematics

Objectives

1. To explain why secular mathematicians may not fully comprehend why so many aspects of the physical universe can be modeled mathematically

2. To relate Greek conic section theory to the universal laws of Kepler and Newton

3. To explain the usefulness of studying and developing pure mathematical theories

Assignments

- **Minimum:** 1–9; 10 (in class)
- **Standard:** 1–9; 10 (in class)
- **Extended:** 1–10

The Utility and Value of Mathematics

How can it be that mathematics, a product of human thought independent of experience, is so admirably adapted to the objects of reality?

...

The eternal mystery of the world is its comprehensibility.

—*Albert Einstein*

Complex numbers provide an interesting example of the connection between pure mathematical theory and real-world applications. The Swiss mathematician Leonhard Euler developed complex number theory in the seventeenth century as a result of his study of polynomial equations with nonreal solutions. He described these abstract solutions as imaginary and represented them using i for $\sqrt{-1}$. Initially there were no practical applications for this field of mathematics. Even now students ask, "When are we ever going to use this?" Over time, complex numbers were applied to circuit analysis in electrical engineering, fractal design in geometry, and the theory of quantum mechanics in physics.

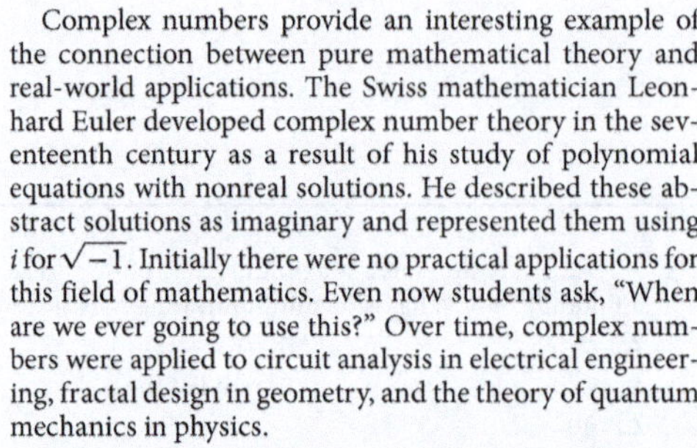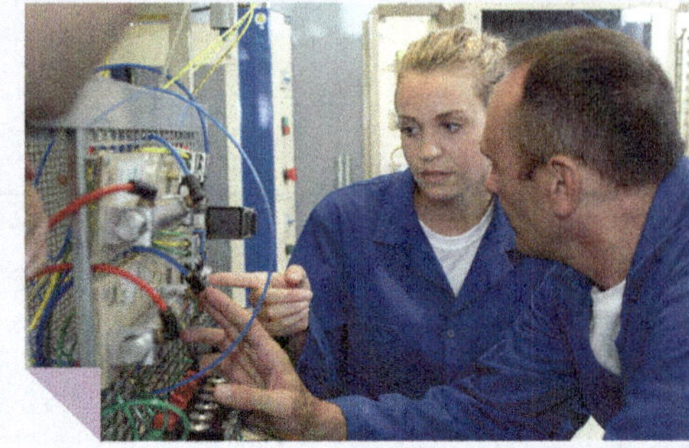

The ability of mathematics to accurately model such a variety of real-world applications often baffles modern mathematicians, many of whom believe that math is solely a product of human thought and do not recognize God as the ultimate source of mathematical truth. Mathematicians with a biblical worldview should be able to provide a clear explanation. As stated by James Nickel, "The mind of man, with its mathematical capabilities, and the physical world, with its observable mathematical order, *cohere* because of a common Creator. Einstein's eternal mystery has a solution. The biblical revelation of the Creator God is the unifying factor that reconciles what is irreconcilable in the humanistic context."

Conic sections provide another famous example of abstract mathematical theory being applied later to solve a practical problem. The ancient Greeks studied conics as early as 350 BC, resulting in its deductive presentation in the work of Apollonius around 200 BC. Archimedes applied conic theory in his development of catapults, delaying Rome's conquest of Syracuse for many years. However, the greatest application of conic section theory did not occur until the early seventeenth century, when Kepler presented his Copernican model of the universe based on elliptical orbits and his three planetary laws. Conic sections are the basis for studying the parabolic terrestrial motion of projectiles, the elliptical orbits of planets and some comets, and the hyperbolic path of several other celestial bodies. Sir Isaac Newton unified terrestrial and celestial motion with his mathematical description of universal gravitation.

Nikolai Lobachevsky, the nineteenth-century Russian founder of a non-Euclidean geometry, observed that "there is no branch of mathematics, however abstract, which may not someday be applied to the phenomena of the real world." Interestingly, his non-Euclidean geometry later became the basis for Albert Einstein's twentieth-century theory of relativity. Special relativity quickly proved valuable in the new fields of atomic and nuclear physics. Coupled with new mathematical measuring techniques, Einstein's general relativity has more recently provided explanations for pulsars and black holes. Classic experiments involving the sun's deflection of light and the gravitational redshift of light have confirmed the accuracy of Einstein's curved spacetime theory. Modern applications of general relativity involve high-precision measurements of both time and global positioning.

PRESENTATION

Briefly review the foundations of mathematics: the origin of mathematics is God, its nature is divine, and its purpose is to glorify God. Mention that both Euler and Kepler maintained a testimony for Christ throughout their lives (see their biographies in BJU Press's *Algebra 2*). Most of their European contemporaries agreed that using mathematics to model the universe reflected God's infinite wisdom and power. However, the Enlightenment in the eighteenth century brought a change in the way people viewed mathematics. According to James Nickel, "Instead of being a monument to God the Creator, mathematics became a monument to the inventiveness of men." Emphasize that because secular mathematicians fail to recognize God as the source of creation, they are sometimes baffled by the ability of abstract mathematics to model real-world applications.

In addition to the introductory quotations by Einstein, read the following quote by Remo Ruffini, a leader in astrophysics: "How a mathematical structure can correspond to nature is a mystery. One way out is just to say that the language in which nature speaks is the language of mathematics. This begs the question. Often we are both shocked and surprised by the correspondence between mathematics and nature, especially when the experiment confirms that our mathematical model describes nature perfectly." Morris Kline reflects the thinking of many secular mathematicians: "Mathematics is a creation of the human mind and yet possesses extraordinary power to represent the processes of nature and to predict the happenings of nature. Why does mathematics work? What is the secret of its power?"

Mention the obvious usefulness of mathematics in fields such as accounting, banking, and engineering. Point out that the utility and value of abstract mathe-

These examples demonstrate the utility and value of mathematics, as well as the importance of a biblical worldview in understanding the usefulness of mathematics. The popular math-puzzle writer Martin Gardner noted that "all mathematicians share . . . a sense of amazement over the infinite depth and the mysterious beauty and usefulness of mathematics." Without acknowledging that God is both the creator of the universe and the source of our ability to describe it, mathematicians will always be mystified by math's usefulness.

❭ Exercises

1. What did Einstein say that mathematics was a product of?

2. List three modern applications of complex numbers.

3. Identify the common source of the physical universe and the human mind.

4. Explain why secular mathematicians are often surprised at the ability of mathematics to accurately model so many aspects of the physical universe.

5. Name the abstract mathematical topic that was used in the development of each application.

  a. fractal geometry design

  b. Kepler's three laws of planetary motion

  c. Einstein's theory of relativity

6. Who unified terrestrial and celestial motion with his work on universal gravitation?

7. Name the conic section used to model the motion of the following.

  a. planetary orbits

  b. terrestrial projectiles such as cannonballs

8. In what fields of physics has special relativity proved especially valuable?

9. Which classic experiments have confirmed the accuracy of Einstein's curved spacetime theory?

10. Discuss: Explain the usefulness of studying and developing abstract mathematical theories. Include several examples.

---

1. human thought

2. circuit analysis, fractal design, and quantum mechanics

3. the Creator, God

4. They fail to recognize God as both the creator of the universe and the source of our ability to describe it.

5a. Euler's complex number theory

5b. conic section theory

5c. non-Euclidean geometry

6. Newton

7a. ellipse

7b. parabola

8. atomic and nuclear physics; astronomy

9. the sun's deflection of light; the gravitational redshift of light

10. Even when there is no readily apparent practical application, mathematicians, scientists, and people from many other fields frequently use previously developed abstract mathematics to model and explain real-world phenomena. Euler developed complex number theory to solve algebraic equations, but centuries later the theory proved useful in the fields of geometry, electricity, and physics. The Greek theory of conic sections waited almost 2000 years before it was applied in the universal laws of Kepler and Newton. Albert Einstein used non-Euclidean geometry developed in the eighteenth century to devise his twentieth-century curved spacetime theory, which is essential to the precise measurements of time and position in the twenty-first century.

---

matics may not be immediately obvious, such as with complex numbers and non-Euclidean geometry. Discuss the classic example of Greek conic section theory used by Kepler and Newton in their universal laws of motion. Students may be interested in the naming of three general views of geometry in terms of conic sections: parabolic (Euclidean) geometry, elliptic (Riemannian) geometry, and hyperbolic (Lobachevskian) geometry. These three views are used to model our universe in terms of a flat surface, a sphere, and a saddle shape, respectively. The hyperbolic model is often used to explain an expanding universe.

Conclude the lesson by explaining that Einstein's laws of relativity depend on non-Euclidian geometry and have practical, modern application in measuring time and location.

The chapter titled "Why Does Mathematics Work?" in James Nickel's *Mathematics: Is God Silent?* provides an excellent extended discussion of this topic.

4.6 Inverse Trigonometric Functions

Objectives

1. To define and graph inverse trigonometric functions

2. To evaluate inverse trigonometric functions

3. To find compositions of trigonometric functions for special angles

Flash

The number of grains of sand on all the beaches and deserts in the world is estimated to be approximately 7.5×10^{18}. With the aid of the Hubble telescope, the number of stars in the observable universe is estimated to be at least 10^{21}, much greater than the number of grains of sand. Sand often appears in Scripture in reference to its innumerable grains. Although we can estimate the number of grains of sand and the number of stars, they are impossible to count, as stated in Hebrews 11:12.

Vocabulary

Arccosine
Arcsine
Arctangent
principal value

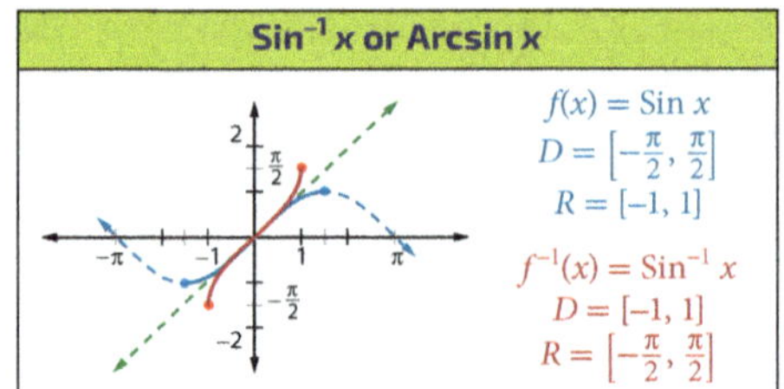

Inverse trigonometric functions are used to find the angle formed by the sides and the base of a pile of sand.

After completing this section, you will be able to

- define and graph inverse trig functions.
- evaluate inverse trig functions.
- evaluate compositions of trig and inverse trig functions.

In Section 4.2 inverse trigonometric functions were defined in terms of acute angles in right triangles. These definitions need to be expanded to include any angle θ and the real numbers. Since the graph of each trigonometric function fails the Horizontal Line Test, these functions are not one-to-one, and their inverses are not functions unless their domains are restricted.

Sin⁻¹ x or Arcsin x

$$f(x) = \operatorname{Sin} x$$
$$D = \left[-\tfrac{\pi}{2}, \tfrac{\pi}{2}\right]$$
$$R = [-1, 1]$$

$$f^{-1}(x) = \operatorname{Sin}^{-1} x$$
$$D = [-1, 1]$$
$$R = \left[-\tfrac{\pi}{2}, \tfrac{\pi}{2}\right]$$

Restricting the domain of $y = \sin x$ to $\left[-\tfrac{\pi}{2}, \tfrac{\pi}{2}\right]$ defines $f(x) = \operatorname{Sin} x$, a one-to-one function with $R = [-1, 1]$. The capital letter indicates that the function's domain has been restricted. This function's inverse, $f^{-1}(x) = \operatorname{Sin}^{-1}x$, or $f^{-1}(x) = \operatorname{Arcsin} x$, is a function that returns a single angle or arc length within the interval $\left[-\tfrac{\pi}{2}, \tfrac{\pi}{2}\right]$. This value is the *principal value* of the inverse sine and is the value returned by a calculator when the $\sin^{-1}$ function is used. As with other inverse functions, the graphs of $f(x)$ and $f^{-1}(x)$ are reflections in the line $y = x$.

Example 1   Evaluating Inverse Sine Functions

Evaluate each expression in radians and degrees. State exact values if possible.

a. $\operatorname{Sin}^{-1}\left(-\tfrac{1}{2}\right)$ b. $\operatorname{Arcsin} 0.75$ c. $\operatorname{Sin}^{-1} \pi$

Answer

a.

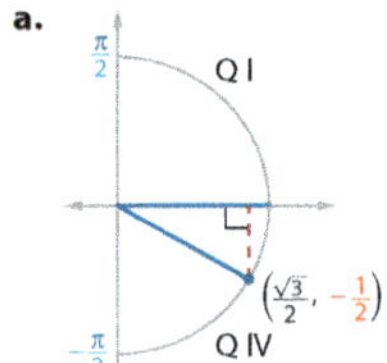

$\operatorname{Sin}^{-1}\left(-\tfrac{1}{2}\right) = -\tfrac{\pi}{6}$ or $-30°$

1. Since the range of the Sin^{-1} function is $\left[-\tfrac{\pi}{2}, \tfrac{\pi}{2}\right]$, locate the point on the right side of the unit circle with a y-coordinate of $-\tfrac{1}{2}$.

2. Use the reference angle $\alpha = 30°$ to state the principal value.

CONTINUED ➡

PRESENTATION

Lesson Opener

Use the unit circle to find all possible angles where $\theta \in [0, 2\pi)$.

1. $\sin \theta = \tfrac{1}{2}$ $\tfrac{\pi}{6}, \tfrac{5\pi}{6}$

2. $\cos \theta = -\tfrac{\sqrt{2}}{2}$ $\tfrac{3\pi}{4}, \tfrac{5\pi}{4}$

3. $\tan \theta = 0$ $0, \pi$

Note: finding multiple solutions for more complicated trigonometric equations and stating both general and principal value solutions are the focus of Section 5.3.

Use the Lesson Opener to demonstrate that these trigonometric functions are not one-to-one functions. Consider asking which parent trigonometric functions pass the horizontal line test. (*none*) Then ask what that indicates about the basic trigonometric functions. (*Their inverses are not functions.*) Ask what needs to be done in order to define inverse functions. (*restricting each function's domain*)

You may also want to generalize the following reasoning for any periodic function $f(x)$: since $f(x)$ has repeated values (by definition), its inverse cannot be a function unless the domain of $f(x)$ is limited.

Some students may not fully appreciate the importance of inverse functions. Explain that unless a relation is one-to-one, it is impossible to know the member of the domain associated with a specific value in the range. For example, if we know that $\sin \theta = \tfrac{1}{2}$, we cannot know whether $\theta = 30°$, $120°$, or any angle coterminal with these angles.

Illustrate how the domain of the sine function is restricted to create the one-to-one function $f(x) = \operatorname{Sin} x$, whose inverse is also a function, $f^{-1}(x) = \operatorname{Sin}^{-1} x$. Use Example 1 to demonstrate evaluating the Sin^{-1} function. While some texts limit the range of inverse trigonometric functions to real numbers (radians), we will not make this distinction so that students can

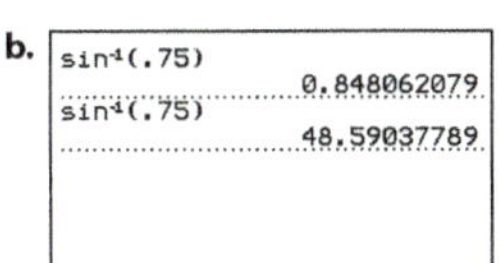

b.

Use your calculator's $\sin^{-1}$ function ([2nd], [SIN]) for x-values not associated with special angles on the unit circle.
Recall that the mode of the calculator determines whether the result is in degrees or radians.

Arcsin $0.75 \approx 0.848$ or $48.6°$

c. $\text{Sin}^{-1}\,\pi$ is not defined.

The domain of $y = \text{Sin}^{-1}$ is $[-1, 1]$ and $\pi > 1$.

SKILL ✔ **EXERCISE 7**

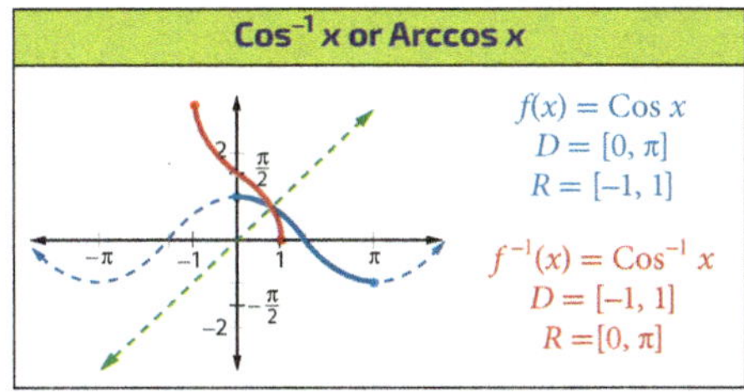

Restricting the domain of $y = \cos x$ to $[0, \pi]$ defines $f(x) = \text{Cos}\,x$, a one-to-one function with $R = [-1, 1]$. This function's inverse, $f^{-1}(x) = \text{Cos}^{-1} x$ or $f^{-1}(x) = \text{Arccos}\,x$, is a function that returns a single angle or arc length within $[0, \pi]$. This principal value of the inverse cosine relation is the value returned by a calculator when its $\cos^{-1}$ function is used.

Example 2 Evaluating Inverse Cosine Functions

Evaluate each expression in radians and degrees. State exact values if possible.

a. $\text{Cos}^{-1}\left(-\frac{\sqrt{2}}{2}\right)$ **b.** Arccos 0   **c.** $\text{Cos}^{-1}\frac{1}{4}$

Answer

a.

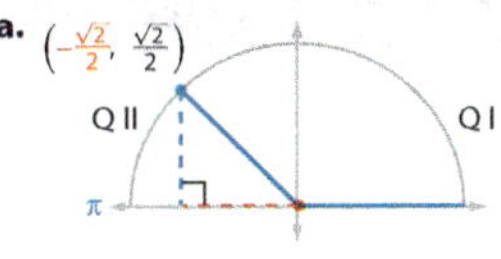

1. Since the range of the Cos^{-1} function is $[0, \pi]$, locate the point on the top half of the unit circle with an x-coordinate of $-\frac{\sqrt{2}}{2}$.

$\text{Cos}^{-1}\left(-\frac{\sqrt{2}}{2}\right) = \frac{3\pi}{4}$ or $135°$

2. Use the reference angle $\alpha = 45°$ to state the principal value.

b.

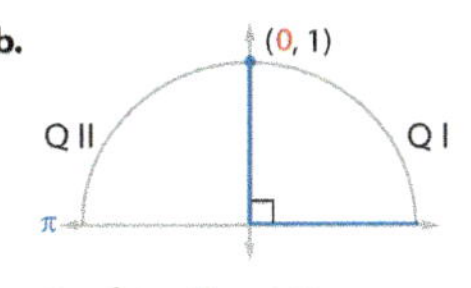

1. Locate the point on the top half of the unit circle with an x-coordinate of 0.

$\text{Cos}^{-1} 0 = \frac{\pi}{2}$ or $90°$

2. Use the quadrantal angle to state the principal value.

CONTINUED ➡

Additional Exercises

Evaluate each expression in degrees.

1. Arccos 1 $0°$

2. $\text{Sin}^{-1}\left(-\frac{\sqrt{3}}{2}\right)$ $-60°$

3. Arctan 11.43 $\approx 85°$

4. State the interval over which each equation is true.

  a. $\cos(\text{Cos}^{-1} x) = x$ $[-1, 1]$

  b. $\text{Tan}^{-1}(\tan x) = x$ $\left(-\frac{\pi}{2}, \frac{\pi}{2}\right)$

Find the exact value for each expression without using a calculator.

5. $\text{Sin}^{-1}\left(\sin\left(-\frac{\pi}{4}\right)\right)$ $-\frac{\pi}{4}$

6. $\text{Tan}^{-1}\left(\tan\frac{4\pi}{3}\right)$ $\frac{\pi}{3}$

7. $\text{Cos}^{-1}\left(\cos\frac{5\pi}{4}\right)$ $\frac{3\pi}{4}$

8. $\cos\left(\text{Cos}^{-1}\frac{3}{4}\right)$ $\frac{3}{4}$

9. $\text{Cos}^{-1}\left(\sin\left(-\frac{\pi}{4}\right)\right)$ $\frac{3\pi}{4}$

10. $\tan\left(\text{Sin}^{-1}\frac{1}{2}\right)$ $\frac{\sqrt{3}}{3}$

11. $\sec\left(\text{Tan}^{-1}\frac{4}{5}\right)$ $\frac{\sqrt{41}}{5}$

Assignments

- **Minimum:** 1–14, 17–18, 20, 22–24, 26, 28–30, 36, 44–45, 47–48, 50–52

- **Standard:** 1–14, 17–22; 24–32 even; 33, 36–39; 42–48 even; 49, 51–53

- **Extended:** 2–16 even; 17–24, 26, 28–29, 32–35, 37–44, 46, 49, 51–53

---

develop an intuitive feeling for the output of the function being the angle in the limited domain with that sine value.

Common Student Error Some students may confuse inverse functions with reciprocal functions. This may be because multiplicative inverses are called reciprocals or because the notation is similar. Stress that multiplicative inverses and inverse functions are not the same thing; the reciprocal of $\sin x = \frac{1}{\sin x} = \csc x$ should not be confused with the inverse sine function $f(x) = \text{Sin}^{-1} x$.

Ask students whether the same restriction works for the cosine function (*no*), and then ask them to propose a restriction that would work. While there are several possibilities, mathematicians have agreed to use $D = [0, \pi]$ to define $f(x) = \text{Cos}\,x$ and its inverse function. Consider having the students first attempt to complete the evaluations in Example 2 without using the solutions in the text.

Use a graph of $y = \tan x$ to discuss how the domain and range of $f(x) = \text{Tan}\,x$ and its inverse function differ from $f(x) = \text{Sin}\,x$ and its inverse. Help the students recognize that they are looking for coordinates where the y-coordinate divided by the x-coordinate gives the desired value. Complete Example 3, emphasizing that we are finding *principal values*.

Introduce the compositions involving an unrestricted function and its related inverse trigonometric function. Be sure students understand why compositions such as $\text{Sin}^{-1}(\sin\theta)$ return a principal value when θ is outside the restricted domain. Use Example 4 to demonstrate these compositions.

Example 5 demonstrates a composition of an unrestricted trigonometric function and a different inverse trigonometric function. This specific case prepares students for the more generalized case presented in Example 6. Explain that $\theta = \text{Cos}^{-1} t$ must be in Q I or Q II and that $y > 0$ in either quadrant.

c. 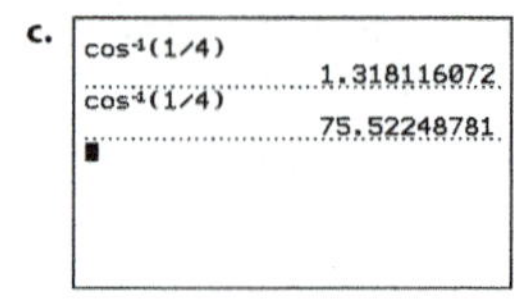

Use your calculator's cos⁻¹ function ([2nd], [COS]) for x-values not associated with special angles on the unit circle.
Switch modes to find radian and degree measures.

Arccos $\frac{1}{4} \approx 1.318$ or $75.5°$

SKILL ✔ EXERCISE 9

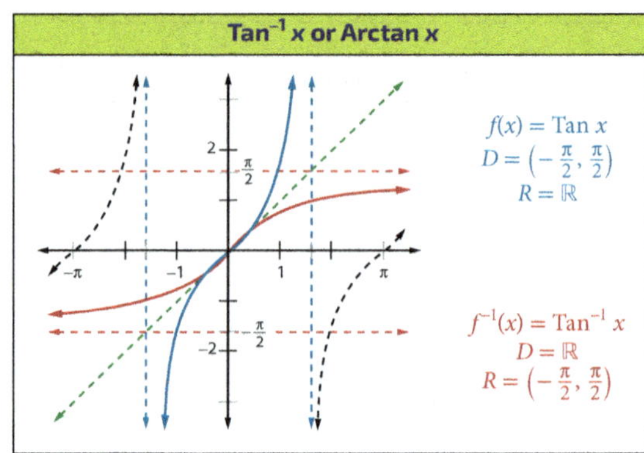

Since $y = \tan x$ is undefined at $-\frac{\pi}{2}$ and $\frac{\pi}{2}$, the function is restricted to $\left(-\frac{\pi}{2}, \frac{\pi}{2}\right)$ to define $f(x) = \text{Tan } x$, a one-to-one function with $R = \mathbb{R}$. This function's inverse, $f^{-1}(x) = \text{Tan}^{-1} x$, or $f^{-1}(x) = \text{Arctan } x$, is a function that returns a single angle or arc length within $\left(-\frac{\pi}{2}, \frac{\pi}{2}\right)$. This principal value of the inverse tangent relation is returned by a calculator when its $\tan^{-1}$ function is used.

Example 3   Evaluating Inverse Tangent Functions

Evaluate each expression in radians and degrees. State exact values if possible.

a. $\text{Tan}^{-1} 1$ **b.** $\text{Arctan}\left(-\sqrt{3}\right)$ **c.** $\text{Tan}^{-1} 1.3764$

Answer

a.

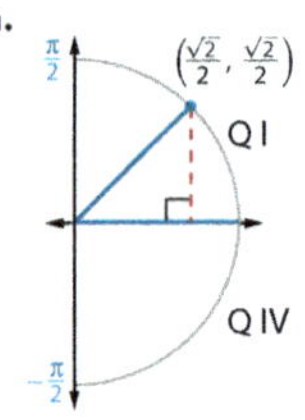

$\text{Tan}^{-1} 1 = \frac{\pi}{4}$ or $45°$

1. Since the range of the Tan^{-1} function is $\left(-\frac{\pi}{2}, \frac{\pi}{2}\right)$, locate the point on the right side of the unit circle where $\frac{y}{x} = 1$.

2. Use the reference angle $\alpha = 45°$ to state the principal value.

CONTINUED ➡

The first known astronomical instrument was a gnomon, a simple vertical rod that allowed an observer to trace the shadow cast by the sun. The time of day, true north, and other compass points could then be determined. The sun's angle of elevation could also be found, using $\theta = \text{Tan}^{-1}\left(\dfrac{\text{gnomon's height}}{\text{shadow's length}}\right)$.

TIPS

Ex. 1, 3, 5 Be sure students understand that the capital letter indicates the restricted domain.

Ex. 21–32 Consider requiring students to show work or reasoning for each evaluation and encouraging them to check their answers with a calculator.

Ex. 36 Similar applications in calculus involve the rate of change in volume and height as piles of substances such as sand are created.

Ex. 37b Consider having the students compare $Y_1 = \tan^{-1}\left(\frac{24}{x}\right) - \tan^{-1}\left(\frac{6}{x}\right)$ and $Y_2 = \tan^{-1}\left(\frac{24}{x} - \frac{6}{x}\right) = \tan^{-1}\frac{18}{x}$. Students should see that $\tan^{-1}$ cannot be factored out of the expression.

b. 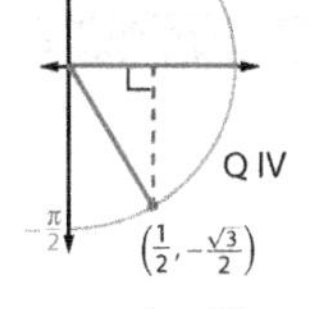

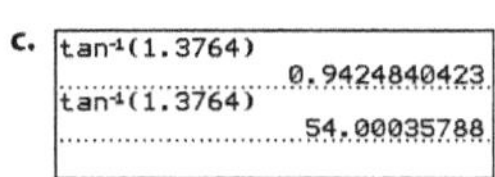

Arctan $\left(-\sqrt{3}\right) = -\frac{\pi}{3}$ or $-60°$

1. Locate the point on the right side of the unit circle where $\frac{y}{x} = -\sqrt{3}$.

2. Use the reference angle $\alpha = 60°$ to state the principal value.

c.

tan⁴(1.3764)	
	0.9424840423
tan⁴(1.3764)	
	54.00035788

Arctan $1.3764 \approx 0.9425$ or $54°$

Use your calculator's tan⁻¹ function ([2nd], [TAN]) for x-values not associated with special angles on the unit circle.
Switch modes to find the angle in radians and degrees.

————————————— SKILL ✓ **EXERCISE 11**

Representing inverse trigonometric functions Sin^{-1}, Cos^{-1}, and Tan^{-1} as Arcsin, Arccos, and Arctan emphasizes that the result is the length of an arc on a unit circle (the radian measure) or an angle (measured in degrees). The inverse functions for the three reciprocal trigonometric functions are explored in exercises 38–40.

Since the trigonometric functions and their related inverse functions are not true inverses of each other, we must be careful when evaluating compositions involving a trigonometric function and its related inverse trigonometric function. While $\sin(\text{Sin}^{-1} x) = x$ for all $x \in [-1, 1]$ and $\text{Sin}^{-1}(\sin x) = x$ for all $x \in \left[-\frac{\pi}{2}, \frac{\pi}{2}\right]$, the expression $\text{Sin}^{-1}(\sin x)$ indicates a principal value when $x < -\frac{\pi}{2}$ or $x > \frac{\pi}{2}$. Similar cautions apply to $\text{Tan}^{-1}(\tan x)$ when $x < -\frac{\pi}{2}$ or $x > \frac{\pi}{2}$ and $\text{Cos}^{-1}(\cos x)$ when $x < 0$ or $x > \pi$.

TIP

Some texts define the inverse trigonometric functions without capital letters, with the assumption of restricted domains for sine, cosine, and tangent functions.

Example 4 Evaluating Compositions

Evaluate each expression.

a. $\sin\left(\text{Sin}^{-1}\frac{1}{2}\right)$ **b.** $\text{Tan}^{-1}(\tan(-45°))$ **c.** $\text{Arccos}\left(\cos\frac{5\pi}{3}\right)$

Answer

a. $\sin\left(\text{Sin}^{-1}\frac{1}{2}\right) = \sin\frac{\pi}{6} = \frac{1}{2}$

The expression represents the sine of the angle within $\left[-\frac{\pi}{2}, \frac{\pi}{2}\right]$ with a sine of $\frac{1}{2}$.

b. $\text{Tan}^{-1}(\tan(-45°)) = \text{Tan}^{-1}(-1) = -45°$

Since $x = -45°$ is within the domain of the Tan^{-1} function, $\text{Tan}^{-1}(\tan x) = x$.

c. $\text{Arccos}\left(\cos\frac{5\pi}{3}\right) = \text{Arccos}\frac{1}{2} = \frac{\pi}{3}$

Since $\frac{5\pi}{3}$ is not in the domain of the Arccos function, the answer represents the principal value with the same cosine as $\frac{5\pi}{3}$.

————————————— SKILL ✓ **EXERCISE 25**

Solutions

❯ A. Exercises

1. 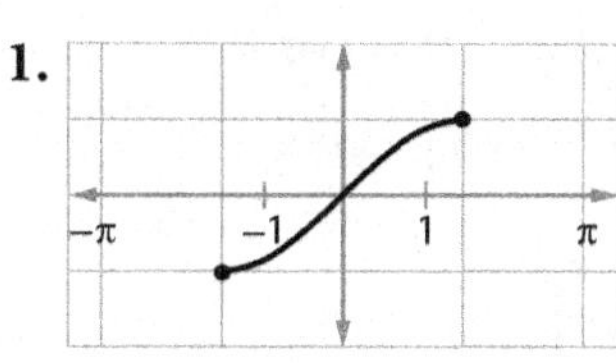

$D = \left[-\frac{\pi}{2}, \frac{\pi}{2}\right]; R = [-1, 1]$

2. 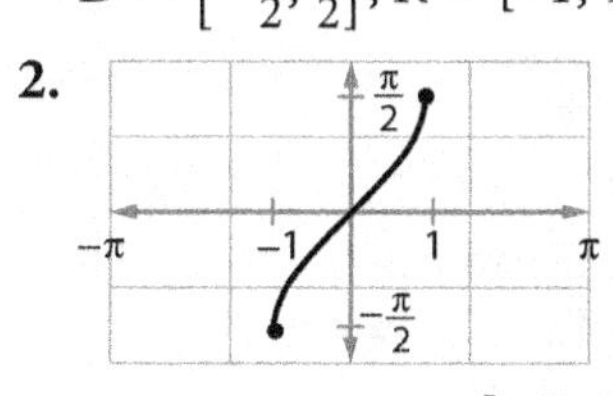

$D = [-1, 1]; R = \left[-\frac{\pi}{2}, \frac{\pi}{2}\right]$

3. 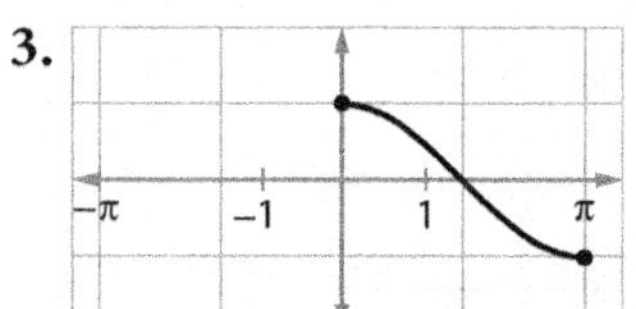

$D = [0, \pi]; R = [-1, 1]$

4. 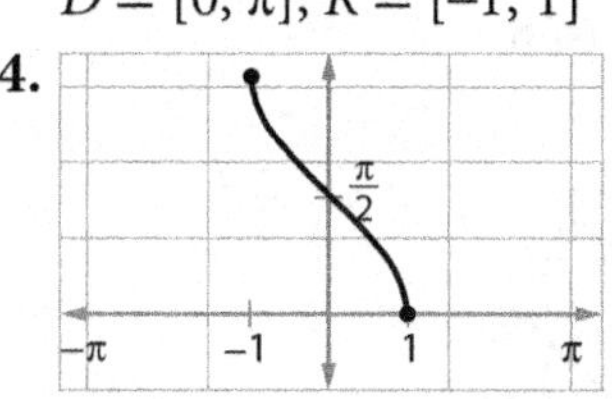

$D = [-1, 1]; R = [0, \pi]$

5. 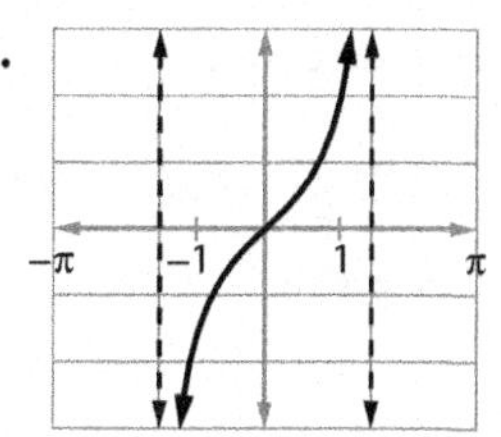

$D = \left(-\frac{\pi}{2}, \frac{\pi}{2}\right); R = (-\infty, \infty)$

6. 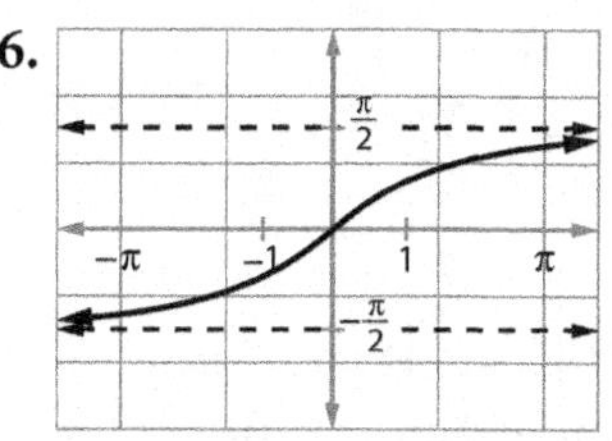

$D = (-\infty, \infty); R = \left(-\frac{\pi}{2}, \frac{\pi}{2}\right)$

7. 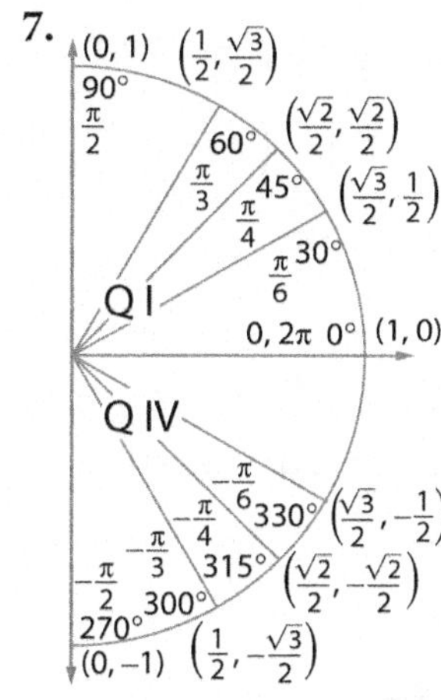

since $\sin\theta = y$, $\text{Sin}^{-1}\frac{\sqrt{3}}{2} = \frac{\pi}{3} = 60°$

8. 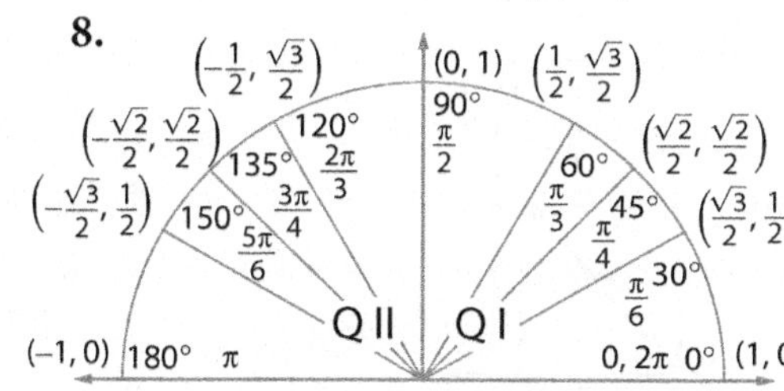

since $\cos\theta = x$, $\text{Cos}^{-1}\frac{1}{2} = \frac{\pi}{3} = 60°$

9. See image from exercise 8.
since $\cos\theta = x$,
$\text{Arccos}\left(-\frac{1}{2}\right) = \frac{2\pi}{3} = 120°$

10. See image from exercise 7.
since $\sin\theta = y$,
$\text{Arcsin}\left(-\frac{\sqrt{2}}{2}\right) = -\frac{\pi}{4} = -45°$

11. See image from exercise 7.
since $\tan\theta = \frac{y}{x}$ and $\frac{-2\sqrt{2}}{2\sqrt{2}} = -1$,
$\text{Tan}^{-1}(-1) = -\frac{\pi}{4} = -45°$

12. See image from exercise 7.

since $\tan\theta = \frac{y}{x}$ and $\dfrac{\frac{1}{2}}{\frac{\sqrt{3}}{2}} = \frac{1}{\sqrt{3}} = \frac{\sqrt{3}}{3}$,

$\text{Arctan}\, \frac{\sqrt{3}}{3} = \frac{\pi}{6} = 30°$

> ### B. Exercises

21. since $x \in [0, \pi]$,

$\text{Cos}^{-1}\left(\cos\frac{5\pi}{6}\right) = \frac{5\pi}{6}$

22. since $x \in \left(-\frac{\pi}{2}, \frac{\pi}{2}\right)$,

$\text{Arctan}\left(\tan\left(-\frac{\pi}{4}\right)\right) = -\frac{\pi}{4}$

23. since $\frac{2\pi}{3} \notin \left[-\frac{\pi}{2}, \frac{\pi}{2}\right]$,

$\text{Arcsin}\left(\sin\frac{2\pi}{3}\right) = \text{Arcsin}\,\frac{\sqrt{3}}{2} = \frac{\pi}{3}$

24. since $-\frac{\pi}{6} \notin [0, \pi]$,

$\text{Cos}^{-1}\left(\cos\left(-\frac{\pi}{6}\right)\right) = \text{Cos}^{-1}\left(\frac{\sqrt{3}}{2}\right) = \frac{\pi}{6}$

25. since $\frac{3}{4} \in [-1, 1]$, $\cos\left(\text{Arccos}\,\frac{3}{4}\right) = \frac{3}{4}$

26. Since $\frac{5}{4} \notin [-1, 1]$, $\text{Sin}^{-1}\,\frac{5}{4}$ is undefined.

27. $\sin\left(\text{Cos}^{-1}\,\frac{\sqrt{2}}{2}\right) = \sin\frac{\pi}{4} = \frac{\sqrt{2}}{2}$

28. $\text{Tan}^{-1}\left(\cos\frac{3\pi}{2}\right) = \text{Tan}^{-1}\,0 = 0$

29.

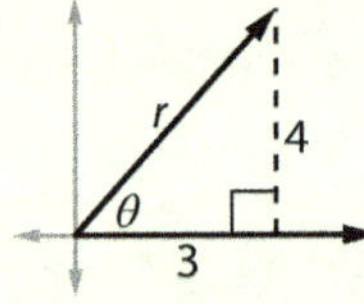

$r^2 = 3^2 + 4^2$
$r = \sqrt{9 + 16} = 5$
$\sin\theta = \frac{y}{r} = \frac{4}{5}$

30.

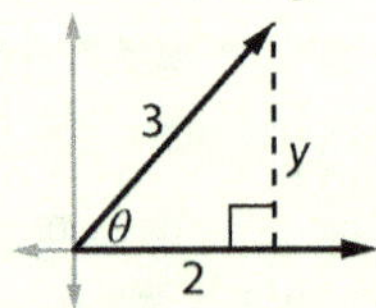

$y^2 + 2^2 = 3^2$
$y = \sqrt{9 - 4} = \sqrt{5}$
$\tan\theta = \frac{y}{x} = \frac{\sqrt{5}}{2}$

31.

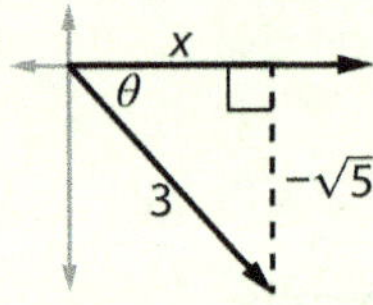

$x^2 + \left(-\sqrt{5}\right)^2 = 3^2$
$x = \sqrt{9 - 5} = 2$
$\cos\theta = \frac{x}{r} = \frac{2}{3}$

A reference triangle can be used to evaluate expressions involving compositions of trigonometric functions and other inverse trigonometric functions.

Example 5 Evaluating Compositions

Evaluate $\sin\left(\text{Cos}^{-1}\left(-\frac{3}{4}\right)\right)$.

Answer

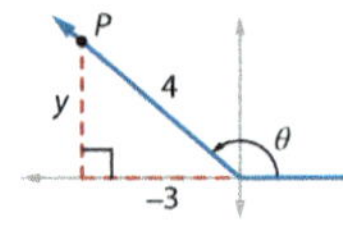

1. Since the range of the Cos^{-1} function is $[0, \pi]$, draw a Q II reference triangle illustrating $\theta = \text{Cos}^{-1}\left(-\frac{3}{4}\right)$.

$(-3)^2 + y^2 = 4^2$
$y = \sqrt{16 - 9} = \sqrt{7}$

2. Calculate the y-coordinate of the illustrated point. Use the positive root since θ is in Q II.

$\therefore \sin\left(\text{Cos}^{-1}\left(-\frac{3}{4}\right)\right) = \sin\theta = \frac{\sqrt{7}}{4}$

3. Use $\sin\theta = \frac{y}{r}$ to evaluate the expression.

SKILL ✔ **EXERCISE 31**

This method allows similar functions composed of trigonometric and inverse trigonometric functions to be expressed algebraically, which is useful in calculus.

Example 6 Evaluating Compositions

Express each function algebraically.

a. $f(t) = \cos\left(\text{Cos}^{-1}\,t\right)$ **b.** $g(t) = \sin\left(\text{Cos}^{-1}\,t\right)$ **c.** $h(t) = \tan\left(\text{Cos}^{-1}\,t\right)$

Answer

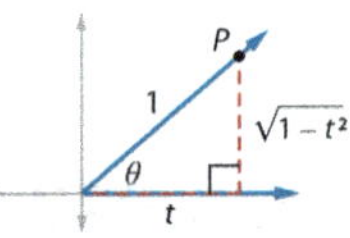

1. Draw a Q I reference triangle illustrating $\theta = \text{Cos}^{-1}\,t$ when $t > 0$ and find an expression for the y-coordinate of P. If $t < 0$, the reference triangle is in Q II, and $y > 0$.

a. $f(t) = \cos\left(\text{Cos}^{-1}\,t\right) = \cos\theta = t$

2. Express each function algebraically.

b. $g(t) = \sin\left(\text{Cos}^{-1}\,t\right) = \sin\theta = \sqrt{1 - t^2}$

c. $h(t) = \tan\left(\text{Cos}^{-1}\,t\right) = \tan\theta = \frac{\sqrt{1 - t^2}}{t}$

SKILL ✔ **EXERCISE 39**

To show King Hezekiah that his prayer for healing had been answered, God moved the sun's shadow backward on a sundial (2 Kings 20:8–11).

32.

$(-5)^2 + y^2 = 8^2$
$y = \sqrt{64 - 25} = \sqrt{39}$
$\csc\theta = \frac{1}{\sin\theta} = \frac{r}{y} = \frac{8}{\sqrt{39}} = \frac{8\sqrt{39}}{39}$

33.

Note that θ is in Q I if $t > 0$ or in Q IV if $t < 0$ and $x > 0$ in both quadrants.

$f(t) = \cos\left(\text{Sin}^{-1}\,t\right) = \cos\theta$
$= \frac{\sqrt{1 - t^2}}{1} = \sqrt{1 - t^2}$

34.

Note that θ is in Q I if $t > 0$ or in Q IV if $t < 0$ and $r > 0$.

$g(t) = \sin\left(\text{Tan}^{-1}\,t\right) = \sin\theta$
$= \frac{t}{\sqrt{t^2 + 1}}$

A. Exercises

Graph each function and state its domain and range.

1. $y = \text{Sin } x$ **2.** $y = \text{Sin}^{-1} x$ **3.** $y = \text{Cos } x$
4. $y = \text{Cos}^{-1} x$ **5.** $y = \text{Tan } x$ **6.** $y = \text{Tan}^{-1} x$

Evaluate each expression in radians and degrees.

7. $\text{Sin}^{-1} \frac{\sqrt{3}}{2}$ $\frac{\pi}{3} = 60°$ **8.** $\text{Cos}^{-1} \frac{1}{2}$ $\frac{\pi}{3} = 60°$

9. $\text{Arccos}\left(-\frac{1}{2}\right)$ $\frac{2\pi}{3} = 120°$ **10.** $\text{Arcsin}\left(-\frac{\sqrt{2}}{2}\right)$ $-\frac{\pi}{4} = -45°$

11. $\text{Tan}^{-1}(-1)$ $-\frac{\pi}{4} = -45°$ **12.** $\text{Arctan} \frac{\sqrt{3}}{3}$ $\frac{\pi}{6} = 30°$

Use a calculator to evaluate each expression in radians and in degrees.

13. $\text{Sin}^{-1} 0.3420$ $\approx 0.3490 \approx 20.0°$ **14.** $\text{Tan}^{-1} 1.732$ $\approx 1.0472 \approx 60.0°$

15. $\text{Cos}^{-1}(-0.7138)$ $\approx 2.3657 \approx 135.5°$ **16.** $\text{Sin}^{-1}(-1.125)$ undefined

B. Exercises

State the interval over which each equation is true.

17. $\sin(\text{Sin}^{-1} x) = x$ $[-1, 1]$ **18.** $\text{Arcsin}(\sin x) = x$ $\left[-\frac{\pi}{2}, \frac{\pi}{2}\right]$
19. $\text{Cos}^{-1}(\cos x) = x$ $[0, \pi]$ **20.** $\tan(\text{Tan}^{-1} x) = x$ $(-\infty, \infty)$

Find the exact value for each expression without using a calculator.

21. $\text{Cos}^{-1}\left(\cos \frac{5\pi}{6}\right)$ $\frac{5\pi}{6}$ **22.** $\text{Arctan}\left(\tan\left(-\frac{\pi}{4}\right)\right)$ $-\frac{\pi}{4}$

23. $\text{Arcsin}\left(\sin \frac{2\pi}{3}\right)$ $\frac{\pi}{3}$ **24.** $\text{Cos}^{-1}\left(\cos\left(-\frac{\pi}{6}\right)\right)$ $\frac{\pi}{6}$

25. $\cos\left(\text{Arccos} \frac{3}{4}\right)$ $\frac{3}{4}$ **26.** $\sin\left(\text{Sin}^{-1} \frac{5}{4}\right)$ undefined

27. $\sin\left(\text{Cos}^{-1} \frac{\sqrt{2}}{2}\right)$ $\frac{\sqrt{2}}{2}$ **28.** $\text{Tan}^{-1}\left(\cos \frac{3\pi}{2}\right)$ 0

29. $\sin\left(\text{Arctan} \frac{4}{3}\right)$ $\frac{4}{5}$ **30.** $\tan\left(\text{Cos}^{-1} \frac{2}{3}\right)$ $\frac{\sqrt{5}}{2}$

31. $\cos\left(\text{Sin}^{-1}\left(-\frac{\sqrt{5}}{3}\right)\right)$ $\frac{2}{3}$ **32.** $\csc\left(\text{Cos}^{-1}\left(-\frac{5}{8}\right)\right)$ $\frac{8\sqrt{39}}{39}$

Express each function algebraically.

33. $f(t) = \cos(\text{Sin}^{-1} t)$ **34.** $g(t) = \sin(\text{Tan}^{-1} t)$
35. $h(t) = \sec(\text{Cos}^{-1} t)$

36. Granular substances forming cone-shaped piles often reach their maximum angle of repose θ when they are poured onto a horizontal surface.

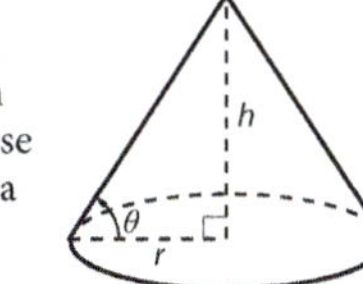

  a. Derive a formula for θ in terms of the cone's height h and radius r.
  b. Find the degree measure of the angle of repose for a 7 in. tall poured pile of sand with a diameter of 2 ft. $\approx 30.3°$

37. A missionary projects a video onto an 18 ft tall screen that is mounted on a wall so its lower edge is 10 ft above the floor. The viewing angle θ of the screen depends on the viewer's distance d from the wall. Assume that the eye level of the viewer is 4 ft above the floor.

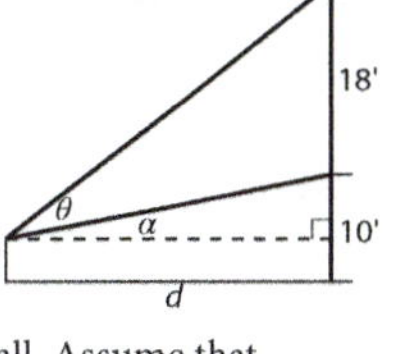

  a. Calculate the viewing angle (to the nearest tenth of a degree) for a viewer 30 ft from the wall and for a second viewer 40 ft from the wall.
  b. Write an equation expressing the viewing angle θ in terms of the viewer's distance from the wall.
  c. Use technology to graph the function and find the maximum viewing angle and the distance from the wall that provides it. $\approx 36.9°$ at 12.0 ft
  d. Use technology to find any distances where the viewing angle is 17°. ≈ 2.6 ft, ≈ 56.3 ft

37a. $\approx 27.4°$; $\approx 22.5°$
37b. $\theta = \tan^{-1}\frac{24}{d} - \tan^{-1}\frac{6}{d}$

C. Exercises

State a restricted domain that makes each reciprocal function one-to-one. Then graph the function and its inverse function on the same set of axes.

38. $f(x) = \text{Sec } x$ **39.** $g(x) = \text{Csc } x$ $D = \left[-\frac{\pi}{2}, 0\right) \cup \left(0, \frac{\pi}{2}\right]$
40. $h(x) = \text{Cot } x$ $D = (0, \pi)$
41. Prove: $\text{Csc}^{-1} t = \text{Sin}^{-1} \frac{1}{t}$ (*Hint:* Let $\text{Csc}^{-1} x = \theta$.)
42. Evaluate each expression (in radians).

  a. $\text{Sec}^{-1} 2$ $\frac{\pi}{3} \approx 1.0472$
  b. $\text{Csc}^{-1} 0.4$ undefined
  c. $\text{Cot}^{-1} -2.819$ ≈ 2.8007

43. The slope of the tangent to a curve, called the *derivative* in calculus, represents the instantaneous rate of change.

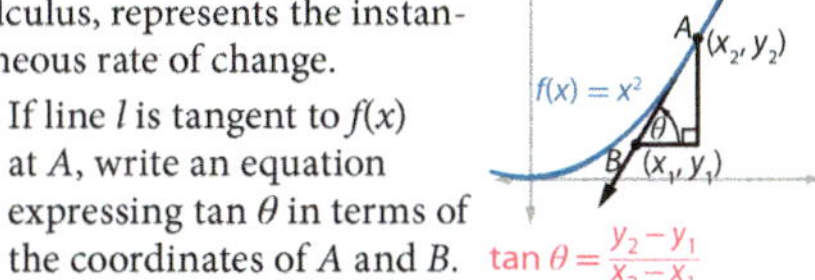

  a. If line l is tangent to $f(x)$ at A, write an equation expressing $\tan \theta$ in terms of the coordinates of A and B. $\tan \theta = \frac{y_2 - y_1}{x_2 - x_1}$
  b. Which characteristic of line l is represented by $\tan \theta$? its slope
  c. Write the equation of the tangent to $y = x^2$ at $x = 3$ if $\tan \theta = 6$.
  d. Find θ to the nearest tenth of a degree.

38. $D = \left[0, \frac{\pi}{2}\right) \cup \left(\frac{\pi}{2}, \pi\right]$

37d.

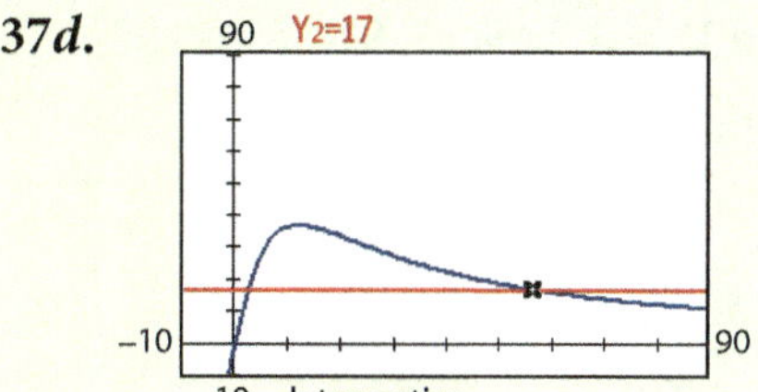

C. Exercises

38.

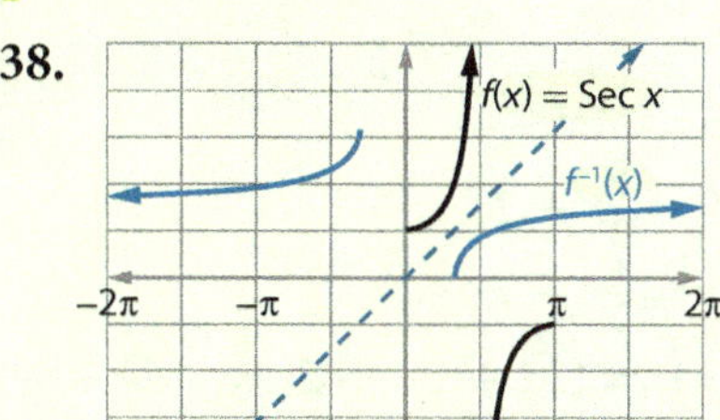

39.

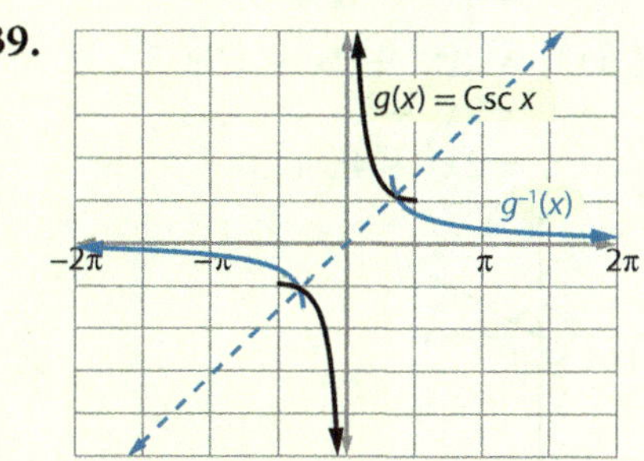

40.

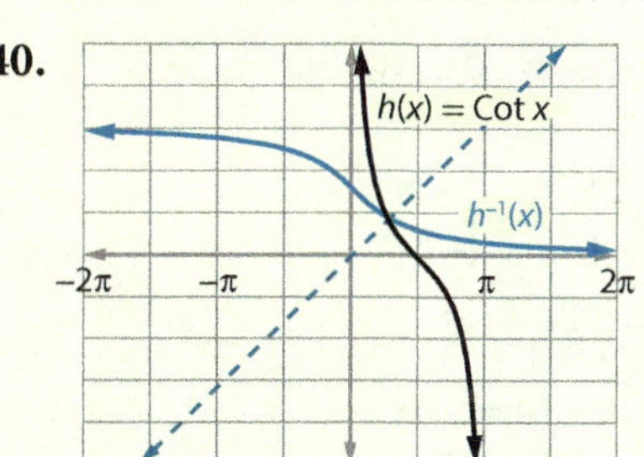

41. Let $\text{Csc}^{-1} x = \theta$.
$$\text{Csc}(\text{Csc}^{-1} t) = \text{Csc } \theta$$
$$t = \text{Csc } \theta$$
$$t = \frac{1}{\text{Sin } \theta}$$
$$\text{Sin } \theta = \frac{1}{t}$$
$$\theta = \text{Sin}^{-1} \frac{1}{t}$$
$$\therefore \text{Csc}^{-1} x = \text{Sin}^{-1} \frac{1}{x}$$

42a. $\text{Sec}^{-1} 2 = \text{Cos}^{-1} \frac{1}{2} = \frac{\pi}{3} \approx 1.0472$

42b. $\text{Csc}^{-1} \frac{2}{5} = \text{Sin}^{-1} \frac{5}{2}$ (undefined)

42c. $\cot \theta < 0$ in Q II and Q IV with a restricted domain: $(0, \pi)$
Therefore, the angle is in Q II.
$$\text{Tan}^{-1}\left(\frac{1}{-2.819}\right) \approx -0.3409$$
$$-0.3409 + \pi \approx 2.8007$$

35.

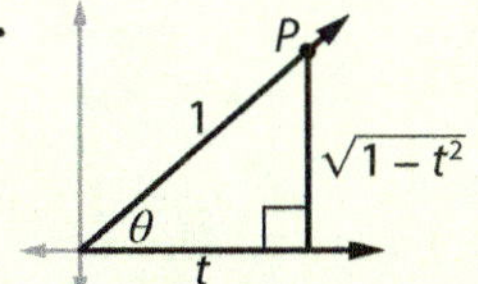

Note that θ is in Q I if $t > 0$ or in Q II if $t < 0$.
$$h(t) = \sec(\text{Cos}^{-1} t) = \sec \theta = \frac{1}{t}$$

36a. $\tan \theta = \frac{h}{r}$
$$\theta = \text{Tan}^{-1} \frac{h}{r}$$

36b. $r = 12$ in.; $\theta = \text{Tan}^{-1} \frac{7}{12} \approx 30.3°$

37a. $\tan(\alpha + \theta) = \frac{18 + 6}{30}$
$$\alpha + \theta = \text{Tan}^{-1} 0.8 \approx 38.7°$$
$$\tan \alpha = \frac{6}{30}$$
$$\alpha = \text{Tan}^{-1} 0.2 \approx 11.3°$$

$$\theta \approx 38.7° - 11.3° = 27.4°$$
$$\tan(\alpha + \theta) = \frac{18 + 6}{40}$$
$$\alpha + \theta = \text{Tan}^{-1} 0.6 \approx 31.0°$$
$$\tan \alpha = \frac{6}{40}$$
$$\alpha = \tan^{-1} \frac{3}{20} \approx 8.5°$$
$$\theta \approx 31.0 - 8.5 = 22.5°$$

37c.

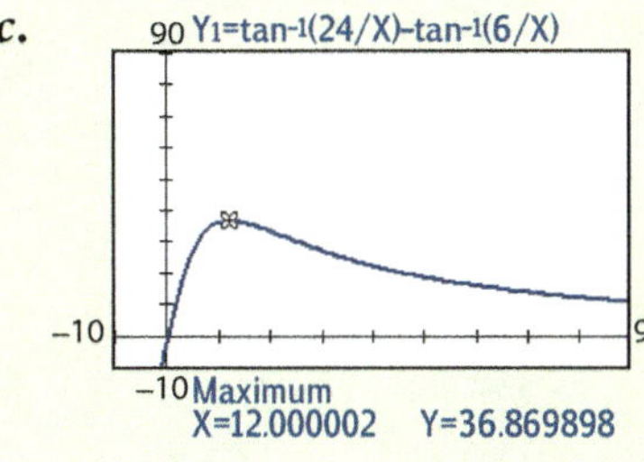

43c. $f(3) = 3^2 = 9; (3, 9)$

$$\tan \theta = \frac{\Delta y}{\Delta x}$$
$$6 = \frac{y - 9}{x - 3}$$
$$y - 9 = 6(x - 3)$$
$$y = 6x - 9$$

43d. $\theta = \text{Tan}^{-1}(6) \approx 80.5°$

❯ Cumulative Review

44. $m = \frac{11 - 5}{8 + 4} = \frac{6}{12} = \frac{1}{2}$

$$y - 5 = \frac{1}{2}(x + 4)$$
$$y - 5 = \frac{1}{2}x + 2$$
$$y = \frac{1}{2}x + 7$$

45. $f(g(x)) = (2x + 1)^2 + 2(2x + 1) + 1$
$$= 4x^2 + 4x + 1 + 4x + 2 + 1$$
$$= 4x^2 + 8x + 4$$

46. $g(f(x)) = 2(x^2 + 2x + 1) + 1$
$$= 2x^2 + 4x + 2 + 1$$
$$= 2x^2 + 4x + 3$$

47.

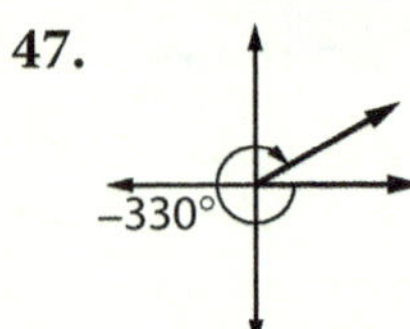

48. $\sin \theta = \frac{x}{17}; \theta = \sin^{-1} \frac{x}{17}$

49. $k = \frac{1}{2}(y_{max} + y_{min})$
$$= \frac{1}{2}(12 + (-6)) = 3$$

50. $\quad 28\left[y = \frac{5}{7}x - \frac{1}{4}\right]$
$$28y = 20x - 7$$
$$20x - 28y = 7$$

51. $\cos \theta = \frac{x + 2}{x^2 - 4} = \frac{x + 2}{(x - 2)(x + 2)}$
$$\theta = \cos^{-1}\left(\frac{1}{x - 2}\right), \text{ where } x \neq \pm 2$$

52. $p = \frac{2\pi}{|b|} = \frac{2\pi}{|4|} = \frac{\pi}{2}$

$$h = -\frac{c}{b} = -\frac{-\frac{\pi}{3}}{4} = \frac{\pi}{12}$$

---

44. Write the slope-intercept form equation of the line passing through the points $(-4, 5)$ and $(8, 11)$. [1.2]

Write the function rule for each composition when $f(x) = x^2 + 2x + 1$ and $g(x) = 2x + 1$. [1.8]

45. $(f \circ g)(x)$ **46.** $(g \circ f)(x)$

47. Draw a $-330°$ angle in standard position. [4.1]

48. Express θ as a function of x. [4.2]

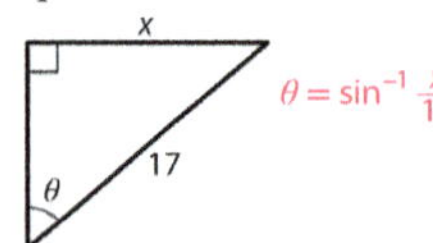

49. State the equation of the midline of a sine function with a relative maximum at $(-1, 12)$ and a relative minimum at $(2, -6)$. [4.4] $y = 3$

50. What is the general form of $f(x) = \frac{5}{7}x - \frac{1}{4}$? [1.2]   C

  A. $28x + 20y = 7$ **D.** $20x + 28y = 7$

  B. $28x - 20y = 7$ **E.** $20x - 28y = -7$

  C. $20x - 28y = 7$

44. $y = \frac{1}{2}x + 7$

45. $f(g(x)) = 4x^2 + 8x + 4$

46. $f(g(x)) = 2x^2 + 4x + 3$

51. Which of the following represents θ as a function of x? [4.1]   D

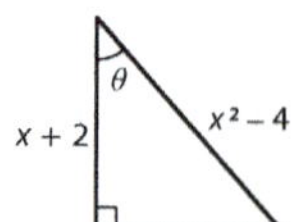

  A. $\theta = \sin^{-1}\left(\frac{x + 2}{x^2 - 4}\right)$

  B. $\theta = \sin^{-1}\left(\frac{1}{x + 2}\right)$ where $x \neq \pm 2$

  C. $\theta = \cos^{-1} \frac{x + 2}{x^2 - 4}$

  D. $\theta = \cos^{-1}\left(\frac{1}{x - 2}\right)$ where $x \neq \pm 2$

  E. none of these

52. Determine the period and phase shift of $f(x) = -2\sin\left(4x - \frac{\pi}{3}\right) - 3$. [4.4]   D

  A. $p = \frac{\pi}{3}, h = 4$ **D.** $p = \frac{\pi}{2}, h = \frac{\pi}{12}$

  B. $p = 4, h = -\frac{\pi}{12}$ **E.** none of these

  C. $p = \frac{\pi}{4}, h = -3$

53. Which of the following is true for the graph of $f(x) = \cot x$? [4.5]   E

  A. vertical asymptotes at $x = n\frac{\pi}{2}, n \in \mathbb{Z}$

  B. symmetric with respect to the y-axis

  C. zeros at $x = (2n + 1)\pi, n \in \mathbb{Z}$

  D. y-intercept at $(0, 1)$

  E. none of these

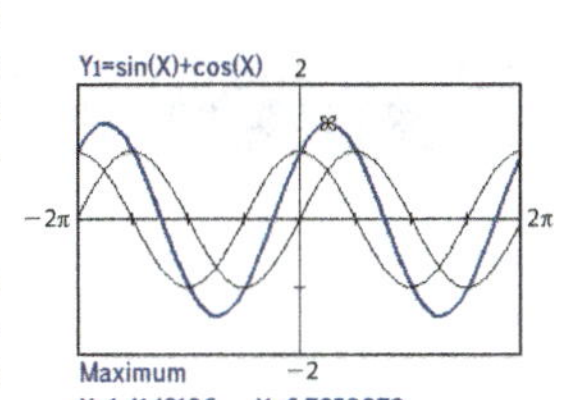

Shock absorbers are designed to dampen the oscillation of a vehicle.

In Chapter 2, we saw that the sum of polynomial functions is a polynomial function. Graphing technology can be used to explore whether the sum of two sinusoids is a sinusoid. Notice that the graph of $f(x) = \sin x + \cos x$ appears to be a sinusoid with a period of 2π, zeros where $\sin x$ and $\cos x$ have opposite values, and relative maximums of $\sqrt{2}$ at $x = \frac{\pi}{4} \pm n2\pi, n \in \mathbb{Z}$.

The sum (or difference) of two sinusoids may not be a sinusoid and frequently has a different period than either of the original sinusoids. Wave phenomena, such as the music of an orchestra, are often combinations of two or more simpler waves. The French mathematician Joseph Fourier (1768–1830) realized that even the most complex waveform could be decomposed into its sinusoidal components.

After completing this section, you will be able to

- analyze sums, products, and compositions of sinusoids and other functions.
- analyze damped oscillations.
- create and apply models of damped harmonic motion.

Example 1 Analyzing the Sum of Two Sinusoids

Determine the period of $f(x) = g(x) + h(x)$. Then use technology to graph two periods of $f(x)$. Does $f(x)$ appear to be a sinusoid?

a. $g(x) = \sin \frac{2}{3}x$ and $h(x) = -2 \cos x$

b. $g(x) = \sin \pi x$ and $h(x) = \frac{1}{2} \cos \left(\pi x - \frac{\pi}{3} \right)$

KEYWORD SEARCH

Fourier series

Answer

1. The period of $g(x)$ is 3π, and the period of $h(x)$ is 2π. The period of $f(x)$ is 6π, the least common multiple of the periods of $g(x)$ and $h(x)$.

1. Since the period of both $g(x)$ and $h(x)$ is 2, the period of $f(x)$ is also 2.

2. Graph $f(x) = \sin \frac{2}{3}x - 2 \cos x$ over $[-6\pi, 6\pi]$.

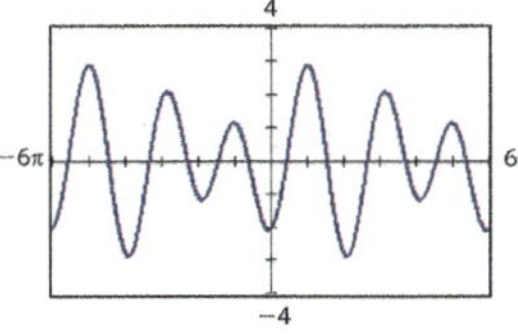

2. Graph $f(x) = \sin \pi x + \frac{1}{2} \cos \left(\pi x - \frac{\pi}{3} \right)$ over $[-4, 4]$.

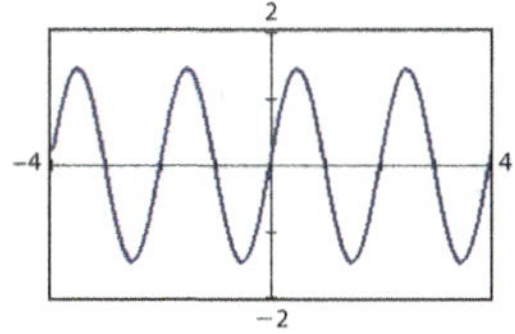

3. The function is not a sinusoid.

3. The function appears to be a sinusoid.

Jules Lissajous studied compound vibrations by attaching small mirrors to two tuning forks with perpendicular planes of vibration. A light beam passed from one mirror to the other and then on to a screen, forming visual patterns. These "Lissajous figures" are still studied in modern physics classes.

SKILL ✔ EXERCISE 3

4.7 Analyzing Combinations of Sinusoidal Functions (Extended)

Objectives

1. To analyze sums and differences of sinusoids

2. To analyze sums, products, and compositions of trigonometric and other functions

3. To analyze damped oscillations

4. To create and apply models of damped harmonic motion

Flash

Worn-out shock absorbers on a vehicle can cause excessive bouncing, which affects tire wear and vehicle control.

Vocabulary

damped harmonic motion
damped oscillation
damped sinusoid
damping constant
damping factor
pseudoperiod
simple harmonic motion

PRESENTATION

Lesson Opener

Write the simplified function rule for each function if $f(x) = 3x - 4$ and $g(x) = -2x + 7$.

1. $f(x) + g(x)$ $f(x) + g(x) = x + 3$
2. $f(x) - g(x)$ $f(x) - g(x) = 5x - 11$
3. $f(x) \cdot g(x)$ $f(x) \cdot g(x) = -6x^2 + 29x - 28$
4. $f(g(x))$ $f(g(x)) = -6x + 17$

This section can be skipped by minimum- and standard-track students, as its content is not required for material appearing later in the text. Consider introducing this lesson with an exploratory approach.

Briefly review the definition, general form, and characteristics of a sinusoid, including the period, maximums, and minimums.

Ask the students whether they think that the sum of two sinusoids will be a sinusoid. (*sometimes*) Then have them graph the following sums of sinusoids using technology.

$$y = \sin x + \cos x$$
$$y = \sin x - 2 \sin x$$
$$y = \sin x + \sin 2x$$

Guide students to the conclusion that the sum of sinusoids is a sinusoid only when the periods are the same.

Use Example 1 to illustrate the use of the LCM to determine the period of the sum of two sinusoids. Point out that both graphs in Example 1 are periodic. Explain that if the ratio of the periods of the functions is rational, the LCM of their periods is the period of the resulting function, as illustrated in Example 1a. Consider the sum of $f(x) = \sin 2x$ and $g(x) = \sin \pi x$. Since the ratio of their periods, $\frac{\pi}{2}$, is irrational, $f(x) + g(x)$ is not periodic. Note that π and 2 are incommensurable since they have no common multiple; that is, there are no positive integers a and b such that $a(\pi) = b(2)$.

Determine the period of $f(x) = g(x) + h(x)$ and state whether $f(x)$ is a sinusoid. Then use technology to graph two periods of $f(x)$.

1. $g(x) = \sin 5x$; $h(x) = -3 \cos (5x - 1)$

$p_f = \dfrac{2\pi}{5}$; sinusoid

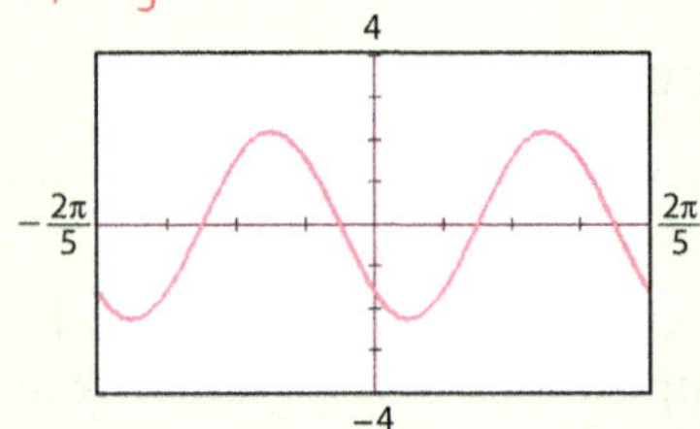

2. $h(x) = 3 \cos 2x - 1$; $g(x) = -4 \sin x$

$p_f = 2\pi$; not a sinusoid

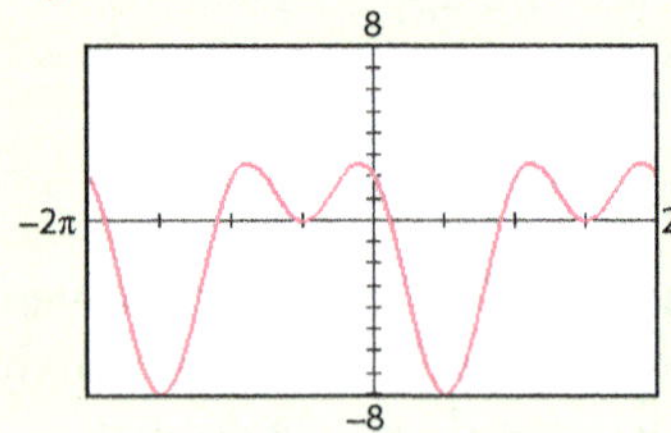

State equations for the bounds of each function; then graph $f(x)$ and its bounding equations. Is $f(x)$ a periodic function?

3. $y = -x + 2 \sin x$

bounds: $y = -x \pm 2$; not periodic

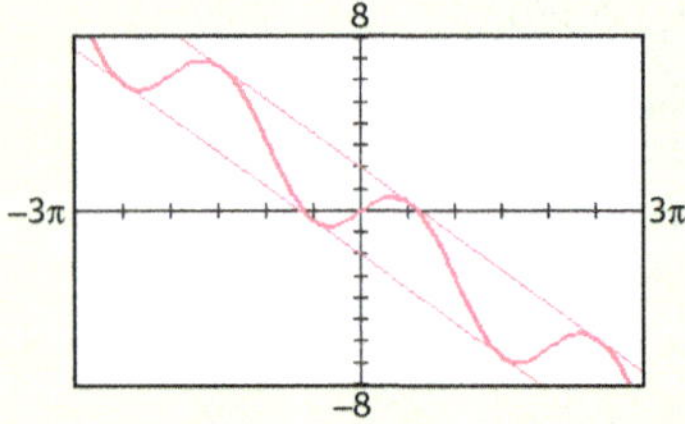

It can be shown that the sum (or difference) of sinusoids with the same period is a sinusoid with that same period. While the sum (or difference) of sinusoids with different periods is not a sinusoid, the resulting function is periodic when the ratio of the addends' periods is rational.

The fact that trigonometric functions are periodic distinguishes them from algebraic, exponential, and logarithmic functions. Combinations of these functions with sinusoidal functions may or may not be periodic.

Consider the sum (or difference) of a polynomial function $p(x)$ and a sinusoidal function. Recall that the graph of $t(x) = a \sin b(x - h)$ or $t(x) = a \cos b(x - h)$ oscillates between $y = \pm a$. Therefore, the graph of $f(x) = p(x) + t(x)$ oscillates between the bounds of $y = p(x) \pm a$.

Example 2 Analyzing a Sum of Polynomial and Sinusoidal Functions

State equations for the bounds of $f(x) = x + 2 \sin 2x$. Then graph $f(x)$ and its bounding equations. Is $f(x)$ a periodic function?

Answer

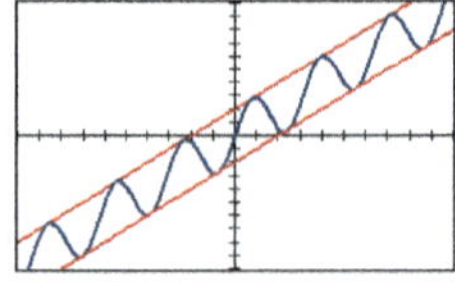

1. Since the value of $2 \sin 2x$ oscillates between ± 2, $f(x)$ oscillates between $y = x + 2$ and $y = x - 2$.

2. Graph $f(x)$ and its bounds.

3. The function $f(x)$ is not periodic since its value does not repeat over a regular interval.

SKILL ✓ EXERCISE 9

The basic sine and cosine functions are frequently composed with basic power functions. Note that $\sin^2 x$ is shortened notation for $(\sin x)^2$ and is different from $\sin x^2$.

Example 3 Analyzing Compositions of a Power and a Sinusoidal Function

Graph each composition of $g(x) = x^2$ and $h(x) = \cos x$. Does $f(x)$ appear to be periodic? If so, state its period.

a. $f(x) = h(g(x)) = \cos x^2$

b. $f(x) = g(h(x)) = \cos^2 x$

Answer

1.
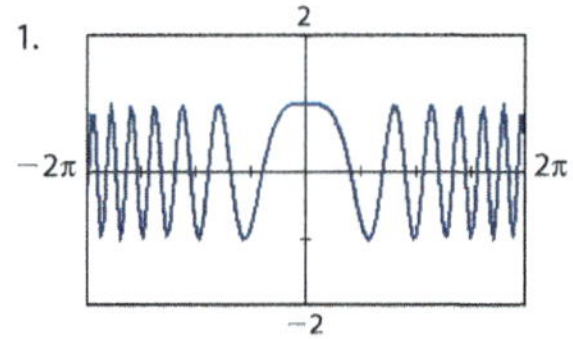

1.
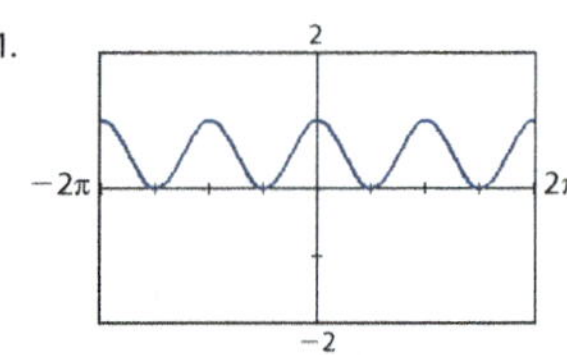

2. The function is not periodic. Note that some graphing calculators have difficulty displaying waves with high frequency.

2. The function appears to have a period of π.

SKILL ✓ EXERCISE 15

The combination of sinusoids can be illustrated by the composition of sound waves created by multiple tuning forks. The Data Analysis note describes how Jules Lissajous (1822–80) was able to "see" sound. The idea of visualizing sound waves eventually led to the invention of the oscilloscope. Use the Internet keyword search *virtual oscilloscope* for sound wave demonstrations.

Introduce sums of a polynomial function and a sinusoidal function. Present Example 2. Explain that the bounds will be $y = p(x) \pm a$, just as the vertical translation in $y = a \sin b(x - h) + k$ or $t(x) = a \cos b(x - h) + k$ is actually a sum of a sinusoid and the constant polynomial function $p(x) = k$ that oscillates between $y = k \pm a$.

One-on-One Write $y = x + \sin x$ in the form $y = \sin x + x$ to compare to $y = \sin x + k$. Point out that k is a constant value that affects vertical translation. Replacing k with x causes a continuous change in the vertical translation and can be thought of as having a "dynamic" vertical translation.

Use Example 3 to illustrate the composition of a polynomial function with a sinusoidal function. Share that some graphing calculators will have difficulty displaying the graph of a wave with high frequency due to limited resolution. Consider viewing smaller intervals on the x-axis or using an online graphing calculator when the graph is not clear.

Interactive Activity Online graphing calculators with sliders provide remarkable opportunities for dynamic demonstrations and interactive explorations for this section.

Introduce the product of a polynomial and sinusoidal function by having students recall the effect of a in the function $y = a \sin x$. (*It affects the amplitude.*) Remind them that a has been a constant, causing the amplitude to be $|a|$. Ask the students how the graph would be affected

When $t(x) = \sin bx$ or $t(x) = \cos bx$ is multiplied by another function $g(x)$, the graph of the product oscillates between $y = \pm g(x)$. You have already seen how the leading constant, a, determines the amplitude of a sinusoid so that $f(x) = 2 \sin x$ oscillates between the bounds of $y = 2$ and $y = -2$. Notice that the graph of $f(x) = x \sin 5x$, the product of $g(x) = x$ and $t(x) = \sin 5x$, is a sine wave with varying amplitude that oscillates between $y = \pm x$.

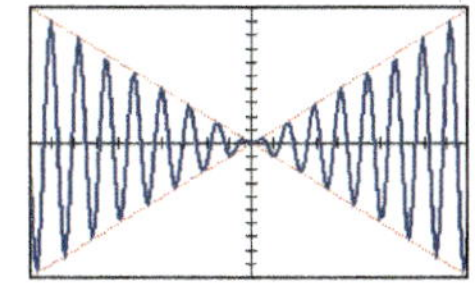

DEFINITIONS

A **damped sinusoid** is a function of the form $f(x) = g(x) \sin bx$ or $f(x) = g(x) \cos bx$. When $g(x)$ reduces the amplitude of the sinusoid model, $g(x)$ is called the **damping factor**, and $f(x)$ has a **damped oscillation**.

The function $f(x) = x \sin 5x$ is damped over $(-\infty, 0]$ with $g(x) = x$ as the damping factor.

Example 4 Analyzing Damped Sinusoids

State the damping factor and equations for the bounds of each function. Then use technology to graph the function and state the interval where a damping effect occurs.

a. $f(x) = \dfrac{x^2}{5} \sin 3x$

b. $f(x) = \dfrac{6}{x} \sin 9x$

Answer

1. damping factor: $\dfrac{x^2}{5}$

   bounds: $y = \pm \dfrac{x^2}{5}$

2. 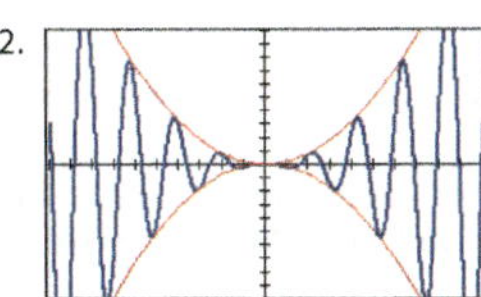

3. $f(x)$ is damped over $(-\infty, 0]$.

1. damping factor: $\dfrac{6}{x}$

   bounds: $y = \pm \dfrac{6}{x}$

2. 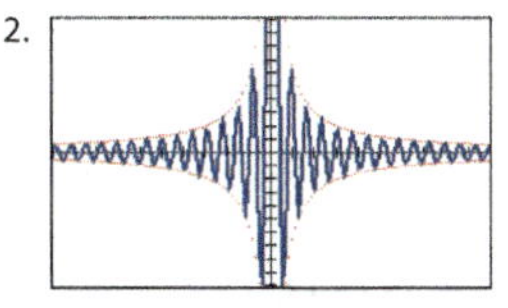

3. $f(x)$ is damped over $(0, \infty)$.

SKILL ✓ EXERCISE 17

When a mass suspended from a spring is pulled down and released, it moves above and below its resting (or equilibrium) position. In the ideal case of no friction, the oscillations would continue with the same amplitude, and *simple harmonic motion* would result. The height above or below the equilibrium position over time t is modeled by the function $y = -a \cos bt$ where $a > 0$ represents the maximum distance from the equilibrium position.

The *damped harmonic motion* that occurs when friction reduces the displacement of an oscillating object can be modeled by $f(t) = ae^{-ct} \cos bt$ or $f(t) = ae^{-ct} \sin bt$ $(c > 0)$, where a is the maximum displacement. Greater values of c, the *damping constant*, indicate a faster decrease in the amplitude of oscillations. The damped function is not periodic since its amplitude varies, but the function's frequency can be used to find b using a *pseudoperiod*:

$$p = \frac{1}{f} = \frac{2\pi}{|b|}.$$

A harmonograph draws patterns by combining the motion of two connected pendulums. These compound pendulums were originally used to model the motion of the earth as seen from the moon. The value of these harmonographs is now mainly aesthetic, exemplifying how mathematical patterns appeal to our God-given sense of beauty.

4. $y = x^3 + 4 \sin 5x$

bounds: $y = x^3 \pm 4$; not periodic

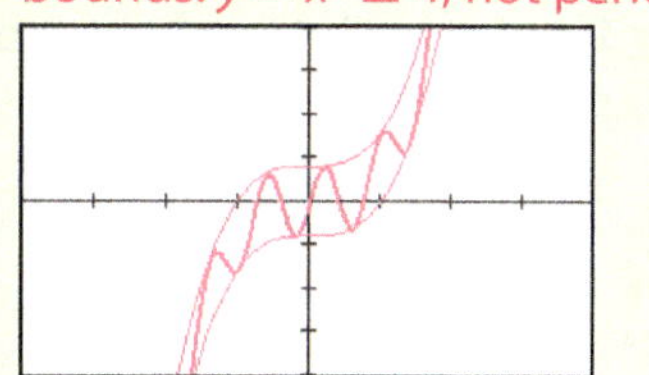

5. $y = 2 \sin x + \ln x$

bounds: $y = \ln x \pm 2$; not periodic

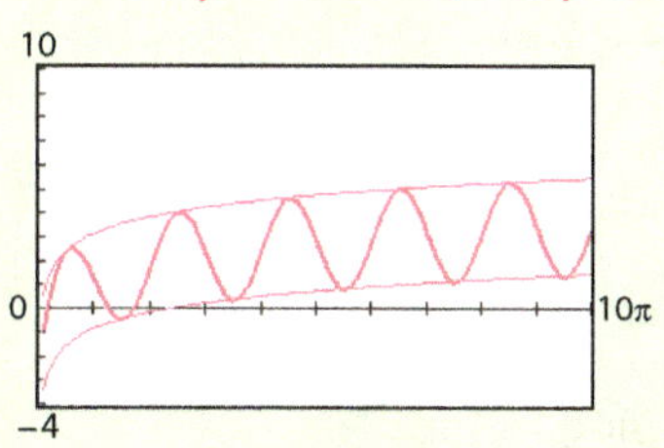

Graph each function. Does the function appear to be periodic? If so, state the period.

6. $y = |\sin^2 x|$ periodic; π

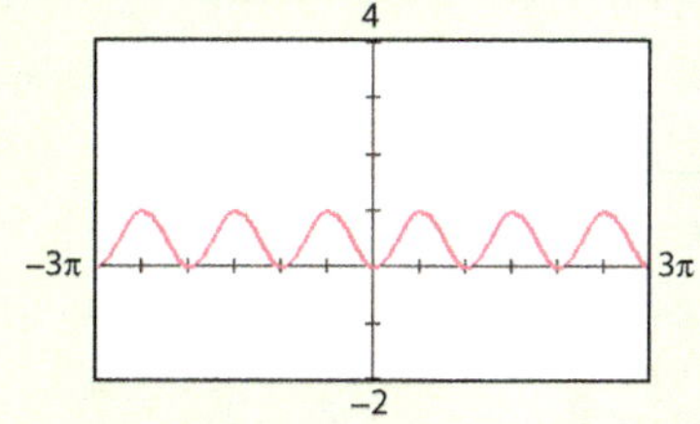

7. $y = \sin x^2$ not periodic

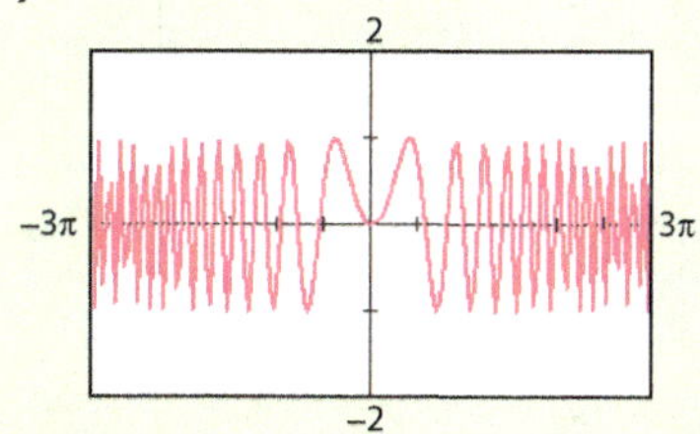

---

if we replaced a with the variable x, giving us the function $y = x \sin x$. (*The amplitude would vary with $|x|$.*) Challenge them to describe the amplitude of $y = x \sin x$. (*It will be 0 when $x = 0$ and then continually increase in both directions from $x = 0$.*)

Define the terms related to *damped* functions. *Damped sinusoids* generally have an amplitude that decreases over a specific interval. In some physics and engineering applications the term refers to sinusoids whose amplitude decreases to 0. Identify the key characteristics of damped sinusoids in Example 4.

Motivational Idea The Internet keyword search *harmonograph* can be used to find videos of harmonographs in action, images of beautiful figures created by them, and tutorials on how to build your own. Use the keyword search *harmonograph simulator* to illustrate one of these beautiful patterns.

Refer to the illustration of the spring and weight motion to describe *simple harmonic motion*. Point out that due to its dynamically changing amplitude, a damped function is not periodic, but because it does have a constant frequency, we can refer to its *pseudoperiod*.

In Example 5, emphasize the choice of a cosine function as opposed to a sine function. *Damped harmonic motion* begins at maximum displacement. In the case of a plucked string or an object on a spring, let $t = 0$ when the object is released from its maximum displacement.

TIPS

Ex. 39 A G_1 note is produced at 49 Hz.

State the damping factor for each function. Then graph the function and its bounds and state the intervals at which a damping effect occurs.

8. $y = \sqrt[3]{x} \sin x$

damping factor: $\sqrt[3]{x}$
bounds: $y = \pm\sqrt[3]{x}$; $(-\infty, 0]$

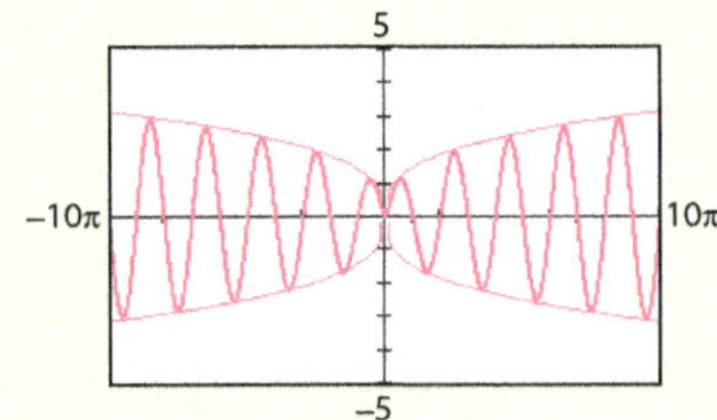

9. $y = \dfrac{2x^2}{3} \sin 2\pi x$

damping factor: $\dfrac{2x^2}{3}$

bounds: $y = \pm\dfrac{2x^2}{3}$; $(-\infty, 0]$

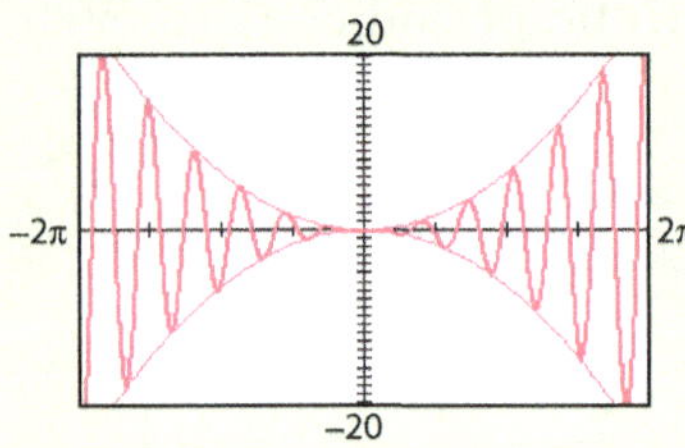

10. $y = -1.5^{-x} \cos 2x$

damping factor: -1.5^{-x}
bounds: $y = \pm1.5^{-x}$; $(-\infty, \infty)$

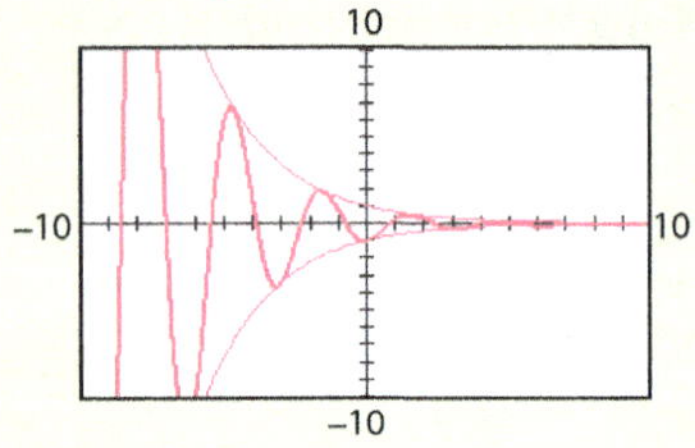

Assignments

- **Minimum:** 1–3, 7, 10–13, 16–18; 21–29 odd; 37–38, 40–41, 45–49, 52, 54
- **Standard:** 2, 4, 6–7, 10, 12–15, 17–19, 21, 23–35, 37–42, 45–48, 50–54
- **Extended:** 2, 5; 8–26 even; 27–35, 37–44, 46–47, 50–54

Assessment

- Quiz 4C covers Sections 4.6–4.7.

Example 5 Modeling Damped Harmonic Motion

A robot that explores tunnels uses springs connected to a steel plate as a shock absorber to protect the robot's electronics in the event of an explosion. In a laboratory test, the steel plate's initial displacement was 5 cm, the damping constant was 0.6, and the plate oscillated with a frequency of 2 Hz. Write a function modeling the plate's damped harmonic motion beginning at maximum displacement. Then use technology to graph the function and determine when the plate's displacement is within ±1 cm.

Answer

$d(t) = ae^{-ct} \cos bt$

$d(t) = 5e^{-0.6t} \cos 4\pi t$

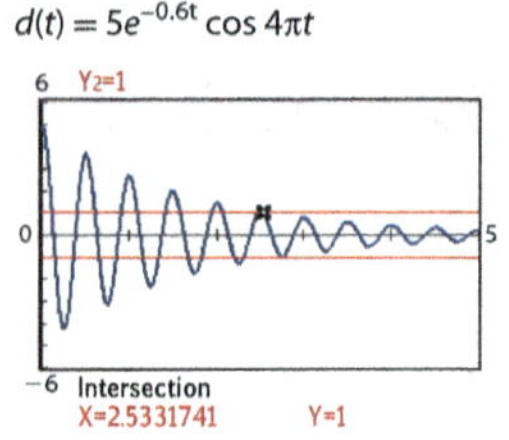

The displacement is within ±1 cm after ≈ 2.53 sec.

1. Substitute values for a, c, and b into the general sinusoid equation that models damped harmonic motion where the maximum displacement occurs when $t = 0$.

   maximum displacement: $a = 5$

   damping constant: $c = 0.6$

   frequency: $f = 2$ Hz; so $p = \dfrac{1}{f} = \dfrac{1}{2} = \dfrac{2\pi}{|b|}$; $b = 4\pi$

2. Graph $Y_1 = 5e^{-0.6x} \cos 4\pi x$, $Y_2 = 1$, and $Y_3 = -1$.

3. A careful examination of the graph indicates that the last intersection of $d(t)$ with $y = \pm1$ occurs at ≈ 2.53 sec.

SKILL ✔ EXERCISE 39

❯ A. Exercises

Determine the period of $f(x) = g(x) + h(x)$ and state whether $f(x)$ is a sinusoid. Then use technology to graph two periods of $f(x)$.

1. $g(x) = \sin x$; $h(x) = 2 \cos x$ 2π; a sinusoid

2. $g(x) = \sin 2x$; $h(x) = -\cos x$ 2π; not a sinusoid

3. $g(x) = \cos \frac{\pi}{2}x$; $h(x) = \sin \pi x + 1$ 4; not a sinusoid

4. $g(x) = \cos (3x + \pi)$; $h(x) = -2 \sin 3x$ $\frac{2\pi}{3}$; a sinusoid

5. $g(x) = 3 \sin \frac{x}{2} + 1$; $h(x) = -2 \cos \frac{x}{2}$ 4π; a sinusoid

6. $g(x) = 2 \cos x - 1$; $h(x) = -\sin (2x - 1)$ 2π; not a sinusoid

State equations for the bounds of each function. Then graph the function and its bounds. Is the function periodic?

7. $y = x - 3 \sin x$

8. $y = \cos 3x - \frac{1}{2}x$

9. $y = (x^2 - 6) + 3 \cos 4x$

10. $y = \dfrac{x^2}{4} \sin 4x$

Use technology to graph each function. Does the function appear to be periodic? If so, state its period.

11. $y = 3 \sin x \cdot \cos x$

12. $y = |\sin x|$

13. $y = x \cdot \cos x$

14. $y = \cos x^3$

15. $y = \cos^3 x$

16. $y = \sin x + \cos \pi x$

State the damping factor for each function. Then graph the function and its bounds and state the interval where a damping effect occurs.

17. $y = \frac{x}{2} \sin 2x$

18. $y = \frac{12}{x} \cos x$

19. $y = 3^{-0.25x} \sin \frac{7x}{2}$

20. $y = e^{0.3x} \cos \pi x$

❯ B. Exercises

Determine the period of each function and state whether the function is a sinusoid. Then use technology to graph two periods of $f(x)$.

21. $f(x) = \sin \pi x - 2 \cos \pi x + \frac{1}{2} \sin \left(\pi x - \frac{\pi}{2}\right)$ $p = 2$; sinusoid

22. $f(x) = \cos (2x - \pi) + 3 \sin \frac{x}{2} - \sin \frac{x}{3}$ $p = 12\pi$; not a sinusoid

Choose the best description for each function in exercises 23–30.

A. a non-periodic function

B. a periodic function

C. a sinusoid

23. $y = \sin x + 3 \cos x$ C

24. $y = \sin x + \cos 2x$ B

25. $y = x + \cos 5x$ A

26. $y = x \cos 5x$ A

27. a sum of sinusoids with the same period   C

Solutions

❯ A. Exercises

1. $p_g = 2\pi$; $p_h = 2\pi$; $p_f = 2\pi$; sinusoid

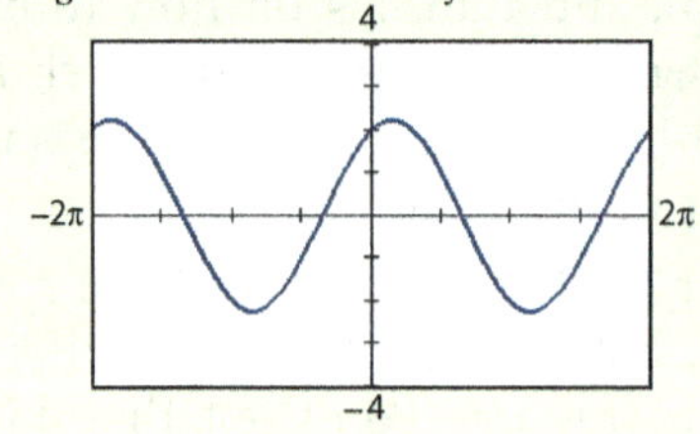

2. $p_g = \pi$; $p_h = 2\pi$; $p_f = 2\pi$; not a sinusoid

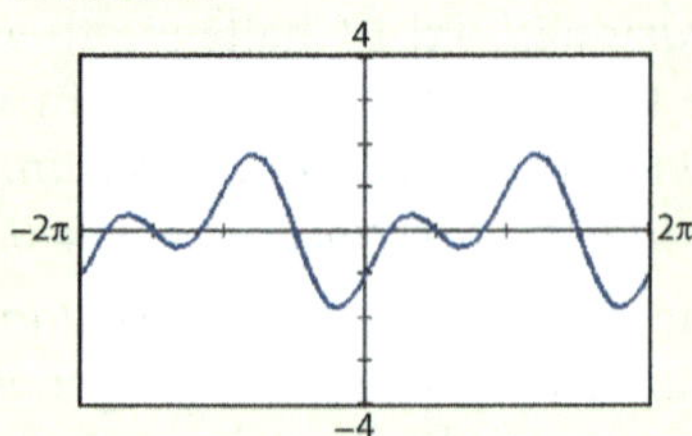

3. $p_g = 4$; $p_h = 2$; $p_f = 4$; not a sinusoid

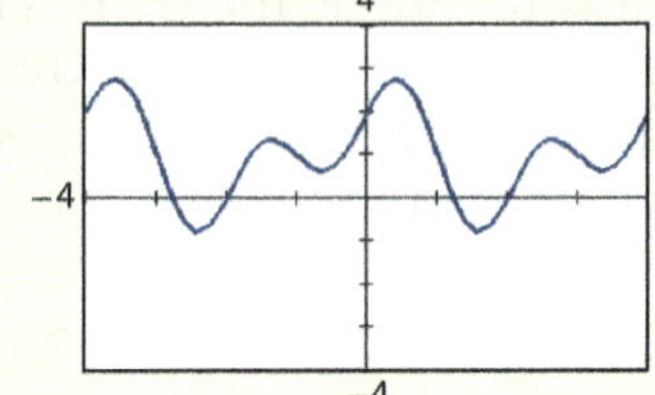

28. a sum of sinusoids whose periods have a common multiple B

29. a product of an exponential function and a sinusoid A

30. a sum of a sinusoid and a non-constant polynomial function A

Match each function with its graph without using technology.

31. $y = x \cos 3x$ C **32.** $y = x + \cos 3x$ D

33. $y = 3 \cos x + \sin x$ B **34.** $y = \cos 3x + 2 \sin x$ E

35. $y = 3 \sin \pi x + 2 \cos x$ A

A. 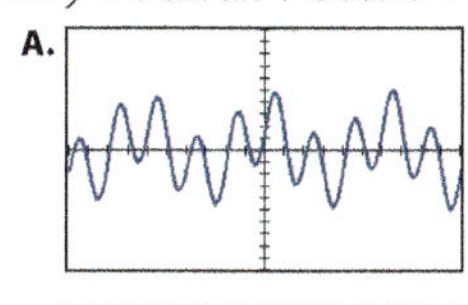

D. 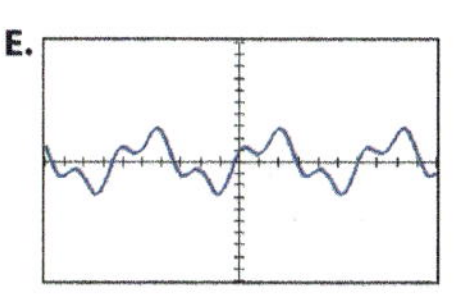

B. 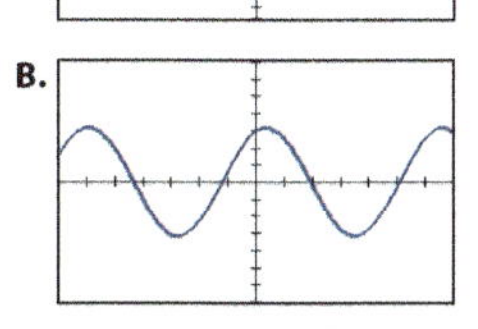

E. 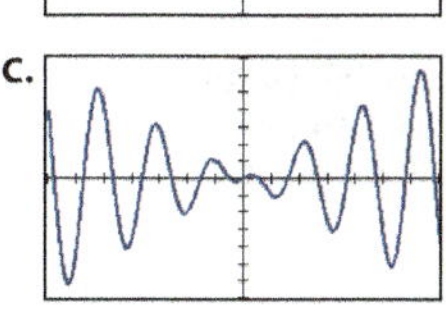

C. 

Write an equation for each damped sine function.

36. damping factor of e^{-2x}; pseudoperiod of π; phase shift of $\frac{\pi}{4}$ units left $y = e^{-2x} \sin 2\left(x + \frac{\pi}{4}\right)$

37. bounds of $y = \pm 0.2x^2$; pseudoperiod of 2; phase shift of π units right $y = 0.2x^2 \sin \pi (x - \pi)$

11. periodic, π

12. periodic; π

13. not periodic

14. not periodic

15. periodic; 2π

16. not periodic

38. Tyler hit a bump on his motorcycle, initiating damped harmonic motion of the front wheel at 3 Hz with an initial displacement of 5 cm and a damping constant of 0.7.

  a. Write a function that models the displacement of the front wheel. $d(t) = 5e^{-0.7t} \cos 6\pi t$

  b. Use a graph of the function to find how much time (to the nearest hundredth of a second) elapses before the front wheel oscillates with a displacement less than 1 cm. 2.19 sec

39. The motion of a point on a plucked violin string can be modeled by $f(x) = ae^{-2.05x} \cos bx$, where a is the initial displacement from the string's resting position in centimeters. A musician displaces a violin string by 0.9 cm, causing damped vibrations at 49 Hz. Determine the displacement of the string one second after it has been plucked. ≈ 0.12 cm

❯ C. Exercises

40. Explore: Graph $f(x) = \sin^2 x + \cos^2 x$ and write the equation suggested by the graph. $\sin^2 x + \cos^2 x = 1$

41. Explore: Graph the product of $g(x) = \sec x$ and $h(x) = \cos x$ and write the equation suggested by the graph. What does the graph reveal about these two functions? sec x cos $x = 1$; Secant and cosine are reciprocal functions.

42. Explore: Graph the product of $g(x) = \tan x$ and $h(x) = \cos x$ and write the equation suggested by the graph. tan $x \cdot \cos x = \sin x$

Use technology to graph each sum of sinusoids. Then use the graph to estimate the altitude and the phase shift to the nearest hundredth and rewrite the sum as a single sinusoid with the given form. Verify your equation by graphing the single sinusoid in the same window as the sum.

43. $f(x) = \sin x + \cos x = a \cos b(x - h)$

44. $f(x) = 2 \cos \pi x - 3 \sin \pi x = a \sin b(x - h)$

8. bounds: $y = -\frac{1}{2}x \pm 1$

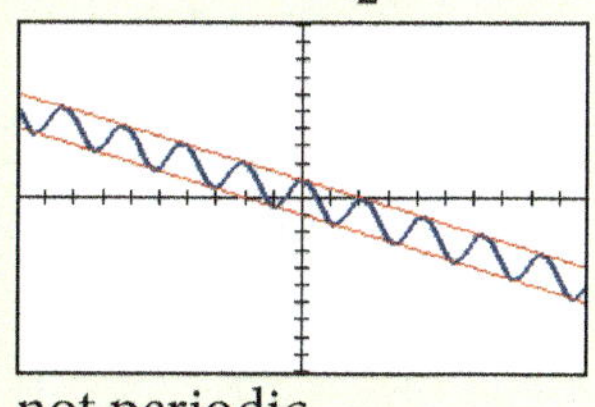

not periodic

9. bounds: $y = (x^2 - 6) \pm 3$
$y = x^2 - 3$ and $y = x^2 - 9$

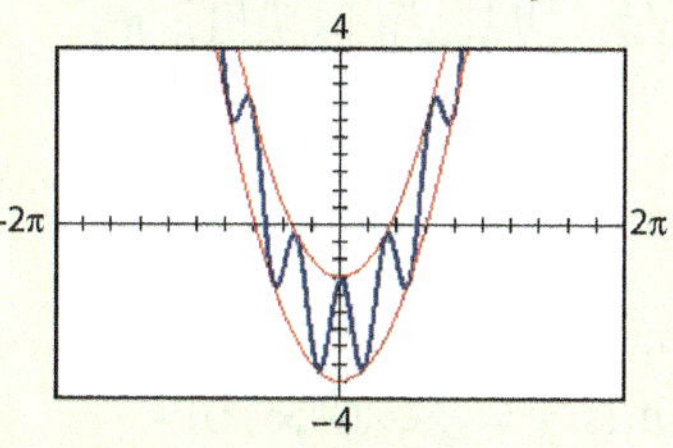

not periodic

10. bounds: $y = \pm \dfrac{x^2}{4}$

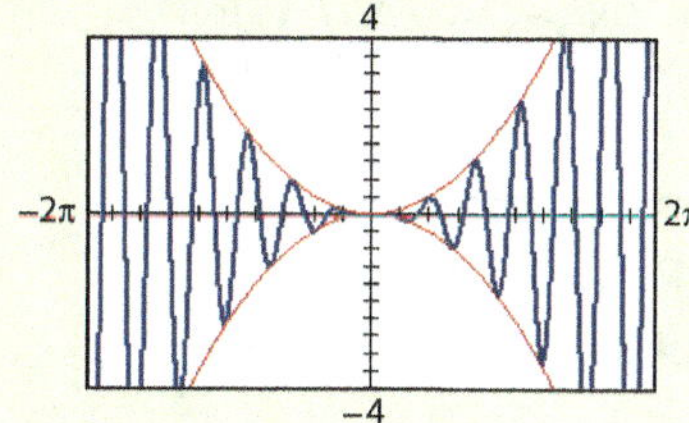

not periodic

11.

12. 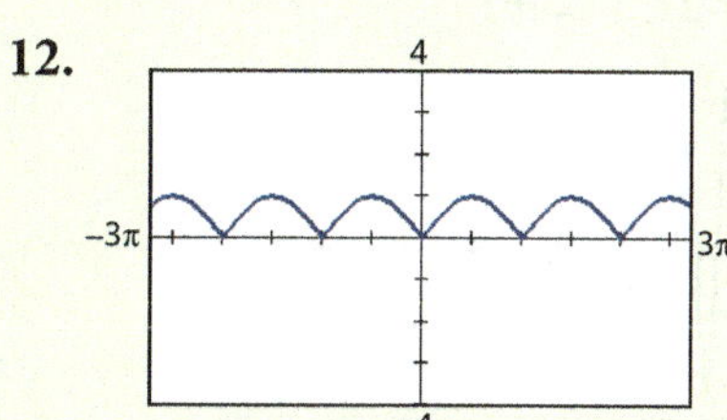

4. $p_g = \dfrac{2\pi}{3}$; $p_h = \dfrac{2\pi}{3}$; $p_f = \dfrac{2\pi}{3}$; sinusoid

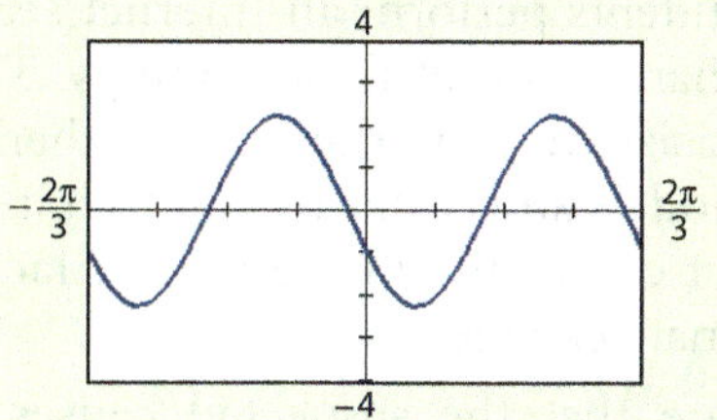

5. $p_g = 4\pi$; $p_h = 4\pi$; $p_f = 4\pi$; sinusoid

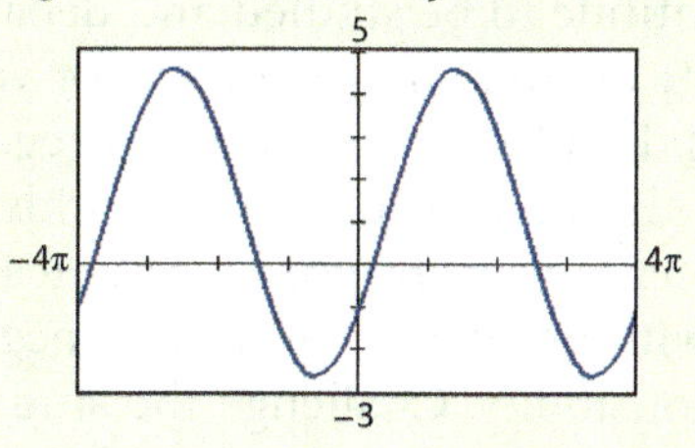

6. $p_g = 2\pi$; $p_h = \pi$; $p_f = 2\pi$; not a sinusoid

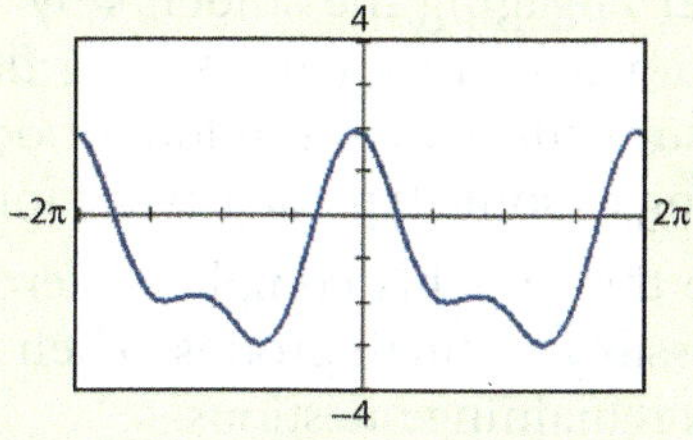

7. bounds: $y = x \pm 3$

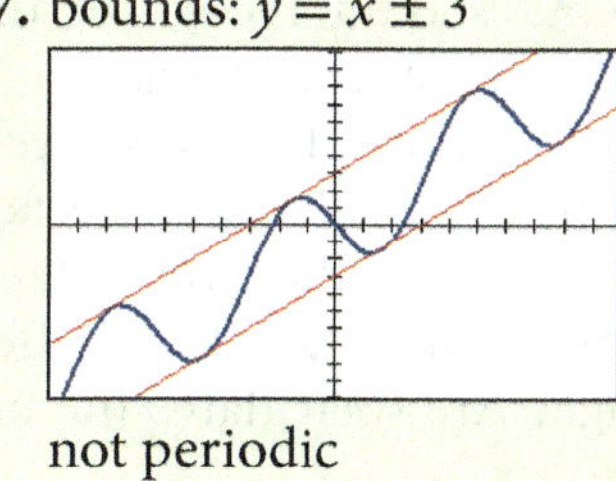

not periodic

13. 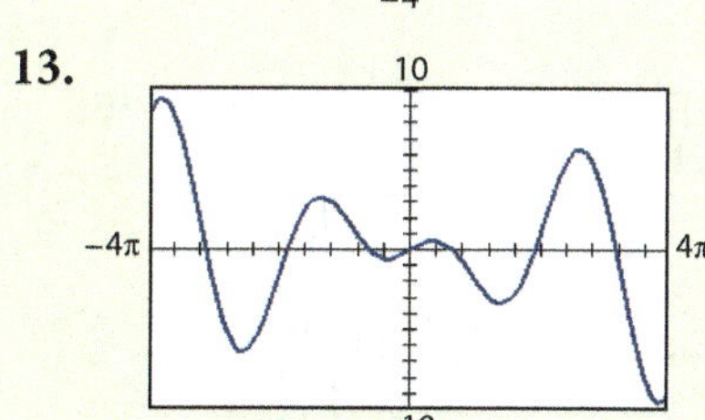

The amplitude varies with $|x|$.

14. 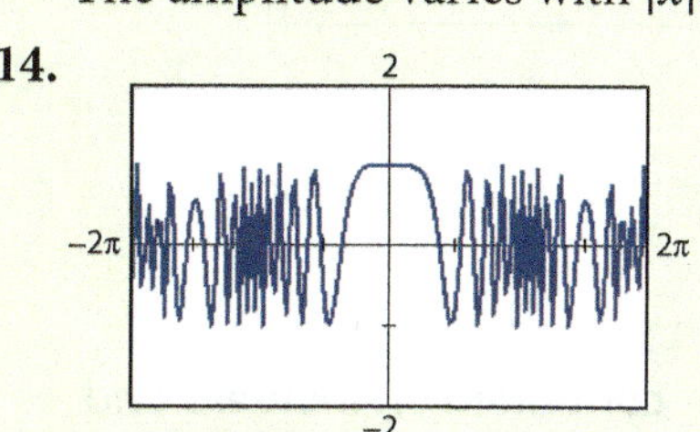

The frequency is not constant.

15.

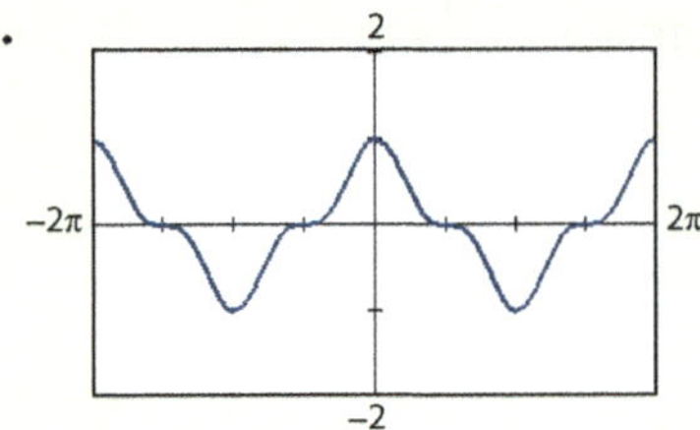

16.

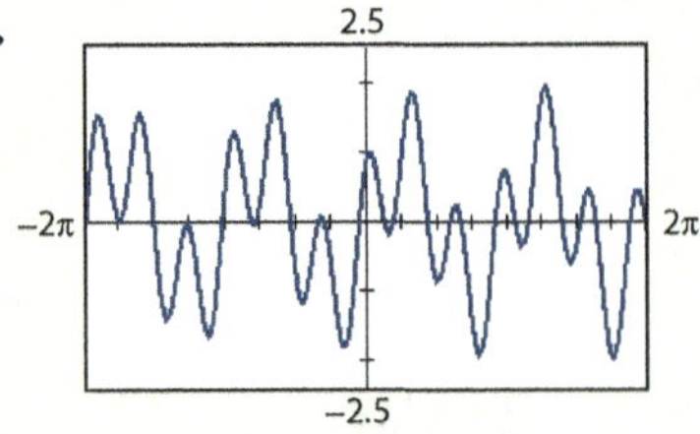

17. damping factor: $\frac{x}{2}$
bounds: $y = \pm\frac{x}{2}$; $(-\infty, 0]$

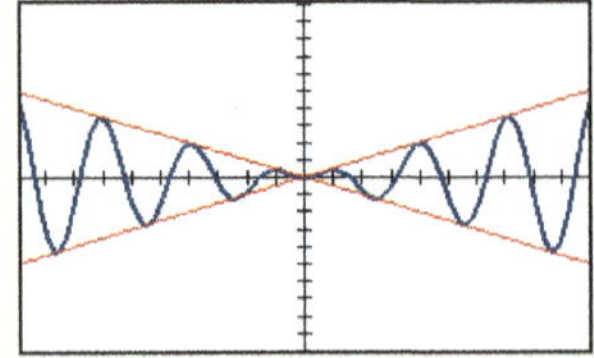

18. damping factor: $\frac{12}{x}$
bounds: $y = \pm\frac{12}{x}$; $(0, \infty)$

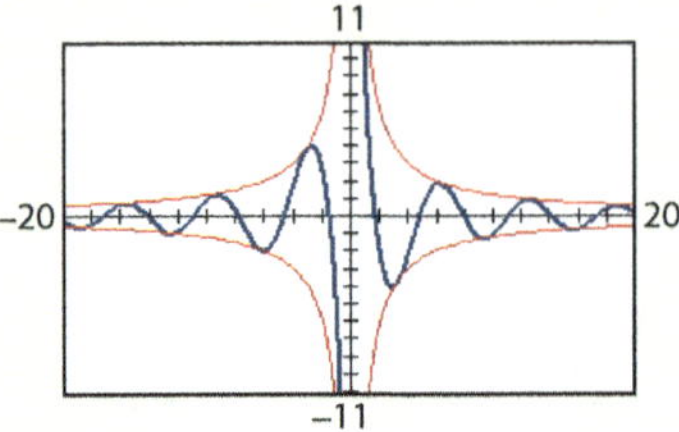

19. damping factor: $3^{-0.25x}$; $(-\infty, \infty)$
bounds: $y = \pm 3^{-0.25x}$

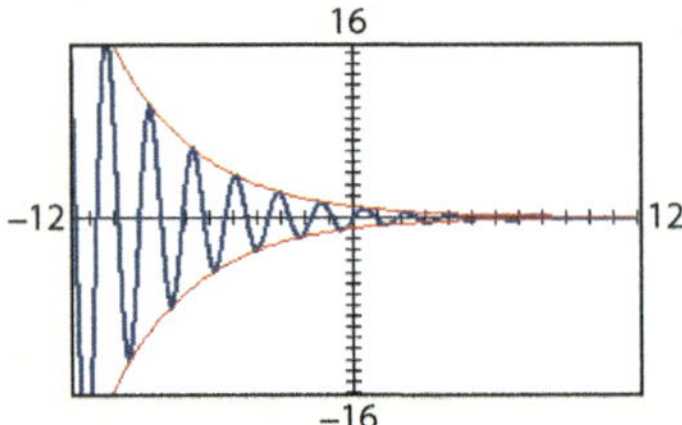

20. damping factor: $e^{0.3x}$; no damping effect

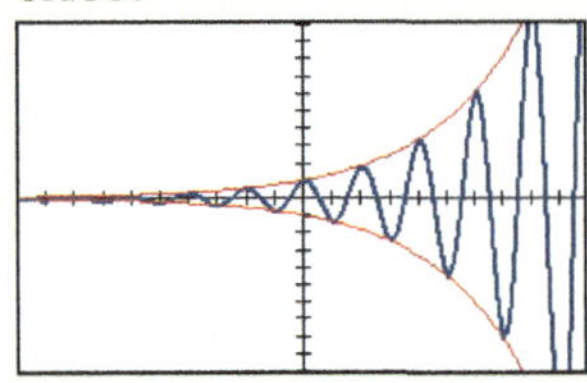

continued in Answers and Solutions Overflow

45. Find the inverse of $f(x) = \frac{2}{3}x - 5$. [1.8] $f^{-1}(x) = \frac{3}{2}x + \frac{15}{2}$

46. Convert $\frac{12\pi}{7}$ radians to degrees. [4.1] $\approx 308.6°$

47. If θ is an angle in standard position whose terminal side passes through $(-7, 2)$, find the exact value of $\sin \theta$. [4.3] $\sin \theta = \frac{2\sqrt{53}}{53}$

48. State the quadrant and reference angle for $\theta = 1000°$. [4.3]   Q IV; 80°

State the zeros of each function. [4.5]

49. $f(x) = \sin x$ [4.4]

50. $g(x) = \cos x$ [4.4]

51. Which function is equivalent to $f(x) = 9 \cdot 3^{2x}$? [3.1]   A

  A. $g(x) = 3^{2x+2}$ **D.** $g(x) = 81^x$

  B. $g(x) = 3^{4x}$ **E.** none of these

  C. $g(x) = 27^{2x}$

49. $x = n\pi, n \in \mathbb{Z}$

50. $x = \frac{\pi}{2} + n\pi, n \in \mathbb{Z}$

52. Which expression is equivalent to $\log \frac{x^2}{y^4}$? [3.3]   C

  A. $\log 2x - \log 4y$ **D.** $\log (x^2 - y^4)$

  B. $2 \log x + 4 \log y$ **E.** none of these

  C. $2 \log x - 4 \log y$

53. Which function is equivalent to $f(x) = \sec\left(\frac{\pi}{2} - x\right)$? D [4.5]

  A. $g(x) = \sin x$ **D.** $g(x) = \csc x$

  B. $g(x) = \cos x$ **E.** $g(x) = \sec x$

  C. $g(x) = \cot x$

54. Evaluate $\text{Sin}^{-1}\left(-\frac{\sqrt{2}}{2}\right)$. [4.6]   C

  A. $\frac{\pi}{4}$ **D.** $-\frac{3\pi}{4}$

  B. $\frac{7\pi}{4}$ **E.** none of these

  C. $-\frac{\pi}{4}$

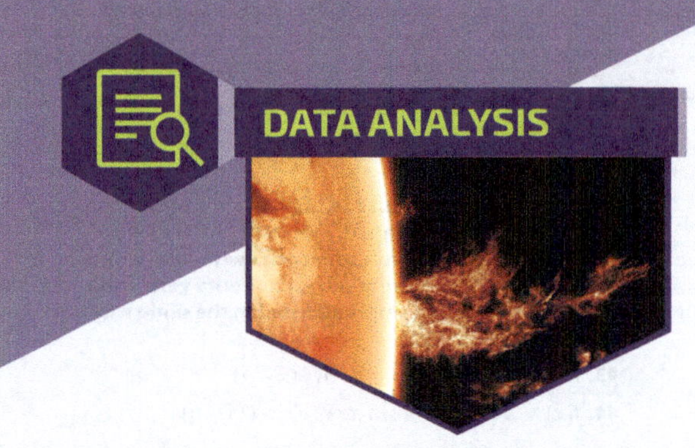

DATA ANALYSIS

Sunspots and Solar Flares

Year	Average Daily Number of Sunspots
1980	149.1
1982	114.8
1984	43.5
1986	11.0
1988	100.9
1990	145.1
1992	93.5
1994	31.0
1996	8.4
1998	61.6
2000	123.3

Sunspot data from the World Data Center SILSO, Royal Observatory of Belgium, Brussels

At about 6300°F, sunspots are dark and relatively cool areas on the sun's surface (average temperature about 10,000°F). They usually appear in pairs having opposite magnetic poles. Solar flares associated with these sunspots are huge eruptions from built-up magnetic energy in the solar atmosphere. These explosions can bombard the earth's atmosphere with magnetic power, interfering with satellite communications and creating radiation haz-ards for astronauts. Some believe that solar flares can also influence our weather.

1. Use technology to complete a scatterplot of the average daily sunspot data where x represents the number of years since 1980. What degree polynomial function would appear to fit the data? Why would this polynomial function be an inappropriate model for predictions outside the data set?

PRESENTATION

Consider engaging the students by asking them whether they think solar flares on the sun's surface are random or occur according to some type of a cyclic form.

Have the students complete exercises 1–8, possibly in small groups. Then discuss the remaining questions.

The students should recognize that the data indicates a cyclic pattern in solar flares. Mention that this cyclic pattern continues decade after decade, century after century. Pose the following questions: "Do solar flares affect climate change? Are solar flares the pri-mary cause of climate change?" Have the students perform an Internet search and share some of their findings. Then emphasize the importance of mathematical models and scientific research in the pursuit of identifying significant factors in climate change.

Share that the suspected causes for climate change, natural and manmade, will continue to be studied and debated. Absolute answers may elude us for years or even decades. Emphasize that regardless of the conclusions, our worldview will affect how we respond—with fear or hope, with despair or faith, with neglect or responsibility. Challenge the students

2. Consider modeling the data with a sinusoid. Using your scatterplot, estimate the following characteristics of such a model.
   a. period ≈ 10 yr
   b. range $\approx [5, 155]$
   c. midline $y \approx 80$
   d. amplitude ≈ 75

3. Within the context of the data, explain what each of the values in exercise 2 means.

4. Use technology to complete a sinusoidal regression and state the resulting sine function. You may be prompted to enter a number of *iterations* (attempts for the calculator to get the best fit) and a guess for the function's period. $f(x) \approx 65.7 \sin(0.615x + 1.47) + 75.6$

5. State the following characteristics of the sine function that models the data.
   a. period ≈ 10.22 yr
   b. midline $y \approx 75.6$
   c. amplitude $|a| \approx 65.7$
   d. range $\approx [9.9, 141.3]$
   e. phase shift $h \approx -2.39$

6. Use your sinusoidal model to make predictions.
   a. Predict the average daily number of sunspots in 2010. 132.7
   b. Would you expect the number of sunspots in 2011 to be higher or lower than in 2010? higher
   c. Compare your results to the recorded numbers of 15.6 in 2010 and 50.1 in 2011. How does this reflect on the accuracy of the model?

According to NASA, worldwide temperatures have been consistently rising since 1980. There is much debate over whether climate change is primarily caused by natural events such as solar flares and volcanic eruptions or by human activity. Many in the scientific community believe that the burning of fossil fuels such as coal and petroleum releases significant amounts of carbon dioxide and may affect the climate. Others are skeptical that human activity can significantly change the global climate. Mathematical modeling is a valuable tool for those on either side of the issue, but a person's worldview also plays a significant role in the way he responds to mathematical models.

7. Does our mathematical model for the average number of sunspots support the claim that they are a significant factor in the recent rise in the earth's temperatures?

8. List several considerations that should be taken into account when analyzing our model.

9. Perform an Internet search to compile a list of factors that can affect the earth's climate. Discuss your results.

10. Many factors affect climate change, and some of these factors may still be unknown. Our response to this issue should be both analytical and biblical. Explain how Genesis 8:20–22, Genesis 1:27–28, Mark 12:31, and Psalm 46:2–3 should shape our view of climate change.

1. fourth-degree; The number of sunspots would not be expected to rapidly approach infinity before 1980 or after 2000.

3. The number of sunspots follows a 10 yr cycle. The minimum number of sunspots per year is about 5 and the maximum is about 160, so the yearly average fluctuates above or below 80 by about 75.

6. Use your sinusoidal model to make predictions.

6c. sample answer: It appears that the period might be closer to 10 yr.

7. No, the pattern of sunspots appears to be cyclical, while the rise in temperature has been consistent.

8. Answers will vary. Since the model is based on a limited amount of data, we could develop more accurate models with more data. We always need to be careful about extrapolating predictions. Many other factors (besides

sunspots and their related solar flares) interact to influence the earth's climate.

9. Possible factors include greenhouse gases (either natural or those caused by man's consumption of fossil fuels), warm El Niño events and cool La Niña events over the tropical Pacific Ocean, the total amount of solar energy absorbed by the earth (which can change by as much as 0.1%), volcanic activity, and the earth's orbit (which is slowly becoming more circular).

10. Christians should not live in fear but should realize that God rules over His creation. Creation and even its seasons will not pass away until ordained by Him. We should seek to be good stewards of God's creation while helping others flourish.

to always seek biblical responses to difficult problems in life.

TIPS

Ex. 4 The larger the number of iterations, the longer it takes for the TI-84 to complete the regression, but the better the fit for the resulting sinusoid.

Data Analysis

Objectives

1. To create, apply, and analyze a sinusoidal model for the average daily number of sunspots

2. To identify several biblical responses to climate change

Assignments

- **Minimum:** 1–6; 7–9 (in class)
- **Standard:** 1–6; 7–9 (in class)
- **Extended:** 1–9

Solutions

1.

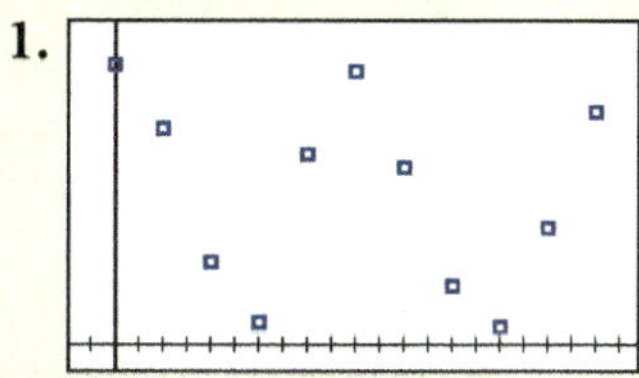

2c. $y \approx \dfrac{5 + 155}{2} = 80$

2d. $|a| \approx \dfrac{155 - 5}{2} = 75$

5a. $p = \dfrac{2\pi}{|0.615|} \approx 10.22$

5d. $75.6 \pm 65.7 = [9.9, 141.3]$

5e. $h \approx -\dfrac{1.47}{0.615} \approx -2.39$ (2.39 units left)

6a. $f(30) \approx 132.7$

6b. The period of $f(x)$ is slightly more than 10 yr, so we would expect the function to not yet reach the third relative maximum.

Objective

To prepare for evaluation

Vocabulary

See Appendix A.

Assignments

- **Minimum:** 1–12, 15–17, 19–23a, 24–25, 28–29, 31, 34–43; 45–51 odd
- **Standard:** 1–51 odd; 2–50 even (in class)
- **Extended:** 1–51 odd; 2–50 even (in class)

Assessment

- Chapter 4 Test

Solutions

1a. $168° + 0.32°\left(\frac{60'}{1°}\right) = 168°19.20'$

$168°19' + 0.20'\left(\frac{60''}{1'}\right) = 168°19'12''$

1b. $29° + 37'\left(\frac{1°}{60'}\right) + 15.6''\left(\frac{1°}{3600''}\right)$
$= 29.621°$

2a. $2°\left(\frac{\pi}{180°}\right) = \frac{\pi}{90}$

2b. $-65°\left(\frac{\pi}{180°}\right) = -\frac{13\pi}{36}$

2c. $13.59°\left(\frac{\pi}{180°}\right) \approx 0.076$

3a. $-\frac{\pi}{10}\left(\frac{180°}{\pi}\right) = -18°$

3b. $-\frac{5\pi}{12}\left(\frac{180°}{\pi}\right) = -75°$

3c. $2.8\left(\frac{180°}{\pi}\right) \approx 160.43°$

4.

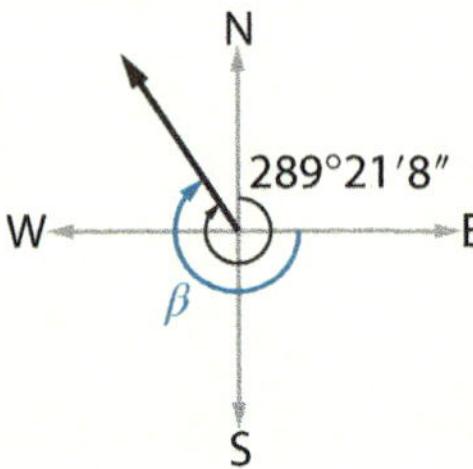

$289° + 21'\left(\frac{1°}{60'}\right) + 8''\left(\frac{1°}{3600''}\right)$
$\approx 289.3522°$
$289.3522 - 90 \approx 199.3522°$
$\beta \approx -199.3522°$

5. $\alpha = -175° \pm n360°$
$\beta = -175 + (1)360 = 185°$
$\theta = -175 - (1)360 = -535°$

---

1. Convert to degrees, minutes, seconds (DMS) or decimal degrees (DD). [4.1]
   a. 168.32° to DMS   b. 29°37'15.6" to DD
      168°19'12"   29.621°
2. Convert each degree measure to radians. [4.1]
   a. 2° $\frac{\pi}{90}$ b. −65° $-\frac{13\pi}{36}$ c. 13.59° ≈ 0.076
3. Convert each radian measure to degrees. [4.1]
   a. $-\frac{\pi}{10}$ −18°   b. $-\frac{5\pi}{12}$ −75°   c. 2.8 $\approx 160.43°$
4. Represent a watercraft's bearing of 289°21'8" relative to a buoy as a standard position angle in DD form. [4.1] −199.3522°
5. Write an expression for all angles coterminal with $\alpha = -175°$. Then specify one positive and one negative coterminal angle. [4.1] $\alpha = -175° \pm n360°$; 185°; −535°
6. Vienna's Giant Ferris Wheel has 15 gondolas and a diameter of approximately 61 m. Find the arc length between consecutive gondolas. Then find the area of the related sector. [4.1] ≈ 12.8 m; ≈ 194.8 m² 
7. A washing machine's spin cycle has a maximum speed of 1300 rpm. Find the angular speed in radians per second. [4.1] ≈ 136.1 radians/sec
8. If the 156 ft long blade of a wind turbine rotates at a maximum speed of 16.9 rpm, find the maximum linear speed of the blade's tip (to the nearest foot per second). [4.1] 276 ft/sec

Find the exact values of the six trigonometric ratios for each angle. [4.2]

9. $\angle A$ 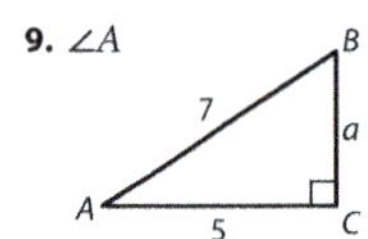
10. β 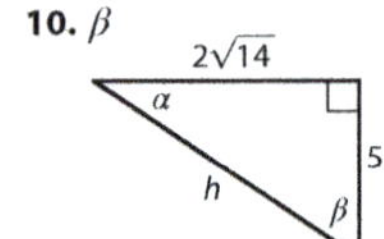

Find the other five trigonometric ratios for each angle θ. Round to the nearest ten thousandth. [4.2]

11. $\tan \theta = 0.6$ 12. $\csc \theta = \frac{21}{19}$

13. Find d (to the nearest tenth). [4.2] $d = 7.1$ 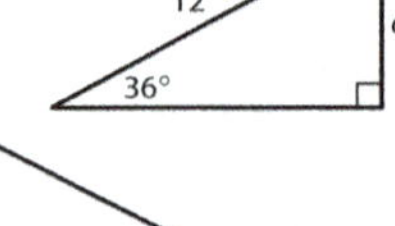

14. Find α (to the nearest tenth of a degree). [4.2] $\alpha = 29.7°$ 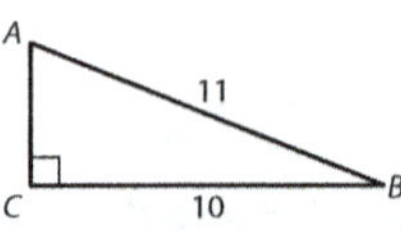

15. Solve $\triangle ABC$, rounding angle measures to the nearest minute. [4.2] 

16. A runaway ramp is used by trucks in the event of brake failure. Calculate the vertical and horizontal distance traveled by a truck on a ramp with a 10% grade (a 5.71° angle of elevation) if the stopping distance after entry was 327 ft. [4.2] ≈ 32.5 ft; ≈ 325.4 ft

17. Find the exact values for the six trigonometric functions of an angle in standard position whose terminal ray passes through (5, −6). [4.3]

18. Sketch $\theta = 315°$ and a reference triangle. Then state the measure of its reference angle α. [4.3] 45°

19. Use a reference triangle to find the exact value of $\tan 210°$. [4.3] $\frac{\sqrt{3}}{3}$

20. If $\cos \theta = -\frac{5}{13}$ and $\sin \theta > 0$, find $\csc \theta$ and $\cot \theta$. [4.3]

Use the unit circle to evaluate the six trigonometric functions for each angle. [4.3]

21. $\theta = 0°$ 22. $\theta = \frac{5\pi}{4}$

20. $\frac{13}{12}$; $-\frac{5}{12}$

6a. $\theta = \left(\frac{360°}{15}\right)\left(\frac{\pi}{180°}\right) = \frac{2\pi}{15}$

$s = 30.5\left(\frac{2\pi}{15}\right) \approx 12.78$ m

6b. $A_{sector} = \frac{1}{2}(30.5)^2\left(\frac{2\pi}{15}\right) \approx 194.83$ m²

7. $\omega = 1300\,\frac{\text{rev}}{\text{min}}\left(\frac{2\pi \text{ radians}}{1 \text{ rev}}\right)\left(\frac{1 \text{ min}}{60 \text{ sec}}\right)$
$\approx 136.136\,\frac{\text{radians}}{\text{sec}}$

8. $\omega = 16.9\,\frac{\text{rev}}{\text{min}}\left(\frac{2\pi \text{ radians}}{1 \text{ rev}}\right)\left(\frac{1 \text{ min}}{60 \text{ sec}}\right)$
$\approx 1.770\,\frac{\text{radians}}{\text{sec}}$
$v \approx (156 \text{ ft})\left(1.770\,\frac{\text{radians}}{\text{sec}}\right)$
$\approx 276.083\,\frac{\text{ft}}{\text{sec}}$

PRESENTATION

Communicate specific elements that the students will be required to know for the chapter test, such as special angle ratios and unit circle values.

23. Use the periodic nature of trigonometric functions to evaluate each expression. [4.3]
a. $\sin\left(-\frac{55\pi}{4}\right)$ $\frac{\sqrt{2}}{2}$ b. $\cos\frac{32\pi}{3}$ $-\frac{1}{2}$

State the amplitude, period, and any phase shift or vertical shift for each function. Then graph the function over [–2π, 2π]. [4.4]

24. $y = -3\cos\frac{x}{2}$

25. $y = \frac{1}{2}\sin x + 2$

26. $y = 2\cos\left(x - \frac{\pi}{3}\right)$

27. $y = \sin(2x + \pi) - 1$

28. Write a sine function rule for the graph. [4.4]
$y = 3\sin\left(x + \frac{\pi}{2}\right)$

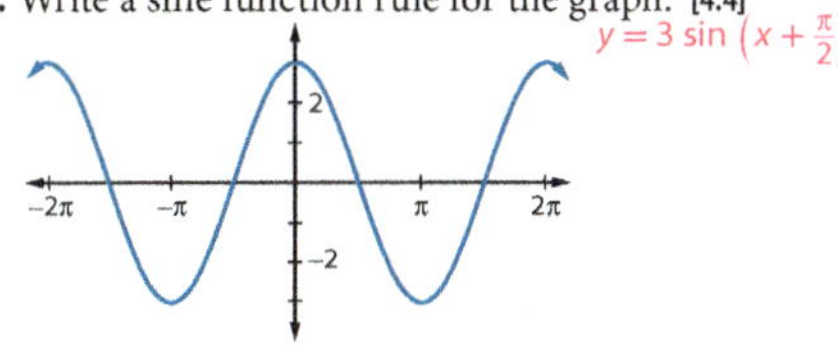

29. Write a function rule for a cosine function with a relative minimum at $(0, -6)$ and a period and amplitude of 2. [4.4] $y = -2\cos\pi x - 4$

30. The loudest animal is the pistol (or snapping) shrimp, whose snapping claw shoots out a high-speed jet of water generating an air bubble that implodes at a noise level of 200 dB. Write a sine function that models the initial behavior of the sound wave over time if the frequency is 200 kHz. [4.4] $y = 200\sin 400{,}000\pi t$

Identify the period, vertical asymptotes, and zeros of each function. [4.5]

31. $f(x) = \sec x$

32. $g(x) = \cot 2x$

Sketch the graph of each function over [–2π, 2π]. [4.5]

33. $y = \csc\left(x - \frac{\pi}{4}\right)$

34. $y = -2\tan\frac{x}{2}$

35. Describe the graph of $g(x) = -2\tan\left(\frac{x}{2} - \pi\right)$ as a transformation of $f(x) = \tan x$. [4.5]

36. Write a function rule for a secant function with a period of $\frac{\pi}{3}$ and a phase shift of $\frac{\pi}{2}$ units left that has been stretched vertically by a factor of 5. [4.5]

37. Match each inverse trigonometric function with its graph. [4.6]
a. $y = \mathrm{Sin}^{-1} x$ II   b. $y = \mathrm{Cos}^{-1} x$ III   c. $y = \mathrm{Tan}^{-1} x$ I

I.

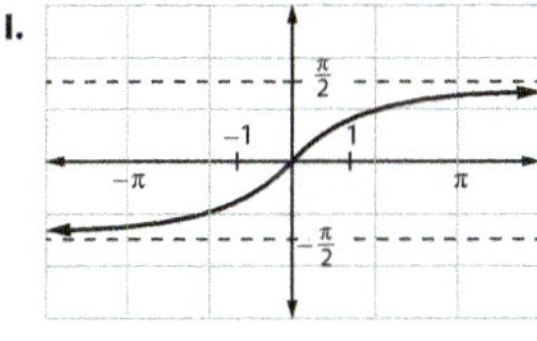

II.

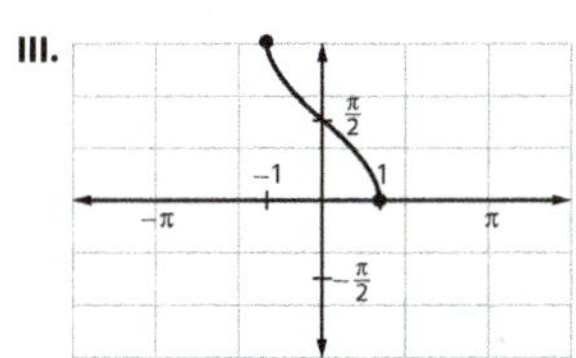

III.

Evaluate each expression in radians and degrees without using technology. [4.6]

38. $\mathrm{Sin}^{-1}\left(-\frac{1}{2}\right)$ $-\frac{\pi}{6} = -30°$

39. $\mathrm{Cos}^{-1}\left(-\frac{\sqrt{2}}{2}\right)$ $\frac{3\pi}{4} = 135°$

40. $\mathrm{Tan}^{-1}\sqrt{3}$ $\frac{\pi}{3} = 60°$

41. Use a calculator to evaluate each expression in radians and degrees. [4.6]
a. $\mathrm{Sin}^{-1}\,0.5976$ $\approx 0.6405 \approx 36.7°$
b. $\mathrm{Cos}^{-1}\,-1.7138$ undefined
c. $\mathrm{Tan}^{-1}\,1.1918$ $\approx 0.8727 \approx 50°$

42. Find the exact value of each composition without using a calculator. [4.6]
a. $\sin\left(\mathrm{Sin}^{-1} 1\right)$ 1
b. $\mathrm{Cos}^{-1}\left(\cos\left(-\frac{5\pi}{4}\right)\right)$ $\frac{3\pi}{4}$

43. Find the exact value of each expression without using a calculator. [4.6]
a. $\mathrm{Sin}^{-1}\left(\cos\frac{2\pi}{3}\right)$ $-\frac{\pi}{6}$
b. $\sec\left(\mathrm{Tan}^{-1}\left(-\frac{\sqrt{5}}{2}\right)\right)$ $\frac{3}{2}$

44. Express $f(t) = \tan\left(\mathrm{Sin}^{-1} t\right)$ algebraically. [4.6]

Determine the period of f(x) and state whether f(x) is a sinusoid. Then use technology to graph two periods of f(x). [4.7]

45. $f(x) = 3\cos x - \sin x$ 2π; sinusoid

46. $f(x) = 2\cos 4x + 2\cos 6x$ π; not a sinusoid

State equations for any bounds of f(x). Then graph the function and its bounds. Is the function periodic? [4.7]

47. $f(x) = \frac{x^2}{3} - 2\sin\pi x$

48. $f(x) = \sin^5 x$

State the damping factor and equations for the bounds of each function. Then use technology to graph the function and its bounds, and state the interval where a damping effect occurs. [4.7]

49. $y = \frac{x^2}{15}\sin x$

50. $y = 3.5e^{-0.12x}\sin 2x$

51. Write an equation for a damped sinusoid with a damping factor of $e^{-0.5x}$, a pseudoperiod of $\frac{\pi}{2}$, and a phase shift of $\frac{\pi}{4}$ to the right. [4.7]
sample answer: $y = e^{-0.5x}\sin 4\left(x - \frac{\pi}{4}\right)$ or $y = e^{-0.5x}\sin(4x - \pi)$

12. $\sin\theta = \dfrac{1}{\csc\theta} = \dfrac{19}{21} = \dfrac{opp.}{hyp.} \approx 0.9048$

$$19^2 + a^2 = 21^2$$
$$a = \sqrt{21^2 - 19^2} = 4\sqrt{5}$$
$$\cos\theta = \frac{4\sqrt{5}}{21} \approx 0.4259$$
$$\sec\theta = \frac{21}{4\sqrt{5}} \approx 2.3479$$
$$\tan\theta = \frac{19}{4\sqrt{5}} \approx 2.1243$$
$$\cot\theta = \frac{4\sqrt{5}}{19} \approx 0.4708$$

13. $\sin 36° = \dfrac{d}{12}$
$$d = 12\sin 36° \approx 7.1$$

14. $\tan\alpha = \dfrac{4}{7}$
$$\alpha = \tan^{-1}\frac{4}{7} \approx 29.7°$$

15. $b = \sqrt{11^2 - 10^2} = \sqrt{21}$
$$B = \cos^{-1}\frac{10}{11} \approx 24.62 \approx 24°37'$$
$$A = \sin^{-1}\frac{10}{11} \approx 65.38 \approx 65°23'$$

16.

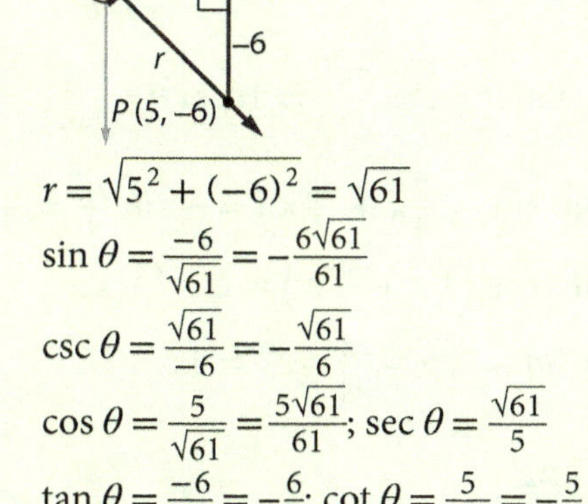

$$\sin 5.71° = \frac{y}{327}$$
$$y = 327\sin 5.71° \approx 32.5\text{ ft}$$
$$\cos 5.71° = \frac{x}{327}$$
$$x = 327\cos 5.71° \approx 325.4\text{ ft}$$

17.

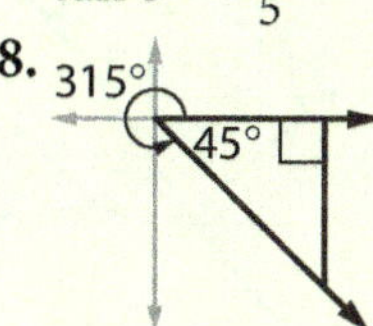

$$r = \sqrt{5^2 + (-6)^2} = \sqrt{61}$$
$$\sin\theta = \frac{-6}{\sqrt{61}} = -\frac{6\sqrt{61}}{61}$$
$$\csc\theta = \frac{\sqrt{61}}{-6} = -\frac{\sqrt{61}}{6}$$
$$\cos\theta = \frac{5}{\sqrt{61}} = \frac{5\sqrt{61}}{61};\ \sec\theta = \frac{\sqrt{61}}{5}$$
$$\tan\theta = \frac{-6}{5} = -\frac{6}{5};\ \cot\theta = \frac{5}{-6} = -\frac{5}{6}$$

18. 315° 45°

9. $a^2 + 5^2 = 7^2$
$$a = \sqrt{7^2 - 5^2} = \sqrt{24} = 2\sqrt{6}$$
$$\sin A = \frac{2\sqrt{6}}{7};\ \csc A = \frac{7}{2\sqrt{6}} = \frac{7\sqrt{6}}{12}$$
$$\cos A = \frac{5}{7};\ \sec A = \frac{7}{5}$$
$$\tan A = \frac{2\sqrt{6}}{5};\ \cot A = \frac{5}{2\sqrt{6}} = \frac{5\sqrt{6}}{12}$$

10. $h^2 = (2\sqrt{14})^2 + 5^2$
$$h = \sqrt{4(14) + 25} = \sqrt{81} = 9$$
$$\sin\beta = \frac{2\sqrt{14}}{9};\ \csc\beta = \frac{9}{2\sqrt{14}} = \frac{9\sqrt{14}}{28}$$
$$\cos\beta = \frac{5}{9};\ \sec\beta = \frac{9}{5}$$
$$\tan\beta = \frac{2\sqrt{14}}{5};\ \cot\beta = \frac{5}{2\sqrt{14}} = \frac{5\sqrt{14}}{28}$$

11. $\tan\theta = \dfrac{opp.}{adj.} = \dfrac{0.6}{1}$

$$h^2 = 1^2 + 0.6^2$$
$$h = \sqrt{1^2 + 0.6^2} \approx 1.1662$$
$$\sin\theta \approx \frac{0.6}{1.1662} = 0.5145$$
$$\csc\theta = \frac{1}{\sin\theta} \approx 1.9437$$
$$\cos\theta \approx \frac{1}{1.1662} = 0.8575$$
$$\sec\theta = \frac{1}{\cos\theta} \approx 1.1662$$
$$\cot\theta = \frac{1}{0.6} \approx 1.6667$$

continued on next page

19.

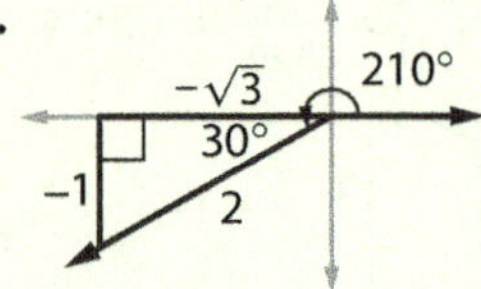

$$\tan 210° = \frac{-1}{-\sqrt{3}} = \frac{\sqrt{3}}{3}$$

20. $\cos\theta = \frac{x}{r} = \frac{-5}{13} = -\frac{5}{13}$

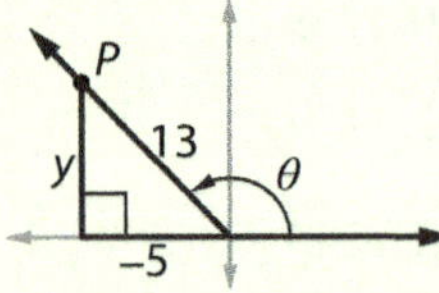

$$y = \sqrt{13^2 - (-5)^2} = 12$$
$$\csc\theta = \frac{r}{y} = \frac{13}{12}$$
$$\cot\theta = \frac{x}{y} = \frac{-5}{12} = -\frac{5}{12}$$

21. Use $P(1, 0)$ on the unit circle.
$$\sin\theta = y = 0$$
$$\csc\theta = \frac{1}{\sin\theta} = \frac{1}{0} \text{ (undefined)}$$
$$\cos\theta = x = 1;\ \sec\theta = \frac{1}{\cos\theta} = \frac{1}{1} = 1$$
$$\tan\theta = \frac{y}{x} = \frac{0}{1} = 0$$
$$\cot\theta = \frac{1}{\tan\theta} = \frac{1}{0} \text{ (undefined)}$$

22.

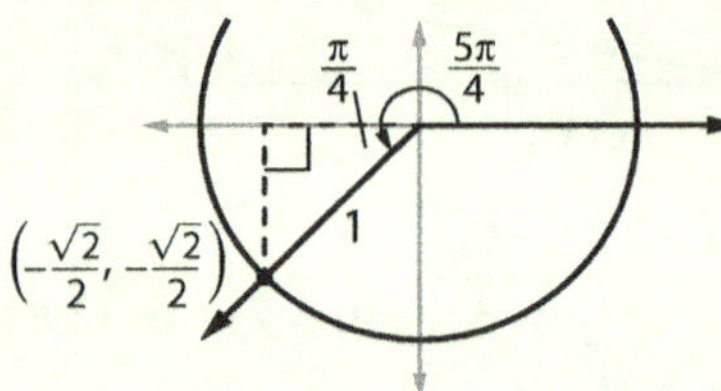

$$\sin\theta = y = -\frac{\sqrt{2}}{2};\ \csc\theta = \frac{1}{\sin\theta} = -\sqrt{2}$$
$$\cos\theta = x = -\frac{\sqrt{2}}{2};\ \sec\theta = \frac{1}{\cos\theta} = -\sqrt{2}$$
$$\tan\theta = \frac{y}{x} = \frac{-\frac{\sqrt{2}}{2}}{-\frac{\sqrt{2}}{2}} = 1;\ \cot\theta = \frac{1}{\tan\theta} = 1$$

23a. $\sin -\left(\frac{7}{4}\pi + \frac{48}{4}\pi\right) = -\sin\frac{7\pi}{4} = \frac{\sqrt{2}}{2}$

23b. $\cos\left(\frac{2}{3}\pi + \frac{30}{3}\pi\right) = \cos\frac{2\pi}{3} = -\frac{1}{2}$

24. $|a| = 3;\ p = \frac{2\pi}{|b|} = \frac{2\pi}{\frac{1}{2}} = 4\pi$

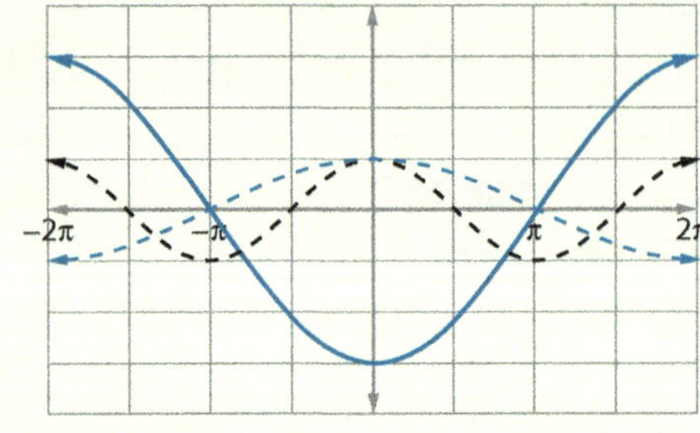

25. $|a| = \frac{1}{2};\ p = \frac{2\pi}{|b|} = \frac{2\pi}{1} = 2\pi;\ k = 2$

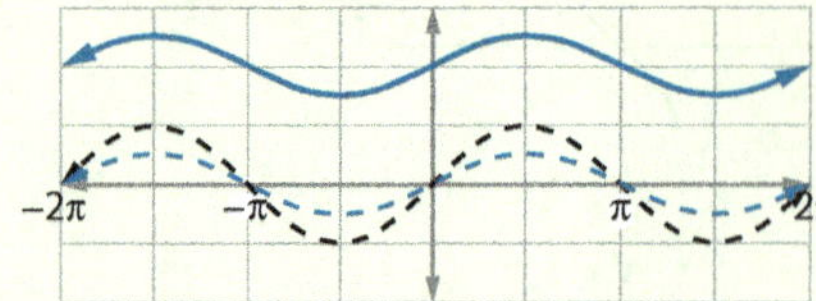

26. $|a| = 2;\ p = \frac{2\pi}{|b|} = \frac{2\pi}{1} = 2\pi;\ h = \frac{\pi}{3}$

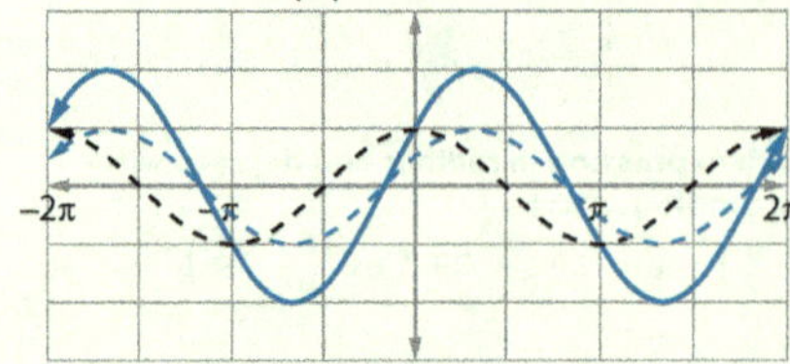

27. $y = \sin 2\left(x + \frac{\pi}{2}\right) - 1$
$$|a| = 1;\ p = \frac{2\pi}{|b|} = \frac{2\pi}{2} = \pi$$
$$h = -\frac{\pi}{2};\ k = -1$$

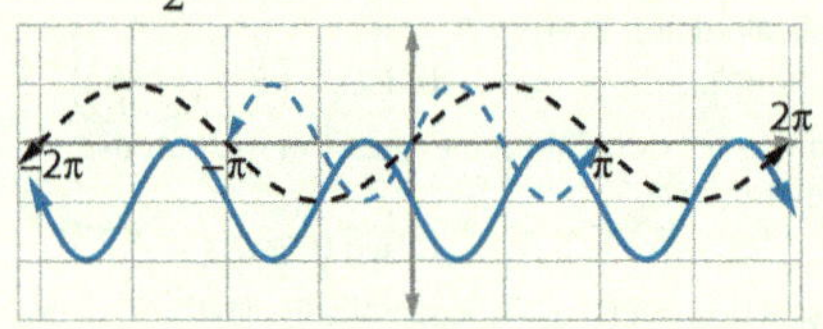

28. sample answer:
$$|a| = 3;\ p = 2\pi = \frac{2\pi}{|b|};\ |b| = 1;\ h = -\frac{\pi}{2}$$
$$y = 3\sin\left(x + \frac{\pi}{2}\right)$$

29. $p = \frac{2\pi}{|b|} = 2;\ |b| = \pi;\ k = -4$
since $f(x) = \cos x$ passes through $(0, 1)$ and the transformed function passes through $(0, -6)$,
$$y = -2\cos\pi x - 4$$

30. $y = 200\sin bt$
$$p = \frac{1}{f} = \frac{1}{200{,}000}$$
$$\frac{1}{200{,}000} = \frac{2\pi}{|b|};\ |b| = 400{,}000\pi$$
$$y = 200\sin 400{,}000\pi t$$

31. $p = 2\pi$
VA: $x = (2n + 1)\frac{\pi}{2},\ n \in \mathbb{Z}$
zeros: none

32. $p = \frac{\pi}{|b|} = \frac{\pi}{2}$
VA: $x = n\frac{\pi}{2},\ n \in \mathbb{Z}$
zeros: $(2n + 1)\frac{\pi}{4},\ n \in \mathbb{Z}$

33. $p = \frac{2\pi}{|b|} = 2\pi;$ shifted right $\frac{\pi}{4}$ units
VA: $n\pi + \frac{\pi}{4},\ n \in \mathbb{Z}$

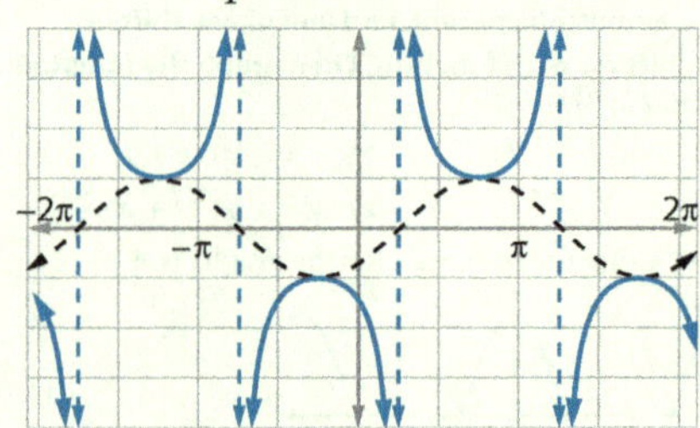

34. $p = \frac{\pi}{|b|} = \frac{\pi}{\frac{1}{2}} = 2\pi$

VA: $x = (2n + 1)\pi,\ n \in \mathbb{Z}$; reflected in the x-axis and stretched vertically and horizontally by a factor of 2

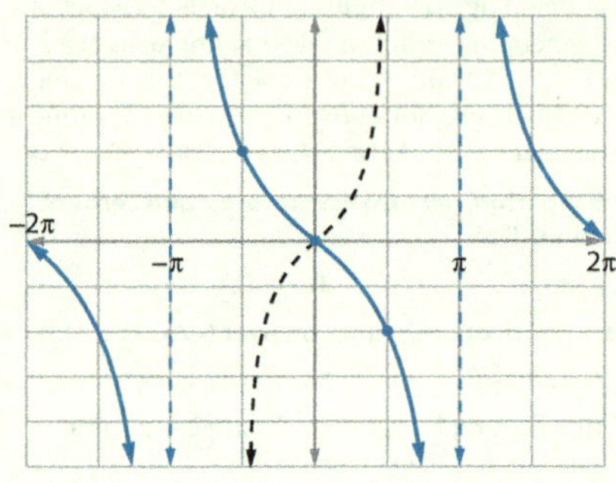

35. reflected in the x-axis, stretched horizontally and vertically by a factor of 2, and translated right 2π units

36. $|a| = 5;\ a = \pm5$
$$p = \frac{2\pi}{|b|} = \frac{\pi}{3};\ |b| = 6;\ h = \frac{\pi}{2}$$
$$g(x) = \pm5\sec 6\left(x + \frac{\pi}{2}\right) \text{ or }$$
$$g(x) = \pm5\sec(6x + 3\pi)$$

38–40.

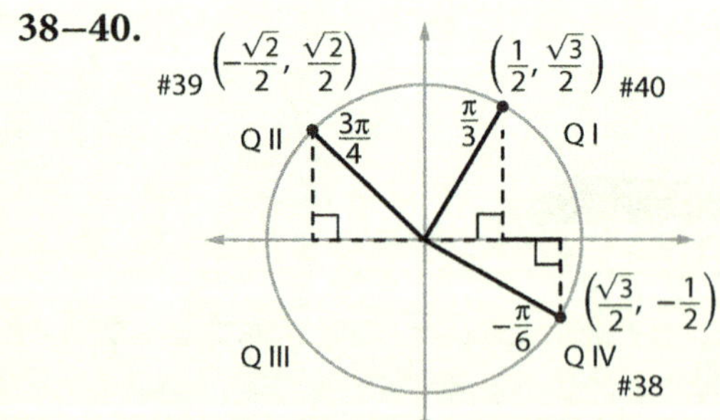

38. $\text{Sin}\left(-\frac{\pi}{6}\right) = -\frac{1}{2};\ \text{Sin}^{-1}\left(-\frac{1}{2}\right) = -\frac{\pi}{6}$

39. $\text{Cos}\frac{3\pi}{4} = -\frac{\sqrt{2}}{2};\ \text{Cos}^{-1}\left(-\frac{\sqrt{2}}{2}\right) = \frac{3\pi}{4}$

40. $\text{Tan}\frac{\pi}{3} = \frac{\frac{\sqrt{3}}{2}}{\frac{1}{2}} = \sqrt{3};\ \text{Tan}^{-1}\sqrt{3} = \frac{\pi}{3}$

42a. since $1 \in [-1, 1],\ \sin(\text{Sin}^{-1} 1) = 1$

42b. since $-\frac{5\pi}{4} \notin [0, \pi]$,

$\text{Cos}^{-1}\left(\cos\left(-\frac{5\pi}{4}\right)\right)$

$= \text{Cos}^{-1}\left(-\frac{\sqrt{2}}{2}\right) = \frac{3\pi}{4}$

43a. $\text{Sin}^{-1}\left(\cos\frac{2\pi}{3}\right)$

$= \text{Sin}^{-1}\left(-\frac{1}{2}\right) = -\frac{\pi}{6}$

43b. $\text{Tan}^{-1}\left(-\frac{\sqrt{5}}{2}\right) = \text{Tan}^{-1}\frac{y}{x}$

$r = \sqrt{2^2 + \left(-\sqrt{5}\right)^2} = 3$ (in Q IV)

$\sec\theta = \frac{r}{x} = \frac{3}{2}$

44.

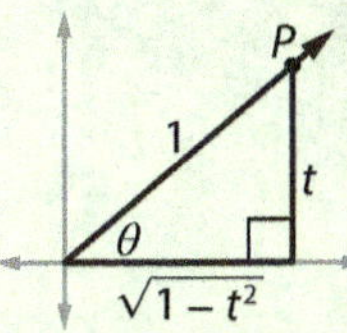

Note that θ is in Q I if $t > 0$ and in Q IV if $t < 0$ and $x > 0$ in both quadrants.

$f(t) = \tan\left(\text{Sin}^{-1} t\right) = \tan\theta$

$= \frac{t}{\sqrt{1-t^2}} = \frac{t\sqrt{1-t^2}}{1-t^2}$ or $\frac{-t\sqrt{1-t^2}}{t^2-1}$

45. $p_g = p_h = p_f = 2\pi$; sinusoid

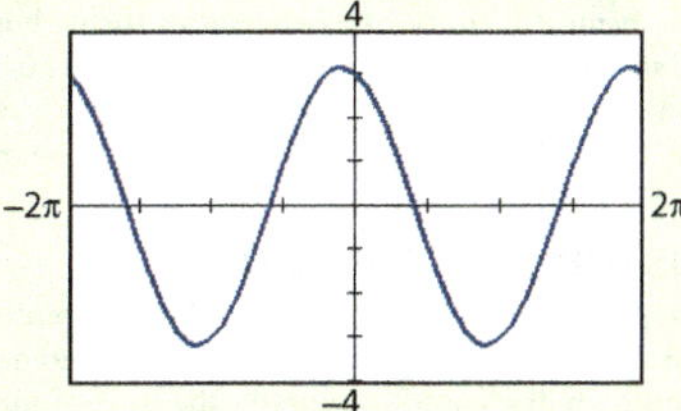

46. $p_g = \frac{\pi}{2}$; $p_h = \frac{\pi}{3}$; $p_f = \pi$; not a sinusoid

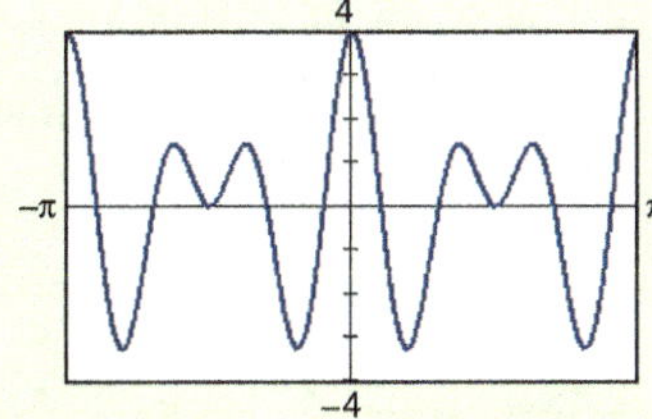

47. bounds: $y = \frac{x^2}{3} \pm 2$

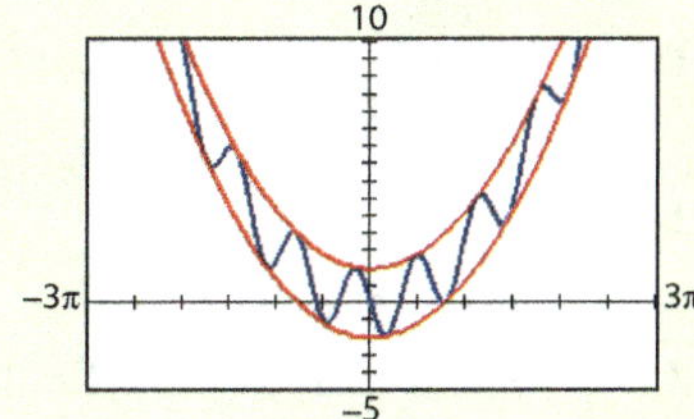

not periodic

48. $f(x) = (\sin x)^5$

bounds: $y = \pm 1$

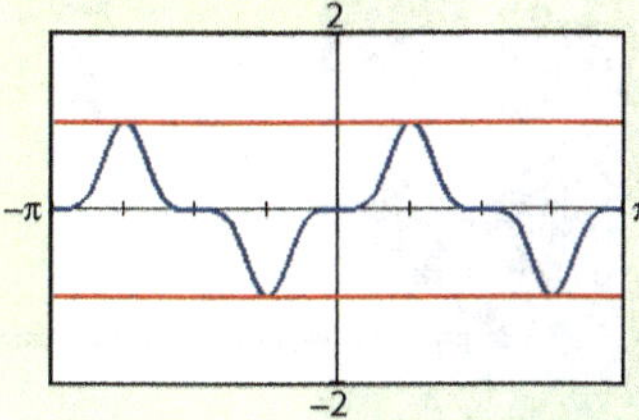

periodic; $p = \pi$

49. damping factor: $\frac{x^2}{15}$

bounds: $y = \pm\frac{x^2}{15}$; $(-\infty, 0]$

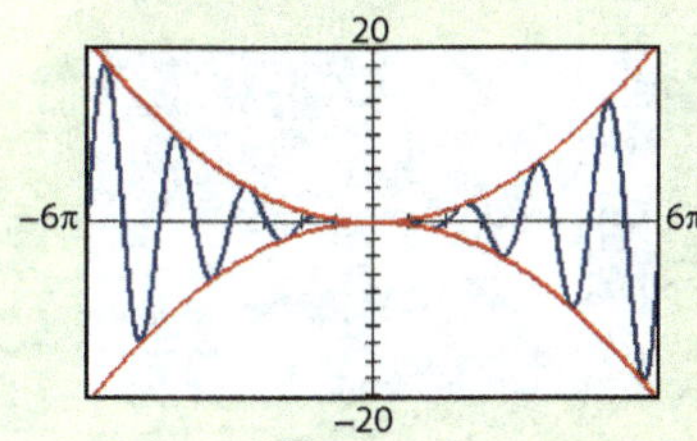

50. damping factor: $3.5e^{-0.12x}$

bounds: $y = \pm 3.5e^{-0.12x}$; $(-\infty, \infty)$

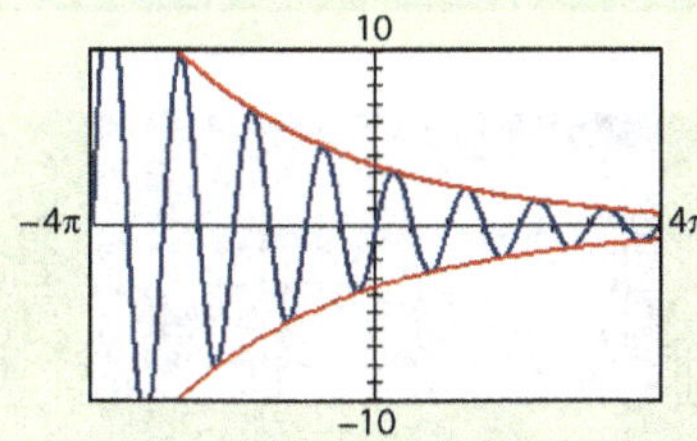

51. $p = \frac{\pi}{2} = \frac{2\pi}{|b|}$; $|b| = 4$

$y = e^{-0.5x}\sin 4\left(x - \frac{\pi}{4}\right)$

$y = e^{-0.5x}\sin(4x - \pi)$

The cosine function would also be acceptable.

Overview

This chapter continues the student's journey from simple computations based on trigonometric ratios to analytic trigonometry. Students will use basic trigonometric identities to prove other identities and solve trigonometric equations. These skills provide many opportunities to apply trigonometry to a broader range of real-world applications.

The Technology Corner illustrates how interactive geometry software can be used to explore trigonometric relationships.

Flash

The tiny glass strands in fiber-optic cables are as small as 8 microns (thousandths of a millimeter). That means the total thickness of 125 strands is about the thickness of a standard paper clip (about 1 mm). Fiber-optic cables' use of light rather than electricity provides more reliable and less expensive service and an environment free from dangerous and damaging heat buildup.

Chapter Objectives

1. To identify trigonometric identities
2. To simplify expressions with trigonometric functions
3. To prove trigonometric identities
4. To apply trigonometric identities to evaluate trigonometric functions
5. To apply trigonometric identities to solve trigonometric equations
6. To solve real-world problems involving oblique triangles using the Law of Sines and the Law of Cosines
7. To identify examples of beauty and unity within mathematics

5 Trigonometric Identities and Equations

5.1 Fundamental Identities
5.2 Verifying Trigonometric Identities
5.3 Solving Trigonometric Equations
Biblical Perspective of Mathematics
5.4 Sum and Difference Identities
5.5 Multiple Angle Identities
Historical Connection
5.6 Law of Sines
5.7 Law of Cosines
Data Analysis

BIBLICAL PERSPECTIVE OF MATHEMATICS
The beautiful unity of mathematics throughout its many diverse applications is evidence of God's creation of both the consistent underlying structure of the physical universe and our ability to describe this structure mathematically.

HISTORICAL CONNECTION
Joseph Fourier's study of heat in the eighteenth century led to the discovery of the power of infinite trigonometric series, harmonic analysis, and eventually the unification of theories of sound and light.

DATA ANALYSIS
Though Ptolemy studied the refraction of light, Willebrord Snell discovered the trigonometric equation that correctly models its refraction. Snell's law played a vital role in the development of fiber-optic networks in the 1970s.

Suggested Teaching Schedule

DAY	
1	5.1
2	5.2
3	5.3
4	Quiz 5A (5.1–5.3) BPM: The Unity and Beauty of Mathematics
5	5.4
6–7	5.5
8	Quiz 5B (5.4–5.5) HC: Harmonic Analysis

DAY	
9–10	5.6 TC: Exploring the Law of Sines
11	5.7
12	Quiz 5C (5.6–5.7) DA: Refraction, Reflection, and Fiber Optics
13	Chapter 5 Review
14	Chapter 5 Test

Fundamental Identities

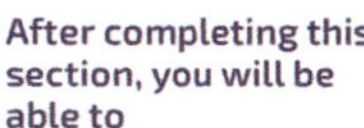

Identities can be used to simplify trigonometric equations that represent voltages across capacitors in an AC generator.

The majority of equations encountered in mathematics are *conditional equations*, equations that are true only for certain values of the variable. Solving a conditional equation, such as $5x + 3 = 13$, involves finding its *solution(s)*, any value of the variable for which the equation is true. Occasionally we encounter *false equations*, such as $x + 3 = x - 2$, that have no solutions. Other equations, such as $x^0 = 1$ or $(a + b)^2 = a^2 + 2ab + b^2$, are true for all values of the variables for which the expressions on each side of the equal sign are defined.

> **DEFINITION**
>
> An **identity** is an equation that is true for all values in the domain of the variable.

Just like theorems in geometry, identities need to be proved before they are accepted as true. In this chapter, you will learn how to use fundamental trigonometric identities to evaluate trigonometric functions, simplify trigonometric expressions, develop other identities, and solve trigonometric equations.

You already know several trigonometric identities that follow directly from the definitions of trigonometric functions.

Reciprocal Identities			Quotient Identities
$\csc \theta = \frac{1}{\sin \theta}$	$\sec \theta = \frac{1}{\cos \theta}$	$\cot \theta = \frac{1}{\tan \theta}$	$\tan \theta = \frac{\sin \theta}{\cos \theta}$
$\sin \theta = \frac{1}{\csc \theta}$	$\cos \theta = \frac{1}{\sec \theta}$	$\tan \theta = \frac{1}{\cot \theta}$	$\cot \theta = \frac{\cos \theta}{\sin \theta}$

The first reciprocal identity can be verified using the definitions $\sin \theta = \frac{y}{r}$ and $\csc \theta = \frac{r}{y}$.

$$\frac{1}{\sin \theta} = \frac{1}{\frac{y}{r}} = \frac{r}{y} = \csc \theta$$

The other reciprocal identities and quotient identities can be similarly proven.

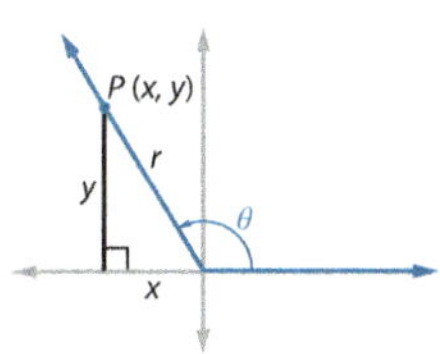

Several other basic trigonometric identities are derived using the Pythagorean Theorem.

$$x^2 + y^2 = r^2 \qquad \text{Pythagorean Theorem}$$
$$\frac{x^2}{r^2} + \frac{y^2}{r^2} = \frac{r^2}{r^2} \qquad \text{Divide each term by } r^2.$$
$$\left(\frac{x}{r}\right)^2 + \left(\frac{y}{r}\right)^2 = 1 \qquad \text{Simplify.}$$
$$\cos^2 \theta + \sin^2 \theta = 1 \qquad \text{Substitute using the definitions of sine and cosine.}$$

Two other versions of this basic Pythagorean identity can also be derived by dividing each term by either $\sin^2 \theta$ or $\cos^2 \theta$.

TIP

Powers of trig functions such as $(\sin \theta)^2$ are notated as $\sin^2 \theta$.

5.1 Fundamental Identities

Objectives

1. To state and verify the reciprocal, quotient, Pythagorean, cofunction, and odd/even identities

2. To use identities to evaluate other trigonometric functions

3. To use trigonometric identities to simplify and rewrite trigonometric expressions

Vocabulary

cofunction
conditional equations
false equations
identity
solutions

Reading and Writing Mathematics

Use the range of $f(x) = \sin x$ to determine the range of $g(x) = \sin^2 x$. Then graph the functions in the same window and algebraically explain the difference in the periods of $f(x)$ and $g(x)$.

The range of $f(x)$ is [−1, 1], so the range of $g(x)$ is [0, 1]. The period of $g(x)$ is half the period of $f(x)$ since squaring its negative values will repeat the squaring of the positive values of its other half-period.

After completing this section, you will be able to

- state and verify the fundamental trig identities.
- use identities to evaluate other trig functions.
- simplify and rewrite trig expressions.

PRESENTATION

Lesson Opener

Simplify each expression.

1. $\frac{3}{y} + \frac{1}{x}$ $\frac{3x + y}{xy}$

2. $\frac{\frac{2}{4}}{\frac{4}{3}} - \frac{\frac{1}{4}}{\frac{1}{2}}$ 1

3. $\frac{x^3 + 2x^2y}{x^2}$ $x + 2y$

4. $\frac{1}{\frac{-\sin \theta}{\cos \theta}}$ $-\cot \theta$

Consider having students graph $y = 1 - \sin^2 x$ and $y = \cos^2 x$ in the same window to illustrate that the two different trigonometric expressions can represent the same function. Then present the definition of an *identity*.

Review the reciprocal and quotient identities. Explain that this section summarizes and presents formal proofs for several identities developed in Chapter 4.

Derive and state the basic Pythagorean identity. This identity can also be derived by substituting the coordinates from $P (\cos \theta, \sin \theta)$ on the unit circle into its equation, $x^2 + y^2 = 1$, to produce $\sin^2 x + \cos^2 x = 1$. Consider deriving one of the other forms of the Pythagorean identity (see exercise 2) together in class.

Example 1 illustrates how to use the Pythagorean identity to evaluate the other trigonometric functions from a given trigonometric value. In Section 4.3 similar evaluations were completed using reference triangles. The students should be able to use either method, finding exact values instead of the decimal approximations produced when using a calculator.

Review the *cofunction* identities introduced in Section 4.2 (see exercise 30*a*) and the even/odd identities introduced in Section 4.3. Consider demonstrating the proof of one of these identities (see exercises 3–4) before presenting Example 2. Then use the periodic identities to remind the students of the period of each trig function.

$$\cos^2\theta + \sin^2\theta = 1 \qquad 1 + \tan^2\theta = \sec^2\theta \qquad \cot^2\theta + 1 = \csc^2\theta$$

Additional Exercises

Figures for Additional Exercises throughout the chapter can be found at TeacherToolsOnline.com.

1. Use the image to prove that
$$\csc\theta = \frac{1}{\sin\theta}.$$

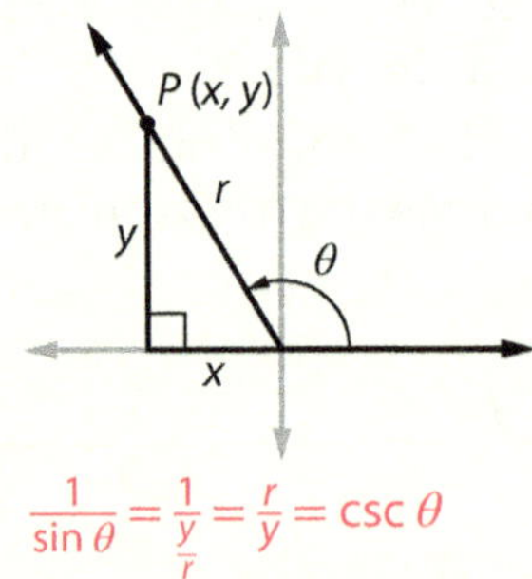

$$\frac{1}{\sin\theta} = \frac{1}{\frac{y}{r}} = \frac{r}{y} = \csc\theta$$

2. Use the image to prove that
$$\cot\left(\frac{\pi}{2} - \theta\right) = \tan\theta.$$

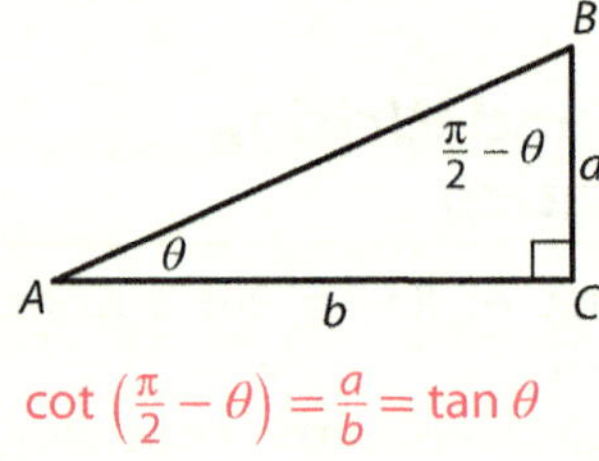

$$\cot\left(\frac{\pi}{2} - \theta\right) = \frac{a}{b} = \tan\theta$$

3. If $\cos\theta = -\frac{1}{3}$ and $\sin\theta > 0$, find the values of the other five trigonometric functions of θ.

$$\sin\theta = \frac{2\sqrt{2}}{3}; \ \tan\theta = -2\sqrt{2}; \ \sec\theta = -3$$
$$\csc\theta = \frac{3\sqrt{2}}{4}; \ \cot\theta = -\frac{\sqrt{2}}{4}$$

Example 1 Using Identities to Evaluate Other Trig Functions

If $\csc\theta = \frac{4}{3}$ and $\cos\theta < 0$, find the other five trigonometric functions of θ.

Answer

$\sin\theta = \frac{1}{\csc\theta} = \frac{1}{\frac{4}{3}} = \frac{3}{4}$

1. Use a reciprocal identity to find $\sin\theta$.

$\cos^2\theta + \sin^2\theta = 1$
$\cos^2\theta = 1 - \sin^2\theta$
$\cos\theta = \pm\sqrt{1 - \left(\frac{3}{4}\right)^2} = -\frac{\sqrt{7}}{4}$

2. Use a Pythagorean identity to find $\cos\theta$. Note that θ terminates in Quadrant II since $\sin\theta > 0$ and $\cos\theta < 0$.

$\tan\theta = \frac{\sin\theta}{\cos\theta} = \frac{\frac{3}{4}}{-\frac{\sqrt{7}}{4}} = -\frac{3}{\sqrt{7}} = -\frac{3\sqrt{7}}{7}$

3. Use a quotient identity to find $\tan\theta$.

$\sec\theta = \frac{1}{\cos\theta} = \frac{1}{-\frac{\sqrt{7}}{4}} = -\frac{4}{\sqrt{7}} = -\frac{4\sqrt{7}}{7}$

4. Use reciprocal identities to find $\sec\theta$ and $\cot\theta$.

$\cot\theta = \frac{1}{\tan\theta} = \frac{1}{-\frac{3\sqrt{7}}{7}} = -\frac{7}{3\sqrt{7}} = -\frac{\sqrt{7}}{3}$

SKILL ✔ EXERCISE 5

In Section 4.2 we illustrated how a right triangle can be used to show that the trigonometric function of an angle is equal to the *cofunction* of the angle's complement. Note that $\sin A = \frac{a}{c} = \cos B$ and $\cos A = \frac{b}{c} = \sin B$. It can also be shown that $\tan A = \cot B$ and $\sec A = \csc B$. Identities developed in Section 5.4 will provide simple proofs for the extension of these identities to non-acute angles and real numbers.

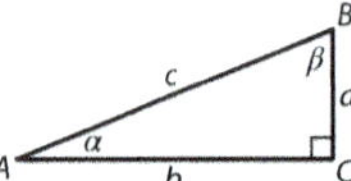

$$\sin\theta = \cos\left(\frac{\pi}{2} - \theta\right) \qquad \tan\theta = \cot\left(\frac{\pi}{2} - \theta\right) \qquad \sec\theta = \csc\left(\frac{\pi}{2} - \theta\right)$$
$$\cos\theta = \sin\left(\frac{\pi}{2} - \theta\right) \qquad \cot\theta = \tan\left(\frac{\pi}{2} - \theta\right) \qquad \csc\theta = \sec\left(\frac{\pi}{2} - \theta\right)$$

In Section 4.3, the unit circle was used to show that cosine is an even function and that sine is an odd function. The other even/odd trigonometric identities can be verified by expressing the functions in terms of sine and cosine.

$$\sin(-\theta) = -\sin\theta \qquad \cos(-\theta) = \cos\theta \qquad \tan(-\theta) = -\tan\theta$$
$$\csc(-\theta) = -\csc\theta \qquad \sec(-\theta) = \sec\theta \qquad \cot(-\theta) = -\cot\theta$$

Explain that identities are used to simplify expressions, prove other trig identities (Section 5.2), and solve trig equations (Section 5.3). Example 3 demonstrates the common strategy of expressing the other trig functions in terms of sine and cosine. Example 4 demonstrates the algebraic technique of factoring, which can be used to help further simplify the expression. Explain that checking your work by comparing the graphs or lists of values for the original and the simplified expressions does not constitute a proof.

Example 5 illustrates simplifying a trigonometric expression by combining terms using a common denominator. Example 6 illustrates the rewriting of a rational trigonometric expression to eliminate fractions. Point out how multiplying the numerator and denominator by the denominator's conjugate in step 1 permits a substitution using a form of a Pythagorean identity in step 2. Step 3 illustrates the technique of expressing a fraction as a sum. Alternatively, we could perform the following in steps 3 and 4:

$$= \frac{1}{\cos^2 x}(1 + \sin x)$$
$$= \sec^2 x\,(1 + \sin x) \text{ or}$$
$$\sec^2 x + \sec^2 x \sin x$$

Challenge students to show that the two answers are equivalent.

TIPS

Ex. 32 While both $2\sin x \sec^2 x$ and $2\csc x \tan^2 x$ are equivalent to the given answer, $2\tan x \sec x$ is simpler since there are no powers of trig functions.

Ex. 37–38 The domain of logarithmic functions excludes negative numbers. Encourage students to evaluate their final answer to see whether absolute value symbols are needed.

Example 2   Using Cofunction and Even/Odd Identities

If $\cot \theta = 2.25$, find $\tan\left(\theta - \frac{\pi}{2}\right)$.

Answer

$$\tan\left(\theta - \frac{\pi}{2}\right) = \tan\left[-\left(\frac{\pi}{2} - \theta\right)\right]$$
$$= -\tan\left(\frac{\pi}{2} - \theta\right)$$
$$= -\cot \theta$$
$$= -2.25$$

1. Factor and apply the fact that cotangent is an odd function.

2. Apply a cofunction identity and substitute.

_______________________________ SKILL ✔ **EXERCISE 11**

In Chapter 4, the unit circle and the graphs of the trigonometric functions illustrated the periodicity of trigonometric functions. This characteristic of each function can also be expressed as an identity. These identities can be used to evaluate functions outside of the interval $[0, 2\pi)$.

Periodic Identities $n \in \mathbb{Z}$		
$\sin(\theta + n2\pi) = \sin\theta$	$\sec(\theta + n2\pi) = \sec\theta$	$\tan(\theta + n\pi) = \tan\theta$
$\cos(\theta + n2\pi) = \cos\theta$	$\csc(\theta + n2\pi) = \csc\theta$	$\cot(\theta + n\pi) = \cot\theta$

The fundamental identities are frequently used to simplify complicated trigonometric expressions. Since the two sides of an identity are equal, one side of an identity may be substituted for the other. It is often helpful to rewrite the expression in terms of sine and cosine. Other typical substitutions involve various forms of the Pythagorean identities. For example, $\sin^2\theta$ can be substituted for $1 - \cos^2\theta$.

Example 3   *Simplifying by Expressing in Terms of Sine and Cosine*

Simplify $\tan\theta - \sec\theta\csc\theta$.

Answer

$$\tan\theta - \sec\theta\csc\theta$$
$$= \frac{\sin\theta}{\cos\theta} - \frac{1}{\cos\theta}\cdot\frac{1}{\sin\theta}$$
$$= \frac{\sin^2\theta}{\cos\theta\sin\theta} - \frac{1}{\cos\theta\sin\theta}$$
$$= \frac{\sin^2\theta - 1}{\cos\theta\sin\theta}$$
$$= \frac{(1 - \cos^2\theta) - 1}{\cos\theta\sin\theta}$$
$$= \frac{-\cos^2\theta}{\cos\theta\sin\theta}$$
$$= -\frac{\cos\theta}{\sin\theta} = -\cot\theta$$

1. Use reciprocal and quotient identities to rewrite the expression in terms of sine and cosine.

2. Combine fractions using the least common denominator.

3. Simplify the numerator using another form of the basic Pythagorean identity $\sin^2\theta = 1 - \cos^2\theta$.

4. Cancel the common factor and express in terms of a single trigonometric function.

Check

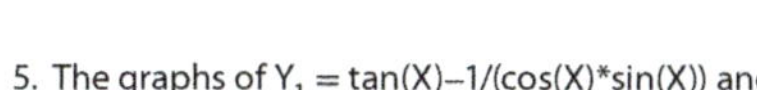

5. The graphs of $Y_1 = \tan(X) - 1/(\cos(X)*\sin(X))$ and $Y_2 = -\cot(X)$ appear to be identical.

_______________________________ SKILL ✔ **EXERCISE 17**

The unity of mathematics was recognized by physicist Richard Feynman in his 1964 lecture series on the laws of the physical universe: "Nature uses only the longest threads to weave her patterns, so each small piece of her fabric reveals the organization of the entire tapestry."

4. Use an identity to evaluate each function.

  a. $\cos(-150°)$ $= \cos(150) = -\frac{\sqrt{3}}{2}$

  b. $\cos\left(\frac{\pi}{2} - \theta\right)$ if $\sin\theta = 0.86$

    $= \sin\theta = 0.86$

  c. $\tan\left(\theta - \frac{\pi}{2}\right)$ if $\cot\theta = 1.52$

    $= -\tan\left(\frac{\pi}{2} - \theta\right) = -\cot\theta = -1.52$

5. Rewrite each expression as a single trigonometric function.

  a. $\dfrac{\cot x}{\csc x}$ $\cos x$

  b. $\dfrac{1 - \csc^2 x}{\cos^2 x}$ $-\csc^2 x$

  c. $\dfrac{1}{1 - \sec x} + \dfrac{1}{1 + \sec x}$ $-2\cot^2 x$

Simplify.

6. $\csc^4 x - 2\csc^2 x + 1$ $\cot^4 x$

7. $\sec^2 x - \csc^2 x - \tan^2 x + \cot^2 x$ 0

8. $\dfrac{\sin^2 x}{\cos x - \cos^2 x}$ $\sec x + 1$

9. $\dfrac{1 - \cos x}{\sin x} + \dfrac{\cos x \sin x}{1 + \cos x}$ $\sin x$

10. Write $\dfrac{1}{\sec x - 1}$ as a simplified expression without fractions.

  $\cos x \csc^2 x + \cot^2 x$ or
  $\cot x \csc x + \cot^2 x$

Assignments

- **Minimum:** 1–5, 8–11, 14–16, 18–19, 21, 23–27; 29–35 odd; 39, 43–44, 46–48, 50
- **Standard:** 1–4, 6–7, 9; 10–22 even; 23–26, 28–30, 32–33, 36–37, 39; 44–52 even
- **Extended:** 1–4, 6, 9–11, 14–15, 18–20, 22–26; 28–36 even; 37–41 odd; 45–47, 50–51

Algebraic manipulations such as factoring are frequently used to simplify an expression.

Example 4 Simplifying by Factoring

Simplify $\cos^2 x \sin\left(\frac{\pi}{2} - x\right) + \cos x \sin^2 x$.

Answer

X	Y1	Y2		
0	1	1		
0.2618	0.9659	0.9659		
0.5236	0.866	0.866		
0.7854	0.7071	0.7071		
1.0472	0.5	0.5		
1.309	0.2588	0.2588		
1.5708	0	0		
1.8326	-0.259	-0.259		
2.0944	-0.5	-0.5		
2.3562	-0.707	-0.707		
2.618	-0.866	-0.866		

X=0

$\cos^2 x \sin\left(\frac{\pi}{2} - x\right) + \cos x \sin^2 x$

$= \cos^2 x \left(\cos x\right) + \cos x \sin^2 x$

1. Substitute using $\sin\left(\frac{\pi}{2} - x\right) = \cos x$.

$= \cos x \left(\cos^2 x + \sin^2 x\right)$

2. Factor $\cos x$ from each term.

$= \cos x \left(1\right)$

$= \cos x$

3. Substitute using the basic Pythagorean identity and simplify.

Check

4. Values for
$Y_1 = \cos(X)^2 * \sin(\pi/2 - X) + \cos(X) * \sin(X)^2$ and
$Y_2 = \cos(X)$ appear to be identical.

_____________________________ SKILL ✔ **EXERCISE 19**

While we may be able to support our simplification graphically or numerically, neither of these methods can be used as a proof since every possible value cannot be verified by these methods. Finding just one member of the domains where the expressions do not have the same value would prove our simplification to be incorrect.

The combining of fractions plays an important part in the simplification of many expressions.

Example 5 Simplifying by Combining Fractions

Simplify $\dfrac{\cos \theta}{\sin \theta} - \dfrac{\sin \theta}{1 - \cos \theta}$.

Answer

$\dfrac{\cos \theta}{\sin \theta} - \dfrac{\sin \theta}{1 - \cos \theta}$

$= \dfrac{\cos \theta}{\sin \theta} \cdot \dfrac{1 - \cos \theta}{1 - \cos \theta} - \dfrac{\sin \theta}{1 - \cos \theta} \cdot \dfrac{\sin \theta}{\sin \theta}$

1. Combine fractions using the least common denominator.

$= \dfrac{\cos \theta - \cos^2 \theta - \sin^2 \theta}{\sin \theta \left(1 - \cos \theta\right)}$

$= \dfrac{\cos \theta - \left(\cos^2 \theta + \sin^2 \theta\right)}{\sin \theta \left(1 - \cos \theta\right)}$

$= \dfrac{\cos \theta - 1}{\sin \theta \left(1 - \cos \theta\right)}$

2. Factor -1 from the last two terms of the numerator and apply the basic Pythagorean identity.

$= \dfrac{-\left(1 - \cos \theta\right)}{\sin \theta \left(1 - \cos \theta\right)}$

3. Factor -1 from the numerator and simplify.

$= -\dfrac{1}{\sin \theta} = -\csc \theta$

_____________________________ SKILL ✔ **EXERCISE 29**

In calculus, it is often desirable to rewrite trigonometric expressions in an equivalent form without fractions. This can be accomplished by using a product of conjugates: $(a + b)(a - b) = a^2 - b^2$.

Solutions

❯ A. Exercises

1a. $\dfrac{1}{\cos \theta} = \dfrac{1}{\frac{x}{r}} = \dfrac{r}{x} = \sec \theta$

1b. $\dfrac{\cos \theta}{\sin \theta} = \dfrac{\frac{x}{r}}{\frac{y}{r}} = \dfrac{x}{y} = \cot \theta$

2a. $\dfrac{\cos^2 \theta}{\sin^2 \theta} + \dfrac{\sin^2 \theta}{\sin^2 \theta} = \dfrac{1}{\sin^2 \theta}$

$\cot^2 \theta + 1 = \csc^2 \theta$

2b. $\dfrac{\cos^2 \theta}{\cos^2 \theta} + \dfrac{\sin^2 \theta}{\cos^2 \theta} = \dfrac{1}{\cos^2 \theta}$

$1 + \tan^2 \theta = \sec^2 \theta$

Rewrite $\dfrac{1}{1-\sin x}$ as an equivalent expression without fractions.

Answer

$$\dfrac{1}{1-\sin x}=\dfrac{1}{1-\sin x}\cdot\dfrac{1+\sin x}{1+\sin x}$$
$$=\dfrac{1+\sin x}{1-\sin^2 x}$$

1. Multiply the numerator and denominator by $1+\sin x$.

$$=\dfrac{1+\sin x}{\cos^2 x}$$

2. Substitute for the denominator using a form of the basic Pythagorean identity $\cos^2 x = 1-\sin^2 x$.

$$=\dfrac{1}{\cos^2 x}+\dfrac{\sin x}{\cos^2 x}$$

3. Write the fraction as a sum.

$$=\dfrac{1}{\cos^2 x}+\dfrac{\sin x}{\cos x}\cdot\dfrac{1}{\cos x}$$
$$=\sec^2 x + \tan x \sec x$$

4. Use reciprocal and quotient identities to eliminate the fractions.

SKILL ✔ EXERCISE 33

5. $\sin\theta=\dfrac{2\sqrt{2}}{3};\ \tan\theta=2\sqrt{2}$

6. $\sec\theta=-\dfrac{13}{12};\ \tan\theta=-\dfrac{5}{12}$

7. $\sec\theta=-\dfrac{17}{8};\ \cos\theta=-\dfrac{8}{17}$

8. $\csc\theta=-\dfrac{4\sqrt{15}}{15};$
$\cot\theta=-\dfrac{\sqrt{15}}{15}$

❯ A. Exercises

1. Use the image to prove each identity.

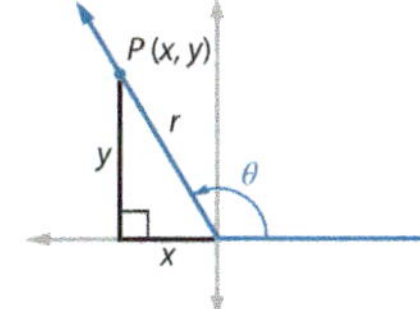

a. $\sec\theta=\dfrac{1}{\cos\theta}$

b. $\cot\theta=\dfrac{\cos\theta}{\sin\theta}$

2. Derive each Pythagorean identity by dividing each term of $\sin^2\theta+\cos^2\theta=1$ by the appropriate quantity.

a. $\cot^2\theta+1=\csc^2\theta$ **b.** $1+\tan^2\theta=\sec^2\theta$

3. Use $\triangle ABC$ to prove each cofunction identity.

a. $\tan\left(\dfrac{\pi}{2}-\theta\right)=\cot\theta$

b. $\sec\left(\dfrac{\pi}{2}-\theta\right)=\csc\theta$

4. Verify each identity.

a. $\tan(-x)=-\tan(x)$ **b.** $\csc(-x)=-\csc x$

Use identities to find the value of each expression.

5. If $\sec\theta=3$ and $\sin\theta>0$, find $\sin\theta$ and $\tan\theta$.

6. If $\csc\theta=\dfrac{13}{5}$ and $\cos\theta<0$, find $\tan\theta$ and $\sec\theta$.

7. If $\tan\theta=-\dfrac{15}{8}$ and $\cos\theta<0$, find $\sec\theta$ and $\cos\theta$.

8. If $\cos\theta=\dfrac{1}{4}$ and $\tan\theta<0$, find $\csc\theta$ and $\cot\theta$.

9. Which of the following is an equivalent form of the basic Pythagorean identity? List all correct answers. B, D

A. $\sin^2 x=\cos^2 x-1$ **C.** $\cos^2 x=\sin^2 x-1$

B. $\sin^2 x=1-\cos^2 x$ **D.** $-\cos^2 x=\sin^2 x-1$

10. Which of the following are equivalent forms of a Pythagorean identity? List all correct answers. A, D

A. $\sec^2 x-\tan^2 x=1$ **C.** $\cot^2 x-\csc^2 x=1$

B. $\tan^2 x=-1-\sec^2 x$ **D.** $\cot^2 x=\csc^2 x-1$

Evaluate each expression.

11. $\sin\left(\dfrac{\pi}{2}-\theta\right)$ if $\cos\theta=0.27$ 0.27

12. $\csc\left(\dfrac{\pi}{2}-\theta\right)$ if $\sec\theta=-1.28$ -1.28

13. $\cos\left(\theta-\dfrac{\pi}{2}\right)$ if $\sin\theta=0.35$ 0.35

14. $\tan\left(\theta-\dfrac{\pi}{2}\right)$ if $\cot\theta=-1.64$ 1.64

Simplify each expression.

15. $\dfrac{\sin\theta-\cos\theta}{\sin\theta}+\cot\theta$ 1

16. $\dfrac{\tan\theta\cos\theta}{\sin\theta}$ 1

17. $\sin\theta\sec\theta-\sin\theta\csc\theta-\tan\theta$ -1

18. $\dfrac{\sec\theta}{\cos\theta}-\dfrac{\tan\theta}{\cot\theta}$ 1

❯ B. Exercises

Simplify by factoring.

19. $\tan^3 x-\cot\left(\dfrac{\pi}{2}-x\right)\sec^2 x$ $-\tan x$

20. $\sin x-\sin^3 x$ $\sin x\cos^2 x$

21. $\tan x-\sin x\sec^3 x$ $-\tan^3 x$

22. $\dfrac{\cos\theta}{\sec\theta+\tan\theta}$ $1-\sin\theta$

---

8. θ in Q IV

$$\sin^2\theta+\left(\dfrac{1}{4}\right)^2=1$$
$$\sin^2\theta=\dfrac{15}{16};\ \sin\theta=-\dfrac{\sqrt{15}}{4}$$
$$\csc\theta=-\dfrac{4\sqrt{15}}{15}$$
$$\cot\theta=-\dfrac{\frac{1}{4}}{\frac{\sqrt{15}}{4}}=-\dfrac{1}{\sqrt{15}}=-\dfrac{\sqrt{15}}{15}$$

11. $\sin\left(\dfrac{\pi}{2}-\theta\right)=\cos\theta=0.27$

12. $\csc\left(\dfrac{\pi}{2}-\theta\right)=\sec\theta=-1.28$

13. $\cos\left(\theta-\dfrac{\pi}{2}\right)=\cos\left(-\left(\dfrac{\pi}{2}-\theta\right)\right)$
$$=\cos\left(\dfrac{\pi}{2}-\theta\right)$$
$$=\sin\theta=0.35$$

14. $\tan\left(\theta-\dfrac{\pi}{2}\right)=\tan\left(-\left(\dfrac{\pi}{2}-\theta\right)\right)$
$$=-\tan\left(\dfrac{\pi}{2}-\theta\right)$$
$$=-\cot\theta=1.64$$

15. $\dfrac{\sin\theta-\cos\theta}{\sin\theta}+\cot\theta$
$$=\dfrac{\sin\theta-\cos\theta}{\sin\theta}+\dfrac{\cos\theta}{\sin\theta}$$
$$=\dfrac{\sin\theta-\cos\theta+\cos\theta}{\sin\theta}=\dfrac{\sin\theta}{\sin\theta}=1$$

16. $\dfrac{\tan\theta\cos\theta}{\sin\theta}=\dfrac{\frac{\sin\theta}{\cos\theta}\cdot\cos\theta}{\sin\theta}=\dfrac{\sin\theta}{\sin\theta}=1$

17. $\sin\theta\sec\theta-\sin\theta\csc\theta-\tan\theta$
$$=\sin\theta\left(\dfrac{1}{\cos\theta}\right)-\sin\theta\left(\dfrac{1}{\sin\theta}\right)-\dfrac{\sin\theta}{\cos\theta}$$
$$=\dfrac{\sin\theta}{\cos\theta}-1-\dfrac{\sin\theta}{\cos\theta}=-1$$

18. $\dfrac{\sec\theta}{\cos\theta}-\dfrac{\tan\theta}{\cot\theta}$
$$=\sec\theta\left(\dfrac{1}{\cos\theta}\right)-\tan\theta\left(\dfrac{1}{\cot\theta}\right)$$
$$=\sec^2\theta-\tan^2\theta=1$$

❯ B. Exercises

19. $\tan^3 x-\cot\left(\dfrac{\pi}{2}-x\right)\sec^2 x$
$$=\tan^3 x-\tan x\sec^2 x$$
$$=\tan x(\tan^2 x-\sec^2 x)$$
$$=-\tan x$$

20. $\sin x-\sin^3 x$
$$=\sin x(1-\sin^2 x)$$
$$=\sin x\cos^2 x$$

21. $\tan x-\sin x\sec^3 x$
$$=\dfrac{\sin x}{\cos x}-\sin x\left(\dfrac{1}{\cos^3 x}\right)$$
$$=\tan x-\left(\dfrac{\sin x}{\cos x}\right)\left(\dfrac{1}{\cos^2 x}\right)$$
$$=\tan x(1-\sec^2 x)$$
$$=\tan x(-\tan^2 x)=-\tan^3 x$$

3a. $\tan\left(\dfrac{\pi}{2}-\theta\right)=\dfrac{b}{a}=\cot\theta$

3b. $\sec\left(\dfrac{\pi}{2}-\theta\right)=\dfrac{c}{a}=\csc\theta$

4a. $\tan(-x)=\dfrac{\sin(-x)}{\cos(-x)}$
$$=\dfrac{-\sin x}{\cos x}=-\tan x$$

4b. $\csc(-x)=\dfrac{1}{\sin(-x)}=\dfrac{1}{-\sin x}=-\csc x$

5. $\cos\theta=\dfrac{1}{3};\ \theta$ in Q I
$$\left(\dfrac{1}{3}\right)^2+\sin^2\theta=1$$
$$\sin^2\theta=\dfrac{8}{9};\ \sin\theta=\dfrac{2\sqrt{2}}{3}$$
$$\tan\theta=\dfrac{\frac{2\sqrt{2}}{3}}{\frac{1}{3}}=2\sqrt{2}$$

6. $\sin\theta=\dfrac{5}{13};\ \theta$ in Q II
$$\left(\dfrac{5}{13}\right)^2+\cos^2\theta=1$$
$$\cos^2\theta=\dfrac{144}{169};\ \cos\theta=-\dfrac{12}{13}$$
$$\sec\theta=-\dfrac{13}{12};\ \tan\theta=\dfrac{\frac{5}{13}}{-\frac{12}{13}}=-\dfrac{5}{12}$$

7. θ in Q III
$$1+\left(-\dfrac{15}{8}\right)^2=\sec^2\theta$$
$$\sec^2\theta=\dfrac{289}{64};\ \sec\theta=-\dfrac{17}{8}$$
$$\cos\theta=-\dfrac{8}{17}$$

22. $\dfrac{\cos\theta}{\sec\theta+\tan\theta}=\dfrac{\cos\theta}{\dfrac{1}{\cos\theta}+\dfrac{\sin\theta}{\cos\theta}}$

$=\dfrac{\cos\theta}{\dfrac{1+\sin\theta}{\cos\theta}}=\dfrac{\cos^2\theta}{1+\sin\theta}=\dfrac{1-\sin^2\theta}{1+\sin\theta}$

$=\dfrac{(1-\sin\theta)(1+\sin\theta)}{1+\sin\theta}=1-\sin\theta$

23. $\csc x\cos\left(\dfrac{\pi}{2}-x\right)=\left(\dfrac{1}{\sin x}\right)\sin x=1$

24. $\sin x\left(\dfrac{\cos x}{\sin x}\right)=\cos x$

25. $\cos x\sec(-x)=\cos x\left(\dfrac{1}{\cos x}\right)=1$

26. $\csc x\sin\left(\dfrac{\pi}{2}-x\right)$

$=\left(\dfrac{1}{\sin x}\right)\cos x=\cot x$

27. $\dfrac{1+\cot^2\theta}{\cot^2\theta}=\dfrac{1}{\cot^2\theta}+\dfrac{\cot^2\theta}{\cot^2\theta}$

$=\tan^2\theta+1=\sec^2\theta$

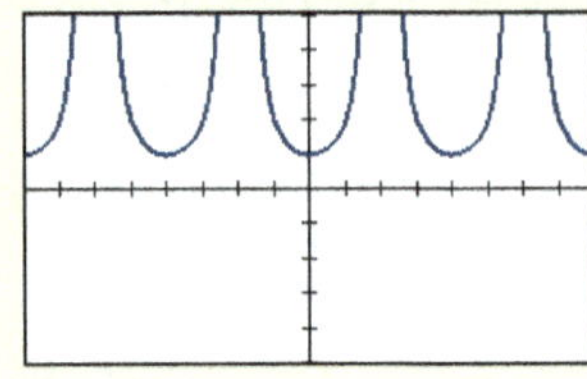

28. $\sin\theta+\cos\theta\cot\theta$

$=\sin\theta+\cos\theta\left(\dfrac{\cos\theta}{\sin\theta}\right)$

$=\dfrac{\sin^2\theta+\cos^2\theta}{\sin\theta}=\dfrac{1}{\sin\theta}=\csc\theta$

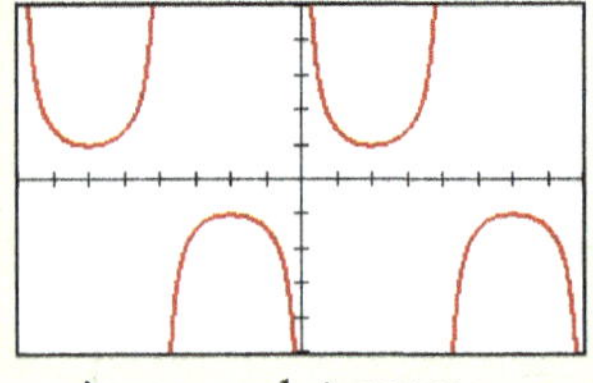

29. $\dfrac{\sin x}{1+\cos x}+\dfrac{1+\cos x}{\sin x}$

$=\dfrac{\sin x}{1+\cos x}\cdot\dfrac{\sin x}{\sin x}+\dfrac{1+\cos x}{\sin x}\cdot\dfrac{1+\cos x}{1+\cos x}$

$=\dfrac{\sin^2 x+1+2\cos x+\cos^2 x}{\sin x(1+\cos x)}$

$=\dfrac{2+2\cos x}{\sin x(1+\cos x)}=\dfrac{2(1+\cos x)}{\sin x(1+\cos x)}$

$=2\csc x$

30. $\cot x-\dfrac{\csc^2 x}{\cot x}$

$=\dfrac{\cot x}{1}\left(\dfrac{\cot x}{\cot x}\right)-\dfrac{\csc^2 x}{\cot x}$

$=\dfrac{\cot^2 x-\csc^2 x}{\cot x}=-\dfrac{1}{\cot x}$

$=-\tan x$

31. $\dfrac{\cos x}{\sec x+1}+\dfrac{\cos x}{\sec x-1}$

$=\dfrac{\cos x}{\sec x+1}\cdot\dfrac{\sec x-1}{\sec x-1}$

$\qquad+\dfrac{\cos x}{\sec x-1}\cdot\dfrac{\sec x+1}{\sec x+1}$

$=\dfrac{\cos x\sec x-\cos x+\cos x\sec x+\cos x}{\sec^2 x-1}$

$=\dfrac{2}{\tan^2 x}=2\cot^2 x$

Match each expression to its equivalent expression. Answers may be used more than once or not at all.

23. $\csc x\cos\left(\dfrac{\pi}{2}-x\right)$ B

24. $\sin x\cot x$ D

25. $\cos x\sec(-x)$ B

26. $\csc x\sin\left(\dfrac{\pi}{2}-x\right)$ E

    A. 0
    B. 1
    C. –1
    D. $\cos x$
    E. $\cot x$
    F. none of these

Simplify the expression. Then support your answer graphically.

27. $\dfrac{1+\cot^2\theta}{\cot^2\theta}$ $\sec^2\theta$ **28.** $\sin\theta+\cos\theta\cot\theta$ $\csc\theta$

Simplify each expression.

29. $\dfrac{\sin x}{1+\cos x}+\dfrac{1+\cos x}{\sin x}$ $2\csc x$

30. $\cot x-\dfrac{\csc^2 x}{\cot x}$ $-\tan x$

31. $\dfrac{\cos x}{\sec x+1}+\dfrac{\cos x}{\sec x-1}$ $2\cot^2 x$

32. $\dfrac{1}{\csc x-1}+\dfrac{1}{\csc x+1}$ $2\tan x\sec x$

Rewrite each expression without fractions.

33. $\dfrac{\cos x}{\sec x-\tan x}$ $1+\sin x$ **35.** $\dfrac{\tan^2 x\sin x}{\sec x-1}$ $\tan x+\sin x$

34. $\dfrac{\cot^2 x}{\csc x+1}$ $\csc x-1$ **36.** $\dfrac{\cos x\cot x}{\sin x+1}$ $\csc x-1$

> **C. Exercises**

Use the properties of the natural log to rewrite each expression as an integer or a single natural logarithm. Support your answers graphically.

37. $\ln|\csc x|-\ln|\sin x|$ $-2\ln|\sin x|$

38. $\ln(\cos^2 x)+\ln(\cot^2 x+1)$ $2\ln|\cot x|$

Write the function rule for $f(g(x))$ and then use a Pythagorean identity to simplify.

39. $f(x)=\sqrt{49-x^2}$ **40.** $f(x)=\sqrt{4x^2+25}$
    $g(x)=7\cos\theta$ $g(x)=\dfrac{5}{2}\tan\theta$
    $f(g(x))=7\,|\sin\theta|$ $f(g(x))=5\,|\sec\theta|$

41. In electronics, an RLC circuit contains a resistor (R), an inductor (L), and a capacitor (C) connected in series across an AC-voltage source. The voltage drop across the capacitor can be modeled by the function $v(t)=V_c\sin\left(\omega t-\dfrac{\pi}{2}\right)$, where V_c is the maximum voltage of the capacitor and ω is the angular frequency. Write a simplified equation for the function.

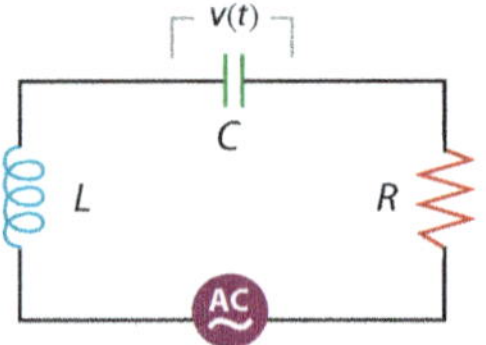

42. The vertical component of the instantaneous velocity of a golf ball that has traveled x ft horizontally after being struck with an initial velocity of v ft/sec at an angle θ above the horizontal can be modeled by the function $v_y(x)=\dfrac{v^2\tan\theta\cos^2\theta-32x}{v\cos\theta}$.

  a. Rewrite the function as the sum or difference of two terms.

  b. Is the function linear, quadratic, or neither?

42a. $v_y(x)=v\sin\theta-\dfrac{32}{v}x\sec\theta$

42b. linear

---

32. $\dfrac{1}{\csc x-1}+\dfrac{1}{\csc x+1}$

$=\dfrac{1}{\csc x+1}\cdot\dfrac{\csc x-1}{\csc x-1}$

$\qquad+\dfrac{1}{\csc x-1}\cdot\dfrac{\csc x+1}{\csc x+1}$

$=\dfrac{\csc x-1+\csc x+1}{\csc^2 x-1}=\dfrac{2\csc x}{\cot^2 x}$

$=\dfrac{2}{\sin x}\cdot\dfrac{\sin^2 x}{\cos^2 x}=\dfrac{2\sin x}{\cos x}\cdot\dfrac{1}{\cos x}$

$=2\tan x\sec x$

33. $\dfrac{\cos x}{\sec x-\tan x}$

$=\dfrac{\cos x}{\sec x-\tan x}\cdot\dfrac{\sec x+\tan x}{\sec x+\tan x}$

$=\dfrac{\cos x\sec x+\cos x\tan x}{\sec^2 x-\tan^2 x}$

$=\dfrac{\cos x\left(\dfrac{1}{\cos x}\right)+\cos x\left(\dfrac{\sin x}{\cos x}\right)}{1}$

$=1+\sin x$

---

34. $\dfrac{\cot^2 x}{\csc x+1}=\dfrac{\cot^2 x}{\csc x+1}\cdot\dfrac{\csc x-1}{\csc x-1}$

$=\dfrac{\cot^2 x(\csc x-1)}{\csc^2 x-1}$

$=\dfrac{\cot^2 x(\csc x-1)}{\cot^2 x}$

$=\csc x-1$

35. $\dfrac{\tan^2 x\sin x}{\sec x-1}=\dfrac{\tan^2 x\sin x}{\sec x-1}\cdot\dfrac{\sec x+1}{\sec x+1}$

$=\dfrac{\tan^2 x\sin x(\sec x+1)}{\sec^2 x-1}$

$=\dfrac{(\sec^2 x-1)(\sin x)(\sec x+1)}{\sec^2 x-1}$

$=\sin x(\sec x+1)\ \text{or}\ \tan x+\sin x$

Simplify. [Algebra]

43. $\dfrac{3x^2 - 11x - 4}{2x^2 - 11x + 12}$ $\dfrac{3x+1}{2x-3}$ **44.** $\dfrac{4a^2 - 20a + 25}{4a^2 - 25}$ $\dfrac{2a-5}{2a+5}$

45. $\dfrac{1}{x+2} + \dfrac{3}{2x}$ $\dfrac{5x+6}{2x^2+4x}$

46. Solve $\dfrac{1}{x+2} + \dfrac{3}{2x} = 1$. [Appendix 2] $x = -\dfrac{3}{2}, 2$

47. State two ways an extraneous root can be introduced when solving a rational equation. [2.6]

48. State the exact value of each expression. [4.3]
 a. $\sin 30°$ $\dfrac{1}{2}$ **b.** $\tan 135°$ -1 **c.** $\cos 210°$ $-\dfrac{\sqrt{3}}{2}$

49. Find the value of x rounded to the nearest hundredth. [4.2]   A

 A. 5.66   **D.** 6.99
 B. 6.03   **E.** none of
 C. 6.68   these

50. Which model best fits growth that eventually slows and approaches a maximum value? [3.5]   D
 A. exponential   **C.** logarithmic
 B. linear   **D.** logistic

51. Which model best fits a rapid initial growth with a slower growth over time? [3.5]   C
 A. exponential   **C.** logarithmic
 B. linear   **D.** logistic

52. Find the value of $i + i^2 + i^3 + \cdots + i^{25}$. [Appendix 5]   A
 A. i **C.** 1   **E.** 0
 B. $-i$ **D.** -1

47. by raising both sides to an even power or by multiplying both sides by a variable expression

---

41. $v_c = V_c \sin\left(\omega t - \dfrac{\pi}{2}\right)$
$= V_c \sin\left(-\left(\dfrac{\pi}{2} - \omega t\right)\right)$
$= -V_c \sin\left(\dfrac{\pi}{2} - \omega t\right)$
$= -V_c \cos(\omega t)$

42a. $v_y(x) = \dfrac{v^2 \tan\theta \cos^2\theta}{v \cos\theta} - \dfrac{32x}{v \cos\theta}$
$= v \tan\theta \cos\theta - \dfrac{32x}{v \cos\theta}$
$= v \dfrac{\sin\theta}{\cos\theta} \cdot \cos\theta - \dfrac{32x}{v \cos\theta}$
$= v \sin\theta - \dfrac{32}{v}x \sec\theta$

42b. linear: $v \sin\theta = b$ and $-\dfrac{32}{v \cos\theta} = m$

> **Cumulative Review**

43. $\dfrac{3x^2 - 11x - 4}{2x^2 - 11x + 12} = \dfrac{(3x+1)(x-4)}{(2x-3)(x-4)}$
$= \dfrac{3x+1}{2x-3}$

44. $\dfrac{4a^2 - 20a + 25}{4a^2 - 25} = \dfrac{(2a-5)(2a-5)}{(2a+5)(2a-5)}$
$= \dfrac{2a-5}{2a+5}$

45. $\dfrac{1}{x+2} + \dfrac{3}{2x} = \dfrac{2x}{2x(x+2)} + \dfrac{3(x+2)}{2x(x+2)}$
$= \dfrac{2x + 3x + 6}{2x(x+2)} = \dfrac{5x+6}{2x^2+4x}$

46. $2x(x+2)\left[\dfrac{1}{x+2} + \dfrac{3}{2x} = 1\right]$
$2x + 3(x+2) = 2x(x+2)$
$5x + 6 = 2x^2 + 4x$
$2x^2 - x - 6 = 0$
$(2x+3)(x-2) = 0$
$x = -\dfrac{3}{2}, 2$

49. $\cos 51° = \dfrac{x}{9}$
$x = 9 \cos 51° \approx 5.66$

52. $i + i^2 + i^3 + i^4 = i - 1 - i + 1 = 0,$
$i^5 + i^6 + i^7 + i^8 = i - 1 - i + 1 = 0,$
etc.
Each group of 4 successive terms has a sum of 0, leaving $0 + i^{25} = i$.

---

36. $\dfrac{\cos x \cot x}{\sin x + 1} = \dfrac{\cos x \cot x}{\sin x + 1} \cdot \dfrac{\sin x - 1}{\sin x - 1}$
$= \dfrac{\cos x \cot x(\sin x - 1)}{\sin^2 x - 1}$
$= \dfrac{\cos x \cot x(\sin x - 1)}{-\cos^2 x}$
$= \dfrac{\cot x \sin x - \cot x}{-\cos x}$
$= \dfrac{\dfrac{\cos x}{\sin x}(\sin x) - \dfrac{\cos x}{\sin x}}{-\cos x} = -1 + \dfrac{1}{\sin x}$
$= \csc x - 1$

> **C. Exercises**

37. $\ln \left|\dfrac{\dfrac{1}{\sin x}}{\sin x}\right| = \ln \left|\dfrac{1}{\sin^2 x}\right|$
$= \ln |\sin^{-2} x| = -2 \ln |\sin x|$

38. $\ln (\cos^2 x (\cot^2 x + 1))$
$= \ln (\cos^2 x \csc^2 x) = \ln \left(\dfrac{\cos^2 x}{\sin^2 x}\right)$
$= \ln (\cot^2 x) = 2 \ln |\cot x|$

39. $f(g(x)) = \sqrt{49 - (7 \cos\theta)^2}$
$= \sqrt{49 - 49 \cos^2\theta}$
$= \sqrt{49(1 - \cos^2\theta)}$
$= \sqrt{49(\sin^2\theta)}$
$= 7 |\sin\theta|$

40. $f(g(x)) = \sqrt{4\left(\dfrac{25}{4} \tan^2\theta\right) + 25}$
$= \sqrt{25(\tan^2\theta + 1)}$
$= \sqrt{25(\sec^2\theta)}$
$= 5 |\sec\theta|$

5.2 Verifying Trigonometric Identities

Objectives

1. To determine whether an equation is an identity
2. To verify identities

Flash

The Rosetta Stone, inscribed in Egypt using ancient writing systems in 196 BC, is now housed in the British Museum in London. It was discovered in Rosetta, Egypt, in 1799 and led to the deciphering of Egyptian hieroglyphs in 1822.

Vocabulary

verifying identities

Reading and Writing Mathematics

Explain how verifying a trigonometric identity is different from solving an algebraic equation. Also explain why properties of equality that involve both sides of the equation cannot be used to prove an identity.

When solving an algebraic equation, the properties of equality can be used because both sides are assumed equal. When proving an identity, the equality properties cannot be used because it cannot be assumed that the identity is an equality.

Additional Exercises

Verify each identity.

1. $\cot\left(\frac{\pi}{2} - \theta\right)\cot\theta = 1$

$\tan\theta\cot\theta = \frac{\tan\theta}{1}\left(\frac{1}{\tan\theta}\right) = 1$

2. $\dfrac{\tan x}{1 - \tan^2 x} = \dfrac{\sin x \cos x}{2\cos^2 x - 1}$

$\dfrac{\frac{\sin x}{\cos x}}{1 - \frac{\sin^2 x}{\cos^2 x}}\left(\frac{\cos^2 x}{\cos^2 x}\right) = \dfrac{\sin x \cos x}{\cos^2 x - \sin^2 x}$

$= \dfrac{\sin x \cos x}{\cos^2 x - (1 - \cos^2 x)} = \dfrac{\sin x \cos x}{2\cos^2 x - 1}$

3. $\dfrac{\csc^2\theta - 1}{\csc\theta} - \dfrac{\sec^2\theta - 1}{\sec\theta}$
$= \csc\theta - \sin\theta - \sec\theta + \cos\theta$

$\dfrac{\csc^2\theta - 1}{\csc\theta}\left(\frac{\sec\theta}{\sec\theta}\right) - \dfrac{\sec^2\theta - 1}{\sec\theta}\left(\frac{\csc\theta}{\csc\theta}\right)$

$= \dfrac{\sec\theta\csc^2\theta - \sec\theta - \sec^2\theta\csc\theta + \csc\theta}{\sec\theta\csc\theta}$

$= \csc\theta - \sin\theta - \sec\theta + \cos\theta$

After completing this section, you will be able to

- determine whether an equation is an identity.
- verify trig identities.

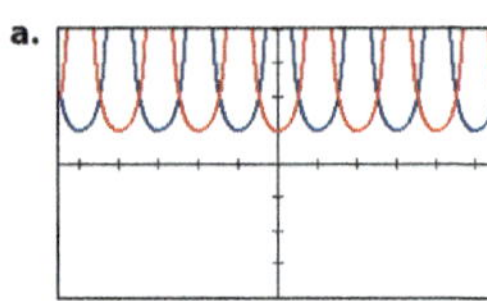

Building on the work of Leonhard Euler, Joseph Fourier made significant contributions to the study of trigonometric series. A Fourier series represents a periodic function as the sum of simple sine waves.

Joseph Fourier accompanied Napoleon Bonaparte on his Egyptian campaign during which the Rosetta Stone was discovered.

In Section 5.1, fundamental identities were used to simplify or rewrite trigonometric expressions. This section focuses on *verifying identities*, showing that both sides of an equation are equal for all values of the variable for which the expressions are defined. Graphing both sides of an equation or examining a table of values can help you decide whether an equation might be an identity.

Example 1   Identifying an Equation That Is Not an Identity

Determine which equation is not an identity.

a. $\dfrac{\tan^2\theta + 1}{\tan^2\theta} = \sec^2\theta$ **b.** $\dfrac{\tan^2\theta + 1}{\tan^2\theta} = \csc^2\theta$

Answer

a.

The graphs of $Y_1 = (\tan^2 X + 1)/\tan^2 X$ and $Y_2 = 1/\cos^2 X$ clearly do not coincide, so the equation is not an identity.

b.

X	Y1	Y3		
0	ERROR	ERROR		
0.2618	14.928	14.928		
0.5236	4	4		
0.7854	2	2		
1.0472	1.3333	1.3333		
1.309	1.0718	1.0718		
1.5708	ERROR	1		
1.8326	1.0718	1.0718		
2.0944	1.3333	1.3333		
2.3562	2	2		
2.618	4	4		

X=0

The values of $Y_1 = (\tan^2 X + 1)/\tan^2 X$ and $Y_3 = 1/\sin^2 X$ appear to be identical wherever both expressions are defined. The fact that $Y_1\left(\frac{\pi}{2}\right)$ is undefined while $Y_3\left(\frac{\pi}{2}\right) = 1$ does not prevent the equation in part *b* from being an identity.

SKILL ✔ EXERCISES 1, 3

Finding just one instance where all the expressions are defined and have different values is sufficient to prove that an equation is not an identity. Using a graph or table of values can indicate the possibility that an equation is an identity, but these methods cannot prove the equation true for all values in its domain.

While you can solve equations by following specific steps, you will see and practice general strategies that can help prove identities. Since an identity cannot be assumed to be true until it has been proved, you cannot apply properties of equality to add or multiply both sides by the same quantity. Instead, begin with one side of the identity and transform it with a sequence of reversible steps using previously proven identities and algebraic manipulations until it matches the other side of the equation. The next example proves that the equation from Example 1*b* is an identity.

PRESENTATION

Lesson Opener

1. Define an identity. an equation that is true for all values of the variable for which the expressions are defined

2. Write the three forms of the Pythagorean identity.
$\cos^2\theta + \sin^2\theta = 1$,
$\tan^2\theta + 1 = \sec^2\theta$, and
$1 + \cot^2\theta = \csc^2\theta$

3. State the conjugate of each expression.

a. $\sqrt{5} + 1$ $\sqrt{5} - 1$

b. $\sec^2 x - 1$ $\sec^2 x + 1$

Review the students' answers to the Lesson Opener exercises. Use Example 1*a* to emphasize that a single counterexample can prove that an equation is not an identity. Example 1*b* demonstrates that both sides of the equation need to be equal only for those values for which each expression is defined. The point discontinuity of Y_1 at $\frac{\pi}{2}$ where Y_2 is defined does not prevent the equation from being an identity.

Remind the students that a visual inspection of graphs or a table of values is not sufficient to prove an identity. While the equation in Example 1*b* appears to be an identity, we have not yet proved that it is an identity.

Verify $\dfrac{\tan^2\theta + 1}{\tan^2\theta} = \csc^2\theta$.

Answer

$\dfrac{\tan^2\theta + 1}{\tan^2\theta} = \dfrac{\sec^2\theta}{\tan^2\theta}$

1. Begin with the more complicated side and substitute using a Pythagorean identity.

$= \dfrac{\frac{1}{\cos^2\theta}}{\frac{\sin^2\theta}{\cos^2\theta}}$

2. Apply reciprocal and quotient identities to express the functions in terms of sine and cosine only.

$= \dfrac{1}{\cos^2\theta} \cdot \dfrac{\cos^2\theta}{\sin^2\theta}$

3. Simplify the complex fraction.

$= \dfrac{1}{\sin^2\theta}$

$= \csc^2\theta$

4. Rewrite the expression using a reciprocal identity.

Check

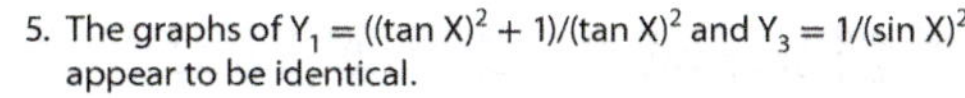

5. The graphs of $Y_1 = ((\tan X)^2 + 1)/(\tan X)^2$ and $Y_3 = 1/(\sin X)^2$ appear to be identical.

SKILL ✔ **EXERCISE 7**

There are often several ways to verify a particular identity. Notice that rewriting the left side of the identity in Example 1 as a sum of fractions produces a shorter proof.

$$\frac{\tan^2\theta + 1}{\tan^2\theta} = \frac{\tan^2\theta}{\tan^2\theta} + \frac{1}{\tan^2\theta} = 1 + \cot^2\theta = \csc^2\theta$$

The proof begins with one side of the equation and lists a series of expressions that are easily seen to be equivalent, ending with the other side of the equation. Adding or subtracting fractions using their least common denominator frequently simplifies the more complicated side.

Example 3   Combining Fractions

Verify $2\sec^2 x = \dfrac{1}{\sin x + 1} - \dfrac{1}{\sin x - 1}$.

Answer

$\dfrac{1}{\sin x + 1} - \dfrac{1}{\sin x - 1} = \dfrac{(\sin x - 1) - (\sin x + 1)}{(\sin x + 1)(\sin x - 1)}$

1. Subtract the fractions on the right side of the equation.

$= \dfrac{-2}{\sin^2 x - 1}$

2. Simplify.

$= \dfrac{2}{1 - \sin^2 x}$

$= \dfrac{2}{\cos^2 x}$

3. Substitute using an alternate form of the basic Pythagorean identity.

$= 2\sec^2 x$

4. Rewrite the expression using a reciprocal identity.

SKILL ✔ **EXERCISE 17**

4. $\dfrac{1}{\sec\theta - \tan\theta} - \dfrac{1}{\sec\theta + \tan\theta} = 2\tan\theta$

$\dfrac{\sec\theta + \tan\theta - (\sec\theta - \tan\theta)}{\sec^2\theta - \tan^2\theta}$

$= \dfrac{2\tan\theta}{(\tan^2\theta + 1) - \tan^2\theta} = 2\tan\theta$

5. $\sin^2 x \sec^4 x - \sec^2 x = \tan^4 x - 1$

$\sec^2 x(\sin^2 x \sec^2 x - 1)$
$= \sec^2 x(\tan^2 x - 1)$
$= (\tan^2 x + 1)(\tan^2 x - 1)$
$= \tan^4 x - 1$

6. $\cot^2 x = \csc^4 x - (\cot^4 x + \csc^2 x)$

$(\csc^4 x - \csc^2 x) - \cot^4 x$
$= \csc^2 x(\csc^2 x - 1) - \cot^4 x$
$= \csc^2 x \cot^2 x - \cot^4 x$
$= \cot^2 x(\csc^2 x - \cot^2 x)$
$= \cot^2 x(1) = \cot^2 x$

7. $\dfrac{\tan\theta}{\sec\theta - 1} = \csc\theta + \cot\theta$

$\dfrac{\tan\theta(\sec\theta + 1)}{(\sec\theta - 1)(\sec\theta + 1)} = \dfrac{\tan\theta(\sec\theta + 1)}{\sec^2\theta - 1}$

$= \dfrac{\tan\theta(\sec\theta + 1)}{\tan^2\theta} = \dfrac{\sec\theta + 1}{\tan\theta}$

$= \dfrac{1}{\cos\theta}\left(\dfrac{\cos\theta}{\sin\theta}\right) + \cot\theta$

$= \csc\theta + \cot\theta$

8. $\dfrac{1}{\cos\theta - 1} + \dfrac{1}{\cos\theta + 1} = -2\cot\theta\csc\theta$

$\dfrac{(\cos\theta + 1) + (\cos\theta - 1)}{\cos^2\theta - 1} = \dfrac{2\cos\theta}{-\sin^2\theta}$

$= -2\cot\theta\csc\theta$

9. $1 - \csc^4 x = -\cot^4 x - 2\cot^2 x$

working on both sides:
$(1 - \csc^2 x)(1 + \csc^2 x)$
$\qquad\qquad = -\cot^2 x(\cot^2 x + 2)$
$-\cot^2 x(2 + \cot^2 x)$
$\qquad\qquad = -\cot^2 x(2 + \cot^2 x)$

10. $\tan^4 x \cos^2 x = \sec^2 x + \cos^2 x - 2$

$\tan^2 x \tan^2 x \cos^2 x$
$= (\sec^2 x - 1)^2 \cos^2 x$
$= (\sec^4 x - 2\sec^2 x + 1)\cos^2 x$
$= \sec^2 x - 2 + \cos^2 x$
$= \sec^2 x + \cos^2 x - 2$

Caution students that while several algebraic strategies are used when *verifying identities*, the properties of equality cannot be used because that would presuppose that the two sides of the equation are equal.

Consider presenting the examples without the students viewing the solutions. Encourage them to identify possible strategies for each example and discuss any invalid strategies that are suggested. Guide the students in identifying strategies that will verify each identity. While the examples list reasons justifying most steps, students do not need to give written explanations as long as they can easily state each reason.

Examples 2 and 3 involve substitutions using Pythagorean identities and basic algebraic operations. Communicate that there may be more than one way to verify an identity.

Factoring either side of the equation in Example 4 enables that side to be transformed into the expression on the other side. Example 5 demonstrates the technique of rationalizing the denominator by multiplying the numerator and denominator by the conjugate of the denominator.

One-on-One Even when the end result is not foreseen, it is important that students test known strategies. While the benefits of rewriting all functions in terms of sine and cosine in step 3 of Example 5 may not be immediately obvious, this strategy allows the expression to be simplified to match the right side of the equation.

Students sometimes wonder how they are going to use what they learn in math. Point out that the development of logical reasoning in math translates into many other areas, including the interpretation of Scripture. Many great mathematicians were also theologians.

At times, it is easier to manipulate both sides of an equation separately until the expressions are the same. Working with both sides is demonstrated in Example 6.

Assignments

- **Minimum:** 1–9, 12; 13–19 odd; 23–24, 27, 43–52
- **Standard:** 1–4, 6, 8–11, 14–15, 18, 22–23, 26, 28–29, 32, 43–48, 50–51
- **Extended:** 2, 4, 6, 9, 12–13, 16, 19, 22, 25, 28, 30–31, 34–35, 37, 39–44, 49–52

Solutions

❯ A. Exercises

1.

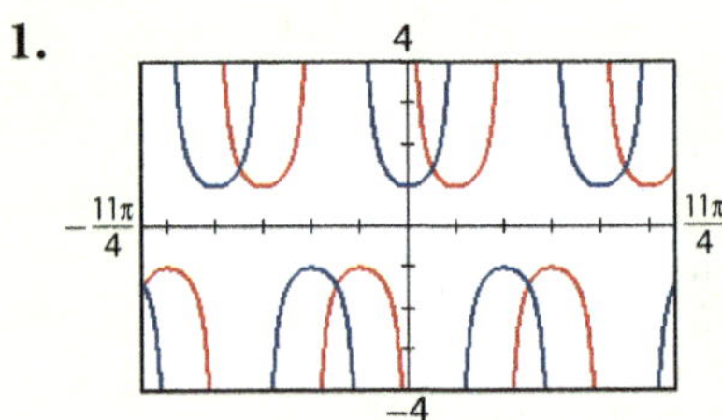

2.

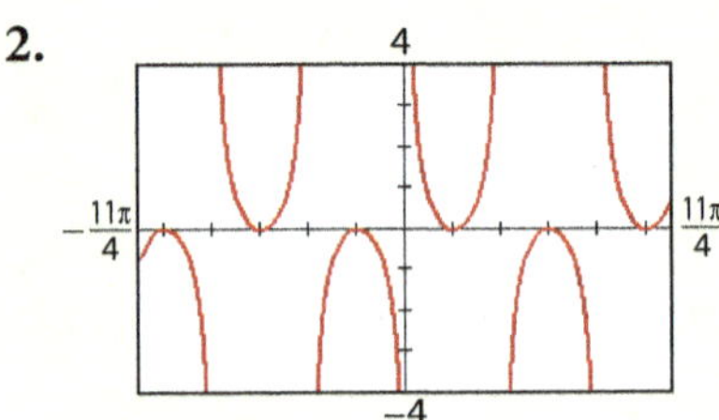

3.

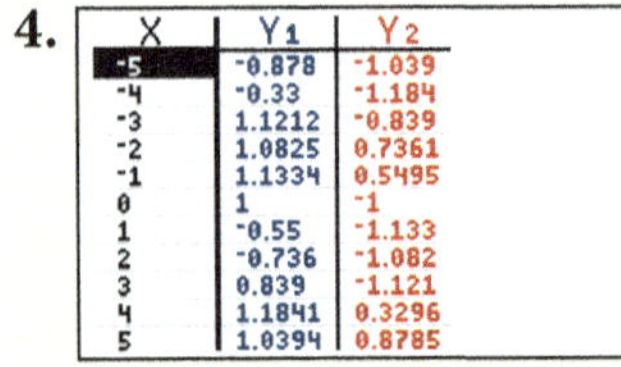

4.

X	Y_1	Y_2
-5	-0.878	-1.039
-4	-0.33	-1.184
-3	1.1212	-0.839
-2	1.0825	0.7361
-1	1.1334	0.5495
0	1	-1
1	-0.55	-1.133
2	-0.736	-1.082
3	0.839	-1.121
4	1.1841	0.3296
5	1.0394	0.8785

5. $\csc x \sin x = \dfrac{1}{\sin x}\left(\dfrac{\sin x}{1}\right) = 1$

6. $\sin^2 x - \cos^2 x$
$= \sin^2 x - (1 - \sin^2 x)$
$= 2\sin^2 x - 1$

7. $\sec\theta - \sin\theta\tan\theta$
$= \dfrac{1}{\cos\theta} - \sin\theta\left(\dfrac{\sin\theta}{\cos\theta}\right)$
$= \dfrac{1 - \sin^2\theta}{\cos\theta} = \dfrac{\cos^2\theta}{\cos\theta} = \cos\theta$

8. $(\csc^2\theta - 1)\sin^2\theta$
$= \cot^2\theta\sin^2\theta$
$= \left(\dfrac{\cos^2\theta}{\sin^2\theta}\right)\dfrac{\sin^2\theta}{1} = \cos^2\theta$

9. $(1 - \cos x)(1 + \sec x)$
$= 1 + \sec x - \cos x - \cos x\sec x$
$= 1 + \sec x - \cos x - \dfrac{\cos x}{1}\left(\dfrac{1}{\cos x}\right)$
$= \sec x - \cos x$

Factoring expressions containing powers of trigonometric functions frequently helps simplify the more complicated side.

Example 4   Using Factoring

Verify $\sec^4\theta - \sec^2\theta = \tan^4\theta + \tan^2\theta$.

Answer

$\tan^4\theta + \tan^2\theta = \tan^2\theta\,(\tan^2\theta + 1)$	1. Factor the expression.
$\qquad = (\sec^2\theta - 1)(\sec^2\theta)$	2. Substitute using two forms of the Pythagorean identity, $\tan^2\theta + 1 = \sec^2\theta$.
$\qquad = \sec^4\theta - \sec^2\theta$	3. Distribute.

— SKILL ✔ **EXERCISE 11**

Utilizing the product of conjugates pattern, $(a - b)(a + b) = a^2 - b^2$, can help simplify fractions with sums or differences in the denominator and set up a substitution using a Pythagorean identity.

Example 5   Using the Product of Conjugates

Verify $\dfrac{\cot x}{\csc x + 1} = \sec x - \tan x$.

Answer

$\dfrac{\cot x}{\csc x + 1} = \dfrac{\cot x}{\csc x + 1}\cdot\dfrac{\csc x - 1}{\csc x - 1}$	1. Multiply the numerator and denominator of the left side by the denominator's conjugate.
$\qquad = \dfrac{\cot x\,(\csc x - 1)}{\csc^2 x - 1}$	
$\qquad = \dfrac{\cot x\,(\csc x - 1)}{\cot^2 x}$	2. Substitute using a Pythagorean identity and simplify.
$\qquad = \dfrac{\csc x - 1}{\cot x}$	
$\qquad = \dfrac{\dfrac{1}{\sin x} - 1}{\dfrac{\cos x}{\sin x}}$	3. Rewrite in terms of sine and cosine and simplify the resulting complex fraction.
$\qquad = \left(\dfrac{1}{\sin x} - 1\right)\dfrac{\sin x}{\cos x}$	
$\qquad = \dfrac{1}{\cos x} - \dfrac{\sin x}{\cos x}$	
$\qquad = \sec x - \tan x$	4. Use reciprocal and quotient identities to rewrite the expression in the form of the right side of the identity.

— SKILL ✔ **EXERCISE 23**

Notice how rewriting the expression in terms of sine and cosine functions helped in the simplification of Example 5. When you struggle to transform one side into the other, you can also try working separately on each side to transform both sides into a common equivalent expression. Remember that the properties of equality cannot be used since the equation has not yet been proved to be true. A more formal proof could then be written starting with either side being transformed into the other.

The identity in Example 4 could also be verified in this way.

$$\sec^2\theta(\sec^2\theta - 1) = \tan^4\theta + \tan^2\theta$$
$$\sec^2\theta\tan^2\theta = (\tan^2\theta + 1)\tan^2\theta$$
$$\sec^2\theta\tan^2\theta = \sec^2\theta\tan^2\theta$$

Finally, Example 7 demonstrates how a power function is written as an expression involving sums that involve lower powers. The more complicated side could be factored and rewritten as $\sin x(1 - \cos^2 x)^2 = \sin x(\sin^2 x)^2 = \sin^5 x$.

Encourage the students to refer to the table on page 243 summarizing the strategies for verifying identities if they experience difficulty when verifying an identity.

One-on-One Success in verifying identities requires familiarity with trigonometric relationships and careful inspection of the expressions. Encourage students to investigate the various strategies used in the examples.

Example 6 Working on Both Sides

Verify $\dfrac{1 - \sin t}{\sin t} = \dfrac{\cot^2 t}{1 + \csc t}$.

Answer

Simplifying the Left Side		Simplifying the Right Side

$$\frac{1 - \sin t}{\sin t} = \frac{\cot^2 t}{1 + \csc t}$$

1. Rewrite as a difference of fractions.

$$\frac{1}{\sin t} - \frac{\sin t}{\sin t} = \frac{\csc^2 t - 1}{1 + \csc t}$$

1. Substitute using a Pythagorean identity.

2. Simplify by substituting using reciprocal and quotient identities.

$$\csc t - 1 = \frac{(\csc t - 1)(\csc t + 1)}{1 + \csc t}$$
$$= \csc t - 1$$

2. Factor the numerator and simplify.

SKILL ✓ **EXERCISE 13**

In order to make certain computations easier in calculus, an expression involving powers of trigonometric functions can be written as a more complicated expression involving sums. Example 7 demonstrates how one such identity can be derived.

Example 7 Rewriting a Power of a Trigonometric Function

Verify $\sin^5 x = \sin x\,(1 - 2\cos^2 x + \cos^4 x)$.

Answer

$$\sin^5 x = \sin x \sin^2 x \sin^2 x$$

1. Factor and apply a form of the basic Pythagorean identity.

$$= \sin x\,(1 - \cos^2 x)(1 - \cos^2 x)$$

$$= \sin x\,(1 - 2\cos^2 x + \cos^4 x)$$

2. Expand the product of binomials.

SKILL ✓ **EXERCISE 15**

We have illustrated how to discern when an equation is not an identity and several strategies that can be used to verify identities. If you are not sure which method of proof will work best, begin by applying one or more of these strategies. While your initial attempt may not be successful, it can provide valuable insights that will lead you to a correct solution.

Strategies for Verifying Identities

1. Begin by trying to transform one side, usually the more complicated side, into the other side.

2. Substitute using previously proven trigonometric identities.
   a. Use various forms of the Pythagorean identities.
   b. Express other trigonometric functions in terms of sine and cosine.

3. Simplify algebraically.
   a. Factor monomial terms, the difference of squares, or trinomials.
   b. Combine, separate, or simplify terms.
   c. Multiply using the Distributive Property, FOIL, or the product of conjugates.

4. Consider working on both sides separately to reach another equivalent expression.

---

10. $\sin^2 x + \sin^2\left(\dfrac{\pi}{2} - x\right)$
$$= \sin^2 x + \cos^2 x = 1$$

11. $\cos\theta - \cos\theta\sin^2\theta$
$$= \cos\theta(1 - \sin^2\theta)$$
$$= \cos\theta(\cos^2\theta) = \cos^3\theta$$

12. $\cot\theta\sec^2\theta - \cot\theta$
$$= \cot\theta(\sec^2\theta - 1)$$
$$= \cot\theta\tan^2\theta$$
$$= \frac{1}{\tan\theta}\left(\frac{\tan^2\theta}{1}\right) = \tan\theta$$

13. working on both sides:
$$\frac{\sin^2 x - 1}{\sin x} = -\cot x\cos x$$
$$-\frac{\cos^2 x}{\sin x} = -\left(\frac{\cos x}{\sin x}\right)\cos\theta$$
$$-\frac{\cos^2 x}{\sin x} = -\frac{\cos^2 x}{\sin x}$$

14. working on both sides:
$$\frac{(1 - \sec\theta)(1 + \sec\theta)}{\sin\theta} = -\tan\theta\sec\theta$$
$$\frac{1 - \sec^2\theta}{\sin\theta} = -\frac{\sin\theta}{\cos\theta}\left(\frac{1}{\cos\theta}\right)$$
$$-\frac{\tan^2\theta}{\sin\theta} = -\frac{\sin\theta}{\cos^2\theta}$$
$$-\frac{\sin^2\theta}{\cos^2\theta}\left(\frac{1}{\sin\theta}\right) = -\frac{\sin\theta}{\cos^2\theta}$$

15. $\cos^4 x = (\cos^2 x)^2 = (1 - \sin^2 x)^2$
$$= 1 - 2\sin^2 x + \sin^4 x$$

16. $\sec^4 x\tan x = \sec^2 x\sec^2 x\tan x$
$$= \sec^2 x(1 + \tan^2 x)\tan x$$
$$= \sec^2 x(\tan x + \tan^3 x)$$

17. $\csc\theta - \cot\theta$
$$= \frac{1}{\sin\theta} - \frac{\cos\theta}{\sin\theta} = \frac{1 - \cos\theta}{\sin\theta}\left(\frac{1 + \cos\theta}{1 + \cos\theta}\right)$$
$$= \frac{1 - \cos^2\theta}{\sin\theta(1 + \cos\theta)} = \frac{\sin^2\theta}{\sin\theta(1 + \cos\theta)}$$
$$= \frac{\sin\theta}{1 + \cos\theta}$$

18. $\dfrac{\csc x + 1}{\cot x} - \dfrac{\cot x}{\csc x - 1}$
$$= \frac{\csc x + 1}{\cot x}\cdot\frac{\csc x - 1}{\csc x - 1} - \frac{\cot x}{\csc x - 1}\cdot\frac{\cot x}{\cot x}$$
$$= \frac{(\csc^2 x - 1) - \cot^2 x}{\cot x(\csc x - 1)}$$
$$= \frac{\cot^2 x - \cot^2 x}{\cot x(\csc x - 1)} = 0$$

19. $\dfrac{1 + \cos\theta}{\sin\theta} + \dfrac{\sin\theta}{1 + \cos\theta}$
$$= \frac{1 + \cos\theta}{\sin\theta}\left(\frac{1 + \cos\theta}{1 + \cos\theta}\right)$$
$$\qquad\qquad + \frac{\sin\theta}{1 + \cos\theta}\left(\frac{\sin\theta}{\sin\theta}\right)$$
$$= \frac{1 + 2\cos\theta + \cos^2\theta + \sin^2\theta}{\sin\theta(1 + \cos\theta)}$$
$$= \frac{1 + 2\cos\theta + 1}{\sin\theta(1 + \cos\theta)} = \frac{2 + 2\cos\theta}{\sin\theta(1 + \cos\theta)}$$
$$= \frac{2(1 + \cos\theta)}{\sin\theta(1 + \cos\theta)} = \frac{2}{\sin\theta}$$
$$= 2\csc\theta$$

20. $\dfrac{\tan^2\theta}{\csc\theta} + \dfrac{\cot^2\theta}{\sec\theta}$
$$= \left(\frac{1}{\csc\theta}\right)\tan^2\theta + \left(\frac{1}{\sec\theta}\right)\cot^2\theta$$
$$= \sin\theta\left(\frac{\sin^2\theta}{\cos^2\theta}\right) + \cos\theta\left(\frac{\cos^2\theta}{\sin^2\theta}\right)$$
$$= \frac{\sin^3\theta}{\cos^2\theta} + \frac{\cos^3\theta}{\sin^2\theta}$$
$$= \frac{\sin^3\theta}{\cos^2\theta}\left(\frac{\sin^2\theta}{\sin^2\theta}\right) + \frac{\cos^3\theta}{\sin^2\theta}\left(\frac{\cos^2\theta}{\cos^2\theta}\right)$$
$$= \frac{\sin^5\theta + \cos^5\theta}{\sin^2\theta\cos^2\theta}$$

21. $\dfrac{\sec^2 x - 3\sec x + 2}{\sec^2 x - 1}$
$$= \frac{(\sec x - 2)(\sec x - 1)}{(\sec x + 1)(\sec x - 1)} = \frac{\sec x - 2}{\sec x + 1}$$

22. $\sec^4\theta + \sec^2\theta\tan^2\theta - 2\tan^4\theta$
$$= (\sec^2\theta + 2\tan^2\theta)(\sec^2\theta - \tan^2\theta)$$
$$= [\sec^2\theta + 2(\sec^2\theta - 1)][\sec^2\theta$$
$$\qquad\qquad\qquad - (\sec^2\theta - 1)]$$
$$= (\sec^2\theta + 2\sec^2\theta - 2)(1)$$
$$= 3\sec^2\theta - 2$$

23. $(\csc x - \cot x)^2 = \left(\dfrac{1}{\sin x} - \dfrac{\cos x}{\sin x}\right)^2$
$$= \left(\frac{1 - \cos x}{\sin x}\right)^2 = \frac{(1 - \cos x)^2}{\sin^2 x}$$
$$= \frac{(1 - \cos x)^2}{1 - \cos^2 x} = \frac{(1 - \cos x)(1 - \cos x)}{(1 - \cos x)(1 + \cos x)}$$
$$= \frac{1 - \cos x}{1 + \cos x}$$

or

working on both sides:

$$\frac{1-\cos x}{1+\cos x} \cdot \frac{1-\cos x}{1-\cos x} = \left(\frac{1}{\sin x} - \frac{\cos x}{\sin x}\right)^2$$

$$\frac{(1-\cos x)^2}{1-\cos^2 x} = \left(\frac{1-\cos x}{\sin x}\right)^2$$

$$\frac{(1-\cos x)^2}{\sin^2 x} = \frac{(1-\cos x)^2}{\sin^2 x}$$

24. working on both sides:

$$\frac{\cos x}{1-\sin x} \cdot \frac{1+\sin x}{1+\sin x} = \sec x + \tan x$$

$$\frac{\cos x(1+\sin x)}{1-\sin^2 x} = \frac{1}{\cos x} + \frac{\sin x}{\cos x}$$

$$\frac{\cos x(1+\sin x)}{\cos^2 x} = \frac{1+\sin x}{\cos x}$$

$$\frac{1+\sin x}{\cos x} = \frac{1+\sin x}{\cos x}$$

25. $\dfrac{\cot x}{\csc x + 1} \cdot \dfrac{\csc x - 1}{\csc x - 1} = \dfrac{\cot x(\csc x - 1)}{\csc^2 x - 1}$

$\qquad = \dfrac{\cot x(\csc x - 1)}{\cot^2 x} = \dfrac{\csc x - 1}{\cot x}$

$\qquad = \dfrac{\csc x}{\cot x} - \dfrac{1}{\cot x}$

$\qquad = \dfrac{1}{\sin x} \cdot \dfrac{\sin x}{\cos x} - \tan x$

$\qquad = \sec x - \tan x$

26. $\dfrac{1}{\sec \theta + 1} \cdot \dfrac{\sec \theta - 1}{\sec \theta - 1}$

$\qquad\qquad + \dfrac{1}{\sec \theta - 1} \cdot \dfrac{\sec \theta + 1}{\sec \theta + 1}$

$\qquad = \dfrac{\sec \theta - 1 + \sec \theta + 1}{\sec^2 \theta - 1} = \dfrac{2 \sec \theta}{\tan^2 \theta}$

$\qquad = \dfrac{2}{\cos \theta} \cdot \dfrac{\cos^2 \theta}{\sin^2 \theta} = 2 \cos \theta \csc^2 \theta$

27. working on both sides:

$$\csc^4 \theta - \cot^4 \theta = \csc^2 \theta + \cot^2 \theta$$

$$(1 + \cot^2 \theta)^2 - \cot^4 \theta$$

$$\qquad\qquad = (1 + \cot^2 \theta) + \cot^2 \theta$$

$$\cot^4 \theta + 2 \cot^2 \theta + 1 - \cot^4 \theta$$

$$\qquad\qquad = 2 \cot^2 \theta + 1$$

$$2 \cot^2 \theta + 1 = 2 \cot^2 \theta + 1$$

28. working on both sides:

$$\tan \theta - \frac{\sin^2 \theta}{\csc \theta \cos \theta} = \frac{\sec^2 \theta - 1}{\sec^2 \theta \tan \theta}$$

$$\frac{\sin \theta}{\cos \theta} \cdot \frac{\csc \theta}{\csc \theta} - \frac{\sin^2 \theta}{\csc \theta \cos \theta} = \frac{\tan^2 \theta}{\sec^2 \theta \tan \theta}$$

$$\frac{\sin \theta \csc \theta - \sin^2 \theta}{\csc \theta \cos \theta} = \frac{\tan \theta}{\sec^2 \theta}$$

$$\frac{1 - \sin^2 \theta}{\csc \theta \cos \theta} = \frac{\sin \theta}{\cos \theta} \cdot \frac{\cos^2 \theta}{1}$$

$$\frac{\cos^2 \theta}{\csc \theta \cos \theta} = \sin \theta \cos \theta$$

$$\frac{\cos \theta}{\frac{1}{\sin \theta}} = \sin \theta \cos \theta$$

$$\sin \theta \cos \theta = \sin \theta \cos \theta$$

29. $\csc^4 x = (\csc^2 x)(\csc^2 x)$

$\qquad = (\cot^2 x + 1) \csc^2 x$

30. $\cos^2 x \sin^5 x = \cos^2 x(\sin^2 x)^2 \sin x$

$\qquad = \cos^2 x(1 - \cos^2 x)^2 \sin x$

$\qquad = \cos^2 x(1 - 2\cos^2 x + \cos^4 x) \sin x$

$\qquad = \sin x(\cos^2 x - 2\cos^4 x + \cos^6 x)$

Use technology to graph each side of the equation and determine whether it appears to be or is not an identity.

1. $\cos x + \sin x \tan x = \csc x$ not an identity

2. $\dfrac{\cot^2 x}{\csc x} = \cos x \cot x$ appears to be an identity

Use technology to make a table of values for each side of the equation and determine whether it appears to be or is not an identity.

3. $\tan^2 x \csc^2 x - \tan^2 x = 1$ appears to be an identity

4. $\cos^2 x - \sin x = \sin^2 x - \sin x - 1$ not an identity

Verify each identity.

5. $\csc x \sin x = 1$

6. $\sin^2 x - \cos^2 x = 2 \sin^2 x - 1$

7. $\sec \theta - \sin \theta \tan \theta = \cos \theta$

8. $(\csc^2 \theta - 1) \sin^2 \theta = \cos^2 \theta$

9. $(1 - \cos x)(1 + \sec x) = \sec x - \cos x$

10. $\sin^2 x + \sin^2 \left(\frac{\pi}{2} - x\right) = 1$

11. $\cos \theta - \cos \theta \sin^2 \theta = \cos^3 \theta$

12. $\cot \theta \sec^2 \theta - \cot \theta = \tan \theta$

13. $\dfrac{\sin^2 x - 1}{\sin x} = -\cot x \cos x$

14. $\dfrac{(1 - \sec \theta)(1 + \sec \theta)}{\sin \theta} = -\tan \theta \sec \theta$

15. $\cos^4 x = 1 - 2\sin^2 x + \sin^4 x$

16. $\sec^4 x \tan x = \sec^2 x (\tan x + \tan^3 x)$

B. Exercises

Verify each identity.

17. $\csc \theta - \cot \theta = \dfrac{\sin \theta}{1 + \cos \theta}$

18. $\dfrac{\csc x + 1}{\cot x} - \dfrac{\cot x}{\csc x - 1} = 0$

19. $\dfrac{1 + \cos \theta}{\sin \theta} + \dfrac{\sin \theta}{1 + \cos \theta} = 2 \csc \theta$

20. $\dfrac{\tan^2 \theta}{\csc \theta} + \dfrac{\cot^2 \theta}{\sec \theta} = \dfrac{\sin^5 \theta + \cos^5 \theta}{\sin^2 \theta \cos^2 \theta}$

21. $\dfrac{\sec^2 x - 3 \sec x + 2}{\sec^2 x - 1} = \dfrac{\sec x - 2}{\sec x + 1}$

22. $\sec^4 \theta + \sec^2 \theta \tan^2 \theta - 2 \tan^4 \theta = 3 \sec^2 \theta - 2$

23. $\dfrac{1 - \cos x}{1 + \cos x} = (\csc x - \cot x)^2$

24. $\dfrac{\cos x}{1 - \sin x} = \sec x + \tan x$

25. $\dfrac{\cot x}{\csc x + 1} = \sec x - \tan x$

26. $\dfrac{1}{\sec \theta + 1} + \dfrac{1}{\sec \theta - 1} = 2 \cos \theta \csc^2 \theta$

27. $\csc^4 \theta - \cot^4 \theta = \csc^2 \theta + \cot^2 \theta$

28. $\tan \theta - \dfrac{\sin^2 \theta}{\csc \theta \cos \theta} = \dfrac{\sec^2 \theta - 1}{\sec^2 \theta \tan \theta}$

29. $\csc^4 x = (\cot^2 x + 1) \csc^2 x$

30. $\cos^2 x \sin^5 x = \sin x (\cos^2 x - 2 \cos^4 x + \cos^6 x)$

Graph each side of the equation to determine whether it appears to be an identity. Then either verify the identity or provide a specific counterexample showing algebraically that the equation is not an identity.

31. $\dfrac{1}{1 - \sin x} + \dfrac{1}{1 + \sin x} = 2 \sec^2 x$

32. $\dfrac{\sin x \cot x + 2 \sin x - \cot x - 2}{\cot x + 2} = \sin x - 1$

33. $\dfrac{\csc x + \cot x}{\csc x - \cot x} = \dfrac{(\cos x + 1)^2}{\sin^2 x}$

34. $\dfrac{\frac{1}{\cos x} + 1}{\frac{1}{\cos x} - 1} = \cos^2 x + 2 \cos x \csc x + \csc^2 x$

C. Exercises

Verify each identity.

35. $\sqrt{\dfrac{\sec x + 1}{\sec x - 1}} = \left|\dfrac{\sec x + 1}{\tan x}\right|$

36. $\sqrt{\dfrac{\csc x - 1}{\csc x + 1}} = \left|\dfrac{\csc x - 1}{\cot x}\right|$

37. $\dfrac{\cos \theta}{\cos \theta - \sin \theta} - \dfrac{\sin \theta}{\sec \theta + \cos \theta} = \dfrac{\tan \theta}{1 + \cos^2 \theta}$

38. $\dfrac{2 \cot x}{1 - \cot^2 x} + \dfrac{1}{2 \sin^2 x - 1} = \dfrac{\sin x + \cos x}{\sin x - \cos x}$

Use the following definitions of hyperbolic trigonometric functions to prove the identities in exercises 39–42.

$\sinh x = \frac{1}{2}(e^x - e^{-x})$ $\operatorname{csch} x = \dfrac{1}{\sinh x}$

$\cosh x = \frac{1}{2}(e^x + e^{-x})$ $\operatorname{sech} x = \dfrac{1}{\cosh x}$

$\tanh x = \dfrac{\sinh x}{\cosh x}$ $\coth x = \dfrac{1}{\tanh x}$

39. $\cosh^2 x - \sinh^2 x = 1$

40. $\coth^2 x - \operatorname{csch}^2 x = 1$

41. $\operatorname{sech}(-x) = \operatorname{sech} x$

42. $\operatorname{csch}(-x) = -\operatorname{csch} x$

31.

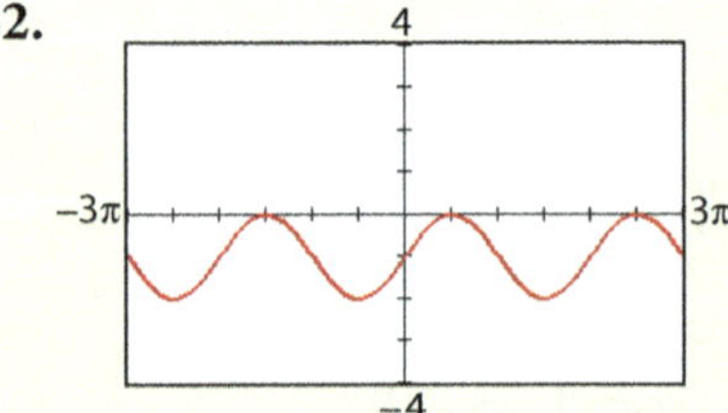

not an identity

sample counterexample:

$$\frac{1}{1 - \sin \frac{\pi}{6}} + \frac{1}{1 + \sin \frac{\pi}{6}} = \frac{1}{1 - \frac{1}{2}} + \frac{1}{1 + \frac{1}{2}}$$

$$= 2 + \frac{2}{3} = \frac{8}{3}$$

$$2 \csc^2 \frac{\pi}{6} = \frac{2}{\sin^2 \frac{\pi}{6}} = \frac{2}{\left(\frac{1}{2}\right)^2} = 8 \neq \frac{8}{3}$$

32.

appears to be an identity

$$\frac{\sin x \cot x + 2 \sin x - \cot x - 2}{\cot x + 2}$$

$$= \frac{\sin x(\cot x + 2) - 1(\cot x + 2)}{(\cot x + 2)}$$

$$= \frac{(\sin x - 1)(\cot x + 2)}{(\cot x + 2)} = \sin x - 1$$

43. Find the inverse of $f(x) = \dfrac{x}{x+3}$. [1.8] $y = -\dfrac{3x}{x-1}$

44. Solve $\ln 3 = -2 + \ln x$. [3.4] $3e^2 \approx 22.2$

45. State the exact value of each expression. [4.3]

  a. $\tan 30°$ $\dfrac{\sqrt{3}}{3}$ **b.** $\cos -135°$ $-\dfrac{\sqrt{2}}{2}$ **c.** $\sin -300°$ $\dfrac{\sqrt{3}}{2}$

46. Express $\dfrac{1}{\sin^2 \theta} - \dfrac{1}{\sin \theta \cos \theta}$ as a single fraction. [5.1] $\dfrac{\cos \theta - \sin \theta}{\sin^2 \theta \cos \theta}$

Express each function in terms of sine or cosine and then simplify. [5.1]

47. $\tan \theta + \sec \theta$ $\dfrac{\sin \theta + 1}{\cos \theta}$ **48.** $\dfrac{\tan \theta \csc \theta}{\sec \theta}$ 1

49. Which term describes the difference between the actual y-value of a data point and the y-value predicted by the regression equation? [1.9] C

  A. coefficient of determination   **C.** residual

  B. regression   **D.** correlation

                     E. none of these

50. Which expression represents the height of a utility pole if a 52 ft cable (guy-wire) attached to the top of the pole is anchored at a 47° angle to the ground? [4.2] C

  A. $\sin 47°$ **D.** $52 \cos 47°$

  B. $\cos 47°$ **E.** $52 \tan 47°$

  C. $52 \sin 47°$

51. Find the value of $\sin\left(\operatorname{Tan}^{-1} \dfrac{3}{4}\right)$. [4.6] D

  A. $\dfrac{4}{3}$ **B.** $\dfrac{3}{4}$ **C.** $\dfrac{4}{5}$ **D.** $\dfrac{3}{5}$ **E.** $\dfrac{5}{3}$

52. Without using technology, determine which function has a wave with varying amplitude. [4.7] A

  A. $y = x^2 \sin x$ **D.** $y = 3 \sin x$

  B. $y = \sin x^2$ **E.** $y = \sin^2 \dfrac{x}{2}$

  C. $y = \sin(3x + \pi)$

$$= \frac{1}{4} + \left(\frac{2\sqrt{3}}{3}\right) + \frac{4}{3} = \frac{19 + 8\sqrt{3}}{12}$$
$$\approx 2.738 \neq 3$$

❯ C. Exercises

35.
$$\sqrt{\frac{\sec x + 1}{\sec x - 1}} = \sqrt{\frac{\sec x + 1}{\sec x - 1}} \cdot \sqrt{\frac{\sec x + 1}{\sec x + 1}}$$
$$= \sqrt{\frac{(\sec x + 1)^2}{\sec^2 x - 1}}$$
$$= \sqrt{\frac{(\sec x + 1)^2}{\tan^2 x}}$$
$$= \left|\frac{\sec x + 1}{\tan x}\right|$$

36.
$$\sqrt{\frac{\csc x - 1}{\csc x + 1}} = \sqrt{\frac{\csc x - 1}{\csc x + 1}} \cdot \sqrt{\frac{\csc x - 1}{\csc x - 1}}$$
$$= \sqrt{\frac{(\csc x - 1)^2}{\csc^2 x - 1}}$$
$$= \sqrt{\frac{(\csc x - 1)^2}{\cot^2 x}}$$
$$= \left|\frac{\csc x - 1}{\cot x}\right|$$

37.
$$\frac{\cos \theta}{\csc \theta - \sin \theta} - \frac{\sin \theta}{\sec \theta + \cos \theta}$$
$$= \frac{\cos \theta}{\dfrac{1}{\sin \theta} - \dfrac{\sin^2 \theta}{\sin \theta}} - \frac{\sin \theta}{\dfrac{1}{\cos \theta} + \dfrac{\cos^2 \theta}{\cos \theta}}$$
$$= \frac{\cos \theta}{\dfrac{1 - \sin^2 \theta}{\sin \theta}} - \frac{\sin \theta}{\dfrac{1 + \cos^2 \theta}{\cos \theta}}$$
$$= \frac{\sin \theta \cos \theta}{\cos^2 \theta} - \frac{\sin \theta \cos \theta}{1 + \cos^2 \theta}$$
$$= \frac{\sin \theta}{\cos \theta} - \frac{\sin \theta \cos \theta}{1 + \cos^2 \theta}$$
$$= \frac{\sin \theta}{\cos \theta}\left(\frac{1 + \cos^2 \theta}{1 + \cos^2 \theta}\right)$$
$$\qquad - \frac{\sin \theta \cos \theta}{1 + \cos^2 \theta}\left(\frac{\cos \theta}{\cos \theta}\right)$$
$$= \frac{\sin \theta + \sin \theta \cos^2 \theta - \sin \theta \cos^2 \theta}{\cos \theta(1 + \cos^2 \theta)}$$
$$= \frac{\sin \theta}{\cos \theta(1 + \cos^2 \theta)} = \frac{\tan \theta}{1 + \cos^2 \theta}$$

33.

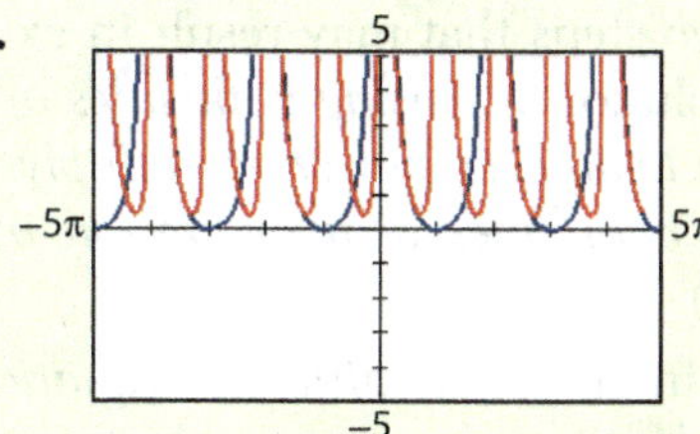

appears to be an identity

working on both sides:

$$\frac{\csc x + \cot x}{\csc x - \cot x} \cdot \frac{\csc x + \cot x}{\csc x + \cot x}$$
$$= \frac{\cos^2 x + 2\cos x + 1}{\sin^2 x}$$
$$\frac{(\csc x + \cot x)^2}{(\csc^2 x - \cot^2 x)}$$
$$\frac{1}{}$$
$$= \frac{\cos^2 x}{\sin^2 x} + \frac{2\cos x}{\sin^2 x} + \frac{1}{\sin^2 x}$$
$$\csc^2 x + 2\cot x \csc x + \cot^2 x$$
$$= \cot^2 x + 2\cot x \csc x + \csc^2 x$$

34.

not an identity

sample counterexample:

$$\frac{\dfrac{1}{\cos \frac{\pi}{3}} + 1}{\dfrac{1}{\cos \frac{\pi}{3}} - 1} = \frac{\dfrac{1}{0.5} + 1}{\dfrac{1}{0.5} - 1} = \frac{2 + 1}{2 - 1} = 3$$

$$\cos^2 \frac{\pi}{3} + 2\cos \frac{\pi}{3}\csc \frac{\pi}{3} + \csc^2 \frac{\pi}{3}$$
$$= \left(\frac{1}{2}\right)^2 + 2\left(\frac{1}{2}\right)\left(\frac{2\sqrt{3}}{3}\right) + \left(\frac{2}{\sqrt{3}}\right)^2$$

continued in Answers and Solutions Overflow

5.3 Solving Trigonometric Equations

Objectives

1. To use standard algebraic techniques to solve trigonometric equations
2. To solve trigonometric equations of the quadratic type
3. To solve trigonometric equations involving multiple angles
4. To use inverse trigonometric functions to find general solutions to trigonometric equations

Flash

Standing waves are a combination of two waves moving in opposite directions. They are frequently created from natural vibrations when an object is struck, bowed, plucked, or strummed.

Vocabulary

conditional trigonometric equations
general solution
primary solutions

The standing wave produced by this machine clearly demonstrates the wave's nodes, points that do not experience any displacement.

After completing this section, you will be able to

- solve trig equations using algebraic techniques and trig identities.

Trigonometric identities are equations that are true for all values of the variable for which the expressions are defined. They are verified by transforming either or both sides independently until the expressions are identical. *Conditional trigonometric equations* may be true for certain values in the domains of the expressions but are not true for all values in their domains. Trigonometric equations are solved using standard algebraic practices such as applying properties of equality, collecting like terms, and factoring along with trigonometric identities. The goal is to isolate the trigonometric function on one side of the equation.

Example 1 — Isolating the Trigonometric Function

Solve $2 \sin \theta + 1 = 0$.

Answer

$$2 \sin \theta + 1 = 0$$
$$2 \sin \theta = -1$$
$$\sin \theta = -\frac{1}{2}$$

1. Use the properties of equality to isolate sin x on the left side of the equation.

$$\theta = \frac{7\pi}{6}, \frac{11\pi}{6}$$

2. Use the unit circle to find the values of θ within $[0, 2\pi)$ for which $\sin \theta = -\frac{1}{2}$.

$$\theta = \frac{7\pi}{6} + n2\pi, \frac{11\pi}{6} + n2\pi$$
where $n \in \mathbb{Z}$

3. Since the period of the sine function is 2π, adding multiples of 2π produces all solutions to the equation.

Check

4. Find the zeros of $y = 2 \sin x + 1$ to verify your solution.

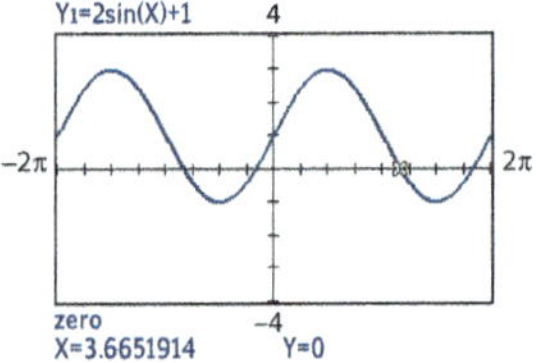

SKILL ✔ EXERCISE 1

Since trigonometric functions are periodic, most trigonometric equations will have an infinite number of solutions. The *general solution* lists all the solutions to the equation. Frequently we are interested only in *primary solutions*, those within $[0, 2\pi)$.

Lesson Opener

Solve each equation.

1. $2x^2 - 12 = 0$ (by taking roots)
   $x = \pm\sqrt{6}$
2. $6x^2 - x - 2 = 0$ (by factoring)
   $x = -\frac{1}{2}, \frac{2}{3}$
3. $\sqrt{x - 8} = -7$ $\varnothing$ (57 is extraneous.)
4. $\frac{1}{a-1} = \frac{2}{a^2 - a}$ $a = 2$, 1 (extraneous)

The Lesson Opener will help the students review methods of solving equations and emphasize the need to check for extraneous roots. Consider asking them to name two steps that may result in extraneous solutions. (*raising both sides of an equation to an even power or multiplying/dividing both sides of an equation by a variable*)

Explain that *conditional equations*, unlike identities, are true only for specific values of the variable. Point out that many of the same strategies used to solve algebraic equations are also employed in the solving of trigonometric equations.

Example 1 illustrates the most basic strategy of isolating the trigonometric function. Explain the difference between the *general solution* and the *primary solutions*.

In Example 2 the square root of each side is taken in order to solve for tan x. Consider asking the students whether this step can produce an extraneous solution. (*no*)

One-on-One A student may struggle with knowing whether to add π or 2π to produce the general solution. Encourage the student to locate the solutions on the unit circle and add π when they are on the same diameter. The student should also recall that the period of trigonometric functions is 2π, except for tangent and cotangent, whose period is π.

Point out how the substitution in Example 3a creates an equation in quadratic

Example 2   Solving by Taking Roots

State primary and general solutions for $6 - \tan^2 x = \tan^2 x$.

Answer

$$6 - \tan^2 x = \tan^2 x$$
$$6 = 2\tan^2 x$$
$$3 = \tan^2 x$$
$$\pm\sqrt{3} = \tan x$$

1. Isolate $\tan x$ on the right side. Be sure to list both the positive and the negative square roots.

$$x = \frac{\pi}{3}, \frac{2\pi}{3}, \frac{4\pi}{3}, \frac{5\pi}{3}$$

2. Use the unit circle to find the primary solutions.

$$x = \frac{\pi}{3} + n\pi, \frac{2\pi}{3} + n\pi$$
where $n \in \mathbb{Z}$

3. Since the tangent function is periodic with a period of π, adding multiples of π to the primary solutions in Q I and Q II produces the general solution.

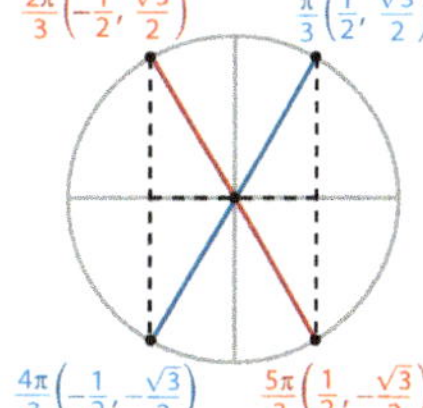

SKILL ✔ EXERCISE 5

Sometimes it is not possible to isolate a trigonometric function on one side of the equation. We can often solve trigonometric equations in quadratic form by factoring and applying the Zero Product Property.

Example 3   Solving by Factoring

Find the general solution for each equation.

a. $\sin^2 x + \cos x + 1 = 0$ **b.** $2\tan x \sin x = \sqrt{2}\tan x$

Answer

a.
$$\sin^2 x + \cos x + 1 = 0$$
$$(1 - \cos^2 x) + \cos x + 1 = 0$$
$$-\cos^2 x + \cos x + 2 = 0$$
$$\cos^2 x - \cos x - 2 = 0$$

1. Substitute using a Pythagorean identity and rewrite the equation as a quadratic in $\cos x$.

$$(\cos x + 1)(\cos x - 2) = 0$$

2. Factor. You can think of the equation as $u^2 - u - 2 = 0$ with $u = \cos x$.

$$\cos x = -1 \quad \text{or} \quad \cos x = 2$$

3. Solve for $\cos x$ using the Zero Product Property.

$$x = \pi$$

4. Identify any primary solutions. Note that the range of the cosine function, $[-1, 1]$, implies that there are no real solutions to $\cos x = 2$.

$$x = \pi + n2\pi = (2n + 1)\pi$$
where $n \in \mathbb{Z}$

5. Write the general solution.

Check

6. Confirm graphically by finding zeros of $y = \sin^2 x + \cos x + 1$.

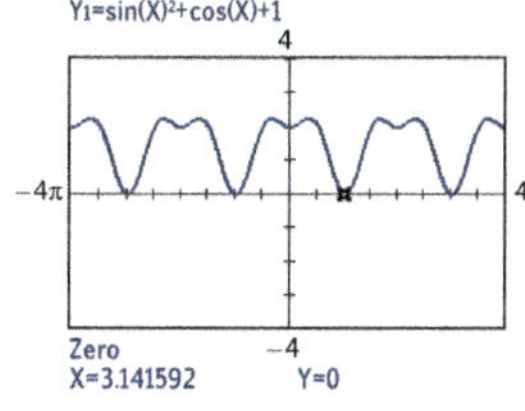

CONTINUED ➡

Additional Exercises

Find all primary solutions to each equation.

1. $2\cos x + 1 = 0$ $x = \frac{2\pi}{3}, \frac{4\pi}{3}$

2. $\sqrt{3}\cot x + 3 = 0$ $x = \frac{5\pi}{6}, \frac{11\pi}{6}$

3. $5\tan^2 x = 4\tan^2 x + 3$
$x = \frac{\pi}{3}, \frac{2\pi}{3}, \frac{4\pi}{3}, \frac{5\pi}{3}$

4. $2\sin x \cos x + \sin x = 0$
$x = 0, \frac{2\pi}{3}, \pi, \frac{4\pi}{3}$

5. $\sin^2 x - \cos x = \cos^2 x$ $x = \frac{\pi}{3}, \pi, \frac{5\pi}{3}$

Find the general solution (in radians) for each equation. Round decimal approximations to the nearest hundredth. Check for extraneous solutions.

6. $\sin x - 2\cos x = 1$ $x = \frac{\pi}{2} + n2\pi,$
$\approx 3.79 + n2\pi, 2.50$ (extraneous)

7. $3\sin x - \cos x = 1$ $x = (2n + 1)\pi,$
$\approx 0.64 + n2\pi, 5.64$ (extraneous)

8. $5\cos x + 2 = 3\sec x$
$x = (2n + 1)\pi, \approx 0.93 + n2\pi,$
$\approx 5.36 + n2\pi$

9. $\cos^2 \frac{x}{2} = \frac{1}{4}$ $\frac{2\pi}{3} + n2\pi, \frac{4\pi}{3} + n2\pi$

10. $\csc 4x - 2 = 0$ $x = \frac{\pi}{24} + \frac{n\pi}{2}, \frac{5\pi}{24} + \frac{n\pi}{2}$

---

form. This equation can then be solved by factoring and applying the Zero Product Property. Quadratic forms that cannot be factored can be solved using the quadratic formula (see exercises 38–39). Example 3b illustrates that there are only two locations where the sum of the sine and cosine values is 1: 0 and $\frac{\pi}{2}$ (with coordinates (1, 0) and (0, 1), respectively). If less rigor is desired, primary solutions may be found without requiring general solutions, as done in the related skill check exercise. Also consider permitting general solutions in a listing format, such as $\{\pm\pi, \pm3\pi, \pm5\pi, \ldots\}$.

This text emphasizes solving algebraically and then checking graphically or algebraically. Exclusively using graphing technology to find zeros or points of intersection can generate correct solutions with little understanding and may not produce exact answers.

Be sure students understand when it is necessary to check for extraneous roots. Since checking each root in the original equation can be time-consuming, help the students understand why each extraneous root fails. Ask the students which step of Example 4a possibly introduces an extraneous root and why. (*step 2, where both sides of the equation are multiplied by a variable expression; This step is not valid if the variable expression equals 0.*) Example 4b illustrates the important technique of squaring both sides so that a Pythagorean identity can be used to express the equation in quadratic form. Ask the students which step could introduce an extraneous root and why. (*step 1, where both sides of the equation are assumed to be equal, although they could be additive inverses of each other; In this case, a false statement is made true by squaring both sides.*)

Why did Viète not find 45 solutions to this equation of the forty-fifth degree? At this time in history the sine of an angle was viewed as the length of a chord of a circle and not as a ratio. Viète disregarded all the negative solutions since a length cannot be negative.

Assignments

- **Minimum:** 1–5, 7, 9, 11–12, 14, 16–19, 21–22, 25, 29, 32–33, 37, 43–47, 51
- **Standard:** 1, 4, 6, 8–9, 11–12, 15, 17–19, 21–22, 25, 27, 29–31, 36–37, 39, 44–47, 50–51
- **Extended:** 3, 6, 9, 12, 15, 17–19, 22, 24, 26–33, 36, 39–40, 42, 45–49, 51

Assessment

- Quiz 5A covers Sections 5.1–5.3.

Solutions

❯ A. Exercises

1. $\sin x = -1$

$$x = \frac{3\pi}{2}$$

2. $\cos x = \frac{\sqrt{3}}{2}$ $\left(\alpha = \frac{\pi}{6} \text{ in Q I and Q IV}\right)$

$$x = \frac{\pi}{6}, \frac{11\pi}{6}$$

3. $\cot x = -1$

(Take the reciprocal of both sides.)

$\tan x = -1$ $\left(\alpha = \frac{\pi}{4} \text{ in Q II and Q IV}\right)$

$$x = \frac{3\pi}{4}, \frac{7\pi}{4}$$

4. $\tan x = \sqrt{3}$ $\left(\alpha = \frac{\pi}{3} \text{ in Q I and Q III}\right)$

$$x = \frac{\pi}{3}, \frac{4\pi}{3}$$

5. $4 \sin^2 x = 3$

$$\sin^2 x = \frac{3}{4}$$

$$\sin x = \pm \frac{\sqrt{3}}{2}$$

$$x = \frac{\pi}{3}, \frac{2\pi}{3}, \frac{4\pi}{3}, \frac{5\pi}{3}$$

6. $4 \cos^2 x = 3$

$$\cos^2 x = \frac{3}{4}$$

$$\cos x = \pm \frac{\sqrt{3}}{2}$$

$$x = \frac{\pi}{6}, \frac{5\pi}{6}, \frac{7\pi}{6}, \frac{11\pi}{6}$$

7. $2 \sin^2 x = 1$

$$\sin^2 x = \frac{1}{2}$$

$$\sin x = \pm \frac{\sqrt{2}}{2}$$

$$x = \frac{\pi}{4}, \frac{3\pi}{4}, \frac{5\pi}{4}, \frac{7\pi}{4}$$

8. $\cos x(2 \cos x - \sqrt{2}) = 0$

$\cos x = 0$ or $\cos x = \frac{\sqrt{2}}{2}$

$x = \frac{\pi}{2}, \frac{3\pi}{2}$ $\qquad x = \frac{\pi}{4}, \frac{7\pi}{4}$

9. $\cot x(\cot x - \sqrt{3}) = 0$

$\cot x = 0$ or $\cot x = \sqrt{3}$

$\cos x = 0$ $\qquad \tan x = \frac{1}{\sqrt{3}} = \frac{\sqrt{3}}{3}$

$x = \frac{\pi}{2}, \frac{3\pi}{2}$ $\qquad x = \frac{\pi}{6}, \frac{7\pi}{6}$

---

b. $2 \tan x \sin x = \sqrt{2} \tan x$ — 1. Collect all terms on one side and factor.

$2 \tan x \sin x - \sqrt{2} \tan x = 0$

$\tan x(2 \sin x - \sqrt{2}) = 0$

$\tan x = 0$ or $\sin x = \frac{\sqrt{2}}{2}$ — 2. Apply the Zero Product Property.

$x = 0, \pi$ $\qquad x = \frac{\pi}{4}, \frac{3\pi}{4}$ — 3. Identify any primary solutions.

$x = n\pi, \frac{\pi}{4} + n2\pi, \frac{3\pi}{4} + n2\pi$ — 4. Write the general solution.

where $n \in \mathbb{Z}$

Check — 5. Confirm graphically by finding zeros of $y = 2 \tan x \sin x - \sqrt{2} \tan x$.

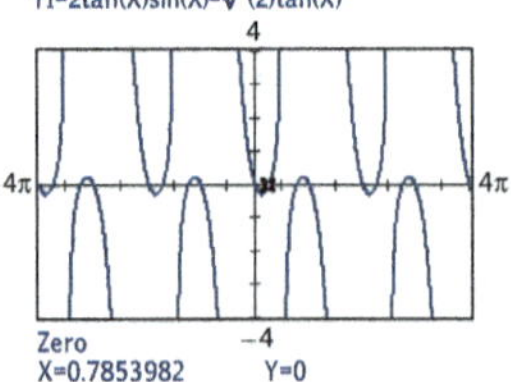

SKILL ✔ EXERCISE 9

Recall that dividing both sides of any equation by an expression containing a variable introduces the possibility of losing solutions to the equation. Dividing both sides of the equation in Example 3*b* by tan *x* would eliminate the solutions of $x = n\pi$ when both sides of the equation are unknowingly divided by 0.

You must check for extraneous solutions when both sides of an equation are multiplied by a trigonometric function or when both sides are squared.

> **Example 4 Checking for Extraneous Solutions**

Find all primary solutions to each equation.

a. $3 \csc x - \sin x = 2$ $\qquad$ **b.** $\cos x + \sin x = 1$

Answer

a. $3 \csc x - \sin x = 2$ — 1. Substitute using the reciprocal identity for cosecant.

$$\frac{3}{\sin x} - \sin x = 2$$

$3 - \sin^2 x = 2 \sin x$ — 2. Eliminate fractions by multiplying both sides by sin *x*.

$\sin^2 x + 2 \sin x - 3 = 0$ — 3. Solve the quadratic by factoring.

$(\sin x - 1)(\sin x + 3) = 0$

$\sin x - 1 = 0$ or $\sin x + 3 = 0$

$\sin x = 1$ or $\sin x = -3$ — Note that the range of sine, $[-1, 1]$, implies that there are no real solutions to $\sin x = -3$.

$$x = \frac{\pi}{2}$$

Check

$3 \csc \frac{\pi}{2} - \sin \frac{\pi}{2} = 3(1) - 1 = 2$ — 4. Substitute to check that $\frac{\pi}{2}$ is not extraneous.

CONTINUED ➡

> **TIP**
>
> Solutions can also be checked on a graphing calculator by finding the intersections of Y₁ (the expression on the left) and Y₂ (the expression on the right).

In Example 5 be sure the students do not confuse finding *primary solutions* of a trig function with finding the *principal value* returned by an inverse function. Note that $\sin^{-1}\left(-\frac{1}{3}\right)$ returns a negative value representing a Q IV angle. The primary solutions can be found using the reference angle $\alpha = \left|\sin^{-1}\left(-\frac{1}{3}\right)\right|$.

Solving a multiple-angle equation is demonstrated in Example 6. Point out that if $n = 3$ in step 3, the solutions would be beyond $[0, 2\pi)$; therefore n must be 1 less than k. If only primary solutions are desired, list solutions for $0 \le 3x < 6\pi$ and then divide each value by 3 so that $0 \le x < 2\pi$.

Use the Internet keyword search *trigonometric equation calculator* to access some powerful teacher tools for quickly checking or creating trig equations.

TIPS

Ex. 3 The solution involves taking the reciprocal of each side.

Ex. 48 A cofunction identity is used to simplify since the difference formulas have not yet been introduced.

b. $\cos x + \sin x = 1$

$\sin x = 1 - \cos x$

$\sin^2 x = 1 - 2\cos x + \cos^2 x$

$(1 - \cos^2 x) = 1 - 2\cos x + \cos^2 x$

1. Since factoring cannot separate $\sin x$ and $\cos x$, solve for $\sin x$ and square both sides to allow a substitution using the Pythagorean identity.

$2\cos^2 x - 2\cos x = 0$

$2\cos x(\cos x - 1) = 0$

$\cos x = 0$ or $\cos x = 1$

$x = 0, \frac{\pi}{2}, \frac{3\pi}{2}$

2. Solve for x.

Check

$\cos 0 + \sin 0 = 1 + 0 = 1$

$\cos \frac{\pi}{2} + \sin \frac{\pi}{2} = 0 + 1 = 1$

$\cos \frac{3\pi}{2} + \sin \frac{3\pi}{2} = 0 + (-1) \neq 1$

3. Substitute to check for extraneous roots.

$\therefore x = 0, \frac{\pi}{2}$

4. Exclude any extraneous solutions.

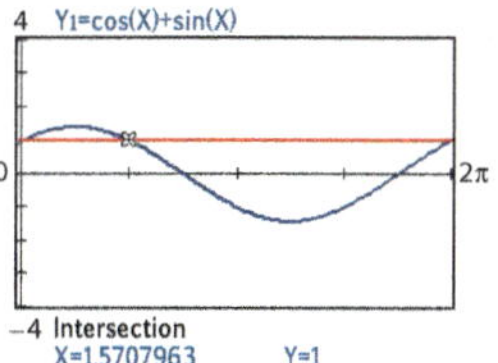

SKILL ✔ EXERCISES 11, 13

Recall that a calculator's inverse trigonometric functions return the principal value within $[0, \pi)$ for $\cos^{-1}$, $\left[-\frac{\pi}{2}, \frac{\pi}{2}\right]$ for $\sin^{-1}$, and $\left(-\frac{\pi}{2}, \frac{\pi}{2}\right)$ for $\tan^{-1}$. These values are used to determine the primary solutions found in other quadrants.

Example 5   Using a Calculator When Solving Equations

Find the primary solutions for $3\sin^2 x - 5\sin x - 2 = 0$ (to the nearest hundredth).

Answer

$3\sin^2 x - 5\sin x - 2 = 0$

$(3\sin x + 1)(\sin x - 2) = 0$

$\sin x = -\frac{1}{3}$ or $\sin x = 2$

1. Factor and apply the Zero Product Property.

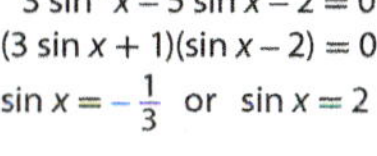
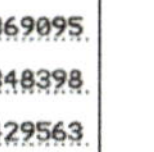

2. Use a calculator (in radian mode) to find the primary solutions.

   a. Find the principal value for $x = \sin^{-1}\left(-\frac{1}{3}\right)$. Note that it is not a primary solution.

   b. Add 2π to find the primary solution in Q IV.

   c. Add π to the reference angle $\alpha = \left|\sin^{-1}\left(-\frac{1}{3}\right)\right|$ to find the primary solution in Q III.

   d. There are no solutions for $\sin x = 2$.

$x \approx 3.48, 5.94$

3. List decimal approximations of primary solutions.

Check

4. Find the 2 zeros of $y = 3\sin^2 x - 5\sin x - 2$ in $[0, 2\pi)$ to verify your solutions.

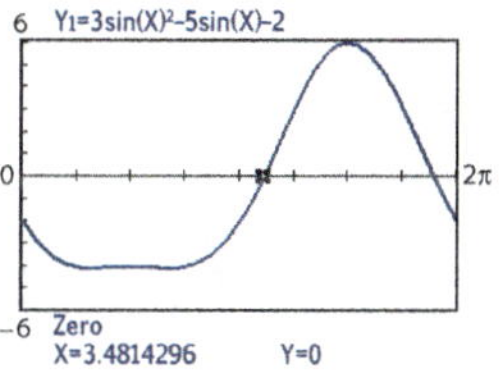

SKILL ✔ EXERCISE 33

In 1593 King Henry IV of France summoned François Viète to respond to a Dutch ambassador's mathematical challenge. When presented with an equation of the forty-fifth degree, Viète used trigonometry to immediately provide one solution and then added twenty-two more solutions the following day.

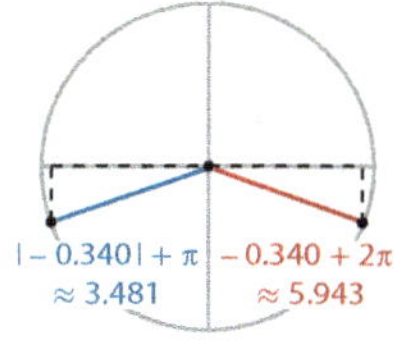

---

10. $\tan x(1 - \tan x) = 0$

$\tan x = 0$ or $\tan x = 1$

$x = 0, \pi$ $\qquad x = \frac{\pi}{4}, \frac{5\pi}{4}$

11. $(1 - \sin x)^2 = \cos^2 x$

$1 - 2\sin x + \sin^2 x = 1 - \sin^2 x$

$2\sin^2 x - 2\sin x = 0$

$2\sin x(\sin x - 1) = 0$

$\sin x = 0$ or $\sin x = 1$

$x = 0, \pi$ $\qquad x = \frac{\pi}{2}$

(π is extraneous.)

12. $\sin x\left[2\sin x = \frac{4}{\sin x} + 7\right]$

$2\sin^2 x = 4 + 7\sin x$

$2\sin^2 x - 7\sin x - 4 = 0$

$(2\sin x + 1)(\sin x - 4) = 0$

$2\sin x = -1$ or $\sin x = 4$

$\sin x = -\frac{1}{2}$

$x = \frac{7\pi}{6}, \frac{11\pi}{6}$

13. $2\cos x + \frac{2}{\cos x} = 5$

$2\cos^2 x + 2 = 5\cos x$

$2\cos^2 x - 5\cos x + 2 = 0$

$(2\cos x - 1)(\cos x - 2) = 0$

$\cos x = \frac{1}{2}$ or $\cos x = 2$

$x = \frac{\pi}{3}, \frac{5\pi}{3}$

14. $\cot^2 x - 2\cot x + 1 = \csc^2 x$

$\cot^2 x - 2\cot x + 1 = 1 + \cot^2 x$

$-2\cot x = 0$

$\cot x = 0$ when $\cos x = 0$

$x = \frac{\pi}{2}, \frac{3\pi}{2}$ (extraneous)

or

$\cot x = \csc x + 1$

$\cot^2 x = \csc^2 x + 2\csc x + 1$

$\csc^2 x - 1 = \csc^2 x + 2\csc x + 1$

$2\csc x = -2$

$\csc x = -1$ when $\sin x = -1$

$x = \frac{3\pi}{2}$

15. $4\sin x \cos x - \sin x = 0$

$\sin x(4\cos x - 1) = 0$

$\sin x = 0$ or $\cos x = \frac{1}{4}$

$x = 0, \pi$ $\qquad \cos^{-1}\frac{1}{4} \approx 1.32$

$x \approx 1.32$ or

$2\pi - 1.32 \approx 4.97$

16. $(3\sin x - 1)(\sin x + 1) = 0$

$\sin x = \frac{1}{3}$ or $\sin x = -1$

$\sin^{-1}\frac{1}{3} \approx 0.34$ $\qquad x = \frac{3\pi}{2} \approx 4.71$

$x \approx 0.34$ or $\pi - 0.34 \approx 2.80$

17. $2\sin 2x = 1$

$\sin 2x = \frac{1}{2}$

$2x = \frac{\pi}{6} + n2\pi, \frac{5\pi}{6} + n2\pi$

$x = \frac{\pi}{12} + n\pi, \frac{5\pi}{12} + n\pi$

using $n = 0, 1$:

$x = \frac{\pi}{12}, \frac{5\pi}{12}, \frac{13\pi}{12}, \frac{17\pi}{12}$

18. $\tan \frac{x}{2} = -\frac{\sqrt{3}}{3}$

$\frac{x}{2} = \frac{5\pi}{6} + n\pi$

$x = \frac{5\pi}{3} + 2n\pi$

using $n = 0$: $x = \frac{5\pi}{3}$

❯ B. Exercises

19–32. Assume $n \in \mathbb{Z}$.

19. $\tan^2 x = 1$

$\tan x = \pm 1$

$x = \frac{\pi}{4}, \frac{3\pi}{4}, \frac{5\pi}{4}, \frac{7\pi}{4}$

$x = \frac{\pi}{4} + \frac{n\pi}{2}$ or $\left(\frac{2n+1}{4}\right)\pi$

20. $(2\cos x + 1)(\cos x + 1) = 0$

$\cos x = -\frac{1}{2}$ or $\cos x = -1$

$x = \frac{2\pi}{3}, \frac{4\pi}{3}$ $\qquad x = \pi$

$x = \frac{2\pi}{3} + n2\pi, (2n+1)\pi, \frac{4\pi}{3} + n2\pi$

21. $\frac{1}{\cos x} - 2\cos x = 1$

$1 - 2\cos^2 x = \cos x$

$2\cos^2 x + \cos x - 1 = 0$

$(2\cos x - 1)(\cos x + 1) = 0$

$\cos x = \frac{1}{2}$ or $\cos x = -1$

$x = \frac{\pi}{3}, \frac{5\pi}{3}$ $\qquad x = \pi$

$x = \frac{\pi}{3} + n2\pi, (2n+1)\pi, \frac{5\pi}{3} + n2\pi$

22.
$$\cos^2 x - 3\sin x = 3$$
$$(1 - \sin^2 x) - 3\sin x = 3$$
$$\sin^2 x + 3\sin x + 2 = 0$$
$$(\sin x + 2)(\sin x + 1) = 0$$
$$\sin x = -2 \quad\text{or}\quad \sin x = -1$$
$$x = \frac{3\pi}{2}$$
$$x = \frac{3\pi}{2} + n2\pi$$

23. $\tan x - 3\cot x = 0$
$$\tan x - \frac{3}{\tan x} = 0$$
$$\tan^2 x - 3 = 0$$
$$\tan x = \pm\sqrt{3}$$
$$x = \frac{\pi}{3}, \frac{2\pi}{3}, \frac{4\pi}{3}, \frac{5\pi}{3}$$
$$x = \frac{\pi}{3} + n\pi, \frac{2\pi}{3} + n\pi$$

24. $\dfrac{\cos^2 x}{\sin^2 x} \cdot \dfrac{\sin x}{1} + \sin x = 2$
$$\frac{\cos^2 x}{\sin x} + \sin x = 2$$
$$\cos^2 x + \sin^2 x = 2\sin x$$
$$1 = 2\sin x$$
$$\sin x = \frac{1}{2}$$
$$x = \frac{\pi}{6}, \frac{5\pi}{6}$$
$$x = \frac{\pi}{6} + n2\pi, \frac{5\pi}{6} + n2\pi$$

25.
$$2(1 - \sin^2 x) = \cos x + 1$$
$$2\cos^2 x - \cos x - 1 = 0$$
$$(2\cos x + 1)(\cos x - 1) = 0$$
$$\cos x = -\frac{1}{2} \quad\text{or}\quad \cos x = 1$$
$$x = \frac{2\pi}{3}, \frac{4\pi}{3} \qquad x = 0$$
$$x = n2\pi, \frac{2\pi}{3} + n2\pi, \frac{4\pi}{3} + n2\pi$$

26.
$$2\cos x + \frac{2}{\cos x} = 5$$
$$2\cos^2 x + 2 = 5\cos x$$
$$2\cos^2 x - 5\cos x + 2 = 0$$
$$(2\cos x - 1)(\cos x - 2) = 0$$
$$\cos x = \frac{1}{2} \quad\text{or}\quad \cos x = 2$$
$$x = \frac{\pi}{3}, \frac{5\pi}{3}$$
$$x = \frac{\pi}{3} + n2\pi, \frac{5\pi}{3} + n2\pi$$

27. $\dfrac{\sin^2 x}{(1 + \cos x)\sin x} + \dfrac{1 + 2\cos x + \cos^2 x}{(1 + \cos x)\sin x}$
$$= -4$$
$$2 + 2\cos x = -4(1 + \cos x)\sin x$$
$$2(1 + \cos x) + 4(1 + \cos x)\sin x = 0$$
$$2(1 + \cos x)(1 + 2\sin x) = 0$$
$$\cos x = -1 \quad\text{or}\quad \sin x = -\frac{1}{2}$$
$$x = \pi \text{ (extraneous)} \quad x = \frac{7\pi}{6}, \frac{11\pi}{6}$$
$$x = \frac{7\pi}{6} + n2\pi, \frac{11\pi}{6} + n2\pi$$

28. $(4\sin^2 x - 3)(2\sin^2 x + 1) = 0$
$$\sin^2 x = \frac{3}{4} \quad\text{or}\quad \sin^2 x = -\frac{1}{2}$$
$$\sin x = \pm\frac{\sqrt{3}}{2} \qquad \sin x = \pm\sqrt{-\frac{1}{2}}$$
$$x = \frac{\pi}{3}, \frac{2\pi}{3}, \frac{4\pi}{3}, \frac{5\pi}{3};\ \text{no } \mathbb{R} \text{ solutions}$$
$$x = \frac{\pi}{3} + n\pi, \frac{2\pi}{3} + n\pi$$

The tones for each number on a phone are composed of sinusoids. The receiving device uses a trigonometric equation known as a Fourier transform to identify the frequencies of the tone and the button that was pressed.

When solving an equation containing a function of a multiple angle, such as $\cos kx$, first solve for the multiple angle kx and then divide the result by k.

Example 6   Solving a Multiple Angle Equation

Find the general and primary solutions for $2\cos 3x = \sqrt{3}$.

Answer

$$2\cos 3x = \sqrt{3}$$
$$\cos 3x = \frac{\sqrt{3}}{2}$$
$$3x = \frac{\pi}{6} \text{ and } \frac{11\pi}{6}$$
$$3x = \frac{\pi}{6} + n2\pi, \frac{11\pi}{6} + n2\pi, n \in \mathbb{Z}$$
$$x = \frac{\pi}{18} + \frac{2n}{3}\pi, \frac{11\pi}{18} + \frac{2n}{3}\pi, n \in \mathbb{Z}$$

$$n = 0;\ x = \frac{\pi}{18}, \frac{11\pi}{18}$$
$$n = 1;\ x = \frac{\pi}{18} + \frac{2}{3}\pi = \frac{13\pi}{18}$$
$$x = \frac{11\pi}{18} + \frac{2}{3}\pi = \frac{23\pi}{18}$$
$$n = 2;\ x = \frac{\pi}{18} + \frac{4}{3}\pi = \frac{25\pi}{18}$$
$$x = \frac{11\pi}{18} + \frac{4}{3}\pi = \frac{35\pi}{18}$$
$$x = \frac{\pi}{18}, \frac{11\pi}{18}, \frac{13\pi}{18}, \frac{23\pi}{18}, \frac{25\pi}{18}, \frac{35\pi}{18}$$

1. Find the general solution for $3x$.
   a. Isolate $\cos 3x$.
   b. Use the unit circle to find primary solutions of $3x$.
   c. Write general solutions for $3x$.
2. Find the general solution for x by dividing both sides by 3.
3. Use $n = 0$, 1, and 2 to generate primary solutions within $[0, 2\pi)$.
   Note that if $n = 3$, $x = \frac{\pi}{18} + 2\pi$ and $x = \frac{11\pi}{18} + 2\pi$ are outside of $[0, 2\pi)$.
   In general, $n \in [0, k - 1]$.

11. $x = 0, \frac{\pi}{2}$

12. $x = \frac{7\pi}{6}, \frac{11\pi}{6}$

13. $x = \frac{\pi}{3}, \frac{5\pi}{3}$

14. $x = \frac{3\pi}{2}$

SKILL ✓ **EXERCISE 29**

❯ A. Exercises

Find the primary solutions of each equation by isolating the trigonometric function.

1. $\sin x + 1 = 0$ $x = \frac{3\pi}{2}$

2. $2\cos x = \sqrt{3}$ $x = \frac{\pi}{6}, \frac{11\pi}{6}$

3. $\cot x + 1 = 0$ $x = \frac{3\pi}{4}, \frac{7\pi}{4}$

4. $\tan x - \sqrt{3} = 0$ $x = \frac{\pi}{3}, \frac{4\pi}{3}$

Find the primary solutions of each equation by taking roots.

5. $\sin^2 x - 3 = -3\sin^2 x$ $x = \frac{\pi}{3}, \frac{2\pi}{3}, \frac{4\pi}{3}, \frac{5\pi}{3}$

6. $2\cos^2 x - 3 = -2\cos^2 x$ $x = \frac{\pi}{6}, \frac{5\pi}{6}, \frac{7\pi}{6}, \frac{11\pi}{6}$

7. $2\sin^2 x - 1 = 0$ $x = \frac{\pi}{4}, \frac{3\pi}{4}, \frac{5\pi}{4}, \frac{7\pi}{4}$

Use factoring to find the primary solutions of each equation.

8. $2\cos^2 x - \sqrt{2}\cos x = 0$ $x = \frac{\pi}{4}, \frac{\pi}{2}, \frac{3\pi}{2}, \frac{7\pi}{4}$

9. $\cot^2 x - \sqrt{3}\cot x = 0$ $x = \frac{\pi}{6}, \frac{\pi}{2}, \frac{7\pi}{6}, \frac{3\pi}{2}$

10. $\tan x - \tan^2 x = 0$ $x = 0, \frac{\pi}{4}, \pi, \frac{5\pi}{4}$

Find the primary solutions of each equation. Be sure to check for extraneous solutions.

11. $1 - \sin x = \cos x$

12. $2\sin x = 4\csc x + 7$

13. $2\cos x + 2\sec x = 5$

14. $\cot x - 1 = \csc x$

Use inverse trigonometric functions to find all primary solutions (to the nearest hundredth).

15. $4\sin x \cos x = \sin x$ $x \approx 0, 1.32, 3.14, 4.97$

16. $3\sin^2 x + 2\sin x - 1 = 0$ $x \approx 0.34, 2.80, 4.71$

Find the primary solutions of each multiple angle equation.

17. $2\sin 2x - 1 = 0$ $x = \frac{\pi}{12}, \frac{5\pi}{12}, \frac{13\pi}{12}, \frac{17\pi}{12}$

18. $\sqrt{3}\tan\frac{x}{2} = -1$ $x = \frac{5\pi}{3}$

❯ B. Exercises

Find the general solutions of each equation.

19. $\tan^2 x - 1 = 0$ $x = \frac{\pi}{4} + \frac{n\pi}{2}$ or $\left(\frac{2n+1}{4}\right)\pi$

20. $2\cos^2 x + 3\cos x + 1 = 0$ $x = \frac{2\pi}{3} + n2\pi, (2n+1)\pi, \frac{4\pi}{3} + n2\pi$

21. $\sec x - 2\cos x = 1$ $x = \frac{\pi}{3} + n2\pi, (2n+1)\pi, \frac{5\pi}{3} + n2\pi$

22. $\cos^2 x - 3\sin x = 3$ $x = \frac{3\pi}{2} + n2\pi$

23. $\tan x - 3\cot x = 0$ $x = \frac{\pi}{3} + n\pi, \frac{2\pi}{3} + n\pi$

24. $\frac{\cot^2 x}{\csc x} + \sin x = 2$ $x = \frac{\pi}{6} + n2\pi, \frac{5\pi}{6} + n2\pi$

25. $2 - 2\sin^2 x = \cos x + 1$ $x = n2\pi, \frac{2\pi}{3} + n2\pi, \frac{4\pi}{3} + n2\pi$

26. $2\cos x + 2\sec x = 5$ $x = \frac{\pi}{3} + n2\pi, \frac{5\pi}{3} + n2\pi$

27. $\frac{\sin x}{1 + \cos x} + \frac{1 + \cos x}{\sin x} = -4$ $x = \frac{7\pi}{6} + n2\pi, \frac{11\pi}{6} + n2\pi$

---

29. $\sin^2 3x = 0$
$$\sin 3x = 0$$
$$3x = 0 + n\pi$$
$$x = \frac{n\pi}{3}$$

30. $\cos\dfrac{x}{2} = \dfrac{1}{2}$
$$\frac{x}{2} = \frac{\pi}{3} + n2\pi, \frac{5\pi}{3} + n2\pi$$
$$x = \frac{2\pi}{3} + n4\pi, \frac{10\pi}{3} + n4\pi$$

31.
$$\sin x(1 + \sin x) = \cos x(1 - \cos x)$$
$$\sin x + \sin^2 x = \cos x - \cos^2 x$$
$$\sin x + (1 - \cos^2 x) = \cos x - \cos^2 x$$
$$\sin x + 1 = \cos x$$
$$\sin^2 x + 2\sin x + 1 = \cos^2 x$$
$$\sin^2 x + 2\sin x + 1 = 1 - \sin^2 x$$
$$2\sin^2 x + 2\sin x = 0$$
$$2\sin x(\sin x + 1) = 0$$

$$\sin x = 0 \quad\text{or}\quad \sin x = -1$$
$$x = 0, \pi \qquad x = \frac{3\pi}{2}$$
$$\varnothing \text{ (All are extraneous.)}$$

32. $\cos 4x(\cos 4x + 1) = 0$
$$\cos 4x = 0$$
$$4x = \frac{\pi}{2} + n2\pi, \frac{3\pi}{2} + n2\pi$$
$$x = \frac{\pi}{8} + \frac{n\pi}{2}, \frac{3\pi}{8} + \frac{n\pi}{2}$$
or
$$\cos 4x = -1$$
$$4x = \pi + n2\pi$$
$$x = \frac{\pi}{4} + \frac{n\pi}{2}$$

33.
$$3 = 1 + 3\sin x$$
$$\frac{2}{3} = \sin x$$
$$\sin^{-1}\frac{2}{3} \approx 0.73$$
$$x \approx 0.73 \text{ or } \pi - 0.73 \approx 2.41$$

28. $8\sin^4 x - 2\sin^2 x - 3 = 0$ $x = \frac{\pi}{3} + n\pi,\ \frac{2\pi}{3} + n\pi$

29. $\sin^2 3x = 0$ $x = \frac{n\pi}{3}$

30. $\cos\frac{x}{2} = \frac{1}{2}$ $x = \frac{2\pi}{3} + n4\pi,\ \frac{10\pi}{3} + n4\pi$

31. $\dfrac{\sin x}{1 - \cos x} = \dfrac{\cos x}{1 + \sin x}$ $\varnothing$ (All are extraneous.)

32. $\cos^2 4x + \cos 4x = 0$ $x = \frac{\pi}{8} + \frac{n\pi}{2},\ \frac{\pi}{4} + \frac{n\pi}{2},\ \frac{3\pi}{8} + \frac{n\pi}{2}$

Find the primary solutions of each equation. Round each radian measure to the nearest hundredth.

33. $\dfrac{3}{1 + 3\sin x} = 1$ $x \approx 0.73,\ 2.41$

34. $5\cos x = 6\sin^2 x$ $x \approx 0.84,\ 5.44$

35. $3\sin^2 x = 2\cos x \sin x$ $x \approx 0,\ 0.59,\ 3.14,\ 3.73$

36. $2\csc^2 x - \cot x - 5 = 0$ $x \approx 0.59,\ 2.36,\ 3.73,\ 5.50$

37. A soccer ball is kicked from the ground at an angle of θ above the horizontal with an initial speed of $v_0 = 85$ ft/sec. Use the formula for the range of a projectile, $d = \dfrac{v_0{}^2 \sin 2\theta}{32}$, to calculate θ (to the nearest tenth of a degree) if the balls lands 210 ft away.   34.2°

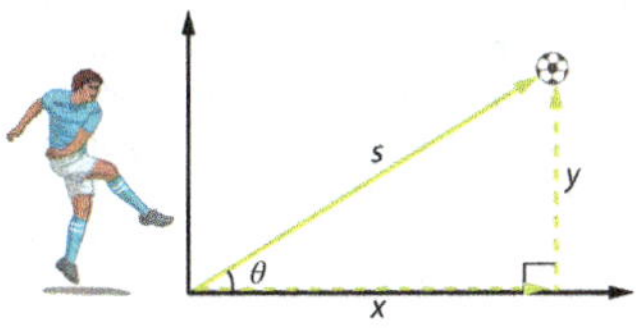

CUMULATIVE REVIEW

Solve each equation.

43. $2x^2 - 5x + 1 = 0$ [1.6]

44. $3 = \dfrac{x}{x+1} + \dfrac{5}{x}$ [1.6]

45. $e^{-2x} = 3$ [3.4]

46. $2x^{-5} = \dfrac{1}{16}$ [3.4] $x = 2$

47. $\sqrt{x - 5} + 3 = 7$ [2.1] $x = 21$

48. Simplify $\dfrac{1}{\sec\left(x - \frac{\pi}{2}\right)}$. [5.1] $\sin x$

49. What is the period of the sum of $f(x) = \sin 2x$ and $g(x) = 3\cos 2x$? [4.7]   C

- **A.** 1
- **B.** 2
- **C.** π
- **D.** 2π
- **E.** cannot be determined

50. Simplify $\cos^3\theta + \sin^2\theta\cos\theta$. [5.1]   D

- **A.** $\tan\theta$
- **B.** $\sec\theta$
- **C.** $\sin\theta$
- **D.** $\cos\theta$
- **E.** $\csc\theta$

51. Which expression is equivalent to $\dfrac{\tan^2 x}{1 + \tan^2 x}$? [5.2]   B

- **A.** $\sin x$
- **B.** $\sin^2 x$
- **C.** $\sin^2 x \cos x$
- **D.** $\sin^2 x \cos^2 x$
- **E.** none of these

52. Which expression is equivalent to $\dfrac{\sin x}{1 - \cos x}$? [5.2]   D

- **A.** $\sin x - \tan x$
- **B.** $\sin x + \tan x$
- **C.** $\csc x - \tan x$
- **D.** $\csc x + \cot x$
- **E.** none of these

43. $x = \dfrac{5 \pm \sqrt{17}}{4}$

44. $x = \dfrac{1 \pm \sqrt{11}}{2}$

45. $x = -\dfrac{\ln 3}{2} \approx -0.55$

C. Exercises

Find the primary solutions of each equation. Round decimal approximations to the nearest hundredth.

38. $8\cos^2 x + 14\cos x - 15 = 0$ $x \approx 0.72,\ 5.56$

39. $\tan^2 x + 8\tan x - 4 = 0$ $x \approx 0.44,\ 1.69,\ 3.58,\ 4.83$

40. $\tan^2 x - \sin^2 x = \tan^2 x \sin^2 x$ $\left\{x \mid x \neq \frac{\pi}{2}, \frac{3\pi}{2}\right\}$

41. $\tan^3 x + \tan^2 x - 3\tan x - 3 = 0$

42. Standing waves result from a combination of waves with the same frequency and amplitude that are traveling at the same speed in opposite directions. A standing wave can be modeled with the function $y = (2A_0 \sin kx)\cos \omega t$. A *node* of a standing wave occurs where its amplitude $2A_0 \sin kx = 0$. Find the position x of the function's first 5 nodes in terms of the wavelength λ if $k = \dfrac{2\pi}{\lambda}$.

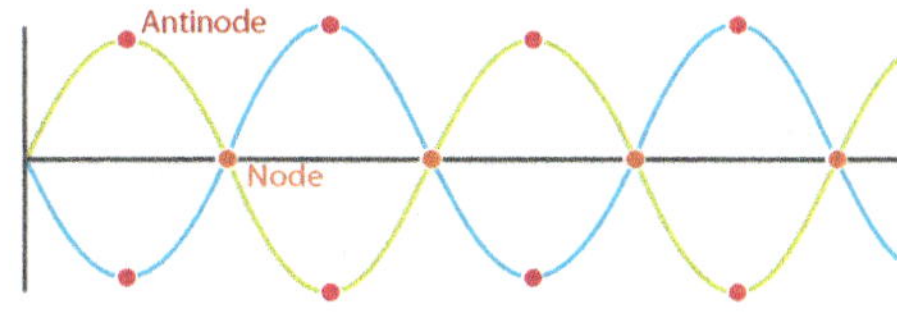

41. $x = \frac{\pi}{3}, \frac{2\pi}{3}, \frac{3\pi}{3}, \frac{4\pi}{3}, \frac{5\pi}{3}, \frac{7\pi}{3}$

42. $x = 0, \frac{\lambda}{2}, \lambda, \frac{3\lambda}{2}, 2\lambda$

37.
$$210 = \frac{(85^2)\sin 2\theta}{32}$$
$$6720 = 7225\sin 2\theta$$
$$2\theta = \sin^{-1}\left(\frac{6720}{7225}\right)$$
$$2\theta \approx 68.45°$$
$$\theta \approx 34.23°$$

C. Exercises

38. Let $u = \cos x$.
$$8u^2 + 14u - 15 = 0$$
$$u = \frac{-14 \pm \sqrt{14^2 - 4(8)(-15)}}{2(8)} = \frac{-14 \pm 26}{16}$$
$$\cos x = \frac{3}{4} \quad \text{or} \quad \cos x = -\frac{5}{2}$$
$$x \approx 0.72 \text{ or } 2\pi - 0.72 \approx 5.56$$

39. Let $u = \tan x$.
$$u^2 + 8u - 4 = 0$$
$$u = \frac{-8 \pm \sqrt{8^2 - 4(1)(-4)}}{2(1)} = -4 \pm 2\sqrt{5}$$
$$\tan x \approx 0.47$$
$$\tan^{-1} 0.47 \approx 0.44$$
$$x \approx 0.44 \text{ or } \pi + 0.44 \approx 3.58$$
or
$$\tan x \approx -8.47$$
$$\tan^{-1}(-8.47) \approx -1.45$$
$$x \approx -1.45 + \pi \approx 1.69 \text{ or}$$
$$-1.45 + 2\pi \approx 4.83$$

40.
$$\frac{\sin^2 x}{\cos^2 x} - \sin^2 x = \frac{\sin^2 x}{\cos^2 x} \cdot \sin^2 x$$
$$\sin^2 x - \sin^2 x \cos^2 x = \sin^4 x$$
$$\sin^2 x(1 - \cos^2 x) = \sin^4 x$$
$$\sin^2 x(\sin^2 x) = \sin^4 x$$
$$\sin^4 x = \sin^4 x$$
$$\text{true } \forall\ x \neq \frac{\pi}{2}, \frac{3\pi}{2}$$

41. $\tan^2 x(\tan x + 1) - 3(\tan x + 1) = 0$
$$(\tan x + 1)(\tan^2 x - 3) = 0$$
$$\tan x = -1 \quad \text{or} \quad \tan^2 x = 3$$
$$x = \frac{3\pi}{4}, \frac{7\pi}{4} \qquad \tan x = \pm\sqrt{3}$$
$$x = \frac{\pi}{3}, \frac{2\pi}{3}, \frac{4\pi}{3}, \frac{5\pi}{3}$$
$$x = \frac{\pi}{3}, \frac{2\pi}{3}, \frac{3\pi}{4}, \frac{4\pi}{3}, \frac{5\pi}{3}, \frac{7\pi}{4}$$

42. $2A_0 \sin\frac{2\pi}{\lambda}x = 0$
$$\sin\frac{2\pi}{\lambda}x = 0$$
$$\frac{2\pi}{\lambda}x = 0 + n\pi$$
$$x = 0 + \left(\frac{\lambda}{2\pi}\right)n\pi = \frac{n\lambda}{2}$$
using $n = 0, 1, 2, 3, 4$:
$$x = 0, \frac{\lambda}{2}, \lambda, \frac{3\lambda}{2}, 2\lambda$$

---

34. $5\cos = 6(1 - \cos^2 x)$
$$6\cos^2 x + 5\cos x - 6 = 0$$
$$(3\cos x - 2)(2\cos x + 3) = 0$$
$$\cos x = \frac{2}{3} \quad \text{or} \quad \cos x = -\frac{3}{2}$$
$$\cos^{-1}\frac{2}{3} \approx 0.84$$
$$x \approx 0.84 \text{ or } 2\pi - 0.84 \approx 5.44$$

35. $3\sin^2 x - 2\cos x \sin x = 0$
$$(3\sin x - 2\cos x)\sin x = 0$$
$$3\sin x - 2\cos x = 0 \quad \text{or} \quad \sin x = 0$$
$$3\sin x = 2\cos x \qquad x = 0, \pi$$
$$\frac{\sin x}{\cos x} = \frac{2}{3} = \tan x$$
$$\tan^{-1}\frac{2}{3} \approx 0.59$$
$$x \approx 0.59 \text{ or}$$
$$\pi + 0.59 \approx 3.73$$
$$\therefore x \approx 0, 0.59, 3.14, 3.73$$

36. $2(1 + \cot^2 x) - \cot x - 5 = 0$
$$2 + 2\cot^2 x - \cot x - 5 = 0$$
$$2\cot^2 x - \cot x - 3 = 0$$
$$(2\cot x - 3)(\cot x + 1) = 0$$
$$\cot x = \frac{3}{2}$$
$$\tan x = \frac{2}{3}$$
$$\tan^{-1}\frac{2}{3} \approx 0.59$$
$$x \approx 0.59 \text{ or } \pi + 0.59 \approx 3.73$$
or
$$\cot x = -1$$
$$\tan x = -1$$
$$\tan^{-1}(-1) \approx -0.79$$
$$x \approx -0.79 + \pi \approx 2.36 \text{ or}$$
$$-0.79 + 2\pi \approx 5.50$$
$$\therefore x \approx 0.59, 2.36, 3.73, 5.50$$

continued in Answers and Solutions Overflow

Biblical Perspective of Mathematics

Objectives

1. To state and analyze Euler's identity
2. To identify examples of the beautiful unity within mathematics involving the irrational numbers π, e, and ϕ
3. To identify examples of the unity of algebra and geometry (both ancient Greek and modern)
4. To explain why mathematics displays such beautiful unity despite its diverse applications

Assignments

- **Minimum:** 1–7; 8–9 (in class)
- **Standard:** 1–7; 8–9 (in class)
- **Extended:** 1–9

The Unity and Beauty of Mathematics

Mathematics compares the most diverse phenomena and discovers the secret analogies that unite them.
— Joseph Fourier (French mathematician and physicist)

Mathematics, rightly viewed, possesses not only truth, but supreme beauty.
— Bertrand Russell (British philosopher and mathematician)

The integers 0 and 1 are quite different from the irrational numbers π and e. The imaginary number i is from another mathematical world. Euler's identity $e^{i\pi} + 1 = 0$ combines these five foundational constants and the three fundamental operations of addition, multiplication, and exponentiation in one of the most beautiful equations in mathematics. The poet Dana Gioia stated that "beauty is the pleasure we get in recognizing the particular manifestation of a broader, universal order." The unity of mathematics across its extreme diversity is one of its most beautiful characteristics.

While π and e are both irrational numbers, each can be expressed as an infinite sum involving only integers.

$$\pi = 4 - \frac{4}{3} + \frac{4}{5} - \frac{4}{7} + \frac{4}{9} - \frac{4}{11} + \cdots \text{ and}$$

$$e = 2 + \frac{1}{2} + \frac{1}{2 \cdot 3} + \frac{1}{2 \cdot 3 \cdot 4} + \frac{1}{2 \cdot 3 \cdot 4 \cdot 5} + \cdots.$$

The number π was used in the solution to the Basel problem, which sought the sum of the reciprocals of all the perfect squares. This problem baffled mathematicians for nearly one hundred years, but Leonard Euler earned his reputation as a world-class mathematician in 1735 by proving that

$$1 + \frac{1}{4} + \frac{1}{9} + \frac{1}{16} + \frac{1}{25} + \cdots = \frac{\pi^2}{6}.$$

The seemingly unrelated constants π and e also appear together in the most important probability distribution function, the normal distribution, whose graph is the bell curve.

James Nickel describes several examples of "unity in diversity": "Basic trigonometric functions relate to the right triangle, music, and the electromagnetic spectrum. The Fibonacci sequence unifies many diverse aspects of creation. The spiral, the hexagon, and the pentagram are just a few of the other beautiful and inherent relationships in creation." Fibonacci discovered that the golden ratio, described much earlier by Euclid, could be estimated to any desired degree of accuracy using a sequence of ratios of successive terms of the Fibonacci sequence: $\frac{3}{2}, \frac{5}{3}, \frac{8}{5}, \frac{13}{8}, \frac{21}{13}, \ldots$. The golden ratio is symbolized by ϕ (phi) and is expressed exactly as $\frac{1 + \sqrt{5}}{2}$, the positive irrational solution to $x^2 - x - 1 = 0$.

The coordinate geometry of René Descartes unified symbolic algebra and the Euclidean geometry that was developed centuries earlier. The various conic sections described in ancient Greece by Apollonius were later represented with second-degree algebraic equations in two variables. While this important connection between geometric figures and abstract algebra is often described by secular mathematicians as "mysterious," this beautiful unity should cause Christian mathematicians to praise the Creator God.

In the nineteenth century, the foundation of Euclidean geometry was shaken by the development of two diverse non-Euclidean geometries. The brilliant German mathematician Carl Gauss applied non-Euclidean theory to the geometric nature of physical space but did not publish his discoveries, perhaps fearing the controversy they would create. In the twentieth century Felix Klein unified the geometries with his beautiful development of projective geometry. As Morris Kline wrote, "It is possible to erect projective geometry on an axiomatic basis in such a way that

PRESENTATION

Begin by asking students to evaluate $e^{i\pi}$ using their calculators (must be in radian mode), verifying Euler's identity since $-1 + 1 = 0$. Show students the origin of Euler's identity by evaluating Euler's more general equation $e^{i\theta} = \cos\theta + i\sin\theta$ for $\theta = \pi$ ($\cos\pi = -1$ and $\sin\pi = 0$). The Internet keyword search *Euler's identity* will provide videos and a wealth of additional information.

After discussing the two series for π and e, mention Euler's solution to the Basel problem, named after the Swiss city where Euler was born in 1707. Several well-known mathematicians such as Gottfried Leibniz, the three Bernoulli brothers, and Abraham de Moivre had worked on the problem before Euler, but they were unable to come up with an exact solution. Euler's solution provided a technique to solve a whole class of problems involving the reciprocals of even powers. He also showed that $1 + \frac{1}{2^4} + \frac{1}{3^4} + \frac{1}{4^4} = \frac{\pi^4}{90}$. Euler was unsuccessful in his attempt to solve $1 + \frac{1}{2^3} + \frac{1}{3^3} + \frac{1}{4^3} + \cdots$, and that problem remains unsolved today.

Euler is consistently ranked as one of the greatest mathematicians of all time. For those interested in a thorough discussion of Euler's mathematical contributions, consult: *Euler: The Master of Us All* (Washington, D.C.: MAA, 1999) by William Dunham.

You may want to show students how the equation $x^2 - x - 1 = 0$ relates to the definition of a golden rectangle. It results from the proportion $\frac{x}{1} = \frac{1}{x-1}$, which compares the length-to-width ratios of a rectangle of length x and width 1 to a rectangle created by cutting a square off one end.

Discuss the examples of the surprising unity of geometry and algebra illustrated by the golden ratio, conic sections, and the relativity theory. Point out that in 1853 Carl Gauss asked his brilliant

the theorems of the other three geometries result as specialized theorems of projective geometry. In other words, the contents of all four geometries are now incorporated in one harmonious whole." Later, Einstein's algebraic relativity formulas fulfilled Gauss's vision by applying non-Euclidean geometry to physical space in yet another example of the beautiful unity of algebra and geometry.

Sets, including geometric sets of points and algebraic sets of numbers, are an essential unifying concept in mathematics. Computers use an iterative equation and complex numbers to produce the amazing Mandelbrot set, an infinitely detailed graph in fractal geometry. Astrophysicist Jason Lisle observed that the fractal's beautifully organized shapes have been "built into mathematics by the Creator of mathematics." Zooming in on the different spirals surrounding the original Mandelbrot set reveals miniature "baby" versions similar to the original. Mathematicians have described portions of the graph with names such as *seahorse valley*, *triple spiral valley*, *elephant valley*, and

The Mandelbrot fractal is a complex plane visualization of the divergent nature of a set of complex numbers.

scepter valley. Analyzing the stems, branches, and bulbs of the Mandelbrot set results in common natural number sequences, including the Fibonacci sequence!

❯ Exercises

1. State Euler's identity and list the fundamental mathematical operations used.

2. Match each constant in Euler's identity with its related foundational concept.

   a. 0   I
   b. 1   III
   c. π V
   d. e IV
   e. i II

      I. additive identity
      II. complex numbers
      III. multiplicative identity
      IV. natural logarithms
      V. radian measure

3. State the exact sum for each infinite series.

   a. $4 - \frac{4}{3} + \frac{4}{5} - \frac{4}{7} + \frac{4}{9} - \frac{4}{11} + \cdots$ π

   b. $2 + \frac{1}{2} + \frac{1}{2 \cdot 3} + \frac{1}{2 \cdot 3 \cdot 4} + \frac{1}{2 \cdot 3 \cdot 4 \cdot 5} + \cdots$ e

   c. $1 + \frac{1}{4} + \frac{1}{9} + \frac{1}{16} + \frac{1}{25} + \cdots$ $\frac{\pi^2}{6}$

4. State the sixth term for the golden ratio sequence given in the text. Then state its decimal value rounded to the nearest thousandth.

5. Solve $x^2 - x - 1 = 0$ using the quadratic formula. Then use a calculator to find a decimal approximation for the golden ratio (rounded to the nearest millionth).

6. Which mathematician's foundational work in applying abstract non-Euclidean theory to physical space led to Einstein's algebraic relativity formulas?

7. State two examples of ancient Greek geometry being later unified with algebraic descriptions.

8. **Discuss:** How does the Mandelbrot set illustrate the beautiful unity of complex numbers and fractal geometry?

9. **Discuss:** From a biblical perspective, explain why mathematics displays such beautiful unity despite its diverse applications.

Solutions

1. $e^{i\pi} + 1 = 0$; addition, multiplication, and exponentiation

4. $\frac{34}{21}$; 1.619

5. $x^2 - x - 1 = 0$
$$x = \frac{-(-1) \pm \sqrt{(-1)^2 - 4(1)(-1)}}{2(1)}$$
$$= \frac{1 \pm \sqrt{1^2 + 4}}{2} = \frac{1 \pm \sqrt{5}}{2}$$
$$\approx 1.618034$$

6. Carl Gauss's

7. The golden ratio described by Euclid was derived later as an exact solution to a quadratic equation. The geometric conic sections studied by Apollonius are now described algebraically by second-degree equations in two variables.

8. The points of the Mandelbrot set are generated using complex numbers in an algebraic equation. The resulting graph involves a shape which repeats itself infinitely in smaller versions under magnification. The number of stems, branches, and bulbs in sections of the graph reveals common natural number sequences.

9. If God is the Creator and if through Jesus all things hold together (Col. 1:17), then it makes sense that the mathematics that models the natural world shows beautiful unity.

student Bernhard Riemann to prepare a lecture on the foundations of geometry. Riemann's lecture on his theory of curved surfaces laid the groundwork for Einstein's algebraic equations of relativity. Gauss's motto was "few, but ripe," perhaps an explanation for why he did not publish his non-Euclidean geometry work.

Share the concluding sentences of Jason Lisle's AIG article: "A biblical creationist expects to find beauty and order in the universe, not only the physical universe, but in the abstract realm of mathematics as well. This order and beauty is possible because there is a logical God who has imparted order and beauty into His universe." You may want to use a brief Mandelbrot set online video for classroom viewing.

5.4 Sum and Difference Identities

Objectives

1. To derive sum and difference identities
2. To apply sum and difference identities when evaluating trigonometric functions
3. To verify other identities using sum and difference identities
4. To solve trigonometric equations using sum and difference identities

Flash

The barrel-shaped components are capacitors. The component with the wire wrappings is an inductor (coil). A capacitor stores an electrical charge and can give a circuit a "boost" of voltage or can serve as an electrical "shock absorber." An inductor also stores an electrical charge, but in the form of a magnetic field of energy. Both components are crucial in circuits involving sinusoidal waves.

Vocabulary

reduction identity

This RLC circuit has a resistor, an inductor, and a capacitor connected in series.

After completing this section, you will be able to
- derive sum and difference identities.
- use sum and difference identities to evaluate functions, verify identities, and solve equations.

The unit circle and relationships within 45-45-90 and 30-60-90 triangles enable us to state exact values for trigonometric functions of multiples of $\frac{\pi}{4}$ and $\frac{\pi}{6}$. In this section, you will derive identities for the sine, cosine, and tangent of a sum or difference of angles. These identities will allow you to calculate other exact trigonometric function values, verify other identities, and solve more trigonometric equations.

A counterexample proves that $\cos(\alpha - \beta) \neq \cos\alpha - \cos\beta$.

$$\cos(90° - 60°) \overset{?}{=} \cos 90° - \cos 60°$$
$$\cos 30° \overset{?}{=} 0 - \frac{1}{2}$$
$$\frac{\sqrt{3}}{2} \neq -\frac{1}{2}$$

An identity for $\cos(\alpha - \beta)$ can be derived from the unit circle where angles α, β, and $\alpha - \beta$ are drawn in standard position. The terminal sides of these angles intersect the unit circle at $A(\cos\alpha, \sin\alpha)$, $B(\cos\beta, \sin\beta)$, and $C(\cos(\alpha - \beta), \sin(\alpha - \beta))$, respectively. Since $m\angle AOB = \alpha - \beta$, $\overline{AB} \cong \overline{CD}$ and their lengths can be found using the distance formula.

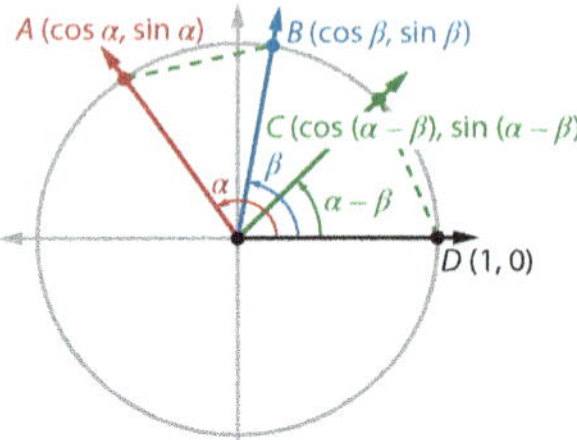

$$CD = AB$$
$$\sqrt{[\cos(\alpha - \beta) - 1]^2 + [\sin(\alpha - \beta) - 0]^2} = \sqrt{(\cos\alpha - \cos\beta)^2 + (\sin\alpha - \sin\beta)^2}$$

Square both sides to remove the radicals and expand the squared binomials.
$$\cos^2(\alpha - \beta) - 2\cos(\alpha - \beta) + 1 + \sin^2(\alpha - \beta)$$
$$= \cos^2\alpha - 2\cos\alpha\cos\beta + \cos^2\beta + \sin^2\alpha - 2\sin\alpha\sin\beta + \sin^2\beta$$

Rearrange the terms and use the Pythagorean identity $\sin^2 x + \cos^2 x = 1$ to simplify.
$$[\cos^2(\alpha - \beta) + \sin^2(\alpha - \beta)] - 2\cos(\alpha - \beta) + 1$$
$$= [\cos^2\alpha + \sin^2\alpha] - 2\cos\alpha\cos\beta - 2\sin\alpha\sin\beta + [\cos^2\beta + \sin^2\beta]$$
$$(1) - 2\cos(\alpha - \beta) + 1 = (1) - 2\cos\alpha\cos\beta - 2\sin\alpha\sin\beta + (1)$$

Subtract 2 from each side and then divide both sides by −2.
$$-2\cos(\alpha - \beta) = -2\cos\alpha\cos\beta - 2\sin\alpha\sin\beta$$
$$\cos(\alpha - \beta) = \cos\alpha\cos\beta + \sin\alpha\sin\beta$$

The identity for the cosine of a sum of two angles is derived using this identity and the odd/even identities for sine and cosine.

$\cos(\alpha + \beta) = \cos[\alpha - (-\beta)]$	1. Apply the definition of subtraction.
$\cos(\alpha + \beta) = \cos\alpha\cos(-\beta) + \sin\alpha\sin(-\beta)$	2. Apply the identity for cosine of a difference.
$\cos(\alpha + \beta) = \cos\alpha\cos\beta - \sin\alpha\sin\beta$	3. Substitute, using $\cos(-\theta) = \cos\theta$ and $\sin(-\theta) = -\sin\theta$.

PRESENTATION

Lesson Opener

1. Evaluate $\cos 30°$, $\cos 90°$, and $\cos 60°$. Does $\cos(90° - 60°) = \cos 90° - \cos 60°$? $\frac{\sqrt{3}}{2}, 0, \frac{1}{2}$; no

2. Evaluate $\sin\pi$, $\sin\frac{\pi}{3}$, and $\sin\frac{2\pi}{3}$. Does $\sin\left(\frac{\pi}{3} + \frac{2\pi}{3}\right) = \sin\frac{\pi}{3} + \sin\frac{2\pi}{3}$? $0, \frac{\sqrt{3}}{2}, \frac{\sqrt{3}}{2}$; no

3. Evaluate $\tan\frac{\pi}{3}$ and $\tan\frac{\pi}{6}$. Does $\tan\frac{\pi}{3} = \tan\frac{\pi}{6} + \tan\frac{\pi}{6}$? $\sqrt{3}, \frac{\sqrt{3}}{3}$; no

Use the Lesson Opener to demonstrate that the trigonometric function of a sum or difference is not equal to the sum or difference of the functions of the addends. The first evaluation and the counterexample in the text show specifically that $\cos(\alpha - \beta) \neq \cos\alpha - \cos\beta$. The second and third evaluations emphasize that the function of a sum is rarely the sum of the functions of the addends. This general fact is also illustrated by the square and square root functions.
$$(a + b)^2 \neq a^2 + b^2 \text{ and}$$
$$\sqrt{a + b} \neq \sqrt{a} + \sqrt{b}$$

Using the figure, derive the identity for $\cos(\alpha - \beta)$. Then show how the odd/even identities are used to derive the identity for $\cos(\alpha + \beta)$. These sum and difference identities are applied in Example 1 to prove two important cofunction identities that are used in the derivation of the sine of a difference identity and other cofunction identities. These may be more familiar to some students when expressed in the degree form, $\cos\theta = \sin(90° - \theta)$ and $\sin\theta = \cos(90° - \theta)$.

Illustrate how the sine of a difference identity is derived using the cosine of a difference identity and the cofunction identities from Example 1. Introduce the other sum and difference identities as presented in the table. The proofs of the

The identity for the cosine of a difference also provides a simple proof for two important cofunction identities. The remaining cofunction identities can be proved using the results of Example 1.

Example 1 Proving Cofunction Identities

Prove the following cofunction identities.

a. $\cos\left(\frac{\pi}{2} - \theta\right) = \sin\theta$ **b.** $\sin\left(\frac{\pi}{2} - \theta\right) = \cos\theta$

Answer

a. $\cos\left(\frac{\pi}{2} - \theta\right) = \cos\frac{\pi}{2}\cos\theta + \sin\frac{\pi}{2}\sin\theta$ 1. Apply the cosine of a difference identity.

$\qquad = (0)\cos\theta + (1)\sin\theta$ 2. Evaluate and simplify.
$\qquad = \sin\theta$

b. $\sin\left(\frac{\pi}{2} - \theta\right) = \cos\left[\frac{\pi}{2} - \left(\frac{\pi}{2} - \theta\right)\right]$ 1. Apply the previous cofunction identity.

$\qquad = \cos(0 + \theta)$ 2. Simplify.
$\qquad = \cos\theta$

—————————————— SKILL ✔ **EXERCISE 15**

These cofunction identities are used to derive the sine of a difference identity.

$$\sin(\alpha - \beta) = \cos\left[\frac{\pi}{2} - (\alpha - \beta)\right]$$
$$= \cos\left[\left(\frac{\pi}{2} - \alpha\right) + \beta\right]$$
$$= \cos\left(\frac{\pi}{2} - \alpha\right)\cos\beta - \sin\left(\frac{\pi}{2} - \alpha\right)\sin\beta$$
$$= \sin\alpha\cos\beta - \cos\alpha\sin\beta$$

The derivations of identities for the sine of a sum, the tangent of a difference, and the tangent of a sum are left as exercises. Because the formats of the sum and difference identities for each function are so similar, the identities are often combined using the $\pm$ (plus or minus) and $\mp$ (minus or plus) symbols, and read either along the top or along the bottom of each symbol.

Sum and Difference Identities
$\cos(\alpha \pm \beta) = \cos\alpha\cos\beta \mp \sin\alpha\sin\beta$
$\sin(\alpha \pm \beta) = \sin\alpha\cos\beta \pm \cos\alpha\sin\beta$
$\tan(\alpha \pm \beta) = \dfrac{\tan\alpha \pm \tan\beta}{1 \mp \tan\alpha\tan\beta}$

These identities can help find exact trigonometric values for many angles other than multiples of $\frac{\pi}{4}$ and $\frac{\pi}{6}$.

Ptolemy used sum and difference identities to construct his table of chords in the second century AD. He used the fact that the product of the diagonals of an inscribed quadrilateral equals the sum of the products of the opposite sides (Ptolemy's Theorem) to prove these identities.

Additional Exercises

Use a sum or difference identity to write each expression in terms of a single trigonometric function.

1. $\sin\frac{5\pi}{6}\cos\frac{\pi}{4} + \cos\frac{5\pi}{6}\sin\frac{\pi}{4}$ $\sin\frac{13\pi}{12}$

2. $\cos\frac{\pi}{4}\cos\frac{\pi}{3} - \sin\frac{\pi}{4}\sin\frac{\pi}{3}$ $\cos\frac{7\pi}{12}$

Find the exact value for each expression.

3. $\cos\left(\frac{\pi}{3} - \frac{\pi}{4}\right)$ $\frac{\sqrt{6} + \sqrt{2}}{4}$

4. $\sin\left(-\frac{7\pi}{12}\right)$ $\frac{-\sqrt{2} - \sqrt{6}}{4}$

5. $\sec\left(\frac{19\pi}{12}\right)$ $\sqrt{6} + \sqrt{2}$

For exercises 6–7, find the exact value for each expression if α is in Q I with $\sin\alpha = \frac{3}{8}$ and β is in Q II with $\sin\beta = \frac{2}{3}$.

6. $\cos(\alpha + \beta)$ $\frac{-5\sqrt{11} - 6}{24}$

7. $\sin(\alpha - \beta)$ $\frac{-3\sqrt{5} - 2\sqrt{55}}{24}$

8. Rewrite $\sin\left(\mathrm{Sin}^{-1}x + \mathrm{Cos}^{-1}\frac{\sqrt{3}}{2}\right)$ as an algebraic expression.

$\frac{x\sqrt{3} + \sqrt{1 - x^2}}{2}$

9. Verify $\cos\left(\theta - \frac{3\pi}{2}\right) = -\sin\theta$.

$\cos\left(\theta - \frac{3\pi}{2}\right)$

$= \cos\theta\cos\frac{3\pi}{2} + \sin\theta\sin\frac{3\pi}{2}$

$= \cos\theta(0) + (-1)\sin\theta = -\sin\theta$

10. State primary solutions for
$\cos\left(x - \frac{4\pi}{3}\right) + \cos\left(x - \frac{2\pi}{3}\right) = \frac{\sqrt{3}}{2}$.

$x = \frac{5\pi}{6}, \frac{7\pi}{6}$

---

remaining sum and difference identities are done in exercises 25–27. Explain that these identities enable us to find exact values of other angles as demonstrated in Example 2. Part *b* reviews the simplification of a complex fraction.

One-on-One A student may think the Distributive Property is valid for $\cos(\alpha - \beta)$, yielding $\cos\alpha - \cos\beta$. Stress that cos here is not a real number value, but rather the name of a trigonometric function; therefore the Distributive Property does not apply.

The first step in Example 3 reviews how to use a Pythagorean identity to evaluate other trig functions of an angle when the sine or cosine of the angle and its quadrant are known, as was done in Section 5.1. This process can also be done with reference triangles as illustrated in Section 4.3. The use of identities is illustrated in the solution for exercises 17–18, while the use of reference triangles is illustrated in Example 4 and exercises 19–24. Note that either method provides exact answers that can be found without determining α or β. Encourage students checking their work with a calculator to store and use unrounded values for α and β. Otherwise, rounding will cause their evaluations to appear different from the exact values.

Example 4 presents a similar problem in more general terms. In these more ab-stract problems, which cannot be solved with a calculator, students generally prefer using reference triangles to determine the ratios represented by the compositions of trig functions and inverse trig functions. Example 4 and associated exercises 21–24 can be viewed as extended material.

One-on-One A student may have difficulty determining the ratio when given an expression such as $\sin\left(\mathrm{Tan}^{-1}\frac{3}{4}\right)$. Using the 3-4-5 right triangle in the text, ask the student to determine $\mathrm{Arcsin}\frac{3}{5}$, $\mathrm{Arccos}\frac{4}{5}$, and $\mathrm{Arctan}\frac{3}{4}$. (*Each of these is α.*) Then ask the student to determine the $\sin\alpha$, $\cos\alpha$, and $\tan\alpha$. $\left(\frac{3}{5}, \frac{4}{5}, and \frac{3}{4}\right)$

- **Minimum:** 1–8, 10–12, 15–18, 25, 28–31, 34, 37, 45–50, 52
- **Standard:** 1–6, 9, 13–18, 22–23, 25–29, 31–32, 35–37, 40–41, 45, 47, 49–53
- **Extended:** 1–4, 7, 10–11, 14–16, 19–21, 24–34, 37–38, 40–45, 49–54

Solutions

❯ A. Exercises

1. $\cos \dfrac{5\pi}{6} \cos \dfrac{\pi}{4} - \sin \dfrac{5\pi}{6} \sin \dfrac{\pi}{4}$

$$= \left(-\frac{\sqrt{3}}{2}\right)\left(\frac{\sqrt{2}}{2}\right) - \left(\frac{1}{2}\right)\left(\frac{\sqrt{2}}{2}\right)$$

$$= \frac{-\sqrt{6} - \sqrt{2}}{4}$$

2. $\sin \dfrac{\pi}{6} \cos \dfrac{\pi}{4} + \cos \dfrac{\pi}{6} \sin \dfrac{\pi}{4}$

$$= \left(\frac{1}{2}\right)\left(\frac{\sqrt{2}}{2}\right) + \left(\frac{\sqrt{3}}{2}\right)\left(\frac{\sqrt{2}}{2}\right) = \frac{\sqrt{2} + \sqrt{6}}{4}$$

3. $\dfrac{\tan \dfrac{5\pi}{6} - \tan \dfrac{\pi}{4}}{1 + \tan \dfrac{5\pi}{6} \tan \dfrac{\pi}{4}} = \dfrac{-\dfrac{\sqrt{3}}{3} - 1}{1 + \left(-\dfrac{\sqrt{3}}{3}\right)(1)}$

$$= \frac{\dfrac{-\sqrt{3} - 3}{3}}{\dfrac{3 - \sqrt{3}}{3}} = \frac{-\sqrt{3} - 3}{3 - \sqrt{3}}\left(\frac{3 + \sqrt{3}}{3 + \sqrt{3}}\right)$$

$$= -2 - \sqrt{3}$$

4. $\sin\left(\dfrac{5\pi}{3} - \dfrac{\pi}{4}\right) = \sin \dfrac{17\pi}{12}$

5. $\cos\left(\dfrac{5\pi}{4} - \dfrac{\pi}{3}\right) = \cos \dfrac{11\pi}{12}$

6. $\tan\left(\dfrac{\pi}{4} + \dfrac{\pi}{6}\right) = \tan \dfrac{5\pi}{12}$

7. $\cos 105° = \cos(45° + 60°)$

$$= \cos 45° \cos 60° - \sin 45° \sin 60°$$

$$= \left(\frac{\sqrt{2}}{2}\right)\left(\frac{1}{2}\right) - \left(\frac{\sqrt{2}}{2}\right)\left(\frac{\sqrt{3}}{2}\right) = \frac{\sqrt{2} - \sqrt{6}}{4}$$

```
(√6+√2)/4
            0.9659258263
sin(7π/12)
            0.9659258263
2-√3
            0.2679491924
tan(15°)
            0.2679491924
```

A calculator can be used to verify your answers.

Find the exact value of each trigonometric expression.

a. $\sin \dfrac{7\pi}{12}$

b. $\tan 15°$

Answer

a. $\sin \dfrac{7\pi}{12} = \sin\left(\dfrac{4\pi}{12} + \dfrac{3\pi}{12}\right)$

$$= \sin\left(\frac{\pi}{3} + \frac{\pi}{4}\right)$$

$$= \sin \frac{\pi}{3} \cos \frac{\pi}{4} + \cos \frac{\pi}{3} \sin \frac{\pi}{4}$$

$$= \left(\frac{\sqrt{3}}{2}\right)\left(\frac{\sqrt{2}}{2}\right) + \left(\frac{1}{2}\right)\left(\frac{\sqrt{2}}{2}\right)$$

$$= \frac{\sqrt{6} + \sqrt{2}}{4}$$

b. $\tan 15° = \tan(45° - 30°)$

$$= \frac{\tan 45° - \tan 30°}{1 + \tan 45° \tan 30°}$$

$$= \frac{1 - \dfrac{\sqrt{3}}{3}}{1 + (1)\left(\dfrac{\sqrt{3}}{3}\right)} \cdot \frac{\dfrac{3}{1}}{\dfrac{3}{1}}$$

$$= \frac{3 - \sqrt{3}}{3 + \sqrt{3}} \cdot \frac{3 - \sqrt{3}}{3 - \sqrt{3}}$$

$$= \frac{9 - 6\sqrt{3} + 3}{9 - 3}$$

$$= 2 - \sqrt{3}$$

SKILL ✓ EXERCISES 7, 11

At times we can evaluate a trigonometric function of a sum or difference of angles without finding decimal approximations of the angles themselves.

Find the exact value for each expression if α is in Q III with $\cos \alpha = -\dfrac{5}{13}$ and β is in Q II with $\sin \beta = \dfrac{3}{5}$.

a. $\cos(\alpha + \beta)$

b. $\sin(\alpha - \beta)$

Answer

$\left(-\dfrac{5}{13}\right)^2 + \sin^2 \alpha = 1$

$\sin^2 \alpha = \dfrac{144}{169}$

$\sin \alpha = -\dfrac{12}{13}$

$(\sin x < 0 \text{ in Q III})$

$\cos^2 \beta + \left(\dfrac{3}{5}\right)^2 = 1$

$\cos^2 \beta = \dfrac{16}{25}$

$\cos \beta = -\dfrac{4}{5}$

$(\cos x < 0 \text{ in Q II})$

1. Use $\cos^2 x + \sin^2 x = 1$ and the quadrant to find $\sin \alpha$ and $\cos \beta$. *Note:* Sketches of reference triangles could also be used to find these values.

a. $\cos(\alpha + \beta) = \cos \alpha \cos \beta - \sin \alpha \sin \beta$

$$= \left(-\frac{5}{13}\right)\left(-\frac{4}{5}\right) - \left(-\frac{12}{13}\right)\left(\frac{3}{5}\right)$$

$$= \frac{20}{65} + \frac{36}{65} = \frac{56}{65}$$

2. Apply the cosine of a sum identity.

b. $\sin(\alpha - \beta) = \sin \alpha \cos \beta - \cos \alpha \sin \beta$

$$= \left(-\frac{12}{13}\right)\left(-\frac{4}{5}\right) - \left(-\frac{5}{13}\right)\left(\frac{3}{5}\right)$$

$$= \frac{48}{65} + \frac{15}{65} = \frac{63}{65}$$

3. Apply the sine of a difference identity.

SKILL ✓ EXERCISE 17

Example 5 introduces two *reduction identities*. Discuss the tip in the margin, explaining that identities are true for all values for which the expressions are defined. In Example 5b we need to choose identities that do not generate undefined values when applying the tangent of a sum identity becomes problematic.

Explain that these new identities expand the number of trig equations that can be solved. Example 6 implements sum and difference identities in solving a trigonometric equation.

Common Student Error The sum and difference identities can be easily confused. Encourage students to check the identities carefully when applying them.

One-on-One Some students are very familiar with functions and terminology in computer programming. These students may quickly relate to the fact that with any function of the form FUNC($a + b$), $a + b$ serves as a single parameter for the function. In general, FUNC($a + b$) $\neq$ FUNC(a) + FUNC(b).

TIPS

Ex. 23 Students may draw two triangles. Recognizing that they are actually the same triangle and that the angles $\alpha = \text{Arcsin } x$ and $\beta = \text{Arccos } x$ are complements would enable an even shorter solution: $\sin(\alpha + \beta) = \sin 90° = 1$.

Ex. 21–24 While rationalizing the denominator is not required when the radical contains variables, students may benefit from the algebraic exercise. Rationalizing the answer in exercise 24 is illustrated here.

Rewrite $\cos\left(\text{Tan}^{-1}\frac{3}{4} - \text{Sin}^{-1} x\right)$ as an algebraic expression.

Answer

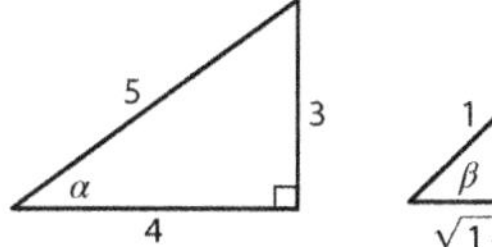

1. Draw reference triangles for $\alpha = \text{Tan}^{-1}\frac{3}{4}$ and $\beta = \text{Sin}^{-1} x$.

$\cos(\alpha - \beta) = \cos\alpha\cos\beta + \sin\alpha\sin\beta$

2. Rewrite the expression in terms of α and β and apply the cosine of a difference identity.

$= \left(\frac{4}{5}\right)\left(\sqrt{1-x^2}\right) + \left(\frac{3}{5}\right)(x)$

3. Use the reference triangles to rewrite each trig function as an algebraic expression.

$= \frac{4\sqrt{1-x^2}}{5} + \frac{3x}{5} = \frac{4\sqrt{1-x^2} + 3x}{5}$

4. Simplify.

SKILL ✓ **EXERCISE 21**

The sum and difference identities are also useful in proving other identities. Methods similar to those used in Example 2a can be used to prove other cofunction and periodic identities listed in Section 5.1. A *reduction identity* simplifies a trigonometric function of a sum or difference where one of the angles is a quadrantal angle.

Verify each reduction identity.

a. $\sin\left(x - \frac{\pi}{2}\right) = -\cos x$ **b.** $\tan(270° + \theta) = -\cot\theta$

Answer

a. $\sin\left(x - \frac{\pi}{2}\right)$

$= \sin x \cos\frac{\pi}{2} - \cos x \sin\frac{\pi}{2}$

$= \sin x \cdot (0) - \cos x \cdot (1)$

$= -\cos x$

b. $\tan(270° + \theta) = \frac{\sin(270° + \theta)}{\cos(270° + \theta)}$

$= \frac{\sin 270°\cos\theta + \cos 270°\sin\theta}{\cos 270°\cos\theta - \sin 270°\sin\theta}$

$= \frac{(-1)\cos\theta + (0)\sin\theta}{(0)\cos\theta - (-1)\sin\theta}$

$= \frac{-\cos\theta}{\sin\theta} = -\cot\theta$

TIP

Express $\tan(\theta + 270°)$ as $\frac{\sin(270° + \theta)}{\cos(270° + \theta)}$ since $\frac{\tan\theta - \tan 270°}{1 + \tan\theta\tan 270°}$ produces undefined values.

SKILL ✓ **EXERCISE 29**

The sum and difference identities extend our ability to solve trigonometric equations.

$\frac{x + x\sqrt{1-x^2}}{\sqrt{1-x^2} - x^2} \cdot \frac{\sqrt{1-x^2} + x^2}{\sqrt{1-x^2} + x^2}$

$= \frac{x\sqrt{1-x^2} + x^3 + x(1-x^2) + x^3\sqrt{1-x^2}}{(1-x^2) - x^4}$

$= \frac{x + (x^3 + x)\sqrt{1-x^2}}{1 - x^2 - x^4}$

Ex. 25, 27 The method used to derive the formula for the cosine of a sum provides a pattern for this derivation.

Ex. 26–27 Example 5b provides a hint for students struggling to complete the proof of exercise 26 and an alternate proof for exercise 27.

Ex. 29a The identity in this exercise is used in a proof in Section 5.6.

Ex. 30 Caution the students that attempting to use $\tan(\alpha - \beta) = \frac{\tan\alpha - \tan\beta}{1 + \tan\alpha\tan\beta}$ to verify this identity produces undefined values. Other methods produce a valid proof.

Ex. 43 Many interesting videos that further explain this concept can be found using the Internet keyword search *standing wave demo.*

---

8. $\sin\frac{11\pi}{12} = \sin\left(\frac{3\pi}{12} + \frac{8\pi}{12}\right)$

$= \sin\frac{\pi}{4}\cos\frac{2\pi}{3} + \cos\frac{\pi}{4}\sin\frac{2\pi}{3}$

$= \left(\frac{\sqrt{2}}{2}\right)\left(-\frac{1}{2}\right) - \left(-\frac{\sqrt{2}}{2}\right)\left(\frac{\sqrt{3}}{2}\right)$

$= \frac{-\sqrt{2} + \sqrt{6}}{4}$

9. $\cos\frac{17\pi}{12} = \cos\left(\frac{8\pi}{12} + \frac{9\pi}{12}\right)$

$= \cos\frac{2\pi}{3}\cos\frac{3\pi}{4} - \sin\frac{2\pi}{3}\sin\frac{3\pi}{4}$

$= \left(-\frac{1}{2}\right)\left(-\frac{\sqrt{2}}{2}\right) - \left(\frac{\sqrt{3}}{2}\right)\left(\frac{\sqrt{2}}{2}\right)$

$= \frac{\sqrt{2} - \sqrt{6}}{4}$

10. $\tan\frac{5\pi}{12} = \tan\left(\frac{2\pi}{12} + \frac{3\pi}{12}\right)$

$= \frac{\tan\frac{\pi}{6} + \tan\frac{\pi}{4}}{1 - \tan\frac{\pi}{6}\tan\frac{\pi}{4}} = \frac{\frac{\sqrt{3}}{3} + 1}{1 - \left(\frac{\sqrt{3}}{3}\right)(1)} \cdot \frac{\frac{3}{1}}{\frac{3}{1}}$

$= \frac{\sqrt{3} + 3}{3 - \sqrt{3}} \cdot \frac{3 + \sqrt{3}}{3 + \sqrt{3}} = \frac{9 + 6\sqrt{3} + 3}{9 - 3}$

$= 2 + \sqrt{3}$

11. $\sin\frac{\pi}{12} = \sin\left(\frac{4\pi}{12} - \frac{3\pi}{12}\right)$

$= \sin\frac{\pi}{3}\cos\frac{\pi}{4} - \cos\frac{\pi}{3}\sin\frac{\pi}{4}$

$= \left(\frac{\sqrt{3}}{2}\right)\left(\frac{\sqrt{2}}{2}\right) - \left(\frac{1}{2}\right)\left(\frac{\sqrt{2}}{2}\right)$

$= \frac{\sqrt{6} - \sqrt{2}}{4}$

12. $\cos(15°) = \cos(45° - 30°)$

$= \cos 45°\cos 30° + \sin 45°\sin 30°$

$= \left(\frac{\sqrt{2}}{2}\right)\left(\frac{\sqrt{3}}{2}\right) + \left(\frac{\sqrt{2}}{2}\right)\left(\frac{1}{2}\right) = \frac{\sqrt{6} + \sqrt{2}}{4}$

13. $\tan 165° = \tan(210° - 45°)$

$= \frac{\tan 120° - \tan 45°}{1 + \tan 120°\tan 45°}$

$= \frac{\frac{\sqrt{3}}{3} - 1}{1 + \frac{\sqrt{3}}{3}(1)} = \frac{\frac{\sqrt{3}}{3} - 1}{\frac{\sqrt{3}}{3} + 1} \cdot \frac{\frac{\sqrt{3}}{3} - 1}{\frac{\sqrt{3}}{3} - 1}$

$= -2 + \sqrt{3}$

14. $\cos\left(-\frac{11\pi}{12}\right) = \cos\left(\frac{3\pi}{12} - \frac{14\pi}{12}\right)$

$= \cos\frac{\pi}{4}\cos\frac{7\pi}{6} + \sin\frac{\pi}{4}\sin\frac{7\pi}{6}$

$= \left(\frac{\sqrt{2}}{2}\right)\left(-\frac{\sqrt{3}}{2}\right) + \left(\frac{\sqrt{2}}{2}\right)\left(-\frac{1}{2}\right)$

$= \frac{-\sqrt{6} - \sqrt{2}}{4}$

15a. $\tan\left(\frac{\pi}{2} - \theta\right) = \frac{\sin\left(\frac{\pi}{2} - \theta\right)}{\cos\left(\frac{\pi}{2} - \theta\right)} = \frac{\cos\theta}{\sin\theta}$

$= \cot\theta$

15b. $\csc\left(\frac{\pi}{2} - \theta\right) = \frac{1}{\sin\left(\frac{\pi}{2} - \theta\right)} = \frac{1}{\cos\theta}$

$= \sec\theta$

16. $\sin\frac{5\pi}{6} = \sin\frac{\pi}{2} + \sin\frac{\pi}{3}$

$\frac{1}{2} \neq 1 + \frac{\sqrt{3}}{2}$

B. Exercises

17–18. using a Pythagorean identity:

$$\left(\frac{7}{25}\right)^2 + \cos^2 \alpha = 1^2$$
$$\cos^2 \alpha = \frac{576}{625}$$
$$\cos \alpha = \frac{24}{25}$$
$$(\cos x > 0 \text{ in Q I})$$

$$\sin^2 \beta + \left(-\frac{5}{13}\right)^2 = 1^2$$
$$\sin^2 \beta = \frac{144}{169}$$
$$\sin \beta = \frac{12}{13}$$
$$(\sin x > 0 \text{ in Q II})$$

17. $\sin \alpha \cos \beta + \cos \alpha \sin \beta$
$$= \left(\frac{7}{25}\right)\left(-\frac{5}{13}\right) + \left(\frac{24}{25}\right)\left(\frac{12}{13}\right)$$
$$= -\frac{35}{325} + \frac{288}{325} = \frac{253}{325}$$

18. $\cos \alpha \cos \beta + \sin \alpha \sin \beta$
$$= \left(\frac{24}{25}\right)\left(-\frac{5}{13}\right) + \left(\frac{7}{25}\right)\left(\frac{12}{13}\right)$$
$$= -\frac{120}{325} + \frac{84}{325} = -\frac{36}{325}$$

19–20. using reference triangles:

$$(-2)^2 + y^2 = 3^2$$
$$y = -\sqrt{9 - 4} = -\sqrt{5}$$
$$\sin \alpha = -\frac{\sqrt{5}}{3}$$
$$\tan \alpha = \frac{\sqrt{5}}{2}$$
$$x^2 + (-4)^2 = 5^2$$
$$x = \sqrt{25 - 16} = 3$$
$$\cos \beta = \frac{3}{5}$$
$$\tan \beta = -\frac{4}{3}$$

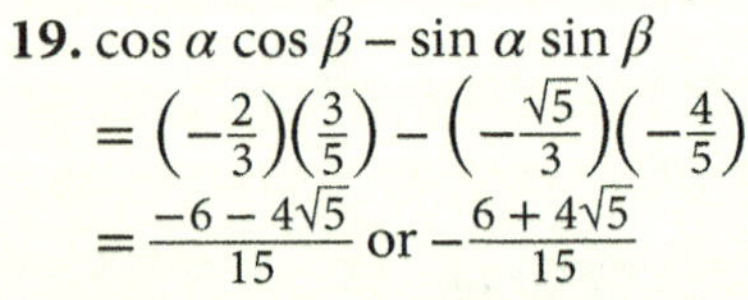

19. $\cos \alpha \cos \beta - \sin \alpha \sin \beta$
$$= \left(-\frac{2}{3}\right)\left(\frac{3}{5}\right) - \left(-\frac{\sqrt{5}}{3}\right)\left(-\frac{4}{5}\right)$$
$$= \frac{-6 - 4\sqrt{5}}{15} \text{ or } -\frac{6 + 4\sqrt{5}}{15}$$

20. $\tan (\alpha + \beta) = \dfrac{\tan \alpha + \tan \beta}{1 - \tan \alpha \tan \beta}$
$$= \frac{\frac{\sqrt{5}}{2} + \left(-\frac{4}{3}\right)}{1 - \left(\frac{\sqrt{5}}{2}\right)\left(-\frac{4}{3}\right)} \cdot \frac{\frac{6}{1}}{\frac{6}{1}} = \frac{-8 + 3\sqrt{5}}{6 + 4\sqrt{5}}$$
$$\cdot \frac{6 - 4\sqrt{5}}{6 - 4\sqrt{5}}$$
$$= \frac{-48 + 50\sqrt{5} - 60}{36 - 80} = \frac{-108 + 50\sqrt{5}}{-44}$$
$$= \frac{54 - 25\sqrt{5}}{22}$$

Example 6   **Solving a Trigonometric Equation**

Find primary solutions for $\sin \left(x + \frac{\pi}{4}\right) = \cos \left(x - \frac{3\pi}{4}\right) + 1$.

Answer

$$\sin x \cos \tfrac{\pi}{4} + \cos x \sin \tfrac{\pi}{4} = \cos x \cos \tfrac{3\pi}{4} + \sin x \sin \tfrac{3\pi}{4} + 1$$

1. Apply the sine of a sum identity and the cosine of a difference identity.

$$(\sin x)\left(\tfrac{\sqrt{2}}{2}\right) + (\cos x)\left(\tfrac{\sqrt{2}}{2}\right) = (\cos x)\left(-\tfrac{\sqrt{2}}{2}\right) + (\sin x)\left(\tfrac{\sqrt{2}}{2}\right) + 1$$

$$\sqrt{2} \cos x = 1$$
$$\cos x = \frac{\sqrt{2}}{2}$$

2. Simplify.

$$x = \frac{\pi}{4}, \frac{7\pi}{4}$$

3. List solutions within $[0, 2\pi)$.

SKILL ✓ **EXERCISE 37**

A. Exercises

Use the sum and difference identities to find the exact value of each expression.

1. $\cos \left(\frac{5\pi}{6} + \frac{\pi}{4}\right)$ $\frac{-\sqrt{6} - \sqrt{2}}{4}$ **2.** $\sin \left(\frac{\pi}{6} + \frac{\pi}{4}\right)$ $\frac{\sqrt{2} + \sqrt{6}}{4}$

3. $\tan \left(\frac{5\pi}{6} - \frac{\pi}{4}\right)$ $-2 - \sqrt{3}$

Use a sum or difference identity to write each expression in terms of a single trigonometric function.

4. $\sin \frac{5\pi}{3} \cos \frac{\pi}{4} - \cos \frac{5\pi}{3} \sin \frac{\pi}{4}$ $\sin \frac{17\pi}{12}$

5. $\cos \frac{5\pi}{4} \cos \frac{\pi}{3} + \sin \frac{5\pi}{4} \sin \frac{\pi}{3}$ $\cos \frac{11\pi}{12}$

6. $\dfrac{\tan \frac{\pi}{4} + \tan \frac{\pi}{6}}{1 - \tan \frac{\pi}{4} \tan \frac{\pi}{6}}$ $\tan \frac{5\pi}{12}$

Use a sum identity to find the exact value of each expression.

7. $\cos 105°$ $\frac{\sqrt{2} - \sqrt{6}}{4}$ **8.** $\sin \frac{11\pi}{12}$ $\frac{-\sqrt{2} + \sqrt{6}}{4}$

9. $\cos \frac{17\pi}{12}$ $\frac{\sqrt{2} - \sqrt{6}}{4}$ **10.** $\tan \frac{5\pi}{12}$ $2 + \sqrt{3}$

Use a difference identity to find the exact value of each expression.

11. $\sin \frac{\pi}{12}$ $\frac{\sqrt{6} - \sqrt{2}}{4}$ **12.** $\cos (15°)$ $\frac{\sqrt{2} + \sqrt{6}}{4}$

13. $\tan 165°$ $-2 + \sqrt{3}$ **14.** $\cos \left(-\frac{11\pi}{12}\right)$ $\frac{-\sqrt{2} - \sqrt{6}}{4}$

15. Use the cofunction identities proved in Example 1 to verify each cofunction identity below.

   a. $\tan \left(\frac{\pi}{2} - \theta\right) = \cot \theta$ **b.** $\csc \left(\frac{\pi}{2} - \theta\right) = \sec \theta$

16. Show that $\sin \left(\frac{\pi}{2} + \frac{\pi}{3}\right) \neq \sin \frac{\pi}{2} + \sin \frac{\pi}{3}$.

B. Exercises

Find the exact value for each expression if α is in Q I with $\sin \alpha = \frac{7}{25}$ and β is in Q II with $\cos \beta = -\frac{5}{13}$.

17. $\sin (\alpha + \beta)$ $\frac{253}{325}$ **18.** $\cos (\alpha - \beta)$ $-\frac{36}{325}$

Find the exact value for each expression if α is in Q III with $\cos \alpha = -\frac{2}{3}$ and β is in Q IV with $\sin \beta = -\frac{4}{5}$.

19. $\cos (\alpha + \beta)$ $\frac{-6 - 4\sqrt{5}}{15}$ **20.** $\tan (\alpha + \beta)$ $\frac{54 - 25\sqrt{5}}{22}$

Rewrite each trigonometric expression as an algebraic expression.

21. $\sin \left(\text{Cos}^{-1} x + \text{Tan}^{-1} \frac{\sqrt{3}}{3}\right)$ $\frac{x + \sqrt{3 - 3x^2}}{2}$

22. $\cos \left(\text{Cos}^{-1} \frac{\sqrt{3}}{2} - \text{Tan}^{-1} x\right)$ $\frac{(x + \sqrt{3})\sqrt{x^2 + 1}}{2x^2 + 2}$

23. $\sin (\text{Arcsin } x + \text{Arccos } x)$ 1

24. $\tan (\text{Sin}^{-1} x + \text{Tan}^{-1} x)$ $\frac{x + x\sqrt{1 - x^2}}{\sqrt{1 - x^2} - x^2}$

25. Derive the sine of a sum identity from the sine of a difference identity.

26. Verify the tangent of a sum identity:
$$\tan (\alpha + \beta) = \frac{\tan \alpha + \tan \beta}{1 - \tan \alpha \tan \beta}.$$

27. Verify the tangent of a difference identity:
$$\tan (\alpha - \beta) = \frac{\tan \alpha - \tan \beta}{1 + \tan \alpha \tan \beta}.$$

28. Use a sum identity to verify each periodic identity.
   a. $\cos (\theta + 2\pi) = \cos \theta$
   b. $\sin (\theta + 2\pi) = \sin \theta$
   c. $\tan (\theta + \pi) = \tan \theta$

21.

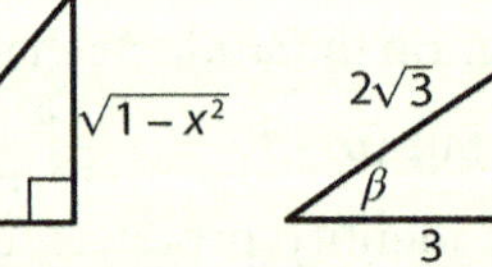

Let $\alpha = \text{Cos}^{-1} x$ and $\beta = \text{Tan}^{-1} \frac{\sqrt{3}}{3}$.

$$\sin (\alpha + \beta) = \sin \alpha \cos \beta + \cos \alpha \sin \beta$$
$$= \sqrt{1 - x^2}\left(\frac{3}{2\sqrt{3}}\right) + x\left(\frac{\sqrt{3}}{2\sqrt{3}}\right)$$
$$= \sqrt{1 - x^2}\left(\frac{\sqrt{3}}{2}\right) + x\left(\frac{1}{2}\right)$$
$$= \frac{x + \sqrt{3 - 3x^2}}{2}$$

22.

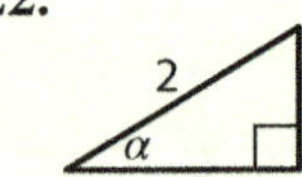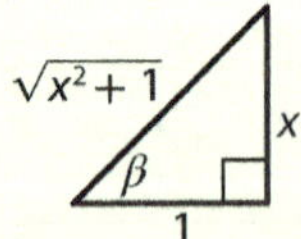

Let $\alpha = \text{Cos}^{-1} \frac{\sqrt{3}}{2} x$ and $\beta = \text{Tan}^{-1} x$.

$$\cos (\alpha - \beta) = \cos \alpha \cos \beta + \sin \alpha \sin \beta$$
$$= \left(\frac{\sqrt{3}}{2}\right)\left(\frac{1}{\sqrt{x^2 + 1}}\right) + \left(\frac{1}{2}\right)\left(\frac{x}{\sqrt{x^2 + 1}}\right)$$
$$= \frac{\sqrt{3}}{2\sqrt{x^2 + 1}} + \frac{x}{2\sqrt{x^2 + 1}}$$
$$= \frac{x + \sqrt{3}}{2\sqrt{x^2 + 1}} \text{ or } \frac{(x + \sqrt{3})\sqrt{x^2 + 1}}{2x^2 + 2}$$

29. Use difference identities to verify the reduction identities for the supplement of an angle.

  a. $\sin(\pi - \theta) = \sin\theta$

  b. $\cos(\pi - \theta) = -\cos\theta$

  c. $\tan(\pi - \theta) = -\tan\theta$

Verify each identity.

30. $\tan\left(x - \frac{\pi}{2}\right) = -\cot x$

31. $\sin\left(\frac{\pi}{2} - \theta\right) = \sin\left(\frac{\pi}{2} + \theta\right)$

32. $\cos(\pi - \alpha) = \cos(\pi + \alpha)$

33. $[\cos(\alpha - \beta)][\cos(\alpha + \beta)] = \cos^2\alpha - \sin^2\beta$

34. $\dfrac{\sin(\alpha - \beta)}{\sin(\alpha + \beta)} = \dfrac{\tan\alpha - \tan\beta}{\tan\alpha + \tan\beta}$

35. $\dfrac{\cos(\alpha - \beta)}{\cos(\alpha + \beta)} = \dfrac{\cot\alpha + \tan\beta}{\cot\alpha - \tan\beta}$

Find the primary solutions for each equation.

36. $\sin(\pi + x) = 1 - \sin(\pi + x)$ $x = \frac{7\pi}{6}, \frac{11\pi}{6}$

37. $\cos\left(x + \frac{\pi}{2}\right) = \sin^2 x$ $x = 0, \pi, \frac{3\pi}{2}$

38. $\sin\left(x - \frac{\pi}{6}\right) + \sin\left(x + \frac{7\pi}{6}\right) = \frac{1}{2}$ $x = \frac{2\pi}{3}, \frac{4\pi}{3}$

39. $\tan(x + \pi) = \sin(\pi - x)$ $x = 0, \pi$

40. In an RLC circuit, the voltage across the inductor is modeled by $v_L = I_m X_L \sin\left(\omega t + \frac{\pi}{2}\right)$. Show that $v_L = V_L \cos\omega t$ is an equivalent model when the maximum voltage across an inductor is $V_L = I_m X_L$.

› C. Exercises

41. Use a sum identity to prove each double-angle identity.

  a. $\sin 2\theta = 2\sin\theta\cos\theta$

  b. $\cos 2\theta = \cos^2\theta - \sin^2\theta$

  c. $\tan 2\theta = \dfrac{2\tan\theta}{1 - \tan^2\theta}$

42. Given $y_1 = m_1 x + b_1$ and $y_2 = m_2 x + b_2$ where $m_2 > m_1$ and both are positive, $\tan\alpha = m_2$ and $\tan\beta = m_1$.

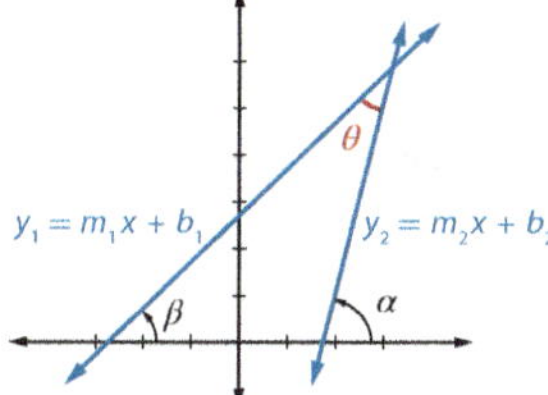

  a. Express $\tan\theta$ in terms of m_1 and m_2.

  b. Use the formula to find (to the nearest degree) the measure of the acute angle between $y_1 = x + 4$ and $y_2 = 4x - 10$.   **31°**

43. A standing (or stationary) wave is formed when two waves having the same frequency, amplitude, and wavelength travel in opposite directions. Verify that the sum of the waves $y_1 = A_0 \sin(kx - \omega t)$ and $y_2 = A_0 \sin(kx + \omega t)$ is the standing wave $y = (2A_0 \sin kx)\cos\omega t$.

44. A function's difference quotient, $\dfrac{f(x + h) - f(x)}{h}$, is the slope of the secant line through two points of the function's graph. In calculus, it is used to examine the function's rate of change. Verify the difference quotient for the sine and cosine functions.

  a. $\dfrac{\sin(x + h) - \sin x}{h}$

     $= \sin x\left(\dfrac{\cos h - 1}{h}\right) + \cos x\left(\dfrac{\sin h}{h}\right)$

  b. $\dfrac{\cos(x + h) - \cos x}{h}$

     $= \cos x\left(\dfrac{\cos h - 1}{h}\right) - \sin x\left(\dfrac{\sin h}{h}\right)$

23.

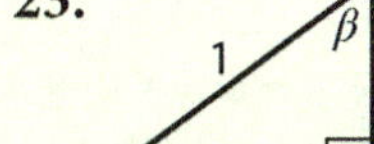

Let $\alpha = \text{Arcsin } x$ and $\beta = \text{Arccos } x$.

$\sin(\alpha + \beta)$

$= \sin\alpha\cos\beta + \cos\alpha\sin\beta$

$= x(x) + \left(\sqrt{1 - x^2}\right)\left(\sqrt{1 - x^2}\right)$

$= x^2 + 1 - x^2 = 1$

24.

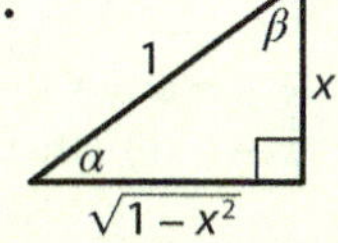

Let $\alpha = \text{Sin}^{-1} x$ and $\beta = \text{Tan}^{-1} x$.

$\tan(\alpha + \beta) = \dfrac{\tan\alpha + \tan\beta}{1 - \tan\alpha\tan\beta}$

$= \dfrac{\frac{x}{\sqrt{1 - x^2}} + x}{1 - \left(\frac{x}{\sqrt{1 - x^2}}\right)x} \cdot \dfrac{\frac{\sqrt{1 - x^2}}{1}}{\frac{\sqrt{1 - x^2}}{1}} = \dfrac{x + x\sqrt{1 - x^2}}{\sqrt{1 - x^2} - x^2}$

25. $\sin(\alpha + \beta) = \sin(\alpha - (-\beta))$

$= \sin\alpha\cos(-\beta) - \cos\alpha\sin(-\beta)$

$= \sin\alpha\cos\beta - \cos\alpha(-\sin\beta)$

$= \sin\alpha\cos\beta + \cos\alpha\sin\beta$

---

26. $\dfrac{\tan\alpha + \tan\beta}{1 - \tan\alpha\tan\beta} = \dfrac{\frac{\sin\alpha}{\cos\alpha} + \frac{\sin\beta}{\cos\beta}}{1 - \frac{\sin\alpha\sin\beta}{\cos\alpha\cos\beta}}$

$= \dfrac{\frac{\sin\alpha\cos\beta + \cos\alpha\sin\beta}{\cos\alpha\cos\beta}}{\frac{\cos\alpha\cos\beta - \sin\alpha\sin\beta}{\cos\alpha\cos\beta}}$

$= \dfrac{\sin\alpha\cos\beta + \cos\alpha\sin\beta}{\cos\alpha\cos\beta - \sin\alpha\sin\beta}$

$= \dfrac{\sin(\alpha + \beta)}{\cos(\alpha + \beta)} = \tan(\alpha + \beta)$

27. $\tan(\alpha - \beta) = \tan(\alpha + (-\beta))$

$= \dfrac{\tan\alpha + \tan(-\beta)}{1 - \tan\alpha\tan(-\beta)}$

$= \dfrac{\tan\alpha - \tan\beta}{1 + \tan\alpha\tan\beta}$

28a. $\cos(\theta + 2\pi)$

$= \cos\theta\cos 2\pi - \sin\theta\sin 2\pi$

$= \cos\theta(1) - \sin\theta(0)$

$= \cos\theta$

28b. $\sin(\theta + 2\pi)$

$= \sin\theta\cos 2\pi + \cos\theta\sin 2\pi$

$= \sin\theta(1) + \cos\theta(0)$

$= \sin\theta$

28c. $\tan(\theta + \pi) = \dfrac{\tan\theta + \tan\pi}{1 - \tan\theta\tan\pi}$

$= \dfrac{\tan\theta + 0}{1 - \tan\theta(0)} = \dfrac{\tan\theta}{1} = \tan\theta$

29a. $\sin(\pi - \theta)$

$= \sin\pi\cos\theta - \cos\pi\sin\theta$

$= (0)\cos\theta - (-1)\sin\theta = \sin\theta$

29b. $\cos(\pi - \theta)$

$= \cos\pi\cos\theta + \sin\pi\sin\theta$

$= (-1)\cos\theta + (0)\sin\theta = -\cos\theta$

29c. $\tan(\pi - \theta) = \dfrac{\tan\pi + \tan\theta}{1 - \tan\pi\tan\theta}$

$= \dfrac{0 - \tan\theta}{1 + (0)\tan\theta} = -\tan\theta$

or

$\tan(\pi - \theta) = \dfrac{\sin(\pi - \theta)}{\cos(\pi - \theta)} = \dfrac{\sin\theta}{-\cos\theta}$

$= -\tan\theta$

30. $\tan\left(x - \frac{\pi}{2}\right) = \tan\left(-\left(\frac{\pi}{2} - x\right)\right)$

$= -\tan\left(\frac{\pi}{2} - x\right) = -\cot x$

or

$\tan\left(x - \frac{\pi}{2}\right) = \dfrac{\sin\left(x - \frac{\pi}{2}\right)}{\cos\left(x - \frac{\pi}{2}\right)}$

$= \dfrac{\sin x\cos\frac{\pi}{2} - \cos x\sin\frac{\pi}{2}}{\cos x\cos\frac{\pi}{2} + \sin x\sin\frac{\pi}{2}}$

$= \dfrac{\sin x(0) - \cos x(1)}{\cos x(0) + \sin x(1)} = \dfrac{-\cos x}{\sin x}$

$= -\cot x$

31. working on both sides:

$\sin\left(\frac{\pi}{2} - \theta\right) = \sin\left(\frac{\pi}{2} + \theta\right)$

$\cos\theta = \sin\frac{\pi}{2}\cos\theta + \cos\frac{\pi}{2}\sin\theta$

$\cos\theta = (1)\cos\theta + (0)\sin\theta$

$\cos\theta = \cos\theta$

32. working on both sides:

$$\cos(\pi - \alpha) = \cos(\pi + \alpha)$$
$$\cos \pi \cos \alpha + \sin \pi \sin \alpha$$
$$\qquad = \cos \pi \cos \alpha - \sin \pi \sin \alpha$$
$$(-1)\cos \alpha + (0)\sin \alpha$$
$$\qquad = (-1)\cos \alpha - 0 \sin \alpha$$
$$-\cos \alpha = -\cos \alpha$$

33. $(\cos(\alpha - \beta))(\cos(\alpha + \beta))$
$$= (\cos \alpha \cos \beta + \sin \alpha \sin \beta)$$
$$\qquad \cdot (\cos \alpha \cos \beta - \sin \alpha \sin \beta)$$
$$= \cos^2 \alpha \cos^2 \beta - \sin^2 \alpha \sin^2 \beta$$
$$= \cos^2 \alpha(1 - \sin^2 \beta)$$
$$\qquad - (1 - \cos^2 \alpha)\sin^2 \beta$$
$$= \cos^2 \alpha - \sin^2 \beta \cos^2 \alpha$$
$$\qquad - \sin^2 \beta + \sin^2 \beta \cos^2 \alpha$$
$$= \cos^2 \alpha - \sin^2 \beta$$

34. $\dfrac{\tan \alpha - \tan \beta}{\tan \alpha + \tan \beta} = \dfrac{\dfrac{\sin \alpha}{\cos \alpha} - \dfrac{\sin \beta}{\cos \beta}}{\dfrac{\sin \alpha}{\cos \alpha} + \dfrac{\sin \beta}{\cos \beta}}$

$$= \dfrac{\dfrac{\sin \alpha \cos \beta - \cos \alpha \sin \beta}{\cos \alpha \cos \beta}}{\dfrac{\sin \alpha \cos \beta + \cos \alpha \sin \beta}{\cos \alpha \cos \beta}}$$

$$= \dfrac{\sin \alpha \cos \beta - \cos \alpha \sin \beta}{\sin \alpha \cos \beta + \cos \alpha \sin \beta} = \dfrac{\sin(\alpha - \beta)}{\sin(\alpha + \beta)}$$

35. $\dfrac{\cot \alpha + \tan \beta}{\cot \alpha - \tan \beta} = \dfrac{\dfrac{\cos \alpha}{\sin \alpha} + \dfrac{\sin \beta}{\cos \beta}}{\dfrac{\cos \alpha}{\sin \alpha} - \dfrac{\sin \beta}{\cos \beta}}$

$$= \dfrac{\dfrac{\cos \alpha \cos \beta + \sin \alpha \sin \beta}{\sin \alpha \cos \beta}}{\dfrac{\cos \alpha \cos \beta - \sin \alpha \sin \beta}{\sin \alpha \cos \beta}}$$

$$= \dfrac{\cos \alpha \cos \beta + \sin \alpha \sin \beta}{\cos \alpha \cos \beta - \sin \alpha \sin \beta} = \dfrac{\cos(\alpha - \beta)}{\cos(\alpha + \beta)}$$

36.
$$2\sin(\pi + x) = 1$$
$$\sin(\pi + x) = \tfrac{1}{2}$$
$$\sin \pi \cos x + \sin x \cos \pi = \tfrac{1}{2}$$
$$(0)\cos x + \sin x(-1) = \tfrac{1}{2}$$
$$\sin x = -\tfrac{1}{2}$$
$$x = \tfrac{7\pi}{6}, \tfrac{11\pi}{6}$$

37. $\cos x \cos \tfrac{\pi}{2} - \sin x \sin \tfrac{\pi}{2} = \sin^2 x$
$$\cos x(0) - \sin x(1) = \sin^2 x$$
$$-\sin x = \sin^2 x$$
$$\sin^2 x + \sin x = 0$$
$$\sin x(\sin x + 1) = 0$$
$$\sin x = 0 \quad \text{or} \quad \sin x = -1$$
$$x = 0, \pi \qquad x = \tfrac{3\pi}{2}$$

45. Evaluate each expression. [3.2]
- **a.** $\log_2 32$ 5
- **b.** $2 \log_3 9$ 4
- **c.** $\log 0.001$ –3
- **d.** $\ln \sqrt[3]{e}$ $\tfrac{1}{3}$

Write each function rule if $f(x) = 2x + 3$ and $g(x) = 2x^2 + x - 3$. [1.7]

46. $(f + g)(x)$ and $(f - g)(x)$ $(f+g)(x) = 2x^2 + 3x$ $(f-g)(x) = -2x^2 + x + 6$

47. $fg(x)$ $(fg)(x) = 4x^3 + 8x^2 - 3x - 9$

48. $\dfrac{f}{g}(x)$ $\dfrac{f}{g}(x) = \dfrac{1}{x-1}$ for $x \neq -\tfrac{3}{2}, x \neq 1$

49. $(f \circ g)(x)$ and $(g \circ f)(x)$ $(f \circ g)(x) = 4x^2 + 2x - 3$ $(g \circ f)(x) = 8x^2 + 26x + 18$

50. Use synthetic division to divide $4x^3 - 3x^2 - 15x - 18$ by $x - 3$. [2.3] $4x^2 + 9x + 12$ R. 18

51. Which of the following is true of $f(x)$ and $g(x)$ if they are inverse functions? [1.8] C
- **a.** $(fg)(x) = (gf)(x) = \tfrac{1}{x}$
- **b.** $(f + g)(x) = (g + f)(x) = -x$
- **c.** $(f \circ g)(x) = (g \circ f)(x) = x$
- **d.** Their graphs are reflections in the x-axis.
- **e.** Their graphs are reflections in the y-axis.

52. Which is not a possible zero of $f(x) = 9x^4 - 16x^3 + 24x^2 - 16$? [2.4] C
- **A.** 2
- **B.** 4
- **C.** $\tfrac{3}{8}$
- **D.** $-\tfrac{4}{3}$
- **E.** $\tfrac{1}{9}$

53. Which function is equivalent to $\cos \theta \cot \theta + \sin \theta$? [5.1] E
- **A.** $\tan \theta$
- **B.** $\sec \theta$
- **C.** $\sin \theta$
- **D.** $\cos \theta$
- **E.** $\csc \theta$

54. Find all primary solutions for $\cos \theta \cot \theta = 2 \cos \theta$. [5.3] E
- **A.** $\theta = 0, \pi$
- **B.** $\theta \approx 0.464, \approx 3.605$
- **C.** $\theta = \tfrac{\pi}{2}, \tfrac{3\pi}{2}$
- **D.** both A and B
- **E.** both B and C

38. $\left(\sin x \cos \tfrac{\pi}{6} - \sin \tfrac{\pi}{6}\cos x\right)$
$$+ \left(\sin x \cos \tfrac{7\pi}{6} + \sin \tfrac{7\pi}{6}\cos x\right) = \tfrac{1}{2}$$
$$\left(\tfrac{\sqrt{3}}{2}\right)\sin x - \left(\tfrac{1}{2}\right)\cos x$$
$$- \left(\tfrac{\sqrt{3}}{2}\right)\sin x - \left(\tfrac{1}{2}\right)\cos x = \tfrac{1}{2}$$
$$-\cos x = \tfrac{1}{2}$$
$$\cos x = -\tfrac{1}{2}$$
$$x = \tfrac{2\pi}{3}, \tfrac{4\pi}{3}$$

39. $\dfrac{\tan x + \tan \pi}{1 - \tan x \tan \pi}$
$$= \sin \pi \cos x - \sin x \cos \pi$$
$$\dfrac{\tan x + 0}{1 - (0)\tan x} = (0)\cos x - (-1)\sin x$$
$$\tan x = \sin x$$
$$\dfrac{\sin x}{\cos x} - \dfrac{\sin x \cos x}{\cos x} = 0$$
$$\dfrac{\sin x(1 - \cos x)}{\cos x} = 0$$
$$\sin x = 0 \quad \text{or} \quad 1 - \cos x = 0$$
$$x = 0, \pi \qquad \cos x = 1$$

40. $v_L = V_L \sin\left(\omega t + \tfrac{\pi}{2}\right)$
$$= V_L \left(\sin \omega t \cos \tfrac{\pi}{2} + \sin \tfrac{\pi}{2}\cos \omega t\right)$$
$$= V_L \left(\sin \omega t(0) + (1)\cos \omega t\right)$$
$$= V_L \cos \omega t$$

continued in Answers and Solutions Overflow

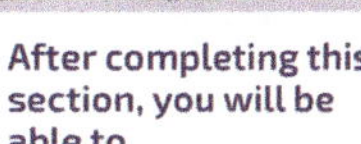

The sinusoidal sound wave of a tuning fork can be viewed on an oscilloscope.

The sum identities examined in the last section are used to develop several other identities that aid in evaluating trigonometric expressions, proving even more identities, and solving trigonometric equations. The *double-angle identities* are derived by replacing α and β with θ.

$$\cos 2\theta = \cos(\theta + \theta)$$ 1. Substitute $(\theta + \theta)$ for 2θ.

$$= \cos\theta\cos\theta - \sin\theta\sin\theta$$ 2. Apply the cosine of a sum identity.

$$= \cos^2\theta - \sin^2\theta$$ 3. Simplify.

Alternate forms of this identity that use a single trigonometric function are derived by applying the Pythagorean identity. Similar derivations of the double-angle identities for sine and cosine are left as exercises.

Double-Angle Identities

$\cos 2\theta = \cos^2\theta - \sin^2\theta$ $= 2\cos^2\theta - 1$ $= 1 - 2\sin^2\theta$	$\sin 2\theta = 2\sin\theta\cos\theta$ $\tan 2\theta = \dfrac{2\tan\theta}{1 - \tan^2\theta}$

Example 1 Applying Double-Angle Identities

If $\tan\theta = \frac{3}{2}$ and $\pi < \theta < \frac{3\pi}{2}$, find exact values for $\cos 2\theta$, $\sin 2\theta$, and $\tan 2\theta$.

Answer

$$r = \sqrt{(-2)^2 + (-3)^2} = \sqrt{13}$$
$$\cos\theta = \frac{-2}{\sqrt{13}}; \ \sin\theta = \frac{-3}{\sqrt{13}}$$

1. Use a reference triangle to determine the ratios for $\cos\theta$ and $\sin\theta$.

$$\cos 2\theta = 2\cos^2\theta - 1 = 2\left(\frac{-2}{\sqrt{13}}\right)^2 - 1 = -\frac{5}{13}$$

2. Apply double-angle identities.

$$\sin 2\theta = 2\sin\theta\cos\theta = 2\left(\frac{-3}{\sqrt{13}}\right)\left(\frac{-2}{\sqrt{13}}\right) = \frac{12}{13}$$

$$\tan 2\theta = \frac{2\tan\theta}{1 - \tan^2\theta} = \frac{2\left(\frac{3}{2}\right)}{1 - \left(\frac{3}{2}\right)^2} = \frac{3}{-\frac{5}{4}} = -\frac{12}{5}$$

SKILL ✔ EXERCISE 3

The sum and double-angle identities can be used to derive other multiple-angle identities, such as $\sin 3\theta$, $\cos 4\theta$, and $\tan 6\theta$.

After completing this section, you will be able to

- derive and apply multiple-angle, power-reducing, and half-angle identities.
- solve equations using trig identities.
- derive and apply product-to-sum and sum-to-product identities.

A *hypocycloid* is a curve traced by a point on a circle as it rolls without slipping inside a circle of larger radius. An *astroid* is a hypocycloid where the ratio of the two radii is 4 : 1. The trigonometric triple angle identity may be used to derive the formula for an astroid.

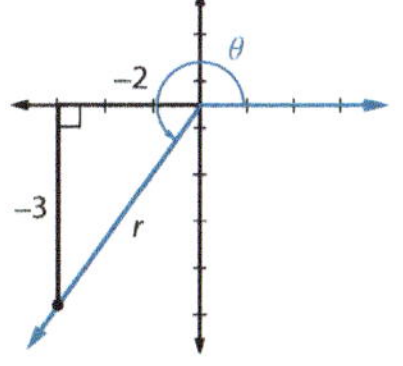

5.5 Multiple Angle Identities

Objectives

1. To derive and apply multiple-angle, power-reducing, half-angle, product-to-sum, and sum-to-product identities

2. To apply trigonometric identities to rewrite and evaluate trigonometric functions

3. To solve equations and real-world problems using trigonometric identities

Flash

An oscilloscope displays an electrical signal's waveform. Using a microphone and an oscilloscope to translate sound waves into an electrical signal provides a visual display of the sinusoidal waves generated by the tuning fork. Consider using a tuning fork and an *online oscilloscope* (Internet keyword search) to display a sinusoidal wave. A low, steady whistle will also create a near-perfect sine wave.

Vocabulary

double-angle identity
half-angle identity
power-reducing identity
product-to-sum identity
sum-to-product identity

PRESENTATION

Lesson Opener

1. If θ is an angle in the given quadrant, state possible quadrants for $\frac{\theta}{2}$ and 2θ.

θ	$\frac{\theta}{2}$	2θ
I	I	I or II
II	I	III or IV
III	II	I or II
IV	II	III or IV

2. Simplify each expression by substituting $1 - \sin^2\theta$ for $\cos^2\theta$.

   a. $\cos^2\theta + 2\sin^2\theta - 1$ $\sin^2\theta$

   b. $\cos^2\theta - \sin^2\theta - 1$ $-2\sin^2\theta$

3. Simplify each expression by substituting $1 - \cos^2\theta$ for $\sin^2\theta$.

   a. $\sin^2\theta - \cos^2\theta - 1$ $-2\cos^2\theta$

   b. $2\cos^2\theta + \sin^2\theta + 1$ $\cos^2\theta + 2$

The first Lesson Opener exercise introduces students to determining quadrants for half and double angles while the other two exercises provide practice in applying the basic Pythagorean identity, a skill used in several proofs in this section.

Convey that the identities presented here are important because they are needed for the development of computer programs and calculators. They also provide exact values, in contrast to the decimal approximations displayed on a calculator. In addition, these identities are needed to derive other useful identities and to solve additional trigonometric equations.

Derive the *double-angle identity* for $\cos 2\theta$. Two other forms of this identity are derived in exercise 2 using substitutions similar to those in the Lesson Opener. The other two double-angle identities are derived in Example 1 using sum identities.

A reference triangle is used to evaluate $\cos\theta$ and $\sin\theta$ in Example 1 since using Pythagorean identities when given $\tan\theta$ can be tedious. Consider having different groups evaluate $\cos 2\theta$ in Example 1

Additional Exercises

Find the exact value of each expression if $\sin\theta = \frac{3}{7}$ and $\frac{\pi}{2} < \theta < \pi$.

1. $\sin 2\theta$, $\cos 2\theta$, and $\tan 2\theta$

$-\frac{12\sqrt{10}}{49}, \frac{31}{49}, -\frac{12\sqrt{10}}{31}$

2. $\sin\frac{\theta}{2}$, $\cos\frac{\theta}{2}$, and $\tan\frac{\theta}{2}$

$\frac{\sqrt{98-28\sqrt{10}}}{14}, \frac{\sqrt{98+28\sqrt{10}}}{14}, \frac{7+2\sqrt{10}}{3}$

3. Use a half-angle identity to find the exact values.

a. $\cos\frac{11\pi}{12}$ $-\frac{\sqrt{2+\sqrt{3}}}{2}$

b. $\cos\frac{11\pi}{24}$ $\frac{\sqrt{2-\sqrt{2+\sqrt{3}}}}{2}$

Verify each statement.

4. $\cos 3\theta = \cos\theta - 4\sin^2\theta\cos\theta$

$\cos 3\theta = \cos(\theta + 2\theta)$
$= \cos\theta\cos 2\theta - \sin\theta\sin 2\theta$
$= \cos\theta(1 - 2\sin^2\theta)$
$\qquad\qquad - \sin\theta(2\sin\theta\cos\theta)$
$= \cos\theta - 2\sin^2\theta\cos\theta - 2\sin^2\theta\cos\theta$
$= \cos\theta - 4\sin^2\theta\cos\theta$

5. $\csc 2\theta = \frac{\cot\theta}{1 + \cos 2\theta}$

$\dfrac{\frac{\cos\theta}{\sin\theta}}{1 + (2\cos^2\theta - 1)} = \dfrac{\frac{\cos\theta}{\sin\theta}}{2\cos^2\theta}$

$= \frac{\cos\theta}{\sin\theta}\left(\frac{1}{2\cos^2\theta}\right)$

$= \frac{1}{2\sin\theta\cos\theta}$

$= \frac{1}{\sin 2\theta} = \csc 2\theta$

Example 2 Deriving a Triple-Angle Identity

Express $\sin 3\theta$ in terms of $\sin\theta$.

Answer

$\sin 3\theta = \sin(2\theta + \theta)$
$\quad = \sin 2\theta\cos\theta + \cos 2\theta\sin\theta$ 1. Express 3θ as a sum and apply the sine of a sum identity.

$\quad = (2\sin\theta\cos\theta)\cos\theta + (1 - 2\sin^2\theta)\sin\theta$ 2. Substitute using double-angle identities and simplify.
$\quad = 2\sin\theta\cos^2\theta + \sin\theta - 2\sin^3\theta$

$\quad = 2\sin\theta(1 - \sin^2\theta) + \sin\theta - 2\sin^3\theta$ 3. Substitute using a Pythagorean identity and simplify.
$\quad = 2\sin\theta - 2\sin^3\theta + \sin\theta - 2\sin^3\theta$
$\quad = -4\sin^3\theta + 3\sin\theta$

Check

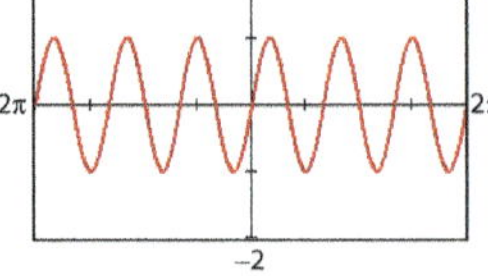

The graphs of $Y_1 = \sin(3X)$ and $Y_2 = -4(\sin(X))^3 + 3\sin(X)$ appear to be identical.

SKILL ✓ **EXERCISE 17**

Power-reducing identities are derived from double-angle identities. For example, solving $\cos 2\theta = 1 - 2\sin^2\theta$ for $\sin^2\theta$ produces $\sin^2\theta = \frac{1 - \cos 2\theta}{2}$.

Power-Reducing Identities		
$\sin^2\theta = \frac{1-\cos 2\theta}{2}$	$\cos^2\theta = \frac{1+\cos 2\theta}{2}$	$\tan^2\theta = \frac{1-\cos 2\theta}{1+\cos 2\theta}$

Example 3 Reducing a Power

Express $\sin^4\theta$ in terms with no power greater than 1.

Answer

$\sin^4\theta = (\sin^2\theta)^2$ 1. Express x^4 as $(x^2)^2$.

$\quad = \left(\frac{1 - \cos 2\theta}{2}\right)^2$ 2. Apply the sine power-reducing identity and expand the power.

$\quad = \frac{1 - 2\cos 2\theta + \cos^2 2\theta}{4}$

$\quad = \frac{1}{4} - \frac{1}{2}\cos 2\theta + \frac{1}{4}\left(\frac{1 + \cos 2(2\theta)}{2}\right)$ 3. Apply the cosine power-reducing identity and simplify.

$\quad = \frac{1}{4} - \frac{1}{2}\cos 2\theta + \frac{1}{8} + \frac{1}{8}\cos 4\theta$

$\quad = \frac{1}{8}(3 - 4\cos 2\theta + \cos 4\theta)$

SKILL ✓ **EXERCISE 21**

Half-angle identities are derived by replacing θ with $\frac{\theta}{2}$ in the power-reducing identities and taking the square root of both sides. The sign of the square root is determined by the quadrant in which $\frac{\theta}{2}$ lies.

$$\sin^2\frac{\theta}{2} = \frac{1 - \cos 2\left(\frac{\theta}{2}\right)}{2}$$

$$\sin\frac{\theta}{2} = \pm\sqrt{\frac{1 - \cos\theta}{2}}$$

---

using each of the three forms of its double-angle identity and then compare their results with those in the example. Rationalizing denominators in the expressions for $\cos\theta$ and $\sin\theta$ is not necessary (they are not answers) and would also create extra work in the next step. Consider having the students confirm the result for $\tan 2\theta$ using $\frac{\sin 2\theta}{\cos 2\theta} = \frac{\frac{12}{13}}{-\frac{5}{13}} = -\frac{12}{5}$.

Example 2 demonstrates how to derive identities for other integral multiples of θ, which can quickly become quite complex.

Power-reducing identities are derived from double-angle identities. For $\cos^2\theta$,

use the cosine double-angle identity that involves just the cosine function. The power-reducing identities are then used to derive the *half-angle identities*. Students should mimic this process for cosine and tangent functions in exercises 7 and 8. The derivation of the other two forms for $\tan\frac{\theta}{2}$ is explored in exercise 47. Demonstrate the application of power-reducing identities in Example 3 and the half-angle identities in Example 4. Consider comparing the use of the three different forms for $\tan\frac{\theta}{2}$.

The double-angle and half-angle identities are used to solve equations in Examples 5 and 6. When finding general solu-

tions for equations involving functions whose period is greater than 2π, students need to extend their list of possible specific solutions. Remind the students that solutions can be checked using graphing technology.

Derivations of the *product-to-sum* and *sum-to-product identities* are intended for the extended track. Consider providing these identities for the students during assessments.

Demonstrate the application of a product-to-sum identity using Example 7. Then demonstrate the verification of the first sum-to-product identity. If students struggle with using the product-to-sum

Half-Angle Identities	
$\sin\frac{\theta}{2} = \pm\sqrt{\frac{1-\cos\theta}{2}}$	$\tan\frac{\theta}{2} = \pm\sqrt{\frac{1-\cos\theta}{1+\cos\theta}}$ $= \frac{1-\cos\theta}{\sin\theta}$
$\cos\frac{\theta}{2} = \pm\sqrt{\frac{1+\cos\theta}{2}}$	$= \frac{\sin\theta}{1+\cos\theta}$

Example 4 Applying Half-Angle Identities

Find the exact value of each expression.

a. $\sin\frac{\pi}{8}$ **b.** $\cos\frac{7\pi}{12}$ **c.** $\tan\frac{5\pi}{12}$

Answer

a. $\sin\frac{\pi}{8}$

$= \sin\frac{\left(\frac{\pi}{4}\right)}{2}$

$= \pm\sqrt{\frac{1-\cos\frac{\pi}{4}}{2}}$

$= \sqrt{\frac{1-\left(\frac{\sqrt{2}}{2}\right)}{2}}$

$= \sqrt{\frac{2-\sqrt{2}}{4}}$

$= \frac{\sqrt{2-\sqrt{2}}}{2}$

b. $\cos\frac{7\pi}{12}$

$= \cos\frac{\left(\frac{7\pi}{6}\right)}{2}$

$= \pm\sqrt{\frac{1+\cos\frac{7\pi}{6}}{2}}$

$= -\sqrt{\frac{1+\left(-\frac{\sqrt{3}}{2}\right)}{2}}$

$= -\sqrt{\frac{2-\sqrt{3}}{4}}$

$= -\frac{\sqrt{2-\sqrt{3}}}{2}$

c. $\tan\frac{5\pi}{12}$

$= \tan\frac{\left(\frac{5\pi}{6}\right)}{2}$

$= \frac{1-\cos\frac{5\pi}{6}}{\sin\frac{5\pi}{6}}$

$= \frac{1-\left(-\frac{\sqrt{3}}{2}\right)}{\left(\frac{1}{2}\right)}$

$= 2+\sqrt{3}$

1. Rewrite the angle as half of a special angle.

2. Apply a half-angle identity. Note that $\sin\frac{\pi}{8} > 0$ since it is in Q I, and $\cos\frac{7\pi}{12} < 0$ since it is in Q II.

3. Simplify.

_______ SKILL ✔ **EXERCISE 9**

These identities provide additional tools for solving trigonometric equations.

Example 5 Solving Equations with a Double-Angle Identity

Find primary solutions for $\sin 2x = \cos x$.

Answer

$\sin 2x = \cos x$

$2\sin x\cos x = \cos x$

1. Use a double-angle identity to substitute for sin 2x.

$2\sin x\cos x - \cos x = 0$

$\cos x(2\sin x - 1) = 0$

2. Collect the terms on one side, factor, and solve using the Zero Product Property.

$\cos x = 0 \quad\text{or}\quad \sin x = \frac{1}{2}$

$x = \frac{\pi}{2}, \frac{3\pi}{2}, \frac{\pi}{6}, \frac{5\pi}{6}$

_______ SKILL ✔ **EXERCISE 13**

---

identity in the initial step, encourage them to substitute using $x = \frac{\alpha+\beta}{2}$ and $y = \frac{\alpha-\beta}{2}$, yielding $2\sin x\cos y = 2\left[\frac{1}{2}(\sin(x+y) + \sin(x-y))\right]$, which is easier to visualize. Then resubstitute to express in terms of α and β.

A sum-to-product identity is used to evaluate a trigonometric expression in Example 8 and to solve a trigonometric equation in Example 9.

TIPS

Ex. 3–6 Encourage students to find tan 2θ using the double-angle formula and then to check their work using $\frac{\sin 2\theta}{\cos 2\theta}$.

Ex. 46 Encourage students to refer to Example 2 and exercise 17.

---

6. $\sin^2\theta\cos^2\theta = \frac{1-\cos^2 2\theta}{4}$

$\sin^2\theta\cos^2\theta = (1-\cos^2\theta)\cos^2\theta$

$= \cos^2\theta - (\cos^2\theta)^2$

$= \frac{1+\cos 2\theta}{2} - \left(\frac{1+\cos 2\theta}{2}\right)^2$

$= \frac{2+2\cos 2\theta}{4} - \frac{1+2\cos 2\theta+\cos^2 2\theta}{4}$

$= \frac{1-\cos^2 2\theta}{4}$

7. Find the primary solutions for $6\sin^2 x + \cos 2x = 4$. $x = \frac{\pi}{3}, \frac{2\pi}{3}, \frac{4\pi}{3}, \frac{5\pi}{3}$

8. Find the general solution for $\sin\frac{x}{2} = 2 - 3\cos x$. $x = \frac{\pi}{3} + n4\pi, \frac{5\pi}{3} + n4\pi, \approx 6.96 + n4\pi, \approx 11.89 + n4\pi$

9. Express each product as a sum.

a. $\cos 5x\cos 3x$ $\frac{1}{2}\cos 2x + \frac{1}{2}\cos 8x$

b. $\sin 4x\cos x$ $\frac{1}{2}\sin 5x + \frac{1}{2}\sin 3x$

10. Evaluate $\sin\frac{7\pi}{12} - \sin\frac{11\pi}{12}$. $\frac{\sqrt{2}}{2}$

11. Find the general solution for $\sin 3x - \sin x = 0$. $x = \frac{\pi}{4} + n\pi, \frac{3\pi}{4} + n\pi, n\pi$

Assignments

- **Minimum:** 1–3, 5, 7–9, 11, 13–14, 16–18, 20–22, 25–26, 41, 49–52, 56–58
- **Standard:** 1–2, 4, 6–9, 12–13, 15, 17–18, 20–21, 23, 25, 29*a*, 30*a*, 31; 35–41 odd; 49–57 odd
- **Extended:** 1–5, 7–8, 10–11, 15–19; 22–44 even; 45, 47, 50, 52–55, 57–58

Assessment

- Quiz 5B covers Sections 5.4–5.5.

Solutions

❯ A. Exercises

1a. $\sin(\theta+\theta) = \sin\theta\cos\theta + \cos\theta\sin\theta$

$= 2\sin\theta\cos\theta$

1b. $\tan(\theta+\theta) = \frac{\tan\theta+\tan\theta}{1-\tan\theta\tan\theta}$

$= \frac{2\tan\theta}{1-\tan^2\theta}$

2a. $\cos 2\theta = \cos^2\theta - \sin^2\theta$

$= \cos^2\theta - (1-\cos^2\theta)$

$= 2\cos^2\theta - 1$

2b. $\cos 2\theta = \cos^2\theta - \sin^2\theta$

$= (1-\sin^2\theta) - \sin^2\theta$

$= 1 - 2\sin^2\theta$

3. $5^2 = (-3)^2 + y^2$

$\qquad y = \pm\sqrt{5^2 - (-3)^2} = 4 \ (\text{Q II})$

$\qquad \sin\theta = \frac{4}{5}; \ \tan\theta = -\frac{4}{3}$

$\qquad \cos 2\theta = 2\left(-\frac{3}{5}\right)^2 - 1 = -\frac{7}{25}$

$\qquad \sin 2\theta = 2\left(\frac{4}{5}\right)\left(-\frac{3}{5}\right) = -\frac{24}{25}$

$\qquad \tan 2\theta = \dfrac{2\left(-\frac{4}{3}\right)}{1 - \left(-\frac{4}{3}\right)^2} = \dfrac{24}{7}$

4. $3^2 = (-2)^2 + x^2$

$\qquad x = \pm\sqrt{3^2 - (-2)^2} = \sqrt{5} \ (\text{Q IV})$

$\qquad \cos\theta = \frac{\sqrt{5}}{3}; \ \tan\theta = -\frac{2}{\sqrt{5}} = -\frac{2\sqrt{5}}{5}$

$\qquad \cos 2\theta = 2\left(\frac{\sqrt{5}}{3}\right)^2 - 1 = \frac{1}{9}$

$\qquad \sin 2\theta = 2\left(-\frac{2}{3}\right)\left(\frac{\sqrt{5}}{3}\right) = -\frac{4\sqrt{5}}{9}$

$\qquad \tan 2\theta = \dfrac{2\left(\frac{-2}{\sqrt{5}}\right)}{1 - \left(\frac{-2}{\sqrt{5}}\right)^2} = -4\sqrt{5}$

5. $r = \pm\sqrt{\left(-\sqrt{3}\right)^2 + (-3)^2}$

$\qquad r = 2\sqrt{3} \ (\text{Q III})$

$\qquad \sin\theta = -\frac{1}{2}; \ \cos\theta = -\frac{\sqrt{3}}{2}$

$\qquad \cos 2\theta = 2\left(-\frac{\sqrt{3}}{2}\right)^2 - 1 = \frac{1}{2}$

$\qquad \sin 2\theta = 2\left(-\frac{1}{2}\right)\left(-\frac{\sqrt{3}}{2}\right) = \frac{\sqrt{3}}{2}$

$\qquad \tan 2\theta = \dfrac{2\left(\frac{\sqrt{3}}{3}\right)}{1 - \left(\frac{\sqrt{3}}{3}\right)^2} = \dfrac{\frac{2\sqrt{3}}{3}}{\frac{2}{3}} = \sqrt{3}$

6. $13^2 = 5^2 + x^2$

$\qquad x = \sqrt{13^2 - 5^2} = -12 \ (\text{Q II})$

$\qquad \cos\theta = -\frac{12}{13}; \ \tan\theta = -\frac{5}{12}$

$\qquad \cos 2\theta = 2\left(-\frac{12}{13}\right)^2 - 1 = \frac{119}{169}$

$\qquad \sin 2\theta = 2\left(\frac{5}{13}\right)\left(-\frac{12}{13}\right) = -\frac{120}{169}$

$\qquad \tan 2\theta = \dfrac{2\left(-\frac{5}{12}\right)}{1 - \left(-\frac{5}{12}\right)^2} = -\frac{120}{119}$

7a. $\quad \cos 2\theta = 2\cos^2\theta - 1$

$\qquad 1 + \cos 2\theta = 2\cos^2\theta$

$\qquad \quad \cos^2\theta = \dfrac{1 + \cos 2\theta}{2}$

7b. $\tan^2\theta = \dfrac{\frac{1 - \cos 2\theta}{2}}{\frac{1 + \cos 2\theta}{2}} = \dfrac{1 - \cos 2\theta}{1 + \cos 2\theta}$

8a. Replace θ with $\frac{\theta}{2}$ in $\cos^2\theta = \dfrac{1 + \cos 2\theta}{2}$.

$\qquad \cos^2\frac{\theta}{2} = \dfrac{1 + \cos 2\left(\frac{\theta}{2}\right)}{2}$

$\qquad \cos\frac{\theta}{2} = \pm\sqrt{\dfrac{1 + \cos\theta}{2}}$

8b. Replace θ with $\frac{\theta}{2}$ in

$\qquad \tan^2\theta = \dfrac{1 - \cos 2\theta}{1 + \cos 2\theta}.$

$\qquad \tan^2\frac{\theta}{2} = \dfrac{1 - \cos 2\left(\frac{\theta}{2}\right)}{1 + \cos 2\left(\frac{\theta}{2}\right)}$

$\qquad \tan\frac{\theta}{2} = \pm\sqrt{\dfrac{1 - \cos\theta}{1 + \cos\theta}}$

9. $\sin\frac{\pi}{8} = \sin\frac{\frac{\pi}{4}}{2}$ (positive in Q I)

$\qquad = \sqrt{\dfrac{1 - \cos\frac{\pi}{4}}{2}} = \sqrt{\frac{1}{2}\left(1 - \frac{\sqrt{2}}{2}\right)}$

$\qquad = \sqrt{\frac{1}{2} - \frac{\sqrt{2}}{4}} = \sqrt{\dfrac{2 - \sqrt{2}}{4}} = \dfrac{\sqrt{2 - \sqrt{2}}}{2}$

10. $\tan\frac{7\pi}{8} = \tan\frac{\frac{7\pi}{4}}{2}$ (negative in Q II)

$\qquad = -\sqrt{\dfrac{1 - \cos\frac{7\pi}{4}}{1 + \cos\frac{7\pi}{4}}} = -\sqrt{\dfrac{1 - \frac{\sqrt{2}}{2}}{1 + \frac{\sqrt{2}}{2}}}$

$\qquad = -\sqrt{\dfrac{2 - \sqrt{2}}{2 + \sqrt{2}} \cdot \dfrac{2 - \sqrt{2}}{2 - \sqrt{2}}} = -\sqrt{\dfrac{6 - 4\sqrt{2}}{2}}$

$\qquad = -\sqrt{3 - 2\sqrt{2}}$

11. $\cos 112.5° = \cos\frac{225°}{2}$

$\qquad$ (negative in Q II)

$\qquad = -\sqrt{\dfrac{1 + \cos 225°}{2}} = -\sqrt{\frac{1}{2}\left(1 + \frac{-\sqrt{2}}{2}\right)}$

$\qquad = -\sqrt{\frac{1}{2} - \frac{\sqrt{2}}{4}} = -\sqrt{\dfrac{2 - \sqrt{2}}{4}}$

$\qquad = -\dfrac{\sqrt{2 - \sqrt{2}}}{2}$

Find the general solution for $\cos\frac{x}{2} = \sin x$.

Answer

$\qquad \cos\frac{x}{2} = \sin x$ — 1. Use a half-angle identity to substitute for $\cos\frac{x}{2}$ and square both sides. Recall that this step may introduce extraneous solutions.

$\qquad \left(\pm\sqrt{\dfrac{1 + \cos x}{2}}\right)^2 = \sin^2 x$

$\qquad \dfrac{1 + \cos x}{2} = \sin^2 x$

$\qquad 1 + \cos x = 2(1 - \cos^2 x)$ — 2. Use a Pythagorean identity to express the equation as a quadratic in terms of $\cos x$. Then collect the terms on one side and factor.

$\qquad 2\cos^2 x + \cos x - 1 = 0$

$\qquad (2\cos x - 1)(\cos x + 1) = 0$

$\qquad \cos x = \frac{1}{2} \ \text{ or } \ \cos x = -1$

$\qquad x = \frac{\pi}{3}, \frac{5\pi}{3}, \frac{7\pi}{3}, \frac{11\pi}{3} \ \text{ or } \ x = \pi, 3\pi$ — 3. Since $\cos\frac{x}{2}$ has period $p = \frac{2\pi}{\left|\frac{1}{2}\right|} = 4\pi$, list all solutions within $[0, 4\pi)$.

$\qquad \cos\frac{\pi}{6} = \sin\frac{\pi}{3} \qquad \cos\frac{5\pi}{6} = \sin\frac{5\pi}{3}$ — 4. Substitute each answer into the original equation to identify the extraneous solutions $\frac{7\pi}{3}$ and $\frac{11\pi}{3}$.

$\qquad \cos\frac{7\pi}{6} \neq \sin\frac{7\pi}{3} \qquad \cos\frac{11\pi}{6} \neq \sin\frac{11\pi}{3}$

$\qquad \cos\frac{\pi}{2} = \sin\pi \qquad \cos\frac{3\pi}{2} = \sin 3\pi$

$\qquad x = \frac{\pi}{3} + n4\pi, \ \frac{5\pi}{3} + n4\pi, \ (2n+1)\pi; \ n \in \mathbb{Z}$ — 5. State the general solution.

Check — 6. Use technology to verify the solution.

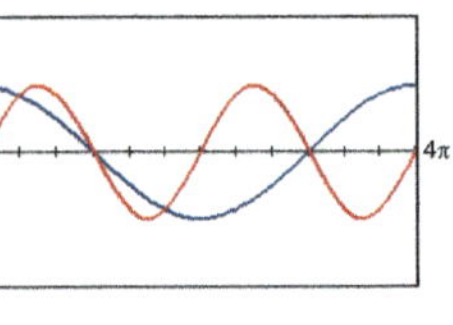

SKILL ✔ **EXERCISE 25**

Product-to-sum and sum-to-product identities are used to rewrite and evaluate certain trigonometric expressions. Each of the following product-to-sum identities can be verified by applying the sum and difference identities to simplify the right side of the equation.

Product-to-Sum Identities	
$\sin\alpha\sin\beta = \frac{1}{2}[\cos(\alpha - \beta) - \cos(\alpha + \beta)]$	$\sin\alpha\cos\beta = \frac{1}{2}[\sin(\alpha + \beta) + \sin(\alpha - \beta)]$
$\cos\alpha\cos\beta = \frac{1}{2}[\cos(\alpha - \beta) + \cos(\alpha + \beta)]$	$\cos\alpha\sin\beta = \frac{1}{2}[\sin(\alpha + \beta) - \sin(\alpha - \beta)]$

Example 7   Expressing a Product as a Sum

Rewrite $\sin 4x \cos 3x$ as a sum or difference.

Answer

$\sin 4x \cos 3x = \frac{1}{2}[\sin(4x + 3x) - \sin(4x - 3x)]$ 1. Apply the appropriate product-to-sum identity.

$= \frac{1}{2}[\sin 7x - \sin x]$ 2. Simplify.

$= \frac{1}{2}\sin 7x - \frac{1}{2}\sin x$

SKILL ✔ EXERCISE 31

Sum-to-Product Identities	
$\sin \alpha + \sin \beta = 2 \sin \frac{\alpha + \beta}{2} \cos \frac{\alpha - \beta}{2}$	$\cos \alpha + \cos \beta = 2 \cos \frac{\alpha + \beta}{2} \cos \frac{\alpha - \beta}{2}$
$\sin \alpha - \sin \beta = 2 \cos \frac{\alpha + \beta}{2} \sin \frac{\alpha - \beta}{2}$	$\cos \alpha - \cos \beta = -2 \sin \frac{\alpha + \beta}{2} \sin \frac{\alpha - \beta}{2}$

Notice how the first sum-to-product identity is verified using the appropriate product-to-sum identity to simplify the right side of the equation.

$$2 \sin \frac{\alpha + \beta}{2} \cos \frac{\alpha - \beta}{2} = 2\left\{ \frac{1}{2}\left[\sin\left(\frac{\alpha + \beta}{2} + \frac{\alpha - \beta}{2} \right) + \sin\left(\frac{\alpha + \beta}{2} - \frac{\alpha - \beta}{2} \right) \right] \right\}$$

$$= \sin\left(\frac{2\alpha}{2} \right) + \sin\left(\frac{2\beta}{2} \right)$$

$$= \sin \alpha + \sin \beta$$

Example 8   Evaluating a Sum

Find the exact value of $\cos \frac{13\pi}{12} - \cos \frac{5\pi}{12}$.

Answer

$\cos \frac{13\pi}{12} - \cos \frac{5\pi}{12} = -2 \sin \frac{\frac{13\pi}{12} + \frac{5\pi}{12}}{2} \sin \frac{\frac{13\pi}{12} - \frac{5\pi}{12}}{2}$ 1. Apply the appropriate sum-to-product identity.

$= -2 \sin \frac{3\pi}{4} \sin \frac{\pi}{3}$ 2. Simplify.

$= -2\left(\frac{\sqrt{2}}{2} \right)\left(\frac{\sqrt{3}}{2} \right) = -\frac{\sqrt{6}}{2}$

SKILL ✔ EXERCISE 35

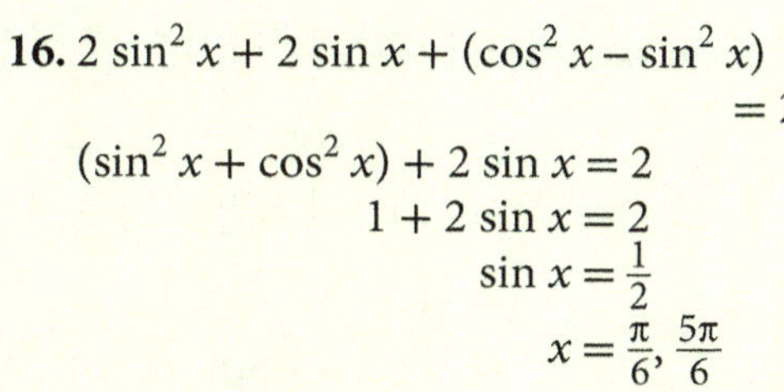

The Spirograph®, a geometric drawing toy developed in the 1960s, made generating cycloidal curves easy and fun. This toy aptly demonstrates our ability to appreciate mathematical beauty even without fully understanding the underlying mathematics.

---

12. $\cos \frac{5\pi}{12} = \cos \frac{\frac{5\pi}{6}}{2}$ (positive in Q I)

$= \pm\sqrt{\frac{1 + \cos \frac{5\pi}{6}}{2}} = \sqrt{\frac{1}{2}\left(1 + \left(-\frac{\sqrt{3}}{2} \right) \right)}$

$= \sqrt{\frac{1}{2} - \frac{\sqrt{3}}{4}} = \sqrt{\frac{2 - \sqrt{3}}{4}} = \frac{\sqrt{2 - \sqrt{3}}}{2}$

13.
$$\sin 2x = \sin x$$
$$2 \sin x \cos x = \sin x$$
$$2 \sin x \cos x - \sin x = 0$$
$$\sin x(2 \cos x - 1) = 0$$
$$\sin x = 0 \quad \text{or} \quad \cos x = \frac{1}{2}$$
$$x = 0, \pi \qquad x = \frac{\pi}{3}, \frac{5\pi}{3}$$

14.
$$\sin 2x + \cos x = 0$$
$$2 \sin x \cos x + \cos x = 0$$
$$\cos x(2 \sin x + 1) = 0$$
$$\cos x = 0 \quad \text{or} \quad \sin x = -\frac{1}{2}$$
$$x = \frac{\pi}{2}, \frac{3\pi}{2} \qquad x = \frac{7\pi}{6}, \frac{11\pi}{6}$$

15.
$$\cos 2x + \cos x + 1 = 0$$
$$(2 \cos^2 x - 1) + \cos x + 1 = 0$$
$$2 \cos^2 x + \cos x = 0$$
$$\cos x(2 \cos x + 1) = 0$$
$$\cos x = 0 \quad \text{or} \quad \cos x = -\frac{1}{2}$$
$$x = \frac{\pi}{2}, \frac{3\pi}{2} \qquad x = \frac{2\pi}{3}, \frac{4\pi}{3}$$

16. $2 \sin^2 x + 2 \sin x + (\cos^2 x - \sin^2 x) = 2$

$(\sin^2 x + \cos^2 x) + 2 \sin x = 2$

$1 + 2 \sin x = 2$

$\sin x = \frac{1}{2}$

$x = \frac{\pi}{6}, \frac{5\pi}{6}$

❯ B. Exercises

17. $\cos 3\theta = \cos(2\theta + \theta)$
$$= \cos 2\theta \cos \theta - \sin 2\theta \sin \theta$$
$$= (2 \cos^2 \theta - 1) \cos \theta - (2 \sin \theta \cos \theta) \sin \theta$$
$$= 2 \cos^3 \theta - \cos \theta - 2 \cos \theta \sin^2 \theta$$
$$= 2 \cos^3 \theta - \cos \theta - 2 \cos \theta(1 - \cos^2 \theta)$$
$$= 2 \cos^3 \theta - \cos \theta - 2 \cos \theta + 2 \cos^3 \theta$$
$$= 4 \cos^3 \theta - 3 \cos \theta$$

18. $\tan 3\theta = \tan(2\theta + \theta)$
$$= \frac{\tan 2\theta + \tan \theta}{1 - \tan 2\theta \tan \theta}$$
$$= \frac{\frac{2 \tan \theta}{1 - \tan^2 \theta} + \tan \theta}{1 - \left(\frac{2 \tan \theta}{1 - \tan^2 \theta} \right)\frac{\tan \theta}{1}} \cdot \frac{\frac{1 - \tan^2 \theta}{1}}{\frac{1 - \tan^2 \theta}{1}}$$
$$= \frac{2 \tan \theta + \tan \theta - \tan^3 \theta}{1 - \tan^2 \theta - 2 \tan^2 \theta}$$
$$= \frac{3 \tan \theta - \tan^3 \theta}{1 - 3 \tan^2 \theta}$$
$$= \frac{\tan^3 \theta - 3 \tan \theta}{3 \tan^2 \theta - 1}$$

19. $\cos 4\theta = \cos(2(2\theta))$
$$= 2(\cos 2\theta)^2 - 1$$
$$= 2(2 \cos^2 \theta - 1)^2 - 1$$
$$= 2(4 \cos^4 \theta - 4 \cos^2 \theta + 1) - 1$$
$$= 8 \cos^4 \theta - 8 \cos^2 \theta + 1$$

20. $\sin 2(2\theta) = 2 \sin 2\theta \cos 2\theta$
$$= 2(2 \sin \theta \cos \theta)(\cos^2 \theta - \sin^2 \theta)$$
$$= 4 \cos^3 \theta \sin \theta - 4 \sin^3 \theta \cos \theta$$

21. $\sin^3 \theta = \sin \theta \sin^2 \theta$
$$= \sin \theta\left(\frac{1 - \cos 2\theta}{2} \right)$$
$$= \frac{1}{2}(\sin \theta - \sin \theta \cos 2\theta)$$

22. $\cos^4 \theta = (\cos^2 \theta)^2 = \left(\frac{1 + \cos 2\theta}{2} \right)^2$
$$= \frac{1}{4}(1 + 2 \cos 2\theta + \cos^2 2\theta)$$
$$= \frac{1}{4}\left(1 + 2 \cos 2\theta + \frac{1}{2}(1 + \cos 4\theta) \right)$$
$$= \frac{1}{4}\left(\frac{3}{2} + 2 \cos 2\theta + \frac{1}{2}\cos 4\theta \right)$$
$$= \frac{1}{8}(3 + 4 \cos 2\theta + \cos 4\theta)$$

23. $\tan^4 \theta = \left(\frac{1 - \cos 2\theta}{1 + \cos 2\theta} \right)^2$
$$= \frac{1 - 2 \cos 2\theta + \cos^2 2\theta}{1 + 2 \cos 2\theta + \cos^2 2\theta}$$
$$= \frac{1 - 2 \cos 2\theta + \left(\frac{1 + \cos 2(2\theta)}{2} \right)}{1 + 2 \cos 2\theta + \left(\frac{1 + \cos 2(2\theta)}{2} \right)} \cdot \frac{2}{2}$$
$$= \frac{3 - 4 \cos 2\theta + \cos 4\theta}{3 + 4 \cos 2\theta + \cos 4\theta}$$

24. $\sin^2\theta\cos^2\theta = \left(\dfrac{1-\cos 2\theta}{2}\right)\left(\dfrac{1+\cos 2\theta}{2}\right)$

$\qquad = \dfrac{1}{4}(1-\cos^2 2\theta)$

$\qquad = \dfrac{1}{4}\left(1 - \dfrac{1+\cos 4\theta}{2}\right)$

$\qquad = \dfrac{1}{4}\left(\dfrac{1-\cos 4\theta}{2}\right)$

$\qquad = \dfrac{1}{8}(1-\cos 4\theta)$

25. $\qquad 2\cos\dfrac{x}{2} = \cos x + 1$

$\left(2\left(\pm\sqrt{\dfrac{1+\cos x}{2}}\right)\right)^2 = (\cos x + 1)^2$

$4\left(\dfrac{1+\cos x}{2}\right) = \cos^2 x + 2\cos x + 1$

$2 + 2\cos x = \cos^2 x + 2\cos x + 1$

$\qquad\qquad \cos^2 x = 1$

$\cos x = 1 \quad$ or $\quad \cos x = -1$

solutions within $[0, 4\pi)$:

$\qquad x = 0, 2\pi \qquad\quad x = \pi, 3\pi$

but $2\cos\dfrac{2\pi}{2} - 1 \neq \cos 2\pi$

$\therefore x = n4\pi, (2n+1)\pi; n \in \mathbb{Z}$

26. $\qquad \left(\pm\sqrt{\dfrac{1-\cos x}{2}}\right)^2 = (\cos x)^2$

$\qquad\qquad \dfrac{1-\cos x}{2} = \cos^2 x$

$2\cos^2 x + \cos x - 1 = 0$

$(2\cos x - 1)(\cos x + 1) = 0$

$\cos x = \dfrac{1}{2} \qquad$ or $\qquad \cos x = -1$

solutions within $[0, 4\pi)$:

$x = \dfrac{\pi}{3}, \dfrac{5\pi}{3}, \dfrac{7\pi}{3}, \dfrac{11\pi}{3} \qquad x = \pi, 3\pi$

but $\sin\dfrac{7\pi}{6} \neq \cos\dfrac{7\pi}{3}$, $\sin\dfrac{11\pi}{6} \neq \cos\dfrac{11\pi}{3}$,

and $\sin\dfrac{\pi}{2} \neq \cos\pi$

$\therefore x = \dfrac{\pi}{3} + n4\pi, \dfrac{5\pi}{3} + n4\pi, 3\pi + n4\pi;$

$n \in \mathbb{Z}$

27. $\left(\pm\sqrt{\dfrac{1-\cos x}{2}}\right)^2 = \sin^2 x$

$\qquad\qquad \dfrac{1-\cos x}{2} = 1 - \cos^2 x$

$\qquad\quad 1 - \cos x = 2 - 2\cos^2 x$

$\quad 2\cos^2 x - \cos x - 1 = 0$

$(\cos x - 1)(2\cos x + 1) = 0$

$\cos x = 1 \ $ or $\ \cos x = -\dfrac{1}{2}$

solutions within $[0, 2\pi)$:

$\quad x = 0 \qquad\quad x = \dfrac{2\pi}{3}, \dfrac{4\pi}{3}$

no extraneous solutions,

$\therefore x = n2\pi, \dfrac{2\pi}{3} + n2\pi, \dfrac{4\pi}{3} + n2\pi; n \in \mathbb{Z}$

28. $\quad \pm\sqrt{\dfrac{1-\cos x}{1+\cos x}} = \left(\pm\sqrt{\dfrac{1-\cos x}{1+\cos x}}\right)^2$

$\qquad \pm\sqrt{\dfrac{1-\cos x}{1+\cos x}} = \dfrac{1-\cos x}{1+\cos x}$

$\left(\pm\sqrt{\dfrac{1-\cos x}{1+\cos x}}\right)^2 = \left(\dfrac{1-\cos x}{1+\cos x}\right)^2$

$\qquad\quad \dfrac{1-\cos x}{1+\cos x} = \dfrac{(1-\cos x)(1-\cos x)}{(1+\cos x)(1+\cos x)}$

$(1-\cos^2 x)(1-\cos x)$

$\qquad = (1-\cos^2 x)(1+\cos x)$

$(\sin^2 x)(1-\cos x)$

$\qquad - (\sin^2 x)(1+\cos x) = 0$

Find the general solution for $\sin 3x - \sin 5x = 0$.

Answer

$\qquad\qquad \sin 3x - \sin 5x = 0$

$2\cos\dfrac{3x+5x}{2}\sin\dfrac{3x-5x}{2} = 0$

$\qquad\quad 2\cos 4x \sin(-x) = 0$

1. Apply the appropriate sum-to-product identity.

$\cos 4x = 0 \qquad$ or $\qquad \sin(-x) = 0$

$\qquad\qquad\qquad\qquad\qquad\quad \sin x = 0$

$\quad 4x = \dfrac{\pi}{2} + n\pi \qquad\qquad\quad x = n\pi$

$\quad x = \dfrac{\pi}{8} + \dfrac{n\pi}{4}$

2. Solve using the Zero Factor Property.

Check

3. Identify the zeros of the graph of $y = \sin 3x - \sin 5x$.

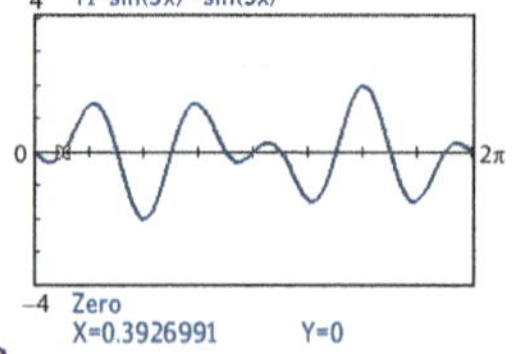

SKILL ✔ **EXERCISE 39**

❯ **A. Exercises**

1. Derive each double-angle identity.

    a. $\sin 2\theta$ **b.** $\tan 2\theta$

2. Use $\cos 2\theta = \cos^2\theta - \sin^2\theta$ to derive each identity.

    a. $\cos 2\theta = 2\cos^2\theta - 1$ **b.** $\cos 2\theta = 1 - 2\sin^2\theta$

Find the exact values of $\cos 2\theta$, $\sin 2\theta$, and $\tan 2\theta$ using the given information.

3. $\cos\theta = -\dfrac{3}{5}$ and $\dfrac{\pi}{2} < \theta < \pi$ $-\dfrac{7}{25}, -\dfrac{24}{25}, \dfrac{24}{7}$

4. $\sin\theta = -\dfrac{2}{3}$ and $270° < \theta < 360°$ $\dfrac{1}{9}, -\dfrac{4\sqrt5}{9}, -4\sqrt5$

5. $\tan\theta = \dfrac{\sqrt3}{3}$ and $180° < \theta < 270°$ $\dfrac{1}{2}, \dfrac{\sqrt3}{2}, \sqrt3$

6. $\sin\theta = \dfrac{5}{13}$ and $\dfrac{\pi}{2} < \theta < \pi$ $\dfrac{119}{169}, -\dfrac{120}{169}, -\dfrac{120}{119}$

7. Derive each power-reducing identity.

    a. $\cos^2\theta$ **b.** $\tan^2\theta$

8. Derive each half-angle identity.

    a. $\cos\dfrac{\theta}{2} = \pm\sqrt{\dfrac{1+\cos\theta}{2}}$ **b.** $\tan\dfrac{\theta}{2} = \pm\sqrt{\dfrac{1-\cos\theta}{1+\cos\theta}}$

Use a half-angle identity to find the exact value of each expression.

9. $\sin\dfrac{\pi}{8}$ $\dfrac{\sqrt{2-\sqrt2}}{2}$ **10.** $\tan\dfrac{7\pi}{8}$ $-\sqrt{3-2\sqrt2}$

11. $\cos 112.5°$ $-\dfrac{\sqrt{2-\sqrt2}}{2}$ **12.** $\cos\dfrac{5\pi}{12}$ $\dfrac{\sqrt{2-\sqrt3}}{2}$

Find the primary solutions of each equation.

13. $\sin 2x = \sin x$ $x = 0, \dfrac{\pi}{3}, \pi, \dfrac{5\pi}{3}$

14. $\sin 2x + \cos x = 0$ $x = \dfrac{\pi}{2}, \dfrac{7\pi}{6}, \dfrac{3\pi}{2}, \dfrac{11\pi}{6}$

15. $\cos 2x + \cos x + 1 = 0$ $x = \dfrac{\pi}{2}, \dfrac{3\pi}{2}, \dfrac{2\pi}{3}, \dfrac{4\pi}{3}$

16. $2\sin^2 x + 2\sin x + \cos 2x = 2$ $x = \dfrac{\pi}{6}, \dfrac{5\pi}{6}$

❯ **B. Exercises**

Derive each multiple-angle identity.

17. Express $\cos 3\theta$ in terms of $\cos\theta$.

18. Express $\tan 3\theta$ in terms of $\tan\theta$.

19. Express $\cos 4\theta$ in terms of $\cos\theta$.

20. Verify that $\sin 4\theta = 4\cos^3\theta\sin\theta - 4\sin^3\theta\cos\theta$.

Use power-reducing identities to rewrite each expression in terms with no power greater than 1.

21. $\sin^3\theta$ **22.** $\cos^4\theta$

23. $\tan^4\theta$ **24.** $\sin^2\theta\cos^2\theta$

Find the general solutions of each equation. Verify your answers graphically.

25. $2\cos\dfrac{x}{2} - 1 = \cos x$ **26.** $\sin\dfrac{x}{2} = \cos x$

27. $\sin^2\dfrac{x}{2} - \sin^2 x = 0$ **28.** $\tan\dfrac{x}{2} = \tan^2\dfrac{x}{2}$

29. Verify each product-to-sum identity.

    a. $\sin\alpha\cos\beta = \dfrac{1}{2}[\sin(\alpha+\beta) + \sin(\alpha-\beta)]$

    b. $\sin\alpha\sin\beta = \dfrac{1}{2}[\cos(\alpha-\beta) - \cos(\alpha+\beta)]$

30. Verify each sum-to-product identity.

    a. $\cos\alpha - \cos\beta = -2\sin\dfrac{\alpha+\beta}{2}\sin\dfrac{\alpha-\beta}{2}$

    b. $\sin\alpha - \sin\beta = 2\cos\dfrac{\alpha+\beta}{2}\sin\dfrac{\alpha-\beta}{2}$

$\sin^2 x((1-\cos x) - (1+\cos x)) = 0$

$\sin^2 x(-2\cos x) = 0$

$\sin^2 x = 0 \ $ or $\ -2\cos x = 0$

solutions within $[0, 2\pi)$:

$\quad x = 0, \pi \qquad\quad x = \dfrac{\pi}{2}, \dfrac{3\pi}{2}$

but $\tan\dfrac{\pi}{2} \neq \tan^2\dfrac{\pi}{2}$ and $\tan\dfrac{3\pi}{4} \neq \tan^2\dfrac{3\pi}{4}$

$\therefore x = n2\pi, \dfrac{\pi}{2} + n2\pi; n \in \mathbb{Z}$

29a. $\dfrac{1}{2}(\sin(\alpha+\beta) + \sin(\alpha-\beta))$

$\qquad = \dfrac{1}{2}(\sin\alpha\cos\beta + \sin\beta\cos\alpha$

$\qquad\qquad + \sin\alpha\cos\beta - \sin\beta\cos\alpha)$

$\qquad = \dfrac{1}{2}(2\sin\alpha\sin\beta) = \sin\alpha\cos\beta$

29b. $\dfrac{1}{2}(\cos(\alpha-\beta) - \cos(\alpha+\beta))$

$\qquad = \dfrac{1}{2}(\cos\alpha\cos\beta + \sin\alpha\sin\beta$

$\qquad\qquad - (\cos\alpha\cos\beta - \sin\alpha\sin\beta))$

$\qquad = \dfrac{1}{2}(2\sin\alpha\sin\beta) = \sin\alpha\sin\beta$

30a. $-2\sin\dfrac{\alpha+\beta}{2}\sin\dfrac{\alpha-\beta}{2}$

$\qquad = -2\left\{\dfrac{1}{2}\left[\cos\left(\dfrac{\alpha+\beta}{2} - \dfrac{(\alpha-\beta)}{2}\right)\right.\right.$

$\qquad\qquad\qquad \left.\left. - \cos\left(\dfrac{\alpha+\beta}{2} + \dfrac{\alpha-\beta}{2}\right)\right]\right\}$

$\qquad = -2\left[\dfrac{1}{2}\left(\cos\dfrac{2\beta}{2} - \cos\dfrac{2\alpha}{2}\right)\right]$

$\qquad = \cos\alpha - \cos\beta$

Rewrite each expression as a sum or difference.

31. $\cos 3x \cos 2x$

32. $\cos 9x \sin (-5x)$

33. $\sin (a - b) \cos (a + b)$

Find the exact value of the expression.

34. $\cos 75° - \cos 15°$ $-\dfrac{\sqrt{2}}{2}$

35. $\sin 105° - \sin 15°$ $\dfrac{\sqrt{2}}{2}$

36. $\sin \dfrac{3\pi}{4} + \sin \dfrac{\pi}{4}$ $\sqrt{2}$

37. A musical instrument can be tuned by listening to the combined sound of a tuning fork and the instrument and eliminating any "beat" caused by slightly different frequencies. If the sound waves produced by the instrument and the tuning fork are modeled by $y_1 = A_0 \cos 2\pi f_1 t$ and $y_2 = A_0 \cos 2\pi f_2 t$, respectively, express the resultant wave $y = y_1 + y_2$ as a product.

Find the general solution of each equation. Verify your answers graphically.

38. $\cos 3\theta - \cos \theta = 0$ $\theta = n\dfrac{\pi}{2}$

39. $\sin \theta - \sin 3\theta = 0$ $\theta = \dfrac{\pi}{4} + n\dfrac{\pi}{2}, \, n\pi$

40. $\dfrac{\sin \theta}{\cos 3\theta - \cos \theta} = 1$ $\theta = \dfrac{7\pi}{12} + n\pi, \dfrac{11\pi}{12} + n\pi$

37. $y = 2A_0 \left[\cos 2\pi \left(\dfrac{f_1 + f_2}{2} \right) t \right] \left[\cos 2\pi \left(\dfrac{f_1 - f_2}{2} \right) t \right]$

Verify each identity.

41. $2 \sin^2 x - 1 = \sin^4 x - \cos^4 x$

42. $\csc 2x = \dfrac{1 + \tan^2 x}{2 \tan x}$

43. $\dfrac{\sin 3x}{\sin x} - \dfrac{\cos 3x}{\cos x} = 2$

44. $\cos^3 x - \sin^3 x = (\cos x - \sin x)\left(1 + \tfrac{1}{2} \sin 2x \right)$

> **C. Exercises**

45. Use half-angle identities to find the exact value of each expression.
   a. $\cos \dfrac{\pi}{12}$ $\dfrac{\sqrt{2 + \sqrt{3}}}{2}$ **b.** $\cos \dfrac{\pi}{24}$ $\dfrac{\sqrt{2 + \sqrt{2 + \sqrt{3}}}}{2}$

46. Express $\sin 5\theta$ in terms of $\sin \theta$.

47. Given $\tan \dfrac{\theta}{2} = \pm\sqrt{\dfrac{1 - \cos \theta}{1 + \cos \theta}}$, derive $\tan \dfrac{\theta}{2} = \dfrac{1 - \cos \theta}{\sin \theta}$.

   a. Show that $\tan \dfrac{\theta}{2} = \pm\left|\dfrac{1 - \cos \theta}{\sin \theta}\right|$.

   b. What is the range of $f(x) = 1 - \cos \theta$? $[0, 2]$

   c. Graph $\sin \theta$ and $\tan \dfrac{\theta}{2}$ in the same window and compare the signs of the functions.

   d. Use the answers to parts b and c to explain why $\pm\left|\dfrac{1 - \cos \theta}{\sin \theta}\right| = \dfrac{1 - \cos \theta}{\sin \theta}$.

48. Solve $\cos 3x + \cos x = \cos 2x$. $x = \dfrac{\pi}{4}, \dfrac{3\pi}{4}, \dfrac{5\pi}{4}, \dfrac{7\pi}{4}, \dfrac{\pi}{3}, \dfrac{5\pi}{3}$

47c. The signs are always the same.

47d. The numerator is never negative and the sign of $\tan \dfrac{\theta}{2}$ always matches the sign of the denominator. Therefore we can omit the $\pm$ sign and the absolute value.

---

49. State the slope and the x and y-intercepts of $3x - 2y = -12$.   [1.2] $\dfrac{3}{2}; (-4, 0); (0, 6)$

50. Find the standard form equation of the line perpendicular to $2x + 5y = 9$ going through $(4, 6)$.   [1.2] $5x - 2y = 8$

51. Given $\triangle ABC$ where $m\angle C = 90°$, $a = 8$, and $b = 15$, find c.   [4.2]   17

52. Find the area of isosceles $\triangle XYZ$ if $m\angle Z = 70°$.   [4.2] ≈ 15.75 in.^{2}

53. State the period and frequency of $f(x) = 3 \cos (-\pi x) + 5$.   [4.4] $p = 2; f = \dfrac{1}{2}$

54. Write the equation of a cosine function with an amplitude of 3, a period of π, and a relative maximum at $(0, 4)$.   [4.4] $f(x) = 3 \cos (2x) + 1$

55. If $P(x) \div (x - 7)$ has a remainder of 0, what can we conclude?   [2.3]   E

   A. $P(7) = 0$ **D.** both B and C

   B. $P(-7) = 0$ **E.** both A and C

   C. $x - 7$ is a factor of $P(x)$.

56. Find $\tan \theta$ if $\sin \theta = -\dfrac{3}{5}$ and $\cos \theta > 0$.   [4.3]   D

   A. $\dfrac{4}{5}$ **C.** $\dfrac{3}{4}$ **E.** $-\dfrac{4}{3}$

   B. $\dfrac{4}{3}$ **D.** $-\dfrac{3}{4}$

57. Find $\sin \theta$ if $\sec \theta = -\dfrac{8}{5}$ and $\tan \theta < 0$.   [5.1]   B

   A. $\dfrac{\sqrt{15}}{4}$ **C.** $\dfrac{3}{8}$ **E.** $\dfrac{39}{64}$

   B. $\dfrac{\sqrt{39}}{8}$ **D.** $\dfrac{\sqrt{255}}{16}$

58. Find the exact value of $\sin 105°$.   [5.4]   D

   A. $\dfrac{\sqrt{3}}{2}$ **C.** $\dfrac{\sqrt{6} - \sqrt{2}}{4}$ **E.** $\dfrac{6 - \sqrt{2}}{4}$

   B. $\dfrac{\sqrt{3}}{4}$ **D.** $\dfrac{\sqrt{6} + \sqrt{2}}{4}$

---

35. $2 \cos \left(\dfrac{105° + 15°}{2} \right) \sin \left(\dfrac{105° - 15°}{2} \right)$

$= 2 \cos 60° \sin 45°$

$= 2\left(\dfrac{1}{2}\right)\left(\dfrac{\sqrt{2}}{2}\right) = \dfrac{\sqrt{2}}{2}$

36. $2 \sin \left(\dfrac{\frac{3\pi}{4} + \frac{\pi}{4}}{2} \right) \cos \left(\dfrac{\frac{3\pi}{4} - \frac{\pi}{4}}{2} \right)$

$= 2 \sin \dfrac{\pi}{2} \cos \dfrac{\pi}{4}$

$= 2(1)\left(\dfrac{\sqrt{2}}{2}\right) = \sqrt{2}$

37. $y = A_0 \cos 2\pi f_1 t + A_0 \cos 2\pi f_2 t$

$= A_0(\cos 2\pi f_1 t + \cos 2\pi f_2 t)$

$= A_0 2\left[\cos \left(\dfrac{2\pi f_1 t + 2\pi f_2 t}{2} \right)\right]$

$\qquad\qquad \cdot \left[\cos \left(\dfrac{2\pi f_1 t - 2\pi f_2 t}{2} \right)\right]$

$= 2A_0$

$\qquad \cdot \left[\cos 2\pi \left(\dfrac{f_1 + f_2}{2}\right)t\right]\left[\cos 2\pi \left(\dfrac{f_1 - f_2}{2}\right)t\right]$

38. $-2 \sin \left(\dfrac{3\theta - \theta}{2} \right) \sin \left(\dfrac{3\theta + \theta}{2} \right) = 0$

$\qquad\qquad -2 \sin 2\theta \sin \theta = 0$

$\sin 2\theta = 0 \quad$ or $\quad \sin \theta = 0$

$\qquad 2\theta = n\pi \qquad\qquad\quad \theta = n\pi$

$\qquad \theta = \dfrac{n\pi}{2}$

39. $2 \cos \left(\dfrac{\theta + 3\theta}{2} \right) \sin \left(\dfrac{\theta - 3\theta}{2} \right) = 0$

$\qquad\qquad 2 \cos 2\theta \sin (-\theta) = 0$

$\cos 2\theta = 0 \quad$ or $\quad \sin (-\theta) = 0$

$\qquad 2\theta = \dfrac{\pi}{2} + n\pi \qquad\quad \sin \theta = 0$

$\qquad \theta = \dfrac{\pi}{4} + \dfrac{n\pi}{2} \qquad\qquad \theta = n\pi$

40. $\sin \theta = \cos 3\theta - \cos \theta$

$\sin \theta = -2 \sin \left(\dfrac{3\theta + \theta}{2} \right) \sin \left(\dfrac{3\theta - \theta}{2} \right)$

$\sin \theta = -2 \sin 2\theta \sin \theta$

$2 \sin \theta \sin 2\theta + \sin \theta = 0$

$\sin \theta(2 \sin 2\theta + 1) = 0$

$\sin \theta = 0 \quad$ or $\quad \sin 2\theta = -\dfrac{1}{2}$

$\qquad \theta = n\pi \qquad\qquad 2\theta = \dfrac{7\pi}{6} + n2\pi,$

$(n\pi$ is extraneous.) $\qquad\qquad \dfrac{11\pi}{6} + n2\pi$

$\qquad\qquad\qquad\qquad\quad \theta = \dfrac{7\pi}{12} + n\pi,$

$\qquad\qquad\qquad\qquad\qquad\quad \dfrac{11\pi}{12} + n\pi$

41. $\sin^4 x - \cos^4 x$

$= (\sin^2 x - \cos^2 x)(\sin^2 x + \cos^2 x)$

$= -(\cos^2 x - \sin^2 x)(1)$

$= -(1 - \sin^2 x - \sin^2 x)$

$= -(1 - 2 \sin^2 x)$

$= 2 \sin^2 x - 1$

42. $\dfrac{1 + \tan^2 x}{2 \tan x} = \dfrac{1 + \dfrac{\sin^2 x}{\cos^2 x}}{2\left(\dfrac{\sin x}{\cos x}\right)} \cdot \dfrac{\dfrac{\cos^2 x}{1}}{\dfrac{\cos^2 x}{1}}$

$= \dfrac{\cos^2 x + \sin^2 x}{2 \cos x \sin x} = \dfrac{1}{\sin 2x} = \csc 2x$

30b. $2 \cos \dfrac{\alpha + \beta}{2} \sin \dfrac{\alpha - \beta}{2}$

$= 2 \sin \dfrac{\alpha - \beta}{2} \cos \dfrac{\alpha + \beta}{2}$

$= 2\left\{\dfrac{1}{2}\left[\sin \left(\dfrac{\alpha - \beta}{2} + \dfrac{\alpha + \beta}{2} \right)\right.\right.$

$\qquad\qquad \left.\left. + \sin \left(\dfrac{\alpha - \beta}{2} - \dfrac{(\alpha + \beta)}{2} \right)\right]\right\}$

$= \sin \dfrac{2\alpha}{2} + \sin \left(-\dfrac{2\beta}{2}\right)$

$= \sin \alpha - \sin \beta$

31. $\dfrac{1}{2}(\cos (3x - 2x) + \cos (3x + 2x))$

$= \dfrac{1}{2}(\cos x + \cos 5x)$

$= \dfrac{1}{2} \cos x + \dfrac{1}{2} \cos 5x$

32. $\dfrac{1}{2}[\sin (9x + (-5x))$

$\qquad\qquad\qquad - \sin (9x - (-5x))]$

$= \dfrac{1}{2}(\sin 4x - \sin 14x)$

$= \dfrac{1}{2} \sin 4x - \dfrac{1}{2} \sin 14x$

33. $\dfrac{1}{2}[\sin ((a - b) + (a + b))$

$\qquad\qquad\qquad + \sin ((a - b) - (a + b))]$

$= \dfrac{1}{2}(\sin 2a + \sin (-2b))$

$= \dfrac{1}{2}(\sin 2a - \sin 2b)$

$= \dfrac{1}{2} \sin 2a - \dfrac{1}{2} \sin 2b$

34. $-2 \sin \left(\dfrac{75° + 15°}{2} \right) \sin \left(\dfrac{75° - 15°}{2} \right)$

$= -2 \sin 45° \sin 30°$

$= -2\left(\dfrac{\sqrt{2}}{2}\right)\left(\dfrac{1}{2}\right) = -\dfrac{\sqrt{2}}{2}$

continued in Answers and Solutions Overflow

Historical Connection

Objectives

1. To identify key contributors to the development of harmonic analysis
2. To identify modern applications of harmonic analysis
3. To evaluate statements to determine worldview bias

Answers

1. the mathematical representation of a vibrating string
2. Joseph Fourier
3. James Clerk Maxwell
4. the existence of radio waves
5. signal processing, computerized music, radio astronomy, or magnetic resonance imaging
6. Viewing mathematics as an exclusively human creation fails to acknowledge a creator and leads to no logical explanation for why mathematical laws are able to accurately model natural phenomena.

HISTORICAL CONNECTION

HARMONIC ANALYSIS

A host of eminent eighteenth-century mathematicians tried to describe the motion of a vibrating string and failed. In 1753 Daniel Bernoulli proposed a solution to the vibrating string problem consisting of an infinite trigonometric series. Leonhard Euler, the most respected mathematics authority of the time, declared him wrong, but in 1807 Bernoulli's solution was definitively proved by Joseph Fourier.

Fourier, who had considered becoming a Benedictine monk in his younger years, spoke of math revealing "an unchangeable order which presides over all natural causes." He was convinced that all complex natural processes are subject to a few simple laws which may be discerned through careful observation and experimentation. Motivated particularly by the success of Isaac Newton's universal laws of motion and gravity, Fourier sought universal laws for heat. His explorations led to the development of what is today called the *Fourier series*. He demonstrated that any function could be described by an infinite trigonometric series—even functions that are discontinuous and not "well behaved." The study of functions that can be represented by the Fourier series is called harmonic analysis.

Fourier's proposal, that simple sine functions could be combined to describe functions of wildly varying properties, seemed so counterintuitive that it took over a quarter of a century to convince the skeptics. During this process, even

the definition of a mathematical function was revised and functions with no possible graphical representation were discovered.

By the middle of the nineteenth century, scientists such as Michael Faraday were beginning to explore the wavelike qualities of electromagnetism. In 1873 James Clerk Maxwell combined the well-known experimentally derived results of Faraday's law and Ampere's law into a set of eight complex mathematical equations that explained how electric and magnetic fields work. He determined that electromagnetic waves travel at the same speed as light—and therefore these waves and visible light were actually two cases of the same phenomenon.

In 1886, seven years after Maxwell's death, Heinrich Hertz first proved the existence of the radio waves predicted by Maxwell. The Italian inventor Guglielmo Marconi then created a commercially successful wireless telegraph. Wireless telegraph operators began to be deployed aboard large cruise liners in the early 1900s. The ship that rescued survivors of the *Titanic* disaster in 1912 was summoned by onboard wireless operators employed by the Marconi Company.

Though some would trace the origins of harmonic analysis back to Pythagoras's musical theories or Ptolemy's planetary cycles and epicycles, its real power was not realized until Fourier discovered the efficacy of an infinite trigonometric series to describe natural phenomena. In her book *The Evolution of Applied Harmonic Analysis: Models of the Real World*, Elena Prestini provides examples of current applications of harmonic analysis in signal processing, computerized music, radio astronomy, and magnetic resonance imaging (MRI). In the preface to this book, Ronald Bracewell states, "Why exact mathematics, a product of the human mind, should have proved so useful in science and technology, remains a mystery."

COMPREHENSION CHECK

1. Which problem baffled many prominent eighteenth-century mathematicians but was solved by Daniel Bernoulli?
2. Who eventually proved Bernoulli's solution to be correct?
3. Who combined previous experimental results into eight equations describing electrical and magnetic fields?
4. What prediction of electromagnetic theory did Heinrich Hertz prove in 1886?
5. Name two current applications of harmonic analysis.
6. **Discuss:** Explain the worldview indicated by the quote from Ronald Bracewell.

PRESENTATION

The French mathematician and physicist Joseph Fourier marveled at the language of mathematics. Fourier stated, "There cannot be a language more universal and more simple, more free from errors and obscurities . . . more worthy to express the invariable relations of all natural things [than mathematics]. [It interprets] all phenomena by the same language, as if to attest the unity and simplicity of the plan of the universe, and to make still more evident that unchangeable order which presides over all natural causes."

The mathematical representation of a vibrating string as a trigonometric series, the application of harmonic analysis in music theory, and mathematical expressions of electromagnetic fields and waves provide a treasure of studies that reveal the presence of mathematics in the world around us. This mathematical presence is a direct result of God's complex and orderly creation.

The Internet keyword search *Fourier series animations* will provide more details and interesting animations.

Oliver Heaviside's 1885 revision of the eight harmonic equations for electric and magnetic fields resulted in the four equations used today.

A minister who visited James Clerk Maxwell in his last days was astonished at his cognitive abilities and remarked regarding his testimony, "his illness drew out the whole heart and soul and spirit of the man: his firm and undoubting faith in the Incarnation and all its results; in the full sufficiency of the Atonement; in the work of the Holy Spirit. He had gauged and fathomed all the schemes and systems of philosophy, and had found them utterly empty and unsatisfying—'unworkable' was his own word about them—and he turned with simple faith to the Gospel of the Saviour."

5.6 Law of Sines

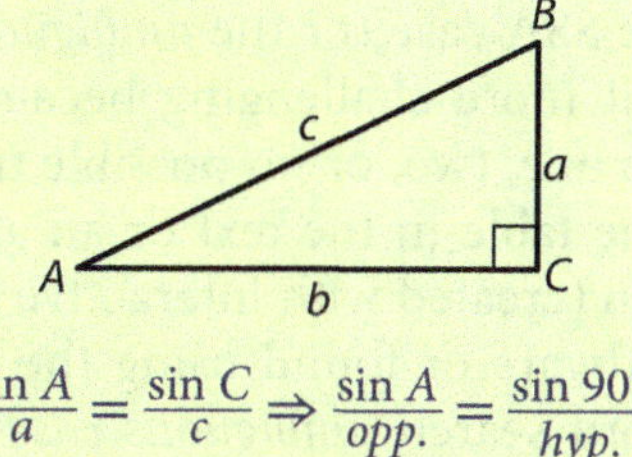

Pilots study trigonometry as an aid to navigation.

In Section 4.2 we solved right triangles. In the last two sections of this chapter, you will learn how to solve *oblique* triangles, those that do not contain a right angle. The *Law of Sines* is used to solve the unique triangle determined by the measures of two angles and the included side (ASA) or the measure of two angles and a side opposite one of the angles (AAS). It is also used to find any possible solutions when you know the measures of two sides and an angle opposite one of the known sides (SSA).

LAW OF SINES

For $\triangle ABC$ with sides of length a, b, and c opposite angles A, B, and C, respectively,
$$\frac{a}{\sin A} = \frac{b}{\sin B} = \frac{c}{\sin C}.$$

To derive this identity, consider the acute and obtuse cases of $\triangle ABC$ with its altitude from B. In both triangles, $\sin A = \frac{h}{c}$ and $h = c \sin A$. In the acute triangle $\sin C = \frac{h}{a}$, and in the obtuse triangle $\sin (\pi - C) = \frac{h}{a}$. But $\sin (\pi - \theta) = \sin \theta$, so $\sin C = \frac{h}{a}$ and $h = a \sin C$ in both cases.

Therefore $c \sin A = a \sin C$, and $\frac{c}{\sin C} = \frac{a}{\sin A}$.

Similar reasoning with altitudes drawn from A produces $\frac{c}{\sin C} = \frac{b}{\sin B}$.

By the Transitive Property, $\frac{a}{\sin A} = \frac{b}{\sin B} = \frac{c}{\sin C}$.

Because the reciprocals of equal ratios are equal, the Law of Sines can also be written as
$$\frac{\sin A}{a} = \frac{\sin B}{b} = \frac{\sin C}{c}.$$

When two angles of a triangle are known, the third angle can be quickly determined. A known side length and the Law of Sines can then be used to find the other two side lengths.

Example 1   Solving the AAS Case

Solve $\triangle ABC$ if $A = 34°$, $B = 76°$, and $a = 9$ cm. Round lengths to the nearest tenth.

Answer

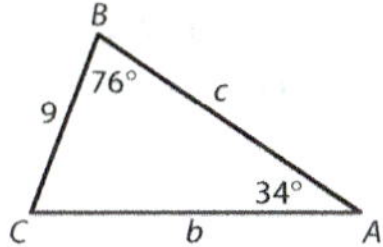

1. Make a sketch of the triangle.

$C = 180° - A - B$
$= 180 - 34 - 76 = 70°$

2. Determine C.

CONTINUED ➡

After completing this section, you will be able to

- prove the Law of Sines.
- solve triangles using the Law of Sines.
- identify the number of possible solutions for a triangle.

Objectives

1. To prove the Law of Sines
2. To solve triangles using the Law of Sines given ASA, AAS, or SSA
3. To identify the number of possible solutions for a triangle

Flash

The operation of an aircraft is highly dependent on trigonometric functions. Trigonometric functions are programmed into computerized systems for take-off, landing, navigation, and communication operations.

Vocabulary

ambiguous case
Law of Sines
oblique triangle

Additional Exercises

Identify each case as ASA, AAS, or SSA, and then determine the number of possible triangles that satisfy the given information.

1. $c = 8$, $B = 35°$, $C = 70°$ AAS; one
2. $a = 9$, $b = 12$, $A = 55°$
   SSA; zero ($h \approx 9.8 > a$)
3. $a = 22$, $B = 18°$, $C = 105°$ ASA; one
4. $b = 13$, $c = 14$, $B = 64°$
   SSA; two ($h \approx 12.6 < b < a$)

PRESENTATION

Lesson Opener

Find each measure of $\triangle ABC$ to the nearest hundredth.

1. b 7.85
2. c 8.81

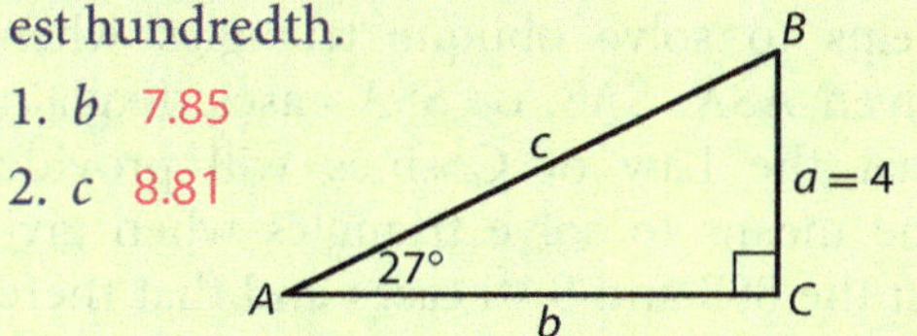

3. Explain why y cannot be found using $\tan 39° = \frac{4}{y}$.

$\triangle XYZ$ is not a right triangle.

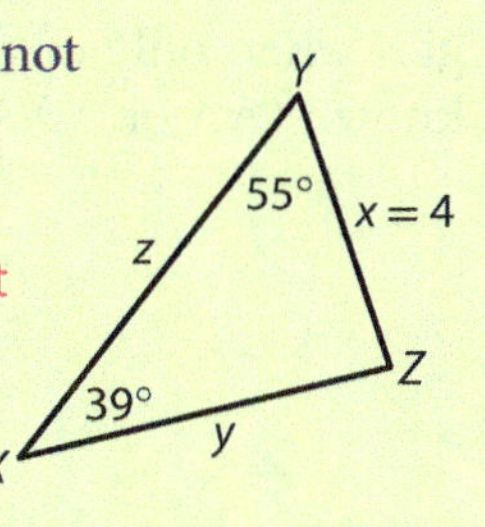

Remind the students that the trigonometric ratios were derived from right triangles, and that we have previously been able to solve only right triangles. These last two sections introduce two important identities which will provide the means to solve *oblique triangles*.

State the *Law of Sines* and refer to the figures in the right margin to show its derivation. The fact that an angle and its supplement have the same sine values was proved in Section 5.4 exercise 29. The Law of Sines can also be stated, "In any triangle, the ratio of the sine of an angle to the length of its opposite side is constant."

Consider having the students apply the Law of Sines to a right triangle to obtain $\sin A = \frac{opp.}{hyp.}$.

$$\frac{\sin A}{a} = \frac{\sin C}{c} \Rightarrow \frac{\sin A}{opp.} = \frac{\sin 90°}{hyp.}$$
$$\Rightarrow \frac{\sin A}{opp.} = \frac{1}{hyp.} \Rightarrow \sin A = \frac{opp.}{hyp.}$$

The application of the Law of Sines when solving triangles given the AAS and ASA cases is demonstrated in the first two examples. Example 2 demon-

5. $a = 12$, $b = 10$, $A = 72°$

SSA; one $(a > b)$

Solve each triangle. Round answers to the nearest tenth.

6. $c = 8$, $B = 35°$, $C = 70°$

$A = 75°$, $a \approx 8.2$, $b \approx 4.9$

7. $a = 22$, $B = 18°$, $C = 105°$

$A = 57°$, $b \approx 8.1$, $c \approx 25.3$

8. $b = 13$, $c = 14$, $B = 64°$

$C \approx 75.5°$, $A \approx 40.5°$, $a \approx 9.4$ or
$C \approx 104.5°$, $A \approx 11.5°$, $a \approx 2.9$

9. $a = 12$, $b = 10$, $A = 72°$

$B \approx 52.4°$, $C \approx 55.6°$, $c \approx 10.4$

10. A triangular piece of property has 2 km of frontage. The other two sides meet the line of the road at 65° and 55°. Find the lengths of the other two sides to the nearest meter.

1892 m, 2093 m

Assignments

- **Minimum:** 1, 3–4, 9–12, 15–22, 25, 27, 31, 35–37; 43–51 odd; 52
- **Standard:** 2, 5–7, 13–20, 23–24, 26, 30, 32–36, 38–39, 44–48, 50–52
- **Extended:** 5–6, 9–10, 13–22, 29–35, 38, 40–44; 46–52 even

Regiomontanus stated the Law of Sines in his major work, *On Triangles*, written in 1464. The geometric proof uses inscribed triangles and demonstrates that the ratio of each side length to the sine of the opposite angle equals twice the radius of the circumscribed circle. (See exercises 41–42.)

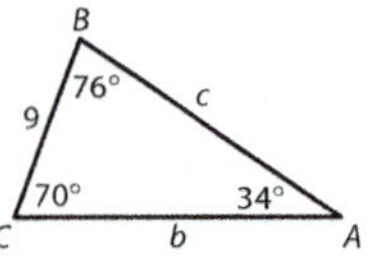

$$\frac{b}{\sin B} = \frac{a}{\sin A}$$

$$\frac{b}{\sin 76°} = \frac{9}{\sin 34°}$$

$$b = \frac{9 \sin 76°}{\sin 34°} \approx 15.6 \text{ cm}$$

$$\frac{c}{\sin C} = \frac{a}{\sin A}$$

$$\frac{c}{\sin 70°} = \frac{9}{\sin 34°}$$

$$c = \frac{9 \sin 70°}{\sin 34°} \approx 15.1 \text{ cm}$$

3. Substitute into the portion of the Law of Sines involving an unknown side length and the known side length.

4. Solve for b.

5. Substitute into the portion of the Law of Sines involving the other unknown side length and the given side length.

SKILL ✔ EXERCISE 9

Example 2 Solving the ASA Case

Johan and Wesley stand 360 ft apart on a level football field to observe the flight of a rocket from opposite sides. If Johan records a 72° angle of elevation to the apex of the flight path and Wesley records an 81° angle of elevation, determine the distance from the rocket to each observer and the height of the rocket's flight.

Answer

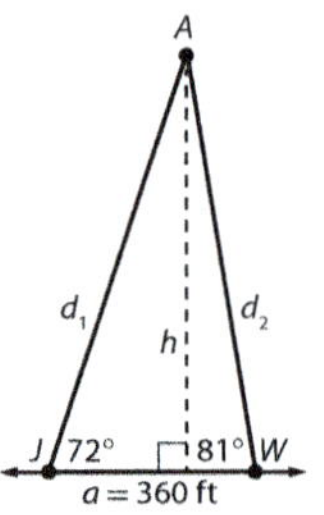

$A = 180 - 72 - 81 = 27°$

1. Make and label a sketch.

2. Determine A.

$$\frac{d_1}{\sin W} = \frac{a}{\sin A} \qquad \frac{d_2}{\sin J} = \frac{a}{\sin A}$$

$$d_1 = \frac{a \sin W}{\sin A} \qquad d_2 = \frac{a \sin J}{\sin A}$$

$$= (360 \text{ ft}) \frac{\sin 81°}{\sin 27°} \qquad = (360 \text{ ft}) \frac{\sin 72°}{\sin 27°}$$

$$\approx 783 \text{ ft} \qquad \approx 754 \text{ ft}$$

3. Use the Law of Sines to find the distance to A from each observer.

$$\sin J = \frac{h}{d_1},$$

so $h = d_1 \sin J \approx 783 \sin 72° \approx 745$ ft

4. Use right-triangle trigonometry to find the height of the rocket's flight.

SKILL ✔ EXERCISE 7

Two side lengths and the measure of an angle opposite one of those sides (SSA) do not determine a unique triangle. Since these measures may describe one, two, or no possible triangles, SSA is known as the *ambiguous case*. Examine the following possibilities when given $\angle A$ and lengths a and b.

---

strates the rearrangement of the equation prior to substituting numerical values and units, a technique that is encouraged in applications.

The SSA case (or the *ambiguous case*) is a bit more challenging because there can be one, two, or no possible triangles. Use the table in the text or an animated version (created with interactive geometric software or found using the Internet keyword search *ambiguous case demonstration*) to explain the reasoning behind each case. Relate the ambiguity to the lack of an SSA congruence postulate or theorem in geometry.

Consistently sketching the angle, the adjacent side, and the opposite side in the same relative position, as done in Example 3, will alleviate confusion when determining the number of possible triangles. Note that when the given angle is acute (Example 3b–c), the opposite side is frequently compared to h, the product of the adjacent side and the sine of a given angle.

Motivational Idea The Internet keyword search *animated SSA case* can be used to illustrate the different number of possible triangles.

Discuss Example 4, noting that the two possible triangles originate from the two possible angles within the interval (0°, 180°) such that $\sin S \approx 0.73$, the principal value returned by the calculator and its supplement.

Review the fact that the Law of Sines helps to solve oblique triangles when given ASA, SAS, or SSA cases. Explain that the Law of Cosines will provide the means to solve triangles when given the SSS and SAS cases and that there are an infinite number of possible triangles when only angles (two or three) are known (AA or AAA).

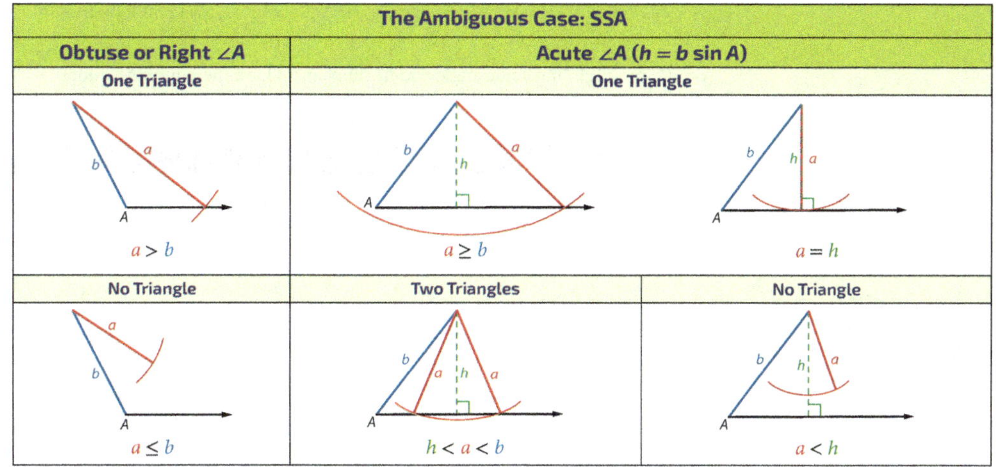

When examining the ambiguous case, make a sketch similar to the illustrated possibilities with the angle, its adjacent side as the side above the horizontal ray, its opposite side, and the possible height h when the angle is acute.

Example 3 Determining the Number of Possible Triangles for SSA

Determine the number of possible triangles in each case.

a. $\triangle JKL$ if $j = 3.5$, $k = 5$, and $J = 97°$

b. $\triangle XYZ$ if $x = 4$, $z = 5$, and $Z = 55°$

c. $\triangle EFG$ if $f = 5$, $g = 2.5$, and $G = 30°$

Answer

a. Sketch obtuse $\angle J$, side k, and side j.

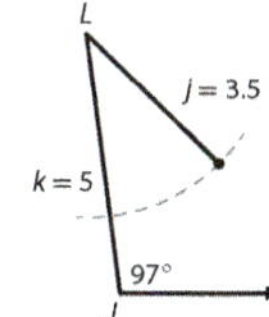

Since $j \le k$, there is no possible $\triangle JKL$.

b. Sketch acute $\angle Z$, side x, and side z.

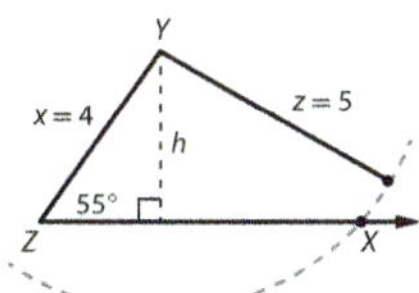

Since $z \ge x$, there is only one $\triangle XYZ$.

c. Sketch acute $\angle G$, side f, and side g.

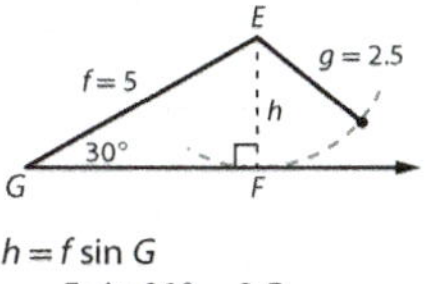

$h = f \sin G$
$\quad = 5 \sin 30° = 2.5$

Since $g = h$, there is only one $\triangle FGH$.

SKILL ✔ **EXERCISES 17, 19**

Attempting to use the Law of Sines to solve $\triangle JKL$ in Example 3a quickly reveals that no solution is possible.

$$\frac{\sin K}{5} = \frac{\sin 97°}{3.5}; \therefore \sin K = \frac{5 \sin 97°}{3.5} \approx 1.4, \text{ but } \sin K \text{ must be within } [-1, 1].$$

Solutions

▶ A. Exercises

3. $B = 180 - 50 - 30 = 100°$

$$\frac{a}{\sin 50°} = \frac{4}{\sin 30°}$$
$$a = \frac{4 \sin 50°}{\sin 30°} \approx 6.1$$
$$\frac{b}{\sin 100°} = \frac{4}{\sin 30°}$$
$$b = \frac{4 \sin 100°}{\sin 30°} \approx 7.9$$

4. $B = 180 - 50 - 44 = 86°$

$$\frac{a}{\sin 50°} = \frac{7}{\sin 86°}$$
$$a = \frac{7 \sin 50°}{\sin 86°} \approx 5.4$$
$$\frac{c}{\sin 44°} = \frac{7}{\sin 86°}$$
$$c = \frac{7 \sin 44°}{\sin 86°} \approx 4.9$$

5. $R = 180 - 68 - 84 = 28°$

$$\frac{s}{\sin 68°} = \frac{8}{\sin 28°}$$
$$s = \frac{8 \sin 68°}{\sin 28°} \approx 15.8$$
$$\frac{t}{\sin 84°} = \frac{8}{\sin 28°}$$
$$t = \frac{8 \sin 84°}{\sin 28°} \approx 16.9$$

6. $K = 180 - 70 - 75 = 35°$

$$\frac{k}{\sin 35°} = \frac{8}{\sin 70°}$$
$$k = \frac{8 \sin 35}{\sin 70°} \approx 4.9$$
$$\frac{j}{\sin 75°} = \frac{8}{\sin 70°}$$
$$j = \frac{8 \sin 75}{\sin 70°} \approx 8.2$$

7. $C = 180 - 52 - 49 = 79°$

$$\frac{a}{\sin 52°} = \frac{16}{\sin 79°}$$
$$a = \frac{16 \sin 52°}{\sin 79°} \approx 12.8$$

$$\frac{b}{\sin 49°} = \frac{16}{\sin 79°}$$
$$b = \frac{16 \sin 49°}{\sin 79°} \approx 12.3$$

8. $A = 180 - 43 - 61 = 76°$

$$\frac{a}{\sin 76°} = \frac{9}{\sin 61°}$$
$$a = \frac{9 \sin 76°}{\sin 61°} \approx 10.0$$
$$\frac{c}{\sin 43°} = \frac{9}{\sin 61°}$$
$$c = \frac{9 \sin 43°}{\sin 61°} \approx 7.0$$

9. $E = 180 - 23 - 42 = 115°$

$$\frac{e}{\sin 115°} = \frac{7}{\sin 23°}$$
$$e = \frac{7 \sin 115°}{\sin 23°} \approx 16.2$$
$$\frac{f}{\sin 42°} = \frac{7}{\sin 23°}$$
$$f = \frac{7 \sin 42°}{\sin 23°} \approx 12.0$$

10. $P = 180 - 26 - 51 = 103°$

$$\frac{q}{\sin 26°} = \frac{9}{\sin 103°}$$
$$q = \frac{9 \sin 26°}{\sin 103°} \approx 4.0$$
$$\frac{r}{\sin 51°} = \frac{9}{\sin 103°}$$
$$r = \frac{9 \sin 51°}{\sin 103°} \approx 7.2$$

11. $$\frac{\sin B}{10} = \frac{\sin 70°}{11}$$
$$\sin B = \frac{10 \sin 70°}{11}$$
$$B \approx 58.7°$$
$$A \approx 180 - 70 - 58.7 = 51.3°$$
$$\frac{a}{\sin 51.3°} = \frac{11}{\sin 70°}$$
$$a = \frac{11 \sin 51.3°}{\sin 70°} \approx 9.1$$

12. $$\frac{\sin X}{17} = \frac{\sin 109°}{20}$$
$$\sin X = \frac{17 \sin 109°}{20}$$
$$X \approx 53.5°$$
$$Y \approx 180 - 109 - 53.5 = 17.5°$$
$$\frac{y}{\sin 17.5°} = \frac{20}{\sin 109°}$$
$$y = \frac{20 \sin 17.5°}{\sin 109°} \approx 6.4$$

13. $$\frac{\sin N}{9} = \frac{\sin 82°}{18}$$
$$\sin N = \frac{9 \sin 82°}{18}$$
$$N \approx 29.7°$$
$$L \approx 180 - 82 - 29.7 = 68.3°$$
$$\frac{l}{\sin 68.3°} = \frac{18}{\sin 82°}$$
$$l = \frac{18 \sin 68.3°}{\sin 82°} \approx 16.9$$

14. $$\frac{\sin B}{8} = \frac{\sin 46°}{11}$$
$$\sin B = \frac{8 \sin 46°}{11}$$
$$B \approx 31.5°$$
$$C \approx 180 - 46 - 31.5 = 102.5°$$
$$\frac{c}{\sin 102.5°} = \frac{11}{\sin 46°}$$
$$c = \frac{11 \sin 102.5°}{\sin 46°} \approx 14.9$$

B. Exercises

15. ASA; one solution

16. SSA; A is obtuse.
no solution when $a < b$

17. SSA; A is obtuse.
one solution when $a > c$

18. SSA; F is acute.
one solution when $f \geq g$

19. SSA; F is acute and $f < g$.
$h = g \sin F = 10 \sin 61 \approx 8.7$
$f < h$; no solution

20. SSA; F is acute and $f < g$.
$h = g \sin F = 10 \sin 61° \approx 8.7$
$h < f < g$; two solutions

21. $\dfrac{\sin B}{6} = \dfrac{\sin 17°}{3}$
$\sin B = \dfrac{6 \sin 17°}{3}$
case 1: $B \approx 35.8°$
$C \approx 180 - 17 - 35.8 = 127.2°$
$\dfrac{c}{\sin 127.2°} = \dfrac{3}{\sin 17°}$
$c = \dfrac{3 \sin 127.2°}{\sin 17°} \approx 8.2$
case 2: $B = 180 - 35.8 = 144.2°$
$C \approx 180 - 17 - 144.2 = 18.8°$
$\dfrac{c}{\sin 18.8°} = \dfrac{3}{\sin 17°}$
$c = \dfrac{3 \sin 18.8°}{\sin 17°} \approx 3.3$
$B \approx 35.8°$, $C \approx 127.2°$, $c \approx 8.2$
or $B \approx 144.2°$, $C \approx 18.8°$, $c \approx 3.3$

22. $\dfrac{\sin C}{8} = \dfrac{\sin 32°}{6}$
$\sin C = \dfrac{8 \sin 32°}{6}$
case 1: $C \approx 45.0°$
$A = 180 - 32 - 45.0 = 103.0°$
$\dfrac{a}{\sin 103.0°} = \dfrac{6}{\sin 32°}$
$a = \dfrac{6 \sin 103.0°}{\sin 32°} \approx 11.0$
case 2: $C = 180 - 45.0 = 135.0°$
$A \approx 180 - 32 - 135.0 = 13.0°$
$\dfrac{a}{\sin 13.0°} = \dfrac{6}{\sin 32°}$
$a = \dfrac{6 \sin 13.0°}{\sin 32°} \approx 2.5$
$A \approx 103.0°$, $C \approx 45.0°$, $a \approx 11.0$
or $A \approx 13.0°$, $C \approx 135.0°$, $a \approx 2.5$

23. $\dfrac{\sin G}{6} = \dfrac{\sin 46°}{5}$
$\sin G = \dfrac{6 \sin 46°}{5}$
case 1: $G \approx 59.7°$
$F \approx 180 - 46 - 59.7 = 74.3°$
$\dfrac{f}{\sin 74.3°} = \dfrac{5}{\sin 46°}$
$f = \dfrac{5 \sin 74.3°}{\sin 46°} \approx 6.7$
case 2: $G = 180 - 59.7 = 120.3°$
$F \approx 180 - 46 - 120.3 = 13.7°$

$\dfrac{f}{\sin 13.7°} = \dfrac{5}{\sin 46°}$
$f = \dfrac{5 \sin 13.7°}{\sin 46°} \approx 1.6$
$F \approx 74.3°$, $G \approx 59.7°$, $f \approx 6.7$
or $F \approx 13.7°$, $G \approx 120.3°$, $f \approx 1.6$

24. $\dfrac{\sin M}{9} = \dfrac{\sin 60°}{8}$
$\sin M = \dfrac{9 \sin 60°}{8}$
case 1: $M \approx 77.0°$
$N \approx 180 - 60 - 77.0 = 43.0°$
$\dfrac{n}{\sin 43.0°} = \dfrac{8}{\sin 60°}$
$n = \dfrac{8 \sin 43.0°}{\sin 60°} \approx 6.3$
case 2: $M = 180 - 77.0 = 103.0°$
$N \approx 180 - 60 - 103.0 = 17.0°$

$\dfrac{n}{\sin 17.0°} = \dfrac{8}{\sin 60°}$
$n = \dfrac{8 \sin 17.0°}{\sin 60°} \approx 2.7$
$M \approx 77.0°$, $N \approx 43.0°$, $n \approx 6.3$
or $M \approx 103.0°$, $N \approx 17.0°$, $n \approx 2.7$

25. AAS; one solution
$B = 180 - 12 - 89 = 79°$
$\dfrac{b}{\sin 79°} = \dfrac{9}{\sin 12°}$
$b = \dfrac{9 \sin 79°}{\sin 12°} \approx 42.5$
$\dfrac{c}{\sin 89°} = \dfrac{9}{\sin 12°}$
$c = \dfrac{9 \sin 89°}{\sin 12°} \approx 43.3$
$B \approx 79°$, $b \approx 42.5$, $c \approx 43.3$

The Law of Sines can also show that there is only one possible $\triangle XYZ$ in Example 3b.

$\dfrac{\sin X}{4} = \dfrac{\sin 55°}{5}$, $\therefore \sin X = \dfrac{4 \sin 55°}{5} \approx 0.6553$ and

$X \approx 41°$ or $\approx 139°$ (an acute angle and its supplement have the same sine value). But $\triangle XYZ$ cannot have $Z = 55°$ and $X = 139°$ since $55° + 139° > 180°$. Example 4 illustrates how the Law of Sines is applied when there are two possible triangles.

Example 4   Solving the Ambiguous Case

Solve $\triangle RST$ if $r = 3$ in., $s = 5$ in., and $R = 26°$.

Answer

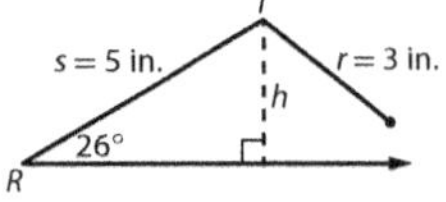

$h = s \sin R = 5 \sin 26° \approx 2.192$

Since $h < r < s$, there are 2 possible triangles.

using $\dfrac{\sin S}{s} = \dfrac{\sin R}{r}$:

$\dfrac{\sin S}{5} = \dfrac{\sin 26°}{3}$

$\sin S = \dfrac{5(\sin 26°)}{3} \approx 0.7306$

$S \approx 46.9°$ or
$S \approx 180 - 46.9 = 133.1°$

If $S \approx 46.9° \ldots$
$T \approx 180 - 26 - 46.9$
$= 107.1°$

$\dfrac{t}{\sin 107.1°} = \dfrac{3}{\sin 26°}$
$t \approx \dfrac{3 \sin 107.1°}{\sin 26°} \approx 6.5$ in.

If $S \approx 133.1° \ldots$
$T \approx 180 - 26 - 133.1$
$= 20.9°$

$\dfrac{t}{\sin 20.9°} = \dfrac{3}{\sin 26°}$
$t \approx \dfrac{3 \sin 20.9°}{\sin 26°} \approx 2.4$ in.

1. Sketch acute $\angle R$, side s, and side r; then determine the height h.

2. Determine the number of possible triangles.

3. Apply the Law of Sines to find the two possible measures for S.

   The calculator's $\sin^{-1}$ function returns the principal value in Q I, which is also the reference angle used to find the obtuse angle (in Q II) with the same sine value, its supplement.

4. Solve both possible triangles.
   a. Find possible measures of T.

   b. Use $\dfrac{t}{\sin T} = \dfrac{r}{\sin R}$ to find the possible values for t.

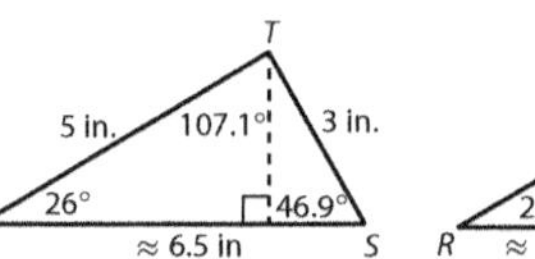

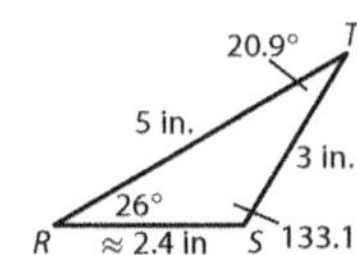

SKILL ✔ EXERCISE 21

You have seen how the Law of Sines can be applied to solve a unique triangle determined by two angle measures and a side length (AAS or ASA) and how it can be used to find all possible solutions for the ambiguous case (SSA). Section 5.7 shows how the Law of Cosines is used to solve the unique triangle described by three side lengths (SSS) or by the measures of two sides and the included angle (SAS). Recall from GEOMETRY that any two angles of a triangle determine the third angle (AA implies AAA) and an infinite number of similar triangles.

> "To those who do not know mathematics it is difficult to get across a real feeling as to the beauty, the deepest beauty, of nature."
>
> – Richard Feynman
> (theoretical physicist)

Interactive geometry software such as The Geometer's Sketchpad® or Geogebra® can be used to explore the Law of Sines and other trigonometric relationships. Use technology to complete the following steps.

1. Open a new file and construct $\triangle ABC$.

2. Measure each side length and each angle.

3. Calculate the ratio of each side length and the sine of the angle opposite that side.

4. Construct the triangle's circumcenter, D, by finding the intersection of perpendicular lines drawn through the midpoints of two sides. Then construct the circumscribed circle.

5. Measure the radius of the circle and calculate its diameter.

6. Compare the ratios to the diameter of the circle.

7. Drag the vertices to change the triangle's shape.
   Does $\dfrac{a}{\sin A} = \dfrac{b}{\sin B} = \dfrac{c}{\sin C} = 2r$ in all cases?

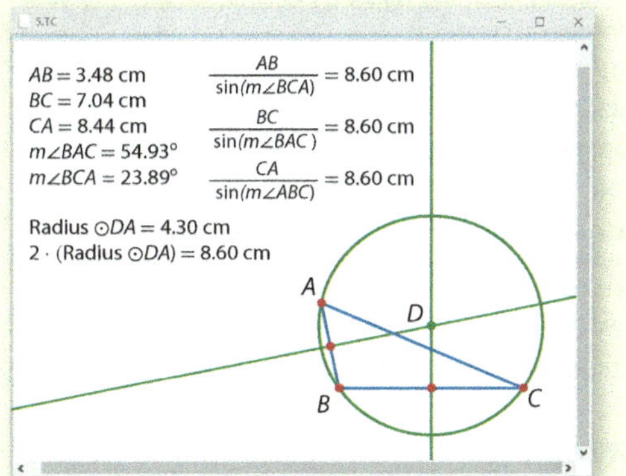

❯ A. Exercises

1. Identify each case as ASA, AAS, or SSA.
   a. $a = 5$, $B = 48°$, $A = 68°$ AAS
   b. $a = 7$, $c = 2$, $\angle C = 12°$ SSA
   c. $R = 12°$, $S = 106°$, $t = 10$ ASA
   d. $x = 2$, $y = 5$, $X = 15°$ SSA

2. Identify each case as ASA, AAS, SSA, or SAS.
   a. $a = 9$, $A = 57°$, $c = 12$ SSA
   b. $A = 70°$, $c = 16$, $C = 37°$ AAS
   c. $j = 3$, $k = 7$, $L = 40°$ SAS
   d. $p = 7$, $\angle Q = 26°$, $\angle R = 42°$ ASA

Solve each triangle. Round answers to the nearest tenth.

3.
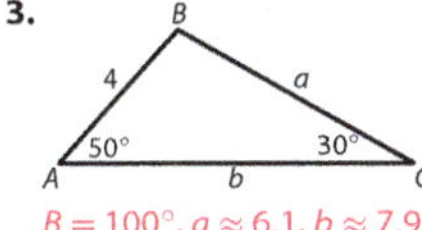
$B = 100°$, $a \approx 6.1$, $b \approx 7.9$

4.
$B = 86°$, $a \approx 5.4$, $c \approx 4.9$

5.
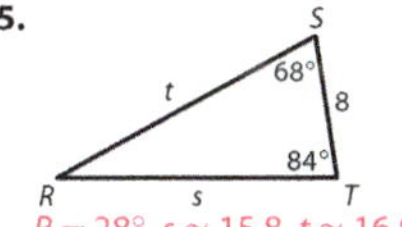
$R = 28°$, $s \approx 15.8$, $t \approx 16.9$

6.
$K = 35°$, $k \approx 4.9$, $j \approx 8.2$

Sketch and solve each triangle. Round answers to the nearest tenth.

7. $\triangle ABC$ with $A = 52°$, $B = 49°$, and $c = 16$

8. $\triangle ABC$ with $C = 43°$, $B = 61°$, and $b = 9$

9. $\triangle DEF$ with $D = 23°$, $F = 42°$, and $d = 7$

10. $\triangle PQR$ with $Q = 26°$, $R = 51°$, and $p = 9$

Solve each triangle. (These SSA cases each have one solution.) Round answers to the nearest tenth.

11.
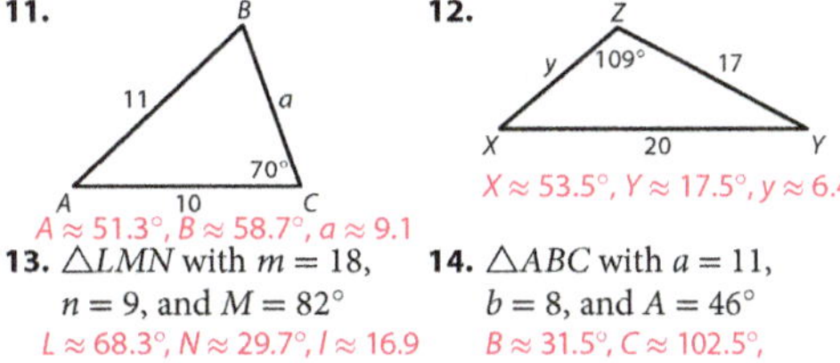
$A \approx 51.3°$, $B \approx 58.7°$, $a \approx 9.1$

12.
$X \approx 53.5°$, $Y \approx 17.5°$, $y \approx 6.4$

13. $\triangle LMN$ with $m = 18$, $n = 9$, and $M = 82°$
$L \approx 68.3°$, $N \approx 29.7°$, $l \approx 16.9$

14. $\triangle ABC$ with $a = 11$, $b = 8$, and $A = 46°$
$B \approx 31.5°$, $C \approx 102.5°$, $c \approx 14.9$

❯ B. Exercises

Determine the number of possible triangles that satisfy the given information.

15. $\triangle ABC$ with $A = 78°$, $B = 46°$, and $c = 11$ one

16. $\triangle ABC$ with $A = 124°$, $a = 8$, and $b = 10$ zero

17. $\triangle ABC$ with $A = 124°$, $a = 13$, and $c = 10$ one

18. $\triangle FGH$ with $F = 61°$, $f = 10$, and $g = 8$ one

26. SSA; B is acute and $b > c$.
one solution
$$\frac{\sin C}{6} = \frac{\sin 82°}{12}$$
$$\sin C = \frac{6 \sin 82°}{12}$$
$$C \approx 29.7°$$
$$A \approx 180 - 82 - 29.7 = 68.3°$$
$$\frac{a}{\sin 68.3°} = \frac{12}{\sin 82°}$$
$$a = \frac{12 \sin 68.3°}{\sin 82°} \approx 11.3$$
$$A \approx 68.3°,\ C \approx 29.7°,\ a \approx 11.3$$

27. SSA; B is acute and $b < a$.
$h = a \sin B = 6 \sin 80° \approx 5.9$
$b < h$; no solution

28. SSA; C is acute and $c < a$.
$h = a \sin C = 14 \sin 15° \approx 3.6$
$h < c < a$; two solutions
$$\frac{\sin A}{14} = \frac{\sin 15°}{12}$$
$$\sin A = \frac{14 \sin 15°}{12}$$
case 1: $A \approx 17.6°$
$B \approx 180 - 15 - 17.6 = 147.4°$
$$\frac{b}{\sin 147.4°} = \frac{12}{\sin 15°}$$
$$b = \frac{12 \sin 147.4°}{\sin 15°} \approx 25.0$$
case 2: $A \approx 180 - 17.6 = 162.4°$
$B \approx 180 - 15 - 162.4 = 2.6°$
$$\frac{b}{\sin 2.6°} = \frac{12}{\sin 15°}$$
$$b = \frac{12 \sin 2.6°}{\sin 15°} \approx 2.1$$
$A \approx 17.6°$, $B \approx 147.4°$, $b \approx 25.0$
or $A \approx 162.4°$, $B \approx 2.6°$, $b \approx 2.1$

29. SSA; E is acute and $e > d$.
one solution
$$\frac{\sin D}{1.2} = \frac{\sin 67°}{4}$$
$$\sin D = \frac{1.2 \sin 67°}{4}$$
$$D \approx 16.0°$$
$$F \approx 180 - 67 - 16.0 = 97.0°$$
$$\frac{f}{\sin 97.0°} = \frac{4}{\sin 67°}$$
$$f = \frac{4 \sin 97.0°}{\sin 67°} \approx 4.3$$
$$D \approx 16.0°,\ F \approx 97.0°,\ f \approx 4.3$$

30. SSA; K is obtuse and $k < g$.
no solution

31. SSA; Q is acute and $q < p$.
$h = p \sin Q = 8 \sin 33° \approx 4.3$
$h < q < p$; two solutions
$$\frac{\sin P}{8} = \frac{\sin 33°}{4.4}$$
$$\sin P = \frac{8 \sin 33°}{4.4}$$
case 1: $P \approx 82.0°$
$R \approx 180 - 33 - 82.0 = 65.0°$
$$\frac{r}{\sin 65.0°} = \frac{4.4}{\sin 33°}$$
$$r = \frac{4.4 \sin 65.0°}{\sin 33°} \approx 7.3$$
case 2: $P \approx 180 - 82.0 = 98.0°$
$R \approx 180 - 33 - 98.0 = 49.0°$
$$\frac{r}{\sin 49.0°} = \frac{4.4}{\sin 33°}$$
$$r = \frac{4.4 \sin 49.0°}{\sin 33°} \approx 6.1$$
$P \approx 82.0°$, $R \approx 65.0°$, $r \approx 7.3$
or $P \approx 98.0°$, $R \approx 49.0°$, $r \approx 6.1$

32. ASA; one solution
$$U = 180 - 62 - 95 = 23°$$
$$\frac{s}{\sin 62°} = \frac{9}{\sin 23°}$$
$$s = \frac{9 \sin 62°}{\sin 23°} \approx 20.3$$
$$\frac{v}{\sin 95°} = \frac{9}{\sin 23°}$$
$$v = \frac{9 \sin 95°}{\sin 23°} \approx 22.9$$
$$U \approx 23°,\ s \approx 20.3,\ v \approx 22.9$$

33.
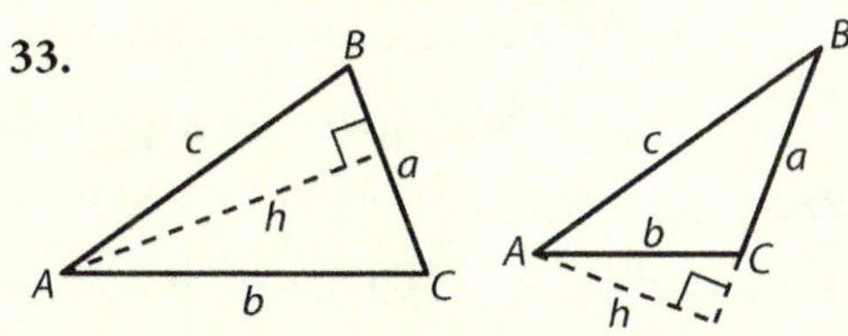

In both triangles, $\sin B = \dfrac{h}{c}$ and $h = c \sin B$.
In the acute triangle $\sin C = \dfrac{h}{b}$, and in the obtuse triangle $\sin (\pi - C) = \dfrac{h}{b}$.
But $\sin (\pi - \theta) = \sin \theta$, so in both cases,
$\sin C = \dfrac{h}{b}$ and $h = b \sin C$.
$\therefore c \sin B = b \sin C$ and $\dfrac{c}{\sin C} = \dfrac{b}{\sin B}$

34.

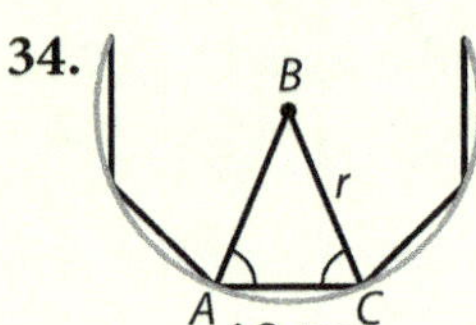

$$B = \frac{360^\circ}{8} = 45^\circ$$
$$A = C = \frac{180^\circ - 45^\circ}{2} = 67.5^\circ$$
$$r = \frac{4.2 \sin 67.5^\circ}{\sin 45^\circ} \approx 5.5 \text{ cm}$$

35. $B = 180 - 37.5 - 17.25 = 125.25^\circ$

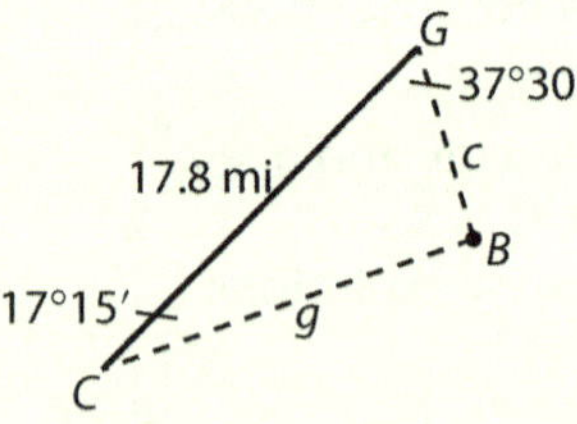

$$c = \frac{b \sin C}{\sin B} = \frac{17.8 \sin 17.25^\circ}{\sin 125.25^\circ}$$
$$\approx 6.5 \text{ mi from Cape Romain}$$
$$g = \frac{b \sin G}{\sin B} = \frac{17.8 \sin 37.5^\circ}{\sin 125.25^\circ}$$
$$\approx 13.3 \text{ mi from Georgetown}$$

36.

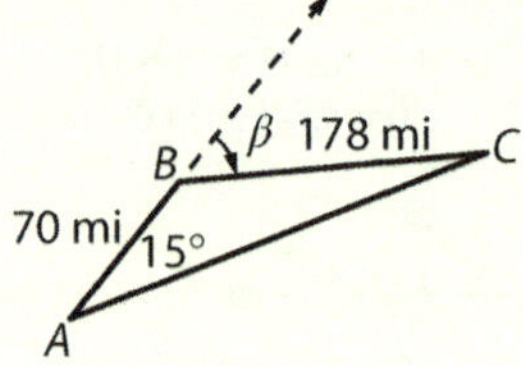

SSA; one solution
$$\frac{\sin C}{70} = \frac{\sin 15^\circ}{178}$$
$$C = \sin^{-1} \frac{70 \sin 15^\circ}{178} \approx 5.8^\circ$$
$$B \approx 180 - 15 - 5.8 = 159.2^\circ$$
$$\beta \approx 180 - 159.2 = 20.8^\circ$$

❯ **C. Exercises**

37. $T = 180 - 20 - 33 = 127^\circ$
$$k = \frac{300 \sin 20^\circ}{\sin 127^\circ} \approx 128.5 \text{ ft}$$
$$\sin 33^\circ \approx \frac{h}{128.5}$$
$$h \approx 128.5 \sin 33^\circ \approx 70.0 \text{ ft}$$

38.

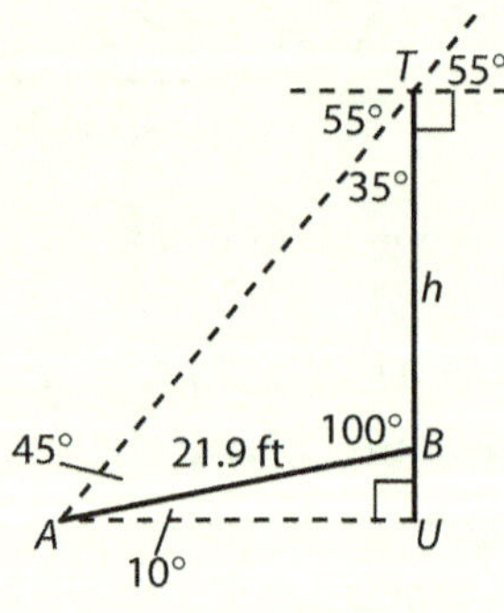

$$\angle ATB = 90 - 55 = 35^\circ$$
$$\angle TAB = 55 - 10 = 45^\circ$$
$$h = \frac{21.9 \sin 45^\circ}{\sin 35^\circ} \approx 27 \text{ ft}$$

---

19. $\triangle FGH$ with $F = 61^\circ$, $f = 8$, and $g = 10$ zero

20. $\triangle FGH$ with $F = 61^\circ$, $f = 9$, and $g = 10$ two

Find both solutions for each SSA case. Round answers to the nearest tenth.

21. $\triangle ABC$ with $A = 17^\circ$, $a = 3$, and $b = 6$

22. $\triangle ABC$ with $B = 32^\circ$, $b = 6$, and $c = 8$

23. $\triangle FGH$ with $H = 46^\circ$, $g = 6$, and $h = 5$

24. $\triangle LMN$ with $L = 60^\circ$, $l = 8$, and $m = 9$

Determine the number of solutions for each case, then solve each possible triangle. Round answers to the nearest tenth.

25. $\triangle ABC$ with $A = 12^\circ$, $C = 89^\circ$, and $a = 9$

26. $\triangle ABC$ with $B = 82^\circ$, $b = 12$, and $c = 6$

27. $\triangle ABC$ with $B = 80^\circ$, $a = 6$, and $b = 2$ no solution

28. $\triangle ABC$ with $C = 15^\circ$, $a = 14$, and $c = 12$

29. $\triangle DEF$ with $E = 67^\circ$, $d = 1.2$, and $e = 4$

30. $\triangle GHK$ with $K = 100^\circ$, $g = 8$, and $k = 6$ no solution

31. $\triangle PQR$ with $Q = 33^\circ$, $p = 8$, and $q = 4.4$

32. $\triangle SUV$ with $S = 62^\circ$, $V = 95^\circ$, and $u = 9$

33. Prove: Copy the acute and obtuse cases of $\triangle ABC$ used in the proof of the Law of Sines on page 269. Then draw the altitude from A in each triangle and show that $\frac{c}{\sin C} = \frac{b}{\sin B}$.

34. If a regular octagon has sides 4.2 cm long, use the Law of Sines to find the distance from its center to a vertex. ≈ 5.5 cm

35. The Georgetown Lighthouse is 17.8 mi northeast of the lighthouse on Cape Romain. If the Georgetown light keeper noticed a boat in distress at 37°30′ from the shoreline between the lighthouses, and the Cape Romain light keeper noticed the same boat at 17°15′ from the same shoreline, find the boat's distance from each lighthouse.

36. To avoid thunderstorms, a pilot flies 70 mi on a path 15° north of the direct route from Atlanta, GA, to Charlotte, NC. How many degrees south must he turn from his current heading if the plane is now 178 mi from its destination? $\approx 20.8^\circ$

❯ **C. Exercises**

37. Calculate the height of a tree if Kenneth measures a 20° angle of elevation from the ground to the top of the tree and Andrew measures a 33° angle of elevation when standing on the opposite side of the tree 300 ft from Kenneth. ≈ 70.0 ft

---

38. Find the height of a windmill if the afternoon sun shining at a 55° angle of elevation casts a 21.9 ft shadow directly down the 10° slope on which the windmill stands. ≈ 27 ft

39. Fire tower 2 is located 19 km from fire tower 1 at a bearing of 96°. A fire is located at a bearing of 157° from tower 1 and 202° from tower 2. Find the distance of the fire from each tower.

40. A 120 ft water tower is set on a vertical cliff next to a river so that the angle of depression from the top of the tower to a boat dock on the opposite side of the river is 29°15′ and the angle of depression from the base of the tower to the dock is 17°22′. Determine the width of the river and the height of the cliff.

41. Prove: Use acute $\triangle ABC$ and circumscribed $\odot M$ to prove that $\frac{a}{\sin A} = \frac{b}{\sin B} = \frac{c}{\sin C} = 2r$.

a. Explain why $m\angle B = \frac{1}{2} m\angle AMC$.

b. Explain why $m\angle B = \theta$.

c. Use $\triangle AMN$ to express $\sin B$ in terms of r and b.

d. Show that $\frac{a}{\sin A} = \frac{b}{\sin B} = \frac{c}{\sin C} = 2r$.

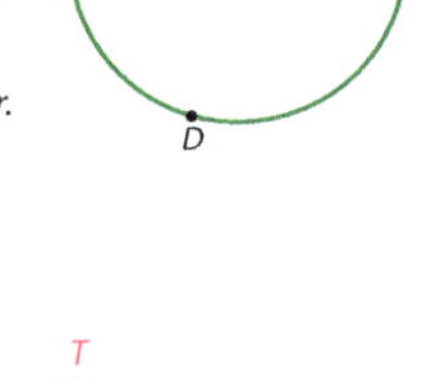

42. Prove: Use obtuse $\triangle ABC$ and circumscribed $\odot M$ to prove $\frac{a}{\sin A} = \frac{b}{\sin B} = \frac{c}{\sin C} = 2r$.

a. Find $m\angle B$ in terms of $m\angle AMC$. Hint: Use $m\widehat{AC}$.

b. Express $m\angle B$ in terms of θ.

c. Use $\triangle AMN$ to express $\sin B$ in terms of r and b.

d. Show that $\frac{a}{\sin A} = \frac{b}{\sin B} = \frac{c}{\sin C} = 2r$.

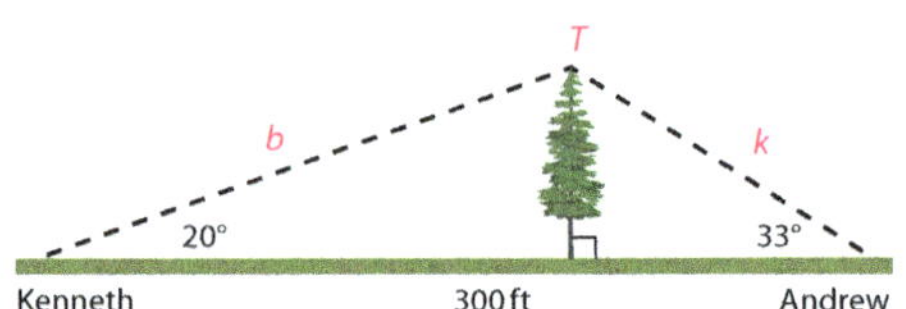

39.

$$\angle T_1 T_2 C = 180 - 96 = 84^\circ$$
(consec. int. $\angle$s)
$$\angle T_1 T_2 F = 360 - 84 - 202 = 74^\circ$$
$$\angle T_2 T_1 F = 157 - 96 = 61^\circ$$
$$\angle F = 180 - 74 - 61 = 45^\circ$$
$$T_1 F = \frac{19 \sin 74^\circ}{\sin 45^\circ} \approx 25.8 \text{ km}$$
$$T_2 F = \frac{19 \sin 61^\circ}{\sin 45^\circ} \approx 23.5 \text{ km}$$

40.

$$\alpha = 17°22′$$
$$\beta = 29°15′ - 17°22′ = 11°53′$$
$$z = \frac{120 \sin 60°45′}{\sin 11°53′} \approx 508.4 \text{ ft}$$
river width: $\cos \alpha = \frac{x}{z}$; $x = z \cos \alpha$
$$\approx 508.4 \cos 17°22′ \approx 485.3 \text{ ft}$$
cliff height: $\sin \alpha = \frac{y}{z}$; $y = z \sin \alpha$
$$\approx 508.4 \sin 17°22′ \approx 151.8 \text{ ft}$$

Simplify each expression. [5.1]

43. $\sin^2\theta\csc\theta\tan\left(\frac{\pi}{2}-\theta\right)$ $\cos\theta$

44. $\dfrac{\cot x + \tan x}{\sec^2 x}$ $\cot x$

45. Verify $\tan\theta\sin\theta + \cos\theta = \sec\theta$. [5.2]

46. State the primary solutions for $3\sin x = \cos x$ (to the nearest hundredth). [5.3] $x = 0.32, 3.46$

47. State the primary solutions for $\sin x \tan x = \sin x$. [5.3] $x = 0, \frac{\pi}{4}, \pi, \frac{5\pi}{4}$

48. Find the exact value of $\cos 15°$. [5.4] $\dfrac{\sqrt{6}+\sqrt{2}}{4}$

49. Solve $\dfrac{2b}{ab+c} = 4$ for b. [2.6]   B

A. $b = \dfrac{2c}{2a-1}$

B. $b = \dfrac{2c}{1-2a}$

C. $b = \dfrac{c}{2a-1}$

D. $b = \dfrac{c}{a-1}$

E. none of these

50. If $\log_b 5 = x$ and $\log_b 4 = y$, what is $\log_b 100$? [3.3]   C

A. $x^2 + y$

B. $x^2 y$

C. $2x + y$

D. $2xy$

E. none of these

51. Describe the measure of central $\angle C$ that subtends an arc whose length is half the circle's diameter. [4.1]   A

A. $m\angle C = 1$

B. $m\angle C = \dfrac{\pi}{4}$

C. $m\angle C = 90°$

D. $m\angle C > 90°$

E. cannot be determined

52. A carpenter braces a wall with a 6 ft board that extends from the floor to a point that is 4 ft 6 in. high on the wall. Which expression represents the angle between the brace and the floor? [4.2]   E

A. $\sin \dfrac{4.6}{6}$

B. $\cos \dfrac{4.5}{6}$

C. $\sin^{-1} \dfrac{4.6}{6}$

D. $\cos^{-1} \dfrac{4.5}{6}$

E. $\sin^{-1} \dfrac{4.5}{6}$

41a. $m\angle B = \frac{1}{2}m\overset{\frown}{AC}$ (inscribed angle measure) and $m\angle AMC = m\overset{\frown}{AC}$ (central angle measure), so $m\angle B = \frac{1}{2}m\angle AMC$ (substitution)

41b. since $\overline{MN} \perp \overline{AC}$, $\triangle AMN \cong \triangle CMN$ (HL $\cong$) with $\angle AMN \cong \angle CMN$, $\therefore \overline{MN}$ bisects $\angle AMC$ and $\frac{1}{2}m\angle AMC = m\angle B = \theta$

41c. $\sin B = \sin\theta = \dfrac{\frac{b}{2}}{r} = \dfrac{b}{2r}$

41d. since $\sin B = \dfrac{b}{2r}$, $2r = \dfrac{b}{\sin B}$ using the Transitive Property of Equality and the Law of Sines: $\dfrac{a}{\sin A} = \dfrac{b}{\sin B} = \dfrac{c}{\sin C} = 2r$

42a. $m\angle B = \frac{1}{2}m\overset{\frown}{ADC} = \frac{1}{2}(360° - m\overset{\frown}{AC})$ $= 180° - \frac{1}{2}m\overset{\frown}{AC}$

since $m\angle AMC = m\overset{\frown}{AC}$, $m\angle B = 180° - \frac{1}{2}m\angle AMC$

42b. since $\overline{MN} \perp \overline{AC}$, $\triangle AMN \cong \triangle CMN$ (HL $\cong$) with $\angle AMN \cong \angle CMN$, $\therefore \overline{MN}$ bisects $\angle AMC$ and $\frac{1}{2}m\angle AMC = \theta$ by substitution, $m\angle B = 180° - \theta$

42c. $\sin B = \sin(180° - \theta)$ $= \sin\theta = \dfrac{\frac{b}{2}}{r} = \dfrac{b}{2r}$

42d. since $\sin B = \dfrac{b}{2r}$, $2r = \dfrac{b}{\sin B}$ using the Transitive Property of Equality and the Law of Sines: $\dfrac{a}{\sin A} = \dfrac{b}{\sin B} = \dfrac{c}{\sin C} = 2r$

> **Cumulative Review**

43. $\sin^2\theta\left(\dfrac{1}{\sin\theta}\right)\cot\theta$ $= \sin^2\theta\left(\dfrac{1}{\sin\theta}\right)\left(\dfrac{\cos\theta}{\sin\theta}\right) = \cos\theta$

44. $\dfrac{\frac{\cos x}{\sin x} + \frac{\sin x}{\cos x}}{\frac{1}{\cos^2 x}} = \left(\dfrac{\cos^2 x + \sin^2 x}{\sin x \cos x}\right)\cos^2 x$ $= \left(\dfrac{1}{\sin x \cos x}\right)\cos^2 x = \dfrac{\cos x}{\sin x} = \cot x$

45. $\tan\theta\sin\theta + \cos\theta$ $= \left(\dfrac{\sin\theta}{\cos\theta}\right)\sin\theta + \cos\theta$ $= \dfrac{\sin^2\theta}{\cos\theta} + \dfrac{\cos^2\theta}{\cos\theta}$ $= \dfrac{\sin^2\theta + \cos^2\theta}{\cos\theta} = \dfrac{1}{\cos\theta} = \sec\theta$

46. $\dfrac{3\sin x}{\cos x} = \dfrac{\cos x}{\cos x}$ $3\tan x = 1$ $\tan x = \dfrac{1}{3}$ $x \approx 0.32, 3.46$

47. $\sin x \tan x - \sin x = 0$ $\sin x(\tan x - 1) = 0$ $\sin x = 0$ or $\tan x = 1$ $x = 0, \pi$ $x = \dfrac{\pi}{4}, \dfrac{5\pi}{4}$

48. $\cos(45° - 30°)$ $= \cos 45° \cos 30° + \sin 45° \sin 30°$ $= \dfrac{\sqrt{2}}{2}\cdot\dfrac{\sqrt{3}}{2} + \dfrac{\sqrt{2}}{2}\cdot\dfrac{1}{2} = \dfrac{\sqrt{6}+\sqrt{2}}{4}$

49. $2b = 4ab + 4c$ $2b - 4ab = 4c$ $2b(1 - 2a) = 4c$ $b = \dfrac{4c}{2(1-2a)} = \dfrac{2c}{1-2a}$

50. $\log_b 100 = \log_b(5^2 \cdot 4)$ $= 2\log_b 5 + \log_b 4$ $= 2x + y$

51. See the definition of radian measure.

52. 4 ft 6 in. = 4.5 ft $\sin\theta = \dfrac{opp.}{hyp.} = \dfrac{4.5}{6}$ $\theta = \sin^{-1}\dfrac{4.5}{6}$

5.7 Law of Cosines

Objectives

1. To prove the Law of Cosines
2. To solve triangles using the Law of Cosines
3. To find the area of oblique triangles using the SAS area formula and Heron's formula

Vocabulary

Heron's formula
Law of Cosines
SAS triangle area formula
semiperimeter

Assessment

- Quiz 5C covers Sections 5.6–5.7.

Reading and Writing Mathematics

Summarize the various methods of solving a triangle.

The summary should include the Pythagorean Theorem and right-triangle trigonometry for right triangles. The Law of Sines can be used for AAS, ASA, and SSA cases and the Law of Cosines for SAS and SSS cases. The fact that the sum of all the angles of any triangle is 180° can also be applied.

5.7 Law of Cosines

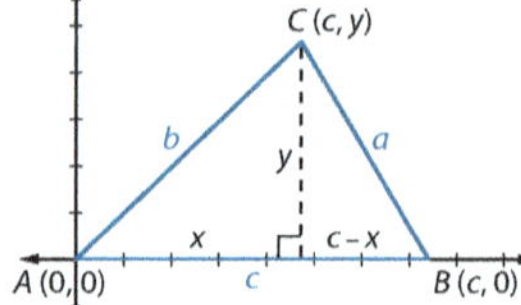

Using Heron's formula is the easiest method for calculating areas of triangular lots.

While the Law of Sines is used to solve triangles when given ASA, AAS, or SSA, the Law of Cosines is used when you know the lengths of all three sides (SSS) or the measures of two sides and the included angle (SAS).

The Law of Cosines can be derived by placing $\triangle ABC$ so that A is at the origin and $\overline{AB}$ is on the positive x-axis. Draw the altitude from C to the x-axis as illustrated and note that the coordinates of C are $x = b \cos A$ and $y = b \sin A$. Then apply the Pythagorean Theorem.

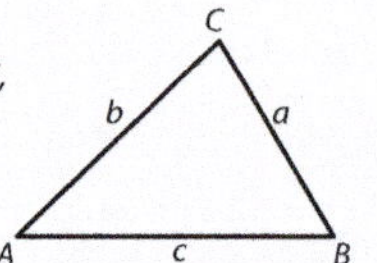

$$a^2 = (c - x)^2 + y^2$$
$$= c^2 - 2cx + x^2 + y^2$$
$$= c^2 - 2c(b \cos A) + (b \cos A)^2 + (b \sin A)^2$$
$$= c^2 - 2cb \cos A + b^2(\cos^2 A + \sin^2 A)$$
$$= c^2 - 2cb \cos A + b^2$$
$$= b^2 + c^2 - 2bc \cos A$$

Similar proofs that place B and C at the origin produce two other forms of this identity.

Before calculators were widely available, the Law of Tangents was often used instead of the Law of Cosines since it required less use of logarithmic tables. Now that calculators are commonplace, the Law of Tangents is rarely included in textbooks.

LAW OF COSINES

For $\triangle ABC$ with side lengths a, b, and c opposite angles A, B, and C, respectively,
$$a^2 = b^2 + c^2 - 2bc \cos A,$$
$$b^2 = a^2 + c^2 - 2ac \cos B, \text{ and}$$
$$c^2 = a^2 + b^2 - 2ab \cos C.$$

Example 1   Solving the SAS Case

Solve $\triangle ABC$ if $A = 23°$, $b = 12$ ft, and $c = 9$ ft.

Answer

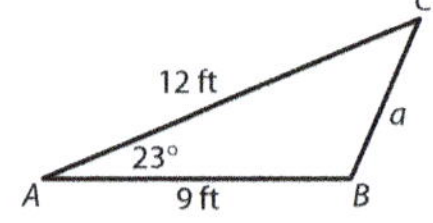

$$a^2 = b^2 + c^2 - 2bc \cos A$$
$$= 12^2 + 9^2 - 2(12)(9)(\cos 23°)$$
$$a^2 \approx 26.17$$
$$a \approx 5.1$$

$$\frac{\sin C}{9} \approx \frac{\sin 23°}{5.1}$$
$$\sin C \approx \frac{9 \sin 23°}{5.1} \approx 0.6874$$
$$C \approx 43.4°$$

$$B \approx 180 - 23 - 43.4 = 113.6°$$

1. Sketch $\triangle ABC$.
2. Find a by applying the Law of Cosines.
3. Use $\dfrac{\sin C}{c} = \dfrac{\sin A}{a}$ to find the smaller of the two remaining angles, which must be acute.
4. Determine the measure of $\angle B$.

SKILL ✔ EXERCISE 1

PRESENTATION

Lesson Opener

1. Explain why the Law of Sines cannot be used to solve the SSS case.
   One variable (the angle measure) will be missing in each ratio, leaving two unknowns in the equation.

2. Explain why the Law of Sines cannot be used to solve the SAS case.
   One variable will be missing in each ratio, either an angle measure in one ratio and a side length in the other, or an angle measure in each ratio, leaving two unknowns in the equation.

Discuss the Lesson Opener. Then engage the students by asking them to name the three cases in which the Law of Sines can be used to solve a triangle. (*ASA, AAS, and SSA*) Share that the *Law of Cosines* enables us to solve the SSS and SAS cases. When deriving the Law of Cosines, ask why the Pythagorean Theorem cannot be directly applied to $\triangle ABC$. (*It is not a right triangle.*) Drawing the altitude creates two right triangles. After applying the Pythagorean Theorem to the triangle on the right, expressing everything in terms of the original oblique $\triangle ABC$ produces the Law of Cosines.

Point out that knowing the pattern of the Law of Cosines is preferred to memorizing the three statements listed in the definition box: a side squared is the sum of the other two sides squared less twice the product of those two sides and the cosine of the angle opposite the first side (which is named with the same letter).

Consider asking the students what the resulting equation for the Law of Cosines would be if the given angle is a right angle. (*the Pythagorean Theorem*)

Use Example 1 to apply the Law of Cosines to solve an oblique triangle. Note that the exact value for a is used in step 3 to obtain the sin C. Encourage the students to store exact values in their calculators for quick retrieval and accurate solutions.

Instead of using the Law of Sines to find the smaller angle in step 3 above, the Law of Cosines can be used to find either angle. Solving $a^2 = b^2 + c^2 - 2bc \cos A$ for $\cos A$ is helpful in this case and when solving the SSS case.

$$2bc \cos A = b^2 + c^2 - a^2$$
$$\cos A = \frac{b^2 + c^2 - a^2}{2bc}$$

Similarly, $\cos B = \frac{a^2 + c^2 - b^2}{2ac}$ and $\cos C = \frac{a^2 + b^2 - c^2}{2ab}$.

Example 2 Solving the SSS Case

Solve $\triangle FDG$ if $f = 8$, $d = 4$, and $g = 10$.

Answer

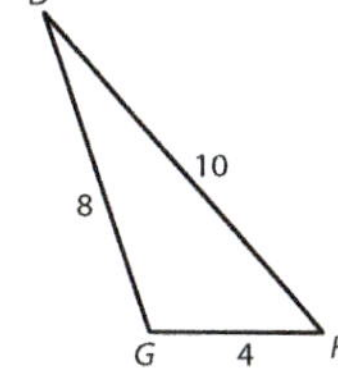

1. Sketch the triangle.

$$\cos G = \frac{f^2 + d^2 - g^2}{2fd}$$
$$= \frac{8^2 + 4^2 - 10^2}{2(8)(4)}$$
$$= -0.3125$$
$$G \approx 108.2°$$

2. Since you are given lengths for three sides of the triangle (SSS), use the Law of Cosines to find the measure of the angle opposite the largest side.

using $\dfrac{\sin F}{f} = \dfrac{\sin G}{g}$,

$$\frac{\sin F}{8} \approx \frac{\sin 108.2°}{10}$$
$$\sin F \approx \frac{8 \sin 108.2°}{10} \approx 0.7599$$
$$F \approx 49.5°$$

3. Find a second angle using either the Law of Cosines or the Law of Sines.

$D \approx 180 - 108.2 - 49.5 = 22.3°$

4. Find the measure of $\angle D$ by subtraction.

SKILL ✔ EXERCISE 3

Many problems can be solved by making a sketch of an oblique triangle, labeling the known parts, and then applying the Law of Sines or the Law of Cosines to solve for the desired parts of the triangular model.

TIP

Initially finding the largest angle with the Law of Cosines guarantees that the other two angles must be acute.

Example 3 Triangulating a Location

An emergency call is made 5.6 mi from one cell tower and 2.7 mi from a second cell tower located 4 mi south of the first. State the two possible headings from the second cell tower for the location of the emergency call.

Answer

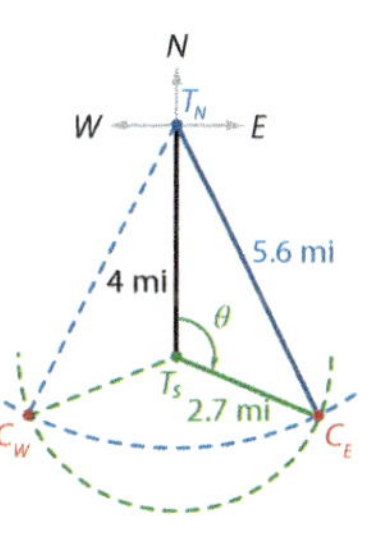

1. Sketch the locations of the towers and the two points 5.6 mi from the northern tower T_N and 2.7 mi from the southern tower T_S.

$$\cos \theta = \frac{4^2 + 2.7^2 - 5.6^2}{2(4)(2.7)} \approx -0.3736$$
$$\theta \approx 111.9°$$

2. Use the Law of Cosines to find θ, the heading to the point southeast of the towers.

$360 - 111.9 = 248.1°$

3. Determine the heading to the second point that is southwest of the towers.

SKILL ✔ EXERCISE 29

Additional Exercises

Solve each $\triangle ABC$. Round answers to the nearest tenth.

1. $a = 3$, $b = 9$, $c = 10$
   $A = 17.1°$, $B = 62.2°$, $C = 100.7°$

2. $a = 125$, $c = 85$, $B = 55°$
   $b = 103.3$, $A = 82.6°$, $C = 42.4°$

3. $a = 38$, $b = 68$, $c = 48$
   $A = 32.9°$, $B = 103.9°$, $C = 43.3°$

4. $a = 250$, $b = 305$, $C = 18°15'$
   $c = 103.4$, $A = 49.2°$, $B = 112.5°$

5. Two boats leave the same dock at the same time. The first boat travels at a bearing of 25° and averages 35 mi/hr and the second averages 20 mi/hr at a bearing of 70°. Find the distance between the boats after 30 min.
   12.6 mi

6. A surveyor measures a triangular lot and determines the following lengths and bearings for two of the sides: 215 m, 38°24′ east of north and 182 m, 28°16′ east of south. Find the length and bearing of the third side.
   332.2 m, 8°12′ east of north

Find the area of each triangle. Round all answers to the nearest tenth.

7.

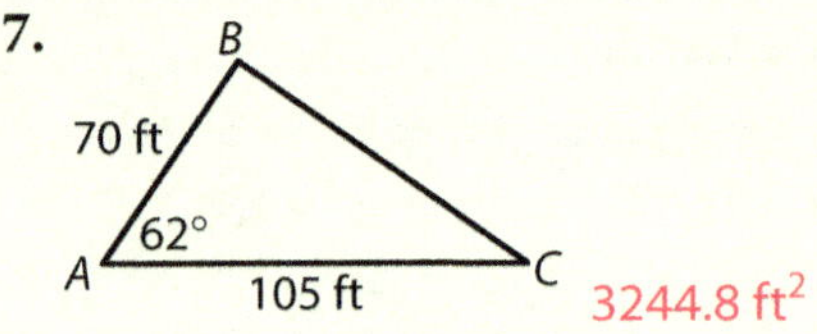

3244.8 ft²

8.

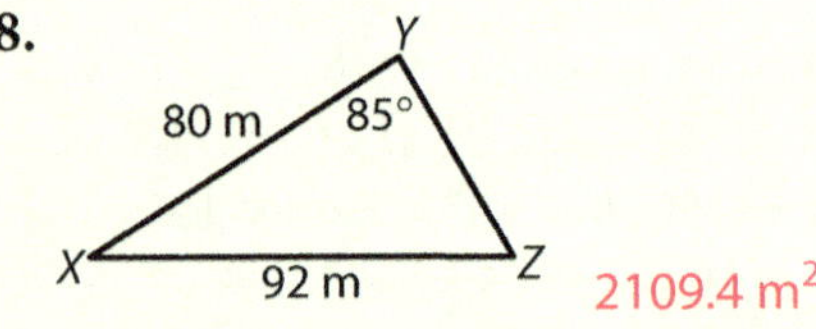

2109.4 m²

9.

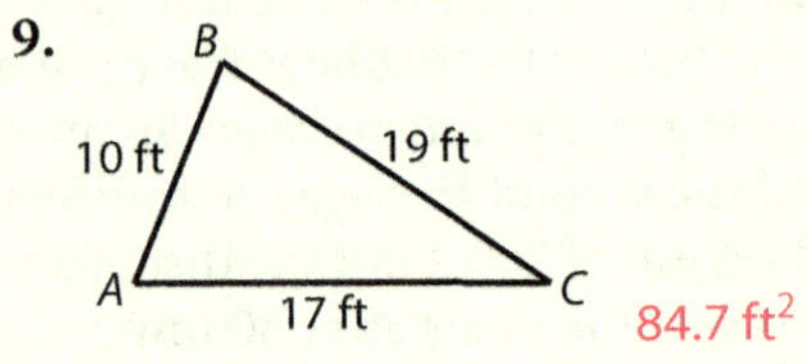

84.7 ft²

10.

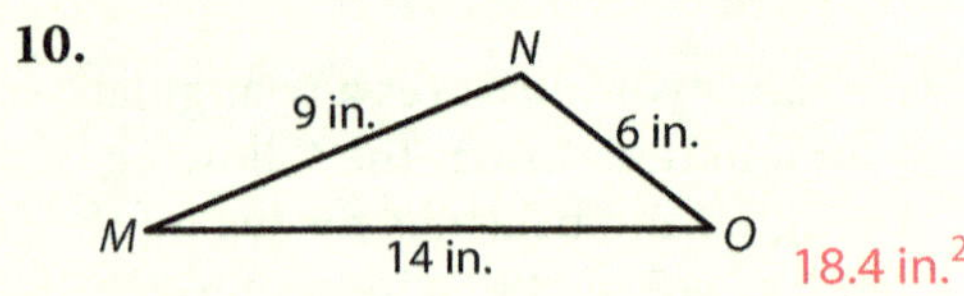

18.4 in.²

Assignments

- **Minimum:** 1, 4, 6–7, 13–20, 23, 25–26, 30–31, 33–35, 37, 43–45, 47, 49, 52
- **Standard:** 2–3, 8, 11, 13–16, 18–20, 22–23, 25, 28–31, 33–37, 39, 44; 45–51 odd
- **Extended:** 2, 5, 8, 11, 14–16, 18, 20–22, 25, 28, 31–33, 37–38, 40–42, 44–48, 50, 51

The distance from a third cell tower that is not on the same north-south line as the first two would be used to determine the exact location of the emergency call.

Formulas can be derived for the area of a triangle in terms of two side lengths and the measure of the included angle (SAS). When $\angle C$ and sides a and b are known, $\sin C = \frac{h}{a}$ and $h = a \sin C$. Substituting for h in $Area = \frac{1}{2}bh$ produces the formula $Area = \frac{1}{2}ab \sin C$. Similar reasoning can be applied to derive any of the three forms of this equation.

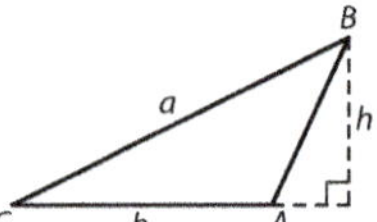

SAS Triangle Area Formulas		
$Area = \frac{1}{2}ab \sin C$	$Area = \frac{1}{2}bc \sin A$	$Area = \frac{1}{2}ac \sin B$

When the given information involves SAA, AAS, or ASA, use the Law of Sines to find the additional side or angle before applying an SAS triangle area formula.

Example 4 Applying the SAS Triangle Area Formula

Find the area of $\triangle PQR$.

Answer

Since $q > r$, there is only one possible triangle.

1. Determine the number of possible triangles.

$$\frac{\sin R}{9} = \frac{\sin 56°}{10}$$

2. Use the Law of Sines to find R.

$$\sin R = \frac{9 \sin 56°}{10} \approx 0.7461$$

$$R \approx 48.3°$$

$$P \approx 180 - 56 - 48.3 = 75.7°$$

3. Find P.

$$A = \frac{1}{2}rq \sin P$$

4. Apply the SAS triangle area formula.

$$\approx \frac{1}{2}(9)(10) \sin 75.7° \approx 43.6 \text{ m}^2$$

SKILL ✓ EXERCISE 13

To find the area when all three side lengths of a triangle are known (SSS), you could use the Law of Cosines to find an angle and then apply an SAS triangle area formula. Heron's formula provides a quick alternative.

HERON'S FORMULA

The area of a triangle having sides of length a, b, and c is given by
$$Area = \sqrt{s(s-a)(s-b)(s-c)} \text{ where the triangle's } semiperimeter \ s = \frac{1}{2}(a+b+c).$$

Explain that many practical applications involving oblique triangles cannot be solved using the Pythagorean Theorem and basic right-triangle trigonometry. The Law of Cosines is used in Example 3 to solve such a real-world problem. Note that the angle opposite the longest side is found first. Consider extending this example by asking students to find the possible headings from the first tower. They could either apply the Law of Cosines a second time or apply the Law of Sines, being confident that this angle must be acute ($\approx 153.4°$, $\approx 206.6°$, respectively).

Motivational Idea Surveyors use triangulation to determine unknown distances, often those that are difficult to measure because of obstacles. The GPS system uses a similar three-dimensional method called trilateration, which determines the intersection of three spheres to identify position, including elevation and speed. Use the Internet keyword search *GPS trilateration video*, or challenge the students to research trilateration using the Internet keyword search *GPS trilateration*.

Present the *SAS triangle area formulas* and present Example 4, in which the Law of Sines must be applied to obtain the required SAS case.

Introduce *Heron's formula* as an easier way to find the area of a triangle when all side lengths are known. More advanced students will appreciate the combination of several identities and unique algebraic manipulations required to prove this powerful theorem. Ask students to explain the following steps of the proof.

3rd: Factor the difference of squares.
4th: Distribute and regroup.
5th: Rewrite using squares of a difference and a sum.
6th: Move $\frac{1}{4}$ under the radical as $\frac{1}{16}$.
7th: Rewrite as a product of fractions.
8th: Add and subtract a, b, or c in the numerator of successive fractions.

A proof of the formula uses an algebraic expression for sin C that is derived from a Pythagorean identity, the Law of Cosines, and some clever algebraic manipulations.

$$\sin C = \sqrt{1 - \cos^2 C} = \sqrt{1 - \left(\frac{a^2 + b^2 - c^2}{2ab}\right)^2} = \frac{\sqrt{4a^2b^2 - (a^2 + b^2 - c^2)^2}}{2ab}$$

substituting into $Area = \frac{1}{2}ab \sin C$:

$$Area = \frac{1}{2}ab \frac{\sqrt{4a^2b^2 - (a^2 + b^2 - c^2)^2}}{2ab}$$

$$= \frac{1}{4}\sqrt{4a^2b^2 - (a^2 + b^2 - c^2)^2}$$

$$= \frac{1}{4}\sqrt{[2ab - (a^2 + b^2 - c^2)][2ab + (a^2 + b^2 - c^2)]}$$

$$= \frac{1}{4}\sqrt{[c^2 - (a^2 - 2ab + b^2)][(a^2 + 2ab + b^2) - c^2]}$$

$$= \frac{1}{4}\sqrt{[c^2 - (a - b)^2][(a + b)^2 - c^2]}$$

$$= \sqrt{\frac{(c - (a - b))(c + (a - b))((a + b) - c)((a + b) + c)}{16}}$$

$$= \sqrt{\frac{(c + b - a)}{2} \cdot \frac{(c + a - b)}{2} \cdot \frac{(a + b - c)}{2} \cdot \frac{(a + b + c)}{2}}$$

noting that $2s = a + b + c$:

$$Area = \sqrt{\frac{(c + b - a + a)}{2} \cdot \frac{(c + a - b - b + b)}{2} \cdot \frac{(a + b - c - c + c)}{2} \cdot \frac{(a + b + c)}{2}}$$

$$= \sqrt{\frac{(2s - 2a)}{2} \cdot \frac{(2s - 2b)}{2} \cdot \frac{(2s - 2c)}{2} \cdot \frac{(2s)}{2}}$$

$$= \sqrt{s(s - a)(s - b)(s - c)}$$

Thankfully, the application of Heron's formula is much easier than its proof!

Example 5   Applying Heron's Formula

A developer desires to purchase a triangular lot with 1100 ft of frontage on Main Street, 820 ft of frontage on Acadia Drive, and 638 ft of frontage on Rutherford Road. Calculate the acreage of the lot. (1 acre = 43,560 ft^2)

Answer

$s = \frac{1}{2}(1100 + 820 + 638) = 1279$ ft

1. Determine the lot's semiperimeter.

$Area = \sqrt{1279(1279 - 1100)(1279 - 820)(1279 - 638)}$
$\approx 259{,}536$ ft^2

2. Apply Heron's formula.

$259{,}536$ ft$^2\left(\dfrac{1 \text{ acre}}{43{,}560 \text{ ft}^2}\right) \approx 5.96$ acres

3. Convert to acres.

————————— SKILL ✔ **EXERCISE 15**

Solutions

❯ A. Exercises

1. $a^2 = 4^2 + 6^2 - 2(4)(6)\cos 55° \approx 24.5$
$a \approx 4.9$
$\dfrac{\sin C}{4} \approx \dfrac{\sin 55°}{4.9}$
$C \approx \sin^{-1}\left(\dfrac{4\sin 55°}{4.9}\right) \approx 41.5°$
$B \approx 180 - 55 - 41.5 = 83.5°$

2. $c^2 = 3^2 + 6^2 - 2(3)(6)\cos 95° \approx 48.1$
$c \approx 6.9$
$\dfrac{\sin B}{3} \approx \dfrac{\sin 95°}{6.9}$
$B \approx \sin^{-1}\left(\dfrac{3\sin 95°}{6.9}\right) \approx 25.5°$
$A \approx 180 - 95 - 25.5 = 59.5°$

3. $Y = \cos^{-1}\left(\dfrac{8^2 + 10^2 - 3^2}{2(8)(10)}\right) \approx 14.4°$
$\dfrac{\sin W}{8} \approx \dfrac{\sin 14.4°}{3}$
$W \approx \sin^{-1}\left(\dfrac{8\sin 14.4°}{3}\right) \approx 41.4°$
$X \approx 180 - 14.4 - 41.4 = 124.2°$

4. $A = \cos^{-1}\left(\dfrac{5^2 + 8^2 - 6^2}{2(5)(8)}\right) \approx 48.5°$
$\dfrac{\sin B}{5} \approx \dfrac{\sin 48.5°}{6}$
$B \approx \sin^{-1}\left(\dfrac{5\sin 48.5°}{6}\right) \approx 38.6°$
$C \approx 180 - 48.5 - 38.6 = 92.9°$

5. $c^2 = 9^2 + 13^2 - 2(9)(13)\cos 28°$
$\approx 43.39; c \approx 6.6$
$\dfrac{\sin A}{9} \approx \dfrac{\sin 28°}{6.59}$
$A \approx \sin^{-1}\left(\dfrac{9\sin 28°}{6.59}\right) \approx 39.9°$
$B \approx 180 - 28 - 39.9 = 112.1°$

9th: Substitute $2s$ for $a + b + c$.
10th: Simplify.

Point out the technique of adding and subtracting the same quantity, which allows the numerators of the fractions to be written using $2s$.

In many real-world applications, as in Example 5, the triangle's side lengths are the easiest measurements to obtain. Emphasize the need to find the *semiperimeter*, s, first when applying Heron's formula.

Interactive Activity Like the Law of Cosines, the Law of Tangents,

$$\frac{a - b}{a + b} = \frac{\tan\frac{A - B}{2}}{\tan\frac{A + B}{2}},$$

is used to solve the SAS case. Some preferred using the Law of Tangents before calculators became common. The solving of a system of equations involving $A + B$ and $A - B$ before applying the Law of Sines to find the missing side length was preferred over the use of logarithmic tables to solve for

$$c = \sqrt{a^2 + b^2 - 2ab \cos C}$$

directly. Using the Internet keyword search *Law of Tangents*, have the students break into small groups and challenge them to find and study proofs for the Law of Tangents.

TIPS

Ex. 34 The Flatiron Building in Times Square is actually the shape of a right triangle, but the given side measurements are rounded to the nearest foot, which determines an angle of about $87°20'$.

Ex. 41 Using the Law of Cosines removes any ambiguity in the SSA case. While the resulting quadratic equation involving the cosine of an angle can be difficult to solve manually, the directness of this solution allows programmers to readily write code that automates the solving of triangles in this case.

6. $a^2 = 26^2 + 18^2 - 2(26)(18)\cos 64°$
$\approx 589.9;\ a \approx 24.3$
$\dfrac{\sin B}{26} \approx \dfrac{\sin 64°}{24.3}$
$B \approx \sin^{-1}\left(\dfrac{26\sin 64°}{24.3}\right) \approx 74.2°$
$C \approx 180 - 64 - 74.2 = 41.8°$

7. $A = \cos^{-1}\left(\dfrac{47^2 + 64^2 - 52^2}{2(47)(64)}\right) \approx 53.2°$
$\dfrac{\sin B}{47} \approx \dfrac{\sin 53.2°}{52}$
$B \approx \sin^{-1}\left(\dfrac{47\sin 53.2°}{52}\right) \approx 46.4°$
$C \approx 180 - 53.2 - 46.4 = 80.4°$

8. $A = \cos^{-1}\left(\dfrac{22^2 + 25^2 - 35^2}{2(22)(25)}\right) \approx 96.1°$
$\dfrac{\sin B}{22} \approx \dfrac{\sin 96.1°}{35}$
$B \approx \sin^{-1}\left(\dfrac{22\sin 96.1°}{35}\right) \approx 38.7°$
$C \approx 180 - 96.1 - 38.7 \approx 45.2°$

9. $z^2 = 6^2 + 10^2 - 2(6)(10)\cos 60° = 76$
$z \approx 8.7$
$\dfrac{\sin X}{6} \approx \dfrac{\sin 60°}{8.7}$
$X \approx \sin^{-1}\left(\dfrac{6\sin 60°}{8.7}\right) \approx 36.6°$
$Y \approx 180 - 60 - 36.6 = 83.4°$

10. $x^2 = 31^2 + 20^2 - 2(31)(20)\cos 42.5°$
$\approx 446.8;\ x \approx 21.1$
$\dfrac{\sin Z}{20} \approx \dfrac{\sin 42.5°}{21.1}$
$Z \approx \sin^{-1}\left(\dfrac{20\sin 42.5°}{21.1}\right) \approx 39.7°$
$Y \approx 180 - 39.7 - 42.5 = 97.8°$

11. $X = \cos^{-1}\left(\dfrac{9^2 + 14^2 - 11^2}{2(9)(14)}\right) \approx 51.8°$
$\dfrac{\sin Y}{9} \approx \dfrac{\sin 51.8°}{11}$
$Y \approx \sin^{-1}\left(\dfrac{9\sin 51.8°}{11}\right) \approx 40.0°$
$Z \approx 180 - 51.8 - 40.0 \approx 88.2°$

12. $X = \cos^{-1}\left(\dfrac{13^2 + 21.2^2 - 17^2}{2(13)(21.2)}\right) \approx 53.3°$
$Z = \cos^{-1}\left(\dfrac{17^2 + 21.2^2 - 13^2}{2(17)(21.2)}\right) \approx 37.8°$
$Y \approx 180 - 53.3 - 37.8 = 88.9°$

13. $Area = \frac{1}{2}(13)(10)\sin 61°$
$\approx 56.85\ \text{cm}^2$

14. $b > c$; one solution
$\dfrac{\sin C}{10} = \dfrac{\sin 76°}{12}$
$C = \sin^{-1}\left(\dfrac{5\sin 76°}{6}\right) \approx 54°$
$A = 180 - (76 + 54) = 50°$
$Area = \frac{1}{2}(12)(10)\sin 50° \approx 45.96\ \text{in.}^2$

15. $s = \frac{1}{2}(11 + 23 + 18) = 26$
$Area =$
$\sqrt{26(26 - 11)(26 - 18)(26 - 23)}$
$\approx 96.7\ \text{ft}^2$

Solve each triangle. Round answers to the nearest tenth.

1.   **2.**

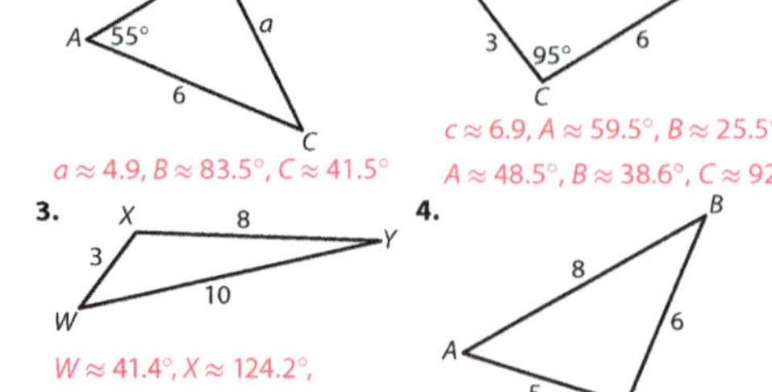

$a \approx 4.9, B \approx 83.5°, C \approx 41.5°$

$c \approx 6.9, A \approx 59.5°, B \approx 25.5°$
$A \approx 48.5°, B \approx 38.6°, C \approx 92.9°$

3.   **4.**

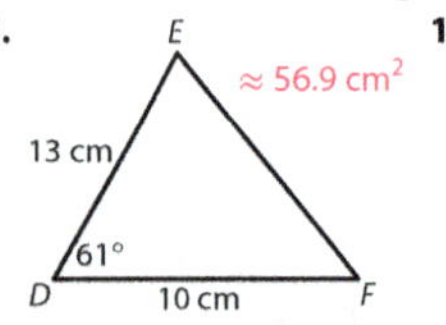 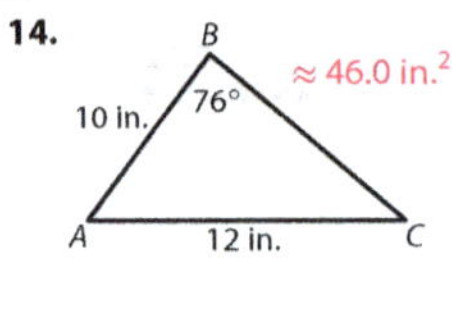

$W \approx 41.4°, X \approx 124.2°,$
$Y \approx 14.4°$

Solve each $\triangle ABC$. Round answers to the nearest tenth.

5. $a = 9, b = 13,$ and $C = 28°$

6. $b = 26, c = 18,$ and $A = 64°$

7. $a = 52, b = 47,$ and $c = 64$

8. $a = 35, b = 22,$ and $c = 25$

Solve each $\triangle XYZ$. Round measures to the nearest tenth.

9. $x = 6, y = 10,$ and $Z = 60°$

10. $y = 31, z = 20,$ and $X = 42.5°$

11. $x = 11, y = 9,$ and $z = 14$

12. $x = 17, y = 21.2,$ and $z = 13$

Find the area of each triangle.

13.   **14.**

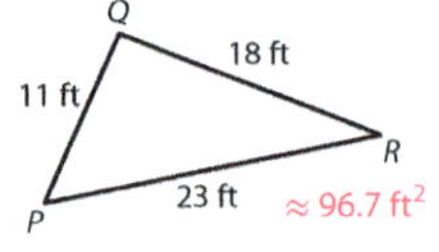

$\approx 56.9\ \text{cm}^2$ $\approx 46.0\ \text{in.}^2$

15. Find the area of $\triangle PQR$ using Heron's formula.

16. Find the area of $\triangle PQR$ using the Law of Cosines and the SAS triangle area formula. $\approx 96.7\ \text{ft}^2$

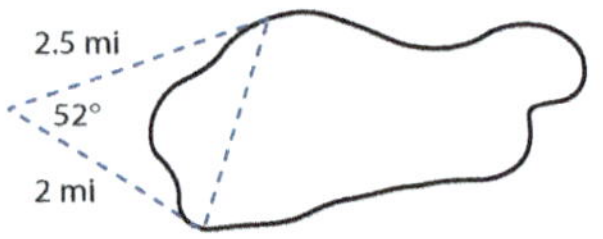

$\approx 96.7\ \text{ft}^2$

17. Use the given measurements to find the indicated distance across the swamp. $\approx 2.0\ \text{mi}$

---

16. $P = \cos^{-1}\left(\dfrac{11^2 + 23^2 - 18^2}{2(11)(23)}\right) \approx 49.9°$
$Area = \frac{1}{2}(11)(23)\sin 49.9° \approx 96.7\ \text{ft}^2$

17. $d^2 = 2^2 + 2.5^2 - 2(2)(2.5)\cos 52°$
$\approx 4.1;\ d \approx 2.0\ \text{mi}$

18. $\dfrac{\sin Y}{218} = \dfrac{\sin 87°}{356}$
$Y = \sin^{-1}\left(\dfrac{218\sin 87°}{356}\right) \approx 37.7°$
$X \approx 180 - 87 - 37.7 = 55.3°$
$\dfrac{x}{\sin 55.3°} \approx \dfrac{356}{\sin 87°}$
$x \approx \dfrac{356\sin 55.3°}{\sin 87°} \approx 293\ \text{ft}$

---

18. An investor knows the lengths of two sides of a triangular lot, but thick woods prevent her from measuring the third side. If she measures θ as 87°, calculate the length of the third side (to the nearest foot). **293 ft**

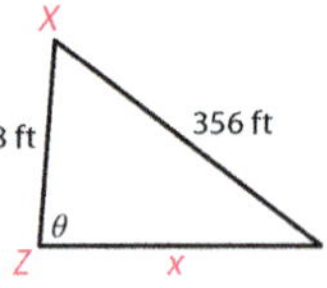

Use the Law of Sines or the Law of Cosines to solve each $\triangle ABC$. Round measures to the nearest tenth.

19. $a = 7, b = 9, c = 10$ $A \approx 42.8°, B \approx 60.9°, C \approx 76.3°$

20. $A = 57°, B = 31°, b = 9$ $C = 92°, a \approx 14.7, c \approx 17.5$

21. $A = 24°, C = 103°, b = 8$ $B = 53°, a \approx 4.1, c \approx 9.8$

22. $C = 38°, a = 13, b = 9$ $A \approx 98.8°, B \approx 43.2°, c \approx 8.1$

23. $C = 43°, a = 12, c = 9$ $A \approx 65.4°, B \approx 71.6°, b \approx 12.5$ or $A \approx 114.6°, B \approx 22.4°, b \approx 5.0$

24. $a = 17.4, b = 15, c = 24.2$ $A \approx 45.6°, B \approx 38°, C \approx 96.4°$

Find the area of each triangle (to the nearest tenth).

25. $\triangle ABC$ with $a = 9$ cm, $b = 14$ cm, and $C = 36°$

26. $\triangle PQR$ with $p = 27$ m, $q = 31$ m, and $r = 40$ m

27. $\triangle XYZ$ with $x = 38$ in., $y = 21$ in., and $z = 47$ in.

28. $\triangle DEF$ with $D = 31°, f = 12$ m, $d = 6.2$ m

29. Find the length of the diagonals of a parallelogram with adjacent sides of 10 cm and 7.2 cm if the measure of its acute angles is 56°. ≈ 8.4 cm and ≈ 15.2 cm

30. A baseball diamond is a square with sides measuring 90 ft. Find the distance of the fielder's throw to third base after catching a ball hit 360 ft from home plate to straightaway center field (directly over second base). ≈ 303 ft

31. A surveyor is positioned 418 ft from one end of a proposed mountain tunnel and 371 ft from the other end. If he measures a 69° angle between the sight lines to the ends of the tunnel, how long will the tunnel be? ≈ 449 ft

32. A radio antenna sits at the edge of a 200 ft tall cliff. A fisherman on the sea below the cliff records a 21° angle of elevation to the top of the antenna while a friend at the bottom of the antenna records a 10° angle of depression to the fisherman. What is the height of the antenna? ≈ 235 ft

33. A boat is located 5 km from one buoy and 14 km from a second buoy anchored 10 km east of the first buoy. Find the boat's two possible bearings from the first buoy. $\approx 225.2°$ and $314.8°$

---

19. $A = \cos^{-1}\left(\dfrac{9^2 + 10^2 - 7^2}{2(9)(10)}\right) \approx 42.8°$
$B = \cos^{-1}\left(\dfrac{7^2 + 10^2 - 9^2}{2(7)(10)}\right) \approx 60.9°$
$C \approx 180 - 42.8 - 60.9 \approx 76.2°$

20. $C = 180 - 57 - 31 = 92°$
$\dfrac{a}{\sin 57°} = \dfrac{9}{\sin 31°}$
$a = \dfrac{9\sin 57°}{\sin 31°} \approx 14.7$
$\dfrac{c}{\sin 92°} = \dfrac{9}{\sin 31°}$
$c = \dfrac{9\sin 92°}{\sin 31°} \approx 17.5$

34. The Flatiron Building in New York City is famous for its triangular shape. Find the area of its base if its sides measure 190 ft, 173 ft, and 87 ft. $\approx 7517.4 \text{ ft}^2$

35. Offices at the tip of the Flatiron Building (described in exercise 34) are much sought after because they offer spectacular views. Calculate the measure of the smallest angle formed by the building's outer walls.

36. Given $\triangle ABC$, find x.
$x \approx 4.7$

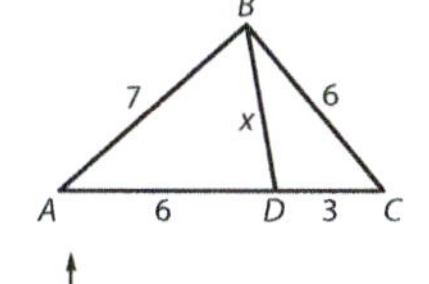

37. **Prove:** Copy the figure and use it to derive the Law of Cosines, $c^2 = a^2 + b^2 - 2ab \cos C$.

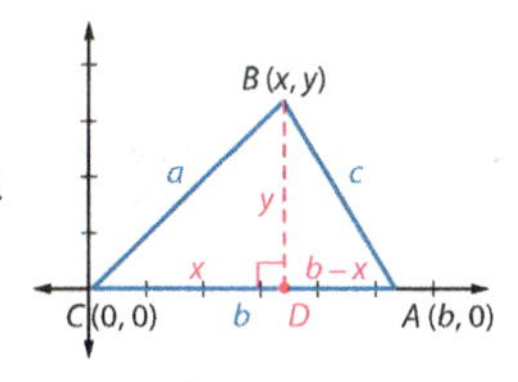

38. **Prove:** Copy the figure and use it to derive the Law of Cosines, $b^2 = a^2 + c^2 - 2ac \cos B$.

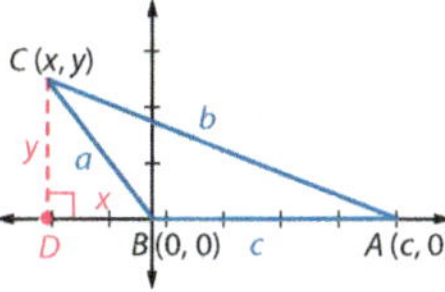

43. Determine the account balance of a $10,000 investment compounded monthly at 5% interest for 7 yr. [3.1] $14,180.36

44. Evaluate $\log_2 \frac{1}{16}$ [3.2] -4

45. Solve $3^{2x+3} = 5^{2x}$. [3.4] $x \approx 3.226$

46. Determine if each equation is an identity. If not, explain. [2.1, 5.1]
a. $\sqrt{x^2} = x$ **b.** $\sqrt[3]{x^3} = x$

47. Solve $\sqrt{2x+3} - 1 = \sqrt{3x-5}$. [2.1] $x = 3$

48. State the period of $f(x) = g(x) + h(x)$ if $g(x) = \sin 3x$ and $h(x) = 2 \cos 3x$. [4.7] $\frac{2\pi}{3}$

49. Find $\cos \theta$. [4.2] C

A. $\frac{3}{7}$ **D.** $\frac{\sqrt{58}}{7}$

B. $\frac{7}{3}$ **E.** $\frac{4\sqrt{10}}{7}$

C. $\frac{2\sqrt{10}}{7}$

46a. not an identity; only true when $x \geq 0$
46b. identity

C. Exercises

39. The illustrated rectangular shipping box is 3 ft wide, 4 ft long, and 2 ft high. Find the measure of $\angle CAB$. $\approx 44.3°$

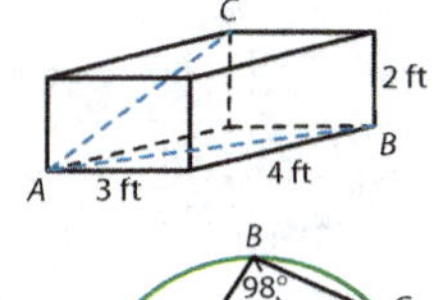

40. Cyclic quadrilaterals can be inscribed in a circle and have supplementary opposite angles. Find the lengths of the diagonals BD and AC of cyclic quadrilateral $ABCD$. ≈ 6.2 and 7.0

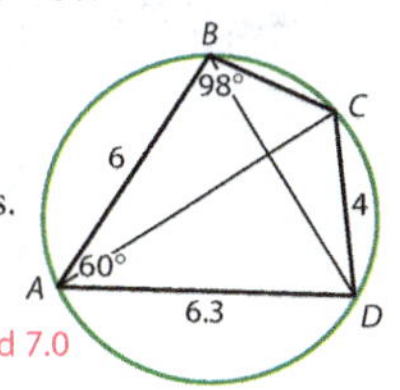

41. While the Law of Sines is commonly used to solve the SSA case, an alternative solution applies the Law of Cosines and solves the resulting quadratic equation. Use the Law of Cosines to find side c in $\triangle ABC$ if $a = 9$, $b = 12$, and $B = 60°$. $c \approx 13.6$

42. Find the exact value of a in $\triangle ABC$ with $A = 60°$, $c = \dfrac{1}{\sqrt{6} + \sqrt{2}}$, and $b = \dfrac{1}{\sqrt{6} - \sqrt{2}}$. $a = \dfrac{\sqrt{3}}{2}$

35. $\approx 27.2°$

50. Which value of x produces a relative minimum for $g(x) = 2 \sin 2x$? [4.4] D
A. $\frac{\pi}{4}$ **C.** $\frac{\pi}{2}$ **E.** π
B. $\frac{3\pi}{8}$ **D.** $\frac{3\pi}{4}$

51. Determine the number of possible triangles that exist for $\triangle ABC$ with $A = 60°$, $a = 7$, and $b = 8$. [5.6] C
A. 0   **C.** 2   **E.** > 3
B. 1   **D.** 3

52. Which expression represents YZ in $\triangle XYZ$? [5.6] B
A. $YZ = \dfrac{33 \sin 56°}{\sin 27°}$
B. $YZ = \dfrac{33 \sin 27°}{\sin 56°}$
C. $YZ = \dfrac{33 \sin 97°}{\sin 56°}$
D. $YZ = \dfrac{33 \sin 56°}{\sin 97°}$
E. cannot be determined

---

21. $B \approx 180 - 24 - 103 = 53°$
$$\frac{a}{\sin 24°} = \frac{8}{\sin 53°}$$
$$a = \frac{8 \sin 24°}{\sin 53°} \approx 4.1$$
$$\frac{8}{\sin 53°} = \frac{c}{\sin 103°}$$
$$c = \frac{8 \sin 103°}{\sin 53°} \approx 9.8$$

22. $c^2 = 13^2 + 9^2 - 2(13)(9) \cos 38°$
≈ 65.6; $c \approx 8.1$
$$\frac{\sin B}{9} \approx \frac{\sin 38°}{8.1}$$
$$B \approx \sin^{-1}\left(\frac{9 \sin 38°}{8.1}\right) \approx 43.2°$$
$$A \approx 180 - 38 - 43.2 = 98.8°$$

23. SSA case where $h = a \sin C$
$\approx 8.18 < c < a$; two solutions
$$\frac{\sin A}{12} = \frac{\sin 43°}{9}$$
$$\sin A = \frac{12 \sin 43°}{9}$$
case 1: $A = \sin^{-1}\left(\frac{12 \sin 43°}{9}\right) \approx 65.4°$
$B \approx 180 - 43 - 65.4 = 71.6°$
$$\frac{b}{\sin 71.6°} \approx \frac{9}{\sin 43°}$$
$$b \approx \frac{9 \sin 71.6°}{\sin 43°} \approx 12.5$$
case 2: $A = 180 - 65.4 \approx 114.6°$
$B \approx 180 - 43 - 114.6 = 22.4°$
$$\frac{b}{\sin 22.4°} \approx \frac{9}{\sin 43°}$$
$$b \approx \frac{9 \sin 22.4°}{\sin 43°} \approx 5.0$$

24. $A \approx \cos^{-1}\left(\dfrac{15^2 + 24.2^2 - 17.4^2}{2(15)(24.2)}\right)$
$\approx 45.6°$
$$\frac{\sin B}{15} \approx \frac{\sin 45.6°}{17.4}$$
$$B \approx \sin^{-1}\left(\frac{15 \sin 45.6°}{17.4}\right) \approx 38.0°$$
$C \approx 180 - 45.6 - 38.0 = 96.4°$

25. $Area = \frac{1}{2}(9)(14) \sin 36° \approx 37.0 \text{ cm}^2$

26. $s = \frac{1}{2}(27 + 31 + 40) = 49$ m
$Area =$
$$\sqrt{49(49 - 27)(49 - 31)(49 - 40)}$$
$\approx 417.9 \text{ m}^2$

27. $s = \frac{1}{2}(38 + 21 + 47) = 53$ in.
$Area =$
$$\sqrt{53(53 - 38)(53 - 21)(53 - 47)}$$
$\approx 390.7 \text{ in.}^2$

28. $d < f$, and $h = 12 \sin 31° \approx 6.18 < d$
two solutions
$$\frac{\sin F}{12} = \frac{\sin 31°}{6.2}$$
$$\sin F = \frac{12 \sin 31°}{6.2}$$
case 1: $F \approx 85.4°$
$E \approx 180 - 85.4 - 31 = 63.6°$
$Area \approx \frac{1}{2}(12)(6.2) \sin 63.6°$
$\approx 33.31 \text{ m}^2$
case 2: $F = 180 - 85.4 \approx 94.6°$
$E \approx 180 - 94.6 - 31 \approx 54.4°$
$Area \approx \frac{1}{2}(12)(6.2) \sin 54.4°$
$\approx 30.27 \text{ m}^2$

29.

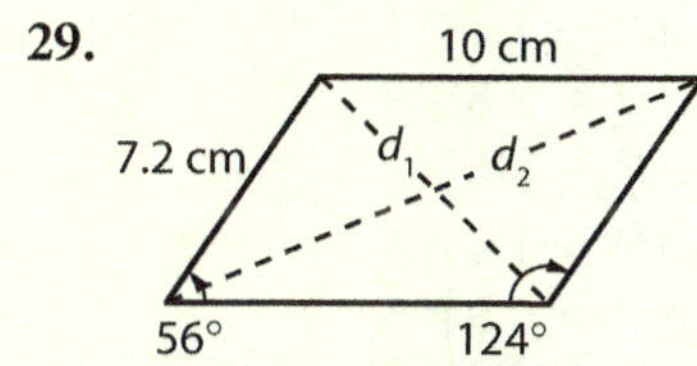

obtuse $\angle s = 180 - 56 = 124°$
$d_1^2 = 7.2^2 + 10^2 - 2(7.2)(10) \cos 56°$
≈ 71.3; $d_1 \approx 8.4$ cm
$d_2^2 = 7.2^2 + 10^2 - 2(7.2)(10) \cos 124°$
≈ 232.4; $d_2 \approx 15.2$ cm

30.

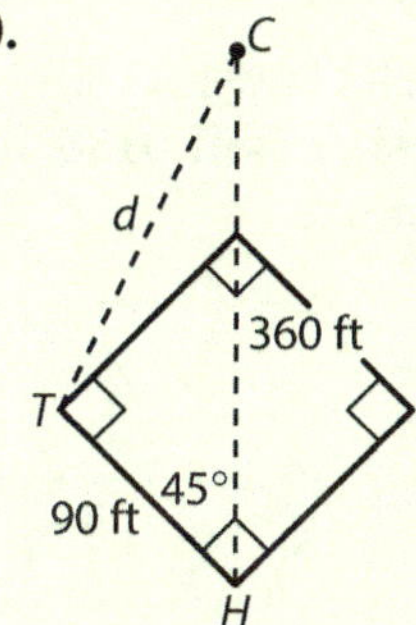

$d^2 = 90^2 + 360^2 - 2(90)(360) \cos 45°$
$\approx 91,879$; $d \approx 303$ ft

continued in Answers and Solutions Overflow

Data Analysis

Objectives

1. To use trigonometry to model and analyze the refraction of light

2. To find critical angles for the refraction of light

3. To acknowledge the Creator as the giver of light

Assignments

- **Minimum:** optional
- **Standard:** 1–9 (in class)
- **Extended:** 1–9

Solutions

1. $\theta_i > \theta_r$ (The light bends toward the normal.)

2. $y = -0.0025x^2 + 0.825x$; 1

3a. $1 \sin 40° = 1.33 \sin \theta_r$

$$\sin \theta_r = \frac{\sin 40°}{1.33} \approx 0.4833$$
$$\theta_r \approx 28.9°$$

3b. $1 \sin 80° = 1.33 \sin \theta_r$

$$\sin \theta_r = \frac{\sin 80°}{1.33} \approx 0.7405$$
$$\theta_r \approx 47.8°$$

4a. $1.33 \sin 40° = 1 \sin \theta_r$

$$\sin \theta_r \approx 0.8549$$
$$\theta_r \approx 58.7°$$

4b. $1.33 \sin 45° = 1 \sin \theta_r$

$$\sin \theta_r \approx 0.9405$$
$$\theta_r \approx 70.1°$$

6a. $1.33 \sin \theta_i = 1 \sin 80°$

$$\theta_i \approx 47.8°$$

6b. $1.33 \sin \theta_i = 1 \sin 90°$

$$\sin \theta_i = \frac{1}{1.33} \approx 0.7519$$
$$\theta_i \approx 48.8°$$

7. According to Snell's law, $1.33 \sin 50° = 1 \sin \theta_r$; $\therefore \sin \theta_r \approx 1.0188$, but there is no such angle.

Refraction, Reflection, and Fiber Optics

Ancient scientists recognized that light generally travels in a straight line but its path bends (or *refracts*) as it crosses a boundary from one medium into another. A classic experiment compares the angle of incidence θ_i, which is measured from the *normal* (the line drawn perpendicular to the boundary), to the angle of refraction θ_r, which is also measured from the normal. Ptolemy is attributed with compiling the following table of data (ca. AD 140) for light crossing the boundary from air into water.

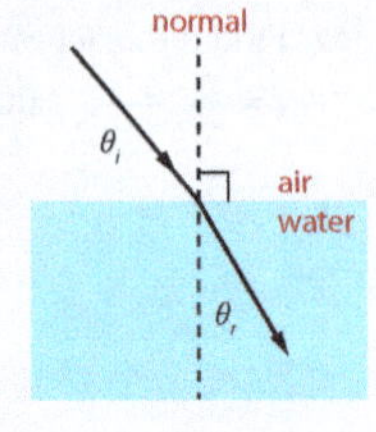

θ_i	10°	20°	30°	40°	50°	60°	70°	80°
θ_r	8.0°	15.5°	22.5°	29.0°	35.0°	40.5°	45.5°	50.0°

1. Examine the data and describe a general relationship between the angles as they pass from air (a less dense medium) to water (a denser medium).

2. Use technology to perform a quadratic regression on the data. State the quadratic function that models the data. What is the correlation coefficient?

Inaccuracies in the actual angles of refraction and the high correlation from the regression suggest that Ptolemy may have modified or created data to fit a quadratic model. In 1621 the Dutch astronomer and mathematician Willebrord Snell provided a more accurate model for the refraction of light.

Snell's law: $n_i \sin \theta_i = n_r \sin \theta_r$ where n is an experimentally determined constant for each medium, called the *index of refraction*.

The refractive index for a vacuum is defined as 1, and every other medium has $n > 1$. Since the refractive index of air ≈ 1.003, it is often rounded to 1. Water has a refractive index of 1.33.

3. Find each angle of refraction θ_r (to the nearest tenth of a degree) when light travels from air into water with the given angle of incidence.
   - **a.** $\theta_i = 40°$ 28.9°
   - **b.** $\theta_i = 80°$ 47.8°

4. Find each angle of refraction θ_r (to the nearest tenth of a degree) when light travels from water (denser) into air (less dense) with the given angle of incidence.
   - **a.** $\theta_i = 40°$ 58.7°
   - **b.** $\theta_i = 45°$ 70.1°

5. When light travels from a medium with a higher refractive index (denser) into a medium of lower index (less dense), does it bend toward or away from the normal?   away from

6. Find each angle of incidence θ_i (to the nearest tenth of a degree) that produces the given angle of refraction when light travels from water into air.
   - **a.** $\theta_r = 80°$ 47.8°
   - **b.** $\theta_r = 90°$ 48.8°

7. Explain why Snell's law cannot predict an angle of refraction into air for an incident light ray in water with $\theta_i = 50°$.

PRESENTATION

Trigonometry is as ancient as the pyramids and as modern as fiber-optic communications. Students will explore how trigonometry is used to model the path of light as it attempts to cross the boundary between media with different refractive indices. René Descartes and Isaac Newton both believed that light traveled faster in denser media, but Willebrord Snell discovered the correct relationship in what is still known as "Snell's law." Snell was a pioneer in the use of triangulation in land surveying, and his familiarity with these techniques no doubt helped him to recognize and quantify the refractive equation.

Encourage students to think critically when discussing Ptolemy's data. Should a correlation coefficient of 1 be questioned when the data is collected with very crude instruments? What are some possible alternative explanations?

The relatively recent development of fiber-optic cables by competing laboratories is a fascinating story. Students interested in technology and engineering may wish to explore NASA's Fiber-Optic Sensing System (FOSS), which finds applications in everything from aerospace and energy to infrastructure and medicine.

When light travels from a medium with a higher refractive index into a medium with a lower refractive index, the *critical angle* is the angle of incidence that produces a 90° angle of refraction. The critical angle θ_c for light rays leaving water and entering air is found by solving $n_{water} \sin \theta_c = n_{air} \sin 90°$. An incident light ray in water at

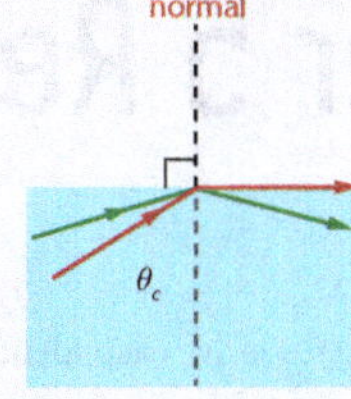

$\theta_c = \sin^{-1} \dfrac{n_{air}}{n_{water}} = \sin^{-1} \dfrac{1}{1.33} \approx 48.8°$ is refracted along the water-air boundary, 90° from the normal.

Incident light rays whose angle of incidence is larger than the critical angle are not refracted but are instead reflected back into the originating medium. In the case of a stream of water surrounded by air, light projected into the stream with $\theta_i > 48°$ will be reflected at the water-air boundary and travel within the path of the water. This phenomenon, known as total internal reflection, was used to create spectacular luminous fountains that amazed visitors at the Universal Exposition in Paris in 1889.

Today, total internal reflection helps to transmit gigabits of data per second through fiber-optic cables—tiny glass strands clad with materials having a lower index of refraction. The first transatlantic fiber-optic cable became operational in 1988. The critical angle θ_c of an optical fiber is calculated using Snell's law and the refractive indices of the fiber's core and cladding.

8. Find the critical angle for each fiber-optic cable.
   a. core with $n_1 = 1.5$ and cladding with $n_2 = 1.485$ 81.9°
   b. core with $n_1 = 1.52$ and cladding with $n_2 = 1.45$ 72.5°

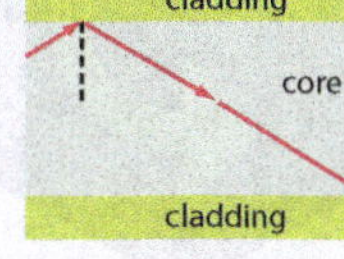

9. Derive a formula for the critical angle of a fiber-optic cable whose core has a refractive index of n_1 and whose cladding has a refractive index of n_2.

"And God said, Let there be light: and there was light. And God saw the light, that it was good" (Gen. 1:3–4). Life on the earth depends on light. Our knowledge has also led to amazing applications of light, such as laser beams and fiber optics. We save money and reduce recovery time by using lasers in medical procedures. We transmit digital information over the Internet at tremendous speeds. It is appropriate to praise the Creator for His good creation. In fact, the consistency of light reflects God's goodness and faithfulness as the Father of Lights, who shows no variation (James 1:17).

8a. $1.5 \sin \theta_c = 1.485 \sin 90°$

$$\sin \theta_c = \frac{1.485}{1.5}$$

$$\theta_c = \sin^{-1} \left(\frac{1.485}{1.5} \right) \approx 81.9°$$

8b. $1.52 \sin \theta_c = 1.45 \sin 90°$

$$\sin \theta_c = \frac{1.45}{1.52}$$

$$\theta_c = \sin^{-1} \left(\frac{1.45}{1.52} \right) \approx 72.5°$$

9. $n_1 \sin \theta_c = n_2 \sin 90°$

$$\sin \theta_c = \frac{n_2}{n_1}$$

$$\theta_c = \sin^{-1} \frac{n_2}{n_1}$$

Objective

To prepare for evaluation

Vocabulary

See Appendix A.

Assignments

- **Minimum:** 1–5, 7, 9–10, 12–14, 17–23, 27, 29–32, 34–35, 37, 39–41, 44–47, 49–50
- **Standard:** 1–49 odd; 2–50 even (in class)
- **Extended:** 2–3, 6–7, 9–10, 12–14, 16, 19–22, 24–41, 43, 45–48, 50

Assessment

- Chapter 5 Test

Solutions

1. $\dfrac{\sin^2 \theta}{\cos^2 \theta} + \dfrac{\cos^2 \theta}{\cos^2 \theta} = \dfrac{1}{\cos^2 \theta}$

$\tan^2 \theta + 1 = \sec^2 \theta$

2. $\cos \theta = \dfrac{2}{3} = \dfrac{x}{r}$; θ in Q IV

$\sin^2 \theta = 1 - \left(\dfrac{2}{3}\right)^2 \theta = \dfrac{5}{9}$; $\sin \theta = -\dfrac{\sqrt{5}}{3}$

$\tan \theta = -\dfrac{\frac{\sqrt{5}}{3}}{\frac{2}{3}} = -\dfrac{\sqrt{5}}{2}$

3a. $\sin (-x) = -\sin x = -0.866$

3b. $\sec (-x) = \dfrac{1}{\cos (-x)} = \dfrac{1}{\cos x} = \dfrac{1}{0.5} = 2$

4a. $\cos \dfrac{5\pi}{12} = \sin \left(\dfrac{\pi}{2} - \dfrac{5\pi}{12}\right) = \sin \dfrac{\pi}{12}$

4b. $\cot \left(\theta - \dfrac{\pi}{2}\right) = -\cot \left(\dfrac{\pi}{2} - \theta\right)$
$= -\tan \theta = 1 - \sqrt{2}$

5. $\sin^2 x - \cos^2 x \sin^2 x$
$= \sin^2 x (1 - \cos^2 x)$
$= \sin^2 x (\sin^2 x) = \sin^4 x$

6. $\dfrac{1 + \cot^2 \theta}{\cot^2 \theta} = \dfrac{\csc^2 \theta}{\cot^2 \theta} = \dfrac{1}{\sin^2 \theta} \cdot \dfrac{\sin^2 \theta}{\cos^2 \theta}$
$= \dfrac{1}{\cos^2 \theta} = \sec^2 \theta$

7. $\cos x + \tan x \sin x$
$= \cos x + \dfrac{\sin^2 x}{\cos x}$
$= \dfrac{\cos^2 x + \sin^2 x}{\cos x} = \dfrac{1}{\cos x} = \sec x$

8. $\dfrac{\sin \theta}{\csc \theta + 1} + \dfrac{\sin \theta}{\csc \theta - 1}$
$= \dfrac{\sin \theta(\csc \theta - 1) + \sin \theta(\csc \theta + 1)}{(\csc \theta + 1)(\csc \theta - 1)}$
$= \dfrac{1 - \sin \theta + 1 + \sin \theta}{\csc^2 \theta - 1} = \dfrac{2}{\cot^2 \theta}$
$= 2 \tan^2 \theta$

9. $\dfrac{\cos x}{1 + \sin x} - \dfrac{\cos x}{1 - \sin x}$
$= \dfrac{\cos x(1 - \sin x)}{1 - \sin^2 x} - \dfrac{\cos x(1 + \sin x)}{1 - \sin^2 x}$
$= \dfrac{\cos x - \cos x \sin x - \cos x - \cos x \sin x}{1 - \sin^2 x}$
$= \dfrac{-2 \cos x \sin x}{\cos^2 x} = \dfrac{-2 \sin x}{\cos x} = -2 \tan x$

10. $\tan^2 \theta \sec^2 \theta - \tan^2 \theta$
$= \tan^2 \theta(\sec^2 \theta - 1)$
$= \tan^2 \theta \tan^2 \theta = \tan^4 \theta$

Chapter 5 Review

2. $-\dfrac{\sqrt{5}}{3}, -\dfrac{\sqrt{5}}{2}$

1. Use the basic Pythagorean identity to derive $\tan^2 \theta + 1 = \sec^2 \theta$. [5.1]

2. If $\sec \theta = \dfrac{3}{2}$ and $\sin \theta < 0$, find $\sin \theta$ and $\tan \theta$. [5.1]

3. Use the even/odd identity to evaluate each expression. [5.1]
  a. If $\sin x = 0.866$, find $\sin (-x)$. -0.866
  b. If $\cos x = 0.5$, find $\sec (-x)$. 2

4. Use a cofunction identity to complete the following. [5.1]
  a. Rewrite $\cos \dfrac{5\pi}{12}$ in terms of sine. $\sin \dfrac{\pi}{12}$
  b. Find $\cot \left(\theta - \dfrac{\pi}{2}\right)$ if $\tan \theta = \sqrt{2} - 1$. $1 - \sqrt{2}$

Write each expression in terms of a single trigonometric function. [5.1]

5. $\sin^2 x - \cos^2 x \sin^2 x$ $\sin^4 x$

6. $\dfrac{1 + \cot^2 \theta}{\cot^2 \theta}$ $\sec^2 \theta$

7. $\cos x + \tan x \sin x$ $\sec x$

8. $\dfrac{\sin \theta}{\csc \theta + 1} + \dfrac{\sin \theta}{\csc \theta - 1}$ $2 \tan^2 \theta$

Verify each identity. [5.2]

9. $\dfrac{\cos x}{1 + \sin x} - \dfrac{\cos x}{1 - \sin x} = -2 \tan x$

10. $\tan^2 \theta \sec^2 \theta - \tan^2 \theta = \tan^4 \theta$

11. $\dfrac{\sin \theta}{1 + \cos \theta} = \dfrac{1 - \cos \theta}{\sin \theta}$

12. $(\sec \theta - \tan \theta)^2 = \dfrac{1 - \sin \theta}{1 + \sin \theta}$

13. $\tan^5 x = \tan x - 2 \tan x \sec^2 x + \tan x \sec^4 x$

Find all primary solutions of each equation. [5.3]

14. $2 \cos \theta + 1 = 0$

15. $4 \sin^2 \theta = 3$

16. $2 \cot x \sin x - \sqrt{3}\cot x = 2 \sin x - \sqrt{3}$

17. $4 \cos^2 x = 7 \cos x + 2$ (to the nearest hundredth)

18. $2 \sin 2x + \sqrt{3} = 0$

Find the general solution for each equation. [5.3]

19. $\tan x + 1 = \sec x$

20. $4 \tan^2 x + 5 \tan x = 6$ (to the nearest hundredth)

22. $\dfrac{\sqrt{2} + \sqrt{6}}{4}$ **23.** $\dfrac{-\sqrt{2} - \sqrt{6}}{4}$ **24.** $2 + \sqrt{3}$

21. Derive each identity. [5.4]
  a. Use the cosine of a difference identity to derive the cosine of a sum identity.
  b. Prove the cofunction identity $\cos \left(\dfrac{\pi}{2} - x\right) = \sin x$.
  c. Derive the sine of the sum of two angles.

Find the exact value of each expression. [5.4]

22. $\sin \left(\dfrac{5\pi}{6} - \dfrac{\pi}{4}\right)$ **23.** $\cos \dfrac{11\pi}{12}$ **24.** $\tan \dfrac{5\pi}{12}$

25. Find the exact value of $\sin (\alpha + \beta)$ if α and β are in Q II with $\sin \alpha = \dfrac{1}{3}$ and $\cos \beta = -\dfrac{3}{4}$. [5.4] $\dfrac{-3 - 2\sqrt{14}}{12}$

26. Rewrite $\sin \left(\text{Tan}^{-1} \dfrac{4}{3} - \text{Cos}^{-1} x\right)$ as an algebraic expression. [5.4] $\dfrac{4x - 3\sqrt{1 - x^2}}{5}$

27. Verify $\tan (\pi - \theta) = -\tan \theta$. [5.4]

28. Find all primary solutions of $\sin \left(x - \dfrac{\pi}{6}\right) = \cos \left(x + \dfrac{\pi}{3}\right) + 1$. [5.4] $\dfrac{\pi}{3}, \pi$

29. Derive the following identities. [5.5]
  a. Derive $\cos 2x = 2 \cos^2 x - 1$.
  b. Then derive the identity for $\cos^2 x$.
  c. Then derive the identity for $\cos \dfrac{x}{2}$.

Find the exact value of each expression if $\tan \theta = -\dfrac{3}{4}$ and $\dfrac{\pi}{2} \leq \theta < \pi$. [5.5]

30. Find $\sin 2\theta$, $\cos 2\theta$, and $\tan 2\theta$.

31. Find $\sin \dfrac{\theta}{2}$, $\cos \dfrac{\theta}{2}$, and $\tan \dfrac{\theta}{2}$.

Find the exact value of each expression. [5.5]

32. $\cos \dfrac{7\pi}{12}$ $-\dfrac{\sqrt{2 - \sqrt{3}}}{2}$ **33.** $\sin \dfrac{\pi}{12} \cos \dfrac{13\pi}{12}$ $-\dfrac{1}{4}$

Find the general solution for each equation. [5.5]

34. $\sin 2x + \sin x = 0$ $x = n\pi, \dfrac{2\pi}{3} + n2\pi, \dfrac{4\pi}{3} + n2\pi$

35. $\cos \dfrac{x}{2} = \sin x$ $x = \dfrac{\pi}{3} + n4\pi, \dfrac{5\pi}{3} + n4\pi, \pi + n2\pi$

36. $\dfrac{\sin \theta}{\sin 3\theta - \sin \theta} = 1$ $x = \dfrac{\pi}{6} + n\pi, \dfrac{5\pi}{6} + n\pi$

Verify each identity. [5.5]

37. $\sin 3x + \sin x = 2 \sin 2x \cos x$

38. $\dfrac{\sin 2x}{1 + \cos 2x} = \dfrac{1 - \cos 2x}{\sin 2x}$

---

11. $\dfrac{\sin \theta}{1 + \cos \theta} = \dfrac{1 - \cos \theta}{\sin \theta}$

PRESENTATION

Be sure to let the students know which identities they are expected to know from memory and which ones can be referenced during an assessment. Consider expecting the students to know all the identities except the reduction, product-to-sum, and sum-to-product identities.

Solve each triangle. Round answers to the nearest tenth. [5.6–5.7]

39. $\triangle ABC$: $B = 25°$, $C = 70°$, $a = 8$
40. $\triangle PQR$: $P = 42°$, $r = 9$, $q = 7$
41. $\triangle WXY$: $Y = 42°$, $y = 2$, $w = 5$
42. $\triangle MNO$: $M = 50°$, $N = 36°$, $n = 6$
43. $\triangle PQR$: $p = 4$, $q = 7$, $r = 10$
44. $\triangle JKL$: $K = 110°$, $j = 7$, $k = 10$
45. $\triangle ABC$: $B = 13°$, $b = 6$, $c = 10$

Find the area of each triangle. [5.7]

46. 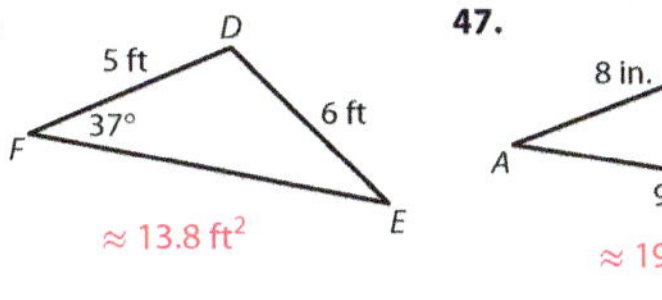
$\approx 13.8 \text{ ft}^2$

47. $\approx 19.9 \text{ in.}^2$

---

14. $\theta = \dfrac{2\pi}{3}, \dfrac{4\pi}{3}$
15. $\theta = \dfrac{\pi}{3}, \dfrac{2\pi}{3}, \dfrac{4\pi}{3}, \dfrac{5\pi}{3}$
16. $x = \dfrac{\pi}{4}, \dfrac{\pi}{3}, \dfrac{2\pi}{3}, \dfrac{5\pi}{4}$
17. $x = 1.82, 4.46$
18. $x = \dfrac{2\pi}{3}, \dfrac{5\pi}{6}, \dfrac{5\pi}{3}, \dfrac{11\pi}{6}$
19. $x = n2\pi$
20. $x = 0.64 + n\pi, 2.03 + n\pi$

30. $\sin 2\theta = -\dfrac{24}{25}$; $\cos 2\theta = \dfrac{7}{25}$; $\tan 2\theta = -\dfrac{24}{7}$
31. $\sin \dfrac{\theta}{2} = \dfrac{3\sqrt{10}}{10}$; $\cos \dfrac{\theta}{2} = \dfrac{\sqrt{10}}{10}$; $\tan \dfrac{\theta}{2} = 3$

39. $A = 85°$, $b \approx 3.4$, $c \approx 7.5$
40. $Q = 51.0°$, $R = 87.0°$, $p = 6.0$
41. no solution
42. $O = 94°$, $o \approx 10.2$, $m \approx 7.8$
43. $P \approx 18.2°$, $Q \approx 33.1°$, $R \approx 128.7°$
44. $J \approx 41.1°$, $L \approx 28.9°$, $l = 5.1$
45. $A \approx 145.0°$, $C \approx 22.0°$, $a \approx 15.3$ or $A \approx 9.0°$, $C \approx 158.0°$, $a \approx 4.2$

---

48. A jungle pilot flies 100 mi west and then makes a 20° adjustment to the north to avoid a storm. After he has flown for 70 mi on the adjusted course, how far is he from the original airstrip? [5.7] ≈ 167.5 mi

49. Grace and Marie both observe a camera-equipped drone hovering over a community event. Find the height of the drone if they are standing 200 ft apart and measure angles of elevation from the ground to the drone as 25° and 12° when the drone is directly between them. [5.6] ≈ 29.2 ft

50. A lost hiker is 2 km from the trailhead and 5 km from a ranger station 4 km due east of the trailhead. Find the hiker's two possible bearings from the trailhead. [5.7] $\approx 198.2°$ and $\approx 341.8°$

---

11. $$\dfrac{\sin \theta}{1 + \cos \theta} \cdot \dfrac{1 - \cos \theta}{1 - \cos \theta} = \dfrac{\sin \theta(1 - \cos \theta)}{1 - \cos^2 \theta}$$
$$= \dfrac{\sin \theta(1 - \cos \theta)}{\sin^2 \theta} = \dfrac{1 - \cos \theta}{\sin \theta}$$

12. working on both sides:
$$(\sec \theta - \tan \theta)^2 = \dfrac{1 - \sin \theta}{1 + \sin \theta}$$
$$\sec^2 \theta - 2 \sec \theta \tan \theta + \tan^2 \theta$$
$$= \dfrac{(1 - \sin \theta)(1 - \sin \theta)}{(1 + \sin \theta)(1 - \sin \theta)}$$
$$\dfrac{1}{\cos^2 \theta} - 2\left(\dfrac{1}{\cos \theta}\right)\left(\dfrac{\sin \theta}{\cos \theta}\right) + \dfrac{\sin^2 \theta}{\cos^2 \theta} = \dfrac{1 - 2 \sin \theta + \sin^2 \theta}{1 - \sin^2 \theta}$$
$$\dfrac{1 - 2 \sin \theta + \sin^2 \theta}{\cos^2 \theta} = \dfrac{1 - 2 \sin \theta + \sin^2 \theta}{\cos^2 \theta}$$

13. $$\tan^5 x = \tan x(\tan^2 x)^2$$
$$= \tan x(\sec^2 x - 1)^2$$
$$= \tan x(1 - 2 \sec^2 x + \sec^4 x)$$
$$= \tan x - 2 \tan x \sec^2 x + \tan x \sec^4 x$$

14. $$2 \cos \theta + 1 = 0$$
$$\cos \theta = -\dfrac{1}{2}$$
$$\theta = \dfrac{2\pi}{3}, \dfrac{4\pi}{3}$$

15. $$\sin^2 \theta = \dfrac{3}{4}$$
$$\sin \theta = \pm\dfrac{\sqrt{3}}{2}$$
$$\theta = \dfrac{\pi}{3}, \dfrac{2\pi}{3}, \dfrac{4\pi}{3}, \dfrac{5\pi}{3}$$

---

16. $$2 \cot x \sin x - 2 \sin x - \sqrt{3} \cot x + \sqrt{3} = 0$$
$$2 \sin x(\cot x - 1) - \sqrt{3}(\cot x - 1) = 0$$
$$(\cot x - 1)(2 \sin x - \sqrt{3}) = 0$$
$$\cot x - 1 = 0 \quad \text{or} \quad 2 \sin x - \sqrt{3} = 0$$
$$\cot x = 1 \qquad \sin x = \dfrac{\sqrt{3}}{2}$$
$$x = \dfrac{\pi}{4}, \dfrac{5\pi}{4} \qquad x = \dfrac{\pi}{3}, \dfrac{2\pi}{3}$$

17. $$4 \cos^2 x - 7 \cos x - 2 = 0$$
$$(4 \cos x + 1)(\cos x - 2) = 0$$
$$\cos x = -\dfrac{1}{4} \quad \text{or} \quad \cos x = 2$$
$$x \approx 1.82$$
$$x \approx 2\pi - 1.82 \approx 4.46$$

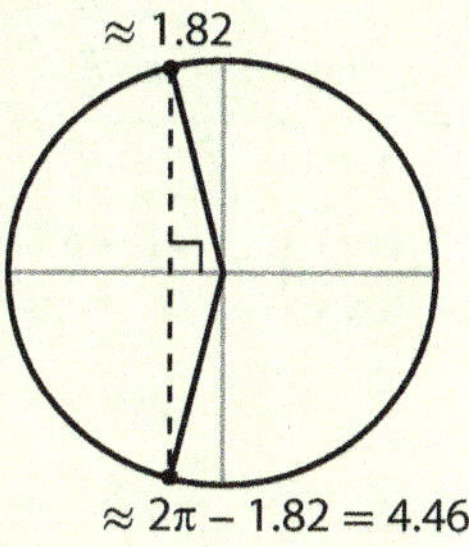

18. $$\sin 2x = -\dfrac{\sqrt{3}}{2}$$
$$2x = \dfrac{4\pi}{3} + n2\pi, \dfrac{5\pi}{3} + n2\pi$$
$$x = \dfrac{2\pi}{3} + n\pi, \dfrac{5\pi}{6} + n\pi$$
using $n = 0, 1$:
$$x = \dfrac{2\pi}{3}, \dfrac{5\pi}{3}, \dfrac{5\pi}{6}, \dfrac{11\pi}{6}$$

19. $$(\tan x + 1)^2 = \sec^2 x$$
$$\tan^2 x + 2 \tan x + 1 = \tan^2 x + 1$$
$$2 \tan x = 0$$
$$\tan x = 0 \text{ when } \sin x = 0$$
$$x = 0, \pi \text{ (extraneous)}$$
$$x = 0 + n2\pi = n2\pi$$

20. $$4 \tan^2 x + 5 \tan x - 6 = 0$$
$$(4 \tan x - 3)(\tan x + 2) = 0$$
$$\tan x = \dfrac{3}{4} \quad \text{or} \quad \tan x = -2$$
$$x \approx 0.64 \qquad x \approx -1.11$$
$$x \approx -1.11 + \pi \approx 2.03$$

21a. $$\cos(x + y) = \cos(x - (-y))$$
$$= \cos x \cos(-y) + \sin x \sin(-y)$$
$$= \cos x \cos y + \sin x(-\sin y)$$
$$= \cos x \cos y - \sin x \sin y$$

21b. $$\cos\left(\dfrac{\pi}{2} - x\right)$$
$$= \cos \dfrac{\pi}{2} \cos x + \sin \dfrac{\pi}{2} \sin x$$
$$= (0) \cos x + (1) \sin x$$
$$= \sin x$$

continued on next page

21c. $\sin(x+y) = \cos\left(\frac{\pi}{2} - (x+y)\right)$

$= \cos\left(\left(\frac{\pi}{2} - x\right) - y\right)$

$= \cos\left(\frac{\pi}{2} - x\right)\cos y$

$\qquad + \sin\left(\frac{\pi}{2} - x\right)\sin y$

$= \sin x \cos y + \cos x \sin y$

22. $\sin\frac{5\pi}{6}\cos\frac{\pi}{4} - \cos\frac{5\pi}{6}\sin\frac{\pi}{4}$

$= \frac{1}{2}\left(\frac{\sqrt{2}}{2}\right) - \left(-\frac{\sqrt{3}}{2}\right)\left(\frac{\sqrt{2}}{2}\right) = \frac{\sqrt{2}+\sqrt{6}}{4}$

23. $\cos\left(\frac{\pi}{4} + \frac{2\pi}{3}\right)$

$= \cos\frac{\pi}{4}\cos\frac{2\pi}{3} - \sin\frac{\pi}{4}\sin\frac{2\pi}{3}$

$= \frac{\sqrt{2}}{2}\left(-\frac{1}{2}\right) - \frac{\sqrt{2}}{2}\left(\frac{\sqrt{3}}{2}\right) = \frac{-\sqrt{2}-\sqrt{6}}{4}$

24. $\tan\left(\frac{3\pi}{4} - \frac{\pi}{3}\right) = \dfrac{\tan\frac{3\pi}{4} - \tan\frac{\pi}{3}}{1 + \tan\frac{3\pi}{4}\left(\tan\frac{\pi}{3}\right)}$

$= \frac{-1 - \sqrt{3}}{1 + (-1)\sqrt{3}}\left(\frac{1 + \sqrt{3}}{1 + \sqrt{3}}\right) = \frac{-1 - 3 - 2\sqrt{3}}{1 - 3}$

$= 2 + \sqrt{3}$

or

$\tan\left(\frac{\frac{5\pi}{6}}{2}\right) = \dfrac{1 - \cos\frac{5\pi}{6}}{\sin\frac{5\pi}{6}} = \dfrac{1 - \left(-\frac{\sqrt{3}}{2}\right)}{\frac{1}{2}}$

$= 2 + \sqrt{3}$

25. $\left(\frac{1}{3}\right)^2 + \cos^2\alpha = 1$

$\cos^2\alpha = \frac{8}{9}$

$\cos\alpha = -\frac{\sqrt{8}}{3}$

$\sin^2\beta + \left(-\frac{3}{4}\right)^2 = 1$

$\sin^2\beta = \frac{7}{16}$

$\sin\beta = \frac{\sqrt{7}}{4}$

$\sin(\alpha+\beta) = \frac{1}{3}\left(-\frac{3}{4}\right) + \left(-\frac{\sqrt{8}}{3}\right)\left(\frac{\sqrt{7}}{4}\right)$

$= \frac{-3 - 2\sqrt{14}}{12}$

26.

$\sin(\alpha - \beta) = \sin\alpha\cos\beta - \cos\alpha\sin\beta$

$= \frac{4}{5}x - \frac{3}{5}\sqrt{1 - x^2} = \frac{4x - 3\sqrt{1 - x^2}}{5}$

27. $\tan(\pi - \theta) = \dfrac{\tan\pi - \tan\theta}{1 + \tan\pi\tan\theta}$

$= \frac{0 - \tan\theta}{1 + (0)\tan\theta} = \frac{-\tan\theta}{1} = -\tan\theta$

28. $\sin\left(x - \frac{\pi}{6}\right) = \cos\left(x + \frac{\pi}{3}\right) + 1$

$\sin x \cos\frac{\pi}{6} - \cos x \sin\frac{\pi}{6}$

$\qquad = \cos x \cos\frac{\pi}{3} - \sin x \sin\frac{\pi}{3} + 1$

$\sin x\left(\frac{\sqrt{3}}{2}\right) - \cos x\left(\frac{1}{2}\right)$

$\qquad = \cos x\left(\frac{1}{2}\right) - \sin x\left(\frac{\sqrt{3}}{2}\right) + 1$

$\sqrt{3}\sin x = \cos x + 1$

$3\sin^2 x = (\cos x + 1)^2$

$3(1 - \cos^2 x) = \cos^2 x + 2\cos x + 1$

$4\cos^2 x + 2\cos x - 2 = 0$

$(2\cos x - 1)(\cos x + 1) = 0$

$\cos x = \frac{1}{2}$ or $\cos x = -1$

$x = \frac{\pi}{3}, \frac{5\pi}{3}$ $\qquad x = \pi$

$\left(\frac{5\pi}{3}$ is extraneous.$\right)$

29a. $\cos 2x = \cos(x + x)$

$= \cos x \cos x - \sin x \sin x$

$= \cos^2 x - \sin^2 x$

$= \cos^2 x - (1 - \cos^2 x)$

$= 2\cos^2 x - 1$

29b. $\cos 2x = 2\cos^2 x - 1$

$2\cos^2 x = \cos 2x + 1$

$\cos^2 x = \frac{\cos 2x + 1}{2}$

29c. $\cos^2 x = \frac{\cos 2x + 1}{2}$

$\cos^2\frac{x}{2} = \frac{\cos 2\left(\frac{x}{2}\right) + 1}{2} = \frac{\cos x + 1}{2}$

$\cos\frac{x}{2} = \pm\sqrt{\frac{\cos x + 1}{2}}$

30–31. $r = \sqrt{(-3)^2 + (4)^2} = 5$

$\cos\theta = -\frac{4}{5}; \sin\theta = \frac{3}{5}$

30. $\sin 2\theta = 2\cos\theta\sin\theta$

$= 2\left(-\frac{4}{5}\right)\left(\frac{3}{5}\right) = -\frac{24}{25}$

$\cos 2\theta = 2\cos^2\theta - 1$

$= 2\left(-\frac{4}{5}\right)^2 - 1$

$= \frac{32}{25} - \frac{25}{25} = \frac{7}{25}$

$\tan 2\theta = \dfrac{2\tan x}{1 - \tan^2 x}$

$= \dfrac{2\left(-\frac{3}{4}\right)}{1 - \left(-\frac{3}{4}\right)^2} = \dfrac{-\frac{3}{2}}{\frac{16 - 9}{16}} = -\frac{24}{7}$

31. $\frac{\theta}{2}$ lies in Q I.

$\sin\frac{\theta}{2} = \sqrt{\frac{1 - \cos\theta}{2}} = \sqrt{\frac{1 - \left(-\frac{4}{5}\right)}{2}}$

$= \sqrt{\frac{9}{10}} = \frac{3\sqrt{10}}{10}$

$\cos\frac{\theta}{2} = \sqrt{\frac{1 + \cos\theta}{2}} = \sqrt{\frac{1 + \left(-\frac{4}{5}\right)}{2}}$

$= \sqrt{\frac{1}{10}} = \frac{\sqrt{10}}{10}$

$\tan\frac{\theta}{2} = \sqrt{\frac{1 - \cos\theta}{1 + \cos\theta}} = \sqrt{\frac{\left(\frac{9}{5}\right)}{\left(\frac{1}{5}\right)}} = 3$

---

32. $\cos\left(\frac{\frac{7\pi}{6}}{2}\right) = -\sqrt{\dfrac{1 + \cos\frac{7\pi}{6}}{2}}$ (Q III)

$= -\sqrt{\dfrac{1 + \left(-\frac{\sqrt{3}}{2}\right)}{2}} = -\frac{\sqrt{2 - \sqrt{3}}}{2}$

33. $\frac{1}{2}\left(\sin\left(\frac{\pi}{12} + \frac{13\pi}{12}\right) + \sin\left(\frac{\pi}{12} - \frac{13\pi}{12}\right)\right)$

$= \frac{1}{2}\left(\sin\frac{7\pi}{6} + \sin(-\pi)\right)$

$= \frac{1}{2}\left(-\frac{1}{2} + 0\right) = -\frac{1}{4}$

34. $2\sin x \cos x + \sin x = 0$

$\sin x(2\cos x + 1) = 0$

$\sin x = 0$ or $2\cos x = -1$

$x = 0, \pi$ $\qquad \cos x = -\frac{1}{2}$

$\qquad\qquad x = \frac{2\pi}{3}, \frac{4\pi}{3}$

35. $\left(\pm\sqrt{\frac{1 + \cos x}{2}}\right)^2 = \sin^2 x$

$1 + \cos x = 2(1 - \cos^2 x)$

$2\cos^2 x + \cos x - 1 = 0$

$(2\cos x - 1)(\cos x + 1) = 0$

$\cos x = \frac{1}{2}$ or $\cos x = -1$

solutions within $[0, 4\pi)$:

$x = \frac{\pi}{3}, \frac{5\pi}{3}, \frac{7\pi}{3}, \frac{11\pi}{3}$ $\quad x = \pi, 3\pi$

$\left(\frac{7\pi}{3}$ and $\frac{11\pi}{3}$ are extraneous.$\right)$

36. $\sin\theta = \sin 3\theta - \sin\theta$

$\sin\theta = 2\cos\left(\frac{3\theta + \theta}{2}\right)\sin\left(\frac{3\theta - \theta}{2}\right)$

$\sin\theta = 2\cos 2\theta\sin\theta$

$\sin\theta = 2(1 - 2\sin^2\theta)\sin\theta$

$\sin\theta - 4\sin^3\theta = 0$

$\sin\theta(1 - 4\sin^2\theta) = 0$

$\sin\theta = 0$ or $1 - 4\sin^2\theta = 0$

$\theta = 0, \pi$ $\qquad \sin^2\theta = \frac{1}{4}$

$\qquad\qquad\qquad \sin\theta = \pm\frac{1}{2}$

$\theta = \frac{\pi}{6}, \frac{5\pi}{6}, \frac{7\pi}{6}, \frac{11\pi}{6}$

(0 and π are extraneous.)

37. $\sin 3x + \sin x = \sin(2x + x) + \sin x$

$= \sin 2x \cos x + \cos 2x \sin x + \sin x$

$= 2\sin x \cos^2 x$

$\qquad + (2\cos^2 x - 1)\sin x + \sin x$

$= \sin x(2\cos^2 x + 2\cos^2 x - 1 + 1)$

$= 4\sin x \cos^2 x$

$= 2(2\sin x \cos x)\cos x$

$= 2\sin 2x \cos x$

or

using a sum-to-product identity:

$\sin 3x + \sin x$

$= 2\sin\left(\frac{3x + x}{2}\right)\cos\left(\frac{3x - x}{2}\right)$

$= 2\sin 2x \cos x$

38. working on both sides:

$$\frac{2\sin x \cos x}{1+(2\cos^2 x - 1)} = \frac{1-(1-2\sin^2 x)}{2\sin x \cos x}$$

$$\frac{2\sin x \cos x}{2\cos^2 x} = \frac{2\sin^2 x}{2\cos x \sin x}$$

$$\frac{\sin x}{\cos x} = \frac{\sin x}{\cos x}$$

39. $A = 180 - 25 - 70 = 85°$

$$\frac{b}{\sin 25°} = \frac{8}{\sin 85°}$$

$$b = \frac{8 \sin 25°}{\sin 85°} \approx 3.4$$

$$\frac{c}{\sin 70°} = \frac{8}{\sin 85°}$$

$$c = \frac{8 \sin 70°}{\sin 85°} \approx 7.5$$

40. $p^2 = 9^2 + 7^2 - 2(9)(7) \cos 42°$
$\approx 36.36;\ p \approx 6.0$

$$\frac{\sin Q}{7} \approx \frac{\sin 42°}{6.0}$$

$$Q \approx \sin^{-1}\left(\frac{7 \sin 42}{6.0}\right) \approx 51.0°$$

$$R \approx 180 - 42 - 51.0 = 87.0°$$

41. SSA; Y is acute and $y < w$.
$h = w \sin Y = 5 \sin 42° \approx 3.3$
$y < h$; no solution

42. $O = 180 - 50 - 36 = 94°$

$$\frac{o}{\sin 94°} = \frac{6}{\sin 36°}$$

$$o = \frac{6 \sin 94°}{\sin 36°} \approx 10.2$$

$$\frac{m}{\sin 50°} = \frac{6}{\sin 36°}$$

$$m = \frac{6 \sin 50°}{\sin 36°} \approx 7.8$$

43. $P = \cos^{-1}\left(\frac{7^2 + 10^2 - 4^2}{2(7)(10)}\right) \approx 18.2°$

$$\frac{\sin Q}{7} \approx \frac{\sin 18.2°}{4}$$

$$Q \approx \sin^{-1}\left(\frac{7 \sin 18.2}{4}\right) \approx 33.1°$$

$$R \approx 180 - 18.2 - 33.1 = 128.7°$$

44. $\dfrac{\sin J}{7} = \dfrac{\sin 110°}{10}$

$$\sin J = \frac{7 \sin 110°}{10}$$

$$J \approx 41.1°$$

$$L \approx 180 - 110 - 41.1 = 28.9°$$

$$\frac{l}{\sin 28.9°} = \frac{10}{\sin 70°}$$

$$l = \frac{10 \sin 28.9°}{\sin 70°} \approx 5.1$$

45. $\dfrac{\sin C}{10} = \dfrac{\sin 13°}{6}$

$$\sin C = \frac{10 \sin 13°}{6}$$

case 1: $C \approx 22.0°$
$A \approx 180 - 13 - 22.0 = 145.0°$

$$\frac{a}{\sin 145.0°} = \frac{6}{\sin 13°}$$

$$a = \frac{6 \sin 145.0°}{\sin 13°} \approx 15.3$$

case 2: $C = 180 - 22.0 = 158.0°$
$A \approx 180 - 13 - 158.0 = 9.0°$

$$\frac{a}{\sin 9.0°} = \frac{6}{\sin 13°}$$

$$a = \frac{6 \sin 9.0°}{\sin 13°} \approx 4.2$$

$A \approx 145.0°,\ C \approx 22.0°,\ a \approx 15.3$ or
$A \approx 9.0°,\ C \approx 158°,\ a \approx 4.2$

46. $\dfrac{\sin E}{5} = \dfrac{\sin 37°}{6}$

$$\sin E = \frac{5 \sin 37°}{6} \approx 0.5015$$

$$E \approx 30.1°$$

$$D \approx 180 - 37 - 30.1 = 112.9°$$

$$\text{Area} = \tfrac{1}{2}(5)(6) \sin 112.9° \approx 13.8 \text{ ft}^2$$

47. $s = \tfrac{1}{2}(8 + 5 + 9) = 11$

$$\text{Area} = \sqrt{11(11-8)(11-5)(11-9)}$$

$$\approx 19.9 \text{ in.}^2$$

48. 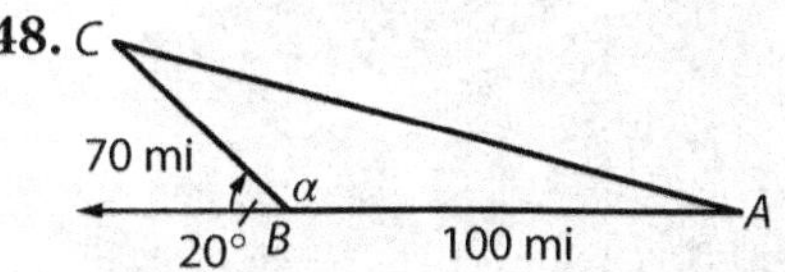

$$\alpha = 180 - 20 = 160°$$

$$b^2 = 70^2 + 100^2 - 2(70)(100) \cos 160°$$

$$\approx 28{,}055.7;\ b \approx 167.5 \text{ mi}$$

49. 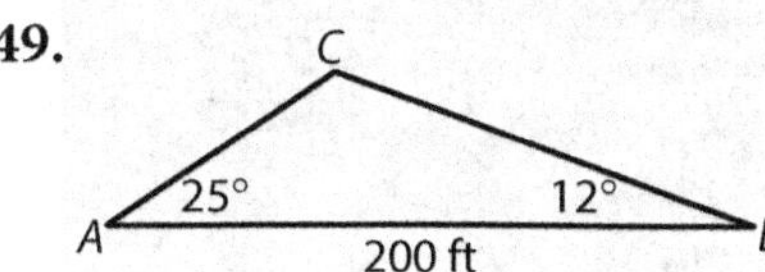

$$C = 180 - 25 - 12 = 143°$$

$$\frac{b}{\sin 12°} = \frac{200}{\sin 143°}$$

$$b \approx 69.1 \text{ ft}$$

$$\sin 25° \approx \frac{h}{69.1}$$

$$h \approx 69.1 \sin 25° \approx 29.2 \text{ ft}$$

50. 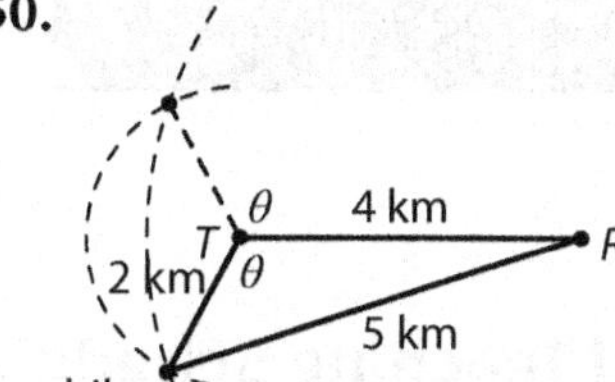

$$\theta = \cos^{-1}\left(\frac{2^2 + 4^2 - 5^2}{2(2)(4)}\right) \approx 108.2°$$

SW bearing $= 90 + 108.2 = 198.2°$
NW bearing $= 360 - (108.2 - 90)$
$$= 341.8°$$

Overview

This chapter gives students a working knowledge of two- and three-dimensional vectors, which is essential for solving real-world problems in physics. The polar coordinate plane is introduced, offering another method of designating points on a plane. Polar equations allow students to generate fascinating graphs. Students will then express complex numbers in polar form and find their products, quotients, powers, and roots.

The Technology Corner provides details for graphing polar equations on the graphing calculator.

Chapter Objectives

1. To express a vector in component form, by its magnitude and direction, and as a sum of unit vectors

2. To perform vector operations using vector components and coordinate geometry

3. To model and solve real-world problems using vectors

4. To graph equations in the polar coordinate plane

5. To convert between polar and rectangular forms of coordinates and equations

6. To determine key characteristics for the graphs of classic polar equations

7. To find products, quotients, powers, and roots of complex numbers using polar form

8. To explain the usefulness of mathematics in light of transcendent mathematical truth.

6 Vectors, Polar Graphs, and Complex Numbers

6.1 Vectors in the Plane
6.2 Dot Products
Biblical Perspective of Mathematics
6.3 Vectors in Space
6.4 Polar Coordinates
6.5 Graphs of Polar Equations
Historical Connection
6.6 Polar Forms of Complex Numbers
Data Analysis

BIBLICAL PERSPECTIVE OF MATHEMATICS
Which types of reasoning are used to discover and verify mathematical truths?

HISTORICAL CONNECTION
While vectors and polar coordinates are often presented before operations with complex numbers, the development of vectors lagged behind the development of complex numbers by about 200 years.

DATA ANALYSIS
While some mathematicians studied operations thought to exist solely in the mind of man, these operations led to fractals, which are used to describe natural patterns occurring all around us and even inside of us.

Suggested Teaching Schedule

DAY	
1	6.1
2–3	6.2
4	Quiz 6A (6.1–6.2) BPM: Truth and Mathematical Proof
5	6.3
6–7	Quiz 6B (6.3) 6.4 TC: Graphing Polar Functions

DAY	
8	6.5
9	Quiz 6C (6.4–6.5) HC: Complex Numbers
10–11	6.6
12	DA: Fractals
13	Quiz 6D (6.6) Chapter 6 Review
14	Chapter 6 Test

Vector diagrams are used extensively in kinematics, the study of motion.

Quantities that can be measured using real numbers are *scalar* quantities. Both a mass of 20 kg and a volume of 250 cm^3 are scalar quantities. Other quantities that are described by both a number and a direction, such as a force applied to an object, its acceleration, or its velocity, are *vector* quantities. A 20 mph wind from the southeast or a 10 lb force pulling up on an object are vector quantities.

These quantities can be modeled with directed line segments called *vectors*. The vector with an *initial point* (or *tail*) at $P(x_1, y_1)$ and a *terminal point* (or *head*) at $Q(x_2, y_2)$ is notated $\overrightarrow{PQ}$ because it has the length of $\overline{PQ}$ and the direction of $\overrightarrow{PQ}$.

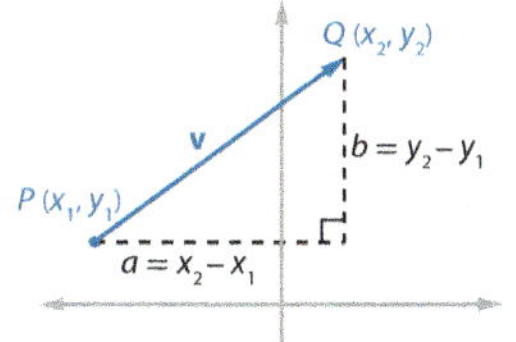

Vectors can also be named in printed text by a bold lowercase letter such as **v**, or in handwritten text by $\vec{v}$.

Vectors are frequently broken down into their horizontal and vertical *components*.

$\mathbf{v} = \langle a, b \rangle$ where $a = x_2 - x_1$ and $b = y_2 - y_1$

This notation allows vectors anywhere in the plane to be translated to an equal vector in *standard position*, one having its initial point at the origin. Two vectors are equal if they have the same magnitude and direction. When expressed in terms of their components, $\mathbf{u} = \langle u_1, u_2 \rangle$ and $\mathbf{v} = \langle v_1, v_2 \rangle$ are equal if and only if $u_1 = v_1$ and $u_2 = v_2$.

Example 1 Expressing a Vector in Component Form

Given $P(4, 2)$ and $O(1, -2)$, express $\overrightarrow{PO}$ in component form and draw an equal vector **v** in standard position.

Answer

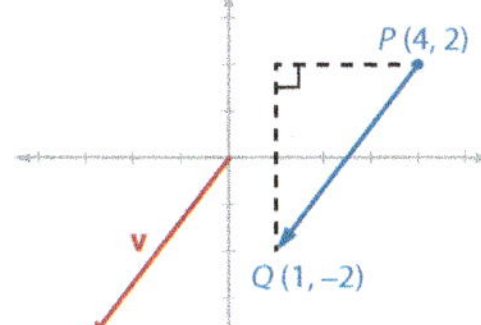

$a = x_2 - x_1$
$= 1 - 4 = -3$
$b = y_2 - y_1$
$= -2 - 2 = -4$
$\overrightarrow{PQ} = \langle -3, -4 \rangle$

1. Draw $\overrightarrow{PQ}$.

2. Express $\overrightarrow{PQ}$ in terms of its horizontal and vertical components.

3. Sketch **v** with its tail at the origin and its head at $(-3, -4)$.

SKILL ✓ EXERCISE 1

The distance formula can be used to find the *magnitude* (length) of a vector. Expressing a vector in component form simplifies the calculation and allows the vector's direction to be found using the reference angle α when the vector is placed in standard position. Unless stated otherwise, the direction of a vector is measured in terms of θ, the directed angle from the positive x-axis.

After completing this section, you will be able to

- describe vectors in the coordinate plane.
- perform vector operations of addition, subtraction, and scalar multiplication.
- write a vector as a linear combination of unit vectors.
- solve real-world problems using vectors.

Objectives

1. To describe vectors in terms of magnitude and direction as well as their horizontal and vertical components

2. To perform arithmetic operations on vectors geometrically and algebraically

3. To write a vector as a linear combination of unit vectors

4. To use vectors to solve real-world problems

Flash

The US Marine Corps Harrier, used extensively in the Gulf War, produces more than 21,000 lb of thrust to hover and perform short takeoffs and vertical landings. The forces permitting hovering, landing, and takeoffs can be described by vectors.

Vocabulary

components
equilibrium
initial point (tail)
linear combination
magnitude
net force
resolved
resultant vector
scalar
scalar product
standard position
standard unit vector

PRESENTATION

Lesson Opener

Find each missing value.

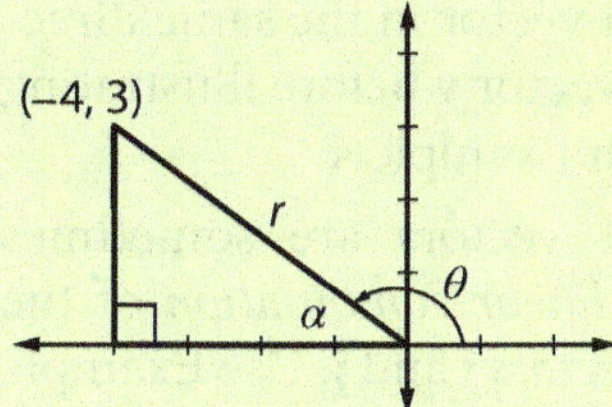

1. r 5
2. $\alpha \approx 36.9°$
3. $\theta \approx 143.1°$

Compare *scalar* and *vector* quantities, asking the students to provide other examples of each. Share that vectors are used extensively in science and engineering. Introduce the terms relating to vectors. Use Example 1 to demonstrate how a vector is described in terms of its horizontal and vertical *components* and placed in *standard position*. Emphasize that the order in which the coordinates are subtracted in step 2 is important. Differences calculated in the opposite order would result in $\langle 3, 4 \rangle$, a vector in the opposite direction.

One-on-One A student may find it difficult to distinguish a vector from a ray. Point out the differences. A ray extends infinitely in a given direction while a vector's length is finite, similar to a segment. The vector's single-sided arrow blends the notation of a segment (with its length) and a ray (with its direction).

Find the *magnitude* of the vector in Example 1 using the distance formula as shown in the table. Mention that some sources use $\|\mathbf{v}\|$ for magnitude of a vector. Discuss how to find the direction angle, θ, using the formulas in the table and demonstrate their application with Example 2. An alternative method of finding this angle is demonstrated later in the chapter. Since the Tan^{-1} function has the range of $\left(-\frac{\pi}{2}, \frac{\pi}{2}\right)$, the following formulas can also be used.

terminal point (head)

unit vector

vector

vector subtraction

vector sum

zero vector

Reading and Writing Mathematics

Explain the difference between speed and velocity in your own words. An Internet search may be helpful.

Speed is a scalar of distance over time without regard to direction, whereas velocity is a vector quantity which involves both speed and direction.

Additional Exercises

Figures for Additional Exercises throughout the chapter can be found at TeacherToolsOnline.com.

1. Express $\overrightarrow{PQ}$ in component form and draw an equal vector in standard position. $\langle -5, 4 \rangle$

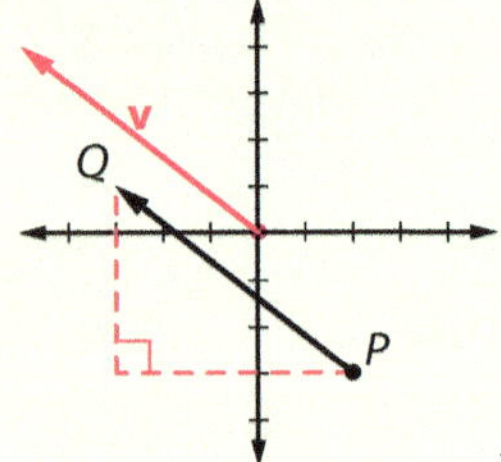

2. Find the magnitude of each vector.
   a. $\mathbf{u} = \langle 6, -8 \rangle$ $|\mathbf{u}| = 10$
   b. $\mathbf{v} = \langle -12, -5 \rangle$ $|\mathbf{v}| = 13$
   c. $\mathbf{w} = \langle -5, 7 \rangle$ $|\mathbf{w}| = \sqrt{74} \approx 8.6$

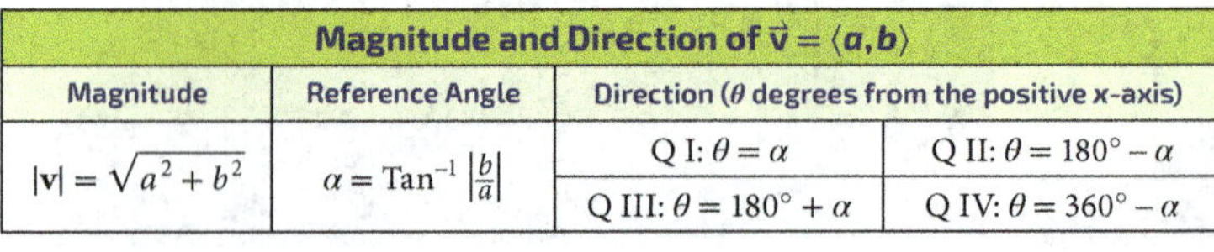

Magnitude and Direction of $\vec{v} = \langle a,b \rangle$		
Magnitude	**Reference Angle**	**Direction (θ degrees from the positive x-axis)**
$|\mathbf{v}| = \sqrt{a^2 + b^2}$	$\alpha = \mathrm{Tan}^{-1}\left|\frac{b}{a}\right|$	Q I: $\theta = \alpha$ Q II: $\theta = 180° - \alpha$ Q III: $\theta = 180° + \alpha$ Q IV: $\theta = 360° - \alpha$

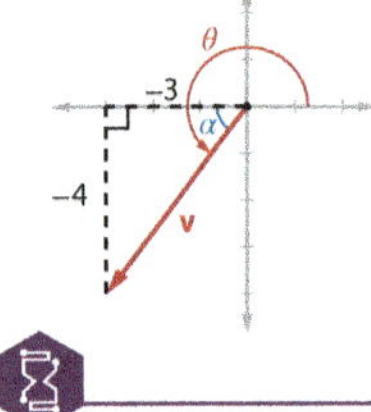

The Norwegian cartographer Caspar Wessel was neither a mathematician nor a scientist, yet in 1699 he became the first to publish the head-to-tail method of adding vectors.

Example 2 Finding the Magnitude and Direction of a Vector

State the magnitude and direction of $\overrightarrow{PQ}$ in Example 1.

Answer

$$|\overrightarrow{PQ}| = \sqrt{a^2 + b^2}$$
$$= \sqrt{(-3)^2 + (-4)^2}$$
$$= \sqrt{25} = 5$$
$$\alpha = \mathrm{Tan}^{-1}\left|\frac{b}{a}\right|$$
$$= \mathrm{Tan}^{-1}\left|\frac{-4}{-3}\right| \approx 53.1°$$

Q III implies that
$\theta \approx 180° + 53.1° = 233.1°$.

1. Apply the formula for a vector's magnitude using $\overrightarrow{PQ} = \langle -3, -4 \rangle$.

2. Find the reference angle when the vector is placed in standard position.

3. Express the direction as a directed angle from the positive x-axis.

SKILL ✔ **EXERCISE 5**

The sum of two vectors can be found by placing the tail of the second at the head of the first and drawing the *resultant vector* from the tail of the first to the head of the second. When the vectors are expressed in component form, their sum is found by adding the corresponding components.

DEFINITION

The **vector sum**, or the **resultant vector**, of $\mathbf{u} = \langle u_1, u_2 \rangle$ and $\mathbf{v} = \langle v_1, v_2 \rangle$ is
$\mathbf{r} = \mathbf{u} + \mathbf{v} = \langle u_1 + v_1, u_2 + v_2 \rangle$.

The product of a real number k (a *scalar*) and a vector $\mathbf{w}$ is defined as a vector with magnitude $k|\mathbf{w}|$. This resulting vector has the same direction as vector $\mathbf{w}$ when $k > 0$, but the direction is reversed when $k < 0$. Section 6.2 discusses the *dot product*, one of two ways to define the product of two vectors.

DEFINITION

The **scalar product** of the real number k and $\mathbf{w} = \langle w_1, w_2 \rangle$ is $k\mathbf{w} = \langle kw_1, kw_2 \rangle$.

Vector subtraction is defined as the addition of the opposite of a vector.
$$\mathbf{u} - \mathbf{v} = \mathbf{u} + (-1)\mathbf{v} = \langle u_1 + (-1)v_1, u_2 + (-1)v_2 \rangle$$
$$= \langle u_1 - v_1, u_2 - v_2 \rangle$$

The sum of a vector and its opposite vector is the *zero vector*, $\mathbf{0} = \langle 0, 0 \rangle$, which has a length of 0 and no direction.

$$\mathrm{Q\ I}\left(\tfrac{a}{b} > 0\right): \theta = \mathrm{Tan}^{-1}\tfrac{b}{a}$$
$$\mathrm{Q\ II}\left(\tfrac{a}{b} < 0\right): \theta = \pi + \mathrm{Tan}^{-1}\tfrac{b}{a}$$
$$\mathrm{Q\ III}\left(\tfrac{a}{b} > 0\right): \theta = \pi + \mathrm{Tan}^{-1}\tfrac{b}{a}$$
$$\mathrm{Q\ IV}\left(\tfrac{a}{b} < 0\right): \theta = 2\pi + \mathrm{Tan}^{-1}\tfrac{b}{a}$$

This method, presented in Section 6.6, will be easily understood after working with reference angles.

Present the vector operations of vector addition, scalar multiplication, and *vector subtraction*. The tail-to-head method of vector addition is presented in Example 3. The alternative parallelogram method for vector addition and vector subtraction is illustrated below.

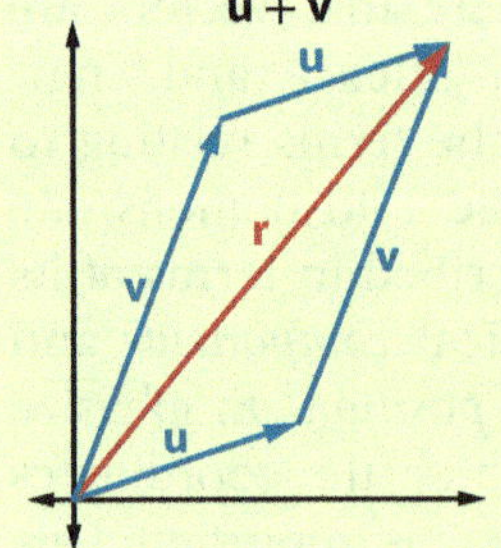

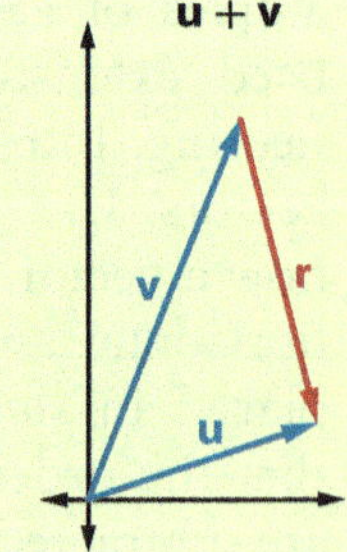

The Internet keyword search *vector parallelogram method* can provide more details, images, and videos explaining these alternative methods.

Interactive Activity Have the students graph the sum of two vectors, $\mathbf{a} + \mathbf{b}$, by placing the tail of $\mathbf{b}$ at the head of $\mathbf{a}$. Then have them graph the sum $\mathbf{b} + \mathbf{a}$ by plac-ing the tail of $\mathbf{a}$ at the head of $\mathbf{b}$. Ask what property of vector addition is demonstrated. (*Commutative*)

Define *unit vector* and describe how to construct a unit vector in the same direction as a given vector $\mathbf{v}$ before illustrating the process with Example 4.

Explain that vectors are sometimes expressed as a *linear combination* of the *standard unit vectors* $\mathbf{i}$ and $\mathbf{j}$. Use Example 5 to acquaint students with this notation and its similarity to the component form for a vector. Caution the students to not confuse the unit vector $\mathbf{i}$ (bold, non-italic) with the imaginary unit i (italic, non-bold).

Given $\mathbf{u} = \langle -2, 1 \rangle$ and $\mathbf{v} = \langle 6, 3 \rangle$, find each resultant vector algebraically and geometrically.

a. $\mathbf{u} + \mathbf{v}$ **b.** $\frac{1}{3}\mathbf{v} - 2\mathbf{u}$

Answer

a. $\mathbf{r} = \langle -2 + 6, 1 + 3 \rangle = \langle 4, 4 \rangle$

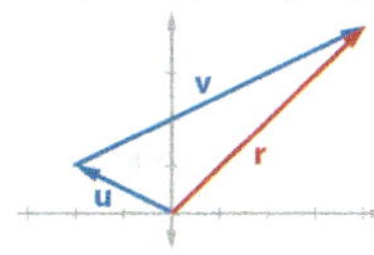

1. Add the corresponding components.
2. Draw $\mathbf{u}$ in standard position, then place the tail of $\mathbf{v}$ at the head of $\mathbf{u}$. Draw the resultant $\mathbf{r}$ from the tail of $\mathbf{u}$ to the head of $\mathbf{v}$.

b. $\mathbf{r} = \frac{1}{3}\langle 6, 3 \rangle - 2\langle -2, 1 \rangle$
$= \langle 2, 1 \rangle + \langle 4, -2 \rangle = \langle 6, -1 \rangle$

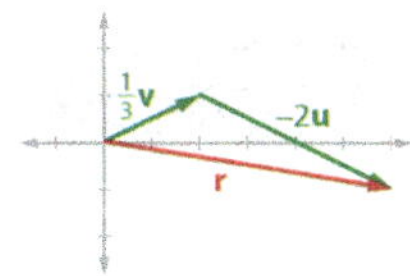

1. Apply the definitions of scalar product and vector addition.
2. Draw $\frac{1}{3}\mathbf{v}$ in standard position, then place the tail of $-2\mathbf{u}$ at the head of $\frac{1}{3}\mathbf{v}$. Draw the resultant $\mathbf{r}$ from the tail of $\frac{1}{3}\mathbf{v}$ to the head of $-2\mathbf{v}$.

SKILL ✔ EXERCISE 9

> **DEFINITION**
>
> Any vector $\mathbf{u}$ with magnitude $|\mathbf{u}| = 1$ is a **unit vector**. The *unit vector in the direction of* $\mathbf{v}$ is found by dividing $\mathbf{v}$ by its magnitude: $\mathbf{u} = \dfrac{\mathbf{v}}{|\mathbf{v}|} = \dfrac{1}{|\mathbf{v}|}\mathbf{v}$.

Example 4   Finding a Unit Vector

Find the unit vector in the direction of $\mathbf{v} = \langle 3, -5 \rangle$ and verify that it has a length of 1.

Answer

$|\mathbf{v}| = \sqrt{3^2 + (-5)^2} = \sqrt{34}$

1. Find the magnitude of $\mathbf{v}$.

$\mathbf{u} = \frac{1}{|\mathbf{v}|}\mathbf{v} = \left(\frac{1}{\sqrt{34}}\right)\langle 3, -5 \rangle$

2. Find $\mathbf{u}$, the unit vector in the direction of $\mathbf{v}$.

$= \left\langle \frac{3}{\sqrt{34}}, -\frac{5}{\sqrt{34}} \right\rangle = \left\langle \frac{3\sqrt{34}}{34}, -\frac{5\sqrt{34}}{34} \right\rangle$

$|\mathbf{u}| = \sqrt{\left(\frac{3}{\sqrt{34}}\right)^2 + \left(-\frac{5}{\sqrt{34}}\right)^2}$

3. Verify that the length of the unit vector is 1.

$= \sqrt{\frac{9}{34} + \frac{25}{34}} = \sqrt{\frac{34}{34}} = 1$

SKILL ✔ EXERCISE 17

The *standard unit vectors* $\mathbf{i} = \langle 1, 0 \rangle$ and $\mathbf{j} = \langle 0, 1 \rangle$ can be used to write a vector $\mathbf{v}$ as a *linear combination* of its horizontal and vertical components.

$\mathbf{v} = \langle a, b \rangle$
$= a\langle 1, 0 \rangle + b\langle 0, 1 \rangle$
$= a\mathbf{i} + b\mathbf{j}$

---

3. Graph each vector and then find its magnitude and direction.
   a. $\langle -5, 5 \rangle$ $5\sqrt{2}, \frac{3\pi}{4}$
   b. $\langle 5\sqrt{3}, -5 \rangle$ $10, \frac{11\pi}{6}$

4. Find the direction of each vector to the nearest tenth of a degree.
   a. $\langle 25, -64 \rangle$ $291.3°$
   b. $\langle -16, 12 \rangle$ $143.1°$

5. Let $\mathbf{u} = \langle 5, 3 \rangle$, $\mathbf{v} = \langle 3, -6 \rangle$, and $\mathbf{w} = \langle -6, 8 \rangle$ to complete the vector operations.
   a. $\mathbf{u} + \mathbf{v}$ $\langle 8, -3 \rangle$
   b. $\mathbf{v} - \mathbf{w}$ $\langle 9, -14 \rangle$
   c. $2\mathbf{u} - \frac{1}{2}\mathbf{w}$ $\langle 13, 2 \rangle$

6. Find the unit vector in the direction of $\mathbf{v} = \langle -2, 4 \rangle$. $\mathbf{u} = \left\langle -\frac{\sqrt{5}}{5}, \frac{2\sqrt{5}}{5} \right\rangle$

7. Write $\overrightarrow{PQ}$ with initial point $P(5, 1)$ and terminal point $Q(-2, 4)$ as a linear combination of standard unit vectors. $-7\mathbf{i} + 3\mathbf{j}$

8. State the component form for $|\mathbf{v}| = 2\sqrt{3}$; $\theta = 120°$. $\langle -\sqrt{3}, 3 \rangle$

9. A plane flying at 320 knots at a bearing of 70° is flying into a 25-knot wind blowing from 10° east of north. Find the resulting speed and bearing of the aircraft.
   304.5 knots at a bearing of 73.6°

10. A 40 lb crate rests on a 30° ramp. The force of friction is up the ramp, and the force of the ramp is perpendicular to the ramp. What is the force of friction on the ramp?   20 lb

---

Example 6 presents a real-life scenario in which a vector is *resolved* into its components. Emphasize that velocity is both speed and direction.

Motivational Idea While vector components are rarely calculated manually in real life, scientists, engineers, and programmers need to understand and apply these concepts to create the programs and systems that automatically perform such calculations.

Discuss Example 7, an application of velocity vectors that can be solved using right-triangle trigonometry. Consider pointing out that this scenario is similar to an airplane flying with a crosswind.

Example 8 presents an application of force vectors and introduces the force of tension. Graphically representing the *vector sum* as a triangle allows the problem to be solved with non-right-triangle trigonometry. Exercise 43 explores the use of equations derived from the component form of each vector and the *resultant*. This method is frequently applied in advanced science classes.

TIPS

Ex. 25–26 Although a radical expression such as $25\sqrt{3}$ is acceptable (and more exact), the decimal approximation often provides more meaning in a real-world problem.

Ex. 41–42 These exercises introduce another scenario for vector problems that will be discussed in Section 6.2.

Assignments

- **Minimum:** 2–3, 5–7, 9–11, 17, 20, 22; 23–31 odd; 32, 35–36, 44–46, 48, 51
- **Standard:** 2–16 even; 17, 20, 22–23, 25–32, 34–37, 39, 41–42, 44–45, 47–51, 53
- **Extended:** 2, 5, 8, 11–14, 17, 20–21, 24–27, 30, 32–45, 47–53

Solutions

❯ A. Exercises

1. $\overrightarrow{PQ} = \langle 1 - (-5), 4 - 1 \rangle = \langle 6, 3 \rangle$

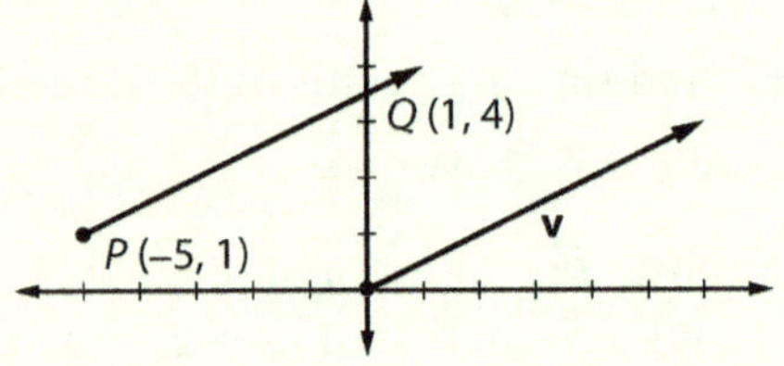

2. $\overrightarrow{PQ} = \langle 2 - 4, -5 - (-1) \rangle = \langle -2, -4 \rangle$

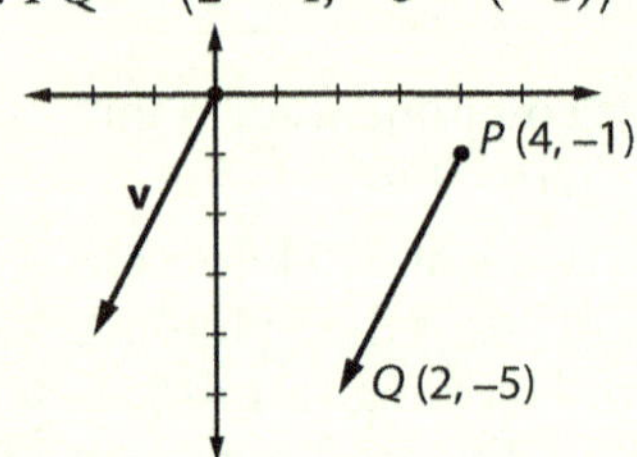

3. $\overrightarrow{PQ} = \langle 2 - 5, 3 - (-1) \rangle = \langle -3, 4 \rangle$

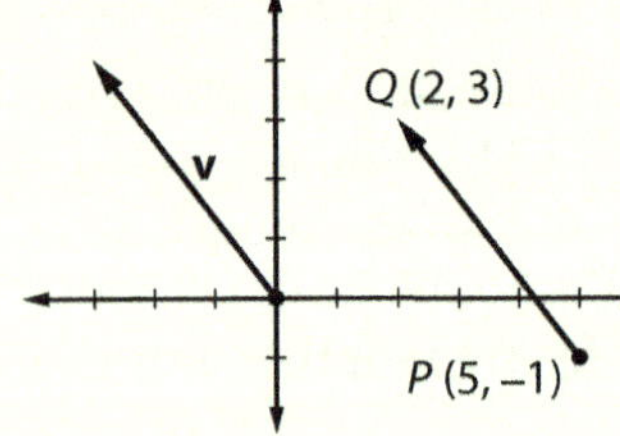

4. $\overrightarrow{PQ} = \langle -1 - (-4), -2 - 1 \rangle$
$= \langle 3, -3 \rangle$

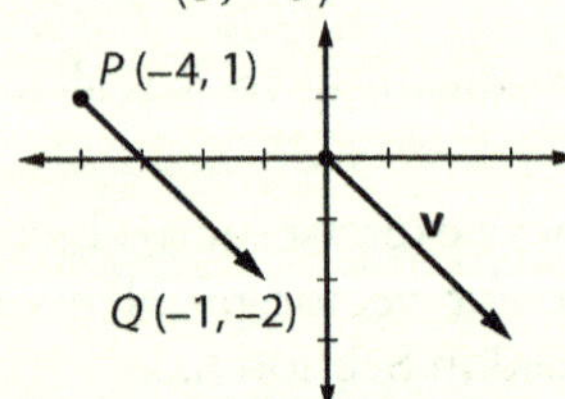

5. $|\mathbf{u}| = \sqrt{4^2 + 4^2} = 4\sqrt{2} \approx 5.7$
$\theta = \alpha = \mathrm{Tan}^{-1} \left| \frac{4}{4} \right| = 45°$

6. $|\mathbf{v}| = \sqrt{(-2)^2 + 6^2} = 2\sqrt{10} \approx 6.3$
$\alpha = \mathrm{Tan}^{-1} \left| \frac{6}{-2} \right| \approx 71.6°$
$\theta = 180 - \alpha \approx 108.4°$

7. $|\mathbf{w}| = \sqrt{5^2 + (-6)^2} = \sqrt{61} \approx 7.8$
$\alpha = \mathrm{Tan}^{-1} \left| \frac{-6}{5} \right| \approx 50.2°$
$\theta = 360 - \alpha \approx 309.8°$

A vector expressed in terms of its magnitude $|\mathbf{v}|$ and its directed angle θ from the positive x-axis can be *resolved* into its horizontal and vertical components with the following formula.

$$\mathbf{v} = \langle a, b \rangle = \langle |\mathbf{v}| \cos \theta, |\mathbf{v}| \sin \theta \rangle$$

An object's velocity vector represents its speed and direction. The velocity of an object is relative to its frame of reference. A blimp's air speed (relative to the air around it) may be 20 mi/hr, but its ground speed (relative to the earth) when flying into a 15 mi/hr headwind is only $20 + (-15) = 5$ mi/hr. The resulting ground velocity of a vessel in moving air (or water) is the sum of its velocity in still air (or still water) and the velocity of the medium through which it travels.

---

8. $|\mathbf{z}| = \sqrt{(-3)^2 + (-8)^2} = \sqrt{73} \approx 8.5$
$\alpha = \mathrm{Tan}^{-1} \left| \frac{-8}{-3} \right| \approx 69.4°$
$\theta = 180 + \alpha \approx 249.4°$

9. $\frac{1}{2}\langle 4, 2 \rangle + \langle -3, 1 \rangle$
$= \langle 2, 1 \rangle + \langle -3, 1 \rangle = \langle -1, 2 \rangle$

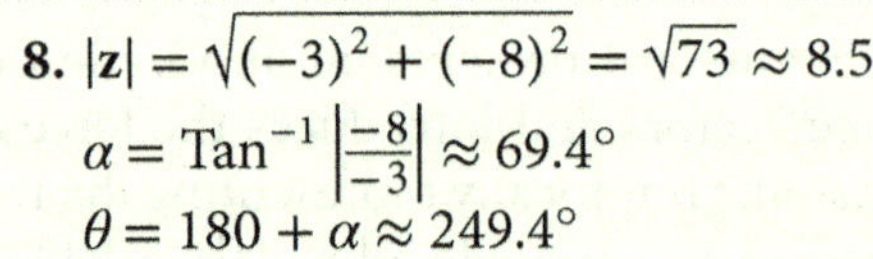

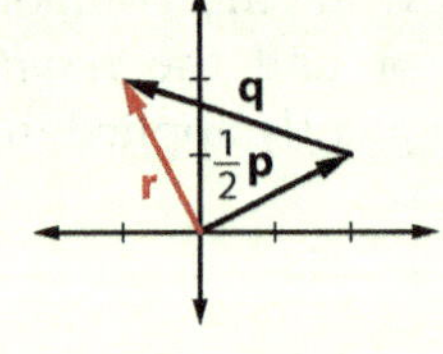

10. $\langle 4, 2 \rangle + 2\langle -3, 1 \rangle$
$= \langle 4, 2 \rangle + \langle -6, 2 \rangle = \langle -2, 4 \rangle$

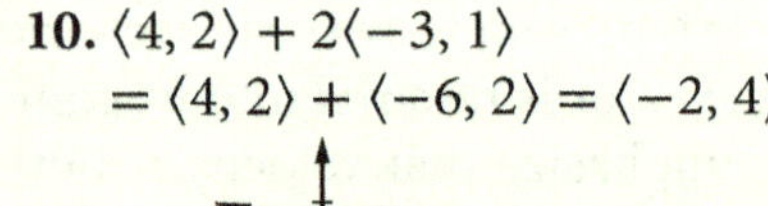

11. $-2\langle 4, 2 \rangle - \langle -3, 1 \rangle$
$= \langle -8, -4 \rangle + \langle 3, -1 \rangle = \langle -5, -5 \rangle$

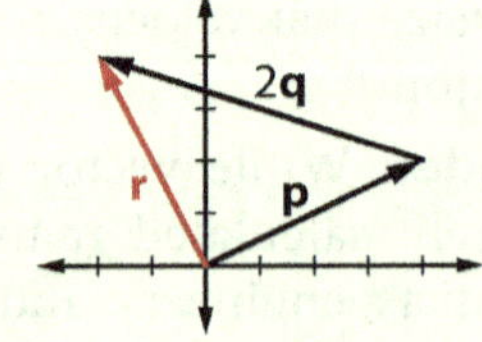

Leigh Anne wants to swim straight across a 300 m wide river flowing at 0.4 m/sec. At what angle upstream (to the nearest degree) should she head if she can swim 1.5 m/sec in still water? How long will it take her to cross the river?

Answer

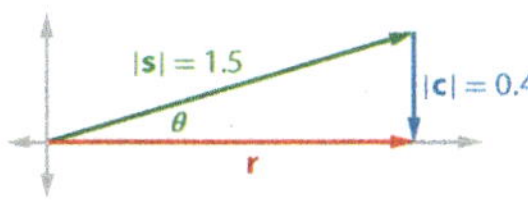

1. Draw vectors representing the swimmer's velocity headed slightly into the current (at θ), the current's velocity headed straight downriver (at 270°), and the resultant velocity heading directly across the river (at 0°).

$\theta = \mathrm{Sin}^{-1} \dfrac{0.4}{1.5} \approx 15°$

$|\mathbf{r}| = \sqrt{1.5^2 - 0.4^2} \approx 1.45 \text{ m/sec}$

2. Since the vector diagram forms a right triangle, determine θ and $|\mathbf{r}|$ using right-triangle trigonometry.

$t \approx \dfrac{300 \text{ m}}{1.45 \text{ m/sec}} \approx 207.5 \text{ sec}$

3. Use rate × time = distance to find the time it takes to cross the river.

SKILL ✔ **EXERCISE 31**

Since forces have both magnitude and direction, they are represented by vectors. The weight of an object is the force gravity exerts on the object and is directed downward, toward the center of the earth. When the sum of all the forces (*net force*) on an object is 0, the object is at rest or has a constant velocity and is said to be at *equilibrium*.

A 10 lb sign hangs motionless from two ropes, each at an angle of 60° above the horizontal. Find the tension in each rope.

Answer

1. Draw vectors in standard position representing the three forces acting on the sign.

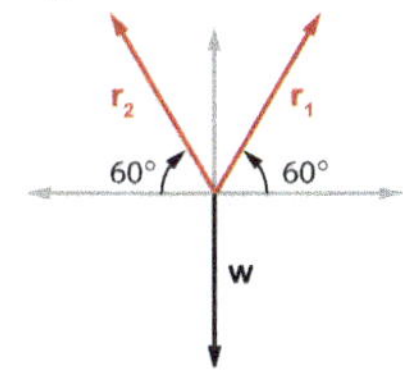

2. Since the sign is at equilibrium, draw a diagram showing $\mathbf{r_1} + \mathbf{r_2} + \mathbf{w} = \mathbf{0}$. The head of the last vector coincides with the tail of the first.

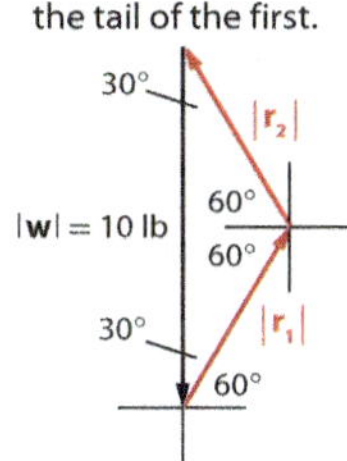

3. Use the Law of Sines to find $|\mathbf{r_1}|$. Note that $|\mathbf{r_2}| = |\mathbf{r_1}|$ since the triangle has two congruent angles.

$\dfrac{|\mathbf{r_1}|}{\sin 30°} = \dfrac{10}{\sin 120°}$

$|\mathbf{r_1}| = \dfrac{10 \sin 30°}{\sin 120°}$

$= \dfrac{10\left(\frac{1}{2}\right)}{\frac{\sqrt{3}}{2}}$

$= \dfrac{10\sqrt{3}}{3} \approx 5.8$

Each rope exerts ≈ 5.8 lb of force on the sign.

SKILL ✔ **EXERCISE 37**

---

12. $-\dfrac{1}{2}\langle 4, 2 \rangle - 2\langle -3, 1 \rangle$
$= \langle -2, -1 \rangle + \langle 6, -2 \rangle = \langle 4, -3 \rangle$

13. $\langle -5, 8 \rangle + \dfrac{1}{2}\langle 4, -3 \rangle$
$= \langle -5, 8 \rangle + \langle 2, -1.5 \rangle$
$= \langle -3, 6.5 \rangle = -3\mathbf{i} + 6.5\mathbf{j}$

14. $\langle -7, 10 \rangle + \langle 4, -3 \rangle - \langle -5, 8 \rangle$
$= \langle 2, -1 \rangle = 2\mathbf{i} - \mathbf{j}$

15. $3\langle 4, -3 \rangle - \dfrac{1}{2}\langle -7, 10 \rangle + 2\langle -5, 8 \rangle$
$= \langle 12, -9 \rangle + \langle 3.5, -5 \rangle + \langle -10, 16 \rangle$
$= \langle 5.5, 2 \rangle = 5.5\mathbf{i} + 2\mathbf{j}$

16. $2\langle -7, 10 \rangle - 5\langle -5, 8 \rangle - \dfrac{3}{2}\langle 4, -3 \rangle$
$= \langle -14, 20 \rangle + \langle 25, -40 \rangle + \langle -6, 4.5 \rangle$
$= \langle 5, -15.5 \rangle = 5\mathbf{i} - 15.5\mathbf{j}$

❭ **B. Exercises**

17. $|\mathbf{v}| = \sqrt{4^2 + (-3)^2} = \sqrt{25} = 5$
$\mathbf{u} = \mathbf{v}\dfrac{1}{|\mathbf{v}|} = \left\langle \dfrac{4}{5}, -\dfrac{3}{5} \right\rangle$

18. $|\mathbf{v}| = \sqrt{(-4)^2 + 1^2} = \sqrt{17}$
$\mathbf{u} = \mathbf{v}\dfrac{1}{|\mathbf{v}|} = \left\langle \dfrac{-4}{\sqrt{17}}, \dfrac{1}{\sqrt{17}} \right\rangle$
$= \left\langle -\dfrac{4\sqrt{17}}{17}, \dfrac{\sqrt{17}}{17} \right\rangle$

19. $|\mathbf{v}| = \sqrt{3^2 + 2^2} = \sqrt{13}$
$\mathbf{u} = \mathbf{v}\dfrac{1}{|\mathbf{v}|} = \left\langle \dfrac{3}{\sqrt{13}}, \dfrac{2}{\sqrt{13}} \right\rangle$
$= \left\langle \dfrac{3\sqrt{13}}{13}, \dfrac{2\sqrt{13}}{13} \right\rangle$

20. $|\mathbf{v}| = \sqrt{2^2 + (-6)^2} = \sqrt{40} = 2\sqrt{10}$
$\mathbf{u} = \mathbf{v}\dfrac{1}{|\mathbf{v}|} = \left\langle \dfrac{2}{2\sqrt{10}}, \dfrac{-6}{2\sqrt{10}} \right\rangle$
$= \left\langle \dfrac{1}{\sqrt{10}}, \dfrac{-3}{\sqrt{10}} \right\rangle = \left\langle \dfrac{\sqrt{10}}{10}, -\dfrac{3\sqrt{10}}{10} \right\rangle$

21. $\langle 5 \cos 20°, 5 \sin 20° \rangle \approx \langle 4.7, 1.7 \rangle$

22. $\left\langle 4\sqrt{3} \cos \dfrac{\pi}{6}, 4\sqrt{3} \sin \dfrac{\pi}{6} \right\rangle$
$= \left\langle 4\sqrt{3}\left(\dfrac{\sqrt{3}}{2}\right), 4\sqrt{3}\left(\dfrac{1}{2}\right) \right\rangle = \langle 6, 2\sqrt{3} \rangle$

23. $\alpha = \mathrm{Tan}^{-1} \left| \dfrac{-5}{2} \right| \approx 68.2°$
$\theta = 360 - \alpha \approx 291.8°$
$\langle 4 \cos 291.8°, 4 \sin 291.8° \rangle$
$\approx \langle 1.5, -3.7 \rangle$

24. $\alpha = \mathrm{Tan}^{-1} \left| \dfrac{15}{-7} \right| \approx 64.98°$
$\theta = 180 - \alpha \approx 115.0°$
$\langle 9 \cos 115.0°, 9 \sin 115.0° \rangle$
$\approx \langle -3.8, 8.2 \rangle$

25. $\theta = 90 - 60 = 30°$
$\mathbf{v} = \langle 50 \cos 30°, 50 \sin 30° \rangle$
$= \langle 25\sqrt{3}, 25 \rangle \approx \langle 43.3, 25 \rangle$

26. $\theta = 90 + 45 = 135°$
$\mathbf{v} = \langle 22 \cos 135°, 22 \sin 135° \rangle$
$= \langle -11\sqrt{2}, 11\sqrt{2} \rangle$
$\approx \langle -15.6, 15.6 \rangle$

27. $\theta = 180°$
$\mathbf{v} = \langle 800 \cos 180°, 800 \sin 180° \rangle$
$= \langle -800, 0 \rangle$

28. $\theta = 270°$
$\mathbf{v} = \langle 3 \cos 270°, 3 \sin 270° \rangle$
$= \langle 0, -3 \rangle$

29. $\theta = 90°$
$\mathbf{v} = \langle 500 \cos 90°, 500 \sin 90° \rangle$
$= \langle 0, 500 \rangle$

30. $\mathbf{v} = \langle 150 \cos 20°, 150 \sin 20° \rangle$
$\approx \langle 141.0, 51.3 \rangle$

31. 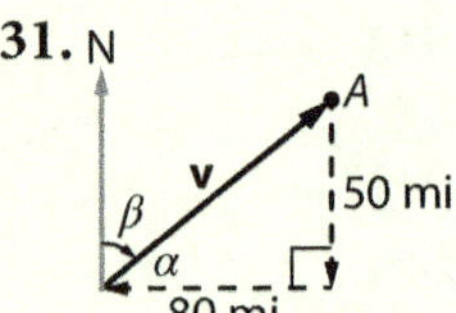

$|\mathbf{v}| = \sqrt{80^2 + 50^2} \approx 94.3 \text{ mi}$

$\alpha = \text{Tan}^{-1}\frac{50}{80} \approx 32°$

$\beta \approx 90 - 32 = 58°$

32. 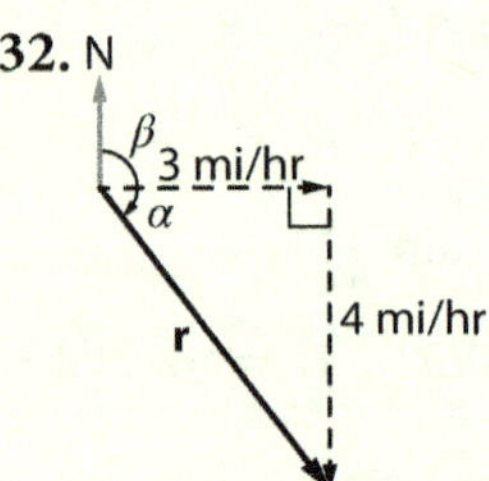

$|\mathbf{r}| = \sqrt{3^2 + 4^2} = 5 \text{ mi/hr}$

$\alpha = \text{Tan}^{-1}\frac{4}{3} \approx 53°$

$\beta \approx 90 + 53 = 143°$

33. 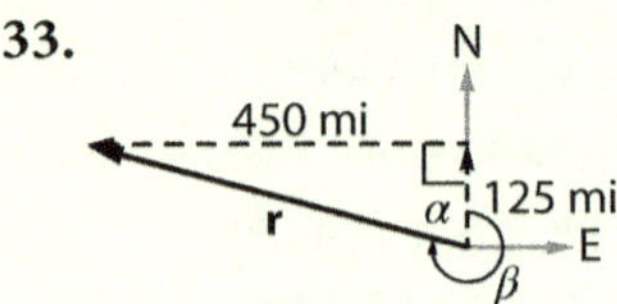

$|\mathbf{r}| = \sqrt{450^2 + 125^2} \approx 467.0 \text{ mi}$

$\alpha = \text{Tan}^{-1}\frac{450}{125} \approx 74°$

$\beta \approx 360 - 74 = 286°$

34. 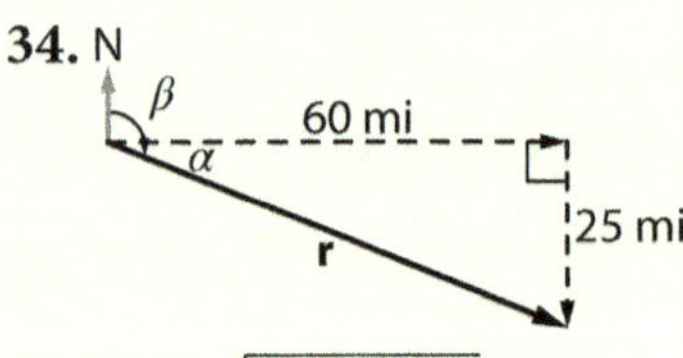

$|\mathbf{r}| = \sqrt{60^2 + 25^2} = 65 \text{ mi}$

$\alpha = \text{Tan}^{-1}\frac{25}{60} \approx 23°$

$\beta \approx 90 + 23 = 113°$

35. 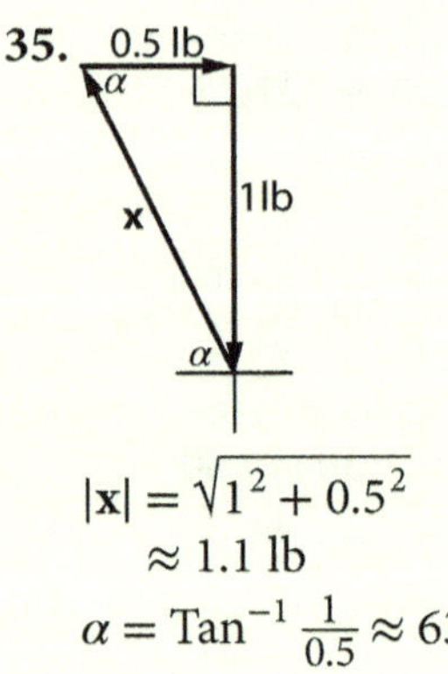

$|\mathbf{x}| = \sqrt{1^2 + 0.5^2}$

$\quad \approx 1.1 \text{ lb}$

$\alpha = \text{Tan}^{-1}\frac{1}{0.5} \approx 63°$

> **A. Exercises**

Draw each vector $\overrightarrow{PQ}$. Then express $\overrightarrow{PQ}$ in component form and draw an equal vector $\vec{v}$ in standard position.

1. $P\,(-5, 1)$ and $Q\,(1, 4)$ **2.** $P\,(4, -1)$ and $Q\,(2, -5)$

3. $P\,(5, -1)$ and $Q\,(2, 3)$ **4.** $P\,(-4, 1)$ and $Q\,(-1, -2)$

Find the magnitude and direction of each vector.

5. $\mathbf{u} = \langle 4, 4 \rangle$ $4\sqrt{2}; 45°$ **6.** $\mathbf{v} = \langle -2, 6 \rangle$ $2\sqrt{10}; \approx 108.4°$

7. $\mathbf{w} = \langle 5, -6 \rangle$ $\sqrt{61}; \approx 309.8°$ **8.** $\mathbf{z} = \langle -3, -8 \rangle$ $\sqrt{73}; \approx 249.4°$

Given $\mathbf{p} = \langle 4, 2 \rangle$ and $\mathbf{q} = \langle -3, 1 \rangle$, find each resultant vector algebraically and geometrically.

9. $\frac{1}{2}\mathbf{p} + \mathbf{q}$ $\langle -1, 2 \rangle$ **10.** $\mathbf{p} + 2\mathbf{q}$ $\langle -2, 4 \rangle$

11. $-2\mathbf{p} - \mathbf{q}$ $\langle -5, -5 \rangle$ **12.** $-\frac{1}{2}\mathbf{p} - 2\mathbf{q}$ $\langle 4, -3 \rangle$

Given $\mathbf{u} = \langle -7, 10 \rangle$, $\mathbf{v} = \langle 4, -3 \rangle$, and $\mathbf{w} = \langle -5, 8 \rangle$, write each resultant vector in component form and as a linear combination of standard unit vectors i and j.

13. $\mathbf{w} + \frac{1}{2}\mathbf{v}$ $\langle -3, 6.5 \rangle;$ **14.** $\mathbf{u} + \mathbf{v} - \mathbf{w}$ $\langle 2, -1 \rangle; 2\mathbf{i} - \mathbf{j}$
$\qquad\qquad -3\mathbf{i} + 6.5\mathbf{j}$

15. $3\mathbf{v} - \frac{1}{2}\mathbf{u} + 2\mathbf{w}$ $\langle 5.5, 2 \rangle;$ **16.** $2\mathbf{u} - 5\mathbf{w} - \frac{3}{2}\mathbf{v}$ $\langle 5, -15.5 \rangle;$
$\qquad\qquad 5.5\mathbf{i} + 2\mathbf{j}$ $\qquad\qquad\qquad 5\mathbf{i} - 15.5\mathbf{j}$

> **B. Exercises**

Find the unit vector in the direction of v.

17. $\mathbf{v} = \langle 4, -3 \rangle$ $\langle \frac{4}{5}, -\frac{3}{5} \rangle$ **18.** $\mathbf{v} = \langle -4, 1 \rangle$ $\langle -\frac{4\sqrt{17}}{17}, \frac{\sqrt{17}}{17} \rangle$

19. $\mathbf{v} = \langle 3, 2 \rangle$ $\langle \frac{3\sqrt{13}}{13}, \frac{2\sqrt{13}}{13} \rangle$ **20.** $\mathbf{v} = \langle 2, -6 \rangle$ $\langle \frac{\sqrt{10}}{10}, -\frac{3\sqrt{10}}{10} \rangle$

State the component form of each vector.

21. $|\mathbf{v}| = 5; \theta = 20°$ $\approx \langle 4.7, 1.7 \rangle$

22. $|\mathbf{u}| = 4\sqrt{3}; \theta = \frac{\pi}{6}$ $\langle 6, 2\sqrt{3} \rangle$

23. $|\mathbf{w}| = 4$ in the direction of $\langle 2, -5 \rangle$ $\approx \langle 1.5, -3.7 \rangle$

24. $|\mathbf{w}| = 9$ in the direction of $\langle -7, 15 \rangle$ $\approx \langle -3.8, 8.2 \rangle$

State the component form of a vector modeling each quantity.

25. a train traveling 60° east of north at 50 mi/hr $\approx \langle 43.3, 25 \rangle$

26. a 22 mi/hr wind from the southeast $\approx \langle -15.6, 15.6 \rangle$

27. a westbound car accelerating at 800 ft/sec² $\langle -800, 0 \rangle$

28. the weight of a 3 lb bird feeder $\langle 0, -3 \rangle$

29. a firework ascending 500 ft straight up $\langle 0, 500 \rangle$

30. a 150 lb force exerted at 20° above the horizontal $\approx \langle 141.0, 51.3 \rangle$

When completing the following exercises, round angles to the nearest degree and lengths to the nearest tenth.

31. A helicopter flies 50 mi due south before turning and flying 80 mi due west. How far and at which bearing must the pilot fly to return to his point of origin?

32. A boy who swims at a speed of 3 mi/hr attempts to swim due east, directly across a river. The river's current flows due south at a speed of 4 mi/hr. Find the boy's resultant velocity as measured from the shore.

33. An airplane flies due north 125 mi from Newark to Albany, then to Detroit, 450 mi due west. What is the plane's distance and bearing from its starting point?

34. A boat sails 60 mi due east from the island of Unus to the island of Dosland, then 25 mi due south to Tres Island. What is the boat's distance and bearing from Unus?   65 mi; 113°

35. A 1 lb mailbox and platform are supported by a horizontal force of 0.5 lb and a diagonal force. What is the magnitude and direction of the diagonal force?   1.1 lb; 63° above horizontal

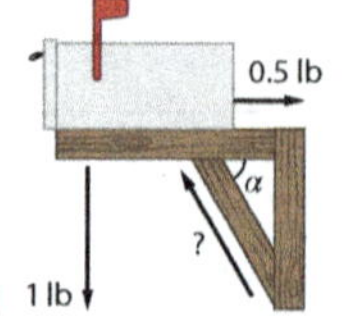

36. A 20 lb sign hangs motionless from two chains, each at a 45° angle to the sign. Find the tension in each chain.   14.1 lb

> **C. Exercises**

37. A plane flies with an air speed of 200 mi/hr at a bearing of 225° through a 50 mi/hr wind from the south. What is the plane's velocity relative to the ground?

38. A wide river flows due south with a current of 10 knots. A boat crosses the river traveling upstream at a bearing of 60° and a speed of 30 knots. Find the resultant bearing and ground speed of the boat.

39. A pilot plots his course from Fernwood to Glendale at a bearing of 295°. If a 40 mi/hr wind is blowing from 60° east of north, at what compass heading and air speed does the pilot need to fly to maintain a ground speed of 350 mi/hr? $\approx 301°; \approx 329$ mi/hr

40. A boat leaves port and sails 50 mi due west to avoid a storm, then turns to a compass heading of 200° and proceeds 120 mi to its destination. How far and at what bearing should the boat travel when headed directly back to its original port? ≈ 144.9 mi; $\approx 39°$

31. 94.3 mi; 58°

32. 5 mi/hr; 143°

33. 467.0 mi; 286°

37. ≈ 168 mi/hr at a bearing of $\approx 237°$

38. $\approx 79°; \approx 26.5$ knots

---

36. 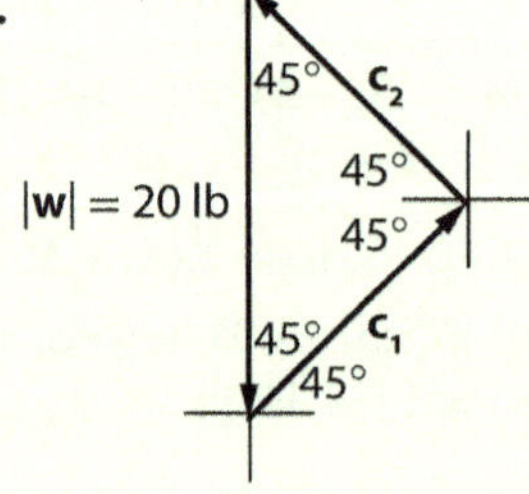

The triangle is an isosceles right triangle.

$|\mathbf{c_1}| = |\mathbf{c_2}| = \frac{20}{\sqrt{2}} = 10\sqrt{2} \approx 14.1 \text{ lb}$

> **C. Exercises**

37. 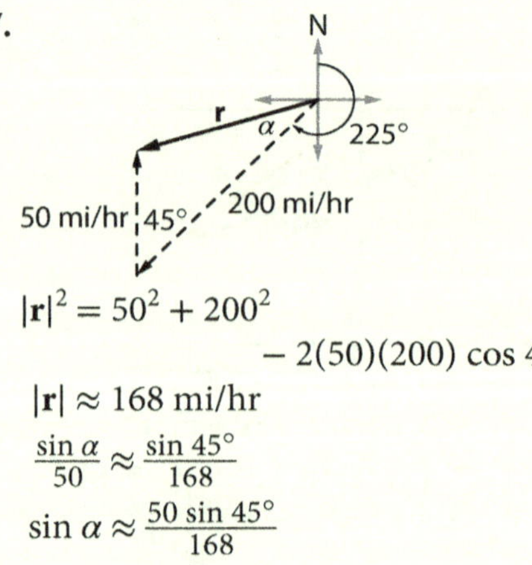

$|\mathbf{r}|^2 = 50^2 + 200^2$
$\qquad\qquad - 2(50)(200)\cos 45°$

$|\mathbf{r}| \approx 168 \text{ mi/hr}$

$\frac{\sin \alpha}{50} \approx \frac{\sin 45°}{168}$

$\sin \alpha \approx \frac{50 \sin 45°}{168}$

$\qquad \alpha \approx 12°$

$\beta \approx 225 + 12 = 237°$

An object resting on a ramp has three forces acting on it: its weight pulling down toward the center of the earth; the force of static friction acting parallel to the ramp; and the normal force, the force exerted by the ramp perpendicular to its surface.

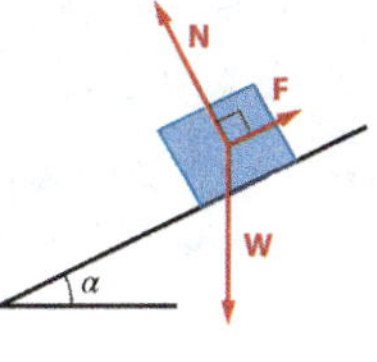

41. Determine the magnitude of the frictional force and the normal force when a 20 lb box of nails rests on a 30° ramp. $|\mathbf{F}| = 10$ lb; $|\mathbf{N}| \approx 17.3$ lb

42. Find a seesaw's angle of incline if a 50 lb child rests on a tilted seesaw and the normal force is 40 lb. $\approx 37°$

43. A problem involving the sum of vectors can also be solved using equations generated by examining the components of the vectors. Complete steps *a–e* to solve the problem from Example 8:

A 10 lb sign hangs motionless from two ropes, each at an angle of 60° above the horizontal. Find the tension in each rope.

   a. Draw the three forces acting on the sign as vectors in standard position.

   b. Write each vector in terms of its horizontal and vertical components.

   c. Using the fact that the net force is 0, write equations for the *x*- and *y*-components.

   d. Solve the *x*-component equation and interpret the result.

   e. Solve the *y*-component equation and interpret the result.

CUMULATIVE REVIEW

44. Convert 1.22 radians to the nearest degree. [4.1] 70°

45. Convert 40° to radian measure. Express your answer in terms of π and as a decimal approximation rounded to the nearest thousandth. [4.1] $\frac{2\pi}{9} \approx 0.698$

Find A in $\triangle ABC$ (to the nearest degree). [5.7]

46. 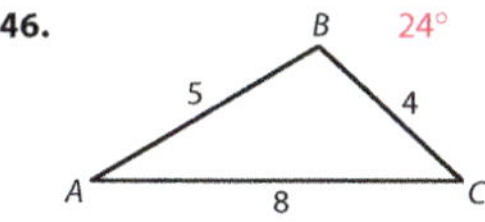 24°

47. $b = 5$, $a = 8$, $C = 40°$ 102°

48. Graph $f(x) = 2 \cos 2x$ over $[0, 2\pi]$. [4.4]

49. Describe how $g(x) = -3 \cos \frac{x}{2} + 1$ is obtained by transforming the parent function $f(x) = \cos x$. [4.4]

50. Which expression is equivalent to i^{-4}? [Appendix] A

   A. 1   **C.** i **E.** none of these

   B. -1 **D.** $-i$

51. Evaluate $\log_5 5^x$. [3.2] D

   A. 1   **C.** 3125   **E.** none of these

   B. 25   **D.** x

52. Which expression represents $m\angle E$? [5.7] B

   A. $\text{Cos}^{-1} \frac{2}{3}$

   B. $\text{Cos}^{-1} \left(-\frac{1}{3}\right)$

   C. $\text{Cos}^{-1} \frac{5}{9}$

   D. $\text{Cos}^{-1} \left(-\frac{2}{3}\right)$

   E. none of these

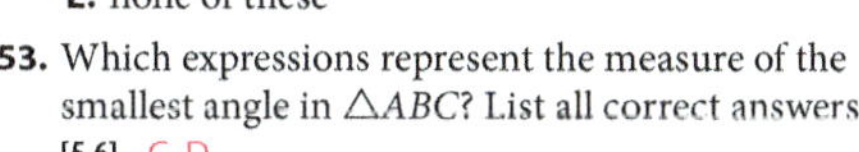

53. Which expressions represent the measure of the smallest angle in $\triangle ABC$? List all correct answers. [5.6] C, D

   A. $\text{Sin}^{-1} \frac{4}{7}$

   B. $\text{Sin}^{-1} \frac{20}{31}$

   C. $\text{Sin}^{-1} \frac{4 \sin A}{7}$

   D. $\text{Sin}^{-1} \frac{20 \sin C}{31}$

   E. none of these

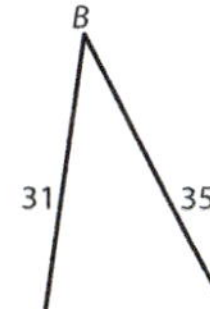

38.

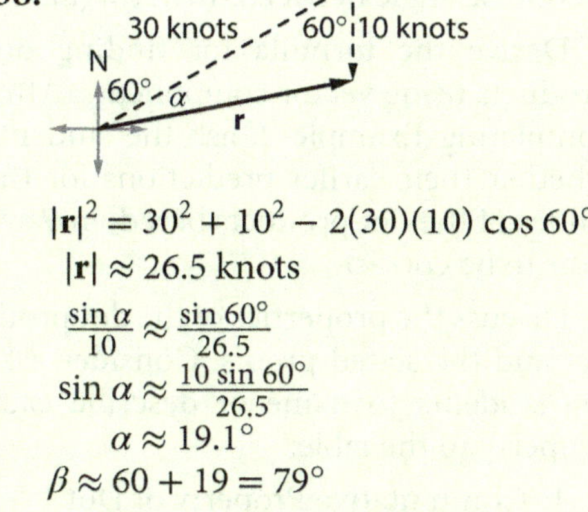

$$|\mathbf{r}|^2 = 30^2 + 10^2 - 2(30)(10) \cos 60°$$
$$|\mathbf{r}| \approx 26.5 \text{ knots}$$
$$\frac{\sin \alpha}{10} \approx \frac{\sin 60°}{26.5}$$
$$\sin \alpha \approx \frac{10 \sin 60°}{26.5}$$
$$\alpha \approx 19.1°$$
$$\beta \approx 60 + 19 = 79°$$

39.

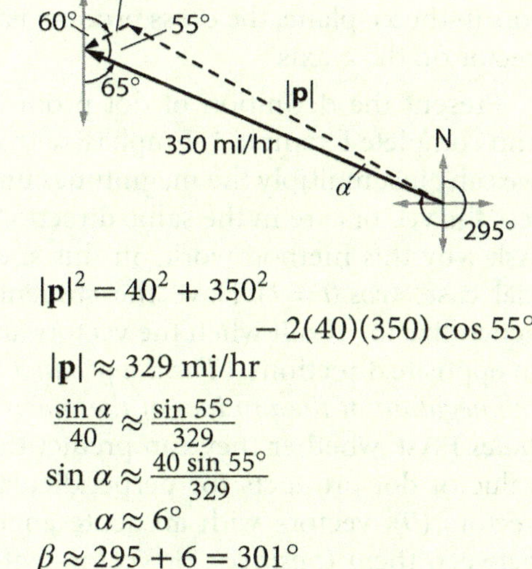

$$|\mathbf{p}|^2 = 40^2 + 350^2 - 2(40)(350) \cos 55°$$
$$|\mathbf{p}| \approx 329 \text{ mi/hr}$$
$$\frac{\sin \alpha}{40} \approx \frac{\sin 55°}{329}$$
$$\sin \alpha \approx \frac{40 \sin 55°}{329}$$
$$\alpha \approx 6°$$
$$\beta \approx 295 + 6 = 301°$$

40.

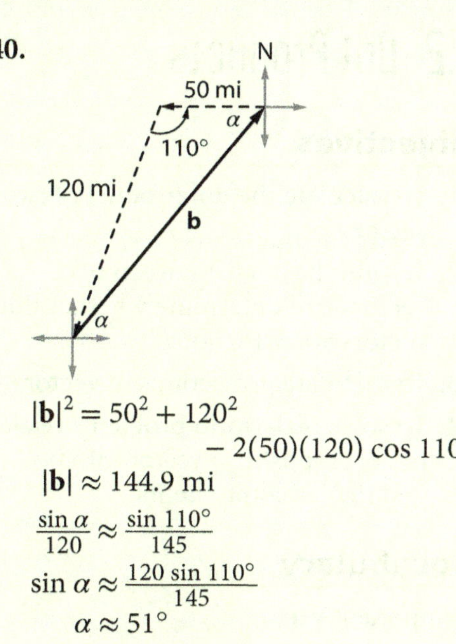

$$|\mathbf{b}|^2 = 50^2 + 120^2 - 2(50)(120) \cos 110°$$
$$|\mathbf{b}| \approx 144.9 \text{ mi}$$
$$\frac{\sin \alpha}{120} \approx \frac{\sin 110°}{145}$$
$$\sin \alpha \approx \frac{120 \sin 110°}{145}$$
$$\alpha \approx 51°$$
$$\beta \approx 90 - 51 = 39°$$

41.

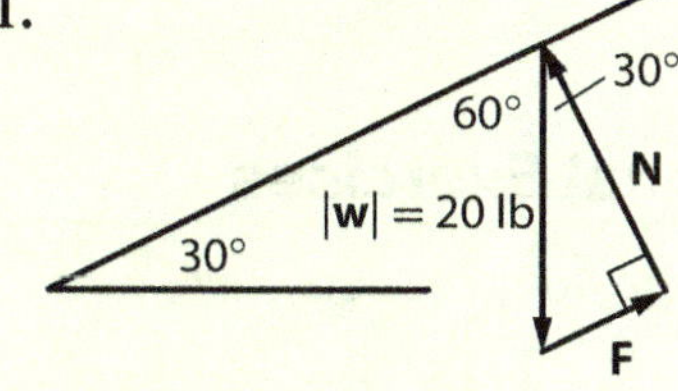

$$\sin 30° = \frac{|\mathbf{F}|}{20}$$
$$|\mathbf{F}| = 20 \sin 30° = 10 \text{ lb}$$
$$\cos 30° = \frac{|\mathbf{N}|}{20}$$
$$|\mathbf{N}| = 20 \cos 30° \approx 17.3 \text{ lb}$$

42.

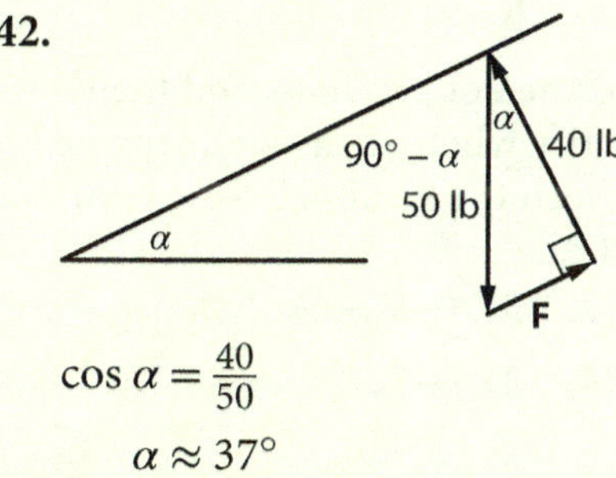

$$\cos \alpha = \frac{40}{50}$$
$$\alpha \approx 37°$$

43a.

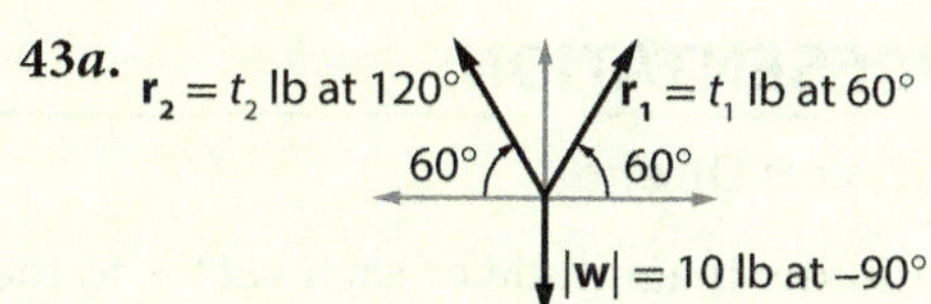

43b. $\mathbf{w} = \langle 10 \cos(-90°), 10 \sin(-90°) \rangle$
$$= \langle 0, -10 \rangle$$
$$\mathbf{r}_1 = \langle t_1 \cos 60°, t_1 \sin 60° \rangle$$
$$= \left\langle \frac{1}{2} t_1, \frac{\sqrt{3}}{2} t_1 \right\rangle$$
$$\mathbf{r}_2 = \langle t_2 \cos 120°, t_2 \sin 120° \rangle$$
$$= \left\langle -\frac{1}{2} t_2, \frac{\sqrt{3}}{2} t_2 \right\rangle$$

continued in Answers and Solutions Overflow

6.2 Dot Products

Objectives

1. To calculate the dot product of two vectors
2. To find the angle between two vectors and determine whether the vectors are orthogonal
3. To calculate projections of vectors
4. To solve real-world problems related to decomposing a vector into perpendicular components

Vocabulary

component vector
dot product
orthogonal vectors
projection

Additional Exercises

1. Find the dot product. $-4\sqrt{3}$

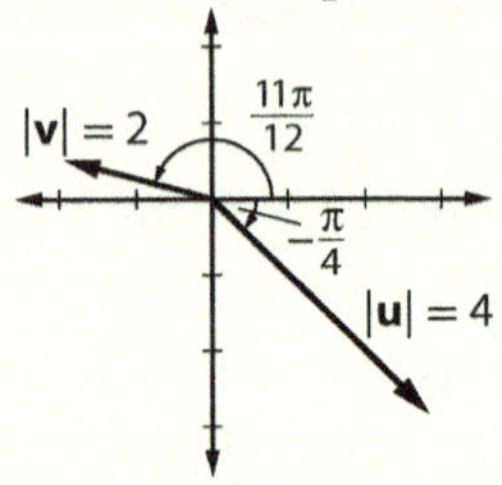

2. Find the dot products and then identify whether the angle formed between the vectors is 90°, < 90°, or > 90°.

   a. $\langle 2, -3 \rangle, \langle -9, -6 \rangle$ 0, 90°

   b. $\langle 4, -3 \rangle, \langle -2, 5 \rangle$ -23, > 90°

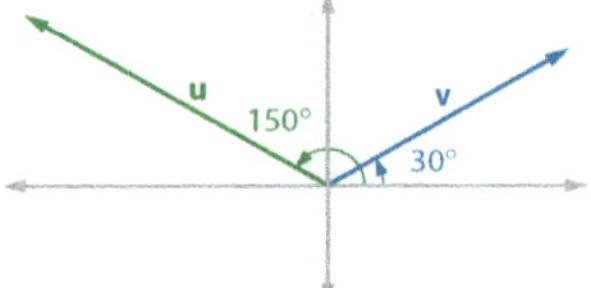

6.2 Dot Products

In physics, work is defined not as the amount of effort expended, but as the dot product of the displacement vector and the amount of force in that direction.

After completing this section, you will be able to

- calculate the dot product of two vectors.
- find the angle between two vectors.
- calculate projections of vectors.
- solve real-world problems requiring the decomposition of a vector.

Well-developed reasoning skills help you make complex decisions based on scriptural principles. Studying proofs can help you develop these skills.

In Section 6.1 you saw that the sum or difference of two vectors is another vector and that the result of multiplying a vector by a scalar (a real number) is a vector. This section explores the *dot product* of two vectors, a multiplication of two vectors that results in a scalar.

DEFINITION

The **dot product** of two vectors **u** and **v** is $\mathbf{u} \cdot \mathbf{v} = |\mathbf{u}|\,|\mathbf{v}| \cos \theta$, where θ is the angle between **u** and **v** when the vectors are placed in standard position ($0 \le \theta \le 180$).

Example 1 Finding a Dot Product

Find $\mathbf{u} \cdot \mathbf{v}$ if $|\mathbf{u}| = 6$ and $|\mathbf{v}| = 4$.

Answer

$\theta = |150° - 30°| = 120°$
$\mathbf{u} \cdot \mathbf{v} = |\mathbf{u}|\,|\mathbf{v}| \cos \theta$
$\quad = 6 \cdot 4 \cos 120°$
$\quad = 24\left(-\frac{1}{2}\right) = -12$

1. Find θ, the angle between **u** and **v**.
2. Substitute into the definition of a dot product.
3. Simplify.

SKILL ✔ EXERCISE 5

The dot product can also be expressed in terms of the vector's horizontal and vertical components.

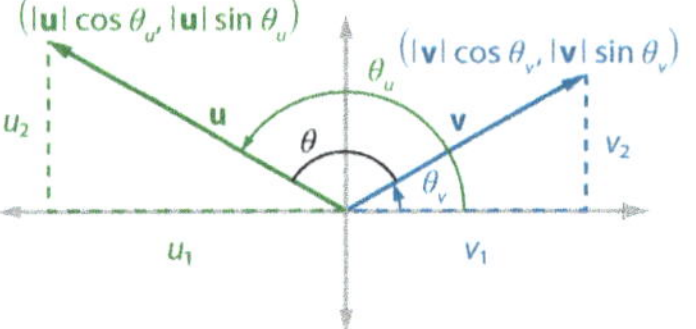

Use $\mathbf{u} = \langle u_1, u_2 \rangle$ where $u_1 = |\mathbf{u}| \cos \theta_u$ and $u_2 = |\mathbf{u}| \sin \theta_u$ and $\mathbf{v} = \langle v_1, v_2 \rangle$ where $v_1 = |\mathbf{v}| \cos \theta_v$ and $v_2 = |\mathbf{v}| \sin \theta_v$.
Then $\mathbf{u} \cdot \mathbf{v} = |\mathbf{u}|\,|\mathbf{v}| \cos (\theta_u - \theta_v)$

$\quad = |\mathbf{u}|\,|\mathbf{v}| (\cos \theta_u \cos \theta_v + \sin \theta_u \sin \theta_v)$

$\quad = |\mathbf{u}| \cos \theta_u\, |\mathbf{v}| \cos \theta_v + |\mathbf{u}| \sin \theta_u\, |\mathbf{v}| \sin \theta_v$

$\quad = u_1 v_1 + u_2 v_2.$

Therefore, if $\mathbf{u} = \langle u_1, u_2 \rangle$ and $\mathbf{v} = \langle v_1, v_2 \rangle$, then $\mathbf{u} \cdot \mathbf{v} = u_1 v_1 + u_2 v_2$.

Find each dot product, given $\mathbf{p} = \langle 0, 3 \rangle$, $\mathbf{q} = \langle 2, 4 \rangle$, $\mathbf{r} = \langle -2, 1 \rangle$, and $\mathbf{s} = \langle 1, -3 \rangle$.

a. $\mathbf{p} \cdot \mathbf{q}$ **b.** $\mathbf{q} \cdot \mathbf{r}$ **c.** $\mathbf{s} \cdot \mathbf{r}$

Answer

Apply the dot product formula that uses vector components:
$\mathbf{u} \cdot \mathbf{v} = \langle u_1, u_2 \rangle \cdot \langle v_1, v_2 \rangle = u_1 v_1 + u_2 v_2$.

a. $\mathbf{p} \cdot \mathbf{q} = \langle 0, 3 \rangle \cdot \langle 2, 4 \rangle$
$= 0(2) + 3(4)$
$= 0 + 12 = 12$

b. $\mathbf{q} \cdot \mathbf{r} = \langle 2, 4 \rangle \cdot \langle -2, 1 \rangle$
$= 2(-2) + 4(1)$
$= -4 + 4 = 0$

c. $\mathbf{s} \cdot \mathbf{r} = \langle 1, -3 \rangle \cdot \langle -2, 1 \rangle$
$= 1(-2) + (-3)(1)$
$= -2 - 3 = -5$

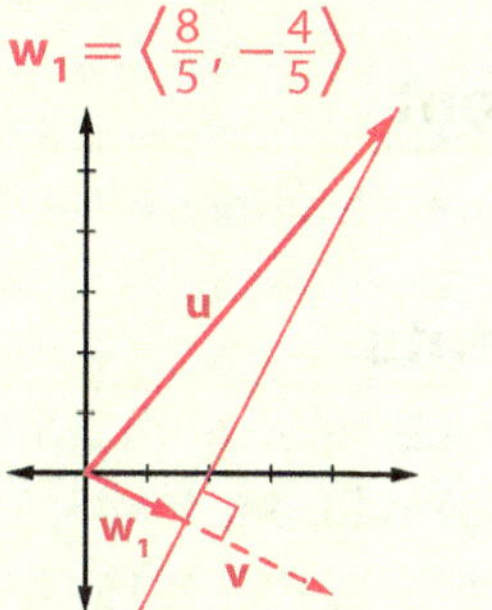

SKILL ✔ EXERCISE 7

It is helpful to know the following properties of dot products. The proofs of the first and last properties are shown below. Proofs of the other properties are left as exercises.

Properties of Dot Products		
Let **u**, **v**, and **w** be vectors and let k be a scalar.		
1. $\mathbf{u} \cdot \mathbf{v} = \mathbf{v} \cdot \mathbf{u}$	3. $k(\mathbf{u} \cdot \mathbf{v}) = k\mathbf{u} \cdot \mathbf{v}$ or $\mathbf{u} \cdot k\mathbf{v}$	5. $\mathbf{u} \cdot \mathbf{u} = |\mathbf{u}|^2$
2. $\mathbf{u} \cdot (\mathbf{v} + \mathbf{w}) = \mathbf{u} \cdot \mathbf{v} + \mathbf{u} \cdot \mathbf{w}$	4. $\mathbf{0} \cdot \mathbf{v} = 0$	

Let $\mathbf{u} = \langle u_1, u_2 \rangle$ and $\mathbf{v} = \langle v_1, v_2 \rangle$.

$\mathbf{u} \cdot \mathbf{v} = u_1 v_1 + u_2 v_2$ and $\mathbf{u} \cdot \mathbf{u} = u_1 u_1 + u_2 u_2$
$= v_1 u_1 + v_2 u_2$ $\qquad\qquad\qquad = u_1^2 + u_2^2$
$= \mathbf{v} \cdot \mathbf{u}$ $\qquad\qquad\qquad\quad = \left(\sqrt{u_1^2 + u_2^2} \right)^2$
$\qquad\qquad\qquad\qquad\qquad = |\mathbf{u}|^2$

Notice that this fifth property provides another way to express the magnitude of a vector.

$|\mathbf{u}| = \sqrt{\mathbf{u} \cdot \mathbf{u}} = \sqrt{u_1^2 + u_2^2}$

The angle θ between two vectors can be found using their dot product. Since $\mathbf{u} \cdot \mathbf{v} = |\mathbf{u}|\,|\mathbf{v}| \cos \theta$, it follows that $\cos \theta = \frac{\mathbf{u} \cdot \mathbf{v}}{|\mathbf{u}|\,|\mathbf{v}|}$.

Find the angle between each pair of vectors.

a. $\mathbf{p} = \langle 0, 3 \rangle$ and $\mathbf{q} = \langle 2, 4 \rangle$ **b.** $\mathbf{r} = \langle -2, 1 \rangle$ and $\mathbf{s} = \langle 1, -3 \rangle$

Answer

Use $\cos \theta = \frac{\mathbf{u} \cdot \mathbf{v}}{|\mathbf{u}|\,|\mathbf{v}|}$.

a. $\mathbf{p} \cdot \mathbf{q} = 0(2) + 3(4) = 12$

$|\mathbf{p}| = \sqrt{0^2 + 3^2} = 3$

$|\mathbf{q}| = \sqrt{2^2 + 4^2} = \sqrt{20}$

$\cos \theta = \frac{12}{3\sqrt{20}} \approx 0.8944$

$\theta \approx \text{Cos}^{-1} 0.8944$
$\approx 26.6°$
or 0.46 radians

b. $\mathbf{r} \cdot \mathbf{s} = -2(1) + 1(-3) = -5$

$|\mathbf{r}| = \sqrt{(-2)^2 + 1^2} = \sqrt{5}$

$|\mathbf{s}| = \sqrt{1^2 + (-3)^2} = \sqrt{10}$

$\cos \theta = \frac{-5}{\sqrt{5} \cdot \sqrt{10}} \approx -0.7071$

$\theta \approx \text{Cos}^{-1} (-0.7071)$
$\approx 135.0°$
or 2.36 radians

1. Find the dot product of the vectors and their magnitudes.

2. Substitute and use the Cos^{-1} function to find θ.

SKILL ✔ EXERCISES 11, 13

---

3. Find the degree measure between the given vectors.

a. $\langle 1, -5 \rangle$, $\langle 4, -2 \rangle$ $\quad \theta \approx 52.1°$

b. $\langle 4, -2 \rangle$, $\langle -2, -3 \rangle$ $\quad \theta \approx 97.1°$

4. If $\mathbf{v} = \langle v_1, v_2 \rangle$ is orthogonal to $\mathbf{u} = \langle -2, 5 \rangle$, find the ratio of v_2 to v_1, and then use this ratio to find two vectors orthogonal to $\mathbf{u}$.

$\frac{v_2}{v_1} = \frac{2}{5}$; any two vectors in the form $\langle 5k, 2k \rangle$, such as $\langle 10, 4 \rangle$ and $\langle -5, -2 \rangle$

5. Find $\mathbf{w}_1$, the projection of $\mathbf{u}$ onto $\mathbf{v}$. Draw vectors $\mathbf{v}$, $\mathbf{u}$, and $\mathbf{w}_1$ to illustrate the projection.

a. $\mathbf{u} = \langle 5, 6 \rangle$, $\mathbf{v} = \langle 4, -2 \rangle$
$\mathbf{w}_1 = \left\langle \frac{8}{5}, -\frac{4}{5} \right\rangle$

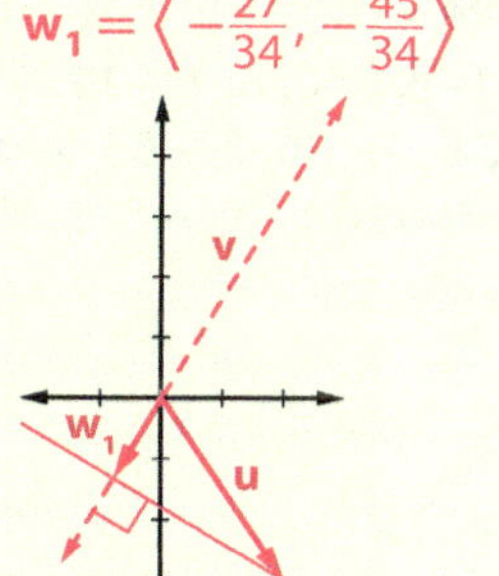

b. $\mathbf{u} = \langle 2, -3 \rangle$, $\mathbf{v} = \langle 3, 5 \rangle$
$\mathbf{w}_1 = \left\langle -\frac{27}{34}, -\frac{45}{34} \right\rangle$

---

3. **Commutative Property of Scalar Multiplication Within Dot Products**

4. **Zero Vector Dot Product Property**

5. The dot product of a vector and itself is the square of the vector's magnitude.

Common Student Error A student may confuse the dot product symbol with the multiplication symbol. Encourage students to pay close attention to the context in order to avoid confusion.

Explain that various applications involving the dot product will be explored in this section, including the determination of the amount of work done by an applied force.

Illustrate how the dot product can be used to find the angle between two vectors in Example 3. Explain that this reasoning will be extended to three-dimensional vectors in Section 6.3.

Common Student Error When using the formula for finding the angle between two vectors and entering an expression such as $\text{Cos}^{-1}\left(\frac{-11}{\sqrt{29} \cdot \sqrt{14}} \right)$ into the calculator, a student may incorrectly enter $\text{Cos}^{-1}(-11/\sqrt{29} * \sqrt{14})$. This error can be avoided by entering $\text{Cos}^{-1}(-11/\sqrt{29 * 14})$, which is also easier to enter.

The paragraphs following Example 3 confirm earlier predictions for dot product values and differentiate the terms *or-thogonal* and *perpendicular*. Example 4 demonstrates how the dot product can be used to find a general expression for all vectors that are orthogonal to a given vector.

Refer the students to the figure to describe resolving the vector $\mathbf{u}$ into $\mathbf{w}_1 = \text{proj}_\mathbf{v} \mathbf{u}$, its component in the direction of vector $\mathbf{v}$, and $\mathbf{w}_2$, another *component vector* orthogonal to $\mathbf{w}_1$. Consider having the students visualize a light source perpendicularly above $\mathbf{v}$, casting a shadow of $\mathbf{u}$ onto $\mathbf{v}$. Discuss the derivation of the formula for $\text{proj}_\mathbf{v} \mathbf{u}$ and explain that $\mathbf{w}_2$ is calculated as a difference of $\mathbf{u}$ and $\text{proj}_\mathbf{v} \mathbf{u}$.

6. Use the vectors from Additional Exercise 5 to find the orthogonal component $\mathbf{w}_2$ such that $\mathbf{u} = \text{proj}_\mathbf{v}\,\mathbf{u} + \mathbf{w}_2$.

a. $\left\langle \frac{17}{5}, \frac{34}{5} \right\rangle$ **b.** $\left\langle \frac{95}{34}, -\frac{57}{34} \right\rangle$

7. An 80 lb wagon is being pushed up a 30° incline. Find the component of its weight that . . .

  a. pulls the wagon down the incline. ≈ 40 lb

  b. pushes into the hill perpendicular to its surface. ≈ 69.3 lb

8. How much work is done by pulling a log for 70 m with a force of 120 N on a rope at a 25° angle above the ground? ≈ 7613 N·m

Assessment

- Quiz 6A covers Sections 6.1–6.2.

Assignments

- **Minimum:** 1–2, 4, 6–9, 13–14, 17, 20, 22–23; 27–37 odd; 44, 46–48, 50, 52
- **Standard:** 1–5, 7, 10, 12, 15–17, 20–22, 24, 27–28, 30, 33, 36, 38, 47–48, 50–52
- **Extended:** 1–2, 5–6, 8, 11–12, 15–17, 20–26, 29, 31–33, 36; 37–53 odd

---

Considering that $\mathbf{u} \cdot \mathbf{v} = |\mathbf{u}|\,|\mathbf{v}| \cos\theta$ and that $|\mathbf{u}|$ and $|\mathbf{v}|$ are always positive, the possible values of $\cos\theta$ are used to see that the dot product must be positive when $0° \leq \theta < 90°$, equal to 0 when $\theta = 90°$, and negative when $90° < \theta \leq 180°$. This reasoning shows that vectors $\mathbf{q}$ and $\mathbf{r}$ in Example 2 are perpendicular.

DEFINITION

Two vectors $\mathbf{u}$ and $\mathbf{v}$ are **orthogonal** if and only if $\mathbf{u} \cdot \mathbf{v} = 0$.

The terms *orthogonal* and *perpendicular* have the same meaning except in the case of the zero vector, **0**. While the zero vector is orthogonal with any other vector (since $\mathbf{0} \cdot \mathbf{v} = 0$), it has no direction and cannot be perpendicular to another vector.

Example 4 Finding and Verifying Orthogonal Vectors

If $\mathbf{v} = \langle v_1, v_2 \rangle$ is orthogonal to $\mathbf{u} = \langle -6, 4 \rangle$, find the ratio of v_2 to v_1. Then find a general expression for all orthogonal vectors and verify that it is perpendicular to $\mathbf{u}$.

Answer

$\mathbf{u} \cdot \mathbf{v} = -6v_1 + 4v_2 = 0$ 1. Set $\mathbf{u} \cdot \mathbf{v} = 0$.

$\quad 4v_2 = 6v_1$

$\quad \dfrac{v_2}{v_1} = \dfrac{3}{2}$ 2. Solve for $\dfrac{v_2}{v_1}$.

$\therefore \mathbf{v} = k\langle 2, 3 \rangle = \langle 2k, 3k \rangle$ where $k \in \mathbb{R}$ 3. Any scalar multiple of $\langle 2, 3 \rangle$ will have $\dfrac{v_2}{v_1} = \dfrac{3}{2}$.

$\mathbf{u} \cdot \mathbf{v} = -6(2k) + 4(3k) = 0$ 4. Use the dot product to show that $\mathbf{u}$ and the general expression for $\mathbf{v}$ are orthogonal.

In Section 6.1 you saw how to resolve (or decompose) a vector into its horizontal and vertical components. However, in many applications it is helpful to resolve a vector, $\mathbf{u}$, into two orthogonal *component vectors*, one of which is parallel to another vector, $\mathbf{v}$.

To resolve the vector $\mathbf{u}$ into $\mathbf{w}_1$ (its component in the direction of vector $\mathbf{v}$) and $\mathbf{w}_2$ (another orthogonal component), draw a perpendicular segment from the head of $\mathbf{u}$ to the line containing $\mathbf{v}$. Then draw the component vectors $\mathbf{w}_1$ and $\mathbf{w}_2$ such that $\mathbf{u} = \mathbf{w}_1 + \mathbf{w}_2$. The vector $\mathbf{w}_1$ is the *projection of u onto v*, notated as $\text{proj}_\mathbf{v}\,\mathbf{u}$. The dot product is used to express $\mathbf{w}_1 = \text{proj}_\mathbf{v}\,\mathbf{u}$ as a scalar multiple of $\mathbf{v}$.

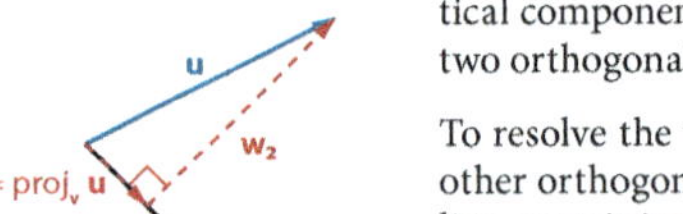

$\mathbf{u} \cdot \mathbf{v} = (\mathbf{w}_1 + \mathbf{w}_2) \cdot \mathbf{v}$ Substitute, using $\mathbf{u} = \mathbf{w}_1 + \mathbf{w}_2$.

$\quad\quad = (\mathbf{w}_1 \cdot \mathbf{v}) + (\mathbf{w}_2 \cdot \mathbf{v})$ Distribute the dot product over vector addition.

$\quad\quad = \mathbf{w}_1 \cdot \mathbf{v} + 0$ Vectors $\mathbf{w}_2$ and $\mathbf{v}$ are orthogonal.

$\quad\quad = k\mathbf{v} \cdot \mathbf{v}$ Express $\mathbf{w}_1$ as a scalar multiple of $\mathbf{v}$, using $\mathbf{w}_1 = k\mathbf{v}$.

$\mathbf{u} \cdot \mathbf{v} = k|\mathbf{v}|^2$ Apply the fifth dot product property.

$\quad k = \dfrac{\mathbf{u} \cdot \mathbf{v}}{|\mathbf{v}|^2}$ Solve for k.

$\therefore \mathbf{w}_1 = \text{proj}_\mathbf{v}\,\mathbf{u} = \left(\dfrac{\mathbf{u} \cdot \mathbf{v}}{|\mathbf{v}|^2} \right)\mathbf{v}$

The component that is orthogonal to $\mathbf{v}$ can then be found using $\mathbf{w}_2 = \mathbf{u} - \text{proj}_\mathbf{v}\,\mathbf{u}$.

Ask how long the "shadow" $\text{proj}_\mathbf{v}\,\mathbf{u}$ is if $\mathbf{u}$ and $\mathbf{v}$ are orthogonal. (*0, since* $\mathbf{u} \cdot \mathbf{v} = 0$) Apply the formula to find the *projection* and $\mathbf{w}_2$ in Example 5. Emphasize the interpretation of the negative projection as described below the example.

When presenting Example 6, emphasize that the weight vector points straight down in these "object-on-a-ramp" problems, making θ the complement of the angle of inclination.

Discuss the derivation of the formula for work, emphasizing that the work accomplished is dependent on the movement of the object. Example 7 demonstrates using this formula.

TIPS

Ex. 39 Encourage students to reason using the definition of dot product: $\mathbf{u} \cdot \mathbf{v} = |\mathbf{u}|\,|\mathbf{v}| \cos\theta$.

Find the projection of $\mathbf{u} = \langle -4, 3 \rangle$ onto $\mathbf{v} = \langle 5, 2 \rangle$. Then find the orthogonal component $\mathbf{w_2}$ such that $\mathbf{u} = \text{proj}_\mathbf{v}\,\mathbf{u} + \mathbf{w_2}$.

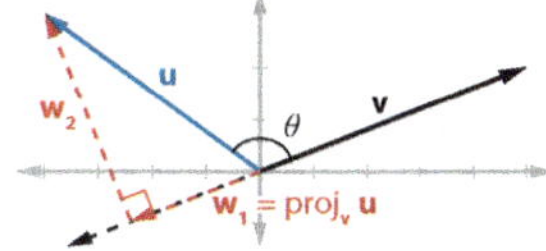

Answer

$$\text{proj}_\mathbf{v}\,\mathbf{u} = \left(\frac{\langle -4, 3 \rangle \cdot \langle 5, 2 \rangle}{|\langle 5, 2 \rangle|^2} \right) \langle 5, 2 \rangle$$

$$= \frac{-4(5) + 3(2)}{\left(\sqrt{5^2 + 2^2} \right)^2} \langle 5, 2 \rangle$$

$$= -\frac{14}{29}\langle 5, 2 \rangle = \left\langle -\frac{70}{29}, -\frac{28}{29} \right\rangle$$

$$\mathbf{w_2} = \langle -4, 3 \rangle - \left\langle -\frac{70}{29}, -\frac{28}{29} \right\rangle$$

$$= \left\langle -\frac{116}{29} + \frac{70}{29}, \frac{87}{29} + \frac{28}{29} \right\rangle = \left\langle -\frac{46}{29}, \frac{115}{29} \right\rangle$$

Check

$$\left\langle -\frac{70}{29}, -\frac{28}{29} \right\rangle \cdot \left\langle -\frac{46}{29}, \frac{115}{29} \right\rangle$$

$$= -\frac{70}{29}\left(-\frac{46}{29} \right) + \left(-\frac{28}{29} \right)\left(\frac{115}{29} \right) = \frac{3220 - 3220}{29^2} = 0$$

1. Find $\mathbf{w_1} = \text{proj}_\mathbf{v}\,\mathbf{u} = \left(\frac{\mathbf{u} \cdot \mathbf{v}}{|\mathbf{v}|^2} \right)\mathbf{v}$.

   The fact that the projection is in the opposite direction of $\mathbf{v}$ is indicated by the negative scalar $\frac{\mathbf{u} \cdot \mathbf{v}}{|\mathbf{v}|^2} = -\frac{14}{29}$.

2. Find $\mathbf{w_2} = \mathbf{u} - \mathbf{w_1}$.

3. Use the dot product to confirm that $\text{proj}_\mathbf{v}\,\mathbf{u}$ and $\mathbf{w_2}$ are orthogonal.

—— SKILL ✔ **EXERCISE 27**

Notice that $\text{proj}_\mathbf{v}\,\mathbf{u} = \left(\frac{\mathbf{u} \cdot \mathbf{v}}{|\mathbf{v}|^2} \right)\mathbf{v}$ is in the same direction as $\mathbf{v}$ when the dot product $\mathbf{u} \cdot \mathbf{v}$ is positive ($0° \leq \theta < 90°$) and that $\text{proj}_\mathbf{v}\,\mathbf{u}$ is in the opposite direction from $\mathbf{v}$ when the dot product is negative ($90° < \theta \leq 180°$).

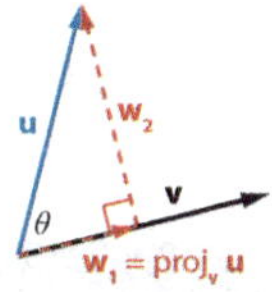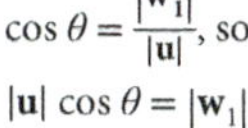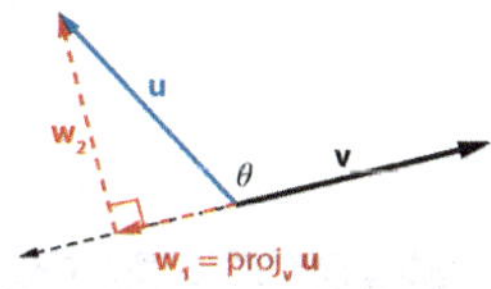

$$\cos \theta = \frac{|\mathbf{w_1}|}{|\mathbf{u}|}, \text{ so}$$
$$|\mathbf{u}| \cos \theta = |\mathbf{w_1}|$$

$$\cos (180° - \theta) = -\cos \theta = \frac{|\mathbf{w_1}|}{|\mathbf{u}|}$$
$$|\mathbf{u}| \cos \theta = -|\mathbf{w_1}|$$

The right-triangle trigonometry demonstrated above shows that $|\mathbf{u}| \cos \theta$ provides the directed magnitude of $\text{proj}_\mathbf{v}\,\mathbf{u}$. Interpret positive values as being in the same direction as $\mathbf{v}$ and negative values as being in the opposite direction from $\mathbf{v}$.

Solutions

❯ A. Exercises

3. $\mathbf{u} \cdot \mathbf{v} = 2 \cdot 3 \cos 135°$
$$= 6\left(-\frac{\sqrt{2}}{2} \right) = -3\sqrt{2}$$

4. $\mathbf{u} \cdot \mathbf{v} = 10 \cdot 5 \cos 180°$
$$= 50(-1) = -50$$

5. expressing $\theta_\mathbf{s}$ as $\frac{5\pi}{3}$:
$$\theta = \frac{5\pi}{3} - \frac{7\pi}{6} = \frac{\pi}{2}$$
$$\mathbf{r} \cdot \mathbf{s} = 4 \cdot \sqrt{2}\,\cos \frac{\pi}{2}$$
$$= 4\sqrt{2}(0) = 0$$

6. $\theta = 50 - (-30) = 80°$
$$\mathbf{f} \cdot \mathbf{g} = 7 \cdot 6 \cos 80° \approx 7.3$$

7–12. If $\mathbf{u} \cdot \mathbf{v} > 0$, $\theta < 90°$;
if $\mathbf{u} \cdot \mathbf{v} = 0$, $\theta = 90°$; and
if $\mathbf{u} \cdot \mathbf{v} < 0$, $\theta > 90°$.

7. $2(5) + 3(7) = 31; \theta < 90°$

8. $(-1)2 + (-4)(-3) = 10; \theta < 90°$

9. $4(4) + 8(-2) = 0; \theta = 90°$

10. $-4(6) + 6(-9) = -78; \theta > 90°$

11. $6\left(\frac{1}{3} \right) + \frac{1}{2}(-12) = -4; \theta > 90°$

12. $\frac{3}{5}\left(\frac{5}{6} \right) + \left(-\frac{4}{7} \right)\frac{7}{8} = 0; \theta = 90°$

13. $\mathbf{u} \cdot \mathbf{v} = 1(-3) + 4(-2) = -11$
$$|\mathbf{u}| = \sqrt{1^2 + 4^2} = \sqrt{17}$$
$$|\mathbf{v}| = \sqrt{(-3)^2 + (-2)^2} = \sqrt{13}$$
$$\cos \theta = \frac{-11}{\sqrt{17} \cdot \sqrt{13}}$$
$$\theta = \text{Cos}^{-1}\left(\frac{-11}{\sqrt{17} \cdot \sqrt{13}} \right) \approx 137.7°$$

14. $\mathbf{r} \cdot \mathbf{s} = 7(2) + 6(9) = 68$
$$|\mathbf{r}| = \sqrt{7^2 + 6^2} = \sqrt{85}$$
$$|\mathbf{s}| = \sqrt{2^2 + 9^2} = \sqrt{85}$$
$$\cos \theta = \frac{68}{85}$$
$$\theta = \text{Cos}^{-1}\frac{68}{85} \approx 36.9°$$

15. $\mathbf{a} \cdot \mathbf{b} = 5(9) + (-8)1 = 37$
$$|\mathbf{a}| = \sqrt{5^2 + (-8)^2} = \sqrt{89}$$
$$|\mathbf{b}| = \sqrt{9^2 + 1^2} = \sqrt{82}$$
$$\cos \theta = \frac{37}{\sqrt{89} \cdot \sqrt{82}}$$
$$\theta = \text{Cos}^{-1}\left(\frac{37}{\sqrt{89} \cdot \sqrt{82}} \right) \approx 64.3°$$

16. $\mathbf{f} \cdot \mathbf{g} = (-3)(-5) + 2(-6) = 3$
$$|\mathbf{f}| = \sqrt{(-3)^2 + 2^2} = \sqrt{13}$$
$$|\mathbf{g}| = \sqrt{(-5)^2 + (-6)^2} = \sqrt{61}$$
$$\cos \theta = \frac{3}{\sqrt{13} \cdot \sqrt{61}}$$
$$\theta = \text{Cos}^{-1}\left(\frac{3}{\sqrt{13} \cdot \sqrt{61}} \right) \approx 83.9°$$

17. $10v_1 - 2v_2 = 0$
$$10v_1 = 2v_2$$
$$\frac{v_2}{v_1} = \frac{5}{1}; k\langle 1, 5\rangle$$

18. $-3v_1 + 8v_2 = 0$
$$8v_2 = 3v_1$$
$$\frac{v_2}{v_1} = \frac{3}{8}; k\langle 8, 3\rangle$$

19. $\frac{1}{3}v_1 + \frac{5}{6}v_2 = 0$
$$5v_2 = -2v_1$$
$$\frac{v_2}{v_1} = -\frac{2}{5}$$
$$k\langle 5, -2\rangle \text{ or } k\langle -5, 2\rangle$$

20. $-2v_1 - \frac{3}{4}v_2 = 0$
$$-8v_1 = 3v_2$$
$$\frac{v_2}{v_1} = -\frac{8}{3}$$
$$k\langle 3, -8\rangle \text{ or } k\langle -3, 8\rangle$$

21. $u_1v_1 + u_2v_2 = 0$
$$u_2v_2 = -u_1v_1$$
$$\frac{v_2}{v_1} = -\frac{u_1}{u_2}$$
$$k\langle u_2, -u_1\rangle \text{ or } k\langle -u_2, u_1\rangle$$

22. $\mathbf{0} \cdot \mathbf{v} = \langle 0, 0\rangle \cdot \langle v_1, v_2\rangle$
$$= 0v_1 + 0v_2 = 0$$

23. $k(\mathbf{u} \cdot \mathbf{v}) = k(u_1v_1 + u_2v_2)$
$$= (ku_1)v_1 + (ku_2)v_2$$
$$= \langle ku_1, ku_2\rangle \cdot \langle v_1, v_2\rangle$$
$$= k\langle u_1, u_2\rangle \cdot \langle v_1, v_2\rangle$$
$$= k\mathbf{u} \cdot \mathbf{v}$$

24. $k(\mathbf{u} \cdot \mathbf{v}) = k(u_1v_1 + u_2v_2)$
$$= u_1(kv_1) + u_2(kv_2)$$
$$= \langle u_1, u_2\rangle \cdot \langle kv_1, kv_2\rangle$$
$$= \langle u_1, u_2\rangle \cdot k\langle v_1, v_2\rangle$$
$$= \mathbf{u} \cdot k\mathbf{v}$$

25. $\mathbf{u} \cdot (\mathbf{v} + \mathbf{w})$
$$= \langle u_1, u_2\rangle \cdot \langle v_1 + w_1, v_2 + w_2\rangle$$
$$= u_1(v_1 + w_1) + u_2(v_2 + w_2)$$
$$= u_1v_1 + u_1w_1 + u_2v_2 + u_2w_2$$
$$= (u_1v_1 + u_2v_2) + (u_1w_1 + u_2w_2)$$
$$= \langle u_1, u_2\rangle \cdot \langle v_1, v_2\rangle + \langle u_1, u_2\rangle \cdot \langle w_1, w_2\rangle$$
$$= \mathbf{u} \cdot \mathbf{v} + \mathbf{u} \cdot \mathbf{w}$$

26. $\text{proj}_{\mathbf{v}}\,\mathbf{u} = \left(\frac{\langle 3, 8\rangle \cdot \langle 1, 2\rangle}{|\langle 1, 2\rangle|^2}\right)\langle 1, 2\rangle$
$$= \left(\frac{3(1) + 8(2)}{\left(\sqrt{1^2 + 2^2}\right)^2}\right)\langle 1, 2\rangle$$
$$= \frac{19}{5}\langle 1, 2\rangle = \left\langle \frac{19}{5}, \frac{38}{5}\right\rangle$$
$$\mathbf{w}_2 = \mathbf{u} - \text{proj}_{\mathbf{v}}\,\mathbf{u}$$
$$= \langle 3, 8\rangle - \left\langle \frac{19}{5}, \frac{38}{5}\right\rangle = \left\langle -\frac{4}{5}, \frac{2}{5}\right\rangle$$

27. $\text{proj}_{\mathbf{v}}\,\mathbf{u} = \left(\frac{\langle -2, 3\rangle \cdot \langle 1, 4\rangle}{|\langle 1, 4\rangle|^2}\right)\langle 1, 4\rangle$
$$= \left(\frac{-2(1) + 3(4)}{\left(\sqrt{1^2 + 4^2}\right)^2}\right)\langle 1, 4\rangle$$
$$= \frac{10}{17}\langle 1, 4\rangle = \left\langle \frac{10}{17}, \frac{40}{17}\right\rangle$$
$$\mathbf{w}_2 = \mathbf{u} - \text{proj}_{\mathbf{v}}\,\mathbf{u}$$
$$= \langle -2, 3\rangle - \left\langle \frac{10}{17}, \frac{40}{17}\right\rangle$$
$$= \left\langle -\frac{44}{17}, \frac{11}{17}\right\rangle$$

28. $\text{proj}_{\mathbf{v}}\,\mathbf{u} = \left(\frac{\langle 4, 1\rangle \cdot \langle 3, 5\rangle}{|\langle 3, 5\rangle|^2}\right)\langle 3, 5\rangle$
$$= \left(\frac{4(3) + 1(5)}{\left(\sqrt{3^2 + 5^2}\right)^2}\right)\langle 3, 5\rangle$$
$$= \frac{1}{2}\langle 3, 5\rangle = \left\langle \frac{3}{2}, \frac{5}{2}\right\rangle$$

$$\mathbf{w}_2 = \mathbf{u} - \text{proj}_{\mathbf{v}}\,\mathbf{u}$$
$$= \langle 4, 1\rangle - \left\langle \frac{3}{2}, \frac{5}{2}\right\rangle = \left\langle \frac{5}{2}, -\frac{3}{2}\right\rangle$$

29. $\text{proj}_{\mathbf{v}}\,\mathbf{u} = \left(\frac{\langle 1, -5\rangle \cdot \langle -2, 3\rangle}{|\langle -2, 3\rangle|^2}\right)\langle -2, 3\rangle$
$$= \left(\frac{1(-2) + (-5)3}{\left(\sqrt{(-2)^2 + 3^2}\right)^2}\right)\langle -2, 3\rangle$$
$$= -\frac{17}{13}\langle -2, 3\rangle = \left\langle \frac{34}{13}, -\frac{51}{13}\right\rangle$$
$$\mathbf{w}_2 = \mathbf{u} - \text{proj}_{\mathbf{v}}\,\mathbf{u}$$
$$= \langle 1, -5\rangle - \left\langle \frac{34}{13}, -\frac{51}{13}\right\rangle$$
$$= \left\langle -\frac{21}{13}, -\frac{14}{13}\right\rangle$$

Example 6　Using a Projection

Determine the downhill component of a 150 lb skier's weight that pulls the skier down a hill with a slope of 30°.

Answer

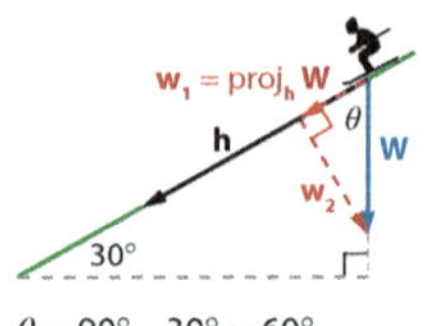

$\theta = 90° - 30° = 60°$

$$\text{proj}_\mathbf{h}\,\mathbf{W} = \mathbf{w}_1 = |\mathbf{W}|\cos\theta$$
$$= |150|\cos 60°$$
$$= 150\left(\frac{1}{2}\right) = 75\text{ lb}$$

1. Sketch a downhill vector $\mathbf{h}$, the force vector $\mathbf{W}$ representing the skier's weight, $\mathbf{w}_1 = \text{proj}_\mathbf{h}\,\mathbf{W}$, and the perpendicular component $\mathbf{w}_2$.

2. Determine θ, the angle between $\mathbf{W}$ and $\mathbf{h}$.

3. Find the magnitude of the projection of $\mathbf{W}$ onto $\mathbf{h}$.

SKILL ✓ **EXERCISE 31**

Vector projections are used to calculate the amount of work done by a constant force as it moves an object from point A to point B. The amount of work done is defined as the product of the component of the force that is parallel to the object's motion and the distance that the object is moved. The typical unit of work in both the British Imperial System and the US customary system is the foot-pound (ft-lb). In the International System of Units, a newton meter (N·m), also called a joule (J), is commonly used.

$$W = \left|\text{proj}_{\overrightarrow{AB}}\,\mathbf{F}\right|\left|\overrightarrow{AB}\right| \qquad \text{definition of work}$$
$$= |\mathbf{F}|\cos\theta\left|\overrightarrow{AB}\right| \qquad \text{definition of proj}_{\overrightarrow{AB}}\,\mathbf{F}$$
$$= |\mathbf{F}|\left|\overrightarrow{AB}\right|\cos\theta \qquad \text{commutative property}$$
$$W = \mathbf{F} \cdot \overrightarrow{AB} \qquad \text{definition of dot product}$$

Example 7　Finding Work

Find the amount of work done by Rebecca as she moves a wagon 300 m down the beach, pulling with 20 N of force on the handle, which is 35° above the horizontal.

Answer

$$W = \mathbf{F} \cdot \overrightarrow{AB} = |\mathbf{F}|\left|\overrightarrow{AB}\right|\cos\theta$$
$$= (20\text{ N})(300\text{ m})\cos 35°$$
$$\approx 4915\text{ N·m or }4915\text{ J}$$

Find the dot product of the vectors representing the applied force and the distance moved.

SKILL ✓ **EXERCISE 37**

⟩ A. Exercises

1. If $\mathbf{r} \cdot \mathbf{s} = 0$, the vectors are said to be __________ and the angle between the vectors (assuming $\mathbf{r} \neq \mathbf{0}$ and $\mathbf{s} \neq \mathbf{0}$) measures _____. *orthogonal; 90° or $\frac{\pi}{2}$*

2. Describe the angle measure between vectors $\mathbf{r}$ and $\mathbf{s}$ in each case.

   a. $\mathbf{r} \cdot \mathbf{s} > 0$ *$0° \leq \theta < 90°$*   **b.** $\mathbf{r} \cdot \mathbf{s} < 0$ *$90° < \theta \leq 180°$*

Find $\mathbf{u} \cdot \mathbf{v}$, given the magnitudes and the angle θ between the vectors.

3. $|\mathbf{u}| = 2, |\mathbf{v}| = 3, \theta = 135°$ *$-3\sqrt{2}$*

4. $|\mathbf{u}| = 10, |\mathbf{v}| = 5, \theta = 180°$ *-50*

Find the dot product of each pair of vectors.

5. 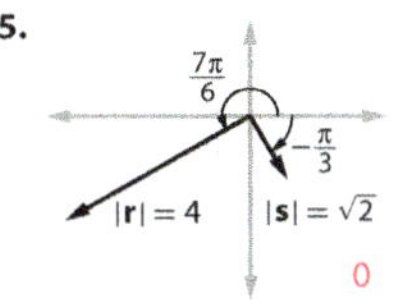

0

6. 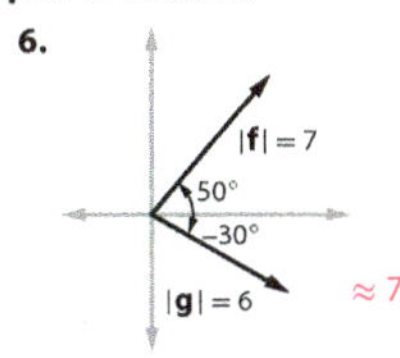

≈ 7.3

Find each dot product. Then state whether the angle between the vectors is $> 90°$, $< 90°$, or $= 90°$.

7. $\langle 2, 3 \rangle \cdot \langle 5, 7 \rangle$ *31; < 90°*   8. $\langle -1, -4 \rangle \cdot \langle 2, -3 \rangle$ *10; < 90°*

9. $\langle 4, 8 \rangle \cdot \langle 4, -2 \rangle$ *0; = 90°*   10. $\langle -4, 6 \rangle \cdot \langle 6, -9 \rangle$ *−78; > 90°*

11. $\left\langle 6, \frac{1}{2} \right\rangle \cdot \left\langle \frac{1}{3}, -12 \right\rangle$ *−4; > 90°*   12. $\left\langle \frac{3}{5}, -\frac{4}{7} \right\rangle \cdot \left\langle \frac{5}{6}, \frac{7}{8} \right\rangle$ *0; = 90°*

Use a dot product to find the angle between each pair of vectors (to the nearest tenth of a degree).

13. $\mathbf{u} = \langle 1, 4 \rangle, \mathbf{v} = \langle -3, -2 \rangle$ *137.7°*

14. $\mathbf{r} = \langle 7, 6 \rangle, \mathbf{s} = \langle 2, 9 \rangle$ *36.9°*

15. $\mathbf{a} = \langle 5, -8 \rangle, \mathbf{b} = \langle 9, 1 \rangle$ *64.3°*

16. $\mathbf{f} = \langle -3, 2 \rangle, \mathbf{g} = \langle -5, -6 \rangle$ *83.9°*

⟩ B. Exercises

If $\langle v_1, v_2 \rangle$ is orthogonal to the given vector, find the ratio of v_2 to v_1. Then find a general expression that represents all the orthogonal vectors.

17. $\langle 10, -2 \rangle$ *$\frac{5}{1}; k\langle 1, 5 \rangle$*   18. $\langle -3, 8 \rangle$ *$\frac{3}{8}; k\langle 8, 3 \rangle$*

19. $\left\langle \frac{1}{3}, \frac{5}{6} \right\rangle$ *$-\frac{2}{5}; k\langle 5, -2 \rangle$*   20. $\left\langle -2, -\frac{3}{4} \right\rangle$ *$-\frac{8}{3}; k\langle -3, 8 \rangle$*

21. If $\mathbf{v} = \langle v_1, v_2 \rangle$ is orthogonal to $\mathbf{u} = \langle u_1, u_2 \rangle$, find the ratio of $\frac{v_2}{v_1}$ in terms of u_1 and u_2. Then find a general expression in terms of u_1 and u_2 for all vectors that are orthogonal to $\mathbf{u}$. *$-\frac{u_1}{u_2}; k\langle u_2, -u_1 \rangle$ or $k\langle -u_2, u_1 \rangle$*

Prove each property of dot products using $\vec{\mathbf{u}} = \langle u_1, u_2 \rangle$, $\vec{\mathbf{v}} = \langle v_1, v_2 \rangle$, $\vec{\mathbf{w}} = \langle w_1, w_2 \rangle$, $\vec{\mathbf{0}} = \langle 0, 0 \rangle$, and scalar k.

22. $\mathbf{0} \cdot \mathbf{v} = 0$

23. $k(\mathbf{u} \cdot \mathbf{v}) = k\mathbf{u} \cdot \mathbf{v}$

24. $k(\mathbf{u} \cdot \mathbf{v}) = \mathbf{u} \cdot k\mathbf{v}$

25. $\mathbf{u} \cdot (\mathbf{v} + \mathbf{w}) = \mathbf{u} \cdot \mathbf{v} + \mathbf{u} \cdot \mathbf{w}$

Find the projection of $\mathbf{u}$ onto $\mathbf{v}$. Then find $\mathbf{w}_2$ so that $\text{proj}_{\mathbf{v}} \mathbf{u} + \mathbf{w}_2 = \mathbf{u}$.

26. $\mathbf{u} = \langle 3, 8 \rangle, \mathbf{v} = \langle 1, 2 \rangle$ 27. $\mathbf{u} = \langle -2, 3 \rangle, \mathbf{v} = \langle 1, 4 \rangle$

28. $\mathbf{u} = \langle 4, 1 \rangle, \mathbf{v} = \langle 3, 5 \rangle$ 29. $\mathbf{u} = \langle 1, -5 \rangle, \mathbf{v} = \langle -2, 3 \rangle$

30. A 200 lb cart is resting on a hill with a slope of 20°.

   a. Determine the down-hill component of the cart's weight that is pulling it down the hill. *≈ 68.4 lb*

   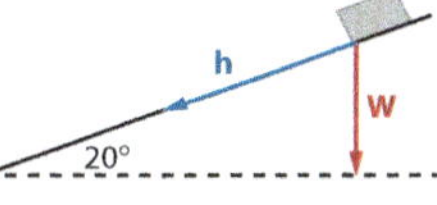

   b. Determine the component of the cart's weight that pushes into the hill perpendicular to the hill's surface. *≈ 187.9 lb*

31. A 50 lb box of gadgets rests on a ramp with a 40° incline.

   a. Determine the component of the box's weight that pulls it down the ramp. *≈ 32.1 lb*

   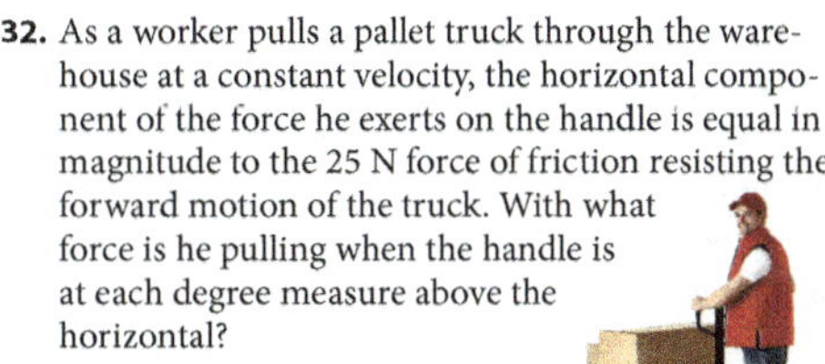

   b. What is the magnitude of the frictional force, F, that prevents the box from sliding down the ramp? *≈ 32.1 lb*

32. As a worker pulls a pallet truck through the warehouse at a constant velocity, the horizontal component of the force he exerts on the handle is equal in magnitude to the 25 N force of friction resisting the forward motion of the truck. With what force is he pulling when the handle is at each degree measure above the horizontal?

   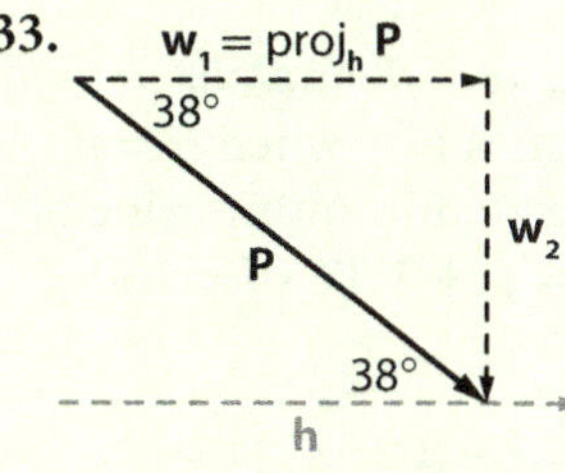

   a. 60° *50 N*

   b. 45° *≈ 35.4 N*

   c. 30° *≈ 28.9 N*

33. Mr. Jones pushes his lawnmower with a 75 lb force directed along its handle, which makes a 38° angle with the ground.

   a. Determine the component of his push that is directed parallel to the ground. *≈ 59.1 lb*

   b. Determine the component of his push that is directed perpendicular to the ground. *≈ 46.2 lb*

32b. $|\mathbf{P}| = \dfrac{25}{\cos 45°} \approx 35.4$ N

32c. $|\mathbf{P}| = \dfrac{25}{\cos 30°} \approx 28.9$ N

33. 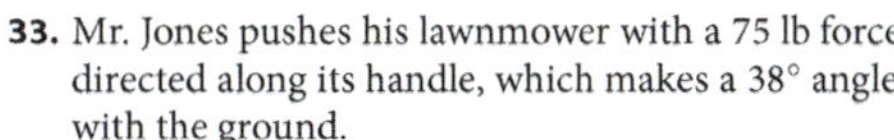

$\theta = 90 - 20 = 70°$

33a. $|\mathbf{w}_1| = |\mathbf{P}| \cos \theta = 75 \cos 38°$
≈ 59.1 lb

33b. $|\mathbf{w}_2| = |\mathbf{P}| \sin \theta = 75 \sin 38°$
≈ 46.2 lb

34. $W = \mathbf{F} \cdot \mathbf{d} = (100 \text{ lb})(6 \text{ ft}) \cos 0°$
$= 600$ ft-lb

35. $W = \mathbf{F} \cdot \mathbf{d} = (30 \text{ N})(15 \text{ m}) \cos 0°$
$= 450$ N·m

36. $W = \mathbf{F} \cdot \mathbf{d} = (70 \text{ N})(100 \text{ m}) \cos 30°$
≈ 6062 N·m

37. $W = \mathbf{F} \cdot \mathbf{d} = (150 \text{ ft})(35 \text{ lb}) \cos 25°$
≈ 4758 ft-lb

---

30. 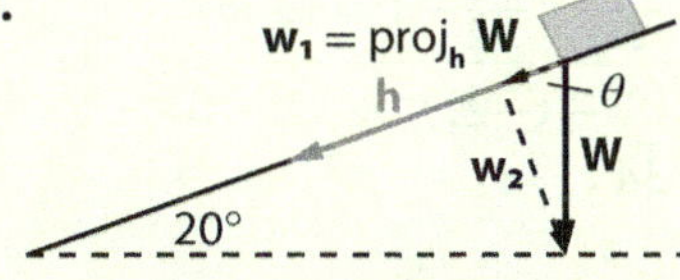

$\theta = 90 - 20 = 70°$

30a. $|\mathbf{w}_1| = |\mathbf{W}| \cos \theta = |200| \cos 70°$
≈ 68.4 lb

30b. $|\mathbf{w}_2| = |\mathbf{W}| \sin \theta = |200| \sin 70°$
≈ 187.9 lb

31. 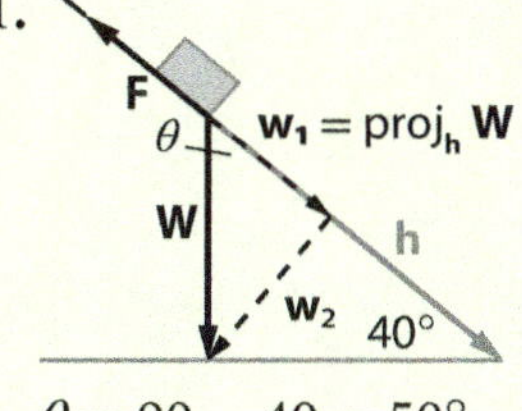

$\theta = 90 - 40 = 50°$

31a. $|\mathbf{w}_1| = |\mathbf{W}| \cos \theta = |50| \cos 50°$
≈ 32.1 lb

31b. If the box is at rest, the frictional force is equal in magnitude to but opposite in direction of the downhill force.

32. 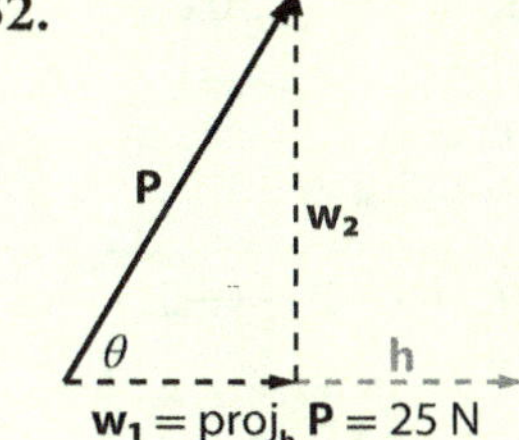

$\cos \theta = \dfrac{|\mathbf{w}_1|}{|\mathbf{P}|}; |\mathbf{P}| = \dfrac{|\mathbf{w}_1|}{\cos \theta}$

32a. $|\mathbf{P}| = \dfrac{25}{\cos 60°} = 50$ N

38. $20(25) + 40(35) = 1900$

39. $[-xy, xy]$

$\mathbf{r} \cdot \mathbf{s} = rs \cos |\alpha - \beta|$ and $\cos |\alpha - \beta|$ has a maximum of 1 when $\alpha = \beta$ $(\theta = 0)$. It has a minimum value of -1 when $\alpha = \beta + 180°$ $(\theta = 180)$.

40. $|\mathbf{u}| = \sqrt{5^2 + 5^2} = 5\sqrt{2}$

$\theta_{\mathbf{u}} = \text{Tan}^{-1}\left|\dfrac{5}{5}\right| = 45°$ (Q I)

$|\mathbf{v}| = \sqrt{(-2)^2 + (-2)^2} = 2\sqrt{2}$

$\theta_{\mathbf{v}} = 180° + \text{Tan}^{-1}\left|\dfrac{-2}{-2}\right|$
$\qquad = 225°$ (Q III)

$\mathbf{u} \cdot \mathbf{v} = 5\sqrt{2} \cdot 2\sqrt{2} \cos |225° - 45°|$
$\qquad = 20 \cos 180 = 20(-1) = -20$

$u_1 v_1 + u_2 v_2 = 5(-2) + 5(-2) = -20$

41. $|\mathbf{u}| = 4; \theta_{\mathbf{u}} = 90°$ (+y-axis)

$|\mathbf{v}| = \sqrt{\left(2\sqrt{3}\right)^2 + 2^2} = 4$

$\theta_{\mathbf{v}} = 180° + \text{Tan}^{-1}\left|\dfrac{2}{2\sqrt{3}}\right| = 30°$ (Q I)

$\mathbf{u} \cdot \mathbf{v} = 4 \cdot 4 \cos |90° - 30°|$
$\qquad = 16 \cos 60° = 8$

$u_1 v_1 + u_2 v_2 = 0(2\sqrt{3}) + 4(2) = 8$

42. $\overrightarrow{AB} = \langle 3, 4 \rangle; \left|\overrightarrow{AB}\right| = 5$

$\overrightarrow{AC} = \langle 4, 1 \rangle; \left|\overrightarrow{AC}\right| = \sqrt{17}$

$\langle 3, 4 \rangle \cdot \langle 4, 1 \rangle = 12 + 4 = 16$

$\cos A = \dfrac{16}{5\sqrt{17}}; A \approx 39.1°$

$\overrightarrow{BA} = \langle -3, -4 \rangle; \left|\overrightarrow{BA}\right| = 5$

$\overrightarrow{BC} = \langle 1, -3 \rangle; \left|\overrightarrow{BC}\right| = \sqrt{10}$

$\langle -3, -4 \rangle \cdot \langle 1, -3 \rangle = -3 + 12 = 9$

$\cos B = \dfrac{9}{5\sqrt{10}}; B \approx 55.3°$

$C \approx 180 - 39.1 - 55.3 = 85.6°$

43. $\overrightarrow{AB} = \langle -3, 6 \rangle; \left|\overrightarrow{AB}\right| = \sqrt{45}$

$\overrightarrow{AC} = \langle -9, 7 \rangle; \left|\overrightarrow{AC}\right| = \sqrt{130}$

$\langle -3, 6 \rangle \cdot \langle -9, 7 \rangle = 27 + 42 = 69$

$\cos A = \dfrac{69}{\sqrt{45} \cdot \sqrt{130}}; A \approx 25.6°$

$\overrightarrow{BA} = \langle 3, -6 \rangle; \left|\overrightarrow{BA}\right| = \sqrt{45}$

$\overrightarrow{BC} = \langle -6, 1 \rangle; \left|\overrightarrow{BC}\right| = \sqrt{37}$

$\langle 3, -6 \rangle \cdot \langle -6, 1 \rangle = -18 - 6 = -24$

$\cos B = \dfrac{-24}{\sqrt{45} \cdot \sqrt{37}}; B \approx 126.0°$

$C \approx 180 - 25.6 - 126.0 = 28.4°$

---

34. How much work is done by lifting a 100 lb weight set up 6 ft? 600 ft-lb

35. How much work is done by lifting an object up 15 m with a force of 30 N? 450 N·m

36. How much work is done by pulling a sled for 100 m with a force of 70 N on a rope at a 30° angle above the ground? $\approx$ 6062 N·m

37. How many foot-pounds of work does Bill do when he pushes an apple cart 50 yd with a force of 35 lb along a handle that is fixed at 25° above the horizontal of the cart? $\approx$ 4758 ft-lb

38. The total cost of purchasing the shorts and jerseys is $1900.

39. $[-xy, xy]$; The maximum occurs when $\alpha = \beta$, and the minimum occurs when $\alpha = \beta + 180°$.

42. $A \approx 39.1°; B \approx 55.3°; C \approx 85.6°$

43. $A \approx 25.6°; B \approx 126.8°; C \approx 28.4°$

CUMULATIVE REVIEW

Solve. [2.6]

44. $\dfrac{x+1}{5-x} = \dfrac{x}{x-5}$ $x = -\dfrac{1}{2}$

45. $\dfrac{2x^2 + 9x - 35}{x^2 - 10x + 1} = 3$ $x = 1, 38$

46. Find the distance between $(5, -2)$ and $(-5, -7)$. [1.2]

47. Find the midpoint between $(-1, 6)$ and $(7, -3)$. [1.2]

48. Find the length of a diameter of the circle that is centered at $(2, 0)$ and contains the point $(6, 3)$. [1.2]

49. Find the center and length of a radius of the circle whose diameter has endpoints $(-2, 0)$ and $(8, 6)$. [1.2]

50. Convert $y = \log_2 x$ to exponential form. [3.2] D

  A. $x = y^2$ **C.** $y = 2^x$ **E.** $x = 2y$

  B. $y = x^2$ **D.** $x = 2^y$

46. $d = 5\sqrt{5}$

47. $\left(3, \dfrac{3}{2}\right)$

48. $d = 10$

49. $C(3, 3); r = \sqrt{34}$

---

38. Analyze: If $\mathbf{n} = \langle 20, 40 \rangle$ represents the number of soccer shorts and jerseys that need to be purchased and $\mathbf{p} = \langle 25, 35 \rangle$ represents the respective costs in dollars, calculate $\mathbf{n} \cdot \mathbf{p}$ and interpret its meaning.

39. Analyze: Given $\mathbf{r}$ with magnitude x at α and $\mathbf{s}$ with magnitude y at β, state the range of possible values for $\mathbf{r} \cdot \mathbf{s}$. Then describe the conditions that generate the maximum and minimum values.

Find the magnitude and argument of each vector and compute the dot product using $\mathbf{u} \cdot \mathbf{v} = |\mathbf{u}|\,|\mathbf{v}| \cos \theta$. Then verify your result by computing the dot product using vector components.

40. $\mathbf{u} = \langle 5, 5 \rangle, \mathbf{v} = \langle -2, -2 \rangle$ -20

41. $\mathbf{u} = \langle 0, 4 \rangle, \mathbf{v} = \langle 2\sqrt{3}, 2 \rangle$ 8

Given the vertices of $\triangle ABC$, use dot products to determine the measure of its angles (to the nearest tenth of a degree).

42. $A(1, 2), B(4, 6),$ and $C(5, 3)$

43. $A(3, -2), B(0, 4),$ and $C(-6, 5)$

51. What are the primary solutions for $2 \sin x = 1$? [5.3] C

  A. $\dfrac{\pi}{3}, \dfrac{2\pi}{3}$ **C.** $\dfrac{\pi}{6}, \dfrac{5\pi}{6}$ **E.** $\dfrac{\pi}{6}, -\dfrac{\pi}{6}$

  B. $\dfrac{\pi}{4}, \dfrac{3\pi}{4}$ **D.** $\dfrac{\pi}{6}, \dfrac{7\pi}{6}$

52. What is the magnitude and direction of $\mathbf{v} = \langle -5, 5 \rangle$? [6.1] E

  A. $50; 45°$ **C.** $25; 135°$ **E.** $5\sqrt{2}; 135°$

  B. $5\sqrt{2}; 45°$ **D.** $2\sqrt{5}; 135°$

53. Which functions have a period of π? List all correct answers. [4.5] C, D

  A. $y = \sin x$ **C.** $y = \tan x$ **E.** $y = \csc x$

  B. $y = \cos x$ **D.** $y = \cot x$ **F.** $y = \sec x$

Cumulative Review

44. $\left[\dfrac{x+1}{-(x-5)} = \dfrac{x}{x-5}\right](x-5)$
$\qquad -x - 1 = x$
$\qquad\qquad -1 = 2x$
$\qquad\qquad\quad x = -\dfrac{1}{2}$

45. $2x^2 + 9x - 35 = 3x^2 - 30x + 3$
$\qquad x^2 - 39x + 38 = 0$
$\qquad (x - 38)(x - 1) = 0$
$\qquad\qquad\qquad x = 38, 1$

46. $d = \sqrt{(-5 - 5)^2 + (-7 - (-2))^2}$
$\qquad = \sqrt{125} = 5\sqrt{5}$

47. $M\left(\dfrac{-1+7}{2}, \dfrac{6+(-3)}{2}\right) = \left(3, \dfrac{3}{2}\right)$

48. $d = 2r = 2\sqrt{(6-2)^2 + (3-0)^2}$
$\qquad = 2(5) = 10$

49. $C\left(\dfrac{-2+8}{2}, \dfrac{0+6}{2}\right) = (3, 3)$

$r = \sqrt{(3 - (-2))^2 + (3 - 0)^2}$
$\quad = \sqrt{34}$

51. $\sin x = \dfrac{1}{2}$

$\qquad x = \dfrac{\pi}{6}, \dfrac{5\pi}{6}$

52. $\sqrt{(-5)^2 + 5^2} = 5\sqrt{2}$

$\qquad \alpha = \text{Tan}^{-1}\left|\dfrac{-5}{5}\right| = -45°$

$\qquad \theta = 180 + \alpha = 135°$

Truth and Mathematical Proof

There exists . . . a world which is the collection of mathematical truths, to which we have access only through our intellects, just as there is the world of physical reality; the one and the other independent of us, both of divine creation.

—*Charles Hermite (nineteenth-century French mathematician)*

Philippians 4:8 admonishes Christians to meditate on those things that are true. But our society values truth less and less. In fact, the Oxford Dictionaries 2016 word of the year was *post-truth*, "relating to or denoting circumstances in which objective facts are less influential in shaping public opinion than appeals to emotion and personal belief." This de-emphasis of truth contrasts with the position taken by great scientists like Galileo, who stated, "I value the discovery of a single even insignificant truth more highly than all the argumentation on the highest questions which fails to reach a truth." Sadly, many modern mathematicians fail to see mathematics as the reflection of an underlying transcendent truth originating with God. In 1940 Edward Kasner and James Newman wrote, "We have overcome the notion that mathematical truths have an existence independent and apart from our own minds. . . . [Mathematics] is man's own handiwork, subject only to the limitations imposed by the laws of thought."

Mathematicians continually seek to discover and verify new aspects of mathematical truth through inductive and deductive reasoning. Mathematical concepts are frequently discovered intuitively when the exploration of specific examples leads to general conclusions. These inductive conclusions are later verified using the logic of deductive reasoning. While some mathematicians such as Archimedes are known for both discovering and proving results, many such as Euclid are famous for the deductive organization and expansion of the discoveries of others. But even Euclid's work accepted many propositions as true without proof, and many major branches of mathematics progressed without logical development. Philip Jourdain wrote, "It is a curious fact that mathematicians have so often arrived at truth by a sort of instinct."

During the explosion of new mathematics in the Renaissance, mathematicians verified each other's inductive discoveries through deductive proof whenever possible. Pierre de Fermat, a seventeenth-century French lawyer and amateur mathematician, corresponded with René

Descartes on calculus, Blaise Pascal on probability, and Marin Mersenne on prime numbers. Fermat is best known for several number theory contentions that he stated without deductive proof. Many mathematicians labored to prove or disprove his statements. Joseph-Louis Lagrange and Leonhard Euler each proved several of his conjectures, but Euler also disproved one of them. Fermat's Last Theorem, his most famous conjecture, eluded proof until the end of the twentieth century, after having gained recognition as the mathematical problem with the greatest number of published incorrect proofs. Carl Gauss observed, "It is characteristic of higher arithmetic that many of the most beautiful theorems can be discovered by induction with the greatest of ease but have proofs that lie anywhere but near at hand and are often found only after many fruitless investigations with the aid of deep analysis and lucky combinations."

Considering this difficulty, it is important to note that the truth of a mathematical statement is not dependent on the existence of a deductive proof. For instance, the famous Goldbach Conjecture (that all positive even integers greater than two can be expressed as the sum of two primes) has yet to be proved deductively, though

Objectives

1. To identify and describe the two main types of mathematical reasoning

2. To explain the role of reasoning in establishing the humanly known body of mathematical truth

3. To explain why transcendent mathematics is a larger body of truth than what has been proved to be true

Assignments

- **Minimum:** 1–7; 9–10 (in class)
- **Standard:** 1–8; 9–11 (in class)
- **Extended:** 1–11

PRESENTATION

John 17:17 states, "Sanctify them through thy truth: thy word is truth." We become more Christlike by meditating on God's Word. Discuss Philippians 4:8 and its contrast to worldly thinking. You may also want to mention that the American Dialect Society's 2005 Word of the Year was "truthiness," defined as "the quality of stating concepts or facts one wishes or believes to be true, rather than concepts or facts known to be true." This word was used in a television comedy to satirize those who attempt to redefine truth to their own liking or benefit. Contrast Galileo's quote and Hermite's quote with the quote from Kasner and Newman.

Spend some time discussing the importance of deductive and inductive reasoning in establishing mathematical truth. The discovery of truth often results from inductive reasoning; the proof of it follows later with deductive reasoning. George Polya in *Mathematics and Plausible Reasoning* (1953) described the relationship as follows. "The result of the mathematician's creative work is demonstrative reasoning, a proof, but the proof is discovered by plausible reasoning, by guessing." Emphasize that inductive reasoning moves from specific examples to

Solutions

8a. $2^{2^2} + 1 = 2^4 + 1 = 17$, which is prime.

8b. $2^{2^5} + 1 = 2^{32} + 1 = 4{,}294{,}967{,}297$ and
$4{,}294{,}967{,}297 \div 641 = 6{,}700{,}417$

9. Inductive reasoning from specific examples can lead to a generalized rule called a conjecture, which may or may not be true. Deductive reasoning may then be applied in an attempt to prove or disprove the truth of the conjecture.

10. Transcendent mathematics includes undiscovered facts that may be proved in the future. Transcendent mathematics also includes math facts that are unprovable but true.

11. Even when we come to mathematical conclusions inductively, we can see how the math truthfully reflects the realities around us. God created the world in an orderly manner. The fact that intuitively derived mathematical concepts are verified deductively demonstrates that we are uncovering transcendent truth.

the inductive evidence in favor of this statement is overwhelming. A computer search in April 2012 verified the conjecture for all even integers less than 4×10^{18}. Such a technique does not provide a satisfactory proof since the search cannot extend to infinity.

Kurt Gödel's incompleteness theorems rattled the mathematical world in 1931. As Larry Zimmerman explains, Gödel demonstrated that it is impossible to "bridge the abyss from finite to infinite, that the consistency of any deductive system which encompasses all of arithmetic can never be proved. Even more astounding, he showed that no set of axioms would be sufficient to account for the mathematical truth in any one branch of mathematics." Mathematicians should not be surprised when they discover truths that they cannot verify deductively. These truths along with mathematical truths that remain to be discovered are all part of transcendent mathematical truth.

❯ Exercises

1. Verify the Goldbach Conjecture for each positive even integer.

   a. 4 $2 + 2$
   b. 8 $3 + 5$
   c. 10 $3 + 7$ or $5 + 5$
   d. 16 $3 + 13$ or $5 + 11$
   e. 30 $7 + 23$ or $11 + 19$ or $13 + 17$

2. Which Bible verse admonishes Christians to think on things that are true? Philippians 4:8

3. State whether inductive or deductive reasoning is being described.

   a. drawing a general principle from specific examples inductive
   b. proving a statement by applying previously proven principles in a logical step-by-step sequence deductive
   c. frequently used by mathematicians when discovering new truths inductive
   d. used by mathematicians to prove new discoveries deductive

4. Which amateur mathematician is famous for stating number theory conjectures? Pierre de Fermat

5. Who disproved one of those conjectures and proved several others? Leonhard Euler

6. Which theorem has had the greatest number of published incorrect proofs? Fermat's Last Theorem

7. Who demonstrated in 1931 that not all mathematical truth can be proved? Kurt Gödel

8. One of Fermat's conjectures stated that $2^{2^n} + 1$ is prime when n is a counting number.

   a. Show that the conjecture is true for $n = 2$.
   b. Use a calculator and divide by 641 to show that the conjecture is false for $n = 5$.

9. **Discuss:** How does the term *conjecture* relate to the types of mathematical reasoning?

10. **Discuss:** Describe transcendent mathematical truth.

11. **Discuss:** Explain how we can know that transcendent mathematical truth exists.

general conclusions, which may or may not be true. On the other hand, deductive reasoning uses logical reasoning to prove that a general statement must be true. As a result, all specific cases of that general statement, now a theorem, must also be true. Use Fermat's conjectures and their related proofs in the years following as an example of the relationship between these two types of reasoning.

Conclude by emphasizing that the mathematical body of transcendent truth is a larger body than humanly proven mathematical truth. Use Goldbach's Conjecture and Godel's proof as you explain this key objective of the lesson. Goldbach's Conjecture, much like Fermat's Last Theorem, points out the frequent difficulty of finding a deductive proof for a statement that seems obviously true by inductive reasoning. Godel's proof of the inherent inconsistencies, or contradictions, in any branch of mathematics highlights the inevitable futility of the efforts made by late nineteenth- and early twentieth-century mathematicians to prove the truth (or in their terminology, the consistency) of all areas of mathematics.

6.3 Vectors in Space

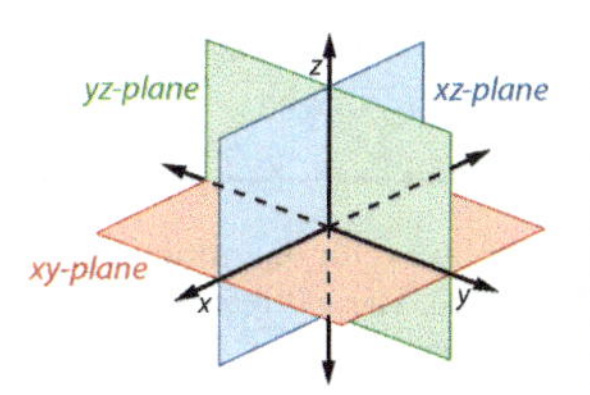

The bearing, ground speed, and angle of inclination of an aircraft in flight can be described by a three-dimensional vector.

The perpendicular x- and y-axes of the Cartesian plane allow each point in the plane to be described by a unique ordered pair. Similarly, three perpendicular axes allow each point in space to be uniquely described by an ordered triple, $P\,(x, y, z)$. The z-axis of our three-dimensional coordinate system passes through the origin of the xy-plane and is perpendicular to both the x- and y-axes. The xy-plane, the yz-plane, and the xz-plane divide space into eight regions called *octants*.

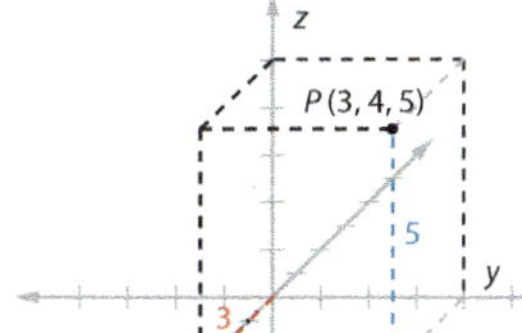

After completing this section, you will be able to

- represent three-dimensional vectors graphically and algebraically.
- perform three-dimensional vector operations.
- find the angle between two vectors in space.
- solve real-world problems involving three-dimensional vectors.

Example 1   Plotting a Point in Space

Plot the point $P\,(3, 4, 5)$.

Answer

1. Sketch a 3D coordinate system.
2. Locate the point (3, 4) in the xy-plane and plot the point 5 units up (parallel to the z-axis).

SKILL ✔ **EXERCISE 5**

TIP

Our 3D coordinate system is "right-handed" since the thumb of the right hand points in the direction of the positive z-axis when its fingers curve from the positive x-axis to the positive y-axis.

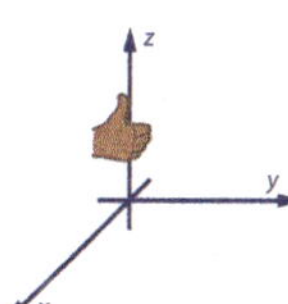

The distance between two points in space and the midpoint of the segment connecting them can be developed by extending the method used in two dimensions to three dimensions.

3D DISTANCE AND MIDPOINT FORMULAS

Given two points in space $P\,(x_1, y_1, z_1)$ and $Q\,(x_2, y_2, z_2)$,

the distance PQ is $d = \sqrt{(x_2 - x_1)^2 + (y_2 - y_1)^2 + (z_2 - z_1)^2}$,

and the midpoint of $\overline{PQ}$ is $M\left(\dfrac{x_1 + x_2}{2}, \dfrac{y_1 + y_2}{2}, \dfrac{z_1 + z_2}{2}\right)$.

6.3 Vectors in Space

Objectives

1. To represent three-dimensional vectors graphically and algebraically
2. To perform vector operations using three-dimensional vectors
3. To find the angle between two vectors in space by using their dot product
4. To solve real-world problems involving three-dimensional vectors

Flash

Commerical jets take off at an average speed of 160–180 mi/hr. Typical climb angles are 15–20°. These numbers vary with factors such as weight and type of aircraft.

Vocabulary

length
midpoint
octant

Reading and Writing Mathematics

Any point in space can be uniquely described on a three-dimensional coordinate system by three ordered magnitudes in rectangular form (x, y, z). Explain how any point in space can be uniquely described in polar form. Any point in space can be uniquely described as a magnitude and two angles in the form (r, θ, φ).

PRESENTATION

Lesson Opener

1. Find the length and midpoint of $\overline{PQ}$.

   a. $P\,(5, 3)$, $Q\,(9, 7)$ $4\sqrt{2}; (7, 5)$

   b. $P\,(2, -5)$, $Q\,(-4, 3)$ $10; (-1, -1)$

   c. $P\,(7, -9)$, $Q\,(4, -3)$ $3\sqrt{5}; \left(\frac{11}{2}, -6\right)$

2. Given $A\,(2, 3)$ and $B\,(4, -6)$, complete the following.

   a. Express $\overrightarrow{AB}$ in component form.
      $(2, -9)$

   b. Find the length of $\overrightarrow{AB}$. $\sqrt{85}$

Use the illustrated three-dimensional coordinate system to introduce plotting points in space. Have the students view it as a two-dimensional coordinate system on a horizontal plane with the positive x-axis pointing toward them. A third axis, z, is then added as a vertical component.

The three planes divide space into eight *octants*, which are numbered in the illustration for exercise 1. Consider completing exercise 1 together before presenting Example 1.

Some students may struggle with the three-dimensional perspective. Encourage them to always move parallel to the corresponding axis when plotting in three-dimensions.

Interactive Activity Use the Internet keyword search *three-dimensional plotter* to find online apps or *three-dimensional plotter app* to find some mobile apps for plotting points in space.

One-on-One If a student is struggling with visualizing the three-dimensional coordinate plane, have the student lay a familiar two-dimensional Cartesian coordinate plane down on their desk with the positive x-axis pointed at them (this places the positive y-axis to the right). Then add a third dimension by modeling the positive z-axis with a pencil placed at the origin pointing straight up.

Extend the two-dimensional distance and midpoint formulas to three-dimensional space and illustrate their application with Example 2.

Additional Exercises

1. Plot each point on a three-dimensional coordinate system.
   a. $P(-4, 2, -3)$
   b. $Q(3, -2, -1)$

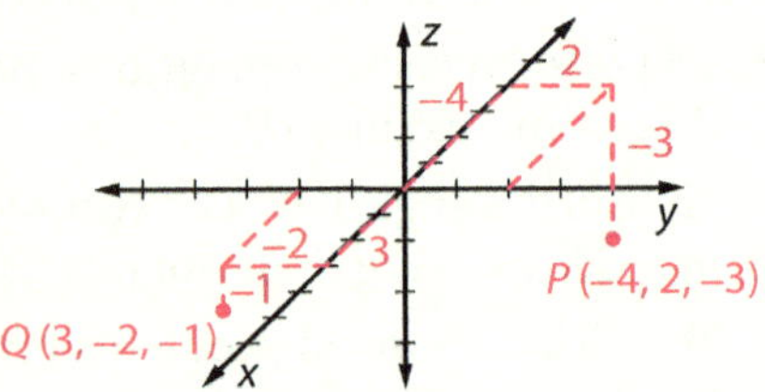

2. Find the length and the midpoint of the segment from $A(1, -3, 0)$ to $B(3, -5, -4)$. $2\sqrt{6}; (2, -4, -2)$

3. Graph each vector in standard position.
   a. $A\langle 1, 4, 2 \rangle$
   b. $B\langle 2, 3, -2 \rangle$

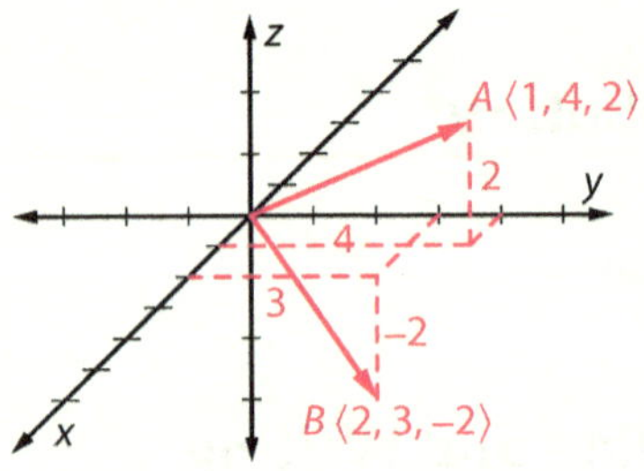

4. Find each resultant if $\mathbf{p} = \langle 3, -1, 4 \rangle$ and $\mathbf{q} = \langle -2, 0, -3 \rangle$.
   a. $2\mathbf{p} - \mathbf{q}$ $\langle 8, -2, 11 \rangle$
   b. $3\mathbf{p} + \frac{1}{2}\mathbf{q}$ $\langle 8, -3, 10.5 \rangle$

5. Find each dot product and use the results to classify each angle as acute, right, or obtuse.
   a. $\mathbf{p} = \langle 3, -1, 4 \rangle$, $\mathbf{q} = \langle -2, 0, -3 \rangle$
      -18; obtuse
   b. $\mathbf{a} = \langle -1, 3, 5 \rangle$, $\mathbf{b} = \langle 6, -3, 3 \rangle$
      0; right

6. Find the measure of the angle between $\mathbf{p} = \langle 2, -7, 1 \rangle$ and $\mathbf{q} = \langle 4, -1, 3 \rangle$.
   $\approx 61.3°$

7. Draw a three-dimensional vector that models a plane taking off heading due south at 155 knots with a climb angle of $17°$. Find the component form for the vector.

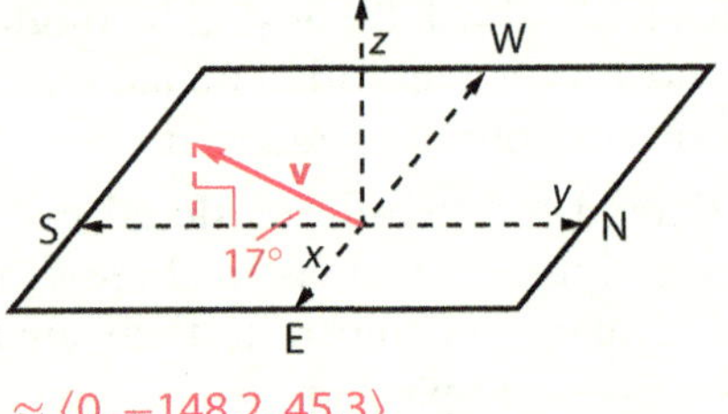

$\approx \langle 0, -148.2, 45.3 \rangle$

Example 2 Finding the Length and Midpoint of a Segment in Space

Find the length and the midpoint of the segment from $A(3, 0, -2)$ to $B(-1, -4, 6)$.

Answer

Length

$$d = \sqrt{(x_2 - x_1)^2 + (y_2 - y_1)^2 + (z_2 - z_1)^2}$$
$$= \sqrt{(-1 - 3)^2 + (-4 - 0)^2 + [6 - (-2)]^2}$$
$$= \sqrt{(-4)^2 + (-4)^2 + 8^2} = \sqrt{96}$$
$$= 4\sqrt{6} \approx 9.80$$

Midpoint

$$\left(\frac{x_1 + x_2}{2}, \frac{y_1 + y_2}{2}, \frac{z_1 + z_2}{2} \right)$$
$$= \left(\frac{3 + (-1)}{2}, \frac{0 + (-4)}{2}, \frac{-2 + 6}{2} \right)$$
$$= (1, -2, 2)$$

SKILL ✔ EXERCISE 11

Many vector quantities, such as forces, velocities, and accelerations, are not confined to a plane. These three-dimensional vectors are often expressed in terms of their components. For example, $\overrightarrow{AB}$ from $A(x_1, y_1, z_1)$ to $B(x_2, y_2, z_2)$ is expressed in component form as
$$\mathbf{v} = \langle v_1, v_2, v_3 \rangle = \langle x_2 - x_1, y_2 - y_1, z_2 - z_1 \rangle.$$

Since $\overrightarrow{AB}$ is equal to the standard position $\mathbf{v}$, its length is $|\mathbf{v}| = \sqrt{v_1^2 + v_2^2 + v_3^2}$ and its direction can be described using the unit vector $\frac{\mathbf{v}}{|\mathbf{v}|}$.

Example 3 Describing a 3D Vector

Given $A(5, 1, -1)$ and $B(3, 6, 2)$, express $\overrightarrow{AB}$ in component form. Then find its length and a unit vector in the direction of $\overrightarrow{AB}$.

Answer

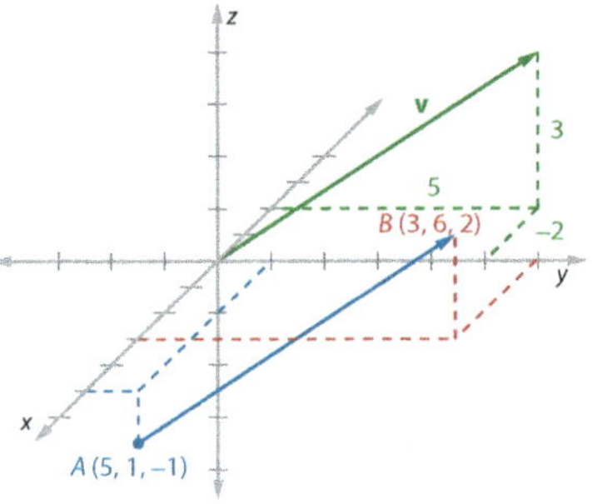

$\mathbf{v} = \langle 3 - 5, 6 - 1, 2 - (-1) \rangle$
$= \langle -2, 5, 3 \rangle$

$|\mathbf{v}| = \sqrt{(-2)^2 + 5^2 + 3^2}$
$= \sqrt{38} \approx 6.16$

$\mathbf{u} = \frac{1}{|\mathbf{v}|}\mathbf{v} = \frac{1}{\sqrt{38}}\langle -2, 5, 3 \rangle$
$= \left\langle -\frac{2\sqrt{38}}{38}, \frac{5\sqrt{38}}{38}, \frac{3\sqrt{38}}{38} \right\rangle$

1. Express $\overrightarrow{AB}$ in component form using $\langle v_1, v_2, v_3 \rangle = \langle x_2 - x_1, y_2 - y_1, z_2 - z_1 \rangle$.

2. Find its length using $|\mathbf{v}| = \sqrt{v_1^2 + v_2^2 + v_3^2}$.

3. Divide the component form of the vector by its length.

SKILL ✔ EXERCISE 19

The last step in the example above utilizes the fact that the properties of vectors extend from two dimensions to any number of dimensions.

The Irish mathematician William Hamilton was the first to develop a method that extended vector operations with complex numbers into three dimensions. Quaternions, as Hamilton called them, are not commutative under multiplication.

Like two-dimensional vectors, three-dimensional vectors are usually illustrated in standard position. Expressing three-dimensional vectors in component form makes drawing the vector and most computations involving the vector easier. Example 3 demonstrates the extension of two-dimensional formulas into three-dimensions. The unit vector in the direction of $\overrightarrow{AB}$ is sometimes referred to as its direction vector.

Discuss how the definitions related to vectors are extended to three-dimensional vectors and how they are applied in Example 4. Explain that the dot product provides a powerful tool for finding the angle between two vectors in space, as demonstrated in Example 5.

When presenting Example 6, remind the students that bearings are measured clockwise from due north (the positive y-axis). Converting to typical mathematical notation for vectors in the xy-plane can help to avoid confusion. Consider completing exercises 38–40 together to illustrate interpreting the component form of the resultant vector.

One-on-One Point out that west, south, and down represent the negative side of the x-, y-, and z-axes, respectively, on a three-dimensional coordinate system.

Let k be a scalar, $\mathbf{v} = \langle v_1, v_2, v_3 \rangle$, $\mathbf{w} = \langle w_1, w_2, w_3 \rangle$, and θ be the angle between $\mathbf{v}$ and $\mathbf{w}$.	
equality	$\mathbf{v} = \mathbf{w}$ if and only if $v_1 = w_1$, $v_2 = w_2$, and $v_3 = w_3$
vector addition	$\mathbf{v} + \mathbf{w} = \langle v_1 + w_1, v_2 + w_2, v_3 + w_3 \rangle$
scalar multiplication	$k\mathbf{v} = \langle kv_1, kv_2, kv_3 \rangle$
vector subtraction	$\mathbf{v} - \mathbf{w} = \mathbf{v} + (-\mathbf{w}) = \langle v_1 - w_1, v_2 - w_2, v_3 - w_3 \rangle$
zero vector	$\mathbf{0} = \langle 0, 0, 0 \rangle$
dot product	$\mathbf{v} \cdot \mathbf{w} = |\mathbf{v}|\,|\mathbf{w}| \cos\theta = v_1 w_1 + v_2 w_2 + v_3 w_3$

Example 4   Operations with 3D Vectors

Use $\mathbf{p} = \langle -5, 2, 0 \rangle$, $\mathbf{q} = \langle 1, -3, -2 \rangle$, and $\mathbf{r} = \langle 6, 4, -3 \rangle$ to complete each vector operation.

a. $2\mathbf{p} + \mathbf{q}$ **b.** $\frac{1}{2}\mathbf{r} - 3\mathbf{q}$ **c.** $\mathbf{q} \cdot \mathbf{r}$

Answer

a. $2\mathbf{p} + \mathbf{q} = 2\langle -5, 2, 0 \rangle + \langle 1, -3, -2 \rangle$ 1. Substitute.
$= \langle -10, 4, 0 \rangle + \langle 1, -3, -2 \rangle$ Complete the scalar multiplication.
$= \langle -9, 1, -2 \rangle$ Complete the vector addition.

b. $\frac{1}{2}\mathbf{r} - 3\mathbf{q} = \frac{1}{2}\langle 6, 4, -3 \rangle - 3\langle 1, -3, -2 \rangle$ 2. Substitute.
$= \langle 3, 2, -\frac{3}{2} \rangle + \langle -3, 9, 6 \rangle$ Complete the scalar multiplication.
    Note that the vector subtraction is done by
$= \langle 0, 11, \frac{9}{2} \rangle$ adding the opposite.

c. $\mathbf{q} \cdot \mathbf{r} = \langle 1, -3, -2 \rangle \cdot \langle 6, 4, -3 \rangle$ 3. Find the dot product.
$= 1(6) + (-3)(4) + (-2)(-3)$
$= 6 - 12 + 6 = 0$

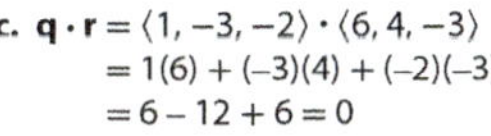 SKILL ✓ EXERCISES 25, 27

Just as in a plane, $\mathbf{v} \cdot \mathbf{w} = |\mathbf{v}|\,|\mathbf{w}| \cos\theta$ can be rearranged to find θ, the measure of the angle formed by $\mathbf{v}$ and $\mathbf{w}$ when placed in standard position: $\cos\theta = \dfrac{\mathbf{v} \cdot \mathbf{w}}{|\mathbf{v}|\,|\mathbf{w}|}$. Note that vectors, such as $\mathbf{q}$ and $\mathbf{r}$ in Example 4, are orthogonal if and only if the dot product is equal to 0. A positive dot product implies $0° \leq \theta < 90°$, and a negative dot product implies $90° < \theta \leq 180°$.

The fact that multiplication was not commutative in Hamilton's algebraic system of quaternions was as surprising to mathematicians as the alternative geometries being developed at that same time. However, we should not assume that any mathematical model completely describes God's infinitely complex creation.

8. A quarterback throws the football due west at a speed of 50 mi/hr and an angle of inclination of 15° with a wind blowing to the southeast at 5 mi/hr. Find a vector that represents the football's resulting initial velocity relative to the point from which it was thrown. What is the football's initial resulting speed?
$\langle -44.8, -3.5, 12.9 \rangle$; 46.7 mi/hr

Assessment

• Quiz 6B covers Section 6.3.

Assignments

• **Minimum (optional):** 1(a–d), 2, 5–7, 9, 11–12, 15, 18–19, 22–24, 26, 28, 31–33, 35, 41–42, 45–48

• **Standard:** 1, 3–4, 7–8, 11–12, 16, 19–20, 24–26, 28, 30, 33–35, 37–38, 42, 44–50

• **Extended:** 3, 6, 9–10, 13, 16–18, 21, 25; 26–36 even; 37–40, 43–44, 46, 48–50

Solutions

❭ A. Exercises

2.

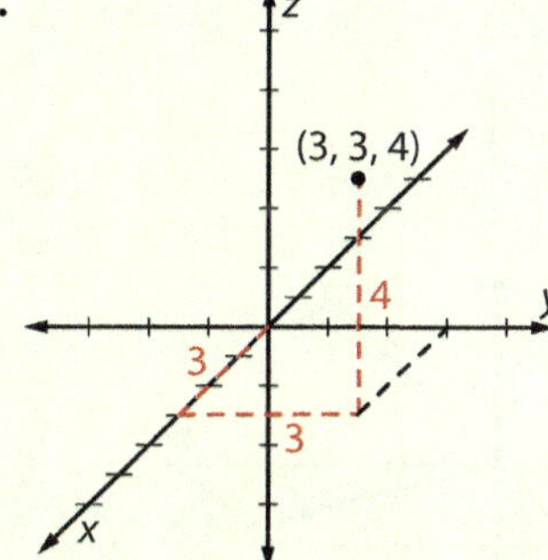

3.

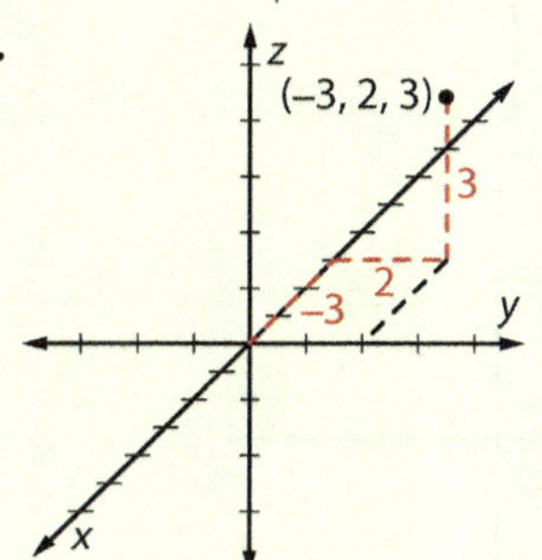

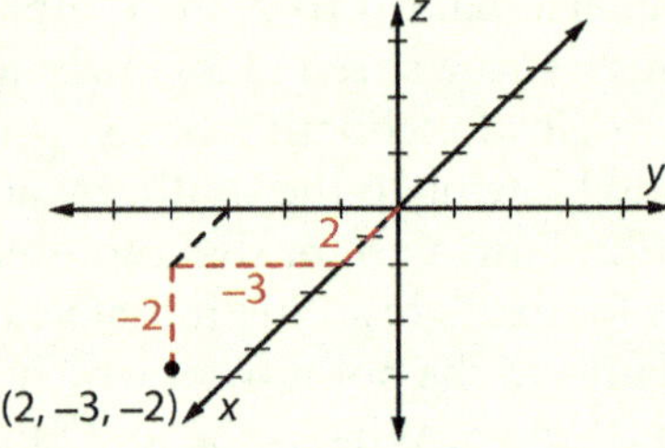

4. $(2, -3, -2)$

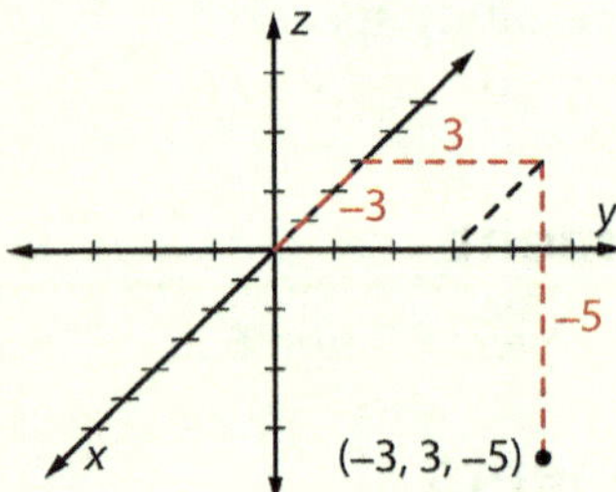

5. $(-3, 3, -5)$

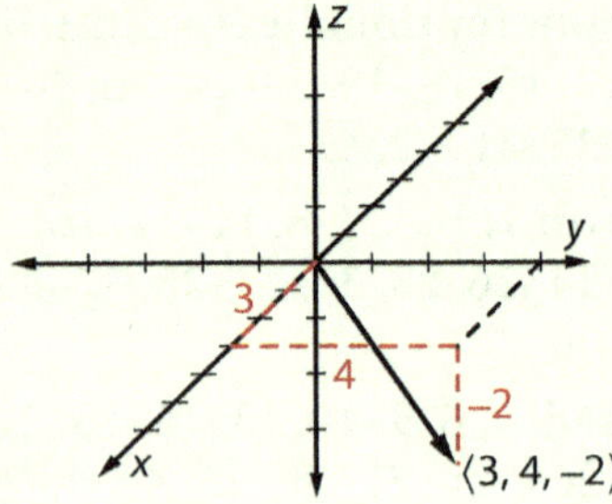

6. $\langle 3, 4, -2 \rangle$

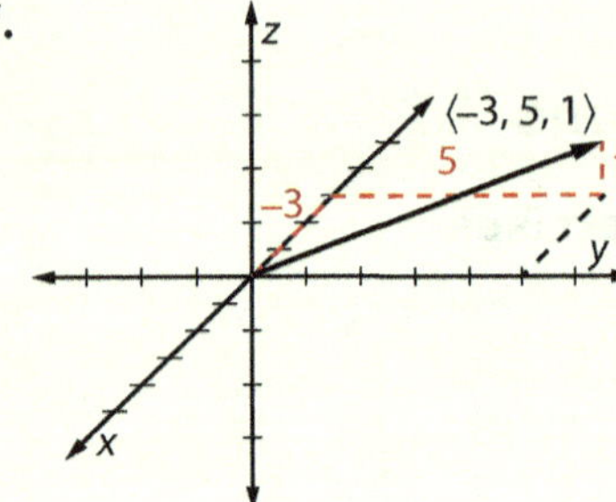

7. $\langle -3, 5, 1 \rangle$

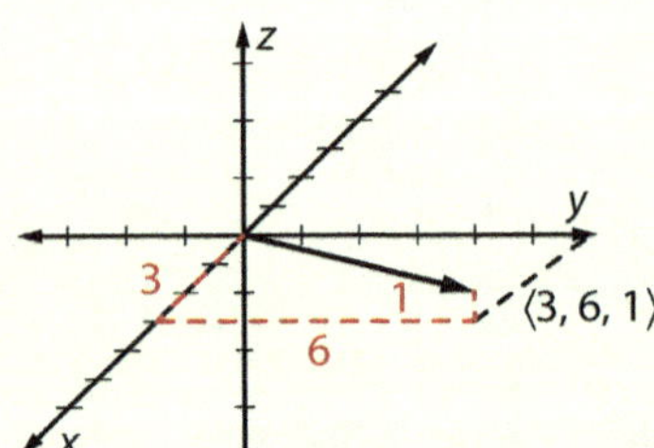

8. $\langle 3, 6, 1 \rangle$

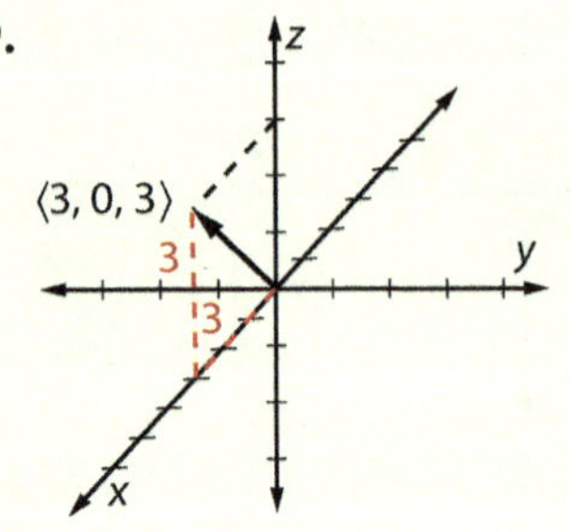

9. $\langle 3, 0, 3 \rangle$

Find the measure of the angle between $\mathbf{p} = \langle -5, 2, 0 \rangle$ and $\mathbf{q} = \langle 1, -3, -2 \rangle$.

Answer

Use $\cos \theta = \dfrac{\mathbf{p} \cdot \mathbf{q}}{|\mathbf{p}|\,|\mathbf{q}|}$.

$\mathbf{p} \cdot \mathbf{q} = -5(1) + 2(-3) + 0(-2) = -11$

$|\mathbf{p}| = \sqrt{(-5)^2 + 2^2 + 0^2} = \sqrt{29}$

$|\mathbf{q}| = \sqrt{1^2 + (-3)^2 + (-2)^2} = \sqrt{14}$

$\cos \theta = \dfrac{-11}{\sqrt{29} \cdot \sqrt{14}}$

$\theta = \text{Cos}^{-1}\left(\dfrac{-11}{\sqrt{29} \cdot \sqrt{14}} \right)$

$\approx 123°$ or 2.15 radians

1. Find the dot product of the vectors and their magnitudes.

2. Substitute and use the Cos^{-1} function to find θ.

— SKILL ✔ EXERCISE 29

Three-dimensional vectors are used in many practical applications.

An airplane taking off at a heading of 180° has an air velocity vector directed 20° above horizontal and an airspeed of 225 knots. There is a 40-knot wind from the southwest. Find a vector that represents the plane's velocity relative to the point of takeoff.

Answer

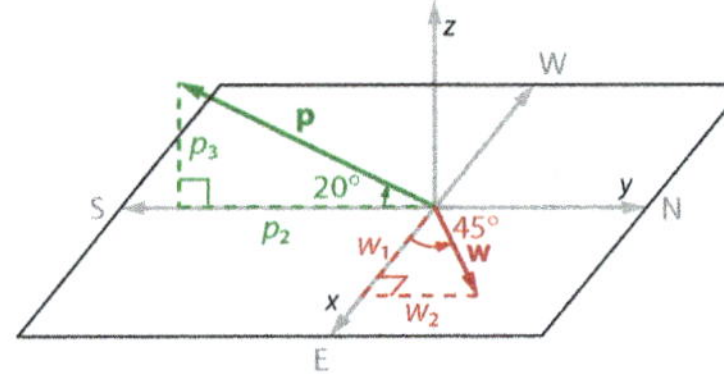

1. Sketch vectors representing the wind and the plane's velocity without the wind.

Letting the positive x-axis represent due east implies the wind is blowing at 45° in the xy-plane and the vector for the plane's heading is directly above the negative y-axis.

$\mathbf{w} = \langle |\mathbf{w}| \cos 45°, |\mathbf{w}| \sin 45°, 0 \rangle$
$= \langle 40 \cos 45°, 40 \sin 45°, 0 \rangle$
$\approx \langle 28.3, 28.3, 0 \rangle$

2. Resolve $\mathbf{w}$ into $\langle w_1, w_2, 0 \rangle$ using $\cos 45° = \dfrac{|w_1|}{|\mathbf{w}|}$ and $\sin 45° = \dfrac{|w_2|}{|\mathbf{w}|}$.

$\mathbf{p} = \langle 0, -|\mathbf{p}| \cos 20°, |\mathbf{p}| \sin 20° \rangle$
$= \langle 0, -225 \cos 20°, 225 \sin 20° \rangle$
$\approx \langle 0, -211.4, 77.0 \rangle$

3. Use right-triangle trigonometry to resolve $\mathbf{p}$ into $\langle 0, p_2, p_3 \rangle$ using $\cos 20° = \dfrac{|p_2|}{|\mathbf{p}|}$ and $\sin 20° = \dfrac{|p_3|}{|\mathbf{p}|}$.

$\mathbf{w} + \mathbf{p} \approx \langle 28.3, 28.3, 0 \rangle + \langle 0, -211.4, 77.0 \rangle$
$\mathbf{r} \approx \langle 28.3, -183.1, 77.0 \rangle$

4. Add the vectors.

— SKILL ✔ EXERCISE 35

Exercises 38–40 explore the interpretation of the component form of the resultant vector in Example 6.

10. $d = \sqrt{(2-2)^2 + (6-4)^2 + (1-5)^2} = \sqrt{0^2 + 2^2 + (-4)^2} = \sqrt{20} = 2\sqrt{5} \approx 4.47$

$M\left(\dfrac{2+2}{2}, \dfrac{4+6}{2}, \dfrac{5+1}{2} \right) = (2, 5, 3)$

11. $d = \sqrt{(4-0)^2 + (-1-7)^2 + (-2-(-4))^2} = \sqrt{4^2 + (-8)^2 + 2^2} = \sqrt{84} = 2\sqrt{21} \approx 9.17$

$M\left(\dfrac{0+4}{2}, \dfrac{7+(-1)}{2}, \dfrac{-4+(-2)}{2} \right) = (2, 3, -3)$

12. $d = \sqrt{(3-0)^2 + (0-0)^2 + (0-4)^2} = \sqrt{3^2 + (-4)^2} = \sqrt{25} = 5$

$M\left(\dfrac{0+3}{2}, \dfrac{0+0}{2}, \dfrac{4+0}{2} \right) = \left(\dfrac{3}{2}, 0, 2 \right)$

13. $d = \sqrt{(2x-2)^2 + (2y-4)^2 + (2z-6)^2}$

$M\left(\dfrac{2x+2}{2}, \dfrac{2y+4}{2}, \dfrac{2z+6}{2} \right) = (x+1, y+2, z+3)$

A. Exercises

1. State the octant or plane in which each point is located.

  a. $(5, 7, -10)$ 5
  b. $(2, -9, -4)$ 8
  c. $(-3, -7, 5)$ 3
  d. $(0, -6, 4)$
  e. $(-9, -12, -4)$ 7
  f. $(-11, 5, -2)$ 6 **1d.** yz-plane (between 3 and 4)

Plot each point on a 3D coordinate system.

2. $(3, 3, 4)$ **3.** $(-3, 2, 3)$

4. $(2, -3, -2)$ **5.** $(-3, 3, -5)$

Graph each vector in standard position.

6. $\langle 3, 4, -2 \rangle$ **7.** $\langle -3, 5, 1 \rangle$

8. $\langle 3, 6, 1 \rangle$ **9.** $\langle 3, 0, 3 \rangle$

Find the length and the midpoint of $\overline{AB}$.

10. $A\,(2, 4, 5),\ B\,(2, 6, 1)$ $2\sqrt{5};\ (2, 5, 3)$

11. $A\,(0, 7, -4),\ B\,(4, -1, -2)$ $2\sqrt{21};\ (2, 3, -3)$

12. $A\,(0, 0, 4),\ B\,(3, 0, 0)$ $5;\ \left(\frac{3}{2}, 0, 2\right)$

13. $A\,(2, 4, 6),\ B\,(2x, 2y, 2z)$

14. An eagle is spotted at $(98, -47, 14)$ and a second eagle is seen at $(37, 122, 78)$. Find the distance between the eagles if the coordinates are stated in feet. ≈ 190.7 ft

15. Lucas's drone is located at $(-1160, 538, 318)$ and Zoey's drone is located at $(455, -212, 270)$ on the same coordinate system. Find the distance between the drones if the coordinates are given in feet. ≈ 1781.3 ft

16. The coordinates of drone A are $(235, -102, 55)$ and the coordinates of drone B are $(-137, 68, 75)$. Find the coordinates of the midpoint between the drones. $(49, -17, 65)$

17. A Navy airplane is located at $(346, 400, c)$ and an Air Force plane is at $(a, b, 3.5)$. Find the value of each variable if the coordinates of the midpoint between the two planes are $(279, 481, 2.5)$. $a = 212;\ b = 562;\ c = 1.5$

13. $\sqrt{(2x-2)^2 + (2y-4)^2 + (2z-6)^2};\ (x+1, y+2, z+3)$

B. Exercises

Express $\overrightarrow{PQ}$ in component form. Then find its length and a unit vector in the direction of $\overrightarrow{PQ}$.

18. $P\,(7, 4, 2),\ Q\,(7, 9, 14)$

19. $P\,(1, 5, -8),\ Q\,(13, 9, -5)$

20. $P\,(-1, 5, 2),\ Q\,(-3, 2, -1)$

21. $P\,(-2, 6, -5),\ Q\,(3, -2, 4)$

Use $\mathbf{u} = \langle 2, -1, 1 \rangle$, $\mathbf{v} = \langle 4, -2, 3 \rangle$, and $\mathbf{w} = \langle 5, 2, -4 \rangle$ to complete each vector operation.

22. $\mathbf{v} + \mathbf{w}$ $\langle 9, 0, -1 \rangle$ **23.** $\mathbf{u} - \mathbf{v}$ $\langle -2, 1, -2 \rangle$

24. $\mathbf{v} - 2\mathbf{u}$ $\langle 0, 0, 1 \rangle$ **25.** $2\mathbf{w} + 3\mathbf{u}$ $\langle 16, 1, -5 \rangle$

26. $\mathbf{u} \cdot \mathbf{v}$ 13   **27.** $\mathbf{v} \cdot \mathbf{w}$ 4

Find the angle between each pair of vectors (to the nearest degree).

28. $\mathbf{v} = \langle 1, 4, 2 \rangle$, $\mathbf{w} = \langle -2, 5, -3 \rangle$ $65°$

29. $\mathbf{v} = \langle 3, 6, -1 \rangle$, $\mathbf{w} = \langle 4, 1, -7 \rangle$ $63°$

30. $\mathbf{p} = \langle 3, 2, 4 \rangle$, $\mathbf{q} = \langle -3, -3, 1 \rangle$ $118°$

31. $\mathbf{p} = \langle 5, 5, -1 \rangle$, $\mathbf{q} = \langle -9, -5, 15 \rangle$ $131°$

In exercises 32–34, draw a three-dimensional vector modeling each quantity and express the vector in component form.

32. driving 45 mi/hr on a level road heading 70° east of north

33. taking off at 140 knots heading due west with a 15° angle of climb

34. hiking southwest up a mountain at 1.5 mi/hr with an angle of inclination of 17.5°

35. An airplane takes off headed due north with an air velocity vector directed 15° above horizontal and an airspeed of 150 knots. There is a 16-knot wind from the northeast. Find a vector that represents the plane's velocity relative to the point of takeoff. $\approx \langle -11.3, 133.6, 38.8 \rangle$

36. A kicker attempting a field goal kicks the football due north at a speed of 83 mi/hr and an initial angle of inclination of 40° with a wind blowing to the southeast at 12 mi/hr. Find a vector that represents the football's resulting initial velocity relative to the point from which it was kicked. What is the football's initial resulting speed? $\approx \langle 8.5, 55.1, 53.4 \rangle;\ \approx 77.2$ mi/hr

14. $d = \sqrt{(37-98)^2 + (122-(-47))^2 + (78-14)^2}$
$= \sqrt{(-61)^2 + 169^2 + 64^2} \approx 190.7$ ft

15. $d = \sqrt{(-1160-455)^2 + (538-(-212))^2 + (318-270)^2}$
$= \sqrt{(-1615)^2 + 750^2 + 48^2} \approx 1781.3$ ft

16. $M\left(\frac{235 + (-137)}{2}, \frac{-102 + 68}{2}, \frac{55 + 75}{2}\right) = (49, -17, 65)$

17. $\frac{346 + a}{2} = 279;\ a = 212$

$\frac{400 + b}{2} = 481;\ b = 562$

$\frac{3.5 + c}{2} = 2.5;\ c = 1.5$

B. Exercises

18. $\mathbf{v} = \langle 7-7, 9-4, 14-2 \rangle$
$= \langle 0, 5, 12 \rangle$

$|\mathbf{v}| = \sqrt{0^2 + 5^2 + 12^2}$
$= \sqrt{169} = 13$

$\frac{\mathbf{v}}{|\mathbf{v}|} = \frac{1}{13}\langle 0, 5, 12 \rangle$
$= \left\langle 0, \frac{5}{13}, \frac{12}{13} \right\rangle$

19. $\mathbf{v} = \langle 13-1, 9-5, -5-(-8) \rangle$
$= \langle 12, 4, 3 \rangle$

$|\mathbf{v}| = \sqrt{12^2 + 4^2 + 3^2} = \sqrt{169} = 13$
$\frac{\mathbf{v}}{|\mathbf{v}|} = \frac{1}{13}\langle 12, 4, 3 \rangle = \left\langle \frac{12}{13}, \frac{4}{13}, \frac{3}{13} \right\rangle$

20. $\mathbf{v} = \langle -3-(-1), 2-5, -1-2 \rangle$
$= \langle -2, -3, -3 \rangle$

$|\mathbf{v}| = \sqrt{(-2)^2 + (-3)^2 + (-3)^2}$
$= \sqrt{22}$

$\frac{\mathbf{v}}{|\mathbf{v}|} = \frac{1}{\sqrt{22}}\langle -2, -3, -3 \rangle$
$= \left\langle -\frac{\sqrt{22}}{11}, -\frac{3\sqrt{22}}{22}, -\frac{3\sqrt{22}}{22} \right\rangle$

21. $\mathbf{v} = \langle 3-(-2), -2-6, 4-(-5) \rangle$
$= \langle 5, -8, 9 \rangle$

$|\mathbf{v}| = \sqrt{5^2 + (-8)^2 + 9^2}$
$= \sqrt{170}$

$\frac{\mathbf{v}}{|\mathbf{v}|} = \frac{1}{\sqrt{170}}\langle 5, -8, 9 \rangle$
$= \left\langle \frac{\sqrt{170}}{34}, -\frac{4\sqrt{170}}{85}, \frac{9\sqrt{170}}{170} \right\rangle$

22. $\langle 4+5, -2+2, 3+(-4) \rangle$
$= \langle 9, 0, -1 \rangle$

23. $\langle 2-4, -1-(-2), 1-3 \rangle$
$= \langle -2, 1, -2 \rangle$

24. $\langle 4, -2, 3 \rangle - 2\langle 2, -1, 1 \rangle$
$= \langle 4, -2, 3 \rangle + \langle -4, 2, -2 \rangle$
$= \langle 0, 0, 1 \rangle$

25. $2\langle 5, 2, -4 \rangle + 3\langle 2, -1, 1 \rangle$
$= \langle 10, 4, -8 \rangle + \langle 6, -3, 3 \rangle$
$= \langle 16, 1, -5 \rangle$

26. $\langle 2, -1, 1 \rangle \cdot \langle 4, -2, 3 \rangle$
$= 2(4) + (-1)(-2) + 1(3)$
$= 8 + 2 + 3 = 13$

27. $\langle 4, -2, 3 \rangle \cdot \langle 5, 2, -4 \rangle$
$= 4(5) + (-2)2 + 3(-4)$
$= 20 - 4 - 12 = 4$

28. $\mathbf{v} \cdot \mathbf{w} = 1(-2) + 4(5) + 2(-3) = 12$
$|\mathbf{v}| = \sqrt{1^2 + 4^2 + 2^2} = \sqrt{21}$
$|\mathbf{w}| = \sqrt{(-2)^2 + 5^2 + (-3)^2} = \sqrt{38}$
$\theta = \text{Cos}^{-1}\left(\frac{12}{\sqrt{21} \cdot \sqrt{38}}\right) \approx 65°$

29. $\mathbf{v} \cdot \mathbf{w} = 3(4) + 6(1) + (-1)(-7) = 25$

$|\mathbf{v}| = \sqrt{3^2 + 6^2 + (-1)^2} = \sqrt{46}$

$|\mathbf{w}| = \sqrt{4^2 + 1^2 + (-7)^2} = \sqrt{66}$

$\theta = \mathrm{Cos}^{-1}\left(\dfrac{25}{\sqrt{46} \cdot \sqrt{66}}\right) \approx 63°$

30. $\mathbf{p} \cdot \mathbf{q} = 3(-3) + 2(-3) + 4(1) = -11$

$|\mathbf{p}| = \sqrt{3^2 + 2^2 + 4^2} = \sqrt{29}$

$|\mathbf{q}| = \sqrt{(-3)^2 + (-3)^2 + 1^2} = \sqrt{19}$

$\theta = \mathrm{Cos}^{-1}\left(\dfrac{-11}{\sqrt{29} \cdot \sqrt{19}}\right) \approx 118°$

31. $\mathbf{p} \cdot \mathbf{q} = 5(-9) + 5(-5) + (-1)15$

$\qquad = -85$

$|\mathbf{p}| = \sqrt{5^2 + 5^2 + (-1)^2} = \sqrt{51}$

$|\mathbf{q}| = \sqrt{(-9)^2 + (-5)^2 + 15^2} = \sqrt{331}$

$\theta = \mathrm{Cos}^{-1}\left(\dfrac{-85}{\sqrt{51} \cdot \sqrt{331}}\right) \approx 131°$

32.

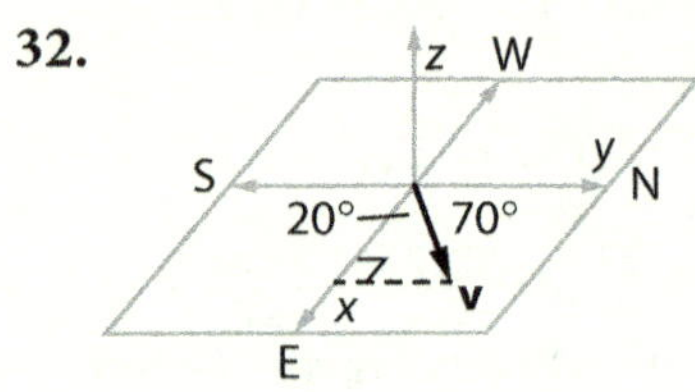

$\mathbf{v} = \langle |45| \cos 20°, |45| \sin 20°, 0 \rangle$

$\quad \approx \langle 42.3, 15.4, 0 \rangle$

33.

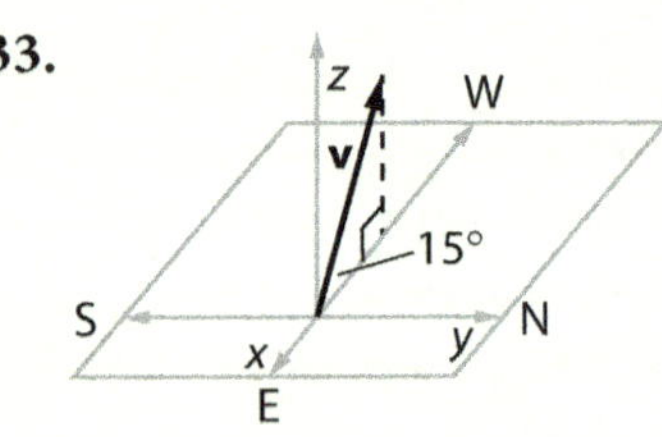

West is on the negative x-axis.

$\mathbf{v} = \langle -|140| \cos 15°, 0, |140| \sin 15° \rangle$

$\quad \approx \langle -135.2, 0, 36.2 \rangle$

34.

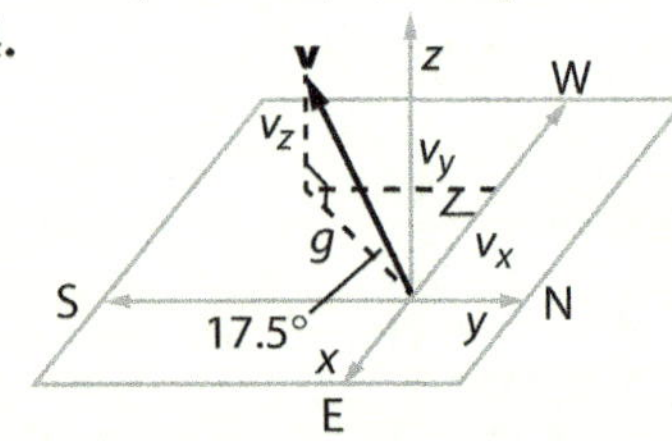

$v_z = 1.5 \sin 17.5° \approx 0.45$

$g = 1.5 \cos 17.5° \approx 1.43$

$v_x = g \cos 225° \approx -1.01$

$v_y = g \sin 225° \approx -1.01$

$\mathbf{v} \approx \langle -1.01, -1.01, 0.45 \rangle$

35.

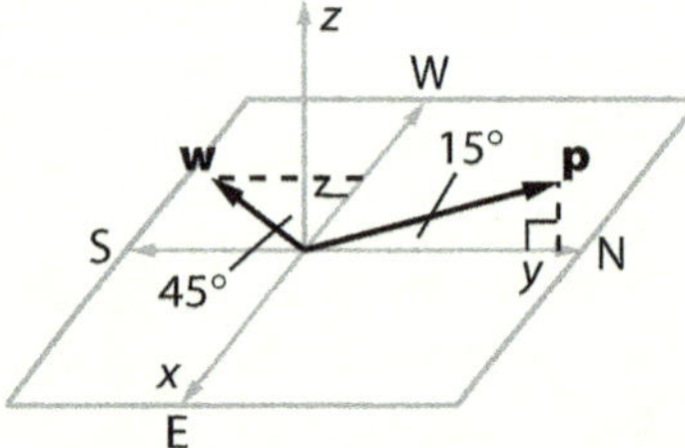

$\mathbf{p} = \langle 0, 150 \cos 15°, 150 \sin 15° \rangle$

$\quad \approx \langle 0, 144.9, 38.8 \rangle$

$\mathbf{w} = \langle 16 \cos 225°, 16 \sin 225°, 0 \rangle$

$\quad \approx \langle -11.3, -11.3, 0 \rangle$

$\mathbf{r} = \mathbf{p} + \mathbf{w} \approx \langle -11.3, 133.6, 38.8 \rangle$

36.

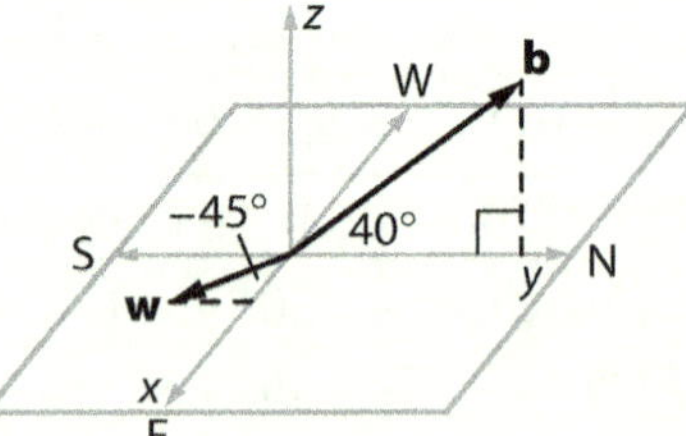

$\mathbf{b} = \langle 0, 83 \cos 40°, 83 \sin 40° \rangle$

$\quad \approx \langle 0, 63.6, 53.4 \rangle$

$\mathbf{w} = \langle 12 \cos(-45°), 12 \sin(-45°), 0 \rangle$

$\quad \approx \langle 8.5, -8.5, 0 \rangle$

$\mathbf{r} = \mathbf{b} + \mathbf{w} \approx \langle 8.5, 55.1, 53.4 \rangle$

$|\mathbf{r}| \approx \sqrt{8.5^2 + 55.1^2 + 53.4^2}$

$\quad \approx 77.2 \text{ mi/hr}$

continued in Answers and Solutions Overflow

> **C. Exercises**

37. A missile is fired southeast at an angle of 50° at a speed of 3000 mi/hr. There is a 22 mi/hr wind from the south. Find a vector that represents the missile's velocity relative to its launch point.

$\approx \langle 1363.6, -1341.6, 2298.1 \rangle$

The airplane in Example 6 has a resultant velocity vector $\mathbf{r} \approx \langle 28.3, -183.1, 77.0 \rangle$.

38. What is the plane's resultant ground speed? $\approx$ 200.6 knots

39. At what bearing is the plane traveling during its take-off (to the nearest degree)? 171°

40. What is the plane's resultant angle of inclination (to the nearest degree)? 23°

CUMULATIVE REVIEW

41. Which quadrant contains the terminal side of each angle? [4.1]

   a. $\dfrac{5\pi}{4}$ III   **b.** $-240°$ II

42. The terminal side of an angle $\theta \in [0, 2\pi)$ passes through the point $(-6, -2)$. Find its reference angle α and the angle θ (to the nearest tenth of a degree). [4.3]   18.4°; 198.4°

43. Find the exact coordinates of point P. [4.3]

$\left(-\dfrac{5}{2}, \dfrac{5\sqrt{3}}{2}\right)$

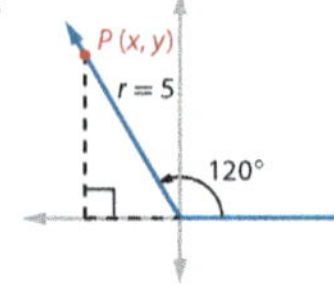

44. A triangular lot has sides of 450 ft and 570 ft. If the included angle is 100°, find the length of the third side to the nearest foot and the area of the lot to the nearest tenth of an acre (1 acre = 43,560 ft²). [5.7]   785 ft; 2.9 acres

Use the definition of a circle and the distance formula to derive an equation for each described circle. [1.2]

45. centered at $(0, 0)$ with a radius of 8 $x^2 + y^2 = 64$

46. centered at $(3, -4)$ with a radius of 5 $(x - 3)^2 + (y + 4)^2 = 25$

47. Find the area of a circle whose diameter has endpoints at $(4, 5)$ and $(-2, -3)$. [1.2]   C

  A. 100 u^2 **D.** $50\pi \text{ u}^2$

  B. 25 u^2 **E.** The area cannot be

  C. $25\pi \text{ u}^2$ determined.

48. Find an expression for the coordinates of A if the point lies on the unit circle. [4.3]   B

  A. $(1, 1)$ **D.** $(\tan \theta, \tan \theta)$

  B. $(\cos \theta, \sin \theta)$ **E.** none of these

  C. $(\sin \theta, \cos \theta)$

49. Which angles are coterminal with $\dfrac{5\pi}{6}$? List all correct answers. [4.3]   B, C, D

  A. $\dfrac{10\pi}{6}$ **C.** $\dfrac{5\pi}{6} \pm 2\pi$ **E.** none of these

  B. $\dfrac{17\pi}{6}$ **D.** $-\dfrac{7\pi}{6}$

50. For which values of θ does $\tan \theta = 1$? List all correct answers. [4.3]   A, C

  A. $\dfrac{\pi}{4}$ **C.** $\dfrac{5\pi}{4}$ **E.** all of these

  B. $\dfrac{3\pi}{4}$ **D.** $\dfrac{7\pi}{4}$

6.4 Polar Coordinates

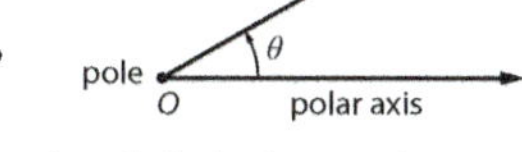
Naval air traffic control systems employ polar coordinates.

Previously, you have graphed points and functions using rectangular coordinates (x, y) in the Cartesian coordinate system defined by the x- and y-axes, which intersect at the origin. Air traffic controllers, the Doppler radar shown in weather reports, flight plans, and sonar positioning systems on submarines all use a polar coordinate system. This system provides an alternative method of describing any position on a plane.

The *pole*, a fixed point O, serves as the origin for the polar coordinate system. The *polar axis* is an initial ray drawn from the pole in the direction of the positive x-axis. Any point P in the plane can be described by the *polar coordinates* (r, θ) where

  r is a directed distance from the pole O to the point P
  and θ is a directed angle from the polar axis.

As usual, counterclockwise rotations are indicated by $\theta > 0$ and clockwise rotations are indicated by $\theta < 0$. When $r > 0$ the point lies on the terminal side of θ, but when $r < 0$ the point lies on the ray that is opposite the terminal side of θ. A *polar grid* can be used when graphing polar coordinates.

Example 1 Plotting Points in the Polar Coordinate System

Graph each point described by its polar coordinates.

a. $A\left(4, \frac{\pi}{3}\right)$ **b.** $B\,(2, -120°)$ **c.** $C\left(-3, \frac{3\pi}{4}\right)$

Answer

a. Draw $\theta = \frac{\pi}{3}$ in Q I and plot A four units from O on the terminal side of θ.

b. Draw $\theta = -120°$ in Q III and plot B two units from O on the terminal side of θ.

c. Draw $\theta = \frac{3\pi}{4}$ in Q II and plot C three units from O on the ray opposite the terminal side of θ (in Q IV).

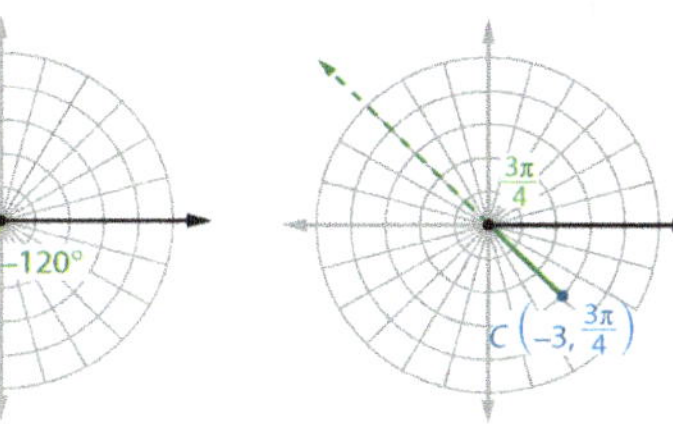

SKILL ✔ **EXERCISE 3**

After completing this section, you will be able to

- graph points using polar coordinates.
- state multiple polar coordinates representing a point in the polar plane.
- find the distance between polar coordinates.
- convert between polar and rectangular coordinates.
- convert from a polar equation to a rectangular equation.

KEYWORD SEARCH

polar graph paper 🔍

Objectives

1. To graph points using polar coordinates

2. To state multiple polar coordinates representing a point in the polar plane

3. To find the distance between two points given the polar coordinates

4. To convert between polar and rectangular coordinates

5. To identify graphs of lines and circles by converting from a polar equation to a rectangular equation

Flash

US Naval ships use navigational systems employing polar coordinates for air traffic control, weather monitoring, and safety alerts.

Vocabulary

polar axis
polar coordinate
polar distance formula
polar grid
pole

PRESENTATION

Lesson Opener

State each general solution in radians.

1. $2 \sin x = 1$ $x = \frac{\pi}{6} + n2\pi,\ \frac{5\pi}{6} + n2\pi$

2. $3 \tan x = \sqrt{3}$ $x = \frac{\pi}{6} + n\pi$

3. Use factoring to find the primary solutions for $2 \cos^2 x - 1 = -\sin x$.
   $x = \frac{\pi}{2}, \frac{7\pi}{6}, \frac{11\pi}{6}$

Engage the students by pointing out that *polar coordinates* have many real-life applications. They are often used in electrical and mechanical engineering, and are very useful in tracking moving objects.

Be sure the students have *polar grids*, which can be easily found using the Internet keyword search *polar graph paper* or using the following image from TeacherToolsOnline.com. The image can also be projected when plotting points and equations in the classroom.

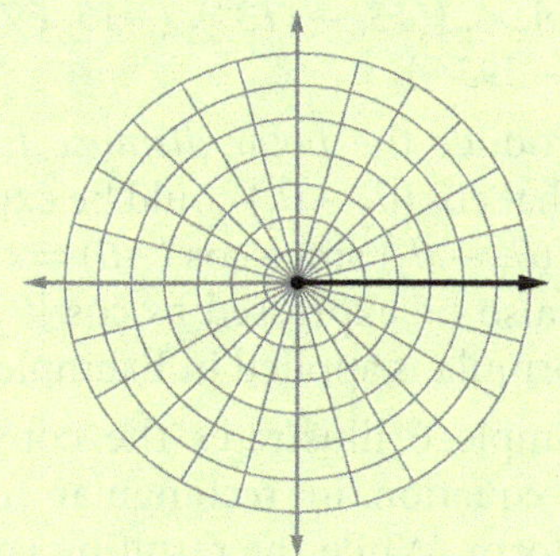

Demonstrate graphing the polar coordinates in Example 1. Note that the angle (the second coordinate) is usually plotted before the directed distance (the first coordinate) is marked off on the line determined by the angle's terminal side. Determining the directed distance on the *polar axis* and then rotating the point would produce the same result, but polar equations usually treat θ as the independent variable and r as the dependent variable.

One-on-One Many videos detailing the graphing of polar coordinates on the TI-84 are available online for a student who needs more instruction. Use the Internet keyword search *TI-84 polar coordinate video*.

Reading and Writing Mathematics

Explain in your own words how to convert rectangular coordinates to polar coordinates.

Find the distance, r, from the origin using the distance formula. Then find the angle, θ, using $\text{Tan}^{-1}\frac{y}{x}$. Express these values in polar form, (r, θ).

Additional Exercises

1. Graph $A\left(4, -\frac{7\pi}{6}\right)$ and $B\left(-1, \frac{5\pi}{4}\right)$.

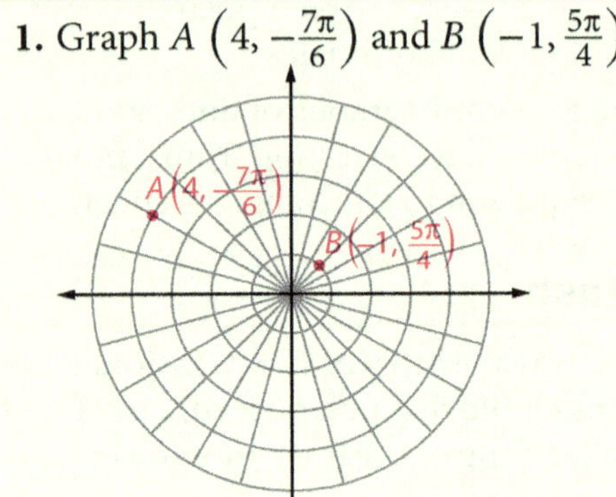

2. State all polar coordinates for each given point if $-360° \le \theta \le 360°$ or $-2\pi \le \theta \le 2\pi$.
   a. $C(-2, 100°)$
   $(-2, -260°)$, $(2, 280°)$, or $(2, -80°)$
   b. $D\left(1, \frac{\pi}{2}\right)$
   $\left(1, -\frac{3\pi}{2}\right)$, $\left(-1, \frac{3\pi}{2}\right)$, $\left(-1, -\frac{\pi}{2}\right)$

3. Write each standard coordinate as a polar coordinate with $0 \le \theta \le 360°$.
   a. $(-3, 4)$ $(5, 127°)$
   b. $(5, -12)$ $(13, 293°)$

In his *Method of Fluxions and Infinite Series* (published in 1736 but written much earlier), Isaac Newton used several different coordinate systems, including Cartesian and polar, to explore the properties of curves.

In the rectangular coordinate system each point is described by exactly one ordered pair, but this is not true in the polar coordinate system. Since there are an infinite number of coterminal angles, $P(r, \theta)$ can also be described by $(r, \theta + n2\pi)$ where $n \in \mathbb{Z}$. The point $P(r, \theta)$ can also be described by $(-r, \theta \pm \pi)$, or more generally $(-r, \theta + (2n + 1)\pi)$ where $n \in \mathbb{Z}$.

Example 2 Multiple Representations of a Point

Plot the point $P(3, 30°)$ and state four additional pairs of polar coordinates for P.

Answer

1. Draw $\theta = 30°$ in Q I and plot P three units from the pole on the terminal side of θ.

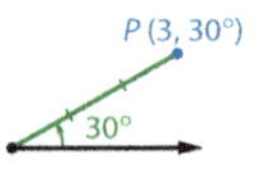

2. Keeping the same value for r, find $\theta \pm 360°$.

$(3, 30° + 360°)$ $(3, 30° - 360°)$
$= (3, 390°)$ $= (3, -330°)$

3. Using the opposite value for r, find $\theta \pm 180°$.

$(-3, 30° + 180°)$ $(-3, 30° - 180°)$
$= (-3, 210°)$ $= (-3, -150°)$

SKILL ✔ EXERCISE 9

To understand the relationship between the polar and rectangular systems, superimpose them so that the pole coincides with the origin and the polar axis lies on the positive x-axis. Sketching the reference triangle related to $P(r, \theta)$ shows that

$$r^2 = x^2 + y^2, \cos\theta = \frac{x}{r}, \text{ and } \sin\theta = \frac{y}{r}.$$

Rearranging the last two equations produces formulas for x and y in terms of r and θ:
$$x = r\cos\theta \text{ and } y = r\sin\theta \text{ so that } (x, y) = (r\cos\theta, r\sin\theta).$$

Example 3 Converting to Rectangular Coordinates

Convert the polar coordinates $\left(2, \frac{2\pi}{3}\right)$ to rectangular coordinates.

Answer

$(x, y) = \left(2\cos\frac{2\pi}{3}, 2\sin\frac{2\pi}{3}\right)$ 1. Substitute into $(x, y) = (r\cos\theta, r\sin\theta)$.

$= \left(2\left(-\frac{1}{2}\right), 2\left(\frac{\sqrt{3}}{2}\right)\right)$ 2. Simplify.

$= (-1, \sqrt{3}) \approx (-1, 1.73)$

SKILL ✔ EXERCISE 13

When converting from rectangular coordinates to polar coordinates, find the reference angle $\alpha = \text{Tan}^{-1}\left|\frac{y}{x}\right|$, then use the appropriate formula from Section 6.1 to find $\theta \in [0, 360°)$ or $[0, 2\pi)$. Other polar coordinates for the point can then be found.

Explaining that any point can be described using an infinite number of coterminal angles provides insight to the nature of polar coordinates. While points are often described using $r > 0$ and $\theta \in [0, 2\pi)$, Example 2 illustrates other equivalent representations of such coordinates. Consider demonstrating that the same point can be located by first plotting the directed distance along the polar axis and then rotating by θ as illustrated.

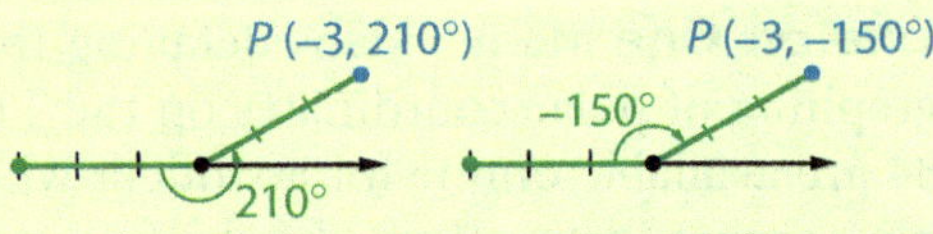

Be sure the students understand the reasoning behind the formulas used in conversions between polar and rectangular coordinates before completing Examples 3 and 4. These formulas will also be used when converting polar and rectangular equations. Consider asking the students for three other equivalent pairs of polar coordinates for point P in Example 4. $((13, -113°), (-13, 67°),$ and $(-13, -293°))$

Introduce the *polar distance formula*. Note that $\cos(\theta_2 - \theta_1)$ could be expressed as $\cos(\theta_1 - \theta_2)$ since $\cos(-\theta) = \cos\theta$. It could also be expressed as $\cos|\theta_2 - \theta_1|$. The formula is applied in Example 5.

Example 6 illustrates the conversion of an equation in rectangular form to polar form. While the resulting equation can be graphed in degree mode, polar equations are typically graphed in radian mode. Consider asking students whether they can predict the polar form of $y = -3$. $(r = -3\csc\theta)$

Consider pointing out that a polar equation for the x-axis, $y = 0$, is $\theta = 0$ and not $r\sin\theta = 0$, which actually plots the origin. Similarly, the polar equation for the y-axis, $x = 0$, is $\theta = \frac{\pi}{2}$ not $r\sin\theta = 0$.

Before converting the polar equation $r = 3$ in Example 7, consider having the students plot $r = 3$ with various values of θ on the polar coordinate system. They should quickly see that the constant distance from the origin determines a circle. Explain that converting from polar

Example 4 Converting to Polar Coordinates

Find polar coordinates for $P(-5, -12)$ where $r > 0$ and $\theta \in [0, 360°)$.

Answer

$r^2 = (-5)^2 + (-12)^2 = 169$

$r = \pm 13$

1. Substitute the known rectangular values to find possible values for r.

$\alpha = \text{Tan}^{-1}\left|\dfrac{-12}{-5}\right| \approx 67°$

$\theta \approx 67° + 180° = 247°$

2. Find the reference angle $\alpha = \text{Tan}^{-1}\left|\dfrac{y}{x}\right|$ and use the fact that P is in Q III to determine $\theta \in [0, 360°)$.

$P(13, 247°)$

3. Express P with coordinates where $r > 0$ and $\theta \in [0°, 360°)$.

SKILL ✔ **EXERCISE 19**

The distance between two points described by polar coordinates can be found using

$$d = \sqrt{r_1^2 + r_2^2 - 2r_1 r_2 \cos(\theta_2 - \theta_1)}.$$

This *polar distance formula* is derived and further investigated in exercise 44.

Example 5 Finding the Distance Between Polar Coordinates

The radar on a battleship detects ships at (15 mi, 140°) and (27 mi, 240°). Find the distance between these two ships.

Answer

1. Sketch a graph.

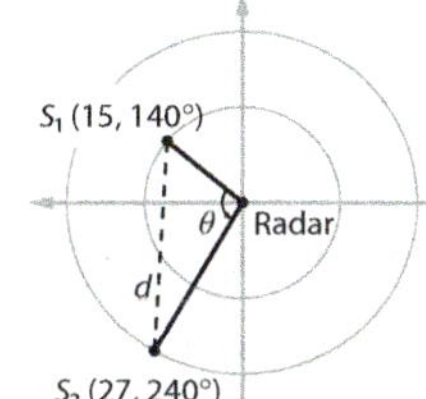

2. Apply the polar distance formula.

$$d = \sqrt{15^2 + 27^2 - 2(15)(27)\cos(240° - 140°)}$$
$$\approx 33.1 \text{ mi}$$

SKILL ✔ **EXERCISE 23**

To convert a rectangular equation to a polar equation, substitute $r \cos \theta$ for x and $r \sin \theta$ for y, then simplify the resulting equation.

Example 6 Converting to a Polar Equation

Write the equation of the line $x = 2$ in polar form where r is expressed in terms of θ.

Answer

$x = 2$

1. Substitute $r \cos \theta$ for x.

$r \cos \theta = 2$

$r = \dfrac{2}{\cos \theta}$

or $r = 2 \sec \theta$

2. Solve for r.

CONTINUED ➡

4. Write each polar coordinate in standard form.

   a. $\left(4, \dfrac{5\pi}{6}\right)$ $(-2\sqrt{3}, 2)$

   b. $(-6, 240°)$ $(3, 3\sqrt{3})$

   c. $(1, 1)$ $(0.54, 0.84)$

5. Find the distance between each pair of polar coordinates.

   a. $(5, 30°), (4, 195°)$ ≈ 8.9

   b. $\left(2, \dfrac{\pi}{3}\right), (-5, \pi)$ ≈ 4.4

6. Write each rectangular equation as a polar equation.

   a. $3x + 2y = 14$ $r = \dfrac{14}{3\cos\theta + 2\sin\theta}$

   b. $(x - 2)^2 + y^2 = 4$ $r = 4\cos\theta$

7. Write $r = 10\sin\theta$ as a rectangular equation. $x^2 + (y - 5)^2 = 25$, a circle centered at $(0, 5)$ with $r = 5$

Assignments

- **Minimum:** 1, 4–5, 7–8, 11–14, 17–18, 21, 23, 25–28, 32–33, 35, 37, 48–50, 52

- **Standard:** 2–8 even; 9, 11–12, 15, 17–19, 21, 24–26, 28, 31–36, 38–39, 41, 46–48, 50–53

- **Extended:** 2, 5, 8–12, 15, 18–19, 23–25, 27–34, 36, 38–41, 43–44, 46, 48–50, 53–54

form requires that each equation first be transformed to an equivalent equation in which substitutions can be made. After completing Example 7a, consider asking the students for the polar equation of the unit circle and of a circle centered at the origin with a radius of 25. $(r = 1; r = 25)$ Ask students to predict the graph of $\theta = \dfrac{\pi}{6}$ in Example 7b before converting to rectangular form. Explain that the graphing calculator cannot graph the polar equation $\theta = \dfrac{\pi}{6}$, as it is not a function in the form of $r(\theta)$.

Present Example 8, emphasizing how the first step transforms the equation to prepare for substitutions. Ask students to predict the graphs of $r = -4\sin\theta$ and $r = 4\cos\theta$. (*circles with radii of 4 centered at $(0, -2)$ and at $(2, 0)$*)

The Technology Corner uses Example 8 to provide basic instruction on graphing in polar mode, a skill that is used extensively in the next section. When defining the graphing WINDOW, θstep determines the number of plotted points that the calculator connects to form the curve. Use integral divisors of $\dfrac{\pi}{12}$, such as $\dfrac{\pi}{36}$ or $\dfrac{\pi}{72}$, to ensure the TRACE function includes typical unit circle values.

Interactive Activity Use the Internet keyword search *polar coordinate applet* to find interactive applications of polar coordinates. The keyword search *polar coordinate game* can be used to locate games that engage student interest. Students may want to explore games using polar coordinates on their own time.

TIPS

Ex. 33 Consider asking whether this equation produces the same graph as $r = 5$. (*no*) While it appears the same on the plane, $r = -5$ has different points. It begins at $(-5, 0)$ and proceeds through Q III to $\left(-5, \dfrac{\pi}{2}\right)$, then through Q IV, etc.

Ex. 36 Caution students that they will have to graph this function manually. The TI-84 Plus cannot plot this function since it is not in the form of $r(\theta)$.

Solutions

❯ A. Exercises

1–3.

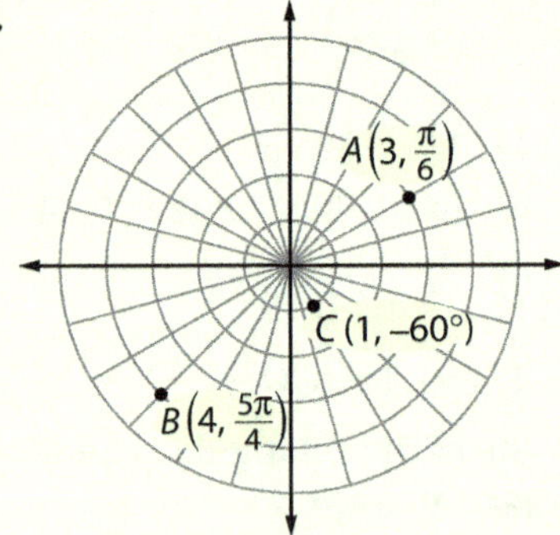

4–6.

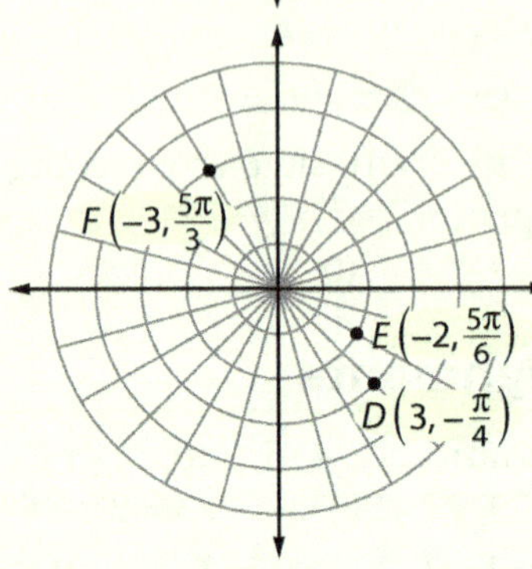

7. $(2, 45° − 360°) = (2, −315°)$
$(−2, 45° + 180°) = (−2, 225°)$
$(−2, 45° − 180°) = (−2, −135°)$

8. $(3, −120° + 360°) = (3, 240°)$
$(−3, −120° + 180°) = (−3, 60°)$
$(−3, −120° − 180°) = (−3, −300°)$

9. $(−2, \pi − 2\pi) = (−2, −\pi)$
$(2, \pi + \pi) = (2, 2\pi)$
$(2, \pi − \pi) = (2, 0)$
$(2, \pi − 3\pi) = (2, −2\pi)$

10. $\left(3, −\frac{\pi}{3} + 2\pi\right) = \left(3, \frac{5\pi}{3}\right)$
$\left(−3, −\frac{\pi}{3} + \pi\right) = \left(−3, \frac{2\pi}{3}\right)$
$\left(−3, −\frac{\pi}{3} − \pi\right) = \left(−3, −\frac{4\pi}{3}\right)$

11. $x = −1 \cos 180° = 1$
$y = −1 \sin 180° = 0$
$(1, 0)$

12. $x = 2 \cos 30° = \sqrt{3} \approx 1.7$
$y = 2 \sin 30° = 1$
$(\sqrt{3}, 1) \approx (1.7, 1)$

13. $x = 5 \cos 45° = \frac{5\sqrt{2}}{2} \approx 3.5$
$y = 5 \sin 45° = \frac{5\sqrt{2}}{2} \approx 3.5$
$\left(\frac{5\sqrt{2}}{2}, \frac{5\sqrt{2}}{2}\right) \approx (3.5, 3.5)$

14. $−\frac{3\pi}{2} + 2\pi = \frac{\pi}{2}$
$x = −2 \cos \frac{\pi}{2} = 0$
$y = −2 \sin \frac{\pi}{2} = −2$
$(0, −2)$

15. $x = 3 \cos \frac{2\pi}{3} = 3(−0.5) = −1.5$
$y = 3 \sin \frac{2\pi}{3} = \frac{3\sqrt{3}}{2} \approx 2.6$
$\left(−\frac{3}{2}, \frac{3\sqrt{3}}{2}\right) \approx (−1.5, 2.6)$

16. $x = − \cos \frac{11\pi}{6} = −\frac{\sqrt{3}}{2} \approx −0.9$
$y = − \sin \frac{11\pi}{6} = −\left(\frac{−1}{2}\right) = \frac{1}{2}$
$\left(−\frac{\sqrt{3}}{2}, \frac{1}{2}\right) \approx (−0.9, 0.5)$

❯ B. Exercises

17. $r = \sqrt{2^2 + 2^2} = \sqrt{8} = 2\sqrt{2}$
$\alpha = \text{Tan}^{-1} \left|\frac{2}{2}\right| = 45°$
Q I: $\theta = \alpha = 45°; \therefore (2\sqrt{2}, 45°)$

18. $r = \sqrt{(−3)^2 + (−4)^2} = \sqrt{25} = 5$
$\alpha = \text{Tan}^{-1} \left|\frac{4}{3}\right| \approx 53°$
Q III: $\theta \approx 53 + 180 = 233°$
$\therefore (5, \approx 233°)$

19. $r = \sqrt{(−15)^2 + 8^2} = \sqrt{289} = 17$
$\alpha = \text{Tan}^{-1} \left|\frac{8}{−15}\right| \approx 28°$
Q II: $\theta \approx 180 − 28 = 152°$
$\therefore (17, \approx 152°)$

20. $r = \sqrt{0^2 + 1^2} = 1$
since S lies on the positive y-axis,
$\theta = 90°; \therefore (1, 90°)$

Check

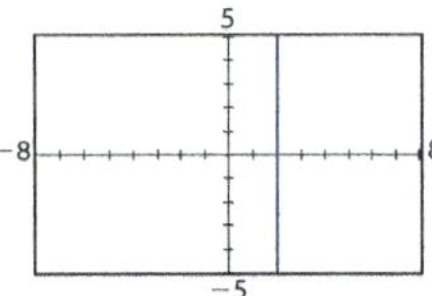

3. Set your calculator to polar mode, select [Y=], and enter 2/cos(θ) for r1 (using [X,T,θ,n] for θ). Then [GRAPH] the polar equation. See the Technology Corner for further instruction on graphing in polar mode.

SKILL ✔ EXERCISE 27

A polar equation can be converted to a rectangular equation by making similar substitutions. Algebraic manipulations may be required to set up substitutions involving the equations used to convert between rectangular and polar coordinates.

Example 7 Converting to a Rectangular Equation

Write each equation in rectangular form and describe its graph.

a. $r = 3$ **b.** $\theta = \frac{\pi}{6}$

Answer

a.
$r = 3$ — 1. Square both sides of the equation.
$r^2 = 9$
$x^2 + y^2 = 9$ — 2. Substitute $x^2 + y^2$ for r^2.

The graph is a circle centered at the origin with a radius of 3.

---

b. $\tan \theta = \tan \frac{\pi}{6}$ — 1. Find the tangent of both sides of the equation.
$\tan \theta = \frac{\sqrt{3}}{3}$
$\frac{y}{x} = \frac{\sqrt{3}}{3}$ — 2. Substitute $\frac{y}{x}$ for $\tan \theta$.
$y = \frac{\sqrt{3}}{3} x$ — 3. Solve for y.

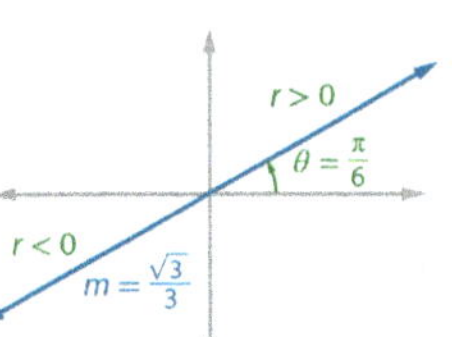

The graph is a line through the origin with a slope of $\frac{\sqrt{3}}{3} \approx 0.58$.

SKILL ✔ EXERCISE 31

Note that any point on the line $\theta = \frac{\pi}{6}$ in Q I has $r > 0$ and that any point on the line in Q IV has $r < 0$.

Example 8 Converting to a Rectangular Equation

Write $r = 4 \sin \theta$ in rectangular form.

Answer

$r = 4 \sin \theta$ — 1. Multiply both sides by r to obtain recognizable forms.
$r^2 = 4r \sin \theta$
$x^2 + y^2 = 4y$ — 2. Substitute using $r^2 = x^2 + y^2$ and $y = r \sin \theta$.
$x^2 + y^2 − 4y = 0$ — 3. Subtract $4y$ from each side; then complete the square.
$x^2 + (y^2 − 4y + 4) = 4$
$x^2 + (y − 2)^2 = 4$

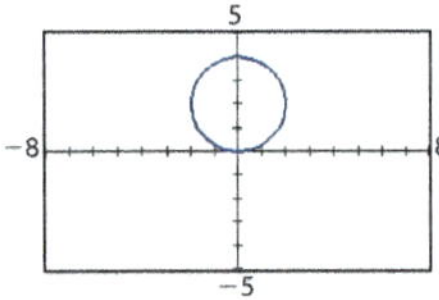

Check — 4. Graphing $r = 4 \sin \theta$ in polar mode confirms that the graph is a circle centered at $(0, 2)$.

SKILL ✔ EXERCISE 33

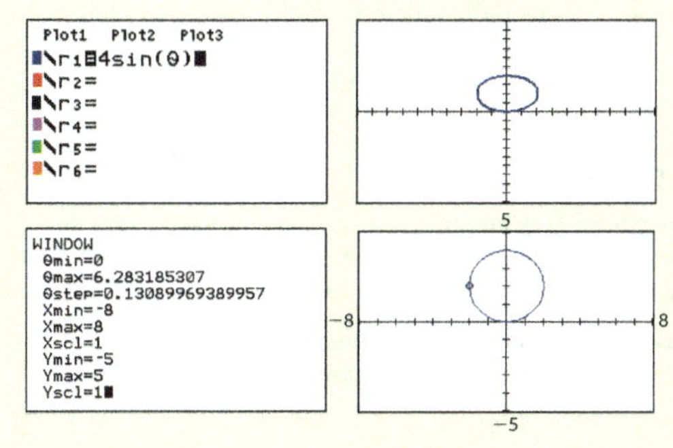

A. Exercises

Graph each point described by its polar coordinates.

1. $A\left(3, \frac{\pi}{6}\right)$ 2. $B\left(4, \frac{5\pi}{4}\right)$

3. $C(1, -60°)$ 4. $D\left(3, -\frac{\pi}{4}\right)$

5. $E\left(-2, \frac{5\pi}{6}\right)$ 6. $F\left(-3, \frac{5\pi}{3}\right)$

State all additional pairs of polar coordinates for each point where $\theta \in [-360°, 360°]$ or $\theta \in [-2\pi, 2\pi]$.

7. $G(2, 45°)$ 8. $H(3, -120°)$

9. $I(-2, \pi)$ 10. $J\left(3, -\frac{\pi}{3}\right)$

Convert the polar coordinates to rectangular coordinates.

11. $(-1, 180°)$ $(1, 0)$ 12. $(2, 30°)$ $(\sqrt{3}, 1)$

13. $(5, 45°)$ $\left(\frac{5\sqrt{2}}{2}, \frac{5\sqrt{2}}{2}\right)$ 14. $\left(-2, -\frac{3\pi}{2}\right)$ $(0, -2)$

15. $\left(3, \frac{2\pi}{3}\right)$ $\left(-\frac{3}{2}, \frac{3\sqrt{3}}{2}\right)$ 16. $\left(-1, \frac{11\pi}{6}\right)$ $\left(-\frac{\sqrt{3}}{2}, \frac{1}{2}\right)$

B. Exercises

Find the polar coordinates for each point where $r > 0$ and $\theta \in [0, 360°)$.

17. $P(2, 2)$ $(2\sqrt{2}, 45°)$ 18. $Q(-3, -4)$ $(5, \approx 233°)$

19. $R(-15, 8)$ $(17, \approx 152°)$ 20. $S(0, 1)$ $(1, 90°)$

29. $r^2 = \dfrac{16}{\cos^2 \theta + 4\sin^2 \theta}$

Find the distance between the given polar coordinates (to the nearest tenth).

21. $(3, 40°), (1, 130°)$ 3.2   22. $(-5, 315°), (3, 50°)$ 5.6

23. $\left(3, \frac{\pi}{3}\right), \left(-1, \frac{\pi}{2}\right)$ 3.9   24. $\left(4, \frac{5\pi}{12}\right), \left(6, \frac{5\pi}{4}\right)$ 9.7

Transform each equation into polar form. Then verify your answer by using technology to graph the polar equation.

25. $y = 3$ $r = 3 \csc \theta$ 26. $x + y = 3$ $r = \dfrac{3}{\cos \theta + \sin \theta}$

27. $2x - y = 4$ $r = \dfrac{4}{2\cos \theta - \sin \theta}$ 28. $x^2 + y^2 = 25$ $r = 5$

29. $x^2 + 4y^2 = 16$ 30. $y = x^2$ $r = \tan \theta \sec \theta$

Graph each polar equation and describe the graph. Then transform the equation into rectangular form.

31. $r = 5 \sec \theta$ $x = 5$ 32. $r = -\csc \theta$ $y = -1$

33. $r = -5$ $x^2 + y^2 = 25$ 34. $r = 8 \cos \theta$ $(x - 4)^2 + y^2 = 16$

35. $r = -6 \sin \theta$ $x^2 + (y + 3)^2 = 9$ 36. $\theta = \frac{5\pi}{6}$ $y = -\frac{\sqrt{3}}{3}x$

37. An airport control tower is tracking the locations of two planes flying at the same altitude. Find the distance between the planes if the first plane is located at (5 km, 155°) and the second is at (6 km, 35°).   ≈ 9.5 km

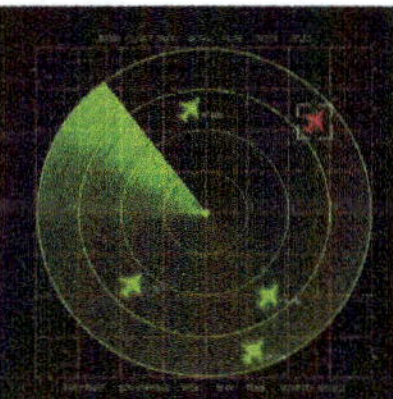

21. $d = \sqrt{3^2 + 1^2 - 2(3)(1)\cos(130° - 40°)} \approx 3.2$

22. $d = \sqrt{(-5)^2 + 3^2 - 2(-5)(3)\cos(50° - 315°)} \approx 5.6$

23. $d = \sqrt{3^2 + (-1)^2 - 2(3)(-1)\cos\left(\frac{\pi}{2} - \frac{\pi}{3}\right)} \approx 3.9$

24. $d = \sqrt{4^2 + 6^2 - 2(4)(6)\cos\left(\frac{5\pi}{4} - \frac{5\pi}{12}\right)} \approx 9.7$

25.
$$y = 3$$
$$r \sin \theta = 3$$
$$r = \frac{3}{\sin \theta} = 3 \csc \theta$$

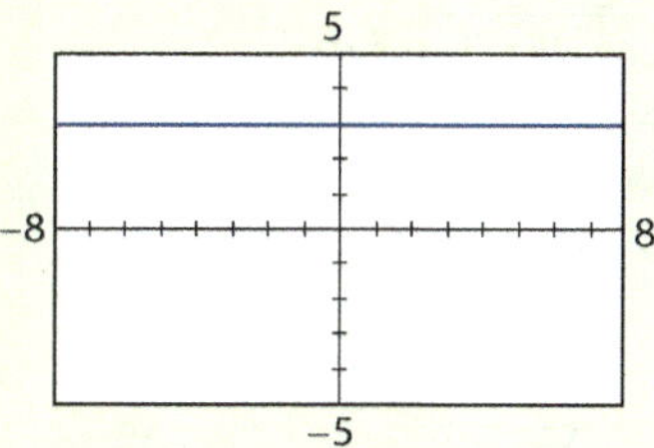

26.
$$x + y = 3$$
$$r \cos \theta + r \sin \theta = 3$$
$$r = \frac{3}{\cos \theta + \sin \theta}$$

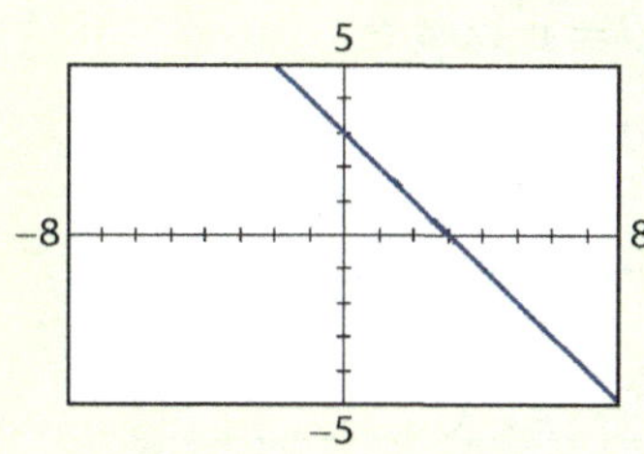

27.
$$2x - y = 4$$
$$2r \cos \theta - r \sin \theta = 4$$
$$r = \frac{4}{2\cos \theta - \sin \theta}$$

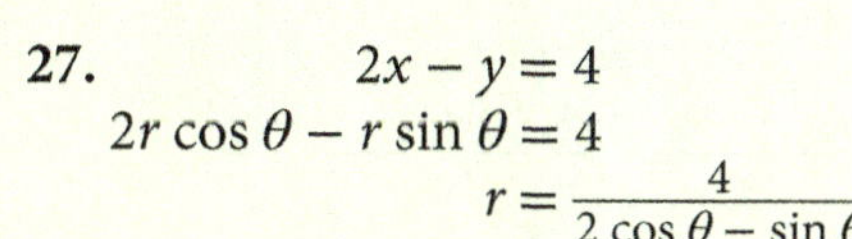

28.
$$x^2 + y^2 = 25$$
$$r^2 \cos^2 \theta + r^2 \sin^2 \theta = 25$$
$$r^2(\cos^2 \theta + \sin^2 \theta) = 25$$
$$r^2 = 25$$
$$r = 5$$

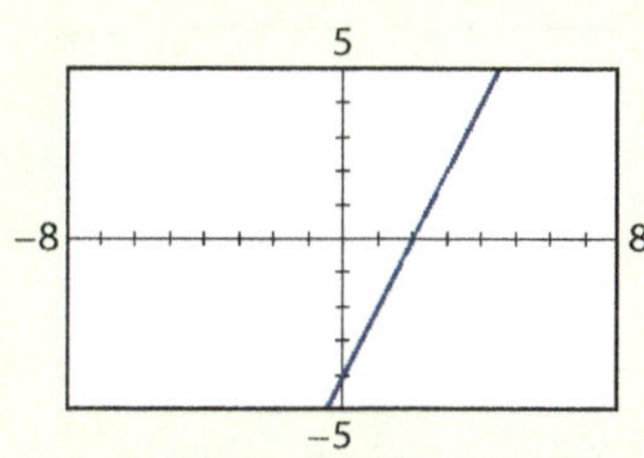

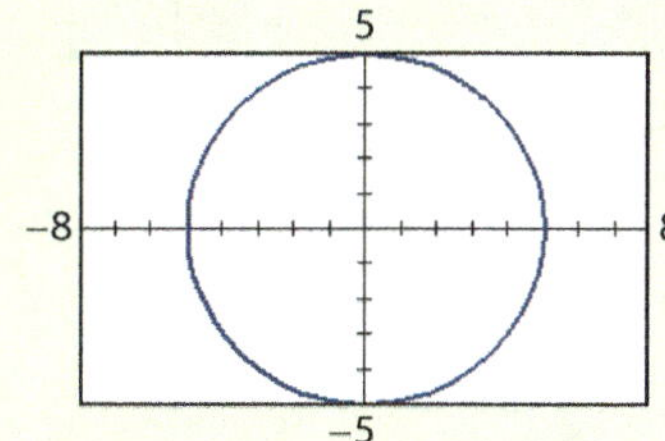

29.
$$x^2 + 4y^2 = 16$$
$$r^2 \cos^2 \theta + 4r^2 \sin^2 \theta = 16$$
$$r^2(\cos^2 \theta + 4 \sin^2 \theta) = 16$$
$$r^2 = \frac{16}{\cos^2 \theta + 4 \sin^2 \theta}$$
$$\text{or } r = \pm \frac{4}{\sqrt{\cos^2 \theta + 4 \sin^2 \theta}}$$

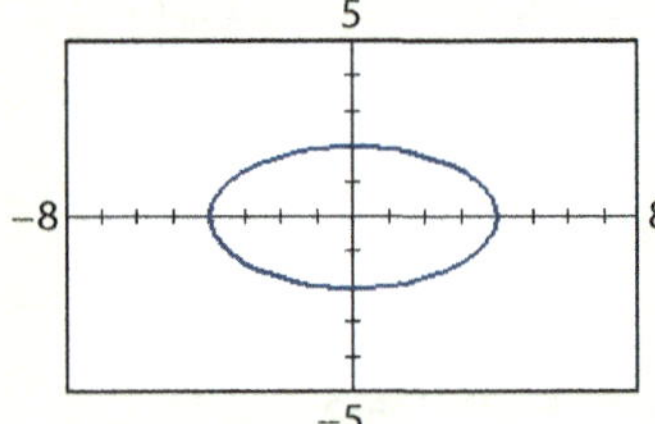

30.
$$r \sin \theta = (r \cos \theta)^2$$
$$r \sin \theta = r^2 \cos^2 \theta$$
$$\frac{\sin \theta}{\cos^2 \theta} = r$$
$$r = \tan \theta \sec \theta$$

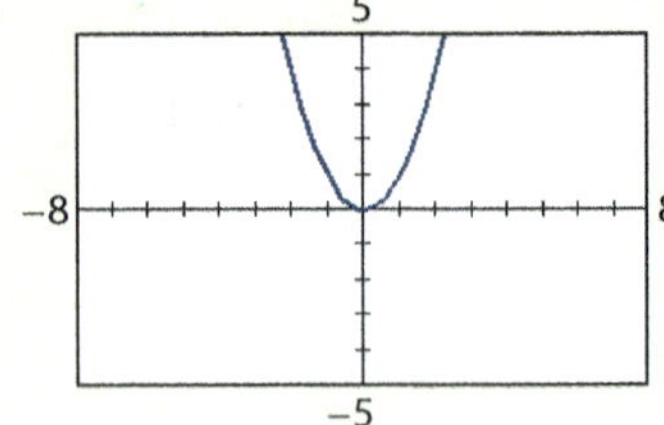

31.

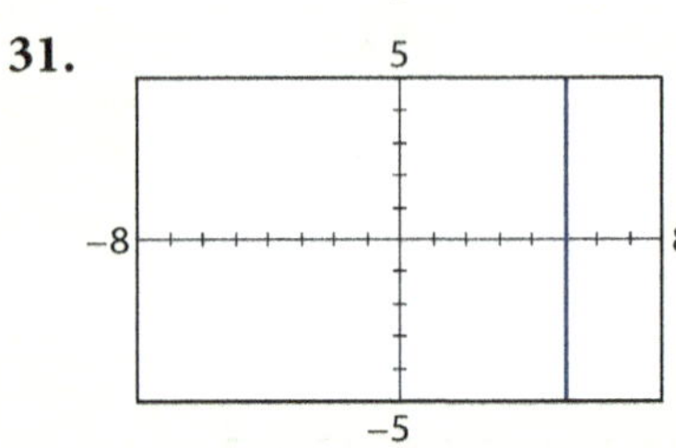

vertical line through $(5, 0)$
$$r = \frac{5}{\cos \theta}$$
$$r \cos \theta = 5$$
$$x = 5$$

32.

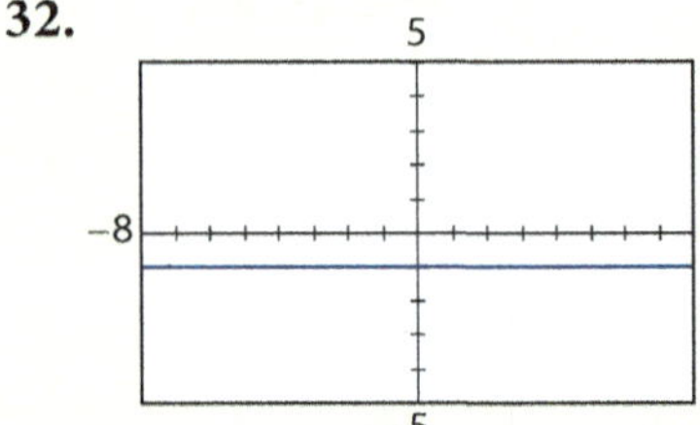

horizontal line through $(0, -1)$
$$r = -\frac{1}{\sin \theta}$$
$$r \sin \theta = -1$$
$$y = -1$$

33.

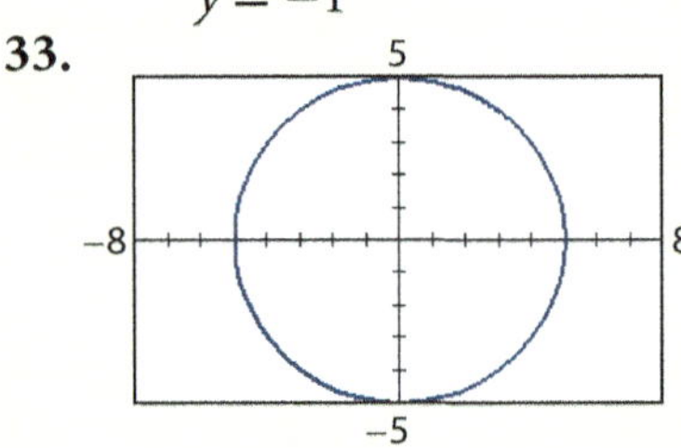

circle centered at $(0, 0)$ with $r = 5$

---

38. A ship's navigator observes ships at (3 nm, 70°) and (7 nm, 110°). Find the distance between the ships (nm = nautical mile). $\approx 5.1 \text{ nm}$

39. Explore: Use technology to graph the four circles: $r = \pm 3 \sin \theta$ and $r = \pm 2 \cos \theta$.

  a. State the radius and center for $r = k \cos \theta$ in terms of k. $\; r = \frac{|k|}{2}, \left(\frac{k}{2}, 0\right)$

  b. State the radius and center for $r = k \sin \theta$ in terms of k. $\; r = \frac{|k|}{2}, \left(\frac{k}{2}, \frac{\pi}{2}\right)$

› C. Exercises

Transform each equation into polar form. Verify your answer by using technology to graph the polar equation.

40. $(x - 2)^2 + (y + 1)^2 = 5$ **41.** $xy = 4$

Graph each polar equation. Then transform the equation into rectangular form and describe the graph.

42. $r = 4 \sin \theta - 6 \cos \theta$ $(x + 3)^2 + (y - 2)^2 = 13$

43. $r = 2 \cos \theta - 6 \sin \theta$ $(x - 1)^2 + (y + 3)^2 = 10$

44. Explore: Investigate the polar distance formula.

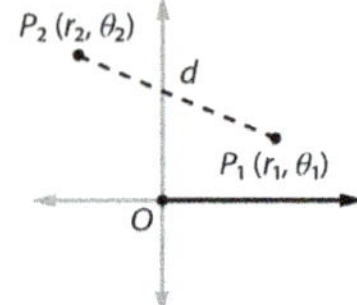

  a. Use the Law of Cosines to show that the distance between $P_1 (r_1, \theta_1)$ and $P_2 (r_2, \theta_2)$ is
$$d = \sqrt{r_1{}^2 + r_2{}^2 - 2r_1 r_2 \cos (\theta_2 - \theta_1)}.$$

  b. Simplify the formula in the case where $\theta_2 = \theta_1$.

  c. Simplify the formula in the case where $\theta_2 = \theta_1 + \pi$.

  d. Simplify the formula in the case where $\theta_2 = \theta_1 + \frac{\pi}{2}$.

40. $r = 4 \cos \theta - 2 \sin \theta$

41. $r = \pm 2\sqrt{\sec \theta \csc \theta}$

44b. $d = |r_1 - r_2|$

44c. $d = |r_1 + r_2|$

44d. $d = \sqrt{r_1{}^2 + r_2{}^2}$

CUMULATIVE REVIEW

Use the sum and difference identities to simplify each expression. [5.4]

45. $\sin (\pi \pm \theta)$ $\mp \sin \theta$ **46.** $\cos (\pi \pm \theta)$ $-\cos \theta$

Graph each function over $\{0, 2\pi\}$. Then list any zeros of the function and any local maximum or minimum points within $[0, 2\pi]$. [4.4]

47. $y = 3 \sin x$ **48.** $y = \cos 2x$

49. Find the general solution for $\cos 2\theta = 1$. [5.3] $\; \theta = n\pi$

50. Write simplified expressions for each power of i from i^0 to i^8. [Appendix 5]

51. If $\sin \theta = \frac{3}{5}$ and $\cos \theta = \frac{4}{5}$, what is $\tan \theta$? [4.2, 5.1] A

  A. $\frac{3}{4}$ **C.** 5   **E.** $\frac{1}{2}$

  B. $\frac{4}{3}$ **D.** $-\frac{4}{3}$

52. Find θ to the nearest degree. [4.2] A

  A. 22°   **D.** 68°

  B. 30°   **E.** none of

  C. 60°   these

53. Evaluate $\text{Tan}^{-1}\left(\tan \frac{3\pi}{4}\right)$. [4.6] D

  A. 1   **C.** $\frac{\pi}{4}$ **E.** $\frac{3\pi}{4}$

  B. -1 **D.** $-\frac{\pi}{4}$

54. Which expression is equivalent to $\frac{\tan^2 x - \sin^2 x}{\sin^2 x \tan^2 x}$? [5.1] B

  A. -1 **D.** $\cot^2 x$

  B. 1   **E.** none of these

  C. $\csc^2 x$

47. $x = 0, \pi, 2\pi$; max: $\left(\frac{\pi}{2}, 3\right)$; min: $\left(\frac{3\pi}{2}, -3\right)$

48. $x = \frac{\pi}{4}, \frac{3\pi}{4}, \frac{5\pi}{4}, \frac{7\pi}{4}$; max: $(0, 1), (\pi, 1), (2\pi, 1)$; min: $\left(\frac{\pi}{2}, -1\right), \left(\frac{3\pi}{2}, -1\right)$

50. $i^0 = 1, i^1 = i, i^2 = -1, i^3 = -i, i^4 = 1, i^5 = i, i^6 = -1, i^7 = -i, i^8 = 1$

$$r = -5$$
$$r^2 = 25$$
$$x^2 + y^2 = 25$$

34.

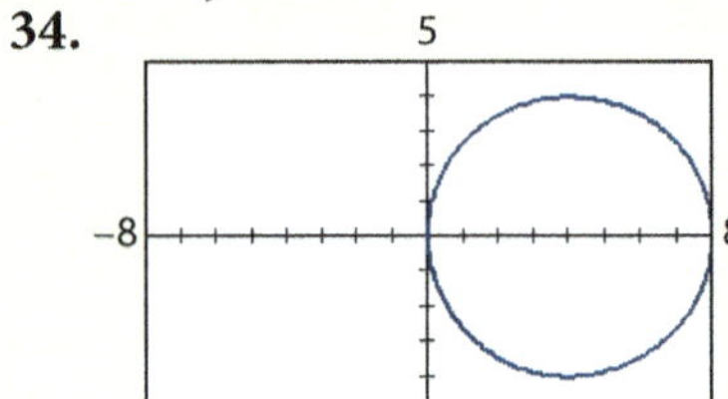

circle centered at $(4, 0)$ with $r = 4$
$$r(r) = r(8 \cos \theta)$$
$$r^2 = 8r \cos \theta$$
$$x^2 + y^2 = 8x$$
$$(x^2 - 8x + 16) + y^2 = 16$$
$$(x - 4)^2 + y^2 = 16$$

35.

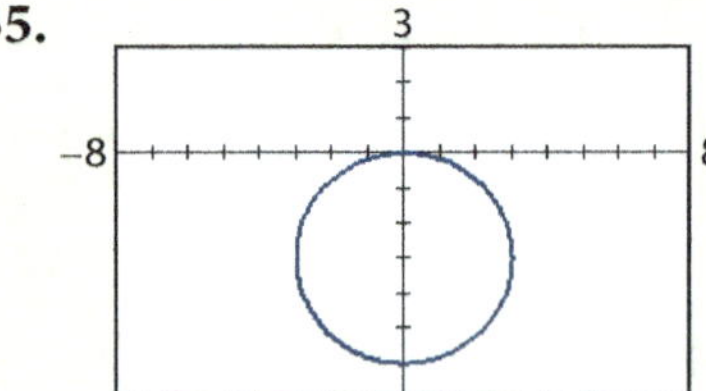

circle centered at $(0, -3)$ with $r = 3$
$$r(r) = r(-6 \sin \theta)$$
$$r^2 = -6(r \sin \theta)$$
$$x^2 + y^2 = -6y$$
$$x^2 + (y^2 + 6y + 9) = 9$$
$$x^2 + (y + 3)^2 = 9$$

continued in Answers and Solutions Overflow

Graphs of Polar Equations

This peacock feather contains shapes that can be modeled with polar equations.

When you first learned to graph equations in the Cartesian coordinate system, you used a table of values to plot a sufficient number of points to illustrate the shape of the graph. You then used general characteristics of each type of graph to accurately sketch the graphs. By converting simple polar equations to rectangular form in the previous section, you saw that the graph of $r = k$ is a circle centered at the origin with radius k and that $\theta = k$ is a line making an angle of θ with the polar axis. The graphs of other polar equations can be found by making a table of values where r is dependent on θ. Each point is plotted as a directed distance r that varies while θ rotates around the pole.

Example 1   Graphing a Spiral by Plotting Points

Graph $r = \theta$ where $\theta \geq 0$.

Answer

1. In order to plot a real number r, θ is expressed in radians rather than degrees. Round values for r to the nearest tenth.

θ	0	$\frac{\pi}{6}$	$\frac{\pi}{3}$	$\frac{\pi}{2}$	$\frac{2\pi}{3}$	$\frac{5\pi}{6}$	π
r	0	0.5	1.0	1.6	2.1	2.6	3.1

θ	$\frac{7\pi}{6}$	$\frac{4\pi}{3}$	$\frac{3\pi}{2}$	$\frac{5\pi}{3}$	$\frac{11\pi}{6}$	2π
r	3.7	4.2	4.7	5.2	5.8	6.3

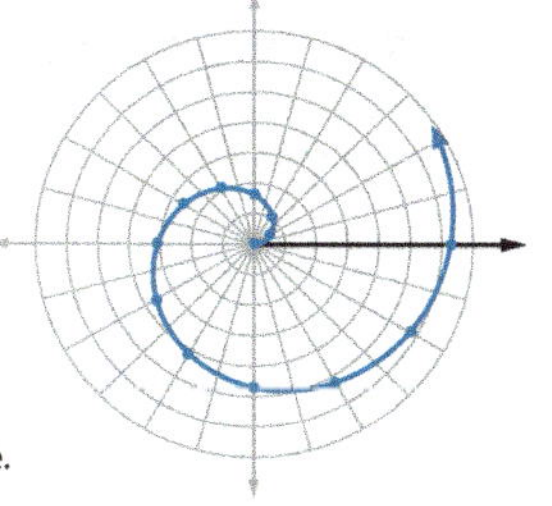

2. Plot the points and connect them with a smooth curve. This type of curve is known as an *Archimedean spiral*.

SKILL ✔ EXERCISE 3

When $\theta \leq 0$, the graph of $r = \theta$ is a reflection of the graph in Example 1 across the line $\theta = \frac{\pi}{2}$.

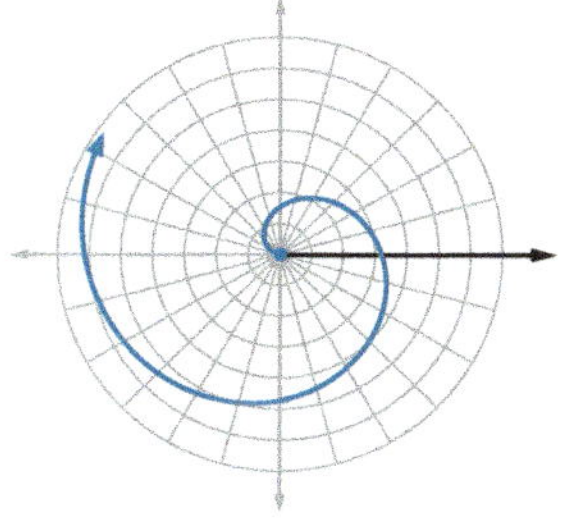

After completing this section, you will be able to

- graph polar equations by plotting points.
- use symmetry, zeros, and maximum values to sketch polar graphs.
- graph and analyze common polar equations.

The study of Archimedes's treatise *On Spirals* during the seventeenth and eighteenth centuries encouraged the development of polar coordinate systems.

6.5 Graphs of Polar Equations

Objectives

1. To graph polar equations by plotting points
2. To use symmetry, zeros, and maximum values to sketch polar graphs
3. To graph and analyze polar equations of circles, lines, spirals, roses, limaçons, and lemniscates

Flash

The bright fluorescent colors of a peacock are from crystal-like structures on their feathers. These crystals reflect varying frequencies of light, providing a visual display of brilliant colors and proclaiming astounding evidence of a great Creator.

Vocabulary

Archimedean spiral
cardioid
lemniscate
limaçon
rose

PRESENTATION

Lesson Opener

State an exact radian measure for each angle.

1. $120°$ $\frac{2\pi}{3}$

2. $150°$ $\frac{5\pi}{6}$

3. $315°$ $\frac{7\pi}{4}$ or $-\frac{\pi}{4}$

Convert the polar coordinates to rectangular coordinates.

4. $\left(5, \frac{3\pi}{2}\right)$ $(0, -5)$

5. $\left(6, \frac{7\pi}{6}\right)$ $(-3\sqrt{3}, -3)$

Explain that polar equations will initially be graphed by plotting points and connecting the points with a smooth curve—just as lines, parabolas, and other graphs were first drawn using rectangular coordinates. This process is demonstrated in Example 1. Note that rounding r to the nearest tenth makes plotting the points reasonable.

Motivational Idea The *Archimedean spiral* is the set of points corresponding to the locations of a point moving away from a fixed point at a constant speed along a line rotating at a constant angular velocity. Use the Internet keyword search *Archimedean spiral animation* to find demonstrations.

In Example 2 the graph of a circle is determined by plotting points from 0 to 2π. Graphing the circle on the calculator with the path icon, ⊕ (selected as explained in the Technology Corner), clearly demonstrates that the points on the circle are plotted twice.

Explain that knowing the symmetry of a graph helps to limit the number of points needed to draw the graph. Present the tests for symmetry in polar graphs with the associated figures. Passing the test for a type of symmetry with either point indicates that the graph exhibits

Reading and Writing Mathematics

Use the Internet keyword search *microphone pickup patterns*. Give the name and a brief description of the application or the characteristics of these different patterns, including the omnidirectional (circular), cardioid, and hypercardioid microphone polar patterns.

sample answer: The omnidirectional pattern picks up sound from all directions in a given radius, making it an ideal choice in events such as weddings. The cardioid is a heart-shaped pickup pattern. Pickup is mostly from the front of the microphone and is common with most handheld microphones. The hypercardioid also has a heart-shaped pickup pattern but significantly diminishes unwanted noise from the sides and rear.

Additional Exercises

List the values of $\theta \in [0, 2\pi)$ for which $r = 0$ and those for which $|r|$ has a maximum value. Then find the maximum length for r.

1. $r = 3 \cos 2\theta$ $\quad \frac{\pi}{4}, \frac{3\pi}{4}, \frac{5\pi}{4}, \frac{7\pi}{4}; 0, \frac{\pi}{2}, \pi, \frac{3\pi}{2}; 3$

2. $r = 4 \sin 2\theta$ $\quad 0, \frac{\pi}{2}, \pi, \frac{3\pi}{2}; \frac{\pi}{4}, \frac{3\pi}{4}, \frac{5\pi}{4}, \frac{7\pi}{4}; 4$

3. $r = -5 \cos 3\theta$ $\quad \frac{\pi}{6}, \frac{\pi}{2}, \frac{5\pi}{6}, \frac{7\pi}{6}, \frac{3\pi}{2}, \frac{11\pi}{6}; 0, \frac{\pi}{3},$ $\frac{2\pi}{3}, \pi, \frac{4\pi}{3}, \frac{5\pi}{3}; 5$

4. $r^2 = 4 \sin 2\theta$ $\quad 0, \frac{\pi}{2}, \pi, \frac{3\pi}{2}; \frac{\pi}{4}, \frac{5\pi}{4}; 2$

Example 2 Graphing a Circle by Plotting Points

Graph $r = 4 \cos \theta$.

Answer

1. Since $\cos \theta$ has a period of 2π, the entire graph can be drawn using $\theta \in [0, 2\pi]$.

θ	0	$\frac{\pi}{6}$	$\frac{\pi}{3}$	$\frac{\pi}{2}$	$\frac{2\pi}{3}$	$\frac{5\pi}{6}$	π
r	4	3.5	2	0	−2	−3.5	−4

θ	$\frac{7\pi}{6}$	$\frac{4\pi}{3}$	$\frac{3\pi}{2}$	$\frac{5\pi}{3}$	$\frac{11\pi}{6}$	2π
r	−3.5	−2	0	2	3.5	4

2. Plot the points and connect them with a smooth curve. Notice that the circle is actually graphed twice.

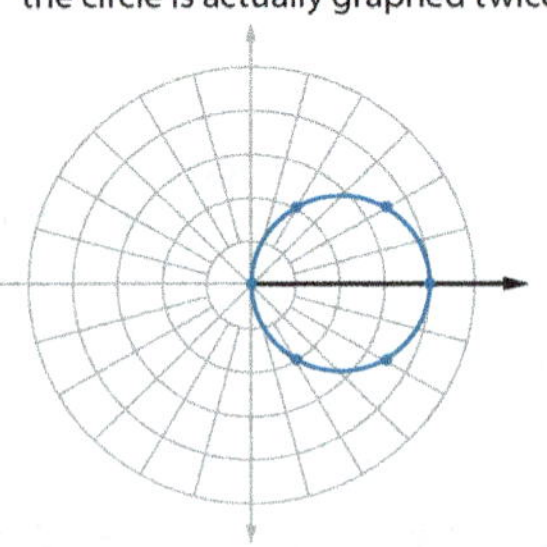

SKILL ✓ EXERCISE 5

Knowing whether the graph is symmetric with respect to the polar axis, the line $\theta = \frac{\pi}{2}$, or the pole reduces the number of points needed to determine the overall shape of the graph.

TESTS FOR SYMMETRY IN POLAR GRAPHS

The graph of a polar equation has the described symmetry if either substitution for (r, θ) produces an equivalent equation.

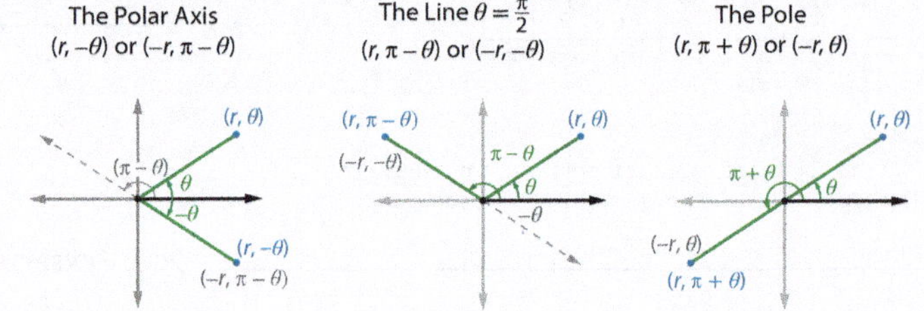

To show that the circle graphed in Example 2 is symmetric with respect to the polar axis, replace (r, θ) in $r = 4 \cos \theta$ with $(r, -\theta)$ and simplify using the fact that cosine is an even function:

$$r = 4 \cos(-\theta) = 4 \cos \theta.$$

The circle graphed in Section 6.4 Example 8 is shown to be symmetric with the line $\theta = \frac{\pi}{2}$ by replacing (r, θ) in $r = 4 \sin \theta$ with $(-r, -\theta)$ and using the fact that sine is an odd function:

$$-r = 4 \sin(-\theta) = 4(-\sin \theta) = -4 \sin \theta; \text{ so } r = 4 \sin \theta.$$

---

that type. Functions that fail a test with both points may still exhibit that type of symmetry, despite the fact that it is not revealed by the tests.

Example 3 applies one of the quick tests to limit the plotted points. Consider sharing that *limaçon* is a French word meaning "cochlea." You may want to demonstrate the more thorough tests for symmetry with respect to $\theta = \frac{\pi}{2}$.

$$3 + 2 \sin(-\theta) = 3 - 2 \sin(\theta) \neq -r,$$

but

$$3 + 2 \sin(\pi - \theta)$$
$$= 3 + 2(\sin \pi \cos \theta - \sin \theta \cos \pi)$$
$$= 3 + 2(0 - \sin \theta(-1))$$
$$= 3 + 2 \sin \theta = r$$

Example 4 demonstrates how knowing the zeros and the maximum values of $|r|$ can help determine key points. Explain that more precise graphs can be obtained by selecting additional points that are multiples of $\frac{\pi}{12}$.

Consider asking the students to demonstrate tests that indicate the graph in Example 4 exhibits the stated symmetries.

the line $\theta = \frac{\pi}{2}$; using $(r, \pi - \theta)$:

$$4 \cos 2(\pi - \theta)$$
$$= 4 \cos(2\pi - 2\theta)$$
$$= 4(\cos 2\pi \cos 2\theta + \sin 2\pi \sin 2\theta)$$
$$= 4((1) \cos 2\theta + (0) \sin 2\theta)$$
$$= 4 \cos 2\theta = r$$

the pole; using $(r, \theta + \pi)$:

$$4 \cos 2(\theta + \pi)$$
$$= 4 \cos(2\theta + 2\pi)$$
$$= 4(\cos 2\theta \cos 2\pi - \sin 2\theta \sin 2\pi)$$
$$= 4((1) \cos 2\theta + (0) \sin 2\theta)$$
$$= 4 \cos 2\theta = r$$

Recognizing general forms of common polar graphs can be very helpful. Encourage the students to become familiar with the most common equations and their graphs. Consider using the Internet keyword search *common polar graphs* for more images and examples of polar graphs.

Referring to the limaçons, point out the correlation between the axis on which

The odd/even identities can be used to verify the following quick tests for symmetry.

> **Quick Tests for Symmetry**
>
> The graph of $r = f(\cos \theta)$ is symmetric with respect to the polar axis.
>
> The graph of $r = f(\sin \theta)$ is symmetric with respect to the line $\theta = \frac{\pi}{2}$.

Example 3   Using Symmetry to Graph a Limaçon

Graph $r = 3 + 2 \sin \theta$.

Answer

1. Since $r = f(\sin \theta)$, the graph is symmetric with respect to $\theta = \frac{\pi}{2}$. The function's period is 2π, so make a table of values using $\theta \in \left[-\frac{\pi}{2}, \frac{\pi}{2}\right]$.

θ	$-\frac{\pi}{2}$	$-\frac{\pi}{3}$	$-\frac{\pi}{6}$	0	$\frac{\pi}{6}$	$\frac{\pi}{3}$	$\frac{\pi}{2}$
r	1	1.3	2	3	4	4.7	5

2. Plot the points over $\left[-\frac{\pi}{2}, \frac{\pi}{2}\right]$ and connect them with a smooth curve.

3. Reflect the curve across the line $\theta = \frac{\pi}{2}$ to complete the graph. The result is a type of limaçon (**lim**-uh-sohn). (There are four types in all.)

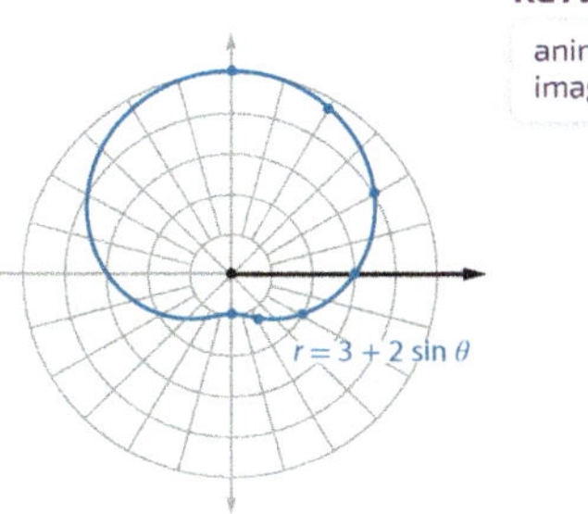

KEYWORD SEARCH

animated limaçon image

—————————————————— SKILL ✔ EXERCISE 13

It is helpful to identify values of θ for which $r = 0$ (where the graph intersects the pole). The fact that the sine and cosine functions range from -1 to 1 can help you identify points where the distance from the pole, $|r|$, is maximized.

Example 4   Using Maximum |r| to Graph a Rose

Graph $r = 4 \cos 2\theta$.

Answer

1. Since $r = f(\cos \theta)$, the graph is symmetric with respect to the polar axis.
2. zeros: when $2\theta = (2n + 1)\frac{\pi}{2}$ or $\theta = (2n + 1)\frac{\pi}{4}$
   maximum $|r|$: when $2\theta = n\pi$ or $\theta = n\frac{\pi}{2}$
3. Make a table of values for $\theta \in [0, \pi]$.

θ	0	$\frac{\pi}{6}$	$\frac{\pi}{4}$	$\frac{\pi}{3}$	$\frac{\pi}{2}$	$\frac{2\pi}{3}$	$\frac{3\pi}{4}$	$\frac{5\pi}{6}$	π
r	4	2	0	-2	-4	-2	0	2	4

4. Plot the points over $[0, \pi]$ and connect them with a smooth curve. Then reflect the curve across the polar axis. The resulting curve is called a rose and has 4 petals.

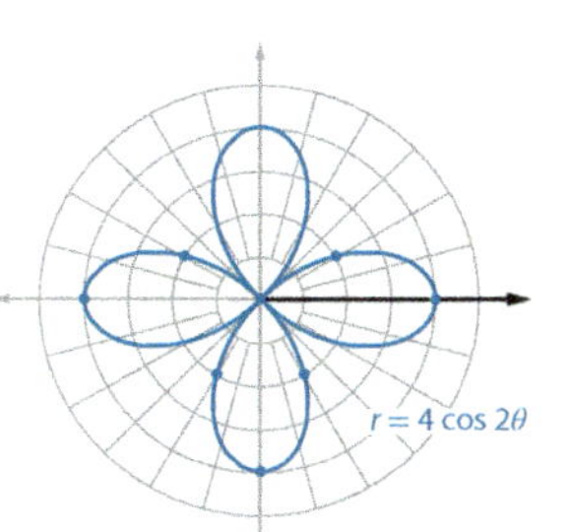

TIP 💡

Notice how the polar graph of $r = 4 \cos 2\theta$ compares to the rectangular coordinate graph of $y = 4 \cos 2x$.

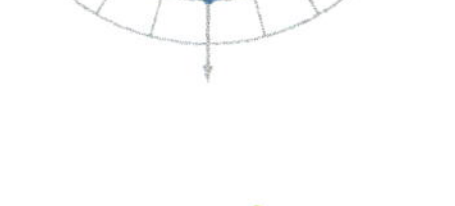

—————————————————— SKILL ✔ EXERCISE 15

List the values of $\theta \in [0, 2\pi)$ for which $r = 0$ and those for which $|r|$ has a maximum value.

5. $r = 2 + 4 \cos \theta$ $\frac{2\pi}{3}, \frac{4\pi}{3}; 0$

6. $r = 4 - 3 \sin \theta$ none; $\frac{3\pi}{2}$

Graph each equation.

7. $r = \sqrt{2}\theta$

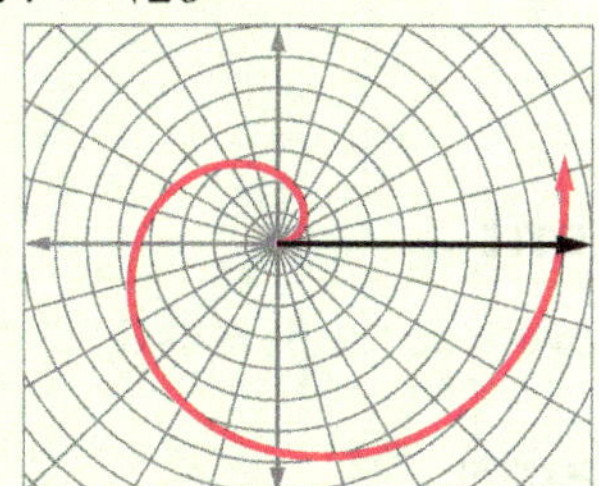

8. $r = -3 \cos \theta$

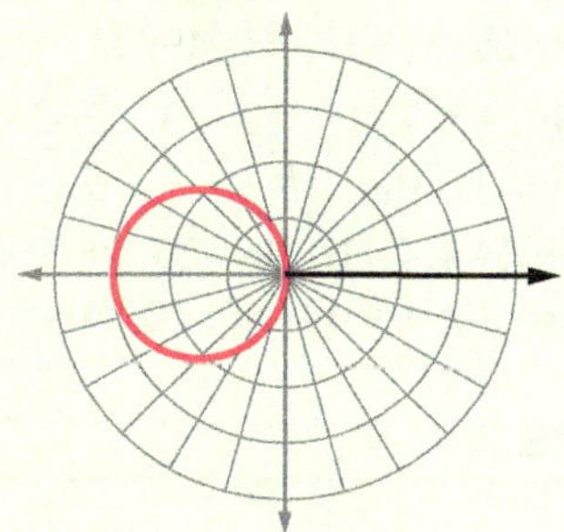

9. $r = 3 - 2 \cos \theta$

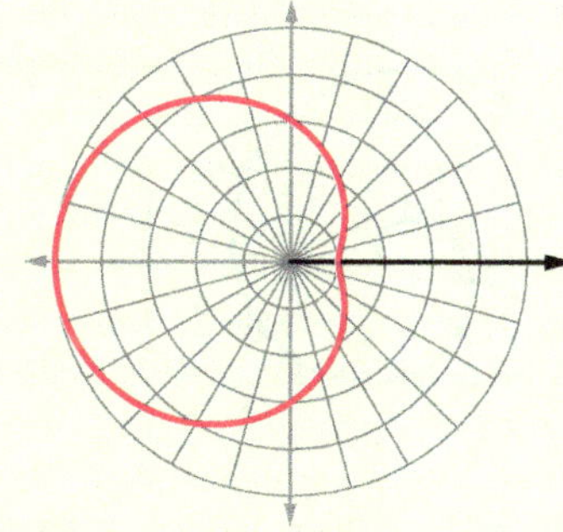

the maximum value $|r|$ occurs with the sign preceding the sine/cosine term. These correspond to sine/cosine circle graphs as well.

Point out how the first petal for $r = k \cos n\theta$ occurs on the polar axis when $k > 0$ and would occur at π if $k < 0$. Consider using graphing technology to explore similar generalizations for $r = k \sin n\theta$.

In Examples 5, 6, and 7, encourage the students to identify the type of graph by observing the general form of the equation before using tests of symmetry and identifying key points such as zeros and maximum values of $|r|$. Point out that the graph of the *rose* is traced twice for $\theta \in [0, 2\pi)$.

Interactive Activity Students will benefit from using technology to explore the graphs of polar equations. A graphing calculator can be used to illustrate the graphing of points as θ moves from 0 to 2π. Other technologies, such as Desmos or The Geometer's Sketchpad, allow the user to define sliders that enable rapid exploration of the effect of key values within a basic equation.

TIPS

Ex. 7d While a more thorough test for symmetry shows the graph to have the desired symmetry, the quick test does not apply since r is not a function of $\cos \theta$. The graph in Example 7 can be used to show that $r^2 = 16 \sin 2\theta$ does not pass the quick test for symmetry across the line $q = \frac{\pi}{2}$.

10. $r = 3 \sin 4\theta$

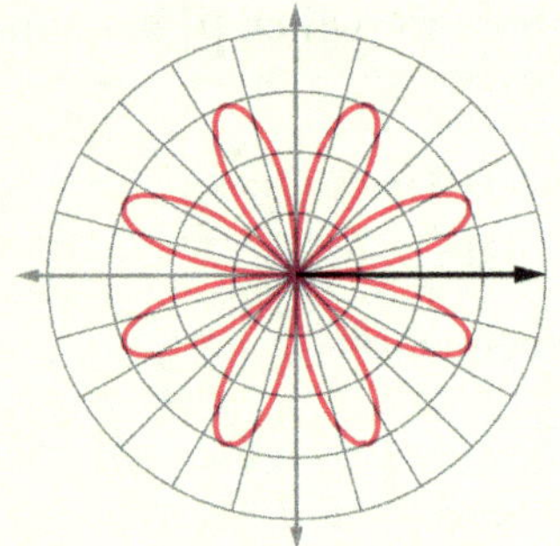

Assessment

- Quiz 6C covers Sections 6.4–6.5.

Assignments

- **Minimum:** 1–3, 5–9, 11–13, 15–18, 22–23, 25, 28, 31–32, 43–49, 51
- **Standard:** 1–5, 7–19, 22–23, 25–26, 28, 31, 33–34, 36, 38(*b*–*c*), 42, 44–51
- **Extended:** 1–4, 7–16, 19–25, 28; 29–35 odd; 36–38, 40, 44–46, 48, 50–51

Solutions

❯ A. Exercises

3.

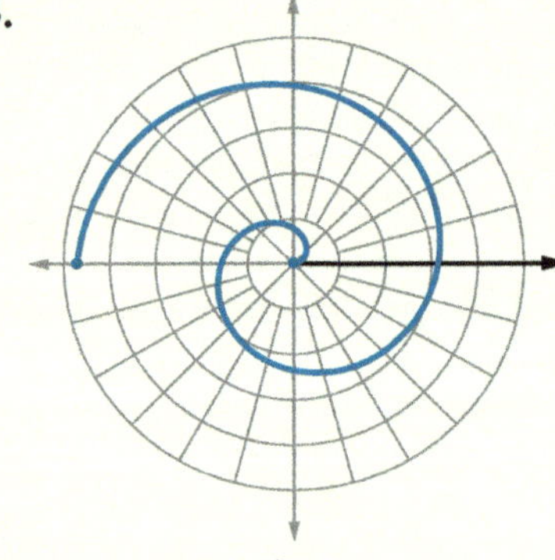

4.

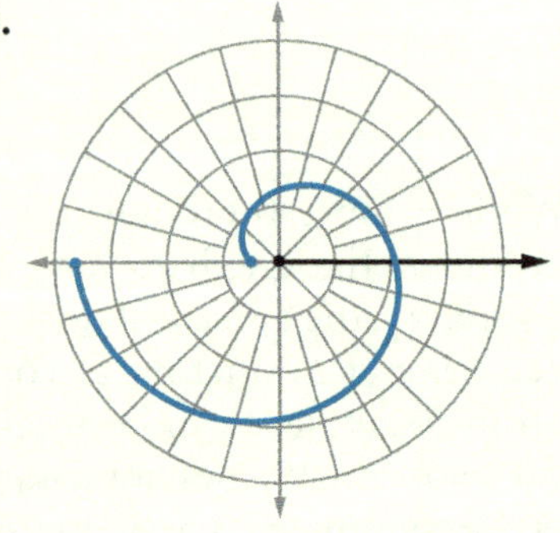

After confirming that the graph in Example 4 is also symmetric with respect to the line $\theta = \frac{\pi}{2}$ and the pole, we could have used $\theta \in \left[0, \frac{\pi}{4}\right]$ and symmetry to draw the rest of the graph.

Although you can always graph polar equations by plotting points, recognizing the general form of the equation for common polar graphs is extremely helpful.

Lines		Circles		
$\theta = k$	$r = k$	$r = k \cos \theta$	$r = k \sin \theta$	

Limaçons
$r = a \pm b \cos \theta$ or $r = a \pm b \sin \theta$ where $a > 0$ and $b > 0$

With Inner Loop $a < b$	Cardioid $a = b$	Dimpled $b < a < 2b$	Convex $2b \leq a$
$r = a + b \sin \theta$	$r = a - b \sin \theta$	$r = a - b \cos \theta$	$r = a + b \cos \theta$

Roses
$r = k \cos n\theta$ or $r = k \sin n\theta$ where $n \in \mathbb{N}$

2*n* Petals if *n* is Even		*n* Petals if *n* is Odd	
$r = k \cos n\theta$	$r = k \sin n\theta$	$r = k \cos n\theta$	$r = k \sin n\theta$

Lemniscates		Archimedean Spirals	
$r^2 = k^2 \cos 2\theta$	$r^2 = k^2 \sin 2\theta$	$r = a\theta + b; \theta \geq 0$	$r = a\theta + b; \theta \leq 0$

KEYWORD SEARCH

animated lemniscates

5.

6.

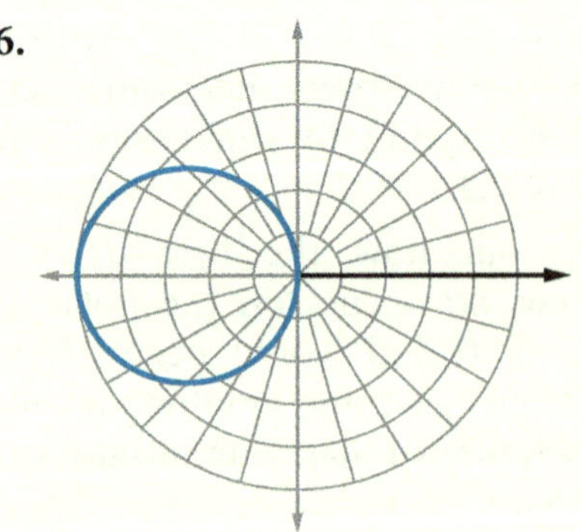

7d. See Tip on p. 317.

9. $\cos \theta = -1$ when $\theta = \pi$

$\cos \theta = 1$ when $\theta = 0, 2\pi$

10. $\sin \theta = 0$ when $\theta = 0, \pi, 2\pi$

$\sin \theta = \pm 1$ when $\theta = \frac{\pi}{2}, \frac{3\pi}{2}$

Example 5   Graphing a Cardioid

Sketch the graph of $r = 2 - 2\cos\theta$.

Answer

1. The general form $r = a - b\cos\theta$ where $a = b$ indicates the graph is a cardioid that is symmetric with respect to the polar axis.
2. Zeros occur when $\cos\theta = 1$ (when $\theta = 0, 2\pi$).
3. Maximum $|r|$ values occur when $\cos\theta = -1$ (when $\theta = \pi$).
4. Make a table of several key points.

θ	0	$\frac{\pi}{4}$	$\frac{\pi}{2}$	$\frac{3\pi}{4}$	π
r	0	0.6	2	3.4	4

5. Plot the points and use symmetry to complete the sketch of the curve.

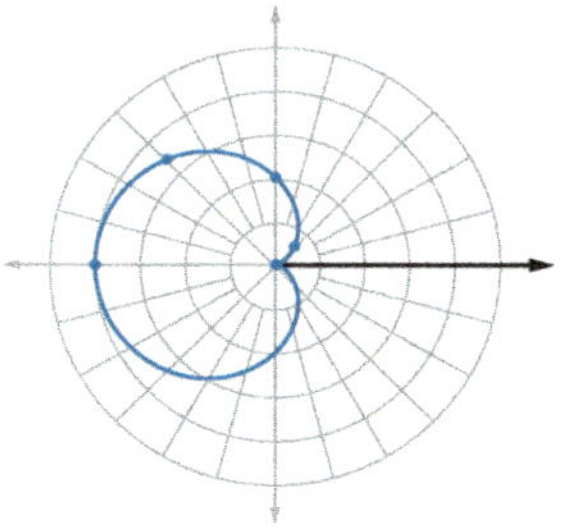

SKILL ✔ EXERCISE 23

Example 4 illustrates that the graph of $r = k\cos n\theta$ or $r = k\sin n\theta$ is a rose with $2n$ petals when n is even. When n is odd the graph contains just n petals, each being traced twice over $[0, 2\pi]$.

Example 6   Graphing a Rose

Sketch the graph of $r = 3\sin 3\theta$.

Answer

1. The general form $r = k\sin n\theta$ where n is odd indicates the graph is a rose with three petals (traced twice from 0 to 2π) that is symmetric with respect to the line $\theta = \frac{\pi}{2}$.
2. Zeros occur when $3\theta = n\pi$ (when $\theta = 0, \frac{\pi}{3}, \frac{2\pi}{3}, \pi, \ldots$).
3. Maximum $|r|$ values occur at $3\theta = (2n+1)\frac{\pi}{2}$ (when $\theta = \frac{\pi}{6}, \frac{\pi}{2}, \frac{5\pi}{6}, \ldots$).
4. Make a table of several key points.

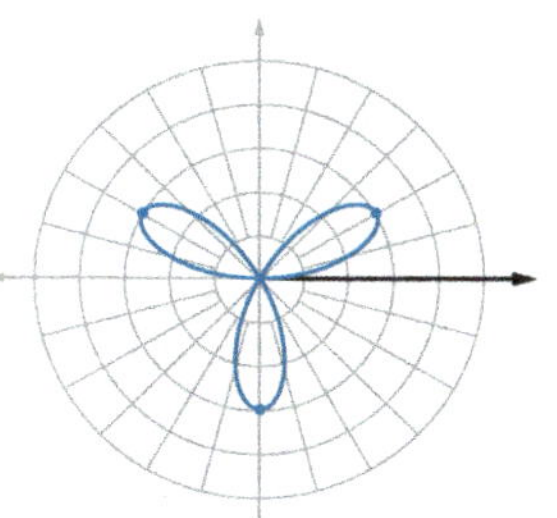

θ	0	$\frac{\pi}{6}$	$\frac{\pi}{3}$	$\frac{\pi}{2}$	$\frac{2\pi}{3}$	$\frac{5\pi}{6}$	π
r	0	3	0	-3	0	3	0

5. Plot the points and sketch the curve. The entire curve is plotted over $[0, \pi]$ and is traced again over $[\pi, 2\pi]$.

SKILL ✔ EXERCISE 25

When $k > 0$, the first maximum $|r|$ value for $r = k\cos n\pi$ occurs on the polar axis (at $\theta = 0$), and the first maximum $|r|$ value for $r = k\sin n\theta$ occurs at $\theta = \frac{\pi}{2} \div n$.

The key characteristic of the equation for a lemniscate (**lem**-nis-keyt) is the squared radius. Taking the square root of both sides reveals that its domain is restricted.

11. $\sin 2\theta = 0$ when
$$2\theta = 0, \pi, 2\pi, 3\pi, 4\pi, \ldots$$
or $\theta = 0, \frac{\pi}{2}, \pi, \frac{3\pi}{2}, 2\pi$
$$\sin 2\theta = \pm 1 \text{ when}$$
$$2\theta = \frac{\pi}{2}, \frac{3\pi}{2}, \frac{5\pi}{2}, \frac{7\pi}{2}, \ldots$$
or $\theta = \frac{\pi}{4}, \frac{3\pi}{4}, \frac{5\pi}{4}, \frac{7\pi}{4}$

12. $r = \pm 5\sqrt{\cos 2\theta}$
$$\cos 2\theta = 0 \text{ when } 2\theta = (2n+1)\frac{\pi}{2}$$
or $\theta = (2n+1)\frac{\pi}{4} = \frac{\pi}{4}, \frac{3\pi}{4}, \frac{5\pi}{4}, \frac{7\pi}{4}$
$$\cos 2\theta = 1 \text{ when } 2\theta = n 2\pi$$
or $\theta = n\pi = 0, \pi, 2\pi$

13.

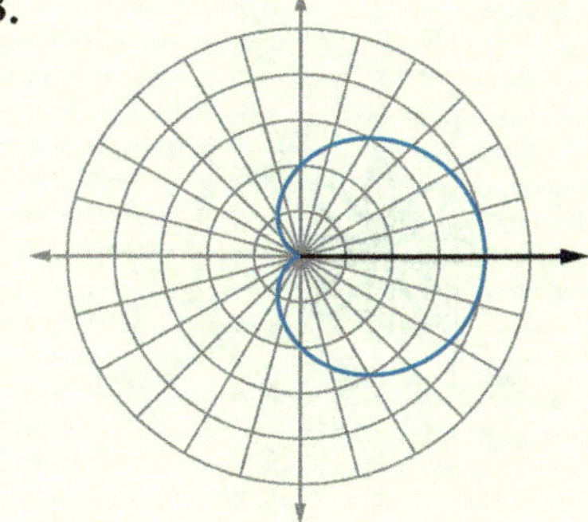

Microphone reception patterns are typically represented by a polar graph, often with overlays of different frequency response curves. The cardioid is a typical reception pattern for microphones designed for singers or announcers.

14.

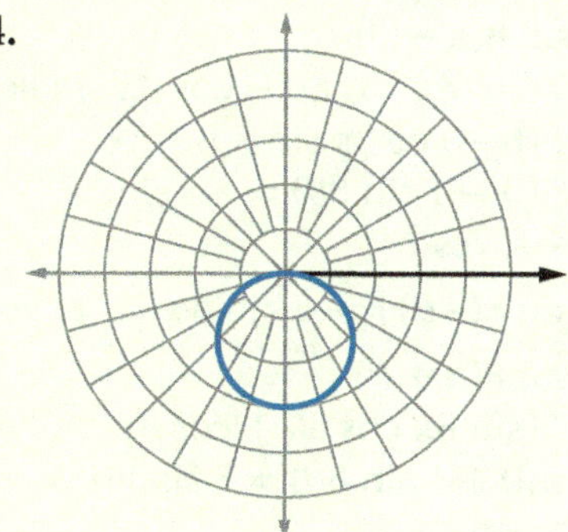

15.

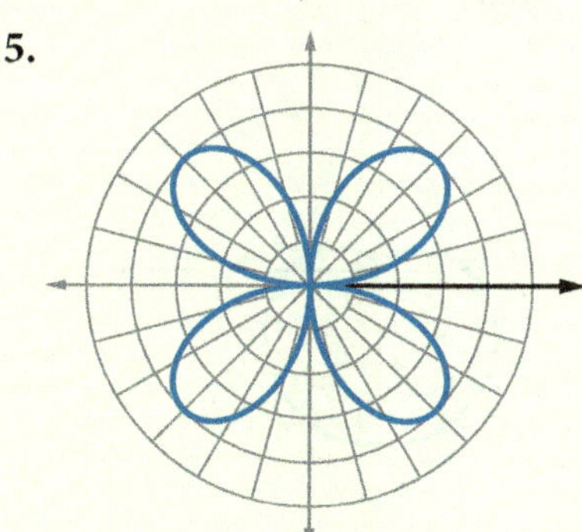

16.

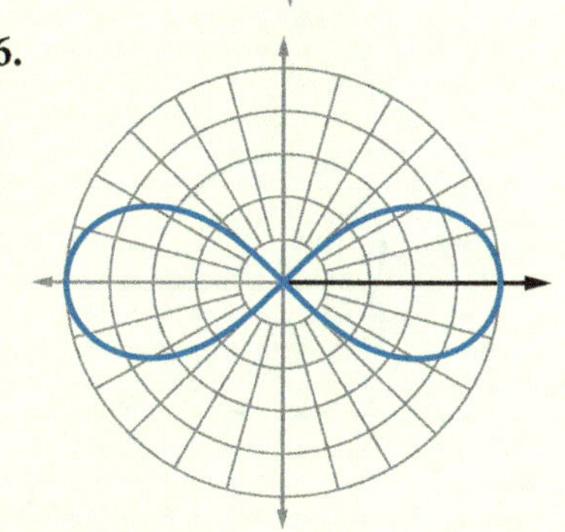

❯ B. Exercises

17. $n = 6$ (even); $2n = 12$ petals
max $|r|$ values at $\cos\theta = 1$ ($\theta = 0$)
$|r| = |2\cos 6(0)| = |2(1)| = 2$

18. $n = 7$ (odd); $n = 7$ petals
max $|r|$ values at $\cos\theta = 1$ ($\theta = 0$)
$|r| = |-6\cos 7(0)| = |-6(1)| = 6$

19a. $1 + \cos(-\theta) = 1 + \cos\theta = r$; yes

19b. $1 + \cos(\pi - \theta) = 1 - \cos\theta \neq -r$
$1 + \cos(-\theta) \neq -r$; no

19c. $1 + \cos(\pi + \theta) = 1 - \cos\theta \neq r$
$1 + \cos\theta \neq -r$; no

20a. $3 + 2\sin(-\theta) = 3 - 2\sin\theta \neq r$
$3 + 2\sin(\pi - \theta) = 3 + 2\sin\theta$
$\neq -r$; no

20b. $3 + 2\sin(\pi - \theta)$
$= 3 + 2(\sin\pi\cos\theta - \cos\pi\sin\theta)$
$= 3 + 2(0 - (-1)\sin\theta)$
$= 3 + 2\sin\theta = r$; yes

20c. $3 + 2\sin(\pi + \theta) = 3 - 2\sin\theta \neq r$
$3 + 2\sin\theta \neq -r$; no

21a. $5 \sin 8(\pi - \theta)$
$= 5(\sin 8\pi \cos 8\theta - \cos 8\pi \sin 8\theta)$
$= 5(0 - \cos 8\pi \sin 8\theta)$
$= 5(0 - 1 \sin 8\theta) = -5 \sin 8\theta$
$= -r;$ yes

21b. $5 \sin 8(-\theta) = -5 \sin 8\theta = -r;$ yes

21c. $5 \sin 8(\pi + \theta)$
$= 5(\sin 8\pi \cos 8\theta + \cos 8\pi \sin 8\theta)$
$= 5(0 + 1 \sin 8\theta) = 5 \sin 8\theta$
$= r;$ yes

22.

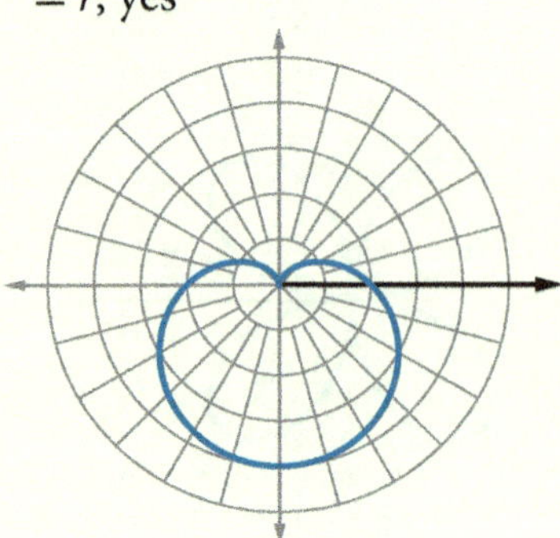

23.

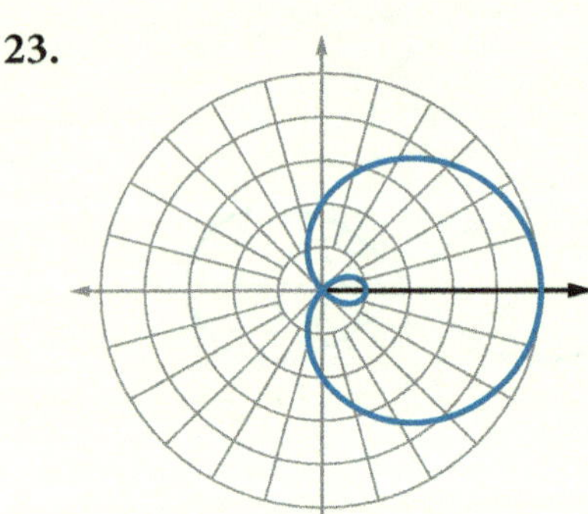

24.

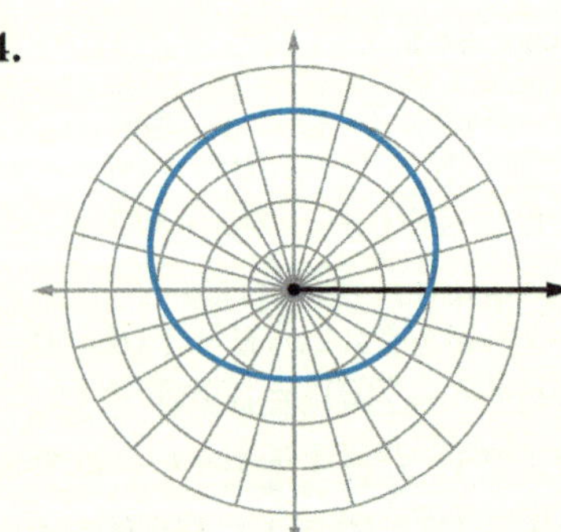

25.

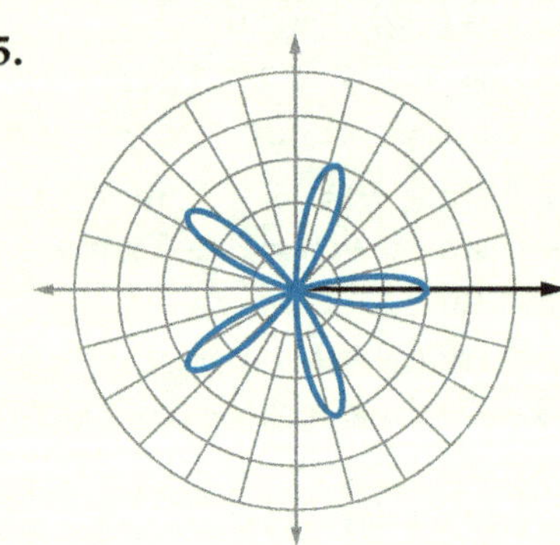

Example 7   Graphing a Lemniscate

Sketch the polar graph of $r^2 = 16 \sin 2\theta$.

Answer

1. The general form indicates the graph is a lemniscate that is symmetric with respect to the pole and the line $\theta = \frac{\pi}{4}$.

Note that $r = \pm 4\sqrt{\sin 2\theta}$ is not defined in $\left(\frac{\pi}{2}, \pi\right)$ or $\left(\frac{3\pi}{2}, 2\pi\right)$.

2. Zeros occur when $2\theta = n\pi$ (when $\theta = 0, \frac{\pi}{2}, \pi, \ldots$).

3. Maximum $|r|$ values for $r = \pm 4\sqrt{\sin 2\theta}$ occur when $\theta = (2n + 1)\frac{\pi}{2}$ is defined (when $\theta = \frac{\pi}{4}, \frac{5\pi}{4}, \ldots$).

4. Make a table of several key points.

θ	0	$\frac{\pi}{4}$	$\frac{\pi}{2}$
r	0	± 4	0

5. Plot the points and sketch the curve. Using positive and negative values of r causes the entire graph to be drawn with $\theta \in \left[0, \frac{\pi}{2}\right]$.

It is then retraced using $\theta \in \left[\pi, \frac{3\pi}{2}\right]$.

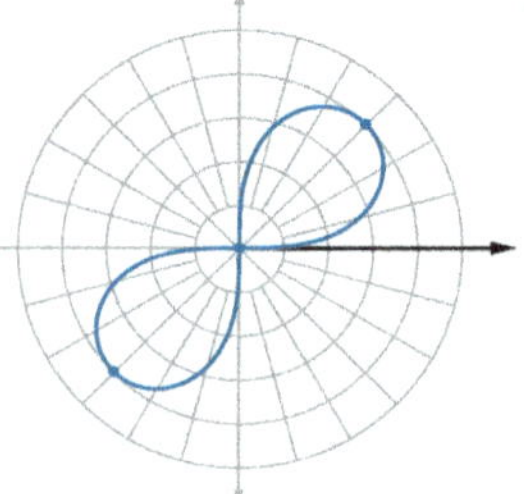

> **TIP**
>
> Notice how the polar graph of $r = 4\sqrt{\sin 2\theta}$ compares to the rectangular coordinate graph of $y = 4\sqrt{\sin 2x}$.

SKILL ✓ EXERCISE 27

❯ **A. Exercises**

1. Match each polar equation to the type of curve.
   a. $r = 3 \cos \theta$ II
   b. $r = \sin 3\theta$ IV
   c. $r = 3\theta$ I
   d. $r^2 = 3 \cos 2\theta$ III

   I. Archimedean spiral
   II. circle
   III. lemniscate
   IV. rose

2. Identify the type of limaçon represented by each polar equation.
   a. $r = 3 - 3 \cos \theta$ I
   b. $r = 3 + 2 \sin \theta$ III
   c. $r = 3 + 4 \sin \theta$ IV
   d. $r = 3 - \cos \theta$ II

   I. cardioid
   II. convex limaçon
   III. dimpled limaçon
   IV. limaçon with inner loop

Sketch the graph of each polar equation by plotting points. Then classify the curve.

3. $r = \frac{1}{2}\theta$ where $\theta \in [0, 3\pi]$ Archimedean spiral
4. $r = \theta - 1$ where $\theta \in [-2\pi, 0]$ Archimedean spiral
5. $r = 4 \sin \theta$ circle
6. $r = -5 \cos \theta$ circle

State whether a quick test for symmetry reveals the stated line of symmetry for each function.

7. with respect to the polar axis
   a. $r = 2 + 2 \cos \theta$ yes
   b. $r = -3 \sin \theta$ no
   c. $r = 4 \sin 2\theta$ no
   d. $r^2 = 25 \cos 2\theta$ no

8. with respect to the line $\theta = \frac{\pi}{2}$
   a. $r = 2 + 2 \cos \theta$ no
   b. $r = -3 \sin \theta$ yes
   c. $r = 4 \sin 2\theta$ yes
   d. $r^2 = 25 \cos 2\theta$ no

List the values of $\theta \in [0, 2\pi]$ for which $r = 0$ and those for which $|r|$ has a maximum value.

9. $r = 2 + 2 \cos \theta$ $\pi; 0, 2\pi$
10. $r = -3 \sin \theta$
11. $r = 4 \sin 2\theta$
12. $r^2 = 25 \cos 2\theta$

Classify each polar curve. Then use the results from exercises 7–12 and plot points as needed to graph each polar equation.

13. $r = 2 + 2 \cos \theta$ cardioid
14. $r = -3 \sin \theta$ circle
15. $r = 4 \sin 2\theta$ rose
16. $r^2 = 25 \cos 2\theta$ lemniscate

State the number of petals in each polar rose and state the maximum value of $|r|$.

17. $r = 2 \cos 6\theta$ 12 petals; 2
18. $r = -6 \cos 7\theta$ 7 petals; 6

26.

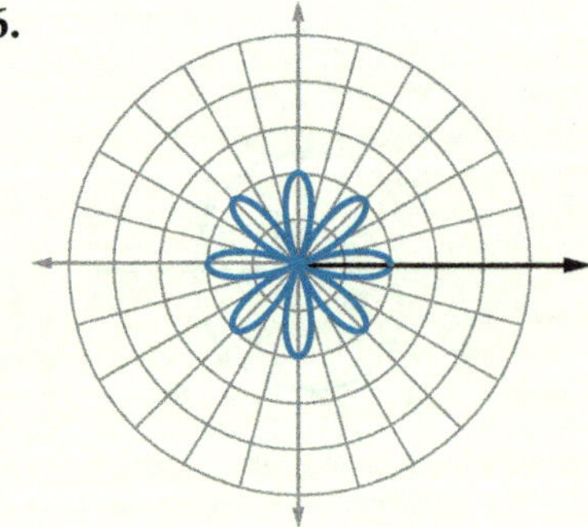

27.

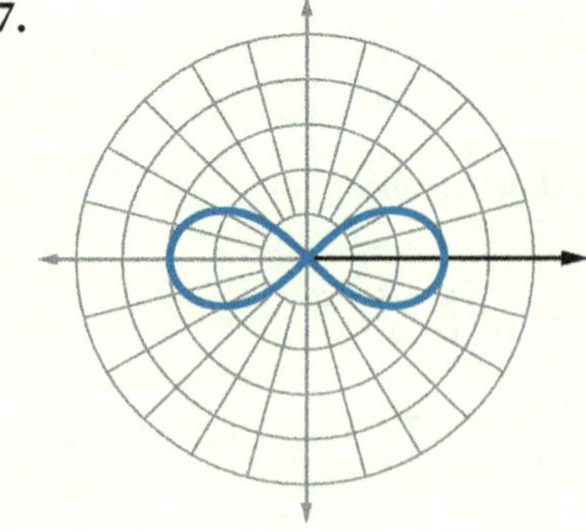

State whether a test reveals that the graph of the polar equation is symmetric with respect to each of the following.

  a. the polar axis   **b.** the line $\theta = \frac{\pi}{2}$ **c.** the pole

19. $r = 1 + \cos\theta$ yes; no; no

20. $r = 3 + 2\sin\theta$ no; yes; no

21. $r = 5\sin 8\theta$ yes; yes; yes

Classify each polar curve and then graph the equation.

22. $r = 4 - 4\sin\theta$ cardioid   **23.** $r = 2 + 3\cos\theta$ limaçon with inner loop

24. $r = -3 + \sin\theta$ **25.** $r = 3\cos 5\theta$ rose

26. $r = 2\cos 4\theta$ rose   **27.** $r^2 = 9\cos 2\theta$ lemniscate

28. $r^2 = -16\sin 2\theta$ **29.** $r = 3\sin 5\theta$ rose

30. $r = 2\cos 2\theta$ rose   **31.** $r = -2\sin 3\theta$ rose

32. $r = 3 - 3\cos\theta$ cardioid   **33.** $r = 3 + 2\sin\theta$ dimpled limaçon

Find the area enclosed by each polar curve.

34. $r = 7$ $49\pi \approx 153.9\ \text{u}^2$ **35.** $r = -5\sin\theta$
                                       $6.25\pi \approx 19.6\ \text{u}^2$

› C. Exercises

36. Explore: Use technology to graph $r = 3\cos\left(\frac{n}{2}\theta\right)$ for several odd values of $n \geq 3$. Explain how these graphs compare to previously graphed roses.

37. Explain: Why is it possible to classify circles of the form $r = \pm b\cos\theta$ or $r = \pm b\sin\theta$ as a rose?

38. Limaçons can be viewed as the sum (or difference) of two circles, $r_1 = \pm a$ and either $r_2 = b\sin\theta$ or $r_2 = b\cos\theta$. The relative magnitudes of a and b define the extent to which each function determines the shape of the limaçon. A similar effect occurs for the graphs of $r = a \pm b\sin n\theta$. Use technology to graph the following functions and explain how the values of a and b affect the petals of the related polar rose.

  a. $r = 2 + 3\cos 2\theta$ **b.** $r = 2 + 2\sin 2\theta$
  c. $r = 3 + 2\sin 2\theta$ **d.** $r = 3 + \cos 2\theta$

24. convex limaçon   **28.** lemniscate

36. There are $2n$ overlapping petals.

37. They have the form of $r = \pm k\cos n\theta$ or $r = \pm k\sin n\theta$, where $b = k$ and $n = 1$. This implies that they are 1-petal (circular) roses with $|r| = b$.

38a. The petals centered on $\theta = 0$ are stretched with $r = 2 + 3 = 5$, and the petals centered on $\theta = \frac{\pi}{2}$ are reduced so that $r = 2 - 3 = -1$.

38b. The petals centered on $\theta = \frac{\pi}{4}$ are stretched with $r = 2 + 2 = 4$ while the petals centered on $\theta = \frac{3\pi}{4}$ are gone since $r = 2 - 2 = 0$.

39. Microphone reception patterns are typically presented on a polar graph that has been rotated 90° where the microphone is at the origin, pointed at 0°. Different frequency response curves are often overlaid on the graph. Classify each polar graph and write an equation that models the microphone's response at each frequency.

  a. 500 Hz

  b. 1000 Hz

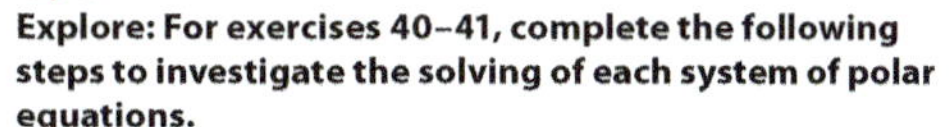
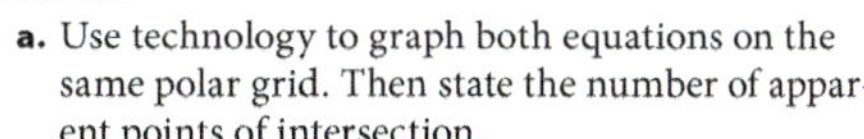

Explore: For exercises 40–41, complete the following steps to investigate the solving of each system of polar equations.

  a. Use technology to graph both equations on the same polar grid. Then state the number of apparent points of intersection.

  b. Use substitution to find the primary solutions of the system algebraically.

  c. Graph each equation in function mode, where $y = r$ and $x = \theta$, and list any primary solutions of the system.

  d. Return to polar mode and graph the equations simultaneously. Explain why some apparent points of intersection on the polar graphs are not solutions.

40. $r = 3\cos\theta$ **41.** $r = \sin\theta$
     $r = 1 + \cos\theta$ $r = 1 - \sin\theta$

38c. The petals centered on $\theta = \frac{\pi}{4}$ are stretched with $r = 3 + 2 = 5$ while "petals" centered on $\theta = \frac{3\pi}{4}$ are now simply dimples with $r = 3 - 2 = 1$.

38d. The petals centered on $\theta = 0$ are stretched with $r = 3 + 1 = 4$ while "petals" centered on $\theta = \frac{\pi}{2}$ are now slight indentations with $r = 3 - 1 = 2$.

39a. cardioid; $r = 12.5 + 12.5\cos\theta$

39b. dimpled limaçon; $r = 15 + 10\cos\theta$

40a. 3   **41a.** 3

40b. $\left(\frac{3}{2}, \frac{\pi}{3}\right), \left(\frac{3}{2}, \frac{5\pi}{3}\right)$ **41b.** $\left(\frac{1}{2}, \frac{\pi}{6}\right), \left(\frac{1}{2}, \frac{5\pi}{6}\right)$

40c. (1.047, 1.5)   **41c.** (0.524, 0.5)
     (5.236, 1.5)   (2.618, 0.5)

30.

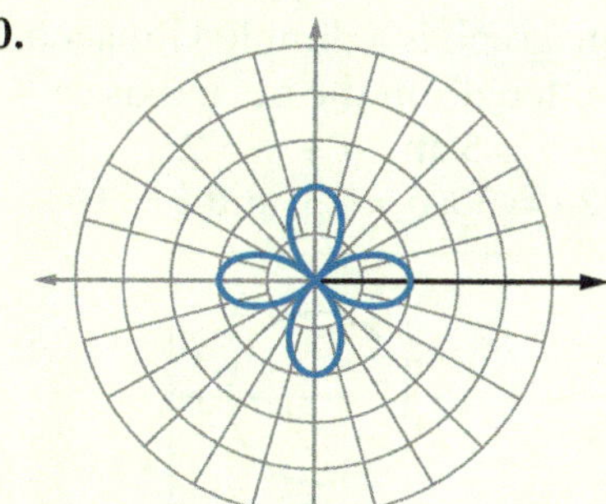

31.

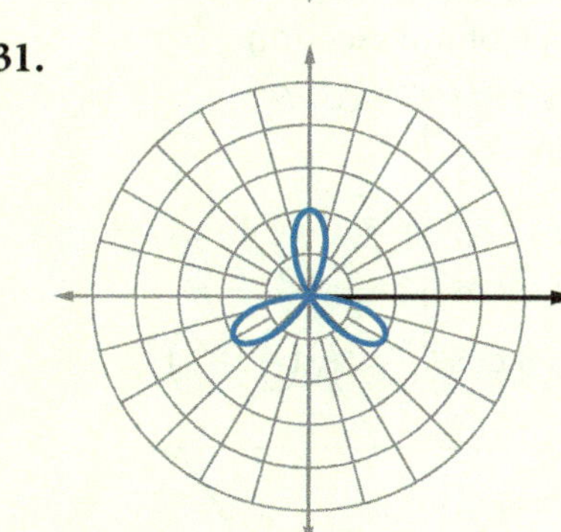

32.

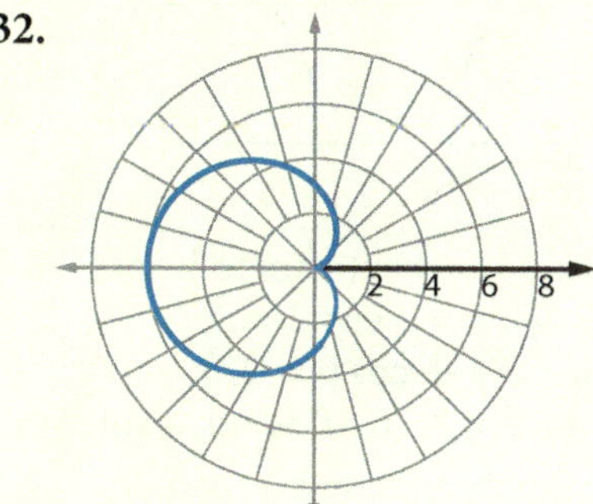

33.

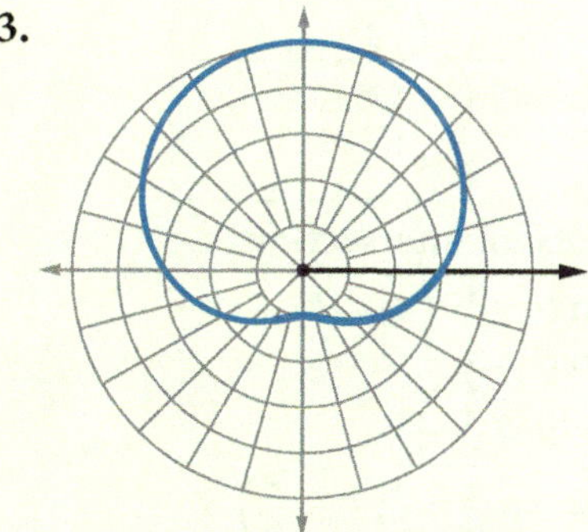

28.

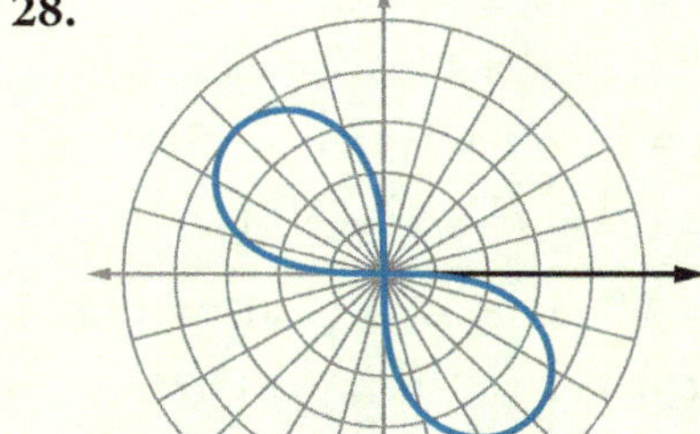

29.

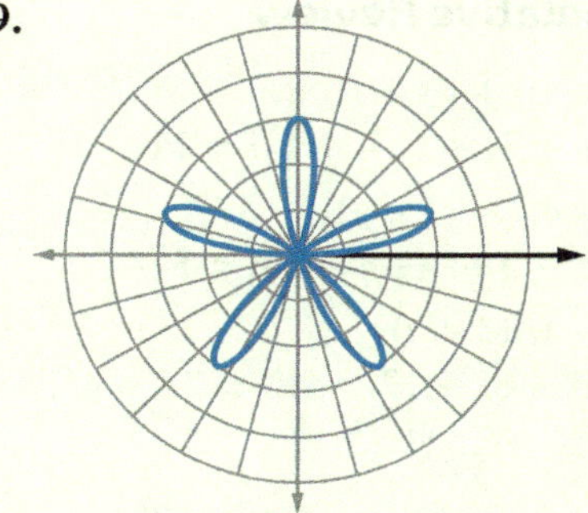

34. The circle is centered at the origin and has a radius of 7.
$A = (7)^2\pi \approx 153.9\ \text{u}^2$

35. The circle is centered at $\left(2.5, \frac{3\pi}{2}\right)$ and is tangent to the polar axis, so its radius is 2.5.
$A = (2.5)^2\pi \approx 19.6\ \text{u}^2$

› C. Exercises

39a. The graph is a cardioid "centered" on the polar axis so $a = b$ and $a + b = 25$.

39b. The graph is a dimpled limaçon "centered" on the polar axis so $a - b = 5$ and $a + b = 25$.
$\therefore 2a = 30$; $a = 15$; and $b = 10$

40a.

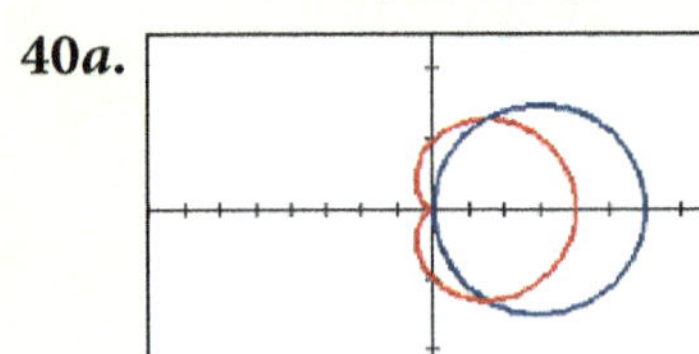

3 pts. of intersection

40b. $3 \cos \theta = 1 + \cos \theta$
$\cos \theta = \frac{1}{2}$
$\theta = \frac{\pi}{3}, \frac{5\pi}{3}$
$r = 3 \cos \frac{\pi}{3} = \frac{3}{2}; \left(\frac{3}{2}, \frac{\pi}{3}\right)$
$r = 3 \cos \frac{5\pi}{3} = \frac{3}{2}; \left(\frac{3}{2}, \frac{5\pi}{3}\right)$

40c.

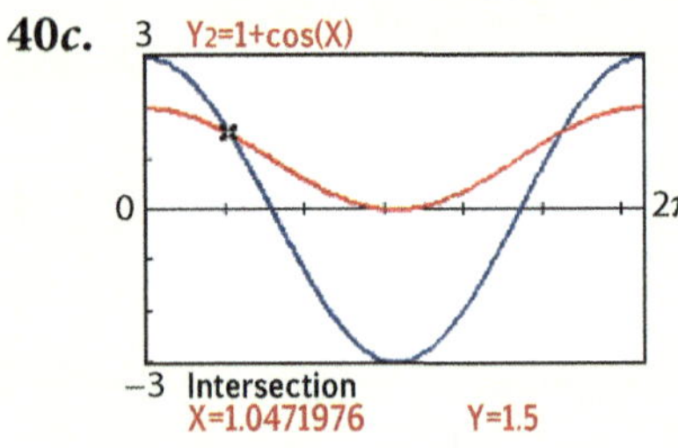

40d. While $(0, 0)$ appears to be an answer, $r = 3 \cos 0 = 3$, and $r = 1 + \cos 0 = 2$.
Since $2 \neq 3$, $(0, 0)$ is not a solution.

41a.

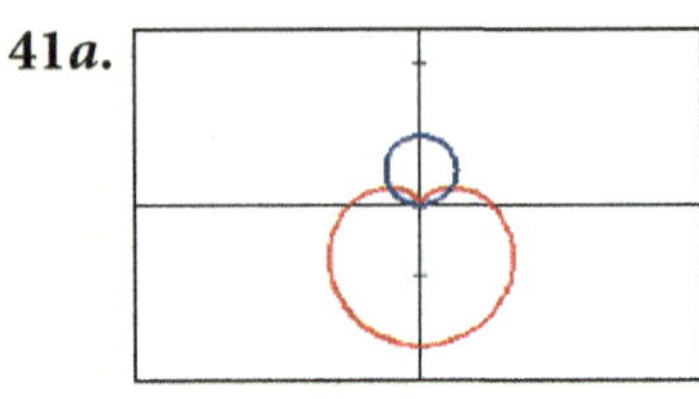

3 pts. of intersection

41b. $\sin \theta = 1 - \sin \theta$
$\sin \theta = \frac{1}{2}$
$\theta = \frac{\pi}{6}, \frac{5\pi}{6}$
$r = \sin \frac{\pi}{6} = \frac{1}{2}; \left(\frac{1}{2}, \frac{\pi}{6}\right)$
$r = \sin \frac{5\pi}{6} = \frac{1}{2}; \left(\frac{1}{2}, \frac{5\pi}{6}\right)$

41c.

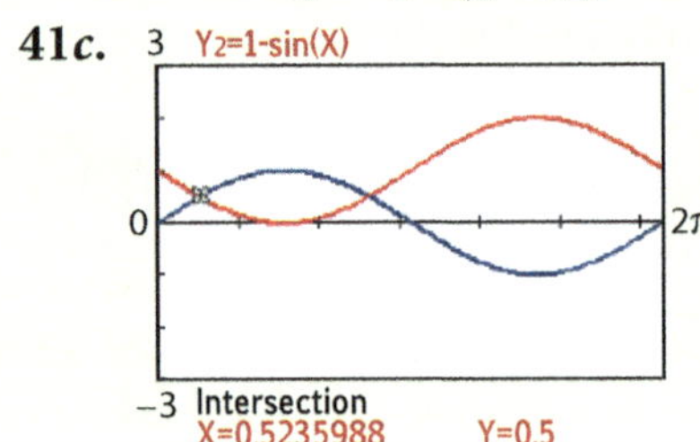

41d. While $(0, 0)$ appears to be an answer, $r = \sin 0 = 0$, and $r = 1 - \sin 0 = 1$.
Since $0 \neq 1$, $(0, 0)$ is not a solution.

CUMULATIVE REVIEW

Use $z_1 = 8 - 4i$ and $z_2 = 3 + i$ to evaluate each expression. [Appendix 5]

42. $z_1 + z_2$ $11 - 3i$

43. $z_1 - z_2$ $5 - 5i$

44. $z_1 \cdot z_2$ $28 - 4i$

45. $z_1 \div z_2$ $2 - 2i$

46. Factor over the set of complex numbers. [Appendix 5]
   a. $x^2 + 9$ $(x - 3i)(x + 3i)$
   b. $x^2 + ix + 12$ $(x - 3i)(x + 4i)$

47. Solve $x^2 + 4 = 0$. [Appendix 5] $x = \pm 2i$
   a. by taking roots
   b. by factoring

48. Simplify $(2 + 2i)^2$. [Appendix 5]   D
   A. 0   **C.** $-4i$ **E.** $-8i$
   B. $4i$ **D.** $8i$

49. Find the angle between $\mathbf{v} = \langle 2, 1, 3 \rangle$ and $\mathbf{w} = \langle -3, 4, -5 \rangle$ (to the nearest degree). [6.3]   B
   A. $50°$ **C.** $55°$ **E.** none of
   B. $130°$ **D.** $143°$ these

50. Convert the polar coordinates $\left(3\sqrt{2}, -\frac{\pi}{4}\right)$ to rectangular coordinates. [6.4]   B
   A. $(3, 3)$ **C.** $(-3, 3)$ **E.** none of
   B. $(3, -3)$ **D.** $(-3, -3)$ these

51. Convert the rectangular coordinates $(\sqrt{3}, 1)$ to polar coordinates. [6.4]   C
   A. $(1, 30°)$ **C.** $(2, 30°)$ **E.** none of
   B. $(1, 60°)$ **D.** $(2, 60°)$ these

❯ Cumulative Review

42. $(8 - 4i) + (3 + i)$
$= (8 + 3) + (-4 + 1)i = 11 - 3i$

43. $(8 - 4i) - (3 + i)$
$= (8 - 3) + (-4 - 1)i = 5 - 5i$

44. $(8 - 4i)(3 + i)$
$= 24 + 8i - 12i - 4i^2 = 28 - 4i$

45. $\frac{8 - 4i}{3 + i} = \frac{8 - 4i}{3 + i} \cdot \frac{3 - i}{3 - i}$
$= \frac{24 - 8i - 12i - 4}{9 - i^2} = \frac{10 - 10i}{5} = 2 - 2i$

47a. $x^2 = -4$
$x = \sqrt{-1 \cdot 4} = \pm 2i$

47b. $x^2 + 4 = 0$
$(x - 2i)(x + 2i) = 0; x = \pm 2i$

48. $(2 + 2i)^2 = 4 + 8i + 4i^2$
$= 4 + 8i - 4 = 8i$

49. $\mathbf{v} \cdot \mathbf{w} = 2(-3) + 1(4) + 3(-5) = -17$
$|\mathbf{v}| = \sqrt{2^2 + 1^2 + 3^2} = \sqrt{14}$
$|\mathbf{w}| = \sqrt{(-3)^2 + 4^2 + (-5)^2} = \sqrt{50}$
$\theta = \text{Cos}^{-1}\left(\frac{-17}{\sqrt{14} \cdot \sqrt{50}}\right) \approx 130°$

50. $x = 3\sqrt{2} \cos\left(-\frac{\pi}{4}\right) = 3\sqrt{2}\left(\frac{\sqrt{2}}{2}\right) = 3$
$y = 3\sqrt{2} \sin\left(-\frac{\pi}{4}\right)$
$= 3\sqrt{2}\left(-\frac{\sqrt{2}}{2}\right) = -3$

51. $r^2 = \left(\sqrt{3}\right)^2 + 1^2 = 4; r = 2$
$\text{Tan}^{-1} \frac{1}{\sqrt{3}} = 30°; \therefore (2, 30°)$

COMPLEX NUMBERS

Impossible. Amphibian. Absurd. Useless. All these words were once used to describe imaginary numbers—by the very people who discovered them! In the sixteenth century, aversion to negative numbers was still the norm even among mathematicians, so the square roots of such numbers were really baffling. Mathematicians understood that any quadratic equation could be written as $ax^2 = bx + c$ and solved by finding the intersections of the parabola $f(x) = ax^2$ and the line $g(x) = bx + c$. When solved algebraically, a negative radicand indicated that the parabola and the line had no points of intersection. However, in the cubic equation $ax^3 + bx^2 = cx + d$ there had to be at least one point of intersection between the graph of $f(x) = ax^3 + bx^2$ and the line $g(x) = cx + d$. The pursuit of a general solution for cubic equations proved the necessity and usefulness of complex numbers.

Early in the sixteenth century Scipione del Ferro discovered the general solution to the depressed cubic, an equation of the form $x^3 + px = q$. It is not hard to show that a depressed cubic where $p > 0$ will always have just one real root, and finding one root was sufficient at that time. Soon, Niccolò Fontana (known as Tartaglia) found the general solution to cubic equations of the form $x^3 + px^2 = q$, and not long after, Gerolamo Cardano developed the method that extended del Ferro's solution to all cubics.

When a cubic with three real roots is evaluated using Cardano's formula, the solution requires manipulation of complex numbers. While Cardano called these cubics "irreducible," Rafael Bombelli had the novel thought that the resulting complex expression actually represents one of the real roots and became the first to demonstrate that complex numbers are subject to the normal rules of mathematical operation. Bombelli's willingness to explore algebraic operations with no obvious geometric interpretations unleashed the power of imaginary numbers. Even so, the search for geometric meaning continued.

In 1799 the Norwegian surveyor Caspar Wessel was the first to use vectors in the complex plane to geometrically represent operations with complex numbers. Wessel's methods received little attention outside of Denmark until 1831 when the highly respected Carl Gauss published the same methods. At that point Wessel's geometric representation of complex numbers, which is still used today, became generally accepted. Extending complex vector operations to three dimensions proved challenging. In 1843 the Irish mathematician William Hamilton discovered quaternions, an extension of complex numbers. A simpler system involving scalars and vectors was later developed by Josiah Gibbs and Oliver Heaviside. This system is the vector analysis still in use today.

Hamilton's quaternions, along with matrices and the complex numbers, are subsets of the *hypercomplex* numbers. Just as the study of conics proved useful in astronomy and infinite series were applied in electrical engineering, unexpected real-world applications of hypercomplex numbers have been discovered in medicine and computer graphics, including virtual reality. This remarkable ability of pure mathematics to model reality provides reasonable evidence of the underlying structure of the universe.

COMPREHENSION CHECK

1. What problem led to the development of imaginary numbers?
2. What is a depressed cubic, and who discovered its solution?
3. Who discovered a general solution for any cubic equation?
4. What was Rafael Bombelli's novel thought about the roots of cubic equations?
5. How did Caspar Wessel represent complex numbers geometrically?
6. Name at least two examples of "useless" mathematics being found to have important real-world applications. What does this imply about our world?

Historical Connection

Objectives

1. To summarize the development of complex numbers
2. To explain how imaginary numbers are represented geometrically
3. To give examples of the unexpected usefulness of mathematics
4. To discuss implications about the nature of our world based on the discovery of useful applications of "useless" mathematics

Answers

1. the search for a general solution to cubic equations
2. a cubic equation with no second-degree term; Scipione del Ferro
3. Gerolamo Cardano
4. that a complex expression in Cardano's formula might represent a real number
5. as vectors in the complex plane
6. sample answers: conics (planetary orbits), infinite series (electrical engineering), complex numbers (medicine)
   The ability of mathematics to model reality implies an underlying structure and design to our world.

PRESENTATION

This Historical Connection chronicles the key events in the exploration and acceptance of imaginary numbers. The well-ingrained geometric traditions in the mathematics of the sixteenth century would allow only those algebraic expressions with a demonstrable counterpart in Euclidean space. It took the creativity and persistence of Rafael Bombelli to break through this boundary and demonstrate the usefulness and necessity of imaginary numbers.

Graphs of the problems mentioned in the first paragraph in the text are illustrated on the chalkboard in the upper-right cartoon.

The graph of the cubic polynomial $x^3 - 15x - 4 = 0$ reveals three real zeros, one of which is 4. Factor the polynomial and use the quadratic formula to find the other real zeros.
$$(x - 4)(x^2 + 4x + 1) = 0$$
$$x = -2 \pm 2\sqrt{3}$$

The del Ferro-Tartaglia-Cardano formula for a solution of the depressed cubic equation $x^3 - px - q = 0$ is
$$x = \sqrt[3]{\frac{q}{2} + \left(\left(\frac{q}{2}\right)^2 - \left(\frac{p}{3}\right)^2\right)} + \sqrt[3]{\frac{q}{2} - \left(\left(\frac{q}{2}\right)^2 - \left(\frac{p}{3}\right)^2\right)}.$$

Therefore, one of the equation's roots is $\sqrt[3]{2 + 11\sqrt{-1}} + \sqrt[3]{2 - 11\sqrt{-1}}$, but which real root does this represent? Bombelli thought this root might be 4 in the form of $(2 + n\sqrt{-1}) + (2 - n\sqrt{-1})$.

Letting $n = 1$ and applying the normal rules of arithmetic to these complex expressions shows that
$$\left(2 + \sqrt{-1}\right)^3 = 2 + 11\sqrt{-1} \text{ and}$$
$$\left(2 - \sqrt{-1}\right)^3 = 2 - 11\sqrt{-1} \text{ so that}$$
$$\sqrt[3]{2 + 11\sqrt{-1}} + \sqrt[3]{2 - 11\sqrt{-1}}$$
$$= (2 + \sqrt{-1}) + (2 - \sqrt{-1}) = 4.$$

continued in Answers and Solutions Overflow

6.6 Polar Forms of Complex Numbers

Objectives

1. To use polar form to express complex numbers, including moduli and arguments

2. To convert between $a + bi$ and $r \operatorname{cis} \theta$ forms

3. To find products, quotients, powers, and roots of complex numbers in polar form

Flash

Mathematician Benoit Mandelbrot coined the term "fractal" in 1975 during his career at IBM.

Vocabulary

argument
complex plane
De Moivre's Theorem
imaginary axis
modulus
nth root of a complex number
nth root of unity
polar form
principal root
real axis
standard form

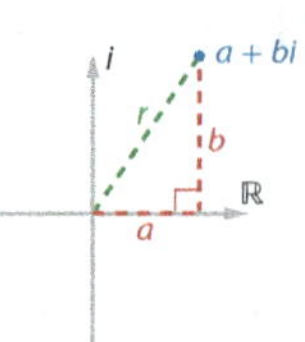

6.6 Polar Forms of Complex Numbers

Fractal art generated by computers utilizes polar graphs of complex numbers.

After completing this section, you will be able to
- graph complex numbers.
- convert between the standard and polar forms of a complex number.
- find products, quotients, powers, and roots of complex numbers in polar form.

Any real number a can be illustrated by its graph on a real number line. Any complex number $a + bi$ can be illustrated by a point in the *complex plane* (as in an Argand diagram). The real component a is plotted along the horizontal *real axis* and the imaginary component bi is plotted along the vertical *imaginary axis* so that the complex number $a + bi$ is represented by the ordered pair (a, b).

Recall that the absolute value of a real number is defined as its distance from the origin. The absolute value of a complex number, $|a + bi|$, is also defined as the distance from the origin to the number's graph. The Pythagorean Theorem implies that $r = |a + bi| = \sqrt{a^2 + b^2}$.

Example 1   Graphing Complex Numbers

Plot each complex number in the complex plane and find its absolute value.

a. $3 - 3i$ **b.** $-\sqrt{3} + i$

Answer

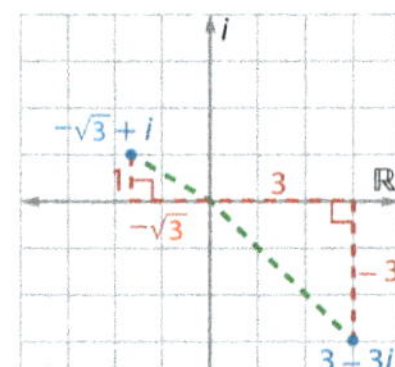

1. Plot each complex number.
2. Apply $|a + bi| = \sqrt{a^2 + b^2}$.

a. $r = \sqrt{3^2 + (-3)^2}$
$= \sqrt{18}$
$= 3\sqrt{2} \approx 4.24$

b. $r = \sqrt{\left(-\sqrt{3}\right)^2 + 1^2}$
$= \sqrt{4}$
$= 2$

SKILL ✔ EXERCISE 1

Defining a directed angle θ from the positive real axis to the terminal ray through the graph of the complex number enables a complex number to be expressed in *polar (or trigonometric) form*. Right-triangle trigonometry indicates four key relationships.

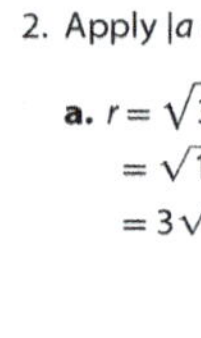

$$r = \sqrt{a^2 + b^2} \qquad a = r \cos \theta \qquad b = r \sin \theta \qquad \tan \theta = \frac{b}{a}$$

$\therefore a + bi = r \cos \theta + i \, r \sin \theta$
$= r (\cos \theta + i \sin \theta)$, which is often abbreviated as $r \operatorname{cis} \theta$.

▶ DEFINITIONS

A complex number in **standard** (or **rectangular**) **form**, $z = a + bi$, can be expressed in **polar** (or **trigonometric**) **form**, $z = r(\cos \theta + i \sin \theta)$ or $z = r \operatorname{cis} \theta$. The absolute value $r = \sqrt{a^2 + b^2}$ is also called the **modulus** (pl. *moduli*). The angle θ is the **argument** of the complex number.

PRESENTATION

Lesson Opener

Convert each to polar coordinates.

1. $(4, 4)$ $\left(4\sqrt{2}, \frac{\pi}{4}\right)$

2. $(\sqrt{3}, 1)$ $\left(2, \frac{\pi}{6}\right)$

3. $(0, -3)$ $\left(3, \frac{3\pi}{2}\right)$

State the number of roots of each.

4. a quadratic equation 2

5. a cubic equation 3

Remind the students that a complex number is expressed in the form $a + bi$, in which a and b are real numbers and i is the square root of -1. Also share that all real numbers are also complex numbers, since they can be expressed as $a + 0i$. Explain that previous graphs were limited to real numbers, but now complex numbers will be plotted on a *complex plane*, which contains the *real axis* and the *imaginary axis*. Example 1 illustrates plotting a complex number and finding its distance from the origin.

Consider having the students plot the sum of the two complex numbers. Demonstrate how this addition is similar to adding vectors expressed in terms of their rectangular components.

Explain that the four equations relating a, b, r, and θ are essential to converting between *standard* and *polar forms* of complex numbers. Define the polar form of complex numbers. Many texts use the shortened format of $r \operatorname{cis} \theta$ extensively so that the concepts covered are not obscured by lengthy notations.

The formulas for the value of θ for each quadrant provide a more direct method than using reference angles, as presented in Section 6.1. Consider asking students to compare the two methods after completing Example 2. Explain that θ can be measured in either radians or degrees and that care should be taken to be sure the correct mode is selected when using a calculator. The text will use deci-

While the standard form for $z = a + bi$ and its absolute value are unique, the existence of coterminal angles for the argument θ implies that its polar form, $z = r$ cis θ, is not. Since the range of the Tan^{-1} function is $\left[-\frac{\pi}{2}, \frac{\pi}{2}\right]$, the following formulas express the argument θ within $[0, 2\pi)$.

$$\text{Q I: } \theta = \text{Tan}^{-1}\frac{b}{a} \qquad \text{Q II \& Q III: } \theta = \pi + \text{Tan}^{-1}\frac{b}{a} \qquad \text{Q IV: } \theta = 2\pi + \text{Tan}^{-1}\frac{b}{a}$$

Example 2   Converting to Polar Form

Convert each complex number to polar form.

a. $z_1 = -2 + 3i$ **b.** $z_2 = -i$

Answer

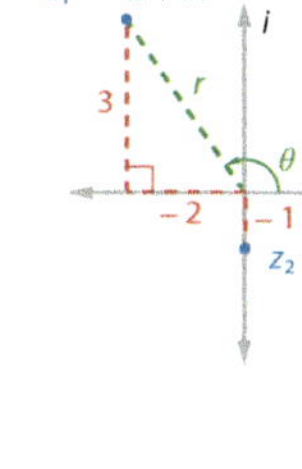

a. $r = \sqrt{(-2)^2 + 3^2}$
  $= \sqrt{13} \approx 3.61$

b. $r = \sqrt{0^2 + (-1)^2} = 1$

1. Determine the modulus.

$\theta = \pi + \text{Tan}^{-1}\left(\frac{3}{-2}\right)$
  $\approx \pi + (-0.98)$
  ≈ 2.16

Since $\text{Tan}^{-1}\left(\frac{-1}{0}\right)$ is undefined, $\theta = \frac{\pi}{2}$ or $\frac{3\pi}{2}$.

2. Since z_1 is in Q II, use $\theta = \pi + \text{Tan}^{-1}\frac{b}{a}$ to determine its argument within $[0, 2\pi)$.

$z_1 \approx 3.61$ cis 2.16

$z_2 = 1$ cis $\frac{3\pi}{2}$

3. Express in polar form.

SKILL ✓ EXERCISE 7

Example 3   Converting to Standard Form

Plot each complex number and convert to standard form.

a. $z_1 = 3$ cis $\frac{\pi}{2}$ **b.** $z_2 = 4$ cis $\frac{11\pi}{6}$

Answer

1. Use a polar grid to plot z_1 and z_2.

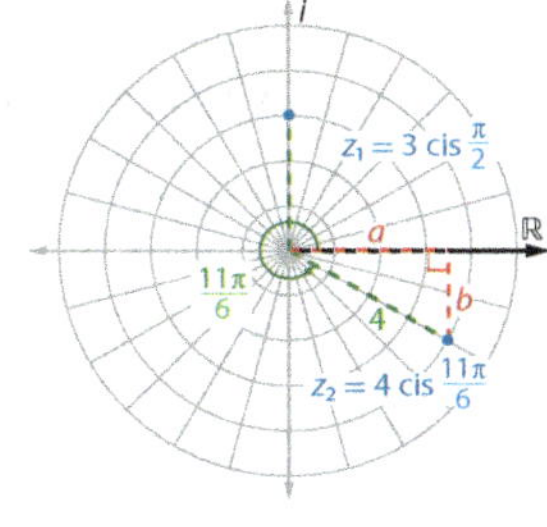

2. Evaluate $r(\cos\theta + i\sin\theta)$.

a. $z_1 = 3\left(\cos\frac{\pi}{2} + i\sin\frac{\pi}{2}\right)$
  $= 3[0 + i(1)] = 3i$

b. $z_2 = 4\left(\cos\frac{11\pi}{6} + i\sin\frac{11\pi}{6}\right)$
  $= 4\left[\frac{\sqrt{3}}{2} + i\left(-\frac{1}{2}\right)\right]$
  $= 2\sqrt{3} - 2i$
  $\approx 3.46 - 2i$

SKILL ✓ EXERCISE 11

Finding the product of two complex numbers in polar form involves finding the product of their moduli and summing their arguments. Finding a quotient involves finding the quotient of the moduli and the difference of the arguments.

Additional Exercises

1. Convert each complex number to polar form.
   a. $u = 6 + 8i$ ≈ 10 cis 0.93
   b. $c = 3 - i$ ≈ 3.16 cis 5.96
   c. $m = -\sqrt{2} - \sqrt{2}i$ 2 cis $\frac{5\pi}{4}$

2. Convert each complex number to rectangular form.
   a. 12 cis $\frac{5\pi}{3}$ $6 - 6\sqrt{3}i$
   b. 5 cis 6 $4.8 - 1.4i$

Complete each operation in polar form and then express your answer in standard form.

3. $\left(3 \text{ cis } \frac{\pi}{6}\right)\left(4 \text{ cis } \frac{7\pi}{6}\right)$
   12 cis $\frac{4\pi}{3}$; $-6 - 6\sqrt{3}i$

4. $\left(3\sqrt{2} \text{ cis } \frac{7\pi}{6}\right) \div \left(\sqrt{2} \text{ cis } \frac{\pi}{3}\right)$
   3 cis $\frac{5\pi}{6}$; $-\frac{3\sqrt{3}}{2} + \frac{3}{2}i$

5. $(-2 - 2i)^5$ $128\sqrt{2}$ cis $\frac{\pi}{3}$; $128 + 128i$

6. Find the cube roots of unity.
   $1, -\frac{1}{2} + \frac{\sqrt{3}}{2}i, -\frac{1}{2} - \frac{\sqrt{3}}{2}i$

7. Find all the fourth roots of i.
   $\approx 0.92 + 0.38i, \approx -0.38 + 0.92i,$
   $\approx -0.92 - 0.38i, \approx 0.38 - 0.92i$

8. Find the square roots of $-1 + \sqrt{3}i$.
   $\frac{\sqrt{2}}{2} + \frac{\sqrt{6}}{2}i; -\frac{\sqrt{2}}{2} - \frac{\sqrt{6}}{2}i$

9. Find all the fifth roots of 4 cis $70°$. ≈ 1.3 cis $14°, \approx 1.3$ cis $86°,$
   ≈ 1.3 cis $158°, \approx 1.3$ cis $230°,$
   ≈ 1.3 cis $302°$

10. Solve $x^3 = 1 + i$. $\approx 1.08 + 0.29i,$
    $\approx -0.79 + 0.79i, \approx -0.29 - 1.08i$

---

mal approximations of the *modulus* when the *argument* is expressed as a decimal approximation.

In Example 3 polar form is converted to standard form. Consider communicating any personal preference of having the students express answers in exact values (radical form) or in decimal approximations. Exact values are important in precise engineering, while decimal values make it easier to appreciate actual lengths.

Go through the derivation of the product formula for complex numbers in polar form and apply the product formula and the quotient formula using Example 4. Introduce *De Moivre's Theorem* and use it

to evaluate the power of a complex number in Example 5. Exercise 38 provides an opportunity to compare De Moivre's Theorem with the FOIL method.

Note that the answer in Example 5 could also be found by squaring the complex number, and then squaring it again, but finding roots is not so easy. This example leads us into a corollary to De Moivre's Theorem that provides a means to find roots of a complex number.

Review that the Fundamental Theorem of Algebra says that an nth degree polynomial function with complex coefficients has exactly n roots in the set of complex numbers.

Present the definitions of an *nth root of a complex number* and an *nth root of unity*. As an example, if v is a sixth root of a complex number z, then $v^6 = z$. Applying the six roots given in the paragraph preceding the definition box, the definition would be interpreted, "if x is a sixth root of 64, then $x^6 = 64$." This can be illustrated by taking any of the six roots to the sixth power, which will yield 64.

The nth roots of unity can be expressed as $v = \sqrt[n]{1}$, in which v is not limited to the principal roots. Introduce the corollary to De Moivre's Theorem.

Guide the students in finding the eighth roots of unity in Example 6. Point

For Graphing Calculators

If $u = a + bi$ and $v = c + di$, write a program for a graphing utility in which the output is $|u + v|$.

Prompt A
Prompt B
Prompt C
Prompt D
Disp $\sqrt{((A + C)^2 + (B + D)^2)}$

Assessment

- Quiz 6D covers Section 6.6.

Assignments

- **Minimum:** 1(a, c), 2–7, 9–11, 15–16, 19, 23–24, 27, 29–31, 44–52
- **Standard:** 2–6, 8–10, 13–14; 17–23 odd; 24, 27; 28–40 even; 44–52
- **Extended:** 2–6, 8, 10; 13–21 odd; 22, 25–26, 29–30, 32, 35–36, 38–39, 41–44, 47–53

Solutions

❯ A. Exercises

1.
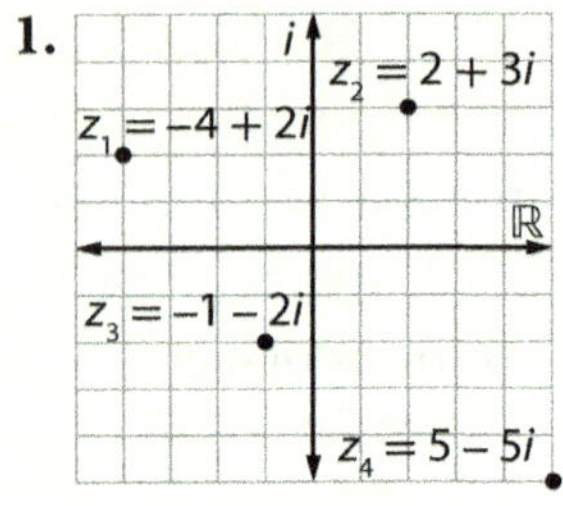

> **PRODUCTS AND QUOTIENTS OF COMPLEX NUMBERS IN POLAR FORM**

If $z_1 = r_1 \operatorname{cis} \theta_1$ and $z_2 = r_2 \operatorname{cis} \theta_2$, then

$$z_1 z_2 = r_1 r_2 \operatorname{cis} (\theta_1 + \theta_2) \text{ and } \frac{z_1}{z_2} = \frac{r_1 \operatorname{cis} \theta_1}{r_2 \operatorname{cis} \theta_2} = \frac{r_1}{r_2} \operatorname{cis} (\theta_1 - \theta_2).$$

The derivation of the product formula utilizes sum identities from Section 5.4.
Let $z_1 = r_1(\cos \theta_1 + i \sin \theta_1)$ and $z_2 = r_2(\cos \theta_2 + i \sin \theta_2)$.
Then
$$
\begin{aligned}
z_1 z_2 &= r_1(\cos \theta_1 + i \sin \theta_1) \, r_2(\cos \theta_2 + i \sin \theta_2) \\
&= r_1 r_2(\cos \theta_1 \cos \theta_2 + i \cos \theta_1 \sin \theta_2 + i \sin \theta_1 \cos \theta_2 + i^2 \sin \theta_1 \sin \theta_2) \\
&= r_1 r_2[(\cos \theta_1 \cos \theta_2 + i^2 \sin \theta_1 \sin \theta_2) + (i \cos \theta_1 \sin \theta_2 + i \sin \theta_1 \cos \theta_2)] \\
&= r_1 r_2[(\cos \theta_1 \cos \theta_2 - \sin \theta_1 \sin \theta_2) + i(\cos \theta_1 \sin \theta_2 + \sin \theta_1 \cos \theta_2)] \\
&= r_1 r_2[\cos (\theta_1 + \theta_2) + i \sin (\theta_1 + \theta_2)] \\
&= r_1 r_2 \operatorname{cis} (\theta_1 + \theta_2)
\end{aligned}
$$

The derivation of the quotient formula, completed in exercise 39, makes similar use of difference identities.

Example 4 Finding Products and Quotients in Polar Form

Find the product and quotient of $z_1 = 5 \operatorname{cis} \frac{4\pi}{3}$ and $z_2 = 2 \operatorname{cis} \frac{\pi}{6}$, expressing each in standard form.

a. $z_1 z_2$ **b.** $\frac{z_1}{z_2}$

Answer

a.
$$
\begin{aligned}
z_1 z_2 &= 5 \cdot 2 \operatorname{cis} \left(\frac{4\pi}{3} + \frac{\pi}{6} \right) \\
&= 10 \operatorname{cis} \frac{3\pi}{2} \\
&= 10(\cos \frac{3\pi}{2} + i \sin \frac{3\pi}{2}) \\
&= 10[0 + i(-1)] \\
&= -10i
\end{aligned}
$$

b.
$$
\begin{aligned}
\frac{z_1}{z_2} &= \frac{5}{2} \operatorname{cis} \left(\frac{4\pi}{3} - \frac{\pi}{6} \right) \\
&= \frac{5}{2} \operatorname{cis} \frac{7\pi}{6} \\
&= \frac{5}{2}(\cos \frac{7\pi}{6} + i \sin \frac{7\pi}{6}) \\
&= \frac{5}{2}\left(-\frac{\sqrt{3}}{2} + i\left(-\frac{1}{2} \right) \right) \\
&= -\frac{5\sqrt{3}}{4} - \frac{5}{4}i
\end{aligned}
$$

1. Apply the product or quotient formula.

2. Convert to standard form.

SKILL ✔ EXERCISES 19, 21

Finding a power or root of a complex number such as $(-1 + i)^7$ or $\sqrt[7]{-1 + i}$ is much easier in polar form. Consider the following powers of $z = r \operatorname{cis} \theta$ to develop a formula for the power of a complex number in polar form.

$$
\begin{aligned}
z^2 &= (r \operatorname{cis} \theta)(r \operatorname{cis} \theta) \\
&= r^2 \operatorname{cis} (\theta + \theta) \\
&= r^2 \operatorname{cis} 2\theta
\end{aligned}
\qquad
\begin{aligned}
z^3 &= z^2 z = (r^2 \operatorname{cis} 2\theta)(r \operatorname{cis} \theta) \\
&= r^3 \operatorname{cis} (2\theta + \theta) \\
&= r^3 \operatorname{cis} 3\theta
\end{aligned}
$$

Can you see why $z^4 = r^4 \operatorname{cis} 4\theta$? De Moivre's Theorem summarizes the general case for the nth power of a complex number.

> **DE MOIVRE'S THEOREM**

If $z = r \operatorname{cis} \theta$, then $z^n = r^n \operatorname{cis} n\theta$, where $n \in \mathbb{Z}$.

---

out that most people would not realize that i and $-i$ were roots, and would probably never guess the radical roots. Consider having different students verify each of the roots w_1 through w_7 in Example 6.

Draw attention to the graph of the roots in Example 6, the circular pattern, and the radius equal to the principal root, $\sqrt[n]{r}$. Then do Example 7 using degrees.

TIPS

Ex. 4–9 Decimal approximations of the modulus are used when the argument is expressed as a decimal approximation.

Ex. 34–37 Ask the students to find the sum of the roots. They can discover that the sum of the n complex roots in each case is 0.

Ex. 39 *Hint*: Multiply the numerator and the denominator by the conjugate of the denominator of $\frac{z_1}{z_2}$, and then apply the basic Pythagorean identity to the denominator.

Ex. 53 Six-petal roses can be generated but not of the form $r = k \cos n\theta$ or $r = k \sin n\theta$.

A formal proof of De Moivre's Theorem requires mathematical induction, which will be studied in Chapter 9.

Example 5 Finding a Power of a Complex Number

Find $(-1 + i)^4$.

Answer

$r = \sqrt{(-1)^2 + 1^2} = \sqrt{2}$

    1. Convert $z = -1 + i$ to polar form. Note that z lies in Q II of the complex plane.

$\theta = \pi + \text{Tan}^{-1}\left(\frac{1}{-1}\right) = \pi + \left(-\frac{\pi}{4}\right) = \frac{3\pi}{4}$

$z = \sqrt{2}\ \text{cis}\ \frac{3\pi}{4}$

$(-1 + i)^4 = z^4 = \left[\sqrt{2}\ \text{cis}\ \frac{3\pi}{4}\right]^4$

    2. Apply De Moivre's Theorem to find the answer in polar form where the argument is within $[0, 2\pi)$.

$= \left(\sqrt{2}\right)^4 \text{cis}\ 4\left(\frac{3\pi}{4}\right)$

$= 4\ \text{cis}\ 3\pi = 4\ \text{cis}\ \pi$

$= 4(\cos \pi + i \sin \pi)$

    3. Express $4\ \text{cis}\ \pi$ in standard form.

$= 4[-1 + i(0)]$

$= -4$

— SKILL ✔ **EXERCISE 23**

How many roots does a complex number have? The principal square root of 4 is 2, but -2 is also a square root of 4. The equation $x^6 = 64$ has two real roots, ± 2. The graph of $y = x^6 - 64$ illustrates those real roots as x-intercepts. Solving $x^6 - 64 = 0$ by factoring and applying the quadratic formula reveals all six complex roots of the equation that are guaranteed by the Fundamental Theorem of Algebra.

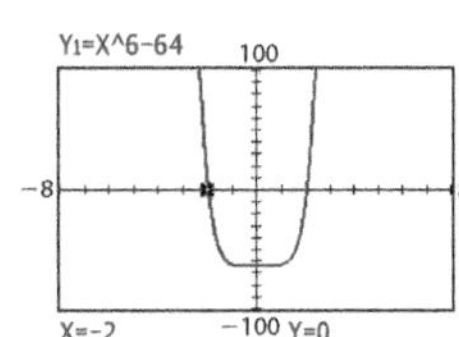

$(x^3 - 8)(x^3 + 8) = 0$

$(x - 2)(x^2 + 2x + 4)(x + 2)(x^2 - 2x + 4) = 0$

$x = \pm 2,\ -1 \pm i\sqrt{3},\ 1 \pm i\sqrt{3}$

In general, there are n complex nth roots of any complex number.

▶ DEFINITIONS

The complex number $w = a + bi$ is an **nth root of the complex number** z if $w^n = z$. If $z = 1$, then w is an **nth root of unity**.

De Moivre's Theorem extends to other rational exponents besides integers. The following corollary uses the fact that $r^{\frac{1}{n}} = \sqrt[n]{r}$ to find the nth roots of a complex number.

▶ COROLLARY TO DE MOIVRE'S THEOREM

If $z = r\ \text{cis}\ \theta$, then the n complex roots of z are $\sqrt[n]{r}\ \text{cis}\ \frac{\theta + k2\pi}{n}$, where $k = 0, 1, 2, \ldots, n - 1$.

8. $r = \sqrt{(-5)^2 + 7^2} = \sqrt{74} \approx 8.60$

Q II: $\theta = \pi + \text{Tan}^{-1}\frac{7}{-5}$

$\approx \pi + (-0.95) \approx 2.19$

$\approx 8.60\ \text{cis}\ 2.19$

9. $r = \sqrt{\left(\frac{\sqrt{3}}{4}\right)^2 + \left(-\frac{1}{4}\right)^2} = \sqrt{\frac{4}{16}} = \frac{1}{2}$

Q IV: $\theta = 2\pi + \text{Tan}^{-1}\left(\frac{-\frac{1}{4}}{\frac{\sqrt{3}}{4}}\right)$

$= 2\pi + \text{Tan}^{-1}\left(-\frac{1}{\sqrt{3}}\right)$

$= 2\pi - \frac{\pi}{6} = \frac{11\pi}{6};\ \frac{1}{2}\ \text{cis}\ \frac{11\pi}{6}$

10.

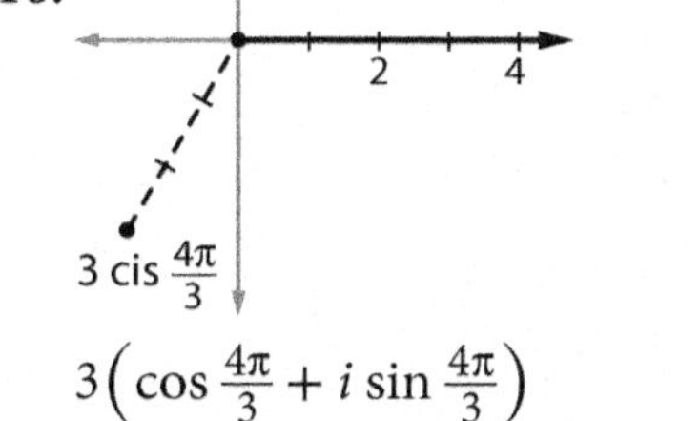

$3\left(\cos \frac{4\pi}{3} + i \sin \frac{4\pi}{3}\right)$

$= 3\left(-\frac{1}{2} + i\left(-\frac{\sqrt{3}}{2}\right)\right) = -\frac{3}{2} - \frac{3\sqrt{3}}{2}i$

11.

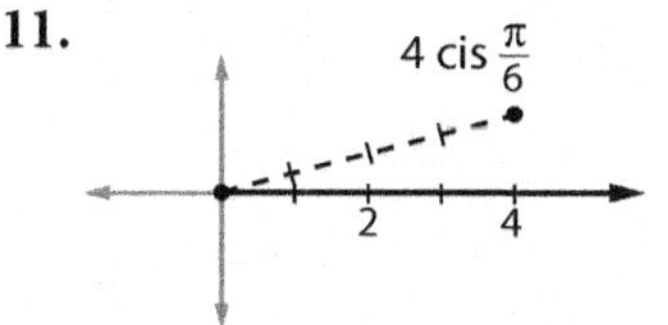

$4\left(\cos \frac{\pi}{6} + i \sin \frac{\pi}{6}\right)$

$= 4\left(\frac{\sqrt{3}}{2} + i\left(\frac{1}{2}\right)\right) = 2\sqrt{3} + 2i$

12.

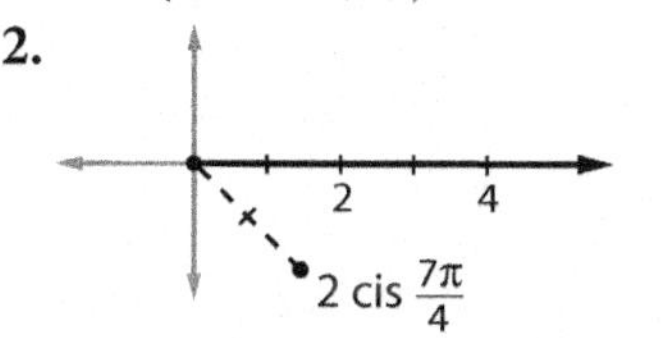

$2\left(\cos \frac{7\pi}{4} + i \sin \frac{7\pi}{4}\right)$

$= 2\left(\frac{\sqrt{2}}{2} - i\frac{\sqrt{2}}{2}\right) = \sqrt{2} - \sqrt{2}\,i$

13.

$2\sqrt{2}\ \text{cis}\ \frac{3\pi}{4}$

$2\sqrt{2}\left(\cos \frac{3\pi}{4} + i \sin \frac{3\pi}{4}\right)$

$= 2\sqrt{2}\left(-\frac{\sqrt{2}}{2} + i\left(\frac{\sqrt{2}}{2}\right)\right)$

$= -2 + 2i$

Abraham de Moivre's biography notes that he remained a faithful Christian until his death. His worldview is reflected in the dedication of his best-known publication, *The Doctrine of Chances*, where he states that the book would actually discourage participation in games of chance.

2.

2a. $r = \sqrt{1^2 + 3^2} = \sqrt{10} \approx 3.16$

2b. $r = \sqrt{2^2 + (-3)^2} = \sqrt{13} \approx 3.61$

2c. $r = \sqrt{0^2 + 4^2} = 4$

2d. $r = \sqrt{(-4)^2 + (-2)^2}$

$= 2\sqrt{5} \approx 4.47$

4. $r = \sqrt{9^2 + 2^2} = \sqrt{85} \approx 9.22$

Q I: $\theta = \text{Tan}^{-1}\frac{2}{9} \approx 0.22$

$\approx 9.22\ \text{cis}\ 0.22$

5. $r = \sqrt{(-7)^2 + 0^2} = \sqrt{49} = 7$

$v = -7$ is at $\theta = \pi$; $7\ \text{cis}\ \pi$

6. $r = |9i| = \sqrt{0^2 + 9^2} = \sqrt{81} = 9$

$9i$ is at $\theta = \frac{\pi}{2}$; $9\ \text{cis}\ \frac{\pi}{2}$

7. $r = \sqrt{(-3)^2 + (-3)^2}$

$= 3\sqrt{2} \approx 4.24$

Q III: $\theta = \pi + \text{Tan}^{-1}\frac{-3}{-3}$

$= \pi + \text{Tan}^{-1} 1 = \pi + \frac{\pi}{4} = \frac{5\pi}{4}$

$3\sqrt{2}\ \text{cis}\ \frac{5\pi}{4}$

14. $7 \cdot 2 \text{ cis} \left(\frac{\pi}{3} + \frac{\pi}{6} \right) = 14 \text{ cis} \frac{\pi}{2}$

$\quad = 14 \left(\cos \frac{\pi}{2} + i \sin \frac{\pi}{2} \right)$

$\quad = 14(0 + i(1)) = 14i$

15. $6 \cdot 1 \text{ cis} (150° + 210°) = 6 \text{ cis } 0$

$\quad = 6(\cos 0 + i \sin 0)$

$\quad = 6(1 + i(0)) = 6$

16. $\frac{9}{3} \text{ cis} \left(\frac{5\pi}{3} - \frac{\pi}{6} \right) = 3 \text{ cis} \frac{3\pi}{2}$

$\quad = 3 \left(\cos \frac{3\pi}{2} + i \sin \frac{3\pi}{2} \right)$

$\quad = 3(0 + i(-1)) = -3i$

17. $6 \text{ cis} (135° - 45°) = 6 \text{ cis } 90°$

$\quad = 6(\cos 90° + i \sin 90°)$

$\quad = 6(0 + i) = 6i$

18. $r_1 = \sqrt{1^2 + 1^2} = \sqrt{2}$

$\quad \text{Q I: } \theta_1 = \text{Tan}^{-1} 1 = \frac{\pi}{4}$

$\quad z_1 = \sqrt{2} \text{ cis} \frac{\pi}{4}$

$\quad r_2 = \sqrt{1^2 + (-1)^2} = \sqrt{2}$

$\quad \text{Q IV: } \theta_2 = 2\pi + \text{Tan}^{-1}(-1) = \frac{7\pi}{4}$

$\quad z_2 = \sqrt{2} \text{ cis} \frac{7\pi}{4}$

$\quad z_1 z_2 = \sqrt{2} \cdot \sqrt{2} \text{ cis} \left(\frac{\pi}{4} + \frac{7\pi}{4} \right)$

$\quad\quad = 2 \text{ cis } 2\pi = 2 \text{ cis } 0 = 2$

$\quad z_1 z_2 = (1 + i)(1 - i)$

$\quad\quad = 1 - i + i - i^2$

$\quad\quad = 1 + 1 = 2$

19. $r_1 = \sqrt{3 + 1^2} = 2$

$\quad \text{Q I: } \theta_1 = \text{Tan}^{-1} \frac{1}{\sqrt{3}} = \frac{\pi}{6}$

$\quad z_1 = 2 \text{ cis} \frac{\pi}{6}$

$\quad r_2 = \sqrt{2^2 + (-2)^2} = 2\sqrt{2}$

$\quad \text{Q II: } \theta_2 = \pi + \text{Tan}^{-1} \frac{-2}{2} = \frac{3\pi}{4}$

$\quad z_2 = 2\sqrt{2} \text{ cis} \frac{3\pi}{4}$

$\quad z_1 z_2 = 2 \cdot 2\sqrt{2} \text{ cis} \left(\frac{\pi}{6} + \frac{3\pi}{4} \right)$

$\quad\quad = 4\sqrt{2} \text{ cis} \frac{11\pi}{12} \approx 5.46 - 1.46i$

$\quad z_1 z_2 = (\sqrt{3} + i)(2 - 2i)$

$\quad\quad = 2\sqrt{3} - 2\sqrt{3}i + 2i - 2i^2$

$\quad\quad = (2 + 2\sqrt{3}) + (2 - 2\sqrt{3})i$

$\quad\quad \approx 5.46 - 1.46i$

20. $r_1 = 2\sqrt{5}; \text{ Q I: } \theta_1 = \text{Tan}^{-1} \frac{2}{4} \approx 0.46$

$\quad z_1 \approx 2\sqrt{5} \text{ cis } 0.46$

$\quad r_2 = 2; \text{ Q IV: } \theta_2 = \text{Tan}^{-1} \frac{-1}{\sqrt{3}} = -\frac{\pi}{6}$

$\quad z_2 = 2 \text{ cis} \left(-\frac{\pi}{6} \right)$

$\quad \frac{z_1}{z_2} \approx \frac{2\sqrt{5}}{2} \text{ cis} \left(0.46 - \left(-\frac{\pi}{6} \right) \right)$

$\quad\quad \approx \sqrt{5} \text{ cis } 0.99 \approx 1.23 + 1.87i$

$\quad \frac{z_1}{z_2} = \frac{4 + 2i}{\sqrt{3} - i} \cdot \frac{\sqrt{3} + i}{\sqrt{3} + i} = \frac{4\sqrt{3} + 4i + 2\sqrt{3}i - 2}{3 - i^2}$

$\quad\quad = \frac{2\sqrt{3} - 1}{2} + \frac{\sqrt{3} + 2}{2} i \approx 1.23 + 1.87i$

Example 6 Finding Roots of Unity

Find the eighth roots of unity.

Answer

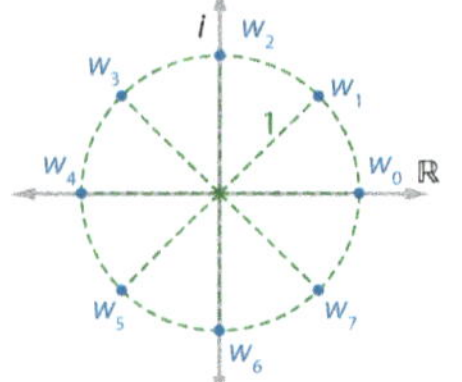

$z = 1 + 0i = 1 \text{ cis } 0$ — 1. Express $z = 1$ in polar form.

$w_k = \sqrt[8]{1} \text{ cis} \frac{0 + k2\pi}{8} = 1 \text{ cis} \left(0 + k\frac{\pi}{4} \right)$
where $k = 0, 1, 2, 3, \ldots, 7$ — 2. Apply the Corollary to De Moivre's Theorem.

$w_0 = \text{cis } 0 = 1 + 0 = 1 \qquad w_4 = \text{cis } \pi = -1 + 0 = -1$ — 3. Express each root in polar and in standard form.

$w_1 = \text{cis} \frac{\pi}{4} = \frac{\sqrt{2}}{2} + \frac{\sqrt{2}}{2} i \qquad w_5 = \text{cis} \frac{5\pi}{4} = -\frac{\sqrt{2}}{2} - \frac{\sqrt{2}}{2} i$

$w_2 = \text{cis} \frac{\pi}{2} = 0 + 1i = i \qquad w_6 = \text{cis} \frac{3\pi}{2} = 0 + (-1i) = -i$

$w_3 = \text{cis} \frac{3\pi}{4} = -\frac{\sqrt{2}}{2} + \frac{\sqrt{2}}{2} i \qquad w_7 = \text{cis} \frac{7\pi}{4} = \frac{\sqrt{2}}{2} - \frac{\sqrt{2}}{2} i$

SKILL ✔ **EXERCISE 27**

The graphs of the complex roots of $z = r \text{ cis } \theta$ lie on the circle centered at the origin with a radius of $\sqrt[n]{r}$. The arguments of its successive roots differ by $\frac{2\pi}{n}$. These facts allow the polar form of other roots to be quickly determined once the *principal root* (the first root using $k = 0$) has been determined.

Example 7 Finding Roots of Complex Numbers

Find the fifth roots of $z = 32 \text{ cis } 50°$, expressing the roots in polar form.

Answer

$w_k = \sqrt[5]{32} \text{ cis} \frac{50° + k360°}{5}$ — 1. Apply the corollary to De Moivre's Theorem.

$\quad = 2 \text{ cis} (10° + k72°)$
where $k = 0, 1, 2, 3, 4$

$w_0 = 2 \text{ cis } 10° \qquad w_3 = 2 \text{ cis } 226°$ — 2. State the principal root and then add 72° four times to find the remaining roots.

$w_1 = 2 \text{ cis } 82° \qquad w_4 = 2 \text{ cis } 298°$

$w_2 = 2 \text{ cis } 154°$

SKILL ✔ **EXERCISE 33**

A calculator can be used to find the standard form decimal approximations.

$w_0 \approx 1.97 + 0.35i \qquad w_3 \approx -1.39 - 1.44i$

$w_1 \approx 0.28 + 1.98i \qquad w_4 \approx 0.94 - 1.77i$

$w_2 \approx -1.80 + 0.88i$

21. $r_1 = 2; \theta_1 = \text{Tan}^{-1} 1 \approx \frac{\pi}{4}$

$\quad z_1 = 2 \text{ cis} \frac{\pi}{4}$

$\quad r_2 = \sqrt{10}$

$\quad \text{Q IV: } \theta_2 \approx \text{Tan}^{-1} \frac{-1}{3} \approx -0.32$

$\quad z_2 \approx \sqrt{10} \text{ cis} (-0.32)$

$\quad \frac{z_1}{z_2} \approx \frac{2}{\sqrt{10}} \text{ cis} \left(\frac{\pi}{4} - (-0.32) \right)$

$\quad\quad \approx \frac{\sqrt{10}}{5} \text{ cis } 1.11 \approx 0.28 + 0.57i$

$\quad \frac{z_1}{z_2} = \frac{\sqrt{2} + \sqrt{2}i}{3 - i} \cdot \frac{3 + i}{3 + i}$

$\quad\quad = \frac{3\sqrt{2} + \sqrt{2}i + 3\sqrt{2}i - \sqrt{2}}{9 - i^2}$

$\quad\quad = \frac{\sqrt{2}}{5} + \frac{2\sqrt{2}}{5} i \approx 0.28 + 0.57i$

22. $z = 1 + i; r = \sqrt{2}$

$\quad \text{Q I: } \theta = \text{Tan}^{-1} 1 = \frac{\pi}{4}$

$\quad z = \sqrt{2} \text{ cis} \frac{\pi}{4}$

$\quad z^3 = (\sqrt{2})^3 \text{ cis} 3 \left(\frac{\pi}{4} \right)$

$\quad\quad = 2\sqrt{2} \text{ cis} \frac{3\pi}{4}$

$\quad\quad = 2\sqrt{2} \left(-\frac{\sqrt{2}}{2} + i \left(\frac{\sqrt{2}}{2} \right) \right)$

$\quad\quad = -2 + 2i$

23. $z = 1 + \sqrt{3}i; r = 2$

$\quad \text{Q I: } \theta = \text{Tan}^{-1} \sqrt{3} = \frac{\pi}{3}$

$\quad z = 2 \text{ cis} \frac{\pi}{3}$

$\quad z^4 = 2^4 \text{ cis} 4 \left(\frac{\pi}{3} \right) = 16 \text{ cis} \frac{4\pi}{3}$

$\quad\quad = 16 \left(-\frac{1}{2} + i \left(-\frac{\sqrt{3}}{2} \right) \right)$

$\quad\quad = -8 - 8\sqrt{3}i$

A. Exercises

1. Graph each complex number in the complex plane.
   a. $z_1 = -4 + 2i$ b. $z_2 = 2 + 3i$
   c. $z_3 = -1 - 2i$ d. $z_4 = 5 - 5i$

2. Graph each complex number in the complex plane and determine its absolute value.
   a. $z_5 = 1 + 3i$ b. $z_6 = 2 - 3i$
   c. $z_7 = 4i$ d. $z_8 = -4 - 2i$

3. State the general formula for the argument θ of the complex number $z = a + bi$ located in the given quadrant.
   a. Q I   b. Q II   c. Q III   d. Q IV

Convert each complex number to polar form.

4. $9 + 2i$ $\approx 9.22 \text{ cis } 0.22$
5. -7 $7 \text{ cis } \pi$
6. $9i$ $9 \text{ cis } \frac{\pi}{2}$
7. $-3 - 3i$ $3\sqrt{2} \text{ cis } \frac{5\pi}{4}$
8. $-5 + 7i$ $\approx 8.60 \text{ cis } 2.19$
9. $\frac{\sqrt{3}}{4} - \frac{1}{4}i$ $\frac{1}{2} \text{ cis } \frac{11\pi}{6}$

Plot each complex number in polar form and then convert to standard form.

10. $3 \text{ cis } \frac{4\pi}{3}$ $-\frac{3}{2} - \frac{3\sqrt{3}}{2}i$
11. $4 \text{ cis } \frac{\pi}{6}$ $2\sqrt{3} + 2i$
12. $2 \text{ cis } \frac{7\pi}{4}$ $\sqrt{2} - \sqrt{2}i$
13. $2\sqrt{2} \text{ cis } \frac{3\pi}{4}$ $-2 + 2i$

B. Exercises

Find each product or quotient in polar form. Then express your answer in standard form.

14. $\left(7 \text{ cis } \frac{\pi}{3}\right)\left(2 \text{ cis } \frac{\pi}{6}\right)$
15. $(6 \text{ cis } 150°)(\text{cis } 210°)$
16. $\dfrac{9 \text{ cis } \frac{5\pi}{3}}{3 \text{ cis } \frac{\pi}{6}}$
17. $3 \text{ cis } 135° \div \frac{1}{2} \text{ cis } 45°$

Convert each number to polar form and then find the product z_1z_2 in both polar form and standard form. Verify your answer by finding the product directly in standard form.

18. $z_1 = 1 + i;\ z_2 = 1 - i$
19. $z_1 = \sqrt{3} + i;\ z_2 = 2 - 2i$

Convert each number to polar form and then find the quotient $\frac{z_1}{z_2}$ in both polar and standard form. Verify your answer by finding the quotient directly in standard form.

20. $z_1 = 4 + 2i;$ $z_2 = \sqrt{3} - i$
21. $z_1 = \sqrt{2} + \sqrt{2}i;$ $z_2 = 3 - i$

3a. $\theta = \text{Tan}^{-1} \frac{b}{a}$
3b. $\theta = \pi + \text{Tan}^{-1} \frac{b}{a}$
3c. $\theta = \pi + \text{Tan}^{-1} \frac{b}{a}$
3d. $\theta = 2\pi + \text{Tan}^{-1} \frac{b}{a}$

14. $14 \text{ cis } \frac{\pi}{2};\ 14i$
15. $6 \text{ cis } 0;\ 6$
16. $3 \text{ cis } \frac{3\pi}{2};\ -3i$
17. $6 \text{ cis } 90°;\ 6i$

Use De Moivre's Theorem to calculate each power. Express your answer in standard form.

22. $(1 + i)^3$ $-2 + 2i$
23. $(1 + \sqrt{3}\,i)^4$ $-8 - 8\sqrt{3}\,i$
24. $(-4 + 4i)^3$ $128 + 128i$
25. $(3 + 2i)^8$ $-239 - 28{,}560i$

Find all nth roots for each complex number, expressing each root in both polar and standard form. Then graph the roots on the complex plane.

26. fourth roots of unity
27. fifth roots of unity
28. cube roots of $-i$
29. fourth roots of -3

Find all nth roots of each complex number in polar form. Label each principal root.

30. square roots of $49 \text{ cis } 50°$
31. cube roots of $125 \text{ cis } \frac{4\pi}{3}$
32. fourth roots of $81 \text{ cis } \frac{\pi}{2}$
33. fifth roots of $\text{cis } 65°$

Use the Corollary to De Moivre's Theorem to find all the roots for each equation. Write your answers in standard form.

34. $x^4 = 16$ $2, 2i, -2, -2i$
35. $x^3 + i = 0$ $i, -\frac{\sqrt{3}}{2} - \frac{1}{2}i, \frac{\sqrt{3}}{2} - \frac{1}{2}i$
36. $x^2 - 2i = 0$ $1 + i, -1 - i$
37. $x^2 + 3i = 3$ $\approx -1.9 + 0.79i, \approx 1.9 - 0.79i$

C. Exercises

38. **Compare:** Expand $(3 + 2i)^8$ using the FOIL method and compare your answer to exercise 25. Which method do you prefer? Why? $-239 - 28{,}560i$; Answers will vary.

39. **Prove:** Given $z_1 = r_1(\cos\theta_1 + i\sin\theta_1)$ and $z_2 = r_2(\cos\theta_2 + i\sin\theta_2)$, show that $\frac{z_1}{z_2} = \frac{r_1}{r_2}[\cos(\theta_1 - \theta_2) + i\sin(\theta_1 - \theta_2)]$.

Simplify.

40. $\dfrac{\left(3 \text{ cis } \frac{\pi}{3}\right)\left(5 \text{ cis } \frac{\pi}{6}\right)}{0.5 \text{ cis } \frac{\pi}{2}}$ 30

41. Compute $\left[(-1 - i)(-4 + 4\sqrt{3}\,i)\right]^2$ in polar form. Then express the result in standard form. $128 \text{ cis } \frac{11\pi}{6};\ 64\sqrt{3} - 64i$

42. Find the cube roots of $\left(25 \text{ cis } \frac{\pi}{6}\right)^2$ in polar form.

43. Find $(2 - 5i)^{100}$ in polar form. $\approx 29^{50} \text{ cis } 0.35$

18. $2 \text{ cis } 0;\ 2$
19. $4\sqrt{2} \text{ cis } \frac{11\pi}{12};\ (2 + 2\sqrt{3}) + (2 - 2\sqrt{3})i \approx 5.46 - 1.46i$
20. $\approx \sqrt{5} \text{ cis } 0.99;\ \frac{2\sqrt{3} - 1}{2} + \frac{\sqrt{3} + 2}{2}i \approx 1.23 + 1.87i$
21. $\approx \frac{\sqrt{10}}{5} \text{ cis } 1.11;\ \frac{\sqrt{2}}{5} + \frac{2\sqrt{2}}{5}i \approx 0.28 + 0.57i$
42. $5\sqrt[3]{5} \text{ cis } \frac{\pi}{9},\ 5\sqrt[3]{5} \text{ cis } \frac{13\pi}{9},\ 5\sqrt[3]{5} \text{ cis } \frac{7\pi}{9}$

27. $z = 1 + 0i = 1 \text{ cis } 0$
$$w_k = \sqrt[5]{1} \text{ cis } \frac{0 + k2\pi}{5} = \text{cis } k\frac{2\pi}{5}$$
$$w_0 = \text{cis } 0 = 1$$
$$w_1 = \text{cis } \frac{2\pi}{5} \approx 0.31 + 0.95i$$
$$w_2 = \text{cis } \frac{4\pi}{5} \approx -0.81 + 0.59i$$
$$w_3 = \text{cis } \frac{6\pi}{5} \approx -0.81 - 0.59i$$
$$w_4 = \text{cis } \frac{8\pi}{5} \approx 0.31 - 0.95i$$

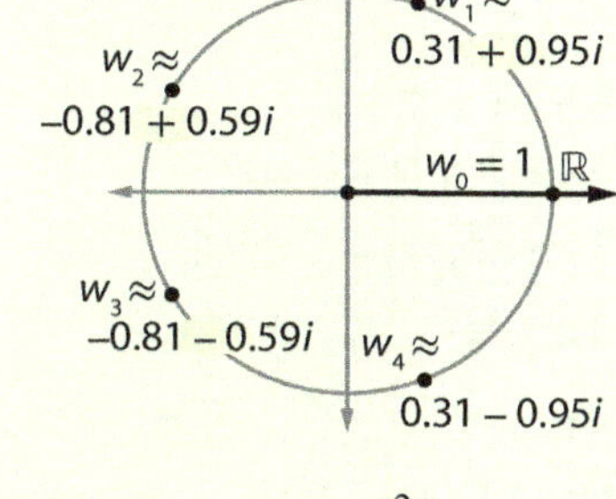

28. $z = 0 - i = 1 \text{ cis } \frac{3\pi}{2}$
$$w_k = \sqrt[3]{1} \text{ cis } \frac{\frac{3\pi}{2} + k2\pi}{3} = \text{cis } \left(\frac{\pi}{2} + k\frac{2\pi}{3}\right)$$
$$w_0 = \text{cis } \frac{\pi}{2} = 0 + 1i = i$$
$$w_1 = \text{cis } \frac{7\pi}{6} = -\frac{\sqrt{3}}{2} - \frac{1}{2}i$$
$$w_2 = \text{cis } \frac{11\pi}{6} = \frac{\sqrt{3}}{2} - \frac{1}{2}i$$

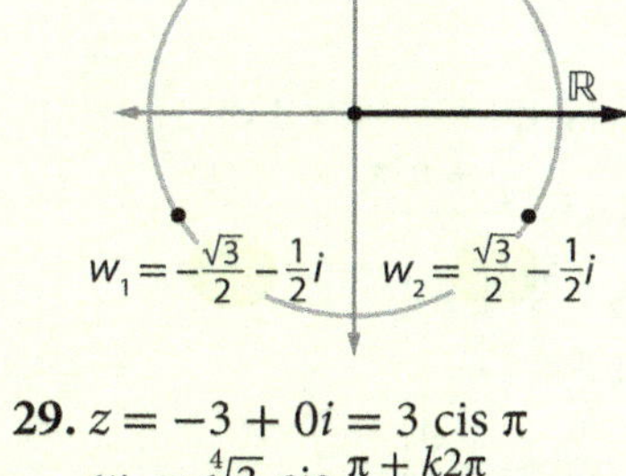

29. $z = -3 + 0i = 3 \text{ cis } \pi$
$$w_k = \sqrt[4]{3} \text{ cis } \frac{\pi + k2\pi}{4} = \sqrt[4]{3} \text{ cis } \left(\frac{\pi}{4} + k\frac{2\pi}{4}\right)$$
$$w_0 = \sqrt[4]{3} \text{ cis } \frac{\pi}{4} \approx 0.93 + 0.93i$$
$$w_1 = \sqrt[4]{3} \text{ cis } \frac{3\pi}{4} \approx -0.93 + 0.93i$$
$$w_2 = \sqrt[4]{3} \text{ cis } \frac{5\pi}{4} \approx -0.93 - 0.93i$$
$$w_3 = \sqrt[4]{3} \text{ cis } \frac{7\pi}{4} \approx 0.93 - 0.93i$$

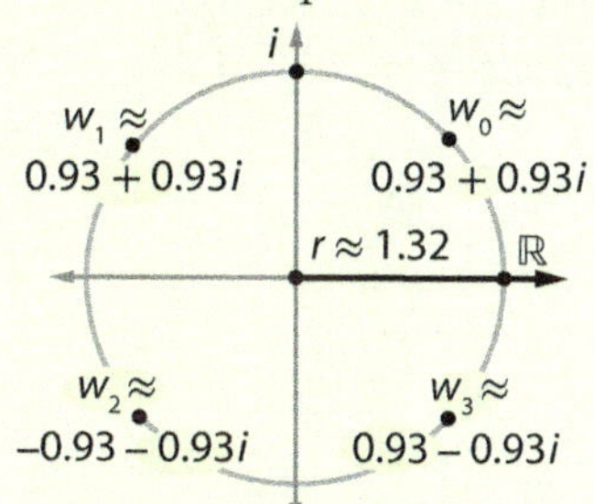

30. $w_k = \sqrt{49} \text{ cis } \frac{50° + k360°}{2} = 7 \text{ cis } (25° + k180°)$

24. $z = -4 + 4i;\ r = 4\sqrt{2}$
$$\text{Q II: } \theta = \pi + \text{Tan}^{-1}(-1) = \frac{3\pi}{4}$$
$$z = 4\sqrt{2} \text{ cis } \frac{3\pi}{4}$$
$$z^3 = \left(4\sqrt{2}\right)^3 \text{ cis } 3\left(\frac{3\pi}{4}\right)$$
$$= 128\sqrt{2} \text{ cis } \frac{9\pi}{4} = 128\sqrt{2} \text{ cis } \frac{\pi}{4}$$
$$= 128\sqrt{2}\left(\frac{\sqrt{2}}{2} + i\left(\frac{\sqrt{2}}{2}\right)\right)$$
$$= 128 + 128i$$

25. $z = 3 + 2i;\ r = \sqrt{13}$
$$\text{Q I: } \theta = \text{Tan}^{-1} \frac{2}{3} \approx 0.59$$
$$z \approx \sqrt{13} \text{ cis } 0.59$$
$$z^8 \approx \left(\sqrt{13}\right)^8 \text{ cis } 8(0.59)$$
$$\approx 28{,}561 \text{ cis } 4.70$$
$$\approx -239 - 28{,}560i$$

26. $z = 1 + 0i = 1 \text{ cis } 0$
$$w_k = \sqrt[4]{1} \text{ cis } \frac{0 + k2\pi}{4} = \text{cis } k\frac{\pi}{2}$$
$$w_0 = \text{cis } 0 = 1 + i(0) = 1$$
$$w_1 = \text{cis } \frac{\pi}{2} = 0 + i(1) = i$$
$$w_2 = \text{cis } \pi = -1 + i(0) = -1$$
$$w_3 = \text{cis } \frac{3\pi}{2} = 0 + i(-1) = -i$$

$w_0 = 7 \text{ cis } 25°$ (principal)
$w_1 = 7 \text{ cis } 205°$

31. $w_k = \sqrt[3]{125} \text{ cis } \dfrac{\frac{4\pi}{3} + k2\pi}{3}$

$\quad = 5 \text{ cis } \left(\dfrac{4\pi}{9} + k\dfrac{6\pi}{9}\right)$

$\quad w_0 = 5 \text{ cis } \dfrac{4\pi}{9}$ (principal)

$\quad w_1 = 5 \text{ cis } \dfrac{10\pi}{9}; \ w_2 = 5 \text{ cis } \dfrac{16\pi}{9}$

32. $w_k = \sqrt[4]{81} \text{ cis } \dfrac{\frac{\pi}{2} + k2\pi}{4}$

$\quad = 3 \text{ cis } \left(\dfrac{\pi}{8} + k\dfrac{4\pi}{8}\right)$

$\quad w_0 = 3 \text{ cis } \dfrac{\pi}{8}$ (principal)

$\quad w_1 = 3 \text{ cis } \dfrac{5\pi}{8}; \ w_2 = 3 \text{ cis } \dfrac{9\pi}{8}$

$\quad w_3 = 3 \text{ cis } \dfrac{13\pi}{8}$

33. $w_k = \sqrt[5]{1} \text{ cis } \dfrac{65° + k360°}{5}$

$\quad = \text{cis } (13° + k72°)$

$\quad w_0 = \text{cis } 13°$ (principal)

$\quad w_1 = \text{cis } 85°; \ w_2 = \text{cis } 157°$

$\quad w_3 = \text{cis } 229°; \ w_4 = \text{cis } 301°$

34. $x = $ 4th rts. of $z = 16 + 0i = 16 \text{ cis } 0$

$\quad w_k = \sqrt[4]{16} \text{ cis } \dfrac{0 + k2\pi}{4}$

$\quad = 2 \text{ cis } \left(0 + k\dfrac{\pi}{2}\right)$

$\quad w_0 = 2 \text{ cis } 0 = 2$

$\quad w_1 = 2 \text{ cis } \dfrac{\pi}{2} = 2i$

$\quad w_2 = 2 \text{ cis } \pi = -2$

$\quad w_3 = 2 \text{ cis } \dfrac{3\pi}{2} = -2i$

35. $x^3 = -i$

$\quad x = $ cu. rts. of $z = -i = 1 \text{ cis } \dfrac{3\pi}{2}$

$\quad w_k = \sqrt[3]{1} \text{ cis } \dfrac{\frac{3\pi}{2} + k2\pi}{3}$

$\quad = \text{cis } \left(\dfrac{3\pi}{6} + k\dfrac{4\pi}{6}\right)$

$\quad w_0 = \text{cis } \dfrac{\pi}{2} = i$

$\quad w_1 = \text{cis } \dfrac{7\pi}{6} = -\dfrac{\sqrt{3}}{2} - \dfrac{1}{2}i$

$\quad w_2 = \text{cis } \dfrac{11\pi}{6} = \dfrac{\sqrt{3}}{2} - \dfrac{1}{2}i$

36. $x^2 = 2i$

$\quad x = $ sq. rts. of $z = 2i = 2 \text{ cis } \dfrac{\pi}{2}$

$\quad w_k = \sqrt{2} \text{ cis } \dfrac{\frac{\pi}{2} + k2\pi}{2} = \text{cis } \left(\dfrac{\pi}{4} + k\pi\right)$

$\quad w_0 = \sqrt{2} \text{ cis } \dfrac{\pi}{4} = 1 + i$

$\quad w_1 = \sqrt{2} \text{ cis } \dfrac{5\pi}{4} = -1 - i$

37. $x^2 = 3 - 3i$

$\quad x = $ sq. rts. of $z = 3 - 3i = \sqrt{18} \text{ cis } \dfrac{7\pi}{4}$

$\quad w_k = \left(\sqrt{18}\right)^{\frac{1}{2}} \text{ cis } \dfrac{\frac{7\pi}{4} + k2\pi}{2} = \sqrt[4]{18} \text{ cis } \left(\dfrac{7\pi}{8} + k\pi\right)$

$\quad w_0 = \sqrt[4]{18} \text{ cis } \dfrac{7\pi}{8} \approx -1.9 + 0.79i$

$\quad w_1 = \sqrt[4]{18} \text{ cis } \dfrac{15\pi}{8} \approx 1.9 - 0.79i$

❯ C. Exercises

38. $(3 + 2i)^8 = \left((3 + 2i)^2\right)^4$

$\quad = (5 + 12i)^4 = \left((5 + 12i)^2\right)^2$

$\quad = (-119 + 120i)^2$

$\quad = -239 - 28{,}560i$

44. State the magnitude and direction for $\mathbf{v} = \langle -5, 7 \rangle$.
[6.1] $\sqrt{74} \approx 8.6; \approx 126°$

45. Find the angle between $\langle 4, -3 \rangle$ and $\langle -5, 7 \rangle$ (to the nearest degree). [6.2] $162°$

46. Find the unit vector that is orthogonal to $\mathbf{v} = \langle 7, -3 \rangle$. [6.2]

47. Given $A\,(-1, 2, 7)$ and $B\,(3, 7, 1)$, express $\overrightarrow{AB}$ in component form. Then find its length and a unit vector in the direction of $\overrightarrow{AB}$. [6.3]

48. Find the angle between $\langle 1, 4, -5 \rangle$ and $\langle -5, 3, 7 \rangle$ (to the nearest degree). [6.3] $118°$

49. Find the distance (to the nearest tenth) between the polar points $(-2, 20°)$, $(5, 280°)$. [6.4] 5.1

50. Which vectors are equal? Choose all correct answers. [6.1] B, C

A. $\mathbf{w}$ and $\mathbf{q}$
B. $\mathbf{p}$ and $\mathbf{v}$
C. $\mathbf{q}$ and $\mathbf{z}$
D. $\mathbf{p}$ and $\mathbf{n}$
E. $\mathbf{z}$ and $\mathbf{n}$

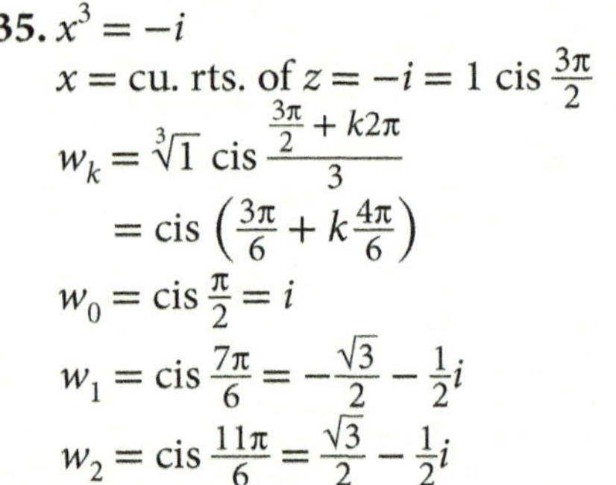

51. Find the angle between vectors $\mathbf{u} = \langle 3, 6 \rangle$ and $\mathbf{v} = \langle 2, -4 \rangle$ (to the nearest degree). [6.2] D

A. $35°$ C. $110°$ E. none of
B. $63°$ D. $127°$ these

52. Which equation is represented by the graph? [6.5] A

A. $r = 4 \cos 2\theta$
B. $r = 4 \sin 2\theta$
C. $r = 4 \cos 4\theta$
D. $r = 4 \sin 4\theta$
E. none of these

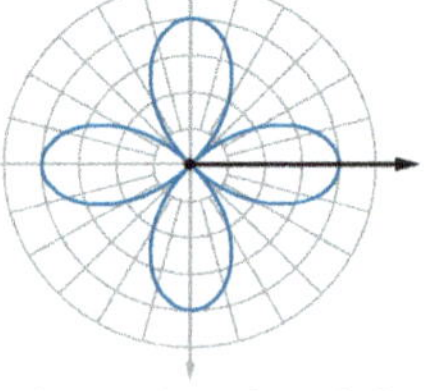

53. Which number cannot be the number of petals for a polar rose? [6.5] B

A. 3   C. 9   E. none of
B. 6   D. 12   these

46. 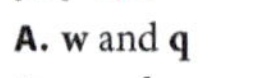 $\left\langle \dfrac{3\sqrt{58}}{58}, \dfrac{7\sqrt{58}}{58} \right\rangle$ or $\left\langle -\dfrac{3\sqrt{58}}{58}, -\dfrac{7\sqrt{58}}{58} \right\rangle$

47. 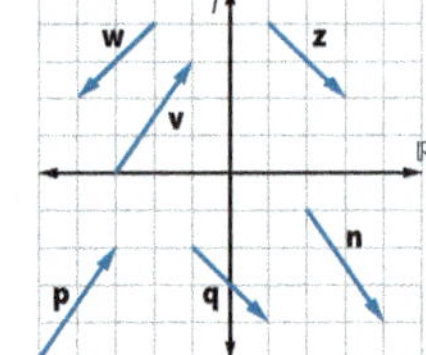 $\overrightarrow{AB} = \langle 4, 5, -6 \rangle; |\overrightarrow{AB}| = \sqrt{77}; \mathbf{u} = \left\langle \dfrac{4\sqrt{77}}{77}, \dfrac{5\sqrt{77}}{77}, -\dfrac{6\sqrt{77}}{77} \right\rangle$

*continued in Answers and
Solutions Overflow*

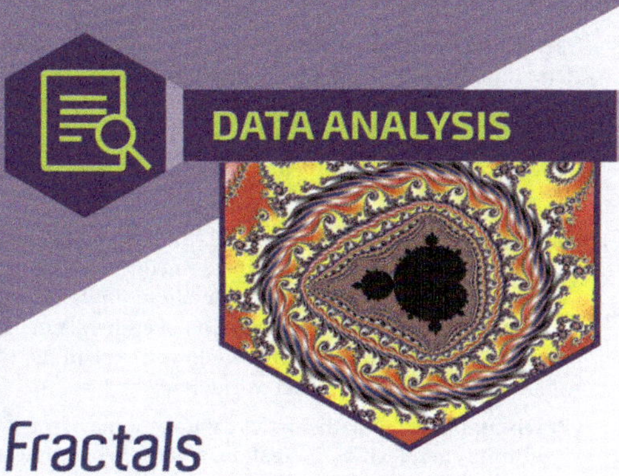

Fractals

In the 1970s the French American mathematician Benoit Mandelbrot (1924–2010) coined the term *fractal* to refer to sets that have fractured or rough edges. Mandelbrot formed beautifully complex self-similar curves (in which each part is a mini-image of the whole) using an iterative process and the relatively simple function $f(z) = z^2 + c$ where c is an imaginary number. The iterative process can be illustrated using $z_{n+1} = z_n^2 + c$ where $z_0 = 0$ and $c = 1$ (a real number).

$$z_1 = 0^2 + 1 = 1 \qquad z_4 = 5^2 + 1 = 26$$
$$z_2 = 1^2 + 1 = 2 \qquad z_5 = 26^2 + 1 = 677$$
$$z_3 = 2^2 + 1 = 5 \qquad z_6 = 677^2 + 1 = 458{,}330$$

Depending on the value of c, as the iterations increase, the points in the complex plane either "escape" to infinity (sometimes very quickly, sometimes more slowly) or are "captured" and remain within a certain distance of the origin. Examine the iterative process using $z_0 = 0$ and a complex number for c.

1. Let $c = i$ and compute z_1 thru z_6 (in terms of i). Does this point escape to infinity?

2. Let $c = 2i$ and compute z_1 thru z_5 (in terms of i). Does this point escape to infinity?

The Mandelbrot set consists of all the points in the complex plane that do not escape to infinity under $f(z) = z^2 + c$. Typically, the points in the set are colored black, while the coloring of other points depends on how quickly they escape to infinity. The most interesting patterns in the set are found near the boundary. It is here that we find self-similar patterns that continue infinitely. As long as you have the computing power to continue to magnify one portion of the curve, you will continue to find fascinating fractals. The study of fractals has contributed to our understanding of complex systems in which a small change in initial conditions creates huge consequential changes, such as weather patterns and the stock market. Fractals are also useful in modeling the human respiratory and circulatory systems.

3. Write and graph a function $r(x)$ representing the ratio of the surface area of a cube to its volume in terms of the cube's edge length x.

  a. Does the ratio increase or decrease as the cube gets bigger?

  b. Which type of function is this?

4. The lungs of a healthy human have a capacity of 305 in.^{3} and a total surface area of 1076 ft^2.

  a. Find the surface area-to-volume ratio (to the nearest in.$^{-1}$) for the lungs.

  b. Use the function from exercise 3 to determine the edge length of a cube having this same ratio.

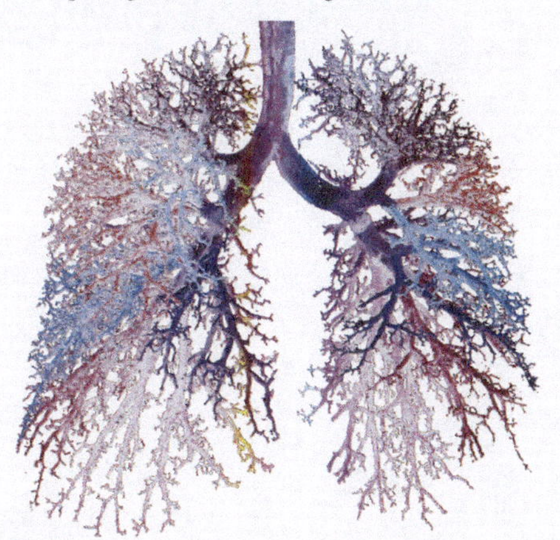

Objectives

1. To use an iterative process to evaluate a fractal function

2. To explain the benefits of fractal patterns in natural systems

3. To evaluate the implications of fractal patterns from a biblical worldview

Assignments

- **Minimum:** optional
- **Standard:** 1–5; 6–7 (in class)
- **Extended:** 1–7

PRESENTATION

Fractals provide an excellent platform to discuss and explore iterative processes, the beauty and complexity of creation, unexpected applications of pure mathematics, chaos theory, and the importance of one's worldview.

In 1958 Benoit Mandelbrot began work with IBM researching turbulence over telephone lines. This position allowed him access to mainframe computers for his exploration of iterative functions in the complex plane. These functions were popularized by Pierre Fatou and Gaston Julia at the turn of the twentieth century, but had soon slipped into obscurity. With the power of primitive computer graphics and a dot matrix printer, Mandelbrot revived interest in and discovered myriad applications of Julia sets. Today, students can produce simple fractals on their programmable calculators or write computer programs to display more complex fractal images.

Exercises 3–5 are designed to explore the marvelous complexity of the human body. Respiration requires a large surface area. An average pair of human lungs provides about 1500 mi of airway and fits about 200 m^2 of surface area into a volume of only about 6 L, which is roughly

Solutions

1. $z_{n+1} = z_n^2 + c$; Let $c = i$.

$z_1 = 0^2 + i = i$

$z_2 = (i)^2 + i = -1 + i$

$z_3 = (-1 + i)^2 + i = 1 - 2i - 1 + i$

$\quad = -i$

$z_4 = (-i)^2 + i = i^2 + i = -1 + i$

$z_5 = (-1 + i)^2 + i = 1 - 2i + i^2 + i$

$\quad = -i$

$z_6 = (-i)^2 + i = -1 + i$; no

2. $z_{n+1} = z_n^2 + c$; Let $c = 2i$.

$z_1 = 0^2 + 2i = 2i$

$z_2 = (2i)^2 + 2i = -4 + 2i$

$z_3 = (-4 + 2i)^2 + 2i$

$\quad = (16 - 16i - 4) + 2i = 12 - 14i$

$z_4 = (12 - 14i)^2 + 2i$

$\quad = (144 - 336i - 196) + 2i$

$\quad = -52 - 334i$

$z_5 = (-52 - 334i)^2 + 2i$

$\quad = (2704 + 34{,}736i - 111{,}556) + 2i$

$\quad = -108{,}852 + 34{,}738i$

3. $r(x) = \dfrac{6x^2}{x^3} = \dfrac{6}{x}$

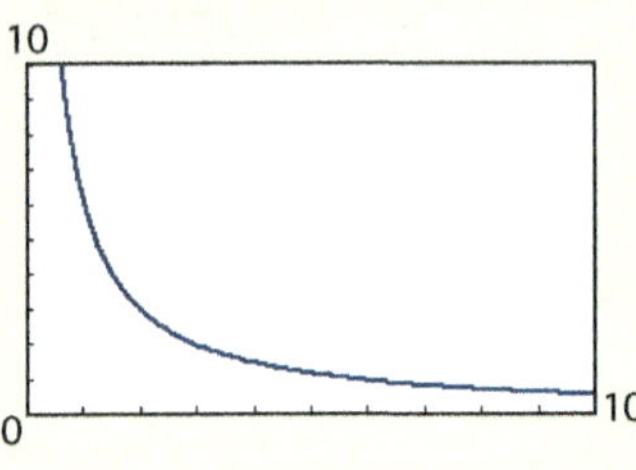

4a. $\dfrac{1076\ \text{ft}^2\left(\frac{12^2\ \text{in.}^2}{1\ \text{ft}^2}\right)}{305\ \text{in.}^3} \approx 508\ \text{in.}^{-1}$

4b. $508 \approx \dfrac{6}{x}; x \approx \dfrac{6}{508} \approx 0.012\ \text{in.}$

5. $67\ \text{L}\left(\dfrac{1000\ \text{cm}^3}{1\ \text{L}}\right)\left(\dfrac{10^3\ \text{mm}^3}{1\ \text{cm}^3}\right)$

$\quad \cdot \left(\dfrac{600\ \text{cap.}}{1\ \text{mm}^3}\right)\left(\dfrac{1\ \text{mm}}{1\ \text{cap.}}\right)\left(\dfrac{1\ \text{m}}{1000\ \text{mm}}\right)$

$\qquad \cdot \left(\dfrac{1\ \text{km}}{1000\ \text{m}}\right) = 40{,}200\ \text{km}$

$4.02 \times 10^4\ \text{km}\left(\dfrac{1000\ \text{m}}{1\ \text{km}}\right)\left(\dfrac{100\ \text{cm}}{1\ \text{m}}\right)$

$\quad \cdot \left(\dfrac{1\ \text{in.}}{2.54\ \text{cm}}\right)\left(\dfrac{1\ \text{ft}}{12\ \text{in.}}\right)\left(\dfrac{1\ \text{mi}}{5280\ \text{ft}}\right)$

$\approx 25{,}000\ \text{mi}$

The circulatory system provides oxygen and nutrients to each of approximately 30 trillion cells in the human body. In order for these transfers to occur, every cell must be within approximately 100 microns of a capillary, the smallest type of blood vessel. Have you ever considered how many meters of blood vessels it takes to spread oxygen throughout your body?

5. Assuming the average adult body has a volume of 67 L and an average of 600 capillaries/mm³ (averaging 1 mm long), calculate an estimate of how many kilometers of capillaries are in the human body. Then express your estimate in miles. (1 L = 1000 cm³)

Systems in the human body are not the only places where we find fractal patterns. Your cell phone's antenna is shaped like a fractal to minimize space and weight while providing greater bandwidth and better reception. Recent patents using fractal patterns involve batteries, invisibility cloaking, and smart clothes. The twentieth-century mathematicians who first investigated the properties of fractal functions called them "pathological" and "monsters." Some believed they were exploring creations that had no possible natural connection and were solely from the mind of man. However, it turns out that we are surrounded by fractal patterns, and their benefits are now obvious.

6. **Discuss:** Consider the structural similarities between tree branches and lung bronchi. Someone with an evolutionary worldview might say these similarities are expected since the function of each system (respiration) is similar. How would you explain the similarities from a biblical worldview?

7. **Discuss:** Fractal mathematics captures some of the infinite complexities of creation. Chaos theory grew out of the study of fractals in which a minor change in initial conditions results in vastly different outcomes. Explain in your own words how fractal mathematics and chaos theory can be in harmony with an orderly creation.

KEYWORD SEARCH

fractals on TI-84

6. Since the trees and the lungs were both created by the same wise God, it is reasonable that they are similar in design.

7. Answers will vary. While processes that appear to be chaotic and unpredictable (such as weather patterns and the functioning of the brain) cannot be perfectly modeled with our current mathematical techniques, their existence does not conflict with God's creative design, purpose, and order. We can model portions of His creation (such as gravitational forces and planetary orbits) with our current mathematics, but more complex mathematical modeling is often required to reflect His infinitely complex creation.

equivalent to packing the surface area of a tennis court into three 2 L soda bottles!

Use exercise 6 to discuss the importance of one's worldview. The "problem" is not with the facts. We all have the same facts, but our worldview determines how we interpret those facts.

The Internet keyword search *Edward Lorenz* will supply a wealth of information for further exploration of chaos theory. In the early 1960s at MIT, Lorenz was running a computer simulation of future weather patterns based on a dozen initial parameters. He was surprised to discover a drastic difference in weather predictions when just one input was rounded from 0.506127 to 0.506. This sensitivity to initial input of some dynamical systems forms the basis of chaos theory and has now been applied in many areas including biology, chemistry, physics, and finance.

TIPS

Ex. 4a The unit label in.$^{-1}$ may be unfamiliar, but it is an appropriate label for the ratio $\dfrac{\text{in.}^2}{\text{in.}^3}$.

Chapter 6 Review

1. Given $P(-2, 5)$ and $Q(3, -4)$, write $\overrightarrow{PQ}$ in component form and as a linear combination of standard unit vectors. [6.1] $\langle 5, -9 \rangle; 5\mathbf{i} - 9\mathbf{j}$

2. Find the magnitude and direction of $\mathbf{v} = \langle -4, 7 \rangle$. [6.1] $\sqrt{65}; \approx 120°$

3. Given $\mathbf{v} = \langle 2, -1 \rangle$ and $\mathbf{w} = \langle 4, 2 \rangle$, find each resultant vector algebraically and geometrically. [6.1]
   a. $\mathbf{v} - \mathbf{w}$ $\langle -2, -3 \rangle$ **b.** $3\mathbf{v} + \frac{1}{2}\mathbf{w}$ $\langle 8, -2 \rangle$

4. Find the unit vector in the direction of each vector. [6.1]
   a. $\langle 7, -3 \rangle$ $\left\langle \frac{7\sqrt{58}}{58}, \frac{-3\sqrt{58}}{58} \right\rangle$ **b.** $\langle -20, 21 \rangle$ $\left\langle \frac{-20}{29}, \frac{21}{29} \right\rangle$

State the component form for each vector. [6.1]

5. $|\mathbf{v}| = 6; \theta = \frac{5\pi}{6}$ $\langle -3\sqrt{3}, 3 \rangle$
6. a southbound car at 45 mi/hr $\langle 0, -45 \rangle$

7. An airplane flies 920 mi due east from Kansas City, MO, to Washington, DC, and then 490 mi due north to Rochester, NY. What is the plane's distance and bearing from Kansas City? [6.1] ≈ 1042 mi; $\approx 62°$

8. Find the normal and frictional forces acting on a 60 lb child sitting on a seesaw whose angle of incline is 25°. [6.1] $|\mathbf{N}| \approx 54.4$ lb; $|\mathbf{F}| \approx 25.4$ lb

9. Find $\mathbf{u} \cdot \mathbf{v}$ if $|\mathbf{u}| = 5$, $|\mathbf{v}| = 8$, and the angle between the vectors is 30°. [6.2] $20\sqrt{3}$

Find each dot product and state whether the angle between the vectors is $> 90°$, $< 90°$, or $= 90°$. [6.2]

10. $\langle 2, -3 \rangle \cdot \langle 7, 1 \rangle$
11. $\langle -4, 7 \rangle \cdot \langle 14, 8 \rangle$

12. Use the dot product to find the angle between the vectors $\mathbf{u} = \langle 5, 3 \rangle$ and $\mathbf{v} = \langle -2, 7 \rangle$ (to the nearest tenth of a degree). [6.2] $75.0°$

13. If $\langle v_1, v_2 \rangle$ is orthogonal to $\langle 5, -3 \rangle$, find the ratio of v_2 to v_1. Then find a general expression that represents all the orthogonal vectors. [6.2] $\frac{5}{3}; k\langle 3, 5 \rangle$

14. Find the projection of $\mathbf{u} = \langle 7, 5 \rangle$ onto $\mathbf{v} = \langle -2, 6 \rangle$. Then find $\mathbf{w}_2$ so that $\text{proj}_\mathbf{v}\, \mathbf{u} + \mathbf{w}_2 = \mathbf{u}$. [6.2]

10. $11; < 90°$
11. $0; = 90°$
14. $\left\langle -\frac{4}{5}, \frac{12}{5} \right\rangle, \left\langle \frac{39}{5}, \frac{13}{5} \right\rangle$

15. An 80 lb box of gadgets rests on a ramp with a 20° incline. [6.2]

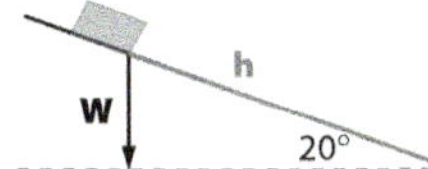

   a. Determine the component of the box's weight that pulls it down the ramp. ≈ 27.4 lb
   b. What is the magnitude of the frictional force directed up along the ramp that prevents the box from sliding down the ramp? ≈ 27.4 lb

16. Find the amount of work done by Jordan as he moves a cart 50 m, pulling with 30 N of force on the handle, which is 50° above the horizontal. [6.2] ≈ 964.2 N·m

17. Plot points $A(5, 1, -2)$ and $B(-2, 4, 3)$. [6.3]

18. Given $A(3, 1, -2)$ and $B(-4, 5, 3)$, find the length and midpoint of $\overline{AB}$. [6.3] $3\sqrt{10}; \left(-\frac{1}{2}, 3, \frac{1}{2}\right)$

19. Given $P(2, 5, -3)$ and $Q(4, -1, 2)$, express $\overrightarrow{PQ}$ in component form, find its length, and find a unit vector in the direction of $\overrightarrow{PQ}$. [6.3]

Find each resultant if $\mathbf{u} = \langle 5, -1, 3 \rangle$, $\mathbf{v} = \langle -3, 4, 1 \rangle$, and $\mathbf{w} = \langle 8, 2, -2 \rangle$. [6.3] $\langle -17, 12, -8 \rangle$

20. $\mathbf{w} - \mathbf{u}$ $\langle 3, 3, -5 \rangle$ 21. $2\mathbf{v} - 3\mathbf{u} + \frac{1}{2}\mathbf{w}$

Use a dot product to find the angle between the given vectors (to the nearest degree). [6.3]

22. $\mathbf{v} = \langle 2, 1, 5 \rangle$, $\mathbf{w} = \langle -2, -3, 7 \rangle$ $50°$
23. $\mathbf{v} = \langle 4, 5, 1 \rangle$, $\mathbf{w} = \langle -2, 3, -7 \rangle$ $90°$

24. An airplane takes off heading due south with a climb angle of 20° and an airspeed of 155 knots. There is a 25-knot wind from the northwest. Find a vector that represents the plane's velocity relative to its point of takeoff. [6.3] $\approx \langle 17.7, -163.3, 53.0 \rangle$

25. Graph each point described by its polar coordinates: $A\left(2, -\frac{3\pi}{4}\right)$ and $B\left(-3, \frac{5\pi}{3}\right)$. [6.4]

19. $\langle 2, -6, 5 \rangle; \sqrt{65}; \left\langle \frac{2\sqrt{65}}{65}, -\frac{6\sqrt{65}}{65}, \frac{\sqrt{65}}{13} \right\rangle$

PRESENTATION

Consider doing exercises 33–35 together in class, promoting discussions on how each type is identified.

Chapter 6 Review

Objective

To prepare for evaluation

Vocabulary

See Appendix A.

Assignments

- **Minimum:** 1–2, 3b, 4a, 5–12, 15–16, 25–28, 31–34, 36–38, 40–41
- **Standard:** 1–47 odd; 2–48 even (in class)
- **Extended:** 1, 2, 4b, 5–10, 12–13, 15–16, 18–19, 21, 23–48

Assessments

- Chapter 10 Test
- Second Quarter Exam covers Chapters 4–6.

Solutions

1. $\overrightarrow{PQ} = \langle 3 - (-2), -4 - 5 \rangle = \langle 5, -9 \rangle = 5\mathbf{i} - 9\mathbf{j}$

2. $|\mathbf{v}| = \sqrt{(-4)^2 + 7^2} = \sqrt{65}$
   $\alpha = \text{Tan}^{-1}\left|\frac{7}{-4}\right| \approx 60°$
   $\theta = 180 - \alpha \approx 120°$

3a. $\langle 2, -1 \rangle + \langle -4, -2 \rangle = \langle -2, -3 \rangle$

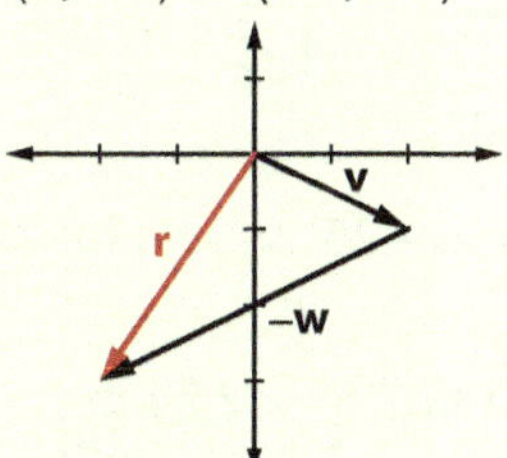

3b. $3\langle 2, -1 \rangle + \frac{1}{2}\langle 4, 2 \rangle = \langle 6, -3 \rangle + \langle 2, 1 \rangle = \langle 8, -2 \rangle$

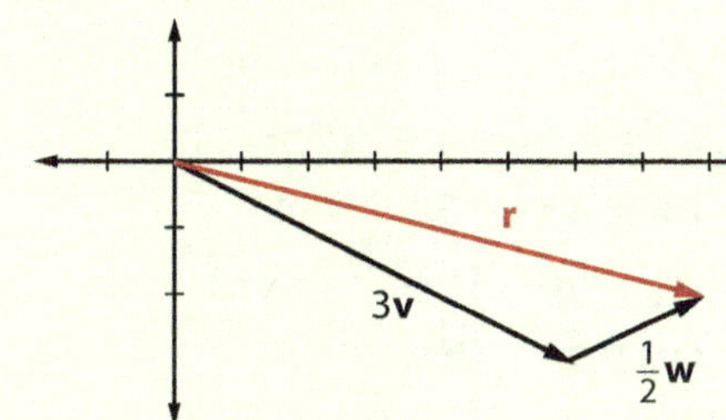

4a. $|\mathbf{v}| = \sqrt{7^2 + (-3)^2} = \sqrt{58}$

$\mathbf{u} = \frac{1}{|\mathbf{v}|}\mathbf{v} = \left\langle \frac{7}{\sqrt{58}}, \frac{-3}{\sqrt{58}} \right\rangle$

$\quad = \left\langle \frac{7\sqrt{58}}{58}, -\frac{3\sqrt{58}}{58} \right\rangle$

4b. $|\mathbf{v}| = \sqrt{(-20)^2 + 21^2} = \sqrt{841} = 29$

$\mathbf{u} = \frac{1}{|\mathbf{v}|}\mathbf{v} = \left\langle -\frac{20}{29}, \frac{21}{29} \right\rangle$

5. $\left\langle 6\cos\frac{5\pi}{6}, 6\sin\frac{5\pi}{6} \right\rangle = \langle -3\sqrt{3}, 3 \rangle$

6.

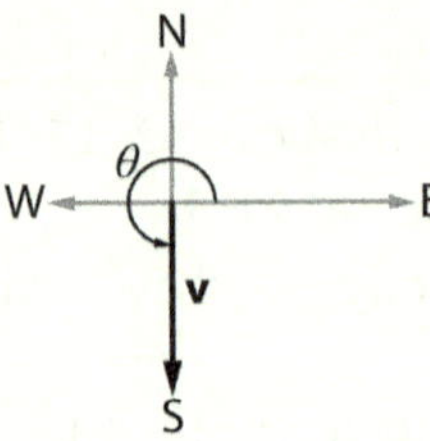

$|\mathbf{v}| = 45;\ \theta = 270°$

$\langle 45\cos 270°, 45\sin 270° \rangle$

$= \langle 0, -45 \rangle$

7.

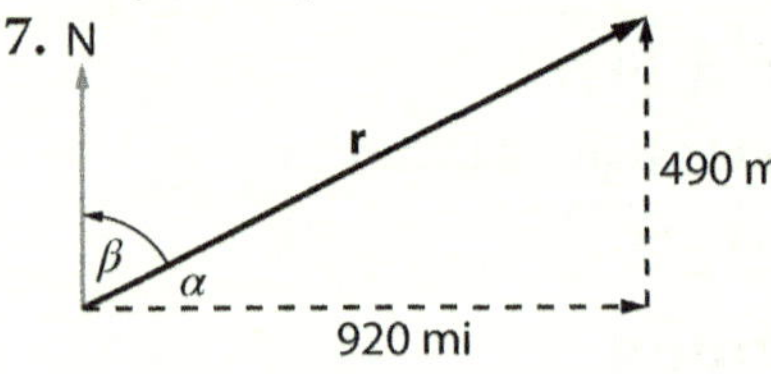

$|\mathbf{r}| = \sqrt{920^2 + 490^2} \approx 1042.4 \text{ mi}$

$\alpha = \text{Tan}^{-1}\frac{490}{920} \approx 28°$

$\beta = 90 - 28 = 62°$

8.

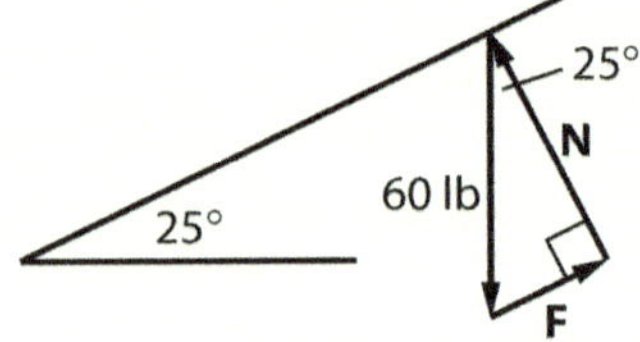

$\cos 25° = \frac{|\mathbf{N}|}{60}$

$\quad |\mathbf{N}| = 60\cos 25° \approx 54.4 \text{ lb}$

$\sin 25° = \frac{|\mathbf{F}|}{60}$

$\quad |\mathbf{F}| = 60\sin 25° \approx 25.4 \text{ lb}$

9. $\mathbf{u} \cdot \mathbf{v} = 5 \cdot 8 \cos 30° = 40\left(\frac{\sqrt{3}}{2}\right)$

$\quad = 20\sqrt{3}$

10. $2(7) + (-3)1 = 11;\ < 90°$

11. $-4(14) + 7(8) = 0;\ = 90°$

12. $\mathbf{u} \cdot \mathbf{v} = 5(-2) + 3(7) = 11$

$\quad |\mathbf{u}| = \sqrt{5^2 + 3^2} = \sqrt{34}$

$\quad |\mathbf{v}| = \sqrt{(-2)^2 + 7^2} = \sqrt{53}$

$\quad \cos\theta = \frac{11}{\sqrt{34} \cdot \sqrt{53}}$

$\quad\quad \theta \approx \text{Cos}^{-1}\left(\frac{11}{\sqrt{34} \cdot \sqrt{53}}\right) \approx 75.0°$

13. $5v_1 - 3v_2 = 0$

$\quad 5v_1 = 3v_2$

$\quad \frac{v_1}{v_2} = \frac{3}{5};\ k\langle 3, 5 \rangle$

---

26. State all the additional pairs of polar coordinates for $(3, 70°)$ if $-360° \le \theta \le 360°$. [6.4]

27. Convert between polar and rectangular coordinates. [6.4]

  a. $\left(-3, \frac{5\pi}{3}\right)$ to rectangular coordinates $\left(-\frac{3}{2}, \frac{3\sqrt{3}}{2}\right)$

  b. $(-2, 7)$ to polar coordinates $\approx (7.3, 105.9°)$

28. Find the distance (to the nearest tenth) between $\left(1, \frac{\pi}{2}\right)$ and $\left(4, \frac{5\pi}{4}\right)$. [6.4] 4.8

Transform each equation into polar form. Then verify your answer by using technology to graph the polar equation. [6.4]

29. $y = -3$ $r = -3\csc\theta$

30. $3x + y = 5$ $r = \frac{5}{3\cos\theta + \sin\theta}$

Transform each equation into rectangular form. Then describe the graph. [6.4]

31. $\theta = 45°$ $y = x$

32. $r = -8\cos\theta$ $(x + 4)^2 + y^2 = 16$

33. Identify the type of curve represented by each polar equation. [6.5]

  a. $r = -2\sin\theta$ III   **I.** line

  b. $r = 3\cos 4\theta$ V   **II.** Archimedean spiral

  c. $\theta = \frac{4\pi}{3}$ I   **III.** circle

  d. $r = -2\theta$ II   **IV.** lemniscate

  e. $r^2 = 8\sin 3\theta$ IV   **V.** rose

  f. $r = 3$ III

34. Identify the type of limaçon represented by each polar equation. [6.5]

  a. $r = 2 + 3\cos\theta$ IV   **I.** cardioid

  b. $r = 3 - 2\sin\theta$ II   **II.** dimpled limaçon

  c. $r = 2 + 2\sin\theta$ I   **III.** convex limaçon

  d. $r = 2 - \cos\theta$ III   **IV.** limaçon with inner loop

35. Identify the type of symmetry in the graph of a polar equation if substituting either ordered pair for (r, θ) produces an equivalent equation. [6.5]

  a. $(r, \pi + \theta)$ or $(-r, \theta)$ III   **I.** the polar axis

  b. $(r, \pi - \theta)$ or $(-r, -\theta)$ II   **II.** the line $\theta = \frac{\pi}{2}$

  c. $(r, -\theta)$ or $(-r, \pi - \theta)$ I   **III.** the pole

26. $(3, -290°),\ (-3, 250°),$ or $(-3, -110°)$

Graph each polar equation. [6.5]

36. $r = 3\sin 2\theta$ **37.** $r = 4\cos 3\theta$

38. $r = 2 - 2\sin\theta$ **39.** $r = 2 - 3\cos\theta$

40. $r^2 = 9\sin 2\theta$

41. Plot each complex number and express it in polar form. [6.6]

  a. $z_1 = 4 - 3i$ $\approx 5\text{ cis }5.64$ **b.** $z_2 = 4i$ $4\text{ cis }\frac{\pi}{2}$

42. Plot $b = 2\text{ cis }1.7$ in polar form and express b in standard form. [6.6] $\approx -0.26 + 1.98i$

Find each product, quotient, or power in polar form. Then express it in standard form. [6.6]

43. $\left(4\text{ cis }\frac{2\pi}{3}\right)\left(-3\text{ cis }\frac{5\pi}{6}\right)$ $-12\text{ cis }\frac{3\pi}{2};\ 12i$

44. $\left(-3\text{ cis }\frac{5\pi}{6}\right) \div \left(4\text{ cis }\frac{2\pi}{3}\right)$ $-\frac{3}{4}\text{ cis }\frac{\pi}{6};\ -\frac{3\sqrt{3}}{8} - \frac{3}{8}i$

45. $(3 + 4i)^4$ $\approx 625\text{ cis }212.5°;\ -527 - 336i$

46. $(1 - i)^3$ $2\sqrt{2}\text{ cis }\frac{5\pi}{4};\ -2 - 2i$

47. Find the cube roots of unity. Then graph the roots on the complex plane. [6.6] $1,\text{ cis }\frac{2\pi}{3},\text{ cis }\frac{4\pi}{3}$

48. Find the fourth roots of $-1 + 2i$, expressing your answers in polar form. [6.6] $\approx \sqrt[8]{5}\text{ cis }29.1°,\ \approx \sqrt[8]{5}\text{ cis }119.1°,\ \approx \sqrt[8]{5}\text{ cis }209.1°,\ \approx \sqrt[8]{5}\text{ cis }299.1°$

---

14. $\text{proj}_\mathbf{v}\,\mathbf{u} = \left(\frac{\langle 7, 5 \rangle \cdot \langle -2, 6 \rangle}{|\langle -2, 6 \rangle|^2}\right)\langle -2, 6 \rangle$

$\quad = \left(\frac{7(-2) + 5(6)}{(-2)^2 + 6^2}\right)\langle -2, 6 \rangle$

$\quad = \frac{2}{5}\langle -2, 6 \rangle = \left\langle -\frac{4}{5}, \frac{12}{5} \right\rangle$

$\mathbf{w}_2 = \mathbf{u} - \text{proj}_\mathbf{v}\,\mathbf{u}$

$\quad = \langle 7, 5 \rangle - \left\langle -\frac{4}{5}, \frac{12}{5} \right\rangle$

$\quad = \left\langle 7 + \frac{4}{5}, 5 - \frac{12}{5} \right\rangle = \left\langle \frac{39}{5}, \frac{13}{5} \right\rangle$

15.

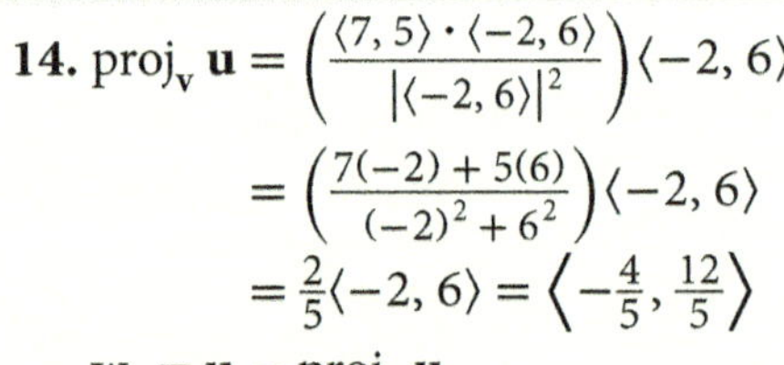

$\theta = 90 - 20 = 70°$

15a. $|\mathbf{w}_1| = |\mathbf{F}|\cos\theta = |80|\cos 70°$

$\quad\quad \approx 27.4 \text{ lb}$

15b. If the box is at rest, the frictional force is equal in magnitude to but opposite in direction of the downhill force.

16. $W = \mathbf{F} \cdot \mathbf{d}$

$\quad = (30 \text{ N})(50 \text{ m})\cos 50°$

$\quad \approx 964.2 \text{ N·m}$

17.

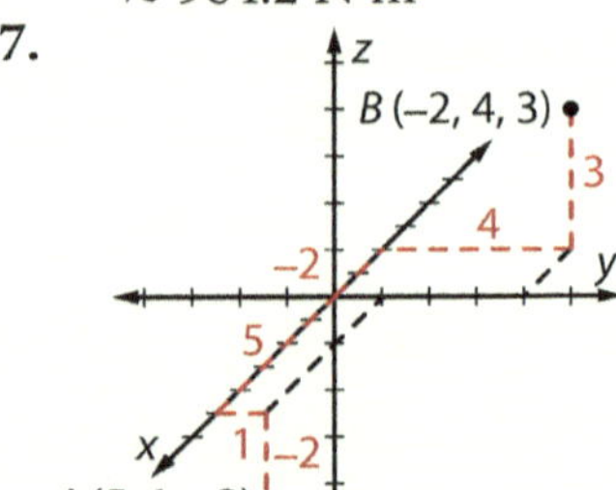

18. $d =$
$$\sqrt{(-4-3)^2 + (5-1)^2 + (3-(-2))^2}$$
$$= \sqrt{(-7)^2 + 4^2 + 5^2} = \sqrt{90}$$
$$= 3\sqrt{10} \approx 9.49$$
$$M\left(\frac{3+(-4)}{2}, \frac{1+5}{2}, \frac{-2+3}{2}\right)$$
$$= \left(-\tfrac{1}{2}, 3, \tfrac{1}{2}\right)$$

19. $\mathbf{v} = \langle 4-2, -1-5, 2-(-3) \rangle$
$$= \langle 2, -6, 5 \rangle$$
$$|\mathbf{v}| = \sqrt{2^2 + (-6)^2 + 5^2}$$
$$= \sqrt{65}$$
$$\frac{\mathbf{v}}{|\mathbf{v}|} = \frac{1}{\sqrt{65}}\langle 2, -6, 5 \rangle$$
$$= \left\langle \frac{2\sqrt{65}}{65}, -\frac{6\sqrt{65}}{65}, \frac{\sqrt{65}}{13} \right\rangle$$

20. $\langle 8-5, 2-(-1), -2-3 \rangle = \langle 3, 3, -5 \rangle$

21. $2\langle -3, 4, 1 \rangle - 3\langle 5, -1, 3 \rangle$
$$+ \tfrac{1}{2}\langle 8, 2, -2 \rangle$$
$$= \langle -6, 8, 2 \rangle + \langle -15, 3, -9 \rangle$$
$$+ \langle 4, 1, -1 \rangle$$
$$= \langle -17, 12, -8 \rangle$$

22. $\mathbf{v} \cdot \mathbf{w} = 2(-2) + 1(-3) + 5(7) = 28$
$$|\mathbf{v}| = \sqrt{2^2 + 1^2 + 5^2} = \sqrt{30}$$
$$|\mathbf{w}| = \sqrt{(-2)^2 + (-3)^2 + 7^2} = \sqrt{62}$$
$$\theta = \mathrm{Cos}^{-1}\left(\frac{28}{\sqrt{30} \cdot \sqrt{62}}\right) \approx 49.5°$$

23. $\mathbf{v} \cdot \mathbf{w} = 4(-2) + 5(3) + 1(-7) = 0$
$$|\mathbf{v}| = \sqrt{4^2 + 5^2 + 1^2} = \sqrt{42}$$
$$|\mathbf{w}| = \sqrt{(-2)^2 + 3^2 + (-7)^2} = \sqrt{62}$$
$$\theta = \mathrm{Cos}^{-1}\left(\frac{0}{\sqrt{42} \cdot \sqrt{62}}\right) = 90°$$

24.

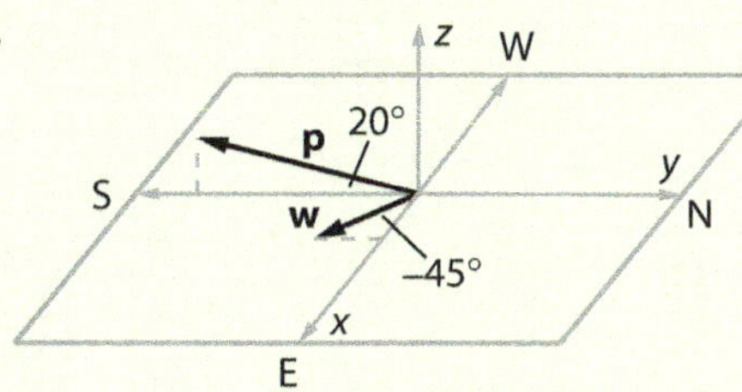

$$\mathbf{w} = \langle 25\cos(-45°), 25\sin(-45°), 0 \rangle$$
$$\approx \langle 17.7, -17.7, 0 \rangle$$
$$\mathbf{p} = \langle 0, -155\cos 20°, 155\sin 20° \rangle$$
$$\approx \langle 0, -145.7, 53.0 \rangle$$
$$\mathbf{r} = \mathbf{w} + \mathbf{p} \approx \langle 17.7, -163.3, 53.0 \rangle$$

25.

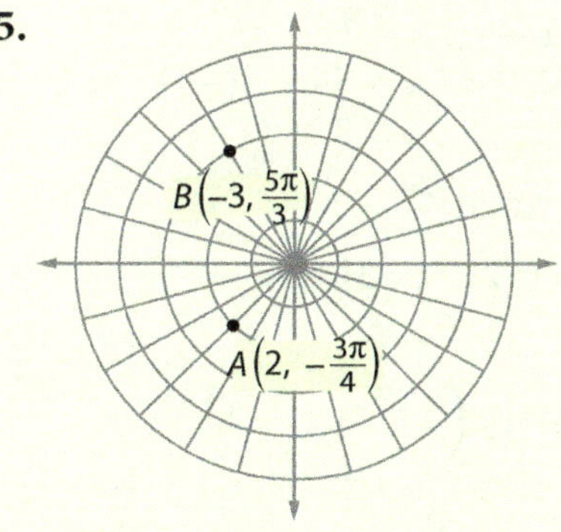

26. $(3, 70° - 360°) = (3, -290°)$
$$(-3, 70° + 180°) = (-3, 250°)$$
$$(-3, 70° - 180°) = (-3, -110°)$$

27a. $\left(-3\cos\frac{5\pi}{3}, -3\sin\frac{5\pi}{3}\right)$
$$= \left(-3\left(\tfrac{1}{2}\right), -3\left(-\tfrac{\sqrt{3}}{2}\right)\right)$$
$$= \left(-\tfrac{3}{2}, \tfrac{3\sqrt{3}}{2}\right) \approx (-1.5, 2.6)$$

27b. $r = \sqrt{(-2)^2 + 7^2} = \sqrt{53} \approx 7.3$
$$\alpha = \mathrm{Tan}^{-1}\left|\frac{7}{-2}\right| \approx 74.1°$$
$$\text{Q II}: = \theta \approx 180 - 74.1 = 105.9°$$
$$\therefore \approx (7.3, 105.9°)$$

28. $d = \sqrt{1^2 + 4^2 - 2(1)(4)\cos\left(\frac{5\pi}{4} - \frac{\pi}{2}\right)}$
$$\approx 4.8$$

29. $r\sin\theta = -3$
$$r = \frac{-3}{\sin\theta} = -3\csc\theta$$

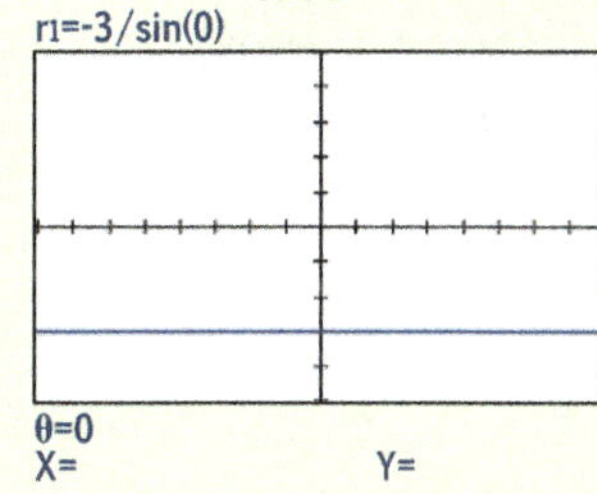

30. $3x + y = 5$
$$3r\cos\theta + r\sin\theta = 5$$
$$r = \frac{5}{3\cos\theta + \sin\theta}$$

31. $\tan\theta = \tan 45°$
$$\tan\theta = 1$$
$$\frac{y}{x} = 1$$
$$y = x$$
line through $(0, 0)$ with $m = 1$

32. $r = -8\cos\theta$
$$r^2 = -8r\cos\theta$$
$$x^2 + y^2 = -8x$$
$$(x^2 + 8x + 16) + y^2 = 16$$
$$(x + 4)^2 + y^2 = 16$$
circle centered at $(-4, 0)$ with $r = 4$

33–34. See the chart on p. 318.

36.

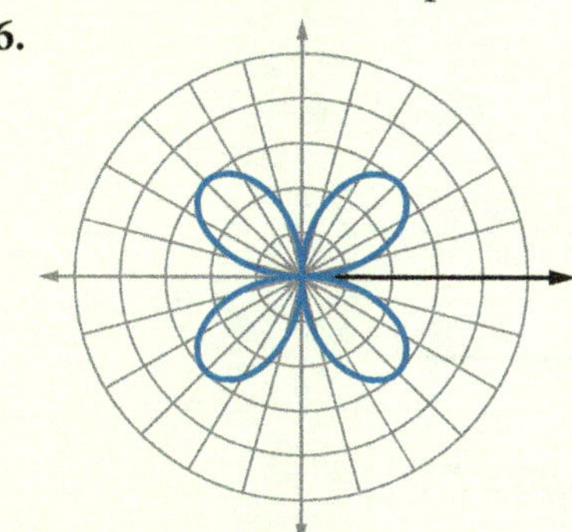

37.

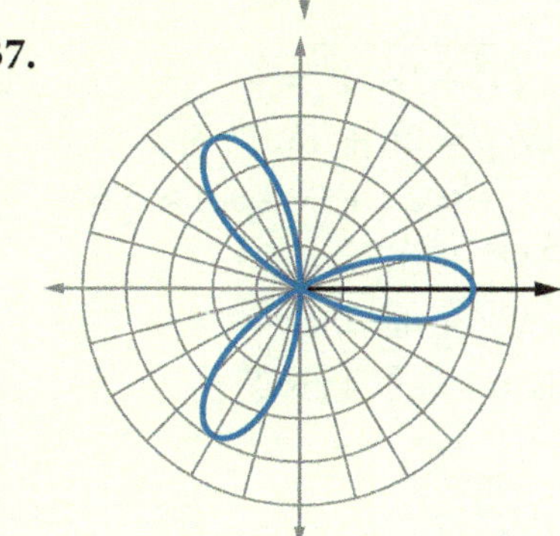

38.

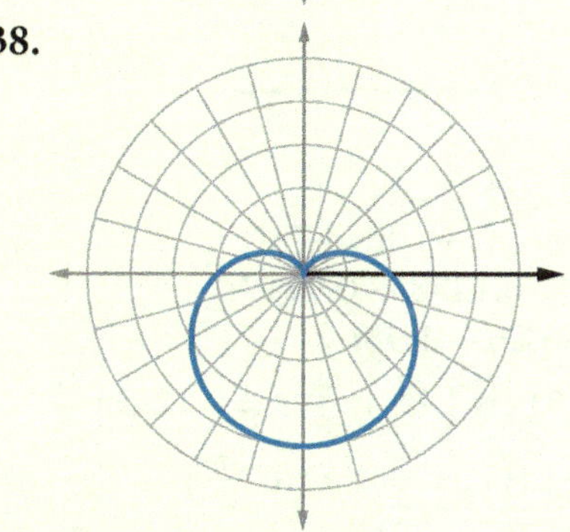

39.

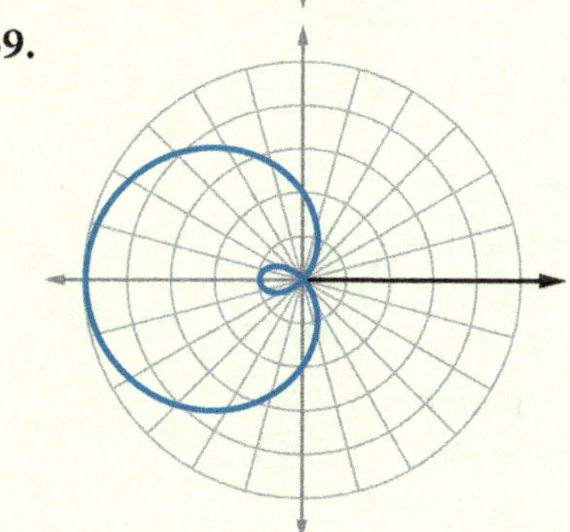

continued on next page

40.

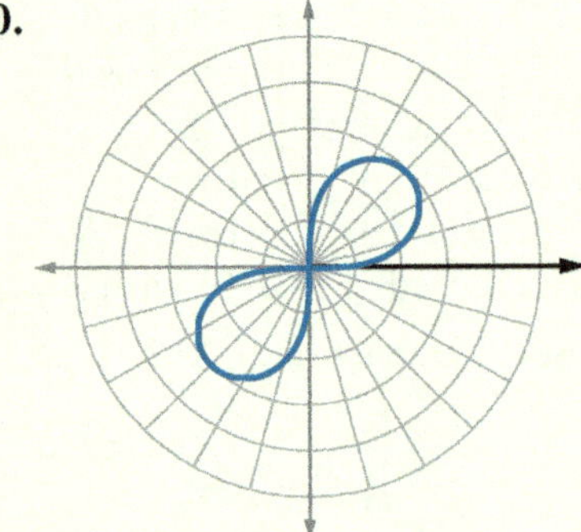

41.

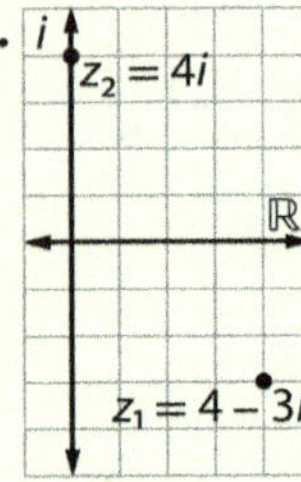

41a. $r = \sqrt{4^2 + (-3)^2} = 5$

Q IV: $\theta = 2\pi + \text{Tan}^{-1}\dfrac{-3}{4}$

$\approx 2\pi + (-0.64) \approx 5.64$

$\approx 5 \text{ cis } 5.64$

41b. $r = \sqrt{0^2 + 4^2} = 4$

$4i$ is at $\theta = \dfrac{\pi}{2}$; $4 \text{ cis } \dfrac{\pi}{2}$

42.

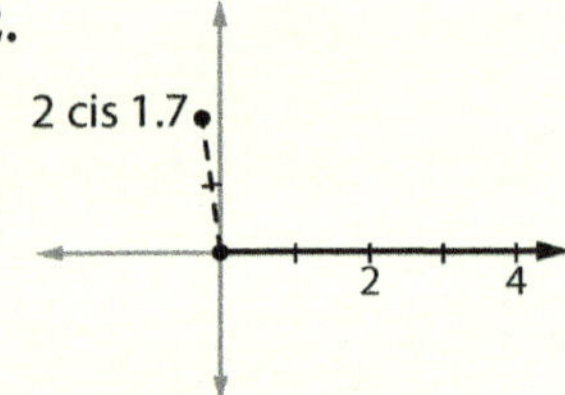

$2(\cos 1.7 + i \sin 1.7)$

$\approx 2(-0.1288 + 0.9917i)$

$\approx -0.26 + 1.98i$

43. $4(-3) \text{ cis } \left(\dfrac{2\pi}{3} + \dfrac{5\pi}{6}\right) = -12 \text{ cis } \dfrac{3\pi}{2}$

$= -12\left(\cos \dfrac{3\pi}{2} + i \sin \dfrac{3\pi}{2}\right)$

$= -12(0) - 12i(-1) = 12i$

44. $-\dfrac{3}{4} \text{ cis } \left(\dfrac{5\pi}{6} - \dfrac{2\pi}{3}\right) = -\dfrac{3}{4} \text{ cis } \dfrac{\pi}{6}$

$= -\dfrac{3}{4}\left(\cos \dfrac{\pi}{6} + i \sin \dfrac{\pi}{6}\right)$

$= -\dfrac{3}{4}\left(\dfrac{\sqrt{3}}{2}\right) - \dfrac{3}{4}\left(\dfrac{1}{2}\right)i$

$= -\dfrac{3\sqrt{3}}{8} - \dfrac{3}{8}i \approx -0.650 - 0.375i$

45. $z = 3 + 4i; r = 5$

Q I: $\theta = \text{Tan}^{-1}\dfrac{4}{3} \approx 53.1°$

$z \approx 5 \text{ cis } 53.1°$

$z^4 \approx 5^4 \text{ cis } 4(53.1°)$

$\approx 625 \text{ cis } 212.5°$

$\approx -527 - 336i$

46. $z = 1 - i; r = \sqrt{2}$

Q IV: $\theta = 2\pi + \text{Tan}^{-1}(-1)$

$= 2\pi - \dfrac{\pi}{4} = \dfrac{7\pi}{4}$

$z = \sqrt{2} \text{ cis } \dfrac{7\pi}{4}$

$z^3 = (\sqrt{2})^3 \text{ cis } 3\left(\dfrac{7\pi}{4}\right)$

$= 2\sqrt{2} \text{ cis } \dfrac{5\pi}{4}$

$= 2\sqrt{2}\left(-\dfrac{\sqrt{2}}{2} + i\left(-\dfrac{\sqrt{2}}{2}\right)\right)$

$= -2 - 2i$

47. $z = 1 + 0i = 1 \text{ cis } 0$

$w_k = \sqrt[3]{1} \text{ cis } \dfrac{0 + k2\pi}{3} = \text{cis } k\dfrac{2\pi}{3}$

$w_0 = \text{cis } 0 = 1 + i(0) = 1$

$w_1 = \text{cis } \dfrac{2\pi}{3} = -\dfrac{1}{2} + \dfrac{\sqrt{3}}{2}i$

$w_2 = \text{cis } \dfrac{4\pi}{3} = -\dfrac{1}{2} - \dfrac{\sqrt{3}}{2}i$

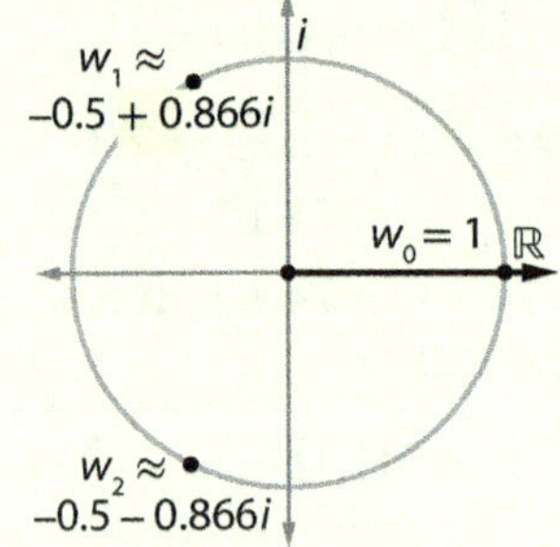

48. $z = -1 + 2i \approx \sqrt{5} \text{ cis } 116.6°$

$w_k \approx (\sqrt{5})^{\frac{1}{4}} \text{ cis } \dfrac{116.6° + k360°}{4}$

$\approx \sqrt[8]{5} \text{ cis } (29.1° + k90°)$

$w_0 \approx \sqrt[8]{5} \text{ cis } 29.1°$ (principal)

$w_1 \approx \sqrt[8]{5} \text{ cis } 119.1°$

$w_2 \approx \sqrt[8]{5} \text{ cis } 209.1°$

$w_3 \approx \sqrt[8]{5} \text{ cis } 299.1°$

Selected Odd Answers

Pages 667–82 in the *Student Edition* contain selected odd answers.

A link to a PDF of these selected odd answers can be found on AfterSchoolHelp.com. Click on "Math" from the "Subjects" menu and then choose "View Overview" for *Precalculus*.

Glossary

absolute value function (p. 18) A function of the form $f(x) = |x|$ where $f(x) = x$ if $x \geq 0$ and $f(x) = -x$ if $x < 0$.

acceleration due to gravity (p. 14) The constant g, where g is approximately 32 ft/sec^2 or 9.8 m/sec^2.

algebraic function (p. 128) A function that involves only the algebraic operations of addition, subtraction, multiplication, division of constants and an independent variable, and raising an independent variable to a rational power.

alternative (research) hypothesis (p. 573) A statement that is the negation of the null hypothesis and contains an inequality such as $>$, $<$, or $\neq$. Denoted H_a.

ambiguous case (p. 270) The SSA case, which does not always uniquely determine a triangle.

amplitude (p. 198) Half the difference of the maximum and minimum values of a periodic function.

anchor step (p. 477) The initial step in mathematical induction that shows $P(n)$ is true for $P(1)$.

angle of depression (p. 183) The angle formed by a horizontal line and the line of sight toward an object that is below the horizontal.

angle of elevation (p. 183) The angle formed by a horizontal line and the line of sight toward an object that is above the horizontal.

angular speed (p. 175) The ratio of the angle of rotation (in radians) per unit of time: $\omega = \frac{\theta}{t}$.

annual percentage rate (APR) (p. 130) The yearly interest rate expressed as a percentage.

annual percentage yield (APY) (p. 134) The equivalent annual rate for an investment at a given annual percentage rate (APR) compounded n times after 1 yr: $P\left(1 + APY\right) = P\left(1 + \frac{APR}{n}\right)^n$.

antiderivative (p. 638) If $F'(x) = f(x)$, the function $F(x)$ is an antiderivative of $f(x)$.

aphelion (pp. 407, 438) The point at which an orbiting object is at its farthest distance from the sun.

Arccosine (p. 217) The inverse function of cosine: $y = \text{Cos}^{-1} x$.

arc length formula (p. 174) A central angle of radian measure θ in a circle with radius r intercepts an arc with length $s = r\theta$.

Arcsine (p. 216) The inverse function of sine: $y = \text{Sin}^{-1} x$.

Arctangent (p. 218) The inverse function of tangent: $y = \text{Tan}^{-1} x$.

argument (p. 324) The angle θ of a complex number in polar form $z = r(\cos\theta + i\sin\theta)$.

arithmetic means (p. 458) The terms between any two given terms of an arithmetic sequence.

arithmetic sequence (p. 456) A sequence in which the difference of any two consecutive terms is a constant.

arithmetic series (p. 459) The sum of the terms of an arithmetic sequence.

asymptote (p. 104) A line that a graph continually approaches.

augmented matrix (p. 353) A combination of the coefficient and constant matrices of a linear system.

average velocity (p. 615) $\dfrac{\text{change in position}}{\text{change in time}} = \dfrac{\Delta d}{\Delta t}$.

axis of symmetry (pp. 39, 396) The line that divides a parabola into mirror-image halves.

bias (p. 585) A factor that systematically causes a sample to differ from the population.

binomial (p. 472) The sum of two unlike terms.

binomial coefficients (p. 472) The coefficients of the terms generated by expanding $(a + b)^n$.

binomial experiment (p. 474) A statistical experiment of n identical independent trials where each trial results in either a success or a failure.

binomial probability (p. 474) The probability of x successes in n trials of a binomial experiment is $P(x) = {}_nC_x p^x q^{n-x}$, where p is the probability of success and q, the probability of failure, is $1 - p$.

binomial probability distribution (p. 549) A table describing all the possible outcomes of a binomial experiment: $P(x) = {}_nC_x p^x q^{n-x}$.

bivariate data (p. 59) Data containing two related variables.

carrying capacity (p. 158) The limit of growth supported by the environment.

center (of a hyperbola) (p. 412) The midpoint of the transverse axis of a hyperbola.

center (of an ellipse) (p. 403) The midpoint of the major axis of an ellipse.

change of base formula (p. 145) $\log_b x = \dfrac{\log_a x}{\log_a b}$ for $a, b \in \mathbb{N}$ when $a \neq 1$ and $b \neq 1$.

circular function (pp. 191–92) A trigonometric function, such as $\tan\theta = \frac{y}{x}$, defined in terms of a point $P(x, y)$ on the unit circle where $x, y,$ and θ are real numbers.

closed interval (p. 4) An interval that includes both endpoints. The interval $\{x \mid a \leq x \leq b\}$ is denoted $[a, b]$.

coefficient of determination (p. 60) (r^2 value) Describes the extent to which the variation in the dependent variable can be explained by variation in the independent variable.

cofactor (p. 362) The cofactor C_{ij} of entry a_{ij} in a square matrix with minor M_{ij} is $C_{ij} = (-1)^{i+j} M_{ij}$.

cofunction (pp. 180, 234) A trigonometric function of an angle's complement.

combination (pp. 473, 492) The number of different ways to choose r out of n items where order is not important, denoted ${}_nC_r$, $C(n, r)$, or $\binom{n}{r}$.

common difference (p. 456) The difference, d, between any two consecutive terms in an arithmetic sequence.

common logarithmic function (p. 137) The function $y = \log x$ where $10^y = x$.

common ratio (p. 463) The quotient, r, between any two consecutive terms in a geometric sequence.

completing the square (p. 40) Adding the constant $\left(\frac{b}{2}\right)^2$ to an expression in the form $x^2 + bx$ to obtain a perfect square trinomial.

complex number (p. 644) Any complex number, z, can be expressed in the standard form $z = a + bi$, where $a, b \in \mathbb{R}$ and $i = \sqrt{-1}$.

complex plane (p. 324) A coordinate plane with real and imaginary axes.

components (p. 287) The horizontal and vertical parts of a vector. Denoted $\mathbf{v} = \langle a, b \rangle$, where a represents the horizontal change $(x_2 - x_1)$ and b represents the vertical change $(y_2 - y_1)$ from the tail to the head of the vector.

component vectors (p. 296) The result of decomposing a vector into two orthogonal vectors, one of which is parallel to another given vector.

composition (p. 47) The composition of two functions f and g, denoted $f \circ g$, is defined as $(f \circ g)(x) = f(g(x))$.

compound inequality (p. 651) The combination of two inequalities into one statement using *and* or *or*.

compound interest (p. 130) The interest on both the principal and any interest accumulated.

conditional equation (pp. 233, 246) An equation that is true only for certain value(s) of the variable.

conditional probability (p. 499) The probability of an event, given the occurrence of a previous event.

confidence interval (p. 567) An interval around the point estimate in which it is reasonably certain that the actual parameter lies.

confidence level (p. 567) The probability that the confidence interval actually contains the population parameter.

conic section (conic) (p. 395) The intersection of a plane with a right circular conical surface.

conjugate axis (p. 412) The segment that passes through a hyperbola's center and is perpendicular to the hyperbola's transverse axis.

conjunction (p. 651) A compound mathematical sentence of the form $a \wedge b$ (or "a and b") that is true when both statements are true (the intersection of their solutions).

constant function (pp. 12, 25–26) A function of the form $f(x) = k$ whose value does not vary.

constant of variation (proportion) (p. 27) The nonzero constant relating the variables in a variation. In the functions $y = kx$ and $y = \frac{k}{x}$, k is the constant of variation.

continuity correction factor (p. 560) A correction applied when a continuous distribution is used to approximate a discrete distribution.

continuous function (pp. 19, 21) A function with no gaps or holes.

continuous on an interval (p. 21) A function is continuous on an interval (a, b) iff it is continuous at every point in (a, b).

continuous random variable (p. 546) A variable representing an uncountable number of outcomes within an interval in a probability experiment.

convergent sequence (p. 449) A sequence whose values approach a finite real number as the position number approaches infinity.

correlation coefficient (p. 60) Denoted r, where $-1 \leq r \leq 1$. Describes the degree to which two variables are related.

cosecant (pp. 179, 188) The trigonometric ratio that is the reciprocal of sine.

cosine (pp. 179, 188) The ratio of the leg adjacent to an acute angle and the hypotenuse of a right triangle. The ratio $\frac{x}{r}$, where $P(x, y)$ is a point on the terminal side of an angle in standard position and $r = \sqrt{x^2 + y^2}$.

cotangent (pp. 179, 188) The trigonometric ratio that is the reciprocal of tangent.

coterminal angles (p. 174) Two angles in standard position with the same terminal side.

co-vertices (pp. 403, 412) The endpoints of an ellipse's minor axis or of a hyperbola's conjugate axis.

critical (rejection) region (p. 575) The area where the difference between the test statistic and the proposed population parameter is enough to reject the null hypothesis.

critical value (p. 567) The z-value that corresponds to a certain confidence level.

damped harmonic motion (p. 225) Motion that occurs when friction reduces the displacement of an oscillating object. Modeled by $f(x) = ae^{-ct} \cos bt$ or $f(x) = ae^{-ct} \sin bt$ ($c > 0$), where a is the maximum displacement.

damped sinusoid (p. 225) A function of the form $f(x) = g(x) \sin bx$ or $f(x) = g(x) \cos bx$.

damping constant (p. 225) The value of $c > 0$ in the equation $f(x) = ae^{-ct} \cos bt$ or $f(x) = ae^{-ct} \sin bt$ that affects the rate of decrease in the amplitude of oscillations.

damping factor (p. 225) A factor $g(x)$ that reduces the amplitude of a sinusoid model $f(x) = g(x) \sin bx$ or $f(x) = g(x) \cos bx$.

decay factor (p. 129) The base, b, in the exponential function $y = ab^x$, given $0 < b < 1$.

decomposition (p. 48) The process of finding simpler functions $g(x)$ and $f(x)$ such that the given function $h(x) = f(g(x))$.

decreasing function (pp. 25–26) A function in which the value of y decreases as x increases.

definite integral (p. 633) The area of a region between the graph of a function $f(x)$ and the interval $[a, b]$ on the x-axis is given by the definite integral $\int_a^b f(x)\, dx = \lim_{n \to \infty} \sum_{i=1}^{n} f(x_i)\, \Delta x$, provided the limit exists.

degenerate conic (p. 395) A point, a line, or two intersecting lines that result when a plane intersecting a conical surface contains the vertex.

dependent events (p. 490) Events in which the probability of one event changes because another event has occurred.

dependent variable (p. 3) The variable representing the values of the range in a relation.

depressed polynomial (p. 88) The quotient of a polynomial divided by one of its binomial factors.

derivative function (p. 613) The derivative of $f(x)$, denoted $f'(x)$, where $f'(x) = \lim_{h \to 0} \frac{f(x+h) - f(x)}{h}$.

descriptive statistics (p. 507) The organization of data and the determination and reporting of its numerical characteristics.

determinant (pp. 361, 363) A characteristic value associated with a square matrix that indicates whether the inverse of the matrix exists.

difference quotient (p. 612) The slope of a secant line through the given points $(x, f(x))$ and $(x + h, f(x + h))$ is $\frac{f(x+h) - f(x)}{h}$.

differentiable function (p. 613) A function that has a derivative at every point of its domain.

differential operator (p. 614) The notation $\frac{d}{dx}$ that indicates to find the derivative of a function.

directrix (p. 435) A fixed line such that all points on a conic are equidistant to a conic's focus and the line.

direct variation (proportion) (p. 27) A function of the form $y = kx$ or $k = \frac{y}{x}$, where k is a nonzero constant of variation.

discontinuity (p. 21) A gap or hole in a relation caused by excluded values.

discrete random variable (p. 546) A variable representing a countable number of outcomes of a probability experiment.

disjunction (p. 651) A compound mathematical sentence of the form $a \vee b$ (or "a or b") that is true when either statement is true (the union of their solutions).

divergent sequence (p. 449) A sequence whose values do not approach a finite real number as the position number approaches infinity.

domain (p. 2) The set of first elements of the ordered pairs in a relation.

dot product (pp. 294–95, 305) For $\mathbf{u}$ and $\mathbf{v}$, $\mathbf{u} \cdot \mathbf{v} = |\mathbf{u}|\,|\mathbf{v}| \cos \theta$, where θ is the angle between $\mathbf{u}$ and $\mathbf{v}$ when the vectors are placed in standard position ($0 \leq \theta \leq 180$). Using vector components: if $\mathbf{u} = \langle u_1, u_2 \rangle$ and $\mathbf{v} = \langle v_1, v_2 \rangle$, then $\mathbf{u} \cdot \mathbf{v} = u_1 v_1 + u_2 v_2$.

eccentricity (pp. 407, 414, 435) The measure of a conic's deviation from a circle.

elementary row operation (p. 354) An operation that transforms an augmented matrix into an equivalent system by interchanging any two rows, multiplying or dividing each entry in a row by a nonzero number, or adding a multiple of one row to another.

eliminating the parameter (p. 52) The process of expressing y in terms of x when given a set of parametric equations.

ellipse (p. 403) The set of points in a plane such that the sum of the distances from two fixed points (the foci) is constant.

empty set (p. 647) A set containing no elements. Denoted $\varnothing$. Also called a *null set*.

end behavior (p. 25) A description of the values of a function as x approaches positive or negative infinity.

equivalent equations (p. 650) Equations that have the same solution set.

even function (p. 26) A function where $f(-x) = f(x)$ for all $x \in D$.

existence theorems (p. 94) Theorems that guarantee the existence of a quantity, but do not provide a means of finding it.

expected value (p. 547) The mean of a probability distribution. Denoted $E(x)$.

experiment (p. 497) A repeatable activity resulting in exactly one of a finite set of outcomes.

experimental (empirical) probability (p. 497) The ratio of the number of times an event occurs to the number of trials in an experiment.

explicit formula (pp. 447, 456) A formula in terms of a_n that allows direct computation of any term in the sequence.

exponential decay (p. 129) The decrease of a quantity by the same percentage per unit of time.

exponential equation (p. 149) An equation in which the variable occurs in the exponent.

exponential function (p. 128) A function of the form $f(x) = ab^x$ where $a \neq 0$, $b > 0$, and $b \neq 1$.

exponential growth (p. 129) The increase of a quantity by the same percentage per unit of time.

extraneous root (p. 74) A derived solution that is invalid because it does not make the original equation true.

extrapolation (p. 60) The process of making predictions outside the range of a data set.

factorial (p. 448) The product of a number, n, and every natural number less than that number. Denoted $n!$.

family of functions (p. 32) A group of functions whose graphs share similar characteristics.

five-number summary (p. 519) The minimum, lower quartile, median, upper quartile, and maximum values of a data set.

focal axis (p. 435) The line that passes through a conic's focus and is perpendicular to the conic's directrix.

focal distance (pp. 403, 412) The distance from the center of an ellipse or a hyperbola to each focus.

focal length (p. 396) The distance from the vertex of a parabola to its focus.

focus (p. 435) A fixed point such that the distances to that point from any point on the conic and from the conic's directrix are in a constant ratio, which is the conic's eccentricity, e.

frequency (p. 199) The reciprocal of the period of a function: $f = \dfrac{1}{p} = \dfrac{|b|}{2\pi}$.

function (p. 2) A relation in which every element of the domain is paired with one and only one element of the range.

future value of an annuity (p. 467) The future value, FV, of an annuity consisting of n periodic payments of P with a periodic interest rate i is given by $FV = P\left[\dfrac{1 - (1+i)^n}{1 - (1+i)}\right]$ or $FV = P\left[\dfrac{(1+i)^n - 1}{i}\right]$.

Gaussian elimination (p. 353) The process of eliminating variables from a system to find an equivalent system in row echelon form.

Gauss-Jordan elimination (p. 355) The process of applying elementary row operations to an augmented matrix until the unique reduced row echelon form is found.

geometric means (p. 465) The terms between any two given terms of a geometric sequence.

geometric sequence (p. 463) A sequence in which the ratio of any two consecutive terms is a constant.

geometric series (p. 465) The sum of the terms of a geometric sequence.

greatest integer function (p. 18) A function that returns the greatest integer less than or equal to x.

growth factor (p. 129) The base, b, in the exponential function $y = ab^x$, given $b > 1$.

growth rate (p. 131) The term r in the exponential growth function, $P(t) = P_0(1 + r)^t$, where $t = $ time.

half-life (p. 132) The time it takes for half a radioactive isotope to decay into a stable isotope of another element.

harmonic sequence (p. 458) The reciprocals of an arithmetic sequence.

histogram (p. 510) A graphic representation of a frequency distribution.

horizontal asymptote (p. 104) The line $y = c$ is a horizontal asymptote of $f(x)$ if $\lim\limits_{x \to -\infty} f(x) = c$ or $\lim\limits_{x \to \infty} f(x) = c$.

hyperbola (p. 412) The set of points in a plane such that the difference of the distances from two fixed points (its foci) is constant.

hypothesis testing (p. 573) The process of using statistics from a sample to examine a claim about the value of a population parameter.

identity (p. 233) An equation that is true for all values in the domain of the variable.

imaginary axis (p. 324) The vertical axis in the complex plane.

imaginary unit (p. 664) The imaginary unit, i, is the square root of -1, so that $i = \sqrt{-1}$ and $i^2 = -1$.

increasing function (pp. 25–26) A function in which the value of y increases as x increases.

indefinite integral (p. 639) The indefinite integral of $f(x)$ is $\int f(x)\, dx = F(x) + C$, where $F(x)$ is the antiderivative of $f(x)$ and C is a constant.

independent events (p. 490) Events in which the occurrence of one has no influence on the probability of the other events occurring.

independent variable (p. 3) The variable representing the values of the domain in a relation.

indeterminate form (p. 606) An expression obtained, such as $\frac{0}{0}$, that implies the function does not contain enough information to determine the limit at that point.

inferential statistics (p. 507) The process of interpreting the statistics of a sample to make predictions about a population's parameters.

infinite discontinuity (pp. 21, 105) An infinite discontinuity exists for $f(x)$ at $x = a$ if a one-sided $\lim\limits_{x \to a^{\pm}} f(x) = \pm\infty$.

infinite limit (p. 600) Describes a function's behavior, but the true limit does not exist.

instantaneous velocity (pp. 615–16) If the position of an object is represented by a function of time $f(t)$, then its instantaneous velocity at time t is $v_{inst}(t) = f'(t) = \lim_{h \to 0} \frac{f(t + h) - f(t)}{h}$.

interpolation (p. 60) The process of making predictions within the range of a data set.

interquartile range (IQR) (p. 519) The difference between the third quartile and the first quartile of a data set.

intersection of sets (p. 647) The set of elements common to the given sets.

interval (p. 4) A continuous set of real numbers.

interval estimate (p. 566) A range of values used to estimate a population's parameter.

inverse functions (p. 56) A relation and its inverse that are both functions: $(f^{-1} \circ f)(x) = (f \circ f^{-1})(x) = x$.

inverse matrix (p. 370) The multiplicative inverse of square matrix A, denoted A^{-1}, is the matrix such that $AA^{-1} = A^{-1}A = I$, the identity matrix.

inverse relation (p. 53) If a relation R contains the ordered pair (a, b), its inverse relation, R^{-1}, contains the ordered pair (b, a).

inverse variation (proportion) (p. 28) A relation in which the product of the two variables is a nonzero constant, k. That is, $xy = k$, where $k \neq 0$.

invertible (nonsingular) matrix (p. 370) A matrix for which an inverse exists.

irrational number (p. 648) A real number that cannot be expressed as a ratio of two integers.

irreducible quadratic factor (p. 99) A second-degree polynomial that cannot be factored over the integers.

leading coefficient (p. 78) The constant factor of the term with the highest degree in a polynomial expression.

least common denominator (LCD) (p. 651) The smallest positive integer that is a multiple of two or more given denominators.

left-hand limit (p. 21) The left-hand limit of $f(x)$ as x approaches a, denoted $\lim_{x \to a^-} f(x)$, is the number to which $f(x)$ gets closer and closer as x approaches a from the left.

level of significance (p. 575) The maximum allowable probability of making a type I error.

limit (pp. 21, 599) The limit of $f(x)$ as x approaches a, denoted $\lim_{x \to a} f(x)$, is L iff $\lim_{x \to a^-} f(x) = \lim_{x \to a^+} f(x) = L$.

linear combination (p. 289) A sum of vectors in the form $a\mathbf{v} + b\mathbf{w}$, where a and b are scalars.

linear correlation (p. 60) Describes how closely data points are clustered along a trend line.

linear programming (p. 377) A technique used to maximize or minimize a particular quantity under a given set of constraints.

linear regression (p. 59) A statistical method of determining the line of best fit (trend line).

linear speed (p. 175) The ratio of arc length, or distance traveled, per unit of time: $v = \frac{s}{t}$.

line of best fit (p. 59) See **trend line**.

logarithmic equation (p. 149) An equation of the form $\log_b x = k$ in which the variable occurs in the argument (input) of a logarithmic expression.

logarithmic function (p. 136) The inverse of the exponential function $b^y = x$, denoted $y = \log_b x$, where x and b are positive and $b \neq 1$: $y = \log_b x$ iff $b^y = x$.

logistic growth function (p. 158) A function that can be written in the form $f(x) = \frac{c}{1 + ae^{-bx}}$ where a, b, and c are positive constants.

magnitude (pp. 287–88) The length of vector $\mathbf{v} = \langle a, b \rangle$ is $|\mathbf{v}| = \sqrt{a^2 + b^2}$.

major axis (p. 403) The chord containing the foci of an ellipse whose endpoints are the vertices of the ellipse.

margin of error (maximum error of the estimate) (p. 567) The greatest acceptable difference between the point estimate and the actual value of the parameter for a given confidence level, c.

mathematical induction (p. 477) A method of proving a proposition $P(n)$ is true for all natural numbers by showing that $P(1)$ is true and that $P(k + 1)$ is true when $P(k)$ is true.

matrix (p. 343) A rectangular array of numbers.

mean (p. 516) The arithmetic average obtained by dividing the sum of the values by the number of values.

median (p. 516) The middle value or the mean of the two middle values in a set of data arranged in numerical order.

midline (p. 201) The horizontal line halfway between the maximum and minimum values of a function: $y = \frac{max - min}{2}$.

minor (p. 362) The minor of entry a_{ij} in square matrix A, notated M_{ij}, is the determinant of the matrix formed by removing the ith row and jth column of A.

minor axis (p. 403) The chord perpendicular to the major axis that passes through the center of an ellipse and whose endpoints are the co-vertices of the ellipse.

mode (p. 516) The most frequently appearing value(s) in a data set.

modulus (p. 324) The absolute value of a complex number (the distance from the complex origin). If $z = a + bi$, then the modulus of $z = \sqrt{a^2 + b^2}$.

monomial functions (p. 78) Power functions of the form $f(x) = kx^n$, where n is a natural number, and constant functions of the form $f(x) = c$, where c is a real number.

multiplicative identity matrix (p. 346) A square matrix in which entries on the main diagonal are 1 and all other entries are 0.

multiplicity (p. 81) A root of a polynomial has a multiplicity of n if the polynomial contains n identical binomial factors.

natural base (p. 130) The number $e = \lim_{x \to \infty} \left(1 + \frac{1}{x}\right)^x \approx 2.71828$.

natural exponential function (p. 130) The function $f(x) = e^x$ where $e \approx 2.71828$.

natural logarithmic function (p. 138) The inverse of the exponential function $e^y = x$, denoted $y = \ln x$, where x is positive: $y = \ln x$ iff $e^y = x$.

nonlinear asymptote (p. 107) Given a rational function $f(x) = \frac{a(x)}{b(x)}$ where degree of $a(x)$ is at least 2 greater than the degree of $b(x)$, the nonlinear asymptote is the quotient (without the remainder) when $a(x)$ is divided by $b(x)$.

nonrigid transformation (p. 33) A transformation that distorts the shape of the graph.

nth partial sum (p. 450) The sum of a sequence's first n terms, denoted S_n.

nth root of a complex number (p. 327) If $w = a + bi$ and z is a complex number such that $w^n = z$, then w is an nth root of z.

nth root of unity (p. 327) If $w = a + bi$ and $w^n = 1$, then w is the nth root of unity.

null hypothesis (p. 573) A statement about a population parameter that contains an equality such as $\geq$, $\leq$, or $=$. Denoted H_0.

objective function (p. 377) A linear function representing the quantity to be optimized in a linear programming problem.

octant (p. 303) One of eight regions created in space by a three-dimensional coordinate system.

odd function (p. 26) $f(-x) = -f(x)$ for all $x \in D$.

one-sided limit (p. 21) See **right-hand limit** or **left-hand limit**.

one-to-one function (p. 55) A function in which each element of the domain corresponds to exactly one element of the range.

open interval (p. 4) An interval that includes all the points between two endpoints but does not include the endpoints. The interval $\{x \mid a < x < b\}$ is denoted (a, b).

orthogonal vectors (p. 296) Two vectors $\mathbf{u}$ and $\mathbf{v}$ are orthogonal iff $\mathbf{u} \cdot \mathbf{v} = 0$. Nonzero orthogonal vectors are also perpendicular.

outlier (pp. 60, 519) A value in a data set that is significantly higher or lower than the other values in the set.

parabola (pp. 39, 396) The set of points in a plane that are equidistant from a fixed line (the directrix) and a fixed point (the focus) not on that line.

parameter (statistical) (p. 507) The numerical value for a characteristic of a population.

parametric curve (p. 429) A set of ordered pairs (x, y) defined by the parametric equations $x = f(t)$ and $y = g(t)$, where $f(t)$ and $g(t)$ are continuous functions of t on the parameter interval, I.

parametric equations (p. 52) A set of functions representing each coordinate in terms of one or more other independent variables (the parameters).

parent function (p. 32) The basic function used as a reference for transformed functions within a family of functions.

partial fraction decomposition (p. 383) The process of expressing a proper rational expression as a sum of simpler rational functions.

percentile (p. 520) When a distribution is divided into 100 equal parts, each percentile represents the amount of the distribution that lies below that percentile's value.

perihelion (pp. 407, 438) The point at which an orbiting object is at its closest distance to the sun.

period (pp. 193, 198) The horizontal length of one cycle in a periodic function.

periodic function (p. 193) A function f with a repeating pattern of y-values so that there exists a constant c such that $f(x + c) = f(x)$ for all x in the domain.

permutation (p. 491) The number of different ways to arrange (or order) r out of n items. Denoted ${}_nP_r$ or $P(n, r)$.

phase shift (p. 200) A horizontal translation in the graph of a trigonometric function.

piecewise function (p. 18) A function whose value is defined by multiple function rules for various intervals of its domain.

point estimate (p. 566) A single-value estimation of a population's parameter.

point (removable) discontinuity (pp. 21, 106) A single point on the graph of $f(x)$ that is not defined.

point-slope form (p. 12) A linear equation of the form $y - y_1 = m(x - x_1)$, where m is the slope of the line and (x_1, y_1) is a point on the line.

polar axis (p. 309) An initial ray drawn from the pole in the direction of the positive x-axis.

polar coordinates (p. 309) Coordinates of the form $P(r, \theta)$, where r is a directed distance from the pole to the point P and θ is a directed angle from the polar axis.

polar (trigonometric) form (p. 324) If $z = a + bi$, then $z = r(\cos \theta + i \sin \theta)$ or $z = r \operatorname{cis} \theta$ where r is the modulus and θ is the argument of z.

pole (p. 309) A fixed point, O, that serves as the origin for the polar coordinate system.

polynomial function (p. 78) A function of the form $p(x) = a_n x^n + a_{n-1} x^{n-1} + a_{n-2} x^{n-2} + \cdots + a_1 x + a_0$, where n is a nonnegative integer and the coefficients are real numbers.

population (p. 507) The larger overall set being considered in a statistical study.

power function (p. 25) A function of the form $f(x) = kx^n$ where k and the power n are nonzero constants.

primary solution (p. 246) A solution to a trigonometric equation that falls within the interval $[0, 2\pi)$.

principal root (p. 328) When $k = 0$ in the complex root of $\sqrt[n]{r} \operatorname{cis} \frac{\theta + k2\pi}{n}$.

principal value (p. 216) A value within a restricted domain of a trigonometric function.

probability distribution (p. 546) A table, graph, or equation that associates each possible value of a random variable, x, with its probability, $P(x)$.

projection (p. 296) If $\mathbf{w}_1$ and $\mathbf{w}_2$ are orthogonal component vectors of $\mathbf{u}$ and $\mathbf{w}_1$ is parallel to vector $\mathbf{v}$, then $\mathbf{w}_1$ is the projection of $\mathbf{u}$ onto $\mathbf{v}$, $\operatorname{proj}_{\mathbf{v}} \mathbf{u}$.

p-value (p. 577) The probability of obtaining a test statistic as extreme as or more extreme than the one found in the sample, assuming the null hypothesis is true.

quadrantal angle (p. 174) An angle in standard position whose terminal side lies along either the x- or y-axis.

quadratic function (p. 40) A function that can be written in the standard form $f(x) = ax^2 + bx + c$, where a, b, and $c \in \mathbb{R}$ and $a \neq 0$.

quartiles (p. 518) Values that divide ordered data into four equal sets.

radian (p. 173) The measure of a central angle that intercepts an arc whose length is the same as the circle's radius: $1 \text{ radian} = \frac{180°}{\pi} \approx 57.3°$.

radian measure (p. 173) The ratio of a central angle's intercepted arc's length, s, to the circle's radius, r: $\theta = \frac{s}{r}$.

radical equation (p. 74) An equation of the form $\sqrt[n]{x} = k$ that contains a radical expression with a variable in the radicand, or an equation of the form $x^{\frac{p}{r}} = k$ that contains a variable with a rational exponent.

radical function (p. 72) A function that contains a radical expression with the independent variable in the radicand.

random sample (pp. 507, 586) A sample for which every member of the population has an equal chance of being selected.

random variable (p. 546) A variable representing all possible numerical outcomes in a probability experiment.

range (of a data set) (pp. 510, 517) The difference between the maximum and minimum values.

range (of a function) (p. 2) The set of second elements of the ordered pairs in a relation.

rational equation (p. 112) An equation containing at least one ratio of polynomials, $\frac{a(x)}{b(x)}$, where $b(x) \neq 0$.

rational function (p. 104) A function that can be expressed as $f(x) = \frac{a(x)}{b(x)}$, where $a(x)$ and $b(x)$ are polynomials and $b(x) \neq 0$.

rational number (p. 648) A number that can be expressed as a ratio of two integers $\frac{a}{b}$, where $b \neq 0$.

real numbers (p. 648) All the numbers that can be graphed on a number line.

recursive formula (pp. 447, 456) A formula that states the first term(s) of a sequence and defines the remaining terms using the previous term(s).

reduced row echelon form (RREF) (p. 355) The form of a coefficient matrix when the first nonzero element in each row is 1 and any other elements in that column are 0.

reference angle (p. 189) The acute angle formed by the x-axis and the terminal side of an angle in standard position.

reference triangle (p. 189) The triangle formed by a perpendicular drawn to the x-axis from the terminal side of an angle in standard position.

reflection (p. 33) A transformation across a reference line that creates a mirror image.

relation (p. 2) Any set of ordered pairs.

relative extrema (p. 80) The maximum or minimum values relative to a local region of points.

residual (p. 60) The difference between the actual y-value of a data point and the value predicted by the modeling function.

resultant vector (vector sum) (p. 288) The vector that represents the sum of two vectors. If $\mathbf{u} = \langle u_1, u_2 \rangle$ and $\mathbf{v} = \langle v_1, v_2 \rangle$, $\mathbf{r} = \mathbf{u} + \mathbf{v} = \langle u_1 + v_1, u_2 + v_2 \rangle$.

right-hand limit (p. 21) The right-hand limit of $f(x)$ as x approaches a, denoted $\lim_{x \to a^+} f(x)$, is the number to which $f(x)$ gets closer and closer as x approaches a from the right.

rigid transformation (p. 33) A transformation that maintains the shape of the graph, such as a translation or reflection.

row echelon form (REF) (p. 353) A system of linear equations written in inverted triangular form in which all leading coefficients are 1.

sample (p. 507) A group selected from a population.

sample space (p. 497) All possible outcomes in an experiment.

sampling error (p. 586) An error that occurs when the sample is not representative of the population.

scalar (pp. 287–88) A real number.

scatterplot (p. 59) A graph of ordered pairs that illustrates the relationship between two sets of data.

secant (pp. 179, 188) The trigonometric ratio that is the reciprocal of cosine.

sector (p. 175) The region of a circle bound by two radii and the intercepted arc.

semi-perimeter (p. 278) One half the sum of a triangle's side lengths.

sequence (p. 446) A function whose domain is the set of natural numbers (infinite sequence) or the first n natural numbers (finite sequence) that is usually written as an ordered list of numbers.

series (p. 449) The sum of the terms in a sequence.

set (p. 647) A group or collection of numbers or objects.

sigma notation (p. 449) See **summation notation**.

simple harmonic motion (p. 225) Oscillations that continue with the same amplitude and period.

sine (pp. 179, 188) The ratio of the leg opposite an acute angle and the hypotenuse of a right triangle. The ratio $\frac{y}{r}$, where $P(x, y)$ is a point on the terminal side of an angle in standard position and $r = \sqrt{x^2 + y^2}$.

singular matrix (p. 370) A matrix for which no inverse exists.

sinusoidal function (p. 198) A function that can be written as $y = a \sin b(x - h) + k$ or $y = a \cos b(x - h) + k$.

slant asymptote (p. 107) Given a rational function $f(x) = \frac{a(x)}{b(x)}$ where the degree of $a(x)$ is one more than the degree of $b(x)$, the slant asymptote is the quotient (without the remainder) when $a(x)$ is divided by $b(x)$.

slope (pp. 10, 14) The ratio of a line's rise (vertical change) to its run (horizontal change). Denoted m.

slope-intercept form (p. 11) A linear equation in the form $y = mx + b$, where m is the slope of a line and $(0, b)$ is the y-intercept.

solution (p. 233) Any value of the variable for which a conditional equation or inequality is true.

square matrix (p. 343) A matrix that has an equal number of rows and columns.

standard deviation (pp. 517, 548) A measure of variance from the mean found by finding the square root of the variance.

standard position (of an angle) (p. 172) An angle with its vertex at the origin of a coordinate plane and its initial side coinciding with the positive x-axis.

standard position (of a vector) (p. 287) A vector with its initial point at the origin.

standard unit vectors (p. 289) $\mathbf{i} = \langle 1, 0 \rangle$ and $\mathbf{j} = \langle 0, 1 \rangle$.

statistic (p. 507) An estimate of a population's parameter based on a sample.

subset (p. 647) A set whose elements are all contained in another set.

summation notation (p. 449) A series represented as $\sum_{i=b}^{c} a_i = a_b + a_{b+1} + a_{b+2} + \cdots + a_c$ where the index (or counter) i has a lower bound of b and an upper bound of c.

sum of the squares (p. 517) The sum of the squares of the difference of each value from the mean.

synthetic division (p. 86) A shortened version of the long-division algorithm used when dividing a polynomial by a linear polynomial of the form $x - c$.

synthetic substitution (p. 89) Using synthetic division to evaluate a function.

tangent (pp. 179, 188) The ratio of the leg opposite an acute angle and the leg adjacent to that angle in a right triangle. The ratio $\frac{y}{x}$, where $P(x, y)$ is a point on the terminal side of an angle in standard position.

terminal side (p. 172) The side of an angle in standard position that rotates around the origin.

theoretical (classical) probability (p. 497) The ratio of the number of favorable outcomes, $n(E)$, to the number of possible outcomes, $n(S)$, where every outcome is equally likely.

transcendental function (p. 128) A function, such as an exponential, logarithmic, or trigonometric function, that goes beyond, or transcends, the basic algebraic operations.

translation (p. 32) A change in the position of a figure without rotation or resizing.

transverse axis (p. 412) The segment containing a hyperbola's vertices whose endpoints are the hyperbola's foci.

trend line (p. 59) A line drawn to model the relationship between two related variables and to make predictions related to other values of either variable.

type I error (p. 574) An error that occurs when the null hypothesis is rejected, but it is true.

type II error (p. 574) An error that occurs when the null hypothesis is not rejected, but it is false.

union of sets (p. 647) The set of all elements that appear in the given sets.

unit circle (p. 191) The circle centered at the origin with a radius of 1 unit: $x^2 + y^2 = 1$.

unit vector (p. 289) A vector with a magnitude of 1.

variance (pp. 517, 548, 550) The average of the sum of the squared deviations from the mean.

vector (p. 287) A quantity $\langle a, b \rangle$ modeled with a directed line segment having both magnitude and direction.

verifying identities (p. 240) Showing that both sides of an equation are equal for all values of the variable for which the expressions are defined.

vertex (of a conic section) (p. 435) Any intersection of a conic and its focal axis.

vertex (of a parabola) (pp. 39, 396) The minimum or maximum point (h, k) of the quadratic function $f(x) = a(x - h)^2 + k$ at the intersection of the parabola and its axis of symmetry.

vertical asymptote (p. 104) The line $x = c$ is a vertical asymptote of $f(x)$ if $\lim_{x \to c^-} f(x) = \pm\infty$ or $\lim_{x \to c^+} f(x) = \pm\infty$.

zero of a function (pp. 41, 656) A value of x such that $f(x) = 0$.

zero vector (pp. 288, 305) A vector having a length of 0 and no direction.

z-score (pp. 527–28) The number of standard deviations a value is from the mean.

Index

Photo Credits

Key: (**t**) top; (**c**) center; (**b**) bottom; (**l**) left; (**r**) right

Cover

tl "Shoreditch, London, United Kingdom (Unsplash gBCHiVSfrYY)" by Tom Eversley/Wikimedia Commons/Public Domain; **tr** Photo by Kevin Crosby on Unsplash; **bl** Photo by Mike Enerio on Unsplash; **br** Photo by Mike Kononov on Unsplash

Front Matter

vt TCD/Prod.DB/Alamy Stock Photo; **vb** naddi/Shutterstock.com; **vit** Vallia/Shutterstock.com; **vicl** xtock/Shutterstock.com; **vicr** Olaf Schulz/Shutterstock.com; **vib** James Cavallini/Science Source; **vii** Guy Corbishley/Alamy Stock Photo

Chapter 1

1 Aspen Photo/Shutterstock.com; **2** Lyu Hu/Shutterstock.com; **10** Stas Vulkanov/Shutterstock.com; **14** alchemist no.9/Shutterstock.com; **25** MikhailSh/Shutterstock.com; **29** HomeArt/Shutterstock.com; **32** © iStock.com/simonbradfield; **37** casadaphoto/Shutterstock.com; **39** photofriday/Shutterstock.com; **46** © iStock.com/Poike; **52** MarcelClemens/Shutterstock.com; **59** Heiko Kiera/Shutterstock.com; **64** WitthayaP/Shutterstock.com; **67** AP Photo/Alex Gallardo

Chapter 2

71 KUPRYNENKO ANDRII/Shutterstock.com; **72** Pavel L Photo and Video/Shutterstock.com; **78** Syda Productions/Shutterstock.com; **80** brovkin/Shutterstock.com; **86** Vicki Vale/Shutterstock.com; **94** Concept Photo/Shutterstock.com; **102** agsandrew/Shutterstock.com; **104** bociek666/Shutterstock.com; **112** chombosan/Shutterstock.com; **114** Andrew Linscott/Shutterstock.com; **116** © iStock.com/Sladic; **118** etnika/Shutterstock.com; **122** "USA-17-flying-cropped" by Pedro de Arechavaleta/Wikimedia Commons/CC BY-SA 3.0; **124** © iStock.com/Daft_Lion_Studio; **126** Odua Images/Shutterstock.com

Chapter 3

127 © iStock.com/FotografiaBasica; **128** © iStock.com/BirdImages; **132** Fotos593/Shutterstock.com; **134** © iStock.com/naturediver; **136** Haje Jan Kamps/EyeEm/Getty Images; **143** AGCuesta/Shutterstock.com; **149** praphab louilarpprasert/Shutterstock.com; **156** f11photo/Shutterstock.com; **166** Michaelstockfoto/Shutterstock.com; **168** © iStock.com/fotoguy22

Chapter 4

171 kdshutterman/Bigstock.com; **172** Design Pics/Bilderbuch/Media Bakery; **177** umnola/Shutterstock.com; **179** © iStock.com/ChuckSchugPhotography; **185** (silhouette) © iStock.com/4x6; **185** (blimp) © iStock.com/luismmolina; **185** (hot air balloon) Miki Studio/Shutterstock.com; **185** (Statue of Liberty) © iStock.com/Samuil_Levich; **188** Alisa24/Shutterstock.com; **197** Debbie Steinhausser/Shutterstock.com; **204** Media Bakery; **206** Nano Calvo/age fotostock/Getty Images; **214** Phovoir/Shutterstock.com; **216** Don Pablo/Shutterstock.com; **223** rommma/Shutterstock.com; **228** Igor Zh./Shutterstock.com; **230l** starfotograf/Bigstock.com; **230r** robert cicchetti/Shutterstock.com

Chapter 5

232, 389 © iStock.com/Henrik5000; **233** © iStock.com/Bosca78; **238** arhendrix/Shutterstock.com; **240** Anthony Lynn/Alamy Stock Photo; **246** Copyright by Educational Innovations, INC. Reprinted by permission.; **253** Visharo/Shutterstock.com; **254** CobraCZ/Shutterstock.com; **261** Charles D. Winters/Science Source; **265** Nor Gal/Shutterstock.com; **269, 394** Johan Swanepoel/Shutterstock.com; **276** © iStock.com/zmeel; **282** BakerJarvis/Shutterstock.com

Chapter 6

286 Vallia/Shutterstock.com; **287** Guy Corbishley/Alamy Stock Photo; **294** mariakraynova/Shutterstock.com; **299** Africa Studio/Shutterstock.com; **301** xtock/Shutterstock.com; **303** sybanto/Shutterstock.com; **309** © iStock.com/FrozenShutter; **313** Photo Melon/Shutterstock.com; **315** Chursina Viktoriia/Shutterstock.com; **324** Desiree Walstra/Shutterstock.com; **331l** Olaf Schulz/Shutterstock.com; **331r** James Cavallini/Science Source

Chapter 7

335, 420 Ezume Images/Shutterstock.com; **336** asharkyu/Shutterstock.com; **343** wavebreakmedia/Shutterstock.com; **349** Andrey Suslov/Shutterstock.com; **351** sokolenok/Shutterstock.com; **353** Yasonya/Shutterstock.com; **361** Bibliothèque de Genève; **370** Wojciech Strozyk/Alamy Stock Photo; **374** Nerthuz/Shutterstock.com; **376** EtiAmmos/Shutterstock.com; **383** Tami Knisely

Chapter 8

395 James Schwabel/Alamy Stock Photo; **400** Xinhua/Alamy Stock Photo; **403** Artem Avetisyan/Shutterstock.com; **409** Jerrie McKenney; **412** John D Sirlin/Shutterstock.com; **417** "Cassegrain.en" by Szőcs Tamás Tamasflex/Wikimedia Commons/CC By-SA 3.0; **418l** kmls/Shutterstock.com; **418r** Michael Rosskothen/Shutterstock.com; **419** Graphical calculator images are powered by Desmos; **425** baimaple/Shutterstock.com; **428** Mick Harper/Shutterstock.com; **429** Sergey Nivens/Shutterstock.com; **433** omnimoney/Shutterstock.com; **434** Air Images/Shutterstock.com; **435** 3Dsculptor/Shutterstock.com; **441tl** Dotted Yeti/Shutterstock.com; **441bl** creativestockexchange/Shutterstock.com; **441br** Golden Sikorka/Shutterstock.com; **443l** LowePhoto/Alamy Stock Photo; **443r** Nomad_Soul/Shutterstock.com

Chapter 9

445 Adam Hart-Davis/Science Source; **446** Organic Matter/Shutterstock.com; **456** Theeraphong/Shutterstock.com; **463** William Potter/Shutterstock.com; **468** canbedone/Shutterstock.com; **472** Science Photo Library/Robert Brook/Media Bakery; **477** © iStock.com/Martin Barraud; **485** Oleksiy Mark/Shutterstock.com

Chapter 10

489 Pirotehnik/iStock/Getty Images Plus/Getty Images; **490** Thomas M Perkins/Shutterstock.com; **492** Cathy Keifer/Shutterstock.com; **494l** © iStock.com/Yevhenii Dubinko; **494r** abxyz/Shutterstock.com; **497t** Believe_In_Me/iStock/Getty Images Plus/Getty Images; **497b** TeddyandMia/Shutterstock.com; **498** Stock Up/Shutterstock.com; **501** Susan Schmitz/Shutterstock.com; **502l** julie deshaies/Shutterstock.com; **502tr** Lantern Works/Shutterstock.com; **502br** Sergiy Kuzmin/Shutterstock.com; **503** Westend61/GmbH/Media Bakery; **505** ALPA PROD/Shutterstock.com; **507** Oksana Kuzmina/Shutterstock.com; **510** cynoclub/Shutterstock.com; **516** © Mbr Images | Dreamstime.com; **523** Preston James Garbe/Shutterstock.com; **527** Basar/Shutterstock.com; **529** Pictorial Press Ltd/Alamy Stock Photo; **534** Mark Conlin/Alamy Stock Photo; **540** Nice_Media_PRODUCTION/Shutterstock.com

Text Acknowledgements

Chapter 1

9 Eves, Howard. *An Introduction to the History of Mathematics*, 6th ed. (Pacific Grove: Brooks/Cole, 1990), 264; **17** Wigner, Eugene. *Symmetries and Reflections: Scientific Essays* (Cambridge, MA: The MIT Press, 1970), 237; **17** Calvin, John. *Institutes on the Christian Religion*, vol. 1, trans. Henry Beveridge (Edinburgh: Calvin Translation Society, 1845), 318–19; **17** Hertz, Heinrich, quoted in E. T. Bell. *Men of Mathematics* (New York: Touchstone Books, 1986), 16; **17** Nickel, James. *Mathematics: Is God Silent?* (Vallecito, CA: Ross House Books, 2001), 223; **49** Fourier, Joseph. *The Analytical Theory of Heat*, trans. Alexander Freeman (Cambridge: The University Press, 1878), 1; **59** Alligator data from Example 2.20, Scheaffer, Richard L. *Probability and Statistics for Engineers*, 5th Edition (Boston: Brooks/Cole, 2011), 97; **64** Average US gas prices data from US Department of Energy; **65** Life expectancy data from the CDC/National Center for Health Statistics; **70** US Census Bureau.

Chapter 2

122 Formula from "International America's Cup Class," Wikipedia.

Chapter 3

134 National Marine Fisheries Service. 2013. Status Review of The Eastern Distinct Population Segment of Steller Sea Lion (Eumetopias jubatus). 144pp + Appendices. Protected Resources Division, Alaska Region, National Marine Fisheries Service, 709 West 9th St, Juneau, Alaska 99802; **134** "Fukushima Radiation: 800 Tera Becquerel of Cesium-137 to Reach West Coast of North America by 2016," Centre for Research on Globalization; **142** Cajori, Florian. *A History of Mathematics*, 2nd ed., rev. (New York: Macmillan, 1919), 149; **155** Kline, Morris. *Mathematics for the Nonmathematician* (New York: Dover Publications, 1967), 206–7; **155** Kepler, Johannes, quoted in James Nickel. *Mathematics: Is God Silent?* (Vallecito, CA: Ross House Books, 2001), 145; **155** Zimmerman, Larry L. "Mathematics: Is God Silent?," Part III, *The Biblical Educator*, 2:3, 2 (1980), n.p.; **155** Leibniz, Gottfried, quoted in Kline, Morris. *Mathematics: The Loss of Certainty* (New York: Oxford University Press, 1980), 60; **155** Fehr, Howard F. "Reorientation in Mathematics Education," *Teachers College Record*, May 1953, vol. 54, no. 8, p. 435; **163** Nesting pairs data from "Chart and Table of Bald Eagle Breeding Pairs in Lower 48 States," US Fish and Wildlife Service.

Chapter 4

214 Einstein, Albert, quoted in Morris Kline. *Mathematics and the Physical World* (New York: Dover, 1981), 464; **214** Einstein, Albert. "Physics and Reality," *Out of My Later Years* (New York: Citadel Press, 1950), 61; **214** Nickel, James. *Mathematics: Is God Silent?* (Vallecito, CA: Ross House Books, 2001), 225; **214** Lobachevsky, Nikolai, quoted in George E. Martin. *The Foundations of Geometry and the Non-Euclidean Plane* (New York: Springer, 1998), 225; **215** Gardner, Martin, quoted in Donald J. Albers and Gerald L. Alexanderson, eds. *Fascinating Mathematical People: Interviews and Memoirs* (Princeton: Princeton University Press, 2011), ix.

Chapter 5

235 Feynman, Richard. *The Character of Physical Law* (Cambridge, MA: The MIT Press, 1985), 34; **252** Fourier, Joseph. *The Analytical Theory of Heat*, trans. Alexander Freeman (Cambridge: The University Press, 1878), 8; **252** Russell, Bertrand. *Mysticism and Logic and Other Essays* (London: George Allen & Unwin Ltd, 1959), 61; **252** Gioia, Dana. "Beauty's Place in the Christian Vision" (chapel sermon), Biola University, February 8, 2012; **252** Nickel, James. *Mathematics: Is God Silent?* (Vallecito, CA: Ross House Books, 2001), 94; **252–53** Kline, Morris. *Mathematics in Western Culture* (New York: Oxford University Press, 1953), 512; **268** Bhatia, Rajendra. *Fourier Series* (Providence, RI: American Mathematical Society, 2004), 116; **268** Bracewell, Ronald, quoted in Prestini, Elena. *The Evolution of Applied Harmonic Analysis: Models of the Real World* (Boston: Birkhäuser, 2004), xii; **282** Data table from Smith, A. Mark. *Ptolemy's Theory of Visual Perception: An English Translation of the "Optics" with Introduction and Commentary* (Philadelphia: American Philosophical Society, 1996), 233.

Chapter 6

301 Hermite, Charles, quoted in Morris Kline. *Mathematics: The Loss of Certainty* (New York: Oxford Univ. Press, 1980), 345; **301** Galilei, Galileo, quoted in Morris Kline. *Mathematics and the Physical World* (New York: Dover, 1981), 206; **301** Kasner, Edward and James Newman. *Mathematics and the Imagination* (London: G. Bell and Sons, 1949), 359; **301** Jourdain, Philip E. B. *The Nature of Mathematics* (London: Jack and Nelson, 1919), 46; **301** Gauss, Carl, quoted in Ioan James. *Remarkable Mathematicians* (New York: Cambridge University Press, 2002), 67; **302** Zimmerman, Larry L. *Truth & the Transcendent* (Florence, KY: Answers in Genesis, 2000), n.p.

Chapter 7

351 Hardy, G. H. *A Mathematician's Apology* (Cambridge: Cambridge University Press, 2004), 94; **352** Bahnsen, Greg L. "The Great Debate: Does God Exist?," University of California, Irvine, CA 1985; **369** Huxley, Thomas, quoted in Cajori, Florian. "A Review of Three Famous Attacks upon the Study of Mathematics as a Training of the Mind," *Popular Science Monthly*, vol. 80, April 1912; **369** Sylvester, James Joseph, ibid.

Chapter 8

427 Wiles, Andrew. "Andrew Wiles on Solving Fermat," NOVA, 1 November 2007, PBS online; **442** Hawking, Stephen. Interview by Kenneth Campbell, *Reality on the Rocks*, 1995.

Chapter 9

454 Hales, Thomas, quoted in George Szpiro. *Kepler's Conjecture* (Hoboken, NJ: Wiley, 2003), 201; **454** Hawking, Stephen, quoted in Rory Cellan-Jones. "Stephen Hawking Warns Artificial Intelligence Could End Mankind.", 2 December 2014, *BBC News* online; **454** Volkers, Cody. "A Christian View of Artificial Intelligence," 1 March 2017, *Ecclesiam Journal* online; **455** ibid; **471** Pascal, Blaise. "Prayer to Ask of God the Proper Use of Sickness," *Minor Works*, trans. O. W. Wight (New York: P. F. Collier & Son, 1910), 371; **481** Van Til, Cornelius. *A Survey of Christian Epistemology* (Phillipsburg, NJ: P&R Publishing, 1969), 19.

Chapter 10

497 First paragraph ideas based on Tagliapietra, Ron. *Math for God's Glory* (Bloomington, IN: Xlibris, 2004); **526** Boyer, Carl B. *A History of Mathematics*, 2nd ed. (New York: John Wiley & Sons, 1992), 492; **526** Poythress, Vern S. *Chance and the Sovereignty of God: A God-Centered Approach to Probability and Random Events* (Wheaton, IL: Crossway, 2014), 101; **540** Hardy, G. H. *A Mathematician's Apology* (Cambridge: Cambridge University Press, 1969), 123–24; **540** Linnebo,

Øystein, "Platonism in the Philosophy of Mathematics," *The Stanford Encyclopedia of Philosophy* (Spring 2018 Edition), Edward N. Zalta (ed.), https://plato.stanford.edu/archives/spr2018/entries/platonism-mathematics/; **540** Kepler, Johannes, quoted in Morris Kline. *Mathematics: The Loss of Certainty* (New York: Oxford University Press), 31; **540** Du Sautoy, Marcus. *The Music of the Primes: Searching to Solve the Greatest Mystery in Mathematics* (New York: HarperCollins, 2003), 33–34.

Chapter 11

559 Graphs generated from the Controlling for Variables tool by SticiGui©. Philip B. Stark, University of California, Berkeley; **560** Teen data from Stage of Life LLC, 24 South Franklin Street, Dallastown, PA 17313. Copyright © 2011 All rights reserved; **562** Bureau of Labor Statistics, American Time Use Study; **562** Molla, Rani. "Why People Are Buying More Expensive Smartphones Than They Have in Years," January 23, 2018, *Recode* online. Vox Media, Inc.; **578** Ioannidis, John P. A. "Why Most Published Research Findings Are False," August 30, 2005, *PLOS Medicine* online; **582** Nickel, James. *Mathematics: Is God Silent?* (Vallecito, CA: Ross House Books, 2001), 59; **582** Eves, Howard. *An Introduction to the History of Mathematics*, 4th ed. (New York: Holt, Rinehart and Winston, 1976), 481; **588** Elbel, Brian, et al. "A Water Availability Intervention in New York City Public Schools: Influence on Youths' Water and Milk Behaviors," *American Journal of Public Health*, vol. 105, no. 2. Washington, DC: American Public Health Association, Feb. 2015; **597** Morning commute data from the US Census Bureau.

Chapter 12

601 Leibniz, Gottfried. *The Philosophical Works of Leibniz*, trans. George Martin Duncan (New Haven, CT: The Tuttle, Morehouse & Taylor Co., 1908), 71; **610** Hilbert, David, quoted in Eli Maor. *To Infinity and Beyond: A Cultural History of the Infinite* (Boston: Birkhäuser Boston, 1987), vii; **610** Weyl, Hermann, quoted in Peter Pesic. *Levels of Infinity* (Mineola, NY: Dover, 2012), 17; **644** Fraser, Alexander Campbell, ed. *The Works of George Berkeley, D.D.*, vol. 3 (Oxford: Clarendon Press, 1871), 301.

TEACHER EDITION

Chapter 3

155 Newton, Isaac, quoted in Florian Cajori. *Sir Isaac Newton's Mathematical Principles of Natural Philosophy and His System of the World*, vol. 2 (Berkeley: University of California Press, 1934), 669; **155** Kepler, Johannes, quoted in Morris Kline. *Mathematics: The Loss of Certainty* (New York: Oxford University Press, 1980), 31.

Chapter 4

214 Nickel, James. *Mathematics: Is God Silent?* (Vallecito, CA: Ross House Books, 2001), 51; **214** Ruffini, Remo, quoted in James Nickel. ibid., 69; **214** Kline, Morris. *Mathematics and the Physical World* (New York: Dover, 1980), 463.

Chapter 5

253 Lisle, Jason. "Fractals: Hidden Beauty Revealed in Mathematics," *Answers* magazine online, January 1, 2007; **268** Fourier, Joseph. *The Analytical Theory of Heat* (Cambridge: The University Press, 1878), 7–8; **268** Campbell, Lewis and William Garnett. *The Life of James Clerk Maxwell* (London: Macmillan, 1882), 416.

Chapter 7

351–52 Wiles, Andrew. "The Proof," NOVA, 28 October 1997, PBS Online.

Chapter 8

428 Wiles, Andrew. "Solving Fermat: Andrew Wiles," NOVA, n.d., PBS Online.

Chapter 9

454 Szpiro, George. *Kepler's Conjecture* (Hoboken, NJ: Wiley, 2003), 212; **454** Stewart, Ian. ibid; **454** Hardy, G. H. *A Mathematician's Apology* (Cambridge: Cambridge University Press, 1969), 113; **455** Volkers, Cody. "A Christian View of Artificial Intelligence," 1 March 2017, *Ecclesiam Journal* online.

Chapter 10

540–41 Kline, Morris. *Mathematics: The Loss of Certainty* (New York: Oxford University Press, 1980), 18; **541** Zimmerman, Larry L. *Truth & the Transcendent* (Florence, KY: Answers in Genesis, 2000), n.p.

Quick Reference

Functions

even	$f(-x) = f(x)$ for all $x \in D$	symmetrical with y-axis
odd	$f(-x) = -f(x)$ for all $x \in D$	symmetrical with origin

Transformations of Functions

	Translation	**Stretch or Shrink**	**Reflection**
vertical	$g(x) = f(x) + k$ up if $k > 0$ down if $k < 0$	$g(x) = af(x)$ stretch by a if $\lvert a \rvert > 1$ shrink by a if $0 \leq \lvert a \rvert < 1$	$g(x) = -f(x)$ reflection across x-axis
horizontal	$g(x) = f(x - h)$ right if $h > 0$ left if $h < 0$	$g(x) = f(bx)$ shrink by a factor of $\frac{1}{\lvert b \rvert}$ if $\lvert b \rvert > 1$ stretch by a factor of $\frac{1}{\lvert b \rvert}$ if $0 \leq \lvert b \rvert < 1$	$g(x) = f(-x)$ reflection across y-axis

Exponents and Logarithms

	of Exponents	**of Logarithms**	
Equality Property	$b^m = b^n$ iff $m = n$	$\log_b m = \log_b n$ iff $m = n$	**change of base**
Product Property	$b^m \cdot b^n = b^{m+n}$	$\log_b mn = \log_b m + \log_b n$	$\log_b x = \dfrac{\log_a x}{\log_a b}$
Quotient Property	$\dfrac{b^m}{b^n} = b^{m-n}$	$\log_b \frac{m}{n} = \log_b m - \log_b n$	
Power Property	$(b^m)^p = b^{mp}$	$\log_b m^p = p \log_b m$	

Vectors

	magnitude	**reference angle**	**Direction (θ degrees from the positive x-axis)**	
	$\lvert \mathbf{v} \rvert = \sqrt{a^2 + b^2}$	$\alpha = \mathrm{Tan}^{-1} \left\lvert \frac{b}{a} \right\rvert$	Q I: $\theta = \alpha$ Q III: $\theta = 180° + \alpha$	Q II: $\theta = 180° - \alpha$ Q IV: $\theta = 360° - \alpha$

dot product $\mathbf{u} \cdot \mathbf{v} = \lvert \mathbf{u} \rvert \, \lvert \mathbf{v} \rvert \cos \theta$
If $\mathbf{u} = \langle u_1, u_2 \rangle$ and $\mathbf{v} = \langle v_1, v_2 \rangle$, then $\mathbf{u} \cdot \mathbf{v} = u_1 v_1 + u_2 v_2$.

unit vector $\mathbf{u} = \frac{\mathbf{v}}{\lvert \mathbf{v} \rvert}$

projection of u onto v $\mathrm{proj}_\mathbf{v} \, \mathbf{u} = \left(\dfrac{\mathbf{u} \cdot \mathbf{v}}{\lvert \mathbf{v} \rvert^2} \right) \mathbf{v} = \mathbf{w}_1;\ \mathbf{w}_2 = \mathbf{u} - \mathrm{proj}_\mathbf{v} \, \mathbf{u}$

Complex Numbers

standard form	$z = a + bi$	**De Moivre's Theorem** $z^n = r^n \operatorname{cis} n\theta$
polar (trig) form	$z = r(\cos \theta + i \sin \theta)$ or $z = r \operatorname{cis} \theta$	
product	$z_1 z_2 = r_1 r_2 \operatorname{cis} (\theta_1 + \theta_2)$	**Corollary to De Moivre's Theorem**
quotient	$\dfrac{z_1}{z_2} = \dfrac{r_1 \operatorname{cis} \theta_1}{r_2 \operatorname{cis} \theta_2} = \dfrac{r_1}{r_2} \operatorname{cis} (\theta_1 - \theta_2)$	The n complex roots of z are $\sqrt[n]{r} \operatorname{cis} \dfrac{\theta + k2\pi}{n}$ where $k = 0, 1, 2, \ldots, n - 1$.

Sequences and Series

	Arithmetic	**Geometric**
recursive formula	$a_n = a_{n-1} + d$	$a_n = a_{n-1} \cdot r$
explicit formula	$a_n = a_1 + (n - 1)d$	$a_n = a_1 r^{n-1}$
finite series	$S_n = \frac{n}{2}(a_1 + a_n)$ or $S_n = \frac{n}{2}[2a_1 + (n - 1)d]$	$S_n = \dfrac{a_1 - a_n r}{1 - r}$ or $S_n = \dfrac{a_1(1 - r^n)}{1 - r}$
infinite series	The series diverges.	If $\lvert r \rvert < 1$, then $S = \dfrac{a_1}{1 - r}$. If $\lvert r \rvert \geq 1$, the series diverges.

Standard Form Equations of a Parabola

with vertex $V(h, k)$ and focal length $|p|$ where $a = \frac{1}{4p}$

$$y = \frac{1}{4p}(x - h)^2 + k \qquad\qquad x = \frac{1}{4p}(y - k)^2 + h$$

focus: $F(h, k + p)$; axis: $x = h$; directrix: $y = k - p$ $\qquad$ focus: $F(h + p, k)$; axis: $y = k$; directrix: $x = h - p$

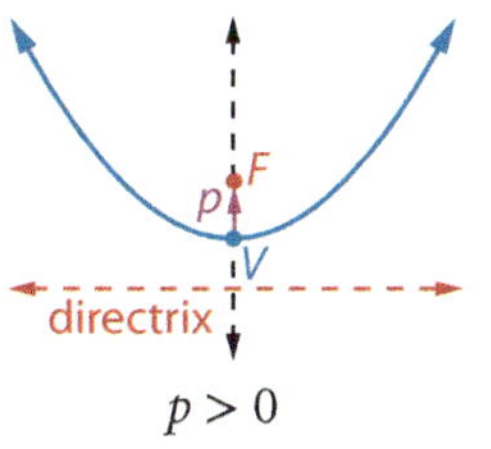

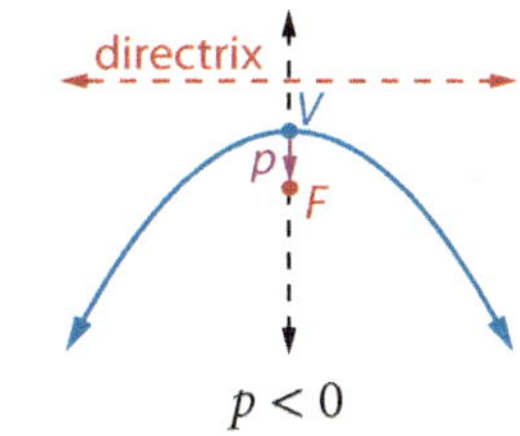

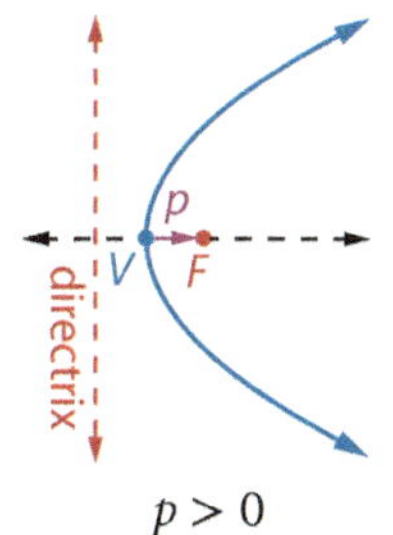

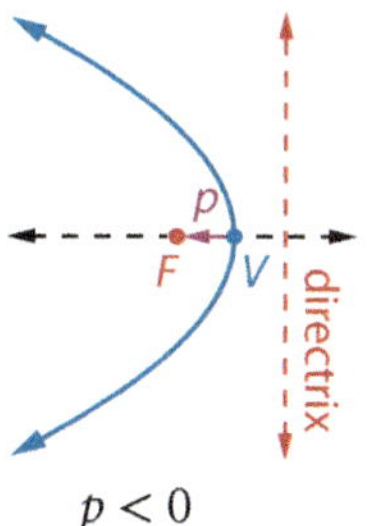

Standard Form Equations of an Ellipse

centered at $C(h, k)$ with $a > b$ and $c^2 = a^2 - b^2$

$$\frac{(x - h)^2}{a^2} + \frac{(y - k)^2}{b^2} = 1 \qquad\qquad \frac{(x - h)^2}{b^2} + \frac{(y - k)^2}{a^2} = 1$$

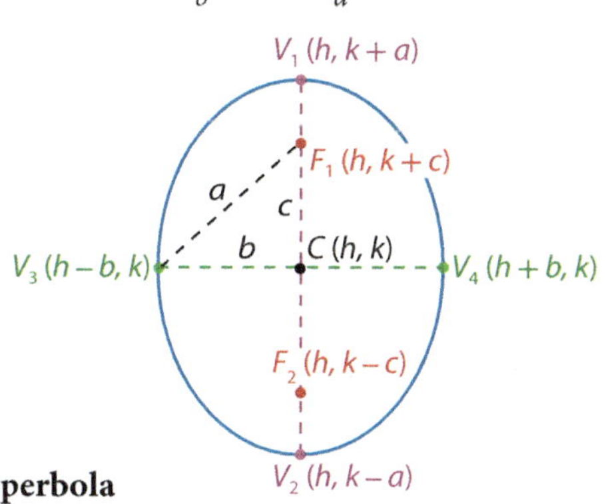

Standard Form Equations of a Hyperbola

centered at $C(h, k)$ with $c^2 = a^2 + b^2$

$$\frac{(x - h)^2}{a^2} - \frac{(y - k)^2}{b^2} = 1 \qquad\qquad \frac{(y - k)^2}{a^2} - \frac{(x - h)^2}{b^2} = 1$$

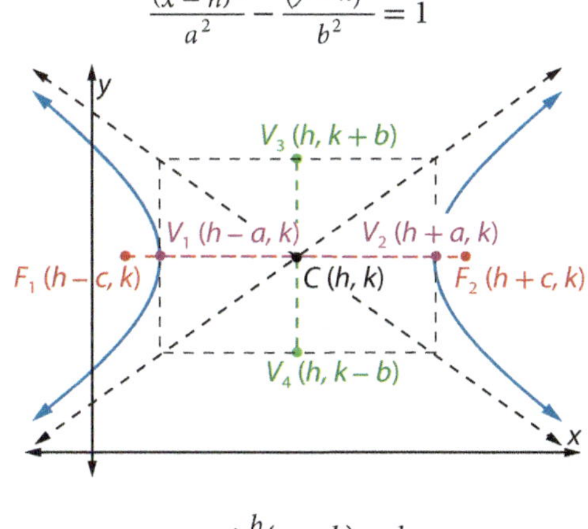

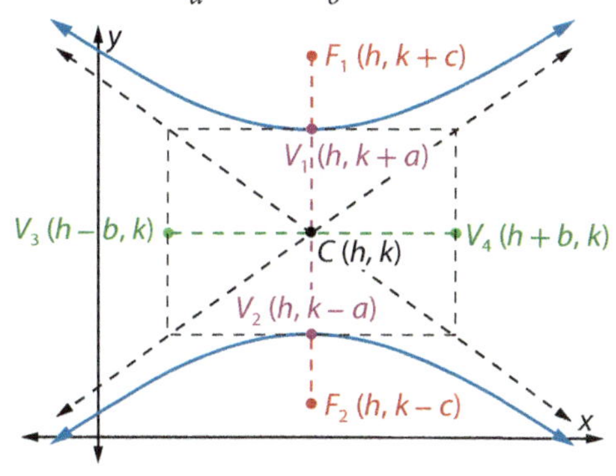

$$y = \pm\frac{b}{a}(x - h) + k \qquad\qquad \textbf{asymptotes} \qquad\qquad y = \pm\frac{a}{b}(x - h) + k$$

General Form Equation of Conics

$$Ax^2 + Bxy + Cy^2 + Dx + Ey + F = 0 \text{ with discriminant } B^2 - 4AC \text{ and eccentricity } e = \frac{c}{a}$$

Circle	Ellipse	Parabola	Hyperbola
$B^2 - 4AC < 0$;	$B^2 - 4AC < 0$;	$B^2 - 4AC = 0$	$B^2 - 4AC > 0$
$B = 0$ and $A = C$	$B \neq 0$ or $A \neq C$	$e = 1$	$e > 1$
$e = 0$	$e < 1$		

Summations

Properties

$$\sum_{i=1}^{n} c = nc \qquad \sum_{i=1}^{n} ca_i = c\sum_{i=1}^{n} a_i \qquad \sum_{i=1}^{n} (a_i + b_i) = \sum_{i=1}^{n} a_i + \sum_{i=1}^{n} b_i \qquad \sum_{i=1}^{n} a_i = \sum_{i=1}^{p} a_i + \sum_{i=p+1}^{n} a_i \ \text{ where } 1 < p < n$$

Partial Sums of Powers of Integers

$$\sum_{k=1}^{n} k = \frac{n^2 + n}{2} = \frac{n(n+1)}{2} \qquad \sum_{k=1}^{n} k^2 = \frac{n(n+1)(2n+1)}{6} \qquad \sum_{k=1}^{n} k^3 = \frac{n^2(n+1)^2}{4} \qquad \sum_{k=1}^{n} k^4 = \frac{n(n+1)(2n+1)(3n^2+3n-1)}{30}$$

Probability

permutation $\quad _nP_r = \dfrac{n!}{(n-r)!}$ $\qquad\qquad$ **combination** $\quad _nC_r = \dfrac{n!}{r!(n-r)!}$

Binomial Theorem

$$(x+y)^n = {}_nC_0\, x^n + {}_nC_1\, x^{n-1}y + {}_nC_2\, x^{n-2}y^2 + {}_nC_3\, x^{n-3}y^3 + \cdots + {}_nC_n\, y^n = \sum_{r=0}^{n} {}_nC_r\, x^{n-r}y^r$$

theoretical $\quad P(E) = \dfrac{n(E)}{n(S)}$ $\qquad$ **experimental** $\quad P(E) = \dfrac{\text{number of times the event occurs}}{\text{number of trials}}$ $\qquad$ **complement of an event** $\quad P(E') = 1 - P(E)$

independent events $\quad P(A \text{ and } B) = P(A) \cdot P(B)$
dependent events $\quad P(A \text{ and } B) = P(A) \cdot P(B|A)$

mutually exclusive events $\quad P(A \text{ or } B) = P(A) + P(B)$
inclusive events $\quad P(A \text{ or } B) = P(A) + P(B) - P(A \text{ and } B)$

Statistics

	Population	**Sample**
mean	$\mu = \dfrac{\sum\limits_{i=1}^{N} x_i}{N}$	$\bar{x} = \dfrac{\sum\limits_{i=1}^{n} x_i}{n}$
variance	$\sigma^2 = \dfrac{\sum\limits_{i=1}^{N} (x_i - \mu)^2}{N} \text{ or } \dfrac{SS}{N}$	$s^2 = \dfrac{\sum\limits_{i=1}^{n} (x_i - \bar{x})^2}{n-1} \text{ or } \dfrac{SS}{n-1}$
standard deviation	$\sigma = \sqrt{\dfrac{\sum\limits_{i=1}^{N} (x_i - \mu)^2}{N}}$	$s = \sqrt{\dfrac{\sum\limits_{i=1}^{n} (x_i - \bar{x})^2}{n-1}}$
z-score	$z = \dfrac{x - \mu}{\sigma}$	$z = \dfrac{x - \bar{x}}{s}$

Binomial Probability Distribution

formula	$P(x) = {}_nC_x\, p^x q^{n-x}$
mean	$\mu = np$
variance	$\sigma^2 = npq$
standard deviation	$\sigma = \sqrt{\sigma^2} = \sqrt{npq}$

Limits

sum $\quad \lim\limits_{x \to a} [f(x) + g(x)] = \lim\limits_{x \to a} f(x) + \lim\limits_{x \to a} g(x)$

product $\quad \lim\limits_{x \to a} [f(x) \cdot g(x)] = \lim\limits_{x \to a} f(x) \cdot \lim\limits_{x \to a} g(x)$

scalar multiple $\quad \lim\limits_{x \to a} cf(x) = c \lim\limits_{x \to a} f(x) \ \text{ where } c \in \mathbb{R}$

quotient $\quad \lim\limits_{x \to a} \dfrac{f(x)}{g(x)} = \dfrac{\lim\limits_{x \to a} f(x)}{\lim\limits_{x \to a} g(x)}$

power $\quad \lim\limits_{x \to a} [f(x)]^n = \left[\lim\limits_{x \to a} f(x)\right]^n$

reciprocal power function at infinity $\quad \lim\limits_{x \to \pm\infty} \dfrac{1}{x^n} = 0 \ \text{ where } n \in \mathbb{N}$

Derivatives

constant	$\frac{d}{dx}[c] = 0$	**product**	$\frac{d}{dx}[u(x) \cdot v(x)] = u'(x) \cdot v(x) + u(x) \cdot v'(x)$
linear	$\frac{d}{dx}[mx + b] = m$	**quotient**	$\frac{d}{dx}\left[\frac{u(x)}{v(x)}\right] = \frac{u'(x) \cdot v(x) - u(x) \cdot v'(x)}{[v(x)]^2}$
power	$\frac{d}{dx}[cx^n] = ncx^{n-1}$	**chain rule**	$\frac{d}{dx}[(u \circ v)(x)] = u'(v(x)) \cdot v'(x)$
sum or difference	$\frac{d}{dx}[u(x) \pm v(x)] = u'(x) \pm v'(x)$	**basic trigonometric**	$\frac{d}{dx}[\sin x] = \cos x$
natural exponential	$\frac{d}{dx}[e^x] = e^x$		$\frac{d}{dx}[\cos x] = -\sin x$

Integrals

constant	$\int k\,dx = kx + C$	**scalar**	$\int cf(x)\,dx = c\int f(x)\,dx$
power	$\int kx^n\,dx = \frac{k}{n+1}x^{n+1} + C$	**sum**	$\int [f(x) \pm g(x)]\,dx = \int f(x)\,dx \pm \int g(x)\,dx$
sine	$\int (\sin x)\,dx = -\cos x + C$	**cosine**	$\int (\cos x)\,dx = \sin x + C$
natural exponential	$\int e^x\,dx = e^x + C$	**Fundamental Theorem of Calculus**	$\int_a^b f(x)\,dx = F(b) - F(a)$
definite integral	$\int_a^b (x)\,dx = \lim_{n \to \infty} \sum_{i=1}^{n} f(x_i)\,\Delta x$		

Other Formulas

distance formula	$d = \sqrt{(x_2 - x_1)^2 + (y_2 - y_1)^2 + (z_2 - z_1)^2}$
midpoint formula	$M = \left(\frac{x_1 + x_2}{2}, \frac{y_1 + y_2}{2}, \frac{z_1 + z_2}{2}\right)$
absolute value of a complex number	If $z = a + bi$, then $|z| = |a + bi| = \sqrt{a^2 + b^2}$.
Heron's formula	$Area = \sqrt{s(s - a)(s - b)(s - c)}$, where $s = \frac{a + b + c}{2}$
compound interest	$A(t) = P\left(1 + \frac{r}{n}\right)^{nt}$
continuously compounded interest	$A(t) = Pe^{rt}$
height of a projectile	$h(t) = -\frac{1}{2}gt^2 + v_0 t + h_0$, where $g = 32$ ft/sec^2 or 9.8 m/sec^2

Trigonometry

Basic Ratios

$$\sin A = \frac{opp.}{hyp.}$$

$$\cos A = \frac{adj.}{hyp.}$$

$$\tan A = \frac{opp.}{adj.}$$

Reciprocal Ratios

$$\csc A = \frac{1}{\sin A} = \frac{hyp.}{opp.}$$

$$\sec A = \frac{1}{\cos A} = \frac{hyp.}{adj.}$$

$$\cot A = \frac{1}{\tan A} = \frac{adj.}{opp.}$$

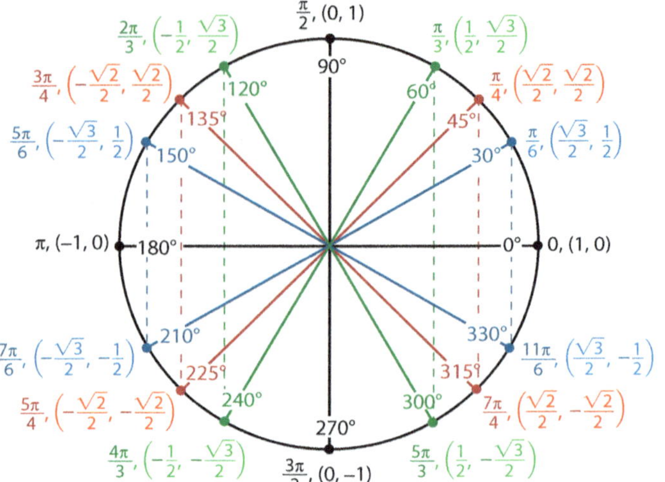

Reciprocal Identities

$$\csc \theta = \frac{1}{\sin \theta} \qquad \sec \theta = \frac{1}{\cos \theta} \qquad \cot \theta = \frac{1}{\tan \theta}$$

Quotient Identities

$$\tan \theta = \frac{\sin \theta}{\cos \theta} \qquad \cot \theta = \frac{\cos \theta}{\sin \theta}$$

Pythagorean Identities

$$\sin^2 \theta + \cos^2 \theta = 1 \qquad \cot^2 \theta + 1 = \csc^2 \theta \qquad \tan^2 \theta + 1 = \sec^2 \theta$$

Even/Odd Identities

$$\sin (-\theta) = -\sin \theta \qquad \cos (-\theta) = \cos \theta \qquad \tan (-\theta) = -\tan \theta$$
$$\csc (-\theta) = -\csc \theta \qquad \sec (-\theta) = \sec \theta \qquad \cot (-\theta) = -\cot \theta$$

Cofunction Identities

$$\sin \theta = \cos \left(\tfrac{\pi}{2} - \theta\right) \qquad \tan \theta = \cot \left(\tfrac{\pi}{2} - \theta\right) \qquad \sec \theta = \csc \left(\tfrac{\pi}{2} - \theta\right)$$
$$\cos \theta = \sin \left(\tfrac{\pi}{2} - \theta\right) \qquad \cot \theta = \tan \left(\tfrac{\pi}{2} - \theta\right) \qquad \csc \theta = \sec \left(\tfrac{\pi}{2} - \theta\right)$$

Sum/Difference Identities

$$\cos (\alpha \pm \beta) = \cos \alpha \cos \beta \mp \sin \alpha \sin \beta$$
$$\sin (\alpha \pm \beta) = \sin \alpha \cos \beta \pm \cos \alpha \sin \beta$$
$$\tan (\alpha \pm \beta) = \frac{\tan \alpha \pm \tan \beta}{1 \mp \tan \alpha \tan \beta}$$

Double-Angle Identities

$$\cos 2\theta = \cos^2 \theta - \sin^2 \theta$$
$$\sin 2\theta = 2 \sin \theta \cos \theta$$
$$\tan 2\theta = \frac{2 \tan \theta}{1 - \tan^2 \theta}$$

angle measure

$$\pi \text{ radians} = 180°$$

arc length

$$s = r\theta \text{ (in radians)}$$

angular speed

$$\omega = \frac{\theta}{t}$$

linear speed

$$v = \frac{s}{t} = r\omega$$

area of a sector
(θ in radians)

$$A_{sector} = \tfrac{1}{2} r^2 \theta$$

Half-Angle Identities

$$\sin \frac{\theta}{2} = \pm \sqrt{\frac{1 - \cos \theta}{2}}$$
$$\cos \frac{\theta}{2} = \pm \sqrt{\frac{1 + \cos \theta}{2}}$$
$$\tan \frac{\theta}{2} = \pm \sqrt{\frac{1 - \cos \theta}{1 + \cos \theta}} = \frac{1 - \cos \theta}{\sin \theta} = \frac{\sin \theta}{1 + \cos \theta}$$

Power-Reducing Identities

$$\sin^2 \theta = \frac{1 - \cos 2\theta}{2}$$
$$\cos^2 \theta = \frac{1 + \cos 2\theta}{2}$$
$$\tan^2 \theta = \frac{1 - \cos 2\theta}{1 + \cos 2\theta}$$

Product-to-Sum Identities

$$\sin \alpha \sin \beta = \tfrac{1}{2}[\cos (\alpha - \beta) - \cos (\alpha + \beta)]$$
$$\cos \alpha \cos \beta = \tfrac{1}{2}[\cos (\alpha - \beta) + \cos (\alpha + \beta)]$$
$$\sin \alpha \cos \beta = \tfrac{1}{2}[\sin (\alpha + \beta) + \sin (\alpha - \beta)]$$
$$\cos \alpha \sin \beta = \tfrac{1}{2}[\sin (\alpha + \beta) - \sin (\alpha - \beta)]$$

Sum-to-Product Identities

$$\sin \alpha + \sin \beta = 2 \sin \frac{\alpha + \beta}{2} \cos \frac{\alpha - \beta}{2}$$
$$\sin \alpha - \sin \beta = 2 \cos \frac{\alpha + \beta}{2} \sin \frac{\alpha - \beta}{2}$$
$$\cos \alpha + \cos \beta = 2 \cos \frac{\alpha + \beta}{2} \cos \frac{\alpha - \beta}{2}$$
$$\cos \alpha - \cos \beta = -2 \sin \frac{\alpha + \beta}{2} \sin \frac{\alpha - \beta}{2}$$

Polar-to-Rectangular Equations

$$r^2 = x^2 + y^2 \qquad x = r \cos \theta \qquad y = r \sin \theta$$

Law of Sines

$$\frac{\sin A}{a} = \frac{\sin B}{b} = \frac{\sin C}{c} \quad \text{or} \quad \frac{a}{\sin A} = \frac{b}{\sin B} = \frac{c}{\sin C}$$

Law of Cosines

$$a^2 = b^2 + c^2 - 2bc \cos A, \quad b^2 = a^2 + c^2 - 2ac \cos B,$$
$$\text{and } c^2 = a^2 + b^2 - 2ab \cos C$$

$\in$	element		$\sum$	sigma, sum (summation notation)
$\cup$	union		$\displaystyle\sum_{i=1}^{\infty} a_i$	infinite series
$\cap$	intersection		$\displaystyle\lim_{n\to\infty} a_n$	limit of a sequence
r	correlation coefficient		$\displaystyle\sum_{i=1}^{n} a_i$	nth partial sum of a sequence (S_n)
r^2	coefficient of determination		$n!$	n factorial
e	the number ≈ 2.71828		$_nP_r$	permutation of r out of n objects
i	imaginary unit, $\sqrt{-1}$		$_nC_r$	combination of r out of n objects
$\theta, \alpha, \beta, \gamma$	theta, alpha, beta, gamma (angles)		$P(E)$	probability of an event
Δ	delta, the change in		$P(A|B)$	probability of A given B
$\{x \mid \ldots\}$	the set of x such that $\ldots$		μ	mu, mean of a population
$\sqrt{x}$	principal square root of x		$\overline{x}$	x-bar, mean of a sample
$\sqrt[n]{x}$	principal nth root of x		σ	sigma, standard deviation of a population
(a, b)	ordered pair		σ^2	sigma squared, variance of a population
(a, b, c)	ordered triple		s	standard deviation of a sample
(a, b)	open interval		s^2	variance of a sample
$[a, b]$	closed interval		x_i	value of random variable x at ith position
$f(x) = |x|$	absolute value function		n	sample size
$f(x) = [x]$	greatest integer function		N	population size
$(f \circ g)(x)$	function composition, $f(g(x))$		Q_n	quartile
$f^{-1}(x)$	inverse function of $f(x)$		$\mu_{\overline{x}}$	mean of the sample means
$f(x, y)$	function of two variables		$\sigma_{\overline{x}}$	standard deviation of the sample means
$f(x) = \dfrac{a(x)}{b(x)}$	rational function		z_c	critical value
$\log_b x$	base b logarithm of x		CI	confidence interval
$\log x$	common logarithm (base 10) of x		α	level of significance
$\ln x$	natural logarithm (base e) of x		p	p-value
ω	omega, angular speed		E	margin of error for the mean
$\mathbf{v}, \vec{v}$	vector $\mathbf{v}$		SS	sum of the squares
$\langle v_1, v_2 \rangle$	component form of vector $\mathbf{v}$		H_0	null hypothesis
$|\mathbf{v}|$	magnitude of $\mathbf{v}$		H_a	alternative (research) hypothesis
$\mathbf{u} \cdot \mathbf{v}$	dot product		$\displaystyle\lim_{x\to a} f(x)$	limit of a function
$\text{proj}_{\mathbf{v}} \mathbf{u}$	projection of $\mathbf{u}$ onto $\mathbf{v}$		$\displaystyle\lim_{x\to a^-} f(x)$	left-hand limit
(r, θ)	polar coordinates		$\displaystyle\lim_{x\to a^+} f(x)$	right-hand limit
$r \operatorname{cis} \theta$	polar form of a complex number		$f'(x), y',$ $\dfrac{dy}{dx}, \dfrac{d}{dx} f(x)$	derivative of a function
$|a + bi|$	modulus of a complex number			
$A_{m \times n}$	matrix dimensions		$\displaystyle\int_a^b f(x)\,dx$	definite integral
a_{mn}	entry in row m and column n			
I	identity matrix		$\displaystyle\int f(x)\,dx$	indefinite integral
$|A|$ or $\det A$	determinant of matrix A			
A^{-1}	inverse matrix			

Appendix A
Vocabulary

CHAPTER 1

abscissa
absolute value function
acceleration due to gravity
axis of symmetry
bivariate data
closed interval
coefficient of determination
completing the square
composition
constant function
constant of variation (proportion)
continuous function
continuous on an interval
correlation coefficient
decomposition
decreasing function
dependent variable
directly proportional
domain
eliminating the parameter
end behavior
even function
extrapolation
family of functions
function
function notation

general linear form
greatest integer function
half-closed interval
half-open interval
height of a projectile
Horizontal Line Test
increasing function
independent variable
infinite discontinuity
interpolation
interval notation
inverse functions
inverse relation
inversely proportional
jump discontinuity
left-hand limit
limit
linear regression
nonrigid transformation
odd function
one-sided limit
one-to-one function
open interval
ordinate
outliers
parabola
parameter

parametric equations
parent function
piecewise functions
point (removable) discontinuity
point-slope form
power
power function
quadratic function
quadratic regression
range
reflection
relation
residuals
right-hand limit
rigid transformation
scatterplot
slope
slope-intercept form
standard form
translation
trend line
vertex
vertex form
Vertical Line Test
zero (root) of a function
Zero Product Property

CHAPTER 2

asymptote
Complete Linear Factorization Theorem
Conjugate Root Theorems
Corollary of the Fundamental
  Theorem of Algebra
depressed polynomial
Descartes's Rule of Signs
existence theorems
extraneous root
Factor Theorem
Fundamental Theorem of Algebra
horizontal asymptote
hyperbola
improper rational function
infinite discontinuity
Intermediate Value Theorem

irreducible quadratic factor
leading coefficient
Leading Term Test
linear factors
lower bound
Lower Bound Test
monic polynomial
monomial functions
multiplicity
nonlinear asymptote
polynomial function
polynomial inequality
proper rational function
radical equation
radical function
rational equation
rational function

rational inequality
Rational Zero Theorem
reduced rational function
relative extrema
relative maximum
relative minimum
Remainder Theorem
sign chart
slant asymptote
standard form
synthetic division
synthetic substitution
test intervals
upper bound
Upper Bound Test
vertical asymptote

CHAPTER 3

algebraic function
annual percentage rate (APR)
annual percentage yield (APY)
carrying capacity
change of base formula
common logarithmic function
compounded interest
decay factor
Equality Property of Exponents

Equality Property of Logarithms
exponential decay
exponential equation
exponential function
exponential growth
growth factor
growth rate
half-life
logarithmic equation
logarithmic function

logistic function
natural base
natural exponential function
natural logarithmic function
transcendental function

CHAPTER 4

acute
amplitude
angle of depression
angle of elevation
angular speed
Arccosine
arc length
arc length formula
Arcsine
Arctangent
circular function
cofunction
cosecant function
cosine function
cotangent function
coterminal angle
damped harmonic motion
damped oscillation
damped sinusoid

damping constant
damping factor
degree
frequency
hertz
inverse trigonometric function
linear speed
midline
minute of a degree
negative angle
obtuse
period
periodic function
phase shift
positive angle
principal value
pseudoperiod
quadrantal angle
radian
radian measure

reference angle
reference triangle
secant function
second of a degree
sector
simple harmonic motion
sine function
sinusoid
sinusoidal function
solving a triangle
standard position
tangent function
terminal side
trigonometric functions
uniform circular motion
unit circle
vertex
wavelength

CHAPTER 5

ambiguous case
cofunction
conditional equations
conditional trigonometric equations
double-angle identity
false equations

general solution
half-angle identity
Heron's formula
identity
Law of Cosines
Law of Sines
oblique triangle

power-reducing identity
primary solutions
reduction identity
SAS triangle area formula
semiperimeter
solutions
verifying identities

CHAPTER 6

Archimedean spiral
argument
cardioid
complex plane
components
component vector
De Moivre's Theorem
dot product
equilibrium
imaginary axis
initial point
lemniscate
length of a segment in space
limaçon
linear combination

magnitude of a vector
midpoint of a segment in space
modulus
net force
nth root of the complex number
nth root of unity
number
octant
orthogonal (perpendicular) vectors
polar axis
polar coordinate
polar distance formula
polar form
polar grid
pole
principal root
projection of a vector

real axis
resolved
resultant vector
rose
scalar
scalar product
standard form
standard position
standard unit vector
terminal point
unit vector
vector
vector subtraction
vector sum
zero vector

Appendix B
Explaining the Gospel

One of the greatest desires of Christian teachers is to see their students repent and believe in Christ. Relying on the Holy Spirit, you should take advantage of the opportunities that arise for presenting the gospel. You may find the following outline helpful, especially when dealing individually with a young person.

1. The Lord God is King over all His creation (Rev. 4:11).

- God created everything that is (Gen. 1–2).
- God created the world with laws about how His world is to work and the way people are to live.

2. All have sinned—including me (Rom. 3:23).

- Since Adam first sinned, all people are born rebels against the rule of God (Rom. 5:12).
- God has made me in His own image so that I might declare His glory by being like Him (Gen. 1:26–27).
- But I am a sinner. I disobey God's Word. The Bible teaches that I am to love God more than anything or anyone (Mark 12:30). It also teaches that I should love other people at all times (Mark 12:31). But I don't enjoy doing what God wants me to do, I don't delight in obeying my parents, and I don't like being kind to other people.
- God will punish me for my rebellion and sin (Rom. 6:23). God hates sin, and there is nothing I can do to get rid of my sin. I can try to change my behavior, but I can never change my heart.

3. Jesus died and rose again for me (Rom. 5:8).

- God loves me even though I am a sinner.
- He sent His Son, Jesus Christ, to live a perfect life and to die on the cross, suffering the punishment for my sin.
- Three days later, God raised Jesus from the dead and made Him the ruler of His eternal kingdom. Jesus is alive today. This is the gospel of Jesus Christ: He died on the cross and was raised up to be God's appointed King (1 Cor. 15:1–4).
- God desires to restore me to bear His image fully by making me like His Son, the perfect image bearer of God (Rom. 8:29).
- God gives a new heart—new loves and desires—to every one of His children (Ezek. 36:26–27).

4. I need to put my trust in Jesus (Rom. 10:9–10).

- I must repent (turn away) from my sin and let Jesus control my life. I must also believe what God has done through Jesus (Mark 1:15).
- If I repent and believe in what Jesus has done, I am putting my trust in Jesus.
- Everyone who is trusting in Jesus is forgiven of sin.
- Everyone who is trusting in Jesus submits to having Him as his King.
- Everyone who has been saved by God will continue trusting Jesus for the rest of his life (Heb. 3:14).

When talking individually with a young person, ask questions to discern sincerity or any misunderstanding. What is sin? Are you a sinner? What is the gospel? What does it mean to repent? Read the verses from your Bible. If the student shows genuine sorrow for sin and a sufficiently accurate understanding of the basics of the gospel, encourage him to call on the Lord as you listen. Perhaps he will pray something like the following:

"God, I know that You hate sin. But I also know that You love me. I believe that Jesus died for me and rose from the dead for me. I now turn away from my sin, and I am trusting in Jesus to forgive me and to be my King forever. I want to follow Him wherever He leads me."

Show the student the Bible's command that believers unite together in regular fellowship (Heb. 10:24–25), and encourage him to get involved right away in a gospel-preaching church. Tell him that whenever he sins, he will be forgiven as he confesses his sins to God (1 John 1:9).

Answers and Solutions Overflow

Chapter 1

1.7 Cumulative Review

49. $y = \dfrac{x+2}{x^2 - x - 6} = \dfrac{x+2}{(x-3)(x+2)} = \dfrac{1}{x-3}$

$x \neq 2$ implies a point discontinuity at $\left(-2, -\frac{1}{5}\right)$ and an infinite discontinuity at $x = 3$.

50. The cost of a stamp has always increased by an integral number of cents.

51. The function must open downward and have a y-intercept of $(0, -2)$.

52. The x-intercepts represent the solutions to $f(x) = 0$.

Data Analysis

3. $y = -0.0015x^2 + 0.719x - 0.2226$
$r^2 \approx 0.9855$

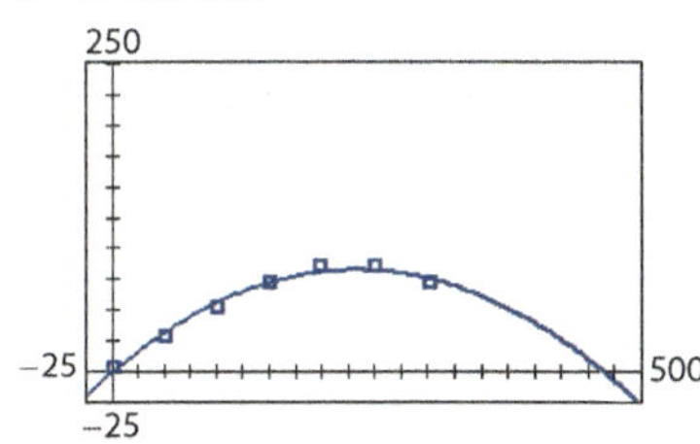

3a. $y = -0.0015x^2 + 0.719x - 0.2226$
$r^2 \approx 0.9855$
Approximately 99% of the variation in height is explained by a variation in horizontal distance.

4a.

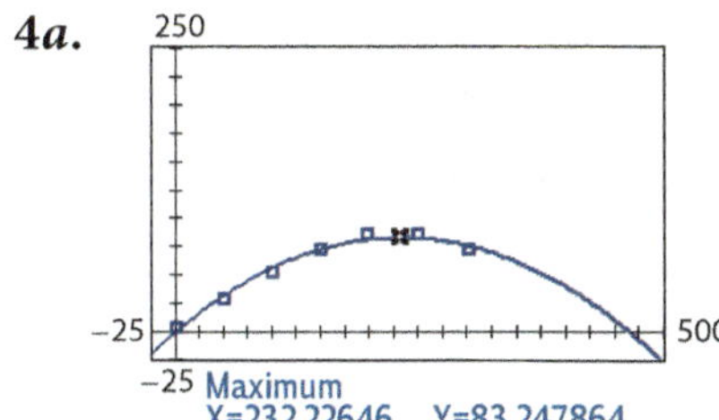

4b. $Y_1(381) \approx 49.0$
yes; The model predicts that the ball will be 49.0 ft high when it is 381 ft from home plate.

4c.

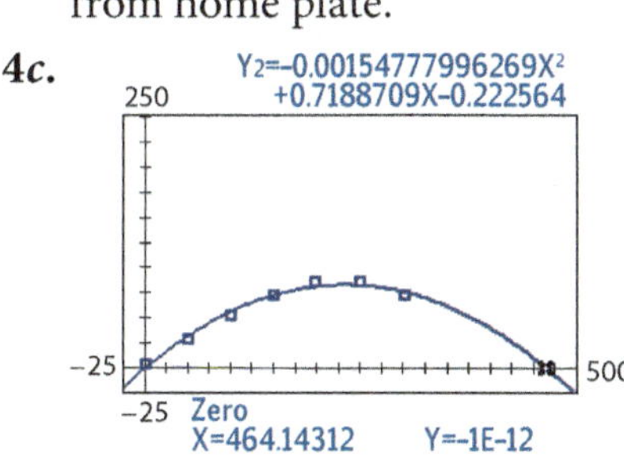

Chapter 1 Review

33. $\dfrac{f}{g}(x) = \dfrac{x^2 - 4x + 4}{x - 2}$ where $x \neq 2$

$\quad = \dfrac{(x-2)^2}{x-2} = x - 2,\ x \neq 2$

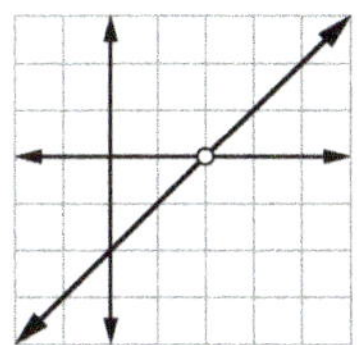

$D = \{x \mid x \neq 2\};\ R = \{y \mid y \neq 0\}$
PD at $x = 2$

34. $f(g(x)) = f(x^2 - 1) = \sqrt{(x^2 - 1) - 3}$
$\quad = \sqrt{x^2 - 4}$
$D_g = \mathbb{R}$, but $x^2 - 4 \geq 0$, so
$D_{f(g(x))} = (-\infty, -2] \cup [2, \infty)$
$D_f = [3, \infty)$
$g(f(x)) = g(\sqrt{x-3})$
$\quad = (\sqrt{x-3})^2 - 1$
$\quad = (x - 3) - 1$
$\quad = x - 4,\ D = [3, \infty)$
$D_{g(f(x))} = D_f = [3, \infty)$

35. sample answers:
$h(x) = -2(x - 6x + 3^2) - 13 - 9(-2)$
$\quad = -2(x - 3)^2 + 5$
$f(x) = -2x^2 + 5,\ g(x) = x - 3$ or
$f(x) = -2x + 5,\ g(x) = (x - 3)^2$

36. $f(g(x)) = \dfrac{(3x + 2)}{3} - 2 = \dfrac{3x - 4}{3} \neq x$ or
$g(f(x)) = 3\left(\dfrac{x}{3} - 2\right) + 2 = x - 4 \neq x$
not inverses

37. $f(g(x)) = \dfrac{\left(\sqrt[5]{4x - 1}\right)^5 + 1}{4} = \dfrac{4x - 1 + 1}{4}$
$\quad = x$ and

$g(f(x)) = \sqrt[5]{4\left(\dfrac{x^5 + 1}{4}\right) - 1}$
$\quad = \sqrt[5]{x^5 + 1 - 1} = \sqrt[5]{x^5} = x$
inverses

38. $x = \dfrac{y - 3}{4}$
$4x = y - 3$
$f^{-1}(x) = 4x + 3$

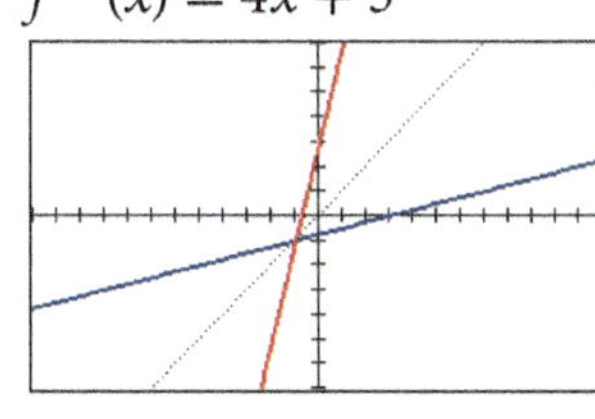

39. $x = \dfrac{1}{y - 3}$
$xy - 3x = 1$
$f^{-1}(x) = \dfrac{3x + 1}{x}$ or $\dfrac{1}{x} + 3$

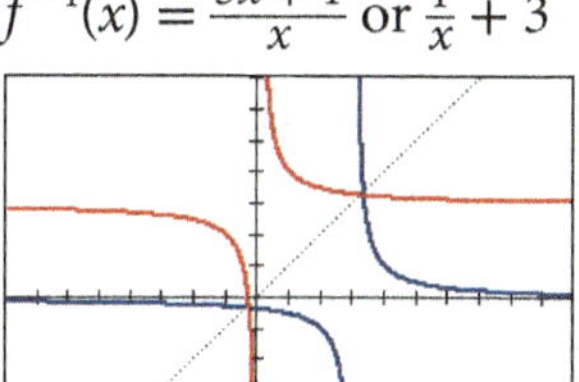

40.

Its inverse is not a function since the function fails the Horizontal Line Test (or the function is not one-to-one).

41a.

41b. $t = x - 4$
$y = (x - 4)^2 - 1$
$\quad = x^2 - 8x + 15$

42a.

using $\boxed{\text{TRACE}}$, $t = 1$: $(45, 47)$;
$t = 2$: $(90, 59)$; $t = 3$: $(135, 39)$

42b. $t = \dfrac{x}{45}$
$y = -16\left(\dfrac{x}{45}\right)^2 + 60\left(\dfrac{x}{45}\right) + 3$
$y = -\dfrac{16}{2025}x^2 + \dfrac{4}{3}x + 3$

42c. $V(h, k)$ is the maximum point.

$h = -\dfrac{\frac{4}{3}}{2\left(-\frac{16}{2025}\right)} = \dfrac{675}{8} = 84.375$ ft

$k = -\dfrac{16}{2025}\left(\dfrac{675}{8}\right)^2 + \dfrac{4}{3}\left(\dfrac{675}{8}\right) + 3$
$\quad = 59.25$ ft

42d. $0 = -\frac{16}{2025}x^2 + \frac{4}{3}x + 3$

$$x = \frac{-\left(\frac{4}{3}\right) \pm \sqrt{\left(\frac{4}{3}\right)^2 - 4\left(-\frac{16}{2025}\right)3}}{2\left(-\frac{16}{2025}\right)}$$

$x \approx -2.22,\ 170.97$ ft

punt distance ≈ 57 yd

43b. $V = \pi\left(\frac{20}{2}\right)^2 H;\ H = \frac{1}{100\pi}V$

43c. $H(t) = H(V(t)) = \frac{1}{100\pi}(4t) = \frac{1}{25\pi}t$

43d. $H(180) = \frac{1}{25\pi}(180) \approx 2.3$ ft

43e. $4 = \frac{1}{25\pi}t$

$t = \frac{4(25)\pi}{1} \approx 314$ min or 5 hr 14 min

44. 6 ft $= 2$ yd and 15 ft $= 5$ yd

$V(35, 5)$; max

$y = a(x - 35)^2 + 5$

$2 = a((20) - 35)^2 + 5$

$2 = 225a + 5$

$a = -\frac{3}{225} = -\frac{1}{75}$

$f(x) = -\frac{1}{75}(x - 35)^2 + 5$

45. $h(t) = -\frac{1}{2}(32)t^2 + 270t + 15$

$= -16t^2 + 270t + 15$

$t = -\frac{270}{2(-16)} = 8.4375$

$k \approx h(8.4375) \approx 1154.1$ ft

46–50.

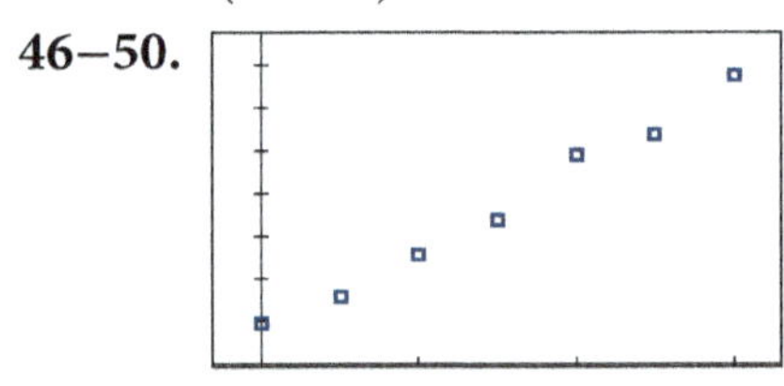

47.

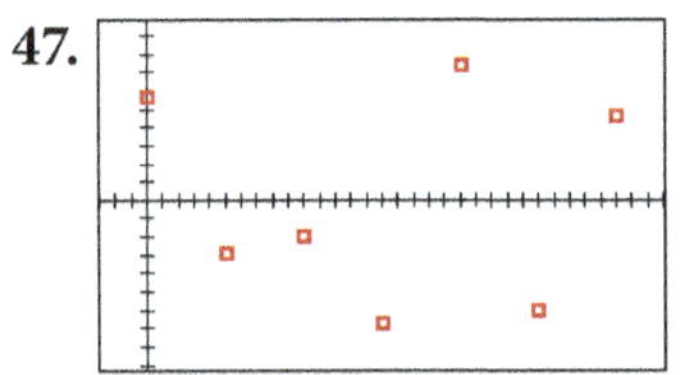

49. $p(13) \approx 4.88799(13) + 17.74 \approx 81.3$

50. $p(40) \approx 4.88799(40) + 17.74 \approx 213.3$

$213{,}300 - 221{,}300 = -8$

$p(40) \approx 0.0297(40)^2 + 3.9979(40)$

$\qquad\qquad + 21.4476 \approx 228.8$

$228.8 - 213.3 = 7.5$

Chapter 2

2.4 Cumulative Review

45. $d = \sqrt{(4 - (-2))^2 + (-1 - 5)^2}$

$= \sqrt{6^2 + (-6)^2} = \sqrt{72} = 6\sqrt{2}$

$M\left(\frac{-2 + 4}{2}, \frac{5 + (-1)}{2}\right) = (1, 2)$

48. $\frac{g}{f}(x) = \frac{7x^2 + 15x + 8}{x + 1}$

$= \frac{(7x + 8)(x + 1)}{x + 1}$

$= 7x + 8,\ x \neq -1$

50. $R = p(4) = 49$

52. $x - 5 \geq 0,\ x \geq 5$

53. $g(x) = \frac{(x - 3)(x + 3)}{x + 3}$

$= x - 3,\ x \neq -3$

54. $-3x^2 - x + 4 = 0$

$3x^2 + x - 4 = 0$

$(3x + 4)(x - 1) = 0$

$x = -\frac{4}{3}, 1$

2.5 B. Exercises

42c. HA: $y = 1$ (or 100%); The percent concentration gets closer and closer to 1 (100%), but it can never be 100%.

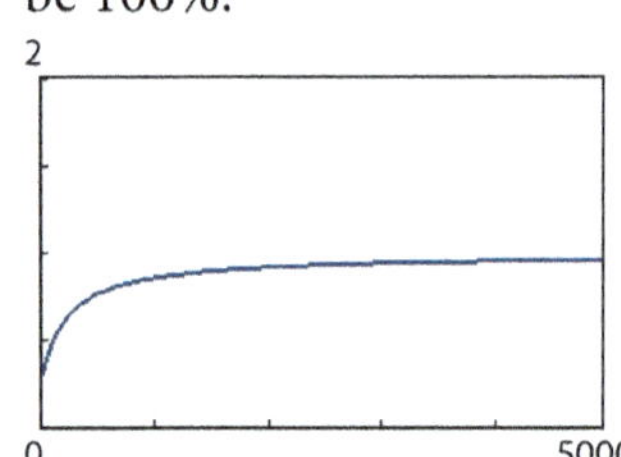

42d. $0.50 = \frac{x + 30}{x + 150}$

$0.5x + 75 = x + 30$

$45 = 0.5x$

$x = 90$ mL

2.5 C. Exercises

43. $f(x) = -\frac{x(x^2 - x + 1)}{1 - x}$

$= -x^2 - 1 + \frac{1}{1 - x}$

NLA: $y = -x^2 - 1$; VA: $x = 1$

y-int.: $f(0) = 0$; $(0, 0)$

x-int.: $x(x^2 - x + 1) = 0$

when $x = 0$; $(0, 0)$

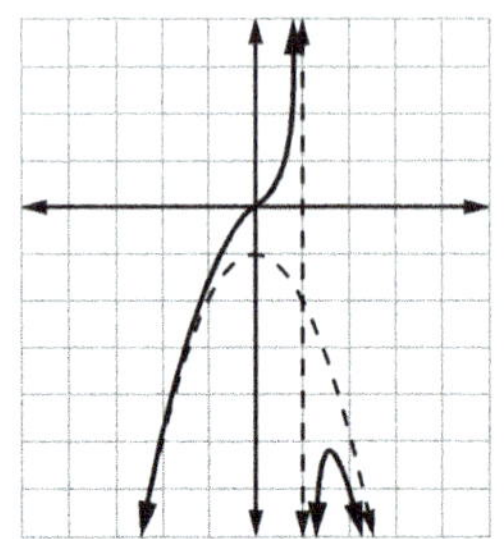

44. $f(x) = \frac{x^2(2x + 1)(x - 1)}{(x - 1)(2x - 1)}$

$= \frac{x^2(2x + 1)}{2x - 1},\ x \neq 1$

$= x^2 + x + 0.5 + \frac{0.5}{2x - 1},\ x \neq 1$

NLA: $y = x^2 + x - 0.5$; VA: $x = \frac{1}{2}$

PD at $x = 1$; y-int.: $f(0) = 0$; $(0, 0)$

x-int.: $x^2(2x + 1) = 0$ when

$x = 0, -\frac{1}{2}$; $(0, 0)$ and $\left(-\frac{1}{2}, 0\right)$

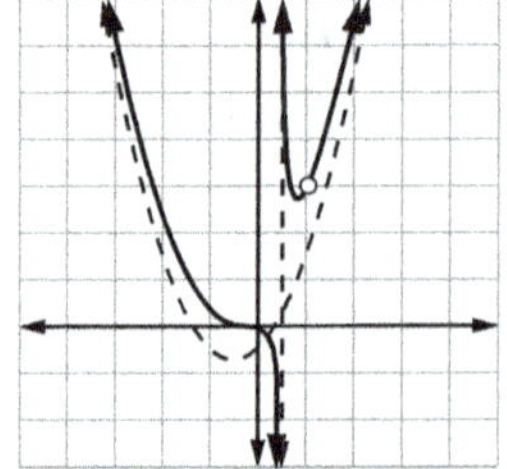

46. $f(x) = \frac{x^n(x^2 + 2x + 4)}{x^{n-1}(x^3 - 8)}$

$= \frac{x^n(x^2 + 2x + 4)}{x^{n-1}(x - 2)(x^2 + 2x + 4)}$

$= \frac{x}{x - 2},\ x \neq 0$

HA: $y = \frac{1}{1} = 1$; VA: $x = 2$; PD at $x = 0$

no intercepts

2.5 Cumulative Review

47. $\frac{a + 3}{(a + 1)(a - 1)} + \frac{2a(a - 1)}{(a + 1)(a - 1)}$

$= \frac{a + 3 + 2a^2 - 2a}{a^2 - 1} = \frac{2a^2 - a + 3}{a^2 - 1}$

48. $\frac{x + 3}{(x + 5)(x - 2)} \cdot \frac{(x + 2)}{(x + 2)}$

$\qquad\qquad - \frac{5x}{(x + 2)(x - 2)} \cdot \frac{(x + 5)}{(x + 5)}$

$= \frac{x^2 + 5x + 6 - 5x^2 - 25x}{(x + 5)(x - 2)(x + 2)}$

$= \frac{-4x^2 - 20x + 6}{(x + 2)(x - 2)(x + 5)}$

49. $2x^2 - 5x + 3 = 0$

$(2x - 3)(x - 1) = 0$

$x = \frac{3}{2}, 1$

50. $x = \frac{-(1) \pm \sqrt{(1)^2 - 4(5)(-3)}}{2(5)} = \frac{-1 \pm \sqrt{61}}{10}$

52. $f(x) = 3x^5 - 23x^4 + 61x^3 - 61x^2$

$\qquad\qquad + 8x + 12$

	3	−23	61	−61	8	12
1	3	−20	41	−20	−12	0
3	3	−11	8	4	0	
2	3	−5	−2	0		

$f(x) = (x - 1)(x - 2)(x - 3)$

$\qquad\qquad \cdot (3x^2 - 5x - 2)$

$f(x) = (x - 1)(x - 2)(x - 3)(3x + 1)$

$\qquad\qquad \cdot (x - 2)$

$f(x) = (x - 1)(x - 2)^2(x - 3)(3x + 1)$

53. $f(-3) = -(-3)^3 + 7 = 34$

54. $y = \dfrac{6}{x^{-3}} = 6x^3$; a polynomial function
$f(-x) = 6(-x)^3 = -6x^3 = -f(x)$

55. $y = x^3$ is the only function that passes the Horizontal Line Test.

56.

2.6 C. Exercises

37b. $A = (11 + 2\sqrt{21})(11 - 2\sqrt{21})$
$= 121 - 4(21) = 37$

38a.

	R	T	D
First Half	r_1	$\dfrac{\left(\frac{d}{2}\right)}{r_1} = \dfrac{d}{2r_1}$	$\dfrac{1}{2}d$
Second Half	r_2	$\dfrac{\left(\frac{d}{2}\right)}{r_2} = \dfrac{d}{2r_2}$	$\dfrac{1}{2}d$
Total Trip	20	$\dfrac{d}{20}$	d

$$\dfrac{d}{2r_1} + \dfrac{d}{2r_2} = \dfrac{d}{20}$$
$$10r_1r_2\left[\dfrac{1}{r_1} + \dfrac{1}{r_2} = \dfrac{1}{10}\right]$$
$$10r_2 + 10r_1 = r_1r_2$$
$$10r_1 = r_1r_2 - 10r_2$$
$$10r_1 = r_2(r_1 - 10)$$
$$r_2 = \dfrac{10r_1}{r_1 - 10}$$

38b. Y1=10X/(X-10)
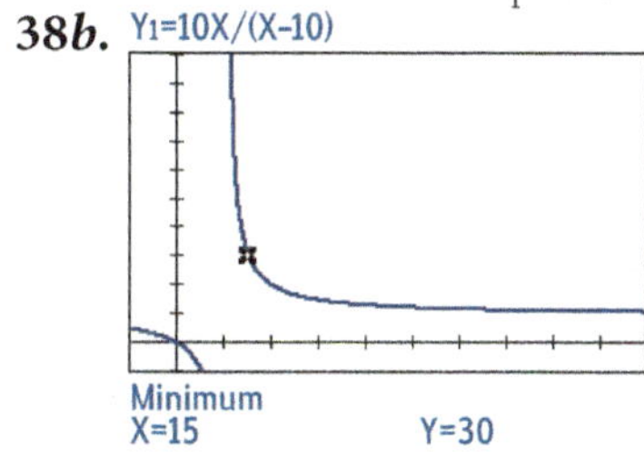

38c. An average speed of 10 mi/hr or less in the first half would use up all the time so that an average overall speed of 20 mi/hr would not be possible.

38d. As the average speed for the first half approaches 10 mi/hr, almost all the time is used up, so the speed for the second half increases, approaching infinity. The horizontal asymptote indicates that the average speed for the second half of the trip approaches 10 mi/hr as the average speed for the first half increases.

2.6 Cumulative Review

39–40. The function's sign changes only when the multiplicity is odd.

43. $\dfrac{1}{(x+3)(x-3)} \cdot \dfrac{x-3}{x-3}$
$+ \dfrac{x-3}{(x-3)(x-3)} \cdot \dfrac{x+3}{x+3}$
$= \dfrac{x-3+x^2-9}{(x+3)(x-3)(x-3)}$
$= \dfrac{x^2+x-12}{(x+3)(x-3)(x-3)}$
$= \dfrac{(x+4)(x-3)}{(x+3)(x-3)(x-3)} = \dfrac{x+4}{x^2-9}$

44. $\dfrac{x}{(x+2)(x+3)} \cdot \dfrac{x-1}{x-1}$
$- \dfrac{6}{(x+2)(x-1)} \cdot \dfrac{x+3}{x+3}$
$= \dfrac{x^2-7x-18}{(x+2)(x+3)(x-1)}$
$= \dfrac{(x+2)(x-9)}{(x+2)(x+3)(x-1)}$
$= \dfrac{x-9}{(x+3)(x-1)}$

45. $y = -x + 3$; $m_G = -1$
$m_\perp = -\dfrac{1}{m_G} = -\dfrac{1}{(-1)} = 1$
$y - 2 = 1(x - 1)$
$y - 2 = x - 1$
$y = x + 1$

47. $x = \dfrac{1}{y+5}$
$y + 5 = \dfrac{1}{x}$
$y = \dfrac{1}{x} - 5 = \dfrac{1-5x}{x}$

2.7 C. Exercises

41. $(2w + 150)w \geq 12(43{,}560)$
$2w^2 + 150w \geq 522{,}720$
$w^2 + 75w \geq 261{,}360$
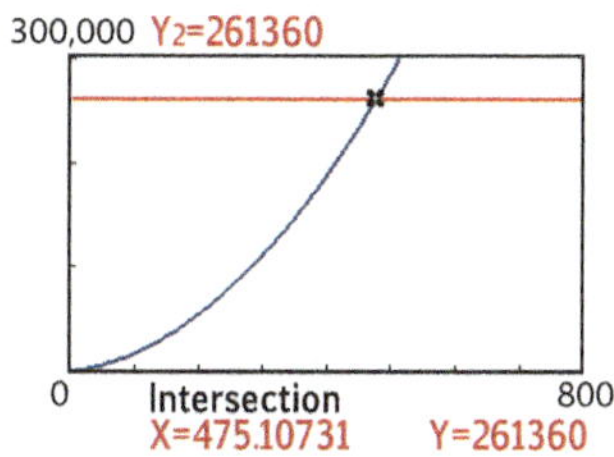

$l \approx 2(475) + 150 \approx 1100$

42. original width: x
original length: $2x$
adjusted length: $2x + 50$
25 acres = 121,000 yd^2
$A(x) = (2x + 50)x$
$2x^2 + 50x \geq 121{,}000$

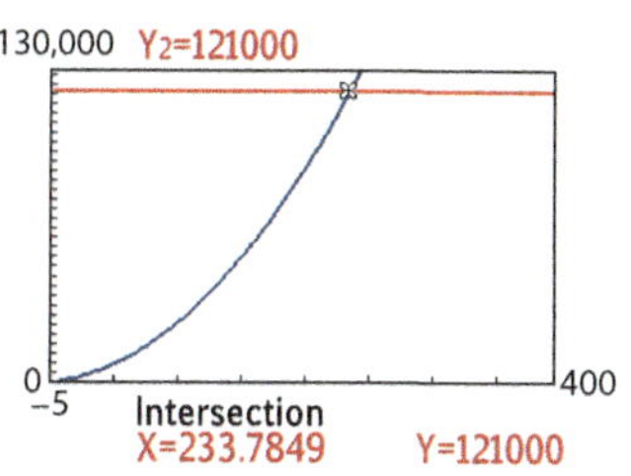

$l \approx 2(233.8) + 50 = 517.6$ yd

2.7 Cumulative Review

43.
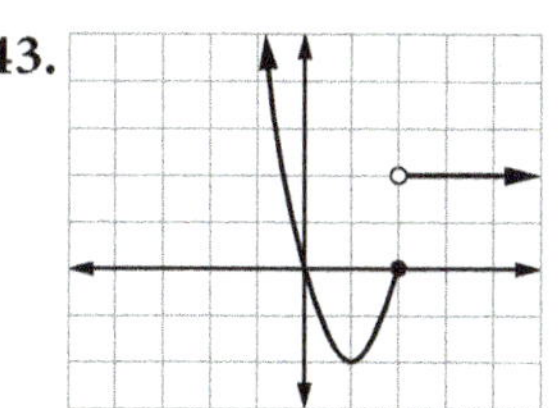

44. $f(x) = -3(x^2 - 4x + 2^2) - 9 - 4(-3)$
$= -3(x - 2)^2 + 3$
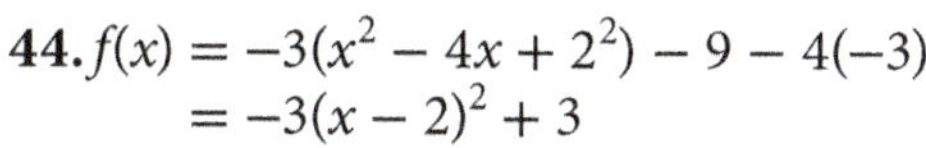

45. $\dfrac{f}{g}(x) = \dfrac{x^3 + 3x^2 + 2x + 6}{x^2 + 2}$
$\dfrac{f}{g}(x) = \dfrac{(x+3)(x^2+2)}{x^2+2}$
$\dfrac{f}{g}(x) = x + 3$; $D_{f/g} = \mathbb{R}$

46b. When a relation passes the Horizontal Line Test, its inverse is a function.

46d. See 46e.

47.
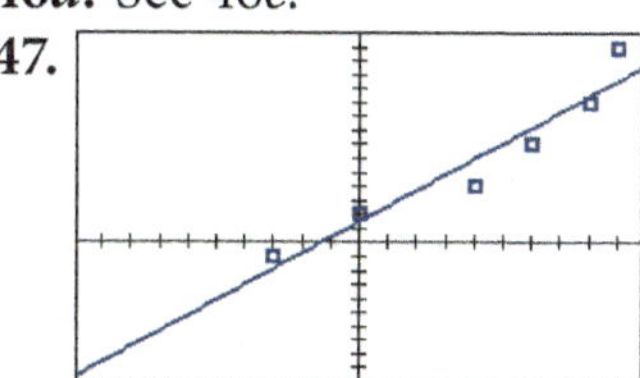

48. The negative leading coefficient implies that the graph will open downward.

49. $V = kT$
$2.86 = k(293)$; $k = \dfrac{2.86}{293}$
$2.72 = \left(\dfrac{2.86}{293}\right)T$
$T = \left(\dfrac{293}{2.86}\right)2.72 \approx 278.66$ K

50. $f(x)$ has been reflected and moved 4 units left and 3 units up.

or
$$f(x) = -(x-3)^2 + 1$$
$$g(x) = -f(x+4) + 3$$
$$= -[-((x+4)-3)^2 + 1] + 3$$
$$= (x+1)^2 - 1 + 3$$
$$= (x+1)^2 + 2$$

51. $D_f = D_g = D = [0, \infty)$
$$(fg)(x) = \sqrt{x} \cdot \sqrt{x(x^2 + 2x + 1)}$$
$$= \sqrt{x^2(x+1)^2}$$
$$= |x(x+1)|, D = [0, \infty)$$
$$= x(x+1), D = [0, \infty)$$

52. $x + 9 = x^2 - 6x + 9$
$$0 = x^2 - 7x$$
$$0 = x(x-7)$$
$$x = 7, 0 \text{ (extraneous)}$$

Data Analysis

3c. 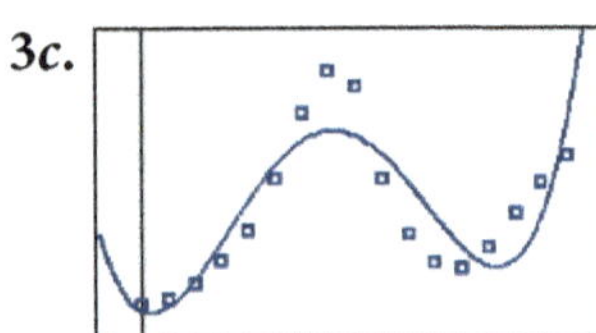

the variation of the real data from the model: 2007 (+75.5), 2008 (+61.8), 2010 (−64.2), or 2011 (−58.6)

4. The function's values (prices) increase sharply toward infinity. The model is likely not reliable beyond the known data set.

5. The sharp increase indicates a boom in housing prices. The sharp decrease indicates a crash in housing prices. Those who had made recent purchases would have found themselves with a home that was worth much less than they paid for it. Additionally, homeowners would find it hard to sell their home if they owed much more than the home was worth.

6. The model would continuously reach relative maximum and minimum points throughout history while the median values gradually increased. This would be a periodic function, similar to a sine wave.

7. The linear model indicates that purchasing a home can be a good investment over a longer period of time. The quartic model indicates that there is some risk, especially with short-term home purchases. The possibility of unsustainable increases in market prices serves as a warning against overextending one's finances when purchasing a home during times of prosperity.

8. Since it is difficult to know when the prices are at a relative minimum, it would be good to buy after the market has experienced a significant price drop. Over time, any further drop in value would likely be recovered, and the long-term investment would be rewarded with an increase in value. However, a buyer's immediate housing needs may not allow him to wait until this point to purchase a home. Any purchase must be carefully weighed against the possibilities of an economic downturn, loss of employment, and health issues.

9. Prudence is important when buying a home because of unpredictable factors such as health issues and inherent risks like market downturns. When a person binds himself to a mortgage, he needs to be able to meet the long-term commitment. He becomes, in a sense, a slave to the loan and the lender until the loan is fully paid.

Chapter 3

3.3 Cumulative Review

49. $(3x-2)(x+6)\left[\dfrac{5}{(3x-2)(x+6)} + \dfrac{3x-2}{x+6} = \dfrac{x}{3x-2}\right]$
$$5 + (3x-2)^2 = x(x+6)$$
$$9x^2 - 12x + 9 = x^2 + 6x$$
$$8x^2 - 18x + 9 = 0$$
$$(2x-3)(4x-3) = 0$$
$$x = \frac{3}{2}, \frac{3}{4}$$

53. $\log_2 2 = 1$ and $\log_2 4 = 2$

54. $f(6) = 74 - 16 \log(6+1) \approx 60$

3.4 Cumulative Review

51. $f(x)$ is translated 3 units up.

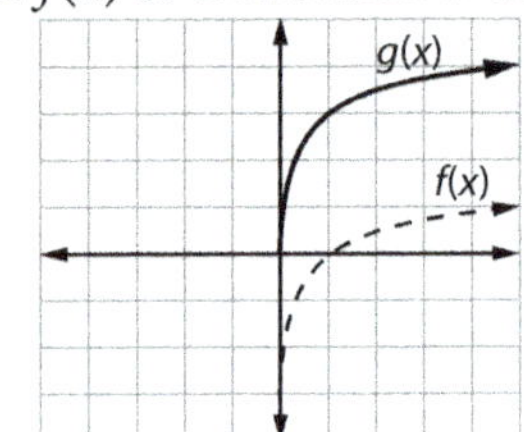

52. $f(x)$ is translated 4 units right and 2 units up.

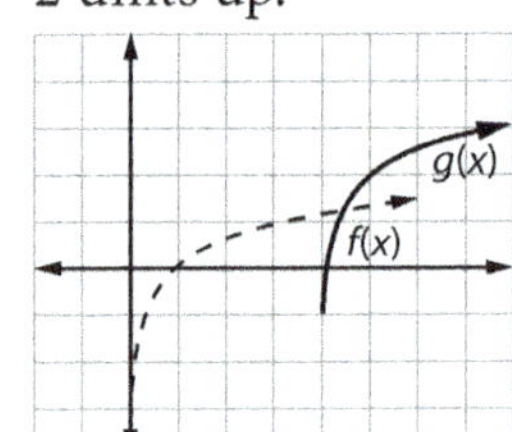

53. $\log 10 + \frac{1}{3} \log x^2 + \frac{1}{3} \log y$
$$1 + \frac{2}{3} \log x + \frac{1}{3} \log y$$

55. $f(x) = \dfrac{x(x+1)(x-2)}{(x-1)(x-2)}$; VA: $x = 1$
$$f(x) = x + 2 + \frac{2x-4}{x^2 - 3x + 2}$$
SA: $y = x + 2$

56. $\dfrac{(x-5)(x+3)}{(2x+3)(x+3)} \geq 0$; $\dfrac{x-5}{2x+3} \geq 0$

$$x = (-\infty, -3) \cup \left(-3, -\frac{3}{2}\right)$$
$$\cup [5, \infty)$$

57. $A(t) = P\left(1 + \dfrac{r}{n}\right)^{nt}$
$$A(t) = 4500\left(1 + \frac{0.065}{2}\right)^{(2)(12)}$$
$$= \$9695.58$$

42.

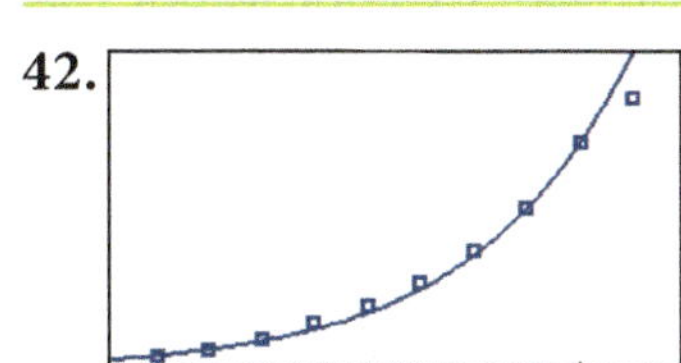

42a. $P(115) = 20.6351(1.0353)^{115}$
≈ 1112

42b.
$$1250 = 20.6351(1.0353)^t$$
$$\frac{1250}{20.6351} = 1.0353^t$$
$$\ln \frac{1250}{20.6351} = t \ln 1.0353$$
$$t = \frac{\ln 1250 - \ln 20.6351}{\ln 1.0353} \approx 118.4$$
or

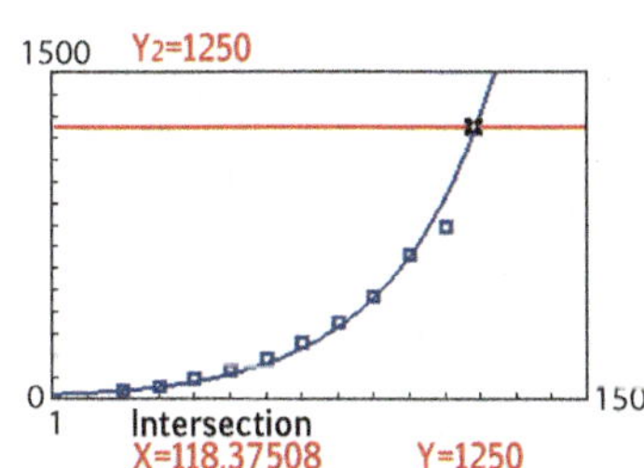

43b. $Y_1=1706.5334502474/(1+95.3535210\ldots$

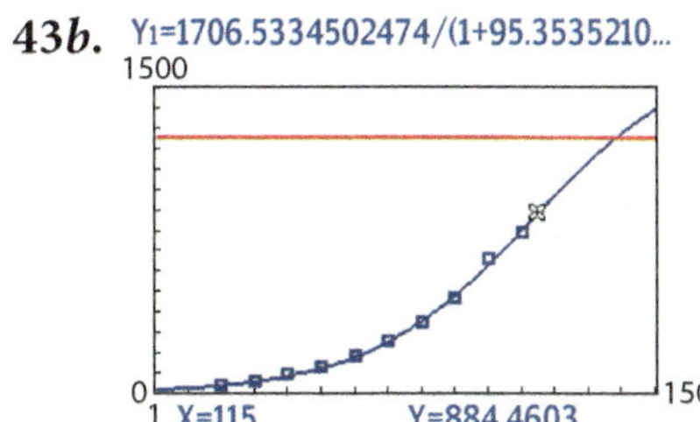

43c.

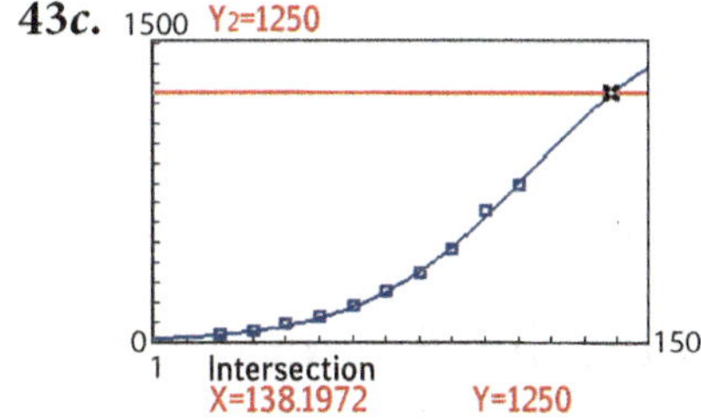

44.

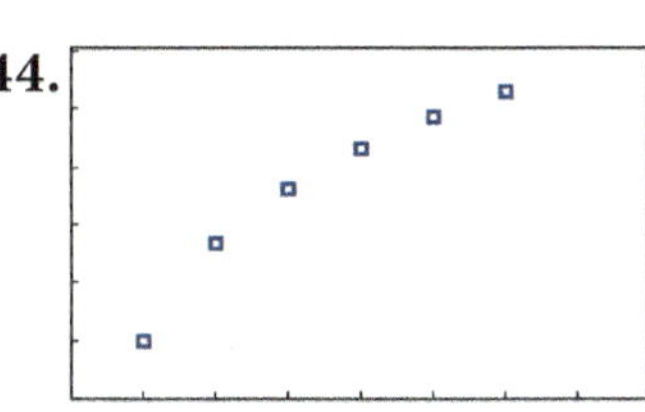

46.

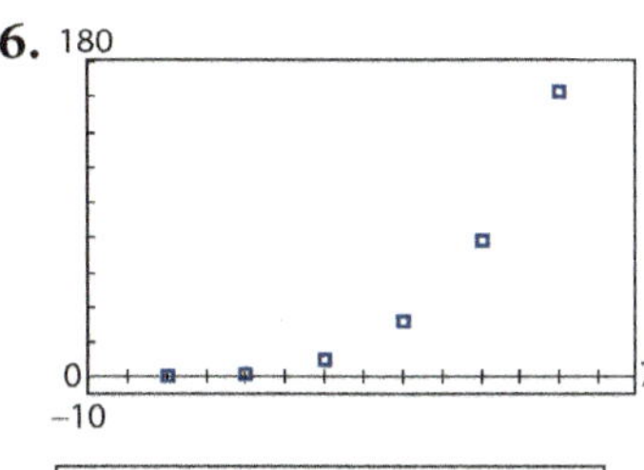

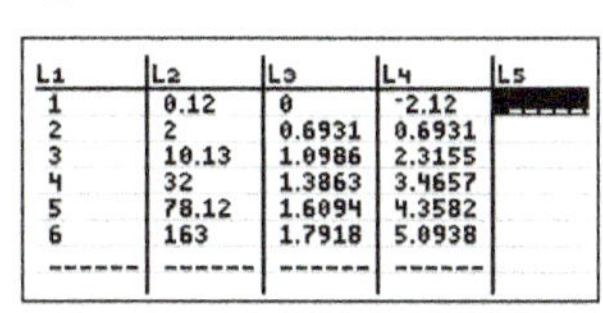

power: $(X, Y) = (\ln x, \ln y)$

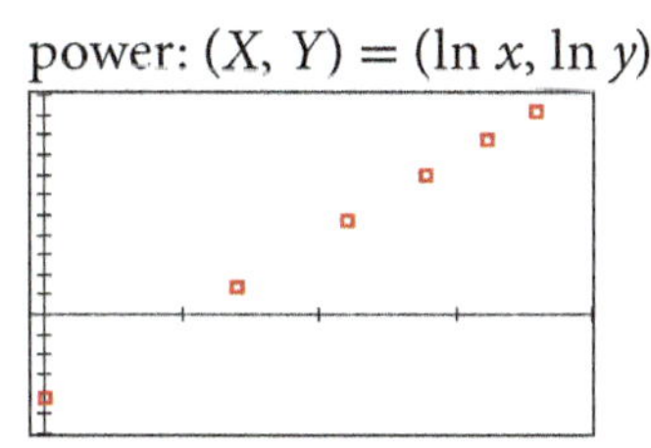

exponential: $(X, Y) = (x, \ln y)$

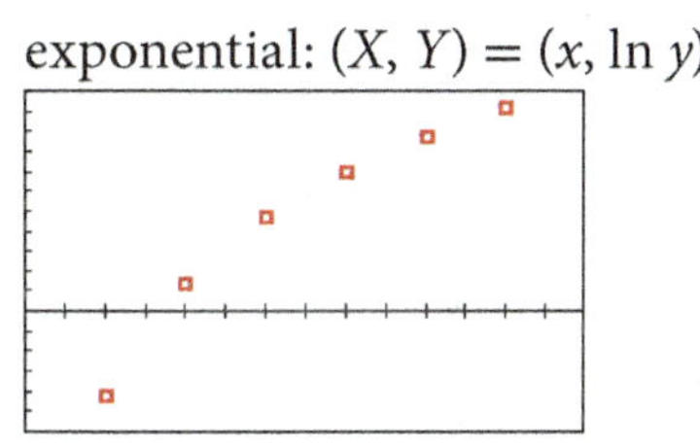

logarithmic: $(X, Y) = (\ln x, y)$

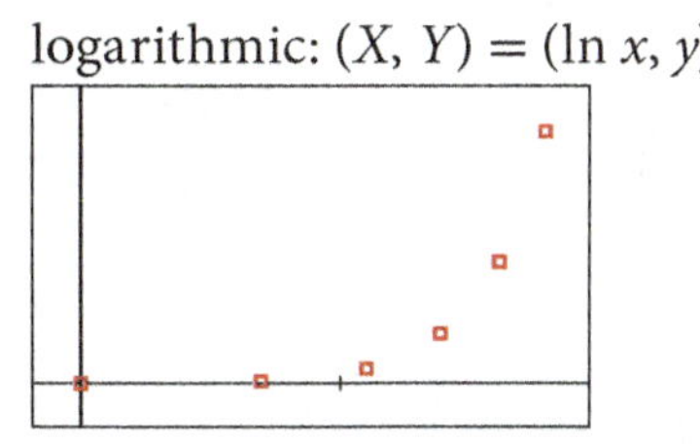

47.
$$Y \approx 4.0223X - 2.1096$$
$$\ln y \approx 4.0223 \ln x - 2.1096$$
$$y \approx e^{4.0223 \ln x + (-2.1096)}$$
$$y \approx (e^{\ln x})^{4.0223} e^{-2.1096}$$
$$y \approx 0.1213x^{4.0223}$$

Chapter 4

4.2 Cumulative Review

46. $f(0) = \frac{0+3}{0^2-4} = -\frac{3}{4}$; y-int.: $\left(0, -\frac{3}{4}\right)$
$x^2 - 4 \neq 0$; $x \neq \pm 2$
$\therefore D = \{x \mid x \neq \pm 2\}$; VA: $x = \pm 2$

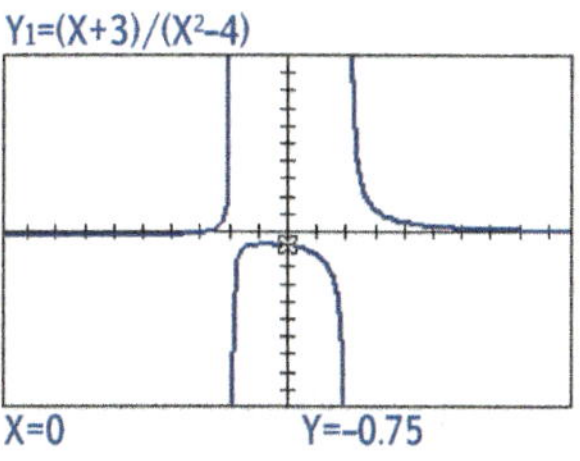

47. $2^1 = x^2 + 3x + 4$
$0 = x^2 + 3x + 2$
$0 = (x + 1)(x + 2)$
$x = -1, -2$

48. $\log (x^2 - 144) - \log (x + 12) = 1$
$\log \frac{(x - 12)(x + 12)}{x + 12} = 1$
$\log (x - 12) = 1$
$10^1 = x - 12$
$x = 22$

49. $36°\left(\frac{\pi}{180°}\right) = \frac{\pi}{5} \approx 0.63$

50. $-148°\left(\frac{\pi}{180°}\right) = -\frac{148\pi}{180} = -\frac{37\pi}{45}$
≈ -2.58

51.
$$\sqrt{x} - 1 = \sqrt{2x - 2}$$
$$x - 2\sqrt{x} + 1 = 2x - 2$$
$$2\sqrt{x} = x - 3$$
$$4x = x^2 - 6x + 9$$
$$x^2 - 10x + 9 = 0$$
$$(x - 1)(x - 9) = 0$$
$$x = 1, 9 \text{ (extraneous)}$$

52. $\ln e^{6x} = \ln 6^2$
$6x = 2 \ln 6$
$x = \frac{2 \ln 6}{6} = \frac{1}{3} \ln 6$

53. $3e^{2x + 1} = 6$
$\ln e^{2x + 1} = \ln 2$
$2x + 1 = \ln 2$
$x = \frac{1}{2}(\ln 2 - 1)$

54. A. $\frac{13\pi}{4} - \frac{5\pi}{4} = 2\pi$
B. $-\frac{3\pi}{4} - \frac{5\pi}{4} = -2\pi$
C. $\frac{9\pi}{4} - \frac{5\pi}{4} = \pi$ (not a multiple of 2π)
D. $-\frac{11\pi}{4} - \frac{5\pi}{4} = -4\pi$
E. $\frac{21\pi}{4} - \frac{5\pi}{4} = 4\pi$

4.4 C. Exercises

44b.

Since both functions are even and symmetric with respect to the y-axis, the upper limit is 1.4. $[-1.4, 1.4]$

4.4 Cumulative Review

45. $f(x) = (x+3)^2(x-4)^2$
$$= (x^2 + 6x + 9)(x^2 - 8x + 16)$$
$$= x^4 - 2x^3 - 23x^2 + 24x + 144$$

46. degree of 3 (the highest power of a term)

47. degree of 5 (Add the highest powers.)

48. degree of 1 (Subtract the highest powers.)

50b. $x^2 + 5x - 14 = 0$
$$(x+7)(x-2) = 0$$
$$x = -7$$

50c. $x^2 + 1 > 0$ for all $x \in \mathbb{R}$

51. $\cos \frac{\pi}{6} = \frac{\sqrt{3}}{2}$

52. A. $\csc \theta = \frac{41}{9} = \frac{r}{y}$
$$x^2 + 9^2 = 41^2$$
$$x = \sqrt{41^2 - 9^2} = 40$$
B. $\tan \theta = \frac{y}{x} = \frac{9}{40}$
$$\cos \theta = \frac{x}{r} = \frac{40}{41}; \sec \theta = \frac{41}{40}$$
C. $\frac{41}{9} > \frac{41}{40}$
D. $\frac{40}{41} > \frac{1}{2}$

53. $\cos(-\theta) + \sec \theta = \cos \theta + \sec \theta$
$$= 0.8 + \frac{10}{8} = 2.05$$

54. in Q II, $\sin \theta > 0$ and $\cos \theta < 0$, so $\tan \theta = \frac{\sin \theta}{\cos \theta} < 0$

4.7 B. Exercises

21. $p = \frac{2\pi}{\pi} = 2$

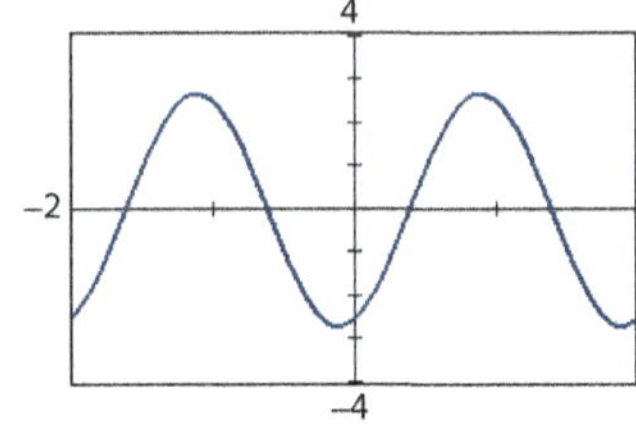

22. $p_1 = \frac{2\pi}{2} = \pi; \; p_2 = \frac{2\pi}{\frac{1}{2}} = 4\pi$
$$p_3 = \frac{2\pi}{\frac{1}{3}} = 6\pi$$

LCM of π, 4π, and 6π is 12π.

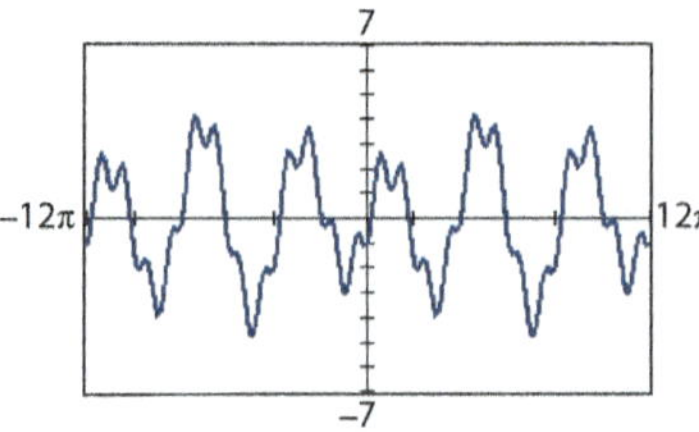

23. Both addends have a period of π.

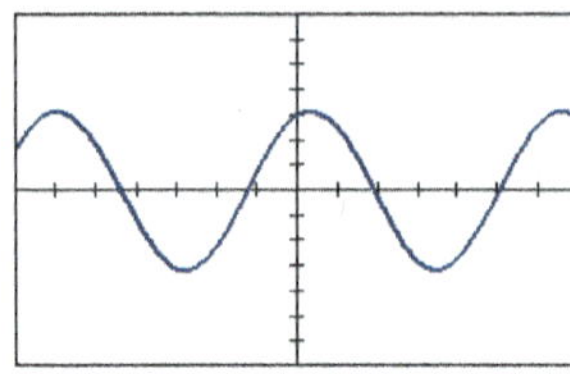

24. The periods of the addends, 2π and π, have an LCM of 2π.

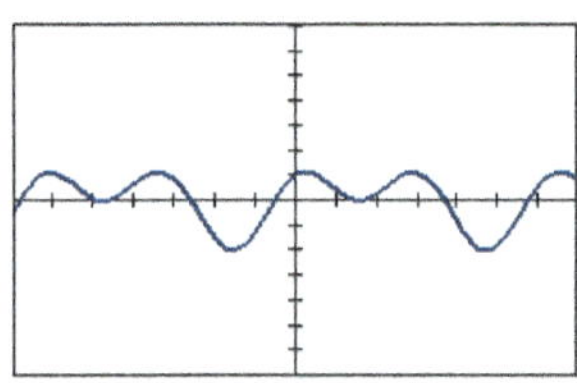

25. The parent sinusoid is vertically translated by a varying amount, x.

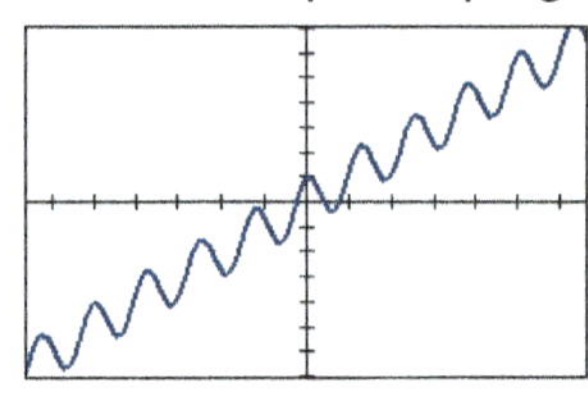

26. The parent sinusoid is stretched vertically by a varying amount, x.

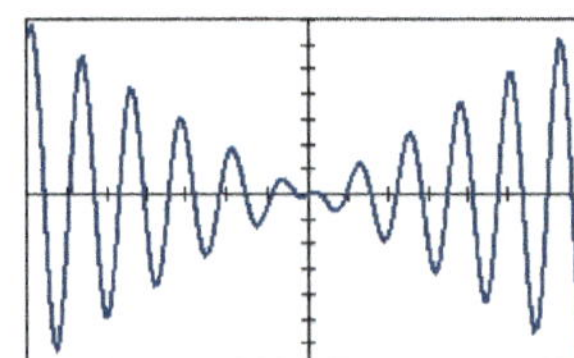

36. $p = \pi = \frac{2\pi}{|b|}; \; b = \pm 2$
$$y = e^{-2x} \sin 2\left(x + \frac{\pi}{4}\right)$$
$$y = e^{-2x} \sin \left(2x + \frac{\pi}{2}\right)$$

37. $p = 2 = \frac{2\pi}{|b|}; \; b = \pm \pi$
$$y = 0.2x^2 \sin \pi(x - \pi)$$
$$y = 0.2x^2 \sin (\pi x - \pi^2)$$

38a. $f = \frac{|b|}{2\pi}; \; |b| = 2\pi(3) = 6\pi; \; a = 5$ cm
$$c = 0.7; \; d(t) = 5e^{-0.7t} \cos 6\pi t$$

38b.

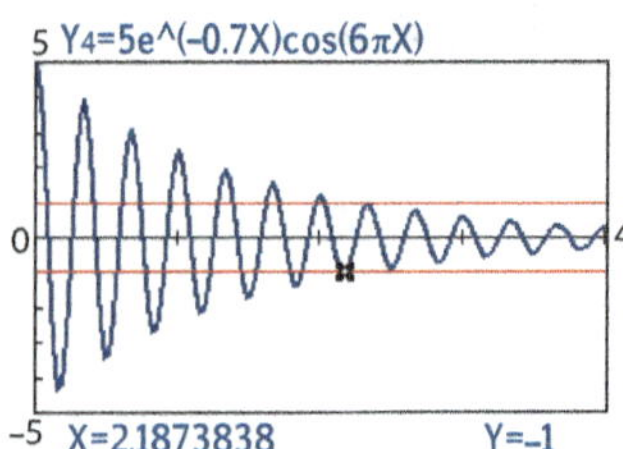

39. $f = \frac{|b|}{2\pi}; \; |b| = 49(2\pi) = 98\pi$
$$a = 0.9 \text{ cm}$$
$$f(x) = 0.9e^{-2.05x} \cos 98\pi x$$
$$f(1) = 0.9e^{-2.05} \cos 98\pi \approx 0.12 \text{ cm}$$

4.7 C. Exercises

43.

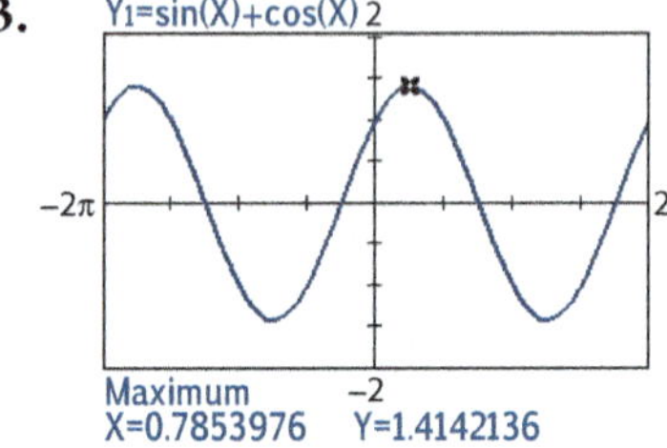

$$p = 2\pi; \; b = 1; \; |a| \approx 1.41$$
$$h \approx 0.79$$
$$\therefore f(x) \approx 1.41 \cos (x - 0.79)$$

44.

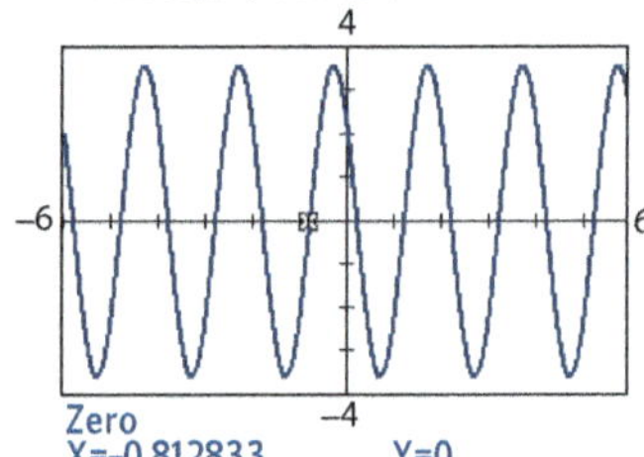

$$p = 2; \; b = \pi; \; |a| \approx 3.61$$
$$h \approx -0.81$$
$$\therefore f(x) \approx 3.61 \sin \pi(x + 0.81)$$

4.7 Cumulative Review

45. $x = \frac{2}{3}y - 5$
$$3x = 2y - 15$$
$$2y = 3x + 15$$
$$f^{-1}(x) = \frac{3}{2}x + \frac{15}{2}$$

46. $\frac{12\pi}{7}\left(\frac{180°}{\pi \text{ radians}}\right) \approx 308.57°$

47. $r = \sqrt{(-7)^2 + 2^2} = \sqrt{53}$
$$\sin \theta = \frac{y}{r} = \frac{2}{\sqrt{53}} = \frac{2\sqrt{53}}{53}$$

48. $1000° - (2)360° = 280°; \therefore$ Q IV
$$360° - 280° = 80°$$

51. $f(x) = 9 \cdot 3^{2x}$
$$= 3^2 \cdot 3^{2x}$$
$$= 3^{2x+2}$$

52. $\log \frac{x^2}{y^4} = \log x^2 - \log y^4$
$$= 2 \log x - 4 \log y$$

53. Shifting the graph of the secant function $\frac{\pi}{2}$ units left and then reflecting it in the y-axis produces the graph of the cosecant function.

Chapter 5

5.2 C. Exercises

38. $\dfrac{2\cot x}{1-\cot^2 x}+\dfrac{1}{2\sin^2 x-1}$

$=\dfrac{2\frac{\cos x}{\sin x}}{1-\frac{\cos^2 x}{\sin^2 x}}\left(\dfrac{\sin^2 x}{\sin^2 x}\right)$

$\qquad+\dfrac{1}{2\sin^2 x-(\sin^2 x+\cos^2 x)}$

$=\dfrac{2\sin x\cos x}{\sin^2 x-\cos^2 x}+\dfrac{1}{\sin^2 x-\cos^2 x}$

$=\dfrac{2\sin x\cos x+(\sin^2 x+\cos^2 x)}{\sin^2 x-\cos^2 x}$

$=\dfrac{\sin^2 x+2\sin x\cos x+\cos^2 x}{\sin^2 x-\cos^2 x}$

$=\dfrac{(\sin x+\cos x)(\sin x+\cos x)}{(\sin x+\cos x)(\sin x-\cos x)}$

$=\dfrac{\sin x+\cos x}{\sin x-\cos x}$

or

working on both sides:

$\dfrac{2\cot x}{1-\cot^2 x}+\dfrac{1}{2\sin^2 x-1}=\dfrac{\sin x+\cos x}{\sin x-\cos x}$

$\dfrac{2\left(\frac{\cos x}{\sin x}\right)\left(\sin^2 x\right)}{\left(1-\frac{\cos^2 x}{\sin^2 x}\right)\left(\sin^2 x\right)}$

$\qquad+\dfrac{1}{2\sin^2 x-(\sin^2 x+\cos^2 x)}$

$=\dfrac{\sin x+\cos x}{\sin x-\cos x}\left(\dfrac{\sin x+\cos x}{\sin x+\cos x}\right)$

$\dfrac{2\sin x\cos x}{\sin x^2-\cos^2 x}+\dfrac{1}{\sin^2 x-\cos^2 x}$

$\qquad=\dfrac{\sin^2 x+2\sin x\cos x+\cos^2 x}{\sin^2 x-\cos^2 x}$

$\dfrac{1+2\sin x\cos x}{\sin^2 x-\cos^2 x}=\dfrac{1+2\sin x\cos x}{\sin^2 x-\cos^2 x}$

39. $\cosh^2 x-\sinh^2 x$

$=\left(\tfrac{1}{2}(e^x+e^{-x})\right)^2-\left(\tfrac{1}{2}(e^x-e^{-x})\right)^2$

$=\tfrac{1}{4}\left[(e^x+e^{-x})^2-(e^x-e^{-x})^2\right]$

$=\tfrac{1}{4}\left[(e^{2x}+2e^0+e^{-2x})\right.$

$\qquad\left.-(e^{2x}-2e^0+e^{-2x})\right]$

$=\tfrac{1}{4}(e^{2x}+2+e^{-2x}-e^{2x}+2-e^{-2x})$

$=\tfrac{1}{4}(4)=1$

40. $\coth^2 x-\operatorname{csch}^2 x$

$=\dfrac{\cosh^2 x}{\sinh^2 x}-\dfrac{1}{\sinh^2 x}$

$=\dfrac{\cosh^2 x-1}{\sinh^2 x}=\dfrac{\left(\frac{1}{2}(e^x+e^{-x})\right)^2-1}{\left(\frac{1}{2}(e^x-e^{-x})\right)^2}$

$=\dfrac{\frac{1}{4}e^{2x}+\frac{1}{2}+\frac{1}{4}e^{-2x}-1}{\frac{1}{4}e^{2x}-\frac{1}{2}+\frac{1}{4}e^{-2x}}$

$=\dfrac{\frac{1}{4}e^{2x}-\frac{1}{2}+\frac{1}{4}e^{-2x}}{\frac{1}{4}e^{2x}-\frac{1}{2}+\frac{1}{4}e^{-2x}}=1$

41. $\operatorname{sech}(-x)$

$=\dfrac{1}{\cosh(-x)}=\dfrac{1}{\frac{1}{2}(e^{(-x)}+e^{-(-x)})}$

$=\dfrac{1}{\frac{1}{2}(e^{-x}+e^x)}=\dfrac{1}{\cosh x}=\operatorname{sech} x$

42. $\operatorname{csch}(-x)=\dfrac{1}{\sinh(-x)}$

$=\dfrac{1}{\frac{1}{2}(e^{(-x)}-e^{-(-x)})}=\dfrac{1}{\frac{1}{2}(e^{-x}-e^x)}$

$=\dfrac{1}{\frac{1}{2}((-1)(e^x-e^{-x}))}=-\dfrac{1}{\frac{1}{2}(e^x-e^{-x})}$

5.2 Cumulative Review

43. $\qquad x=\dfrac{y}{y+3}$

$x(y+3)=y$

$xy+3x=y$

$xy-y=-3x$

$y(x-1)=-3x$

$\qquad y=-\dfrac{3x}{x-1}$

44. $2=\ln x-\ln 3$

$2=\ln\dfrac{x}{3}$

$e^2=\dfrac{x}{3}$

$x=3e^2\approx 22.2$

46. $\dfrac{1}{\sin^2\theta}-\dfrac{1}{\sin\theta\cos\theta}$

$=\dfrac{\cos\theta}{\sin^2\theta(\cos\theta)}-\dfrac{\sin\theta}{\sin\theta\cos\theta(\sin\theta)}$

$=\dfrac{\cos\theta-\sin\theta}{\sin^2\theta\cos\theta}$

47. $\dfrac{\sin\theta}{\cos\theta}+\dfrac{1}{\cos\theta}=\dfrac{\sin\theta+1}{\cos\theta}$

48. $\dfrac{\frac{\sin\theta}{\cos\theta}\cdot\frac{1}{\sin\theta}}{\frac{1}{\cos\theta}}=\dfrac{\frac{1}{\cos\theta}}{\frac{1}{\cos\theta}}=1$

50. $\sin 47°=\dfrac{h}{52}$

$h=52\sin 47°$

51.

$r=\sqrt{3^2+4^2}=5$

$\sin\left(\operatorname{Tan}^{-1}\dfrac{3}{4}\right)=\sin\alpha=\dfrac{3}{5}$

52. $y=x^2\sin x$ is the only function for which the value of a is a variable.

5.3 Cumulative Review

43. $x=\dfrac{5\pm\sqrt{25-4(2)}}{2(2)}=\dfrac{5\pm\sqrt{17}}{4}$

---

44. $x(x+1)\left[3=\dfrac{x}{x+1}+\dfrac{5}{x}\right]$

$3x(x+1)=x^2+5(x+1)$

$3x^2+3x=x^2+5x+5$

$2x^2-2x-5=0$

$x=\dfrac{2\pm\sqrt{4-4(-10)}}{2(2)}=\dfrac{2\pm\sqrt{44}}{4}$

$=\dfrac{1\pm\sqrt{11}}{2}$

45. $\ln e^{-2x}=\ln 3$

$-2x=\ln 3$

$x=-\dfrac{\ln 3}{2}\approx -0.55$

46. $x^{-5}=\dfrac{1}{32}=2^{-5};\ x=2$

47. $\sqrt{x-5}=4$

$x-5=16;\ x=21$

48. $\dfrac{1}{\sec\left(x-\frac{\pi}{2}\right)}=\cos\left(x-\dfrac{\pi}{2}\right)$

$=\cos\left(-\left(\dfrac{\pi}{2}-x\right)\right)$

$=\cos\left(\dfrac{\pi}{2}-x\right)$

$=\sin x$

49. Both sinusoids have a period of $\dfrac{2\pi}{2}=\pi$, so the sum has the same period.

50. $\cos^3\theta+\sin^2\theta\cos\theta$

$=\cos\theta(\cos^2\theta+\sin^2\theta)$

$=\cos\theta(1)=\cos\theta$

51. $\dfrac{\tan^2 x}{1+\tan^2 x}=\dfrac{\tan^2 x}{\sec^2 x}=\dfrac{\frac{\sin^2 x}{\cos^2 x}}{\frac{1}{\cos^2 x}}=\sin^2 x$

52. $\dfrac{\sin x}{1-\cos x}\cdot\dfrac{1+\cos x}{1+\cos x}$

$=\dfrac{\sin x(1+\cos x)}{1-\cos^2 x}=\dfrac{\sin x(1+\cos x)}{\sin^2 x}$

$=\dfrac{1+\cos x}{\sin x}=\dfrac{1}{\sin x}+\dfrac{\cos x}{\sin x}$

$=\csc x+\cot x$

5.4 C. Exercises

41a. $\sin 2\theta=\sin(\theta+\theta)$

$\qquad=\sin\theta\cos\theta+\cos\theta\sin\theta$

$\qquad=2\sin\theta\cos\theta$

41b. $\cos 2\theta=\cos(\theta+\theta)$

$\qquad=\cos\theta\cos\theta-\sin\theta\sin\theta$

$\qquad=\cos^2\theta-\sin^2\theta$

41c. $\tan 2\theta=\tan(\theta+\theta)$

$\qquad=\dfrac{\tan\theta+\tan\theta}{1-\tan\theta\cdot\tan\theta}=\dfrac{2\tan\theta}{1-\tan^2\theta}$

42a. $\alpha=\beta+\theta;\therefore\theta=\alpha-\beta$ and

$\tan\theta=\tan(\alpha-\beta)=\dfrac{\tan\alpha-\tan\beta}{1+\tan\alpha\tan\beta}$

$\therefore\tan\theta=\dfrac{m_2-m_1}{1+m_2 m_1}$

42b. $\tan\theta=\dfrac{4-1}{1+(4)(1)}=\dfrac{3}{5}$

$\theta=\tan^{-1}\dfrac{3}{5}\approx 31°$

43. $y = y_1 + y_2$
$$= A_0 \sin(kx - \omega t) + A_0 \sin(kx + \omega t)$$
$$= A_0(\sin kx \cos \omega t - \sin \omega t \cos kx$$
$$\quad + \sin kx \cos \omega t + \sin \omega t \cos kx)$$
$$= (2A_0 \sin kx) \cos \omega t$$

44a. $\dfrac{\sin(x+h) - \sin x}{h}$
$$= \frac{\sin x \cos h + \cos x \sin h - \sin x}{h}$$
$$= \frac{(\sin x \cos h - \sin x) + \cos x \sin h}{h}$$
$$= \sin x\left(\frac{\cos h - 1}{h}\right) + \cos x\left(\frac{\sin h}{h}\right)$$

44b. $\dfrac{\cos(x+h) - \cos x}{h}$
$$= \frac{\cos x \cos h - \sin x \sin h - \cos x}{h}$$
$$= \frac{(\cos x \cos h - \cos x) - \sin x \sin h}{h}$$
$$= \cos x\left(\frac{\cos h - 1}{h}\right) - \sin x\left(\frac{\sin h}{h}\right)$$

5.4 Cumulative Review

45a. $\log_2 2^5 = 5$

45b. $\log_3 (3^2)^2 = \log_3 3^4 = 4$

45c. $\log_{10} 10^{-3} = -3$

45d. $\ln e^{\frac{1}{3}} = \frac{1}{3}$

46. $(f+g)(x) = (2x+3) + (2x^2 + x - 3)$
$$= 2x^2 + 3x$$
$$(f-g)(x) = (2x+3) - (2x^2 + x - 3)$$
$$= 2x + 3 - 2x^2 - x + 3$$
$$= -2x^2 + x + 6$$

47. $(fg)(x) = (2x+3)(2x^2 + x - 3)$
$$= 4x^3 + 2x^2 - 6x + 6x^2 + 3x - 9$$
$$= 4x^3 + 8x^2 - 3x - 9$$

48. $\dfrac{f}{g}(x) = \dfrac{2x+3}{2x^2 + x - 3} = \dfrac{2x+3}{(2x+3)(x-1)}$
$$= \frac{1}{x-1}; x \neq -\frac{3}{2}, 1$$

49. $(f \circ g)(x) = 2(2x^2 + x - 3) + 3$
$$= 4x^2 + 2x - 6 + 3$$
$$= 4x^2 + 2x - 3$$
$$(g \circ f)(x) = 2(2x+3)^2 + (2x+3) - 3$$
$$= 2(4x^2 + 12x + 9) + 2x$$
$$= 8x^2 + 24x + 18 + 2x$$
$$= 8x^2 + 26x + 18$$

50.

$$\begin{array}{r|rrrr} 3 & 4 & -3 & -15 & -18 \\ & \downarrow & 12 & 27 & 36 \\ \hline & 4 & 9 & 12 & 18 \end{array}$$

$4x^2 + 9x + 12$ R. 18

52. any possible zero in the form of $\pm \dfrac{\text{factor of } 16}{\text{factor of } 9}$; $\dfrac{3}{8}$ does not fit this requirement.

53. $\cos \theta \cot \theta + \sin \theta$
$$= \cos \theta \frac{\cos \theta}{\sin \theta} + \sin \theta$$
$$= \frac{\cos^2 \theta}{\sin \theta} + \sin \theta$$
$$= \frac{\cos^2 \theta + \sin^2 \theta}{\sin \theta} = \frac{1}{\sin \theta} = \csc \theta$$

54. $\cos \theta \cot \theta - 2 \cos \theta = 0$
$$\cos \theta(\cot \theta - 2) = 0$$
$$\cos \theta = 0 \quad \text{or} \quad \cot \theta = 2$$
$$\theta = \frac{\pi}{2}, \frac{3\pi}{2} \quad \tan \theta = \frac{1}{2}$$
$$\theta \approx 0.464 \text{ or}$$
$$\pi + 0.464 \approx 3.605$$

5.5 B. Exercises

43. $\dfrac{\sin 3x}{\sin x} - \dfrac{\cos 3x}{\cos x}$
$$= \frac{\sin(2x+x)}{\sin x} - \frac{\cos(2x+x)}{\cos x}$$
$$= \frac{\sin 2x \cos x + \cos 2x \sin x}{\sin x}$$
$$\quad - \frac{\cos 2x \cos x - \sin 2x \sin x}{\cos x}$$
$$= \frac{2 \sin x \cos^2 x + (2\cos^2 x - 1)\sin x}{\sin x}$$
$$\quad - \frac{(1 - 2\sin^2 x)\cos x - 2\sin^2 x \cos x}{\cos x}$$
$$= \frac{\sin x(4\cos^2 x - 1)}{\sin x} - \frac{\cos x(1 - 4\sin^2 x)}{\cos x}$$
$$= 4\cos^2 x - 1 - 1 + 4\sin^2 x$$
$$= 4(\cos^2 x + \sin^2 x) - 2$$
$$= 4(1) - 2 = 2$$

44. $\cos^3 x - \sin^3 x$
$$= (\cos x - \sin x)$$
$$\quad \cdot (\cos^2 x + \cos x \sin x + \sin^2 x)$$
$$= (\cos x - \sin x)(1 + \cos x \sin x)$$
$$= (\cos x - \sin x)\left(1 + \frac{2\sin x \cos x}{2}\right)$$
$$= (\cos x - \sin x)\left(1 + \frac{1}{2}\sin 2x\right)$$

5.5 C. Exercises

45a. $\cos \dfrac{\pi}{12} = \cos \dfrac{\frac{\pi}{6}}{2}$ (positive in Q I)
$$= \sqrt{\frac{1 + \cos \frac{\pi}{6}}{2}} = \sqrt{\frac{1}{2}\left(1 + \frac{\sqrt{3}}{2}\right)}$$
$$= \sqrt{\frac{1}{2} + \frac{\sqrt{3}}{4}} = \frac{\sqrt{2 + \sqrt{3}}}{2}$$

45b. $\cos \dfrac{\pi}{24} = \cos \dfrac{\frac{\pi}{12}}{2}$ (positive in Q I)
$$= \sqrt{\frac{1}{2}\left(1 + \cos \frac{\pi}{12}\right)}$$
$$= \sqrt{\frac{1}{2}\left(1 + \frac{\sqrt{2 + \sqrt{3}}}{2}\right)}$$
$$= \sqrt{\frac{1}{2} + \frac{\sqrt{2 + \sqrt{3}}}{4}} = \frac{\sqrt{2 + \sqrt{2 + \sqrt{3}}}}{2}$$

46. $\sin 5\theta = \sin(3\theta + 2\theta)$
$$= \sin 3\theta \cos 2\theta + \cos 3\theta \sin 2\theta$$
$$= (-4\sin^3 \theta + 3\sin \theta)(1 - 2\sin^2 \theta)$$
$$\quad + (4\cos^3 \theta - 3\cos \theta)(2\sin \theta \cos \theta)$$
$$= 8\sin^5 \theta - 10\sin^3 \theta + 3\sin \theta$$
$$\quad + (4\cos^2 \theta - 3)(2\sin \theta \cos^2 \theta)$$
$$= 8\sin^5 \theta - 10\sin^3 \theta + 3\sin \theta$$
$$\quad + (4(1 - \sin^2 \theta) - 3)$$
$$\quad \cdot (2\sin \theta(1 - \sin^2 \theta))$$
$$= 8\sin^5 \theta - 10\sin^3 \theta + 3\sin \theta$$
$$\quad + (1 - 4\sin^2 \theta)(2\sin \theta - 2\sin^3 \theta)$$
$$= 8\sin^5 \theta - 10\sin^3 \theta + 3\sin \theta$$
$$\quad + 8\sin^5 \theta - 10\sin^3 \theta + 2\sin \theta$$
$$= 16\sin^5 \theta - 20\sin^3 \theta + 5\sin \theta$$

47a. $\tan \dfrac{\theta}{2} = \pm\sqrt{\dfrac{1 - \cos \theta}{1 + \cos \theta}}$
$$= \pm\sqrt{\frac{1 - \cos \theta}{1 + \cos \theta} \cdot \frac{1 - \cos \theta}{1 - \cos \theta}}$$
$$= \pm\sqrt{\frac{(1 - \cos \theta)^2}{1 - \cos^2 \theta}}$$
$$= \pm\sqrt{\frac{(1 - \cos \theta)^2}{\sin^2 \theta}}$$
$$= \pm\left|\frac{1 - \cos \theta}{\sin \theta}\right|$$

47b. The range of $\cos \theta$ is $[-1, 1]$, so the range of $1 - \cos \theta$ is $[1 - (-1), 1 - 1] = [2, 0]$ or $[0, 2]$.

47c.

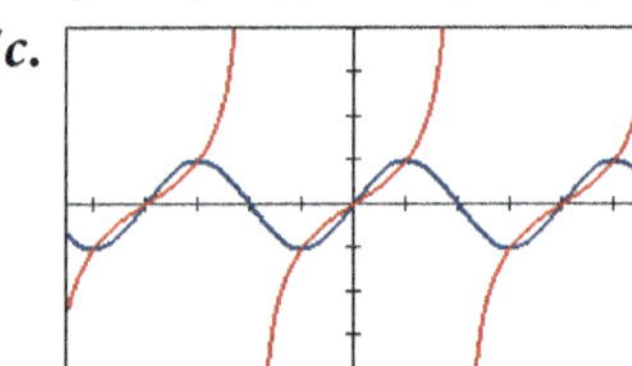

The signs are the same.

47d. Since $\tan \dfrac{\theta}{2}$ always has the same sign as $\sin \theta$ and $1 - \cos \theta$ is never negative, we can omit the $\pm$ sign and the absolute value.

48. $\cos 3x + \cos x = \cos 2x$
$$\cos(2x + x) + \cos x - \cos 2x = 0$$
$$\cos 2x \cos x - \sin 2x \sin x + \cos x$$
$$\quad - \cos 2x = 0$$
$$(2\cos^2 x - 1)\cos x$$
$$\quad - (2\sin x \cos x)\sin x + \cos x$$
$$\quad - (2\cos^2 x - 1) = 0$$
$$2\cos^3 x - \cos x - 2\sin^2 x \cos x$$
$$\quad + \cos x - 2\cos^2 x + 1 = 0$$
$$2\cos^3 x - 2\cos x(1 - \cos^2 x)$$
$$\quad - 2\cos^2 x + 1 = 0$$

$$2\cos^3 x - 2\cos x + 2\cos^3 x$$
$$- 2\cos^2 x + 1 = 0$$
$$4\cos^3 x - 2\cos^2 x - 2\cos x + 1 = 0$$

$$\frac{1}{2}\begin{array}{r|rrrr} & 4 & -2 & -2 & 1 \\ & \downarrow & 2 & 0 & -1 \\ \hline & 4 & 0 & -2 & 0 \end{array}$$

$$\left(\cos x - \tfrac{1}{2}\right)(4\cos^2 x - 2) = 0$$
$$\cos x = \tfrac{1}{2} \quad \text{or} \quad \cos^2 x = \tfrac{1}{2}$$
$$x = \tfrac{\pi}{3}, \tfrac{5\pi}{3} \qquad \cos x = \pm\tfrac{\sqrt{2}}{2}$$
$$x = \tfrac{\pi}{4}, \tfrac{3\pi}{4}, \tfrac{5\pi}{4}, \tfrac{7\pi}{4}$$

5.5 Cumulative Review

49. $m = -\dfrac{A}{B} = -\dfrac{3}{-2} = \dfrac{3}{2}$

x-int.: $\left(\dfrac{C}{A}, 0\right) = \left(\dfrac{-12}{3}, 0\right)$

y-int.: $\left(0, \dfrac{C}{B}\right) = \left(0, \dfrac{-12}{-2}\right)$

50. $m = -\dfrac{A}{B} = -\dfrac{2}{5}; \; m_\perp = \dfrac{5}{2}$

using $y = mx + b$:
$$6 = \tfrac{5}{2}(4) + b; \; b = -4$$
$$2\left(y = \tfrac{5}{2}x - 4\right)$$
$$5x - 2y = 8$$

51. $c = \sqrt{8^2 + 15^2} = 17$

52. $\cos 70° = \dfrac{\tfrac{1}{2}YZ}{7}$

$$\tfrac{1}{2}YZ = 7\cos 70° \approx 2.39$$
$$h = \sqrt{7^2 - \left(\tfrac{1}{2}YZ\right)^2} \approx 6.58$$
$$Area = \tfrac{1}{2}YZ \cdot h \approx 15.75 \text{ in.}^2$$

53. $p = \dfrac{2\pi}{|-\pi|} = 2; \; f = \dfrac{1}{p} = \dfrac{1}{2}$

54. $|a| = 3; \; \pi = \dfrac{2\pi}{|b|}; \; b = \pm 2$

Since $y = 3\cos 2x$ has a relative maximum at $(0, 3)$, let $k = 1$.

56. $y = -3; \; r = 5; \; x = \sqrt{5^2 - (-3)^2} = 4$

$$\tan\theta = \tfrac{y}{x} = \tfrac{-3}{4} = -\tfrac{3}{4}$$

57. $\cos\theta = -\dfrac{5}{8}; \; \theta$ in Q II

$$\sin^2\theta + \left(-\tfrac{5}{8}\right)^2 = 1$$
$$\sin^2\theta = 1 - \tfrac{25}{64} = \tfrac{39}{64}$$
$$\sin\theta = \tfrac{\sqrt{39}}{8}$$

58. $\sin(60 + 45°)$
$$= \sin 60°\cos 45° + \cos 60°\sin 45°$$
$$= \tfrac{\sqrt{3}}{2}\cdot\tfrac{\sqrt{2}}{2} + \tfrac{1}{2}\cdot\tfrac{\sqrt{2}}{2} = \tfrac{\sqrt{6}+\sqrt{2}}{4}$$

5.7 B. Exercises

31. $x^2 = 418^2 + 371^2$
$$- 2(418)(371)\cos 69°$$
$$\approx 201{,}215; \; x \approx 448.6 \text{ ft}$$

32.
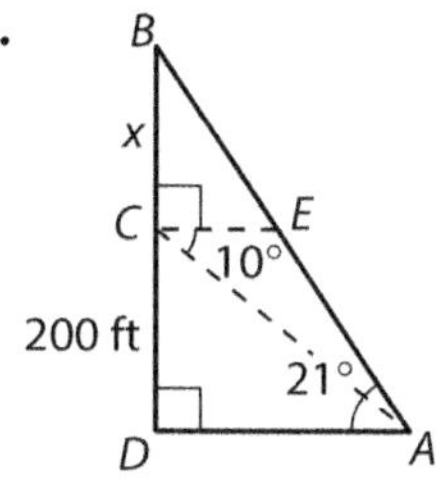

$$m\angle B = 180 - 90 - 21 = 69°$$
$$m\angle CAD = m\angle ECA = 10°$$
$$\text{(alt. int. } \angle\text{s)}$$
$$m\angle EAC = 21 - 10 = 11°$$
$$\sin 10° = \dfrac{200}{CA}$$
$$CA = \dfrac{200}{\sin 10°} \approx 1152$$
$$\dfrac{x}{\sin 11°} \approx \dfrac{1152}{\sin 69°}$$
$$x \approx \dfrac{1152\sin 11°}{\sin 69°} \approx 235 \text{ ft}$$

33.
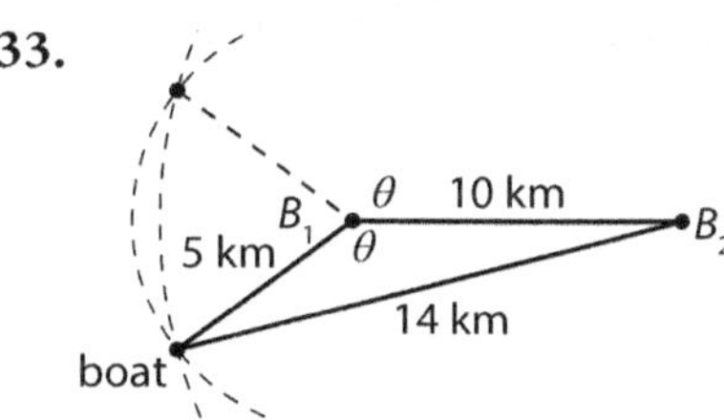

$$\theta = \cos^{-1}\left(\dfrac{5^2 + 10^2 - 14^2}{2(5)(10)}\right) \approx 135.2°$$
$$\text{SW bearing} = 90 + 135.2 = 225.2°$$
$$\text{NW bearing} = 360 - (135.2 - 90)$$
$$= 314.8°$$

34. $s = \tfrac{1}{2}(190 + 173 + 87) = 225$

$$Area =$$
$$\sqrt{225(225 - 190)(225 - 173)(225 - 87)}$$
$$\approx 7517.4 \text{ ft}^2$$

35. $A = \cos^{-1}\left(\dfrac{190^2 + 173^2 - 87^2}{2(190)(173)}\right) \approx 27.2°$

36. $\cos A = \dfrac{9^2 + 7^2 - 6^2}{2(9)(7)} = \dfrac{94}{126} = \dfrac{47}{63}$

$$x^2 = 6^2 + 7^2 - 2(6)(7)\left(\tfrac{47}{63}\right) = 22.\overline{3}$$
$$x \approx 4.7$$

37. Draw the altitude from B to the x-axis.

$x = a\cos C$ and $y = a\sin C$
Apply the Pythagorean Theorem to $\triangle BDA$.
$$c^2 = (b - x)^2 + y^2$$
$$= b^2 - 2bx + x^2 + y^2$$
$$= b^2 - 2b(a\cos C)$$
$$\quad + (a\cos C)^2 + (a\sin C)^2$$

$$= b^2 - 2ab\cos C$$
$$\quad + a^2(\cos^2 C + \sin^2 C)$$
$$= b^2 - 2ab\cos C + a^2$$
$$= a^2 + b^2 - 2ab\cos C$$

38. Draw the altitude from C to the x-axis.

$x = a\cos B$ (a negative number) and
$y = a\sin B$
$AD = c - x$ (since x is negative)
Apply the Pythagorean Theorem to $\triangle CDA$.
$$b^2 = (c - x)^2 + y^2$$
$$= c^2 - 2cx + x^2 + y^2$$
$$= c^2 - 2c(a\cos B)$$
$$\quad + (a\cos B)^2 + (a\sin B)^2$$
$$= c^2 - 2ac\cos B$$
$$\quad + a^2(\cos^2 B + \sin^2 B)$$
$$= c^2 - 2ac\cos B + a^2$$
$$= a^2 + c^2 - 2ac\cos B$$

5.7 C. Exercises

39. $AB = \sqrt{3^2 + 4^2} = 5$
$$AC = \sqrt{2^2 + 4^2} = \sqrt{20}$$
$$BC = \sqrt{2^2 + 3^2} = \sqrt{13}$$
$$A \approx \cos^{-1}\left(\dfrac{5^2 + (\sqrt{20})^2 - (\sqrt{13})^2}{2(5)(\sqrt{20})}\right)$$
$$\approx 44.3°$$

40. $BD^2 = 6^2 + 6.3^2 - 2(6)(6.3)\cos 60°$
$$BD^2 = 37.89; \; BD \approx 6.2$$
$$\angle D = 180 - 98 = 82°$$
$$AC^2 = 4^2 + 6.3^2 - 2(4)(6.3)\cos 82°$$
$$AC^2 \approx 48.7; \; AC \approx 7.0$$

41.
$$b^2 = a^2 + c^2 - 2ac\cos B$$
$$12^2 = 9^2 + c^2 - 2(9)c\cos 60°$$
$$144 = 81 + c^2 - 9c$$
$$c^2 - 9c - 63 = 0$$
$$c = \dfrac{9 \pm \sqrt{81 - 4(1)(-63)}}{2} = \dfrac{9 \pm \sqrt{333}}{2}$$
$$c \approx 13.62 \text{ or } -4.62; \; c \approx 13.62$$

42. $a^2 = \left(\dfrac{1}{\sqrt{6} - \sqrt{2}}\right)^2 + \left(\dfrac{1}{\sqrt{6} + \sqrt{2}}\right)^2$
$$- 2\left(\dfrac{1}{\sqrt{6} - \sqrt{2}}\right)\left(\dfrac{1}{\sqrt{6} + \sqrt{2}}\right)\cos 60°$$
$$= \dfrac{1}{8 + 4\sqrt{3}} + \dfrac{1}{8 - 4\sqrt{3}} - 2\left(\tfrac{1}{4}\right)\left(\tfrac{1}{2}\right)$$
$$= \dfrac{8 - 4\sqrt{3}}{16} + \dfrac{8 + 4\sqrt{3}}{16} - \tfrac{1}{4}$$
$$= \dfrac{16 - 4}{16} = \dfrac{3}{4}$$
$$a = \dfrac{\sqrt{3}}{2}$$

43. $A(t) = P\left(1 + \frac{r}{n}\right)^{nt}$

$$= 10{,}000\left(1 + \frac{0.05}{12}\right)^{12(7)}$$

$$= \$14{,}180.36$$

44. $2^x = \frac{1}{16} = 2^{-4}; \; x = -4$

45. $\quad (2x + 3)\ln 3 = 2x \ln 5$

$$2x \ln 3 + 3 \ln 3 = 2x \ln 5$$

$$2x \ln 3 - 2x \ln 5 = -3 \ln 3$$

$$x(2 \ln 3 - 2 \ln 5) = -3 \ln 3$$

$$x = -\frac{3 \ln 3}{2(\ln 3 - \ln 5)} \approx 3.226$$

47. $\quad (\sqrt{2x+3} - 1)^2 = (\sqrt{3x-5})^2$

$$2x + 4 - 2\sqrt{2x+3} = 3x - 5$$

$$(-2\sqrt{2x+3})^2 = (x-9)^2$$

$$8x + 12 = x^2 - 18x + 81$$

$$x^2 - 26x + 69 = 0$$

$$(x-3)(x-23) = 0$$

$$x = 3, \; 23 \text{ (extraneous)}$$

48. Since the period of both $g(x)$ and $h(x)$ is $\frac{2\pi}{|b|} = \frac{2\pi}{3}$, the period of $f(x)$ is also $\frac{2\pi}{3}$.

49. $b = \sqrt{7^2 - 3^2} = 2\sqrt{10}$

$$\cos \theta = \frac{adj.}{hyp.} = \frac{2\sqrt{10}}{7}$$

50. $f(x) = \sin x$ has a relative minimum at $\frac{3\pi}{2}$, and $g(x)$ has a horizontal shrink by a factor of $\frac{1}{2}$: $\frac{1}{2}\left(\frac{3\pi}{2}\right) = \frac{3\pi}{4}$.

51. $h = 8 \sin 60° \approx 6.92 < a < b$

two solutions

52. $X = 180 - 97 - 56 = 27°$

$$\frac{x}{\sin 27°} = \frac{33}{\sin 56°}$$

$$x = \frac{33 \sin 27°}{\sin 56°}$$

6.1 C. Exercises

43c. x-components: $0 + \frac{1}{2}t_1 - \frac{1}{2}t_2 = 0$

y-components:

$$-10 + \frac{\sqrt{3}}{2}t_1 + \frac{\sqrt{3}}{2}t_2 = 0$$

43d. $0 + \frac{1}{2}t_1 - \frac{1}{2}t_2 = 0$

$$\frac{1}{2}t_1 = \frac{1}{2}t_2$$

$$t_1 = t_2$$

The tension in the ropes is the same.

43e. $-10 + \frac{\sqrt{3}}{2}t_1 + \frac{\sqrt{3}}{2}t_1 = 0$

$$\sqrt{3}\,t_1 + \sqrt{3}\,t_1 = 20$$

$$2\sqrt{3}\,t_1 = 20$$

$$t_1 = \frac{10\sqrt{3}}{3} \approx 5.8 \text{ lb}$$

The tension in each rope is about 5.8 lb.

6.1 Cumulative Review

44. $1.22\left(\frac{180°}{\pi}\right) \approx 70°$

45. $40°\left(\frac{\pi}{180°}\right) = \frac{2\pi}{9} \approx 0.698$

46. $\cos A = \frac{5^2 + 8^2 - 4^2}{2(5)(8)}$

$$A = \text{Cos}^{-1} \frac{73}{80} \approx 24°$$

47. $c^2 = 5^2 + 8^2 - 2(5)(8)\cos 40°$

$$c \approx 5.26$$

$$\cos A \approx \frac{5^2 + 5.26^2 - 8^2}{2(5)(5.26)}$$

$$A \approx \text{Cos}^{-1}(-0.2154) \approx 102°$$

48.

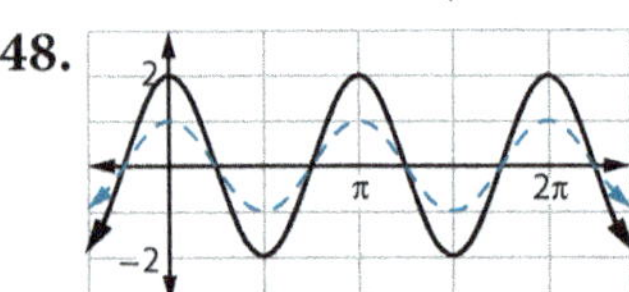

49. The graph of $f(x)$ is stretched horizontally by a factor of 2, stretched vertically by a factor of 3, reflected in the x-axis, and translated 1 unit up.

50. $\dfrac{1}{\sqrt{-1} \cdot \sqrt{-1} \cdot \sqrt{-1} \cdot \sqrt{-1}} = \dfrac{1}{-1(-1)} = 1$

51. $5^? = 5^x$

52. $\cos E = \dfrac{d^2 + f^2 - e^2}{2df}$

$$= \dfrac{5^2 + 6^2 - 9^2}{2(5)(6)}$$

$$E = \text{Cos}^{-1}\left(-\frac{1}{3}\right)$$

53. $\angle B$ is opposite the shortest side.

$$\frac{\sin B}{20} = \frac{\sin A}{35}$$

$$\sin B = \frac{20 \sin A}{35} = \frac{4 \sin A}{7}$$

$$B = \text{Sin}^{-1}\left(\frac{4 \sin A}{7}\right)$$

$$\frac{\sin B}{20} = \frac{\sin C}{31}$$

$$\sin B = \frac{20 \sin C}{31}$$

$$B = \text{Sin}^{-1}\left(\frac{20 \sin C}{31}\right)$$

6.3 C. Exercises

37.

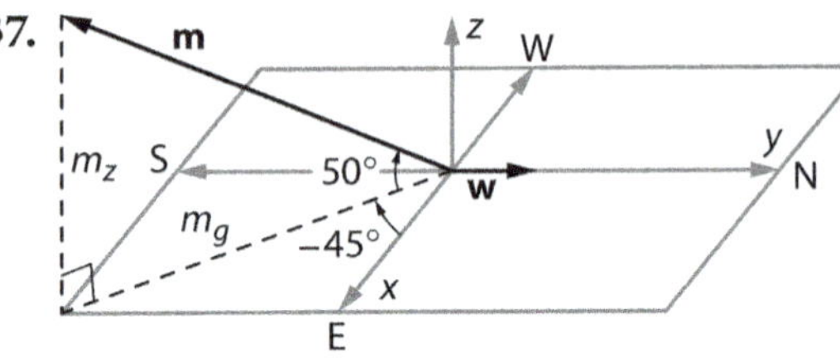

$m_z = 3000 \sin 50° \approx 2298.1$

$m_g = 3000 \cos 50° \approx 1928.4$

$m_x \approx 1928.4 \cos(-45°) \approx 1363.6$

$m_y \approx 1928.4 \sin(-45°) \approx -1363.6$

$\mathbf{m} \approx \langle 1363.6, -1363.6, 2298.1 \rangle$

$\mathbf{w} = \langle 22 \cos 90°, 22 \sin 90°, 0 \rangle$

$\quad \approx \langle 0, 22, 0 \rangle$

$\mathbf{r} = \mathbf{w} + \mathbf{m}$

$\quad \approx \langle 1363.6, -1341.6, 2298.1 \rangle$

38. $|\mathbf{r}| \approx \sqrt{28.3^2 + (-183.1)^2 + 77.0^2}$

$\quad \approx 200.6$ knots

39. $\mathbf{r}_{xy} = \langle 28.3, -183.1 \rangle$

$$\alpha = \text{Tan}^{-1}\left|\frac{-183.1}{28.3}\right| \approx 81°$$

$$\beta \approx 90 + 81 = 171°$$

40. $|\mathbf{r}_{xy}| \approx \sqrt{28.3^2 + (-183.1)^2}$

$\quad \approx 185.3$

$$\alpha = \text{Cos}^{-1}\left|\frac{185.3}{200.6}\right| \approx 23°$$

(above the xy-plane)

6.3 Cumulative Review

42. $\alpha = \text{Tan}^{-1}\left|\frac{-2}{-6}\right| \approx 18.4°$

$$\theta \approx 180 + 18.4 = 198.4°$$

43. $\cos \theta = \frac{x}{r}$ and $\sin \theta = \frac{y}{r}$

$$\cos 120° = \frac{x}{5}$$

$$x = 5 \cos 120° = 5\left(-\frac{1}{2}\right) = -\frac{5}{2}$$

$$\sin 120° = \frac{y}{5}$$

$$y = 5 \sin 120° = 5\left(-\frac{\sqrt{3}}{2}\right) = -\frac{5\sqrt{3}}{2}$$

$$\left(-\frac{5}{2}, \frac{5\sqrt{3}}{2}\right)$$

44. $x^2 = 450^2 + 570^2$
$\qquad - 2(450)(570) \cos 100°$
$\quad = 616{,}481.5151$
$\quad x \approx 785$ ft
$\quad A = \frac{1}{2}(450)(570) \sin 100°$
$\qquad \approx 126{,}311 \text{ ft}^2 \left(\dfrac{1 \text{ acre}}{43{,}560 \text{ ft}^2}\right)$
$\qquad \approx 2.9$ acres

45. $8 = \sqrt{(x - 0)^2 + (y - 0)^2}$
$\quad x^2 + y^2 = 64$

46. $5 = \sqrt{(x - 3)^2 + (y - (-4))^2}$
$\quad (x - 3)^2 + (y + 4)^2 = 25$

47. $d = \sqrt{(-2 - 4)^2 + (-3 - 5)^2}$
$\quad = \sqrt{100} = 10$
$\quad r = 5; A = \pi(5)^2 = 25\pi \text{ u}^2$

50. $\tan \theta = 1$ where $\sin \theta = \cos \theta$
$\quad \sin \theta = \cos \theta$ at $\frac{\pi}{4}$ and $\frac{5\pi}{4}$

6.4 B. Exercises

36.

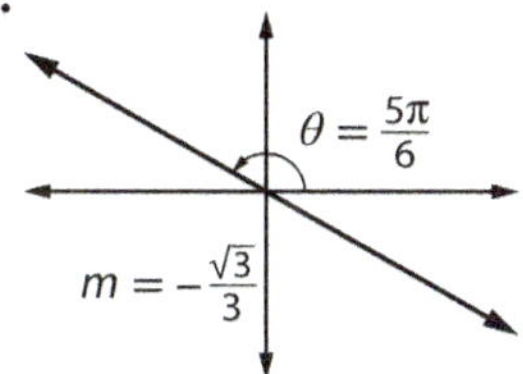

line through $(0, 0)$ with $m = -\dfrac{\sqrt{3}}{3}$
$$\tan \theta = \tan \frac{5\pi}{6}$$
$$\tan \theta = -\frac{\sqrt{3}}{3}$$
$$\frac{y}{x} = -\frac{\sqrt{3}}{3}$$
$$y = -\frac{\sqrt{3}}{3}x$$

37. $d =$
$\quad \sqrt{5^2 + 6^2 - 2(5)(6) \cos (35° - 155°)}$
$\quad \approx 9.5$ km

38. $d =$
$\quad \sqrt{3^2 + 7^2 - 2(3)(7) \cos (110° - 70°)}$
$\quad \approx 5.1$ nm

6.4 C. Exercises

40. $x^2 - 4x + 4 + y^2 + 2y + 1 = 5$
$\qquad x^2 + y^2 = 4x - 2y$
$\quad r^2 = 4r \cos \theta - 2r \sin \theta$
$\quad r = 4 \cos \theta - 2 \sin \theta$

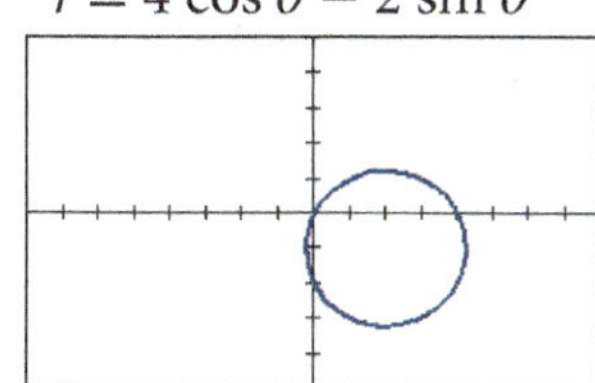

41. $\qquad xy = 4$
$\quad r^2 \cos \theta \sin \theta = 4$
$\quad r^2 = \dfrac{4}{\cos \theta \sin \theta} = 4 \sec \theta \csc \theta$ or
$\quad r = \pm\dfrac{2}{\sqrt{\cos \theta \sin \theta}} = \pm 2\sqrt{\sec \theta \csc \theta}$

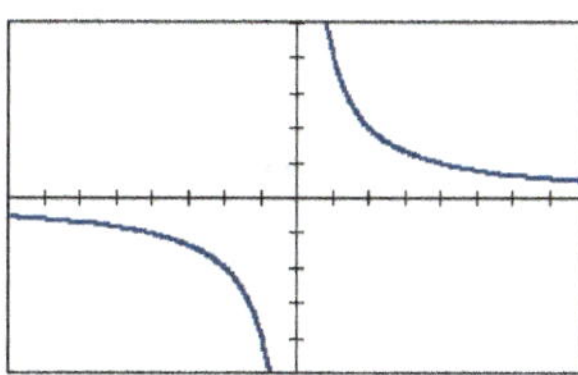

42.

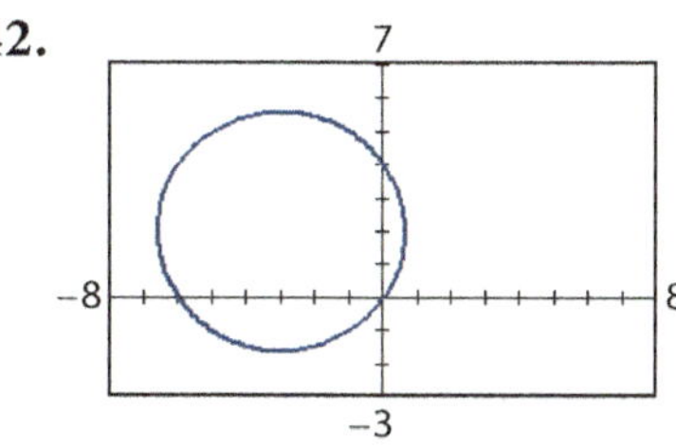

$r(r) = 4r \sin \theta - 6r \cos \theta$
$r^2 = 4y - 6x$
$x^2 + y^2 = 4y - 6x$
$(x^2 + 6x + 9) + (y^2 - 4y + 4) = 9 + 4$
$\qquad (x + 3)^2 + (y - 2)^2 = 13$
circle centered at $(-3, 2)$
with $r = \sqrt{13} \approx 3.6$

43.

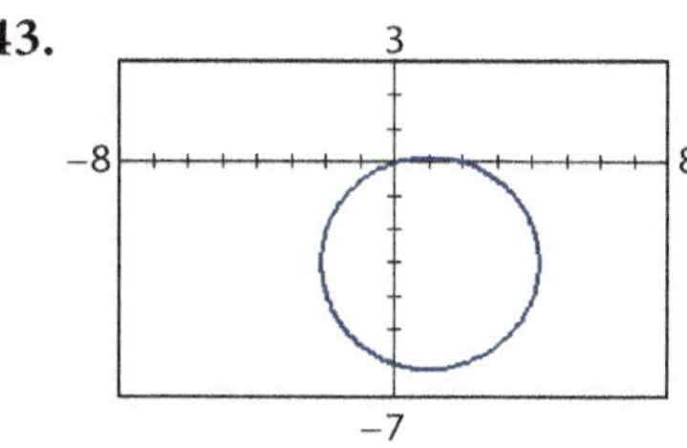

$r(r) = 2r \cos \theta - 6r \sin \theta$
$r^2 = 2x - 6y$
$x^2 + y^2 = 2x - 6y$
$(x^2 - 2x + 1) + (y^2 + 6y + 9) = 1 + 9$
$\qquad (x - 1)^2 + (y + 3)^2 = 10$
circle centered at $(1, -3)$
with $r = \sqrt{10} \approx 3.2$

44.
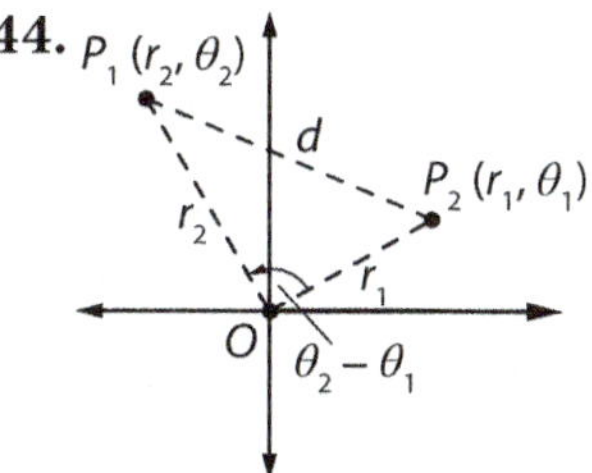

44a. In $\triangle P_1 O P_2$, $m\angle O = \theta_2 - \theta_1$.
applying the Law of Cosines,
$$d^2 = r_1{}^2 + r_2{}^2 - 2r_1 r_2 \cos (\theta_2 - \theta_1)$$
$$d = \sqrt{r_1{}^2 + r_2{}^2 - 2r_1 r_2 \cos (\theta_2 - \theta_1)}$$

44b. $d^2 = r_1{}^2 + r_2{}^2 - 2r_1 r_2 \cos 0$
$\quad = r_1{}^2 - 2r_1 r_2 + r_2{}^2$
$\quad = (r_1 - r_2)^2$
$\quad d = |r_1 - r_2|$

44c. $d^2 = r_1{}^2 + r_2{}^2 - 2r_1 r_2 \cos \pi$
$\quad = r_1{}^2 + 2r_1 r_2 + r_2{}^2$
$\quad = (r_1 + r_2)^2$
$\quad d = |r_1 + r_2|$

44d. $d^2 = r_1{}^2 + r_2{}^2 - 2r_1 r_2 \cos \frac{\pi}{2}$
$\quad = r_1{}^2 + r_2{}^2$
$\quad d = \sqrt{r_1{}^2 + r_2{}^2}$

6.4 Cumulative Review

45. $\sin (\pi \pm \theta) = \sin \pi \cos \theta \pm \cos \pi \sin \theta$
$\qquad = (0) \cos \theta \pm (-1) \sin \theta$
$\qquad = \mp \sin \theta$

46. $\cos (\pi \pm \theta) = \cos \pi \cos \theta \mp \sin \pi \sin \theta$
$\qquad = (-1) \cos \theta \mp (0) \sin \theta$
$\qquad = - \cos \theta$

47.
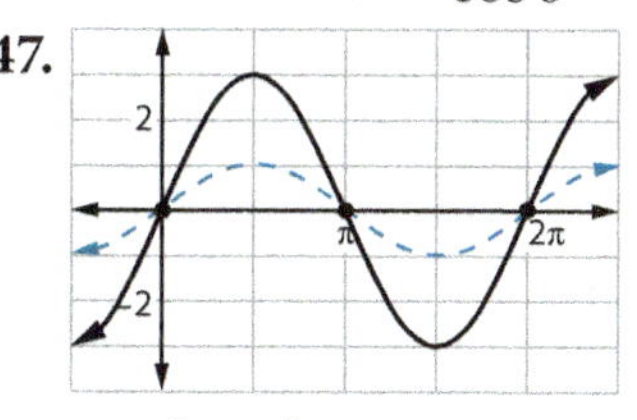

$x = 0, \pi, 2\pi$
max: $\left(\frac{\pi}{2}, 3\right)$; min: $\left(\frac{3\pi}{2}, -3\right)$

48.
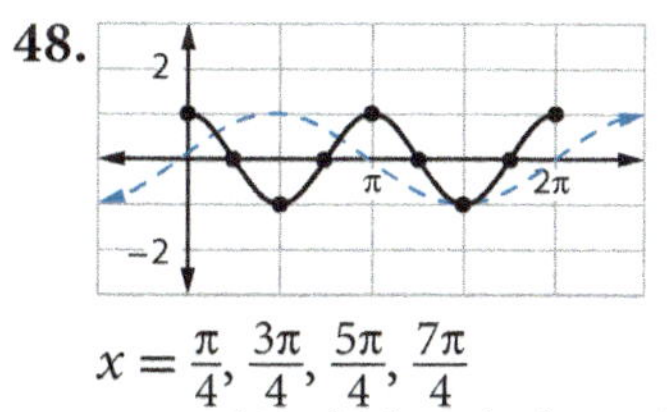

$x = \dfrac{\pi}{4}, \dfrac{3\pi}{4}, \dfrac{5\pi}{4}, \dfrac{7\pi}{4}$
max: $(0, 1), (\pi, 1), (2\pi, 1)$
min: $\left(\frac{\pi}{2}, -1\right), \left(\frac{3\pi}{2}, -1\right)$

49. $2\theta = n2\pi; \theta = n\pi$

50. $i = i$
$i^1 = i$
$i^2 = (\sqrt{-1})^2 = -1$
$i^3 = i \cdot i^2 = i(-1) = -i$
$i^4 = i^2 \cdot i^2 = -1(-1) = 1$
$i^5 = i \cdot i^4 = i$
$i^6 = i^2 \cdot i^4 = i^2 = -1$
$i^7 = i^3 \cdot i^4 = i^3 = -i$
$i^8 = i^4 \cdot i^4 = 1$

51. $\tan \theta = \dfrac{\sin \theta}{\cos \theta} = \dfrac{\frac{3}{5}}{\frac{4}{5}} = \dfrac{3}{4}$

52. $\tan \theta = \frac{2}{5}$; $\mathrm{Tan}^{-1} \frac{2}{5} \approx 22°$

53. $\mathrm{Tan}^{-1} (-1) = -\dfrac{\pi}{4}$

54. $\dfrac{\tan^2 x - \sin^2 x}{\sin^2 x \tan^2 x}$
$\quad = \dfrac{\tan^2 x}{\sin^2 x \tan^2 x} - \dfrac{\sin^2 x}{\sin^2 x \tan^2 x}$
$\quad = \dfrac{1}{\sin^2 x} - \dfrac{1}{\tan^2 x} = \csc^2 x - \cot^2 x = 1$

The geometrical method of representing imaginary numbers as vectors in the complex plane, when finally developed 200 yr later, came not from a mathematician but from the surveyor Caspar Wessel. His solution grew from the very practical attempt to simplify calculations required in his work. Generalizing this result to three dimensions was eventually accomplished by William Hamilton, who worked on the problem for 15 yr. According to Hamilton, the long-sought breakthrough came to him suddenly while out walking. He was so excited by the thought that he carved the formula, $i^2 + j^2 + k^2 = ijk = -1$, into a nearby stone bridge. Hamilton's quaternions essentially comprised a new algebra in which the commutative property of multiplication does not hold. The method of Gibbs and Heaviside pulled the quaternion apart into what we would call dot product and cross product. While cross product is beyond the scope of this text, its history is included here as a natural and necessary result of extending vector operations to three dimensions and the vector analysis that arose from it.

Jean Robert Argand described the complex plane in an 1806 paper published in France. The complex plane is sometimes called the "Argand plane" since Argand's paper was more widely known than Wessel's.

6.6 C. Exercises

39. $\dfrac{z_1}{z_2} = \dfrac{r_1(\cos\theta_1 + i\sin\theta_1)}{r_2(\cos\theta_2 + i\sin\theta_2)} \cdot \dfrac{\cos\theta_2 - i\sin\theta_2}{\cos\theta_2 - i\sin\theta_2}$

$= \dfrac{r_1}{r_2}\left[\dfrac{\cos\theta_1\cos\theta_2 - i\cos\theta_1\sin\theta_2 + i\sin\theta_1\cos\theta_2 - i^2\sin\theta_1\sin\theta_2}{\cos^2\theta_2 - i^2\sin^2\theta_2}\right]$

$= \dfrac{r_1}{r_2}\left[\dfrac{(\cos\theta_1\cos\theta_2 + \sin\theta_1\sin\theta_2) + i(\sin\theta_1\cos\theta_2 - \cos\theta_1\sin\theta_2)}{\cos^2\theta_2 + \sin^2\theta_2}\right]$

$= \dfrac{r_1}{r_2}[\cos(\theta_1 - \theta_2) + i\sin(\theta_1 - \theta_2)]$

40. $\dfrac{\left(3\text{ cis }\frac{\pi}{3}\right)\left(5\text{ cis }\frac{\pi}{6}\right)}{0.5\text{ cis }\frac{\pi}{2}} = \dfrac{3(5)}{0.5}\text{ cis }\left(\frac{\pi}{3} + \frac{\pi}{6} - \frac{\pi}{2}\right) = 30\text{ cis }0 = 30$

41. $(-1 - i) = \sqrt{2}\text{ cis }\dfrac{5\pi}{4}$

$(-4 + 4\sqrt{3}i) = 8\text{ cis }\dfrac{2\pi}{3}$

$\left[\left(\sqrt{2}\text{ cis }\frac{5\pi}{4}\right)\left(8\text{ cis }\frac{2\pi}{3}\right)\right]^2 = \left(8\sqrt{2}\text{ cis }\frac{23\pi}{12}\right)^2$

$= \left((8\sqrt{2})^2\text{ cis }2\left(\frac{23\pi}{12}\right)\right) = 128\text{ cis }\dfrac{23\pi}{6}$

$= 128\text{ cis }\dfrac{11\pi}{6} = 128\left(\frac{\sqrt{3}}{2} + i\left(-\frac{1}{2}\right)\right) = 64\sqrt{3} - 64i$

42. $w_k = \left(25\text{ cis }\frac{\pi}{6}\right)^{\frac{2}{3}} = \sqrt[3]{5^4}\text{ cis }2\left(\dfrac{\frac{\pi}{6} + k2\pi}{3}\right)$

$= 5\sqrt[3]{5}\text{ cis }\dfrac{\pi}{9} + k\dfrac{4\pi}{3}$

$w_0 = 5\sqrt[3]{5}\text{ cis }\dfrac{\pi}{9}$

$w_1 = 5\sqrt[3]{5}\text{ cis }\dfrac{13\pi}{9}$

$w_2 = 5\sqrt[3]{5}\text{ cis }\dfrac{25\pi}{9} = 5\sqrt[3]{5}\text{ cis }\dfrac{7\pi}{9}$

43. $z = 2 - 5i \approx \sqrt{29}\text{ cis }5.09$

Q IV: $\theta = 2\pi + \text{Tan}^{-1}\left(-\frac{5}{2}\right) \approx 5.09$

$z^{100} \approx \left(29^{\frac{1}{2}}\right)^{100}\text{ cis }100(5.0928)$

$\approx 29^{50}\text{ cis }(509.28 - 81(2\pi))$

$\approx 29^{50}\text{ cis }0.35$

6.6 Cumulative Review

44. $|\mathbf{v}| = \sqrt{(-5)^2 + 7^2} = \sqrt{74} \approx 8.6$

$\alpha = \text{Tan}^{-1}\left|\frac{7}{-5}\right| \approx 54°$

$\theta = 180 - \alpha \approx 126°$

45. $\mathbf{u} \cdot \mathbf{v} = 4(-5) + (-3)7 = -41$

$|\mathbf{u}| = \sqrt{4^2 + (-3)^2} = 5$

$|\mathbf{v}| = \sqrt{(-5)^2 + 7^2} = \sqrt{74}$

$\cos\theta = \dfrac{-41}{5\sqrt{74}}$

$\theta = \text{Cos}^{-1}\left(\dfrac{-41}{5\sqrt{74}}\right) \approx 162°$

46. $\mathbf{u} \cdot \mathbf{v} = 7a - 3b = 0$

$7a = 3b$

$a = \frac{3}{7}b$

unit vector for $\langle 3, 7\rangle$:

$\dfrac{1}{\sqrt{3^2 + 7^2}}\langle 3, 7\rangle = \left\langle\dfrac{3\sqrt{58}}{58}, \dfrac{7\sqrt{58}}{58}\right\rangle$

47. $\mathbf{v} = \langle 3 - (-1), 7 - 2, 1 - 7\rangle$

$= \langle 4, 5, -6\rangle$

$|\mathbf{v}| = \sqrt{4^2 + 5^2 + (-6)^2} = \sqrt{77}$

$\dfrac{\mathbf{v}}{|\mathbf{v}|} = \dfrac{1}{\sqrt{77}}\langle 4, 5, -6\rangle$

$= \left\langle\dfrac{4\sqrt{77}}{77}, \dfrac{5\sqrt{77}}{77}, -\dfrac{6\sqrt{77}}{77}\right\rangle$

48. $\mathbf{u} \cdot \mathbf{v} = 1(-5) + 4(3) + (-5)7 = -28$

$|\mathbf{u}| = \sqrt{1^2 + 4^2 + (-5)^2} = \sqrt{42}$

$|\mathbf{v}| = \sqrt{(-5)^2 + 3^2 + 7^2} = \sqrt{83}$

$\theta = \text{Cos}^{-1}\left(\dfrac{-28}{\sqrt{42}\cdot\sqrt{83}}\right) \approx 118°$

49. $d = \sqrt{(-2)^2 + 5^2 - 2(-2)(5)\cos(280° - 20°)}$

≈ 5.1

50. $\mathbf{w} = \langle -2, -2\rangle;\ \mathbf{q} = \mathbf{z} = \langle 2, -2\rangle$

$\mathbf{p} = \mathbf{v} = \langle 2, 3\rangle;\ \mathbf{n} = \langle 2, -3\rangle$

51. $\mathbf{u} \cdot \mathbf{v} = 3(2) + 6(-4) = -18$

$|\mathbf{u}| = \sqrt{3^2 + 6^2} = \sqrt{45}$

$|\mathbf{v}| = \sqrt{2^2 + (-4)^2} = \sqrt{20}$

$\cos\theta = \dfrac{-18}{\sqrt{45}\cdot\sqrt{20}}$

$\theta = \text{Cos}^{-1}\left(\dfrac{-18}{\sqrt{45}\cdot\sqrt{20}}\right) \approx 127°$

52. The general form $r = k\cos n\theta$ is a rose with $2n = 4$ petals centered along the polar axis and the line.

53. If n is even, there are $2n$ petals. If $2n = 6$, then $n = 3$ which is odd (implying only 3 petals).

PRECALCULUS

Second Edition

Teacher Edition

Part 2

bju **press**®

Greenville, South Carolina

Note:

The fact that materials produced by other publishers may be referred to in this volume does not constitute an endorsement of the content or theological position of materials produced by such publishers. Any references and ancillary materials are listed as an aid to the student or the teacher and in an attempt to maintain the accepted academic standards of the publishing industry.

PRECALCULUS Teacher Edition

Second Edition

Coordinating Writer
Mark Wetzel, MEd

Writers
Ben Adams, MEd
Gene Bucholtz, MEd
Jeffrey A. Jaeger, MEd
Timothy King
Tamera Knisely, MEd
Steve McKisic, MEd
Kathy Pilger, EdD
Sarah Ream, MPA

Academic Oversight
Jeff Heath, EdD

Editors
Heather Lonaberger, MA
Abigail Sivyer

Biblical Worldview
Vern Poythress, PhD
Bryan Smith, PhD
Tyler Trometer, MDiv

Previous Edition Writers
Larry Hall, MS
Kathy Pilger, EdD
Ron Tagliapietra, EdD

Consultants
Steve McKisic, MEd
Kathy Pilger, EdD

Permissions
Sylvia Gass
Carrie Hanna
Ashleigh Schieber
Carrie Walker

Project Coordinator
Kyla J. Smith

Page Layout
Dzign Associates,
   Patricia Tirado

Designers and Illustrators
Chris Barnhart
Dzign Associates,
   Patricia Tirado
Drew Fields
Josh Frederick
Emily Heinz
Pixel Mouse House,
   David Casas

Cover Designers
Chris Barnhart
Drew Fields
Josh Frederick

p. 471 "Triangulo Pascal" by Blaise Pascal/Wikimedia Commons/Public Domain
Photograph credits appear on pages 696–97.
Text acknowledgements appear on pages 698–99.

Contents

Precalculus Standard Lesson Plan Overview

The Standard Lesson Plan allocates 90 days for each semester.

Key:

- Assessments
- *Student Edition* features:
  - **TC:** Technology Corner
  - **HC:** Historical Connection
  - **BPM:** Biblical Perspective of Mathematics
  - **DA:** Data Analysis
- Biblical worldview topics in the *Teacher Edition*

DAY	TOPIC	PAGES	ASSESSMENT	BIBLICAL WORLDVIEW
	Chapter 7 Systems and Matrices			
91	**7.1** Solving Systems of Equations	336–42		
92	**7.2** Matrices	343–50		• using math for the benefit of mankind and the glory of God
93	**BPM:** Direct and Indirect Deductive Reasoning	351–52	Quiz 7A (7.1–7.2)	• examining direct and indirect reasoning used in Scripture and in the Stein-Bahnsen debate on the existence of God (Acts 17:2; 18:4,19; 24:25; Heb. 5:14; 1 Pet. 3:15) • identifying transcendent mathematical truth as originating from God and His nature
94–95	**7.3** Gaussian Elimination **TC:** Matrix Operations	353–60 358		
96–97	**7.4** Determinants	361–68		
98	**HC:** Matrices and Determinants	369	Quiz 7B (7.3–7.4)	
99	**7.5** Inverse Matrices	370–75		
100–101	**7.6** Systems of Inequalities	376–82		• understanding the importance of biblical principles in decision making • using linear programming to make good decisions, maximize profit, and serve others (Luke 10:33–37; 14:28–30; Prov. 11:24–26; Matt. 5:16)
102–3	**7.7** Partial Fractions	383–88	Quiz 7C (7.5–7.6)	
104	**DA:** Cryptology	389–90	Quiz 7D (7.7)	• applying biblical principles in the context of privacy rights and the need for lawful investigation
105	Chapter 7 Review	391–93		
106	Chapter 7 Test			

DAY	TOPIC	PAGES	ASSESSMENT	BIBLICAL WORLDVIEW
	Chapter 10 Descriptive Statistics			
135	**10.1** Counting Principles	490–96		• recognizing the high risk of the lottery by using combinatory mathematics
136	**10.2** Basic Probability	497–504		• learning how God's Word and the Holy Spirit provide guidance (Ps. 37:23; James 4:13–15) • affirming that mathematical laws are discovered, not manmade • responding biblically to statistical indicators (Eph. 4:32; Phil. 2:3; James 2:1–7)
137	**DA:** Medical Screenings and Probability	505–6	Quiz 10A (10.1–10.2)	• upholding the sanctity of life and caring for others
138	**10.3** Representing Data Graphically	507–15		• planning while God directs our steps (Prov. 16:9)
139–40	**10.4** Describing Data Numerically	516–25		• using research to present truth
141	**HC:** Political Arithmetic and Probability	526	Quiz 10B (10.3–10.4)	• reflecting on the wisdom and character of God (Rom. 1:19–20)
142–43	**10.5** Data Distributions	527–33		
144–45	**10.6** Normal Distributions	534–39		• avoiding unethical use of descriptive statistics
146	**BPM:** Mathematical Platonism	540–41	Quiz 10C (10.5–10.6)	• comparing Platonism and a biblical perspective of mathematical truth • viewing mathematics as a direct result of God's creation
147	Chapter 10 Review	542–44		
148	Chapter 10 Test			

DAY	TOPIC	PAGES	ASSESSMENT	BIBLICAL WORLDVIEW
	Chapter 11 Inferential Statistics			
149–50	**11.1** Probability Distributions	546–54		• understanding that statistical interpretation is affected by worldview
151	**11.2** Central Limit Theorem	555–64		• understanding that theorems reflect natural laws
152	**HC:** Mathematical Expectation	565		• applying biblical principles to shape worldview
153–54	**11.3** Confidence Intervals	566–72	Quiz 11A (11.1–11.2)	
155–56	**11.4** Hypothesis Testing	573–81		• avoiding bias in research
157	**BPM:** Logicism, Intuitionism, and Formalism	582–83	Quiz 11B (11.3–11.4)	• examining modern philosophies of mathematics (1 Cor. 3:11) • understanding that Christ is the foundation of all mathematical truth (Col. 1:15–17)
158	**11.5** Research Studies	584–92		• realizing the role of moral values and judgment in inferential statistics
159	**DA:** Markov Chains	593–94	Quiz 11C (11.5)	• promoting honesty and integrity in computer programming
160	Chapter 11 Review	595–97		
161	Chapter 11 Test			

DAY	TOPIC	PAGES	ASSESSMENT	BIBLICAL WORLDVIEW
	Chapter 12 Limits, Derivatives, and Integrals			
162	**12.1** Evaluating Limits	599–603		• viewing God as the author of infinity • recognizing the limits on our lives (Ps. 90:10)
163	**12.2** Properties of Limits	604–9		• comparing mortal lifespan to eternity (James 4:14; Col. 3:2)
164	**BPM:** Mathematics and Infinity	610–11	Quiz 12A (12.1–12.2)	• acknowledging God's infinite wisdom (Isa. 55:9; Rom. 11:33) • discovering order in creation (Gen. 1:1; Isa. 66:1) • worshipping and praising our infinite God (Ps. 90:2; 139:1–12) • relating the mathematical concept of infinity to the infinite characteristics of God (Ps. 147:5; 139:7–10; 1 John 3:20; Deut. 33:27; Isa. 40:28; 1 Tim. 1:17; Matt. 19:26; Rev. 19:6; Prov. 15:3)
165–66	**12.3** Tangents and Derivatives	612–17		
167	**12.4** Basic Derivative Theorems	618–23		
168–69	**12.5** Product, Quotient, and Chain Rules	624–28		
170	**DA:** Marginal Analysis in Economics	629–30	Quiz 12B (12.3–12.5)	• applying biblical principles in business (Lev. 25:14; Ps. 112:1–3; Prov. 11:1; Eccl. 5:19; Isa. 10:1–2; Matt. 25:14–23; 1 Tim. 6:10)
171	**12.6** Area Under a Curve and Integration	631–37		
172–73	**12.7** The Fundamental Theorem of Calculus	638–43		• demonstrating the unity and beauty in God's created order
174	**HC:** *The Analyst*	644	Quiz 12C (12.6–12.7)	• reflecting the glory of God and His attributes
175	Chapter 12 Review	645–46		
176	Chapter 12 Test			
177–79	Review for Fourth Quarter Exam (Chapters 7–9) or Second Semester Exam (Chapters 7–12)			
180	Fourth Quarter Exam or Second Semester Exam			

Systems and Matrices

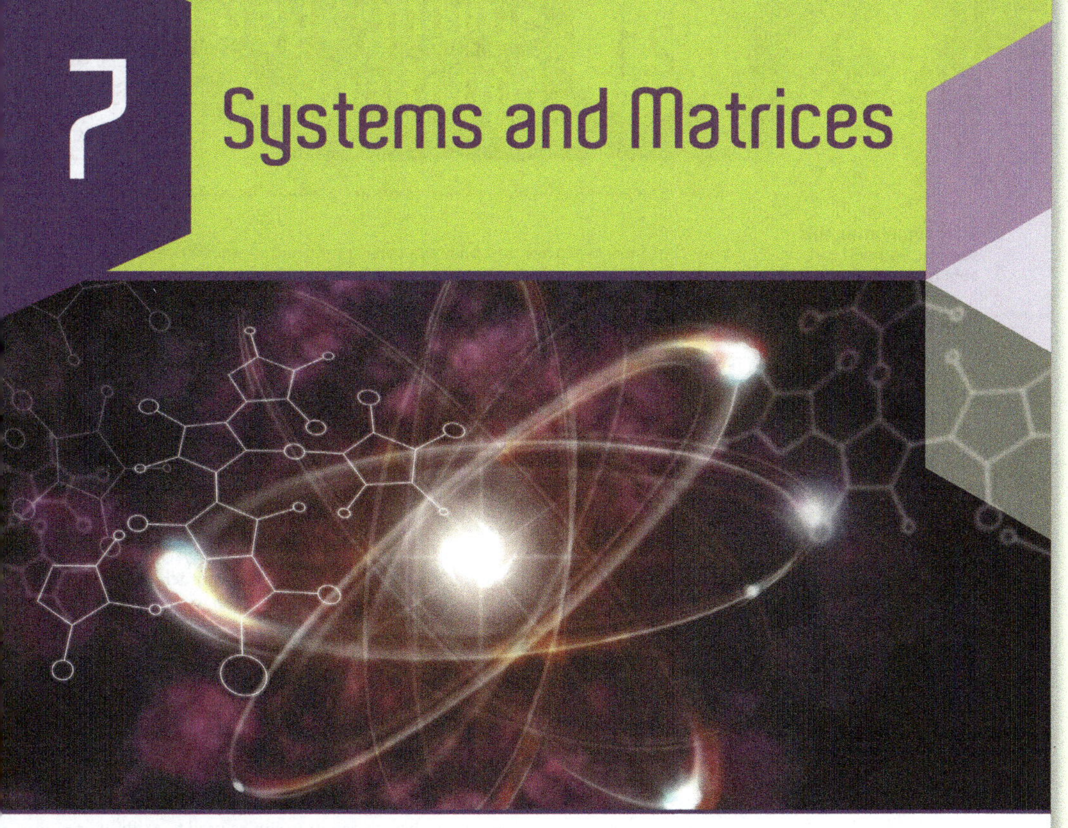

Overview

This chapter focuses on solving linear systems of equations. Algebraic methods of solving systems are reviewed before matrices are introduced. Three different solving methods using matrices are developed: using row operations on an augmented matrix, using determinants described by Cramer's rule, and using an inverse matrix to solve a matrix equation. Although the Gaussian elimination and matrix inverse methods are the hardest to do manually, they are the best methods for calculators and computers, especially when the systems are 3×3 or larger.

The Technology Corner gives details and illustrations of matrix operations on the TI-84 Plus calculator.

Chapter Objectives

1. To solve systems of equations and inequalities graphically and algebraically
2. To perform matrix operations
3. To solve a system of linear equations using row operations on an augmented matrix
4. To solve matrix equations using determinants and inverse matrices
5. To model and solve real-world problems using systems of equations, matrices, and systems of inequalities
6. To compare and contrast conflicting arguments using sound reasoning and a biblical worldview

BIBLICAL PERSPECTIVE OF MATHEMATICS
Is it possible to prove the existence of God? After examining direct and indirect deductive proofs, you will analyze an indirect argument presented by Dr. Greg Bahnsen.

HISTORICAL CONNECTION
From astronomy to subatomic physics, matrices and determinants continue to play an important role in pure and applied mathematics.

DATA ANALYSIS
While secret codes and ciphers date to antiquity, the American mathematician Lester Hill first published a block cipher that used matrix operations for encryption in 1929.

Suggested Teaching Schedule

DAY	
1	7.1
2	7.2
3	Quiz 7A (7.1–7.2) BPM: Direct and Indirect Deductive Reasoning
4–5	7.3; TC: Matrix Operations
6–7	7.4
8	Quiz 7B (7.3–7.4) HC: Matrices and Determinants

DAY	
9	7.5
10–11	7.6
12–13	Quiz 7C (7.5–7.6) 7.7
14	Quiz 7D (7.7) DA: Cryptology
15	Chapter 7 Review
16	Chapter 7 Test

7.1 Solving Systems of Equations

Objectives

1. To classify systems of equations as inconsistent, dependent consistent, or independent consistent

2. To solve independent consistent systems of equations in two or three variables using elimination and substitution

3. To find general solutions of dependent systems

4. To model and solve real-world problems with systems of equations

Flash

Robotics in smartphone production continues to eliminate massive numbers of manual labor jobs around the globe, while competition necessitates automation for survival.

Vocabulary

dependent consistent system
inconsistent system
independent consistent system
solution
system of equations

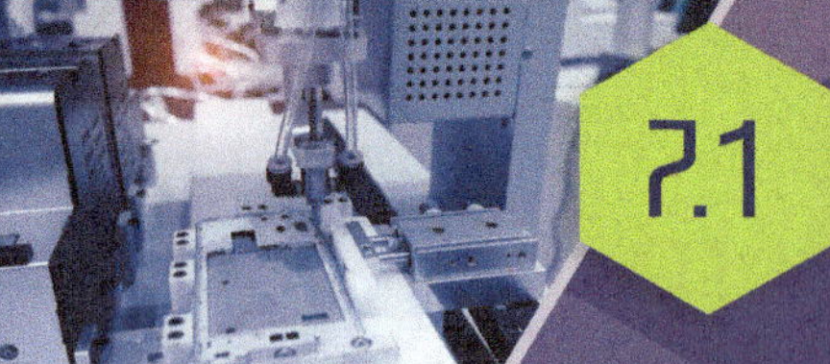

7.1 Solving Systems of Equations

Automating the production of cell phones has helped meet the demand for popular smartphones.

After completing this section, you will be able to

- classify systems of equations.
- solve systems of equations.
- model and solve real-world problems using systems of equations.

Many problems in science and business can be modeled mathematically using a set of equations, each having more than one variable. A *system of equations* consists of several such equations that are solved simultaneously.

A linear equation can be written in the form $a_1 x_1 + a_2 x_2 + \cdots + a_n x_n = b$, where the real coefficients $a_1, a_2, \ldots, a_n$ are not all 0 and the constant b is real. Note the following examples.

Linear Equations	Nonlinear Equations
$2x + 3y = 7$	$4x^2 + y^2 = 16$
$5x - 6y + 3z = -37$	$xy = 12$
$x = 5$	$y = x^3 - 1$

When all of the equations are linear, the system is referred to as a system of linear equations. A *solution* to a system of equations in two variables is an ordered pair satisfying each equation in the system. The ordered pair $(5, -4)$ is a solution to the following system because the values $x = 5$ and $y = -4$ satisfy both equations.

Linear System	$(5, -4)$ is a solution.
$2x + 3y = -2$	$2(5) + 3(-4) = -2$
$6x - y = 34$	$6(5) - (-4) = 34$

Recall from previous courses how the algebraic methods of substitution and elimination were used to find the solutions to a system. Substitution is frequently used to solve linear or nonlinear systems in two variables when an equation can be used to easily express one of the variables in terms of the other.

Example 1 Solving a Linear System by Substitution

Use substitution to solve the system.

$$2x - y = 8$$
$$x + 3y = -3$$

Answer

$x + 3y = -3$
$\quad x = -3 - 3y$

1. Solve the second equation for x.

$2(-3 - 3y) - y = 8$
$\quad -6 - 7y = 8$
$\quad -7y = 14$
$\quad y = -2$

2. Substitute the result into the first equation and solve for y.

$x = -3 - 3y$
$\quad = -3 - 3(-2) = 3$

3. Substitute $y = -2$ back into the equation from the first step to find the value of x.

$(3, -2)$

4. Write the solution as an ordered pair.

CONTINUED ➡

PRESENTATION

Lesson Opener

Identify any ordered pair that is a solution to both equations. List all correct answers.

1. $3x - 2y = -4$
   $2x + 3y = -7$ B
   A. $(1, -3)$
   B. $(-2, -1)$
   C. $\left(-1, \frac{1}{2}\right)$

2. $y = x^2 + 5x - 4$
   $y = x + 1$ B, C
   A. $(-2, -1)$
   B. $(-5, -4)$
   C. $(1, 2)$

3. $4x - 2y = -10$
   $y - 2x = 5$ A, B, C
   A. $(-2, 1)$
   B. $(-1, 3)$
   C. $(1, 7)$

Most students should be familiar with solving *systems of equations* from previous courses. The review of solving systems in this section will prepare the students for solving systems of equations using matrices. You may want to ask them to recall the different methods used to solve a system of equations. (*substitution, elimination, graphing*)

Define a system of equations and distinguish between the linear and nonlin-

ear equations and systems. In Example 1 consider solving the first equation for y and substituting the resulting expression into the second equation to show the students that it doesn't matter which variable is chosen. Explain that choosing to solve the first equation for x or the second equation for y would introduce fractions and make solving the system more complicated. In step 3 the value of x could also be found by substituting $y = -2$ into either original equation. Communicate that the *solution* can be checked graphically or by substituting into the original equations.

Check

$2(3) - (-2) = 8$
$(3) + 3(-2) = -3$

5. Substitute into the original equations to verify your solution.

Using several variables and a system of equations can simplify the solution to many problems.

Example 2   Solving a Nonlinear System by Substitution

Find the dimensions of a rectangle whose perimeter is 48 ft and whose area is 108 ft^2.

Answer

$2x + 2y = 48$
$xy = 108$

1. Write equations for the perimeter and area where $x =$ the rectangle's width and $y =$ its length.

$x + y = 24$
$\quad y = 24 - x$

2. Solve the first equation for y.

$x(24 - x) = 108$
$-x^2 + 24x = 108$
$x^2 - 24x + 108 = 0$
$(x - 6)(x - 18) = 0$
$\quad x = 6 \quad$ or $\quad x = 18$

3. Substitute the result into the second equation and solve for x.

$6y = 108 \quad 18y = 108$
$\quad y = 18 \qquad y = 6$

4. Substitute both values back into one of the original equations and solve for y.

The rectangle is 6 ft by 18 ft.

5. Interpret the results.

Check

$2(6) + 2(18) = 48; 6(18) = 108$

6. Confirm the rectangle's perimeter and area.

SKILL ✔ EXERCISE 3

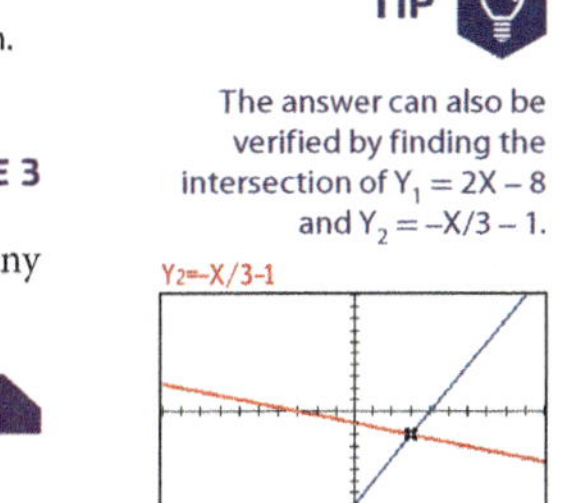

The answer can also be verified by finding the intersection of $Y_1 = 2X - 8$ and $Y_2 = -X/3 - 1$.

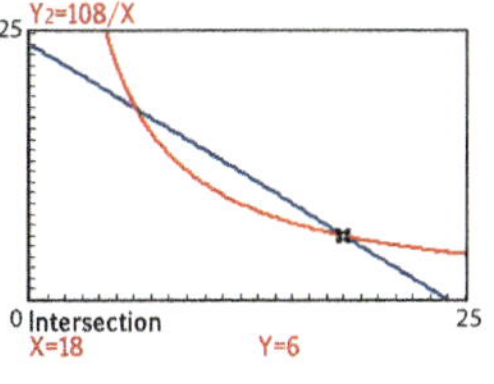

SKILL ✔ EXERCISE 7

Linear systems of two or more equations are frequently solved by elimination. The key is to rewrite the equations as equivalent equations where the coefficients of one of the variables are opposites. The equations can then be added to eliminate that variable.

Example 3   Solving a Linear System by Elimination

Solve the system using elimination.

$6x - 5y = -43$
$7x + 2y = -11$

Answer

$2(6x - 5y = -43) \Rightarrow 12x - 10y = -86$
$5(7x + 2y = -11) \Rightarrow 35x + 10y = -55$

1. Multiply the first equation by 2 and the second by 5 to obtain opposite coefficients for y.

$\qquad 47x = -141$
$\qquad\quad x = -3$

2. Add the equations to eliminate the y-terms and solve for x.

$6(-3) - 5y = -43$
$-18 - 5y = -43$
$\quad -5y = -25$
$\qquad y = 5$

3. Substitute $x = -3$ back into either original equation and solve for y.

$(-3, 5)$

4. Write the solution as an ordered pair.

CONTINUED ➡

Additional Exercises

Figures for Additional Exercises throughout the chapter can be found at TeacherToolsOnline.com.

Solve each system and classify it according to its number of solutions.

1. $x + 3y = 10$
   $2x - 5y = -24$
   $(-2, 4)$; independent consistent

2. $3x - 4y = 9$
   $7x - 11y = 16$
   $(7, 3)$; independent consistent

3. $4x + 3y = -9$
   $y = -\frac{4}{3}x - 3$ $\quad \{(x, y) \mid y = -\frac{4}{3}x - 3\}$
   dependent consistent

4. $4x + 13y = 7$
   $12x + 9y = -69$
   $(-8, 3)$; independent consistent

5. $2x - 3y = 16$
   $9y = 6x + 7$ $\quad \varnothing$; inconsistent

6. $\frac{3}{x} - \frac{5}{y} = -\frac{1}{12}$
   $\frac{6}{x} + \frac{3}{y} = \frac{7}{10}$
   (*Hint*: Let $a = \frac{1}{x}$ and $b = \frac{1}{y}$.)
   $(12, 15)$; independent consistent

7. Find the dimensions of a rectangle whose perimeter is 173 ft and whose area is 1755 ft^2.   32.5 ft $\times$ 54 ft

8. Use technology to solve the system.
   $y = -1.8x^3 + 4x^2 + 5x - 3.4$
   $y = -2x^2 + 3x + 5$ $\quad \approx (-1.15, -1.11),$
   $\approx (1.25, 5.62), \approx (3.23, -6.18)$

---

9. Find the equilibrium point for the given supply and demand equations.
supply: $p = 250 + 0.8x$
demand: $p = 500 - 0.2x$ (250, 450)

Solve each system.

10. $3x + 4y - 7z = 11$
$2x - 5y + z = 0$
$x + 3y - 2z = -1$ $(-3, -2, -4)$

11. $2x + y - 4z = 10$
$3x - 2y + z = 1$ $(t + 3, 2t + 4, t)$

Assignments

- **Minimum:** 1–3, 6–8, 11, 13–14, 18, 21–23, 26, 29, 31–32, 35, 42–44, 46, 48–50

- **Standard:** 2, 4, 6–8; 10–20 even; 21, 23; 24–30 even; 33–35, 40; 43–51 odd

- **Extended:** 5–8, 10–12; 14–20 even; 21, 23–25, 27–28, 32–35, 38, 40–41, 48–49, 51

Solutions

> **A. Exercises**

1. $3(-2) - 2(-1) = -4$
$2(-2) + 3(-1) = -7$

2. $(-5)^2 + 5(-5) - 4 = -4$
$(-5) + 1 = -4$
$(1)^2 + 5(1) - 4 = 2$
$(1) + 1 = 2$

3. $2(-2y + 6) - 5y = 3$
$-9y + 12 = 3$
$-9y = -9$
$y = 1$
$x = -2(1) + 6 = 4; \ (4, 1)$

René Descartes pioneered the graphical representation of functions on a coordinate system in an appendix to his 1637 philosophical treatise *Discourse on Method*. In Part IV he attempted to prove the existence of God and the human soul.

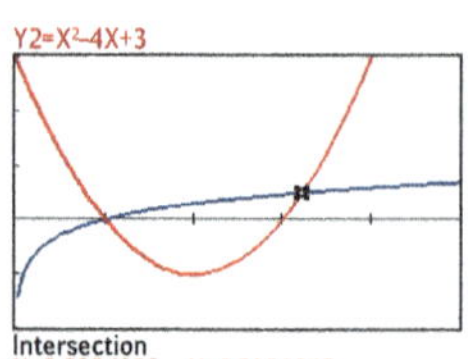

Check

$7(-3) + 2(5) = -21 + 10 = -11$ 5. Substitute into the other original equation to verify your solution.

_______ SKILL ✔ **EXERCISE 11**

Graphical solutions for systems in two variables were also studied in Algebra. Exact solutions for many systems are difficult to obtain graphically. The use of technology enables this method to provide decimal approximations for solutions of systems when known algebraic methods cannot be employed.

Example 4   Solving a Nonlinear System Graphically

Use technology to solve the system graphically. $y = \log x$
$y = x^2 - 4x + 3$

Answer

1. Graph $Y_1 = \log X$ and $Y_2 = X^2 - 4X + 3$ and find the two points of intersection.

$(1, 0)$ and $\approx (3.23, 0.51)$

Check

$\log(1) = 0$ and $(1)^2 - 4(1) + 3 = 0$ 2. The first point $(1, 0)$ is easily verified. A calculator could be used to verify the decimal approximations for the second point.

_______ SKILL ✔ **EXERCISE 17**

Examining the graphs of the equations in a system provides insight into the number of solutions. A unique solution to a system of linear equations in two variables represents the point of intersection of two lines with different slopes. Note how linear systems can also have an infinite number of solutions or no solution.

Types of Linear Systems in Two Variables			
Classification	independent consistent	dependent consistent	inconsistent
Solutions	one	an infinite number	no solution
Graphs	intersecting lines	coinciding lines	parallel lines

In general, consistent systems have one or more solutions, while inconsistent systems have no solution. The graphs of the equivalent equations in a dependent system coincide.

solving systems using matrices later in this chapter. You may want to reserve discussing the last part of this lesson as an introduction to Section 7.3. The following illustrations show the three possibilities for a three-variable, three-condition system.

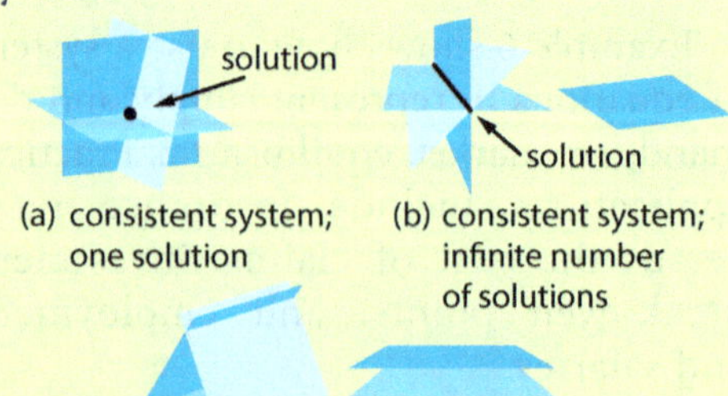

(a) consistent system; one solution
(b) consistent system; infinite number of solutions
(b) inconsistent system; no solution

Because the three-variable system in Example 8 has only two equations, a unique solution cannot be determined. The general solution can be expressed in terms of one of the variables, in this case z, or a parameter, such as t. This solution represents all the points on the three-dimensional line of intersection of the two planes. Particular solutions can be found by substituting for t (or z) as shown in the text and can be used to help verify the solution.

TIPS

Ex. 35–36 Stating the general solution in terms of x or y instead of z produces a different ordered triple. The solution stated for exercise 36 results from solving for y and z in terms of x. Solving for x and y in terms of z results in a generalized solution of $\left(-\frac{5}{9}t - \frac{5}{9}, \frac{2}{9}t - \frac{34}{9}, t\right)$.

Solve each linear system algebraically and classify the system.

a. $2x + 3y = 6$
$y = -\frac{2}{3}x + 2$

b. $3x - 4y = -8$
$6x - 8y = 12$

Answers

a. Solve by substitution.

$2x + 3\left(-\frac{2}{3}x + 2\right) = 6$
$2x - 2x + 6 = 6$
$6 = 6$

Since the statement is true regardless of the value of x, there is an infinite number of solutions.

The solution to this dependent consistent system is a line:

$\{(x, y) \mid y = -\frac{2}{3}x + 2\}$.

b. Solve by elimination.

$-2(3x - 4y = -8) \Rightarrow -6x + 8y = 16$
$ \underline{6x - 8y = 12}$
$ 0 = 28$

The false statement indicates there is no solution.

The solution to this inconsistent system is the empty set, $\varnothing$.

—————— SKILL ✓ **EXERCISE 21**

Businesses use supply and demand curves to describe their willingness to produce more of a product and consumers' declining interest in purchasing the product as prices increase. The intersection of these two curves represents the equilibrium price, the point at which there is a stable price and quantity of goods in the marketplace.

Example 6   **Finding the Equilibrium Price**

A phone manufacturer has determined the supply and demand curves for a cell phone where p is the price (in dollars) and x is the supply (in thousands of phones).

supply: $p = 400 + 0.8x$
demand: $p = 1000 - 0.4x$

Find the equilibrium price and the number of phones that should be produced.

Answer

$400 + 0.8x = 1000 - 0.4x$ 1. Solve by substitution.
$1.2x = 600$
$x = 500$

$p = 400 + 0.8(500) = 800$

The equilibrium price is $800 and 500,000   2. Interpret the results.
phones should be manufactured.

—————— SKILL ✓ **EXERCISE 29**

Algebraic solutions to systems of three linear equations typically eliminate the same variable from two different pairs of equations. The two resulting equations form a new system in two variables. After solving this smaller system, the values of the two variables are substituted into one of the original equations to find the value of the third variable.

8. $x = 2y + 4$
$y = (2y + 4)^2 + 2(2y + 4) - 3$
$y = 4y^2 + 16y + 16 + 4y + 8 - 3$
$4y^2 + 19y + 21 = 0$
$(4y + 7)(y + 3) = 0$
$\qquad\qquad y = -\frac{7}{4}, -3$
$x = 2\left(-\frac{7}{4}\right) + 4 = \frac{1}{2}; (0.5, -1.75)$
$x = 2(-3) + 4 = -2; (-2, -3)$

9. $\qquad 2x^2 + x - 2 = -6x^2 - x + 4$
$\qquad\quad 8x^2 + 2x - 6 = 0$
$\qquad 2(4x - 3)(x + 1) = 0$
$\qquad\qquad\qquad x = \frac{3}{4}, -1$
$y = 2\left(\frac{3}{4}\right)^2 + \frac{3}{4} - 2 = -\frac{1}{8}; \left(\frac{3}{4}, -\frac{1}{8}\right)$
$y = 2(-1)^2 + (-1) - 2 = -1$
$(-1, -1)$

10. $y = 4x - 20$
$4x - 20 = x^3 - 4x^2 - 7x + 10$
$\qquad 0 = x^3 - 4x^2 - 11x + 30$
$\qquad 0 = (x - 2)(x - 5)(x + 3)$
$\qquad\quad x = 2, 5, -3$
$y = (2)^3 - 4(2)^2 - 7(2) + 10 = -12$
$(2, -12)$
$y = (5)^3 - 4(5)^2 - 7(5) + 10 = 0$
$(5, 0)$
$y = (-3)^3 - 4(-3)^2 - 7(-3) + 10$
$= -32; (-3, -32)$

11. $-2(\text{1st}) \Rightarrow -2x - 6y = -14$
$\text{2nd} \qquad \Rightarrow \underline{2x - 5y = -8}$
$\qquad\qquad\qquad\quad -11y = -22$
$\qquad\qquad\qquad\qquad y = 2$
$\text{1st: } x + 3(2) = 7$
$\qquad\qquad x = 7 - 6 = 1; (1, 2)$

12. $\text{1st} \qquad \Rightarrow \quad 6x - 4y = 60$
$2(\text{2nd}) \Rightarrow \underline{6x + 4y = 36}$
$\qquad\qquad\qquad 12x \qquad = 96$
$\qquad\qquad\qquad\qquad x = 8$
$\text{2nd: } 3(8) + 2y = 18$
$\qquad\qquad\quad 2y = -6$
$\qquad\qquad\qquad y = -3; (8, -3)$

13. $2(\text{1st}) \qquad \Rightarrow \quad 4x + 6y = -4$
$-3(\text{2nd}) \Rightarrow \underline{-15x - 6y = -18}$
$\qquad\qquad\qquad -11x \qquad\quad = -22$
$\qquad\qquad\qquad\qquad\quad x = 2$
$\text{1st: } 2(2) + 3y = -2$
$\qquad\qquad\quad 3y = -6$
$\qquad\qquad\qquad y = -2; (2, -2)$

14. $4(\text{1st}) \qquad \Rightarrow \quad 32x + 12y = 154$
$-3(\text{2nd}) \Rightarrow \underline{-21x - 12y = -99}$
$\qquad\qquad\qquad\quad 11x \qquad\quad = 55$
$\qquad\qquad\qquad\qquad\quad x = 5$
$\text{2nd: } 7(5) + 4y = 33$
$\qquad\qquad\quad 4y = -2$
$\qquad\qquad\qquad y = -0.5; (5, -0.5)$

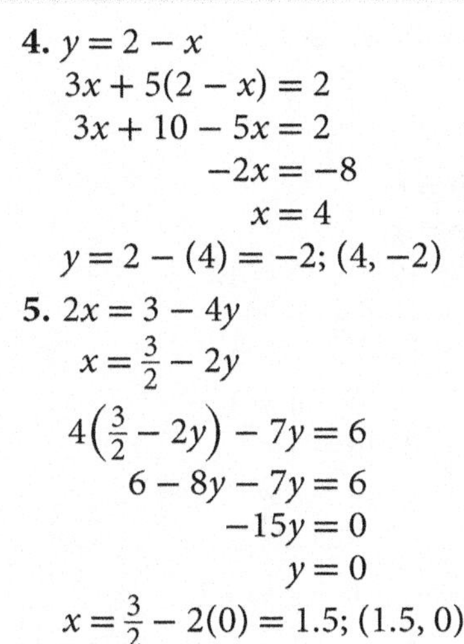

4. $y = 2 - x$
$3x + 5(2 - x) = 2$
$3x + 10 - 5x = 2$
$\qquad\quad -2x = -8$
$\qquad\qquad x = 4$
$y = 2 - (4) = -2; (4, -2)$

5. $2x = 3 - 4y$
$x = \frac{3}{2} - 2y$
$4\left(\frac{3}{2} - 2y\right) - 7y = 6$
$6 - 8y - 7y = 6$
$\qquad -15y = 0$
$\qquad\qquad y = 0$
$x = \frac{3}{2} - 2(0) = 1.5; (1.5, 0)$

6. $5y = -4x - 10$
$y = -\frac{4}{5}x - 2$
$3x + 2\left(-\frac{4}{5}x - 2\right) = 3$
$3x - \frac{8}{5}x - 4 = 3$
$\qquad \frac{7}{5}x = 7$
$\qquad\quad x = 5$
$y = -\frac{4}{5}(5) - 2 = -6; (5, -6)$

7. $\qquad x(2x - 9) = 5$
$\qquad 2x^2 - 9x - 5 = 0$
$\qquad (2x + 1)(x - 5) = 0$
$\qquad\qquad x = -\frac{1}{2}, 5$
$y = 2\left(-\frac{1}{2}\right) - 9 = -10; \left(-\frac{1}{2}, -10\right)$
$y = 2(5) - 9 = 1; (5, 1)$

15.
$$18(\text{1st}) \Rightarrow \quad 9x + 6y = 15$$
$$6(\text{2nd}) \Rightarrow \quad \underline{4x - 6y = -15}$$
$$13x = 0$$
$$x = 0$$
$$\text{2nd: } \tfrac{2}{3}(0) - y = -\tfrac{5}{2}$$
$$y = \tfrac{5}{2}; \ \left(0, \tfrac{5}{2}\right)$$

16.
$$10(\text{1st}) \Rightarrow \quad 5x - 30y = 26$$
$$30(\text{2nd}) \Rightarrow \quad \underline{24x + 30y = -84}$$
$$29x = -58$$
$$x = -2$$
$$\text{1st: } \tfrac{1}{2}(-2) - 3y = \tfrac{13}{5}$$
$$-3y = \tfrac{18}{5}$$
$$y = -\tfrac{6}{5}; \ \left(-2, -\tfrac{6}{5}\right)$$

17. Y₂=3X+4

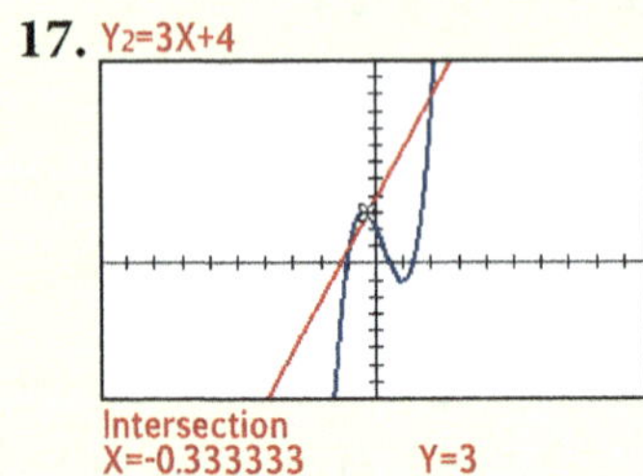

18. Y₂=-4X²+3X+5   10

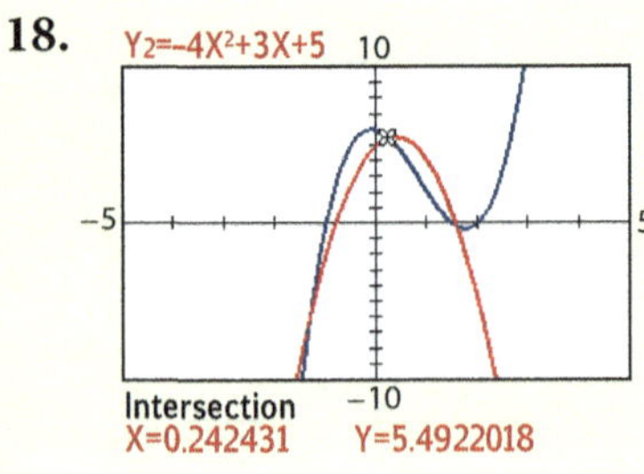

19. Y₂=3sin(X) 5

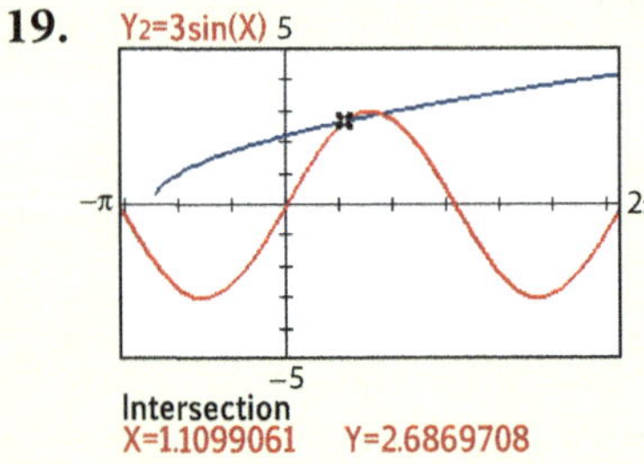

20. Y₂=-2X²+11X-6

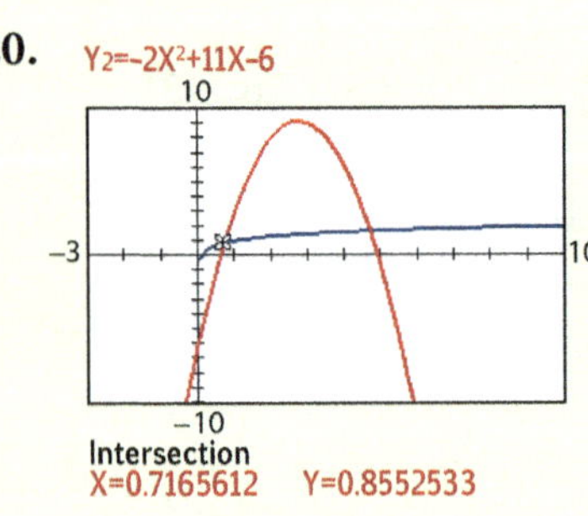

Solve the system.
$$-2x + 5y + 2z = 23$$
$$4x - y + 2z = -13$$
$$x - 3y + 6z = -17$$

Answer

$$-2x + 5y + 2z = 23$$
$$2(x - 3y + 6z = -17) \Rightarrow \underline{2x - 6y + 12z = -34}$$
$$\text{(a)} \qquad -y + 14z = -11$$

1. Eliminate the x-term from the first and third equations by adding the first equation to 2 times the third equation.

$$-4(x - 3y + 6z = -17) \Rightarrow -4x + 12y - 24z = 68$$
$$\underline{4x - y + 2z = -13}$$
$$\text{(b)} \qquad 11y - 22z = 55$$

2. Eliminate the x-term from the second and third equations by adding -4 times the third equation and the second equation.

$$-y + 14z = -11$$
$$\tfrac{1}{11}(11y - 22z = 55) \Rightarrow \underline{y - 2z = 5}$$
$$12z = -6$$
$$z = -\tfrac{1}{2}$$
$$y - 2\left(-\tfrac{1}{2}\right) = 5$$
$$y = 4$$

3. Solve the two-variable system formed by equations (a) and (b) using elimination.

$$x - 3(4) + 6\left(-\tfrac{1}{2}\right) = -17$$
$$x - 15 = -17$$
$$x = -2$$

4. Substitute the values of y and z into one of the original equations and solve for x.

$$\left(-2, 4, -\tfrac{1}{2}\right)$$

5. Write the solution as an ordered triple.

Check
$$-2(-2) + 5(4) + 2\left(-\tfrac{1}{2}\right) = 23$$
$$4(-2) - (4) + 2\left(-\tfrac{1}{2}\right) = -13$$

6. Substitute the coordinates into the other two equations to verify your solution.

SKILL ✔ EXERCISE 31

The graph of each linear equation in three variables represents a plane. The unique solution in Example 7 represents the point of intersection of three distinct planes in space. In order for a linear system to have a unique solution, there must be as many equations as variables.

Solve the system.
$$2x + 3y - 7z = 8$$
$$6x + 8y - 11z = 4$$

Answer

$$-3(2x + 3y - 7z = 8) \Rightarrow -6x - 9y + 21z = -24$$
$$\underline{6x + 8y - 11z = 4}$$
$$-y + 10z = -20$$

1. Eliminate x by multiplying the first equation by -3 and adding the result to the second equation.

There are infinitely many solutions to this dependent system.

2. Since there are no other equations, a second equation in y and z cannot be derived.

$$y = 10z + 20$$

3. Express y in terms of z.

CONTINUED ➡

> **B. Exercises**

21.
$$\text{1st} \Rightarrow 0.6x + 2y = 3.66$$
$$4(\text{2nd}) \Rightarrow \underline{5.2x - 2y = 8.52}$$
$$5.8x = 12.18$$
$$x = 2.1$$
$$\text{1st: } 0.6(2.1) + 2y = 3.66$$
$$2y = 2.4$$
$$y = 1.2$$

22.
$$3(\text{1st}) \Rightarrow \quad 15x + 6y = 60$$
$$-2(\text{2nd}) \Rightarrow \underline{-15x - 6y = -60}$$
$$0 = 0$$

23.
$$y = -2x + 3$$
$$4x + 2(-2x + 3) = 7$$
$$4x - 4x + 6 = 7$$
$$6 \neq 7$$

24.
$$2\left(\tfrac{7}{2} + \tfrac{1}{2}y\right) - y = 7$$
$$7 + y - y = 7$$
$$7 = 7$$

25.
$$5a - 2b = 6$$
$$3a + 4b = 1$$
$$2(\text{1st}) \Rightarrow 10a - 4b = 12$$
$$(\text{2nd}) \Rightarrow \underline{3a + 4b = 1}$$
$$13a = 13$$
$$a = 1$$
$$\text{2nd: } 3(1) + 4b = 1$$
$$4b = -2$$
$$b = -\tfrac{1}{2}$$
$$a = \tfrac{1}{x} = 1 \qquad b = \tfrac{1}{y} = -\tfrac{1}{2}$$
$$x = 1 \qquad y = -2$$

$2x + 3(10z + 20) - 7z = 8$
$2x + 23z + 60 = 8$
$2x = -23z - 52$
$x = -\frac{23}{2}z - 26$

$(x, y, z) = \left(-\frac{23}{2}t - 26, 10t + 20, t\right)$

4. Substitute into either original equation to express x in terms of z.

5. The general solution can be expressed as an ordered triple where z is represented by the arbitrary parameter t.

SKILL ✔ EXERCISE 35

You can find particular solutions to the system in Example 8 by substituting any real number for t. For example, if $t = -2$, the solution is

$$\left(-\frac{23}{2}(-2) - 26, 10(-2) + 20, -2\right) = (-3, 0, -2).$$

The ordered triple $(-3, 0, -2)$ is just one of the points on the line in space formed by the intersection of the two planes.

❭ A. Exercises

Identify any ordered pair that is a solution to the system. List all correct answers.

1. $3x - 2y = -4$
$2x + 3y = -7$ **B**
A. $(1, -3)$
B. $(-2, -1)$
C. $\left(-1, \frac{1}{2}\right)$

2. $y = x^2 + 5x - 4$
$y = x + 1$ **B, C**
A. $(-2, -1)$
B. $(-5, -4)$
C. $(1, 2)$

Use substitution to solve each system.

3. $2x - 5y = 3$
$x = -2y + 6$ **(4, 1)**

4. $3x + 5y = 2$
$x + y = 2$ **(4, -2)**

5. $2x + 4y = 3$
$4x - 7y = 6$ **(1.5, 0)**

6. $3x + 2y = 3$
$4x + 5y = -10$ **(5, -6)**

Use substitution to solve each system. Then verify your solution(s) graphically.

7. $y = 2x - 9$
$xy = 5$

8. $y = x^2 + 2x - 3$
$x - 2y = 4$

9. $y = 2x^2 + x - 2$
$y = -6x^2 - x + 4$

10. $4x - y = 20$
$y = x^3 - 4x^2 - 7x + 10$

Solve each system using elimination.

11. $x + 3y = 7$
$2x - 5y = -8$ **(1, 2)**

12. $6x - 4y = 60$
$3x + 2y = 18$ **(8, -3)**

13. $2x + 3y = -2$
$5x + 2y = 6$ **(2, -2)**

14. $8x + 3y = 38.5$
$7x + 4y = 33$ **(5, -0.5)**

15. $\frac{1}{2}x + \frac{1}{3}y = \frac{5}{6}$
$\frac{2}{3}x - y = -\frac{5}{2}$ **$\left(0, \frac{5}{2}\right)$**

16. $\frac{1}{2}x - 3y = \frac{13}{5}$
$\frac{4}{5}x + y = -\frac{14}{5}$ **$\left(-2, -\frac{6}{5}\right)$**

Solve each system using technology. Round answers to the nearest tenth if necessary.

17. $y = 3x^3 - 2x^2 - 4x + 2$
$y = 3x + 4$

18. $y = 2x^3 - 5x^2 - x + 6$
$y = -4x^2 + 3x + 5$

19. $y = \sqrt{2x + 5}$
$y = 3 \sin x$
(1.1, 2.7), (1.8, 2.9)

20. $y = \log 10x$
$y = -2x^2 + 11x - 6$
(0.7, 0.9), (4.7, 1.7)

❭ B. Exercises

Solve each system. Then classify the system as independent consistent, dependent consistent, or inconsistent.

21. $0.6x + 2y = 3.66$
$1.3x - 0.5y = 2.13$

22. $5x + 2y = 20$
$\frac{15}{2}x + 3y = 30$

23. $2x + y = 3$
$4x + 2y = 7$

24. $x = \frac{7}{2} + \frac{1}{2}y$
$2x - y = 7$

25. Solve the system. (*Hint*: Let $a = \frac{1}{x}$ and $b = \frac{1}{y}$.)
$$\frac{5}{x} - \frac{2}{y} = 6$$
$$\frac{3}{x} + \frac{4}{y} = 1 \quad (1, -2)$$

Find the dimensions of a rectangle with the given perimeter and area.

26. perimeter of 52 m and area of 144 m^2 **8 m × 18 m**

27. perimeter of 427 ft and area of 11,025 ft^2 **87.5 ft × 126 ft**

17. $(-1, 1), \left(-\frac{1}{3}, 3\right), (2, 10)$

18. $(-1.3, -5.8), (0.2, 5.5), (1.6, -0.2)$

21. $(2.1, 1.2)$; independent consistent

22. $\{(x, y) \mid 5x + 2y = 20\}$; dependent consistent

23. $\varnothing$; inconsistent

24. $\{(x, y) \mid 2x - y = 7\}$; dependent consistent

7. $\left(-\frac{1}{2}, -10\right), (5, 1)$

8. $(0.5, -1.75), (-2, -3)$

9. $\left(\frac{3}{4}, -\frac{1}{8}\right), (-1, -1)$

10. $(2, -12), (5, 0), (-3, -32)$

---

26. $2l + 2w = 52; l = 26 - w$
$lw = 144$
$(26 - w)w = 144$
$w^2 - 26w + 144 = 0$
$(w - 8)(w - 18) = 0$
$w = 8, 18$
$l = 26 - (8) = 18; l = 26 - (18) = 8$
8 m × 18 m

27. $2l + 2w = 427$
$lw = 11,025; l = \dfrac{11,025}{w}$
$\left(2\left(\dfrac{11,025}{w}\right) + 2w = 427\right)w$
$2w^2 - 427w + 22,050 = 0$
$(2w - 175)(w - 126) = 0$
$w = 87.5, 126$
$l(87.5) = 11,025; l = 126$
$l(126) = 11,025; l = 87.5$
87.5 ft × 126 ft

28. $50 + 0.5x = 950 - 2.5x$
$3x = 900$
$x = 300$ seats
$p = 50 + 0.5(300) = \$200$

29. $14,900 + 0.2x = 32,120 - 0.3x$
$0.5x = 17,220$
$x = 34,440$ cars
$p = 14,900 + 0.2(34,440) = \$21,788$

30.
$$\begin{aligned}
\text{1st} &\Rightarrow 2x + 3y + 4z = 3 \\
3(\text{2nd}) &\Rightarrow 9x - 3y - 15z = 30 \\
\hline
&\quad 11x \quad\;\; - 11z = 33
\end{aligned}$$
(a) $\Rightarrow x - z = 3$
$$\begin{aligned}
\text{3rd} &\Rightarrow 5x - 2y - 3z = 5 \\
-2(\text{2nd}) &\Rightarrow -6x + 2y + 10z = -20 \\
\hline
\text{(b)} &\quad -x + 7z = -15
\end{aligned}$$
$$\begin{aligned}
\text{(a)} &\Rightarrow x - z = 3 \\
\text{(b)} &\Rightarrow -x + 7z = -15 \\
\hline
&\quad 6z = -12 \\
&\quad z = -2
\end{aligned}$$
(a) $x - (-2) = 3$
$x = 1$
1st: $2(1) + 3y + 4(-2) = 3$
$3y - 6 = 3$
$y = 3$
$(1, 3, -2)$

31.
$$\begin{aligned}
-6(\text{1st}) &\Rightarrow -12x - 18y + 6z = -54 \\
\text{2nd} &\Rightarrow 4x - 5y - 6z = 1 \\
\hline
\text{(a)} &\quad -8x - 23y = -53
\end{aligned}$$
$$\begin{aligned}
-3(\text{1st}) &\Rightarrow -6x - 9y + 3z = -27 \\
\text{3rd} &\Rightarrow 5x + 2y - 3z = 8 \\
\hline
\text{(b)} &\quad -x - 7y = -19
\end{aligned}$$
$$\begin{aligned}
-1(\text{a}) &\Rightarrow 8x + 23y = 53 \\
8(\text{b}) &\Rightarrow -8x - 56y = -152 \\
\hline
&\quad -33y = -99 \\
&\quad y = 3
\end{aligned}$$
$(-a)\ 8x + 23(3) = 53$
$8x = -16$
$x = -2$
1st: $2(-2) + 3(3) - z = 9$
$-z + 5 = 9$
$z = -4$
$(-2, 3, -4)$

32.
$$\begin{aligned}
-3(\text{1st}) &\Rightarrow -3x - 6y - 6z = 3 \\
\text{2nd} &\Rightarrow 3x - 2y + z = -2 \\
\hline
\text{(a)} &\quad -8y - 5z = 1
\end{aligned}$$
$$\begin{aligned}
\text{(a)} &\Rightarrow -8y - 5z = 1 \\
\text{3rd} &\Rightarrow 3y + 5z = -16 \\
\hline
&\quad -5y = -15 \\
&\quad y = 3
\end{aligned}$$
(a) $-8(3) - 5z = 1$
$-5z = 25$
$z = -5$
1st: $x + 2(3) + 2(-5) = -1$
$x - 4 = -1$
$x = 3$
$(3, 3, -5)$

33.

$$\begin{array}{rl} \text{1st} \Rightarrow & 4x - 6y + 6z = -21 \\ -4(\text{3rd}) \Rightarrow & \underline{-4x + 8y - 16z = 32} \\ \text{(a)} & \phantom{-4x + {}} 2y - 10z = 11 \\ \text{2nd} \Rightarrow & 3x + 4y - z = -6 \\ -3(\text{3rd}) \Rightarrow & \underline{-3x + 6y - 12z = 24} \\ \text{(b)} & \phantom{-3x + {}} 10y - 13z = 18 \\ -5(\text{a}) \Rightarrow & -10y + 50z = -55 \\ \text{(b)} \Rightarrow & \underline{10y - 13z = 18} \\ & \phantom{-10y + {}} 37z = -37 \\ & z = -1 \end{array}$$

(a) $2y - 10(-1) = 11$

$ 2y = 1$

$ y = \dfrac{1}{2}$

3rd: $x - 2\left(\dfrac{1}{2}\right) + 4(-1) = -8$

$\phantom{\text{3rd: }} x - 5 = -8$

$\phantom{\text{3rd: }} x = -3$

$\left(-3, \dfrac{1}{2}, -1\right)$

34a. for $t = -1$: $(3(-1) + 4, -1 - 6, -1)$

$\phantom{\textbf{34a.} \text{ for } t = -1:} = (1, -7, -1)$

$\phantom{\textbf{34a.} } \text{for } t = 2$: $(3(2) + 4, 2 - 6, 2)$

$\phantom{\textbf{34a.} \text{ for } t = 2:} = (10, -4, 2)$

34b. $3t + 4 = -8$; $t = -4$

34c. $3t + 4 = 13$; $t = 3$

$\phantom{\textbf{34c.} } (3(3) + 4, 3 - 6, 3) = (13, -3, 3)$

35.

$$\begin{array}{rl} -2(\text{1st}) \Rightarrow & -2x - 10y + 2z = -16 \\ \text{2nd} \Rightarrow & \underline{2x - 4y + 4z = 2} \\ & \phantom{-2x - {}} -14y + 6z = -14 \\ & y = \dfrac{3}{7}z + 1 \end{array}$$

1st: $x + 5\left(\dfrac{3}{7}z + 1\right) - z = 8$

$\phantom{\text{1st: }} x + \dfrac{8}{7}z + 5 = 8$

$\phantom{\text{1st: }} x = -\dfrac{8}{7}z + 3$

letting $z = t$: $\left(-\dfrac{8}{7}t + 3, \dfrac{3}{7}t + 1, t\right)$

for $t = 7$: $(-5, 4, 7)$

1st: $-5 + 20 - 7 = 8$

2nd: $-10 - 16 + 28 = 2$

36.

$$\begin{array}{rl} -1(\text{1st}) \Rightarrow & -x + 2y - z = -7 \\ 2(\text{2nd}) \Rightarrow & \underline{10x - 2y + 6z = 2} \\ & 9x \phantom{- 2y {}} + 5z = -5 \\ & 9x z = -\dfrac{9}{5}x - 1 \end{array}$$

1st: $x - 2y + \left(-\dfrac{9}{5}x - 1\right) = 7$

$\phantom{\text{1st: }} -2y - \dfrac{4}{5}x - 1 = 7$

$\phantom{\text{1st: }} y = -\dfrac{2}{5}x - 4$

letting $x = r$: $\left(r, -\dfrac{2}{5}r - 4, -\dfrac{9}{5}r - 1\right)$

for $r = 5$: $(5, -6, -10)$

1st: $5 + 12 - 10 = 7$

2nd: $25 + 6 - 30 = 1$

Find the equilibrium price.

28. An airline determines the supply and demand curves for a flight where p is the price (in dollars) and x is the number of seats booked. Find the equilibrium price and the number of seats that should be provided.

supply: $p = 50 + 0.5x$

demand: $p = 950 - 2.5x$ 300 seats at \$200 each

29. An automobile manufacturer determines the supply and demand curves for a base model car where p is the price (in dollars) and x is the number of cars sold. Find the equilibrium price and the number of cars that should be produced.

supply: $p = 14{,}900 + 0.2x$

demand: $p = 32{,}120 - 0.3x$ 34,440 cars at \$21,788 each

Solve each system.

30. $\begin{aligned} 2x + 3y + 4z &= 3 \\ 3x - y - 5z &= 10 \\ 5x - 2y - 3z &= 5 \end{aligned}$ (1, 3, −2)

31. $\begin{aligned} 2x + 3y - z &= 9 \\ 4x - 5y - 6z &= 1 \\ 5x + 2y - 3z &= 8 \end{aligned}$ (−2, 3, −4)

32. $\begin{aligned} x + 2y + 2z &= -1 \\ 3x - 2y + z &= -2 \\ 3y + 5z &= -16 \end{aligned}$ (3, 3, −5)

33. $\begin{aligned} 4x - 6y + 6z &= -21 \\ 3x + 4y - z &= -6 \\ x - 2y + 4z &= -8 \end{aligned}$ $\left(-3, \dfrac{1}{2}, -1\right)$

34. A two-equation system in three variables has a general solution of $(3t + 4, t - 6, t)$.   (1, −7, −1) and (10, −4, 2)

   a. Find particular solutions for $t = -1$ and $t = 2$.

   b. What value of t corresponds to the point with an x-coordinate of −8? $t = -4$

   c. What solution has an x-coordinate of 13?   (13, −3, 3)

CUMULATIVE REVIEW

42. State the x-intercept, y-intercept, and slope of the line $5x + y = -10$.   [1.2]   (−2, 0); (0, −10); $m = -5$

43. Find the slope of the line containing $(2, 7)$ and $(4, -5)$.   [1.2] $m = -6$

Describe how the graph of $f(x) = x^2$ is transformed into the graph of $g(x)$.   [1.3]

44. $g(x) = (x + 4)^2 - 3$

45. $g(x) = -|x - 3| + 1$

Find the zeros for each polynomial function, including the multiplicity of any repeated zeros.   [2.2]

46. $p(x) = (x^2 - 25)(x^2 - 9)$

47. $p(x) = (x - 7)^4$

48. Which general form equation is equivalent to $y = \dfrac{2}{5}x - 3$?   [1.2]   D

   A. $5x - 2y = 15$

   B. $-2x + 5y = 15$

   C. $2x - 5y = -15$

   D. $2x - 5y = 15$

   E. none of these

Find a general solution for each system. Then state one particular solution and use it to verify your solution.

35. $\begin{aligned} x + 5y - z &= 8 \\ 2x - 4y + 4z &= 2 \end{aligned}$

36. $\begin{aligned} x - 2y + z &= 7 \\ 5x - y + 3z &= 1 \end{aligned}$

❯ C. Exercises

37. Let $f(x) = ax + 4$, $g(x) = cx - 1$, $2f(x) + g(x) = 12x + 7$, and $f(x) - 3g(x) = -x + 7$. Find a and c. $a = 5, c = 2$

38. Given $\triangle ABC$ with $A\,(-2, 5)$, $B\,(-6, 1)$, and $C\,(4, -1)$, find the intersection of side BC and the altitude from A. $\left(-\dfrac{38}{13}, \dfrac{5}{13}\right)$

39. Find all (x, y) within $[0, 2\pi)$ such that $2 \sin x + 4 \cos y = 4$ and $\sin x = 2 \cos y$. $\left(\dfrac{\pi}{2}, \dfrac{\pi}{3}\right)$, $\left(\dfrac{\pi}{2}, \dfrac{5\pi}{3}\right)$

40. Use the general equation for a circle, $x^2 + y^2 + Ax + By + C = 0$, to find A, B, and C for the circle passing through $(3, -6)$, $(-1, 2)$, and $(6, 3)$.

41. **Explore:** Solve the following system graphically. Then solve the system algebraically, listing all possible solutions. Reconcile the seeming contradiction caused by solving with different methods.

$$3x + 2y = 6$$
$$xy = 12$$

35. $\left(-\dfrac{8}{7}t + 3, \dfrac{3}{7}t + 1, t\right)$ for $t = 7$: $(-5, 4, 7)$

36. $\left(r, -\dfrac{2}{5}r - 4, -\dfrac{9}{5}r - 1\right)$ for $r = 5$: $(5, -6, -10)$

40. $A = -6, B = 2, C = -15$

41. no solution; $\left(1 + \sqrt{7}\,i, \dfrac{3}{2} - \dfrac{3\sqrt{7}}{2}\,i\right)$, $\left(1 - \sqrt{7}\,i, \dfrac{3}{2} + \dfrac{3\sqrt{7}}{2}\,i\right)$; Solving graphically reveals only real solutions but solving algebraically reveals the complex solutions.

49. Find the area of the triangle.   [5.7]   D

   A. 11 ft^2

   B. 20 ft^2

   C. 29.8 ft^2

   D. 12.9 ft^2

   E. none of these

50. Find the midpoint of $\overline{AB}$ with $A\,(-3, 1, 2)$ and $B\,(7, -4, -2)$.   [6.3]   B

   A. $\left(5, -\dfrac{5}{2}, -2\right)$

   B. $\left(2, -\dfrac{3}{2}, 0\right)$

   C. $(4, -3, 0)$

   D. $(11, -5, -4)$

   E. none of these

51. Which of the following equations passes a symmetry test for the polar axis? List all correct answers.   [6.5]   A, C, D

   A $r = -\sin 3$

   B. $r = -5 \sin \theta$

   C. $r = 2 \sin 2\theta$

   D. $r^2 = 9 \cos 2\theta$

   E. none of these

❯ C. Exercises

37. $2(ax + 4) + (cx - 1) = 12x + 7$

$\phantom{\textbf{37.} } 2ax + cx = 12x$

(a) $\phantom{2ax + {}} 2a + c = 12, x \neq 0$

$\phantom{\textbf{37.} } (ax + 4) - 3(cx - 1) = -x + 7$

$\phantom{\textbf{37.} } ax - 3cx = -x$

(b) $\phantom{2ax + {}} a - 3c = -1, x \neq 0$

$$\begin{array}{rl} \text{(a)} \Rightarrow & 2a + c = 12 \\ -2(\text{b}) \Rightarrow & \underline{-2a + 6c = 2} \\ & \phantom{-2a + {}} 7c = 14 \\ & c = 2 \end{array}$$

(b) $a - 3(2) = -1$

$\phantom{\text{(b) }} a = 5$

continued in Answers and Solutions Overflow

7.2 Matrices

Apps are available to help coaches record statistics for their teams.

Numerical data is often presented in a table or stored in the rows and columns of a spreadsheet. A *matrix* is a rectangular array of numbers and is designated with an upper-case letter. A matrix with m rows and n columns has *dimensions* of $m \times n$ (read "m by n"). Each number in the array is an *entry* (or *element*) of the matrix. The double-subscript notation a_{ij} denotes the entry in row i and column j of matrix A.

$$\underset{\downarrow}{\text{Column 2}}$$

$$A = [a_{ij}] = \begin{bmatrix} a_{11} & a_{12} & a_{13} & \cdots & a_{1n} \\ a_{21} & a_{22} & a_{23} & \cdots & a_{2n} \\ a_{31} & a_{32} & a_{33} & \cdots & a_{3n} \\ \vdots & \vdots & \vdots & & \vdots \\ a_{m1} & a_{m2} & a_{m3} & \cdots & a_{mn} \end{bmatrix} \leftarrow \text{Row 3}$$

Example 1 Identifying Matrix Dimensions and Entries

State the dimensions of each matrix. Then state the value of the designated entry.

a. $A = \begin{bmatrix} 8 & -5 & 1 \\ 3 & 0 & 2 \end{bmatrix}$ **b.** $B = [-7 \quad 2]$ **c.** $C = \begin{bmatrix} 3 \\ -1 \\ 5 \end{bmatrix}$ **d.** $D = \begin{bmatrix} a & b & c \\ r & s & t \\ x & y & z \end{bmatrix}$

$a_{21} = ?$ $b_{12} = ?$ $c_{21} = ?$ $d_{32} = ?$

Answer

a. 2×3; $a_{21} = 3$ **b.** 1×2; $b_{12} = 2$ **c.** 3×1; $c_{21} = -1$ **d.** 3×3; $d_{32} = y$

SKILL ✓ EXERCISE 1

In Example 1, B is a *row matrix* and C is a *column matrix*. Because matrix D has the same number of rows and columns, it is a *square matrix* and is said to have an *order* of 3.

Two matrices $A = [a_{ij}]$ and $B = [b_{ij}]$ are *equal* if they have the same dimensions ($m \times n$) and their corresponding entries are equal, that is, $a_{ij} = b_{ij}$ for all $i = 1, 2, \ldots, m$ and $j = 1, 2, \ldots, n$.

The first basic matrix operation is *matrix addition*. Matrices are added by adding corresponding entries.

▎MATRIX ADDITION

If $A = [a_{ij}]$ and $B = [b_{ij}]$ are both $m \times n$ matrices, then $A + B = [a_{ij} + b_{ij}]$ for all $i = 1, 2, \ldots, m$ and $j = 1, 2, \ldots, n$.

After completing this section, you will be able to

- define *matrix* and related terms.
- perform matrix operations.
- compare and contrast the properties of matrices with the properties of real numbers.
- use matrices to represent real-world data.

Benjamin Peirce, one of the earliest American mathematicians, included multiplication tables in matrix form for all complex associative algebras up to 6 dimensions in his 1870 text *Linear Associative Algebra*.

7.2 Matrices

Objectives

1. To define a matrix and related terms
2. To perform matrix operations of addition, subtraction, scalar multiplication, and matrix multiplication
3. To compare and contrast the properties of matrices with the properties of real numbers

Flash

There are hundreds of apps designed to track team stats. These apps use rectangular grids of numbers, which are mathematical matrices.

Vocabulary

additive identity matrix
additive inverse matrix
column matrix
dimensions
entry (element)
equal matrices
main diagonal
matrix
matrix addition
matrix product
matrix subtraction
multiplicative identity matrix
multiplicative inverse matrix
order
row matrix
scalar product
square matrix
zero matrix

PRESENTATION

Lesson Opener

A double subscript is used to indicate the location of data within a *matrix*. Data entered in row 2, column C in a spreadsheet would be designated as e_{23} (*element two three*) in a matrix. Indicate the location in a matrix for each *entry* in the spreadsheet.

	A	B	C
1	P_{11}		
2	45		P_{23}
3		81	
4			56

1. 45 e_{21} 2. 81 e_{32} 3. 56 e_{43}

Use the Lesson Opener to help students begin to think in terms of rows and columns. Define a matrix and the related terms, emphasizing the use of double-subscript notation. Use Example 1 to assess student understanding of matrix *dimensions* and the notation for the location of an element. Define *row*, *column*, and *square matrices*.

Define *matrix addition* and illustrate the operation with Example 2. Have the students explain why matrices of different dimensions cannot be added. (*They will lack corresponding entries.*)

Then define a *scalar product*, the *zero matrix*, 0 (the *additive identity matrix*), an *additive inverse matrix*, and *matrix subtraction*. Example 3 illustrates both *scalar multiplication* and matrix subtraction.

Motivational Idea Explain that the hue of each pixel on a display can be represented collectively by a large matrix of numbers. A scalar multiplication of the matrix would create an instant change in the hue of the entire display.

Direct students to the properties of matrix addition and scalar multiplication. While many of these will be obvious to the students, explain that proving a property usually involves the following steps.

  a. Apply a definition to get to real number expressions.

Reading and Writing Mathematics

Describe how you would find C_{23} if $C = AB$. **Find the product of the first element of row 2 in A and the first element of column 3 in B. Then find the product of the second element of row 2 in A and the second element of column 3 in B. Continue for each corresponding pair of elements in row 2 of A and column 3 of B. C_{23} will be the sum of the products.**

Additional Exercises

Find C if $A = \begin{bmatrix} 4 & 2 & -5 \\ -1 & 3 & 1 \end{bmatrix}$ and

$B = \begin{bmatrix} -7 & 4 & 6 \\ -1 & -2 & 0 \end{bmatrix}$.

1. $C = A + 3B$ $\quad \begin{bmatrix} -17 & 14 & 13 \\ -4 & -3 & 1 \end{bmatrix}$

2. $C = 2A - B$ $\quad \begin{bmatrix} 15 & 0 & -16 \\ -1 & 8 & 2 \end{bmatrix}$

Perform the following matrix operations.

3. $\begin{bmatrix} -7 & -2 \\ 8 & 10 \end{bmatrix} + \begin{bmatrix} 2 & 6 \\ 3 & -3 \end{bmatrix}$ $\quad \begin{bmatrix} -5 & 4 \\ 11 & 7 \end{bmatrix}$

4. $\begin{bmatrix} 8 & -2 & -3 \\ 5 & 6 & 2 \end{bmatrix} - 2\begin{bmatrix} 4 & 1 & 8 \\ 7 & \frac{3}{2} & -6 \end{bmatrix}$

$\begin{bmatrix} 0 & -4 & -19 \\ -9 & 3 & 14 \end{bmatrix}$

In his 1853 presidential address to the American Association for the Advancement of Science, Benjamin Peirce cautioned that mathematics should not become a toy or an idol, but should be used "for the benefit of mankind and the glory of his Creator."

Example 2   Matrix Addition

Find each sum if $A = \begin{bmatrix} 3 & -2 & 1 \\ -4 & 5 & 7 \end{bmatrix}$, $B = \begin{bmatrix} -11 & 0 & -4 \\ 5 & 9 & -3 \end{bmatrix}$, and $C = \begin{bmatrix} 1 & -4 \\ 0 & 3 \end{bmatrix}$.

a. $A + B$ $\qquad\qquad$ **b.** $B + C$

Answer

a. $A + B = \begin{bmatrix} 3-11 & -2+0 & 1-4 \\ -4+5 & 5+9 & 7-3 \end{bmatrix}$

$\quad = \begin{bmatrix} -8 & -2 & -3 \\ 1 & 14 & 4 \end{bmatrix}$

b. $B + C$ is undefined since the matrices do not have the same dimensions.

SKILL ✔ EXERCISE 5

The second basic matrix operation is *scalar multiplication*. To multiply a matrix A by a real number k, multiply each entry in A by k.

DEFINITION

The **scalar product** of the real number k and the $m \times n$ matrix $A = [a_{ij}]$ is $kA = [ka_{ij}]$, the $m \times n$ matrix obtained by multiplying each entry of A by k.

An $m \times n$ matrix with every entry equal to 0 is a *zero matrix*, 0. This matrix is the *additive identity matrix* since $A + 0 = A$. Every $m \times n$ matrix A has an *additive inverse (or opposite) matrix*, designated as $-A$, where each entry is the opposite of the corresponding entry of A. Note that $-A = (-1)A$ and that $A + (-A) = -A + A = 0$.

Matrix subtraction is defined as the addition of the opposite matrix.

$$A - B = A + (-B) = [a_{ij}] + [-b_{ij}] = [a_{ij} - b_{ij}]$$

This means that matrix subtraction is defined only for matrices with the same dimensions and that corresponding entries are subtracted.

Example 3   Scalar Multiplication and Subtraction

Find $2B - A$ if $A = \begin{bmatrix} -9 & 4 \\ 24 & 16 \end{bmatrix}$ and $B = \begin{bmatrix} 11 & 2 \\ -13 & -8 \end{bmatrix}$.

Answer

$2\begin{bmatrix} 11 & 2 \\ -13 & -8 \end{bmatrix} - \begin{bmatrix} -9 & 4 \\ 24 & 16 \end{bmatrix}$

Multiply each entry in B by 2, then add the opposite of each entry in A.

$= \begin{bmatrix} 22 & 4 \\ -26 & -16 \end{bmatrix} + \begin{bmatrix} 9 & -4 \\ -24 & -16 \end{bmatrix} = \begin{bmatrix} 31 & 0 \\ -50 & -32 \end{bmatrix}$

SKILL ✔ EXERCISE 7

Because matrices are arrays of real numbers, they share some of their properties.

Properties of Matrix Addition and Scalar Multiplication	
The following properties apply to the $m \times n$ matrices A, B, and C and real scalars g, h, and k.	
Commutative Property of Matrix Addition	$A + B = B + A$
Associative Property of Matrix Addition	$(A + B) + C = A + (B + C)$
Commutative Property of Scalar Multiplication	$kA = Ak$
Associative Property of Scalar Multiplication	$(gh)A = g(hA)$
Distributive Properties	$k(A + B) = kA + kB$ and $(g + h)A = gA + hA$
Additive Identity Property	$A + 0 = A$
Additive Inverse Property	$A + (-A) = 0$
Scalar Identity Property	$1A = A$

The fact that matrix addition is commutative is easily demonstrated using the Commutative Property of Addition for real numbers.
$$A + B = [a_{ij} + b_{ij}] = [b_{ij} + a_{ij}] = B + A$$

Just as the subtraction of real numbers is not commutative, neither is the subtraction of matrices.
$$A - B \neq B - A; \text{ instead, } A - B = -(B - A)$$

The multiplication of two matrices is defined only when the number of columns in the first matrix is equal to the number of rows in the second matrix. The product of $A_{m \times n}$ and $B_{n \times p}$ is an $m \times p$ matrix. Each entry c_{ij} in the product AB is the sum of the products of the entries in row i of A and corresponding entries in column j of B.

$$A \quad \cdot \quad B \quad = \quad AB$$
$$m \times n \quad n \times p \quad m \times p$$
$$\text{equal}$$
$$\text{dimensions of } AB$$

If $A_{2 \times 2} = \begin{bmatrix} a & b \\ c & d \end{bmatrix}$ and $B_{2 \times 3} = \begin{bmatrix} u & v & w \\ x & y & z \end{bmatrix}$, then the product AB is a 2×3 matrix with the following entries.

$$c_{11} = [a \quad b]\begin{bmatrix} u \\ x \end{bmatrix} \qquad c_{12} = [a \quad b]\begin{bmatrix} v \\ y \end{bmatrix} \qquad c_{13} = [a \quad b]\begin{bmatrix} w \\ z \end{bmatrix}$$
$$= au + bx \qquad\qquad = av + by \qquad\qquad = aw + bz$$

$$c_{21} = [c \quad d]\begin{bmatrix} u \\ x \end{bmatrix} \qquad c_{22} = [c \quad d]\begin{bmatrix} v \\ y \end{bmatrix} \qquad c_{23} = [c \quad d]\begin{bmatrix} w \\ z \end{bmatrix}$$
$$= cu + dx \qquad\qquad = cv + dy \qquad\qquad = cw + dz$$

The product BA is undefined since the number of columns in $B_{2 \times 3}$ is not equal to the number of rows in $A_{2 \times 2}$.

Example 4 Multiplying Matrices

Find AB if $A = \begin{bmatrix} 2 & -3 \\ 1 & 5 \end{bmatrix}$ and $B = \begin{bmatrix} 4 & 6 & 2 \\ -3 & 1 & -5 \end{bmatrix}$.

Answer

The product of $A_{2 \times 2}$ and $B_{2 \times 3}$ is a 2×3 matrix.

$$AB = \begin{bmatrix} 2(4) + (-3)(-3) & 2(6) + (-3)1 & 2(2) + (-3)(-5) \\ 1(4) + 5(-3) & 1(6) + 5(1) & 1(2) + 5(-5) \end{bmatrix} = \begin{bmatrix} 8 + 9 & 12 - 3 & 4 + 15 \\ 4 - 15 & 6 + 5 & 2 - 25 \end{bmatrix}$$

$$= \begin{bmatrix} 17 & 9 & 19 \\ -11 & 11 & -23 \end{bmatrix}$$

SKILL ✓ EXERCISE 13

State the dimension of each matrix product, then find the product.

5. $[5 \quad -9 \quad 4]\begin{bmatrix} 3 & 4 \\ 10 & 0 \\ -2 & 5 \end{bmatrix}$

$1 \times 2; [-83 \quad 40]$

6. $\begin{bmatrix} -3 & 2 & 20 & -10 \\ 0 & 5 & 2 & 3 \end{bmatrix}\begin{bmatrix} 2 \\ -4 \\ -1 \\ 4 \end{bmatrix}$

$2 \times 1; \begin{bmatrix} -74 \\ -10 \end{bmatrix}$

7. $\begin{bmatrix} 5 & 2 & -4 \\ 6 & -1 & 2 \end{bmatrix}\begin{bmatrix} -1 & 4 & 1 \\ 2 & 3 & -2 \\ -5 & 4 & 1 \end{bmatrix}$

$2 \times 3; \begin{bmatrix} 19 & 10 & -3 \\ -18 & 29 & 10 \end{bmatrix}$

Evaluate each matrix, given
$A = \begin{bmatrix} 7 & 5 \\ -4 & -2 \end{bmatrix}$ and $B = \begin{bmatrix} -5 & -5 \\ -3 & 0 \end{bmatrix}$.

8. BA $\begin{bmatrix} -15 & -15 \\ -21 & -15 \end{bmatrix}$

9. A^2 $\begin{bmatrix} 29 & 25 \\ -20 & -16 \end{bmatrix}$

Multiply.

10. $\begin{bmatrix} -4 & 1 & 4 \\ 0 & -2 & -1 \\ -2 & 3 & 2 \end{bmatrix}\begin{bmatrix} -1 & -3 & 2 \\ -1 & 0 & 4 \\ 3 & 3 & -2 \end{bmatrix}$

$\begin{bmatrix} 15 & 24 & -12 \\ -1 & -3 & -6 \\ 5 & 12 & 4 \end{bmatrix}$

Assignments

- **Minimum:** 1, 3–6, 9–10, 12–13, 17–21, 24; 27–31 odd; 34, 42, 44–45, 47–49

- **Standard:** 2–4, 6, 10, 13–14, 16, 19–23, 27, 29, 31, 34, 36, 38, 42; 45–51 odd

- **Extended:** 1, 4, 7, 10, 13–16, 19–22, 24, 26, 30–35, 37–39; 41–49 odd; 50–51

Assessment

- Quiz 7A covers Sections 7.1–7.2.

Illustrate how the matrix equation $AX = C$ can be used to represent a system of linear equations in standard form. Students will use these *coefficient*, *variable*, and *constant matrices* when learning three different approaches to solving linear systems in Sections 7.3–7.5.

Example 6 provides a real-world application of matrix multiplication. Be sure the students fully understand the meaning of the values of a resulting matrix product in a real-world application.

Motivational Idea Explain that businesses frequently use rectangular arrays of numbers within spreadsheets. A single Excel worksheet has 1,048,576 (2^{20}) rows and 16,384 (2^{14}) columns. That is one big matrix with 17,179,869,184 (2^{34}) possible entries.

Note the formal definition of matrix multiplication.

> ### DEFINITION
>
> The **matrix product** AB of the $m \times n$ matrix A and the $n \times p$ matrix B is the $m \times p$ matrix $AB = [c_{ij}]$ where $c_{ij} = a_{i1}b_{1j} + a_{i2}b_{2j} + \cdots + a_{in}b_{nj}$.

Example 5 Multiplying Matrices

If $A = \begin{bmatrix} 2 & 3 \\ -1 & 4 \\ 5 & -2 \end{bmatrix}$ and $B = \begin{bmatrix} -3 & 1 & 3 \\ 6 & -2 & -1 \end{bmatrix}$, state the dimensions and find each product.

a. AB

b. BA

Answer

a. $A_{3 \times 2} \cdot B_{2 \times 3}$ is a 3×3 matrix.

$$AB = \begin{bmatrix} 2 & 3 \\ -1 & 4 \\ 5 & -2 \end{bmatrix}\begin{bmatrix} -3 & 1 & 3 \\ 6 & -2 & -1 \end{bmatrix}$$

$$= \begin{bmatrix} -6+18 & 2+(-6) & 6+(-3) \\ 3+24 & -1+(-8) & -3+(-4) \\ -15+(-12) & 5+4 & 15+2 \end{bmatrix}$$

$$= \begin{bmatrix} 12 & -4 & 3 \\ 27 & -9 & -7 \\ -27 & 9 & 17 \end{bmatrix}$$

b. $B_{2 \times 3} \cdot A_{3 \times 2}$ is a 2×2 matrix.

$$BA = \begin{bmatrix} -3 & 1 & 3 \\ 6 & -2 & -1 \end{bmatrix}\begin{bmatrix} 2 & 3 \\ -1 & 4 \\ 5 & -2 \end{bmatrix}$$

$$= \begin{bmatrix} -6-1+15 & -9+4-6 \\ 12+2-5 & 18-8+2 \end{bmatrix}$$

$$= \begin{bmatrix} 8 & -11 \\ 9 & 12 \end{bmatrix}$$

SKILL ✓ **EXERCISE 15**

Noting that $AB \neq BA$ in Example 5 indicates that multiplication of matrices is not commutative. In fact, the dimensions of AB and BA are not the same.

While every matrix has an additive identity matrix *0*, only square matrices have a *multiplicative identity matrix*, *I*, such that $AI = IA = A$. The multiplicative identity matrix of order n, I_n, is an $n \times n$ matrix with entries on the *main diagonal* (entries e_{ij} such that $i = j$) equal to 1 and all other entries equal to 0. Exercises 32–33 further investigate multiplicative identity matrices I_n. Section 7.5 discusses the *multiplicative inverse matrix*.

While matrix multiplication is generally not commutative, some of the other properties of real numbers do extend to matrix multiplication.

$$I_n = \begin{bmatrix} 1 & 0 & 0 & \cdots & 0 \\ 0 & 1 & 0 & \cdots & 0 \\ 0 & 0 & 1 & \cdots & 0 \\ \vdots & \vdots & \vdots & & \vdots \\ 0 & 0 & 0 & \cdots & 1 \end{bmatrix}$$

Properties of Matrix Multiplication	
Given matrices A, B, and C and scalar k; these properties apply if the products are defined.	
Associative Property of Matrix Multiplication	$(AB)C = A(BC)$
Associative Property of Scalar Multiplication	$k(AB) = (kA)B = A(kB)$
Left Distributive Property	$A(B + C) = AB + AC$
Right Distributive Property	$(B + C)A = BA + CA$
Multiplicative Identity for a Square Matrix of Order n	$AI_n = I_nA = A$

Because matrix multiplication is not commutative, $A(B + C) \neq (B + C)A$ and $AB + AC \neq BA + CA$.

Matrices provide several powerful techniques for solving systems of linear equations that will be explored in Sections 7.3–7.5. One of these methods involves writing the system as a matrix equation in the form of $AX = C$ where A is the *coefficient matrix* representing the coefficients when the equations are written in standard form, X is a column matrix containing the variables, and C is a column matrix containing the constants (the *constant matrix*). Notice how matrix multiplication is used to create a matrix equation that is equivalent to the following system of equations.

System of Equations in General Form

$$2x - 3y - 4z = -4$$
$$5x + 4y - z = -7$$
$$-8x + 0y + 3z = 0$$

Matrix Equation: $A \cdot X = C$

$$\begin{bmatrix} 2 & -3 & -4 \\ 5 & 4 & -1 \\ -8 & 0 & 3 \end{bmatrix} \begin{bmatrix} x \\ y \\ z \end{bmatrix} = \begin{bmatrix} -4 \\ -7 \\ 0 \end{bmatrix}$$

Example 6   Applying Matrices to Real-World Data

A high school basketball coach records individual statistics on a spreadsheet. The illustrated portion shows the shots made by his top 3 players during the first 10 games of the season. Write matrix S representing each player's made shots and matrix P representing the point value for each type of shot. Then write and evaluate a matrix expression representing the total points for each player.

	A	B	C	D
1			**Shots Made**	
2		2-Pointers	3-Pointers	Free Throws
3	Freeman	80	21	19
4	Gonzalez	59	15	12
5	Rodgers	55	5	9
6			**Point Value**	
7		2	3	1

Answer

$$S = \begin{bmatrix} 80 & 21 & 19 \\ 59 & 15 & 12 \\ 55 & 5 & 9 \end{bmatrix}; P = \begin{bmatrix} 2 \\ 3 \\ 1 \end{bmatrix}$$

$$SP = \begin{bmatrix} 80 & 21 & 19 \\ 59 & 15 & 12 \\ 55 & 5 & 9 \end{bmatrix} \begin{bmatrix} 2 \\ 3 \\ 1 \end{bmatrix} = \begin{bmatrix} 160 + 63 + 19 \\ 118 + 45 + 12 \\ 110 + 15 + 9 \end{bmatrix} = \begin{bmatrix} 242 \\ 175 \\ 134 \end{bmatrix} \begin{matrix} \text{Freeman} \\ \text{Gonzalez} \\ \text{Rodgers} \end{matrix}$$

SKILL ✓ **EXERCISE 35**

❯ A. Exercises

1. Given $A = \begin{bmatrix} 12 & 4 & -3 \\ 7 & 0 & 2 \\ 19 & -1 & 6 \\ 5 & -9 & 7 \end{bmatrix}$, state the following.

  a. dimensions of A 4×3

  b. value of a_{21} 7

  c. value of a_{12} 4

  d. element whose value is 2 a_{23}

  e. element whose value is -1 a_{32}

2. Given $B = \begin{bmatrix} 14 & 2 & 6 & -1 & 19 \\ 0 & -3 & 5 & 2 & 7 \\ 1 & 9 & 16 & -11 & 3 \end{bmatrix}$, state the following.

  a. dimensions of B 3×5

  b. value of b_{24} 2

  c. value of b_{31} 1

  d. element whose value is 0 b_{21}

  e. element whose value is 6 b_{13}

14. $[12 - 6 - 35 \quad 36 + 4 + 14]$
$$= [-29 \quad 54]$$

15. $\begin{bmatrix} 6 + 10 - 6 - 6 \\ 4 + 10 - 12 - 4 \end{bmatrix} = \begin{bmatrix} 4 \\ -2 \end{bmatrix}$

18. $\begin{bmatrix} 14 & 37 & 79 \\ 9 & 15 & 26 \\ 6 & 10 & 13 \\ 8 & 11 & 9 \\ 1 & 4 & 7 \\ 3 & 12 & 18 \end{bmatrix} + \begin{bmatrix} 14 & 31 & 79 \\ 9 & 9 & 22 \\ 6 & 5 & 9 \\ 8 & 14 & 16 \\ 1 & 5 & 7 \\ 3 & 4 & 17 \end{bmatrix}$

$$= \begin{bmatrix} 28 & 68 & 158 \\ 18 & 24 & 48 \\ 12 & 15 & 22 \\ 16 & 25 & 25 \\ 2 & 9 & 14 \\ 6 & 16 & 35 \end{bmatrix}$$

❯ B. Exercises

19. $[2(4) + (-6)(-3) + 3(8)]$
$$= [8 + 18 + 24] = [50]$$

21. $\begin{bmatrix} 4 & -2 \\ 7 & 1 \end{bmatrix} \begin{bmatrix} 5 & 1 \\ -4 & -1 \end{bmatrix}$
$$= \begin{bmatrix} 20 + 8 & 4 + 2 \\ 35 - 4 & 7 - 1 \end{bmatrix} = \begin{bmatrix} 28 & 6 \\ 31 & 6 \end{bmatrix}$$

22. $\begin{bmatrix} 5 & 1 \\ -4 & -1 \end{bmatrix} \begin{bmatrix} 4 & -2 \\ 7 & 1 \end{bmatrix}$
$$= \begin{bmatrix} 20 + 7 & -10 + 1 \\ -16 - 7 & 8 - 1 \end{bmatrix}$$
$$= \begin{bmatrix} 27 & -9 \\ -23 & 7 \end{bmatrix}$$

23. $\begin{bmatrix} 4 & -2 \\ 7 & 1 \end{bmatrix} \begin{bmatrix} 4 & -2 \\ 7 & 1 \end{bmatrix}$
$$= \begin{bmatrix} 16 - 14 & -8 - 2 \\ 28 + 7 & -14 + 1 \end{bmatrix} = \begin{bmatrix} 2 & -10 \\ 35 & -13 \end{bmatrix}$$

24. $\begin{bmatrix} 5 & 1 \\ -4 & -1 \end{bmatrix} \begin{bmatrix} 5 & 1 \\ -4 & -1 \end{bmatrix}$
$$= \begin{bmatrix} 25 - 4 & 5 - 1 \\ -20 + 4 & -4 + 1 \end{bmatrix} = \begin{bmatrix} 21 & 4 \\ -16 & -3 \end{bmatrix}$$

Solutions

❯ A. Exercises

4. $\begin{bmatrix} 13 & -6 & -2 \\ -7 & 3 & 12 \end{bmatrix}$

5. $\begin{bmatrix} -5 & -8 & 4 \\ 3 & 7 & 4 \end{bmatrix}$

6. $\begin{bmatrix} 12 & -21 & 3 \\ -6 & 15 & 24 \end{bmatrix} + \begin{bmatrix} -18 & -2 & 6 \\ 10 & 4 & -8 \end{bmatrix}$
$$= \begin{bmatrix} -6 & -23 & 9 \\ 4 & 19 & 16 \end{bmatrix}$$

7. $\begin{bmatrix} 8 & -14 & 2 \\ -4 & 10 & 16 \end{bmatrix} + \begin{bmatrix} 27 & 3 & -9 \\ -15 & -6 & 12 \end{bmatrix}$
$$= \begin{bmatrix} 35 & -11 & -7 \\ -19 & 4 & 28 \end{bmatrix}$$

10. $\begin{bmatrix} -16 \\ 14 \\ 6 \end{bmatrix} + \begin{bmatrix} 8 \\ -24 \\ 28 \end{bmatrix} = \begin{bmatrix} -8 \\ -10 \\ 34 \end{bmatrix}$

11. $\begin{bmatrix} 13 & -9 & 14 \\ 8 & 7 & -5 \end{bmatrix} + \begin{bmatrix} -9 & -21 & 18 \\ -12 & -1 & -30 \end{bmatrix}$
$$= \begin{bmatrix} 4 & -30 & 32 \\ -4 & 6 & -35 \end{bmatrix}$$

12. $\begin{bmatrix} -6 - 20 & 9 - 35 \\ -12 + 16 & 18 + 28 \end{bmatrix}$
$$= \begin{bmatrix} -26 & -26 \\ 4 & 46 \end{bmatrix}$$

13. $\begin{bmatrix} -9 + 0 + 4 & 6 - 2 + 16 & 18 + 4 + 8 \\ -15 + 0 + 2 & 10 + 0 + 8 & 30 + 0 + 4 \end{bmatrix}$
$$= \begin{bmatrix} -5 & 20 & 30 \\ -13 & 18 & 34 \end{bmatrix}$$

25. $\begin{bmatrix} -2+3+10 & 0+12+0 & 3+6-2 \\ -4-1+0 & 0-4+0 & 6-2+0 \\ -8+0+5 & 0+0+0 & 12+0-1 \end{bmatrix}$

$= \begin{bmatrix} 11 & 12 & 7 \\ -5 & -4 & 4 \\ -3 & 0 & 11 \end{bmatrix}$

26. $\begin{bmatrix} 0+0+2 & 0-4+1 & 0-8+6 \\ 16+0+12 & 20+2+6 & 12+4+36 \\ 12+0-4 & 15-10-2 & 9-20-12 \end{bmatrix}$

$= \begin{bmatrix} 2 & -3 & -2 \\ 28 & 28 & 52 \\ 8 & 3 & -23 \end{bmatrix}$

27. $\begin{bmatrix} 0+4+0 & 0+2-6 & 0+12-10 \\ -4+14+0 & 0+7-3 & 1+42-5 \\ 20+2+0 & 0+1+9 & -5+6+15 \end{bmatrix}$

$= \begin{bmatrix} 4 & -4 & 2 \\ 10 & 4 & 38 \\ 22 & 10 & 16 \end{bmatrix}$

28. $\begin{bmatrix} ad+0+0 & 0+0+0 & 0+0+0 \\ 0+0+0 & 0+be+0 & 0+0+0 \\ 0+0+0 & 0+0+0 & 0+cf+0 \end{bmatrix}$

$= \begin{bmatrix} ad & 0 & 0 \\ 0 & be & 0 \\ 0 & 0 & cf \end{bmatrix}$

31a. $\begin{bmatrix} 1 & 0 \\ 0 & 1 \end{bmatrix}$ **31b.** $\begin{bmatrix} 1 & 0 & 0 \\ 0 & 1 & 0 \\ 0 & 0 & 1 \end{bmatrix}$

32a. $AI = \begin{bmatrix} -2 & 5 \\ 9 & 6 \end{bmatrix}\begin{bmatrix} 1 & 0 \\ 0 & 1 \end{bmatrix}$

$= \begin{bmatrix} -2+0 & 0+5 \\ 9+0 & 0+6 \end{bmatrix} = \begin{bmatrix} -2 & 5 \\ 9 & 6 \end{bmatrix} = A$

$IA = \begin{bmatrix} 1 & 0 \\ 0 & 1 \end{bmatrix}\begin{bmatrix} -2 & 5 \\ 9 & 6 \end{bmatrix}$

$= \begin{bmatrix} -2+0 & 5+0 \\ 0+9 & 0+6 \end{bmatrix} = \begin{bmatrix} -2 & 5 \\ 9 & 6 \end{bmatrix} = A$

32b. $AI = \begin{bmatrix} a & b & c \\ r & s & t \\ u & v & w \end{bmatrix}\begin{bmatrix} 1 & 0 & 0 \\ 0 & 1 & 0 \\ 0 & 0 & 1 \end{bmatrix}$

$= \begin{bmatrix} a+0+0 & 0+b+0 & 0+0+c \\ r+0+0 & 0+s+0 & 0+0+t \\ u+0+0 & 0+v+0 & 0+0+w \end{bmatrix}$

$= \begin{bmatrix} a & b & c \\ r & s & t \\ u & v & w \end{bmatrix} = A$

$IA = \begin{bmatrix} 1 & 0 & 0 \\ 0 & 1 & 0 \\ 0 & 0 & 1 \end{bmatrix}\begin{bmatrix} a & b & c \\ r & s & t \\ u & v & w \end{bmatrix}$

$= \begin{bmatrix} a+0+0 & b+0+0 & c+0+0 \\ 0+r+0 & 0+s+0 & 0+t+0 \\ 0+0+u & 0+0+v & 0+0+w \end{bmatrix}$

$= \begin{bmatrix} a & b & c \\ r & s & t \\ u & v & w \end{bmatrix} = A$

3. Let $D = \begin{bmatrix} 5 & -1 & 9 \\ 2 & 0 & -4 \end{bmatrix}$, $E = \begin{bmatrix} 0 & 1 \\ 1 & 0 \end{bmatrix}$,

and $F = \begin{bmatrix} 1 & 3 \\ -7 & 5 \\ -2 & 1 \end{bmatrix}$.

  a. Which matrix is a 2×3 matrix? D

  b. Which matrix is a square matrix? E

  c. Which two matrices could be added? none

Find D if $A = \begin{bmatrix} 4 & -7 & 1 \\ -2 & 5 & 8 \end{bmatrix}$ and $B = \begin{bmatrix} -9 & -1 & 3 \\ 5 & 2 & -4 \end{bmatrix}$.

4. $D = A - B$ **5.** $D = A + B$

6. $D = 3A + 2B$ **7.** $D = 2A - 3B$

Perform the following matrix operations.

8. $\begin{bmatrix} 4 & -7 & 1 \\ -2 & 5 & 8 \end{bmatrix} + \begin{bmatrix} -9 & -1 & 3 \\ 5 & 2 & -4 \end{bmatrix}$ $\begin{bmatrix} -5 & -8 & 4 \\ 3 & 7 & 4 \end{bmatrix}$

9. $\begin{bmatrix} 3 & -6 \\ 8 & -1 \end{bmatrix} - \begin{bmatrix} -4 & 9 \\ 6 & -13 \end{bmatrix}$ $\begin{bmatrix} 7 & -15 \\ 2 & 12 \end{bmatrix}$

10. $-2\begin{bmatrix} 8 \\ -7 \\ -3 \end{bmatrix} + 4\begin{bmatrix} 2 \\ -6 \\ 7 \end{bmatrix}$ $\begin{bmatrix} -8 \\ -10 \\ 34 \end{bmatrix}$

11. $\begin{bmatrix} 13 & -9 & 14 \\ 8 & 7 & -5 \end{bmatrix} - 3\begin{bmatrix} 3 & 7 & -6 \\ 4 & \frac{1}{3} & 10 \end{bmatrix}$ $\begin{bmatrix} 4 & -30 & 32 \\ -4 & 6 & -35 \end{bmatrix}$

State the dimensions of each matrix product, then find the product.

12. $\begin{bmatrix} 3 & -5 \\ 6 & 4 \end{bmatrix}\begin{bmatrix} -2 & 3 \\ 4 & 7 \end{bmatrix}$ $2 \times 2;\ \begin{bmatrix} -26 & -26 \\ 4 & 46 \end{bmatrix}$

13. $\begin{bmatrix} 3 & -1 & 4 \\ 5 & 0 & 2 \end{bmatrix}\begin{bmatrix} -3 & 2 & 6 \\ 0 & 2 & -4 \\ 1 & 4 & 2 \end{bmatrix}$ $2 \times 3;\ \begin{bmatrix} -5 & 20 & 30 \\ -13 & 18 & 34 \end{bmatrix}$

14. $[6 \ \ -2 \ \ 7]\begin{bmatrix} 2 & 6 \\ 3 & -2 \\ -5 & 2 \end{bmatrix}$ $1 \times 2;\ [-29 \ \ 54]$

15. $\begin{bmatrix} 3 & 2 & -1 & 6 \\ 2 & 2 & -2 & 4 \end{bmatrix}\begin{bmatrix} 2 \\ 5 \\ 6 \\ -1 \end{bmatrix}$ $2 \times 1;\ \begin{bmatrix} 4 \\ -2 \end{bmatrix}$

Write each linear system of equations as a matrix equation in the form $AX = C$.

16. $2x + 3y - 5z = 0$
$\quad -x + 2y + z = 4$
$\quad 6x - y = -7$

17. $-3x + y + 2z = 1$
$\quad y = x + 2z$
$\quad 3z = -6$

16. $\begin{bmatrix} 2 & 3 & -5 \\ -1 & 2 & 1 \\ 6 & -1 & 0 \end{bmatrix}\begin{bmatrix} x \\ y \\ z \end{bmatrix} = \begin{bmatrix} 0 \\ 4 \\ -7 \end{bmatrix}$

17. $\begin{bmatrix} -3 & 1 & 2 \\ -1 & 1 & -2 \\ 0 & 0 & 3 \end{bmatrix}\begin{bmatrix} x \\ y \\ z \end{bmatrix} = \begin{bmatrix} 1 \\ 0 \\ -6 \end{bmatrix}$

18. A class tracked the cookie sales of its two competing groups in a spreadsheet. Write matrix A representing sales from Group A and matrix B representing sales from Group B. Then write and evaluate a matrix expression representing the total boxes of each type of cookie sold by the class on each day.

	A	B	C	D	E	F	G
1		Group A			Group B		
2		Thur	Fri	Sat	Thur	Fri	Sat
3	Mint	14	37	79	14	31	79
4	Caramel	9	15	26	9	9	22
5	Peanut Butter	6	10	13	6	5	9
6	Toffee	8	11	9	8	14	16
7	Shortbread	1	4	7	1	5	7
8	S'more	3	12	18	3	4	17

❯ B. Exercises

Given $A = \begin{bmatrix} 4 & -2 \\ 7 & 1 \end{bmatrix}$, $B = \begin{bmatrix} 5 & 1 \\ -4 & -1 \end{bmatrix}$, $C = [2 \ \ -6 \ \ 3]$,

and $D = \begin{bmatrix} 4 \\ -3 \\ 8 \end{bmatrix}$, evaluate each matrix expression if possible. If not possible, explain why.

19. CD $[50]$ **20.** DB not possible; the number of columns in $D \neq$ the number of rows in B

21. AB $\begin{bmatrix} 28 & 6 \\ 31 & 6 \end{bmatrix}$ **22.** BA $\begin{bmatrix} 27 & -9 \\ -23 & 7 \end{bmatrix}$

23. A^2 $\begin{bmatrix} 2 & -10 \\ 35 & -13 \end{bmatrix}$ **24.** B^2 $\begin{bmatrix} 21 & 4 \\ -16 & -3 \end{bmatrix}$

Multiply.

25. $\begin{bmatrix} 1 & 3 & 2 \\ 2 & -1 & 0 \\ 4 & 0 & 1 \end{bmatrix}\begin{bmatrix} -2 & 0 & 3 \\ 1 & 4 & 2 \\ 5 & 0 & -1 \end{bmatrix}$ $\begin{bmatrix} 11 & 12 & 7 \\ -5 & -4 & 4 \\ -3 & 0 & 11 \end{bmatrix}$

26. $\begin{bmatrix} 0 & 2 & 1 \\ 4 & -1 & 6 \\ 3 & 5 & -2 \end{bmatrix}\begin{bmatrix} 4 & 5 & 3 \\ 0 & -2 & -4 \\ 2 & 1 & 6 \end{bmatrix}$ $\begin{bmatrix} 2 & -3 & -2 \\ 28 & 28 & 52 \\ 8 & 3 & -23 \end{bmatrix}$

27. $\begin{bmatrix} 0 & 2 & -2 \\ -1 & 7 & -1 \\ 5 & 1 & 3 \end{bmatrix}\begin{bmatrix} 4 & 0 & -1 \\ 2 & 1 & 6 \\ 0 & 3 & 5 \end{bmatrix}$ $\begin{bmatrix} 4 & -4 & 2 \\ 10 & 4 & 38 \\ 22 & 10 & 16 \end{bmatrix}$

28. $\begin{bmatrix} a & 0 & 0 \\ 0 & b & 0 \\ 0 & 0 & c \end{bmatrix}\begin{bmatrix} d & 0 & 0 \\ 0 & e & 0 \\ 0 & 0 & f \end{bmatrix}$ $\begin{bmatrix} ad & 0 & 0 \\ 0 & be & 0 \\ 0 & 0 & cf \end{bmatrix}$

Use technology to find the following matrix products.

29. $\begin{bmatrix} 47 & 68 \\ -25 & 39 \end{bmatrix}\begin{bmatrix} 22 & -31 \\ 75 & 27 \end{bmatrix}$ $\begin{bmatrix} 6134 & 379 \\ 2375 & 1828 \end{bmatrix}$

30. $\begin{bmatrix} 239 & 332 \\ 112 & 159 \end{bmatrix}\begin{bmatrix} 99 & 51 \\ 187 & 307 \end{bmatrix}$ $\begin{bmatrix} 85{,}745 & 114{,}113 \\ 40{,}821 & 54{,}525 \end{bmatrix}$

33. $\begin{bmatrix} 3 & 3 \\ 2 & 2 \end{bmatrix}\begin{bmatrix} -1 & 1 \\ 1 & -1 \end{bmatrix}$

$= \begin{bmatrix} -3+3 & 3-3 \\ -2+2 & 2-2 \end{bmatrix} = \begin{bmatrix} 0 & 0 \\ 0 & 0 \end{bmatrix}$

Zero Product Property

34. $SP = \begin{bmatrix} 29 & 11 & 19 & 1 & 0 \\ 22 & 7 & 14 & 0 & 0 \\ 37 & 9 & 26 & 1 & 1 \end{bmatrix}\begin{bmatrix} 6 \\ 3 \\ 1 \\ 2 \\ 2 \end{bmatrix}$

$= \begin{bmatrix} 174+33+19+2+0 \\ 132+21+14+0+0 \\ 222+27+26+2+2 \end{bmatrix}$

$= \begin{bmatrix} 228 \\ 167 \\ 279 \end{bmatrix}$

35. $C = [120 \ \ 28 \ \ 85]$

$S = [159 \ \ 35 \ \ 112]$

$N = \begin{bmatrix} 15 \\ 29 \\ 11 \end{bmatrix}$

$R = SN = [159 \ \ 35 \ \ 112]\begin{bmatrix} 15 \\ 29 \\ 11 \end{bmatrix}$

$= [2385 + 1015 + 1232] = [4632]$

$P = R - CN$

$= [4632] - [120 \ \ 28 \ \ 85]\begin{bmatrix} 15 \\ 29 \\ 11 \end{bmatrix}$

$= [4632] - [1800 + 812 + 935]$

$= [4632] - [3547] = [1085]$

31. State the identity matrix for each order n.

   a. I_2 **b.** I_3

32. For each matrix, show that $AI = A$ and $IA = A$.

   a. $A = \begin{bmatrix} -2 & 5 \\ 9 & 6 \end{bmatrix}$ **b.** $A = \begin{bmatrix} a & b & c \\ r & s & t \\ u & v & w \end{bmatrix}$

33. Find the product AB if $A = \begin{bmatrix} 3 & 3 \\ 2 & 2 \end{bmatrix}$ and $B = \begin{bmatrix} -1 & 1 \\ 1 & -1 \end{bmatrix}$. Which property of real numbers is shown not to extend to matrix multiplication by this counterexample? $\begin{bmatrix} 0 & 0 \\ 0 & 0 \end{bmatrix}$; Zero Product Property

34. The spreadsheet shows seasonal data for three high school football teams. Write matrices S and P representing the scoring plays and the point values, respectively. Then write and evaluate a matrix expression representing the total points scored by each team.

▲	A	B	C	D	E	F
1		TD	FG	PAT	2PT-Conv	Safety
2	Defenders	29	11	19	1	0
3	Eagles	22	7	14	0	0
4	Minutemen	37	9	26	1	1
5				**Point Values**		
6		6	3	1	2	2

35. Hunter purchases three types of smart home devices at wholesale prices and sells them at retail for a profit as shown on the spreadsheet. Write 1×3 matrices C and S representing the devices' wholesale costs and retail prices, respectively, and a 3×1 matrix N representing the number of each device sold. Then write and evaluate matrix expressions representing his total revenue R and total profit P. $R = [4632]; P = [1085]$

▲	A	B	C	D
1			**Smart Home Devices**	
2		Wholesale Cost	Retail Price	Number Sold
3	Smart Entrance	$120	$159	15
4	Leak Detector	$28	$35	29
5	Garage Door Tender	$85	$112	11

36. Use $k = 3$, $A = \begin{bmatrix} 8 & -6 \\ 1 & 7 \end{bmatrix}$, and $B = \begin{bmatrix} -5 & 4 \\ 3 & 11 \end{bmatrix}$ to illustrate the Distributive Property of Scalar Multiplication over Matrix Addition.

37. Use $A = \begin{bmatrix} 1 & -3 \\ 4 & 2 \end{bmatrix}$, $B = \begin{bmatrix} 2 & 3 & 2 \\ 1 & 0 & 4 \end{bmatrix}$, and $C = \begin{bmatrix} 5 \\ -2 \\ 6 \end{bmatrix}$ to illustrate the Associative Property of Matrix Multiplication.

38. Write the linear system as a matrix equation in the form of $AX = C$. Then use $X = \begin{bmatrix} 6 \\ 3 \\ -2 \end{bmatrix}$ to show that $(6, 3, -2)$ is a solution to the system.
$$3x - y + 4z = 7$$
$$2x - y + 6z = -3$$
$$-x + 2y - 3z = 6$$

39. **Analyze:** Steven claims that $(A + B)(A - B) = A^2 - B^2$ since $(a + b)(a - b) = a^2 - b^2$ is true for real numbers.

   a. Use $A = \begin{bmatrix} 1 & 2 \\ 3 & -1 \end{bmatrix}$ and $B = \begin{bmatrix} 0 & 1 \\ 2 & -1 \end{bmatrix}$ to show that $(A + B)(A - B) \neq A^2 - B^2$.

   b. Expand $(A + B)(A - B)$ and explain why $(A + B)(A - B) \neq A^2 - B^2$.

40. Let $A = \begin{bmatrix} 0 & i \\ i & 0 \end{bmatrix}$.

   a. Find A^2, A^3, and A^4.

   b. Use the results above to find A^5, A^6, A^7, and A^8.

41. Matrix multiplication can be used to rotate figures in the Cartesian plane. Represent the figure with n vertices in an $n \times 2$ matrix where each row contains the coordinates of a vertex. Then right-multiply by the rotation matrix below to produce a matrix containing the coordinates of the image after a clockwise rotation by the angle α. Use R_α to find the vertices of each rotation of $\triangle ABC$. Then graph each image.

$$R_\alpha = \begin{bmatrix} \cos \alpha & -\sin \alpha \\ \sin \alpha & \cos \alpha \end{bmatrix}$$

   a. $\triangle DEF$, a $90°$ rotation

   b. $\triangle JKL$, a $180°$ rotation

   c. $\triangle PQR$, a $270°$ rotation

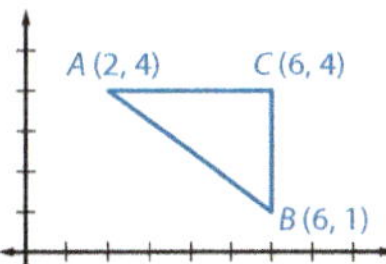

39b. $(A + B)(A - B)$
$$= (A + B)A - (A + B)B$$
$$= A^2 + BA - AB - B^2$$
But $AB \neq BA$, so the middle terms do not sum to 0.

40a. $A^2 = \begin{bmatrix} 0 & i \\ i & 0 \end{bmatrix}\begin{bmatrix} 0 & i \\ i & 0 \end{bmatrix}$
$$= \begin{bmatrix} 0 + i^2 & 0 + 0 \\ 0 + 0 & i^2 + 0 \end{bmatrix} = \begin{bmatrix} -1 & 0 \\ 0 & -1 \end{bmatrix}$$
$$A^3 = A^2 A = \begin{bmatrix} -1 & 0 \\ 0 & -1 \end{bmatrix}\begin{bmatrix} 0 & i \\ i & 0 \end{bmatrix}$$
$$= \begin{bmatrix} 0 + 0 & -i + 0 \\ 0 - i & 0 + 0 \end{bmatrix} = \begin{bmatrix} 0 & -i \\ -i & 0 \end{bmatrix}$$
$$A^4 = A^2 A^2 = \begin{bmatrix} -1 & 0 \\ 0 & -1 \end{bmatrix}\begin{bmatrix} -1 & 0 \\ 0 & -1 \end{bmatrix}$$
$$= \begin{bmatrix} 1 + 0 & 0 + 0 \\ 0 + 0 & 0 + 1 \end{bmatrix} = \begin{bmatrix} 1 & 0 \\ 0 & 1 \end{bmatrix}$$

40b. using $A^4 = I$:
$$A^5 = AA^4 = AI = A$$
$$A^6 = A^2 A^4 = A^2 I = A^2$$
$$A^7 = A^3 A^4 = A^3 I = A^3$$
$$A^8 = A^4 A^4 = I^2 = I$$

41. $T = \begin{bmatrix} 2 & 4 \\ 6 & 1 \\ 6 & 4 \end{bmatrix}$

41a. $T \begin{bmatrix} \cos 90° & -\sin 90° \\ \sin 90° & \cos 90° \end{bmatrix}$
$$= \begin{bmatrix} 2 & 4 \\ 6 & 1 \\ 6 & 4 \end{bmatrix}\begin{bmatrix} 0 & -1 \\ 1 & 0 \end{bmatrix} = \begin{bmatrix} 4 & -2 \\ 1 & -6 \\ 4 & -6 \end{bmatrix}$$

41b. $T \begin{bmatrix} \cos 180° & -\sin 180° \\ \sin 180° & \cos 180° \end{bmatrix}$
$$= \begin{bmatrix} 2 & 4 \\ 6 & 1 \\ 6 & 4 \end{bmatrix}\begin{bmatrix} -1 & 0 \\ 0 & -1 \end{bmatrix} = \begin{bmatrix} -2 & -4 \\ -6 & -1 \\ -6 & -4 \end{bmatrix}$$

41c. $T \begin{bmatrix} \cos 270° & -\sin 270° \\ \sin 270° & \cos 270° \end{bmatrix}$
$$= \begin{bmatrix} 2 & 4 \\ 6 & 1 \\ 6 & 4 \end{bmatrix}\begin{bmatrix} 0 & 1 \\ -1 & 0 \end{bmatrix} = \begin{bmatrix} -4 & 2 \\ -1 & 6 \\ -4 & 6 \end{bmatrix}$$

36. $k(A + B) = 3\begin{bmatrix} 3 & -2 \\ 4 & 18 \end{bmatrix} = \begin{bmatrix} 9 & -6 \\ 12 & 54 \end{bmatrix}$
$$kA + kB = \begin{bmatrix} 24 & -18 \\ 3 & 21 \end{bmatrix} + \begin{bmatrix} -15 & 12 \\ 9 & 33 \end{bmatrix}$$
$$= \begin{bmatrix} 9 & -6 \\ 12 & 54 \end{bmatrix}$$
$$\therefore k(A + B) = kA + kB$$

37. $(AB)C = \begin{bmatrix} -1 & 3 & -10 \\ 10 & 12 & 16 \end{bmatrix}\begin{bmatrix} 5 \\ -2 \\ 6 \end{bmatrix}$
$$= \begin{bmatrix} -71 \\ 122 \end{bmatrix}$$
$$A(BC) = \begin{bmatrix} 1 & -3 \\ 4 & 2 \end{bmatrix}\begin{bmatrix} 16 \\ 29 \end{bmatrix} = \begin{bmatrix} -71 \\ 122 \end{bmatrix}$$
$$\therefore (AB)C = A(BC)$$

38. $\begin{bmatrix} 3 & -1 & 4 \\ 2 & -1 & 6 \\ 1 & 2 & -3 \end{bmatrix}\begin{bmatrix} x \\ y \\ z \end{bmatrix} = \begin{bmatrix} 7 \\ -3 \\ 6 \end{bmatrix}$
$$\begin{bmatrix} 3 & -1 & 4 \\ 2 & -1 & 6 \\ 1 & 2 & -3 \end{bmatrix}\begin{bmatrix} 6 \\ 3 \\ -2 \end{bmatrix}$$
$$= \begin{bmatrix} 18 - 3 - 8 \\ 12 + 3 - 12 \\ 6 - 6 + 6 \end{bmatrix} = \begin{bmatrix} 7 \\ -3 \\ 6 \end{bmatrix}$$

39a. $A^2 = \begin{bmatrix} 7 & 0 \\ 0 & 7 \end{bmatrix}$; $B^2 = \begin{bmatrix} 2 & -1 \\ -2 & 3 \end{bmatrix}$
$$A^2 - B^2 = \begin{bmatrix} 5 & 1 \\ 2 & 4 \end{bmatrix}$$
$$A + B = \begin{bmatrix} 1 & 3 \\ 5 & -2 \end{bmatrix}; A - B = \begin{bmatrix} 1 & 1 \\ 1 & 0 \end{bmatrix}$$
$$(A + B)(A - B) = \begin{bmatrix} 4 & 1 \\ 3 & 5 \end{bmatrix}$$

41. 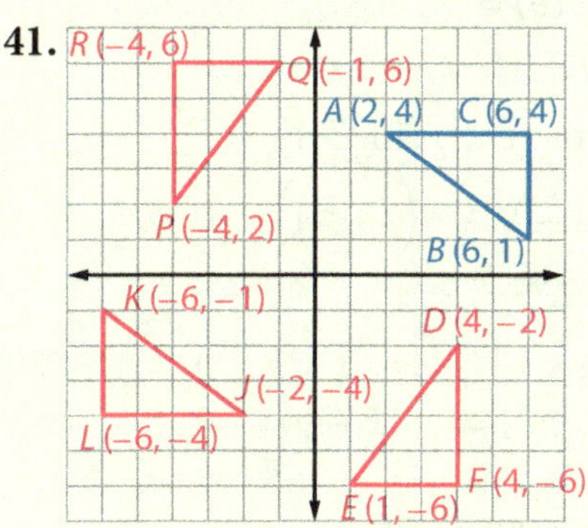

42. possible roots: $\pm 1, \pm 2, \pm 3, \pm 6$

$$
\begin{array}{c|cccc}
-1 & 1 & -4 & 1 & 6 \\
 & \downarrow & -1 & 5 & -6 \\
\hline
 & 1 & -5 & 6 & 0
\end{array}
$$

$(x+1)(x^2 - 5x + 6)$
$(x+1)(x-2)(x-3)$

43. $\tan x \csc 2x = \dfrac{\sin x}{\cos x} \cdot \dfrac{1}{\sin 2x}$

$= \dfrac{\sin x}{\cos x} \cdot \dfrac{1}{2 \sin x \cos x} = \dfrac{1}{2 \cos^2 x}$

$= \dfrac{1}{2} \sec^2 x$

44. $\quad x(3x - 6) = 9$
$3x^2 - 6x - 9 = 0$
$x^2 - 2x - 3 = 0$
$(x + 1)(x - 3) = 0$
$\qquad\qquad x = -1, 3$
$y = 3(-1) - 6 = -9; \; (-1, -9)$
$y = 3(3) - 6 = 3; \; (3, 3)$

45. $x^2 + 3x + 6 = 2x^2 + 3x + 5$
$\qquad\quad x^2 - 1 = 0$
$\qquad\qquad x = \pm 1$
$y = (-1)^2 + 3(-1) + 6 = 4; \; (-1, 4)$
$y = 1^2 + 3(1) + 6 = 10; \; (1, 10)$

46. $1\text{st} \quad\Rightarrow\quad x + 4y = 32$
$2\text{nd} \div 2 \Rightarrow \dfrac{-x + \; y = \; 8}{5y = 40}$
$\qquad\qquad\qquad\quad y = 8$

$1\text{st: } x + 4(8) = 32$
$\qquad\qquad x = 0; \; (0, 8)$

47. $2(1\text{st}) \;\Rightarrow\; 6x + 10y = 26$
$3(2\text{nd}) \Rightarrow \dfrac{-6x + 21y = 36}{31y = 62}$
$\qquad\qquad\qquad\quad y = 2$

$1\text{st: } 3x + 5(2) = 13$
$\qquad\qquad 3x = 3$
$\qquad\qquad\; x = 1; \; (1, 2)$

48. $6(4) + 7(0) = 24$
$4 - 4(0) = 4$

49. $3(7) + 1(-2) = 19$

50. $5\left(\cos \dfrac{5\pi}{4} + i \sin \dfrac{5\pi}{4}\right)$

$= 5\left(-\dfrac{\sqrt{2}}{2} + i\left(-\dfrac{\sqrt{2}}{2}\right)\right)$

$= -\dfrac{5\sqrt{2}}{2} - \dfrac{5\sqrt{2}}{2} i$

51. $4(1\text{st}) \quad\Rightarrow -3x + y = \;\; 2$
$-1(2\text{nd}) \Rightarrow \dfrac{3x - y = -8}{0 \neq -6}$

$\varnothing$; inconsistent

---

42. Find the zeros of $x^3 - 4x^2 + x + 6$. [2.3] $-1, 2, 3$

43. Prove $\frac{1}{2} \sec^2 x = \tan x \csc 2x$. [5.7]

Solve each system using substitution. Then verify graphically. [7.1]

44. $xy = 9$
$y = 3x - 6$ $(-1, -9), (3, 3)$

45. $y = 2x^2 + 3x + 5$
$y = x^2 + 3x + 6$
$(-1, 4), (1, 10)$

Solve each system using elimination. [7.1]

46. $x + 4y = 32$
$-2x + 2y = 16$ $(0, 8)$

47. $3x + 5y = 13$
$-2x + 7y = 12$ $(1, 2)$

48. Which ordered pair is a solution of $6x + 7y = 24$ and $x - 4y = 4$? [7.1] C

A. $(0, -1)$ **D.** $(-4, -2)$
B. $(11, -6)$ **E.** none of these
C. $(4, 0)$

49. Find $\mathbf{u} \cdot \mathbf{v}$ if $\mathbf{u} = \langle 3, 1 \rangle$ and $\mathbf{v} = \langle 7, -2 \rangle$. [6.2] B

A. 8   **D.** 23
B. 19   **E.** none of these
C. 1

50. Convert $z = 5 \operatorname{cis} \dfrac{5\pi}{4}$ to standard form. [6.6] D

A. $-\dfrac{5\sqrt{2}}{2} + \dfrac{5\sqrt{2}}{2} i$ **D.** $-\dfrac{5\sqrt{2}}{2} - \dfrac{5\sqrt{2}}{2} i$

B. $\dfrac{5\sqrt{2}}{2} + \dfrac{5\sqrt{2}}{2} i$ **E.** none of these

C. $\dfrac{5\sqrt{2}}{2} - \dfrac{5\sqrt{2}}{2} i$

51. Which of the following describes the number of solutions for the system? [7.1] A

$-\dfrac{3x}{4} + \dfrac{y}{4} = \dfrac{1}{2}$
$-3x + y = 8$

A. 0   **D.** infinitely many
B. 1   **E.** none of these
C. 2

Direct and Indirect Deductive Reasoning

No human investigation can be called real science if it cannot be demonstrated mathematically.

—*Leonardo da Vinci*

In stating that real science must be demonstrated mathematically, Leonardo da Vinci asserted that science should be supported by mathematical reasoning. He made this statement before the scientific laws of Galileo, Kepler, and Newton were supported by mathematical deductive reasoning. Scientific theories are often based on inductive reasoning, which draws general conclusions from specific examples and cannot guarantee the truth of the results. As science seeks to verify unproven theories, those theories that are based on mathematical reasoning are more reliable than those that are not. As Einstein declared, "It is mathematics that offers the exact natural sciences a certain measure of security which, without mathematics, they could not attain." The ability to verify mathematical concepts through various kinds of proofs has led to the universal acceptance of a large body of mathematical truth.

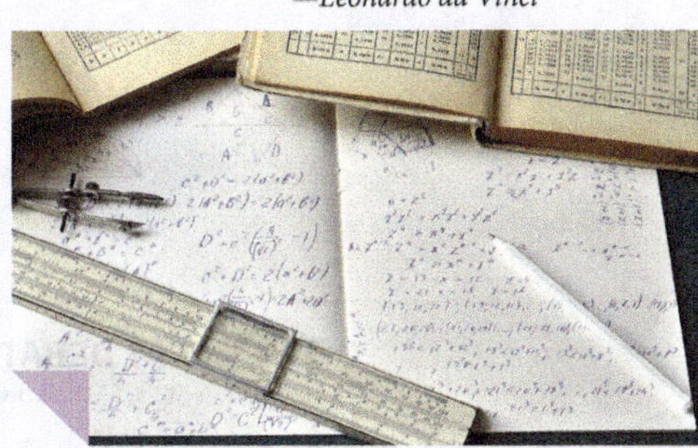

Mathematical proofs commonly use direct deductive reasoning, in which step-by-step logical reasoning establishes the truth of a statement. You did many direct proofs in Geometry. These proofs established the theorems that proved the many formulas and techniques you applied to real-world problems. While some direct proofs can be short and easy, others can be long and difficult. Andrew Wiles's proof of Fermat's Last Theorem is over 150 pages long and required seven years of research.

An indirect deductive proof assumes the negation of the statement that is to be proved and then applies logical reasoning to reach a contradiction. The contradiction shows that the negation is false and that the original statement must then be true. British mathematician G. H. Hardy described indirect proofs as "one of a mathematician's finest weapons," saying, "It is a far finer gambit than any chess gambit: a chess player may offer the sacrifice of a pawn or even a piece, but a mathematician offers *the game*" (the assumption that the negation might be true). Consider an indirect proof for there being no smallest rational number greater than 0. Start by assuming that there is a smallest rational number x greater than 0. But the rational number $\frac{x}{2}$ would be less than x and greater than 0.

Therefore the assumption that x is the smallest rational number greater than 0 is false, and the original statement must be true. This indirect reasoning proves there is no smallest rational number greater than 0. Examples of classic indirect proofs include proving that $\sqrt{2}$ is irrational and that the number of prime numbers is infinite.

Scripture often uses the word *prove* or *proof*. While these words often refer to a testing of God's chosen people or a particular person, other passages refer to a reasoned argument used to show the truthfulness of a fact. While these proofs can be inductive, the Holy Spirit can use direct and indirect deductive reasoning to strengthen our belief in spiritual truths and convince others of these truths (Acts 17:2; 18:4,19; 24:25; Heb. 5:14; 1 Pet. 3:15).

❯ Exercises

1. State whether the illustrated reasoning is deductive or inductive.
   a. $1^2 \geq 1, 2^2 \geq 2, 3^2 \geq 3, 4^2 \geq 4, 5^2 \geq 5$ inductive
   Therefore, $n^2 \geq n$ for all natural numbers n.
   b. For $n \in \mathbb{Z}, n^2 + n = n(n + 1)$, the product of one odd and one even integer.
   Any even integer can be expressed as $2c$, with $c \in$ natural numbers.
   Therefore, $n^2 + n$ is divisible by 2.   deductive
2. Which type of deductive proof is the most common?   direct

Biblical Perspective of Mathematics

Objectives

1. To distinguish between direct and indirect deductive proofs
2. To compare and contrast mathematical proofs and a proof from Scripture
3. To identify the reasoning of an indirect deductive proof offered in a debate on the existence of God

Assignments

- **Minimum:** 1–7; 8–10 (in class)
- **Standard:** 1–7; 8–10 (in class)
- **Extended:** 1–10

PRESENTATION

Read Leonardo da Vinci's quote and consider how mathematics helped establish much of the science of the Renaissance. Discuss the use of mathematics in establishing scientific truth and the limitations of mathematical reasoning's role in modern scientific theories. Contrast modern science and its many unproven theories with modern mathematics and its proven mathematical truth.

Explain the techniques of establishing truth through two common types of deductive proofs, direct and indirect. Then consider the examples of each type of proof. Math induction proofs, which will be studied in Section 9.5, are actually direct deductive proofs rather than a form of inductive reasoning.

Discuss Fermat's Last Theorem as an example of a difficult direct deductive proof and the long road that is sometimes necessary to establish the truth of a mathematical statement, promoting it from a conjecture to a theorem. Andrew Wiles must have experienced great frustration and disappointment when the review of his original incorrect proof exposed a critical flaw. Three years after his second failed approach, Wiles commented on his discovery of a remedy, saying, "It was the most—the most important

Answers

4. After assuming the negation of what you are trying to prove, the goal in an indirect proof is to reason logically until you arrive at a contradiction.

5a. Assume that the set of prime numbers is finite.

5b. Assume that $\sqrt{2}$ is rational.

6. the resurrection of Jesus Christ

7. Jesus' multiple appearances to the disciples over a forty-day period

8. Direct and indirect proofs both use logical step-by-step reasoning to prove a statement but begin at different places. Direct reasoning begins with a given or known fact, while indirect reasoning begins with the negation of the statement to be proved and reasons logically to a contradiction of a given or known fact.

9. Deductive proofs establish the truth of a statement using a logical series of step-by-step statements. The inductive proof in Acts 1:3 uses the combined weight of evidence from multiple witnesses to support the statement.

10. inductive; Assuming there is no God leads to the impossibility of the existence of laws in logic, science, and morality. This obvious contradiction of the nature of our universe and human nature itself disproves the assumption of atheists and verifies God's existence.

---

3. Use the figure to complete the proof of the Pythagorean Theorem.

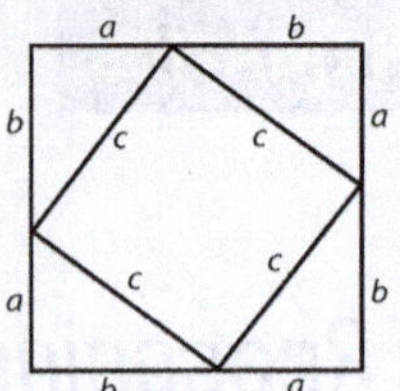

$$A_{Large\ Square} = A_{Small\ Square} + 4A_{Right\ Triangle}$$
$$(a+b)^2 = c^2 + 4(0.5ab)$$
$$a^2 + 2ab + b^2 = c^2 + 2ab$$
$$a^2 + b^2 = c^2$$

4. Why is an indirect proof often called proof by contradiction?

5. Write the assumption that is the first step for each indirect proof.
   a. Prove that the set of prime numbers is infinite.
   b. Prove that $\sqrt{2}$ is irrational.

6. Read Acts 1:3. What is being proved?

7. What evidence is used as proof in Acts 1:3? (See also Matt. 28:16–17; Mark 16:14; Luke 24:33–37; John 20:19–20; and 1 Cor. 15:3–8.)

8. Discuss: Compare and contrast direct and indirect deductive proofs.

9. Discuss: Compare and contrast deductive proofs with the proof in Acts 1:3.

10. Discuss: Read the following excerpt from the Bahnsen-Stein debate. Identify the type of proof that Dr. Bahnsen employs and summarize his argument in your own words.

THE GREAT DEBATE: DOES GOD EXIST?

Dr. Gordon Stein (Atheist) v. Dr. Greg Bahnsen (Christian)

❯ From Bahnsen's Opening Case

I suggest *we can prove the existence of God from the impossibility of the contrary. The transcendental proof for God's existence is that without Him it is impossible to prove anything.* . . . The atheist world view cannot allow for laws of logic, the uniformity of nature, the ability for the mind to understand the world, and moral absolutes. . . .

❯ From Bahnsen's Closing Statement

Why should [Dr. Stein] feed the poor? He says they want to do that. I grant that. My argument has never been that atheists are the lousiest people in the world. That's not the point. Some Christians can be pretty lousy, too. But why is it that I can call atheists or Christians lousy when they act in the ways we're thinking of? [It's] because I have absolute standards of morality to judge. Dr. Stein does not.

Therefore, from a transcendental standpoint the atheistic view cannot account for this debate tonight; because this debate has assumed that we're going to use the laws of logic as standards of reasoning, or else we're irrational; that we're going to use laws of science. . . . It's assumed in a moral sense that we're not going to be dishonest and try to lie or just try to deceive you. . . .

How, in a material, naturalistic outlook on life and man his place in the world, can you account for the laws of logic, science, and morality?

The atheist world view cannot do it, and therefore I feel justified concluding . . . that the proof of the Christian God is the impossibility of the contrary. Without the Christian worldview this debate wouldn't make sense.

---

moment of my working life. It was so indescribably beautiful; it was so simple and so elegant." For further information about Fermat's Last Theorem, consult the following texts.

Aczel, Amir. *Fermat's Last Theorem: Unlocking the Secret of an Ancient Mathematical Problem.* New York: Four Walls, 1996.

Singh, Simon. *Fermat's Enigma: The Epic Quest to Solve the World's Greatest Mathematical Problem.* New York: Walker and Company, 1997.

Conclude with a discussion of the references to "prove" and "proof" in Scripture. Emphasize the relationship between truth as a key characteristic of God and mathematical truth as a result of human proof. Remind students that transcendent mathematical truth is not dependent on humans but instead originates from God and His nature.

TIPS

Ex. 10 Transcripts and recordings of this debate can be found with the Internet keyword search *Bahnsen–Stein debate video.*

Gaussian Elimination

Solving linear systems of more than 2 variables can be challenging. Using elimination to solve Example 7 of Section 7.1 converted the original system into an *equivalent system* (one with the same solution set) that was easier to solve.

$$\begin{aligned} -2x + 5y + 2z &= 23 \\ 4x - y + 2z &= -13 \\ x - 3y + 6z &= -17 \end{aligned} \quad \Rightarrow \quad \begin{aligned} x - 3y + 6z &= -17 \\ y - 2z &= 5 \\ z &= -\tfrac{1}{2} \end{aligned}$$

When a system with this inverted triangular form has all leading coefficients of 1, it is said to be in *row echelon form* (REF). The process of eliminating variables from the system to find an equivalent system in REF is called *Gaussian elimination*. While this process is similar to solving by elimination, a more formal notation illustrates how an equivalent system in REF is generated.

$$\begin{aligned} x - 3y + 6z &= -17 \\ 4x - y + 2z &= -13 \\ -2x + 5y + 2z &= 23 \end{aligned}$$

1. Interchange the first and third equations so the coefficient of the *x*-term in the first equation is 1.

$$\begin{aligned} x - 3y + 6z &= -17 \\ 11y - 22z &= 55 \\ -y + 14z &= -11 \end{aligned}$$

2. Replace the second equation with the sum of the second and -4 times the first to eliminate the *x*-term.

   Then replace the third equation with the sum of the third and 2 times the first to eliminate the *x*-term.

$$\begin{aligned} x - 3y + 6z &= -17 \\ y - 2z &= 5 \\ -y + 14z &= -11 \end{aligned}$$

3. Divide each term in the second equation by 11.

$$\begin{aligned} x - 3y + 6z &= -17 \\ y - 2z &= 5 \\ 12z &= -6 \end{aligned}$$

4. Replace the third equation with the sum of the second and third equations to eliminate the *y*-term.

$$\begin{aligned} x - 3y + 6z &= -17 \\ y - 2z &= 5 \\ z &= \tfrac{1}{2} \end{aligned}$$

5. Divide the last equation by 12 to get an equivalent REF system.

Once each equation in a linear system is written in standard form, Gaussian elimination uses operations that affect only the coefficients of the variables and the constant term. An *augmented matrix* keeps the coefficients and constants organized when converting the original system to an equivalent REF. Notice how an augmented matrix is made from the standard form equations by combining the system's coefficient matrix and its constant matrix.

$$\begin{array}{cc} \text{System} & \text{Augmented Matrix} \\ \begin{aligned} -2x + 5y + 2z &= 23 \\ 4x - y + 2z &= -13 \\ x - 3y + 6z &= -17 \end{aligned} \Rightarrow & \left[\begin{array}{ccc|c} -2 & 5 & 2 & 23 \\ 4 & -1 & 2 & -13 \\ 1 & -3 & 6 & -17 \end{array}\right] \end{array}$$

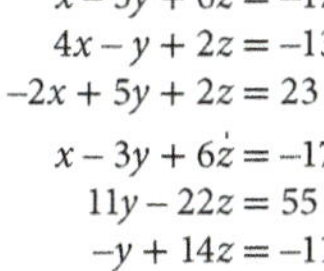

After completing this section, you will be able to

- list the elementary row operations.
- solve systems of equations using Gaussian elimination or Gauss-Jordan elimination.
- use matrices to model and solve real-world problems.

The German mathematician Carl Friedrich Gauss (1777–1855) helped develop the branch of mathematics called linear algebra.

7.3 Gaussian Elimination

Objectives

1. To apply the three elementary row operations
2. To solve a system of equations using an augmented matrix and Gaussian elimination
3. To solve systems of equations using Gauss-Jordan elimination (RREF)
4. To model and solve real-world problems using matrices

Vocabulary

augmented matrix
equivalent system
Gaussian elimination
Gauss-Jordan elimination
reduced row echelon form (RREF)
row echelon form (REF)

Reading and Writing Mathematics

Compare and contrast REF and RREF. Both are elimination methods for systems of equations in which elementary row operations are performed on augmented matrices. The strategy for REF is to get 1s across the main diagonal and 0s to the left of the main diagonal. The strategy for RREF is to get 0s on both sides of the main diagonal of 1s except for the last column. RREF provides the values of the variables directly from the matrix, while REF requires back substitution.

PRESENTATION

Lesson Opener

Describe all possible intersections and state the number of solutions for a system represented by each case.

1. two lines   a point (one solution), a line (infinite number of solutions), or no intersection (∅)

2. three planes   a point (one solution), a line (infinite number of solutions), a plane (infinite number of solutions), or no intersection (∅)

Solve each system by elimination.

3. $2x - 3y = 16$
   $-x + 4y = -13$ (5, −2)

4. $2x + y - z = 2$
   $x - 2y + z = 3$
   $x + y + z = 6$ (2, 1, 3)

Introduce *Gaussian elimination* as a different way of notating the solving of a system of linear equations by elimination. Show how the solving of Example 7 from Section 7.1 converted the system to an *equivalent system* in the triangular *row echelon form* (REF) and how the notation of Gaussian elimination emphasizes the equivalent system throughout the process.

Explain that matrices can be used to simplify the process, then introduce *augmented matrices* and the elementary row operations. When presenting Example 1, emphasize the need to notate the steps used during Gaussian elimination. Explain that there are many possible row echelon forms and that checking work will be difficult if each step is not recorded. Consider reviewing the back substitutions that lead to the solution $\left(-2, 4, -\tfrac{1}{2}\right)$.

In Example 2 Gaussian elimination is used to obtain a matrix in REF. Explain the strategies involved in Gaussian elimination. The basic strategies are as follows.

1. Divide row 1 by the leading coefficient to obtain 1 in the first column.

Additional Exercises

Convert each matrix into row echelon form and use back substitution to find the solution.

1. $\begin{bmatrix} 3 & 4 & | & 13 \\ 0 & -1 & | & 2 \end{bmatrix}$ $(7, -2)$

2. $\begin{bmatrix} -2 & 3 & 1 & | & 6 \\ 0 & 4 & -2 & | & 14 \\ 0 & 0 & 1 & | & 3 \end{bmatrix}$ $(6, 5, 3)$

Solve each system using an augmented matrix and Gauss-Jordan elimination.

3. $6x + 3y = 4$
$2x + 3y = 1$ $\left(\frac{3}{4}, -\frac{1}{6}\right)$

4. $3x - 4y = 6$
$2x + 3y = -13$ $(-2, -3)$

5. $x + y + z = -2$
$3x + 2z = 0$
$5y - 3z = 4$ $(2, -1, -3)$

6. $3x - 6y + 6z = -7$
$2x + 3y - 5z = 18$
$9x - 6y + 2z = 19$ $\left(3, \frac{2}{3}, -2\right)$

7. $x + 4y + 3z = 8$
$2x + 10y + 8z = 19$
$x - y - 2z = 9$ $\varnothing$

8. Find the general solution and a specific solution for the following system.
$2x - 3y + z = -15$
$-4x + y + 5z = 13$
$\left(\frac{8t - 12}{5}, \frac{7t + 17}{5}, t\right); (-4, 2, -1)$

Each operation that produced an equivalent system has a corresponding *elementary row operation* used to transform the augmented matrix into an equivalent REF matrix.

Elementary Row Operation
1. Interchange any two rows.
2. Multiply (or divide) each entry in a row by any nonzero number.
3. Add a multiple of one row to another row.

Notice how each step of the Gaussian elimination in Example 1 is notated using elementary row operations on the augmented matrix.

Example 1 Illustrating Gaussian Elimination with an Augmented Matrix

Illustrate the row operations used to convert the augmented matrix representing the system to an equivalent REF.
$\begin{bmatrix} -2 & 5 & 2 & 23 \\ 4 & -1 & 2 & -13 \\ 1 & -3 & 6 & -17 \end{bmatrix}$

Answer

$r_3 \Leftrightarrow r_1 \begin{bmatrix} 1 & -3 & 6 & | & -17 \\ 4 & -1 & 2 & | & -13 \\ -2 & 5 & 2 & | & 23 \end{bmatrix}$

1. Interchange the first and third rows.

$\begin{matrix} -4r_1 + r_2 \Rightarrow r_2 \\ 2r_1 + r_3 \Rightarrow r_3 \end{matrix} \begin{bmatrix} 1 & -3 & 6 & | & -17 \\ 0 & 11 & -22 & | & 55 \\ 0 & -1 & 14 & | & -11 \end{bmatrix}$

2. Replace the second row with the sum of the second and −4 times the first.

   Then replace the third row with the sum of the third and 2 times the first.

$\frac{1}{11}r_2 \Rightarrow r_2 \begin{bmatrix} 1 & -3 & 6 & | & -17 \\ 0 & 1 & -2 & | & 5 \\ 0 & -1 & 14 & | & -11 \end{bmatrix}$

3. Divide all terms in the second row by 11.

$r_2 + r_3 \Rightarrow r_3 \begin{bmatrix} 1 & -3 & 6 & | & -17 \\ 0 & 1 & -2 & | & 5 \\ 0 & 0 & 12 & | & -6 \end{bmatrix}$

4. Replace the third row with the sum of the second and third rows.

$\frac{1}{12}r_2 \Rightarrow r_2 \begin{bmatrix} 1 & -3 & 6 & | & -17 \\ 0 & 1 & -2 & | & 5 \\ 0 & 0 & 1 & | & -\frac{1}{2} \end{bmatrix}$

5. Divide the last row by 12 to get an equivalent system in REF.

SKILL ✔ **EXERCISE 7**

Once the augmented matrix is in REF, the corresponding system is relatively easy to solve. Use the value of the last variable, $z = -\frac{1}{2}$, to back-substitute and find the value of the other variables as illustrated in Section 7.1.

---

2. Obtain 0s in the first column of the succeeding rows by adding multiples of rows and replacing with the resulting row.

3. Divide row 2 to obtain a leading coefficient of 1 in column 2.

4. Continue by obtaining 0s in the second column of the succeeding rows.

5. Continue the process until the major diagonal contains only 1s.

Point out that there is more than one way to obtain an REF matrix. In step 4, some might prefer to divide row 2 by 10 so the first nonzero coefficient is 1. While this introduces fractions, back substitut-ing into $y - \frac{7}{10}z = -\frac{43}{10}$ is relatively easy with $z = -1$, yielding $y = -5$.

One-on-One A student may prefer to think in terms of subtracting equations instead of multiplying by a negative and adding. Also, in some cases it is easier to subtract the upper equation from the lower instead of subtracting the lower from the upper.

Discuss the *reduced row echelon form* (RREF). This modification of the REF reflects the efforts of the German geodesist Wilhelm Jordan (1842–99). Geodesy is the study of measurement relating to the earth's shape, size, orientation in space, and gravitational fields.

Use Example 3 to reinforce the process of obtaining the REF in steps 1–4 and to introduce the process of obtaining the RREF in steps 5–6.

Point out that obtaining all 0s in row 3 of Example 4 indicates that two of the equations (the first and third) represent the same plane. The general solution then represents a line in space: the intersection of that plane and the plane represented by the second equation.

Ask why the matrix in step 3 is in RREF when the left portion contains −2 and −8. (*The first nonzero element in each row is 1 and all other elements in that column are 0.*) The infinite number of solu-

Write an augmented matrix representing the system. Then use Gaussian elimination to find an equivalent REF matrix and solve the system.

$$3x + 4y + 5z = -13$$
$$x - 2y = 14$$
$$2y + 10 = 4z + x$$

Answer

$$\begin{aligned} 3x + 4y + 5z &= -13 \\ x - 2y + 0z &= 14 \\ -x + 2y - 4z &= -10 \end{aligned} \qquad \begin{bmatrix} 3 & 4 & 5 & \vline & -13 \\ 1 & -2 & 0 & \vline & 14 \\ -1 & 2 & -4 & \vline & -10 \end{bmatrix}$$

1. Write each equation in standard form before representing the system with an augmented matrix.

$$r_2 \Leftrightarrow r_1 \begin{bmatrix} 1 & -2 & 0 & \vline & 14 \\ 3 & 4 & 5 & \vline & -13 \\ -1 & 2 & -4 & \vline & -10 \end{bmatrix}$$

2. Interchange rows to make the leading coefficient of the first equation equal to 1.

$$\begin{aligned} 3r_3 + r_2 &\Rightarrow r_2 \\ r_1 + r_3 &\Rightarrow r_3 \end{aligned} \begin{bmatrix} 1 & -2 & 0 & \vline & 14 \\ 0 & 10 & -7 & \vline & -43 \\ 0 & 0 & -4 & \vline & 4 \end{bmatrix}$$

3. Use elementary row operations to produce 0s under the leading 1 in the first column.

$$-\tfrac{1}{4}r_3 \Rightarrow r_3 \begin{bmatrix} 1 & -2 & 0 & \vline & 14 \\ 0 & 10 & -7 & \vline & -43 \\ 0 & 0 & 1 & \vline & -1 \end{bmatrix}$$

4. Divide row 3 by -4 to obtain a 1 along the major diagonal.

$$7r_3 + r_2 \Rightarrow r_2 \begin{bmatrix} 1 & -2 & 0 & \vline & 14 \\ 0 & 10 & 0 & \vline & -50 \\ 0 & 0 & 1 & \vline & -1 \end{bmatrix}$$

5. Since dividing row 2 by 10 would introduce fractions, add the multiple of row 3 that eliminates the z term in row 2.

$$\tfrac{1}{10}r_2 \Rightarrow r_2 \begin{bmatrix} 1 & -2 & 0 & \vline & 14 \\ 0 & 1 & 0 & \vline & -5 \\ 0 & 0 & 1 & \vline & -1 \end{bmatrix}$$

6. Divide the second row by 10 to finish converting to REF. The last row implies that $z = -1$ and the second row implies that $y = -5$.

$$\begin{aligned} 1x - 2(-5) + 0 &= 14 \\ x &= 4 \end{aligned}$$

7. Substitute $z = -1$ and $y = -5$ into the equation indicated by the first row to find x.

$$(4, -5, -1)$$

8. Write the solution as an ordered triple.

TIP

The REF of an augmented matrix is not unique. While the result using a calculator may differ from your work, the system's solution will be the same.

```
[A]
        [ 3   4   5  -13 ]
        [ 1  -2   0   14 ]
        [ -1  2  -4  -10 ]
ref([A])▸Frac
        [ 1  4/3  5/3  -13/3 ]
        [ 0   1   1/2  -11/2 ]
        [ 0   0    1    -1   ]
```

_______ SKILL ✔ **EXERCISE 23**

Performing the additional row operation of $2r_2 + r_1 \Rightarrow r_1$ on the last matrix in Example 2 produces a coefficient matrix equal to the multiplicative identity matrix I_3. The system's solution can then be read directly: $x = 4$, $y = -5$, and $z = -1$.

$$\begin{bmatrix} 1 & 0 & 0 & \vline & 4 \\ 0 & 1 & 0 & \vline & -5 \\ 0 & 0 & 1 & \vline & -1 \end{bmatrix}$$

A matrix is in _reduced row echelon form_ (RREF) when the first nonzero element in each row is 1 and any other elements in that column are 0. Solving a system by applying elementary row operations to an augmented matrix until the unique RREF is found is called _Gauss-Jordan elimination_.

Example 3   Using Gauss-Jordan Elimination

Write an augmented matrix representing the system. Then solve the system using Gauss-Jordan elimination.

$$\begin{aligned} 2x + y - z &= -7 \\ -2x + 3y - 4z &= -18 \\ -x - 2y + 2z &= 11 \end{aligned}$$

Answer

$$\begin{bmatrix} 2 & 1 & -1 & \vline & -7 \\ -2 & 3 & -4 & \vline & -18 \\ -1 & -2 & 2 & \vline & 11 \end{bmatrix}$$

1. Write an augmented matrix representing the system.

$$\begin{aligned} -r_3 &\Rightarrow r_1 \\ r_1 + r_2 &\Rightarrow r_2 \\ r_1 + 2r_3 &\Rightarrow r_3 \end{aligned} \begin{bmatrix} 1 & 2 & -2 & \vline & -11 \\ 0 & 4 & -5 & \vline & -25 \\ 0 & -3 & 3 & \vline & 15 \end{bmatrix}$$

2. Use elementary row operations to make $e_{11} = 1$, $e_{21} = 0$, and $e_{31} = 0$.

CONTINUED ➡

tions can be expressed in terms of one of the variables or in terms of an arbitrary parameter t, as shown in step 5. Encourage students to use one or more particular solutions to check general solutions for dependent systems.

Emphasize that solving a dependent consistent system produces a true equation ($0 = 0$ in Example 4) and solving an inconsistent system produces a false equation ($0 = -\tfrac{3}{10}$ in Example 5). The inconsistency of the system in Example 5 might be seen at the end of step 2, where rows 2 and 3 represent equations with the same coefficients but different constants: $y - z = \tfrac{1}{2}$ and $y - z = \tfrac{1}{5}$. Since there is no solution to this smaller system, a false statement is generated in step 3.

Example 6 presents a real-world application in which the system can be solved quickly by using technology to find the RREF of the augmented matrix. Focus on helping students determine the equations in the system.

Motivational Idea   Applications for solving systems of equations can be found on the Internet using the Internet keyword search _Gaussian elimination calculator_. Once the students understand and become proficient with the process, you may want to introduce them to some of these applications.

For Graphing Calculators

Use matrices and a calculator to solve each system. (See the Technology Corner on p. 358.)

1. $5.5x - 6.7y = 13.8$
   $1.3x + 2.6y = 12.4$ (5.169, 2.184)

2. $0.08x - 0.13y = 2.75$
   $-0.14x + 0.58y = 1.98$ (65.69, 19.27)

Assignments

- **Minimum:** 1–7, 9–11, 13–15; 17–21 odd; 25–27, 31, 34, 37, 45, 50–54
- **Standard:** 2–12 even; 13–15; 17–25 odd; 26, 28–33, 38–40, 42, 45–49, 53–54
- **Extended:** 5, 8–9, 12–16, 18–19, 23, 25–27, 29–30, 32, 35; 36–48 even; 53–54

Solutions

❯ A. Exercises

1. $5x + 3(4) = 2$
   $$5x = -10$$
   $$x = -2; \ (-2, 4)$$

2. $2y + (-1) = -5$
   $$2y = -4; \ y = -2$$
   $$x + 3(-2) - 4(-1) = 1$$
   $$x - 6 + 4 = 1$$
   $$x = 3; \ (3, -2, -1)$$

3. $\begin{bmatrix} 3 & -4 & \vline & 6 \\ 2 & 5 & \vline & -19 \end{bmatrix}$

4. $\begin{bmatrix} 2 & 4 & -3 & \vline & 9 \\ 3 & 0 & -2 & \vline & 9 \\ 4 & -1 & 0 & \vline & 18 \end{bmatrix}$

5. $\begin{bmatrix} 2 & 5 & \vline & -16 \\ 1 & -1 & \vline & 6 \end{bmatrix}$

5a. $\begin{aligned} x - y &= 6 \\ 2x + 5y &= -16 \end{aligned} \quad \begin{bmatrix} 1 & -1 & \vline & 6 \\ 2 & 5 & \vline & -16 \end{bmatrix}$

5b. $\begin{aligned} x - y &= 6 \\ 7y &= -28 \end{aligned} \quad \begin{bmatrix} 1 & -1 & \vline & 6 \\ 0 & 7 & \vline & -28 \end{bmatrix}$

5c. $\begin{aligned} x - y &= 6 \\ y &= -4 \end{aligned} \quad \begin{bmatrix} 1 & -1 & \vline & 6 \\ 0 & 1 & \vline & -4 \end{bmatrix}$

6. $\begin{bmatrix} 3 & -6 & \vline & 24 \\ 4 & -3 & \vline & 7 \end{bmatrix}$

6a. $\begin{aligned} x - 2y &= 8 \\ 4x - 3y &= 7 \end{aligned} \quad \begin{bmatrix} 1 & -2 & \vline & 8 \\ 4 & -3 & \vline & 7 \end{bmatrix}$

6b. $\begin{aligned} x - 2y &= 8 \\ 5y &= -25 \end{aligned} \quad \begin{bmatrix} 1 & -2 & \vline & 8 \\ 0 & 5 & \vline & -25 \end{bmatrix}$

6c. $\begin{aligned} x - 2y &= 8 \\ y &= -5 \end{aligned} \quad \begin{bmatrix} 1 & -2 & \vline & 8 \\ 0 & 1 & \vline & -5 \end{bmatrix}$

7. $\begin{bmatrix} 1 & 1 & 1 & | & 4 \\ 4 & 3 & 1 & | & 4 \\ 3 & -1 & -2 & | & -1 \end{bmatrix}$

7a. $\begin{aligned} x + y + z &= 4 \\ -y - 3z &= -12 \\ -4y - 5z &= -13 \end{aligned}$ $\qquad \begin{bmatrix} 1 & 1 & 1 & | & 4 \\ 0 & -1 & -3 & | & -12 \\ 0 & -4 & -5 & | & -13 \end{bmatrix}$

7b. $\begin{aligned} x + y + z &= 4 \\ y + 3z &= 12 \\ -4y - 5z &= -13 \end{aligned}$ $\qquad \begin{bmatrix} 1 & 1 & 1 & | & 4 \\ 0 & 1 & 3 & | & 12 \\ 0 & -4 & -5 & | & -13 \end{bmatrix}$

7c. $\begin{aligned} x + y + z &= 4 \\ y + 3z &= 12 \\ 7z &= 35 \end{aligned}$ $\qquad \begin{bmatrix} 1 & 1 & 1 & | & 4 \\ 0 & 1 & 3 & | & 12 \\ 0 & 0 & 7 & | & 35 \end{bmatrix}$

7d. $\begin{aligned} x + y + z &= 4 \\ y + 3z &= 12 \\ z &= 5 \end{aligned}$ $\qquad \begin{bmatrix} 1 & 1 & 1 & | & 4 \\ 0 & 1 & 3 & | & 12 \\ 0 & 0 & 1 & | & 5 \end{bmatrix}$

8. $\begin{bmatrix} 1 & -1 & -1 & | & 9 \\ 2 & 1 & 3 & | & 1 \\ 5 & -6 & 4 & | & 8 \end{bmatrix}$

8a. $\begin{aligned} x - y - z &= 9 \\ 3y + 5z &= -17 \\ -y + 9z &= -37 \end{aligned}$ $\qquad \begin{bmatrix} 1 & -1 & -1 & | & 9 \\ 0 & 3 & 5 & | & -17 \\ 0 & -1 & 9 & | & -37 \end{bmatrix}$

8b. $\begin{aligned} x - y - z &= 9 \\ y - 9z &= 37 \\ 3y + 5z &= -17 \end{aligned}$ $\qquad \begin{bmatrix} 1 & -1 & -1 & | & 9 \\ 0 & 1 & -9 & | & 37 \\ 0 & 3 & 5 & | & -17 \end{bmatrix}$

8c. $\begin{aligned} x - y - z &= 9 \\ y - 9z &= 37 \\ 32z &= -128 \end{aligned}$ $\qquad \begin{bmatrix} 1 & -1 & -1 & | & 9 \\ 0 & 1 & -9 & | & 37 \\ 0 & 0 & 32 & | & -128 \end{bmatrix}$

8d. $\begin{aligned} x - y - z &= 9 \\ y - 9z &= 37 \\ z &= -4 \end{aligned}$ $\qquad \begin{bmatrix} 1 & -1 & -1 & | & 9 \\ 0 & 1 & -9 & | & 37 \\ 0 & 0 & 1 & | & -4 \end{bmatrix}$

9. $\begin{aligned} y &= -4 \\ x - (-4) &= 6 \\ x &= 2; \ (2, -4) \end{aligned}$

10. $\begin{aligned} y &= -5 \\ x - 2(-5) &= 8 \\ x &= -2; \ (-2, -5) \end{aligned}$

11. $\begin{aligned} z &= 5 \\ y + 3(5) &= 12; \ y = -3 \\ x + (-3) + (5) &= 4 \\ x &= 2; \ (2, -3, 5) \end{aligned}$

12. $\begin{aligned} z &= -4 \\ y - 9(-4) &= 37; \ y = 1 \\ x - (1) - (-4) &= 9 \\ x &= 6; \ (6, 1, -4) \end{aligned}$

13. $\begin{aligned} z &= 4 \\ y + 2(4) &= -1; \ y = -9 \\ x + 2(-9) + 3(4) &= 5; \ x = 11 \end{aligned}$
$(11, -9, 4)$; independent consistent

14. $0 + 0 + 0 = 6$ is false.
$\varnothing$; inconsistent

$-\frac{1}{3}r_3 \Rightarrow r_2$ $\begin{bmatrix} 1 & 2 & -2 & | & -11 \\ 0 & 1 & -1 & | & -5 \\ 0 & 0 & -3 & | & -15 \end{bmatrix}$
$3r_2 + 4r_3 \Rightarrow r_3$

3. Use elementary row operations to make $e_{22} = 1$. Notice how fractions can be avoided by combining multiples of two rows when making $e_{32} = 0$.

$-\frac{1}{3}r_3 \Rightarrow r_3$ $\begin{bmatrix} 1 & 2 & -2 & | & -11 \\ 0 & 1 & -1 & | & -5 \\ 0 & 0 & 1 & | & 5 \end{bmatrix}$

4. Finish converting to REF.

$2r_3 + r_1 \Rightarrow r_1$ $\begin{bmatrix} 1 & 2 & 0 & | & -1 \\ 0 & 1 & 0 & | & 0 \\ 0 & 0 & 1 & | & 5 \end{bmatrix}$
$r_3 + r_2 \Rightarrow r_2$

5. Use elementary row operations to make $e_{13} = 0$ and $e_{23} = 0$.

$-2r_2 + r_1 \Rightarrow r_1$ $\begin{bmatrix} 1 & 0 & 0 & | & -1 \\ 0 & 1 & 0 & | & 0 \\ 0 & 0 & 1 & | & 5 \end{bmatrix}$

6. Finish converting to RREF.

$(-1, 0, 5)$

7. Write the solution as an ordered triple.

SKILL ✓ **EXERCISE 27**

Consistent independent systems like those in Examples 1–3 produce a unique solution. A dependent consistent system has an infinite number of solutions that can be expressed using one or more arbitrary parameters.

Example 4 Finding a General Solution

Solve the system of linear equations. $\begin{aligned} x - 2y - 4z &= 2 \\ 3x - 10y - 4z &= -10 \\ -2x + 4y + 8z &= -4 \end{aligned}$

Answer

$\begin{bmatrix} 1 & -2 & -4 & | & 2 \\ 3 & -10 & -4 & | & -10 \\ -2 & 4 & 8 & | & -4 \end{bmatrix}$

1. Write the system as an augmented matrix.

$-3r_1 + r_2 \Rightarrow r_2$ $\begin{bmatrix} 1 & -2 & -4 & | & 2 \\ 0 & -4 & 8 & | & -16 \\ 0 & 0 & 0 & | & 0 \end{bmatrix}$
$2r_1 + r_3 \Rightarrow r_3$

2. Use elementary row operations to convert to REF.

$-\frac{1}{4}r_2 \Rightarrow r_2$ $\begin{bmatrix} 1 & -2 & -4 & | & 2 \\ 0 & 1 & -2 & | & 4 \\ 0 & 0 & 0 & | & 0 \end{bmatrix}$

$2r_2 + r_1 \Rightarrow r_1$ $\begin{bmatrix} 1 & 0 & -8 & | & 10 \\ 0 & 1 & -2 & | & 4 \\ 0 & 0 & 0 & | & 0 \end{bmatrix}$

3. Convert to RREF.

$\begin{aligned} x - 8z &= 10 & \qquad y - 2z &= 4 \\ x &= 8z + 10 & y &= 2z + 4 \end{aligned}$

4. Use the equations from the first two rows to solve for x and y in terms of z.

$(10 + 8t, 4 + 2t, t)$

5. Write the general solution to the dependent system as an ordered triple where z is replaced by the arbitrary parameter t.

SKILL ✓ **EXERCISE 31**

15. 3 variables with only 2 equations
dependent consistent
$\begin{aligned} y + 3z &= 2 \\ y &= 2 - 3z \\ 2x - 3(2 - 3z) + z &= 3 \\ 2x - 6 + 10z &= 3 \\ 2x &= 9 - 10z \\ x &= \frac{9}{2} - 5z \end{aligned}$
letting $z = t$: $\left(\frac{9}{2} - 5t, 2 - 3t, t \right)$

16. $r_2 \Leftrightarrow r_1$ $\begin{bmatrix} 1 & 6 & -4 \\ 0 & -10 & 10 \end{bmatrix}$
$-3r_2 + r_1 \Rightarrow r_2$

$-\frac{1}{10}r_2 \Rightarrow r_2$ $\begin{bmatrix} 1 & 6 & -4 \\ 0 & 1 & -1 \end{bmatrix}$

17. $\frac{1}{2}r_1 \Rightarrow r_1$ $\begin{bmatrix} 1 & \frac{7}{2} & -2 \\ 0 & -9 & 18 \end{bmatrix}$
$-2r_1 + r_2 \Rightarrow r_2$

$-\frac{1}{9}r_2 \Rightarrow r_2$ $\begin{bmatrix} 1 & \frac{7}{2} & -2 \\ 0 & 1 & -2 \end{bmatrix}$

18. $-2r_1 + r_2 \Rightarrow r_2$ $\begin{bmatrix} 1 & 1 & -1 & 3 \\ 0 & 1 & 2 & -8 \\ 0 & 6 & 1 & -26 \end{bmatrix}$

$-6r_2 + r_3 \Rightarrow r_3$ $\begin{bmatrix} 1 & 1 & -1 & 3 \\ 0 & 1 & 2 & -8 \\ 0 & 0 & -11 & 22 \end{bmatrix}$

$-\frac{1}{11}r_3 \Rightarrow r_3$ $\begin{bmatrix} 1 & 1 & -1 & 3 \\ 0 & 1 & 2 & -8 \\ 0 & 0 & 1 & -2 \end{bmatrix}$

The system has an infinite number of solutions where every real value of t corresponds to an ordered triple that is a particular solution to the system. The particular solution corresponding to $t = 1$ is $(18, 6, 1)$, which can be checked in each equation of the original system.

An inconsistent system produces a row representing a false equation with all the coefficients being 0 but with a nonzero constant.

Example 5   Finding No Solution

Solve the system of linear equations.

$$3x - 4y + z = 7$$
$$-3x + 3z = -9$$
$$-x - 2y + 3z = -3$$

Answer

$$\begin{bmatrix} 3 & -4 & 1 & | & 7 \\ -3 & 0 & 3 & | & -9 \\ -1 & -2 & 3 & | & -3 \end{bmatrix}$$

1. Write the system as an augmented matrix.

$$-r_3 \Leftrightarrow r_1 \begin{bmatrix} 1 & 2 & -3 & | & 3 \\ -3 & 0 & 3 & | & -9 \\ 3 & -4 & 1 & | & 7 \end{bmatrix}$$

2. Use elementary row operations to convert to REF.

$$\begin{matrix} r_3 + r_2 \Rightarrow r_2 \\ -3r_1 + r_3 \Rightarrow r_3 \end{matrix} \begin{bmatrix} 1 & 2 & -3 & | & 3 \\ 0 & -4 & 4 & | & -2 \\ 0 & -10 & 10 & | & -2 \end{bmatrix}$$

$$\begin{matrix} -\frac{1}{4}r_2 \Rightarrow r_2 \\ -\frac{1}{10}r_3 \Rightarrow r_3 \end{matrix} \begin{bmatrix} 1 & 2 & -3 & | & 3 \\ 0 & 1 & -1 & | & \frac{1}{2} \\ 0 & 1 & -1 & | & \frac{1}{5} \end{bmatrix}$$

$$-r_2 + r_3 \Rightarrow r_3 \begin{bmatrix} 1 & 2 & -3 & | & 3 \\ 0 & 1 & -1 & | & \frac{1}{2} \\ 0 & 0 & 0 & | & -\frac{3}{10} \end{bmatrix}$$

3. Note that the system is inconsistent since the last row represents a false equation:
$$0 + 0 + 0 = -\frac{3}{10}.$$

no solution (or $\varnothing$)

SKILL ✓ **EXERCISE 31**

Many problems related to mixtures can be modeled by a system of equations. Once a system of equations is written, technology is frequently used to find the RREF of the related augmented matrix.

Example 6   Finding Profit

To raise funds, Sweet Valley Christian School sells boxes of varying combinations of candies, as shown. Find the profit from each type of candy if the school makes a profit of $6.60 on each bronze box, $8.35 on each silver box, and $15.15 on each gold box.

Box	Caramel	Almond	Truffle
Bronze	6	0	3
Silver	4	3	3
Gold	5	5	8

Answer

$$6c + 0a + 3t = 6.60$$
$$4c + 3a + 3t = 8.35$$
$$5c + 5a + 8t = 15.15$$

1. Write a system of equations representing the profit from each box. Use c, a, and t to represent the profit from each caramel, almond, and truffle candy, respectively.

CONTINUED ➡

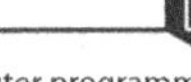

In computer programming, Gauss-Jordan elimination is frequently used to find the multiplicative inverse of a square matrix. This procedure is illustrated in Section 7.4.

21.
$$\begin{bmatrix} 3 & 5 & | & 9 \\ 6 & 7 & | & 9 \end{bmatrix}$$

$$\frac{1}{3}r_1 \Rightarrow r_1 \begin{bmatrix} 1 & \frac{5}{3} & | & 3 \\ 0 & -3 & | & -9 \end{bmatrix}$$
$$-2r_1 + r_2 \Rightarrow r_2$$

$$-\frac{1}{3}r_2 \Rightarrow r_2 \begin{bmatrix} 1 & \frac{5}{3} & | & 3 \\ 0 & 1 & | & 3 \end{bmatrix}$$

$$x + \frac{5}{3}(3) = 3$$
$$x = 3 - 5 = -2;\ (-2, 3)$$

22.
$$\begin{bmatrix} 4 & 1 & 3 & | & 5 \\ 0 & 2 & 5 & | & 6 \\ 0 & 0 & 3 & | & 1 \end{bmatrix}$$

$$\frac{1}{4}r_1 \Rightarrow r_1 \begin{bmatrix} 1 & \frac{1}{4} & \frac{3}{4} & | & \frac{5}{4} \end{bmatrix}$$
$$\frac{1}{2}r_2 \Rightarrow r_2 \begin{bmatrix} 0 & 1 & \frac{5}{2} & | & 3 \end{bmatrix}$$
$$\frac{1}{3}r_3 \Rightarrow r_3 \begin{bmatrix} 0 & 0 & 1 & | & \frac{1}{3} \end{bmatrix}$$

$$y + \frac{5}{2}\left(\frac{1}{3}\right) = 3$$
$$y = 3 - \frac{5}{6} = \frac{13}{6}$$
$$x + \frac{1}{4}\left(\frac{13}{6}\right) + \frac{3}{4}\left(\frac{1}{3}\right) = \frac{5}{4}$$
$$x = \frac{5}{4} - \frac{13}{24} - \frac{1}{4}$$
$$x = \frac{11}{24};\ \left(\frac{11}{24}, \frac{13}{6}, \frac{1}{3}\right)$$

23.
$$\begin{bmatrix} 1 & -1 & -1 & | & 10 \\ 3 & 0 & 5 & | & 9 \\ 4 & -9 & 0 & | & 23 \end{bmatrix}$$

$$\begin{matrix} -3r_1 + r_2 \Rightarrow r_2 \\ -4r_1 + r_3 \Rightarrow r_3 \end{matrix} \begin{bmatrix} 1 & -1 & -1 & | & 10 \\ 0 & 3 & 8 & | & -21 \\ 0 & -5 & 4 & | & -17 \end{bmatrix}$$

$$\frac{1}{3}r_2 \Rightarrow r_2 \begin{bmatrix} 1 & -1 & -1 & | & 10 \\ 0 & 1 & \frac{8}{3} & | & -7 \\ 0 & -5 & 4 & | & -17 \end{bmatrix}$$

$$r_2 + r_1 \Rightarrow r_1 \begin{bmatrix} 1 & 0 & \frac{5}{3} & | & 3 \\ 0 & 1 & \frac{8}{3} & | & -7 \\ 0 & 0 & \frac{52}{3} & | & -52 \end{bmatrix}$$
$$5r_2 + r_3 \Rightarrow r_3$$

$$\frac{3}{52}r_3 \Rightarrow r_3 \begin{bmatrix} 1 & 0 & \frac{5}{3} & | & 3 \\ 0 & 1 & \frac{8}{3} & | & -7 \\ 0 & 0 & 1 & | & -3 \end{bmatrix}$$

$$y + \frac{8}{3}(-3) = -7;\ y = 1$$
$$x + \frac{5}{3}(-3) = 3$$
$$x = 8;\ (8, 1, -3)$$

19.
$$\begin{matrix} 5r_1 \Rightarrow r_1 \\ -4r_2 \Rightarrow r_2 \end{matrix} \begin{bmatrix} 20 & 15 & -5 \\ -20 & -4 & -28 \end{bmatrix}$$

$$r_1 + r_2 \Rightarrow r_2 \begin{bmatrix} 20 & 15 & -5 \\ 0 & 11 & -33 \end{bmatrix}$$

$$\frac{1}{20}r_1 \Rightarrow r_1 \begin{bmatrix} 1 & \frac{3}{4} & -\frac{1}{4} \\ 0 & 1 & -3 \end{bmatrix}$$
$$\frac{1}{11}r_2 \Rightarrow r_2$$

$$-\frac{3}{4}r_2 + r_1 \Rightarrow r_1 \begin{bmatrix} 1 & 0 & 2 \\ 0 & 1 & -3 \end{bmatrix}$$

20.
$$\begin{matrix} -3r_1 + r_2 \Rightarrow r_2 \\ -4r_1 + r_3 \Rightarrow r_3 \end{matrix} \begin{bmatrix} 1 & -2 & 1 & -7 \\ 0 & 7 & -5 & 33 \\ 0 & 5 & 0 & 20 \end{bmatrix}$$

$$\frac{1}{5}r_3 \Leftrightarrow r_2 \begin{bmatrix} 1 & -2 & 1 & -7 \\ 0 & 1 & 0 & 4 \\ 0 & 7 & -5 & 33 \end{bmatrix}$$

$$-7r_2 + r_3 \Rightarrow r_3 \begin{bmatrix} 1 & -2 & 1 & -7 \\ 0 & 1 & 0 & 4 \\ 0 & 0 & -5 & 5 \end{bmatrix}$$

$$-\frac{1}{5}r_3 \begin{bmatrix} 1 & -2 & 1 & -7 \\ 0 & 1 & 0 & 4 \\ 0 & 0 & 1 & -1 \end{bmatrix}$$

$$2r_2 + r_1 \Rightarrow r_1 \begin{bmatrix} 1 & 0 & 1 & 1 \\ 0 & 1 & 0 & 4 \\ 0 & 0 & 1 & -1 \end{bmatrix}$$

$$-r_3 + r_1 \Rightarrow r_1 \begin{bmatrix} 1 & 0 & 0 & 2 \\ 0 & 1 & 0 & 4 \\ 0 & 0 & 1 & -1 \end{bmatrix}$$

24.
$$\begin{bmatrix} 8 & -7 & | & -32 \\ 4 & 1 & | & 2 \end{bmatrix}$$

$\frac{1}{4}r_2 \Leftrightarrow r_1 \begin{bmatrix} 1 & \frac{1}{4} & | & \frac{1}{2} \\ 0 & -9 & | & -36 \end{bmatrix}$
$-2r_2 + r_1 \Rightarrow r_2$

$-\frac{1}{9}r_2 \Rightarrow r_2 \begin{bmatrix} 1 & \frac{1}{4} & | & \frac{1}{2} \\ 0 & 1 & | & 4 \end{bmatrix}$

$-\frac{1}{4}r_2 + r_1 \Rightarrow r_1 \begin{bmatrix} 1 & 0 & | & -\frac{1}{2} \\ 0 & 1 & | & 4 \end{bmatrix}$

$\left(-\frac{1}{2}, 4\right)$

25.
$$\begin{bmatrix} 15 & 2 & | & -15 \\ 6 & -5 & | & 23 \end{bmatrix}$$

$\frac{1}{15}r_1 \Rightarrow r_1 \begin{bmatrix} 1 & \frac{2}{15} & | & -1 \\ 0 & -29 & | & 145 \end{bmatrix}$
$-2r_1 + 5r_2 \Rightarrow r_2$

$-\frac{1}{29}r_2 \Rightarrow r_2 \begin{bmatrix} 1 & \frac{2}{15} & | & -1 \\ 0 & 1 & | & -5 \end{bmatrix}$

$-\frac{2}{15}r_2 + r_1 \Rightarrow r_1 \begin{bmatrix} 1 & 0 & | & -\frac{1}{3} \\ 0 & 1 & | & -5 \end{bmatrix}$

$\left(-\frac{1}{3}, -5\right)$

26.
$$\begin{bmatrix} 8 & -1 & | & 9 \\ 2 & -\frac{1}{4} & | & 5 \end{bmatrix}$$

$\frac{1}{8}r_1 \Rightarrow r_1 \begin{bmatrix} 1 & -\frac{1}{8} & | & \frac{9}{8} \\ 0 & 0 & | & -11 \end{bmatrix}$
$-4r_2 + r_1 \Rightarrow r_2$

$\varnothing$; inconsistent

27.
$$\begin{bmatrix} 4 & -3 & 2 & | & -7 \\ 6 & 2 & 3 & | & 9 \\ -8 & 3 & -5 & | & 3 \end{bmatrix}$$

$-3r_1 + 2r_2 \Rightarrow r_2 \begin{bmatrix} 4 & -3 & 2 & | & -7 \\ 0 & 13 & 0 & | & 39 \\ 0 & -3 & -1 & | & -11 \end{bmatrix}$
$2r_1 + r_3 \Rightarrow r_3$

$\frac{1}{13}r_2 \Rightarrow r_2 \begin{bmatrix} 4 & -3 & 2 & | & -7 \\ 0 & 1 & 0 & | & 3 \\ 0 & -3 & -1 & | & -11 \end{bmatrix}$

$3r_2 + r_1 \Rightarrow r_1 \begin{bmatrix} 4 & 0 & 2 & | & 2 \\ 0 & 1 & 0 & | & 3 \\ 0 & 0 & -1 & | & -2 \end{bmatrix}$
$3r_2 + r_3 \Rightarrow r_3$

$2r_3 + r_1 \Rightarrow r_1 \begin{bmatrix} 4 & 0 & 0 & | & -2 \\ 0 & 1 & 0 & | & 3 \\ 0 & 0 & -1 & | & -2 \end{bmatrix}$

$\frac{1}{4}r_1 \Rightarrow r_1 \begin{bmatrix} 1 & 0 & 0 & | & -\frac{1}{2} \\ 0 & 1 & 0 & | & 3 \\ 0 & 0 & 1 & | & 2 \end{bmatrix}$
$-r_3 \Rightarrow r_3$

$\left(-\frac{1}{2}, 3, 2\right)$

28.
$$\begin{bmatrix} 3 & -2 & -5 & | & -9 \\ 3 & 4 & -4 & | & 1 \\ -2 & 3 & 5 & | & 6 \end{bmatrix}$$

$r_1 - r_2 \Rightarrow r_2 \begin{bmatrix} 3 & -2 & -5 & | & -9 \\ 0 & -6 & -1 & | & -10 \\ 0 & 5 & 5 & | & 0 \end{bmatrix}$
$2r_1 + 3r_3 \Rightarrow r_3$

$\frac{1}{5}r_3 \Leftrightarrow r_2 \begin{bmatrix} 3 & -2 & -5 & | & -9 \\ 0 & 1 & 1 & | & 0 \\ 0 & -6 & -1 & | & -10 \end{bmatrix}$

$2r_2 + r_1 \Rightarrow r_1 \begin{bmatrix} 3 & 0 & -3 & | & -9 \\ 0 & 1 & 1 & | & 0 \\ 0 & 0 & 5 & | & -10 \end{bmatrix}$
$6r_2 + r_3 \Rightarrow r_3$

$\frac{1}{3}r_1 \Rightarrow r_1 \begin{bmatrix} 1 & 0 & -1 & | & -3 \\ 0 & 1 & 1 & | & 0 \\ 0 & 0 & 1 & | & -2 \end{bmatrix}$
$\frac{1}{5}r_3 \Rightarrow r_3$

$r_1 + r_3 \Rightarrow r_1 \begin{bmatrix} 1 & 0 & 0 & | & -5 \\ 0 & 1 & 0 & | & 2 \\ 0 & 0 & 1 & | & -2 \end{bmatrix}$
$r_2 - r_3 \Rightarrow r_2$

$(-5, 2, -2)$

29.
$$\begin{bmatrix} 1 & -1 & -17 & | & -2 \\ 0 & 1 & 4 & | & 2 \\ 2 & 6 & -2 & | & 8 \end{bmatrix}$$

$-2r_1 + r_3 \Rightarrow r_3 \begin{bmatrix} 1 & -1 & -17 & | & -2 \\ 0 & 1 & 4 & | & 2 \\ 0 & 8 & 32 & | & 12 \end{bmatrix}$

$-8r_2 + r_3 \Rightarrow r_3 \begin{bmatrix} 1 & -1 & -17 & | & -2 \\ 0 & 1 & 4 & | & 2 \\ 0 & 0 & 0 & | & -4 \end{bmatrix}$

$\varnothing$; inconsistent

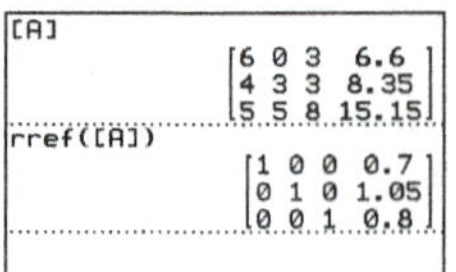

There is a profit of $0.70 on each caramel candy, $1.05 on each almond candy, and $0.80 on each truffle candy.

2. Create an augmented matrix representing the system and use technology to find its equivalent RREF.

3. Interpret the results.

TECHNOLOGY CORNER (TI-84 PLUS FAMILY)

The MATRIX menu ([2nd], [x⁻¹]) can be used to complete many matrix operations, such as multiplying matrices, finding the equivalent row echelon form and reduced row echelon form, calculating the determinant, and finding inverse matrices.

To enter matrix A from Example 3, select the EDIT submenu from the MATRIX menu, select 1: [A], and then enter the dimensions and each individual element. Select QUIT ([2nd], [MODE]) to return to the home screen.

To find the row echelon form use [MATRIX], MATH, A:ref(, and then enter the matrix using [MATRIX], 1:[A] 3×4, [ENTER]. The reduced row echelon form can be found using the B:rref(command, one of many matrix operations found in the [MATRIX], MATH submenu.

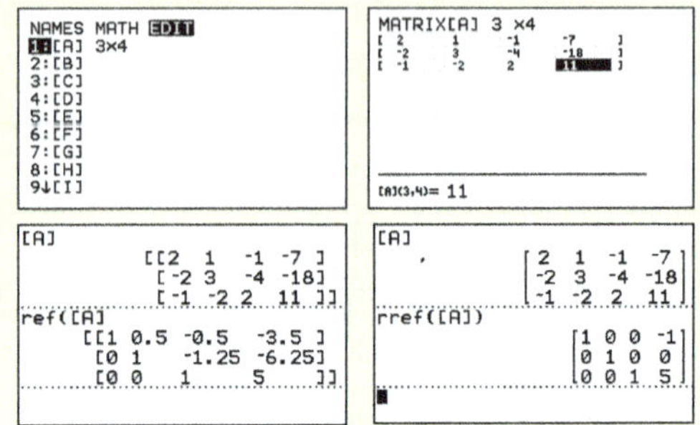

❯ **A. Exercises**

Use back substitution to solve each system.

1. $5x + 3y = 2$
$y = 4$ $(-2, 4)$

2. $x + 3y - 4z = 1$
$2y + z = -5$
$z = -1$ $(3, -2, -1)$

Write each system of equations as an augmented matrix.

3. $3x - 4y = 6$
$2x + 5y = -19$

4. $2x + 4y - 3z = 9$
$3x - 2z = 9$
$4x - y = 18$

Write each system as an augmented matrix. Then complete the series of elementary row operations on both the system and the augmented matrix.

5. $2x + 5y = -16$
$x - y = 6$
a. $r_1 \Leftrightarrow r_2$
b. $-2r_1 + r_2 \Rightarrow r_2$
c. $\frac{1}{7}r_2 \Rightarrow r_2$

6. $3x - 6y = 24$
$4x - 3y = 7$
a. $\frac{1}{3}r_1 \Rightarrow r_1$
b. $-4r_1 + r_2 \Rightarrow r_2$
c. $\frac{1}{5}r_2 \Rightarrow r_2$

7. $x + y + z = 4$
$4x + 3y + z = 4$
$3x - y - 2z = -1$
a. $-4r_1 + r_2 \Rightarrow r_2$
  $-3r_1 + r_3 \Rightarrow r_3$
b. $-r_2 \Rightarrow r_2$
c. $4r_2 + r_3 \Rightarrow r_3$
d. $\frac{1}{7}r_3 \Rightarrow r_3$

8. $x - y - z = 9$
$2x + y + 3z = 1$
$5x - 6y + 4z = 8$
a. $-2r_1 + r_2 \Rightarrow r_2$
  $-5r_1 + r_3 \Rightarrow r_3$
b. $-r_3 \Leftrightarrow r_2$
c. $-3r_2 + r_3 \Rightarrow r_3$
d. $\frac{1}{32}r_3 \Rightarrow r_3$

Use the results of exercises 5–8 and back substitution to solve each system.

9. $2x + 5y = -16$
$x - y = 6$ $(2, -4)$

10. $3x - 6y = 24$
$4x - 3y = 7$ $(-2, -5)$

11. $x + y + z = 4$
$4x + 3y + z = 4$
$3x - y - 2z = -1$ $(2, -3, 5)$

12. $x - y - z = 9$
$2x + y + 3z = 1$
$5x - 6y + 4z = 8$ $(6, 1, -4)$

Write the solution for the system represented by each matrix and classify the system as independent consistent, dependent consistent, or inconsistent.

13. $\begin{bmatrix} 1 & 2 & 3 & 5 \\ 0 & 1 & 2 & -1 \\ 0 & 0 & 1 & 4 \end{bmatrix}$

14. $\begin{bmatrix} 3 & 5 & 2 & -7 \\ 0 & -2 & 8 & 4 \\ 0 & 0 & 0 & 6 \end{bmatrix}$

15. $\begin{bmatrix} 2 & -3 & 1 & 3 \\ 0 & 1 & 3 & 2 \\ 0 & 0 & 0 & 0 \end{bmatrix}$

13. $(11, -9, 4)$; independent consistent
14. $\varnothing$; inconsistent
15. $\left(\frac{9}{2} - 5t, 2 - 3t, t \right)$; dependent consistent

Write each matrix in equivalent row echelon form.

16. $\begin{bmatrix} 3 & 8 & -2 \\ 1 & 6 & -4 \end{bmatrix}$

17. $\begin{bmatrix} 2 & 7 & -4 \\ 4 & 5 & 10 \end{bmatrix}$

18. $\begin{bmatrix} 1 & 1 & -1 & 3 \\ 2 & 3 & 0 & -2 \\ 0 & 6 & 1 & -26 \end{bmatrix}$

Write each matrix in reduced row echelon form.

19. $\begin{bmatrix} 4 & 3 & -1 \\ 5 & 1 & 7 \end{bmatrix}$

20. $\begin{bmatrix} 1 & -2 & 1 & -7 \\ 3 & 1 & -2 & 12 \\ 4 & -3 & 4 & -8 \end{bmatrix}$

B. Exercises

Solve each system using Gaussian elimination.

21. $3x + 5y = 9$
$6x + 7y = 9$ $(-2, 3)$

22. $4x + y + 3z = 5$
$2y + 5z = 6$
$3z = 1$ $\left(\frac{11}{24}, \frac{13}{6}, \frac{1}{3} \right)$

23. $x - y - z = 10$
$3x + 5z = 9$
$4x - 9y = 23$ $(8, 1, -3)$

Use Gauss-Jordan elimination to solve each system. Classify any inconsistent or dependent systems.

24. $8x - 7y = -32$
$4x + y = 2$

25. $15x + 2y = -15$
$6x - 5y = 23$

26. $8x - y = 9$
$2x - \frac{1}{4}y = 5$

27. $4x - 3y + 2z = -7$
$6x + 2y + 3z = 9$
$-8x + 3y - 5z = 3$

28. $3x - 2y - 5z = -9$
$3x + 4y - 4z = 1$
$-2x + 3y + 5z = 6$

29. $x - y - 17z = -2$
$y + 4z = 2$
$2x + 6y - 2z = 8$

30. $-2x + 3y + 5z = -1$
$6x - 2y - 3z = 5$

31. $3x + 4y + 5z = 4$
$x + 2y - 2z = 2$

Solve by using technology to find the reduced row echelon form of an augmented matrix.

32. $\frac{1}{8}x + \frac{1}{4}y = \frac{7}{5}$
$\frac{1}{2}x - \frac{1}{5}y = -\frac{7}{8}$ $\left(\frac{49}{120}, \frac{259}{48} \right)$

33. $1.5x - 3.3y + 0.7z = 9.9$
$0.2x + 4.1y - 1.3z = 10.5$
$2.4x + 0.4y - 7.5z = 2.2$
(Round to the nearest hundredth.) $(11.81, 3.14, 3.65)$

34. $2x_1 + 3x_2 + x_3 + 2x_4 = 1$
$4x_1 - x_2 + 3x_3 = 4$
$-6x_1 + 4x_2 + 2x_3 + 3x_4 = -2$
$-4x_1 - 2x_2 + 3x_3 + 2x_4 = 8$ $\left(\frac{1}{2}, -2, 0, 3 \right)$

35. $6x_1 + x_2 - 4x_3 - 5x_4 = 1$
$2x_1 + 3x_2 - 2x_3 - 2x_4 = 6$
$4x_1 - 2x_2 - 2x_3 - 3x_4 = -5$
$4x_1 + 2x_2 - 3x_3 - 5x_4 = -1$ $\left(\frac{18 + 5t}{8}, \frac{10 + t}{4}, t, 3 \right)$

36. Auto Mart sells a full-strength antifreeze. Tom already has a brand that is a 50% antifreeze-water mixture. How many gallons of each should be mixed together to get 3 gal of a 70% antifreeze-water mixture? 1.2 gal 100%; 1.8 gal 50%

37. Suzanne sells three different trail mixes at a local fair. Write and solve a system of equations to find the profit per cup for each ingredient.

Mix	Peanuts	Pretzels	Chocolate Pieces	Profit
Twisty	3 c	4 c	1 c	$5.40
Nutty	4 c	3 c	1 c	$5.90
Snack	1 c	2 c	0.5 c	$2.35

38. Joey sold the listed quantities of concessions at a local tournament and made $138.75, $106.25, and $188.75 in profits on Thursday, Friday, and Saturday, respectively. Write and solve a system of equations to find the profit per item for each type of concession.

Day	Popcorn	Drinks	Candy Bars
Thursday	30	60	15
Friday	25	40	20
Saturday	40	75	35

39. Josh has 72 quarters, dimes, and nickels, totaling $9.60. Write and solve a system of equations to find how many of each coin Josh has if he has 10 more quarters than dimes. 26 quarters, 16 dimes, 30 nickels

37. peanuts: $0.70; pretzels: $0.20; chocolate pieces: $2.50
38. popcorn: $1.25; drinks: $1.50; candy bars: $0.75

---

30. $\begin{bmatrix} -2 & 3 & 5 & | & -1 \\ 6 & -2 & -3 & | & 5 \end{bmatrix}$

$3r_1 + r_2 \Rightarrow r_2 \begin{bmatrix} -2 & 3 & 5 & | & -1 \\ 0 & 7 & 12 & | & 2 \end{bmatrix}$

$-7r_1 + 3r_2 \Rightarrow r_1 \begin{bmatrix} 14 & 0 & 1 & | & 13 \\ 0 & 7 & 12 & | & 2 \end{bmatrix}$

$\frac{1}{14}r_1 \Rightarrow r_1 \begin{bmatrix} 1 & 0 & \frac{1}{14} & | & \frac{13}{14} \\ 0 & 1 & \frac{12}{7} & | & \frac{2}{7} \end{bmatrix}$
$\frac{1}{7}r_2 \Rightarrow r_2$

$y + \frac{12}{7}z = \frac{2}{7}; y = \frac{2 - 12z}{7}$

$x + \frac{1}{14}z = \frac{13}{14}; x = \frac{13 - z}{14}$

letting $z = t$: $\left(\frac{13 - t}{14}, \frac{2 - 12t}{7}, t \right)$
dependent

31. $r_1 \Leftrightarrow r_2 \begin{bmatrix} 1 & 2 & -2 & | & 2 \\ 3 & 4 & 5 & | & 4 \end{bmatrix}$

$-3r_1 + r_2 \Rightarrow r_2 \begin{bmatrix} 1 & 2 & -2 & | & 2 \\ 0 & -2 & 11 & | & -2 \end{bmatrix}$

$r_1 + r_2 \Rightarrow r_1 \begin{bmatrix} 1 & 0 & 9 & | & 0 \\ 0 & 1 & -\frac{11}{2} & | & 1 \end{bmatrix}$
$-\frac{1}{2}r_2 \Rightarrow r_2$

$y - \frac{11}{2}z = 1; y = \frac{2 + 11z}{2}$

$x + 9z = 0; x = -9z$

letting $z = t$: $\left(-9t, \frac{2 + 11t}{2}, t \right)$
dependent

32. ```
rref([A])
      [1 0 0.4083333333]
      [0 1 5.395833333]
Ans▶Frac
              [1 0 49/120]
              [0 1 259/48]
```

33. ```
rref([A])
      [[1 0 0 11.8102966 ]
       [0 1 0 3.143323867]
       [0 0 1 3.65360551811]
```

34. ```
rref([A])
      [[1 0 0 0 0.5]
       [0 1 0 0 -2]
       [0 0 1 0 0]
       [0 0 0 1 3]]
■
```

35. ```
rref([A])▶Frac
      [1 0 -5/8 0 9/4]
      [0 1 -1/4 0 5/2]
      [0 0 0 1 3]
      [0 0 0 0 0]
■
```

$x_1 - \frac{5}{8}x_3 = \frac{9}{4}; x_1 = \frac{18 + 5x_3}{8}$

$x_2 - \frac{1}{4}x_3 = \frac{5}{2}; x_2 = \frac{10 + x_3}{4}$

letting $x_3 = t$: $\left(\frac{18 + 5t}{8}, \frac{10 + t}{4}, t, 3 \right)$

36.

Gallons of Mixtures	Strength	Gallons of Antifreeze
x	1	x
y	0.5	$0.5y$
3	0.7	2.1

$x + y = 3; x + 0.5y = 2.1$

$\begin{bmatrix} 1 & 1 & | & 3 \\ 1 & 0.5 & | & 2.1 \end{bmatrix}$

$r_1 - r_2 \Rightarrow r_2 \begin{bmatrix} 1 & 1 & | & 3 \\ 0 & 0.5 & | & 0.9 \end{bmatrix}$

$2r_2 \Rightarrow r_2 \begin{bmatrix} 1 & 1 & | & 3 \\ 0 & 1 & | & 1.8 \end{bmatrix}$

$-r_2 + r_1 \Rightarrow r_1 \begin{bmatrix} 1 & 0 & | & 1.2 \\ 0 & 1 & | & 1.8 \end{bmatrix}$

1.2 gal 100%; 1.8 gal 50%

37. $3n + 4p + c = 5.40$
$4n + 3p + c = 5.90$
$n + 2p + 0.5c = 2.35$

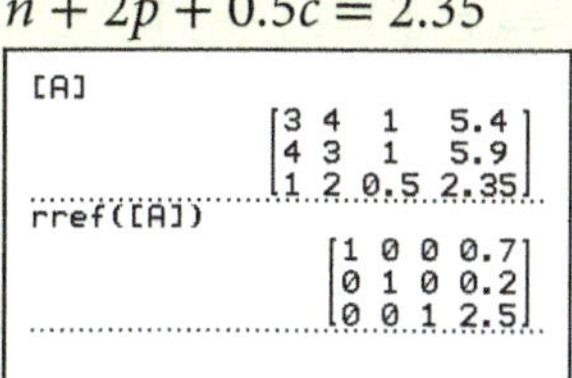
```
[A]
      [3 4  1   5.4 ]
      [4 3  1   5.9 ]
      [1 2 0.5 2.35]
rref([A])
      [1 0 0 0.7]
      [0 1 0 0.2]
      [0 0 1 2.5]
```

38. $30p + 60d + 15c = 138.75$
$25p + 40d + 20c = 106.25$
$40p + 75d + 35c = 188.75$

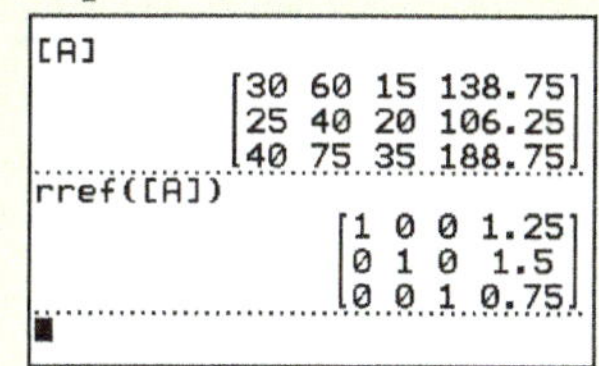
```
[A]
      [30 60 15 138.75]
      [25 40 20 106.25]
      [40 75 35 188.75]
rref([A])
      [1 0 0 1.25]
      [0 1 0 1.5 ]
      [0 0 1 0.75]
■
```

39. $q + d + n = 72$

$\quad 25q + 10d + 5n = 960$ (in cents)

$\quad q = d + 10$ or $q - d = 10$

```
[A]
        [ 1   1  1 |  72 ]
        [25  10  5 | 960 ]
        [ 1  -1  0 |  10 ]
rref([A])
        [1 0 0 | 26]
        [0 1 0 | 16]
        [0 0 1 | 30]
■
```

❯ C. Exercises

40.

$$\begin{bmatrix} 5 & -2 & 4 & | & -9 \\ 3 & -4 & 2 & | & -2 \\ 2 & 6 & 5 & | & -7 \end{bmatrix}$$

$$\begin{matrix} \\ -5r_2 + 3r_1 \Rightarrow r_2 \\ -2r_1 + 5r_3 \Rightarrow r_3 \end{matrix} \begin{bmatrix} 5 & -2 & 4 & | & -9 \\ 0 & 14 & 2 & | & -17 \\ 0 & 34 & 17 & | & -17 \end{bmatrix}$$

$$\tfrac{1}{14}r_2 \Rightarrow r_2 \begin{bmatrix} 5 & -2 & 4 & | & -9 \\ 0 & 1 & \frac{1}{7} & | & -\frac{17}{14} \\ 0 & 34 & 17 & | & -17 \end{bmatrix}$$

$$\begin{matrix} 2r_2 + r_1 \Rightarrow r_1 \\ \\ -34r_2 + r_3 \Rightarrow r_3 \end{matrix} \begin{bmatrix} 5 & 0 & \frac{30}{7} & | & -\frac{80}{7} \\ 0 & 1 & \frac{1}{7} & | & -\frac{17}{14} \\ 0 & 0 & \frac{85}{7} & | & \frac{170}{7} \end{bmatrix}$$

$$\begin{matrix} \tfrac{1}{5}r_1 \Rightarrow r_1 \\ \\ \tfrac{7}{85}r_3 \Rightarrow r_3 \end{matrix} \begin{bmatrix} 1 & 0 & \frac{6}{7} & | & -\frac{16}{7} \\ 0 & 1 & \frac{1}{7} & | & -\frac{17}{14} \\ 0 & 0 & 1 & | & 2 \end{bmatrix}$$

$$\begin{matrix} -\tfrac{6}{7}r_3 + r_1 \Rightarrow r_1 \\ -\tfrac{1}{7}r_3 + r_2 \Rightarrow r_2 \end{matrix} \begin{bmatrix} 1 & 0 & 0 & | & -4 \\ 0 & 1 & 0 & | & -\frac{3}{2} \\ 0 & 0 & 1 & | & 2 \end{bmatrix}$$

41.

$$\begin{bmatrix} 12 & 6 & | & 9 \\ -2 & 1 & | & -\frac{3}{2} \\ 4 & 1 & | & 4 \end{bmatrix}$$

$$\begin{matrix} 6r_2 + r_1 \Rightarrow r_2 \\ -3r_3 + r_1 \Rightarrow r_3 \end{matrix} \begin{bmatrix} 12 & 6 & | & 9 \\ 0 & 12 & | & 0 \\ 0 & 3 & | & -3 \end{bmatrix}$$

$$\begin{matrix} \tfrac{1}{12}r_1 \Rightarrow r_1 \\ \tfrac{1}{12}r_2 \Rightarrow r_2 \\ \tfrac{1}{3}r_3 \Rightarrow r_3 \end{matrix} \begin{bmatrix} 1 & \frac{1}{2} & | & \frac{3}{4} \\ 0 & 1 & | & 0 \\ 0 & 1 & | & -1 \end{bmatrix}$$

$$r_2 - r_3 \Rightarrow r_3 \begin{bmatrix} 1 & \frac{1}{2} & | & \frac{3}{4} \\ 0 & 1 & | & 0 \\ 0 & 0 & | & 1 \end{bmatrix}$$

❯ C. Exercises

Use row operations on an augmented matrix to solve each system.

40. $5x - 2y + 4z = -9$
$\quad 3x - 4y + 2z = -2$
$\quad 2x + 6y + 5z = -7$
$\quad \left(-4, -\frac{3}{2}, 2\right)$

41. $12x + 6y = 9$
$\quad -2x + y = -\frac{3}{2}$
$\quad 4x + y = 4$
$\quad \varnothing$, inconsistent

The equation of the parabola passing through three given points can be found by solving the system of equations created by substituting each point into $y = ax^2 + bx + c$.

Use matrices to find $f(x) = ax^2 + bx + c$, whose graph passes through the given points.

42. $(1, 2), (2, 5), (3, 14)$ $y = 3x^2 - 6x + 5$

43. $(-3, -8), (1, 4), (2, -3)$ $y = -2x^2 - x + 7$

44. A Christian school earned \$150, \$400, and \$750, respectively, in the first three weeks of a paper drive. Find the quadratic function $f(x) = ax^2 + bx + c$ modeling the earnings during these weeks and use it to predict the fourth week's earnings.
$f(x) = 50x^2 + 100x; f(4) = \$1200$

CUMULATIVE REVIEW

45. Find the height of a building if a 6 ft tall observer 500 ft from the building views the top at a 52° angle of elevation. [4.2] ≈ 646 ft

46. Write $y = 3x^2 - 7x + 1$ in vertex form. State the coordinates of the vertex and identify it as a maximum or minimum point. [1.6]

47. Write the equation of $g(x)$ as a transformation of $f(x) = \cos x$ with an amplitude of 3 and a period of $\frac{\pi}{2}$, having been translated $\frac{\pi}{3}$ units right. [4.4]

48. Convert $z = 2 - 2i$ to polar form. [6.4] $2\sqrt{2} \operatorname{cis} \frac{7\pi}{4}$

49. Convert the polar equation $r = \frac{\sin\theta}{\cos^2\theta}$ to rectangular form. [6.6] $y = x^2$

50. Find the measure of each angle (to the nearest tenth of a degree) in a triangle with sides $AC = 10$ cm, $BC = 9$ cm, and $AB = 14$ cm. [5.7] $39.8°, 45.4°, 94.8°$

51. Identify any solutions to the system. List all correct answers. [7.1]   A, D
$y = -x^2 + 2x - 15$
$y = 3x - 21$

A. $(2, -15)$ **C.** $(3, 30)$ **E.** none of
B. $(-2, 15)$ **D.** $(-3, -30)$ these

46. $y = 3\left(x - \frac{7}{6}\right)^2 - \frac{37}{12}; \left(\frac{7}{6}, -\frac{37}{12}\right)$; minimum

47. $f(x) = 3\cos 4\left(x - \frac{\pi}{3}\right)$

52. Use graphing technology to find any solutions to the system. List all correct answers [7.1]   C
$y = x^3 - x^2 - 2x$
$y = x^2 - 4x + 1$

A. $(-2, 1)$ **C.** $(1, -2)$ **E.** none of
B. $(1, 2)$ **D.** $(2, -1)$ these

53. Which of the following is true of matrix operations? List all correct answers. [7.2]   B, D

A. $A_{2\times3}$ times $B_{3\times2}$ is a 2×3 matrix.
B. $AI_n = I_nA$ **D.** $A(B + C) = AB + AC$
C. $AB = BA$ **E.** none of these

54. Find entry e_{23} in AB, given
$A = \begin{bmatrix} -4 & 1 \\ 2 & -3 \end{bmatrix}$ and $B = \begin{bmatrix} -3 & 2 & 6 \\ 1 & 0 & -2 \end{bmatrix}$. [7.2]   B

A. 8   **C.** 6   **E.** none of
B. 18   **D.** -6   these

42.

$$\begin{bmatrix} 1 & 1 & 1 & | & 2 \\ 4 & 2 & 1 & | & 5 \\ 9 & 3 & 1 & | & 14 \end{bmatrix}$$

$$\begin{matrix} 4r_1 - r_2 \Rightarrow r_2 \\ 9r_1 - r_3 \Rightarrow r_3 \end{matrix} \begin{bmatrix} 1 & 1 & 1 & | & 2 \\ 0 & 2 & 3 & | & 3 \\ 0 & 6 & 8 & | & 4 \end{bmatrix}$$

$$\begin{matrix} -\tfrac{1}{2}r_2 + r_1 \Rightarrow r_1 \\ \\ 3r_2 - r_3 \Rightarrow r_3 \end{matrix} \begin{bmatrix} 1 & 0 & -\frac{1}{2} & | & \frac{1}{2} \\ 0 & 2 & 3 & | & 3 \\ 0 & 0 & 1 & | & 5 \end{bmatrix}$$

$$\begin{matrix} \tfrac{1}{2}r_3 + r_1 \Rightarrow r_1 \\ -3r_3 + r_2 \Rightarrow r_2 \end{matrix} \begin{bmatrix} 1 & 0 & 0 & | & 3 \\ 0 & 2 & 0 & | & -12 \\ 0 & 0 & 1 & | & 5 \end{bmatrix}$$

$$\tfrac{1}{2}r_2 \Rightarrow r_2 \begin{bmatrix} 1 & 0 & 0 & | & 3 \\ 0 & 1 & 0 & | & -6 \\ 0 & 0 & 1 & | & 5 \end{bmatrix}$$

continued in Answers and Solutions Overflow

7.4 Determinants

Gabriel Cramer (1704–52)

Many applications of matrices, including the solving of systems and the calculation of polygonal areas, use the *determinant* of a matrix. The determinant of a square matrix $A_{n \times n}$ is a real number denoted as $|A|$ or det (A). The determinant of a 1×1 matrix $A = [a]$ is equal to its only entry, a.

> **DEFINITION**
>
> The **determinant of a 2×2 matrix** $A = \begin{bmatrix} a & b \\ c & d \end{bmatrix}$ is given by
>
> $$|A| = \det(A) = \begin{vmatrix} a & b \\ c & d \end{vmatrix} = ad - cb.$$

The illustrated pattern emphasizes that the determinant of a 2×2 matrix is the difference of the products of the major and minor diagonals.

$$\begin{vmatrix} a & b \\ c & d \end{vmatrix} = ad - cb$$

Example 1 Finding the Determinant of a 2×2 Matrix

Find $|A|$ if $A = \begin{bmatrix} 6 & -2 \\ 5 & 8 \end{bmatrix}$.

Answer

$|A| = 6(8) - 5(-2) = 48 + 10 = 58$

SKILL ✔ EXERCISE 3

A pattern of products of diagonals can also be used to find the determinant of a 3×3 matrix. Begin by rewriting the first two columns to the right of the matrix.

$$\begin{bmatrix} a & b & c \\ d & e & f \\ g & h & i \end{bmatrix} \Rightarrow \begin{bmatrix} a & b & c \\ d & e & f \\ g & h & i \end{bmatrix}\begin{matrix} a & b \\ d & e \\ g & h \end{matrix}$$

Then find the sum of the products of each set of diagonals.

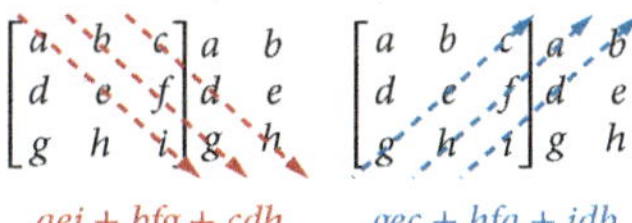

$aei + bfg + cdh \qquad gec + hfa + idb$

The determinant is the difference of the first and second sums.

> **DEFINITION**
>
> The **determinant of a 3×3 matrix** A is given by
>
> $$\det(A) = |A| = \begin{vmatrix} a & b & c \\ d & e & f \\ g & h & i \end{vmatrix} = (aei + bfg + cdh) - (gec + hfa + idb).$$

After completing this section, you will be able to

- calculate determinants.
- use Cramer's rule to solve systems of equations.
- use determinants to find polygonal areas.

Determinants were actually developed before matrices, but as matrices were explored, their importance and usefulness surpassed determinants to the point that determinants are now defined in terms of square matrices.

7.4 Determinants

Objectives

1. To calculate determinants of square matrices
2. To solve systems of equations using Cramer's rule
3. To find polygonal areas using determinants

Flash

Gabriel Cramer received his doctorate from the University of Geneva, Switzerland, at 18 yr old. Cramer's rule, named after him, was published in 1750.

Vocabulary

cofactor
Cramer's rule
determinant of a matrix
minor

Reading and Writing Mathematics

Explain how you would create a linear system of equations with a specific (x, y) solution. Then give an example for $(5, -7)$. Choose coefficients for the *x*- and *y*-terms, then use the coefficients and the values of *x* and *y* to find the constant term. Repeat for the second equation with at least one coefficient that is different from the ones used in the first equation.

$3(5) + 2(-7) = 1;\ 3x + 2y = 1$
$-5(5) - 3(-7) = -4;\ -5x - 3y = -4$

PRESENTATION

Lesson Opener

Solve the following system using an augmented matrix.

$$2x - y + 3z = 8$$
$$x + y + z = 2$$
$$-x + 3z = 1 \quad (2, -1, 1)$$

Introduce the *determinant* as a real number associated with any square matrix. Explain that it has several different applications, including finding the area of a polygonal region (Example 8), determining whether a matrix has an inverse (Section 7.5), and finding the inverse of a 2×2 matrix (Section 7.5). The area of a parallelogram is the absolute value of

the determinant of a 2×2 matrix. Determinants can also be viewed as a scaling factor of a transformation of a region described by a matrix. In calculus, changes in area or volume are represented by Jacobian determinants.

While the determinant of a 1×1 matrix may seem trivial, its definition is foundational for the determinant of an $n \times n$ matrix. Present the formula for the *determinant of a 2×2 matrix* and illustrate its application with Example 1.

One-on-One Because of the similarity of the determinant notation $|A|$ and absolute value notation, some may choose to use det(A). Encourage students to use context to avoid confusion.

Present the formula for the *determinant of a 3×3 matrix*, which uses products of diagonals. The difference of the sums of the products is demonstrated in Example 2. Crossing out each diagonal as its product is calculated may help some students. Students may recall using products of diagonals to find determinants of 2×2 and 3×3 matrices in ALGEBRA 2.

Example 3 highlights the formal definitions of *minors* and *cofactors*. Be sure students understand that the cofactor is equal to the minor when the entry is in an *even position* and that it is the opposite of the minor when the entry is in an *odd position*. Example 4 emphasizes that any row or column can be used to find the

Additional Exercises

Use a formula to find the determinant of each matrix.

1. $\begin{bmatrix} -2 & -3 \\ -5 & -8 \end{bmatrix}$ 1

2. $\begin{bmatrix} -6 & 9 & -3 \\ 2 & -3 & 1 \\ -4 & 3 & 1 \end{bmatrix}$ 0

Use cofactors of the given row or column to find the determinant of each matrix.

3. $\begin{bmatrix} 1 & 1 & 2 \\ 2 & -3 & 3 \\ 2 & 1 & 5 \end{bmatrix}$

  a. row 1   −6

  b. column 2   −6

4. $\begin{bmatrix} -1 & 3 & 2 \\ -3 & 1 & 4 \\ 2 & -1 & 1 \end{bmatrix}$

  a. row 3   30

  b. column 2   30

5. Expand by cofactors to find the determinant.

$$\begin{bmatrix} 2 & -6 & 1 & 4 \\ 3 & 0 & 6 & 0 \\ 1 & 4 & 0 & -1 \\ 5 & 2 & -1 & 2 \end{bmatrix}$$ 120

Use Cramer's rule to solve each system of equations.

6. $2x - 3y = -16$
$5x + 11y = 34$ $(-2, 4)$

7. $5x + 6y = 4$
$10x + 9y = 7$ $\left(\frac{2}{5}, \frac{1}{3}\right)$

8. $3x - 10y + 4z = -3$
$-x + 5y + 3z = 26$
$4x - 2y + 5z = 17$ $(-1, 2, 5)$

Use determinants to find the area of each shaded region.

9.
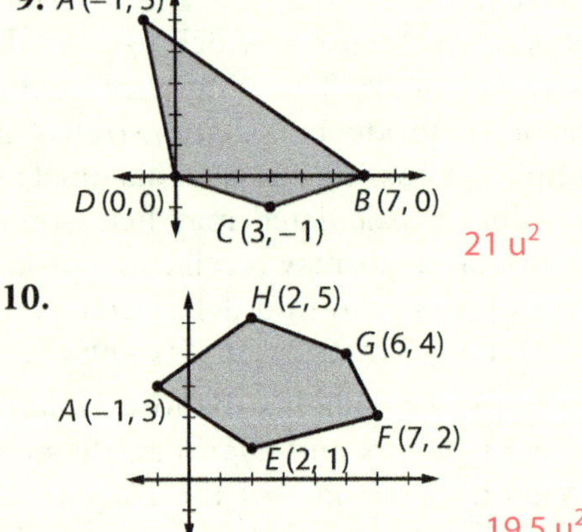

21 u²

10.

19.5 u²

Find $|B|$ if $B = \begin{bmatrix} 2 & -3 & 4 \\ 0 & 5 & -1 \\ 4 & 6 & 1 \end{bmatrix}$.

Answer

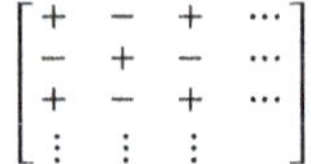

$|A| = [2(5)(1) + (-3)(-1)(4) + 4(0)(6)] - [4(5)(4) + 6(-1)(2) + 1(0)(-3)]$
$= [10 + 12 + 0] - [80 + (-12) + 0]$
$= 22 - 68 = -46$

Check

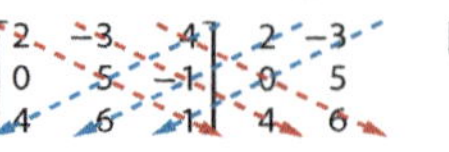

A graphing calculator or Internet app can be used to verify $|B|$.

_______________________ SKILL ✔ **EXERCISE 7**

Determinants of square matrices of an order higher than 3 require the use of *minors* and *cofactors*.

▶ DEFINITIONS

The **minor** of entry a_{ij} in square matrix A, notated M_{ij}, is the determinant of the matrix formed by removing the ith row and jth column of A.
The **cofactor** C_{ij} of entry a_{ij} is $C_{ij} = (-1)^{i+j}M_{ij}$.

$$\begin{bmatrix} + & - & + & \cdots \\ - & + & - & \cdots \\ + & - & + & \cdots \\ \vdots & \vdots & \vdots & \end{bmatrix}$$

When finding cofactors, the sign of the minor is not changed in *even positions* (where $i + j$ is even) but is negated in *odd positions* (where $i + j$ is odd).

Find the minors and cofactors for each entry in row 1 of $B = \begin{bmatrix} 2 & -3 & 4 \\ 0 & 5 & -1 \\ 4 & 6 & 1 \end{bmatrix}$.

Answer

$M_{11} = \begin{vmatrix} 2 & -3 & 4 \\ 0 & 5 & -1 \\ 4 & 6 & 1 \end{vmatrix}$ $M_{12} = \begin{vmatrix} 2 & -3 & 4 \\ 0 & 5 & -1 \\ 4 & 6 & 1 \end{vmatrix}$ $M_{13} = \begin{vmatrix} 2 & -3 & 4 \\ 0 & 5 & -1 \\ 4 & 6 & 1 \end{vmatrix}$

$= 5(1) - 6(-1) = 11$ $= 0(1) - 4(-1) = 4$ $= 0(6) - 4(5) = -20$

Use the sign pattern and the minors to state the cofactors.

$C_{11} = (+1)M_{11} = 11$ $C_{12} = (-1)M_{12} = -4$ $C_{13} = (+1)M_{13} = -20$

_______________________ SKILL ✔ **EXERCISE 11**

The use of minors and cofactors allows the determinant of a square matrix of order n to be defined inductively in terms of the determinants of square matrices of order $n - 1$.

determinant. Point out that the power of -1 from the cofactor is often applied to the entry before multiplying by the minor.

Example 5 illustrates how determinants of matrices with order higher than 3 are calculated using expansion by cofactors.

Discuss the pros and cons of finding the determinant of a 3 × 3 matrix and higher-order matrices using cofactors, the products of diagonals, and technology. Clearly communicate how you will expect students to calculate determinants for these larger-order matrices. Consider allowing the use of calcula-tors or Internet apps (Internet keyword search *matrix calculator*) when checking work and solving matrix equations.

Introduce *Cramer's rule* as another way to solve a system of linear equations. Discuss the derivation of the rule for 2 × 2 matrices before stating the generalized theorem. Use Example 6 to illustrate Cramer's rule. Point out that while systems of equations with fractional solutions can sometimes be challenging, they can be relatively easy using Cramer's rule.

Have the students recall that an in-consistent system has no solution and a dependent system has infinite solutions.

DEFINITION

The **determinant of an $n \times n$ matrix** $A = [a_{ij}]$ (where $n \geq 2$) is the sum of the product of entries in any row or column and their respective cofactors. For example, expanding by the ith row:

$$\det(A) = |A| = a_{i1}C_{i1} + a_{i2}C_{i2} + \ldots + a_{in}C_{in}.$$

Example 4   Expanding by Cofactors

If $B = \begin{bmatrix} 2 & -3 & 4 \\ 0 & 5 & -1 \\ 4 & 6 & 1 \end{bmatrix}$ find $|B|$ expanding along each of the following.

a. row 1   **b.** column 1

Answers

$|B| = b_{11}C_{11} + b_{12}C_{12} + b_{13}C_{13}$
$ = 2(11) + (-3)(-4) + 4(-20)$
$ = 22 + 12 - 80 = -46$

a. Use the entries in the first row and their cofactors found in Example 3.

$|B| = b_{11}C_{11} + b_{21}C_{21} + b_{31}C_{31}$
$ = 2(M_{11}) - 0(M_{21}) + 4(M_{31})$
$ = 2[5(1) - 6(-1)] + 4[-3(-1) - 5(4)]$
$ = 2(11) + 4(-17) = -46$

b. Notice that the sign from the cofactor's power of -1 is often applied to the entry before calculating the minor.

SKILL ✔ **EXERCISE 13**

Notice that the value of the determinant is the same no matter which row or column you use in the expansion. Expanding along a row or column with entries of 0 reduces the number of cofactors that need to be calculated since the product of an entry of 0 and its cofactor is always 0.

Example 5   Finding the Determinant of a 4×4 Matrix

Find $|D|$ if $D = \begin{bmatrix} 1 & 2 & 0 & -1 \\ -3 & 2 & 1 & 4 \\ -2 & -1 & 0 & 2 \\ 0 & 4 & 3 & 1 \end{bmatrix}$.

Answer

$|D| = 0C_{13} + 1C_{23} + 0C_{33} + 3C_{43}$

$ = 1(-1)\begin{vmatrix} 1 & 2 & -1 \\ -2 & -1 & 2 \\ 0 & 4 & 1 \end{vmatrix} + 3(-1)\begin{vmatrix} 1 & 2 & -1 \\ -3 & 2 & 4 \\ -2 & -1 & 2 \end{vmatrix}$

1. Only two cofactors need to be calculated when expanding along the third column.

$|D| = -1[0 - 4(2 - 2) + (-1 + 4)] - 3[(4 + 4) - 2(-6 + 8) - (3 + 4)]$

2. Evaluate M_{23} along row 3 and M_{43} along row 1.

$ = -1[0 + 3] - 3[8 - 4 - 7]$

3. Simplify.

$ = -3 + 9 = 6$

SKILL ✔ **EXERCISE 21**

For Graphing Calculators

To find a determinant on your calculator, first enter the matrix as instructed on page 358. Next, return to the MATRIX menu, arrow over to the MATH menu, and press 1:det(. Without pressing ENTER, go back into the MATRIX menu and choose your matrix.

Find each determinant.

1. $\begin{bmatrix} 3.5 & 8.7 \\ 1.4 & -2.6 \end{bmatrix}$ -21.28

2. $\begin{bmatrix} \frac{1}{2} & \frac{3}{4} & -\frac{9}{16} \\ \frac{3}{8} & -\frac{4}{15} & \frac{6}{7} \\ \frac{15}{16} & -\frac{7}{8} & \frac{5}{6} \end{bmatrix}$ 0.676

3. $\begin{bmatrix} 235 & 862 & -419 \\ 521 & 786 & -512 \\ 921 & -486 & 1024 \end{bmatrix}$ $-326{,}280{,}824$

Assignments

- **Minimum:** 1, 3, 6–7, 10, 13, 17a, 19, 25, 27, 31, 35, 38, 46–48, 52–54
- **Standard:** 1, 4, 7, 10–11, 15, 17b, 18b, 20, 27–29, 34–35, 38–39, 41–42, 46–47, 49, 52–54
- **Extended:** 2, 5, 8, 11, 14, 18a; 19–29 odd; 32, 37, 40–47, 49–50, 52–55

Assessment

- Quiz 7B covers Sections 7.3–7.4.

Then use Example 7 to demonstrate applying Cramer's rule to solve these types of systems. In each case the x- and y-value ratios are undefined. Their values are expressed as $\frac{0}{0}$ in the dependent system.

Be sure students understand that the $\pm$ sign in the formula for the area of a triangle is used to ensure that the area is positive. Cofactors are used to calculate the determinants in Example 8, but products of diagonals or technology could also be used.

Motivational Idea The stated formula for the area of a triangle uses the *transpose* of the matrix given in BJU Press's ALGEBRA 2 (Section 4.4). Consider introducing students to this operation in which each row is written as the corresponding column. Given $m \times n$ matrix $A = [a_{ij}]$, its transpose is the $n \times m$ matrix $A^T = [a_{ji}]$. The fact that the determinant can be found using rows or columns implies that $|A| = |A^T|$ and either area formula provides the same result.

TIPS

Ex. 25–34 You may want to simplify some of these problems by asking students to find the value of just one of the variables in the system's solution.

Solutions

A. Exercises

3. $8(9) - (-3)5 = 72 + 15 = 87$

4. $-2(-5) - (-1)(-4) = 10 - 4 = 6$

5. $11(-9) - 7(-13) = -99 + 91 = -8$

6. $\frac{1}{2}(8) - 5(6) = 4 - 30 = -26$

7. $\begin{bmatrix} 2 & 1 & -4 \\ 3 & 2 & -5 \\ 6 & 2 & 1 \end{bmatrix}\begin{matrix} 2 & 1 \\ 3 & 2 \\ 6 & 2 \end{matrix}$
$(4 - 30 - 24) - (-48 - 20 + 3)$
$= -50 + 65 = 15$

8. $\begin{bmatrix} 2 & -1 & 4 \\ 1 & 0 & -2 \\ 3 & 1 & 5 \end{bmatrix}\begin{matrix} 2 & -1 \\ 1 & 0 \\ 3 & 1 \end{matrix}$
$(0 + 6 + 4) - (0 - 4 - 5)$
$= 10 + 9 = 19$

9. $\begin{bmatrix} 1 & 0 & 2 \\ 1 & 0 & 3 \\ 0 & 1 & -2 \end{bmatrix}\begin{matrix} 1 & 0 \\ 1 & 0 \\ 0 & 1 \end{matrix}$
$(0 + 0 + 2) - (0 + 3 + 0)$
$= 2 - 3 = -1$

10. $\begin{bmatrix} 5 & -3 & 4 \\ 2 & 3 & -2 \\ 4 & -2 & 4 \end{bmatrix}\begin{matrix} 5 & -3 \\ 2 & 3 \\ 4 & -2 \end{matrix}$
$(60 + 24 - 16) - (48 + 20 - 24)$
$= 68 - 44 = 24$

11. $M_{11} = 2(-2) - 3(1) = -7$
$M_{12} = 4(-2) - 1(1) = -9$
$M_{13} = 4(3) - 1(2) = 10$
$C_{11} = (-1)^{1+1}M_{11} = (1)(-7) = -7$
$C_{12} = (-1)^{1+2}M_{12} = (-1)(-9) = 9$
$C_{13} = (-1)^{1+3}M_{13} = (1)(10) = 10$

12. $M_{12} = 4(-2) - 1(1) = -9$
$M_{22} = 2(-2) - 1(3) = -7$
$M_{32} = 2(1) - 4(3) = -10$
$C_{12} = (-1)^{1+2}M_{12} = (-1)(-9) = 9$
$C_{22} = (-1)^{2+2}M_{22} = (1)(-7) = -7$
$C_{23} = (-1)^{3+2}M_{32} = (-1)(-10) = 10$

13. $2(-4 - 3) + 1(-8 - 1) + 3(12 - 2)$
$= -14 - 9 + 30 = 7$

14. $-2(6 - 21) - 1(30 - 3) + 0$
$= 30 - 27 = 3$

15. $0 + 5(-1 + 6) + 0 = 25$

16. $4(5 - 14) - 3(-30 + 4) - 1(-42 + 2)$
$= -36 + 78 + 40 = 82$

B. Exercises

19a. $4(9 + 2) + 0 - 1(-4 - 3)$
$= 44 + 7 = 51$

19b. $-3(-1 - 12) + 2(2 + 4)$
$= 39 + 12 = 51$

20a. $-1(-4 - 5) - 1(8 - 3) - 3(20 + 6)$
$= 9 - 5 - 78 = -74$

20b. $4(-15 - 2) - 3(6 - 1) - 1(-4 - 5) = -68 - 15 + 9 = -74$

21. $2\begin{vmatrix} 2 & -2 & 3 \\ 3 & -1 & 2 \\ 4 & 3 & 1 \end{vmatrix} - 1\begin{vmatrix} 0 & 3 & 1 \\ 3 & -1 & 2 \\ 4 & 3 & 1 \end{vmatrix} + 0 + 0$
using row 1 in each 3×3 matrix:
$= 2[2(-1 - 6) + 2(3 - 8) + 3(9 + 4)] - [0 - 3(3 - 8) + 1(9 + 4)]$
$= 2(-14 - 10 + 39) - (15 + 13) = 30 - 28 = 2$

22. $0 - 2\begin{vmatrix} 1 & 2 & 2 \\ -1 & 1 & 3 \\ -2 & 3 & 2 \end{vmatrix} + 0 - 1\begin{vmatrix} 1 & 0 & 2 \\ -1 & 2 & 1 \\ -2 & 1 & 3 \end{vmatrix}$
using row 1 in each 3×3 matrix:
$= -2[1(2 - 9) - 2(-2 + 6) + 2(-3 + 2)] - [1(6 - 1) + 0 + 2(-1 + 4)]$
$= -2(-7 - 8 - 2) - (5 + 6) = 34 - 11 = 23$

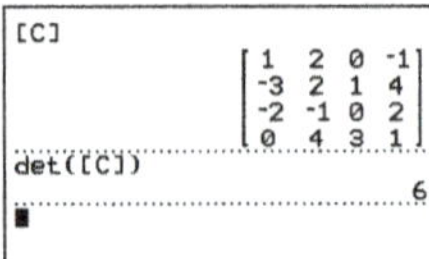

While expansion by cofactors applies to square matrices of any order, technology is frequently used to calculate determinants beyond an order of 2 since the arithmetic often becomes laborious.

Determinants provide another method of solving systems of equations. Examine the following solution to the generalized system of linear equations in two variables.

$$ax + by = e$$
$$cx + dy = f$$

finding x by eliminating y:
$$d(ax + by = e) \Rightarrow adx + bdy = ed$$
$$-b(cx + dy = f) \Rightarrow -bcx - bdy = -fb$$
$$\overline{(ad - bc)x = de - bf}$$
$$x = \frac{ed - fb}{ad - bc}$$

finding y by eliminating x:
$$-c(ax + by = e) \Rightarrow -acx - bcy = -ce$$
$$a(cx + dy = f) \Rightarrow acx + ady = af$$
$$\overline{(ad - bc)y = af - ce}$$
$$y = \frac{af - ce}{ad - bc}$$

Recall that this system can be written as the matrix equation $AX = C$ where A is the coefficient matrix, X is a column matrix containing the variables, and C is a column matrix containing the constants.

$$\begin{bmatrix} a & b \\ c & d \end{bmatrix}\begin{bmatrix} x \\ y \end{bmatrix} = \begin{bmatrix} e \\ f \end{bmatrix}$$

Notice that the denominator of both x and y is the determinant of the coefficient matrix. The numerators can be expressed as $|A_x|$ where A_x is the matrix formed by replacing the first column in A with C, and $|A_y|$ where A_y is the matrix formed by replacing the second column in A with C.

$$x = \frac{|A_x|}{|A|} = \frac{\begin{vmatrix} e & b \\ f & d \end{vmatrix}}{\begin{vmatrix} a & b \\ c & d \end{vmatrix}} = \frac{ed - fb}{ad - bc} \text{ and } y = \frac{|A_y|}{|A|} = \frac{\begin{vmatrix} a & e \\ c & f \end{vmatrix}}{\begin{vmatrix} a & b \\ c & d \end{vmatrix}} = \frac{af - ce}{ad - bc}$$

Unique solutions to linear systems with more variables can be found in a similar manner using the theorem named after Gabriel Cramer (1704–52), who popularized the use of matrices and introduced this method of solving systems.

Real-world problems in engineering often include square matrices of an order greater than 1000. Understanding the math behind the formulas allows computer programmers to maximize efficiency and minimize errors.

CRAMER'S RULE

If a system of n linear equations in n variables has the coefficient matrix A with $|A| \neq 0$, its unique solution is given by

$$x_1 = \frac{|A_1|}{|A|}, x_2 = \frac{|A_2|}{|A|}, \ldots, x_n = \frac{|A_n|}{|A|}$$

where A_i is formed by replacing the ith column in A with a column of the system's constants.

Example 6   Using Cramer's Rule

Use Cramer's rule to solve the system of equations.
$$x + 3z = 2$$
$$3x - 2y = 6$$
$$2y - z = 8$$

Answer

$$\begin{bmatrix} 1 & 0 & 3 \\ 3 & -2 & 0 \\ 0 & 2 & -1 \end{bmatrix}\begin{bmatrix} x \\ y \\ z \end{bmatrix} = \begin{bmatrix} 2 \\ 6 \\ 8 \end{bmatrix}$$

1. Write the system as a matrix equation $AX = C$.

CONTINUED ➡

$$|A| = \begin{vmatrix} 1 & 0 & 3 \\ 3 & -2 & 0 \\ 0 & 2 & -1 \end{vmatrix} \qquad |A_x| = \begin{vmatrix} 2 & 0 & 3 \\ 6 & -2 & 0 \\ 8 & 2 & -1 \end{vmatrix}$$

2. Find $|A|$, $|A_x|$, $|A_y|$, and $|A_z|$.

$$= 0 - 2[0 - 9] - 1[-2 - 0] \qquad = 2[2 - 0] - 0 + 3[12 + 16]$$

a. using row 3 to find $|A|$

$$= 18 + 2 = 20 \qquad = 4 + 84 = 88$$

b. using row 1 to find $|A_x|$

$$|A_y| = \begin{vmatrix} 1 & 2 & 3 \\ 3 & 6 & 0 \\ 0 & 8 & -1 \end{vmatrix} \qquad |A_z| = \begin{vmatrix} 1 & 0 & 2 \\ 3 & -2 & 6 \\ 0 & 2 & 8 \end{vmatrix}$$

c. using row 3 to find $|A_y|$

$$= 0 - 8[0 - 9] - 1[6 - 6] \qquad = 1[-16 - 12] - 0 + 2[6 - 0]$$

d. using row 1 to find $|A_z|$

$$= 72 \qquad = -28 + 12 = -16$$

$$x = \frac{|A_x|}{|A|} \qquad y = \frac{|A_y|}{|A|} \qquad z = \frac{|A_z|}{|A|}$$

3. Apply Cramer's rule.

$$= \frac{88}{20} = 4.4 \qquad = \frac{72}{20} = 3.6 \qquad = \frac{-16}{20} = -0.8$$

The solution is $(4.4, 3.6, -0.8)$.

SKILL ✔ EXERCISE 35

If the determinant of the coefficient matrix is 0, the system does not have a unique solution and is either dependent or inconsistent. Cramer's rule does not produce the solution.

Example 7 Using Cramer's Rule

Apply Cramer's rule to each system of equations.

a. $3x + 8y = 5$
   $6x + 16y = 7$

b. $3x + 8y = 5$
   $6x + 16y = 10$

Answer

a. $|A| = \begin{vmatrix} 3 & 8 \\ 6 & 16 \end{vmatrix} = 0$

$$|A_x| = \begin{vmatrix} 5 & 8 \\ 7 & 16 \end{vmatrix} \qquad |A_y| = \begin{vmatrix} 3 & 5 \\ 6 & 7 \end{vmatrix}$$
$$= 24 \qquad\qquad = -9$$
$$x = \frac{|A_x|}{|A|} = \frac{24}{0} \qquad y = \frac{|A_y|}{|A|} = \frac{-9}{0}$$
$$\text{undefined} \qquad\qquad \text{undefined}$$

Since the equations represent parallel lines, the system is inconsistent and its solution is $\varnothing$.

b. $|A| = \begin{vmatrix} 3 & 8 \\ 6 & 16 \end{vmatrix} = 0$

$$|A_x| = \begin{vmatrix} 5 & 8 \\ 10 & 16 \end{vmatrix} \qquad |A_y| = \begin{vmatrix} 3 & 5 \\ 6 & 10 \end{vmatrix}$$
$$= 0 \qquad\qquad = 0$$
$$x = \frac{|A_x|}{|A|} = \frac{0}{0} \qquad y = \frac{|A_y|}{|A|} = \frac{0}{0}$$
$$\text{undefined} \qquad\qquad \text{undefined}$$

Since the equations represent the same line, the system is dependent, and its solution is $\{(x, y) \mid 3x + 8y = 5\}$.

SKILL ✔ EXERCISE 29

Determinants can also be used to find the area of a triangle. Other polygonal areas can be estimated by dividing the region into several triangular areas.

▍AREA OF TRIANGLE

The area of a triangle with vertices (x_1, y_1), (x_2, y_2), and (x_3, y_3) can be found using

$$\text{Area} = \pm\frac{1}{2}\begin{vmatrix} x_1 & y_1 & 1 \\ x_2 & y_2 & 1 \\ x_3 & y_3 & 1 \end{vmatrix} \text{ where the area must be positive.}$$

23. using row 2:

$$-3\begin{vmatrix} -3 & 1 & 2 \\ 1 & 3 & 4 \\ 2 & -1 & -2 \end{vmatrix} + 0 + 1\begin{vmatrix} 2 & -3 & 2 \\ -1 & 1 & 4 \\ 0 & 2 & -2 \end{vmatrix} + 0$$

using column 1 in each 3×3 matrix:

$$= -3[-3(-6 + 4) - 1(-2 + 2) + 2(4 - 6)] + [2(-2 - 8) + 1(6 - 4) + 0]$$
$$= -3(6 - 0 - 4) + (-20 + 2) = -6 - 18 = -24$$

24. using column 2:

$$0 + 3\begin{vmatrix} 4 & 2 & 3 \\ -3 & -1 & 2 \\ -6 & -4 & -2 \end{vmatrix} + 0 + 0$$

using row 1 of the 3×3 matrix:

$$= 3[4(2 + 8) - 2(6 + 12) + 3(12 - 6)] = 3(40 - 36 + 18) = 66$$

25. $|A| = \begin{vmatrix} 2 & 2 \\ 4 & -1 \end{vmatrix} = -10$

$$|A_x| = \begin{vmatrix} 10 & 2 \\ 5 & -1 \end{vmatrix} = -20$$

$$|A_y| = \begin{vmatrix} 2 & 10 \\ 4 & 5 \end{vmatrix} = -30$$

$$x = \frac{-20}{-10} = 2; \ y = \frac{-30}{-10} = 3$$

26. $|A| = \begin{vmatrix} 3 & -2 \\ -3 & 4 \end{vmatrix} = 6$

$$|A_x| = \begin{vmatrix} 3 & -2 \\ -4 & 4 \end{vmatrix} = 4$$

$$|A_y| = \begin{vmatrix} 3 & 3 \\ -3 & -4 \end{vmatrix} = -3$$

$$x = \frac{4}{6} = \frac{2}{3}; \ y = \frac{-3}{6} = -\frac{1}{2}$$

27. $y = \frac{1}{3}x + \frac{4}{3}; \ -x + 3y = 4$

$$|A| = \begin{vmatrix} 5 & 6 \\ -1 & 3 \end{vmatrix} = 21$$

$$|A_x| = \begin{vmatrix} -13 & 6 \\ 4 & 3 \end{vmatrix} = -63$$

$$|A_y| = \begin{vmatrix} 5 & -13 \\ -1 & 4 \end{vmatrix} = 7$$

$$x = \frac{-63}{21} = -3; \ y = \frac{7}{21} = \frac{1}{3}$$

28. $y = -\frac{3}{2}x + 1; \ 3x + 2y = 2$

$$|A| = \begin{vmatrix} 3 & 2 \\ 3 & 2 \end{vmatrix} = 0$$

$$|A_x| = \begin{vmatrix} 2 & 2 \\ -2 & 2 \end{vmatrix} = 8$$

$$|A_y| = \begin{vmatrix} 3 & 2 \\ 3 & -2 \end{vmatrix} = -12$$

$$x = \frac{8}{0}; \ y = \frac{-12}{0} \text{ (both undefined)}$$

$\varnothing$; The equations represent parallel lines.

29. $|A| = \begin{vmatrix} 8 & -12 \\ -6 & 9 \end{vmatrix} = 0$

$$|A_x| = \begin{vmatrix} 20 & -12 \\ -15 & 9 \end{vmatrix} = 0$$

$$|A_y| = \begin{vmatrix} 8 & 20 \\ -6 & -15 \end{vmatrix} = 0$$

$$x = \frac{0}{0}; \ y = \frac{0}{0} \text{ (both undefined)}$$

The equations represent the same line: $\{(x, y) \mid 2x - 3y = 5\}$.

30. $|A| = \begin{vmatrix} 6 & -2 \\ 14 & 5 \end{vmatrix} = 58$

$$|A_x| = \begin{vmatrix} 11 & -2 \\ -13 & 5 \end{vmatrix} = 29$$

$$|A_y| = \begin{vmatrix} 6 & 11 \\ 14 & -13 \end{vmatrix} = -232$$

$$x = \frac{29}{58} = \frac{1}{2}; \ y = \frac{-232}{58} = -4$$

31. $|A| = \begin{vmatrix} 1 & -1 & 2 \\ 2 & 1 & -1 \\ 1 & 1 & 1 \end{vmatrix} = 7$

$|A_x| = \begin{vmatrix} 4 & -1 & 2 \\ -4 & 1 & -1 \\ 3 & 1 & 1 \end{vmatrix} = -7$

$|A_y| = \begin{vmatrix} 1 & 4 & 2 \\ 2 & -4 & -1 \\ 1 & 3 & 1 \end{vmatrix} = 7$

$|A_z| = \begin{vmatrix} 1 & -1 & 4 \\ 2 & 1 & -4 \\ 1 & 1 & 3 \end{vmatrix} = 21$

$x = \dfrac{-7}{7} = -1; \ y = \dfrac{7}{7} = 1; \ z = \dfrac{21}{7} = 3$

32. $|A| = \begin{vmatrix} 2 & -1 & -3 \\ 3 & 1 & -1 \\ -1 & 1 & 2 \end{vmatrix} = -1$

$|A_x| = \begin{vmatrix} -4 & -1 & -3 \\ 2 & 1 & -1 \\ 3 & 1 & 2 \end{vmatrix} = -2$

$|A_y| = \begin{vmatrix} 2 & -4 & -3 \\ 3 & 2 & -1 \\ -1 & 3 & 2 \end{vmatrix} = 1$

$|A_z| = \begin{vmatrix} 2 & -1 & -4 \\ 3 & 1 & 2 \\ -1 & 1 & 3 \end{vmatrix} = -3$

$x = \dfrac{-2}{-1} = 2; \ y = \dfrac{1}{-1} = -1$

$z = \dfrac{-3}{-1} = 3$

33. $|A| = \begin{vmatrix} 2 & -3 & 4 \\ -1 & 4 & -10 \\ 4 & 6 & -1 \end{vmatrix} = 147$

$|A_x| = \begin{vmatrix} 16 & -3 & 4 \\ -13 & 4 & -10 \\ 8 & 6 & -1 \end{vmatrix} = 735$

$|A_y| = \begin{vmatrix} 2 & 16 & 4 \\ -1 & -13 & -10 \\ 4 & 8 & -1 \end{vmatrix} = -294$

$|A_z| = \begin{vmatrix} 2 & -3 & 16 \\ -1 & 4 & -13 \\ 4 & 6 & 8 \end{vmatrix} = 0$

$x = \dfrac{735}{147} = 5; \ y = \dfrac{-294}{147} = -2$

$z = \dfrac{0}{147} = 0$

34. $|A| = \begin{vmatrix} 3 & -1 & -1 \\ 21 & -3 & 1 \\ -9 & 2 & -7 \end{vmatrix} = -96$

$|A_a| = \begin{vmatrix} -5 & -1 & -1 \\ -7 & -3 & 1 \\ 0 & 2 & -7 \end{vmatrix} = -32$

$|A_b| = \begin{vmatrix} 3 & -5 & -1 \\ 21 & -7 & 1 \\ -9 & 0 & -7 \end{vmatrix} = -480$

$|A_c| = \begin{vmatrix} 3 & -1 & -5 \\ 21 & -3 & -7 \\ -9 & 2 & 0 \end{vmatrix} = -96$

$a = \dfrac{-32}{-96} = \dfrac{1}{3}; \ b = \dfrac{-480}{-96} = 5$

$c = \dfrac{-96}{-96} = 1$

Example 8 Calculating the Area of a Polygonal Region

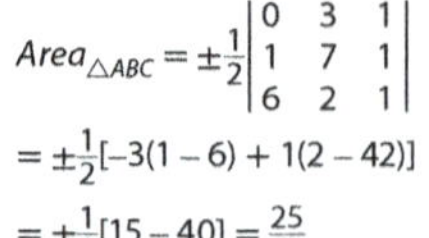

Use determinants to find the area of the shaded region.

Answer

$Area = Area_{\triangle ABC} + Area_{\triangle ADC}$

1. Divide the area into two triangular areas.

$Area_{\triangle ABC} = \pm\dfrac{1}{2}\begin{vmatrix} 0 & 3 & 1 \\ 1 & 7 & 1 \\ 6 & 2 & 1 \end{vmatrix}$ $Area_{\triangle ADC} = \pm\dfrac{1}{2}\begin{vmatrix} 0 & 3 & 1 \\ 4 & 0 & 1 \\ 6 & 2 & 1 \end{vmatrix}$

2. Find the area of each triangle. Expanding along row 1 requires the calculation of only two minors.

$= \pm\dfrac{1}{2}[-3(1-6) + 1(2-42)]$ $= \pm\dfrac{1}{2}[-3(4-6) + 1(8-0)]$

$= \pm\dfrac{1}{2}[15 - 40] = \dfrac{25}{2}$ $= \pm\dfrac{1}{2}[6+8] = 7$

$Area = 12.5 + 7 = 19.5 \ u^2$

3. Calculate the sum.

SKILL ✔ **EXERCISE 37**

❯ A. Exercises

Find the determinant of each matrix.

1. $[7]$ 7   **2.** $[-2]$ −2

3. $\begin{bmatrix} 8 & -3 \\ 5 & 9 \end{bmatrix}$ 87   **4.** $\begin{bmatrix} -2 & -4 \\ -1 & -5 \end{bmatrix}$ 6

5. $\begin{bmatrix} 11 & 7 \\ -13 & -9 \end{bmatrix}$ −8   **6.** $\begin{bmatrix} \frac{1}{2} & 5 \\ 6 & 8 \end{bmatrix}$ −26

Use the products of diagonals to find the determinant of each matrix.

7. $\begin{bmatrix} 2 & 1 & -4 \\ 3 & 2 & -5 \\ 6 & 2 & 1 \end{bmatrix}$ 15   **8.** $\begin{bmatrix} 2 & -1 & 4 \\ 1 & 0 & -2 \\ 3 & 1 & 5 \end{bmatrix}$ 19

9. $\begin{bmatrix} 1 & 0 & 2 \\ 1 & 0 & 3 \\ 0 & 1 & -2 \end{bmatrix}$ −1   **10.** $\begin{bmatrix} 5 & -3 & 4 \\ 2 & 3 & -2 \\ 4 & -2 & 4 \end{bmatrix}$ 24

Given $A = \begin{bmatrix} 2 & -1 & 3 \\ 4 & 2 & 1 \\ 1 & 3 & -2 \end{bmatrix}$, **find each minor and cofactor.**

11. in row 1   **12.** in column 2

Use cofactors from the given row or column to find the determinant of each matrix.

13. $\begin{bmatrix} 2 & -1 & 3 \\ 4 & 2 & 1 \\ 1 & 3 & -2 \end{bmatrix}$ **14.** $\begin{bmatrix} 5 & 1 & 3 \\ 2 & -1 & 0 \\ 1 & 7 & 6 \end{bmatrix}$

   row 1   7   row 2   3

15. $\begin{bmatrix} -1 & 0 & -3 \\ 3 & 5 & -2 \\ 2 & 0 & 1 \end{bmatrix}$ **16.** $\begin{bmatrix} -6 & -2 & 4 \\ 1 & 7 & 3 \\ 2 & 5 & -1 \end{bmatrix}$

   column 2   25   column 3   82

17. Use technology to find the determinant of each 3×3 matrix.

  a. $A = \begin{bmatrix} 2 & -3 & 2 \\ 8 & 5 & 7 \\ -2 & 1 & 1 \end{bmatrix}$ $|A| = 98$

  b. $B = \begin{bmatrix} -8 & -1 & 3 \\ 1 & 6 & 5 \\ 4 & -5 & 2 \end{bmatrix}$ $|B| = -401$

18. Use technology to find the determinant of each 4×4 matrix.

  a. $A = \begin{bmatrix} 1 & 3 & 0 & -2 \\ 4 & 0 & 5 & 5 \\ 2 & -3 & -4 & 1 \\ 2 & 1 & -1 & 3 \end{bmatrix}$ $|A| = 323$

  b. $B = \begin{bmatrix} 9 & 0 & 2 & 3 \\ 11 & -4 & 1 & 6 \\ 3 & 1 & 7 & -1 \\ -1 & 5 & -2 & 8 \end{bmatrix}$ $|B| = -2236$

❯ B. Exercises

Use cofactors from the given row or column to find the determinant of each matrix.

19. $\begin{bmatrix} -2 & 3 & -1 \\ 1 & 2 & 3 \\ 4 & 0 & -1 \end{bmatrix}$ **20.** $\begin{bmatrix} 4 & -2 & 1 \\ 3 & 5 & 2 \\ -1 & 1 & -3 \end{bmatrix}$

  a. row 3   51   **a.** row 3   −74

  b. column 2   51   **b.** column 1   −74

21. $\begin{bmatrix} 2 & 0 & 3 & 1 \\ 1 & 2 & -2 & 3 \\ 0 & 3 & -1 & 2 \\ 0 & 4 & 3 & 1 \end{bmatrix}$ **22.** $\begin{bmatrix} 1 & 0 & 2 & 2 \\ -1 & 2 & 1 & 3 \\ 0 & 2 & 0 & 1 \\ -2 & 1 & 3 & 2 \end{bmatrix}$

  column 1   2   row 3   23

---

35. $Area = \pm\dfrac{1}{2}\begin{vmatrix} 0 & 2 & 1 \\ 5 & 6 & 1 \\ 10 & 1 & 1 \end{vmatrix}$

using row 1:

$= \pm\dfrac{1}{2}[0 - 2(5-10) + 1(5-60)]$

$= \pm\dfrac{1}{2}(-45) = 22.5 \ u^2$

36. $Area = \pm\dfrac{1}{2}\begin{vmatrix} -5 & -3 & 1 \\ 1 & 6 & 1 \\ 7 & -2 & 1 \end{vmatrix}$

using column 3:

$= \pm\dfrac{1}{2}[(-2 - 42) - (10 + 21)$

$\qquad\qquad + (-30 + 3)]$

$= \pm\dfrac{1}{2}(-102) = 51 \ u^2$

37. $Area_{\triangle ABC} = \pm\dfrac{1}{2}\begin{vmatrix} 0 & 0 & 1 \\ -3 & 4 & 1 \\ 3 & 5 & 1 \end{vmatrix}$

using row 1:

$= \pm\dfrac{1}{2}(1(-15 - 12)) = 13.5 \ u^2$

$Area_{\triangle ADC} = \pm\dfrac{1}{2}\begin{vmatrix} 0 & 0 & 1 \\ 9 & 3 & 1 \\ 3 & 5 & 1 \end{vmatrix}$

using row 1:

$= \pm\dfrac{1}{2}(1(45 - 9)) = 18 \ u^2$

$Area_{poly} = 13.5 + 18 = 31.5 \ u^2$

Expand by cofactors to find the determinant of each matrix.

23. $\begin{bmatrix} 2 & -3 & 1 & 2 \\ 3 & 0 & -1 & 0 \\ -1 & 1 & 3 & 4 \\ 0 & 2 & -1 & -2 \end{bmatrix}$ -24 **24.** $\begin{bmatrix} 4 & 0 & 2 & 3 \\ 1 & 3 & -5 & 0 \\ -3 & 0 & -1 & 2 \\ -6 & 0 & -4 & -2 \end{bmatrix}$ 66

Use Cramer's rule to solve each system of equations.

25. $2x + 2y = 10$
$4x - y = 5$ $(2, 3)$

26. $3x - 2y = 3$
$-3x + 4y = -4$ $\left(\frac{2}{3}, -\frac{1}{2}\right)$

27. $5x + 6y = -13$
$y = \frac{1}{3}x + \frac{4}{3}$ $\left(-3, \frac{1}{3}\right)$

28. $y = -\frac{3}{2}x + 1$
$3x + 2y = -2$ $\varnothing$

29. $8x - 12y = 20$
$-6x + 9y = -15$

30. $6x - 2y = 11$
$14x + 5y = -13$ $\left(\frac{1}{2}, -4\right)$

31. $x - y + 2z = 4$
$2x + y - z = -4$
$x + y + z = 3$ $(-1, 1, 3)$

32. $2x - y - 3z = -4$
$3x + y - z = 2$
$-x + y + 2z = 3$ $(2, -1, 3)$

33. $2x - 3y + 4z = 16$
$-x + 4y - 10z = -13$
$4x + 6y - z = 8$ $(5, -2, 0)$

34. $3a - b - c = -5$
$21a - 3b + c = -7$
$-9a + 2b - 7c = 0$ $\left(\frac{1}{3}, 5, 1\right)$

Find the area of each triangle.

35. $\triangle ABC$ with $A(0, 2)$, $B(5, 6)$, and $C(10, 1)$. 22.5 u^2

36. $\triangle PQR$ with $P(-5, -3)$, $Q(1, 6)$, and $R(7, -2)$. 51 u^2

Use determinants to find the area of each shaded region.

37. 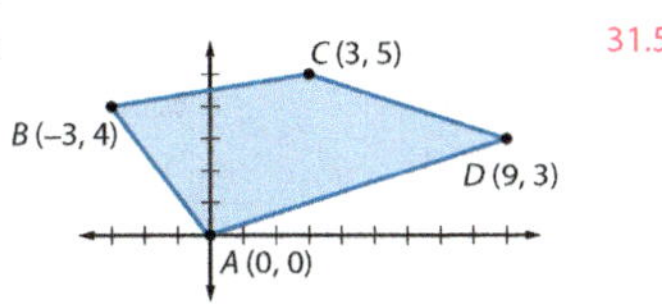 31.5 u^2

38. 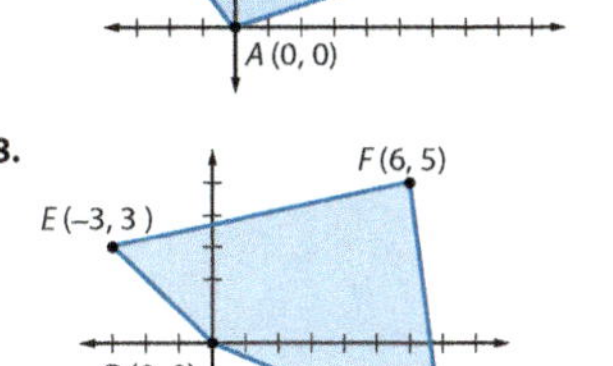 43 u^2

39. Three utility poles define a triangular area. If a utility worker standing at one of the poles walks 250 ft east and then 312 ft north, he arrives at the second pole. If instead he walks 140 ft west and 78 ft north, he arrives at the third pole. Find the area of the triangular region defined by the three poles. $31,590 \text{ ft}^2$

29. $\{(x, y) \mid 2x - 3y = 5\}$

40. Carter and Charlotte set up four 50 gal barrels for a disc golf practice field. Starting at the red barrel they walk 60 ft west and 320 ft north and set the green barrel. The blue barrel is set 600 ft north of the red barrel, and the yellow barrel is set 210 ft east and 50 ft south of the red barrel. Find the area of the region defined by the barrels in square feet and in acres (1 acre = 43,560 ft^2). $81,000 \text{ ft}^2 \approx 1.86 \text{ acres}$

> **C. Exercises**

41. Verify the formulas for determinants that use products of diagonals by using cofactors to derive the same result.

a. $A = \begin{bmatrix} a & b \\ c & d \end{bmatrix}$ **b.** $A = \begin{bmatrix} a & b & c \\ d & e & f \\ g & h & i \end{bmatrix}$

42. Explore: Investigate a method of determining whether three points are collinear.

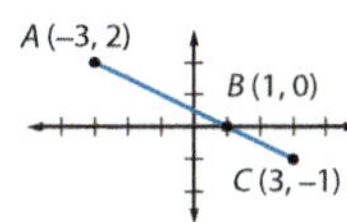

a. What is the result of applying the formula for the area of the triangle to points A, B, and C? $Area = 0$

b. Will this result occur whenever the points are collinear? Explain your reasoning.

c. Are $D(-5, -4)$, $E(2, 1)$, and $F(6, 4)$ collinear? Explain your reasoning.

d. Are $X(14, 0)$, $Y(0, 6)$, and $Z(-7, 9)$ collinear? Explain your reasoning.

43. Extend: The results of exercise 42 imply that the equation of a line passing through distinct points $P_1(x_1, y_1)$ and $P_2(x_2, y_2)$ can be expressed as

$$\begin{vmatrix} x & y & 1 \\ x_1 & y_1 & 1 \\ x_2 & y_2 & 1 \end{vmatrix} = 0.$$

Use this equation to find the equation of the line passing through each pair of points.

a. $(0, 2)$ and $(3, 0)$ $2x + 3y = 6$

b. $(-3, 2)$ and $(3, -1)$ $x + 2y = 1$

c. (a, k) and (b, k) $y = k$

> **C. Exercises**

41a. using row 1: $a|d| - b|c| = ad - bc$

41b. using row 1:

$a(ei - hf) - b(di - gf) + c(dh - ge)$
$= aei - ahf - bdi + bgf + cdh - cge$
$= aei + bfg + cdh - gec - hfa - idb$
$= (aei + bfg + cdh)$
$\qquad\qquad - (gec + hfa + idb)$

42a. $Area = \pm\frac{1}{2} \begin{vmatrix} -3 & 2 & 1 \\ 1 & 0 & 1 \\ 3 & -1 & 1 \end{vmatrix} = 0$

42b. yes; Three collinear points could be considered a degenerate triangle with a height of 0.

42c. no; $Area = \pm\frac{1}{2} \begin{vmatrix} -5 & -4 & 1 \\ 2 & 1 & 1 \\ 6 & 4 & 1 \end{vmatrix} = \pm\frac{1}{2} \neq 0$

42d. yes; $Area = \pm\frac{1}{2} \begin{vmatrix} 14 & 0 & 1 \\ 0 & 6 & 1 \\ -7 & 9 & 1 \end{vmatrix} = 0$

43a. $\begin{vmatrix} x & y & 1 \\ 0 & 2 & 1 \\ 3 & 0 & 1 \end{vmatrix} = 0$

using row 1:
$x(2 + 0) - y(0 - 3) + 1(-6) = 0$
$\qquad\qquad 2x + 3y = 6$

43b. $\begin{vmatrix} x & y & 1 \\ -3 & 2 & 1 \\ 3 & -1 & 1 \end{vmatrix} = 0$

using row 1:
$x(2 + 1) - y(-3 - 3) + 1(3 - 6) = 0$
$\qquad\qquad 3x + 6y = 3$

43c. $\begin{vmatrix} x & y & 1 \\ a & k & 1 \\ b & k & 1 \end{vmatrix} = 0$

using row 1:
$x(k - k) - y(a - b) + 1(ak - bk) = 0$
$-(a - b)y = -(a - b)k$
$\qquad\qquad y = k$

38. $Area_{\triangle DEF} = \pm\frac{1}{2} \begin{vmatrix} 0 & 0 & 1 \\ -3 & 3 & 1 \\ 6 & 5 & 1 \end{vmatrix}$

using row 1:
$= \pm\frac{1}{2}(1(-33)) = 16.5 \text{ u}^2$

$Area_{\triangle DFG} = \pm\frac{1}{2} \begin{vmatrix} 0 & 0 & 1 \\ 7 & -3 & 1 \\ 6 & 5 & 1 \end{vmatrix}$

using row 1:
$= \pm\frac{1}{2}(1(53)) = 26.5 \text{ u}^2$
$Area_{poly} = 16.5 + 26.5 = 43 \text{ u}^2$

39. $Area = \pm\frac{1}{2} \begin{vmatrix} 0 & 0 & 1 \\ 250 & 312 & 1 \\ -140 & 78 & 1 \end{vmatrix}$

$= \pm\frac{1}{2}(63,180) = 31,590 \text{ ft}^2$

40. letting $R = (0, 0)$: $G = (-60, 320)$
$B = (0, 600)$; $Y = (210, -50)$

$Area_{\triangle RGB} = \pm\frac{1}{2} \begin{vmatrix} 0 & 0 & 1 \\ -60 & 320 & 1 \\ 0 & 600 & 1 \end{vmatrix}$

$= 18,000 \text{ ft}^2$

$Area_{\triangle RYB} = \pm\frac{1}{2} \begin{vmatrix} 0 & 0 & 1 \\ 210 & -50 & 1 \\ 0 & 600 & 1 \end{vmatrix}$

$= 63,000 \text{ ft}^2$

$Area_{poly} = 18,000 + 63,000$
$\qquad\qquad = 81,000 \text{ ft}^2$

$81,000 \text{ ft}^2 \left(\dfrac{1 \text{ acre}}{43,560 \text{ ft}^2}\right) \approx 1.86 \text{ acres}$

44a. $Area = \pm \begin{vmatrix} 4 & -3 \\ 3 & 4 \end{vmatrix} = \pm(25) = 25 \text{ u}^2$

44b. $Area = \pm \begin{vmatrix} 3 & 5 \\ 7 & 1 \end{vmatrix} = \pm(-32) = 32 \text{ u}^2$

45a. $V = \pm \begin{vmatrix} 4 & 0 & 0 \\ 0 & 7 & 0 \\ 0 & 0 & 5 \end{vmatrix} = \pm(140) = 140 \text{ u}^3$

45b. $V = \pm \begin{vmatrix} 7 & 0 & 2 \\ 1 & 5 & 0 \\ 2 & 2 & 8 \end{vmatrix} = \pm(264) = 264 \text{ u}^3$

❯ Cumulative Review

46.
$$5^{2x} = 129$$
$$\log 5^{2x} = \log 129$$
$$2x \log 5 = \log 129$$
$$x = \frac{\log 129}{2 \log 5} \approx 1.5098$$

47. $\ln(-2x) = 8.5$
$$-2x = e^{8.5}$$
$$x = \frac{e^{8.5}}{-2} \approx -2457.38$$

48.

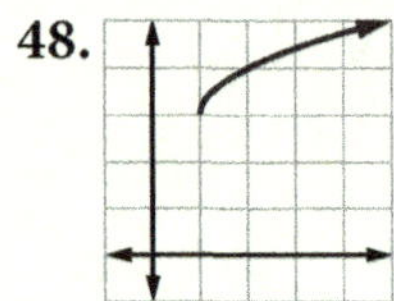

49.

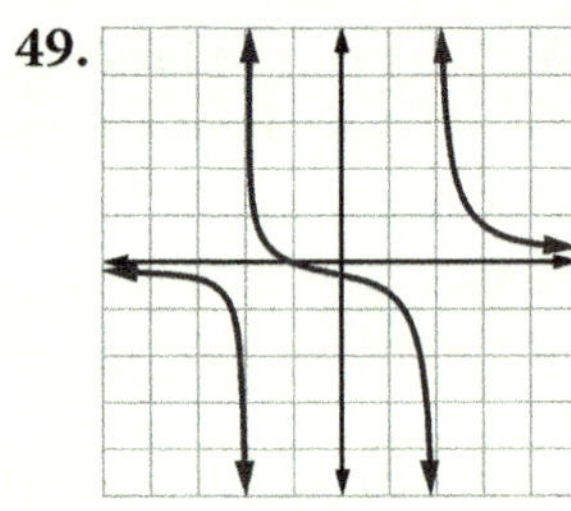

50. $\cos^4 x - \sin^4 x$
$$= (\cos^2 x - \sin^2 x)(\cos^2 x + \sin^2 x)$$
$$= (\cos^2 x - \sin^2 x)(1)$$
$$= \cos^2 x - \sin^2 x$$

52. $p = \frac{2\pi}{\left|\frac{\pi}{3}\right|} = 6$

53. $\dfrac{\sin A}{10} = \dfrac{\sin 55°}{12}$
$$A = \sin^{-1}\left(\frac{10 \sin 55°}{12}\right) \approx 43.05°$$

55. When $m_1 = m_2$, the graphs are parallel lines and the system is inconsistent.
$$m = -\frac{A}{B}; \ m_1 = -\frac{2}{7} \text{ and } m_2 = -\frac{6}{B}$$
$$-\frac{2}{7} = -\frac{6}{B}; \ -2B = -42; \ B = 21$$

---

44. If the sides of a parallelogram are represented by $\mathbf{u} = \langle a, b \rangle$ and $\mathbf{v} = \langle c, d \rangle$, then its area is $\pm \begin{vmatrix} a & b \\ c & d \end{vmatrix}$. Find the area of the parallelogram represented by each pair of vectors.

  a. $\mathbf{u} = \langle 4, -3 \rangle$ and $\mathbf{v} = \langle 3, 4 \rangle$ 25 u²

  b. $\mathbf{u} = \langle 3, 5 \rangle$ and $\mathbf{v} = \langle 7, 1 \rangle$ 32 u²

45. If the sides of a parallelepiped are represented by vectors $\mathbf{u}$, $\mathbf{v}$, and $\mathbf{w}$ as illustrated, its volume is
$$\pm \begin{vmatrix} u_1 & u_2 & u_3 \\ v_1 & v_2 & v_3 \\ w_1 & w_2 & w_3 \end{vmatrix}.$$

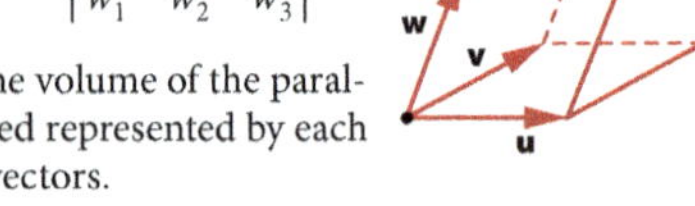

Find the volume of the parallelepiped represented by each set of vectors.

  a. $\mathbf{u} = \langle 4, 0, 0 \rangle$, $\mathbf{v} = \langle 0, 7, 0 \rangle$, $\mathbf{w} = \langle 0, 0, 5 \rangle$ 140 u³

  b. $\mathbf{u} = \langle 7, 1, 2 \rangle$, $\mathbf{v} = \langle 0, 5, 2 \rangle$, $\mathbf{w} = \langle 2, 0, 8 \rangle$ 264 u³

CUMULATIVE REVIEW

Solve each equation.

46. $5^{2x} = 129$ [3.4] $x \approx 1.5098$

47. $\ln(-2x) = 8.5$ [3.4] $x \approx -2457.38$

Graph each function.

48. $f(x) = 3 + \sqrt{x - 1}$ [2.1]

49. $g(x) = \dfrac{x + 1}{x^2 - 4}$ [2.5]

50. Verify $\cos^4 x - \sin^4 x = \cos^2 x - \sin^2 x$.   [5.2]

51. Write the system as an augmented matrix.   [7.3]
$$\begin{aligned} 2x + 3y - z &= -6 \\ 3x - 2z &= -1 \\ 4x - 2y &= 2 \end{aligned} \qquad \begin{bmatrix} 2 & 3 & -1 & -6 \\ 3 & 0 & -2 & -1 \\ 4 & -2 & 0 & 2 \end{bmatrix}$$

52. Find the period of $y = \frac{3}{2} \cos \frac{\pi x}{3}$.   [4.4]   D

  A. $\frac{\pi}{3}$ **C.** $\frac{1}{6}$ **E.** none of these

  B. $\frac{3}{2}$ **D.** 6

53. Find $m\angle A$ (to the nearest degree) in a triangle with sides $AC = 12$ cm and $BC = 10$ cm and $m\angle B = 55°$.   [5.6]   B

  A. 82°   **C.** 35°   **E.** none of these

  B. 43°   **D.** 39°

54. Find entry e_{22} in AB if
$$A = \begin{bmatrix} d & e \\ f & g \end{bmatrix} \text{ and } B = \begin{bmatrix} t & v & w \\ x & y & z \end{bmatrix}. \quad [7.2] \quad C$$

  A. $dv + ey$ **C.** $fv + gy$ **E.** none of these

  B. $ft + gx$ **D.** $fw + gz$

55. Find B so that the system is inconsistent.
$$\begin{aligned} 2x + 7y &= 6 \\ 6x + By &= 6 \end{aligned} \quad [7.3] \quad D$$

  A. 6   **C.** 14   **E.** none of these

  B. 7   **D.** 21

MATRICES AND DETERMINANTS

How many planets are in our solar system? The Italian astronomer Giuseppe Piazzi discovered a very small one on January 1, 1801. He recorded its position until the second week of February, when it was lost behind the sun. Piazzi appealed for help in determining the planet's orbit so that it could be reacquired. Many attempted this nearly impossible feat, but it was the 24-yr-old German mathematician Carl Gauss who accurately determined the planet's orbit by solving systems of equations involving more than 80 variables in three different coordinate systems. Using Gauss's results, the dwarf planet Ceres was reacquired nearly a year later on the first attempt. Gauss published his method in 1809 and it became the standard in astronomy.

In an 1888 textbook on geodesy, German surveyor Wilhelm Jordan published an improved version of Gauss's method, now known as Gauss-Jordan elimination. Many other mathematicians made contributions to the processes and notation used to solve systems, including Colin MacLaurin, Gabriel Cramer, and Augustin-Louis Cauchy. Arthur Cayley (1821–95)

published the first comprehensive treatise on matrix theory in 1857. Working closely with Cayley was another British mathematician, James Joseph Sylvester (1814–97), who had given up on mathematics until he met Cayley while both of them were studying law.

When an influential British biologist stated that mathematics is "that study which knows nothing of observation, nothing of induction, nothing of experiment, nothing of causation," Sylvester replied, "I think no statement could have been made more opposite to the undoubted facts of the case, that mathematical analysis . . . affords a boundless scope for the exercise of the highest efforts of imagination and invention." Sylvester's emphasis on the role of creativity in mathematics is evidenced in his own work on matrices. In fact, he coined the term *matrix* from the Latin word for "womb." An Internet search for "Sylvester matrix" supplies many current research papers that extend Sylvester's work and provide practical applications over a century later.

Twentieth-century physicists wrestled with experimental results at the atomic level that could not be explained by Bohr's planetary atomic model. Werner Heisenberg (1901–76) was concerned that the use of models to portray unobservable physical phenomena was an obstacle to understanding. In 1925 he used matrices to develop a system that explained observed results without a physical representation. The merging of Heisenberg's matrix mechanics with Erwin Schrödinger's wave mechanics became quantum mechanics. It may seem unlikely that a mathematical concept could be useful in the vastness of space, on the surface of the earth, and in submicroscopic physics. Even Newton's laws, which work well in our everyday experience and in astronomy, are insufficient to explain experimental results at the atomic level. But determinants and matrices continue to play a crucial role in our understanding of our world at each of these levels.

COMPREHENSION CHECK

1. By whom and in what year was the first comprehensive treatise on matrices published?
   Arthur Cayley; 1857

2. Who coined the term *matrix* and from what Latin word?
   James Sylvester; womb

3. What connection does matrix theory have to quantum mechanics?
   Heisenberg used matrices to develop matrix mechanics, which merged with Schrödinger's wave mechanics to become quantum mechanics.

4. **Discuss:** Both quotations in the third paragraph have some merit. What positive insights can you see in each person's view?

Historical Connection

Objectives

1. To summarize the major events in the history of matrices and determinants

2. To explain how matrices and determinants have been used to solve real-world problems

3. To evaluate the role of imagination in mathematical discovery

Answers

4. sample answer:
Mathematicians use creativity and experimentation to solve problems. Bombelli's success with imaginary numbers, Hamilton's quaternions, and the non-Euclidean geometries are masterpieces of creativity. Mathematicians also seek a deductive proof building on definitions and accepted postulates and theorems. Once a concept has been proved deductively, experimental tests are not needed to establish truth. Manipulation of symbols in branches of mathematics without known applications frequently generate mathematics with real-world applications.

PRESENTATION

The study of matrices began in the mid-1800s, about 150 yr after the study of determinants began (in the context of linear systems). Matrices soon outpaced determinants in both development and usefulness.

Ceres was eventually downgraded to an asteroid when the asteroid belt between Mars and Jupiter was discovered. Today it is considered a dwarf planet, the only such planet in the inner solar system.

Gauss referred to the method of elimination he used as the "common method," indicating he probably was not the originator. His claim to the method of least squares was disputed by French mathematician Adrien-Marie Legendre, who was a contemporary of Gauss. The Internet keyword search *Gauss Ceres orbit* provides documentation describing the mathematics used by Gauss to determine an elliptical orbit from three observed positions.

Further research and discussion of James Sylvester is recommended. Although an Englishman, Sylvester had a major influence on the development of mathematics in the United States during the nineteenth century. In addition to teaching at Johns Hopkins University, he founded the first mathematical journal in the United States, the *American Journal of Mathematics*, in 1878.

Heisenberg is perhaps most famous for his uncertainty principle, which has been interpreted as a challenge to a deterministic worldview. Use the Internet keyword search *uncertainty principle* for more details. In his book, *The Physical Principles of the Quantum Theory*, Heisenburg says that this principle must result in the renunciation of "the idea that natural phenomena obey exact laws—the principle of causality." In *Mathematics: Is God Silent?*, James Nickel provides a biblical worldview response to Heisenberg's claim (pp. 204–5).

Objectives

1. To define and determine the existence of an inverse matrix

2. To determine the inverse of a 2×2 matrix

3. To use an augmented matrix and row operations to find the inverse of a 3×3 matrix

4. To use an inverse matrix to solve a matrix equation of the form $AX = C$

Flash

Technology significantly impacted the outcome of World War II. This technology included Colossus, a British computer that broke encrypted codes sent from Enigma cipher machines. Some historians estimate that this shortened the war by two or more years, saving at least 14 million lives.

Vocabulary

inverse matrix
invertible (nonsingular) matrix
singular matrix

7.5 Inverse Matrices

The Allies' ability to decode messages encrypted by Enigma cipher machines helped turn the tide of World War II.

After completing this section, you will be able to

- determine the inverse of a square matrix.
- solve systems of equations using inverse matrices.
- model and solve real-world problems using matrix equations.

The product of a nonzero real number a and its multiplicative inverse (or reciprocal) $a^{-1} = \frac{1}{a}$ is 1, the multiplicative identity. Similarly, many square matrices A have a unique *multiplicative inverse matrix* A^{-1}, defined so that the product of A and A^{-1} is the identity matrix of the same order, I_n.

$$a \cdot \frac{1}{a} = \frac{1}{a} \cdot a = 1 \quad \text{and} \quad A \cdot A^{-1} = A^{-1} \cdot A = I_n$$

Example 1 Verifying Inverse Matrices

Show that $A = \begin{bmatrix} 2 & -1 \\ 5 & -3 \end{bmatrix}$ and $B = \begin{bmatrix} 3 & -1 \\ 5 & -2 \end{bmatrix}$ are inverse matrices.

Answer

$$AB = \begin{bmatrix} 2 & -1 \\ 5 & -3 \end{bmatrix}\begin{bmatrix} 3 & -1 \\ 5 & -2 \end{bmatrix} = \begin{bmatrix} 6-5 & -2+2 \\ 15-15 & -5+6 \end{bmatrix} = \begin{bmatrix} 1 & 0 \\ 0 & 1 \end{bmatrix} = I_2$$

$$BA = \begin{bmatrix} 3 & -1 \\ 5 & -2 \end{bmatrix}\begin{bmatrix} 2 & -1 \\ 5 & -3 \end{bmatrix} = \begin{bmatrix} 6-5 & -3+3 \\ 10-10 & -5+6 \end{bmatrix} = \begin{bmatrix} 1 & 0 \\ 0 & 1 \end{bmatrix} = I_2$$

Since $AB = BA = I$, A and B are inverse matrices ($B = A^{-1}$ and $A = B^{-1}$).

SKILL ✓ **EXERCISE 5**

Any matrix A for which A^{-1} exists is said to be *invertible* (or *nonsingular*) while a matrix for which no inverse exists is called *singular*. If $|A| = 0$, the matrix is singular. Notice that while matrix multiplication is not generally commutative, this property holds for the products of a matrix and its inverse.

The inverse of a 2×2 matrix can be found using a formula that includes its determinant.

🔆 **TIP**

Note that entries on the main diagonal are swapped, and entries on the minor diagonal are negated.

▶ **INVERSE OF A 2 × 2 MATRIX**

If $A = \begin{bmatrix} a & b \\ c & d \end{bmatrix}$ and $|A| \neq 0$, then $A^{-1} = \frac{1}{|A|}\begin{bmatrix} d & -b \\ -c & a \end{bmatrix}$. If $|A| = 0$, then A is singular.

Example 2 Finding the Inverse of a 2 × 2 Matrix

Find the inverse of each matrix.

a. $A = \begin{bmatrix} 4 & 1 \\ 5 & 2 \end{bmatrix}$

b. $B = \begin{bmatrix} 4 & -2 \\ 6 & -3 \end{bmatrix}$

Answer

a. $|A| = 8 - 5 = 3$  1. Find the determinant of A.

$$A^{-1} = \frac{1}{3}\begin{bmatrix} 2 & -1 \\ -5 & 4 \end{bmatrix} = \begin{bmatrix} \frac{2}{3} & -\frac{1}{3} \\ -\frac{5}{3} & \frac{4}{3} \end{bmatrix}$$

2. Apply the formula for the inverse of a 2×2 matrix.

CONTINUED ➡

PRESENTATION

Lesson Opener

State the following.

1. the additive identity for real numbers and for $m \times n$ matrices $0; 0_{m \times n}$

2. the additive inverse for the real number a and for the $m \times n$ matrix A $-a; -A$

3. the multiplicative identity for the real number a and for the $n \times n$ matrix A $1; I_n$

4. the multiplicative inverse for the real number a $\frac{1}{a}$

Use the Lesson Opener to help students connect the additive and multiplicative identities and inverses for real numbers with those covered in Section 7.2 and highlight the need to define the inverse for matrix multiplication.

Explain that the product of a matrix and its inverse must be the multiplicative identity. Then demonstrate how matrices A and B in Example 1 fulfill this requirement. After emphasizing the fact that the product of a matrix and its inverse is commutative, ask students why any non-square matrix must be *singular*. (*The dimensions prohibit* $A \cdot A^{-1} = A^{-1} \cdot A = I_n$.)

Present the formula for the inverse of a 2×2 matrix and Example 2. Consider having half the students verify that $A \cdot A^{-1} = I_n$ and the other half verify that $A^{-1} \cdot A = I_n$. Ask students to explain why any square matrix with a determinant of 0 is singular. (*An undefined fraction is generated.*)

Use the matrix from Example 2a to illustrate how to use elementary row operations to transform the augmented $[A|I]$ to $[I|A^{-1}]$. Explain that this method is used for higher-order square matrices and demonstrate the process with Example 3. Explain that technology is usually used to find inverses for square matri-

b. $|B| = -12 + 12 = 0$; B is singular.   Since $|B| = 0$, the matrix has no inverse.

____ SKILL ✔ **EXERCISE 9**

The reasoning that supports the formula for the inverse of a 2×2 matrix is illustrated by solving the systems generated by the matrix equation $AA^{-1} = I$. Using A from Example 2,

$$\begin{bmatrix} 4 & 1 \\ 5 & 2 \end{bmatrix} \begin{bmatrix} r & s \\ v & w \end{bmatrix} = \begin{bmatrix} 1 & 0 \\ 0 & 1 \end{bmatrix} \quad \text{implies} \quad \begin{matrix} 4r + v = 1 & 4s + w = 0 \\ 5r + 2v = 0 & 5s + 2w = 1. \end{matrix}$$

Since both systems have the same coefficient matrix, they can be solved simultaneously by augmenting the coefficient matrix with both constant column matrices to form $[A|I]$. Elementary row operations can then be used to transform this matrix to reduced row echelon form, $[I|A^{-1}]$.

$$\begin{bmatrix} 4 & 1 & | & 1 & 0 \\ 5 & 2 & | & 0 & 1 \end{bmatrix} \qquad \begin{matrix} \frac{1}{4}r_1 \Rightarrow r_1 \\ \frac{1}{3}r_2 \Rightarrow r_2 \end{matrix} \begin{bmatrix} 1 & \frac{1}{4} & | & \frac{1}{4} & 0 \\ 0 & 1 & | & -\frac{5}{3} & \frac{4}{3} \end{bmatrix}$$

$$-5r_1 + 4r_2 \Rightarrow r_2 \begin{bmatrix} 4 & 1 & | & 1 & 0 \\ 0 & 3 & | & -5 & 4 \end{bmatrix} \qquad -\frac{1}{4}r_2 + r_1 \Rightarrow r_1 \begin{bmatrix} 1 & 0 & | & \frac{2}{3} & -\frac{1}{3} \\ 0 & 1 & | & -\frac{5}{3} & \frac{4}{3} \end{bmatrix}$$

This method can be extended to find the inverse of square matrices of an order higher than 2. As in 2×2 matrices, any $n \times n$ matrix is singular if its determinant is 0. When $\det(A) = 0$, the augmented matrix $[A|I]$ cannot be converted to $[I|A^{-1}]$.

Example 3 Finding the Inverse of a 3×3 Matrix

Find A^{-1} if $A = \begin{bmatrix} 1 & 0 & 2 \\ 2 & -1 & 1 \\ 1 & 1 & 1 \end{bmatrix}$.

Answer

$$[A|I] = \begin{bmatrix} 1 & 0 & 2 & | & 1 & 0 & 0 \\ 2 & -1 & 1 & | & 0 & 1 & 0 \\ 1 & 1 & 1 & | & 0 & 0 & 1 \end{bmatrix} \qquad \begin{matrix} \\ \\ -\frac{1}{4}r_3 \Rightarrow r_3 \end{matrix} \begin{bmatrix} 1 & 0 & 2 & | & 1 & 0 & 0 \\ 0 & 1 & 3 & | & 2 & -1 & 0 \\ 0 & 0 & 1 & | & \frac{3}{4} & -\frac{1}{4} & -\frac{1}{4} \end{bmatrix}$$

$$\begin{matrix} -2r_1 + r_2 \Rightarrow r_2 \\ -r_1 + r_3 \Rightarrow r_3 \end{matrix} \begin{bmatrix} 1 & 0 & 2 & | & 1 & 0 & 0 \\ 0 & -1 & -3 & | & -2 & 1 & 0 \\ 0 & 1 & -1 & | & -1 & 0 & 1 \end{bmatrix} \qquad \begin{matrix} -2r_3 + r_1 \Rightarrow r_1 \\ -3r_3 + r_2 \Rightarrow r_2 \end{matrix} \begin{bmatrix} 1 & 0 & 0 & | & -\frac{1}{2} & \frac{1}{2} & \frac{1}{2} \\ 0 & 1 & 0 & | & -\frac{1}{4} & -\frac{1}{4} & \frac{3}{4} \\ 0 & 0 & 1 & | & \frac{3}{4} & -\frac{1}{4} & -\frac{1}{4} \end{bmatrix}$$

$$\begin{matrix} -r_2 \Rightarrow r_2 \\ r_2 + r_3 \Rightarrow r_3 \end{matrix} \begin{bmatrix} 1 & 0 & 2 & | & 1 & 0 & 0 \\ 0 & 1 & 3 & | & 2 & -1 & 0 \\ 0 & 0 & -4 & | & -3 & 1 & 1 \end{bmatrix} \qquad A^{-1} = \begin{bmatrix} -\frac{1}{2} & \frac{1}{2} & \frac{1}{2} \\ -\frac{1}{4} & -\frac{1}{4} & \frac{3}{4} \\ \frac{3}{4} & -\frac{1}{4} & -\frac{1}{4} \end{bmatrix}$$

____ SKILL ✔ **EXERCISE 19**

Since the arithmetic involved in this technique can be tedious and formulas for the inverse of a square matrix of an order higher than 2 are quite complicated, technology is frequently used. The $\boxed{x^{-1}}$ key is used to find the inverse and the ▸Frac function from the $\boxed{\text{MATH}}$ menu is used to convert decimals to fractions.

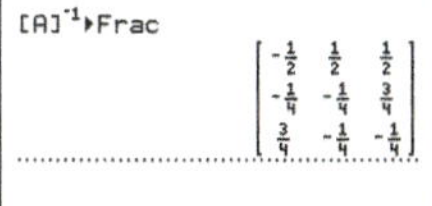

ces of order $n \geq 3$ and have the students complete Example 3 on their calculators.

Common Student Error Explain that using row operations to transform $[A|I]$ to $[I|A^{-1}]$ does not imply $[A|I] = [I|A^{-1}]$.

Illustrate the similarity of solving the linear equation $ax = c$ and the matrix equation $AX = C$ that represents a system of linear equations. Emphasize the necessity of left-multiplying both sides of the matrix equation since matrix multiplication is generally not commutative, then complete Example 4.

Interactive Activity Have the students create either an inconsistent or a dependent linear system in two variables and state its related coefficient matrix. Then verify that $|A| = 0$. In either case $|A| = 0$ and A^{-1} does not exist, indicating that there is no unique solution.

Illustrate how to use matrix equations to solve linear systems with Example 5. Focus on deriving the equations within the system, emphasizing the need to have as many equations as there are variables in order to have a unique solution to the system.

Consider explaining that technology enables efficient solutions of complex systems by using matrix equations or by finding the RREF of the augmented matrix. Cramer's rule and algebraic

Reading and Writing Mathematics

State the method you prefer for solving a system of equations (algebraically, solving the related matrix equation, Cramer's rule, Gaussian elimination) and why that is your preferred method.   _Answers will vary._

Additional Exercises

Determine whether A and B are inverse matrices.

1. $A = \begin{bmatrix} 0 & -1 \\ 1 & 2 \end{bmatrix}, B = \begin{bmatrix} 2 & 1 \\ -1 & 0 \end{bmatrix}$ _inverses_

2. $A = \begin{bmatrix} 1 & 0 & -1 \\ 0 & -1 & 2 \\ 3 & 0 & -4 \end{bmatrix}$,

$\quad B = \begin{bmatrix} 4 & 0 & -1 \\ 6 & -1 & -2 \\ 3 & 0 & -1 \end{bmatrix}$ _inverses_

Use the determinant to determine whether each matrix is invertible or singular.

3. $\begin{bmatrix} 4 & -3 \\ 7 & 10 \end{bmatrix}$ _61; yes_

4. $\begin{bmatrix} 3 & 2 & -5 \\ 7 & -9 & 3 \\ -6 & -4 & 10 \end{bmatrix}$ _0; no_

Use the formula to find the inverse of each matrix.

5. $\begin{bmatrix} 8 & 2 \\ 6 & 7 \end{bmatrix}$ $\begin{bmatrix} \frac{7}{44} & -\frac{1}{22} \\ -\frac{3}{22} & \frac{2}{11} \end{bmatrix}$

6. $\begin{bmatrix} 5 & -1 \\ -3 & -4 \end{bmatrix}$ $\begin{bmatrix} \frac{4}{23} & -\frac{1}{23} \\ -\frac{3}{23} & -\frac{5}{23} \end{bmatrix}$

Use row operations on $[A|I]$ to find the inverse of each matrix.

7. $\begin{bmatrix} -3 & 5 \\ 1 & -2 \end{bmatrix}$ $\begin{bmatrix} -2 & -5 \\ -1 & -3 \end{bmatrix}$

8. $\begin{bmatrix} 3 & -2 & 2 \\ 0 & 1 & -2 \\ 1 & -1 & 1 \end{bmatrix}$ $\begin{bmatrix} 1 & 0 & -2 \\ 2 & -1 & -6 \\ 1 & -1 & -3 \end{bmatrix}$

Write and solve a matrix equation for each system.

9. $5x - 2y = 2$
$\quad x + 2y = 10$ _(2, 4)_

10. $x - 2y + 2z = 0$
$\quad 3x - 4y - 4z = 28$
$\quad 5x - 2y + 10z = -8$ _(4, −1, −3)_

Solutions

❯ A. Exercises

1. $3(16) - (-4)(-12) = 0$; singular

2. $10(-36) - 4(90) = -720$; invertible

3. using column 2:
$$-(0)M_{12} + (0)M_{22} - (0)M_{32} = 0$$
singular

4. $\begin{vmatrix} 5 & 2 & 14 \\ 1 & -2 & -2 \\ 6 & -4 & 4 \end{vmatrix} \begin{matrix} 5 & 2 \\ 1 & -2 \\ 6 & -4 \end{matrix}$
$= (-40 - 24 - 56) - (-168 + 40 + 8)$
$= -120 + 120 = 0$; singular

5. $AB = \begin{bmatrix} -2 & 3 \\ -3 & 4 \end{bmatrix} \begin{bmatrix} 4 & -3 \\ 3 & -2 \end{bmatrix}$
$= \begin{bmatrix} -8+9 & 6-6 \\ -12+12 & 9-8 \end{bmatrix} = \begin{bmatrix} 1 & 0 \\ 0 & 1 \end{bmatrix} = I_2$

$BA = \begin{bmatrix} 4 & -3 \\ 3 & -2 \end{bmatrix} \begin{bmatrix} -2 & 3 \\ -3 & 4 \end{bmatrix}$
$= \begin{bmatrix} -8+9 & 12-12 \\ -6+6 & 9-8 \end{bmatrix} = \begin{bmatrix} 1 & 0 \\ 0 & 1 \end{bmatrix} = I_2$

6. $AB = BA = \begin{bmatrix} 0 & 1 \\ 1 & 0 \end{bmatrix} \begin{bmatrix} 0 & 1 \\ 1 & 0 \end{bmatrix}$
$= \begin{bmatrix} 0+1 & 0+0 \\ 0+0 & 1+0 \end{bmatrix} = \begin{bmatrix} 1 & 0 \\ 0 & 1 \end{bmatrix} = I_2$

7. $AB = \begin{bmatrix} 1 & 0 & 2 \\ 2 & 1 & -2 \\ 1 & 0 & 1 \end{bmatrix} \begin{bmatrix} -1 & 0 & 2 \\ 4 & 1 & -6 \\ 1 & 0 & -1 \end{bmatrix}$
$= \begin{bmatrix} -1+0+2 & 0+0+0 & 2+0-2 \\ -2+4-2 & 0+1+0 & 4-6+2 \\ -1+0+1 & 0+0+0 & 2+0-1 \end{bmatrix}$
$= \begin{bmatrix} 1 & 0 & 0 \\ 0 & 1 & 0 \\ 0 & 0 & 1 \end{bmatrix} = I_3$

$BA = \begin{bmatrix} -1 & 0 & 2 \\ 4 & 1 & -6 \\ 1 & 0 & -1 \end{bmatrix} \begin{bmatrix} 1 & 0 & 2 \\ 2 & 1 & -2 \\ 1 & 0 & 1 \end{bmatrix}$
$= \begin{bmatrix} -1+0+2 & 0+0+0 & -2+0+2 \\ 4+2-6 & 0+1+0 & 8-2-6 \\ 1+0-1 & 0+0+0 & 2+0-1 \end{bmatrix}$
$= \begin{bmatrix} 1 & 0 & 0 \\ 0 & 1 & 0 \\ 0 & 0 & 1 \end{bmatrix} = I_3$

If a system of linear equations has the same number of independent equations as variables, its unique solution can be found using the matrix equation $AX = C$ where A is the square coefficient matrix, X is the variable matrix, and C is the constant column matrix. Notice how left-multiplying each side of the equation by A^{-1} produces the solution $X = A^{-1}C$.

$$\begin{aligned} AX &= C \\ A^{-1}(AX) &= A^{-1}C \quad &&\text{Multiplication Property of Equality} \\ (A^{-1}A)X &= A^{-1}C \quad &&\text{Associative Property of Matrix Multiplication} \\ (I)X &= A^{-1}C \quad &&\text{definition of multiplicative inverse} \\ X &= A^{-1}C \quad &&\text{definition of multiplicative identity} \end{aligned}$$

If the determinant of the coefficient matrix is 0, the system is either dependent or inconsistent and a matrix equation cannot be used to solve the system.

Example 4 Using an Inverse Matrix to Solve a 2 × 2 System

Solve the system using an inverse matrix. $\quad \begin{aligned} 3x - 2y &= 12 \\ 4x + 3y &= -1 \end{aligned}$

Answer

$\begin{bmatrix} 3 & -2 \\ 4 & 3 \end{bmatrix} \begin{bmatrix} x \\ y \end{bmatrix} = \begin{bmatrix} 12 \\ -1 \end{bmatrix}$

1. Write the system as a matrix equation in the form $AX = C$.

$A^{-1} = \frac{1}{17} \begin{bmatrix} 3 & 2 \\ -4 & 3 \end{bmatrix}$

2. Use the formula to find the inverse of the 2 × 2 coefficient matrix.

$X = A^{-1}C = \frac{1}{17} \begin{bmatrix} 3 & 2 \\ -4 & 3 \end{bmatrix} \begin{bmatrix} 12 \\ -1 \end{bmatrix}$

3. Find the matrix product $X = A^{-1}C$.

$= \frac{1}{17} \begin{bmatrix} 34 \\ -51 \end{bmatrix} = \begin{bmatrix} 2 \\ -3 \end{bmatrix}$

The solution is the ordered pair $(2, -3)$.

SKILL ✔ EXERCISE 21

Inverse matrices can greatly simplify the solution of real-life problems modeled by a system of linear equations.

Example 5 Using an Inverse Matrix to Solve a 3 × 3 System

Mr. Jones wants to invest a $100,000 inheritance in an aggressive-growth mutual fund with a projected return of 10% APY, a municipal bond fund with a projected return of 3% APY, and CDs with a 2% APY. How much should he invest in each fund if he hopes to earn $6300 and wants to invest as much in the mutual fund as he does in municipal bonds and CDs combined?

Answer

investment: $g + b + c = 100{,}000$

return: $0.10g + 0.03b + 0.02c = 6300$

risk limit: $g = b + c$

1. Write a system of equations modeling the scenario, letting $c =$ the amount in CDs, $b =$ the amount in bonds, and $g =$ the amount in the growth mutual funds.

$\begin{aligned} g + b + c &= 100{,}000 \\ 10g + 3b + 2c &= 630{,}000 \\ g - b - c &= 0 \end{aligned}$

2. Write the equations in standard form. Multiplying the second equation by 100 eliminates the decimals.

$\begin{bmatrix} 1 & 1 & 1 \\ 10 & 3 & 2 \\ 1 & -1 & -1 \end{bmatrix} \begin{bmatrix} g \\ b \\ c \end{bmatrix} = \begin{bmatrix} 100{,}000 \\ 630{,}000 \\ 0 \end{bmatrix}$

3. Express the system as a matrix equation in the form of $AX = C$.

CONTINUED ➡

---

methods are often more cumbersome. Finding the RREF of an augmented matrix has the benefits of clearly indicating inconsistent systems and providing the means to state a general solution for dependent systems.

Many students will enjoy seeing how inverse matrices are used to decode encrypted matrices. You may want to show students how to store the coded matrix C on their calculator so it can be used in later steps. This material lays the foundation for the chapter's Data Analysis feature.

Consider summarizing the three methods of solving linear systems that have been presented: using row operations on an augmented matrix, using the determinants described in Cramer's rule, and solving the related matrix equation $AX = C$.

TIPS

Ex. 5–8 Finding any single entry that differs from the identity matrix is sufficient to show the matrices are not inverses.

Ex. 42 Initially express the variation with k as the constant of proportionality and solve for k.

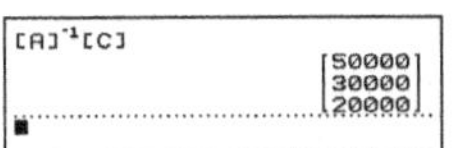

He should invest \$50,000 in the aggressive-growth mutual fund, \$30,000 in municipal bonds, and \$20,000 in CDs.

4. Use a calculator to find $X = A^{-1}C$.

5. Interpret the result.

SKILL ✔ **EXERCISE 31**

Matrices and their inverses are used in cryptography to encrypt and decode messages. Each letter or character is assigned a number, as in the example below.

A	B	C	D	E	F	G	H	I	J	K	L	M
0	1	2	3	4	5	6	7	8	9	10	11	12
N	O	P	Q	R	S	T	U	V	W	X	Y	Z
13	14	15	16	17	18	19	20	21	22	23	24	25

Spaces between words are frequently omitted when the message is converted into numbers and partitioned into several $1 \times n$ row matrices. (In this example, $n = 3$.) Extra Xs may be added to the end of a message if necessary.

G	O	D	I	S	M	Y	J	O	Y	X	X
6	14	3	8	18	12	24	9	14	24	23	23

To encrypt a message, create an $m \times n$ uncoded matrix U by stacking the row matrices and then right-multiply by an invertible $n \times n$ encrypting matrix E to form the coded matrix C.

$$
\begin{array}{ccccc}
U & \cdot & E & = & C
\end{array}
$$

$$
\begin{bmatrix} 6 & 14 & 3 \\ 8 & 18 & 12 \\ 24 & 9 & 14 \\ 24 & 23 & 23 \end{bmatrix}
\begin{bmatrix} 1 & 3 & 3 \\ 1 & 4 & 3 \\ 1 & 3 & 4 \end{bmatrix}
=
\begin{bmatrix} 23 & 83 & 72 \\ 38 & 132 & 126 \\ 47 & 150 & 155 \\ 70 & 233 & 233 \end{bmatrix}
$$

The message can then be sent as a string of values: 23, 83, 72, 38, 132, ... The recipient reconstructs the coded matrix and multiplies it by E^{-1}, the decoding matrix, to reveal the uncoded matrix: $CE^{-1} = (UE)E^{-1} = U(EE^{-1}) = U(I) = U$.

$$
\begin{array}{ccccc}
C & \cdot & E^{-1} & = & U
\end{array}
$$

$$
\begin{bmatrix} 23 & 83 & 72 \\ 38 & 132 & 126 \\ 47 & 150 & 155 \\ 70 & 233 & 233 \end{bmatrix}
\begin{bmatrix} 7 & -3 & -3 \\ -1 & 1 & 0 \\ -1 & 0 & 1 \end{bmatrix}
=
\begin{bmatrix} 6 & 14 & 3 \\ 8 & 18 & 12 \\ 24 & 9 & 14 \\ 24 & 23 & 23 \end{bmatrix}
$$

The resulting string of numbers can then be converted back to the original message.

Encryption is a valuable tool for protecting information, but it can also be used maliciously. In 2017 the WannaCry virus infected over 300,000 computers, encrypting each user's files and demanding a bitcoin ransom. Several affected hospitals in the United Kingdom had to cancel operations and turn away patients.

14.

15.

16.

B. Exercises

17.
$$[A\,|\,I] = \begin{bmatrix} 3 & 7 & | & 1 & 0 \\ 4 & 9 & | & 0 & 1 \end{bmatrix}$$

$$\frac{1}{3}r_1 \Rightarrow r_1 \begin{bmatrix} 1 & \frac{7}{3} & | & \frac{1}{3} & 0 \\ & & & & \end{bmatrix}$$
$$4r_1 - 3r_2 \Rightarrow r_2 \begin{bmatrix} 1 & \frac{7}{3} & | & \frac{1}{3} & 0 \\ 0 & 1 & | & 4 & -3 \end{bmatrix}$$
$$-\frac{7}{3}r_2 + r_1 \Rightarrow r_1 \begin{bmatrix} 1 & 0 & | & -9 & 7 \\ 0 & 1 & | & 4 & -3 \end{bmatrix}$$
$$= [I\,|\,A^{-1}]$$

18.
$$[A\,|\,I] = \begin{bmatrix} 2 & 4 & | & 1 & 0 \\ -3 & -1 & | & 0 & 1 \end{bmatrix}$$

$$\frac{1}{2}r_1 \Rightarrow r_1 \begin{bmatrix} 1 & 2 & | & \frac{1}{2} & 0 \\ & & & & \end{bmatrix}$$
$$3r_1 + 2r_2 \Rightarrow r_2 \begin{bmatrix} 1 & 2 & | & \frac{1}{2} & 0 \\ 0 & 10 & | & 3 & 2 \end{bmatrix}$$
$$-\frac{1}{5}r_2 + r_1 \Rightarrow r_1 \begin{bmatrix} 1 & 0 & | & -\frac{1}{10} & -\frac{2}{5} \\ & & & & \end{bmatrix}$$
$$\frac{1}{10}r_2 \Rightarrow r_2 \begin{bmatrix} 1 & 0 & | & -\frac{1}{10} & -\frac{2}{5} \\ 0 & 1 & | & \frac{3}{10} & \frac{1}{5} \end{bmatrix}$$
$$= [I\,|\,A^{-1}]$$

19.
$$[A\,|\,I] = \begin{bmatrix} 1 & 2 & 0 & | & 1 & 0 & 0 \\ 2 & 1 & -3 & | & 0 & 1 & 0 \\ 0 & 3 & 2 & | & 0 & 0 & 1 \end{bmatrix}$$

$$-2r_1 + r_2 \Rightarrow r_2 \begin{bmatrix} 1 & 2 & 0 & | & 1 & 0 & 0 \\ 0 & -3 & -3 & | & -2 & 1 & 0 \\ 0 & 3 & 2 & | & 0 & 0 & 1 \end{bmatrix}$$

$$-\frac{1}{3}r_2 \Rightarrow r_2 \begin{bmatrix} 1 & 2 & 0 & | & 1 & 0 & 0 \\ 0 & 1 & 1 & | & \frac{2}{3} & -\frac{1}{3} & 0 \\ 0 & 3 & 2 & | & 0 & 0 & 1 \end{bmatrix}$$

$$-2r_2 + r_1 \Rightarrow r_1 \begin{bmatrix} 1 & 0 & -2 & | & -\frac{1}{3} & \frac{2}{3} & 0 \\ & 0 & 1 & 1 & | & \frac{2}{3} & -\frac{1}{3} & 0 \\ 0 & 0 & -1 & | & -2 & 1 & 1 \end{bmatrix}$$
$$-3r_2 + r_3 \Rightarrow r_3$$

$$-2r_3 + r_1 \Rightarrow r_1 \begin{bmatrix} 1 & 0 & 0 & | & \frac{11}{3} & -\frac{4}{3} & -2 \\ 0 & 1 & 0 & | & -\frac{4}{3} & \frac{2}{3} & 1 \\ 0 & 0 & 1 & | & 2 & -1 & -1 \end{bmatrix}$$
$$r_3 + r_2 \Rightarrow r_2$$
$$-r_3 \Rightarrow r_3$$
$$= [I\,|\,A^{-1}]$$

8. $AB = \begin{bmatrix} 0 & 1 & 1 \\ 2 & 1 & 1 \\ 1 & 2 & 0 \end{bmatrix} \begin{bmatrix} 1 & 0 & 3 \\ 1 & 2 & 0 \\ 0 & -2 & -4 \end{bmatrix}$

$$= \begin{bmatrix} 0+1+0 & 0+2-2 & 0+0-4 \\ 2+1+0 & 0+2-2 & 6+0-4 \\ 1+2+0 & 0+4+0 & 3+0+0 \end{bmatrix}$$

$$= \begin{bmatrix} 1 & 0 & -4 \\ 3 & 0 & 2 \\ 3 & 4 & 3 \end{bmatrix} \neq I_3$$

9. $|A| = 16 - 12 = 4$

$$A^{-1} = \frac{1}{4}\begin{bmatrix} 2 & -3 \\ -4 & 8 \end{bmatrix} = \begin{bmatrix} \frac{1}{2} & -\frac{3}{4} \\ -1 & 2 \end{bmatrix}$$

10. $|A| = -27 + 35 = 8$

$$A^{-1} = \frac{1}{8}\begin{bmatrix} -9 & -7 \\ 5 & 3 \end{bmatrix} = \begin{bmatrix} -\frac{9}{8} & -\frac{7}{8} \\ \frac{5}{8} & \frac{3}{8} \end{bmatrix}$$

11. $|A| = \frac{4}{3} - 1 = \frac{1}{3}$

$$A^{-1} = 3\begin{bmatrix} \frac{2}{3} & -1 \\ -1 & 2 \end{bmatrix} = \begin{bmatrix} 2 & -3 \\ -3 & 6 \end{bmatrix}$$

12. $|A| = \frac{-35}{9} + \frac{32}{9} = -\frac{1}{3}$

$$A^{-1} = -3\begin{bmatrix} \frac{7}{3} & -\frac{4}{3} \\ \frac{8}{3} & -\frac{5}{3} \end{bmatrix} = \begin{bmatrix} -7 & 4 \\ -8 & 5 \end{bmatrix}$$

13.

20.

$$[A|I] = \begin{bmatrix} 3 & 5 & 6 & | & 1 & 0 & 0 \\ 1 & 2 & 2 & | & 0 & 1 & 0 \\ 2 & 4 & 5 & | & 0 & 0 & 1 \end{bmatrix}$$

$$r_1 \Leftrightarrow r_2 \begin{bmatrix} 1 & 2 & 2 & | & 0 & 1 & 0 \\ 3 & 5 & 6 & | & 1 & 0 & 0 \\ 2 & 4 & 5 & | & 0 & 0 & 1 \end{bmatrix}$$

$$\begin{matrix} -3r_1 + r_2 \Rightarrow r_2 \\ -2r_1 + r_3 \Rightarrow r_3 \end{matrix} \begin{bmatrix} 1 & 2 & 2 & | & 0 & 1 & 0 \\ 0 & -1 & 0 & | & 1 & -3 & 0 \\ 0 & 0 & 1 & | & 0 & -2 & 1 \end{bmatrix}$$

$$\begin{matrix} -2r_3 + r_1 \Rightarrow r_1 \\ -r_2 \Rightarrow r_2 \end{matrix} \begin{bmatrix} 1 & 2 & 0 & | & 0 & 5 & -2 \\ 0 & 1 & 0 & | & -1 & 3 & 0 \\ 0 & 0 & 1 & | & 0 & -2 & 1 \end{bmatrix}$$

$$-2r_2 + r_1 \Rightarrow r_1 \begin{bmatrix} 1 & 0 & 0 & | & 2 & -1 & -2 \\ 0 & 1 & 0 & | & -1 & 3 & 0 \\ 0 & 0 & 1 & | & 0 & -2 & 1 \end{bmatrix}$$

$$= [I \,|\, A^{-1}]$$

21. $\begin{bmatrix} 1 & -5 \\ 2 & -1 \end{bmatrix} \begin{bmatrix} x \\ y \end{bmatrix} = \begin{bmatrix} -30 \\ 3 \end{bmatrix}$

$$A^{-1} = \frac{1}{9} \begin{bmatrix} -1 & 5 \\ -2 & 1 \end{bmatrix}$$

$$X = A^{-1}C = \frac{1}{9} \begin{bmatrix} -1 & 5 \\ -2 & 1 \end{bmatrix} \begin{bmatrix} -30 \\ 3 \end{bmatrix}$$

$$= \frac{1}{9} \begin{bmatrix} 45 \\ 63 \end{bmatrix} = \begin{bmatrix} 5 \\ 7 \end{bmatrix}$$

22. $\begin{bmatrix} 2 & 1 \\ 5 & 2 \end{bmatrix} \begin{bmatrix} x \\ y \end{bmatrix} = \begin{bmatrix} 8 \\ 19 \end{bmatrix}$

$$A^{-1} = -1 \begin{bmatrix} 2 & -1 \\ -5 & 2 \end{bmatrix}$$

$$X = A^{-1}C = \begin{bmatrix} -2 & 1 \\ 5 & -2 \end{bmatrix} \begin{bmatrix} 8 \\ 19 \end{bmatrix} = \begin{bmatrix} 3 \\ 2 \end{bmatrix}$$

23. $\begin{bmatrix} 2 & -5 \\ 1 & 4 \end{bmatrix} \begin{bmatrix} x \\ y \end{bmatrix} = \begin{bmatrix} -39 \\ 26 \end{bmatrix}$

$$A^{-1} = \frac{1}{13} \begin{bmatrix} 4 & 5 \\ -1 & 2 \end{bmatrix}$$

$$X = A^{-1}C = \frac{1}{13} \begin{bmatrix} 4 & 5 \\ -1 & 2 \end{bmatrix} \begin{bmatrix} -39 \\ 26 \end{bmatrix}$$

$$= \frac{1}{13} \begin{bmatrix} -26 \\ 91 \end{bmatrix} = \begin{bmatrix} -2 \\ 7 \end{bmatrix}$$

24. $\begin{bmatrix} 4 & 2 \\ 1 & 6 \end{bmatrix} \begin{bmatrix} x \\ y \end{bmatrix} = \begin{bmatrix} -8 \\ 1 \end{bmatrix}$

$$A^{-1} = \frac{1}{22} \begin{bmatrix} 6 & -2 \\ -1 & 4 \end{bmatrix}$$

$$X = A^{-1}C = \frac{1}{22} \begin{bmatrix} 6 & -2 \\ -1 & 4 \end{bmatrix} \begin{bmatrix} -8 \\ 1 \end{bmatrix}$$

$$= \frac{1}{22} \begin{bmatrix} -50 \\ 12 \end{bmatrix} = \begin{bmatrix} -\frac{25}{11} \\ \frac{6}{11} \end{bmatrix}$$

25. $\begin{bmatrix} 1 & 3 & -4 \\ 2 & -1 & 5 \\ 4 & 2 & 4 \end{bmatrix} \begin{bmatrix} x \\ y \\ z \end{bmatrix} = \begin{bmatrix} 41 \\ 0 \\ 44 \end{bmatrix}$

$$X = A^{-1}C = \begin{bmatrix} 9 \\ 8 \\ -2 \end{bmatrix}$$

❯ A. Exercises

Use the determinant to decide whether each matrix is invertible or singular.

1. $\begin{bmatrix} 3 & -4 \\ -12 & 16 \end{bmatrix}$ singular   **2.** $\begin{bmatrix} 10 & 90 \\ 4 & -36 \end{bmatrix}$ invertible

3. $\begin{bmatrix} -37 & 0 & 96 \\ 11 & 0 & 118 \\ 52 & 0 & 75 \end{bmatrix}$ singular   **4.** $\begin{bmatrix} 5 & 2 & 14 \\ 1 & -2 & -2 \\ 6 & -4 & 4 \end{bmatrix}$ singular

Determine whether A and B are inverse matrices.

5. $A = \begin{bmatrix} -2 & 3 \\ -3 & 4 \end{bmatrix}, B = \begin{bmatrix} 4 & -3 \\ 3 & -2 \end{bmatrix}$ inverses

6. $A = \begin{bmatrix} 0 & 1 \\ 1 & 0 \end{bmatrix}, B = \begin{bmatrix} 0 & 1 \\ 1 & 0 \end{bmatrix}$ inverses

7. $A = \begin{bmatrix} 1 & 0 & 2 \\ 2 & 1 & -2 \\ 1 & 0 & 1 \end{bmatrix}, B = \begin{bmatrix} -1 & 0 & 2 \\ 4 & 1 & -6 \\ 1 & 0 & -1 \end{bmatrix}$ inverses

8. $A = \begin{bmatrix} 0 & 1 & 1 \\ 2 & 1 & 1 \\ 1 & 2 & 0 \end{bmatrix}, B = \begin{bmatrix} 1 & 0 & 3 \\ 1 & 2 & 0 \\ 0 & -2 & -4 \end{bmatrix}$ not inverses

Use the formula to find the inverse of each 2 × 2 matrix.

9. $\begin{bmatrix} 8 & 3 \\ 4 & 2 \end{bmatrix}$ $\begin{bmatrix} \frac{1}{2} & -\frac{3}{4} \\ -1 & 2 \end{bmatrix}$ **10.** $\begin{bmatrix} 3 & 7 \\ -5 & -9 \end{bmatrix}$ $\begin{bmatrix} -\frac{9}{8} & -\frac{7}{8} \\ \frac{5}{8} & \frac{3}{8} \end{bmatrix}$

11. $\begin{bmatrix} 2 & 1 \\ 1 & \frac{2}{3} \end{bmatrix}$ $\begin{bmatrix} 2 & -3 \\ -3 & 6 \end{bmatrix}$ **12.** $\begin{bmatrix} -\frac{5}{3} & \frac{4}{3} \\ -\frac{8}{3} & \frac{7}{3} \end{bmatrix}$ $\begin{bmatrix} -7 & 4 \\ -8 & 5 \end{bmatrix}$

Use technology to find the inverse of each matrix. Express answers in fractional form.

13. $\begin{bmatrix} 16 & 4 \\ -7 & -2 \end{bmatrix}$ $\begin{bmatrix} \frac{1}{2} & 1 \\ -\frac{7}{4} & -4 \end{bmatrix}$ **14.** $\begin{bmatrix} 2 & 8 \\ -1 & 4 \end{bmatrix}$ $\begin{bmatrix} \frac{1}{4} & -\frac{1}{2} \\ \frac{1}{16} & \frac{1}{8} \end{bmatrix}$

15. $\begin{bmatrix} 7 & 0 & 4 \\ -5 & 1 & -2 \\ 2 & 2 & 3 \end{bmatrix}$ **16.** $\begin{bmatrix} 5 & -2 & 4 \\ 0 & 2 & 1 \\ 1 & 3 & 3 \end{bmatrix}$

❯ B. Exercises

Use row operations on $[A \,|\, I]$ to find the inverse of each matrix.

17. $\begin{bmatrix} 3 & 7 \\ 4 & 9 \end{bmatrix}$ $\begin{bmatrix} -9 & 7 \\ 4 & -3 \end{bmatrix}$ **18.** $\begin{bmatrix} 2 & 4 \\ -3 & -1 \end{bmatrix}$ $\begin{bmatrix} -\frac{1}{10} & -\frac{2}{5} \\ \frac{3}{10} & \frac{1}{5} \end{bmatrix}$

19. $\begin{bmatrix} 1 & 2 & 0 \\ 2 & 1 & -3 \\ 0 & 3 & 2 \end{bmatrix}$ **20.** $\begin{bmatrix} 3 & 5 & 6 \\ 1 & 2 & 2 \\ 2 & 4 & 5 \end{bmatrix}$

Write and solve a matrix equation to solve each system.

21. $x - 5y = -30$
$2x - y = 3$ (5, 7)

22. $2x + y = 8$
$5x + 2y = 19$ (3, 2)

23. $2x - 5y = -39$
$x + 4y = 26$ (−2, 7)

24. $4x + 2y = -8$
$x + 6y = 1$ $\left(-\frac{25}{11}, \frac{6}{11}\right)$

26. $\left(-\frac{29}{2}, -3, 14\right)$ **28.** $\left(\frac{139}{3}, \frac{62}{3}, -\frac{56}{3}\right)$

27. $\left(\frac{99}{10}, \frac{3}{2}, \frac{33}{10}\right)$

Write and use technology to solve a matrix equation for each system. Express answers in fractional form.

25. $x + 3y - 4z = 41$
$2x - y + 5z = 0$
$4x + 2y + 4z = 44$ (9, 8, −2)

26. $2x + y + 3z = 10$
$2x - 3y + 4z = 36$
$-2x + 3y + 2z = 48$

27. $2x + 3y - z = 21$
$x + y + 2z = 18$
$x - 3y + 2z = 12$

28. $6y + 3z = 68$
$4x + 5z = 92$
$7x - 9y + 5z = 45$

29. The following systems cannot be solved by solving a related matrix equation. Classify each system and explain why its solution cannot be found using this method.

   a. $2x + y = 8$
       $4x + 2y = 16$

   b. $3a - 5b = 11$
       $-3a + 5b = 7$

Write a system of equations and solve the related matrix equation for each problem.

30. How old are Jenn and Dan if three times Jenn's age is the same as twice Dan's age and the sum of their ages is 60?   Jenn: 24; Dan: 36

31. On Tuesday, Grayson filled up his ATV at $3.95/gal. On Friday, he filled up at $4.12/gal. Together the 5.7 gal of gasoline cost $22.94. How much gas did Grayson put into the tank each day?   3.2 gal Mon.; 2.5 gal Fri.

32. Find two numbers whose sum is 20 if twice the first is 2 more than the opposite of the second.   −18, 38

33. Find three numbers whose sum is 35 if the least is 126 less than twice the greatest, and the sum of the least and the greatest is 6 times the other number.   −22, 5, 52

34. Jackie wants to earn $500 in 1 yr by investing in a savings account earning 2% APY, a CD earning 3.5% APY, and a bond earning 4.7% APY. She plans on investing twice as much in the bond as she does in the CD and putting half of the amount invested in the CD and bond combined into savings. How much (to the nearest dollar) should Jackie deposit in each?   savings: $4717; CD: $3145; bond: $6289 or $6290

Use the encoding matrix E and the letter assignments from p. 373 to make a coded matrix for each message.

$$E = \begin{bmatrix} 1 & 2 & -1 \\ 0 & -2 & 1 \\ 1 & 1 & 0 \end{bmatrix}$$

35. PLEASE CALL

$$\begin{bmatrix} 19 & 12 & -4 \\ 4 & -32 & 18 \\ 13 & 15 & -2 \\ 34 & -1 & 12 \end{bmatrix}$$

36. MEET ME AT NOON

$$\begin{bmatrix} 16 & 20 & -8 \\ 23 & 18 & -7 \\ 13 & -25 & 19 \\ 27 & 13 & 0 \end{bmatrix}$$

26. $\begin{bmatrix} 2 & 1 & 3 \\ 2 & -3 & 4 \\ -2 & 3 & 2 \end{bmatrix} \begin{bmatrix} x \\ y \\ z \end{bmatrix} = \begin{bmatrix} 10 \\ 36 \\ 48 \end{bmatrix}$

$$X = A^{-1}C = \begin{bmatrix} -\frac{29}{2} \\ -3 \\ 14 \end{bmatrix}$$

27. $\begin{bmatrix} 2 & 3 & -1 \\ 1 & 1 & 2 \\ 1 & -3 & 2 \end{bmatrix} \begin{bmatrix} x \\ y \\ z \end{bmatrix} = \begin{bmatrix} 21 \\ 18 \\ 12 \end{bmatrix}$

$$X = A^{-1}C = \begin{bmatrix} \frac{99}{10} \\ \frac{3}{2} \\ \frac{33}{10} \end{bmatrix}$$

28. $\begin{bmatrix} 0 & 6 & 3 \\ 4 & 0 & 5 \\ 7 & -9 & 5 \end{bmatrix} \begin{bmatrix} x \\ y \\ z \end{bmatrix} = \begin{bmatrix} 68 \\ 92 \\ 45 \end{bmatrix}$

$$X = A^{-1}C = \begin{bmatrix} \frac{139}{3} \\ \frac{62}{3} \\ -\frac{56}{3} \end{bmatrix}$$

29a. dependent; It has an infinite number of solutions.
$\{(x, y) \mid y = -2x + 8\}$

29b. inconsistent; It has no solution.

C. Exercises

37. Find B given A and AB.

$A = \begin{bmatrix} 2 & 1 \\ 4 & -1 \end{bmatrix}$, $AB = \begin{bmatrix} 2 & 7 \\ 4 & 5 \end{bmatrix}$ $\begin{bmatrix} 1 & 2 \\ 0 & 3 \end{bmatrix}$

38. Find A given B and $(AB)^{-1}$.

$B = \begin{bmatrix} 4 & 3 \\ 0 & -\frac{1}{4} \end{bmatrix}$, $(AB)^{-1} = \begin{bmatrix} -5 & 3 \\ 7 & -4 \end{bmatrix}$ $\begin{bmatrix} 1 & 0 \\ \frac{7}{4} & 1 \end{bmatrix}$

Decode each matrix that was encrypted using the letter assignments from the lesson and the encoding matrix

$$E = \begin{bmatrix} 1 & 2 & -1 \\ 0 & -2 & 1 \\ 1 & 1 & 0 \end{bmatrix}.$$

39. $C = \begin{bmatrix} 31 & 41 & -14 \\ 17 & -4 & 9 \\ 37 & 42 & -9 \end{bmatrix}$ **40.** $C = \begin{bmatrix} 8 & 1 & 3 \\ 18 & 6 & -1 \\ 19 & -1 & 1 \\ 12 & -4 & 6 \\ 36 & 43 & -10 \end{bmatrix}$

SEND MONEY

BE HONEST BE KIND

41. Verify the formula for the inverse of a 2×2 matrix by using row operations on $[A|I]$ to find the inverse of $A = \begin{bmatrix} a & b \\ c & d \end{bmatrix}$.

CUMULATIVE REVIEW

42. The quantity p varies jointly with q and r and inversely with t. If $p = 3q$ when $r = 4t$, find q when $p = 8$, $r = 12$, and $t = 20$. [Algebra] $q = \frac{160}{9}$

43. Rewrite the expression $4 \sin\theta \cos\theta$ in terms of a single trigonometric function. [5.5] $2 \sin 2\theta$

44. Write an augmented matrix representing the system. [7.3]

$3x + y = 3$
$4x + 3z = -16$
$y - 2z = 14$ $\begin{bmatrix} 3 & 1 & 0 & | & 3 \\ 4 & 0 & 3 & | & -16 \\ 0 & 1 & -2 & | & 14 \end{bmatrix}$

45. Sketch the graph of $p(x) = x^3 - x$. [2.2]

46. Solve the system. [7.3]
$5x + 2y - 3z = 9$
$4y + z = 19$
$z = 3$ $(2, 4, 3)$

47. If matrix A has a row of 0s, what is $|A|$? Explain your reasoning. [7.4] $|A| = 0$; The product of each entry and its cofactor is 0.

48. Evaluate $\cos\frac{5\pi}{12}$. [5.4] A

A. $\dfrac{\sqrt{6} - \sqrt{2}}{4}$ **C.** $\dfrac{\sqrt{6} - \sqrt{2}}{2}$ **E.** none of these

B. $\dfrac{\sqrt{6} + \sqrt{2}}{4}$ **D.** $\dfrac{\sqrt{3}}{4}$

49. Find the area of $\triangle ABC$ given $a = 3.2$ mi, $b = 4.8$ mi, and $m\angle C = 75°$. [5.7] C

A. 2 mi^2 **C.** 7.4 mi^2 **E.** none of these
B. 7.1 mi^2 **D.** 3 mi^2

50. Find $m\angle A$ (to the nearest degree). [4.2] D
A. 70°
B. 20°
C. 68°
D. 22°
E. none of these

51. At which angle would the plots of consecutive 12th roots of a complex number differ on the polar coordinate plane? [6.6] A

A. $\frac{\pi}{6}$ **C.** $\frac{\pi}{3}$ **E.** none of these
B. $\frac{\pi}{12}$ **D.** $\frac{\pi}{24}$

15. $\begin{bmatrix} 7 & 8 & -4 \\ 11 & 13 & -6 \\ -12 & -14 & 7 \end{bmatrix}$ **16.** $\begin{bmatrix} \frac{3}{5} & \frac{18}{5} & -2 \\ \frac{1}{5} & \frac{11}{5} & -1 \\ -\frac{2}{5} & -\frac{17}{5} & 2 \end{bmatrix}$ **19.** $\begin{bmatrix} \frac{11}{3} & -\frac{4}{3} & -2 \\ -\frac{4}{3} & \frac{2}{3} & 1 \\ 2 & -1 & -1 \end{bmatrix}$ **20.** $\begin{bmatrix} 2 & -1 & -2 \\ -1 & 3 & 0 \\ 0 & -2 & 1 \end{bmatrix}$

30. $3j = 2d$; $3j - 2d = 0$
$j + d = 60$

$\begin{bmatrix} 3 & -2 \\ 1 & 1 \end{bmatrix}\begin{bmatrix} j \\ d \end{bmatrix} = \begin{bmatrix} 0 \\ 60 \end{bmatrix}$

$X = A^{-1}C = \begin{bmatrix} 24 \\ 36 \end{bmatrix}$

31. gallons: $m + f = 5.7$
cost: $3.95m + 4.12f = 22.94$

$\begin{bmatrix} 1 & 1 \\ 3.95 & 4.12 \end{bmatrix}\begin{bmatrix} m \\ f \end{bmatrix} = \begin{bmatrix} 5.7 \\ 22.94 \end{bmatrix}$

$X = A^{-1}C = \begin{bmatrix} 3.2 \\ 2.5 \end{bmatrix}$

32. $x + y = 20$
$2x = -y + 2$; $2x + y = 2$

$\begin{bmatrix} 1 & 1 \\ 2 & 1 \end{bmatrix}\begin{bmatrix} x \\ y \end{bmatrix} = \begin{bmatrix} 20 \\ 2 \end{bmatrix}$

$X = A^{-1}C = \begin{bmatrix} -18 \\ 38 \end{bmatrix}$

33. Let $x =$ least number,
$y =$ other number, and
$z =$ greatest number.
$$x + y + z = 35$$
$x = 2z - 126$; $x - 2z = -126$
$x + z = 6y$; $x - 6y + z = 0$

$\begin{bmatrix} 1 & 1 & 1 \\ 1 & 0 & -2 \\ 1 & -6 & 1 \end{bmatrix}\begin{bmatrix} x \\ y \\ z \end{bmatrix} = \begin{bmatrix} 35 \\ -126 \\ 0 \end{bmatrix}$

$X = A^{-1}C = \begin{bmatrix} -22 \\ 5 \\ 52 \end{bmatrix}$

34. $0.02s + 0.035c + 0.047b = 500$
$20s + 35c + 47b = 500{,}000$
$b = 2c$; $2c - b = 0$
$s = \frac{1}{2}(c + b)$; $2s - c - b = 0$

$\begin{bmatrix} 20 & 35 & 47 \\ 0 & 2 & -1 \\ 2 & -1 & -1 \end{bmatrix}\begin{bmatrix} x \\ y \\ z \end{bmatrix} = \begin{bmatrix} 500{,}000 \\ 0 \\ 0 \end{bmatrix}$

$X = A^{-1}C \approx \begin{bmatrix} 4717 \\ 3145 \\ 6289 \end{bmatrix}$

35. $U = \begin{bmatrix} 15 & 11 & 4 \\ 0 & 18 & 4 \\ 2 & 0 & 11 \\ 11 & 23 & 23 \end{bmatrix}$

$C = UE = \begin{bmatrix} 19 & 12 & -4 \\ 4 & -32 & 18 \\ 13 & 15 & -2 \\ 34 & -1 & 12 \end{bmatrix}$

36. $U = \begin{bmatrix} 12 & 4 & 4 \\ 19 & 12 & 4 \\ 0 & 19 & 13 \\ 14 & 14 & 13 \end{bmatrix}$

$C = UE = \begin{bmatrix} 16 & 20 & -8 \\ 23 & 18 & -7 \\ 13 & -25 & 19 \\ 27 & 13 & 0 \end{bmatrix}$

37. Let $B = \begin{bmatrix} a & b \\ c & d \end{bmatrix}$, then

$\begin{bmatrix} 2 & 1 \\ 4 & -1 \end{bmatrix}\begin{bmatrix} a & b \\ c & d \end{bmatrix} = \begin{bmatrix} 2 & 7 \\ 4 & 5 \end{bmatrix}.$

$2a + c = 2 \qquad\qquad 2b + d = 7$
$\underline{4a - c = 4} \qquad\qquad \underline{4b - d = 5}$
$6a \quad\ = 6 \qquad\qquad 6b \quad\ = 12$
$\qquad a = 1 \qquad\qquad\qquad b = 2$
$2(1) + c = 2 \qquad 2(2) + d = 7$
$\qquad c = 0 \qquad\qquad\qquad d = 3$

38. $\left((AB)^{-1}\right)^{-1} = AB$

$\dfrac{1}{-1}\begin{bmatrix} -4 & -3 \\ -7 & -5 \end{bmatrix} = \begin{bmatrix} 4 & 3 \\ 7 & 5 \end{bmatrix}$

Let $A = \begin{bmatrix} a & b \\ c & d \end{bmatrix}$, then

$\begin{bmatrix} a & b \\ c & d \end{bmatrix}\begin{bmatrix} 4 & 3 \\ 0 & -\frac{1}{4} \end{bmatrix} = \begin{bmatrix} 4 & 3 \\ 7 & 5 \end{bmatrix}.$

$4a + 0b = 4 \qquad\qquad 3a - \frac{1}{4}b = 3$
$\qquad a = 1 \qquad\qquad 3(1) - \frac{1}{4}b = 3$
$\qquad\qquad\qquad\qquad\qquad\qquad b = 0$

$4c + 0d = 7 \qquad\qquad 3c - \frac{1}{4}d = 5$
$\qquad c = \frac{7}{4} \qquad 3\left(\frac{7}{4}\right) - \frac{1}{4}d = 5$
$\qquad\qquad\qquad\qquad\qquad -\frac{1}{4}d = -\frac{1}{4}$
$\qquad\qquad\qquad\qquad\qquad\qquad d = 1$

continued in Answers and Solutions Overflow

7.6 Systems of Inequalities

Objectives

1. To solve a system of inequalities in two variables

2. To maximize or minimize an objective function for a set of constraints

3. To model and solve real-world problems using linear programming

Vocabulary

boundary
constraints
feasible region
linear programming
objective function
system of inequalities
Vertex Principle of Linear
  Programming

Reading and Writing Mathematics

Explain why you cannot solve $\frac{3x+7}{x-1} < 10$ by multiplying both sides of the inequality by $x - 1$.

Since we cannot know whether $x - 1$ is positive or negative, we do not know whether we should change the order of the inequality. In solving this inequality we do not know whether the solution is $x < \frac{17}{7}$ or $x > \frac{17}{7}$. *Note:* Substituting 0 in the original inequality would confirm that $x < \frac{17}{7}$ (where $x \neq 1$).

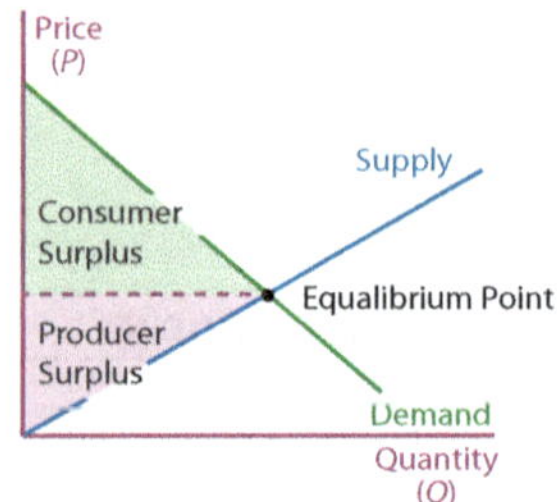

After completing this section, you will be able to

- solve a system of inequalities in two variables.

- maximize or minimize an objective function for a set of linear constraints.

- model and solve real-world problems using linear programming.

Businesses examine supply and demand graphs to find the equilibrium price and the regions representing consumer surplus (a measure of consumer willingness to pay more for a product) and producer surplus (a measure of producer welfare). These regions can be described by a *system of inequalities.* Use the Internet keyword searches *consumer surplus* and *producer surplus* to find detailed definitions and descriptions of these business terms.

To solve a system of inequalities, graph each inequality and find the intersection of the graphs. Recall that the graph of each inequality's related equation (its *boundary*) is drawn as a dashed line for inequalities using > or <, or as a solid line for inequalities using ≤ or ≥. The set of points that satisfies all the inequalities is the system's solution.

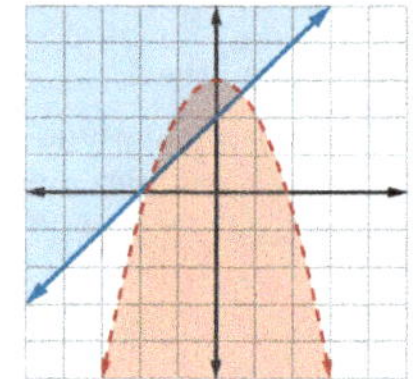

Example 1 Solving a System of Inequalities

Solve the system.
$$x - y \leq -2$$
$$y < -x^2 + 4$$

Answer

The solution is represented by the intersection of the two graphs.

1. Graph the linear inequality.
   a. Convert to slope-intercept form.
      $$-y \leq -x - 2$$
      $$y \geq x + 2$$
   b. Graph $y = x + 2$, using a solid line to indicate that the line is part of the solution.
   c. Using (0, 0) as a test point, $0 - 0 \leq -2$ indicates that the half-plane below the line is not the solution. Shade the half-plane above the line.

2. Graph the quadratic inequality.
   a. Reflect the graph of $y = x^2$ across the x-axis and shift the result 4 units up, using a dashed curve to indicate that the parabola is not part of the solution.
   b. Shade the region below the parabola.

SKILL ✔ EXERCISE 3

It is often important to know the vertices of the solution region when solving a system of linear inequalities. You may need to solve a system of equations when the exact coordinates of a vertex are not apparent from the graph.

PRESENTATION

Lesson Opener

Solve each inequality for y and describe its graph.

1. $2x + 3y \geq 3$ $y \geq -\frac{2}{3}x + 1$; the half-plane above the solid line

2. $4x - y > -10$ $y < 4x + 10$; the half-plane below the dashed line

3. For which inequalities is the origin a solution? List all correct answers. A, D

   A. $x - y < 2$

   B. $x + y \geq 3$

   C. $x > 5$

   D. $y \leq 3$

Use the final Lesson Opener exercise and the linear and quadratic inequalities in Example 1 to review using a test point to identify the solution region.

Review how transformations are used to determine the graph of the quadratic inequality. Emphasize that any point at which the graphs intersect satisfies the inequalities in the system.

Common Student Error Students may forget to reverse the inequality when dividing both sides of an inequality by a negative value, or they may mistakenly reverse the sign when dividing by a positive value results in a negative value.

The rest of the lesson focuses on linear systems. Example 2 demonstrates how a matrix equation can be used to find the exact coordinates of a vertex of a polygonal solution region. Ask students to describe other possible methods of solving the system.

Define the terms related to *linear programming* and present the *Vertex Principle of Linear Programming.* Discuss Example 3 and the following paragraph, which explains the reasoning behind this principle. Consider asking students whether the vertex generating the maximum value would change if $f(x, y) = 2x + 5y$. (yes; *The slope of the family of lines would be* $-\frac{2}{5}$

Example 2 Solving a System of Linear Inequalities

Solve the system and state the coordinates of each vertex of the solution region.

$$x \geq -3 \qquad x - y \leq 1$$
$$y \geq -1 \qquad 2x + 3y \leq 6$$

Answer

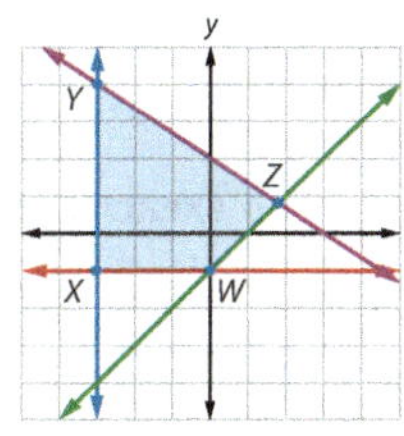

1. Graph each inequality and shade the common region.
   a. to the right of $x = -3$
   b. above $y = -1$
   c. above $y = x - 1$
   d. below $y = -\frac{2}{3}x + 2$

2. Vertices at $(0, -1)$, $(-3, -1)$, and $(-3, 4)$ can be determined from the graph.

3. Find the last vertex, Z, by solving the matrix equation $AX = C$ that represents the following system of linear equations.
   $$x - y = 1$$
   $$2x + 3y = 6$$

$$\begin{bmatrix} 1 & -1 \\ 2 & 3 \end{bmatrix} \begin{bmatrix} x \\ y \end{bmatrix} = \begin{bmatrix} 1 \\ 6 \end{bmatrix}$$

$$X = A^{-1}C = \frac{1}{3 - (-2)} \begin{bmatrix} 3 & 1 \\ -2 & 1 \end{bmatrix} \begin{bmatrix} 1 \\ 6 \end{bmatrix}$$

$$= \frac{1}{5} \begin{bmatrix} 9 \\ 4 \end{bmatrix} = \begin{bmatrix} 1.8 \\ 0.8 \end{bmatrix}$$

The coordinates of Z are $(1.8, 0.8)$.

SKILL ✔ EXERCISE 21

Optimization problems in management science strive to maximize or minimize the value of a quantity within given limitations, or *constraints*. In a two-dimensional *linear programming* problem, the quantity to be optimized is represented by a linear *objective function*, $f(x, y) = ax + by + c$, and the constraints are a system of linear inequalities. The solution to the system of constraints is known as the *feasible region*.

In such an optimization problem, any optimized value of the objective function occurs at a vertex of the feasible region.

VERTEX PRINCIPLE OF LINEAR PROGRAMMING

If a maximum or minimum value of the objective function in a linear programming problem occurs, it occurs at one or more of the vertices of the feasible region.

Example 3 Solving a Linear Programming Problem

Maximize and minimize the objective function $f(x, y) = 5x + 2y$ under the given constraints.

$$x \geq 0 \qquad y \geq \frac{1}{2}x - 1$$
$$1 \leq y \leq 4 \qquad x + y \leq 8$$

Answer

1. Graph each constraint to determine the feasible region.
   $x \geq 0$
   $1 \leq y \leq 4$
   $y \geq \frac{1}{2}x - 1$
   $x + y \leq 8$

2. Find the coordinates of the vertices.

CONTINUED ➡

Solve each system.

1. $y > -3|x + 2| + 3$
$x + y \geq -2$

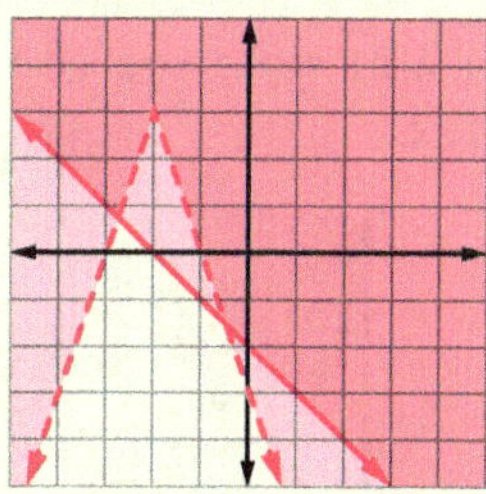

2. $-2x + y \geq 4$
$y \leq -(x - 1)^2 + 2$

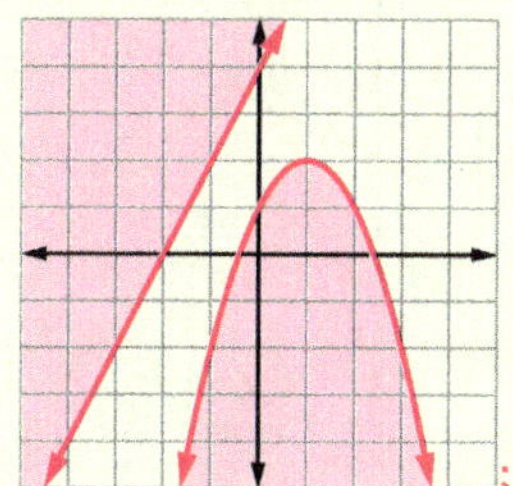

; ∅

Write a system of inequalities representing each graph.

3.

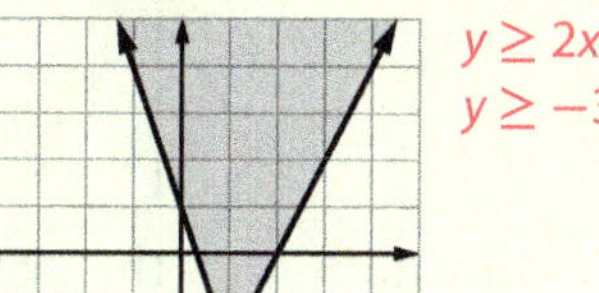

$y \geq 2x - 4$
$y \geq -3x + 1$

---

and the line farthest from the origin would pass through (4, 4) *with a maximum value* $z = 28$.) Ask whether the vertex generating the minimum value would change. (*no; The line closest to the origin would still pass through* (0, 1).)

Motivational Idea Expose students to linear programming with constraints involving more than two variables. Use the Internet keyword search *George Dantzig linear programming* to find additional information and related visuals.

Summarize and demonstrate the steps in solving a linear programming problem with Example 4, which illustrates how a business could determine and achieve

a maximum profit within a given set of constraints.

Motivational Idea Share that linear programming methods can enable business managers to better serve customers and provide for the needs of employees by helping them make informed decisions that favor maximum profit (Luke 14:28–30). The profits of a successful Christian business can also be used to advance the cause of Christ (Prov. 11:24–26; Matt. 5:16; Luke 10:33–37).

Example 5 illustrates minimizing cost over an unbounded region. Consider exploring the conditions that would cause an objective function to not have an op-

timized value. Then discuss the implications of the optimal value occurring at a vertex with decimal coordinates in a context requiring integral answers.

Interactive Activity Use the Internet keyword search *linear programming applet* to find interactive linear programming tools.

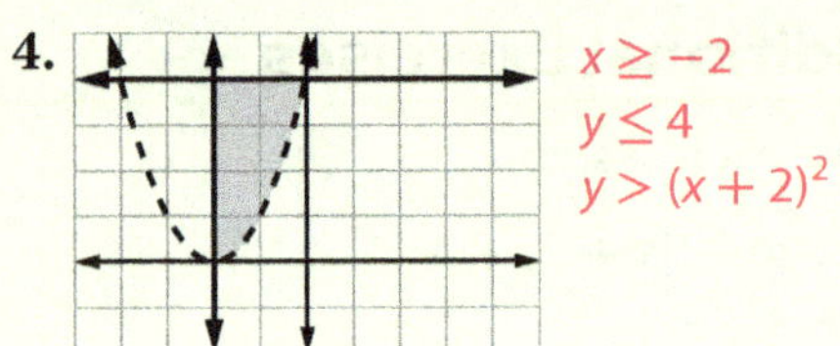

4.
$$x \geq -2$$
$$y \leq 4$$
$$y > (x + 2)^2$$

Solve the system and state the coordinates of each vertex.

5. $x + 3y \geq 6$
$-x + 3y \leq 6$
$x + y \leq 6$

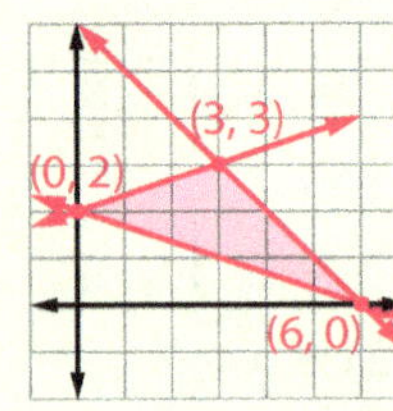

6. $-x + y \geq -1$
$-x + 2y \leq 4$
$-3x + y \leq -3$

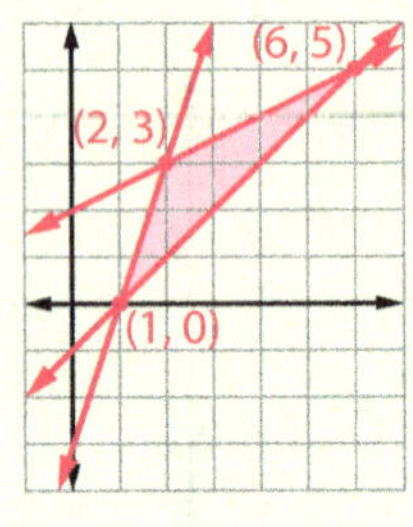

7. $3x + y \leq 6$
$-2x + y \leq 3$
$x \geq 0; y \geq 0$

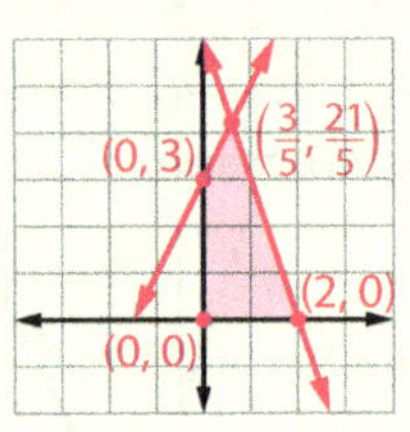

Maximize and minimize the objective function under the given constraints.

8. $f(x, y) = -3x + 5y$
$3x + y \geq 3; x + y \leq 5$
$x \geq 0; y \geq 0$
max: $f(0, 5) = 25$; min: $f(5, 0) = -15$

9. $f(x, y) = 2x - y; 5x + 3y \leq 15$
$-x + y \leq 3; x + y \geq 3$
max: $f(3, 0) = 6$; min: $f(0, 3) = -3$

10. Use a graph of a supply and demand curve for a limited edition basketball shoe where p is the price (in dollars) and x is the number of pairs of basketball shoes sold (in thousands) to find the equilibrium point, consumer surplus, and producer surplus.
supply: $p = 0.80x + 20$
demand: $p = -2x + 160$
equilibrium: (50, 60); consumer surplus: \$2500; producer surplus: \$1000

(x, y)	$5x + 2y$	$f(x, y)$
(0, 4)	$5(0) + 2(4)$	8
(0, 1)	$5(0) + 2(1)$	2
(4, 1)	$5(4) + 2(1)$	22
(6, 2)	$5(6) + 2(2)$	34
(4, 4)	$5(4) + 2(4)$	28

maximum: $f(6, 2) = 34$
minimum: $f(0, 1) = 2$

SKILL ✔ EXERCISE 25

3. Evaluate the objective function at each vertex to find the maximum and minimum values of $f(x, y)$.

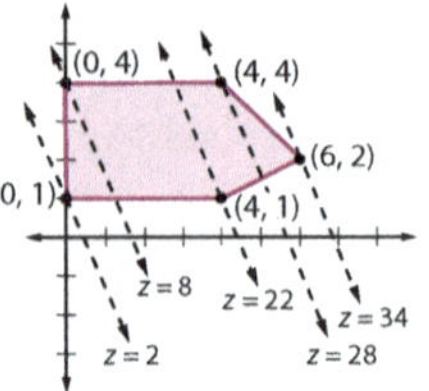

To understand why the Vertex Principle of Linear Programming works in Example 3, view the graphs of the objective function $f(x, y) = z = 5x + 2y$ as a family of parallel lines $y = -\frac{5}{2}x + \frac{z}{2}$. As the value of z increases from 0, the lines move away from the origin. The first line intersecting the feasible region does so at the vertex of the feasible region that generates the minimum value of $z = f(x, y)$ and the last line intersecting the region does so at the vertex that generates the maximum value of $z = f(x, y)$.

This reasoning also implies that when an optimal value occurs at two vertices of the feasible region, the objective function has that optimal value at any point on the connecting segment.

Businesses often seek to minimize costs and maximize profits within the market's constraints. While we will limit our study of linear programming problems to objective functions of two variables, the technique can be expanded to more complex problems involving many more variables.

Solving a Linear Programming Problem
1. Write an objective function for the quantity being maximized or minimized.
2. Write a system of inequalities modeling the constraints on the variables.
3. Graph the feasible region and find its vertices.
4. Substitute the coordinates of each vertex into the objective function to determine the maximum or minimum value.

Example 4   Maximizing Profit

Gaming Inc. manufactures the Vintage-X and the YnotPlay gaming consoles. The company budgets \$150,000 for labor and \$600,000 for components. The components cost \$100 for each Vintage-X and \$150 for each YnotPlay. Labor costs are \$40 for each Vintage-X and \$30 for each YnotPlay. If the profit for each Vintage-X is \$45 and the profit for each YnotPlay is \$60, how many of each console should be made to maximize profit? What is the maximum profit?

Answer

$P(x, y) = 45x + 60y$ where
$x = $ the number of Vintage-X consoles and
$y = $ the number of YnotPlay consoles.

1. Define the objective function representing profit.

$$x \geq 0$$
$$y \geq 0$$
$$100x + 150y \leq 600,000 \text{ (components)}$$
$$40x + 30y \leq 150,000 \text{ (labor)}$$

2. Write the system of inequalities representing the constraints.

CONTINUED ➡

Assignments

- **Minimum:** 1–2, 5, 7–8, 10–11, 14–17, 20, 22, 25, 28, 35, 37, 42, 46, 48
- **Standard:** 3, 6, 9, 12–13, 15–17, 18, 20, 24, 26, 28, 31, 34–35, 38–39, 44, 47, 49
- **Extended:** 5, 8, 11, 14–16, 19, 21–23, 27, 30–35, 38–40, 45, 49–50

Assessment

- Quiz 7C covers Sections 7.5–7.6.

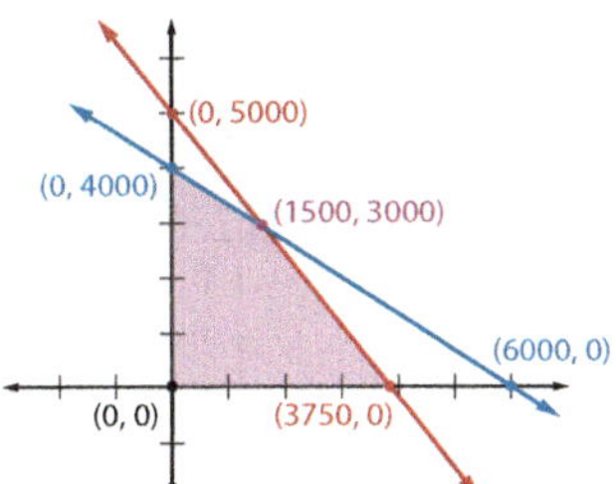

3. Graph the constraints, shade the feasible region, and find the vertices of the region.

4. Evaluate $P(x, y) = 45x + 60y$ at each vertex.

A maximum profit of $247,500 can be achieved by producing 1500 Vintage-X consoles and 3000 YnotPlay consoles.

5. Interpret the results.

The vertex formed by the component and labor constraints can be found graphically using
$$y = -\frac{2}{3}x + 4000 \text{ and}$$
$$y = -\frac{4}{3}x + 5000.$$

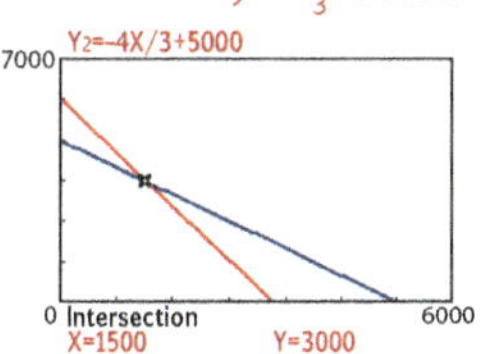

SKILL ✔ EXERCISES 31–33

Example 5   Minimizing Cost

A transportation company wants to acquire at least 10 vehicles, including at least two 24-passenger minibuses, in order to accommodate at least 204 additional passengers. They can purchase a 15-passenger van for $60,000 and a 24-passenger minibus for $85,000. What is the minimum cost at which the company can expand their fleet to meet these needs?

Answer

$C(x, y) = 60x + 85y$ where x = the number of vans and y = the number of minibuses.

1. Define the objective function to represent cost in thousands of dollars.

$$\begin{aligned} x &\geq 0 \quad \text{(vans)} \\ y &\geq 2 \quad \text{(minibuses)} \\ x + y &\geq 10 \quad \text{(total vehicles)} \\ 15x + 24y &\geq 204 \quad \text{(passengers)} \end{aligned}$$

2. Write the system of inequalities representing the constraints.

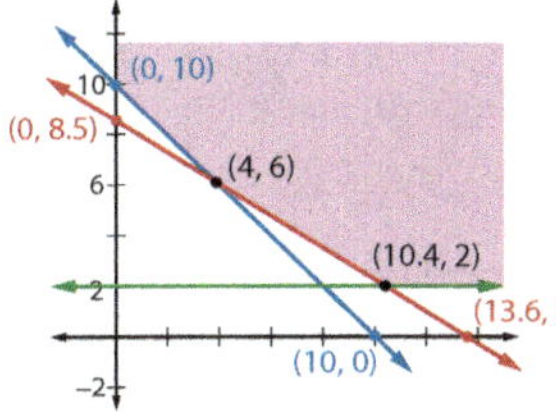

3. Graph the constraints, shade the feasible region, and find the vertices of the region.
4. Evaluate $C(x, y) = 60x + 85y$ at each vertex.

(x, y)	$C(x, y)$
$(0, 10)$	850
$(4, 6)$	750
$(10.4, 2)$	794

The company can purchase 4 vans and 6 minibuses for a minimum cost of $750,000.

SKILL ✔ EXERCISES 34–36

The objective function clearly has no maximum value over the unbounded region in Example 5. Since the vertex with decimal coordinates does not produce a minimum value, nearby points in the feasible region with integral coordinates would also not generate the minimum cost.

Even the most carefully crafted models using the most accurate data cannot answer ethical questions raised when making business decisions. While mathematics may help determine whether something can be done, biblical principles should be applied to determine whether it should be done.

5.
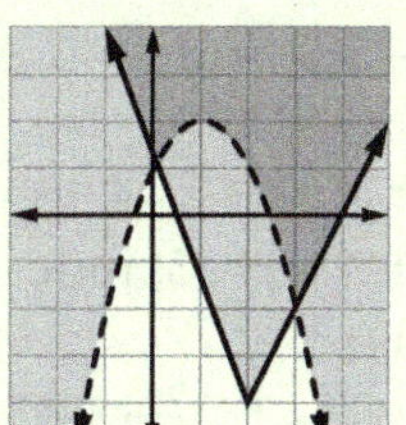

6. $-6y \leq -3x - 24;\ y \geq \frac{x}{2} + 4$
   $4y \leq 2x + 2;\ y \leq \frac{x}{2} + \frac{1}{2}$

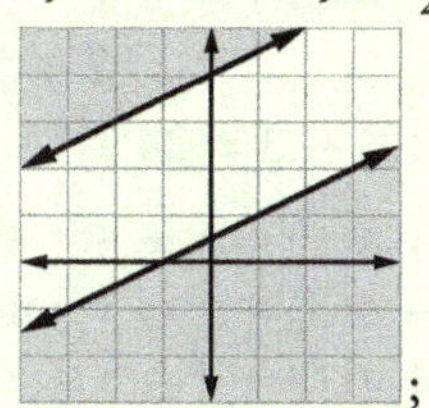
$;\ \varnothing$

7. $m_1 = -\frac{3}{2};\ y\text{-int.: } b_1 = -2$
   $m_2 = 2;\ b_2 = 1$
   $y < -\frac{3}{2}x - 2;\ y > 2x + 1$

8. $m = -\frac{3}{4};\ b = 3$
   $3x + 4y \geq 12;\ x \leq 4;\ y \leq 3$

9. $y < -(x - 2)^2 + 4$
   $y \geq |x - 2|$

10. $-4 \leq x \leq 0$
    $y \geq |x + 2| - 2$
    $y \leq -|x + 2| + 4$

11.
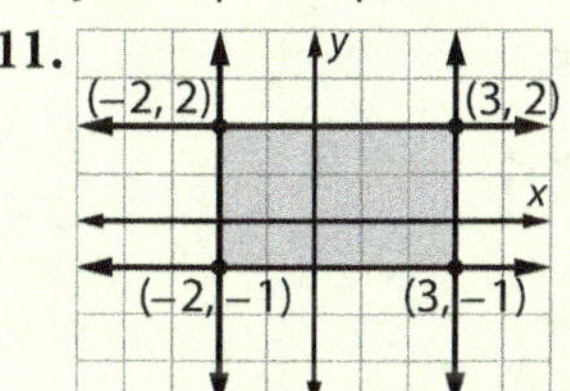

12.
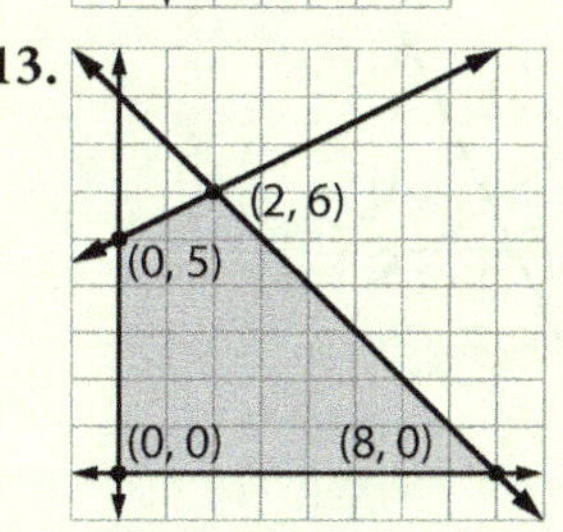

13.
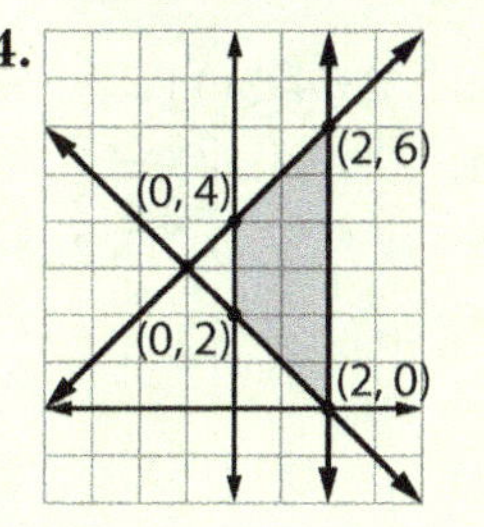

Solutions

❯ A. Exercises

1.
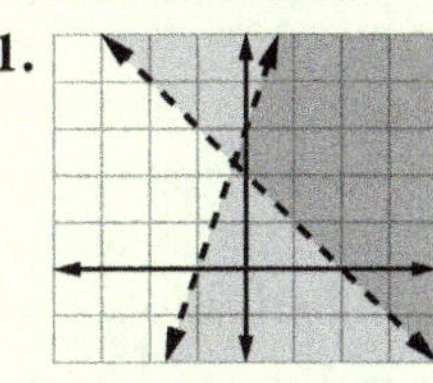

2. $4y \geq -x - 8;\ y \geq -\frac{x}{4} - 2$
   $-3y < -x + 9;\ y > \frac{x}{3} - 3$

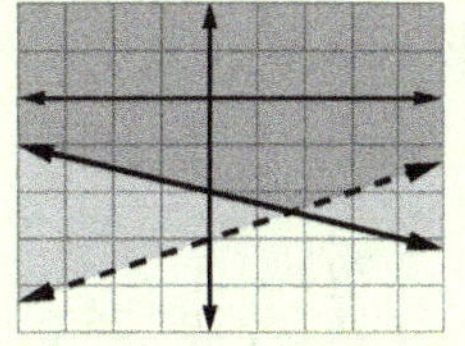

3.
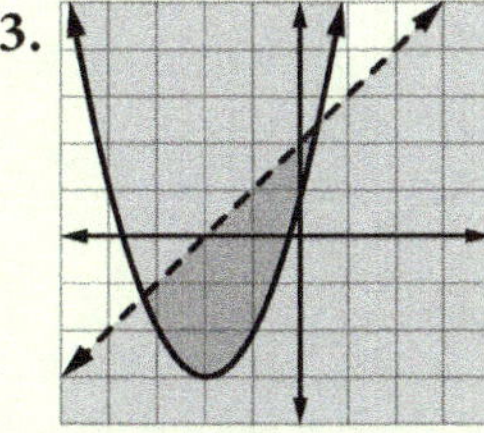

4.
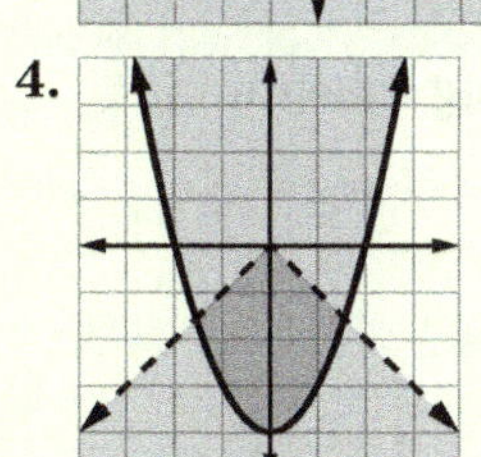

15. $f(0, 0) = -2(0) + 5(0) = 0$
 $f(0, 6) = -2(0) + 5(6) = 30$; max
 $f(6, 0) = -2(6) + 5(0) = -12$; min

16. $f(0, 3) = 8(0) + 11(3) = 33$; min
 $f(0, 7) = 8(0) + 11(7) = 77$
 $f(6, 7) = 8(6) + 11(7) = 125$; max

17. $f(0, 0) = 8(0) + (0) - 3 = -3$; min
 $f(0, 4) = 8(0) + (4) - 3 = 1$
 $f\left(\frac{15}{4}, 0\right) = 8\left(\frac{15}{4}\right) + (0) - 3$
 $\qquad\qquad = 27$; max
 $f(3, 3) = 8(3) + (3) - 3 = 24$

18. $f(0, 1) = -3(0) + 4(1) + 1 = 5$
 $f(0, 3) = -3(0) + 4(3) + 1 = 13$; max
 $f(2, 4) = -3(2) + 4(4) + 1 = 11$
 $f(2, 1) = -3(2) + 4(1) + 1 = -1$; min
 $f(4, 3) = -3(4) + 4(3) + 1 = 1$

19. $f(0, 10) = -6(0) + (10) = 10$
 $f(0, 8) = -6(0) + (8) = 8$
 $f(3, 4) = -6(3) + (4) = -14$
 $f(6, 3) = -6(6) + (3) = -33$
 $f(10, 3) = -6(10) + (3) = -57$

20. $f(0, 0) = 3(0) + 2(0) + 5 = 5$; min
 $f(0, 1) = 3(0) + 2(1) + 5 = 7$
 $f\left(2, \frac{3}{2}\right) = 3(2) + 2\left(\frac{3}{2}\right) + 5 = 14$
 $f(8, 0) = 3(8) + 2(0) + 5 = 29$; max

> ## B. Exercises

21. $6 \leq x + 3y;\ y \geq -\frac{x}{3} + 2$
 $x + 3y \leq 21;\ y \leq -\frac{x}{3} + 7$

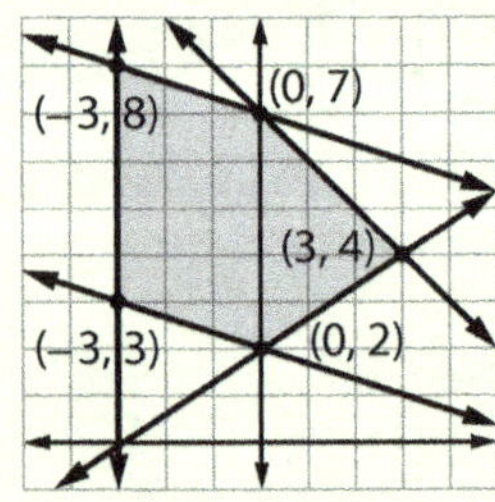

22.
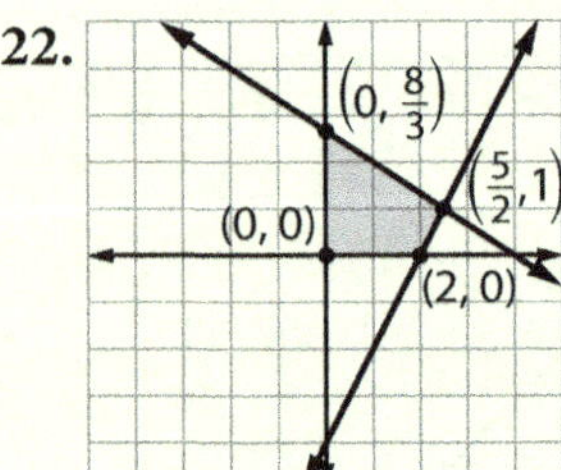

3rd and 4th:
$$\begin{array}{rl} 2x + 3y = & 8 \\ -2x + \ \ y = & -4 \\ \hline 4y = & 4 \\ y = & 1 \end{array}$$
$2x + 3(1) = 8$
$2x = 5$
$x = \frac{5}{2}$; $\left(\frac{5}{2}, 1\right)$

> ## A. Exercises

Solve each system.

1. $y < 3x + 3$
 $y > -x + 2$

2. $x + 4y \geq -8$
 $x - 3y < 9$

3. $y \geq (x + 2)^2 - 3$
 $y < x + 2$

4. $y \geq x^2 - 4$
 $y < -|x|$

5. $y \geq 2|x - 2| - 4$
 $y > -(x - 1)^2 + 2$

6. $3x - 6y \leq -24$
 $-2x + 4y \leq 2$ $\varnothing$

Write a system of inequalities representing each graph.

7.
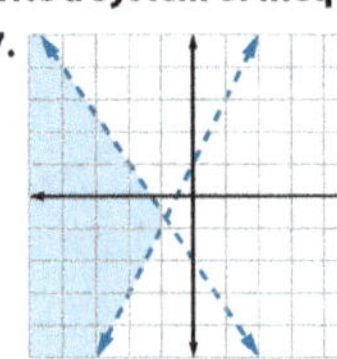

8.
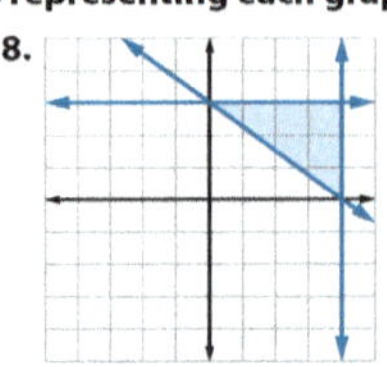

9.
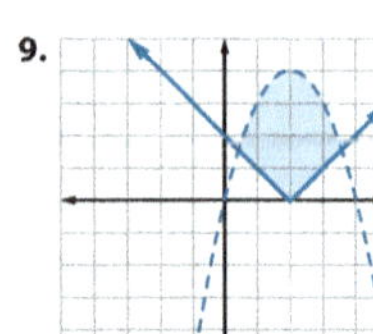

10.
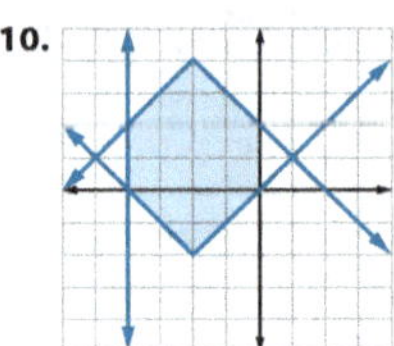

Solve the system and state the coordinates of each vertex of the solution region.

11. $-2 \leq x \leq 3$
 $-1 \leq y \leq 2$

12. $y \geq 0$
 $x \geq 0$
 $3x + 5y \leq 15$

13. $x \geq 0$
 $y \geq 0$
 $y \leq 8 - x$
 $y \leq \frac{1}{2}x + 5$

14. $0 \leq x \leq 2$
 $x + y \geq 2$
 $-x + y \leq 4$

Maximize and minimize the objective function under the given constraints.

15. $f(x, y) = -2x + 5y$
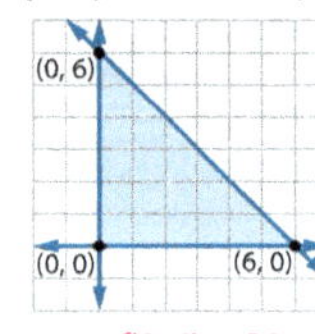
max: $f(0, 6) = 30$
min: $f(6, 0) = -12$

16. $f(x, y) = 8x + 11y$
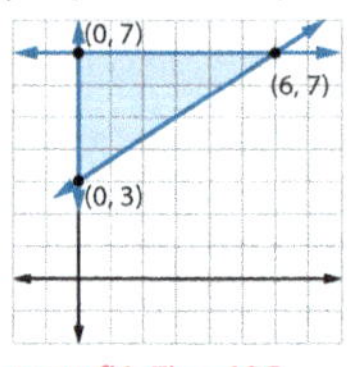
max: $f(6, 7) = 125$
min: $f(0, 3) = 33$

17. $f(x, y) = 8x + y - 3$
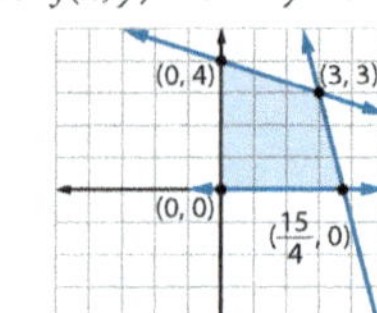

18. $f(x, y) = -3x + 4y + 1$
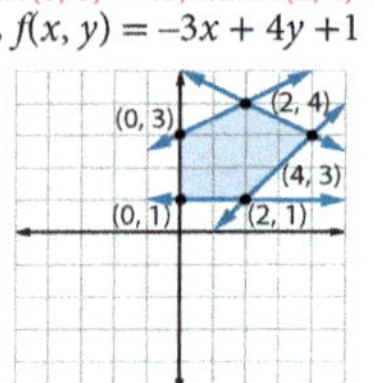

19. $f(x, y) = -6x + y$
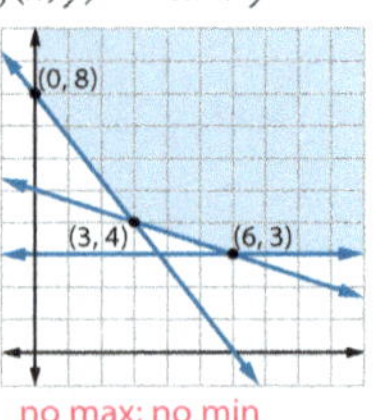
no max; no min

20. $f(x, y) = 3x + 2y + 5$
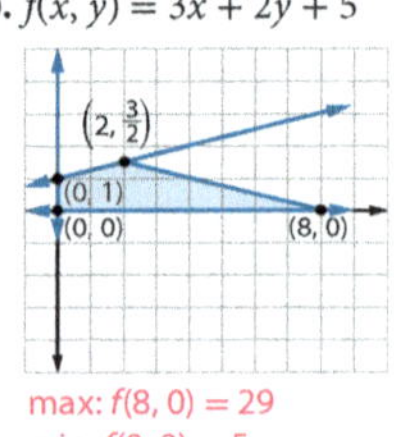
max: $f(8, 0) = 29$
min: $f(0, 0) = 5$

> ## B. Exercises

Graph each system and find the vertices of the region representing the solution.

21. $x \geq -3$
 $x + y \leq 7$
 $2x - 3y \leq -6$
 $6 \leq x + 3y \leq 21$

22. $x \geq 0$
 $y \geq 0$
 $2x + 3y \leq 8$
 $-2x + y \geq -4$

23. $x \geq -1$
 $x + 2y \leq 6$
 $-2x + y \geq -2$
 $6x + 2y \geq -6$

24. $y \geq 1$
 $2x + 5y \leq 20$
 $3x + 2y \geq 12$

Maximize and minimize the objective function under the given constraints.

25. $f(x, y) = 9x + 3y - 4$
 $x \geq 0$
 $y \geq 0$
 $3x + 2y \leq 15$
 $2x + 3y \leq 15$
 $2x + y \geq 2$

26. $f(x, y) = 4x + 5y$
 $x \leq 3$
 $-2x + 3y \leq 18$
 $6x + 5y \geq 30$
 max: $f(3, 8) = 52$
 min: $f\left(3, \frac{12}{5}\right) = 24$

27. $f(x, y) = -4x + 4y$
 $x \geq 1$
 $y \geq x$
 $y \leq 6$
 $y \leq -2x + 12$

28. $f(x, y) = 2x + 2y$
 $0 \leq x \leq 4$
 $y \geq 0$
 $2x + 2y \leq 12$
 $x + 2y \leq 10$

29. $f(x, y) = 7x - 6y$
 $x + y \leq 7$
 $3x + y \leq 11$
 $-x + 3y \geq 3$
 max: $f(3, 2) = 9$; no min

30. $f(x, y) = -x + 6y$
 $x \geq 0$
 $y \geq 0$
 $17x + 10y \leq 200$
 $-3x + 5y \geq -15$
 $4x + 5y \geq 20$
 max: $f(0, 20) = 120$
 min: $f(5, 0) = -5$

23.
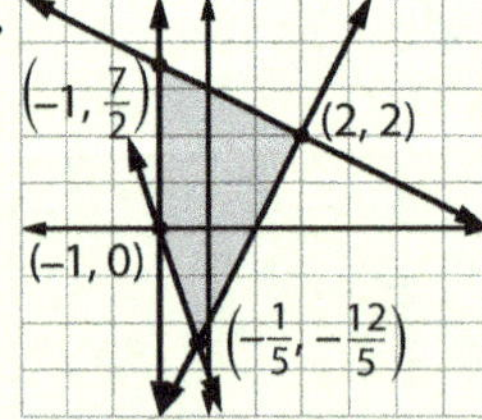

1st and 2nd, using substitution:
$(-1) + 2y = 6;\ 2y = 7;\ y = \frac{7}{2}$
$\left(-1, \frac{7}{2}\right)$
3rd and 4th:
$$\begin{bmatrix} -2 & 1 \\ 6 & 2 \end{bmatrix}^{-1} \begin{bmatrix} -2 \\ -6 \end{bmatrix} = \begin{bmatrix} -\frac{1}{5} \\ -\frac{12}{5} \end{bmatrix}$$
$\left(-\frac{1}{5}, -\frac{12}{5}\right)$

24.
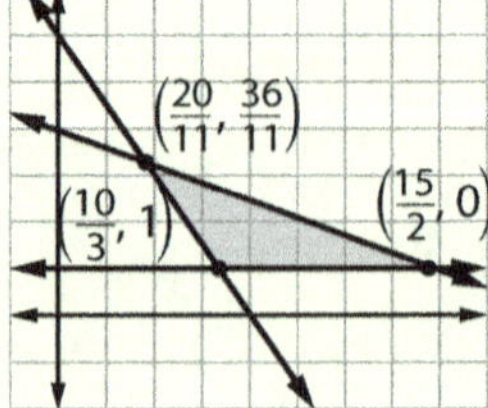

1st and 2nd, using substitution:
$2x + 5(1) = 20;\ 2x = 15;\ x = \frac{15}{2}$
$\left(\frac{15}{2}, 1\right)$
1st and 3rd, using substitution:
$3x + 2(1) = 12;\ 3x = 10;\ x = \frac{10}{3}$
$\left(\frac{10}{3}, 1\right)$
2nd and 3rd:
$$\begin{bmatrix} 2 & 5 \\ 3 & 2 \end{bmatrix}^{-1} \begin{bmatrix} 20 \\ 12 \end{bmatrix} = \begin{bmatrix} \frac{20}{11} \\ \frac{36}{11} \end{bmatrix};\ \left(\frac{20}{11}, \frac{36}{11}\right)$$

The juniors are baking coconut pies and macaroons for the Southwest Christian Academy bake-off. They have 51 c of shredded coconut and 21 c of sugar. Each coconut pie requires 0.5 c of sugar and 2 c of shredded coconut, and each batch of macaroons requires 3.5 c of sugar and 3 c of coconut. They can sell each coconut pie for $15 and each batch of coconut macaroons for $26, and they desire to maximize their revenue.

31. Write a revenue function and the system of constraints.

32. Graph the feasible region and find its vertices.

33. Determine the number of coconut pies and batches of coconut macaroons the juniors should make to maximize their revenue. What is the maximum revenue?

Due to higher-than-normal demand, Acme Inc. needs to assemble at least 3000 more gizmos each day. Regular employees can assemble 20 gizmos/day, while temporary employees can assemble 15. Management needs the ratio of temporary to regular employees to be at least 3 : 4 and at most 4 : 3. It costs $196/day to employ a regular assembler and $98/day to hire a temporary assembler.

34. Write a cost function for the additional assemblers and the system of constraints.

35. Graph the feasible region and find its vertices.

36. Determine the number of regular and temporary employees that will minimize the cost. How much will the additional employees cost?
75 regular and 100 temporary employees; $24,500/day

> **C. Exercises**

Review the introduction to this section. The area under the demand curve and above the equilibrium price represents the consumer surplus. The area under the equilibrium price and above the supply curve represents the producer surplus.

Use a graph of the supply and demand curves to find the equilibrium price, the consumer surplus, and the producer surplus.

37. A martial arts center determines the supply and demand curves for enrollment where p is the price (in dollars) and x is the number of students enrolled.
supply: $p = 0.2x + 30$ (125, $55); $937.50;
demand: $p = -0.12x + 70$ $1562.50

38. Vons Sporting Goods determines the supply and demand curves for kayaks where p is the price (in dollars) and x is the number of kayaks sold.
supply: $p = 0.25x + 300$ (750, $487.50); $42,187.50;
demand: $p = -0.15x + 600$ $70,312.50

Write and solve a system of inequalities for each problem.

39. Stephanie earns $10/hr working at an afterschool daycare and $15/hr working on Saturdays as a personal trainer. The gym is open 8:00 AM–4:00 PM, and Stephanie can work at most 15 hr/week in addition to her studies. How can Stephanie earn the most money and how much can she make?

40. A wireless carrier outlet obtains the Zphone 7 for $300 and the Zphone 8 for $500 and sells them for $475 and $775, respectively. Sales data predicts that the outlet will sell at most 200 Zphones per month and the budget allocates a maximum of $75,000 to acquire Zphones. Determine the outlet's maximum profit under these conditions and the number of each phone that should be ordered to obtain that profit.

41. Todd wants to increase his protein intake by adding protein powder to his pancake mix. He wants at least 60 g of protein and 90 g of carbohydrates in his breakfast. A serving of pancakes costs $0.75 and a serving of protein powder costs $1.00. How many servings of pancake mix and protein powder should Todd eat to achieve his nutritional goals at the lowest cost?

Per Serving	Protein (g)	Carbs (g)	Cost ($)
pancakes	10	45	$0.75
protein powder	30	15	$1.00

25. max: $f(5, 0) = 41$; min: $f(0, 2) = 2$

27. max: $f(1, 6) = 20$; min: 0 on segment with endpoints (1, 1) and (4, 4)

28. max: 12 on segment with endpoints (2, 4) and (4, 2) min: $f(0, 0) = 0$

31. $R(x, y) = 15x + 26y$ where x = the number of pies and y = the batches of macaroons; $x \geq 0, y \geq 0, 2x + 3y \leq 51, 0.5x + 3.5y \leq 21$

32. (0, 0), (0, 6), (21, 3), (25.5, 0)

33. 21 pies and 3 batches of coconut macaroons; $393

34. $C(x, y) = 196x + 98y$, where x = regular employees and y = temporary employees; $20x + 15y \geq 3000, \frac{y}{x} \geq \frac{3}{4}, \frac{y}{x} \leq \frac{4}{3}$

39. by working 7 hr at the daycare and 8 hr at the gym; $190/week

40. $42,500 profit; Zphone 7: 125; Zphone 8: 75

41. 1.5 servings of both pancakes and protein powder

27.
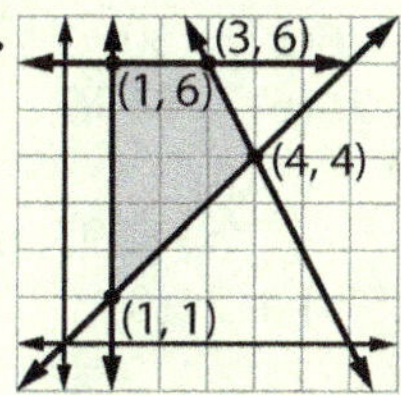

$f(1, 1) = -4(1) + 4(1) = 0$; min
$f(1, 6) = -4(1) + 4(6) = 20$; max
$f(3, 6) = -4(3) + 4(6) = 12$
$f(4, 4) = -4(4) + 4(4) = 0$; min

28.
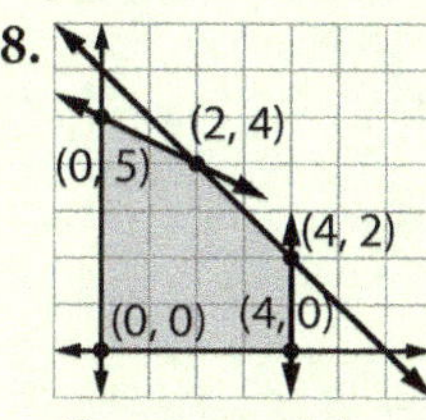

$f(0, 5) = 2(0) + 2(5) = 10$
$f(0, 0) = 2(0) + 2(0) = 0$; min
$f(2, 4) = 2(2) + 2(4) = 12$; max
$f(4, 2) = 2(4) + 2(2) = 12$; max
$f(4, 0) = 2(4) + 2(0) = 8$

29.
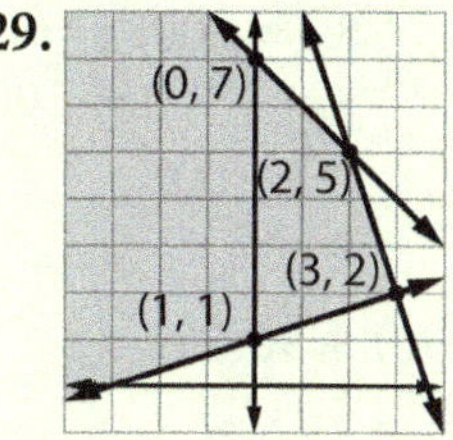

$f(0, 7) = 7(0) - 6(7) = -42$
$f(2, 5) = 7(2) - 6(5) = -16$
$f(3, 2) = 7(3) - 6(2) = 9$; max
$f(1, 1) = 7(1) - 6(1) = 1$

30.
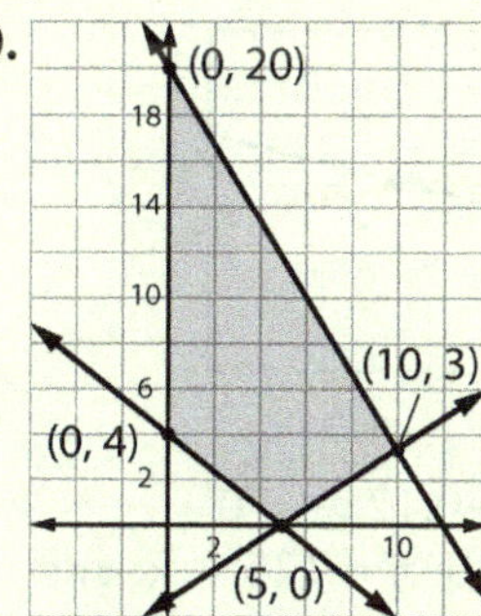

$f(0, 4) = -(0) + 6(4) = 24$
$f(0, 20) = -(0) + 6(20) = 120$; max
$f(5, 0) = -(5) + 6(0) = -5$; min
$f(10, 3) = -(10) + 6(3) = 8$

32. $2x + 3y \leq 51; y \leq -\frac{2}{3}x + 17$

$0.5x + 3.5y \leq 21; y \leq -\frac{1}{7}x + 6$

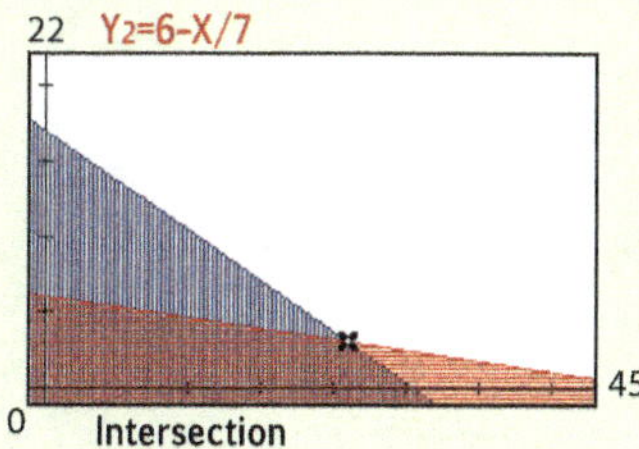

25.
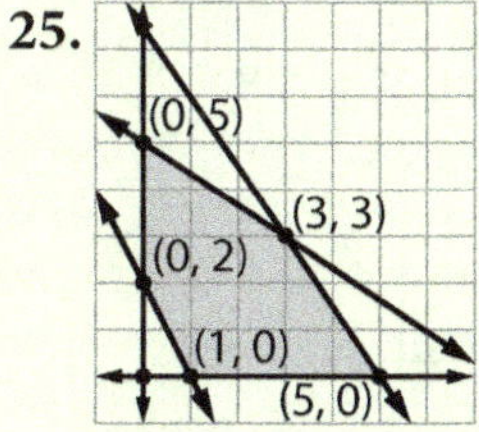

$f(0, 2) = 9(0) + 3(2) - 4 = 2$; min
$f(1, 0) = 9(1) + 3(0) - 4 = 5$
$f(0, 5) = 9(0) + 3(5) - 4 = 11$
$f(3, 3) = 9(3) + 3(2) - 4 = 29$
$f(5, 0) = 9(5) + 3(0) - 4 = 41$; max

26.
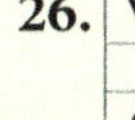
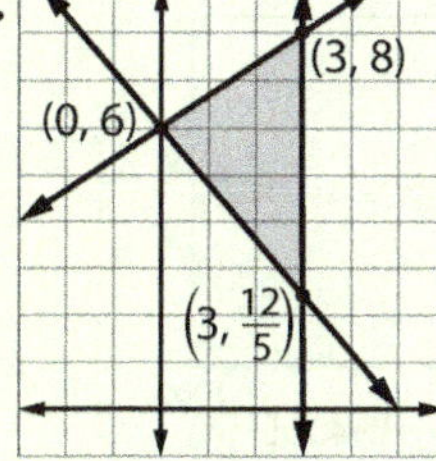

1st and 3rd, using substitution:
$6(3) + 5y = 30; 5y = 12; y = \frac{12}{5}$
$\left(3, \frac{12}{5}\right)$
$f(0, 6) = 4(0) + 5(6) = 30$
$f(3, 8) = 4(3) + 5(8) = 52$; max
$f\left(3, \frac{12}{5}\right) = 4(3) + 5\left(\frac{12}{5}\right) = 24$; min

33. $R(0, 0) = 15(0) + 26(0) = \$0$
$R(0, 6) = 15(0) + 26(6) = \$156$
$R(21, 3) = 15(21) + 26(3)$
$$= \$393; \text{ max}$$
$R(25.5, 0) = 15(25.5) + 26(0)$
$$= \$382.5$$

35. $4x + 3y \geq 600; \; y = -\frac{4}{3}x + 200$

$\frac{y}{x} \geq \frac{3}{4}; \; y \geq \frac{3}{4}x$
$\frac{y}{x} \leq \frac{4}{3}; \; y \leq \frac{4}{3}x$

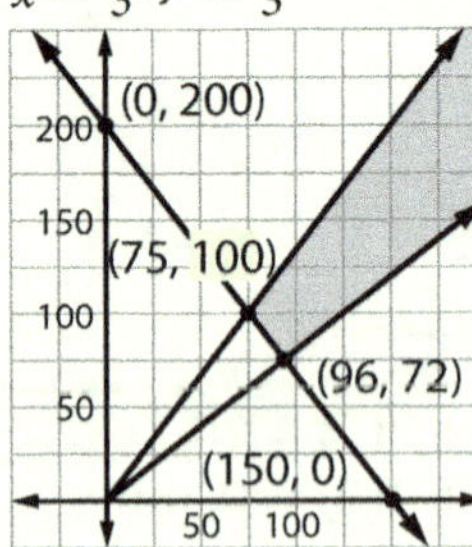

1st and 2nd:
$4x + 3\left(\frac{3}{4}x\right) = 600$
$\frac{25}{4}x = 600; \; x = 96$
$y = \frac{3}{4}(96) = 72$
$(96, 72)$

1st and 3rd:
$4x + 3\left(\frac{4}{3}x\right) = 600$
$8x = 600; \; x = 75$
$y = \frac{4}{3}(75) = 100$
$(75, 100)$

36. $C(96, 72) = 196(96) + 98(72)$
$$= \$25{,}872$$
$C(75, 100) = 196(75) + 98(100)$
$$= \$24{,}500; \text{ min}$$

› C. Exercises

37.

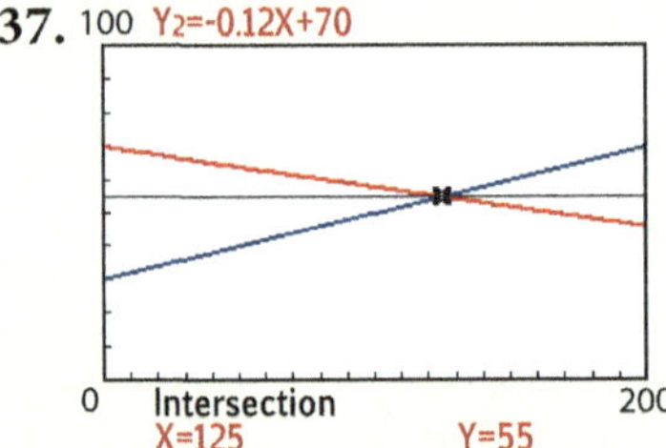

$0.2x + 30 = -0.12x + 70$
$0.32x = 40; \; x = 125$ students
$p = 0.2(125) + 30 = \$55; \; (125, \$55)$

consumer surplus:
$\frac{1}{2}(125)(70 - 55) = \$937.50$

producer surplus:
$\frac{1}{2}(125)(55 - 30) = \$1562.50$

42. Use laws of logarithms to evaluate $\log \frac{a^2 b^3}{c}$ if $\log a = 2.3$, $\log b = 1.4$, and $\log c = 0.9$. [3.3] 7.9

43. Use a double-angle identity to rewrite the expression $10 \sin^2 \theta - 5$. [5.5] $-5 \cos 2\theta$

44. State the number of petals and the maximum value of $|r|$ in the polar rose $r = 7 \sin 9\theta$. [6.5] 9 petals; 7

45. Convert $z = -4 - 6i$ to polar form. [6.6] 7.21 cis 4.12

46. Solve the system by elimination. [7.1] (6, 0)
$4x - 6y = 24$
$5x + 6y = 30$

47. Find the inverse of $A = \begin{bmatrix} 0 & 3 \\ -2 & 7 \end{bmatrix}$. [7.5] $\begin{bmatrix} \frac{7}{6} & -\frac{1}{2} \\ \frac{1}{3} & 0 \end{bmatrix}$

48. What is the force required to prevent a 234 lb dirt bike from rolling down a 40° incline (to the nearest tenth of a pound)? [6.2] A
A. 150.4 lb **C.** 225.8 lb **E.** none of
B. 179.3 lb **D.** 120.0 lb these

49. Which of the following is an approximate polar form representation of $z = 10 - 8i$? [6.6] D
A. 4.24 cis 5.61 **D.** 12.81 cis 5.61
B. 10.77 cis 7.18 **E.** none of these
C. 12.81 cis 6.95

50. Find the determinant of $A = \begin{bmatrix} 7 & -2 \\ 4 & -8 \end{bmatrix}$. [7.4] B
A. 64 **C.** 62 **E.** none of
B. -48 **D.** 50 these

51. Solve $\log_4 (x^2 - x - 2) - \log_4 (x + 1) = 2$. [3.4] B
A. $x = 14$ **C.** $x = 20$ **E.** none of
B. $x = 18$ **D.** $x = 12$ these

38.

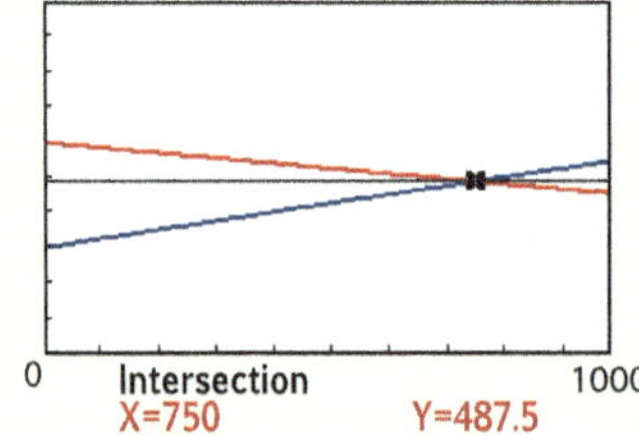

$0.25x + 300 = -0.15x + 600$
$0.4x = 300; \; x = 750$ kayaks
$p = 0.25(750) + 300 = \$487.50$
$(750, \$487.50)$

consumer surplus:
$\frac{1}{2}(750)(600 - 487.5) = \$42{,}187.50$

producer surplus:
$\frac{1}{2}(750)(487.5 - 300) = \$70{,}312.50$

39. $I(x, y) = 10x + 15y$
$x + y \leq 15; \; x \geq 0; \; 0 \leq y \leq 8$

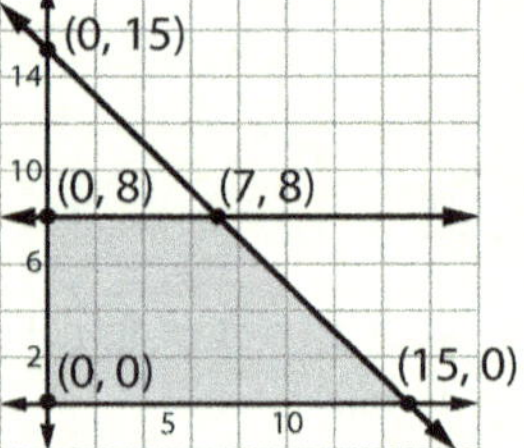

$I(0, 0) = 10(0) + 15(0) = \$0$
$I(0, 8) = 10(0) + 15(8) = \$120$
$I(7, 8) = 10(7) + 15(8) = \$190; \text{ max}$
$I(15, 0) = 10(15) + 15(0) = \$150$

continued in Answers and Solutions Overflow

7.7 Partial Fractions

Polynomials were factored into a product of linear and irreducible quadratic factors in Chapter 2. You will now learn how to express a proper rational expression as a sum of simpler rational functions whose denominators are powers of linear or irreducible quadratic factors. This process of finding the *partial fraction decomposition* of a rational expression (the equivalent sum of *partial fractions*) is used in calculus when finding the area under the graph of a rational function.

If the rational expression can be factored into the form $\dfrac{px+q}{(x+a)(x+b)}$ with distinct linear factors in the denominator, the partial fraction decomposition has the form $\dfrac{A}{x+a} + \dfrac{B}{x+b}$.

After completing this section, you will be able to

- decompose rational expressions into partial fractions.

partial fraction decomposition

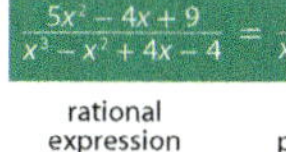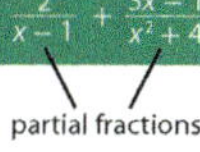
$$\underbrace{\frac{5x^2 - 4x + 9}{x^3 - x^2 + 4x - 4}}_{\text{rational expression}} = \underbrace{\frac{2}{x-1} + \frac{3x-1}{x^2+4}}_{\text{partial fractions}}$$

Example 1   Decomposing with Distinct Linear Factors

Find the partial fraction decomposition of $\dfrac{2x+10}{x^2 - 2x - 8}$.

Answer

$$\frac{2x+10}{(x-4)(x+2)} = \frac{A}{x-4} + \frac{B}{x+2}$$

1. When the denominator factors into distinct linear factors, the numerator of each partial fraction is a constant.

$$\left[\frac{2x+10}{(x-4)(x+2)} = \frac{A}{x-4} + \frac{B}{x+2}\right](x-4)(x+2)$$
$$2x+10 = A(x+2) + B(x-4)$$

2. Multiply the equation by the LCD to clear fractions.

$$2x+10 = Ax + 2A + Bx - 4B$$
$$2x+10 = (A+B)x + (2A-4B)$$
$$\therefore \ A + B = 2$$
$$2A - 4B = 10$$

3. Rearrange the terms on the right side and equate the coefficients of like terms to form a system of equations.

$$\begin{bmatrix} 1 & 1 \\ 2 & -4 \end{bmatrix}\begin{bmatrix} A \\ B \end{bmatrix} = \begin{bmatrix} 2 \\ 10 \end{bmatrix}$$
$$\begin{bmatrix} A \\ B \end{bmatrix} = -\frac{1}{6}\begin{bmatrix} -4 & -1 \\ -2 & 1 \end{bmatrix}\begin{bmatrix} 2 \\ 10 \end{bmatrix}$$
$$= -\frac{1}{6}\begin{bmatrix} -18 \\ 6 \end{bmatrix} = \begin{bmatrix} 3 \\ -1 \end{bmatrix}$$

4. Solve the system by writing a matrix equation $CX = D$ and left-multiplying by the inverse of the coefficient matrix to produce $X = C^{-1}D$.

$$\frac{2x+10}{x^2 - 2x + 8} = \frac{3}{x-4} - \frac{1}{x+2}$$

5. Write the partial fraction decomposition as a difference.

Check

$$\frac{3}{(x-4)} \cdot \frac{(x+2)}{(x+2)} - \frac{1}{(x+2)} \cdot \frac{(x-4)}{(x-4)}$$
$$= \frac{(3x+6) - (x-4)}{(x-4)(x+2)} = \frac{2x+10}{x^2 - 2x - 8}$$

6. Combine the terms to verify the result.

SKILL ✔ EXERCISE 7

Objectives

1. To find partial fraction decompositions of rational expressions with linear factors

2. To extend partial fraction decompositions to include prime quadratic factors

Vocabulary

partial fraction
partial fraction decomposition

Additional Exercises

Find the partial fraction decomposition of each expression.

1. $\dfrac{11 - 2x}{2x^2 + 3x - 2}$ $\dfrac{4}{2x-1} - \dfrac{3}{x+2}$

2. $\dfrac{13x - 7}{3x^2 + x - 2}$ $\dfrac{1}{3x-2} + \dfrac{4}{x+1}$

3. $\dfrac{4x - 7}{x^2 - 6x + 9}$ $\dfrac{4}{x-3} + \dfrac{5}{(x-3)^2}$

4. $\dfrac{15x + 4}{25x^2 + 20x + 4}$ $\dfrac{3}{5x+2} - \dfrac{2}{(5x+2)^2}$

5. $\dfrac{5x^2 + x - 15}{x^3 + 3x^2}$ $\dfrac{2}{x} - \dfrac{5}{x^2} + \dfrac{3}{x+3}$

6. $\dfrac{3x^2 + 2x + 2}{x^3 + 3x^2 - 2x - 2}$ $\dfrac{1}{x-1} + \dfrac{2x}{x^2 + 4x + 2}$

7. $\dfrac{x^2 - 8x - 8}{x^3 - 2x^2 - 14x + 3}$ $\dfrac{1}{x+3} - \dfrac{3}{x^2 - 5x + 1}$

8. $\dfrac{x^3 + 5x^2 + x - 11}{x^4 - 4x^2 + 4}$ $\dfrac{x+5}{x^2 - 2} + \dfrac{3x - 1}{(x^2 - 2)^2}$

PRESENTATION

Lesson Opener

Write each fraction as the sum of two fractions.

1. $\dfrac{3}{4}$ $\dfrac{1}{2} + \dfrac{1}{4}$

2. $\dfrac{x+2}{3}$ $\dfrac{x}{3} + \dfrac{2}{3}$

3. $\dfrac{x+5}{5x}$ $\dfrac{1}{5} + \dfrac{1}{x}$

Write the fraction as a sum of three fractions.

4. $\dfrac{x^2 + 2x + 3}{6x}$ $\dfrac{x}{6} + \dfrac{1}{3} + \dfrac{1}{2x}$

The Lesson Opener provides insight into decomposing expressions into *partial fractions*. Explain that difficult decompositions such as those presented in this section are used in calculus. The students should be able to decompose the expression being integrated after completing Example 3.

Ask the students to explain the difference between a proper and an improper rational expression. (*In a proper rational expression, the degree of the numerator's polynomial is less than the degree of the denominator's polynomial.*)

Explain that our decompositions will involve rewriting a sum in terms of fractions whose denominators are factors of the original denominator. In Example 1 the partial fractions contain unknown constants in the numerators and the two distinct linear factors as the denominators. Discuss the strategy of rearranging the terms and equating the coefficients of like terms. While the resulting system of equations can be solved using a variety of methods, the text reviews solving the system with a matrix equation. Discuss how the alternative method of substituting convenient values of x can be used to find the values of A and B in this case.

It is customary to use uppercase letters for the unknown constants in the numerators. This distinguishes them from other variables in the expressions; they do not indicate a matrix in this context.

9. $\dfrac{7x^4 + 19x^2 - 2x + 12}{x\left(x^2 + 2\right)^2}$

$\dfrac{3}{x} + \dfrac{4x}{x^2 + 2} - \dfrac{x + 2}{\left(x^2 + 2\right)^2}$

10. $\dfrac{x^3 + 4x^2 - 5x - 9}{x^2 + x - 6}$

$x + 3 - \dfrac{3}{x + 3} + \dfrac{1}{x - 2}$

Assignments

- **Minimum:** optional
- **Standard:** 1, 3–8, 12–13, 15, 18–19, 21, 39–43, 45–48
- **Extended:** 3–7, 10–11, 14–17, 20–21; 24–30 even; 34, 37–39, 42, 44–45, 47

Assessment

- Quiz 7D covers Section 7.7.

Solutions

❯ A. Exercises

1. $\dfrac{A}{x - 2} + \dfrac{B}{x - 12}$

2. $\dfrac{Ax + B}{x^2 + 6x + 3} + \dfrac{Cx + D}{\left(x^2 + 6x + 3\right)^2}$

3. $\dfrac{A}{x - 1} + \dfrac{Bx + C}{x^2 - 2x - 5}$

4. $\dfrac{A}{x} + \dfrac{B}{x^2} + \dfrac{C}{x^3} + \dfrac{D}{x^4}$

5. $\dfrac{A}{x - 5} + \dfrac{B}{x + 6}$

6. $\dfrac{A}{x - 4} + \dfrac{B}{(x - 4)^2}$

7. $\left[\dfrac{7x + 11}{(x - 2)(x + 3)} = \dfrac{A}{x - 2} + \dfrac{B}{x + 3}\right]$
$\cdot (x - 2)(x + 3)$

$7x + 11 = A(x + 3) + B(x - 2)$
$ = (A + B)x + (3A - 2B)$

$A + B = 7$

$3A - 2B = 11$

$\begin{bmatrix} 1 & 1 \\ 3 & -2 \end{bmatrix}\begin{bmatrix} A \\ B \end{bmatrix} = \begin{bmatrix} 7 \\ 11 \end{bmatrix}$

$\begin{bmatrix} A \\ B \end{bmatrix} = -\dfrac{1}{5}\begin{bmatrix} -2 & -1 \\ -3 & 1 \end{bmatrix}\begin{bmatrix} 7 \\ 11 \end{bmatrix}$

$\phantom{\begin{bmatrix} A \\ B \end{bmatrix}} = -\dfrac{1}{5}\begin{bmatrix} -25 \\ -10 \end{bmatrix} = \begin{bmatrix} 5 \\ 2 \end{bmatrix}$

$\therefore \dfrac{5}{x - 2} + \dfrac{2}{x + 3}$

An alternative method of solving for A and B is frequently employed when there are no repeated factors. Substituting zeros of the original denominator into the equation after the fractions have been cleared produces equations that can be quickly solved for A and B.

Use $2x + 10 = A(x + 2) + B(x - 4)$.

letting $x = -2$:
$2(-2) + 10 = A(-2 + 2) + B(-2 - 4)$
$6 = -6B$
$B = -1$

letting $x = 4$:
$2(4) + 10 = A(4 + 2) + B(4 - 4)$
$18 = 6A$
$A = 3$

When a linear factor of the expression's denominator is repeated n times, include a partial fraction with a distinct constant numerator for each power of the factor from 1 to n.

$$\frac{px^2 + qx + r}{(x + a)^3} = \frac{A}{x + a} + \frac{B}{(x + a)^2} + \frac{C}{(x + a)^3}$$

> **Example 2** **Decomposing with a Repeated Linear Factor**

Find the partial fraction decomposition of $\dfrac{7x^2 - 21x + 16}{x^3 - 4x^2 + 4x}$.

Answer

$\dfrac{7x^2 - 21x + 16}{x(x - 2)(x - 2)} = \dfrac{A}{x} + \dfrac{B}{x - 2} + \dfrac{C}{(x - 2)^2}$

1. Include partial fractions with a distinct constant numerator for each power of the repeated linear factors.

$\left[\dfrac{7x^2 - 21x + 16}{x(x - 2)^2} = \dfrac{A}{x} + \dfrac{B}{x - 2} + \dfrac{C}{(x - 2)^2}\right]x(x - 2)^2$

2. Multiply by the LCD to clear fractions.

$7x^2 - 21x + 16 = A(x - 2)^2 + Bx(x - 2) + Cx$

$ = Ax^2 - 4Ax + 4A + Bx^2 - 2Bx + Cx$

$ = (A + B)x^2 + (-4A - 2B + C)x + 4A$

$\therefore\ A + B = 7$

$ -4A - 2B + C = -21$

$ 4A = 16$

3. Rearrange the terms on the right side and equate the coefficients of like terms to form a system of equations.

3rd equation: $A = 4$

1st equation: $(4) + B = 7; B = 3$

2nd equation: $-4(4) - 2(3) + C = -21; C = 1$

4. Since the system is in triangular form, it can be solved quickly using back substitution.

$\left[\dfrac{7x^2 - 21x + 16}{x^3 - 4x^2 + 4x} = \dfrac{4}{x} + \dfrac{3}{x - 2} + \dfrac{1}{(x - 2)^2}\right]$

5. Write the partial fraction decomposition as a sum.

Check

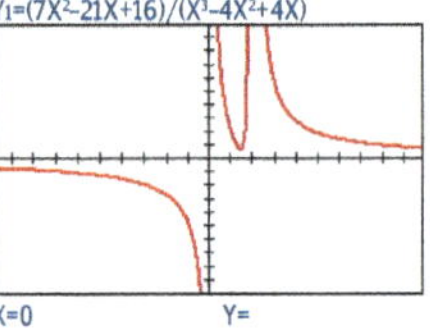

6. The graphs of 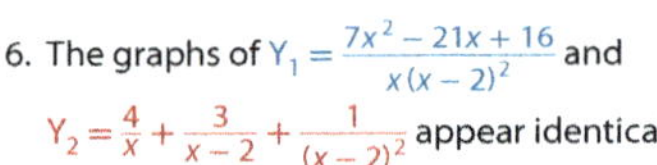 $Y_1 = \dfrac{7x^2 - 21x + 16}{x(x - 2)^2}$ and $Y_2 = \dfrac{4}{x} + \dfrac{3}{x - 2} + \dfrac{1}{(x - 2)^2}$ appear identical.

———— SKILL ✓ **EXERCISE 13**

An alternative solution involves substituting $x = 0$ and $x = 2$ into the results of Step 2 to find A and C.

$7x^2 - 21x + 16 = A(x - 2)^2 + Bx(x - 2) + Cx$

letting $x = 0$:
$0 - 0 + 16 = 4A + 0 + 0$
$16 = 4A$
$A = 4$

letting $x = 2$:
$7(4) - 21(2) + 16 = 0 + 0 + 2C$
$2 = 2C$
$C = 1$

Decomposing with repeated linear factors is demonstrated in Example 2. Emphasize using a separate term for each power of any repeated linear factor, as shown in step 1. Be sure students understand the equating of coefficients of like terms in step 3. Encourage students to check their answers by graphing the original expression and the sum of the partial fractions.

Point out that convenient substitutions produce just two of the unknown constants directly—the third can be found by substituting the first two constants and another value for x.

Example 3 introduces decomposing with an irreducible quadratic factor. Emphasize the use of a linear numerator with the quadratic denominator in step 1. The system in step 3 could also be solved by solving the first equation for B, solving the third equation for C, and then substituting the resulting expressions into the second equation to find A. Discussing various methods of solving the system provides helpful review.

Explain that the decomposition in Example 4 lists a separate term for each power of any repeated quadratic factor (as was done with repeated linear factors in Example 2) and that each numerator

Substituting these values and any other convenient value for x (such as $x = 1$) produces an equation that can then be solved to find B.

$$7(1) - 21(1) + 16 = 4(1) + B(-1) + 1(1)$$
$$2 = 5 - B$$
$$B = 3$$

When the original expression's denominator contains an irreducible quadratic factor, include a partial fraction with a linear numerator.

$$\frac{px^2 + qx + r}{(x+a)(x^2 + bx + c)} = \frac{A}{x+a} + \frac{Bx + C}{x^2 + bx + c}$$

Example 3 Decomposing with an Irreducible Quadratic Factor

Find the partial fraction decomposition of $\frac{5x^2 - 4x - 3}{x^3 + 1}$.

Answer

$\dfrac{5x^2 - 4x - 3}{(x+1)(x^2 - x + 1)} = \dfrac{A}{x+1} + \dfrac{Bx + C}{x^2 - x + 1}$

1. Include a partial fraction with a linear numerator and the irreducible quadratic denominator.

$\left[\dfrac{5x^2 - 4x - 3}{(x+1)(x^2 - x + 1)} = \dfrac{A}{x+1} + \dfrac{Bx + C}{x^2 - x + 1} \right](x+1)(x^2 - x + 1)$
$5x^2 - 4x - 3 = A(x^2 - x + 1) + (Bx + C)(x + 1)$

2. Multiply by the LCD to clear fractions.

$\quad = (Ax^2 - Ax + A) + (Bx^2 + Bx + Cx + C)$
$\quad = (A + B)x^2 + (-A + B + C)x + (A + C)$
$\therefore\ A + B = 5$
$\quad -A + B + C = -4$
$\quad A + C = -3$

3. Rearrange the terms on the right side and equate the coefficients of like terms to form a system of equations.

```
[A]
         [ 1  1  0   5]
         [-1  1  1  -4]
         [ 1  0  1  -3]
rref([A])
         [1  0  0   2]
         [0  1  0   3]
         [0  0  1  -5]
```

4. The resulting system can be quickly solved by using a calculator to find the equivalent RREF of the augmented matrix.

$\dfrac{5x^2 - 4x - 3}{x^3 + 1} = \dfrac{2}{x+1} + \dfrac{3x - 5}{x^2 - x + 1}$

5. Write the partial fraction decomposition as a sum.

SKILL ✔ **EXERCISE 17**

The alternative method of substituting zeros of the original denominator does not work well when there is an irreducible quadratic factor since there are not enough convenient values of x.

is linear (as in Example 3). The students should recognize that back substitution will quickly solve the resulting system in step 4.

Discuss the approach to decomposing an improper rational fraction and then refer to the table for a review of the methods discussed in this section.

One-on-One Encourage students who struggle with these concepts to use the Internet keyword search *partial fraction decomposition* to find videos and other presentations of this process.

TIPS

Ex. 27, 30 Encourage students to identify possible rational roots and apply the remainder theorem and synthetic division to factor the denominator.

Ex. 38 This exercise requires students to recall summation notation from *ALGEBRA 1* and *ALGEBRA 2*. Explain that this notation will be reviewed and further explored in Chapter 9.

8. $\left[\dfrac{3x - 6}{(x+4)(x+1)} = \dfrac{A}{x+4} + \dfrac{B}{x+1} \right]$
$$\qquad\qquad\qquad\qquad \cdot (x+4)(x+1)$$
$$3x - 6 = A(x+1) + B(x+4)$$
$$\quad = (A + B)x + (A + 4B)$$
$$A + B = 3$$
$$A + 4B = -6$$
$$\begin{bmatrix} 1 & 1 \\ 1 & 4 \end{bmatrix}\begin{bmatrix} A \\ B \end{bmatrix} = \begin{bmatrix} 3 \\ -6 \end{bmatrix}$$
$$\begin{bmatrix} A \\ B \end{bmatrix} = \frac{1}{3}\begin{bmatrix} 4 & -1 \\ -1 & 1 \end{bmatrix}\begin{bmatrix} 3 \\ -6 \end{bmatrix}$$
$$\quad = \frac{1}{3}\begin{bmatrix} 18 \\ -9 \end{bmatrix} = \begin{bmatrix} 6 \\ -3 \end{bmatrix}$$
$$\therefore\ \frac{6}{x+4} - \frac{3}{x+1}$$

9. $\left[\dfrac{20x - 17}{(2x+1)(x-4)} = \dfrac{A}{2x+1} + \dfrac{B}{x-4} \right]$
$$\qquad\qquad\qquad\qquad \cdot (2x+1)(x-4)$$
$$20x - 17 = A(x-4) + B(2x+1)$$
$$\quad = (A + 2B)x + (-4A + B)$$
$$A + 2B = 20$$
$$-4A + B = -17$$
$$\begin{bmatrix} 1 & 2 \\ -4 & 1 \end{bmatrix}\begin{bmatrix} A \\ B \end{bmatrix} = \begin{bmatrix} 20 \\ -17 \end{bmatrix}$$
$$\begin{bmatrix} A \\ B \end{bmatrix} = \frac{1}{9}\begin{bmatrix} 1 & -2 \\ 4 & 1 \end{bmatrix}\begin{bmatrix} 20 \\ -17 \end{bmatrix}$$
$$\quad = \frac{1}{9}\begin{bmatrix} 54 \\ 63 \end{bmatrix} = \begin{bmatrix} 6 \\ 7 \end{bmatrix}$$
$$\therefore\ \frac{6}{2x+1} + \frac{7}{x-4}$$

10. $\left[\dfrac{5x - 38}{(3x-2)(x+8)} = \dfrac{A}{3x-2} + \dfrac{B}{x+8} \right]$
$$\qquad\qquad\qquad\qquad \cdot (3x-2)(x+8)$$
$$5x - 38 = A(x+8) + B(3x-2)$$
$$\quad = (A + 3B)x + (8A - 2B)$$
$$A + 3B = 5$$
$$8A - 2B = -38$$
$$\begin{bmatrix} 1 & 3 \\ 8 & -2 \end{bmatrix}\begin{bmatrix} A \\ B \end{bmatrix} = \begin{bmatrix} 5 \\ -38 \end{bmatrix}$$
$$\begin{bmatrix} A \\ B \end{bmatrix} = -\frac{1}{26}\begin{bmatrix} -2 & -3 \\ -8 & 1 \end{bmatrix}\begin{bmatrix} 5 \\ -38 \end{bmatrix}$$
$$\quad = -\frac{1}{26}\begin{bmatrix} 104 \\ -78 \end{bmatrix} = \begin{bmatrix} -4 \\ 3 \end{bmatrix}$$
$$\therefore\ -\frac{4}{3x-2} + \frac{3}{x+8}$$

11. $\left[\dfrac{4x + 3}{x^2} = \dfrac{A}{x} + \dfrac{B}{x^2} \right]x^2$
$$4x + 3 = Ax + B$$
$$A = 4;\ B = 3$$
$$\therefore\ \frac{4}{x} + \frac{3}{x^2}$$

12. $\left[\dfrac{1 - 3x}{x^2} = \dfrac{A}{x} + \dfrac{B}{x^2} \right]x^2$
$$1 - 3x = Ax + B$$
$$A = -3;\ B = 1$$
$$\therefore\ -\frac{3}{x} + \frac{1}{x^2}$$

13. $\left[\dfrac{2x+1}{(x-1)^2} = \dfrac{A}{x-1} + \dfrac{B}{(x-1)^2}\right](x-1)^2$

$2x + 1 = A(x-1) + B$

$\qquad = Ax + (-A+B)$

$A = 2; -A + B = 1$

$B = 1 + (2) = 3$

$\therefore \dfrac{2}{x-1} + \dfrac{3}{(x-1)^2}$

14. $\left[\dfrac{4x+11}{(x+3)^2} = \dfrac{A}{x+3} + \dfrac{B}{(x+3)^2}\right](x+3)^2$

$4x + 11 = A(x+3) + B$

$\qquad = Ax + (3A+B)$

$A = 4; 3A + B = 11$

$3(4) + B = 11; B = -1$

$\therefore \dfrac{4}{x+3} - \dfrac{1}{(x+3)^2}$

15. $\left[\dfrac{7x^2 + 9x + 5}{x(x+1)^2} = \dfrac{A}{x} + \dfrac{B}{x+1} + \dfrac{C}{(x+1)^2}\right]$
$\qquad\qquad\qquad\qquad\qquad \cdot x(x+1)^2$

$7x^2 + 9x + 5$

$\quad = A(x+1)^2 + Bx(x+1) + Cx$

$\quad = (A+B)x^2 + (2A + B + C)x + A$

$A + B = 7; 2A + B + C = 9; A = 5$

$B = 7 - (5) = 2$

$C = 9 - 2(5) - (2) = -3$

$\therefore \dfrac{5}{x} + \dfrac{2}{x+1} - \dfrac{3}{(x+1)^2}$

16. $\left[\dfrac{2x^2 - 7x + 16}{x(x-2)^2} = \dfrac{A}{x} + \dfrac{B}{x-2} + \dfrac{C}{(x-2)^2}\right]$
$\qquad\qquad\qquad\qquad\qquad \cdot x(x-2)^2$

$2x^2 - 7x + 16$

$\quad = A(x-2)^2 + Bx(x-2) + Cx$

$\quad = (A+B)x^2 + (-4A - 2B + C)x + 4A$

$A + B = 2; -4A - 2B + C = -7$

$4A = 16; A = 4; B = 2 - (4) = -2$

$C = -7 + 4(4) + 2(-2) = 5$

$\therefore \dfrac{4}{x} - \dfrac{2}{x-2} + \dfrac{5}{(x-2)^2}$

❯ B. Exercises

17. $\left[\dfrac{7x^2 - 3x + 14}{(x-2)(x^2 + 2x + 4)} = \dfrac{A}{x-2}\right.$

$\qquad \left. + \dfrac{Bx + C}{x^2 + 2x + 4}\right](x-2)(x^2 + 2x + 4)$

$7x^2 - 3x + 14$

$= A(x^2 + 2x + 4) + (Bx + C)(x-2)$

$= (A+B)x^2 + (2A - 2B + C)x$
$\qquad\qquad\qquad\qquad\qquad + (4A - 2C)$

$A + B = 7; 2A - 2B + C = -3$

$4A - 2C = 14$

```
[A]
      ⎡1  1   0   7⎤
      ⎢2  -2  1  -3⎥
      ⎣4  0  -2  14⎦
rref([A])
      ⎡1  0  0   3⎤
      ⎢0  1  0   4⎥
      ⎣0  0  1  -1⎦
■
```

$\therefore \dfrac{3}{x-2} + \dfrac{4x - 1}{x^2 + 2x + 4}$

When an irreducible quadratic factor of the expression's denominator is repeated n times, include a partial fraction with a distinct linear numerator for each power of the factor from 1 to n.

$$\dfrac{mx^3 + px^2 + qx + r}{(x^2 + bx + c)^2} = \dfrac{Ax + B}{x^2 + bx + c} + \dfrac{Cx + D}{(x^2 + bx + c)^2}$$

Example 4   Decomposing with a Repeated Irreducible Quadratic Factor

Find the partial fraction decomposition of $\dfrac{5x^4 + 21x^2 - x + 18}{x^5 + 6x^3 + 9x}$.

Answer

$\dfrac{5x^4 + 21x^2 - x + 18}{x(x^2 + 3)^2} = \dfrac{A}{x} + \dfrac{Bx + C}{x^2 + 3} + \dfrac{Dx + E}{(x^2 + 3)^2}$

1. Include partial fractions with a linear numerator for each power of the irreducible quadratic denominator.

$\left[\dfrac{5x^4 + 21x^2 - x + 18}{x(x^2 + 3)^2} = \dfrac{A}{x} + \dfrac{Bx + C}{x^2 + 3} + \dfrac{Dx + E}{(x^2 + 3)^2}\right]x(x^2 + 3)^2$

2. Multiply by the LCD to clear fractions.

$5x^4 + 21x^2 - 1x + 18$

$= A(x^2 + 3)^2 + (Bx + C)x(x^2 + 3) + (Dx + E)x$

$= (Ax^4 + 6Ax^2 + 9A) + (Bx^4 + Cx^3 + 3Bx^2 + 3Cx)$
$\qquad\qquad\qquad\qquad\qquad\qquad + (Dx^2 + Ex)$

$= (A + B)x^4 + Cx^3 + (6A + 3B + D)x^2 + (3C + E)x + 9A$

$\therefore\ A + B = 5$

$\quad C = 0$

$\quad 6A + 3B + D = 21$

$\quad 3C + E = -1$

$\quad 9A = 18$

3. Rearrange the terms on the right side and equate the coefficients of like terms to form a system of equations.

5th equation: $A = 2$

1st equation: $(2) + B = 5; B = 3$

3rd equation: $6(2) + 3(3) + D = 21; D = 0$

4th equation: $3(0) + E = -1; E = -1$

4. Solve the system using back substitution.

$\dfrac{5x^4 + 21x^2 - x + 18}{x^5 + 6x^3 + 9x} = \dfrac{2}{x} + \dfrac{3x}{x^2 + 3} - \dfrac{1}{(x^2 + 3)^2}$

5. Write the partial fraction decomposition as a sum.

SKILL ✓ EXERCISE 25

Improper rational fractions such as $\dfrac{x^3}{x^2 - 1}$ can be decomposed by using polynomial division to obtain the quotient $x + \dfrac{x}{x^2 - 1}$. A remainder with a nonlinear denominator can then be decomposed into partial fractions using the techniques from this section.

$\dfrac{x}{x^2 - 1} = \dfrac{A}{x-1} + \dfrac{B}{x+1}$

$\left[\dfrac{x}{x^2 - 1} = \dfrac{A}{x-1} + \dfrac{B}{x+1}\right](x-1)(x+1)$

$x = A(x+1) + B(x-1)$

$\quad = Ax + A + Bx - B$

$\quad = (A+B)x + (A-B)$

$\therefore A + B = 1$

$A - B = 0; A = B$

and $A + (A) = 1$

$2A = 1$

$A = \dfrac{1}{2} = B$

$\therefore \dfrac{x}{x^2 - 1} = \dfrac{\frac{1}{2}}{x-1} + \dfrac{\frac{1}{2}}{x+1}$

$\qquad\qquad = \dfrac{1}{2(x-1)} + \dfrac{1}{2(x+1)}$

and

$\dfrac{x^3}{x^2 - 1} = x + \dfrac{1}{2(x-1)} + \dfrac{1}{2(x+1)}$

---

18. $\left[\dfrac{4x^2 + 4x + 17}{(x-1)(x^2 + x + 3)} = \dfrac{A}{x-1}\right.$

$\qquad \left. + \dfrac{Bx + C}{x^2 + x + 3}\right](x-1)(x^2 + x + 3)$

$4x^2 + 4x + 17$

$= A(x^2 + x + 3) + (Bx + C)(x-1)$

$= (A+B)x^2 + (A - B + C)x$
$\qquad\qquad\qquad\qquad\qquad + (3A - C)$

$A + B = 4; A - B + C = 4$

$3A - C = 17$

```
[A]
      ⎡1   1   0   4⎤
      ⎢1  -1   1   4⎥
      ⎣3   0  -1  17⎦
rref([A])
      ⎡1  0  0   5⎤
      ⎢0  1  0  -1⎥
      ⎣0  0  1  -2⎦
```

$\therefore \dfrac{5}{x-1} + \dfrac{-x-2}{x^2 + x + 3}$

$\qquad = \dfrac{5}{x-1} - \dfrac{x+2}{x^2 + x + 3}$

19. $\left[\dfrac{8x^2 + 55x + 12}{(x+1)(x^2 + 8x + 2)} = \dfrac{A}{x+1}\right.$

$\qquad \left. + \dfrac{Bx + C}{x^2 + 8x + 2}\right](x+1)(x^2 + 8x + 2)$

$8x^2 + 55x + 12$

$= A(x^2 + 8x + 2) + (Bx + C)(x+1)$

$= (A+B)x^2 + (8A + B + C)x$
$\qquad\qquad\qquad\qquad\qquad + (2A + C)$

$A + B = 8; 8A + B + C = 5$

$2A + C = 12$

Partial Fraction Decomposition for Proper Rational Expressions		
Denominator	Rational Expression	Partial Fractions
distinct linear factors	$\dfrac{px + q}{(x + a)(x + b)}$	$\dfrac{A}{x + a} + \dfrac{B}{x + b}$
repeated linear factors	$\dfrac{px^{n-1} + qx^{n-2} + \cdots + rx + s}{(x + a)^n}$	$\dfrac{A_1}{x + a} + \dfrac{A_2}{(x + a)^2} + \cdots + \dfrac{A_n}{(x + a)^n}$
distinct irreducible quadratic factors	$\dfrac{px^2 + qx + r}{(x + a)(x^2 + bx + c)}$	$\dfrac{A}{x + a} + \dfrac{Bx + C}{x^2 + bx + c}$
repeated irreducible quadratic factors	$\dfrac{mx^{n+1} + px^n + \cdots + qx + r}{(x^2 + bx + c)^n}$	$\dfrac{A_1 x + B_1}{x^2 + bx + c} + \dfrac{A_2 x + B_2}{(x^2 + bx + c)^2} + \cdots + \dfrac{A_n x + B_n}{(x^2 + bx + c)^n}$

A. Exercises

Write the form for each partial fraction decomposition. Do not solve.

1. $\dfrac{5x - 30}{(x - 2)(x - 12)}$

2. $\dfrac{x^3 + 6x^2 + 3x + 3}{(x^2 + 6x + 3)^2}$

3. $\dfrac{5x^2 - 8x + 11}{(x - 1)(x^2 - 2x - 5)}$

4. $\dfrac{5x^3 - 2}{x^4}$

5. $\dfrac{2x + 23}{x^2 + x - 30}$

6. $\dfrac{3x - 7}{x^2 - 8x + 16}$

Write the partial fraction decomposition for each rational expression. Use a matrix equation to solve the generated system.

7. $\dfrac{7x + 11}{(x - 2)(x + 3)}$

8. $\dfrac{3x - 6}{(x + 4)(x + 1)}$

9. $\dfrac{20x - 17}{2x^2 - 7x - 4}$

10. $\dfrac{5x - 38}{3x^2 + 22x - 16}$

Write the partial fraction decomposition for each rational expression.

11. $\dfrac{4x + 3}{x^2}$

12. $\dfrac{1 - 3x}{x^2}$

13. $\dfrac{2x + 1}{(x - 1)^2}$

14. $\dfrac{4x + 11}{(x + 3)^2}$

15. $\dfrac{7x^2 + 9x + 5}{x(x + 1)^2}$

16. $\dfrac{2x^2 - 7x + 16}{x(x - 2)^2}$

B. Exercises

Write the partial fraction decomposition for each rational expression. Use technology to solve the generated system with an augmented matrix.

17. $\dfrac{7x^2 - 3x + 14}{x^3 - 8}$

18. $\dfrac{4x^2 + 4x + 17}{(x - 1)(x^2 + x + 3)}$

19. $\dfrac{8x^2 + 55x + 12}{(x + 1)(x^2 + 8x + 2)}$

20. $\dfrac{3x^2 + 7x + 10}{(x + 2)(x^2 + 3x + 4)}$

Write the partial fraction decomposition for each rational expression. Use technology to confirm your solution graphically.

21. $\dfrac{5x^3 + 5x + 3}{(x^2 + 1)^2}$

22. $\dfrac{7x^3 - 14x - 1}{(x^2 - 2)^2}$

23. $\dfrac{5x^4 + 25x^2 + x + 27}{x(x^2 + 3)^2}$

24. $\dfrac{4x^4 + x^3 - 4x^2 - 2x + 4}{x(x^2 - 2)^2}$

Write the partial fraction decomposition for each rational expression.

25. $\dfrac{9x - 8}{9x^2 - 12x + 4}$

26. $\dfrac{23 - x}{x^2 - x - 20}$

27. $\dfrac{3x^2 + x - 2}{x^3 - 6x^2 + 7x - 2}$

28. $\dfrac{x^3 + 5x - 1}{x^4 + 4x^2 + 4}$

29. $\dfrac{5x - 23}{2x^2 - 17x + 30}$

30. $\dfrac{3x^2 - 2x + 2}{x^3 + x^2 - 10x - 6}$

31. $\dfrac{3x^2 - 24x + 32}{x^3 - 8x^2 + 16x}$

32. $\dfrac{10x^4 - 32x^2 + 2x + 27}{4x^5 - 12x^3 + 9x}$

C. Exercises

Divide each improper rational expression and then find the partial fraction decomposition of the rational remainder. Write your answer as the sum of the quotient and the expanded remainder.

33. $\dfrac{4x^2 + 3}{x^2 - x}$

34. $\dfrac{2x^2 - x + 3}{x^2 + 3x}$

35. $\dfrac{x^3 - 8x + 11}{x^2 + 2x - 3}$

36. $\dfrac{3x^3 - 3x^2 - 33x - 19}{x^2 - x - 12}$

$$A + B = 3; \quad 3A + 2B + C = 7$$
$$4A + 2C = 10$$

[A]
$$\begin{bmatrix} 1 & 1 & 0 & 3 \\ 3 & 2 & 1 & 7 \\ 4 & 0 & 2 & 10 \end{bmatrix}$$
rref([A])
$$\begin{bmatrix} 1 & 0 & 0 & 4 \\ 0 & 1 & 0 & -1 \\ 0 & 0 & 1 & -3 \end{bmatrix}$$

$$\therefore \frac{4}{x + 2} + \frac{-x - 3}{x^2 + 3x + 4}$$
$$= \frac{4}{x + 2} - \frac{x + 3}{x^2 + 3x + 4}$$

[A]
$$\begin{bmatrix} 1 & 1 & 0 & 8 \\ 8 & 1 & 1 & 55 \\ 2 & 0 & 1 & 12 \end{bmatrix}$$
rref([A])
$$\begin{bmatrix} 1 & 0 & 0 & 7 \\ 0 & 1 & 0 & 1 \\ 0 & 0 & 1 & -2 \end{bmatrix}$$

$$\therefore \frac{7}{x + 1} + \frac{x - 2}{x^2 + 8x + 2}$$

20. $\left[\dfrac{3x^2 + 7x + 10}{(x + 2)(x^2 + 3x + 4)} = \dfrac{A}{x + 2} \right.$
$\left. + \dfrac{Bx + C}{x^2 + 3x + 4}\right](x + 2)(x^2 + 3x + 4)$

$$3x^2 + 7x + 10$$
$$= A(x^2 + 3x + 4) + (Bx + C)(x + 2)$$
$$= (A + B)x^2 + (3A + 2B + C)x$$
$$\qquad + (4A + 2C)$$

---

21. $\left[\dfrac{5x^3 + 5x + 3}{(x^2 + 1)^2} = \dfrac{Ax + B}{(x^2 + 1)} + \dfrac{Cx + D}{(x^2 + 1)^2}\right]$
$$\cdot (x + 1)^2$$

$$5x^3 + 5x + 3$$
$$= (Ax + B)(x^2 + 1) + Cx + D$$
$$= Ax^3 + Bx^2 + (A + C)x + (B + D)$$
$$A = 5; \; B = 0; \; A + C = 5$$
$$C = 5 - (5) = 0$$
$$B + D = 3; \; D = 3 - (0) = 3$$
$$\therefore \frac{5x + 0}{(x^2 + 1)} + \frac{0x + 3}{(x^2 + 1)^2}$$
$$= \frac{5x}{x^2 + 1} + \frac{3}{(x^2 + 1)^2}$$

22. $\left[\dfrac{7x^3 - 14x - 1}{(x^2 - 2)^2} = \dfrac{Ax + B}{(x^2 - 2)} + \dfrac{Cx + D}{(x^2 - 2)^2}\right]$
$$\cdot (x - 2)^2$$

$$7x^3 - 14x - 1$$
$$= (Ax + B)(x^2 - 2) + Cx + D$$
$$= Ax^3 + Bx^2 + (-2A + C)x$$
$$\qquad + (-2B + D)$$

$$A = 7; \; B = 0$$
$$-2A + C = -14; \; C = 2(7) - 14 = 0$$
$$-2B + D = -1; \; D = -1 + 2(0) = -1$$
$$\therefore \frac{7x + 0}{x^2 - 2} + \frac{0x + (-1)}{(x^2 - 2)^2}$$
$$= \frac{7x}{x^2 - 2} - \frac{1}{(x^2 - 2)^2}$$

23. $\left[\dfrac{5x^4 + 25x^2 + x + 27}{x(x^2 + 3)^2} = \dfrac{A}{x} + \dfrac{Bx + C}{x^2 + 3}\right.$
$\left. + \dfrac{Dx + E}{(x^2 + 3)^2}\right]x(x^2 + 3)^2$

$$5x^4 + 25x^2 + x + 27$$
$$= A(x^2 + 3)^2 + (Bx + C)(x^3 + 3x)$$
$$\qquad + (Dx + E)x$$
$$= (A + B)x^4 + Cx^3 + (6A + 3B + D)x^2$$
$$\qquad + (3C + E)x + 9A$$

$$A + B = 5; \; C = 0$$
$$6A + 3B + D = 25$$
$$3C + E = 1; \; 9A = 27$$
$$A = 3; \; B = 5 - (3) = 2$$
$$D = 25 - 6(3) - 3(2) = 1$$
$$E = 1 - 3(0) = 1$$
$$\therefore \frac{3}{x} + \frac{2x}{x^2 + 3} + \frac{x + 1}{(x^2 + 3)^2}$$

24. $\left[\dfrac{4x^4 + x^3 - 4x^2 - 2x + 4}{x(x^2 - 2)^2} = \dfrac{A}{x}\right.$
$+ \dfrac{Bx + C}{x^2 - 2} + \dfrac{Dx + E}{(x^2 - 2)^2}\left.\right]x(x^2 - 2)^2$

$$4x^4 + x^3 - 4x^2 - 2x + 4$$
$$= A(x^2 - 2)^2 + (Bx + C)(x^3 - 2x)$$
$$\qquad + (Dx + E)x$$
$$= (A + B)x^4 + Cx^3$$
$$\qquad + (4A - 2B + D)x^2 + (-2C + E)x$$
$$\qquad + 4A$$

$$A + B = 4; \; C = 1$$
$$-4A - 2B + D = -4$$
$$-2C + E = -2; \; 4A = 4$$

$$A = 1; B = 4 - (1) = 3$$
$$D = -4 + 4(1) + 2(3) = 6$$
$$E = -2 + 2(1) = 0$$
$$\therefore \frac{1}{x} + \frac{3x + 1}{x^2 - 2} + \frac{6x}{\left(x^2 - 2\right)^2}$$

25. $\left[\dfrac{9x - 8}{(3x - 2)^2} = \dfrac{A}{3x - 2} + \dfrac{B}{(3x - 2)^2}\right]$
$$\cdot (3x - 2)^2$$
$$9x - 8 = A(3x - 2) + B$$
$$= 3Ax + (-2A + B)$$
$$3A = 9; A = 3$$
$$-2A + B = -8; B = -8 + 2(3) = -2$$
$$\therefore \frac{3}{3x - 2} - \frac{2}{(3x - 2)^2}$$

26. $\left[\dfrac{23 - x}{(x - 5)(x + 4)} = \dfrac{A}{x - 5} + \dfrac{B}{x + 4}\right]$
$$\cdot (x - 5)(x + 4)$$
$$23 - x = A(x + 4) + B(x - 5)$$
$$= (A + B)x + (4A - 5B)$$
$$A + B = -1; 4A - 5B = 23$$
$$4A - 5(-1 - A) = 23$$
$$9A = 18; A = 2$$
$$B = -1 - (2) = -3$$
$$\therefore \frac{2}{x - 5} - \frac{3}{x + 4}$$

27. $\left[\dfrac{3x^2 + x - 2}{(x - 1)(x^2 - 5x + 2)} = \dfrac{A}{x - 1}\right.$
$$\left. + \frac{Bx + C}{x^2 - 5x + 2}\right](x - 1)(x^2 - 5x + 2)$$
$$3x^2 + x - 2$$
$$= A(x^2 - 5x + 2) + (Bx + C)(x - 1)$$
$$= (A + B)x^2 + (-5A - B + C)x$$
$$+ (2A - C)$$
$$A + B = 3$$
$$-5A - B + C = 1$$
$$2A - C = -2$$

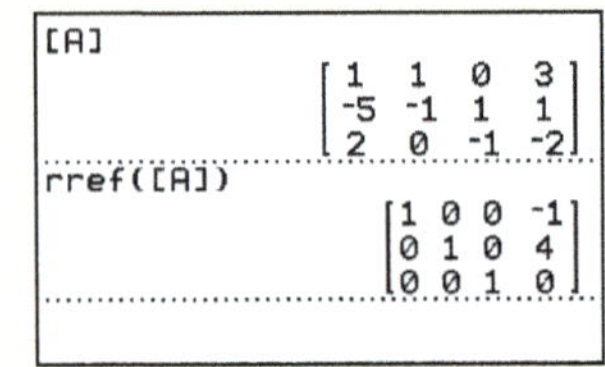

$$\therefore \frac{-1}{x - 1} + \frac{4x + 0}{x^2 - 5x + 2}$$
$$= \frac{4x}{x^2 - 5x + 2} - \frac{1}{x - 1}$$

---

37. Analyze: Use technology to analyze $f(x) = \frac{4x + 5}{x^2 + 5x}$.

  a. Find the partial fraction decomposition. Verify your answer by graphing the original expression as Y_1 and the sum of the partial fractions as Y_2. $\frac{1}{x} + \frac{3}{x + 5}$

  b. Graph the partial fractions separately as Y_3 and Y_4 along with Y_2, the graph of their sum. Do the three graphs coincide? no

  c. Compare the asymptotes of each partial fraction and their sum.

    All have the same horizontal asymptote of $y = 0$. The vertical asymptote in each partial fraction determines the asymptotes for their sum, $f(x)$.

38. Explore: Consider how a partial fraction decomposition can help evaluate the sum $\displaystyle\sum_{n=1}^{\infty} \frac{12}{16n^2 + 8n - 3}$.

  a. Find the partial fraction decomposition of $\frac{12}{16n^2 + 8n - 3}$. $\frac{3}{4n - 1} - \frac{3}{4n + 3}$

  b. Use the partial fraction decomposition to list the sum of the first four terms in the series.

  c. Make a conjecture about what happens to the sum when the series is extended infinitely. Explain your reasoning. sum = 1; All the fractions have a sum of

38b. $\left(1 - \frac{3}{7}\right) + \left(\frac{3}{7} - \frac{3}{11}\right) + \left(\frac{3}{11} - \frac{3}{15}\right) + \left(\frac{3}{15} - \frac{3}{19}\right)$

CUMULATIVE REVIEW

39. Find the angle between the vectors $\mathbf{p} = \langle -1, 2\rangle$ and $\mathbf{q} = \langle 3, -3\rangle$ (to the nearest degree). [6.2] 162°

40. State all equivalent polar coordinates for $(r, \theta) = \left(3, -\frac{\pi}{3}\right)$ if $-2\pi \leq \theta \leq 2\pi$. [6.4]

41. Find the inverse of $f(x) = 3e^x$. [3.3] $g(x) = \ln x - \ln 3$

42. Verify that $A = \begin{bmatrix} 1 & -3 \\ 2 & 4 \end{bmatrix}$ and $B = \begin{bmatrix} \frac{2}{5} & \frac{3}{10} \\ -\frac{1}{5} & \frac{1}{10} \end{bmatrix}$ are inverse matrices. [7.5]

43. Without graphing, specify all asymptotes and the domain and range for each function. [2.5]

  a. $f(x) = \frac{1}{x}$

  b. $g(x) = \frac{1}{x^2}$

  c. $h(x) = \frac{x + 1}{x^2 + 4x + 3}$

44. What is the determinant of any identity matrix? [7.4] $|I| = 1$

45. Which of the following is *not* an equivalent pair of polar coordinates for $(r, \theta) = \left(-4, \frac{5\pi}{6}\right)$? [6.4] D

  A. $\left(-4, -\frac{7\pi}{6}\right)$ **D.** $\left(4, -\frac{2\pi}{3}\right)$

  B. $\left(4, \frac{11\pi}{6}\right)$ **E.** none of these

  C. $\left(4, -\frac{\pi}{6}\right)$

46. Which of the following is the determinant of $\begin{bmatrix} -14 & 3.5 \\ -42 & -5.6 \end{bmatrix}$? [7.4] A

  A. 225.4   **C.** −225.4   **E.** none of

  B. 68.6   **D.** −68.6   these

47. Which of the following is the approximate length of a in $\triangle ABC$ if $A = 47°$, $b = 86$, and $c = 79$? [5.7] A

  A. 66   **C.** 11   **E.** 4370

  B. 123   **D.** 87

48. Which points are vertices of the region defined by the given constraints? List all correct answers. [7.6]
$$x \geq 0$$
$$y \geq 1$$
$$x + y \leq 5$$
  B, C, D

  A. (0, 0)   **C.** (0, 1)   **E.** all of these

  B. (0, 5)   **D.** (4, 1)

---

28. $\left[\dfrac{x^3 + 5x - 1}{\left(x^2 + 2\right)^2} = \dfrac{Ax + B}{x^2 + 2} + \dfrac{Cx + D}{\left(x^2 + 2\right)^2}\right]$
$$\cdot \left(x^2 + 2\right)^2$$
$$x^3 + 5x - 1$$
$$= (Ax + B)(x^2 + 2) + Cx + D$$
$$= Ax^3 + Bx^2 + (2A + C)x + (2B + D)$$
$$A = 1; B = 0$$
$$2A + C = 5; C = 5 - 2(1) = 3$$
$$2B + D = -1; D = -1 - 2(0) = -1$$
$$\therefore \frac{1x + 0}{x^2 + 2} + \frac{3x + (-1)}{\left(x^2 + 2\right)^2}$$
$$= \frac{x}{x^2 + 2} + \frac{3x - 1}{\left(x^2 + 2\right)^2}$$

29. $\left[\dfrac{5x - 23}{2x^2 - 17x + 30} = \dfrac{A}{2x - 5} + \dfrac{B}{x - 6}\right]$
$$\cdot (2x - 5)(x - 6)$$
$$5x - 23 = A(x - 6) + B(2x - 5)$$
$$= (A + 2B)x + (-6A - 5B)$$
$$A + 2B = 5$$
$$-6A - 5B = -23$$
$$-6(5 - 2B) - 5B = -23$$
$$7B = 7; B = 1$$
$$A = 5 - 2(1) = 3$$
$$\therefore \frac{3}{2x - 5} + \frac{1}{x - 6}$$

continued in Answers and Solutions Overflow

Cryptology

In 2017 over 147 million people had their personal data exposed to hackers due to a security breach at Equifax, a leading US credit reporting company. Governments and military forces have always been keenly interested in information security, but in the modern era of digital communications and online commerce, the secure storage and transmission of data is now extremely important to everyone. In 2001 the National Institute of Standards and Technology published the Advanced Encryption Standard, an algorithm that relies on the difficulty of factoring products of very large prime numbers to encrypt messages.

Cryptology includes both cryptography (creating and using ciphers) and cryptanalysis (breaking ciphers). Simple monoalphabetic ciphers pair a letter with a different letter or symbol based on a key. For example, the Caesar cipher key adds three to the original letter so that A becomes D, B becomes E, and so on. Even if the key is more complex, these ciphers are easily broken by analyzing how often each symbol occurs and comparing its frequency to the known frequency of occurrence of letters in normal usage.

In 1929 Lester Hill, a teacher at Hunter College in New York, published a polyalphabetic block cipher using an algebraic alphabet. While never widely used, the Hill cipher introduced mathematics into cryptology and is still commonly taught in undergraduate linear algebra and computer science classes. The Hill cipher pairs each letter of the alphabet with a whole number from 0 to 25.

A	B	C	D	E	F	G	H	I	J	K	L	M
0	1	2	3	4	5	6	7	8	9	10	11	12

N	O	P	Q	R	S	T	U	V	W	X	Y	Z
13	14	15	16	17	18	19	20	21	22	23	24	25

The plaintext message is then converted into an uncoded $m \times n$ matrix U and encrypted by right-multiplying by an $n \times n$ invertible matrix E as described in Section 7.5.

Additionally, each number in the coded matrix C can be replaced with its remainder when divided by 26 (known as modulus 26, or mod 26). This allows the message "SEND FOOD" to be sent as a string of letters.

$$U \cdot E = C \Rightarrow C_{mod\,26}$$

$$\begin{bmatrix} 18 & 4 \\ 13 & 3 \\ 5 & 14 \\ 14 & 3 \end{bmatrix} \cdot \begin{bmatrix} 2 & 3 \\ 1 & 4 \end{bmatrix} = \begin{bmatrix} 40 & 70 \\ 29 & 51 \\ 24 & 71 \\ 31 & 54 \end{bmatrix} \Rightarrow \begin{bmatrix} 14 & 18 \\ 3 & 25 \\ 24 & 19 \\ 5 & 2 \end{bmatrix} = \text{OSDZYTFC}$$

While the product CE^{-1} decodes matrix C, multiplying $C_{mod\,26}$ by E^{-1} does not. To decipher the message, the recipient recreates $C_{mod\,26}$ and multiplies by the mod 26 inverse of the encryption matrix, $E^{-1}{}_{mod\,26}$. Only matrices where $|E|$ mod 26 = 1, 3, 5, 7, 9, 11, 15, 17, 19, 21, 23, or 25 have a mod 26 inverse matrix.

1. Find $|E|$, $|E|_{mod\,26}$, and E^{-1} for $E = \begin{bmatrix} 2 & 3 \\ 1 & 4 \end{bmatrix}$.

To find $E^{-1}{}_{mod\,26}$, each fraction in E^{-1} must be converted into a whole number by multiplying the numerator by the modular multiplicative inverse of the denominator (listed in the table) and writing the result in mod 26. Since e_{12} in $E^{-1} = -\frac{3}{5}$, e_{12} in $E^{-1}{}_{mod\,26}$ is the remainder when $-3(21)$ is divided by 26.

$$e_{21} = -11$$

Convert a negative mod 26 value to an equivalent mod 26 whole number by adding 26 or a multiple of 26.

$$-11 \bmod 26$$
$$= (-11 + 26) \bmod 26$$
$$= 15 \bmod 26$$

denominator	1	3	5	7	9	11	15	17	19	21	23	25
mod 26 mult. inv.	1	9	21	15	3	19	7	23	11	5	17	25

2. Convert E^{-1} to $E^{-1}{}_{mod\,26}$. $\begin{bmatrix} 6 & 15 \\ 5 & 16 \end{bmatrix}$

3. Decode OSDZYTFC using $E^{-1}{}_{mod\,26}$. SEND FOOD

Objectives

1. To encrypt and decode messages using mod 26 matrices
2. To apply a biblical perspective to conflicting concerns for privacy rights and the need of law enforcement to access digital information

Assignments

- **Minimum:** 1–10 (in class)
- **Standard:** 1–8; 9–10 (in class)
- **Extended:** 1–10

Solutions

1. $|E| = 8 - 3 = 5$; $5 \bmod 26 = 5$

$$E^{-1} = \frac{1}{5}\begin{bmatrix} 4 & -3 \\ -1 & 2 \end{bmatrix} = \begin{bmatrix} \frac{4}{5} & -\frac{3}{5} \\ -\frac{1}{5} & \frac{2}{5} \end{bmatrix}$$

2. $E^{-1}{}_{mod\,26} = 21\begin{bmatrix} 4 & -3 \\ -1 & 2 \end{bmatrix}$

$$= \begin{bmatrix} 84 & -63 \\ -21 & 42 \end{bmatrix} = \begin{bmatrix} 6 & -11 \\ -21 & 16 \end{bmatrix}$$

$$= \begin{bmatrix} 6 & 15 \\ 5 & 16 \end{bmatrix}$$

3. $U_{mod\,26} = C_{mod\,26}E^{-1}{}_{mod\,26}$

$$= \begin{bmatrix} 14 & 18 \\ 3 & 25 \\ 24 & 19 \\ 5 & 2 \end{bmatrix}\begin{bmatrix} 6 & 15 \\ 5 & 16 \end{bmatrix} = \begin{bmatrix} 174 & 498 \\ 143 & 445 \\ 239 & 664 \\ 40 & 107 \end{bmatrix}$$

$$= \begin{bmatrix} 18 & 4 \\ 13 & 3 \\ 5 & 14 \\ 14 & 3 \end{bmatrix}$$

PRESENTATION

Identity theft, malware, spyware, and ransomware are all too familiar topics in our technological, data-driven age. But the emphasis on user-friendly interfaces often masks the herculean efforts required to protect digital information. This Data Analysis helps students appreciate the essential role of math in one of its many behind-the-scenes applications.

Government interest in cryptology has increased as technology has increased. The National Security Agency is commonly listed as the largest employer of mathematicians.

The Advanced Encryption Standard adopted by the National Institute of Standards and Technology (NIST) is a symmetric block cipher developed by Belgian cryptographers Joan Daemen and Vincent Rijmen. It replaced the Data Encryption Standard, published in 1977, as the specification used by the US government to protect sensitive data.

Consider asking students to guess the ten most-used letters in the English alphabet. (*In order of occurrence they are E, A, T, O, I, N, S, H, R, D.*) Frequency analysis methods also identify common multi-letter patterns such as *th* and *ing*. Bletchley Park cryptanalysts used common greetings and closings in military letters to help them break German codes during World War II.

Hill's cipher was more secure since he assigned numbers to letters in random order and then added a key matrix after multiplying. Hill also patented a machine that would encrypt messages with keys up to $n = 6$.

The first widely used polyalphabetic cipher, the Vigenère cipher, dates back to the seventeenth century. Alphabetic ciphers are popular for transmitting codes manually, since a string of letters is more user-friendly to send or receive than a long string of numbers.

4. $|E| = [2(2)(3) + 0(4)(1) + 1(3)(5)]$
$\qquad - [1(2)(1) + 4(5)(2) + 3(0)(3)]$
$\quad = (12 + 0 + 15) - (2 + 40 + 0)$
$\quad = 27 - 42 = -15$

$-15 \bmod 26$
$= (-15 + 26) \bmod 26 = 11$

5. $U = \begin{bmatrix} 0 & 11 & 11 \\ 8 & 18 & 22 \\ 4 & 11 & 11 \end{bmatrix}$

$UE = \begin{bmatrix} 44 & 77 & 77 \\ 92 & 146 & 146 \\ 52 & 77 & 81 \end{bmatrix} = C$

$C_{mod\,26} = \begin{bmatrix} 18 & 25 & 25 \\ 14 & 16 & 16 \\ 0 & 25 & 3 \end{bmatrix}$
or SZZOQQAZD

6. $E^{-1} = \begin{bmatrix} \frac{14}{15} & -\frac{1}{3} & \frac{2}{15} \\ \frac{1}{3} & -\frac{1}{3} & \frac{1}{3} \\ -\frac{13}{15} & \frac{2}{3} & -\frac{4}{15} \end{bmatrix}$

$E^{-1}{}_{mod\,26} = \begin{bmatrix} 14(7) & -1(9) & 2(7) \\ 1(9) & -1(9) & 1(9) \\ -13(7) & 2(9) & -4(7) \end{bmatrix}$

$= \begin{bmatrix} 98 & -9 & 14 \\ 9 & -9 & 9 \\ -91 & 18 & -28 \end{bmatrix} = \begin{bmatrix} 20 & 17 & 14 \\ 9 & 17 & 9 \\ 13 & 18 & 24 \end{bmatrix}$

7. $C_{mod\,26}E^{-1}{}_{mod\,26}$

$= \begin{bmatrix} 18 & 25 & 25 \\ 14 & 16 & 16 \\ 0 & 25 & 3 \end{bmatrix}\begin{bmatrix} 20 & 17 & 14 \\ 9 & 17 & 9 \\ 13 & 18 & 24 \end{bmatrix}$

$= \begin{bmatrix} 910 & 1181 & 1077 \\ 632 & 798 & 724 \\ 264 & 479 & 297 \end{bmatrix} = \begin{bmatrix} 0 & 11 & 11 \\ 8 & 18 & 22 \\ 4 & 11 & 11 \end{bmatrix}$

$\Rightarrow$ ALL IS WELL

continued in Answers and Solutions Overflow

If the mod 26 determinant of E shares a common factor with 26, $E^{-1}{}_{mod\,26}$ does not exist and a mod 26 encryption cannot be decoded.

Consider the 3×3 encryption matrix $E = \begin{bmatrix} 2 & 0 & 1 \\ 3 & 2 & 4 \\ 1 & 5 & 3 \end{bmatrix}$.

4. Find $|E|$ and the whole number $|E|$ mod 26. −15; 11

5. Use $E_{3\times3}$ to find the alphabetic encryption of "ALL IS WELL." SZZOQQAZD

6. Find the decrypting matrix $E^{-1}{}_{mod\,26}$. $\begin{bmatrix} 20 & 17 & 14 \\ 9 & 17 & 9 \\ 13 & 18 & 24 \end{bmatrix}$

7. Use $E^{-1}{}_{mod\,26}$ to decode your encrypted message from exercise 5. ALL IS WELL

8. Decode this message encrypted with $E_{3\times3}$: CALKKXTDLTUA. BY GRACE ALONE

Cryptology plays an essential role in preserving private information, intellectual property, and international security. Cryptology can also be controversial. Law enforcement agencies claim that the surge in encryption creates a major impediment to law enforcement as it is more difficult to access digitally stored information when investigating criminal activity. Tech companies and privacy advocates claim that providing "back doors" to law enforcement would violate the user's right to privacy. How much privacy should be relinquished in order to apprehend criminals?

9. Discuss: Which biblical principles should be considered in the cryptology debate between law enforcement and privacy advocates?

10. Discuss: Evaluate the following statement: "In criminal cases, law enforcement officials should be able to obtain a search warrant granting access to all information stored on related electronic devices."

9. Answers will vary. God clearly states that people can own property and that it is sinful to take another person's property. God also gives guidance about how to carry out justice in criminal cases. The preservation of the rights of the accused is illustrated by the protection offered within Old Testament cities of refuge.

10. Answers may include an evaluation that takes either side of the issue. Students may emphasize that corruption exists within law enforcement so that granting access to private information could be dangerous. Students could also emphasize that law enforcement should have some means of quickly accessing information to catch criminals, deter criminal activity, and protect the public.

Notice how matrix multiplication hides patterns required for frequency analysis. The two Ds in the original message come out as different letters when encoded and the OO in *food* turns into *TF*.

Consider working the first three exercises together since many students may need some assistance with modulo 26 arithmetic. Due to the cyclic nature of modular arithmetic, x mod y can be found by subtracting y from x until a value less than y is obtained. For example, you can find 70 mod 26 by subtracting 26 twice from 70, yielding 18. The Internet keyword search *modular arithmetic* provides additional instructional aids, including online modular calculators.

The expression $|E|$ mod 26 may be confused with $E_{mod\,26}$. $|E|$ mod 26 is the expression for the mod 26 value of the determinant of E, while $E_{mod\,26}$ refers to the matrix E whose elements have been converted to mod 26.

The two discussion exercises provide an excellent opportunity to model the practical application of a biblical worldview. You and your students will be affected by how these privacy issues are resolved by our lawmakers. Answering these questions clearly requires knowledge as well as discernment and wisdom from God.

TIPS

Ex. 1–8 The TI-84 remainder(a, b) function (found in the MATH, NUM submenu) returns the remainder when the whole number a is divided by the natural number b. You can create a modular arithmetic calculator by using TRACE, on a graph of Y1 = remainder(X, 26). The Windows® scientific calculator has a Mod function that accepts negative integers as the argument.

Ex. 8 Decimal approximations of the modulus are used when the argument is expressed as a decimal approximation.

Chapter 7 Review

Solve each system. Then classify each system as independent consistent, dependent consistent, or inconsistent. [7.1]

1. $3x - y = 1$
$5x + 2y = -24$

2. $2x + y = 15$
$4x + 2y = 19$

3. $5x - 3y = 10$
$1.5y - 2.5x = -5$

4. $2x + 5y = 14$
$3x - 2y = -17$

5. Find the dimensions of a rectangle whose perimeter is 156 ft and area is 1296 ft. [7.1]   24 ft × 54 ft

6. A cell phone manufacturer has determined the supply and demand curves for a cell phone where p is the price (in dollars) and x is the supply in hundred thousands of phones. Find the equilibrium price and the number of phones that should be produced. [7.1]
supply: $p = 100x - 460$ 1,100,000 phones
demand: $p = 915 - 25x$ at $640 each

Solve each system. [7.1]

7. $2x - 4y + 5z = -7$
$5x + 3y + 2z = 1$
$7x - y + 3z = 6$ $(2, -1, -3)$

8. $3x + 2y - z = 9$
$9x - 2y + 2z = 3$
$\left(1 - \frac{1}{12}t, 3 + \frac{5}{8}t, t\right)$

9. Given $A = \begin{bmatrix} 5 & 12 & -2 & 0 \\ 9 & 7 & 14 & 2 \\ -1 & 3 & 6 & 8 \end{bmatrix}$, state the following. [7.2]

   a. dimensions of A 3×4
   b. value of a_{32} 3
   c. value of a_{23} 14
   d. element whose value is 2 a_{24}
   e. element whose value is -1 a_{31}

10. If $A = \begin{bmatrix} 5 & -2 & 10 \\ 1 & 7 & 9 \end{bmatrix}$ and $B = \begin{bmatrix} 3 & 1 & -5 \\ -4 & 6 & 2 \end{bmatrix}$, find $D = 3A - 2B$. [7.2] $\begin{bmatrix} 9 & -8 & 40 \\ 11 & 9 & 23 \end{bmatrix}$

11. State the dimensions of each matrix product, or state *not defined.* [7.2]

   a. $A_{3 \times 2} \times B_{2 \times 3}$ 3×3
   b. $A_{3 \times 2} \times C_{3 \times 2}$ not defined
   c. $D_{4 \times 1} \times E_{1 \times 2}$ 4×2
   d. $F_{2 \times 2} \times G_{2 \times 5}$ 2×5

Multiply. [7.2]

12. $\begin{bmatrix} 3 & 5 \\ -1 & 2 \end{bmatrix}\begin{bmatrix} 4 & 0 \\ 2 & -1 \end{bmatrix}$ $\begin{bmatrix} 22 & -5 \\ 0 & -2 \end{bmatrix}$

13. $\begin{bmatrix} 6 & -2 & 1 \\ 3 & 4 & -2 \end{bmatrix}\begin{bmatrix} -5 & 4 \\ -1 & 3 \\ 6 & 3 \end{bmatrix}$ $\begin{bmatrix} -22 & 21 \\ -31 & 18 \end{bmatrix}$

14. Write each 3×3 matrix. [7.2]
   a. multiplicative identity matrix
   b. additive identity matrix

15. The school athletic director needs to order tennis equipment for the new season. Write matrix $N_{2 \times 3}$ representing the team's needs and matrix C representing the costs. Then write and evaluate a matrix expression to find the cost for each group. [7.2]

Team Equipment Needs			
	Balls	Rackets	Uniforms
Boys	45	12	16
Girls	40	14	18
Cost			
Ball	Racket	Uniform	
$1.50	$42.00	$57.00	

16. Write $\begin{bmatrix} 2 & 5 & 4 \\ 3 & 6 & 3 \end{bmatrix}$ in equivalent REF. [7.3] $\begin{bmatrix} 1 & 2 & 1 \\ 0 & 1 & 2 \end{bmatrix}$

17. Write $\begin{bmatrix} 2 & 1 & 2 \\ 3 & -2 & 17 \end{bmatrix}$ in RREF. [7.3] $\begin{bmatrix} 1 & 0 & 3 \\ 0 & 1 & -4 \end{bmatrix}$

18. Solve using Gaussian elimination. [7.3]
$2x + 4y + 2z = -10$
$3x - 2y + 8z = 17$
$6x + 2y - 3z = 10$ $(3, -4, 0)$

Use Gauss-Jordan elimination to solve each system. Classify any inconsistent or dependent systems. [7.3]

19. $2x - 3y + z = 3$
$y + 3z = 2$
$-4x + 6y - 2z = -6$

20. $x + y + z = 7$
$3x + 4y = 6$
$2x + z = 2$ $(-2, 3, 6)$

21. $x + 4y + 2z = 7$
$2x + 9y + 7z = 7$
$y + 3z = -5$ $\varnothing$; inconsistent

19. $\left(\frac{9 - 10t}{2}, 2 - 3t, t\right)$; dependent

PRESENTATION

Be sure to communicate your expectations for calculator use on the chapter review and test.

Objective

To prepare for evaluation

Vocabulary

See Appendix A.

Assignments

- **Minimum:** 1–3, 5–12, 14–15, 17, 19–24, 27, 29, 31–35, 37–38, 41–43
- **Standard:** 1–47 odd; 2–46 even (in class)
- **Extended:** 2–8, 11, 13–19, 22–27, 30–31, 33–35, 37, 39–40, 42–45, 47–48

Assessment

- Chapter 7 Test

Solutions

1. $y = 3x - 1$
$5x + 2(3x - 1) = -24$
$11x = -22$
$x = -2$
$y = 3(-2) - 1 = -7; (-2, -7)$
independent consistent

2. $-2(\text{1st}) \Rightarrow -4x - 2y = -30$
$\text{2nd} \Rightarrow \underline{4x + 2y = 19}$
$0 \neq -11$
$\varnothing$; inconsistent

3. $2(\text{2nd}) \Rightarrow -5x + 3y = -10$
$\text{1st} \Rightarrow \underline{5x - 3y = 10}$
$0 = 0$
$\left\{(x, y) \mid y = \frac{5}{3}x - \frac{10}{3}\right\}$
dependent consistent

4. $3(\text{1st}) \Rightarrow 6x + 15y = 42$
$-2(\text{2nd}) \Rightarrow \underline{-6x + 4y = 34}$
$19y = 76$
$y = 4$
$2x + 5(4) = 14$
$2x = -6$
$x = -3; (-3, 4)$
independent consistent

5. $2l + 2w = 156; l = 78 - w$
$lw = 1296$
$(78 - w)w = 1296$
$w^2 - 78w + 1296 = 0$
$(w - 24)(w - 54) = 0$
$w = 24, 54$
$l = 78 - (24) = 54; l = 78 - (54) = 24$
24 ft × 54 ft

6. $100x - 460 = 915 - 25x$
$$125x = 1375$$
$$x = 11$$
$$p = 100(11) - 460 = \$640$$

7. 1st $\Rightarrow 2x - 4y + 5z = -7$
$-4(\text{3rd}) \Rightarrow -28x + 4y - 12z = -24$
(a) $\qquad -26x \qquad - 7z = -31$

2nd $\Rightarrow 5x + 3y + 2z = 1$
$3(\text{3rd}) \Rightarrow 21x - 3y + 9z = 18$
(b) $\qquad 26x \qquad + 11z = 19$

(a) $\Rightarrow -26x - 7z = -31$
(b) $\Rightarrow 26x + 11z = 19$
$$4z = -12$$
$$z = -3$$

(b) $26x + 11(-3) = 19$
$$26x = 52$$
$$x = 2$$

2nd: $5(2) + 3y + 2(-3) = 1$
$$3y + 4 = 1$$
$$y = -1$$

$(2, -1, -3)$

8. $-3(\text{1st}) \Rightarrow -9x - 6y + 3z = -27$
2nd $\Rightarrow 9x - 2y + 2z = 3$
$$-8y + 5z = -24$$
$$y = \tfrac{5}{8}z + 3$$

1st: $3x + 2\left(\tfrac{5}{8}z + 3\right) - z = 9$
$$3x + \tfrac{1}{4}z = 3$$
$$x = -\tfrac{1}{12}z + 1$$

letting $z = t$: $\left(1 - \tfrac{1}{12}t,\; 3 + \tfrac{5}{8}t,\; t\right)$

10. $\begin{bmatrix} 15 & -6 & 30 \\ 3 & 21 & 27 \end{bmatrix} + \begin{bmatrix} -6 & -2 & 10 \\ 8 & -12 & -4 \end{bmatrix}$
$$= \begin{bmatrix} 9 & -8 & 40 \\ 11 & 9 & 23 \end{bmatrix}$$

12. $\begin{bmatrix} 12 + 10 & 0 - 5 \\ -4 + 4 & 0 - 2 \end{bmatrix} = \begin{bmatrix} 22 & -5 \\ 0 & -2 \end{bmatrix}$

13. $\begin{bmatrix} -30 + 2 + 6 & 24 + (-6) + 3 \\ -15 + (-4) + (-12) & 12 + 12 - 6 \end{bmatrix}$
$$= \begin{bmatrix} -22 & 21 \\ -31 & 18 \end{bmatrix}$$

14a. $\begin{bmatrix} 1 & 0 & 0 \\ 0 & 1 & 0 \\ 0 & 0 & 1 \end{bmatrix}$ 14b. $\begin{bmatrix} 0 & 0 & 0 \\ 0 & 0 & 0 \\ 0 & 0 & 0 \end{bmatrix}$

15. $NC = \begin{bmatrix} 45 & 12 & 16 \\ 40 & 14 & 18 \end{bmatrix} \begin{bmatrix} 1.5 \\ 42 \\ 57 \end{bmatrix}$
$$= \begin{bmatrix} 1483.5 \\ 1674 \end{bmatrix}$$
boys: \$1483.50; girls: \$1674.00

16. $\tfrac{1}{3}r_2 \Leftrightarrow r_1 \begin{bmatrix} 1 & 2 & 1 \\ 2 & 5 & 4 \end{bmatrix}$

$-2r_1 + r_2 \Rightarrow r_2 \begin{bmatrix} 1 & 2 & 1 \\ 0 & 1 & 2 \end{bmatrix}$

---

22. John sold the listed quantities of art pieces at a craft fair and made \$690, \$470, and \$1325 in profits on Thursday, Friday, and Saturday, respectively. Write and solve a system of equations to find the profit per item for each type of art piece. [7.3]

Day	Ornaments	Paintings	Prints
Thursday	12	1	4
Friday	10	0	8
Saturday	15	2	10

Find the determinant of each matrix using the product of the diagonals. [7.4]

23. $\begin{bmatrix} 7 & 9 \\ -3 & 2 \end{bmatrix}$ 41

24. $\begin{bmatrix} 5 & -2 & 0 \\ 2 & -4 & 7 \\ -3 & 1 & 1 \end{bmatrix}$ -9

25. Find the determinant of the matrix using minors and cofactors for each given row or column. [7.4]
$\begin{bmatrix} 1 & 2 & 0 \\ 3 & -1 & -2 \\ 1 & 3 & 2 \end{bmatrix}$
  a. row 1 -12
  b. column 1 -12

26. Evaluate the determinant of
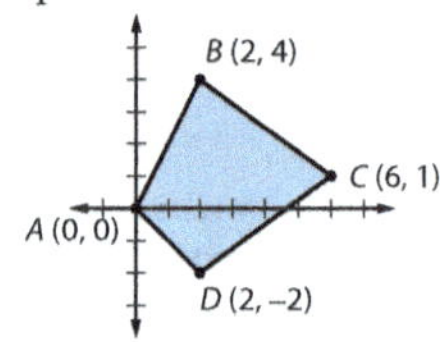
$\begin{bmatrix} -1 & 1 & -4 & 2 \\ 2 & 0 & 0 & 3 \\ 5 & 3 & 1 & 0 \\ 0 & -2 & 3 & -2 \end{bmatrix}$ [7.4] 50

Solve each system using Cramer's rule. [7.4]

27. $2x - y = 13$
$4x + 2y = 6$ (4, -5)

28. $2x - y = 1$
$x + 2z = -1$
$y + 3z = -1$ (3, 5, -2)

Find the area of each polygonal region. [7.4]

29. $\triangle ABC$ with $A\,(0, 0)$, $B\,(-2, 5)$, and $C\,(4, 3)$ 13 u²

30. quadrilateral $ABCD$ 18 u²

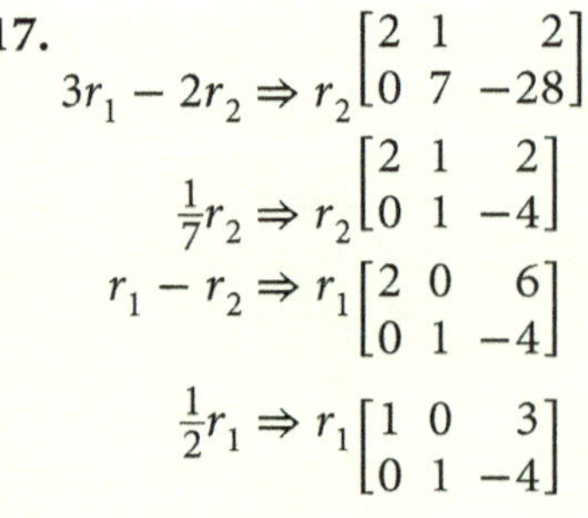

31. Determine whether each pair of matrices are inverse matrices. [7.5]
  a. $A = \begin{bmatrix} 2 & -3 \\ -1 & 2 \end{bmatrix}$, $B = \begin{bmatrix} 2 & 3 \\ 1 & 2 \end{bmatrix}$ inverses

  b. $C = \begin{bmatrix} 4 & 3 \\ 1 & 1 \end{bmatrix}$, $D = \begin{bmatrix} 1 & -3 \\ -1 & 3 \end{bmatrix}$ not inverses

32. Use the formula to find the inverse of $A = \begin{bmatrix} -3 & 4 \\ -2 & 2 \end{bmatrix}$. [7.5]

---

33. Use row operations on $[A \,|\, I]$ to find the inverse of
$B = \begin{bmatrix} 2 & -1 & 3 \\ 6 & -1 & 9 \\ -4 & 3 & -11 \end{bmatrix}$. [7.5]

34. Use a matrix equation to solve the system. [7.5]
$2x + 5y = 4$
$-3x - 4y = 1$ (-3, 2)

35. Solve the system by writing and solving a matrix equation with technology. [7.5]
$3x + y - z = 16$
$x + y + 2z = 12$ $\left(6, \tfrac{2}{3}, \tfrac{8}{3}\right)$
$x - y + z = 8$

36. Write and solve a matrix equation to find three numbers whose sum is 4 if the difference of the first and the third is 1 more than twice the second, and the sum of the first two is 23 less than twice the third. [7.5] (0, -5, 9)

37. Joseph wants to place \$42,000 for his children's college fund into a savings account earning 3.5% APY, a CD earning 4% APY, and a bond earning 5.5% APY. He wants to earn \$1700 in interest next year while investing twice as much in the CD as the other two investments. Write and solve a matrix equation to find the amount that should be placed in each investment. [7.5] savings: \$9500; CD: \$28,000; bond: \$4500

Solve each system of inequalities. [7.6]

38. $y < -x + 5$
$y < \tfrac{1}{2}x - 2$

39. $y < -3|x - 4| + 5$
$y > (x - 4)^2 - 5$

Graph each system and state the coordinates of the vertices of the solution region. [7.6]

40. $1 \le y \le 2$
$x + 3y \le 10$
$-x + y \le 1$

41. $x - y \ge -3$
$2x + y \le 3$
$2x + 3y \ge -6$

42. Maximize and minimize $f(x, y) = 7x - 2y$ with the given constraints. [7.6]
$x \le 5 \qquad 2x + y \ge 6$ max: $f(5, 2) = 31$
$y \le 4 \qquad x - y \le 3$ min: $f(1, 4) = -1$

43. A grocer makes \$5.00 profit on the larger Tropical fruit basket and \$3.00 on the smaller Heartland basket. Research indicates the store will sell at most 60 fruit baskets and sales of the Heartland will be at least double the sales of the Tropical. This week the grocer has only 50 of the smaller baskets to use for the Heartland. Write the profit function and a system of constraints. Then graph the feasible region, find how many of each basket should be made to maximize the possible profit this week, and find the maximum profit. [7.6] 20 Tropical and 40 Heartland; \$220

17.
$3r_1 - 2r_2 \Rightarrow r_2 \begin{bmatrix} 2 & 1 & 2 \\ 0 & 7 & -28 \end{bmatrix}$

$\tfrac{1}{7}r_2 \Rightarrow r_2 \begin{bmatrix} 2 & 1 & 2 \\ 0 & 1 & -4 \end{bmatrix}$

$r_1 - r_2 \Rightarrow r_1 \begin{bmatrix} 2 & 0 & 6 \\ 0 & 1 & -4 \end{bmatrix}$

$\tfrac{1}{2}r_1 \Rightarrow r_1 \begin{bmatrix} 1 & 0 & 3 \\ 0 & 1 & -4 \end{bmatrix}$

18.
$\tfrac{1}{2}r_1 \Rightarrow r_1 \begin{bmatrix} 1 & 2 & 1 & -5 \\ 3 & -2 & 8 & 17 \\ 6 & 2 & -3 & 10 \end{bmatrix}$

$3r_1 - r_2 \Rightarrow r_2 \begin{bmatrix} 1 & 2 & 1 & -5 \\ 0 & 8 & -5 & -32 \\ 6r_1 - r_3 \Rightarrow r_3 & 0 & 10 & 9 & -40 \end{bmatrix}$

$\tfrac{1}{8}r_2 \Rightarrow r_2 \begin{bmatrix} 1 & 2 & 1 & -5 \\ 0 & 1 & -\tfrac{5}{8} & -4 \\ 0 & 10 & 9 & -40 \end{bmatrix}$

$-2r_2 + r_1 \Rightarrow r_1 \begin{bmatrix} 1 & 0 & \tfrac{9}{4} & 3 \\ 0 & 1 & -\tfrac{5}{8} & -4 \end{bmatrix}$

$-10r_2 + r_3 \Rightarrow r_3 \begin{bmatrix} 0 & 0 & \tfrac{61}{4} & 0 \end{bmatrix}$

$\begin{bmatrix} 1 & 0 & \tfrac{9}{4} & 3 \\ 0 & 1 & -\tfrac{5}{8} & -4 \end{bmatrix}$

$\tfrac{4}{61}r_3 \Rightarrow r_3 \begin{bmatrix} 0 & 0 & 1 & 0 \end{bmatrix}$

$-\tfrac{9}{4}r_3 + r_1 \Rightarrow r_1 \begin{bmatrix} 1 & 0 & 0 & 3 \\ \tfrac{5}{8}r_3 + r_2 \Rightarrow r_2 & 0 & 1 & 0 & -4 \\ & 0 & 0 & 1 & 0 \end{bmatrix}$

Find the partial fraction decomposition for each rational expression. [7.7]

44. $\dfrac{3x-7}{x^2-6x+9}$

45. $\dfrac{7x-32}{x^2-10x+24}$

46. $\dfrac{2x^2-x+2}{4x^3-4x^2+x}$

47. $\dfrac{5x^2-16x-12}{(x-2)(x^2-3x-2)}$

48. $\dfrac{4x^4+x^3+3x^2+x+2}{x^5+2x^3+x}$

44. $\dfrac{3}{x-3}+\dfrac{2}{(x-3)^2}$

45. $\dfrac{5}{x-6}+\dfrac{2}{x-4}$

46. $\dfrac{2}{x}-\dfrac{3}{2x-1}+\dfrac{4}{(2x-1)^2}$

47. $\dfrac{6}{x-2}-\dfrac{x}{x^2-3x-2}$

48. $\dfrac{2}{x}+\dfrac{2x+1}{x^2+1}-\dfrac{3x}{(x^2+1)^2}$

32. $\begin{bmatrix} 1 & -2 \\ 1 & -\frac{3}{2} \end{bmatrix}$

33. $\begin{bmatrix} \frac{4}{5} & \frac{1}{10} & \frac{3}{10} \\ -\frac{3}{2} & \frac{1}{2} & 0 \\ -\frac{7}{10} & \frac{1}{10} & -\frac{1}{5} \end{bmatrix}$

19.
$$\begin{bmatrix} 2 & -3 & 1 & 3 \\ 0 & 1 & 3 & 2 \\ -4 & 6 & -2 & -6 \end{bmatrix}$$
$3r_2+r_1\Rightarrow r_1\ \begin{bmatrix} 2 & 0 & 10 & 9 \\ 0 & 1 & 3 & 2 \\ & & & \end{bmatrix}$
$2r_1+r_3\Rightarrow r_3\ \begin{bmatrix} & & & \\ & & & \\ 0 & 0 & 0 & 0 \end{bmatrix}$
$\frac{1}{2}r_1\Rightarrow r_1\ \begin{bmatrix} 1 & 0 & 5 & \frac{9}{2} \\ 0 & 1 & 3 & 2 \\ 0 & 0 & 0 & 0 \end{bmatrix}$
$x+5z=\dfrac{9}{2}$
$x=\dfrac{9-10z}{2}$
$y+3z=2;\ y=2-3z$
letting $z=t$: $\left(\dfrac{9-10t}{2},\, 2-3t,\, t\right)$

20.
$$\begin{bmatrix} 1 & 1 & 1 & 7 \\ 3 & 4 & 0 & 6 \\ 2 & 0 & 1 & 2 \end{bmatrix}$$
$\begin{aligned} -3r_1+r_2&\Rightarrow r_2 \\ -2r_1+r_3&\Rightarrow r_3 \end{aligned}\ \begin{bmatrix} 1 & 1 & 1 & 7 \\ 0 & 1 & -3 & -15 \\ 0 & -2 & -1 & -12 \end{bmatrix}$
$\begin{aligned} -r_2+r_1&\Rightarrow r_1 \\ 2r_2+r_3&\Rightarrow r_3 \end{aligned}\ \begin{bmatrix} 1 & 0 & 4 & 22 \\ 0 & 1 & -3 & -15 \\ 0 & 0 & -7 & -42 \end{bmatrix}$
$-\frac{1}{7}r_3\Rightarrow r_3\ \begin{bmatrix} 1 & 0 & 4 & 22 \\ 0 & 1 & -3 & -15 \\ 0 & 0 & 1 & 6 \end{bmatrix}$
$\begin{aligned} -4r_3+r_1&\Rightarrow r_1 \\ 3r_3+r_2&\Rightarrow r_2 \end{aligned}\ \begin{bmatrix} 1 & 0 & 0 & -2 \\ 0 & 1 & 0 & 3 \\ 0 & 0 & 1 & 6 \end{bmatrix}$

21.
$$\begin{bmatrix} 1 & 4 & 2 & 7 \\ 2 & 9 & 7 & 7 \\ 0 & 1 & 3 & -5 \end{bmatrix}$$
$-2r_1+r_2\Rightarrow r_2\ \begin{bmatrix} 1 & 4 & 2 & 7 \\ 0 & 1 & 3 & -7 \\ 0 & 1 & 3 & -5 \end{bmatrix}$
$-r_2+r_3\Rightarrow r_3\ \begin{bmatrix} 1 & 4 & 2 & 7 \\ 0 & 1 & 3 & -7 \\ 0 & 0 & 0 & 2 \end{bmatrix}$

22. $12x+y+4z=690$
$10x+8z=470$
$15x+2y+10z=1325$

```
[A]
            [[12 1 4   690 ]
             [10 0 8   470 ]
             [15 2 10 1325]]
rref([A])
            [[1 0 0 15 ]
             [0 1 0 350]
             [0 0 1 40 ]]
■
```

23. $2(7)-(-3)9=41$

24. $\begin{vmatrix} 5 & -2 & 0 \\ 2 & -4 & 7 \\ -3 & 1 & 1 \end{vmatrix}\begin{matrix} 5 & -2 \\ 2 & -4 \\ -3 & 1 \end{matrix}$
$(-20+42+0)-(0+35-4)$
$=22-31=-9$

25a. $1(-2+6)-2(6+2)+0(9+1)$
$=4-16=-12$

25b. $1(-2+6)-3(4-0)+1(-4+0)$
$=4-12-4=-12$

26. using row 2:
$$-2\begin{vmatrix} 1 & -4 & 2 \\ 3 & 1 & 0 \\ -2 & 3 & -2 \end{vmatrix}+0+0$$
$$+3\begin{vmatrix} -1 & 1 & -4 \\ 5 & 3 & 1 \\ 0 & -2 & 3 \end{vmatrix}$$
using row 2 and column 1, respectively:
$=-2[-3(8-6)+1(-2+4)+0]$
$\qquad +3[-1(9+2)-5(3-8)+0]$
$=-2(-6+2)+3(-11+25)$
$=8+42=50$

27. $|A|=\begin{vmatrix} 2 & -1 \\ 4 & 2 \end{vmatrix}=8$

$|A_x|=\begin{vmatrix} 13 & -1 \\ 6 & 2 \end{vmatrix}=32$

$|A_y|=\begin{vmatrix} 2 & 13 \\ 4 & 6 \end{vmatrix}=-40$

$x=\dfrac{32}{8}=4;\ y=\dfrac{-40}{8}=-5$

continued on next page

28. $|A| = \begin{bmatrix} 2 & -1 & 0 \\ 1 & 0 & 2 \\ 0 & 1 & 3 \end{bmatrix}$

$= 2(-2) + 1(3) + 0 = -1$

$|A_x| = \begin{vmatrix} 1 & -1 & 0 \\ -1 & 0 & 2 \\ -1 & 1 & 3 \end{vmatrix}$

$= 1(-2) + 1(-1) + 0 = -3$

$|A_y| = \begin{vmatrix} 2 & 1 & 0 \\ 1 & -1 & 2 \\ 0 & -1 & 3 \end{vmatrix}$

$= 2(-1) - 1(3) + 0 = -5$

$|A_z| = \begin{vmatrix} 2 & -1 & 1 \\ 1 & 0 & -1 \\ 0 & 1 & -1 \end{vmatrix}$

$= 2(1) + 1(-1) + 1(1) = 2$

$x = \frac{-3}{-1} = 3;\ y = \frac{-5}{-1} = 5;\ z = \frac{2}{-1} = -2$

29. $Area = \pm\frac{1}{2} \begin{vmatrix} 0 & 0 & 1 \\ -2 & 5 & 1 \\ 4 & 3 & 1 \end{vmatrix}$

using row 1:

$= \pm\frac{1}{2}(0 + 0 + 1(-26))$

$= \pm\frac{1}{2}(-26) = 13 \text{ u}^2$

30. $Area_{\triangle ABC} = \pm\frac{1}{2} \begin{vmatrix} 0 & 0 & 1 \\ 2 & 4 & 1 \\ 6 & 1 & 1 \end{vmatrix}$

using row 1:

$= \pm\frac{1}{2}(1(-22)) = 11 \text{ u}^2$

$Area_{\triangle ACD} = \pm\frac{1}{2} \begin{vmatrix} 0 & 0 & 1 \\ 2 & -2 & 1 \\ 6 & 1 & 1 \end{vmatrix}$

using row 1:

$= \pm\frac{1}{2}(1(2 + 12)) = 7 \text{ u}^2$

$Area_{poly} = 11 + 7 = 18 \text{ u}^2$

31a. $AB = \begin{bmatrix} 2 & -3 \\ -1 & 2 \end{bmatrix}\begin{bmatrix} 2 & 3 \\ 1 & 2 \end{bmatrix}$

$= \begin{bmatrix} 4-3 & 6-6 \\ -2+2 & -3+4 \end{bmatrix} = \begin{bmatrix} 1 & 0 \\ 0 & 1 \end{bmatrix} = I_2$

$BA = \begin{bmatrix} 2 & 3 \\ 1 & 2 \end{bmatrix}\begin{bmatrix} 2 & -3 \\ -1 & 2 \end{bmatrix}$

$= \begin{bmatrix} 4-3 & -6+6 \\ 2-2 & -3+4 \end{bmatrix} = \begin{bmatrix} 1 & 0 \\ 0 & 1 \end{bmatrix} = I_2$

31b. $CD = \begin{bmatrix} 4 & 3 \\ 1 & 1 \end{bmatrix}\begin{bmatrix} 1 & -3 \\ -1 & 3 \end{bmatrix}$

$= \begin{bmatrix} 4-3 & -12+9 \\ 1-1 & -3+3 \end{bmatrix} = \begin{bmatrix} 1 & -3 \\ 0 & 0 \end{bmatrix}$

32. $|A| = -6 + 8 = 2$

$A^{-1} = \frac{1}{2}\begin{bmatrix} 2 & -4 \\ 2 & -3 \end{bmatrix} = \begin{bmatrix} 1 & -2 \\ 1 & -\frac{3}{2} \end{bmatrix}$

33. $[A|I] = \begin{bmatrix} 2 & -1 & 3 & | & 1 & 0 & 0 \\ 6 & -1 & 9 & | & 0 & 1 & 0 \\ -4 & 3 & -11 & | & 0 & 0 & 1 \end{bmatrix}$

$\begin{matrix} -3r_1 + r_2 \Rightarrow r_2 \\ 2r_1 + r_3 \Rightarrow r_3 \end{matrix} \begin{bmatrix} 2 & -1 & 3 & | & 1 & 0 & 0 \\ 0 & 2 & 0 & | & -3 & 1 & 0 \\ 0 & 1 & -5 & | & 2 & 0 & 1 \end{bmatrix}$

$\begin{matrix} r_1 + r_3 \Rightarrow r_1 \\ \\ -2r_3 + r_2 \Rightarrow r_3 \end{matrix} \begin{bmatrix} 2 & 0 & -2 & | & 3 & 0 & 1 \\ 0 & 2 & 0 & | & -3 & 1 & 0 \\ 0 & 0 & 10 & | & -7 & 1 & -2 \end{bmatrix}$

$\begin{matrix} \frac{1}{2}r_1 \Rightarrow r_1 \\ \frac{1}{2}r_2 \Rightarrow r_2 \\ \frac{1}{10}r_3 \Rightarrow r_3 \end{matrix} \begin{bmatrix} 1 & 0 & -1 & | & \frac{3}{2} & 0 & \frac{1}{2} \\ 0 & 1 & 0 & | & -\frac{3}{2} & \frac{1}{2} & 0 \\ 0 & 0 & 1 & | & -\frac{7}{10} & \frac{1}{10} & -\frac{1}{5} \end{bmatrix}$

$r_3 + r_1 \Rightarrow r_1 \begin{bmatrix} 1 & 0 & 0 & | & \frac{4}{5} & \frac{1}{10} & \frac{3}{10} \\ 0 & 1 & 0 & | & -\frac{3}{2} & \frac{1}{2} & 0 \\ 0 & 0 & 1 & | & -\frac{7}{10} & \frac{1}{10} & -\frac{1}{5} \end{bmatrix}$

$= [I|A^{-1}]$

34. $\begin{bmatrix} 2 & 5 \\ -3 & -4 \end{bmatrix}\begin{bmatrix} x \\ y \end{bmatrix} = \begin{bmatrix} 4 \\ 1 \end{bmatrix}$

$A^{-1} = \frac{1}{7}\begin{bmatrix} -4 & -5 \\ 3 & 2 \end{bmatrix}$

$X = A^{-1}C = \frac{1}{7}\begin{bmatrix} -4 & -5 \\ 3 & 2 \end{bmatrix}\begin{bmatrix} 4 \\ 1 \end{bmatrix}$

$= \frac{1}{7}\begin{bmatrix} -21 \\ 14 \end{bmatrix} = \begin{bmatrix} -3 \\ 2 \end{bmatrix}$

35. $\begin{bmatrix} 3 & 1 & -1 \\ 1 & 1 & 2 \\ 1 & -1 & 1 \end{bmatrix}\begin{bmatrix} x \\ y \\ z \end{bmatrix} = \begin{bmatrix} 16 \\ 12 \\ 8 \end{bmatrix}$

$X = A^{-1}C = \begin{bmatrix} 6 \\ \frac{2}{3} \\ \frac{8}{3} \end{bmatrix}$

36. $x + y + z = 4$

$x - z = 2y + 1;\ x - 2y - z = 1$

$x + y = 2z - 23;\ x + y - 2z = -23$

$\begin{bmatrix} 1 & 1 & 1 \\ 1 & -2 & -1 \\ 1 & 1 & -2 \end{bmatrix}\begin{bmatrix} x \\ y \\ z \end{bmatrix} = \begin{bmatrix} 4 \\ 1 \\ -23 \end{bmatrix}$

$X = A^{-1}C = \begin{bmatrix} 0 \\ -5 \\ 9 \end{bmatrix}$

37. $s + c + b = 42{,}000$

$0.035s + 0.04c + 0.055b = 1700$

$3.5s + 4c + 5.5b = 170{,}000$

$c = 2(s + b);\ 2s - c + 2b = 0$

$\begin{bmatrix} 1 & 1 & 1 \\ 3.5 & 4 & 5.5 \\ 2 & -1 & 2 \end{bmatrix}\begin{bmatrix} x \\ y \\ z \end{bmatrix} = \begin{bmatrix} 42{,}000 \\ 170{,}000 \\ 0 \end{bmatrix}$

$X = A^{-1}C = \begin{bmatrix} 9500 \\ 28{,}000 \\ 4500 \end{bmatrix}$

38. 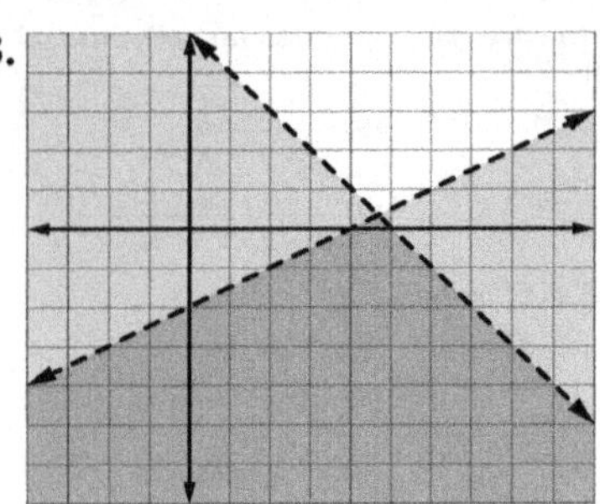

39. 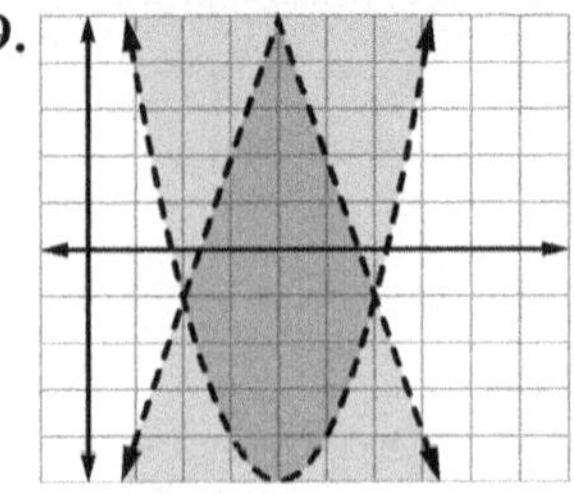

40. 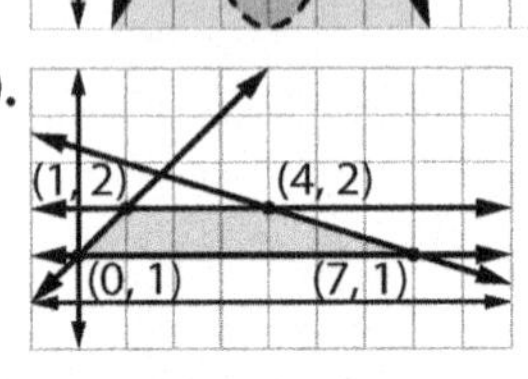

41. 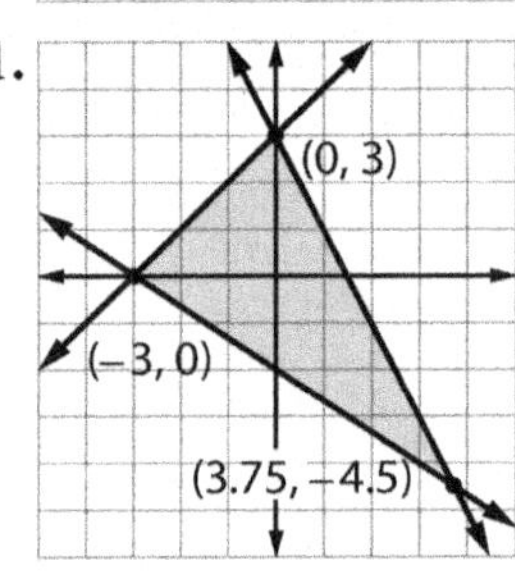

2nd and 3rd, using substitution:

$2x + 3(3 - 2x) = -6$

$-4x = -15;\ x = \frac{15}{4}$

$y = 3 - 2\left(\frac{15}{4}\right) = -\frac{9}{2}$

$\left(\frac{15}{4}, -\frac{9}{2}\right)$

42. 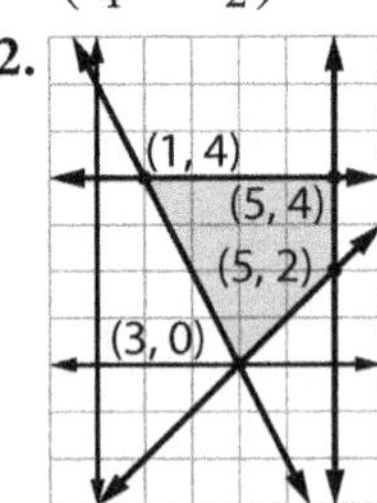

$f(3, 0) = 7(3) - 2(0) = 21$

$f(1, 4) = 7(1) - 2(4) = -1;\ \text{min}$

$f(5, 4) = 7(5) - 2(4) = 27$

$f(5, 2) = 7(5) - 2(2) = 31;\ \text{max}$

43. $P(x, y) = 5x + 3y$,
where $x =$ Tropical baskets and
$y =$ Heartland baskets
$x + y \le 60$; $y \ge 2x$; $y \le 50$

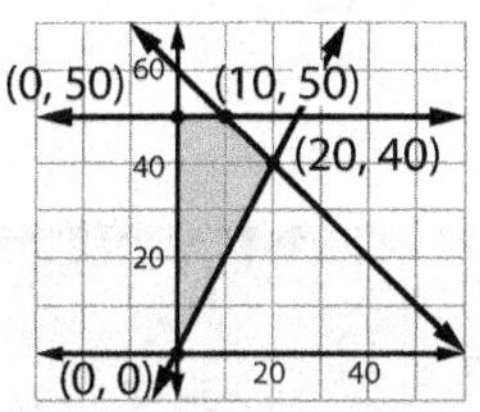

$P(0, 0) = 5(0) + 3(0) = \$0$
$P(0, 50) = 5(0) + 3(50) = \$150$
$P(10, 50) = 5(10) + 3(50) = \$200$
$P(20, 40) = 5(20) + 3(40)$
$$= \$220; \text{ max}$$

44. $\left[\dfrac{3x - 7}{x^2 - 6x + 9} = \dfrac{A}{x - 3} + \dfrac{B}{(x - 3)^2}\right](x - 3)^2$
$3x - 7 = A(x - 3) + B$
$\qquad = Ax + (-3A + B); A = 3$
$-3A + B = -7$
$-3(3) + B = -7; B = 2$

45. $\left[\dfrac{7x - 32}{(x - 6)(x - 4)} = \dfrac{A}{x - 6} + \dfrac{B}{x - 4}\right]$
$$\cdot (x - 6)(x - 4)$$
$7x - 32 = A(x - 4) + B(x - 6)$
$\qquad = Ax - 4A + Bx - 6B$
$\qquad = (A + B)x + (-4A - 6B)$
$A + B = 7$; $-4A - 6B = -32$
$4A + 6(7 - A) = 32$
$\quad -2A + 42 = 32; A = 5$
$B = 7 - (5) = 2$

46. $\left[\dfrac{2x^2 - x + 2}{4x^3 - 4x^2 + x} = \dfrac{A}{x} + \dfrac{B}{2x - 1}\right.$
$$\left. + \dfrac{C}{(2x - 1)^2}\right]x(2x - 1)^2$$
$2x^2 - x + 2$
$= A(2x - 1)^2 + B(2x^2 - x) + Cx$
$= 4Ax^2 - 4Ax + A + 2Bx^2 - Bx + Cx$
$= (4A + 2B)x^2 + (-4A - B + C)x + A$
$4A + 2B = 2$; $-4A - B + C = -1$
$A = 2$; $2B = 2 - 4(2)$; $B = -3$
$C = -1 + 4(2) + (-3) = 4$

47. $\left[\dfrac{5x^2 - 16x - 12}{(x - 2)(x^2 - 3x - 2)} = \dfrac{A}{x - 2}\right.$
$$\left. + \dfrac{Bx + C}{x^2 - 3x - 2}\right](x - 2)(x^2 - 3x - 2)$$
$5x^2 - 16x - 12$
$= A(x^2 - 3x - 2) + (Bx + C)(x - 2)$
$= Ax^2 - 3Ax - 2A + Bx^2 - 2Bx$
$$+ Cx - 2C$$
$= (A + B)x^2 + (-3A - 2B + C)x$
$$+ (-2A - 2C)$$
$A + B = 5$; $-3A - 2B + C = -16$
$-2A - 2C = -12$

```
[A]
            [[1   1   0   5  ]
             [-3  -2  1   -16]
             [-2  0   -2  -12]]
rref([A])
            [[1  0  0  6 ]
             [0  1  0  -1]
             [0  0  1  0 ]]
```

48. $\left[\dfrac{4x^4 + x^3 + 3x^2 + x + 2}{x(x^2 + 1)^2} = \dfrac{A}{x}\right.$
$$\left. + \dfrac{Bx + C}{(x^2 + 1)} + \dfrac{Dx + E}{(x^2 + 1)^2}\right]x(x^2 + 1)^2$$
$4x^4 + x^3 + 3x^2 + x + 2$
$= A(x^2 + 1)^2 + (Bx + C)(x^3 + x)$
$$+ x(Dx + E)$$
$= Ax^4 + 2Ax^2 + A + Bx^4 + Bx^2 + Cx^3$
$$+ Cx + Dx^2 + Ex$$
$= (A + B)x^4 + Cx^3 + (2A + B + D)x^2$
$$+ (C + E)x + A$$
$A + B = 4$; $C = 1$
$2A + B + D = 3$
$C + E = 1$; $A = 2$
$B = 4 - (2) = 2$
$D = 3 - 2(2) - (2) = -3$
$E = 1 - (1) = 0$

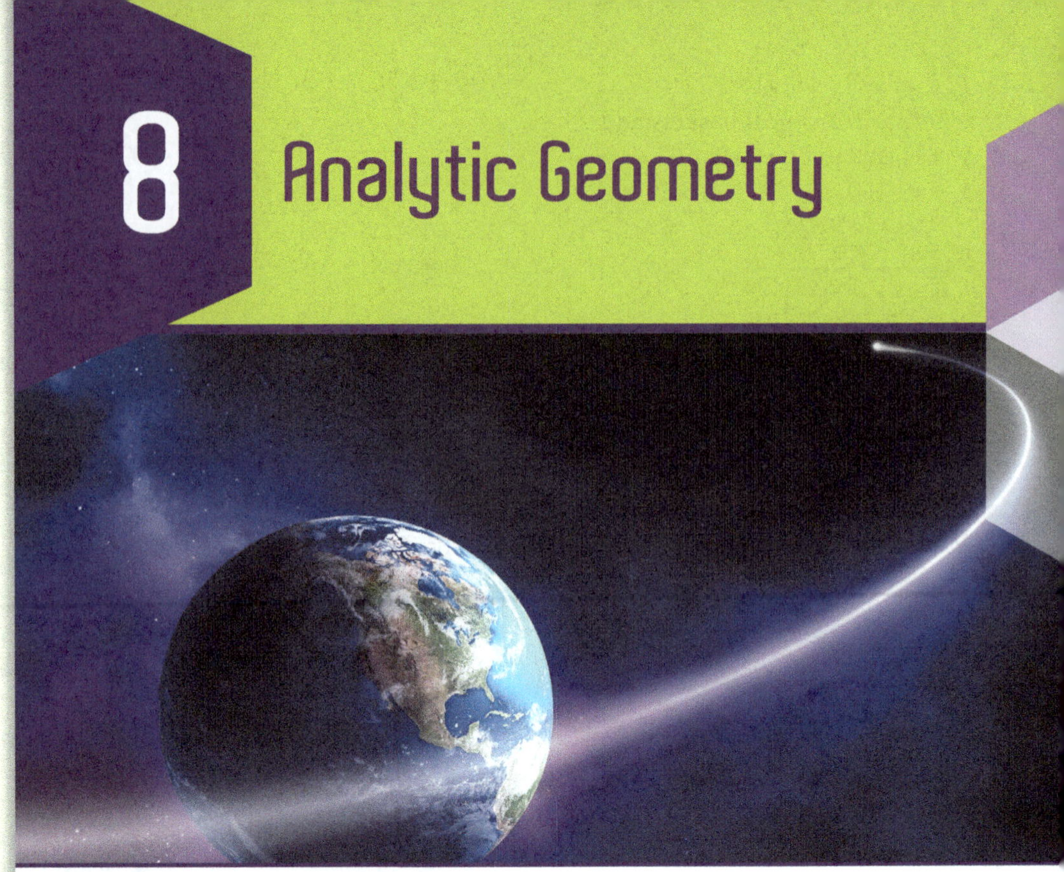

Overview

The orbits of celestial bodies can be modeled by equations representing conic sections. Analytic geometry provides a way to describe their motion through God's created universe. Students will learn to identify the standard form equations for parabolas, ellipses, circles, and hyperbolas. A solid foundation in analytic geometry is established through identifying key characteristics of the graphs and equations of conics throughout the chapter. This chapter also introduces parametric and polar equations of conics.

Flash

Our temporary home, the earth, is traveling at an average speed of approximately 67,000 mi/hr. We travel about 584 million mi/yr, and we are just one of the countless bodies constantly orbiting throughout the universe. The orbits of celestial bodies provide evidence of God's majesty and power (Ps. 19:1).

Chapter Objectives

1. To define each conic section and related terminology

2. To write the equation of a conic section given its key characteristics

3. To graph a conic section (quadratic relation) using key characteristics of its equation

4. To classify a conic section by analyzing its equation

5. To represent a conic section using parametric equations

6. To write and graph the polar equation of a conic section

7. To model and solve real-world problems involving conic sections

8. To evaluate statements about the vastness of our universe

 **HISTORICAL CONNECTION**
Like most ideas in the history of mathematics, the advent of analytic geometry required prior advancements in other areas, such as symbolic notation and algebra. Unlike most other developments, the two individuals most responsible for the origin of analytic geometry were not professional mathematicians.

 **BIBLICAL PERSPECTIVE OF MATHEMATICS**
All worldviews ultimately require faith in foundational axioms. Explore how counterexamples and existence proofs are related to a biblical worldview.

 **DATA ANALYSIS**
Conic sections are used to represent the orbits of the planets and comets within our solar system, satellites of the earth, and even interstellar objects.

Suggested Teaching Schedule

DAY	
1–2	8.1
3	8.2
4	Quiz 8A (8.1–8.2) HC: Analytic Geometry
5	8.3 TC: Graphing with Desmos®
6–7	8.4

DAY	
8	Quiz 8B (8.3–8.4) BPM: Counterexamples and Existence Proofs
9–10	8.5
11	8.6
12	Quiz 8C (8.5–8.6) DA: Conics in Space
13	Chapter 8 Review
14	Chapter 8 Test

8.1 Conics and Parabolas

The Frederick Douglass–Susan B. Anthony Memorial Bridge in Rochester, NY, has a nearly parabolic triple arch that spans 433 ft.

Apollonius unified the classical Greek knowledge of *conic sections* in his eight-volume treatise, *Conics*. The combining of geometry and algebra into analytic geometry by René Descartes and Pierre Fermat in the seventeenth century laid the foundations for the use of conics in the modeling of projectile motion by Galileo and planetary orbits by Johannes Kepler. The analytic geometry of conic sections also paved the way for Isaac Newton and Gottfried Leibniz to develop the foundations of calculus. Since Christ "is before all things, and by him all things consist" (Col. 1:17), He is the controlling and unifying force in nature. We can expect to find wonderful order in His creation. Mathematics provides many models of Christ's creative order.

A conic section (or *conic*) is formed by the intersection of a plane with a *right circular conical surface*. A *conical surface* is the union of all lines (*elements*) through the points of a *generating curve* and the *vertex*, a point not in the plane of the curve. The vertex divides the surface into two *nappes*. The generating curve of a right circular conical surface is a circle, and the surface's axis (the line connecting the circle's center to the vertex) is perpendicular to the plane of the circle.

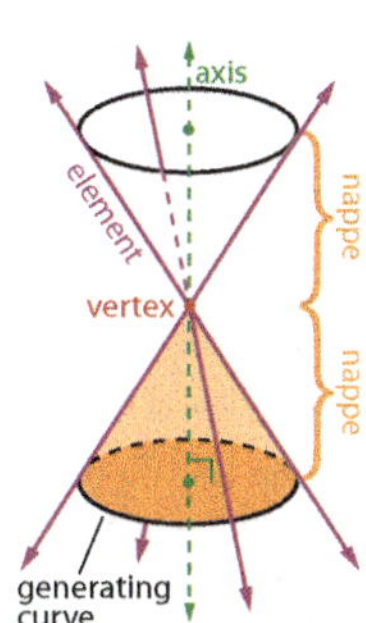

A plane can intersect this conical surface in seven ways. If the plane does not contain the vertex, the intersection is a circle, ellipse, parabola, or hyperbola. A *degenerate conic* (a point, a line, or two intersecting lines) is formed when the plane passes through the vertex.

Basic Conics

Parabola	Ellipse	Circle	Hyperbola
The plane is parallel to an element of the conical surface.	The plane intersects one nappe and is not parallel to an element.	The plane is perpendicular to the axis of the conical surface.	The plane intersects both nappes of the conical surface.

Objectives

1. To model conic sections as a section of a conical surface
2. To define a parabola as a locus of points
3. To write the equation of a parabola using its vertex, focus, and directrix
4. To graph a parabola given its equation
5. To solve real-world problems related to parabolas

After completing this section, you will be able to

- describe conics as a section of a conical surface.
- define a parabola as a locus of points.
- write the equation of a parabola.
- graph a parabola, given its equation.
- solve real-world problems related to parabolas.

Flash

Construction started on the Frederick Douglass–Susan B. Anthony Memorial Bridge in April 2004 and was completed in June 2007. The bridge carries more than 70,000 vehicles per day.

Vocabulary

chord
conical surface
conic section
directrix
element
focal length
focal width
focus
general form

generating curve
locus
nappe
parabola
right circular conical surface
tangent
vertex

PRESENTATION

Lesson Opener

Identify the type of graph that each equation represents.

1. $2x - 3y = 17$ line
2. $y = 2(x - 5)^2 + 3$ parabola
3. $x^2 + y^2 = 6$ circle

The Lesson Opener can be used to help the students recall the equations of some of the *conic sections* that they will be encountering in this chapter.

Introduce the terms related to conics. When most people think of a cone, they envision a finite object with a single *nappe*, like an ice cream cone or a traffic cone. A conic section is formed from a *conical surface* with two nappes that extend infinitely.

Use the figures to discuss the seven possible conics. Point out that a circle is a special case of an ellipse. The plane that intersects the two nappes to form a hyperbola may or may not be parallel to axis of the surface. Explain that a degeneracy occurs when a complex object can be reduced to a simpler object. A point is a degenerate circle in which the radius is 0.

Introduce the general quadratic equation that represents any conic section.

This equation will be referenced throughout the chapter. Then use the definition of a circle to introduce the *locus* definition of a conic.

Discuss the locus definition of a *parabola* and derive the standard form equation of the illustrated parabola. Some texts use $(x - h)^2 = 4p(y - k)$ as the standard form in terms of p, h, and k. Use Example 1 to illustrate how to graph a parabola in *vertex* form and how to find its vertex, *focus*, and *directrix*.

Define the *focal width* of a parabola and then extend the definition to parabolas of the form $x = a(y - k)^2 + h$, which have a horizontal axis of symmetry as

Reading and Writing Mathematics

Is the intersection of a plane and a conic surface *always*, *sometimes*, or *never* an infinite number of points? Explain.

sometimes; The intersection will always be an infinite number of points except when the plane intersects the vertex (apex).

Additional Exercises

Figures for Additional Exercises throughout the chapter can be found at TeacherToolsOnline.com.

Graph each parabola, its directrix, and its focus.

1. $y = \frac{1}{8}(x+3)^2 - 1$

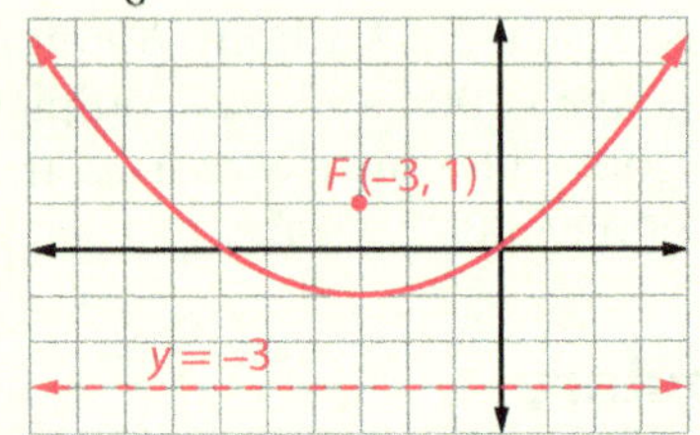

2. $x^2 - 10x + 2y + 19 = 0$

$y = -\frac{1}{2}(x-5)^2 + 3$

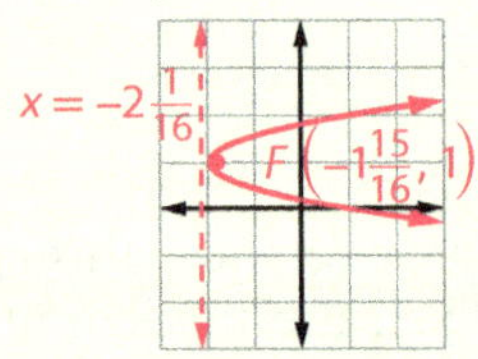

3. $x = 4(y-1)^2 - 2$

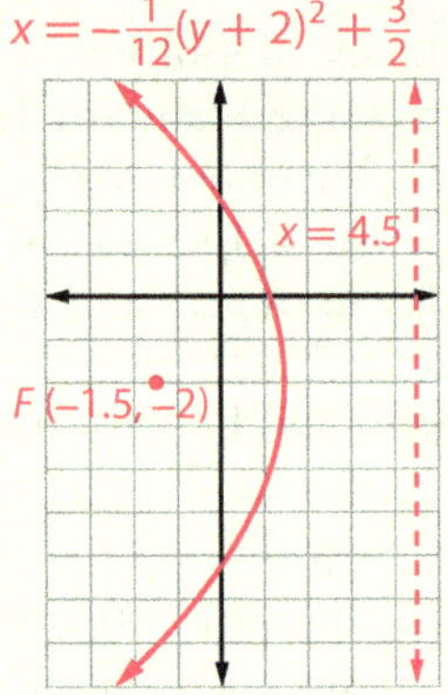

4. $y^2 + 12x + 4y - 14 = 0$

$x = -\frac{1}{12}(y+2)^2 + \frac{3}{2}$

Degenerate Conics		
Point	**Line**	**Intersecting Lines**
The plane intersects just the vertex.	The plane contains an element.	The plane contains the axis.

KEYWORD SEARCH

interactive conic sections

In analytic geometry conic sections are also described algebraically with second-degree (quadratic) equations in the form of $Ax^2 + Bxy + Cy^2 + Dx + Ey + F = 0$. If $A = B = C = 0$, the equation is a first-degree (linear) equation representing one of the degenerate conics. If $B \neq 0$, the equation represents a rotation of one of the nondegenerate conic sections. The study of rotated conic sections will be introduced in Section 8.4. The values of D, E, and F translate the curve.

A conic can also be defined as a set of points that satisfy a certain condition, or a *locus* of points. A circle has been defined as the set of points in a plane that are equidistant from a given point. This definition was used to derive the standard equation for a circle, $(x - h)^2 + (y - k)^2 = r^2$. In this chapter we will define each basic conic section as a locus of points and derive standard equations representing them, beginning with parabolas.

In Section 1.6 you graphed quadratic functions of the form $y = ax^2 + bx + c$ using the vertex form $y = a(x - h)^2 + k$. These parabolas have a vertical axis of symmetry. We will see how a parabola's locus definition is used to derive other standard form equations for parabolas with either a vertical or a horizontal axis of symmetry.

In orbital mechanics the parabola represents the path followed by a satellite which has reached escape velocity, that is, the minimum speed required to escape the gravitational pull of the attracting body.

DEFINITIONS

A **parabola** is the set of points in a plane that are equidistant from a fixed line (its **directrix**) and a fixed point (its **focus**) not on that line.

The distance d_1 from any point $P(x, y)$ on the parabola to the focus F is equal to d_2, the point's perpendicular distance to the directrix.

The parabola's axis of symmetry passes through its focus and is perpendicular to its directrix. The parabola's vertex, V, is the midpoint of the segment on the axis connecting the focus and the directrix. If p represents the directed distance from the vertex to the focus, then $|p|$ is the parabola's *focal length*.

If the vertex of the parabola is at $V(h, k)$ and represents a minimum point, the parabola opens upward, $p > 0$, the focus is $F(h, k + p)$, and the directrix is the line $y = k - p$. An equation for the parabola in terms of h, k, and p can be derived using the fact that $d_1 = d_2$.

illustrated in the table summarizing the standard forms. Have the students note that the sign of *a* will indicate which direction the parabola opens.

Encourage students to sketch the parabola from the given characteristics when writing its equation in Example 2. Emphasize that this graph is formed by the union of two functions. Complete Example 3, reviewing how to complete the square in order to obtain the standard form equation.

Common Student Error Some students may confuse *h* and *k* when working with the $x = a(y - k)^2 + h$ form. Reinforce that although *h* and *k* are interchanged, *h* is always the *x*-value of the vertex and *k* is always the *y*-value.

Introduce a *tangent* to a conic and the isosceles triangle formed by a tangent to a parabola. Then use Example 4 to demonstrate finding the equation of a line tangent to a parabola. Consider sharing that calculus provides another way of finding the equation of this tangent using derivatives.

Discuss the reflective property of a parabola. You may want to use an Internet keyword search such as *parabolic reflector animation*. Use Example 5 to show how these reflective properties are applied in radio telescopes. Ask students to list other similar applications.

$$d_1 = d_2$$

$$\sqrt{(x-h)^2 + [y-(k+p)]^2} = \sqrt{(x-x)^2 + [y-(k-p)]^2}$$

Square both sides and simplify.

$$(x-h)^2 + [y-k-p]^2 = [y-k+p]^2$$

Apply the Associative Property to group y and k.

$$(x-h)^2 + [(y-k)-p]^2 = [(y-k)+p]^2$$

Expand, keeping $(x-h)^2$ and $(y-k)$ intact.

$$(x-h)^2 + (y-k)^2 - 2p(y-k) + p^2 = (y-k)^2 + 2p(y-k) + p^2$$

Subtract $(y-k)^2$ and p^2 from each side and solve for y.

$$(x-h)^2 - 2p(y-k) = 2p(y-k)$$
$$(x-h)^2 = 4p(y-k)$$
$$\frac{1}{4p}(x-h)^2 = y-k$$
$$y = \frac{1}{4p}(x-h)^2 + k$$

This equation is equivalent to the vertex form $y = a(x-h)^2 + k$ if $a = \frac{1}{4p}$ or $p = \frac{1}{4a}$.

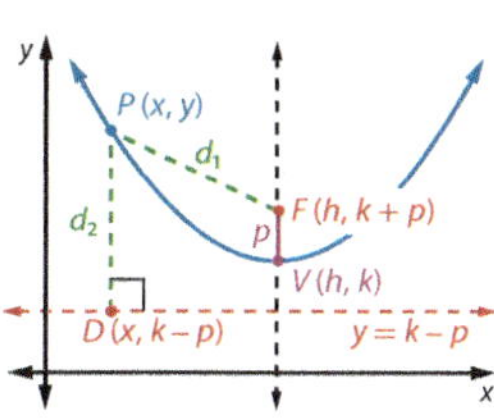

Example 1   Analyzing and Graphing a Parabola

Graph $y = -\frac{1}{2}(x+3)^2 + 2$ and its vertex, focus, and directrix.

Answer

$V(-3, 2)$

$p = \dfrac{1}{4\left(-\frac{1}{2}\right)} = -\dfrac{1}{2}$

$F\left(-3, 2 - \frac{1}{2}\right) = (-3, 1.5)$

directrix: $y = 2 + \frac{1}{2}$ or $y = 2.5$

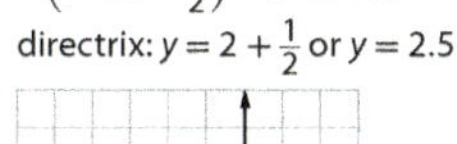
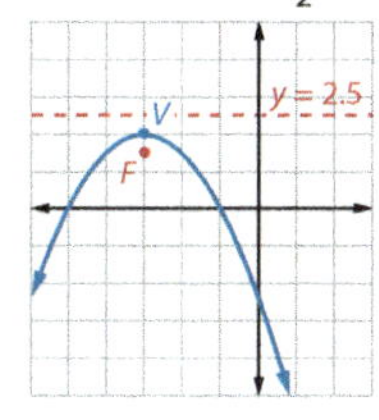

1. Identify (h, k) from the vertex form $y = a(x-h)^2 + k$.

2. Use $p = \frac{1}{4a}$ to find the directed focal length.

3. For parabolas with a vertical axis, the focus is $F(h, k+p)$, and the directrix is the line $y = k - p$.

4. Graph the parabola, using its vertex and $a = -\frac{1}{2}$. Then plot $F(-3, 1.5)$ and graph the directrix $y = 2.5$.

   Notice that the parabola opens toward its focus and away from its directrix.

SKILL ✔ EXERCISES 7, 11

A *chord of a curve* is a segment with both endpoints on the curve. The length of the chord through the parabola's focus and perpendicular to its axis defines the parabola's *focal width*, which is $|4p|$. The focal width of the parabola in Example 1 is $|4p| = \left|4\left(-\frac{1}{2}\right)\right| = 2$, the length of the chord with endpoints $(-4, 1.5)$ and $(-2, 1.5)$.

The inverse relation of a parabola of the form $y = \frac{1}{4p}x^2$ is $x = \frac{1}{4p}y^2$, a parabola with a horizontal axis and a vertex at the origin. If this parabola is translated so its vertex is at $V(h, k)$, its standard form equation is $x = \frac{1}{4p}(y-k)^2 + h$. This formula is derived in exercise 42.

Write a standard form equation for each parabola with the following characteristics. Use technology to check your answer.

5. $V(-2, 4); F(-2, 0)$
$y = -\frac{1}{16}(x+2)^2 + 4$

6. $V(-3, -2)$; directrix: $x = -2.5$
$x = -\frac{1}{2}(y+2)^2 - 3$

Find the equation of the line tangent to each parabola at the point given.

7. $x^2 + 2y - 4 = 0, P(-2, 0)$ $y = 2x + 4$

8. $y^2 + 6x - 30 = 0, P(-1, 6)$
$y = -\frac{1}{2}x + \frac{11}{2}$

9. Find the equation for a parabola passing through $(13, 6)$ with a horizontal axis and a vertex of $(-5, 3)$.
$x = 2(y-3)^2 - 5$

10. The reflectors in flashlights are in the shape of a paraboloid. If the reflector has a diameter of 6 cm and a depth of 2 cm, how far should the bulb be positioned in front of the vertex to give the most concentrated beam of light? 1.125 cm

For Graphing Calculators

Graph each parabola and its directrix.

1. $y = 0.2518(x - 2.44)^2 - 1.36$
$y = -2.35$

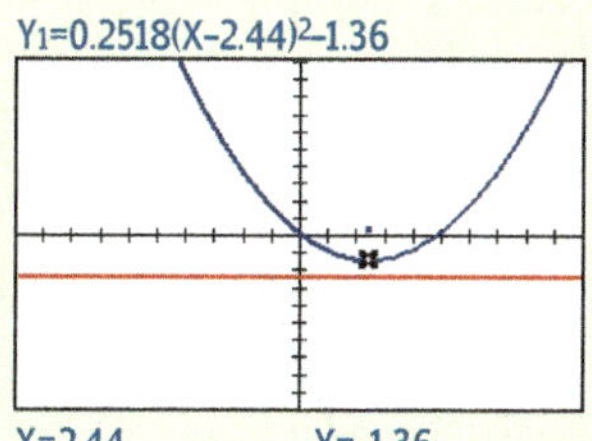

To graph horizontal parabolas, follow the instructions found in the check of Example 2. To draw a vertical directrix, from the home screen enter Vertical 1.5 by accessing the draw menu (2nd, PRGM), press 4, enter (1.5) and then press ENTER.

2. $x = 0.125(y - 5.8)^2 + 3.5$ $x = 1.5$

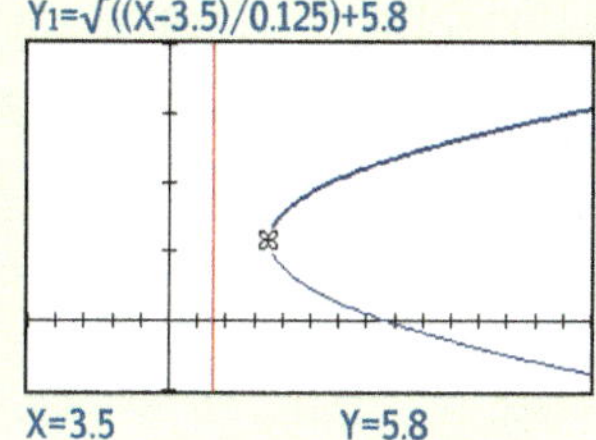

Inequalities such as those found in exercises 36–39 on page 401 can be graphed on a calculator. Enter the boundary's equation in the equation editor, move the cursor to the left of the function (Y1 or Y2), and press $\boxed{\text{ENTER}}$ to select the line style ▜ for greater than or ▟ for less than.

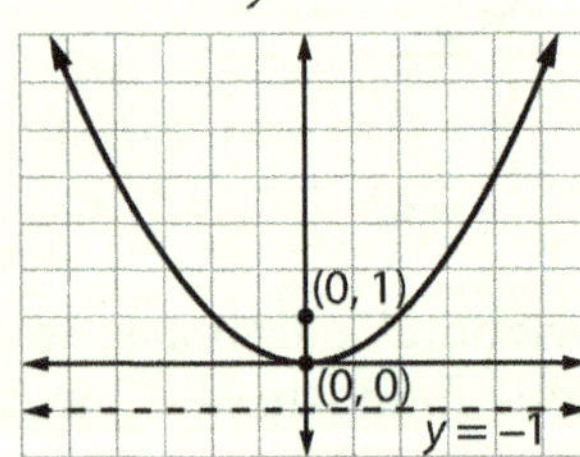

Assignments

- **Minimum:** 1–6, 9–11, 13–15, 17, 20–21; 29–33 odd; 44–46, 48, 51, 53
- **Standard:** 1–2, 4, 7, 9–12, 15, 17–18, 20, 23, 25; 28–36 even; 42–50 even; 53
- **Extended:** 1, 4, 7–9, 12, 15, 18, 20, 23, 26, 29–31, 33–35, 38–40, 42–43, 45; 48–52 even; 53

Solutions

❯ A. Exercises

2. $V(0, 0)$; opens up

3. $V(0, 0)$; opens left

4. $V(1, 2)$; opens right

5. $V(-1, 2)$; opens down

6. $V(0, 0)$; $p = \dfrac{1}{4\left(\frac{1}{4}\right)} = 1$

$F(0, 0 + 1) = (0, 1)$

directrix: $y = 0 - 1 = -1$

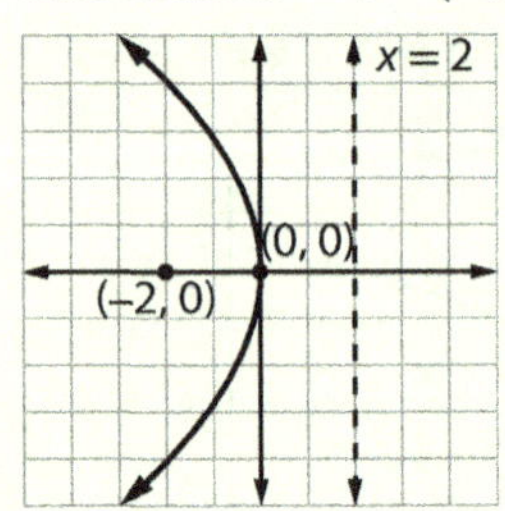

7. $V(0, 0)$; $p = \dfrac{1}{4\left(-\frac{1}{8}\right)} = -2$

$F(0 - 2, 0) = (-2, 0)$

directrix: $x = 0 - (-2) = 2$

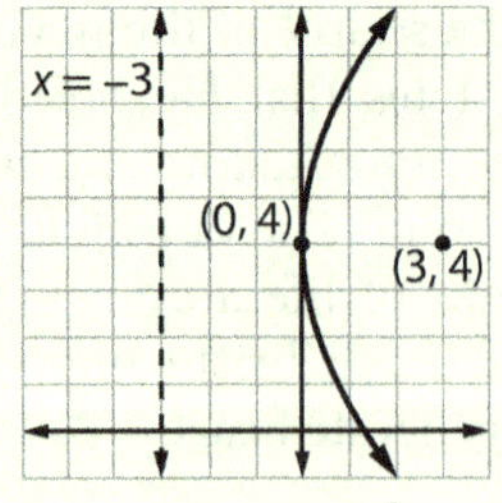

8. $V(0, 4)$; $p = \dfrac{1}{4\left(\frac{1}{12}\right)} = 3$

$F(0 + 3, 4) = (3, 4)$

directrix: $x = 0 - 3 = -3$

9. $V(1, 6)$; $p = \dfrac{1}{4\left(-\frac{1}{2}\right)} = -\dfrac{1}{2}$

$F\left(1, 6 - \dfrac{1}{2}\right) = (1, 5.5)$

directrix: $y = 6 - \left(-\dfrac{1}{2}\right) = 6.5$

10. $V(-2, 3)$; $p = \dfrac{1}{4\left(-\frac{1}{4}\right)} = -1$

$F(-2 + 1, 3) = (-3, 3)$

directrix: $x = -2 - (-1) = -1$

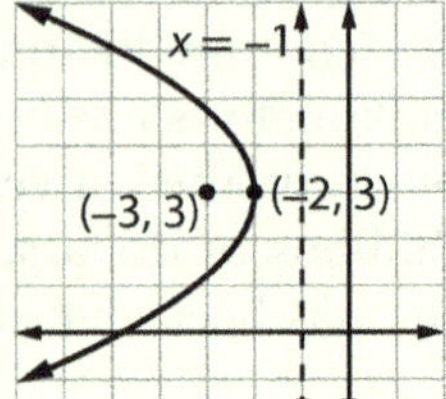

Standard Form Equations of a Parabola
with vertex $V(h, k)$ and focal length $|p|$ where $a = \dfrac{1}{4p}$

$y = \dfrac{1}{4p}(x - h)^2 + k$	$x = \dfrac{1}{4p}(y - k)^2 + h$
focus: $F(h, k + p)$; axis: $x = h$; directrix: $y = k - p$	focus: $F(h + p, k)$; axis: $y = k$; directrix: $x = h - p$

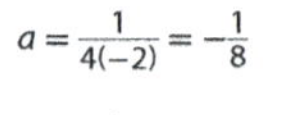

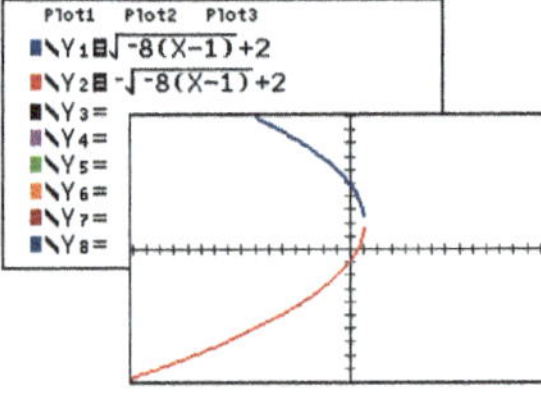

Example 2   Writing the Equation of a Parabola

Write the standard form equation of the parabola with focus $(-1, 2)$ and directrix $x = 3$.

Answer

$V\left(\dfrac{-1 + 3}{2}, \dfrac{2 + 2}{2}\right) = (1, 2)$

$p = -2$

$a = \dfrac{1}{4(-2)} = -\dfrac{1}{8}$

$x = -\dfrac{1}{8}(y - 2)^2 + 1$

Check

$x - 1 = -\dfrac{1}{8}(y - 2)^2$

$-8(x - 1) = (y - 2)^2$

$y - 2 = \pm\sqrt{-8(x - 1)}$

$y = \pm\sqrt{-8(x - 1)} + 2$

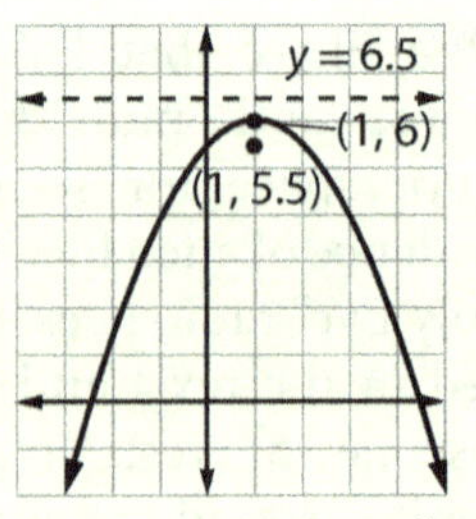

1. Plot the focus and draw the directrix.
2. Locate the vertex halfway between the focus and the directrix.
3. Sketch the parabola curving around the focus.
4. Use the sketch to determine p, and use $a = \dfrac{1}{4p}$ to find the value of a.
5. Substitute the values of a, k, and h into the standard form equation $x = a(y - k)^2 + h$.
6. Solve the equation for y.
7. Since the parabola is not a function, enter the two functions whose union forms the parabola, and graph them in a square window to avoid distortion. Your calculator may not plot the nearly vertical portions of the graphs.

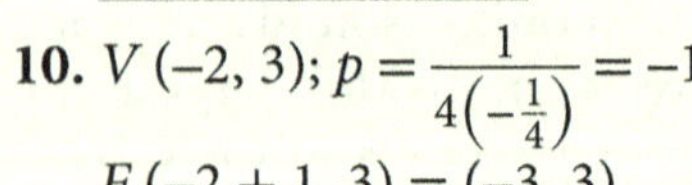

Apollonius's definition of a parabola is a one-sentence paragraph of over 200 words. When manipulating formulas seems difficult, imagine trying to perform the same operations using only words, with no symbols at all.

SKILL ✔ **EXERCISE 15**

When the equation of a parabola is written in the *general form* for conics,

$$Ax^2 + Bxy + Cy^2 + Dx + Ey + F = 0,$$

parabolas with a vertical axis have the form $Ax^2 + Dx + Ey + F = 0$ (with $B = C = 0$), and parabolas with a horizontal axis have the form $Cy^2 + Dx + Ey + F = 0$ (with $A = B = 0$). Converting these equations to a standard form for parabolas frequently involves the process of completing a square.

Example 3 Analyzing and Graphing a Parabola

Graph $2y^2 - x + 4y - 1 = 0$ and its focus and directrix.

Answer

$x = 2y^2 + 4y - 1$

   1. Since the equation has a y^2-term but no x^2-term, solve for x.

$x = 2(y^2 + 2y) - 1$

   2. Complete the square for the y-terms.

$\quad = 2(y^2 + 2y + 1^2) - 1 - 2(1)$
$x = 2(y + 1)^2 - 3$

$p = \dfrac{1}{4(2)} = \dfrac{1}{8}$

   3. Use $p = \dfrac{1}{4a}$ to find the directed focal length.

   4. Use the vertex $V(-3, -1)$ and $a = 2$ to graph the parabola.

   5. Plot the focus $\dfrac{1}{8}$ unit to the right of the vertex and the directrix $\dfrac{1}{8}$ unit to its left.

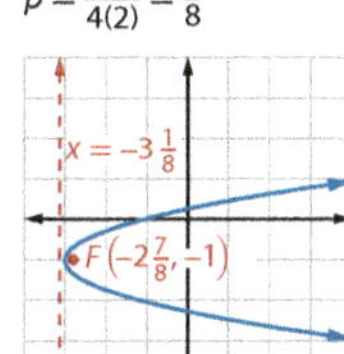
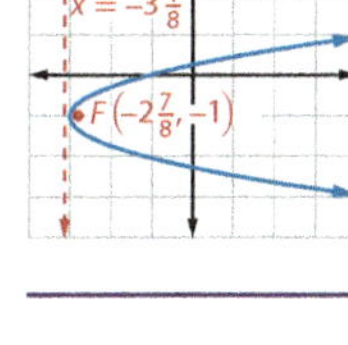

SKILL ✔ EXERCISE 21

A tangent to a conic intersects the conic at just one point. The tangent to a parabola at point P defines an isosceles triangle whose vertex is the parabola's focus F and base is $\overline{PA}$ where A is the intersection of the tangent and the parabola's axis. This property allows us to find the equation of the tangent drawn at a given point on the parabola.

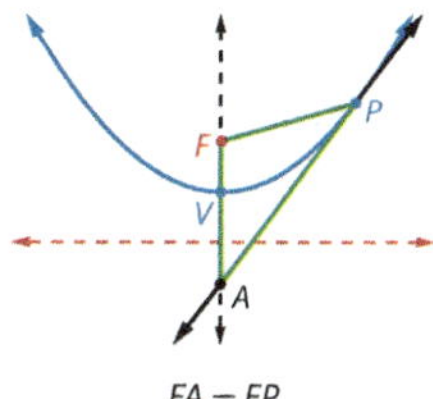

Example 4 Finding a Tangent to a Parabola

Find the equation of the line tangent to $y = -\dfrac{1}{2}x^2$ at $P(-2, -2)$.

Answer

$p = \dfrac{1}{4\left(-\frac{1}{2}\right)} = -\dfrac{1}{2}, \therefore F\left(0, -\dfrac{1}{2}\right)$

   1. Note that the vertex is at the origin and use $p = \dfrac{1}{4a}$ to find the parabola's focus.

$FP = \sqrt{(-2 - 0)^2 + \left(-2 + \dfrac{1}{2}\right)^2}$

   2. Find the distance from F to P.

$\quad = \sqrt{4 + \dfrac{9}{4}} = \sqrt{\dfrac{25}{4}} = \dfrac{5}{2}$

$A\left(0, -\dfrac{1}{2} + \dfrac{5}{2}\right) = (0, 2)$

   3. Use $FA = FP$ to locate A on the parabola's axis.

$m = \dfrac{-2 - 2}{-2 - 0} = 2$

   4. Find the slope of the tangent, using $P(-2, -2)$ and $A(0, 2)$.

$y = 2x + 2$

   5. Use $y = mx + b$ to write the equation of the tangent.

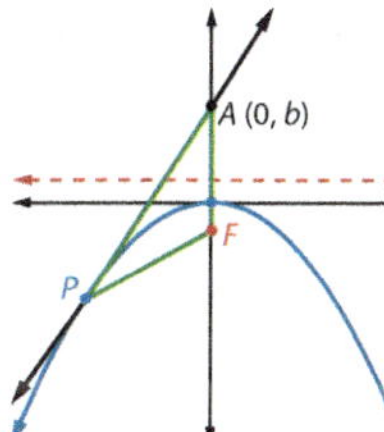

SKILL ✔ EXERCISE 25

11. $V(-3, 2); \; p = \dfrac{1}{4\left(\frac{1}{16}\right)} = 4$

$F(-3 + 4, 2) = (1, 2)$

directrix: $x = -3 - 4 = -7$

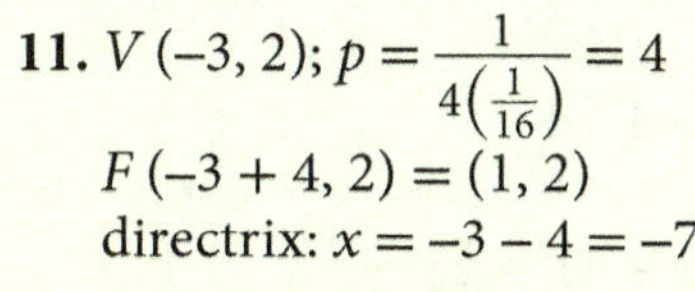
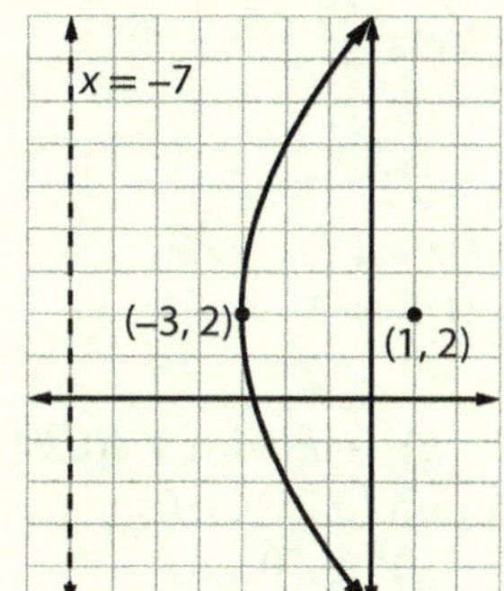

12. $V(2, 1); \; p = \dfrac{1}{4\left(\frac{1}{6}\right)} = \dfrac{3}{2}$

$F\left(2, 1 + \dfrac{3}{2}\right) = (2, 2.5)$

directrix: $y = 1 - \dfrac{3}{2} = -0.5$

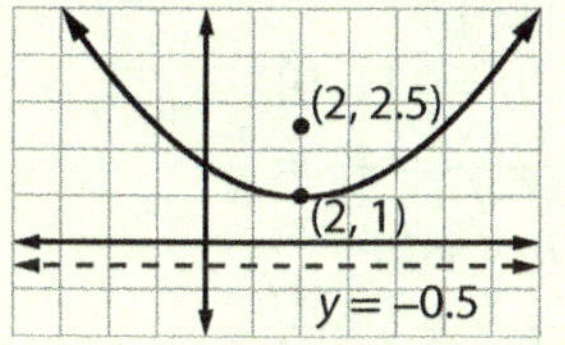

13. $V(2, -2); \; p = \dfrac{1}{4\left(\frac{1}{8}\right)} = 2$

$F(2, -2 + 2) = (2, 0)$

directrix: $y = -2 - 2 = -4$

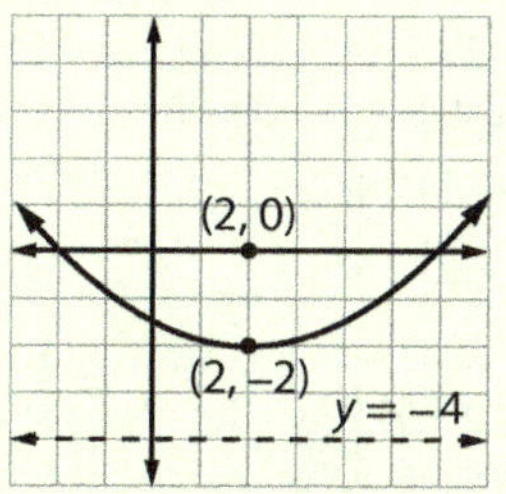

14. $p = -4; \; a = \dfrac{1}{4(-4)} = -\dfrac{1}{16}$

$x = -\dfrac{1}{16}y^2$

15. $p = 5; \; a = \dfrac{1}{4(5)} = \dfrac{1}{20}$

$y = \dfrac{1}{20}(x - 4)^2 - 6$

16. $p = -\dfrac{1}{2}; \; a = \dfrac{1}{4\left(-\frac{1}{2}\right)} = -\dfrac{1}{2}$

$y = -\dfrac{1}{2}(x + 1)^2 - 6$

17. $p = 3; \; a = \dfrac{1}{4(3)} = \dfrac{1}{12}$

$x = \dfrac{1}{12}(y - 3)^2 + 5$

18. $V(-1, 3)$

$p = -2; \; a = \dfrac{1}{4(-2)} = -\dfrac{1}{8}$

$y = -\dfrac{1}{8}(x + 1)^2 + 3$

19. $V\left(\dfrac{-\frac{5}{4} + \left(-\frac{3}{4}\right)}{2}, \dfrac{-3 + (-3)}{2}\right) = (-1, -3)$

$p = \dfrac{1}{4}; \; a = \dfrac{1}{4\left(\frac{1}{4}\right)} = 1$

$x = (y + 3)^2 - 1$

❯ B. Exercises

20. $x = (y^2 - 8y) + 18$

$x = (y^2 - 8y + 16) + 18 - 16$

$x = (y - 4)^2 + 2$

$p = \dfrac{1}{4(1)} = \dfrac{1}{4}$

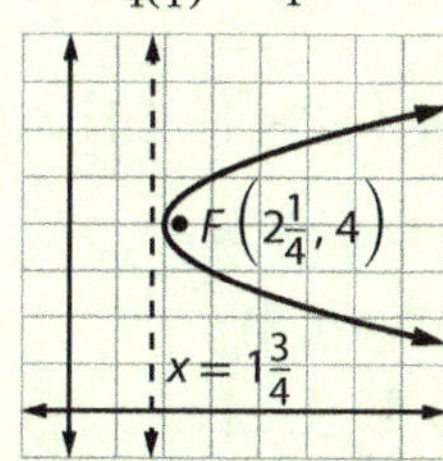

21. $-8y = (x^2 - 10x) + 9$

$-8y = (x^2 - 10x + 25) + 9 - 25$

$-8y = (x - 5)^2 - 16$

$y = -\dfrac{1}{8}(x - 5)^2 + 2$

$p = \dfrac{1}{4\left(-\frac{1}{8}\right)} = -2$

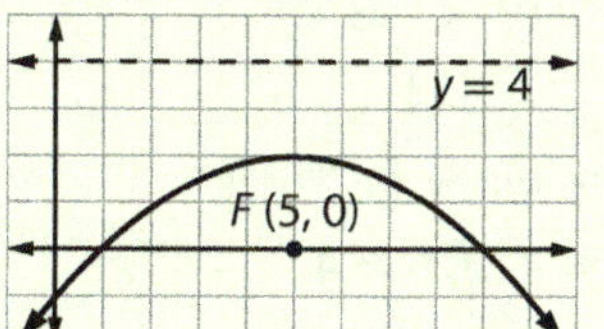

8.1 CONICS AND PARABOLAS   399

22. $-2x = (y^2 - 2y) - 6$

$-2x = (y^2 - 2y + 1) - 6 - 1$

$x = -\frac{1}{2}(y - 1)^2 + \frac{7}{2}$

$p = \dfrac{1}{4\left(-\frac{1}{2}\right)} = -\dfrac{1}{2}$

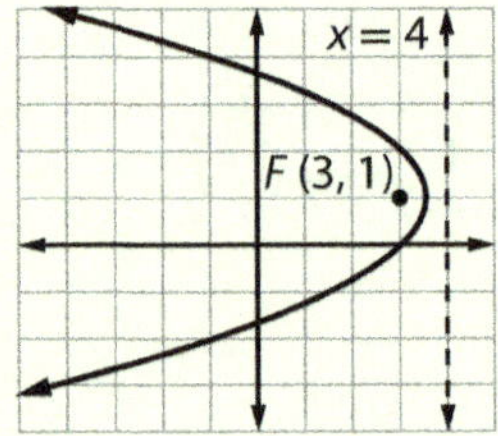

23. $4y = (x^2 + 8x) + 20$

$4y = (x^2 + 8x + 16) + 20 - 16$

$y = \frac{1}{4}(x + 4)^2 + 1$

$p = \dfrac{1}{4\left(\frac{1}{4}\right)} = 1$

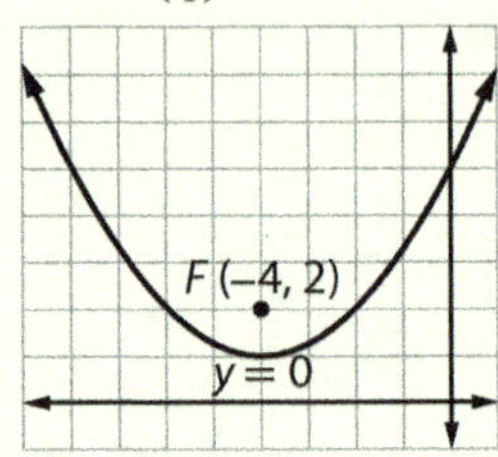

24. $y = \frac{1}{4}x^2$

$V(0, 0); \ p = \dfrac{1}{4\left(\frac{1}{4}\right)} = 1$

$F(0, 1); \ FP = 2$

$A(0, 1 - 2) = (0, -1)$

$m_{PA} = \dfrac{-1 - 1}{0 - (-2)} = \dfrac{-2}{2} = -1$

$y = -x - 1$

25. $y = -\frac{1}{2}x^2; \ V(0, 0)$

$p = \dfrac{1}{4\left(-\frac{1}{2}\right)} = -\dfrac{1}{2}; \ F\left(0, -\dfrac{1}{2}\right)$

$FP = \sqrt{(4 - 0)^2 + \left(-8 - \left(-\frac{1}{2}\right)\right)^2}$

$= \sqrt{16 + \dfrac{225}{4}} = \sqrt{\dfrac{289}{4}} = \dfrac{17}{2}$

$A\left(0, -\frac{1}{2} + \frac{17}{2}\right) = (0, 8)$

$m_{PA} = \dfrac{8 - (-8)}{0 - 4} = \dfrac{16}{-4} = -4$

$y = -4x + 8$

26. $V(0, 0)$

$p = \dfrac{1}{4\left(\frac{1}{8}\right)} = 2; \ F(2, 0)$

$FP = \sqrt{(2 - 8)^2 + (0 - 8)^2}$

$= \sqrt{100} = 10$

$A(2 - 10, 0) = (-8, 0)$

$m_{PA} = \dfrac{0 - 8}{-8 - 8} = \dfrac{1}{2}$

using $y = mx + b$:

$0 = \frac{1}{2}(-8) + b; \ b = 4$

$y = \frac{1}{2}x + 4$

Satellite dishes, automobile headlights, and radio telescopes all take advantage of another property of parabolas. Signals emitted from a parabola's focus are reflected parallel to its axis, and all the signals entering the parabola parallel to the axis are reflected to the focus.

Example 5 Modeling with Parabolas

The "Heavenly Eye" radio telescope in Guizhou province, China, has a diameter of 500 m. The paraboloid reflects incoming radio waves to a receiver 140 m above the reflective surface. Write an equation modeling a cross section through the center of the telescope and use it to find the height of the reflective surface.

Answer

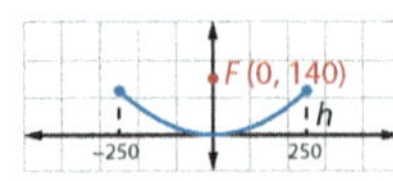

$a = \dfrac{1}{4(140)} = \dfrac{1}{560}$

$y = \dfrac{1}{560}x^2$

$y = \dfrac{1}{560}(\pm 250)^2 \approx 111.6$ m

1. Sketch a cross section of the telescope and its focus. Place the vertex at the origin to simplify the problem and use $p = 140$. Since the diameter is 500 m, the parabola's domain is [−250, 250].

2. Use $a = \dfrac{1}{4p}$ to find a, and then write the equation modeling the cross section.

3. Find y when $x = \pm 250$.

_______________ SKILL ✔ EXERCISE 33

❯ A. Exercises

1. Match each description of the plane intersecting a right circular conical surface to its basic conic section.

  a. intersects both nappes of the conical surface III

  b. parallel to an element of the conical surface IV

  c. intersects one nappe and is not parallel to an element II

  d. perpendicular to the axis of the conical surface I

  I. circle

  II. ellipse

  III. hyperbola

  IV. parabola

Match each equation with its corresponding graph.

2. $y = \frac{1}{2}x^2$ C

3. $x = -\frac{1}{2}y^2$ E

4. $x = \frac{1}{2}(y - 2)^2 + 1$ B

5. $y = -\frac{1}{4}(x + 1)^2 + 2$ D

A. 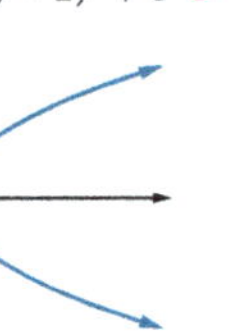

B. 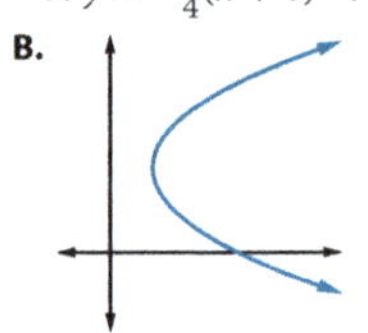

C.

D.

E.

F.

Graph each parabola and its vertex, focus, and directrix.

6. $y = \frac{1}{4}x^2$

7. $x = -\frac{1}{8}y^2$

8. $x = \frac{1}{12}(y - 4)^2$

9. $y = -\frac{1}{2}(x - 1)^2 + 6$

10. $x = -\frac{1}{4}(y - 3)^2 - 2$

11. $x = \frac{1}{16}(y - 2)^2 - 3$

12. $y = \frac{1}{6}(x - 2)^2 + 1$

13. $y = \frac{1}{8}(x - 2)^2 - 2$

---

27. $x = -\frac{1}{4}y^2 + 2; \ V(2, 0)$

$p = \dfrac{1}{4\left(-\frac{1}{4}\right)} = -1; \ F(1, 0)$

$FP = \sqrt{(-7 - 1)^2 + (-6 - 0)^2}$

$= \sqrt{100} = 10$

$A(1 + 10, 0) = (11, 0)$

$m_{PA} = \dfrac{0 - (-6)}{11 - (-7)} = \dfrac{6}{18} = \dfrac{1}{3}$

using $y = mx + b$:

$0 = \frac{1}{3}(11) + b; \ b = -\dfrac{11}{3}$

$y = \frac{1}{3}x - \dfrac{11}{3}$

28. $x = a(y - 5)^2 + 4$

$0 = a(9 - 5)^2 + 4$

$-4 = 16a$

$a = -\dfrac{1}{4}$

29. $y = a(x - 4)^2 + 1$

$4 = a(5 - 4)^2 + 1$

$a = 3$

30. $y = a(x + 1)^2 - 2$

$-5 = a(5 + 1)^2 - 2$

$3 = 36a$

$a = -\dfrac{1}{12}$

31. $x = a(y - 3)^2 - 4$

$-2 = a(2 - 3)^2 - 4$

$a = 2$

32. Let $y = a(x - h)^2 + k$ with $V(0, 70)$ passing through $(\pm 216.5, 0)$.

$0 = a(216.5 - 0)^2 + 70$

$a = \dfrac{-70}{(216.5)^2} \approx -0.0015$

$y = -0.0015x^2 + 70$

Write a standard form equation for each parabola with the given characteristics. Check your answer by using technology to graph the parabola.

14. $V(0, 0); F(-4, 0)$ $x = -\frac{1}{16}y^2$

15. $V(4, -6); F(4, -1)$ $y = \frac{1}{20}(x - 4)^2 - 6$

16. $V(-1, -6)$; directrix: $y = -5.5$ $y = -\frac{1}{2}(x + 1)^2 - 6$

17. $V(5, 3)$; directrix: $x = 2$ $x = \frac{1}{12}(y - 3)^2 + 5$

18. $F(-1, 1)$; directrix: $y = 5$ $y = -\frac{1}{8}(x + 1)^2 + 3$

19. $F\left(-\frac{3}{4}, -3\right)$; directrix: $x = -\frac{5}{4}$ $x = (y + 3)^2 - 1$

> **B. Exercises**

Write each equation in standard form. Then graph the parabola, its focus, and its directrix.

20. $y^2 - x - 8y + 18 = 0$ **21.** $x^2 - 10x + 8y + 9 = 0$

22. $y^2 + 2x - 2y - 6 = 0$ **23.** $x^2 + 8x - 4y + 20 = 0$

Write the equation of the tangent to the parabola at the given point P.

24. $x^2 = 4y, P(-2, 1)$ **25.** $x^2 + 2y = 0, P(4, -8)$

26. $x = \frac{1}{8}y^2, P(8, 8)$ **27.** $y^2 + 4x = 8, P(-7, -6)$

Use the vertex and the other point on the parabola to find a, then write the standard form equation of the parabola.

28.

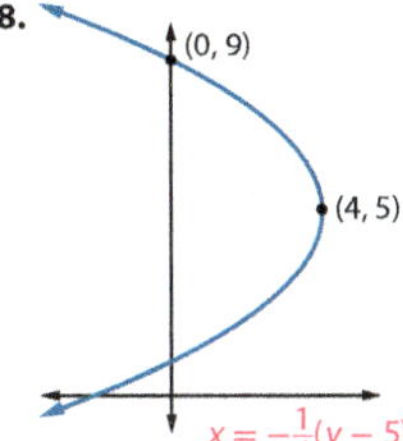

29.
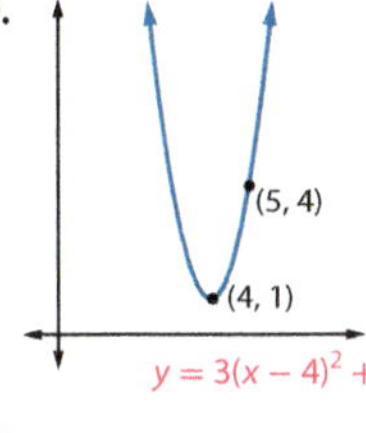

30. the parabola passing through $(5, -5)$ with a vertical axis and a vertex at $(-1, -2)$ $y = -\frac{1}{12}(x + 1)^2 - 2$

31. the parabola passing through $(-2, 2)$ with a horizontal axis and a vertex at $(-4, 3)$ $x = 2(y - 3)^2 - 4$

32. The 433 ft center span of the Frederick Douglass–Susan B. Anthony Memorial Bridge in Rochester, NY, is supported by triple arches that rise 70 ft above the highway. Write an equation of a parabola modeling the arches. $y = -0.0015x^2 + 70$

33. A parabolic satellite dish for a cable company has a diameter of 8 ft and a depth of 3 ft. Find the parabolic equation that models a cross section through the center of the satellite and determine the location of the receiver (its focal point). $y = \frac{3}{16}x^2$; 16 in. in front of the center

---

34. A solar furnace uses a parabolic reflector to direct the light to the focal point. If the reflector is 10 m high and has a focal length of 3.2 m, find the depth of the reflector. ≈ 1.95 m

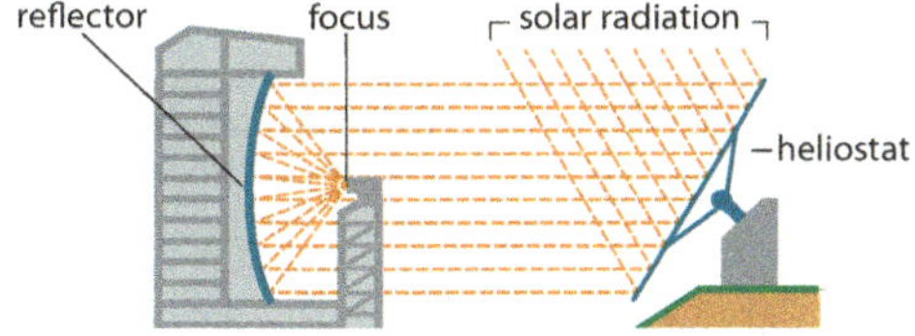

35. Which of the following statements are true? List all correct answers.   B, E

  A. A tangent to a conic can intersect that conic at more than one point.

  B. Conic sections are described algebraically with second-degree equations in the form of $Ax^2 + Bxy + Cy^2 + Dx + Ey + F = 0$.

  C. A conic section is formed by the intersection of a line and a conical surface.

  D. The focal width is the distance from the vertex to the focus.

  E. Degenerate conics include a point, a line, and two intersecting lines.

> **C. Exercises**

Graph the solution of each system of inequalities.

36. $y \leq -(x - 5)^2 + 5$
$y \geq (x - 5)^2 - 3$

37. $y \leq -2(x - 1)^2 + 5$
$x \leq -(y - 1)^2 + 5$

38. $x^2 + y^2 \leq 36$
$x \geq \frac{1}{2}y^2 - 8$
$3y - x^2 + 12 \geq 0$

39. $x^2 + 12x + y^2 + 10y \leq -52$
$y^2 - x + 10y + 19 \geq 0$
$y^2 + x + 10y + 31 \geq 0$

Find the equation of each tangent to the parabola at the given point P.

40. $x^2 + 16x + 8y + 8 = 0, P(-4, 5)$ $y = -x + 1$

41. $y^2 - 4x - 4y - 12 = 0, P(5, 8)$ $y = \frac{1}{3}x + \frac{19}{3}$

24. $y = -x - 1$ **26.** $y = \frac{1}{2}x + 4$

25. $y = -4x + 8$ **27.** $y = \frac{1}{3}x - \frac{11}{3}$

---

39. $(x^2 + 12x) + (y^2 + 10y) \leq -52$
$(x + 6)^2 + (y + 5)^2 \leq -52 + 36 + 25$
$(x + 6)^2 + (y + 5)^2 \leq 9$

$(y^2 + 10y) + 19 \geq x$
$(y + 5)^2 - 25 + 19 \geq x$
$x \leq (y + 5)^2 - 6$

$(y^2 + 10y) + 31 \geq -x$
$(y + 5)^2 - 25 + 31 \geq -x$
$x \geq -(y + 5)^2 - 6$

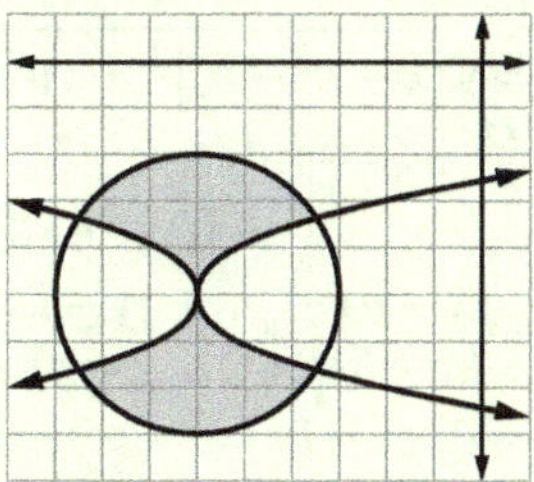

40. $8y = -(x^2 + 16x) - 8$
$8y = -(x^2 + 16x + 64) - 8 + 64$
$8y = -(x + 8)^2 + 56$
$y = -\frac{1}{8}(x + 8)^2 + 7$
$V(-8, 7); F(-8, 5)$

$FP = \sqrt{(-4 - (-8))^2 + (5 - 5)^2}$
$\quad = \sqrt{16 + 0} = 4$
$A(-8, 5 + 4) = (-8, 9)$

$m_{PA} = \frac{9 - 5}{-8 - (-4)} = \frac{4}{-4} = -1$

using $y = mx + b$:
$5 = -(-4) + b; b = 1$
$y = -x + 1$

41. $(y^2 - 4y) = 4x + 12$
$4x + 12 = (y^2 - 4y + 4) - 4$
$4x = (y - 2)^2 - 16$
$x = \frac{1}{4}(y - 2)^2 - 4$
$V(-4, 2); F(-3, 2)$

$FP = \sqrt{(5 - (-3))^2 + (8 - 2)^2}$
$\quad = \sqrt{64 + 36} = 10$
$A(-3 - 10, 2) = (-13, 2)$

$m_{PA} = \frac{2 - 8}{-13 - 5} = \frac{-6}{-18} = \frac{1}{3}$

using $y = mx + b$:
$8 = \frac{1}{3}(5) + b; b = \frac{19}{3}$
$y = \frac{1}{3}x + \frac{19}{3}$

---

33. Let $y = ax^2$ with $V(0, 0)$.
$3 = a(4)^2; a = \frac{3}{16}$
$y = \frac{3}{16}x^2$
$p = \frac{1}{4\left(\frac{3}{16}\right)} = \frac{4}{3}$ ft $= 16$ in.

34. Let $x = \frac{1}{4p}y^2$ with $V(0, 0)$.
$x = \frac{1}{4(3.2)}(5^2) \approx 1.95$ m

> **C. Exercises**

36.
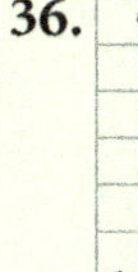
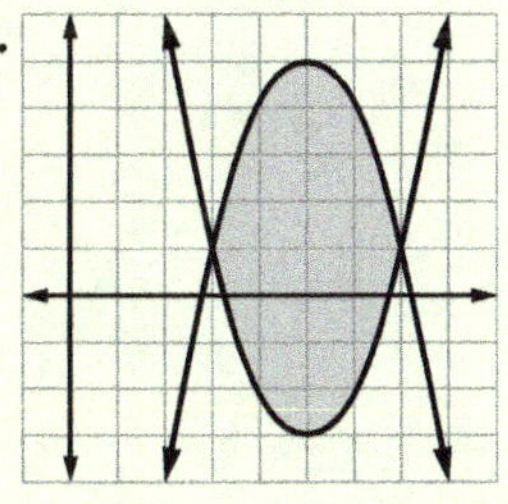

37.
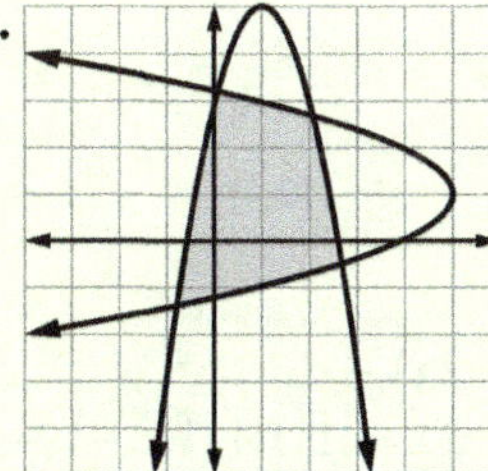

38. $3y \geq x^2 - 12$
$y \geq \frac{1}{3}x^2 - 4$

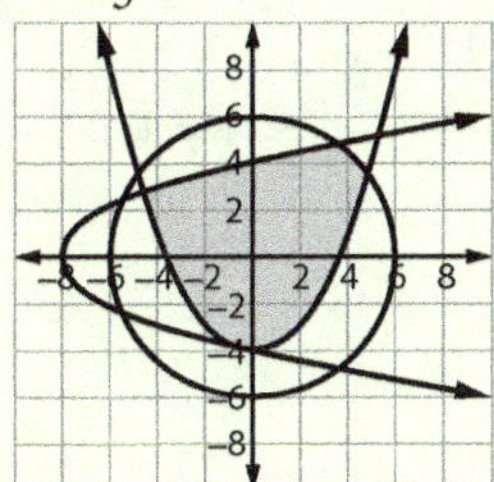

42.

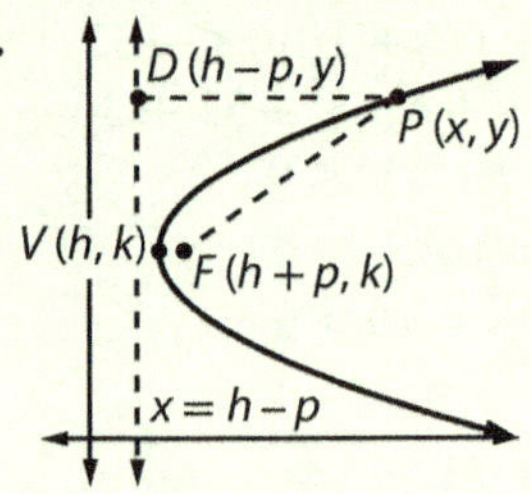

$$PF = PD$$

$$\sqrt{[x-(h+p)]^2 + (y-k)^2}$$
$$= \sqrt{[x-(h-p)]^2 + (y-y)^2}$$

$$[(x-h)-p]^2 + (y-k)^2$$
$$= [(x-h)+p]^2$$

$$(x-h)^2 - 2p(x-h) + p^2 + (y-k)^2$$
$$= (x-h)^2 + 2p(x-h) + p^2$$

$$(y-k)^2 = 4p(x-h)$$

$$\frac{1}{4p}(y-k)^2 = x-h$$

$$x = \frac{1}{4p}(y-k)^2 + h$$

43.

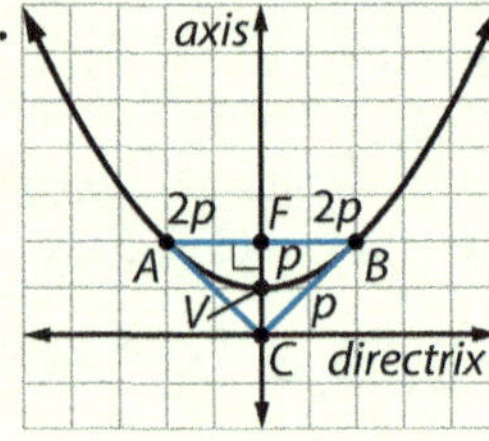

$\overline{AB} \perp \overline{FC}$ (def. of focal width)
$\angle AFC$ and $\angle BFC$ are right $\angle$s and
$\triangle AFC$ and $\triangle BFC$ are right $\triangle$s.
$\overline{FC} \cong \overline{FC}$ (Refl. Prop.)
$\overline{AF} \cong \overline{FB}$ (def. of axis of sym.)
$\triangle AFC \cong \triangle BFC$ (LL congruence)
$\overline{AC} \cong \overline{BC}$ (CPCTC)
$FC = 2p$ (distance from focus to
directrix)
$AF + FB = 4p \Rightarrow AF = FB = 2p$
$\overline{AF} \cong \overline{FC}; \overline{BF} \cong \overline{FC}$
$\triangle AFC$ and $\triangle BFC$ are isos. rt. $\triangle$s.
$m\angle FCA = m\angle FCB = 45°$
$m\angle ACB = 90°$ ($\angle$ Add. Post)

❯ **Cumulative Review**

45. $d = \sqrt{(2-(-1))^2 + (1-(-3))^2}$
$$= \sqrt{3^2 + 4^2} = 5$$
$$(x+1)^2 + (y+3)^2 = 25$$

46.

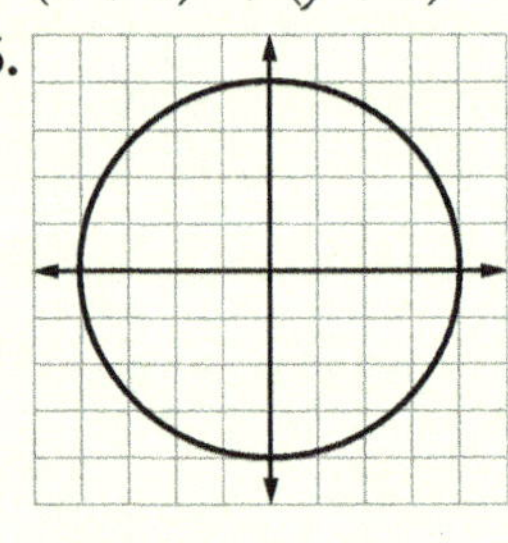

---

42. Derive: Sketch a parabola with vertex (h, k) and opening to the right. Label the focus and the directrix in terms of h, k, and p. Sketch $\overline{PF}$ from a point $P(x, y)$ on the curve to the focal point and $\overline{PD}$ representing the perpendicular distance from P to the directrix. Label the coordinates of D in terms of y, h, and p. Use the figure and the locus definition of a parabola to derive the standard form equation of a parabola with a horizontal axis: $x = \frac{1}{4p}(y-k)^2 + h$.

43. Prove: Prove that the triangle defined by the endpoints of the chord representing the focal width of a parabola and the point of intersection of the parabola's axis and directrix is an isosceles right triangle.

CUMULATIVE REVIEW

Write the equation for each circle. [1.2]

44. a circle centered at the origin with a radius of 8 $x^2 + y^2 = 64$

45. a circle centered at $(-1, -3)$ and containing the point $(2, 1)$ $(x+1)^2 + (y+3)^2 = 25$

Graph each circle. [1.2]

46. $x^2 + y^2 = 16$

47. $(x-2)^2 + (y+3)^2 = 4$

Write each quadratic function in vertex form. [1.6]

48. $f(x) = -3x^2 + 6x + 13$

49. $g(x) = 6x^2 + 48x + 90$

50. Which of the following is not a zero of the equation $f(x) = 2x^3 - 5x^2 - 11x - 4$? [2.4]   D

  A. −1   **C.** $-\frac{1}{2}$ **E.** These are
  B. 4   **D.** 1   all zeros.

48. $f(x) = -3(x-1)^2 + 16$ **49.** $g(x) = 6(x+4)^2 - 6$

51. Which values of x are zeros of $f(x) = 2x^2 + x - 10$? [2.4] A

  A. $-\frac{5}{2}, 2$ **C.** $-\frac{1}{4}, 2$ **E.** none of these
  B. $-2, \frac{5}{2}$ **D.** $-2, \frac{1}{4}$

52. Which function has an amplitude of 3, period of π, and phase shift of π? [4.4] B

  A. $f(x) = 3 \sin x + \pi$ **D.** $f(x) = 3 \sin 2x + \pi$
  B. $f(x) = 3 \sin 2(x - \pi)$ **E.** none of these
  C. $f(x) = 3 \sin (x - \pi)$

53. From the top of a 210 ft cliff, the angle of depression of a ship at sea is 4°. Find the distance from the foot of the cliff to the ship (to the nearest foot). [4.2] C

  A. 209 ft   **C.** 3003 ft   **E.** none of these
  B. 15 ft   **D.** 50 ft

47.

48. $f(x) = -3(x^2 - 2x) + 13$
$$= -3(x^2 - 2x + 1^2) + 13 - 1(-3)$$
$$= -3(x-1)^2 + 16$$

49. $g(x) = 6(x^2 + 8x) + 90$
$$= 6(x^2 + 8x + 4^2) + 90 - 6(16)$$
$$= 6(x+4)^2 - 6$$

50. $f(1) = 2 - 5 + 11 - 4 = 4 \neq 0$

51. $0 = (x-2)(2x+5)$
$$x = 2, -\frac{5}{2}$$

52. using $f(x) = a \sin b(x - h)$:
$$p \frac{2\pi}{|b|} = \pi; \ |b| = 2; \ h = \pi$$
$$\therefore f(x) = 3 \sin 2(x - \pi)$$

53. $\tan 86° = \frac{x}{210}$
$$x = 210 \tan 86°$$
$$x \approx 3003 \text{ ft}$$

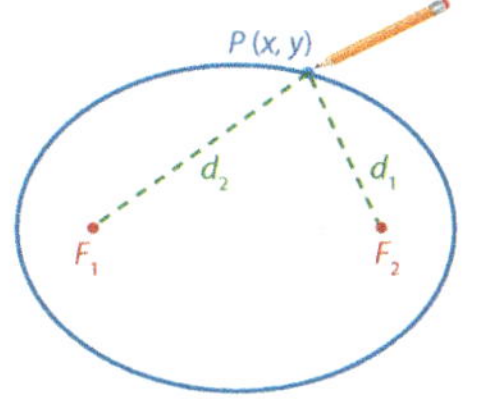

National Statuary Hall in the US Capitol functions as a whispering gallery.

A parabola has been defined as a set of points equidistant from a given point (its focus) and a given line (its directrix). The definition of an ellipse involves the distances from two given points. Ellipses are formed by intersections of one nappe of the conical surface and a plane that is not parallel to an element. Compare the locus definition of an ellipse to the definition of a parabola.

> **DEFINITION**
>
> An **ellipse** is the set of points $P(x, y)$ in a plane such that the sum of the distances from two fixed points (its foci) is constant.

An ellipse can be formed by attaching a string to the two foci and using a pencil to draw the points of the ellipse while keeping the string taut. The section of string on each side of the pencil illustrates the distance from a point to each focus, and the sum of both sections (the string's length) remains constant.

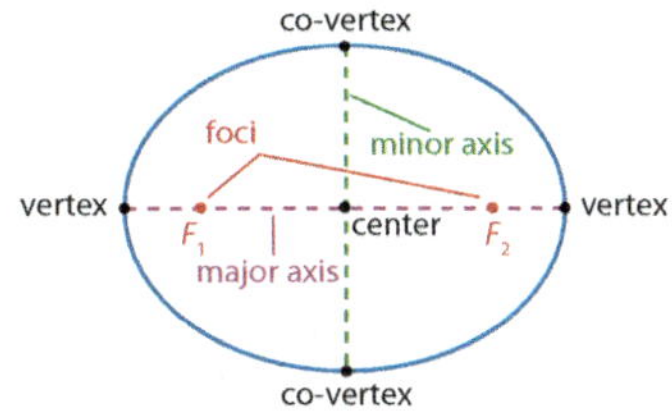

The chord of the ellipse containing the foci (F_1 and F_2) is its *major axis* and the chord's midpoint is the *center* of the ellipse, typically notated as $C(h, k)$. The distance from the center to each focus is represented by c, the *focal distance*. The distance from the center to the ellipse's *vertices* (the endpoints of the major axis) is notated as a, the length of the *semimajor axis*. For an ellipse that is elongated horizontally, the major axis is horizontal, while an ellipse elongated vertically has a vertical major axis. The *minor axis* is the chord through the center and perpendicular to the major axis. Its endpoints are sometimes called *co-vertices*, and the distance between a co-vertex and the center is represented by b, the length of the *semiminor axis*.

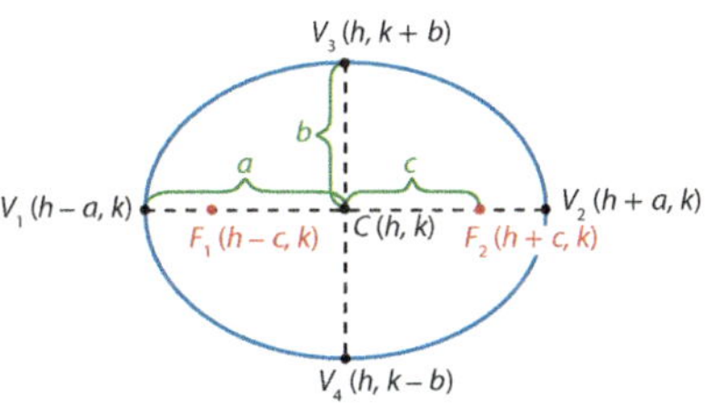

The sum of the distances from the foci to any point $P(x, y)$ on the ellipse, $d_1 + d_2$, can be determined using V_1 on the major axis. Since $V_1F_1 = a - c$ and $V_1F_2 = a + c$, the sum $d_1 + d_2 = (a - c) + (a + c) = 2a$. Therefore, the constant sum in the definition is the length of the major axis.

Objectives

1. To define an ellipse as a locus of points
2. To analyze and graph an ellipse given its equation
3. To write the equation of an ellipse using its center, vertices, and foci
4. To identify a circle as a special case of an ellipse
5. To solve real-world problems related to ellipses

After completing this section, you will be able to

- define an ellipse.
- analyze and graph an ellipse, given its equation.
- write the equation of an ellipse.
- identify a circle as a special case of an ellipse.
- solve real-world problems related to ellipses.

The first known reference to the now-familiar method of creating an ellipse by using two fixed points and a loop of string was made by the Greek mathematician Anthemius of Tralles in the sixth century.

Flash

National Statuary Hall was initially the home for the House of Representatives. However, the acoustics of the room made it unsuitable to conduct business. Currently the room is used to host special events and is home to 38 statues, each donated by a different state.

Vocabulary

aphelion	major axis
center	minor axis
co-vertices	perihelion
eccentricity	semimajor axis
ellipse	semiminor axis
focal distance	vertices

PRESENTATION

Lesson Opener

Solve by completing the square.

1. $x^2 + 2x = 3$ $\quad x = -3, 1$
2. $y^2 + 2y = 1$ $\quad y = -1 \pm \sqrt{2}$
3. $x^2 + 3x = 10$ $\quad x = -5, 2$

Consider reviewing the basic conic sections using an interactive illustration. (Use the Internet keyword search *interactive conic sections*.) Then present the locus (set of points) definition of an *ellipse* in terms of the sum of distances from two fixed points. If time permits, have students construct ellipses and explore various placements of the two points

and lengths of the string. Define the key points and terms in the diagrams. Some texts refer to the line through the foci of an ellipse or hyperbola as the focal axis.

Derive the standard form equation of an ellipse centered at (h, k) with a horizontal *major axis*. Discuss standard forms of ellipses centered at the origin before summarizing the results with the standard forms and characteristics given in the table. Exercise 43 asks students to derive the standard form when the major axis is vertical.

Demonstrate how the standard form is used to identify characteristics and graph the ellipses in Example 1. In Example 2

emphasize how the process of completing the square is used twice to convert the general form equation for a conic to the standard form for an ellipse. Remind the students to add the appropriate values to both sides of the equation when completing the square.

Example 3 demonstrates the process of finding the equation of an ellipse when certain values are known. Students will find this process helpful when creating mathematical models in real-world scenarios.

Discuss characteristics of the general second degree equation in two variables when it represents an ellipse. This topic

Additional Exercises

Write each equation in the standard form for an ellipse and graph each ellipse. Then find the foci and the eccentricity for each ellipse.

1. $x^2 + 2y^2 - 4x - 4y - 10 = 0$

$\dfrac{(x-2)^2}{16} + \dfrac{(y-1)^2}{8} = 1; e = \dfrac{\sqrt{2}}{2}$

$F: (2 \pm 2\sqrt{2}, 1) \approx (4.83, 1), \approx (-0.83, 1)$

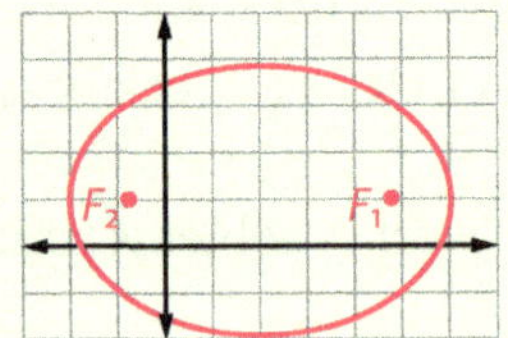

2. $81x^2 + 25y^2 + 486x + 100y$
$$- 1196 = 0$$

$\dfrac{(x+3)^2}{25} + \dfrac{(y+2)^2}{81} = 1; e = \dfrac{2\sqrt{14}}{9}$

$F: (-3, -2 \pm 2\sqrt{14}) \approx (-3, 5.48),$
$\approx (-3, -9.48)$

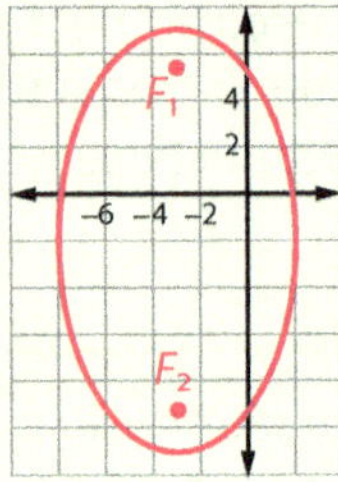

Write the standard form equation of each ellipse with the given characteristics.

3. $F: \left(2 \pm 3\sqrt{3}, -1\right); e = \dfrac{\sqrt{3}}{2}$

$\dfrac{(x-2)^2}{36} + \dfrac{(y+1)^2}{9} = 1$

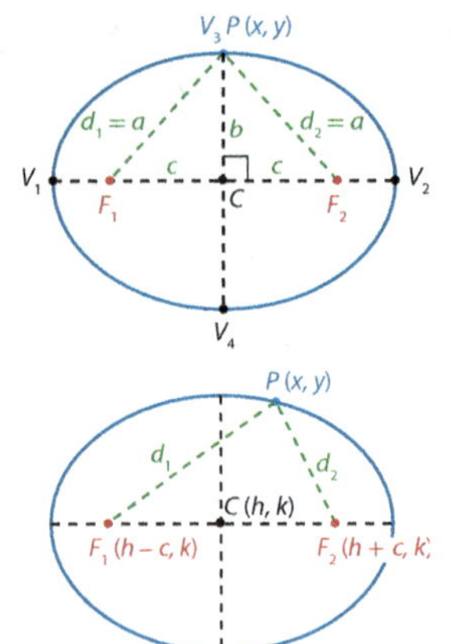

The relationships between a, b, and c can be determined by examining V_3 on the minor axis. Since the two right triangles formed are congruent and the sum of the distances from the foci is $2a$, the distance from V_3 to each focus is a. The Pythagorean Theorem indicates that $b^2 + c^2 = a^2$ or $c^2 = a^2 - b^2$. This implies $a > b$ and $a > c$.

The definition of an ellipse can be used to derive a standard form equation of an ellipse with a horizontal major axis centered at $C(h, k)$ whose foci $F(h \pm c, k)$ are c units from the center.

Applying the distance formula to any point $P(x, y)$ on the ellipse:
$$PF_1 + PF_2 = 2a.$$
$$\sqrt{[x-(h-c)]^2 + (y-k)^2} + \sqrt{[x-(h+c)]^2 + (y-k)^2} = 2a$$

Regroup the terms using the Associative Property of Addition.
$$\sqrt{[(x-h)+c]^2 + (y-k)^2} + \sqrt{[(x-h)-c]^2 + (y-k)^2} = 2a$$

Isolate a radical and square both sides.
$$\sqrt{[(x-h)+c]^2 + (y-k)^2} = 2a - \sqrt{[(x-h)-c]^2 + (y-k)^2}$$
$$[(x-h)+c]^2 + (y-k)^2 = 4a^2 - 4a\sqrt{[(x-h)-c]^2 + (y-k)^2} + [(x-h)-c]^2 + (y-k)^2$$

Subtract $(y-k)^2$ from both sides and expand the binomials outside the radicals.
$$(x-h)^2 + 2c(x-h) + c^2 = 4a^2 - 4a\sqrt{[(x-h)-c]^2 + (y-k)^2} + (x-h)^2 - 2c(x-h) + c^2$$

Isolate the radical term and simplify.
$$4a\sqrt{[(x-h)-c]^2 + (y-k)^2} = 4a^2 - 4c(x-h)$$
$$a\sqrt{[(x-h)-c]^2 + (y-k)^2} = a^2 - c(x-h)$$

Square both sides and expand the left side, keeping $(x-h)$ and $(y-k)$ intact.
$$a^2\left\{[(x-h)-c]^2 + (y-k)^2\right\} = a^4 - 2a^2c(x-h) + c^2(x-h)^2$$
$$a^2\left[(x-h)^2 - 2c(x-h) + c^2 + (y-k)^2\right] = a^4 - 2a^2c(x-h) + c^2(x-h)^2$$
$$a^2(x-h)^2 - 2a^2c(x-h) + a^2c^2 + a^2(y-k)^2 = a^4 - 2a^2c(x-h) + c^2(x-h)^2$$

Collect all terms containing x or y on the left side and the other terms on the right.
$$a^2(x-h)^2 - c^2(x-h)^2 + a^2(y-k)^2 = a^4 - a^2c^2$$

Factor $(x-h)^2$ from the first two terms on the left and a^2 on the right side.
$$(a^2 - c^2)(x-h)^2 + a^2(y-k)^2 = a^2(a^2 - c^2)$$

Since $b^2 + c^2 = a^2$, substitute using $b^2 = a^2 - c^2$.
$$b^2(x-h)^2 + a^2(y-k)^2 = a^2b^2$$

Finally, divide both sides by a^2b^2.
$$\frac{b^2(x-h)^2}{a^2b^2} + \frac{a^2(y-k)^2}{a^2b^2} = \frac{a^2b^2}{a^2b^2}$$
$$\frac{(x-h)^2}{a^2} + \frac{(y-k)^2}{b^2} = 1$$

An ellipse with a horizontal major axis, center at $(0, 0)$, and foci at $(\pm c, 0)$ has the standard form equation $\dfrac{x^2}{a^2} + \dfrac{y^2}{b^2} = 1$. Its inverse, $\dfrac{y^2}{a^2} + \dfrac{x^2}{b^2} = 1$, is an ellipse with a vertical major axis, center at $(0, 0)$, and foci at $(0, \pm c)$. Translations of these vertically elongated ellipses to a center at $C(h, k)$ could be written as $\dfrac{(y-k)^2}{a^2} + \dfrac{(x-h)^2}{b^2} = 1$ or in the standard form stated in the following table.

will be discussed further in Section 8.4. This chapter's Technology Corner illustrates how technology can be used to effectively visualize the effects of changing the values of constants in the general form of conics or standard form for a particular conic. In the first three sections, the conics are not rotated ($B = 0$). In this case, if A and C have the same sign ($AC > 0$), the conic is an ellipse. Additionally, $A = C$, so the ellipse is the special case of a circle and could be graphed using the standard form for a circle or as a circular ellipse where $a = b = r$ and $c = 0$, so the foci have the same coordinates as the *center*.

Present the formula for *eccentricity* and apply it in Example 5. Emphasize that eccentricity is a measure of deviation from a circle.

Common Student Error Students who are unclear on the definition of eccentricity may use the ratio $a : b$ and state the eccentricity of a circle as 1. Emphasize that eccentricity is the ratio $c : a$, the distance from the center to a focus and to a vertex. Since the foci are at the center of a circle, $c = 0$ and $e = 0$.

An elliptical equation modeling Mercury's orbit is determined in Example 6. Check that student graphs are in square windows so they accurately reflect the eccentricity of its orbit. Other graphing technologies can also be used to model the orbit.

The paragraph following Example 6 introduces the reflective properties of an ellipse. An ellipsoid formed by rotating an ellipse about its major axis is a *prolate spheroid*, while one formed by rotating about the *minor axis* is an *oblate spheroid*.

Motivational Idea Consider having the students look up the definition of *eccentric*. This term is used in several areas of life. In mechanics, an eccentric rotation is one that has a non-circular effect or off-center rotation, such as the lobs on a camshaft in an engine. In geometry,

Standard Form Equations of an Ellipse
centered at $C(h, k)$ with $a > b$ and $c^2 = a^2 - b^2$

$\dfrac{(x-h)^2}{a^2} + \dfrac{(y-k)^2}{b^2} = 1$		$\dfrac{(x-h)^2}{b^2} + \dfrac{(y-k)^2}{a^2} = 1$
horizontal on $y = k$	**major axis**	vertical on $x = h$
$(h \pm a, k)$	**vertices**	$(h, k \pm a)$
vertical on $x = h$	**minor axis**	horizontal on $y = k$
$(h, k \pm b)$	**co-vertices**	$(h \pm b, k)$
$(h \pm c, k)$	**foci**	$(h, k \pm c)$

Example 1   Graphing an Ellipse

Graph each ellipse and its foci.

a. $\dfrac{(x+3)^2}{25} + \dfrac{(y-1)^2}{9} = 1$ **b.** $4x^2 + 3y^2 = 48$

Answer

a.
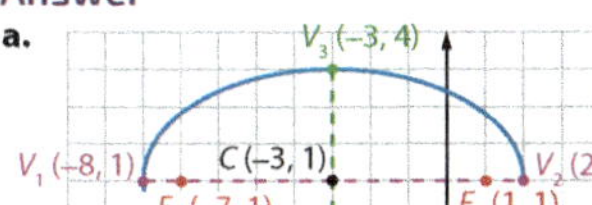

1. Identify and plot the center: $C(h, k) = (-3, 1)$.
2. Since $a^2 = 25$, use $a = 5$ to plot the vertices: $(-3 \pm 5, 1)$.
3. Since $b^2 = 9$, use $b = 3$ to plot the co-vertices: $(-3, 1 \pm 3)$.
4. Sketch the ellipse.
5. Since $c^2 = 25 - 9 = 16$, use $c = 4$ to plot the foci: $(-3 \pm 4, 1)$.

b. $\dfrac{4x^2}{48} + \dfrac{3y^2}{48} = \dfrac{48}{48}, \dfrac{x^2}{12} + \dfrac{y^2}{16} = 1$

1. Convert to the standard form for an ellipse by dividing both sides by 48.
2. The ellipse has a vertical major axis and is centered at the origin.
3. Since $a^2 = 16$, use $a = 4$ to plot the vertices: $(0, \pm 4)$.
4. Since $b^2 = 12$, use $b \approx 3.5$ to plot the co-vertices: $(\pm 2\sqrt{3}, 0)$.
5. Sketch the ellipse.
6. Since $c^2 = 16 - 12 = 4$, use $c = 2$ to plot the foci: $(0, \pm 2)$.

SKILL ✔ **EXERCISE 7**

---

4. major axis length: 12
   minor axis length: 8
   F: $(-2 \pm 2\sqrt{5}, 4)$
   $\dfrac{(x+2)^2}{36} + \dfrac{(y-4)^2}{16} = 1$

5. foci lie on the line $x = -4$
   major axis length: 16
   $C(-4, -4)$; $e = 0.5$
   $\dfrac{(x+4)^2}{48} + \dfrac{(y+4)^2}{64} = 1$

6. F: $(\pm 6, 0)$
   semimajor axis length: 9
   $\dfrac{x^2}{81} + \dfrac{y^2}{45} = 1$

7. If a whisper room can be modeled with the given equation of an ellipse, find the two locations in the coordinate plane at which you could whisper and be heard.
   $49x^2 + 4y^2 + 294x + 8y = -249$
   $(-3, -1 \pm 3\sqrt{5})$

8. The formula for the area of an ellipse is $A = \pi ab$. Find the area of an ellipse with semimajor axis of 12 units and eccentricity of 0.75.
   $c = 9$; $b = \sqrt{63}$; $A = \pi(12)\sqrt{63} \approx 299 \text{ u}^2$

9. Graph $f(x) = \dfrac{3\sqrt{25 - (x+1)^2}}{5}$.

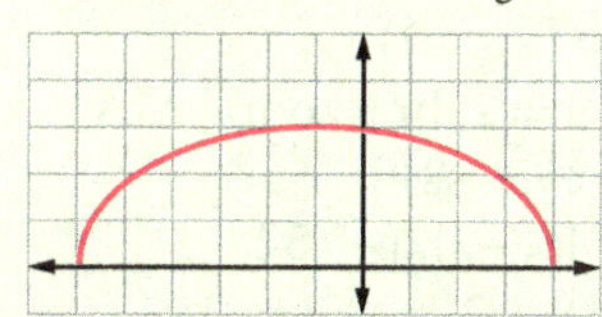

10. Write an equation modeling the orbit of Saturn if its eccentricity is 0.0542 and its semimajor axis is 9.537 AU. Then graph the ellipse using technology and find its perihelion and aphelion distances.
    $\dfrac{x^2}{90.954} + \dfrac{y^2}{90.687} = 1$
    ≈ 9.02 AU; ≈ 10.1 AU

Assignments

- **Minimum:** 1, 4–5, 7–8, 11–13, 15–19, 21, 24–25, 27, 31, 41, 45, 47, 51, 53

- **Standard:** 2–10 even; 11, 14–15, 18, 20–21, 24–25, 27, 29, 32–35, 37, 41; 45–53 odd

- **Extended:** 3, 6, 9–10, 13–14, 17, 20, 22, 24–25; 28–36 even; 39–43; 44–50 even; 53

---

the word refers to a deviation from the center or circular form. The term eccentric is also commonly used to describe a person or behavior that is odd or unconventional, deviating from the norm. God has established moral and behavioral standards, and worldly behaviors that many may consider normal are actually eccentric, deviating from God's will (Titus 2:14) and the way God intended the world to function. Consider discussing with students some areas in which the world has attempted to deviate the norm from God's will.

TIPS

Ex. 15–20 Encourage students to make a sketch of the given information and add information such as the center and key distances a, b, and c, as they are determined.

Assessment

- Quiz 8A covers Sections 8.1–8.2.

Solutions

A. Exercises

1. $C(h, k) = (0, 0)$; $a = 5$, $b = 3$
   major axis (MA): $2a = 10$
   minor axis: $2b = 6$

2. $C(h, k) = (2, -1)$; $a = 4$, $b = 1$
   major axis: $2a = 8$
   minor axis: $2b = 2$

3. $C(h, k) = (0, 0)$
   $a = 7$, $b = 5$; vertical MA
   $V: (0, 0 \pm 7) = (0, \pm 7)$
   $cV: (0 \pm 5, 0) = (\pm 5, 0)$

4. $C(h, k) = (3, -5)$
   $a = 6$, $b = 3$; horizontal MA
   $V: (3 \pm 6, -5) = (-3, -5), (9, -5)$
   $cV: (3, -5 \pm 3) = (3, -2), (3, -8)$

5. $C(h, k) = (-1, 3)$
   $a = 5$, $b = 3$; vertical MA
   $V: (-1, 3 \pm 5) = (-1, -2), (-1, 8)$
   $c^2 = 5^2 - 3^2 = 16$; $c = 4$
   $F: (-1, 3 \pm 4) = (-1, -1), (-1, 7)$
   $e = \frac{c}{a} = \frac{4}{5} = 0.8$

6. $C(h, k) = (-6, -2)$
   $a = 2\sqrt{3}$, $b = 2$; horizontal MA
   $V: (-6 \pm 2\sqrt{3}, -2)$
   $\qquad \approx (-9.46, -2), (-2.54, -2)$
   $c^2 = (2\sqrt{3})^2 - 2^2 = 8$; $c = 2\sqrt{2}$
   $F: (-6 \pm 2\sqrt{2}, -2)$
   $\qquad \approx (-8.83, -2), (-3.17, -2)$
   $e = \frac{c}{a} = \frac{2\sqrt{2}}{2\sqrt{3}} \approx 0.82$

7. $C(0, 0)$
   $a = 4$, $b = 2$; horizontal MA
   $c^2 = 4^2 - 2^2 = 12$; $c = 2\sqrt{3}$
   $F: (\pm 2\sqrt{3}, 0) \approx (\pm 3.5, 0)$

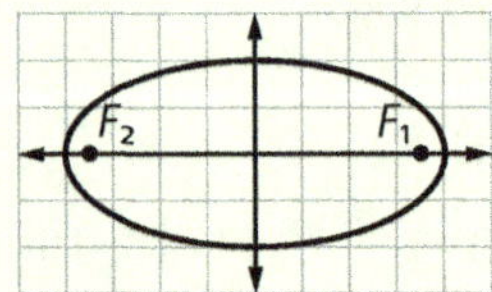

8. $C(0, 0)$
   $a = 3$, $b = 2$; vertical MA
   $c^2 = 3^2 - 2^2 = 5$; $c = \sqrt{5}$
   $F: (0, \pm\sqrt{5}) \approx (0, \pm 2.24)$

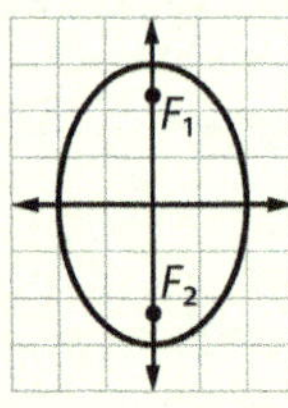

9. $C(-1, 5)$
   $a = 2$, $b = 1$; vertical MA
   $c^2 = 2^2 - 1^2 = 3$; $c = \sqrt{3}$
   $F: (-1, 5 \pm \sqrt{3})$
   $\qquad \approx (-1, 6.73), (-1, 3.3)$

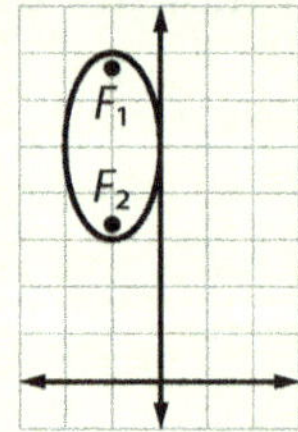

10. $C(2, 2)$
    $a = 3$, $b = 2$; horizontal MA
    $c^2 = 3^2 - 2^2 = 5$; $c = \sqrt{5}$
    $F: (2 \pm \sqrt{5}, 2) \approx (4.24, 2), (-0.24, 2)$

When the equation of an ellipse is written in general form for conics, convert the equation to a standard form for an ellipse to reveal the center and the lengths of the semimajor and semiminor axes. You will likely need to complete the square for both the x-terms and the y-terms.

Convert $4x^2 + 16y^2 - 8x + 64y + 4 = 0$ to a standard form for an ellipse.

Answer

$(4x^2 - 8x) + (16y^2 + 64y) = -4$ 1. Group the x-terms and the y-terms on the left side and the constant on the right.

$4(x^2 - 2x) + 16(y^2 + 4y) = -4$ 2. Complete the square for both variables.
$4(x^2 - 2x + 1) + 16(y^2 + 4y + 4) = -4 + 4 + 64$
$4(x - 1)^2 + 16(y + 2)^2 = 64$
$\frac{(x-1)^2}{16} + \frac{(y+2)^2}{4} = 1$ 3. Divide both sides by 64.

SKILL ✔ EXERCISE 13

You can also use characteristics of an ellipse to determine its standard form equation.

Write the equation of the ellipse with foci at $(2, \pm\sqrt{21})$ and a minor axis that is 4 units long.

Answer

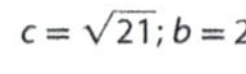

$c = \sqrt{21}$; $b = 2$ 1. Use a sketch to determine the center $(2, 0)$ and the values of c and b.

$21 = a^2 - 4$ 2. Use $c^2 = a^2 - b^2$ to find the value of a.
$25 = a^2$
$a = 5$

$\frac{(x-2)^2}{2^2} + \frac{(y-0)^2}{5^2} = 1$ 3. Since the major axis is vertical, substitute into
$\frac{(x-2)^2}{4} + \frac{y^2}{25} = 1$ $\frac{(x-h)^2}{b^2} + \frac{(y-k)^2}{a^2} = 1$ and simplify.

SKILL ✔ EXERCISE 17

The equation for the ellipse in Example 3 can be written in the general form for conics, $Ax^2 + Bxy + Cy^2 + Dx + Ey + F = 0$ (by multiplying by 100 and simplifying) as $25x^2 + 4y^2 - 100x = 0$. From this form you can see that the ellipse has not been rotated (since $B = 0$) or translated vertically (since $E = 0$). The fact that A and C have the same sign (and therefore $AC > 0$) indicates that the equation represents an ellipse.

When $A = C$, the equation represents a circle, a special case of an ellipse where both foci are at the center. In this case $c = 0$ and $a = b = r$. Dividing the standard form equation of a circle $(x - h)^2 + (y - k)^2 = r^2$ by r^2 converts the equation into the standard form for an ellipse: $\frac{(x-h)^2}{r^2} + \frac{(y-k)^2}{r^2} = 1$.

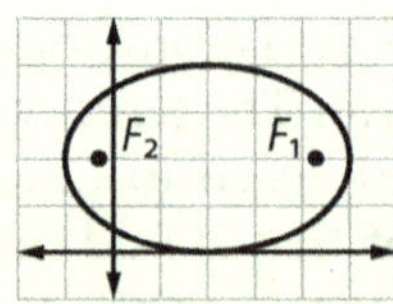

11. $\frac{1}{64}(16x^2 + 4y^2 = 64)$
    $\frac{x^2}{4} + \frac{y^2}{16} = 1$; ellipse

12. $(x^2 - 6x + 9) + (y^2 - 2y + 1)$
    $\qquad\qquad\qquad = -1 + 9 + 1$
    $(x - 3)^2 + (y - 1)^2 = 9$
    $\frac{(x-3)^2}{9} + \frac{(y-1)^2}{9} = 1$; circle

Convert $x^2 + y^2 + 4x - 6y - 3 = 0$ to a standard form of an ellipse and graph the conic section.

Answer

$(x^2 + 4x) + (y^2 - 6y) = 3$
$(x^2 + 4x + 4) + (y^2 - 6y + 9) = 3 + 4 + 9$
$(x + 2)^2 + (y - 3)^2 = 16$

$$\dfrac{(x + 2)^2}{16} + \dfrac{(y - 3)^2}{16} = 1$$

$C(-2, 3); a = b = 4 = r$
$c^2 = 16 - 16 = 0$
$\therefore$ foci at $(-2 \pm 0, 3 \pm 0) = (-2, 3)$

1. Complete the square for the x- and y-terms.

2. Divide by 16 to write in the standard form of an ellipse.

3. Since $a = b$, the ellipse is a circle, and both foci are located at the center.

4. Graph the circle centered at $(-2, 3)$ with a radius of 4.

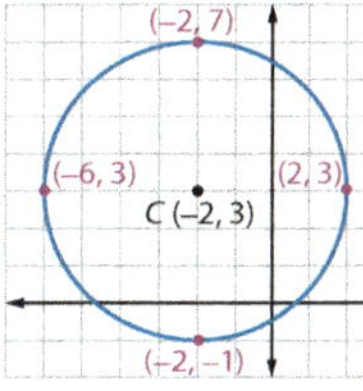

An ellipse's amount of elongation from a circle is measured by its *eccentricity*, defined as $e = \dfrac{c}{a}$.

Find the eccentricity of each ellipse.

a.

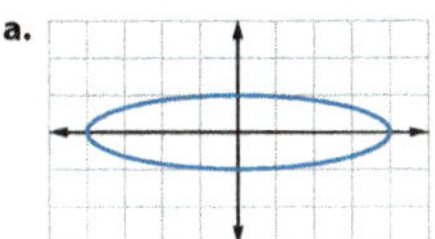

b.

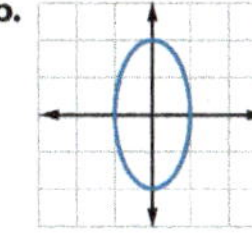

Answer

a. $a = 4, b = 1$

$c^2 = 4^2 - 1^2 = 15$
$c = \sqrt{15}$

$e = \dfrac{c}{a}$
$= \dfrac{\sqrt{15}}{4} \approx 0.968$

b. $a = 2, b = 1$

$c^2 = 2^2 - 1^2 = 3$
$c = \sqrt{3}$

$e = \dfrac{c}{a}$
$= \dfrac{\sqrt{3}}{2} \approx 0.866$

1. Find the lengths of the semimajor and semiminor axes.
2. Use $c^2 = a^2 - b^2$ to find the value of c.

3. Apply the definition of eccentricity.

TIP 💡

Note that e is used for both the irrational number $e \approx 2.718$ and the eccentricity of a conic. The symbol's context is used to distinguish its meaning.

The more elongated ellipse has greater eccentricity since the foci's distance from the center, c, is greater and the foci are closer to the vertices. In any ellipse $0 \le c < a$ and $0 \le e < 1$. The eccentricity of a circle is 0 since $c = 0$.

Kepler's first law of planetary motion states that the orbit of a planet is an ellipse with the sun at one of the foci. The point at which an orbiting object is closest to the sun is the *perihelion*, while its farthest point from the sun is the *aphelion*. These distances are frequently measured in astronomical units (AU), which are multiples of the earth's average distance from the sun. The average eccentricity of the sun's planets is a nearly circular $e \approx 0.0601$.

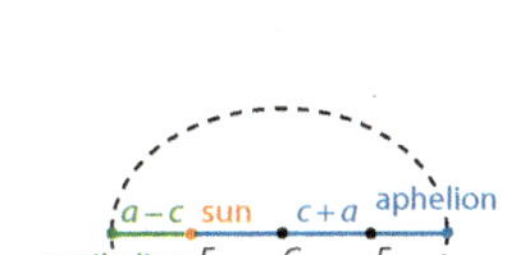

18. vertical MA

$C\left(\dfrac{-8 + (-2)}{2}, 3\right) = (-5, 3)$
$b = -5 - (-8) = 3; b^2 = 9$
$2a = 12; a = 6; a^2 = 36$
$$\dfrac{(x + 5)^2}{9} + \dfrac{(y - 3)^2}{36} = 1$$

19. vertical MA

$C\left(-2, \dfrac{5 + (-3)}{2}\right) = (-2, 1)$
$c = 5 - 1 = 4; c^2 = 16$
$b = 1 - (-2) = 3; b^2 = 9$
$a^2 = c^2 + b^2 = 16 + 9 = 25$
$$\dfrac{(x + 2)^2}{9} + \dfrac{(y - 1)^2}{25} = 1$$

20. horizontal MA

$2a = 34; a = 17; a^2 = 289$
$C\left(\dfrac{-11 + 5}{2}, -2\right) = (-3, -2)$
$c = 5 - (-3) = 8; c^2 = 64$
$b^2 = a^2 - c^2 = 289 - 64 = 225$
$$\dfrac{(x + 3)^2}{289} + \dfrac{(y + 2)^2}{225} = 1$$

❭ B. Exercises

21. $a = 5; 0.6 = \dfrac{c}{5}; c = 3$
$b^2 = 5^2 - 3^2 = 16$
$$\dfrac{x^2}{25} + \dfrac{y^2}{16} = 1$$

22. $a = 25; 0.28 = \dfrac{c}{25}; c = 7$
$b^2 = 25^2 - 7^2 = 576$
$$\dfrac{(x - 4)^2}{576} + \dfrac{(y - 1)^2}{625} = 1$$

23. $\dfrac{1}{100}(x^2 + 25y^2 = 100)$
$$\dfrac{x^2}{100} + \dfrac{y^2}{4} = 1$$
$C(0, 0);$ horizontal MA; $a = 10, b = 2$
$c^2 = 10^2 - 2^2 = 96; c = 4\sqrt{6}$
$F: (\pm 4\sqrt{6}, 0) \approx (\pm 9.80, 0)$
$e = \dfrac{c}{a} = \dfrac{4\sqrt{6}}{10} \approx 0.98$

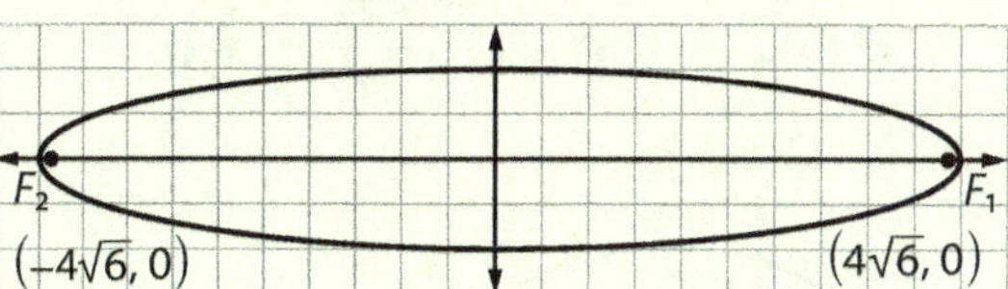

24. $\dfrac{1}{16}(4x^2 + y^2 = 16)$ $(0, 2\sqrt{3})$
$$\dfrac{x^2}{4} + \dfrac{y^2}{16} = 1$$
$C(0, 0);$ vertical MA
$a = 4, b = 2$
$c^2 = 4^2 - 2^2 = 12$
$c = 2\sqrt{3}$
$F: (0, \pm 2\sqrt{3})$
$\approx (0, \pm 3.46)$
$e = \dfrac{c}{a} = \dfrac{2\sqrt{3}}{4} \approx 0.87$

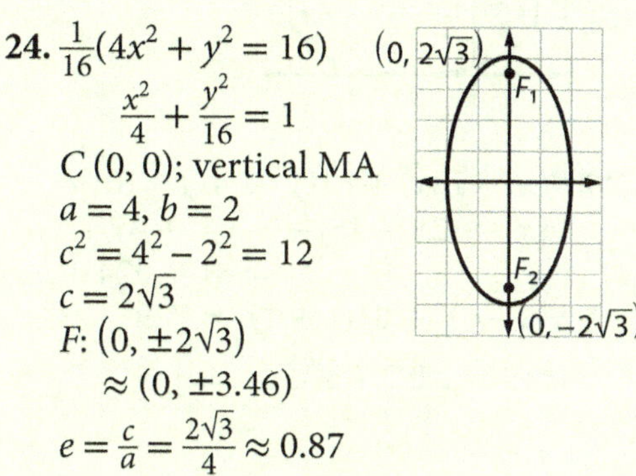

13. $25x^2 + 9(y^2 - 6y) = 144$
$25x^2 + 9(y^2 - 6y + 9) = 144 + 81$
$25x^2 + 9(y - 3)^2 = 225$
$$\dfrac{x^2}{9} + \dfrac{(y - 3)^2}{25} = 1;\ \text{ellipse}$$

14. $16(x^2 + 8x) + 9(y^2 - 6y) = -193$
$16(x^2 + 8x + 16) + 9(y^2 - 6y + 9)$
$\qquad = -193 + 256 + 81$
$16(x + 4)^2 + 9(y - 3)^2 = 144$
$$\dfrac{(x + 4)^2}{9} + \dfrac{(y - 3)^2}{16} = 1;\ \text{ellipse}$$

15. $C(-3, 5);$ vertical MA
$a = \dfrac{9 - 1}{2} = 4; a^2 = 16$
$b = \dfrac{-1 - (-5)}{2} = 2; b^2 = 4$
$$\dfrac{(x + 3)^2}{4} + \dfrac{(y - 5)^2}{16} = 1$$

16. $C(2, -4);$ horizontal MA
$a = \dfrac{7 - (-3)}{2} = 5; a^2 = 25$
$b = \dfrac{-1 - (-7)}{2} = 3; b^2 = 9$
$$\dfrac{(x - 2)^2}{25} + \dfrac{(y + 4)^2}{9} = 1$$

17. $C\left(\dfrac{-3 + 11}{2}, 2\right) = (4, 2)$
horizontal MA
$a = 11 - 4 = 7; a^2 = 49$
$2b = 4; b = 2; b^2 = 4$
$$\dfrac{(x - 4)^2}{49} + \dfrac{(y - 2)^2}{4} = 1$$

25. $(x^2 - 2x) + (y^2 + 4y) = -1$

$(x^2 - 2x + 1) + (y^2 + 4y + 4)$
$$= -1 + 1 + 4$$

$(x - 1)^2 + (y + 2)^2 = 4$

$\dfrac{(x-1)^2}{4} + \dfrac{(y+2)^2}{4} = 1$

$C\,(1, -2)$

$a = b = 2 = r$

$c^2 = 2^2 - 2^2 = 0;\ c = 0$

$F\,(1, -2)$

$e = \dfrac{c}{a} = \dfrac{0}{2} = 0$

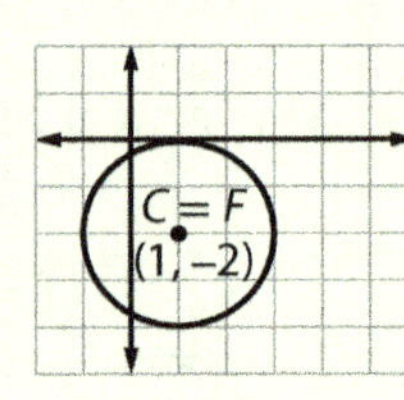

26. $25(x^2 + 6x) + 16(y^2 + 8y) = -81$

$25(x^2 + 6x + 9) + 16(y^2 + 8y + 16)$
$$= -81 + 25(9) + 16(16)$$

$25(x + 3)^2 + 16(y + 4)^2 = 400$

$\dfrac{(x+3)^2}{16} + \dfrac{(y+4)^2}{25} = 1$

$C\,(-3, -4);$ vertical MA; $a = 5,\ b = 4$

$c^2 = 5^2 - 4^2 = 9;\ c = 3$

$F:\ (-3, -4 \pm 3) = (-3, -1),\ (-3, -7)$

$e = \dfrac{c}{a} = \dfrac{3}{5} = 0.6$

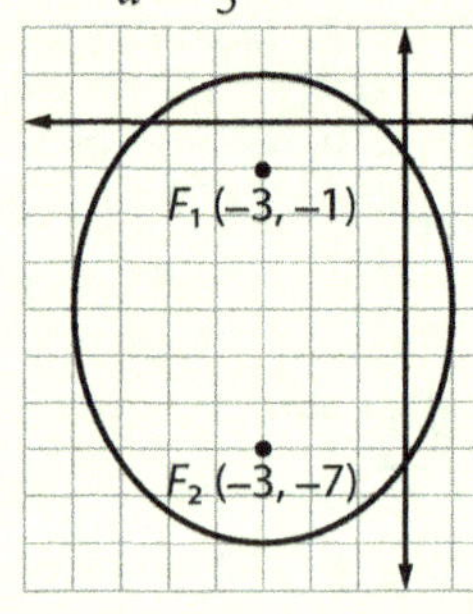

27. $9(x^2 - 4x) + 36(y^2 - 4y) = -36$

$9(x^2 - 4x + 4) + 36(y^2 - 4y + 4)$
$$= -36 + 9(4) + 36(4)$$

$9(x - 2)^2 + 36(y - 2)^2 = 144$

$\dfrac{(x-2)^2}{16} + \dfrac{(y-2)^2}{4} = 1$

$C\,(2, 2);$ horizontal MA; $a = 4,\ b = 2$

$c^2 = 4^2 - 2^2 = 12;\ c = 2\sqrt{3}$

$F:\ (2 \pm 2\sqrt{3}, 2) \approx (5.46, 2),\ (-1.46, 2)$

$e = \dfrac{c}{a} = \dfrac{2\sqrt{3}}{4} \approx 0.87$

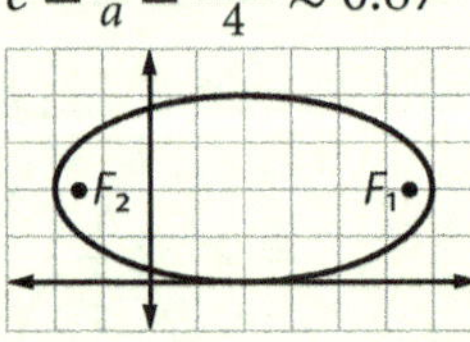

28. $9(x^2 - 12x) + 81(y^2 + 4y) = 81$

$9(x^2 - 12x + 36) + 81(y^2 + 4y + 4)$
$$= 81 + 324 + 324$$

$9(x - 6)^2 + 81(y + 2)^2 = 729$

$\dfrac{(x-6)^2}{81} + \dfrac{(y+2)^2}{9} = 1$

$C\,(6, -2);$ horizontal MA

$a = 9,\ b = 3$

$c^2 = 9^2 - 3^2 = 72;\ c = 6\sqrt{2}$

$F:\ (6 \pm 6\sqrt{2}, -2)$
$$\approx (14.49, -2),\ (-2.49, -2)$$

$e = \dfrac{c}{a} = \dfrac{6\sqrt{2}}{9} \approx 0.94$

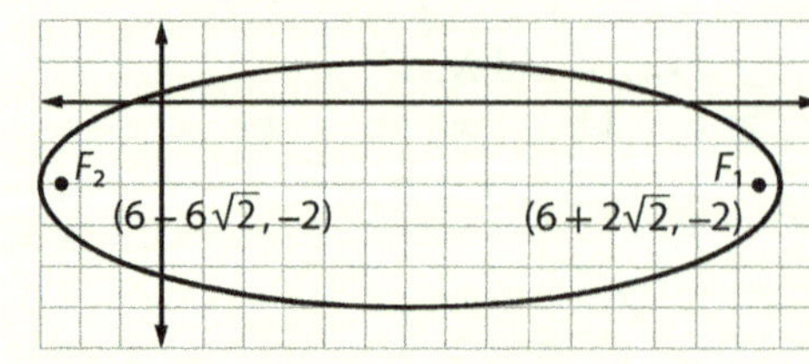

Write an equation modeling the orbit of Mercury if its aphelion is 0.467 AU from the sun and its perihelion is 0.307 AU away. Then graph the ellipse using technology and find its eccentricity, the highest of any of the sun's planets.

Answer

$2a = 0.307 + 0.467 = 0.774$

$a = 0.387$

1. Use the fact that the length of the major axis is the sum of the distances from the sun to the perihelion and the aphelion to find a.

$c = a - (a - c) = 0.387 - 0.307 = 0.08$

2. Determine the value of c by subtracting the perihelion distance from a.

$b^2 = a^2 - c^2 = 0.387^2 - 0.08^2 \approx 0.143$

3. Determine the value of b^2 using $c^2 = a^2 - b^2$.

$\dfrac{x^2}{0.387^2} + \dfrac{y^2}{0.143} \approx 1$

$\dfrac{x^2}{0.150} + \dfrac{y^2}{0.143} \approx 1$

4. Substitute into the standard form equation $\dfrac{x^2}{a^2} + \dfrac{y^2}{b^2} = 1$ and simplify.

$\dfrac{y^2}{0.143} \approx 1 - \dfrac{x^2}{0.150}$

5. Solve the equation for y.

$y^2 \approx 0.143\left(1 - \dfrac{x^2}{0.150}\right)$

$y \approx \pm\sqrt{0.143\left(1 - \dfrac{x^2}{0.150}\right)}$

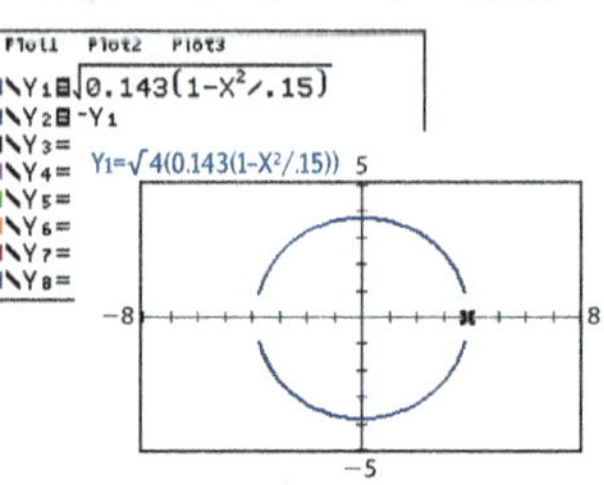

6. Graph the two functions that model the ellipse in a square window to avoid distortion. Remember that parts of the graphs may not display. Use [TRACE] or [TABLE] to verify points that are not displayed.

$e = \dfrac{c}{a} = \dfrac{0.08}{0.387} \approx 0.207$

7. Find the eccentricity of the orbit.

Kepler's laws are specifically stated in terms of planets orbiting the sun, but they also model the orbit of natural and manmade satellites about the earth and other planets.

SKILL ✓ **EXERCISE 35**

Rotating an ellipse about the major axis forms an ellipsoid in which signals emitted from one focus are reflected off the ellipsoid and pass through its other focus. Exercise 32 investigates how a lithotripter applies this property with ultra high frequency (UHF) shockwaves to break up kidney stones without surgery.

❯ **A. Exercises**

State the coordinates of the center and the length of the major and the minor axis for each ellipse.

1. $\dfrac{x^2}{9} + \dfrac{y^2}{25} = 1$ $C\,(0, 0);\ 10;\ 6$

2. $\dfrac{(x-2)^2}{16} + (y + 1)^2 = 1$ $C\,(2, -1);\ 8;\ 2$

State the coordinates of the center, the vertices, and the co-vertices for each ellipse.

3. $\dfrac{x^2}{25} + \dfrac{y^2}{49} = 1$
$C\,(0, 0);\ V:\ (0, \pm 7);\ cV:\ (\pm 5, 0)$

4. $\dfrac{(x-3)^2}{36} + \dfrac{(y+5)^2}{9} = 1$
$C\,(3, -5);\ V:\ (-3, -5),\ (9, -5);$
$cV:\ (3, -2),\ (3, -8)$

State the coordinates for the vertices and the foci for each ellipse, then calculate its eccentricity.

5. $\dfrac{(x+1)^2}{9} + \dfrac{(y-3)^2}{25} = 1$

6. $\dfrac{(x+6)^2}{12} + \dfrac{(y+2)^2}{4} = 1$

Graph each ellipse and its foci.

7. $\dfrac{x^2}{16} + \dfrac{y^2}{4} = 1$

8. $\dfrac{x^2}{4} + \dfrac{y^2}{9} = 1$

9. $(x + 1)^2 + \dfrac{(y-5)^2}{4} = 1$

10. $\dfrac{(x-2)^2}{9} + \dfrac{(y-2)^2}{4} = 1$

5. $V:\ (-1, -2),\ (-1, 8);\ F:\ (-1, -1),\ (-1, 7);\ e = 0.8$

6. $V:\ (-6 \pm 2\sqrt{3}, -2);\ F:\ (-6 \pm 2\sqrt{2}, -2);\ e \approx 0.82$

29. $(y - 2)^2 = 9\left(1 - \dfrac{(x+1)^2}{4}\right)$

$$y = 2 \pm 3\sqrt{1 - 0.25(x + 1)^2}$$

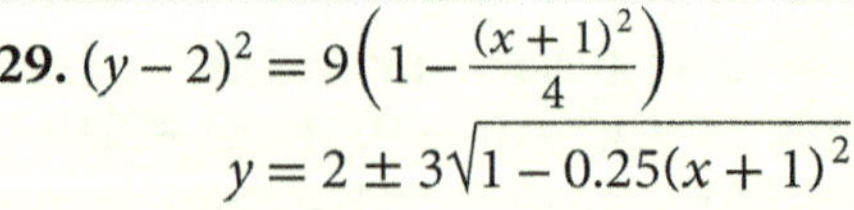
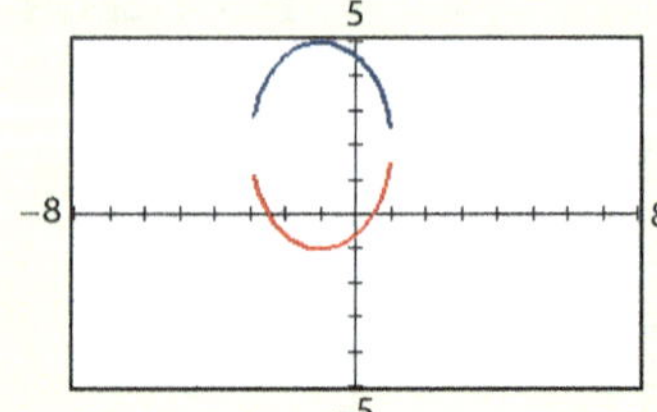

Write each equation in standard form. Then identify the conic as an ellipse or a circle.

11. $16x^2 + 4y^2 = 64$ $\dfrac{x^2}{4} + \dfrac{y^2}{16} = 1$; ellipse

12. $x^2 + y^2 - 6x - 2y = -1$ $\dfrac{(x-3)^2}{9} + \dfrac{(y-1)^2}{9} = 1$; circle

13. $25x^2 + 9y^2 - 54y = 144$ $\dfrac{x^2}{9} + \dfrac{(y-3)^2}{25} = 1$; ellipse

14. $16x^2 + 9y^2 + 128x - 54y = -193$ $\dfrac{(x+4)^2}{9} + \dfrac{(y-3)^2}{16} = 1$; ellipse

Write the standard form equation of each described ellipse.

15. vertices: $(-3, 9)$, $(-3, 1)$
co-vertices: $(-5, 5)$, $(-1, 5)$ $\dfrac{(x+3)^2}{4} + \dfrac{(y-5)^2}{16} = 1$

16. vertices: $(7, -4)$, $(-3, -4)$
co-vertices: $(2, -1)$, $(2, -7)$ $\dfrac{(x-2)^2}{25} + \dfrac{(y+4)^2}{9} = 1$

17. vertices: $(-3, 2)$, $(11, 2)$
length of minor axis: 4 $\dfrac{(x-4)^2}{49} + \dfrac{(y-2)^2}{4} = 1$

18. co-vertices: $(-8, 3)$, $(-2, 3)$
length of major axis: 12 $\dfrac{(x+5)^2}{9} + \dfrac{(y-3)^2}{36} = 1$

19. co-vertices: $(-5, 1)$, $(1, 1)$
foci: $(-2, 5)$, $(-2, -3)$ $\dfrac{(x+2)^2}{9} + \dfrac{(y-1)^2}{25} = 1$

20. foci: $(-11, -2)$, $(5, -2)$
length of major axis: 34 $\dfrac{(x+3)^2}{289} + \dfrac{(y+2)^2}{225} = 1$

❯ B. Exercises

21. Write the standard form equation of an ellipse centered at the origin with an eccentricity of 0.6, a semimajor axis length of 5, and a horizontal major axis.

22. Write the standard form equation of an ellipse centered at $(4, 1)$ with an eccentricity of 0.28, a semimajor axis length of 25, and a vertical major axis.

Graph each ellipse and its foci. State the eccentricity of the ellipse.

23. $x^2 + 25y^2 = 100$ $e \approx 0.98$

24. $4x^2 + y^2 = 16$ $e \approx 0.87$

25. $x^2 + y^2 - 2x + 4y + 1 = 0$ $e = 0$

26. $25x^2 + 16y^2 + 150x + 128y + 81 = 0$ $e = 0.6$

27. $9x^2 + 36y^2 - 36x - 144y + 36 = 0$ $e \approx 0.87$

28. $9x^2 + 81y^2 - 108x + 324y = 81$ $e \approx 0.94$

Use technology to graph each conic.

29. $\dfrac{(x+1)^2}{4} + \dfrac{(y-2)^2}{9} = 1$

30. $(x - 4.2)^2 + (y + 1.6)^2 = 5.75$

21. $\dfrac{x^2}{25} + \dfrac{y^2}{16} = 1$

22. $\dfrac{(x-4)^2}{576} + \dfrac{(y-1)^2}{625} = 1$

$\dfrac{x^2}{81} + \dfrac{y^2}{256} = 1$; $F: (0, \pm 5\sqrt{7})$; 32 in.

31. A mirror frame is being cut so the opening is an 18 in. wide and 32 in. high ellipse. Write the standard form equation modeling the ellipse if it is centered at the origin. Then find the foci and the string length needed to draw the ellipse.

32. A lithotripter is a medical device used to break up kidney stones. Shock waves generated by an electrode at one focus of a half ellipsoidal chamber are reflected toward a kidney stone positioned at the second focus. If the semimajor axis measures 50 mm and the minor axis measures 15 mm, find the distance the electrode should be positioned from the kidney stone.   98.9 mm

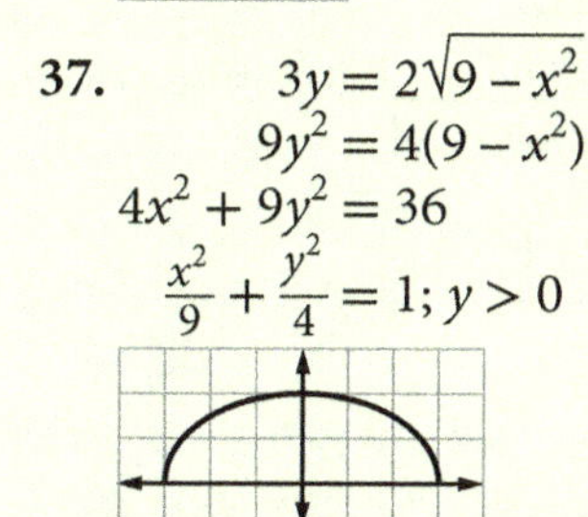

33. Due to its semi-elliptical design, National Statuary Hall in the US Capitol is known as a whispering gallery. If the room is 96 ft long and 46 ft wide, find the distance the foci are from the wall, and then find the eccentricity.   34.3 ft; $e \approx 0.29$

34. The earth travels around the sun in an elliptical path with the sun at one focus. The semimajor axis of the ellipse is about 1.49×10^8 km. If the eccentricity of its orbit is 0.017, find the perihelion and the aphelion. $\approx 1.46 \times 10^8$ km; $\approx 1.52 \times 10^8$ km

35. If the orbit of Mars has a semimajor axis of 1.52 AU and an eccentricity of 0.0934, write a standard form equation modeling its orbit, and then find the perihelion and aphelion. $\dfrac{x^2}{2.31} + \dfrac{y^2}{2.29} = 1$; ≈ 1.38 AU; ≈ 1.66 AU

Graph without the aid of technology.

36. $f(x) = 3\sqrt{1 - x^2}$ 37. $f(x) = \dfrac{2\sqrt{9 - x^2}}{3}$

❯ C. Exercises

Graph the solution to each system of inequalities.

38. $x^2 + y^2 - 6x - 2y + 1 \le 0$
$4x^2 + 25y^2 - 24x - 50y - 39 \le 0$

39. $4x^2 + 9y^2 + 48x - 90y + 225 \le 0$
$x^2 + y^2 + 8x - 10y + 40 \ge 0$

---

34. $0.017 = \dfrac{c}{1.49 \times 10^8}$
$c = 2{,}533{,}000$
perihelion: $a - c$
$= 1.49 \times 10^8 - 2{,}533{,}000$
$\approx 1.46 \times 10^8$ km
aphelion: $a + c$
$= 1.49 \times 10^8 + 2{,}533{,}000$
$\approx 1.52 \times 10^8$ km

35. $e = \dfrac{c}{a}$; $c = ae = 1.52(0.0934) \approx 0.14$
$b^2 \approx 1.52^2 - 0.14^2 \approx 2.29$
$\dfrac{x^2}{2.31} + \dfrac{y^2}{2.29} = 1$
perihelion: $a - c$
$= 1.52 - 0.14 = 1.38$ AU
aphelion: $a + c$
$= 1.52 + 0.14 = 1.66$ AU

36. $y^2 = 9(1 - x^2)$
$9x^2 + y^2 = 9$
$\dfrac{x^2}{1} + \dfrac{y^2}{9} = 1$; $y > 0$

37. $3y = 2\sqrt{9 - x^2}$
$9y^2 = 4(9 - x^2)$
$4x^2 + 9y^2 = 36$
$\dfrac{x^2}{9} + \dfrac{y^2}{4} = 1$; $y > 0$

38. $x^2 - 6x + y^2 - 2y \le -1$
$(x^2 - 6x + 9) + (y^2 - 2y + 1) \le -1 + 9 + 1$
$(x - 3)^2 + (y - 1)^2 \le 9$
circle: $C(3, 1)$; $r = 3$
$4x^2 - 24x + 25y^2 - 50y \le 39$
$4(x^2 - 6x + 9) + 25(y^2 - 2y + 1) \le 39 + 36 + 25$
$4(x - 3)^2 + 25(y - 1)^2 \le 100$
$\dfrac{(x - 3)^2}{25} + \dfrac{(y - 1)^2}{4} \le 1$
ellipse: $C(3, 1)$; $a = 5$, $b = 2$

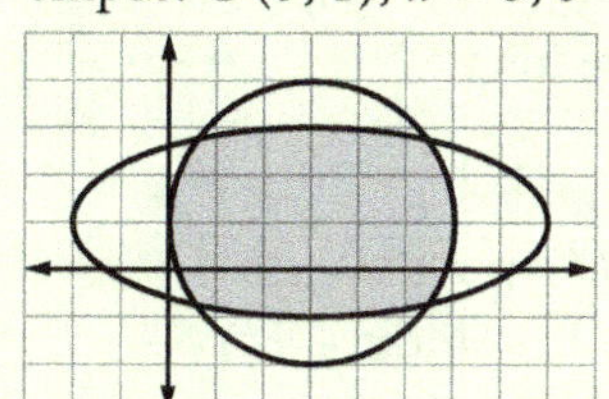

---

30. $(y + 1.6)^2 = 5.75 - (x - 4.2)^2$
$y + 1.6 = \pm\sqrt{5.75 - (x - 4.2)^2}$
$y = \pm\sqrt{5.75 - (x - 4.2)^2} - 1.6$

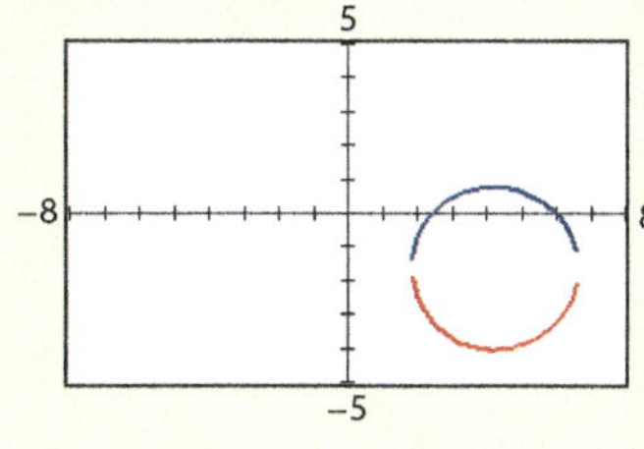

31. $a = 16$, $b = 9$
$\dfrac{x^2}{9^2} + \dfrac{y^2}{16^2} = 1$
$\dfrac{x^2}{81} + \dfrac{y^2}{256} = 1$
$c^2 = 256 - 81 = 175$; $c = 5\sqrt{7}$
$F: (0, 0 \pm 5\sqrt{7}) \approx (0, \pm 13.23)$
string length: $2a = 32$ in.

32. $a = 50$, $2b = 15$; $b = 7.5$
$c^2 = 50^2 - 7.5^2 = 2443.75$
$c \approx 49.434$; $2c \approx 98.87$ mm

33. $a = 48$, $b = 46$
$c^2 = 48^2 - 46^2 = 188$
$c \approx 13.7$
$48 - 13.7 = 34.3$ ft
$e = \dfrac{13.7}{48} \approx 0.29$

39. $4x^2 + 9y^2 + 48x - 90y \leq -225$

$4(x^2 + 12x + 36) + 9(y^2 - 10y + 25)$
$$\leq -225 + 4(36) + 9(25)$$

$4(x + 6)^2 + 9(y - 5)^2 \leq 144$

$$\dfrac{(x + 6)^2}{36} + \dfrac{(y - 5)^2}{16} \leq 1$$

$x^2 + y^2 + 8x - 10y \geq -40$

$(x^2 + 8x + 16) + (y^2 - 10y + 25)$
$$\geq -40 + 16 + 25$$

$(x + 4)^2 + (y - 5)^2 \geq 1$

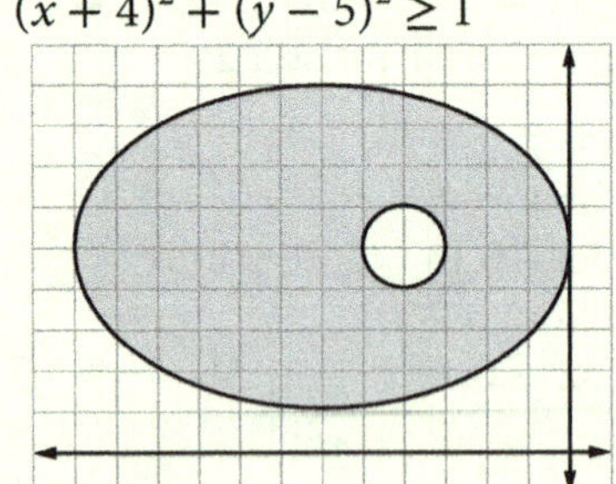

40a. $\dfrac{x^2}{a^2} + \dfrac{y^2}{b^2} = 1$

$$y^2 = b^2\left(1 - \dfrac{x^2}{a^2}\right) = \dfrac{b^2}{a^2}(a^2 - x^2)$$

$$y = \pm\dfrac{b}{a}\sqrt{a^2 - x^2}$$

40b. $x^2 + y^2 = a^2$

$$y^2 = a^2 - x^2$$

$$y = \pm\sqrt{a^2 - x^2}$$

40c. $\dfrac{\pm\frac{b}{a}\sqrt{a^2 + x^2}}{\pm\sqrt{a^2 + x^2}} = \dfrac{b}{a}$

40d. $\dfrac{A_{ellipse}}{A_{circle}} = \dfrac{b}{a}$

$$A_{ellipse} = \dfrac{b}{a}A_{circle} = \dfrac{b}{a}(\pi r^2) = \dfrac{b}{a}(\pi a^2)$$
$$= ab\pi$$

41. $\dfrac{x^2}{27.5^2} + \dfrac{y^2}{16^2} = 1$

$256x^2 + 756.25y^2 - 193{,}600 = 0$

$$y = \sqrt{\dfrac{193{,}600 - 256x^2}{756.25}}$$

$$A = \tfrac{1}{2}\pi ab = \tfrac{1}{2}\pi\left(\dfrac{55}{2}\right)(16) = 220\pi$$
$$\approx 691 \text{ in.}^2$$

42. In any ellipse, $c^2 = a^2 - b^2$.

In P_1 and P_2, $x = c$.

Substitute $x^2 = c^2 = a^2 - b^2$
and solve for y.

$$\dfrac{a^2 - b^2}{a^2} + \dfrac{y^2}{b^2} = 1$$

$b^2(a^2 - b^2) + a^2y^2 = a^2b^2$

$a^2y^2 = a^2b^2 - b^2(a^2 - b^2)$

$y^2 = \dfrac{b^4}{a^2}, \therefore y_1 = \dfrac{b^2}{a}$ and $y_2 = -\dfrac{b^2}{a}$

$P_1P_2 = \dfrac{b^2}{a} - \left(-\dfrac{b^2}{a}\right) = 2\dfrac{b^2}{a}$

---

40. Derive: Complete each step to derive the formula for the area of an ellipse: $A_{ellipse} = \pi ab$.

a. Write the equation for the standard form of an ellipse centered at the origin and solve for y. $y = \pm\dfrac{b}{a}\sqrt{a^2 - x^2}$

b. Write the equation for the standard form of a circle centered at the origin with $r = a$ and solve for y. $y = \pm\sqrt{a^2 - x^2}$

c. What is the ratio of corresponding vertical chords of the circle and the ellipse? $\dfrac{b}{a}$

d. Write a proportion using the fact that the areas of the ellipse and circle also have this ratio (Cavalieri's principle) and solve to derive the formula for the area of an ellipse. $A_{ellipse} = \pi ab$

41. Write a function that models the top curve of the window. Then use the formula for the area of an ellipse, $A_{ellipse} = \pi ab$, to find the area of the window to the nearest square inch.

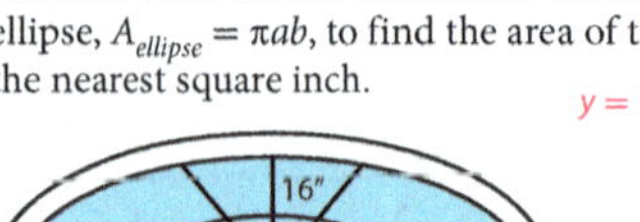

$y = \sqrt{\dfrac{193{,}600 - 256x^2}{756.25}}$; 691 in.^{2}

42. Prove: The focal width of an ellipse is defined as the length of the chord through a focus and perpendicular to the major axis. Show that the ellipse $\dfrac{x^2}{a^2} + \dfrac{y^2}{b^2} = 1$ has a focal width of $P_1P_2 = 2\left(\dfrac{b^2}{a}\right)$.

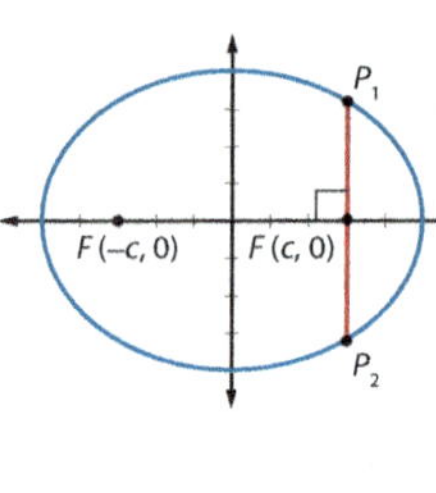

43. Prove: Sketch an ellipse with a vertical major axis centered at $C(h, k)$ with foci $F(h, k \pm c)$ and semimajor and semiminor axis lengths of a and b, respectively, where $c^2 = a^2 - b^2$. Then use the definition of an ellipse and the distances to any point $P(x, y)$ to derive its standard form equation.

Graph each function as a transformation of a parent function. [1.5]

44. $g(x) = \dfrac{1}{x - 3}$

45. $h(x) = \dfrac{1}{2}|x + 2| - 2$

Graph the parabola, its focus, and its directrix. [8.1]

46. $y = \dfrac{1}{2}(x - 2)^2 - 3$

47. $x = -\dfrac{1}{4}(y + 1)^2 - 2$

State the maximum number of real zeros and relative extrema for each function. Then find the real zeros. [2.2]

48. $y = 4x^4 - 36x^3 + 32x^2$

49. $y = 5x^6 - 25x^5 - 70x^4$

50. Classify the graph of $r = 2\cos\theta$. [6.4] D

A. hyperbola

B. parabola

C. ellipse

D. circle centered on the polar axis

E. circle centered on $\theta = \dfrac{\pi}{2}$

48. 4; 3; $x = 0, 1, 8$

49. 6; 5; $x = 0, -2, 7$

51. Shawna makes an initial investment of $5000. How much does she have after 10 yr if the interest is 4.5% compounded continuously? [3.1] A

A. \$7841.56

B. \$7764.85

C. \$5230.14

D. \$450,085.66

E. none of these

52. A 52° central angle of a circle intercepts an arc whose length is 358 ft. What is the radius of the circle (to the nearest foot)? [4.1] D

A. 539 ft

B. 788 ft

C. 160 ft

D. 394 ft

E. none of these

53. Find xy if $x = \dfrac{\sqrt{3}u - v}{2}$ and $y = \dfrac{u + \sqrt{3}v}{2}$. [Algebra] B

A. $\dfrac{\sqrt{3}u^2 - 2uv - \sqrt{3}v^2}{4}$

B. $\dfrac{\sqrt{3}u^2 + 2uv - \sqrt{3}v^2}{4}$

C. $\dfrac{\sqrt{3}u^2 - 2uv + \sqrt{3}v^2}{4}$

D. $\dfrac{\sqrt{3}u^2 + 2uv - \sqrt{3}v^2}{2}$

E. none of these

continued in Answers and Solutions Overflow

ANALYTIC GEOMETRY

Geometry has a rich history that is thousands of years old and includes observations by astronomers and mathematicians dating back to before the time of Moses. Analytic geometry, on the other hand, is not yet 400 yr old and was pioneered by a philosopher and a lawyer. René Descartes (1596–1650), the philosopher, was primarily concerned with geometry as a vehicle to demonstrate the validity of his method for discovering truth and published just one mathematical work. Pierre de Fermat (1607–65), the lawyer, was really more interested in number theory than geometry. Together these men are considered the cofounders of analytic geometry.

Descartes's *The Geometry* was the third appendix to his philosophical treatise, *Discourse on the Method*. According to Ian Maclean, professor emeritus at the Univerisity of Oxford, the work's publication in 1637 "marks the moment … at which geometry and algebra ceased being separate." While geometers had previously used algebra in a limited way, Descartes described how to convert any geometric problem into an algebraic equation or system of equations. Once solved, the answer could be reinterpreted with a geometric construction. The key to Descartes's method was a system of coordinates that allowed the easy location of any point in the plane

(analytic geometry is sometimes called *coordinate geometry*). The modern practices of using a lowercase letter to denote the length of a line, using the first letters of the alphabet to denote constants and the last letters to denote variables, and consistently using exponents all had their origins in *The Geometry*.

Only one of Fermat's many mathematical works was published in his lifetime ("Concerning the Comparison of Curved Lines with Straight Lines"), but through frequent written correspondence with French mathematician Marin Mersenne, his papers were widely circulated among mathematicians. One of Fermat's projects was restoring an ancient Greek manuscript on the loci of plane curves, written by Apollonius. This is apparently where Fermat developed the idea that all equations can be represented as curves in the plane. His algebraic representations of parabolas, hyperbolas, and spirals are still used today. In correspondence with French mathematician Gilles Personne de Roberval in 1636, Fermat detailed the results of his analytic geometry that were eventually published posthumously as "Introduction to Plane and Solid Loci."

Isaac Newton added an extensive appendix on analytic geometry in his *Opticks* (1704) that included 72 "species" of cubic equations, with a sketch of each. Euler's 1748 *Introduction to the Analysis of the Infinite* contained a "systematic treatise on analytic geometry in the sense of Fermat," according to mathematics historian Carl Boyer, who calls Euler's book "probably the most influential textbook of modern times." In the nineteenth century Julius Plucker devoted a lifetime of study to analytic geometry, paving the way for modern breakthroughs in specialized branches of this subject that originated with the work of Descartes and Fermat.

COMPREHENSION CHECK

1. List the cofounders of analytic geometry and the primary occupation of each.

2. Describe the motivation of each cofounder of analytic geometry.

3. Name three other scientist-mathematicians who made major contributions to analytic geometry.

4. In *Principles of Philosophy* Descartes claims, "in order to seek truth, it is necessary once in the course of our life, to doubt, as far as possible, of all things." Is this statement consistent with a biblical worldview? Why or why not?

Historical Connection

Objectives

1. To summarize the development of analytic geometry

2. To explain the motivation behind the work of Descartes and Fermat

3. To evaluate the worldview indicated by statements of one of the founders of analytic geometry

Answers

1. Descartes, a philosopher; Fermat, a lawyer

2. Descartes sought proof for his "method" of determining truth, and Fermat sought to reconstruct the work of Apollonius.

3. Newton, Euler, and Plucker

4. Answers will vary. Scripture presents the Bible as true (John 17:17) and even more trustworthy than our experience (2 Pet. 1:16–21). All worldviews choose a foundation. Descartes made reason his foundation for determining truth (I think, therefore I am.) His assumption that reason is the path to truth governed all his actions. A Christian pursues truth with the assumption that Scripture is the source and path to truth.

PRESENTATION

Students are now introduced to concepts of analytic geometry (coordinate geometry) early in their education. This Historical Connection explores how geometry and algebra became combined into one familiar and useful discipline.

René Descartes's method provides an interesting way to introduce and discuss worldviews that are based on human reasoning. For a helpful exposition of this topic, see Appendix A in *Redeeming Sociology: A God-Centered Approach* by Vern Poythress.

8.3 Hyperbolas

Objectives

1. To define a hyperbola as a locus of points
2. To analyze and graph a hyperbola given its equation
3. To write the equation of a hyperbola using its center, vertices, and foci
4. To solve real-world problems related to hyperbolas

Flash

The distance from a lightning bolt can be estimated by counting the seconds between when the lightning is seen and when the thunder is heard, estimating 1 mi for every 5 sec or 1 km for every 3 sec.

Vocabulary

branches
center
conjugate axis
co-vertices
focal distance
hyperbola
semiconjugate axis
semitransverse axis
transverse axis
vertices

Reading and Writing Mathematics

Describe a parabola, an ellipse, and a hyperbola in terms of the distance from any point on the conic to its foci or directrix. In a parabola the distances from each point to the focus and directrix are the same. In an ellipse the sum of the distances to the two foci is constant, but in a hyperbola it is the difference of the distances to the two foci that is constant.

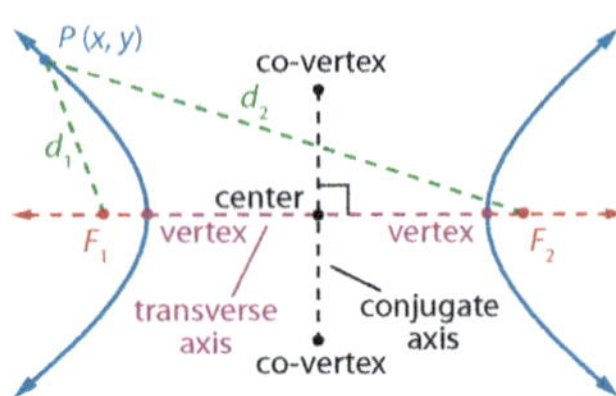

8.3 Hyperbolas

The equation of a hyperbola can be written to model the possible positions of a lightning strike.

After completing this section, you will be able to

- define a hyperbola.
- analyze and graph a hyperbola.
- write the equation of a hyperbola.
- solve real-world problems related to hyperbolas.

Recall that parabolas and ellipses intersect a single nappe of a right circular conical surface, while a hyperbola intersects both nappes of the surface. Compare the definitions and characteristics of a hyperbola and an ellipse and note their similarities and differences.

▌ **DEFINITION**

A **hyperbola** is the set of points $P(x, y)$ in a plane such that the difference of the distances from two fixed points (its foci) is constant.

A hyperbola consists of two disjoint, symmetrical curves called *branches*. The *vertices* are the intersections of each branch and the line containing its foci (F_1 and F_2). The segment connecting the vertices is the hyperbola's *transverse axis* and its midpoint is the hyperbola's *center*. The *conjugate axis* is a segment through the center and perpendicular to the transverse axis. While the conjugate axis does not intersect the hyperbola, the endpoints are the *co-vertices*.

The characteristic distances a, b, and c are similar to those of the ellipse, but they are related differently. The hyperbola's *focal distance* is the distance from the center to each focus, c. The length of the *semitransverse axis* is the distance from the hyperbola's center to either vertex, a. The length of the *semiconjugate axis* is the distance from the center to each co-vertex, b, where $b^2 = c^2 - a^2$ (or $c^2 = a^2 + b^2$). This implies $c > a$, $c > b$, and that a may be greater than, less than, or equal to b. The vertices and co-vertices can be used to draw a rectangle whose diagonals are asymptotes of the hyperbola's branches.

The difference of the distances from the foci can be determined using V_1 on the transverse axis. Since $V_1F_2 = a + c$ and $V_1F_1 = c - a$, the difference is $(a + c) - (c - a) = 2a$, the length of the transverse axis. By definition, the difference is constant for any point on the hyperbola.

The definition of a hyperbola can then be used to derive a standard form equation for a hyperbola with a horizontal transverse axis centered at $C(h, k)$ whose foci $F(h \pm c, k)$ are c units from the center.

Applying the distance formula to the point $P(x, y)$ on the hyperbola:
$$PF_1 - PF_2 = 2a.$$
$$\sqrt{[x - (h - c)]^2 + (y - k)^2} - \sqrt{[x - (h + c)]^2 + (y - k)^2} = 2a$$

Regroup the terms, using the Associative Property of Addition.
$$\sqrt{[(x - h) + c]^2 + (y - k)^2} - \sqrt{[(x - h) - c]^2 + (y - k)^2} = 2a$$

PRESENTATION

Lesson Opener

Solve for y.

1. $\dfrac{x^2}{8} + \dfrac{y^2}{7} = 1$ $\quad y = \pm\sqrt{7 - \dfrac{7x^2}{8}}$

2. $\dfrac{(y - 3)^2}{5} - \dfrac{(x + 4)^2}{10} = 1$

   $y = \pm\sqrt{5 + \dfrac{(x + 4)^2}{2}} + 3$

3. $\dfrac{(x - 1)^2}{6} - \dfrac{(y + 2)^2}{2} = 1$

   $y = \pm\sqrt{-2 + \dfrac{(x - 1)^2}{3}} - 2$

The Lesson Opener reviews skills often required for entering conic equations on a graphing calculator.

Present the definition of a *hyperbola*. Use the figure in the text to introduce the terms and distances associated with a hyperbola. Compare and contrast definitions and equations of ellipses and hyperbolas. Emphasize that in a hyperbola, a^2 is the denominator of the positive squared term, while a^2 is the larger denominator in an ellipse. Point out that $c^2 = a^2 + b^2$ in a hyperbola while $c^2 = a^2 - b^2$ in an ellipse. Note that a could be $<$, $>$, or $= b$.

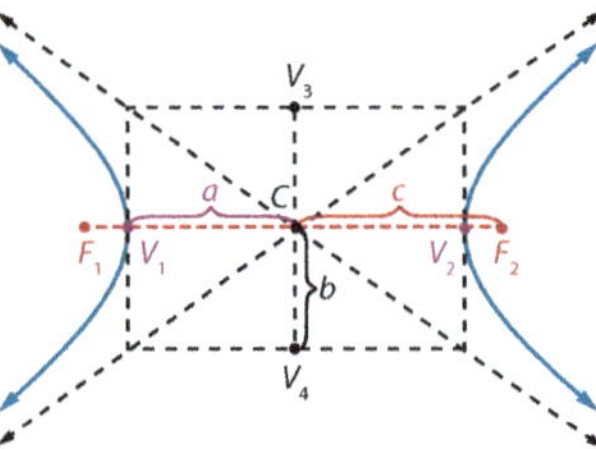

Isolate a radical and square both sides.

$$\sqrt{[(x-h)+c]^2+(y-k)^2}=2a+\sqrt{[(x-h)-c]^2+(y-k)^2}$$
$$[(x-h)+c]^2+(y-k)^2=4a^2+4a\sqrt{[(x-h)-c]^2+(y-k)^2}+[(x-h)-c]^2+(y-k)^2$$

Subtract $(y-k)^2$ from both sides and expand the binomials outside the radical.

$$(x-h)^2+2c(x-h)+c^2=4a^2+4a\sqrt{[(x-h)-c]^2+(y-k)^2}+(x-h)^2-2c(x-h)+c^2$$

Isolate the radical term and simplify, keeping $(x-h)$ and $(y-k)$ intact.

$$4c(x-h)-4a^2=4a\sqrt{[(x-h)-c]^2+(y-k)^2}$$
$$c(x-h)-a^2=a\sqrt{[(x-h)-c]^2+(y-k)^2}$$

Square both sides and use the Distributive Property to expand the right side.

$$c^2(x-h)^2-2a^2c(x-h)+a^4=a^2\{[(x-h)-c]^2+(y-k)^2\}$$
$$c^2(x-h)^2-2a^2c(x-h)+a^4=a^2[(x-h)^2-2c(x-h)+c^2+(y-k)^2]$$
$$c^2(x-h)^2-2a^2c(x-h)+a^4=a^2(x-h)^2-2a^2c(x-h)+a^2c^2+a^2(y-k)^2$$

Collect all terms containing x or y on the left side and all other terms on the right.

$$c^2(x-h)^2-a^2(x-h)^2-a^2(y-k)^2=a^2c^2-a^4$$

Factor $(x-h)^2$ from the first two terms on the left and a^2 from the right side.

$$(c^2-a^2)(x-h)^2-a^2(y-k)^2=a^2(c^2-a^2)$$

Since $b^2=c^2-a^2$ in a hyperbola, substitute and divide both sides by a^2b^2.

$$b^2(x-h)^2-a^2(y-k)^2=a^2b^2$$
$$\frac{b^2(x-h)^2}{a^2b^2}-\frac{a^2(y-k)^2}{a^2b^2}=\frac{a^2b^2}{a^2b^2}$$
$$\frac{(x-h)^2}{a^2}-\frac{(y-k)^2}{b^2}=1$$

The central rectangle can be used to find the hyperbola's asymptotes. Consider the hyperbola $\frac{x^2}{a^2}-\frac{y^2}{b^2}=1$, which is centered at the origin and has a horizontal transverse axis. Clearing fractions produces $b^2x^2-a^2y^2=a^2b^2$, and solving for y produces $y^2=\frac{b^2(x^2-a^2)}{a^2}$ and $y=\pm\frac{b}{a}\sqrt{x^2-a^2}$. As $x\to\pm\infty$, the subtraction of a^2 becomes insignificant and $\sqrt{x^2-a^2}\to\sqrt{x^2}$ or x, but is never equal to x. Therefore, the hyperbola $y=\pm\frac{b}{a}\sqrt{x^2-a^2}$ approaches its asymptotes $y=\pm\frac{b}{a}x$, lines intersecting at the hyperbola's center $(0,0)$ with slopes of $\frac{b}{a}$ and $-\frac{b}{a}$.

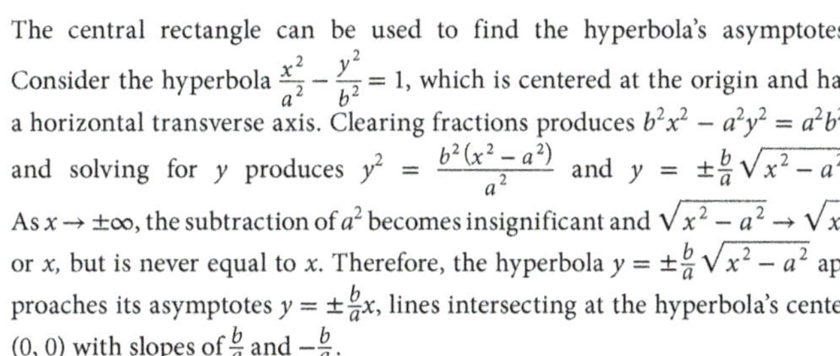

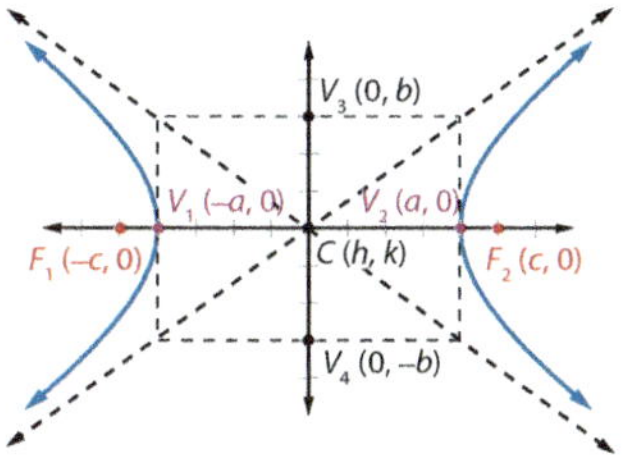

A similar derivation for the inverse hyperbola $\frac{y^2}{a^2}-\frac{x^2}{b^2}=1$, which is centered at the origin with a vertical transverse axis, yields oblique asymptotes of $y=\pm\frac{a}{b}x$. When a hyperbola is translated, the asymptotes experience the same translation.

Notice that the signs of the squared terms in the standard form determine the orientation of the transverse axis.

In his seventeenth-century work "Introduction to Plane and Solid Loci," Fermat stated that a hyperbola was the locus of points represented by the second-degree equation "A in E aequetur Z plano" (or $xy=k^2$ in modern notation).

Additional Exercises

Identify the center, the vertices, and the equations of the asymptotes for each hyperbola. Then graph the hyperbola and its foci.

1. $\dfrac{(x-2)^2}{16}-\dfrac{(y-4)^2}{16}=1$

$C\,(2,4)$; $V:\,(6,4),\,(-2,4)$

$y=x+2$ and $y=-x+6$

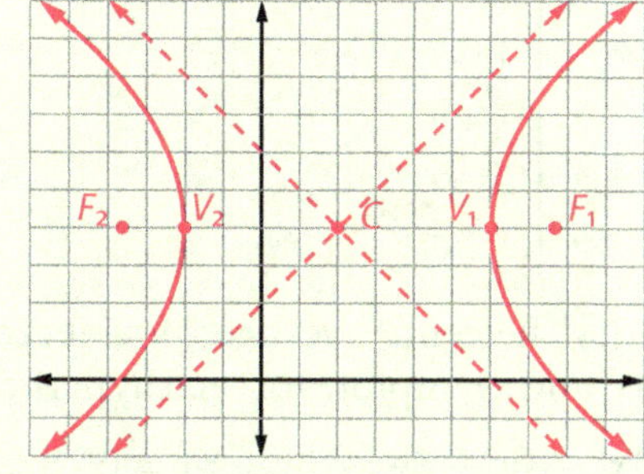

2. $\dfrac{(x+2)^2}{4}-y^2=1$

$C\,(-2,0)$; $V:\,(0,0),\,(-4,0)$

$y=\frac{1}{2}x+1$ and $y=-\frac{1}{2}x-1$

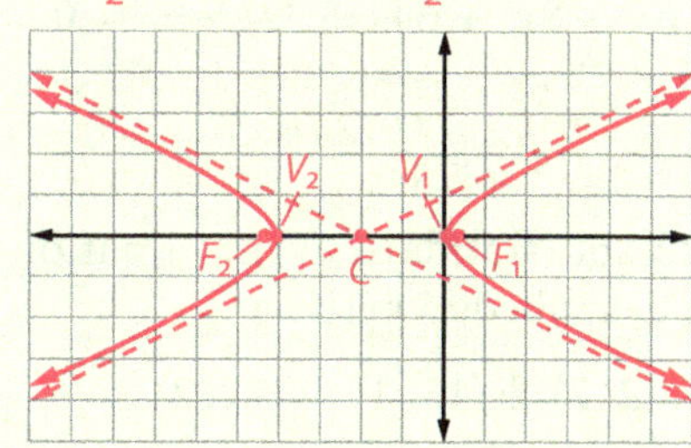

Common Student Error A common misconception is that a curve can never intersect an asymptote. This is true for vertical asymptotes, but not for slant or horizontal asymptotes. A rational function may touch or even cross a slant or horizontal asymptote. Encourage students to use the Internet keyword search *crossing an asymptote*.

The derivation of the standard form equation for a hyperbola presented in the text is similar to the derivation of the formula for an ellipse. The difference in distances is found by subtracting the shorter distance from the longer distance. While reversing the order produces a difference of $-2a$, squaring both sides produces $4a^2$ regardless of the order used to find the difference.

Although the diagonals of the central rectangle can be used to sketch the asymptotes, students will want to use the generalized formulas to determine their equations. Show how hyperbolas centered on the origin approach their asymptotes. Students are challenged to formally derive the standard form equation for a hyperbola with a *vertical transverse axis* in exercise 43.

The information in the table for the standard form equations of a hyperbola may seem overwhelming at first. Consider examining all the values related to a hyperbola with a horizontal transverse axis first, then carefully compare to those values related to a hyperbola with a vertical transverse axis. Note that a is associated with the positive term in a hyperbola and b is associated with the negative term. Contrast this with an ellipse where both fractions are positive and $a>b$.

Interactive Activity Use the Internet keyword search *desmos interactive hyperbola* to find interactive demonstrations.

You may want to have the students sketch the graph for Example 1. Begin by asking the students what they know directly from looking at the equation. (*a*

3. $\dfrac{(y+1)^2}{25} - \dfrac{(x-3)^2}{9} = 1$

$C\,(3, -1);\ V:\,(3, 4),\,(3, -6)$

$y = -\dfrac{5}{3}x + 4$ and $y = \dfrac{5}{3}x - 6$

Identify the center, vertices, foci, eccentricity, and asymptotes of each hyperbola.

4. $9x^2 - 4y^2 + 54x - 8y + 41 = 0$

$C\,(-3, -1);\ V:\,(-5, -1),\,(-1, -1)$

$F: \approx (-6.6, -1),\,(0.6, -1);\ e \approx 1.8$

$y = \dfrac{3}{2}x + \dfrac{7}{2}$ and $y = -\dfrac{3}{2}x - \dfrac{11}{2}$

5. $16x^2 - 8y^2 + 64x + 32y - 96 = 0$

$C\,(-2, 2);\ V: \approx (-4.8, 2),\,(0.8, 2)$

$F: \approx (-6.9, 2),\,(2.9, 2);\ e \approx 1.7$

$y \approx 1.4x + 4.8$ and $y \approx -1.4x - 0.8$

Write the standard form equation of each described hyperbola.

6. $C\,(2, 2),\ F\,(15, 2),\ V\,(-3, 2)$

$\dfrac{(x-2)^2}{25} - \dfrac{(y-2)^2}{144} = 1$

7. $e = \dfrac{5}{4};\ V:\,(-2, 3),\,(-2, -2)$

$\dfrac{4\left(y - \frac{1}{2}\right)^2}{25} - \dfrac{64(x+2)^2}{225} = 1$

or $\dfrac{(y - 0.5)^2}{6.25} - \dfrac{(x+2)^2}{3.515625} = 1$

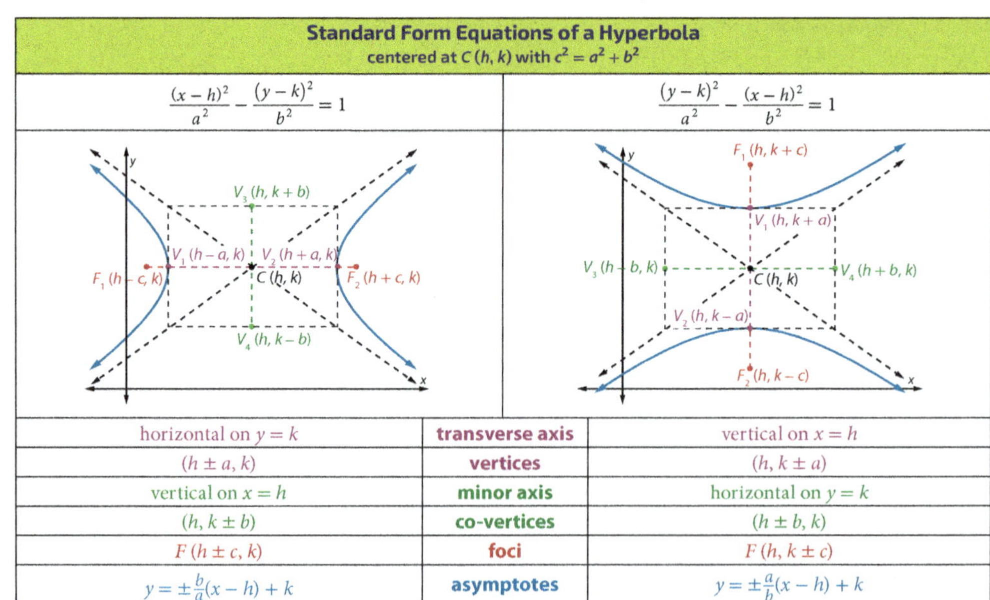

Standard Form Equations of a Hyperbola
centered at $C\,(h, k)$ with $c^2 = a^2 + b^2$

$$\dfrac{(x-h)^2}{a^2} - \dfrac{(y-k)^2}{b^2} = 1 \qquad\qquad \dfrac{(y-k)^2}{a^2} - \dfrac{(x-h)^2}{b^2} = 1$$

horizontal on $y = k$	**transverse axis**	vertical on $x = h$
$(h \pm a, k)$	**vertices**	$(h, k \pm a)$
vertical on $x = h$	**minor axis**	horizontal on $y = k$
$(h, k \pm b)$	**co-vertices**	$(h \pm b, k)$
$F\,(h \pm c, k)$	**foci**	$F\,(h, k \pm c)$
$y = \pm\frac{b}{a}(x - h) + k$	**asymptotes**	$y = \pm\frac{a}{b}(x - h) + k$

Example 1   Graphing a Hyperbola

Graph $\dfrac{x^2}{9} - \dfrac{(y-2)^2}{16} = 1$ and its foci. Then state the equations of its asymptotes.

Answer

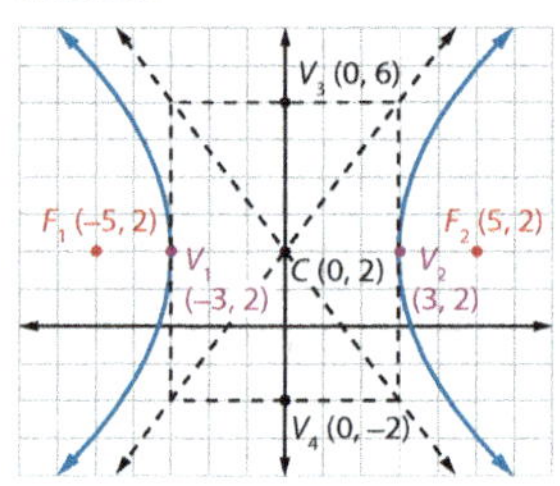

asymptotes: $y = \pm\dfrac{4}{3}(x - 0) + 2$ or

$\qquad\qquad y = \pm\dfrac{4}{3}x + 2$

1. Note that the hyperbola has a horizontal transverse axis at $y = 2$ since the x^2-term is positive.
2. Identify and plot the center: $C\,(0, 2)$.
3. Sketch the central rectangle using $a = 3$ and $b = 4$ and draw the asymptotes.
4. Plot the vertices: $(0 \pm 3, 2)$.
5. Sketch the hyperbola through its vertices and approaching its asymptotes.
6. Since $c^2 = 9 + 16 = 25$, use $c = 5$ to plot the foci: $(\pm 5, 2)$.
7. Substitute into $y = \pm\dfrac{b}{a}(x - h)^2 + k$ and simplify.

SKILL ✔ **EXERCISE 13**

Like an ellipse, the eccentricity of a hyperbola is defined as $e = \frac{c}{a}$. In any hyperbola $c > a$, so its eccentricity is always greater than 1. When e is close to 1, the foci are near the vertices and the branches are near the transverse axis. As e increases, the foci are farther from the vertices and the hyperbola's branches are wider. In Section 8.6 you will see that the eccentricity of any parabola is 1.

---

horizontal TA, $C\,(0, 2)$, $a = 3$, and $b = 4$)
Technology is often used to graph hyperbolas, but sketching them will help the students appreciate some of the related characteristics and values.

Introduce the eccentricity formula for a hyperbola. Use the illustrations to compare eccentricities of conic sections.

The equation given in Example 2 is in the general form for conics. Converting the equation to standard form is a key step in analyzing the hyperbola. The equation must be solved for y before it can be entered into most graphing calculators. Consider using tables on the calculator to explore how the hyperbola's *branches* approach its asymptotes.

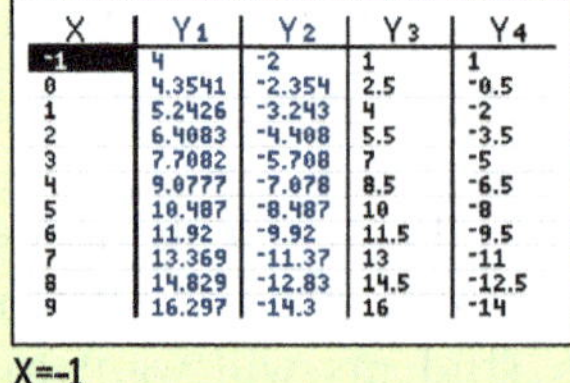

X	Y₁	Y₂	Y₃	Y₄
-1	4	-2	1	1
0	4.3541	-2.354	2.5	-0.5
1	5.2426	-3.243	4	-2
2	6.4083	-4.408	5.5	-3.5
3	7.7082	-5.708	7	-5
4	9.0777	-7.078	8.5	-6.5
5	10.487	-8.487	10	-8
6	11.92	-9.92	11.5	-9.5
7	13.369	-11.37	13	-11
8	14.829	-12.83	14.5	-12.5
9	16.297	-14.3	16	-14

$X = -1$

Common Student Error Mistakes can easily be made when transforming conic equations. Encourage students to use graphing technology to confirm their work and to take time to find their mistake if graphing indicates an error.

Sketching the given characteristics is extremely helpful when writing an equation of the hyperbola in Example 3. Encourage the students to focus on using the

known values to find unknown values. Point out that the signs of the quadratic terms in the final equation in step 5 are different, which indicates a hyperbola.

In Example 4 the constant difference in distances from the foci, $\Delta d = 2a$, is calculated in step 2. The possible locations lie on the resulting hyperbola. The intersection of two hyperbolas is often used to find an exact location, as in exercise 41.

The use of hyperbolic mirrors in the Cassegrain telescope is an interesting application of hyperbolas. Use the Internet keyword search *Cassegrain video* to find examples of its detailed operation.

Notice how a conic's eccentricity (the measure of its deviation from a circle) increases as the intersecting plane deviates from being perpendicular to the axis of the conical surface.

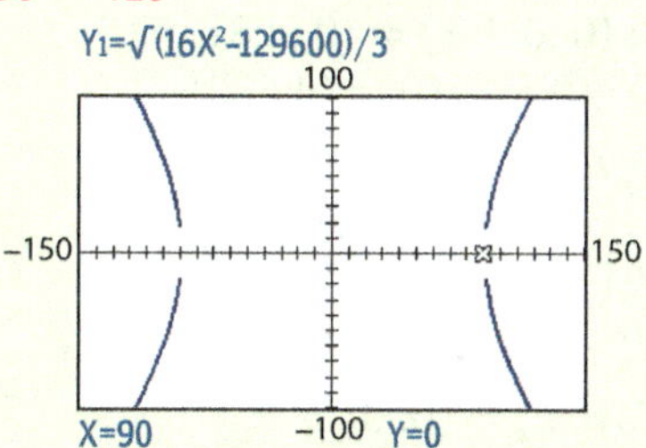

Circle	Ellipse	Parabola	Hyperbola
$e = 0$	$0 \le e < 1$	$e = 1$	$e > 1$

Example 2 Analyzing a Hyperbola

Given the hyperbola $9x^2 - 4y^2 + 18x + 8y + 41 = 0$, identify its center, vertices, co-vertices, foci, eccentricity, and asymptotes.

Answer

$$9(x^2 + 2x) - 4(y^2 - 2y) = -41$$
$$9(x^2 + 2x + 1) - 4(y^2 - 2y + 1) = -41 + 9 - 4$$
$$9(x + 1)^2 - 4(y - 1)^2 = -36$$
$$\frac{(y - 1)^2}{9} - \frac{(x + 1)^2}{4} = 1$$

1. Write the equation in standard form by completing the square for the x- and y-terms and dividing both sides by the constant, -36. Note that the hyperbola has a vertical transverse axis.

$C\,(-1, 1)$
$a = 3; b = 2$

2. Use $\frac{(y - k)^2}{a^2} - \frac{(x - h)^2}{b^2} = 1$ to identify the center and the values of a and b.

vertices: $(-1, 1 \pm 3) = (-1, 4)$ and $(-1, -2)$
co-vertices: $(-1 \pm 2, 1) = (1, 1)$ and $(-3, 1)$

3. Identify the vertices $(h, k \pm a)$ and co-vertices $(h \pm b, k)$.

$c^2 = 9 + 4 = 13; c = \sqrt{13}$
$F\,(-1, 1 \pm \sqrt{13}) \approx (-1, 4.6)$ and $(-1, -2.6)$

4. Use $c^2 = a^2 + b^2$ to find c and identify the foci: $F\,(h, k \pm c)$.

$e = \frac{\sqrt{13}}{3} \approx 1.2$

5. Calculate the eccentricity using $e = \frac{c}{a}$.

$y = \pm\frac{3}{2}(x + 1) + 1$
$y = \frac{3}{2}x + \frac{5}{2}$ and $y = -\frac{3}{2}x - \frac{1}{2}$

6. Use $y = \pm\frac{a}{b}(x - h) + k$ to identify the asymptotes.

Check

$$(y - 1)^2 = 9\left(1 + \frac{(x + 1)^2}{4}\right)$$
$$y = 1 \pm 3\sqrt{1 + \frac{(x + 1)^2}{4}}$$

7. Solve the standard form equation for y and graph the two functions that form the hyperbola. Then graph the asymptotes.

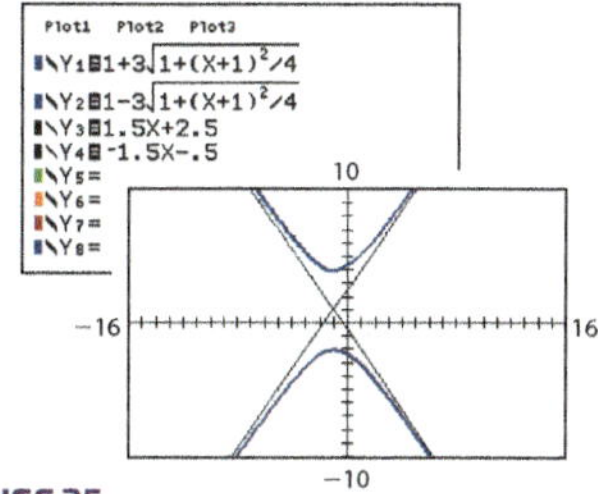

SKILL ✔ **EXERCISE 25**

The APPS menu on the TI-84 graphing calculator includes a conics application. This is an easy way to graph a conic, but this method has limited functions.

TIPS

Ex. 21–26 Encourage students to be careful in identifying h and k, because finding the *center, vertices, co-vertices*, foci, and equations of asymptotes all depends on these values.

Ex. 41 Encourage the students to consider which signal arrives first to determine the branch of the hyperbola that contains the possible locations for the ship.

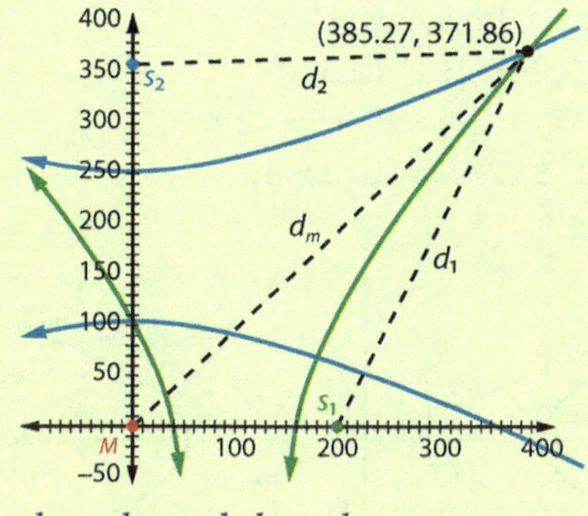

$d_1 < d_M$ and $d_2 < d_M$

8. A ship receives signals from two stations 300 mi apart and determines that it is 180 mi closer to the first station than it is to the second. Write the equation that models the hyperbolic path that the ship must be on. Then use technology to graph the hyperbola.

$$\frac{x^2}{90^2} - \frac{y^2}{120^2} = 1$$

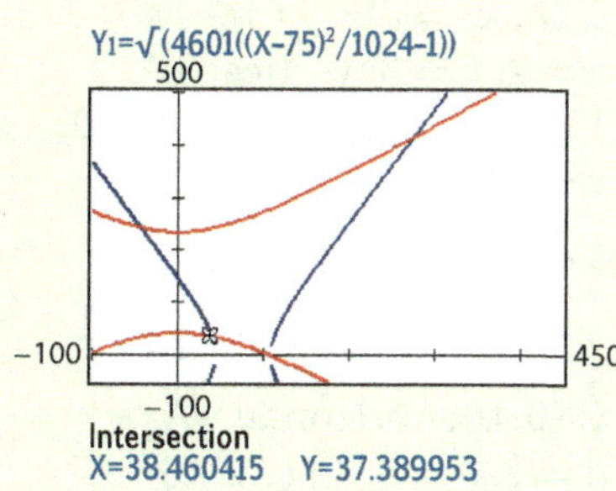

9. Assume that transmitters at $M\,(0, 0)$, $S_1\,(150\text{ km}, 0)$, and $S_2\,(0, 280\text{ km})$ simultaneously broadcast radio signals that travel at 320 km/sec. Write the equation of the hyperbola modeling a ship's possible locations if the signal from…

   a. S_1 arrives 200 ms later than the signal from M.

   $$\frac{(x - 75)^2}{1024} - \frac{y^2}{4601} = 1$$

   b. S_2 arrives 600 ms later than the signal from M.

   $$\frac{(y - 140)^2}{9216} - \frac{x^2}{10{,}384} = 1$$

   c. Use graphing technology to find the coordinates of the ship to the nearest km. (38, 37)

Assignments

- **Minimum:** 1–7, 10–13, 15, 18–19, 22, 25, 29, 35, 44–45, 50–52

- **Standard:** 1–5, 8–9, 12–13, 15–16, 19, 22; 27–31 odd; 35, 39, 44–46, 50–53

- **Extended:** 1–4; 6–18 even; 19, 22, 25, 28–29, 32, 35–36, 38, 41, 43; 46–52 even

Solutions

▶ A. Exercises

5. $C(h, k) = (0, 0)$; $a = 12$, $b = 5$
TA: $2a = 24$; CA: $2b = 10$

6. $C(h, k) = (-7, -2)$; $a = 3$, $b = 6$
TA: $2a = 6$; CA: $2b = 12$

7. $C(h, k) = (0, 0)$
$a = 8$, $b = 6$; vertical TA
$V: (0, 0 \pm 8) = (0, \pm 8)$
$cV: (0 \pm 6, 0) = (\pm 6, 0)$

8. $C(h, k) = (1, 5)$
$a = 4$, $b = 6$; horizontal TA
$V: (1 \pm 4, 5) = (5, 5), (-3, 5)$
$cV: (1, 5 \pm 6) = (1, 11), (1, -1)$

9. $C(h, k) = (0, 0)$
$a = 9$, $b = 12$; vertical TA
$V: (0, 0 \pm 9) = (0, \pm 9)$
$c^2 = 9^2 + 12^2 = 225$; $c = 15$
$F: (0, 0 \pm 15) = (0, \pm 15)$
$e = \dfrac{c}{a} = \dfrac{15}{9} = \dfrac{5}{3} \approx 1.67$

10. $C(h, k) = (2, -4)$
$a = 8$, $b = 15$; horizontal TA
$V: (2 \pm 8, -4) = (10, -4), (-6, -4)$
$c^2 = 8^2 + 15^2 = 289$; $c = 17$
$F: (2 \pm 17, -4) = (19, -4), (-15, -4)$
$e = \dfrac{c}{a} = \dfrac{17}{8} = 2.125$

11. $C(h, k) = (0, 0)$
$a = 3$, $b = 4$; horizontal TA
$V: (0 \pm 3, 0) = (\pm 3, 0)$
$y = \pm\dfrac{b}{a}(x - h) + k = \pm\dfrac{b}{a}(x - 0) + 0$
$y = \pm\dfrac{4}{3}x$

12. $C(h, k) = (-2, 1)$
$a = 2$, $b = 3$; vertical TA
$V: (-2, 1 \pm 2) = (-2, 3), (-2, -1)$
$y = \pm\dfrac{a}{b}(x - h) + k$
$\quad = \pm\dfrac{2}{3}(x - (-2)) + 1$
$y = \dfrac{2}{3}x + \dfrac{7}{3}$ and $y = -\dfrac{2}{3}x - \dfrac{1}{3}$

13. $C(0, 0)$; horizontal TA
$a = 2$, $b = 3$; $V: (\pm 2, 0)$
$c^2 = 2^2 + 3^2 = 13$; $c = \sqrt{13}$
$F: \left(0, \pm\sqrt{13}\right) \approx (0, \pm 3.6)$
$y = \pm\dfrac{3}{2}x$

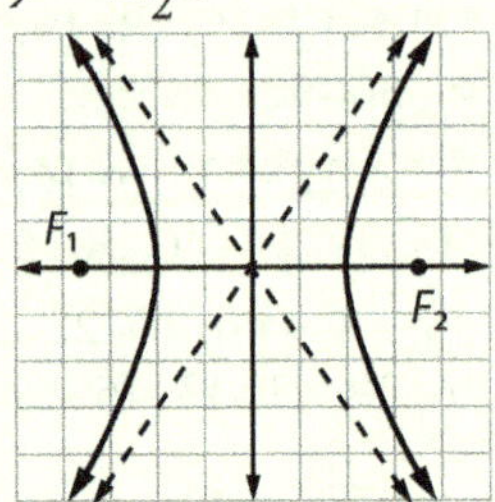

To write the standard form equation of a hyperbola, determine its center and the lengths a and b. The standard form equation can then be changed to the general quadratic equation used for conics, $Ax^2 + Bxy + Cy^2 + Dx + Ey + F = 0$.

Write a standard form equation for the hyperbola with vertices at (4, 1) and (4, 5) and an eccentricity of $\frac{3}{2}$. Then write an equivalent equation in the general quadratic form for conics.

Answer

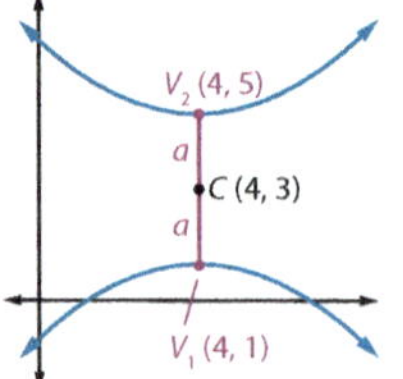

$C(h, k) = \left(\dfrac{4 + 4}{2}, \dfrac{1 + 5}{2}\right) = (4, 3)$

$a = 2$
$\dfrac{c}{2} = \dfrac{3}{2}$; $c = 3$
$b^2 = c^2 - a^2 = 3^2 - 2^2 = 5$; $b = \sqrt{5}$

$\dfrac{(y - 3)^2}{4} - \dfrac{(x - 4)^2}{5} = 1$

$5(y - 3)^2 - 4(x - 4)^2 = 20$
$5y^2 - 30y + 45 - 4x^2 + 32x - 64 = 20$
$-4x^2 + 5y^2 + 32x - 30y - 39 = 0$
$4x^2 - 5y^2 - 32x + 30y + 39 = 0$

1. Sketch the hyperbola and find the center, the midpoint of the transverse axis.
2. Use the center and vertices to find a.
3. Use $e = \frac{c}{a}$ to find c and $c^2 = a^2 + b^2$ to find b.
4. Substitute into $\dfrac{(y - h)^2}{a^2} - \dfrac{(x - k)^2}{b^2} = 1$.
5. Multiply both sides by the LCD and expand to convert to the general form for conics. Multiplying by -1 allows the lead coefficient to be positive, which is a common practice.

SKILL ✔ **EXERCISES 27, 33**

Note that $B = 0$ in the general form equation for Example 3, and that A and C have different signs ($AC < 0$). This indicates that the conic is a hyperbola with a horizontal or vertical transverse axis.

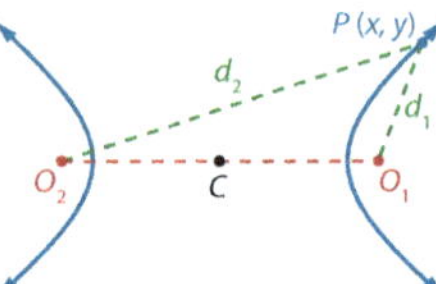

By examining the difference of the distances between a lightning strike and two observers (O_1 and O_2), the equation of a hyperbola can be written to model all possible positions of the strike, the points where this difference of distances $\Delta d = d_1 - d_2 = 2a$. Comparing this hyperbola with a second hyperbola derived from the positions of another set of observers allows the exact location to be found.

Eric observes a lightning strike and hears the thunder 2 sec earlier than Abby, who is standing 1 km due east of Eric. If sound travels at 340 m/sec, model the possible locations of the lightning strike by writing the equation of a hyperbola centered at the origin and with the observers at the foci.

Answer

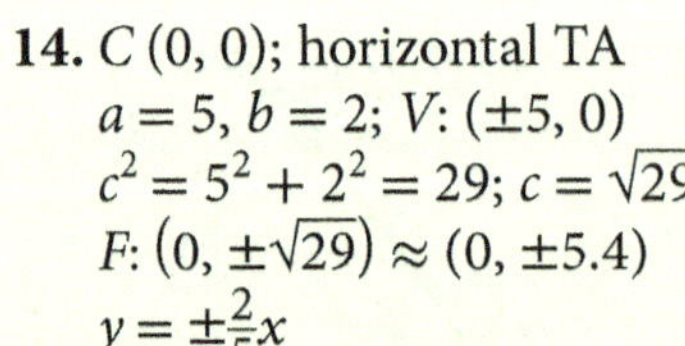

Using $C(0, 0)$, $c = 500$ m.
$\Delta d = \left(340\,\frac{m}{sec}\right)(2 \text{ sec}) = 680 \text{ m} = 2a$
$a = 340$ m; $a^2 = 115,600$
$b^2 = 500^2 - 340^2 = 134,400$

$\dfrac{x^2}{115,600} - \dfrac{y^2}{134,400} = 1$

$\dfrac{x^2}{115,600} - 1 = \dfrac{y^2}{134,400}$

$y = \pm\sqrt{134,400\left(\dfrac{x^2}{115,600} - 1\right)}$

1. The observers are at the foci (± 500, 0).
2. Use $d = rt$ and the fact that the difference of the distances is $2a$ to find a and a^2.
3. Use $c^2 = a^2 + b^2$ to find b^2.
4. Substitute into $\dfrac{(x - k)^2}{a^2} - \dfrac{(y - h)^2}{b^2} = 1$.
5. Solve for y and use technology to plot the possible locations, which are any point on the hyperbola's branches.

SKILL ✔ **EXERCISE 35**

---

14. $C(0, 0)$; horizontal TA
$a = 5$, $b = 2$; $V: (\pm 5, 0)$
$c^2 = 5^2 + 2^2 = 29$; $c = \sqrt{29}$
$F: \left(0, \pm\sqrt{29}\right) \approx (0, \pm 5.4)$
$y = \pm\dfrac{2}{5}x$

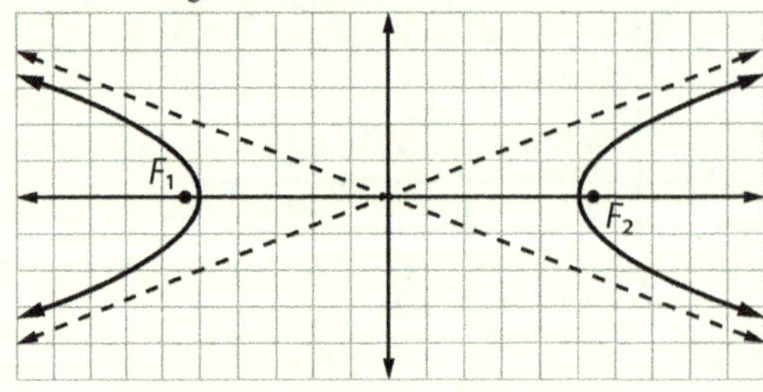

15. $C(0, 0)$; vertical TA
$a = 4$, $b = 2$; $V: (0, \pm 4)$
$c^2 = 4^2 + 2^2 = 20$; $c = 2\sqrt{5}$
$F: \left(0, \pm 2\sqrt{5}\right) \approx (0, \pm 4.5)$
$y = \pm 2x$

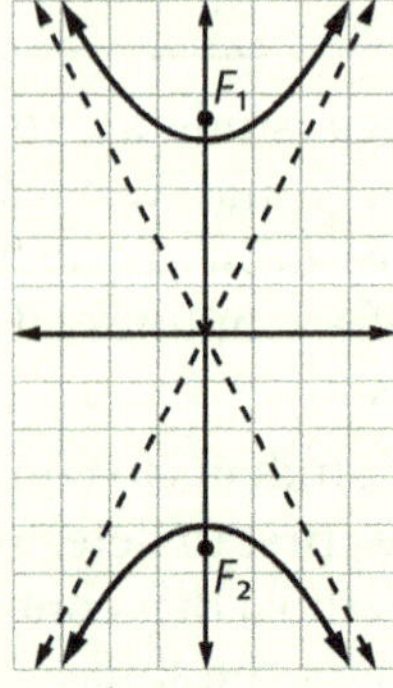

Light rays headed toward one focus of a hyperbolic mirror are reflected to its other focus. In a Cassegrain reflecting telescope, a small hyperbolic mirror is positioned in front of a larger parabolic mirror along the same optical axis so that the paraboloid's focus and one of the hyperboloid's foci coincide. Signals reflected to the parabola's focus reflect off the hyperbolic mirror through a hole in the center of the primary mirror to the hyperbola's other focus.

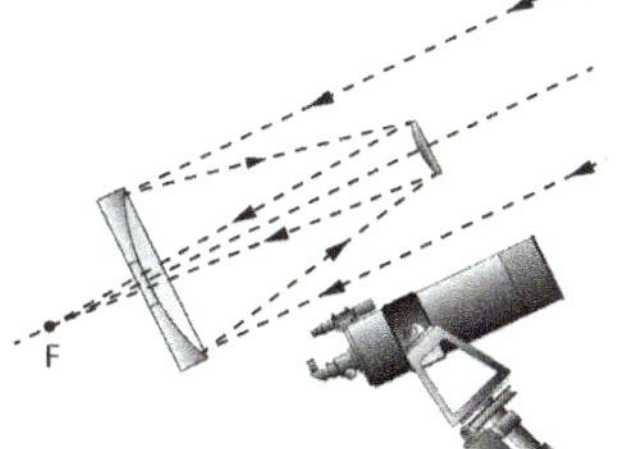

Cassegrain antennas are also used for sending and receiving radio waves. NASA's Goldstone Deep Space Communications Complex in the Mojave Desert includes a Cassegrain antenna that is 24 stories tall, has a diameter of 230 ft, and weighs 16 million lb.

❭ A. Exercises

1. A hyperbola represents all the points in a plane such that the _______ of the distances from two fixed points is _______. difference; constant

2. The _______ is the segment connecting the vertices of a hyperbola, while the _______ is the segment connecting its co-vertices. transverse axis; conjugate axis

3. Match each formula or expression with its description.

   I. length of the transverse axis
   II. length of the conjugate axis
   III. distance from the center to a focal point
   IV. equation of a hyperbola with a horizontal transverse axis
   V. equation of a hyperbola with a vertical transverse axis

   a. $\dfrac{(y-k)^2}{a^2} - \dfrac{(x-h)^2}{b^2} = 1$ V

   b. $\dfrac{(x-h)^2}{a^2} - \dfrac{(y-k)^2}{b^2} = 1$ IV

   c. c III
   d. $2a$ I
   e. $2b$ II

4. Which of the following is true for a hyperbola? E
   A. The difference of the distances from any point on the hyperbola to the foci is $2a$.
   B. The distance from the center to a co-vertex is b.
   C. The distance from a vertex to a co-vertex is c.
   D. $c^2 = a^2 + b^2$
   E. All are true statements.

State the coordinates of the center and the lengths of the transverse and conjugate axes for each hyperbola.

5. $\dfrac{x^2}{144} - \dfrac{y^2}{25} = 1$ $C\,(0, 0)$; 24; 10

6. $\dfrac{(y+2)^2}{9} - \dfrac{(x+7)^2}{36} = 1$
   $C\,(-7, -2)$; 6; 12

State the coordinates of the center, the vertices, and the co-vertices for each hyperbola.

7. $\dfrac{y^2}{64} - \dfrac{x^2}{36} = 1$

8. $\dfrac{(x-1)^2}{16} - \dfrac{(y-5)^2}{36} = 1$

State the coordinates of the center, the vertices, and the foci for each hyperbola. Then calculate its eccentricity.

9. $\dfrac{y^2}{81} - \dfrac{x^2}{144} = 1$

10. $\dfrac{(x-2)^2}{64} - \dfrac{(y+4)^2}{225} = 1$

State the coordinates of the center, the vertices, and the equations of the asymptotes for each hyperbola.

11. $\dfrac{x^2}{9} - \dfrac{y^2}{16} = 1$

12. $\dfrac{(y-1)^2}{4} - \dfrac{(x+2)^2}{9} = 1$

Graph each hyperbola and its foci, stating the equations of its asymptotes.

13. $\dfrac{x^2}{4} - \dfrac{y^2}{9} = 1$ $y = \pm\frac{3}{2}x$

14. $\dfrac{x^2}{25} - \dfrac{y^2}{4} = 1$ $y = \pm\frac{2}{5}x$

15. $\dfrac{y^2}{16} - \dfrac{x^2}{4} = 1$ $y = \pm 2x$

16. $y^2 - x^2 = 4$ $y = \pm x$

17. $(y+2)^2 - (x-1)^2 = 1$
    $y = x - 3,\ y = -x - 1$

18. $x^2 - \dfrac{(y-2)^2}{9} = 1$
    $y = \pm 3x + 2$

❭ B. Exercises

Give the center, vertices, and equations of the asymptotes for each hyperbola. Then use technology to graph each hyperbola and its asymptotes.

19. $\dfrac{(x-3)^2}{9} - \dfrac{(y+2)^2}{25} = 1$

20. $\dfrac{(y+1)^2}{4} - \dfrac{(x-4)^2}{9} = 1$

Identify the center, vertices, foci, eccentricity, and asymptotes for each hyperbola.

21. $25(x-6)^2 - 9y^2 = 225$

22. $9(y+7)^2 - 16(x-12)^2 = 144$

23. $16x^2 - 25y^2 - 150y = 625$

24. $y^2 - 9x^2 + 4y - 18x = 14$

25. $4x^2 - y^2 - 40x + 4y = -92$

26. $4y^2 - 49x^2 + 24y - 294x = 601$

7. $C\,(0, 0)$; $V\!: (0, \pm 8)$;
   $cV\!: (\pm 6, 0)$

8. $C\,(1, 5)$; $V\!: (5, 5), (-3, 5)$;
   $cV\!: (1, 11), (1, -1)$

9. $C\,(0, 0)$; $V\!: (0, \pm 9)$;
   $F\!: (0, \pm 15)$; $e = \frac{5}{3}$

10. $C\,(2, -4)$; $V\!: (10, -4), (-6, -4)$;
    $F\!: (19, -4), (-15, -4)$; $e = 2.125$

---

18. $C\,(0, 2)$; horizontal TA
    $a = 1,\ b = 3$; $V\!: (\pm 1, 2)$
    $c^2 = 1^2 + 3^2 = 10$; $c = \sqrt{10}$
    $F\!: \left(0 \pm \sqrt{10}, 2\right) \approx (\pm 3.2, 2)$
    $y = \pm 3x + 2$

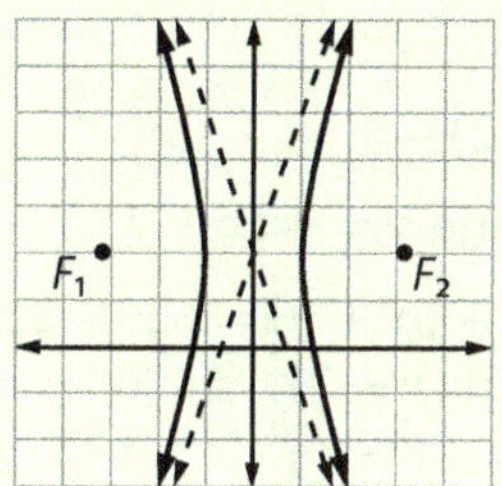

❭ B. Exercises

19. $C\,(3, -2)$; horizontal TA
    $a = 3,\ b = 5$; $V\!: (0, -2), (6, -2)$
    $y = \pm\frac{5}{3}(x-3) - 2$
    $y = \frac{5}{3}x - 7$ and $y = -\frac{5}{3}x + 3$
    $f(x) = \pm\sqrt{-25\left(1 - \dfrac{(x-3)^2}{9}\right)} - 2$

20. $C\,(4, -1)$; vertical TA; $a = 2,\ b = 3$
    $V\!: (4, -3), (4, 1)$
    $y = \pm\frac{2}{3}(x-4) - 1$
    $y = \frac{2}{3}x - \frac{11}{3}$ and $y = -\frac{2}{3}x + \frac{5}{3}$
    $f(x) = \pm\sqrt{4\left(1 + \dfrac{(x-4)^2}{9}\right)} - 1$

21. $\dfrac{(x-6)^2}{9} - \dfrac{y^2}{25} = 1$
    $C\,(6, 0)$; horizontal TA; $a = 3,\ b = 5$
    $V\!: (3, 0), (9, 0)$
    $c^2 = 3^2 + 5^2 = 34$; $c = \sqrt{34}$
    $F\!: \left(6 \pm \sqrt{34}, 0\right) \approx (11.8, 0), (0.2, 0)$
    $e = \dfrac{\sqrt{34}}{3} \approx 1.94$
    $y = \pm\frac{5}{3}(x-6)$
    $y = \frac{5}{3}x - 10$ and $y = -\frac{5}{3}x + 10$

---

16. $\dfrac{y^2}{4} - \dfrac{x^2}{4} = 1$
    $C\,(0, 0)$; vertical TA
    $a = b = 2$; $V\!: (0, \pm 2)$
    $c^2 = 2^2 + 2^2 = 8$; $c = 2\sqrt{2}$
    $F\!: \left(0, \pm 2\sqrt{2}\right) \approx (0, \pm 2.8)$
    $y = \pm x$

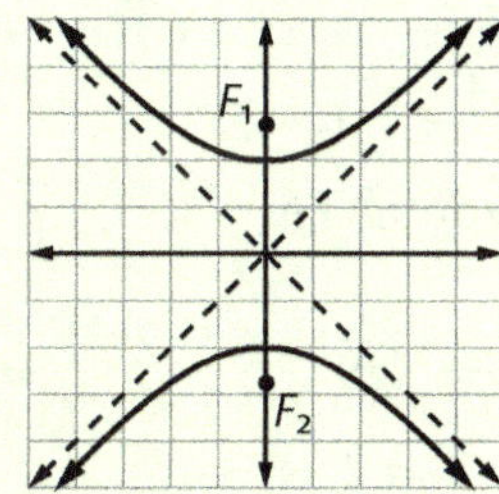

17. $C\,(1, -2)$; vertical TA; $a = b = 1$
    $V\!: (1, -1), (1, -3)$
    $c^2 = 1^2 + 1^2 = 2$; $c = \sqrt{2}$
    $F\!: \left(1, -2 \pm \sqrt{2}\right) \approx (1, -0.6), (1, -3.4)$
    $y = \pm(x-1) - 2$
    $y = x - 3$ and $y = -x - 1$

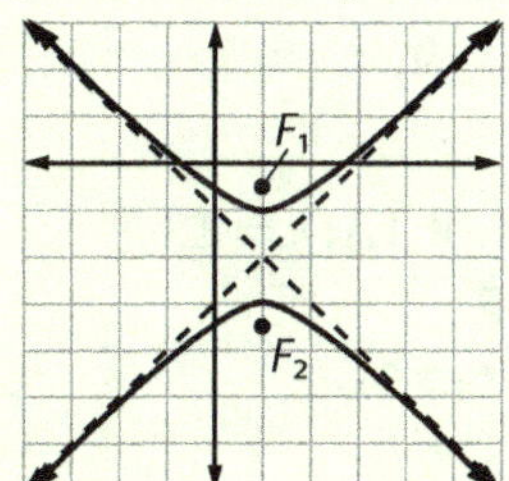

22. $\dfrac{(y+7)^2}{16} - \dfrac{(x-12)^2}{9} = 1$

$C\,(12, -7)$; vertical TA; $a = 4$, $b = 3$

$V:\ (12, -11),\ (12, -3)$

$c^2 = 4^2 + 3^2 = 25;\ c = 5$

$F:\ (12, -7 \pm 5) = (12, -2),\ (12, -12)$

$e = \dfrac{5}{4} = 1.25$

$y = \pm\dfrac{4}{3}(x - 12) - 7$

$y = \dfrac{4}{3}x - 23$ and $y = -\dfrac{4}{3}x + 9$

23. $16x^2 - 25(y^2 + 6y + 9) = 625 - 225$

$16x^2 - 25(y + 3)^2 = 400$

$\dfrac{x^2}{25} - \dfrac{(y+3)^2}{16} = 1$

$C\,(0, -3)$; horizontal TA

$a = 5,\ b = 4;\ V:\ (\pm5, -3)$

$c^2 = 5^2 + 4^2 = 41;\ c = \sqrt{41}$

$F:\ \left(\pm\sqrt{41}, -3\right) \approx (\pm6.4, -3)$

$e = \dfrac{\sqrt{41}}{5} \approx 1.28$

$y = \pm\dfrac{4}{5}x - 3$

24. $(y^2 + 4y + 4) - 9(x^2 + 2x + 1)$

$\qquad\qquad = 14 + 4 - 9$

$(y + 2)^2 - 9(x + 1)^2 = 9$

$\dfrac{(y+2)^2}{9} - \dfrac{(x+1)^2}{1} = 1$

$C\,(-1, -2)$; vertical TA; $a = 3$, $b = 1$

$V:\ (-1, 1),\ (-1, -5)$

$c^2 = 3^2 + 1^2 = 10;\ c = \sqrt{10}$

$F:\ \left(-1, -2 \pm \sqrt{10}\right)$

$\qquad \approx (-1, 1.2),\ (-1, -5.2)$

$e = \dfrac{\sqrt{10}}{3} \approx 1.05$

$y = \pm3(x + 1) - 2$

$y = 3x + 1$ and $y = -3x - 5$

25. $4(x^2 - 10x + 25) - (y^2 - 4y + 4)$

$\qquad\qquad = -92 + 100 - 4$

$4(x - 5)^2 - (y - 2)^2 = 4$

$\dfrac{(x-5)^2}{1} - \dfrac{(y-2)^2}{4} = 1$

$C\,(5, 2)$; horizontal TA; $a = 1$, $b = 2$

$V:\ (4, 2),\ (6, 2)$

$c^2 = 1^2 + 2^2 = 5;\ c = \sqrt{5}$

$F:\ \left(5 \pm \sqrt{5}, 2\right) \approx (7.2, 2),\ (2.8, 2)$

$e = \dfrac{\sqrt{5}}{1} \approx 2.24$

$y = \pm2(x - 5) + 2$

$y = 2x - 8$ and $y = -2x + 12$

26. $4(y^2 + 6y + 9) - 49(x^2 + 6x + 9)$

$\qquad\qquad = 601 + 36 - 441$

$4(y + 3)^2 - 49(x + 3)^2 = 196$

$\dfrac{(y+3)^2}{49} - \dfrac{(x+3)^2}{4} = 1$

$C\,(-3, -3)$; vertical TA; $a = 7$, $b = 2$

$V:\ (-3, -10),\ (-3, 4)$

$c^2 = 7^2 + 2^2 = 53;\ c = \sqrt{53}$

$F:\ \left(-3, -3 \pm \sqrt{53}\right)$

$\qquad \approx (-3, 4.3),\ (-3, -10.3)$

Write the standard form equation for each hyperbola with the given characteristics.

27. foci at $(\pm5, 0)$ and $e = \dfrac{5}{3}$ $\quad \dfrac{x^2}{9} - \dfrac{y^2}{16} = 1$

28. foci at $(0, \pm17)$ and $e = \dfrac{17}{15}$ $\quad \dfrac{y^2}{225} - \dfrac{x^2}{64} = 1$

29. centered at $(2, 4)$, a vertex at $(6, 4)$, and an asymptote $y = x + 2$ $\quad \dfrac{(x-2)^2}{16} - \dfrac{(y-4)^2}{16} = 1$

30. centered at $(-4, -3)$, a focus at $(-4, 2)$, and a vertex at $(-4, 1)$ $\quad \dfrac{(y+3)^2}{16} - \dfrac{(x+4)^2}{9} = 1$

Write a general form equation for each hyperbola with the given characteristics.

31. vertices at $(\pm2, 0)$ and $e = \dfrac{\sqrt{29}}{2}$ $\quad 25x^2 - 4y^2 - 100 = 0$

32. vertices at $(-4, 0)$ and $(2, 0)$ and $e = \sqrt{5}$

33. vertices at $(7, 1)$ and $(7, 9)$ and $e = \dfrac{\sqrt{65}}{4}$

34. vertices at $(10, -21)$ and $(10, -15)$ and $e = \dfrac{\sqrt{13}}{3}$

35. Jayden hears the firing of a cannon 3 sec before Olivia. If Olivia is 2000 m south of Jayden and the sound traveled at 340 m/sec, write the equation of a hyperbola centered at the origin and modeling the possible locations of the cannon. Then use technology to plot the possible locations. $\quad \dfrac{y^2}{260{,}100} - \dfrac{x^2}{739{,}900} = 1$

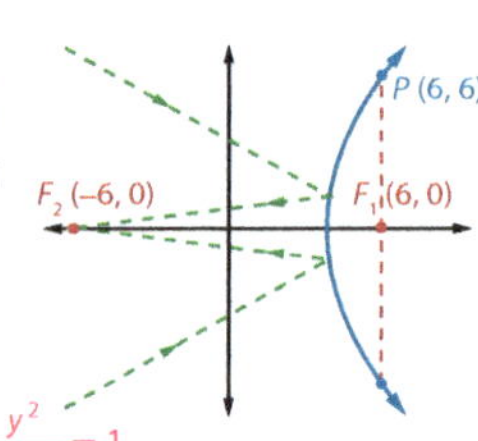

36. The focal width of a hyperbola is defined as the length of a chord through a focus and perpendicular to the line containing its foci. A proof similar to exercise 42 in Section 8.2 shows that the hyperbola $\dfrac{x^2}{a^2} - \dfrac{y^2}{b^2} = 1$ has a focal width of $2\left(\dfrac{b^2}{a}\right)$.

Use this fact to find the equation of a hyperbola modeling the illustrated mirror that reflects light rays headed toward its focus at $(0, 6)$ to its other focus at $(-6, 0)$ if its focal width is 12.

C. Exercises $\quad \dfrac{x^2}{13.75} - \dfrac{y^2}{22.25} = 1$

37. State the center, vertices, and equations of the asymptotes for $\dfrac{(y+1)^2}{5} - \dfrac{(x-4)^2}{8} = 1$. Then use technology to graph the hyperbola and its asymptotes.

38. Graph each degenerate hyperbola.

a. $x^2 - 4y^2 = 0$ $\qquad$ **b.** $\dfrac{y^2}{16} - \dfrac{(x-3)^2}{9} = 0$

Graph each hyperbolic inequality.

39. $\dfrac{x^2}{9} - \dfrac{(y-1)^2}{4} \ge 1$ $\qquad$ **40.** $y^2 - x^2 < 9$

41. The LORAN (long range navigation) network was developed during World War II to guide ships and airplanes across the ocean. The difference in the times required to receive radio signals broadcasted simultaneously from a master station and each of two substations was used to determine two hyperbolic equations modeling the receiver's possible location. The vessel's exact location was then determined using an intersection of the two hyperbolas. Assume that transmitters located at $M\,(0, 0)$, $S_1\,(200\text{ km}, 0)$, and $S_2\,(0, 360\text{ km})$ simultaneously broadcast radio signals that travel at 300 km/sec.

a. Write the equation of a hyperbola modeling a ship's possible locations if the signal from S_1 arrives at the ship 400 ms earlier than the signal from M. (1 ms = 10^{-3} sec) $\quad \dfrac{(x-100)^2}{3600} - \dfrac{y^2}{6400} = 1$

b. Write the equation of a hyperbola modeling the ship's possible locations if the signal from S_2 arrives at the ship 500 ms earlier than the signal from M. $\quad \dfrac{(y-180)^2}{5625} - \dfrac{x^2}{26{,}775} = 1$

c. Use graphing technology to find the coordinates of the ship (to the nearest km). $\quad (385, 372)$

42. A Cassegrain telescope is constructed by placing a hyperbolic mirror 12 in. in front of a parabolic mirror with a focal length of 14 in. (see the figure on p. 417). If the surface of the hyperbolic mirror is modeled by $13x^2 - 36y^2 = 468$, how far behind the parabolic mirror are the reflected light rays focused? $\quad$ 1 in.

43. Prove: Sketch a hyperbola centered at $C\,(h, k)$ and foci $(h, k \pm c)$ with semitransverse and semiconjugate axes with lengths of a and b, respectively, where $c^2 = a^2 + b^2$. Then use the definition of a hyperbola to derive its standard form equation.

$e = \dfrac{\sqrt{53}}{7} \approx 1.04$

$y = \pm\dfrac{7}{2}(x + 3) - 3$

$y = \dfrac{7}{2}x + \dfrac{15}{2}$ and $y = -\dfrac{7}{2}x - \dfrac{27}{2}$

27. $C\,(0, 0)$; horizontal TA; $c = 5$

$e = \dfrac{5}{3};\ a = 3$

$b^2 = 5^2 - 3^2 = 16$

$\dfrac{x^2}{9} - \dfrac{y^2}{16} = 1$

28. $C\,(0, 0)$; vertical TA; $c = 17$

$e = \dfrac{17}{15};\ a = 15$

$b^2 = 17^2 - 15^2 = 64$

$\dfrac{y^2}{225} - \dfrac{x^2}{64} = 1$

29. horizontal TA; $a = 6 - 2 = 4$

$m = \dfrac{b}{a} = \dfrac{b}{4} = 1;\ b = 4$

$\dfrac{(x-2)^2}{16} - \dfrac{(y-4)^2}{16} = 1$

30. vertical TA; $a = 1 - (-3) = 4$

$c = 2 - (-3) = 5$

$b^2 = 5^2 - 4^2 = 9$

$\dfrac{(y+3)^2}{16} - \dfrac{(x+4)^2}{9} = 1$

31. $C\,(0, 0)$; horizontal TA

$a = 2 - 0 = 2$

$e = \dfrac{c}{2} = \dfrac{\sqrt{29}}{2};\ c = \sqrt{29}$

$b^2 = \left(\sqrt{29}\right)^2 - 2^2 = 25$

$\dfrac{x^2}{4} - \dfrac{y^2}{25} = 1$

$25x^2 - 4y^2 = 100$

$25x^2 - 4y^2 - 100 = 0$

Graph each conic and any foci.

44. $9x^2 + 16y^2 + 128y + 112 = 0$ [8.2]

45. $y^2 - 2x - 8y + 16 = 0$ [8.1]

Find each if $\theta \in \left[0, \frac{\pi}{2}\right]$. [4.6]

46. $\cot 2\theta = \frac{\sqrt{3}}{3}$ $\frac{\pi}{6}$ **47.** $\cot 2\theta = 0$ $\frac{\pi}{4}$

Use reference triangles and the half-angle formula to find the exact value of $\cos \theta$. [5.5]

48. $\cot 2\theta = \frac{5}{12}$ $\frac{3\sqrt{13}}{13}$ **49.** $\cot 2\theta = \frac{7}{24}$ $\frac{4}{5}$

50. Find x^2 if $x = \frac{\sqrt{3}\,u - v}{2}$. [Algebra]   D

   A. $\dfrac{3u^2 - v^2}{4}$ **D.** $\dfrac{3u^2 - 2\sqrt{3}\,uv + v^2}{4}$

   B. $\dfrac{2\sqrt{3}\,u^2 - 2\sqrt{3}\,uv + v^2}{4}$ **E.** none of these

   C. $\dfrac{3u^2 + 2\sqrt{3}\,uv + v^2}{4}$

51. Find xy if $x = \dfrac{\sqrt{2}\,u - \sqrt{2}\,v}{2}$ and $y = \dfrac{\sqrt{2}\,u + \sqrt{2}\,v}{2}$. [Algebra]   C

   A. $\dfrac{u^2 - \sqrt{2}\,uv + v^2}{2}$ **D.** $\dfrac{u^2 - \sqrt{2}\,uv - v^2}{2}$

   B. $\dfrac{u^2 + \sqrt{2}\,uv + v^2}{2}$ **E.** none of these

   C. $\dfrac{u^2 - v^2}{2}$

52. Solve $9^{x+1} = 27^{1-2x}$. [3.4]   A

   A. $x = \frac{1}{8}$ **C.** $x = \frac{1}{4}$ **E.** $x = 4$

   B. $x = -\frac{1}{8}$ **D.** $x = -\frac{1}{4}$

53. Solve $\log_4 2 + \log_4 x = 2$. [3.3]   D

   A. $x = 2$ **C.** $x = 4$ **E.** $x = 16$

   B. $x = \frac{1}{2}$ **D.** $x = 8$

TECHNOLOGY CORNER (DESMOS® GRAPHING CALCULATOR)

Many interactive graphing applications such as the Desmos online graphing calculator allow equations to be entered in a variety of forms and allow the use of sliders to define constants. These features enable the user to quickly graph a variety of rectangular and polar equations and can be extremely helpful when studying conic sections.

Notice how the hyperbola and asymptotes from Example 2 can be quickly verified using the general form equation of the hyperbola. The asymptotes and key points can also be easily plotted on the same graph. Click and hold the icon to the left of the equation to adjust the color and format of the graphs. This is especially helpful when checking graphs of rotated conics in Section 8.4.

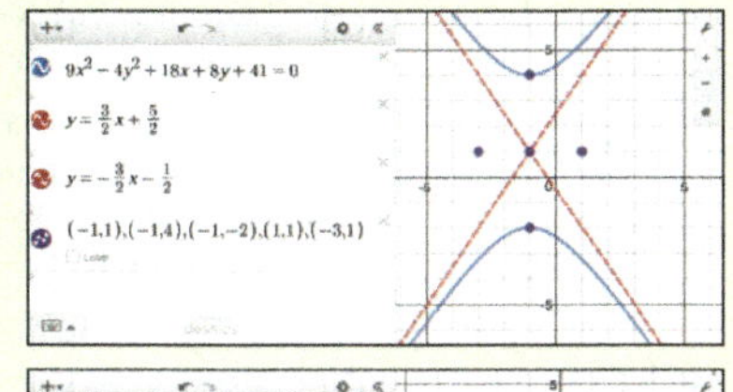

Enter parametric equations from Section 8.5 Example 1a as an ordered pair and define the bounds of the parameter's interval. The domain of the rectangular equation can be limited by writing the conjunction within curly braces, { }. Clicking the icon to the left of the equation turns that graph on and off.

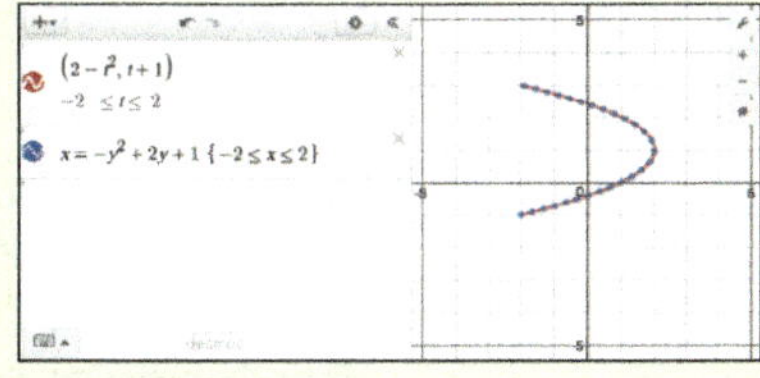

Polar equations for conics from Section 8.6 are easily explored. Use E to represent the eccentricity since e is reserved for the constant $e \approx 2.718$. Type pi or theta to enter these Greek letters.

Entering more than 2 variables in an equation prompts the user to define sliders that are used to quickly change the values of the other variables that act as constants, such as E and d. Clicking on the minimum or maximum value of the slider allows you to define those values and the step for that slider.

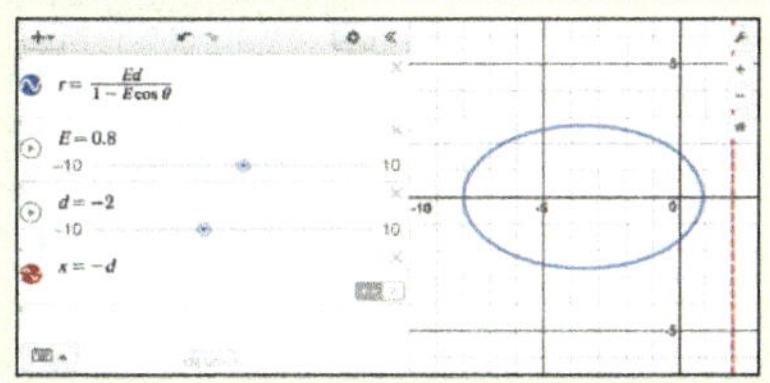

32. $C\left(\frac{-4+2}{2}, 0\right) = (-1, 0)$

horizontal TA; $a = 2 - (-1) = 3$

$e = \frac{c}{3} = \sqrt{5}; \; c = 3\sqrt{5}$

$b^2 = \left(3\sqrt{5}\right)^2 - 3^2 = 36$

$\dfrac{(x+1)^2}{9} - \dfrac{y^2}{36} = 1$

$4(x+1)^2 - y^2 = 36$

$4x^2 + 8x + 4 - y^2 - 36 = 0$

$4x^2 - y^2 + 8x - 32 = 0$

33. $C\left(\frac{7+7}{2}, \frac{1+9}{2}\right) = (7, 5);$ vertical TA

$a = 9 - 5 = 4$

$e = \frac{c}{4} = \frac{\sqrt{65}}{4}; \; c = \sqrt{65}$

$b^2 = \left(\sqrt{65}\right)^2 - 4^2 = 49$

$\dfrac{(y-5)^2}{16} - \dfrac{(x-7)^2}{49} = 1$

$49(y-5)^2 - 16(x-7)^2 = 784$

$49(y^2 - 10y + 25)$
$\qquad - 16(x^2 - 14x + 49) = 784$

$-16x^2 + 49y^2 + 224x - 490y$
$\qquad\qquad\qquad\qquad - 343 = 0$

$16x^2 - 49y^2 - 224x + 490y + 343 = 0$

34. $C\left(\frac{10+10}{2}, \frac{-21-15}{2}\right) = (10, -18)$

vertical TA; $a = -15 - (-18) = 3$

$e = \frac{c}{3} = \frac{\sqrt{13}}{3}; \; c = \sqrt{13}$

$b^2 = \left(\sqrt{13}\right)^2 - 3^2 = 4$

$\dfrac{(y+18)^2}{9} - \dfrac{(x-10)^2}{4} = 1$

$4(y+18)^2 - 9(x-10)^2 = 36$

$4(y^2 + 36y + 324)$
$\qquad - 9(x^2 - 20x + 100) = 36$

$-9x^2 + 4y^2 + 180x + 144y + 360 = 0$
$9x^2 - 4y^2 - 180x - 144y - 360 = 0$

35. $C(0, 0);$ vertical TA; $c = 1000$ m

$\Delta d = 340 \frac{\text{m}}{\text{sec}}(3 \text{ sec}) = 1020 \text{ m} = 2a$

$a = 510; \; a^2 = 260{,}100$

$b^2 = 1000^2 - 510^2 = 739{,}900$

$\dfrac{y^2}{260{,}100} - \dfrac{x^2}{739{,}900} = 1$

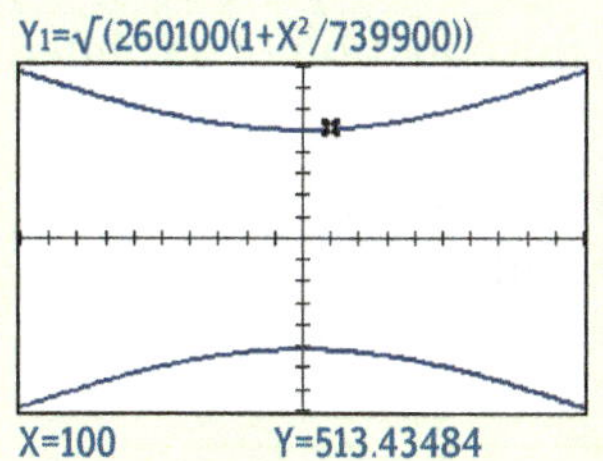

36. $2\frac{b^2}{a} = 12; \; b^2 = 6a$

$\left[\dfrac{6^2}{a^2} - \dfrac{6^2}{(6a)^2} = 1\right]a^2$

$36 - 6a = a^2$

$0 = a^2 + 6a - 36$

$a = \dfrac{-6 \pm \sqrt{6^2 - 4(-36)}}{2} \approx 3.71$

$a^2 \approx 13.75$

$b^2 = c^2 - a^2 \approx 36 - 13.75 = 22.25$

$\dfrac{x^2}{13.75} - \dfrac{y^2}{22.25} = 1$

❯ C. Exercises

37. $C(4, -1);$ vertical TA

$a = \sqrt{5}; \; V: \left(4, -1 \pm \sqrt{5}\right)$

$b = 2\sqrt{2}$

$y = \pm\dfrac{\sqrt{5}}{2\sqrt{2}}(x-4) - 1$

$y = \pm\dfrac{\sqrt{10}}{4}(x-4) - 1$

$y = \dfrac{\sqrt{10}}{4}x - \left(\sqrt{10} + 1\right)$ and

$y = -\dfrac{\sqrt{10}}{4}x + \left(\sqrt{10} - 1\right)$

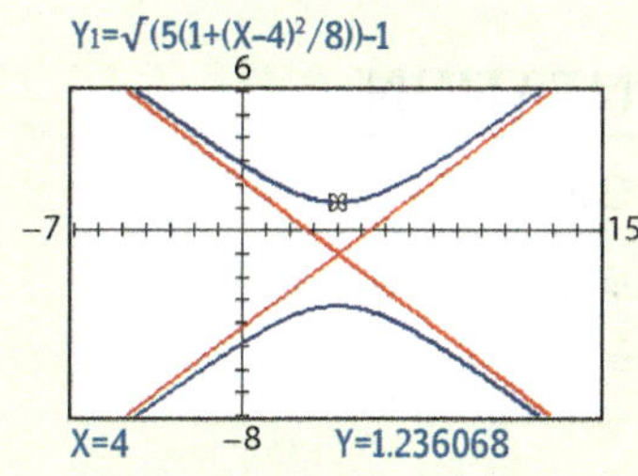

38a. $y = \pm\frac{1}{2}x$

continued in Answers and Solutions Overflow

8.4 Rotated Conics (Extended)

Objectives

1. To determine the angle of rotation and sketch the graph of a rotated conic section

2. To write the equation of a rotated conic section

3. To classify conics using the discriminant

Flash

Saturn, the second-largest planet and the sixth planet from the sun, is known for its ring system, which is comprised of mostly ice particles mixed with dust and rocks.

Vocabulary

discriminant
invariant under rotation
rotation formulas

Saturn's axial tilt makes its rings take on the appearance of rotated ellipses.

After completing this section, you will be able to

- determine the angle of rotation and graph a rotated conic.
- write the equation of a rotated conic.
- classify conics using the discriminant test.

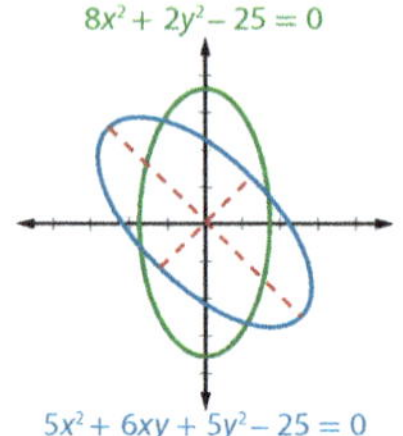

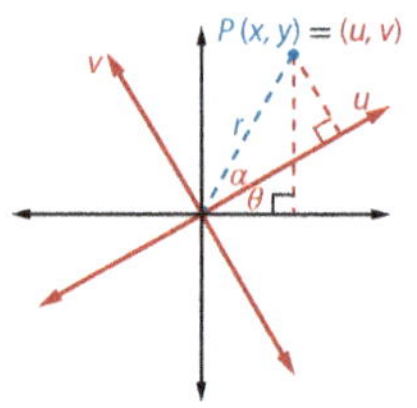

Recall that the equation of a conic section can be written using the general form of a quadratic (second-degree) equation: $Ax^2 + Bxy + Cy^2 + Dx + Ey + F = 0$. In prior sections, the axes of the conics were horizontal or vertical since $B = 0$. When $B = 0$ the conic is a parabola if there is no x^2- or y^2-term ($AC = 0$), a hyperbola if A and C have different signs ($AC < 0$), and an ellipse if A and C have the same sign ($AC > 0$), including the special case of a circle where $A = C$. Conics whose axes are rotated so they are not parallel to the x- or y-axis are represented by the general form equations which contain an xy-term where $B \neq 0$. This section introduces a method of quickly identifying any conic, given its general form equation.

The previous methods used to analyze conic sections can be employed if the x- and y-axes are rotated about the origin by an acute angle θ so that the new u- and v-axes are parallel to the axes of the conic section. The conic's equation in the new uv-coordinate plane is then $A'u^2 + C'v^2 + D'u + E'v + F' = 0$ where there is no uv-term ($B' = 0$).

The equations relating the (x, y) and (u, v) coordinates of any point P can be derived by examining the right triangles drawn to each axis. Note that $(u, v) = (r \cos \alpha, r \sin \alpha)$ and
$(x, y) = (r \cos (\theta + \alpha), r \sin (\theta + \alpha))$
$$\therefore x = r(\cos \theta \cos \alpha - \sin \theta \sin \alpha) \qquad \text{and} \qquad y = r(\sin \theta \cos \alpha + \cos \theta \sin \alpha)$$
$$x = (r \cos \alpha) \cos \theta - (r \sin \alpha) \sin \theta \qquad\qquad y = (r \cos \alpha) \sin \theta + (r \sin \alpha) \cos \theta$$
$$x = u \cos \theta - v \sin \theta \qquad\qquad\qquad\qquad y = u \sin \theta + v \cos \theta$$

To determine θ, the angle of rotation that causes the x- and y-axes to be parallel to the conic's axes, substitute the *rotation formulas* into the general form of a conic. After expanding and regrouping the terms (see exercise 35), set the resulting coefficient of the uv-term, B', equal to 0 and solve.

$$2(C - A) \sin \theta \cos \theta + B(\cos^2 \theta - \sin^2 \theta) = 0$$
$$(C - A) \sin 2\theta + B \cos 2\theta = 0$$
$$B \cos 2\theta = (A - C) \sin 2\theta$$
$$\cot 2\theta = \frac{A - C}{B}$$

Solving the system consisting of the two rotation formulas above (see exercise 36) produces equations for u and v.
$$u = x \cos \theta + y \sin \theta \quad \text{and} \quad v = y \cos \theta - x \sin \theta$$

▍ROTATION OF CONICS

A general quadratic equation $Ax^2 + Bxy + Cy^2 + Dx + Ey + F = 0$ where $B \neq 0$ can be rewritten as $A'u^2 + C'v^2 + D'u + E'v + F' = 0$ in the uv-plane by rotating the xy-plane by the acute angle θ such that $\cot 2\theta = \frac{A - C}{B}$.

The equations in the uv-plane and the xy-plane are related using the following *rotation formulas*.
$$x = u \cos \theta - v \sin \theta \quad \text{and} \quad y = u \sin \theta + v \cos \theta$$
$$u = x \cos \theta + y \sin \theta \quad \text{and} \quad v = y \cos \theta - x \sin \theta$$

PRESENTATION

Lesson Opener

Identify each characteristic of $\dfrac{(x - h)^2}{a^2} - \dfrac{(y - k)^2}{b^2} = 1$.

1. center (h, k)

2. vertices $(h \pm a, k)$

3. foci $(h \pm c, k)$

4. asymptotes $y = \pm \dfrac{b}{a}(x - h) + k$

The study of rotated conics should be reserved as a challenge for advanced students. For the standard class you may want to discuss the *discriminant* test for rotated conics and introduce how to find the angle of rotation. Consider graphing rotated conics with an application that does not require equations to be input as a function, such as the Desmos online graphing calculator.

Review the general form equation of a conic and how the conic can be classified when $B = 0$. Use the equation of the rotated ellipse where $A = C$ to show that these generalizations do not apply to rotated conics ($B \neq 0$). Explain that this lesson develops a test that works for any conic section (see Example 4) and that the previous guidelines are special cases of this more general test.

Introduce the formula for the angle of rotation and related *rotation formulas*. Allowing students to reference these formulas during assessments permits them to focus on the process of rotating the conic instead of memorizing four very similar formulas.

Apply the rotation formulas in Example 1 and illustrate how the standard form equation can be used to sketch the graph. Notice how these formulas are used to find the (x, y) coordinates of the vertices and foci from their (u, v) coordinates. Explain that either ordered pair can be used when sketching the graph. Applying the second set of rotation formulas allows the asymptotes to be described in terms of x and y.

In step 1 of Example 2, explain that 2θ is in Q II because θ is defined to be acute.

Given the equation $xy - 4 = 0$, find the angle θ that rotates the xy-plane so that there is no uv-term. Then write the general form equation in terms of the uv-plane and classify the conic. Finally, sketch a graph of the conic and any foci.

Answer

$A = 0, B = 1, C = 0$

$\cot 2\theta = \frac{0 - 0}{1} = 0; 2\theta = \frac{\pi}{2}; \theta = \frac{\pi}{4}$

$x = u \cos \frac{\pi}{4} - v \sin \frac{\pi}{4}$ and $y = u \sin \frac{\pi}{4} + v \cos \frac{\pi}{4}$

$x = \frac{\sqrt{2}}{2}u - \frac{\sqrt{2}}{2}v \qquad y = \frac{\sqrt{2}}{2}u + \frac{\sqrt{2}}{2}v$

$$xy - 4 = 0$$
$$\left(\frac{\sqrt{2}}{2}u - \frac{\sqrt{2}}{2}v \right)\left(\frac{\sqrt{2}}{2}u + \frac{\sqrt{2}}{2}v \right) - 4 = 0$$
$$\frac{u^2}{2} - \frac{v^2}{2} - 4 = 0$$
$$u^2 - v^2 - 8 = 0$$

The equation represents a hyperbola.

$$u^2 - v^2 = 8$$
$$\frac{u^2}{8} - \frac{v^2}{8} = 1$$

$C\,(0, 0); a = 2\sqrt{2}; b = 2\sqrt{2}$

$c^2 = a^2 + b^2 = 8 + 8 = 16; c = 4$

$\therefore V'\,(\pm 2\sqrt{2}, 0)$, and $F'\,(\pm 4, 0)$

$V'\,(\pm 2\sqrt{2}, 0)$:

$x = (\pm 2\sqrt{2})\frac{\sqrt{2}}{2} - (0)\frac{\sqrt{2}}{2} = \pm 2$

$y = (\pm 2\sqrt{2})\frac{\sqrt{2}}{2} + (0)\frac{\sqrt{2}}{2} = \pm 2$

V: (2, 2) and (−2, −2)

1. Use $\cot 2\theta = \frac{A - C}{B}$ to determine the angle of rotation.
2. Determine the rotation formulas.

3. Rewrite the equation as a general quadratic in terms of u and v. Note that $A' > 0$ and $B' < 0$.

4. Convert to the standard form for a hyperbola with a transverse axis along the u-axis.
5. Identify the center, vertices, and foci in the uv-plane.

6. The rotation formulas
   $x = u \cos \theta - v \sin \theta$ and
   $y = u \sin \theta + v \cos \theta$
   can be used to locate the vertices in the xy-plane.

7. Sketch the graph of the hyperbola.

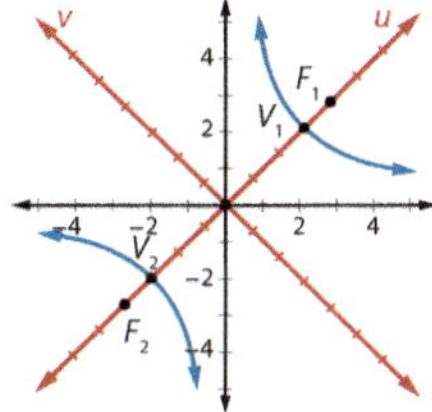

SKILL ✓ EXERCISE 11

The formulas for u and v in terms of x and y can be used to confirm that the asymptotes of the hyperbola rotated by $\theta = \frac{\pi}{4}$ in Example 1, $v = \pm u$, are the x-axis and y-axis.

$u = x \cos \theta + y \sin \theta = \frac{\sqrt{2}}{2}x + \frac{\sqrt{2}}{2}y$ and $v = -x \sin \theta + y \cos \theta = -\frac{\sqrt{2}}{2}x + \frac{\sqrt{2}}{2}y$

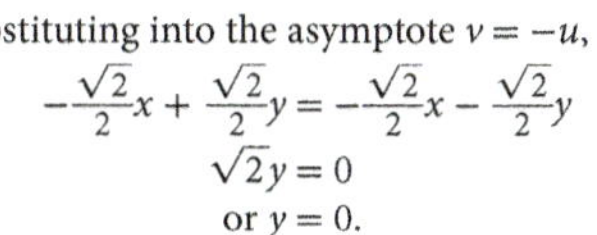

Substituting into the asymptote $v = u$,

$$-\frac{\sqrt{2}}{2}x + \frac{\sqrt{2}}{2}y = \frac{\sqrt{2}}{2}x + \frac{\sqrt{2}}{2}y$$
$$0 = \sqrt{2}x$$
or $x = 0$.

Substituting into the asymptote $v = -u$,

$$-\frac{\sqrt{2}}{2}x + \frac{\sqrt{2}}{2}y = -\frac{\sqrt{2}}{2}x - \frac{\sqrt{2}}{2}y$$
$$\sqrt{2}y = 0$$
or $y = 0$.

Fermat's treatise on conics used substitutions to reduce more complicated second-degree equations to a standard form. These substitutions resulted in a transformation of the axes.

Reading and Writing Mathematics

Explain how the guidelines given in the first three sections for classifiying a general form equation for a conic section with $B = 0$ are special cases of the discriminant test presented in this lesson. In a non-rotated parabola, either A or $C = 0$ and the discriminant $B^2 - 4AC = 0^2 - 4(0)$. In a non-rotated ellipse (or circle), both A and C have the same sign so the discriminant $B^2 - 4AC = 0^2 - 4$ (a positive product), which is always negative. In a non-rotated hyperbola A and B have opposite signs, so the discriminant $B^2 - 4AC = 0^2 - 4$ (a negative product), which is always positive.

Additional Exercises

State the angle of rotation and the formulas that will rotate each general quadratic equation to a uv-plane without a uv-term.

1. $3x^2 + 10\sqrt{3}xy - 7y^2 = 72$

$\frac{\pi}{6}; x = \frac{\sqrt{3}}{2}u - \frac{1}{2}v; y = \frac{1}{2}u + \frac{\sqrt{3}}{2}v$

2. $3x^2 - 6xy + 3y^2 - \sqrt{2}x + \sqrt{2}y + 2 = 0$

$\frac{\pi}{4}; x = \frac{\sqrt{2}}{2}u - \frac{\sqrt{2}}{2}v; y = \frac{\sqrt{2}}{2}u + \frac{\sqrt{2}}{2}v$

Using $2\theta = \mathrm{Cos}^{-1}\left(-\frac{3}{5}\right)$ produces the correct value directly, while 180° would need to be added to the Q IV value obtained using $\mathrm{Tan}^{-1}\left(-\frac{4}{3}\right)$, or $\mathrm{Sin}^{-1}\frac{4}{5}$ would need to be subtracted from 180°. Converting to the standard form allows the ellipse to be quickly sketched on the rotated axis.

In Example 3 the uv-equation and the angle of rotation for the uv-plane are used to write the equation in terms of x and y by substituting with the rotation formulas for u and v. Draw attention to the equation in the second line of step 3, where the equation is written as a quadratic in terms of y: $ay^2 + by + c = 0$. The quadratic formula can then be used to solve for y, and the equations can be entered into a graphing calculator. Consider the simpler alternative of using an Internet application, such as the Desmos online graphing calculator, that allows the general form equation to be graphed without solving for y. You may want to conclude by discussing how to find the location of other key characteristics in the xy-plane.

Explain the reasoning behind the discriminant ($B^2 - 4AC$) test. Then illustrate its application using Example 4. Note that the equation in part a represents a hyperbola despite $A = C$. Use the equation in part b to point out that only A, C, and the absolute value of B affect the type of conic, while D and E affect its position and eccentricity. While $A = C$, the equation is an ellipse since $B \neq 0$. The equation in part c is a parabola despite having both x^2- and y^2-terms. The same is true for the equation in Example 3.

Interactive Activity Use the Internet keyword search *rotating ellipse* or *rotating conic applet* to locate a video or interactive conic applet.

Find the angle of rotation of the *xy*-plane that causes the *u*- or *v*-axis to be parallel to the conic's axes. Then write the standard form equation in terms of *u* and *v* and sketch a graph of the conic and any foci.

3. $x^2 + 2xy + y^2 + \sqrt{2}x - \sqrt{2}y - 4 = 0$

$\theta = 45°; v = u^2 - 2$

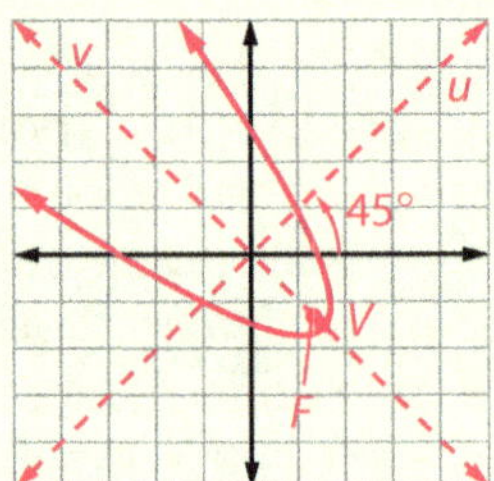

4. $32x^2 + 60xy + 7y^2 + 52 = 0$

$\theta = 33.7°; \dfrac{v^2}{4} - u^2 = 1$

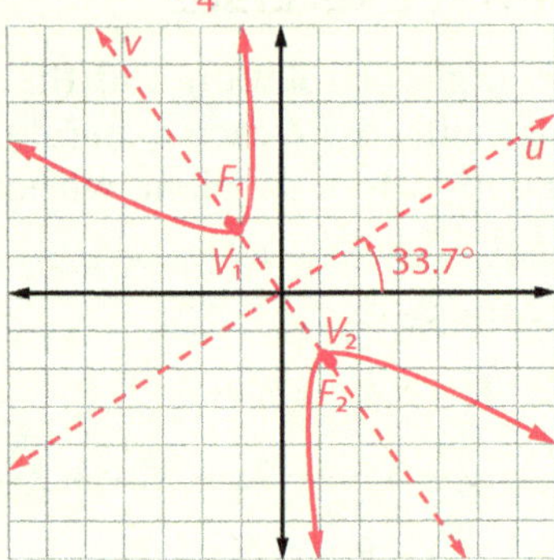

5. $21x^2 - 8\sqrt{3}xy + 13y^2 - 225 = 0$

$\theta = 60°; \dfrac{u^2}{25} + \dfrac{v^2}{9} = 1$

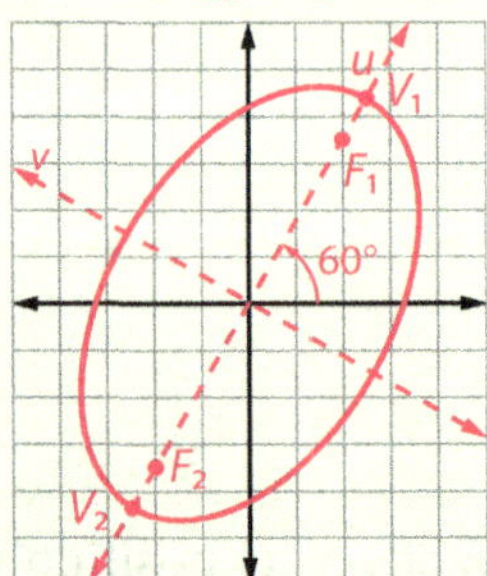

Write a general form equation for each conic in terms of x and y given its angle of rotation and equation in terms of u and v.

6. $\theta = \dfrac{\pi}{3}; v^2 + 2u + 4 = 0$

$3x^2 - 2\sqrt{3}xy + y^2 + 4x + 4\sqrt{3}y + 16 = 0$

7. $\theta = \dfrac{\pi}{6}; 9u^2 + 4v^2 - 36 = 0$

$31x^2 + 10\sqrt{3}xy + 21y^2 - 144 = 0$

Given the equation $8x^2 + 12xy + 17y^2 = 20$, find the angle θ that rotates the *xy*-plane so that there is no *uv*-term. Then write the general form equation in terms of the *uv*-plane and classify the conic. Finally, sketch a graph of the conic.

Answer

$\cot 2\theta = \dfrac{8 - 17}{12} = -\dfrac{3}{4}$

$2\theta \approx 126.9°; \theta \approx 63.4°$

$\cos 2\theta = -\dfrac{3}{5}$

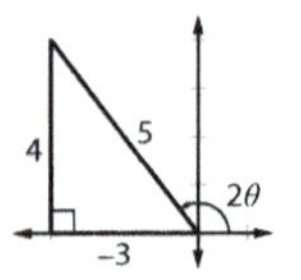

1. Since $\cot 2\theta = \dfrac{A - C}{B}$ does not produce an exact value for θ, use a reference triangle to find $\cos 2\theta$.

$\cos\theta = \sqrt{\dfrac{1 + \cos 2\theta}{2}}$ $\qquad \sin\theta = \sqrt{\dfrac{1 - \cos 2\theta}{2}}$

$= \sqrt{\dfrac{1 + \left(-\frac{3}{5}\right)}{2}}$ $\qquad = \sqrt{\dfrac{1 - \left(-\frac{3}{5}\right)}{2}}$

$= \sqrt{\dfrac{1}{5}} = \dfrac{1}{\sqrt{5}}$ $\qquad = \sqrt{\dfrac{4}{5}} = \dfrac{2}{\sqrt{5}}$

2. Use half-angle identities to find $\cos\theta$ and $\sin\theta$ for the acute angle θ.

$x = u\cos\theta - v\sin\theta$ and $y = u\sin\theta + v\cos\theta$

$x = \dfrac{1}{\sqrt{5}}u - \dfrac{2}{\sqrt{5}}v$ $\qquad y = \dfrac{2}{\sqrt{5}}u + \dfrac{1}{\sqrt{5}}v$

3. Determine the rotation formulas for the conic and substitute.

$8x^2 + 12xy + 17y^2 = 20$

$8\left(\dfrac{u - 2v}{\sqrt{5}}\right)^2 + 12\left(\dfrac{u - 2v}{\sqrt{5}}\right)\left(\dfrac{2u + v}{\sqrt{5}}\right) + 17\left(\dfrac{2u + v}{\sqrt{5}}\right)^2 = 20$

Then expand, multiply both sides by 5, and simplify.

$8(u^2 - 4uv + 4v^2) + 12(2u^2 - 3uv - 2v^2) + 17(4u^2 + 4uv + v^2) = 100$

$(8 + 24 + 68)u^2 + (-32 - 36 + 68)uv + (32 - 24 + 17)v^2 = 100$

$100u^2 + 25v^2 = 100$

$4u^2 + v^2 - 4 = 0$

The equation is an ellipse.

$\dfrac{u^2}{1} + \dfrac{v^2}{4} = 1$

4. A' and $B' > 0$

5. Convert to the standard form for an ellipse.

$a' = 2; b' = 1$

$(c')^2 = (a')^2 - (b')^2 = 4 - 1 = 3; c' = \sqrt{3} \approx 1.7$

$\therefore V'(0, \pm 2)$ and $F'(0, \pm\sqrt{3})$

6. Rotate the axes by $\theta \approx 63.4°$ and graph an ellipse centered at the origin with its major axis on the *v*-axis.

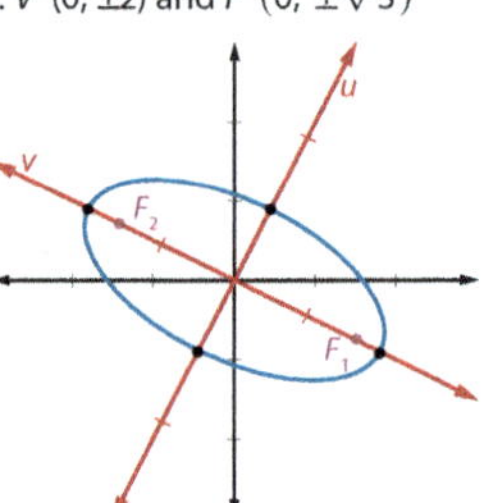

SKILL ✓ EXERCISE 15

State whether the conic is rotated. Then use the discriminant test to classify the conic.

8. $3x^2 - 2xy + y^2 + x - 3y - 2 = 0$

yes; ellipse

9. $4x^2 + 40x - y + 97 = 0$ no; parabola

10. $3x^2 + 9xy + y^2 + 88 = 0$

yes; hyperbola

Assignments

- **Minimum:** optional

- **Standard:** 1–7, 9, 11, 14–15, 17, 21, 23, 27, 43–45, 47–48, 52

- **Extended:** 1–2; 5–17 odd; 18, 22–24, 26, 28–29, 31–32, 34–35, 37, 41, 43–44, 46, 48, 51–52

Assessment

- Quiz 8B covers Sections 8.3–8.4.

The formulas for u and v in terms of x and y can be used to write the equation of a rotated conic.

After being rotated by 30°, the equation for a parabola in the uv-plane is $v = -\frac{1}{4}u^2 + 3$. Write a general form equation for the parabola in the xy-plane. Then use technology to graph the parabola.

Answer

$u = x\cos 30° + y\sin 30°$ and $v = y\cos 30° - x\sin 30°$

$u = \frac{\sqrt{3}}{2}x + \frac{1}{2}y$ $\qquad\qquad$ $v = \frac{\sqrt{3}}{2}y - \frac{1}{2}x$

1. Express u and v in terms of x and y.

$-4v = u^2 - 12$

2. Multiply both sides of $v = -\frac{1}{4}u^2 + 3$ by -4 before substituting.

Then expand and simplify.

$-4\left(\frac{\sqrt{3}y - x}{2}\right) = \left(\frac{\sqrt{3}x + y}{2}\right)^2 - 12$

$-2\sqrt{3}y + 2x = \frac{3x^2 + 2\sqrt{3}xy + y^2}{4} - 12$

$-8\sqrt{3}y + 8x = 3x^2 + 2\sqrt{3}xy + y^2 - 48$

$3x^2 + 2\sqrt{3}xy + y^2 - 8x + 8\sqrt{3}y - 48 = 0$

$y^2 + 2\sqrt{3}xy + 8\sqrt{3}y + 3x^2 - 8x - 48 = 0$

$y^2 + (2\sqrt{3}x + 8\sqrt{3})y + (3x^2 - 8x - 48) = 0$

3. Solve the equation for y.
   a. Express in quadratic form.
   b. Use the quadratic formula.

$y = \dfrac{-(2\sqrt{3}x + 8\sqrt{3}) \pm \sqrt{(2\sqrt{3}x + 8\sqrt{3})^2 - 4(3x^2 - 8x - 48)}}{2}$

4. Graph both functions whose union forms the parabola.

Check

Since $\tan 30° = \frac{\sqrt{3}}{3}$, the u-axis is at $y = \frac{\sqrt{3}}{3}x$, and the v-axis is at $y = -\sqrt{3}x$.

5. Graph the u- and v-axes using the slope of the u-axis $= \tan\theta$ and $m_v = -\frac{1}{m_u}$.

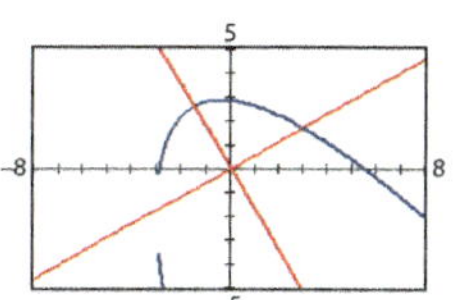

SKILL ✓ EXERCISE 27

The location of the focus or other characteristic points of the parabola in Example 3 can be found from their position in the uv-plane. Since $V'(0, 3)$ and $p = \frac{1}{4a} = -1$, the focus is $F'(0, 2)$ in the uv-plane.

using the rotation formulas:

$x = u\cos\theta - v\sin\theta$ and $y = u\sin\theta + v\cos\theta$

$x = (0)\frac{\sqrt{3}}{2} - (2)\frac{1}{2}$ $\qquad$ $y = (0)\frac{1}{2} + (2)\frac{\sqrt{3}}{2}$

Therefore, the parabola's focus is $F\left(-1, \sqrt{3}\right)$ in the xy-plane.

When using the quadratic formula, $x = \dfrac{-b \pm \sqrt{b^2 - 4ac}}{2a}$, to solve a quadratic equation of the form $ax^2 + bx + c = 0$, the *discriminant*, $b^2 - 4ac$, is used to identify the type and number of roots. A similar expression $B^2 - 4AC$, which is also called the discriminant, can be used to identify the type of conic section represented by a general form quadratic equation in two variables.

Solutions

❯ A. Exercises

1. $B = 0;\ (0)^2 - 4(4)(-9) = 144 > 0$

2. $B \neq 0;\ (1)^2 - 4(1)(2) = -7 < 0$

3. $B = 0;\ (0)^2 - 4(5)(5) = -100 < 0$
   ellipse but $A = C = 5$ and $B = 0$, so a circle

4. $B = 0;\ (0)^2 - 4(4)(9) = -144 < 0$

5. $B \neq 0;\ (-2)^2 - 4(1)(1) = 0$

6. $B \neq 0;\ (-12)^2 - 4(3)(2) = 120 > 0$

7. $A = 0;\ B = 1;\ C = 0$

$\cot 2\theta = \frac{0-0}{1} = 0;\ 2\theta = \frac{\pi}{2};\ \theta = \frac{\pi}{4}$

$x = u\cos\frac{\pi}{4} - v\sin\frac{\pi}{4} = \frac{\sqrt{2}}{2}u - \frac{\sqrt{2}}{2}v$

$y = u\sin\frac{\pi}{4} + v\cos\frac{\pi}{4} = \frac{\sqrt{2}}{2}u + \frac{\sqrt{2}}{2}v$

8. $A = 2;\ B = 1;\ C = 2$

$\cot 2\theta = \frac{2-2}{1} = 0;\ 2\theta = \frac{\pi}{2};\ \theta = \frac{\pi}{4}$

$x = u\cos\frac{\pi}{4} - v\sin\frac{\pi}{4} = \frac{\sqrt{2}}{2}u - \frac{\sqrt{2}}{2}v$

$y = u\sin\frac{\pi}{4} + v\cos\frac{\pi}{4} = \frac{\sqrt{2}}{2}u + \frac{\sqrt{2}}{2}v$

9. $A = 2;\ B = \sqrt{3};\ C = 1$

$\cot 2\theta = \frac{2-1}{\sqrt{3}} = \frac{1}{\sqrt{3}};\ 2\theta = \frac{\pi}{3};\ \theta = \frac{\pi}{6}$

$x = u\cos\frac{\pi}{6} - v\sin\frac{\pi}{6} = \frac{\sqrt{3}}{2}u - \frac{1}{2}v$

$y = u\sin\frac{\pi}{6} + v\cos\frac{\pi}{6} = \frac{1}{2}u + \frac{\sqrt{3}}{2}v$

10. $A = 0;\ B = -\sqrt{3};\ C = 1$

$\cot 2\theta = \frac{0-1}{-\sqrt{3}} = \frac{1}{\sqrt{3}};\ 2\theta = \frac{\pi}{3};\ \theta = \frac{\pi}{6}$

$x = u\cos\frac{\pi}{6} - v\sin\frac{\pi}{6} = \frac{\sqrt{3}}{2}u - \frac{1}{2}v$

$y = u\sin\frac{\pi}{6} + v\cos\frac{\pi}{6} = \frac{1}{2}u + \frac{\sqrt{3}}{2}v$

11. $\cot 2\theta = \frac{0-0}{1} = 0;\ 2\theta = \frac{\pi}{2};\ \theta = \frac{\pi}{4}$

$x = \frac{\sqrt{2}u - \sqrt{2}v}{2};\ y = \frac{\sqrt{2}u + \sqrt{2}v}{2}$

$\left(\frac{\sqrt{2}u - \sqrt{2}v}{2}\right)\left(\frac{\sqrt{2}u + \sqrt{2}v}{2}\right) - 8 = 0$

$\frac{u^2}{2} - \frac{v^2}{2} = 8;\ \frac{u^2}{16} - \frac{v^2}{16} = 1$

$(c')^2 = 16 + 16 = 32;\ c' = 4\sqrt{2}$

$V': (\pm 4, 0);\ F': (\pm 4\sqrt{2}, 0)$

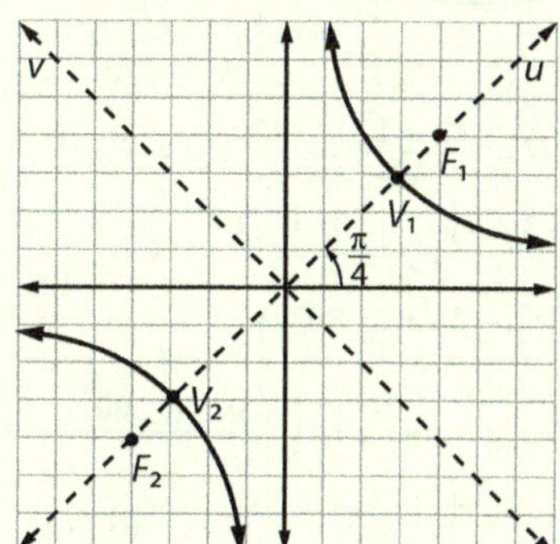

12. $\cot 2\theta = \frac{0-0}{1} = 0;\ 2\theta = \frac{\pi}{2};\ \theta = \frac{\pi}{4}$

$x = \frac{\sqrt{2}u - \sqrt{2}v}{2};\ y = \frac{\sqrt{2}u + \sqrt{2}v}{2}$

$\left(\frac{\sqrt{2}u - \sqrt{2}v}{2}\right)\left(\frac{\sqrt{2}u + \sqrt{2}v}{2}\right) + 6 = 0$

$\frac{u^2}{2} - \frac{v^2}{2} = -6;\ \frac{v^2}{12} - \frac{u^2}{12} = 1$

$(c')^2 = 12 + 12 = 24;\ c' = 2\sqrt{6}$

$V': (0, \pm 2\sqrt{3});\ F': (0, \pm 2\sqrt{6})$

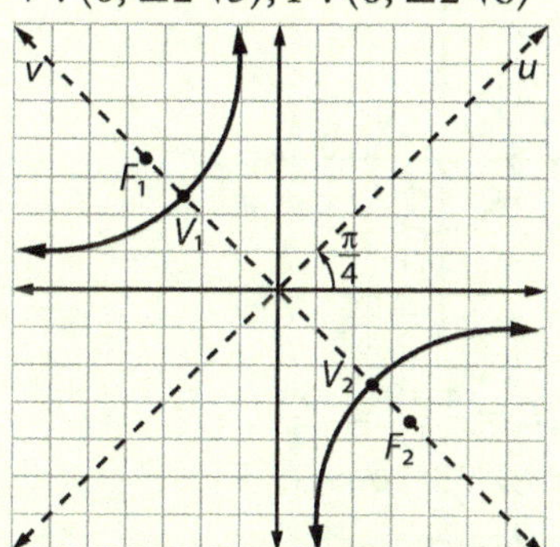

13. $\cot 2\theta = \frac{9-15}{-6\sqrt{3}} = \frac{1}{\sqrt{3}};\ 2\theta = \frac{\pi}{3};\ \theta = \frac{\pi}{6}$

$x = \frac{\sqrt{3}u - v}{2};\ y = \frac{u + \sqrt{3}v}{2}$

$9\left(\frac{3u^2 - 2\sqrt{3}uv + v^2}{4}\right)$

$\qquad - 6\sqrt{3}\left(\frac{\sqrt{3}u^2 + 2uv - \sqrt{3}v^2}{4}\right)$

$\qquad + 15\left(\frac{u^2 + 2\sqrt{3}uv + 3v^2}{4}\right) = 288$

$27u^2 - 18\sqrt{3}uv + 9v^2 - 18u^2$

$\qquad - 12\sqrt{3}uv + 18v^2 + 15u^2$

$\qquad + 30\sqrt{3}uv + 45v^2 = 288(4)$

$\qquad\qquad 24u^2 + 72v^2 = 1152$

$\qquad\qquad \frac{u^2}{48} + \frac{v^2}{16} = 1$

$(c')^2 = 48 - 16 = 32;\ c' = 8$

$V': (\pm 4\sqrt{3}, 0);\ cV': (0, \pm 4)$

$F': (\pm 4\sqrt{2}, 0)$

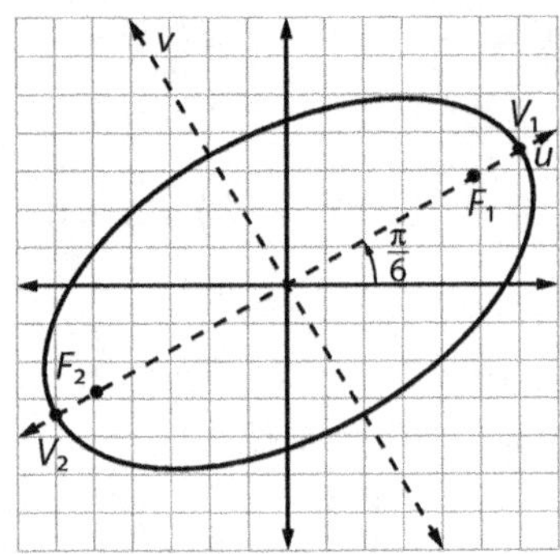

For the parabola in Example 3, $B^2 - 4AC = (2\sqrt{3})^2 - 4(3)(1) = 0$. Examining its general form in the uv-plane, $u^2 - 4v - 12 = 0$, the discriminant $(B')^2 - 4A'C' = (0)^2 - 4(1)(0) = 0$ as well. In fact, the discriminant of any conic is *invariant under rotation*, so that $B^2 - 4AC = (B')^2 - 4A'C'$. When the conic is rotated so there is no uv-term, $B' = 0$ and $B^2 - 4AC = -4A'C'$. Notice how the characteristics for conics with horizontal and vertical axes reviewed at the beginning of this section are special cases for the following general characteristics of conic sections.

Identifying a Conic by Its Discriminant			

The graph of a basic conic section whose equation is of the form
$$Ax^2 + Bxy + Cy^2 + Dx + Ey + F = 0$$
can be classified using the value of the discriminant: $B^2 - 4AC$.

Circle	Ellipse	Parabola	Hyperbola
$B^2 - 4AC < 0$ $B = 0$ and $A = C$	$B^2 - 4AC < 0$ $B \neq 0$ or $A \neq C$	$B^2 - 4AC = 0$	$B^2 - 4AC > 0$

Example 4 Using the Discriminant Test

Use the discriminant to classify each conic section.

a. $3x^2 + 9xy + 3y^2 - 10 = 0$ **b.** $3x^2 - 4xy + 3y^2 + 2x - 6y - 10 = 0$

c. $x^2 + 6y - 2x = 6xy - 9y^2 + 22$ **d.** $6x^2 + 6y^2 = 10 - 3x - 2y$

Answer

a. $3x^2 + 9xy + 3y^2 - 10 = 0$ Evaluate the discriminant.
$B^2 - 4AC = (9)^2 - 4(3)(3) = 45 > 0$
$\therefore$ a hyperbola

b. $3x^2 - 4xy + 3y^2 + 2x - 6y - 10 = 0$ Evaluate the discriminant.
$B^2 - 4AC = (-4)^2 - 4(3)(3) = -20 < 0$
$\therefore$ an ellipse (which is not a circle, since $B \neq 0$)

c. $x^2 + 6y - 2x = 6xy - 9y^2 + 22$ Express the equation in general form
$x^2 - 6xy + 9y^2 - 2x + 6y - 22 = 0$ before evaluating the discriminant.
$B^2 - 4AC = (-6)^2 - 4(1)(9) = 0$
$\therefore$ a parabola

d. $6x^2 + 6y^2 = 10 - 3x - 2y$ This is a special case of an ellipse.
$\therefore$ a circle, since $B = 0$ and $A = C$ $B^2 - 4AC = 0^2 - 4(6)(6) = -144 < 0$

SKILL ✔ **EXERCISE 5**

---

14. $\cot 2\theta = \dfrac{15 - 13}{2\sqrt{3}} = \dfrac{1}{\sqrt{3}};\ 2\theta = \dfrac{\pi}{3};\ \theta = \dfrac{\pi}{6}$

$x = \dfrac{\sqrt{3}u - v}{2};\ y = \dfrac{u + \sqrt{3}v}{2}$

$15\left(\dfrac{3u^2 - 2\sqrt{3}uv + v^2}{4}\right)$
$\qquad + 2\sqrt{3}\left(\dfrac{\sqrt{3}u^2 + 2uv - \sqrt{3}v^2}{4}\right)$
$+ 13\left(\dfrac{u^2 + 2\sqrt{3}uv + 3v^2}{4}\right) - 48 = 0$
$45u^2 - 30\sqrt{3}uv + 15v^2 + 6u^2$
$\qquad + 4\sqrt{3}uv - 6v^2 + 13u^2$
$\qquad + 26\sqrt{3}uv + 39v^2 = 48(4)$
$\qquad\qquad 64u^2 + 48v^2 = 192$
$\qquad\qquad\qquad \dfrac{u^2}{3} + \dfrac{v^2}{4} = 1$

$(c')^2 = 4 - 3 = 1;\ c' = 1$
$V': (0, \pm 2);\ cV': (0, \pm\sqrt{3});\ F': (0, \pm 1)$

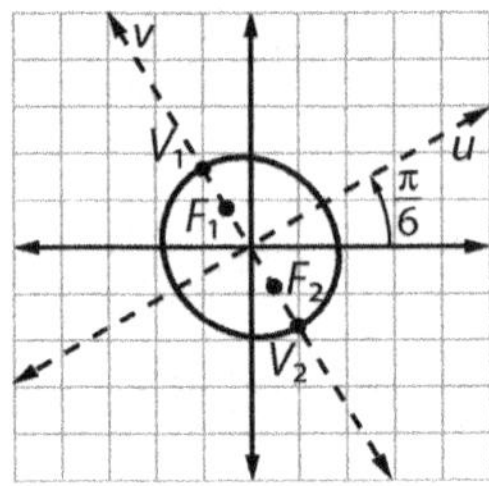

15. $\cot 2\theta = \dfrac{4 - 9}{-12} = \dfrac{5}{12}$

$2\theta \approx 67.38°;\ \theta \approx 33.7°$

$\cos 2\theta = \dfrac{5}{13}$

$\cos\theta = \sqrt{\dfrac{1 + \left(\frac{5}{13}\right)}{2}} = \dfrac{3}{\sqrt{13}}$

$\sin\theta = \sqrt{\dfrac{1 - \left(\frac{5}{13}\right)}{2}} = \dfrac{2}{\sqrt{13}}$

$x = \dfrac{3u - 2v}{\sqrt{13}};\ y = \dfrac{2u + 3v}{\sqrt{13}}$

$4\left(\dfrac{9u^2 - 12uv + 4v^2}{13}\right)$
$\qquad - 12\left(\dfrac{6u^2 + 5uv - 6v^2}{13}\right)$
$\qquad + 9\left(\dfrac{4u^2 + 12uv + 9v^2}{13}\right)$
$\qquad - 9\sqrt{13}\left(\dfrac{3u - 2v}{\sqrt{13}}\right)$
$\qquad - 6\sqrt{13}\left(\dfrac{2u + 3v}{\sqrt{13}}\right) = 0$

$\left(\dfrac{36}{13} - \dfrac{72}{13} + \dfrac{36}{13}\right)u^2$
$\qquad + \left(-\dfrac{48}{13} - \dfrac{60}{13} + \dfrac{108}{26}\right)uv$
$\qquad + \left(\dfrac{16}{13} + \dfrac{72}{13} + \dfrac{81}{13}\right)v^2 - 27u$
$\qquad + 18v - 12u - 18v = 0$
$\qquad\qquad 13v^2 - 39u = 0$
$\qquad\qquad\qquad u = \dfrac{1}{3}v^2$

$V'(0, 0)$
$F'\left(\dfrac{3}{4}, 0\right)$

16. $\cot 2\theta = \dfrac{1 - 1}{2} = 0;\ 2\theta = \dfrac{\pi}{2};\ \theta = \dfrac{\pi}{4}$

$x = \dfrac{\sqrt{2}u - \sqrt{2}v}{2};\ y = \dfrac{\sqrt{2}u + \sqrt{2}v}{2}$

$\left(\dfrac{u^2 - \sqrt{2}uv + v^2}{2}\right) + 2\left(\dfrac{u^2 - v^2}{2}\right)$
$\qquad + \left(\dfrac{u^2 + \sqrt{2}uv + v^2}{2}\right) + \dfrac{\sqrt{2}}{2}\left(\dfrac{\sqrt{2}u - \sqrt{2}v}{2}\right)$
$\qquad - \dfrac{\sqrt{2}}{2}\left(\dfrac{\sqrt{2}u + \sqrt{2}v}{2}\right) = 0$

$\left(\dfrac{1}{2} + 1 + \dfrac{1}{2}\right)u^2 + \left(\dfrac{-\sqrt{2}}{2} + \dfrac{\sqrt{2}}{2}\right)uv$
$\qquad + \left(\dfrac{1}{2} - 1 + \dfrac{1}{2}\right)v^2 + \left(\dfrac{1}{2} - \dfrac{1}{2}\right)u$
$\qquad + \left(-\dfrac{1}{2} - \dfrac{1}{2}\right)v = 0$
$\qquad\qquad 2u^2 - v = 0$
$\qquad\qquad\qquad v = 2u^2$

A. Exercises

State whether the conic is rotated. Then use the discriminant test to classify the graph as a circle, ellipse, parabola, or hyperbola.

1. $4x^2 - 9y^2 - 8x + 54y - 113 = 0$ no; hyperbola
2. $x^2 + xy + 2y^2 - 6 = 0$ yes; ellipse
3. $5x^2 + 5y^2 - 30x - 6y + 5 = 0$ no; circle
4. $4x^2 + 9y^2 - 8x + 126y - 131 = 0$ no; ellipse
5. $x^2 - 2xy + y^2 - 5x - 15y = 0$ yes; parabola
6. $3x^2 - 12xy + 2y^2 - 144 = 0$ yes; hyperbola

State the rotation formulas that will rotate each general quadratic equation to a uv-plane without a uv-term.

7. $xy = -2$ $x = \frac{\sqrt{2}}{2}u - \frac{\sqrt{2}}{2}v,\ y = \frac{\sqrt{2}}{2}u + \frac{\sqrt{2}}{2}v$
8. $2x^2 + xy + 2y^2 - 1 = 0$ $x = \frac{\sqrt{2}}{2}u - \frac{\sqrt{2}}{2}v,\ y = \frac{\sqrt{2}}{2}u + \frac{\sqrt{2}}{2}v$
9. $2x^2 + \sqrt{3}xy + y^2 + y + x = 0$ $x = \frac{\sqrt{3}}{2}u - \frac{1}{2}v,\ y = \frac{1}{2}u + \frac{\sqrt{3}}{2}v$
10. $3x - \sqrt{3}xy + y^2 + 5 = 0$ $x = \frac{\sqrt{3}}{2}u - \frac{1}{2}v,\ y = \frac{1}{2}u + \frac{\sqrt{3}}{2}v$

Find the angle of rotation of the xy-plane that causes the u- or v-axis to be parallel to the conic's axes. Then write the standard form equation in terms of u and v and sketch a graph of the conic and any foci.

11. $xy - 8 = 0$ $\theta = \frac{\pi}{4};\ \frac{u^2}{16} - \frac{v^2}{16} = 1$
12. $xy + 6 = 0$ $\theta = \frac{\pi}{4};\ \frac{v^2}{12} - \frac{u^2}{12} = 1$
13. $9x^2 - 6\sqrt{3}xy + 15y^2 - 288 = 0$ $\theta = \frac{\pi}{6};\ \frac{u^2}{48} + \frac{v^2}{16} = 1$
14. $15x^2 + 2\sqrt{3}xy + 13y^2 - 48 = 0$ $\theta = \frac{\pi}{6};\ \frac{u^2}{3} + \frac{v^2}{4} = 1$
15. $4x^2 - 12xy + 9y^2 - 9\sqrt{13}x - 6\sqrt{13}y = 0$ $\theta \approx 33.7°;\ u = \frac{1}{3}v^2$
16. $x^2 + 2xy + y^2 + \frac{\sqrt{2}}{2}x - \frac{\sqrt{2}}{2}y = 0$ $\theta = \frac{\pi}{4};\ v = 2u^2$

B. Exercises

Write a general form equation for each conic in terms of x and y, given its angle of rotation and equation in terms of u and v.

17. $\theta = \frac{\pi}{4};\ 2u^2 + v^2 = 9$
18. $\theta = \frac{\pi}{6};\ 3u = v^2$
19. $\theta = \frac{\pi}{3};\ 4u^2 - 3v^2 = 12$
20. $\theta = \frac{\pi}{4};\ \frac{u^2}{25} - \frac{v^2}{16} = 1$
21. $\theta = \frac{\pi}{6};\ \frac{u^2}{9} + \frac{v^2}{4} = 1$
22. $\theta = \frac{\pi}{3};\ \frac{u^2}{28} + \frac{v^2}{7} = 1$

17. $3x^2 + 2xy + 3y^2 - 18 = 0$
18. $x^2 - 2\sqrt{3}xy + 3y^2 - 6\sqrt{3}x - 6y = 0$
19. $-5x^2 + 14\sqrt{3}xy + 9y^2 - 48 = 0$
20. $9x^2 - 82xy + 9y^2 + 800 = 0$
21. $21x^2 - 10\sqrt{3}xy + 31y^2 - 144 = 0$
22. $13x^2 - 6\sqrt{3}xy + 7y^2 - 112 = 0$

Find the angle of rotation of the xy-plane that causes the u- or v-axis to be parallel to the conic's axes. Then write the standard form equation in terms of u and v and sketch a graph of the conic and any foci.

23. $6x^2 - 6xy + 14y^2 = 90$ $\theta \approx 18.4°;\ \frac{u^2}{18} + \frac{v^2}{6} = 1$
24. $16x^2 + 24xy + 9y^2 + 30x - 40y = 0$ $\theta \approx 36.9°;\ v = \frac{u^2}{2}$
25. $3x^2 + 2\sqrt{3}xy + y^2 + 4x - 4\sqrt{3}y + 24 = 0$ $\theta = \frac{\pi}{6};\ v = \frac{1}{2}u^2 + 3$
26. $621x^2 + 290xy - 75y^2 = 2600$ $\theta = 11.3°;\ \frac{u^2}{4} - \frac{v^2}{25} = 1$

Write a general form equation for each conic in terms of x and y, given its angle of rotation and equation in terms of u and v. Then use technology to graph the conic.

27. $\theta = \frac{\pi}{4},\ 3u^2 + v^2 = 4$
28. $\theta = 30°,\ v^2 = 6u$
29. $\theta = \frac{\pi}{4},\ \frac{u^2}{25} - v^2 = 1$
30. $\theta = 60°,\ \frac{u^2}{4} + \frac{v^2}{5} = 1$

Recall that a point is a degenerate ellipse, a line is a degenerate parabola, and two intersecting lines are a degenerate hyperbola. Use the discriminant test to classify each conic section. Then solve for y and describe the degenerate conic.

31. $4x^2 + 12xy + 9y^2 = 0$ parabola; $y = \frac{2}{3}x$; a line
32. $16x^2 - 25y^2 = 0$ hyperbola; $y = \pm\frac{4}{5}x$; 2 intersecting lines

33. A satellite dish with a parabolic cross section is represented (in feet) by the equation
$x^2 - 2xy + y^2 - 10\sqrt{2}x - 10\sqrt{2}y = 0$.

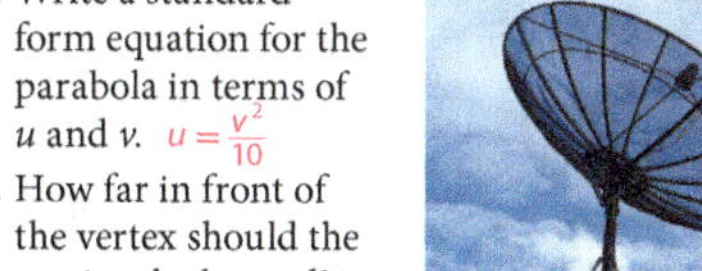

  a. Write a standard form equation for the parabola in terms of u and v. $u = \frac{v^2}{10}$

  b. How far in front of the vertex should the receiver be located?   2.5 ft

34. **Prove:** Show that a rotation has no effect on a circle with the equation $x^2 + y^2 = r^2$.

35. **Prove:** Substitute the rotation formulas $x = u\cos\theta - v\sin\theta$ and $y = u\sin\theta - v\cos\theta$ into the general quadratic equation $Ax^2 + Bxy + Cy^2 + Dx + Ey + F = 0$ and show that B', the coefficient of the uv-term, is $2(C - A)\sin\theta\cos\theta + B(\cos^2\theta - \sin^2\theta)$.

36. **Prove:** Solve the system of equations $x = u\cos\theta - v\sin\theta$ and $y = u\sin\theta + v\cos\theta$ to derive the rotation formulas expressing u and v in terms of x and y.

27. $x^2 + xy + y^2 - 2 = 0$
28. $x^2 - 2\sqrt{3}xy + 3y^2 - 12\sqrt{3}x - 12y = 0$
29. $12x^2 - 26xy + 12y^2 + 25 = 0$
30. $17x^2 + 2\sqrt{3}xy + 19y^2 - 80 = 0$

$$4(x + \sqrt{3}y)^2 - 3(y - \sqrt{3}x)^2 = 12(4)$$
$$4(x^2 + 2\sqrt{3}xy + 3y^2)$$
$$- 3(y^2 - 2\sqrt{3}xy + 3x^2) = 48$$
$$-5x^2 + 14\sqrt{3}xy + 9y^2 - 48 = 0$$

20. $u = x\cos\frac{\pi}{4} + y\sin\frac{\pi}{4} = \dfrac{\sqrt{2}x + \sqrt{2}y}{2}$

$v = y\cos\frac{\pi}{4} - x\sin\frac{\pi}{4} = \dfrac{\sqrt{2}y - \sqrt{2}x}{2}$

$$16u^2 - 25v^2 = 400$$
$$16\left(\frac{\sqrt{2}(x+y)}{2}\right)^2 - 25\left(\frac{\sqrt{2}(y-x)}{2}\right)^2 = 400$$
$$16(x + y)^2 - 25(y - x)^2 = 400(2)$$
$$16(x^2 + 2xy + y^2)$$
$$- 25(y^2 - 2xy + x^2) = 800$$
$$-9x^2 + 82xy - 9y^2 - 800 = 0$$
$$\text{or } 9x^2 - 82xy + 9y^2 + 800 = 0$$

21. $u = x\cos\frac{\pi}{6} + y\sin\frac{\pi}{6} = \dfrac{\sqrt{3}x + y}{2}$

$v = y\cos\frac{\pi}{6} - x\sin\frac{\pi}{6} = \dfrac{\sqrt{3}y - x}{2}$

$$4u^2 + 9v^2 = 36$$
$$4\left(\frac{\sqrt{3}x + y}{2}\right)^2 + 9\left(\frac{\sqrt{3}y - x}{2}\right)^2 = 36$$
$$4(3x^2 + 2\sqrt{3}xy + y^2)$$
$$+ 9(3y^2 - 2\sqrt{3}xy + x^2) = 36(4)$$
$$21x^2 - 10\sqrt{3}xy + 31y^2 - 144 = 0$$

22. $u = x\cos\frac{\pi}{3} + y\sin\frac{\pi}{3} = \dfrac{x + \sqrt{3}y}{2}$

$v = y\cos\frac{\pi}{3} - x\sin\frac{\pi}{3} = \dfrac{y - \sqrt{3}x}{2}$

$$u^2 + 4v^2 = 28$$
$$\left(\frac{x + \sqrt{3}y}{2}\right)^2 + 4\left(\frac{y - \sqrt{3}x}{2}\right)^2 = 28$$
$$(x + \sqrt{3}y)^2 + 4(y - \sqrt{3}x)^2 = 28(4)$$
$$(x^2 + 2\sqrt{3}xy + 3y^2)$$
$$+ 4(y^2 - 2\sqrt{3}xy + 3x^2) = 112$$
$$13x^2 - 6\sqrt{3}xy + 7y^2 - 112 = 0$$

23. $\cot 2\theta = \frac{4}{3};\ \cos 2\theta = \frac{4}{5};\ \theta \approx 18.4°$

$$\cos\theta = \sqrt{\frac{1 + \left(\frac{4}{5}\right)}{2}} = \frac{3}{\sqrt{10}}$$
$$\sin\theta = \sqrt{\frac{1 - \left(\frac{4}{5}\right)}{2}} = \frac{1}{\sqrt{10}}$$
$$x = \frac{3u - v}{\sqrt{10}};\ y = \frac{u + 3v}{\sqrt{10}}$$
$$6\left(\frac{3u - v}{\sqrt{10}}\right)^2 - 6\left(\frac{3u - v}{\sqrt{10}}\right)\left(\frac{u + 3v}{\sqrt{10}}\right)$$
$$+ 14\left(\frac{u + 3v}{\sqrt{10}}\right)^2 - 90 = 0$$
$$6(9u^2 - 6uv + v^2) - 6(3u^2 + 8uv - 3v^2)$$
$$+ 14(u^2 + 6uv + 9v^2) = 90(10)$$
$$50u^2 + 150v^2 = 900$$
$$\frac{u^2}{18} + \frac{v^2}{6} = 1$$
$V': \approx (\pm 4.2, 0);\ cV': \approx (\pm 2.4, 0)$
$F': \approx (\pm 3.5, 0)$

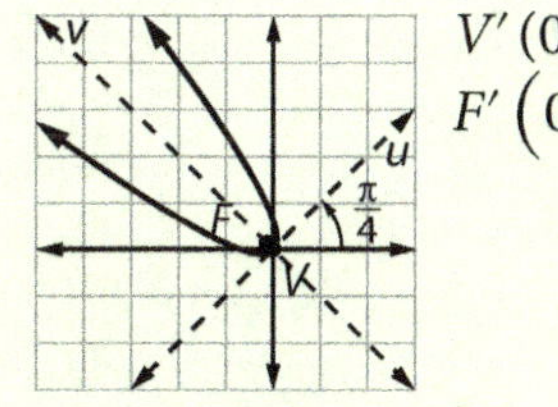

$V'(0, 0)$
$F'\left(0, \frac{1}{8}\right)$

B. Exercises

17. $u = x\cos\frac{\pi}{4} + y\sin\frac{\pi}{4} = \dfrac{\sqrt{2}x + \sqrt{2}y}{2}$

$v = y\cos\frac{\pi}{4} - x\sin\frac{\pi}{4} = \dfrac{\sqrt{2}y - \sqrt{2}x}{2}$

$$2\left(\frac{\sqrt{2}x + \sqrt{2}y}{2}\right)^2 + \left(\frac{\sqrt{2}x - \sqrt{2}y}{2}\right)^2 = 9$$
$$2\left(\frac{1}{2}x^2 + xy + \frac{1}{2}y^2\right)$$
$$+ \left(\frac{1}{2}y^2 - xy + \frac{1}{2}x^2\right) = 9$$
$$x^2 + 2xy + y^2 + \frac{1}{2}y^2 - xy + \frac{1}{2}x^2 = 9$$
$$\frac{3}{2}x^2 + xy + \frac{3}{2}y^2 - 9 = 0$$
$$\text{or } 3x^2 + 2xy + 3y^2 - 18 = 0$$

18. $u = x\cos\frac{\pi}{6} + y\sin\frac{\pi}{6} = \dfrac{\sqrt{3}x + y}{2}$

$v = y\cos\frac{\pi}{6} - x\sin\frac{\pi}{6} = \dfrac{\sqrt{3}y - x}{2}$

$$3\left(\frac{\sqrt{3}x + y}{2}\right) = \left(\frac{\sqrt{3}y - x}{2}\right)^2$$
$$6(\sqrt{3}x + y) - (\sqrt{3}y - x)^2 = 0(4)$$
$$6\sqrt{3}x + 6y - (3y^2 - 2\sqrt{3}xy + x^2) = 0$$
$$-x^2 + 2\sqrt{3}xy - 3y^2 + 6\sqrt{3}x + 6y = 0$$
$$x^2 - 2\sqrt{3}xy + 3y^2 - 6\sqrt{3}x - 6y = 0$$

19. $u = x\cos\frac{\pi}{3} + y\sin\frac{\pi}{3} = \dfrac{x + \sqrt{3}y}{2}$

$v = y\cos\frac{\pi}{3} - x\sin\frac{\pi}{3} = \dfrac{y - \sqrt{3}x}{2}$

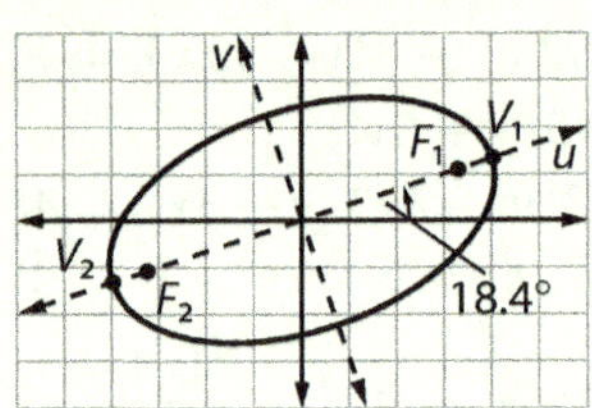

24. $\cot 2\theta = \frac{7}{24}$; $\cos 2\theta = \frac{7}{25}$; $\theta \approx 36.9°$

$$\cos\theta = \sqrt{\frac{1 + \left(\frac{7}{25}\right)}{2}} = \frac{4}{5}$$

$$\sin\theta = \sqrt{\frac{1 - \left(\frac{7}{25}\right)}{2}} = \frac{3}{5}$$

$$x = \frac{4u - 3v}{5};\ y = \frac{3u + 4v}{5}$$

$$16\left(\frac{4u - 3v}{5}\right)^2 + 24\left(\frac{4u - 3v}{5}\right)\left(\frac{3u + 4v}{5}\right)$$
$$+ 9\left(\frac{3u + 4v}{5}\right)^2 + 30\left(\frac{4u - 3v}{5}\right)$$
$$- 40\left(\frac{3u + 4v}{5}\right) = 0$$

$$16(16u^2 - 24uv + 9v^2)$$
$$+ 24(12u^2 + 7uv - 12v^2)$$
$$+ 9(9u^2 + 24uv + 16v^2)$$
$$+ 150(4u - 3v) - 200(3u + 4v)$$
$$= 0(25)$$

$$625u^2 - 1250v = 0$$
$$v = \frac{u^2}{2}$$

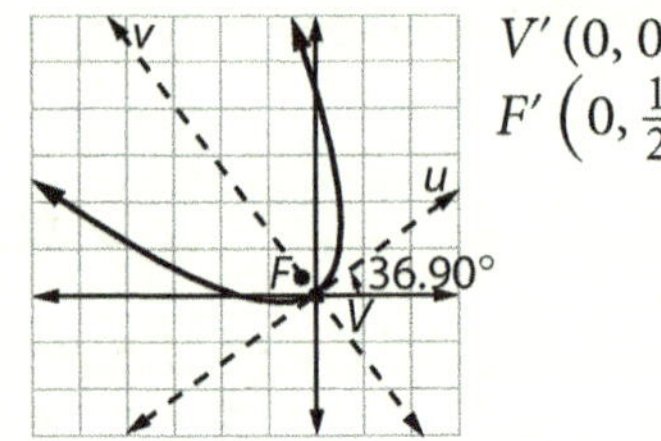

$V'(0, 0)$

$F'\left(0, \frac{1}{2}\right)$

25. $\cot 2\theta = \frac{3 - 1}{2\sqrt{3}} = \frac{1}{\sqrt{3}}$; $2\theta = \frac{\pi}{3}$; $\theta = \frac{\pi}{6}$

$$x = \frac{\sqrt{3}u - v}{2};\ y = \frac{u + \sqrt{3}v}{2}$$

$$3\left(\frac{3u^2 - 2\sqrt{3}uv + v^2}{4}\right)$$
$$+ 2\sqrt{3}\left(\frac{\sqrt{3}u^2 + 2uv - \sqrt{3}v^2}{4}\right)$$
$$+ \left(\frac{u^2 + 2\sqrt{3}uv + 3v^2}{4}\right) + 4\left(\frac{\sqrt{3}u - v}{2}\right)$$
$$- 4\sqrt{3}\left(\frac{u + \sqrt{3}v}{2}\right) = -24$$

$$\left(\frac{9}{4} + \frac{6}{4} + \frac{1}{4}\right)u^2$$
$$+ \left(-\frac{6\sqrt{3}}{4} + \frac{4\sqrt{3}}{4} + \frac{2\sqrt{3}}{4}\right)uv$$
$$+ \left(\frac{3}{4} - \frac{6}{4} + \frac{3}{4}\right)v^2 + (2\sqrt{3} - 2\sqrt{3})u$$
$$+ (-2 - 6)v = -24$$
$$4u^2 - 8v = -24$$
$$v = \frac{1}{2}u^2 + 3$$

$V'(0, 3)$

$F'\left(0, 3\tfrac{1}{2}\right)$

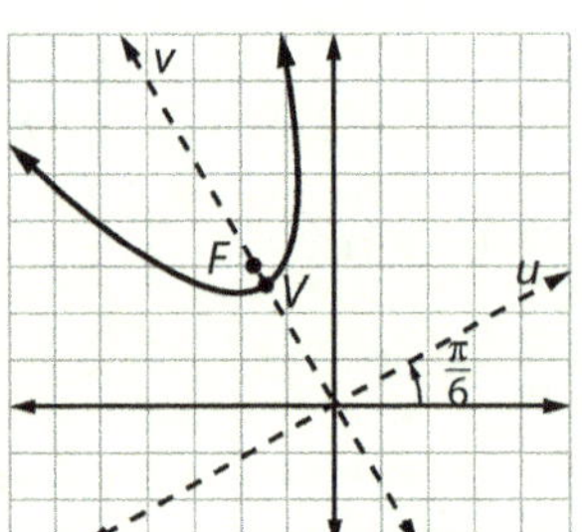

Find the angle of rotation of the *xy*-plane that causes the *u*- or *v*-axis to be parallel to the conic's axes. Then write the standard form equation in terms of *u* and *v* and sketch a graph of the conic and any foci.

37. $4x^2 - 16xy + 16y^2 + 20\sqrt{5}y = 0$ $\theta \approx 26.6°$ $u = -(v + 1)^2 + 1$

38. $3x^2 + 6xy + 3y^2 + 3\sqrt{2}x + \sqrt{2}y = 0$ $\theta = \frac{\pi}{4}$; $v = 3u^2 + 2u$

Write the general form equation for each conic in terms of *x* and *y*, given its angle of rotation and equation in terms of *u* and *v*. Then use technology to graph the conic.

39. $\theta = \frac{\pi}{6}$, $v = 4u^2 + 2u$ $6x^2 + 4\sqrt{3}xy + 2y^2 + (2\sqrt{3} + 1)x + (2 - \sqrt{3})y = 0$

40. $\theta = \frac{\pi}{4}$, $(u - 2\sqrt{2})^2 - (v - \sqrt{2})^2 = 1$ $2xy - 6x - 2y + 5 = 0$

41. Prove: Substitute the rotation formulas into general quadratic form for conics to find each coefficient of the rotated formula in the (u, v) plane having $B' = 0$:
$A'u^2 + C'v^2 + D'u + E'v + F' = 0$.

42. Explain: Use the results of exercise 41 to explain why each expression is invariant under rotation.

  a. F **b.** $A + C$

CUMULATIVE REVIEW

43. Find the ordered pair for the parametric equations $x = 2t - 1$ and $y = t - 4$ when $t = 2$.   [1.8] $(3, -2)$

44. Eliminate the parameter for the equations $x = t + 5$ and $y = 3t + 1$.   [1.8] $y = 3x - 14$

Simplify.   [5.1]

45. $2\cos^2\theta + 2\sin^2\theta$ 2   **46.** $4\tan^2\theta - 4\sec^2\theta$ -4

47. Write the standard form equation of each conic with its center not at the origin.

  a. a parabola with a vertical axis of symmetry   [8.1]

  b. an ellipse with a horizontal major axis   [8.2]

  c. a hyperbola with a horizontal transverse axis   [8.3]

48. Express $3x^2 + 6y^2 + 18x - 24y + 33 = 0$ in standard form.   [8.2]

47a. $y = (x - h)^2 + k$

47b. $\dfrac{(x - h)^2}{a^2} + \dfrac{(y - k)^2}{b^2} = 1$

47c. $\dfrac{(x - h)^2}{a^2} - \dfrac{(y - k)^2}{b^2} = 1$

48. $\dfrac{(x + 3)^2}{6} + \dfrac{(y - 2)^2}{3} = 1$

49. What is the sum of the measures of the acute angles α and β if $\tan\alpha = \cot\beta$?   [4.3]   C

  A. $45°$ **C.** $90°$ **E.** $180°$

  B. $60°$ **D.** $120°$

50. What is $\sin x \csc x + \sec x \tan x$ if $x = 30°$?   [4.3, 5.1]   B

  A. $\frac{2}{3}$ **C.** 2 **E.** none of these

  B. $\frac{5}{3}$ **D.** $\frac{\sqrt{3}}{2}$

51. What is $\dfrac{\sin^2\theta + \cos^2\theta + \tan\theta}{\sec^2\theta - \tan^2\theta}$ if $\theta = \frac{\pi}{4}$?   [5.1]   C

  A. 0 **C.** 2 **E.** $-\dfrac{\sqrt{3}}{3}$

  B. -2 **D.** $\dfrac{\sqrt{3}}{3}$

52. Which set of polar coordinates represents the rectangular coordinates $(6, 8)$?   [6.4]   E

  A. $(6\sin\theta, 8\cos\theta)$ **D.** $(10, 36.9°)$

  B. $(6\cos\theta, 8\sin\theta)$ **E.** $(10, 53.1°)$

  C. $(10\cos\theta, 10\sin\theta)$

continued in Answers and Solutions Overflow

Counterexamples and Existence Proofs

Some mathematics problems look simple, and you try them for a year or so, and then you try them for a hundred years, and it turns out that they're extremely hard to solve. There's no reason why these problems shouldn't be easy, and yet they turn out to be extremely intricate. Fermat's Last Theorem is the most beautiful example of this.

—Andrew Wiles (British mathematician who proved Fermat's Last Theorem)

A counterexample proves that a statement is false. The statement "there are no solutions to $a^2 + b^2 = c^2$ where a, b, and c are consecutive even integers" is easily disproved by doubling the most famous Pythagorean triple (3, 4, 5) to get the counterexample $6^2 + 8^2 = 10^2$. This example also provides an existence proof for the related statement "there is a solution to $a^2 + b^2 = c^2$ where a, b, and c are consecutive even integers."

Several special cases of the general equation $a^x + b^y = c^z$ in which all variables represent natural numbers have generated famous problems in mathematics. The first special case is the Pythagorean Theorem, in which $x = y = z = 2$. The equation $a^2 + b^2 = c^2$ has an infinite number of solutions (a, b, c), including the frequently occurring Pythagorean triples (3, 4, 5), (5, 12, 13), (7, 24, 25), and (8, 15, 17) and all their integral multiples.

Fermat's Last Theorem addresses a second special case, stating there are no solutions to $a^n + b^n = c^n$ where $n > 2$. While no counterexample could be found to this 1637 conjecture, its proof became the most famous unsolved problem in mathematics until it was finally proven in 1994 by Andrew Wiles.

Another special case states that "no solution exists to $a^x + b^y = c^z$ with unique x, y, and z where a, b, and c are relatively prime." At the beginning of the millennium, there were ten known counterexamples. The first five could be referred to as "small" solutions. The missing values for x are found in exercise 5. Notice that each of the five "large" solutions contains 2 as one of the exponents.

$$1 + 2^x = 3^2$$
$$2^5 + 7^2 = 3^x$$
$$7^x + 13^2 = 2^9$$
$$2^x + 17^3 = 71^2$$
$$3^x + 11^4 = 122^2$$

$$17^7 + 76{,}271^3 = 21{,}063{,}928^2$$
$$1414^3 + 2{,}213{,}459^2 = 65^7$$
$$43^8 + 96{,}222^3 = 30{,}042{,}907^2$$
$$33^8 + 1{,}549{,}034^2 = 15{,}613^3$$
$$9262^3 + 15{,}312{,}283^2 = 113^7$$

Andrew Beal, a Texas banker and number theory enthusiast, presented a related conjecture in 1993 after extensive computer testing. The Beal Conjecture states there are no solutions to $a^x + b^y = c^z$ where a, b, and c are *relatively prime* (share no common integral factor other than 1) for $x \geq 3$, $y \geq 3$, and $z \geq 3$. Currently, the American Mathematical Society is offering a $1,000,000 prize for either a correct proof or a counterexample for the Beal Conjecture.

Counterexamples are frequently used to disprove statements in other areas of reasoning. While these proofs may not be as definitive as those found in mathematics, counterexamples can be used to arrive at reasonable conclusions in many fields of study.

Consider the claim "there are no miracles." Instead of accepting the possibility of a miracle, many believe that people accidentally or intentionally misinterpret an explainable event as supernatural. To prove this statement to be false, many counterexamples have been cited. The resurrection of Christ provides the supreme counterexample.

For a Christian, experiencing the love of Christ and the guidance of the Holy Spirit provides a counterexample

Biblical Perspective of Mathematics

Objectives

1. To identify the role of counterexamples in proofs
2. To explore examples and counterexamples related to the general case of $a^x + b^y = c^z$
3. To evaluate the use of counterexamples in evangelism

Assignments

- **Minimum:** 1–7; 8–10 (in class)
- **Standard:** 1–8; 9–12 (in class)
- **Extended:** 1–12

PRESENTATION

Explain that the purpose of a counterexample is to invalidate a statement. Contrast that with an existence proof, which uses an example to prove that a statement is true. Consider the illustrations using the Pythagorean triple 3, 4, 5 for both a counterexample and an example proving existence. Then relate a Christian's relationship to God as a counterexample to the fool's assertion in Psalms 14:1 and 53:1.

Mention that creation offers proof of the existence of God. Isaac Newton asserted, "In the absence of any other proof, the thumb alone would convince me of God's existence." He also said, "This most beautiful System of the Sun, Planets and Comets, could only proceed from the counsel and dominion of an intelligent and powerful being." Romans 1 and 2 makes clear that unbelievers suppress the knowledge of God and their own consciences when they choose to deny God and reject Jesus. The miracles of Christ recorded throughout Scripture and the statements Jesus made about Himself in John, specifically, prove that He was God in the flesh. Encourage students to be bold in expressing their faith as they witness to others.

Use the Pythagorean Theorem to introduce the most familiar case of the general statement $a^x + b^y = c^z$. Then discuss the famous Fermat's Last Theorem and its lack of a counterexample. Discuss the ten known counterexamples to the statement that $a^x + b^y = c^z$ has no solutions with x, y, and z being unique where a, b, and c are relatively prime. Then discuss the Beal Conjecture and the million-dollar prize available for its successful proof or for a counterexample disproving it.

In a page margin of Diophantus's *Arithmetica*, where Fermat wrote his final theorem, Fermat also wrote that he had discovered a truly wonderful proof of his theorem, but that the margin did not

to the statement "there is no God." Anyone who has not seen the reality of God in his life may not accept these examples as proof. God wants us to accept His existence by faith (Heb. 11:6). Just as it is impossible to mathematically prove God's existence, it is also impossible to prove that there is no God.

❭ Exercises

1. What is the purpose of a counterexample?

2. Which type of statement can be proved by an example?

Consider $a^x + b^y = c^z$ where all variables represent natural numbers.

3. What additional condition is required by the Pythagorean Theorem?

4. What additional condition is required by Fermat's Last Theorem?

5. Find x in each "small solution" $a^x + b^y = c^z$, where a, b, and c are relatively prime.
   a. $1 + 2^x = 3^2$ 3
   b. $2^5 + 7^2 = 3^x$ 4
   c. $7^x + 13^2 = 2^9$ 3
   d. $2^x + 17^3 = 71^2$ 7
   e. $3^x + 11^4 = 122^2$ 5

6. Explain why each equation is not a counterexample to the Beal Conjecture.
   a. $2^9 + 8^3 = 4^5$ 2 is a common factor.
   b. $19^4 + 38^3 = 57^3$ 19 is a common factor.
   c. $34^5 + 51^4 = 85^4$ 17 is a common factor.
   d. $1414^3 + 2{,}213{,}459^2 = 65^7$ One of the exponents is 2.

7. How could you reasonably refute the statement "all men are lazy"?

8. Name and describe the common powerful counterexample frequently used in court to reasonably disprove the possibility that someone committed a crime at a given time and location.

9. The complexity of brain functionality required for vision would render it useless in any primitive form. What theory can be reasonably proven false with this counterexample?

10. How does our personal relationship with Christ relate to the statement "there is no God"?

11. **Discuss:** Christ appeared to Mary, the disciples, and many other witnesses after His resurrection. List several statements that these counterexamples disprove.

12. **Discuss:** How should the truths found in John 12:37 and Acts 1:3 affect how we share the gospel with others?

have enough space for him to write it all. Ask students for their thoughts in light of Wiles's quote at the beginning of the feature. Share with students that Andrew Wiles said, "I don't believe Fermat had a proof. I think he fooled himself into thinking he had a proof."

In Chapter 7 students examined a portion of the famous debate between Dr. Greg Bahnsen and Dr. Gordon Stein. In the final question of the debate, Stein was asked what he would consider adequate evidence for the existence of God. Responses to that question can be found by using the Internet keyword search *Stein Bahnsen debate* and searching for *final question* in the text or by locating the discussion near the end of a video of the debate.

8.5 Parametric Representations

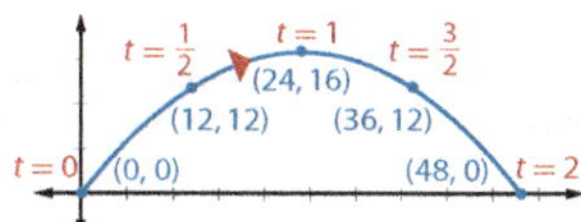

Parametric equations can be used to model the path of a kicked ball.

Parametric equations were introduced in Section 1.8. The introduction of the parameter t, a third variable representing time, provided a more complete model for the path of a projectile. The trajectory of a ball kicked at a speed of 40 ft/sec with an angle of elevation of approximately 53° can be modeled by $y = -\frac{1}{36}x^2 + \frac{4}{3}x$. However, this model provides no information regarding the times associated with the ball's flight. Expressing the horizontal displacement and height as functions of time allows the x- and y-coordinates to be determined at any time during the ball's flight. An arrowhead is drawn to indicate the curve's *orientation*, the order of the points as the parameter t increases.

$x(t) = 24t$ and $y(t) = -16t^2 + 32t$

t	0	$\frac{1}{2}$	1	$\frac{3}{2}$	2
x	0	12	24	36	48
y	0	12	16	12	0

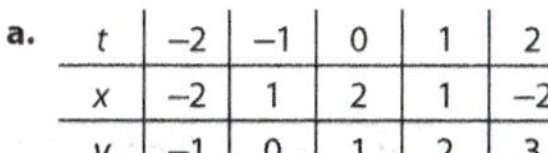

DEFINITIONS

A **parametric curve** is a set of ordered pairs (x, y) defined by the **parametric equations** $x = f(t)$ and $y = g(t)$ where $f(t)$ and $g(t)$ are continuous functions of t on the *parameter interval, I*.

If the parameter interval is not specified, the implied interval is all values of t that produce real-number values of x and y.

Example 1   Graphing Parametric Equations

Draw the curve described by each set of parametric equations.

a. $x = 2 - t^2$ and $y = t + 1; t \in [-2, 2]$ **b.** $x = 2 - \frac{t^2}{4}$ and $y = 1 - \frac{t}{2}$

Answer

a.

t	-2	-1	0	1	2
x	-2	1	2	1	-2
y	-1	0	1	2	3

1. Evaluate x and y for several values of t within the interval.

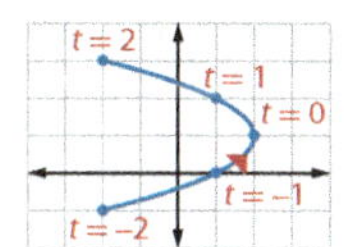

2. Plot the resulting points and connect them with a smooth curve. Add an arrowhead to indicate the curve's orientation.

CONTINUED ➡

8.5 Parametric Representations

Objectives

1. To graph curves represented by parametric equations

2. To eliminate the parameter to rewrite parametric equations as a rectangular equation

3. To write parametric equations for a curve

4. To solve projectile motion problems using parametric equations

Flash

Professional soccer players typically shoot the ball at about 60 mi/hr. The Brazilian professional player Ronny Heberson is generally credited with the fastest recorded kick at about 131 mi/hr, which can be viewed online using the Internet keyword search *Heberson free kick video*.

Vocabulary

parameter interval
parametric curve
parametric equations
orientation

PRESENTATION

Lesson Opener

1. Find a set of five ordered pairs defined by the parametric equations $x = t + 2$ and $y = t^2 + 1$ if $t \in [-2, 2]$.
   (0, 5), (1, 2), (2, 1), (3, 2), (4, 5)

2. Graph the set of parametric equations $x = t + 2$ and $y = t^2 + 1$ for $-2 \le t \le 2$ by plotting points.

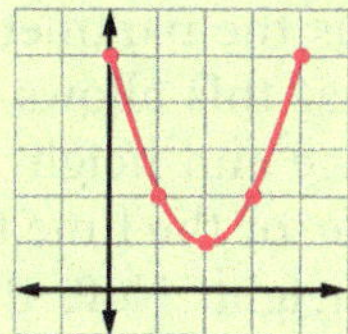

The Lesson Opener provides a quick review of *parametric equations* from Section 1.8. Consider asking the students which component is added when parametric equations are used to model the path of a projectile. (*time*)

Review the plotting of a set of parametric equations using the table of values and graph shown in the text. Engage the students by briefly discussing the importance of time as a factor in the flight of an object, then present the definitions related to parametric equations.

Example 1 involves finding x and y values for integral values of t within the specified interval and then plotting points. The graphs can be checked using a graphing calculator in parametric mode or other graphing technology with parametric capabilities, such as the Desmos online graphing calculator.

Challenge the students to explain how the *orientation* can be identified. (*by comparing the relative position of the points in relation to the value of the parameter*) The orientation can be demonstrated using the TRACE function on a graphing calculator. Consider reminding students to use WINDOW to define Tmin and Tmax.

The general rectangular form of a parabolic curve is found by eliminating the parameter, t. This process is demonstrated

Reading and Writing Mathematics

Explain the advantage of using a parametric model for a trajectory instead of a rectangular model. The rectangular model provides two variables, often the horizontal and vertical displacements, enabling the height to be determined at any horizontal position and vice versa. Parametric models provide another variable, often time, in addition to the vertical and horizontal displacements. Given a value for any one of these three variables, the other two values can be determined.

Additional Exercises

Draw the curve described by each set of parametric equations and describe the graph's orientation.

1. $x = t$ and $y = t + 2$; $t \in [-2, 2]$

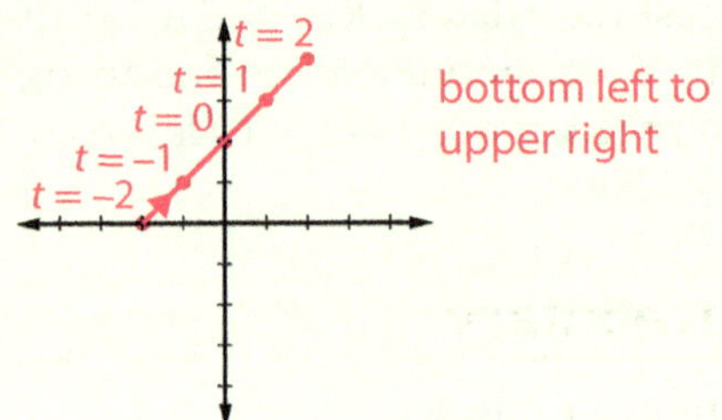

2. $x = t^2 - 2$ and $y = t + 1$; $t \in [-3, 3]$

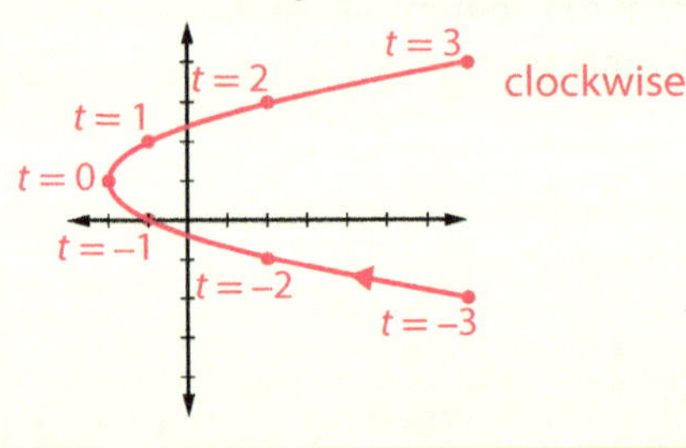

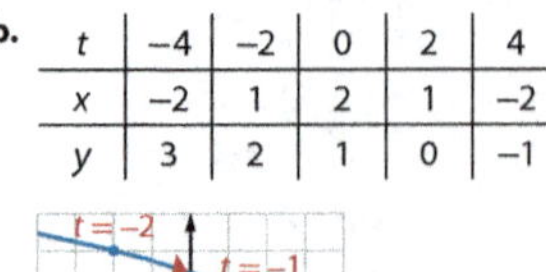

Use the trace function to find x- and y-values for each value of t and verify the orientation. (See Section 1.8.)

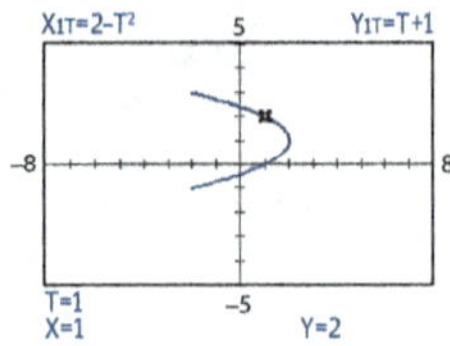

---

b.

t	-4	-2	0	2	4
x	-2	1	2	1	-2
y	3	2	1	0	-1

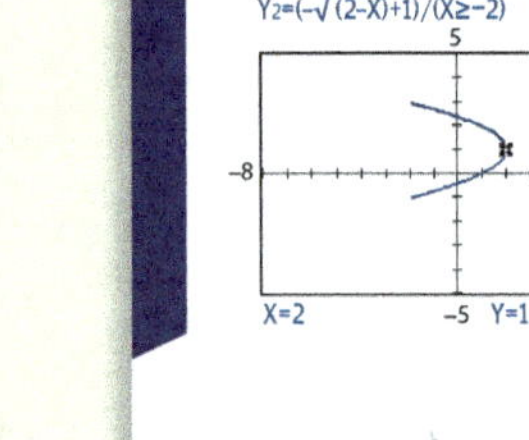

1. Evaluate $x = 2 - \frac{t^2}{4}$ and $y = 1 - \frac{t}{2}$ for several values of t.

2. Plot the resulting points and connect them with a smooth curve. Add an arrowhead to indicate the curve's orientation.

SKILL ✓ **EXERCISE 3**

Parametric equations can be used to represent curves that fail the Vertical Line Test and are not functions. Examples 1*a* and 1*b* define the same curve over different intervals and different orientations. Notice the first curve is traced more rapidly (in terms of t) than the second curve.

The parameter can be eliminated by solving either equation for t and substituting into the other equation. This process frequently simplifies the graphing process by revealing a familiar form of the curve's equation in terms of x and y (rectangular form). The domain may need to be restricted to ensure the same set of points is generated.

Example 2 Eliminating the Parameter

Given $x = 2 - t^2$ and $y = t + 1$ with $t \in [-2, 2]$, eliminate the parameter and restrict the domain to generate the same curve as in Example 1.

Answer

$y = t + 1$
$t = y - 1$
$x = 2 - (y - 1)^2$
$x = -(y - 1)^2 + 2$
$x = -y^2 + 2y + 1$

1. Solving for t in terms of y and substituting into the other equation confirms the graph of a parabola with vertex $(2, 1)$ that opens to the left.

$t^2 = 2 - x$ and $|t| \le 2$
$0 \le 2 - x \le 4$
$-2 \le -x$ and $-x \le 2$
$\quad x \le 2 \qquad x \ge -2$
$\therefore D = [-2, 2]$

2. Solve for t^2 in terms of x and examine its minimum value of 0 and maximum value of 4 from the interval $t \in [-2, 2]$ to find the appropriate restriction on the domain.

SKILL ✓ **EXERCISE 11**

The parameter does not always represent time. A Pythagorean identity is often used to eliminate parameters that represent angles.

Example 3 Eliminating the Parameter for an Ellipse

Given $x = 5 \cos \theta$ and $y = 3 \sin \theta$, eliminate the parameter and write the rectangular equation in standard form. Then sketch the graph.

Answer

$\cos \theta = \frac{x}{5}$ and $\sin \theta = \frac{y}{3}$

$\left(\frac{x}{5}\right)^2 + \left(\frac{y}{3}\right)^2 = 1$

$\frac{x^2}{25} + \frac{y^2}{9} = 1$

1. Solve for $\cos \theta$ and $\sin \theta$ and then substitute into $\cos^2 \theta + \sin^2 \theta = 1$.

CONTINUED ➡

---

in Example 2. The interval for x is determined by using the value of t in terms of x and the interval for t.

The identity $\sin^2 \theta + \cos^2 \theta = 1$ is used in Example 3 to eliminate the parameter in the equation of an ellipse. You may want to preview the trigonometric identity used in Example 5 for a hyperbola by asking the students which Pythagorean identity could be used if the equation was a difference instead of a sum. ($\sec^2 \theta - \tan^2 \theta = 1$)

Example 4 demonstrates the process of introducing a parameter into a standard equation. Explain that the domain of t is found by substituting the endpoint of the domain of x into the parametric equation. Emphasize that the curve remains the same in the different parameterizations, but the values of t along the curve and the orientation differ. Consider asking the students to identify the value of t at $x = -1$ in each parameterization, noting that each parameterization produces a different value of t for each point on the curve. ((a.) $t = -2$, (b.) $t = 0$, (c.) $t = 7$) The different parameterizations can indicate different speeds of an object following the same path.

In step 1 of Example 5, a is the distance from the center to a vertex; that is, $5 - 2 = 3$. The value of c is the distance from the center to the focus; that is, $7 - 2 = 5$. Point out that either vertex or focus can be used to find the distances. In step 2 the Pythagorean identity $\sec^2 \theta - \tan^2 \theta = 1$ is used because the equation is a difference instead of a sum.

When graphing a set of parametric equations with θ as the parameter, use the $\boxed{X,T,\Theta,n}$ key to enter the variable T.

In Example 6 the path of a projectile is modeled with time (t) as the parameter. Remind the students that this allows us to determine the distance and height of the ball at any given time, or the time for any given distance or height. Note that the ball in this example will be too high,

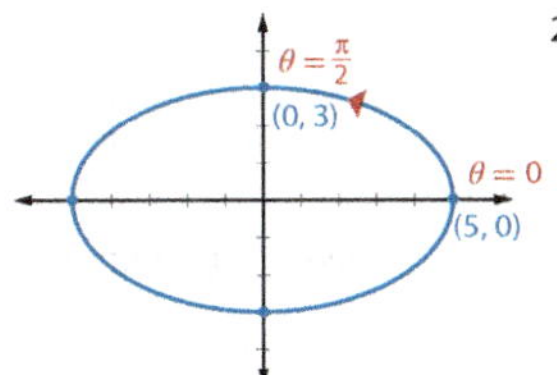

2. Sketch the resulting ellipse centered at the origin with a horizontal major axis where $a = 5$ and $b = 3$. Evaluating when $\theta = 0$ and $\frac{\pi}{2}$ produces $(5, 0)$ and $(0, 3)$, which indicates a counterclockwise orientation.

_______________ SKILL ✔ **EXERCISE 15**

How can you find a set of parametric equations for a given graph or rectangular equation? Recall from Example 1 that the parametric representations for a curve are not unique. The orientation and speed of the parametric curve depend on the definition of the parameter.

Example 4 Writing Parametric Equations for a Parabola

Use each parameter to write parametric equations for the graph of $y = x^2 - 1$; $x \in [-2, 2]$.

a. $t = x$ **b.** $t = \frac{x}{2} + 1$ **c.** $t = 1 - 3x$

Answer

a. using $t = x$; $t \in [-2, 2]$
$$y = t^2 - 1$$

b. using $t = \frac{x}{2} + 1$; $t \in [0, 2]$
$$x = 2t - 2$$
$$y = (2t - 2)^2 - 1$$
$$= 4t^2 - 8t + 3$$

c. using $t = 1 - 3x$; $t \in [-5, 7]$
$$x = \frac{1 - t}{3}$$
$$y = \left(\frac{1 - t}{3}\right)^2 - 1 = \frac{t^2 - 2t - 8}{9}$$

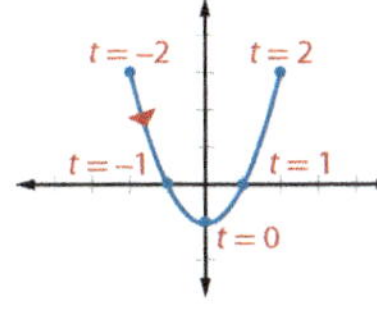
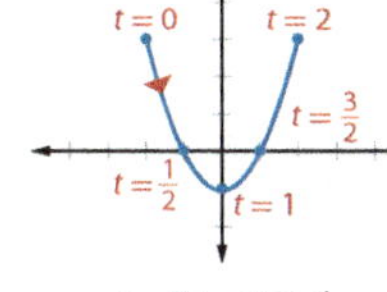
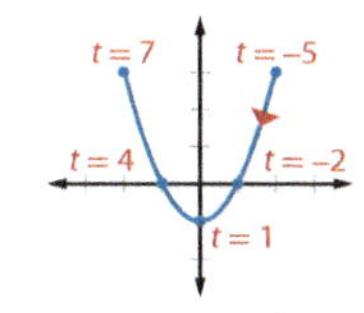

$x = t$ and $y = t^2 - 1$; $t \in [-2, 2]$

$x = 2t - 2$ and $y = 4t^2 - 8t + 3$; $t \in [0, 2]$

$x = \frac{1 - t}{3}$ and $y = \frac{t^2 - 2t - 8}{9}$; $t \in [-5, 7]$

_______________ SKILL ✔ **EXERCISE 25**

Example 5 Writing Parametric Equations for a Hyperbola

Write a set of parametric equations for the hyperbola with vertices at $(1, 5)$ and $(1, -1)$ and foci at $(1, 7)$ and $(1, -3)$.

Answer

transverse axis on $x = 1$
$C\left(1, \frac{-1 + 5}{2}\right) = (1, 2)$
$a = 3; c = 5$
$b^2 = c^2 - a^2 = 25 - 9 = 16; b = 4$
$\therefore \frac{(y - 2)^2}{3^2} - \frac{(x - 1)^2}{4^2} = 1$

Since $\sec^2 \theta - \tan^2 \theta = 1$, let
$\sec \theta = \frac{y - 2}{3}$ and $\tan \theta = \frac{x - 1}{4}$.
$\therefore x = 1 + 4 \tan \theta$ and $y = 2 + 3 \sec \theta$

1. Write the standard form equation for the hyperbola: $\frac{(y - k)^2}{b^2} - \frac{(x - h)^2}{a^2} = 1$.

Recall that $c^2 = a^2 + b^2$ in a hyperbola.

2. Use a Pythagorean identity to create parametric equations for x and y.

CONTINUED ➡

---

and therefore it will miss the target even if it is vertically aligned with the center of the target.

Motivational Idea While time is a commonly used parameter, it is important to recognize that there are other parameters that affect mathematical models. Challenge the students to name some other possible parameters. (*temperature, humidity, light intensity, pressure or force, friction, speed, and oxygen level*)

Interactive Activity Use the Internet keyword search *parametric equation grapher* for several interactive graphing options.

TIPS

Ex. 23–28 Encourage students to check their work by using technology to graph both parameterized and rectangular equations in the same window of a graphing application such as the Desmos online graphing calculator.

Eliminate the parameter and restrict the domain to find the rectangular equation generating the same curve as the parametric equations. Use graphing technology to verify your answer.

3. $x = t + 2$ and $y = 2t$; $t \in [0, 4]$
$y = 2x - 4$; $D = [2, 6]$

4. $x = t + 2$ and $y = 1 - \frac{1}{2}t^2$; $t \in [-2, 2]$
$y = -\frac{1}{2}x^2 + 2x - 1$; $D = [0, 4]$

5. $x = 2 \cos t$ and $y = 20 \sin t$; $t \in [0, 2\pi]$
$\frac{x^2}{4} + \frac{y^2}{400} = 1$; $D = [-2, 2]$

6. $x = 2 \sec t + 1$ and $y = 2 \tan t - 1$
$t \in [0, 2\pi]$
$\frac{(x - 1)^2}{4} - \frac{(y + 1)^2}{4} = 1$

7. Use each parameter to write parametric equations for the graph of $y = x^2 - 1$; $x \in [-2, 2]$.
  a. $t = x$ $x = t$; $y = t^2 - 1$; $t \in [-2, 2]$
  b. $t = x + 1$
  $x = t - 1$; $y = t^2 - 2t$; $t \in [-1, 3]$
  c. $t = 3 - \frac{x}{2}$ $x = 6 - 2t$
  $y = 4t^2 - 24t + 35$; $t \in [2, 4]$

Write a set of parametric equations for each conic with the given characteristics.

8. circle; $C(1, -3)$; $r = 5$
$x = 5 \sin \theta + 1$ and $y = 5 \cos \theta - 3$

9. hyperbola; $C(1, 3)$; $V(-2, 3)$
$F(-3, 3)$
$x = 3 \sec \theta + 1$ and $y = \sqrt{7} \tan \theta + 3$

10. Abigail serves a volleyball from one corner of the 9 m × 18 m court toward the opposite corner from a height of 2 m with an initial speed of 18 m/sec at a 12° angle of elevation.
  a. Write a set of parametric equations modeling the ball's trajectory. $x = 18 \cos(12°)t$ and
  $y = -4.9t^2 + 18 \sin(12°)t + 2$
  b. Find the horizontal distance the ball is from the server when the ball is directly over the net, the height of the ball at that point, and the time it took to get there. If the top of the net is at a height of 2.24 m, will the ball clear the net?
  $d \approx 10.1$ m; 2.54 m; $t \approx 0.57$ sec; yes
  c. Will the ball stay inbounds if no one touches it after the serve? Explain your reasoning.
  yes; The ball will land ≈ 19.8 m from the server, which is less than the 20.1 m to the opposite corner.

Assignments

- **Minimum:** 1–2, 5, 7, 10–11, 15, 19, 25, 31, 33, 43, 46–49, 51
- **Standard:** 1, 3, 4, 6–7, 9, 12, 15, 18, 21, 23, 25, 28–29, 32, 35, 43–45, 47, 49–52
- **Extended:** 2–10 even; 14–15, 18, 20–21, 24, 26, 30, 34, 37–40; 44–52 even

Solutions

❯ A. Exercises

1.

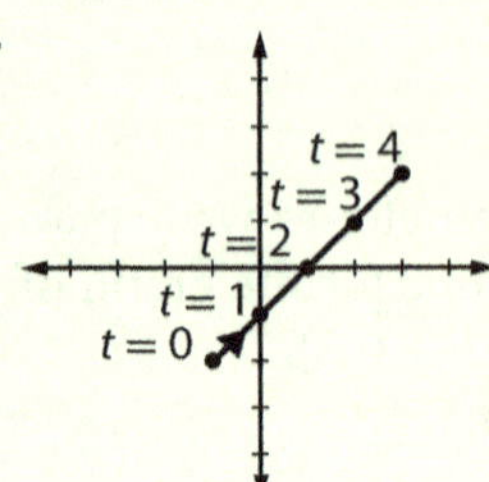

2.

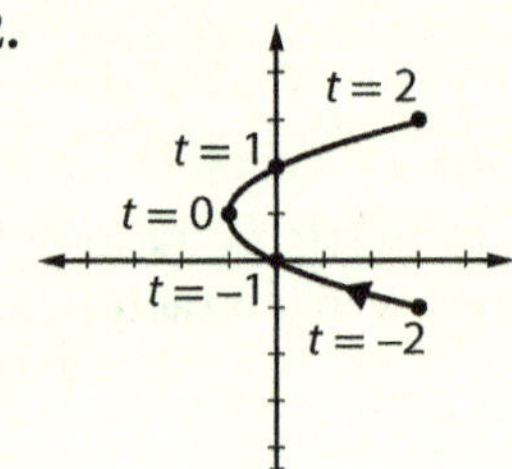

3.

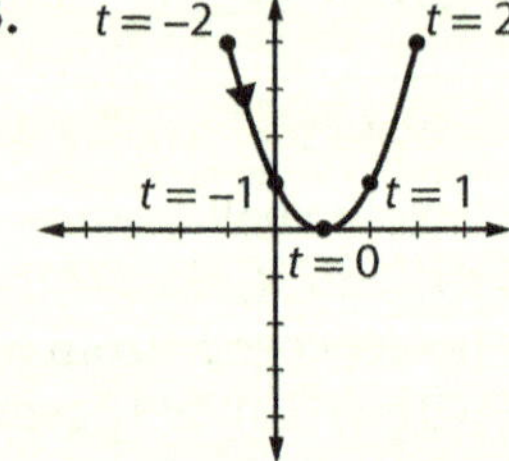

4.

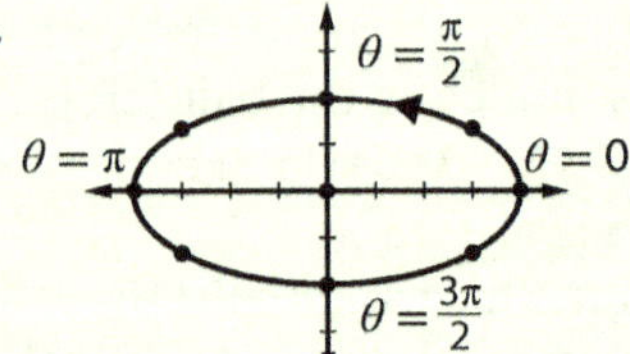

5.

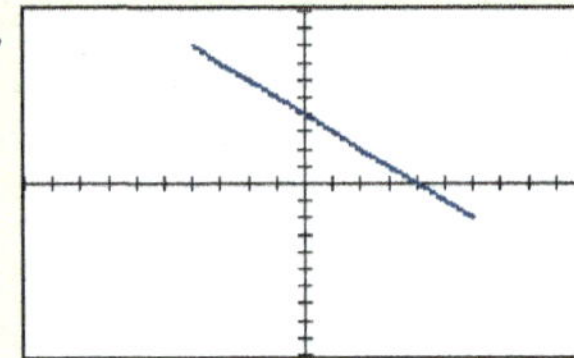

lower right to upper left

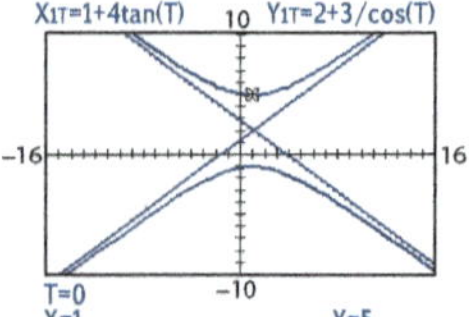
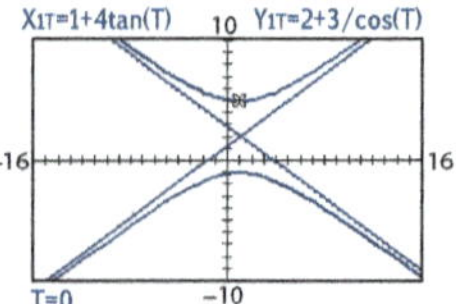

Check

Architects use parametric equations to model the effects of earthquakes on structures in order to enhance public safety.

TIP

If $t \in [0, 1]$, the target can be represented by $x = 30$ and $y = 3.5 + t$.

3. Graph the equations in parametric mode using $\theta \in [0, 2\pi]$. Extending the interval for θ retraces the hyperbola.

SKILL ✔ EXERCISE 29

Parametric equations are frequently used to model motion within a plane. You have already seen how the parabolic path of a projectile can be modified. In general, the position of an object launched from an initial height h_0 at an angle of θ above the horizontal with an initial speed of v_0 after t seconds is modeled by the parametric equations

$$x = (v_0 \cos \theta)t \quad \text{and} \quad y = -\tfrac{1}{2}gt^2 + (v_0 \sin \theta)t + h_0$$

where the acceleration due to gravity, g, is approximately 32 ft/sec² or 9.8 m/sec².

Example 6 Modeling 2D Motion with Parametric Equations

Julie throws a ball at a 1 ft wide circular dunk tank target 30 ft away whose center is 4 ft above the ground. She releases the ball from a height of 6 ft with a 10° angle of elevation and a speed of 50 ft/sec. Write a set of parametric equations modeling the ball's trajectory and determine whether the 4 in. diameter ball hits the target.

Answer

$x = (50 \cos 10°)t$ and
$y = -16t^2 + (50 \sin 10°)t + 6$

1. Write the parametric equations for projectile motion.

$30 = (50 \cos 10°)t_{30}$
$t_{30} = \dfrac{30}{50 \cos 10°} \approx 0.61$ sec

2. Determine the time it takes for the ball to reach a horizontal displacement of 30 ft.

$y = -16t_{30}^2 + (50 \sin 10°)t_{30} + 6$
≈ 5.35 ft or ≈ 5 ft 4 in.

3. Find the height of the ball at that time.

The balls misses the target by $\approx (16 - 6 - 2) \approx 8$ in.

4. The center of the ball (2 in. radii) passes ≈ 16 in. above the center of the target (6 in. radii).

Check

5. Graph the equations in parametric mode. Using the trace function with $t \approx 0.61$ confirms that the ball is too high.

SKILL ✔ EXERCISE 35

❯ A. Exercises

Complete the table and draw the graph described by each set of parametric equations.

1. $x = t - 1$ and $y = t - 2$; $t \in [0, 4]$

t	0	1	2	3	4
x	−1	0	1	2	3
y	−2	−1	0	1	2

2. $x = t^2 - 1$ and $y = t + 1$; $t \in [-2, 2]$

t	−2	−1	0	1	2
x	3	0	−1	0	3
y	−1	0	1	2	3

3. $x = t + 1$ and $y = t^2$; $t \in [-2, 2]$

t	−2	−1	0	1	2
x	−1	0	1	2	3
y	4	1	0	1	4

4. $x = 4 \cos \theta$ and $y = 2 \sin \theta$; $\theta \in [0, 2\pi]$

θ	0	$\frac{\pi}{4}$	$\frac{\pi}{2}$	$\frac{3\pi}{4}$	π	$\frac{5\pi}{4}$	$\frac{3\pi}{2}$	$\frac{7\pi}{4}$	2π
x	4	2.8	0	−2.8	−4	−2.8	0	2.8	4
y	0	1.4	2	1.4	0	−1.4	−2	−1.4	2

6.

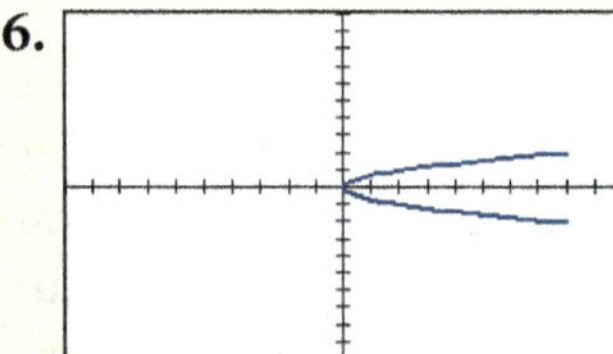

clockwise

7.

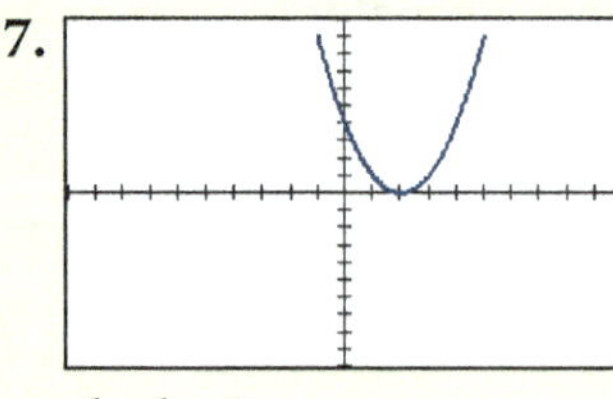

clockwise

8.

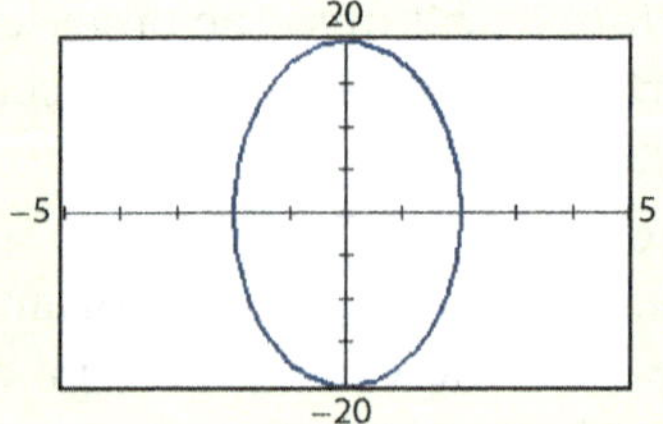

counterclockwise

9. $t = -x + 1;\ y = -x + 1$
$-2 \le -x + 1 \le 2$
$-3 \le -x$ and $-x \le 1$
$x \le 3 \qquad x \ge -1$
$\therefore D = [-1, 3]$

Use a graphing calculator in parametric mode to graph each set of parametric equations. Then explore the graph with the trace function and describe the graph's orientation.

5. $x = 1 - t$ and $y = 3 + t; t \in [-5, 5]$ lower right to upper left

6. $x = 2t^2$ and $y = t; t \in [-2, 2]$ clockwise

7. $x = 2 - t$ and $y = t^2; t \in [-3, 3]$ clockwise

8. $x = 2 \cos t$ and $y = 20 \sin t; t \in [0, 2\pi]$ counterclockwise

Eliminate the parameter and restrict the domain to find the rectangular equation generating the same curve as the parametric equations. Use graphing technology to verify your answer.

9. $x = 1 - t$ and $y = t; t \in [-2, 2]$

10. $x = t - 2$ and $y = 1 + 2t; t \in [-1, 2]$

11. $x = t - 1$ and $y = -t^2 + 4; t \in [-2, 2]$

12. $x = -t + 1$ and $y = t^2 - 5; t \in [-3, 3]$

13. $x = -1 + t^2$ and $y = t - 1; t \in [-2, 2]$

14. $x = 1 - t^2$ and $y = t - 2; t \in [-2, 2]$

▶ B. Exercises

Eliminate the parameter $\theta \in [0, 2\pi]$ to write the standard form equation for each conic section. Then sketch the graph with its orientation.

15. $x = 3 \sin \theta$ and $y = 4 \cos \theta$ $\dfrac{x^2}{9} + \dfrac{y^2}{16} = 1$

16. $x = 5 \cos \theta$ and $y = 2 \sin \theta$ $\dfrac{x^2}{25} + \dfrac{y^2}{4} = 1$

17. $x = 3 \sin \theta$ and $y = 3 \cos \theta$ $x^2 + y^2 = 9$

18. $x = 2 \sec \theta$ and $y = 2 \tan \theta$ $\dfrac{x^2}{4} - \dfrac{y^2}{4} = 1$

19. $x = 3 \tan \theta$ and $y = \sec \theta$ $y^2 - \dfrac{x^2}{9} = 1$

20. $x = \cot \theta$ and $y = 2 \csc \theta$ $\dfrac{y^2}{4} - x^2 = 1$

Eliminate the parameter $\theta \in [0, 2\pi]$ to write a standard form equation for each conic.

21. $x = 2 \cos \theta + 1$ and $y = 4 \sin \theta + 3$

22. $x = 10 \cos \theta - 2$ and $y = 10 \sin \theta + 4$

23. $x = 6 \csc \theta - 4$ and $y = 5 \cot \theta - 8$

24. $x = 3 \sec \theta + 5$ and $y = 9 \tan \theta - 5$

Use the given parameter to write a set of parametric equations for each parabolic curve.

25. $y = x^2 + 1; x \in [-2, 2]$
$t = x + 1$ $x = t - 1, y = t^2 - 2t + 2; t \in [-1, 3]$

26. $y = x^2 - 3; x \in [-2, 2]$
$t = x - 2$ $x = t + 2, y = t^2 + 4t + 1; t \in [-4, 0]$

27. $y = 2x^2 - 3; x \in [-2, 1]$
$t = x + 4$ $x = t - 4, y = 2t^2 - 16t + 29; t \in [2, 5]$

28. $y = -4x^2 + 7; x \in [-1, 2]$
$t = 2x - 6$ $x = \frac{1}{2}t + 3, y = -t^2 - 12t - 29; t \in [-8, -2]$

Write a set of parametric equations for each conic section.

29. a hyperbola with vertices at $(-2, -2)$ and $(-2, 4)$ and foci at $(-2, -4)$ and $(-2, 6)$

30. a hyperbola with vertices at $(-2, 3)$ and $(4, 3)$ and foci at $(-4, 3)$ and $(6, 3)$

31. an ellipse with vertices at $(1, -12)$ and $(1, 8)$ and foci at $(1, -8)$ and $(1, 4)$

32. an ellipse with vertices at $(-6, 3)$ and $(4, 3)$ and co-vertices at $(-1, 0)$ and $(-1, 6)$

33. A baseball player throws a ball with a $15°$ angle of elevation from a height of 6.5 ft at a speed of 70 ft/sec.

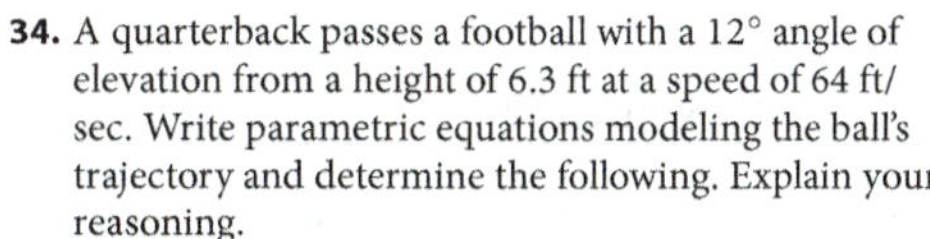

   a. Write a set of parametric equations modeling the ball's path.

   b. What is the height of the ball 1 sec after release? ≈ 8.6 ft

   c. How far will the ball have traveled across the field after 1.2 sec? ≈ 81.1 ft

34. A quarterback passes a football with a $12°$ angle of elevation from a height of 6.3 ft at a speed of 64 ft/sec. Write parametric equations modeling the ball's trajectory and determine the following. Explain your reasoning.

   a. Does a defensive player whose maximum vertical reach is 7.5 ft have the ability to block the pass if it takes 0.5 sec for the ball to reach his position?

   b. Will the football be catchable for a wide receiver 30 yd down the field?

35. Autumn hits her golf ball at 120 ft/sec with a $15°$ launch angle. Write and graph parametric equations modeling the ball's flight and determine the following.

   a. How long is the ball in flight, and how far does the ball travel down the fairway (to the nearest yard)? ≈ 1.94 sec; 75 yd

   b. By how many feet does her drive clear the top of a 10 ft tree that is 70 ft away? 3 ft

36. A kicked soccer ball leaves the ground at a $30°$ angle of elevation and an initial speed of 85 ft/sec. Write a set of parametric equations modeling the ball's path. Determine the amount of time (to the nearest hundredth of a second) the ball is in the air and the percentage of the length of the 100 yd soccer field it travels if it was kicked straight down the field.
$x = 85 \cos (30°)t$ and $y = -16t^2 + 85 \sin (30°)t$; ≈ 2.66 sec; 65%

15. $\sin \theta = \frac{x}{3}; \cos \theta = \frac{y}{4}$

$$\left(\frac{x}{3}\right)^2 + \left(\frac{y}{4}\right)^2 = 1$$
$$\frac{x^2}{9} + \frac{y^2}{16} = 1$$

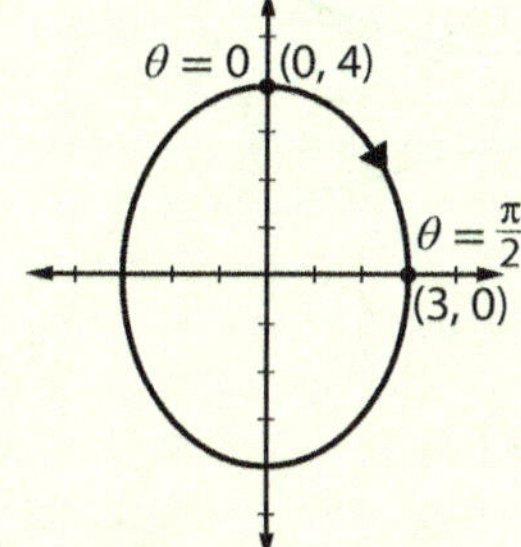

16. $\cos \theta = \frac{x}{5}; \sin \theta = \frac{y}{2}$

$$\left(\frac{x}{5}\right)^2 + \left(\frac{y}{2}\right)^2 = 1$$
$$\frac{x^2}{25} + \frac{y^2}{4} = 1$$

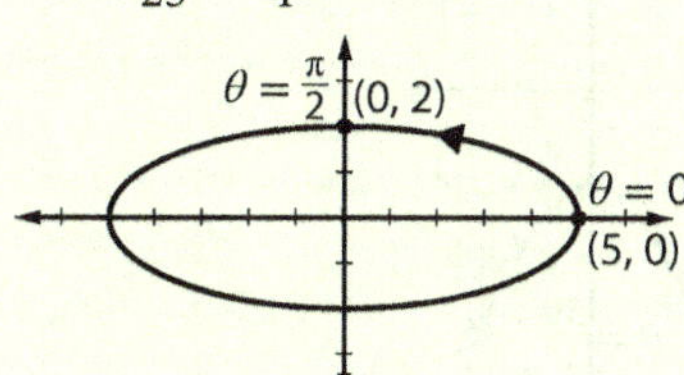

17. $\sin \theta = \frac{x}{3}; \cos \theta = \frac{y}{3}$

$$\left(\frac{x}{3}\right)^2 + \left(\frac{y}{3}\right)^2 = 1$$
$$\frac{x^2}{9} + \frac{y^2}{9} = 1; x^2 + y^2 = 9$$

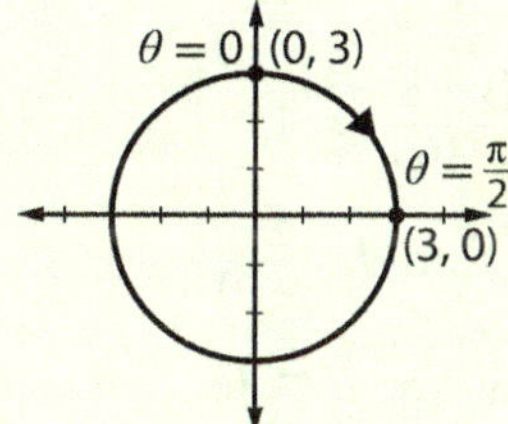

18. $\sec \theta = \frac{x}{2}; \tan \theta = \frac{y}{2}$

$$\left(\frac{y}{2}\right)^2 + 1 = \left(\frac{x}{2}\right)^2$$
$$\frac{x^2}{4} - \frac{y^2}{4} = 1$$

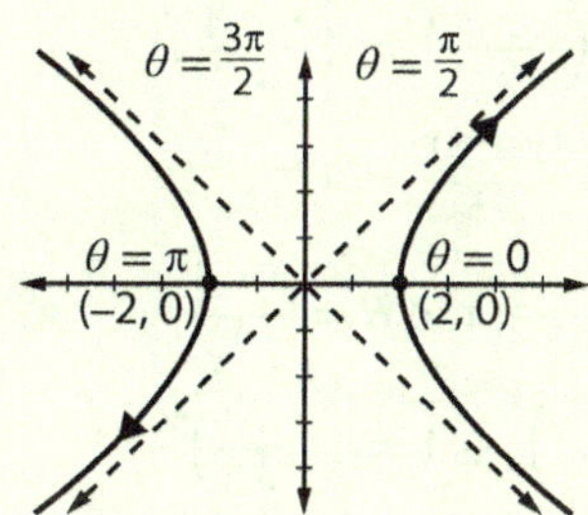

10. $t = x + 2$
$y = 1 + 2(x + 2) = 2x + 5$
$-1 \leq x + 2 \leq 2$
$-3 \leq x$ and $x \leq 0$
$\therefore D = [-3, 0]$

11. $t = x + 1$
$y = -(x + 1)^2 + 4 = -x^2 - 2x + 3$
$t = x + 1$ and $t \in [-2, 2]$
$-2 \leq x + 1 \leq 2$
$-3 \leq x$ and $x \leq 1$
$\therefore D = [-3, 1]$

12. $t = -x + 1$
$y = (-x + 1)^2 - 5 = x^2 - 2x - 4$
$x = -t + 1$ and $t \in [-3, 3]$
$-3 \leq -x + 1 \leq 3$
$x \leq 4$ and $x \geq -2$
$\therefore D = [-2, 4]$

13. $t = y + 1$
$x = (y + 1)^2 - 1 = y^2 + 2y$
$t^2 = x + 1$ and $|t| \leq 2$
$0 \leq x + 1 \leq 4$
$x \geq -1$ and $x \leq 3$
$\therefore D = [-1, 3]$

14. $t = y + 2$
$x = 1 - (y + 2)^2 = -y^2 - 4y - 3$
$t^2 = -x + 1$ and $|t| \leq 2$
$0 \leq -x + 1 \leq 4$
$x \leq 1$ and $x \geq -3$
$\therefore D = [-3, 1]$

19. $\tan \theta = \frac{x}{3}$; $\sec \theta = y$

$$\left(\frac{x}{3}\right)^2 + 1 = y^2$$

$$y^2 - \frac{x^2}{9} = 1$$

20. $\cot \theta = x$; $\csc \theta = \frac{y}{2}$

$$1 + x^2 = \left(\frac{y}{2}\right)^2$$

$$\frac{y^2}{4} - x^2 = 1$$

> **B. Exercises**

21. $\cos \theta = \frac{x-1}{2}$; $\sin \theta = \frac{y-3}{4}$

$$\left(\frac{x-1}{2}\right)^2 + \left(\frac{y-3}{4}\right)^2 = 1$$

$$\frac{(x-1)^2}{4} + \frac{(y-3)^2}{16} = 1$$

22. $\cos \theta = \frac{x+2}{10}$; $\sin \theta = \frac{y-4}{10}$

$$\left(\frac{x+2}{10}\right)^2 + \left(\frac{y-4}{10}\right)^2 = 1$$

$$\frac{(x+2)^2}{100} + \frac{(y-4)^2}{100} = 1$$

$$(x+2)^2 + (y-4)^2 = 100$$

23. $\csc \theta = \frac{x+4}{6}$; $\cot \theta = \frac{y+8}{5}$

$$1 + \left(\frac{y+8}{5}\right)^2 = \left(\frac{x+4}{6}\right)^2$$

$$\frac{(x+4)^2}{36} - \frac{(y+8)^2}{25} = 1$$

24. $\sec \theta = \frac{x-5}{3}$; $\tan \theta = \frac{y+5}{9}$

$$\left(\frac{y+5}{9}\right)^2 + 1 = \left(\frac{x-5}{3}\right)^2$$

$$\frac{(x-5)^2}{9} - \frac{(y+5)^2}{81} = 1$$

25. $t \in [-1, 3]$; $x = t - 1$

$$y = (t-1)^2 + 1 = t^2 - 2t + 2$$

> **C. Exercises**

37. Eliminate the parameter to express each set of parametric equations in the standard rectangular form of a circle.

  a. $x = r \cos \theta$ and $y = r \sin \theta$ $x^2 + y^2 = r^2$

  b. $x = r \cos \theta + h$ and $y = r \sin \theta + k$ $(x - h)^2 + (y - k)^2 = r^2$

38. Write a set of parametric equations for the circle with the given center and radius.

  a. $C\,(0, 0)$; $r = 7$ **b.** $C\,(3, -2)$; $r = 6$

39. Describe the graph of the parametric equations $x = a \cos \theta$ and $y = b \sin \theta$ with both a and $b > 0$ (a) if $a = b$, (b) if $a < b$, and (c) if $a > b$. Explain your reasoning.

38a. $x = 7 \cos \theta$, $y = 7 \sin \theta$

38b. $x = 6 \cos \theta + 3$, $y = 6 \sin \theta - 2$

40. $x = h + a \cos \theta$, $y = k + b \sin \theta$

41. $x = h + a \sec \theta$, $y = k + b \tan \theta$

Write a set of parametric equations for each standard form of a conic.

40. $\dfrac{(x-h)^2}{a^2} + \dfrac{(y-k)^2}{b^2} = 1$ **41.** $\dfrac{(x-h)^2}{a^2} - \dfrac{(y-k)^2}{b^2} = 1$

42. Explore: Analyze various launch angles for a field goal attempt taken 40 yd from a goal post whose horizontal crossbar is 10 ft high. Use parametric equations to find the height of a ball kicked with an initial speed of 70 ft/sec when it reaches the goal post if kicked with launch angles of 15°, 30°, 45°, and 60°. Which of these launch angles results in a field goal?
−18.2 ft, 6.6 ft, 26.0 ft, 19.8 ft; 45° and 60°

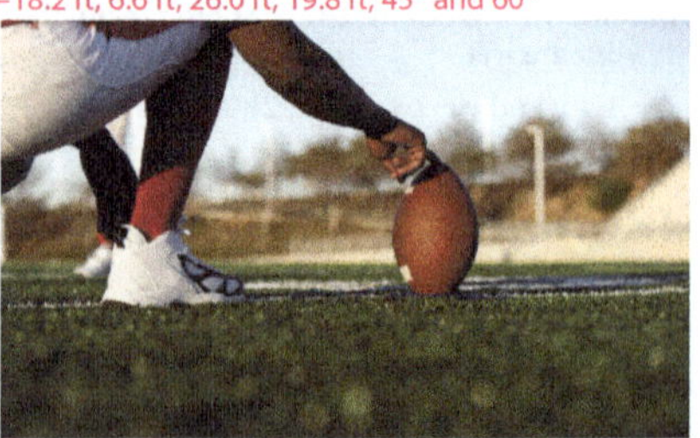

CUMULATIVE REVIEW

Write each polar equation in rectangular form and identify the conic. [6.6]

43. $r = 8$ $x^2 + y^2 = 64$; circle

44. $r \sin \theta = 5$ $y = 5$; horizontal line

45. $r \sin \theta = r^2 \cos^2 \theta$ $y = x^2$; parabola

Identify each conic and then state the coordinates of its vertices and foci and its eccentricity.

46. $\dfrac{(x-2)^2}{36} + \dfrac{y^2}{16} = 1$ [8.2]

47. $9x^2 - 4y^2 = 36$ [8.3]

48. $x + (y-1)^2 = 3$ [8.1]

49. For which θ does $\cos \theta = -\dfrac{\sqrt{3}}{2}$? List all correct answers. [4.1]   C, D

  A. $\dfrac{\pi}{6}$ **C.** $\dfrac{5\pi}{6}$ **E.** $300°$

  B. $\dfrac{2\pi}{3}$ **D.** $210°$

50. What is the smallest positive value of x for which $y = \sin 2x$ has a relative minimum? [4.4]   C

  A. $\dfrac{\pi}{4}$ **C.** $\dfrac{3\pi}{4}$ **E.** $\dfrac{3\pi}{2}$

  B. $\dfrac{\pi}{2}$ **D.** π

51. Which equation represents a parabola with a vertex at $(2, 6)$ and a focus at $(2, 3)$? [8.1]   E

  A. $y = \frac{1}{12}(x+2)^2 + 6$ **D.** $y = -\frac{1}{12}(x+2)^2 + 6$

  B. $y = \frac{1}{12}(x-2)^2 + 6$ **E.** $y = -\frac{1}{12}(x-2)^2 + 6$

  C. $y = \frac{1}{12}(x-2)^2 - 6$

52. Use the discriminant test to classify the conic $4x^2 + 12xy = -9y^2 + 16$. [8.4]   C

  A. circle   **C.** parabola

  B. ellipse   **D.** hyperbola

46. ellipse; V: $(-4, 0)$, $(8, 0)$
F: $(2 \pm 2\sqrt{5}, 0) \approx (-2.47, 0), (6.47, 0)$;
$e = \frac{\sqrt{5}}{3} \approx 0.75$

47. hyperbola; V: $(\pm 2, 0)$
F: $(\pm \sqrt{13}, 0) \approx (\pm 3.61, 0)$;
$e = \frac{\sqrt{13}}{2} \approx 1.8$

48. parabola; $V\,(1, 3)$
$F\left(\frac{3}{4}, 3\right)$; $e = 1$

---

26. $t \in [-4, 0]$; $x = t + 2$
$$y = (t+2)^2 - 3 = t^2 + 4t + 1$$

27. $t \in [2, 5]$; $x = t - 4$
$$y = 2(t-4)^2 - 3 = 2t^2 - 16t + 29$$

28. $t \in [-8, -2]$; $x = \frac{1}{2}t + 3$
$$y = -4\left(\frac{1}{2}t + 3\right)^2 + 7$$
$$= -4\left(\frac{1}{4}t^2 + 3t + 9\right) + 7$$
$$= -t^2 - 12t - 29$$

29. vertical TA on $x = -2$
$$C\left(-2, \frac{-2+4}{2}\right) = (-2, 1)$$
$$a = 3;\ c = 5$$
$$b^2 = 5^2 - 3^2 = 16;\ b = 4$$
$$\frac{(y-1)^2}{3^2} - \frac{(x+2)^2}{4^2} = 1$$
Using $\sec^2 \theta - \tan^2 \theta = 1$,
let $\sec \theta = \frac{y-1}{3}$ and $\tan \theta = \frac{x+2}{4}$.
$$x = -2 + 4 \tan \theta;\ y = 1 + 3 \sec \theta$$

continued in Answers and Solutions Overflow

8.6 Polar Equations of Conics

Engineers calculate the precise amount of thrust that generates the hyperbolic path used by spacecraft as they escape Earth's gravitational pull.

Earlier in this chapter a parabola was defined as the set of points equidistant from a given point (the focus) and a given line (the directrix). Ellipses and hyperbolas were defined as the set of points such that the sum or difference of the distances from two points (the foci) is constant. Defining parabolas, ellipses, and hyperbolas in terms of the distance from each point to a focus and to its related directrix provides a unified definition for these conics.

▶ DEFINITIONS

A **conic section** is the set of points in a plane whose distances from a given line (its **directrix**) and a given point (its **focus**) not on that line have a constant ratio. This ratio, $\frac{PF}{PD} = e$, is the **eccentricity** of the conic.

> **After completing this section, you will be able to**
> - identify polar equations of conics in terms of eccentricity.
> - write and graph the polar equation of a conic, given its eccentricity and directrix.
> - analyze planetary orbits with polar conics.

The line through the focus that is perpendicular to the directrix is the conic's *focal axis*, and any intersection of the conic and its focal axis is a *vertex*. If the center of a conic is located at the origin, the resulting rectangular equation is relatively simple. When representing a conic section by a polar equation, it is often simpler to position a focus at the pole (or origin) with the directrix perpendicular to or parallel to the polar axis. While a parabola has one focus and one directrix, either of the two focus-directrix pairs in an ellipse or a hyperbola can be used to define the conic.

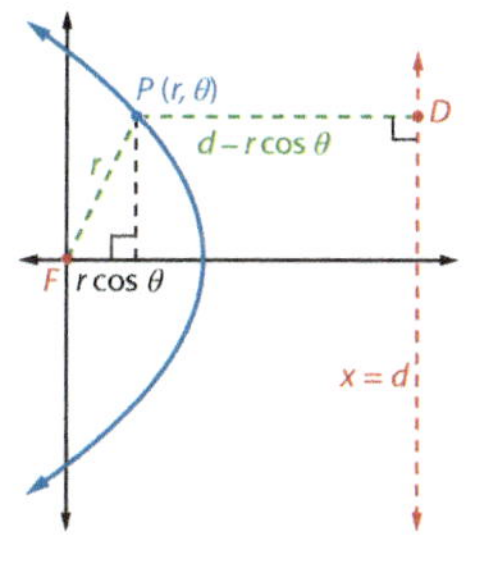

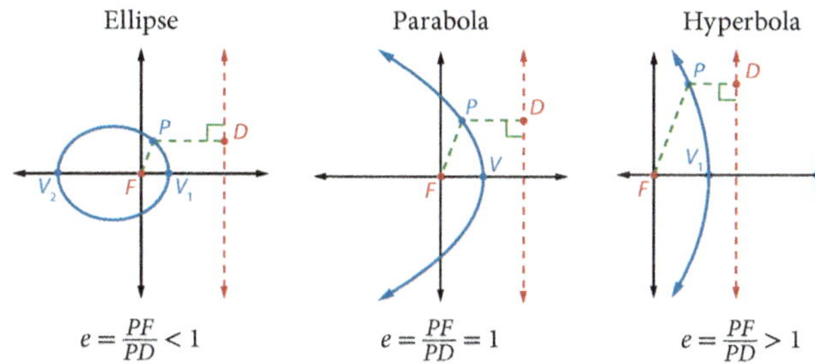

$$e = \frac{PF}{PD} < 1 \qquad e = \frac{PF}{PD} = 1 \qquad e = \frac{PF}{PD} > 1$$

To find a unified polar equation for conics, consider the point $P\,(r, \theta)$ on the conic whose focus is at the pole and whose directrix is d units to the right of the pole perpendicular to the polar axis. From the figure, note that $PF = r$ and $PD = d - r \cos \theta$. Solving $\frac{PF}{PD} = e$ for PF implies that $PF = e \cdot PD$. Substitute and solve for r.

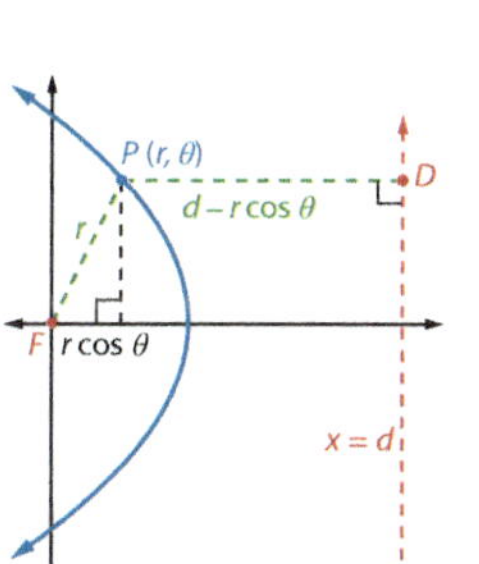

$$PF = e \cdot PD$$
$$r = e(d - r \cos \theta)$$
$$r = ed - er \cos \theta$$
$$r + er \cos \theta = ed$$
$$r(1 + e \cos \theta) = ed$$
$$r = \frac{ed}{1 + e \cos \theta}$$

8.6 Polar Equations of Conics

Objectives

1. To identify polar equations of conics in terms of eccentricity and directrix
2. To write and graph the polar equation of a conic, given its eccentricity and directrix
3. To analyze planetary orbits with polar conics

Flash

Spacecraft are launched vertically and then turn on a parabolic path to enter a circular orbit around the earth. This orbit allows the spacecraft to use the earth's gravitational pull and travel at 18,000 mi/hr (the orbital escape velocity from the earth) using minimum fuel consumption to escape into space. Launching the spaceship straight into space would require much more fuel and would create the high risk of falling straight back to earth like a rock if a failure were to occur.

Vocabulary

conic section
directrix
eccentricity
focal axis
focus
vertex

PRESENTATION

Lesson Opener

Express each polar coordinate in rectangular form. Round to the nearest tenth if necessary.

1. $(7, 90°)$ $(0, 7)$
2. $(12, 40°)$ $(9.2, 7.7)$

Identify the graph of each polar equation.

3. $r = 2 \csc \theta$ horizontal line $(y = 2)$
4. $r = -5 \sec \theta$ vertical line $(x = -5)$

Discuss the Lesson Opener, reminding students that $x = r \cos \theta$ and $y = r \sin \theta$ are used to convert polar coordinates to rectangular coordinates and that $r = a \csc \theta$ and $r = a \sec \theta$ represent horizontal and vertical lines, respectively.

In this section distances to a *focus* and a related *directrix* are used to present a unified polar definition of conics. Emphasize that a focus is located at the pole. Reference the chart illustrating ellipses, parabolas, and hyperbolas and point out how the conic wraps around the focus with a *vertex* between the focus and directrix. Remind students that the vertex of a parabola is the midpoint of the segment from the focus perpendicular to its directrix.

As an ellipse becomes more elongated ($e \Rightarrow 1$), the second vertex (and center) moves farther away from the first vertex and directrix. A parabola ($e = 1$) could then be viewed as an elongated ellipse whose second vertex (and center) is infinitely far from its focal point. When $e > 1$ the second vertex (and center) of the hyperbola is on the same half-axis as the directrix and gets closer to the directrix as e increases.

Consider illustrating this progression using interactive graphing technology with a slider for the *eccentricity* of a *conic section*.

Guide the students through the steps for deriving the unified polar equation for a conic with a focus on the pole and a

Additional Exercises

Determine the eccentricity, type of conic, and directrix for each polar equation.

1. $r = \dfrac{6}{3 - 3\sin\theta}$ $\quad e = 1$; parabola; $y = -2$

2. $r = \dfrac{5}{5 + \cos\theta}$ $\quad e = 0.2$; ellipse; $x = 5$

3. $r = \dfrac{-12}{4\cos\theta - 4}$ $\quad e = 1$; parabola; $x = -3$

4. Given a conic having $e = 0.4$ and a directrix $x = -4$, write a polar equation, classify the conic, and then verify your answer by using technology to graph the curve and its directrix.

$r = \dfrac{8}{5 - 2\cos\theta}$; ellipse

Write a polar equation for each conic section with a focus at the pole. Then sketch a graph of the conic and its directrix.

5. $e = 1$; $V(3, 0)$ $\quad r = \dfrac{6}{1 + \cos\theta}$

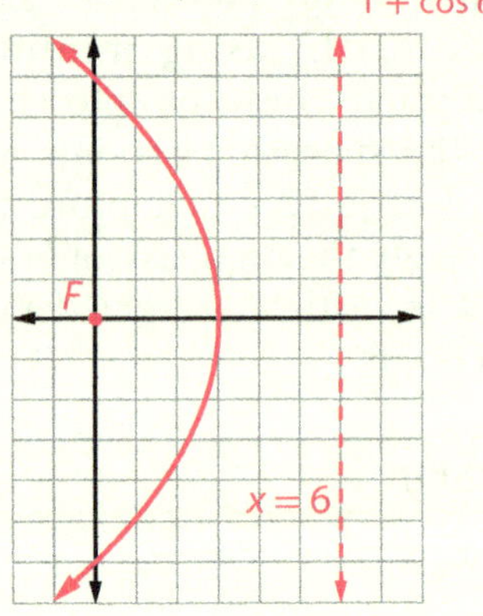

Similar derivations can be done when the directrix is d units to the left of the pole and when it is d units above or below the pole.

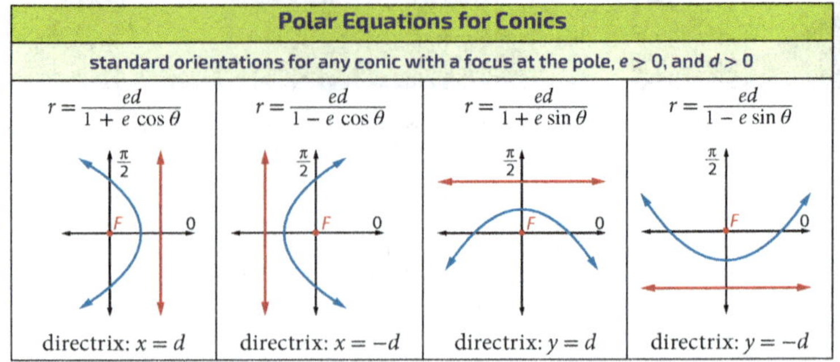

Explore the graphs of these conics as described in the Technology Corner on p. 419. Circles and degenerate conics cannot be represented by these polar equations. The type of conic and its orientation can be determined by analyzing its polar equation.

Example 1 Analyzing Polar Equations of Conics

Determine the eccentricity, type of conic, and directrix for each polar equation.

a. $r = \dfrac{6}{4 + 3\sin\theta}$ $\qquad$ b. $r = \dfrac{-9}{6\cos\theta - 3}$

Answer

a. $r = \dfrac{6 \div 4}{(4 + 3\sin\theta) \div 4}$

$r = \dfrac{1.5}{1 + 0.75\sin\theta}$

$e = 0.75;$ $\quad ed = 1.5$
$\qquad\qquad d = \dfrac{1.5}{0.75} = 2$

The conic is an ellipse with directrix $y = 2$.

Check

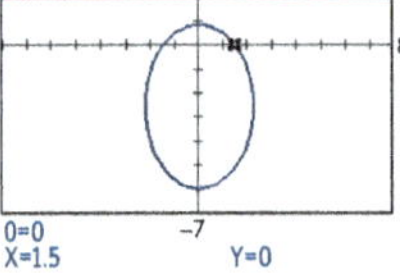

1. Divide the numerator and denominator by 4 to rewrite the equation in the standard form $r = \dfrac{ed}{1 + e\sin\theta}$.

2. Identify the eccentricity, e, and the distance of the directrix from the pole, d.

3. Use the eccentricity, $e < 1$, and the standard form to identify the conic and locate the directrix.

4. Use technology to graph the ellipse and its directrix, $r = 2\csc\theta$, in polar mode.

b. $r = \dfrac{-9 \div (-3)}{(6\cos\theta - 3) \div (-3)}$

$r = \dfrac{3}{1 - 2\cos\theta}$

$e = 2;$ $\quad ed = 3$
$\qquad\qquad d = \dfrac{3}{2} = 1.5$

The conic is a hyperbola with directrix $x = -1.5$.

Check

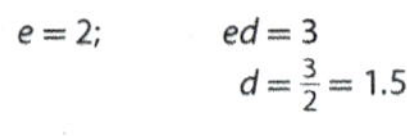
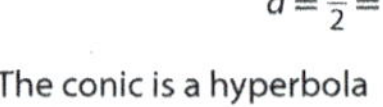

1. Divide the numerator and denominator by -3 to rewrite the equation in the standard form $r = \dfrac{ed}{1 - e\cos\theta}$.

2. Identify the eccentricity, e, and the distance of the directrix from the pole, d.

3. Use the eccentricity, $e > 1$, and the standard form to identify the conic and locate the directrix.

4. Use technology to graph the hyperbola and its directrix, $r = -1.5\sec\theta$, in polar mode.

SKILL ✔ EXERCISE 3

---

directrix that is perpendicular to the polar (or positive x-) axis. Explain the table illustrating the four basic forms of polar equations for conics, which apply to parabolas, ellipses, and hyperbolas. Emphasize how the operator in the denominator and the trigonometric function relate to the location of the directrix.

Use Example 1 to illustrate how the type of conic and location of the directrix are determined from the conic's polar equation. To graph the directrix, use $x = r\cos\theta$ or $y = r\sin\theta$ to convert the rectangular form to an equivalent polar equation. For $y = 2$ in Example 1a, substitute to obtain $2 = r\sin\theta$, so that $r = \dfrac{2}{\sin\theta}$ or $r = 2\csc\theta$.

Ask the students how the characteristics in Example 2a reveal the type and orientation of the conic section. ($e = 1$ indicates a parabola. *The directrix is at $y = -3$ and the focus is on the pole, so the parabola opens upward.*) Students should apply similar reasoning for Example 2b.

The illustrated polar graph for Example 3 includes the directrix, $x = 1.8\sec\theta$, which can be found as illustrated in Example 1. While the diagonal lines appear to be the asymptotes, they are actually lines connecting consecutive evaluated points at the extremes of the branches.

Since the vertices are on the polar axis, their coordinates can be identified by finding the value of r at $\theta = 0$ and π, as

shown in step 2 of Example 3. Reinforce that the polar coordinates $(-9, \pi)$ are equivalent to the rectangular coordinates $(0, 9)$. The $\boxed{\text{TRACE}}$ function can be used to verify the vertices found algebraically. You may want to ask the students to state which values of θ would be used if the vertices were along the y-axis. $\left(\dfrac{\pi}{2} \text{ and } \dfrac{3\pi}{2}\right)$ In step 4, remind the students that because the standard polar equations are based on a conic with a focus at the origin, we know that $c = 5$.

Explain that hyperbolic comets are interstellar, that is, they do not belong to our solar system. They enter and leave again. Orbits of objects within the solar system are elliptical. The orbit of Comet

The characteristics of a conic can be used to write the polar equation of a conic.

Example 2 Writing the Polar Equation of a Conic

Write a polar equation for each conic section with a focus at the origin.

a. $e = 1$; directrix: $y = -3$
b. $e = \frac{1}{3}$; vertices at $(-4, 0)$ and $(2, 0)$

Answer

a. $r = \dfrac{(1)(3)}{1 - (1)\sin\theta}$, $\therefore r = \dfrac{3}{1 - \sin\theta}$

1. Substitute into the standard form $r = \dfrac{ed}{1 - e\sin\theta}$ where $d = 3$, since the directrix crosses the negative y-axis.

Check

2. Use technology to graph the parabola and its directrix, $r = -3\csc\theta$.

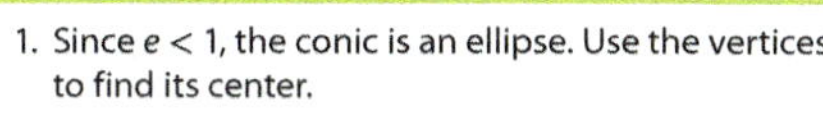

b. $C\left(\dfrac{-4+2}{2}, \dfrac{0+0}{2}\right) = (-1, 0)$

1. Since $e < 1$, the conic is an ellipse. Use the vertices to find its center.

$2 = \dfrac{\left(\frac{1}{3}\right)d}{1 + \left(\frac{1}{3}\right)\cos(0)} \cdot \dfrac{3}{3}$

$2 = \dfrac{d}{3 + 1}$; $d = 8$

2. Since the center is on the negative x-axis, the directrix has form $x = d$. Use the standard form $r = \dfrac{ed}{1 + e\cos\theta}$, $e = \frac{1}{3}$, and a point on the ellipse, $(r, \theta) = (2, 0)$, to find d.

$r = \dfrac{\frac{1}{3}(8)}{1 + \frac{1}{3}\cos\theta}$ or $r = \dfrac{8}{3 + \cos\theta}$

3. Substitute e and d into the standard form and simplify the complex fraction.

Check

4. Use technology to graph the ellipse and its directrix, $r = 8\sec\theta$.

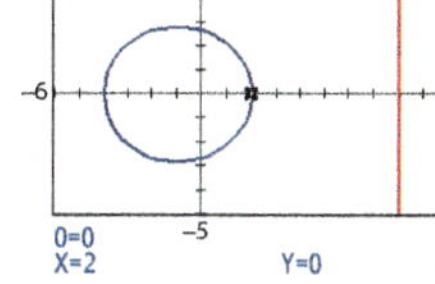

— SKILL ✔ EXERCISE 15

The relationships between the center, foci, vertices, and co-vertices learned in earlier sections of this chapter can be used to rewrite polar equations of conics as equivalent rectangular equations.

Example 3 Writing the Rectangular Equation of a Conic

Write a rectangular equation for the conic represented by $r = \dfrac{9}{4 + 5\cos\theta}$.

Answer

$r = \dfrac{9 \div 4}{(4 + 5\cos\theta) \div 4}$

$r = \dfrac{2.25}{1 + 1.25\cos\theta}$; $e = 1.25$

1. Rewrite the equation in the standard form $r = \dfrac{ed}{1 + e\cos\theta}$ and identify the eccentricity which indicates that the conic is a hyperbola with vertices on the polar axis.

$r(0) = \dfrac{9}{4 + 5\cos(0)} = \dfrac{9}{4 + 5} = 1$

$r(\pi) = \dfrac{9}{4 + 5\cos(\pi)} = \dfrac{9}{4 - 5} = -9$

2. Evaluate the polar function at 0 and π to find the vertices in polar form.

3. Express the vertices using rectangular coordinates.

$V_1: x = 1\cos 0 = 1; y = 1\sin 0 = 0$
$V_2: x = -9\cos\pi = 9; y = -9\sin\pi = 0$

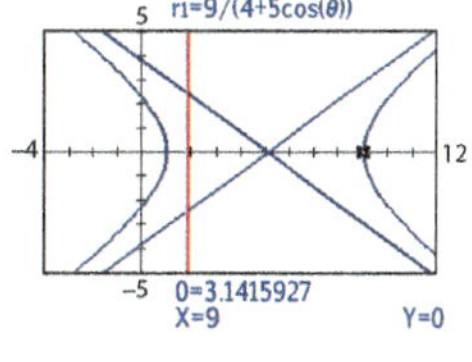

	(r, θ)	(x, y)
V_1	$(1, 0)$	$(1, 0)$
V_2	$(-9, \pi)$	$(9, 0)$

$C\left(\dfrac{1+9}{2}, \dfrac{0+0}{2}\right) = (5, 0)$; $c = 5$

4. Find the center (h, k) and use the fact that the focus is at the origin to find the value of c.

$a = \dfrac{c}{e} = \dfrac{5}{1.25} = 4$

$b^2 = c^2 - a^2 = 25 - 16 = 9$

5. Find the values of a and b^2 using $e = \dfrac{c}{a}$ and $c^2 = a^2 + b^2$.

CONTINUED ➡

6. $e = 0.75$; V: $\left(1, \dfrac{\pi}{2}\right)$, $\left(7, \dfrac{3\pi}{2}\right)$

$r = \dfrac{7}{4 + 3\sin\theta}$

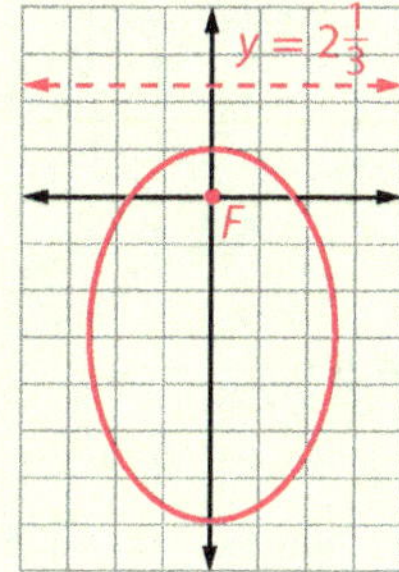

Write a rectangular equation for each conic.

7. $r = \dfrac{3}{1 - \cos\theta}$ $x = \dfrac{1}{6}y^2 - \dfrac{3}{2}$

8. $r = \dfrac{16}{5 + 3\sin\theta}$ $\dfrac{x^2}{16} + \dfrac{(y + 3)^2}{25} = 1$

9. $r = \dfrac{36}{4 + 5\cos\theta}$ $\dfrac{(x - 20)^2}{256} - \dfrac{y^2}{144} = 1$

10. Write a polar equation modeling the orbit of the dwarf planet Pluto given its eccentricity of 0.248 and its semimajor axis of 39.5294 AU. Then find the distances to its perihelion and aphelion. $r = \dfrac{39.5294(1 - 0.248^2)}{1 + 0.248\cos\theta}$

perihelion: ≈ 29.726 AU

aphelion: ≈ 49.333 AU

Assignments

- **Minimum:** 1, 3–4, 9–12; 15–23 odd; 27, 43–46, 48

- **Standard:** 1, 4, 8–11, 15–17; 19–29 odd; 30, 33, 41, 43–46, 49

- **Extended:** 1–2, 6–7, 12–15, 18–19, 22–24, 26; 29–35 odd; 36; 39–49 odd

Borrelly, discussed in Example 4, is elliptical.

You may want to share that the asteroid Oumuamua, referenced in the Data Analysis margin note, was the first and only object to date to be identified as having a hyperbolic orbit. Discovered in 2017, Oumuamua is a word of Hawaiian origin that roughly translates "first to arrive from afar."

Interactive Activity The Technology Corner on page 83 provides an excellent interactive opportunity.

TIPS

Ex. 19–22 Students can check their work by graphing the polar and rectangular equations with technology, such as the Desmos online graphing calculator, that permits simultaneous graphing of polar and rectangular graphs.

Ex. 23–28 Encourage students to check their work by graphing polar and rectangular equations in the same coordinate plane.

Ex. 32–36 These exercises can be solved using the techniques from this section (see solutions for exercises 32 and 35) or using the techniques from Section 6.4 with some creative algebra (see solutions for exercises 33–34). The solution to exercise 36 is similar to those for exercises 33–34, but the final polar form of exercise 36 differs from those presented in this lesson.

- Quiz 8C covers Sections 8.5–8.6.

Solutions

⟩ A. Exercises

1. $e = 1$; parabola
$ed = 4 = (1)d$; $d = 4$
directrix: $y = 4$

2. $r = \dfrac{27}{1 + 3\cos\theta}$
$e = 3$; hyperbola
$ed = 27 = (3)d$; $d = 9$
directrix: $x = 9$

3. $r = \dfrac{18 \div 2}{(2 - 12\cos\theta) \div 2} = \dfrac{9}{1 - 6\cos\theta}$
$e = 6$; hyperbola
$ed = 9 = (6)d$; $d = 1.5$
directrix: $x = -1.5$

4. $r = \dfrac{12 \div 8}{(8 + 2\sin\theta) \div 8} = \dfrac{1.5}{1 + 0.25\sin\theta}$
$e = 0.25$; ellipse
$ed = 1.5 = (0.25)d$; $d = 6$
directrix: $y = 6$

5. $r = \dfrac{-6\left(-\frac{4}{3}\right)}{\left(\cos\theta - \frac{3}{4}\right)\left(-\frac{4}{3}\right)} = \dfrac{8}{1 - \frac{4}{3}\cos\theta}$
$e = \frac{4}{3}$; hyperbola
$ed = 8 = \left(\frac{4}{3}\right)d$; $d = 6$
directrix: $x = -6$

6. $r = \dfrac{49 \div 2}{(2 - 7\sin\theta) \div 2} = \dfrac{24.5}{1 - 3.5\sin\theta}$
$e = 3.5$; hyperbola
$ed = 24.5 = (3.5)d$; $d = 7$
directrix: $y = -7$

7. $r = \dfrac{7 \div 5}{(5 - 3\cos\theta) \div 5} = \dfrac{1.4}{1 - 0.6\cos\theta}$
$e = 0.6$; ellipse
$ed = 1.4 = (0.6)d$; $d = \frac{7}{3}$
directrix: $x = -\frac{7}{3}$

8. $r = \dfrac{10 \div 2}{(2 - 5\sin\theta) \div 2} = \dfrac{5}{1 - 2.5\sin\theta}$
$e = 2.5$; hyperbola
$ed = 5 = (2.5)d$; $d = 2$
directrix: $y = -2$

9. $e = 2$; hyperbola
$1 + 2\sin\theta \Rightarrow$ directrix: $y = d$

10. $e = 1$; parabola
$1 + \cos\theta \Rightarrow$ directrix: $x = d$

11. $e = 0.5$; ellipse
$2 - \cos\theta \Rightarrow$ directrix: $x = -d$

12. $r = \dfrac{2}{1 + 0.5\sin\theta}$; $e = \frac{1}{2}$; ellipse
$1 + 0.5\sin\theta \Rightarrow$ directrix: $y = d$

13. $r = \dfrac{2.5}{1 - \sin\theta}$; $e = 1$; parabola
$1 - \sin\theta \Rightarrow$ directrix: $y = -d$

14. $r = \dfrac{2}{1 - 2\cos\theta}$; $e = 2$; hyperbola
$1 - 2\cos\theta \Rightarrow$ directrix: $x = -d$

15. $r = \dfrac{1(5)}{1 + (1)\sin\theta} = \dfrac{5}{1 + \sin\theta}$; parabola

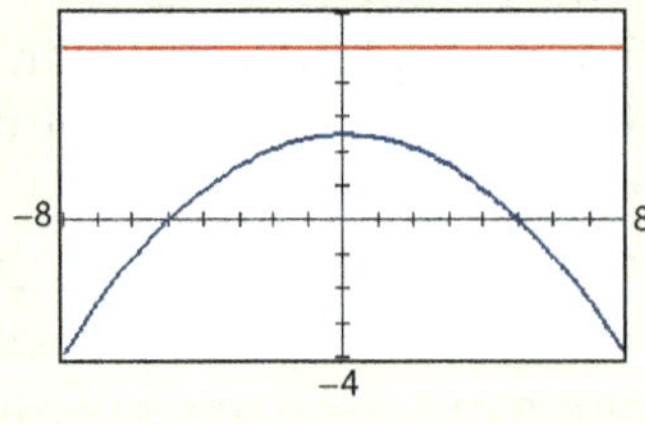

16. $r = \dfrac{\frac{1}{2}(3)}{1 - \left(\frac{1}{2}\right)\cos\theta} = \dfrac{1.5}{1 - 0.5\cos\theta}$; ellipse

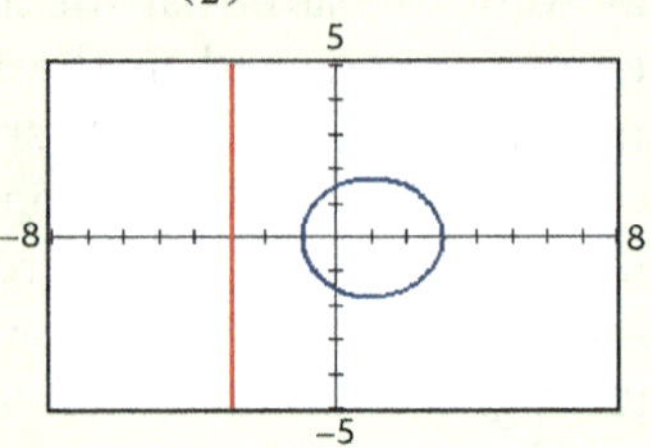

17. $r = \dfrac{2(1)}{1 - (2)\sin\theta} = \dfrac{2}{1 - 2\sin\theta}$; hyperbola

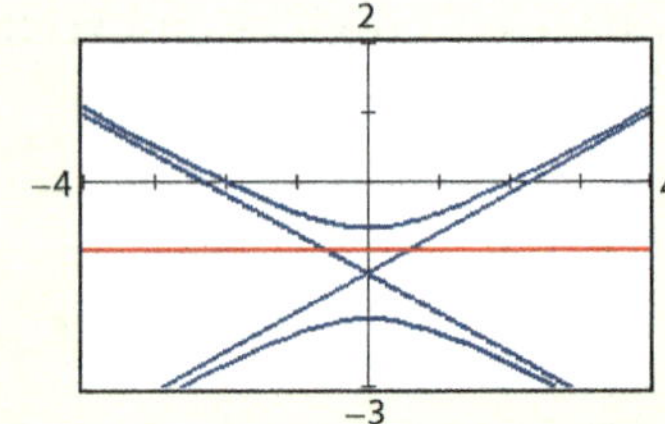

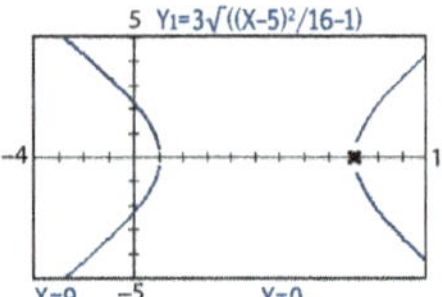

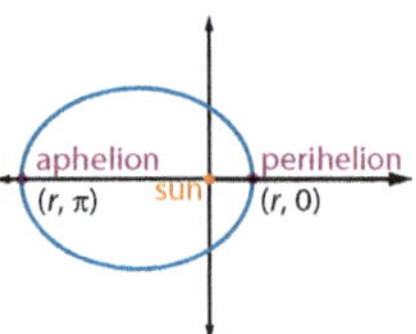

The eccentricity of the hyperbolic path of Oumuamua, the first interstellar object ever observed, is 1.20. From its perihelion it is approximately 0.25 AU (23,700,000 mi) from the sun.

$$\frac{(x - 5)^2}{16} - \frac{y^2}{9} = 1$$

Check

6. Write the standard rectangular form for a hyperbola with a horizontal transverse axis:
$$\frac{(x - h)^2}{a^2} - \frac{(y - k)^2}{b^2} = 1.$$

7. Graph the functions $y = \pm 3\sqrt{\dfrac{(x - 5)^2}{16} - 1}$ and compare their graphs to the polar graph.

SKILL ✔ **EXERCISE 23**

The orbits of planets, comets, and satellites are frequently modeled by polar equations in which the much more massive central body is placed at the focus. This expresses the distance to the orbiting object as a function of the angle traveled from the polar axis.

Modeling an elliptical orbit with $r = \dfrac{ed}{1 + e\cos\theta}$ causes the perihelion (the point closest to the sun) to occur when $\theta = 0$ and the aphelion (the point farthest from the sun) to occur when $\theta = \pi$. The length of the major axis is the sum of the r-values for these two vertices:

$$2a = \frac{ed}{1 + e\cos 0} + \frac{ed}{1 + e\cos\pi} = \frac{ed}{1 + e} + \frac{ed}{1 - e} = \frac{ed(1 - e) + ed(1 + e)}{1 - e^2} = \frac{2ed}{1 - e^2}.$$
$$\therefore ed = a(1 - e^2).$$

The elliptical orbit can then be modeled in terms of the semimajor axis and the eccentricity:

$$r = \frac{a(1 - e^2)}{1 + e\cos\theta}.$$

Example 4 Modeling an Orbit with a Polar Equation

Write a polar equation modeling the orbit of Comet Borrelly, given its eccentricity of 0.624 and its semimajor axis of 3.61 AU. Then find the distances to its perihelion and aphelion.

Answer

$r = \dfrac{3.61(1 - 0.624^2)}{1 + 0.624\cos\theta}$

$r(0) = \dfrac{3.61(1 - 0.624^2)}{1 + 0.624} \approx 1.36$ AU

$r(\pi) = \dfrac{3.61(1 - 0.624^2)}{1 - 0.624} \approx 5.86$ AU

Check

$1.36 + 5.86 = 2(3.61) = 7.22$ AU

1. Write the equation in the standard form $r = \dfrac{a(1 - e^2)}{1 + e\cos\theta}$.

2. Evaluate $r(0)$ to determine the perihelion distance and $r(\pi)$ to determine the aphelion distance.

3. The sum of the perihelion and aphelion is the length of the major axis.

SKILL ✔ **EXERCISE 31**

A. Exercises

Determine the eccentricity, type of conic, and directrix for each polar equation.

1. $r = \dfrac{4}{1 + \sin\theta}$ **2.** $r = \dfrac{27}{3\cos\theta + 1}$

3. $r = \dfrac{18}{2 - 12\cos\theta}$ **4.** $r = \dfrac{12}{2\sin\theta + 8}$

5. $r = \dfrac{-6}{\cos\theta - 0.75}$ **6.** $r = \dfrac{49}{2 - 7\sin\theta}$

7. $r = \dfrac{7}{5 - 3\cos\theta}$ **8.** $r = \dfrac{10}{2 - 5\sin\theta}$

Match each equation with its graph.

A. 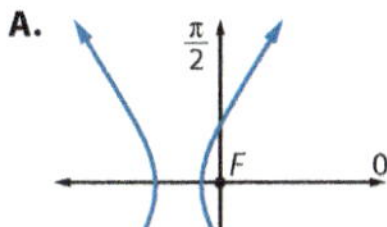 **D.** 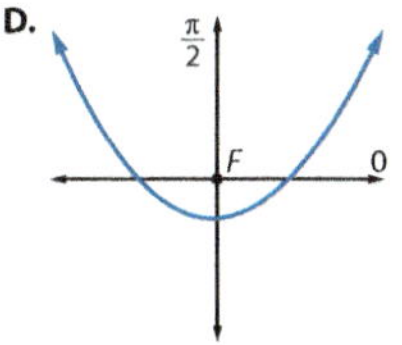

B. 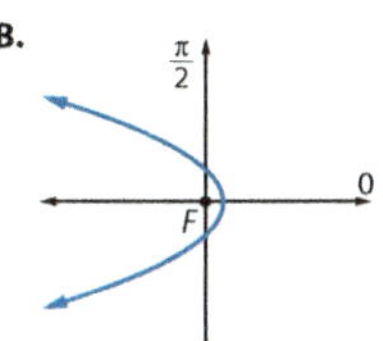 **E.** 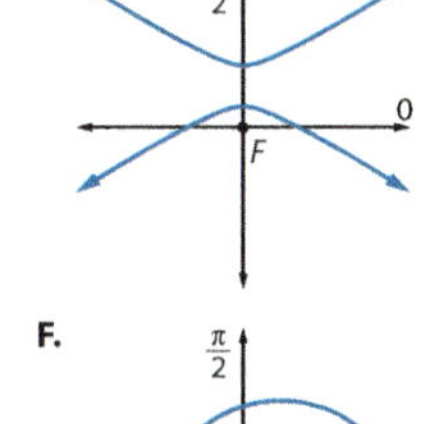

C. 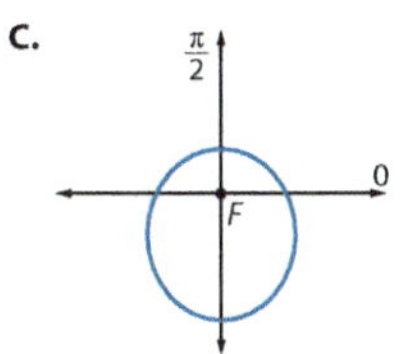 **F.** 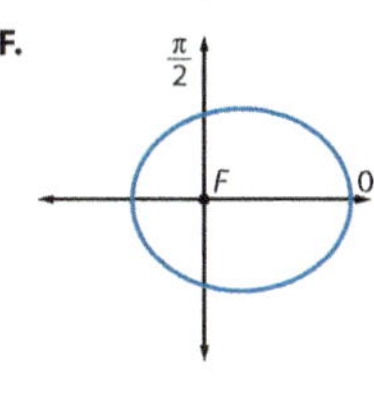

9. $r = \dfrac{2}{1 + 2\sin\theta}$ E   **10.** $r = \dfrac{1}{1 + \cos\theta}$ B

11. $r = \dfrac{2.5}{1 - 0.5\cos\theta}$ F   **12.** $r = \dfrac{4}{2 + \sin\theta}$ C

13. $r = \dfrac{5}{2 - 2\sin\theta}$ D   **14.** $r = \dfrac{6}{3 - 6\cos\theta}$ A

Write a polar equation for each conic section. Then classify the conic and verify your answer by using technology to graph the curve and its directrix.

15. $e = 1$; directrix: $y = 5$ **16.** $e = \frac{1}{2}$; directrix: $x = -3$

17. $e = 2$; directrix: $y = -1$ **18.** $e = \frac{5}{2}$; directrix: $x = 2$

15. $r = \dfrac{5}{1 + \sin\theta}$; parabola   **16.** $r = \dfrac{1.5}{1 - 0.5\cos\theta}$; ellipse

17. $r = \dfrac{2}{1 - 2\sin\theta}$; hyperbola   **18.** $r = \dfrac{5}{1 + 2.5\cos\theta}$; hyperbola

B. Exercises

Write a polar equation for each conic section with a focus at the pole. Then use technology to graph the conic and its directrix.

19. $e = 0.5$; $V(x, y)$: $(-3, 0)$, $(1, 0)$ $r = \dfrac{1.5}{1 + 0.5\cos\theta}$

20. $e = \frac{3}{2}$; $V(x, y)$: $(3, 0)$, $(15, 0)$ $r = \dfrac{15}{2 + 3\cos\theta}$

21. $e = 2$; $V(r, \theta)$: $\left(2, \frac{\pi}{2}\right)$, $\left(-6, \frac{3\pi}{2}\right)$ $r = \dfrac{6}{1 + 2\sin\theta}$

22. $e = 1$; $V(r, \theta)$: $\left(\frac{5}{2}, \frac{3\pi}{2}\right)$ $r = \dfrac{5}{1 - \sin\theta}$

Write a rectangular equation for each conic.

23. $r = \dfrac{30}{4 + \sin\theta}$ **24.** $r = \dfrac{10}{2 - 3\cos\theta}$

25. $r = \dfrac{4}{1 + \cos\theta}$ **26.** $r = \dfrac{5}{2 - 2\sin\theta}$

27. $r = \dfrac{24}{1 - 3\sin\theta}$ **28.** $r = \dfrac{42}{5 + 2\cos\theta}$

29. Elliptical orbits can be modeled by $r = \dfrac{a(1 - e^2)}{1 + e\cos\theta}$. Use this equation to find expressions for the perihelion distance and the aphelion distance in terms of the semimajor axis and eccentricity. $a(1 - e)$; $a(1 + e)$

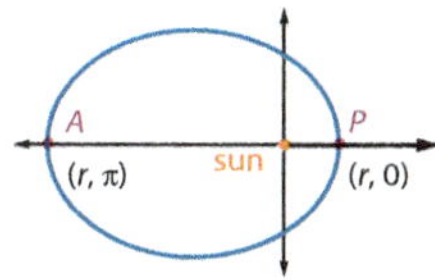

Write the equation modeling each planet's orbit, given its eccentricity and semimajor axis length. Then find its perihelion and aphelion distances.

30. Neptune: $e = 0.009$, $a = 30.06$ AU

31. Mars: $e = 0.093$, $a = 1.524$ AU

Write the polar form equation for each conic section.

32. $x^2 = 6y + 9$ **33.** $y^2 - 8x - 16 = 0$

34. $3(x + 2)^2 + 4y^2 = 48$ **35.** $\dfrac{(y - 5)^2}{16} - \dfrac{x^2}{9} = 1$

C. Exercises

36. Write the polar form equation for $\dfrac{x^2}{64} - \dfrac{y^2}{36} = 1$.

Make a sketch of the given directrix ($d > 0$) and the conic with its focus at the origin (similar to the last figure on p. 435). Then use the sketch and the definition of a conic to derive the standard form polar equation for the conic.

37. $x = -d$; $r = \dfrac{ed}{1 - e\cos\theta}$

38. $y = d$; $r = \dfrac{ed}{1 + e\sin\theta}$

39. $y = -d$; $r = \dfrac{ed}{1 - e\sin\theta}$

18. $r = \dfrac{\frac{5}{2}(2)}{1 + \left(\frac{5}{2}\right)\cos\theta} = \dfrac{5}{1 + 2.5\cos\theta}$

hyperbola

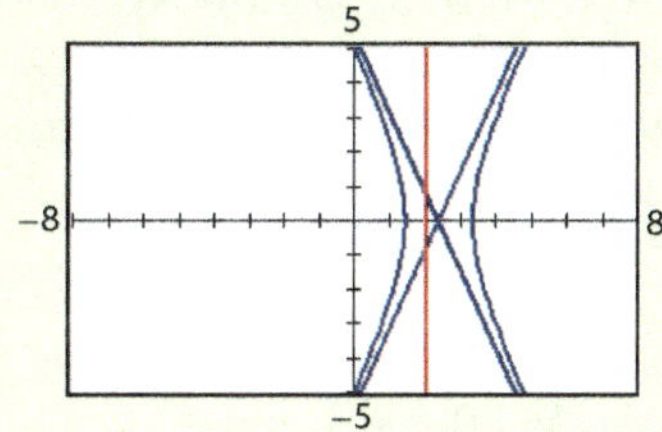

B. Exercises

19. ellipse with $C(-1, 0)$; $\therefore r = \dfrac{ed}{1 + e\cos\theta}$

using $(1, 0$ radians$)$:

$$1 = \dfrac{0.5d}{1 + (0.5)\cos 0}; \quad 1 = \dfrac{0.5d}{1.5}; \quad d = 3$$

$$r = \dfrac{1.5}{1 + 0.5\cos\theta}; \quad \text{directrix: } x = 3$$

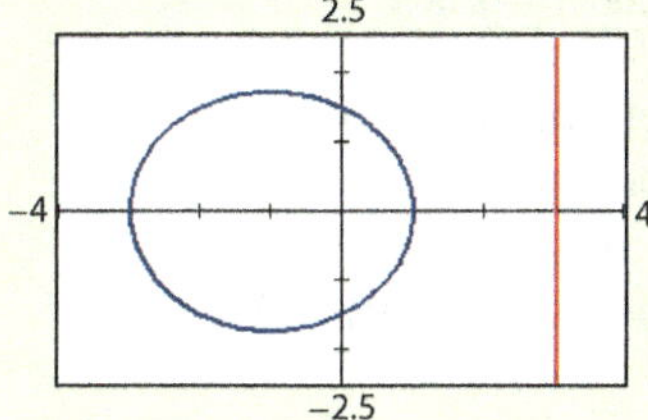

20. hyperbola with $C\left(\dfrac{15 + 3}{2}, 0\right) = (9, 0)$

$\therefore r = \dfrac{ed}{1 + e\cos\theta}$; using $(3, 0$ radians$)$:

$$3 = \dfrac{\left(\frac{3}{2}\right)d}{1 + \left(\frac{3}{2}\right)\cos 0} \cdot \dfrac{2}{2} = \dfrac{3d}{2 + 3(1)}; \quad d = 5$$

$$r = \dfrac{\frac{3}{2}(5)}{1 + \frac{3}{2}\cos\theta} \cdot \dfrac{2}{2} = \dfrac{15}{2 + 3\cos\theta}$$

directrix: $x = 5$

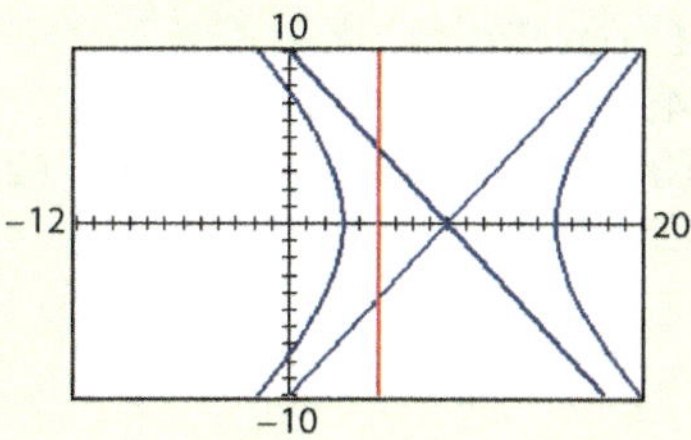

21. $(r\cos\theta, r\sin\theta) \Rightarrow V: (0, 6), (0, 2)$

hyperbola with $C\left(0, \dfrac{6 + 2}{2}\right) = (0, 4)$

$\therefore r = \dfrac{ed}{1 + e\sin\theta}$; using $\left(2, \frac{\pi}{2}\right)$:

$$2 = \dfrac{2d}{1 + (2)\sin\frac{\pi}{2}} = \dfrac{2d}{1 + 2(1)}; \quad d = 3$$

$$r = \dfrac{2(3)}{1 + 2\sin\theta} = \dfrac{6}{1 + 2\sin\theta}$$

directrix: $y = 3$

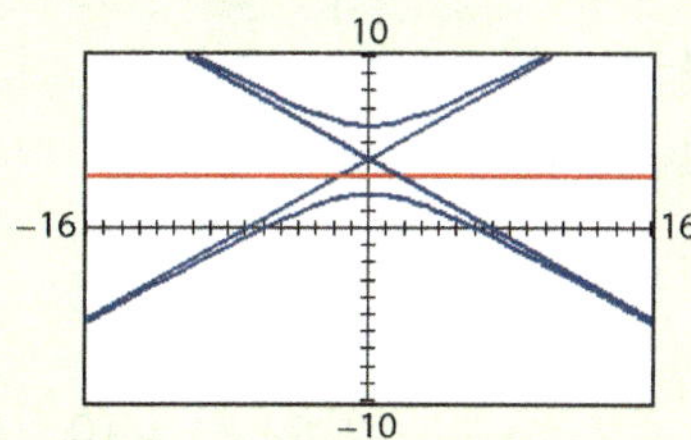

22. parabola with $y = -d$; $\therefore r = \dfrac{ed}{1 - e\sin\theta}$

$d = 2|p| = 2(2.5) = 5$

$$r = \dfrac{1(5)}{1 - (1)\sin\theta} = \dfrac{5}{1 - \sin\theta}$$

directrix: $y = -5$

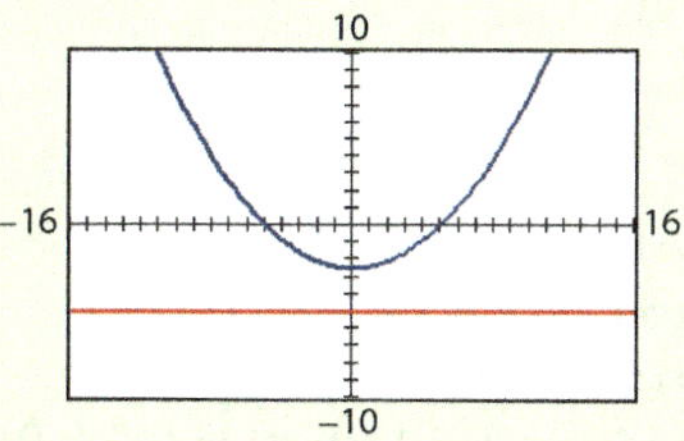

23. $r = \dfrac{7.5}{1 + 0.25\sin\theta}$; $e = 0.25$ (ellipse)

$V_1: r\left(\dfrac{\pi}{2}\right) = 6$

$\left(6\cos\dfrac{\pi}{2}, 6\sin\dfrac{\pi}{2}\right) = (0, 6)$

$V_2: r\left(\dfrac{3\pi}{2}\right) = 10$

$\left(10\cos\dfrac{3\pi}{2}, 10\sin\dfrac{3\pi}{2}\right) = (0, -10)$

$C(0, -2)$; $c = 2$

$a = \dfrac{c}{e} = \dfrac{2}{0.25} = 8$; $b^2 = 8^2 - 2^2 = 60$

$$\dfrac{x^2}{60} + \dfrac{(y + 2)^2}{64} = 1$$

24. $r = \dfrac{5}{1 - 1.5 \cos \theta}$; $e = 1.5$ (hyperbola)

$V_1: r(0) = -10$

$(-10 \cos 0, -10 \sin 0) = (-10, 0)$

$V_2: r(\pi) = 2$

$(2 \cos \pi, 2 \sin \pi) = (-2, 0)$

$C(-6, 0); c = 6$

$a = \dfrac{c}{e} = \dfrac{6}{1.5} = 4; \ b^2 = 6^2 - 4^2 = 20$

$\dfrac{(x+6)^2}{16} - \dfrac{y^2}{20} = 1$

25. $e = 1$ (parabola)

$d = 4$; directrix: $x = 4$

$V: r(0) = 2; \ (2 \cos 0, 2 \sin 0) = (2, 0)$

$p = -2; \ a = \dfrac{1}{4(-2)} = -\dfrac{1}{8}$

$x = -\dfrac{1}{8}(y - 0)^2 + 2$

$x = -\dfrac{1}{8}y^2 + 2$

26. $r = \dfrac{2.5}{1 - \sin \theta}$; $e = 1$ (parabola)

$d = 2.5$; directrix: $y = -2.5$

$V: r\left(\dfrac{3\pi}{2}\right) = 1.25$

$\left(1.25 \cos \dfrac{3\pi}{2}, 1.25 \sin \dfrac{3\pi}{2}\right)$

$\qquad\qquad = (0, -1.25)$

$p = 1.25; \ a = \dfrac{1}{4(1.25)} = \dfrac{1}{5}$

$y = \dfrac{1}{5}(x - 0)^2 - 1.25$

$y = \dfrac{1}{5}x^2 - 1.25$

27. $e = 3$ (hyperbola)

$V_1: r\left(\dfrac{\pi}{2}\right) = -12$

$\left(-12 \cos \dfrac{\pi}{2}, -12 \sin \dfrac{\pi}{2}\right) = (0, -12)$

$V_2: r\left(\dfrac{3\pi}{2}\right) = 6$

$\left(6 \cos \dfrac{3\pi}{2}, 6 \sin \dfrac{3\pi}{2}\right) = (0, -6)$

$C(0, -9); c = 9$

$a = \dfrac{c}{e} = \dfrac{9}{3} = 3; \ b^2 = 9^2 - 3^2 = 72$

$\dfrac{(y+9)^2}{9} - \dfrac{x^2}{72} = 1$

28. $r = \dfrac{8.4}{1 + 0.4 \cos \theta}$; $e = 0.4$ (ellipse)

$V_1: r(0) = 6$

$(6 \cos 0, 6 \sin 0) = (6, 0)$

$V_2: r(\pi) = 14$

$(14 \cos \pi, 14 \sin \pi) = (-14, 0)$

$C(-4, 0); c = 4$

$a = \dfrac{c}{e} = \dfrac{4}{0.4} = 10; \ b^2 = 10^2 - 4^2 = 84$

$\dfrac{(x+4)^2}{100} + \dfrac{y^2}{84} = 1$

State the rotation formulas that will rotate each general quadratic equation to a *uv*-plane without a *uv*-term. [8.4]

40. $-x^2 + 2xy - y^2 + \sqrt{2}x + \sqrt{2}y + 144 = 0$

41. $3x^2 - 4\sqrt{3}xy + 7y^2 + 120 = 0$

42. Graph $(x - 4)^2 + \dfrac{(y-1)^2}{25} = 1$ and its foci. Then state its eccentricity. [8.2]

State the next three terms of each sequence. [Algebra]

43. 5, 10, 15, 20…   25, 30, 35

44. 4, 16, 64, 256…   1024, 4096, 16,384

45. Find the first, second, and tenth term of the sequence $a_n = 4n$. [Algebra] $a_1 = 4, a_2 = 8, a_{10} = 40$

46. Which polar coordinate expression is equivalent to the rectangular coordinates $(1, \sqrt{3})$? [6.4]   D

   A. $(1, 30º)$ **C.** $(2, 30º)$ **E.** $(4, 30º)$

   B. $(1, 60º)$ **D.** $(2, 60º)$

47. Find AB if $A = \begin{bmatrix} 2 & 1 \\ 4 & -2 \end{bmatrix}$ and $B = \begin{bmatrix} 0 & -1 \\ 3 & 1 \end{bmatrix}$. [7.2]   D

   A. $\begin{bmatrix} 2 & -1 \\ 12 & -2 \end{bmatrix}$ **C.** $\begin{bmatrix} 3 & -1 \\ 12 & -6 \end{bmatrix}$ **E.** $\begin{bmatrix} 0 & -1 \\ 12 & -2 \end{bmatrix}$

   B. $\begin{bmatrix} 6 & 0 \\ 12 & 0 \end{bmatrix}$ **D.** $\begin{bmatrix} 3 & -1 \\ -6 & -6 \end{bmatrix}$

48. Find the domain for the graph of $x = t + 1$ and $y = 1 - t^2$ where $t \in [-2, 2]$. [8.5]   B

   A. $x \in [-2, 2]$ **C.** $x \in [-3, 1]$ **E.** $x \in \mathbb{R}$

   B. $x \in [-1, 3]$ **D.** $x \in [0, 4]$

49. Find the domain for the graph of $x = t^2 - 3$ and $y = t - 1$ where $t \in [-2, 2]$. [8.5]   E

   A. $x \in [-2, 2]$ **C.** $x \in [1, 7]$ **E.** $x \in [-3, 1]$

   B. $x \in [1, 9]$ **D.** $x \in [3, 7]$

1. $e = 1$; parabola; $y = 4$

2. $e = 3$; hyperbola; $x = 9$

3. $e = 6$; hyperbola; $x = -1.5$

4. $e = 0.25$; ellipse; $y = 6$

5. $e = \dfrac{4}{3}$; hyperbola; $x = -6$

6. $e = 3.5$; hyperbola; $y = -7$

7. $e = 0.6$; ellipse; $x = -\dfrac{7}{3}$

8. $e = 2.5$; hyperbola; $y = -2$

23. $\dfrac{x^2}{60} + \dfrac{(y+2)^2}{64} = 1$

24. $\dfrac{(x+6)^2}{16} - \dfrac{y^2}{20} = 1$

25. $x = -\dfrac{1}{8}y^2 + 2$

26. $y = \dfrac{1}{5}x^2 - 1.25$

27. $\dfrac{(y+9)^2}{9} - \dfrac{x^2}{72} = 1$

28. $\dfrac{(x+4)^2}{100} + \dfrac{y^2}{84} = 1$

30. $r = \dfrac{30.06(1 - 0.009^2)}{1 + 0.009 \cos \theta}$; ≈ 29.79 AU; ≈ 30.33 AU

31. $r = \dfrac{1.524(1 - 0.093^2)}{1 + 0.093 \cos \theta}$; ≈ 1.38 AU; ≈ 1.67 AU

32. $r = \dfrac{3}{1 - \sin \theta}$

33. $r = \dfrac{4}{1 - \cos \theta}$

34. $r = \dfrac{-3}{1 - \frac{1}{2} \cos \theta}$

35. $r = \dfrac{9}{4 + 5 \sin \theta}$

36. $r^2 = \dfrac{576}{9 - 25 \sin^2 \theta}$

40. $x = \dfrac{\sqrt{2}u - \sqrt{2}v}{2}, y = \dfrac{\sqrt{2}u + \sqrt{2}v}{2}$

41. $x = \dfrac{\sqrt{3}u - v}{2}, y = \dfrac{u + \sqrt{3}v}{2}$

29. perihelion: $r(0) = \dfrac{a(1 - e^2)}{1 + e \cos 0}$

$= \dfrac{a(1-e)(1+e)}{1+e} = a(1 - e)$

aphelion: $r(\pi) = \dfrac{a(1 - e^2)}{1 + e \cos \pi}$

$= \dfrac{a(1-e)(1+e)}{1-e} = a(1 + e)$

30. $r = \dfrac{30.06(1 - 0.009^2)}{1 + 0.009 \cos \theta}$

perihelion: $a(1 - e)$

$= 30.06(1 - 0.009) \approx 29.79$ AU

aphelion: $a(1 + e)$

$= 30.06(1 + 0.009) \approx 30.33$ AU

31. $r = \dfrac{1.524(1 - 0.093^2)}{1 + 0.093 \cos \theta}$

perihelion: $a(1 - e)$

$= 1.524(1 - 0.093) \approx 1.38$ AU

aphelion: $a(1 + e)$

$= 1.524(1 + 0.093) \approx 1.67$ AU

32. $6y = x^2 - 9$

$y = \dfrac{1}{6}x^2 - \dfrac{3}{2}$

parabola; $V\left(0, -\dfrac{3}{2}\right); p = \dfrac{3}{2}$

$e = 1; d = |2p| = 3$; directrix: $y = -3$

$r = \dfrac{3(1)}{1 - (1) \sin \theta} = \dfrac{3}{1 - \sin \theta}$

continued in Answers and Solutions Overflow

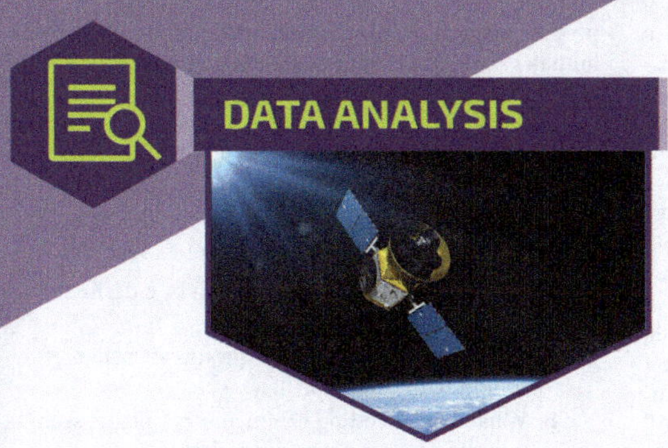

Conics in Space

In his State of the Union address on January 25, 1984, President Ronald Reagan directed NASA to construct an international space station (ISS). The first crew arrived in November 2000, and the ISS has been continuously occupied ever since. Reusable American space shuttles completed more than 30 missions delivering and assembling components of the space station. Rendezvousing and docking with a platform traveling at nearly 5 mi/sec requires careful planning.

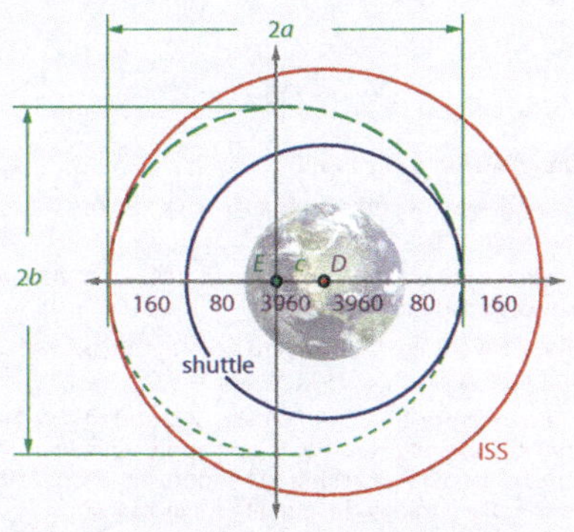

The diagram provides a simplified overview of how the properties of conic sections are used to complete a successful rendezvous. The shuttle initially assumes a circular orbit closer to the earth than the space station so that it can "catch up" to the station. The shuttle fires its rockets to complete half of an elliptical orbit (dashed line) that takes it to the orbit of the ISS.

The radius of the earth is 3960 mi. Assume that the ISS is in a circular orbit 240 mi above the surface of the earth, and the initial orbit of the shuttle is 80 mi above the surface of the earth. Let D represent the center of the earth and E represent the center of the elliptical orbit. Use E as the origin of your coordinate system and assume the orbits are coplanar.

1. State the coordinates of D. (80, 0)

2. Write standard form equations that model the circular orbits of the ISS and the shuttle.

3. Write a standard form equation modeling the elliptical orbit and state the orbit's eccentricity.

While the touchdown of Atlantis on July 21, 2011, marked the end of the space shuttle program, the essential role of conics in space exploration continues. In April 2018, NASA used a commercial rocket to launch the Transiting Exoplanet Survey Satellite (TESS) into an elliptical high-earth orbit. TESS is designed to conduct an exhaustive search for exoplanets, planets that orbit a star other than our sun.

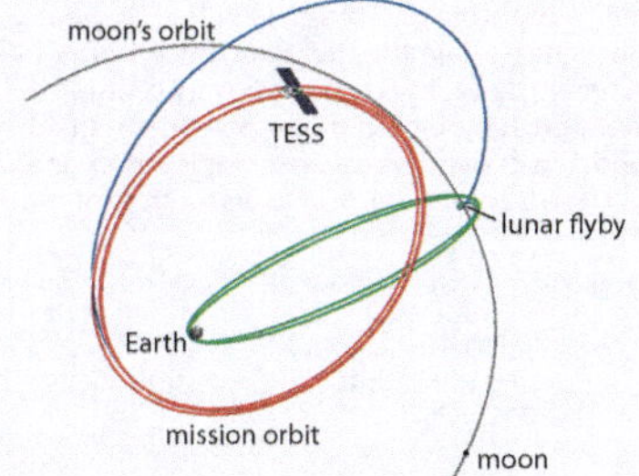

PRESENTATION

Space flight provides a good context to explore real-world applications of conic sections. Since the planets in our solar system are basically spherical, the gravitational forces can be calculated as if the mass was concentrated at the center point of the sphere. This Data Analysis explores circular, elliptical, parabolic, and hyperbolic orbits (in a simplified form) that are actually required to conduct space flight.

Data Analysis

Objectives

1. To describe the usefulness of conics in space flight

2. To write the equations of conics that meet given parameters

3. To evaluate from a biblical worldview the lessons we should learn from space exploration

Assignments

- **Minimum:** 1–9; 10 (in class)
- **Standard:** 1–9; 10 (in class)
- **Extended:** 1–10

Solutions

1. $2a = 240 + 2(3960) + 80 = 8240$
   $a = 4120$
   $x_D = 4120 - (3960 + 80) = 80$

   or

   The center of the ellipse, E, is half the difference between the radii of the circular orbits, $\frac{4200 - 4040}{2} = 80$, from D.

2. ISS: $(x - 80)^2 + y^2 = (3960 + 240)^2$
   $(x - 80)^2 + y^2 = 4200^2$

   shuttle: $(x - 80)^2 + y^2 = (3960 + 80)^2$
   $(x - 80)^2 + y^2 = 4040^2$

3. $a = \frac{4040 + 4200}{2} = 4120$

   Since the central body is at a focus of an elliptical orbit, $c = 80$.
   $b^2 = a^2 - c^2$
   $ = 4120^2 - 80^2 = 16{,}968{,}000$
   $\frac{x^2}{16{,}974{,}400} + \frac{y^2}{16{,}968{,}000} = 1$
   $e = \frac{c}{a} = \frac{80}{4120} \approx 0.019$

4.

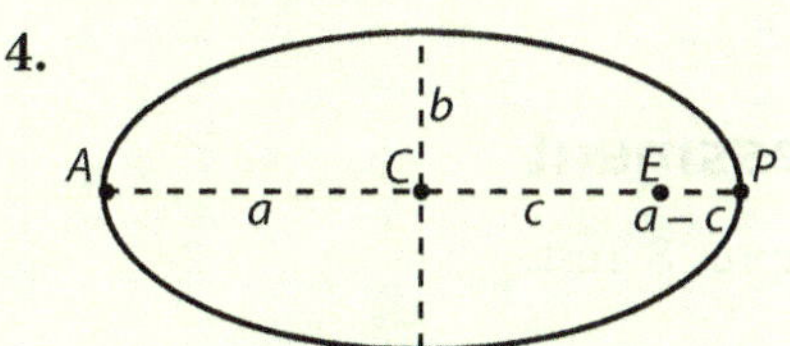

   $AE = a + c$
   $235{,}960 = a + 82{,}500$
   $a = 153{,}460$
   $b^2 = a^2 - c^2 = 1.675 \times 10^{10}$
   $b \approx 129{,}400$
   $\frac{x^2}{153{,}460^2} + \frac{y^2}{129{,}400^2} = 1$

5. $EP = a - c = 70{,}960$

   $EP - r_E = 70{,}960 - 3960 = 67{,}000 \text{ mi}$

6. $e = \dfrac{c}{a} = \dfrac{82{,}500}{153{,}460} \approx 0.5376$

7. using $r = \dfrac{ed}{1 + e \cos \theta}$ and $(0,\, 70{,}960)$

   or $r(0) = 70{,}960$

$$70{,}960 \approx \frac{(0.5376)d}{1 + (0.5376)}$$

$$d = 70{,}960\left(\frac{1.5376}{0.5376}\right) \approx 203{,}000$$

$$\therefore r \approx \frac{(0.5376)(203{,}000)}{1 + (0.5376)\cos \theta}$$

$$\approx \frac{109{,}100}{1 + 0.5376 \cos \theta}$$

or

since $r = \dfrac{ed}{1 + e \cos \theta} = \dfrac{a(1 - e^2)}{1 + e \cos \theta}$,

$ed = a(1 - e^2)$

$$\approx 153{,}460(1 - 0.5376^2)$$

$$\approx 109{,}100$$

$$\therefore r \approx \frac{109{,}100}{1 + 0.5376 \cos \theta}$$

Chapter 8 Review

Objective

To prepare for evaluation

Vocabulary

See Appendix A.

Assignments

- **Minimum:** 1–3, 5, 8; 9–15 odd; 16–19, 22–23, 25, 34; 35–43 odd; 44–48 even

- **Standard:** 1–49 odd; 2–50 even (in class)

- **Extended:** 1, 4; 5–11 odd; 14–20 even; 21–29; 31–39 odd; 40–42, 45–46, 49–50

Assessment

- Chapter 8 Test

Solutions

1. $y = \frac{1}{8}(x - 3)^2 - 1$

   $V(3, -1); \ p = \dfrac{1}{4\left(\frac{1}{8}\right)} = 2$

   $F(3, -1 + 2) = (3, 1)$

   directrix: $y = -1 - (2) = -3$

Its unique orbit requires a gravitational assist from a flyby of the moon that propels it into a hyperbolic trajectory toward its final orbit. To reach the precise path required for the flyby, TESS will make a series of "phasing" orbits whose apogee (farthest point from the earth) increases from about 150,000 mi to over 240,000 mi. Once established in its final orbit, TESS will complete one orbit every 13.7 days, two orbits for every one moon orbit. Combining this period with an orbital plane orthogonal to the moon's orbit virtually eliminates perturbations (orbital disruptions) caused by the moon's gravitational pull. The orbit is also designed so that each perigee (closest point to the earth) is well above the narrow band occupied by geostationary satellites. At each perigee TESS will download approximately 10 billion pixels of data to NASA's Pleiades supercomputer.

Assume that TESS's final elliptical orbit has an apogee of 235,960 mi and a focal length of 82,500 mi.

4. Write a standard form rectangular equation that models this orbit with an ellipse centered at the origin and Earth at the focus on the positive x-axis.

5. How close is TESS to Earth's surface at perigee?

6. Find the eccentricity of TESS's final elliptical orbit.

7. Write a polar equation modeling TESS's final orbit.

Scientists expect to identify tens of thousands of new exoplanets from this data. As more planets are discovered and the vastness of space becomes more apparent, many people question our place in the universe, or whether the human race is significant at all. Despair would be a logical conclusion in a godless universe. A biblical worldview, however, allows us to marvel and rejoice as the Creator God's majesty and power are increasingly displayed.

8. Psalm 8 records David's musings as he considered God's vast creation. Use Psalm 8 to list some appropriate biblical worldview responses to a vast universe.

9. Much of space exploration is driven by a desire to find life on other planets.

   a. The answer to what major question is being sought with this goal?

   b. What would finding life on another planet seem to indicate from a secular worldview?

10. In a 1995 television interview Stephen Hawking said, "The human race is just a chemical scum on a moderate-sized planet, orbiting around a very average star in the outer suburb of one among a hundred billion galaxies. We are so insignificant that I can't believe the whole universe exists for our benefit." Evaluate each part of this statement from a biblical worldview.

   a. The human race is a chemical scum.

   b. The earth is very small in a vast universe.

   c. The human race is insignificant.

   d. The universe exists for our benefit.

4. $\dfrac{x^2}{153{,}460^2} + \dfrac{y^2}{129{,}400^2} = 1$

5. 67,000 mi

6. $e \approx 0.54$

7. $r \approx \dfrac{109{,}100}{1 + 0.5376 \cos \theta}$

8. Answers will vary. We should have a sense of awe and thankfulness that God has given humans such special attention. We should thank Him for our role as rulers under His greater rule and praise Him for His great creation. The heavens declare the glory of God. A better understanding of the vastness of the universe simply helps us to have a proper opinion of the power and majesty of our God. We know that we are significant to God—He proved it by sending His Son as a sacrifice for our sins.

9a. Where did life come from?

9b. It would seem to indicate that life can begin spontaneously.

10a. Humans are created in the image of God, have eternal souls, and are loved by God.

10b. The heavens are designed to declare the glory of an infinite God.

10c. Humans were created to please God and to give Him glory. Humanity is so significant and God loves it so much that He was willing to become a man and suffer in our place to pay the penalty for all our sin.

10d. God's creation is to please Him. His thoughts and His ways are beyond our understanding. We can't know all that He knows, but we can know what He has revealed to us.

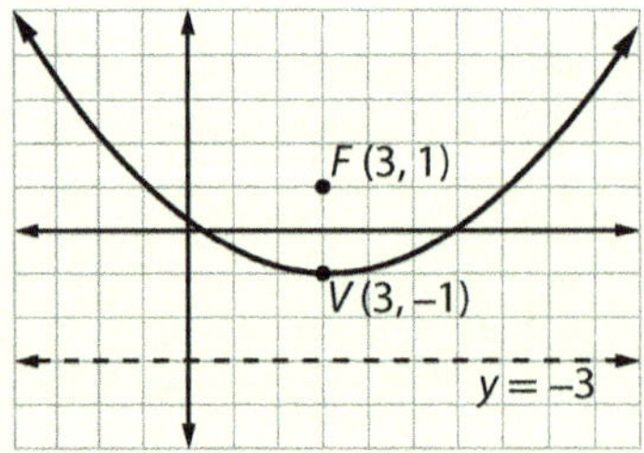

2. $x = -\frac{1}{4}y^2 + 1$

   $V(1, 0); \ p = \dfrac{1}{4\left(-\frac{1}{4}\right)} = -1$

   $F(1 - 1, 0) = (0, 0)$

   directrix: $x = 1 - (-1) = 2$

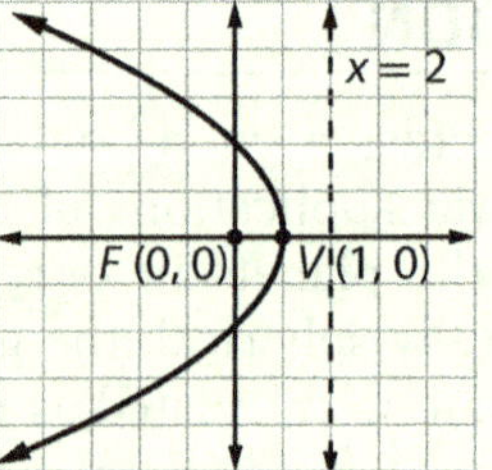

3. $p = \frac{1}{2}; \ a = \dfrac{1}{4\left(\frac{1}{2}\right)} = \dfrac{1}{2}$

   $y = \frac{1}{2}(x - 3)^2 - 3$

4. $V\left(\dfrac{\frac{5}{4} + \frac{3}{4}}{2},\, 2\right) = (1, 2)$

   $p = \frac{1}{4}; \ a = \dfrac{1}{4\left(\frac{1}{4}\right)} = 1$

   $x = (y - 2)^2 + 1$

Chapter 8 Review

Graph the parabola and its vertex, focus, and directrix. [8.1]

1. $(x - 3)^2 = 8(y + 1)$ **2.** $y^2 = -4(x - 1)$

Write the equation of the parabola with the following characteristics. [8.1]

3. $V(3, -3)$; $F(3, -2.5)$ $y = \frac{1}{2}(x - 3)^2 - 3$

4. $F\left(\frac{5}{4}, 2\right)$; directrix: $x = \frac{3}{4}$ $x = (y - 2)^2 + 1$

Graph the parabola and its focus and directrix. [8.1]

5. $-\frac{1}{4}x^2 - y + 3x - 5 = 0$ **6.** $-y^2 + 8x = 4y - 20$

7. Find the equation of the line tangent to $x^2 + 2y = 0$ at $(2, -2)$. [8.1] $y = -2x + 2$

8. The deck of the Hulme Arch Bridge in Manchester, England, is supported by cables attached to a parabolic arch that is 25 m high and that is 52 m wide at its base. Write a standard form equation that models the arch with one base at the origin and the other on the positive x-axis. [8.1]

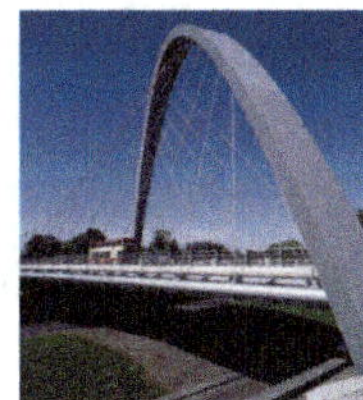

$y = -\frac{25}{676}(x - 26)^2 + 25$

9. Find the vertices, the foci, and the eccentricity for $\frac{x^2}{9} + \frac{y^2}{16} = 1$. Then graph the ellipse and its foci. [8.2]

Write each equation in the standard form for an ellipse. Then find its foci and eccentricity and graph the conic. [8.2]

10. $x^2 + y^2 - 2x + 4y = 4$ $\frac{(x - 1)^2}{9} + \frac{(y + 2)^2}{9} = 1$; $e = 0$

11. $x^2 + 4y^2 + 6x - 8y + 9 = 0$ $\frac{(x + 3)^2}{4} + \frac{(y - 1)^2}{1} = 1$; $e = \frac{\sqrt{3}}{2} \approx 0.866$

Write the standard form equation of each described ellipse. [8.2]

12. vertices: $(7, -4)$, $(-3, -4)$ co-vertices: $(2, -1)$, $(2, -7)$ $\frac{(x - 2)^2}{25} + \frac{(y + 4)^2}{9} = 1$

13. co-vertices: $(-3, -3)$, $(7, -3)$ foci: $(2, 9)$, $(2, -15)$ $\frac{(x - 2)^2}{25} + \frac{(y + 3)^2}{169} = 1$

14. $C(0, 0)$, $e = 0.75$ vertices: $(0, \pm 4)$ $\frac{x^2}{7} + \frac{y^2}{16} = 1$

15. $F(3, 3)$ and $(3, -5)$ a minor axis length of 6 $\frac{(x - 3)^2}{9} + \frac{(y + 1)^2}{25} = 1$

16. A whispering room is designed with a focus 2 ft from the northern wall and 16 ft from the southern wall. Write a standard form equation modeling the elliptical room, using the northern focus as the origin. Verify your equation by using technology to graph the ellipse. [8.2] $\frac{x^2}{32} + \frac{(y + 7)^2}{81} = 1$

17. Graph $\frac{x^2}{9} - \frac{y^2}{4} = 1$ and state the equations of its asymptotes. [8.3]

18. State the center, vertices, and equations of the asymptotes for $\frac{(x - 4)^2}{16} - \frac{(y + 1)^2}{4} = 1$. [8.3]

State the center, vertices, co-vertices, foci, and eccentricity for each hyperbola. [8.3]

19. $16(y + 9)^2 - 9(x - 15)^2 = 144$

20. $y^2 - 4x^2 - 2y - 16x = 31$

Write a standard form equation for each hyperbola with the given characteristics. [8.3]

21. foci at $(0, \pm 4)$; $e = \frac{4}{3}$ $\frac{x^2}{9} - \frac{y^2}{7} = 1$

22. center at $(-1, 3)$, vertex at $(5, 3)$, an asymptote of $y = 2x + 5$ $\frac{(x + 1)^2}{36} - \frac{(y - 3)^2}{144} = 1$

Write a general quadratic form equation for each hyperbola with the given characteristics. [8.3]

23. vertices at $(6, 1)$ and $(-4, 1)$; $e = \frac{\sqrt{34}}{5}$

24. vertices at $(5, -3)$ and $(5, 5)$; $e = \frac{\sqrt{5}}{2}$

25. Frank is 3000 ft due west of Rodger. Rodger hears a dynamite blast 1 sec before Frank hears it. If the sound traveled at 1126 ft/sec, model the possible locations of the blast with the equation of a hyperbola in which the locations of Frank and Rodger are foci on the horizontal axis. [8.3]

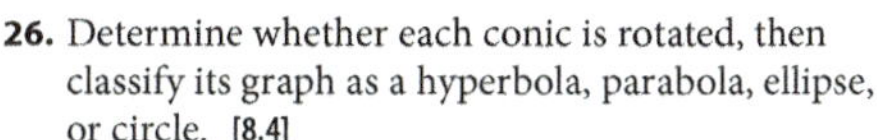

26. Determine whether each conic is rotated, then classify its graph as a hyperbola, parabola, ellipse, or circle. [8.4]

  a. $3x^2 - 2\sqrt{3}xy + y^2 + 8\sqrt{3}x + 8y - 8x - 8\sqrt{3}y + 96 = 0$ yes; parabola

  b. $x^2 + 4y^2 - 4x + 40y + 96 = 0$ no; ellipse

  c. $7x^2 + 2xy - 3y^2 + 5x + 11y - 7 = 0$ yes; hyperbola

18. $C(4, -1)$; V: $(8, -1)$, $(0, -1)$; $y = \frac{1}{2}x - 3$, $y = -\frac{1}{2}$

PRESENTATION

Students should know the standard and general forms of each conic section and should be able to identify the type of conic and its key characteristics given its equation. They will also be expected to write the equation of a conic given its key characteristics. You may want to encourage the students to use the Quick Reference and/or the Internet keyword search *conic study sheet* for reviewing equations for each conic.

Rotated, polar, and parametric equations present a wealth of concepts that can be challenging. Consider limiting their review to just those concepts that will be assessed.

---

5. $y = -\frac{1}{4}x^2 + 3x - 5$

$ = -\frac{1}{4}(x^2 - 12x) - 5$

$ = -\frac{1}{4}(x^2 - 12x + 6^2) - 5 + \frac{1}{4}(36)$

$y = -\frac{1}{4}(x - 6)^2 + 4$

$p = \dfrac{1}{4\left(-\frac{1}{4}\right)} = -1$

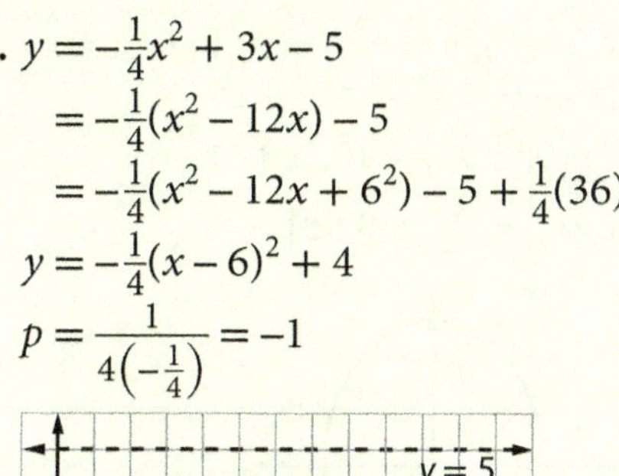

6. $8x + 20 = y^2 + 4y$

$8x + 20 + 4 = y^2 + 4y + 4$

$ 8x = (y + 2)^2 - 24$

$ x = \frac{1}{8}(y + 2)^2 - 3$

$p = \dfrac{1}{4\left(\frac{1}{8}\right)} = 2$

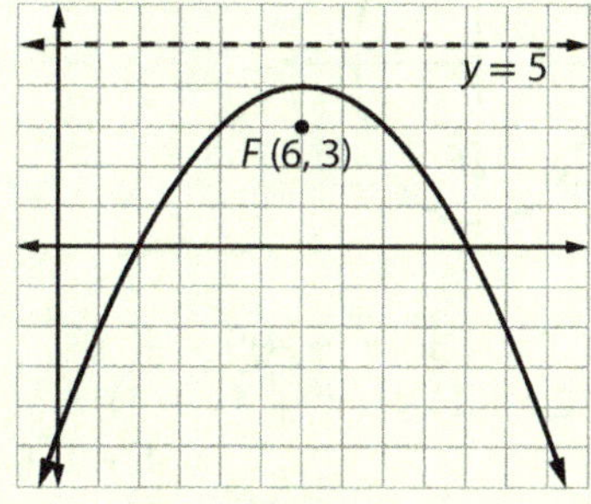

7. $y = -\frac{1}{2}x^2$; $V(0, 0)$

$p = \dfrac{1}{4\left(-\frac{1}{2}\right)} = -\frac{1}{2}$; $F\left(0, -\frac{1}{2}\right)$

$FP = \sqrt{(0 - 2)^2 + \left(-\frac{1}{2} - (-2)\right)^2}$

$ = \sqrt{4 + \frac{9}{4}} = \sqrt{\frac{25}{4}} = \frac{5}{2}$

$A\left(0, -\frac{1}{2} + \frac{5}{2}\right) = (0, 2)$

$m_{PA} = \dfrac{-2 - 2}{2 - 0} = -2$

$y = -2x + 2$

8. passing through $(52, 0)$ with $V(26, 25)$

$y = a(x - 26)^2 + 25$

$0 = a(52 - 26)^2 + 25$

$-25 = 676a$; $a = -\frac{25}{676}$

$y = -\frac{25}{676}(x - 26)^2 + 25$

9. $C\,(0,0);\ a = 4,\ b = 3;\ V:\ (0,\pm 4)$
$c^2 = 4^2 - 3^2 = 7;\ c = \sqrt{7}$
$F:\ \left(0,\ \pm\sqrt{7}\right) \approx (0,\ \pm 2.6)$
$e = \dfrac{c}{a} = \dfrac{\sqrt{7}}{4} \approx 0.661$

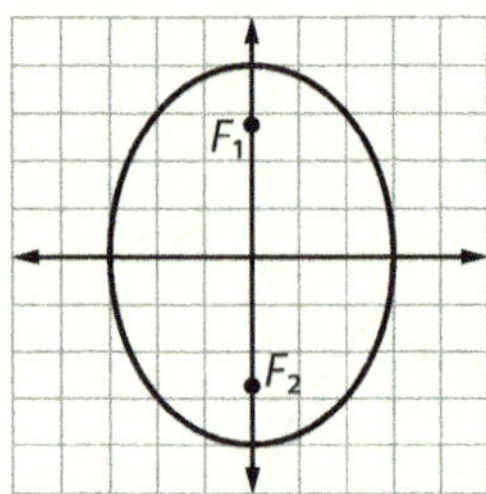

10. $\quad (x^2 - 2x) + (y^2 + 4y) = 4$
$(x^2 - 2x + 1) + (y^2 + 4y + 4)$
$\qquad\qquad\qquad = 4 + 1 + 4$
$(x - 1)^2 + (y + 2)^2 = 9$
$\dfrac{(x - 1)^2}{9} + \dfrac{(y + 2)^2}{9} = 1;\ \text{circle}$
$c^2 = 9 - 9 = 0;\ c = 0$
$F = C\,(1, -2);\ e = \dfrac{c}{a} = \dfrac{0}{3} = 0$

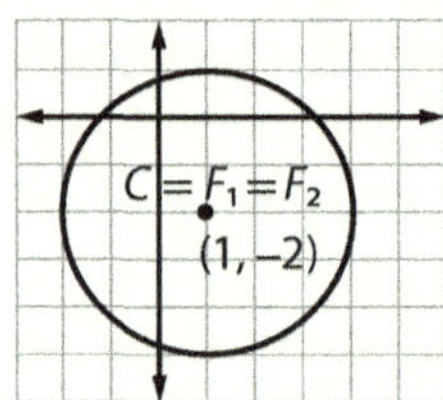

11. $(x^2 + 6x) + 4(y^2 - 2y) = -9$
$(x^2 + 6x + 9) + 4(y^2 - 2y + 1)$
$\qquad\qquad\qquad = -9 + 9 + 1(4)$
$(x + 3)^2 + 4(y - 1)^2 = 4$
$\dfrac{(x + 3)^2}{4} + \dfrac{(y - 1)^2}{1} = 1;\ C\,(-3, 1)$
$c^2 = 4 - 1 = 3;\ c = \sqrt{3}$
$F:\ \left(-3 \pm \sqrt{3},\ 1\right) \approx (-4.73, 1),\ (-1.27, 1)$
$e = \dfrac{c}{a} = \dfrac{\sqrt{3}}{2} \approx 0.866$

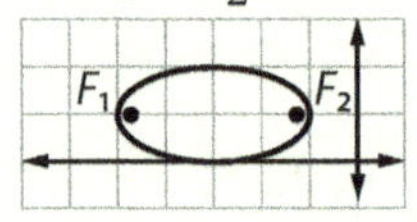

12. horizontal MA; $C\,(2, -4)$
$a = \dfrac{7 - (-3)}{2} = 5;\ a^2 = 25$
$b = \dfrac{-1 - (-7)}{2} = 3;\ b^2 = 9$
$\dfrac{(x - 2)^2}{25} + \dfrac{(y + 4)^2}{9} = 1$

13. vertical MA; $C\,(2, -3)$
$b = 7 - (-2) = 5;\ b^2 = 25$
$c = \dfrac{9 - (-15)}{2} = 12;\ c^2 = 144$
$a^2 = c^2 + b^2 = 144 + 25 = 169$
$\dfrac{(x - 2)^2}{25} + \dfrac{(y + 3)^2}{169} = 1$

State the rotation formulas that will rotate the equation to the uv-plane without a uv-term. [8.4]

27. $x^2 + 8xy + y^2 + 9 = 0 \quad x = \dfrac{\sqrt{2}}{2}u - \dfrac{\sqrt{2}}{2}v,\ y = \dfrac{\sqrt{2}}{2}u + \dfrac{\sqrt{2}}{2}v$

28. $13x^2 + 6\sqrt{3}xy + 7y^2 - 15x + 8y - 221 = 0$

Find the angle of rotation of the xy-plane that causes the u- or v-axis to be parallel to the conic's axis. Then write the standard form equation in terms of u and v and sketch a graph of the conic and any foci. [8.4]

29. $5x^2 + 26xy + 5y^2 - 72 = 0 \quad \theta = \dfrac{\pi}{4};\ \dfrac{u^2}{4} - \dfrac{v^2}{9} = 1$

30. $3x^2 - 4\sqrt{3}xy + 7y^2 - 9 = 0 \quad \theta = \dfrac{\pi}{6};\ \dfrac{u^2}{9} + v^2 = 1$

31. $9x^2 + 12xy + 4y^2 + 8\sqrt{13}x + \sqrt{13}y + 13 = 0$
$\theta \approx 33.7°;\ v = (u + 1)^2$

Write a general form equation for each conic in terms of x and y, given its angle of rotation and equation in terms of u and v. Then use technology to graph the conic. [8.4]

32. $v^2 - 4u = 0;\ \theta = 45°$ **33.** $u^2 + 4v^2 = 25;\ \theta = 30°$

34. Graph $x = t + 1$ and $y = t^2 - 2;\ t \in [-2, 2]$ and indicate the orientation. [8.5]

35. Given $x = 1 + t^2$ and $y = t + 2;\ t \in [-2, 2]$, eliminate the parameter and restrict the domain to find the rectangular equation generating the same curve. Use graphing technology to verify your answer. [8.5]

Eliminate the parameter for each conic and write the resulting equation in standard form. [8.5]

36. $x = 10 \sin\theta$ and $y = 10 \cos\theta \quad x^2 + y^2 = 100$

37. $x = 8 \sin\theta + 5$ and $y = 3 \cos\theta - 2 \quad \dfrac{(x - 5)^2}{64} + \dfrac{(y + 2)^2}{9} = 1$

38. Write a set of parametric equations for $y = 2x^2 - 1$ where $x \in [-2, 2]$ given $t = x - 2$.

Write a set of parametric equations for each described conic. [8.5]

39. an ellipse with vertices at $(-3, 1)$ and $(9, 1)$ and a minor axis length of 4 $x = 6\cos\theta + 3,\ y = 2\sin\theta + 1$

40. a hyperbola with vertices at $(1, 10)$ and $(1, -6)$ and foci at $(1, 12)$ and $(1, -8)$ $x = 1 + 6\tan\theta,\ y = 2 + 8\sec\theta$

28. $x = \dfrac{\sqrt{3}}{2}u - \dfrac{1}{2}v,\ y = \dfrac{1}{2}u + \dfrac{\sqrt{3}}{2}v$

32. $x^2 - 2xy + y^2 - 4\sqrt{2}x - 4\sqrt{2}y = 0$

33. $7x^2 - 6\sqrt{3}xy + 13y^2 - 100 = 0$

35. $x = y^2 - 4y + 5;\ D = [1, 5]$

38. $x = t + 2,\ y = 2t^2 + 8t + 7;\ t \in [-4, 0]$

---

41. A quarterback throws a football from a height of 6.3 ft with a 20° angle of elevation and speed of 60 ft/sec. [8.5] $x = (60 \cos 20°)t,$
$y = -16t^2 + (60 \sin 20°)t + 6.3$
  a. Write a set of parametric equations modeling the ball's trajectory.
  b. Find the height of the ball 1.2 sec after release. ≈ 7.9 ft
  c. How far has the ball traveled down the field after 1.4 sec? ≈ 78.9 ft or 26.3 yd

Determine the eccentricity, type of conic, and directrix for each polar equation. [8.6]

42. $r = \dfrac{6}{2 + 3\cos\theta}$ **43.** $r = \dfrac{5}{1 - 0.2\sin\theta}$

Write a polar equation for each conic section. Then use technology to graph the conic and its directrix. [8.6]

44. $e = 1$; directrix: $x = -2$ **45.** $e = \dfrac{2}{3}$; directrix: $y = -3$

Write a polar equation for each conic section with a focus at the pole. Then use technology to graph the conic and its directrix. [8.6]

46. $e = 2;\ V_1\,(-3, 0),\ V_2\,(-1, 0) \quad r = \dfrac{3}{1 - 2\cos\theta};\ x = -1.5$

47. $e = 1;\ V\,(0, 1) \quad r = \dfrac{2}{1 + \sin\theta};\ y = 2$

Write a rectangular equation for each conic. [8.6]

48. $r = \dfrac{6}{1 - \sin\theta}$ **49.** $r = \dfrac{3}{1 + 2\cos\theta}$

50. Write a polar equation modeling the orbit of Venus, given its eccentricity of 0.007 and semimajor axis length of 0.723 AU. Then find its perihelion and aphelion distances. [8.6]
$r = \dfrac{0.723(1 - 0.007^2)}{1 + 0.007\cos\theta};\ \approx 0.718$ AU; ≈ 0.728 AU

42. $e = 1.5$; hyperbola; $x = 2$

43. $e = 0.2$; ellipse; $y = -25$

44. $r = \dfrac{2}{1 - \cos\theta}$

45. $r = \dfrac{2}{1 - \frac{2}{3}\sin\theta}$

48. $y = \dfrac{1}{12}x^2 - 3$

49. $\dfrac{(x - 2)^2}{1} - \dfrac{y^2}{3} = 1$

---

14. vertical MA; $a = 4;\ a^2 = 16$
$e = 0.75 = \dfrac{c}{4};\ c = 3$
$b^2 = a^2 - c^2 = 16 - 9 = 7$
$\dfrac{x^2}{7} + \dfrac{y^2}{16} = 1$

15. vertical MA; $C\left(3, \dfrac{3 + (-5)}{2}\right) = (3, -1)$
$b = 3;\ b^2 = 9$
$c = \dfrac{3 - (-5)}{2} = 4;\ c^2 = 16$
$a^2 = b^2 + c^2 = 16 + 9 = 25$
$\dfrac{(x - 3)^2}{9} + \dfrac{(y + 1)^2}{25} = 1$

16. $V_N\,(0, 2);\ V_S\,(0, -16);\ 2a = 18;\ a = 9$
$C\left(0, \dfrac{2 + (-16)}{2}\right) = (0, -7)$
$c = 7;\ c^2 = 49$
$b^2 = a^2 - c^2 = 81 - 49 = 32$
$\dfrac{x^2}{32} + \dfrac{(y + 7)^2}{81} = 1$
$y = \pm 9\sqrt{\left(1 - \dfrac{x^2}{32}\right)} - 7$

17. horizontal TA; $C(0, 0)$
$a = 3$; $V: (\pm 3, 0)$; $b = 2$
$y = \pm \frac{2}{3}x$

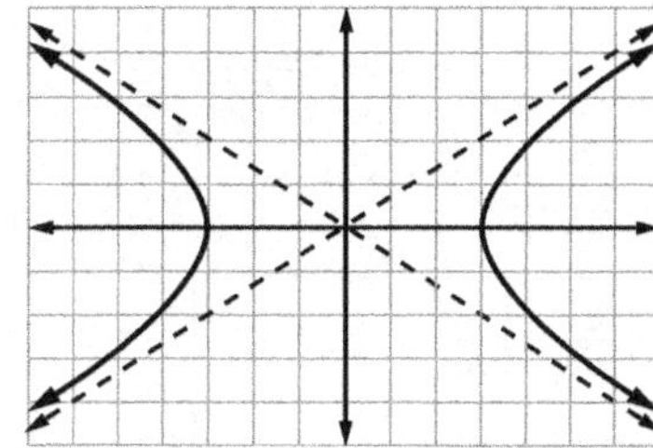

18. horizontal TA; $C(4, -1)$; $a = 4$, $b = 2$
$V: (8, -1), (0, -1)$
$y = \pm \frac{1}{2}(x - 4) - 1$
$y = \frac{1}{2}x - 3$ and $y = -\frac{1}{2}x + 1$

19. $\frac{(y + 9)^2}{9} - \frac{(x - 15)^2}{16} = 1$
vertical TA; $C(15, -9)$, $a = 3$, $b = 4$
$V: (15, -6), (15, -12)$
$cV: (19, -9), (11, -9)$
$c^2 = 9 + 16 = 25$; $c = 5$
$F: (15, -14), (15, -4)$; $e = \frac{5}{3} \approx 1.67$

20. $y^2 - 2y - 4x^2 - 16x = 31$
$(y^2 - 2y + 1) - 4(x^2 + 4x + 4)$
$\qquad = 31 + 1 - 16$
$(y - 1)^2 - 4(x + 2)^2 = 16$
$\frac{(y - 1)^2}{16} - \frac{(x + 2)^2}{4} = 1$
vertical TA; $C(-2, 1)$; $a = 4$, $b = 2$
$V: (-2, 5), (-2, -3)$; $cV: (0, 1), (-4, 1)$
$c^2 = 16 + 4$; $c = 2\sqrt{5}$
$F: (-2, 1 \pm 2\sqrt{5})$
$\quad \approx (-2, 5.47), (-2, -3.47)$
$e = \frac{\sqrt{5}}{2} \approx 1.12$

21. vertical TA; $C(0, 0)$; $c = 4$
$a = \frac{c}{e} = \frac{4}{\frac{4}{3}} = 3$; $b^2 = 4^2 - 3^2 = 7$
$\frac{x^2}{9} - \frac{y^2}{7} = 1$

22. horizontal TA; $a = 6$
$m = \frac{b}{a} = \frac{b}{6} = 2$; $b = 12$
$\frac{(x + 1)^2}{36} - \frac{(y - 3)^2}{144} = 1$

23. horizontal TA
$C\left(\frac{6 + (-4)}{2}, 1\right) = (1, 1)$
$a = 6 - 1 = 5$; $e = \frac{c}{5} = \frac{\sqrt{34}}{5}$; $c = \sqrt{34}$
$b^2 = c^2 - a^2 = 34 - 25 = 9$
$\frac{(x - 1)^2}{25} - \frac{(y - 1)^2}{9} = 1$
$9(x - 1)^2 - 25(y - 1)^2 = 225$
$9(x^2 - 2x + 1) - 25(y^2 - 2y + 1)$
$\qquad\qquad\qquad = 225$
$9x^2 - 25y^2 - 18x + 50y - 241 = 0$

24. vertical TA; $C\left(5, \frac{-3 + 5}{2}\right) = (5, 1)$
$a = 5 - 1 = 4$; $e = \frac{c}{4} = \frac{\sqrt{5}}{2}$; $c = 2\sqrt{5}$
$b^2 = c^2 - a^2 = 20 - 16 = 4$
$\frac{(y - 1)^2}{16} - \frac{(x - 5)^2}{4} = 1$
$(y - 1)^2 - 4(x - 5)^2 = 16$
$y^2 - 2y + 1 - 4(x^2 - 10x + 25) = 16$
$-4x^2 + y^2 + 40x - 2y - 115 = 0$

25. $C(0, 0)$; $c = 1500$
$\Delta d = 1126 \frac{\text{ft}}{\text{sec}}(1 \text{ sec}) = 1126 \text{ ft} = 2a$
$a = 563$; $a^2 = 316{,}969$
$b^2 = 1500^2 - 563^2 = 1{,}933{,}031$
$\frac{x^2}{316{,}969} - \frac{y^2}{1{,}933{,}031} = 1$

26. If $B \neq 0$, then yes. If $B = 0$, then no.
Use $B^2 - 4AC$.

26a. $\left(2\sqrt{3}\right)^2 - 4(3)(1) = 0$; parabola

26b. $(0)^2 - 4(1)(4) = -16 < 0$; ellipse

26c. $(2)^2 - 4(7)(-3) = 88 > 0$; hyperbola

27. $A = 1$; $B = 8$; $C = 1$
$\cot 2\theta = \frac{1 - 1}{8} = 0$; $2\theta = \frac{\pi}{2}$; $\theta = \frac{\pi}{4}$
$x = u \cos \frac{\pi}{4} - v \sin \frac{\pi}{4} = \frac{\sqrt{2}}{2}u - \frac{\sqrt{2}}{2}v$
$y = u \sin \frac{\pi}{4} + v \cos \frac{\pi}{4} = \frac{\sqrt{2}}{2}u + \frac{\sqrt{2}}{2}v$

28. $A = 13$; $B = 6\sqrt{3}$; $C = 7$
$\cot 2\theta = \frac{13 - 7}{6\sqrt{3}} = \frac{1}{\sqrt{3}}$; $2\theta = \frac{\pi}{3}$; $\theta = \frac{\pi}{6}$
$x = u \cos \frac{\pi}{6} - v \sin \frac{\pi}{6} = \frac{\sqrt{3}}{2}u - \frac{1}{2}v$
$y = u \sin \frac{\pi}{6} + v \cos \frac{\pi}{6} = \frac{1}{2}u + \frac{\sqrt{3}}{2}v$

29. $\cot 2\theta = \frac{0 - 0}{1} = 0$; $2\theta = \frac{\pi}{2}$; $\theta = \frac{\pi}{4}$
$x = \frac{\sqrt{2}u - \sqrt{2}v}{2}$; $y = \frac{\sqrt{2}u + \sqrt{2}v}{2}$
$5\left(\frac{u^2 - 2uv + v^2}{2}\right) + 26\left(\frac{u^2 - v^2}{2}\right)$
$\qquad + 5\left(\frac{u^2 + 2uv + v^2}{2}\right) - 72 = 0$
$\left(\frac{5}{2} + 13 + \frac{5}{2}\right)u^2 + (-5uv + 5uv)$
$\qquad\qquad + \left(\frac{5}{2} - 13 + \frac{5}{2}\right)v^2 = 72$
$18u^2 - 8v^2 = 72$
$\frac{u^2}{4} - \frac{v^2}{9} = 1$

$V': (\pm 2, 0)$; $F': \left(\pm\sqrt{13}, 0\right)$

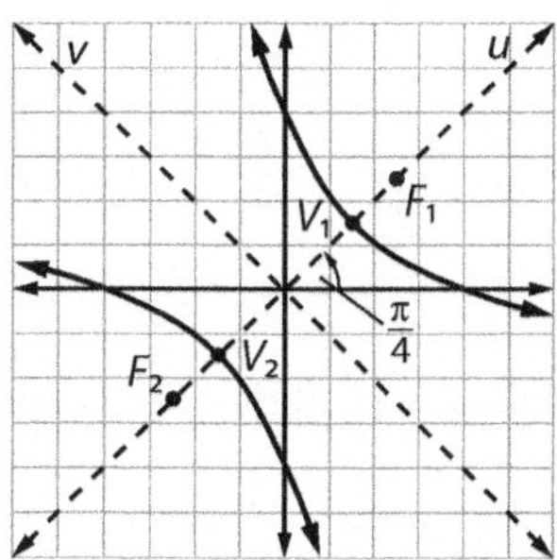

30. $\cot 2\theta = \frac{3 - 1}{2\sqrt{3}} = \frac{1}{\sqrt{3}}$; $2\theta = \frac{\pi}{3}$; $\theta = \frac{\pi}{6}$
$x = \frac{\sqrt{3}u - v}{2}$; $y = \frac{u + \sqrt{3}v}{2}$
$3\left(\frac{3u^2 - 2\sqrt{3}uv + v^2}{4}\right)$
$\qquad - 4\sqrt{3}\left(\frac{\sqrt{3}u^2 + 2uv - \sqrt{3}v^2}{4}\right)$
$\qquad + 7\left(\frac{u^2 + 2\sqrt{3}uv + 3v^2}{4}\right) = 9$
$\left(\frac{9}{4} - 3 + \frac{7}{4}\right)u^2$
$\qquad - \left(-\frac{3\sqrt{3}}{2} - 2\sqrt{3} + \frac{7\sqrt{3}}{2}\right)uv$
$\qquad + \left(\frac{3}{4} + 3 + \frac{21}{4}\right)v^2 = 9$
$u^2 + 9v^2 = 9$
$\frac{u^2}{9} + v^2 = 1$

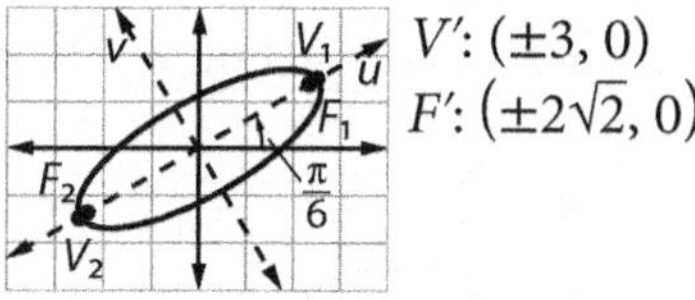

$V': (\pm 3, 0)$
$F': (\pm 2\sqrt{2}, 0)$

31. $\cot 2\theta = \frac{5}{12}$; $\cos 2\theta = \frac{5}{13}$; $\theta \approx 33.7°$
$\cos \theta = \sqrt{\frac{1 + \left(\frac{5}{13}\right)}{2}} = \frac{3}{\sqrt{13}}$
$\sin \theta = \sqrt{\frac{1 - \left(\frac{5}{13}\right)}{2}} = \frac{2}{\sqrt{13}}$
$x = \frac{3u - 2v}{\sqrt{13}}$; $y = \frac{2u + 3v}{\sqrt{13}}$
$9\left(\frac{9u^2 - 12uv + 4v^2}{13}\right)$
$\qquad + 12\left(\frac{6u^2 + 5uv - 6v^2}{13}\right)$
$\qquad + 4\left(\frac{4u^2 + 12uv + 9v^2}{13}\right)$
$\qquad + 8(3u - 2v) + (2u + 3v) + 13 = 0$
$13u^2 + 26u - 13v + 13 = 0$
$13u^2 + 26u + 13 = 13v$
$v = (u + 1)^2$
$V'(-1, 0)$
$F'\left(-1, \frac{1}{4}\right)$

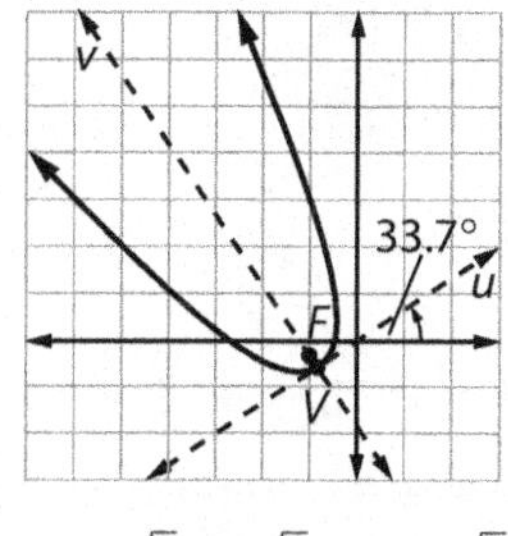

32. $u = \frac{\sqrt{2}x + \sqrt{2}y}{2}$; $v = \frac{\sqrt{2}y - \sqrt{2}x}{2}$
$\frac{y^2 - 2xy + x^2}{2} - 4\left(\frac{\sqrt{2}x + \sqrt{2}y}{2}\right) = 0$
$\frac{1}{2}y^2 - xy + \frac{1}{2}x^2 - 2\sqrt{2}x - 2\sqrt{2}y = 0$
$x^2 - 2xy + y^2 - 4\sqrt{2}x - 4\sqrt{2}y = 0$
$y^2 - (2x + 4\sqrt{2})y + (x^2 - 4\sqrt{2}x) = 0$
$y = \frac{2x + 4\sqrt{2} \pm \sqrt{(2x + 4\sqrt{2})^2 - 4(x^2 - 4\sqrt{2}x)}}{2}$

continued on next page

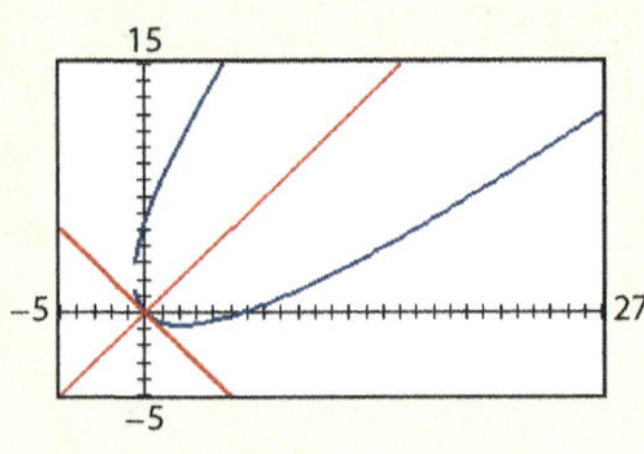

33. $u = \dfrac{\sqrt{3}x + y}{2}$; $v = \dfrac{\sqrt{3}y - x}{2}$

$$\dfrac{3x^2 + 2\sqrt{3}xy + y^2}{4} + 4\left(\dfrac{3y^2 - 2\sqrt{3}xy + x^2}{4}\right)$$
$$= 25$$
$$\dfrac{7x^2 - 6\sqrt{3}xy + 13y^2}{4} = 25$$
$$7x^2 - 6\sqrt{3}xy + 13y^2 - 100 = 0$$
$$13y^2 - (6\sqrt{3}\,x)y + (7x^2 - 100) = 0$$
$$y = \dfrac{6\sqrt{3}x \pm \sqrt{(6\sqrt{3}x)^2 - 4(13)(7x^2 - 100)}}{2(13)}$$

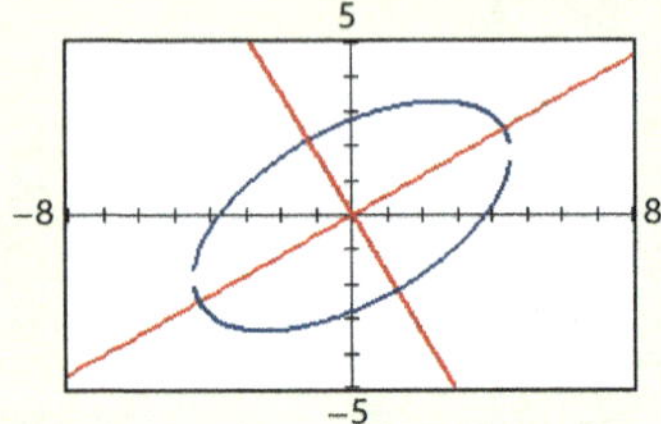

34.

t	-2	-1	0	1	2
x	-1	0	1	2	3
y	2	-1	-2	-1	2

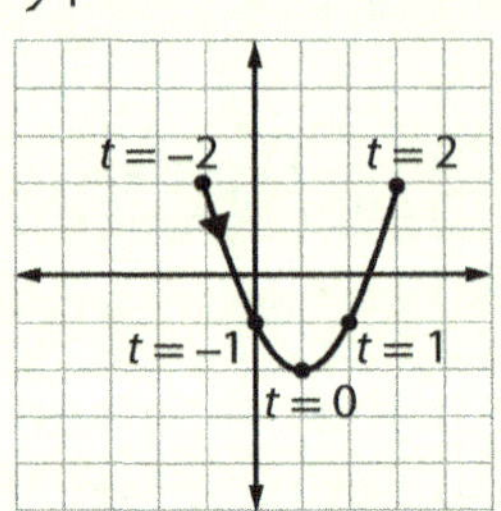

35. $t = y - 2$
$x = (y - 2)^2 + 1 = y^2 - 4y + 5$
$x = 1 + t^2$ and $t \in [-2, 2]$
$t^2 = x - 1$ and $|t| \le 2$
$0 \le x - 1 \le 4$
$x \ge 1$ and $x \le 5$
$\therefore D = [1, 5]$

36. $\sin \theta = \dfrac{x}{10}$; $\cos \theta = \dfrac{y}{10}$
$$\left(\dfrac{x}{10}\right)^2 + \left(\dfrac{y}{10}\right)^2 = 1$$
$$\dfrac{x^2}{100} + \dfrac{y^2}{100} = 1; \ x^2 + y^2 = 100$$

37. $\sin \theta = \dfrac{x - 5}{8}$; $\cos \theta = \dfrac{y + 2}{3}$
$$\left(\dfrac{x - 5}{8}\right)^2 + \left(\dfrac{y + 2}{3}\right)^2 = 1$$
$$\dfrac{(x - 5)^2}{64} + \dfrac{(y + 2)^2}{9} = 1$$

38. $t \in [-4, 0]$; $x = t + 2$
$y = 2(t + 2)^2 - 1 = 2t^2 + 8t + 7$

39. horizontal TA; $C\left(\dfrac{-3 + 9}{2}, 1\right) = (3, 1)$
$a = 9 - 3 = 6$; $2b = 4$; $b = 2$
$$\left(\dfrac{x - 3}{6}\right)^2 + \left(\dfrac{y - 1}{2}\right)^2 = 1$$
$\cos \theta = \dfrac{x - 3}{6}$; $\sin \theta = \dfrac{y - 1}{2}$
$x = 6 \cos \theta + 3$; $y = 2 \sin \theta + 1$

40. TA on $x = 1$; $C\left(1, \dfrac{-6 + 10}{2}\right) = (1, 2)$
$a = 10 - 2 = 8$; $c = 12 - 2 = 10$
$b^2 = c^2 - a^2 = 100 - 64 = 36$; $b = 6$
$$\left(\dfrac{y - 2}{8}\right)^2 - \left(\dfrac{x - 1}{6}\right)^2 = 1$$
Using $\sec^2 \theta - \tan^2 \theta = 1$,
let $\sec \theta = \dfrac{y - 2}{8}$ and $\tan \theta = \dfrac{x - 1}{6}$.
$x = 1 + 6 \tan \theta$; $y = 2 + 8 \sec \theta$

41b. $y = -16(1.2)^2$
$$\qquad + (60 \sin 20°)(1.2) + 6.3$$
$$\approx 7.9 \text{ ft}$$

41c. $x = (60 \cos 20°)(1.4)$
$$\approx 78.9 \text{ ft or } 26.3 \text{ yd}$$

42. $r = \dfrac{6 \div 2}{(2 + 3 \cos \theta) \div 2} = \dfrac{3}{1 + 1.5 \cos \theta}$
$e = 1.5$; hyperbola
$(1.5)d = 3$; $d = 2$; directrix: $x = 2$

43. $e = 0.2$; ellipse
$(0.2)d = 5$; $d = 25$; directrix: $y = -25$

44. $r = \dfrac{2(1)}{1 - (1) \cos \theta} = \dfrac{2}{1 - \cos \theta}$

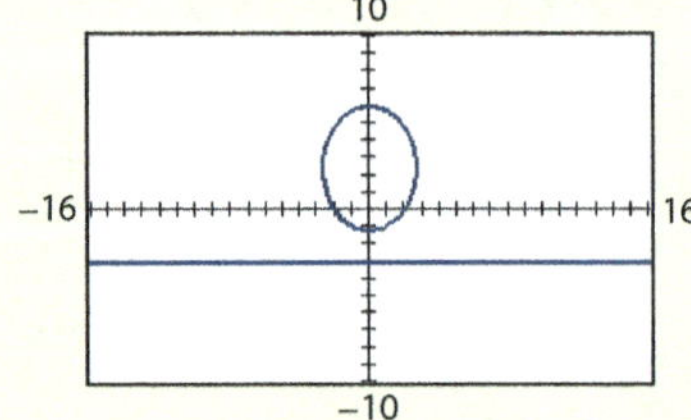

45. $r = \dfrac{3\left(\frac{2}{3}\right)}{1 - \left(\frac{2}{3}\right) \sin \theta} = \dfrac{2}{1 - \frac{2}{3} \sin \theta}$

46. hyperbola with $C\left(\dfrac{-3 + (-1)}{2}, 0\right)$
$$= (-2, 0)$$
$\therefore r = \dfrac{ed}{1 - e \cos \theta}$; using $(1, \pi)$:
$1 = \dfrac{(2)d}{1 - (2) \cos \pi}$; $1 = \dfrac{2d}{3}$; $d = \dfrac{3}{2}$
$r = \dfrac{3}{1 - 2 \cos \theta}$; directrix: $x = -1.5$

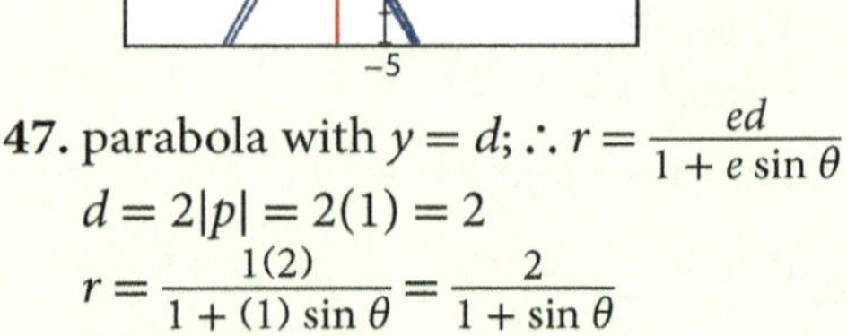

47. parabola with $y = d$; $\therefore r = \dfrac{ed}{1 + e \sin \theta}$
$d = 2|p| = 2(1) = 2$
$r = \dfrac{1(2)}{1 + (1) \sin \theta} = \dfrac{2}{1 + \sin \theta}$
directrix: $y = 2$

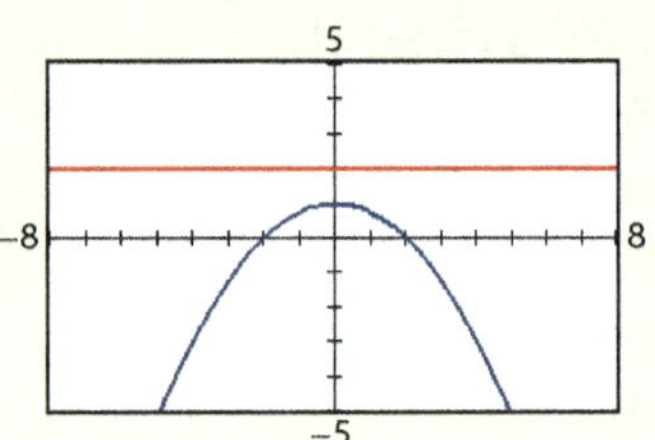

48. $e = 1$ (parabola)
$d = 6$; directrix: $y = -6$
$V: r\left(\dfrac{3\pi}{2}\right) = 3$
$$\left(3 \cos \dfrac{3\pi}{2}, 3 \sin \dfrac{3\pi}{2}\right) = (0, -3)$$
$p = 3$; $a = \dfrac{1}{4(3)} = \dfrac{1}{12}$
$y = \dfrac{1}{12}(x - 0)^2 - 3$
$y = \dfrac{1}{12}x^2 - 3$

49. $e = 2$ (hyperbola)
$V_1: r(0) = 1$; $(1 \cos 0, 1 \sin 0) = (1, 0)$
$V_2: r(\pi) = -3$
$$(-3 \cos \pi, -3 \sin \pi) = (3, 0)$$
$C(2, 0)$; $c = 2$
$a = \dfrac{c}{e} = \dfrac{2}{2} = 1$; $b^2 = 2^2 - 1^2 = 3$
$$\dfrac{(x - 2)^2}{1} - \dfrac{y^2}{3} = 1$$

50. $r = \dfrac{0.723(1 - 0.007^2)}{1 + 0.007 \cos \theta}$
$r(0) \approx \dfrac{0.723(1 - 0.007^2)}{1 + 0.007 \cos (0)} \approx 0.718 \text{ AU}$
$r(\pi) \approx \dfrac{0.723(1 - 0.007^2)}{1 + 0.007 \cos (\pi)} \approx 0.728 \text{ AU}$

9 Sequences and Series

Overview

Whether estimating population growth, determining future values related to compound interest, or studying patterns in music, students will benefit from and enjoy learning about sequences and series. Series such as the one in the Binomial Theorem will open new avenues in the study of combinations and probabilities.

Flash

The vertical distances traveled by a bouncing ball can be represented by a geometric sequence. Rebounding heights of a bouncing ball are explored in Section 9.3.

Chapter Objectives

1. To determine recursive and explicit formulas for arithmetic, geometric, and other sequences
2. To find the nth term and sums of arithmetic, geometric, and other sequences
3. To prove statements using mathematical induction
4. To use the Binomial Theorem to expand powers of a binomial
5. To approximate values using summations
6. To model and solve real-world problems using sequences and series
7. To articulate a biblical worldview of automation and artificial intelligence

BIBLICAL PERSPECTIVE OF MATHEMATICS

From Siri® to chatbots to self-driving cars, artificial intelligence (AI) seems to be the unstoppable wave of the future. Bill Gates and Stephen Hawking have warned of its potential dangers. How does a biblical worldview inform our opinions of AI?

HISTORICAL CONNECTION

While Isaac Newton is credited with the Binomial Theorem and Blaise Pascal with his eponymous triangle, the first written expansion of a binomial can be traced to Euclid, and charts equivalent to Pascal's triangle can be found in a 1303 Chinese publication.

DATA ANALYSIS

Leonhard Euler's solution to a popular puzzle concerning bridges crossed while taking a stroll resulted in the birth of graph theory—the mathematics at the heart of navigation applications that provide optimal routing in near real time.

Suggested Teaching Schedule

DAY	
1–2	9.1 TC: Graphing Sequences
3–4	BPM: Computer Proofs and Artificial Intelligence 9.2
5–6	Quiz 9A (9.1–9.2) 9.3 HC: Binomial Expansion and Pascal's Triangle

DAY	
7	9.4
8–9	Quiz 9B (9.3–9.4) 9.5
10	Quiz 9C (9.5) DA: Graph Theory and Networks
11	Chapter 9 Review
12	Chapter 9 Test

9.1 Introduction to Sequences and Series

Objectives

1. To represent sequences numerically, graphically, and algebraically with recursive and explicit formulas

2. To evaluate expressions containing factorial notation

3. To determine whether a sequence is divergent or convergent

4. To use sigma notation to find the sum of terms in a sequence

5. To use sequences and series to model and solve real-world problems

Vocabulary

convergent
divergent
explicit formula
factorial
finite sequence
general term
infinite sequence
infinite series
limit of the sequence
nth partial sum
recursive formula
sequence
series
summation (sigma) notation
term

9.1 Introduction to Sequences and Series

The Fibonacci sequence can be used to model a population of rabbits.

After completing this section, you will be able to

- represent sequences numerically, algebraically, and graphically.
- evaluate expressions containing factorial notation.
- determine whether a sequence is divergent or convergent.
- use sigma notation to find the sum of terms in a sequence.
- model and solve real-world problems using sequences.

The golden ratio can be represented using the Fibonacci sequence, $\lim_{n \to \infty} \frac{a_n}{a_{n-1}} = \phi$ (phi). This number can frequently be found in God's creation.

KEYWORD SEARCH

Fibonacci sequence

In his 1202 book *Liber Abaci*, Leonardo Fibonacci theorized regarding the maximum number of pairs of rabbits resulting from a newborn pair of rabbits after one year. He assumed that each pair produces a pair of rabbits after two months and then another pair each month following. A simple chart revealed that the maximum number of fully grown pairs of rabbits at each monthly interval is given by the *sequence* 1, 1, 2, 3, 5, 8, 13, 21, ... Each value (such as 8) is associated with its position in the sequence (month 6). Unlike in a set, the order of elements is important in a sequence.

DEFINITIONS

An **infinite sequence** is a function whose domain is the natural numbers, $\mathbb{N}$, and whose range contains the function values $a_1, a_2, a_3, \ldots, a_n, \ldots$, which are the *terms* of the sequence. The nth term is called the *general term*. The domain of a **finite sequence** of n terms consists of the first n natural numbers.

Since functions are sets of ordered pairs, Fibonacci's sequence above is a shortened notation for $\{(1, 1), (2, 1), (3, 2), (4, 3), (5, 5), (6, 8), (7, 13), (8, 21), \ldots\}$, and $f(6) = a_6 = 8$. Set notation is rarely used since the simple list is more convenient.

Example 1 Writing the Terms of a Sequence

Find the first 4 terms and the tenth term of each sequence.

a. $a_n = 5n + 3$ **b.** $a_n = n^2 + 1$ **c.** $a_n = \frac{(-1)^{n-1}}{2n}$

Answer

a. $a_1 = 5(1) + 3 = 8$
$a_2 = 5(2) + 3 = 13$
$a_3 = 5(3) + 3 = 18$
$a_4 = 5(4) + 3 = 23$
$\vdots$
$a_{10} = 5(10) + 3 = 53$

b. $a_1 = (1)^2 + 1 = 2$
$a_2 = (2)^2 + 1 = 5$
$a_3 = (3)^2 + 1 = 10$
$a_4 = (4)^2 + 1 = 17$
$\vdots$
$a_{10} = (10)^2 + 1 = 101$

c. $a_1 = \frac{(-1)^{1-1}}{2(1)} = \frac{1}{2}$
$a_2 = \frac{(-1)^{2-1}}{2(2)} = -\frac{1}{4}$
$a_3 = \frac{(-1)^{3-1}}{2(3)} = \frac{1}{6}$
$a_4 = \frac{(-1)^{4-1}}{2(4)} = -\frac{1}{8}$
$\vdots$
$a_{10} = \frac{(-1)^{10-1}}{2(10)} = -\frac{1}{20}$

SKILL ✓ EXERCISE 5

PRESENTATION

Lesson Opener

Use end behavior to determine each limit.

1. $\lim_{x \to \infty} \frac{1}{x}$ 0

2. $\lim_{x \to \infty} \frac{x - 1}{5}$ ∞

3. $\lim_{x \to \infty} \frac{3x - 1}{2x + 1}$ $\frac{3}{2}$

4. $\lim_{x \to \infty} \frac{x^2 + 3}{2x}$ ∞

5. $\lim_{x \to \infty} (-2x + 9)$ $-\infty$

This section reviews many concepts covered in ALGEBRA 2. Some students may need to spend two days in this section if they are unfamiliar with the material.

Introduce *sequences* using the famous Fibonacci sequence. Consider mentioning that the title of Leonardo Fibonacci's book, *Liber Abaci*, is Latin for "Book of Calculation." Define *infinite* and *finite sequences*. A sequence can also be described as an ordered list or an ordered progression.

Explicit formulas are used to find specific *terms* of sequences in Example 1. This easy substitution process leads into the more challenging process of finding an explicit formula, as done in Example 2.

One-on-One Writing the position number under each term can help students see a pattern between the term and its position.

In Example 3 a *recursive formula* is used to find terms that are described by an expression involving previous terms. After discussing Example 3, compare the characteristics of explicit and recursive formulas. Stress that an explicit formula can be used to directly calculate any nth term, but a recursive formula is dependent on other terms.

After finding and applying the recursive formulas in Example 4, you may want to have the students continue the

Mathematicians frequently study the patterns found in sequences. Examining the relationship between successive terms and the position of each term can help determine the unlisted terms in a sequence. Stating an *explicit formula* for the nth term as a function of n clearly defines the sequence.

Example 2 Finding an Explicit Formula for a Sequence

Write an explicit formula for the value of a_n and use it to find the tenth term.

a. 17, 20, 23, 26, ...   **b.** −5, 10, −20, 40, ...   **c.** 0, 3, 8, 15, ...

Answer

a. Each successive term increases by 3.
$$a_2 = 17 + 3$$
$$a_3 = (17 + 3) + 3$$
$$= 17 + 2(3)$$
$$a_4 = [17 + 2(3)] + 3$$
$$= 17 + 3(3)$$
$$\vdots$$
$$a_n = 17 + 3(n - 1)$$
$$= 3n + 14$$
$$a_{10} = 3(10) + 14 = 44$$

b. Each term is the product of the previous term and −2.
$$a_2 = -5(-2)$$
$$a_3 = [-5(-2)](-2) = -5(-2)^2$$
$$a_4 = [-5(-2)^2](-2) = -5(-2)^3$$
$$\vdots$$
$$a_n = -5(-2)^{n-1}$$
$$a_{10} = -5(-2)^{10-1} = 2560$$

c. Each term is 1 less than the square of its position.
$$a_1 = 1^2 - 1$$
$$a_2 = 2^2 - 1$$
$$a_3 = 3^2 - 1$$
$$a_4 = 4^2 - 1$$
$$\vdots$$
$$a_n = n^2 - 1$$
$$a_{10} = 10^2 - 1 = 99$$

SKILL ✓ **EXERCISE 15**

You may have noticed in Example 2c that the differences between consecutive terms are consecutive odd numbers. While this fact allows the next term to be determined, it does not easily lead to an explicit formula. A *recursive formula* for a sequence states the first term(s) and defines the remaining terms using the previous term(s).

Example 3 Writing the Terms of a Sequence

Write the first 6 terms of the sequence defined by $b_1 = 2$, $b_2 = 4$, $b_k = b_{k-1} + b_{k-2}$ where $k \geq 3$.

Answer

by definition:
$$b_1 = 2$$
$$b_2 = 4$$

using the recursive formula:
$$b_3 = b_2 + b_1 = 4 + 2 = 6$$
$$b_4 = b_3 + b_2 = 6 + 4 = 10$$
$$b_5 = b_4 + b_3 = 10 + 6 = 16$$
$$b_6 = b_5 + b_4 = 16 + 10 = 26$$

Therefore, the first 6 terms are: 2, 4, 6, 10, 16, 26.

SKILL ✓ **EXERCISE 7**

Comparing consecutive terms is especially helpful when finding the pattern used in a recursive formula.

TIP 

While any variable can be used to represent the sequence or the subscripts, i, j, k, and n are frequently used for the subscripts. If the general term is b_k, the previous term is b_{k-1} and the next term is b_{k+1}.

Reading and Writing Mathematics

Use the Internet keyword search *Fibonacci metric conversion* and explain which conversion the Fibonacci sequence can be used for and why it works. The Fibonacci sequence can be used to estimate the conversion between miles and kilometers. Letting any number in the Fibonacci sequence represent miles, the next number will estimate the kilometers and vice versa. This works because the geometric ratio for the Fibonacci sequence is approximately 1.618, while the mile-to-kilometer ratio is approximately 1.609.

Additional Exercises

1. Find the first 4 terms and the tenth term of the sequence defined by $a_n = n^3 - n^2$.
0, 4, 18, 48; $a_{10} = 900$

2. Write an explicit formula for the value of $a_n = \frac{1}{2}, \frac{4}{3}, \frac{9}{4}, \frac{16}{5}, \ldots$ and use it to find the tenth term.
$a_n = \frac{n^2}{n+1}$; $a_{10} = \frac{100}{11}$

3. Write the first 6 terms of the recursively defined sequence: $a_1 = 1$, $a_2 = 2$; $a_n = (a_{n-1})^2 - (a_{n-2})^2$.
1, 2, 3, 5, 16, 231

Write a recursive formula for each sequence and use it to find the next 3 terms.

4. 42, 38, 32, 24, ... $a_1 = 42$, $a_n = a_{n-1} - 2n$ where $n \geq 2$; 14, 2, −12

---

sequences to verify the values of a_{10} found in Example 2.

Discuss how a calculator or computer program can quickly calculate the values of a sequence recursively.

Motivational Idea Challenge the students to write a calculator or computer program that generates sequences recursively. Students may want to expand the program illustrated in the Data Analysis margin note to show each term in the sequence. The following program lists the first ten terms of the sequence.

```
PROGRAM:EXRECUR
:1→A
:Disp "TERM 1 IS ",A
:For(N,2,10)
:A+N-1→A
:Disp "PRESS ENTER"
:Pause :ClrHome
:Disp "TERM ",N," IS ",A
:End
```

The following BASIC program generates the first 300 terms of the sequence 1, 4, 7, 10, 13, ...

```
DIM A(310)
A(1) = 1: PRINT A(1)
FOR n = 2 TO 300
A(n) = A(n − 1) + 3
PRINT A(n)
NEXT n
```

Many free online programming compilers are available. Use an Internet keyword search such as *BASIC online compiler* or *C++ online compiler*. Encourage students with programming knowledge to use the language of their choice.

Present the definition of *n factorial* (first presented in the C. Exercises of Section 4.4). Many students will already be familiar with factorial and its notation. Stress that 0! is defined as 1. Notice that the factorial function grows very rapidly.

Motivational Idea Expressions such as x^0 and 0! may cause some confusion for students. Have the students verify the recursive formula $n! = \frac{(n+1)!}{n+1}$ with a few

5. 2, 3, 5, 9, 17, . . .

$a_1 = 2, a_n = 2a_{n-1} - 1$; 33, 65, 129

6. Find $a_0, a_1, a_2, a_3,$ and a_4 for the sequence $a_n = \dfrac{3n-1}{n!}$.

$-1, 2, \dfrac{5}{2}, \dfrac{4}{3}, \dfrac{11}{24}$

7. Determine whether the sequence $a_n = \dfrac{n+1}{2n-1}$ converges or diverges. State the limit if it converges.

converges; $\dfrac{1}{2}$

8. Evaluate the finite series $\displaystyle\sum_{k=3}^{5} kj + 2$.

$12j + 6$

9. Find the partial sums $S_5, S_6,$ and S_7 for the sequence $a_n = \dfrac{2^{n-1}}{2^n - 1}$.

$\approx 3.2876, \approx 3.7955, \approx 4.2994$

10. Evaluate the infinite series $\displaystyle\sum_{n=1}^{\infty} \dfrac{n!}{n^2}$ and describe it as convergent or divergent. ∞; divergent

Assignments

- **Minimum:** 1–2, 4, 6; 7–13 odd; 14, 17–19, 22, 25–26, 29, 33, 37–38, 44–46, 49, 52

- **Standard:** 1, 3–4, 6, 8–10, 13, 15–16, 19–20, 23, 27–30, 33–34, 39–40, 44–46, 49, 51, 53

- **Extended:** 1, 4–5, 8–9, 12–13, 16–17, 20–21, 23–24; 26–32 even; 38, 40–43, 45, 48, 50, 52–53

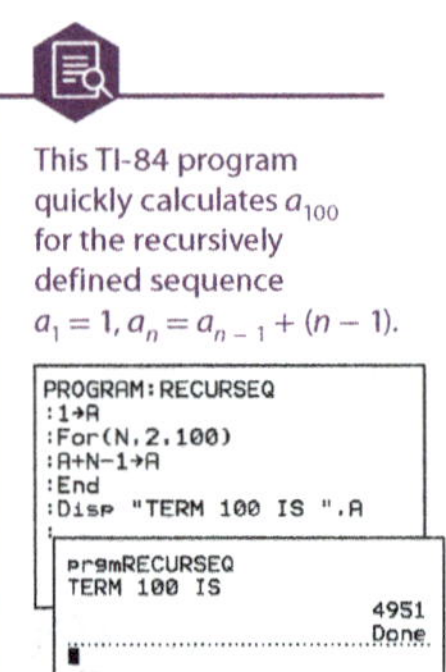

This TI-84 program quickly calculates a_{100} for the recursively defined sequence $a_1 = 1, a_n = a_{n-1} + (n-1)$.

Example 4 Finding a Recursive Formula for a Sequence

Write a recursive formula for each sequence and use it to find $a_5, a_6,$ and a_7.

a. 17, 20, 23, 26, . . .   **b.** $-5, 10, -20, 40, . . .$ **c.** 0, 3, 8, 15, . . .

Answer

a. After 17, each term increases by 3.

$a_1 = 17, a_n = a_{n-1} + 3$ where $n \geq 2$

$a_5 = 26 + 3 = 29$
$a_6 = 29 + 3 = 32$
$a_7 = 32 + 3 = 35$

b. After -5, each term is the product of the previous term and -2.

$a_1 = -5, a_n = -2a_{n-1}$ where $n \geq 2$

$a_5 = -2(40) = -80$
$a_6 = -2(-80) = 160$
$a_7 = -2(160) = -320$

c. After 0 and 3, the difference between a term and the preceding term is two more than the difference between the two previous terms.

$a_1 = 0, a_2 = 3$
$a_n = a_{n-1} + (a_{n-1} - a_{n-2}) + 2$
$\quad = 2a_{n-1} - a_{n-2} + 2, n \geq 3$
$a_5 = 2(15) - (8) + 2 = 24$
$a_6 = 2(24) - (15) + 2 = 35$
$a_7 = 2(35) - (24) + 2 = 48$

SKILL ✔ **EXERCISE 19**

Be aware that a listing of the first several terms is not sufficient to uniquely define a sequence. For example, 1, 2, 4, . . . can be the first three terms of either of the following sequences:

1, 2, 4, 8, 16, . . . with an explicit formula of $a_n = 2^n$ or
1, 2, 4, 7, 11, . . . with a recursive formula of $a_1 = 1, a_n = a_{n-1} + (n-1), n \geq 2$.

It is clearly easier to find a_{100} with an explicit formula. While using a recursive formula to find the value of this term would require the calculation of all previous terms of the sequence, a calculator or computer can be programmed to perform these calculations. Numerous relationships can only be modeled recursively.

Many sequences involve a special product called a *factorial*. Note the explicit formula used to define this product.

▶ **DEFINITION**

The product of n and every natural number less than n is notated **n!**, which is read as **n factorial**. Symbolically, $n! = n(n-1)(n-2)(n-3) . . . (1)$. Note that 0! is defined as 1.

In some sequences it is convenient to use the set of whole numbers as the domain: $a_0, a_1, a_2, a_3, ...$ A sequence of factorials can then be generated using the recursive definition $a_n = na_{n-1}$ with $a_0 = 1$.

$a_0 = 1$
$a_1 = 1(1) = 1$
$a_2 = 2(1) = 2$
$a_3 = 3(2) = 6$
$a_4 = 4(6) = 24$
$a_5 = 5(24) = 120$
$a_6 = 6(120) = 720$
⋮

Example 5 Using Factorials to Find Terms of a Sequence

Given $a_n = \dfrac{2n!}{(n+1)^2}$, find $a_0, a_1, a_2, a_3,$ and a_4.

Answer

$a_0 = \dfrac{2 \cdot 0!}{(0+1)^2}$ $a_1 = \dfrac{2 \cdot 1!}{(1+1)^2}$ $a_2 = \dfrac{2 \cdot 2!}{(2+1)^2}$ $a_3 = \dfrac{2 \cdot 3!}{(3+1)^2}$ $a_4 = \dfrac{2 \cdot 4!}{(4+1)^2}$

$= \dfrac{2(1)}{1^2} = 2$ $= \dfrac{2(1)}{2^2} = \dfrac{1}{2}$ $= \dfrac{2(2 \cdot 1)}{3^2} = \dfrac{4}{9}$ $= \dfrac{2(3 \cdot 2 \cdot 1)}{4^2} = \dfrac{3}{4}$ $= \dfrac{2(4 \cdot 3 \cdot 2 \cdot 1)}{5^2} = \dfrac{48}{25}$

SKILL ✔ **EXERCISE 11**

values such as $7! = \dfrac{8!}{8}$. Then have them apply the formula to determine a value for 0! $(0! = \dfrac{1!}{1} = 1)$. Share that an empty product is defined as a product that contains no factors. The empty product convention states that an empty product is equal to 1, the identity element for multiplication. This supports that $x^0 = 1$ and $0! = 1$. An empty sum is defined as 0, the identity element for addition. Consider challenging the students to use the Internet keyword search *empty product* for reinforcement and to recognize important applications such as computer programming and logic.

In Example 5 students use factorials in an explicit formula. Like exponents, the factorial symbol applies only to the factor immediately preceding the term. $\therefore 2n! \neq (2n)!$

The *convergence* or *divergence* of the sequences in Example 6 can be determined by analyzing the graph of the sequence or by examining the function's end behavior (see the chart summarizing rational functions on p. 107). Note that a divergent sequence may approach infinity or the limit may not exist.

Define a *series* and the related terms. Use Example 7 to demonstrate evaluating a series expressed in *summation notation*. *Partial sums* can be expressed in summa-

tion notation and evaluated as shown in Example 8.

Finally, the concepts of convergent and divergent *infinite series* are introduced. Be sure that the students clearly distinguish between a convergent sequence (whose terms approach a certain number) and a convergent series (whose sequence of partial sums approaches a certain number). Emphasize that the infinite series of any sequence in which $\displaystyle\lim_{x \to \infty} a_n \neq 0$ is divergent. In Example 9 point out that the sequence is examined first and then the series.

Common Student Error Students may assume that the converse of the test for

The end behavior of functions has been described using limits. If the values of a sequence approach a finite real number L as the position number approaches infinity, the sequence is *convergent* and $\lim_{n\to\infty} a_n = L$ is the *limit of the sequence*. If not, the sequence is *divergent*. Note that finding the limit of rational sequences is similar to finding the horizontal asymptote of a rational function, which was described in Section 2.5.

Example 6   Finding the Limit of an Infinite Sequence

Determine whether each sequence converges or diverges and state the limit of any convergent sequences.

a. $a_n = \dfrac{5n-1}{n+1}$ **b.** $a_n = -2n+4$ **c.** $a_n = 3(-1)^n$

Answer

a. converges; since
$$\lim_{n\to\infty} \frac{5n-1}{n+1} = 5$$

b. diverges; since
$$\lim_{n\to\infty}(-2n+4) = -\infty$$

c. diverges; The limit does not exist since the terms alternate between 3 and -3.

Check

The Technology Corner at the end of this section explains how the sequences can be graphed.

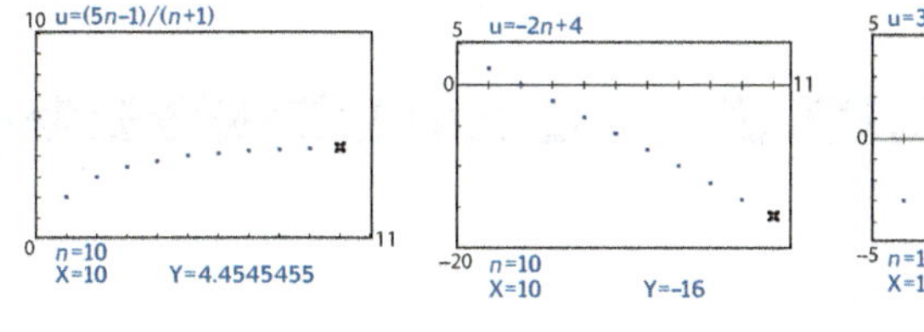

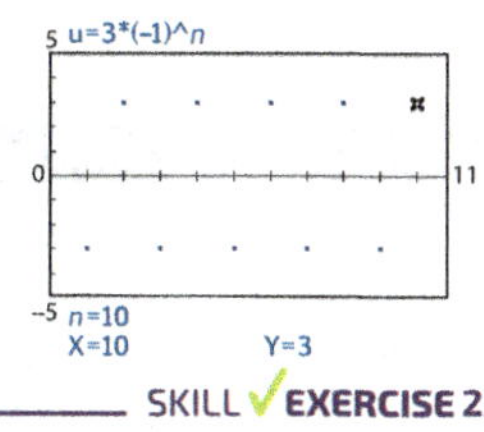

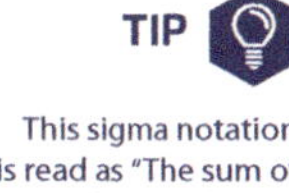

SKILL ✔ EXERCISE 25

A sum of the terms in a sequence is a *series*. The Greek capital letter sigma, Σ, is frequently used to represent a series.

DEFINITION

A **series** of a given sequence $a_1, a_2, a_3, a_4, \ldots$ can be represented in *summation* (or *sigma*) *notation* as

$$\sum_{i=b}^{c} a_i = a_b + a_{b+1} + a_{b+2} + \cdots + a_c$$

where the index (or counter) i has a lower bound of b and an upper bound of c. An *infinite series* has an upper bound of ∞.

TIP

This sigma notation is read as "The sum of a_i from $i = b$ to c."

Example 7   Evaluating Summation Notation

Evaluate $\sum_{i=3}^{7} (i^2 - 2i)$.

Answer

The bounds of the index indicate that the series is the sum of the terms from a_3 to a_7. Find each term by substituting the values 3, 4, 5, 6, and 7 for i, and then find the sum.

$$\sum_{i=3}^{7}(i^2 - 2i) = [3^2 - 2(3)] + [4^2 - 2(4)] + [5^2 - 2(5)] + [6^2 - 2(6)] + [7^2 - 2(7)]$$
$$= 3 + 8 + 15 + 24 + 35$$
$$= 85$$

SKILL ✔ EXERCISE 29

---

divergence mentioned in the text is true, but the fact that the terms approach 0 does not guarantee that their sum converges. Harmonic sequences (introduced in the next section) can be used to demonstrate this fact. The discussion of most tests for convergence are beyond the scope of this text.

Consider working through the Technology Corner with the students. When using SEQUENCE mode, some calculators may provide prompts for u(1) and u(2) instead of u(nMin). The linear sequence $v(n) = 4n + 3$ can be defined recursively as $v(n) = v(n-1) + 4$ where $v(n\text{Min}) = 7$.

TIPS

Ex. 6 Sequences in which the only variable expression is a_{n-1} can be calculated by entering a_0, pressing [ENTER], then entering the recursive formula using [ANS] ([2nd], [(-)]) for a_{n-1}, and then pressing [ENTER] to calculate the successive terms.

Ex. 33–36 These partial-sum sequences quickly converge to the values of e, $\frac{1}{e}$, $-\frac{1}{e}$, and π.

Ex. 42–43 These exercises revisit concepts presented in the C. Exercises of Section 4.4, using partial sums to show the quick convergence to sine and cosine values.

Solutions

❯ A. Exercises

2. $a_1 = 2(1) - 5 = -3$
$a_2 = 2(2) - 5 = -1$
$a_3 = 2(3) - 5 = 1; a_4 = 2(4) - 5 = 3$
$a_{10} = 2(10) - 5 = 15$

3. $a_1 = \dfrac{3}{(1^2 + 1)} = \dfrac{3}{2}; a_2 = \dfrac{3}{(2^2 + 1)} = \dfrac{3}{5}$
$a_3 = \dfrac{3}{(3^2 + 1)} = \dfrac{3}{10}; a_4 = \dfrac{3}{(4^2 + 1)} = \dfrac{3}{17}$
$a_{10} = \dfrac{3}{(10^2 + 1)} = \dfrac{3}{101}$

4. $a_1 = \dfrac{1(1-1)}{2} = 0; a_2 = \dfrac{2(2-1)}{2} = 1$
$a_3 = \dfrac{3(3-1)}{2} = 3; a_4 = \dfrac{4(4-1)}{2} = 6$
$a_{10} = \dfrac{10(10-1)}{2} = 45$

5. $a_1 = 1 + \dfrac{1}{1} = 2; a_2 = 1 + \dfrac{1}{2} = \dfrac{3}{2}$
$a_3 = 1 + \dfrac{1}{3} = \dfrac{4}{3}; a_4 = 1 + \dfrac{1}{4} = \dfrac{5}{4}$
$a_{10} = 1 + \dfrac{1}{10} = \dfrac{11}{10}$

6. $a_2 = 3(4) - 9 = 3$
$a_3 = 3(3) - 9 = 0$
$a_4 = 3(0) - 9 = -9$
$a_5 = 3(-9) - 9 = -36$
$a_6 = 3(-36) - 9 = -117$

7. $a_2 = 2^1(3) = 6$
$a_3 = 2^2(6) = 24$
$a_4 = 2^3(24) = 192$
$a_5 = 2^4(192) = 3072$
$a_6 = 2^5(3072) = 98{,}304$

8. $a_3 = [(4) - (1)]^2 = 9$
$a_4 = [(9) - (4)]^2 = 25$
$a_5 = [(25) - (9)]^2 = 256$
$a_6 = [(256) - (25)]^2 = 53{,}361$

9. $a_3 = 3(10) + 2(2) = 34$
$a_4 = 3(34) + 2(10) = 122$
$a_5 = 3(122) + 2(34) = 434$
$a_6 = 3(434) + 2(122) = 1546$

10. $a_0 = \dfrac{4}{0!} = \dfrac{4}{1} = 4$
$a_1 = \dfrac{4}{1!} = \dfrac{4}{1} = 4$
$a_2 = \dfrac{4}{2!} = \dfrac{4}{2 \cdot 1} = 2$
$a_3 = \dfrac{4}{3!} = \dfrac{4}{3 \cdot 2 \cdot 1} = \dfrac{2}{3}$
$a_4 = \dfrac{4}{4!} = \dfrac{4}{4 \cdot 3 \cdot 2 \cdot 1} = \dfrac{1}{6}$

11. $a_0 = \dfrac{3(0!)}{2(0) + 1} = \dfrac{3(1)}{1} = 3$
$a_1 = \dfrac{3(1!)}{2(1) + 1} = \dfrac{3(1)}{3} = 1$
$a_2 = \dfrac{3(2!)}{2(2) + 1} = \dfrac{3(2 \cdot 1)}{5} = \dfrac{6}{5}$
$a_3 = \dfrac{3(3!)}{2(3) + 1} = \dfrac{3(3 \cdot 2 \cdot 1)}{7} = \dfrac{18}{7}$
$a_4 = \dfrac{3(4!)}{2(4) + 1} = \dfrac{3(4 \cdot 3 \cdot 2 \cdot 1)}{9} = 8$

12. $a_0 = \dfrac{0!}{(0+2)!} = \dfrac{1}{2!} = \dfrac{1}{2}$

$a_1 = \dfrac{1!}{(1+2)!} = \dfrac{1}{3!} = \dfrac{1}{6}$

$a_2 = \dfrac{2!}{(2+2)!} = \dfrac{2!}{4!} = \dfrac{1}{12}$

$a_3 = \dfrac{3!}{(3+2)!} = \dfrac{3!}{5!} = \dfrac{1}{20}$

$a_4 = \dfrac{4!}{(4+2)!} = \dfrac{4!}{6!} = \dfrac{1}{30}$

13. $a_0 = \dfrac{(-1)^0}{2(0!)} = \dfrac{1}{2(1)} = \dfrac{1}{2}$

$a_1 = \dfrac{(-1)^1}{2(1)!} = \dfrac{-1}{2(1)} = -\dfrac{1}{2}$

$a_2 = \dfrac{(-1)^2}{2(2!)} = \dfrac{1}{2(2)} = \dfrac{1}{4}$

$a_3 = \dfrac{(-1)^3}{2(3!)} = \dfrac{-1}{2(6)} = -\dfrac{1}{12}$

$a_4 = \dfrac{(-1)^4}{2(4!)} = \dfrac{1}{2(24)} = \dfrac{1}{48}$

❭ B. Exercises

14. $a_1 = \sqrt{1} = 1;\ a_2 = \sqrt{2};\ a_3 = \sqrt{3}$
$a_4 = \sqrt{4} = 2;\ a_n = \sqrt{n};\ a_{10} = \sqrt{10}$

15. $a_1 = 1 + 2(1) = 3;\ a_2 = 1 + 2(2) = 5$
$a_3 = 1 + 2(3) = 7;\ a_4 = 1 + 2(4) = 9$
$a_n = 1 + 2n;\ a_{10} = 1 + 2(10) = 21$

16. $a_1 = (-1)^{1-1} = 1;\ a_2 = (-1)^{1-2} = -1$
$a_3 = (-1)^{3-1} = 1;\ a_4 = (-1)^{4-1} = -1$
$a_n = (-1)^{n-1}$ or $a_n = (-1)^{n+1}$
$a_{10} = (-1)^{10-1} = -1$

17. $a_1 = \dfrac{1-1}{1} = 0;\ a_2 = \dfrac{2-1}{2} = \dfrac{1}{2}$
$a_3 = \dfrac{3-1}{3} = \dfrac{2}{3};\ a_4 = \dfrac{4-1}{4} = \dfrac{3}{4}$
$a_n = \dfrac{n-1}{n};\ a_{10} = \dfrac{10-1}{10} = \dfrac{9}{10}$

18. $a_1 = 3,\ a_n = a_{n-1} + 3$ where $n \geq 2$
$a_5 = 12 + 3 = 15;\ a_6 = 15 + 3 = 18$
$a_7 = 18 + 3 = 21$

19. $a_1 = 3,\ a_2 = 6$
$a_n = a_{n-1} + a_{n-2}$ where $n \geq 3$
$a_5 = 15 + 9 = 24;\ a_6 = 24 + 15 = 39$
$a_7 = 39 + 24 = 63$

20. $a_1 = 3,\ a_n = a_{n-1} + n$ where $n \geq 2$
$a_5 = 12 + 5 = 17;\ a_6 = 17 + 6 = 23$
$a_7 = 23 + 7 = 30$

21. $a_1 = 8,\ a_n = 0.1a_{n-1} + 1$ where $n \geq 2$
$a_5 = 0.1(1.118) + 1 = 1.1118$
$a_6 = 0.1(1.1118) + 1 = 1.11118$
$a_7 = 0.1(1.11118) + 1 = 1.111118$

22a. $A = P(1 + rt)$
$\quad = P(1 + 0.05(1))$
$a_n = a_{n-1}(1.05)$

22b. $A = P(1 + rt)$
$\quad = P(1 + 0.07(1))$
$a_n = a_{n-1}(1.07)$

The sum of a sequence's first n terms, $\displaystyle\sum_{i=1}^{n} a_i$, is called the *nth partial sum* and is abbreviated as S_n.

Example 8   Evaluating Partial Sums

Given the sequence $2, 4, 6, 8, \ldots, 2n$; find S_5, S_6, and S_7.

Answer

$S_5 = \displaystyle\sum_{i=1}^{5} a_i = 2 + 4 + 6 + 8 + 10 = 30$ — 1. Find S_5 by adding the first 5 terms.

$S_6 = S_5 + a_6 = 30 + 12 = 42$ — 2. Successive partial sums can be found by adding the next term.

$S_7 = S_6 + a_7 = 42 + 14 = 56$

SKILL ✔ **EXERCISE 33**

The series in Examples 7 and 8 are finite series. Can an infinite series be calculated when a person cannot add an infinite number of terms? Given a sequence of terms a_k, you can generate a related sequence of partial sums: $S_1, S_2, S_3, \ldots, S_n$. If the sequence of partial sums converges to a real number S, then the infinite series $\displaystyle\sum_{i=1}^{\infty} a_i = S$ and the series is *convergent*. If the terms of the sequence a_i do not approach 0, the sequence of partial sums S_n approaches infinity and the series is *divergent*. Be careful to distinguish the limit of the sequence from the limit of the partial sums.

Example 9   Evaluating Infinite Series

Evaluate each infinite series and describe the series as convergent or divergent.

a. $3 + 3 + 3 + 3 + \cdots$

b. $0.3 + 0.03 + 0.003 + 0.0003 + \cdots$

Answer

a. Since $\displaystyle\lim_{i \to \infty} 3 = 3$, the sequence of partial sums $3, 6, 9, 12, 15, \ldots$ has no finite limit.

Therefore, $\displaystyle\sum_{i=1}^{\infty} 3 = \infty$ and the series is divergent.

b. Note that $\displaystyle\lim_{n \to \infty} \dfrac{3}{10^n} = 0$ and that the sequence of partial sums $0.3, 0.33, 0.333, 0.3333, \ldots$ approaches $0.\overline{3}$ or $\dfrac{1}{3}$.

Therefore, $\displaystyle\sum_{n=1}^{\infty} \dfrac{3}{10^n} = \dfrac{1}{3}$ and the series is convergent.

SKILL ✔ **EXERCISE 37**

❭ A. Exercises

1. Identify each expression.

- **a.** $2, 4, 6, 8, \ldots$ IV
- **b.** $3 + 6 + 9 + 12 + \cdots + 30$ I
- **c.** $1, 4, 7, \ldots, 25$ III
- **d.** $\displaystyle\sum_{i=1}^{15} 5i$ I
- **e.** $\displaystyle\sum_{i=9}^{\infty} 0.2^i$ II

   I. finite series
   II. infinite series
   III. finite sequence
   IV. infinite sequence

2. $-3, -1, 1, 3; a_{10} = 15$
3. $\dfrac{3}{2}, \dfrac{3}{5}, \dfrac{3}{10}, \dfrac{3}{17}; a_{10} = \dfrac{3}{101}$
4. $0, 1, 3, 6; a_{10} = 45$
5. $2, \dfrac{3}{2}, \dfrac{4}{3}, \dfrac{5}{4}; a_{10} = \dfrac{11}{10}$

Find the first 4 terms and the tenth term of each explicitly defined sequence.

2. $a_n = 2n - 5$
3. $a_n = \dfrac{3}{(n^2 + 1)}$
4. $a_n = \dfrac{n(n-1)}{2}$
5. $a_n = 1 + \dfrac{1}{n}$

Write the first 6 terms of each recursively defined sequence.

6. $a_1 = 4;\ a_n = 3(a_{n-1}) - 9$ where $n \geq 2$
$\quad$ 4, 3, 0, −9, −36, −117
7. $a_1 = 3;\ a_n = 2^{n-1}(a_{n-1})$ where $n \geq 2$
$\quad$ 3, 6, 24, 192, 3072, 98,304
8. $a_1 = 1,\ a_2 = 4;\ a_n = (a_{n-1} - a_{n-2})^2$ where $n \geq 3$
$\quad$ 1, 4, 9, 25, 256, 53,361
9. $a_1 = 2,\ a_2 = 10;\ a_n = 3a_{n-1} + 2a_{n-2}$ where $n \geq 3$
$\quad$ 2, 10, 34, 122, 434, 1546

23a. $a_7 = 13,\ a_8 = 21,\ a_9 = 34$
$a_{10} = 55,\ a_{11} = 89,\ a_{12} = 144$
$a_{13} = 233,\ a_{14} = 377,\ a_{15} = 610$
$a_{16} = 987,\ a_{17} = 1597,\ a_{18} = 2584$

23b. $a_1 = 1,\ a_2 = 1$
$a_3 = 1 + 1 = a_1 + a_2 = 2$
$a_4 = 1 + 2 = a_2 + a_3 = 3$
$a_5 = 2 + 3 = a_3 + a_4 = 5$
$a_n = a_{n-2} + a_{n-1}$, where $n \geq 3$

29. $4^2 + 5^2 + 6^2 + 7^2 + 8^2 = 190$

30. $(2^2 - 1) + (3^2 - 1) + (4^2 - 1)$
$\quad = 3 + 8 + 15 = 26$

31. $\dfrac{4}{1} + \dfrac{4}{2} + \dfrac{4}{3} + \dfrac{4}{4} + \dfrac{4}{5}$
$\quad = 4 + 2 + \dfrac{4}{3} + 1 + \dfrac{4}{5}$
$\quad = \dfrac{137}{15} = 9.1\overline{3}$

32. $i^0 + i^1 + i^2 + i^3 + i^4 + i^5$
$\quad = 1 + i + (-1) + (-i) + 1 + i$
$\quad = 1 + i$

33. $S_5 = \dfrac{1}{0!} + \dfrac{1}{1!} + \dfrac{1}{2!} + \dfrac{1}{3!} + \dfrac{1}{4!} = 2.708\overline{3}$
$S_6 = S_5 + a_6 = 2.708\overline{3} + \dfrac{1}{5!} = 2.71\overline{6}$
$S_7 = S_6 + a_7 = 2.71\overline{6} + \dfrac{1}{6!} = 2.7180\overline{5}$

Find a_0, a_1, a_2, a_3, and a_4 for each sequence.

10. $a_n = \dfrac{4}{n!}$ 4, 4, 2, $\dfrac{2}{3}$, $\dfrac{1}{6}$

11. $a_n = \dfrac{3n!}{2n+1}$ 3, 1, $\dfrac{6}{5}$, $\dfrac{18}{7}$, 8

12. $a_n = \dfrac{n!}{(n+2)!}$ $\dfrac{1}{2}$, $\dfrac{1}{6}$, $\dfrac{1}{12}$, $\dfrac{1}{20}$, $\dfrac{1}{30}$

13. $a_n = \dfrac{(-1)^n}{2n!}$ $\dfrac{1}{2}$, $-\dfrac{1}{2}$, $\dfrac{1}{4}$, $-\dfrac{1}{12}$, $\dfrac{1}{48}$

❯ B. Exercises

Write an explicit formula for the value of a_n and use it to find the tenth term.

14. 1, $\sqrt{2}$, $\sqrt{3}$, 2, …

15. 3, 5, 7, 9, …

16. 1, −1, 1, −1, …

17. 0, $\dfrac{1}{2}$, $\dfrac{2}{3}$, $\dfrac{3}{4}$, …

Write a recursive formula for each sequence and use it to find a_5, a_6, and a_7.

18. 3, 6, 9, 12, …

19. 3, 6, 9, 15, …

20. 3, 5, 8, 12, …

21. 8, 1.8, 1.18, 1.118, …

22. Assume a home's value appreciates by the same percentage each year. Write a sequence with a_0 representing the home's original value and a_1, a_2, a_3, and a_4 representing the value of the home (rounded to the nearest dollar) in each of the next 4 years. Then write a recursive formula for the sequence.

  a. $a_0 = \$225{,}000$; 5% APR

  b. $a_0 = \$380{,}000$; 7% APR

23. The Fibonacci sequence 1, 1, 2, 3, 5, 8, … models the number of mature rabbit pairs after n months (assuming no rabbits die).

  a. Find the number of rabbit pairs after 1 yr and after 18 mo.   144; 2584

  b. Write the recursive formula for the Fibonacci sequence.

24. The golden rectangles commonly found in art and architecture have a length-to-width ratio that aproximates the golden ratio, ϕ. This ratio can be approximated by the converging sequence $b_n = \dfrac{a_n}{a_{n-1}}$ where a_n is the Fibonacci sequence with $n > 1$. Write the first 10 terms of b_n as fractions and then evaluate b_9 and b_{10} to three decimal places.

Determine whether each sequence converges or diverges and state the limit of any convergent sequences.

25. 1, $\dfrac{1}{2}$, $\dfrac{1}{3}$, $\dfrac{1}{4}$, …

26. 1, 1, 2, 3, 5, …

27. $a_1 = 4$, $a_n = a_{n-1} - 7$

28. $a_n = \dfrac{n-1}{n+1}$

14. $a_n = \sqrt{n}$; $a_{10} = \sqrt{10}$

15. $a_n = 1 + 2n$; $a_{10} = 21$

16. $a_n = (-1)^{n-1}$ or $a_n = (-1)^{n+1}$; $a_{10} = -1$

17. $a_n = \dfrac{n-1}{n}$; $a_{10} = \dfrac{9}{10}$

18. $a_1 = 3$, $a_n = a_{n-1} + 3$ where $n \geq 2$; 15, 18, 21

19. $a_1 = 3$, $a_2 = 6$, $a_n = a_{n-1} + a_{n-2}$ where $n \geq 3$; 24, 39, 63

20. $a_1 = 3$, $a_n = a_{n-1} + n$ where $n \geq 2$; 17, 23, 30

21. $a_1 = 8$, $a_n = 0.1a_{n-1} + 1$, where $n \geq 2$; 1.1118, 1.11118, 1.111118

Evaluate each finite series.

29. $\displaystyle\sum_{n=4}^{8} n^2$ 190

30. $\displaystyle\sum_{k=2}^{4} (k^2 - 1)$ 26

31. $\displaystyle\sum_{i=1}^{5} \dfrac{4}{i}$ $\dfrac{137}{15}$

32. $\displaystyle\sum_{j=0}^{5} i^j$, where $i = \sqrt{-1}$ $1 + i$

Find the partial sums S_5, S_6, and S_7 for each sequence. Round answers to the hundred-thousandth place if necessary.

33. $a_n = \dfrac{1}{(n-1)!}$

34. $a_n = \dfrac{(-1)^{n-1}}{(n-1)!}$

35. $a_n = \dfrac{(-1)^n}{(n+1)!}$

36. $a_n = \dfrac{4(-1)^{n-1}}{2n-1}$

Evaluate each infinite series and describe it as convergent or divergent.

37. $\displaystyle\sum_{k=1}^{\infty} 2\,(0.1)^k$ $0.\overline{2}$ or $\dfrac{2}{9}$ convergent

38. $\displaystyle\sum_{k=1}^{\infty} \dfrac{k}{3}$ ∞; divergent

39. $\displaystyle\sum_{i=1}^{\infty} (-5i)$ $-\infty$; divergent

40. $\displaystyle\sum_{i=1}^{\infty} 45\,(0.01)^i$ $0.\overline{45}$ or $\dfrac{5}{11}$ convergent

❯ C. Exercises

41. Express the number of seconds in 6 weeks in factorial notation.   10!

42. Calculators determine sine and cosine values using the following infinite series:

$$\sin x = \sum_{k=0}^{n} \dfrac{(-1)^k x^{2k+1}}{(2k+1)!} \quad \text{and} \quad \cos x = \sum_{k=0}^{n} \dfrac{(-1)^k x^{2k}}{(2k)!}.$$

Estimate each trigonometric value by finding the partial sums S_0, S_1, S_2, and S_3 (to five decimal places).

  a. $\sin \dfrac{\pi}{6}$

  b. $\cos \dfrac{\pi}{4}$

  c. $\cos\left(-\dfrac{\pi}{6}\right)$

  d. $\sin\left(-\dfrac{\pi}{3}\right)$

43. Use technology to graph $Y_1 = \sin x$ and $Y_2 = x - \dfrac{x^3}{3!} + \dfrac{x^5}{5!} - \dfrac{x^7}{7!}$. State the interval of x for which the two graphs appear to have identical values. $[-\pi, \pi]$

22a. $\$225{,}000$, $\$236{,}250$, $\$248{,}063$, $\$260{,}466$, $\$273{,}489$; $a_n = 1.05a_{n-1}$

22b. $\$380{,}000$, $\$406{,}600$, $\$435{,}062$, $\$465{,}516$, $\$498{,}102$; $a_n = 1.07a_{n-1}$

23b. $a_1 = 1$, $a_2 = 1$, $a_n = a_{n-2} + a_{n-1}$ where $n \geq 3$

24. $\dfrac{1}{1}$, $\dfrac{2}{1}$, $\dfrac{3}{2}$, $\dfrac{5}{3}$, $\dfrac{8}{5}$, $\dfrac{13}{8}$, $\dfrac{21}{13}$, $\dfrac{34}{21}$, $\dfrac{55}{34}$, $\dfrac{89}{55}$; $b_9 \approx 1.618 \approx b_{10}$

25. converges; 0

26. diverges

27. diverges

28. converges; 1

33. $2.708\overline{3}$; $2.71\overline{6}$; 2.71805

34. 0.375; $0.3\overline{6}$; 0.36805

35. $-0.3680\overline{5}$; -0.36786; -0.36788

36. 3.33968; 2.97605; 3.28374

42a. 0.52360; 0.49967; 0.50000; 0.50000

42b. 1; 0.69157; 0.70743; 0.70710

42c. 1; 0.86292; 0.86605; 0.86603

42d. -1.04720; -0.85580; -0.86630; -0.86602

❯ C. Exercises

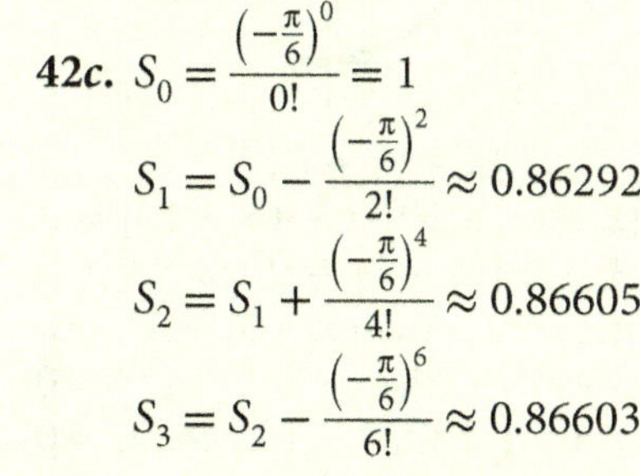

41. $6 \text{ weeks } \dfrac{7 \text{ days}}{1 \text{ week}} \cdot \dfrac{24 \text{ hr}}{1 \text{ day}} \cdot \dfrac{60 \text{ min}}{1 \text{ hr}} \cdot \dfrac{60 \text{ sec}}{1 \text{ min}}$

$= 6 \cdot 7 \cdot 2(3)(4) \cdot 2(5)(2)(3) \cdot 2(5)(2)(3)$

$= 6 \cdot 7 \cdot 2(3)(4) \cdot 2(5)(2)(3) \cdot 2(5)(2)(3)$

$= 7 \cdot 6 \cdot 4 \cdot 3 \cdot 2 \cdot 5 \cdot 2(2)(2) \cdot 3(3) \cdot 2(5)$

$= 10 \cdot 9 \cdot 8 \cdot 7 \cdot 6 \cdot 5 \cdot 4 \cdot 3 \cdot 2 = 10!$

42a. $S_0 = \dfrac{\left(\frac{\pi}{6}\right)^1}{1!} \approx 0.52360$

$S_1 = S_0 - \dfrac{\left(\frac{\pi}{6}\right)^3}{3!} \approx 0.49967$

$S_2 = S_1 + \dfrac{\left(\frac{\pi}{6}\right)^5}{5!} \approx 0.50000$

$S_3 = S_2 - \dfrac{\left(\frac{\pi}{6}\right)^7}{7!} \approx 0.50000$

42b. $S_0 = \dfrac{\left(\frac{\pi}{4}\right)^0}{0!} = 1$

$S_1 = S_0 - \dfrac{\left(\frac{\pi}{4}\right)^2}{2!} \approx 0.69157$

$S_2 = S_1 + \dfrac{\left(\frac{\pi}{4}\right)^4}{4!} \approx 0.70743$

$S_3 = S_2 - \dfrac{\left(\frac{\pi}{4}\right)^6}{6!} \approx 0.70710$

42c. $S_0 = \dfrac{\left(-\frac{\pi}{6}\right)^0}{0!} = 1$

$S_1 = S_0 - \dfrac{\left(-\frac{\pi}{6}\right)^2}{2!} \approx 0.86292$

$S_2 = S_1 + \dfrac{\left(-\frac{\pi}{6}\right)^4}{4!} \approx 0.86605$

$S_3 = S_2 - \dfrac{\left(-\frac{\pi}{6}\right)^6}{6!} \approx 0.86603$

42d. $S_0 = \dfrac{\left(-\frac{\pi}{3}\right)^1}{1!} \approx -1.04720$

$S_1 = S_0 - \dfrac{\left(-\frac{\pi}{3}\right)^3}{3!} \approx -0.85580$

$S_2 = S_1 + \dfrac{\left(-\frac{\pi}{3}\right)^5}{5!} \approx -0.86630$

$S_3 = S_2 - \dfrac{\left(-\frac{\pi}{3}\right)^7}{7!} \approx -0.86602$

43.

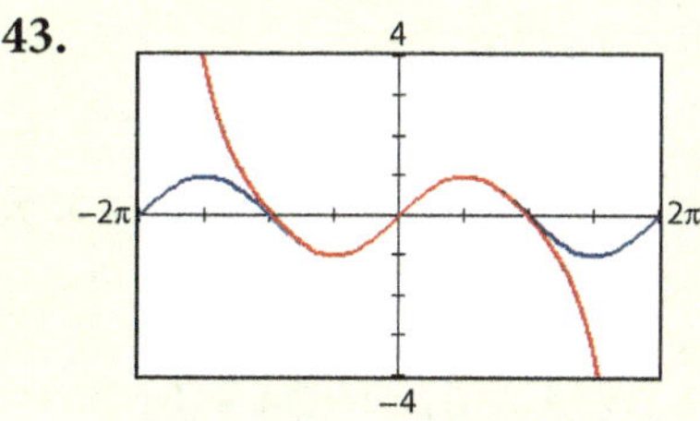

---

34. $S_5 = \dfrac{1}{0!} - \dfrac{1}{1!} + \dfrac{1}{2!} - \dfrac{1}{3!} + \dfrac{1}{4!} = 0.375$

$S_6 = S_5 + a_6 = 0.375 - \dfrac{1}{5!} = 0.3\overline{6}$

$S_7 = S_6 + a_7 = 0.3\overline{6} + \dfrac{1}{6!} = 0.36805$

35. $S_5 = -\dfrac{1}{2!} + \dfrac{1}{3!} - \dfrac{1}{4!} + \dfrac{1}{5!} - \dfrac{1}{6!}$

$= -0.3680\overline{5}$

$S_6 = S_5 + a_6 = -0.3680\overline{5} + \dfrac{1}{7!}$

≈ -0.36786

$S_7 = S_6 + a_7 \approx -0.36786 - \dfrac{1}{8!}$

≈ -0.36788

36. $S_5 = \dfrac{4}{1} - \dfrac{4}{3} + \dfrac{4}{5} - \dfrac{4}{7} + \dfrac{4}{9} \approx 3.33968$

$S_6 = S_5 + a_6 \approx 3.34 - \dfrac{4}{11} \approx 2.97605$

$S_7 = S_6 + a_7 \approx 2.98 + \dfrac{4}{13} \approx 3.28374$

37. $0.2 + 0.02 + 0.002 + \cdots = 0.\overline{2}$ or $\dfrac{2}{9}$ convergent

38. $\dfrac{1}{3} + \dfrac{2}{3} + 1 + \cdots = \infty$; divergent

39. $(-5) + (-10) + (-15) + \cdots = -\infty$ divergent

40. $0.45 + 0.0045 + 0.000045 + \cdots = 0.\overline{45}$ or $\dfrac{5}{11}$; convergent

44. $m = \dfrac{-2-3}{8-0} = -\dfrac{5}{8}$

$y - 3 = -\dfrac{5}{8}(x - 0)$

$y = -\dfrac{5}{8}x + 3$

45. $m_G = \dfrac{5-4}{-3-0} = -\dfrac{1}{3}$; $m_\perp = 3$

$y - 5 = 3(x - (-1))$

$y = 3x + 8$

46. $D = \mathbb{R}$; $R = (0, \infty)$; growth

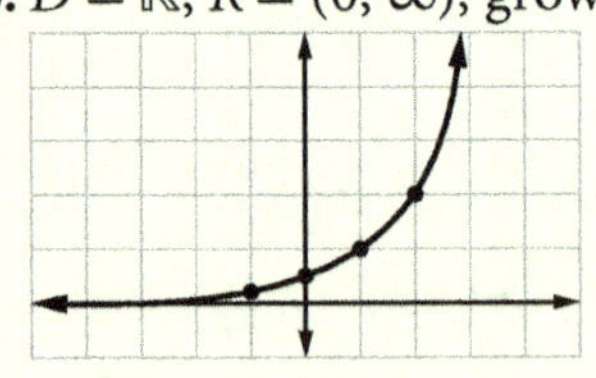

47. $D = \mathbb{R}$; $R = (0, \infty)$; decay

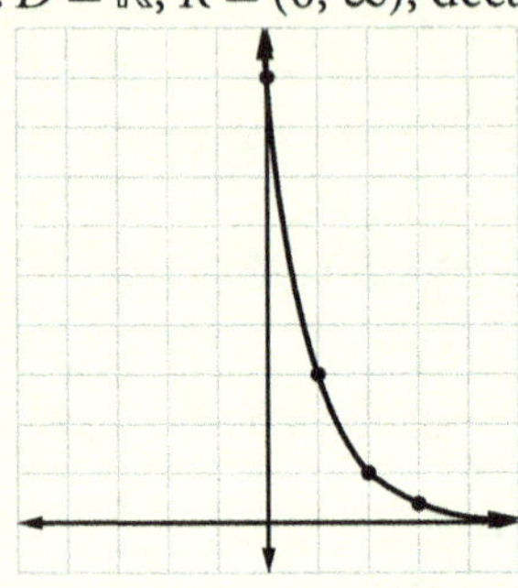

48. $[x - (3 - i)][x - (3 + i)]$
$\quad \cdot [x - (5 + 2i)][x - (5 - 2i)] = 0$
$[(x - 3) + i][(x - 3) - i]$
$\quad \cdot [(x - 5) - 2i][(x - 5) + 2i] = 0$
$[(x - 3)^2 - i^2][(x - 5)^2 - 4i^2] = 0$
$(x^2 - 6x + 10)(x^2 - 10x + 29) = 0$
$x^4 - 16x^3 + 99x^2 - 274x + 290 = 0$

49. $c^2 = 6^2 + 10^2$

$c = \sqrt{136} = 2\sqrt{34}$

(The smaller acute $\angle A$ is opposite the shortest side.)

$\sin A = \dfrac{6}{2\sqrt{34}} = \dfrac{3}{\sqrt{34}} = \dfrac{3\sqrt{34}}{34}$

$\cos A = \dfrac{10}{2\sqrt{34}} = \dfrac{5}{\sqrt{34}} = \dfrac{5\sqrt{34}}{34}$

$\tan A = \dfrac{6}{10} = \dfrac{3}{5}$

51. $\left[\dfrac{2x + 3}{(x + 1)(x + 2)} = \dfrac{A}{x + 1} + \dfrac{B}{x + 2} \right]$
$\qquad\qquad\qquad \cdot (x + 1)(x + 2)$

$2x + 3 = A(x + 2) + B(x + 1)$
$\qquad = Ax + 2A + Bx + B$
$\qquad = (A + B)x + (2A + B)$

$A + B = 2$
$- (2A + B = 3)$
$\overline{ -A = -1}$
$\qquad A = 1; B = 1$

Write the slope-intercept form equation of each described line. [1.2]

44. passing through $(0, 3)$ and $(8, -2)$ $y = -\frac{5}{8}x + 3$

45. passing through $(-1, 5)$ and perpendicular to the line containing $(0, 4)$ and $(-3, 5)$ $y = 3x + 8$

Graph each exponential function and state its domain and range. Then classify the function as exponential growth or decay. [3.1]

46. $y = \frac{1}{2} \cdot 2^x$ **47.** $y = 9\left(\frac{1}{3}\right)^x$

48. Find a fourth-degree polynomial equation with roots of $3 - i$ and $5 + 2i$. [2.4]

49. Right $\triangle ABC$ has legs 6 cm and 10 cm long. Find the exact value of the sine, cosine, and tangent of the smaller acute $\angle A$. [4.2]

50. Which region(s) represent the solution of the system?
$-3x + 5y \le 7$
$y \ge -2x^2 - 8x + 5$ [7.6] E

A. region 1
B. region 4
C. region 5
D. regions 2 and 4
E. regions 3 and 5

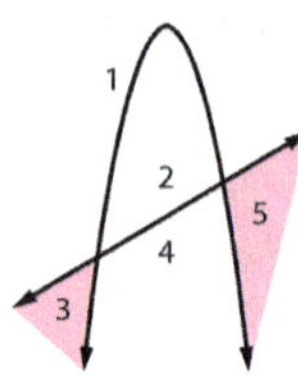

48. $x^4 - 16x^3 + 99x^2 - 274x + 290 = 0$

49. $\sin A = \frac{3\sqrt{34}}{34}$; $\cos A = \frac{5\sqrt{34}}{34}$; $\tan A = \frac{3}{5}$

51. Find the partial fraction decomposition for $\dfrac{2x + 3}{x^2 + 3x + 2}$. [7.7] A

A. $\frac{1}{x+1} + \frac{1}{x+2}$ **D.** $\frac{2}{x+1} + \frac{3}{x+2}$

B. $\frac{6}{x+1} + \frac{1}{x+2}$ **E.** none of these

C. $\frac{3}{x+1} + \frac{2}{x+2}$

52. Convert $x^2 + 3y^2 - 6x + 6y = 0$ to a standard form for an ellipse. [8.2] B

A. $\frac{(x-3)^2}{12} - \frac{(y+1)^2}{4} = 1$ **D.** $\frac{(x-3)^2}{3} + \frac{(y+1)^2}{4} = 1$

B. $\frac{(x-3)^2}{12} + \frac{(y+1)^2}{4} = 1$ **E.** none of these

C. $\frac{(x+3)^2}{3} + \frac{(y+1)^2}{4} = 1$

53. Use the discriminant test to classify the conic $x^2 + y^2 - 6x + 8y + 7 = 0$. [8.4] D

A. parabola **C.** ellipse **E.** none of these
B. hyperbola **D.** circle

52. $(x^2 - 6x) + 3(y^2 + 2) = 1$
$(x^2 - 6x + 9) + 3(y^2 + 2y + 1)$
$\qquad\qquad\qquad = 1 + 9 + 3(1)$
$(x - 3)^2 + 3(y + 1)^2 = 12$
$\dfrac{(x - 3)^2}{12} + \dfrac{(y + 1)^2}{4} = 1$

53. $B^2 - 4AC = (0)^2 - 4(1)(1) = -4 < 0$
an ellipse but $A = C = 1$ and $B = 0$,
so the special case of a circle

A sequence is a function whose domain contains natural numbers. It can be graphed in FUNCTION mode as a scatter-plot using its explicit formula. It can also be graphed in SEQUENCE mode using either its recursive or explicit formula.

To graph a sequence while in FUNCTION mode, select option 5 from the LIST ([2nd], [STAT]) OPS menu. To store the first ten natural numbers in L_1, store seq(X, X, 1, 10, 1) in L_1, where the first X is the expression and the second X is the variable that starts at 1 and ends at 10 with a step of 1. The explicit formula for a sequence can then be used to store the first ten terms of the sequence in L_2. Given $a_n = \dfrac{2n-1}{4n}$, store $(2L_1-1) \div (4L_1)$ as L_2. The answer can be converted to fractions using 1:▶ from the [MATH] menu.

Clear any equations in the [Y=] editor and then use STAT PLOT ([2nd], [Y=]) to turn on Plot1 and [GRAPH] the scatterplot. [TRACE] can be used to identify the ordered pairs of the sequence. Notice how the graph illustrates the sequence as it approaches its $\lim\limits_{n\to\infty} \dfrac{2n-1}{4n} = 0.5$.

Selecting SEQUENCE mode allows the sequence to be entered using either its explicit or recursive formula. Press [Y=] and turn Plot1 off. To graph the Fibonacci sequence, enter 1 for nMin, then enter the recursive formula u(n)=u(n−1)+u(n−2) and the first term (or in this case a list of the first two initial terms) of the sequence for u(nMin). Enter the explicit formula for the linear sequence v(n)=4n+3. Note that v(nMin) is left blank when the sequence is defined explicitly. Use [WINDOW] to define nMax and other characteristics of the graphing window. [TRACE] can again be used to identify ordered pairs of the sequence. The sequences can also be viewed using TABLE ([2nd], [GRAPH]). Use the TBLSET menu ([2nd], [WINDOW]) to be sure that TblStart = 1 and ΔTbl = 1.

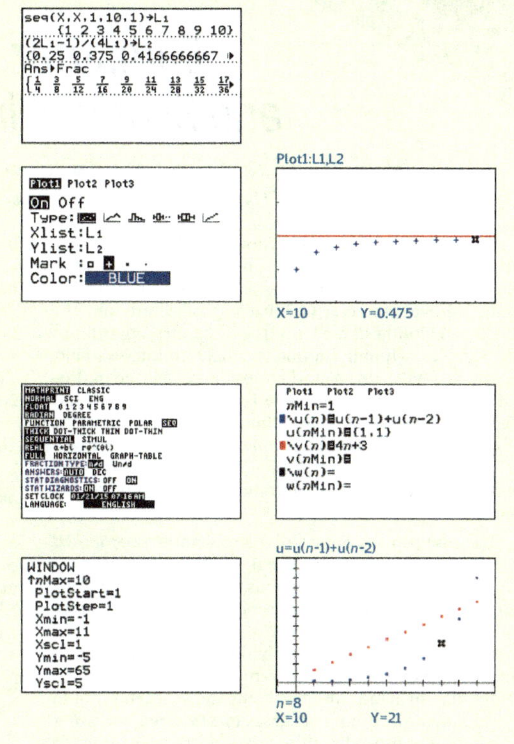

Biblical Perspective of Mathematics

Objectives

1. To describe how computers have advanced the quest for mathematical truth

2. To describe and apply the technique of proof by enumeration of cases (also know as proof by exhaustion)

3. To evaluate artificial intelligence from a biblical worldview

Assignments

- **Minimum:** 1–8; 9–10 (in class)
- **Standard:** 1–8; 9–10 (in class)
- **Extended:** 1–10

Computer Proofs and Artificial Intelligence

[My proof] differs significantly from earlier approaches by making extensive use of computers.

—*Thomas Hales (American mathematician who proved the Kepler Conjecture)*

Over the last six decades computers have been very helpful in attacking famous unsolved problems in mathematics. For example, computers have supplied extensive inductive evidence that Goldbach's Conjecture and the Riemann Hypothesis are both true by confirming trillions of examples without finding a single counterexample. Computers have also played an instrumental role in providing deductive proofs for the Four-Color Problem and Kepler's Conjecture using a technique called enumeration of cases (or proof by exhaustion). This technique establishes the conclusion by dividing the problem into a finite number (often a very large number) of cases and proving each case separately.

According to the Four-Color Problem, first proposed in 1852, four colors at most are needed to color any separation of a plane into contiguous regions so that no two adjacent regions have the same color. The problem's proof by exhaustion in 1976 was controversial since it was the first major proof to heavily rely on computer assistance; the majority of its 1936 cases were checked by a computer rather than by hand. Another computer-assisted proof in 1996 reduced the number of cases to 633. The Four-Color Theorem was proved with generalized theorem-proving software in 2005. The field of automated theorem proving (ATP) is still growing in application and acceptability.

The Kepler Conjecture was first analyzed in the late-sixteenth century as a cannonball packing problem. It is now a mathematical theorem stating that the density of equal-sized spheres packed in space is never greater than $\frac{\pi}{3\sqrt{2}} \approx 74.05\%$. In the early 1990s Thomas Hales and Samuel Ferguson, one of Hales's graduate students at the University of Michigan, developed a proof by exhaustion that systematically applied linear programming methods to over 5000 different configurations of spheres. Hales used Mathematica® software to complete the symbolic math, and his work was verified by Ferguson using Maple® software. After independently writing key portions of the programs twice and checking the results on different types of computers, Hales declared their success in an email to colleagues around the world in 1998. After a referee's panel declared it could not certify the correctness of the computer calculations in 2003, a collaborative project led by Hales sought to create a formal proof verified by automated proof-checking software. The project was completed in 2014 and the formal proof of the Kepler Conjecture was accepted by a leading research journal in 2017.

ATP is a subfield of artificial intelligence (AI). From playing chess and other games at higher and higher levels, AI has improved to the point that an IBM® computer named Watson defeated two champions from the popular game show *Jeopardy!* in 2011. *Forbes* magazine reported in 2013 that the technology is now being used to improve successful patient diagnosis rates in hospitals, where nurses using Watson as a resource take the AI program's advice 90% of the time. AI is now nearly ubiquitous with applications such as Apple's Siri® and GPS navigation.

In 2014 the world-renowned physicist Stephen Hawking told a BBC interviewer, "The development of full artificial intelligence could spell the end of the human race. It would take off on its own, and re-design itself at an ever-increasing rate . . . . Humans, who are limited by slow biological evolution, couldn't compete, and would be superseded."

In his 2017 magazine article entitled "A Christian View of Artificial Intelligence," Cody Volkers said, "Technology has always been about making life better for the human race. In its best forms, AI continues this tradition. But in its more dangerous iterations, AI aspires to attain omniscience, something reserved only for God." We should realize that real-world applications of AI rely on imperfect algorithms for sifting imperfect data sets which do not represent all the information related to the problem. It is crucial that humans understand that they are made in God's image, retain their own sense of humanity, and have the courage to overrule AI output. Despite the hype, computers don't think; they merely perform algorithms written by people who do think.

PRESENTATION

John Casti's 2001 text *Mathematical Mountaintops: The Five Most Famous Problems of All Time* describes five famous mathematical problems that were resolved in the last half of the twentieth century. The Four-Color Problem and the Kepler Conjecture were solved as computer proofs in 1976 and 1998, respectively. Explain that computers helped complete the proofs by exhaustion for these two mathematical truths.

Computer-assisted proofs were not universally accepted at first. Until the twentieth century it was assumed that any proof could, in principle, be quickly confirmed for validity by a competent mathematician. Some mathematicians were concerned that the possibility of a programming error or a run-time error would make a computer-assisted proof unreliable. Another objection to computer-assisted proofs was the lack of beauty and clarity in the proofs. In his book *Kepler's Conjecture* George Szpiro praises Andrew Wiles's proof of Fermat's Last Theorem, saying, "*That* was a classy proof. It was beautiful, it had elegance, it had style." In contrast, English mathematician Ian Stewart compared Hales's computer-assisted proof to a telephone directory, saying, "Short it is not. Elegant it is not." G. H. Hardy remarked in his *A Mathematician's Apology* that "'enumeration of cases', indeed, is one of the duller forms of mathematical argument. A mathematical proof should resemble a simple and clear-cut constellation, not a scattered cluster in the Milky Way."

Discuss automated theorem proving (ATP) as a subfield of artificial intelligence (AI). Consider having the students name positive and negative examples of AI, such as medical diagnosis, market analysis, individualized tutoring, robotic control, training simulators, and Internet hacking tools.

Direct attention to the apocalyptic warning in the feature and guide the stu-

1. Name two famous unproven math statements for which computers have provided extensive inductive evidence.

2. Name two famous math problems that were solved in the latter half of the twentieth century using deductive computer proofs.

3. Name and describe the type of proof used in solving these two famous math problems.

4. To verify the conjecture that there are no 4-digit numbers that are integral multiples of their "reversals" using the proof-by-exhaustion technique, how many cases would need to be checked?

5. Show that each number is a counterexample for the conjecture in exercise 4.

   a. 8172   **b.** 9801

6. A computer can verify that the other 4-digit numbers are not integral multiples of their reversals. Revise the conjecture in exercise 4 to state a theorem that is proved by enumeration of cases.

7. How are ATP and AI related?

8. Which two popular mathematic software packages were utilized by Hales and Ferguson in their proof of the Kepler Conjecture?

9. **Discuss:** Describe several ways that computers have advanced the quest for mathematical truth.

10. **Discuss:** In his 2017 article Cody Volkers continues, "Some AI pioneers believe they can create a machine with boundless potential for learning, a machine that can conquer all knowledge. This kind of AI would go beyond mere usefulness and amount to a modern Tower of Babel." How should a Christian view artificial intelligence?

1. Goldbach's Conjecture and the Riemann Hypothesis

2. the Four-Color Problem and the Kepler Conjecture

3. proof by exhaustion (or enumeration of cases); This technique splits the problem into a finite number of cases and proves each one separately.

4. 9000—all the 4-digit numbers individually from 1000 to 9999

5a. $8712 = 2178 \times 4$

5b. $9801 = 1089 \times 9$

6. There are exactly 2 four-digit numbers that are integral multiples of their reversals.

7. ATP is a subfield of AI.

8. Mathematica and Maple

9. Computers have provided inductive evidence for some famous problems and deductive proofs for others. Software has been developed to aid in tedious calculations for proving theorems.

10. Christians can embrace AI as a tool that can be used to better fulfill the Creation Mandate and to glorify God by helping others. If the pursuit of AI becomes a search for omniscience or a tool that we rely upon instead of seeking God's wisdom, then we are destined for failure and judgment.

dents to a biblical response recognizing God's sovereignty. Read Cody Volkers's quote concerning those who would seek to usurp God's authority and characteristics. In the same article Volkers wrote, "AI is quickly advancing, and its moral trajectory is at odds with biblical morality, values, and assertions about reality. AI aspires to replicate human minds, but we know only real humans bear the image and likeness of God. AI's atheistic proponents assume that human beings can be reduced to a set of cognitive patterns, a numerical code, a mathematical formula."

For a deeper look at the relationship between AI and religion, read Brandon Withrow's article "The New Religions Obsessed with A.I." His discussion of transhumanism is enlightening and sobering. One specific new religion mentioned in the article is Way of the Future. Withrow writes that "according to WIRED, the mission of the new religion is to 'develop and promote the realization of a Godhead based on Artificial Intelligence,' and 'through understanding and worship of the Godhead, [to] contribute to the betterment of society.'"

9.2 Arithmetic Sequences and Series

Objectives

1. To represent arithmetic sequences using recursive and explicit formulas
2. To find the *n*th terms and arithmetic means of arithmetic sequences
3. To find the *n*th partial sums of arithmetic sequences
4. To model and solve real-world problems using arithmetic sequences

Flash

A water well can cost a few thousand to tens of thousands of dollars to drill. Water wells have been used for thousands of years. The first drilled wells in the United States were drilled in the early nineteenth century, but most wells were dug by hand until the twentieth century.

Vocabulary

arithmetic mean
arithmetic sequence
arithmetic series
common difference
harmonic sequence

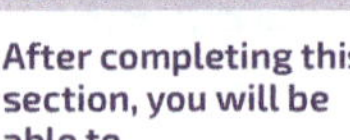

9.2 Arithmetic Sequences and Series

The cost of drilling a well is modeled by an arithmetic series in exercise 38.

After completing this section, you will be able to

- represent arithmetic sequences using recursive and explicit formulas.
- find *n*th terms and arithmetic means of arithmetic sequences.
- find *n*th partial sums of arithmetic sequences.
- model and solve real-world problems using arithmetic sequences.

Using an arithmetic sequence to model mortality rates was a key simplification that allowed early actuaries to easily compute the cost of an annuity.

If \$1000 is invested at 8% simple interest for 5 yr, it earns \$400 in interest during that time. The \$80 annual return can be seen in the sequence of annual balances: 1080, 1160, 1240, 1320, 1400. Notice that the difference between consecutive terms is the same.

> **DEFINITION**
>
> In an **arithmetic sequence**, the difference between any two consecutive terms is a constant, *d*, which is called the *common difference*.

This definition implies that the common difference can be found by subtracting any two consecutive terms and that arithmetic sequences can be described by the *recursive formula*

$$a_n = a_{n-1} + d.$$

Example 1   Identifying an Arithmetic Sequence

Determine whether each sequence is arithmetic. If so, state the common difference and list the next 3 terms. If not, explain why.

a. 9, 13, 17, 21, 25, . . .   **b.** 2, 5, 10, 17, 26, . . .

Answer

a. arithmetic; $d = 4$;
$a_6 = 29$; $a_7 = 33$; $a_8 = 37$

b. not arithmetic; The differences are not constant: 3, 5, 7, 9, . . .

SKILL ✔ EXERCISE 3

The *explicit formula* for an arithmetic sequence can be derived using the recursive formula, which states that each term is found by adding *d* to the previous term.

$$a_1 = a_1$$
$$a_2 = a_1 + d$$
$$a_3 = a_1 + 2d$$
$$a_4 = a_1 + 3d$$
$$\vdots$$
$$a_n = a_1 + (n-1)d$$

Example 2   Finding and Applying Recursive and Explicit Formulas

Write both the recursive formula and a simplified explicit formula for each arithmetic sequence. Then find a_{20}.

a. 19, 16, 13, 10, 7, . . .   **b.** ln 2, ln 6, ln 18, ln 54, . . .

Answer

a. $a_1 = 19$; $a_n = a_{n-1} - 3$

1. Use the common difference $d = -3$ to write the recursive formula.

$$a_n = 19 + (n-1)(-3)$$
$$a_n = -3n + 22$$

2. Substitute into the explicit formula $a_n = a_1 + (n-1)d$ and simplify.

$$a_{20} = -3(20) + 22 = -38$$

3. Substitute and simplify to find a_{20}.

CONTINUED ➡

PRESENTATION

Lesson Opener

Write each expression as a single term.

1. $\ln 3 + \ln x$ ln 3*x*
2. $\log x - \log y$ $\log \frac{x}{y}$
3. $3 \ln y - \ln x$ $\ln \frac{y^3}{x}$

Classify each formula as recursive or explicit.

4. $a_n = a_{n-1} + a_{n-2}$ recursive
5. $a_n = dn + a_0$ explicit

The Lesson Opener can be used to review properties of logarithms that are used in Example 2. The students may also need to review recursive versus explicit formulas.

Although compound interest is significantly more common, some CDs will pay simple interest. An ordered list of annual balances for the simple interest example given in the text represents an *arithmetic sequence.*

Present the definition of an arithmetic sequence and the recursive formula. Emphasize the need for a *common difference* and have students determine whether the sequences in Example 1 are arithmetic.

One-on-One The terms of the sequences can be thought of as pickets in a fence and the common difference as a uniform gap between any pair of pickets. Between any *n* pickets there are $n - 1$ gaps.

Show how the explicit formula for an arithmetic sequence is derived and then derive the recursive and explicit formulas in Example 2. In Example 2*b*, emphasize the fact that the common difference can be found using any two consecutive terms. Encourage some students to find $d = a_2 - a_1$ and the others to find $d = a_4 - a_3$. Then compare their results to the text's use of $d = a_3 - a_2$. The simplification in step 3 emphasizes the linear (arithmetic) nature of the sequence. An alternate simplification follows.

$$a_n = \ln 2 + (n-1)(\ln 3)$$
$$= \ln 2 + n \ln 3 - \ln 3$$
$$= \ln 2 + \ln 3^n - \ln 3$$
$$= \ln 2 + \ln 3^{n-1} = \ln 2 \cdot 3^{n-1}$$

b. $d = \ln 18 - \ln 6 = \ln \frac{18}{6} = \ln 3$

$a_1 = \ln 2;\ a_n = a_{n-1} + \ln 3$

$a_n = \ln 2 + (n-1)(\ln 3)$
$ = n\ln 3 + \ln 2 - \ln 3$

$a_n = n\ln 3 + \ln \frac{2}{3}$

$a_{20} = 20\ln 3 + \ln \frac{2}{3}$

$\phantom{a_{20}} = \ln \frac{2 \cdot 3^{20}}{3} = \ln(2 \cdot 3^{19})$

1. Determine the common difference.

2. Write the recursive formula.

3. Substitute into the explicit formula
   $a_n = a_1 + (n-1)d$ and simplify.

4. Substitute and simplify to find a_{20}.

_____________________ SKILL ✔ **EXERCISES 9, 13**

The graph of the sequence in Example 2a shows that the function is linear with a slope of -3, the common difference. In general, the explicit formula can also be stated as $a_n = dn + a_0$ where $a_0 = a_1 - d$.

The fact that arithmetic sequences are linear with a slope of d can be used to derive a generalized explicit formula given any two terms (k, a_k) and (n, a_n).

$$\frac{a_n - a_k}{n - k} = d$$
$$a_n - a_k = (n - k)d$$
$$a_n = a_k + (n - k)d$$

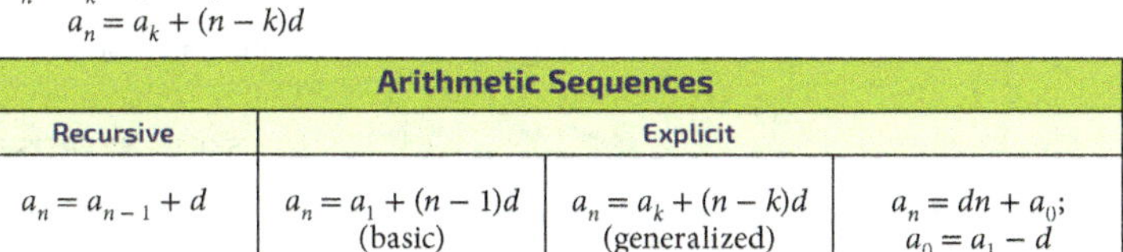

Arithmetic Sequences			
Recursive	**Explicit**		
$a_n = a_{n-1} + d$	$a_n = a_1 + (n-1)d$ (basic)	$a_n = a_k + (n-k)d$ (generalized)	$a_n = dn + a_0;$ $a_0 = a_1 - d$ (simplified)

Example 3 Writing and Using Explicit Formulas

Given the arithmetic sequence with $b_3 = 8$ and $b_7 = -6$, write a simplified explicit formula for b_n and then use the formula to find b_{20}.

Answer

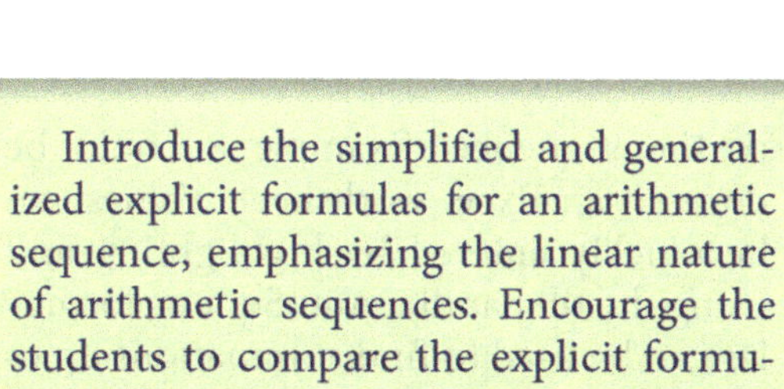

$b_7 = b_3 + (7-3)d$
$-6 = 8 + 4d$
$-14 = 4d$
$d = -\frac{7}{2}$

1. Find d by substituting both terms into $b_n = b_k + (n-k)d$ and solving.

$b_n = 8 + (n-3)\left(-\frac{7}{2}\right)$
$b_n = 8 - \frac{7}{2}n + \frac{21}{2}$
$b_n = -\frac{7}{2}n + \frac{37}{2}$

2. Substitute d and $b_3 = 8$ (with $k = 3$) into $b_n = b_k + (n-k)d$ and simplify.

$b_{20} = -\frac{7}{2}(20) + \frac{37}{2}$
$\phantom{b_{20}} = -\frac{103}{2}$ or -51.5

3. Evaluate the explicit formula to find b_{20}.

_____________________ SKILL ✔ **EXERCISE 15**

Additional Exercises

1. Determine whether the sequence 17, 14, 11, 8, 5, . . . is arithmetic. If so, state the common difference and list the next 3 terms. If not, explain why. yes; $d = -3;\ 2, -1, -4$

2. Write the recursive formula and a simplified explicit formula for the arithmetic sequence 3, 7, 11, . . . , and then find a_{20}. $a_n = a_{n-1} + 4,$ $a_1 = 3;\ a_n = 4n - 1;\ a_{20} = 79$

3. Write the simplified explicit formula when $a_5 = 5$ and $a_{11} = 17$. Then find a_{20}. $a_n = 2n - 5;\ 35$

4. Find the arithmetic means between $a_{10} = 17$ and $a_{14} = 5$. 14, 11, 8

5. Write the first 5 terms of the sequence $a_n = \dfrac{1}{n^2 + 2}$. Then classify the sequence as arithmetic, harmonic, or neither and state whether it is convergent or divergent. $\dfrac{1}{3}, \dfrac{1}{6}, \dfrac{1}{11}, \dfrac{1}{18}, \dfrac{1}{27};$ neither; convergent

Evaluate each arithmetic series.

6. $\displaystyle\sum_{k=1}^{20} 5k - 3$ 990

7. $9 + 18 + 27 + \cdots + 243$ 3402

8. Write the simplified explicit formula when $a_1 = 7$ and $S_{30} = 1950$. $a_n = 4n + 3$

Introduce the simplified and generalized explicit formulas for an arithmetic sequence, emphasizing the linear nature of arithmetic sequences. Encourage the students to compare the explicit formulas in the summary table and to describe the distinctions of each. Point out that the basic formula references a_1, the generalized form a_k, and the simplified form a_0.

Students should recognize that the generalized formula is applied in Example 3 because a_1 and d are not known.

Define *arithmetic means* and demonstrate how they are found with Example 4. After using an explicit formula to find the common difference, the recursive formula is used to find each arithmetic mean. Point out how the arithmetic means can be found for the terms in Example 3.

One-on-One While the single arithmetic mean is also called the average, there are three commonly used measures of central tendency for describing the typical or average value: median, mode, and mean.

Define *harmonic sequence* and then present Example 5. Point out that any arithmetic sequence (except the degenerate case of $d = 0$) is divergent. Note that the pattern for the sequence in part b can be seen when it is rewritten as

$$\frac{2}{1}, \frac{2}{2}, \frac{2}{3}, \frac{2}{4}, \frac{2}{5}, \dots$$

Define an *arithmetic series* and derive the related formulas. This concept is introduced in BJU Press's *ALGEBRA 2* Section 12.4 by the narrative of Gauss using a similar technique to derive the sum of the first 100 integers. At the age of ten Carl Gauss astounded his teacher and classmates by responding to his teacher's challenge to find the sum. The teacher expected that the challenge would keep the students busy for some time, but Gauss quickly and accurately found the sum by recognizing a pattern. He saw that $100 + 1 = 101, 99 + 2 = 101, \dots, 51 + 50 = 101$, and that the sum of these fifty pairs of numbers would be $50(101) = 5050$.

9. Daniel wants to participate in a missions project by giving his $25/week allowance and increasing the amount he gives by $2 each week after the first week. Write a simplified explicit formula for the arithmetic sequence representing his giving. What would be Daniel's total giving for 10 weeks? for 18 weeks?

$a_n = 2n + 23; \$340; \$756$

Assignments

- **Minimum:** 1, 3–5, 7–9, 14–15, 19, 22, 26(*a–b*), 27, 30, 33, 37, 45–46, 49–51, 53–54
- **Standard:** 1–7 odd; 10, 13, 15, 17–18, 23–24, 26(*b, d*); 27–29 odd; 46, 48–49, 51–54
- **Extended:** 4–7, 11–12, 14, 16–17, 20–22, 25, 26(*c–d*), 28–29, 32, 35, 38, 40, 42–44, 47–48, 52, 54

Assessment

- Quiz 9A covers Sections 9.1–9.2.

The terms between two given terms of an arithmetic sequence are called *arithmetic means*.

Example 4   Finding Arithmetic Means

Write an arithmetic sequence with the following arithmetic means.

a. three arithmetic means between 1.7 and 7.3

b. one arithmetic mean between −8 and 15

Answer

a. Let $a_1 = 1.7$ and $a_5 = 7.3$.
$$7.3 = 1.7 + (5 - 1)d$$
$$5.6 = 4d$$
$$d = 1.4$$
$$a_2 = 1.7 + 1.4 = 3.1$$
$$a_3 = 3.1 + 1.4 = 4.5$$
$$a_4 = 4.5 + 1.4 = 5.9$$

1. The three arithmetic means are a_2, a_3, and a_4.
2. Substitute a_5 and a_1 into $a_n = a_1 + (n - 1)d$ and solve for d.
3. Use the recursive formula to find the three arithmetic means.
   Find $a_5 = 5.9 + 1.4 = 7.3$ to verify your work.

b. Let $a_1 = -8$ and $a_3 = 15$.
$$15 = -8 + (3 - 1)d$$
$$23 = 2d$$
$$d = 11.5$$
$$a_2 = -8 + 11.5 = 3.5$$

1. The arithmetic mean is a_2.
2. Substitute a_3 and a_1 into $a_n = a_1 + (n - 1)d$ and solve for d.
3. Use the recursive formula to find the arithmetic mean.
   Find $a_3 = 3.5 + 11.5 = 15$ to verify your work.

_________________________ SKILL ✔ **EXERCISE 19**

Note that finding a single arithmetic mean between two numbers is the same as finding the average of the numbers: $\frac{-8 + 15}{2} = \frac{7}{2} = 3.5$.

The reciprocals of the terms of an arithmetic sequence form a *harmonic sequence*.

Example 5   Classifying a Sequence

Classify each sequence as arithmetic, harmonic, or neither. Then describe each as convergent or divergent.

a. $a_n = 7n - 2$

b. $2, 1, \frac{2}{3}, \frac{1}{2}, \frac{2}{5}, \ldots$

Answer

a. arithmetic
   divergent

1. The explicit formula is linear.
2. Each term in the sequence 5, 12, 19, … is 7 more than the previous term.

b. harmonic

   convergent

1. The term's reciprocals $\frac{1}{2}, 1, \frac{3}{2}, 2, \frac{5}{2}, \ldots$ form the arithmetic sequence $a_n = \frac{1}{2}n$.
2. $\lim\limits_{n \to \infty} \frac{2}{n} = 0$

_________________________ SKILL ✔ **EXERCISE 23**

Common Student Error Students can easily confuse sequences with series.

Demonstrate evaluating the series given in Example 6. Note that the second formula could be used in part *a* after noting that $a_1 = 2$ and $d = 2$:

$$S_{50} = \frac{50}{2}[2(2) + 49(2)] = 2550.$$

The first formula could be used in part *b* after finding $a_n = 11(500) - 6 = 5494$:

$$S_{50} = \frac{500}{2}(5 + 5494) = 1{,}374{,}750.$$

Have the students note that they need to find d to write an explicit formula in Example 7. Because a_1, S_n, and n are known, we can plug those values into the arithmetic series formula to find the common difference, d. Some of the students may recall from past courses that the sum of the first n counting numbers is $\frac{n(n + 1)}{2}$.

Applications of sequences and series, such as amortizations for loans, are common in the financial world. Example 8 illustrates one such application.

Interactive Activity A spreadsheet works well for replicating terms of an arithmetic sequence and finding partial summations. This tool allows the students to create long sequences with little effort and provides a good visualization of the sequence. Example 8 could be done in an Excel worksheet. Values can be quickly entered by highlighting and using the fill handle. AutoSum or formulas can be used to find summations up to a specific term. Graphs of sequences can then be easily created on the worksheet.

DEFINITION

An **arithmetic series** is a sum of the terms of an arithmetic sequence. The sum of the first n terms of $a_1, a_2, a_3, \ldots$ is the nth partial sum, $S_n = a_1 + a_2 + a_3 + \cdots + a_n$.

Given the arithmetic sequence $2, 4, 6, \ldots$, the 50th partial sum $S_{50} = 2 + 4 + 6 + \cdots + 100$ could be evaluated by completing each addition. A better approach is to derive and apply a general formula for S_n in any arithmetic sequence. The key is to use two expressions for $S_n = a_1 + a_2 + \cdots + a_{n-1} + a_n$, the first in terms of a_1 and the second in terms of a_n.

$S_n = a_1 + (a_1 + d) + (a_1 + 2d) + \cdots + [a_1 + (n-1)d]$ 1. Each term after a_1 is found by repeatedly adding d.

$S_n = a_n + (a_n - d) + (a_n - 2d) + \cdots + [a_n - (n-1)d]$ 2. Reverse the order of the terms, beginning with a_n, and repeatedly subtract d.

$S_n + S_n = (a_1 + a_n) + (a_1 + a_n) + \cdots + (a_1 + a_n)$ 3. Add the two equations for S_n.

$2S_n = n(a_1 + a_n)$ 4. There are n identical terms.

$S_n = \frac{n}{2}(a_1 + a_n)$ 5. Solve for S_n.

When a_n is not known, substitute using $a_n = a_1 + (n-1)d$ and simplify.

$$S_n = \frac{n}{2}[a_1 + (a_1 + (n-1)d)] = \frac{n}{2}[2a_1 + (n-1)d]$$

Arithmetic Series
$S_n = \frac{n}{2}(a_1 + a_n)$ or $S_n = \frac{n}{2}[2a_1 + (n-1)d]$

Example 6 Evaluating an Arithmetic Series

Evaluate each sum.

a. $2 + 4 + 6 + \cdots + 100$

b. $\displaystyle\sum_{k=1}^{500} (11k - 6)$

Answer

a. $100 = 2 + (n-1)(2)$
$100 = 2n$
$n = 50$

1. Determine the value of n using $a_n = a_1 + (n-1)d$.

$S_n = \frac{50}{2}(2 + 100)$
$= 25(102) = 2550$

2. Substitute $a_1 = 2$, $a_n = 100$, and $n = 50$ into $S_n = \frac{n}{2}(a_1 + a_n)$ and simplify.

b. $a_1 = 11(1) - 6 = 5$
$d = 11$

1. Recognize the linear expression, $11k - 6$, as the explicit formula for an arithmetic sequence, $a_k = dk + a_0$.

$S_{500} = \frac{500}{2}[2(5) + (500 - 1)11]$
$= 1,374,750$

2. Substitute into $\frac{n}{2}[2a_1 + (n-1)d]$ and simplify.

SKILL ✔ EXERCISE 27

TIPS

Ex. 42–44 Each partial sum of the harmonic sequence $a_n = \frac{1}{n}$ is a harmonic number.

Ex. 42 Manual and calculator solutions are provided. Calculator functions cumSum and seq(are found on the [LIST], OPS menu.

Ex. 43 The proof that the growth of harmonic numbers is logarithmic is beyond the scope of this text. (Since $S_1 - \ln(1) = 1$ while $S_{30} - \ln(30) \approx 0.6$, students might understandably question the claim in this exercise.)

Ex. 44 You may want to encourage students to use these functions to check answers in other exercises.

Solutions

❯ A. Exercises

6. $d = -1$; $a_2 = 4$, $a_3 = 3$, $a_4 = 2$, $a_5 = 1$

7. $m = 6 = d$; $a_5 = 6(5) - 8 = 22$

8. $d = -5 - (-1) = -4$
$a_1 = -1$, $a_n = a_{n-1} - 4$
$a_n = -1 + (n-1)(-4) = -4n + 3$
$a_{20} = -4(20) + 3 = -77$

9. $d = 19 - 7 = 12$
$a_1 = 7$, $a_n = a_{n-1} + 12$
$a_n = 7 + (n-1)(12) = 12n - 5$
$a_{20} = 12(20) - 5 = 235$

10. $d = 8$; $a_1 = 8$, $a_n = a_{n-1} + 8$
$a_n = 8 + (n-1)(8) = 8n$
$a_{20} = 8(20) = 160$

11. $d = 2$; $a_1 = 1$, $a_n = a_{n-1} + 2$
$a_n = 1 + (n-1)(2) = 2n - 1$
$a_{20} = 2(20) - 1 = 39$

12. $d = \ln 12 - \ln 3 = \ln \frac{12}{3} = \ln 4$
$a_1 = \ln 3$, $a_n = a_{n-1} + \ln 4$
$a_n = \ln 3 + (n-1)(\ln 4)$
$\quad = n \ln 4 + \ln 3 - \ln 4$
$\quad = n \ln 4 + \ln \frac{3}{4}$
$a_{20} = (20) \ln 4 + \ln \frac{3}{4}$
$\quad = \ln \frac{3 \cdot 4^{20}}{4} = \ln (3 \cdot 4^{19})$

13. $d = \log 1000 - \log 10 = 3 - 1 = 2$
$a_1 = \log 10$, $a_n = a_{n-1} + 2$
$a_n = \log 10 + (n-1)(2)$
$\quad = 2n + 1 - 2$
$\quad = 2n - 1$
$a_{20} = 2(20) - 1 = 39$

14. $a_1 = 9$; $d = 4$
$a_n = 9 + (n-1)4 = 4n + 5$
$a_{20} = 4(20) + 5 = 85$

15. $a_{23} = a_9 + (23 - 9)d$
$-49 = 35 + 14d$
$\quad d = -6$
$a_n = a_9 + (n - 9)(-6)$
$\quad = 35 - 6n + 54 = -6n + 89$
$a_{20} = -6(20) + 89 = -31$

16. $a_{15} = a_7 + (15 - 7)d$
$90 = 52 + 8d$
$\quad d = 4.75$
$a_n = a_7 + (n - 7)4.75$
$\quad = 52 + 4.75n - 33.25$
$\quad = 4.75n + 18.75$
$a_{20} = 4.75(20) + 18.75 = 113.75$

17. The arithmetic sequence of denominators has $a_1 = 5$ and $d = 7$.

$$a_n = 5 + (n-1)7 = 7n - 2$$

The harmonic sequence is the reciprocal sequence.

$$a_{20} = \frac{1}{7n-2} = \frac{1}{7(20)-2} = \frac{1}{138}$$

B. Exercises

18. $a_{21} = a_{18} + (21 - 18)d$
$$30 = 9 + 3d$$
$$d = 7$$
$$a_{19} = a_{18} + 7 = 9 + 7 = 16$$
$$a_{20} = a_{19} + 7 = 16 + 7 = 23$$

19. $a_{14} = a_{10} + (14 - 10)d$
$$71 = 32 + 4d$$
$$d = 9.75$$
$$a_{11} = a_{10} + 9.75 = 32 + 9.75 = 41.75$$
$$a_{12} = a_{11} + 9.75 = 41.75 + 9.75 = 51.5$$
$$a_{13} = a_{12} + 9.75 = 51.5 + 9.75 = 61.25$$

20. The single arithmetic mean between two terms is their average.

$$a_7 = \frac{a_6 + a_8}{2} = \frac{25.83 + 94.05}{2} = 59.94$$

21. $a + d = x$ and $x + d = b$
$$a = x - d$$
$$+\ (b = x + d)$$
$$\overline{a + b = 2x}$$
$$x = \frac{a+b}{2}$$

26a.

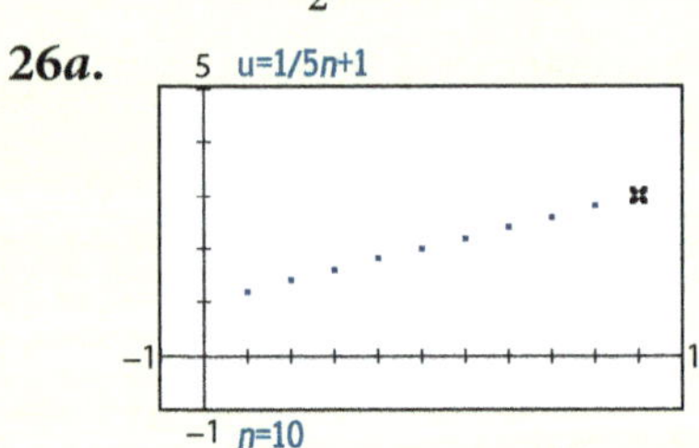

26b.

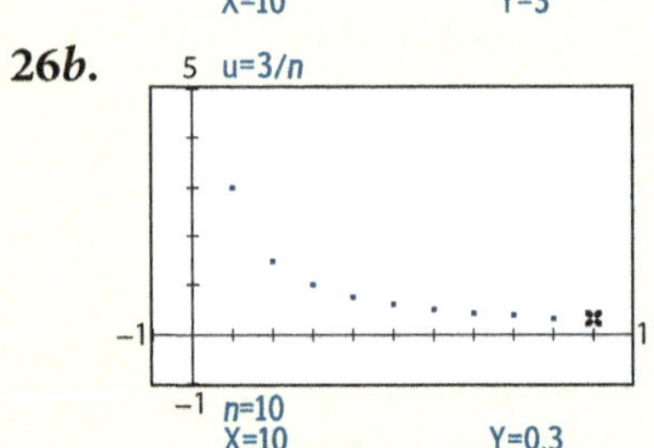

26c.

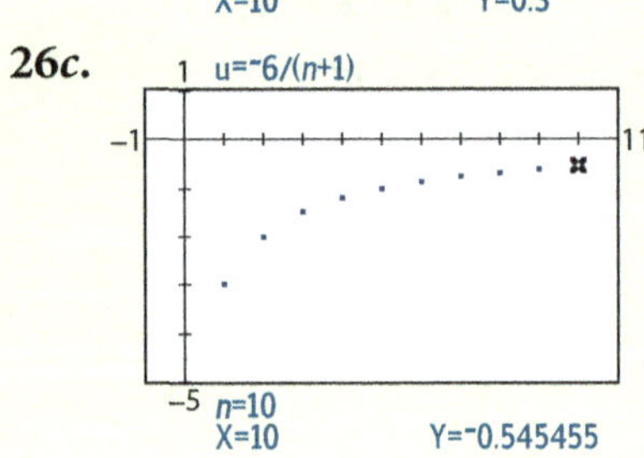

Example 7 — Using a Partial Sum to Find the Explicit Formula

Write a simplified explicit formula for the arithmetic sequence with $a_1 = 10$ if $S_{100} = 10{,}900$.

Answer

$$10{,}900 = \frac{100}{2}[2(10) + (100 - 1)d]$$
$$10{,}900 = 50(20 + 99d)$$
$$218 = 20 + 99d$$
$$198 = 99d$$
$$d = 2$$

1. Substitute into $S_n = \frac{n}{2}[2a_1 + (n-1)d]$ and solve for d, the common difference.

$$a_n = 10 + (n-1)2$$
$$a_n = 2n + 8$$

2. Substitute into $a_n = a_1 + (n-1)d$ and simplify.

SKILL ✔ EXERCISE 35

Example 8 — Applying Arithmetic Sequences and Series

Mr. and Mrs. King decided to pay off their mortgage early by adding an extra \$25 to their \$800 mortgage payment in January and then increasing their additional payment by another \$5 each successive month. Write a simplified explicit formula for the arithmetic sequence representing their payments. Then find the amount paid in December of the second year and the total amount paid during those 2 yr.

Answer

$$a_1 = 825, d = 5$$
$$a_n = 825 + (n-1)5$$
$$= 5n + 820$$

1. Substitute values for a_1 and d into $a_n = a_1 + (n-1)d$ and simplify.

$$a_{24} = 5(24) + 820 = \$940$$

2. In December of the second year, $n = 24$.

$$S_{24} = \frac{24}{2}(825 + 940) = \$21{,}180$$

3. Substitute values for n, a_1, and a_n into $a_n = a_1 + (n-1)d$ and simplify.

SKILL ✔ EXERCISE 39

A. Exercises

1. Identify each formula related to arithmetic sequences.

a. $S_n = \frac{n(a_1 + a_n)}{2}$ VI
b. $a_n - a_{n-1} = d$ V
c. $a_n = a_{n-1} + d$ I
d. $a_n = a_k + (n - k)d$ III
e. $a_n = a_1 + (n-1)d$ II
f. $a_n = dn + a_0$ IV

I. recursive
II. basic explicit
III. generalized explicit
IV. simplified explicit
V. common difference
VI. finite series

If the sequence is arithmetic, state the common difference and list the next 3 terms. If it is not arithmetic, explain why.

2. $-14, -10, -6, -2, 2, \ldots$
3. $66, 77, 88, 99, 110, \ldots$
4. $37, 28, 19, 10, 1, \ldots$
5. $28, 35, 43, 54, 67, \ldots$

Identify the common difference for each arithmetic sequence and then find the fifth term.

6. $a_n = a_{n-1} - 1, a_1 = 5$ $d = -1; a_5 = 1$
7. $a_n = 6n - 8$ $d = 6; a_5 = 22$

Write the recursive formula and a simplified explicit formula for each arithmetic sequence. Then find a_{20}.

8. $-1, -5, -9, \ldots$
9. $7, 19, 31, \ldots$
10. positive multiples of 8
11. odd natural numbers
12. $\ln 3, \ln 12, \ln 48, \ln 192, \ldots$
13. $\log 10, \log 1000, \log 100{,}000, \log 10{,}000{,}000, \ldots$

Write the simplified explicit formula for the following sequences, then find a_{20}.

14. $9, 13, 17, 21, \ldots$ $a_n = 4n + 5; a_{20} = 85$
15. a_n is arithmetic with $a_9 = 35$ and $a_{23} = -49$.
16. a_n is arithmetic with $a_7 = 52$ and $a_{15} = 90$.
17. $\frac{1}{5}, \frac{1}{12}, \frac{1}{19}, \frac{1}{26}, \ldots$ $a_n = \frac{1}{7n-2}; a_{20} = \frac{1}{138}$

15. $a_n = -6n + 89; a_{20} = -31$
16. $a_n = 4.75n + 18.75; a_{20} = 113.75$

2. $d = 4; 6, 10, 14$
3. $d = 11; 121, 132, 143$
4. $d = -9; -8, -17, -26$
5. It is not arithmetic since the differences are not constant: 7, 8, 11, 13.
8. $a_1 = -1, a_n = a_{n-1} - 4;$ $a_n = -4n + 3; a_{20} = -77$
9. $a_1 = 7, a_n = a_{n-1} + 12;$ $a_n = 12n - 5; a_{20} = 235$
10. $a_1 = 8, a_n = a_{n-1} + 8;$ $a_n = 8n; a_{20} = 160$
11. $a_1 = 1, a_n = a_{n-1} + 2;$ $a_n = 2n - 1; a_{20} = 39$
12. $a_1 = \ln 3, a_n = a_{n-1} + \ln 4;$ $a_n = n \ln 4 + \ln \frac{3}{4};$ $a_{20} = \ln(3 \cdot 4^{19})$
13. $a_1 = \log 10, a_n = a_{n-1} + 2;$ $a_n = 2n - 1; a_{20} = 39$

26d.

27. $d = 95 - 100 = -5$
$$5 = 100 + (n-1)(-5)$$
$$5 = -5n + 105; n = 20$$
$$S_{20} = \frac{20}{2}(100 + 5) = 1050$$

28. $d = 5 - 3 = 2$
$$31 = 3 + (n-1)2$$
$$31 = 2n + 1; n = 15$$
$$S_{15} = \frac{15}{2}(3 + 31) = 255$$

29. $a_1 = 2(1) + 3 = 5$
$$a_{500} = 2(500) + 3 = 1003$$
$$S_{500} = \frac{500}{2}(5 + 1003) = 252{,}000$$

30. $a_1 = \frac{1 + 6}{3} = \frac{7}{3}; a_{60} = \frac{60 + 6}{3} = 22$
$$S_{60} = \frac{60}{2}\left(\frac{7}{3} + 22\right) = 30\left(\frac{7 + 66}{3}\right)$$
$$= 10(73) = 730$$

31. arithmetic with $d = 9 - 5 = 4$
$$117 = 5 + (n-1)4$$
$$117 = 4n + 1; n = 29$$
$$a_0 = a_1 - d = 5 - 4 = 1$$

32. arithmetic with $d = -6 - (-3) = -3$
$$-300 = -3 + (n-1)(-3)$$
$$-300 = -3n; n = 100$$
$$a_0 = a_1 - d = -3 - (-3) = 0$$

B. Exercises

Find the arithmetic means between the given terms.

18. $a_{18} = 9$ and $a_{21} = 30$ 16, 23

19. $a_{10} = 32$ and $a_{14} = 71$ 41.75, 51.5, 61.25

20. $a_6 = 25.83$ and $a_8 = 94.05$ 59.94

21. Prove that the single arithmetic mean x between a and b is the average of a and b.

Write the first 5 terms of each sequence. Then classify the sequence as arithmetic, harmonic, or neither and state whether it converges or diverges.

22. $a_1 = 11, a_n = a_{n-1} - 4$ 11, 7, 3, −1, −5; arithmetic; diverges

23. $a_n = -2 + 3(n-1)$ −2, 1, 4, 7, 10; arithmetic; diverges

24. $a_n = \dfrac{1}{3 + 2n}$ $\frac{1}{5}, \frac{1}{7}, \frac{1}{9}, \frac{1}{11}, \frac{1}{13}$; harmonic; converges

25. $a_1 = \frac{1}{2}, a_n = \dfrac{1}{\frac{1}{a_{n-1}} + 2}$ $\frac{1}{2}, \frac{1}{4}, \frac{1}{6}, \frac{1}{8}, \frac{1}{10}$; harmonic; converges

26. Classify each sequence as arithmetic or harmonic. Graph the first 10 terms and determine whether the sequence converges or diverges and state the limit of any convergent sequences.

   a. $a_n = \frac{1}{5}n + 1$ **b.** $a_n = \frac{3}{n}$

   c. $a_n = -\dfrac{6}{n-1}$ **d.** $a_n = -\frac{n}{3} + 2$

Evaluate each arithmetic series.

27. $100 + 95 + 90 + \cdots + 5$ 1050

28. $3 + 5 + 7 + \cdots + 31$ 255

29. $\displaystyle\sum_{k=1}^{500} (2k + 3)$ 252,000

30. $\displaystyle\sum_{j=1}^{60} \frac{j+6}{3}$ 730

Use sigma notation to write each sum.

31. $5 + 9 + 13 + 17 + \cdots + 117$ $\displaystyle\sum_{n=1}^{29} (4n + 1)$

32. $-3 - 6 - 9 - 12 - \cdots - 300$ $\displaystyle\sum_{n=1}^{100} (-3n)$

Find the requested value for each arithmetic sequence.

33. Find d if $a_1 = 43$ and $a_{19} = -137$.   −10

34. Find a_1 if $a_{25} = -1$ and $d = 14$.   −337

35. Find a_{15} if $S_{15} = 30$ and $a_1 = 3$.   1

36. Find S_{10} if $a_{10} = 30$ and $a_{15} = 70$.   −60

26a. arithmetic; diverges

26b. harmonic; converges; 0

26c. harmonic; converges; 0

26d. arithmetic; diverges

37. A church borrows \$50,000 at 3% simple interest for 4 yr with one balloon payment at the end. Use a sequence to express the balance due at the end of each year.   \$51,500, \$53,000, \$54,500, \$56,000

38. A well-drilling company charges \$2 for the first foot, \$3 for the second foot, \$4 for the third foot and so on. What is the cost of drilling a 75 ft well?   \$2925

39. A construction crew designs a trapezoidal stone patio for a customer. If the first row requires 11 stones and each additional row requires two more stones, how many total stones will be needed for the patio if it will have 18 total rows?   504 stones

40. Prove that $a_n = a_1 + (n-1)d$ simplifies to a linear function with slope of d and y-intercept $a_0 = a_1 - d$.

C. Exercises

41. Prove that the explicit formula for a general harmonic sequence, $a_n = \dfrac{b}{dn + c}$, simplifies to the reciprocal of a simplified arithmetic sequence.

Partial sums of the harmonic sequence $a_n = \frac{1}{n}$ have applications in many areas including music, physics, business, and theoretical mathematics.

42. Express the first 5 partial sums of $a_n = \frac{1}{n}$ in fractional form. $1, \frac{3}{2}, \frac{11}{6}, \frac{25}{12}, \frac{137}{60}$

43. Use a graphing calculator to graph the sequence of partial sums for $a_n = \frac{1}{n}$ and use $\boxed{\text{TRACE}}$ to find the tenth partial sum (rounded to the nearest thousandth). Then graph a second sequence $v(n) = \ln(n)$ in the same window and use the graphs to support the conjecture that the sequence of partial sums diverges.

44. Use the "sum" and "seq" functions of a graphing calculator to find S_{10}, S_{100}, S_{400}, and S_{900} (rounded to the nearest thousandth).   2.929; 5.187; 6.570; 7.380

43. 2.929; Since $S_n > \ln(n)$ for all illustrated values of n, it appears that $\lim\limits_{n \to \infty} S_n = \infty$ just as $\lim\limits_{n \to \infty} \ln(n) = \infty$.

---

38. $a_1 = 2; d = 1$
$$S_{75} = \frac{75}{2}[2(2) + (75 - 1)1]$$
$$= 37.5(4 + 74) = \$2925$$

39. $a_1 = 11; d = 2$
$$S_{18} = \frac{18}{2}[2(11) + (18 - 1)2]$$
$$= 9(22 + 34) = 504 \text{ stones}$$

40. $a_n = a_1 + (n-1)d$
$$= a_1 + dn - d$$
$$= dn + (a_1 - d)$$
$$f(n) = dn + a_0$$
$$f(x) = mx + b \text{ with } x \in \text{ natural}$$
numbers, $m = d$, and $b = a_0$

C. Exercises

41. The reciprocal of $\dfrac{b}{dn + c}$ is
$$\frac{1}{\frac{b}{dn+c}} = \frac{dn + c}{b} = \frac{d}{b}n + \frac{c}{b},$$
which is linear and therefore arithmetic.

42. $a_n = 1, \frac{1}{2}, \frac{1}{3}, \frac{1}{4}, \frac{1}{5}, \ldots$
$$S_1 = 1; \quad S_2 = 1 + \frac{1}{2} = \frac{3}{2}$$
$$S_3 = \frac{3}{2} + \frac{1}{3} = \frac{11}{6}$$
$$S_4 = \frac{11}{6} + \frac{1}{4} = \frac{25}{12}$$
$$S_5 = \frac{25}{12} + \frac{1}{5} = \frac{137}{60}$$

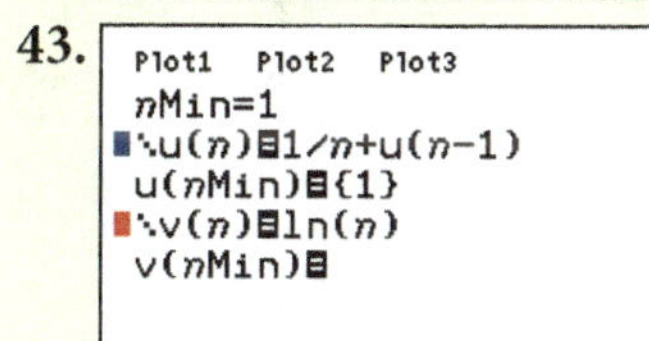

43. 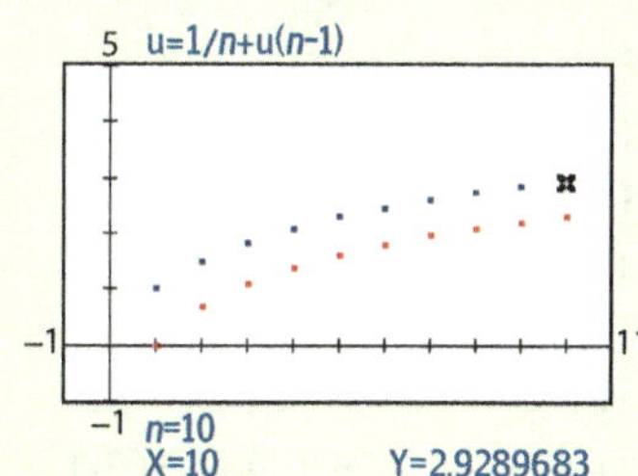

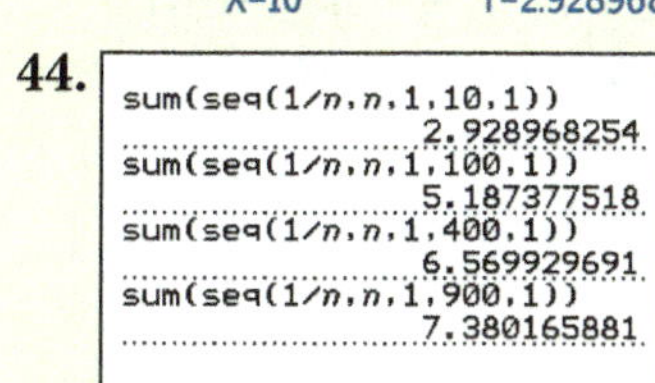

44.

---

33. $a_{19} = a_1 + (19 - 1)d$
$$-137 = 43 + 18d$$
$$18d = -180; \ d = -10$$

34. $a_{25} = a_1 + (25 - 1)14$
$$-1 = a_1 + 336; \ a_1 = -337$$

35. $S_{15} = 30, \ a_1 = 3$
$$S_{15} = \frac{15}{2}(a_1 + a_{15})$$
$$30 = \frac{15}{2}(3 + a_{15})$$
$$60 = 45 + 15a_{15}; \ a_{15} = 1$$

36. $a_{15} = a_{10} + (15 - 10)d$
$$70 = 30 + 5d; \ d = 8$$
$$a_n = a_1 + (n-1)d$$
$$a_{10} = a_1 + 9(8)$$
$$30 = a_1 + 72; \ a_1 = -42$$
$$S_{10} = \frac{10}{2}(-42 + 30) = -60$$

37. $I = Prt = 50,000(0.03)1$
$$= \$1500 \text{ annual interest}$$
$$A_1 = 50,000 + 1500 = \$51,500$$
$$A_2 = 51,500 + 1500 = \$53,000$$
$$A_3 = 53,000 + 1500 = \$54,500$$
$$A_4 = 54,500 + 1500 = \$56,000$$

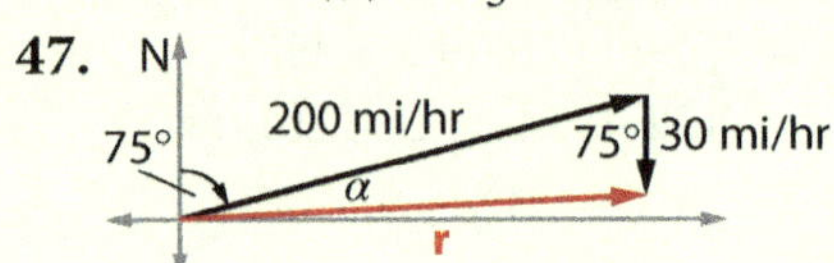

45. $f(-2) = 4^{-2} = \dfrac{1}{4^2} = \dfrac{1}{16}$

46. $p(4) = 3\left(\dfrac{2}{3}\right)^4 = \dfrac{2^4}{3^3} = \dfrac{16}{27}$

47.

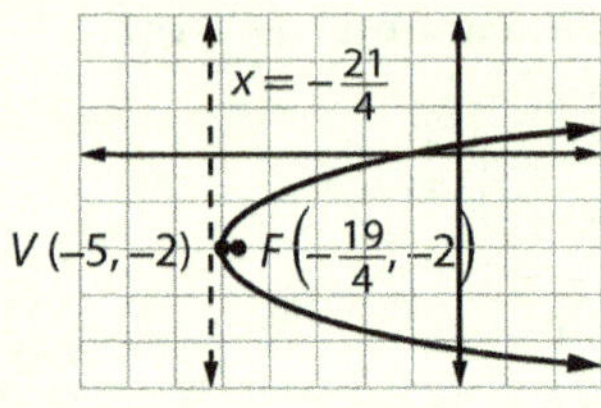

$|r|^2 = 30^2 + 200^2 - 2(30)(200)\cos 75°$

$|r| \approx 194$ mi/hr

$\dfrac{\sin \alpha}{30} \approx \dfrac{\sin 75°}{194}$

$\sin \alpha \approx \dfrac{30 \sin 75°}{194}$

$\alpha \approx 8.6°$

$\beta \approx 75 + 8.6 = 83.6°$

48. $x = y^2 + 4y - 1$
$x = (y+2)^2 - 5$

$x = -\dfrac{21}{4}$

$V(-5, -2)$ $F\left(-\dfrac{19}{4}, -2\right)$

49. $a_1 = 7(1) - 2 = 5$
$a_2 = 7(2) - 2 = 12$
$a_3 = 7(3) - 2 = 19$
$a_4 = 7(4) - 2 = 26$
$a_{10} = 7(10) - 2 = 68$

50. $a_1 = \dfrac{1^2 - 2}{1} = -1; a_2 = \dfrac{2^2 - 2}{2} = 1$

$a_3 = \dfrac{3^2 - 2}{3} = \dfrac{7}{3}; a_4 = \dfrac{4^2 - 2}{4} = \dfrac{7}{2}$

$a_{10} = \dfrac{10^2 - 2}{10} = \dfrac{49}{5}$

51. $B^2 - 4AC = (10)^2 - 4(13)(13)$
$= -576 < 0$ and $B \neq 0$

52. $e = 2 > 1$

53. $a_2 = 4(-2)$
$a_3 = 4(-2)(-2) = 4(-2)^2$
$a_4 = (4(-2)^2)(-2) = 4(-2)^3$
$a_n = 4(-2)^{n-1}$
$a_{10} = 4(-2)^{10-1} = -2048$

54. $a_3 = 4 + 2 = 6; a_4 = 6 + 4 = 10$
$a_5 = 10 + 6 = 16; a_6 = 16 + 10 = 26$

Evaluate. [3.1]

45. $f(-2)$ when $f(x) = 4^x$ $\dfrac{1}{16}$

46. $p(4)$ when $p(x) = 3\left(\dfrac{2}{3}\right)^x$ $\dfrac{16}{27}$

47. Find a plane's resultant velocity and heading if it flies at a bearing of 75° with an air speed of 200 mi/hr and the wind is blowing from the north at 30 mi/hr. [6.1]   194 mi/hr; 83.6°

48. Graph the parabola $x = y^2 + 4y - 1$, its vertex, focus, and directrix. [8.1]

Find the first 4 terms and the tenth term of each explicitly defined sequence. [9.1]

49. $a_n = 7n - 2$ 5, 12, 19, 26; $a_{10} = 68$

50. $a_n = \dfrac{n^2 - 2}{n}$ $-1, 1, \dfrac{7}{3}, \dfrac{7}{2}; a_{10} = \dfrac{49}{5}$

51. Classify the conic $13x^2 + 10xy + 13y^2 - 72 = 0$. [8.4]   B
- **A.** circle
- **B.** ellipse
- **C.** parabola
- **D.** hyperbola
- **E.** none of these

52. Classify the conic $r = \dfrac{-3}{2\cos\theta - 1}$. [8.6]   D
- **A.** circle
- **B.** ellipse
- **C.** parabola
- **D.** hyperbola
- **E.** none of these

53. Find a_{10} in the sequence $4, -8, 16, -32, \ldots$. [9.1]   C
- **A.** 1024
- **B.** -1024
- **C.** -2048
- **D.** 4096
- **E.** none of these

54. Find a_6 in the sequence $a_1 = 2, a_2 = 4, a_n = a_{n-1} + a_{n-2}$. [9.1]   C
- **A.** 12
- **B.** 16
- **C.** 26
- **D.** 64
- **E.** none of these

9.3 Geometric Sequences and Series

Compound interest can multiply early investments toward your retirement.

If \$1000 is invested at 8% APR compounded yearly, the balance at the end of each of the next 5 yr can be represented by the sequence 1080.00, 1166.40, 1259.71, 1360.49, 1469.33. This sequence is not arithmetic because the difference between consecutive terms is not constant. However, the ratio of any two consecutive terms is approximately 1.08.

DEFINITION

In a **geometric sequence**, the quotient of any two consecutive terms is a constant, $r = \frac{a_{n+1}}{a_n}$, which is called the *common ratio*.

The definition shows that the common ratio can be found by dividing any term by its previous term and that a geometric sequence can be described by its recursive formula $a_n = a_{n-1} r$.

Example 1   Identifying a Geometric Sequence

Determine whether each sequence is geometric. If so, state the common ratio and list the next 3 terms. If not, explain why.

a. $1, 4, 9, 16, \ldots$

b. $36, -24, 16, -\frac{32}{3}, \ldots$

Answer

a. $\frac{a_2}{a_1} = 4; \frac{a_3}{a_2} = \frac{9}{4} = 2.25$

not geometric; The ratio of consecutive terms is not constant.

b. $\frac{a_2}{a_1} = \frac{-24}{36} = -\frac{2}{3}; \frac{a_3}{a_2} = \frac{16}{-24} = -\frac{2}{3}; \frac{a_4}{a_3} = \frac{-\frac{32}{3}}{16} = -\frac{2}{3}$

geometric $\left(\text{with a common ratio of } -\frac{2}{3}\right)$

$a_5 = a_4\left(-\frac{2}{3}\right) = -\frac{32}{3}\left(-\frac{2}{3}\right) = \frac{64}{9}$

$a_6 = a_5\left(-\frac{2}{3}\right) = \frac{64}{9}\left(-\frac{2}{3}\right) = -\frac{128}{27}$

$a_7 = a_6\left(-\frac{2}{3}\right) = -\frac{128}{27}\left(-\frac{2}{3}\right) = \frac{256}{81}$

SKILL ✓ EXERCISE 3

The explicit formula for a geometric sequence is derived from its recursive formula, which states that each term is found by multiplying the previous term by r.

$a_1 = a_1$

$a_2 = a_1 \cdot r$

$a_3 = a_1 \cdot r^2$

$a_4 = a_1 \cdot r^3$

$\vdots$

$a_n = a_1 \cdot r^{n-1}$

After completing this section, you will be able to

- represent geometric sequences using recursive and explicit formulas.
- find nth terms and geometric means of geometric sequences.
- find sums of finite and infinite geometric sequences.
- model and solve real-world problems using geometric sequences.

The mathematical knowledge that allows insurance companies to provide a cost-effective service for those desiring to wisely prepare for unforeseen events also enables governments to profit from lottery schemes that promise easy money at astronomical odds.

9.3 Geometric Sequences and Series

Objectives

1. To represent geometric sequences using recursive and explicit formulas
2. To find the nth terms and geometric means in geometric sequences
3. To find the sums of finite and infinite geometric sequences
4. To model and solve compound interest and other real-world problems using geometric sequences

Flash

Online compound interest calculators can be used to demonstrate the power of early and regular retirement fund investments.

Vocabulary

common ratio
future value of an annuity
geometric mean
geometric sequence
geometric series

PRESENTATION

Lesson Opener

1. Show that $48\left(\frac{3}{2}\right)^{x-1} = 32\left(\frac{3}{2}\right)^x$.

$48\left(\frac{3}{2}\right)^x \left(\frac{3}{2}\right)^{-1} = 48\left(\frac{3}{2}\right)^x \left(\frac{2}{3}\right) = 32\left(\frac{3}{2}\right)^x$

2. Show that $18\left(\frac{3}{2}\right)^{x-2} = 8\left(\frac{3}{2}\right)^x$.

$18\left(\frac{3}{2}\right)^x \left(\frac{3}{2}\right)^{-2} = 18\left(\frac{3}{2}\right)^x \left(\frac{4}{9}\right) = 8\left(\frac{3}{2}\right)^x$

Use the Lesson Opener to review properties of exponents. Engage the students with the \$1000 investment example. Point out that compound interest is much more common than simple interest, which was used to introduce arithmetic sequences in the previous section. Explain how the balance of an account earning compound interest can be represented by a *geometric sequence* in which the ratio of consecutive terms is constant.

Present the definition of a geometric sequence and its recursive formula. Then determine whether the sequences in Example 1 are geometric.

Develop the explicit formula for a geometric sequence. Then find the recursive and explicit formulas for the sequences in Example 2.

Explain that the explicit formulas are actually exponential functions. The first Lesson Opener exercise illustrates how the sequence function $a_n = 48\left(\frac{3}{2}\right)^{n-1}$ in Example 2a can be expressed as the exponential function $f(x) = 32\left(\frac{3}{2}\right)^x$. The simplified exponential form $a_n = a_0 r^n$ and the generalized form $a_n = a_k r^{n-k}$ are derived in exercises 40–41. Discuss the summary of formulas for geometric sequences. Explain that it may be easier to memorize a few key formulas and then quickly derive the others as needed.

The basic explicit formula reveals two possible values for r in Example 3. The generalized formula is then used to find the simplified formula for each case. Explain that there will be only one pos-

Reading and Writing Mathematics

Suppose 3 people tag 3 others on the first day of April. Then each of those 3 people tag 3 others on the second day of April. Write a series that represents the total number of people tagged if this continues until the last day of the month, and explain how we know that someone gets tagged more than once. $3 + 3^2 + 3^3 + \cdots + 3^{30}$; The number of people tagged on the last day would be 3^{30}, which is more than the world's population, so some have to be tagged more than once.

Additional Exercises

1. Determine whether the sequence $\frac{3}{16}$, $\frac{3}{4}$, 3, 12, … is geometric. If so, state the common ratio and list the next three terms. If not, explain why.
   geometric; $r = 4$; 48, 192, 768

2. Write the recursive and explicit formulas for the geometric sequence $\frac{3}{16}, \frac{3}{4}, 3, 12, \ldots$. Then find a_{10}.
   $a_n = 4a_{n-1}$; $a_n = \frac{3}{16}(4)^{n-1}$; 49,152

3. Given the terms of a geometric sequence, write a simplified explicit formula for a_n. Then use the formula to find the specified term(s).
   a. $a_2 = \frac{2}{9}$; $a_5 = 6$; a_{10}
      $a_n = \frac{2}{27}(3)^{n-1}$ 1458
   b. $a_1 = 4$; $a_5 = 5184$; a_7 and a_8
      $a_n = 4(\pm 6)^{n-1}$
      186,624; ±1,119,744

n	$u(n)$		
0	32		
1	48		
2	72		
3	108		
4	162		
5	243		
6	364.5		
7	546.75		
8	820.13		
9	1230.2		
10	1845.3		

$n=0$

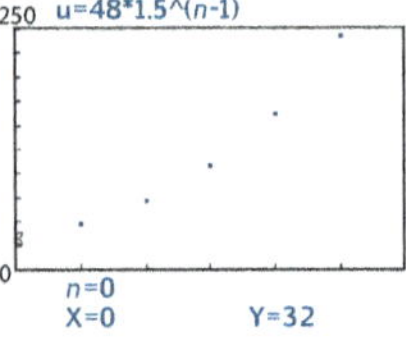

Example 2 Finding and Applying Recursive and Explicit Formulas

Write the recursive formula and an explicit formula for each geometric sequence. Then find a_{10}.

a. 48, 72, 108, 162, …

b. $10^7, 10^4, 10^1, 10^{-2}, \ldots$

Answer

a. $a_1 = 48$, $a_n = \frac{3}{2}a_{n-1}$

$a_n = 48\left(\frac{3}{2}\right)^{n-1}$

$a_{10} = 48\left(\frac{3}{2}\right)^{10-1} = \frac{3 \cdot 2^4 \cdot 3^9}{2^9} = \frac{3^{10}}{2^5}$
$= \frac{59{,}049}{32} \approx 1845.28$

1. Use the common ratio $r = \frac{72}{48} = \frac{3}{2}$ to write the recursive formula.
2. Substitute into the explicit formula $a_n = a_1 r^{n-1}$.
3. Substitute and simplify to find a_{10}.

b. $a_1 = 10^7$, $a_n = 10^{-3}a_{n-1}$

$a_n = 10^7(10^{-3})^{n-1}$

$a_n = 10^7(10^{-3})^{10-1}$
$= 10^7(10^{-27}) = 10^{-20}$

1. Use the common ratio $r = \frac{10^4}{10^7} = 10^{-3}$ to write the recursive formula.
2. Substitute into $a_n = a_1 r^{n-1}$.
3. Substitute and simplify to find a_{10}.

SKILL ✔ EXERCISE 7

The graph of the sequence in Example 2a shows that the function contains points on the graph of the exponential function $f(x) = 48\left(\frac{3}{2}\right)^{x-1}$ or $f(x) = 32\left(\frac{3}{2}\right)^{x}$. Notice that the explicit formula $a_n = a_1 r^{n-1}$ can be rewritten as $a_n = a_0 r^n$ where $a_0 = \frac{a_1}{r}$.

Observing the pattern of obtaining a_n from a_k (where $k < n$) by repeated multiplications of r produces a more generalized explicit formula: $a_n = a_k r^{n-k}$.

Geometric Sequences			
Recursive	**Explicit**		
$a_n = a_{n-1} \cdot r$	$a_n = a_1 r^{n-1}$ (basic)	$a_n = a_k r^{n-k}$ (generalized)	$a_n = a_0 r^n$; $a_0 = \frac{a_1}{r}$ (simplified)

Example 3 Writing and Using Explicit Formulas

Given the geometric sequence with $a_3 = 6$ and $a_7 = 96$, write a simplified explicit formula for a_n and use the formula to find a_{10}.

Answer

$a_7 = a_3 r^{7-3}$
$96 = 6r^4$
$16 = r^4$
$r = \pm 2$

1. Find r by substituting both terms into $a_n = a_k r^{n-k}$ and solving. Note that there are two possible values for r in this case.

CONTINUED ➡

---

sible value for r when finding an even number of *geometric means* since the power of r will be odd.

The geometric mean in Example 4 also has two possibilities since $r^2 = 5$ and $r = \pm\sqrt{5}$. You may want to review the right triangle theorems that use the geometric mean to describe the lengths of segments formed by an altitude from the right angle.

Define a *geometric series* and then derive the formulas for the sum of a finite geometric series. These two formulas are used in Example 5. Using $S_n = \frac{a_1(1 - r^n)}{1 - r}$ to evaluate series similar to the series in Example 5a may

require the use of logarithms to find the value of n.

using $a_n = a_1 r^{n-1}$:

$$2 = 1458\left(\frac{1}{3}\right)^{n-1}$$
$$\frac{1}{729} = \left(\frac{1}{3}\right)^{n-1}$$
$$729 = 3^{n-1}$$
$$\log_3 729 = n - 1$$
$$n = \frac{\log 729}{\log 3} + 1 = 7$$

Discuss the illustrated annuity payments, emphasizing that the values of the individual investments form a geometric sequence where $r = 1.005$. The first investment is compounded every month (480 times), while the last investment

earns no interest. Explain that the total value of the annuity is a geometric series, and then calculate the sum.

Present the formulas for the *future value of an annuity* and ask students to explain how the second formula is derived from the first. (*Simplify the denominator and distribute the resulting negative sign across the numerator.*) Complete Example 6 and direct attention to the power of compound interest over a prolonged period.

Consider explaining that an infinite series is not a true sum, as it would require the impossible task of adding an infinite number of terms. If the limit of

Case 1: $r = 2$

$a_n = 6(2)^{n-3}$
$a_n = 6(2)^n(2)^{-3}$
$a_n = \frac{3}{4}(2)^n$

$a_{10} = \frac{3}{4}(2)^{10} = 768$

2. Substitute $r = 2$ and $a_3 = 6$ (with $k = 3$) into $a_n = a_k r^{n-k}$ and simplify.

3. Evaluate the explicit formula to find a_{10}.

Case 2: $r = -2$

$a_n = 6(-2)^{n-3}$
$a_n = 6(-2)^n(-2)^{-3}$
$a_n = -\frac{3}{4}(-2)^n$

$a_{10} = -\frac{3}{4}(-2)^{10} = -768$

4. Substitute $r = -2$ and $a_3 = 6$ (with $k = 3$) into $a_n = a_k r^{n-k}$ and simplify.

5. Evaluate the explicit formula to find a_{10}.

SKILL ✓ EXERCISE 11

The terms between two given terms of a geometric sequence are called the *geometric means*. The geometric means between a_3 and a_7 in the sequence from Example 3 can be found using the recursive formula once r has been determined. If $r = 2$, the geometric means are 12, 24, and 48. If $r = -2$, they are -12, 24, and -48.

Example 4   Finding a Geometric Mean

Find the single geometric mean between 10 and 50.

Answer

Let $a_1 = 10$ and $a_3 = 50$.
$50 = 10r^{3-1}$
$5 = r^2$
$r = \pm\sqrt{5}$
$a_2 = 10(\pm\sqrt{5}) = \pm10\sqrt{5}$

1. The geometric mean is a_2.
2. Substitute a_3 and a_1 into $a_n = a_1 r^{n-1}$ and solve for r.

3. Use the recursive formula, $a_n = a_{n-1}r$, to find the geometric mean.

Check

4. Find $a_3 = \pm10\sqrt{5}(\pm\sqrt{5}) = 50$.

SKILL ✓ EXERCISE 17

Notice that the geometric mean between 10 and 50 is $\pm\sqrt{10 \cdot 50} = \pm10\sqrt{5}$. The fact that the geometric mean between two numbers a and c is $\pm\sqrt{ac}$ is used in geometry and proved in exercise 42.

▎ DEFINITION

A **geometric series** is a sum of the terms of a geometric sequence. The sum of the first n terms of $a_1, a_2, a_3, \ldots$ is the nth partial sum, $S_n = a_1 + a_2 + a_3 + \cdots + a_n$.

4. Find the geometric means between $a_3 = 8$ and $a_6 = 27$.   12, 18

5. Evaluate each geometric series.

   a. $\sum_{n=1}^{9} \frac{5}{6}(3)^n$ 24,602.5

   b. $1.25 + 2.5 + 5 + \cdots + 80$ 158.75

6. Madison invested \$100 monthly in an annuity earning 6% APR compounded monthly for 30 yr. Addison started 5 yr later investing at these same terms. How much would Addison need to invest each month for 25 yr to have the same amount as Madison?   \$144.95

7. Evaluate each infinite geometric series or state that the series is divergent.

   a. $\sum_{n=1}^{\infty} \frac{1}{2}\left(\frac{2}{3}\right)^{n-1}$ $\frac{3}{2}$

   b. $\frac{5}{24} + \frac{5}{12} + \frac{5}{6} + \cdots$ divergent

8. A harmonic sequence is the reciprocal of an arithmetic sequence. What is the reciprocal of a geometric sequence? Explain your reasoning.
   a geometric sequence;
   given $a_n = a_{n-1}(r_1)$ where $r_1 = \frac{a_n}{a_{n-1}}$,
   $\frac{1}{a_n} = \frac{1}{a_{n-1}r}$ or $\frac{1}{a_{n-1}} \cdot \frac{1}{r}$
   $= \frac{1}{a_{n-1}}(r_2)$ where $r_2 = \frac{a_{n-1}}{a_n}$

Assignments

- **Minimum:** 1–3, 6–7, 9, 11–12, 15–16, 19, 23–25, 29, 33, 35, 37, 47–49, 53
- **Standard:** 1, 3, 5–7; 9–19 odd; 20, 24–26, 29–30, 34–35, 38–39, 48, 50, 52–54
- **Extended:** 1, 4–6, 8, 10, 14, 16, 18, 21–22, 25; 28–38 even; 39–41, 44–45, 51, 54–56

the sequence of the partial sums is a real number, that limit is said to be the sum of the infinite series.

Discuss why an infinite geometric series has a sum only when $|r| < 1$ and derive the related formula for that case before evaluating the series given in Example 7.

The distance traveled by a bouncing ball in Example 7 is modeled by a geometric series.

the values of the two given terms, the terms are between the given terms in the sequence.

Ex. 21 Attempting to evaluate an expression containing such a large exponent returns a value beyond the range that many calculators can display. Many prefer the expression given as the answer since both the numerator and denominator represent positive values.

TIPS

Ex. 15–18 While the value of a particular geometric mean may not lie between

For the geometric sequence $36, -24, 16, \ldots$, the seventh partial sum, $S_7 = 36 - 24 + 16 - \cdots + \frac{256}{81}$, could be evaluated by completing each addition. Instead, we will derive and apply a general formula for S_n in any geometric sequence by examining the sums S_n and $-rS_n$.

$$S_n = a_1 + a_1 r + a_1 r^2 + \cdots + a_1 r^{n-2} + a_1 r^{n-1}$$

1. Write S_n in terms of a_1 and multiples of r.

$$-rS_n = -a_1 r - a_1 r^2 - a_1 r^3 - \cdots - a_1 r^{n-1} - a_1 r^n$$

2. Multiply both sides by $-r$.

$$S_n - rS_n = a_1 - a_1 r^n$$

3. Add the equations. Note that most of the terms cancel.

$$S_n(1 - r) = a_1(1 - r^n)$$

4. Solve for S_n.

$$S_n = \frac{a_1(1 - r^n)}{1 - r}$$

When given the last term a_n instead of n, you can solve $a_n = a_1 r^{n-1}$ for n, a process that may require the use of logarithms. Alternately, substituting $a_n r$ for $a_1 r^n$ in step 3 above produces a second formula.

$$S_n - rS_n = a_1 - a_n r$$
$$S_n(1 - r) = a_1 - a_n r$$
$$S_n = \frac{a_1 - a_n r}{1 - r}$$

Finite Geometric Series

$$S_n = \frac{a_1(1 - r^n)}{1 - r} \quad \text{or} \quad S_n = \frac{a_1 - a_n r}{1 - r}$$

Example 5 Evaluating Finite Geometric Series

Evaluate each geometric series.

a. $1458 + 486 + 162 + \cdots + 2$

b. $\displaystyle\sum_{k=3}^{10} \frac{3}{4}(2)^n$

Answer

a. $\dfrac{a_2}{a_1} = \dfrac{486}{1458} = \dfrac{1}{3} = r$

1. Determine the common ratio.

$$S_n = \frac{1458 - 2\left(\frac{1}{3}\right)}{1 - \frac{1}{3}} = 2186$$

2. Evaluate the sum using $S_n = \frac{a_1 - a_n r}{1 - r}$.

b. $a_1 = \dfrac{3}{4}(2)^3 = 6$
$r = 2$
$n = (10 - 3) + 1 = 8$

1. Recognize the sum as a geometric series and identify a_1, r, and n.

$$S_n = \frac{a_1(1 - r^n)}{1 - r} = \frac{6(1 - 2^8)}{1 - 2} = \frac{6(-255)}{-1} = 1530$$

2. Evaluate the sum using $S_n = \frac{a_1(1 - r^n)}{1 - r}$.

SKILL ✓ **EXERCISE 19**

Suppose a 25-yr-old invests \$100 at the end of each month into an annuity that pays 6% APR compounded monthly (0.5% per month) until retiring at the age of 65. The value of each of these $40(12) = 480$ monthly investments at retirement can be determined using the compound interest formula. The total value of the annuity can be calculated as a geometric series with $a_1 = 100$, $r = 1.005$, and $n = 480$.

$$S_n = \frac{a_1(1 - r^n)}{1 - r} = \frac{100(1 - 1.005^{480})}{1 - 1.005} = \$199{,}149.07$$

Payment	Amount	Value
1	100	$100(1.005)^{480}$
2	100	$100(1.005)^{479}$
3	100	$100(1.005)^{478}$
⋮	⋮	⋮
478	100	$100(1.005)^{2}$
479	100	$100(1.005)^{1}$
480	100	100

Note that the contributions of $100(480) = \$48{,}000$ earned \$151,149.07 in interest over 40 yr.

FUTURE VALUE OF AN ANNUITY

The future value, *FV*, of an annuity consisting of n periodic payments of *P* with a periodic interest rate *i* is given by $FV = P\left[\dfrac{1 - (1 + i)^n}{1 - (1 + i)}\right]$ or $FV = P\left[\dfrac{(1 + i)^n - 1}{i}\right]$.

Example 6 Finding the Future Value of an Annuity

Find the future value and interest earned in an annuity with \$500 quarterly payments after 45 yr if the account earns 8% APR compounded quarterly.

Answer

$i = \dfrac{0.08}{4} = 0.02;\ n = 45(4) = 180$

1. Determine the quarterly interest rate and number of payments.

$FV = 500\left[\dfrac{1.02^{180} - 1}{0.02}\right] = \$858{,}020.78$

2. Substitute into $FV = P\left[\dfrac{(1 + i)^n - 1}{i}\right]$ and evaluate.

$\$858{,}020.78 - \$90{,}000 = \$768{,}020.78$

3. Subtract the original contributions of $180(\$500)$ to determine the amount of interest earned.

SKILL ✓ **EXERCISE 37**

Recall from Section 9.1 that the sum of an infinite series is defined if the related sequence of partial sums $S_1, S_2, S_3, \ldots S_n$ converges to a real number S as $n \to \infty$. The sum of an infinite arithmetic series cannot be calculated since a_n does not approach 0 as $n \to \infty$ and the sequence of partial sums is divergent. Similar reasoning shows infinite geometric series with $|r| \geq 1$ cannot be evaluated. However, if $|r| < 1$, $r^n \to 0$ as $n \to \infty$ and $S_n = \dfrac{a_1(1 - r^n)}{1 - r}$ converges to $S = \dfrac{a_1}{1 - r}$.

The nineteenth-century Norwegian mathematician Niels Abel referred to divergent series as "the work of the Devil" since they led to many contradictions and paradoxes.

Infinite Geometric Series

An infinite geometric series with $|r| < 1$ *converges* to $S = \dfrac{a_1}{1 - r}$. If $|r| \geq 1$, the series *diverges*.

Solutions

> ### A. Exercises

2. $r = \dfrac{1}{2} \div \dfrac{1}{8} = 2 \div \dfrac{1}{2} = 8 \div 2 = 4$

$a_5 = 8(4) = 32$

$a_6 = 32(4) = 128$

$a_7 = 128(4) = 512$

3. $r = \dfrac{0.53}{0.053} = \dfrac{5.3}{0.53} = \dfrac{53}{5.3} = 10$

$a_5 = 53(10) = 530$

$a_6 = 530(10) = 5300$

$a_7 = 5300(10) = 53{,}000$

4. $\dfrac{9}{-3} = -3$, but $\dfrac{-36}{9} = -4$ and

$\dfrac{-81}{9} = -\dfrac{9}{4}$; The ratio of consecutive terms is not constant.

5. $r = \dfrac{4}{9} \div \left(-\dfrac{8}{27}\right) = -\dfrac{2}{3} \div \dfrac{4}{9}$

$\quad = 1 \div \left(-\dfrac{2}{3}\right) = -\dfrac{3}{2}$

$a_5 = 1\left(-\dfrac{3}{2}\right) = -\dfrac{3}{2}$

$a_6 = -\dfrac{3}{2}\left(-\dfrac{3}{2}\right) = \dfrac{9}{4}$

$a_7 = \dfrac{9}{4}\left(-\dfrac{3}{2}\right) = -\dfrac{27}{8}$

6. B is quadratic, D is linear (arithmetic); only A and C are exponential (geometric).

7. $r = -2;\ a_1 = -7;\ a_n = -2a_{n-1}$

$a_n = -7(-2)^{n-1}$ or $a_n = \dfrac{7}{2}(-2)^n$

$a_{10} = -7(-2)^{10-1} = 3584$

8. $r = -\dfrac{1}{5};\ a_1 = 625;\ a_n = -\dfrac{1}{5}a_{n-1}$

$a_n = 625\left(-\dfrac{1}{5}\right)^{n-1}$

$a_{10} = 625\left(-\dfrac{1}{5}\right)^{10-1} = -\dfrac{1}{3125}$

9. $r = 2.5;\ a_1 = 4;\ a_n = 2.5a_{n-1}$

$a_n = 4(2.5)^{n-1}$ or $a_n = 1.6(2.5)^n$

$a_{10} = 4(2.5)^{10-1} \approx 15{,}259$

10. $r = \dfrac{1}{x};\ a_1 = \dfrac{1}{x};\ a_n = \dfrac{1}{x}a_{n-1}$

$a_n = \dfrac{1}{x}\left(\dfrac{1}{x}\right)^{n-1} = \dfrac{1}{x^n}$

$a_{10} = \left(\dfrac{1}{x}\right)^{10} = \dfrac{1}{x^{10}}$

11. $\quad a_8 = a_3 r^{8-3}$

$9375 = 3r^5$

$r^5 = 3125;\ r = 5$

$a_n = 3(5)^{n-3} = 3(5)^n(5)^{-3}$

$\quad = \dfrac{3}{125}(5)^n$

$a_5 = \dfrac{3}{125}(5)^5 = 75$

12. $a_6 = a_2 r^{6-2}$
$486 = 6r^4$
$r^4 = 81; r = \pm 3$
$a_n = 6(\pm 3)^{n-2} = 6(\pm 3)^n (\pm 3)^{-2}$
$\quad = \frac{2}{3}(\pm 3)^n$
$a_{10} = \frac{2}{3}(\pm 3)^{10} = 39{,}366$

13. $a_5 = a_3 r^{5-3}$
$\frac{1}{2} = 2r^2$
$r^2 = \frac{1}{4}; r = \pm \frac{1}{2}$
case 1: $r = \frac{1}{2}$
$a_n = 2\left(\frac{1}{2}\right)^{n-3} = 2\left(\frac{1}{2}\right)^n \left(\frac{1}{2}\right)^{-3}$
$\quad = 2\left(\frac{1}{2}\right)^n (8) = 16\left(\frac{1}{2}\right)^n$
$a_{12} = 16\left(\frac{1}{2}\right)^{12} = \frac{1}{256}$

case 2: $r = -\frac{1}{2}$
$a_n = 2\left(-\frac{1}{2}\right)^{n-3} = 2\left(-\frac{1}{2}\right)^n \left(-\frac{1}{2}\right)^{-3}$
$\quad = -16\left(-\frac{1}{2}\right)^n$
$a_{12} = -16\left(-\frac{1}{2}\right)^{12} = -\frac{1}{256}$

14. $a_4 = a_2 r^{4-2}$
$12 = 3r^2$
$r^2 = 4; r = \pm 2$
$a_n = 3(\pm 2)^{n-2}$
$\quad = 3(\pm 2)^n (\pm 2)^{-2}$
$\quad = \frac{3}{4}(\pm 2)^n$
$a_9 = \frac{3}{4}(\pm 2)^9 = \pm 384$

15. $a_{10} = a_7 r^{10-7}$
$250 = -2r^3$
$-125 = r^3; r = -5$
$a_8 = a_7(-5) = -2(-5) = 10$
$a_9 = a_8(-5) = 10(-5) = -50$

16. $a_{23} = a_{19} r^{23-19}$
$768 = 3r^4$
$256 = r^4; r = \pm 4$
$a_{20} = a_{19}(\pm 4) = 3(\pm 4) = \pm 12$
$a_{21} = a_{20}(\pm 4) = \pm 12(\pm 4) = 48$
$a_{22} = a_{21}(\pm 4) = 48(\pm 4) = \pm 192$

17. $a_{40} = a_{38} r^{40-38}$
$196 = 4r^2$
$49 = r^2; r = \pm 7$
$a_{39} = a_{38}(\pm 7) = 4(\pm 7) = \pm 28$

18. $a_{32} = a_{29} r^{32-29}$
$-832 = 13r^3$
$-64 = r^3; r = -4$
$a_{30} = a_{29}(-4) = 13(-4) = -52$
$a_{31} = a_{30}(-4) = -52(-4) = 208$

Evaluate each infinite geometric series if possible.

a. $\sum_{k=1}^{\infty} 3(0.9)^k$ **b.** $\sum_{n=1}^{\infty} \left(\frac{e}{2}\right)^{n-1}$ **c.** $\frac{1}{2} - \frac{1}{4} + \frac{1}{8} - \frac{1}{16} + \cdots$

Answer

a. $S = \frac{3(0.9)^1}{1-(0.9)} = \frac{2.7}{0.1} = 27$ Since $|r| = |0.9| < 1$, substitute into $S = \frac{a_1}{1-r}$ and evaluate.

b. The sum does not exist.   The series is divergent since $|r| = \left|\frac{e}{2}\right| \approx 1.36 > 1$.

c. $S = \frac{\frac{1}{2}}{1-\left(-\frac{1}{2}\right)} = \frac{1}{2}\left(\frac{2}{3}\right) = \frac{1}{3}$ Since $|r| = \left|-\frac{1}{2}\right| < 1$, substitute into $S = \frac{a_1}{1-r}$ and evaluate.

SKILL ✓ EXERCISE 23

Geometric sequences and series can be used to model and solve real-world problems.

Suppose a rubber ball is dropped from a height of 20 ft and it rebounds to 80% of its previous height on each bounce.

a. How high will the ball bounce on the eighth bounce?

b. How far will the ball have traveled vertically before it finally comes to rest?

Answer

a. $b_1 = 20(0.8) = 16$ ft
$b_2 = 16(0.8) = 12.8$ ft
$b_3 = 12.8(0.8) = 10.24$ ft
$\vdots$
$b_n = 20(0.8)^n$
$b_8 = 20(0.8)^8 \approx 3.36$ ft

1. Compute the height of the first 3 bounces.

2. Recognize that the heights can be represented as a geometric sequence with $a_0 = 20$ and $r = 0.8$.

3. Use the simplified explicit formula to find the height of the eighth bounce.

b. $\sum_{n=1}^{\infty} 20(0.8)^n = \frac{20(0.8)^1}{1-0.8} = \frac{16}{0.2} = 80$

the total distance traveled
$= 20 + 2(80) = 180$ ft

1. Model the sum of the heights with an infinite geometric series and evaluate the sum using $S = \frac{a_1}{1-r}$.

2. The ball falls from the initial height of 20 ft and then rises to and falls from each bounce's subsequent height.

SKILL ✓ EXERCISE 39

❯ B. Exercises

19. $a_1 = 2; r = 2$
$$S_{16} = \frac{2(1-2^{16})}{1-2} = 2^{17} - 2 = 131{,}070$$

20. $a_1 = -\frac{3}{2}; r = -\frac{1}{2}$
$$S_{15} = \frac{-\frac{3}{2}\left[1-\left(-\frac{1}{2}\right)^{15}\right]}{1-\left(-\frac{1}{2}\right)}$$
$$= -\left[1-\left(-\frac{1}{2}\right)^{15}\right]$$
$$= -\left(1+\frac{1}{32{,}768}\right)$$
$$= -\frac{32{,}769}{32{,}768}$$

21. $a_1 = e; r = e$
$$S_{1000} = \frac{e(1-e^{1000})}{1-e}$$
$$= \frac{e-e^{1001}}{1-e} \text{ or } \frac{e^{1001}-e}{e-1}$$

22. $a_1 = 5; r = 0.2$
$$S_{14} = \frac{5(1-0.2^{14})}{1-0.2} = \frac{25}{4}(1-0.2^{14})$$
$$= \frac{25}{4}\left(\frac{6{,}103{,}515{,}624}{6{,}103{,}515{,}625}\right) = \frac{1{,}525{,}878{,}906}{244{,}148{,}625}$$
$$\approx 6.25$$

A. Exercises

1. Identify each formula related to geometric sequences.

a. $S = \dfrac{a_1}{1-r}$ VI   **I.** recursive

b. $S_n = \dfrac{a_1(1-r^n)}{1-r}$ V   **II.** basic explicit

c. $a_n = r \cdot a_{n-1}$ I   **III.** generalized explicit

d. $a_n = a_k r^{n-k}$ III   **IV.** common ratio

e. $a_n = a_1 r^{n-1}$ II   **V.** finite series

f. $\dfrac{a_n}{a_{n-1}} = r$ IV   **VI.** infinite series

If the sequence is geometric, state the common ratio and list the next 3 terms. If the sequence is not geometric, explain why.

2. $\frac{1}{8}, \frac{1}{2}, 2, 8, \ldots$ **3.** $0.053, 0.53, 5.3, 53, \ldots$

4. $-3, 9, -36, 81, \ldots$ **5.** $-\frac{8}{27}, \frac{4}{9}, -\frac{2}{3}, 1, \ldots$

6. Which of the following is an explicit formula for a geometric sequence? (Select all that apply.) A, C

A. $a_n = 4^{n-1}$ **C.** $a_n = \left(\frac{2}{3}\right)^n$

B. $a_n = -5n^2$ **D.** $a_n = \dfrac{7n+8}{19}$

Write the recursive formula and an explicit formula for each geometric sequence. Then find a_{10}.

7. $-7, 14, -28, 56, \ldots$ **8.** $625, -125, 25, -5, \ldots$

9. $4, 10, 25, 62.5, \ldots$ **10.** $\frac{1}{x}, \frac{1}{x^2}, \frac{1}{x^3}, \frac{1}{x^4}, \ldots$

Given the terms of a geometric sequence, write a simplified explicit formula for a_n. Then use the formula to find the specified term.

11. $a_3 = 3; a_8 = 9375; a_5$ **12.** $a_2 = 6; a_6 = 486; a_{10}$

13. $a_3 = 2; a_5 = \frac{1}{2}; a_{12}$ **14.** $a_2 = 3; a_4 = 12; a_9$

Find the geometric means between the given terms.

15. $a_7 = -2$ and $a_{10} = 250$ $10, -50$

16. $a_{19} = 3$ and $a_{23} = 768$ $\pm 12, 48, \pm 192$

17. $a_{38} = 4$ and $a_{40} = 196$ ± 28

18. $a_{29} = 13$ and $a_{32} = -832$ $-52, 208$

B. Exercises

Evaluate each finite geometric series.

19. $\displaystyle\sum_{r=1}^{16} 2^r$ $131,070$ **20.** $\displaystyle\sum_{n=1}^{15} 3\left(-\frac{1}{2}\right)^n$ $-\dfrac{32,769}{32,768}$

21. $\displaystyle\sum_{n=1}^{1000} e^n$ $\dfrac{e^{1001}-e}{e-1}$ **22.** $\displaystyle\sum_{n=1}^{14} 5(0.2)^{n-1}$ $\dfrac{1,525,878,906}{244,148,625} \approx 6.25$

35. $a_6 = \$53,906; S_6 = \$300,781$

$b_1 = \$54,880; S_6 = \$292,482$

Evaluate each infinite geometric series or state that the series is divergent.

23. $4 + 1 + \frac{1}{4} + \cdots$ $\frac{16}{3}$ **24.** $1 - \frac{2}{5} + \frac{4}{25} - \cdots$ $\frac{5}{7}$

25. $\frac{1}{16} + \frac{1}{8} + \frac{1}{4} + \cdots$ divergent   **26.** $\displaystyle\sum_{n=1}^{\infty} 2\left(\frac{2}{5}\right)^{n-1}$ $\frac{10}{3}$

27. $\displaystyle\sum_{n=1}^{\infty} \left(\frac{1}{2}\right)^n$ 1 **28.** $\displaystyle\sum_{n=1}^{\infty} 18\left(\frac{1}{3}\right)^n$ 9

Find the requested value for each geometric sequence or series.

29. Find n if $a_2 = 36$, $r = 9$, and $a_n = 2916$. 4

30. Find n if $a_5 = -\frac{1}{3}$, $r = 6$, and $a_n = -432$. 9

31. Find n if $a_2 = 30$, $r = 6$, and $S_n = 215$. 3

32. Find r if $a_1 = 8$ and $S = \frac{32}{7}$. $-\frac{3}{4}$

33. A church borrows \$50,000 at 3% APR compounded annually with one balloon payment at the end of 4 yr. Write a sequence representing the balance due at the end of each year. $\$51,500.00, \$53,045.00, \$54,636.35, \$56,275.44$

34. A logging company purchased a piece of equipment for \$269,000. If the depreciation rate is 30% per year, what will be the value of the equipment after 4 yr? $\$64,586.90$

35. One job offers a beginning salary of \$46,500 with an annual raise of 3% while a second job starts at \$43,000 with a 5% raise each year. Find the annual salary and total amount earned (to the nearest dollar) for each job offer at the end of 6 yr.

36. A software engineer has earned \$1,340,325 over the last 12 yr at his current position. If his salary has increased by 4% each year, how much (to the nearest dollar) did he earn during his first year? during his twelfth year? $\$89,235; \$137,373$

37. Calculate the amount of interest earned over 5 yr if you invest \$1000 at the beginning of each year in an annuity that earns 5% annually. $\$525.63$

38. When he was born, Sam's parents began to deposit \$200 at the end of each month in a mutual fund for his college education. Estimate (to the nearest dollar) the fund's balance after 18 yr if the fund is projected to grow at an annual rate of 6%. $\$77,471$

39. A rubber ball is thrown to a height of 80 ft and then rebounds to 50% of its previous height on each bounce.

a. Find the height of the fifth bounce.   2.5 ft

b. Find the total distance the ball has traveled vertically when it hits the ground the sixth time.   315 ft

c. Find the total distance the ball has traveled vertically when it finally comes to rest.   320 ft

---

31. $a_1 = \dfrac{a_2}{r} = \dfrac{30}{6} = 5$

$$S_n = \frac{5(1-6^n)}{1-6}$$
$$215 = -1 + 6^n$$
$$216 = 6^n$$
$$6^3 = 6^n$$
$$n = 3$$

32. $S = \dfrac{32}{7} = \dfrac{8}{1-r}$

$$32 - 32r = 56$$
$$-32r = 24$$
$$r = -\frac{3}{4}$$

33. $A_n = A_{n-1}(1+r)$

$A_1 = 50{,}000(1.03) = \$51{,}500.00$
$A_2 = 51{,}500(1.03) = \$53{,}045.00$
$A_3 = 53{,}045(1.03) = \$54{,}636.35$
$A_4 = 54{,}636.35(1.03) = \$56{,}275.44$

34. $a_0 = 269{,}000; r = 1 - 0.3 = 0.7; n = 4$

$a_4 = a_0 r^n = 269{,}000(0.7)^4$
$\quad = \$64{,}586.90$

35. first: $a_6 = 46{,}500(1.03)^{6-1} \approx \$53{,}906$

$$S_6 = \frac{46{,}500(1-1.03^6)}{1-1.03} \approx \$300{,}781$$

second: $a_6 = 43{,}000(1.05)^{6-1}$
$\qquad \approx \$54{,}880$

$$S_6 = \frac{43{,}000(1-1.05^6)}{1-1.05}$$
$\qquad \approx \$292{,}482$

36. $r = 1.04; n = 12$

$$S_{12} = 1{,}340{,}825 = \frac{a_1(1-1.04^{12})}{1-1.04}$$
$$1{,}340{,}825 \approx 15.0258 a_1$$
$$a_1 \approx \$89{,}234.82$$
$$a_{12} = 89{,}234.82(1.04)^{12-1}$$
$$\approx \$137{,}372.90$$

37. $i = 0.05; n = 5$

$$FV = 1000\left(\frac{1.05^5 - 1}{0.05}\right) \approx \$5525.63$$
$$5525.63 - 5000.00 = \$525.63$$

38. $i = \dfrac{0.06}{12} = 0.005; n = 18(12) = 216$

$$FV = 200\left(\frac{1.005^{216} - 1}{0.005}\right) \approx \$77{,}470.64$$

39a. $r = 0.5; n = 5; b_1 = 80(0.5) = 40$

$b_5 = 40(0.5)^{5-1} = 2.5$ ft

39b. $n = 5; b_1 = 40$

$$S_5 = \frac{40(1-0.5^5)}{1-0.5} \approx 77.5$$
$$2(b_0 + S_5) = 2(80 + 77.5) = 315 \text{ ft}$$

39c. $S = \dfrac{40}{1-0.5} \approx 80$

$$2(b_0 + S) = 2(80 + 80) = 320 \text{ ft}$$

---

23. $a_1 = 4; r = \frac{1}{4}; S = \dfrac{4}{1-\frac{1}{4}} = \dfrac{4}{\frac{3}{4}} = \dfrac{16}{3}$

24. $a_1 = 1; r = -\frac{2}{5}$

$$S = \frac{1}{1-\left(-\frac{2}{5}\right)} = \frac{1}{\frac{7}{5}} = \frac{5}{7}$$

25. $r = 2$; divergent

26. $a_1 = 2; r = \frac{2}{5}; S = \dfrac{2}{1-\frac{2}{5}} = \dfrac{10}{3}$

27. $a_1 = \frac{1}{2}; r = \frac{1}{2}; S = \dfrac{\frac{1}{2}}{1-\frac{1}{2}} = \dfrac{\frac{1}{2}}{\frac{1}{2}} = 1$

28. $a_1 = 6; r = \frac{1}{3}; S = \dfrac{6}{1-\frac{1}{3}} = \dfrac{6}{\frac{2}{3}} = 9$

29.
$$a_n = a_2(9)^{n-2}$$
$$2916 = 36(9)^{n-2}$$
$$81 = 9^{n-2}$$
$$9^2 = 9^{n-2}$$
$$2 = n - 2$$
$$n = 4$$

30.
$$a_n = a_5(6)^{n-5}$$
$$-432 = -\frac{1}{3}(6)^{n-5}$$
$$1296 = 6^{n-5}$$
$$\log 1296 = (n-5)\log 6$$
$$n - 5 = \frac{\log 1296}{\log 6}$$
$$n = \frac{\log 1296}{\log 6} + 5 = 4 + 5 = 9$$

40. $a_n = a_1 r^{n-1} = a_1 r^n r^{-1} = \left(\frac{a_1}{r}\right) r^n$
$= a_0 r^n$

❯ C. Exercises

41. $a_n = a_1 \cdot r^{n-1}$ and $a_k = a_1 \cdot r^{k-1}$
$\dfrac{a_n}{a_k} = \dfrac{a_1 r^{n-1}}{a_1 r^{k-1}} = r^{(n-1)-(k-1)}$

$\therefore \dfrac{a_n}{a_k} = r^{n-k}$ and $a_n = a_k \cdot r^{n-k}$

42. Let $a_1 = a$, $a_2 = b$, and $a_3 = c$.
$a_3 = a_1 r^{3-1}$
$c = ar^2;\ r = \pm\sqrt{\dfrac{c}{a}}$
$a_2 = a_1 r$
$b = a\left(\pm\sqrt{\dfrac{c}{a}}\right) = \left(\pm\sqrt{\dfrac{a^2 c}{a}}\right) = \pm\sqrt{ac}$

43. $a_n = r a_{n-1}$
$r = \dfrac{a_n}{a_{n-1}}$
$r^2 = \dfrac{a_n^{\,2}}{a_{n-1}^{\,2}}$
$a_n^{\,2} = r^2 (a_{n-1})^2$
Since r^2 is a common ratio,
$a_n^{\,2}$ is a geometric sequence.

44. $i = \dfrac{0.06}{12} = 0.005;\ n = 35(12) = 420$
$1{,}000{,}000 = P\left[\dfrac{(1+0.005)^{420} - 1}{0.005}\right]$
$P = 1{,}000{,}000\left[\dfrac{0.005}{(1+0.005)^{420} - 1}\right]$
$\approx \$701.90$

45. $i = \dfrac{0.048}{12} = 0.004$
$1{,}000{,}000 = 200\left[\dfrac{(1+0.004)^n - 1}{0.004}\right]$
$20 = (1.004)^n - 1$
$21 = (1.004)^n$
$\ln 21 = n \ln 1.004$
$n = \dfrac{\ln 21}{\ln 1.004} \approx 762.65$ mo
or 63.6 yr

46. $n = 20(12)$
$1{,}000{,}000 = 500\left[\dfrac{(1+i)^{240} - 1}{i}\right]$

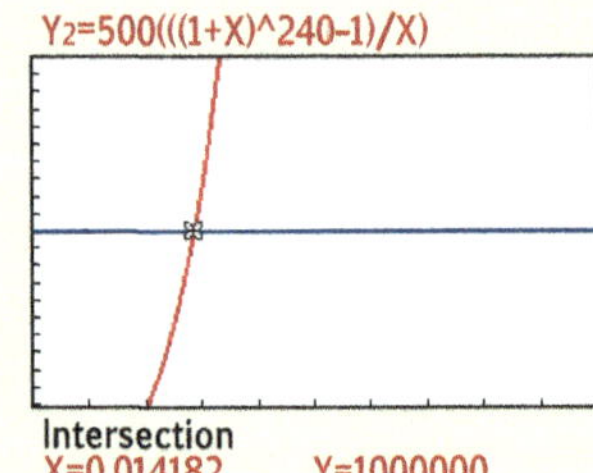

$i = \dfrac{APR}{12}$
$APR \approx 12(0.014182) \approx 0.1701$

❯ Cumulative Review

47. $a_2 = 3^1\left(\frac{1}{9}\right) = \frac{1}{3};\ a_3 = 3^2\left(\frac{1}{3}\right) = 3$
$a_4 = 3^3(3) = 81;\ a_5 = 3^4(81) = 6561$
$a_6 = 3^5(6561) = 1{,}594{,}323$

---

40. Use the basic explicit formula $a_n = a_1 r^{n-1}$ to derive the simplified explicit formula $a_n = a_0 r^n$.

❯ C. Exercises

41. Prove: If a_k is the kth term of a geometric sequence with a common ratio r, then $a_n = a_k r^{n-k}$.

42. Prove: The geometric mean between two numbers a and c is $\pm\sqrt{ac}$.

43. Prove: Show that $a_n^{\,2}$ is a geometric sequence if a_n is a geometric sequence.

CUMULATIVE REVIEW

Write the first 6 terms of each recursively defined sequence. [9.1]

$\frac{1}{9}, \frac{1}{3}, 3, 81, 6561, 1{,}594{,}323$

47. $a_1 = \frac{1}{9};\ a_n = 3^{n-1}(a_{n-1})$ where $n \geq 2$

48. $a_1 = 1,\ a_2 = 3;\ a_n = 4a_{n-1} - a_{n-2}$ where $n \geq 3$

$1, 3, 11, 41, 153, 571$

Determine whether each sequence is arithmetic. If so, state the common difference and list the next 3 terms. If not, explain why. [9.2]

49. $1, 10, 19, 28, 37, \ldots$ **50.** $39, 33, 26, 20, 14, \ldots$

Write a simplified polar equation for each conic section. Then classify the conic and verify your answer by using technology to graph the curve and its directrix. [8.6]

51. $e = 2.5$; directrix: $y = 3$ $r = \dfrac{7.5}{1 + 2.5 \sin\theta}$; hyperbola

52. $e = \frac{1}{3}$; directrix: $x = -6$ $r = \dfrac{6}{3 - \cos\theta}$; ellipse

53. Find A such that $\sin(A - \theta) = \cos\theta$. [5.4]   B

  A. π **C.** $\frac{\pi}{4}$ **E.** none of
  B. $\frac{\pi}{2}$ **D.** 0 these

54. Find the direction of $\mathbf{z} = \langle 3, -7\rangle$. [6.1]   D

  A. $\theta \approx 66.8°$ **C.** $\theta \approx 246.8°$ **E.** $\theta \approx 336.8°$
  B. $\theta \approx 113.2°$ **D.** $\theta \approx 293.2°$

49. yes; $d = 9$; 46, 55, 64

50. no; The differences are not constant.

---

44. Ada would like to retire in 35 yr with \$1,000,000. How much should she deposit at the end of each month into an account paying 6% APR compounded monthly?   \$701.90

45. How many years will it take Bob to save \$1,000,000 if he contributes \$200 at the end of each month in an account that pays 4.8% APR compounded monthly? ≈ 63.6 yr

46. Use technology to calculate the APR (to the nearest tenth of a percent) needed for \$500 monthly payments into an annuity over 20 yr to grow to \$1,000,000.   17.0%

55. Find the partial fraction decomposition for $\dfrac{2x+1}{x^2 - 2x + 1}$. [7.7]   C

  A. $\dfrac{2x}{x-1} + \dfrac{1}{(x-1)^2}$ **D.** $\dfrac{2}{x+1} + \dfrac{3}{(x+1)^2}$

  B. $\dfrac{3}{x-1} + \dfrac{2}{(x-1)^2}$ **E.** none of these

  C. $\dfrac{2}{x-1} + \dfrac{3}{(x-1)^2}$

56. Eliminate the parameter θ in $x = 4\sec\theta$ and $y = \frac{1}{4}\tan\theta$ to write a rectangular equation for the conic. [8.5]   D

  A. $\dfrac{x^2}{4} - \dfrac{y^2}{4} = 1$ **D.** $\dfrac{x^2}{16} - 16y^2 = 1$

  B. $\dfrac{x^2}{4} + 4y^2 = 1$ **E.** none of these

  C. $\dfrac{x^2}{16} - \dfrac{y^2}{16} = 1$

---

48. $a_3 = 4(3) - 1 = 11$
$a_4 = 4(11) - 3 = 41$
$a_5 = 4(41) - 11 = 153$
$a_6 = 4(153) - 41 = 571$

49. $d = 10 - 1 = 9$
$37 + 9 = 46$

50. no; $d = 6$ but $33 - 26 = 7$

51. $r = \dfrac{2.5(3)}{1 + (2.5)\sin\theta} = \dfrac{7.5}{1 + 2.5\sin\theta}$
Since $e > 1$, r represents a hyperbola.

52. $r = \dfrac{\frac{1}{3}(6)}{1 - \left(\frac{1}{3}\right)\cos\theta} = \dfrac{2}{1 - \frac{1}{3}\cos\theta}$
$= \dfrac{6}{3 - \cos\theta}$
Since $e < 1$, r represents an ellipse.

54. $\alpha = \text{Tan}^{-1}\left|\dfrac{-7}{3}\right| \approx 66.8°$
$\theta = 360° - \alpha \approx 293.2°$

55. $\left[\dfrac{2x+1}{(x-1)^2} = \dfrac{A}{x-1} + \dfrac{B}{(x-1)^2}\right](x-1)^2$
$2x + 1 = A(x-1) + B$
$= Ax + (B - A)$
$A = 2;\ B - A = 1;\ B = 3$

56. $\sec\theta = \dfrac{x}{4};\ \tan\theta = 4y$
$\sec^2\theta - \tan^2\theta = 1$
$\left(\dfrac{x}{4}\right)^2 - (4y)^2 = 1$
$\dfrac{x^2}{16} - 16y^2 = 1$

BINOMIAL EXPANSION AND PASCAL'S TRIANGLE

Consider Proposition 4 from Book II of Euclid's *Elements* (ca. 300 BC): When a segment is randomly divided, the square drawn on the whole segment is equal to the squares drawn on the pieces plus twice the rectangle formed by the pieces. Can you express this statement in modern algebraic notation? If the cut pieces have lengths of x and y, then $(x + y)^2 = x^2 + y^2 + 2xy$. Euclid used geometric figures to prove his propositions, including Proposition 7, which is equivalent to $(x - y)^2 = x^2 + y^2 - 2xy$ in modern notation. While Euclid did not address equations beyond those of degree two, this is the first written evidence of a binomial expansion.

Over 1000 yr later the Persian scholar al-Khwārizmī (ca. AD 800) used geometric figures to show how to solve second-degree equations and then explained how to multiply binomials that we would write as $(a + bx)(c + dx)$. The Persian mathematician and engineer al-Karajī (953–1029) created a table similar to Pascal's triangle with columns that listed binomial expansion coefficients up to $(a + b)^5$. The Chinese mathematician Zhu Shijie included a chart like Pascal's triangle in a 1303 publication and explained how to use it to solve equations up to degree 14. He referred to this method as "ancient," indicating it had been known in China for some time.

Several hundred years later, a chart similar to those found in Persian and Chinese documents was developed independently in Europe. Michael Stifel's *Arithmetica Integra* (1544) contained a version of Pascal's triangle in columnar form. Finally, in 1654, Blaise Pascal (1623–62) published his *Treatise on the Arithmetical Triangle*. This comprehensive work explained how to construct the triangle and explored several patterns and applications beyond binomial coefficients. It became the basis of mathematical probability theory and explicitly introduced proof by mathematical induction.

Pascal was a child prodigy born into a wealthy family and educated at home by his father. At the age of 15 he discovered a theorem of projective geometry that he called the "Theorem of the Mystic Hexagon," known today as Pascal's Theorem (the extended opposite sides of a hexagon formed by six random points on a conic will intersect in three collinear points). Pascal published a treatise on conic sections at 16 yr of age, and at the age of 22 he completed a 3-yr endeavor of inventing a calculating machine to aid his father in collecting taxes. This became the first commercially produced desktop calculating machine. Pascal also studied physics and was the first to publish a system of hydrostatics in *Treatise on the Equilibrium of Liquids*, which contains Pascal's principle of pressure. His later treatise on the measurement of the weight of the atmosphere led to the invention of the barometric altimeter.

With all of Pascal's privilege and accomplishments, you might never suspect the illnesses and pain he struggled with throughout his life, especially in his last few years. He devoted a large portion of his time to studying and writing on theology. In 1659 he wrote a 15-paragraph treatise entitled "Prayer, to Ask of God the Proper Use of Sickness," published in *Minor Works*. In the third paragraph he states, "I praise thee, my God, and I will bless thee all the days of my life, that it has pleased thee to reduce me to the incapacity of enjoying the sweets of health and the pleasures of the world."

COMPREHENSION CHECK

1. Express Euclid's Book II Proposition 7, the expansion of $(x - y)^2$, in words with no symbols. *Hint*: write all terms as positive so that $x^2 + y^2 = 2xy$.
2. Name two cultures where "Pascal's triangle" was used hundreds of years prior to Pascal.
3. What 1654 publication by Blaise Pascal introduced mathematical probability theory and proof by induction?
4. List several ways that Pascal's statement about his suffering reflects a biblical worldview.

Objectives

1. To summarize the development of binomial expansion and Pascal's triangle
2. To list mathematical concepts introduced in Pascal's *Treatise on the Arithmetical Triangle*
3. To evaluate the worldview indicated by Pascal's writings on suffering

Answers

1. If a line is cut at random, the sum of the squares of the whole and one of the pieces is equal to twice the rectangle contained by the whole and the said piece plus the square on the remaining piece.
2. Persian and Chinese
3. *Treatise on the Arithmetical Triangle*
4. sample answer: Pascal's statement indicates a confidence in God's goodness regardless of circumstances. It acknowledges God's sovereignty in our lives. It seems to express a sense of humility and an understanding that the pleasures of the world are fleeting but joys in heaven are eternal.

PRESENTATION

Binomial expansion is another example of a mathematical idea that was, to varying degrees, developed independently in several different cultures. We can even trace the origin of our English word *algebra* to the Arabic word *al-jabr*, which first appeared in the title of a book by al-Khwārizmī.

The triangular diagram originally published by Pascal looks different than most modern examples. Pascal constructed his triangle starting with two intersecting perpendicular lines, one horizontal and one vertical.

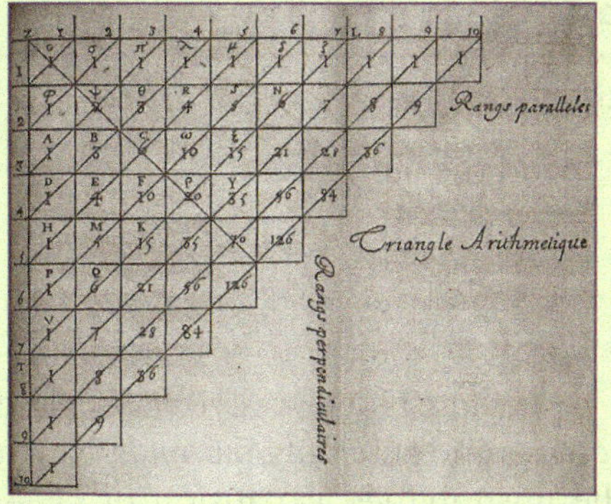

Encourage students to research the life and work of Pascal, who biographer Donald Adamson described as a "mathematician, physicist, and thinker about God." Pascal's triangle, Pascal's Theorem, and Pascal's Principle (or Law) can easily be explored online. Pascal built twenty more mechanical calculators, nine of which still exist today. They can be viewed online using the Internet keyword search *Pascal calculator image*.

In the last two years of his life, Pascal voluntarily rid himself of all but essential possessions and at one point allowed a destitute family to take shelter in his house. Consider discussing Pascal's treatise on sickness or his "Memorial," a poetic account of his 1654 conversion experience, both of which can be found online by title search.

9.4 The Binomial Theorem

Objectives

1. To expand powers of a binomial using Pascal's triangle
2. To expand powers of a binomial using the combinatorial notation of the Binomial Theorem
3. To find probabilities within a binomial experiment

Vocabulary

binomial
binomial coefficient
binomial experiment
Binomial Theorem
combination

Additional Exercises

1. Use Pascal's triangle to expand each binomial power.
   a. $(3x + 2y)^5$
   $243x^5 + 810x^4y + 1080x^3y^2 + 720x^2y^3 + 240xy^4 + 32y^5$
   b. $(5e - 3f)^3$
   $125e^3 - 225e^2f + 135ef^2 - 27f^3$

2. Evaluate $_7C_5$ and $_7C_3$.   21; 35

3. Use the Binomial Theorem to expand each binomial.
   a. $(5x - 2y)^4$
   $625x^4 - 1000x^3y + 600x^2y^2 - 160xy^3 + 16y^4$
   b. $(7z + 2)^5$
   $16{,}807z^5 + 24{,}010z^4 + 13{,}720z^3 + 3920z^2 + 560z + 32$
   c. $(2m^3 + 3n^2)^5$
   $32m^{15} + 240m^{12}n^2 + 720m^9n^4 + 1080m^6n^6 + 810m^3n^8 + 243n^{10}$

4. Find the given term in the expansion of $(2x + y)^8$.
   a. third term $1792x^6y^2$
   b. seventh term $112x^2y^6$
   c. x^4y^4 term $1120x^4y^4$

5. Suppose a standard die is rolled 10 times. Find the probability (to the nearest tenth of a percent) of each result.
   a. exactly 4 threes   5.4%
   b. Exactly half the results are odd.   24.6%
   c. Fewer than 3 of the rolls result in a 5 or a 6.   29.9%

After completing this section, you will be able to

- expand powers of a binomial using Pascal's triangle.
- expand powers of a binomial using the combinatorial notation of the Binomial Theorem.
- find probabilities within a binomial experiment.

This section is devoted to studying the series generated when expanding powers of a *binomial*, the sum of two unlike terms. As discussed in the Biblical Perspective of Mathematics feature in Chapter 6, many mathematical concepts are discovered intuitively by studying patterns found in specific examples. Examine the following expansions of powers of $(x + y)^n$.

$$(x + y)^0 = 1$$
$$(x + y)^1 = 1x + 1y$$
$$(x + y)^2 = 1x^2 + 2xy + 1y^2$$
$$(x + y)^3 = 1x^3 + 3x^2y + 3xy^2 + 1y^3$$
$$(x + y)^4 = 1x^4 + 4x^3y + 6x^2y^2 + 4xy^3 + 1y^4$$
$$(x + y)^5 = 1x^5 + 5x^4y + 10x^3y^2 + 10x^2y^3 + 5xy^4 + 1y^5$$

Notice the following patterns.
Each expansion contains $n + 1$ terms, each having a degree of n.
The exponents of x decrease by 1 from n in the first term to 0 in the last term.
The exponents of y increase by 1 from 0 in the first term to n in the last term.
The coefficients exhibit symmetry, with the first two coefficients being 1 and n.

Pascal's Triangle

1		row 0
1 1		row 1
1 2 1		row 2
1 3 3 1		row 3
1 4 6 4 1		row 4
1 5 10 10 5 1		row 5

Extracting the *binomial coefficients* produces Pascal's triangle, an array of numbers that can be used to determine the other coefficients for expansions of higher binomial powers. The top line containing 1 is row 0 since it corresponds to $(x + y)^0$. Row 1 contains the coefficients of $(x + y)^1$. The remaining rows begin and end with coefficients of 1, and the other coefficients are the sum of the two numbers above it in the previous row. Row n provides the coefficients for the corresponding expansion of $(x + y)^n$.

KEYWORD SEARCH

Sierpinski in Pascal's Triangle

> **Example 1   Using Pascal's Triangle**

Use Pascal's triangle to expand $(a + b)^6$.

Answer

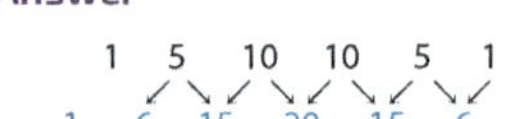

$a^6 + 6a^5b + 15a^4b^2 + 20a^3b^3 + 15a^2b^4 + 6ab^5 + b^6$

1. Derive row 6 of Pascal's triangle from row 5.

2. Use the descending powers of a, the ascending powers of b, and coefficients from row 6 of Pascal's triangle to expand the binomial power.

SKILL ✓ EXERCISE 1

PRESENTATION

Lesson Opener

Expand the binomials.

1. $(x + y)^2$ $x^2 + 2xy + y^2$
2. $(x + y)^3$ $x^3 + 3x^2y + 3xy^2 + y^3$
3. $(2x - y)^3$ $8x^3 - 12x^2y + 6xy^2 - y^3$

The Lesson Opener reviews expanding squares and cubes of *binomials*. Note that the unsimplified expansion of $(x + y)^2$ has 4 terms and the unsimplified expansion of $(x + y)^3$ has 8 terms. In general, the unsimplified expansion of $(x + y)^n$ will generate 2^n terms.

Common Student Error You may want to remind students that $(x + y)^2 \neq x^2 + y^2$.

Guide the students in identifying the stated patterns in the binomial expansions of the powers of $(x + y)^n$. Then demonstrate how the coefficients of the binomial expansions form Pascal's triangle. Show how each number is derived using the sum of the numbers above it. Use Example 1 to demonstrate how Pascal's triangle can be used to quickly write a binomial expansion.

The numerical coefficient and the subtraction sign make Example 2 a bit more challenging. Encourage the students to view the process as $(a + b)^4$,

To expand a binomial power of the form $(x - y)^n$, consider the equivalent expression $(x + (-y))^n$.

Example 2 Using Pascal's Triangle

Use Pascal's triangle to expand $(2x - y)^4$.

Answer

$1(2x)^4 + 4(2x)^3(-y) + 6(2x)^2(-y)^2 + 4(2x)(-y)^3 + 1(-y)^4$

1. Expand the binomial power using row 4 of Pascal's triangle, descending powers of the first term, $2x$, and ascending powers of the second term.

$= 16x^4 - 32x^3y + 24x^2y^2 - 8xy^3 + y^4$

2. Simplify each term. Notice that the signs of the terms alternate when expanding powers of $(x - y)^n$.

SKILL ✓ **EXERCISE 5**

Using Pascal's triangle to recursively determine the coefficients for the expansion of a binomial power can become tedious. Examining the coefficients in the expansion of $(x + y)^3$ can lead to a more efficient explicit process.

$$(x + y)^3 = (x + y)(x + y)(x + y) = x^3 + 3x^2y + 3xy^2 + y^3$$

There is only one x^3 term and it is the product of the x-term in each of the three factors. The expansion generates three separate terms of x^2y. Each is obtained using y from one of the factors and x from the other two factors. The three ways of choosing the factor that contributes the y (and by default, choosing the other two factors to contribute an x) is often notated as $_3C_1$ which can be read "3 choose 1." There are also three ways to select two factors to contribute a y (and by default, the one factor contributing an x) which generate the $_3C_2 = 3$ separate terms of xy^2. There is only one way to multiply a y from each factor, so there is only $_3C_3 = 1$ term of y^3.

The number of different ways to choose r out of n items is given by the formula $_nC_r = \frac{n!}{r!(n-r)!}$, where $_nC_r$ can be read "n choose r." The number of ways to choose 1 factor from the 3 factors above can be calculated using $_3C_1 = \frac{3!}{1!(3-1)!} = \frac{3 \cdot 2!}{1 \cdot 2!} = 3$. The expression $_nC_r$ is called a *combination* and may also be notated as $C(n, r)$ or $\binom{n}{r}$.

Example 3 Evaluating Combinations

Evaluate each combination.

a. $_6C_2$ **b.** $_6C_4$ **c.** $_nC_0$ **d.** $_nC_1$

Answer

a. $_6C_2 = \frac{6!}{2!(6-2)!}$

$= \frac{6 \cdot 5 \cdot 4!}{2 \cdot 1 \cdot 4!} = 15$

b. $_6C_4 = \frac{6!}{4!(6-4)!}$

$= \frac{6 \cdot 5 \cdot 4!}{4! \cdot 2 \cdot 1} = 15$

c. $_nC_0 = \frac{n!}{0!(n-0)!}$

$= \frac{n!}{1 \cdot n!} = 1$

d. $_nC_1 = \frac{n!}{1!(n-1)!}$

$= \frac{n \cdot (n-1)!}{1 \cdot (n-1)!} = n$

SKILL ✓ **EXERCISE 9**

TIP

Most calculators have an nCr function (found in the [MATH], PROB menu on the TI-84 Plus) that can be used to calculate combinations.

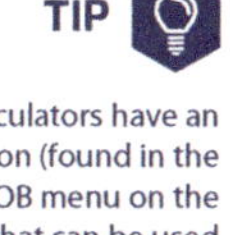

where $a = 2x$ and $b = -y$.

The recursive method of finding the *binomial coefficients* in the nth row of Pascal's triangle from row $(n - 1)$ becomes tedious as the binomial power increases. Introduce the explicit method of obtaining the coefficients of a binomial expansion using *combinations*. This method is especially helpful when finding only a specific term.

Consider reviewing the difference between permutations, in which the order of the selections is important, and combinations, in which the order is not considered. These topics will be thoroughly reviewed in the next chapter. Example 3 demonstrates the evaluation and simplification of several basic combinations. Handheld or online calculators can also be used to evaluate a combination. Have the students evaluate a combination using their calculator's *nCr* function, as shown in the margin tip.

Introduce the *Binomial Theorem* and demonstrate its use by expanding the binomial power in Example 4. Note how the entire row of coefficients can be obtained by entering a list of values for r in the *nCr* function or by using the alternate method described in the margin tip.

Example 5 demonstrates finding specific terms in the expansion of a binomial power. This skill is essential for calculating probabilities within a *binomial experiment*, a topic introduced here as an application of the Binomial Theorem and explored further in Chapters 10–11.

Define a binomial experiment and review the basic definition of the probability of an event: the ratio of the number of successes to the total number of possibilities. Use Example 6 to demonstrate finding the probability of x successes in a binomial experiment. Emphasize that calculating the probability for 3 or more correct answers requires adding the probabilities for 3, 4, and 5 correct answers.

7. $(2a)^4 + 4(2a)^3(-3b) + 6(2a)^2(-3b)^2$
$$+ 4(2a)(-3b)^3 + (-3b)^4$$
$$= 16a^4 - 96a^3b + 216a^2b^2 - 216ab^3$$
$$+ 81b^4$$

8. $(4m)^5 + 5(4m)^4(5n) + 10(4m)^3(5n)^2$
$$+ 10(4m)^2(5n)^3 + 5(4m)(5n)^4 + (5n)^5$$
$$= 1024m^5 + 6400m^4n + 16{,}000m^3n^2$$
$$+ 20{,}000m^2n^3 + 12{,}500mn^4$$
$$+ 3{,}125n^5$$

9. $\dfrac{6!}{4!2!} = \dfrac{6 \cdot 5 \cdot 4!}{4! \cdot 2 \cdot 1} = 15$

$\dfrac{6!}{2!4!} = \dfrac{6 \cdot 5 \cdot 4!}{2 \cdot 1 \cdot 4!} = 15$

10. $\dfrac{10!}{7!3!} = \dfrac{10 \cdot 9 \cdot 8 \cdot 7!}{7! \cdot 3 \cdot 2 \cdot 1} = 120$

$\dfrac{10!}{3!7!} = \dfrac{10 \cdot 9 \cdot 8 \cdot 7!}{3 \cdot 2 \cdot 1 \cdot 7!} = 120$

11. $\dfrac{5!}{4!1!} = \dfrac{5 \cdot 4!}{4!} = 5;\ \dfrac{5!}{1!4!} = \dfrac{5 \cdot 4!}{4!} = 5$

12. $\dfrac{9!}{9!0!} = 1;\ \dfrac{9!}{0!9!} = 1$

13. $\dfrac{n!}{0!(n-0)!} = \dfrac{n!}{n!} = 1;\ \dfrac{n!}{n!(n-n)!} = \dfrac{n!}{n!} = 1$

14. $\dfrac{n!}{1!(n-1)!} = \dfrac{n(n-1)!}{(n-1)!} = n$

$\dfrac{n!}{(n-1)!1!} = \dfrac{n(n-1)!}{(n-1)!} = n$

❯ B. Exercises

15. $a^3 + 3a^2b + 3ab^2 + b^3$

16. $1x^5 + 5x^4(2) + 10x^3(2)^2$
$$+ 10x^2(2)^3 + 5x(2)^4 + 1(2)^5$$
$$= x^5 + 10x^4 + 40x^3 + 80x^2 + 80x + 32$$

17. $1(2c)^3 + 3(2c)^2(-1)$
$$+ 3(2c)(-1)^2 + 1(-1)^3$$
$$= 8c^3 - 12c^2 + 6c - 1$$

18. $1(3m)^4 + 4(3m)^3(4n) + 6(3m)^2(4n)^2$
$$+ 4(3m)(4n)^3 + 1(4n)^4$$
$$= 81m^4 + 432m^3n + 864m^2n^2$$
$$+ 768mn^3 + 256n^4$$

19. $1(4)^5 + 5(4)^4z + 10(4)^3(z)^2$
$$+ 10(4)^2(z)^3 + 5(4)(z)^4 + 1z^5$$
$$= 1024 + 1280z + 640z^2$$
$$+ 160z^3 + 20z^4 + z^5$$

Gilbert and Sullivan's 1879 comic opera *The Pirates of Penzance* mentions the Binomial Theorem in its popular "Major-General's Song." The inference is that a properly educated person would be quite familiar with this theorem.

TIP

The list of coefficients can also be generated by entering Y1 = 4 nCr X and displaying the [TABLE] after using [TBLSET] to set Tblstart = 0 and ΔTbl = 1.

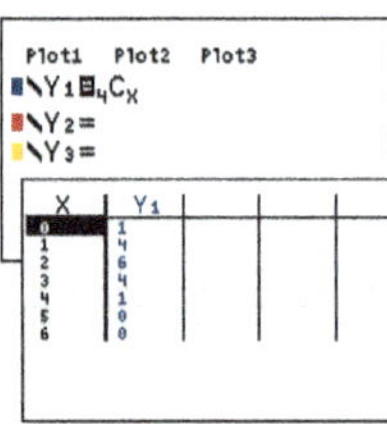

In the expansion of $(a + b)^6$ in Example 1, the coefficient of a^4b^2 is $_6C_2 = 15$ and the coefficient of a^2b^4 is $_6C_4 = 15$. The fact that $_6C_2 = _6C_4 = 15$ illustrates that $_nC_r = _nC_{n-r}$, which accounts for the symmetry in the expansion's coefficients.

In general, $_nC_r$ is the coefficient of $x^{n-r}y^r$ in the expansion of $(x + y)^n$.

▌ THE BINOMIAL THEOREM

For any positive integer n,
$$(x + y)^n = _nC_0x^n + _nC_1x^{n-1}y + _nC_2x^{n-2}y^2 + _nC_3x^{n-3}y^3 + \cdots + _nC_ny^n$$
$$= \sum_{r=0}^{n} {_nC_r}\,x^{n-r}y^r \text{ where } _nC_r = \dfrac{n!}{r!(n-r)!}.$$

Example 4   Applying the Binomial Theorem

Use the Binomial Theorem to expand $(c^3 + 2d)^4$.

Answer

1. The binomial coefficients can be found using the calculator's nCr function and a list of values for r. Enter 4 nCr {0,1,2,3,4}.

$(c^3)^4 + 4(c^3)^3(2d) + 6(c^3)^2(2d)^2$
$$+ 4(c^3)(2d)^3 + (2d)^4$$

2. Use the list of coefficients, descending powers of c^3, and ascending powers of $2d$ to write the expansion.

$c^{12} + 8c^9d + 24c^6d^2 + 32c^3d^3 + 16d^4$

3. Simplify.

SKILL ✔ EXERCISE 17

Using the explicit formula for $_nC_r$ to find the coefficient is especially helpful when determining a specific term in the expansion of a binomial power.

Example 5   Finding Specific Terms of a Binomial Expansion

Given $(2x - 1)^8$, find the x^5 term and the seventh term of its expansion.

Answer

$_8C_3(2x)^5(-1)^3 = 56(32x^5)(-1) = -1792x^5$

1. The term containing x^5 has $r = 3$.

$_8C_6(2x)^2(-1)^6 = 28(4x^2)(1) = 112x^2$

2. Since r begins at 0, the seventh term has $r = 6$.

SKILL ✔ EXERCISE 25

The Binomial Theorem can be used to model probabilities associated with a *binomial experiment* consisting of n independent trials where each trial results in either a success or a failure. For example, rolling three dice looking for how many times a four results is a binomial experiment consisting of three trials where the probability of success is $p = \frac{1}{6}$ and the probability of failure is $q = 1 - p = \frac{5}{6}$.

▌ DEFINITION

The **probability** of x successes in a binomial experiment with n independent trials is $P(x) = _nC_x p^x q^{n-x}$, where p is the probability of success and q, the probability of failure, is $1 - p$.

Motivational Idea Probabilities are sometimes associated with gambling, but probability has many applications in everyday life. Actuarial scientists and statisticians, who hold two of the top-rated professions in the United States, use probabilities to make predictions, set prices, assess reliability of products, and make environmental and financial decisions—all vital parts of wise stewardship.

TIPS

Ex. 25–27, 30–31 Consider reminding students that the kth term in the expansion uses $r = k - 1$ since the first term uses $r = 0$.

Ex. 44–45 These exercises will prepare students for binomial probability distributions in Section 11.1.

Jeff randomly selects answers to 5 multiple-choice questions, each having 4 choices. Find the probability of each event.

a. He gets exactly 3 questions correct.   **b.** He gets at least 3 questions correct.

Answer

a. $p = \frac{1}{4}; q = \frac{3}{4}$

1. Determine the probability of a success and a failure for each question.

$$_5C_3\left(\frac{1}{4}\right)^3\left(\frac{3}{4}\right)^2 = 10\left(\frac{1}{64}\right)\left(\frac{9}{16}\right) = \frac{45}{512}$$
$$\approx 0.0879 \text{ or } 8.79\%$$

2. Determine the probability of exactly 3 successes in 5 independent trials.

b. $_5C_3\left(\frac{1}{4}\right)^3\left(\frac{3}{4}\right)^2 + {_5C_4}\left(\frac{1}{4}\right)^4\left(\frac{3}{4}\right)^1 + {_5C_5}\left(\frac{1}{4}\right)^5\left(\frac{3}{4}\right)^0$

1. Find the sum of the probabilities of getting 3, 4, or all 5 correct.

$$= 10\left(\frac{1}{64}\right)\left(\frac{9}{16}\right) + 5\left(\frac{1}{256}\right)\left(\frac{3}{4}\right) + 1\left(\frac{1}{1024}\right)(1)$$
$$= \frac{90}{1024} + \frac{15}{1024} + \frac{1}{1024} = \frac{53}{512}$$
$$\approx 0.1035 \text{ or } 10.35\%$$

SKILL ✓ **EXERCISE 33**

❯ A. Exercises

Use Pascal's triangle to expand each binomial power.

1. $(x + y)^4$ **2.** $(p - q)^3$

3. $(w + v)^8$ **4.** $(a - 4)^6$

5. $(3a - 1)^5$ **6.** $(x + 2y)^5$

7. $(2a - 3b)^4$ **8.** $(4m + 5n)^5$

Evaluate each pair of combinations without using technology.

9. $_6C_4$ and $_6C_2$ 15; 15   **10.** $_{10}C_7$ and $_{10}C_3$ 120; 120

11. $_5C_4$ and $_5C_1$ 5; 5   **12.** $_9C_9$ and $_9C_0$ 1; 1

13. $_nC_0$ and $_nC_n$ 1; 1   **14.** $_nC_1$ and $_nC_{n-1}$ $n; n$

❯ B. Exercises

Use the Binomial Theorem to expand each binomial.

15. $(a + b)^3$ **16.** $(x + 2)^5$

17. $(2c - 1)^3$ **18.** $(3m + 4n)^4$

19. $(4 + z)^5$ **20.** $(c + 10)^5$

21. $(2a - 3b)^4$ **22.** $(6v - 5w)^3$

23. $(5a + 4z)^5$ **24.** $(3c + 2d^2)^4$

Find the given term in the expansion of each binomial power.

25. seventh term in $(x + y)^8$ $28x^2y^6$

26. fifth term in $(a - 3b)^{10}$ $17{,}010a^6b^4$

27. m^4n^3 term in $(2m - 5n)^7$ $-70{,}000m^4n^3$

28. v^5 term in $(4v + 2)^{11}$ $30{,}277{,}632v^5$

29. c^8d^7 term in $(c - d)^{15}$ $-6435c^8d^7$

30. fourth term in $(5x - 3y)^9$ $-35{,}437{,}500x^6y^3$

31. third term in $(w + 7)^{12}$ $3234w^{10}$

32. c^4d^3 term in $(6c + \frac{1}{3}d)^7$ $1680c^4d^3$

Find each probability to the nearest tenth of a percent.

33. getting exactly 4 twos in 10 tosses of a die   5.4%

34. getting exactly 2 threes in 12 tosses of a die   29.6%

35. getting heads exactly 4 times in 8 tosses of a coin   27.3%

36. getting tails exactly 3 times in 11 tosses of a coin   8.1%

37. getting 5 or more correct on a 7-question multiple-choice quiz if random guessing is used on each of the 4-choice questions   1.3%

Prove each statement involving combinations.

38. $_nC_r = {_nC_{n-r}}$ **39.** $_nC_2 + {_{n+1}C_2} = n^2$

❯ C. Exercises

Expand each binomial.

40. $\left(\sqrt{3} + 5t\right)^4$ **41.** $\left(\sqrt{x} - 2\right)^5$

42. $\left(m^{\frac{2}{3}} + 2n^{\frac{4}{3}}\right)^3$ **43.** $\left(4i - \sqrt{7}\right)^4$

44. A coin is tossed 3 times.
 a. Find $P(0 \text{ heads})$ and $P(3 \text{ heads})$.   12.5%; 12.5%
 b. Find $P(1 \text{ head})$ and $P(2 \text{ heads})$.   37.5%; 37.5%
 c. Find the sum of the 4 probabilities above.   100%

---

31. $_{12}C_2w^{10}(7)^2 = 66w^{10}(49) = 3234w^{10}$

32. $_7C_3(6c)^4\left(\frac{1}{3}d\right)^3 = 35(1296c^4)\left(\frac{1}{27}d^3\right)$
$$= 1680c^4d^3$$

33. $_{10}C_4\left(\frac{1}{6}\right)^4\left(\frac{5}{6}\right)^6 = 210\left(\frac{5^6}{6^{10}}\right) \approx 5.4\%$

34. $_{12}C_2\left(\frac{1}{6}\right)^2\left(\frac{5}{6}\right)^{10} = 66\left(\frac{5^{10}}{6^{12}}\right) \approx 29.6\%$

35. $_8C_4\left(\frac{1}{2}\right)^4\left(\frac{1}{2}\right)^4 = 70\left(\frac{1}{2^8}\right) \approx 27.3\%$

36. $_{11}C_3\left(\frac{1}{2}\right)^3\left(\frac{1}{2}\right)^8 = 165\left(\frac{1}{2^{11}}\right) \approx 8.1\%$

37. $_7C_5\left(\frac{1}{4}\right)^5\left(\frac{3}{4}\right)^2 + {_7C_6}\left(\frac{1}{4}\right)^6\left(\frac{3}{4}\right)^1$
$$+ {_7C_7}\left(\frac{1}{4}\right)^7\left(\frac{3}{4}\right)^0$$
$$= 21\left(\frac{3^2}{4^7}\right) + 7\left(\frac{3}{4^7}\right) + 1\left(\frac{1}{4^7}\right)$$
$$\approx 1.15 + 0.13 + 0.01 \approx 1.3\%$$

38. $_nC_r = \frac{n!}{r!(n-r)!} = \frac{n!}{(n-r)!r!} = {_nC_{n-r}}$

39. $_nC_2 + {_{n+1}C_2} = \frac{n!}{2!(n-2)!} + \frac{(n+1)!}{2!(n+1-2)!}$
$$= \frac{n(n-1)n - 2!}{2(n-2)!} + \frac{(n+1)(n)(n-1)!}{2(n-1)!}$$
$$= \frac{n^2 - n}{2} + \frac{n^2 + n}{2} = \frac{2n^2}{2} = n^2$$

❯ C. Exercises

40. $\left(\sqrt{3}\right)^4 + 4\left(\sqrt{3}\right)^3(5t) + 6\left(\sqrt{3}\right)^2(5t)^2$
$$+ 4\left(\sqrt{3}\right)(5t)^3 + (5t)^4$$
$$= 9 + 60\sqrt{3}t + 450t^2$$
$$+ 500\sqrt{3}t^3 + 625t^4$$

41. $\left(\sqrt{x}\right)^5 + 5\left(\sqrt{x}\right)^4(-2)$
$$+ 10(\sqrt{x})^3(-2)^2 + 10(\sqrt{x})^2(-2)^3$$
$$+ 5(\sqrt{x})(-2)^4 + (-2)^5$$
$$= x^2\sqrt{x} - 10x^2 + 40x\sqrt{x} - 80x$$
$$+ 80\sqrt{x} - 32$$

42. $\left(m^{\frac{2}{3}}\right)^3 + 3\left(m^{\frac{2}{3}}\right)^2\left(2n^{\frac{4}{3}}\right)$
$$+ 3\left(m^{\frac{2}{3}}\right)\left(2n^{\frac{4}{3}}\right)^2 + \left(2n^{\frac{4}{3}}\right)^3$$
$$= m^2 + 6m^{\frac{4}{3}}n^{\frac{4}{3}} + 12m^{\frac{2}{3}}n^{\frac{8}{3}} + 8n^4$$

43. $(4i)^4 + 4(4i)^3\left(-\sqrt{7}\right) + 6(4i)^2\left(-\sqrt{7}\right)^2$
$$+ 4(4i)\left(-\sqrt{7}\right)^3 + \left(-\sqrt{7}\right)^4$$
$$= 256 + 256\sqrt{7}i - 672$$
$$- 112\sqrt{7}i + 49$$
$$= -367 + 144\sqrt{7}i$$

44a. $_3C_0\left(\frac{1}{2}\right)^0\left(\frac{1}{2}\right)^3 = 1\left(\frac{1^3}{2^3}\right) = 12.5\%$

$_3C_3\left(\frac{1}{2}\right)^3\left(\frac{1}{2}\right)^0 = 1\left(\frac{1^3}{2^3}\right) = 12.5\%$

44b. $_3C_1\left(\frac{1}{2}\right)^1\left(\frac{1}{2}\right)^2 = 3\left(\frac{1^3}{2^3}\right) = 37.5\%$

$_3C_2\left(\frac{1}{2}\right)^2\left(\frac{1}{2}\right)^1 = 3\left(\frac{1^3}{2^3}\right) = 37.5\%$

44c. $2(12.5) + 2(37.5) = 100\%$

---

20. $1c^5 + 5c^4(10) + 10c^3(10)^2$
$$+ 10c^2(10)^3 + 5c(10)^4 + 1(10)^5$$
$$= c^5 + 50c^4 + 1000c^3 + 10{,}000c^2$$
$$+ 50{,}000c + 100{,}000$$

21. $1(2a)^4 + 4(2a)^3(-3b) + 6(2a)^2(-3b)^2$
$$+ 4(2a)(-3b)^3 + 1(-3b)^4$$
$$= 16a^4 - 96a^3b + 216a^2b^2 - 216ab^3$$
$$+ 81b^4$$

22. $1(6v)^3 + 3(6v)^2(-5w) + 3(6v)(-5w)^2$
$$+ 1(-5w)^3$$
$$= 216v^3 - 540v^2w + 450vw^2 - 125w^3$$

23. $1(5a)^5 + 5(5a)^4(4z) + 10(5a)^3(4z)^2$
$$+ 10(5a)^2(4z)^3 + 5(5a)(4z)^4 + 1(4z)^5$$
$$= 3125a^5 + 12{,}500a^4z + 20{,}000a^3z^2$$
$$+ 16{,}000a^2z^3 + 6400az^4 + 1024z^5$$

24. $1(3c)^4 + 4(3c)^3(2d^2) + 6(3c)^2(2d^2)^2$
$$+ 4(3c)(2d^2)^3 + 1(2d^2)^4$$
$$= 81c^4 + 216c^3d^2 + 216c^2d^4$$
$$+ 96cd^6 + 16d^8$$

25. $_8C_6x^2y^6 = 28x^2y^6$

26. $_{10}C_4a^6(-3b)^4 = 210a^6(81b^4)$
$$= 17{,}010a^6b^4$$

27. $_7C_3(2m)^4(-5n)^3$
$$= 35(16m^4)(-125n^3)$$
$$= -70{,}000m^4n^3$$

28. $_{11}C_6(4v)^5(2)^6 = 462(1024v^5)(64)$
$$= 30{,}277{,}632v^5$$

29. $_{15}C_7c^8(-d)^7 = -6435c^8d^7$

30. $_9C_3(5x)^6(-3y)^3$
$$= 84(15{,}625x^6)(-27y^3)$$
$$= -35{,}437{,}500x^6y^3$$

45a. ${}_4C_0\left(\frac{1}{6}\right)^0\left(\frac{5}{6}\right)^4 = 1\left(\frac{5^4}{6^4}\right) \approx 48.2\%$

$${}_4C_4\left(\frac{1}{6}\right)^4\left(\frac{5}{6}\right)^0 = 1\left(\frac{1}{6^4}\right) \approx 0.1\%$$

45b. ${}_4C_1\left(\frac{1}{6}\right)^1\left(\frac{5}{6}\right)^3 = 4\left(\frac{5^3}{6^4}\right) \approx 38.6\%$

$${}_4C_3\left(\frac{1}{6}\right)^3\left(\frac{5}{6}\right)^1 = 4\left(\frac{5}{6^4}\right) \approx 1.5\%$$

45c. ${}_4C_2\left(\frac{1}{6}\right)^2\left(\frac{5}{6}\right)^2 = 6\left(\frac{25}{6^4}\right) \approx 11.6\%$

45d. $48.3 + 40.1 + 11.6 = 100\%$

46. ${}_{n-1}C_{r-1} + {}_{n-1}C_r$

$$= \frac{(n-1)!}{(r-1)![(n-1)-(r-1)]!}$$
$$+ \frac{(n-1)!}{r![(n-1)-r]!}$$
$$= \frac{r}{r}\cdot\frac{(n-1)!}{(r-1)!(n-r)!}$$
$$+ \frac{(n-1)!}{r!(n-r-1)!}\cdot\frac{(n-r)}{(n-r)}$$
$$= \frac{r(n-1)!}{r!(n-r)!} + \frac{n(n-1)! - r(n-1)!}{r!(n-r)!}$$
$$= \frac{r(n-1)!}{r!(n-r)!} + \frac{n!}{r!(n-r)!} - \frac{r(n-1)!}{r!(n-r)!}$$
$$= \frac{n!}{r!(n-r)!} = {}_nC_r$$

Cumulative Review

47. $\mathbf{v} = \langle 4-2, -4-7, 2-5\rangle$

$$= \langle 2, -11, -3\rangle$$
$$|\mathbf{v}| = \sqrt{2^2 + (-11)^2 + (-3)^2}$$
$$= \sqrt{134}$$
$$\frac{\mathbf{v}}{|\mathbf{v}|} = \frac{1}{\sqrt{134}}\langle 2, -11, -3\rangle$$
$$= \left\langle \frac{\sqrt{134}}{67}, -\frac{11\sqrt{134}}{134}, -\frac{3\sqrt{134}}{134}\right\rangle$$

48. $\mathbf{v} = \langle 8-(-2), -2-3, 1-1\rangle$

$$= \langle 10, -5, 0\rangle$$
$$|\mathbf{v}| = \sqrt{10^2 + (-5)^2 + 0^2}$$
$$= \sqrt{125} = 5\sqrt{5}$$
$$\frac{\mathbf{v}}{|\mathbf{v}|} = \frac{1}{\sqrt{125}}\langle 10, -5, 0\rangle$$
$$= \left\langle \frac{2\sqrt{5}}{5}, -\frac{\sqrt{5}}{5}, 0\right\rangle$$

51. $r = \frac{3}{2}$; $a_1 = 4$; $a_n = \frac{3}{2}a_{n-1}$

$$a_n = 4\left(\frac{3}{2}\right)^{n-1} \text{ or } a_n = \frac{8}{3}\left(\frac{3}{2}\right)^n$$
$$a_{10} = 4\left(\frac{3}{2}\right)^{10-1} = \frac{19{,}683}{128}$$

52. $r = -\frac{1}{3}$; $a_1 = 729$; $a_n = -\frac{1}{3}a_{n-1}$

$$a_n = 729\left(-\frac{1}{3}\right)^{n-1}$$
$$a_{10} = 729\left(-\frac{1}{3}\right)^{10-1} = -\frac{1}{27}$$

---

45. Four dice are rolled.
  a. Find $P(0\ \text{ones})$ and $P(4\ \text{ones})$. 48.2%; 0.1%
  b. Find $P(1\ \text{one})$ and $P(3\ \text{ones})$. 38.6%; 1.5%
  c. Find $P(2\ \text{ones})$. 11.6%
  d. Find the sum of the 5 probabilities above. 100%

46. Verify the rule for determining numbers in Pascal's triangle by proving that ${}_nC_r = {}_{n-1}C_{r-1} + {}_{n-1}C_r$.

Pascal's Triangle	
$1 \quad (n-1)\ldots\ {}_{n-1}C_{r-1}\ \ {}_{n-1}C_r\ldots(n-1)\ \ 1$	row $n-1$
$1 \qquad\qquad n\ldots\ldots\ldots\ {}_nC_r\ \ldots\ldots\ldots\ n \qquad 1$	row n

Express $\overrightarrow{PQ}$ in component form. Then find its length and a unit vector in the direction of $\overrightarrow{PQ}$. **[6.3]**

47. $P(2, 7, 5)$, $Q(4, -4, 2)$ **48.** $P(-2, 3, 1)$, $Q(8, -2, 1)$

If the sequence is geometric, state the common ratio and list the next 3 terms. If the sequence is not geometric, explain why. **[9.3]**

49. $1, 4, 9, 16, 25, \ldots$ **50.** $25, -10, 4, -\frac{8}{5}, \frac{16}{25}, \ldots$

Write both the recursive formula and an explicit formula for each geometric sequence. Then find a_{10}. **[9.3]**

51. $4, 6, 9, \frac{27}{2}, \ldots$ **52.** $729, -243, 81, -27, \ldots$

53. Find the value of $\frac{9!}{5!4!}$. **[9.1]** C

  A. 220   **C.** 126   **E.** none of
  B. 756   **D.** 3024   these

54. Find the maximum value of $f(x, y) = 2x + y$ under the given constraints.
  $x \geq 0$, $y \geq 0$, $x + 2y \leq 8$, and $3x + y \leq 9$ **[7.6]** B
  A. 16   **C.** 6   **E.** none of
  B. 7   **D.** 4   these

55. Find the sum of the first 4 terms if $a_1 = 5$ and $a_n = a_{n-1} - 2$. **[9.2]** A
  A. 8   **C.** 13   **E.** none of
  B. 5   **D.** −3   these

56. Find an expression for the nth term of the following sequence: triangle, pentagon, heptagon, ... **[9.2]** C
  A. $2(n+1)$-gon   **D.** $2(n-1)$-gon
  B. $(2n-1)$-gon   **E.** none of these
  C. $(2n+1)$-gon

47. $\mathbf{v} = \langle 2, -11, -3\rangle$; $|\mathbf{v}| = \sqrt{134} \approx 11.58$
  $\frac{\mathbf{v}}{|\mathbf{v}|} = \left\langle \frac{\sqrt{134}}{67}, -\frac{11\sqrt{134}}{134}, -\frac{3\sqrt{134}}{134}\right\rangle$

48. $\mathbf{v} = \langle 10, -5, 0\rangle$; $|\mathbf{v}| = 5\sqrt{5} \approx 11.18$
  $\frac{\mathbf{v}}{|\mathbf{v}|} = \left\langle \frac{2\sqrt{5}}{5}, -\frac{\sqrt{5}}{5}, 0\right\rangle$

49. It is not geometric since $\frac{a_n}{a_{n-1}}$ is not constant: $4, \frac{9}{4}, \frac{16}{9}, \frac{25}{16}$.

50. $r = -\frac{2}{5}$; $-\frac{32}{125}, \frac{64}{625}, -\frac{128}{3125}$

51. $a_1 = 4$, $a_n = \frac{3}{2}a_{n-1}$
  $a_n = 4\left(\frac{3}{2}\right)^{n-1}$; $a_{10} = \frac{19{,}683}{128}$

52. $a_1 = 729$, $a_n = -\frac{1}{3}a_{n-1}$
  $a_n = 729\left(-\frac{1}{3}\right)^{n-1}$; $a_{10} = -\frac{1}{27}$

53. $\frac{9!}{5!4!} = \frac{9\cdot 8\cdot 7\cdot 6\cdot 5!}{5!\cdot 4!} = \frac{9\cdot 8\cdot 7\cdot 6}{4\cdot 3\cdot 2\cdot 1} = 126$

54.

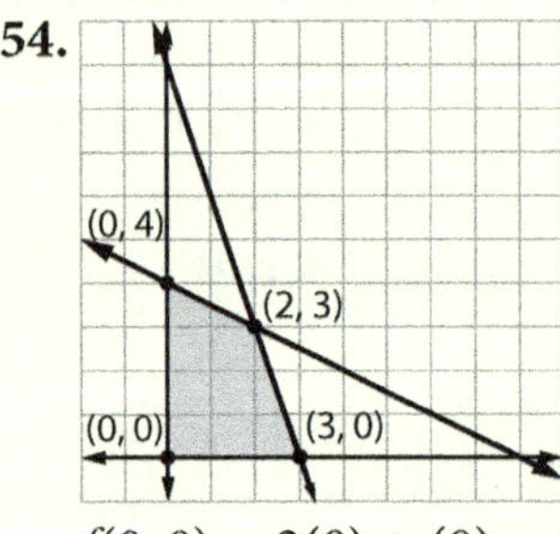

$$f(0, 0) = 2(0) + (0) = 0$$
$$f(0, 4) = 2(0) + (4) = 4$$
$$f(2, 3) = 2(2) + (3) = 7;\ \text{max}$$
$$f(3, 0) = 2(3) + (0) = 6$$

55. $a_1 = 5$; $a_2 = 5 - 2 = 3$
  $a_3 = 3 - 2 = 1$; $a_4 = 1 - 2 = -1$
  $S_4 = 5 + 3 + 1 - 1 = 8$

56. The number of sides forms an arithmetic sequence with $a_1 = 3$ and $d = 2$: $a_n = 3 + (n-1)2 = 2n+1$.

9.5 Mathematical Induction

If the first domino falls and each domino knocks down the next, all the dominoes will fall.

Consider the statement: $n^2 - n + 11$ is prime for all natural numbers. This statement is true for:

$n = 1;\ 1^2 - 1 + 11 = 11$ (prime)	$n = 5;\ 5^2 - 5 + 11 = 31$ (prime)
$n = 2;\ 2^2 - 2 + 11 = 13$ (prime)	$n = 6;\ 6^2 - 6 + 11 = 41$ (prime)
$n = 3;\ 3^2 - 3 + 11 = 17$ (prime)	$n = 7;\ 7^2 - 7 + 11 = 53$ (prime)
$n = 4;\ 4^2 - 4 + 11 = 23$ (prime)	$n = 8;\ 8^2 - 8 + 11 = 67$ (prime)

Can we conclude that the statement is true for all natural numbers? No, in fact, we can prove that the statement is false by using the counterexample of $n = 11;\ 11^2 - 11 + 11 = 121$ (not prime). Showing that a statement is true for a few or even many values of n does not constitute a mathematical proof. In this section we will study *mathematical induction*, a common method of proving statements concerning natural numbers.

PRINCIPLE OF MATHEMATICAL INDUCTION

The proposition $P(n)$ is true for all natural numbers n if
(1) $P(1)$ is true (the *anchor step*), and
(2) $P(k + 1)$ is true when $P(k)$ is true (the *inductive step*).

In other words, you first show that the proposition is true when $n = 1$. Then you prove the induction step: if the proposition is true for k, it must be true for $k + 1$. The reasoning behind mathematical induction is similar to a sequence of falling dominoes. If the first domino falls ($P(1)$ is true), and if any falling domino knocks down the next domino ($P(k + 1)$ is true whenever $P(k)$ is true), then all the dominoes fall ($P(n)$ is true for all $\mathbb{N}$).

It may seem like the inductive step assumes the proposition that is being proved, but this step does not prove the proposition for any given value of n. Instead it shows only that every proposition follows from the previous proposition. It may help to write both the *inductive hypothesis*, $P(k)$, and the *inductive conclusion*, $P(k + 1)$, before attempting to prove the inductive step.

Example 1 Preparing for a Math Induction Proof

Consider the proposition $P(n)$: $\sum_{k=1}^{n} 8k = 8 + 16 + 24 + \cdots + 8n = 4n(n + 1)$ for all $n \in \mathbb{N}$.

a. Verify $P(1)$, the anchor step.
b. State the assumption of $P(k)$, the inductive hypothesis.
c. State $P(k + 1)$, the inductive conclusion.

Answer

a. $P(1)$: $8 = 4(1)(1 + 1) = 4(2) = 8$
b. $P(k)$: $8 + 16 + 24 + \cdots + 8k = 4k(k + 1)$
c. $P(k + 1)$: $8 + 16 + 24 + \cdots + 8k + 8(k + 1) = 4(k + 1)[(k + 1) + 1]$

SKILL ✔ **EXERCISE 5**

After completing this section, you will be able to

- use mathematical induction to prove statements involving natural numbers.
- prove and apply summation properties and formulas for sums of powers of integers.
- determine and prove a formula for the nth term of a sequence.

Fermat developed a form of proof similar to induction that he called "infinite descent." Use the Internet keyword search *Fermat infinite descent* to find more information and examples.

Objectives

1. To use mathematical induction to prove statements involving natural numbers

2. To prove and apply summation properties and formulas for the sums of powers of integers

3. To determine and prove a formula for the nth term of a sequence using pattern recognition and mathematical induction

Flash

The concept of math induction is often compared to toppling dominoes. The Internet keyword search *domino topple video* can provide a possible introduction to mathematical induction.

Vocabulary

anchor step
inductive conclusion
inductive hypothesis
inductive step
mathematical induction

PRESENTATION

Lesson Opener

Identify each expression as true or false.

1. $5^{x + 1 - 1} = 5(5^{x - 1})$ true
2. $7^{x + 2} = 7(7^{x + 1})$ true
3. $3^x = 3(3^{x - 1})$ true
4. $2^x + 2^x = 2^{x + 1}$ true
5. $x^2 + x + 2x + 2 = (x + 1)^2 + (x + 1)$ true

The Lesson Opener will help students recognize and verify equivalent forms of expressions similar to those that will be encountered in this section.

Consider taking two days for this section, covering the sequence and summation statements in Examples 1–4 on the first day and covering Examples 5–8 on the second day.

Use the introductory paragraph to illustrate the fallacy of using a limited number of examples to mathematically prove a statement (see exercises 2–4).

Present the principle of *mathematical induction* and explain that mathematical induction shows that the statement is true by proving an infinite number of examples. The first example is verified in the *anchor step*, and all the following examples are simultaneously proved deductively in the *inductive step*. Consider referring the students back to the discussion of deductive and inductive reasoning in the Biblical Perspective of Mathematics feature on page 303.

One-on-One While the domino effect is commonly used to illustrate mathematical induction, some students may benefit from the analogy of climbing an infinitely tall ladder. Explain that the first (anchor) step is relatively easy. If you can always get to the next step from the step you are on, then you can climb all the steps on the ladder.

Mathematical induction is similar to a recursive formula. If you state the first term as a_1 and a_{n+1} is defined in terms of a_n, then a_n can be calculated for every value of n.

Reading and Writing Mathematics

Explain the principle of mathematical induction. Showing that a statement is true for the first case (P_1) and that the general case (P_k) implies the next case (P_{k+1}) proves that it is true for any case in the set of natural numbers. Math induction is often illustrated with the climbing of a ladder or the toppling of dominoes.

Additional Exercises

1. For proposition $P(n) = 7 + 14 + 21 + \cdots + 7n = \dfrac{7n(n+1)}{2}$, complete the following.

a. Verify the anchor step.

$P(1) = \dfrac{7(1)((1)+1)}{2} = \dfrac{7(2)}{2} = 7$

b. State the inductive hypothesis.

Assume $P(k)$:
$7 + 14 + 21 + \cdots + 7k = \dfrac{7k(k+1)}{2}$.

c. State the inductive conclusion.

$P(k+1) = 7 + 14 + 21 + \cdots + 7(k+1)$
$= \dfrac{7(k+1)((k+1)+1)}{2}$

2. Prove the summation formula using mathematical induction: $P(n) = 3 + 7 + 11 + \cdots + (4n-1) = n(2n+1)$.

$P(1) = 4(1) - 1 = 1(2(1)+1) = 3$
Assume $P(k)$:
$3 + 7 + 11 + \cdots + (4k-1) = k(2k+1)$.
$P(k+1) = k(2k+1) + 4(k+1) - 1$
$\quad = 2k^2 + k + 4k + 4 - 1$
$\quad = 2k^2 + 5k + 3$
$\quad = (k+1)(2k+3)$
$\quad = (k+1)(2(k+1)+1)$

3. Use the partial sum formulas to evaluate $\displaystyle\sum_{k=1}^{19} (k^3 - 5k + 8)$. 35,302

Use mathematical induction to prove the following.

4. 8 is a factor of $3^{2n} - 1$ for $n \in \mathbb{N}$.

$P(1)$: $3^{2(1)} - 1 = 8$
Assume $P(k)$: $3^{2k} - 1 = 8c, c \in \mathbb{N}$.
$P(k+1) = 3^{2(k+1)} - 1 = 3^{2k+2} - 1$
$\quad = 3^{2k} \cdot 3^2 - 1 = 9 \cdot 3^{2k} - 9 + 8$
$\quad = 9(3^{2k} - 1) + 8 = 9(8c) + 8$
$\quad = 8(9c + 1) = 8d, d \in \mathbb{N}$

Mathematical induction is actually a deductive method of proof that can be written in the "if-then" format. The fact that all possible cases are proved makes it similar to inductive reasoning, which derives a general result from the pattern seen in several different instances.

The inductive step for the statement in Example 1 is often stated in "if-then" form as "If $8 + 16 + 24 + \cdots + 8k = 4k(k+1)$, then $8 + 16 + 24 + \cdots + 8(k+1) = 4(k+1)(k+2)$."

Math induction can be used to prove theorems about sequences and series that we have already studied.

Example 2 Proving a Sequence's Explicit Formula

Prove the explicit formula $a_n = a_1 r^{n-1}$ for a geometric sequence.

Answer

$P(1)$: $a_1 = a_1 r^{1-1} = a_1 r^0 = a_1$

Assume $P(k)$: $a_k = a_1 r^{k-1}$.

$a_k \cdot r = a_1 r^{k-1} \cdot r$

$a_{k+1} = a_1 r^{k-1+1}$

$P(k+1)$: $a_{k+1} = a_1 r^{(k+1)-1}$

$\therefore a_n = a_1 r^{n-1}$ for all $n \in \mathbb{N}$

1. Show that $P(n)$ is true when $n = 1$.
2. Begin the inductive step by assuming $P(k)$ is true for k.
3. Show that $P(k)$ implies $P(k+1)$.
   a. Since the sequence is geometric, multiply both sides by the common ratio r.
   b. Simplify using the recursive definition on the left side and a rule for exponents on the right side.
   c. Use the Commutative Property within the exponent to express the equation in the form of $P(k+1)$.
4. Conclude that $P(n)$ is true for all $n \in \mathbb{N}$.

SKILL ✔ EXERCISE 19

Many of the properties of operations of real numbers can be generalized using summation notation.

Properties of Summations	
a_i and b_i are sequences; c and p are constants.	
1. $\displaystyle\sum_{i=1}^{n} c = nc$ (repeated additions of c defined as multiplication)	2. $\displaystyle\sum_{i=1}^{n} ca_i = c \sum_{i=1}^{n} a_i$ (Distributive Property)
3. $\displaystyle\sum_{i=1}^{n} (a_i + b_i) = \sum_{i=1}^{n} a_i + \sum_{i=1}^{n} b_i$	4. $\displaystyle\sum_{i=1}^{n} a_i = \sum_{i=1}^{p} a_i + \sum_{i=p+1}^{n} a_i$ where $1 < p < n$.

(Commutative and Associative Properties of Addition)

Mathematical induction can also be used to prove formulas for other frequently occurring summations.

Example 3 Proving a Summation Formula

Prove $\displaystyle\sum_{k=1}^{n} k = 1 + 2 + 3 + \cdots + n = \dfrac{n^2 + n}{2}$.

Answer

$P(1)$: $\displaystyle\sum_{k=1}^{1} k = 1 = \dfrac{1^2 + 1}{2} = \dfrac{2}{2} = 1$

Assume $P(k)$: $1 + 2 + 3 + \cdots + k = \dfrac{k^2 + k}{2}$.

1. Show that $P(n)$: $\displaystyle\sum_{k=1}^{n} k = \dfrac{n^2 + n}{2}$ is true for $n = 1$.
2. Begin the inductive step by assuming $P(k)$ for any specific k value.

CONTINUED ➡

Example 1 introduces the key parts of the mathematical induction process. Explain that the inductive step begins with the *inductive hypothesis* and ends with the *inductive conclusion*, and that these parts can be combined into a single if-then statement (see exercises 9–11). Emphasize that n is used to represent any natural number, while k is used to indicate a specific value in the sequence of statements.

Example 2 demonstrates a proof in which multiplying both sides of the statement of $P(k)$ by r leads to the statement of $P(k+1)$. Point out that r^{k-1+1} in step $2b$ could be simplified to r^k, but the goal is to express the exponent in terms of $k+1$, so it is written as $r^{(k+1)-1}$.

Discuss the properties of summations. Direct deductive proofs for the second and third properties are illustrated in Section 9.3 C. Exercises.

Example 3 proves the commonly used summation formula for the first n integers. Note that $k+1$ is added to both sides of the statement of $P(k)$, and then the right side is expressed in the form of $P(k+1)$. Note that simplifying $k^2 + k + 2k + 2$ in step 3c does not lead to the desired expression of $P(k+1)$. Instead, regrouping and factoring permits the expression to be written in the desired form.

$$1 + 2 + 3 + \cdots + k + (k+1) = \frac{k^2 + k}{2} + (k+1)$$

$$= \frac{k^2 + k}{2} + \frac{2(k+1)}{2}$$

$$= \frac{(k^2 + k + 2k + 2)}{2}$$

$$= \frac{(k^2 + 2k + 1) + (k+1)}{2}$$

$$P(k+1): \ 1 + 2 + 3 + \cdots + (k+1) = \frac{(k+1)^2 + (k+1)}{2}$$

Conclude that $P(n)$ is true for all $n \in \mathbb{N}$.

3. Show that $P(k)$ implies $P(k+1)$.
   a. Add the next term $(k+1)$ to both sides.
   b. Obtain a common denominator on the right side.
   c. Add the fractions, then consider $P(k+1)$ below.
   d. Group in terms of $k+1$.
   e. Factor the perfect square trinomial to achieve $P(k+1)$.
4. Apply the principle of math induction.

SKILL ✓ EXERCISE 21

Explicit formulas for sums of the first n powers of integers can be proved using math induction. The proofs of the other formulas in the table below are left as exercises.

Partial Sums of Powers of Integers	
1. $\displaystyle\sum_{k=1}^{n} k = \frac{n^2 + n}{2} = \frac{n(n+1)}{2}$	2. $\displaystyle\sum_{k=1}^{n} k^2 = \frac{n(n+1)(2n+1)}{6}$
3. $\displaystyle\sum_{k=1}^{n} k^3 = \frac{n^2(n+1)^2}{4}$	4. $\displaystyle\sum_{k=1}^{n} k^4 = \frac{n(n+1)(2n+1)(3n^2 + 3n - 1)}{30}$

The properties of summation and these formulas are used to evaluate partial sums of polynomial sequences. These techniques will also be used in the introduction to calculus found in Chapter 12.

Example 4 Using Properties to Evaluate Partial Sums

Evaluate.

a. $\displaystyle\sum_{k=1}^{17} (4k^2 - 9)$

b. $\displaystyle\sum_{k=8}^{20} k^3$

Answer

a. $\displaystyle\sum_{k=1}^{17} (4k^2 - 9) = 4\sum_{k=1}^{17} k^2 + \sum_{k=1}^{17} (-9)$

$$= 4\left(\frac{17(17+1)[2(17)+1]}{6}\right) + 17(-9)$$

$$= 4(1785) - 153 = 6987$$

b. $\displaystyle\sum_{k=8}^{20} k^3 = \sum_{k=1}^{20} k^3 - \sum_{k=1}^{7} k^3$

$$= \frac{20^2(20+1)^2}{4} - \frac{7^2(7+1)^2}{4}$$

$$= 44{,}100 - 784 = 43{,}316$$

SKILL ✓ EXERCISES 13, 15

You can prove many divisibility statements using math induction. Saying that 7 is a factor of an expression is equivalent to saying that the expression is equal to $7c$ where c is an integer. For example, saying 7 is a factor of -42 is equivalent to saying that $-42 = 7c$ where $c = -6$.

Discuss how the difficulty of converting $k^2 + k + 2k + 2$ to $(k+1)^2 + (k+1)$ emphasizes the advantage of stating $P(k+1)$ and reasoning backward toward $P(k)$ in addition to reasoning forward from $P(k)$.

Present the summary of the formulas for partial sums of the first four powers of integers. These formulas and the properties of summations are applied in Example 4.

Math induction is commonly used to prove divisibility, the fact that a number is a factor of another expression (or that the expression is evenly divisible by the number). In step $3b$ of Example 5, consider asking which properties justify the statement $d = 6c - 4$ where d is an integer. (*closure of integers under multiplication and subtraction*)

Mathematical induction is also used to prove statements of inequality, as demonstrated in Example 6. Emphasize reasoning backward from the inductive conclusion $P(k+1)$ as well as forward from the inductive hypothesis $P(k)$ to "bridge the gap."

Example 7 illustrates a typical mathematical process of using patterns to inductively discover a mathematical concept and then verifying the results with a proof.

---

5. $n^2 - 1 > (n-1)^2;\ n \geq 2$

$P(2) = 2^2 - 1 > (2-1)^2;\ 3 > 1$
Assume $P(k)$: $k^2 - 1 > (k-1)^2$
$\qquad\qquad$ or $k^2 - 1 > k^2 - 2k + 1$.
Add $2k + 1$ to both sides.
$k^2 + 2k > k^2 + 2$ and $k^2 + 2 > k^2$,
so $k^2 + 2k > k^2$
or $(k^2 + 2k + 1) - 1 > ((k+1) - 1)^2$
$P(k+1)$: $(k+1)^2 - 1 > ((k+1) - 1)^2$
$\therefore P(n)$ is true for all $\mathbb{N} \geq 2$.

6. The sum of all interior angles in a convex n-gon is $S_n = 180(n-2)$ where n is the number of interior angles.

$P(3) = 180(3-2) = 180$
Assume $P(k)$: $S_k = 180(k-2)$.
$P(k+1)$: $S_{k+1} = 180(k-2) + 180$
$\qquad\qquad\quad = 180k - 360 + 180$
$\qquad\qquad\quad = 180k + 180 - 360$
$\qquad\qquad\quad = 180(k + 1 - 2)$
$\qquad\qquad\quad = 180((k+1) - 2)$
$\therefore P(n)$ is true for all $n \in \mathbb{N}$.

Find each partial sum formula, then use mathematical induction to prove your result.

7. $5 + 7 + 9 + 11 + \cdots + (2n+3)$

$S_n = n(n+4);\ S_1 = 1(1+4) = 5$
Assume P_k: $S_k = k(k+4)$.
$S_k + [2(k+1) + 3]$
$\qquad\qquad = k(k+4) + [2(k+1) + 3]$
$S_{k+1} = k^2 + 4k + 2k + 5$
$\qquad\quad = k^2 + 6k + 5$
$\qquad\quad = (k+1)(k+5)$
$\qquad\quad = (k+1)[(k+1) + 4]$
$\therefore \displaystyle\sum_{k=1}^{n} 2n + 3 = n(n+4)$ for all $n \in \mathbb{N}$

8. $3 + 15 + 75 + 375 + \cdots + 3(5)^{n-1}$

$S_n = \frac{3(5^n - 1)}{4}$

$S_1 = \frac{3(5^1 - 1)}{4} = \frac{3(4)}{4} = 3$

Assume P_k: $S_k = \frac{3(5^k - 1)}{4}$.

$S_k + 3(5)^k = \frac{3(5^k - 1)}{4} + 3(5)^k$

$S_{k+1} = \frac{3(5^k - 1) + 12(5^k)}{4}$

$\qquad\quad = \frac{3(5^k) + 12(5^k) - 3}{4} = \frac{15(5^k) - 3}{4}$

$\qquad\quad = \frac{3(5^{k+1}) - 3}{4} = \frac{3(5^{k+1} - 1)}{4}$

$\therefore S_n = \frac{3(5^n - 1)}{4}$ for all $n \in \mathbb{N}$

Assignments

- **Minimum:** 1–5, 7–8, 10, 13–14, 21, 23, 45, 47–49, 52
- **Standard:** 1, 3–5, 7–9, 13–16, 19–20, 24, 27, 31, 34, 43–44, 46–52
- **Extended:** 1–2; 6–12 even; 13, 15–18, 20, 22, 27, 30, 32, 35, 37–38, 48, 50–52

Assessment

- Quiz 9C covers Section 9.5.

Solutions

A. Exercises

2. $P(1)\colon 5 = \dfrac{1(1+9)}{2}$

$P(2)\colon 5 + 8 = \dfrac{2(2+9)}{2}$ is false.

3. Since $n \geq 2$, skip $P(1)$.
$P(2)\colon 3$ is a factor of $2(2+1) = 6$.
$P(3)\colon 3$ is a factor of $3(3+1) = 12$.
$P(4)\colon 3$ is not a factor of
$4(4+1) = 20$.

4. $P(1)\colon \left(1 + \dfrac{1}{1}\right)^{1} = 2 > 1$

$P(2)\colon \left(1 + \dfrac{1}{2}\right)^{2} = \dfrac{9}{4} > 2$

$P(3)\colon \left(1 + \dfrac{1}{3}\right)^{3} = \dfrac{64}{27} > 3$ is false.

5. $P(1)\colon 8 = 1(1 + 7)$ is true.
$P(k)\colon 8 + 10 + 12 + \cdots + 2(k + 3)$
$\qquad\qquad\qquad\qquad = k(k + 7)$

$P(k + 1)\colon$
$8 + 10 + 12 + \cdots + 2((k + 1) + 3)$
$\qquad = (k + 1)((k + 1) + 7)$
$\qquad\qquad$ or $(k + 1)(k + 8)$

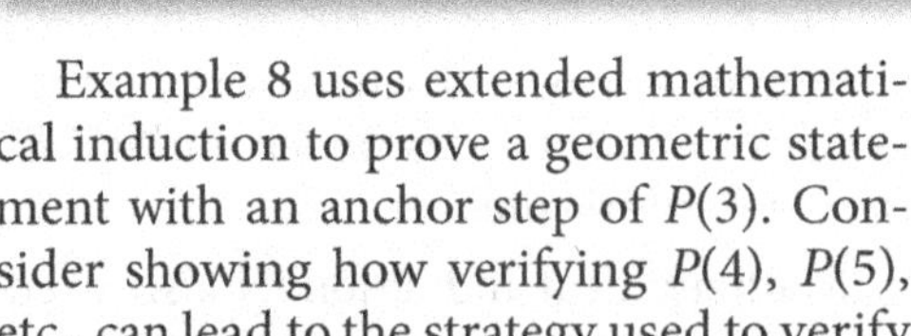

Example 5 Proving a Divisibility Statement with Math Induction

Prove that 5 is a factor of $6^{n-1} + 4$ for all $n \in \mathbb{N}$.

Answer

$P(1)\colon 6^{1-1} + 4 = 1 + 4 = 5$ $= 5c$ where $c = 1$	1. Verify the anchor step, $P(n)$ is true when $n = 1$.
Assume $P(k)\colon 6^{k-1} + 4 = 5c$ where c is an integer. $\therefore 6^{k-1} = 5c - 4$	2. Write the inductive hypothesis and subtract 4 from both sides.
$6^{(k+1)-1} + 4 = 6^{k-1+1} + 4$ $= 6(6^{k-1}) + 4$	3. Begin the statement of the inductive conclusion by substituting $k + 1$ for n and rearranging the exponential expression.
$= 6(5c - 4) + 4$ $= 30c - 20$	a. Substitute using the result of the inductive hypothesis in step 2 and simplify.
$= 5(6c - 4) = 5d$	b. Factor and substitute the integer $d = 6c - 4$.
$P(k + 1)\colon 6^{(k+1)-1} = 5d$ where d is an integer.	c. Note that the inductive conclusion has been derived from the inductive hypothesis.
$\therefore 5$ is a factor of $6^{n-1} + 4$ for all $n \in \mathbb{N}$.	4. Conclude that $P(n)$ is true for all $n \in \mathbb{N}$.

SKILL ✓ **EXERCISE 23**

You can also prove inequalities using math induction. Recall that an Addition Property of Inequality states that if $a \leq b$ and $c \leq d$, then $a + c \leq b + d$.

Example 6 Proving an Inequality with Math Induction

Prove that $n + 1 \leq 2^n$ for all $n \in \mathbb{N}$.

Answer

$P(1)\colon 1 + 1 \leq 2^1$	1. Verify the anchor step.
Assume $P(k)\colon k + 1 \leq 2^k$.	2. Write the inductive hypothesis and reason to the inductive conclusion.
Since $1 \leq 2^k$ for $k \in \mathbb{N}$, $(k + 1) + 1 \leq 2^k + 2^k$.	a. Apply the Addition Property of Inequality that states if $a \leq b$ and $c \leq d$, then $a + c \leq b + d$, where $c = 1$ and $d = 2^k$ lead to the inductive conclusion.
$\leq 2 \cdot 2^k$ $\leq 2^{k+1}$	b. Simplify the right side.
$P(k + 1)\colon (k + 1) + 1 \leq 2^{k+1}$	c. State the inductive conclusion.
$\therefore n + 1 \leq 2^n$ for all $n \in \mathbb{N}$	3. Conclude that $P(n)$ is true for all $n \in \mathbb{N}$.

SKILL ✓ **EXERCISE 27**

Consider expanding on the Van Til quote in the Historical Connection margin note. Van Til argued that induction and deduction are both part of a spiral reasoning process since all reasoning is a "method of implication into the truth of God." This transcendental argument maintains that apart from the existence of God, no reasoning or law would make sense (see the Biblical Perspective of Mathematics feature on p. 351). Van Til addresses the seemingly circular nature of this argument in the first chapter of the referenced book, which can be downloaded for free.

Example 8 uses extended mathematical induction to prove a geometric statement with an anchor step of $P(3)$. Consider showing how verifying $P(4)$, $P(5)$, etc., can lead to the strategy used to verify $P(k + 1)$ using $P(k)$.

One-on-One It may take multiple presentations before a student becomes comfortable with the mathematical induction process. A student can use the Internet keyword search *mathematical induction* to find various resources to clarify or reinforce this concept.

TIPS

Ex. 22–25 These exercises assume $n \in \mathbb{N}$.

Ex. 40–41 Enter the number of each term in L_1 and the terms in L_2 using STAT, EDIT menu option 1. Complete regressions using STAT, CALC menu options 4:LinReg, 5:QuadReg, and 0:ExpReg until the result has r^2 or R^2 of 1. For linear and exponential regressions (discussed in Section 3.5), use $r^2 = 1$, but for quadratic regressions (and cubic regressions, introduced in exercise 41), use $R^2 = 1$.

Math induction frequently allows us to prove results that we arrive at inductively.

Example 7 Finding and Proving a Partial-Sum Formula

Find an explicit formula for $\sum_{j=1}^{n} \frac{1}{j(j+1)}$.

Then use mathematical induction to verify the formula for all $n \in \mathbb{N}$.

Answer

$S_1 = \frac{1}{1(2)} = \frac{1}{2}$

1. Find the first several partial sums, looking for a pattern.

$S_2 = \frac{1}{2} + \frac{1}{2(3)} = \frac{3}{6} + \frac{1}{6} = \frac{2}{3}$

$S_3 = \frac{2}{3} + \frac{1}{3(4)} = \frac{8}{12} + \frac{1}{12} = \frac{9}{12} = \frac{3}{4}$

$S_4 = \frac{3}{4} + \frac{1}{4(5)} = \frac{15}{20} + \frac{1}{20} = \frac{16}{20} = \frac{4}{5}$

$S_n = \frac{n}{n+1}$

2. Write an explicit formula for S_n.

$P(n)$: $\sum_{j=1}^{n} \frac{1}{j(j+1)} = \frac{n}{n+1}$

3. State the proposition $P(n)$.

$P(1)$: $\frac{1}{1(2)} = \frac{1}{1+1}$ is true.

4. Verify the anchor step.

Assume $P(k)$: $S_k = \frac{k}{k+1}$.

5. Write the inductive hypothesis, using S_k for the partial sum and reason to the inductive conclusion.

$S_k + \frac{1}{(k+1)[(k+1)+1]} = \frac{k}{k+1} + \frac{1}{(k+1)[(k+1)+1]}$

   a. Add the $(k+1)$ term to both sides.

$S_{k+1} = \frac{k(k+2)}{(k+1)(k+2)} + \frac{1}{(k+1)(k+2)}$

   b. Rename the left side and combine the fractions on the right side.

$= \frac{k^2 + 2k + 1}{(k+1)(k+2)}$

$= \frac{(k+1)(k+1)}{(k+1)(k+2)}$

   c. Factor the numerator and cancel.

$P(k+1)$: $S_{k+1} = \frac{(k+1)}{(k+1)+1}$

   d. State the inductive conclusion.

$\sum_{j=1}^{n} \frac{1}{j(j+1)} = \frac{n}{n+1}$ for all $n \in \mathbb{N}$

6. Conclude that $P(n)$ is true for all $n \in \mathbb{N}$.

SKILL ✓ **EXERCISE 31**

A statement $P(n)$ may be false for $k = 1$ but true for all $n \geq j$ where $j > 1$. Extended mathematical induction is used to prove such a statement. The anchor step verifies $P(j)$ instead of $P(1)$ and the inductive step verifies all the other propositions as done previously.

16. $3 \sum_{k=1}^{38} k^2 + 7 \sum_{k=1}^{38} k - \sum_{k=1}^{38} 4$

$= 3\left(\frac{38(38+1)[2(38)+1]}{6} \right) + 7\left(\frac{38(38+1)}{2} \right) - 4(38)$

$= 3(19{,}019) + 7(741) - 152$

$= 62{,}092$

17. $\sum_{k=15}^{43} k^4 = \sum_{k=1}^{43} k^4 - \sum_{k=1}^{14} k^4$

$= \frac{43(43+1)[2(43)+1][3(43)^2 + 3(43) - 1]}{30}$

$- \frac{14(14+1)[2(14)+1][3(14)^2 + 3(14) - 1]}{30}$

$= \frac{43(44)(87)(5675)}{30} - \frac{14(15)(29)(629)}{30}$

$= 31{,}137{,}590 - 127{,}687$

$= 31{,}009{,}903$

❯ B. Exercises

18. $P(1)$: $a_1 = a_1 + (1-1)d = a_1$
Assume $P(k)$: $a_k = a_1 + (k-1)d$.
$a_k + d = a_1 + (k-1)d + d$
$a_{k+1} = a_1 + (k - 1 + 1)d$
$P(k+1)$: $a_{k+1} = a_1 + ((k+1) - 1)d$
$\therefore P(n)$ is true for all $n \in \mathbb{N}$.

19. $S_1 = \frac{a_1(1 - r^1)}{1 - r} = a_1$

Assume $P(k)$: $S_k = \frac{a_1(1 - r^k)}{1 - r}$.

$a_1 + a_2 + \cdots + a_1 r^{k-1} = \frac{a_1(1 - r^k)}{1 - r}$
Add $a_{k+1} = a_1 r^{(k+1)-1}$ or $a_1 r^k$ to both sides.
$a_1 + a_2 + \cdots + a_1 r^{k-1} + a_1 r^k$

$= \frac{a_1(1 - r^k)}{1 - r} + a_1 r^k$

$S_{k+1} = \frac{a_1(1 - r^k) + a_1 r^k(1 - r)}{1 - r}$

$= \frac{a_1 - a_1 r^k + a_1 r^k - a_1 r^k r}{1 - r}$

$= \frac{a_1 - a_1 r^{k+1}}{1 - r}$

$P(k+1)$: $S_{k+1} = \frac{a_1(1 - r^{k+1})}{1 - r}$
$\therefore P(n)$ is true for all $n \in \mathbb{N}$.

6. $P(1)$: $1 = \frac{1(1+1)(1+2)}{6} = \frac{1(2)(3)}{6} = 1$

$P(k)$: $1 + 3 + 6 + \cdots + \frac{k(k+1)}{2}$

$= \frac{k(k+1)(k+2)}{6}$

$P(k+1)$: $1 + 3 + 6 + \cdots$

$+ \frac{(k+1)((k+1)+1)}{2}$

$= \frac{(k+1)((k+1)+1)((k+1)+2)}{6}$

or $P(k+1)$: $1 + 3 + 6 + \cdots$

$+ \frac{(k+1)(k+2)}{2}$

$= \frac{(k+1)(k+2)(k+3)}{6}$

7. $P(1)$: 3 is a factor of $7^1 - 4^1 = 3$.
$P(k)$: 3 is a factor of $7^k - 4^k$.
$P(k+1)$: 3 is a factor of
$7^{k+1} - 4^{k+1}$.

8. $P(1)$: $3^1 \geq 2(1) + 1 = 3$
$P(k)$: $3^k \geq 2k + 1$
$P(k+1)$: $3^{k+1} \geq 2(k+1) + 1$
       or $\geq 2k + 3$

12. $3 \sum_{k=1}^{200} a_k = 3(20) = 60$

13. $4 \sum_{k=1}^{200} c_k + \sum_{k=1}^{200} 3$
$= 4(404) + 3(200) = 2216$

14. $5 \sum_{k=1}^{200} b_k - 5 \sum_{k=1}^{3} b_k$
$= 5(67) - 5(17) = 250$

15. $\frac{20(20+1)[2(20)+1][3(20)^2 + 3(20) - 1]}{30}$
$= 2(7)(41)(1259) = 722{,}666$

20. $P(1)$: $\displaystyle\sum_{k=1}^{1} k^2 = 1^2 = \frac{1(1+1)[2(1)+1]}{6}$

$$= \frac{2(3)}{6} = 1$$

Assume $P(k)$:

$$1^2 + 2^2 + \cdots + k^2 = \frac{k(k+1)(2k+1)}{6}.$$

Add $(k+1)^2$ to both sides.

$$1^2 + 2^2 + \cdots + k^2 + (k+1)^2$$
$$= \frac{k(k+1)(2k+1)}{6} + (k+1)^2$$
$$= \frac{k(k+1)(2k+1) + 6(k+1)^2}{6}$$
$$= \frac{(k+1)[k(2k+1) + 6(k+1)]}{6}$$
$$= \frac{(k+1)(2k^2 + 7k + 6)}{6}$$
$$= \frac{(k+1)(k+2)(2k+3)}{6}$$
$$= \frac{(k+1)[(k+1)+1][2(k+1)+1]}{6}$$
$$= P(k+1)$$

$\therefore P(n)$ is true for all $n \in \mathbb{N}$.

21. $P(1)$:

$$\sum_{k=1}^{1} k^3 = 1^3 = 1 = \frac{1^2(1+1)^2}{4} = \frac{2^2}{4} = 1$$

Assume $P(k)$:

$$1^3 + 2^3 + \cdots + k^3 = \frac{k^2(k+1)^2}{4}.$$

Add $(k+1)^3$ to both sides.

$$1^3 + 2^3 + \cdots + k^3 + (k+1)^3$$
$$= \frac{k^2(k+1)^2}{4} + (k+1)^3$$
$$= \frac{k^2(k+1)^2 + 4(k+1)^3}{4}$$
$$= \frac{(k+1)^2[k^2 + 4(k+1)]}{4}$$
$$= \frac{(k+1)^2(k^2 + 4k + 4)}{4}$$
$$= \frac{(k+1)^2(k+2)^2}{4}$$
$$= \frac{(k+1)^2[(k+1)+1]^2}{4} = P(k+1)$$

$\therefore P(n)$ is true for all $n \in \mathbb{N}$.

22. $P(1)$: $1(1+1) = 1(2)$; 2 is a factor.

Assume $P(k)$: $k(k+1) = 2c$, $c \in \mathbb{Z}$.

$P(k+1)$: $(k+1)((k+1)+1)$
$$= (k+1)(k+2) = k^2 + k + 2k + 2$$
$$= k(k+1) + 2(k+1)$$
$$= 2c + 2(k+1)$$
$$= 2(c + k + 1) = 2d,\ d \in \mathbb{Z}$$

$\therefore P(n)$ is true for all $n \in \mathbb{N}$.

23. $P(1)$: $1^3 - 1 + 12 = 12 = 6(2)$

6 is a factor.

Assume $P(k)$: $k^3 - k + 12 = 6c$, $c \in \mathbb{Z}$.

$P(k+1)$: $(k+1)^3 - (k+1) + 12$
$$= k^3 + 3k^2 + 3k + 1 - k - 1 + 12$$
$$= (k^3 - k + 12) + 3k^2 + 3k$$
$$= 6c + 3k(k+1)$$
$$= 6c + 3(2d) = 6(c + d)$$
$$= 6e,\ e \in \mathbb{Z}$$

$\therefore P(n)$ is true for all $n \in \mathbb{N}$.

24. $P(1)$: $4^1 - 1 = 3 = 3(1)$; 3 is a factor.

Assume $P(k)$: $4^k - 1 = 3c$, $c \in \mathbb{Z}$.

$P(k+1)$: $4^{k+1} - 1$
$$= 4 \cdot 4^k - 1$$
$$= 4(4^k - 1) - 1 + 4$$
$$= 4(3c) + 3$$
$$= 3(4c + 1) = 3d,\ d \in \mathbb{Z}$$

$\therefore P(n)$ is true for all $n \in \mathbb{N}$.

25. $P(1)$: $8^1 - 3^1 = 5 = 5(1)$; 5 is a factor.

Assume $P(k)$: $8^k - 3^k = 5c$, $c \in \mathbb{Z}$.

$P(k+1)$: $8^{k+1} - 3^{k+1}$
$$= 8(8^k) - (3)3^k$$
$$= 8(8^k) - 8(3^k) + 5(3^k)$$
$$= 8(8^k - 3^k) + 5(3^k)$$
$$= 8(5c) + 5(3^k)$$
$$= 5(8c + 3^k) = 5d,\ d \in \mathbb{Z}$$

$\therefore P(n)$ is true for all $n \in \mathbb{N}$.

26. $P(1)$: $(1+1)^2 \geq 1^2 + 1$

$$4 \geq 2$$

Assume $P(k)$: $(k+1)^2 \geq k^2 + 1$.

$$k^2 + 2k + 1 \geq k^2 + 1$$

Add $2k + 3$ to both sides.

$$k^2 + 4k + 4 \geq k^2 + 2k + 4$$
$$(k+2)^2 \geq (k^2 + 2k + 1) + 3$$
$$(k+2)^2 \geq (k+1)^2 + 1 + 2$$

$\therefore P(k+1)$:

$$((k+1)+1)^2 \geq (k+1)^2 + 1 \text{ and }$$
$$P(n) \text{ is true for all } n \in \mathbb{N}.$$

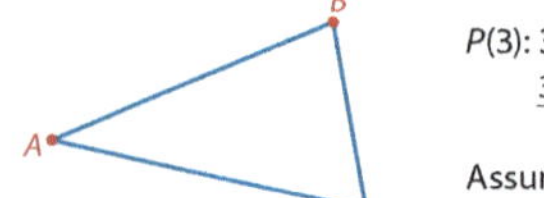

Example 8 — Applying Math Induction for Anchor $j > 1$

Use math induction to prove that any set of n coplanar points, no three of which are collinear, determines $\frac{n(n-1)}{2}$ line segments.

Answer

$P(3)$: 3 noncollinear points determine $\frac{3(3-1)}{2} = 3$ segments.

1. Since any set of noncollinear points has at least 3 points, verify the anchor step $P(3)$.

Assume $P(k)$: A set of k coplanar points, no three of which are collinear, determines $\frac{k(k-1)}{2}$ line segments.

2. Write the inductive hypothesis and reason to the inductive conclusion.

$P(k+1)$: $\frac{k(k-1)}{2} + k$

a. Adding another point that is not collinear with any two previous points allows k additional segments to be drawn, one to each of the previous k points.

$$= \frac{k^2 - k}{2} + \frac{2k}{2} = \frac{k^2 + k}{2}$$

b. Combine the fractions.

$$= \frac{(k+1)k}{2} = \frac{(k+1)[(k+1)-1]}{2}$$

c. Factor and rewrite the numerator to match $P(k+1)$.

$P(k+1)$: A set of $k+1$ coplanar points, no three of which are collinear, determines $\frac{(k+1)[(k+1)-1]}{2}$ line segments.

d. State the inductive conclusion.

Therefore n coplanar points, no three of which are collinear, determine $\frac{n(n-1)}{2}$ line segments.

3. Conclude that $P(n)$ is true for all integers $n \geq 3$.

SKILL ✔ EXERCISE 35

❯ A. Exercises

1. Identify each part in a mathematical induction proof.

   a. $P(n)$ is true for all $n \in \mathbb{N}$.   I
   b. $P(k+1)$ is true when $P(k)$ is true.   III
   c. $P(k+1)$ V
   d. $P(k)$ IV
   e. $P(1)$ is true.   II

   I. induction proposition
   II. anchor step
   III. inductive step
   IV. inductive hypothesis
   V. inductive conclusion

Find the smallest natural number n that provides a counterexample for each proposition $P(n)$.

2. $P(n)$: $5 + 8 + 11 + \cdots + (3n+2) = \frac{n(n+9)}{2}$ 2

3. $P(n)$: 3 is a factor of $n(n+1)$ for $n \geq 2$.   4

4. $P(n)$: $\left(1 + \frac{1}{n}\right)^n > n$ 3

For each proposition $P(n)$ state the anchor step, the inductive hypothesis, and the inductive conclusion.

5. $8 + 10 + 12 + \cdots + 2(n+3) = n(n+7)$

6. $1 + 3 + 6 + \cdots + \frac{j(j+1)}{2} = \frac{n(n+1)(n+2)}{6}$

7. 3 is a factor of $7^n - 4^n$.

8. $3^n \geq 2n + 1$

State (do not prove) the "if-then" form of the inductive step for each proposition $P(n)$ for all $n \in \mathbb{N}$.

9. $1 + 5 + 9 + \cdots + (4n - 3) = 2n^2 - n$

10. 8 is a factor of $3^{2n} - 1$.

11. $3n < 2^n + 3$ If $3k < 2^k + 3$, then $3(k+1) < 2^{(k+1)} + 3$.

9. If $1 + 5 + 9 + \cdots + (4k - 3) = 2k^2 - k$, then $1 + 5 + 9 + \cdots + [4(k+1) - 3] = 2(k+1)^2 - (k+1)$.

10. If 8 is a factor of $3^{2k} - 1$, then 8 is a factor of $3^{2(k+1)} - 1$.

If $\sum_{k=1}^{200} a_k = 20$, $\sum_{k=1}^{200} b_k = 67$, and $\sum_{k=1}^{200} c_k = 404$, use proper-

ties of summations to evaluate each sum.

12. $\sum_{k=1}^{200} 3a_k$ 60

13. $\sum_{k=1}^{200} (4c_k + 3)$ 2216

14. $\sum_{k=4}^{200} 5b_k$ if $b_1 + b_2 + b_3 = 17$ 250

Evaluate each partial sum.

15. $\sum_{k=1}^{20} k^4$ 722,666

16. $\sum_{k=1}^{38} (3k^2 + 7k - 4)$ 62,092

17. $\sum_{k=15}^{43} k^4$ 31,009,903

❯ B. Exercises

Use math induction to prove each formula.

18. $P(n)$: given the arithmetic sequence a_n with common difference d, $a_n = a_1 + (n-1)d$

19. $P(n)$: given the geometric sequence a_n with common ratio r, $S_n = \dfrac{a_1(1 - r^n)}{1 - r}$

Use math induction to prove each summation formula.

20. $\sum_{k=1}^{n} k^2 = \dfrac{n(n+1)(2n+1)}{6}$

21. $\sum_{k=1}^{n} k^3 = \dfrac{n^2(n+1)^2}{4}$

Use math induction to prove each divisibility statement.

22. 2 is a factor of $n(n+1)$.

23. 6 is a factor of $n^3 - n + 12$.

24. 3 is a factor of $4^n - 1$.

25. 5 is a factor of $8^n - 3^n$.

Use math induction to prove each inequality.

26. $(n+1)^2 \geq n^2 + 1$ for all $n \in \mathbb{N}$

27. $2^n > n$ for all $n \in \mathbb{N}$

28. $\left(1 + \frac{1}{n}\right)^n < n$ for $n \geq 3$

29. $n! > 2^n$ for $n \geq 4$

Find an explicit formula for each partial sum. Then use math induction to prove your result.

30. $1 + 3 + 5 + 7 + 9 + \cdots + (2n - 1)$ $S_n = n^2$

31. $1 + 2 + 4 + 8 + \cdots + 2^{n-1}$ $S_n = 2^n - 1$

32. $\frac{1}{2} + \frac{1}{2^2} + \frac{1}{2^3} + \cdots + \frac{1}{2^n}$ $S_n = 1 - \frac{1}{2^n}$

33. Use the summation formulas to show that
$$\sum_{k=1}^{n} k^3 = \left(\sum_{k=1}^{n} k\right)^2.$$

34. A recreational league plans a round-robin volleyball tournament (each team plays every other team once). $a_n = \frac{n(n-1)}{2}$

---

a. Write the sequence representing the total number of games played by n teams.   0, 1, 3, 6, 10, …

b. Determine an explicit formula for a_n and explain your reasoning.

c. State the inductive proposition $P(n)$ for your formula.

d. Use math induction to prove $P(n)$.   180, 360, 540, 720, …

e. Use the explicit formula to determine the total number of games if 9 teams sign up.   36 games

35. A convex n-gon has n vertices.

a. Write the sequence representing the sum of the interior angle degree measures in a convex n-gon.

b. Determine the explicit formula for a_n and explain your reasoning. $a_n = 180(n-2)$; $n \geq 3$

c. State the induction proposition $P(n)$ for your formula.   1440°

d. Use math induction to prove $P(n)$.

e. Use the explicit formula to find the sum of the interior angle degree measures in a convex decagon.

❯ C. Exercises

Use math induction to prove the following partial-sum formulas.

36. given the arithmetic sequence a_n with common difference d, $S_n = \dfrac{n(2a_1 + (n-1)d)}{2}$ $S_n = \dfrac{n}{2(n+1)}$

37. $\sum_{k=1}^{n} k^4 = \dfrac{n(n+1)(2n+1)(3n^2 + 3n - 1)}{30}$

38. Find an explicit formula for
$$S_n = \frac{1}{4} + \frac{1}{12} + \frac{1}{24} + \frac{1}{40} + \frac{1}{60} + \cdots + \frac{1}{2n(n+1)}.$$
Then use math induction to prove your formula.

39. Use math induction to prove that $\cos n\pi = (-1)^n$.

40. While previous sections emphasized arithmetic and geometric sequences, this section includes quadratic and other polynomial sequences. Using technology, complete a regression to determine whether each sequence is arithmetic (linear), quadratic, geometric (exponential), or none of these. State the explicit formula for any sequence if the coefficient of determination, r^2, is 1.

a. $\frac{5}{32}, \frac{1}{4}, \frac{2}{5}, \frac{16}{25}, \ldots$ geometric; $a_n = 0.09765625(1.6)^n$

b. $\frac{3}{8}, \frac{5}{4}, \frac{19}{8}, \frac{15}{4}, \ldots$ quadratic; $a_n = 0.125n^2 + 0.5n - 0.25$

c. $\frac{1}{4}, \frac{3}{8}, \frac{1}{2}, \frac{5}{8}, \ldots$ arithmetic; $a_n = 0.125n + 0.125$

d. $\frac{5}{32}, \frac{51}{32}, \frac{185}{32}, \frac{479}{32}, \ldots$ none

---

$$1 + 3 + \cdots + (2k - 1) + 2(k+1) - 1 = k^2 + (2k + 1)$$
$P(k+1)$:
$$1 + 3 + \cdots + 2(k+1) - 1 = (k+1)^2$$
$\therefore P(n)$ is true for all $n \in \mathbb{N}$.

31. $S_1 = 1$; $S_2 = 3$; $S_3 = 7$
$S_4 = 15$; $S_5 = 31$; $S_n = 2^n - 1$
$P(n)$: $1 + 2 + 4 + 8 + \cdots + 2^{n-1} = 2^n - 1$
$P(1)$: $1 = 2^1 - 1 = 1$
Assume $P(k)$:
$$1 + 2 + \cdots + 2^{k-1} = 2^k - 1.$$
$$1 + 2 + \cdots + 2^{k-1} + 2^k = 2^k - 1 + 2^k = 2(2^k) - 1$$
$P(k+1)$: $1 + 2 + \cdots + 2^k = 2^{k+1} - 1$
$\therefore P(n)$ is true for all $n \in \mathbb{N}$.

32. $S_1 = \frac{1}{2}$; $S_2 = \frac{3}{4}$; $S_3 = \frac{7}{8}$
$S_4 = \frac{15}{16}$; $S_5 = \frac{31}{32}$; $S_n = 1 - \frac{1}{2^n}$
$P(n)$: $\frac{1}{2} + \frac{1}{2^2} + \frac{1}{2^3} + \cdots + \frac{1}{2^n} = 1 - \frac{1}{2^n}$
$P(1)$: $\frac{1}{2} = 1 - \frac{1}{2^1} = \frac{1}{2}$
Assume $P(k)$:
$$\frac{1}{2} + \frac{1}{2^2} + \cdots + \frac{1}{2^k} = 1 - \frac{1}{2^k}.$$
$$\frac{1}{2} + \frac{1}{2^2} + \cdots + \frac{1}{2^k} + \frac{1}{2^{k+1}}$$
$$= 1 - \frac{1}{2^k} + \frac{1}{2^{k+1}}$$
$$= 1 - \frac{1(2)}{2^k(2)} + \frac{1}{2^{k+1}}$$
$P(k+1)$:
$$\frac{1}{2} + \frac{1}{2^2} + \cdots + \frac{1}{2^{k+1}} = 1 - \frac{1}{2^{k+1}}$$
$\therefore P(n)$ is true for all $n \in \mathbb{N}$.

33. $\sum_{k=1}^{n} k^3 = \dfrac{n^2(n+1)^2}{4}$
$$= \left(\frac{n(n+1)}{2}\right)^2 = \left(\sum_{k=1}^{n} k\right)^2$$

34b. $a_n = \dfrac{n(n-1)}{2}$; Each of the n teams plays each of the other $n - 1$ teams. Since 2 teams are playing in each game, divide this total of $n(n-1)$ games by 2 to eliminate double counting.

34c. $P(n)$: The total number of games in a round-robin tournament with n teams is $a_n = \dfrac{n(n-1)}{2}$.

---

27. $P(1)$ is true.; $2^1 > 1$
Assume $P(k)$: $2^k > k$.
since $2^k > 1$ for $k > 0$,
$2^k + 2^k > k + 1$
$2(2^k) > k + 1$
$\therefore P(k+1)$: $2^{k+1} > k + 1$
and $P(n)$ is true for all $n \in \mathbb{N}$.

28. $P(3)$: $\left(1 + \frac{1}{3}\right)^3 = \frac{64}{27} < 3$
Assume $P(k)$: $\left(1 + \frac{1}{k}\right)^k < k$.
$$\left(1 + \frac{1}{k}\right)^k \left(1 + \frac{1}{k}\right) < k\left(1 + \frac{1}{k}\right)$$
$$\left(1 + \frac{1}{k}\right)^{k+1} < k + 1$$
since $\frac{1}{k+1} < \frac{1}{k}$ for $n \geq 3$,
$$\left(1 + \frac{1}{k+1}\right)^{k+1} < \left(1 + \frac{1}{k}\right)^{k+1}$$

by the Transitive Property,
$$P(k+1): \left(1 + \frac{1}{k+1}\right)^{k+1} < k + 1$$
$\therefore P(n)$ is true for all $n \geq 3$, $n \in \mathbb{N}$.

29. $P(4)$: $4! \; (= 24) > 2^4 \; (= 16)$
Assume $P(k)$: $k! > 2^k$ when $k \geq 4$.
since $k + 1 > 2$ when $k \geq 4$,
$(k+1)k! > (2)2^k$
$\therefore P(k+1)$: $(k+1)! > 2^{k+1}$ and
$P(n)$ is true for all $n \geq 4$, $n \in \mathbb{N}$.

30. $S_1 = 1$; $S_2 = 4$; $S_3 = 9$
$S_4 = 16$; $S_5 = 25$; $S_n = n^2$
$P(n)$: $1 + 3 + 5 + \cdots + (2n - 1) = n^2$
$P(1)$: $1 = 1^2$
Assume $P(k)$:
$1 + 3 + \cdots + (2k - 1) = k^2$.

34d. $P(1): a_1 = \dfrac{1(1-1)}{2} = 0$

Assume $P(k): a_k = \dfrac{k(k-1)}{2}$.

Any additional team plays each of the previous k teams adding k more games.

$\dfrac{k(k-1)}{2} + k = \dfrac{k^2 - k}{2} + \dfrac{2k}{2} = \dfrac{k^2 + k}{2}$

$= \dfrac{(k+1)k}{2} = \dfrac{(k+1)(k+1-1)}{2}$

$P(k+1): a_{k+1} = \dfrac{(k+1)((k+1)-1)}{2}$

$\therefore P(n)$ is true for all $n \in \mathbb{N}$.

34e. $a_9 = \dfrac{9(9-1)}{2} = 36$ games

35.

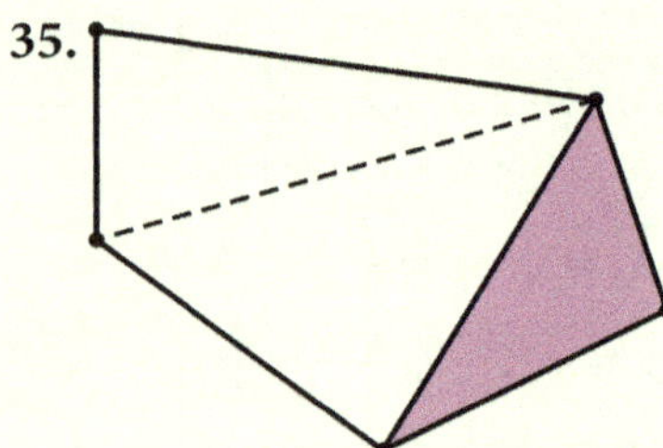

35a. beginning with $n = 3$:
180, 360, 540, 720, ...

35b. $a_n = 180(n-2)$; $n \geq 3$
In a triangle, the sum is $180°$. A quadrilateral can be divided into 2 triangles, so the sum is $180(2) = 360°$. A pentagon can be divided into 3 triangles, so its sum is $180(3) = 540°$. Each additional vertex adds another triangle, so a convex n-gon can be divided into $n - 2$ triangles and has a sum of $180(n-2)$.

35c. $P(n)$: For n vertices, the number of degrees for all interior angles is $180(n-2)$.

35d. $P(3): a_3 = 180(3-2) = 180$, and the sum of the degree measures in any triangle is $180°$.
Assume $P(k): a_k = 180(k-2)$.
An additional vertex adds another triangle and another $180°$.
$P(k+1): a_{k+1} = 180(k-2) + 180$
$\qquad\qquad = 180(k-2+1)$
$\qquad\qquad = 180((k+1)-2)$
$\therefore P(n)$ is true for $n \geq 3,\, n \in \mathbb{N}$.

35e. $a_{10} = 180(10-2) = 1440°$

---

41. A line is uniquely determined by two points, a parabola by three points, and a cubic function by four points. Use your calculator and a cubic regression to find the unique cubic explicit formula for each set of four points given in exercise 40.

a. $\dfrac{5}{32}, \dfrac{1}{4}, \dfrac{2}{5}, \dfrac{16}{25}, \dots$ $a_n = 0.005625n^3 - 0.005625n^2 + 0.07125n + 0.085$

b. $\dfrac{3}{8}, \dfrac{5}{4}, \dfrac{19}{8}, \dfrac{15}{4}, \dots$ $a_n = 0n^3 + 0.125n^2 + 0.5n - 0.25$

c. $\dfrac{1}{4}, \dfrac{3}{8}, \dfrac{1}{2}, \dfrac{5}{8}, \dots$ $a_n = 0n^3 + 0n^2 + 0.125n + 0.125$

d. $\dfrac{5}{32}, \dfrac{51}{32}, \dfrac{185}{32}, \dfrac{479}{32}, \dots$ $a_n = 0.375n^3 - 0.875n^2 + 1.4375n - 0.78125$

42. Use mathematical induction to prove the Binomial Theorem.

CUMULATIVE REVIEW

Consider the following functions.
a. $f(x) = x^4 - 3x^2 + 2$ **b.** $g(x) = x^3 - x^2$
c. $h(x) = 2x^3 - x$ **d.** $k(x) = 2\cos\dfrac{x}{2}$

43. Classify each function as even, odd, or neither.   [3.1]

44. Which function has a zero at $x = \sqrt{2}$?   [2.6]   f(x)

Determine whether A and B are inverse matrices.   [7.5]

45. $A = \begin{bmatrix} -3 & -6 \\ 1 & 3 \end{bmatrix}, B = \begin{bmatrix} -1 & -2 \\ \frac{1}{3} & 1 \end{bmatrix}$ inverses

46. $A = \begin{bmatrix} -1 & 0 \\ -3 & 1 \end{bmatrix}, B = \begin{bmatrix} -1 & 0 \\ -3 & 1 \end{bmatrix}$ inverses

Find the geometric means between the given terms.   [9.3]

47. $a_8 = 3$ and $a_{10} = 507$ ± 39

48. $a_{38} = -5$ and $a_{41} = 3645$ 45; −405

49. Atlanta, Georgia, had a population of 416,474 in 2000. The city's population in 2010 was 418,981. Find the city's average annual rate of growth from 2000 to 2010.   [3.5]   D
A. 1%   **C.** 0.6%   **E.** none of
B. 6%   **D.** 0.06%   these

50. Which of the following represents the period, vertical asymptotes, and zeros of the cotangent function?   [4.5]   B
A. 2π, none, $(2n+1)\dfrac{\pi}{2}$ **D.** π, $x = (2n+1)\dfrac{\pi}{2}$, $n\pi$
B. π, $x = n\pi$, $(2n+1)\dfrac{\pi}{2}$ **E.** none of these
C. 2π, $x = n\pi$, none

51. Write the partial fraction decomposition for the rational expression $\dfrac{5x-12}{x^2-6x+9}$.   [7.7]   A
A. $\dfrac{5}{x-3} + \dfrac{3}{(x-3)^2}$ **D.** $\dfrac{4}{x-3} - \dfrac{1}{(x-3)^2}$
B. $\dfrac{4}{x-3} + \dfrac{1}{(x-3)^2}$ **E.** none of these
C. $\dfrac{5}{x-3} - \dfrac{3}{(x-3)^2}$

52. Find a_{98} in the sequence $-12, -15, -18, -21, \dots$ [9.2]   B
A. −300   **C.** −306   **E.** none of
B. −303   **D.** −309   these

43a. even
43b. neither
43c. odd
43d. even

> ### C. Exercises

36. $P(1): S_1 = \dfrac{1(2a_1 + (1-1)d)}{2} = \dfrac{2a_1}{2} = a_1$

Assume $P(k)$:
$a_1 + a_2 + \cdots + (a_1 + (k-1)d)$
$\qquad\qquad = \dfrac{k(2a_1 + (k-1)d)}{2}.$

Add $a_{k+1} = a_1 + ((k+1)-1)d$ or $a_1 + kd$ to both sides.
$a_1 + a_2 + \cdots + (a_1 + (k-1)d)$
$\qquad\qquad\qquad + (a_1 + kd)$
$\qquad = \dfrac{k(2a_1 + (k-1)d)}{2} + (a_1 + kd)$

$S_{k+1} = \dfrac{k(2a_1 + kd - d)}{2} + \dfrac{2(a_1 + kd)}{2}$

$= \dfrac{2a_1 k + k^2 d - kd + 2a_1 + 2kd}{2}$

$= \dfrac{2a_1 k + 2a_1 + k^2 d + kd}{2}$

$= \dfrac{2a_1(k+1) + kd(k+1)}{2}$

$= \dfrac{(k+1)(2a_1 + kd)}{2}$

$= \dfrac{(k+1)(2a_1 + (k+1-1)d)}{2}$

$\therefore P(n)$ is true for all $n \in \mathbb{N}$.

continued in Answers and Solutions Overflow

Graph Theory and Networks

In 1736 Leonhard Euler published his solution to a well-known problem from Prussia: can the seven bridges in Königsberg all be crossed exactly one time in a single walk? Euler generalized the problem to a system of points and lines that could answer the question for any number of bridges and islands. This was the genesis of graph theory.

Consider the following 4×4 grid (or *network*). How many different paths are there from A to any given *node* (point of intersection) if you can only move to the right or down along the segments (or *edges*)? The figure shows the number of possible paths from A to each node.

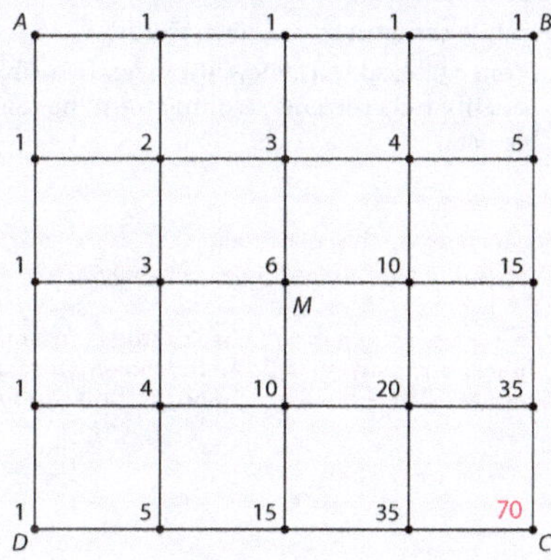

1. What pattern can be seen diagonally in the numbers on this figure? Pascal's triangle

2. Sketch the 6 possible paths to the middle node, M.

3. How many ways are there to go from A to C? 70

4. If you cannot pass through M, how many paths are there from A to C? 34

Graph theory led to the study of networks, a vital component of myriad real-world applications in many diverse fields. Navigation programs like Google Maps are especially helpful applications. Each location is a node (or vertex), and the roads are edges. In a *weighted* graph, the decisions become more difficult as each edge is assigned a nonnegative number indicating the time (or cost) required to traverse that segment. The goal is to find the shortest (or least expensive) route between two nodes—not simply the number of different ways to get there.

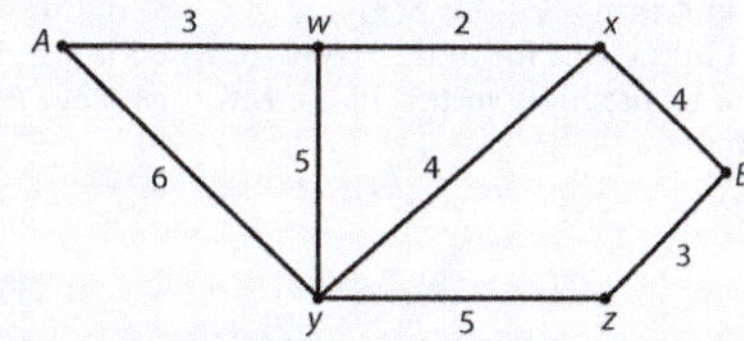

Finding the shortest path from A to B is a fairly simple task in this small network, but more complicated scenarios can be daunting. A basic algorithm to find the shortest distance between two locations on a weighted graph was published by E. W. Dijkstra in 1959.

PRESENTATION

You may wish to research the Königsberg bridge problem and use it to introduce this Data Analysis feature. The problem is easy to illustrate and demonstrates how the effort to find a general solution to a simple problem can lead to a completely new branch of mathematics (in this case, graph theory and topology). In this context a graph refers to a network of points and lines.

The grid problem is an interesting way to present another appearance of Pascal's triangular numbers. Also notice that to go from A to C on any path will require 4 moves to the right and 4 moves down.

If your students are familiar with combinations, explain that this is the same as finding $_8C_4$. In general, any such $n \times n$ grid will have $_{2n}C_n$ possible paths. Chapter 10 discusses how the Binomial Theorem (and thus Pascal's triangle) applies to probability and provides an easier method of answering these questions on the number of possible paths. You may also wish to explore rectangular grids.

Graph theory provides another example of how pure mathematics, developed with no practical application in mind, has become indispensable in several fields. In fact, it is difficult to overstate the number of real-world applications.

Objectives

1. To describe the origins of graph theory

2. To find the number of possible paths between two points on a grid

3. To find the lowest-cost path on a weighted graph

4. To evaluate the effects of artificial intelligence and big data from a biblical worldview

Assignments

- **Minimum:** 1–8; 9–12 (in class)
- **Standard:** 1–8; 9–12 (in class)
- **Extended:** 1–12

Solutions

2. 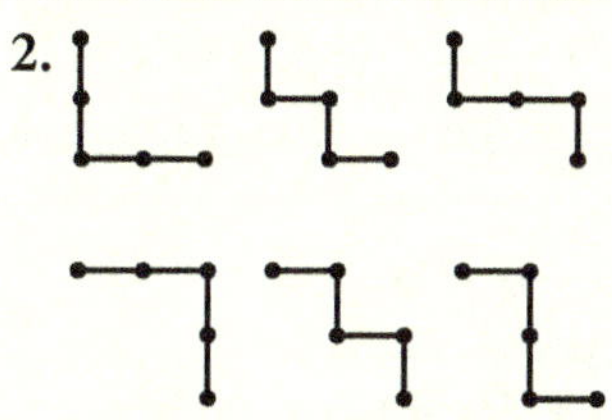

3. With 35 paths to the node left of C and 35 paths to the node above C, there are 70 paths to C.

4. There are 6 paths to M and 6 paths from M to C, so there are 36 paths from A to C containing M, $70 - 36 = 34$.

   Alternatively, renumber the vertices, letting $M = 0$, and add numbers diagonally to get the numbers for the next node.

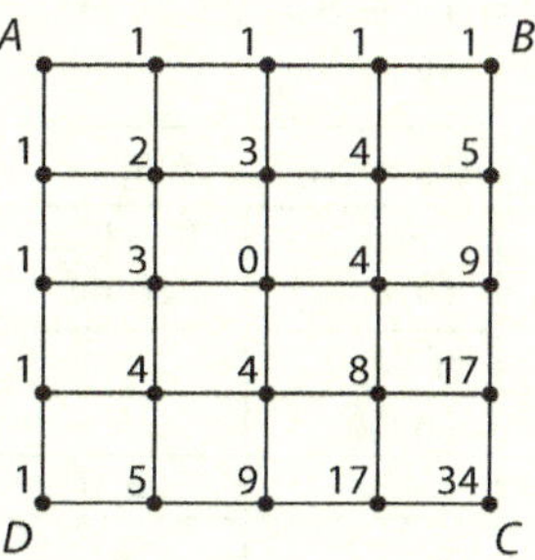

5. 3rd iteration (from x):

A-x-$B = 5 + 4 = 9$

Change row B to 9, x.

A-x-$y = 5 + 4 = 9$

Leave row y as is.

4th iteration (from y):

A-y-$z = 6 + 5 = 11$

Change row z to 11, y.

5th iteration (from B):

A-x-B-$z = 9 + 3 = 12$

Leave row z as is.

Not Visited: $\{A, w, x, y, z, B\}$		
Node	Cost (from A)	Previous Node
A	0	
w	∞ 3	A
x	∞ 5	w
y	∞ 6	A
z	∞ 11	y
B	∞ 9	x
Visited: $\{A, w, x, y, B\}$		

6. 1st iteration (from S):

S-$a = 10$; Change row a to 9, S.

S-$c = 12$; Change row c to 12, S.

2nd iteration (from a):

S-a-$b = 10 + 6 = 16$

Change row b to 16, a.

S-a-$d = 10 + 3 = 13$

Change row d to 13, a.

3rd iteration (from c):

S-c-$d = 12 + 5 = 17$

Leave row d as is.

4th iteration (from d):

S-a-d-$b = 13 + 2 = 15$

Change row b to 15, d.

S-a-d-$F = 13 + 9 = 22$

Change row F to 22, d.

5th iteration (from b):

S-a-d-b-$F = 15 + 8 = 23$

Leave row F as is.

Not Visited: $\{S, a, b, c, d, F\}$		
Node	Cost (from A)	Previous Node
S	0	
a	∞ 10	S
b	∞ 16 15	a d
c	∞ 12	S
d	∞ 13	a
F	∞ 22	d
Visited: $\{S, a, c, d, b\}$		

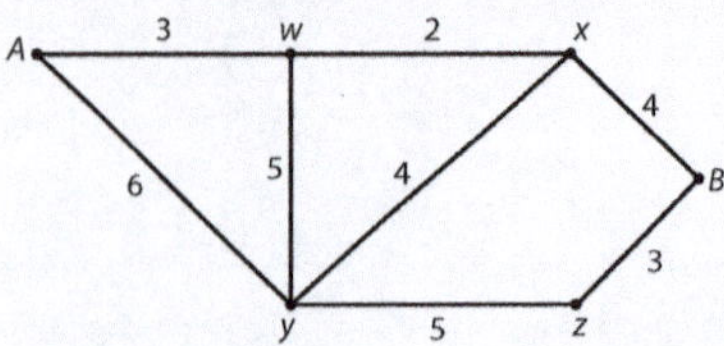

The iterative algorithm can be explained with a table. Begin by listing the vertices as a set of points that have not been visited. Then set the cost of the starting point to 0 and the cost of all other points as infinite. The first iteration begins by moving from the node with the least cost (A) to a set of visited nodes and then finding the cost to reach the connected vertices. (A to $w = 3$ and A to $y = 6$.) If these costs are less than the entry currently listed in that row, its cost and the "previous node" value are updated.

Not Visited: $\{a, w, x, y, z, B\}$		
Node	Cost (from A)	Previous Node
A	0	
w	∞ 3	A
x	∞ 5	w
y	∞ 6	A
z	∞	
B	∞	
Visited: $\{A, w,$ $\}$		

initial condition

1st iteration, 2nd iteration

The algorithm then starts over by moving the lowest-cost node in the not-visited set (w this time) to the visited set and finding the cost to reach each connected node (w to $x = 2$ so A to $x = 3 + 2 = 5$, and w to $y = 5$ so A to $y = 3 + 5 = 8$.) In the row for node x, replace ∞ with 5 and record w for its previous vertex. In the row for y leave 6 and A since $6 < 8$.

5. Copy the table and continue the iterations until the lowest cost to each node has been determined.

  a. List the lowest-cost route from A to B and state the route's cost. A–w–x–B; 9

  b. List the lowest-cost route from A to z and state the route's cost. A–y–z; 11

6. Make a table and use Dijkstra's algorithm to find the lowest-cost route from S to F in the following network and state its cost. S–a–d–F; 22

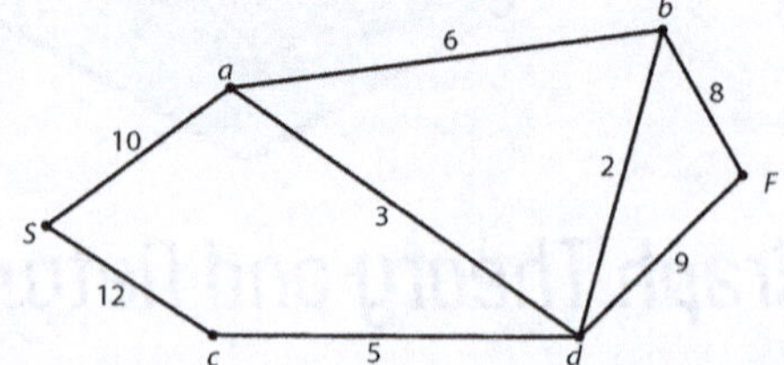

Modern mapping applications have improved upon Dijkstra's algorithm and are able to consider constantly changing conditions such as traffic volume, road construction, and weather delays. While the benefits of mapping programs are obvious, there are concerns about how the vast amounts of data may be used. If you've ever visited a restaurant or store and later received an unsolicited message on your phone asking you to rate the location, you may have wondered who was watching. The implications of big data are worth considering.

7. Discuss: Research the term *big data*.

  a. Define the term in your own words.

  b. From a biblical worldview, list several possible benefits and problems resulting from the use of big data.

sample answers:

7a. Big data refers to the collection and analysis of large amounts of many forms of data collected from a variety of sources.

7b. Beneficial uses might include health studies, navigation programs, and marketing. Possible problems include illegal surveillance, bank fraud, and invasion of privacy.

Most students will be familiar with navigation programs like Google Maps. The algorithm presented in the text was used in pioneering mapping programs and is still taught in computer science classes even though many improvements have been made since its initial publication.

One model of the San Francisco Bay area consists of more than 330,000 vertices. The Dijkstra algorithm required only about 80 ms to find the shortest path between two random vertices. The best algorithm required less than half a millisecond.

Big data, privacy issues, and artificial intelligence all suggest several possible questions you may wish to discuss with students from a biblical worldview.

Chapter 9 Review

1. Classify each as a finite sequence, infinite sequence, finite series, or infinite series. [9.1]

   a. 1, 2, 3, 4, 5   finite sequence

   b. $2 + 6 + 18 + 54 + \cdots$ infinite series

2. Find the first 4 terms and the tenth term of the explicitly defined sequence $a_n = \frac{n(n+1)}{4}$. [9.1]

3. Given 1, 4, 9, 16, . . . , write an explicit formula for the value of a_n and use it to find the tenth term. [9.1]

4. Write the first 6 terms of each recursively defined sequence. [9.1]

   a. $a_1 = 3$; $a_n = (a_{n-1} - 1)^2$, where $n \geq 2$

   b. $a_1 = 5$; $a_n = \frac{1}{a_{n-1} + 2}$, where $n \geq 2$

5. Write a recursive formula for $1, \frac{1}{2}, \frac{1}{4}, \frac{1}{8}, \ldots$. Then use the formula to find a_5, a_6, and a_7. [9.1]

6. Given $a_n = \frac{n!}{2^n}$, find a_0, a_1, a_2, a_3, and a_4. [9.1]

7. Determine whether the sequence $a_n = \frac{2n+3}{n-2}$ converges or diverges and state the limit if it is convergent. [9.1]   converges; 2

8. Evaluate $\sum\limits_{k=6}^{7} 5k$. [9.1]   65

9. Given $a_n = \frac{(n+1)!}{(n-1)!}$, find the partial sums S_5, S_6, and S_7. [9.1]   70, 112, 168

10. Evaluate the infinite series $\sum\limits_{i=1}^{\infty} 10\,(0.01)^i$ and classify it as convergent or divergent. [9.1]

11. Determine whether 41, 26, 11, -4, -19, . . . is an arithmetic sequence. If so, state the common difference and list the next 3 terms. If not, explain why. [9.2]   yes; $d = -15$; $-34, -49, -64$

12. Identify the common difference d for each arithmetic sequence and then find the fifth term. [9.2]

   a. $a_1 = 2$, $a_n = a_{n-1} + 3$ $d = 3$; $a_5 = 14$

   b. $a_n = -5n + 9$ $d = -5$; $a_5 = -16$

13. Write the recursive and simplified explicit formulas for the sequence of whole numbers: 0, 1, 2, ... Then find a_{20}. [9.2] $a_1 = 0$, $a_n = a_{n-1} + 1$; $a_n = n - 1$; $a_{20} = 19$

14. Given an arithmetic sequence with $a_1 = 20$ and $d = 39$, find a_{58}. [9.2]   2243

15. Find the arithmetic means between $a_5 = 54$ and $a_9 = 18$. [9.2]   45, 36, 27

16. Write the simplified explicit formula for the sequence $\frac{1}{8}, \frac{1}{13}, \frac{1}{18}, \frac{1}{23}, \ldots$, then find a_{20}. [9.2]

Evaluate each arithmetic series. [9.2]

17. $\sum\limits_{k=1}^{50} 4n$ 5100

18. $\sum\limits_{j=12}^{30} \frac{1}{2}(k+3)$ $\frac{555}{2}$

19. Write a simplified explicit formula for the arithmetic sequence with $a_1 = 1$ if $S_{250} = 62,500$. [9.2]

20. If you borrow \$7595 from your parents to buy a used car, how many months will it take to pay back the loan if you pay \$200 the first month and increase your payment by \$3 each month? [9.2]   31 mo

Determine whether each sequence is geometric. If so, state the common ratio and list the next 3 terms. If not, explain why. [9.3]

21. 2, 3, 4.5, 6, . .

22. 18, 6, 2, $\frac{2}{3}$, . . .

Write both the recursive and an explicit formula for each geometric sequence. Then find a_{10}. [9.3]

23. 216, 72, 24, 8, . . .

24. 4, -6, 9, $-\frac{27}{2}$, . . .

25. Given a geometric sequence with $a_2 = \frac{5}{2}$ and $a_7 = 80$, write a simplified explicit formula for a_n. Then find a_{10}. [9.3] $a_n = \frac{5}{8}(2)^n$; $a_{10} = 640$

26. Find the geometric means between $a_8 = 16$ and $a_{11} = \frac{1}{4}$. [9.3]   4, 1

Evaluate each geometric series. [9.3]

27. $\sum\limits_{n=1}^{10} -2\left(\frac{3}{2}\right)^{n-1}$ ≈ -226.66

28. $-\frac{1}{3} + \frac{1}{6} - \frac{1}{12} + \frac{1}{24}, \cdots$ $-\frac{2}{9}$

29. Find the interest earned in an annuity with \$3000 quarterly deposits for 15 yr if the account earns 3% APR compounded quarterly. [9.3]   \$46,272.41

Objective

To prepare for evaluation

Vocabulary

See Appendix A.

Assignments

- **Minimum:** 1–9, 11–15, 17, 19–23, 26, 29, 31, 34–36, 41, 43

- **Standard:** 1–49 odd; 2–50 even (in class)

- **Extended:** 3, 4a, 5, 7–22, 24–25, 27–30, 32, 34, 37–38, 40–41, 43–45, 49–50

Assessments

- Chapter 9 Test

- Third Quarter Exam covers Chapters 7–9.

Solutions

2. $a_1 = \frac{1(1+1)}{4} = \frac{1}{2}$; $a_2 = \frac{2(2+1)}{4} = \frac{3}{2}$

   $a_3 = \frac{3(3+1)}{4} = 3$; $a_4 = \frac{4(4+1)}{4} = 5$

   $a_{10} = \frac{10(10+1)}{4} = \frac{55}{2}$

3. $a_1 = 1^2 = 1$; $a_2 = 2^2 = 4$

   $a_3 = 3^2 = 9$; $a_4 = 4^2 = 16$

   $a_n = n^2$; $a_{10} = 10^2 = 100$

4a. $a_2 = (a_1 - 1)^2 = (3 - 1)^2 = 4$

   $a_3 = (a_2 - 1)^2 = (4 - 1)^2 = 9$

   $a_4 = (a_3 - 1)^2 = (9 - 1)^2 = 64$

   $a_5 = (a_4 - 1)^2 = (64 - 1)^2 = 3969$

   $a_6 = (a_5 - 1)^2 = (3969 - 1)^2$

   $\quad = 15{,}745{,}024$

4b. $a_2 = \frac{1}{a_1 + 2} = \frac{1}{5 + 2} = \frac{1}{7}$

   $a_3 = \frac{1}{a_2 + 2} = \frac{1}{\frac{1}{7} + 2} = \frac{7}{15}$

   $a_4 = \frac{1}{a_3 + 2} = \frac{1}{\frac{7}{15} + 2} = \frac{15}{37}$

   $a_5 = \frac{1}{a_4 + 2} = \frac{1}{\frac{15}{37} + 2} = \frac{37}{89}$

   $a_6 = \frac{1}{a_5 + 2} = \frac{1}{\frac{37}{89} + 2} = \frac{89}{215}$

5. $a_1 = 1$; $a_n = \frac{1}{2}a_{n-1}$ where $n \geq 2$

   $a_5 = \frac{1}{2}\left(\frac{1}{8}\right) = \frac{1}{16}$; $a_6 = \frac{1}{2}\left(\frac{1}{16}\right) = \frac{1}{32}$

   $a_7 = \frac{1}{2}\left(\frac{1}{32}\right) = \frac{1}{64}$

6. $a_0 = \frac{0!}{2^0} = \frac{1}{1} = 1$

$a_1 = \frac{1!}{2^1} = \frac{1}{2}$

$a_2 = \frac{2!}{2^2} = \frac{2 \cdot 1}{4} = \frac{1}{2}$

$a_3 = \frac{3!}{2^3} = \frac{3 \cdot 2 \cdot 1}{8} = \frac{3}{4}$

$a_4 = \frac{4!}{2^4} = \frac{4 \cdot 3 \cdot 2 \cdot 1}{16} = \frac{3}{2}$

7. $\lim\limits_{n\to\infty} \frac{2n+3}{1n-2} = \frac{2}{1} = 2$; converges

8. $5(6) + 5(7) = 65$

9. $S_5 = \frac{2!}{0!} + \frac{3!}{1!} + \frac{4!}{2!} + \frac{5!}{3!} + \frac{6!}{4!}$

$= 2 + 6 + 12 + 20 + 30 = 70$

$S_6 = S_5 + \frac{7!}{5!} = 70 + 42 = 112$

$S_7 = S_3 + \frac{8!}{6!} = 112 + 56 = 168$

10. $0.1 + 0.001 + 0.00001 + \cdots = 0.\overline{10}$

or $\frac{10}{99}$; convergent

11. $d = 26 - 41 = -15$

$a_6 = -19 - 15 = -34$

$a_7 = -34 - 15 = -49$

$a_6 = -49 - 15 = -64$

12a. $d = 3; a_n = 2, 5, 8, 11, 14, \ldots$

so $a_5 = 14$

12b. $d = m = -5; a_5 = -5(5) + 9 = -16$

13. $a_1 = 0; d = 1$

$a_n = 0 + (n-1)(1) = n - 1$

14. $a_{58} = 20 + (58 - 1)39$

$= 20 + 2223 = 2243$

15. $a_9 = a_5 + (9 - 5)d$

$18 = 54 + 4d$

$-36 = 4d; d = -9$

$a_6 = a_5 - 9 = 54 - 9 = 45$

$a_7 = a_6 - 9 = 45 - 9 = 36$

$a_8 = a_7 - 9 = 36 - 9 = 27$

16. The denominators form an arithmetic sequence with $a_1 = 8$ and $d = 5$.

$a_n = 8 + (n-1)5 = 5n + 3$

The harmonic sequence is the reciprocal sequence.

$\frac{1}{a_n} = \frac{1}{5n+3}; \frac{1}{a_{20}} = \frac{1}{5(20)+3} = \frac{1}{103}$

17. $a_1 = 4; a_{50} = 200$

$S_{50} = \frac{50}{2}(4 + 200) = 5100$

18. $a_1 = 2; a_2 = 2.5; d = \frac{1}{2}$

$S_{30} = \frac{30}{2}\left(2(2) + (30-1)\left(\frac{1}{2}\right)\right)$

$= 15\left(4 + \frac{29}{2}\right) = 15\left(\frac{37}{2}\right) = \frac{555}{2}$

or

$a_1 = 2; a_{30} = \frac{33}{2}$

$S_{50} = \frac{30}{2}\left(2 + \frac{33}{2}\right) = 15\left(\frac{37}{2}\right) = \frac{555}{2}$

19. $S_n = \frac{n}{2}(a_1 + a_n)$

$S_{250} = 62,500$

$= \frac{250}{2}(2(1) + (250 - 1)d)$

$500 = 2 + 249d$

$498 = 249d; d = 2$

$a_n = 1 + (n-1)2 = 2n - 1$

20. $a_1 = 200; d = 3$

$a_n = 200 + (n-1)3 = 197 + 3n$

$S_n = \frac{n}{2}(200 + (197 + 3n)) = 7595$

$n(397 + 3n) = 15,194$

$3n^2 + 397n - 15,194 = 0$

$n = \dfrac{-397 \pm \sqrt{397^2 - 4(3)(15,194)}}{2(3)}$

≈ 31 mo

21. no; $\frac{3}{2} = 1.5$, but $\frac{6}{4.5} = \frac{4}{3}$

The ratio is not constant.

22. yes; $r = \frac{6}{18} = \frac{2}{6} = \frac{\frac{2}{3}}{2} = \frac{1}{3}$

$a_5 = \frac{2}{3}\left(\frac{1}{3}\right) = \frac{2}{9}$

$a_6 = \frac{2}{9}\left(\frac{1}{3}\right) = \frac{2}{27}$

$a_7 = \frac{2}{27}\left(\frac{1}{3}\right) = \frac{2}{81}$

23. $r = \frac{1}{3}; a_1 = 216; a_n = \frac{1}{3}a_{n-1}$

$a_n = 216\left(\frac{1}{3}\right)^{n-1}$

$a_{10} = 216\left(\frac{1}{3}\right)^{10-1} = \frac{8}{729}$

---

30. A rubber ball is thrown to a height of 50 ft and then rebounds to 75% of its previous height on each bounce. [9.3]

  a. Find the height of the seventh bounce. ≈ 6.7 ft

  b. Find the total distance the ball has traveled vertically when it hits the ground the eighth time. ≈ 360 ft

  c. Find the total distance the ball has traveled vertically when it finally comes to rest. ≈ 400 ft

Use Pascal's triangle to expand. [9.4]

31. $(a - 3)^5$ **32.** $(3m + 4n)^4$ **33.** $(2 - 2z)^6$

Evaluate. [9.4]

34. $_7C_3$ and $_7C_5$ 35; 21   **35.** $_4C_2$ and $_4C_4$ 6; 1

Use the Binomial Theorem to expand each binomial. [9.4]

36. $(3c + 1)^3$ **37.** $\left(\sqrt{2} - 3t\right)^4$

Find the given term in the expansion of each binomial power. [9.4]

38. a^6b^3 term of $(3a - 2b)^9$ $-489{,}888a^6b^3$

39. eighth term of $(w + 5)^{11}$ $25{,}781{,}250w^4$

40. What is the probability (to the nearest tenth of a percent) of getting exactly 3 fours on 12 tosses of a die? [9.4]   19.7%

31. $a^5 - 15a^4 + 90a^3 - 270a^2 + 405a - 243$

32. $81m^4 + 432m^3n + 864m^2n^2 + 768mn^3 + 256n^4$

33. $64 - 384z + 960z^2 - 1280z^3 + 960z^4 - 384z^5 + 64z^6$

36. $27c^3 + 27c^2 + 9c + 1$

37. $4 - 24\sqrt{2}t + 108t^2 - 108\sqrt{2}t^3 + 81t^4$

For each proposition $P(n)$,

  a. verify $P(1)$, the anchor step,

  b. state the inductive hypothesis, $P(k)$, and

  c. state the inductive conclusion, $P(k + 1)$. [9.5]

41. 4 is a factor of $9^n - 5^n$.

42. $2^2 + 4^2 + 6^2 + \cdots + (2n)^2 = \dfrac{2n(n+1)(2n+1)}{3}$

43. Given $\sum\limits_{k=1}^{100} a_k = 18$, $\sum\limits_{k=1}^{100} b_k = 54$, and $\sum\limits_{k=1}^{100} c_k = 243$, use summation properties to evaluate each sum. [9.5]

  a. $\sum\limits_{k=1}^{100} 5a_k$ 90   **b.** $\sum\limits_{k=1}^{100} \left(2b_k + 3c_k\right)$ 837

44. Evaluate each sum. [9.5]

  a. $\sum\limits_{k=1}^{15} \left(2k^3 - 2\right)$ 28,770   **b.** $\sum\limits_{k=6}^{20} k^4$ 721,687

Use math induction to prove each statement. [9.5]

45. 2 is a factor of $n^2 + n + 4$.

46. 5 is a factor of $6^n - 1$.

47. $3^n > 3n$ for $n > 1$

48. $1^2 + 3^2 + 5^2 + \cdots + (2n-1)^2 = \dfrac{n(2n-1)(2n+1)}{3}$

49. A convex n-gon has $\dfrac{n(n-3)}{2}$ diagonals.

50. Find an explicit formula for $2 + 4 + 6 + \cdots + 2n$, and then verify the formula for all $n \in \mathbb{N}$. [9.5] $S_n = n(n + 1)$

continued in Answers and Solutions Overflow

10 Descriptive Statistics

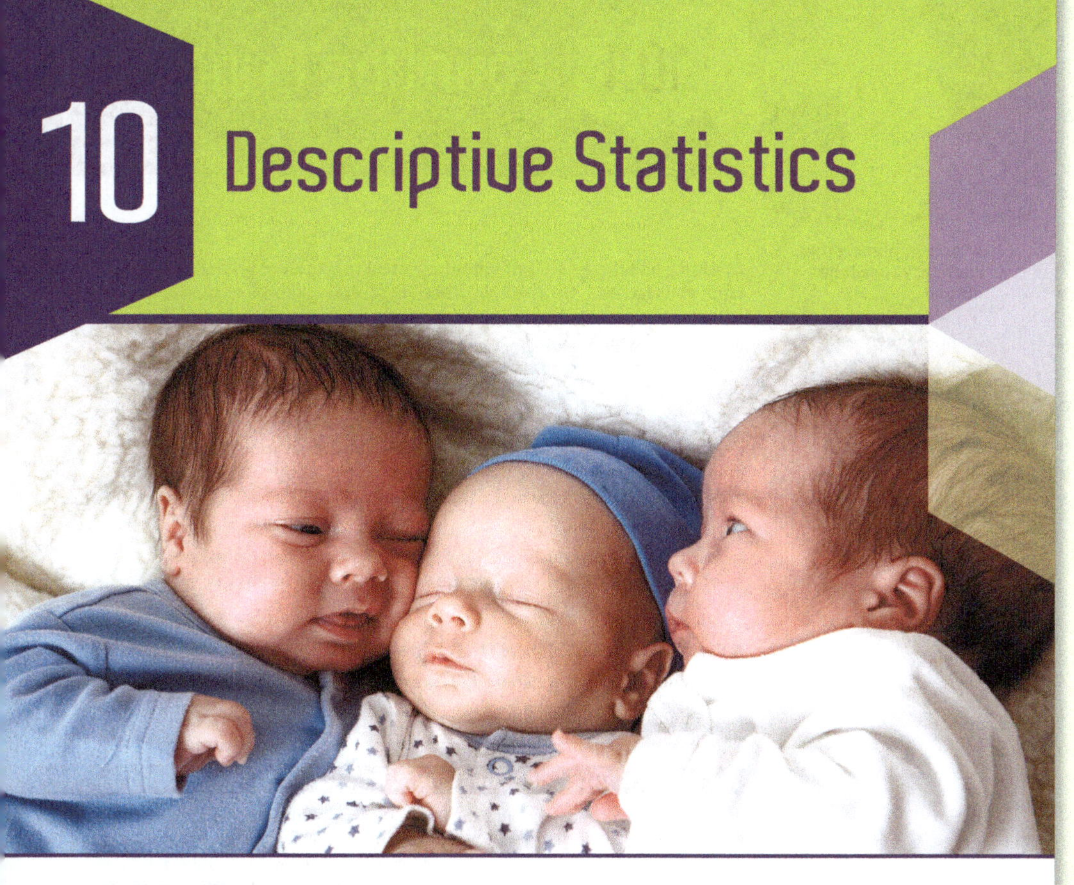

10.1 Counting Principles
10.2 Basic Probability
Data Analysis
10.3 Representing Data Graphically
10.4 Describing Data Numerically
Historical Connection
10.5 Data Distributions
10.6 Normal Distributions
Biblical Perspective of Mathematics

DATA ANALYSIS
If you test positive in a medical screening that correctly identifies people with a given condition ninety-eight percent of the time, how likely is it that you actually have the condition?

HISTORICAL CONNECTION
You may be surprised to learn that Florence Nightingale and the Crimean War played important roles in the development of descriptive statistics.

BIBLICAL PERSPECTIVE OF MATHEMATICS
In *Celestial Mechanics*, a foundational work on probability theory, Pierre-Simon de Laplace supported a Platonic view of the universe. Should a Christian endorse mathematical Platonism?

Overview

From weather to sports to finances to disease control, statistics are used to make decisions that affect us every day. This chapter provides students with the foundational terminology and organizational tools related to the study of statistics. Descriptive statistics set the foundation for the study of inferential statistics in the next chapter.

The use of calculators and spreadsheets is directly integrated into this chapter's lessons instead of a separate Technology Corner.

Flash

The probability of having twins naturally is about 1.5 in 100, or 1.5%. The probability of having triplets naturally is about 1 in 9000, or 0.01%.

Chapter Objectives

1. To use counting principles, permutations, and combinations to calculate basic probabilities

2. To describe univariate data graphically and numerically

3. To use standard normal distributions to compare standardized scores of raw numerical data

4. To recognize the sovereignty of God as well as the uncertainty that exists from a human perspective and to apply a biblical worldview to probability and statistics

Suggested Teaching Schedule

DAY	
1	10.1
2	10.2
3	Quiz 10A (10.1–10.2) DA: Medical Screenings and Probability
4	10.3
5–6	10.4
7	Quiz 10B (10.3–10.4) HC: Political Arithmetic and Probability

DAY	
8–9	10.5
10–11	10.6
12	Quiz 10C (10.5–10.6) BPM: Mathematical Platonism
13	Chapter 10 Review
14	Chapter 10 Test

10.1 Counting Principles

10.1 Counting Principles

Objectives

1. To distinguish between independent and dependent events

2. To solve counting problems using tree diagrams or the Fundamental Counting Principle

3. To apply permutations and combinations to solve counting problems

Flash

Before cream became the base for ice cream in the eighteenth century, flavored ice was a popular frozen dessert. Ancient Greeks are believed to have mixed snow with honey and fruit.

Vocabulary

combination
dependent events
distinguishable permutations
Fundamental Counting Principle
independent events
permutation

After completing this section, you will be able to

- distinguish between independent and dependent events.
- solve counting problems using tree diagrams or the Fundamental Counting Principle.
- apply permutations and combinations to solve counting problems.

Counting techniques are extremely important in a study of probability and statistics. For example, there are several options for fall and spring sports. If the playing of a particular fall sport does not affect the playing of any particular spring sport, those sports are *independent* events. If the selection of a fall sport affects the selection of a spring sport, the choices are *dependent* events.

The tree diagram represents all the possibilities if soccer, cross-country, and volleyball are the fall sports and baseball and track are the spring sports. With 3 fall sports and 2 spring sports, there are 6 ways to choose a fall and a spring sport.

Tree diagrams can become cumbersome as the number of choices increases. The *Fundamental Counting Principle* summarizes the results and is usually used to determine the number of possible ways to complete a process.

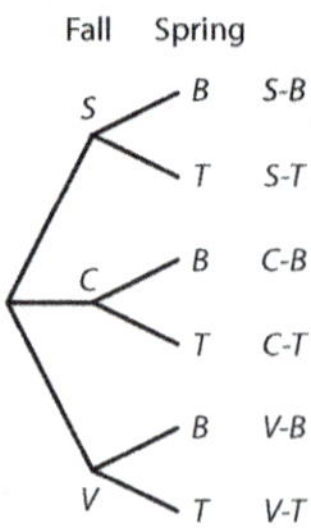

FUNDAMENTAL COUNTING PRINCIPLE

In a multi-event process where there are m_1 possible outcomes for the first event, m_2 possible outcomes for the second event, and so on, there are $m_1 \cdot m_2 \cdot \ldots \cdot m_n$ total possible outcomes.

Example 1   Applying the Fundamental Counting Principle

A school fundraiser poster claims to offer 12 meal combinations. Participants can choose a fried chicken plate or a spaghetti plate as the main course; a dessert of chocolate cake, banana pudding, or pecan pie; and either tea or coffee to drink.

a. Use the Fundamental Counting Principle to determine whether the claim of offering 12 meal combinations is correct.

b. Are the choices of main course, dessert, and drink independent or dependent events?

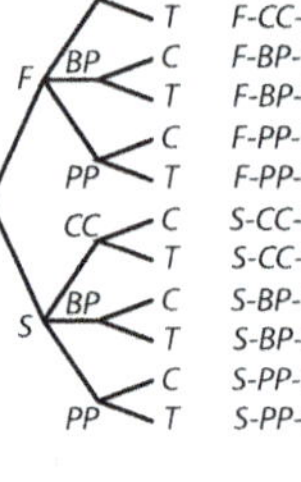

Answer

a. $2(3)(2) = 12$

The claim is correct.

There are 2 possible main courses, 3 possible desserts, and 2 possible beverages.

b. The choices are independent events.

Each choice does not affect the other choices.

Check

A tree diagram can be used to verify the number of possible outcomes.

SKILL ✓ **EXERCISE 5**

Example 2 Applying the Fundamental Counting Principle

A local deli offers triple-scoop ice-cream cones. For each case, find the number of different arrangements (distinct orders) of flavors and whether the arrangements are independent or dependent.

a. The deli has 5 flavors and the cone can contain repeated flavors.

b. The deli has 5 flavors and the cone cannot contain repeated flavors.

c. The deli has 28 flavors and the cone cannot contain repeated flavors.

Answer

a. $5(5)(5) = 125$
independent — There are 5 different choices for each scoop.

b. $5(4)(3) = 60$
dependent — There are 5 choices for the first scoop, 4 choices for the second scoop, and 3 choices for the third scoop.

c. $28(27)(26) = 19,656$
dependent — There are 28 choices for the first scoop, 27 choices for the second scoop, and 26 choices for the third scoop.

SKILL ✓ EXERCISE 7

Example 3 Applying the Fundamental Counting Principle

A computer programming team consists of 8 members.

a. How many ways can they line up for a yearbook photo?

b. How many ways can a president, vice president, and secretary be selected?

Answer

a. $8(7)(6)(5)(4)(3)(2)(1) = 40,320$ — There are 8 choices for the first spot, 7 choices for the second, 6 choices for the third, and so on until there is only 1 choice left for the last spot.

b. $8(7)(6) = 336$ — Any of the 8 can be selected as president, then 7 choices remain for vice president, and then 6 choices are left for secretary.

SKILL ✓ EXERCISE 15

Applying the Fundamental Counting Principle to determine the number of possible arrangements of n objects produces $n(n - 1)(n - 2) \ldots 1 = n!$ arrangements, as seen in Example 3a. When arranging only r of the n objects, the ways of arranging the remaining $(n - r)$ objects are not included. The possible arrangements of 3 of the 8 computer programming team members in Example 3b can be calculated using

$$\frac{8!}{(8 - 3)!} = \frac{8(7)(6)5!}{5!} = 8(7)(6) = 336.$$

> ### DEFINITION
>
> A **permutation** is an arrangement of objects in which order is important. The number of permutations of n objects taken r at a time is $_nP_r = \dfrac{n!}{(n - r)!}$.

The notation $_nP_r$ can be read as "arrange r out of n objects" or simply "n arrange r." Note that this definition implies $_nP_n = \dfrac{n!}{(n - n)!} = \dfrac{n!}{0!} = \dfrac{n!}{1} = n!$, as stated previously.

TIP

The factorial function, found in the MATH, PROB submenu, can be used to calculate the answers to Example 3.

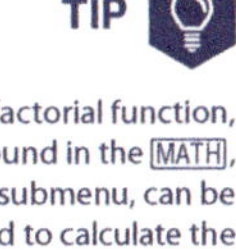

Reading and Writing Mathematics

Show algebraically that $_nC_n = 1$ and $_nC_0 = 1$. Then illustrate why this is true using a real-life scenario.

$$_nC_n = \frac{n!}{r!(n - r)!} = \frac{n!}{n!(n - n)!} = \frac{1}{0!} = \frac{1}{1} = 1$$

$$_nC_n = \frac{n!}{r!(n - r)!} = \frac{n!}{0!(n - 0)!} = \frac{1}{0!} = \frac{1}{1} = 1$$

sample answer: There is only 1 way to choose a team of 10 players from 10 students who try out. There is also only 1 way to choose none of those who try out.

Additional Exercises

1. How many different outfits can Thomas make if he can choose from 4 shirts, 3 pairs of pants, and 2 pairs of shoes? 24

2. Three marbles will be chosen from 8 different bags of marbles each containing a different color. For each case find the number of possible arrangements and whether the arrangements are independent or dependent.

   a. A marble color can be repeated.
   512; independent

   b. A marble color cannot be repeated. 336; dependent

3. A committee of student body officers consists of 10 different students.

   a. How many different ways can they sit in a row? 3,628,800

   b. How many ways can a president, vice president, and secretary be selected? 720

players to start a baseball game $(_{20}C_9)$ and creating the batting order $(_{20}P_9)$. Define a combination and apply the formula in Example 6.

Consider asking the students to explain why $_nC_n$ will logically always equal 1, as seen in Example 6b. (*Only one set of n objects can be created from n objects.*)

Example 7 is a scenario with two independent combinations, the possible selections of the women and of the men. The total number of possible committees is then found by applying the Fundamental Counting Principle.

Refer the students to the table summarizing the counting formulas from this section. The Internet keyword search *permutation combination calculator* can provide apps for calculating these functions.

One-on-One A student may confuse permutations and combinations. Reinforce that permutations involve different possible arrangements or orderings, while combinations involve different possible sets or groups.

Motivational Idea It has been said that the lottery is for the greedy and for the mathematically challenged. Understanding combinations supports the latter. The odds of picking 5 correct numbers from the numbers 1 to 59 is $1 : {}_{59}C_5 = 1 :$ 5,006,386. To have a 50% chance of winning the lottery, a person would need to buy a ticket every hour (24/7) for 286 yr, spending over $5 million at $2 a ticket.

TIPS

Ex. 38–39 These exercises prepare the students for the proof of the formula for distinguishable permutations.

4. How many ways can 4 names be drawn from a hat containing 6 names if names are not replaced? 360

5. Find the number of permutations of the letters in each word.
   a. *SUGGEST* 1260
   b. *COMMITTEE* 45,360

6. Calculate each combination.
   a. $_{10}C_4$ 210   b. $_7C_3$ 35

7. A decorating committee of 4 women and 3 men will be chosen from a group of 8 women and 5 men. How many different committees can be selected? 700

Assignments

- **Minimum:** 1–5, 7, 9–10, 12–13, 16–17, 21, 23–24, 26, 31, 41, 45–46, 48–49
- **Standard:** 1–10, 13–15; 16–26 even; 27–28, 31–36, 41–42, 45–47, 49–50
- **Extended:** 1–4; 6–16 even; 17–23 odd; 24–27, 29–30, 32–33, 35–41; 42–50 even

TIP

The nPr function, found in the MATH, PROB submenu, can be used to calculate permutations.

```
7 nPr 4
                      840
```

Some mathematicians were not in favor of using an exclamation point for factorial. British mathematician Augustus De Morgan called it a "barbarism" and suggested those who used it were expressing "surprise and admiration" that such results were to be found in mathematics.

Example 4   Applying the Permutation Formula

How many 4-letter arrangements can be made from $\{a, b, c, d, e, f, g\}$ if letters cannot be repeated?

Answer

$$_7P_4 = \frac{7!}{(7-4)!} = \frac{7(6)(5)(4)3!}{3!}$$

Apply the formula $_nP_r = \frac{n!}{(n-r)!}$.

$$= 7(6)(5)(4) = 840$$

Note that $_7P_4$ is often read as "7 arrange 4."

SKILL ✓ EXERCISE 17

How many distinguishable arrangements can be made using the letters in the word *SCHOOL*? If the two *O*s were distinguishable (using O_1 and O_2), there would be $6! = 720$ arrangements; but this number includes otherwise undistinguishable arrangements such as $SCHO_1O_2L$ and $SCHO_2O_1L$. Dividing 720 by 2!, the number of ways O_1 and O_2 can be arranged, produces the correct number of distinguishable permutations—360.

> **DEFINITION**
>
> Given n objects with n_1 identical objects of one type, n_2 identical objects of a second type, and so on where $n = n_1 + n_2 + \cdots + n_k$, the number of **distinguishable permutations** is $\frac{n!}{n_1!\,n_2! \ldots n_k!}$.

Example 5   Finding Permutations with Identical Objects

Find the number of permutations of the letters in each word.
a. *BUTTERFLY*
b. *CATERPILLAR*

Answer

a. $\frac{9!}{2!} = 181,440$

There are nine letters in *BUTTERFLY*. All are distinguishable except the 2 *T*s.

b. $\frac{11!}{2!2!2!} = 4,989,600$

There are 11 letters in *CATERPILLAR*. Three letters (*A*, *R*, and *L*) are each used twice.

SKILL ✓ EXERCISE 23

In permutations, order is important. Selections of objects where order is not important are called *combinations*. There are $3! = 6$ permutations of the three letters *abc*: *abc, acb, bac, bca, cab*, and *cba*. The number of ways to select r items from a group of n objects can be found by dividing its number of permutations by $r!$.

$$_nC_r = \frac{_nP_r}{r!} = \frac{\frac{n!}{(n-r)!}}{r!} = \frac{n!}{r!(n-r)!}$$

> **DEFINITION**
>
> A **combination** is a collection of objects in which order is not important. The number of combinations of n objects taken r at a time is written as $_nC_r = \frac{n!}{r!(n-r)!}$.

The notation $_nC_r$ can be read as "choose r out of n objects" or simply "n choose r."

Example 6 Calculating Combinations

Calculate each combination.

a. $_8C_3$

b. $_{12}C_{12}$

Answer

a. $_8C_3 = \dfrac{8!}{3!(8-3)!}$
$= \dfrac{8(7)(6)5!}{3(2)5!}$
$= 8(7) = 56$

b. $_{12}C_{12} = \dfrac{12!}{12!(12-12)!}$
$= \dfrac{12!}{12!(1)}$
$= 1$

1. Substitute into $_nC_r = \dfrac{n!}{r!(n-r)!}$ and simplify.

Check

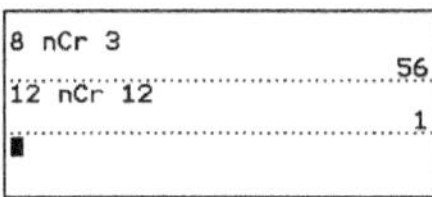

2. Use the nCr function from the MATH, PROB submenu to evaluate each expression.

——————— SKILL ✓ **EXERCISE 13**

Example 7 Applying the Combination Formula

A committee of 2 women and 4 men will be selected from 10 students, of which 4 are women and 6 are men. How many different committees can be selected?

Answer

women: $_4C_2 = \dfrac{4!}{2!(4-2)!} = \dfrac{4!}{2!2!} = 6$

men: $_6C_4 = \dfrac{6!}{4!(6-4)!} = \dfrac{6!}{4!2!} = 15$

$6(15) = 90$ different committees

1. Calculate the number of combinations for each of the two independent events: selecting 2 of 4 women and selecting 4 of 6 men.

2. Apply the Fundamental Counting Principle to find the number of possible committees.

——————— SKILL ✓ **EXERCISE 27**

Counting Formulas	
Fundamental Counting Principle	In a multi-event process where there are m_1 possible outcomes for the first event, m_2 possible outcomes for the second event, and so on, there are $m_1 \cdot m_2 \cdot \ldots \cdot m_n$ total possible outcomes.
permutations	$_nP_r = \dfrac{n!}{(n-r)!}$ ways to arrange r out of n objects
distinguishable permutations	Given n objects with n_1 identical objects of one type, n_2 identical objects of a second type, and so on where $n = n_1 + n_2 + \cdots + n_k$, the number of distinguishable permutations is $\dfrac{n!}{n_1! \, n_2! \, \ldots \, n_k!}$.
combinations	$_nC_r = \dfrac{n!}{r!(n-r)!}$ ways to choose r out of n objects

Solutions

❯ A. Exercises

3.

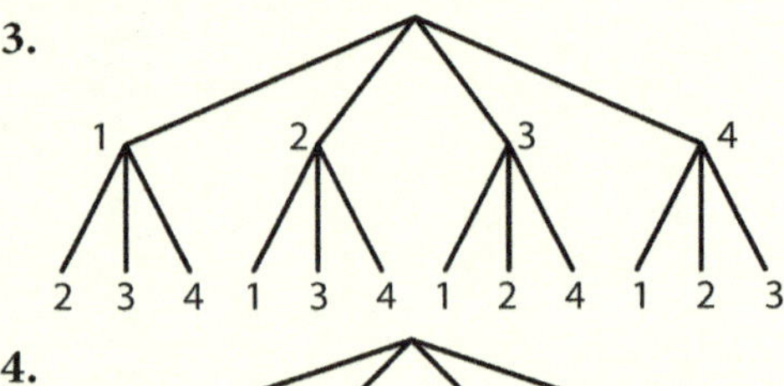

4.

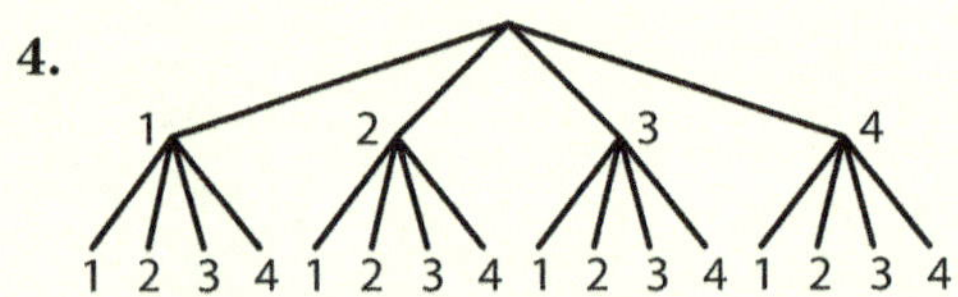

5. $4(5)(3)(6) = 360$

6. $2(10^{13}) = 20{,}000{,}000{,}000{,}000$

7. $9(8)(7)(6)(5)(4)(3) = 181{,}440$

8. $26^3(10^4) = 175{,}760{,}000$

9. $\dfrac{9!}{(9-3)!} = \dfrac{9 \cdot 8 \cdot 7 \cdot 6!}{6!} = 504$

10. $\dfrac{8!}{(8-4)!} = \dfrac{8 \cdot 7 \cdot 6 \cdot 5 \cdot 4!}{4!} = 1680$

11. $\dfrac{14!}{(14-3)!} = \dfrac{14 \cdot 13 \cdot 12 \cdot 11!}{11!} = 2184$

12. $\dfrac{12!}{9!(12-9)!} = \dfrac{12 \cdot 11 \cdot 10 \cdot 9!}{9! \cdot 3 \cdot 2 \cdot 1} = 220$

13. $\dfrac{7!}{5!(7-5)!} = \dfrac{7 \cdot 6 \cdot 5!}{5! \cdot 2 \cdot 1} = 21$

14. $\dfrac{100!}{98!(100-98)!} = \dfrac{100 \cdot 99 \cdot 98!}{98! \cdot 2 \cdot 1} = 4950$

15. $_6P_6 = 6! = 6(5)(4)(3)(2)(1) = 720$

16. $(3)(6)(2) = 36$

17. $_9P_6 = \dfrac{9!}{(9-6)!} = \dfrac{9 \cdot 8 \cdot 7 \cdot 6 \cdot 5 \cdot 4 \cdot 3!}{3!}$
$= 60{,}480$

18. $_{20}P_3 = \dfrac{20!}{(20-3)!} = \dfrac{20 \cdot 19 \cdot 18 \cdot 17!}{17!}$
$= 6840$

19. $_{24}C_5 = \dfrac{24!}{5!(24-5)!}$
$= \dfrac{24 \cdot 23 \cdot 22 \cdot 21 \cdot 20 \cdot 19!}{5 \cdot 4 \cdot 3 \cdot 2 \cdot 1 \cdot 19!} = 42{,}504$

20. $_{100}C_2 = \dfrac{100!}{2!(100-2)!} = \dfrac{100 \cdot 99 \cdot 98!}{2 \cdot 1 \cdot 98!}$
$= 4950$

❯ B. Exercises

21. $_7P_7 = 7! = 5040$

22. $\dfrac{8!}{2!} = 20{,}160$

23. $\dfrac{7!}{2!2!} = 1260$

24. $\dfrac{15!}{3!5!4!3!} = 12{,}612{,}600$

25a. $6(5)(4)(3)(2)(1) = {_6P_6} = 6! = 720$

25b. $6(1)(4)(1)(2)(1) = 48$

26a. $_{15}C_9 = 5005$

26b. $_{15}P_9 = 1{,}816{,}214{,}400$

27a. $_{18}C_2 \cdot {_{22}C_3} = 235{,}620$

❯ A. Exercises

1. Events whose possible outcomes are affected by other events are said to be __dependent__, while events whose possible outcomes have no effect on each other are called __independent__.

2. Selections of objects in which order is not important are called __combinations__, while arrangements of objects in which order is important are __permutations__.

A bag contains four cards numbered 1 through 4. Isaiah draws a card from the bag, records the first digit, and then draws a second card and records the second digit. Use a tree diagram to determine how many two-digit numbers are possible in each case below. Then state whether the two draws are independent or dependent.

3. Isaiah does not replace the first card before drawing the second card. 12; dependent

4. Isaiah replaces the first card before drawing the second card. 16; independent

Apply the Fundamental Counting Principle to answer the following questions.

5. Downtown Catering Service offers box lunches that include a sandwich, a bag of chips, a fruit, and a cookie. How many different box lunches are possible if customers can choose from 4 kinds of sandwiches, 5 types of chips, 3 different fruits, and 6 kinds of cookies? 360

6. How many different credit card numbers are possible if a company's card numbers are 15 digits long and start with either 34 or 37? 20 trillion

7. How many different ways can 7 out of 9 different chocolates be distributed to 7 students? 181,440

8. How many different license plates can be made containing 3 letters followed by 4 numbers? 175,760,000

Evaluate each expression without using technology, and then check your answer with technology.

9. $_9P_3$ 504

10. $_8P_4$ 1680

11. $_{14}P_3$ 2184

12. $_{12}C_9$ 220

13. $_7C_5$ 21

14. $_{100}C_{98}$ 4950

Use the appropriate counting technique.

15. How many ways are there to arrange 6 books on a bookshelf? 720

16. Jessica has chosen 3 possible paint colors, 6 possible rugs, and 2 possible sets of curtains for her dining room. How many different ways can she redecorate the room? 36

---

17. Josh's family is planning a Holy Land tour but only has time to see 6 out of 9 historic sites outside of the Jerusalem area. How many different itineraries can be planned for those sites? 60,480

18. If there are 20 rising seniors vying to be a class officer, how many different ways can a president, vice president, and secretary/treasurer be selected? 6840

19. How many different 5-person committees can be chosen from a group of 24 people? 42,504

20. Out of the first 100 people who enter a new grocery store, 2 people will be selected to win a $500 shopping spree. How many different pairs of shoppers could win the prizes? 4950

❯ B. Exercises

Determine how many ways the letters in each word can be arranged.

21. *LEOPARD* 5040

22. *CALENDAR* 20,160

23. *BALLOON* 1260

24. A bag of marbles contains 3 blue, 5 red, 4 yellow, and 3 green marbles. In how many different ways can all the marbles be lined up in a row? 12,612,600

25. Three couples have tickets in the same row for a football game. Determine the different seating arrangements in each case.
   a. They don't mind who sits next to whom. 720
   b. Each couple sits together. 48

26. Coach Grant has 15 boys on his baseball team.
   a. How many groups of 9 starting players can he have? 5005
   b. How many different 9-person batting orders can he make? 1,816,214,400

27. There are 18 boys and 22 girls in the church youth group. How many different banquet committees can be formed with the given conditions?
   a. 2 boys and 3 girls 235,620
   b. 2 boys and 2 girls 35,343
   c. 2 boys and 2 or 3 girls 270,963

28. The 1947 North American Number Plan (NANP) standardized area codes and phone numbers. Originally the 3-digit area code was of the form NPX where N could be any number 2 through 9, P could be only 0 or 1, and X could be any digit 0 through 9.
   a. How many different area codes were possible with the original format? 160

27b. $_{18}C_2 \cdot {_{22}C_2} = 35{,}343$

27c. $235{,}620 + 35{,}343 = 270{,}963$

28a. $8(2)(10) = 160$

b. In 1995 the area code numbering plan was updated to the form NXX. How many different codes of this form are possible? 800

c. When the second and third digits of an area code are the same, that code is called an easily recognizable code (ERC). ERCs designate special services, e.g. 888 for toll-free service. How many ERCs are available? 80

d. Codes of the form N11 are used as service codes instead of area codes. How many of these codes are there? Give a common example of one of these codes. 8; sample answer: 911

e. Area codes of the form N9X are reserved as expansion codes and will be used when the current 10-digit number formatting needs to be expanded. How many expansion codes are there? 80

29. How many different 4- to 6-character passcodes are possible using letters, the digits 0–9, or the symbols # or $?

a. if no character is repeated 2,049,694,920

b. if characters can be repeated 3,092,256,688

30. If 8 sprinters compete in a semifinal heat, how many possible groups of 2, 3, or 4 runners could advance to the final from this race? 154

31. Simplify each permutation.

a. $_nP_n$ $n!$ **b.** $_nP_{n-r}$ $\frac{n!}{r!}$

c. $_nP_1$ n **d.** $_nP_0$ 1

32. Simplify each combination.

a. $_nC_n$ 1 **b.** $_nC_{n-r}$ $\frac{n!}{(n-r)!r!}$

c. $_nC_1$ n **d.** $_nC_0$ 1

> **C. Exercises**

33. The high school choir consists of 8 seniors, 6 juniors, 5 sophomores, and 3 freshmen. In how many ways can 8 choir members be selected so that there are 2 students from each grade represented? 12,600

34. Jesse needs to read 3 novels and 2 biographies from a summer reading list that includes 10 novels and 7 biographies.

a. How many different combinations of books can Jesse read to fulfill the requirement? 2520

b. In how many different orders can Jesse read 5 books from the list to fulfill the requirement? 302,400

35. The information technology department is exploring minimum security requirements for 6-character passwords used within their company. Determine the number of different possible passwords for each set of requirements.

a. One character must be a number 0 through 9, and the rest are lowercase letters. 712,882,560

b. One character must be a number, one an uppercase letter, and the rest are lowercase letters. 3,564,412,800

c. One character must be a number, one an uppercase letter, another is 1 of 10 special characters, and the rest are lowercase letters. 5,483,712,000

d. Why is a password containing lowercase and uppercase letters, numbers, and symbols safer?

36. Since the two illustrated circular arrangements are considered to be identical, the number of circular arrangements differs from the number of linear arrangements of the same items. Consider linear and circular arrangements of the letters A, B, C, D, and E to derive an expression for the number of circular arrangements of n items.

a. How many different linear arrangements are possible for the 5 letters? $5! = 120$

b. List all the linear arrangements that are identical to $ABCDE$ when the letters are arranged in a circle.

c. How many linear arrangements are equivalent to each circular arrangement of the 5 letters? 5

d. How many different circular arrangements can be made with the 5 letters? 24

e. Describe how the number of circular arrangements of n items can be found if there are l linear arrangements. $\frac{n!}{n}$ or $(n-1)!$

37. Prove: Show that $n(_{n-1}C_{r-1}) = (n-r+1)\,_nC_{r-1}$.

38. Complete the following steps to determine the number of distinguishable permutations of the letters in the word *MALL*.

a. How many ways are there to select 2 of the 4 positions in which to place the Ls? $_4C_2 = 6$

b. After placing the 2 Ls, how many ways are there to select a position for the M? $_2C_1 = 2$

c. Finally, how many ways are there to select the position for the A? $_1C_1 = 1$

d. Use the Fundamental Counting Principle to calculate the total number of ways the 4 letters can be arranged. 12

> **C. Exercises**

33. $_8C_2 \cdot _6C_2 \cdot _5C_2 \cdot _3C_2$
$= \frac{8!}{6!2!} \cdot \frac{6!}{4!2!} \cdot \frac{5!}{3!2!} \cdot \frac{3!}{1!2!}$
$= 28 \cdot 15 \cdot 10 \cdot 3 = 12{,}600$

34a. $_{10}C_3 \cdot _7C_2 = \frac{10!}{7!3!} \cdot \frac{7!}{5!2!}$
$= 120 \cdot 21 = 2520$

34b. $2520 \cdot 5! = 2520 \cdot 120 = 302{,}400$

35a. $6(10) \cdot 26^5 = 712{,}882{,}560$

35b. $6(10) \cdot 5(26) \cdot 26^4 = 3{,}564{,}412{,}800$

35c. $6(10) \cdot 5(26) \cdot 4(10) \cdot 26^3$
$= 5{,}483{,}712{,}000$

36b. $BCDEA$, $CDEAB$, $DEABC$, and $EABCD$

36d. $\frac{120}{5} = 24$

36e. $\frac{n!}{n} = \frac{n(n-1)!}{n} = (n-1)!$

37. working on both sides:
$$n(_{n-1}C_{r-1}) \overset{?}{=} (n-r+1)\,_nC_{r-1}$$
$$n\left[\frac{(n-1)!}{(r-1)!(n-1-(r-1))!}\right]$$
$$\overset{?}{=} (n-r+1)\left[\frac{n!}{(r-1)![n-(r-1)!]}\right]$$
$$\frac{n(n-1)!}{(r-1)!(n-r)!}$$
$$\overset{?}{=} (n-r+1)\left[\frac{n!}{(r-1)!(n-r+1)!}\right]$$
$$\frac{n!}{(r-1)!(n-r)!}$$
$$= \left(\frac{n-r+1}{n-r+1}\right)\left[\frac{n!}{(r-1)!(n-r)!}\right]$$

38d. $_4C_2 \cdot _2C_1 \cdot _1C_1 = 6 \cdot 2 \cdot 1 = 12$

28b. $8(10)(10) = 800$

28c. $8(10)(1) = 80$

28d. $8(1)(1) = 8$

28e. $(8)(1)(10) = 80$

29a. $_{38}P_4 + _{38}P_5 + _{38}P_6 = 2{,}049{,}694{,}920$

29b. $38^4 + 38^5 + 38^6 = 3{,}092{,}256{,}688$

30. $_8C_2 + _8C_3 + _8C_4 = 28 + 56 + 70$
$= 154$

31a. $\frac{n!}{(n-n)!} = \frac{n!}{0!} = \frac{n!}{1}$

31b. $\frac{n!}{(n-(n-r))!} = \frac{n!}{r!}$

31c. $\frac{n!}{(n-1)!} = \frac{n \cdot (n-1)!}{(n-1)!} = n$

31d. $\frac{n!}{(n-0)!} = \frac{n!}{n!} = 1$

32a. $\frac{n!}{n!(n-n)!} = \frac{n!}{n!0!} = \frac{1}{1} = 1$

32b. $\frac{n!}{(n-r)!(n-(n-r))!} = \frac{n!}{(n-r)!r!}$

32c. $\frac{n!}{1!(n-1)!} = \frac{n(n-1)!}{1 \cdot (n-1)!} = n$

32d. $\frac{n!}{0!(n-n)!} = \frac{n!}{0!n!} = \frac{1}{1} = 1$

39d. $_7C_3 \cdot {_4C_2} \cdot {_2C_1} \cdot {_1C_1} = 35 \cdot 6 \cdot 2 \cdot 1$
$= 420$

40d. $_nC_a \cdot {_{(n-a)}C_b} \cdot {_{(n-a-b)}C_c} \cdot \cdots \cdot {_zC_z}$

$$= \frac{n!}{a!(n-a)!} \cdot \frac{(n-a)!}{b!(n-a-b)!}$$
$$\cdot \frac{(n-a-b)!}{c!(n-a-b-c)!} \cdot \cdots$$
$$\cdot \frac{(n-a-b-c-\cdots-y)!}{z!(n-a-b-c-\cdots-y-z)!}$$

$$= \frac{n!}{a!(n-a)!} \cdot \frac{(n-a)!}{b!(n-a-b)!}$$
$$\cdot \frac{(n-a-b)!}{c!(n-a-b-c)!} \cdot \cdots$$
$$\cdot \frac{(n-a-b-c-\cdots-y)!}{z!(n-n)!}$$

$$= \frac{n!}{a!b!c!\ldots z!0!} = \frac{n!}{a!b!c!\ldots z!}$$

❯ Cumulative Review

41. $f(x) = \dfrac{x+1}{(x-3)(x-4)}$
$n < m$; HA: $y = 0$
zeros of $a(x)$: -1
zeros of $b(x)$: $3, 4$
VA: $x = 3$, $x = 4$

42. $h(x) = \dfrac{x(x^2+2)}{(x-1)(x+1)}$
$n = m + 1$ and $\dfrac{x^3+2x}{x^2-1} = x + \dfrac{3x}{x^2-1}$
SA: $y = x$
zeros of $a(x)$: 0
zeros of $b(x)$: ± 1
VA: $x = \pm 1$

44. $\left[\dfrac{9x-24}{(x+2)(x-5)} = \dfrac{A}{x+2} + \dfrac{B}{x-5} \right]$
$\qquad\qquad\qquad \cdot (x+2)(x-5)$
$9x - 24 = Ax - 5A + Bx + 2B$
$9x - 24 = (A+B)x + (-5A+2B)$
$\qquad\qquad A + B = 9$
$\qquad\quad -5A + 2B = -24$
$-5(9 - B) + 2B = -24$
$\qquad -45 + 7B = -24;\ B = 3$
$A + 3 = 9;\ A = 6$
$\therefore \dfrac{6}{x+2} + \dfrac{3}{x-5}$

45. $_3C_0 c^3 d^0 + {_3C_1} c^2 d^1 + {_3C_2} c^1 d^2 + {_3C_3} c^0 d^3$
$= c^3 + 3c^2 d + 3cd^2 + d^3$

46. $_3C_0 (3x)^3 (-1)^0 + {_3C_1}(3x)^2(-1)^1$
$\quad + {_3C_2}(3x)^1(-1)^2 + {_3C_3}(3x)^0(-1)^3$
$= 27x^3 - 27x^2 + 9x - 1$

47. $r = \sqrt{(-3)^2 + 4^2} = \sqrt{25} = 5$
Q II: $\theta = \pi + \text{Tan}^{-1}\dfrac{4}{-3} \approx 2.21$
$z \approx 5 \text{ cis } 2.21$

---

39. Complete the following steps to determine the number of distinguishable permutations of the letters in the word *SUCCESS*.

  a. How many ways are there to select the positions in which to place the *S*s? $_7C_3 = 35$

  b. After placing the *S*s, how many ways are there to select the positions for the *C*s? $_4C_2 = 6$

  c. How many ways are there to then place the *U*? and then the *E*? $_2C_1 = 2;\ {_1C_1} = 1$

  d. Use the Fundamental Counting Principle to calculate the total number of ways to place all the letters. 420

CUMULATIVE REVIEW

Classify each rational function as proper or improper. Then identify all asymptotes and any point discontinuities. [2.5]

41. $f(x) = \dfrac{x+1}{x^2-7x+12}$

42. $h(x) = \dfrac{x^3+2x}{x^2-1}$

43. Write the general form for the partial fraction decomposition of $\dfrac{x^2}{x^3+x^2}$. Do not solve. [7.7]

44. Find the partial fraction decomposition of $\dfrac{9x-24}{x^2-3x-10}$. [7.7]

Use the Binomial Theorem to expand each binomial. [9.4]

$c^3 + 3c^2 d + 3cd^2 + d^3 \qquad 27x^3 - 27x^2 + 9x - 1$

45. $(c+d)^3$ **46.** $(3x-1)^3$

47. Convert $z = -3 + 4i$ to polar form. [6.6] C

  A. 5 cis 0.93   **D.** 5 cis -2.21

  B. 5 cis -0.93 **E.** -5 cis -2.21

  C. 5 cis 2.21

41. proper; HA: $y = 0$; VA: $x = 4$, $x = 3$

42. improper; SA: $y = x$; VA: $x = \pm 1$

43. $\dfrac{A}{x} + \dfrac{B}{x^2} + \dfrac{C}{x+1}$

44. $\dfrac{6}{x+2} + \dfrac{3}{x-5}$

---

40. **Prove:** Use the reasoning illustrated in exercises 38–39 to show that $\dfrac{n!}{a!b!c!\ldots z!}$ is the number of distinguishable permutations of n objects with a identical objects of one type, b identical objects of a second type, and so on where $n = a + b + c + \cdots + z$.

  a. Write an expression for the number of ways to select positions for the first a objects. $_nC_a$

  b. Write an expression for the number of ways to then place the next b objects. $_{n-a}C_b$

  c. Write an expression for the number of ways to place the next $c, \ldots, z$ objects. $_{n-(a+b)}C_c \cdot \cdots \cdot {_zC_z}$

  d. Use the Fundamental Counting Principle to calculate the total number of ways to place all the letters.

48. Which of the following is true for an ellipse with vertices $(0, -2)$ and $(10, -2)$ and eccentricity of 0.75? List all correct answers. [8.2] B, D

  A. $a = 10$ **D.** $b^2 = 10.9375$

  B. $a = 5$ **E.** none of these

  C. $c = 7.5$

49. What is the sixth term of $a_n = 5a_{n-1} + 3$, $a_1 = 2$ where $n \geq 2$? [9.1] D

  A. 33   **C.** 3128   **E.** none of

  B. 1718   **D.** 8593   these

50. What is the sequence of steps in a math induction proof? [9.5] C

  A. I, II, III   **I.** Assume $P(k)$.

  B. I, III, II   **II.** Verify $P(1)$.

  C. II, I, III   **III.** Prove $P(k+1)$.

  D. II, III, I

  E. III, II, I

---

48. $C(5, -2)$; horizontal MA
$a = 5$; $e = \dfrac{c}{a}$
$c = ea = \left(\dfrac{3}{4}\right)(5) = 3.75$
$b^2 = a^2 - c^2 = 5^2 - 3.75^2 = 10.9375$
$\dfrac{(x-5)^2}{25} + \dfrac{(y+2)^2}{10.9375} = 1$

49. $a_2 = 5(2) + 3 = 13$
$a_3 = 5(13) + 3 = 68$
$a_4 = 5(68) + 3 = 343$
$a_5 = 5(343) + 3 = 1718$
$a_6 = 5(1718) + 3 = 8593$

10.2 Basic Probability

Comparing the omniscience and sovereignty of God to our finite and fallen human nature highlights His greatness and our inadequacies. God knows and is in control of the future, but even our best predictions of future events are frequently flawed. Thankfully, God's Word and the Holy Spirit provide guidance as we plan for our future (Ps. 37:23). The study of probability and statistics can also help us make wise decisions related to education, business, healthcare, and other areas. Due to the limited knowledge of humans, "chance" seems present all around, but in reality God is sovereign over all things. James 4:13–15 reminds us that all we do depends on God's will.

In the study of probability, an *experiment* consists of one or more actions whose outcome may be uncertain. An experiment can be as simple as a single trial where a coin is flipped, a die is rolled, or a name is drawn from a hat. The set S containing all the possible outcomes is the experiment's *sample space*. An *event* is any subset of a sample space and is denoted by a capital letter such as E. The outcomes are said to be *equally likely* when every outcome in the experiment has the same chance of occurring.

> **After completing this section, you will be able to**
> - define terms related to basic probabilities.
> - calculate basic probabilities.
> - calculate probabilities of mutually exclusive and inclusive events.
> - calculate probabilities of independent and conditional events.

Example 1   Identifying a Sample Space and an Event

A spinner has 7 equally spaced numbers. List the sample space and the event of getting an odd number on a spin.

Answer

$S = \{1, 2, 3, 4, 5, 6, 7\}$ 1. The sample space is the set of all possible outcomes.

$E = \{1, 3, 5, 7\}$ 2. The event is the set of odd numbers.

SKILL ✔ **EXERCISE 1**

Theoretical probabilities describe long-term behaviors. If an experiment is literally conducted, the result may be quite different from the theoretical probability. These results determine the *experimental probability*. As the number of trials increases, the experimental probability approaches the theoretical probability.

DEFINITIONS

The **theoretical** (or **classical**) **probability** of an event, $P(E)$, is $n(E)$, the ratio of the number of favorable outcomes, to $n(S)$, the number of equally likely possible outcomes:

$$P(E) = \frac{n(E)}{n(S)}.$$

The **experimental** (or **empirical**) **probability** of an event, $P(E)$, is the ratio of the number of times the event occurs to the number of trials in an experiment:

$$P(E) = \frac{\text{number of times the event occurs}}{\text{number of trials}}.$$

10.2 Basic Probability

Objectives

1. To define terms related to basic probabilities
2. To calculate theoretical and experimental probabilities
3. To calculate probabilities of mutually exclusive and mutually inclusive events
4. To calculate probabilities of independent and dependent events

Flash

Six in ten Americans believe the Bible has transformed their life. Early morning is the most popular time for reading the Bible.

Vocabulary

certain event
complement
compound event
conditional probability
equally likely
event
experiment
experimental probability
impossible event
mutually exclusive
mutually inclusive
sample space
simple event
theoretical probability

PRESENTATION

Lesson Opener

Complete each formula.

1. $_nP_r = \dfrac{n!}{(n-r)!}$ 2. $_nC_r = \dfrac{n!}{r!(n-r)!}$

3. How many ways can you arrange the letters in the word *FAITH*? 120

4. How many ways can a team of 4 be selected from 10 students? 210

Consider taking some time to discuss the concepts of chance and probability from a biblical perspective. There is some debate regarding whether true randomness exists. A computer cannot generate a truly random number; the random number generators apply some type of algorithm to one or more initial values. Results are sometimes referred to as pseudo-random. Definitions of *chance* often allude to an event that happens with no foreknowledge or assignable cause. Emphasize that while some events may appear coincidental, random, accidental, unpredictable, or by chance, these events are actually part of God's design and plan.

Define *experiment*, *sample space*, and *event*—key terms in probability. Use Example 1 to identify these terms.

Present the definitions of *theoretical probability* and *experimental probability* and illustrate them with Example 2. Point out that as the trials increase, the experimental probabilities approach theoretical probabilities.

Interactive Activity Challenge students to complete exercise 44, either individually or as a group. Performing the calculations will validate that the experimental probabilities approach theoretical probabilities as the number of rolls increases. The Internet keyword search *interactive coin toss* can also be used to find programs that generate pseudo-random outcomes.

Summarize the terms and rules for basic probabilities, emphasizing that the probability of the *complement* is simply

Additional Exercises

1. List the sample space and the event of rolling a prime number on a regular 6-sided die.
   $S = \{1, 2, 3, 4, 5, 6\}; E = \{2, 3, 5\}$

2. Find the experimental and theoretical probabilities of rolling a four 10 times in an experiment of 50 tosses of a regular 6-sided die. $\frac{1}{5}, \frac{1}{6}$

3. What is the probability of not pulling a double-digit card from a stack of 15 cards numbered 0 through 14?
   $1 - \frac{5}{15} = \frac{2}{3}$

4. An experiment consists of flipping a coin and rolling an 8-sided die. Find the probability of each event.
   a. $P(T \text{ and } 7)$ $\frac{1}{2} \cdot \frac{1}{8} = \frac{1}{16}$

   b. $P(H \text{ and prime})$ $\frac{1}{2} \cdot \frac{1}{2} = \frac{1}{4}$

5. A bag contains 4 squares and 6 triangles. Find the probability of drawing 2 squares under the given conditions.
   a. The first figure is replaced before the second is drawn. $\frac{4}{25}$
   b. The first figure is not replaced before the second is drawn. $\frac{2}{15}$

6. Use the following probabilities to determine whether the events of being a junior and being in calculus class are independent or dependent.
   $P(J) = 0.2$
   $P(C) = 0.15$
   $P(J \text{ and } C) = 0.05$ dependent

7. Using the information from the previous exercise, what is the probability that a student chosen at random is a junior or is in calculus class? 0.3

8. If the remaining calculus students mentioned in Additional Exercise 6 are seniors and the probability of being a senior is 0.35, calculate the probability of being a junior or a senior who does not take calculus. 0.4

Number	Freq.
1	0
2	2
3	1
4	3
5	2
6	2

Attempts to reconcile differing astronomical observations, not gaming applications, motivated much of the early research on the theory of probability. Experiments conducted with coins and dice were often studied since these were easy to perform and provided significant insight.

The table displays the results of Marcie's 10 rolls of a die. State the theoretical and experimental probabilities for each event.

a. rolling a 4

b. rolling an odd number

Answer

a. theoretical:
   $P(4) = \frac{1}{6} = 0.1\overline{6}$ or $16.\overline{6}\%$

   experimental:
   $P(4) = \frac{3}{10} = 0.3$ or 30%

   1. Let E = rolling a 4.
   Use $P(4) = \frac{n(4)}{n(S)}$.

   2. $P(4) = \frac{\text{number of times a 4 was rolled}}{\text{number of trials}}$

b. theoretical:
   $P(\text{odd}) = \frac{3}{6} = 0.5$ or 50%

   experimental:
   $P(\text{odd}) = \frac{3}{10} = 0.3$ or 30%

   1. Let E = rolling an odd number.
   Use $P(\text{odd}) = \frac{n(\text{odd})}{n(S)}$.

   2. $P(\text{odd}) = \frac{\text{number of times an odd number was rolled}}{\text{number of trials}}$

SKILL ✔ EXERCISE 9

Since the die in Example 2 has no 7s, the probability of selecting a 7, an *impossible event*, is 0. The probability of a *certain event* (such as rolling a number less than 7) is 1. Other probabilities range from 0 to 1. If E_k is an outcome of an experiment and the sample space is $S = \{E_1, E_2, E_3, \dots, E_n\}$, then the sum of the probabilities of all the possible events is 1.

$$P(S) = \sum_{k=1}^{n} P(E_i) = 1$$

The set of all events in the sample space that are not in E is the *complement* of E, denoted as E'. Together E and E' form the sample space.

$$P(E) + P(E') = P(S)$$
$$P(E) + P(E') = 1$$
$$P(E') = 1 - P(E)$$

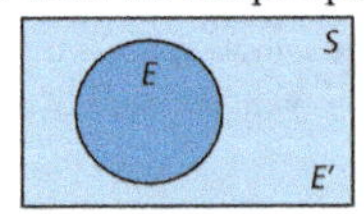

Basic Probabilities	
theoretical	$P(E) = \frac{n(E)}{n(S)}$
experimental	$P(E) = \frac{\text{number of times the event occurs}}{\text{number of trials}}$
event	$0 \leq P(E) \leq 1$
complement of an event	$P(E') = 1 - P(E)$
certain event	$P(C) = 1$
impossible event	$P(\varnothing) = 0$

1 minus the probability, as shown in Example 3.

Discuss *simple events* versus *compound events*, asking students to describe the type of event found in Example 4. (*independent, compound*) Explain that the probability of independent compound events is the product of the probability of their individual events.

Example 5 contrasts the probabilities of independent and dependent events. Demonstrate how a tree diagram can be used to find probabilities for dependent events.

In Example 6, the fact that $P(B|H) \neq P(B)$ indicates that events B and H are not independent. Point out that this indicates a higher income level had an effect on reported bullying.

Motivational Idea Consider addressing the fact that mathematics can produce numeric indicators, but our worldview will affect our response to those indicators. Ask the students what would be a biblical response to the indicators in Example 6. Possible responses include being kind to others (Eph. 4:32), esteeming others above ourselves (Phil. 2:3), not treating people differently because of their income level, and if we have more than others, not being boastful or

If there were 3,978,497 live births reported in the United States in 2015 and 139,862 of these were multiple births (twins, triplets, etc.), what was the probability of not having a multiple birth?

Answer

Let M = having a multiple birth.

$P(M) = \frac{139{,}862}{3{,}978{,}497} \approx 0.035$ or 3.5%

$P(M') \approx 1 - 0.035 = 0.965$ or 96.5%

1. Find the probability of having a multiple birth.

2. Apply $P(E') = 1 - P(E)$.

SKILL ✓ EXERCISE 15

An event that has a single outcome, such as rolling a die, is a *simple event*. A *compound event*, such as flipping a coin several times, is made of two or more simple events. Section 10.1 introduced independent events, where the outcome of one event has no influence on the outcomes of the other event(s).

PROBABILITY OF INDEPENDENT EVENTS

If A and B are independent events, $P(A \text{ and } B) = P(A) \cdot P(B)$.

An experiment consists of flipping a coin followed by rolling a 6-sided die. Find the probability of each event.

a. getting heads and a 5

b. getting tails and an even number

Answer

a. $P(H \text{ and } 5) = P(H) \cdot P(5) = \frac{1}{2} \cdot \frac{1}{6} = \frac{1}{12}$

b. $P(T \text{ and even}) = P(T) \cdot P(\text{even}) = \frac{1}{2} \cdot \frac{3}{6} = \frac{1}{4}$

Since the compound events consist of two independent events, apply $P(A \text{ and } B) = P(A) \cdot P(B)$.

SKILL ✓ EXERCISE 25

When the events are dependent, the probability of one event changes the probability of the other event(s). The Fundamental Counting Principle implies that the formula for independent events then needs to be changed to $P(A \text{ and } B) = P(A) \cdot P(B|A)$, where $P(B|A)$ is the *conditional probability* of B, given that A has happened.

PROBABILITY OF DEPENDENT EVENTS

If A and B are dependent events, $P(A \text{ and } B) = P(A) \cdot P(B|A)$.

Pierre-Simon de Laplace was able to show mathematically that small deviations from Newton's predicted paths of celestial objects were due to effects of gravitational forces after all. This finding supported a Platonic view of the universe.

Assignments

- **Minimum:** 1, 4, 7, 10, 12–21, 23, 25–26, 31–32, 35, 38, 46–52
- **Standard:** 1–2, 7–16, 19, 21–22, 25–28, 31–32, 35, 38–39, 41–42, 47–49, 51, 54
- **Extended:** 5, 9–10, 11*b*, 12, 15, 20, 23–31, 33–34, 36, 38–41, 43, 47–48, 50–53

Assessment

- Quiz 10A covers Sections 10.1–10.2.

arrogant, inciting others to bully (James 2:1–7).

Describe *mutually inclusive* and *mutually exclusive* events. Use the Venn diagrams and the definition for the probability of A or B to promote understanding of these terms.

Demonstrate how the Venn diagram in Example 7 helps determine probabilities of the mutually inclusive events.

Example 8 illustrates finding a probability involving mutually exclusive events and the events' complement. Use the table to summarize the formulas for compound probabilities.

TIPS

Ex. 5–6 Point out that order is important in exercise 5; *CST* and *CTS* are different events. But order is not important in exercise 6, in which A & B is the same event as B & A.

Ex. 42 Alternate solutions would involve finding $P(B \text{ and } G|B) + P(G \text{ and } B|G)$ for each bag.

Ex. 43*b*–*c* The solutions consider ways the teacher can choose the questions Jon knows and those questions he doesn't know.

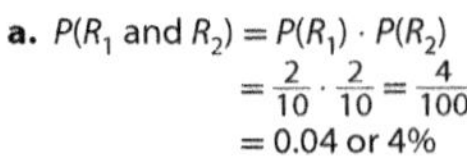

Example 5 Probabilities of Independent and Dependent Events

A hat contains 2 red marbles and 8 blue marbles. What is the probability of drawing 2 red marbles from the hat under the following conditions?

a. The first marble drawn is replaced before the second is drawn.

b. The first marble drawn is not replaced before the second is drawn.

Answer

a. $P(R_1 \text{ and } R_2) = P(R_1) \cdot P(R_2)$

$\qquad = \frac{2}{10} \cdot \frac{2}{10} = \frac{4}{100}$

$\qquad = 0.04 \text{ or } 4\%$

The events are independent because the first marble is replaced.

b. $P(R_1 \text{ and } R_2) = P(R_1) \cdot P(R_2 | R_1)$

$\qquad = \frac{2}{10} \cdot \frac{1}{9} = \frac{2}{90} = \frac{1}{45}$

$\qquad = 0.0\overline{2} \text{ or } 2.\overline{2}\%$

1. The events are dependent because the first marble is not replaced.

Check

$P(R_1 \text{ and } R_2) = \frac{2}{10} \cdot \frac{1}{9} = \frac{2}{90} = \frac{1}{45}$

2. Notice how a tree diagram can be used to verify the probability of dependent events.

——————————————— SKILL ✔ **EXERCISE 21**

When events A and B are independent, the probability of B is not affected by the fact that A has happened and $P(B|A) = P(B)$.

Example 6 Applying Conditional Probabilities

In a 2014–15 school-year survey of 12–18-yr-old students, 20.8% of the students reported being bullied. The survey indicated that 53.8% of the students were from households with annual incomes of $50,000 or more. If 10.2% of the students reported being bullied and being from households with higher incomes, are the events of being bullied and being in a household with a higher income independent or dependent?

Answer

Let B = bullied student
and H = higher-income household student.

1. Define the events.

$P(B|H) = \frac{0.102}{0.538} \approx 0.190 \text{ or } 19.0\%$

$\neq P(B) = 20.8\%$

2. Find $P(B|H) = \frac{P(B \text{ and } H)}{P(H)}$ and compare it to $P(B)$.

Events B and H are dependent.

3. Since $P(B|H) \neq P(B)$, the events are not independent.

——————————————— SKILL ✔ **EXERCISE 37**

To find the probability of event A or event B happening, consider whether the occurrence of one event prevents the other from occurring. The events are *mutually exclusive* or *disjoint* if both events cannot happen. If events A and B can both happen, they are *mutually inclusive* events.

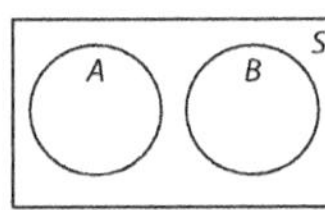

mutually exclusive

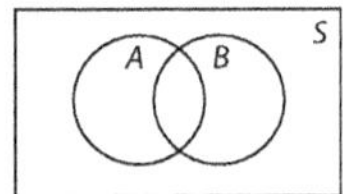

mutually inclusive

PROBABILITY OF *A* OR *B*

If *A* and *B* are mutually inclusive events, $P(A \text{ or } B) = P(A) + P(B) - P(A \text{ and } B)$.
If *A* and *B* are mutually exclusive, $P(A \text{ and } B) = 0$ and $P(A \text{ or } B) = P(A) + P(B)$.

Example 7 Calculating Combined Probabilities

A survey of 371 teens shows that 266 own a dog and 157 own a cat. If 108 own both a dog and a cat, what is the probability that a randomly selected teen owns either a dog or a cat?

Answer

Let D = owns a dog
and C = owns a cat.

$P(D \text{ or } C) = P(D) + P(C) - P(D \text{ and } C)$

$\quad = \frac{266}{371} + \frac{157}{371} - \frac{108}{371}$

$\quad = \frac{315}{371} \approx 0.85 \text{ or } 85\%$

The probability that a randomly selected teen owns a dog or cat is 85%.

The Venn diagram shows that the events are mutually inclusive.

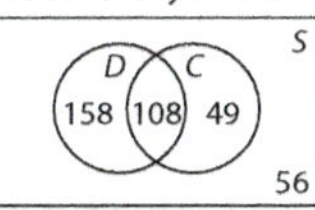

—————————————— SKILL ✔ **EXERCISE 35**

Example 8 Calculating Combined Probabilities

The table summarizes the results of a survey that asked 434 students how many books they read last summer. Determine the probability that a randomly selected student read each number of books.

a. 4 or fewer books

b. 5 or more books

Number of Books	Percent of Students
0	3.9%
1–2	14.1%
3–4	19.8%
5–6	16.1%
7–8	8.1%
9+	38.0%

Answer

a. $P(\leq 4) = P(0) + P(1-2) + P(3-4)$
$\quad\quad\quad = 0.039 + 0.141 + 0.198$
$\quad\quad\quad = 0.378 \text{ or } 37.8\%$

Find $P(\leq 4) = P(0) + P(1-2) + P(3-4)$ since these events are mutually exclusive.

b. $P(\geq 5) = 1 - P(\leq 4) = 1 - 0.378$
$\quad\quad\quad\quad\quad\quad = 0.622 \text{ or } 62.2\%$

Apply the Complement Property.

—————————————— SKILL ✔ **EXERCISE 39**

Compound Probabilities	
independent events	$P(A \text{ and } B) = P(A) \cdot P(B)$
dependent events	$P(A \text{ and } B) = P(A) \cdot P(B|A)$
inclusive events	$P(A \text{ or } B) = P(A) + P(B) - P(A \text{ and } B)$
mutually exclusive events	$P(A \text{ or } B) = P(A) + P(B)$

Solutions

❯ A. Exercises

1. $S = \{1, 2, 3, 4, 5, 6\}$; $E = \{5, 6\}$

2. $S = \{H1, H2, H3, H4, H5, H6, T1, T2, T3, T4, T5, T6\}$
$E = \{T1, T3, T5\}$

3. $S = \{HH, HT, TH, TT\}$; $E = \{HT, TH\}$

4. $S = \{HL, HW, CL, CW, TL, TW\}$
$E = \{CL\}$

5. $S = \{CST, CTS, SCT, STC, TCS, TSC\}$
$E = \{STC\}$

6. $S = \{A \& B, A \& C, A \& D, A \& E, B \& C, B \& D, B \& E, C \& D, C \& E, D \& E\}$; $E = \{A \& B\}$

7. $T: P(1) = \frac{1}{6}$

$\quad E: P(1) = \frac{10}{60} = \frac{1}{6}$

8. $T: P(2 \text{ or } 3 \text{ or } 5) = \frac{3}{6} = \frac{1}{2}$

$\quad E: P(2 \text{ or } 3 \text{ or } 5) = \frac{12}{60} + \frac{10}{60} + \frac{6}{60}$

$\quad\quad\quad\quad\quad\quad\quad\quad = \frac{28}{60} = \frac{7}{15}$

List the sample space and the event for each outcome.

1. rolling a number greater than 4 on a 6-sided die

2. getting tails and an odd number when flipping a coin and rolling a 6-sided die

3. getting heads once and tails once when flipping two coins simultaneously

4. A randomly selected student chooses a chicken sandwich and lemonade from a menu offering hamburgers, chicken sandwiches, and turkey wraps to eat and lemonade and water to drink.

5. Given the options of cola, a sports drink, or tea, a randomly selected student ranks his preferred beverages as (1) sports drink, (2) tea, and (3) cola.

6. Out of the 5 starting players for the basketball team, Adam and Ben are randomly chosen to carry the equipment. The other starting players are Chris, Dave, and Ethan.

The table displays the results from 60 rolls of a die. State the theoretical and experimental probabilities of each event.

Number	Times Rolled
1	10
2	12
3	10
4	9
5	6
6	13

7. a 1 $\frac{1}{6}, \frac{1}{6}$

8. a prime number $\frac{1}{2}; \frac{7}{15}$

9. a number greater than 4 $\frac{1}{3}, \frac{19}{60}$

10. an even number $\frac{1}{2}; \frac{17}{30}$

11. Examine the table of results.

   a. For which outcome(s) did the experimental probability equal the theoretical probability?

   b. If the experiment was increased to 1000 rolls, would you expect to continue getting a 2 in 20% of the trials and a 5 in 10% of the trials? Explain your reasoning.

A standard deck of 108 Uno® cards has four regular "Wild" cards and four "Wild Draw 4" cards. The remaining 100 cards are evenly divided into red, blue, yellow, and green cards. Each color has one 0 and two each of the numbers 1 through 9, "Reverse," "Draw 2," and "Skip" cards. Determine each probability.

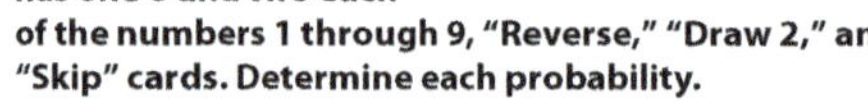

12. $P(\text{red})$ $\frac{25}{108}$

13. $P(\text{"Skip"})$ $\frac{2}{27}$

14. $P(\text{purple})$ 0

15. $P(\text{not blue})$ $\frac{83}{108}$

16. $P(\text{"Reverse" or regular "Wild"})$ $\frac{1}{9}$

17. $P(\text{"Wild Draw 4" or blue "Reverse"})$ $\frac{1}{18}$

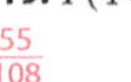

18. $P(\text{red or blue})$ $\frac{25}{54}$

19. $P(\text{4 or yellow})$ $\frac{31}{108}$

20. $P(\text{odd number or green})$ $\frac{55}{108}$

B. Exercises

Using the standard deck of Uno cards described above, find the probability of drawing the stated cards consecutively under the following conditions.
a. The first drawn card is replaced.
b. The first drawn card is not replaced.

21. 5, 8 $\frac{4}{729}; \frac{16}{2889}$

22. "Reverse," regular "Wild" $\frac{2}{729}; \frac{8}{2889}$

23. yellow 5, yellow $\frac{25}{5832}; \frac{4}{963}$

24. numbered card less than 5, "Wild Draw 4" $\frac{1}{81}; \frac{4}{321}$

Find each probability when a 6-sided die and an 8-sided die are rolled.

25. an odd number on the 6-sided die and a number greater than 2 on the 8-sided die $\frac{3}{8}$

26. a number greater than 2 on the 6-sided die and a number less than 4 on the 8-sided die $\frac{1}{4}$

Make a table listing the sums for all possible outcomes when a 6-sided die and an 8-sided die are rolled. Then find the probability for each event.

27. a sum of 11 $\frac{1}{12}$

28. a sum less than 7 $\frac{5}{16}$

29. a sum greater than 10 $\frac{5}{24}$

30. a sum that is prime $\frac{19}{48}$

Emily and Tim each bought a bag of chocolate candies from King's Kandy Shop and recorded how many of each color of candy were in their bag. Assume candies are randomly drawn without replacement and express probabilities to the nearest tenth of a percent.

Emily	Color	Tim
6	purple	3
13	green	8
2	red	9
20	orange	14
9	yellow	14
13	blue	18
63	**Total**	66

31. What is the probability that the first candy drawn from Emily's bag is purple and the first candy drawn from Tim's bag is red? 1.3%

32. What is the probability that none of the first three candies drawn from Emily's bag are orange? 31.1%

---

9. $T: P(5 \text{ or } 6) = \frac{2}{6} = \frac{1}{3}$
$E: P(5 \text{ or } 6) = \frac{6}{60} + \frac{13}{60} = \frac{19}{60}$

10. $T: P(\text{even}) = \frac{3}{6} = \frac{1}{2}$
$E: P(2 \text{ or } 4 \text{ or } 6) = \frac{12}{60} + \frac{9}{60} + \frac{13}{60}$
$= \frac{34}{60} = \frac{17}{30}$

11a. Both rolling a 1 and rolling a 3 occurred 10 times out of the 60 trials and have a theoretical probability of $\frac{1}{6}$.

11b. no; As the number of trials increases, the experimental probabilities tend to approach the theoretical probability of $\approx 17\%$.

13. $P(\text{"Skip"}) = \frac{4(2)}{108} = \frac{2}{27}$

14. impossible event

15. $P(\text{not blue}) = 1 - P(\text{blue})$
$= 1 - \frac{25}{108} = \frac{83}{108}$

16. $P(\text{Rev}) + P(\text{W})$
$= \frac{4(2)}{108} + \frac{4}{108} = \frac{12}{108} = \frac{1}{9}$

17. $P(\text{WD4}) + P(\text{blue Rev})$
$= \frac{4}{108} + \frac{2}{108} = \frac{6}{108} = \frac{1}{18}$

18. $P(\text{red}) + P(\text{blue}) = \frac{25}{108} + \frac{25}{108} = \frac{25}{54}$

19. $P(4) + P(\text{yellow}) - P(\text{yellow } 4)$
$= \frac{8}{108} + \frac{25}{108} - \frac{2}{108} = \frac{31}{108}$

20. $P(\text{odd}) + P(\text{green}) - P(\text{odd green})$
$= \frac{4 \cdot 2 \cdot 5}{108} + \frac{25}{108} - \frac{10}{108} = \frac{55}{108}$

B. Exercises

21a. $P(5) \cdot P(8) = \frac{8}{108} \cdot \frac{8}{108} = \frac{4}{729}$

21b. $P(5) \cdot P(8|5) = \frac{8}{108} \cdot \frac{8}{107} = \frac{16}{2889}$

22a. $P(\text{Rev}) \cdot P(\text{reg. W})$
$= \frac{8}{108} \cdot \frac{4}{108} = \frac{2}{729}$

22b. $P(\text{Rev}) \cdot P(\text{W}|\text{Rev})$
$= \frac{8}{108} \cdot \frac{4}{107} = \frac{8}{2889}$

23a. $P(\text{yellow } 5) \cdot P(\text{yellow})$
$= \frac{2}{108} \cdot \frac{25}{108} = \frac{25}{5832}$

23b. $P(\text{yellow } 5) \cdot P(\text{yellow } 5|\text{yellow})$
$= \frac{2}{108} \cdot \frac{24}{107} = \frac{4}{963}$

24a. $P(< 5) \cdot P(\text{WD4})$
$= \frac{4(9)}{108} \cdot \frac{4}{108} = \frac{1}{3} \cdot \frac{1}{27} = \frac{1}{81}$

24b. $P(< 5) \cdot P(\text{WD4}|< 5)$
$= \frac{4(9)}{108} \cdot \frac{4}{107} = \frac{1}{3} \cdot \frac{4}{107} = \frac{4}{321}$

25. $\frac{3}{6} \cdot \frac{6}{8} = \frac{3}{8}$

26. $\frac{4}{6} \cdot \frac{3}{8} = \frac{1}{4}$

27–30.

	1	2	3	4	5	6
1	2	3	4	5	6	7
2	3	4	5	6	7	8
3	4	5	6	7	8	9
4	5	6	7	8	9	10
5	6	7	8	9	10	11
6	7	8	9	10	11	12
7	8	9	10	11	12	13
8	9	10	11	12	13	14

27. $\frac{4}{48} = \frac{1}{12}$

28. $\frac{15}{48} = \frac{5}{16}$

29. $\frac{10}{48} = \frac{5}{24}$

30. primes (with occurances):
2 (1), 3 (2), 5 (4), 7 (6), 11 (4), 13 (2)
19 total prime results: $\frac{19}{48}$

31. $\frac{6}{63} \cdot \frac{9}{66} = \frac{1}{77} \approx 1.3\%$

32. $\frac{43}{63} \cdot \frac{42}{62} \cdot \frac{41}{61} = \frac{1763}{5673} \approx 31.1\%$

33. What is the probability that Lily randomly selects Tim's bag and then draws a red candy? 6.8%

34. If three identical bags are placed beside the original two, what is the probability that Lily randomly chooses an orange or a yellow candy from Emily's bag? 9.2%

35. Of the church youth group's 78 teens, 48 teens went to camp last year. Of these 48, 38 teens will be going back to camp this summer, while 15 teens who did not go last year plan to join them. Draw a Venn diagram illustrating this scenario and then express the probability of each random selection as a fraction.

  a. a teen attending last summer but not this summer $\frac{5}{39}$

  b. a teen attending just one summer $\frac{25}{78}$

  c. a teen not attending either summer $\frac{5}{26}$

  d. a teen attending at least one summer $\frac{21}{26}$

36. Of Eastside Christian High School's 120 students, 72 students play fall sports, 70 students play spring sports, and 42 of these students play both. Draw a Venn diagram and label each region with its corresponding number of students. Express each probability for a randomly chosen student as a reduced fraction.

  a. The student plays in the fall but not in the spring. $\frac{1}{4}$

  b. The student plays in either season. $\frac{5}{6}$

  c. The student does not play in either season. $\frac{1}{6}$

  d. Are playing fall sports and spring sports independent events or dependent events? Explain your reasoning. independent

37. Jim estimates that he experiences pain during 8% of his working hours. Draw a Venn diagram illustrating the time he doesn't feel well if his neck hurts 7% of the time, and his hand is sore 4% of the time.

  a. What percent of the time does he experience both neck and hand pain? 3%

  b. Are his neck pain and hand pain independent events or dependent events? Explain your reasoning. dependent

38. Apple's iPhone® sales accounted for 216.8 million of the 1536.5 million smartphones sold worldwide in 2017. The table records the quarterly iPhone sales (in millions) for that year. Find each probability to the nearest tenth of a percent.

2017 iPhone Sales	
Qtr	Millions
1st	78.3
2nd	50.8
3rd	41.0
4th	46.7

  a. A person who purchased a smartphone in 2017 bought an iPhone. 14.1%

  b. An iPhone bought in 2017 was purchased in the last quarter. 21.5%

  c. An iPhone bought in 2017 was purchased in one of the first three quarters. 78.5%

39. Use the 2017 world population of 7.6 billion people and the smartphone information from excersise 38 to find each probability to the nearest tenth of a percent.

  a. A randomly chosen person did not purchase a smartphone in 2017. 79.8%

  b. A randomly chosen person purchased an iPhone in the first half of 2017. 1.7%

> **C. Exercises**

Exercises 40–42 refer to the table used for exercises 31–34.

40. How much more likely is Emily to first draw a purple candy than Tim? ≈ 5% or twice as likely

41. Whose bag should Lily choose from if she would like to randomly draw a blue and then a green candy? How much higher (to the nearest tenth of a percent) are her chances? Emily's bag; 1.0%

42. Whose bag should Lily choose from if she would like to randomly draw one blue and one green candy simultaneously? How much higher (to the nearest tenth of a percent) are her chances? Emily's bag; 1.9%

36d. $P(S|F) = \frac{42}{72} = \frac{7}{12}$ or

$$= \frac{P(F \text{ and } S)}{P(F)} = \frac{\frac{42}{120}}{\frac{72}{120}} = \frac{7}{12}$$

and $P(S) = \frac{70}{120} = \frac{7}{12}$

Since $P(S|F) = P(S)$, the events are independent.
Similarly, $P(F|S) = P(F)$ can be confirmed.

37a.

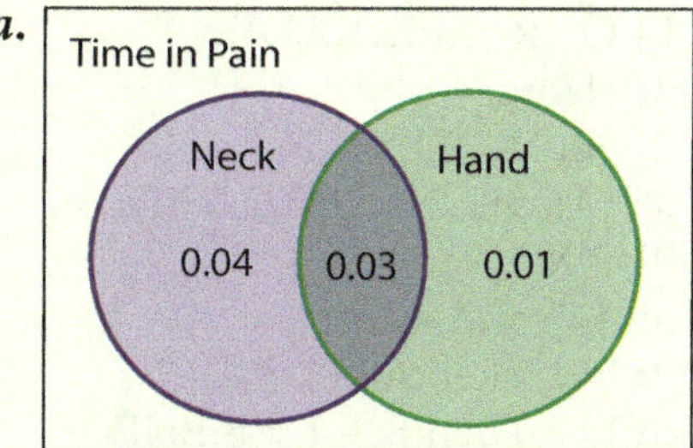

$P(N \text{ and } H) = 0.03$ or 3%

37b. $P(H|N) = \frac{P(N \text{ and } H)}{P(N)} = \frac{0.03}{0.07} = 0.43$

and $P(H) = 0.04$

Since $P(H|N) \neq P(H)$, the two events are dependent.
Similarly, $P(N|H) \neq P(N)$ can be confirmed.

38a. $\frac{216.8}{1536.5} \approx 14.1\%$

38b. $\frac{46.7}{216.8} \approx 21.5\%$

38c. $P(\text{not } 4) = 1 - P(4)$
$\approx 1 - 0.215 = 78.5\%$

39a. $1 - \frac{1.5365}{7.6} \approx 79.8\%$

39b. $\frac{78.3 + 50.8}{7600} \approx 1.7\%$

> **C. Exercises**

40. Emily: $\frac{6}{63} \approx 9.5\%$

Tim: $\frac{3}{66} \approx 4.5\%$

$9.5\% - 4.5\% = 5\%$ or $\frac{9.5\%}{4.5\%} \approx 2.1$

41. Emily's bag: $\frac{13}{63} \cdot \frac{13}{62} = \frac{169}{3906} \approx 4.33\%$

Tim's bag: $\frac{18}{66} \cdot \frac{8}{65} = \frac{144}{4290} \approx 3.36\%$

$4.33\% - 3.36\% = 0.97\%$

42. Emily's bag:

$$\frac{_{13}C_1 \cdot {}_{13}C_1}{_{63}C_2} = \frac{13 \cdot 13}{\frac{63 \cdot 62}{2}} = 2\left(\frac{169}{3906}\right)$$

$\approx 8.65\%$

Tim's bag:

$$\frac{_{18}C_1 \cdot {}_8C_1}{_{66}C_2} = \frac{18 \cdot 8}{\frac{66 \cdot 65}{2}} = 2\left(\frac{144}{4290}\right)$$

$\approx 6.71\%$

$8.65\% - 6.71\% = 1.94\%$

$P(B \text{ and } G|B) + P(G \text{ and } B|G)$

33. $\frac{1}{2} \cdot \frac{9}{66} = \frac{3}{44} \approx 6.8\%$

34. $\frac{1}{5} \cdot \frac{29}{63} = \frac{29}{315} \approx 9.2\%$

35.

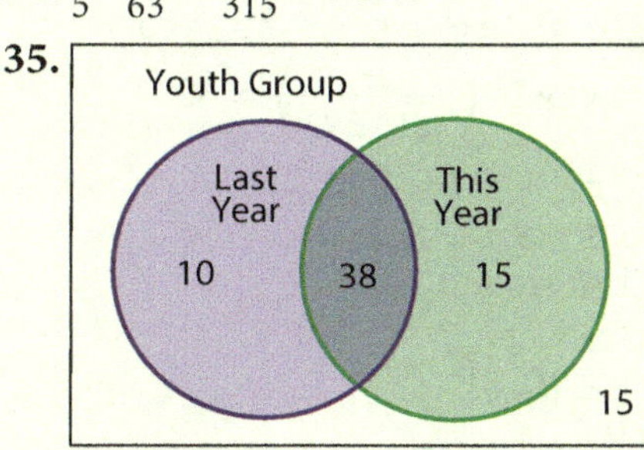

35a. $\frac{10}{78} = \frac{5}{39}$

35b. non-overlapping regions
$10 + 15 = 25; \frac{25}{78}$

35c. $\frac{15}{78} = \frac{5}{26}$

35d. $1 - P(\text{not attending}) = 1 - \frac{5}{26} = \frac{21}{26}$

36.

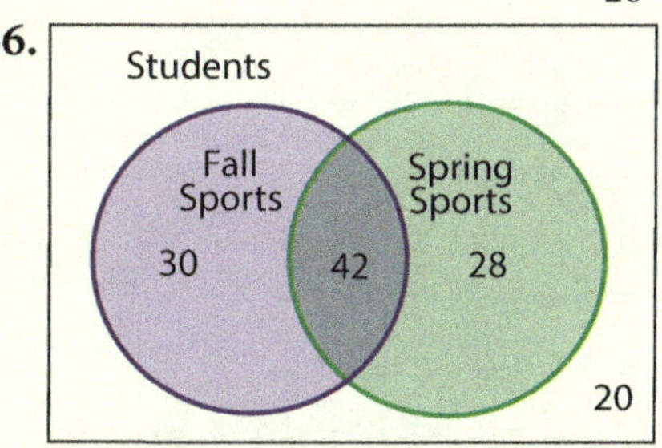

36a. $\frac{72 - 42}{120} = \frac{30}{120} = \frac{1}{4}$

36b. $\frac{72}{120} + \frac{70}{120} - \frac{42}{120} = \frac{100}{120} = \frac{5}{6}$

36c. $1 - \frac{5}{6} = \frac{1}{6}$

43a. $\dfrac{{}_8C_6}{{}_{12}C_6} = \dfrac{28}{924} = \dfrac{1}{33} \approx 3\%$

43b. $\dfrac{{}_8C_3 \cdot {}_4C_3}{{}_{12}C_6} = \dfrac{224}{924} = \dfrac{8}{33} \approx 24\%$

43c. $\dfrac{{}_8C_4 \cdot {}_4C_2}{{}_{12}C_6} + \dfrac{{}_8C_5 \cdot {}_4C_1}{{}_{12}C_6} + \dfrac{{}_8C_6}{{}_{12}C_6}$

$= \dfrac{420}{924} + \dfrac{224}{924} + \dfrac{28}{924}$

$= \dfrac{5}{11} + \dfrac{8}{33} + \dfrac{1}{33} = \dfrac{8}{11} \approx 73\%$

44. (simple) sample program:

```
PROGRAM:ROLLDIE
:ClrHome
:Input "ROLLS=",N
:0→A:0→B:0→C:0→D:0→E:0→F
:For(X,1,N)
:int(6*rand)+1→I
:If I=1:Then:A + 1 →A:END
:If I=2:Then:B + 1 →B:END
:If I=3:Then:C + 1 →C:END
:If I=4:Then:D + 1 →D:END
:If I=5:Then:E + 1 →E:END
:If I=6:Then:F + 1 →F:END
:End
:Disp "{1S 2S 3S 4S 5S 6S} ="
:Disp "{A, B, C, D, E, F}"
```

sample results:

	$n = 10$		$n = 20$		$n = 100$		$n = 1000$	
	#	%	#	%	#	%	#	%
1s	2	20	3	15	19	19	189	19
2s	2	20	1	5	16	16	168	17
3s	2	20	4	20	12	12	171	17
4s	0	0	2	10	15	15	167	17
5s	2	20	6	30	14	14	156	16
6s	2	20	4	20	24	24	149	15

Cumulative Review

45.

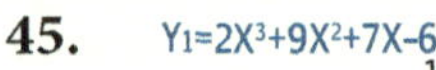
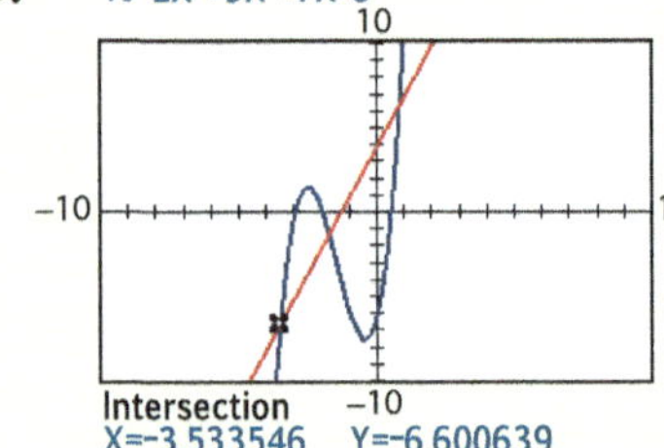

46.

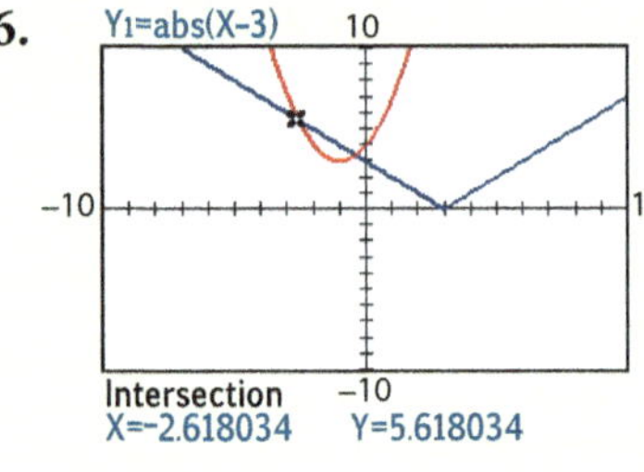

---

43. Mrs. Ream randomly selects 6 essay questions from a list of 12 that she has given to her class for the final exam. If Jon knows 8 of the answers, find the probability (to the nearest percent) that he will answer each number of questions correctly.

  a. all 6 questions 3%

  b. exactly 3 24%

  c. at least 4 73%

44. To demonstrate that experimental probabilities approach theoretical probabilities, write a calculator program that randomly generates n numbers from 1 to 6 and tallies the frequency of each outcome. The program's output should display the number of times each number was "rolled." Run the program for $n = 10, 20, 100,$ and 1000 and then calculate the experimental probability for each outcome.

CUMULATIVE REVIEW

Solve each system using technology. Round answers to the nearest tenth if necessary. [7.1]

45. $y = 2x^3 + 9x^2 + 7x - 6$
$y = 3x + 4$ $(-3.5, -6.6), (-1.8, -1.3), (0.8, 6.4)$

46. $y = |x - 3|$
$y = x^2 + 2x + 4$ $(-2.6, 5.6), (-0.4, 3.4)$

Write the standard form equation of a conic with the given characteristics.

47. hyperbola with vertices at $(\pm 6, 0)$ and perpendicular asymptotes [8.3] $\dfrac{x^2}{36} - \dfrac{y^2}{36} = 1$

48. ellipse with vertices at $(3, 5)$ and $(3, -9)$ and an eccentricity of $\dfrac{\sqrt{13}}{7}$ [8.2] $\dfrac{(x-3)^2}{36} + \dfrac{(y+2)^2}{49} = 1$

If $\displaystyle\sum_{k=1}^{200} a_k = 45$ and $\displaystyle\sum_{k=1}^{200} c_k = 505$, **use properties of summations to evaluate each sum.** [9.5]

49. $\displaystyle\sum_{k=1}^{200} 7a_k$ 315

50. $\displaystyle\sum_{k=1}^{200} (5c_k - 6)$ 1325

51. What is the angle (to the nearest degree) between the vectors $\mathbf{p} = \langle 3, 4, 6 \rangle$ and $\mathbf{q} = \langle -4, 5, 3 \rangle$? [6.3] E

  A. 25°   **C.** 61°   **E.** none of these

  B. 39°   **D.** 118°

52. Which matrix is in reduced row echelon form? [7.3] D

  A. $\begin{bmatrix} 3 & 5 & 2 & -7 \\ 0 & -2 & 8 & 4 \\ 0 & 0 & 0 & 6 \end{bmatrix}$ **D.** $\begin{bmatrix} 1 & 0 & 0 & 3 \\ 0 & 1 & 0 & 2 \\ 0 & 0 & 1 & 0 \end{bmatrix}$

  B. $\begin{bmatrix} 1 & 0 & 0 & -7 \\ 0 & 1 & 0 & 2 \\ 0 & 0 & 3 & 3 \end{bmatrix}$ **E.** none of these

  C. $\begin{bmatrix} 1 & 5 & 2 & 4 \\ 0 & 1 & 8 & 4 \\ 0 & 0 & 1 & 2 \end{bmatrix}$

53. What is the domain for the graph of the parametric equations $x = 2 - t$ and $y = t$; $t \in [-3, 2]$? [8.5] B

  A. $x \in [-3, 2]$ **C.** $x \in [-5, 0]$ **E.** $x \in \mathbb{R}$

  B. $x \in [0, 5]$ **D.** $x \in [-1, 4]$

54. How many ways can the letters in the word *MISSISSIPPI* be arranged? [10.1] C

  A. 24   **C.** 34,650   **E.** none of these

  B. 330   **D.** 39,916,800

---

47. horizontal TA; $C(0, 0)$
$a = 6 = b$
$\dfrac{x^2}{36} - \dfrac{y^2}{36} = 1$

48. vertical MA
$C\left(\dfrac{3+3}{2}, \dfrac{5+(-9)}{2}\right) = (3, -2)$
$a = 5 - (-2) = 7$
$e = \dfrac{c}{a} = \dfrac{\sqrt{13}}{7}; c = \sqrt{13}$
$b^2 = a^2 - c^2 = 49 - 13 = 36$
$\dfrac{(x-3)^2}{36} + \dfrac{(y+2)^2}{49} = 1$

49. $7\displaystyle\sum_{k=1}^{200} a_k = 7(45) = 315$

50. $5\displaystyle\sum_{k=1}^{200} c_k - \sum_{k=1}^{200} 6$
$= 5(505) - 6(200) = 1325$

51. $\mathbf{p} \cdot \mathbf{q} = 3(-4) + 4(5) + 6(3) = 26$
$|\mathbf{p}| = \sqrt{3^2 + 4^2 + 6^2} = \sqrt{61}$
$|\mathbf{q}| = \sqrt{(-4)^2 + 5^2 + 3^2} = 5\sqrt{2}$
$\theta = \text{Cos}^{-1}\left(\dfrac{26}{\sqrt{61} \cdot 5\sqrt{2}}\right) \approx 62°$

53. $t = -x + 2; t \in [-3, 2]$
$-3 \leq -x + 2 \leq 2$
$-5 \leq -x$ and $-x \leq 0$
$x \leq 5 \qquad x \geq 0$
$\therefore x \in [0, 5]$

54. $\dfrac{11!}{2!4!4!} = 34{,}650$

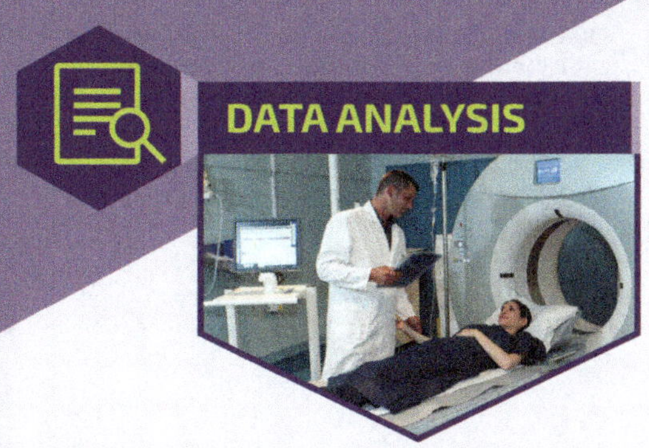

Medical Screenings and Probability

Suppose that, during a routine physical examination, you test positive for a serious but rare (only 1 in every 200,000) medical condition. The test accurately identifies an individual with the condition 95% of the time and gives a false positive only 2% of the time. What is the probability that you actually have this condition? Many of us would assume it is 98%, 95%, or maybe only 93%. Doing the math provides surprisingly counterintuitive results.

Let C represent having the condition and let T represent testing positive.

1. Which conditional probability is represented by the 95% effectiveness of the test, $P(T|C)$ or $P(C|T)$? $P(T|C)$

2. Which conditional probability represents the likelihood that a person who tests positive actually has the condition? $P(C|T)$

Recall that the probability for these conditional events is found using $P(C|T) = \frac{P(C \text{ and } T)}{P(T)}$. Because C and T are dependent events, the probability that you have the condition and test positive is $P(C \text{ and } T) = P(C) \cdot P(T|C)$.

3. State $P(C)$ if 1 person out of 200,000 has the condition. 5×10^{-6} or 0.0005%

4. Find $P(C \text{ and } T)$ in this scenario. 4.75×10^{-6} or 0.000475%

To find the probability of a positive test, $P(T)$, consider the number of true positives from those who have the condition and the number of false positives from those who do not. In a sample of 200,000 people, the one person with the condition tests positive 95% of the time and the 199,999 who do not will test positive 2% of the time for a total of $0.95(1) + 0.02(199,999) = 0.95 + 3999.98 = 4000.93$ positive tests. Therefore, $P(T) = \frac{4000.93}{200,000} \approx 0.02$.

5. Find $P(C|T) = \frac{P(C \text{ and } T)}{P(T)}$. 2.375×10^{-4} or 0.02375%

6. What is the probability that you actually have the condition? $\approx 0.02\%$

In this hypothetical scenario, the effective rate of the test, $P(T|C)$, was given along with the percentage of false positives. To find the likelihood of having the condition knowing that you have tested positive, $P(C|T)$ was found using

$$P(C|T) = \frac{P(C \text{ and } T)}{P(T)} = \frac{P(C) \cdot P(T|C)}{P(T)}.$$

Breast cancer is the second leading cause of cancer deaths in women in the United States. In 2017 over 40,000 women died from the disease. In 2009 the United States Preventive Services Task Force made headlines when they revised previously issued guidelines for when and how often women with no unusual risk factors should be screened for breast cancer. In particular, the task force found that the risks outweighed the benefits of mammograms for women 40–50 yr old, and changed the recommended age for women to begin routine screenings from 40 to 50. Risks noted in the task force report include overdiagnosis and overtreatment of non-life-threatening cancers, false-positive results leading to unnecessary and sometimes invasive follow-up testing, psychological harm, and false reassurance from false-negative tests.

Assume that screening mammograms are 80% effective and give false positives 10% of the time. Let C represent a woman who has breast cancer and T represent a positive result from a screening.

7. Find each probability for a woman in her 40s if on average 14 out of 1000 women will develop breast cancer in their 40s.

   a. $P(T)$ 10.98%   b. $P(C|T)$ $\approx 10.20\%$

Data Analysis

Objectives

1. To find conditional probabilities

2. To find probabilities of dependent events

3. To find the probability that a given event is the result of a particular cause

4. To evaluate the benefits and risks of medical screenings from a biblical worldview

Assignments

- **Minimum:** optional
- **Standard:** 1–6; 7–10 (in class)
- **Extended:** 1–10

Solutions

4. $P(C \text{ and } T) = P(C) \cdot P(T|C)$
   $= (5 \times 10^{-6})(0.95) = 4.75 \times 10^{-6}$

5. $P(C|T) = \dfrac{P(C \text{ and } T)}{P(T)} = \dfrac{P(C) \cdot P(T|C)}{P(T)}$
   $= \dfrac{4.75 \times 10^{-6}}{0.02} = 2.375 \times 10^{-4}$

6. The probability is an interpretation of the result in exercise 5.

7a. $P(T) = \dfrac{0.80(14) + 0.10(986)}{1000}$
   $= \dfrac{11.2 + 98.6}{1000} = 0.1098$

7b. $P(C|T) = \dfrac{P(C) \cdot P(T|C)}{P(T)}$
   $= \dfrac{(0.014)(0.80)}{0.1098} = \dfrac{0.0112}{0.1098} \approx 0.1020$

PRESENTATION

Summarize the hypothetical situation and the procedure for finding the probability that you actually have a rare disease, given a positive result on a routine test. The second given scenario relates to a real-world application of probability theory that resulted in significant changes to recommended screenings for breast cancer.

Emphasize that in both cases, the condition is rare in the general population, there are no pre-disposing factors, and the positive result occurred during a routinely scheduled physical examination. Changes to any of these assumptions may radically change the probabilities.

Point out that in the hypothetical scenario even a low rate of false positives (2%) makes a big difference when the condition is very rare. Notice that the probability of a positive test for any reason, $P(T)$, is only slightly higher than the probability of a false positive.

Since women are more likely to develop breast cancer as they age, the increase in actual occurrence of the disease increases the probability that a positive test indicates a true positive. According to the American Cancer Society, the likelihood that an American woman will develop breast cancer is approximately 0.5% in her 30s, 1.5% in her 40s, 2.3% in her 50s, and 3.4% in her 60s. The Preventative Services Task Force estimates that lives saved by annual mammograms per 10,000 women screened are 3 for women in their 40s, 8 for women in their 50s, and 21 for women in their 60s.

If you choose to discuss questions 9–10, be sensitive to students who may have friends or relatives affected by breast cancer. While governments must prioritize apportionment of scarce resources, "only 3" lives saved in 10,000 has a very different meaning if 1 of those 3 is close to you.

8. $P(C|T) = \dfrac{P(C) \cdot P(T|C)}{P(T)}$

$\quad = \dfrac{(0.004)(0.80)}{\frac{0.8(4) + 0.10(996)}{1000}}$

$\quad = \dfrac{0.0032}{0.1028} \approx 0.0311$

sample answers:

9. Since we are made in God's image, every life is sacred. Just as Christ seeks to redeem fallen humankind and will eventually redeem His creation, we should seek to reduce the effects of the Fall in this world. We can be thankful that God has enabled people to develop tests for diseases and use them wisely. Ultimately, we know that God is sovereign in all creation and the length of our lives is in His hands. We should encourage and support refinement of screening methods— particularly reduction of the 1 in 5 false positive results.

10. increasing women's awareness of other cancer indicators so that screenings of younger women are more meaningful; using the re- sources conserved by doing fewer screenings to improve existing screenings, increasing the rate of effectiveness and reducing the rate of false positives

8. Find the probability that a woman in her 30s who has tested positive actually has breast cancer if on average 4 out of 1000 women will develop breast cancer in their 30s. ≈ 3.11%

In 2016, after reviewing multiple studies involving over 600,000 women, the task force issued another report supporting their 2009 conclusions. This report noted that routine screening of women 39–49 yr old would likely prevent 3 deaths per 10,000 women screened over a 10 yr period.

9. **Discuss:** While statistical analysis is obviously helpful, it cannot prescribe solutions to ethical questions. How should a biblical worldview influence the recommendations on when and how often screenings should take place?

10. **Discuss:** In lieu of more screenings, what could be done to reduce the number of deaths due to breast cancer?

Remind the students that in all our afflictions, we can still find joy and hope. Our joy, hope, comfort, and peace are in the Lord, Who gave Himself for us. Even in suffering we an look forward to the hope we have in Christ (2 Cor. 1:3–4; Rom. 8:18).

10.3 Representing Data Graphically

Our ultimate goal in the study of mathematics should be to glorify God and to help others reach their God-given potential. The study of statistics plays an important part in that goal. God created humans in His image as reasoning beings and expects us to use these abilities as we plan and make decisions for the future. While God directs our steps, He expects us to plan our ways (Prov. 16:9). Careful planning can be facilitated by information and predictions derived from statistics.

The study of statistics seeks to predict (or *infer*) outcomes based on the results of similar situations. Mathematicians have developed techniques that permit the study of a manageable number of trials to reasonably predict outcomes. For example, a statistician may predict the average weight of one-year-old boys in the United States based on a sample of 300 one-year-old boys. The larger overall set being considered is called the *population*. The *sample* is a carefully selected subset of the population. The sample is *random* if all members of the population have an equal chance of being included. If the sample is biased in any way, the inferences made will not be reliable.

DEFINITIONS

A **parameter** is the actual value for a characteristic of the population. A **statistic** is an estimate of the population's parameter based on a sample.

In the example above, the average weight of all one-year-old boys in the United States is a parameter of the population and the average weight of the 300 boys in the study is a statistic. In larger populations, the actual value of a parameter is difficult or even impossible to determine. The results of studying a sample are summarized with *descriptive statistics*. The use of statistics from a sample to make predictions about the population's parameters is called *inferential statistics*.

Example 1 Identifying Parameters and Statistics

State whether each characteristic is a parameter or a statistic. Explain your reasoning.

a. In a study of the health of cats in humane societies, the average weight of cats at the local humane society was found.

b. When reviewing the performance of students in the university's Calculus I course, Dr. Holcomb found the range of homework averages for all the students in the six sections of the class.

Answer

a. statistic; The limited number of cats represents a sample of the larger population.

b. parameter; This is the value for the entire population being studied.

SKILL ✔ EXERCISE 5

It is difficult to estimate probabilities for events that have no history. For example, how could the probability of losing a space shuttle be predicted before any shuttles had been launched? Bayesian statistics provide an alternative form of analysis for these situations.

KEYWORD SEARCH

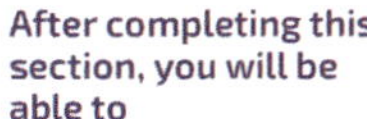

Bayesian statistics

10.3 Representing Data Graphically

Objectives

1. To distinguish between a parameter and a statistic

2. To determine the level of measurement for a characteristic

3. To create bar graphs and pie charts of categorical data

4. To create dot plots, stem plots, frequency distribution tables, and histograms for individualized data and intervals of data

Flash

The first pediatric hospital was founded in 1802 in Paris, France. Becoming a pediatrician typically involves 8 yr of post-secondary education followed by a 3 yr residency program.

Vocabulary

descriptive statistics
frequency distribution table
frequency polygon
grouped frequency distribution table
histogram
infer
inferential statistics
interval
line graph
nominal
ordinal
parameter
percentile graph (ogive)
population
random
range
ratio
sample
statistic
Sturges's formula

Reading and Writing Mathematics

Write a brief comparison of a bar graph and a histogram. Bar graphs are used for categorical (nominal) data, but histograms are used for quantitative values. The horizontal axis of a histogram contains values of the variable being examined or intervals (classes) for these values.

PRESENTATION

Lesson Opener

Find each value.

1. What percent of 60 is 12? 20%

2. What is 32% of 140? 44.8

3. 70 is 28% of what number? 250

4. What is 17% of 360°? 61.2°

Take time to fully explain and discuss the terms in this section. One of the first steps in a statistical study is to define the *population*. Explain that the population is often too large to practically obtain the information needed for the study; in these cases information is collected from a representative portion (or *sample*) of the larger population.

Be sure students understand the difference between a *parameter* (a measure of the population) and a *statistic* (a measure of the sample). Have the students consider the task of finding the average income of 20-yr-olds in Europe, or in the United States, or in just the state of Michigan. Discuss the impossibility of polling every person in any of these populations. Use Example 1 to reinforce these terms.

Discuss the different levels of measurements. The difference between the *interval* and *ratio* levels can be illustrated with temperature scales. Note that 0°F

Additional Exercises

1. Identify each of the following as a parameter or a statistic.
   a. heights of all the seventh-grade boys at the local middle school
      parameter
   b. political affiliation of 20 students chosen at random statistic
2. State the level of measurement and explain your reasoning.
   a. amount (in inches) of snowfall last winter in Denver ratio; absolute zero with a fixed interval
   b. favorite season of the year
      nominal; an unordered category
   c. order of the top three contenders in the national spelling bee
      ordinal; an ordered category without a fixed scale
3. Create a bar graph and pie chart illustrating the history test scores.

Grade	A	B	C	D	F
Frequency	5	8	9	2	1

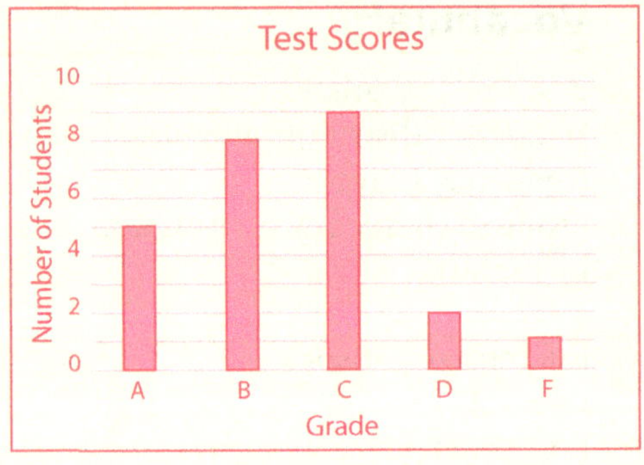

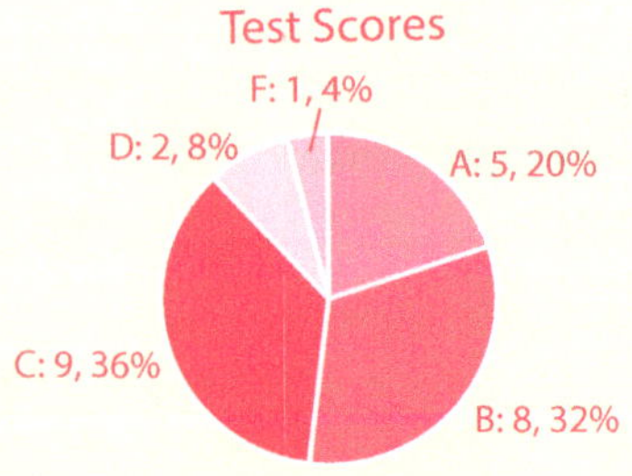

Data collected for a statistical study falls into one of four levels of measurement. *Nominal* data represents unordered groups or classifications. *Ordinal* data describes ordered categories within the set without indicating the interval between values. *Interval* data measures a characteristic on a scale having a fixed width without a meaningful zero. *Ratio* data is measured on a scale with a fixed width and an absolute zero. The mathematical significance of the measurement increases at each level.

Measurement Levels	
nominal	unordered categories examples: political party, eye color, gender
ordinal	ordered categories without distances examples: customer satisfaction rating, class rank
interval	distances without an absolute zero examples: Fahrenheit temperature, time of day
ratio	distances with an absolute zero examples: test score, precipitation, income, height

Example 2   Identifying Levels of Measurement

State the level of measurement for each characteristic. Explain your reasoning.
a. grade point average   b. place in a race   c. favorite professional baseball team

Answer

a. ratio; GPA has an absolute zero and a scale with a fixed width.
b. ordinal; The data describes an ordering without a fixed scale.
c. nominal; The categories are unordered classifications.

SKILL ✓ EXERCISE 9

A statistical graph organizes the data and illustrates its distribution. Categorical data is usually illustrated with a bar graph or a pie chart. Spreadsheet programs can be used to quickly produce accurate graphs.

KEYWORD SEARCH

online statistical graph generator 🔍

Example 3   Illustrating Categorical Data

Create a bar graph and a pie chart illustrating the grades on a Statistics test.

Grade	A	B	C	D	F
Frequency	4	7	10	3	1

Answer

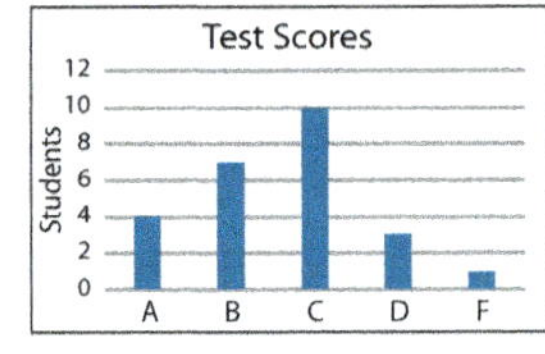

1. Create a bar graph with the vertical axis representing the frequencies and the horizontal axis representing the grades.

	A	B
1	Grade	Freq.
2	A	4
3	B	7
4	C	10
5	D	3
6	F	1

or

In a spreadsheet, enter grades in one column and the frequency of each grade in a second column. Select both columns and insert a bar graph. Then enter a title and label the axes as needed.

CONTINUED ➡

does not indicate the absence of heat. Also, 40°F is not twice a warm as 20°F. Compare the Fahrenheit scale to the Kelvin scale (ratio level) where 0 K is a meaningful zero representing the absence of heat (absolute zero).

Interval and ratio are often classified as quantitative while *nominal* and *ordinal* are classified as categorical. The mathematical significance of each level increases from nominal to ratio, which provides the most useful measurements.

Use Example 2 to help the students recognize the level of measurement for a characteristic. There are several ways to make statistical graphs. While students can create hand-drawn graphs, using technology is often quicker and produces a more accurate and polished graph. Note that the grades in Example 3 consist of 5 categories. Explain that bar graphs and pie charts are well suited for categorical data. *Histograms* can be created on the TI-84, but bar graphs cannot be. Graphs can also be easily created on a spreadsheet or with an online graph generator.

Motivational Idea Consider allowing students the freedom of making their statistical graphs with the tool of their choice. Use Internet keyword searches for specific types of graph generators, such as *online histogram*.

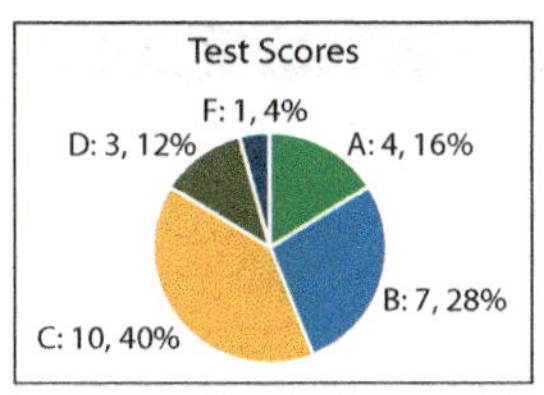

Test Scores

F: 1, 4%
D: 3, 12%
A: 4, 16%
B: 7, 28%
C: 10, 40%

2. Calculate the percentage of students who received each letter grade and multiply each percentage by 360°. Create a pie chart demonstrating the distribution of the letter grades. Label each sector appropriately.

	A	B	C	D
1	Grade	Freq	Percent	Angle (°)
2	A	4	16%	57.6
3	B	7	28%	100.8
4	C	10	40%	144
5	D	3	12%	43.2
6	F	1	4%	14.4
7	Total	25		

or

In a spreadsheet, select the grade labels and frequency data and then insert a pie chart. Add a title and labels and adjust the format as needed.

SKILL ✓ EXERCISE 25

Dot plots and stem plots are used to organize and illustrate interval and ratio data. Relatively small sets of data, such as the following heights (in inches) of choir members, can be represented by a dot plot.

{58, 61, 65, 72, 70, 66, 62, 70, 66, 64, 71, 70, 62, 66, 60, 71, 73, 68, 66, 59, 60, 73, 71, 64}

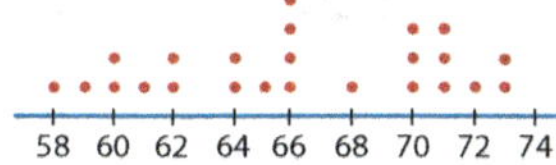

From this dot plot it is easy to see that the heights range from 58 in. to 73 in. and that the majority of heights range from 62 in. to 71 in.

This data can also be illustrated with a stem plot (or a stem-and-leaf plot). The largest digit that adequately groups the data acts as the stem and the remaining digit(s) are the leaves. The tens digit of a choir member's height is the stem and the ones digit is the leaf.

Stem	Leaves
5	8 9
6	0 0 1 2 2 4 4 5 6 6 6 6 8
7	0 0 0 1 1 1 2 3 3

Occasionally it is helpful to split a stem plot into smaller categories. The first row with the same stem contains leaves 0–4 and the second row with that stem contains leaves 5–9.

Two similar data sets can be compared using back-to-back stem plots. The leaves of one data set are placed to the right of the shared stem and the leaves of the second data set are placed to its left.

Stem	Leaves
5	
5	8 9
6	0 0 1 2 2 4 4
6	5 6 6 6 6 8
7	0 0 0 1 1 1 2 3 3
7	

Florence Nightingale made polar graphs that represented the number of Crimean War deaths due to combat, disease, and other causes. Each sector represented a different month and different colors indicated different causes of death.

4. Create a back-to-back stem plot summarizing the ages of fans sitting in section E at a high school football game. Then compare the two data sets.

Visitors: 12, 17, 19, 22, 24, 33, 37, 39, 41, 41, 48, 56, 58, 62, 63

Home: 14, 14, 15, 19, 20, 21, 27, 32, 36, 40, 42, 48, 48, 50, 51, 51, 58, 59, 61, 62, 62

Visitors		Home
9 7 2	1	4 4 5 9
4 2	2	0 1 7
9 7 3	3	2 6
8 1 1	4	0 2 8 8
8 6	5	0 1 1 8 9
3 2	6	1 2 2

Although the visitors had a wider range of ages, the home crowd was larger.

5. Construct a frequency distribution table and histogram to illustrate the following heights (in inches) of 17-yr-old girls.

60, 61, 61, 62, 62, 62, 64, 64, 64, 64, 64, 64, 64, 65, 65, 65, 65, 65, 65, 66, 66, 66, 68, 68, 70

Height (inches)	Tally	Frequency
60	I	1
61	II	2
62	III	3
64	ꟷꟷꟷ II	7
65	ꟷꟷꟷ I	6
66	III	3
68	II	2
70	I	1

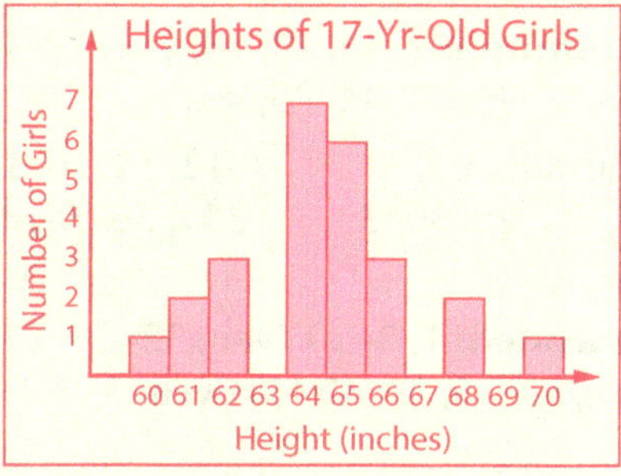

Heights of 17-Yr-Old Girls

Number of Girls

Height (inches)

Introduce dot plots and stem plots, and then discuss the back-to-back stem plot in Example 4.

Present the *frequency distribution table* and the histogram in Example 5. Having the students create a histogram on the graphing calculator is encouraged. All lists can be cleared using 4:ClrAllLists from the MEM ([2nd], [+]) menu.

When there is a large number of data elements, grouping the data into classes will makes it easier to analyze and visualize. Use *Sturges's formula* to create a *grouped frequency distribution table* in Example 6. Consider having the students create the histogram on a spreadsheet if they have already created one on a graphing calculator.

Example 7 uses additional columns in the frequency distribution table to create a *percentile graph*. Point out that the cumulative frequency column can also be used to find the median and other quartiles of the data; it will be used in Section 10.4 to find the variance and standard deviation of data.

6. Complete the following steps to organize and illustrate the ages of visitors to the art museum.

42, 50, 38, 40, 66, 51, 49, 31, 50, 64, 55, 46, 36, 44, 60, 44, 42, 32, 58, 70, 52, 59, 36, 47, 29, 51, 47, 32, 52, 57

a. Use Sturges's formula to find the number of classes for a grouped frequency distribution table and their corresponding widths.

6 classes

b. Create a frequency distribution table.

◢	A	B	C
1	Class	Tally	Freq
2	29–35	IIII	4
3	36–42	ЖI	6
4	43–49	ЖI	6
5	50–56	ЖII	7
6	57–63	IIII	4
7	64–70	III	3
8	Total		30

c. Use technology to create a histogram illustrating the frequency distribution.

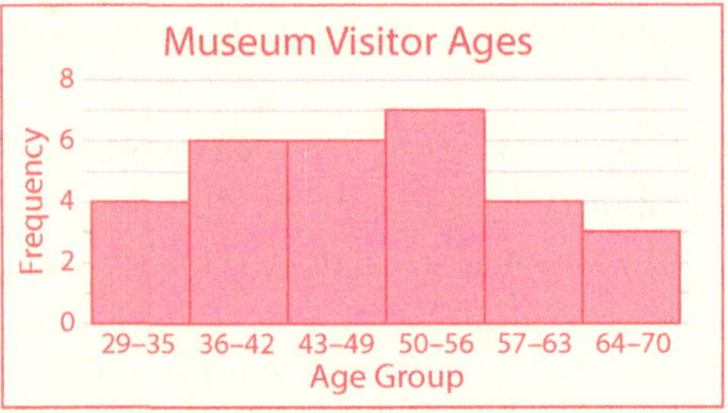

7. Create a percentile graph for the data from Additional Exercise 6.

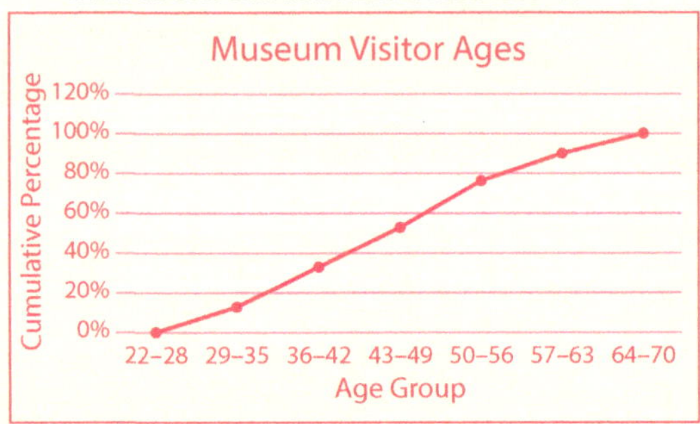

Assignments

- **Minimum:** 1–13, 16, 18, 21–22, 27, 32–33, 36, 42, 44–45, 48
- **Standard:** 1, 2, 4, 6, 8–12, 14, 16, 18, 21–22, 28, 30–31, 33, 35*a*, 36, 38, 42, 44–48
- **Extended:** 1; 2–20 even; 25–26, 29, 31, 34, 35*b*, 36, 39; 40–48 even

---

Create a back-to-back stem plot summarizing the ages of US presidents when they took office in the 1800s and the 1900s. Then compare the two data sets.

1800s: 47, 61, 46, 64, 48, 49, 51, 55, 57, 58, 56, 57, 54, 54, 50, 49, 57, 54, 51, 52, 55

1900s: 42, 51, 56, 55, 51, 54, 51, 60, 62, 43, 55, 56, 61, 52, 69, 64, 46

Answer

1800s	Stem	1900s
9 9 8 7 6	4	2 3 6
8 7 7 7 6 5 5 4 4 4 2 1 1 0	5	1 1 1 2 4 5 5 6 6
4 1	6	0 1 2 4 9

Use the tens digit as the stem for both data sets and the ones digits as the leaves. Notice that the digits for the leaves are arranged to increase as they extend farther from the stem.

There is a wider range of ages in the 1900s and the average age when assuming office seems to be higher in the 1900s.

SKILL ✔ **EXERCISES 15, 27**

Interval and ratio data can also be illustrated by a *histogram*. Begin by organizing the data in a *frequency distribution table*. Ungrouped frequency distribution tables are made for data sets whose *range* (the difference between the maximum and minimum values) is relatively small (often ≤ 15).

Construct a frequency distribution table and histogram to illustrate the following cockatiel weights (in grams).

$$W = \left\{ \begin{array}{l} 87, 83, 92, 86, 88, 86, 90, 85, 86, 85, \\ 88, 84, 87, 86, 89, 86, 90, 87, 86, 87 \end{array} \right\}$$

Answer

1. Since the range of the data is $92 - 83 = 9$, use an ungrouped frequency distribution table. List all the possible weights from 82 to 94 and tally their occurrences to determine the frequency for each weight.

Data x_i	Tally	Frequency f_i
83	I	1
84	I	1
85	II	2
86	ЖI	6
87	IIII	4
88	II	2
89	I	1
90	II	2
91		0
92	I	1

$$\sum f_i = 20$$

2. Construct a histogram placing the weights along the horizontal axis and frequency (or percent) along the vertical axis. Draw a bar to the frequency for each data value. Clearly label the axes and title the graph.

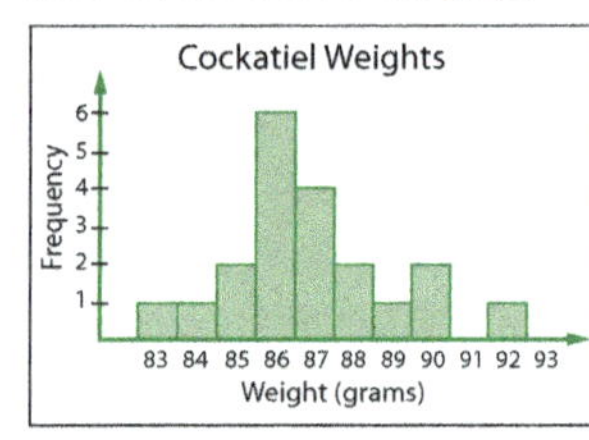

SKILL ✔ **EXERCISE 19**

To draw the histogram on your calculator, enter the 20 original data values in L₁. Define Plot1 and the [WINDOW] as illustrated with Xscl = 1.

```
Plot1 Plot2 Plot3
On  Off
Type: ⊯ ⬩ ⊞ ⬚ ⬚ ⬩
Xlist:L₁
Freq :1
```

Solutions

❯ A. Exercises

13.

14.

15.

4	9
5	
6	0
7	3 6
8	2 4 4 5
9	0 2

16.

2	6 7 9
3	1 2 4 7 8
4	6 8 9
5	1 2 5 7 8
6	1 2

A *grouped frequency distribution table* is usually used if the range of the data is greater than 15. Every data value is placed within classes that have the same width. Sturges's formula is often used to find an appropriate number of classes.

STURGES'S FORMULA

The number of classes, k, for a sample with n values is $k = 1 + 3.3(\log n)$, in which k is always rounded up to the next integer.

The width of each class, w, is then found using $w = \dfrac{\text{range}}{k}$, where w is also rounded up.

Example 6   Creating and Using a Grouped Frequency Distribution

Construct a frequency distribution table and histogram to illustrate the SAT® math scores received by students at Pinbrook Christian School.

291, 312, 321, 329, 365, 365, 382, 402, 417, 418, 421, 424, 428, 439, 443, 445, 463, 464, 465, 471, 473, 487, 492, 494, 501, 507, 510, 510, 511, 515, 518, 519, 521, 529, 538, 552, 557, 567, 572, 582, 584, 593, 607, 609, 612, 615, 641, 665, 682, 705

Answer

$1 + 3.3 \log n = 1 + 3.3 \log 50 \approx 6.6$
$\therefore k = 7$

1. Use Sturges's formula with $n = 50$ to determine the number of classes, rounding up to the next integer.

$\dfrac{\text{range}}{k} = \dfrac{705 - 291}{7} = 59.1; \therefore w = 60$

2. Find the width of each class.

3. A spreadsheet can be used to construct a frequency distribution table with 7 classes of width 60.

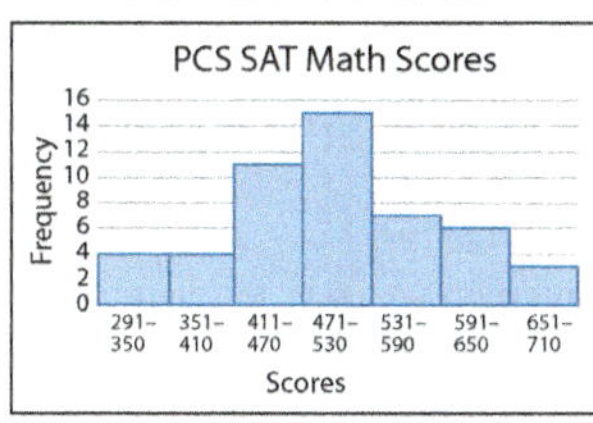

	A	B	C
1	Class	Tally	Freq
2	291–350	IIII	4
3	351–410	IIII	4
4	411–470	JHT JHT I	11
5	471–530	JHT JHT JHT	15
6	531–590	JHT II	7
7	591–650	JHT I	6
8	651–710	III	3
9	Total	$\Sigma f_i =$	50

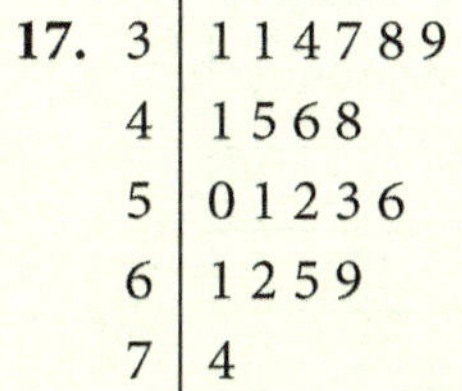

PCS SAT Math Scores

4. Draw the related histogram.

In a spreadsheet, highlight the class limits and frequency data and insert a bar graph. Add a title and label the axes.
Click on one of the bars to access the Series Options; then set the gap width to 0 and draw borders to separate the bars.

SKILL ✔ EXERCISE 33

A *percentile graph* or *ogive* (pronounced ō'jīve) is a line graph that represents the cumulative frequencies of the data as percentages.

TIP

To draw a similar histogram on the TI-84, use the midpoints of each class stored in L1 for the XList and the frequencies stored in L2 for the Freq. Use the class width as the Xscl.

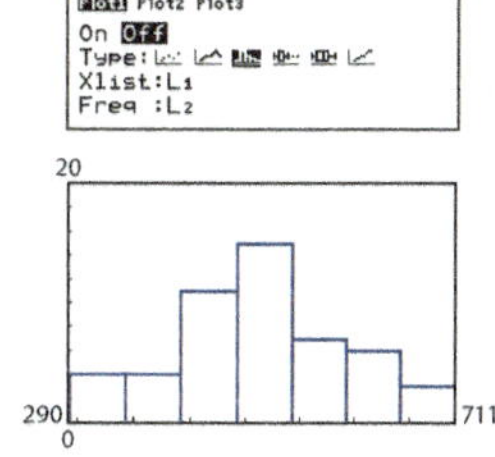

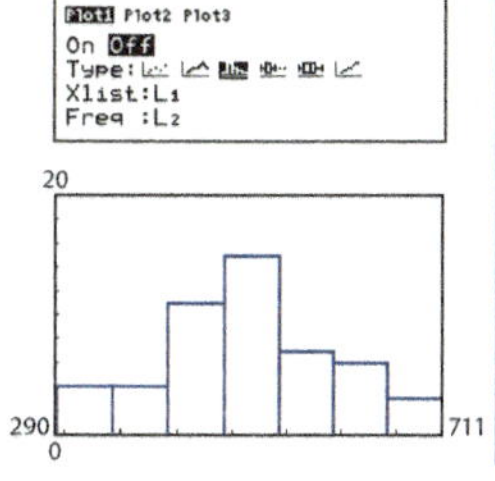

---

21. A: $0.2(360°) = 72°$
B: $0.35(360°) = 123°$
C: $0.25(360°) = 90°$
D: $0.15(360°) = 54°$
F: $0.05(360°) = 18°$

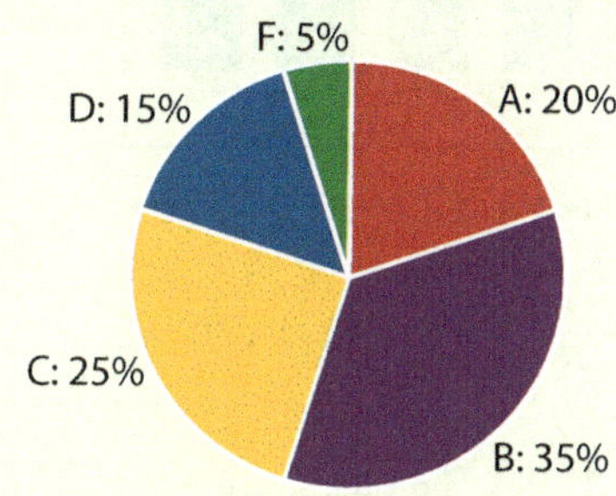

22. A: $0.2(20) = 4$
B: $0.35(20) = 7$
C: $0.25(20) = 5$
D: $0.15(20) = 3$
F: $0.05(20) = 1$

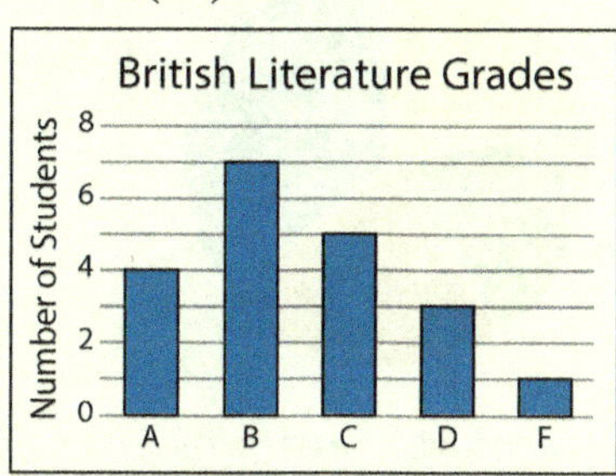

23.

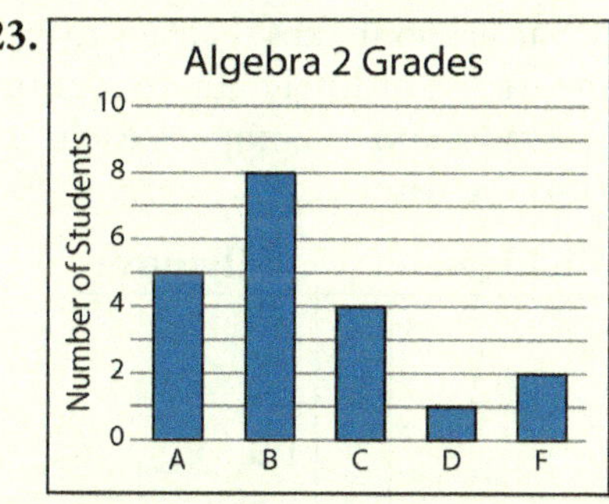

24. $n = 5 + 8 + 4 + 1 + 2 = 20$

A: $\dfrac{5}{20} = 0.25; \approx 90°$

B: $\dfrac{8}{20} = 0.4; \approx 144°$

C: $\dfrac{4}{20} = 0.2; \approx 72°$

D: $\dfrac{1}{20} = 0.05; \approx 18°$

F: $\dfrac{2}{20} = 0.1; \approx 36°$

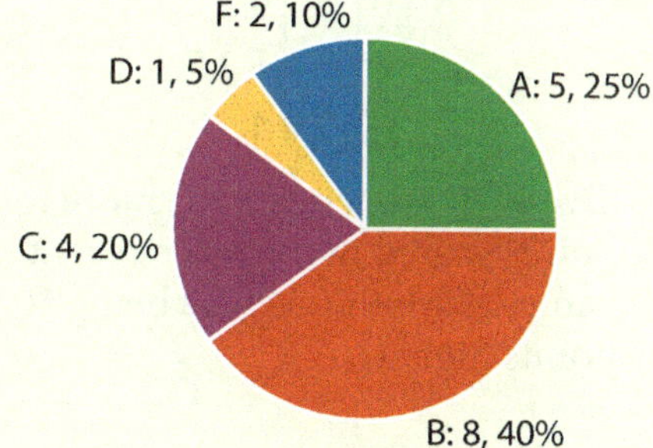

17.

3	1 1 4 7 8 9
4	1 5 6 8
5	0 1 2 3 6
6	1 2 5 9
7	4

18a. $n = 60 + 18 + 22 + 20 + 20$
$= 140$

$\dfrac{60}{140} \approx 42.9\%$

18b. $\dfrac{20}{140} \approx 14.3\%$

> B. Exercises

19.

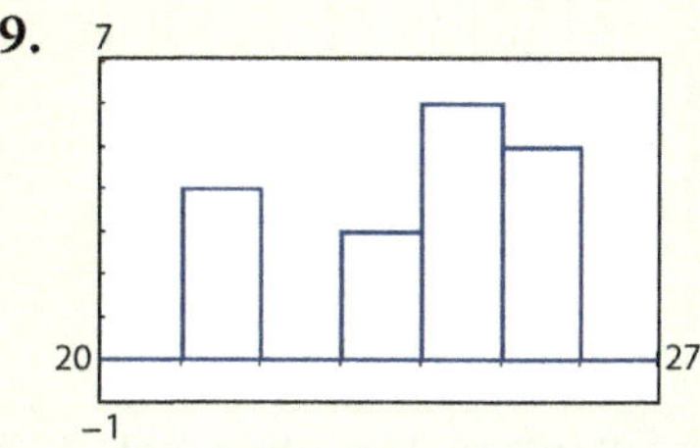

20.

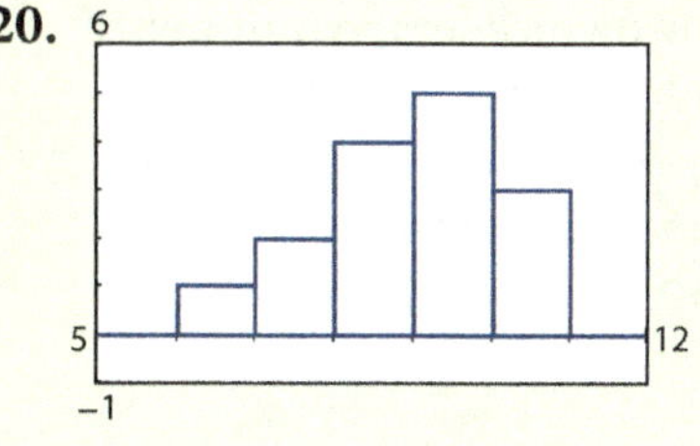

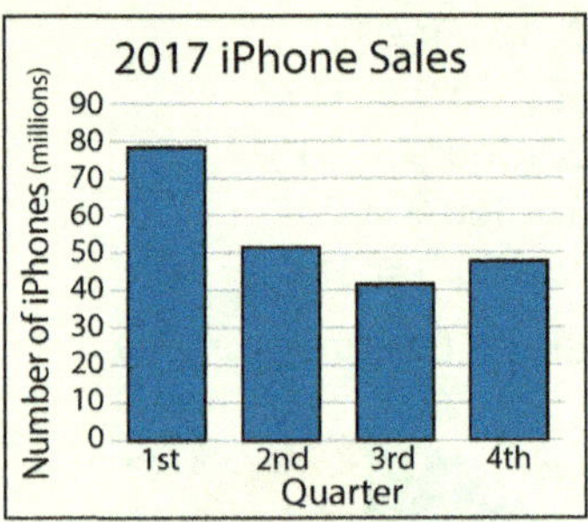

25.

$$n = 78.3 + 50.8 + 41.0 + 46.7$$
$$= 216.8$$

Q1: $\frac{78.3}{216.8} \approx 36.1\%; \approx 130°$

Q2: $\frac{50.8}{216.8} \approx 23.4\%; \approx 84°$

Q3: $\frac{41.0}{216.8} \approx 18.9\%; \approx 68°$

Q4: $\frac{46.7}{216.8} \approx 21.5\%; \approx 77°$

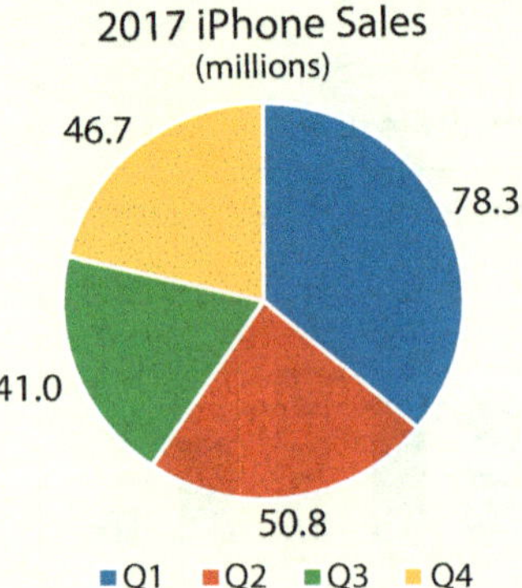

26. The bar graph makes it easier to compare the ordinal data, to identify any trends, and to compare each part to the whole.

27.

Michigan		Villanova
8	5	
1 2 4 9	6	
	7	1 9
	8	1 7
9	9	0 5

Answers will vary. Villanova generally scored more than Michigan.

28.

SB 1–20		SB 33–52
9 7 3 2 1	2	
8 7 7 1 0	3	1 1 4 7 8 9
7 7 7 6 5 4	4	1 5 6 8
6 4 0	5	0 1 2 3 6
6	6	1 2 5 9
	7	4

Overall there tends to be more total points scored in the more recent games (with the average being 10 points higher).

Example 7 Constructing a Percentile Graph

Construct a percentile graph for the data in Example 6.

Answer

1. Add Cumulative Frequency and Cumulative Percentage columns to the frequency distribution table and calculate the percentages.

	A	B	C	D
1	Class	Freq	Cumulative Frequency	Cumulative Percentage
2	291–350	4	4	8%
3	351–410	4	8	16%
4	411–470	11	19	38%
5	471–530	15	34	68%
6	531–590	7	41	82%
7	591–650	6	47	94%
8	651–710	3	50	100%

2. Plot ordered pairs representing each class's upper limit and its cumulative percentage. Add a point at (290, 0) to show that there are no scores below 291. Connect the points to make the line graph.

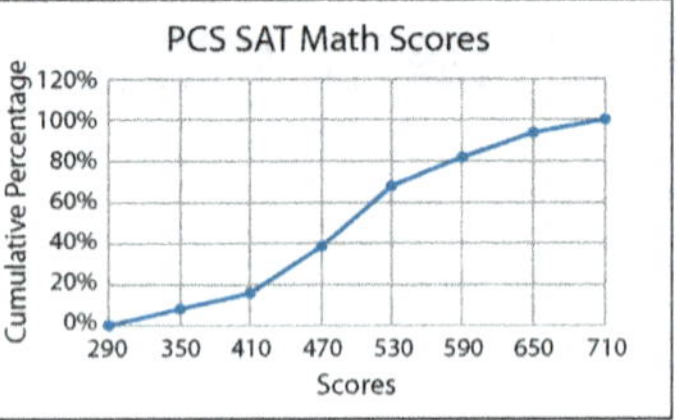

SKILL ✔ EXERCISE 37

A. Exercises

1. A __parameter__ represents a characteristic of a population while a __statistic__ measures a characteristic of a sample.

Identify each group as a population or a sample.

2. all the students taking the SAT in the spring of 2020   population

3. 100 randomly selected 16-yr-olds in Ohio   sample

4. the students in 1 of 6 randomly selected sections of Calculus I   sample

State whether each of the following is most likely a parameter or a statistic.

5. the favorite sport of 25 randomly selected students from Riverside High   statistic

6. the average height of a 17-yr-old boy in America   parameter

7. the median household income of residents in California   parameter

8. the range of scores on 1 of the 10 Precalculus quizzes   statistic

Identify the level of measurement.

9. the day of the week   interval

10. the number of questions answered correctly on a math test   ratio

11. class salutatorian   ordinal

12. blood type   nominal

Create a dot plot for each data set.

13. heights of students in a fifth-grade class:
51, 55, 52, 54, 55, 54, 53, 52, 55,
54, 54, 52, 56, 55, 54, 55, 56, 55

14. number of pets owned by classmates:
1, 0, 3, 2, 2, 1, 6, 2, 1, 0, 3,
2, 0, 5, 4, 2, 1, 1, 3, 0, 4

Create a stem-and-leaf plot for each data set.

15. grades from an AP History test:
85, 92, 76, 49, 84, 82, 73, 60, 84, 90

16. ages of employees:
37, 32, 49, 27, 62, 57, 46, 31, 26,
48, 51, 55, 52, 34, 61, 58, 38, 29

17. point totals in Super Bowls XXXIII through LII:
53, 39, 41, 37, 69, 61, 45, 31, 46, 31
50, 48, 56, 38, 65, 51, 52, 34, 62, 74

18. Use the survey results represented by the bar graph to find each percentage of the sample.

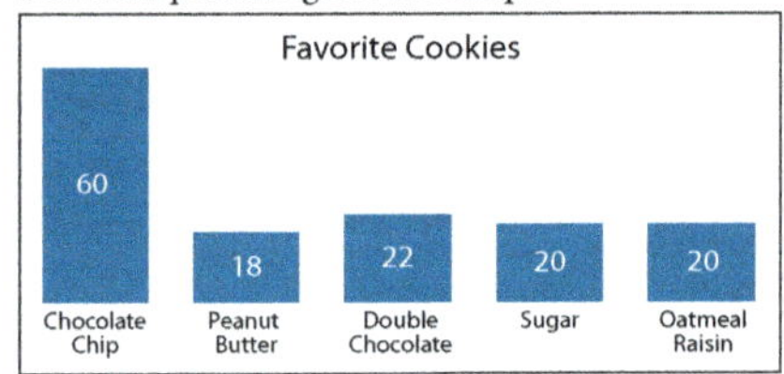

a. those preferring chocolate chip cookies   42.9%

b. those preferring oatmeal raisin cookies   14.3%

29.

Payton		Brown
	1	0	
5 7 7 8 8	0	9
1 1 1 1 3	1	0 0 1 4
6 6	1	5 8 8
	2	1

While Payton played more seasons, Brown's average number of touchdowns per season is higher (14 vs. 9.6). The fact that he accomplished this in fewer games only adds to his accomplishment.

31a.

Number of Vehicles	Frequency
0	2
1	18
2	11
3	4
4	3
5	2
Total	40

Draw a histogram representing each frequency distribution table.

19.

Students in a Class	Freq.
21	4
22	0
23	3
24	6
25	5

20.

Quiz Scores	Freq.
10	3
9	5
8	4
7	2
6	1

Represent the British Literature grades of 20 students with the indicated graph.

Grades	Percent of Students
A	20%
B	35%
C	25%
D	15%
F	5%

21. pie chart

22. bar graph showing the number of students per grade

Represent the Algebra 2 grades with the indicated graph.

Grades	Number of Students
A	5
B	8
C	4
D	1
F	2

23. bar graph

24. pie chart

25. Make a bar graph and a pie chart to illustrate the number of 2017 iPhone sales.

Quarter	Millions
1st	78.3
2nd	50.8
3rd	41.0
4th	46.7

26. Which graph from exercise 25 better illustrates the quarterly sales? Explain your reasoning.

27. Make a back-to-back stem plot for the top two teams in the 2018 NCAA basketball tournament. Then compare the two teams's scoring.

Michigan: 61, 64, 99, 58, 69, 62
Villanova: 87, 81, 90, 71, 95, 79

28. Make a back-to-back stem plot for the point totals in Super Bowls I–XX (listed below) and the point totals in Super Bowls XXXIII–LII (listed in exercise 17). Compare the two sets of data.
45, 47, 23, 30, 29, 27, 21, 31, 22, 38, 46, 37, 66, 50, 37, 47, 44, 47, 54, 56

29. Make a split back-to-back stem plot for each player's total touchdowns scored per season played. The first stem should represent 0–4, the second should represent 5–9, and so on. Then compare their statistics. How does the fact that Brown played, on average, one fewer game per season affect your comparison?
Walter Payton: 7, 13, 16, 11, 16, 7, 8, 1, 8, 11, 11, 11, 5
Jim Brown: 10, 18, 14, 11, 10, 18, 15, 9, 21

30. Describe the error in each set of classes for a frequency distribution table.

a.

Ages	The classes are not the same width.
0–5	
6–10	
11–15	
16–20	
21–25	
26–30	
31–35	

b.

Ages	The classes overlap.
20–30	
30–40	
40–50	
50–60	
60–70	
70–80	

31b.

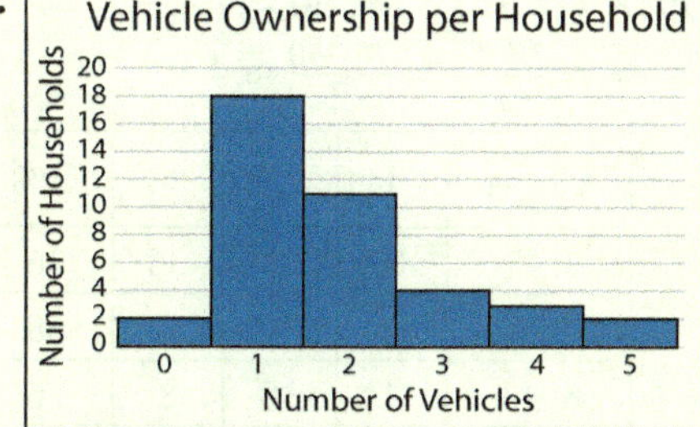

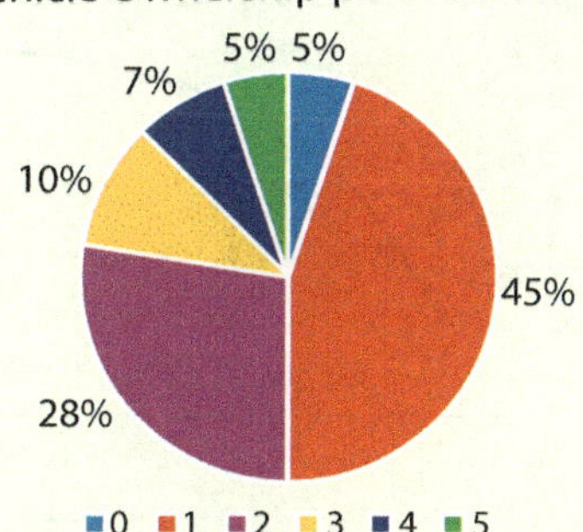

31c. $0: \dfrac{2}{40} = 0.05$

$1: \dfrac{18}{40} = 0.45$

$2: \dfrac{11}{40} = 0.275$

$3: \dfrac{4}{40} = 0.1$

$4: \dfrac{3}{40} = 0.075$

$5: \dfrac{2}{40} = 0.05$

31d. The histogram makes it easy to compare the amounts, while the pie chart allows comparison values as percentages.

32.

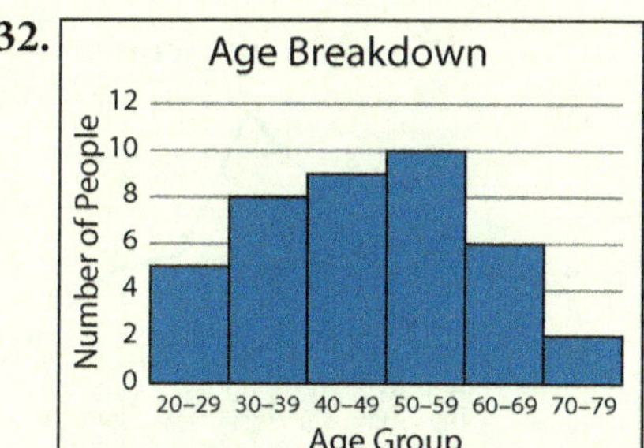

33a. $k = 1 + 3.3 \log 25 \approx 6$

$w = \dfrac{range}{k} = \dfrac{42 - 16}{6} \approx 5$

33b.

Class	Frequency
15–19	1
20–24	1
25–29	11
30–34	8
35–39	3
40–44	1
Total	25

33c.

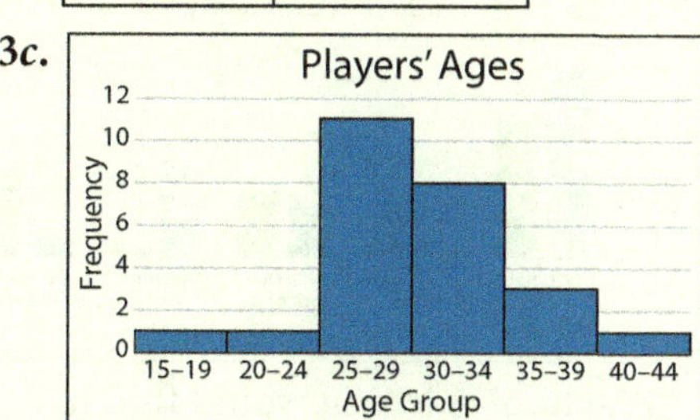

34a. $k = 1 + 3.3 \log 30 \approx 6$

$w = \dfrac{range}{k} = \dfrac{48 - 7}{6} \approx 7$

34b.

Bridges	Frequency
7–13	11
14–20	10
21–27	7
28–34	1
35–41	0
42–48	1
Total	30

34c.

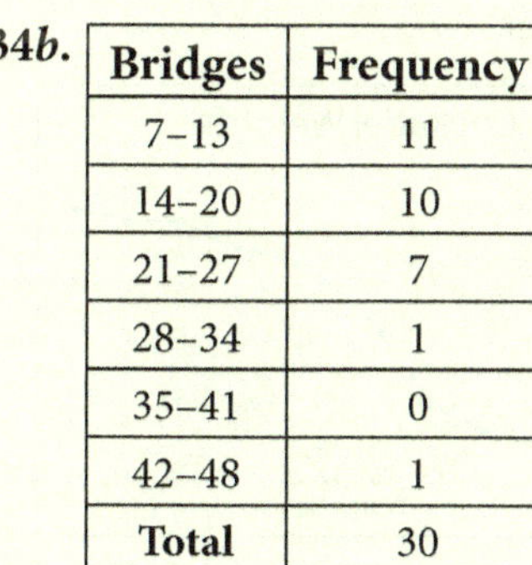

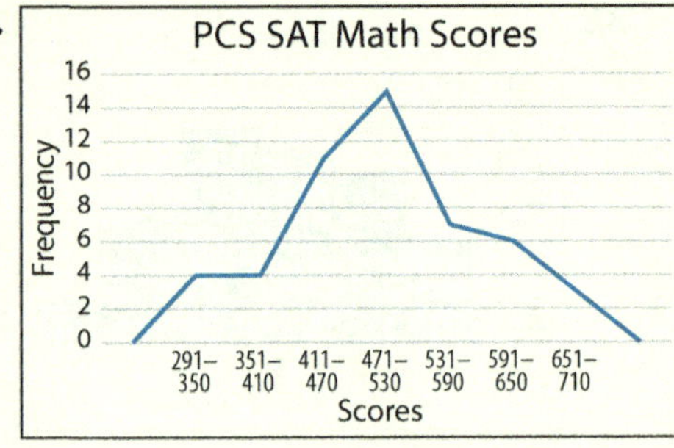

35a.

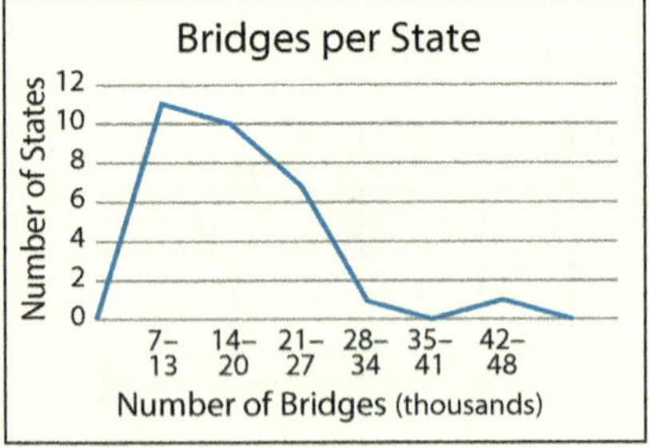

35b.

> ## C. Exercises

36.

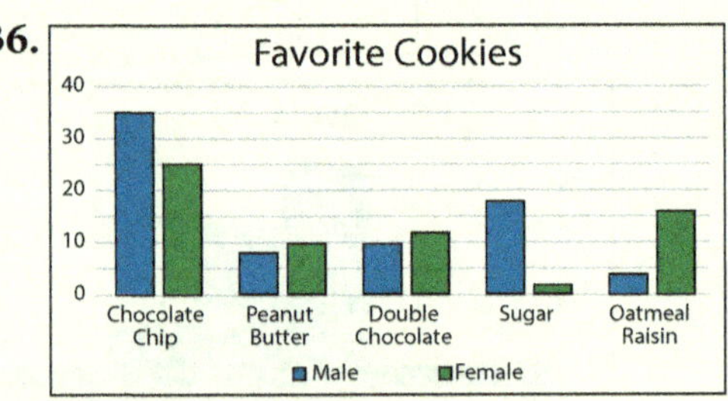

A school cafeteria would find it helpful to plan their dessert menu. A class could use this survey to determine which cookies to sell for a fundraiser.

37.

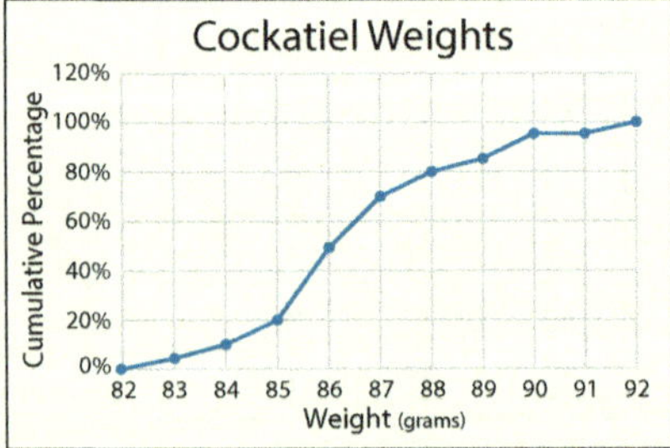

38.

x_i	f_i	Σf_i	Σp_i
0	2	2	5%
1	18	20	50%
2	11	31	77.5%
3	4	35	87.5%
4	3	38	95%
5	2	40	100%
Total	40		

31. The numbers of vehicles owned as reported in a random sampling of 40 households are listed below.

5, 1, 1, 2, 0, 1, 1, 2, 1, 1,
1, 3, 3, 0, 2, 5, 1, 2, 3, 4,
2, 1, 2, 2, 1, 2, 2, 1, 1, 1,
4, 2, 1, 1, 2, 1, 1, 4, 1, 3

a. Make a frequency distribution table to organize the data.

b. Draw a histogram representing the data.

c. Draw a pie chart representing the data.

d. Which graph better illustrates the data? Explain your reasoning.

32. Use technology to create a histogram illustrating the grouped frequency distribution table.

Ages	Freq.
20–29	5
30–39	8
40–49	9
50–59	10
60–69	6
70–79	2
Total	**40**

33. Complete the following steps to organize and illustrate the following ages of players in a recreational league soccer game.

16, 26, 26, 35, 28,
29, 22, 30, 28, 31,
33, 35, 42, 37, 27,
30, 32, 29, 26, 31,
34, 27, 31, 26, 28   6 classes; $w = 5$

a. Use Sturges's formula to determine the appropriate number of classes for a grouped frequency distribution table and their corresponding widths.

b. Create a frequency distribution table with the first class having a lower limit of 15.

c. Use technology to create a histogram illustrating the frequency distribution.

34. The approximate numbers of bridges (in thousands) in 30 different states are listed below.

7, 11, 14, 18, 24, 24, 14, 14, 17, 28,
8, 7, 13, 22, 26, 13, 13, 7, 19, 17, 12,
26, 48, 16, 23, 11, 17, 9, 25, 15   6 classes; $w = 7$

a. Use Sturges's formula to determine the appropriate number of classes for a grouped frequency distribution table and their corresponding widths.

b. Make a frequency table from the information provided and the results from part a.

c. Use technology to create a histogram illustrating the distribution.

35. Frequency polygons are line graphs that are made by connecting the midpoints of the top of each bar in a histogram. Create a frequency polygon for the data from the following.

a. Example 6   **b.** exercise 34

> ## C. Exercises

36. Create a double bar graph from the favorite cookie survey data. Then explain how the data from this survey might be valuable to a school.

Cookie Flavor	Males	Females
chocolate chip	35	25
peanut butter	8	10
double chocolate	10	12
sugar	18	2
oatmeal raisin	4	16

37. Create a percentile graph for the cockatiel weight distribution given in Example 5.

38. Create a percentile graph for the data from the survey of vehicles owned per household in exercise 31.

39. The numbers of NCAA Division I schools per state are listed below. Using Sturges's formula, create a frequency distribution table, histogram, and cumulative percentile graph for the data. Use Sturges's formula to find the appropriate number of groups and their corresponding widths.

9, 0, 3, 5, 24, 5, 7, 2, 13, 7,
1, 3, 13, 10, 4, 3, 7, 12, 1, 9,
6, 7, 1, 6, 5, 2, 3, 2, 2, 8, 2,
22, 18, 2, 13, 4, 4, 14, 4, 12,
2, 12, 21, 6, 1, 14, 5, 2, 4, 1

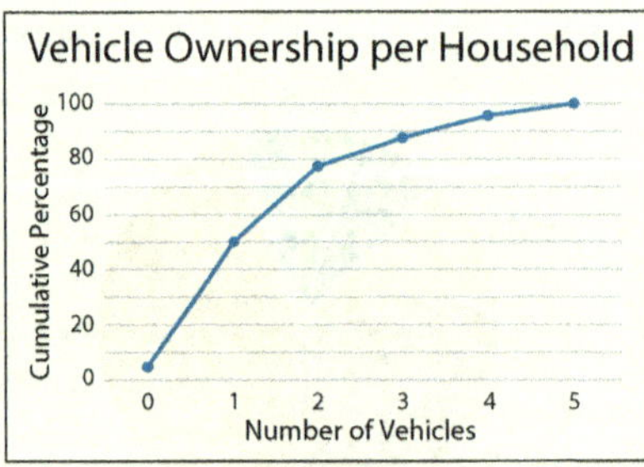

39. $k = 1 + 3.3 \log 50 \approx 7$

$$w = \frac{range}{k} = \frac{24 - 0}{7} \approx 4$$

Class	f	Σf	%	Σ%
0–3	18	18	36	36
4–7	16	34	32	68
8–11	4	38	8	76
12–15	8	46	16	92
16–19	1	47	2	94
20–23	2	49	4	98
24–27	1	50	2	100
Total	50		100	

Use a half-angle identity to find the exact value of each expression. [5.5]

40. $\sin \frac{7\pi}{12}$ $\frac{\sqrt{2+\sqrt{3}}}{2}$

41. $\tan \frac{7\pi}{12}$ $-2-\sqrt{3}$

42. Write the recursive formula and a simplified explicit formula for the sequence 7, 10, 13, 16, 19, ... [9.2]

43. Write an explicit formula for the geometric sequence with $a_3 = 45$ and $a_6 = 1215$ and use the formula to find a_{10}. [9.3] $a_n = \frac{5}{3}(3)^n;\ 98,415$

Marbles are drawn from a bag that contains 5 red, 7 blue, and 8 green marbles. Express each probability below to the nearest tenth of a percent. [10.2]

44. What is the probability that the first 3 marbles drawn (without replacement) are red? 0.9%

45. What is the probability that the first 3 marbles drawn (with replacement) are all blue or green? 42.2%

46. Which of the following is NOT true for $y = \cot x$? [4.5]
A. period: $p = \pi$ D
B. VA: $x = n\pi, n \in \mathbb{Z}$
C. zeros: $x = (2n + 1)\frac{\pi}{2}, n \in \mathbb{Z}$
D. It is an increasing function.
E. all of these

42. $a_1 = 7;\ a_n = a_{n-1} + 3;\ a_n = 3n + 4$

47. Which describes the graph of $r = \dfrac{5}{2 + 3\cos\theta}$? [8.6] E
A. line **C.** ellipse **E.** hyperbola
B. circle **D.** parabola

48. How many different ways can 5 of 7 pictures be arranged on a shelf? [10.1] D
A. 21 **C.** 120 **E.** 5040
B. 25 **D.** 2520

49. Which of the following is a trigonometric identity? [5.5] A
A. $\cos(\alpha - \beta) = \sin\alpha\sin\beta + \cos\alpha\cos\beta$
B. $\cos(\alpha + \beta) = \sin\alpha\sin\beta - \cos\alpha\cos\beta$
C. $\sin(\alpha + \beta) = \sin\alpha\cos\beta - \cos\alpha\sin\beta$
D. $\sin(\alpha - \beta) = \cos\alpha\sin\beta - \sin\alpha\cos\beta$
E. all of these

40. $\sin \dfrac{7\pi}{12} = \sin \dfrac{\frac{7\pi}{6}}{2}$ (positive in Q II)

$$= \sqrt{\frac{1 - \cos\frac{7\pi}{6}}{2}} = \sqrt{\frac{1}{2}\left(1 + \frac{\sqrt{3}}{2}\right)}$$

$$= \sqrt{\frac{1}{2} + \frac{\sqrt{3}}{4}} = \sqrt{\frac{2+\sqrt{3}}{4}} = \frac{\sqrt{2+\sqrt{3}}}{2}$$

41. $\tan \dfrac{7\pi}{12} = \tan \dfrac{\frac{7\pi}{6}}{2}$

$$= \frac{1 - \cos\frac{7\pi}{6}}{\sin\frac{7\pi}{6}} = \frac{1 - \left(-\frac{\sqrt{3}}{2}\right)}{-\frac{1}{2}}$$

$$= \frac{\frac{2+\sqrt{3}}{2}}{-\frac{1}{2}} = -2 - \sqrt{3}$$

42. arithmetic with $d = 10 - 7 = 3$
$a_1 = 7;\ a_n = a_{n-1} + 3$
$a_n = a_1 + (n - 1)d$
$\quad = 7 + (n - 1)(3)$
$\quad = 7 + 3n - 3 = 3n + 4$

43. $a_6 = a_3 r^{6-3}$
$1215 = 45r^3$
$\quad r^3 = 27;\ r = 3$
$a_n = 45(3)^{n-3} = 45(3)^n(3)^{-3}$
$\quad = \frac{5}{3}(3)^n$ or $5(3)^{n-1}$
$a_{10} = \frac{5}{3}(3)^{10} = 98,415$

44. $\dfrac{5}{20} \cdot \dfrac{4}{19} \cdot \dfrac{3}{18} = \dfrac{1}{114} \approx 0.9\%$

45. $\dfrac{3}{4} \cdot \dfrac{3}{4} \cdot \dfrac{3}{4} = \dfrac{27}{64} \approx 42.2\%$

47. $r = \dfrac{5 \div 2}{(2 + 3\cos\theta) \div 2} = \dfrac{2.5}{1 + 1.5\cos\theta}$
$e = 1.5; \therefore$ a hyperbola

48. $_7P_5 = \dfrac{7!}{2!} = 2520$

49. $\cos(\alpha \pm \beta) = \cos\alpha\cos\beta \mp \sin\alpha\sin\beta$
$\sin(\alpha \pm \beta) = \sin\alpha\cos\beta \pm \cos\alpha\sin\beta$

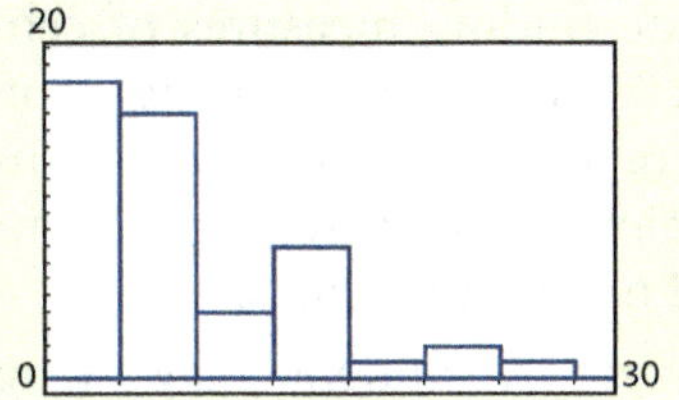

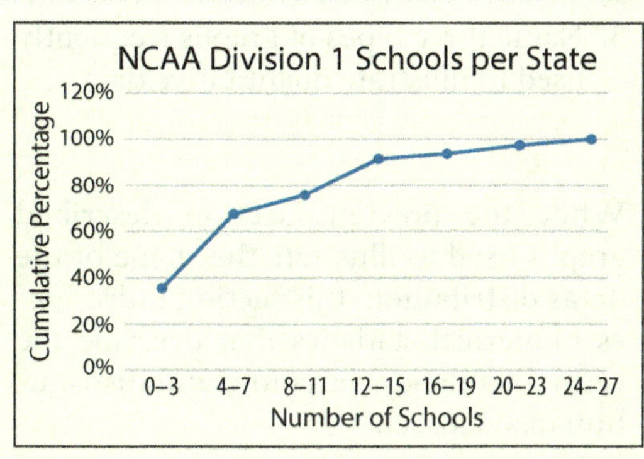

10.4 Describing Data Numerically

Objectives

1. To calculate measures of central tendency: median, mode, mean
2. To calculate measures of variability for samples and populations: range, interquartile range, variance, and standard deviation
3. To use a boxplot to illustrate the distribution of data
4. To describe the distribution of data as symmetric, skewed right, or skewed left
5. To calculate quartiles and percentiles
6. To choose appropriate descriptive statistics based on the data distribution

Flash

The professional baseball league right fielder (1954–76) Hank Aaron had a career batting average of 0.305. He passed Babe Ruth's record with 755 runs (second only to Barry Bonds). He held the home run record for 33 yr. In 1977 Hank Aaron's 44 jersey number was retired by the Atlanta Braves.

Vocabulary

box plot
deviation

Statistics play an ever-increasing role in sports.

<table>
<tr><td>After completing this section, you will be able to</td></tr>
</table>

After completing this section, you will be able to

- calculate measures of central tendency.
- calculate measures of variability.
- illustrate the distribution of data using a box plot.
- calculate quartiles and percentiles.
- choose appropriate descriptive statistics based on the data's distribution.

10.4 Describing Data Numerically

The previous section presented several ways to organize and illustrate a set of data. This section discusses several statistics that help describe the overall shape of the distribution of data by numerically describing its central tendency and variation. The mean, median, and mode are measures of central tendency that describe the typical or middle value of a set of data.

DEFINITIONS

The **mode** is the most frequently appearing value(s) in a data set.
The **median** is the middle value when the data is arranged in ascending or descending order (or the average of the two middle values when there is an even number of values).
The **mean** is the sum of the values divided by the number of values.

Population Mean	Sample Mean
$\mu = \dfrac{\sum\limits_{i=1}^{N} x_i}{N}$	$\bar{x} = \dfrac{\sum\limits_{i=1}^{n} x_i}{n}$

Notice that the Greek letter μ (mu) represents the population's mean (or average), while $\bar{x}$ (x bar) represents the sample's mean. The bar over the x distinguishes the sample mean from an individual value of the variable x. The total number of values is notated as N for a population and n for a sample. In general, parameters are represented by Greek letters, while statistics are represented by Latin letters.

Example 1 Finding Measures of Central Tendency

Find the sample mean, median, and mode of the following scores.
$S = \{24, 15, 20, 17, 25, 19, 20, 21, 24, 17, 24\}$

Answer

mean: $\bar{x} = \dfrac{\sum\limits_{i=1}^{11} x_i}{n} = \dfrac{226}{11} \approx 20.5$

$\{15, 17, 17, 19, 20, 20, 21, 24, 24, 24, 25\}$
The median is the 6th value, 20.

mode: 24

1. Find the mean by dividing the sum of the scores by the number of scores, $n = 11$.
2. Arrange the scores in ascending order and find the middle (the $\dfrac{n+1}{2}$ th) value.
3. The most frequently occurring value is 24.

SKILL ✔ EXERCISE 3

Multiple modes can occur. If Example 1 had one less value of 24, there would be three modes: 17, 20, and 24. If no two data points are the same, then no mode exists.

PRESENTATION

Lesson Opener

Write a brief statement comparing each set of terms.

1. population, sample A sample is a representative subset of a larger population.
2. parameter, statistic A parameter is a characteristic value for a population, and a statistic is a characteristic value from a sample of the population that is used to estimate corresponding parameters of the population.
3. Name three types of graphs frequently used to illustrate quantitative data.
dot plot, stem plot, histogram, and line graph (any three)

While the previous section described graphs used to illustrate the shape of the data's distribution, this section indroduces numerical statistics that describe the central tendency, variation, and distribution of a data set.

Define *mode*, *median*, and *mean*, and then state the formulas for the population mean and sample mean. Point out that, in general, parameters are represented with Greek letters, while statistics are represented with Latin letters. Use Example 1 to illustrate finding measures of central tendency. Consider introducing another measure of central tendency, the midrange, which is the average of the minimum and maximum values.

One-on-One A student may struggle with the difference between the mean and the median. Review that the mean is the average and the median is the middle value.

Introduce the terms related to variability and the definitions for *variance* and *standard deviation*. Point out that the division by $n - 1$ in the sample variance formula provides a more accurate estimate. The processes for finding variance

While the mean and median of each data set A, B, and C are both 6, the values are spread out in different ways around the mean. The *range, variance, standard deviation,* and *interquartile range* describe how far the values are spread apart in the distribution. The simplest measure of variability, the range, was defined in Section 10.3 as the difference between the maximum and minimum values. The range for A is 4 while the range for both B and C is 8. While this shows that the data in A is less spread out, it does not reveal that the data in C is more clustered at the mean than the data in B.

$$A = \{4, 5, 6, 7, 8\}$$
$$B = \{2, 4, 6, 8, 10\}$$
$$C = \{2, 6, 6, 6, 10\}$$

The *deviation* of each data value is its difference from the mean, $x_i - \bar{x}$ or $x_i - \mu$. Since the sum of the deviations is always 0 (see exercise 35), an average of the deviations is meaningless. Instead, an average of the squares of the deviations is calculated and the square root of this average, the standard deviation, provides a measure of variableness for the data. The sum of the squares of the deviations, either $\sum_{i=1}^{n} (x_i - \bar{x})^2$ or $\sum_{i=1}^{N} (x_i - \mu)^2$, is often called the *sum of the squares* and is often abbreviated as *SS*.

DEFINITIONS

The **variance** is an average of the squared deviations.
The **standard deviation** is the square root of the variance.

Population Variance	Sample Variance
$\sigma^2 = \dfrac{\sum_{i=1}^{N} (x_i - \mu)^2}{N}$ or $\dfrac{SS}{N}$	$s^2 = \dfrac{\sum_{i=1}^{n} (x_i - \bar{x})^2}{n-1}$ or $\dfrac{SS}{n-1}$
Population Standard Deviation	**Sample Standard Deviation**
$\sigma = \sqrt{\dfrac{\sum_{i=1}^{N} (x_i - \mu)^2}{N}}$	$s = \sqrt{\dfrac{\sum_{i=1}^{n} (x_i - \bar{x})^2}{n-1}}$

Note that the sample variance and sample standard deviation are divided by $n - 1$ instead of n. The fact that this allows the statistic to more accurately predict the actual variance and standard deviation of the population is proved in higher-level statistics classes.

Example 2 Finding the Sample Variance and Standard Deviation

Calculate the sample variance and standard deviation for data set $B = \{2, 4, 6, 8, 10\}$.

Answer
$\bar{x} = 6$

Values	Deviation^2
x_i	$(x_i - \bar{x})^2$
10	$(10 - 6)^2 = 16$
8	$(8 - 6)^2 = 4$
6	$(6 - 6)^2 = 0$
4	$(4 - 6)^2 = 4$
2	$(2 - 6)^2 = 16$

$SS = \sum_{i=1}^{5} (x_i - \bar{x})^2 = 40$

1. Find the mean.

2. Make a table to organize the calculation and recording of the squared deviations.

3. Find the sum of the squared deviations.

CONTINUED ➡

In the 1930s Walter A. Shewhart of Bell Labs developed process-control charts that are still widely used in many manufacturing processes. The relationship between the mean of a random sample and the mean of the population is used to determine whether a process is within normal variance limits or is out of control.

---

five-number summary
interquartile range
mean
median
mode
outlier
percentile
quartile
range
resistant
standard deviation
sum of the squares
variance

Additional Exercises

Use the following test scores for Additional Exercises 1–5.

87, 52, 87, 78, 65, 91, 77, 84, 93, 87, 89, 72

1. Find the mean, median, and mode.
80.2; 85.5; 87

2. Find the sample variance and standard deviation.   147.2; 12.1

3. Identify the outlier. Find the mean, median, and interquartile range and draw a box plot for the data with the outlier included and a box plot with the outlier excluded.
52; with: 80.2, 85.5, 13.5
without: 82.7, 87, 12

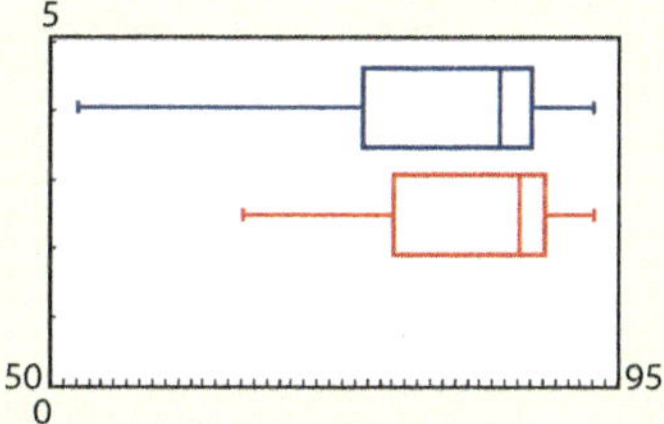

---

and standard deviation are presented in Example 2. Consider demonstrating how to use a spreadsheet to complete the example.

One-on-One Have the student create a summary of the steps in Example 2 that can serve as a guide for the exercises.

1. Find the mean.

2. Find the deviation of each value from the mean and square it.

3. Find *SS*, the *sum of the squares*.

4. Find the variance by dividing by $n - 1$ for a sample or by N for a population.

5. Find the standard deviation by taking the square root of the variance.

Use the Historical Connection margin note to emphasize the importance of statistics in areas such as manufacturing and engineering.

A spreadsheet is a great tool for data analysis and manipulation. The statistics in Example 3 can be found with a graphing calculator, a spreadsheet, or an online app. Having the students create a Microsoft Excel or Google Sheets spreadsheet allows them to focus on the relationships without being distracted by arithmetic calculations and provides valuable experience using these powerful tools.

Introduce *box plot, five-number summary,* and associated terms. Explain what

an *outlier* is and consider briefly discussing how it is often desirable to omit outliers because they usually do not represent the typical values. For example, including the salary of a professional athlete in a sample when finding the average income of 24-yr-old males would significantly affect the sample mean. The fact that outliers can be intentionally included or excluded to change the distribution in a desired direction emphasizes the need for mathematicians and statisticians with a biblical worldview.

Explain that outliers are often easy to identify without doing any math, but sometimes they are not obvious. Use

4. Find a student's score if the score is at the 90th percentile. **92.4**

5. How is the distribution of scores skewed after the outlier is removed from the data set? Explain.
negatively; The mean is less than the median and there is a left tail.

6. Complete the frequency distribution table for the sample heights (in inches) and then find the mean and the standard deviation. $\bar{x} = 66.125$ **$s = 4.62$**

x_i	f_i	$f_i x_i$	$f_i(x_i - \bar{x})^2$
58	1	58	66.02
60	3	180	112.55
61	1	61	26.27
62	2	124	34.03
64	3	192	13.55
66	4	264	0.06
68	1	68	3.52
70	3	210	45.05
71	4	284	95.06
73	2	146	94.53

$$s^2 = \frac{\sum_{i=1}^{n}(x_i - \bar{x})^2}{n - 1} = \frac{40}{5 - 1} = 10$$

$$s = \sqrt{10} \approx 3.16$$

4. Find the sample variance by dividing the total by $(n - 1)$.

5. The standard deviation is the square root of the variance.

Similar calculations for set $C = \{2, 6, 6, 6, 10\}$ produce $SS = 16 + 0 + 0 + 0 + 16 = 32$, $s^2 = \frac{32}{4} = 8$, and $s \approx 2.83$. This lower standard deviation indicates that more data values are closer to the mean in set C than in set B.

Since the calculations can become tedious, technology is often used to automate the computations.

Example 3   Using a Frequency Distribution

Use a spreadsheet to find the mean and population standard deviation of the class's test scores.

Answer

	A	B	C	D
1	Score (x_i)	Freq (f_i)	$x_i f_i$	$f_i(x_i - x_{bar})^2$
2	73	1	73	157.75
3	75	2	150	223.03
4	78	1	78	57.15
5	80	3	240	92.74
6	83	5	415	32.77
7	85	2	170	0.63
8	88	3	264	17.86
9	90	2	180	39.43
10	93	2	186	110.71
11	95	3	285	267.34
12	98	1	98	154.75
13		n	Sum	Sum
14		25	2139	1154.16
15		Mean:	85.56	
16			Variance:	46.1664
17			Standard Deviation:	6.79

Test Scores

73, 75, 75, 78, 80, 80, 80, 83, 83, 83, 83, 83, 85, 85, 88, 88, 88, 90, 90, 93, 93, 95, 95, 95, 98

1. Enter the headings in row 1, the data values in cells A2 through A12, the frequencies in cells B2 through B12, and the headings in row 13.

2. Find each product $x_i f_i$ by entering the formula =B1*A1 in cell C2 and then dragging the fill handle (small square in the lower right corner of the cell) down to fill in the formulas for C3 through C12.

3. Find the mean by entering =SUM(B2:B12) in cell B14, =SUM(C2:C12) in cell C14, and then =C14/B14 in cell C15. Enter a label for the mean in cell B15.

4. Calculate the squares of the deviations by entering =B2*(A2−C15)^2 in cell D2 and then dragging the fill handle down to fill in the formulas for D3 through D12.

5. Calculate the sum of the squares by entering =SUM(D2:D12) in cell D14.

6. Find the sample variance, s^2, by entering =D14/B14 in cell D16 and find the standard deviation, s, by entering =SQRT(D16) in cell D17; label these values appropriately.

The median value in Example 3, a test score of 85, divides the ordered list of data into two parts, each containing the same amount of data. *Quartiles* divide the ordered data into four sets containing the same amount of data. After finding the median (the second quartile, Q_2), find Q_1 and Q_3 by finding the median of the lower and upper halves of the data. When finding quartiles for data sets with an odd number of values, do not include the median in either the upper or lower half of the data. For the test scores in Example 3, $Q_1 = 80$ and $Q_3 = 91.5$. A *box plot* (often called a box-and-whisker plot) is used to illustrate

---

> **TIP**
>
> The \$ symbols prevent the reference to B12 from changing when the formula is copied. Otherwise cell references are copied relative to the cell's position in the spreadsheet.

Example 4 to find measures of central tendency and variability, and to identify an outlier. While removing an outlier reduces the *range* of the data and often decreases the *interquartile range (IQR)*, it may actually increase the *IQR* in some data sets.

Discuss the Data Analysis margin note and communicate the extent to which students may use technology on their assessments. Functions returning specific statistics for a list of data can be found in the [LIST] ([2nd], [STAT]), MATH submenu.

```
median(L1)
                         61.5
```

Engage the students by asking if they can explain what an 85th *percentile* ranking means on a college entrance exam. (*85% of those taking the test scored lower.*) Present the steps for finding a percentile and use Example 5 to demonstrate the process.

Explain that unimodal data has only one peak, as illustrated in the three graphs. Discuss the left-skewed, normal, and right-skewed distributions and their characteristics. Point out that the tail and the mean are to the left on the left-skewed graph, and to the right on the right-skewed graph. Bimodal distributions are illustrated in Section 10.5.

Common Student Error Students focusing on the peak of the data may confuse left- and right-skewed distributions. Despite the fact that a left-skewed figure often appears to slant or weigh heavily to the right (positive direction), the skew (distortion) is actually the long tail on the left side. Encourage the students to focus on the long-tailed skew rather than the peak.

Have students refer to the illustrated figures to classify the data in the graph for Example 6.

Interactive Activity Some students may not understand the importance of the median or the significance of calculating the *five-number summary* of a set of data which includes the minimum, Q_1, Q_2 (the median), Q_3, and the maximum values.

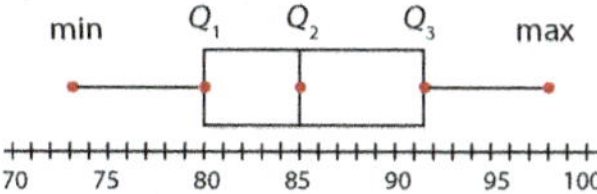

The *interquartile range* (IQR), the difference between Q_3 and Q_1, is another measure of variability. The *IQR* "box" contains the middle half (50%) of the data, while the "whiskers" represent the lower and upper fourths of the data.

An *outlier* is a value that is significantly higher or lower than the rest of the data. Any value more than 1.5(IQR) below Q_1 or above Q_3 is usually considered an outlier. It may be an extreme, nonrepresentative value or the result of an error in collecting the data. Because the value does not fit the data's overall distribution, it is sometimes excluded from the data set.

Example 4   Comparing Data Sets with Box-and-Whisker Plots

Identify the outlier in the following heights. Find the population mean, median, and interquartile range, and then draw box plots when the outlier is included and when it is excluded. Then compare the resulting means, medians, and interquartile ranges.

$H = \{64, 58, 61, 68, 74, 25, 56, 62, 70, 64, 61, 59\}$

Answer

The value of 25 appears to be an outlier.	
With	**Without**
{25, 56, 58, 59, 61, 61, 62, 64, 64, 68, 70, 74}	{56, 58, 59, 61, 61, 62, 64, 64, 68, 70, 74}
$\mu = \dfrac{\sum x_i}{N} = \dfrac{722}{12} \approx 60.2$	$\mu = \dfrac{\sum x_i}{N} = \dfrac{697}{11} \approx 63.4$
median: $\dfrac{61 + 62}{2} = 61.5$	median: 62
Q_1: $\dfrac{58 + 59}{2} = 58.5$	Q_1: 59; Q_3: 68
Q_3: $\dfrac{64 + 68}{2} = 66$	
IQR = 66 − 58.5 = 7.5	IQR = 68 − 59 = 9

$Q_1 - 1.5(IQR) = 58.5 - 1.5(7.5) = 47.25$
$Q_3 + 1.5(IQR) = 66 + 1.5(7.5) = 77.25$
25 is an outlier.

1. Arranging the data in ascending order helps to identify potential outliers.

2. Find the mean of each data set.

3. Find the median of each data set.

4. Find Q_1 and Q_3 for each data set.

5. Use $IQR = Q_3 - Q_1$ to find each interquartile range.

6. Since 25 is the only value outside the interval [47.25, 77.25], it is the only outlier.

7. Enter the first data set as L1 and the second as L2. Then use [STAT PLOT] to graph box plots ([STAT]) of L1 with Plot1 and L2 with Plot2.

8. Compare the means, medians, and interquartile ranges.

Including the outlier lowers the mean by ≈ 3.2 but lowers the median by just 0.5. While the *IQR*s with and without the outlier were about the same, the *IQR* actually decreases by 1.5 when the outlier is included.

SKILL ✓ **EXERCISES 17, 27, 31**

Excluding outliers is one way researchers may either intentionally or unintentionally bias their results. Stanford professor John Ioannidis cites "serendipitous inclusion or exclusion" as one factor that contributes to his claim that most published research findings are false.

Your calculator can also be used to find statistics related to a univariate (one-variable) set of data. Enter the data in L1. Then select [STAT], CALC, 1:1-Var Stats, [2nd], L1, [ENTER].

```
1-Var Stats
x̄=60.16666667
Σx=722
Σx²=45084
Sx=12.22392091
σx=11.70351324
n=12
minX=25
Q₁=58.5
Med=61.5
Q₃=66
maxX=74
```

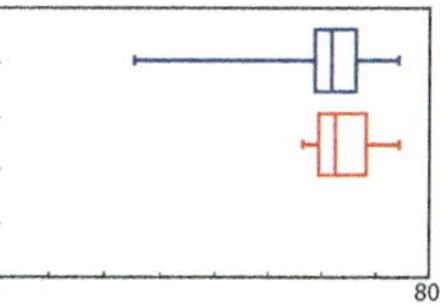

Assignments

- **Minimum:** 1–3, 5–9, 11, 13–17, 19, 21, 23–25, 29, 31, 36, 41–42, 44–47, 51
- **Standard:** 1–2, 4–9, 11, 14–16, 18–19, 22–25, 29, 32, 34, 36–37, 41–42, 44–47, 50–51
- **Extended:** 1–2, 5; 6–14 even; 15, 18, 21–22, 26–28, 30, 32–35, 38, 43–44, 48–51

Assessment

- Quiz 10B covers Sections 10.3–10.4.

the median and the mean. Consider having the students use the Internet to find values for mean and median net worth, annual income, or home value. It is difficult to obtain accurate numbers for these values, so the numbers may vary significantly, but the relative differences should be interesting. For example, one estimate for the 2016 average household net worth in the United States was about $692,000, while median net worth was about $97,000. The net worth of the very rich right-skews the mean significantly, but has less effect on the median, which better defines the majority of Americans.

The results of standardized tests are frequently reported in terms of *percentiles*. It is important to understand that getting a score in the 70th percentile is not the same as getting 70% of the possible points on the test. Scoring in the 70th percentile means that your score was higher than 70% of all the scores. In general $P\%$ of the data is less than the number at the Pth percentile. The first quartile Q_1 is the 25th percentile, the median is the 50th percentile, and the third quartile Q_3 is the 75th percentile.

Finding Percentiles from an Ordered List of Data

1. Find the rank, R, of the Pth percentile using $R = \frac{P}{100}(n + 1)$
   and identify the rank's integer part (IR) and its fractional part (FR).
   (For example, if $R = 3.25$, then $IR = 3$ and $FR = 0.25$.)

2. Let A equal the data value at the IR position and B equal the next value in the ordered list.

3. Apply the formula: Pth percentile $= A + FR(B - A)$.

Example 5   Determining a Percentile

Find the 87th percentile of the following weights.
$W = \{41, 42, 46, 47, 47, 48, 48, 50, 51, 52, 53, 56, 61\}$

Answer

$R = \frac{87}{100}(13 + 1) = 12.18$ 1. Use $n = 13$ and $R = \frac{P}{100}(n + 1)$ to find R;

$IR = 12;\ FR = 0.18$ then identify IR and FR.

Let $A = 56$ and $B = 61$.   2. Let $A = x_{12}$ and $B = x_{13}$ in the ordered data.

87th percentile $= 56 + 0.18(61 - 56) = 56.9$ 3. Use Pth percentile $= A + FR(B - A)$ to find the 87th percentile.

SKILL ✓ EXERCISE 19

The shape of a distribution can often be classified as negatively skewed, normal, or positively skewed by examining the overall shape of the histogram or box plot and the relationship between the mean and median.

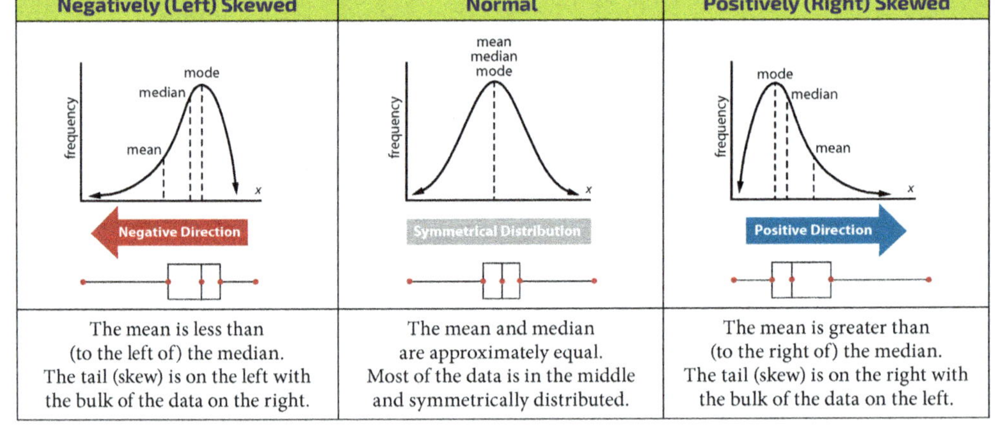

Negatively (Left) Skewed	Normal	Positively (Right) Skewed
The mean is less than (to the left of) the median. The tail (skew) is on the left with the bulk of the data on the right.	The mean and median are approximately equal. Most of the data is in the middle and symmetrically distributed.	The mean is greater than (to the right of) the median. The tail (skew) is on the right with the bulk of the data on the left.

Classify the shape of the illustrated distribution and predict the relationship between the data's mean and median.

Answer

The distribution of scores is negatively (left) skewed, so the mean is less than the median.

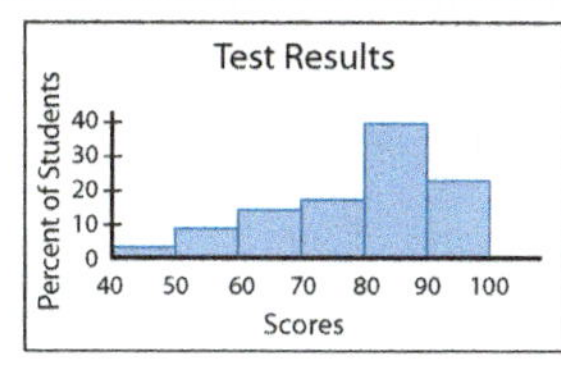

SKILL ✓ EXERCISE 13

Which measures of central tendency and variation should be used to describe a distribution of data? The median, interquartile range, and five-number summary are frequently used to describe skewed distributions and those with significant outliers. The median and interquartile range are said to be *resistant* since they are not as easily affected by the presence of extreme values. Note in Example 4 how the outlier had a greater effect on the mean than on the median and caused the distribution to be negatively skewed. The outlier also had a much greater effect on the sample standard deviation ($s \approx 12.2$ with the outlier and $s \approx 5.4$ without) than on the interquartile range.

The mean and standard deviation are used to describe normally distributed data. Many naturally occurring phenomena produce normal distributions, so these measures are the most commonly used statistics. They will be explored further in Sections 10.5 and 10.6.

❯ A. Exercises

1. Match each definition to the correct term.
   a. the middle value in a sorted list of data   III
   b. the quotient of the sum of the data values and the number of values   II
   c. the most frequently occurring data value   IV
   d. the difference between a data value and the mean   I

   I. deviation
   II. mean
   III. median
   IV. mode
   V. standard deviation
   VI. variance

2. Match each definition to the correct term.
   a. the average of the squared deviations   V
   b. the square root of the variance   IV
   c. one of the values that separate an ordered data set into four equal sets   III
   d. a value that is significantly higher or lower than the rest of the data   II

   I. deviation
   II. outlier
   III. quartile
   IV. standard deviation
   V. variance

Find the mean, median, and mode for each set of data.

3. 96, 87, 91, 90, 88, 83   89.2; 89; none

4. 28, 28, 31, 35, 35, 36, 39, 40, 43, 43, 43, 43, 47   37.8; 39; 43

5. heights of 24 choir members   66.2; 66; 66

6. ages of US presidents when they took office in the 1900s   54.6; 55; 51

```
58 60 62 64 66 68 70 72 74
```

Stem	Leaves
4	2 3 6
5	1 1 1 2 4 5 5 6 6
6	0 1 2 4 9

Use a table to find the mean, variance, and standard deviation (to the nearest tenth).

7. sample:
   96, 87, 91, 90, 88, 83
   $\bar{x} = 89.2; s^2 = 19.0; s = 4.4$

8. population:
   28, 31, 35, 35, 36, 39, 40, 43, 43, 47
   $\mu = 37.7; \sigma^2 = 30.6; \sigma = 5.5$

8. $\mu = \dfrac{377}{10} = 37.7$

x_i	$(x_i - \bar{x})^2$
28	$(28 - 37.7)^2 = 94.09$
31	$(31 - 37.7)^2 = 44.89$
35	$(35 - 37.7)^2 = 7.29$
35	$(35 - 37.7)^2 = 7.29$
36	$(36 - 37.7)^2 = 2.89$
39	$(39 - 37.7)^2 = 1.69$
40	$(40 - 37.7)^2 = 5.29$
43	$(43 - 37.7)^2 = 28.09$
43	$(43 - 37.7)^2 = 28.09$
47	$(47 - 37.7)^2 = 86.49$

$$SS = \sum_{i=1}^{10} (x_i - \mu)^2 = 306.1$$

$$\sigma^2 = \frac{306.1}{10} = 30.61; \sigma \approx 5.5$$

9. $\mu = \dfrac{1362}{20} = 68.1$

$$\sigma^2 \approx \frac{553.80}{20} = 27.69; \sigma \approx 5.3$$

10. $\bar{x} = \dfrac{656}{16} = 41$

$$s^2 = \frac{510}{16 - 1} = 34; s \approx 5.8$$

11. $IQR = 8 - 5 = 3$

$$1.5(IQR) = 1.5(3) = 4.5$$
$$Q_1 - 4.5 = 0.5; Q_3 + 4.5 = 12.5$$
$$14 \notin [0.5, 12.5]$$

12. $Q_1 = \dfrac{x_2 + x_3}{2} = 4.5; Q_3 = \dfrac{x_7 + x_8}{2} = 8$

$$IQR = 8 - 4.5 = 3.5$$
$$1.5(IQR) = 1.5(3.5) = 5.25$$
$$Q_1 - 5.25 = -0.75$$
$$Q_3 + 5.25 = 13.25$$
$$2, 9 \in [-0.75, 13.25]$$

❯ B. Exercises

17. 6, 7, 7, 8, 8, 9, 9, 9, 10
   median: $x_5 = 8$
   $Q_1: \dfrac{x_2 + x_3}{2} = 7; Q_3: \dfrac{x_8 + x_9}{2} = 9$

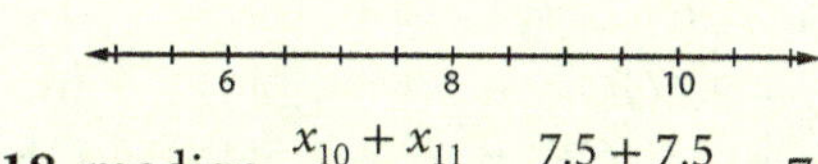

18. median: $\dfrac{x_{10} + x_{11}}{2} = \dfrac{7.5 + 7.5}{2} = 7.5$
   $Q_1: \dfrac{x_5 + x_6}{2} = 7; Q_3: \dfrac{x_{15} + x_{16}}{2} = 8.5$

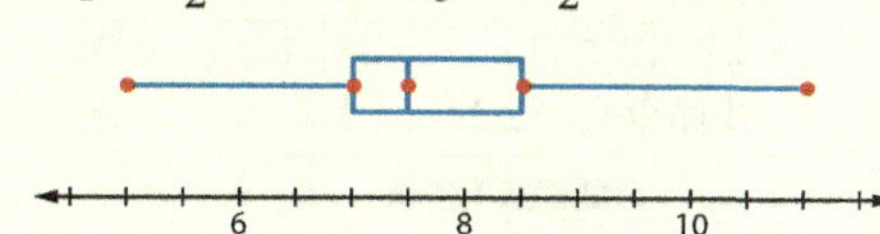

Solutions

❯ A. Exercises

3. $\bar{x} = \dfrac{\sum_{i=1}^{6} x_i}{n} = \dfrac{535}{6} \approx 89.2$

   median: $\dfrac{x_3 + x_4}{2} = \dfrac{88 + 90}{2} = 89$

   mode: none

4. $\bar{x} = \dfrac{\sum_{i=1}^{13} x_i}{n} = \dfrac{491}{13} \approx 37.8$

   median: $x_7 = 39$; mode: 43

5. $\bar{x} = \dfrac{1588}{24} \approx 66.2$

   median: $\dfrac{x_{12} + x_{13}}{2} = \dfrac{66 + 66}{2} = 66$

   mode: 66

6. $\bar{x} = \dfrac{928}{17} \approx 54.6$

   median: $x_9 = 55$; mode: 51

7. $\bar{x} = \dfrac{535}{6} \approx 89.17$

x_i	$(x_i - \bar{x}^2)$
96	$(96 - 89.17)^2 \approx 46.69$
91	$(91 - 89.17)^2 \approx 3.36$
90	$(90 - 89.17)^2 \approx 0.69$
88	$(88 - 89.17)^2 \approx 1.36$
87	$(87 - 89.17)^2 = 4.69$
83	$(83 - 89.17)^2 \approx 38.03$

$$SS = \sum_{i=1}^{6} (x_i - \bar{x})^2 \approx 94.83$$

$$s^2 = \frac{94.83}{6 - 1} = 18.97; s \approx 4.35$$

19. $R = \frac{50}{100}(20 + 1) = 10.5$

$x_{10} = 86; x_{11} = 87$

$86 + (0.5)(87 - 86) = 86.5$

$R = \frac{20}{100}(20 + 1) = 4.2$

$x_4 = 85; x_5 = 86$

$85 + (0.2)(86 - 85) = 85.2$

20. $R = \frac{75}{100}(18 + 1) = 14.25$

$x_{14} = 92; x_{15} = 95$

$92 + (0.25)(95 - 92) = 92.75$

$R = \frac{90}{100}(18 + 1) = 17.1$

$x_{17} = 96; x_{18} = 97$

$96 + (0.1)(97 - 96) = 96.1$

21. $R = \frac{30}{100}(24 + 1) = 7.5$

$x_7 = 62; x_8 = 64$

$62 + 0.5(64 - 62) = 63$

$R = \frac{75}{100}(24 + 1) = 18.75$

$x_{18} = 70; x_{19} = 71$

$70 + 0.75(71 - 70) = 70.75$

22. $R = \frac{10}{100}(30 + 1) = 3.1$

$x_3 = 402; x_4 = 417$

$402 + (0.1)(417 - 402) = 403.5$

$R = \frac{80}{100}(30 + 1) = 24.8$

$x_{24} = 582; x_{28} = 607$

$582 + (0.8)(607 - 582) = 602$

23–24.

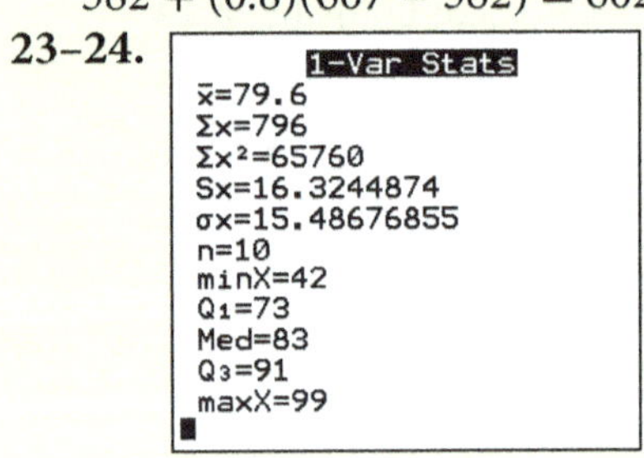

24.

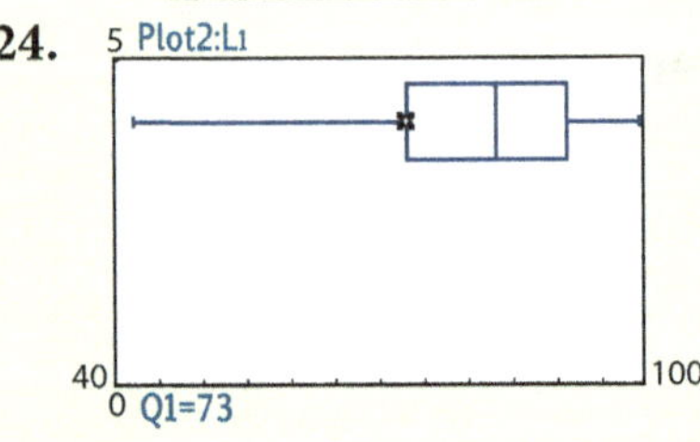

$IQR = 91 - 73 = 18$

$Q_1 - 1.5(IQR) = 73 - 1.5(18) = 46$

$Q_3 + 1.5(IQR) = 91 + 1.5(18) = 118$

$42 \notin [46, 118]$

25. *Hint:* On the home screen, store L1 as L2 ([2nd], [1], [STO▸], [2nd], [2]) and delete 42 from L2.

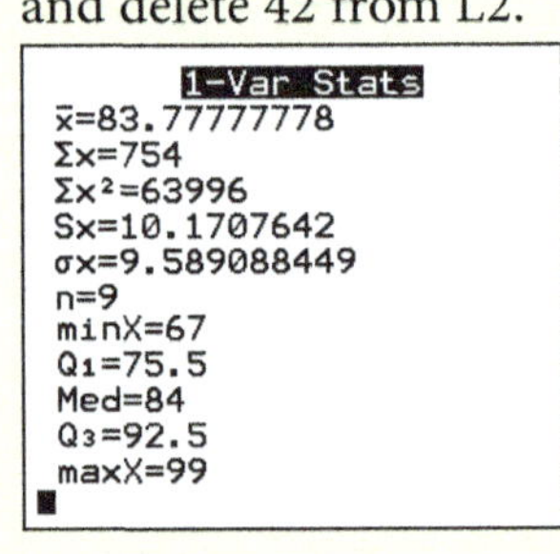

Complete the frequency distribution table for the sample data to find the mean and the standard deviation.

9. population data $\mu = 68.1; \sigma \approx 5.3$

x_i	f_i	$x_i f_i$	$f_i(x_i - \mu)^2$
55	2	110	343.22
63	2	126	52.02
67	2	134	2.42
69	5	345	4.05
70	3	210	10.83
72	3	216	45.63
73	2	146	48.02
75	1	75	47.61
Total	**20**	**1362**	**553.80**

10. sample data $\bar{x} = 41; s \approx 5.8$

x_i	f_i	$x_i f_i$	$f_i(x_i - \bar{x})^2$
33	3	99	192
35	1	35	36
38	2	76	18
39	2	78	8
42	2	84	2
46	3	138	75
48	2	96	98
50	1	50	81
Total	**16**	**656**	**510**

Determine whether each minimum or maximum value is an outlier.

11. min: 1, Q_1: 5, Q_2: 6.5, Q_3: 8, max: 14

14 is the only outlier.

12. 2, 4, 5, 5, 6, 6, 7, 9, 9 neither

13. Classify the shape of the illustrated distribution as negatively skewed, normal, or positively skewed.

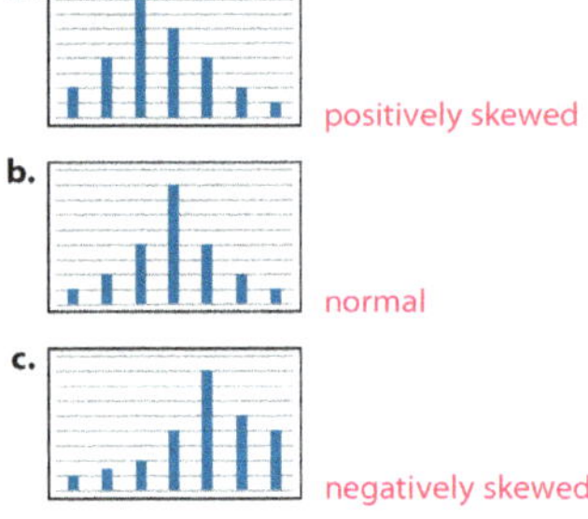

a. positively skewed

b. normal

c. negatively skewed

---

14. Classify the data represented by each box plot as left skewed, normal, or right skewed.

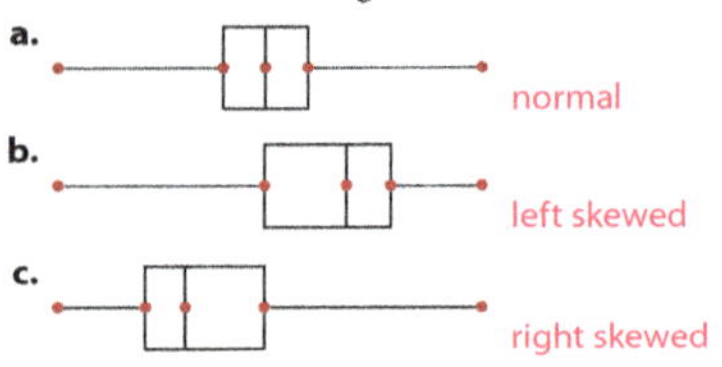

a. normal

b. left skewed

c. right skewed

15. Classify each described data distribution.

a. approximately equal mean and median II

b. mean significantly to the right of the median III

c. tail of data to the left I

d. symmetric distribution II

  I. negatively skewed

  II. normal

  III. positively skewed

16. Which set of statistics should be used to describe each type of data distribution?

a. negatively skewed II

b. normally distributed I

c. data set with outliers II

  I. mean and standard deviation

  II. five-number summary

> **B. Exercises**

State the five-number summary for each data set. Then sketch a box plot illustrating the data.

17. quiz scores: 10, 7, 8, 7, 9, 9, 6, 8, 9 min: 6; Q_1: 7; median: 8; Q_3: 9; max: 10

18. sizes of women's shoes sold on Friday: 5, 6, 6, 6.5, 7, 7, 7, 7.5, 7.5, 7.5, 7.5, 8, 8, 8, 8.5, 8.5, 9, 9, 10, 11 min: 5; Q_1: 7; median: 7.5; Q_3: 8.5; max: 11

Determine the indicated percentiles.

19. 50th and 20th percentiles for these cockatiel weights: 82, 83, 84, 85, 86, 86, 86, 86, 86, 86, 87, 87, 87, 87, 88, 88, 89, 90, 92, 94 86.5; 85.2

20. 75th and the 90th percentiles for these test scores: 62, 65, 72, 75, 79, 80, 82, 85, 88, 88, 89, 90, 90, 92, 95, 95, 96, 97 92.75; 96.1

21. 30th and 75th percentiles for these heights: 58, 59, 60, 60, 61, 62, 62, 64, 64, 65, 66, 66, 66, 68, 70, 70, 70, 71, 71, 71, 72, 73, 73 63; 70.75

22. 10th and 80th percentiles for these SAT scores: 312, 382, 402, 417, 418, 424, 428, 439, 463, 465, 487, 491, 492, 501, 507, 510, 510, 511, 515, 518, 538, 557, 567, 582, 607, 612, 615, 641, 665, 682 403.5; 602.0

The outlier of 42 lowers the mean by more than 4 points (from 83.8 to 79.6) and increases the standard deviation by more than 6 (from 10.2 to 16.3).

26–27.

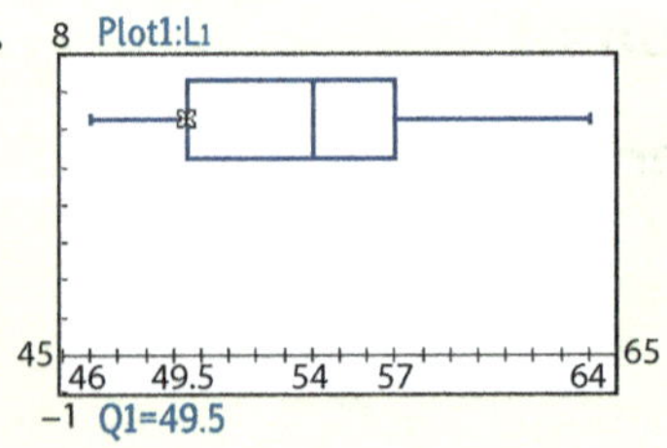

27.

$IQR = 57 - 49.5 = 7.5$

$Q_1 - 1.5(IQR) = 49.5 - 1.5(7.5)$
$\qquad = 38.25$

$Q_3 + 1.5(IQR) = 57 + 1.5(7.5)$
$\qquad = 68.25$

All ages fall between [38.5, 68.25].

28a. While the mean of 53.6 is slightly below the median of 54, the box plot indicates a slight tail to the right. The best classification is likely a normal distribution.

Use the sample AP History test scores to complete exercises 23–25.

$T = \{73, 42, 67, 78, 99, 84, 91, 82, 86, 94\}$

23. Use technology to find the mean and sample standard deviation. $\bar{x} = 79.6; s \approx 16.3$

24. Use technology to create a box plot. Then identify any outliers. 42

25. Use technology to find the mean and sample standard deviation without the outlier(s). Compare these statistics to those found in exercise 24.

Use the ages of presidents when they took office during the 1800s to complete exercises 26–28.

$A = \{47, 61, 46, 64, 48, 49, 51,$
$\quad 55, 57, 58, 56, 57, 54, 54,$
$\quad 50, 49, 57, 54, 51, 52, 55\}$

26. Use technology to find the mean and the population standard deviation. $\mu \approx 53.6; \sigma \approx 4.5$

27. Use technology to create a box plot. Then identify any outliers. none

28. Analyze the distribution of the data.
    a. Classify the distribution as negatively skewed, normal, or postively skewed. Explain your reasoning. normal
    b. Which set of statistics better describes the data set—the mean and standard deviation or a five-number summary? Explain your reasoning.

29. The number of games won by each NFL team during the 2016–17 season is summarized in the frequency distribution table.

Wins	0	3	4	5	6	7	8	9
Teams	1	1	2	4	3	3	1	6
Wins	10	11	12	13	14	15	16	
Teams	2	3	2	1	1	1	1	

    a. Does the data represent a population or a sample? population
    b. Find the data's mean and standard deviation.
    c. Draw a box plot illustrating the five-number summary. $\mu \approx 8.3; \sigma \approx 3.6$
    d. Draw a histogram and classify the shape of the distribution. Explain your reasoning. normal

30. An average of 1224 tornadoes occurred annually in the United States from 1991 to 2015. The list displays how 12 different states contributed to that number.
146.7, 49.5, 54.6, 92.4, 65.4, 54.6,
46.7, 54.0, 49.2, 41.9, 45.1, 47.1
    a. Find the data's mean and sample standard deviation. $\bar{x} \approx 62.3; s \approx 29.8$
    b. Draw a box plot illustrating the five-number summary and identify any outliers. 92.4 and 146.7
    c. Classify the shape of the distribution and explain your reasoning. positively skewed
    d. Which set of statistics better summarizes the data? five-number summary

Create box plots and write a brief statement comparing the two distributions.

31. **Super Bowl Total Points**

1–20		33–52
9 7 3 2 1	2	
8 7 7 1 0	3	1 1 4 7 8 9
7 7 7 6 5 4	4	1 5 6 8
6 4 0	5	0 1 2 3 6
6	6	1 2 5 9
	7	4

32.

2003–17 Hurricane Occurrences	
Eastern Pacific	7, 6, 7, 10, 4, 7, 7, 3, 10, 10, 9, 15, 13, 13, 9
Atlantic	7, 9, 15, 5, 6, 8, 3, 12, 7, 10, 2, 6, 4, 7, 10

33. The midrange of a data set, the average of the minimum and maximum values, is another measure of central tendency. Find the midrange for each set of data in exercise 32. Which measure is most resistant to the presence of outliers—the midrange, mean, median, or mode?
Eastern Pacific = 9, Atlantic = 8.5; median

30b. 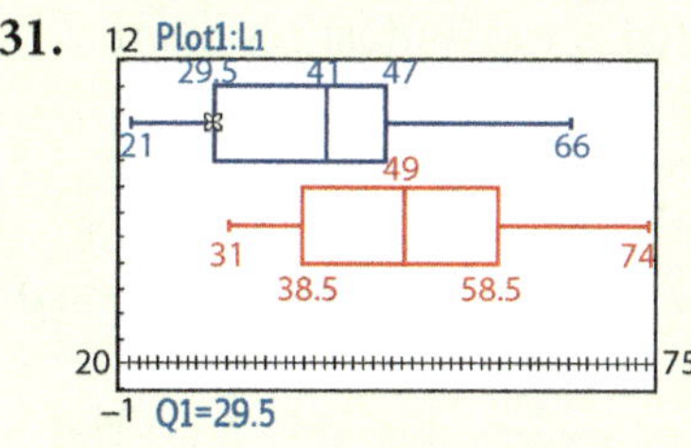

$1.5(IQR) = 1.5(60 - 46.9) = 19.65$
$Q_1 - 19.65 = 27.25$
$Q_3 + 19.65 = 79.65$
$92.4, 146.7 \notin [27.25, 79.65]$

30c. positively skewed; The mean is significantly above the median and there is a tail to the right.

31. 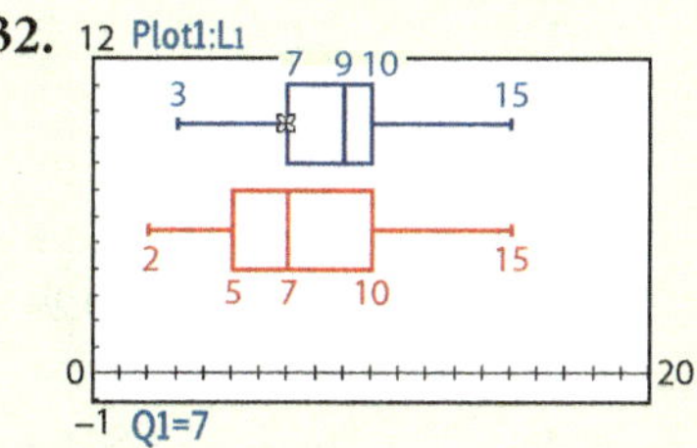

More points have been scored in recent Super Bowls than in the first 20.

32. 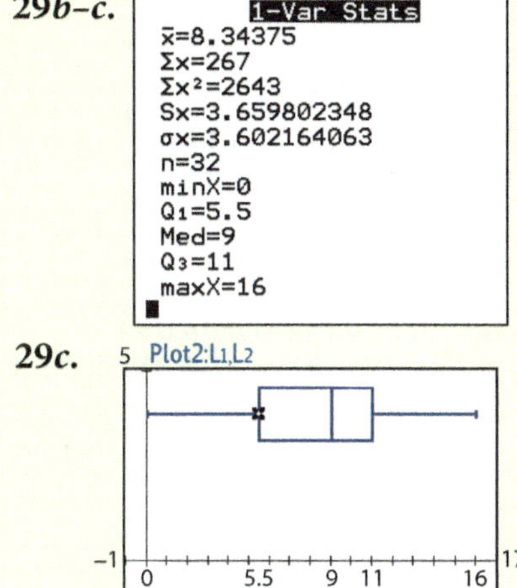

Despite similar maximums and third quartiles, the lower first quartile and median of hurricanes in the Atlantic indicate a greater variation in the yearly numbers. It appears that hurricanes typically occur more frequently in the Eastern Pacific than in the Atlantic.

33. EP: $\dfrac{3 + 15}{2} = 9$; A: $\dfrac{2 + 15}{2} = 8.5$

> **C. Exercises**

35. using the properties from Section 9.5 (p. 478):

$$\sum_{i=1}^{n} (x_i - \bar{x}) = \sum_{i=1}^{n} x_i - \sum_{i=1}^{n} \bar{x} \quad \text{(3rd property)}$$
$$= \sum_{i=1}^{n} x_i - n\bar{x} \quad \text{(1st property)}$$
$$= \sum_{i=1}^{n} x_i - n\left(\dfrac{\sum_{i=1}^{n} x_i}{n}\right) \quad \text{(def. of mean)}$$
$$= \sum_{i=1}^{n} x_i - \sum_{i=1}^{n} x_i = 0$$

28b. mean and standard deviation; These are used to describe normally distributed data.

29b–c.
```
            1-Var Stats
x̄=8.34375
Σx=267
Σx²=2643
Sx=3.659802348
σx=3.602164063
n=32
minX=0
Q₁=5.5
Med=9
Q₃=11
maxX=16
```

29c.

29d. normal; The mean is slightly below the median but the tail appears to be on the right.

30a–b.
```
            1-Var Stats
x̄=62.26666667
Σx=747.2
Σx²=56273.94
Sx=29.76924384
σx=28.50188103
n=12
minX=41.9
Q₁=46.9
Med=51.75
Q₃=60
maxX=146.7
```

36a. AR: $\mu = \dfrac{696}{22} \approx 31.6$

$\sigma = \sqrt{\dfrac{5769.09}{22}} \approx 16.2$

HA: $\mu = \dfrac{755}{23} \approx 32.8$

$\sigma = \sqrt{\dfrac{2751.30}{23}} \approx 10.9$

36b. While the average number of home runs was about the same, Rodriguez's yearly totals must have fluctuated more, implying that Aaron's yearly totals were more consistent. Even though Rodriguez's average was lower, he likely had some years that exceeded Aaron's most productive year.

37.

	A	B	C	D
1	**Age (x_i)**	f_i	$x_i f_i$	$f_i(x_i - x_{bar})^2$
2	42	1	42	158.46
3	43	1	43	134.29
4	46	1	46	73.76
5	51	3	153	38.63
6	52	1	52	6.70
7	54	1	54	0.35
8	55	2	110	0.34
9	56	2	112	3.99
10	60	1	60	29.29
11	61	1	61	41.11
12	62	1	62	54.93
13	64	1	64	88.58
14	69	1	69	207.70
15		**n**	**Sum**	**Sum**
16		17	928	838.12
17		**Mean:**	54.59	
18			**Variance:**	49.30
19			**Standard Deviation:**	7.02

B16: =SUM(B2:B14)
C2: =A2*B2
Use C2 to fill from C3 to C14.
Use B16 to fill C16.
C17: =C16/B16
D2: =B2*(A2-C17)^2
Use D2 to fill from D3 to D14.
Use C16 to fill D16.
D18: =D16/B16
D19: =SQRT(D18)

❯ **C. Exercises**

34. Generalize: The ordered data set $x_1, x_2, x_3, \ldots, x_{10}$ has the five-number summary: $x_1, x_3, \dfrac{x_5 + x_6}{2}, x_8, x_{10}$. Write a generalized five-number summary for each data set with n values.

  a. $n = 11$ **b.** $n = 12$ **c.** $n = 13$

35. Prove: Show that the sum of the deviations from the mean is 0: $\displaystyle\sum_{i=1}^{n}(x_i - \bar{x}) = 0$.

36. Find the indicated statistics for each player from the information provided.

	Alex Rodriguez	Hank Aaron
seasons played	22	23
total home runs	696	755
sum of the squared deviations, SS	5769.09	2751.30

  a. Find the average number of home runs per season and the population standard deviation for each player. $\mu_{AR} \approx 31.6; \sigma_{AR} \approx 16.2; \mu_{HA} \approx 32.8; \sigma_{HA} \approx 10.9$

  b. Use the results of part a to compare the players' home run hitting careers.

37. Use a spreadsheet to create a frequency distribution and calculate the mean and the population standard deviation. $\mu \approx 54.6; \sigma \approx 7.02$

Ages of Presidents Who Took Office in the 1900s

Stem	Leaves
4	2 3 6
5	1 1 1 2 4 5 5 6 6
6	0 1 2 4 9

38. Given the ages from a random sample of employees, use Sturges's formula and a spreadsheet to create a grouped frequency distribution for the data. Using each class's median age as its value, calculate the data's mean and standard deviation.

37, 32, 49, 62, 57, 46, 31, 48, 51, 55, 52, 34, 61, 58, 38 $\bar{x} \approx 47.5; s \approx 10.1$

34a. $x_1, x_3, x_6, x_9, x_{11}$

34b. $x_1, \dfrac{x_3 + x_4}{2}, \dfrac{x_6 + x_7}{2}, \dfrac{x_9 + x_{10}}{2}, x_{12}$

34c. $x_1, \dfrac{x_3 + x_4}{2}, x_7, \dfrac{x_{10} + x_{11}}{2}, x_{13}$

A *weighted mean* is often used to ensure that various types of assessments within a class contribute the desired percentage (or weight) toward the final grade. After the mean is calculated for each category, the percentage is multiplied by each mean, and the final grade is the sum of those products.

A: 94.8%; B: 81.65%

39. Find the weighted average for each student.

	Exam	Tests	Projects	Quizzes	Home-work
weight	15%	50%	10%	15%	10%
student A	98	95	97	92	91
student B	78	80	90	83	85

40. The semester grade for Ralph's Precalculus class is determined by six chapter tests and one final exam, which is weighted as 20% of the final grade. His six chapter test scores were 85, 89, 90, 86, 92, and 89. Assuming that an 89.5 rounds up to an A-, what minimum grade would he need on his final exam to receive an A- for the semester? *93.5*

41. Use the percentile graph to estimate each value.

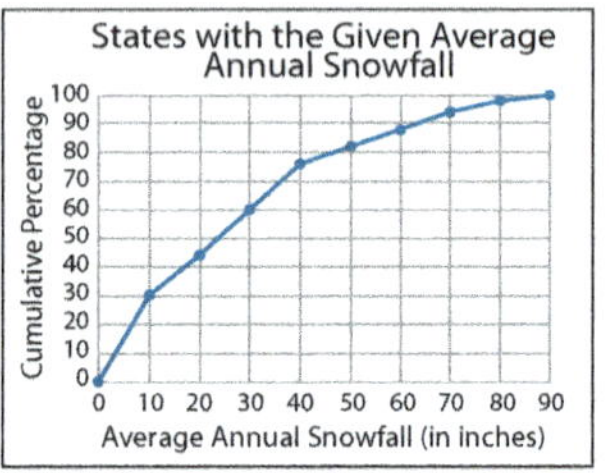

  a. What is the percentile rank of a state with an annual snowfall of 10 in.? *≈ 30th*

  b. What is the percentile rank of a state with an annual snowfall of 35 in.? *≈ 68th*

  c. What annual snowfall is at the 80th percentile? *≈ 47 in.*

  d. Use the Internet to find your state's average annual snowfall and then determine your state's approximate percentile rank. *Answers will vary.*

---

38. $1 + 3.3 \log 15 \approx 4.89; k = 5$

$\dfrac{62 - 31}{5} \approx 6.2; w = 7$

	A	B	C	D	E
1	**Class**	M_i	f_i	$M_i f_i$	$f_i(M_i - x_{bar})^2$
2	31–37	34	4	136	732.60
3	38–44	41	1	41	42.68
4	45–51	48	4	192	0.87
5	52–58	55	4	220	223.00
6	59–65	62	2	124	418.57
7			**n**	**Sum**	**Sum**
8			15	713	1417.73
9				**Mean:**	47.53
10				**Variance:**	101.27
11				**Standard Deviation:**	10.06

C8: =SUM(D2:C6)
D2: =B2*C2
Use D2 to fill from D3 to D6.
Use C8 to fill D8.
D9: =D8/C8
E2: =C2*(B2-D9)^2
Use E2 to fill from E3 to E6.
Use D9 to fill E9.
E10: =E8/(C8-1)
E11: =SQRT(E10)

39. A: $0.15(98) + 0.5(95) + 0.1(97)$
$+ 0.15(92) + 0.1(91) = 94.8$

B: $0.15(78) + 0.5(80) + 0.1(90)$
$+ 0.15(83) + 0.1(85) = 81.65$

Solve each system of equations using Cramer's rule. [7.4]

42. $x + y = 1$
$3x + 2y = -2$ $\quad (-4, 5)$

43. $4x - 5y = 8$
$3x + 2y = 4$ $\quad \left(\frac{36}{23}, -\frac{8}{23}\right)$

44. Write a simplified explicit formula for the arithmetic sequence with $b_6 = 21$ and $b_{10} = 37$. Then use the formula to find b_{20}. [9.2] $\quad b_n = 4n - 3; b_{20} = 77$

45. Find n in the geometric sequence if $a_3 = 112$, $r = 4$, and $a_n = 7168$. [9.3] $\quad n = 6$

Use the table to complete exercises 46–47. [10.3]

Goals Scored	Number of Players
0	2
1	5
2	8
3	10
4	6
5	4
6	1

46. Construct a histogram representing the data.

47. Construct a pie chart representing the data.

48. Solve $x^3 > x$. [2.7] E

A. $x \in (-\infty, -1)$ **D.** $x \in (1, \infty)$

B. $x \in (-\infty, 1)$ **E.** none of these

C. $x \in (-1, \infty)$

49. Which equation represents an asymptote of $f(x) = \frac{8x - 5}{9x + 10}$? [2.5] C

A. $y = 0$ **C.** $y = \frac{8}{9}$ **E.** $y = -\frac{10}{9}$

B. $x = -\frac{1}{2}$ **D.** $x = \frac{5}{8}$

50. Find the standard form of a parabola with a focus $F(-2, 3)$ and a directrix $x = 2$. [8.1] B

A. $x = -\frac{1}{8}(y + 3)^2$ **D.** $y = -\frac{1}{8}(x - 3)^2$

B. $x = -\frac{1}{8}(y - 3)^2$ **E.** none of these

C. $y = \frac{1}{8}(x + 3)^2$

51. The administration can nominate up to 3 of the school's 6 eligible seniors for a local award. How many different sets of nominations can the school submit if at least 1 senior is nominated? [10.2] B

A. 20 **C.** 120 **E.** none of these
B. 41 **D.** 156

45. $a_n = a_3 \cdot 4^{n-3}$
$7168 = 112(4)^{n-3}$
$64 = 4^{n-3}$
$4^3 = 4^{n-3}$
$3 = n - 3; n = 6$

46. 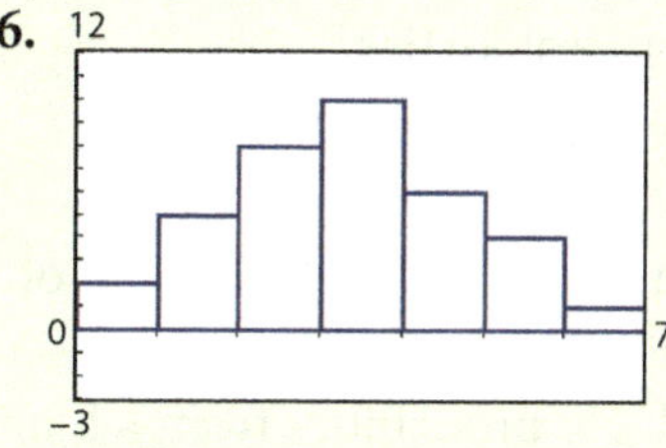

47. zero: $\frac{2}{36}(360°) = 20°$

one: $\frac{5}{36}(360°) = 50°$

two: $\frac{8}{36}(360°) = 80°$

three: $\frac{10}{36}(360°) = 100°$

four: $\frac{6}{36}(360°) = 60°$

five: $\frac{4}{36}(360°) = 40°$

six: $\frac{1}{36}(360°) = 10°$

Goals Scored

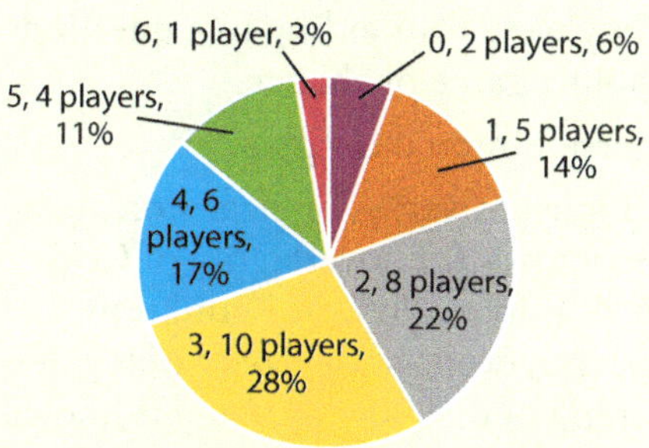

(Rounding causes the sum of the percentages to exceed 100%).

48. Examine $f(x) = x^3 - x$
$= x(x + 1)(x - 1) > 0.$
zeros: $-1, 0, 1$
using the end behavior of $f(x)$ and the multiplicity of 1 for each zero:
$f(x) < 0$ for $x \in (-\infty, -1)$
$f(x) > 0$ for $x \in (-1, 0)$
$f(x) < 0$ for $x \in (0, 1)$
$f(x) > 0$ for $x \in (1, \infty)$
$\therefore x \in (-1, 0) \cup (1, \infty)$

49. $n = m$; HA: $y = \frac{8}{9}$
VA: $x = -\frac{10}{9}$

50. $V\left(\frac{-2 + 2}{2}, 3\right) = (0, 3)$
opens to the left
$p = -2; a = \frac{1}{4(-2)} = -\frac{1}{8}$
$x = -\frac{1}{8}(y - 3)^2$

51. $_6C_3 + {_6}C_2 + {_6}C_1 = 20 + 15 + 6 = 41$

40. $\frac{85 + 89 + 90 + 86 + 92 + 89}{6} = 88.5$
$0.8(88.5) + 0.2x \geq 89.5$
$70.8 + 0.2x \geq 89.5$
$x \geq 93.5$

> **Cumulative Review**

42. $|A| = \begin{vmatrix} 1 & 1 \\ 3 & 2 \end{vmatrix} = -1$

$|A_x| = \begin{vmatrix} 1 & 1 \\ -2 & 2 \end{vmatrix} = 4$

$|A_y| = \begin{vmatrix} 1 & 1 \\ 3 & -2 \end{vmatrix} = -5$

$x = \frac{4}{-1} = -4; y = \frac{-5}{-1} = 5$

43. $|A| = \begin{vmatrix} 4 & -5 \\ 3 & 2 \end{vmatrix} = 23$

$|A_x| = \begin{vmatrix} 8 & -5 \\ 4 & 2 \end{vmatrix} = 36$

$|A_y| = \begin{vmatrix} 4 & 8 \\ 3 & 4 \end{vmatrix} = -8$

$x = \frac{36}{23}; y = \frac{-8}{23} = -\frac{8}{23}$

44. $b_{10} = b_6 + (10 - 6)d$
$37 = 21 + 4d$
$16 = 4d; d = 4$
$b_n = 21 + (n - 6)(4) = 4n - 3$
$b_{20} = 4(20) - 3 = 77$

Historical Connection

Objectives

1. To summarize the development of descriptive statistics
2. To state the origin of the word *statistics*
3. To summarize the development of probability theory
4. To evaluate probability from a biblical worldview

Answers

1. Florence Nightingale used polar area charts to emphasize the much greater loss of British soldiers from preventable diseases than from battle.
2. *Statistics* comes from the Latin word for nation-state.
3. Blaise Pascal and Pierre de Fermat; They were responding to a question about a game of chance.
4. Pierre-Simon de Laplace
5. sample answer: A biblical worldview recognizes the sovereignty of God in all things. Nothing is unknown to Him; even the outcome of the rolling of dice (or "the lot") is known to Him (Prov. 16:33). Man does not know as God knows, therefore some things seem to happen randomly. Even these events are under God's control.

POLITICAL ARITHMETIC AND PROBABILITY

Half a league, half a league,
Half a league onward,
All in the valley of Death
Rode the six hundred.

These lines open Alfred Tennyson's "The Charge of the Light Brigade," a poem memorializing a doomed British cavalry charge during the Crimean War. The descriptive statistics recorded by Florence Nightingale reveal that the army's casualties due to preventable diseases were nearly four times greater than those due to battle in the same month of October 1854. Her polar area graphs present the data in a clear and compelling visual format. Statistics summarize data in a form that aids understanding. The derivation of the word *statistics* from the Latin word for a nation-state indicates the origin of statistics as a mathematical aid to governments in carrying out their responsibilities. In fact, the term *political arithmetic* was frequently used before the term *statistics* was adopted. Governments have kept various records from antiquity, but John Graunt, a seventeenth-century British merchant, is generally credited as the first to systematically study and interpret collected data. His 1662 publication *Natural and Political Observations . . . Made Upon the Bills of Mortality* analyzed the reported deaths published weekly by local parishes. "Men of great experience" estimated London's population to be in the millions, possibly as high as six or seven million. Graunt used the available data to more accurately estimate the population as 384,000.

By the twentieth century, descriptive statistics were important enough to the governing process that one of Winston Churchill's first actions upon assuming control of the British Royal Navy was to establish his own department of statistics. Churchill credited this department with allowing him to form a "just and comprehensible view of the innumerable facts and figures" that otherwise would have been overwhelming.

The first attempt to create a mathematical theory of probability occurred in correspondence between Blaise Pascal and Pierre de Fermat in 1654. Their discussion was prompted by a gambler's question regarding a dice game. They developed a theory of probability for games of chance but did not attempt to apply the theory to other areas or to infer properties of the games based on outcomes. Dutch mathematician Christiaan Huygens published the first book on probability in 1657. In the early 1800s Pierre-Simon de Laplace published the comprehensive treatise *The Analytical Theory of Probabilities* and a simplified version called *A Philosophical Essay on Probabilities*. The historian Carl Boyer claims that probability theory "owes more to Laplace than to any other mathematician."

Understanding probability is important since we see both regularity and chance events in our world. Physical determinism says that nothing occurs by chance; if we could at one instant know the state of every particle, then we would know both the future and the past. Others claim that chance events are beyond even the control of God. A biblical worldview recognizes the sovereignty of God as well as the uncertainty that exists from a human perspective. As American theologian Vern Poythress states, "God controls small, unpredictable events as well as big patterns and regularities in history. . . . God in his wisdom has given us the whole tapestry of regularities and unpredictabilities and their connections with one another. This tapestry . . . reflects his wisdom and his character, as Romans 1:19–20 indicates."

COMPREHENSION CHECK

1. Describe Florence Nightingale's use of descriptive statistics.
2. What is the origin of the word *statistics*?
3. Which two Frenchmen first developed a theory of probability, and what initiated their efforts?
4. According to historian Carl Boyer, which mathematician made the greatest contribution to probability theory?
5. Describe a biblical worldview of probability.

PRESENTATION

Most students will be familiar with Florence Nightingale and her tremendous influence in improving nursing and sanitary conditions in hospitals. Most will not realize that her use of descriptive statistics played an important role in her ability to bring about these changes. Nightingale studied mathematics (she was elected the first female member of the Royal Statistical Society in 1858) and her charts were carefully crafted so that the relative frequency of each cause corresponded to the area in the graph. Point out that simply plotting raw data with radii would be deceptive, since the eye sees area, not length, as the indicator of frequency in a pie chart.

Nightingale's graphs present data for one year, so each sector corresponds to one month and has a central angle of 30°. She divided the number of deaths due to a particular cause by the total troop strength for that month to obtain a percentage. She then multiplied that number by 1,000 to find deaths per 1,000 due to this cause, and then multiplied by 12 to annualize the result. Finally, she used the square root of this number to plot a radius in the graph. For example, her data for October, 1854, lists troop strength as 30,643, deaths due to disease as 503, deaths due to wounds as 132, and deaths due to other causes as 128. Thus the radius representing deaths due to disease (14.0) is nearly twice as long as the radius representing deaths due to wounds (7.2), since the number of deaths due to disease is nearly four times that of deaths due to wounds.

The British merchant John Graunt made noteworthy contributions to the study of statistics. He gathered data of annual birth rates and burials as well as estimations of family size and women of child-bearing age to estimate the true population of London at about 6% of the previous assumed estimates.

Technology such as a graphing calculator or a spreadsheet enables us to quickly explore how transformations of a data set affect its mean and standard deviation. Enter $L_1 = \{4, 6, 7, 9, 9, 10\}$ in your calculator and use the 1-Var Stats function from the [STAT], CALC submenu to determine the mean $\bar{x} = 7.5$ and the sample standard deviation $s \approx 2.26$. Then enter four transformations of the data set: $L_2 = L_1 + 5$, $L_3 = L_1 - 4$, $L_4 = L_1 \times 3$, and $L_5 = L_1 \div 2$ and find the mean and sample deviation of each transformed data set.

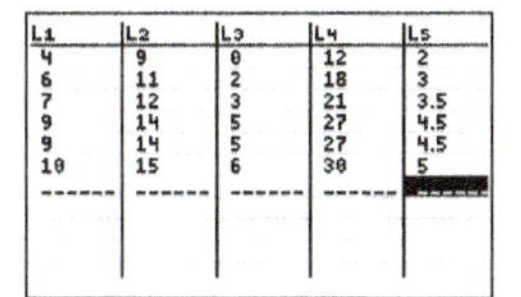

L_1	$L_2 = L_1 + 5$	$L_3 = L_1 - 4$	$L_4 = L_1 \times 3$	$L_5 = L_1 \div 2$
$\bar{x} = 7.5$	$\bar{x} = 12.5$	$\bar{x} = 3.5$	$\bar{x} = 22.5$	$\bar{x} = 3.75$
$s \approx 2.26$	$s \approx 2.26$	$s \approx 2.26$	$s \approx 6.77$	$s \approx 1.13$

Can you describe how the mean and standard deviation are affected when the same number is added to (or subtracted from) each value? What if each data value is multiplied (or divided) by the same number? The following theorems summarize these results.

TRANSLATED DATA THEOREM

If $A = \{x_1, x_2, \ldots, x_n\}$ and $B = \{y_1, y_2, \ldots, y_n\}$ where $y_i = x_i + k$, then $\bar{y} = \bar{x} + k$ and $s_y = s_x$.

Apply the properties of summations to prove this theorem.

$$\bar{y} = \frac{\sum_{i=1}^{n} y_i}{n} = \frac{\sum_{i=1}^{n} (x_i + k)}{n} = \frac{\sum_{i=1}^{n} x_i + \sum_{i=1}^{n} k}{n} = \frac{\sum_{i=1}^{n} x_i + nk}{n} = \frac{\sum_{i=1}^{n} x_i}{n} + k = \bar{x} + k$$

$$\therefore \bar{y} = \bar{x} + k$$

$$s_y^2 = \frac{\sum_{i=1}^{n} (y_i - \bar{y})^2}{n-1} = \frac{\sum_{i=1}^{n} [(x_i + k) - (\bar{x} + k)]^2}{n-1} = \frac{\sum_{i=1}^{n} (x_i - \bar{x})^2}{n-1} = s_x^2$$

$$\therefore s_y = s_x$$

SCALED DATA THEOREM

If $A = \{x_1, x_2, \ldots, x_n\}$ and $B = \{y_1, y_2, \ldots, y_n\}$ where $y_i = kx_i$, then $\bar{y} = k\bar{x}$ and $s_y = ks_x$.

The proof of this theorem is left as an exercise.

Raw data values are frequently transformed into z-scores. The z-scores measure the deviation of each individual score from the mean in units of standard deviation.

After completing this section, you will be able to

- describe how the mean and variance are affected when data is translated or scaled.
- calculate z-scores and other transformed scores for data.
- apply Chebyshev's Theorem and the Empirical Rule to describe variability.

Objectives

1. To describe how the mean and variance are affected when data is translated or scaled

2. To calculate z-scores and other transformed scores for data

3. To apply Chebyshev's Theorem and the Empirical Rule to describe variability in a set of data

Flash

Reading the data on a bubble sheet is done with light reflection and optical mark recognition (OMR) software. This technology is very similar to the old technology for reading computer punch cards or punched tape, and on a smaller scale, the more modern technology of CD, DVD, and Blu-ray Disc™ optical drives, in which data is read by reflecting a laser beam.

Vocabulary

bimodal
Chebyshev's Theorem
Empirical Rule (68-95-99.7 Rule)
Scaled Data Theorem
Translated Data Theorem
z-score

PRESENTATION

Lesson Opener

Without using technology, find the mean, mode(s), and median of each set.

1. $\{5, 10, 10, 10, 15\}$ 10; 10; 10

2. $\{2, 6, 8, 16\}$ 8; none; 7

3. $\{1, 2, 4, 5, 5, 6, 6, 7\}$ 4.5; 5 and 6; 5

Engage the students by explaining that statistical terms and concepts are becoming increasingly more common in society. Also explain that many people do not understand statistical terms relating to standardized test scores and that a good understanding of the terms in this section will provide greater insight and appreciation of such numeric data.

Begin this section by having the students enter the data from the text into a list (L1) on their calculators. Then have them create the described lists of translated data. Students should then use the 1-Var Stats function to find the mean and standard deviation of each data set. These lists illustrate the *Translated Data Theorem* and the *Scaled Data Theorem*. State the theorems and ask the students to supply the reason for each step in the proofs.

Define the *z-score* and explain that it is used to compare scores from two different distributions by describing the scores in terms of their standard deviations from the means. Many students will encounter z-scores in the future. For example, z-scores are used to simplify test results such as echocardiographic measurements, which have a mean of 0 and a normal range of -2 to $+2$.

Use Example 1 to encourage students to explore z-scores from original and transformed data. This example shows how z-scores standardize values from different scales. Note that using rounded values to calculate $\bar{z}$ and s will result in values that are not exactly 0 and 1.

Finding the z-score provides a way to evaluate how a score compares to the

Additional Exercises

If the mean of a set of data values is 75 and the standard deviation is 10, find the mean and standard deviation for the data transformed as follows.

1. $y = x + 7$ $\overline{y} = 82; s = 10$

2. $y = \frac{1}{5}x$ $\overline{y} = 15; s = 2$

3. Find Kim's z-score if she scored an 80 on a test having a mean of 78 and standard deviation of 4. Then interpret the meaning of the z-score. $z = 0.5$; Kim's test score was 0.5 standard deviations above the mean.

4. Jen received a 47 on a quiz with a mean of 44 and standard deviation of 2.4, and she earned 113 on an English paper with a mean of 106 and standard deviation of 6. Use z-scores to compare Jen's relative performance on the two assessments. $z_Q = 1.25; z_P \approx 1.17$; The quiz score was better.

5. Find Jack's z-score if he scored a 37 on a quiz with a mean of 43.3 and standard deviation of 3.54. $z = -1.78$

6. Use Chebyshev's Theorem to determine the smallest percentage of data that must lie within 3 standard deviations of the mean, then confirm the theorem using the following sample of 15 quiz scores: 6, 6, 7, 7, 7, 8, 8, 9, 9, 9, 9, 9, 10, 10. 88.9%; $\overline{x} = 8.2$, $\sigma = 1.3$, 100% of the given data is between $8.2 \pm 3(1.3) = (4.24, 12.16)$.

Given $A = \{x_1, x_2, \ldots, x_n\}$ with a mean of $\overline{x}$ and standard deviation of s, the **z-score** for each data value is $z = \frac{x - \overline{x}}{s}$.

Both theorems on page 527 are used in the definition of a z-score. It is first translated by the subtraction of x and then scaled through the division by s.

Example 1   Exploring z-scores

Find the z-score for each value in $L_1 = \{4, 6, 7, 9, 9, 10\}$ and for each transformed value in $L_4 = \{12, 18, 21, 27, 27, 30\}$. Then compare the z-scores, the mean $\overline{z}$, and the sample standard deviation s for each standardized set of data.

Answer

$L_2 = (L_1 - 7.5)/2.26$	$L_5 = (L_4 - 22.5)/6.77$
$z_1 \approx \frac{4 - 7.5}{2.26} \approx -1.55$	$z_1 \approx \frac{12 - 22.5}{6.77} \approx -1.55$
$z_2 \approx \frac{6 - 7.5}{2.26} \approx -0.66$	$z_2 \approx \frac{18 - 22.5}{6.77} \approx -0.66$
$z_3 \approx \frac{7 - 7.5}{2.26} \approx -0.22$	$z_3 \approx \frac{21 - 22.5}{6.77} \approx -0.22$
$z_4 \approx \frac{9 - 7.5}{2.26} \approx 0.66$	$z_4 \approx \frac{27 - 22.5}{6.77} \approx 0.66$
$z_5 \approx \frac{9 - 7.5}{2.26} \approx 0.66$	$z_5 \approx \frac{27 - 22.5}{6.77} \approx 0.66$
$z_6 \approx \frac{10 - 7.5}{2.26} \approx 1.11$	$z_6 \approx \frac{30 - 22.5}{6.77} \approx 1.11$

In each set of z-scores, $\overline{z} = 0$ and $s \approx 1.00$.

1. Use $z = \frac{x - \overline{x}}{s}$ and the mean and standard deviation from each set.

   This can be done quickly by redefining L_2 and L_5.

   The z-scores are the same for both sets.

2. Use 1-Var Stats on L_2 and L_5 to determine $\overline{z}$ and s.

SKILL ✓ EXERCISE 7

The mean of the z-score distribution is always 0 and its standard deviation is always 1. The distribution of the z-scores reflects the distribution of the raw scores.

Example 2   Finding and Interpreting a z-score

Find the z-score for a student who scored a 79 on a test having a mean of 84 and a standard deviation of 4. Then interpret the meaning of the z-score.

Answer

$z = \frac{x - \overline{x}}{s} = \frac{79 - 84}{4} = -1.25$ The student's score is 1.25 standard deviations below the mean.

SKILL ✓ EXERCISE 9

Using z-scores is the best way to compare individual scores from different normal distributions.

average (mean) score in terms of standard deviations above or below the mean, as shown in Example 2. Example 3 shows how this same comparison can be made with two scores from different tests or quizzes.

Example 4 shows how a z-score can be translated into an actual test score. Point out that this requires the mean and standard deviation to be known.

Discuss unimodal and *bimodal* distributions and *Chebyshev's Theorem*. Apply Chebyshev's Theorem with Example 5. Then present the *Empirical Rule* and apply it using Example 6. Students should know the percentages of the data within the first 3 standard deviations of a normal distribution.

Images for normal distributions can be easily found using the Internet keyword search *normal distribution image*. Note that z-scores are described by standard normal distribution, which has a mean of 0 and a standard deviation of 1.

Interactive Activity Use the Internet keyword search *interactive normal distribution* to find an interactive graph displaying areas and probabilities for various intervals.

TIPS

Ex. 17a An answer of 0 should indicate to the student that Chebyshev's Theorem does not apply for $k = 1$.

Mary earned 9 points on a quiz for which the class average was 7 and the standard deviation was 1.5. She then earned 15 points on a second quiz for which the class average was 13 and the standard deviation was 2.1. Find the z-score for each quiz grade and use them to decide which quiz Mary did better on compared to her classmates.

Answer

$z_1 = \frac{9-7}{1.5} \approx 1.3$; $z_2 = \frac{15-13}{2.1} \approx 0.95$

1. Use $z = \frac{x-\bar{x}}{s}$ to find the number of standard deviations from the mean for each score.

Mary's score on the first quiz is more standard deviations above the mean than her score on the second quiz.

2. Compare the z-scores.

_______________________________ SKILL ✓ **EXERCISE 19**

A known z-score can be transformed to a distribution with any desired mean and standard deviation while maintaining the distribution's shape and the score's relative position.

The 2010 ACT® English test had a mean of 20.5 and a standard deviation of 6.4. Find Brenton's ACT English score if his z-score was 0.43.

Answer

$sz = x - \bar{x}$
$x = sz + \bar{x}$

1. Solve $z = \frac{x-\bar{x}}{s}$ for x.

$x = 6.4(0.43) + 20.5 = 23.252$

2. Substitute and simplify.

His ACT English score was 23.

3. Round to find the integral raw score.

_______________________________ SKILL ✓ **EXERCISE 23**

Distributions of data can have many different shapes. The shape of the distribution is generally determined by looking at the data's histogram. Section 10.4 examined unimodal distributions that were skewed and normally distributed. In other cases, the distribution may be *bimodal*, having two clusters of data as illustrated below. Bimodal distributions can represent polarized opinions or sample data from overlapping distributions.

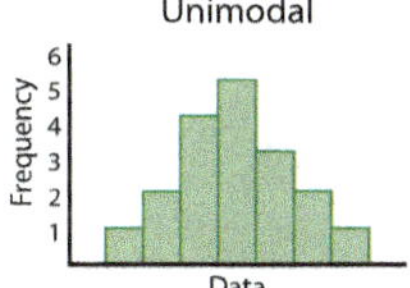

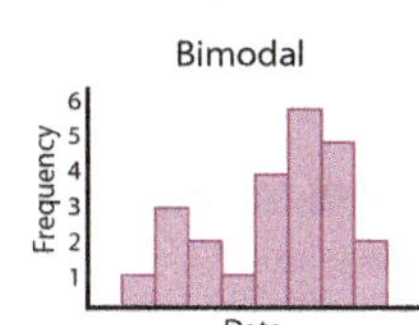

The Russian mathematician Pafnuty Chebyshev (1821–94) proved the following theorem describing the fraction of data values that falls within a given interval, regardless of the shape of the data distribution.

CHEBYSHEV'S THEOREM

When $k > 1$, at least $1 - \frac{1}{k^2}$ of the data values are within k standard deviations of the mean.

7. A normally distributed set of test scores has a mean of 84 and a standard deviation of 6. What percent of the data is within each interval?
   a. 72 to 84   47.5%
   b. 78 to 90   68%
   c. 90 or greater   16%

Assignments

- **Minimum:** 1–9, 12, 15–16, 19–20, 23–24, 27, 30–31, 34, 43–47
- **Standard:** 1–10, 12–14, 17, 19–20, 23–24, 27–29, 30–32, 35, 37, 41–42, 44–50
- **Extended:** 1–8, 10, 12–14, 17; 18–30 even; 31–33; 36–42 even; 46–50

In 1995 the SAT recentered scaled scores for the verbal (reading and writing) and math tests so that average scores would be closer to the midpoint of the scale at 500. Average math scores had fallen to 475 and verbal scores to 425.

This implies that less than $\frac{1}{k^2}$ of the data is outside the interval $[\bar{x} - ks, \bar{x} + ks]$.

Example 5 Applying Chebyshev's Theorem

Use Chebyshev's Theorem to find the smallest percentage of data that must always lie within 2 standard deviations of the mean. Confirm the theorem using the following sample of 18 quiz scores.

$Q = \{5, 6, 7, 8, 8, 8, 9, 9, 9, 10, 10, 10, 10, 10, 10, 11, 11, 11\}$

Answer

$1 - \frac{1}{2^2} = \frac{3}{4} = 0.75$

At least 75% of the data lies within 2 standard deviations of the mean.

1. Evaluate $1 - \frac{1}{k^2}$ with $k = 2$ and interpret the results.

$\bar{x} = 9$ and $s \approx 1.71$

2. Use technology to find $\bar{x}$ and s.

$[\bar{x} - 2s, \bar{x} + 2s]$
$[9 - 2(1.71), 9 + 2(1.71)] = [5.58, 12.42]$

3. Find the interval of values within 2 standard deviations of the mean.

outside: $\frac{1}{18} \approx 5.6\%$; inside: $\frac{17}{18} \approx 94.4\%$

4. Determine the percentages of values within and outside the interval.

At least 75% of the data falls within 2 standard deviations of the mean.
(Less than 25% of the data is more than 2 standard deviations from the mean.)

5. Chebyshev's Theorem accurately describes the data.

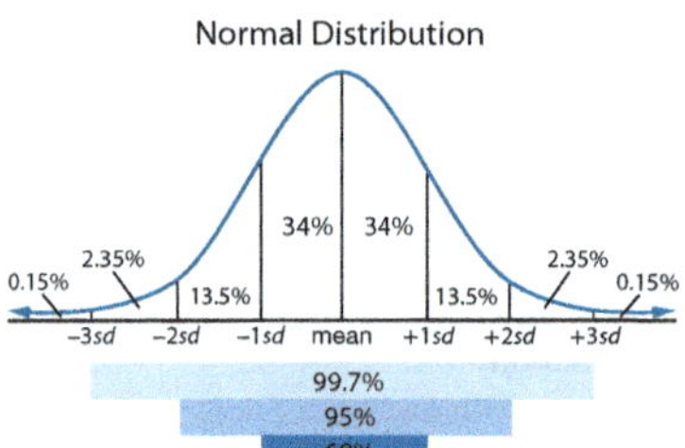

Since Chebyshev's Theorem applies to every set of data, it makes a very broad statement. More specific statements can be made for normally distributed data where the shape is described by the function

$$y = \frac{1}{\sqrt{2\pi}} e^{-\frac{1}{2}x^2}.$$

Recall that the mean, median, and mode are all located at the center of a normal distribution. The x-axis is the asymptote of the symmetric bell-shaped curve, and the area under the curve represents 100% of the data.

▌ THE EMPIRICAL RULE (OR THE 68-95-99.7 RULE)

In a normal distribution, practically all the data is within four standard deviations of the mean:

(1) about 68% of the data lies within one standard deviation of the mean, $\bar{x} \pm 1s$,
(2) about 95% of the data lies within two standard deviations of the mean, $\bar{x} \pm 2s$, and
(3) about 99.7% of the data lies within three standard deviations of the mean, $\bar{x} \pm 3s$.

When data is distributed normally, the Empirical Rule can be used to estimate the amount of data within certain intervals. Remember that the z-score represents the number of standard deviations that the individual data value is above or below the mean.

Example 6   Applying the Empirical Rule

A normally distributed set of test scores has a mean of 86 and a standard deviation of 4. What percentage of the data is within each interval?

a. from 82 to 90   **b.** 94 or less   **c.** from 78 to 90

Answer

a.

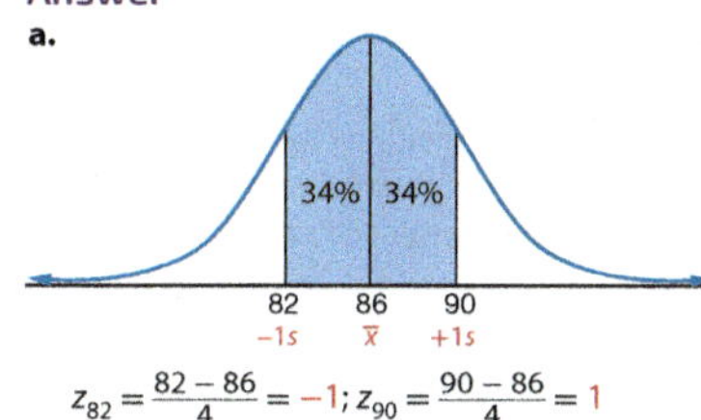

$$z_{82} = \frac{82 - 86}{4} = -1; \, z_{90} = \frac{90 - 86}{4} = 1$$

68% of the data is within [82, 90].

1. Sketch a normal curve centered at $\bar{x} = 86$ with the interval [82, 90]. Then shade the area under the curve within the interval.

2. Find the z-score for the lower and upper limits.

3. Apply the Empirical Rule.

b.

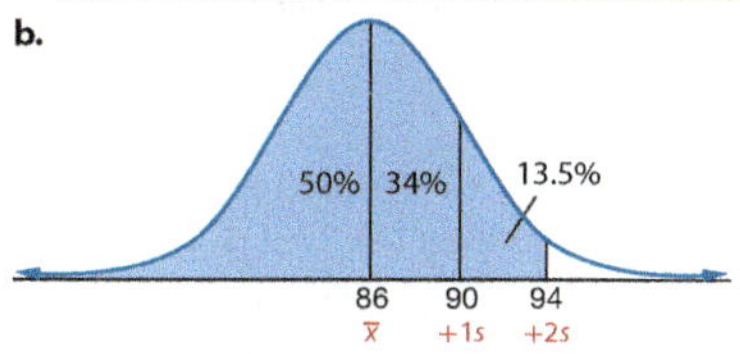

$$z_{94} = \frac{94 - 86}{4} = 2$$

$50\% + 34\% + 13.5\% = 97.5\%$

97.5% of the scores are 94 or less.

1. Sketch the normal curve with the interval $(-\infty, 94]$. Then shade the area under the curve within the interval.

2. Find the z-score for the upper limit.

3. Apply the Empirical Rule.

c.

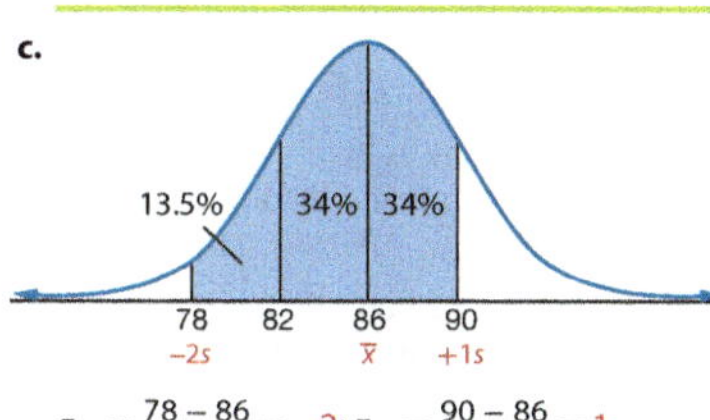

$$z_{78} = \frac{78 - 86}{4} = -2; \, z_{90} = \frac{90 - 86}{4} = 1$$

$13.5\% + 68\% = 81.5\%$

81.5% of the data is within [78, 90].

1. Sketch the normal curve with the interval [78, 90]. Then shade the area between the curve and the interval.

2. Find the z-scores for the lower and upper limits.

3. Apply the Empirical Rule.

SKILL ✔ EXERCISE 35

Even more specific statements regarding normal distributions will be investigated in the next section.

12. $z = \dfrac{83 - 83}{7} = 0$

The score is the mean.

13. $0.27 + (-0.16) + (0.75)$
$\quad + 1.33 + (-1.39) + z_6 = 0$
$\quad\quad\quad 0.80 + z_6 = 0$
$\quad\quad\quad\quad\quad z_6 = -0.80$

14a. $1 - \dfrac{1}{2^2} = \dfrac{3}{4} = 75\%$

15a. $1 - \dfrac{1}{3^2} = \dfrac{8}{9} = 88.9\%$

16a. $1 - \dfrac{1}{4^2} = \dfrac{15}{16} = 93.75\%$

B. Exercises

18. $z_1 = \dfrac{15 - 12}{2.2} = 1.36$

$z_2 = \dfrac{12 - 10}{1.5} = 1.33$

Her Quiz 1 grade was relatively better than her Quiz 2 grade.

19. $z_T = \dfrac{82 - 74}{10} = 0.8$

$z_P = \dfrac{162 - 150}{15} = 0.8$

His test grade was relatively as good as his project grade.

20. $z_T = \dfrac{79 - 83}{5} = -0.8$

$z_E = \dfrac{120 - 127}{9} \approx -0.78$

Her essay grade was relatively better than her test grade.

21. $z_A = \dfrac{25 - 22}{1.8} = 1.67$

$z_B = \dfrac{26 - 22}{2.5} = 1.60$

Her Quiz A grade was relatively better than her Quiz B grade.

22. $z = \dfrac{x - 41.3}{3.2}$
$x \approx 3.2(1.47) + 41.3 \approx 46$

23. $z = \dfrac{x - 20}{2.7}$
$x \approx 2.7(-1.1) + 20 \approx 17$

24.

```
         1-Var Stats
x̄=5.8
Σx=58
Σx²=370
Sx=1.932183566
σx=1.833030278
```

$\bar{x} = 5.8; s \approx 1.93; x = sz + \bar{x}$

24a. $x \approx 1.93(0.62) + 5.8 \approx 7$

24b. $x \approx 1.93(1.14) + 5.8 \approx 8$

24c. $x \approx 1.93(-0.41) + 5.8 \approx 5$

25.

```
         1-Var Stats
x̄=78.3
Σx=783
Σx²=62431
Sx=11.16592037
σx=10.59292217
```

$\bar{x} = 78.3; s \approx 11.17; x = sz + \bar{x}$

25a. $x \approx 11.17(-1.01) + 78.3 \approx 67$

25b. $x \approx 11.17(0.15) + 78.3 \approx 80$

A. Exercises

1. The z-scores measure the <u>deviation</u> of each individual score from the mean in units of <u>standard deviation</u>.

2. In the distribution of z-scores, the mean is <u>0</u> and the standard deviation is <u>1</u>.

3. The <u>mean</u>, <u>median</u>, and <u>mode</u> are all located at the center of a normal distribution.

4. When data is distributed normally, the <u>Empirical Rule</u> can be used to estimate the amount of data within certain intervals.

5. What percentage of data is represented under the normal curve? 100%

6. Explain when Chebyshev's Theorem and the Empirical Rule apply.

7. Describe the effect on each value if 5 points are added to each student's test score.
   a. mean increases by 5
   b. standard deviation remains the same
   c. z-scores remain the same

8. Describe the effect on each value if each test score is doubled.
   a. mean doubled
   b. standard deviation doubled
   c. z-scores remain the same

If a set of test scores has a mean of 83 and standard deviation of 7, find the z-score for each test score and interpret its meaning.

9. 88   **10.** 96   **11.** 79   **12.** 83

13. Find the missing z-score from a set of six sample data values.
$z_1 = 0.27, z_2 = -0.16, z_3 = 0.75, z_4 = 1.33, z_5 = -1.39,$
$z_6 = ?$ −0.80

State the percentage of data that must lie within each stated interval.
a. when using Chebyshev's Theorem
b. when using the Empirical Rule

14. within 2 standard deviations of the mean
   a. ≈ 75%; b. ≈ 95%
15. within 3 standard deviations of the mean
   a. ≈ 88.9%; b. ≈ 99.7%
16. within 4 standard deviations of the mean
   a. ≈ 93.75%; b. ≈ 100%
17. within 1 standard deviation of the mean
   a. does not apply; b. ≈ 68%

6. Chebyshev's Theorem applies to all distributions (if $k > 1$), while the Empirical Rule applies only to normal distributions.

B. Exercises

Use z-scores to compare the students' relative performances on two different assessments.

18. Sarah's grade of 15 on Quiz 1 with a class average of 12 and standard deviation of 2.2 versus her grade of 12 on Quiz 2 with a class average of 10 and standard deviation of 1.5

19. Adam's grade of 82 on a test with a class average of 74 and standard deviation of 10 versus his grade of 162 on a project with a class average of 150 and standard deviation of 15

20. Jessi's grade of 79 on a test with a class average of 83 and standard deviation of 5 versus her grade of 120 on an essay with a class average of 127 and standard deviation of 9

21. Grace's grade of 25 on Quiz A with a class average of 22 and standard deviation of 1.8 versus her grade of 26 on Quiz B with a class average of 22 and standard deviation of 2.5

Use the data to find the integral raw score.

22. $z \approx 1.47; \sigma = 3.2; \mu = 41.3$ 46
23. $z \approx -1.1; s = 2.7; \bar{x} = 20$ 17
24. sample scores: 3, 4, 4, 5, 5, 6, 7, 7, 8, 9
   a. $z \approx 0.62$ 7   **b.** $z \approx 1.14$ 8   **c.** $z \approx -0.41$ 5
25. sample scores: 62, 67, 68, 72, 75, 80, 86, 88, 90, 95
   a. $z \approx -1.01$ 67   **b.** $z \approx 0.15$ 80   **c.** $z \approx 1.50$ 95

A fisherman recorded the following lengths (in inches) of the fish caught during his trip.
$L = \{10, 11, 11, 11, 11, 12, 15, 15, 15, 15, 15, 16, 16, 17, 20\}$

26. Find the interval $\bar{x} \pm 2s$. [8.3, 19.7]
27. Find the interval $\bar{x} \pm 3s$. [5.4, 22.6]
28. What percentage of the data actually lies within 2 standard deviations of the mean? Is this consistent with Chebyshev's Theorem? 93%; Yes, it exceeds 75%.
29. What percentage of the data actually lies within 3 standard deviations of the mean? Is this consistent with Chebyshev's Theorem? 100%; Yes, it exceeds 88.9%.

SAT scores are calculated from z-scores using the transformation SAT $= 100z + 500$.

30. State the mean score on the SAT. 500
31. State the standard deviation of the SAT. 100
32. State the SAT score that is 2.5 standard deviations above the mean. 750
33. Determine the number of standard deviations each score is above or below the mean.
   a. 563   **b.** 381

25c. $x \approx 11.17(1.50) + 78.3 \approx 95$

26–27. $\bar{x} = \dfrac{210}{15} = 14; SS = 114$

$s = \sqrt{\dfrac{114}{14}} \approx 2.85$

26. $[14 - 2(2.85), 14 + 2(2.85)]$
$= [8.3, 19.7]$

27. $[14 - 3(2.85), 14 + 3(2.85)]$
$= [5.4, 22.6]$

28. There are 14 out of 15 data values in the interval [8.3, 19.7].
$\dfrac{14}{15} = 93\% \geq 75\%$

30–31. using $x_i = \sigma z + \mu, \sigma = 100$:
$\mu = 500$

32. $x = 100(2.5) + 500 = 750$

33a. $100z + 500 = 563; z = 0.63$
0.63 standard deviations above the mean

33b. $100z + 500 = 381; z = -1.19$
1.19 standard deviations below the mean

34a. $100\% - (0.15\% + 0.15\%) = 99.7\%$
34b. $13.5\% + 34\% + 34\% = 81.5\%$
34c. $34\% + 13.5\% = 47.5\%$
34d. $2.35\% + 0.15\% = 2.5\%$
35a. $[70, 100] = [\mu - 2\sigma, \mu]$
$13.5\% + 34\% = 47.5\%$
35b. $[85, 130] = [\mu - \sigma, \mu + 2\sigma]$
$34\% + 34\% + 13.5\% = 81.5\%$

34. Use the Empirical Rule to find the percentage of data within each described interval.
  - **a.** $\mu \pm 3\sigma$ 99.7%
  - **b.** $[\mu - 2\sigma, \mu + \sigma]$ 81.5%
  - **c.** $[\mu, \mu + 2\sigma]$ 47.5%
  - **d.** $x_i \geq \mu + 2\sigma$ 2.5%

35. A set of normally distributed test scores has $\mu = 100$ and $\sigma = 15$. Find the percentage of scores within each interval.
  - **a.** [70, 100] 47.5%
  - **b.** [85, 130] 81.5%
  - **c.** [0, 115] 84%
  - **d.** $x_i \geq 115$ 16%

A normally distributed set of test scores has a mean of 88 and a standard deviation of 3.

36. What percentage of the data is in each interval?
  - **a.** [88, 94] 47.5% **b.** [85, 94] 81.5% **c.** $x_i \leq 85$ 16%

37. Which interval contains each described percentage of scores?
  - **a.** the middle 68% [85, 91]
  - **b.** the top 2.5% $x_i \geq 94$
  - **c.** scores below the 84th percentile $x_i \leq 91$

❯ C. Exercises

Prove the following relations for sets of transformed data $y_i = kx_i$.

38. $\bar{y} = k\bar{x}$

39. $s_y = ks_x$

40. The normally distributed 2016 ACT scores have a mean of 20.8 and a standard deviation of 4.7.
  - **a.** What percentage of scores were from 16.1 to 25.5? 68%
  - **b.** What score represents the 84th percentile? 25.5
  - **c.** If students whose scores are in the top 2.5% are eligible for a scholarship, what is the minimum score needed to earn the scholarship? 30.2

41. The normally distributed 2017 SAT scores have a mean of 1060 and a standard deviation of 195.
  - **a.** What scores make up the middle 95% of the data? [670, 1450]
  - **b.** What is the percentile rank for a score of 865? 16th
  - **c.** If 1.7 million students took the test, estimate how many students scored above 1450. 42,500 Students

42. Jeff received a 28 on the 2016 ACT and a 1355 on the 2017 SAT. Based on the means and standard deviations given in exercises 40–41, which score was better? Explain your reasoning.
The ACT score was better since $z_A \approx 1.53$ while $z_S \approx 1.51$.

40b. 84th percentile: 50% + 34%
$$x_i \leq \mu + 1\sigma \text{ or } 20.8 + 4.7 = 25.5$$

40c. $x_i \geq 97.5\%$ or $\mu + 2\sigma$
$$= 20.8 + 2(4.7) = 30.2$$

41a. $\mu \pm 2\sigma = 1060 \pm 2(195)$
$$= [670, 1450]$$

41b. $z = \dfrac{865 - 1060}{195} = -1; \ x_i \leq \mu - \sigma$
$$13.5\% + 2.35\% + 0.15\% = 16\%$$

41c. $z = \dfrac{1450 - 1060}{195} = 2; \ x_i \geq \mu + 2\sigma$
$$2.35\% + 0.15\% = 2.5\%$$
$$0.025(1{,}700{,}000) = 42{,}500 \text{ students}$$

42. $z_A = \dfrac{28 - 20.8}{4.7} \approx 1.53$
$$z_S = \dfrac{1355 - 1060}{195} \approx 1.51$$

❯ Cumulative Review

43. 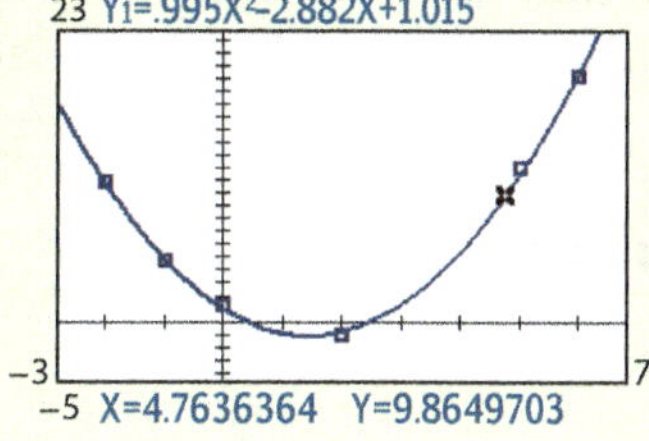

44. 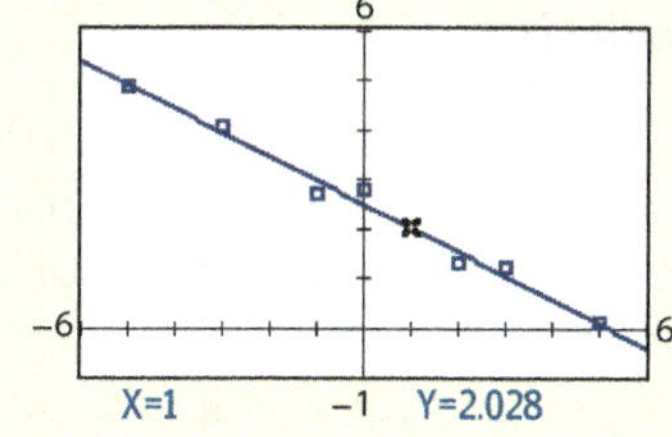

46. $E = \begin{bmatrix} 2 + 3(3) & 6 + 3(-5) \\ 4 + 3(-2) & 1 + 3(0) \end{bmatrix}$
$$= \begin{bmatrix} 11 & -9 \\ -2 & 1 \end{bmatrix}$$

47. $r = \sqrt{(5 - 2)^2 + (0 - (-4))^2}$
$$= \sqrt{9 + 16} = 5$$
$$(x - 2)^2 + (y + 4)^2 = 5^2$$

48. $B^2 - 4AC = 0 - 4(4)(9) = -144 < 0$
∴ an ellipse
$$4x^2 + 9y^2 - 16x + 54y + 61 = 0$$
$$4(x^2 - 4x + 4) + 9(y^2 + 6y + 9)$$
$$= -61 + 4(4) + 9(9)$$
$$4(x - 2)^2 + 9(y + 3)^2 = 36$$
$$\frac{(x - 2)^2}{9} + \frac{(y + 3)^2}{4} = 1$$

50. $_{10}C_5(2a)^5(-b)^5 = 252(32a^5)(-b)^5$
$$= -8064a^5b^5$$

51. $1 + 3.3 \log 55 \approx 6.74; \ k = 7$
$$w = \frac{range}{k} = \frac{533 - 166}{7} \approx 53$$

CUMULATIVE REVIEW

Use technology to create a scatterplot of each set of data and determine the linear or quadratic function that best models the data. Round coefficients to the nearest thousandth. [1.9]

43. $f(x) = 0.995x^2 - 2.882x + 1.015$

x	-2	-1	0	2	5	6
y	10.8	4.8	1.2	-1.1	12.0	19.2

44. $f(x) = -0.484x + 2.512$

x	-5	-3	-1	0	2	3	5
y	4.9	4.1	2.7	2.8	1.3	1.2	0.1

Given $A = \begin{bmatrix} 2 & 6 \\ 4 & 1 \end{bmatrix}$ and $B = \begin{bmatrix} 3 & -5 \\ -2 & 0 \end{bmatrix}$, find D and E. [7.2]

45. $D = -\frac{3}{4}B$

46. $E = A + 3B$

47. Write the standard form equation of a circle centered at $(2, -4)$ and passing through point $(5, 0)$. [8.2]

48. Classify the conic $4x^2 + 9y^2 - 16x + 54y + 61 = 0$ and then write the equation in standard form. [8.2]

49. Which strategies can be used to verify trigonometric identities? List all correct answers. [5.2] A, B, C, D
  - **A.** factoring
  - **B.** combining fractions
  - **C.** simplifying each side separately
  - **D.** applying fundamental identities
  - **E.** squaring both sides

50. Find the sixth term of $(2a - b)^{10}$. [9.4] C
  - **A.** $-13{,}440a^4b^6$
  - **B.** $-64a^5b^5$
  - **C.** $-8064a^5b^5$
  - **D.** $-128a^4b^6$
  - **E.** none of these

51. A set of 55 data values ranges from 166 to 533. What class width results from Sturges's formula for the grouped frequency distribution table or histogram representing the data? [10.3] D
  - **A.** 6.7
  - **B.** 7
  - **C.** 52.4
  - **D.** 53
  - **E.** none of these

52. Which is the formula for the sample standard deviation? [10.4] A
  - **A.** $s = \sqrt{\dfrac{SS}{n - 1}}$
  - **B.** $s^2 = \dfrac{SS}{n - 1}$
  - **C.** $\sigma = \sqrt{\dfrac{SS}{N}}$
  - **D.** $\sigma^2 = \dfrac{SS}{N}$
  - **E.** none of these

45. $\begin{bmatrix} -\frac{9}{4} & \frac{15}{4} \\ \frac{3}{2} & 0 \end{bmatrix}$

46. $\begin{bmatrix} 11 & -9 \\ -2 & 1 \end{bmatrix}$

47. $(x - 2)^2 + (y + 4)^2 = 25$

48. ellipse; $\dfrac{(x - 2)^2}{9} + \dfrac{(y + 3)^2}{4} = 1$

35c. $x_i \leq 115 = x_i \leq \mu + \sigma$
$$50\% + 34\% = 84\%$$

35d. using part c: $100\% - 84\% = 16\%$

36a. $[88, 94] = [\mu, \mu + 2\sigma]$
$$34\% + 13.5\% = 47.5\%$$

36b. $[82, 91] = [\mu - 2\sigma, \mu + \sigma]$
$$13.5\% + 34\% + 34\% = 81.5\%$$

36c. $x_i \leq 85 = x_i \leq \mu - \sigma$
$$50\% - 34\% = 16\%$$

37a. $[\mu - 1\sigma, \mu + 1\sigma] = [88 - 3, 88 + 3]$
$$= [85, 91]$$

37b. $x_i \geq \mu + 2\sigma$ or $88 + 2(3); \ x_i \geq 94$

37c. 84th percentile: 50% + 34%
$$x_i \leq \mu + 1\sigma \text{ or } 88 + 3; \ x_i \leq 91$$

❯ C. Exercises

38. $\bar{y} = \dfrac{\sum\limits_{i=1}^{n} y_i}{n} = \dfrac{\sum\limits_{i=1}^{n} kx_i}{n} = \dfrac{k\sum\limits_{i=1}^{n} x_i}{n}$
$$= k\bar{x}$$

39. $s_y = \sqrt{\dfrac{\sum\limits_{i=1}^{n}(y_i - \bar{y})^2}{n - 1}} = \sqrt{\dfrac{\sum\limits_{i=1}^{n}(kx_i - k\bar{x})^2}{n - 1}}$

$$= \sqrt{\dfrac{\sum\limits_{i=1}^{n}[k(x_i - \bar{x})]^2}{n - 1}} = \sqrt{\dfrac{\sum\limits_{i=1}^{n}k^2(x_i - \bar{x})^2}{n - 1}}$$

$$= \sqrt{\dfrac{k^2\sum\limits_{i=1}^{n}(x_i - \bar{x})^2}{n - 1}} = k\sqrt{\dfrac{\sum\limits_{i=1}^{n}(x_i - \bar{x})^2}{n - 1}}$$

$$= ks_x$$

40a. $[16.1, 25.5] = \mu \pm \sigma; \ 68\%$

10.6 Normal Distributions

Objectives

1. To use *z*-scores and standard normal distribution tables or technology to find the percentage of data within a range of normally distributed values

2. To find percentiles and percentile ranks for given values in a normal distribution

Flash

The Chinook salmon is the biggest of the Pacific salmons. Their God-given navigational ability to return back to the river or stream in which they were born to spawn after spending 1 to 8 yr maturing in the ocean continues to amaze us.

Vocabulary

percentile rank

Reading and Writing Mathematics

Compare and contrast normal distributions and the standard normal distribution. Then explain the advantage of the standard normal distribution. All normal distributions exhibit a bell-shaped curve, but they can have different means and standard deviations; the standard normal distribution has a mean of 0 and a standard deviation of 1. Representing different normal distributions in terms of the standard normal distribution permits valuable comparisons.

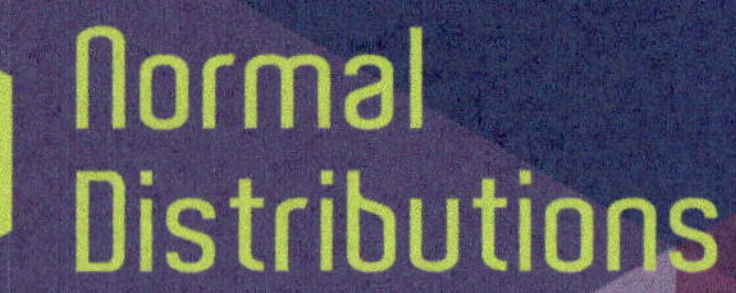
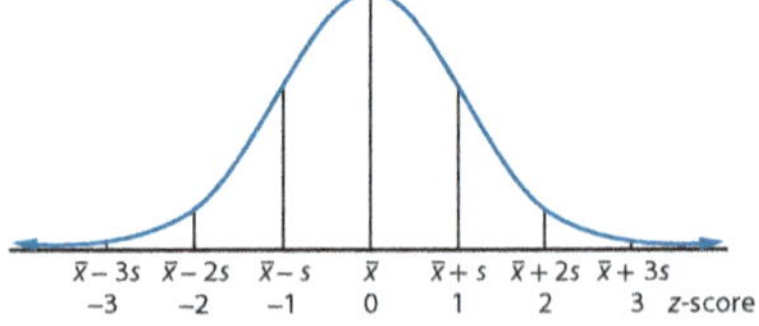

Some Chinook salmon have reached more than 1.5 m in length.

After completing this section, you will be able to

- use *z*-scores to find the percentage of data within a range of normally distributed values.
- find percentiles and percentile ranks for given values in a normal distribution.

z	0.00	0.01
0.0	0.0000	0.0040
0.1	0.0398	0.0438
0.2	0.0793	0.0832
0.3	0.1179	0.1217
0.4	0.1554	0.1591
0.5	0.1915	0.1950
0.6	0.2257	0.2291
0.7	0.2580	0.2611
0.8	0.2881	0.2910
0.9	0.3159	0.3186
1.0	0.3413	0.3438
1.1	0.3643	0.3665
1.2	0.3849	0.3869
1.3	0.4032	0.4049
1.4	0.4192	0.4207
1.5	0.4332	0.4345

KEYWORD SEARCH

interactive normal distribution

Remember that the area under the normal curve represents 100% of the data values. Since the normal curve is symmetric with respect to the mean, half of the data is represented by positive *z*-scores and the other half is represented by negative *z*-scores. Recall that practically all the area under the curve lies within 4 standard deviations from the mean and that the Empirical (or 64-95-99.7) Rule describes the percentages of data values within 1, 2, and 3 standard deviations from the mean. The probability that a randomly selected value in a normal distribution is within a given interval is found by determining the percentage of the area under the curve for that interval.

Standard normal distribution tables (available online) often list the portion of the area from the mean to a positive *z*-score. Notice that the table shows that the portion of the area from the mean to a *z*-score of 1.00 is 0.3413 or 34.13%. Other tables may list the portion of the area at or below the listed positive or negative *z*-score. Those tables indicate an area of 0.8413 or 84.13% at or below a *z*-score of 1.

The portion of the area can also be found using your calculator.

Example 1   Finding an Area Under the Normal Curve

Find the portion of the area under the normal curve within the interval $-0.75 \leq z \leq 0$.

Answer

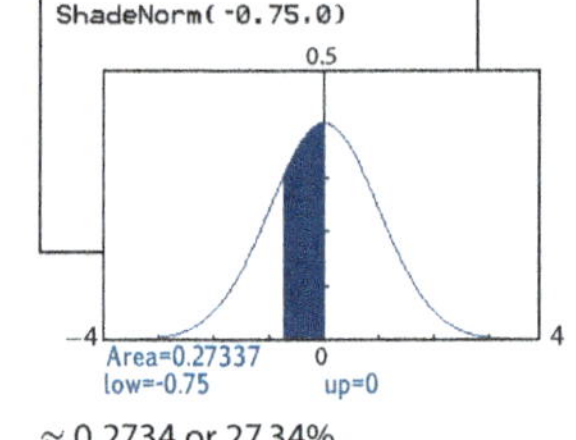

≈ 0.2734 or 27.34%

Symmetry implies that ≈ 0.2734 or 27.34% of the area under the curve lies within [−0.75, 0].

Use DRAW ([2nd], [PRGM]), 1:ClrDraw to clear the graphing window. Then enter DISTR ([2nd], [VARS]), select the DRAW submenu, 1:ShadeNorm, and then enter the *lower z-score*, and the *upper z-score*.

Note that the default values of 0 and 1 for the mean and standard deviation could be added to the list of the function's parameters. A [WINDOW] of [−4, 4] × [0, 0.5] is recommended when graphing standard normal distributions.

or

Mean-centered standard normal distribution tables state that ≈ 0.2734 of the area lies between the mean and $z = 0.75$.

SKILL ✔ EXERCISE 3

PRESENTATION

Lesson Opener

State the percentage of values within each interval of the standard normal distribution.

1. $[-1\sigma, \mu]$ 34%   2. $[-1\sigma, 1\sigma]$ 68%

3. $(-\infty, \mu]$ 50%   4. $[-2\sigma, 2\sigma]$ 95%

The students may have heard the normal distribution curve informally referred to as the bell curve. A normal distribution is a special type of bell curve with specific portions of values in each interval of the distribution. Re-emphasize that a normal distribution with a mean of 0 and a standard deviation of 1 is a standard normal distribution.

The procedure for finding the areas under the normal curve in Example 1 and Example 2 is given using the ShadeNorm function of the TI-84 calculator and a mean-centered standard normal distribution table. This type of table and several other types of standard normal distribution tables (typically left-tailed tables) are readily available online. The text relies on the ShadeNorm, normalcdf, and invNorm functions found on the calculator, but the related exercises can also be solved using tables. Be sure to communicate how you expect students to solve these types of exercises.

Examples 3 and 4 continue to use the ShadeNorm function. Note that turning ON the STAT WIZARDS option (found in [MODE] options), causes this command to produce the following screen, which by default uses *z*-scores and the standard normal distribution.

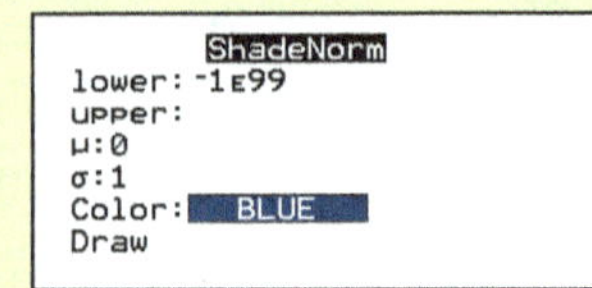

Entering the population's mean and standard deviation allows the probabilities to be found using the raw scores instead of *z*-scores, but determining a good

Find the area under the normal curve lying within 1.5 standard deviations of the mean.

Answer

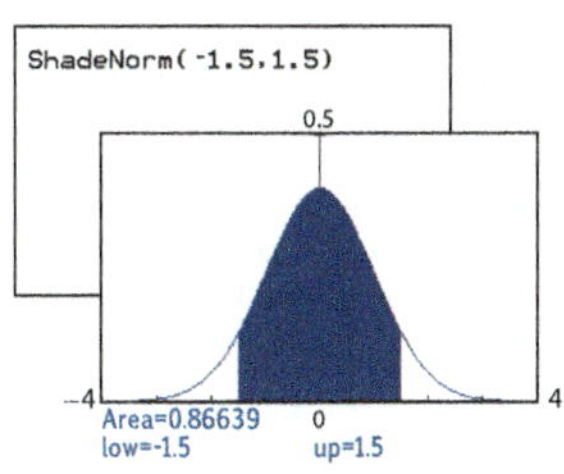

Use 1:ClrDraw from the DRAW menu to clear any previously drawn graphs.

Then select 1:ShadeNorm from the DISTR menu and enter *lower limit, upper limit*.

≈ 0.8664 or 86.64%

or

Symmetry implies that $2(0.4332) = 0.8664$ or $\approx 86.64\%$ of the data lies in the interval $-1.5 \le z \le 1.5$.

Mean-centered standard normal distribution tables state that ≈ 0.4332 of the area lies between the mean and $z = 1.5$.

SKILL ✓ EXERCISE 7

Recall that the area under the normal curve represents the percentage of values under the normal curve in a normally distributed set of data. Therefore, the area also represents the probability of a randomly selected outcome falling within a given interval.

Example 3   Using Technology to Find a Probability

Find the probability that a data value has $z \ge 0.98$.

Answer

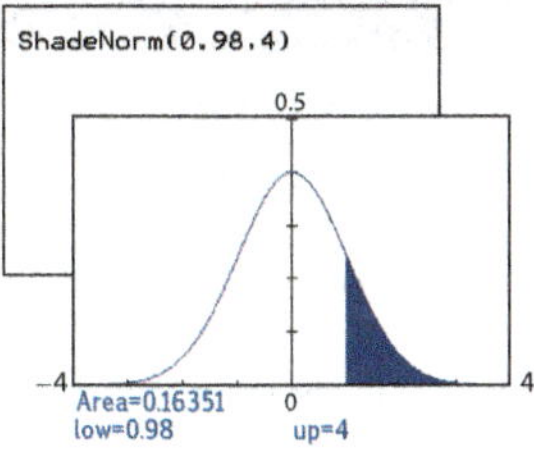

Use 1:ClrDraw to clear any previously drawn graphs.

Then use 1:ShadeNorm with 0.98 as the lower limit and 4 as the upper limit (since almost all the data is within 4 standard deviations of the mean in a standard normal distribution).

The probability that $z \ge 0.98 \approx 16.35\%$.

or

$50\% - 33.65\% = 16.35\%$
16.35% of the data has $z \ge 0.98$.

Mean-centered standard normal distribution tables state that 0.3365 of the area lies between the mean and $z = 0.98$.

SKILL ✓ EXERCISE 15

While the following examples are solved using technology, they could also be solved using standard normal distribution tables.

viewing window becomes much more complicated.

Common Student Error Students may forget to change window settings when using ShadeNorm if a display is desired when the mean is not 0 and the data range increases.

Using -4 and 4 for lower and upper limits produces very reasonable approximations, but some may prefer to use values such as the default lower limit value of $-1\text{E}99$ and an upper limit value of $1\text{E}99$ to ensure greater accuracy.

The use of the normalcdf function is demonstrated in Example 5. The example utilizes z–scores and the standard normal distribution. Most statistics software allows probabilities to be found directly without z-scores by entering the population's mean and standard deviation after the interval limits. We encourage the use of z-scores, which standardize different distributions, enabling quick and meaningful comparisons.

Remind the students of the difference between percentile and percentage to avoid confusion. In Example 6 the normalcdf function is used to find the *percentile rank* after the raw score is converted to a z-score.

There are several accepted definitions for percentile, all of which round

English statisticians Francis Galton and Karl Pearson made many contributions in the field of descriptive statistics. However, they also used their results to champion eugenics, an attempt to improve the gene pool by sterilizing persons of the lower class and encouraging procreation in the upper class. In 1927 the US Supreme Court ruled that involuntary sterilization of disabled persons did not violate the Constitution. The ruling was overturned in 1942.

Additional Exercises

1. Find the area under the normal curve in the interval $-0.88 \le z \le 0$. 31.1%

2. Find the area under the normal curve lying within 1.2 standard deviations of the mean. 77.0%

3. Find the probability that a data value has $z \ge 1.8$. 3.6%

4. Find the probability that a randomly selected value in a normal distribution with a mean of 51 and standard deviation of 5.2 is within the interval [48, 60]. 67.6%

5. Find the probability that a randomly selected student scored between 25 and 30 on the ACT with a mean of 20.9 and a standard deviation of 5.6. 18.0%

6. Find the percentile rank of Jon's ACT score of 31 if the mean is 21 and the standard deviation is 5.4. 97th percentile

7. Using the information from the previous problem, what score would be at the 80th percentile? ≈ 26

8. A normally distributed set of data has a mean of 120 and a standard deviation of 15. Find the interval of values centered at the mean that contains 50% of the scores. [110, 130]

Assignments

- **Minimum:** 1–11, 14, 17, 19, 21, 24–25, 28, 31–32, 35–36, 45, 49–51, 53–54

- **Standard:** 1–7, 10, 12, 15, 17, 20–21, 24–25, 28, 30, 33–34, 37, 39–40; 46–52 even; 53–54

- **Extended:** 5–7, 9; 10–18 even; 24–40 even; 42–44, 48, 51–54

Assessment

- Quiz 10C covers Sections 10.5–10.6.

Example 4   Using Technology to Find a Probability

Find the probability that a randomly selected value in a normal distribution with a mean of 43 and standard deviation of 6.25 is within the interval [36, 51].

Answer

$$z_L = \frac{36 - 43}{6.25} = -1.12; z_U = \frac{51 - 43}{6.25} = 1.28$$

1. Find the z-scores for the upper and lower bounds of the interval.

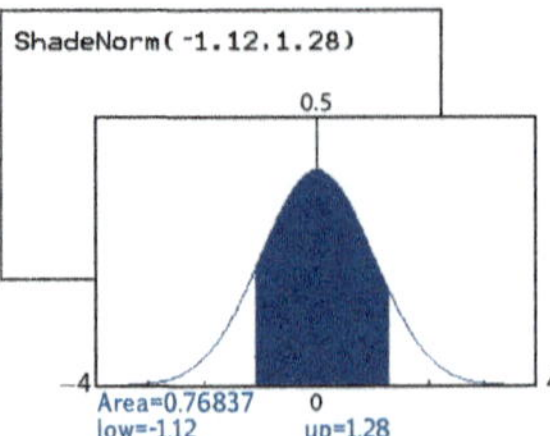

2. Clear any drawings and use 1:ShadeNorm to find the portion of the area under the standard normal distribution curve within [−1.12, 1.28].

```
normalcdf(-1.12,1.28)
             0.7683704396
```

≈ 0.768 or 76.8%

or

Using the normal cumulative density function, 2:normalcdf, from the DISTR menu and entering *lower z-score, upper z-score* returns the portion of the area under the curve within the stated *z*-scores.

SKILL ✔ **EXERCISE 21**

If the population's mean and standard deviation are known, the individual scores are standardized relative to the entire population by $z = \frac{x - \mu}{\sigma}$. The z-score formula, $z = \frac{x - \bar{x}}{s}$, standardizes individual scores in relation to the other scores in the sample data.

Example 5   Finding Probabilities

The Chinook salmon is the largest Pacific salmon species. If a study found that the lengths of the Chinook salmon caught were normally distributed with a mean of 78.5 cm and a standard deviation of 8.2 cm, find the probability that a randomly caught Chinook salmon has each of the following lengths.

a. 95 cm or more

b. from 70 cm to 80 cm

Answer

a. $z = \frac{95 - 78.5}{8.2} \approx 2.01$

1. Find the z-score for the lower limit, 95.

```
normalcdf(2.01,4)
           0.0221838387
```

$P(l \geq 95) \approx 0.022$ or 2.2%

2. Use 2:normalcdf from the [DISTR] menu to find the portion of the area under the standard normal distribution curve within [2.01, 4].

b. $z_L = \frac{70 - 78.5}{8.2} \approx -1.04; z_U = \frac{80 - 78.5}{8.2} \approx 0.18$

1. Find the z-scores for the upper and lower bounds of the interval.

```
normalcdf(-1.04,0.18)
           0.4222537389
```

$P(70 < l < 80) \approx 0.42$ or 42%

2. Use 2:normalcdf to find the portion of the area under the standard normal distribution curve within [−1.04, 0.18].

SKILL ✔ **EXERCISE 27**

---

percentile rankings to a whole number. Some round to the nearest whole number while others always round up.

In Example 7 invNorm is used to find the *z*-score; then the raw score is found by solving for *x* in the *z*-score formula.

Be sure students understand why the *z*-score for the 72nd percentile (22% above the mean) in Example 8 is found using invNorm when looking for the middle 44% of the data. Emphasize that the *z*-score is again converted to a raw score using the formula for *z*-scores before the raw score is added to and subtracted from the mean to get the raw score interval.

Interactive Activity There are many statistical functions in spreadsheets, statistical mobile apps, and websites with interactive normal distributions. Consider encouraging students to explore these resources, or select one of these resources to explore together in class.

TIPS

Ex. 1–4 Consider having students use tables to complete these exercises and then use technology to verify their work.

The *percentile rank* of a particular data value within a normally distributed set of data can be found by finding its z-score and finding the percent of data in the area to its left.

Example 6 Finding the Percentile Rank of a Data Value

Find the percentile rank of a student whose quiz score was 29 in a class whose scores were approximately normally distributed with a mean of 27 and a standard deviation of 4.

Answer

$z = \dfrac{x - \bar{x}}{s} = \dfrac{29 - 27}{4} = 0.5$

```
normalcdf(-4,0.5)
              0.6914307818
```

1. Convert to a z-score.

2. Find the portion of scores below $z = 0.5$ in the standard normal distribution.

The student scored at the 69th percentile.

— SKILL ✔ **EXERCISE 31**

A calculator's invNorm (inverse Normal) function or standard normal deviation tables can be used to find the z-value representing a specific percentile when data is normally distributed.

Example 7 Finding the Data Value for a Percentile Rank

Find the value at the 75th percentile in a normally distributed set of data whose mean is 19.7 and whose standard deviation is 2.8.

Answer

```
invNorm(0.75)
              0.6744897495
```

$\dfrac{x - 19.7}{2.8} \approx 0.674$
$x \approx 2.8(0.674) + 19.7 \approx 21.6$
$\therefore$ 21.6 is at the 75th percentile.

1. Selecting 3:invNorm from the DISTR menu and entering the percentage of data (or area) to the left of the stated percentile produces the related z-score.

2. Find the data value associated with that z-score.

TIP

Many of your calculator's distribution functions allow you to enter the mean and standard deviation as parameters and will then return the related data value instead of the z-score.

```
invNorm(.75,19.7,2.8)
              21.5885713
```

— SKILL ✔ **EXERCISE 17**

Example 8 Finding an Interval of Scores

A normally distributed set of data has a mean of 1200 and a standard deviation of 150. Find the interval of values centered at the mean that contains 44% of the scores.

Answer

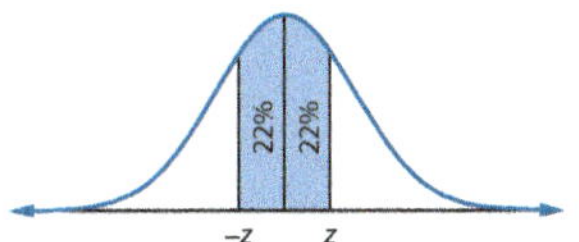

1. Sketch the desired area.

   The upper boundary of the interval is at the 72nd percentile because 50% of the data is to the left of the mean, leaving 22% between the mean and the upper boundary.

   The lower boundary is the 28th percentile.

```
invNorm(.72)
              0.582841502
```

$1200 \pm (0.5828)(150) \approx 1113, 1287$
$\therefore$ [1113, 1287] contains the middle 44% of the data.

2. Use 3:invNorm to find the z-score for the 72nd percentile.

3. Symmetry implies that the interval's bounds are ≈ 0.5828 standard deviations above and below the mean.

— SKILL ✔ **EXERCISE 35**

Solutions

❭ A. Exercises

1. normalcdf$(0, 1.7) \approx 0.4554$

2. normalcdf$(-2.47, 0) \approx 0.4932$

The first line of each solution in exercises 3–16 uses a standard normal distribution table.

3. $0.5 - 0.3554 = 0.1446$ or
normalcdf$(1.06, 4) \approx 0.1445$

4. $0.3485 + 0.0987 = 0.4472$ or
normalcdf$(-1.03, 0.25) \approx 0.4472$

5. normalcdf$(0, 0.69) \approx 0.2549$

6. normalcdf$(-1.21, 0) \approx 0.3869$

7. $2(0.2704) = 0.5408$ or
normalcdf$(-0.74, 0.74) \approx 0.5407$

8. $2(0.4678) = 0.9356$ or
normalcdf$(-1.85, 1.85) \approx 0.9357$

9. $0.5 - 0.4911 = 0.0089$ or
normalcdf$(2.37, 4) \approx 0.0090$

10. $0.5 - 0.4406 = 0.0594$ or
normalcdf$(-4, -1.56) \approx 0.05935$

11. $0.5 - 0.3212 = 0.1788$ or
normalcdf$(-4, -0.92) \approx 0.1788$

12. $0.4192 + 0.5 = 0.9192$ or
normalcdf$(-1.4, 4) \approx 0.9192$

13. $0.5 + 0.4772 = 0.9772$ or
normalcdf$(-4, 2) \approx 0.9772$

14. $2(0.1179) = 0.2358$ or
normalcdf$(-0.3, 0.3) \approx 0.2358$

15. $0.4192 + 0.4893 = 0.9085$ or
normalcdf$(-1.4, 2.3) \approx 0.9085$

16. $0.4977 + 0.3665 = 0.8642$ or
normalcdf$(-2.83, 1.11) = 0.8642$

❭ B. Exercises

17. $z_L = \dfrac{79 - 83}{5.9} \approx -0.678$
$z_U = \dfrac{90 - 83}{5.9} \approx 1.186$
normalcdf$(-0.678, 1.186) \approx 0.633$

18. $z_L = \dfrac{6.5 - 8.375}{1.04} \approx -1.803$
$z_U = \dfrac{9 - 8.375}{1.04} \approx 0.601$
normalcdf$(-1.803, 0.601) \approx 0.690$

19. $z_L = \dfrac{22 - 20}{7.9} \approx 0.253$
$z_U = \dfrac{25 - 20}{7.9} \approx 0.633$
normalcdf$(0.253, 0.633) \approx 0.137$

20. $z_L = \dfrac{35 - 45}{9.26} \approx -1.080$
$z_U = \dfrac{40 - 45}{9.26} \approx -0.540$
normalcdf$(-1.080, -0.540) \approx 0.155$

21. $z_L = \dfrac{67 - 68.36}{2.72} = -0.5$
$z_U = \dfrac{70 - 68.36}{2.72} \approx 0.603$
normalcdf$(-0.5, 0.603) \approx 0.418$

22. $z_L = \dfrac{18 - 16.44}{1.95} = 0.8$
$z_U = \dfrac{20 - 16.44}{1.95} \approx 1.826$
normalcdf$(0.8, 1.826) \approx 0.178$

23. $z_L = \dfrac{32 - 20.9}{5.6} \approx 1.982$
$z_U = \dfrac{36 - 20.9}{5.6} \approx 2.696$
normalcdf$(1.982, 2.696) \approx 0.020$

24. $z_L = \dfrac{400 - 527}{107} \approx -1.187$
$z_U = \dfrac{500 - 527}{107} \approx -0.252$
normalcdf$(-1.187, -0.252) \approx 0.283$

25. $0.5 + 0.3577 = 0.8577$ or
normalcdf$(-4, 1.07) \approx 0.8577$

26. $0.5 - 0.1026 = 0.3974$ or
normalcdf$(-4, -0.26) \approx 0.3974$

27. $z = \dfrac{90 - 85}{3} \approx 1.67$
$0.5 + 0.4525 = 0.9525$ or
normalcdf$(-4, 1.67) \approx 0.9525$

28. $z = \dfrac{64 - 69}{2.94} \approx -1.7$
$0.5 - 0.4554 = 0.0446$ or
$\text{normalcdf}(-4, -1.7) \approx 0.0445$

29. $z = \dfrac{8.5 - 8.83}{1.6} \approx -0.21$
$0.5 - 0.0832 = 0.4168$ or
$\text{normalcdf}(-4, -0.21) \approx 0.4168$

30. $z = \text{invNorm}(0.6) \approx 0.2533$
$\dfrac{x - 40}{7.2} \approx 0.2533;\ x \approx 41.8$

31. $z = \text{invNorm}(0.45) \approx -0.1257$
$\dfrac{x - 8.2}{1.2} \approx -0.1257;\ x \approx 8.05$

32. $z = \text{invNorm}(0.05) \approx -1.6449$
$\dfrac{x - 59}{3.1} \approx -1.6449;\ x \approx 53.9$

33. $z = \text{invNorm}(0.75) \approx 0.6744$
$\dfrac{x - 79}{6.8} \approx 0.6744;\ x \approx 83.6$

34. $z = \text{invNorm}(0.2) \approx -0.842$

35. 90th percentile or better
$z = \text{invNorm}(0.9) \approx 1.282$

36. $\dfrac{0.34}{2} = 0.17$
$0.5 \pm 0.17 = [0.33, 0.67]$
$z = \text{invNorm}(0.33) \approx -0.4399$
$z = \text{invNorm}(0.67) \approx 0.4399$

37. $\dfrac{0.97}{2} = 0.485$
$0.5 \pm 0.485 = [0.015, 0.985]$
$z = \text{invNorm}(0.015) \approx -2.17$
$z = \text{invNorm}(0.985) \approx 2.17$

38. $z = \text{invNorm}(0.75) \approx 0.6745$
$7.5 + 0.6745(1.5) \approx 8.5$

39. $z = \text{invNorm}(0.33) \approx -0.4399$
$81 - 0.4399(9.6) \approx 76.77$

40. $\dfrac{0.2}{2} = 0.1;\ [0.4, 0.6]$
$z = \text{invNorm}(0.6) \approx 0.2533$
$75 \pm 0.2533(7.8) \approx 73.02, 76.98$

41. $\dfrac{0.5}{2} = 0.25;\ [0.25, 0.75]$
$z = \text{invNorm}(0.75) \approx 0.6745$
$50 \pm 0.6745(6.1) = 45.89, 54.11$

C. Exercises

42. $1 + 3.3 \log 50 \approx 6.6;\ k = 7$
$w = \dfrac{98 - 66}{7} \approx 5$

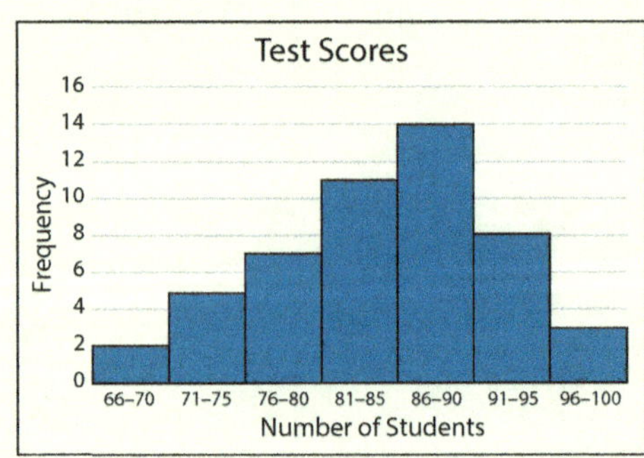

A. Exercises

Find the portion of the area of the shaded region under the normal distribution curve using the z-score tables and technology.

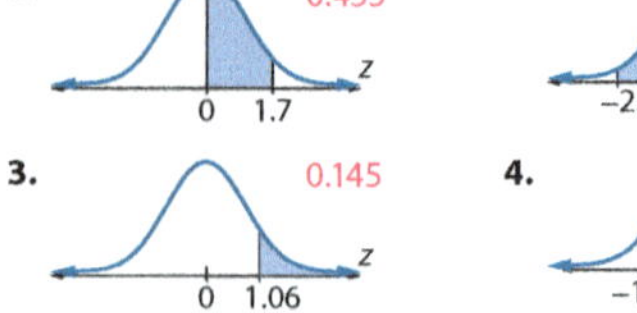

Find the percentage of data values in each interval.

5. $0 \le z \le 0.69$ 25.5%
6. $-1.21 \le z \le 0$ 38.7%
7. $-0.74 \le z \le 0.74$ 54.1%
8. $-1.85 \le z \le 1.85$ 93.6%
9. $z \ge 2.37$ 0.9%
10. $z \le -1.56$ 5.9%

Find the probability that a randomly chosen score falls within the given interval.

11. $P(z \le -0.92)$ 17.9%
12. $P(z \ge -1.4)$ 91.9%
13. $P(z \le 2.0)$ 97.7%
14. $P(-0.3 \le z \le 0.3)$ 23.6%
15. $P(-1.4 \le z \le 2.3)$ 90.9%
16. $P(-2.83 \le z \le 1.11)$ 86.4%

B. Exercises

Find the probability that a randomly selected value in a normal distribution is within the given interval.

17. $[79, 90]$ with $\mu = 83$ and $\sigma = 5.9$ 63.3%
18. $[6.5, 9]$ with $\mu = 8.375$ and $\sigma = 1.04$ 69.0%
19. $[22, 25]$ with $\mu = 20$ and $\sigma = 7.9$ 13.7%
20. $[35, 40]$ with $\mu = 45$ and $\sigma = 9.26$ 15.5%
21. A randomly selected 16-yr-old male will have a height between 67 in. and 70 in. if the mean is 68.36 in. and the standard deviation is 2.72. 41.8%
22. A randomly selected 6-mo-old baby girl will weigh between 18 lb and 20 lb if the mean is 16.44 lb and the standard deviation is 1.95. 17.8%
23. A randomly selected student taking the ACT will score between 32 and 36 if the mean score is 20.9 and the standard deviation is 5.6. 2.0%
24. A randomly selected student taking the SAT mathematics test in 2017 would have a score between 400 and 500 if the mean score is 527 and the standard deviation is 107. 28.3%

Find the percentile rank for each student.

25. Joe's z-score of 1.07 on his college entrance exam 86th percentile
26. Mary's z-score of -0.26 on her economics test 40th percentile
27. Leesha's score of 90 on a test with a mean of 85 and a standard deviation of 3 95th percentile
28. Marcus's 64 in. height in an age group with a mean of 69 in. and a standard deviation of 2.94 4th percentile
29. Thomas's birth weight of 8.5 lb, given that the average male birth weight is 8.83 lb with a standard deviation of 1.6 42nd percentile

Find the data value for the specified percentile in a normally distributed set of data.

30. 60th percentile if $\mu = 40$ and $\sigma = 7.2$ ≈ 41.8
31. 45th percentile if $\mu = 8.2$ and $\sigma = 1.2$ ≈ 8.0
32. Find the height of a 12-yr-old boy at the 5th percentile if the mean is 59 in. and the standard deviation is 3.1. ≈ 53.9 in.
33. Find the history test score that is at the 75th percentile if the mean is 79 and the standard deviation is 6.8. ≈ 83.6

Find the z-scores for the given interval.

34. lower 20% $z \le -0.84$
35. upper 10% $z \ge 1.28$
36. middle 34% $-0.44 \le z \le 0.44$
37. middle 97% $-2.17 \le z \le 2.17$

Find the interval of the data given the mean and the standard deviation.

38. upper 25% of data when $\mu = 7.5$ and $\sigma = 1.5$ $x_i \ge 8.5$
39. lower 33% of data when $\mu = 81$ and $\sigma = 9.6$ $x_i \le 76.8$
40. middle 20% of scores with $\mu = 75$ and $\sigma = 7.8$ [73.0, 77.0]
41. middle 50% of scores with $\mu = 50$ and $\sigma = 6.1$ [45.9, 54.1]

C. Exercises

Use technology and the test grades below to complete exercises 42–44.
66, 70, 71, 71, 74, 75, 75, 76, 76, 77,
78, 79, 80, 80, 81, 81, 81, 82, 83, 83,
84, 84, 85, 85, 85, 86, 88, 88, 88, 88,
89, 89, 89, 89, 90, 90, 90, 90, 90, 91,
91, 92, 92, 93, 94, 95, 95, 96, 97, 98

42. Use Sturges's formula to divide the data set into intervals and create a histogram representing the data.
43. Complete a grouped frequency distribution table and use the midpoint of each class to find the estimated mean, μ, and standard deviation of the population, σ. $\mu = 84.6;\ \sigma \approx 7.5$

43.

Class	M_i	f_i	f_iM_i	$f_i(M_i - \bar{x})^2$
66–70	68	2	136	551.12
71–75	73	5	365	672.8
76–80	78	7	546	304.92
81–85	83	11	913	28.16
86–90	88	14	1232	161.84
91–95	93	8	744	564.48
96–100	98	3	294	538.68
Sum		50	4230	2822
Mean	84.6		**Var.**	56.44
			SD	7.51

$\text{variance} = \dfrac{2822}{50} = 56.44$
$\sigma = \sqrt{56.44} \approx 7.51$

44a. 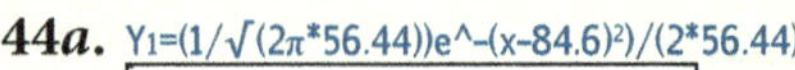

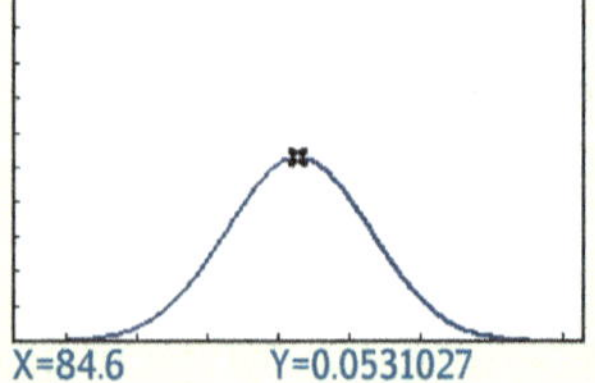

44b–c. $[62.1, 69.6];\ \dfrac{1}{50} = 2\%$ vs. 2.35%
$[69.6, 77.1];\ \dfrac{9}{50} = 18\%$ vs. 13.5%
$[77.1, 84.6];\ \dfrac{12}{50} = 24\%$ vs. 34%
$[84.6, 92.1];\ \dfrac{21}{50} = 42\%$ vs. 34%
$[92.1, 99.6];\ \dfrac{7}{50} = 14\%$ vs. 13.5%
$[99.6, 107.1];\ \dfrac{0}{50} = 0\%$ vs. 2.35%

44. Write a normal distribution function for your data by substituting the estimated mean, μ, and variance, σ^2, into $y = \frac{1}{\sqrt{2\pi\sigma^2}} e^{-\frac{(x-\mu)^2}{2\sigma^2}}$. $\quad y = \frac{1}{\sqrt{2\pi(56.44)}} e^{-\frac{(x-84.6)^2}{2(56.44)}}$

  a. Graph the data's normal distribution function in the window $x \in [\mu - 4\sigma, \mu + 4\sigma]$ with Xscl $= \sigma$ and $y \in [0, 0.1]$.

  b. Express each interval in terms of the data values.

    $[\mu - 3\sigma, \mu - 2\sigma]$ $[\mu, \mu + \sigma]$
    $[\mu - 2\sigma, \mu - \sigma]$ $[\mu + \sigma, \mu + 2\sigma]$
    $[\mu - \sigma, \mu]$ $[\mu + 2\sigma, \mu + 3\sigma]$

  c. Find the percentage of test scores that lie within each of the above intervals. Compare your results with the Empirical Rule to determine whether the data is normally distributed. Explain why or why not.

CUMULATIVE REVIEW

Use technology to complete a regression and write the exponential function modeling the data. Round coefficients to the nearest thousandth. [3.5]

45.

x	y
1	6
2	9
3	12
4	16
5	22
6	30

$y \approx 4.579(1.370)^x$

46.

x	y
0	62.2
2	57.9
4	54.8
6	49.9
8	47.5
10	44.0

$y \approx 62.221(0.966)^x$

Use technology to solve each system of inequalities graphically. [7.6]

47. $x + 2y > 4$
$y < 2x^2 - 1$

48. $y \geq |x + 2| - 2$
$y \leq -|x + 2| + 4$

State the five-number summary and construct a box plot for each set of data. [10.4]

49. $B = \{58, 60, 57, 73, 64, 65, 70, 71, 63, 65\}$

50. $H = \{130, 116, 133, 131, 121, 127, 123, 125, 126, 126,$
    $127, 124, 128, 130, 112, 134, 137, 119, 129\}$

49. min: 57; Q_1: 60; median: 64.5; Q_3: 70; max: 73

50. min: 112; Q_1: 123; median: 127; Q_3: 130; max: 137

51. Use technology to estimate solutions of
$4.3x^3 - 2.45x^2 - 4.55$
    $= 4.72x^4 + 5.21x^3 - 12.2x^2 - 8.22$
to the nearest hundredth. List all correct answers. **[Appendix 3]** B, C

  A. -1.64 **C.** 1.46 **E.** none of these
  B. -1.63 **D.** 3.71

52. Which of the following is true of $f(x) = g(x) + h(x)$ if $g(x) = \sin 2x$ and $h(x) = 2\cos 2x$? List all correct answers. **[4.7]** A, C

  A. period: $p = \pi$ **D.** It is not a sinusoid.
  B. period: $p = 2$ **E.** none of these
  C. It is a sinusoid.

53. Find the sum $\sum_{n=1}^{5} 64\left(\frac{3}{4}\right)^{n-1}$. **[9.3]** B

  A. 210.4375   **C.** 192   **E.** 101.25
  B. 195.25   **D.** 175

54. Find the z-score for a score of 33 on an exam with $\mu = 23$ and $\sigma = 6$. **[10.4]** D

  A. 10.00   **C.** 5.50   **E.** none of these
  B. 3.83   **D.** 1.67

48. 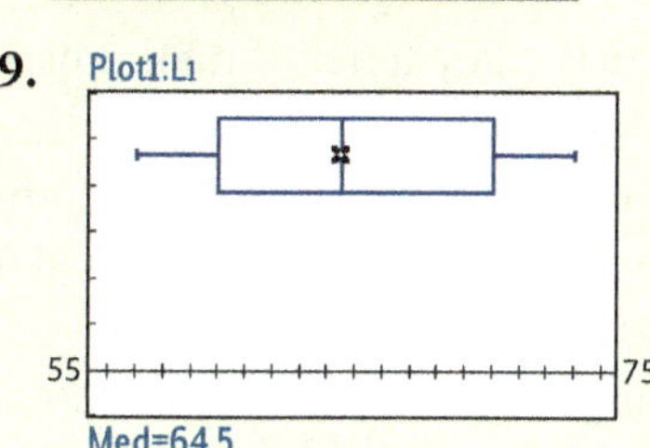

49. 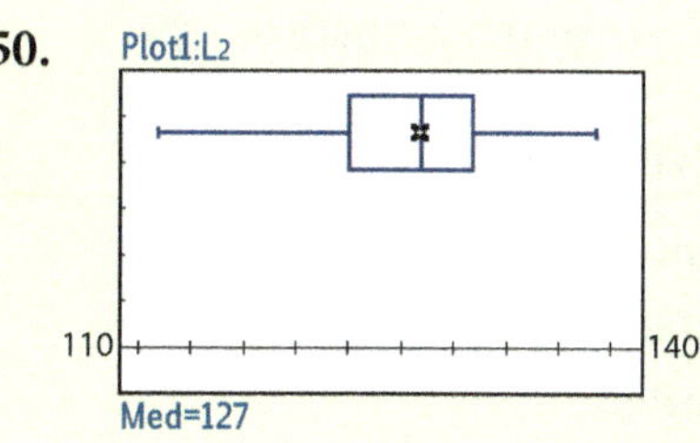

50. 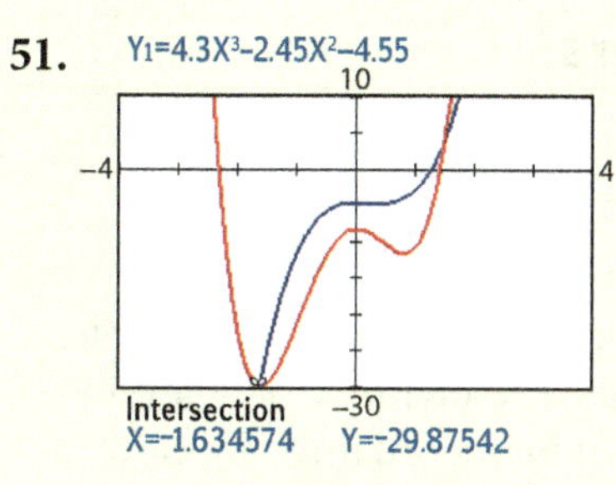

51. 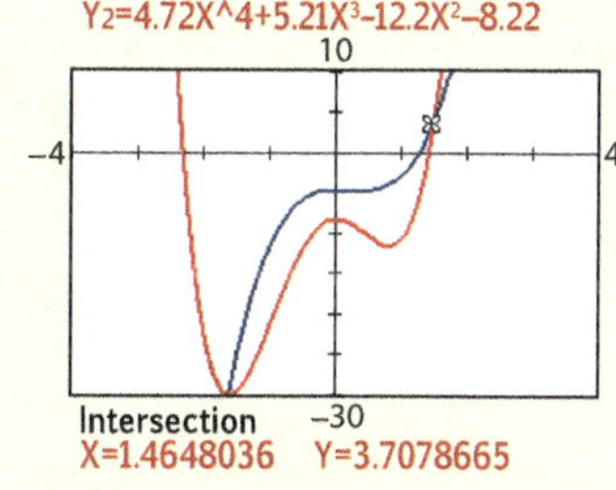

52. Both $g(x)$ and $h(x)$ are sinusoids with $p = \frac{2\pi}{2} = \pi$.

53. $S_n = \dfrac{a_1(1 - r^n)}{1 - r}$

$S_5 \approx \dfrac{64\left(1 - \left(\frac{3}{4}\right)^5\right)}{1 - \frac{3}{4}} = 195.25$

54. $z = \dfrac{33 - 23}{6} \approx 1.67$

44c. While the distribution is close to the normal curve, it is not an exact fit. 68% of the data is within $\pm 1\sigma$ of the mean, but the distribution is not symmetric. The scores above the mean represent 56% of the data, which has a slight negative skew (median of 85.5).

> **Cumulative Review**

45. 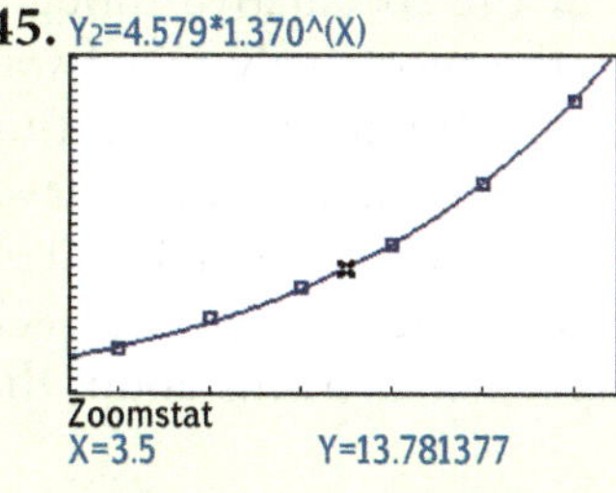

46. 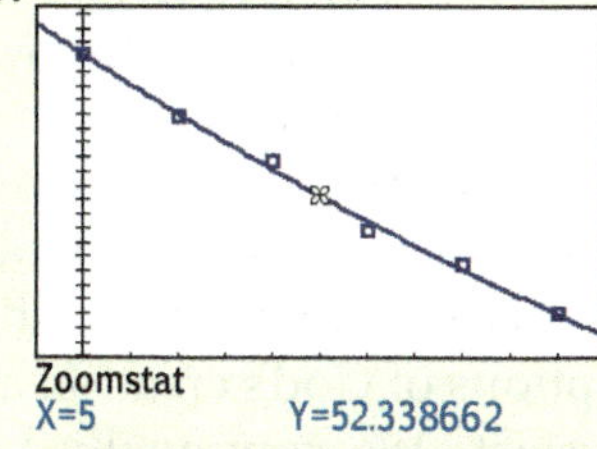

47. 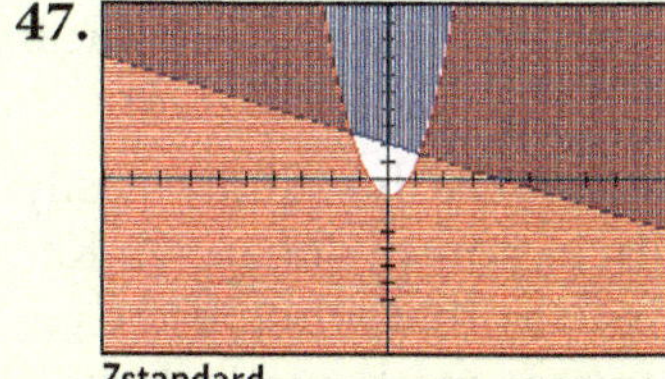

Biblical Perspective of Mathematics

Objectives

1. To state the key tenet of mathematical Platonism
2. To explain why many secular mathematicians have become skeptical of mathematical Platonism
3. To compare and contrast mathematical Platonism and a biblical perspective of mathematics

Assignments

- **Minimum:** 1–7; 8–9 (in class)
- **Standard:** 1–7; 8–9 (in class)
- **Extended:** 1–9

Answers

3. Kepler
4. Platonists believe mathematical truths were established by God.
5. Universal mathematical principles exist independently from God instead of being ordained by God.
6. the development of three seemingly contradictory systems of valid geometry and the discovery of paradoxes involving set theory

Mathematical Platonism

I believe that mathematical reality lies outside us, that our function is to discover or observe it, and that the theorems which we prove, and which we describe grandiloquently as our "creations," are simply the notes of our observations.

—G. H. Hardy (English mathematician)

Almost everyone has heard of Plato, but not many know much about him. Plato wrote about morality, philosophy, theology, politics, aesthetics, and mathematics. While he considered mathematics his most important field of study, he is most remembered for his philosophy, which has become known as Platonism. He thought that the things we see in this world are only representations of a reality that actually exists in forms or ideas. So how did this philosophy view mathematics?

The *Stanford Encyclopedia of Philosophy* states that "*mathematical Platonism . . .* is the metaphysical view that there are abstract mathematical objects whose existence is independent of us and our language, thought, and practices. Just as electrons and planets exist independently of us, so do numbers and sets. . . . Mathematical truths are therefore discovered, not invented." The belief that mathematics is discovered dominated mathematics for nearly two millennia. Mathematical Platonism views abstract mathematical objects and truths as existing eternally, independent of God.

A biblical worldview acknowledges God as the source of all truth and humans as discoverers of mathematical truths derived from God's self-consistent character. Being created in God's image, humans are able to create a mathematical language that describes some of the underlying mathematical principles governing our universe. Many prominent mathematicians of the past gladly acknowledged the supernatural Creator God as the source of transcendent truth. For example, Johannes Kepler stated that "the chief aim of all investigations of the external world should be to discover the rational order and harmony which has been imposed on it by God and which he revealed to us in the language of mathematics."

Many modern mathematicians maintain that mathematics is simply a result of the human mind and is therefore invented or created. But this belief cannot account for math's intuitive impression of immutable truth or the striking instances of coherence between mathematics and the physical universe. British mathematician Marcus du Sautoy wrote, "Many mathematicians fluctuate between feeling they are being creative and a sense they are discovering absolute scientific truths. Mathematical ideas can often appear very personal and dependent on the creative mind that conceived them. Yet that is balanced by the belief that its logical character means that every mathematician is living in the same mathematical world that is full of immutable truths. These truths are simply waiting to be unearthed, and no amount of creative thinking will undermine their existence."

In his 2008 essay "Mathematical Platonism and its Opposites," Harvard mathematician Barry Mazur states, "If we adopt the Platonic view that mathematics is discovered, we are suddenly in surprising territory, for this is a full-fledged theistic position." He confuses Platonism and a theistic perspective since Platonism does not necessitate theism, seeing universal mathematical truths as existing independently from God instead.

Mathematical Platonism lost some credibility when three seemingly contradictory valid geometries were developed in the nineteenth century and when paradoxes involving set theory were discovered in the early twenti-

PRESENTATION

Ask students to comment on whether mathematics is created or discovered. Refer to the Biblical Perspective of Mathematics feature in Chapter 1 (p. 17) for more ideas on this topic. The conclusion in that feature was that man is both a discoverer and a creator. Different civilizations explored many of the same math concepts but with vocabulary and notation specific to each civilization. Our creative acts are possible because we are created in God's image. This feature focuses on man as a discoverer of mathematics.

Discuss the G. H. Hardy quote, pointing out his emphasis on how we discover mathematical truth through observation rather than creating it. Stress that mathematics is a direct result of God's creation. Also discuss how our mathematical descriptions of God's creation are not always perfect. We continually test theorems and question our mathematical understanding of the universe.

A discussion of Hardy's quote easily leads into a discussion of Johannes Kepler's quote. Explain that while the Creator is the source of mathematical truth with all its order and harmony, we continually develop mathematical symbols and conventions to express our understanding of the universe.

According to Morris Kline, the ancient Greeks believed that man's role in the study of math was to discover math in nature. In his book *Mathematics: The Loss of Certainty,* Kline states that the Greeks believed nature functioned by rational design. Kline concluded, "Of all the triumphs of the speculative thought of the Greeks, the most truly novel was their conception of the cosmos operating in accordance with mathematical laws discoverable by human thought. The Greeks, then, were determined to seek truths and in particular truths about the

eth century. Several mathematicians then embarked on a quest to restore consistency to the foundations of our mathematics. Although those efforts never achieved their ultimate goal, they did produce new philosophies, including several modifications of traditional Platonism.

From a biblical point of view, the three different geometries reveal a variety of ways that mathematics can describe the physical world that God created. Understanding our finite and fallen nature allows us to accept new human insights and current inadequacies of our mathematical descriptions of our universe. Christians can have complete confidence in the infallibility of our infinite God and His system of transcendent mathematics (Ps. 147:5; Isa. 40:28).

Exercises

1. What major tenet of mathematical Platonism was the dominant view of mathematicians for the last two millennia? Our role in mathematics is to discover.

2. What two mathematical objects are compared to electrons and planets in that they exist independently of humans? numbers and sets

3. Which famous astronomer stated that the rational order and harmony of the world was due to God?

4. What misconception of Platonism is illustrated by Barry Mazur's quote?

5. What common Platonist view is problematic for Christian mathematicians?

6. Name the key events that damaged the credibility of mathematical Platonism in the nineteenth and twentieth centuries.

7. How is God's understanding described in Psalm 147:5? infinite

8. **Discuss:** Compare and contrast a biblical worldview of mathematics with mathematical Platonism.

9. **Discuss:** Is mathematics discovered or invented?

8. Both maintain that the primary role of man in mathematics is to discover rather than to create and that mathematical truths exist universally whether or not the human mind is aware of them. These universal mathematical truths necessitate a supernatural source as the origin for mathematics. A biblical worldview of mathematics recognizes the God of the Bible as this source—mathematical Platonism does not.

9. Mathematics is both discovered and created. God has created us in His image. Even in our fallen and finite state, we can discover some of the universal principles underlying His created universe. People have created mathematical concepts, vocabulary, and notation that seek to describe God's universal truths and abstract principles.

mathematical design of nature." Many of the Greeks were schooled in the philosophy of Plato, which did not acknowledge the Christian or Hebrew God.

Thus, strict mathematical Platonism is an anti-biblical philosophy. Consider sharing the following excerpt from Larry Zimmerman's book *Truth & the Transcendent.*

Though there is an element of similarity between the platonic and the theistic view of the nature of mathematics, the differences are significant. The latter posits the existence of the infinite; an abnormal world, due to sin; the importance of mathematics in obeying God's command to take dominion over the earth, subduing, and replenishing it; and most important, a personal and ethical deity who creates ex nihilo. These concepts are foreign to Platonism, which allows for categories but not for meaning. As [Morris] Kline observes, Plato "wished not merely to understand mathematics but to substitute mathematics for nature herself."

Discuss the quote by Barry Mazur, which inaccurately views Platonism as a full-fledged theistic position. Conclude the lesson with a discussion of the reasons for the loss of credibility of mathematical Platonism and the resulting development of new mathematical philosophies in the early twentieth century. The Biblical Perspective of Mathematics feature in Chapter 11 addresses the three best known of these new humanistic philosophies.

Objective

To prepare for evaluation

Vocabulary

See Appendix A.

Assignments

- **Minimum:** 1–3, 5*a*, 6–8, 9*b*–12, 14–16, 18, 21, 23, 25, 27, 29*a*, 30–31, 33, 36, 39, 41, 43, 45–47
- **Standard:** 1–47 odd; 2–46 even (in class)
- **Extended:** 2–4, 5*b*, 6–7, 10–14, 16–18, 20, 22, 24–29, 31, 32*b*, 33, 35–37, 40, 42, 44–47

Assessment

- Chapter 10 Test

Solutions

❯ A. Exercises

1. $2(3)(2) = 12$

2a. $20(19)(18) = 6840$

2b. $20(20)(20) = 8000$

3a. $8! = 40,320$

3b. $8(7)(6) = 336$

4. $_{15}P_4 = \dfrac{15!}{(15-4)!} = \dfrac{15 \cdot 14 \cdot 13 \cdot 12 \cdot 11!}{11!}$
$= 32,760$

5a. $\dfrac{7!}{2!} = 2520$

5b. $\dfrac{8!}{3!2!2!} = 1680$

6a. $_9C_4 = \dfrac{9!}{5!4!} = 126$

6b. $_{10}C_4 = \dfrac{10!}{6!4!} = 210$

7. $_8C_2 \cdot {}_4C_2 = 28 \cdot 6 = 168$

9a. T: $\dfrac{1}{6}$; E: $\dfrac{2}{20} = \dfrac{1}{10}$

9b. T: $P(1 \text{ or } 3 \text{ or } 5) = \dfrac{3}{6} = \dfrac{1}{2}$
E: $P(1 \text{ or } 3 \text{ or } 5) = \dfrac{5+3+4}{20} = \dfrac{3}{5}$

10. $P(\text{mult. of } 3) = \dfrac{3+3}{20} = \dfrac{3}{10}$
$P(\text{not mult. of } 3) = 1 - \dfrac{3}{10} = \dfrac{7}{10}$

11a. $P(H \text{ and even}) = \dfrac{1}{2} \cdot \dfrac{1}{2} = \dfrac{1}{4}$

11b. $P(T \text{ and } < 3) = \dfrac{1}{2} \cdot \dfrac{1}{3} = \dfrac{1}{6}$

12a. $\dfrac{4}{15} \cdot \dfrac{4}{15} = \dfrac{16}{225} = 7.1\%$

Chapter 10 Review

1. Caleb likes to wear dress pants, a dress shirt, and a tie to church on Sunday. How many different outfits can he make from his 2 pairs of dress pants, 3 dress shirts, and 2 ties? **[10.1]** 12

2. Determine the number of different ways 3 flavors of lollipops from a store offering 20 different flavors can be arranged in each case below. Then state whether such arrangements are independent or dependent. **[10.1]**
 a. Flavors cannot be repeated. 6840; dependent
 b. Flavors can be repeated. 8000; independent

3. Eight contestants compete in the high school track meet. **[10.1]**
 a. How many ways can they line up for a group photo? 40,320
 b. Assuming there are no ties, how many possible ways could 1st, 2nd, and 3rd be secured? 336

4. In how many different orders can the top 4 of 15 contestants be announced as advancing to the final round of a Scripture memory competition? **[10.1]** 32,760

5. Find the number of permutations of the letters in each word. **[10.1]**
 a. *CONTEST* 2520 **b.** *LOLLIPOP* 1680

6. Evaluate each combination. **[10.1]**
 a. $_9C_4$ 126 **b.** $_{10}C_4$ 210

7. How many 4-person teams can be formed from a group of 8 boys and 4 girls if each gender will have 2 representatives? **[10.1]** 168

8. A regular 6-sided die is rolled. List the sample space and the event of getting a prime number. **[10.2]** $S = \{1, 2, 3, 4, 5, 6\}; E = \{2, 3, 5\}$

Use the results of 20 rolls of a 6-sided die listed below for exercises 9–10. **[10.2]**

Number	1	2	3	4	5	6
Frequency	5	3	3	2	4	3

9. State the theoretical and experimental probability for each event.
 a. $P(4)$ $\dfrac{1}{6}, \dfrac{1}{10}$ **b.** $P(\text{odd})$ $\dfrac{1}{2}, \dfrac{3}{5}$

10. Find the experimental probability of not getting a multiple of 3. $\dfrac{7}{10}$

11. Find the probability of each event when a coin is flipped and a regular 6-sided die is rolled. **[10.2]**
 a. heads and a multiple of 2 $\dfrac{1}{4}$
 b. tails and a number less than 3 $\dfrac{1}{6}$

12. A bag contains 3 red marbles, 4 blue marbles, and 8 green marbles. Find the probability of drawing 2 blue marbles consecutively in each case. **[10.2]**
 a. The first marble is replaced. $\dfrac{16}{225} \approx 7.1\%$
 b. The first marble is not replaced. $\dfrac{2}{35} \approx 5.7\%$

In a recent survey of high school juniors, 40% played fall sports, 30% earned an A in their math class that fall, and 20% earned an A while playing fall sports.

13. Determine whether the two events are independent or dependent. **[10.2]** dependent

14. Find the probability that a randomly selected student fits each description. **[10.2]**
 a. played fall sports or earned an A, but not both 0.3
 b. did not play fall sports and did not earn an A 0.5

15. The table summarizes the results of a survey that asked 434 students how many books they read the previous summer. Find the probability that a randomly selected student read each number of books. **[10.2]**

Number of Books	Percent of Students
0	3.9%
1–2	14.1%
3–4	19.8%
5–6	16.1%
7–8	8.1%
9+	38.0%

 a. more than 8 or fewer than 3 books 56%
 b. between 2 and 9 books 44%

16. Label the following as a parameter or a statistic. **[10.3]**
 a. The average speed of all vehicles traveling on Interstate 10 is 69 mi/hr. parameter
 b. The average speed of black cars traveling on Interstate 10 between 8:05 AM and 8:10 AM is 69 mi/hr. statistic

17. Identify the level of measurement. **[10.3]**
 a. student body president ordinal
 b. memberships in volunteer groups categorical
 c. the number of votes received ratio
 d. Celsius temperatures interval

PRESENTATION

If exercise 26 is assigned, ensure that the students have access to a spreadsheet, or permit the exercise to be completed without a spreadsheet. Be sure to communicate what technology will be permitted or required for the chapter test.

TIPS

Ex. 17*a* If the offices are not considered to be ordered, then this could be nominal level.

18. Create a pie chart illustrating the Sunday school attendance data. **[10.3]**

nursery	14
preschool	23
primary	43
juniors	47
teens	97
adults	380

19. Create a back-to-back stem plot summarizing the ages of newcomers attending the local church. **[10.3]**
2017: 1, 8, 35, 12, 16, 40, 41, 14, 16, 56, 62, 66
2018: 1, 27, 28, 7, 4, 41, 41, 48, 29, 29, 50, 57, 58

20. Determine the number of classes and the class width for a grouped frequency distribution table summarizing 20 data values having a minimum value of 5 and a maximum value of 28. **[10.3]** 6 classes; $w = 4$

21. Draw a histogram representing the ages of the cousins in the Smith family. **[10.3]**

Ages	Freq.
1–3	2
4–6	2
7–9	4
10–12	3
13–15	2

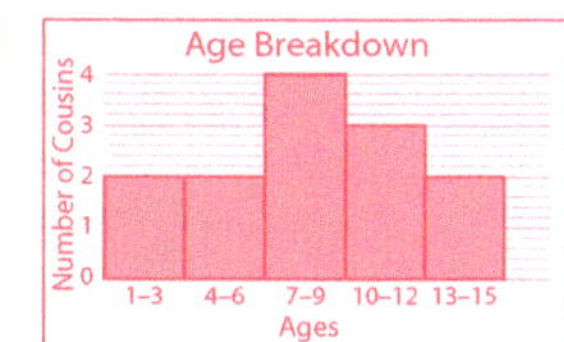

22. Extend the frequency distribution table in exercise 21 by adding Cumulative Frequency and Cumulative Percentage columns and use the table to draw a percentile graph of the data. **[10.3]**

23. Find the sample mean, median, and mode of the average monthly temperatures for Charlotte, North Carolina. **[10.4]** mean = 70.83; median = 72; mode = 72
$T = \{51, 55, 63, 72, 79, 86, 89, 87, 81, 72, 62, 53\}$

24. Find the variance and standard deviation of sample data containing 13 values in which the sum of the squares $SS = 60$. **[10.4]** $s^2 = 5; s \approx 2.24$

25. Calculate the mean and sample standard deviation for the data set. **[10.4]** $\bar{x} = 53.6; s \approx 6.3$
$D = \{47, 61, 46, 64, 48, 49, 51, 55, 57, 58\}$

26. Use a spreadsheet to find the mean and sample standard deviation for the data in exercise 21. **[10.4]**
$\bar{x} \approx 8.23; s \approx 3.81$

27. Analyze the data set $Q = \{34, 27, 28, 25, 26, 21, 24, 25, 29, 23, 28, 27\}$. **[10.4]**
   a. Find the mean, median, and interquartile range.
   b. Identify any outliers and then draw box plots that include and exclude any outliers.

28. Use technology to analyze the data set. **[10.4]**
$D = \{13, 27, 21, 18, 3, 15, 29, 21, 19, 24\}$
   a. State the five-number summary and draw the box plot. 3, 15, 20, 24, and 29
   b. State the mean and the population standard deviation. $\mu = 19; \sigma \approx 7.11$
   c. Which provides a better summary for the data: the five-number summary or the mean and standard deviation? Explain your reasoning.

29. Find the data value at each percentile for the points scored by a sample of basketball players. **[10.4]**
$P = \{13, 15, 18, 19, 21, 21, 24, 27, 29, 31\}$
   a. 43rd percentile 20.46   **b.** 78th percentile 28.16

30. Classify a data distribution in which the bulk of the data is to the left with a tail to the right and the mean is greater than the median. **[10.4]** C
   a. negatively skewed   **b.** normal distribution
   c. positively skewed

31. The data in L_1 has $\mu = 6$ and $\sigma = 2$. State the mean and standard deviation of each transformed set of data. **[10.5]**
   a. $L_2 = 5L_1$ $\mu = 30$ $\sigma = 10$ **b.** $L_3 = L_1 + 4$ $\mu = 10$ $\sigma = 2$

32. Find the z-score for each data value in a distribution with $\mu = 85$ and $\sigma = 5$. **[10.5]**
   a. $x_i = 100$ $z = 3$ **b.** $x_i = 78$ $z = -1.4$

33. Andrew scored a 1206 on the SAT where the mean was 1060 and the standard deviation was 195. He also received a 26 on the ACT where the mean was 21 and the standard deviation was 5.4. Find the z-score for each test and determine the test on which he received the better score. **[10.5]** ≈ 27.0

34. Find Mikayla's raw score if she had a z-score of 1.32 on an ACT test with $\mu = 20.8$ and $\sigma = 4.7$. **[10.5]**

35. Use Chebyshev's Theorem to find the smallest percentage of data that must always lie within 2 standard deviations of the mean. Confirm your results using the set of quiz scores $Q = \{1, 5, 6, 7, 7, 8, 8, 9, 9, 9\}$. **[10.5]** 75%

27a. $\mu \approx 26.4$; median = 26.5; IQR = 3.5

28c. With the possible outlier of 3 and apparent left skew of the distribution, the five-number summary would likely provide a better description.

33. $z_{SAT} \approx 0.75$; $z_{ACT} \approx 0.93$; His ACT score was better.

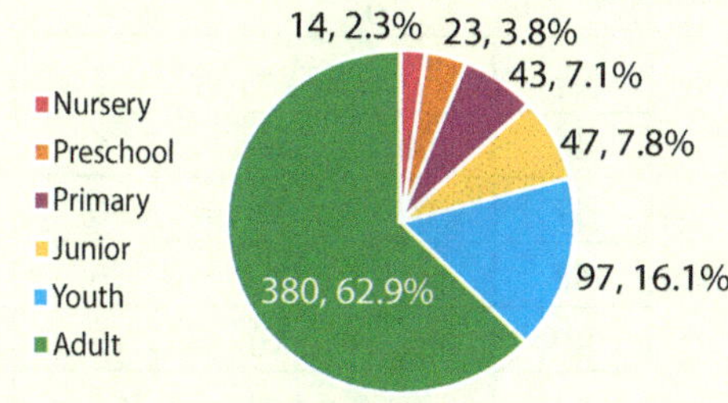

19.

2017		2018
8 1	0	1 4 7
6 6 4 2	1	
	2	7 8 9 9
5	3	
1 0	4	1 1 8
6	5	0 7 8
6 2	6	

20. $1 + 3.3 \log 20 \approx 5.37; k = 6$
$$w = \frac{28 - 5}{6} \approx 4$$

22.

Ages	f_i	Σf	$\Sigma\%$
1–3	2	2	15.4%
4–6	2	4	30.8%
7–9	4	8	61.5%
10–12	3	11	84.6%
13–15	2	13	100%
Total	13		

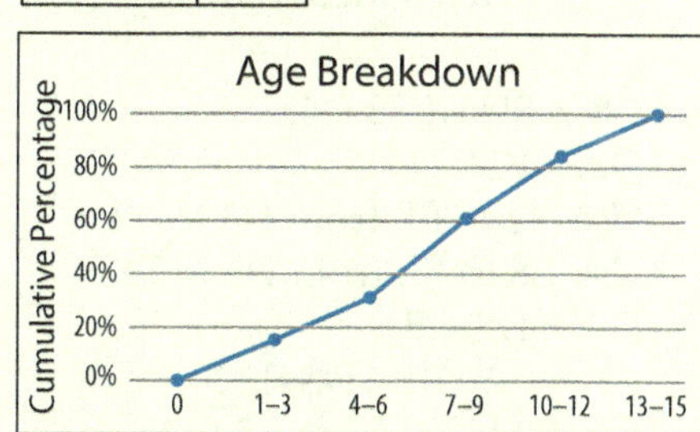

12b. $\dfrac{4}{15} \cdot \dfrac{3}{14} = \dfrac{2}{35} \approx 5.7\%$

13. $P(A|S) = \dfrac{P(A \text{ and } S)}{P(S)} = \dfrac{0.2}{0.4} = 0.5$
and $P(A) = 0.3$
Since $P(A|S) \neq P(A)$, the two events are dependent.
Similarly, $P(S|A) \neq P(S)$ can be confirmed.

14a. $0.4 - 0.2 + 0.3 - 0.2 = 0.3$ or 30%

14b.

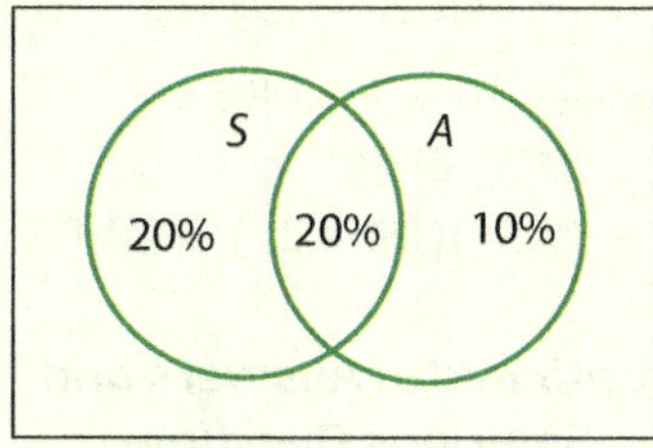

$P((S \text{ or } A)') = 1 - 0.5 = 0.5$ or 50%

15a. $P(9+) + P(0) + P(1–2)$
   $= 38.0\% + 3.9\% + 14.1\% = 56\%$

15b. $P(3–4) + P(5–6) + P(7–8)$
   $= 19.8\% + 16.1\% + 8.1\% = 44\%$

18. $n = 14 + 23 + 43 + 47 + 97 + 380$
   $= 604$

nursery: $\dfrac{14}{604}(360°) \approx 8°$

preschool: $\dfrac{23}{604}(360°) \approx 14°$

primary: $\dfrac{43}{604}(360°) \approx 26°$

junior: $\dfrac{47}{604}(360°) \approx 28°$

teen: $\dfrac{97}{604}(360°) \approx 58°$

adult: $\dfrac{380}{604}(360°) \approx 226°$

23. $\bar{x} = \dfrac{\sum\limits_{i=1}^{n} x_i}{n} = \dfrac{850}{12} \approx 70.83$
51, 53, 55, 62, 63, 72, 72, 79, 81, 86, 87, 89

24. $s^2 = \dfrac{60}{13 - 1} = 5; s \approx 2.24$

25. $\bar{x} = \dfrac{\sum\limits_{i=1}^{n} x_i}{n} = \dfrac{536}{10} \approx 53.6$

x_i	$(x_i - \bar{x})^2$
46	$(46 - 53.6)^2 = 57.76$
47	$(47 - 53.6)^2 = 43.56$
48	$(48 - 53.6)^2 = 31.36$
49	$(49 - 53.6)^2 = 21.16$
51	$(51 - 53.6)^2 = 6.76$
55	$(55 - 53.6)^2 = 1.96$
57	$(57 - 53.6)^2 = 11.56$
58	$(58 - 53.6)^2 = 19.36$
61	$(61 - 53.6)^2 = 54.76$
64	$(64 - 53.6)^2 = 108.16$

$SS = \sum\limits_{i=1}^{10} (x_i - \bar{x})^2 = 356.4$

$s^2 = \dfrac{356.4}{10 - 1} = 39.6; \; s \approx 6.3$

26.

	A	B	C	D	E
1	**Class**	M_i	f_i	M_if_i	$f_i(M_i-\mu)^2$
2	1–3	2	2	4	77.64
3	4–6	5	2	10	20.88
4	7–9	8	4	32	0.21
5	10–12	11	3	33	23.01
6	13–15	14	2	28	66.57
7			n	**Sum**	**Sum**
8			13	107	188.31
9			**Mean:**	8.23	
10			**Variance:**		14.49
11		**Standard Deviation:**			3.81

C8: =SUM(C2:C6)
D2: =B2*C2
Use D2 to fill from D3 to D6.
Use C8 to fill from D8 to E8.
D9: =D8/C8
E2: =C2*(B2-D9)^2
Use E2 to fill from E3 to E6.
E10: =E8/(C8-1)
E11: =SQRT(E10)

27a. $Q = \{21, 23, 24,|\ 25, 25, 26,|\ 27, 27, 28,|\ 28, 29, 34\}$

$\mu = \dfrac{317}{12} \approx 26.4$; median $= 26.5$

$IQR = Q_3 - Q_1 = 28 - 24.5 = 3.5$

27b. $Q_1 - 1.5(IQR) = 24.5 - 5.25$
$\qquad\qquad\qquad = 19.25$
$Q_3 + 1.5(IQR) = 28 + 5.25 = 33.25$
$34 \notin [19.25, 33.25]$

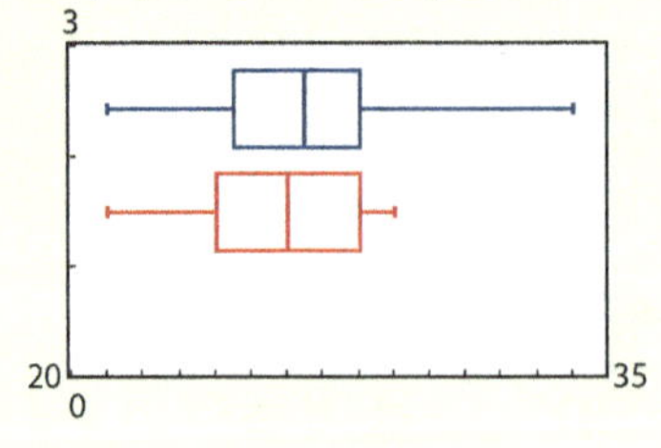

28.

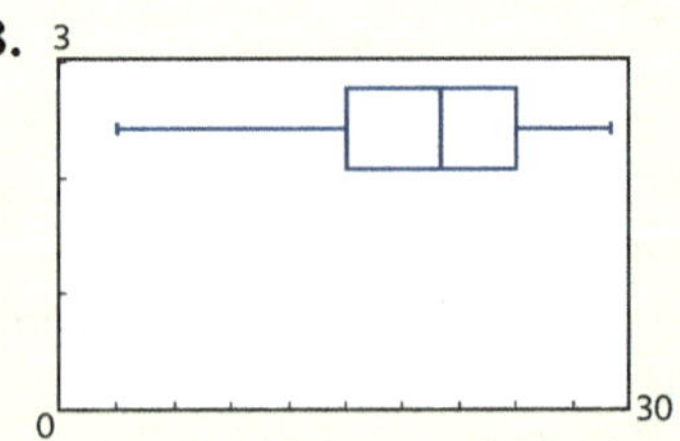

29a. $R = \dfrac{43}{100}(10 + 1) = 4.73$

$x_4 = 19; \; x_5 = 21$

$19 + (0.73)(21 - 19) = 20.46$

29b. $R = \dfrac{78}{100}(10 + 1) = 8.58$

$x_8 = 27; \; x_9 = 29$

$27 + (0.58)(29 - 27) = 28.16$

---

36. A normally distributed set of test scores has a mean of 41 and a standard deviation of 5. What percentage of the data is in each interval? [10.5]

  a. from 36 to 46 68%   **b.** from 26 to 51 97.35%

  c. greater than 46 16%

Find the portion of the area under the normal curve within each interval. [10.6]

37. $-1.2 \le z \le 0$ 0.385 or 38.5%

38. $-0.75 \le z \le 0.75$ 0.547 or 54.7%

39. $-0.9 \le z \le 1.85$ 0.784 or 78.4%

Find the probability that a data value chosen at random meets the following criteria. [10.6]

40. $z \le -2.1$ 1.8%

41. $-0.2 \le z \le 1.35$ 49.1%

Find the probability that a randomly selected value in a normal distribution is within the given interval. [10.6]

42. [90, 95] with $\mu = 82$ and $\sigma = 7.8$ 10.5%

43. [70, 89] with $\mu = 78$ and $\sigma = 8.6$ 72.3%

Use the normally distributed sample of scores that has a mean of 58 and a standard deviation of 4 to solve the following exercises. [10.6]

44. Find the probability that a randomly selected score is in the given interval.

  a. $x_i \in [53, 62]$ 73.6%   **b.** $x_i \ge 66$ 2.3%

45. Find the percentile rank for a score of 63. 89th percentile

46. Find the score at the 60th percentile. ≈ 59.0

47. Find the interval of values that contains the middle 50% of data. [55.3, 60.7]

continued in Answers and Solutions Overflow

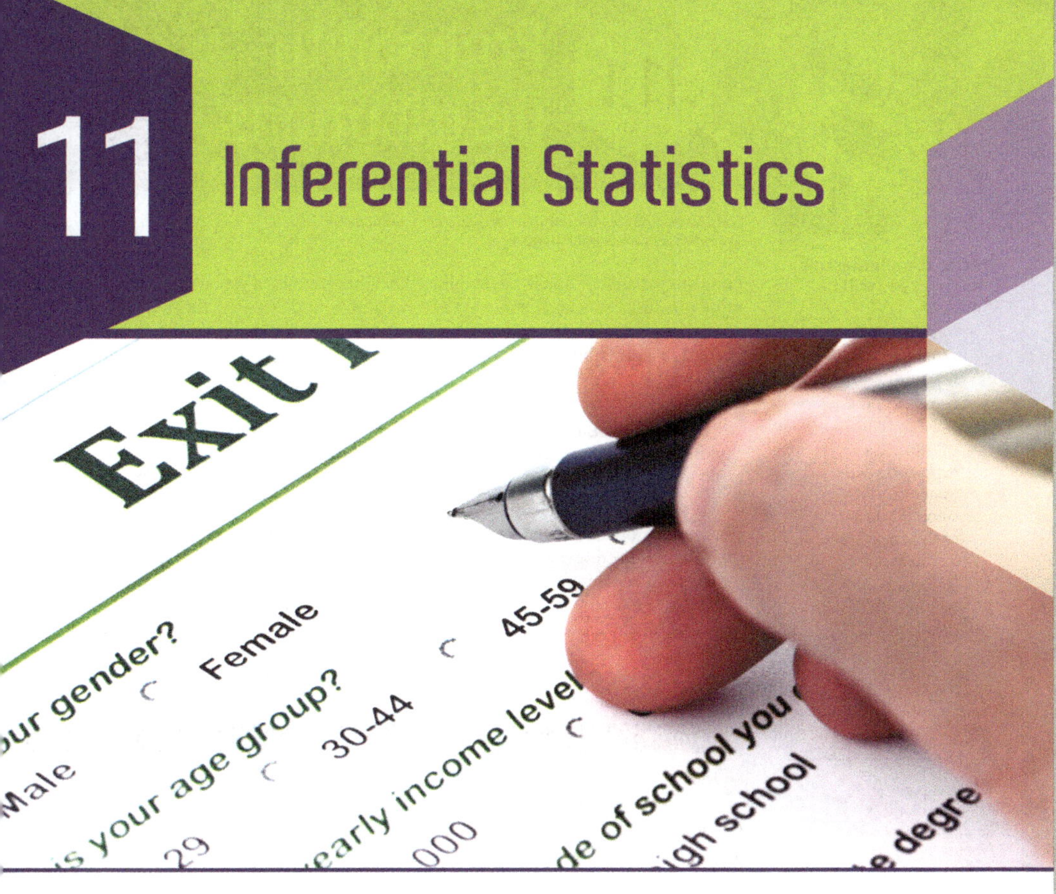

11 Inferential Statistics

11.1 Probability Distributions
11.2 Central Limit Theorem
Historical Connection
11.3 Confidence Intervals
11.4 Hypothesis Testing
Biblical Perspective of Mathematics
11.5 Research Studies
Data Analysis

HISTORICAL CONNECTION
A theorem from the seventeenth century finds new life and expansive applications in the power of twenty-first century computing.

BIBLICAL PERSPECTIVE OF MATHEMATICS
As Platonism fell out of favor in the nineteenth century, three competing schools of thought emerged in an effort to establish a foundation and philosophical approach for mathematics.

DATA ANALYSIS
Internet search engine results owe a lot to a Russian mathematician and his probability experiments of the early twentieth century.

Suggested Teaching Schedule

DAY	
1–2	11.1
3	11.2
4	Quiz 11A (11.1–11.2) HC: Mathematical Expectation
5–6	11.3
7–8	11.4

DAY	
9	Quiz 11B (11.3–11.4) BPM: Logicism, Intuitionism, and Formalism
10	11.5
11	Quiz 11C (11.5) DA: Markov Chains
12	Chapter 11 Review
13	Chapter 11 Test

Overview

Inferential statistics involves collecting random samples of data to describe and make conclusions (inferences) about the population.

This chapter provides students with the terminology and tools to understand, calculate, and interpret the results of statistical studies. Students will also be challenged to recognize the impact of one's worldview on the decision-making process and the need for godly wisdom and discernment when making such decisions.

Detailed instruction related to the use of technology is directly integrated into the lessons in lieu of a Technology Corner.

Flash

Many people believe that exit polls are an accurate predictor of election results and encourage their use. Others argue for caution since early reporting may discourage some from voting and affect the outcome of the election.

Chapter Objectives

1. To construct probability distributions for experimental, binomial, and other theoretical probabilities
2. To calculate the expected value and measures of variability for probability distributions
3. To find the maximum error and corresponding confidence intervals for estimates of a population's mean
4. To perform a hypothesis test to determine significant differences in the means of a population
5. To evaluate strengths and weaknesses of a statistical experiment and its reported results
6. To design and conduct a statistical experiment to answer a research question
7. To identify how a person's worldview affects the decisions made based on statistical tests
8. To compare and contrast three modern philosophical theories of mathematics with a biblical philosophy of mathematics

Objectives

1. To classify random variables as discrete or continuous
2. To construct probability distributions using theoretical and experimental probabilities
3. To calculate the expected value (mean), variance, and standard deviation of a probability distribution
4. To define a binomial experiment
5. To construct a binomial probability distribution and calculate its expected value (mean), variance, and standard deviation

Flash

In 2018 marketing research revenues reached 23 billion dollars in the United States. Researchers use surveys, in-store traffic, and consumer purchases to gather data about consumers.

Vocabulary

binomial experiment
binomial probability distribution
binomial probability distribution formula
continuous random variable
discrete random variable
expected value
mean of a probability distribution

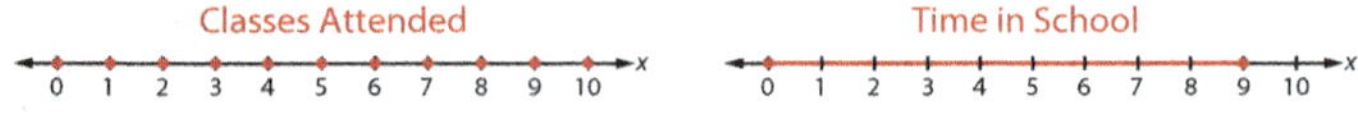

11.1 Probability Distributions

Consumers rank comfortable seats and dual climate controls as key features when purchasing a car.

After completing this section, you will be able to

- classify random variables as discrete or continuous.
- construct probability distributions.
- calculate the expected value (mean) and standard deviation of a probability distribution.
- construct and apply binomial probability distributions.

The 1937 publication of Harald Cramér's *Random Variables and Probability Distributions* was the first book to use *random variable* in its title and helped standardize the term.

Frequency distributions were introduced in Chapter 10 as a way to organize data, determine measures of central tendency and variability, and visualize the overall shape of the distribution. This section relates these concepts to the outcomes for the *random variable* studied in a probability experiment.

▶ **DEFINITIONS**

A **random variable** *x* represents all possible numerical outcomes of a probability experiment. A *discrete random variable* represents a countable number of outcomes. A *continuous random variable* represents an uncountable number of outcomes within an interval.

The number of classes that a randomly selected student attends during a school day represents a discrete outcome, while the time actually spent at school during that day represents a continuous outcome.

Considering whether the data is counted or measured can help distinguish discrete and continuous variables.

Example 1   Classifying a Random Variable

Classify each random variable as discrete or continuous and justify your answer.

a. Let *x* represent the length of newborn babies at birth.

b. Let *x* represent the days that it rains in a given week.

Answer

a. continuous; The value can be any reasonable real number.

b. discrete; The value is a whole number from 0 to 7.

SKILL ✔ **EXERCISE 1**

Different statistical techniques are used to analyze discrete and continuous random variables. This section examines the shape, center, and variability of probability distributions for discrete random variables.

	1	2	3	4	5	6
1	2	3	4	5	6	7
2	3	4	5	6	7	8
3	4	5	6	7	8	9
4	5	6	7	8	9	10
5	6	7	8	9	10	11
6	7	8	9	10	11	12

▶ **DEFINITION**

A **probability distribution** for the random variable, *x*, is a table, graph, or equation that associates each possible value of a random variable, x_i, and its probability, $P(x_i)$.

Recall that the probability of any event is $0 \leq P(x_i) \leq 1$ and that $\sum_{i=1}^{n} P(x_i) = 1$.

Consider the theoretical probability experiment examining the sum when a pair of dice is rolled. The sample space consists of the outcomes {2, 3, 4, 5, 6, 7, 8, 9, 10, 11, 12}, but each result is not equally likely. The table listing all the possible results can be used to

---

PRESENTATION

Lesson Opener

Expand each power of a binomial.

1. $(a - b)^3$ $a^3 - 3a^2b + 3ab^2 - b^3$
2. $(p + q)^4$ $p^4 + 4p^3q + 6p^2q^2 + 4pq^3 + q^4$

Consider briefly reviewing how repetitive multiplication, Pascal's triangle, and the Binomial Theorem can each be used to expand the above binomials.

Engage the students by explaining that while the study of probability can be negatively associated with gambling, it plays a major role in many businesses, such as insurance, healthcare, sports analytics, and computer software. Even in texting software probability is used to generate suggestions for the words that you probably want to type.

Define a *random variable* using the figures for the number of classes attended and time in school to differentiate *discrete random variables* and *continuous random variables*. Then use Example 1 to assess student understanding. Emphasize that this section focuses on discrete random variables.

Define a *probability distribution* and use the table of possible sums for a roll of a pair of dice to illustrate how to create a table and histogram representing its probability distribution.

Motivational Idea Point out that in some gambling games, the most likely result of rolling a sum of 7 is considered a win. While the $\frac{1}{6}$ chance of winning may not sound bad when compared to the chances of winning a lottery, you are still 5 times more likely to lose than win. Understanding probabilities is an important part of wise stewardship of what God has provided.

Create the probability distribution table from the frequency table in Example 2. Stress the importance of identifying the random variable, even when it seems to be obvious. Point out that the sum of the probabilities equals 1. Consider hav-

determine the probability of each outcome. The distribution can then be represented with a table or a graph.

sum, x_i	2	3	4	5	6	7	8	9	10	11	12
probability, $P(x_i)$	$\frac{1}{36}$	$\frac{1}{18}$	$\frac{1}{12}$	$\frac{1}{9}$	$\frac{5}{36}$	$\frac{1}{6}$	$\frac{5}{36}$	$\frac{1}{9}$	$\frac{1}{12}$	$\frac{1}{18}$	$\frac{1}{36}$

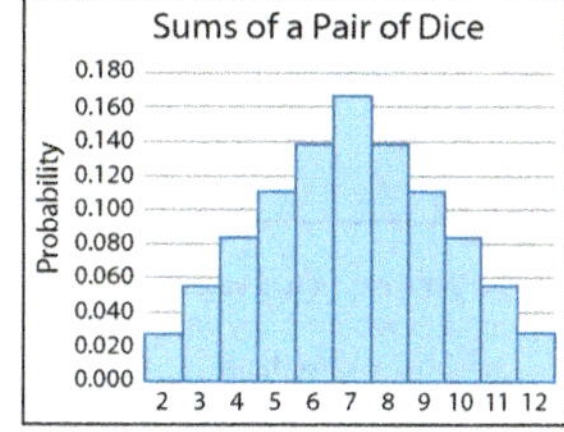

Probability distributions can also be constructed to represent experimental or observed data.

Example 2 Creating Probability Distributions

The frequency distribution table shows the results of a customer survey in which car comfort is rated from 1, very uncomfortable, to 5, very comfortable. Create numerical and graphical probability distributions for the data.

Rating	Freq.
1	2
2	3
3	12
4	20
5	13

Answer

Let x = rating.

$P(1) = \frac{2}{50} = 0.04$

$P(2) = \frac{3}{50} = 0.06$

$P(3) = \frac{12}{50} = 0.24$

$P(4) = \frac{20}{50} = 0.40$

$P(5) = \frac{13}{50} = 0.26$

x_i	$P(x_i)$
1	0.04
2	0.06
3	0.24
4	0.40
5	0.26

1. Define the random variable.

2. Determine the probability of each rating. Begin by finding the frequency total $n = 50$.

3. Verify that $0 \le P(x_i) \le 1$ for each x_i and that

$$\sum_{i=1}^{n} P(x_i) = 0.04 + 0.06 + 0.24 + 0.40 + 0.26 = 1.$$

4. Create a histogram for the probability distribution. Notice that the probability can be expressed as a percent.

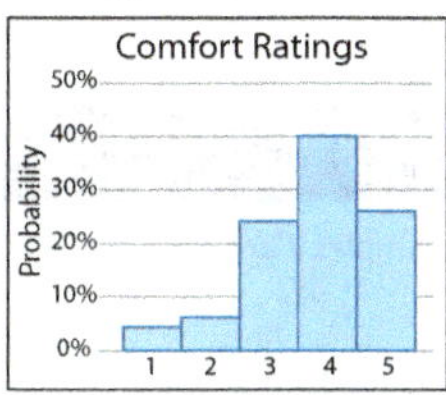

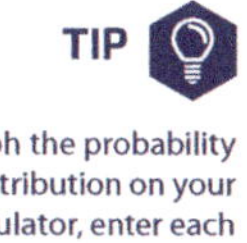
TIP

To graph the probability distribution on your calculator, enter each outcome x_i in L1 and their probabilities $P(x_i)$ in L2. Then use STAT PLOT to draw a histogram with List: L1 and FreqList: L2.

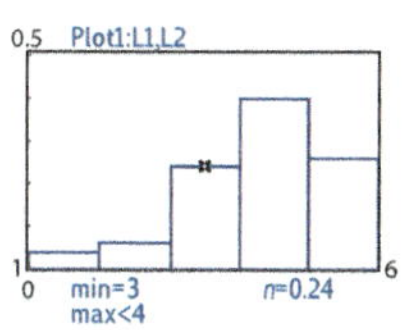

SKILL ✔ EXERCISE 27

The graph of a probability distribution illustrates its overall shape. Since probability distributions include all possible outcomes, they represent the experiment's population. The mean and standard deviation are the typical measures of central tendency and variation in a discrete probability distribution. Different formulas for μ and σ must be used since the probability distribution describes what we expect to happen and N, a total number of values, is unknown.

▌ DEFINITION

The **mean of a probability distribution**, μ, is the sum of the products of each outcome x_i of the random variable and its associated probability $P(x_i)$.

$$\mu = \sum_{i=1}^{n} x_i P(x_i)$$

The mean of a probability distribution is also called the *expected value*, denoted as $E(x)$. The expected value is a weighted average that estimates the average result after many

ing the students create the histogram on their calculators or on a spreadsheet. Emphasize that the histogram illustrates the overall shape of the distribution and note the need for different formulas for the central tendency and variation of a probability distribution.

Introduce the definition for the *mean of a probability distribution*, which is also called the *expected value*. Explain that this value is frequently used to make predictions. Use the values in Example 2 and the formula for the expected value to find the expected value of the probability distribution in Example 3. Share that the mean is the average value a person might

expect if the experiment contained many more surveys.

Consider discussing the Biblical Perspective of Mathematics margin note concerning the expected probability of male/female births. Some evolutionists have argued that natural selection has accounted for higher male mortality rates by increasing the number of male births. The fact of a higher male birth rate is interpreted through the lens of a person's worldview. Scottish physician John Arbuthnot (1667–1735) studied gender ratios at birth and argued that divine providence, not chance, was the governing factor. Arbuthnot was the first to use

probability distribution
random variable
standard deviation
variance of a probability distribution

Additional Exercises

State whether each random variable is discrete or continuous and justify your answer.

1. the number of days last week that students were given homework
   discrete; a whole number from 0 to 5

2. the amount of time students spent last night studying for their history test   continuous; a real number greater than or equal to 0

Use the following results from rolling a die to complete Additional Exercises 3–5.

x_i	1	2	3	4	5	6
Freq.	9	7	7	9	13	5

3. Create a probability distribution for the data.

x_i	$P(x_i)$	$x_i P(x_i)$	$(x_i - \mu)^2 P(x_i)$
1	0.18	0.18	1.125
2	0.14	0.28	0.315
3	0.14	0.42	0.035
4	0.18	0.72	0.045
5	0.26	1.30	0.585
6	0.10	0.60	0.625
Sum	1.00	$\mu = 3.5$	$\sigma^2 = 2.73$
			$\sigma \approx 1.65$

4. Find the expected value of the probability distribution. $E(x) = \mu = 3.5$

5. Find the standard deviation for the probability distribution. $\sigma \approx 1.65$

6. Find the probability of each outcome by rolling a regular die.
   a. 2 fives out of 3 rolls   0.069
   b. 2 fives out of 6 rolls   0.201

7. Find the mean and standard deviation of the binomial probability distribution that describes the number of fives obtained by rolling 4 dice.
   $\mu \approx 0.667; \sigma \approx 0.745$

Assignments

- **Minimum:** 1–6, 9, 11–13, 15, 17, 19–21; 23–27 odd; 41, 43–46, 48
- **Standard:** 2, 4–6, 8, 10–11, 14–15, 17, 19–20, 22, 24, 27, 29–31, 36, 39, 42, 48
- **Extended:** 3–4, 6; 12–20 even; 21, 24–28, 32, 34–35, 37, 40, 45, 47–48

Is 0.5 the expected value that a newborn baby will be a boy? World Bank statistics show there are 105 male babies born in the United States for every 100 females. Your worldview will determine how you interpret this statistic.

trials of the experiment. The expected value may not be equal to any of the outcomes for the random variable.

Example 3 Finding the Expected Value of a Probability Distribution

Find the expected value of the probability distribution from Example 2.

Answer

x_i	$P(x_i)$	$x_i \cdot P(x_i)$
1	0.04	$1(0.04) = 0.04$
2	0.06	$2(0.06) = 0.12$
3	0.24	$3(0.24) = 0.72$
4	0.40	$4(0.40) = 1.60$
5	0.26	$5(0.26) = 1.30$

$$E(x) = \sum x_i P(x_i) = 3.78$$

1. Extend the table to include a column containing the products of $x_i \cdot P(x_i)$ for each possible outcome.

2. Find $\mu = E(x) = \sum_{i=1}^{n} x_i P(x_i)$.

SKILL ✓ **EXERCISE 27**

The formulas for the variance and standard deviation of a discrete probability distribution are different from the formulas for populations and samples given in Chapter 10.

DEFINITIONS

The **variance of a probability distribution**, σ^2, is the sum of the products of the squared deviation of each outcome from the mean and the outcome's probability.
The **standard deviation**, σ, is the square root of the variance.

$$\sigma^2 = \sum_{i=1}^{n} (x_i - \mu)^2 P(x_i) \text{ and } \sigma = \sqrt{\sigma^2}$$

Example 4 Finding the Standard Deviation of a Probability Distribution

Find the standard deviation of the car comfort rating data from Examples 2 and 3.

Answer

x_i	$P(x_i)$	$x_i \cdot P(x_i)$	$(x_i - \mu)^2 P(x_i)$
1	0.04	0.04	$(1 - 3.78)^2(0.04) \approx 0.309$
2	0.06	0.12	$(2 - 3.78)^2(0.06) \approx 0.190$
3	0.24	0.72	$(3 - 3.78)^2(0.24) \approx 0.146$
4	0.40	1.60	$(4 - 3.78)^2(0.40) \approx 0.019$
5	0.26	1.30	$(5 - 3.78)^2(0.26) \approx 0.387$
		$\mu = 3.78$	$\sigma^2 \approx 1.05$
			$\sigma \approx 1.03$

1. Extend the table to include a column containing the product of $(x_i - \mu)^2 P(x_i)$ for each outcome.

2. Find $\sigma^2 = \sum_{i=1}^{n} (x_i - \mu)^2 P(x_i)$.

3. Find $\sigma = \sqrt{\sigma^2}$.

4. Enter each possible outcome x_i in L1 and their probabilities $P(x_i)$ in L2. Select 1-Var Stats from the STAT, CALC menu and enter L1 for the List: and L2 for the FreqList:. Note that μ is given as $\bar{x}$ and σ is given as σx.

Check

```
1-Var Stats
x̄=3.78
Σx=3.78
Σx²=15.34
Sx=
σx=1.025475499
n=1
minX=1
↓Q₁=3
```

SKILL ✓ **EXERCISE 27**

a formal test of statistical significance in his research. Use the Internet keyword search *Arbuthnot birth rate* for more interesting facts about his research.

Present the definitions of *variance of a probability distribution* and its *standard deviation*. Explain that while values describing central tendencies such as the average are important, the amount that values typically deviate from the average is also important when making predictions and, ultimately, decisions. Demonstrate finding these values using Example 4. It is important that the students obtain these values step-by-step, as in exercises 26–30, before relying on calculator-generated values. Going through each step and then using the calculator to check the values promotes understanding of how these values are found and provides valuable calculator experience.

Interactive Activity Having students use a spreadsheet to generate these values can help them focus on the relationships expressed by the formulas without getting bogged down in the arithmetic.

Review the requirements of a *binomial experiment* (see exercises 15–18) and discuss the characteristics of the resulting *binomial probability distribution*. Rolling dice and flipping coins are commonly used experiments that meet these char-

acteristics. Review how $_nC_x$ provides the number of ways to get x successes in n trials and formally define the *binomial probability distribution formula*.

Use Example 5 to demonstrate how to find single and multiple binomial probabilities. Then refer the students to the margin tip for related calculator functions.

Point out how the Binomial Theorem is an equation representing a binomial probability distribution. Then present the simpler formulas for its mean, variance, and standard deviation. Example 6 demonstrates that the simpler formulas produce the same result as the more gen-

Section 9.4 introduced a *binomial experiment*, which consists of *n* identical, independent trials in which each trial results in either a success or a failure. A *binomial probability distribution* can be constructed to describe all the possible outcomes of such an experiment.

CHARACTERISTICS OF A BINOMIAL PROBABILITY DISTRIBUTION

1. The experiment consists of *n* identical, independent trials in which each trial results in either a success (S) or a failure (F).
2. The probabilities of success and failure are $P(S) = p$ and $P(F) = q = 1 - p$.
3. The binomial random variable *x* is the number of successes in *n* trials of the experiment.

Flipping a coin three times is a simple binomial experiment. Each of the 3 trials has only two possible outcomes: heads or tails. The experiment's sample space is {*HHH, HHT, HTH, HTT, THH, THT, TTH, TTT*}. If the random variable represents the number of heads on three flips of the coin, it can have values of 0, 1, 2, or 3 and has the following binomial probability distribution.

heads, x_i	0	1	2	3
frequency, f	1	3	3	1
probability, $P(x_i)$	$\frac{1}{8} = 0.125$	$\frac{3}{8} = 0.375$	$\frac{3}{8} = 0.375$	$\frac{1}{8} = 0.125$

Recall from Section 9.4 that $_nC_x$ provides the number of ways to select *x* successes in *n* trials and that the *binomial probability distribution formula*

$$P(x) = {_nC_x}\,p^x q^{n-x}$$

gives the probability of *x* successes in *n* independent trials in which the probability of success is *p* and the probability of failure is $q = 1 - p$.

Example 5   Finding Binomial Probabilities

Find the probability of a randomly selected golfer getting each number of holes in one on a 9-hole miniature golf course if the probability of an average golfer getting a hole in one on each hole is 0.15.

a. 3 holes in one

b. fewer than 3 holes in one

c. 3 or more holes in one

Answer

x = number of holes in one
$n = 9$ holes; $p = 0.15$
$q = 1 - 0.15 = 0.85$

Define the variables for the binomial experiment.

a. $P(3) = {_9C_3}(0.15)^3(0.85)^{9-3}$
$= 84(0.15)^3(0.85)^6$
≈ 0.11 or 11%

Substitute into $P(x) = {_nC_x}\,p^x q^{n-x}$ and evaluate.

b. $P(2) = {_9C_2}(0.15)^2(0.85)^7 \approx 0.26$
$P(1) = {_9C_1}(0.15)^1(0.85)^8 \approx 0.37$
$P(0) = {_9C_0}(0.15)^0(0.85)^9 \approx 0.23$
$P(x_i < 3) \approx 0.26 + 0.37 + 0.23$
≈ 0.86 or 86%

1. Note that $P(x < 3) = P(2) + P(1) + P(0)$ and use the binomial distribution probability formula to find these 3 probabilities.

2. Calculate $P(x < 3) = P(2) + P(1) + P(0)$.

CONTINUED ➡

TIP

Your calculator's binomial probability distribution function can be used to determine the probability of each outcome, $P(x_i)$. Select DISTR, A: binompdf, and enter the parameters n, p, x.

```
binompdf(9,0.15,3)
          0.1069218877
```

The binomial cumulative distribution function calculates $P(\le x_i)$.

```
binomcdf(9,0.15,2)
          0.8591465972
```

Solutions

❯ A. Exercises

1. continuous; The value can be any real number greater than 0 up to the maximum miles per gallon for your car.

2. discrete; The value can be any whole number 0 to 50.

3. discrete; The value can only be 0, 1, 2, 3, or 4.

4. continuous; The value can be any real number between 0 and the maximum pressure the tire can sustain.

9a. $0.39 + 0.22 + 0.25 = 0.86$

9b. $0.39 + 0.25 = 0.64$

9c. $0.14 + 0.39 = 0.53$

10a. $0.75 + 0.10 = 0.85$

10b. $0.10 + 0.05 = 0.15$

11–14. $P(x) = {_nC_x}\,p^x q^{n-x}$
where $q = 1 - p$
$\mu = np$ and $\sigma^2 = npq$

11. $P(8) = {_{12}C_8}(0.75)^8(0.25)^4 \approx 0.19$
$\mu = (12)(0.75) = 9$
$\sigma = \sqrt{12(0.75)(0.25)} = 1.5$

12. $P(9) = {_{16}C_9}(0.33)^9(0.67)^7 \approx 0.03$
$\mu = (16)(0.33) = 5.28$
$\sigma = \sqrt{16(0.33)(0.67)} \approx 1.88$

13. $P(3) = {_9C_3}(0.6)^3(0.4)^6 \approx 0.07$
$\mu = (9)(0.6) = 5.4$
$\sigma = \sqrt{9(0.6)(0.4)} \approx 1.47$

14. $P(5) = {_{22}C_5}(0.45)^5(0.55)^{17} \approx 0.02$
$\mu = 22(0.45) = 9.9$
$\sigma = \sqrt{22(0.45)(0.55)} \approx 2.33$

15. yes; $p = \frac{32}{42} \approx 0.76$; $q = \frac{10}{42} \approx 0.24$
$n = 42$

16. no; There are more than two possible outcomes for each trial and the random variable is continuous.

17. yes; $p = \frac{13}{18} \approx 0.72$; $q = \frac{5}{18} \approx 0.28$
$n = 18$

18. yes; $p = \frac{4}{60} \approx 0.07$; $q = \frac{56}{60} \approx 0.93$
$n = 60$

19. $n = 5$; $p = \frac{1}{3}$; $q = \frac{2}{3}$

$x = $ number of correct responses

$$P(x_i \geq 3) = {}_5C_3\left(\frac{1}{3}\right)^3\left(\frac{2}{3}\right)^2$$
$$+ {}_5C_4\left(\frac{1}{3}\right)^4\left(\frac{2}{3}\right)^1 + {}_5C_5\left(\frac{1}{3}\right)^5\left(\frac{2}{3}\right)^0$$
$$\approx 0.165 + 0.041 + 0.004 = 0.210$$

20. $n = 12$; $p = 0.1$; $q = 0.9$

$x = $ number of left-handed people
$$P(x_i > 3) = 1 - P(x_i \leq 3)$$
$$= 1 - ({}_{12}C_0(0.1)^0(0.9)^{12}$$
$$+ {}_{12}C_1(0.1)^1(0.9)^{11}$$
$$+ {}_{12}C_2(0.1)^2(0.9)^{10}$$
$$+ {}_{12}C_3(0.1)^3(0.9)^9)$$
$$\approx 1 - (0.282 + 0.377 + 0.230 + 0.085)$$
$$= 1 - 0.974 = 0.026$$

21. $n = 20$; $p = 0.08$; $q = 0.92$

$x = $ number of blue-eyed people

21a. $P(x_i = 2) = {}_{20}C_2(0.08)^2(0.92)^{18}$
$$\approx 0.271$$

21b. $P(x_i \leq 3) = {}_{20}C_0(0.08)^0(0.92)^{20}$
$$+ {}_{20}C_1(0.08)^1(0.92)^{19}$$
$$+ {}_{20}C_2(0.08)^2(0.92)^{18}$$
$$+ {}_{20}C_3(0.08)^3(0.92)^{17}$$
$$\approx 0.189 + 0.328 + 0.271 + 0.141$$
$$= 0.929$$

21c. $P(x_i > 3) = 1 - P(x_i \leq 3)$
$$\approx 1 - 0.929 = 0.071$$

22a. $x = $ number of red candies

$x_i = 12$; $p = 0.2$; $q = 0.8$; $n = 60$

$$P(12) = {}_{60}C_{12}(0.2)^{12}(0.8)^{48} \approx 0.128$$

22b. $P(10 \leq x_i \leq 14)$
$$= P(x_i \leq 14) - P(x_i \leq 9)$$
$$= \text{binomcdf}(60, 1/5, 14)$$
$$\qquad - \text{binomcdf}(60, 1/5, 9)$$
$$\approx 0.7935 - 0.2132 = 0.580$$

22c. $P(0) = \text{binompdf}(60, 1/5, 0)$
$$\approx 1.5 \times 10^{-6}$$

22d. $P(60) = \text{binompdf}(60, 1/5, 60)$
$$\approx 1.15 \times 10^{-42}$$

22e. There was a malfunction in the production line or an intentional change in packaging.

23. $n = 8$; $p = 0.41$; $q = 0.59$

$x = $ number of times light is green
$$E(x) = \mu = np = 8(0.41) = 3.28$$

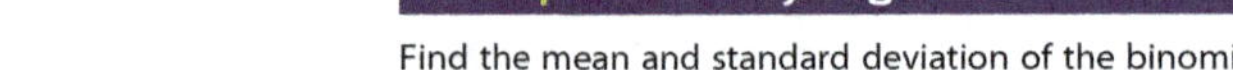

TIP

The Microsoft Excel function
=BINOM.DIST(x_i,n,p,false)
returns $P(x_i)$, while
=BINOM.DIST(x_i,n,p,true)
returns $P(\leq x_i)$.
For Example 6,
=BINOM.DIST(2,4,1/6,false)
returns 0.1157, and
=BINOM.DIST(2,4,1/6,true)
returns 0.9838.

---

c. $P(x_i \geq 3) = 1 - P(x_i < 3)$
$$\approx 1 - 0.86 = 0.14 \text{ or } 14\%$$

Find $P(x \geq 3)$ using the complement rule for probabilities.

SKILL ✓ EXERCISE 21

The binomial probability distribution describes the probability of each possible outcome in a binary experiment. If 4 dice are rolled with the random variable of getting a six on a die being a success, $p = \frac{1}{6}$ and $q = \frac{5}{6}$. The probabilities of the outcomes $x_i = \{0, 1, 2, 3, 4\}$ are the terms in the expansion of $(q + p)^4$, which can be found using the Binomial Theorem.

$$(q + p)^4 = \sum_{x=0}^{4} {}_nC_x\, p^x q^{n-x} = {}_4C_0q^4 + {}_4C_1pq^3 + {}_4C_2p^2q^2 + {}_4C_3p^3q + {}_4C_4p^4$$
$$= 1\left(\frac{5}{6}\right)^4 + 4\left(\frac{1}{6}\right)\left(\frac{5}{6}\right)^3 + 6\left(\frac{1}{6}\right)^2\left(\frac{5}{6}\right)^2 + 4\left(\frac{1}{6}\right)^3\left(\frac{5}{6}\right) + 1\left(\frac{1}{6}\right)^4$$
$$\approx 0.4823 + 0.3858 + 0.1157 + 0.0154 + 0.0008$$

sixes, x_i	0	1	2	3	4
probability, $P(x_i)$	0.4823	0.3858	0.1157	0.0154	0.0008

The sum of these probabilities is 1 since $(q + p)^4 = \left(\frac{5}{6} + \frac{1}{6}\right)^4 = 1^4 = 1$.

While the mean and standard deviation of a binomial probability distribution can be found using the formulas given earlier in this section, simpler formulas have been derived using the properties of the binomial distribution.

Mean, Variance, and Standard Deviation of Binomial Probability Distributions		
mean: $\mu = np$	variance: $\sigma^2 = npq$	standard deviation: $\sigma = \sqrt{\sigma^2} = \sqrt{npq}$

Example 6   Analyzing a Binomial Distribution

Find the mean and standard deviation of the binomial probability distribution describing the number of sixes obtained by rolling 4 dice.

Answer

using the formulas for a binomial probability distribution:

$$\mu = np = 4\left(\frac{1}{6}\right) = \frac{2}{3} \approx 0.6667$$

$$\sigma = \sqrt{npq} = \sqrt{4\left(\frac{1}{6}\right)\left(\frac{5}{6}\right)} = \sqrt{\frac{5}{9}} \approx 0.745$$

using a spreadsheet and the general formulas for probability distributions:

	A	B	C	D
1	Sixes	Probability		
2	x_i	$P(x_i)$	$x_i \cdot P(x_i)$	$(x_i - \mu)^2 P(x_i)$
3	0	0.4823	0	0.2143
4	1	0.3858	0.3858	0.0429
5	2	0.1157	0.2315	0.2058
6	3	0.0154	0.0463	0.0840
7	4	0.0008	0.0031	0.0086
8		Sum	Sum	Sum
9		1	0.6667	0.5556
10			Std Dev:	0.7454

SKILL ✓ EXERCISE 23

Since the expected value (or mean) is two-thirds, we would expect a six in about 2 out of 3 rolls of the 4 dice. Any roll of the dice producing 3 sixes ($\mu + 3.13\sigma$) would be unlikely, since $P(3 \text{ sixes}) \approx 1.5\%$, and 4 sixes ($\mu + 4.47\sigma$) would be extremely rare, since $P(4 \text{ sixes}) \approx 0.08\%$.

24. $x = $ number of boys

x_i	$P(x_i)$	$x_i P(x_i)$
0	0.058	0
1	0.240	0.240
2	0.375	0.749
3	0.260	0.780
4	0.068	0.271
Sum	1	2.04

$$E(x) = \mu = np = 4(0.51) = 2.04$$

25. $x = $ number of points scored

x_i	$P(x_i)$	$x_i P(x_i)$
0	$\frac{5}{25} = 0.20$	0
1	$\frac{17}{25} = 0.68$	0.68
2	$\frac{3}{25} = 0.12$	0.24
Sum	1.00	0.92

$$E(x) = 0(0.20) + 1(0.68) + 2(0.12)$$
$$= 0.92$$

Classify each random variable as discrete or continuous and justify your answer.

1. Let x represent the miles per gallon your car gets on a 200 mi interstate trip.

2. Let x represent the number of questions answered correctly on a test consisting of 50 multiple-choice questions.

3. Let x represent the number of red marbles pulled on 4 successive tries (with replacement) from a bag containing 3 blue, 2 white, and 6 red marbles.

4. Let x represent the tire pressure in a car tire.

Determine whether each table could represent a probability distribution and justify your answer.

5.

x_i	3	5	7
$P(x_i)$	0.5	−0.3	0.2

no; A probability cannot be negative.

6.

x_i	−2	0	2	4
$P(x_i)$	0.03	0.28	0.61	0.08

yes; $\sum P(x_i) = 1$ and $0 \le P(x) \le 1$

7.

x_i	4	9	15	22	30
$P(x_i)$	0.21	0.07	0.42	0.16	0.09

no; $\sum P(x_i) = 0.95 \ne 1$

8.

x_i	−5	−10	−15
$P(x_i)$	0.58	0.29	0.41

no; $\sum P(x_i) = 1.28 \ne 1$

Use the probability distribution to find each probability.

9.

x_i	1	2	3	4
$P(x_i)$	0.14	0.39	0.22	0.25

a. $P(x_i > 1)$ 0.86
b. $P(\text{an even number})$ 0.64
c. $P(x_i \le 2)$ 0.53

10.

x_i	−100	−50	50	100
$P(x_i)$	0.75	0.10	0.10	0.05

a. $P(x_i < 0)$ 0.85
b. $P(x_i > 0)$ 0.15

Use the binomial probability distribution formula to find $P(x_i)$, the probability of x successes in n independent trials where p is the probability of a success in a trial. Then find the expected value and standard deviation of the related binomial probability distribution. Round answers to the nearest hundredth if necessary.

11. $x = 8$, $n = 12$, $p = 0.75$ $P(8) = 0.19$; $\mu = 9$; $\sigma = 1.5$

12. $x = 9$, $n = 16$, $p = 0.33$ $P(9) = 0.03$; $\mu = 5.28$; $\sigma = 1.88$

13. $x = 3$, $n = 9$, $p = 0.6$ $P(3) = 0.07$; $\mu = 5.4$; $\sigma = 1.47$

14. $x = 5$, $n = 22$, $p = 0.45$ $P(5) = 0.02$; $\mu = 9.9$; $\sigma = 2.33$

Determine whether each of the following describes a binomial experiment. If so, state the values of p, q, and n, and if not, explain why.

15. Forty-two seniors at a high school are asked if they regularly eat breakfast before coming to school. Thirty-two of them affirm that they do. The random variable represents the number of seniors who eat breakfast.

16. A survey of 115 recent ER patients at a particular hospital asked how many minutes patients had to wait before a nurse talked with them. The random variable represents the number of minutes they waited.

17. A penny is flipped 18 times. Heads lands up 13 of those times. The random variable represents the number of times a flip resulted in heads.

18. In a group of 60 randomly selected donors, 27 have type O blood, 24 have type A, 5 have type B, and 4 have type AB. The random variable represents the number of donors having type AB blood.

≈ 0.210 or 21.0%

19. Jarnel randomly guesses on the 5 multiple-choice questions on his biology quiz. If his chance of guessing the correct answer on each question is $\frac{1}{3}$, what is the probability that he gets at least 60% correct?

20. Approximately 10% of people are left-handed. What is the probability that in a group of 12 children more than 3 are left-handed? ≈ 0.026 or 2.6%

21. If 8% of the world's population has blue eyes, find the probability of each number of people with blue eyes in a group of 20 randomly chosen individuals.
a. 2 blue-eyed people ≈ 0.271 or 27.1%
b. at most 3 blue-eyed people ≈ 0.929 or 92.9%
c. more than 3 blue-eyed people ≈ 0.071 or 7.1%

22. A manufacturer claims that a regular bag of its candies has an equal distribution of its 5 colors. Use a binomial distribution in which x represents the number of red candies in a 2.17 oz bag containing 60 candies to find the following probabilities.

a. $P(x_i = 12)$ ≈ 0.128
b. $P(10 \le x_i \le 14)$ ≈ 0.580
c. $P(x_i = 0)$ ≈ 1.5×10^{-6}
d. $P(x_i = 60)$ ≈ 1.15×10^{-42}
e. Give a possible explanation for a bag with either no red or all red candies.

x_i	$P(x_i)$	$x_iP(x_i)$	$(x_i - \mu)^2 P(x_i)$
2	$\frac{1}{12} \approx 0.083$	0.167	0.521
3	$\frac{2}{12} \approx 0.167$	0.500	0.375
4	$\frac{3}{12} = 0.250$	1.000	0.063
5	$\frac{3}{12} = 0.250$	1.250	0.063
6	$\frac{2}{12} \approx 0.167$	1.000	0.375
7	$\frac{1}{12} \approx 0.083$	0.583	0.521
Sum	1	$\mu = 4.5$	$\sigma^2 \approx 1.917$
			$\sigma \approx 1.384$

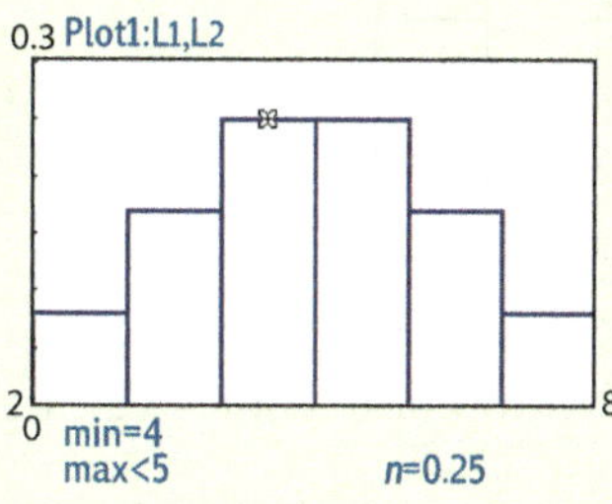

28. $x = $ sum of the two dice

x_i	$P(x_i)$	$x_iP(x_i)$	$(x_i - \mu)^2 P(x_i)$
2	$\frac{1}{16} = 0.0625$	0.1250	0.5625
3	$\frac{2}{16} = 0.1250$	0.3750	0.5000
4	$\frac{3}{16} = 0.1875$	0.7500	0.1875
5	$\frac{4}{16} = 0.250$	1.2500	0.0000
6	$\frac{3}{16} = 0.1875$	1.1250	0.1875
7	$\frac{2}{16} = 0.1250$	0.8750	0.5000
8	$\frac{1}{16} = 0.0625$	0.5000	0.5625
Sum	1	$\mu = 5.0$	$\sigma^2 = 2.5000$
			$\sigma \approx 1.5811$

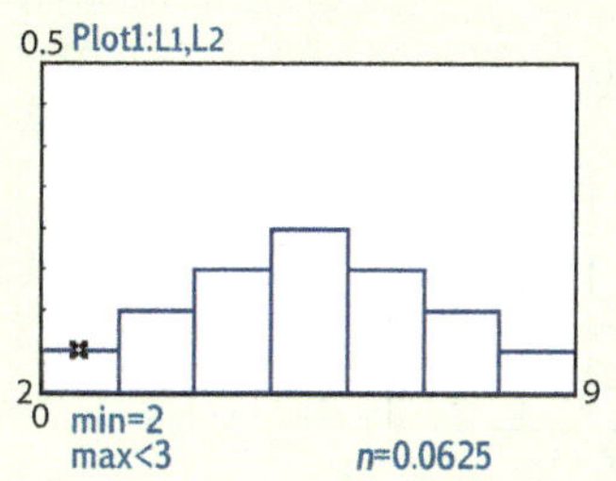

26. $x = $ number of heads

x_i	$P(x_i)$	$x_iP(x_i)$	$(x_i - \mu)^2 P(x_i)$
0	$\frac{1}{8}$	0	0.2813
1	$\frac{3}{8}$	0.375	0.0938
2	$\frac{3}{8}$	0.75	0.0938
3	$\frac{1}{8}$	0.375	0.2813
Sum	1	$\mu = 1.5$	$\sigma^2 \approx 0.75$
			$\sigma \approx 0.866$

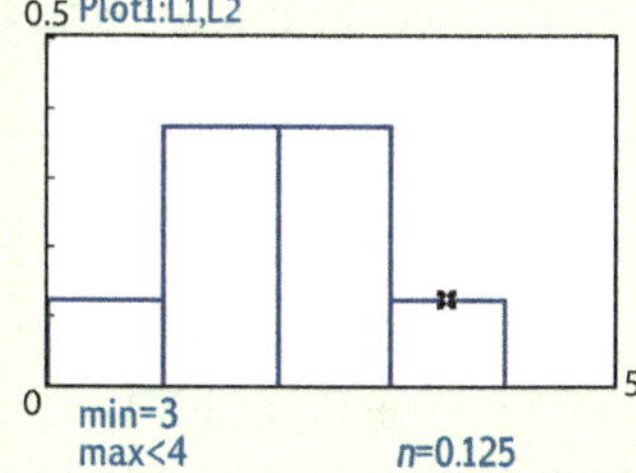

$$E(x) = \mu = np = 3(0.5) = 1.5$$
$$\sigma^2 \approx npq = 3(0.5)(0.5) = 0.75$$
$$\sigma \approx 0.866$$

27. $x = $ sum of the spins

	1	2	3	4
1	2	3	4	5
2	3	4	5	6
3	4	5	6	7

29. $x =$ sum of the two dice

x_i	$P(x_i)$	$x_i P(x_i)$	$(x_i - \mu)^2 P(x_i)$
2	$\frac{9}{36} = 0.25$	0.500	0.444
3	$\frac{12}{36} \approx 0.333$	1.000	0.037
4	$\frac{10}{36} \approx 0.278$	1.111	0.123
5	$\frac{4}{36} \approx 0.111$	0.556	0.309
6	$\frac{1}{36} \approx 0.028$	0.167	0.198
Sum	1	$\mu \approx 3.333$	$\sigma^2 \approx 1.11$
			$\sigma \approx 1.05$

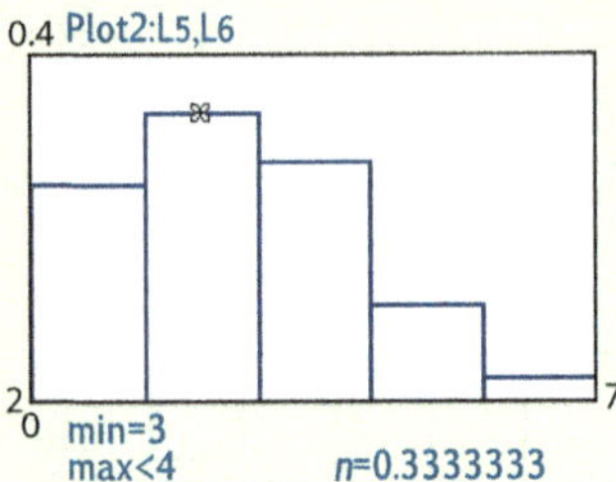

30a–b. $x =$ number of Bible-reading days

x_i	$P(x_i)$	$x_i P(x_i)$	$(x_i - \mu)^2 P(x_i)$
0	$\frac{4}{50} = 0.08$	0	0.9577
1	$\frac{2}{50} = 0.04$	0.04	0.2421
2	$\frac{12}{50} = 0.24$	0.48	0.5116
3	$\frac{9}{50} = 0.18$	0.54	0.0381
4	$\frac{8}{50} = 0.16$	0.64	0.0467
5	$\frac{7}{50} = 0.14$	0.70	0.3320
6	$\frac{3}{50} = 0.06$	0.36	0.3871
7	$\frac{5}{50} = 0.10$	0.70	1.2532
Sum	1	$\mu = 3.46$	$\sigma^2 \approx 3.7684$
			$\sigma \approx 1.9412$

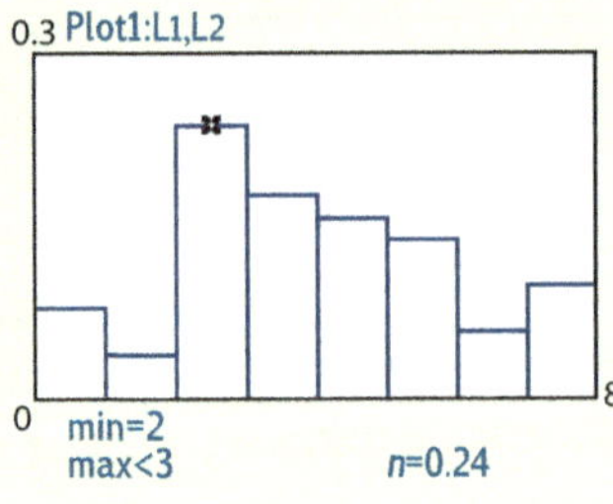

23. A traffic light on the way to church is green 41% of the time, yellow 12% of the time, and red 47% of the time. How many times would you expect it to be green during 8 trips to church? $E(x) = \mu \approx 3.28$

24. If 51% of the babies born in the United States are boys, create a probability distribution for the number of boys born into a 4-child family and determine the expected number of boys. $E(x) = \mu = 2.04$

25. After scoring a touchdown, a football team may attempt an extra point or a 2-point conversion. The table lists the results for such plays by a high school football team over the past 6 games. Create a probability distribution for these scoring plays and use it to find the expected number of points for the team after scoring a touchdown. $E(x) = \mu = 0.92$

Extra Points	2	1	0
Frequency	3	17	5

Create a table and a histogram representing the probability distribution for each experiment. Then use the probability distribution to find the expected value and the standard deviation.

26. A penny is flipped 3 times and the number of heads is recorded. $E(x) = \mu = 1.5; \sigma \approx 0.87$

27. The illustrated spinners are each spun once and the resulting sum is noted. $E(x) = \mu = 4.5; \sigma \approx 1.38$

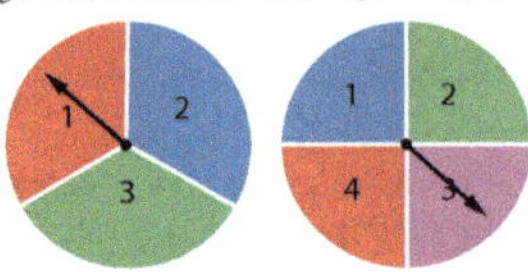

28. Two tetrahedral dice (each producing a result of 1, 2, 3, or 4) are rolled. $E(x) = \mu = 5; \sigma \approx 1.58$

29. Two 6-sided dice, each having three faces with a 1, two faces with a 2, and one face with a 3, are rolled. $E(x) = \mu \approx 3.33; \sigma \approx 1.05$

30. Members of the church youth group were asked the number of days per week they read their Bibles outside of church. The results are summarized in the table below.

Number of Days	0	1	2	3	4	5	6	7
Number of Members	4	2	12	9	8	7	3	5

a. Create a table and a histogram to represent the probability distribution for this data.

b. Find the mean and the standard deviation of the distribution. $E(x) = \mu = 3.46; \sigma \approx 1.94$

c. Interpret the results.

Find the mean of a probability distribution to determine whether the warranty will likely save money for the purchaser. $E(x) = \mu = -304$
On average, the purchaser loses $304.

31. A used car dealer offers a 1-yr extended warranty for $600 that covers 100% of all repair costs. During the next year there is a 22% chance the car will need an $800 repair, a 12% chance it will need a $1000 repair, and a 66% chance that no repairs will be needed.

32. An electronics store offers a 2-yr extended warranty covering all repairs for $10 with the purchase of a $520 camera. In the first 2 yr, 6% of these cameras require repairs costing $100 and 1.5% require repairs costing $200.
$E(x) = \mu = -1$; On average, the purchaser loses $1.

> **C. Exercises**

33. Use a spreadsheet to create a probability distribution of the sums obtained by rolling a pair of dodecahedral dice. Which sum is most likely to occur? Which sum is least likely to occur? What is the expected value of the sum for this probability distribution?
most likely: 13; least likely: 2 and 24; $E(x) \approx 13$

30c. The average number of days is 3.46 and the most common answer is 2 days.

31. $x =$ amount saved by purchasing the warranty

x_i	−600	200	400
$P(x_i)$	0.66	0.22	0.12

$E(x) = 0.66(-600) + 0.22(200)$
$\qquad\qquad + 0.12(400) = -304$
On average, the purchaser loses $304.

32. $x =$ amount saved by purchasing the warranty

x_i	−10	90	190
$P(x_i)$	0.925	0.060	0.015

$E(x) = 0.925(-10) + 0.06(90)$
$\qquad\qquad + 0.015(190) = -1$
On average, the purchaser loses $1.

> **C. Exercises**

33. $x =$ sum of the two dice

x_i	$P(x_i)$	$x_i P(x_i)$	x_i	$P(x_i)$	$x_i P(x_i)$
2	$\frac{1}{144}$	0.014	14	$\frac{11}{144}$	1.069
3	$\frac{2}{144}$	0.042	15	$\frac{10}{144}$	1.042
4	$\frac{3}{144}$	0.083	16	$\frac{9}{144}$	1.000
5	$\frac{4}{144}$	0.139	17	$\frac{8}{144}$	0.944
6	$\frac{5}{144}$	0.208	18	$\frac{7}{144}$	0.875
7	$\frac{6}{144}$	0.292	19	$\frac{6}{144}$	0.792
8	$\frac{7}{144}$	0.389	20	$\frac{5}{144}$	0.694

34. A phone company offers two plans for a 2-yr extended warranty on broken screens. Plan A costs $130 and covers up to two screen replacements, each with a $30 copay. Plan B costs $115 and covers just one screen replacement with no copay. Screen replacement with no warranty costs $175. If 18% of screens need replacement once during the 2-yr period and 4% of screens need replacement twice during the 2-yr period, find the expected value of probability distributions for each plan (A, B, or none) to determine which would likely cost the consumer the least.

35. A CMO Impact Study completed by K. A. Whitler researched the time that chief marketing officers (CMOs) had held their current position. Let the random variable x be the number of years a CMO has held that position.

Years	$P(x_i)$
1	0.22
2	0.19
3	0.16
4	0.10
5	0.10
6	0.05
7	0.04
8	0.02
9	0.01
10+	0.11

   a. Does the table satisfy the properties of a probability distribution? Explain why or why not.

   b. Draw a histogram representing the data and describe the shape of the distribution. skewed right

   c. Find the probability of a CMO holding that position for more than 5 yr. 0.23

   d. Find the probability of a CMO holding that position less than 4 yr. 0.57

   e. Use the data to determine the expected number of years a CMO has been in that position and interpret the result. 3.91 yr

36. According to the CDC, about 1 out of 700 babies is born with Down syndrome. According to the US Census Bureau, about 10,800 babies are born each day in the United States. Let x represent the number of children born with Down syndrome each day in the United States.

   a. Explain why a binomial probability distribution is an appropriate model for studying the probabilities of the number of children born with Down syndrome.

   b. What is the expected number of babies born with Down syndrome on a single day in the United States? ≈ 15

   c. Use technology to find $P(12)$. ≈ 0.076 or 7.6%

   d. Use technology to find $P(x \le 10)$. ≈ 0.099 or 9.9%

34. $E_A(x) \approx -\$137.80$; $E_B(x) \approx -\$122$; $E_C(x) \approx -\$45.50$; no warranty

35a. yes; $\Sigma P(x_i) = 1$ and $0 \le P(x) \le 1$

37. Experiment: Use an Internet coin toss simulator to flip 5 pennies and record the number of heads. Repeat this 50 times.

   a. Construct an experimental probability distribution from the trials using the number of heads as the random variable and use it to find $P(3)$ and $P(\ge 2)$. sample answer: $P_E(x = 3) = 0.38$; $P_E(x \ge 2) = 0.82$

   b. Construct a theoretical probability distribution for this experiment and use it to find $P(3)$ and $P(\ge 2)$. $P_T(3) \approx 0.313$; $P_T(x \ge 2) \approx 0.813$

   c. Discuss the differences beteeen and the advantages or disadvantages of theoretical and experimental probabilities.

   d. When would it be necessary to use an experimental probability distribution?

38. Use an Internet keyword such as *wages by education* to find approximate median salaries for US workers according to their level of education. Create a histogram representing the data. Then determine the expected median salary for a randomly selected worker.

Level of Education	Percent of Workers	Median Annual Salary
no high school diploma	8%	$25,636
high school diploma or GED	30%	$35,256
some college or associate's degree	29%	$41,496
bachelor's degree	23%	$59,124
advanced degree	10%	$75,597

sample answer using 2018 data: Expected median income is $45,819.74.

35b. 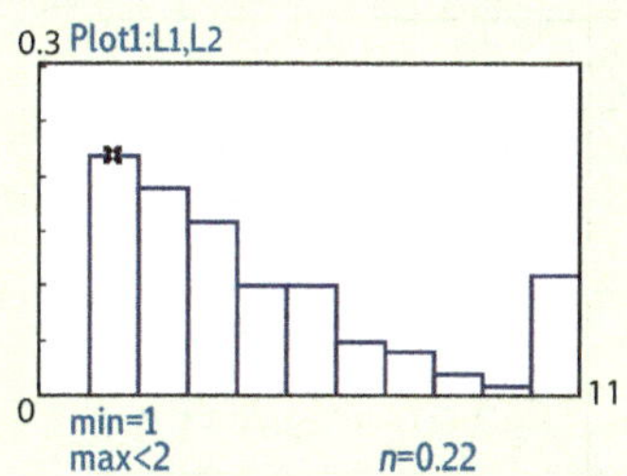

35c. $P(x_i > 5) = 0.05 + 0.04 + 0.02$
$+ 0.01 + 0.11 = 0.23$

35d. $P(x_i < 4) = 0.22 + 0.19 + 0.16$
$= 0.57$

35e. $E(x) = 1(0.22) + 2(0.19) + 3(0.16)$
$+ 4(0.10) + 5(0.10) + 6(0.05)$
$+ 7(0.04) + 8(0.02) + 9(0.01)$
$+ 10(0.11) = 3.91$

CMOs, on average, have been in their position for almost 4 yr.

36a. The outcome is one of two choices, with Down Syndrome or without Down Syndrome. The births are independent of each other.

36b. $E(x) = np$
$= 10,800(0.001429) \approx 15.4$

36c–d. 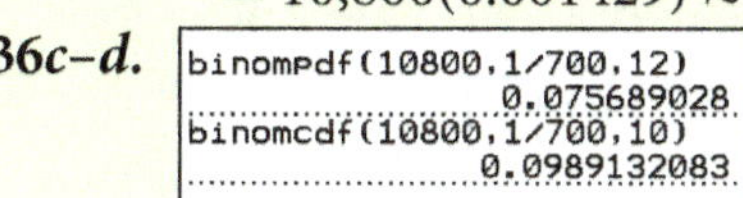

or

36c. $P(x_i = 12)$
$= \text{binompdf}(10800, 1/700, 12)$
≈ 0.076

36d. $P(x_i \le 10)$
$= \text{binomcdf}(10800, 1/700, 10n)$
≈ 0.099

37. $x = $ number of heads

37a. sample answer:

x_i	f_i	$P(x_i)$
0	0	0.00
1	9	0.18
2	11	0.22
3	19	0.38
4	11	0.22
5	0	0.00

$P_E(x_i = 3) = 0.38$
$P_E(x_i \ge 2) = 0.22 + 0.38 + 0.22$
$= 0.82$

9	$\frac{8}{144}$	0.500	21	$\frac{4}{144}$	0.583
10	$\frac{9}{144}$	0.625	22	$\frac{3}{144}$	0.458
11	$\frac{10}{144}$	0.764	23	$\frac{2}{144}$	0.319
12	$\frac{11}{144}$	0.917	24	$\frac{1}{144}$	0.167
13	$\frac{12}{144}$	1.083	**Sum**	1	$\mu = 13.0$

The most likely sum is 13, and the least likely sums are 2 and 24.

$$E(x) = \sum_{i=2}^{24} x_i P(x_i) \approx 13$$

34. $x = $ cost of the plan

Claims	Plan A		Plan B		No Plan	
	x_i	$P(x_i)$	x_i	$P(x_i)$	x_i	$P(x_i)$
0	−130	0.78	−115	0.78	0	0.78
1	−160	0.18	−115	0.18	−175	0.18
2	−190	0.04	−290	0.04	−350	0.04

$E_A(x) \approx (-130)(0.78) + (-160)(0.18)$
$+ (-190)(0.04) = -\$137.80$
$E_B(x) \approx (-115)(0.78) + (-115)(0.18)$
$+ (-290)(0.04) = -\$122$
$E_C(x) \approx (0)(0.78) + (-175)(0.18)$
$+ (-350)(0.04) = -\$45.50$; none

37b.

x_i	$P(x_i)$
0	$_5C_0(0.5)^5 \approx 0.031$
1	$_5C_1(0.5)^5 \approx 0.156$
2	$_5C_2(0.5)^5 \approx 0.313$
3	$_5C_3(0.5)^5 \approx 0.313$
4	$_5C_4(0.5)^5 \approx 0.156$
5	$_5C_5(0.5)^5 \approx 0.031$

$$P_T(3) \approx 0.313$$
$$P_T(x_i \geq 2) \approx 2(0.313) + 0.156$$
$$+ 0.031 = 0.813$$

37c. Theoretical distributions can be created quickly without having to accumulate large amounts of data. Experimental distributions may better reflect realities such as an unfair die or coin.

37d. when the theoretical p-value is unknown

38. sample answer based on 2018 data:

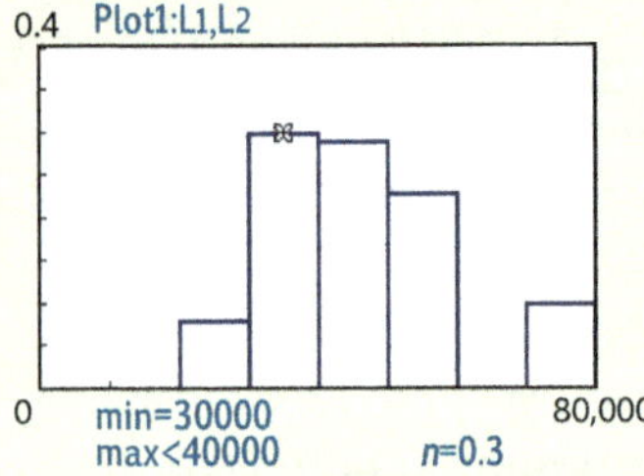

$$E(x) = 0.08(25{,}636) + 0.30(35{,}256)$$
$$+ 0.29(41{,}496) + 0.23(59{,}124)$$
$$+ 0.1(75{,}597) \approx \$45{,}819.74$$

❯ Cumulative Review

39. $4 \sin^2 x = 3$
$$\sin^2 x = \frac{3}{4}$$
$$\sin x = \pm\frac{\sqrt{3}}{2}$$
$$x = \frac{\pi}{3}, \frac{2\pi}{3}, \frac{4\pi}{3}, \frac{5\pi}{3}$$

40. $(\cos x + 1)(2\cos x - 1) = 0$
$$\cos x = -1 \quad \text{or} \quad \cos x = \frac{1}{2}$$
$$x = \pi \qquad\qquad x = \frac{\pi}{3}, \frac{5\pi}{3}$$

41. $\begin{bmatrix} 2 & -5 \\ 3 & -1 \end{bmatrix}\begin{bmatrix} x \\ y \end{bmatrix} = \begin{bmatrix} -16 \\ 2 \end{bmatrix}$
$$A^{-1} = \frac{1}{13}\begin{bmatrix} -1 & 5 \\ -3 & 2 \end{bmatrix}$$
$$X = A^{-1}C = \frac{1}{13}\begin{bmatrix} -1 & 5 \\ -3 & 2 \end{bmatrix}\begin{bmatrix} -16 \\ 2 \end{bmatrix}$$
$$= \frac{1}{13}\begin{bmatrix} 26 \\ 52 \end{bmatrix} = \begin{bmatrix} 2 \\ 4 \end{bmatrix}; \ (2, 4)$$

Find the primary solutions of each equation. [5.3]

39. $2\sin^2 x = 3 - 2\sin^2 x$ $\quad x = \frac{\pi}{3}, \frac{2\pi}{3}, \frac{4\pi}{3}, \frac{5\pi}{3}$

40. $2\cos^2 x + \cos x - 1 = 0$ $\quad x = \frac{\pi}{3}, \pi, \frac{5\pi}{3}$

Write and solve a matrix equation to solve each system. [7.5]

41. $2x - 5y = -16$
$\quad 3x - y = 2$ $\ (2, 4)$

42. $4x + 7y = -1$
$\quad 6x + 2y = -10$ $\ (-2, 1)$

Find each probability for the score of a randomly selected student from a normally distributed population with a mean score of 82 and standard deviation of 6.2. [10.6]

43. $P(x_i < 85)$ $\ \approx 68\%$

44. $P(x_i > 80)$ $\ \approx 63\%$

45. What is the remainder when $p(x) = 5x^5 - 4x^4 + 3x^3 - 2x^2 + x$ is divided by $x - 2$? [2.3] A
- **A.** 114
- **B.** 57
- **C.** 2
- **D.** −258
- **E.** none of these

46. Points D and F are 19 m apart on one side of a river. If point E lies on the opposite bank with $\angle EDF = 60°$ and $\angle EFD = 50°$, which of the following represents the measure of ED? [5.6] D
- **A.** $\dfrac{(19 \text{ m}) \sin 60°}{\sin 70°}$
- **B.** $\dfrac{(19 \text{ m}) \sin 70°}{\sin 50°}$
- **C.** $\dfrac{\sin 70°}{(19 \text{ m}) \sin 60°}$
- **D.** $\dfrac{(19 \text{ m}) \sin 50°}{\sin 70°}$
- **E.** none of these

47. Which of the following represents the fifth term of $(3x - 4y)^9$? [9.4] C
- **A.** $_9C_4(3x)^4(-4y)^5$
- **B.** $_9C_5(3x)^4(-4y)^5$
- **C.** $_9C_5(3x)^5(-4y)^4$
- **D.** $_9C_5(3x)^5(-4y)^5$
- **E.** none of these

48. A sports league has an average salary of \$2.1 million with a standard deviation of \$1.6 million. Assuming a normal distribution, what is the probability (to the nearest percent) that a player will make between \$3 million and \$4 million? [10.6] D
- **A.** 63%
- **B.** 56%
- **C.** 34%
- **D.** 17%
- **E.** none of these

---

42. $\begin{bmatrix} 4 & 7 \\ 6 & 2 \end{bmatrix}\begin{bmatrix} x \\ y \end{bmatrix} = \begin{bmatrix} -1 \\ -10 \end{bmatrix}$
$$A^{-1} = -\frac{1}{34}\begin{bmatrix} 2 & -7 \\ -6 & 4 \end{bmatrix}$$
$$X = A^{-1}C = -\frac{1}{34}\begin{bmatrix} 2 & -7 \\ -6 & 4 \end{bmatrix}\begin{bmatrix} -1 \\ -10 \end{bmatrix}$$
$$= -\frac{1}{34}\begin{bmatrix} 68 \\ -34 \end{bmatrix} = \begin{bmatrix} -2 \\ 1 \end{bmatrix}; \ (-2, 1)$$

43. $z = \dfrac{x - \mu}{\sigma} = \dfrac{85 - 82}{6.2} \approx 0.48$
$$P(x_i < 85) = P(z < 0.48)$$
$$\text{normalcdf}(-4, 0.48) \approx 0.68$$

44. $z = \dfrac{x - \mu}{\sigma} = \dfrac{80 - 82}{6.2} \approx -0.32$
$$P(x_i > 80) = P(z > -0.32)$$
$$\text{normalcdf}(-0.32, 4) \approx 0.63$$

45. using the Remainder Theorem:
$$p(2) = 5(2)^5 - 4(2)^4 + 3(2)^3 - 2(2)^2 + (2) = 114$$
or

$2 \rfloor$	5	−4	3	−2	1	0
↓		10	12	30	56	114
	5	6	15	28	57	114

continued in Answers and Solutions Overflow

11.2 Central Limit Theorem

Approximately 27.5 million people watched prime time programming for the 2016 Summer Olympic Games in Rio.

The actual parameter for a population, such as the average height of 18-yr-old males in the United States, can be difficult or even impossible to determine. In these cases, statistics from a sample of the population are used to make inferences about the population's parameters. A probability distribution can be created to describe any statistic from the sample.

> **DEFINITION**
>
> A **sampling distribution** is the probability distribution of a statistic from all possible samples of a given sample size from a population.

We will focus on sampling distributions for the means, $\bar{x}$, of samples with a size of n from a given population.

Example 1 Analyzing a Sampling Distribution of Sample Means

Make a table listing all possible samples of size 2 (taken with replacement) from the population {7, 9, 11, 13} and their means. Use the table to create a sampling distribution for the means. Then find the mean and standard deviation of the sample means and compare them to the population's mean and standard deviation.

Answer

1. List all (4 · 4 = 16) possible samples of size n = 2 and their means.

Sample; $\bar{x}$	
{7, 7}; 7	{11, 7}; 9
{7, 9}; 8	{11, 9}; 10
{7, 11}; 9	{11, 11}; 11
{7, 13}; 10	{11, 13}; 12
{9, 7}; 8	{13, 7}; 10
{9, 9}; 9	{13, 9}; 11
{9, 11}; 10	{13, 11}; 12
{9, 13}; 11	{13, 13}; 13

2. Find the probability of each mean to make the sampling distribution.

Sampling Distribution	
$\bar{x}$	Probability
7	1 ÷ 16 = 0.0625
8	2 ÷ 16 = 0.1250
9	3 ÷ 16 = 0.1875
10	4 ÷ 16 = 0.2500
11	3 ÷ 16 = 0.1875
12	2 ÷ 16 = 0.1250
13	1 ÷ 16 = 0.0625

3. Enter the sampling distribution in L1 and L2 and the population in L3.

L1	L2	L3	L4
7	0.0625	7	---
8	0.125	9	
9	0.1875	11	
10	0.25	13	
11	0.1875	------	
12	0.125		
13	0.0625		
------	------		

CONTINUED ➡

After completing this section, you will be able to

- create a sampling distribution and find its mean and standard deviation.
- apply the Central Limit Theorem to find probabilities.
- approximate binomial probabilities using the Central Limit Theorem.

Objectives

1. To apply the Central Limit Theorem to calculate the standard deviation of a sample's mean
2. To find real-world probabilities using the Central Limit Theorem
3. To identify when a binomial distribution can be approximated with a normal distribution
4. To approximate binomial probabilities by applying the Central Limit Theorem

Flash

The 2016 Olympics in Rio were the first Olympics hosted by a South American country and cost over \$13 billion. Around 85,000 security forces made up of police and military were used, twice as many as employed by the Olympic Games in London 4 yr earlier.

Vocabulary

Central Limit Theorem
continuity correction factor
normal approximation
 of a binomial distribution
sampling distribution
standard error of the mean

PRESENTATION

Lesson Opener

A police radar unit measures speeds along an interstate highway. The average speed is 65 mi/hr with a standard deviation of 8.7 mi/hr.

1. Find the probability that a randomly selected motor vehicle is registered traveling more than 70 mi/hr.
   ≈ 0.283 or 28.3%

2. Find the probability that a randomly selected motor vehicle is registered traveling more than 75 mi/hr.
   ≈ 0.125 or 12.5%

One-on-One The math in this section is basic and the real-world applications are valuable, but the terminology will be unfamiliar to most students and can be confusing. Encourage students to pay special attention to the definitions and to discern differences between similar terms.

Introduce *sampling distributions*, emphasizing that the text will examine just one type—the distribution for the means of samples of a given size, including those that involve replacement. Then use Example 1 to create a sampling distribution, stressing the need to find the mean of all samples of size n. Note that the mean of the sampling distribution is equal to the mean of the population, but the standard deviation of the means is less than the standard deviation of the population. The table "Properties of Sampling Distributions of Sample Means" describes the relationships of these values.

Examining the illustrated sampling distributions for both uniform and skewed population distributions leads into the *Central Limit Theorem* (CLT). This theorem defines the mean and standard deviation of (1) the normal sampling distribution of the sample means from a normally distributed population or (2) a normal distribution approximating the sampling distribution from

Reading and Writing Mathematics

As a class, the eleventh graders at Calvary Christian School scored in the 90th percentile on the math portion of a standard achievement test. Explain why this does not indicate that the average student in the eleventh grade at the school had a 90th percentile ranking on the math portion of the test.

Group percentile rankings differ from individual percentile rankings. Individual percentile rankings are relative comparisons to individual scores, while group rankings (all eleventh graders at each school) are relative comparisons to group scores. The eleventh graders at a school could have an individual average percentile ranking of 78 and still be in the top 10% (90th percentile) of all eleventh-grade groups in the state or nation.

Additional Exercises

1. Make a table listing all the possible samples of size 2 (taken with replacement) from the population $\{2, 4, 6\}$ and their means.

Sample; $\bar{x}$	
$\{2, 2\}$; 2	$\{4, 6\}$; 5
$\{2, 4\}$; 3	$\{6, 2\}$; 4
$\{2, 6\}$; 4	$\{6, 4\}$; 5
$\{4, 2\}$; 3	$\{6, 6\}$; 6
$\{4, 4\}$; 4	

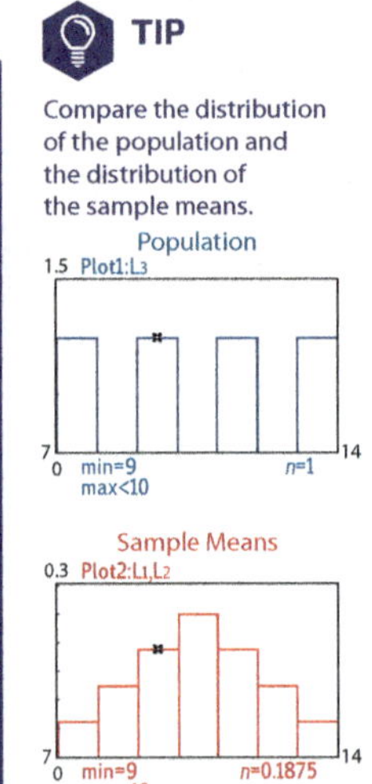

4. Use 1-Var Stats to determine the mean and standard deviation for the sampling distribution and the population.

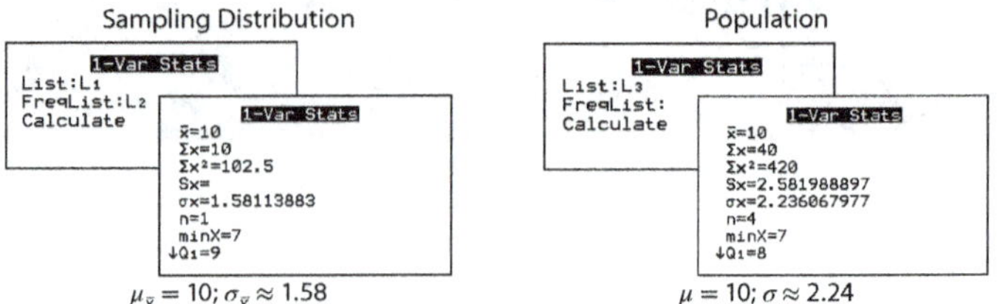

$$\mu_{\bar{x}} = 10; \sigma_{\bar{x}} \approx 1.58 \qquad \mu = 10; \sigma \approx 2.24$$

5. While the means are the same, the standard deviation of the sample means is less than the standard deviation of the population.

SKILL ✔ EXERCISE 21

Properties of Sampling Distributions of Sample Means	
$\mu_{\bar{x}} = \mu$	The mean of the sample means, $\mu_{\bar{x}}$, is equal to the population's mean, μ.
$\sigma_{\bar{x}} = \dfrac{\sigma}{\sqrt{n}}$	The standard deviation of the sample means, $\sigma_{\bar{x}}$, is equal to the population standard deviation, σ, divided by the square root of the sample size, n.

Notice that the histogram illustrating the sampling distribution in Example 1 approximates a normal distribution. What effect does increasing n, the size of the sample, have on the shape of the sampling distributions from the uniform and skewed populations illustrated below?

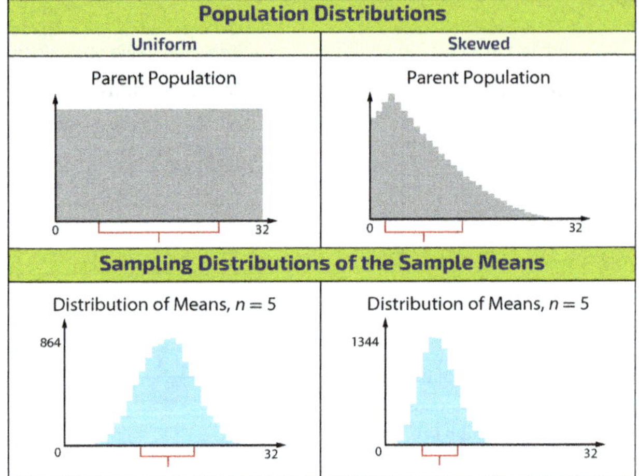

CONTINUED ➡

populations with other distributions (if $n \geq 30$). Consider asking the students what will make the sampling distribution from a non-normally distributed population become more normally distributed. (*increasing n, the sample size*) Explain that data from an entire population is often very difficult and sometimes impossible to gather. Therefore statistics from a large enough random sample can often accurately estimate the parameters of the population.

Emphasize that the sampling distribution does not describe individuals, but a statistic from all possible samples of the same size taken from the population. Using sampling distributions to predict the population distribution is a key concept in inferential statistics.

Example 2 uses a population with a known mean and standard deviation to demonstrate the difference between the probability for an individual (using the population's distribution) and for the mean of a sample (using the sampling distribution). The use of the ShadeNorm command without calculating z-scores is illustrated in the margin tip, which shows the different normal distributions for the population and the sampling distribution. Since students frequently have difficulty determining an appropriate window when using this command, it is usually easier to use z-scores and the standard normal distribution.

One-on-One A student may forget that the normalcdf function on the TI-84 calculator is found under the DISTR menu and that -4 and 4 are commonly used as the lower and upper ends of the z-scores.

The second case of the CLT applies in Example 3 and the first case applies in Example 4 ($n < 30$, but the weights of the candy bars are normally distributed). These can be solved without z-scores using normalcdf(-1E99, 18000, 19250, 3650) and normalcdf(47.5, 48.75, 48, 0.617), but it is important that the students understand

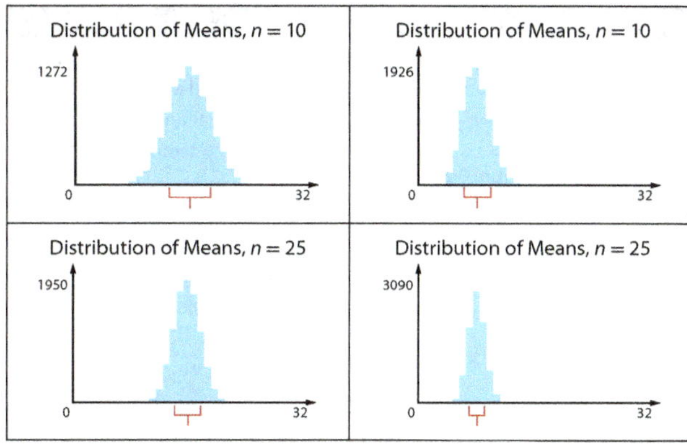

sampling distributions generated from 10,000 random samples of size *n*

Increasing *n* causes the sampling distribution of the means for both the uniform and the skewed populations to become more normally distributed with less variance. The Central Limit Theorem summarizes this idea and lays the foundation for much of the study of inferential statistics by describing the relationship between a population and its sampling distribution.

CENTRAL LIMIT THEOREM

1. In a normally distributed population, the sampling distribution of sample means is normally distributed for any sample size *n*.
2. Random samples of size $n \geq 30$ from other populations with a mean μ and standard deviation σ have sampling distributions of the sample means that approximate normal distributions. Better approximations result with larger sample sizes. In both cases $\mu_{\bar{x}} = \mu$ and $\sigma_{\bar{x}} = \frac{\sigma}{\sqrt{n}}$. The standard deviation of the sample means, $\sigma_{\bar{x}}$, is also called the *standard error of the mean*.

Pierre-Simon de Laplace developed the concepts of what would come to be called the Central Limit Theorem over the course of almost 40 yr, culminating with the publication of *Théorie Analytique des Probabilités* in 1810.

The population distribution describes the data from the entire population and is the most accurate but hardest information to gather. A sample distribution includes the data from one sample and is the least accurate but easiest information to gather. The sampling distribution does not describe individuals but a statistic from all possible samples of the same size taken from the population. Inferential statistics uses sampling distributions to predict the population distribution.

In Chapter 10, individual scores from a population were standardized within the population's distribution using $z = \frac{x - \mu}{\sigma}$, and individual scores from a single sample were standardized within the sample's distribution using $z = \frac{x - \bar{x}}{s}$. We will now use the Central Limit Theorem and $z = \frac{\bar{x} - \mu_{\bar{x}}}{\sigma_{\bar{x}}}$ to standardize the distribution of the means of all possible samples of the same size taken from the population, the sampling distribution of the means.

From Laplace's era until the twentieth century, the Central Limit Theorem was generally viewed as the expression of an underlying natural law.

Be careful to determine which distribution you are using when solving problems. Different *z*-score formulas are used depending on the distribution.

---

2. Use the table to create a sampling distribution for the means.

Sampling Distribution	
Mean	**Probability**
2	$1 \div 9 = 0.111$
3	$2 \div 9 = 0.222$
4	$3 \div 9 = 0.333$
5	$2 \div 9 = 0.222$
6	$1 \div 9 = 0.111$

3. Find the mean and standard deviation of the sample means and compare them to the population's mean and standard deviation.

sample: $\mu_{\bar{x}} = 4$; $\sigma_{\bar{x}} \approx 1.15$
population: $\mu = 4$; $\sigma \approx 1.63$
While the means are the same, the standard deviation of the sample means is less than the standard deviation of the population.

Find the following probabilities for randomly selected 5-yr-old girls, given that a normally distributed population of 5-yr-old girls has an average weight of 40 lb with a standard deviation of 5 lb.

4. A girl weighs between 35 and 42 lb.
   ≈ 0.50 or 50%

5. The average weight of 30 girls is more than 42 lb. ≈ 0.014 or 1.4%

6. The average weight of 45 girls is less than 39.2 lb. ≈ 0.14 or 14%

Use a normal distribution to estimate the following probabilities for randomly selected Americans given that a recent survey found that 58.6% of Americans vote.

7. At least 58 out of 100 Americans vote.
   ≈ 0.589 or 58.9%

8. Between 19 and 29 out of 50 Americans vote. ≈ 0.407 or 40.7%

9. Exactly 29 out of 50 Americans vote.
   ≈ 0.114 or 11.4%

10. Exactly 58 out of 100 Americans vote.
    ≈ 0.080 or 8.0%

---

how to use *z*-scores and the standard normal curve.

Interactive Activity The Internet keyword search *normal distribution calculator* can provide statistical calculators that will display graphs and calculate the probabilities without having to determine the *z*-scores. These can be used for teacher tools, interactive activities, or student use after the concepts are well understood.

Examine the effect of increasing the number of trials in the illustrated binomial distributions on page 559 where $p = 0.5, 0.3$, and 0.1. Explain that when both $np \geq 5$ and $nq \geq 5$, statisticians generally accept the use of a continuous normal distribution to estimate discrete binomial probabilities that would be tedious to calculate on a case-by-case basis. Emphasize the use of $\mu = np$ and $\sigma = \sqrt{npq}$ when converting to a *z*-score, which allows for the use of the standard normal distribution in these cases.

Explain the continuity corrections shown in the table. Emphasize how the discrete binomial probability distributions in Example 5 and Example 6 are approximated with a continuous normal probability distribution using *z*-scores of values that have had the *continuity correction factor* applied. After completing

Assignments

- **Minimum:** 1–2; 3–9 odd; 12–13, 22–25, 29, 33; 34–38 even; 39–47 odd; 50
- **Standard:** 2, 4–6, 10, 13–14, 18, 21, 24–27, 30, 32, 35–36, 42, 46, 48–50
- **Extended:** 5–6, 8, 10, 15–16, 20–21, 24, 28; 31–35 odd; 37–39, 42, 46, 50

Assessment

- Quiz 11A covers Sections 11.1–11.2.

TIP

The probability for the individual is found using the population's distribution: ShadeNorm (140, 1E99, 128, 41.5).

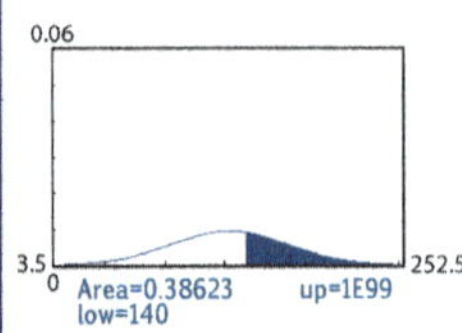

The probability for the sample's mean is found using the sampling distribution: ShadeNorm (140, 1E99, 128, 7.01).

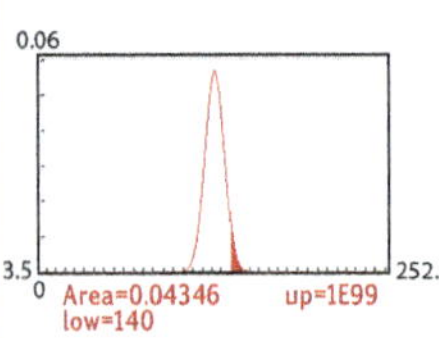

On average, 18–24-yr-olds in the United States send and receive 128 texts per day. Find each probability if the standard deviation for this normally distributed population is 41.5.

a. A randomly selected 18–24-yr-old sends and receives more than 140 texts per day.

b. The average number of texts sent and received per day for a sample of 35 randomly selected 18–24-yr-olds is more than 140.

Answer

a. $z = \dfrac{x - \mu}{\sigma} = \dfrac{140 - 128}{41.5} \approx 0.29$

1. Find the z-score for an individual using the population's mean and standard deviation.
2. Use technology to find the probability.

```
normalcdf(0.29,4)
              0.3858764957
```

$P(x > 140) = P(z > 0.29)$
≈ 0.386 or 38.6%

b. $\mu_{\bar{x}} = \mu = 128$

$\sigma_{\bar{x}} = \dfrac{\sigma}{\sqrt{n}} = \dfrac{41.5}{\sqrt{35}} \approx 7.01$

$z = \dfrac{\bar{x} - \mu_{\bar{x}}}{\sigma_{\bar{x}}} = \dfrac{140 - 128}{7.01} \approx 1.71$

1. Since $n = 35 \geq 30$, use $\mu_{\bar{x}}$ and the standard error of the mean, $\sigma_{\bar{x}}$, to find the z-score for the mean in a sampling distribution.

```
normalcdf(1.71,4)
              0.0436012169
```

2. Use technology to find the probability.

$P(\bar{x} > 140) = P(z > 1.71)$
≈ 0.044 or 4.4%

SKILL ✓ EXERCISE 23

Example 2 demonstrates that it is not unusual for individuals to deviate from the mean. It is much more unusual for the mean of a larger sample to deviate significantly from the population's mean.

In this section we will study the means of samples from populations whose mean and standard deviation are known. In later sections we will use the Central Limit Theorem to make inferences about a population's parameters.

If the average cost of a wedding in central Illinois is $19,250 with a standard deviation of $3650, find the probability that the average cost of 40 randomly selected weddings in central Illinois is less than $18,000.

Answer

$\mu_{\bar{x}} = \mu = 19{,}250$

$\sigma_{\bar{x}} = \dfrac{\sigma}{\sqrt{n}} = \dfrac{3650}{\sqrt{40}} \approx 577.1$

1. Since $n \geq 30$, apply the Central Limit Theorem to find the mean of the approximately normal sampling distribution, $\mu_{\bar{x}}$, and the standard error of the mean, $\sigma_{\bar{x}}$.

$z = \dfrac{\bar{x} - \mu_{\bar{x}}}{\sigma_{\bar{x}}} \approx \dfrac{18{,}000 - 19{,}250}{577.1} \approx -2.17$

$P(\bar{x} < 18{,}000) = P(z < -2.17)$

2. Let $\bar{x}$ = average cost of a wedding from a sample. Find the z-score associated with 18,000.

```
normalcdf(-4,-2.17)
              0.0149716833
```

3. Find the portion of the area under the normal curve (the probability) within the interval $-4 \leq z \leq -2.17$.

CONTINUED ➡

Example 6, consider having students use the binomial distribution formula to find the probability and compare the results. ($P(x = 120) = {}_{250}C_{120}(0.464)^{120}(0.536)^{130} = binompdf(250, 0.464, 120) = 0.044$ or 4.4%)

Many applets for statistics are available online. Use the Internet keyword search *inferential statistics applets*.

TIPS

Ex. 22–35 Encourage students to use exact results of calculations until they arrive at the final answer. Storing key values such as $\sigma_{\bar{x}}$, z_L, and z_U (as illustrated below for exercise 23*b*) simplifies these calculations.

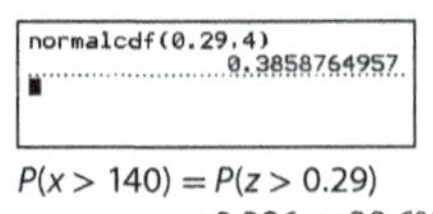
```
.28/√35→S
              0.0473286383
(3.2−3.15)/S→L
              1.056442818
(3.8−3.15)/S→H
              13.73375664
normalcdf(L,H)
              0.1453830027
```

Ex. 36–37 Consider using these exercises for group work or take-home projects to allow for the numerous calculations and tables needed.

$P(\bar{x} < 18,000) = P(z < -2.17)$
≈ 0.015 or 1.5%

4. The probability that the average cost of the 40 weddings is less than \$18,000 is very unlikely.

SKILL ✔ **EXERCISE 27**

Example 4   Applying the Central Limit Theorem

Factory specifications require the weight of a candy bar coming off the production line to be 48 g with a standard deviation of 1.95 g. Weights of the candy bars are normally distributed. What is the probability that the average weight of a random sample of 10 candy bars is between 47.5 g and 48.75 g?

Answer

$\mu_{\bar{x}} = \mu = 48$

$\sigma_{\bar{x}} = \dfrac{\sigma}{\sqrt{n}} = \dfrac{1.95}{\sqrt{10}} \approx 0.617$

$z_{lower} = \dfrac{\bar{x} - \mu_{\bar{x}}}{\sigma_{\bar{x}}} = \dfrac{47.5 - 48}{0.617} \approx -0.81$

$z_{upper} = \dfrac{\bar{x} - \mu_{\bar{x}}}{\sigma_{\bar{x}}} = \dfrac{48.75 - 48}{0.617} \approx 1.22$

$P(47.5 < \bar{x} < 48.75) \approx P(-0.81 < z < 1.22)$
$=$ normalcdf$(-0.81, 1.22) \approx 0.68$ or 68%

1. Since the population is normally distributed, all sampling distributions are normally distributed with $\mu_{\bar{x}} = \mu$ and $\sigma_{\bar{x}} = \dfrac{\sigma}{\sqrt{n}}$.

2. Find the lower and upper z-scores for the interval of sample means.

3. Use technology to find the probability.

SKILL ✔ **EXERCISE 29**

The Central Limit Theorem justifies the use of the normal distribution to approximate a sampling distribution when $n \geq 30$ or when the population is normally distributed. Binomial probability distributions that represent discrete outcomes can also be approximated by a normal distribution under certain conditions. The table illustrates how binomial distributions become more normally distributed as n increases.

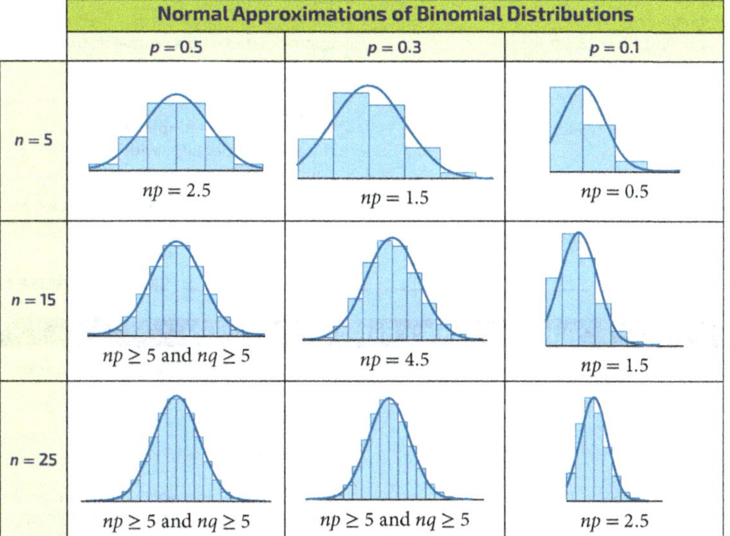

Normal Approximations of Binomial Distributions		
$p = 0.5$	$p = 0.3$	$p = 0.1$
$n = 5$ $np = 2.5$	$np = 1.5$	$np = 0.5$
$n = 15$ $np \geq 5$ and $nq \geq 5$	$np = 4.5$	$np = 1.5$
$n = 25$ $np \geq 5$ and $nq \geq 5$	$np \geq 5$ and $nq \geq 5$	$np = 2.5$

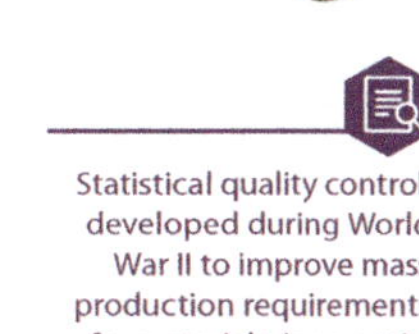

Statistical quality control, developed during World War II to improve mass production requirements for materials, is essential to success in modern manufacturing of everything from food products to computers.

Solutions

❭ A. Exercises

5. $\mu = \displaystyle\sum_{i=1}^{5} x_i = \dfrac{25}{5} \approx 5$

$\sigma = \sqrt{\dfrac{\displaystyle\sum_{i=1}^{5}(x_i - \bar{x})^2}{N}}$

$= \sqrt{\dfrac{4^2 + 2^2 + 0^2 + 4^2 + 2^2}{5}} = \sqrt{\dfrac{40}{5}}$

$= 2\sqrt{2} \approx 2.83$

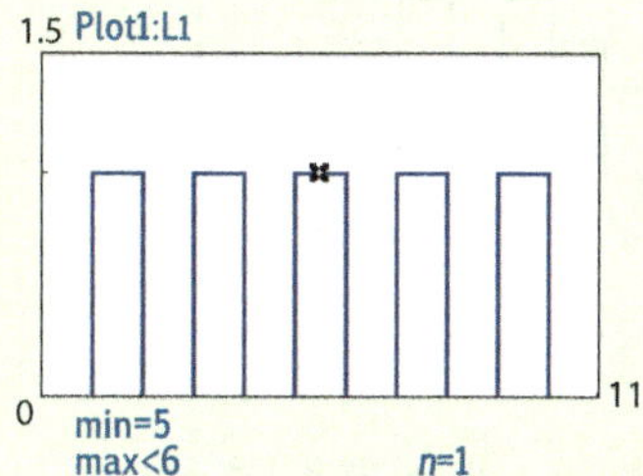

6.

Sample; $\bar{x}$	
{1, 1}; 1	{5, 7}; 6
{1, 3}; 2	{5, 9}; 7
{1, 5}; 3	{7, 1}; 4
{1, 7}; 4	{7, 3}; 5
{1, 9}; 5	{7, 5}; 6
{3, 1}; 2	{7, 7}; 7
{3, 3}; 3	{7, 9}; 8
{3, 5}; 4	{9, 1}; 5
{3, 7}; 5	{9, 3}; 6
{3, 9}; 6	{9, 5}; 7
{5, 1}; 3	{9, 7}; 8
{5, 3}; 4	{9, 9}; 9
{5, 5}; 5	

7.

$\bar{x}$	Probability	$\bar{x}$	Probability
Sampling Distribution			
1	$1 \div 25 = 0.04$	6	$4 \div 25 = 0.16$
2	$2 \div 25 = 0.08$	7	$3 \div 25 = 0.12$
3	$3 \div 25 = 0.12$	8	$2 \div 25 = 0.08$
4	$4 \div 25 = 0.16$	9	$1 \div 25 = 0.04$
5	$5 \div 25 = 0.2$		

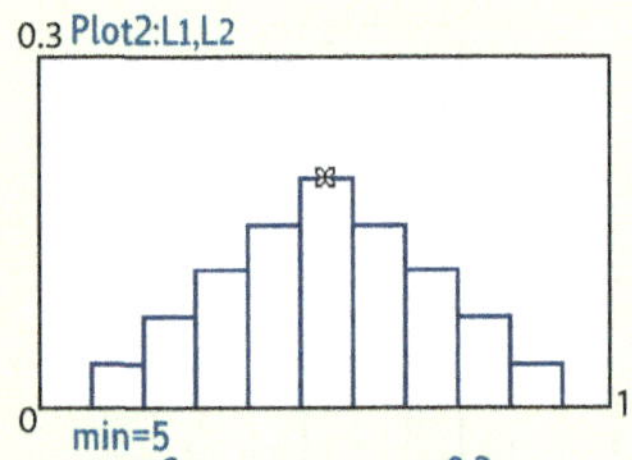

8. $\mu_{\bar{x}} = \sum\limits_{i=1}^{9} \bar{x}_i P(\bar{x}_i) = 5;\ \mu_{\bar{x}} = \mu$

$\sigma_{\bar{x}} = \dfrac{\sigma}{\sqrt{n}} = \dfrac{2\sqrt{2}}{\sqrt{2}} = 2;\ \sigma_{\bar{x}} < \sigma$

9. $\mu_{\bar{x}} = \mu = 86;\ \sigma_{\bar{x}} = \dfrac{\sigma}{\sqrt{n}} = \dfrac{8.2}{\sqrt{15}} \approx 2.1$

10. $\mu_{\bar{x}} = \mu = 86;\ \sigma_{\bar{x}} = \dfrac{\sigma}{\sqrt{n}} = \dfrac{8.2}{\sqrt{50}} \approx 1.2$

11. (1) when using samples of any size from a normally distributed population
(2) when using a sample size of $n \geq 30$ from any population

12. no; The population is not normally distributed and $n \ngeq 30$.

13. yes; $\mu_{\bar{x}} = \mu = 126;\ \sigma_{\bar{x}} = \dfrac{18}{\sqrt{8}} \approx 6.36$

14. yes; $\mu_{\bar{x}} = \mu = 9;\ \sigma_{\bar{x}} = \dfrac{0.8}{\sqrt{38}} \approx 0.13$

15. no; The population's distribution is unknown and $n \ngeq 30$.

17. $np = 10(0.15) = 1.5 \ngeq 5$

18. $np = 200(0.82) = 164 \geq 5$
$nq = 200(0.18) = 36 \geq 5$

19. $np = 32(0.78) = 24.96 \geq 5$
$nq = 32(0.22) = 7.04 \geq 5$

20. $nq = 16(0.03) = 0.48 \ngeq 5$

❯ B. Exercises

21a.

Sample; $\bar{x}$	
$\{3, 3\}$; 3	$\{8, 3\}$; 5.5
$\{3, 6\}$; 4.5	$\{8, 6\}$; 7
$\{3, 8\}$; 5.5	$\{8, 8\}$; 8
$\{3, 10\}$; 6.5	$\{8, 10\}$; 9
$\{6, 3\}$; 4.5	$\{10, 3\}$; 6.5
$\{6, 6\}$; 6	$\{10, 6\}$; 8
$\{6, 8\}$; 7	$\{10, 8\}$; 9
$\{6, 10\}$; 8	$\{10, 10\}$; 10

21b.

$\bar{x}$	$P(\bar{x})$	$\bar{x}$	$P(\bar{x})$
3.0	1/16	7.0	2/16
3.5	0/16	7.5	0/16
4.0	0/16	8.0	3/16
4.5	2/16	8.5	0/16
5.0	0/16	9.0	2/16
5.5	2/16	9.5	0/16
6.0	1/16	10.0	1/16
6.5	2/16		

Using a normal approximation for the binomial distribution when $np \geq 5$ and $nq \geq 5$ allows binomial probabilities to be found more easily. Recall that the mean and standard deviation in a binomial distribution are given by $\mu = np$ and $\sigma = \sqrt{npq}$.

▍ NORMAL APPROXIMATION OF A BINOMIAL DISTRIBUTION

A binomial distribution can be approximated by a normal distribution if $np \geq 5$ and $nq \geq 5$ where n is the number of trials, p is the probability of success, and q is the probability of failure. The distribution can be converted to standard normal distribution using $z = \dfrac{x - \mu}{\sigma} = \dfrac{x - np}{\sqrt{npq}}$.

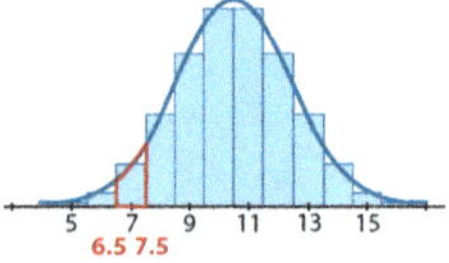

When the discrete binomial distribution is approximated by a continuous normal distribution, a *continuity correction factor* of 0.5 is subtracted from the lowest value and added to the highest value to obtain accurate results. To find $P(x = 7)$ in the discrete binomial distribution, find $P(6.5 \leq x \leq 7.5)$ in the continuous normal distribution.

Corrections for Continuous Probability Distributions				
discrete	equal to 7	at least 11	less than 7	between 7 and 11
meaning	$x = 7$	$x \geq 11$	$x \leq 6$	$8 \leq x \leq 10$
correction for continuity	$6.5 \leq x \leq 7.5$	$x \geq 10.5$	$x \leq 6.5$	$7.5 \leq x \leq 10.5$

Example 5 Approximating a Binomial Probability

Sixty-one percent of teens volunteer to help others during the Christmas season. If 60 randomly selected teens are asked if they have volunteered to help others during the Christmas season, what is the probability that less than 35 of them answer yes?

Answer

$n = 60, p = 0.61, q = 0.39$ — 1. Define n, p, and q for the binomial experiment.

$np = 60(0.61) = 36.6 \geq 5$
$nq = 60(0.39) = 23.4 \geq 5$ — 2. Verify that the binomial distribution can be approximated with a normal distribution.

$\mu = np = 36.6$
$\sigma = \sqrt{npq} = \sqrt{60(0.61)(0.39)} \approx 3.78$ — 3. Find μ and σ.

$P(x < 35) \approx P(x \leq 34.5)$ — 4. Write the discrete probability and restate the probability with its continuity correction factor.

$z = \dfrac{x - \mu}{\sigma} \approx \dfrac{34.5 - 36.6}{3.78} \approx -0.56$ — 5. Convert to a z-score and use technology to find the probability.
$P(z \leq -0.56) \approx 0.288$ or 28.8%

SKILL ✔ EXERCISE 17

Example 6 Approximating a Binomial Probability

A survey found that 46.4% of teens like to watch the Summer Olympics. Estimate the probability that 120 out of 250 randomly selected teens like watching the Summer Olympics.

Answer

$n = 250, p = 0.464, q = 0.536$ — 1. Define n, p, and q for the binomial experiment.

$np = 250(0.464) = 116 \geq 5$
$nq = 250(0.536) = 134 \geq 5$ — 2. Verify that the binomial distribution can be approximated with a normal distribution.

CONTINUED ➡

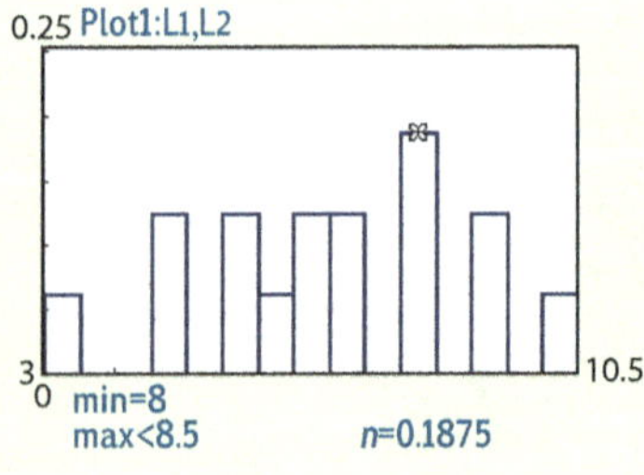

21c. $\mu_{\bar{x}} = 6.75;\ \sigma_{\bar{x}} = \dfrac{2.586}{\sqrt{2}} \approx 1.83$
$\mu = \mu_{\bar{x}}$, but $\sigma > \sigma_{\bar{x}}$

22a. $z_L = \dfrac{x - \mu}{\sigma} = \dfrac{9 - 8}{2.1} \approx 0.48$
$z_U = \dfrac{12 - 8}{2.1} \approx 1.90$
$P(9 < x_i < 12) \approx P(0.48 < z < 1.90)$
$\text{normalcdf}(0.48, 1.90) \approx 0.29$

$\mu = np = 116$

$\sigma = \sqrt{npq} = \sqrt{250(0.464)(0.536)} \approx 7.89$

$P(x = 120) \approx P(119.5 \le x \le 120.5)$

$z_{low} = \frac{119.5 - 116}{7.89} \approx 0.44$

$z_{high} = \frac{120.5 - 116}{7.89} \approx 0.57$

$P(0.44 \le z \le 0.57) \approx 0.046$ or 4.6%

3. Find μ and σ.

4. Write the discrete probability and restate the probability with its continuity correction factor.

5. Convert to a z-score and use technology to find the probability.

❯ A. Exercises

For exercises 1–4, identify the described distribution as a (I) population, (II) sample, or (III) sampling distribution.

1. Which distribution represents data from each of the following?

   a. a group within the population II

   b. all possible samples of the same size III

   c. an entire group I

2. Which distribution has values that are standardized by each z-score?

   a. $z = \frac{\overline{x} - \mu_{\overline{x}}}{\sigma_{\overline{x}}}$ III   **b.** $z = \frac{x - \overline{x}}{s}$ II   **c.** $z = \frac{x - \mu}{\sigma}$ I

3. When making predictions about a population's parameter, which distribution has each given characteristic?

   a. most accurate I   **b.** least accurate II

4. Which distribution describes each of the following in a study of Texas high school basketball team members?

   a. a randomly selected group of 20 basketball team members from Texas high schools II

   b. all basketball team members from Texas high schools I

   c. the collection of all possible groupings of 20 basketball team members from Texas high schools III

Use the population {1, 3, 5, 7, 9} for exercises 5–8.

5. Find the mean and standard deviation of the population, and then draw a histogram representing the population's distribution. $\mu \approx 5; \sigma \approx 2.83$

6. Create a table of all possible samples of size 2 (taken with replacement) from this population and find the mean for each sample.

7. Create a table and a histogram representing the sampling distribution for the means.

$\mu_{\overline{x}} \approx 5; \sigma_{\overline{x}} \approx 2$; The means are the same, and $\sigma_{\overline{x}} < \sigma$.

8. Find the mean and standard deviation of the sampling distribution and compare them to the population's mean and standard deviation.

A population has a mean of 86 and a standard deviation of 8.2. Find the mean and the standard deviation of a sampling distribution of the means having the given sample size.

9. $n = 15$ $\mu_{\overline{x}} = 86; \sigma_{\overline{x}} \approx 2.1$ 10. $n = 50$ $\mu_{\overline{x}} = 86; \sigma_{\overline{x}} \approx 1.2$

11. State the two conditions for which the Central Limit Theorem implies that the sampling distribution of the means approximates a normal distribution.

Determine whether the Central Limit Theorem can be applied to each population distribution with the given mean and standard deviation using a sample size of n. If so, find $\mu_{\overline{x}}$ and $\sigma_{\overline{x}}$. If not, explain why.

12. skewed left with $\mu = 52$, $\sigma = 3.5$, and $n = 12$

13. normal with $\mu = 126$, $\sigma = 18$, and $n = 8$

14. skewed right with $\mu = 9$, $\sigma = 0.8$, and $n = 38$

15. unknown with $\mu = 37$, $\sigma = 4.9$, and $n = 19$

16. State the continuity correction factor used when using a normal approximation to find each binomial probability.

   a. $P(x = 12)$ $11.5 \le x \le 12.5$ **b.** $P(\text{at least } 9)$ $x \ge 8.5$

   c. $P(\text{between 3 and 7})$ **d.** $P(\text{less than } 6)$ $x \le 5.5$

   $3.5 \le x \le 6.5$

Determine whether the normal distribution can be used to approximate the binomial distribution for an experiment with n trials, each having a probability of success of p. Explain your reasoning.

17. $n = 10$, $p = 0.15$ 18. $n = 200$, $p = 0.82$

19. $n = 32$, $p = 0.78$ 20. $n = 16$, $p = 0.97$

17. no; $np \not\ge 5$ 18. yes; $np \ge 5$ and $nq \ge 5$

19. yes; $np \ge 5$ and $nq \ge 5$ 20. no; $nq \not\ge 5$

---

27. $\mu_{\overline{x}} = 363; \sigma_{\overline{x}} = \frac{28.2}{\sqrt{36}} = 4.7$

   $z = \frac{350 - 363}{4.7} \approx -2.77$

   $P(x_i > 350) = P(z > -2.77) \approx 0.997$

28. $\mu_{\overline{x}} = 82; \sigma_{\overline{x}} = \frac{6.2}{\sqrt{20}} \approx 1.39$

   $z = \frac{85 - 82}{1.39} \approx 2.16$

   $P(\overline{x} > 85) = P(z > 2.16) \approx 0.015$

29. $\mu_{\overline{x}} = 40; \sigma_{\overline{x}} = \frac{5}{\sqrt{35}} \approx 0.845$

   $z_L = \frac{39 - 40}{0.845} \approx -1.18$

   $z_U = \frac{41 - 40}{0.845} \approx 1.18$

   $P(39 < \overline{x} < 41)$
   $= P(-1.18 < z < 1.18) \approx 0.76$

30. $\mu_{\overline{x}} = 1040; \sigma_{\overline{x}} = \frac{3.7}{\sqrt{15}} \approx 0.96$

   $z_L = \frac{1038 - 1040}{0.96} \approx -2.08$

   $z_U = \frac{1041 - 1040}{0.96} \approx 1.04$

   $P(1038 < \overline{x} < 1041)$
   $= P(-2.08 < z < 1.04) \approx 0.83$

31. $\mu_{\overline{x}} = \mu = 5; \sigma_{\overline{x}} = \frac{2.5}{\sqrt{10}} \approx 0.79$

   $z = \frac{3 - 5}{0.79} = -2.5$

   $P(\overline{x} < 3) = P(z < -2.5) \approx 0.006$

32. $p = 0.5; q = 0.5$

   $\mu = np = 10(0.5) = nq = 10(0.5) = 5$

   $\sigma = \sqrt{npq} = \sqrt{10(0.5)(0.5)} \approx 1.58$

32a. $z = \frac{6.5 - 5}{1.58} \approx 0.95$

   $P(x_i > 6.5) = P(z > 0.95) \approx 0.17$

32b. $z_L = \frac{3.5 - 5}{1.58} \approx -0.95$

   $z_U = \frac{6.5 - 5}{1.58} \approx 0.95$

   $P(3.5 < x_i < 6.5)$
   $= P(-0.95 < z < 0.95) \approx 0.66$

32c. $P(\text{at most } 3)$
   $= 1 - P(\text{at least } 7) - P(4 < x_i < 6)$
   $\approx 1 - 0.17 - 0.66 = 0.17$

33. $p = 0.46; q = 0.54$
   $\mu = np = 60(0.46) = 27.6$
   $nq = 60(0.54) = 32.4$
   $\sigma = \sqrt{npq} = \sqrt{60(0.46)(0.54)}$
   ≈ 3.86

33a. $z = \frac{29.5 - 27.6}{3.86} \approx 0.49$
   $P(x_i > 29.5) = P(z > 0.49) \approx 0.31$

33b. $z_L = \frac{20.5 - 27.6}{3.86} \approx -1.84$
   $z_U = \frac{29.5 - 27.6}{3.86} \approx 0.49$
   $P(20.5 < x_i < 29.5)$
   $= P(-1.84 < z < 0.49) \approx 0.66$

34. $p = 0.62; q = 0.38$
   $\mu = np = 30(0.62) = 18.6$
   $nq = 30(0.38) = 11.4$
   $\sigma = \sqrt{npq} = \sqrt{30(0.62)(0.38)} \approx 2.659$

22b. $\mu_{\overline{x}} = 8; \sigma_{\overline{x}} = \frac{2.1}{\sqrt{40}} \approx 0.332$

   $z = \frac{\overline{x} - \mu_{\overline{x}}}{\sigma_{\overline{x}}} = \frac{7.5 - 8}{0.332} \approx -1.51$

   $P(\overline{x} > 7.5) \approx P(z > -1.51)$

   normalcdf$(-1.51, 4) \approx 0.93$

23a. $z = \frac{3 - 3.15}{0.28} \approx -0.54$

   $P(x_i < 3.0) \approx P(z < -0.54)$

   normalcdf$(-4, -0.54) \approx 0.29$

23b. $\mu_{\overline{x}} = 3.15; \sigma_{\overline{x}} = \frac{0.28}{\sqrt{35}} \approx 0.047$

   $z_L = \frac{3.2 - 3.15}{0.047} \approx 1.06$

   $z_U = \frac{3.8 - 3.15}{0.047} \approx 13.73$

   $P(3.2 < \overline{x} < 3.8)$
   $\approx P(1.06 < z < 13.73)$
   normalcdf$(1.06, 13.73) \approx 0.15$

24a. $z = \frac{8 - 9.8}{1.86} \approx -0.97$

   $P(x_i < 8) \approx P(z < -0.97) \approx 0.17$

24b. $\mu_{\overline{x}} = 9.8; \sigma_{\overline{x}} = \frac{1.86}{\sqrt{40}} \approx 0.29$

   $z = \frac{8 - 9.8}{0.29} \approx -6.12$

   $P(\overline{x} < 8) \approx P(z < -6.12) \approx 0$

25a. $z = \frac{64 - 66}{4.8} \approx -0.42$

   $P(x_i > 64) \approx P(z > -0.42) \approx 0.66$

25b. $\mu_{\overline{x}} = 66; \sigma_{\overline{x}} = \frac{4.8}{\sqrt{15}} \approx 1.24$

   $z = \frac{64 - 66}{1.24} \approx -1.61$

   $P(\overline{x} > 64) = P(z > -1.61) \approx 0.95$

26. $\mu_{\overline{x}} = 42.2; \sigma_{\overline{x}} = \frac{4.7}{\sqrt{5}} \approx 2.1$

   $z = \frac{45 - 42.2}{2.1} \approx 1.33$

   $P(x_i > 45) \approx P(z > 1.33) \approx 0.09$

34a. $z_L = \dfrac{13.5 - 18.6}{2.659} \approx -1.918$

$z_U = \dfrac{14.5 - 18.6}{2.659} \approx -1.542$

$P(13.5 < x_i < 14.5)$
$= P(-1.918 < z < -1.542) \approx 0.034$

34b. $z_L = \dfrac{20.5 - 18.6}{2.659} \approx 0.715$

$z_U = \dfrac{24.5 - 18.6}{2.659} \approx 2.219$

$P(20.5 < x_i < 24.5)$
$= P(0.715 < z < 2.219) \approx 0.22$

34c. $z = \dfrac{14.5 - 18.6}{2.659} \approx -1.542$

$P(x_i \geq 14.5) = P(z \geq -1.542) \approx 0.94$

35. $p = 0.25;\ q = 0.75$

$\mu = np = 25(0.25) = 6.25$

$nq = 25(0.75) = 18.75$

$\sigma = \sqrt{npq} = \sqrt{25(0.25)(0.75)} \approx 2.17$

35a. $z = \dfrac{4.5 - 6.25}{2.17} \approx -0.81$

$P(x_i < 9.5) = P(z < -0.81) \approx 0.21$

35b. $z = \dfrac{9.5 - 6.25}{2.17} \approx 1.50$

$P(x_i > 9.5) = P(z > 1.50) \approx 0.07$

❭ C. Exercises

36a. Enter the digits 0–9 into L1 and use 1-VarStats to find $\mu = 4.5$ and $\sigma \approx 2.87$.

36b.

Sample; $\bar{x}$	Distribution	
	$\bar{x}$	$P(\bar{x})$
$A;\ 5$		
$B;\ 5.2$	1.8	0.167
$C;\ 1.8$	3.6	0.167
$D;\ 5$	5	0.333
$E;\ 3.6$	5.2	0.167
$F;\ 6.6$	6.6	0.167

no; All possible samples of size 5 are not included.

36c. Enter the means into L2 and $P(\bar{x})$ into L3 and use 1-VarStats with List:L2 and FreqList:L3 to find $\mu_{\bar{x}} \approx 4.53$ and $\sigma_{\bar{x}} \approx 1.50$. The mean of the sampling distribution is the mean of the population, $\mu = 4.5$, and the standard deviation of the sampling distribution is $\dfrac{\sigma}{\sqrt{n}} \approx \dfrac{2.87}{\sqrt{5}} \approx 1.28$. The mean and standard deviation of the means of these selected samples are not equal to the mean and standard deviation of the sampling distribution of the means.

❭ B. Exercises

21. Complete the following steps to create and analyze a sampling distribution of the means for the population $\{3, 6, 8, 10\}$ in which $\mu = 6.75$ and $\sigma \approx 2.59$.
   a. List all possible samples of size 2 and their corresponding means, $\bar{x}$.
   b. Create a table and a histogram representing the sampling distribution for the means.
   c. Find the mean and standard deviation for the sampling distribution and compare them to the population's mean and standard deviation.

22. Find each probability if the average weight of an adult rainbow trout is normally distributed with a mean of 8 lb and a standard deviation of 2.1 lb.

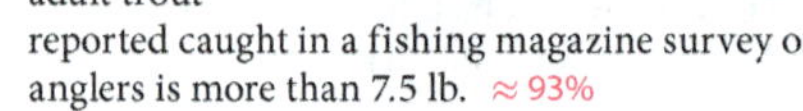
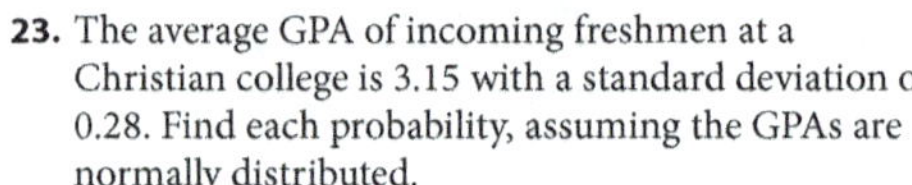

   a. The adult rainbow trout you catch weighs between 9 lb and 12 lb. ≈ 29%
   b. The average weight of 40 adult trout reported caught in a fishing magazine survey of anglers is more than 7.5 lb. ≈ 93%

23. The average GPA of incoming freshmen at a Christian college is 3.15 with a standard deviation of 0.28. Find each probability, assuming the GPAs are normally distributed.
   a. A randomly selected freshman has a GPA less than 3.0. ≈ 29%
   b. The average GPA of 35 randomly selected freshmen is between 3.2 and 3.8. ≈ 15%

24. According to the Bureau of Labor Statistics, teenagers aged 15–19 get an average of 9.8 hr of sleep per night. Find each probability, assuming that the hours of sleep for teens are normally distributed and the standard deviation is 1.86 hr.
   a. A randomly chosen 17-yr-old averages less than 8 hr of sleep. ≈ 17%
   b. Forty randomly selected 15–19-yr-olds average less than 8 hr of sleep. ≈ 0%

25. The average annual snowfall in Bangor, Maine, is normally distributed with a mean of 66 in. Find each probability, assuming a standard deviation of 4.8 in.
   a. more than 64 in. of snow in one year ≈ 66%
   b. an average annual snowfall of more than 64 in. over 15 yr ≈ 95%

26. If a town has an average January high temperature of 42.2° with a standard deviation of 4.7°, what is the probability that 5 randomly chosen days in January have a high temperature greater than 45°? ≈ 9%

27. The average smartphone in 2017 cost $363. If the standard deviation is $28.20, what is the probability that the mean cost for 36 randomly selected consumers was more than $350? ≈ 99.7%

28. If a normally distributed set of test scores has a mean of 82 and standard deviation of 6.2, find the probability that a normally distributed sample of 20 scores will average greater than 85. ≈ 1.5%

29. A normally distributed population of 5-yr-old girls has an average weight of 40 lb with a standard deviation of 5 lb. Find the probability that the average weight of 35 randomly selected 5-yr-old girls is between 39 lb and 41 lb. ≈ 76%

30. If frozen chicken pot pies contain an average of 1040 mg of sodium and the sodium level is normally distributed with a standard deviation of 3.7 mg, what is the probability that 15 pot pies coming off the production line contain an average of 1038 mg to 1041 mg of sodium? ≈ 83%

31. The time that a randomly selected student waits for a school bus is normally distributed with a mean of 5 min and a standard deviation of 2.5 min. Find the probability that the average wait time for a random sample of 10 students is less than 3 minutes. ≈ 0.6%

Use a normal approximation for the binomial distribution and continuity correction factor to estimate each binomial probability.

32. Mark randomly guesses at each of the 10 true-false questions on a test. Find the probability that he gets each number of these questions correct.
   a. $P(\text{at least } 7)$ ≈ 17%
   b. $P(4 \leq x \leq 6)$ ≈ 66%
   c. $P(\text{at most } 3)$ ≈ 17%

33. According to AAA, 46% of teenagers admit to texting while driving. If 60 teen drivers are randomly selected, find each probability.
   a. more than 30 admit to texting while driving ≈ 31%
   b. between 20 and 30 admit to texting while driving ≈ 66%

37a.

Sample; $\bar{x}$	Sample; $\bar{x}$	Sample; $\bar{x}$	Sample; $\bar{x}$	Sample; $\bar{x}$
$\{3, 3, 3\};\ 3$	$\{3, 10, 6\};\ 6.33$	$\{6, 8, 8\};\ 7.33$	$\{8, 6, 10\};\ 8$	$\{10, 6, 3\};\ 6.33$
$\{3, 3, 6\};\ 4$	$\{3, 10, 8\};\ 7$	$\{6, 8, 10\};\ 8$	$\{8, 8, 3\};\ 6.33$	$\{10, 6, 6\};\ 7.33$
$\{3, 3, 8\};\ 4.67$	$\{3, 10, 10\};\ 7.67$	$\{6, 10, 3\};\ 6.33$	$\{8, 8, 6\};\ 7.33$	$\{10, 6, 8\};\ 8$
$\{3, 3, 10\};\ 5.33$	$\{6, 3, 3\};\ 4$	$\{6, 10, 6\};\ 7.33$	$\{8, 8, 8\};\ 8$	$\{10, 6, 10\};\ 8.67$
$\{3, 6, 3\};\ 4$	$\{6, 3, 6\};\ 5$	$\{6, 10, 8\};\ 8$	$\{8, 8, 10\};\ 8.67$	$\{10, 8, 3\};\ 7$
$\{3, 6, 6\};\ 5$	$\{6, 3, 8\};\ 5.67$	$\{6, 10, 10\};\ 8.67$	$\{8, 10, 3\};\ 7$	$\{10, 8, 6\};\ 8$
$\{3, 6, 8\};\ 5.67$	$\{6, 3, 10\};\ 6.33$	$\{8, 3, 3\};\ 4.67$	$\{8, 10, 6\};\ 8$	$\{10, 8, 8\};\ 8.67$
$\{3, 6, 10\};\ 6.33$	$\{6, 6, 3\};\ 5$	$\{8, 3, 6\};\ 5.67$	$\{8, 10, 8\};\ 8.67$	$\{10, 8, 10\};\ 9.33$
$\{3, 8, 3\};\ 4.67$	$\{6, 6, 6\};\ 6$	$\{8, 3, 8\};\ 6.33$	$\{8, 10, 10\};\ 9.33$	$\{10, 10, 3\};\ 7.67$
$\{3, 8, 6\};\ 5.67$	$\{6, 6, 8\};\ 6.67$	$\{8, 3, 10\};\ 7$	$\{10, 3, 3\};\ 5.33$	$\{10, 10, 6\};\ 8.67$
$\{3, 8, 8\};\ 6.33$	$\{6, 6, 10\};\ 7.33$	$\{8, 6, 3\};\ 5.67$	$\{10, 3, 6\};\ 6.33$	$\{10, 10, 8\};\ 9.33$
$\{3, 8, 10\};\ 7$	$\{6, 8, 3\};\ 5.67$	$\{8, 6, 6\};\ 6.67$	$\{10, 3, 8\};\ 7$	$\{10, 10, 10\};\ 10$
$\{3, 10, 3\};\ 5.33$	$\{6, 8, 6\};\ 6.67$	$\{8, 6, 8\};\ 7.33$	$\{10, 3, 10\};\ 7.67$	

34. Benson's Garage advertises a 20 min oil change. Their records indicate that this actually occurs 62% of the time. Find the probability that each number of customers from a random survey of 30 customers will have had their oil changed in less than 20 minutes.

  a. $P(x = 14)$ ≈ 3.4%

  b. $P(20 < x < 25)$ ≈ 22%

  c. P(at least half) ≈ 94%

35. A score of 1200 on a recent standardized test represents the 75th percentile. Find each probability if 25 test scores are randomly selected.

  a. less than 5 have a score of 1200 or more ≈ 21%

  b. at least 10 have a score of 1200 or more ≈ 7%

> **C. Exercises**

36. Six sets of 5 numbers are created by randomly drawing (with replacement) from the digits 0 through 9. Use technology and these samples with $n = 5$ to complete the following.

$A = \{8, 4, 1, 9, 3\}$; $B = \{4, 4, 6, 9, 3\}$
$C = \{2, 0, 2, 0, 5\}$; $D = \{9, 1, 1, 8, 6\}$
$E = \{5, 2, 5, 6, 0\}$; $F = \{8, 4, 9, 7, 5\}$

  a. Compute the population's mean and standard deviation. $\mu = 4.5$; $\sigma_{\bar{x}} \approx 2.87$

  b. Find the mean of each sample and create a table representing the distribution of these means. Does the table represent a sampling distribution of the means? Explain your reasoning.

  c. Use technology to find the mean and standard deviation for the means of the samples, and then compare them to the mean and standard deviation of a sampling distribution of the means.

37. Using technology, complete the following steps to create and analyze a sampling distribution of the sample means when $n = 3$ for the population $\{3, 6, 8, 10\}$ with $\mu = 6.75$ and $\sigma \approx 2.59$.

  a. List all possible samples of size $n = 3$ and their corresponding means, $\bar{x}$.

  b. Create a table and a histogram representing the sampling distribution for the means.

  c. Find the mean and standard deviation for the sampling distribution. $\mu_{\bar{x}} = 6.75$; $\sigma_{\bar{x}} \approx 1.49$

36c. $\mu_{\bar{x}} \approx 4.53 \neq \mu = 4.5$; $\sigma_{\bar{x}} \approx 1.50 \neq \frac{\sigma}{\sqrt{n}} = \frac{2.87228}{\sqrt{5}} \approx 1.28$

38. The mean systolic blood pressure for men in a recent study was 124.71 millimeters of mercury (mm Hg). Assume the population is normally distributed with a standard deviation of 14.53 mm Hg.

  a. What is the probability that the average systolic blood pressure of a sample of 30 men selected randomly from the population is between 122.0 and 130.4? ≈ 83%

  b. If an individual from the population is randomly selected and his blood pressure taken, what is the probability that his systolic blood pressure is between 122.0 and 130.4? ≈ 23%

  c. Are the probabilities from parts a and b the same? Explain why or why not.

39. A telemarketer who earns $25 per sale is told that he can expect to make a sale on only 8% of his calls. During his first week, he makes 588 calls and gets 25 sales.

  a. How many sales are expected with the given number of calls? ≈ 47

  b. What is the probability of making 25 sales or fewer? ≈ 0.05%

  c. What is the probability of making the expected number of sales or fewer? 50%

  d. Determine his expected weekly earnings after several weeks when he can make 550 calls a week and makes a sale on 6% of his calls. $825

  e. What is the probability that he will make at least $900 per week? ≈ 33%

40. A vendor selling hot dogs and hamburgers at a baseball game knows that 55% of his customers buy hot dogs. If 20 customers want to make a purchase before he restocks and he only has 12 hotdogs and 12 hamburgers in his case, what is the probability every customer gets the item he wants? ≈ 69%

38a. $\mu = 124.71$

$$\sigma_{\bar{x}} = \frac{14.53}{\sqrt{30}} \approx 2.653$$

$$z_L = \frac{122.0 - 124.71}{2.653} \approx -1.021$$

$$z_U = \frac{130.4 - 124.71}{2.653} \approx 2.145$$

$$P(122 < \bar{x} < 130.4)$$
$$= P(-1.021 < z < 2.145) \approx 0.83$$

38b. $z_L = \frac{122 - 124.71}{14.53} \approx -0.187$

$$z_U = \frac{130.4 - 124.71}{14.53} \approx 0.392$$

$$P(122 < x_i < 130.4)$$
$$= P(-0.187 < z < 0.392) \approx 0.23$$

38c. no; Part a is an average of the sample of 30 men and the standard deviation for the sample means, $\sigma_{\bar{x}} = \frac{\sigma}{\sqrt{n}}$, is much smaller than that of individuals within the population, σ. (It is much more likely for an average of a sample to be "average" than for an individual to be "average.")

39a. $n = 588$; $p = 0.08$; $q = 0.92$
$\mu = np = 588(0.08) = 47.04$
$nq = 588(0.92) = 540.96$
$\sigma = \sqrt{npq} = \sqrt{588(0.08)(0.92)}$
≈ 6.58

39b. $z = \frac{25.5 - 47.04}{6.58} \approx -3.27$
$P(x_i \leq 25.5) \approx P(z \leq -3.27)$
≈ 0.0005

39c. $P(x_i \leq 47.04) = 0.50$

39d. $n = 550$; $p = 0.06$; $np = 33$
$E(x) = \mu = 33$, so $33(\$25) = \$825$

39e. sales needed $= \frac{900}{25} = 36$; $q = 0.94$
$\sigma = \sqrt{npq} = \sqrt{550(0.06)(0.94)}$
≈ 5.57
$z = \frac{35.5 - 33}{5.57} \approx 0.45$
$P(x_i \geq 35.5) = P(z \geq 0.45) \approx 0.33$

40. All customers are satisfied if any number from 8 to 12 want hot dogs. Let p = probability customer wants a hot dog.
$n = 20$; $p = 0.55$; $q = 0.45$
$\mu = np = 20(0.55) = 11$
$nq = 20(0.45) = 9$
$\sigma = \sqrt{20(0.55)(0.45)} \approx 2.225$
$P(8 \leq x_i \leq 12) \approx P(7.5 \leq x_i \leq 12.5)$
$z_L = \frac{7.5 - 11}{2.225} \approx -1.57$
$z_U = \frac{12.5 - 11}{2.225} \approx 0.67$
$P(-1.57 \leq z \leq 0.67) \approx 0.69$

37b.

$\bar{x}$	$P(\bar{x})$	$\bar{x}$	$P(\bar{x})$
3.00	1/64	6.67	3/64
3.33	0/64	7.00	6/64
3.67	0/64	7.33	6/64
4.00	3/64	7.67	3/64
4.33	0/64	8.00	7/64
4.67	3/64	8.33	0/64
5.00	3/64	8.67	6/64
5.33	3/64	9.00	0/64
5.67	6/64	9.33	3/64
6.00	1/64	9.67	0/64
6.33	9/64	10	1/64

0.15 Plot1:L1,L3

3     $10\frac{1}{3}$
0   min=6.3333333
    max<6.6666667   n=0.140625

37c. $\mu_{\bar{x}} = 6.75$; $\sigma_{\bar{x}} = \frac{2.586}{\sqrt{3}} \approx 1.49$

$\mu = \mu_{\bar{x}}$; $\sigma > \sigma_{\bar{x}}$

Cumulative Review

41. $(x - 8)(x - 3)(x + 3)$

$$\cdot \left[\frac{3x + 1}{(x - 8)(x - 3)} + \frac{5}{(x - 3)(x + 3)} \right.$$
$$\left. = \frac{3x}{(x - 8)(x + 3)} \right]$$

$(x + 3)(3x + 1) + 5(x - 8)$
$$= 3x(x - 3)$$
$3x^2 + 10x + 3 + 5x - 40 = 3x^2 - 9x$
$$24x = 37$$
$$x = \frac{37}{24}$$

42. $(x + 5)(x - 2)(x + 2)$

$$\cdot \left[\frac{3x + 5}{(x - 2)(x + 2)} + \frac{4}{(x + 2)(x + 5)} \right.$$
$$\left. = \frac{4x}{(x - 2)(x + 5)} \right]$$

$(x + 5)(3x + 5) + 4(x - 2)$
$$= 4x(x + 2)$$
$3x^2 + 20x + 25 + 4x - 8 = 4x^2 + 8x$
$$x^2 - 16x - 17 = 0$$
$$(x - 17)(x + 1) = 0$$
$$x = 17, -1$$

43. $2\langle -5, 8 \rangle + \langle 12, -5 \rangle$
$$= \langle -10, 16 \rangle + \langle 12, -5 \rangle$$
$$= \langle 2, 11 \rangle = 2\mathbf{i} + 11\mathbf{j}$$

44. $4\langle 12, -5 \rangle - \frac{1}{2}\langle 6, -12 \rangle + 3\langle -5, 8 \rangle$

$$= \langle 48, -20 \rangle + \langle -3, 6 \rangle + \langle -15, 24 \rangle$$
$$= \langle 30, 10 \rangle = 30\mathbf{i} + 10\mathbf{j}$$

45. $z = \frac{4564 - 4021}{421} \approx 1.29$

normalcdf$(-4, 1.29) \approx 0.9014$

90th percentile

46. $z = \frac{28 - 20.6}{5.3} \approx 1.40$

normalcdf$(-4, 1.40) \approx 0.9186$

92nd percentile

47.

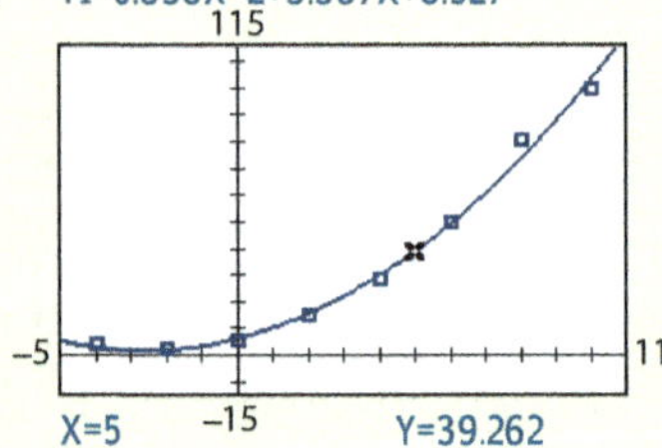

48. $\tan (\mathrm{Sin}^{-1}(-1)) = \tan \left(-\frac{\pi}{2} \right)$, which is undefined.

50. $P(9) = {}_{15}C_9 (0.73)^9 (0.27)^6 \approx 0.11$

Solve each rational equation. [2.6]

41. $\dfrac{3x + 1}{x^2 - 11x + 24} + \dfrac{5}{x^2 - 9} = \dfrac{3x}{x^2 - 5x - 24}$ $x = \frac{37}{24}$

42. $\dfrac{3x + 5}{x^2 - 4} + \dfrac{4}{x^2 + 7x + 10} = \dfrac{4x}{x^2 + 3x - 10}$ $x = 17, -1$

Given $\mathbf{u} = \langle 6, -12 \rangle$, $\mathbf{v} = \langle 12, -5 \rangle$, **and** $\mathbf{w} = \langle -5, 8 \rangle$, **write each resultant vector in component form and as a linear combination of standard unit vectors i and j. [6.1]**

43. $2\mathbf{w} + \mathbf{v}$ $\langle 2, 11 \rangle$; $2\mathbf{i} + 11\mathbf{j}$

44. $4\mathbf{v} - \frac{1}{2}\mathbf{u} + 3\mathbf{w}$ $\langle 30, 10 \rangle$; $30\mathbf{i} + 10\mathbf{j}$

Find the percentile rank for each value (to the nearest percent). [10.6]

45. an automobile weighing 4564 lb if the mean is 4021 lb and standard deviation is 421 lb   90th

46. a student with a score of 28 on an assignment with a mean of 20.6 and standard deviation of 5.3   92nd

47. Use technology to create a scatterplot of the data. Then determine the linear, quadratic, or exponential function that best models the data. Round coefficients to the nearest thousandth. [1.9] B

x	y
-4	4.4
-2	2.1
0	5.3
2	15.1
4	28.7
6	49.7
8	80.7
10	99.7

A. $f(x) = 0.749x^2 + 3.267x + 4.787$
B. $f(x) = 0.636x^2 + 3.367x + 6.527$
C. $f(x) = 7.184x + 14.161$
D. $f(x) = 7.467(1.326)^x$
E. none of these

48. Find the exact value of tan $(\mathrm{Sin}^{-1}(-1))$ without using technology. [4.6] E

A. 1   **C.** 0   **E.** none of these
B. -1 **D.** ∞

49. Which statement is the anchor step in the inductive proof of $14 + 18 + 22 + \cdots + (4n + 10) = 2n(n + 6)$? [9.5] A

A. $P(1) = 14 = 2(1)(1 + 6)$
B. $P(k) = 2k(k + 6)$
C. $P(k + 1) = 2(k + 1)((k + 1) + 6)$
D. $P(k) = P(k + 1)$
E. none of these

50. Use the binomial probability distribution formula to find the probability (to the nearest percent) of 9 successes in 15 independent trials if the probability of success in each trial is 0.73. [11.1] B

A. 4%   **C.** 12%   **E.** none of these
B. 11%   **D.** 27%

PRESENTATION

Eighteenth-century debates over smallpox inoculations illustrate the role of statistics in informing public policy. Until the nineteenth century, inoculation actually meant variolation, the process of scraping material from smallpox pustules and exposing uninfected people to the virus through skin lesions or through inhalation. Variolation often resulted in a mild case of smallpox but was sometimes fatal. Jean Le Rond d'Alembert argued that common sense must not be abandoned when interpreting statistical data. After the immunization process was improved and better data was available, d'Alembert then supported vaccinations for smallpox.

The industrial revolution of the late eighteenth and early nineteenth centuries brought assembly lines and the need for interchangeable parts (a huge departure from the days when craftsmen fashioned each part by hand). Effective methods for sampling parts and finished products became essential to profitability (and the war effort in the 1940s).

R. A. Fisher's book *The Design of Experiments* (1935) included the "lady tasting tea" thought experiment. Fisher addressed the statistical reasoning behind the experiment, such as how many cups

MATHEMATICAL EXPECTATION

Imagine a lottery in which the chances of winning are one in two. Winners are guaranteed a lifespan of 180 yr—and losers are immediately put to death. If the government requires participation of all citizens, then the average lifespan will be 90 yr. Would you volunteer to migrate to this country to enjoy the enviable longevity of its citizens? A similar thought experiment was used by French mathematician Jean Le Rond d'Alembert (1717–83) to argue against probability theory findings by Swiss mathematician Daniel Bernoulli (1700–82) that supported smallpox inoculations. Even if not very likely, the large immediate risk (death from purposeful exposure to smallpox) outweighed the small future benefit (living for an additional two or three years) for the majority. D'Alembert pointed out that what is optimal for the state may not be optimal for the individual.

Pierre-Simon de Laplace (1749–1827), the preeminent French mathematician and scientist of his time, laid out ten principles concerning the calculation of probabilities in the 1814 edition of his *Analytic Theory of Probability*. The first principle was the definition of probability as the ratio of favorable to possible outcomes. After nearly 200 yr, these principles have not changed. Laplace is also responsible for two major developments in probability theory that are now known as the Central Limit Theorem and Bayesian inference (known at the time as inverse probability).

English mathematician Karl Pearson (1857–1936) developed the chi-squared test to determine the reasonableness of fit between observed and expected frequencies in large data sets, while another Englishman, R. A. Fisher (1890–1962), was interested in how to make sound deductions from small data sets. Fisher was a researcher at an agricultural experimental center and published a book in 1925 entitled *Statistical Methods for Research Workers* that statistics author John Tabak claims "may well be the most successful book on statistics ever written." Fisher was particularly interested in experimental design, and one of his later books includes a discussion on how to design an experiment to test the claim that a person could tell by the taste whether milk was added to a cup before or after tea was added. Fisher's work, published initially in agricultural journals, contains a wealth of pioneering mathematics that continues to be explored and used in modern times.

In the beginning of the twentieth century, the Russian mathematician Andrey Kolmogorov (1903–87) axiomatized the theory of probability. This tremendous breakthrough brought mathematical rigor to the field and allowed mathematicians to apply the full weight of analysis to probability theory. Even so, controversy still surrounds the application of statistical findings to real-world problems. Although algorithms are written by authors who have their own worldview and prejudices, many people view algorithms as unbiased and unquestionable because the results come from a computer and are presented in terms of probabilities. Statistics-based algorithms are now common in finance (determining who gets a loan and at what rate), education, criminal justice (sentencing), and advertising. As the use of algorithms rapidly increases, an increasing number of people are warning of pitfalls and calling for experiential judgment and wisdom in interpreting and implementing algorithmic metrics.

COMPREHENSION CHECK

1. What public policy did Daniel Bernoulli's probability research support? inoculations for smallpox

2. Who published ten principles of probability theory in 1814 that remain valid today? Pierre-Simon de Laplace

3. What English mathematician wrote the "most successful book on statistics ever written"? R. A. Fisher

4. What was the major twentieth century contribution of Andrey Kolmogorov to probability theory?

5. **Discuss:** One problem frequently noted with applying quantitative statistics to social issues (sometimes called Campbell's law) is that inevitably the measures will succumb to corruption. Use the Internet keyword search *CompStat* to research a common modern policing tool. List some pros and cons of the system. What biblical principles can help shape our worldview concerning the use of systems like these?

---

of tea should be included in the taste test, whether the before or after conditions should be predetermined or random, and whether the taster should know how many of each option are in the test. In 2002 David Salsburg expounded on Fisher's influence in his book *The Lady Tasting Tea: How Statistics Revolutionized Science in the Twentieth Century.*

The axiomatization of probability was specifically called for by David Hilbert in his famous 1900 address to the International Congress of Mathematicians. Hilbert proposed 23 problems for the coming century, and problem number six included this goal.

Applying pure mathematics to real-world situations is a difficult task. This modeling process is the focus of much effort in modern statistical research. One promising approach that owes much to the advent of computer processing power dates back to Thomas Bayes's and Simon Laplace's method of inverse probability. According to Sharon McGrayne in *The Theory that Would Not Die*, Bayes's theorem has made prolific contributions "from high finance to e-commerce, from sociology to machine learning, and from astronomy to neurophysiology."

Historical Connection

Objectives

1. To summarize the development of the theory of probability
2. To list key publications in probability theory
3. To evaluate uses of probability theory from a biblical worldview

Assignments

- **Minimum:** 1–4; 5 (in class)
- **Standard:** 1–4; 5 (in class)
- **Extended:** 1–5

Answers

4. He used measurement theory to axiomatize probability theory.

5. CompStat was developed by the NYPD in the early 1990s in an effort to reduce crime and increase public safety. It uses near real-time crime statistics to help prevent crime instead of simply reporting it. Most sources agree there was a real decrease in crime after implementation. However, credible reports indicate that data is purposefully misreported so that the results are more favorable. Some crimes may be downgraded, and CompStat may make it more difficult for citizens to report crime. Corruption is part of our fallen nature. We should rejoice when people exercise good judgment and honesty in evaluating and verifying reports.

11.3 Confidence Intervals

Objectives

1. To calculate point estimates for the mean of a population
2. To find the critical value and margin of error for a given confidence level
3. To construct and interpret confidence intervals for the mean of a population
4. To calculate the minimum sample size

Flash

The African bush elephant is the world's largest land animal, roaming in 37 African countries. Poaching and environmental changes put these elephants at risk. Statistical studies aid in protecting this enormous animal.

Vocabulary

confidence interval
confidence level
critical value
interval estimate
margin of error
maximum error of the estimate
point estimate
unbiased

Reading and Writing Mathematics

Give an example of a study in which 99% confidence would be crucial and one in which it would not be necessary.

A 99% or higher confidence level would be crucial when performing a study on the safety of a commercial airliner. A 99% or higher confidence level would not be necessary when performing a study on the average weight of cows.

There are approximately 300,000 African bush elephants in the world, some weighing over 11 tons.

> **After completing this section, you will be able to**
> - calculate point estimates for the mean of a population.
> - find the critical value and margin of error for a given confidence level.
> - construct and interpret confidence intervals for the mean of a population.
> - calculate the minimum sample size.

What is the average weight of a male African bush elephant? You would have to find and weigh every one of them to calculate the population's mean weight, but this is not possible. How can you find a statistic to estimate the average weight, and how confident can you be of the accuracy of your estimate?

In the last section, the Central Limit Theorem was used to find the probability of a sample mean being a certain value when the mean and the standard deviation of the population were known. However, this information is rarely known since the population being studied is usually very large or inaccessible. Instead, statistics such as the mean and standard deviation of a sample are determined and used to estimate (or infer) the related parameters for the population.

There are two types of estimates: the *point estimate* and the *interval estimate*.

> **DEFINITIONS**
>
> A **point estimate** is a single-value estimation of a population's parameter. An **interval estimate** is a range of values used to estimate a population's parameter.

We will limit our discussion to estimates of the mean for populations in which the Central Limit Theorem applies. When the population is normally distributed or the sample size $n \geq 30$, the mean of the sampling distribution of the sample means is equal to the mean of the population ($\mu_{\bar{x}} = \mu$). In this case, a sample's mean, $\bar{x}$, is the best point estimate of the population's mean, μ.

Estimates from small samples ($n < 30$) of populations that are not normally distributed and estimates for other parameters are reserved for later courses in statistics. Any estimate needs to be *unbiased*, which means the estimate is a value that does not consistently underestimate or overestimate the parameter.

Example 1   Finding a Point Estimate

Find the best point estimate for μ using the following random sample of values from a population.

{51, 49, 56, 81, 44, 65, 69, 72, 57, 48, 55, 64, 73, 57, 63, 64, 71, 69, 73, 58, 70, 67, 74, 86, 75, 69, 83, 67, 77, 58, 74, 66, 84, 75, 71, 68, 64, 76, 81, 77}

Answer

$$\bar{x} = \frac{\sum_{i=1}^{40} x_i}{40} = \frac{2701}{40} = 67.525$$

67.525 is the point estimate of μ.

Since $n = 40$, the Central Limit Theorem applies and the sample's mean is the best point estimate of μ.

SKILL ✔ **EXERCISE 15a**

It is extremely unlikely that the mean μ of the population in Example 1 is exactly $\bar{x} = 67.525$, but there is a good chance it is fairly close to the sample's mean. Statisticians typically report an interval around the point estimate in which they are reasonably sure

PRESENTATION

Lesson Opener

1. List the two cases in which the Central Limit Theorem (CLT) states that a sampling distribution of the means can be modeled by a normal distribution. **(1) a normally distributed population, (2) any population with sample size $n \geq 30$**

2. State the formula for the standard deviation for sample means.
$$\sigma_{\bar{x}} = \frac{\sigma}{\sqrt{n}}$$

Engage the students by discussing the difficulty or impossibility of knowing the parameters of various populations, such

as African bush elephants. In addition to the impossible task of rounding up every elephant, the mean weight is constantly changing. Explain that this section shows how inferential statistics can be used to find reliable estimates of a parameter, the population's mean.

Introduce the two types of estimations that will be used in this section, the *point estimate* and the *interval estimate*. Our estimates of the mean will be limited to populations to which the CLT applies. Review student responses to the Lesson Opener to assess their understanding of the CLT.

Example 1 illustrates that the best

the actual parameter lies. The larger the interval, the more confident we can be that it includes the parameter. The normal distribution predicted by the Central Limit Theorem allows us to quantify the *confidence level, c*, as the probability that this interval estimate (called a *confidence interval*) contains the population parameter.

The level of confidence is the area under the standard normal distribution curve between the *critical values*, $\pm z_c$. To find the critical values for an 80% confidence level, find the z-scores, which define the interval containing the middle 80% of the area under the standard normal distribution curve. In this case, 10% of the area is in each tail, so 90% of the data lies below $z_c \approx 1.28$.

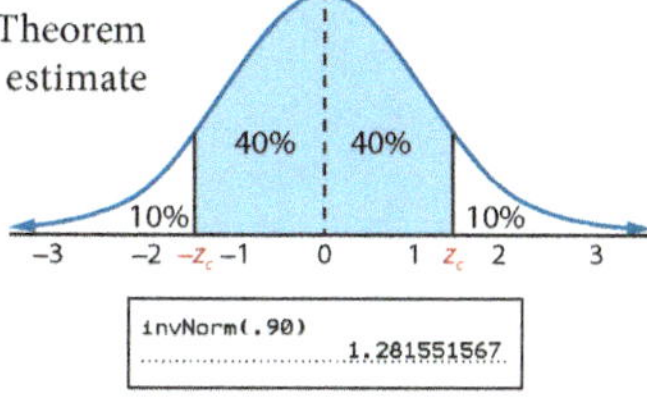

Example 2　Determining the Critical Value

Use technology to find the critical value, z_c, for a 95% confidence level.

Answer

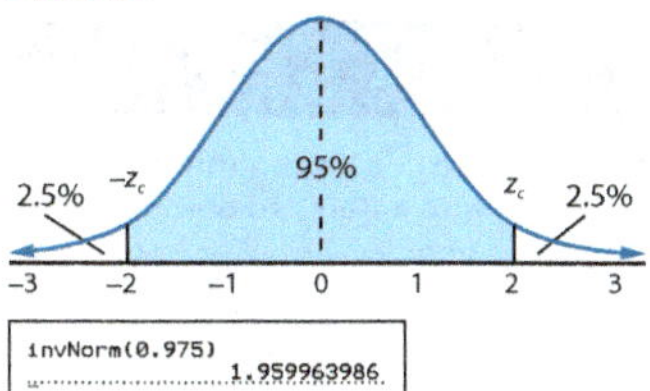

$z_c \approx 1.96$

1. Since $c = 0.95$ is the portion of the area under the curve within $\pm z_c$, each tail contains $\frac{1-c}{2} = 0.025$ or 2.5% of the area.

2. Use technology to find z_c such that 97.5% of the data lies to its left.

SKILL ✓ EXERCISE 15b

To construct a confidence interval (CI) for the population mean, a *margin of error* (sometimes called the *maximum error of the estimate*) is added to and subtracted from the sample mean, $\bar{x}$. This margin of error is determined by the critical value, z_c, for the desired confidence level, the standard deviation of the population (either estimated or known), and the sample size.

▌ DEFINITIONS

For a given confidence level, c, the **confidence interval for the mean** is $\bar{x} \pm E$ where E, the **margin of error for the mean**, is found using

$$E = z_c\left(\frac{s}{\sqrt{n}}\right),$$

where z_c is the critical value for the confidence level and s is the standard deviation of a randomized sample of size $n \geq 30$. If the population standard deviation is known, substitute σ for s.

Example 3　Finding a Margin of Error and Confidence Interval

A random sample of 35 male African bush elephants has a mean weight of 13,483 lb with a standard deviation of 1263 lb. Find the 90% margin of error and 90% confidence interval for the population mean weight and interpret the meaning of the confidence interval.

Answer

Note that $\bar{x} = 13{,}483$ and $s = 1263$.

$$E = z_c\left(\frac{s}{\sqrt{n}}\right) = 1.645\left(\frac{1263}{\sqrt{35}}\right) \approx 351$$

1. Since $n = 35$, the Central Limit Theorem applies. Use the critical value $z_c = 1.645$ for a 90% confidence level.

CONTINUED ➡

TIP 

Remembering the critical values for these frequently used confidence levels will save time and effort in the future.

Confidence Level	Critical Value
c	z_c
90%	1.645
95%	1.96
99%	2.58

In 1937 Polish mathematician Jerzy Neyman published the theory behind confidence intervals. He also developed the experimental method that the FDA uses to test medicines today.

Additional Exercises

Find the best point estimate μ for each set of random sample values.

1. {57, 48, 55, 64, 73, 57, 63, 64, 71, 69, 73, 58, 70, 67, 74, 86, 75, 69, 83, 67, 77, 58, 74, 66 84, 75, 71, 68, 64, 76, 81, 77} 69.2

2. {50, 32, 16, 5, 50, 10, 20, 6, 13, 43, 19, 30, 31, 4, 1, 35, 36, 1, 45, 19, 6, 23, 50, 35, 28, 32, 47, 34, 40, 28} 26.3

3. Use technology to find the critical values for each confidence level.
   a. 98% ≈ 2.326　　b. 75% ≈ 1.150

4. Given the following values, find the margin of error and the confidence interval for the mean.
   a. 95% CL; $n = 45$; $\bar{x} = 85$; $s = 6.9$ ≈ 2.016; ≈ (82.98, 87.02)
   b. 90% CL; $n = 60$; $\bar{x} = 7.5$; $s = 1.2$ ≈ 0.255; ≈ (7.25, 7.75)

5. A random sample of 500 teens was asked how many hours they spent studying or doing homework each week. The mean was 17 hr with a standard deviation of 5.9 hr. Find the 95% confidence interval and interpret its meaning.
   (16.5, 17.5); We are 95% confident that the mean of the population is between 16.5 and 17.5 hr.

point estimate for the population mean is the mean of a randomly selected sample, but this estimate is insufficient. Using an interval (range) of values allows us to state a level of confidence that we have included the actual population mean in our estimate.

The first step in constructing an interval estimate is to determine the *critical values* (*z*-scores) required for the desired *confidence level* by finding the values for the interval. The critical values can be found using the invNorm function for an 80% confidence level, as illustrated in the top figure, and for the 95% confidence level in Example 2. Recall from Section 10.6 that the invNorm function is the inverse of the cumulative normal distribution function and it returns the *z*-score with a given percentage of the data to the left of that value.

Consider leading the students through a generalized explanation of the invNorm function. Since each tail contains $\frac{1-c}{2}$ of the area, $1 - \frac{1-c}{2} = \frac{2-(1-c)}{2} = \frac{1+c}{2}$ of the data lies to the left of z_c. Therefore invNorm $\left(\frac{1+c}{2}\right) = z_c$. In Example 2, invNorm $\left(\frac{1.95}{2}\right) = $ invNorm$(0.975) \approx 1.96$. Remembering the critical *z*-score values for the common confidence levels listed in the margin tip can save a significant amount of time and effort.

Formally define a *confidence interval for the mean* and *the margin of error for the mean* and state the formula of the latter. Then use Example 3 to demonstrate constructing a confidence interval using the margin of error.

Example 4 combines the steps to construct a confidence interval. This provides a good review of each step and details for checking the work on a calculator. Work can also be checked online using the Internet keyword search *confidence interval calculator*.

Students should see in Example 5 that the higher confidence levels require wider confidence intervals (less precise estimates). They should also see that a larger

6. Two random months from hospital records are chosen for a sample of birth weights.

sample A:
$n = 550; \bar{x} = 7.7$ lb; $s = 1.4$ lb
sample B:
$n = 600; \bar{x} = 7.5$ lb; $s = 1.25$ lb

  a. Construct the 85% CI and the 95% CI from sample A.
  $\approx (7.61, 7.79); \approx (7.58, 7.82)$

  b. Construct the 95% CI from sample B and compare it to the 95% CI from sample A. $\approx (7.40, 7.60)$
  Sample B had a smaller margin of error, resulting in a smaller confidence interval and a more precise estimate.

  c. Which confidence level and sample size resulted in the widest confidence interval? 95% CL from sample A, in which $n = 550$

7. A sample of 39 parents reported they spent $796 per year, per student on school lunches with a standard deviation of $8.37.

  a. Find the 90% and 98% confidence intervals. $\approx (793.80, 798.20)$
  $\approx (792.88, 799.12)$

  b. Assume the mean and standard deviation remain the same, but the sample size is increased to 100. Compare the 98% confidence intervals found using the two samples sizes.
  $\approx (794.05, 797.95)$; The larger sample produced a smaller margin of error and smaller confidence interval.

  c. What would the sample size need to be if the researcher wanted a margin of error of about 1 and a CI of 98%? $n \geq 380$

Assignments

- **Minimum:** 1–6; 9–25 odd; 28–29, 31, 34(*a–b*); 43–51 odd

- **Standard:** 1–8; 12–22 even; 25–28, 30, 32, 33, 36, 39; 42–46 even; 51

- **Extended:** 1–7, 10, 13–14, 16, 18, 22, 24, 27–28, 32–35; 37–43 odd; 48, 51

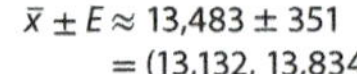

$\bar{x} \pm E \approx 13{,}483 \pm 351$
$= (13{,}132, 13{,}834)$

We are 90% confident that this interval contains μ.

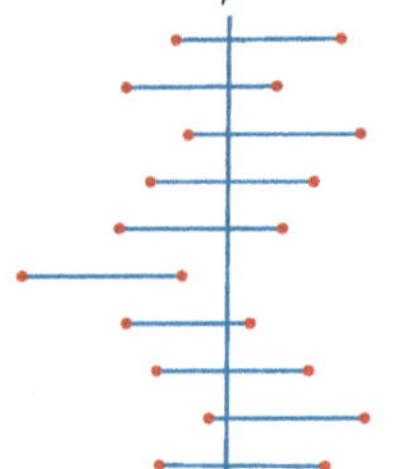

In the 1980s many research journals, particularly in the medical field, began to require authors to include confidence intervals in the presentation of their results.

2. The margin of error is subtracted from and added to the sample mean to construct the 90% confidence interval.

3. Interpret the confidence interval.

— SKILL ✔ **EXERCISE 15c**

Confidence intervals are often misunderstood. The mean of the population, μ, is a fixed number for a given population. Different samples will produce varying confidence intervals. In Example 3, 90% of these intervals would contain μ, but 10% of them would not. We cannot be sure that μ is in our calculated confidence interval. The figure illustrates ten different 90% confidence intervals for the population mean. Notice that nine contain the population mean and one does not.

Example 4   Constructing and Interpreting a 97% Confidence Interval

Students studying earthquakes took a random sample of Richter scale values from 50 earthquakes over the past 25 yr. The mean of the values was 1.1842 with a standard deviation of 0.3449. Find the 97% confidence interval and interpret its meaning.

Answer

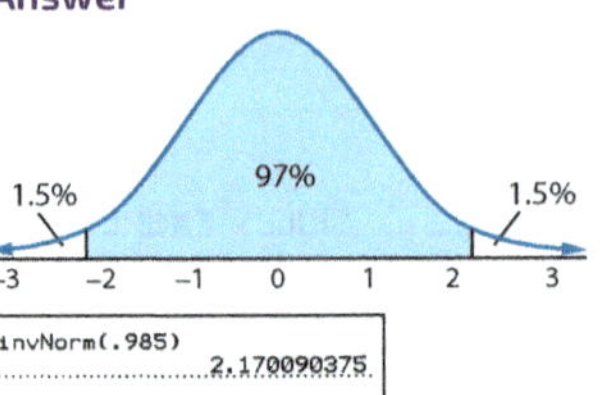

$s = 0.3449, n = 50$
$$E = z_c\left(\frac{s}{\sqrt{n}}\right) = 2.17\left(\frac{0.3449}{\sqrt{50}}\right) \approx 0.1058$$

$\bar{x} \pm E \approx 1.1842 \pm 0.1058 = (1.0784, 1.2900)$

We are 97% confident that the population mean μ of the Richter scale values is in the interval (1.0784, 1.2900).

Check

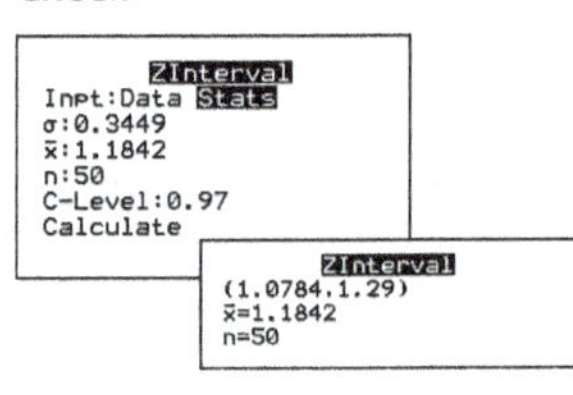

1. Since 97% of the area is within $\pm z_c$, each tail contains 1.5% of the area.

Use technology to find z_c such that 98.5% of the data lies to its left.

2. Substitute z_c, s, and n into the formula for the margin of error of the mean and evaluate.

3. Calculate the 97% CI.

4. Interpret the confidence interval.

5. The STAT, TESTS submenu contains a ZInterval function that can be used to calculate confidence intervals. Select the Stats option for Inpt: and enter the standard deviation (using the value of s for σ), the sample mean, and the confidence level.

The interval is listed in the first line.

— SKILL ✔ **EXERCISE 17**

The next example examines the effect of the confidence level and sample size on the precision of the resulting confidence interval.

sample size reduces the width of the confidence interval (more precise estimates). Emphasize that the only effective way to reduce the confidence interval while maintaining the same level of confidence is to increase the sample size. Consider completing exercise 27 together to derive the formula for the minimum sample size. Demonstrate its application in Example 6. Ensuring a sufficient sample size is an important factor in the validation of a study and its conclusions.

Consider sharing John 20:30–31. While John did not record all of Christ's miracles, he recorded a large enough sample for us to confidently believe that Jesus is the Son of God.

TIPS

Ex. 24–26 Encourage the students to recall the related critical values in the table on page 567.

Example 5   Analyzing Confidence Intervals

Researchers recorded the body temperatures of individuals from two samples of randomly selected adults and calculated each sample's mean and standard deviation.

	Sample A	Sample B
	$n = 40$	$n = 95$
	$\bar{x} = 98.165°F$	$\bar{x} = 98.204°F$
	$s = 0.687°F$	$s = 0.631°F$

a. Construct and compare the 90% CI and the 99% CI from Sample A.

b. Construct the 90% CI from Sample B and compare it to the 90% CI from Sample A.

Answer

a. Sample A, 90% CI

$$E = z_c\left(\frac{s}{\sqrt{n}}\right)$$

$$= 1.645\left(\frac{0.687}{\sqrt{40}}\right)$$

$$\approx 0.179$$

$$\bar{x} \pm E \approx 98.165 \pm 0.179$$
$$= (97.986°F, 98.344°F)$$

Sample A, 99% CI

$$E = z_c\left(\frac{s}{\sqrt{n}}\right)$$

$$= 2.58\left(\frac{0.687}{\sqrt{40}}\right)$$

$$\approx 0.280$$

$$\bar{x} \pm E \approx 98.165 \pm 0.280$$
$$= (97.885°F, 98.445°F)$$

1. Use $z_{90} = 1.645$ and $z_{99} = 2.58$ to determine each margin of error, the 90% CI, and the 99% CI from Sample A.

The higher (99%) level of confidence is created using a larger margin of error, which provides a less precise estimate.

2. Compare the results.

b. Sample A, 90% CI

$$E \approx 0.179$$

$$\bar{x} \pm E \approx (97.986°F, 98.344°F)$$

Sample B, 90% CI

$$E = z_c\left(\frac{s}{\sqrt{n}}\right)$$

$$= 1.645\left(\frac{0.631}{\sqrt{95}}\right)$$

$$\approx 0.106$$

$$\bar{x} \pm E \approx 98.204 \pm 0.106$$
$$= (98.098°F, 98.31°F)$$

1. Use $z_{90} = 1.645$ and $n = 95$ to find the 90% CI from Sample B.

The larger sample size of Sample B reduced the margin of error and the width of the confidence interval, allowing a more precise estimate. The slightly smaller value for s also made the 90% confidence interval slightly smaller.

2. Compare the results.

SKILL ✓ **EXERCISES 29–31**

Researchers desire the narrowest possible confidence interval while maintaining a high level of confidence. But the margin of error formula, $E = z_c\left(\frac{s}{\sqrt{n}}\right)$, implies that higher confidence levels generate larger margins of error and wider confidence intervals. The sample standard deviations, s, cannot be controlled by the researcher and will likely be close to the population standard deviation, σ. Therefore, increasing the sample size, n, is the only way to effectively lower the margin of error and obtain a more precise confidence interval with the desired level of confidence. Solving the margin of error formula for n (see exercise 27) results in a formula that can be used to find the minimum sample size required for the desired confidence level and margin of error.

$$n = \left(\frac{z_c s}{E}\right)^2$$

A researcher studying body mass index (BMI) desires a 95% confidence level and the margin of error for the mean to be no more than 2. A small pilot study determined that $s = 7.4$ is a reasonable standard deviation for the population. Determine the minimum sample size for the research study.

Answer

$$n = \left(\frac{z_c s}{E}\right)^2 = \left(\frac{1.96(7.4)}{2}\right)^2 \approx 52.6$$

1. Use the sample size formula with $z_{95} = 1.96$.

The minimum sample size is 53.

2. Always round up since the formula yields the minimum number required.

SKILL ✓ EXERCISE 23

It would be wise to include several more people in the study referenced in Example 6 since the actual sample standard deviation, s, will likely differ somewhat from 7.4 in the study.

❭ A. Exercises

Match each term with its description.
A. confidence interval
B. confidence level
C. critical value
D margin of error
E. point estimate

1. a single value that estimates a population's parameter   E

2. a range of values that estimates a population's parameter   A

3. the probability that a confidence interval contains the population's parameter   B

4. the maximum difference between the point estimate and the actual parameter value   D

5. What is the only way to effectively lower the margin of error?   Increase the sample size.

6. Researchers desire the (*smallest* or *largest*) possible CI while maintaining a (*low* or *high*) level of confidence.   smallest; high

Use technology to find the critical value z_c (to the nearest thousandth) for each confidence level.

7. 90% confidence level   1.645

8. 99% confidence level   2.576

9. 85% confidence level   1.440

10. 78% confidence level   1.227

Find the margin of error for the mean, E, for the given confidence level, c.

11. $c = 95\%$ if $n = 42$ and $s = 2.6$ ≈ 0.79

12. $c = 99\%$ if $n = 64$ and $s = 158$ ≈ 50.87

13. $c = 80\%$ if $n = 120$ and $s = 5.9$ ≈ 0.69

14. $c = 72\%$ if $n = 30$ and $s = 71$ ≈ 14.00

Use each random sample to find (a) the point estimate for the population's mean, (b) the critical value for the stated confidence level, and (c) the related confidence interval.

15. {4, 9, 1, 8, 6, 8, 5, 5, 4, 0, 6, 1, 1, 1, 1, 7, 3, 3, 5, 6, 3, 7, 7, 7, 1, 9, 2, 3, 9, 8}
    $c = 95\%$ $\bar{x} \approx 4.67$; $z_c \approx 1.96$; CI: $\approx (3.64, 5.70)$

16. {42, 18, 93, 32, 92, 3, 95, 51, 48, 54, 81, 7, 62, 78, 13, 99, 49, 8, 32, 34, 34, 81, 30, 9, 22, 97, 49, 43, 6, 8}
    $c = 98\%$ $\bar{x} \approx 45.7$; $z_c \approx 2.33$; CI: $\approx (32.3, 59.0)$

❭ B. Exercises

Find each confidence interval from the given sample statistics and interpret its meaning.

17. $c = 90\%$, $n = 52$, $\bar{x} = 10{,}372$, $s = 1018$

18. $c = 95\%$, $n = 115$, $\bar{x} = 38$, $s = 4.7$

19. $c = 98\%$, $n = 32$, $\bar{x} = 79$, $s = 8.2$

20. $c = 97\%$, $n = 30$, $\bar{x} = 2.81$, $s = 0.17$

17. $\approx (10{,}140, 10{,}604)$; We are 90% confident that μ is within the interval (10,140, 10,604).

18. $\approx (37.14, 38.86)$; We are 95% confident that μ is within the interval (37.14, 38.86).

19. $\approx (75.6, 82.4)$; We are 98% confident that μ is within the interval (75.6, 82.4).

20. $\approx (2.74, 2.88)$; We are 97% confident that μ is within the interval (2.74, 2.88).

Solutions

❭ A. Exercises

7. $z_c = \text{invNorm}(0.95) \approx 1.645$

8. $z_c = \text{invNorm}(0.995) \approx 2.576$

9. $z_c = \text{invNorm}(0.925) \approx 1.440$

10. $z_c = \text{invNorm}(0.89) \approx 1.227$

11. $E = z_c\left(\frac{s}{\sqrt{n}}\right) = 1.96\left(\frac{2.6}{\sqrt{42}}\right) \approx 0.79$

12. $E = z_c\left(\frac{s}{\sqrt{n}}\right) = 2.58\left(\frac{158}{\sqrt{64}}\right) \approx 50.87$

13. $z_c = \text{invNorm}(0.90) \approx 1.28$
    $E = z_c\left(\frac{s}{\sqrt{n}}\right) \approx 1.28\left(\frac{5.9}{\sqrt{120}}\right) \approx 0.69$

14. $z_c = \text{invNorm}(0.86) \approx 1.08$
    $E = z_c\left(\frac{s}{\sqrt{n}}\right) \approx 1.08\left(\frac{71}{\sqrt{30}}\right) \approx 14.00$

15. $\bar{x} \approx 4.67$; $s \approx 2.87$; $z_c \approx 1.96$
    $E = z_c\left(\frac{s}{\sqrt{n}}\right) \approx 1.96\left(\frac{2.87}{\sqrt{30}}\right) \approx 1.03$
    CI: $4.67 \pm 1.03 = (3.64, 5.70)$

16. $\bar{x} \approx 45.7$; $s \approx 31.4$
    $z_c = \text{invNorm}(0.99) \approx 2.33$
    $E = z_c\left(\frac{s}{\sqrt{n}}\right) \approx 2.33\left(\frac{31.4}{\sqrt{30}}\right) \approx 13.3$
    CI: $45.7 \pm 13.3 = (32.3, 59.0)$

❭ B. Exercises

17. $E = 1.645\left(\frac{1018}{\sqrt{52}}\right) \approx 232$
    CI: $10{,}372 \pm 232 \approx (10{,}140, 10{,}604)$

18. $E = 1.96\left(\frac{4.7}{\sqrt{115}}\right) \approx 0.86$
    CI: $38 \pm 0.86 \approx (37.14, 38.86)$

19. $z_c = \text{invNorm}(0.99) \approx 2.326$
    $E \approx 2.326\left(\frac{8.3}{\sqrt{32}}\right) \approx 3.4$
    CI: $79 \pm 3.4 \approx (75.6, 82.4)$

20. $z_c = \text{invNorm}(0.985) \approx 2.17$
    $E \approx 2.17\left(\frac{0.17}{\sqrt{30}}\right) \approx 0.07$
    CI: $2.81 \pm 0.07 \approx (2.74, 2.88)$

21. $\bar{x} = 533.8$; $s \approx 252.4$
    $E \approx 1.645\left(\frac{252.4}{\sqrt{30}}\right) \approx 75.8$
    CI: $533.8 \pm 75.8 \approx (458.0, 609.6)$

22. $\bar{x} \approx 423.5$; $s \approx 202.5$
    $E \approx 1.645\left(\frac{202.5}{\sqrt{30}}\right) \approx 60.8$
    CI: $423.5 \pm 60.8 \approx (362.7, 484.3)$

23. $z_c = \text{invNorm}(0.9) \approx 1.28$
    $n = \left(\frac{z_c s}{E}\right)^2 \approx \left(\frac{1.28(7.4)}{1.5}\right)^2 \approx 39.9$
    $n \geq 40$

24. $n = \left(\frac{z_c s}{E}\right)^2 = \left(\frac{1.645(38.5)}{11}\right)^2 \approx 33.1$
    $n \geq 34$

25. $n = \left(\frac{z_c s}{E}\right)^2 = \left(\frac{2.58(0.47)}{0.2}\right)^2 \approx 36.6$
    $n \geq 37$

Use each random sample to find the point estimate and the 90% confidence interval for the population's mean.

21. {403, 906, 387, 519, 285, 691, 592, 482,
905, 350, 987, 186, 872, 436, 362, 571,
790, 301, 815, 283, 693, 109, 363, 531, $\bar{x} = 533.8$
364, 314, 780, 799, 794, 144}   CI: $\approx$ (458.0, 609.6)

22. {110, 465, 489, 284, 434, 557, 656, 694,
598, 343, 180, 346, 315, 170, 103, 477,
412, 406, 860, 632, 620, 284, 492, 391, $\bar{x} \approx 423.5$
882, 112, 386, 328, 284, 394}   CI: $\approx$ (362.7, 484.3)

Find the minimum sample size, n, for the given margin of error in a normally distributed population with the given standard deviation.

23. 80% CI, $E = 1.5$, $\sigma = 7.4$ 40

24. 90% CI, $E = 11$, $\sigma = 38.5$ 34

25. 99% CI, $E = 0.2$, $\sigma = 0.47$ 37

26. 95% CI, $E = 38$, $\sigma = 896$ 2136

27. Derive the formula for minimum sample size from the formula for the margin of error of the mean.

A researcher compares different-sized samples from a population. Construct the 90% confidence interval and the 99% confidence interval from each sample if $\bar{x} = 800$ and $s = 50$.

28. $n = 50$ **29.** $n = 500$ **30.** $n = 5000$

31. In exercises 28–30, which sample size and confidence level provide the narrowest confidence interval?

32. In 2015 a sample of 2600 teens spent an average of 81 min/day playing video games with $s = 21.4$ min/day. Find the 95% confidence interval for the population's mean. What causes the confidence interval to be so narrow? $\approx$ (80.2, 81.8); the large sample size

33. The maximum drop on 55 randomly selected roller coasters in the United States was determined. If the average was 132.37 ft with a standard deviation of 64.90 ft, find a 95% confidence interval for the average maximum drop for US roller coasters. $\approx$ (115.22, 149.52)

34. The average distance of 32 randomly selected home runs hit during the 2016 MLB season was 406.81 ft with a standard deviation of 25.75 ft.
a. Use your calculator's ZInterval function to find the 97% confidence interval for the mean distance of all the home runs hit that season. $\approx$ (396.93, 416.69)
b. State the point estimate for the mean and the margin of error. $\bar{x} \approx 406.81$; $E \approx 9.88$
c. Would a population mean distance of 400.3 ft be surprising? Why or why not?
no; It is within the 97% confidence interval.

35. An ornithologist seeks to determine the average heart rate of hummingbirds in flight. Preliminary research suggests a standard deviation of 115 bpm. What is the minimum sample size for this study if he selects a 95% confidence interval and the acceptable margin of error is 30 bpm? $n \geq 57$

36. If a study is conducted to determine the average lifespan of a certain brand of microwaves, how many microwaves should be tested if the accepted margin of error is 3 mo, the standard deviation based on past lifespans of the brand's microwaves is 0.8 yr, and a 99% confidence interval is desired? $n \geq 69$

> C. Exercises

37. A random sample of 44 GPAs was selected from the seniors in a city's high schools. The 90% confidence interval for the mean was found to be (3.17, 3.43). Determine whether each statement is a correct interpretation of the result and explain your reasoning.
a. 90% of all the students have a GPA between 3.17 and 3.43.
b. 90% of the confidence intervals constructed from random samples of this population will contain the population's mean GPA.
c. 90% of the sampled students have a GPA between 3.17 and 3.43.
d. 90% of all samples of students will have average GPAs between 3.17 and 3.43.

38. A study in *Developmental Medicine & Child Neurology* reported the mean age at which 72 babies learned to roll from their back to their stomach and vice versa. Data from the study is displayed in the following table.

	Back to Stomach	Stomach to Back
mean age	4 mo	4.8 mo
95% CI	3.7 to 4.3	4.4 to 5.2

a. State the point estimate and E for both rolling back to stomach and rolling stomach to back.
b. Would you be surprised to hear that the population's mean age of rolling back to stomach was truly 4.4 mo? Why or why not?

28. 90% CI: $\approx$ (788.4, 811.6); 99% CI: $\approx$ (781.8, 818.2)
29. 90% CI: $\approx$ (796.3, 803.7); 99% CI: $\approx$ (794.2, 805.8)
30. 90% CI: $\approx$ (798.8, 801.2); 99% CI: $\approx$ (798.2, 801.8)
31. the lower confidence level (90%) and the largest sample size ($n = 5000$)

33. $E = 1.96\left(\dfrac{64.9}{\sqrt{55}}\right) \approx 17.15$
CI: 132.37 ± 17.15
  $= (115.22, 149.52)$

34a.
```
             ZInterval
Inpt:Data Stats
σ:25.75
x̄:406.81
n:32
C-Level:0.97
Calculate
```
```
             ZInterval
(396.93,416.69)
x̄=406.81
n=32
```

34b. $\bar{x} = 406.81$
  $E \approx \dfrac{416.69 - 396.93}{2} = 9.88$

35. $n = \left(\dfrac{z_c s}{E}\right)^2 = \left(\dfrac{1.96(115)}{30}\right)^2 \approx 56.45$
  $n \geq 57$

36. $n = \left(\dfrac{z_c s}{E}\right)^2 = \left(\dfrac{2.58[0.8(12)]}{3}\right)^2 \approx 68.16$
  $n \geq 69$

> C. Exercises

37a. no; The confidence interval does not describe individual GPAs in the population. Rather, it describes the population mean.

37b. yes; The confidence interval is used to infer the value of the population mean.

37c. no; The confidence interval does not describe individual GPAs in the sample. Rather, it describes the population mean.

37d. no; The confidence interval does not describe sample means. Rather, it describes the population mean.

38a. back to stomach:
  $\bar{x} = \dfrac{4.3 + 3.7}{2} = 4$; $E = \dfrac{4.3 - 3.7}{2} = 0.3$
stomach to back:
  $\bar{x} = \dfrac{5.2 + 4.4}{2} = 4.8$
  $E = \dfrac{5.2 - 4.4}{2} = 0.4$

38b. yes; It is only 5% likely that the sample caused the confidence interval to be the 1 out of 20 that did not include the mean. It would be more likely that the sample was not representative of the population.

26. $n = \left(\dfrac{z_c s}{E}\right)^2 = \left(\dfrac{1.96(896)}{38}\right)^2 \approx 2135.7$
  $n \geq 2136$

27. $E = z_c\left(\dfrac{s}{\sqrt{n}}\right)$
  $E\sqrt{n} = z_c s$
  $\sqrt{n} = \dfrac{z_c s}{E}$
  $n = \left(\dfrac{z_c s}{E}\right)^2$

28. $E_{90} = 1.645\left(\dfrac{50}{\sqrt{50}}\right) \approx 11.6$
  CI_{90}: $800 \pm 11.6 = (788.4, 811.6)$
  $E_{99} = 2.58\left(\dfrac{50}{\sqrt{50}}\right) \approx 18.2$
  CI_{99}: $800 \pm 18.21 = (781.8, 818.2)$

29. $E_{90} = 1.645\left(\dfrac{50}{\sqrt{500}}\right) \approx 3.7$
  CI_{90}: $800 \pm 3.7 = (796.3, 803.7)$
  $E_{95} = 2.58\left(\dfrac{50}{\sqrt{500}}\right) \approx 5.8$
  CI_{99}: $800 \pm 5.8 = (794.2, 805.8)$

30. $E_{90} = 1.645\left(\dfrac{50}{\sqrt{5000}}\right) \approx 1.2$
  CI_{90}: $800 \pm 1.16 = (798.8, 801.2)$
  $E_{99} = 2.58\left(\dfrac{50}{\sqrt{5000}}\right) \approx 1.8$
  CI_{99}: $800 \pm 1.8 = (798.2, 801.8)$

32. $E = 1.96\left(\dfrac{21.4}{\sqrt{2600}}\right) \approx 0.8$
  CI: $81 \pm 0.8 = (80.2, 81.8)$

39a. CI_{sun}: $55.82 \pm 1.64\left(\dfrac{7.26}{\sqrt{100}}\right)$

$\approx (54.63, 57.01)$

CI_{shade}: $47.67 \pm 1.64\left(\dfrac{3.96}{\sqrt{30}}\right)$

$\approx (46.48, 48.86)$

39b. Although not likely, it could be if the sample caused the confidence interval to be the 1 out of 10 that did not include the true mean of the population. It would be best to take another sample and follow additional steps to be sure the sample is representative of the entire population.

40. $z_c = \text{invNorm}(0.99) \approx 2.33$

$E \approx 2.33\left(\dfrac{3.76}{\sqrt{40}}\right) \approx 1.38$

CI: $72.38 \pm 1.38 = (71.00, 73.76)$

Since the CI describes the average height of the women in the conference, not individual heights, it would not be unusual for a 66 in. woman to be in the conference.

41a. using technology:

$\bar{x} \approx 189.3; \; s \approx 34.7$

41b. CI_{99}: $189.3 \pm 2.58\left(\dfrac{34.7}{\sqrt{30}}\right)$

$\approx (173.0, 205.6)$

CI_{95}: $189.3 \pm 1.96\left(\dfrac{34.7}{\sqrt{30}}\right)$

$\approx (176.9, 201.7)$

CI_{90}: $189.3 \pm 1.645\left(\dfrac{34.7}{\sqrt{30}}\right)$

$\approx (178.9, 199.7)$

41c. Higher confidence levels have wider confidence intervals, and lower confidence levels have narrower confidence intervals.

> **Cumulative Review**

42. $p = 4\pi = \dfrac{2\pi}{b}; \; b = \dfrac{2\pi}{4\pi} = \dfrac{1}{2}$

$g(x) = \csc\dfrac{x}{2} + 2$

43. $p = 2 = \dfrac{2\pi}{b}; \; b = \pi$

$g(x) = \dfrac{1}{4}\sec \pi x$

44. $\tan^2\theta + 1 = \sec^2\theta$

$\tan^2\theta + 1 = \dfrac{1}{\cos^2\theta}$

$\tan^2\theta = \dfrac{1}{(0.8764)^2} - 1$

$\tan^2\theta \approx 0.30195$

$\tan\theta \approx -0.5495 \; (\theta \text{ in Q IV})$

45. $\cos\theta = \dfrac{1}{-1.236} \approx -0.8091$

$\sin^2\theta = 1 - \cos^2\theta$

$\approx 1 - (-0.8091)^2 \approx 0.3454$

$\sin\theta \approx 0.5877 \; (\theta \text{ in Q II})$

---

39. A study compared the leaf widths from samples of camellia plants grown in the sun and those grown in the shade.

	Sun	Shade
n	100	30
$\bar{x}$	55.82 mm	47.67 mm
s	7.26 mm	3.96 mm

  a. Construct the 90% confidence interval for the mean leaf width of each population.

  b. Based on the confidence interval for the shrubs growing in the shade, do you think the mean leaf width of all camellia plants growing in the shade could be 50 mm? Why or why not?

39a. sun: $\approx (54.63, 57.01)$; shade: $\approx (46.48, 48.86)$

41a. 189.3 cal/serving

41b. 99% CI: $\approx (173.0, 205.6)$
95% CI: $\approx (176.9, 201.7)$
90% CI: $\approx (178.9, 199.7)$

40. A sample of the heights of 40 players in the Eastern Conference of the WNBA during the 2014 season revealed the mean to be 72.38 in. with a standard deviation of 3.76 in. Create a 98% confidence interval for the mean height of all women in the conference that year. Would it be unusual for a 5 ft 6 in. woman to be in the conference? Why or why not?

41. The number of calories per serving from 30 different brands of chocolate ice cream was determined.
{130, 260, 185, 172, 220, 214, 190, 175, 204, 218, 158, 217, 138, 159, 184, 195, 289, 174, 149, 163, 184, 173, 210, 221, 167, 180, 205, 148, 176, 221}

  a. Find the point estimate for the population mean.

  b. Use the sample to find 99%, 95%, and 90% confidence intervals for the population mean.

  c. Describe the relationship between the level of confidence and the width of the confidence interval for a given sample.

CUMULATIVE REVIEW

Write the rule for $g(x)$, the described transformation of $f(x)$. [4.5]

42. The function $f(x) = \csc x$ is horizontally stretched so its period is 4π and it is shifted up 2 units.

43. The function $f(x) = \sec x$ is horizontally shrunk so its period is 2 and it is vertically shrunk by a factor of $\frac{1}{4}$.

Use identities to find each value. [5.1]

44. Find $\tan\theta$ if $\cos\theta = 0.8764$ and $\sin\theta < 0$. ≈ -0.5495

45. Find $\sin\theta$ if $\sec\theta = -1.236$ and $\tan\theta < 0$. ≈ 0.5877

Write the simplified explicit formula for each arithmetic sequence and then find a_{20}. [9.2]

46. $2, 9, 16, 23, \ldots$ **47.** $12, 4, -4, -12, \ldots$

48. Select the number of petals in the graph of the polar equation $r = 6\cos 3\theta$ and the maximum value of $|r|$. [6.5] B

  A. 6; 6   **C.** 3; 3   **E.** none of these

  B. 3; 6   **D.** 3; 2

42. $g(x) = \csc\frac{x}{2} + 2$

43. $g(x) = \frac{1}{4}\sec \pi x$

46. $a_n = 7n - 5; a_{20} = 135$

47. $a_n = -8n + 20; a_{20} = -140$

49. Classify the graph of $4x^2 - y + 40x + 97 = 0$. [8.4] D

  A. circle   **C.** hyperbola   **E.** none of these

  B. ellipse   **D.** parabola

50. Identify the value whose standardized score is $z \approx 1.47$. [10.5] D
{61, 68, 69, 73, 75, 81, 85, 87, 90, 94}

  A. 61   **C.** 90   **E.** 95

  B. 68   **D.** 94

51. Which quantity is described by discrete data? List all correct answers. [11.1] A, D

  A. oranges in a box

  B. orange juice in a container

  C. height of a tree

  D. students in a class

  E. none of these

---

46. $d = 9 - 2 = 7$

$a_n = 2 + (n-1)7 = 7n - 5$

$a_{20} = 7(20) - 5 = 135$

47. $d = 12 - 4 = 8$

$a_n = 12 + (n-1)(-8) = -8n + 20$

$a_{20} = -8(20) + 20 = -140$

48. If n is odd, then $r = k\cos n\theta$ implies n petals and the maximum $|r| = k$.

49. $D = 0^2 - 4(4)(0) = 0$

50. using technology:

$\bar{x} \approx 78.3; \; s \approx 10.8; \; 1.47 \approx \dfrac{x - 78.3}{10.8}$

$x \approx 10.8(1.47) + 78.3 \approx 94$

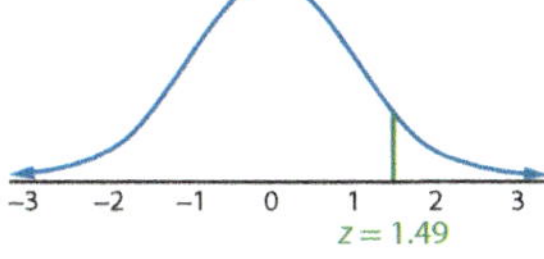

A university reports that the average SAT math score for incoming students is 663. A survey of 50 randomly selected freshmen found they had an average SAT math score of 672 with a standard deviation of 42.8. Do the statistics from this sample provide enough evidence to support a claim that the average SAT math score is higher than 663?

If the reported mean of $\mu = 663$ is accurate, the Central Limit Theorem implies that the sampling distribution of the sample means (with $n = 50$) is normally distributed with $\mu_{\bar{x}} = \mu = 663$ and $\sigma_{\bar{x}} = \frac{s}{\sqrt{n}} = \frac{42.8}{\sqrt{50}} \approx 6.1$. The sample mean, $\bar{x} = 672$, would have a z-score of $z = \frac{\bar{x} - \mu}{\sigma_{\bar{x}}} = \frac{672 - 663}{6.1} \approx 1.49$. Plotting this z-score on the standard normal distribution curve indicates that the amount of deviation is not that unusual, so this sample would not seem to provide enough evidence to reject the reported score of 663.

> **After completing this section, you will be able to**
> - write null and alternative hypotheses.
> - identify type I and type II errors.
> - perform hypothesis tests and interpret the results.

DEFINITION

The process of using statistics from a sample to examine a *claim* about the value of a population parameter is called **hypothesis testing**.

Hypothesis testing does not determine the truth of the claim; instead, it considers whether the evidence from a sample is strong enough to reach a conclusion related to a proposed value for a population parameter. Two hypotheses are used to test the claim, which is written as a mathematical statement such as $\mu = 663$. The claim is one of the hypotheses, and its negation (or complement), $\mu \neq 663$ in this case, is the other hypothesis.

R. A. Fisher popularized the use of hypothesis testing with the publication of *Statistical Methods for Research Workers* in 1925, but it was Karl Pearson and Jerzy Neyman who added the concept of an alternative hypothesis.

DEFINITIONS

A **null hypothesis, H_0,** is a statement about a population parameter that contains an equality such as =, ≤, or ≥.

An **alternative** (or **research**) **hypothesis, H_a,** is a statement that is the negation of the null hypothesis and contains an inequality such as ≠, <, or >.

The university's claim that $\mu = 663$ is the null hypothesis, and the alternative hypothesis is $\mu \neq 663$. Note the alternative hypothesis for each null hypothesis listed below.

If H_0: $\mu = k$, then H_a: $\mu \neq k$.   If H_0: $\mu \geq k$, then H_a: $\mu < k$.   If H_0: $\mu \leq k$, then H_a: $\mu > k$.

11.4 Hypothesis Testing

Objectives

1. To list the steps used to test a hypothesis

2. To write null and alternative hypotheses of a research question

3. To perform hypothesis testing using critical values and using *p*-values

4. To interpret the results of hypothesis testing in a research question

5. To perform a hypothesis test to determine significant differences in the means of different populations

Flash

Standardized test scores act as a good indicator of a student's future success in college, but many other factors influence a student's performance. These scores can also be an indicator of the overall academic strength of a school's student body.

Vocabulary

alternative (research) hypothesis
claim
critical (rejection) region
hypothesis testing
level of significance
null hypothesis
p-value
statistically significant difference
test statistic
type I error
type II error

PRESENTATION

Lesson Opener

State the critical value for each confidence level.

1. 90% 1.645

2. 95% 1.96

3. 97.5% 2.24

If students have memorized the critical values from the previous section, the first two Lesson Opener exercises will not take much time. The third exercise assesses whether the students recall the method of determining other critical values.

A scenario involving SAT math scores of incoming college freshmen is used throughout this section to illustrate the process of *hypothesis testing*, a process that determines which of two mutually exclusive statements about a population's parameter is best supported by a statistic from a sample. Note that the test cannot prove either statement to be true, but it can reject the truth of one of the statements with a given degree of certainty.

Present the SAT scores scenario and define the *null hypothesis* and the *alternative hypothesis*. In this section we focus on testing null hypotheses of the form $\mu = k$, $\mu \geq k$, or $\mu \leq k$ where k is a constant, such as H_0: $\mu = 663$ in the SAT math scores scenario. Explain that the null hypothesis utilizes the equality when z-scores are calculated using the assumption that $\mu = k$.

Example 1 illustrates defining the null and alternative hypotheses that would be used to test two other *claims*. Emphasize that the test can have only two results. We can either reject the null hypothesis or not reject it. Since the test for the SAT math scores does not provide enough evidence to prove the null hypothesis, we can only decide to not reject it. Consider explaining that researchers generally try to show that the statistic supports rejecting the null hypothesis, which in turn supports the validity of the mutually exclusive alternative hypothesis.

Reading and Writing Mathematics

Explain the logic of hypothesis testing. Then explain how the reasoning behind hypothesis testing applies to the US legal system.

The logic of hypothesis testing is similar to the logic used in indirect proofs. We assume the null hypothesis is true and then examine the evidence in the sample to see whether it reasonably contradicts our assumption. If so, the null hypothesis is rejected and the alternative (research) hypothesis is shown to be true. If the evidence does not reasonably show the null hypothesis to be false, we do not reject it, but we also cannot accept it as being true.

In the US legal system, the null hypothesis is that the defendant is innocent. When there is enough evidence to reasonably contradict this assumption, the null hypothesis is rejected, and when the defendant is found guilty, the alternative hypothesis is accepted. If there is not enough evidence, the person is acquitted but is not declared innocent.

Additional Exercises

Write the null and alternative hypotheses for each statement, and identify which hypothesis represents the claim.

1. The oxygen content for pollution control is at least 5 parts per million.
   $H_0: \mu \geq 5$ (claim); $H_a: \mu < 5$

In 2016 the US Department of Health and Human Services issued a rule requiring submission of all results from clinical trials of FDA-regulated drug, biological, and device products, regardless of whether results were statistically significant. This rule addressed the well-documented bias of underreporting negative results.

Write the null and alternative hypotheses for each statement, and identify which hypothesis represents the claim.

a. A gas station states its gasoline contains up to 10% ethanol.

b. A physical education teacher states that her students can do an average of more than 5 pull-ups.

Answer

	Null Hypotheses	Alternative Hypotheses	
a.	$H_0: \mu \leq 0.10$ (claim)	$H_a: \mu > 0.10$	Up to 10% of each gallon implies $\mu \leq 0.10$.
b.	$H_0: \mu \leq 5$	$H_a: \mu > 5$ (claim)	More than 5 implies $\mu > 5$. Since this claim does not involve an equality, it is the alternative hypothesis.

SKILL ✔ **EXERCISE 15**

Regardless of which hypothesis represents the claim, the hypothesis test begins by assuming the null hypothesis is true. This limits the logical outcome of the test to one of two decisions:

1. Reject the null hypothesis.
2. Do not reject the null hypothesis.

Since this decision is based on a limited sample, there is always a chance that the decision is wrong. You might reject a null hypothesis that is actually true or you might fail to reject a null hypothesis that is actually false.

> **DEFINITIONS**
>
> A **type I error** occurs when the null hypothesis is rejected, but it is true.
> A **type II error** occurs when the null hypothesis is not rejected, but it is false.

The table summarizes the four possible outcomes of a hypothesis test.

	True H_0	False H_0
Reject H_0.	type I error	correct decision
Do not reject H_0.	correct decision	type II error

Hypothesis testing is similar to the US legal system in that a defendant is assumed innocent (H_0) until there is enough evidence to show beyond reasonable doubt that this assumption is incorrect. Convicting the innocent is a type I error, and acquitting the guilty is a type II error. In statistics the problem being studied determines whether rejecting a true null hypothesis or accepting a false null hypothesis would be a more serious error.

Be sure students understand that there is always a chance that limited samples may cause the researcher to make the wrong conclusion. The analogy to the US legal system can help students understand the definitions of *type I* and *type II* errors. Consider having the students fill in the following chart.

	Actually Innocent	Actually Guilty
conviction	type I error	correct decision
acquittal	correct decision	type II error

The consequences of making an error depends on the claim that is being tested. For example, errors in medical studies could significantly affect many lives. Consider asking the students which type of error would be worse in the justice system. (*In most instances, a type I error would probably be considered worse, although a type II error in context of a very serious crime could jeopardize public safety.*)

Example 2 illustrates that errors in test conclusions can have significant consequences. Point out that costs and safety are commonly involved in our decision making, but our worldview also affects our conclusions. Contrast this with the consequences of an error in the hypothesis testing of the SAT scores.

While there is always some uncertainty in making a decision in statistics, more important decisions should have less uncertainty. To help avoid rejecting a true hypothesis, a *level of significance* is used. Explain the level of significance using the illustrated graphs with the *critical regions*. Emphasize that the null hypothesis is rejected if the *z*-score is within a critical region. Use the summary box to review the steps in hypothesis testing and apply these steps with Example 3.

Define *p-values* and discuss their use in hypothesis testing. Review how the normalcdf function can be used to find

An airbag manufacturer determines their airbags fail in at most 1% of crashes in which they are designed to deploy. Researchers examine a random sample of crashes to test whether the company's claim is accurate.

a. State the null and alternative hypotheses and identify which represents the claim.

b. State the possible type I and type II errors.

c. Which error would be more serious?

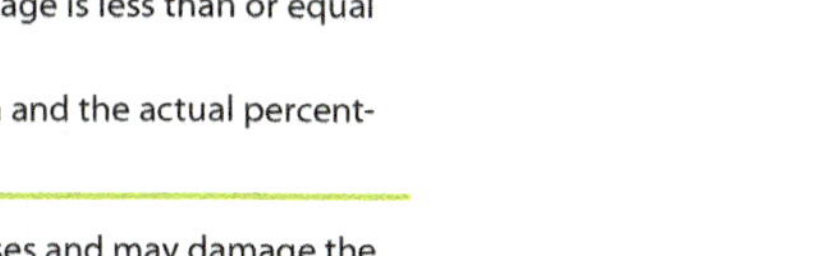

Answer

a. $H_0: \mu \leq 0.01$ (claim); $H_a: \mu > 0.01$

b. type I error: Researchers reject the company's claim, saying the percentage is actually more than 1% when the actual percentage is less than or equal to 1%.

  type II error: Researchers fail to reject the company's claim and the actual percentage of failure is greater than 1%.

c. A type I error may result in unnecessary recalls and extra expenses and may damage the company's reputation. A type II error may result in more serious injuries or fatalities during automobile crashes. A person's worldview, not mathematics, determines which set of consequences is more serious.

SKILL ✓ EXERCISE 17

In this section's introduction we saw that the mean SAT math score for the sample of freshmen was somewhat higher than the university's claimed mean of 663, but is it significantly higher? The probability of rejecting a true null hypothesis is lowered by using a lower level of significance.

> **DEFINITION**
>
> The **level of significance**, α, is the maximum allowable probability of making a type I error.

The level of significance is related to the desired confidence level for the test using $\alpha = 1 - c$. The level of significance is determined before data is collected and hypothesis testing is begun. Selecting $\alpha = 0.05$ implies there is a 5% chance that a true null hypothesis will be rejected (a type I error) and that there is a 95% chance that a correct decision will be made. Commonly used levels of significance are $\alpha = 0.10$, $\alpha = 0.05$, and $\alpha = 0.01$.

The *test statistic* is the statistic from a sample that is compared to a population parameter in the null hypothesis. This section focuses on using the sample mean, $\bar{x}$, to test claims about the population mean, μ. When the sample size $n \geq 30$ or the population is normally distributed, the Central Limit Theorem states that the sampling distribution of the means can be modeled by a normal distribution in which $\mu_{\bar{x}} = \mu$ and $\sigma_{\bar{x}} = \frac{\sigma}{\sqrt{n}}$.

The level of significance and the alternative hypothesis determine the *critical (or rejection) region*(s) where the differences between the test statistic and the proposed population parameter would be enough to justify rejecting the null hypothesis.

TIP 💡

The table lists the critical values, z_c, that define the critical regions for the commonly used levels of significance.

α	Tails	z_c
0.10	left	−1.28
	two	±1.645
	right	1.28
0.05	left	−1.645
	two	±1.96
	right	1.645
0.01	left	−2.33
	two	±2.576
	right	2.33

2. The average height of a certain plant after 5 weeks grown without fertilizer is 78.5 mm.
$H_0: \mu = 78.5$ (claim); $H_a: \mu \neq 78.5$

Label the error made as a type I or type II error.

3. A researcher rejects a claim that he should not have rejected. type I error

4. A researcher fails to reject a claim he should have rejected. type II error

5. A researcher fails to reject $H_0: \mu = 3$ with a given p-value of 0.04 at the 95% level of confidence. type II error

If 30 water samples taken downstream of Kurtwood Manufacturing yield a mean oxygen content of 4.9 parts per million and a standard deviation of 0.17, is the oxygen content significantly below the state standard of at least 5 parts per million? Should the state fine the company for breaking pollution laws? For stronger support, use $\alpha = 0.01$.

6. State the hypotheses and identify the claim. $H_0: \mu \geq 5$ (claim); $H_a: \mu < 5$

7. Establish the level of significance.
$\alpha = 0.01$

8. Calculate the z-score. $z = -3.23$

9. Calculate the p-value. $p = 0.0006$

10. Reject or fail to reject the null hypothesis.
Since $p < \alpha$, H_0 is rejected and H_a is accepted. The manufacturing company is in violation and should be fined.

Assignments

- **Minimum:** 1–6; 9–15 odd; 16, 23, 27, 31, 41, 43–45, 47, 50
- **Standard:** 1–10, 12, 15–16, 19–20, 22, 25–26; 29–33 odd; 42–46 even; 49–50
- **Extended:** 9, 11, 14, 16–18, 20–21, 24, 28–29, 32, 34–35, 38–39, 46, 48–50

Assessment

- Quiz 11B covers Sections 11.3–11.4.

the p-value and summarize the steps for this alternative method of testing the null hypothesis. Since the p-value in Example 4 is less than the level of significance, the null hypothesis is rejected. Your calculator's Z-Test function can be used to check the students' work. An online calculator may also be used. Use the Internet keyword search *hypothesis testing calculator*.

One-on-One A student may wonder why all the steps are needed when the Z-Test function is available. Performing each step reinforces the reasoning related to hypothesis testing.

Motivational Idea Use the Internet keyword search *hypothesis testing video* to find several short videos related to this section.

TIPS

Ex. 33–40 Be sure students understand that these exercises all utilize the z-score described on page 578.

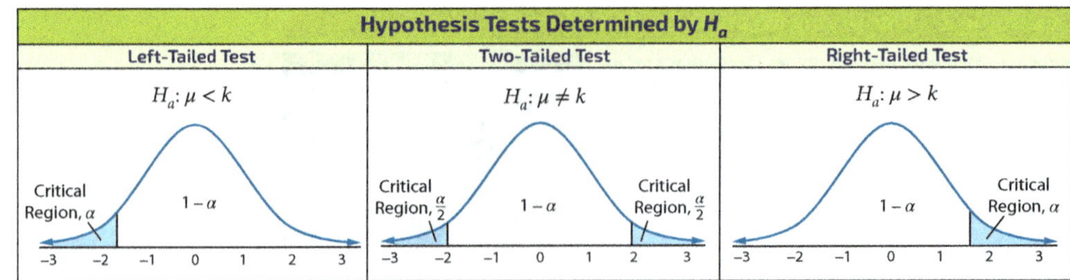

The standardized test statistic, $z = \dfrac{\bar{x} - \mu}{\sigma_{\bar{x}}}$, in which $\sigma_{\bar{x}} = \dfrac{s}{\sqrt{n}}$, states the difference between the sample mean and the hypothesized population mean in terms of standard deviations. The null hypothesis is rejected if this z-score falls in the critical region, and the alternative hypothesis is accepted. If the z-score does not fall in the critical region, the null hypothesis is not rejected.

If the researchers studying the SAT scores of incoming freshmen discussed in this section's introduction choose a level of significance of $\alpha = 0.10$, the critical regions for $H_a: \mu \neq 663$ would be defined by $|z| \geq 1.645$. Since the standardized test statistic $z \approx 1.49$ does not fall in either critical region, they would fail to reject the null hypothesis. The sample's mean did not provide enough evidence to allow the researchers to be 90% certain that the claim made by the null hypothesis was false.

Hypothesis Testing for Means
using critical regions

1. State the null and alternative hypotheses and identify the claim.
2. Establish the level of significance.
3. Determine the critical value(s) and region(s).
4. Calculate the standardized test statistic, $z = \dfrac{\bar{x} - \mu}{\sigma_{\bar{x}}}$.
5. Reject or fail to reject the null hypothesis.

Example 3 Hypothesis Testing with Critical Regions

Milltown Foods manufactures granola bars with a stated net weight of 31 g, assuming the average granola bar weighs at least 31 g. A random sample of 50 bars has a mean of 30.75 g and standard deviation of 0.75 g. Complete a hypothesis test for the company's assumption about the average net weight of its granola bars at a 0.05 level of significance.

Answer

$H_0: \mu \geq 31$ (claim): $H_a: \mu < 31$

1. State the null and alternative hypotheses and identify which represents the claim.

$\alpha = 0.05$

2. Note the chosen level of significance.

$\therefore z_c = -1.645$ and a critical region $z < -1.645$

3. Determine the critical value and critical region for the left-tailed test.

CONTINUED ➡

$n = 50$, $\bar{x} = 30.75$, $s = 0.75$

$$\sigma_{\bar{x}} = \frac{s}{\sqrt{n}} = \frac{0.75}{\sqrt{50}} \approx 0.106$$

$$z = \frac{\bar{x} - \mu}{\sigma_{\bar{x}}} = \frac{30.75 - 31}{0.106} \approx -2.357$$

Reject the claim that the average weight of the granola bars is at least 31 g.

Check

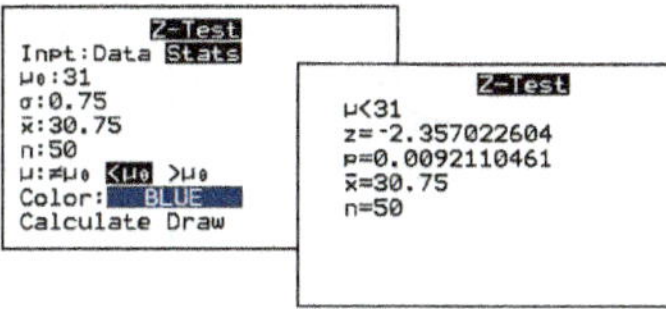

4. Since $n \geq 30$ and the Central Limit Theorem applies, calculate the standardized test statistic, $z = \dfrac{\bar{x} - \mu}{\frac{s}{\sqrt{n}}}$.

5. The standardized z-score is in the critical region.

The Z-Test function from the $\boxed{\text{STAT}}$, TESTS submenu can be used to verify the standardized z-score.

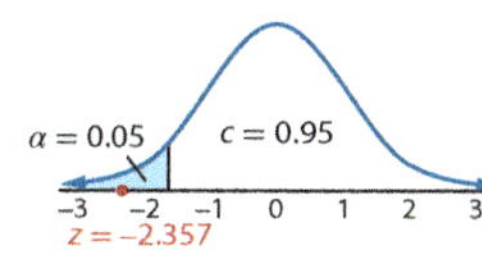

_______ SKILL ✔ EXERCISE 25

Notice that the calculator also reports a *p-value* of ≈ 0.009 for the sample mean in Example 3. *P*-values provide another way to decide whether to reject the null hypothesis.

DEFINITION

> The ***p*-value** is the probability of obtaining a test statistic as extreme as or more extreme than the one found in the sample assuming the null hypothesis is true.

The *p*-value in Example 3 can be calculated directly using the normalcdf function and the *z*-score, as shown in Section 11.2.

The *p*-value is calculated by finding the area under the standard normal curve in the tail(s) defined by the *z*-score (left, right, or in both depending on the inequality in the alternative hypothesis). Compare the *p*-value to the level of significance to decide whether to reject the null hypothesis. If $p \leq \alpha$, then there is a statistically significant difference between the sample statistic and the population parameter at the α level of significance.

If $p \leq \alpha$, then reject the null hypothesis.
If $p > \alpha$, then fail to reject the null hypothesis.

Since the *p*-value for the sample mean in Example 3 of $p \approx 0.009$ is lower than the established level of significance $\alpha = 0.05$, the null hypothesis is rejected. In general, if $p \leq \alpha$, there is a *statistically significant difference* between the sample statistic and the population parameter at the α level of significance. The *p*-value can indicate the strength of the evidence against the null hypothesis. The smaller the *p*-value from the sample data, the more confident the researcher is in the decision to reject H_0.

<table>
<tr><td colspan="1">Hypothesis Testing for Means
using p-values</td></tr>
</table>

1. State the null and alternative hypotheses and identify the claim.

2. Establish the level of significance.

3. Calculate the standardized test statistic, $z = \dfrac{\bar{x} - \mu}{\sigma_{\bar{x}}}$.

4. Calculate the *p*-value of the standardized test statistic.

5. Reject or fail to reject the null hypothesis.

According to the National Institutes of Health (NIH), the average level of high-density lipoprotein (HDL) cholesterol, the good cholesterol, is 50 mg/dL. A sample of 88 patients has a mean HDL cholesterol level of 45.77 mg/dL with a standard deviation of 16.40 mg/dL. If the level of significance is 0.05, is the mean of this sample significantly different from the recommended cholesterol level?

Answer

$H_0: \mu = 50; H_a: \mu \neq 50$

1. To prove a significant difference, use equality for the null hypothesis and not equal to for the alternative hypothesis.

$\alpha = 0.05$

2. Record the stated significance level.

$n = 88, \bar{x} = 45.77, s = 16.40$

$\sigma_{\bar{x}} = \dfrac{s}{\sqrt{n}} = \dfrac{16.40}{\sqrt{88}} \approx 1.748$

$z = \dfrac{\bar{x} - \mu}{\frac{s}{\sqrt{n}}} \approx \dfrac{45.77 - 50}{1.748} \approx -2.42$

3. Since $n \geq 30$ and the Central Limit Theorem applies, calculate the standardized test statistic, $z = \dfrac{\bar{x} - \mu}{\frac{s}{\sqrt{n}}}$.

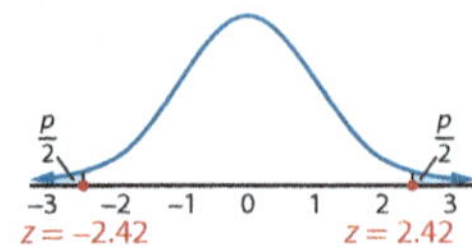

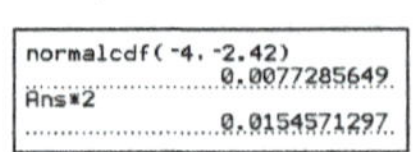

$p \approx 0.015$

4. Find the *p*-value in a two-tailed test. Entering normalcdf($-4, -2.42$) returns the area to the left of $z \approx -2.42$, which must be doubled to account for the area to the right of $z \approx 2.42$.

$p \approx 0.015 < 0.05 = \alpha$
Reject H_0, which implies accepting H_a.
The sample mean HDL is significantly different from 50 mg/dL at the 0.05 level of significance.

5. Compare p with α to make and interpret the decision.

Check

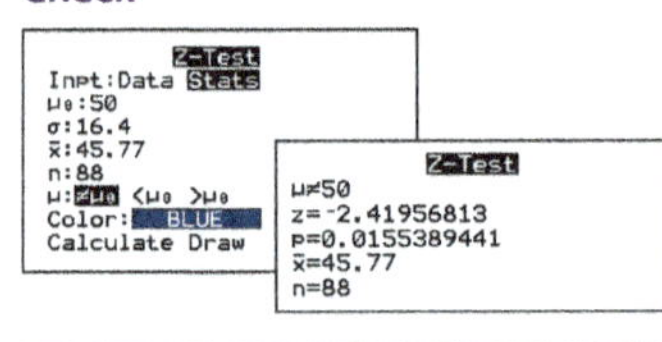

6. The Z-Test function from the STAT, TESTS submenu can be used to verify the *p*-value of the test statistic.

SKILL ✓ **EXERCISE 27**

Bias in design, sample selection, reporting, and even in publication is a real problem in medical research. In fact, physician and Stanford professor John Ioannidis asserted, "It can be proven that most claimed research findings are false."

Another type of statistical test determines whether the means of two different populations are significantly different. The null hypothesis assumes the population means are the same: $H_0: \mu_1 = \mu_2, H_a: \mu_1 \neq \mu_2$. The test statistic is the difference of the sample means and the z-score is found using $z = \dfrac{\bar{x}_1 - \bar{x}_2}{\sqrt{\frac{s_1^2}{n_1} + \frac{s_2^2}{n_2}}}$. This formula is used in exercises 33–40.

The logic of hypothesis testing is similar to the logic used in an indirect proof. Assume the null hypothesis value for the parameter, then compare the test statistic with the proposed value. If the sample statistic is then very unlikely (its z-score is in a critical region, or $p \leq \alpha$), the null hypothesis is rejected and the alternative hypothesis, its negation, must be true. However, if the test statistic is not that unusual (its z-score is not in a critical region, or $p > \alpha$), the null hypothesis is not rejected since there is not enough evidence from the sample to show that the null hypothesis is false. This is not the same as accepting H_0; we only know that we fail to reject H_0. We need to accurately interpret and communicate the results of our statistical tests so that the results can be used to God's glory and to the benefit of others.

Solutions

▶ A. Exercises

14. $60.2 \times 10^6 \dfrac{\text{candies}}{\text{day}} \left(\dfrac{1 \text{ day}}{24 \text{ hr}} \right) \left(\dfrac{1 \text{ hr}}{60 \text{ min}} \right)$
$\approx 41{,}800$ candies/min
Since $p > \alpha$, they cannot reject H_0.

15. right-tailed test
$z = \dfrac{\bar{x} - \mu_{\bar{x}}}{\frac{s}{\sqrt{n}}} = \dfrac{48 - 42}{\frac{12.4}{\sqrt{30}}} \approx \dfrac{6}{2.263}$
≈ 2.650

16. two-tailed test
$z = \dfrac{\bar{x} - \mu_{\bar{x}}}{\frac{s}{\sqrt{n}}} = \dfrac{0.55 - 0.61}{\frac{0.09}{\sqrt{34}}} \approx \dfrac{-0.06}{0.0154}$
≈ -3.887

17. left-tailed test
$z = \dfrac{\bar{x} - \mu_{\bar{x}}}{\frac{s}{\sqrt{n}}} = \dfrac{78 - 82}{\frac{11.9}{\sqrt{45}}} \approx \dfrac{-4}{1.774}$
≈ -2.2548

18. $z = \dfrac{\bar{x} - \mu_{\bar{x}}}{\frac{s}{\sqrt{n}}} \approx \dfrac{73 - 71}{\frac{5.91}{\sqrt{42}}} \approx \dfrac{2}{0.912} \approx 2.193$

19. $z = \dfrac{\bar{x} - \mu_{\bar{x}}}{\frac{s}{\sqrt{n}}} = \dfrac{2006 - 1843}{\frac{497}{\sqrt{58}}} \approx \dfrac{163}{65.26}$
≈ 2.498

▶ B. Exercises

23a. $H_0: \mu = 355$ (claim); $H_a: \mu \neq 355$

23b. $\alpha = 0.05$

23c. two-tailed test with $z_c \approx \pm 1.96$
critical regions: $z < -1.96$ and $z > 1.96$

23d. $z = \dfrac{\bar{x} - \mu_{\bar{x}}}{\frac{s}{\sqrt{n}}} = \dfrac{364 - 355}{\frac{38.3}{\sqrt{58}}} \approx \dfrac{9}{5.03}$
≈ 1.790

23e. Since z is not within a critical region, fail to reject H_0. Based on this sample, the average weight does not differ significantly from 355 lb at the 95% level of confidence. The feed company could use these statistics to support their claim.

A. Exercises

In exercises 1–5, match each term to its definition.

A. alternative hypothesis   **D.** null hypothesis

B. confidence level   **E.** type I error

C. level of significance   **F.** type II error

1. A true null hypothesis is rejected.   E

2. A false null hypothesis is not rejected.   F

3. a statement about a population parameter that involves an equality   D

4. a statement about a population that involves only an inequality   A

5. the maximum probability of making a type I error allowed by the researcher   C

6. The process of using statistics from a sample to examine a claim about the value of a population parameter is called <u> hypothesis testing </u>.

7. The statistic from a sample that is compared to the hypothetical population parameter is called a <u> test statistic </u>.

8. The probability of obtaining a statistic as extreme as or more extreme than the one found in the sample if the null hypothesis is true is the <u> p-value </u>.

9. Order the steps (I–V) in the hypothesis testing process when using p-values.

   a. Calculate the standardized test statistic (z-score).   III

   b. Calculate the p-value of the standardized test statistic.   IV

   c. Establish the level of significance.   II

   d. Reject or fail to reject the null hypothesis.   V

   e. State the hypotheses and identify the claim.   I

Write the null and alternative hypotheses for each statement, and identify which hypothesis represents the claim.

10. Jimmy claims that he can usually eat more than six slices of pizza. $H_0: \mu \leq 6; H_a: \mu > 6$ (claim)

11. A car manufacturer claims their new SUV's highway fuel economy is 35 mpg. $H_0: \mu = 35$ (claim); $H_a: \mu \neq 35$

Determine whether the decision is a type I error, a type II error, or a correct decision.

12. A researcher fails to reject $H_0: \mu = 187.3$ when in reality the population's mean is significantly less than 187.3.   type II error

13. After testing the hypothesis, the decision is made to reject $H_0: \mu = 42$. Later it is found that μ truly is 42.   type I error

15. right-tailed; $z \approx 2.65$

16. two-tailed; $z \approx -3.89$

17. left-tailed; $z \approx -2.25$

14. After choosing a 5% level of significance, a one-minute test of a company's 24-hr-a-day production line finds that approximately 42,000 candies came off the line and indicates a p-value of 0.38. The company maintains its claim that they produce 60.2 million candies each day.   correct decision

State whether the hypothesis test is a left-, right-, or two-tailed test. Then calculate the z-score of the test statistic.

15. $H_0: \mu \leq 42$, $H_a: \mu > 42$, $\bar{x} = 48$, $s = 12.4$, $n = 30$

16. $H_0: \mu = 0.61$, $H_a: \mu \neq 0.61$, $\bar{x} = 0.55$, $s = 0.09$, $n = 34$

17. $H_0: \mu \geq 82$, $H_a: \mu < 82$, $\bar{x} = 78$, $s = 11.9$, $n = 45$

Use technology to calculate the p-value.

18. $H_0: \mu \leq 71$, $H_a: \mu > 71$
$\bar{x} = 73$, $s = 5.91$, $n = 42$ $p \approx 0.014$

19. $H_0: \mu = 1843$, $H_a: \mu \neq 1843$
$\bar{x} = 2006$, $s = 497$, $n = 58$ $p \approx 0.012$

State the decision that should be made for the given level of significance and p-value of the test statistic.

20. $\alpha = 0.01$, $p = 0.20$ Fail to reject H_0.

21. $\alpha = 0.05$, $p = 0.006$ $p < \alpha$; Reject H_0.

22. $\alpha = 0.001$, $p = 0.0005$ $p < \alpha$; Reject H_0.

B. Exercises

In exercises 23–25, complete a five-step hypothesis test using critical regions.

   a. State the hypotheses and identify the claim.

   b. Establish the level of significance.

   c. Determine the critical value(s) and region(s).

   d. Calculate the standardized test statistic.

   e. Make a decision and interpret it relative to the problem.

23. A feed company claims their llama food keeps animals healthy and trim at an average weight of 355 lb. A sample of 58 adult llamas that eat this food have an average weight of  364 lb with a standard deviation of 38.3 lb. Can the company use these statistics to support their claim at the 95% confidence level?

24. An airline calculates that a certain flight must have at least 53 passengers to be profitable. After 60 flights, it has an average of 51 passengers with a standard deviation of 9. Is this sufficient evidence at a 95% confidence level to imply that the flight is unprofitable and should be discontinued?

24a. $H_0: \mu \geq 53$ (claim); $H_a: \mu < 53$

24b. $\alpha = 0.05$

24c. left-tailed test with $z_c \approx -1.645$
critical region: $z < -1.645$

24d. $z = \dfrac{\bar{x} - \mu_{\bar{x}}}{\frac{s}{\sqrt{n}}} = \dfrac{51 - 53}{\frac{9}{\sqrt{60}}} \approx \dfrac{-2}{1.16}$
≈ -1.721

24e. Since z is in the critical region, reject H_0 and accept H_a. According to this sample, the average number of passengers per flight is significantly less than 53 at the 95% level of confidence. The airline should discontinue this flight.

25a. $H_0: \mu \leq 128$; $H_a: \mu > 128$ (claim)

25b. $\alpha = 0.05$

25c. right-tailed test with $z_c \approx 1.645$
critical region: $z > 1.645$

25d. $z = \dfrac{\bar{x} - \mu_{\bar{x}}}{\frac{s}{\sqrt{n}}} = \dfrac{142 - 128}{\frac{38.7}{\sqrt{37}}} \approx \dfrac{14}{6.36} \approx 2.20$

25e. Since z is in the critical region, reject H_0 and accept H_a. According to this sample, there was a significant increase in texting by young adults between 2016 and 2018 at the 0.05 level of significance.

26a. $H_0: \mu \leq 500$; $H_a: \mu > 500$ (claim)

26b. $\alpha = 0.02$

26c. $z = \dfrac{\bar{x} - \mu_{\bar{x}}}{\frac{s}{\sqrt{n}}} = \dfrac{537 - 500}{\frac{100}{\sqrt{47}}} \approx \dfrac{37}{14.59}$
≈ 2.537

26d. right-tailed test

```
normalcdf(2.536592202,4)
                0.005565199
```

$p \approx 0.006$

26e. Since $p < \alpha$, reject H_0 and accept H_a. Based on this sample, the seniors at Cloverdale Christian High School scored significantly higher than the national average at a 98% confidence level.

27a. $H_0: \mu \geq 90$; $H_a: \mu < 90$ (claim)

27b. $\alpha = 0.05$

27c. $z = \dfrac{\bar{x} - \mu_{\bar{x}}}{\frac{s}{\sqrt{n}}} = \dfrac{88 - 90}{\frac{7.9}{\sqrt{52}}} \approx \dfrac{-2}{1.10}$
≈ -1.826

27d. left-tailed test

```
normalcdf(-4,-1.825595583)
                0.0339238568
```

$p \approx 0.03$

27e. Since $p < \alpha$, reject H_0 and accept H_a. The depression inventory average score was significantly lower after 12 weeks of exercise at the 95% confidence level, based on this sample.

28a. $H_0: \mu \geq 2500$; $H_a: \mu < 2500$ (claim)

28b. $\alpha = 0.10$

28c. $z = \dfrac{\bar{x} - \mu_{\bar{x}}}{\frac{s}{\sqrt{n}}} = \dfrac{2462 - 2500}{\frac{426}{\sqrt{41}}} \approx \dfrac{-38}{66.53}$
≈ -0.571

28d. left-tailed test

```
normalcdf(-4,-.5711707066)
                0.2839102387
```

$p \approx 0.28$

28e. Since $p > \alpha$, fail to reject H_0. This sample of 41 rural teens does not spend significantly less than the average of \$2500/yr on purchases at the 90% confidence level.

29. using critical regions:

$H_0: \mu \geq 8.75$; $H_a: \mu < 8.75$ (claim)

$\alpha = 0.05$; left-tailed test with

$z_c \approx -1.645$

critical region: $z < -1.645$

$$z = \frac{\overline{x} - \mu_{\overline{x}}}{\frac{s}{\sqrt{n}}} = \frac{7.9 - 8.75}{\frac{2.3}{\sqrt{38}}} \approx \frac{-0.85}{0.373}$$

$$\approx -2.28$$

Since z is in the critical region, reject H_0 and accept H_a. This sample of 38 teenagers took vitamin C fewer days than the national average at the 93% confidence level.

30. using p-values:

$H_0: \mu = 128.1$; $H_a: \mu \neq 128.1$ (claim)

$\alpha = 0.05$

$$z = \frac{\overline{x} - \mu_{\overline{x}}}{\frac{s}{\sqrt{n}}} = \frac{108.5 - 128.1}{\frac{52.6}{\sqrt{18}}} \approx \frac{-19.6}{12.40}$$

$$\approx -1.58$$

two-tailed test

```
normalcdf(-4,-1.58)
               0.057021751
Ans*2
               0.1140435019
```

Since $p \approx 0.11 > \alpha$, fail to reject H_0. The plants treated with fertilizer did not grow to a significantly different height after five weeks at a 95% level of confidence.

31. $E = 1.96 \left(\frac{3.7}{\sqrt{30}} \right) \approx 1.32$

CI: $58.6 \pm 1.3 = (57.3, 59.9)$

32. $H_0: \mu = 60$; $H_a: \mu \neq 60$; $\alpha = 0.05$

two-tailed test with $z_c \approx \pm 1.96$

critical regions:

$z < -1.96$ and $z > 1.96$

$$z = \frac{\overline{x} - \mu_{\overline{x}}}{\frac{s}{\sqrt{n}}} = \frac{58.6 - 60}{\frac{3.7}{\sqrt{30}}} \approx \frac{-1.4}{0.68} \approx -2.06$$

Since z falls in a critical region, reject H_0 and accept H_a.

or

```
normalcdf(-4,-2.06)
               0.0196675174
Ans*2
               0.0393350349
```

Since $p \approx 0.039 < \alpha$, reject H_0 and accept H_a.

33. $z = \dfrac{\overline{x}_1 - \overline{x}_2}{\sqrt{\frac{s_1^2}{n_1} + \frac{s_2^2}{n_2}}} = \dfrac{72.3 - 75.8}{\sqrt{\frac{4.7^2}{41} + \frac{8.2^2}{38}}} \approx \dfrac{-3.5}{1.52}$

≈ -2.30

34. $z = \dfrac{492 - 483}{\sqrt{\frac{17.8^2}{44} + \frac{9.5^2}{52}}} \approx \dfrac{9}{2.99} \approx 3.01$

---

25. In 2016 Experian Marketing Services reported that young adults send and receive an average of 128 text messages per day. In 2018 a random sample of 37 young adults averaged 142 texts with a standard deviation of 38.7. Can the claim be made at a 95% level of confidence that there was a significant increase in the number of text messages sent and received by young adults from 2016 to 2018?

Conduct a five-step hypothesis test using p-values.
  a. State the hypotheses and identify the claim.
  b. Establish the level of significance.
  c. Calculate the standardized test statistic.
  d. Calculate the p-value.
  e. Make a decision and interpret it relative to the problem.

26. The average SAT score of the 47 seniors at Cloverdale Christian High School was 537 with a standard deviation of 100. Can they claim at the 98% confidence level that they exceed the national mean of 500?

27. The average score on a depression inventory for young adults is 90. After a sample of 52 young adults participated in an aerobic exercise routine for 20–30 min each day for 12 weeks, the average score improved to 88 with a standard deviation of 7.9. Does this research support, at the 95% level of confidence, the claim that daily aerobic exercise can improve depression?

28. According to a 2017 survey, teenagers spend about $2500 each year. A small-town survey of 41 randomly selected teens resulted in a mean of $2462 spent each year with a standard deviation of $426. Does this data support the claim that rural teens spend significantly less money per year at the 90% confidence level?

Conduct a hypothesis test to examine each claim.

29. The average duration of the common cold for a sample of 38 teenagers who took vitamin C daily after being diagnosed was 7.9 days with a standard deviation of 2.3 days. Can it be claimed that at the 95% level of confidence the sample's mean was less than the national average duration of 8.75 days?

30. Students testing a plant fertilizer find that after 5 weeks their 18 plants have a mean height of 108.5 mm with a standard deviation of 52.6 mm. Research journals report that after 5 weeks similar plants without the fertilizer have normally distributed heights with a mean of 128.1 mm. Did the fertilizer make a significant difference in the plants' growth at the 95% confidence level?

Complete exercises 31–32 using a sample with $n = 30$, $\overline{x} = 58.6$, and $s = 3.7$ to explore the relationship between confidence intervals and hypothesis testing.

31. Find the 95% confidence interval for the mean. Would a hypothesized mean of $\mu = 60$ be in the confidence interval? $\approx (57.3, 59.9)$; no

32. Complete a hypothesis test with $\alpha = 0.05$ to decide if $H_0: \mu = 60$ should be rejected. yes

Use $z = \dfrac{\overline{x}_1 - \overline{x}_2}{\sqrt{\frac{s_1^2}{n_1} + \frac{s_2^2}{n_2}}}$ to determine the z-score used when comparing the following samples from two different populations.

33. $z \approx -2.30$

	Sample 1	Sample 2
$\overline{x}$	72.3	75.8
s	4.7	8.2
n	41	38

34. $z \approx 3.01$

	Sample 1	Sample 2
$\overline{x}$	492	483
s	17.8	9.5
n	44	52

Determine whether the means of the samples from different populations are different at the given level of significance.

35. $\alpha = 0.05$ yes

	Sample 1	Sample 2
$\overline{x}$	16.9	18.4
s	1.69	3.18
n	30	30

36. $\alpha = 0.10$ no

	Sample 1	Sample 2
$\overline{x}$	56.01	55.9
s	3.6	2.1
n	100	50

> C. Exercises

Conduct a hypothesis test to determine whether the sample means from different populations are significantly different.

37. Mrs. D's class of 33 students had a mean score of 85 with a standard deviation of 6, while Mr. C's class of 37 students had a mean of 79 with a standard deviation of 10. Are the means statistically different at the 0.05 level?

---

35. $z = \dfrac{16.9 - 18.4}{\sqrt{\frac{1.69^2}{30} + \frac{3.18^2}{30}}} \approx \dfrac{-1.5}{0.66} \approx -2.28$

two-tailed test

```
normalcdf(-4,-2.28)
               0.011272125
Ans*2
               0.0225442501
```

Since $p \approx 0.02 < \alpha$, reject H_0 and accept H_a.

36. $z = \dfrac{56.01 - 55.9}{\sqrt{\frac{3.6^2}{100} + \frac{2.1^2}{50}}} \approx \dfrac{0.11}{0.47} \approx 0.24$

two-tailed test

```
normalcdf(-4,-0.24)
               0.4051334889
Ans*2
               0.8102669779
```

Since $p \approx 0.81 > \alpha$, fail to reject H_0.

> C. Exercises

37. $H_0: \mu_1 = \mu_2$; $H_a: \mu_1 \neq \mu_2$; $\alpha = 0.05$

$$z = \frac{79 - 85}{\sqrt{\frac{10^2}{37} + \frac{6^2}{33}}} \approx -3.08$$

two-tailed test

```
normalcdf(-4,-3.08)
               0.0010033854
Ans*2
               0.0020067709
```

Since $p \approx 0.002 < \alpha$, reject H_0 and accept H_a. The means of the two classes are statistically different at the 0.05 level.

38. Sharon and Jodi took a word processing test involving 30 samples. Sharon averaged 94 with a standard deviation of 5 and Jodi averaged 96 with a standard deviation of 6. Test for a significant difference using $\alpha = 0.025$.

39. A 30-day contest included 2 groups of 45 dieters. The first group lost an average of 10.3 lb with a standard deviation of 4.2 lb, while the second group lost an average of 8.2 lb with a standard deviation of 5.9 lb. Does the evidence suggest that the difference in effectiveness between the two diet groups is significant at the 95% level?

40. At Calvary Christian School the average SAT math score for 35 male students is 521.9 with a standard deviation of 4.597, and for 41 female students the mean is 518.7 with a standard deviation of 7.312. Is there a significant difference ($\alpha = 0.05$) in the SAT math scores of the male and female students?

CUMULATIVE REVIEW

Expand each logarithmic expression. Assume all variables are positive values. [3.3]

41. $\log_5 \dfrac{25\sqrt{y}}{x^2}$

42. $\log_3 \dfrac{\sqrt{3x}}{81y}$

Use the appropriate counting technique. [10.1]

43. How many different ways can five activities be chosen out of eight? 56

44. How many different ways can the letters in *MARSHMALLOW* be arranged? 4,989,600

Find each confidence interval from the given sample statistics. [11.3]

45. 95% CI, $n = 37$, $\bar{x} = 84$, $s = 9.7$ $\approx (80.9, 87.1)$

46. 90% CI, $n = 109$, $\bar{x} = 39$, $s = 5.3$ $\approx (38.2, 39.8)$

47. Which function represents $f(x) = x^2$ translated 3 units left and vertically shrunk by a factor of $\frac{1}{4}$? [1.5] B

A. $g(x) = \frac{1}{4}(x^2 + 3)$

B. $g(x) = \frac{1}{4}(x + 3)^2$

C. $g(x) = \frac{1}{4}(x^2 - 3)$

D. $g(x) = \frac{1}{4}(x - 3)^2$

E. $g(x) = 4(x + 3)^2$

41. $\frac{1}{2}\log_5 y - 2\log_5 x + 2$

42. $\frac{1}{2}\log_3 x - \log_3 y - \frac{7}{2}$

48. Identify the angle (to the nearest degree) between $\mathbf{u} = \langle 3, 4 \rangle$ and $\mathbf{v} = \langle -5, -2 \rangle$. [6.2] A

A. 149° **C.** 90° **E.** none of these

B. 119° **D.** 31°

49. Write a simplified explicit formula for a_n in 5, 8, 11, 14, 17, … and find the value of the tenth term. [9.1] A

A. $a_n = 2 + 3n$; $a_{10} = 32$

B. $a_n = a_{n-1} + 3$; $a_{10} = 29$

C. $a_n = 5 + (n - 1)3$; $a_{10} = 32$

D. $a_{n+1} = a_n + 3$; $a_{10} = 29$

E. none of these

50. If 26 people are randomly selected from a normal population having a mean of 97 and a standard deviation of 7.9, compute the probability that the sample mean is between 98 and 100. [11.2] D

A. 97% **C.** 26% **E.** 2.6%

B. 74% **D.** 23%

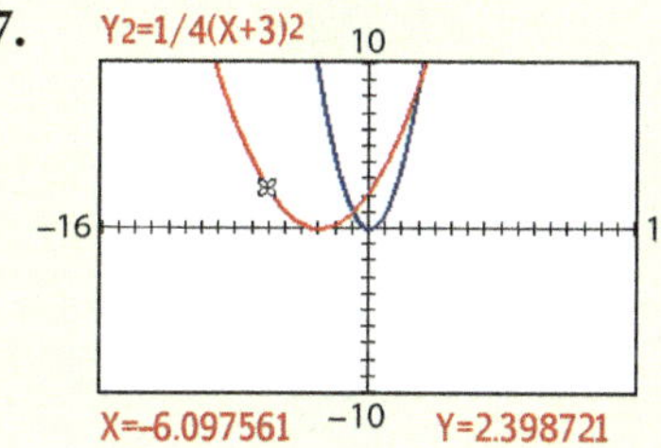

Since $p \approx 0.02 < \alpha$, reject H_0 and accept H_a. There is a significant difference at the 0.05 level of significance between the SAT scores of the male and female students.

› Cumulative Review

41. $\log_5 25 + \log_5 \sqrt{y} - \log_5 x^2$
$= \frac{1}{2}\log_5 y - 2\log_5 x + 2$

42. $\log_3 (3x)^{\frac{1}{2}} - \log_3 3^4 y$
$= \frac{1}{2}(\log_3 3 + \log_3 x)$
$\qquad\qquad - (\log_3 3^4 + \log_3 y)$
$= \frac{1}{2} + \frac{1}{2}\log_3 x - 4 - \log_3 y$
$= \frac{1}{2}\log_3 x - \log_3 y - \frac{7}{2}$

43. $_8C_5 = \dfrac{8!}{(8-5)!5!} = \dfrac{8 \cdot 7 \cdot 6 \cdot 5!}{3 \cdot 2 \cdot 1 \cdot 5!} = 56$

44. $\dfrac{11!}{2!2!2!} = 4{,}989{,}600$

45. $E = 1.96\left(\dfrac{9.7}{\sqrt{37}}\right) \approx 3.1$
CI: $84 \pm 3.13 \approx (80.9, 87.1)$

46. $E = 1.645\left(\dfrac{5.3}{\sqrt{109}}\right) \approx 0.8$
CI: $39 \pm 0.8 = (38.2, 39.8)$

47.

48. $\mathbf{u} \cdot \mathbf{v} = 3(-5) + 4(-2) = -23$
$|\mathbf{u}| = \sqrt{3^2 + 4^2} = 5$
$|\mathbf{v}| = \sqrt{(-5)^2 + (-2)^2} = \sqrt{29}$
$\cos \theta = \dfrac{-23}{5\sqrt{29}}$
$\theta = \text{Cos}^{-1}\left(-\dfrac{23}{5\sqrt{29}}\right) \approx 149°$

49. $a_1 = 2 + 3(1) = 5$; $a_2 = 2 + 3(2) = 8$
$a_3 = 2 + 3(3) = 11$
$a_4 = 2 + 3(4) = 14$
$a_n = 2 + 3n$; $a_{10} = 2 + 3(10) = 32$

50. $\mu_{\bar{x}} = 97$; $\sigma_{\bar{x}} = \dfrac{7.9}{\sqrt{26}} \approx 1.55$
$z_L = \dfrac{98 - 97}{1.55} \approx 0.65$
$z_U = \dfrac{100 - 97}{1.55} \approx 1.94$
$P(98 < \bar{x} < 100)$
$= P(0.65 < z < 1.94)$
normalcdf(0.65, 1.94) ≈ 0.233

---

38. $H_0: \mu_1 = \mu_2$; $H_a: \mu_1 \neq \mu_2$; $\alpha = 0.025$
$z = \dfrac{94 - 96}{\sqrt{\frac{5^2}{30} + \frac{6^2}{30}}} \approx -1.40$

two-tailed test

Since $p \approx 0.16 > \alpha$, fail to reject H_0. There is no statistical difference in the word processing skills between the two candidates at the 0.025 significance level.

39. $H_0: \mu_1 = \mu_2$; $H_a: \mu_1 \neq \mu_2$; $\alpha = 0.05$
$z = \dfrac{10.3 - 8.2}{\sqrt{\frac{4.2^2}{45} + \frac{5.9^2}{45}}} \approx 1.95$

two-tailed test

Since $p \approx 0.051 > \alpha$, fail to reject H_0. There is no significant difference in the weight lost between the two groups at the 95% confidence level.

40. $H_0: \mu_1 = \mu_2$; $H_a: \mu_1 \neq \mu_2$; $\alpha = 0.05$
$z = \dfrac{521.9 - 518.7}{\sqrt{\frac{4.597^2}{35} + \frac{7.312^2}{41}}} \approx 2.32$

two-tailed test

Biblical Perspective of Mathematics

Objectives

1. To name and describe three common modern philosophies of mathematics

2. To identify the key founders of and major flaws in logicism, intuitionism, and formalism

3. To contrast logicism, intuitionism, and formalism with a Christian worldview of mathematics

Assignments

- **Minimum:** 1–7; 8 (in class)
- **Standard:** 1–7; 8 (in class)
- **Extended:** 1–8

Logicism, Intuitionism, and Formalism

To the biblical Christian, Gödel's results confirmed the truth that Scripture had always proclaimed. Autonomy belongs to the biblical God alone. Whenever man tries to construct any system of thought without reference to this God, it will ultimately fall short.

—James Nickel (Christian mathematician and author)

After contradictions arose in mathematics in the nineteenth century, many mathematicians viewed Platonism and its ideal realm of abstract ideas with skepticism. Scientists and some traditional mathematicians who saw mathematics as a branch of science turned to the philosophy of empiricism and its reliance on the senses and experience. In the early twentieth century, other mathematicians attempted to repair the cracks in the foundations of math as they developed three new philosophical schools: logicism, intuitionism, and formalism.

In 1910–13, British mathematicians Bertrand Russell and Alfred North Whitehead produced the three-volume work *Principia Mathematica*, which presented classical mathematics as a branch of logic. Logicism seeks to deduce all of mathematics from a few simple, undeniable axioms using valid logical steps. In an effort to reconcile such an attempt by the German philosopher Gottlob Frege, Russell discovered troubling paradoxes concerning infinite sets. Like Frege, Russell and Whitehead were also unsuccessful in developing mathematics exclusively from logic axioms. Two of their nine set theory axioms, the axiom of infinity and the axiom of choice, were not axioms of logic. Instead, they were accepted without proof based on mathematicians' daily experience with sets. Thus, modern mathematical logic is one of the branches of math, rather than math being considered a branch of logic.

Intuitionism, originated by the Dutch mathematician and philosopher L. E. J. Brouwer, views mathematics as an activity of mental construction. After observing the contradictions in classical mathematics, intuitionists attempted to rebuild mathematics from the ground up, beginning with the natural numbers. Intuitionists, like the ancient Greeks, accept that the set of natural numbers is potentially infinite since there is no largest natural number, but would reject the natural numbers as actually infinite, since the set does not exist as one finished set. Intuitionists also reject non-constructive indirect proofs as a demonstration of existence since these proofs do not provide a means of generating an example. Math historian Howard Eves concludes, "So far, intuitionist mathematics has turned out to be considerably less powerful than classical mathematics, and in many ways it is much more complicated to develop. This is the fault found with the intuitionist approach—too much that is dear to most mathematicians is sacrificed."

The German mathematician David Hilbert was instrumental in developing formalism. Formalists say that mathematics is its own foundation rather than human intuition, nature, or logic. Math is simply a game in which one manipulates symbols according to well-defined rules. It has and needs no other basis other than itself. Hilbert's lofty goal was to produce a complete and consistent axiomatization of all of mathematics. However, mathematical logician Kurt Gödel proved in 1931 that for any robust mathematical system, deriving all true mathematics from a list of axioms is impossible. Hilbert's goal was crushed, but the formalist school of extreme axiomatization survives.

These three new schools of mathematics believe that man alone, with no need for God, can provide reliability

PRESENTATION

Explain that the prevailing mathematical philosophy for thousands of years was the Greek philosophy of Platonism, which corresponds in many ways to the modern general philosophy of idealism. The modern mathematical school of logicism is compatible with both Platonism and idealism, but it failed to reduce mathematics to logic.

Intuitionism contrasts with logicism in that it requires any abstract entities to be constructed in the human mind, a philosophy sometimes referred to as conceptualism. Point out the parts of classical mathematics such as actual infinity and indirect proofs that are not acceptable to the intuitionist. On pages 157–58 of his book *Redeeming Mathematics*, Vern Poythress presents an interesting discussion of the limitations of intuitionism in declaring propositions as true or false, since in doing so the human mind would take precedence over God's knowledge. Intuitionism failed to develop much of the body of classical mathematical truth.

Formalism does not depend on an ideal realm as logicism does, and it does not depend on the human mind for concepts as intuitionism does. But since formalism is like a game that is subject only to its own rules, it does not lead to practical applications evident in the world around us. Kurt Gödel's work showed that David Hilbert's goal of completeness and consistency in mathematics was impossible.

Read the introductory quote by James Nickel and 1 Corinthians 3:11, pointing out that logicism, intuitionism, and formalism are each incompatible with a biblical worldview of mathematics.

and consistency to the foundation of an orderly mathematical system. A biblical worldview differs from all these approaches. It acknowledges that we need our created universe (empirical input), logic (axiomatic development), the human mind (intuitions), and the symbols and rules of language (formal structure). But it attributes God as the source of all these foundational elements rather than any of these being the source itself. Colossians 1:15–17 states that Jesus Christ created all things, both visible and invisible, and that by Him all things consist. Any humanistic philosophy for the foundations of mathematics is doomed to failure.

❯ Exercises

1. Match each mathematical philosophy with its view of math.

   I. empiricism   **IV.** logicism

   II. formalisim   **V.** Platonism

   III. intuitionism

   a. Math is a game of symbols played according to arbitrary rules. II

   b. Math exists in an ideal realm of ideas independent of human minds. V

   c. Math is part of symbolic logic rather than logic being a subset of math. IV

   d. Math is similar to science in its reliance on senses and experience. I

   e. Math comes from and is dependent on the mind of man. III

2. Name the mathematical philosophy associated with each mathematician.

   a. L. E. J. Brouwer intuitionism

   b. David Hilbert formalism

   c. Bertrand Russell logicism

3. Name the mathematical philosophy that is challenged by each key problem.

   a. Gödel's proof that completeness with consistency is impossible formalism

   b. much of classical mathematics sacrificed intuitionism

   c. two of its necessary axioms not the type required logicism

4. Which modern mathematical philosophy starts with the natural numbers? intuitionism

5. Which modern mathematical philosophy views math as its own foundation? formalism

6. List two key mathematical concepts rejected by logicism. actual infinity and indirect proofs

7. According to Colossians 1:15–17, who is the only sure foundation for a man's philosophy? Jesus Christ

8. **Discuss:** Explain how a Christian philosophy of mathematics differs from the three new philosophies of the early twentieth century.

   sample answer: A Christian philosophy of mathematics acknowledges the role of the human mind and the axiomatic nature and formal structures of mathematics as derivatives of the character of God, while the new philosophies deify one of these foundational elements and ignore God as their source.

11.5 Research Studies

Objectives

1. To describe the process of making a statistical inference

2. To compare and contrast sample surveys, experimental studies, and observational studies

3. To identify sample types and evaluate possible biases

4. To evaluate strengths and weaknesses of a statistical experiment and its reported results

5. To design and conduct a statistical experiment to answer a research question

Flash

Research and data analysis are powerful tools in decision-making, but data and conclusions can be adjusted, compromised, and reported in various ways. Sometimes researchers adjust their data to support a desired conclusion. Encourage students to strive to always be honest and upright (Prov. 16:8, 21:21).

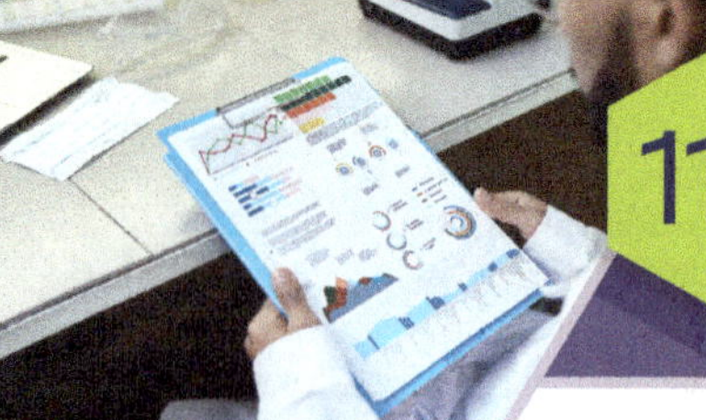

After completing this section, you will be able to

- describe the process of making a statistical inference.
- compare and contrast surveys, observational studies, and experimental studies.
- identify the type of sample used and evaluate possible biases.
- evaluate strengths and weaknesses of a research study and its reported results.
- design and conduct a study to answer a research question.

Game theory, a branch of mathematics involving the analysis of strategies in competitive decision-making, had its modern genesis in the 1944 book *Theory of Games and Economic Behavior* coauthored by mathematician John von Neumann and economist Oskar Morgenstern.

Mikayla spends a lot of time preparing for her church's Bible quiz competitions. After hearing that teens spend an average of 18.9 hr/week viewing social media, she decides to research whether this statistic is true for the teens in her national quizzing association. Considering there are approximately 2700 teenagers in this organization, how would she go about investigating this question?

Any research project begins with a compelling question or concern. A research question that clearly identifies the population and the exact characteristics (parameters) to be studied is stated. The null and alternative (research) hypotheses are then stated and the level of significance is determined.

The design of the study depends on the study's goal and the methods of data collection. The goal might be to compare parameters of two different groups, to determine whether significant relationships exist between variables, or to predict values of parameters based on other variables in the study. The sampling techniques are extremely important to the outcome of the research.

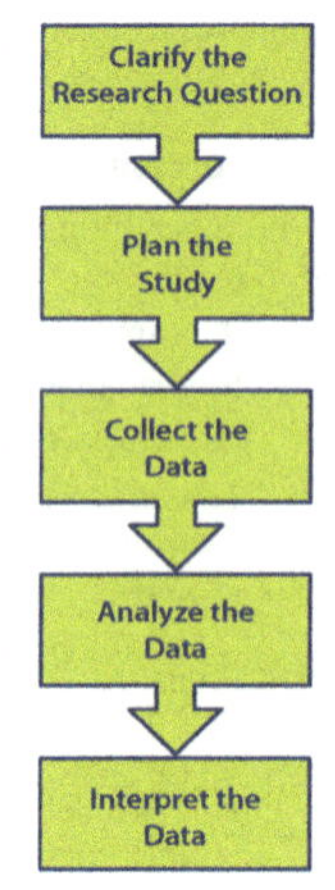

RESEARCH DESIGNS

experimental	Data is created by recording the results of treating an experimental group differently from a control group.
observational	Data is collected without the sample being treated by the researcher.
survey	Data is selected from responses given by members of the sample.

Experimental studies seek to establish a *cause and effect* relationship by comparing the effect of a specific treatment on a randomly selected experimental group to the effect of a placebo or no treatment on a similar, randomly selected control group. The cause and effect relationship can be difficult to define due to the many other factors, or *confounding variables*, that can influence the results.

Observational studies collect data without defining and applying a treatment to experimental and control groups. These studies often use preexisting data. The goal of observational studies is to determine a correlation between variables instead of a causation. A correlation between variables (one variable increases or decreases as the second variable increases) does not guarantee that a change in the first variable *causes* a change in the second variable (or vice versa). Another variable could actually cause changes in one or both variables being studied. Survey design is a form of observational design that gathers data using a survey of a sample from the population.

PRESENTATION

Lesson Opener

1. A manufacturer claims that a vehicle gets at least 35 mi to a gallon of gas. State the null and alternative hypotheses and identify the claim.
$H_0: \mu \geq 35$ (claim); $H_a: \mu < 35$

2. What type of error occurs when a false null hypothesis is not rejected?
type II error

3. Calculate the standardized test statistic for testing $H_0: \mu \geq 20$ if a random sample of 36 has a mean of 21 and a standard deviation of 2. $z = 3$

Inform students that a research study is often the key process in using data to make predictions and decisions. Decisions made from research studies often have significant financial or physical ramifications related to the health and safety of many people. A research study offers a measure of assurance that would not be present otherwise. Final decisions often involve both statistical and moral reasoning. Applying a biblical worldview to the mathematical analysis provides the means for making good decisions.

Introduce the steps involved in a research project and emphasize the need to have a clearly defined research question. Consider having the students expand the diagram to include the details involved in each step. Discuss the characteristics and purposes of each type of research design. Use Example 1 to demonstrate identifying the population and research design of a given study.

Since data collection is often a major undertaking involving significant amounts of time and money, the process must be carefully planned so that the sample represents the entire population and the data is *unbiased*.

The Data Analysis margin note provides an opportunity to discuss the role one's worldview plays in making decisions based on statistical data. Artificial intelligence, often based on statistical

Identify the population and research design for each study.

a. A study analyzed the composite ACT scores from 62 randomly selected high school seniors (31 boys and 31 girls) from schools in the Keystone Christian School Association to determine whether there was any significant difference between the girls' and boys' scores.

b. The Milwaukee Chamber of Commerce seeks to compare the commute time of workers in the Milwaukee metro area to the average commute time for US citizens of 25.5 min (as reported by the US Census Bureau). They analyzed commute-time data from questionnaires sent to 300 randomly selected workers in the metro area.

c. Before investing in a new reading curriculum, a school district randomly selects 5 first-grade classes to be taught with ABC Reading curriculum and 5 other first-grade classes to be taught with First Primer curriculum. The students' standardized test scores for reading comprehension and speed are compared at the end of the school year.

Answer

a. **population:** senior boys and girls in the school association

design: observational | Scores were reported from data that had been gathered without different treatments being applied.

b. **population:** workers in the Milwaukee metro area

design: survey | The questionnaire asks for information that will answer the question of concern.

c. **population:** first graders in the school district

design: experimental | Different treatments were applied to similar, randomly selected groups before data was collected and compared.

SKILL ✔ **EXERCISE 7**

After the research design is selected, a detailed plan for collecting data is developed. If the inferences made are to be reliable, the sample must be large enough ($n \geq 30$ if applying the Central Limit Theorem) and *unbiased* (representing every member of the population fairly). A sample is *biased* if a factor exists that could systematically cause the sample to differ from the population being studied in a way that would cause the population's parameter to be overestimated or underestimated. Bias may be introduced into a study by the type of sample used, the conduct of the researcher, or an insufficient sample size.

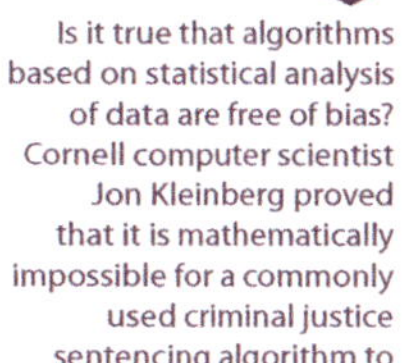

Vocabulary

abstract
biased sample
cause and effect
causes
confounding variables
experimental
observational
research design
researcher bias
sampling error
survey
unbiased sample

Reading and Writing Mathematics

Use the Internet to research and explain the meaning of each term related to statistical studies: lurking variable, retrospective study, and prospective study.

A lurking variable is a factor that affects the statistical experiment but has not been taken into account by the study. Lurking variables must be identified and eliminated or controlled within the study. An example would be if the texting while driving variable was ignored in a statistical study on the causes of motor vehicle accidents.

A retrospective study is based on existing data of situations in which the outcome has already been determined.

A prospective study is one in which ongoing data is generated and collected during the study, such as medical studies on the effects of new drugs on volunteers.

---

data, falls far short in making judgments in the criminal justice system. While mathematical algorithms are valuable tools, seeking justice requires more than data, statistics, and artificial intelligence. Human reasoning, moral values, and spiritual discernment are essential in making critical decisions.

Present the various types of samples, and distinguish between *sampling error* and *researcher bias*. Have the students identify each type of sample in Example 2 and discuss how the sample in part *a* could be *biased*. For example, those who brought lunches may be the first to leave because they do not have to stand in the tray return line, or students who eat more may leave later.

Explain that the students should then be able to apply the inferential statistics techniques from Chapters 10 and 11 to the collected data in order to test the hypotheses and make conclusions. The rest of this section focuses on analyzing the design and conclusions of other studies.

Interactive Activity Provide a handout of only the left column of Example 3 for a project or group discussion. Consider having several groups complete the analysis portion and then share the results with the rest of the class. Guide the students to any conclusions in the text that they omitted.

Example 4 also provides a good opportunity for a group activity. Consider providing a handout of the abstract and the 5 steps without the answers. For a quicker activity, have each of 5 groups analyze one of the steps.

TIPS

Ex. 37–38 These exercises could be used as class or small group projects. Be sure to communicate teacher expectations on the extent of the research study along with grading guidelines for the project.

Additional Exercises

Identify the population and research design for each study.

1. One hundred college football fans were asked to record their commute time to the game. *fans at the game; survey*

2. A Korean pizza parlor sells American-style pizzas and Korean-style pizzas. To determine which is more popular, the owner randomly records the third customer's choice and every tenth customer's choice thereafter. *local customer base; observational*

For Additional Exercises 3–4, identify the sampling technique and discuss any potential sampling errors or researcher bias.

3. The temperatures of 30 randomly selected babies from a medical clinic in Uganda were taken to determine the average body temperature of babies less than 2 yr old. *convenience; Sample error is likely since temperatures of those at the clinic will probably be higher than the rest of the population.*

4. The senior class is randomly chosen to participate in a survey to determine a new name for the school mascot. *cluster; Sample error is likely unless the sample's opinions closely match those of the rest of the student body.*

5. Find an abstract of a research study by performing an Internet search on a topic of your interest. Use the flow chart found at the beginning of Section 11.5 and your chosen abstract to describe each step of the research. *Answers will vary.*

Assignments

- **Minimum:** 1–14, 19–21; 25–29 odd; 33; 39–47 odd; 48
- **Standard:** 1–12, 14, 16, 18, 22, 24, 28, 30–33, 35, 40–41; 44–48 even
- **Extended:** 2–6; 8–12 even; 13–14, 16–17, 20, 23, 26, 29–35, 37, 40, 42, 46, 48

TYPES OF SAMPLES

random	Every member of the population has an equal chance of being selected.
stratified random	The population is divided into groups based on a characteristic and a random sample is taken from each group.
cluster	The population is divided into groups and then an entire group is randomly selected to be in the overall sample.
systematic	The first member is selected randomly from an ordered list of the population, and then every nth person on the list is selected.
convenience	The members are selected from the population based on accessibility.
self-selected	The members are volunteers from the population.

Random and systematic samples are usually unbiased. Convenience and self-selected samples should be avoided because of the likelihood of introducing bias. If the sample is not representative of the population, a *sampling error* has occurred. *Researcher bias* occurs when the researcher influences the respondents, either intentionally or unintentionally, by the way a question is worded or by nonverbal signals such as facial expressions during an interview. The design and execution of a research study should demonstrate precautions taken to avoid bias.

Example 2 Analyzing Sampling Techniques

Identify each type of sample and determine whether the sampling technique is likely to introduce bias.

a. The first 50 students leaving the cafeteria reported what they ate for lunch and whether they brought their lunch from home or purchased a school lunch. Researchers then calculate the calories consumed by each student and compare the mean calories consumed by those who brought lunch from home and those who ate a school lunch.

b. The 80 students taking Statistics were placed in one of two sections of the class taught by the same professor by flipping a coin. One section did not use laptops in class but the other section used them regularly. At the end of the semester, researchers compared the averages of the final grades of the students in the two sections.

c. All the third graders line up along their school gym's sidelines. A teacher randomly selects the number 3 from a stack of five index cards numbered 1–5. The third child in line and every fifth child after that is selected to be in a study.

d. The student body officers send a survey regarding food preferences for an upcoming activity to 10% of the students in each class whose names are randomly selected.

Answer

a. convenience sample; likely to be biased
b. random sample; not biased
c. systematic sample; not biased
d. stratified random sample; not biased

SKILL ✔ **EXERCISE 13**

Assessment

- Quiz 11C covers Section 11.5.

The data analysis techniques found in Chapters 10 and 11 form the heart of the research study. Descriptive statistics are calculated to describe the sample's shape, center, and variability. These statistics are then used to infer estimates of the population's parameter of interest and to test hypotheses about the parameters. Proper inferential statistical techniques allow conclusions to be made about the population.

The researcher then has the responsibility to communicate these conclusions, usually in a written report and often in an oral presentation as well. A thorough knowledge of the entire research process enables the researcher to present the results effectively and credibly and allows others to carefully examine the validity of the conclusions.

Example 3   Evaluating a Research Project

Evaluate each step in Mikayla's research project.

Answer

Mikayla's Research Project	Analysis
Clarify the Research Question.	
Do the Bible quizzers in Mikayla's association spend an average of 18.9 hr/week on social media?	The question is clearly stated, identifying the population and parameter.
She writes the null and alternative hypotheses and determines the level of significance.	The null hypothesis states equality so the alternative hypothesis states not equal to.
$H_0: \mu = 18.9$; $H_a: \mu \neq 18.9$; $\alpha = 0.05$	A 95% level of certainty seems to be appropriate for her study.
Plan the Study.	
Mikayla creates a short, carefully worded questionnaire that has been reviewed by several students and teachers to help eliminate bias. It asks, "How many hours do you spend each week on social media?" She will ask all 5 members of each of the 10 teams competing at the regional tournament to complete the anonymous survey.	The cluster sample may not be representative of the 2700 members of the national association, since only the top teams compete at a regional tournament and her region may not be representative of quizzers from the entire nation.
Collect the Data.	
Mikayla numbers the returned surveys and then randomly selects 30 cards from a stack of similarly numbered index cards to choose the surveys to include in her research.	Her selection of 30 surveys from the 50 possible surveys allows for some surveys to not be completed while keeping a sufficient number to apply the Central Limit Theorem. The random selection of the surveys has made some provision for eliminating bias.
Analyze the Data.	
The 30 surveys indicate $\bar{x} = 16.65$ and $s = 5.68$. She calculates a standardized test statistic $z = -2.17$ with a p-value of 0.03 in a two-tailed test.	$\sigma_{\bar{x}} = \dfrac{s}{\sqrt{n}} = \dfrac{5.68}{\sqrt{30}} \approx 1.037$ $z = \dfrac{\bar{x} - \mu}{\sigma_{\bar{x}}} \approx \dfrac{16.65 - 18.9}{1.037} \approx -2.17$ $2 \times \text{normalcdf}(-4, -2.17) = 0.030$ Her calculations are correct.

CONTINUED ➡

Interpret the Data.

Since $p < \alpha$, she rejects the null hypothesis and states that the time spent on social media by quizzers in her association is significantly different from the average of 18.9 hr/week.	Mikayla correctly rejects the null hypothesis. The possible bias in the sample causes some concern about the validity of generalizing her results to all the quizzers in the national association.
Wondering if she could have shown that the quizzers spend significantly less time on social media, Mikayla does a second hypothesis test. She rewrites her hypotheses. $H_0: \mu \geq 18.9$; $H_a: \mu < 18.9$ (claim) Then she compares the p-value for the standardized test statistic in a left-tailed test to the level of significance and concludes that the quizzers spend significantly less time on social media than the average teenager at the 0.05 level of significance.	The p-value for a z-score of -2.17 in a left-tailed test, $p = 0.015$, is less than the previously established level of significance, $\alpha = 0.05$, so her conclusion is valid.

SKILL ✔ **EXERCISES 29–33**

Reading and evaluating research studies is an important skill in many careers. An *abstract* is a brief summary that precedes a formal research report. When evaluating a study, refer to the diagram at the beginning of this section and determine whether all the necessary steps are included and whether the conclusions are correct.

Example 4 Analyzing an Abstract

Use the flow chart from the beginning of this section to analyze the following abstract from "A Water Availability Intervention in New York City Public Schools: Influence on Youths' Water and Milk Behaviors," a study published in the *American Journal of Public Health*.

Objectives. We determined the influence of "water jets" [drinking water dispensers] on observed water and milk taking and self-reported fluid consumption in New York City public schools.

Methods. From 2010 to 2011, before and 3 months after water jet installation in 9 schools, we observed water and milk taking in cafeterias (mean 1000 students per school) and surveyed students in grades 5, 8, and 11 (n = 2899) in the 9 schools that received water jets and 10 schools that did not. We performed an observation 1 year after implementation (2011–2012) with a subset of schools. We also interviewed cafeteria workers regarding the intervention.

Results. Three months after implementation we observed a 3-fold increase in water taking (increase of 21.63 events per 100 students; $P < .001$) and a much smaller decline in milk taking (-6.73 events per 100 students; $P = .012$), relative to comparison schools. At 1 year, relative to baseline, there was a similar increase in water taking and no decrease in milk taking. Cafeteria workers reported that the water jets were simple to clean and operate.

Conclusions. An environmental intervention in New York City public schools increased water taking and was simple to implement.

CONTINUED ➡

Answer

Clarify the Research Question.

The research question (whether the installation of the water jets would benefit students by encouraging them to take in more fluids) and the population (students in the New York City public school system) are clearly communicated.

Plan the Study.

The study is mainly observational but also includes two surveys. Data was to be collected by observing students in schools with newly installed water jets and schools without them. The study compared milk and water intake by students prior to, 3 mo after, and 1 yr after the installation of the water jets. The null and alternative hypotheses and the level of significance are not stated.

Collect the Data.

The sample includes fifth-, eighth-, and eleventh-grade students in schools having water jets installed and students in the same grades in schools without water jets. It is difficult to determine whether a sampling error has occurred, since the methods of choosing which grades were studied or which schools without water jets were included are not discussed. The reported sample sizes appear to be more than adequate for the study.

Analyze the Data.

While the abstract states significant *p*-values, hypothesis testing procedures are not reported in the abstract.

Interpret the Data.

The abstract's conclusion clearly interprets the results of the study relative to New York City public schools.

SKILL ✔ **EXERCISE 35**

❯ **A. Exercises**

1. List the order of the steps in a research study.   E, D, B, A, C

A. Analyze the data.   **D.** Plan the study.

B. Collect the data.   **E.** Clarify the research question.

C. Interpret the data.

In which stage of the research study (listed in exercise 1) should each of the following occur?

2. Perform an experiment, gather data, or conduct a survey.   B

3. Define the population and parameters related to a compelling concern.   E

4. Calculate test statistic(s).   A

5. Decide whether to reject the null hypothesis.   C

6. Determine the research design that best fits the goal of the research question.   D

Identify the population and research design for each study.

7. People at a local restaurant are asked which college team they support.   the local community; survey

8. Preexisting test score data is collected from the Pickens County school district to determine the math aptitude of their high school graduates.   high school graduates in Pickens County; observational

9. The relationship between drinking coffee and risk of death was studied by analyzing data from a NIH-AARP Diet and Health Study that included 229,119 men and 173,141 women ranging in age from 50 to 71.   50–71-yr-old men and women; observational

10. The average yield of oyster mushrooms in the control group was compared to the average yield of groups in which a nutritional supplement of alfalfa meal, soybean meal, or vermicompost was added to the substrate.   oyster mushrooms; experimental

11. The school nutritionist asks students from one class in each grade to record what they ate for breakfast each day in an attempt to determine whether the school's students are getting enough fiber in their breakfast.   the students in the school; survey

12. A pharmaceutical company randomly selects half the lactose-intolerant patients who are volunteering to participate in a study to receive an experimental drug while the other half is given a placebo. They compare symptom reduction rates of the two groups during a 3 yr period.   lactose-intolerant patients; experimental

Identify each type of sample. Then determine whether the sampling technique is likely to introduce bias.

13. Mrs. King chooses a banquet committee by randomly placing all the junior girls on a numbered list, rolling a die to choose the first girl, and then selecting every eighth girl on the list. systematic; Bias is not likely.

14. A cat food company testing its claim to prevent iron-deficiency anemia asks cat owners to volunteer their cats for participation in the study. self-selected; Bias is likely.

15. Mr. Siebert wonders if students' grades are affected by whether the students choose to sit in the front, middle, or back of the classroom. At the end of the semester, he compares the grades of students sitting in the front, middle, and back of his classroom.

16. A research company documenting the effect of a stair-climbing-machine workout on the resting heart rates of males randomly selects 45 males from major US cities to exercise 20 min on the machine 5 times a week for 6 weeks. random; Bias is not likely.

> **B. Exercises**

17. Explain the difference between a sampling error and researcher bias.

Describe how you might select a representative sample from the population for each described study. Answers will vary.

18. the names on voter registration lists for a poll regarding an upcoming election

19. a study examining the time spent on community service by students attending Edwards Academy

Identify the sampling technique and discuss any potential sampling errors or researcher bias.

20. Forty teens walking in a mall, some of whom are accompanied by their mother, are selected to rate their mother's cooking on a scale of 1 to 10.

21. When studying the average tenure of teachers in Arkansas, researchers found that 15% of its teachers belong to a minority group and randomly selected 15% of their sample from minority teachers.

22. A magazine advertisement states that the first 100 hearing-impaired people who call a given number will be able to participate in a study of micro hearing aids at no cost.

23. A reading coach tests a new reading program by randomly selecting 5 of the 32 second-grade classes in the district, which include several classes for advanced students.

24. State officials conducting a yearlong study investigating the impact of weekly multiplication speed drills on fourth-grade achievement test scores select 15 fourth-grade classes, each composed of a wide range of student abilities.

25. When attempting to determine whether gas prices in their state have increased significantly in the last month, students record the current gas price at 30 gas stations within 15 mi of the school.

26. In order to study the average time Americans watch the news each day, researchers purchase data describing the viewing habits of 4300 American households that was gathered by a television ratings company by using viewing diaries. The number 7 was randomly selected and the researchers selected data from the 7th, 57th, 107th, . . . households as their sample.

27. Statistics students use a random number generator to select 40 students from a list of the 723 students in their school to determine whether the average number of pairs of shoes by students at their school differs significantly from the national average of 13.28 pairs.

28. A researcher using data from the US Census Bureau finds that housing expenses in Ohio are fairly typical of the housing expenses in the United States, so he selects the homeowners in Ohio to be the sample for his study.

15. convenience; Bias is likely.

17. A sampling error occurs when the sample is not representative of the population; researcher bias occurs when the researcher intentionally or unintentionally influences the responses.

Solutions

> **B. Exercises**

18. Use a systematic sample in which a random number generator determines the first choice and then continue by choosing every 100th name thereafter.

19. Use a random sample in which each name is given a number and a random number generator is used to pick 50 numbers (without replacement).

20. convenience; Researcher bias is likely since those walking with their mother would likely feel pressured to elevate their ranking.

21. stratified random; Bias is not likely.

22. self-selected; Sampling error is likely since the most motivated individuals will compose the sample.

23. random; Sampling error is likely unless the percentage of advanced classes in the sample reflects that of the population. A stratified random sample would be appropriate.

24. cluster; Bias is not likely.

25. convenience; Sampling error is likely since gas prices may vary significantly in other regions of the state.

26. systematic; Some sample bias may occur, since those who report in the diaries may not be typical American households, they may not report accurately, and those who don't watch will not be included.

27. random; Bias is not likely.

28. cluster; Bias is not likely.

29a. 12–17-yr-olds from similar churches; the time spent watching TV each week

29b. $H_0: \mu = 18.1$; $H_a: \mu \neq 18.1$

29c. $\alpha = 0.10$ or 0.05 is reasonable.

30a. How many hours per week do you watch television?

31. He could use a random number generator to choose a number 1–12 to determine the first form selected from the stack of forms that included responses and then choose every twelfth survey thereafter.

32. convenience, self-selected, and random; Since those who respond may not be representative of the entire population, there is a good chance of sampling error. Unintended researcher bias may occur since the survey is not anonymous and some teens may underreport their average viewing time.

33. $\sigma_{\bar{x}} = \dfrac{s}{\sqrt{n}} = \dfrac{3.4}{\sqrt{50}} \approx 0.481$

$z = \dfrac{\bar{x} - \mu}{\sigma_{\bar{x}}} \approx \dfrac{17.2 - 18.1}{0.481} \approx -1.87$

two-tailed test:
$p \approx 2 \times \text{normalcdf}(-4, -1.87)$
≈ 0.0612
Since $p > \alpha$, fail to reject H_0.

Complete exercises 29–33 to plan and evaluate a research project.

29. Jack reads that 12–17-yr-olds spend an average of 18.1 hr/week watching television. He wonders if this average is accurate for 12–17-yr-olds who attend churches similar to his church. State the following.
   a. the population and parameter
   b. the null and alternative hypotheses
   c. an appropriate level of significance

30. Jack gets permission to put a question on the sign-up form for the regional church youth rally.
   a. How should he word the question?
   b. Which research design is he using? *survey*

31. Of the 1200 12–17-yr-olds signed up for the rally, 600 teens answer Jack's question. Describe a method he can use to randomly select a sample of 50 responses for his research.

32. Identify the sampling techniques used and discuss the possibilities for sample error and researcher bias.

33. If his sample has a mean of 17.2 hr with a standard deviation of 3.4 hr, find the standardized test score and determine whether he should reject the null hypothesis at a 0.05 level of significance. Explain your reasoning. $z = -1.87$; Since $p \approx 0.06 > \alpha = 0.05$, H_0 should not be rejected.

> **C. Exercises**

34. A math professor is interested in finding a method of quizzing that helps her students learn better. She institutes daily quizzes in her courses for a semester and compares the average final grade of her 105 students during this semester with the average final grade of students in her previous courses, an 81.4. The mean grade of her students during this semester is 83.5 with a standard deviation of 15.2.
   a. Clarify the objective by writing a research question.
   b. Identify the population and parameters.
   c. Write null and alternative hypotheses.
   d. Identify the research design. *experimental*
   e. Identify the sampling technique and discuss possible sample errors and researcher bias.
   f. Calculate the standardized test statistic. $z \approx 1.42$
   g. Should null hypotheses be rejected at a level of significance of 0.10? of 0.05? *yes; no*

34c. $H_0: \mu \leq 81.4;\ H_a: \mu > 81.4$

In exercises 35–36, use the flow chart at the beginning of the lesson and the given abstract to evaluate each study.

35. *The Effectiveness of Implementing the Classic Reading Program*

   Objective: to determine whether implementing the Classic Reading program as an intervention for struggling students increases their reading comprehension and speed of reading
   Methods: From 2014 to 2015 struggling readers from first and second grade took part in a 90-day trial implementing the Classic Reading program. Students were taken from 30 different classrooms in 5 different Nashville-area schools and were considered to be at least 8 mo behind in their reading level. Similar reading comprehension tests were given to these students before and after the program.
   Results: While their speed of reading showed slight improvement, significant improvement was observed in the students' reading comprehension. According to the posttest results, 80% of the students improved to no more than 4 mo behind grade level.
   Conclusions: The Classic Reading program allows struggling readers to improve their skills, quickly preventing them from falling further behind. The earlier the implementation of this reading program, the better the results.

36. *The Ability of Al's Feed to Control Weight in Alpacas*

   Objective: to test the effectiveness of the alpaca food supplied by the feed company to keep the weight of alpacas at its average of 150 lb with 97% confidence
   Methods: From a farm of 75 alpacas, 30 are provided the special feed in an area with no other significant food source. After one year the average weight for the special feed group will be compared to the average weight of the other alpacas.
   Results: One year after implementation, the average weight of the test group was 158 lb with a standard deviation of 21.2 lb ($p = 0.039$).
   Conclusion: The food produced by the feed company helps alpacas maintain their average weight.

37. Research: Use the Internet to find a research report on a topic you find interesting. Use the flow chart from the beginning of the lesson to analyze the study.

38. Research: Use the flow chart from the beginning of the lesson to design and complete a research study. Write a report summarizing your study and presenting your interpretation of the results.

tified. The level of significance and the null and alternative hypotheses are not discussed in the abstract.
Plan the Study: The study is experimental since data was collected before and after the intervention.
Collect the Data: The method of selecting which struggling readers to include in the sample was not discussed, so the possibility of sampling error should be further investigated in the research report. While a specific sample size is not mentioned, it appears to be sufficient since students are chosen from 30 different classrooms.
Analyze the Data: No hypothesis testing is mentioned, nor is significant descriptive data provided.
Interpret the Data: The conclusion clearly interprets the results of the study relative to Nashville-area schools, but the abstract provides no support for the statement regarding earlier intervention leading to better results.

36. *Clarify the Research Question*: The research question (effectiveness of the feed for weight maintenance) and population (alpacas) are clearly stated. The null and alternative hypotheses were not stated, but the level of significance is 0.03.
Plan the Study: The study is experimental since it compares the mean weight of a group to a known mean after applying a treatment to the experimental group.
Collect the Data: The sample size is 30 and is compared to the overall alpaca average weight. There was no mention of the alpacas being weighed before the experiment.
Analyze the Data: The p-value and level of significance are mentioned along with other helpful descriptive data.
Interpret the Data: The abstract's conclusion clearly interprets the results of the study. The study would carry more meaning if the average weight for each group was found before the experiment began.

37–38. Answers will vary.

> **C. Exercises**

34a. Will daily quizzes help my students attain higher grades in the courses I teach?

34b. The population is the professor's students, and the parameter is their final grades.

34e. cluster; Sampling error occurs if the students in her classes this semester are not representative of previous and future students. Researcher bias may occur if she makes other changes in her courses that influence the students' grades.

34f. $\sigma_{\bar{x}} = \dfrac{s}{\sqrt{n}} = \dfrac{15.2}{\sqrt{105}} \approx 1.483$

$z = \dfrac{\bar{x} - \mu}{\sigma_{\bar{x}}} \approx \dfrac{83.5 - 81.4}{1.483} \approx 1.42$

34g. right-tailed test
$p \approx \text{normalcdf}(1.42, 4) \approx 0.078$
Reject if $\alpha = 0.10$, and fail to reject if $\alpha = 0.05$.
(if $\alpha = 0.10$, $z_c \approx 1.28$;
 if $\alpha = 0.05$, $z_c \approx 1.645$)

35. *Clarify the Research Question*: The research question (effectiveness of the Classic Reading program in increasing reading comprehension and speed) and population (struggling first- and second-grade readers in the Nashville area) are clearly iden-

39.

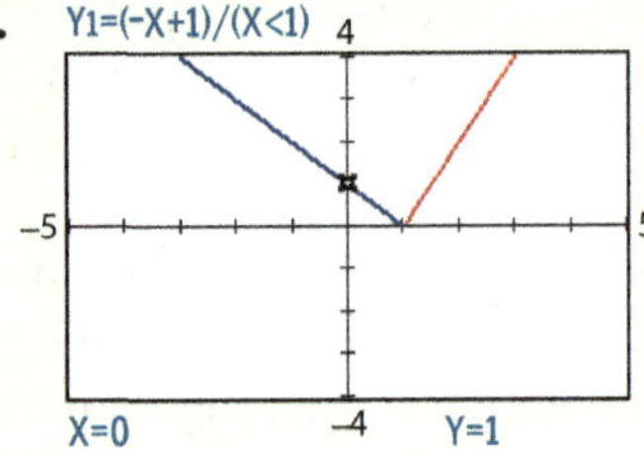

40.

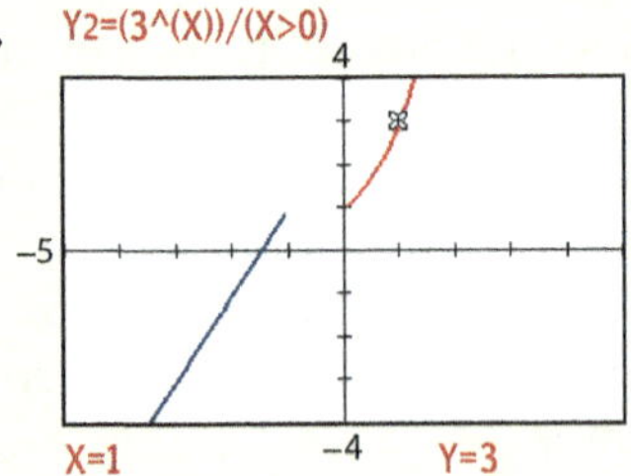

41. $a^2 = 24^2 + 17^2 - 2(24)(17)\cos 57°$
$\approx 420.6; a \approx 20.5$
$\dfrac{\sin B}{24} = \dfrac{\sin 57°}{20.5}$
$B \approx \text{Sin}^{-1}\left(\dfrac{24 \sin 57°}{20.5}\right) \approx 79.0°$
$C \approx 180 - 79.1 - 57 = 44.0°$

42. $A = \text{Cos}^{-1}\left(\dfrac{42^2 - 53^2 - 70^2}{-2(53)(70)}\right) \approx 36.8°$
$\dfrac{\sin B}{53} \approx \dfrac{\sin 36.8°}{42}$
$B \approx \text{Sin}^{-1}\left(\dfrac{53 \sin 36.8°}{42}\right) \approx 49.0°$
$C \approx 180 - 36.8 - 49.0 = 94.2°$

43. $IQR = 8 - 4.5 = 3.5$
$1.5(IQR) = 1.5(3.5) = 5.25$
$Q_1 - 5.25 = -0.75$
$Q_3 + 5.25 = 13.25$
$2, 9 \in [-0.75, 13.25]$; no outliers

44. min: 1; Q_1: 5; Q_2: 6.5; Q_3: 8; max: 13
$IQR = 8 - 5 = 3$
$1.5(IQR) = 1.5(3) = 4.5$
$Q_1 - 4.5 = 0.5; Q_3 + 4.5 = 12.5$
$13 \notin [0.5, 12.5]$

45. $\ln e^{\frac{4}{3}} = \dfrac{4}{3}$

46. $\cos\theta = \dfrac{-9}{41} = \dfrac{x}{r}; (-9)^2 + y^2 = 41^2$
$y = \sqrt{41^2 - 9^2} = 40$
$\cot\theta = \dfrac{x}{y} = \dfrac{-9}{40} = -\dfrac{9}{40}$

47. $r = \sqrt{3^2 + 6^2} = \sqrt{45} = 3\sqrt{5}$
$\alpha = \text{Tan}^{-1}\left|\dfrac{6}{3}\right| \approx 63°$
Q I: $\theta = \alpha \approx 63°; \therefore (3\sqrt{5}, 63°)$
$(3\sqrt{5}, 63° - 360°) = (3\sqrt{5}, -297°)$
$(-3\sqrt{5}, 63° + 180°) = (-3\sqrt{5}, 243°)$

48. $n = \left(\dfrac{z_c s}{E}\right)^2 = \left(\dfrac{1.96(8.4)}{2.1}\right)^2 \approx 61.47$
$n \geq 62$

Graph each function. Then state its domain, range, and whether it is continuous or discontinuous. [1.3]

39. $g(x) = \begin{cases} -x + 1 & \text{if } x < 1 \\ 2x - 2 & \text{if } x \geq 1 \end{cases}$ $D = \mathbb{R}; R = [0, \infty)$; continuous

40. $f(x) = \begin{cases} 2x + 3 & \text{if } x < -1 \\ 3^x & \text{if } x > 0 \end{cases}$ $D = (-\infty, -1) \cup (0, \infty)$
$R = (-\infty, 1) \cup (1, \infty)$; discontinuous

Solve each $\triangle ABC$. Round answers to the nearest tenth. [5.7]

41. $b = 24, c = 17, A = 57°$ **42.** $a = 42, b = 53, c = 70$

Determine whether each minimum or maximum value is an outlier. [10.4]

43. min: 2, Q_1: 4.5, Q_2: 6, Q_3: 8, max: 9   neither

44. 1, 3, 5, 5, 6, 7, 8, 8, 10, 13   13 is the only outlier.

45. Simplify $\ln \sqrt[3]{e^4}$. [3.2]   D

   A. 0.2877   **C.** $\dfrac{3}{4}$ **E.** none of these

   B. $\sqrt[3]{4}$ **D.** $\dfrac{4}{3}$

41. $a = 20.5; B = 79.0°; C = 44.0°$
42. $A = 36.8°; B = 49.0°; C = 94.2°$

46. If $\cos\theta = -\dfrac{9}{41}$ and $\tan\theta < 0$, what is $\cot\theta$? [4.3]   A

   A. $-\dfrac{9}{40}$ **C.** $-\dfrac{40}{41}$ **E.** $\dfrac{9}{41}$

   B. $\dfrac{9}{40}$ **D.** $-\dfrac{40}{9}$

47. Which polar coordinates are equivalent (to the nearest degree) to the rectangular coordinates $(3, 6)$? List all correct answers. [6.4]   A, B, C

   A. $(3\sqrt{5}, -297°)$ **D.** $(3\sqrt{5}, 243°)$

   B. $(3\sqrt{5}, 63°)$ **E.** $(3\sqrt{5}, 603°)$

   C. $(-3\sqrt{5}, 243°)$

48. Find the minimum sample size, n, for a 95% confidence interval with a maximum margin of error of 2.1 in a population with a standard deviation of 8.4. [11.3]   E

   A. 8   **C.** 44   **E.** 62

   B. 16   **D.** 61

Data Analysis

Objectives

1. To describe the origin and some of the modern uses of Markov chains
2. To interpret the information presented in a transition graph
3. To use a matrix to find probabilities of future states
4. To evaluate the reliability and effectiveness of mathematical models in determining the truth from a biblical worldview

Assignments

- **Minimum:** 1–9
- **Standard:** 1–9
- **Extended:** 1–9

Markov Chains

In 1903 Orville Wright ushered in the aircraft era with a flight that lasted little more than 10 sec and covered only 120 ft. Just 3 yr later Andrey Markov, a student of Pafnuty Chebyshev, published the first results in his study of probabilistic processes that are now called Markov chains. In June 2017 the first operational F-35A Lightning II squadron of stealth fighter jets received its final aircraft. It's hard to say who would be more surprised to see this jet—the Wright brothers, who pioneered the aerodynamics, or Markov, whose work was crucial to the development of the voice-controlled cockpit.

Applications of Markov chains are found in biological science, business, and engineering in addition to voice recognition algorithms. Unlike traditional probability that considers independent events, Markov chains model processes in which the current state depends only on the previous state. Consider a token on an integer number line. You flip a coin and move the token one number to the right for heads and one to the left for tails. If the token starts on 5 and you flip a fair coin, the probability of the token landing on 4 is 50% and the probability of landing on 6 is 50%. It doesn't matter how the token got to 5, but the fact that it is at 5 determines the possible outcomes after the next coin flip. Since the current state depends only on the state immediately preceding it, this is an example of a Markov chain.

A transition graph is a useful tool in exploring Markov chains. The graph consists of nodes that represent the states and arrows between the nodes that represent transitions. Arrows are labeled with the appropriate probability, and the sum of probabilities from each node must equal one.

Suppose every day is either calm, breezy, or windy. The following transition graph shows that if it is calm today, there is a 40% chance it will be breezy tomorrow.

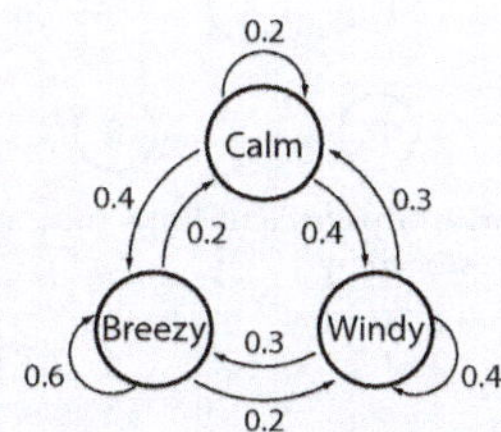

1. If it is windy today, what is the chance it will be calm tomorrow? 30%

2. What is the probability it will be calm two days in a row? 20%

3. In our model, does the condition yesterday affect the probability that it will be calm tomorrow? Why or why not? no; Tomorrow's state depends only on today's.

Transition graphs with n states can be written as an $n \times n$ transition matrix T in which entry t_{ij} represents the probability of transitioning from state i to state j. Using the order of calm (c), breezy (b), and windy (w) for both the rows and the columns:

$$T = \begin{bmatrix} 0.2_{cc} & 0.4_{cb} & 0.4_{cw} \\ 0.2_{bc} & 0.6_{bb} & 0.2_{bw} \\ 0.3_{wc} & 0.3_{wb} & 0.4_{ww} \end{bmatrix}.$$

The product of a row matrix C representing the current condition and the transition matrix T gives the probabilities for the next day's condition. For example, using $C = [1 \quad 0 \quad 0]$ to represent a calm day, the product $CT = [0.2\ 0.4\ 0.4]$ represents the probabilities that tomorrow will be a calm, breezy, or windy day. Raising the transition matrix to the nth power gives the probability for transitioning to each state in n days. Therefore, the probability a certain state will exist after n iterations is the product

---

PRESENTATION

This Data Analysis feature allows students to experiment with some simple examples of Markov chains. In order to demonstrate how to construct and use transition graphs and matrices, we use an imaginary climate in which the wind condition today depends only on the wind condition yesterday. Point out this distinction to students, since real-world weather predictions obviously include many other factors.

Proficiency in basic matrix operations is required to complete Markov chains. You may wish to briefly review these procedures before assigning these questions. Ask students what must be true about the entries in any row of a transition matrix and if this is also true for every column. (*They must add up to 1; no*)

Consider sharing that Google's page-ranking algorithm was developed by its cofounders Larry Page and Sergey Brin. The trademarked name of this search algorithm is PageRank, a play on web pages and Larry Page's name.

4. $C_w T^3$

$$= [0 \ 0 \ 1] \begin{bmatrix} 0.2_{cc} & 0.4_{cb} & 0.4_{cw} \\ 0.2_{bc} & 0.6_{bb} & 0.2_{bw} \\ 0.3_{wc} & 0.3_{wb} & 0.4_{ww} \end{bmatrix}^3$$

$$= [0.234 \ 0.45 \ 0.316]$$

5.

```
[A]⁴
        [0.231 0.460 0.309]
        [0.230 0.464 0.306]
        [0.232 0.458 0.310]
[A]⁵
        [0.231 0.461 0.308]
        [0.231 0.462 0.307]
        [0.231 0.461 0.308]
```

```
[A]⁶
        [0.231 0.461 0.308]
        [0.231 0.462 0.308]
        [0.231 0.461 0.308]
[A]⁷
        [0.231 0.462 0.308]
        [0.231 0.462 0.308]
        [0.231 0.461 0.308]
```

```
[A]⁸
        [0.231 0.462 0.308]
        [0.231 0.462 0.308]
        [0.231 0.462 0.308]
[A]⁹
        [0.231 0.462 0.308]
        [0.231 0.462 0.308]
        [0.231 0.462 0.308]
```

6. in rows with 2 links,

$$0.5 - \frac{3(0.05)}{2} = 0.425$$

in the row with 1 link,

$$1 - 4(0.05) = 0.80$$

7.

```
[B]¹⁰⁰
[0.18 0.27 0.22 0.13 0.20]
[0.18 0.27 0.22 0.13 0.20]
[0.18 0.27 0.22 0.13 0.20]
[0.18 0.27 0.22 0.13 0.20]
[0.18 0.27 0.22 0.13 0.20]
```

---

of the initial condition matrix and the nth power of the transition matrix, CT^n.

4. If it is windy today, what is the probability it will be windy 3 days from today? 31.6%

5. Using technology set to display results to the nearest thousandth, find T^4, T^5, T^6, T^7, T^8, and T^9. Describe the result as the power of T increases.

Google® search results are ranked by an algorithm based on a Markov chain. The Internet's vast size and continual growth make the important task of ranking results difficult. The page-ranking algorithm rates importance by the probability that the searcher will choose a particular page based on hyperlinks as well as the amount of time spent on each page.

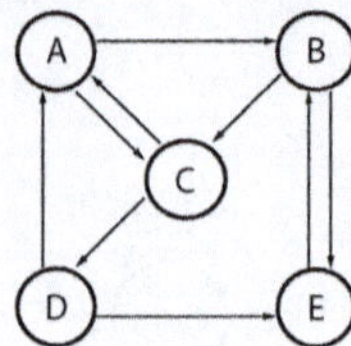

This web with 5 pages and links indicated by arrows would be represented by

$$T = \begin{bmatrix} 0 & 0.5 & 0.5 & 0 & 0 \\ 0 & 0 & 0.5 & 0 & 0.5 \\ 0.5 & 0 & 0 & 0.5 & 0 \\ 0.5 & 0 & 0 & 0 & 0.5 \\ 0 & 1 & 0 & 0 & 0 \end{bmatrix}.$$

5. The entries in each row converge to probabilities of 0.231 (c), 0.462 (b), and 0.308 (w).

6.
$$\begin{bmatrix} 0.05 & 0.425 & 0.425 & 0.05 & 0.05 \\ 0.05 & 0.05 & 0.425 & 0.05 & 0.425 \\ 0.425 & 0.05 & 0.05 & 0.425 & 0.05 \\ 0.425 & 0.05 & 0.05 & 0.05 & 0.425 \\ 0.05 & 0.8 & 0.05 & 0.05 & 0.05 \end{bmatrix}$$

The page-ranking algorithm also assumes that the searcher may access a page directly without linking from another site.

6. To simulate the matrix obtained by the page-ranking algorithm, replace each zero in the transition matrix with 0.05 and then reduce the other original entries by the same amount so that the sum of each row still equals one.

7. Using technology set to display results to the nearest hundredth, find the probability of each state after 100 iterations. 0.18, 0.27, 0.22, 0.13, 0.20

8. Use the results to rank the web pages. B, C, E, A, D

9. Discuss: The word *google* now appears in dictionaries as a verb. The vast amount of information at our fingertips makes information and expertise easily accessible. How can understanding the process of ranking search results affect the confidence that should be placed in this information?

sample answer: Knowing that the results are based on algorithms that are people's attempts to guide others to the best information helps us realize that the rankings are subjective. The algorithms are simply models created by fallible people. Scandals related to rigged algorithms raise concerns about the integrity of some search engines.

Chapter 11 Review

Classify each random variable as discrete or continuous and justify your answer. [11.1]

1. Let x represent the number of pizzas sold each day over a two-week period at Pizza Pizzaz.

2. Let x represent the actual number of ounces in a gallon of milk coming off a dairy's processing line.

3. The table lists results of a survey of 158 people that asked the number of days per week they use a cell phone. Create a table and a histogram representing the probability distribution. Then use the distribution to find the expected value and standard deviation. [11.1] $\mu \approx 5.18; \sigma \approx 1.96$

Days Used	Frequency
0	7
1	4
2	8
3	10
4	17
5	23
6	38
7	51

4. Use the probability distribution from exercise 3 to find the probability of a survey participant using a cell phone each number of days. [11.1]

   a. P(at least 6) ≈ 0.563 **b.** P(fewer than 6) ≈ 0.437

Use the binomial probability distribution formula to find the probability of x successes in n independent trials where p is the probability of a success in a trial. Then find the mean and standard deviation of the binomial probability distribution. Round answers to the nearest hundredth if necessary. [11.1]

5. $x = 8, n = 15, p = 0.6$ 6. $x = 4, n = 10, p = 0.25$

7. An insurance company's data shows that 33% of men in a given age group receive a traffic ticket each year. Create a table and a histogram representing the probability distribution for the number of men from a randomly selected sample of 5 men in that age group who received a traffic ticket in the past year. Then find the expected number of traffic tickets for this sample and the distribution's standard deviation. [11.1] $\mu = 1.65; \sigma \approx 1.05$

   5. $P(8) \approx 0.18; \mu = 9; \sigma \approx 1.90$

   6. $P(4) \approx 0.15; \mu = 2.5; \sigma \approx 1.37$

8. Use the probability distribution from exercise 7 to find the probability that each number of men in the sample received a ticket. [11.1] $\approx 79.5\%$ $\approx 53.3\%$
   a. P(all 5) $\approx 0.4\%$ **b.** P(at most 2)   **c.** P(at least 2)

9. A die with 3 twos, 2 threes, and 1 four is rolled and a nickel is flipped with heads recorded as 1 and tails recorded as 2. Create a table and a histogram representing the probability distribution for the sums of this two-step experiment. Then find the distribution's expected value and standard deviation. [11.1]

10. The number of months that 4 patients lived after being diagnosed with a rare disease was recorded. {18, 21, 16, 23} [11.2]

    a. Create a table containing all possible samples of size 2 (taken with replacement) from this population.

    b. Create a table and a histogram representing the sampling distribution for the means.

    c. Find and compare the means and standard deviations for the population and the sampling distribution.

11. Determine whether the Central Limit Theorem can be applied to each population distribution with the given mean and standard deviation using a sample size of n. If so, find $\mu_{\bar{x}}$ and $\sigma_{\bar{x}}$. If not, explain why. [11.2]

    a. normal with $\mu = 5200, \sigma = 1450, n = 12$

    b. $\mu = 16, \sigma = 3.92, n = 48$

12. State the continuity correction factor for using a normal approximation to find each binomial probability. [11.2]

    a. P(between 52 and 81) $52.5 \leq x \leq 80.5$

    b. P(at most 19) $x \leq 19.5$

13. A survey of 15–24-yr-olds reveals their average Christmas spending is $243. Assume the standard deviation of the amount spent was $328.25. Find each probability for the average spending of 50 randomly selected people in this age bracket. [11.2]

    a. The average is less than $200. $\approx 18\%$

    b. The average is between $125 and $150. $\approx 1.7\%$

9. $E(x) = 4.17; \sigma \approx 0.90$

PRESENTATION

Exercises 1–2, 29–33, and 36–45 could be done as class discussion exercises.

Objective

To prepare for evaluation

Vocabulary

See Appendix A.

Assignments

- **Minimum:** 1–5, 7–8, 11–12, 14–15, 19–21, 23, 27–33, 35–36, 38–41, 44, 45a, 46a

- **Standard:** 1–45 odd; 2–46 even (in class)

- **Extended:** 4, 6–12, 14, 16–19; 22–28 even; 29–31, 35, 37–42, 44–46

Assessment

- Chapter 11 Test

Solutions

1. discrete; The value can only be a whole number.

2. continuous; The value can be a real number within a range of values.

3. $x =$ number of days cell phone used

x_i	$P(x_i)$	$x_i P(x_i)$	$(x_i - \mu)^2 P(x_i)$
0	$\frac{7}{158} \approx 0.044$	0	1.187
1	$\frac{4}{158} \approx 0.025$	0.025	0.442
2	$\frac{8}{158} \approx 0.051$	0.101	0.511
3	$\frac{10}{158} \approx 0.063$	0.190	0.300
4	$\frac{17}{158} \approx 0.108$	0.430	0.149
5	$\frac{23}{158} \approx 0.146$	0.728	0.005
6	$\frac{38}{158} \approx 0.241$	1.443	0.163
7	$\frac{51}{158} \approx 0.323$	2.259	1.073
Sum	1	$\mu \approx 5.177$	$\sigma^2 \approx 3.829$
			$\sigma \approx 1.957$

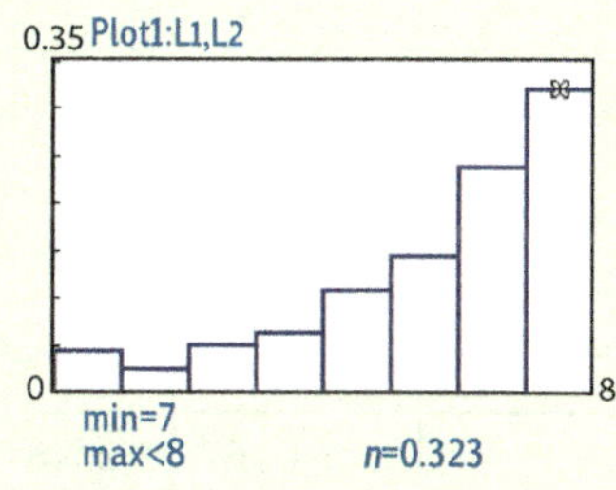

4a. $P(6 \text{ or } 7) = \frac{38}{158} + \frac{51}{158} = \frac{89}{158} \approx 0.563$

4b. $P(<6) = 1 - P(\geq 6)$
$$= 1 - \frac{89}{158} = \frac{69}{158} \approx 0.437$$

5. $P(8) = {}_{15}C_8(0.6)^8(0.4)^7 \approx 0.18$
   $\mu = np = 15(0.6) = 9$
   $\sigma = \sqrt{npq} = \sqrt{15(0.6)(0.4)} \approx 1.90$

6. $P(4) = {}_{10}C_4(0.25)^4(0.75)^6 \approx 0.15$
   $\mu = np = 10(0.25) = 2.5$
   $\sigma = \sqrt{npq} = \sqrt{10(0.25)(0.75)} \approx 1.37$

7. x = number of men in the sample receiving a traffic ticket

x_i	$P(x_i)$
0	${}_5C_0(0.33)^0(0.67)^5 \approx 0.135$
1	${}_5C_1(0.33)^1(0.67)^4 \approx 0.332$
2	${}_5C_2(0.33)^2(0.67)^3 \approx 0.328$
3	${}_5C_3(0.33)^3(0.67)^2 \approx 0.161$
4	${}_5C_4(0.33)^4(0.67)^1 \approx 0.040$
5	${}_5C_5(0.33)^5(0.67)^0 \approx 0.004$

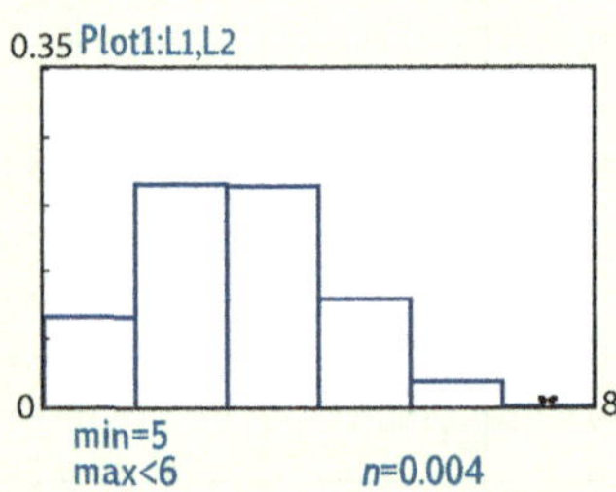

$$\mu = np = 5(0.33) = 1.65$$
$$\sigma = \sqrt{npq} = \sqrt{5(0.33)(0.37)} \approx 1.05$$

8b. $P(0, 1, \text{or } 2)$
$$\approx 0.135 + 0.332 + 0.328 = 0.795$$

8c. $P(\text{at least } 2) = 1 - P(0 \text{ or } 1)$
$$\approx 1 - (0.135 + 0.332) = 0.533$$

9. x = sum of die and nickel

x_i	$P(x_i)$	$x_i P(x_i)$	$(x_i - \mu)^2 P(x_i)$
3	$\frac{3}{12} = 0.25$	0.75	0.340
4	$\frac{5}{12} \approx 0.417$	1.667	0.012
5	$\frac{3}{12} = 0.25$	1.25	0.174
6	$\frac{1}{12} \approx 0.083$	0.5	0.280
Sum	1	$\mu \approx 4.167$	$\sigma^2 \approx 0.806$
			$\sigma \approx 0.898$

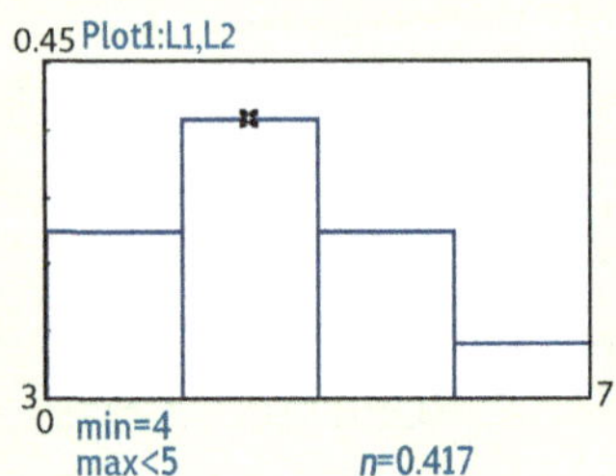

14. Find each probability if the height of mature sunflowers is normally distributed with an average height of 64 in. and a standard deviation of 3.5 in. **[11.2]**
  a. A randomly selected mature sunflower is between 58 in. and 62 in. tall. $\approx 24\%$
  b. The average height of 150 mature sunflower plants is more than 64.5 in. $\approx 4\%$

15. A manufacturer found that the lifespan of its 40-watt light bulbs is normally distributed with a mean of 10,000 hr and a standard deviation of 1150 hr. What is the probability that the average lifespan of a random sample of 15 bulbs is less than 9315 hr? **[11.2]** $\approx 1.0\%$

16. A cannery packs 32 oz of tomato sauce in cans with an acceptable standard deviation of 0.77 oz. If the variation is normally distributed, what is the probability that a random sample of 16 cans has a mean weight of 32.5 oz? **[11.2]** $\approx 0.5\%$

17. Pew Research Center reports that 31% of adult Internet users use Pinterest. Find each probability for a sample of 200 randomly selected adult Internet users. **[11.2]**
  a. at least 55 use Pinterest $\approx 87\%$
  b. more than 75 use Pinterest $\approx 1.9\%$

18. A manufacturer recalls a car part that has a 28% failure rate. In a randomly selected sample of 70 cars, what is the probability of having a faulty part in each number of cars? **[11.2]**
  a. $P(\text{fewer than 20 cars})$ $\approx 49\%$ **b.** $P(\text{from 20 to 25 cars})$ $\approx 45\%$

19. In June 2018, Scooter Gennett of the Cincinnati Reds led the National League in batting averages at 0.332, meaning that his probability of getting a hit is 33.2%. How many hits is Gennett expected to get in his next 60 at bats? Find the probability of Gennett getting at most 20 hits in the next 60 at bats. **[11.2]** 20 hits; $\approx 56\%$

20. Use technology to find the critical value, z_c, for each confidence level. **[11.3]**
  a. 80% ≈ 1.28 **b.** 85% ≈ 1.44

30. $H_0: \mu \leq 2500$; $H_a: \mu > 2500$ (claim)

21. Find the 99% confidence interval for a sample with $n = 50$, $\bar{x} = 19.6$, and $s = 3.87$ and interpret its meaning. **[11.3]** $\approx (18.19, 21.01)$

Use each random sample to find the point estimate for the population's mean, the margin of error for the stated confidence level, and the related confidence interval. [11.3]

22. {46, 33, 13, 20, 45, 80, 77, 64, 71, 2, 37, 10, 40, 47, 75, 84, 39, 10, 52, 6, 55, 23, 80, 49, 63, 33, 54, 64, 0, 8}
$c = 90\%$
$\bar{x} \approx 42.67$
$E \approx 7.78$
CI: $\approx (34.89, 50.45)$

23. {51, 49, 56, 81, 44, 65, 69, 72, 57, 48, 55, 64, 73, 57, 63, 64, 71, 69, 73, 58, 70, 67, 74, 86, 75, 69, 83, 67, 77, 58, 74, 66}
$c = 95\%$
$\bar{x} \approx 65.78$
$E \approx 3.52$
CI: $\approx (62.21, 69.35)$
90% CI: $\approx (14.34, 16.66)$; 95% CI: $\approx (14.11, 16.89)$

24. A test was given to predict success in a Calculus 1 course. Normally distributed results with $s = 3$ show that a mean score of 15.5 predicted an A in the course. Find the 90% and 95% confidence intervals for an A if 18 students take the test. **[11.3]**

25. Using the information from exercise 24, find the 90% and 95% confidence intervals if 180 students take the test. **[11.3]** 90% CI: $\approx (15.13, 15.87)$; 95% CI: $\approx (15.06, 15.94)$

26. A sample of 70 students report spending an average of $267.40 per semester on books. If the standard deviation is $185.67, find the 90% confidence interval for the mean and interpret its meaning. **[11.3]**

27. Find the minimum sample size for an 88% confidence interval with a margin of error of 2.3 if the population standard deviation is 7.64. **[11.3]** $n \geq 27$

28. A researcher studying the price of comic books desires a 95% confidence level and the margin of error for the mean to be no more than $0.27. A previous study determined that $s = \$1.39$ is a reasonable standard deviation for the population. Determine the minimum sample size for the research study. **[11.3]** $n = 102$

Write the null and alternative hypotheses for each statement and identify which hypothesis represents the claim. [11.4]

29. A manufacturer tests the 250 mg dose of its pain reliever. $H_0: \mu = 250$ (claim); $H_a: \mu \neq 250$

30. A local clothing store predicts they will have more than 2500 customers on Black Friday this year.

For exercises 31–33, classify each decision as a type I error, type II error, or correct decision given $H_0: \mu \geq 50$ and $H_a: \mu < 50$. [11.4]

31. Researchers reject H_0 and μ is found to be 70.
type I error

10a.

Sample; $\bar{x}$	
{16, 16}; 16	{21, 16}; 18.5
{16, 18}; 17	{21, 18}; 19.5
{16, 21}; 18.5	{21, 21}; 21
{16, 23}; 19.5	{21, 23}; 22
{18, 16}; 17	{23, 16}; 19.5
{18, 18}; 18	{23, 18}; 20.5
{18, 21}; 19.5	{23, 21}; 22
{18, 23}; 20.5	{23, 23}; 23

10b. using technology:
$$\bar{x} = L1; \ P(\bar{x}) = L2$$

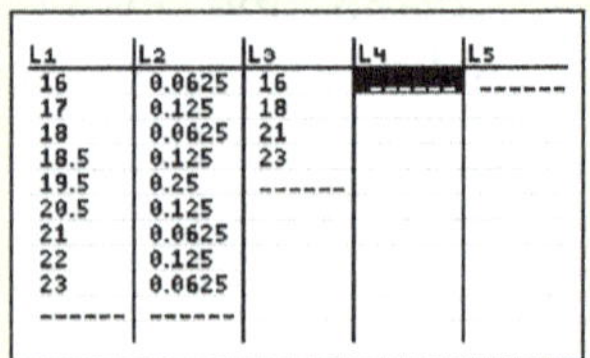

L1	L2	L3	L4	L5
16	0.0625	16		------
17	0.125	18		
18	0.0625	21		
18.5	0.125	23		
19.5	0.25	------		
20.5	0.125			
21	0.0625			
22	0.125			
23	0.0625			
------	------			

L4(1)=

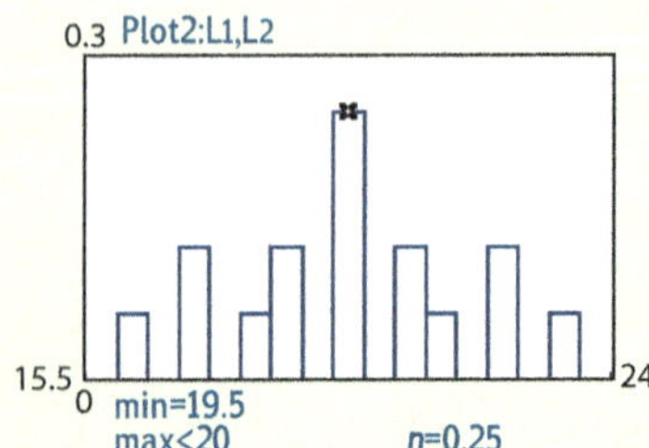

32. Researchers fail to reject H_0 and μ is found to be 40.

33. Using a sample of classmates in which the average height is 61.5 in., Joshua fails to reject that the average height of all high school seniors is 60.75 in. Later studies reveal that the average height of high school seniors is 66.8 in.

34. Researchers use 52 randomly selected teens to test the claim that the average weekly earnings for 16–19-yr-olds is at least \$420. The mean of the sample's weekly earnings is \$415/week with a standard deviation of \$24.18. Perform a hypothesis test to determine whether the data supports their claim at a 0.01 level of significance. [11.4]

35. The average morning commute time for US residents is 26.1 min. Mrs. Krackenbush's statistics class wanted to know if this was true for the commuters in their city. They found the average commute time for 40 randomly selected commuters from their city to be 30.4 min with a standard deviation of 8.2 min. Perform a hypothesis test at the 95% confidence level to answer their question. [11.4]

36. State the 5 major steps involved in a research project. [11.5]

37. Explain the differences between an observational study and an experimental study. [11.5]

38. Identify each sampling technique. [11.5]
  a. The population is divided into representative groups and an entire group is randomly selected to be in the sample.
  b. The first member of the sample is randomly selected from a list and then every nth member of the list is selected.
  c. The researcher selects members of the population that are easily accessible.

Identify the population and research design for each study. [11.5]

39. A researcher randomly selects 60 eighth graders from area schools and divides them into two groups. The members of the first group continue to study using methods of their own choice. The members of the second group are coached in specific methods of studying. Average test scores from the two groups are then compared.

40. A researcher collects information about Asian US citizens from the US Census Bureau to see how their level of education relates to their rate of moving to a different state.

Identify the sampling technique and discuss potential sampling error or researcher bias. [11.5]

41. A 3-question survey is distributed to customers in a store. Participants will be added into a drawing for a gift card.

42. A researcher studying recreational spending examines median home prices in neighborhoods within the city and randomly selects participants in proportion to the number of homes in each price range.

43. Fifty dirt bike owners at a dirt bike rally are randomly selected to fill out a survey that includes the cost of their bike.

Complete exercises 44–46 to analyze a research study. [11.5]

44. A social researcher reads that the average age women in the United States get married is 27.4 yr but thinks that men are significantly older when they marry.
  a. Identify the population and parameter.
  b. State an appropriate research question for the study.
  c. State the null and alternative hypotheses.

45. The researcher decides to obtain data from the 360 weddings recorded at the county courthouse last year.
  a. Which type of research design is used?
  b. Describe a method of selecting a representative sample of size $n = 60$.
  c. Discuss any possible sampling error or research bias.

46. The researcher finds the average age of the men in the sample from exercise 45 is 29.1 yr with a standard deviation of 1.59 yr.
  a. Calculate the standardized test statistic and determine whether the null hypothesis should be rejected at a 0.05 level of significance.
  b. Interpret the results of the study.

14b. $\sigma_{\bar{x}} = \dfrac{3.5}{\sqrt{150}} \approx 0.286$

$z = \dfrac{64.5 - 64}{0.286} \approx 1.75$

$P(\bar{x} > 64.5) = P(z > 1.75)$

normalcdf$(1.75, 4) \approx 0.04$

15. $n = 15;\ \mu_{\bar{x}} = 9315;\ \mu = 10{,}000$

$\sigma = 1150;\ \sigma_{\bar{x}} = \dfrac{1150}{\sqrt{15}} \approx 297$

$z = \dfrac{9315 - 10{,}000}{297} \approx -2.31$

$P(\bar{x} < 9315) = P(z < -2.31)$

normalcdf$(-4, -2.31) \approx 0.01$

16. $n = 16;\ \mu_{\bar{x}} = 32.5;\ \mu = 32;\ \sigma = 0.77$

$\sigma_{\bar{x}} = \dfrac{0.77}{\sqrt{16}} = 0.1925$

$z = \dfrac{32.5 - 32}{0.1925} \approx 2.60$

$P(\bar{x} > 32.5) = P(z > 2.60)$

normalcdf$(2.60, 4) \approx 0.0047$

17. $n = 200;\ p = 0.31;\ q = 0.69$

$\mu = np = 200(0.31) = 62$

$nq = 200(0.69) = 138$

$\sigma = \sqrt{npq} = \sqrt{200(0.31)(0.69)} \approx 6.54$

17a. $z = \dfrac{54.5 - 62}{6.54} \approx -1.147$

$P(x_i \geq 54.5) = P(z \geq -1.15)$

normalcdf$(-1.15, 4) \approx 0.87$

17b. $z = \dfrac{75.5 - 62}{6.54} \approx 2.06$

$P(x_i > 75.5) = P(z > 2.06)$

normalcdf$(2.06, 4) \approx 0.019$

18. $n = 70;\ p = 0.28;\ q = 0.72$

$\mu = np = 70(0.28) = 19.6$

$nq = 70(0.72) = 50.4$

$\sigma = \sqrt{npq} = \sqrt{70(0.28)(0.72)} \approx 3.76$

18a. $z = \dfrac{19.5 - 19.6}{3.76} \approx -0.0266$

$P(x_i < 19.5) \approx P(z < -0.0266)$

normalcdf$(-4, -0.0266) \approx 0.49$

18b. $z_L = \dfrac{19.5 - 19.6}{3.76} \approx -0.03$

$z_U = \dfrac{25.5 - 19.6}{3.76} \approx 1.57$

$P(19.5 < x_i < 25.5)$

$= P(-0.03 < z < 1.57)$

normalcdf$(-0.03, 1.57) \approx 0.45$

19. $n = 60;\ p = 0.332;\ q = 0.668$

$\mu = np = 60(0.332) = 19.92$

$nq = 60(0.668) = 40.08;$ both ≥ 5

$\sigma = \sqrt{npq} = \sqrt{60(0.332)(0.668)} \approx 3.65$

$z = \dfrac{20.5 - 19.92}{3.65} \approx 0.16$

$P(x_i \leq 20.5) \approx P(z \leq 0.159)$

normalcdf$(-4, 0.159) \approx 0.56$

20a. $z_c = \text{invNorm}(0.9) \approx 1.28$

20b. $z_c = \text{invNorm}(0.925) \approx 1.44$

10c. population using 1-Var Stats, L3:

$\mu = 19.5;\ \sigma \approx 2.69$

sampling distribution using 1-Var Stats, L1, L2:

$\mu_{\bar{x}} = 19.5;\ \sigma_{\bar{x}} \approx 1.90$

$\mu_{\bar{x}} = \mu;\ \sigma_{\bar{x}} < \sigma$

11a. yes; The sampling distribution of the means is also normally distributed.

$\mu_{\bar{x}} = 5200;\ \sigma_{\bar{x}} = \dfrac{1450}{\sqrt{12}} \approx 418.58$

11b. yes; Since $n = 48 \geq 30$, the Central Limit Theorem applies.

$\mu_{\bar{x}} = 16;\ \sigma_{\bar{x}} = \dfrac{3.92}{\sqrt{48}} \approx 0.57$

13. $\mu = 243;\ \sigma_{\bar{x}} = \dfrac{328.25}{\sqrt{50}} \approx 46.42$

13a. $\mu_{\bar{x}} = 200;\ z = \dfrac{200 - 243}{46.42} \approx -0.93$

$P(\bar{x} < 200) \approx P(z < -0.93)$

normalcdf$(-4, -0.93) \approx 0.18$

13b. $z_L = \dfrac{125 - 243}{46.42} \approx -2.54$

$z_U = \dfrac{150 - 243}{46.42} \approx -2.00$

$P(125 < \bar{x} < 150)$

$\approx P(-2.54 < z < -2.00)$

normalcdf$(-2.54, -2) \approx 0.017$

14a. $z_L = \dfrac{58 - 64}{3.5} \approx -1.71$

$z_U = \dfrac{62 - 64}{3.5} \approx -0.57$

$P(58 < x_i < 62)$

$\approx P(-1.71 < z < -0.57)$

normalcdf$(-1.71, -0.57) \approx 0.24$

continued in Answers and Solutions Overflow

Overview

This chapter provides an introduction to calculus, one of the foremost branches of mathematics. Limits are evaluated and used to define the two major concepts in calculus, differentiation and integration. Differentiation involves finding the slope of a curve at any point and integration involves finding the area between a curve and an interval on the x-axis. Until this chapter, students have solved problems in which variables, such as speed, had an unchanging value. Calculus allows students to model problems with variables whose values vary.

Working with limits, the concept of infinity, and instantaneous rates of change through the study of calculus will extend the students' understanding and appreciation of God's creation from a biblical worldview.

Detailed instruction in the use of technology is integrated directly into the lessons of this chapter rather than given in a Technology Corner.

Flash

The typical speedboat can travel at speeds in excess of 90 mi/hr, with the fastest exceeding 300 mi/hr. Experience, weather conditions, maintenance, and physical limits all play an important role in safe boating speeds.

Chapter Objectives

1. To evaluate limits of functions and sequences

2. To determine instantaneous rates of change

3. To find the derivative of algebraic and trigonometric functions

4. To calculate the area under a curve using approximation methods and integration

5. To use the Fundamental Theorem of Calculus to evaluate integrals and to represent antiderivatives

6. To solve real-world problems involving limits, derivatives, and integrals

7. To identify influences of a biblical worldview of infinity on the development of mathematics

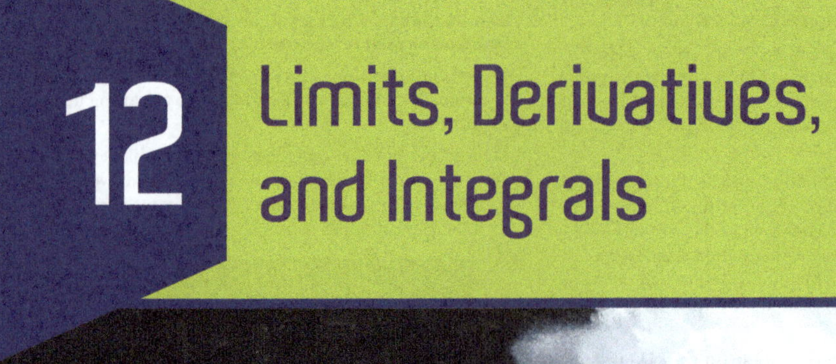

12 Limits, Derivatives, and Integrals

12.1 Evaluating Limits
12.2 Properties of Limits
Biblical Perspective of Mathematics
12.3 Tangents and Derivatives
12.4 Basic Derivative Theorems
12.5 Product, Quotient, and Chain Rules
Data Analysis
12.6 Area Under a Curve and Integration
12.7 The Fundamental Theorem of Calculus
Historical Connection

BIBLICAL PERSPECTIVE OF MATHEMATICS
From simple arithmetic to the most complex mathematics, there is no avoiding the concept of infinity. How should we think biblically about this idea we cannot fully comprehend?

DATA ANALYSIS
Whenever a process can be modeled mathematically, derivatives provide insight into instantaneous rates of change. In business and economics this is known as marginal analysis.

HISTORICAL CONNECTION
When you hear the name Berkeley, you probably think of a research university in the city of Berkeley, California. Both the university and the city were named after an Irish bishop who motivated mathematicians to place calculus on a firm foundation.

Suggested Teaching Schedule

DAY	
1	12.1
2	12.2
3	Quiz 12A (12.1–12.2) BPM: Mathematics and Infinity
4–5	12.3
6	12.4
7–8	12.5

DAY	
9	Quiz 12B (12.3–12.5) DA: Marginal Analysis in Economics
10	12.6
11–12	12.7
13	Quiz 12C (12.6–12.7) HC: *The Analyst*
14	Chapter 12 Review
15	Chapter 12 Test

12.1 Evaluating Limits

Limits govern our lives in many ways (Ps. 90:10).

The study of calculus focuses on two foundational problems related to the graph of a function: finding the equation of the line tangent to the curve and finding the area between the graph and the x-axis. The concept of a limit is essential in the solutions of both problems. Section 1.4 defined $\lim_{x \to a} f(x) = L$ if the function value $f(x)$ approaches a unique real number L as x approaches a from both sides.

Examining the function's graph over a small interval around $x = a$ will help you evaluate limits and prepare you for the more technical definition of a limit used in calculus.

Example 1   Finding Limits Graphically

Find each limit of the greatest integer function $f(x) = [x]$.

a. $\lim_{x \to 1} f(x)$
b. $\lim_{x \to -2.5} f(x)$

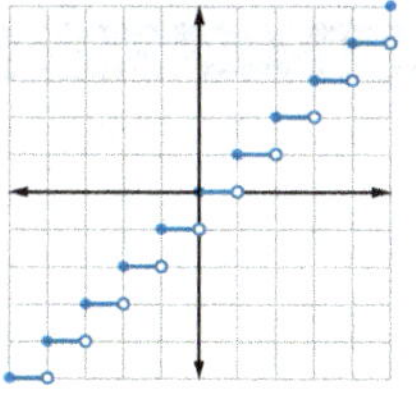

Answer

a. Consider the one-sided limits.
$\lim_{x \to 1^-} f(x) = 0$ and $\lim_{x \to 1^+} f(x) = 1$
$\therefore \lim_{x \to 1} f(x)$ does not exist.

Since the one-sided limits are not equal at $x = 1$, the function does not have a limit as $x \to 1$.

b. Consider the one-sided limits.
$\lim_{x \to -2.5^-} f(x) = -3$ and $\lim_{x \to -2.5^+} f(x) = -3$
$\therefore \lim_{x \to -2.5} f(x) = -3$

Since both one-sided limits are equal to -3, the two-sided limit exists.

SKILL ✓ EXERCISE 5

Recall from Section 1.4 that if the function is discontinuous at $x = a$, the limit exists only if both one-sided limits are equal to a unique real number L. You also learned that $\lim_{x \to a} f(x) = f(a)$ if the function is continuous over an interval of x containing a. Therefore the limits of polynomial, exponential, logarithmic, sinusoidal, or any other continuous functions can be found using direct substitution.

12.1 Evaluating Limits

Objectives

1. To evaluate limits of functions at points of continuity using direct substitution

2. To use limits to describe the end behavior of a function

3. To use limits to describe the behavior of functions at points of discontinuity

4. To find limits of sequences

5. To solve real-world problems involving limits

Flash

In many countries, signs with a red circled number indicate the speed limit. 80 km/hr is about 50 mi/hr. In 1652 the first speed limit in America was imposed in what is now New York City. Wagons and carts were restricted from being driven at a gallop. In 1901 Connecticut established the first speed limits for motor vehicles as 12 mi/hr in cities and 15 mi/hr on country roads.

Vocabulary

infinite limit

PRESENTATION

Lesson Opener

Describe the end behavior of each function using limit notation.

1. $f(x) = x^2$ $\lim_{x \to -\infty} f(x) = \infty$; $\lim_{x \to \infty} f(x) = \infty$

2. $f(x) = x^3$ $\lim_{x \to -\infty} f(x) = -\infty$; $\lim_{x \to \infty} f(x) = \infty$

3. Identify and classify any discontinuity of $g(x) = \dfrac{3x}{x^2 - 4}$.   VA: $x = \pm 2$

Limits are foundational in the development of the two main concepts of calculus, derivatives and integrals. This section reviews and extends concepts of limits presented in Section 1.4.

Consider discussing limits on our lives, using Psalm 90:10 to remind the students of the need to be wise stewards of the limited time God has given us.

Review the definition of a limit. The students should recognize that, by definition, the limits found in the Lesson Opener are not true limits because ∞ is not a real number.

The greatest integer function is used in Example 1 to provide a function in which the left-hand limit is often not equal to the right-hand limit. This function also provides multiple discontinuities not restricted to limits that approach infinity.

Note that when both one-sided limits are not equal, $\lim_{x \to a} f(x)$ does not exist. If the function is continuous at a, then $\lim_{x \to a} f(x)$ will always equal $f(a)$.

Common Student Error A student may think that the limit as a function approaches a is always the value of the function at a, but this is only true when the function is continuous at $x = a$. If the function is discontinuous at $x = a$, the limit of the function may not exist or may be different from $f(a)$. Emphasize that a limit is a value approached that is "in the neighborhood" of a, not necessarily the value at $f(a)$.

Reading and Writing Mathematics

Explain why $\lim\limits_{x\to 0}\ [x] \neq 0$ even though $f(0) = 0$.

$\lim\limits_{x\to 0^-}[x] = -1$, but $\lim\limits_{x\to 0^+}[x] = 0$. Because these limits are not equal, $\lim\limits_{x\to 0}[x]$ does not exist.

Additional Exercises

Graph each function and evaluate the limit.

1. $\lim\limits_{x\to 1} 2\ln(x+3)\ \approx 2.77$

2. $\lim\limits_{x\to \frac{\pi}{4}} 2\tan x\quad 2$

3. $\lim\limits_{x\to -2}\sqrt{x+2}$
does not exist since $D = [-2, \infty)$

4. Use limits to describe the end behavior of each function.
   a. $g(x) = x^3 + x$
   $\lim\limits_{x\to \infty} g(x) = \infty;\ \lim\limits_{x\to -\infty} g(x) = -\infty$
   b. $h(x) = -x^2 + 4x - 2$
   $\lim\limits_{x\to \pm\infty} h(x) = -\infty$

5. Find each limit of $f(x) = \dfrac{3x^2}{(x+3)(x-2)}$.
   a. $\lim\limits_{x\to \infty} f(x)\quad 3$
   b. $\lim\limits_{x\to -3} f(x)\quad$ does not exist

6. Find each limit of
   $f(x) = \dfrac{2x^2}{(x+3)^2(3x-5)}$.
   a. $\lim\limits_{x\to -\infty} f(x)\quad 0$
   b. $\lim\limits_{x\to -3} f(x)\quad \infty$

7. Find the limit of each sequence.
   a. $a_n = \dfrac{2a+3}{3a-4}\quad \dfrac{2}{3}$
   b. $b_n = (0.8)^n\quad 0$

8. A car has an initial value of \$38,000 and depreciates 15% per year.
   a. Write a function $f(x)$ modeling the value of the car after x years.
   $f(x) = 38{,}000(0.85)^x$
   b. Use technology to find the value of the car after 8 yr. \$10,354.64
   c. During which year will the car's value become less than \$2500?
   seventeenth year

The formula modeling continuously compounded interest, $P = e^{rt}$, is derived by evaluating the limit as $n \to \infty$ of the compound interest formula, $A = P\left(1 + \frac{r}{n}\right)^{nt}$.

Find each limit.

a. $\lim\limits_{x\to -8}(x^2 - 6x + 17)$

b. $\lim\limits_{x\to 4} 5\ln(x+3)$

c. $\lim\limits_{x\to \frac{\pi}{6}}\cos 2x$

Answer

a. $\begin{aligned}\lim\limits_{x\to -8}(x^2 - 6x + 17) &= (-8)^2 - 6(-8) + 17\\ &= 64 + 48 + 17 = 129\end{aligned}$ — Since quadratic functions are continuous, use direct substitution.

b. $\lim\limits_{x\to 4} 5\ln(x+3) = 5\ln(4+3) \approx 9.73$ — Since logarithmic functions are continuous, use direct substitution.

c. $\lim\limits_{x\to \frac{\pi}{6}}\cos 2x = \cos 2\left(\frac{\pi}{6}\right) = \cos\frac{\pi}{3} = \frac{1}{2}$ — Since sinusoidal functions are continuous, use direct substitution.

SKILL ✓ EXERCISE 9

In Chapters 1 and 2, limit notation was used to describe the end behavior of a function and the function's behavior at vertical asymptotes and point discontinuities. Recall that the end behavior of a polynomial function can be determined by examining the leading term, since other terms become relatively insignificant as $x \to \pm\infty$.

Use limits to describe the end behavior of $g(x) = x^3 - 8x - 57$ and $h(x) = -x^4 + 3x^2 + 9x$.

Answer

a. $\lim\limits_{x\to \infty} g(x) = \infty$ and $\lim\limits_{x\to -\infty} g(x) = -\infty$ — The end behavior of $g(x)$ matches that of $f(x) = x^3$.

b. $\lim\limits_{x\to \pm\infty} h(x) = -\infty$ — The end behavior of $h(x)$ matches that of a reflection of $f(x) = x^4$ across the x-axis.

SKILL ✓ EXERCISE 15

Note that *infinite limits* such as $\lim\limits_{x\to \infty} g(x) = \infty$ describe a function's behavior, but they are not true limits since $g(x)$ does not approach a unique real number L.

Evaluate each limit for $f(x) = \dfrac{2x^4}{x(x-3)(x+1)^2}$.

a. $\lim\limits_{x\to \pm\infty} f(x)$

b. $\lim\limits_{x\to -1} f(x)$

c. $\lim\limits_{x\to 3} f(x)$

d. $\lim\limits_{x\to 0} f(x)$

Answer

a. $\lim\limits_{x\to \pm\infty} f(x) = 2$ — Since the degrees of the polynomials in the numerator and the denominator are the same, $f(x)$ has a horizontal asymptote at $y = 2$.

b. $\lim\limits_{x\to -1^-} f(x) = \infty$ and $\lim\limits_{x\to -1^+} f(x) = \infty$
$\therefore \lim\limits_{x\to -1} f(x) = \infty$ — Since the function approaches infinity on both sides of $x = -1$, its behavior as $x \to -1$ is described with an infinite limit.

c. $\lim\limits_{x\to 3^-} f(x) = -\infty$ and $\lim\limits_{x\to 3^+} f(x) = \infty$
$\therefore \lim\limits_{x\to 3} f(x)$ does not exist. — Since the function approaches $-\infty$ on one side of $x = 3$ and ∞ on the other side, its behavior as $x \to 3$ cannot be described using a limit.

CONTINUED ➡

Example 2 enforces that limits of continuous functions can be found by substitution. Remind students that logarithmic and some trigonometric functions have limited domains. A limited domain can prevent the existence of a left- or right-hand limit, as in exercise 30 and Additional Exercise 3.

The limits in Example 3 can be determined numerically by mentally substituting progressively larger values as $x \to \pm\infty$ or by recalling the end behavior of the parent functions. Consider sketching or projecting the graphs of $f(x) = x^3$ and $f(x) = -x^4$ or having the students graph these functions using technology. The limits in Example 3 are not true limits and are often referred to as *infinite limits* since ∞ is not a real number.

Limits are found by analyzing the expression and its graph in Example 4. Consider reviewing the table describing end behavior of rational functions on page 107 after completing part a. Part d illustrates that $\lim\limits_{x\to a} f(x)$ can exist even though $f(a)$ is undefined.

A graphing calculator or a spreadsheet can be used to help visualize the limits of the sequences in Example 5. These limits could also be viewed as the end behavior as $x \to \infty$ of the related function. Note that the values in Y2 would alternate

d. $\lim\limits_{x\to 0^-} f(x) = 0$ and $\lim\limits_{x\to 0^+} f(x) = 0$

$\therefore \lim\limits_{x\to 0} f(x) = 0$

Since the function approaches the same real number from both sides of $x = 0$, the limit as $x \to 0$ exists.

_______________________ SKILL ✔ **EXERCISE 29**

In Example 4d the fact that $f(x)$ is not defined at $x = 0$ but $\lim\limits_{x\to 0} f(x)$ is a real number indicates a point discontinuity. The fact that $\lim\limits_{x\to 0} f(x) \neq f(0)$ illustrates an important property of limits: a limit describes what occurs to the function's values as the function approaches (or is in the neighborhood of) $x = a$. If a function is discontinuous at $x = a$, then $\lim\limits_{x\to a} f(x) \neq f(a)$.

When finding the limit of a sequence as $n \to \infty$, consider the effect of substituting large numbers for the variable.

Example 5 Finding Limits of Sequences

Find the limit of each sequence.

a. $a_n = \dfrac{2n + 7}{3n - 4}$ **b.** $b_n = (-1)^n$ **c.** $c_n = \left(-\dfrac{2}{3}\right)^n$ **d.** $d_n = -\dfrac{2}{3}n$

Answer

X	Y1	Y2	Y3	Y4
1	-9	-1	-0.667	-0.667
2	5.5	1	0.4444	-1.333
3	2.6	-1	-0.296	-2
4	1.875	1	0.1975	-2.667
5	1.5455	-1	-0.132	-3.333
100	0.6993	1	2E-18	-66.67
1000	0.6699	1	0	-666.7
10000	0.667	1	0	-6667
100000	0.6667	1	0	-66667

$Y_1=(2X+7)/(3X-4)$

Define $Y_1 = a_n$, $Y_2 = b_n$, $Y_3 = c_n$, and $Y_4 = d_n$. In the TABLE SETUP menu set Indpnt: to Ask and Depend: to Auto. Then Select Table and enter increasingly larger values for X.

a. $\lim\limits_{n\to\infty} a_n = \lim\limits_{n\to\infty} \dfrac{2n + 7}{3n - 4} = \dfrac{2}{3}$

As n increases, a_n converges to $\dfrac{2}{3}$.

b. $\lim\limits_{n\to\infty} b_n = \lim\limits_{n\to\infty} (-1)^n$ does not exist.

As n increases, b_n alternates between 1 and −1 and the sequence is divergent.

c. $\lim\limits_{n\to\infty} c_n = \lim\limits_{n\to\infty} \left(-\dfrac{2}{3}\right)^n = 0$

As n increases, c_n converges to 0.

d. $\lim\limits_{n\to\infty} d_n = \lim\limits_{n\to\infty} -\dfrac{2}{3}n = -\infty$

As n increases, d_n continues to decrease and the sequence is divergent.

_______________________ SKILL ✔ **EXERCISE 33**

You can use limits and graphing calculators to model many real-world situations.

Example 6 Applying Limits to Depreciation

Philip buys a new car for $45,000. The estimated depreciation is 20% per year.

a. Write a function $f(x)$ modeling the value of the car after x years.

b. Use technology to graph the function and find $\lim\limits_{x\to\infty} f(x)$.

c. Find the value of the car after 10 yr.

d. When will the car's estimated value drop below $1000?

Answer

a. $f(x) = 45{,}000(0.8)^x$

The modeling function will be exponential with a rate of $100\% - 20\% = 80\%$.

CONTINUED ➡

if the X values were sequential, but the last four X values are even terms whose values are all positive. Point out that c_n is a geometric sequence with $|r| < 1$ so it must converge to 0, and that d_n is an arithmetic sequence so it must be divergent.

Example 6 explores a real-world application of limits. Discuss algebraic and graphical methods for answering parts c and d. Options include evaluating $f(10)$, using [TRACE] on the graph of the function, solving the equation $45{,}000(0.8)^x = 10$, and adding the graph of $y = 1000$ and finding the x-value of the intersection of the graphs. Although it may be obvious that the car's estimated value will approach \$0, this is an example of the important concept of limits.

TIPS

Ex. 8–31 Consider encouraging the students to verify their answers graphically.

Ex. 30 The limit does not exist because the domain is not defined for the function when $x < 5$.

A table can also be used to examine the values of the function near $x = 0$.

X	Y1
-0.5	0.2857
-0.4	0.1046
-0.3	0.0334
-0.2	0.0078
-0.1	8E-4
0	ERROR
0.1	-6E-4
0.2	-0.004
0.3	-0.012
0.4	-0.025
0.5	-0.044

$Y_1=2X^4/(X(X-3)(X+1)^2)$

In a 1693 letter Leibniz wrote, "I am so much in favor of the actual infinite that instead of admitting that nature abhors it, as is commonly said, I hold that it assumes it everywhere, in order to better show the perfections of its author."

Assignments

- **Minimum:** 1–5, 9–11, 15, 17–18, 20, 23, 25, 28, 31–32, 35–36, 41, 46–49, 52, 55

- **Standard:** 1–5, 10; 11–15 odd; 16–20 even; 26, 28, 31; 32–40 even; 44–54 even; 55

- **Extended:** 2, 5–7, 11–12, 14; 18–24 even; 28–36 even; 39–40, 42–44; 46–50 even; 54–55

Solutions

❯ A. Exercises

8. A linear function is continuous.
$7(2) - 5 = 9$

9. An absolute value function is continuous.
$|-1 - 4| = |-5| = 5$

10. A cubic function is continuous.
$(-2)^3 + 2(-2) - 6 = -18$

11. $\lim\limits_{x\to -3^-} [x + 1] = -3$
$\lim\limits_{x\to -3^+} [x + 1] = -2$; does not exist

12. A sine function is continuous.
$4\sin\dfrac{7\pi}{6} = 4(-0.5) = -2$

13. A radical function is continuous.
$\sqrt{17 - 8} = \sqrt{9} = 3$

14. an even degree polynomial with a positive leading coefficient

15. an odd degree polynomial with a negative leading coefficient

❯ B. Exercises

16. $D = \mathbb{R}$; $\lim\limits_{x\to -3} |x| = |-3| = 3$

17. $D = \mathbb{R}$
$\lim\limits_{x\to -1} (5x^4 + 2x - 7)$
$= 5(-1)^4 + 2(-1) - 7 = -4$

18. $x - 8 > 0$; $x > 8$; $D = (8, \infty)$
$\lim\limits_{x\to 9} \ln (x - 8) = \ln (9 - 8) = \ln 1 = 0$

19. $D = \{x \mid x \neq 0\}$
$x \neq 0$ (vertical asymptote)
$\lim\limits_{x\to 0^+} \dfrac{3}{x^2} = \infty$; $\lim\limits_{x\to 0^-} \dfrac{3}{x^2} = \infty$

20. $2x + 6 \geq 0$; $x \geq -3$; $D = [-3, \infty)$
$\lim\limits_{x\to 5} \sqrt{2x + 6} = \sqrt{2(5) + 6} = 4$

21. $\cos x \neq 0$ so
$D = \left\{x \mid x \neq (2n+1)\frac{\pi}{2}, n \in \mathbb{Z}\right\}$
$x \neq (2n+1)\frac{\pi}{2}$ (vertical asymptotes)
$\lim\limits_{x \to \frac{\pi}{2}^-} \tan x = \infty;\ \lim\limits_{x \to \frac{\pi}{2}^+} \tan x = -\infty$
The limit does not exist since the one-sided limits are not equal.

22. $x^2 - 49 \neq 0;\ x = \pm 7$
$D = \{x \mid x \neq \pm 7\}$
$\lim\limits_{x \to 7^-} \dfrac{x+5}{x^2-49} = -\infty;\ \lim\limits_{x \to 7^+} \dfrac{x+5}{x^2-49} = \infty$
The limit does not exist since the one-sided limits are not equal.

23. $x^2 + 1 > 0$; true; $D = \mathbb{R}$
$\lim\limits_{x \to -1} \dfrac{4}{x^2+1} = \dfrac{4}{(-1)^2+1} = 2$

28. $\lim\limits_{x \to 4^-} \dfrac{2}{(x-4)^2} = \infty;\ \lim\limits_{x \to 4^+} \dfrac{2}{(x-4)^2} = \infty$

29. $\lim\limits_{x \to 2^-} \dfrac{3}{x-2} = -\infty;\ \lim\limits_{x \to 2^+} \dfrac{3}{x-2} = \infty$
The limit does not exist since the one-sided limits are not equal.

30. Since $D = [5, \infty)$, $\lim\limits_{x \to 5^-} \sqrt{x-5}$ does not exist.

31. $\lim\limits_{x \to 0} \sqrt{x+4} = \sqrt{0+4} = \sqrt{4} = 2$

32. $a_{1000} = \dfrac{3 - 2(1000)}{1000} = \dfrac{-1997}{1000} \approx -2$
$a_{1,000,000} = \dfrac{3 - 2(1,000,000)}{1,000,000}$
$= \dfrac{-1,999,997}{1,000,000} \approx -2$
$\lim\limits_{n \to \infty} a_n = \lim\limits_{n \to \infty} \dfrac{3 - 2n}{n} = -2$

33. $a_{1000} = \dfrac{1001}{3000} \approx \dfrac{1}{3}$
$a_{1,000,000} = \dfrac{1,000,001}{3,000,000} \approx \dfrac{1}{3}$
$\lim\limits_{n \to \infty} a_n = \lim\limits_{n \to \infty} \dfrac{n+1}{3n} = \dfrac{1}{3}$

34. $a_n = \dfrac{8}{4n^3} = \dfrac{2}{n^3}$
$a_{1000} = \dfrac{2}{1000^3} \approx 0$
$a_{1,000,000} = \dfrac{2}{1,000,000^3} \approx 0$
$\lim\limits_{n \to \infty} a_n = \lim\limits_{n \to \infty} \dfrac{8}{4n^3} = 0$

35. $a_{1000} = 3 + \dfrac{1}{1000} = 3.001$
$a_{1,000,000} = 3 + \dfrac{1}{1,000,000} \approx 3$
$\lim\limits_{n \to \infty} a_n = \lim\limits_{n \to \infty} \left(3 + \dfrac{1}{n}\right) = 3$

36. Since this is a constant sequence, all terms equal 5.
$\lim\limits_{n \to \infty} a_n = \lim\limits_{n \to \infty} 5 = 5$

37. $a_{1000} = 2(1000) = 2000$
$a_{1,000,000} = 2(1,000,000) = 2,000,000$
$\therefore$ the limit $\to \infty$

---

b. 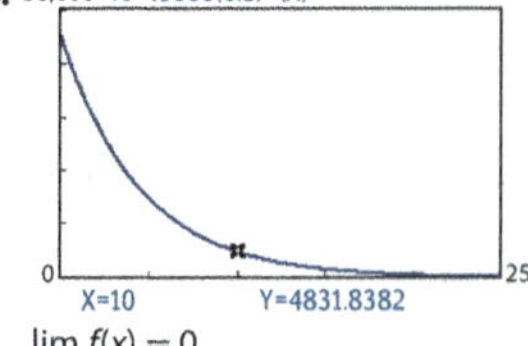

$\lim\limits_{x \to \infty} f(x) = 0$

The graph of $f(x)$ has a horizontal asymptote at $y = 0$.

c. $f(10) = \$4831.84$

d. During the 18th year the estimated value will drop below \$1000.

A table of values can also be helpful.

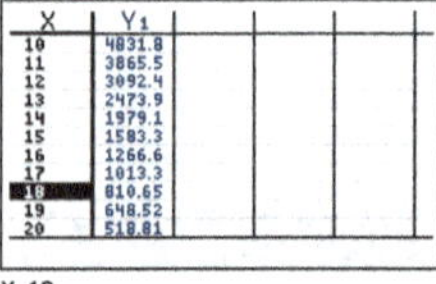

X	Y₁
10	4831.8
11	3865.5
12	3092.4
13	2473.9
14	1979.1
15	1583.3
16	1266.6
17	1013.3
18	810.65
19	648.52
20	518.81

X=18

The key concepts in solving fundamental problems in calculus, the derivative and the integral, are both defined in terms of limits. Limit theorems that make the deductive development of the derivative possible are discussed in the following sections.

❯ A. Exercises

Use the graph of
$f(x) = \dfrac{x(x-3)(x+1)}{x(x+2)(x-1)^2}$
to find each limit.

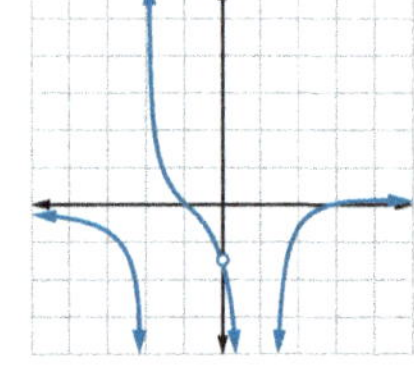

1. $\lim\limits_{x \to \infty} f(x)$ 0
2. $\lim\limits_{x \to 0} f(x)$ $-\frac{3}{2}$
3. $\lim\limits_{x \to -2^+} f(x)$ ∞
4. $\lim\limits_{x \to -2^-} f(x)$ $-\infty$
5. $\lim\limits_{x \to -2} f(x)$ does not exist
6. $\lim\limits_{x \to 1} f(x)$ $-\infty$
7. $\lim\limits_{x \to 3} f(x)$ 0

Evaluate each limit, if it exists.

8. $\lim\limits_{x \to 2} (7x - 5)$ 9
9. $\lim\limits_{x \to -1} |x - 4|$ 5
10. $\lim\limits_{x \to -2^+} (x^3 + 2x - 6)$ -18
11. $\lim\limits_{x \to -3} [x + 1]$ does not exist
12. $\lim\limits_{x \to \frac{7\pi}{6}} (4 \sin x)$ -2
13. $\lim\limits_{x \to 17} \sqrt{x - 8}$ 3

Use limits to describe the end behavior of each polynomial function.

14. $g(x) = x^6 - 3x^3 + 94$
$\lim\limits_{x \to \pm\infty} g(x) = \infty$
15. $f(x) = -x^3 - 7x + 5$
$\lim\limits_{x \to -\infty} f(x) = \infty$
$\lim\limits_{x \to \infty} f(x) = -\infty$

❯ B. Exercises

Use technology to graph each function. Then state the function's domain and evaluate the limit of the function at the given value of x.

16. $f(x) = |x|; x \to -3$ $D = \mathbb{R}; 3$
17. $g(x) = 5x^4 + 2x - 7; x \to -1$ $D = \mathbb{R}; -4$
18. $h(x) = \ln(x - 8); x \to 9$ $D = (8, \infty); 0$
19. $n(x) = \dfrac{3}{x^2}; x \to 0$ $D = \{x \mid x \neq 0\}; \infty$
20. $m(x) = \sqrt{2x + 6}; x \to 5$ $D = [-3, \infty); 4$
21. $j(x) = \tan x; x \to \dfrac{\pi}{2}$ $D = \left\{x \mid x \neq (2n+1)\frac{\pi}{2}, n \in \mathbb{Z}\right\}$ does not exist
22. $p(x) = \dfrac{x+5}{x^2-49}; x \to 7$ $D = \{x \mid x \neq \pm 7\}$; does not exist
23. $q(x) = \dfrac{4}{x^2+1}; x \to -1$ $D = \mathbb{R}; 2$

Write the equation of the asymptote in the graph of $f(x)$ that is implied by each limit.

24. $\lim\limits_{x \to -\infty} f(x) = 7$ $y = 7$
25. $\lim\limits_{x \to -2} f(x) = -\infty$ $x = -2$
26. $\lim\limits_{x \to 5} f(x) = \infty$ $x = 5$
27. $\lim\limits_{x \to \infty} f(x) = -4$ $y = -4$

Use technology to graph each function. Then state the limit, if it exists. If the limit does not exist, explain why.

28. $\lim\limits_{x \to 4} \dfrac{2}{(x-4)^2}$ ∞
29. $\lim\limits_{x \to 2} \dfrac{3}{x-2}$ does not exist
30. $\lim\limits_{x \to 5} \sqrt{x - 5}$ does not exist
31. $\lim\limits_{x \to 0} \sqrt{x + 4}$ 2

---

38. exponential growth function,
$\therefore$ the limit $\to \infty$

39. The sign will alternate, $\therefore$ the limit does not exist.

40c. $f(x) = 359,000(0.7)^x$

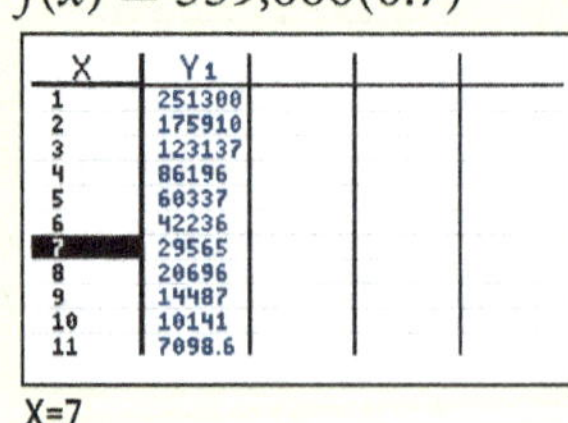

X	Y₁
1	251300
2	175910
3	123137
4	86196
5	60337
6	42236
7	29565
8	20696
9	14487
10	10141
11	7098.6

X=7

41c. $f(x) = 4,575,000(1.06)^x$

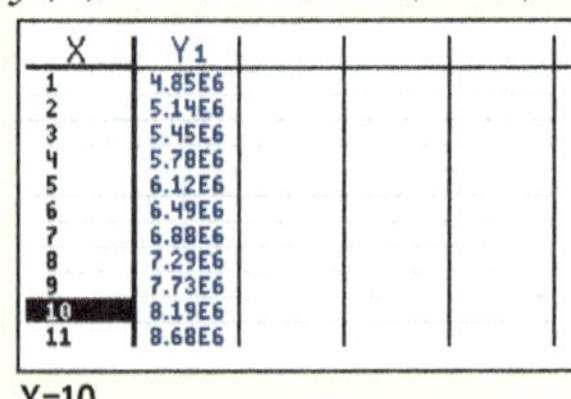

X	Y₁
1	4.85E6
2	5.14E6
3	5.45E6
4	5.78E6
5	6.12E6
6	6.49E6
7	6.88E6
8	7.29E6
9	7.73E6
10	8.19E6
11	8.68E6

X=10

Find the limit of the each sequence.

32. $a_n = \frac{3 - 2n}{n}$ -2 **33.** $a_n = \frac{n+1}{3n}$ $\frac{1}{3}$

34. $a_n = \frac{8}{4n^3}$ 0 **35.** $a_n = 3 + \frac{1}{n}$ 3

36. $a_n = 5$ 5 **37.** $a_n = 2n$ ∞ (divergent)

38. $a_n = 4(1.5)^n$ ∞ (divergent) **39.** $a_n = (-2)^n$ does not exist

40. A construction company buys a new piece of equipment for \$359,000. The estimated depreciation is 30% per year.

  a. Write a function $f(x)$ modeling the value of the equipment after x years. $f(x) = 359{,}000(0.7)^x$

  b. Find $\lim\limits_{x \to \infty} f(x)$. 0

  c. In which year will the equipment's value become less than \$30,000? 7th year

41. A real estate investor buys a new property for \$4.575 million. The estimated appreciation is 6% per year.

  a. Write a function $f(x)$ modeling the value of the property after x years. $f(x) = 4{,}575{,}000(1.06)^x$

  b. Find $\lim\limits_{x \to \infty} f(x)$. ∞

  c. In which year will the property's value become greater than \$8 million? tenth year

CUMULATIVE REVIEW

45. Write the slope-intercept form equation of a line passing through $(3, 7)$ and parallel to $y = \frac{4}{3}x + 2$. [1.2] $y = \frac{4}{3}x + 3$

46. If $f(x) = 3x + 7$, find the simplified expression for $\frac{f(x+h) - f(x)}{h}$. [1.7] 3

Evaluate each infinite geometric series. [9.3]

47. $5 + 1 + \frac{1}{5} + \cdots$ $\frac{25}{4}$ **48.** $\sum\limits_{n=1}^{\infty} 6\left(\frac{1}{4}\right)^n$ 2

Determine the p-value and then use it to determine whether to reject the null hypothesis at a 0.05 level of significance. [11.4]

49. $H_0: \mu \geq 84;\ H_a: \mu < 84$

  $\bar{x} = 82,\ s = 5.92,\ n = 32$ $p \approx 0.028$; Reject H_0.

50. $H_0: \mu = 850;\ H_a: \mu \neq 850$

  $\bar{x} = 878,\ s = 123,\ n = 55$ $p \approx 0.091$; Fail to reject H_0.

51. Find the length of side a (to the nearest tenth) in $\triangle ABC$ with $\angle A = 51°$, $b = 12$, and $c = 18$. [5.7] E

  A. 196.1   **C.** 14.2   **E.** none of these

  B. 21.6   **D.** 6

C. Exercises

42. Graph the function $f(x) = e^x + x^e + \ln(x + e)$.

  a. State the domain. $[0, \infty)$ **b.** Find $f(0)$. 2

  c. Find $\lim\limits_{x \to 0^+} f(x)$. 2 **d.** Find $\lim\limits_{x \to 0} f(x)$. does not exist

  e. Why should the graph of an unfamiliar function be considered when finding limits?

43. Find $\lim\limits_{x \to -\infty} r(x)$ and $\lim\limits_{x \to \infty} r(x)$ for each case of the rational function $r(x) = \frac{ax^m}{bx^n}$.

  a. $m = n$ $\frac{a}{b}; \frac{a}{b}$

  b. $m < n$ $0; 0$

  c. $m - n > 0$ and even, $ab > 0$ $\infty; \infty$

  d. $m - n > 0$ and even, $ab < 0$ $-\infty; -\infty$

  e. $m - n > 0$ and odd, $ab < 0$ $\infty; -\infty$

  f. $m - n > 0$ and odd, $ab > 0$ $-\infty; \infty$

44. Find $\lim\limits_{n \to \infty} a_n$ for each sequence.

  a. $a_n = a_0 + nd$ and $d < 0$ $-\infty$

  b. $a_n = a_0 + nd$ and $d > 0$ ∞

  c. $a_n = a_0 r^n$ and $r < 0$ does not exist

  d. $a_n = a_0 r^n$ and $0 < r < 1$ 0

  e. $a_n = a_0 r^n$ and $r = 1$ a_0

  f. $a_n = a_0 r^n$, $a_0 > 0$, and $r > 1$ ∞

52. Use technology to find the probability that a randomly selected value in a normal distribution with $\mu = 83$ and $\sigma = 12.7$ is within $[69, 92]$. [10.6] C

  A. 13.6%   **C.** 62.6%   **E.** none of these

  B. 23.9%   **D.** 86.4%

53. Which sampling technique is used when the population is divided into distinguishable groups and members of each group are randomly selected for the sample? [11.5] D

  A. random   **D.** stratified random

  B. convenience   **E.** none of these

  C. cluster

54. Evaluate $\sum\limits_{i=1}^{n} c$. [9.5] C

  A. c **C.** nc **E.** none of these

  B. i **D.** n

Cumulative Review

45. $m_{\parallel} = m_G = \frac{4}{3}$

$7 = \frac{4}{3}(3) + b$

$7 = 4 + b$

$b = 3;\ y = \frac{4}{3}x + 3$

46. $\dfrac{3(x+h) + 7 - (3x + 7)}{h}$

$= \dfrac{3x + 3h + 7 - 3x - 7}{h} = \dfrac{3h}{h} = 3$

47. $a_1 = 5;\ r = \frac{1}{5};\ S = \dfrac{5}{1 - \frac{1}{5}} = \dfrac{5}{\frac{4}{5}} = \dfrac{25}{4}$

48. $a_1 = \frac{3}{2};\ r = \frac{1}{4};\ S = \dfrac{\frac{3}{2}}{1 - \frac{1}{4}} = \dfrac{\frac{3}{2}}{\frac{3}{4}} = 2$

49. $z = \dfrac{\bar{x} - \mu}{\frac{s}{\sqrt{n}}} = \dfrac{82 - 84}{\frac{5.92}{\sqrt{32}}} \approx -1.911$

$p = \text{normalcdf}(-4, -1.911) \approx 0.028$

50. $z = \dfrac{\bar{x} - \mu}{\frac{s}{\sqrt{n}}} = \dfrac{878 - 850}{\frac{123}{\sqrt{55}}} \approx 1.688$

$p = 2 \times \text{normalcdf}(1.688, 4) \approx 0.091$

51.

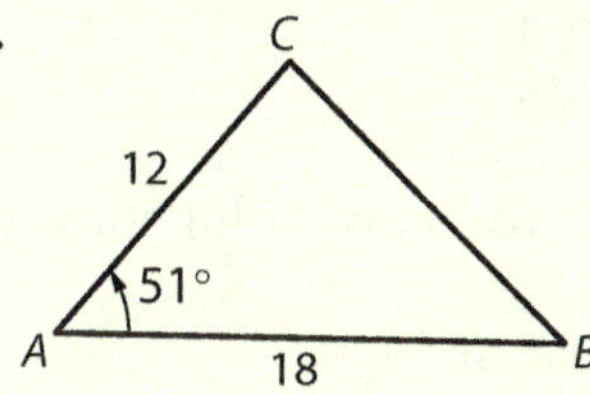

$a^2 = 12^2 + 18^2 - 2(12)(18)\cos 51°$

$\approx 196.1;\ a = 14.0$

52. $z_L = \dfrac{69 - 83}{12.7} = -1.10$

$z_U = \dfrac{92 - 83}{12.7} = 0.71$

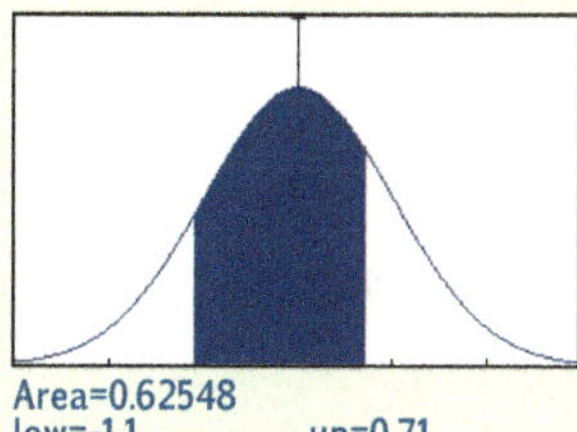

$p \approx 0.626$

54. Multiplication is repeated addition. (See the table on p. 478.)

C. Exercises

42a–b. e^x has $D = \mathbb{R}$; x^e has $D = [0, \infty)$

$\ln(x + e)$ has $D = (-e, \infty)$

$D_f = [0, \infty)$ (the intersection)

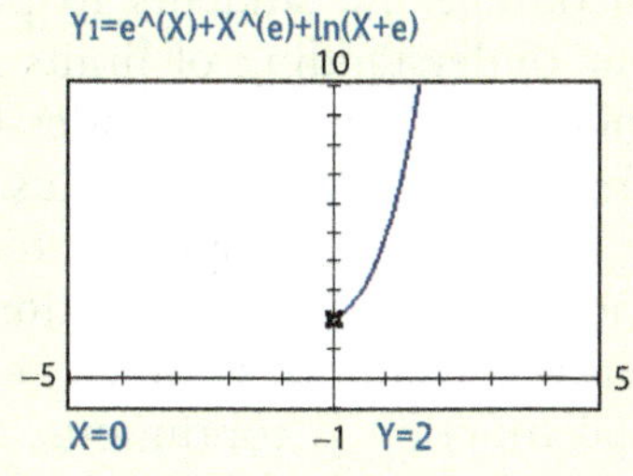

42b. $e^0 + 0^e + \ln(0 + e) = 1 + 0 + 1 = 2$

42d. The limit does not exist since the left-hand limit does not exist.

42e. The function value may not be the limit. Check to make sure the function is continuous and that the right- and left-hand limits are equal.

12.2 Properties of Limits

Objectives

1. To apply properties of limits to evaluate limits of polynomial and rational functions algebraically

2. To evaluate limits at points of discontinuity algebraically using canceling, rationalizing the numerator, or trigonometric identities

Flash

Encourage your students to spend their lives investing in the lives of others (Heb. 6:10; 1 Pet. 4:10; Col. 3:23–24; Gal. 5:13).

Vocabulary

indeterminate form
limit of a constant
limit of a power
limit of a product
limit of a quotient
limit of a reciprocal power function at
  infinity
limit of a scalar multiple
limit of a sum
limit of the identity function

Reading and Writing Mathematics

If $f(x) = a_n x^n + a_{n-1} x^{n-1} + \cdots + a_1 x + a_0$ and $g(x) = b_n x^m + b_{n-1} x^{m-1} + \cdots + b_1 x + b_0$, state the possible values for $\lim\limits_{x \to \infty} \dfrac{f(x)}{g(x)}$, when each occurs, and why that is the limit.

0 when $n < m$, since the higher degree polynomial in the denominator grows faster than the polynomial in the numerator.
$\dfrac{a_n}{b_n}$ when $n = m$, since the leading terms grow in this ratio and the other terms eventually become insignificant.
$\pm\infty$ when $n > m$, since the higher degree polynomial in the numerator grows faster than the polynomial in the denominator.

Investing in the lives of others produces eternal rewards.

After completing this section, you will be able to

- apply properties of limits to evaluate limits algebraically.
- evaluate limits by canceling, rationalizing the numerator, and using trig identities.

The French mathematician Augustin-Louis Cauchy (1789–1857) proved many of these properties and established the theoretical foundation of calculus. He wrote so much that the periodical in which many of his proofs were published, *Comptes Rendus*, decided not to accept any article exceeding four pages, a rule still in use today.

The concept of limits can be used to illustrate an important truth. If you lived 80 yr and there was no life after death, your life on the earth would be $\frac{80}{80}$, or 100%, of your existence. If your life after death was 80 yr long, your earthly life would be $\frac{80}{160}$, or 50%, of your entire existence. If your life after death was 720 yr, your life here would be only $\frac{80}{80+720}$, or 10%, of your existence. Extending this pattern to an eternal life after death, $\lim\limits_{x \to \infty} \frac{80}{80+x} = 0$, demonstrates that the duration of life spent on earth is insignificant in comparison to the duration spent in eternity. In light of eternity, earthly life and its current joys and troubles are like a vapor, appearing for a little time and then vanishing away (James 4:14). Any selfish achievements are also worthless. Only what you do for the Lord becomes important to the infinite, eternal God whom you serve. With the little time we have here on earth, we should focus more on things with eternal, heavenly benefits than on those with temporary, earthly benefits (Col. 3:2).

In the previous section we intuitively evaluated limits using graphs and tables of values. Algebraic methods of evaluating limits are necessary if limits are to serve as the foundation of calculus. The following properties of limits (some of which simply state what we have already learned about continuous functions and direct substitution) are the basis of these computational techniques.

LIMIT OF A CONSTANT

If $f(x) = c$, where c is a constant, then $\lim\limits_{x \to a} f(x) = c$ for all a.

The graph of $y = c$ is a horizontal line, which implies that the function's value remains constant for all values of x.

LIMIT OF THE IDENTITY FUNCTION

If $f(x) = x$ for all x, then $\lim\limits_{x \to a} f(x) = a$.

Since the graph of $y = x$ is continuous, if $x = a$, then $y = a$ also.

LIMIT OF A SCALAR MULTIPLE

If f is a function such that $\lim\limits_{x \to a} f(x) = L$, then $\lim\limits_{x \to a} cf(x) = c \lim\limits_{x \to a} f(x) = cL$, where $c \in \mathbb{R}$.

In other words, the limit of the product of a constant and a function is equal to the constant times the limit of the function.

LIMIT OF A SUM

If f and g are functions such that $\lim\limits_{x \to a} f(x) = L_1$ and $\lim\limits_{x \to a} g(x) = L_2$, then $\lim\limits_{x \to a} [f(x) + g(x)] = \lim\limits_{x \to a} f(x) + \lim\limits_{x \to a} g(x) = L_1 + L_2$.

PRESENTATION

Lesson Opener

Use technology to graph
$$f(x) = \frac{2x^3}{x(x-2)(x+3)^2}$$
and use the graph to evaluate each limit.

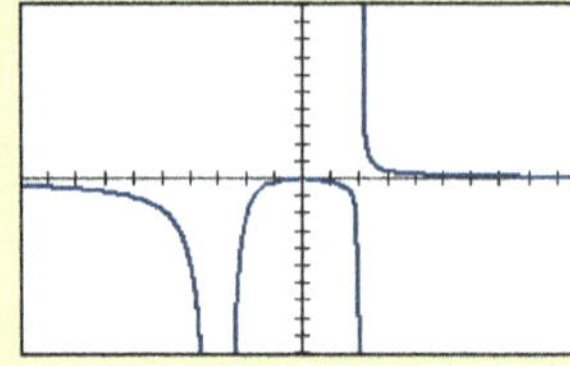

1. $\lim\limits_{x \to 0} f(x)$ 0

2. $\lim\limits_{x \to -3} f(x)$ $-\infty$

3. $\lim\limits_{x \to 2} f(x)$ does not exist

Consider following the Lesson Opener with the thoughts relating to the concept of limits and the Scripture verses found in the first paragraph of the text.

Encourage the students to pursue a deeper understanding of limits in light of their importance in understanding the main concepts of calculus. Introduce the first five properties for limits and emphasize their application when presenting Example 1. Since the polynomial function is continuous, its limit could easily be found using substitution, as evident in step 4. Explain that using substitution will not always be possible and applying the properties may be the only way to find a limit algebraically.

Simply put, the limit of a sum is the sum of the limits. This powerful property allows you to consider the limit of each term of a function separately. It also applies to the limit of a difference since subtraction is equivalent to adding the opposite.

LIMIT OF A POWER

If f is a function such that $\lim\limits_{x \to a} f(x) = L$, then $\lim\limits_{x \to a} [f(x)]^n = \left[\lim\limits_{x \to a} f(x)\right]^n = L^n$.

Example 1 Finding a Limit Using Properties

Evaluate $\lim\limits_{x \to 2} (2x^2 - 3x + 5)$.

Answer

$\lim\limits_{x \to 2} (2x^2 - 3x + 5)$
$= \lim\limits_{x \to 2} 2x^2 - \lim\limits_{x \to 2} 3x + \lim\limits_{x \to 2} 5$ 1. Apply the limit of a sum.

$= 2 \lim\limits_{x \to 2} x^2 - 3 \lim\limits_{x \to 2} x + \lim\limits_{x \to 2} 5$ 2. Apply the limit of a scalar multiple.

$= 2 \left[\lim\limits_{x \to 2} x\right]^2 - 3 \lim\limits_{x \to 2} x + \lim\limits_{x \to 2} 5$ 3. Apply the limit of a power.

$= 2(2)^2 - 3(2) + 5$ 4. Apply the limit of an identity and the limit of a constant
$= 7$ and simplify.

SKILL ✔ EXERCISE 7

Example 1 validates finding the limit of the continuous quadratic function by direct substitution. Since all polynomials are continuous, direct substitution can be used to find limits of any polynomial function.

LIMIT OF A PRODUCT

If f and g are functions such that $\lim\limits_{x \to a} f(x) = L_1$ and $\lim\limits_{x \to a} g(x) = L_2$, then $\lim\limits_{x \to a} [f(x) \cdot g(x)] = \lim\limits_{x \to a} f(x) \cdot \lim\limits_{x \to a} g(x) = L_1 \cdot L_2$.

LIMIT OF A QUOTIENT

If f and g are functions such that $\lim\limits_{x \to a} f(x) = L_1$ and $\lim\limits_{x \to a} g(x) = L_2 \neq 0$, then $\lim\limits_{x \to a} \dfrac{f(x)}{g(x)} = \dfrac{\lim\limits_{x \to a} f(x)}{\lim\limits_{x \to a} g(x)} = \dfrac{L_1}{L_2}$.

The limit of a quotient allows us to find limits for rational functions.

Example 2 Finding the Limit of a Rational Function

Find $\lim\limits_{x \to -3} \dfrac{x^3 - 8x^2 + 4}{5x - 10}$.

Answer

$\lim\limits_{x \to -3} \dfrac{x^3 - 8x^2 + 4}{5x - 10} = \dfrac{\lim\limits_{x \to -3} (x^3 - 8x^2 + 4)}{\lim\limits_{x \to -3} (5x - 10)}$ 1. Apply the limit of a quotient.

$= \dfrac{\lim\limits_{x \to -3} x^3 - \lim\limits_{x \to -3} 8x^2 + \lim\limits_{x \to -3} 4}{\lim\limits_{x \to -3} 5x - \lim\limits_{x \to -3} 10}$ 2. Apply the limit of a sum.

CONTINUED ➡

Additional Exercises

Find each limit using the properties of limits.

1. $\lim\limits_{x \to -2} 3x^2 - 2$ 10

2. $\lim\limits_{x \to 2} \dfrac{x^2 + 3x - 2}{x - 5}$ $-\dfrac{8}{3}$

Find each limit.

3. $\lim\limits_{x \to 3} \dfrac{x^2 - 4x + 3}{x - 3}$ 2

4. $\lim\limits_{x \to 0} \dfrac{\sqrt{x + 9} - 3}{x^2 + x}$ $\dfrac{1}{6}$

5. $\lim\limits_{x \to \frac{\pi}{2}} \dfrac{\sin 2x}{\cos x}$ 2

6. $\lim\limits_{x \to \frac{\pi}{4}} \dfrac{\sec^2 x - 1}{\sin^2 x}$ 2

7. $\lim\limits_{x \to \infty} \dfrac{3x^2 - 4x + 3}{2x^2 - 3}$ $\dfrac{3}{2}$

8. $\lim\limits_{x \to \infty} \dfrac{5x^2 + 2x + 7}{3x^3 - 5}$ 0

Assignments

- **Minimum:** 1–5, 8–9, 12–13, 15, 17, 22–23, 25, 27, 30, 34, 36–37; 43–51 odd; 52
- **Standard:** 3, 6, 8, 10–13, 15, 20–21, 23, 25–29, 31, 33, 35, 39, 44–45, 48, 50–52
- **Extended:** 7, 10, 13–16, 18–19, 23–25, 27, 29–30, 33, 36, 38, 40–42; 44–52 even

Assessment

- Quiz 12A covers Sections 12.1–12.2.

Present the *limit of a product* and the *limit of a quotient* properties. Consider asking the students what they think the limit for the expression in Example 2 might be. Emphasize that it would be nearly impossible to guess the limit of $\frac{19}{5}$. Have the students apply the properties to find the limit in Example 2. If students appear to need practice, have them complete exercise 29.

Interactive Activity Consider having the students verify the answer for Example 2 by entering the function into a calculator and examining the table of values as $x \to -3$.

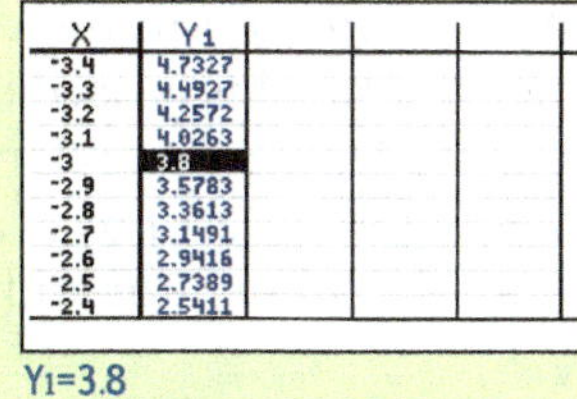

Explain that limits can be found by substitution for points within continuous intervals of a discontinuous function, but substitution at a point discontinuity produces $\frac{0}{0}$, an *indeterminate form*. The limit must then be found by factoring and canceling, as shown in Example 3.

Example 4 demonstrates rationalizing the numerator by multiplying the numerator and denominator by the conjugate, which prepares the students for finding the derivative of a radical function in Section 12.3.

The indeterminate form in Example 5 indicates a point discontinuity at $x = \pi$. You may want to discuss the difficulty of identifying this point discontinuity by using the graph or by using a table of values since π is irrational.

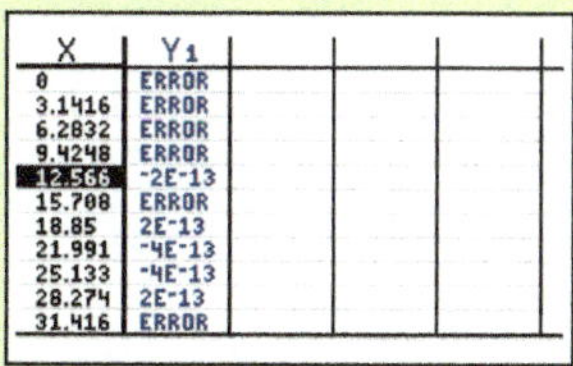

$$= \frac{\left(\lim\limits_{x \to -3} x\right)^3 - 8\left(\lim\limits_{x \to -3} x\right)^2 + \lim\limits_{x \to -3} 4}{5 \lim\limits_{x \to -3} x - \lim\limits_{x \to -3} 10}$$

3. Apply the limit of a power and limit of a scalar multiple.

$$= \frac{(-3)^3 - 8(-3)^2 + 4}{5(-3) - 10} = \frac{-27 - 72 + 4}{-15 - 10} = \frac{-95}{-25} = \frac{19}{5}$$

4. Apply the limit of an identity and limit of a constant and simplify.

SKILL ✔ **EXERCISE 9**

Since the only discontinuity in the function in Example 2 occurs where $5x - 10 = 0$ (at $x = 2$), direct substitution can be used to find the limit approaching any point except $x = 2$.

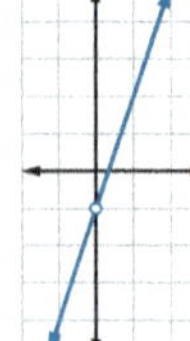

Using direct substitution to find $\lim\limits_{x \to 0} \frac{3x^2 - x}{x}$ produces $\frac{0}{0}$, which is called an *indeterminate form* since it does not contain enough information to determine the limit. In these cases, the limit may not exist, it may diverge to ∞ or $-\infty$, or it may be a real number. Factoring and reducing $f(x) = \frac{3x^2 - x}{x} = \frac{(3x - 1)x}{x} = 3x - 1$ (where $x \neq 0$) reveals that the original expression represents a line with a point discontinuity at $x = 0$. Therefore, $\lim\limits_{x \to 0} \frac{3x^2 - x}{x} = \lim\limits_{x \to 0} (3x - 1) = 3(0) - 1 = -1$.

Example 3   Canceling to Find a Limit

Evaluate $\lim\limits_{x \to 2} \frac{x - 2}{x^2 - x - 2}$.

Answer

$$\lim\limits_{x \to 2} \frac{x - 2}{x^2 - x - 2} = \frac{(2) - 2}{(2)^2 - (2) - 2} = \frac{0}{0}$$

1. Using direct substitution produces an indeterminate form.

$$\lim\limits_{x \to 2} \frac{x - 2}{x^2 - x - 2} = \lim\limits_{x \to 2} \frac{x - 2}{(x - 2)(x + 1)} = \lim\limits_{x \to 2} \frac{1}{x + 1}$$

2. Factor and cancel. The result implies the original function has a point discontinuity at $x = 2$.

$$= \frac{1}{(2) + 1} = \frac{1}{3}$$

3. Use direct substitution to find the limit.

Check

4. A graph of the function can be used to verify the limit.

SKILL ✔ **EXERCISE 21**

Since $\lim\limits_{x \to 2} f(x)$ considers only values approaching $x = 2$ without reaching it, the reduced form clearly shows that both one-sided limits are equal to $\frac{1}{3}$ for $x = 2$, the location of the hole in the graph.

Some limits with indeterminate forms can be evaluated by rationalizing the numerator or the denominator.

Present the *limit of a reciprocal power function at infinity*. This is used in Example 6. Some students may prefer viewing step 1, dividing each term by x^3, as multiplying both the numerator and denominator by $\frac{1}{x^3}$.

Stress that the limit of a rational function with a numerator and denominator of the same degree is the ratio of the leading coefficients only when evaluating the limit as $x \to \pm\infty$ (end behavior).

TIPS

Ex. 35, 41 Additional Exercise 8 can be used to help students prepare for these exercises.

Ex. 37 The limit of $-\infty$ should be obvious because as $x \to \infty$, the numerator grows much faster than the denominator.

Find $\lim\limits_{x \to 0} \dfrac{\sqrt{x+3} - \sqrt{3}}{x^2 - x}$.

Answer

$\lim\limits_{x \to 0} \dfrac{\sqrt{x+3} - \sqrt{3}}{x^2 - x} = \dfrac{\sqrt{0+3} - \sqrt{3}}{0^2 - 0} = \dfrac{0}{0}$

1. Using direct substitution produces an indeterminate form.

$\lim\limits_{x \to 0} \dfrac{(\sqrt{x+3} - \sqrt{3})(\sqrt{x+3} + \sqrt{3})}{(x^2 - x)(\sqrt{x+3} + \sqrt{3})}$

2. Multiply both the numerator and the denominator by the conjugate of the numerator.

$= \lim\limits_{x \to 0} \dfrac{(x+3) - 3}{x(x-1)(\sqrt{x+3} + \sqrt{3})}$

$= \lim\limits_{x \to 0} \dfrac{x}{x(x-1)(\sqrt{x+3} + \sqrt{3})}$

3. Simplify the numerator and cancel.

$= \lim\limits_{x \to 0} \dfrac{1}{(x-1)(\sqrt{x+3} + \sqrt{3})}$

$= \dfrac{1}{(0-1)(\sqrt{0+3} + \sqrt{3})} = -\dfrac{1}{2\sqrt{3}}$

4. Use direct substitution to find the limit.

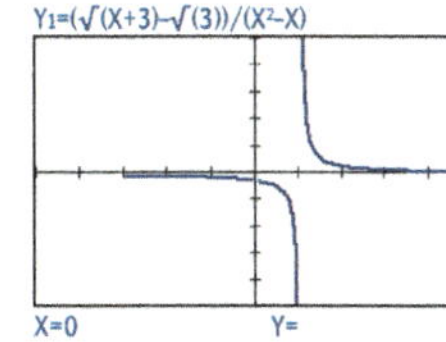

SKILL ✔ EXERCISE 25

A graph of the function in Example 4 and the canceling of x in step 3 indicate that the limit of the function exists despite a point discontinuity at $x = 0$.

You may need to use trigonometric identities to rewrite an expression whose limit is an indeterminate form.

Evaluate $\lim\limits_{x \to \pi} \dfrac{1 - \cos^2 x}{\sin x}$.

Answer

$\lim\limits_{x \to \pi} \dfrac{1 - \cos^2 x}{\sin x} = \dfrac{1 - \cos^2 \pi}{\sin \pi} = \dfrac{1 - (-1)^2}{0} = \dfrac{0}{0}$

1. Using direct substitution produces an indeterminate form.

$\lim\limits_{x \to \pi} \dfrac{1 - \cos^2 x}{\sin x} = \lim\limits_{x \to \pi} \dfrac{\sin^2 x}{\sin x}$
$= \lim\limits_{x \to \pi} \sin x$

2. Use a Pythagorean identity to rewrite the expression in the numerator and reduce. The result implies the original function has point discontinuities where $\sin x = 0$ (at $x = n\pi$, $n \in \mathbb{Z}$).

$= \sin \pi = 0$

3. Use direct substitution to find the limit.

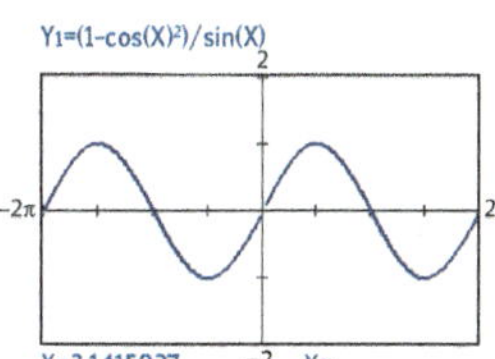

SKILL ✔ EXERCISE 27

A graph of the function in Example 5 and the canceling of $\sin x$ in step 2 indicate that the limit of the function exists despite a point discontinuity at $x = \pi$ (where $\sin x = 0$).

A final property of limits is used to find limits of rational functions at infinity.

LIMIT OF A RECIPROCAL POWER FUNCTION AT INFINITY

$\lim\limits_{x \to \pm\infty} \dfrac{1}{x^n} = 0$ when n is a natural number.

Solutions

❯ A. Exercises

7. $\lim\limits_{x \to 0} 5x = 5 \lim\limits_{x \to 0} x = 5(0) = 0$

8. $\lim\limits_{x \to -1} \left(x^2 - 3x\right) = \lim\limits_{x \to -1} x^2 - \lim\limits_{x \to -1} 3x$
$= \left(\lim\limits_{x \to -1} x\right)^2 - 3 \lim\limits_{x \to -1} x$
$= (-1)^2 - 3(-1) = 1 + 3 = 4$

9. $\lim\limits_{x \to 1} \dfrac{-2x - 9}{x + 1} = \dfrac{\lim\limits_{x \to 1} (-2x - 9)}{\lim\limits_{x \to 1} (x + 1)}$

$= \dfrac{\lim\limits_{x \to 1} (-2x) - \lim\limits_{x \to 1} 9}{\lim\limits_{x \to 1} x + \lim\limits_{x \to 1} 1}$

$= \dfrac{-2 \lim\limits_{x \to 1} x - \lim\limits_{x \to 1} 9}{\lim\limits_{x \to 1} x + \lim\limits_{x \to 1} 1} = \dfrac{-2(1) - 9}{1 + 1} = -\dfrac{11}{2}$

10. $\dfrac{\lim\limits_{x \to 2} (x^4 - x^2 + 4x)}{\lim\limits_{x \to 2} (2x^2 - 3x - 9)}$

$= \dfrac{\lim\limits_{x \to 2} x^4 - \lim\limits_{x \to 2} x^2 + \lim\limits_{x \to 2} 4x}{\lim\limits_{x \to 2} 2x^2 - \lim\limits_{x \to 2} 3x - \lim\limits_{x \to 2} 9}$

$= \dfrac{\left(\lim\limits_{x \to 2} x\right)^4 - \left(\lim\limits_{x \to 2} x\right)^2 + 4 \lim\limits_{x \to 2} x}{2\left(\lim\limits_{x \to 2} x\right)^2 - 3 \lim\limits_{x \to 2} x - \lim\limits_{x \to 2} 9}$

$= \dfrac{(2)^4 - (2)^2 + 4(2)}{2(2)^2 - 3(2) - 9}$

$= \dfrac{16 - 4 + 8}{8 - 6 - 9} = \dfrac{20}{-7} = -\dfrac{20}{7}$

11. $\lim\limits_{x \to 4} (x + 3)^2 = (4 + 3)^2 = 7^2 = 49$

12. $\lim\limits_{x \to 2} \dfrac{3}{x^2} = \dfrac{3}{(2)^2} = \dfrac{3}{4}$

13. $\lim\limits_{x \to -1} \dfrac{x^2 - 1}{x + 1} = \dfrac{(-1)^2 - 1}{(-1) + 1} = \dfrac{0}{0}$

14. $\lim\limits_{x \to 3} x^3 = (3)^3 = 27$

15. $\lim\limits_{x \to -1} \dfrac{x^2 + 2x - 1}{x + 1} = \dfrac{(-1)^2 + 2(-1) - 1}{(-1) + 1}$
$= \dfrac{-2}{0}$

16. $\lim\limits_{x \to -1} \dfrac{2x^2 + 5x + 3}{x + 1} = \dfrac{2(-1)^2 + 5(-1) + 3}{(-1) + 1}$
$= \dfrac{0}{0}$

17. $\dfrac{1}{0 - 1} = \dfrac{1}{-1} = -1$

18. $\lim\limits_{h \to 0} \dfrac{3h^2 - h + 1}{h} = \dfrac{3(0)^2 - 0 + 1}{0} = \dfrac{1}{0}$

19. $\sin \dfrac{\pi}{3} = \dfrac{\sqrt{3}}{2}$

20. $\sec \pi = \dfrac{1}{\cos \pi} = \dfrac{1}{-1} = -1$

B. Exercises

21. $\lim\limits_{x\to 1}\dfrac{(x+1)(x-1)}{x-1}=\lim\limits_{x\to 1}(x+1)$
$= (1)+1 = 2$

22. $\lim\limits_{x\to 1}\dfrac{(x-1)(x^2+x+1)}{x-1}$
$= \lim\limits_{x\to 1}(x^2+x+1)$
$= (1)^2+(1)+1 = 3$

23. $\lim\limits_{x\to 2}\dfrac{-(x-2)}{(x-2)(x+2)}=\lim\limits_{x\to 2}\dfrac{-1}{x+2}$
$= \dfrac{-1}{2+2}=-\dfrac{1}{4}$

24. $\lim\limits_{t\to 0}\dfrac{(\sqrt{t+1}-1)(\sqrt{t+1}+1)}{t(\sqrt{t+1}+1)}$
$= \lim\limits_{t\to 0}\dfrac{t+1-1}{t(\sqrt{t+1}+1)}=\lim\limits_{t\to 0}\dfrac{t}{t(\sqrt{t+1}+1)}$
$= \lim\limits_{t\to 0}\dfrac{1}{\sqrt{t+1}+1}=\dfrac{1}{\sqrt{1}+1}=\dfrac{1}{2}$

25. $\lim\limits_{x\to 0}\dfrac{(\sqrt{x+2}-\sqrt{2})(\sqrt{x+2}+\sqrt{2})}{x(\sqrt{x+2}+\sqrt{2})}$
$= \lim\limits_{x\to 0}\dfrac{x+2-2}{x(\sqrt{x+2}+\sqrt{2})}$
$= \lim\limits_{x\to 0}\dfrac{x}{x(\sqrt{x+2}+\sqrt{2})}$
$= \lim\limits_{x\to 0}\dfrac{1}{\sqrt{x+2}+\sqrt{2}}$
$= \dfrac{1}{\sqrt{2}+\sqrt{2}}=\dfrac{1}{2\sqrt{2}}=\dfrac{\sqrt{2}}{4}$

26. $\lim\limits_{a\to 0}\left(\dfrac{\sqrt{a+4}-2}{a}\cdot\dfrac{\sqrt{a+4}+2}{\sqrt{a+4}+2}\right)$
$= \lim\limits_{a\to 0}\dfrac{a+4-4}{a(\sqrt{a+4}+2)}$
$= \lim\limits_{a\to 0}\dfrac{a}{a(\sqrt{a+4}+2)}=\lim\limits_{a\to 0}\dfrac{1}{\sqrt{a+4}+2}$
$= \dfrac{1}{\sqrt{4}+2}=\dfrac{1}{4}$

27. $\lim\limits_{\theta\to 0}\dfrac{-\sin^2\theta}{\sin\theta}=\lim\limits_{\theta\to 0}(-\sin\theta)$
$= -\sin 0 = 0$

28. $\lim\limits_{\theta\to\frac{\pi}{2}}\left(\dfrac{\tan\theta}{\sec\theta}-\dfrac{1}{\sec\theta}\right)$
$= \lim\limits_{\theta\to\frac{\pi}{2}}\left(\dfrac{\frac{\sin\theta}{\cos\theta}}{\frac{1}{\cos\theta}}-\cos\theta\right)$
$= \lim\limits_{\theta\to\frac{\pi}{2}}(\sin\theta-\cos\theta)$
$= \sin\dfrac{\pi}{2}-\cos\dfrac{\pi}{2}=1-0=1$

29. $\lim\limits_{x\to 3}\dfrac{(x-3)(x+1)}{4x(x-3)}=\lim\limits_{x\to 3}\dfrac{x+1}{4x}$
$= \dfrac{3+1}{4(3)}=\dfrac{4}{12}=\dfrac{1}{3}$

30. $\dfrac{5(0)^2-8}{4(0)^2-3(0)+1}=\dfrac{-8}{1}=-8$

31. Since $D=[4,\infty)$, $\lim\limits_{x\to -4^-}\sqrt{x+4}$ does not exist.

32. right-hand limit = 7,
but left-hand limit = 6

33. $\lim\limits_{h\to 0}\dfrac{16+8h+h^2-16}{h}=\lim\limits_{h\to 0}\dfrac{h(h+8)}{h}$
$= \lim\limits_{h\to 0}(h+8)=0+8=8$

34. $\lim\limits_{h\to 0}\dfrac{x^2+2xh+h^2-x^2}{h}=\lim\limits_{h\to 0}\dfrac{h(2x+h)}{h}$
$= \lim\limits_{h\to 0}(2x+h)=2x+0=2x$

35. $\lim\limits_{x\to\infty}\dfrac{\frac{x^2}{x^3}+\frac{4x}{x^3}-\frac{7}{x^3}}{\frac{x^3}{x^3}-\frac{5x^2}{x^3}+\frac{8}{x^3}}=\dfrac{\lim\limits_{x\to\infty}\left(\frac{1}{x}+\frac{4}{x^2}-\frac{7}{x^3}\right)}{\lim\limits_{x\to\infty}\left(1-\frac{5}{x}+\frac{8}{x^3}\right)}$
$= \dfrac{\lim\limits_{x\to\infty}\frac{1}{x}+4\lim\limits_{x\to\infty}\frac{1}{x^2}-7\lim\limits_{x\to\infty}\frac{1}{x^3}}{\lim\limits_{x\to\infty}1-5\lim\limits_{x\to\infty}\frac{1}{x}+8\lim\limits_{x\to\infty}\frac{1}{x^3}}$
$= \dfrac{0+4(0)-7(0)}{1-5(0)+8(0)}=\dfrac{0}{1}=0$

36. $\lim\limits_{x\to\infty}\dfrac{\frac{-6x^4}{x^4}-\frac{x}{x^4}+\frac{15}{x^4}}{\frac{8x^4}{x^4}+\frac{11x^2}{x^4}-\frac{2}{x^4}}$
$= \dfrac{\lim\limits_{x\to\infty}\left(-6-\frac{1}{x^3}+\frac{15}{x^4}\right)}{\lim\limits_{x\to\infty}\left(8+\frac{11}{x^2}-\frac{2}{x^4}\right)}$
$= \dfrac{\lim\limits_{x\to\infty}(-6)-\lim\limits_{x\to\infty}\frac{1}{x^3}+15\lim\limits_{x\to\infty}\frac{1}{x^4}}{\lim\limits_{x\to\infty}8+11\lim\limits_{x\to\infty}\frac{1}{x^2}-2\lim\limits_{x\to\infty}\frac{1}{x^4}}$
$= \dfrac{-6-0+15(0)}{8+11(0)-2(0)}=\dfrac{-6}{8}=-\dfrac{3}{4}$

37. $\lim\limits_{n\to\infty}\dfrac{8n-1}{2n}=\lim\limits_{n\to\infty}\left(4-\dfrac{1}{2n}\right)$
$= \lim\limits_{n\to\infty}4-\dfrac{1}{2}\lim\limits_{n\to\infty}\dfrac{1}{n}=4-\dfrac{1}{2}(0)=4$

38. $\lim\limits_{n\to\infty}\dfrac{4n-1}{7n^2-2n+3}=\dfrac{\lim\limits_{n\to\infty}\left(\frac{4}{n}-\frac{1}{n^2}\right)}{\lim\limits_{n\to\infty}\left(7-\frac{2}{n}+\frac{3}{n^2}\right)}$
$= \dfrac{4\lim\limits_{n\to\infty}\frac{1}{n}-\lim\limits_{n\to\infty}\frac{1}{n^2}}{\lim\limits_{n\to\infty}7-2\lim\limits_{n\to\infty}\frac{1}{n}+3\lim\limits_{n\to\infty}\frac{1}{n^2}}$
$= \dfrac{4(0)-0}{7-2(0)+3(0)}=\dfrac{0}{7}=0$

Example 6   Finding the Limit of a Rational Function at Infinity

Find $\lim\limits_{x\to\infty}\dfrac{6x^3-7x-11}{2x^3+x+3}$.

Answer

$\lim\limits_{x\to\infty}\dfrac{6x^3-7x-11}{2x^3+x+3}=\lim\limits_{x\to\infty}\dfrac{\frac{6x^3}{x^3}-\frac{7x}{x^3}-\frac{11}{x^3}}{\frac{2x^3}{x^3}+\frac{x}{x^3}+\frac{3}{x^3}}$

1. Divide each term by x^d, where d is the degree of the denominator.

$= \lim\limits_{x\to\infty}\dfrac{6-\frac{7}{x^2}-\frac{11}{x^3}}{2+\frac{1}{x^2}+\frac{3}{x^3}}$

2. Simplify by reducing each term in the numerator and denominator.

$= \dfrac{\lim\limits_{x\to\infty}6-7\lim\limits_{x\to\infty}\frac{1}{x^2}-11\lim\limits_{x\to\infty}\frac{1}{x^3}}{\lim\limits_{x\to\infty}2+\lim\limits_{x\to\infty}\frac{1}{x^2}+3\lim\limits_{x\to\infty}\frac{1}{x^3}}$

3. Apply the limit of a sum and limit of a scalar multiple.

$= \dfrac{6-7(0)-11(0)}{2+0+3(0)}=\dfrac{6}{2}=3$

4. Apply the limit of a constant and limit of a reciprocal power function, and simplify.

Check

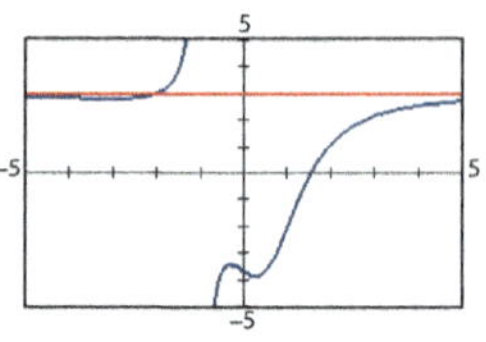

5. The horizontal asymptote of the function's graph at $y=3$ illustrates that the limit as $x\to\pm\infty$ of a rational function with a numerator and a denominator of the same degree is the ratio of leading coefficients.

$Y_1 = 3/(X-2)$

$X=2 \qquad Y=$

$Y_2 = 3/(X-2)^2$

$X=2 \qquad Y=$

SKILL ✓EXERCISE 35

While algebraic methods of finding limits are preferred when applicable, graphs and tables can provide confirmation of algebraic answers or valuable insight when direct substitution produces an undefined expression $\frac{k}{0}$, where k is a constant. Notice that evaluating the limit as $x\to 2$ of both $f(x)=\dfrac{3}{x-2}$ and $g(x)=\dfrac{3}{(x-2)^2}$ by direct substitution produces the result of $\frac{3}{0}$. Such a result indicates a vertical asymptote at $x=2$, and graphing the functions illustrates that $\lim\limits_{x\to 2}\dfrac{3}{x-2}$ does not exist, while $\lim\limits_{x\to 2}\dfrac{3}{(x-2)^2}=\infty$. Graphs should also be considered for limits such as $\lim\limits_{x\to 3}[x]$ and $\lim\limits_{x\to 3}\sqrt{x-3}$ in which the one-sided limits differ.

3. limit of a scalar multiple
4. limit of a reciprocal power function at infinity

A. Exercises

5. limit of a product

Name the property of limits that justifies each statement.

1. $\lim\limits_{x\to -4}(-3)=-3$ limit of a constant

2. $\lim\limits_{x\to a}x^4=\left(\lim\limits_{x\to a}x\right)^4$ limit of a power

3. $\lim\limits_{x\to 15}9f(x)=9\lim\limits_{x\to 15}f(x)$

4. $\lim\limits_{x\to\infty}\dfrac{1}{x^5}=0$

5. $\lim\limits_{x\to 7}[8x]\sqrt{2x+1}=\lim\limits_{x\to 7}[8x]\cdot\lim\limits_{x\to 7}\sqrt{2x+1}$

6. $\lim\limits_{x\to 7}(5x^2-4x)=\lim\limits_{x\to 7}5x^2-\lim\limits_{x\to 7}4x$ limit of a sum

Evaluate each limit by applying the properties of limits.

7. $\lim\limits_{x\to 0}5x$ 0

8. $\lim\limits_{x\to -1}(x^2-3x)$ 4

9. $\lim\limits_{x\to 1}\dfrac{-2x-9}{x+1}$ $-\dfrac{11}{2}$

10. $\lim\limits_{x\to 2}\dfrac{x^4-x^2+4x}{2x^2-3x-9}$ $-\dfrac{20}{7}$

Evaluate each limit using direct substitution or explain why this method fails.

11. $\lim\limits_{x\to 4}(x+3)^2$ 49

12. $\lim\limits_{x\to 2}\dfrac{3}{x^2}$ $\dfrac{3}{4}$

13. $\lim\limits_{x\to -1}\dfrac{x^2-1}{x+1}$ indeterminate form

14. $\lim\limits_{x\to 3}x^3$ 27

15. $\displaystyle\lim_{x\to-1}\frac{x^2+2x-1}{x+1}$

16. $\displaystyle\lim_{x\to-1}\frac{2x^2+5x+3}{x+1}$

17. $\displaystyle\lim_{x\to0}\frac{1}{x-1}$ -1

18. $\displaystyle\lim_{h\to0}\frac{3h^2-h+1}{h}$

19. $\displaystyle\lim_{\theta\to\frac{\pi}{3}}\sin\theta$ $\dfrac{\sqrt{3}}{2}$

20. $\displaystyle\lim_{\theta\to\pi}\sec\theta$ -1

B. Exercises

Evaluate each limit.

21. $\displaystyle\lim_{x\to1}\frac{x^2-1}{x-1}$ 2

22. $\displaystyle\lim_{x\to1}\frac{x^3-1}{x-1}$ 3

23. $\displaystyle\lim_{x\to2}\frac{2-x}{x^2-4}$ $-\dfrac{1}{4}$

24. $\displaystyle\lim_{t\to0}\frac{\sqrt{t+1}-1}{t}$ $\dfrac{1}{2}$

25. $\displaystyle\lim_{x\to0}\frac{\sqrt{x+2}-\sqrt{2}}{x}$ $\dfrac{\sqrt{2}}{4}$

26. $\displaystyle\lim_{a\to0}\frac{\sqrt{a+4}-2}{a}$ $\dfrac{1}{4}$

27. $\displaystyle\lim_{\theta\to0}\frac{\cos^2\theta-1}{\sin\theta}$ 0

28. $\displaystyle\lim_{\theta\to\frac{\pi}{2}}\frac{\tan\theta-1}{\sec\theta}$ 1

29. $\displaystyle\lim_{x\to3}\frac{x^2-2x-3}{4x^2-12x}$ $\dfrac{1}{3}$

30. $\displaystyle\lim_{x\to0}\frac{5x^2-8}{4x^2-3x+1}$ -8

31. $\displaystyle\lim_{x\to-4}\sqrt{x+4}$

32. $\displaystyle\lim_{x\to7}[x]$ 6

33. $\displaystyle\lim_{h\to0}\frac{(4+h)^2-4^2}{h}$ 8

34. $\displaystyle\lim_{h\to0}\frac{(x+h)^2-x^2}{h}$ $2x$

Apply the properties of limits to find the limit of each rational function at infinity.

35. $\displaystyle\lim_{x\to\infty}\frac{x^2+4x-7}{x^3-5x^2+8}$ 0

36. $\displaystyle\lim_{x\to\infty}\frac{-6x^4-x+15}{8x^4+11x^2-2}$ $-\dfrac{3}{4}$

Apply the properties of limits to find the limits of each sequence.

37. $a_n=\dfrac{8n-1}{2n}$ 4

38. $a_n=\dfrac{4n-1}{7n^2-2n+3}$ 0

C. Exercises

39. When does $\displaystyle\lim_{x\to a}\frac{\sec x}{\sin x}=0$? never

40. When does $\displaystyle\lim_{x\to a}\frac{e^{x^2}-e^x}{e^x}=e^{20}-1$? $a=5$ or $a=-4$

Given the general rational function
$$f(x)=\frac{a_mx^m+a_{m-1}x^{m-1}+\cdots+a_2x^2+a_1x+a_0}{b_nx^n+b_{n-1}x^{n-1}+\cdots+b_2x^2+b_1x+b_0},$$
use the properties of limits to derive the formulas for asymptotes found on page 107.

41. $m<n$

42. $m=n$

15. undefined expression
16. indeterminate form
18. undefined expression
31. does not exist
41. $y=0$ is a horizontal asymptote.
42. $y=\dfrac{a_m}{b_n}$ is a horizontal asymptote.

CUMULATIVE REVIEW

43. If $f(x)=x^2-4x$, find the simplified expression for $\dfrac{f(x+h)-f(x)}{h}$. [1.7] $h+2x-4$

44. State functions $u(x)$ and $v(x)$ such that $f(x)=e^{6x}=(u\circ v)(x)$. [1.7] $u(x)=e^x;\ v(x)=6x$

45. Verify the identity $\dfrac{\tan x}{1-\tan^2 x}=\dfrac{\sin x\cos x}{2\cos^2 x-1}$. [5.2]

46. Use the explicit formula for the sum of the first 50 perfect squares to evaluate $\displaystyle\sum_{k=1}^{50}k^2$. [9.5] 42,925

Classify each conic. Then state its center, vertices, and focal length. [8.2–8.3]

47. $\dfrac{(x-1)^2}{25}-\dfrac{y^2}{144}=1$

48. $\dfrac{(x-9)^2}{9}+\dfrac{(y+2)^2}{16}=1$

49. State the equation for the bounds of $y=2\sin 3x+x$. [4.7] C

  A. $y=\pm2$ **C.** $y=x\pm2$ **E.** none of these

  B. $y=x$ **D.** $y=2x\pm2$

50. Given $P(2,3,-1)$ and $Q(6,5,2)$, find the length of $\overrightarrow{PQ}$ to the nearest tenth. [6.3] B

  A. 4.5   **C.** 9   **E.** none of these

  B. 5.4   **D.** 29

51. According to AAA, 21% of fatal accidents involving teen drivers are caused by cell phone distraction. If a random sample of 30 fatal accidents involving teens is investigated, list all the reasons that a normal approximation of a binomial distribution can be used. [11.2] B, C

  A. The population size is 30.

  B. If p is the probability of success, $np=6.3$.

  C. If q is the probability of failure, $nq=23.7$.

  D. The continuity correction factor is 0.5.

  E. none of these

52. A food processing company buys a new piece of equipment for \$312,000. The estimated depreciation is 15% per year. In which year will the equipment's value become less than \$80,000? [12.1] D

  A. first year   **C.** eighth year   **E.** none of these

  B. fourth year   **D.** ninth year

47. hyperbola; $C(1,0)$; $V:(-4,0),(6,0)$; $c=13$
48. ellipse; $C(9,-2)$; $V:(9,-6),(9,2)$; $c=\sqrt{7}$

$$\lim_{n\to\infty}\frac{a_mx^m+\cdots+a_2x^2+a_1x+a_0}{b_nx^n+\cdots+b_2x^2+b_1x+b_0}$$
$$=\frac{\displaystyle\lim_{n\to\infty}\left(a_m+\cdots+\frac{a_2}{x^{n-2}}+\frac{a_1}{x^{n-1}}+\frac{a_0}{x^n}\right)}{\displaystyle\lim_{n\to\infty}\left(b_n+\cdots+\frac{b_2}{x^{n-2}}+\frac{b_1}{x^{n-1}}+\frac{b_0}{x^n}\right)}$$
$$=\frac{a_m}{b_n}$$
$$\therefore\ y=\frac{a_m}{b_n}\ \text{is a horizontal asymptote.}$$

Cumulative Review

43.
$$\frac{(x+h)^2-4(x+h)-(x^2-4x)}{h}$$
$$=\frac{x^2+2xh+h^2-4x-4h-x^2+4x}{h}$$
$$=\frac{h^2+2xh-4h}{h}=h+2x-4$$

45.
$$\frac{\tan x}{1-\tan^2 x}=\frac{\dfrac{\sin x}{\cos x}}{1-\dfrac{\sin^2 x}{\cos^2 x}}\cdot\frac{\cos^2 x}{\cos^2 x}$$
$$=\frac{\sin x\cos x}{\cos^2 x-\sin^2 x}=\frac{\sin x\cos x}{\cos^2 x-(1-\cos^2 x)}$$
$$=\frac{\sin x\cos x}{2\cos^2 x-1}$$

46.
$$\sum_{k=1}^{n}k^2=\frac{n(n+1)(2n+1)}{6}$$
$$\sum_{k=1}^{50}k^2=\frac{50(50+1)(2(50)+1)}{6}=42{,}925$$

47. hyperbola (with horizontal TA)
$$C(h,k)=(1,0)$$
$$a=5;\ V:(6,0),(-4,0)$$
$$c^2=25+144;\ c=13$$

48. ellipse (with vertical MA)
$$C(h,k)=(9,-2)$$
$$a=4;\ V:(9,-6),(9,2)$$
$$c^2=16-9;\ c=\sqrt{7}$$

49. bounds: $y=x\pm2$

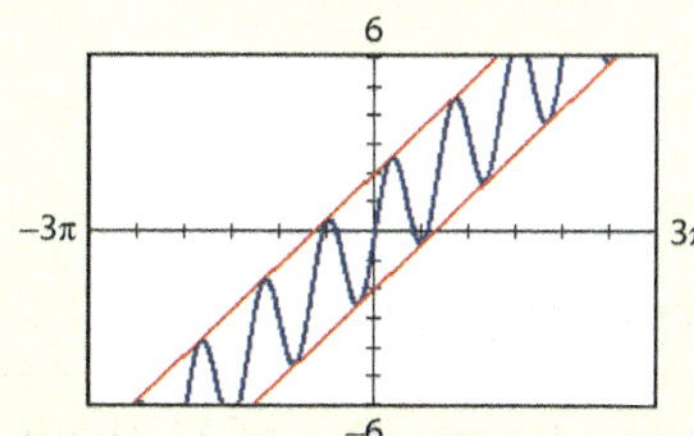

50. $\mathbf{v}=\langle 6-2,\ 5-3,\ 2-(-1)\rangle$
$$=\langle 4,2,3\rangle$$
$$|\mathbf{v}|=\sqrt{4^2+2^2+3^2}=\sqrt{29}\approx5.4$$

51. See the definition of a normal approximation to the binomial distribution (p. 560).

52. $f(x)=312{,}000(0.85)^x$

X	Y1
1	265200
2	225420
3	191607
4	162866
5	138436
6	117671
7	100020
8	85017
9	72264
10	61425
11	52211

Y1=72264.487240358

C. Exercises

39. $\displaystyle\lim_{x\to a}\frac{\sec x}{\sin x}=\lim_{x\to a}\frac{1}{\cos x\sin x}$; The limit is 0 only if the denominator increases without bound, but both trig functions have maxima at 1, so their product can never exceed 1. Therefore, the limit will never be 0.

40. The graph of $f(x)=\dfrac{e^{x^2}-e^x}{e^x}$ is continuous.

Let $\displaystyle\lim_{x\to a}\frac{e^{x^2}-e^x}{e^x}=e^{20}-1$.
$$e^{a^2-a}-1=e^{20}-1$$
$$e^{a^2-a}=e^{20}$$
$$a^2-a=20$$
$$a^2-a-20=0$$
$$(a-5)(a+4)=0$$
$$a=5,\ -4$$

41. Divide the numerator and denominator by x^n where $m<n$.
$$\lim_{n\to\infty}\frac{a_mx^m+\cdots+a_2x^2+a_1x+a_0}{b_nx^n+\cdots+b_2x^2+b_1x+b_0}$$
$$=\frac{\displaystyle\lim_{n\to\infty}\left(\frac{a_m}{x^{n-m}}+\cdots+\frac{a_2}{x^{n-2}}+\frac{a_1}{x^{n-1}}+\frac{a_0}{x^n}\right)}{\displaystyle\lim_{n\to\infty}\left(b_n+\cdots+\frac{b_2}{x^{n-2}}+\frac{b_1}{x^{n-1}}+\frac{b_0}{x^n}\right)}$$
$$=\frac{0}{b_n}=0$$
$$\therefore\ y=0\ \text{is a horizontal asymptote.}$$

42. Divide the numerator and denominator by x^n where $m=n$.

Biblical Perspective of Mathematics

Objectives

1. To recognize the importance of the concept of infinity to different branches of mathematics
2. To explain how the concept of infinity has been a source of frustration to mathematicians
3. To relate the mathematical concept of infinity to our infinite God

Assignments

- **Minimum:** 1–8; 9 (in class)
- **Standard:** 1–8; 9 (in class)
- **Extended:** 1–9

Mathematics and Infinity

The infinite! No other question has ever moved so profoundly the spirit of man; no other idea has so fruitfully stimulated his intellect; yet no other concept stands in greater need of clarification than that of the infinite.

—*David Hilbert (German mathematician; key proponent of formalism)*

In 1930 the German mathematician and philosopher Hermann Weyl opened a speech with the statement, "Mathematics is the science of the infinite." Few mathematicians would place that great of an emphasis on the term "infinite," but the concept of infinity does permeate all of mathematics.

Even the simple skill of counting leads quickly to the idea that natural numbers continue on and on without end. In college-level mathematics natural numbers are referred to as "countably infinite." Surprisingly, rational numbers are also countably infinite because they can be placed in one-to-one correspondence with natural numbers. However, the subset of real numbers in the interval between zero and one is referred to as "uncountably infinite" due to the irrational numbers involved.

In 1900 David Hilbert declared the greatest unsolved modern mathematical problem to be the continuum hypothesis, which states that there is not a level of infinity between the countably infinite natural numbers and the uncountably infinite real numbers. In the 1960s American mathematician Paul Cohen proved that the continuum hypothesis is independent of the axiomatic framework of set theory. In other words, the continuum hypothesis is impossible to either prove or disprove. Cohen's work demonstrates that as powerful as axiomatic systems are, they still have limitations.

Beginning geometry students quickly encounter infinite sets when they contemplate the basic definition of space as the set of all points and the three undefined terms *point*, *line*, and *plane*. A line contains an infinite number of points, a plane contains an infinite number of lines, and space contains an infinite number of planes. Non-Euclidean geometries consider the infinite sets of points on a sphere, on a saddle shape, and on many other unusual surfaces.

Algebra students describe infinite sets when they use set builder notation to represent a line as $\{(x, y) \mid y = mx + b\}$, any conic section as $\{(x, y) \mid Ax^2 + Bxy + Cy^2 + Dx + Ey + F = 0\}$, a basic trigonometric curve as $\{(x, y) \mid y = a \cos bx\}$, or a sphere as $\{(x, y, z) \mid (x - h)^2 + (y - k)^2 + (z - j)^2 = r^2\}$. Domains and ranges of algebraic and transcendental functions are also described with infinite set notation. For example, the domain of the logarithmic function $f(x) = \log_2 x$ is $\{x \mid x > 0\}$ and the range of the trigonometric function $f(x) = \sin x$ is $\{f(x) \mid -1 \leq f(x) \leq 1\}$.

Calculus uses the concept of infinity to define its three big ideas: limits, derivatives, and integrals. Defining the limit requires *infinitesimals*, infinitely small quantities. The derivative is defined as the limit of the slope of a segment with endpoints on the graph of a function, which makes it possible to find instantaneous velocities or to find maximums in algebraic and geometric applications. The integral is defined as the limit of the sum of rectangle areas under a function curve with the number of rectangles approaching infinity. The integral allows the calculation of the amount of work done in physics or the volume of a solid of revolution.

Although the concept of infinity is pervasive in mathematics, the mystery surrounding its use has been a source of irritation for mathematicians throughout history. Irrational numbers, which cannot be expressed as ratios of integers and whose decimal expansions are infinite, shattered the cult-like natural number philosophy of the Pythagoreans. The ancient Greek Zeno stated many paradoxes involving infinity. Contradictions involving infinity occurred early in the development of calculus before the introduction of the formal limit resolved those contradictions. The development of set theory gave rise to paradoxes such as "the set of all sets."

The concept of infinity has been a source of inspiration and amazement for Christian mathematicians, as it points to our infinite God. One reason the Renaissance mathematicians and scientists made such great advances was due to their belief in an infinite God who controls the universe in an orderly manner. They realized that God's infinite wisdom was far above their finite knowledge (Isa. 55:9; Rom. 11:33). So Kepler, Galileo, Newton, and others searched for and discovered principles of order in God's creation (Gen. 1:1; Isa. 66:1), successfully inventing mathematical descriptions for those discoveries.

Mathematical infinity is quantitative, while God's infinity is qualitative. By this we mean that God cannot be reduced to His smallest parts to be understood better. God is essentially different from us in His being, not just infinitely more intelligent or present. God is unlimited yet retains His oneness and simplicity. The famous theologian Herman Bavinck describes God's infinity as "not limited by anything finite and creaturely." God's eternality transcends time and He is not bound by it. If the concept of infinity in mathematics creates wonder and awe, how much more should we worship and praise our limitless and infinite God (Ps. 90:2; Ps. 139:1–12)!

❯ Exercises

1. Classify each infinite set of numbers as countable or uncountable.

   a. integer   countable
   b. irrational   uncountable
   c. natural   countable
   d. rational   countable
   e. real   uncountable

2. Match each mathematical concept with its algebraic description.

   I. $\{x \mid x \in \mathbb{R}\}$
   II. $\{f(x) \mid f(x) > 0\}$
   III. $\{(x, y) \mid y = mx + b\}$
   IV. $\{(x, y) \mid Ax^2 + Bxy + Cy^2 + Dx + Ey + F = 0\}$
   V. $\{(x, y, z) \mid (x - h)^2 + (y - k)^2 + (z - j)^2 = r^2\}$

   a. line on a plane   III
   b. range of the exponential function   II
   c. sphere   V
   d. domain of the sine function   I
   e. conic section   IV

3. How did Herman Wyle define mathematics?

4. Which famous statement asserts there is not a level of infinity between the infinities of the natural and real numbers?   continuum hypothesis

5. How do the geometric concepts of a line, a plane, and space relate to the concept of infinity?

6. Name each of the described big ideas of calculus.

   a. represents the slope of the tangent to a function's graph   derivative
   b. represents the area under a function's graph   integral
   c. used to define the other two big ideas   limit

7. Which big idea of calculus can be used to find maximums in algebra and geometry?   derivative

8. **Discuss:** Give several examples of how the concept of infinity has been a source of frustration to mathematicians over the history of mathematics.

9. **Discuss:** Explain how a belief in an infinite God helped make Renaissance mathematicians successful in discovering mathematical laws in the universe.

Answers

3. science of the infinite

5. A line contains an infinite number of points, a plane contains an infinite number of lines, and space contains an infinite number of planes.

8. irrational numbers, which cannot be expressed as ratios of integers and whose decimal expansions are infinite; the paradoxes of Zeno; contradictions in calculus before the introduction of the formal limit theorems; paradoxes in set theory such as "the set of all sets"; the failed efforts to prove the continuum hypothesis

9. Their belief in an infinite God made them realize that God's infinite wisdom was far above their finite knowledge and that He controls the universe in an orderly manner. They were willing to consider the concept of infinity even if they could not fully explain it. As a result, they were successful in discovering mathematical descriptions for the order they were searching for in God's infinite creation.

12.3 Tangents and Derivatives

Objectives

1. To find the slope of a line tangent to a curve at a given point
2. To find the derivative of a function
3. To find average and instantaneous velocities in real-world applications

Flash

The Tower of Hanoi problem was believed to have been introduced by French mathematician Édouard Lucas in 1883. The game starts with all the disks arranged on a single peg from the largest on the bottom to the smallest on top. The minimum number of moves is $2^n - 1$; n = number of disks. Legend has it that priests in a Vietnamese or Indian temple have been moving 64 disks from one peg to another using the same procedure; when they finish, the world will end. Moving one disk per second results in $2^{64} - 1 = 1.8 \times 10^{19}$ sec, or about 585 billion yr. Try the game with your students with fewer than 64 disks.

Vocabulary

average velocity
derivative
derivative function
difference quotient
differentiable function
differential operator
instantaneous velocity

Reading and Writing Mathematics

Explain why using only the difference quotient $\dfrac{f(x + h) - f(x)}{h}$ for the slope of a tangent of a polynomial will never produce the exact value of the slope.

As the value of h progressively decreases, the slope of the secant line approaches the slope of the tangent line; but when h becomes equal to 0, the quotient is indeterminate.

To solve the Tower of Hanoi problem, move only one disk at a time to rebuild the tower in the correct order on a different peg without placing a larger disk on top of a smaller disk.

After completing this section, you will be able to

- find the slope of the tangent to a curve at a given point.
- determine the derivative of a function.
- find average and instantaneous velocities in real-world applications.

Classic problems, such as the Tower of Hanoi, have challenged mathematicians for centuries. One such problem in calculus was finding the slope of the line tangent to a curve at a given point. The solution to the "tangent problem" involves the slopes of secants through the point. To find the slope of the tangent to $f(x) = \frac{1}{4}x^2$ at $P(2, 1)$, begin by examining the slopes of several secants that are successively better approximations for the tangent.

using secant $\overleftrightarrow{PQ}$ with $Q\,(6, 9)$:	using secant $\overleftrightarrow{PQ}$ with $Q\,(4, 4)$:	using secant $\overleftrightarrow{PQ}$ with $Q\left(3, \frac{9}{4}\right)$:
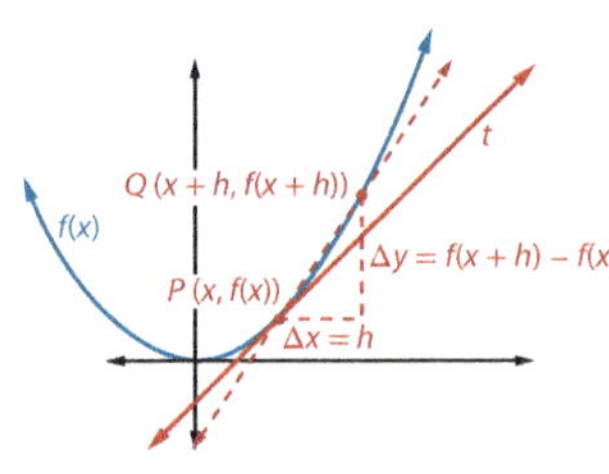		
$m = \dfrac{\Delta y}{\Delta x} = \dfrac{8}{4} = 2$	$m = \dfrac{\Delta y}{\Delta x} = \dfrac{3}{2} = 1.5$	$m = \dfrac{\Delta y}{\Delta x} = \dfrac{\frac{5}{4}}{1} = \dfrac{5}{4} = 1.25$

The challenge is finding the exact slope of the tangent to the curve at P. When Q reaches P, the secant becomes a tangent (intersecting the curve at just one point), but applying the formula for slope produces $\frac{0}{0}$, an indeterminate form. This difficulty is solved by finding the limit of the secant line's slope as $\Delta x \to 0$.

Generalizing the problem, let $P\,(x, f(x))$ represent the point of tangency on the graph, and let the second point be $Q\,(x + h, f(x + h))$ where $h = \Delta x$, the difference between the x-coordinates. The slope of secant $\overleftrightarrow{PQ}$ is $m = \dfrac{\Delta y}{\Delta x}$

$$= \dfrac{f(x + h) - f(x)}{(x + h) - x} = \dfrac{f(x + h) - f(x)}{h},$$ an expression called the *difference quotient*.

Finding $\lim\limits_{h \to 0} \dfrac{f(x + h) - f(x)}{h}$ produces the exact slope of the tangent line t.

▶ DEFINITION

The **derivative** of the function f at the point where $x = a$, denoted by $f'(a)$ and read "f prime of a," is

$$f'(a) = \lim_{h \to 0} \dfrac{f(a + h) - f(a)}{h},$$ if the limit exists.

PRESENTATION

Lesson Opener

Simplify the expression $\dfrac{f(x + h) - f(x)}{h}$ for each function.

1. $f(x) = -2x$ -2
2. $f(x) = x^2$ $2x + h$
3. $f(x) = 3x^2 + 5x$ $6x + 3h + 5$

Engage the students by explaining that the concepts of limits introduced in the previous section are used to define one of the foundational concepts of calculus, the *derivative*. Use the figures in the text to describe how we can approach the value for the slope of a tangent line by moving point Q closer and closer to the given point P. Because the slope becomes the indeterminate form $\frac{0}{0}$ when Q is at P, the concept of a limit is used to determine the slope of the tangent line at P.

Emphasize that the tangent intersects the graph at a given point and its slope (or rate of change) is the slope of the graph at that point.

Common Student Error A student may think that a tangent can intersect a graph at only one point. This is true for many curves, including conics, but a tangent line can intersect a graph at more than one point. Consider a tangent line to a curve such as $f(x) = \sin x$. The graph will

The slope of the line tangent to $f(x)$ at $x = a$ is $f'(a)$.

If $f'(a)$ exists, the function is said to be *differentiable* at $x = a$. All three of the following conditions must be true in order for a function to be differentiable at $x = a$.

1. The function must be continuous at $x = a$ (no asymptote, jump, or point discontinuities).

2. The function cannot have a sharp point at $x = a$.

3. The function cannot have a vertical tangent at the point. (The slope would be undefined.)

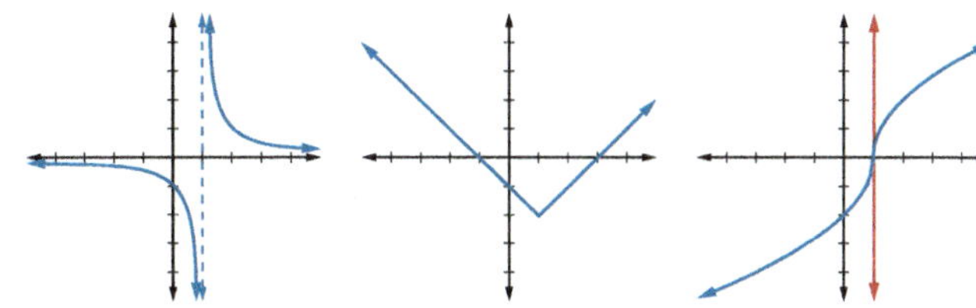

None of these graphed functions are differentiable at $x = 1$. At this value of x, the first is discontinuous, the second has a sharp point, and the third has a vertical tangent.

Example 1   Finding the Derivative at a Point

Find $f'(1)$ if $f(x) = 2x^2 - 4$.

Answer

$f'(1) = \lim\limits_{h \to 0} \dfrac{f(1+h) - f(1)}{h}$

    1. Substitute for a in the definition of the derivative at a point: $f'(a) = \lim\limits_{h \to 0} \dfrac{f(a+h) - f(a)}{h}$.

$= \lim\limits_{h \to 0} \dfrac{[2(1+h)^2 - 4] - [2(1)^2 - 4]}{h}$

    2. Rewrite the difference quotient using substitutions for $f(1+h)$ and $f(1)$.

$= \lim\limits_{h \to 0} \dfrac{2(1 + 2h + h^2) - 4 - (2 - 4)}{h}$

    3. Simplify the difference quotient. Note that division by h is allowed since $h \neq 0$; h is only approaching 0.

$= \lim\limits_{h \to 0} \dfrac{2 + 4h + 2h^2 - 4 + 2}{h}$

$= \lim\limits_{h \to 0} \dfrac{4h + 2h^2}{h} = \lim\limits_{h \to 0} (4 + 2h)$

$= 4 + 2(0) = 4$

    4. Evaluate the limit of the polynomial by substituting 0 for h.

SKILL ✔ EXERCISE 13

Notice that the slope of the tangent to the parabola $f(x) = 2x^2 - 4$ at $x = 1$ is 4.

The definition of a derivative can be used to find a function for the slope of the tangent to the graph of $f(x)$ at any differentiable point.

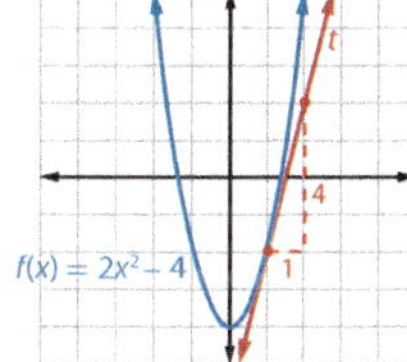

DEFINITION

The **derivative function** represents the derivative for any value of x where the limit exists. It is denoted $f'(x)$ where $f'(x) = \lim\limits_{h \to 0} \dfrac{f(x+h) - f(x)}{h}$.

<hr>

have only one tangent to the curve at a given point, but a tangent line can intersect the curve many times.

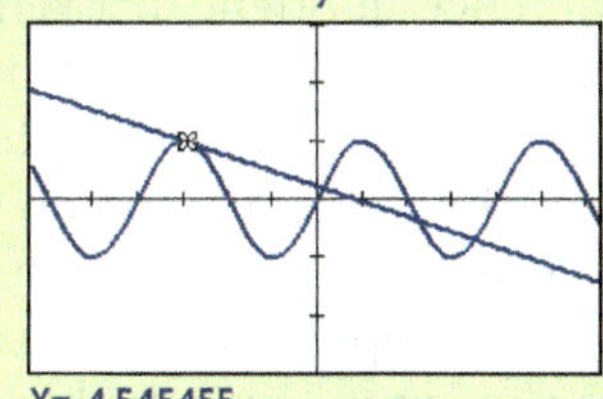

Introduce the *difference quotient* and the definition of the derivative of a function. The students should memorize this formula.

Using the figures, discuss the conditions in which a function is not *differentiable*. Use the middle figure to demonstrate why a function is not differentiable at a sharp point. (*The limit does not exist because the left-hand limit is -1 and the right-hand limit is 1.*) Zooming in on any portion of a smooth, continuous curve eventually produces a graph that appears to be a straight line. This will not happen when zooming in on the sharp point at $x = 1$ in the middle figure, so the function is not differentiable at $x = 1$.

Similarly, a function is not differentiable at a point where its graph terminates since one of the one-sided limits does not exist at that point.

The derivative of a function at a given point is found in Example 1. Consider asking the students what the 4 means in step 4. (*It is the slope of the line tangent to the graph at $x = 1$.*)

Share that a *derivative function* can be found that describes the slope of the tangent to the graph at any point where the function is differentiable (not just at a specific point as in Example 1). Present the derivative function definition and use Example 2 to demonstrate the benefit of using the derivative function. Ask the students what the function in step 4 represents. (*It is the equation for the slope of*

Solutions

A. Exercises

1. $\dfrac{(2(x+h)+3)-(2x+3)}{h}$

$=\dfrac{2x+2h+3-2x-3}{h}=\dfrac{2h}{h}=2$

2. $\dfrac{(-5(x+h)-7)-(-5x-7)}{h}$

$=\dfrac{-5x-5h-7+5x+7}{h}=\dfrac{-5h}{h}=-5$

3. $\dfrac{((x+h)^2+3(x+h))-(x^2+3x)}{h}$

$=\dfrac{(x^2+2xh+h^2+3x+3h)-(x^2+3x)}{h}$

$=\dfrac{x^2+2xh+h^2+3x+3h-x^2-3x}{h}$

$=\dfrac{2xh+h^2+3h}{h}=2x+3+h$

4. $\dfrac{(3(x+h)^2-4(x+h))-(3x^2-4x)}{h}$

$=\dfrac{3(x^2+2xh+h^2)-4x-4h-(3x^2-4x)}{h}$

$=\dfrac{3x^2+6xh+3h^2-4x-4h-3x^2+4x}{h}$

$=\dfrac{6xh+3h^2-4h}{h}=6x-4+3h$

5b. $f(x)$ is not defined for $x<0$.

7d.

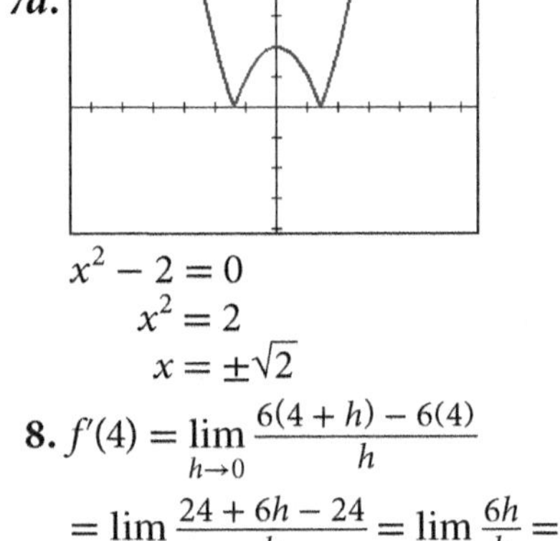

$x^2-2=0$

$x^2=2$

$x=\pm\sqrt{2}$

8. $f'(4)=\lim\limits_{h\to0}\dfrac{6(4+h)-6(4)}{h}$

$=\lim\limits_{h\to0}\dfrac{24+6h-24}{h}=\lim\limits_{h\to0}\dfrac{6h}{h}=6$

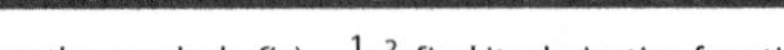

Example 2 Finding and Applying a Derivative Function

Given the parabola $f(x)=\frac{1}{4}x^2$, find its derivative function $f'(x)$.
Then use $f'(x)$ to find the slope of the tangent to the parabola at $x=-2$ and $x=1$.

Answer

$f'(x)=\lim\limits_{h\to0}\dfrac{f(x+h)-f(x)}{h}$ — 1. State the definition of the derivative function.

$=\lim\limits_{h\to0}\dfrac{\frac{1}{4}(x+h)^2-\frac{1}{4}x^2}{h}$ — 2. Rewrite the difference quotient using substitutions for $f(x+h)$ and $f(x)$.

$=\lim\limits_{h\to0}\dfrac{\frac{1}{4}(x^2+2xh+h^2-x^2)}{h}$ — 3. Simplify the difference quotient.

$=\lim\limits_{h\to0}\dfrac{2xh+h^2}{4h}=\lim\limits_{h\to0}\left(\dfrac{x}{2}+\dfrac{h}{4}\right)$

$=\dfrac{x}{2}+\dfrac{0}{4}=\dfrac{x}{2}$ — 4. Evaluate the limit by substituting for h.

$\therefore f'(x)=\dfrac{x}{2}$

$f'(-2)=\dfrac{(-2)}{2}=-1$ $\qquad f'(1)=\dfrac{(1)}{2}=\dfrac{1}{2}$ — 5. Substitute $x=-2$ and $x=1$ into $f'(x)$ to find the slope of the tangent to the parabola at those points.

SKILL ✔ **EXERCISE 31**

Since the slope of any tangent to the function $f(x)=\frac{1}{4}x^2$ is given by $f'(x)=\frac{1}{2}x$, the slope varies and depends on the x-coordinate of the point. To find the exact slope of the tangent to $f(x)=\frac{1}{4}x^2$ at $x=2$ as sought in the introduction of the lesson, evaluate $f'(2)=\frac{1}{2}(2)=1$. The ability to find the slope of the tangent to the curve at multiple points using the single $f'(x)$ function demonstrates the power of the derivative function.

Alternative notations for the derivative function, $f'(x)$, include $\frac{dy}{dx}$ (read "the derivative of y with respect to x"), y' (read "y prime"), and $\frac{d}{dx}f(x)$ (where $\frac{d}{dx}$ is the *differential operator* indicating the derivative of the following function).

Example 3 Finding the Derivative Function

Find $\dfrac{dy}{dx}$ if $y=3x+5$.

Answer

$\dfrac{dy}{dx}=f'(x)=\lim\limits_{h\to0}\dfrac{f(x+h)-f(x)}{h}$ — 1. State the definition in function form.

$=\lim\limits_{h\to0}\dfrac{[3(x+h)+5]-(3x+5)}{h}$ — 2. Rewrite the difference quotient using substitutions for $f(x+h)$ and $f(x)$.

$=\lim\limits_{h\to0}\dfrac{3x+3h+5-3x-5}{h}$ — 3. Simplify the difference quotient.

$=\lim\limits_{h\to0}3=3$ — 4. Find the limit, recalling that the limit of a constant is the constant.

$\therefore f'(x)=3$

SKILL ✔ **EXERCISE 17**

To indicate a derivative, Leibniz used $\frac{dy}{dx}$ notation but Newton used a dot, as in $\dot y$. When Charles Babbage, a founding member of the Analytical Society (1812) of Cambridge, stated that one of the society's purposes was to promote "The Principles of pure D-ism in opposition to the Dot-age of the University," he was referring to a preference for Leibniz's notation.

---

the *line tangent to the graph at any differentiable point.*) In step 5, the slope of two tangents to the graph is found. Point out that this function provides the solution to the "tangent problem" posed at the beginning of the lesson.

Share the alternate notations for the derivative function, $f'(x)$: $\frac{dy}{dx}$ and $\frac{d}{dx}f(x)$. The $\frac{dy}{dx}$ notation is commonly used and is read "the derivative of y with respect to x." This may be the best way for students to think of the notation until they take calculus. In the expression $\frac{d}{dx}f(x)$, the notation $\frac{d}{dx}$ is viewed as the *differential operator*, indicating that the derivative of $f(x)$ should be found.

One-on-One A student may want to know why the d's in $\frac{dy}{dx}$ can't be canceled. Explain that it would be similar to canceling the Δ's in $\frac{\Delta y}{\Delta x}$. Further explain that $\frac{dy}{dx}$ is a single notation and is not a fraction like $\frac{\Delta y}{\Delta x}$ (change in y over change in x).

Before presenting Example 3, consider asking the students which characteristic of a linear function the derivative function provides. (*the slope of the line*) Complete Example 3 and be sure students

understand why $\frac{d}{dx}(mx+b)$ is the constant m, which is proved in exercise 31.

The difference quotient in Example 4 is simplified by multiplying the numerator by its conjugate. Point out that the derivative of this radical function is not a constant, implying that its graph is not a line but has a slope that varies as indicated by the independent variable in the derivative function.

Example 5 illustrates finding the *average velocity*, which is simply the change in position divided by the change in time. Using the derivative allows us to find the *instantaneous velocity* at any given point along the graph, as shown in Example 6.

Since $f(x) = 3x + 5$ is a linear function, its slope at every point is the coefficient of the first degree term, $m = 3$. For other types of functions, the derivative function will not be a constant.

Example 4   Finding the Derivative of a Radical Function

Find $\frac{d}{dx}\sqrt{x}$.

Answer

$\frac{d}{dx}f(x) = \lim_{h\to 0}\frac{f(x+h) - f(x)}{h}$

$\frac{d}{dx}\sqrt{x} = \lim_{h\to 0}\frac{\sqrt{x+h} - \sqrt{x}}{h}$

1. Using the definition of the derivative, rewrite the difference quotient using substitutions for $f(x+h)$ and $f(x)$.

$= \lim_{h\to 0}\frac{(\sqrt{x+h} - \sqrt{x})(\sqrt{x+h} + \sqrt{x})}{h(\sqrt{x+h} + \sqrt{x})}$

2. Rationalize the numerator, then simplify and cancel to eliminate the common factor h.

$= \lim_{h\to 0}\frac{x + h - x}{h(\sqrt{x+h} + \sqrt{x})}$

$= \lim_{h\to 0}\frac{h}{h(\sqrt{x+h} + \sqrt{x})} = \lim_{h\to 0}\frac{1}{\sqrt{x+h} + \sqrt{x}}$

3. Simplify.

$= \frac{1}{\sqrt{x+0} + \sqrt{x}} = \frac{1}{2\sqrt{x}}$

4. Evaluate the limit by substitution. Note that the function can be expressed using rational exponents.

$\therefore \frac{d}{dx}\sqrt{x} = \frac{1}{2\sqrt{x}}$ or $\frac{1}{2}x^{-\frac{1}{2}}$

SKILL ✓ EXERCISE 27

Geometrically the derivative represents the slope of the tangent to the curve at a given point. In real-world applications the derivative is used to model rates of change. The following examples illustrate how the derivative can be used to find *instantaneous velocities*. The *average velocity* of an object is defined as $\frac{\text{change in position}}{\text{change in time}} = \frac{\Delta d}{\Delta t}$. If the object's position is given by $f(t)$, its average velocity from time a to time b can be calculated using $v_{avg} = \frac{\Delta d}{\Delta t} = \frac{f(b) - f(a)}{b - a}$.

Example 5   Finding Average Velocity

The height (in meters) of a ball t seconds after it is shot straight up at 20 m/sec can be modeled by the equation $f(t) = -10t^2 + 20t$. Find the ball's average velocity from 0.5 sec to 1.0 sec.

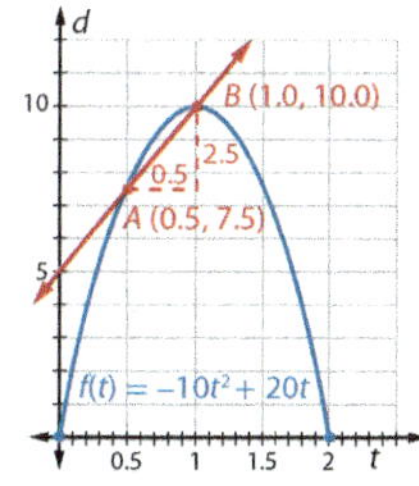

Answer

$f(0.5) = -10(0.5)^2 + 20(0.5) = 7.5$ m
$f(1.0) = -10(1.0)^2 + 20(1.0) = 10.0$ m

1. Find the ball's height at $t = 0.5$ sec and $t = 1.0$ sec.

$v_{avg} = \frac{\Delta d}{\Delta t} = \frac{f(1.0) - f(0.5)}{1.0 - 0.5}$

$= \frac{10.0 - 7.5}{0.5} = 5$ m/sec

2. Substitute into the formula for average velocity and evaluate.

SKILL ✓ EXERCISE 21

The average velocity of the ball in Example 5 is represented by the slope of the secant connecting the two points on the distance-time graph of $f(t)$. Its instantaneous velocity is represented by the slope of the tangent to the curve. To find the instantaneous velocity of the ball at $t = 0.5$ sec, take the limit of the average velocity as $\Delta t \to 0$. In the following definition, $h = \Delta t$.

Consider comparing these last two examples. Example 5 is similar to exercises students have done in algebra using the formula $r = \frac{d}{t}$ to find the average speed for a given trip. Example 6 would be similar to taking a look at the speedometer at a given time along the way. Note that the negative velocity implies the object is falling.

Interactive Activity A variety of interactive tools for illustrating and discussing tangents and derivatives can be found with the Internet keyword search *tangent derivative applet* or *applets for calculus*.

TIPS

Ex. 5–7 Encourage students to use technology to graph the function if needed.

Ex. 28–29 Consider reminding students that they need the slope of the tangent and the point of tangency to determine the tangent's equation.

Ex. 31, 36 These exercises are also used as assessments in Quiz 12B and the Chapter 12 Test.

Ex. 38 Accept piecewise functions that explicitly state that the limit does not exist when $x = 0$.

9. $f'(-1)$

$= \lim_{h\to 0}\frac{2(-1 + h) - 60 - (2(-1) - 60)}{h}$

$= \lim_{h\to 0}\frac{-2 + 2h - 60 + 2 + 60}{h} = \lim_{h\to 0}\frac{2h}{h}$

$= 2$

10. $f'(3) = \lim_{h\to 0}\frac{(3 + h)^2 - 3^2}{h}$

$= \lim_{h\to 0}\frac{3^2 + 6h + h^2 - 9}{h} = \lim_{h\to 0}\frac{h(6 + h)}{h}$

$= \lim_{h\to 0}(6 + h) = 6$

11. $f'(-8) = \lim_{h\to 0}\frac{7 - 7}{h} = \lim_{h\to 0}\frac{0}{h} = 0$

12. $f'(x) = \lim_{h\to 0}\frac{(x + h) - x}{h} = \lim_{h\to 0}\frac{h}{h} = 1$

13. $f'(x) = \lim_{h\to 0}\frac{3 - 3}{h} = \lim_{h\to 0}\frac{0}{h} = 0$

14. $f'(x) = \lim_{h\to 0}\frac{-2(x + h) + 5 - (-2x + 5)}{h}$

$= \lim_{h\to 0}\frac{-2x - 2h + 5 + 2x - 5}{h} = \lim_{h\to 0}\frac{-2h}{h}$

$= -2$

15. $f'(x) = \lim_{h\to 0}\frac{\left(\frac{x + h}{5} - 3\right) - \left(\frac{x}{5} - 3\right)}{h}$

$= \lim_{h\to 0}\frac{\left(\frac{x + h - 15}{5}\right) - \left(\frac{x - 15}{5}\right)}{h}$

$= \lim_{h\to 0}\frac{\frac{x + h - 15 - x + 15}{5}}{h} = \lim_{h\to 0}\frac{\frac{h}{5}}{h} = \frac{1}{5}$

16. $f'(x) = \lim_{h\to 0}\frac{3(x + h)^2 - 3x^2}{h}$

$= \lim_{h\to 0}\frac{3x^2 + 6xh + 3h^2 - 3x^2}{h}$

$= \lim_{h\to 0}\frac{h(6x + 3h)}{h} = \lim_{h\to 0}(6x + 3h) = 6x$

17. $f'(x) = \lim_{h\to 0}\frac{(x + h)^3 - x^3}{h}$

$= \lim_{h\to 0}\frac{x^3 + 3x^2h + 3xh^2 + h^3 - x^3}{h}$

$= \lim_{h\to 0}\frac{h(3x^2 + 3xh + h^2)}{h}$

$= \lim_{h\to 0}(3x^2 + 3xh + h^2) = 3x^2$

❯ B. Exercises

18. $f(0) = -1.6(0)^2 - 5(0) = 0$
$f(3) = -1.6(3)^2 - 5(3) = -29.4$
$\frac{f(b) - f(a)}{b - a} = \frac{-29.4 - 0}{3 - 0} = \frac{-29.4}{3}$

$= -9.8$ m/sec

19. $f(1) = 0.9(1)^2 + 1.1(1) = 2$
$f(2.5) = 0.9(2.5)^2 + 1.1(2.5) = 8.375$
$\frac{f(b) - f(a)}{b - a} = \frac{8.375 - 2}{2.5 - 1} = 4.25$ m/sec

20. $f(0.4) = -9.8(0.4)^2 + 7(0.4) = 1.232$
$f(0.5) = 21(0.5)^2 + 7(0.5) = 1.05$
$\dfrac{f(b) - f(a)}{b - a} = \dfrac{1.05 - 1.232}{0.5 - 0.4}$
$= -1.82$ m/sec

21. $f(0.25) = 48(0.25)^2 = 3$
$f(0.75) = 48(0.75)^2 = 27$
$\dfrac{f(b) - f(a)}{b - a} = \dfrac{27 - 3}{0.75 - 0.25} = 48$ m/sec

22. $v_{inst}(t) = f'(t) = \lim\limits_{h \to 0} \dfrac{19(t + h) - (19t)}{h}$
$= \lim\limits_{h \to 0} \dfrac{19t + 19h - 19t}{h} = \lim\limits_{h \to 0} \dfrac{19h}{h} = 19$
$v_{inst}(t) = 19$

23. $v_{inst}(t) = f'(t)$
$= \lim\limits_{h \to 0} \dfrac{-12(t + h) + 15 - (-12t + 15)}{h}$
$= \lim\limits_{h \to 0} \dfrac{-12t - 12h + 15 + 12t - 15}{h}$
$= \lim\limits_{h \to 0} \dfrac{-12h}{h} = -12; \; v_{inst}(t) = -12$

24. $v_{inst}(t) = f'(t)$
$= \lim\limits_{h \to 0} \dfrac{19(t + h)^2 - 6(t + h) - (19t^2 - 6t)}{h}$
$= \lim\limits_{h \to 0} \dfrac{19(t^2 + 2ht + h^2) - 6t - 6h - 19t^2 + 6t}{h}$
$= \lim\limits_{h \to 0} \dfrac{19t^2 + 38ht + 19h^2 - 6h - 19t^2}{h}$
$= \lim\limits_{h \to 0} \dfrac{h(38t + 19h - 6)}{h}$
$= \lim\limits_{h \to 0} (38t + 19h - 6) = 38t - 6$
$v_{inst}(t) = 38t - 6$

25. $v_{inst}(t) = f'(t)$
$= \lim\limits_{h \to 0} \dfrac{8(t + h)^2 - 13(t + h) - (8t^2 - 13t)}{h}$
$= \lim\limits_{h \to 0} \dfrac{8(t^2 + 2ht + h^2) - 13t - 13h - 8t^2 + 13t}{h}$
$= \lim\limits_{h \to 0} \dfrac{8t^2 + 16ht + 8h^2 - 13h - 8t^2}{h}$
$= \lim\limits_{h \to 0} \dfrac{h(16t + 8h - 13)}{h}$
$= \lim\limits_{h \to 0} (16t + 8h - 13) = 16t - 13$
$v_{inst}(t) = 16t - 13$

26. $f'(x) = \lim\limits_{h \to 0} \left[\dfrac{2\sqrt{x + h} - 2\sqrt{x}}{h} \right.$
$\left. \cdot \dfrac{2\sqrt{x + h} + 2\sqrt{x}}{2\sqrt{x + h} + 2\sqrt{x}} \right]$
$= \lim\limits_{h \to 0} \dfrac{4(x + h) - 4x}{2h(\sqrt{x + h} + \sqrt{x})}$
$= \lim\limits_{h \to 0} \dfrac{4h}{2h(\sqrt{x + h} + \sqrt{x})}$
$= \lim\limits_{h \to 0} \dfrac{2}{\sqrt{x + h} + \sqrt{x}}$
$= \dfrac{2}{\sqrt{x} + \sqrt{x}} = \dfrac{2}{2\sqrt{x}} = \dfrac{1}{\sqrt{x}}$

DEFINITION

If the position of an object is represented by a function of time $f(t)$, then its **instantaneous velocity** at time t is $v_{inst}(t) = f'(t) = \lim\limits_{h \to 0} \dfrac{f(t + h) - f(t)}{h}$.

Example 6 Finding Instantaneous Velocity

The height of an object dropped from a bridge 150 ft above a river is modeled by $f(t) = -16t^2 + 150$. Find the object's instantaneous velocity and height after falling for 3 sec.

Answer

$v_{inst}(t) = f'(t) = \lim\limits_{h \to 0} \dfrac{f(t + h) - f(t)}{h}$ — 1. Apply the limit definition and simplify to find the instantaneous velocity function $f'(t)$.

$= \lim\limits_{h \to 0} \dfrac{-16(t + h)^2 + 150 - (-16t^2 + 150)}{h}$

$= \lim\limits_{h \to 0} \dfrac{-16(t^2 + 2th + h^2) + 150 + 16t^2 - 150}{h}$

$= \lim\limits_{h \to 0} \dfrac{-32th - 16h^2}{h}$

$= \lim\limits_{h \to 0} (-32t - 16h) = -32t - 16(0)$

$\therefore v_{inst}(t) = -32t$

$v_{inst}(3) = -32(3) = -96$ ft/sec — 2. Substitute 3 into $v_{inst}(t)$.
 or 96 ft/sec down

$f(3) = -16(3)^2 + 150 = 6$ ft above the ground — 3. Substitute 3 into $f(t)$ to find its height.

SKILL ✓ EXERCISE 23

The fact that the derivative represents the slope of the tangent to a curve (its rate of change at that instant) is a foundational concept in calculus.

❯ A. Exercises

Find a simplified expression for the difference quotient, $\dfrac{f(x + h) - f(x)}{h}$, of each function.

1. $f(x) = 2x + 3$ $\;$ 2
2. $f(x) = -5x - 7$ $\;$ -5
3. $f(x) = x^2 + 3x$ $\;$ $2x + 3 + h$
4. $f(x) = 3x^2 - 4x$ $\;$ $6x - 4 + 3h$

5. State whether each function is differentiable at the given point. If not, explain why.
 a. $f(x) = 5x^2 + 6; x = -2$ $\;$ yes
 b. $g(x) = \sqrt{x}; x = 0$ $\;$ no; discontinuous
 c. $h(x) = \cos x; x = \pi$ $\;$ yes

6. State whether each function is differentiable at the given point. If not, explain why.
 a. $f(x) = |x|; x = 1$ $\;$ yes
 b. $f(x) = \dfrac{6}{x^2 - 2x - 15}; x = -3$ $\;$ no; discontinuous (vertical asymptote)
 c. $f(x) = \tan x; x = \dfrac{\pi}{2}$ $\;$ no; discontinuous (vertical asymptote)

7. State whether each function is differentiable over its entire domain. If not, explain why.
 a. $f(x) = [x]$
 b. $f(x) = \log x$ $\;$ yes
 c. $f(x) = \sin x$ $\;$ yes
 d. $f(x) = |x^2 - 2|$

Find the derivative of each function at the given point.

8. $f(x) = 6x; x = 4$ $\;$ $f'(4) = 6$
9. $f(x) = 2x - 60; x = -1$ $\;$ $f'(-1) = 2$
10. $f(x) = x^2; x = 3$ $\;$ $f'(3) = 6$
11. $f(x) = 7; x = -8$ $\;$ $f'(-8) = 0$

Find $f'(x)$ for each function.

12. $f(x) = x$ $\;$ $f'(x) = 1$
13. $f(x) = 3$ $\;$ $f'(x) = 0$
14. $f(x) = -2x + 5$ $\;$ $f'(x) = -2$
15. $f(x) = \dfrac{x}{5} - 3$ $\;$ $f'(x) = \dfrac{1}{5}$
16. $f(x) = 3x^2$ $\;$ $f'(x) = 6x$
17. $f(x) = x^3$ $\;$ $f'(x) = 3x^2$

❯ B. Exercises

The height (in meters) of an object after t seconds is modeled by the equation $f(t)$. Find the average velocity of the object between a and b seconds.

18. $f(t) = -1.6t^2 - 5t; a = 0, b = 3$ $\;$ -9.8 m/sec
19. $f(t) = 0.9t^2 + 1.1t; a = 1, b = 2.5$ $\;$ 4.25 m/sec
20. $f(t) = -9.8t^2 + 7t; a = 0.4, b = 0.5$ $\;$ -1.82 m/sec
21. $f(t) = 48t^2; a = 0.25, b = 0.75$ $\;$ 48 m/sec

7a. no; jumps at $x \in$ integers
7d. no; sharp points at $x = \pm\sqrt{2}$

---

27. $f'(x) = \lim\limits_{h \to 0} \left[\dfrac{\sqrt{3(x + h)} - \sqrt{3x}}{h} \right.$
$\left. \cdot \dfrac{\sqrt{3(x + h)} + \sqrt{3x}}{\sqrt{3(x + h)} + \sqrt{3x}} \right]$
$= \lim\limits_{h \to 0} \dfrac{3(x + h) - 3x}{h(\sqrt{3(x + h)} + \sqrt{3x})}$
$= \lim\limits_{h \to 0} \dfrac{3h}{h(\sqrt{3(x + h)} + \sqrt{3x})}$
$= \lim\limits_{h \to 0} \dfrac{3}{\sqrt{3(x + h)} + \sqrt{3x}}$
$= \dfrac{3}{\sqrt{3x} + \sqrt{3x}} = \dfrac{3}{2\sqrt{3x}}$ or $\dfrac{\sqrt{3x}}{2x}$

28. $f'(x) = \lim\limits_{h \to 0} \dfrac{(x + h)^2 - 4 - (x^2 - 4)}{h}$
$= \lim\limits_{h \to 0} \dfrac{x^2 + 2xh + h^2 - 4 - x^2 + 4}{h}$
$= \lim\limits_{h \to 0} \dfrac{2xh + h^2}{h} = \lim\limits_{h \to 0} \dfrac{h(2x + h)}{h}$
$= \lim\limits_{h \to 0} (2x + h) = 2x$
$m = f'(-1) = 2(-1) = -2$
$f(-1) = (-1)^2 - 4 = -3; \; (-1, -3)$
$y - (-3) = -2(x - (-1))$
$y + 3 = -2x - 2$
$y = -2x - 5$

Find the instantaneous velocity function $v_{inst}(t)$ for each distance function $f(t)$.

22. $f(t) = 19t$

23. $f(t) = -12t + 15$

24. $f(t) = 19t^2 - 6t$

25. $f(t) = 8t^2 - 13t$

Find $f'(x)$ for each function.

26. $f(x) = 2\sqrt{x}$

27. $f(x) = \sqrt{3x}$

Find $f'(x)$ for each function and then determine the equation of the tangent at the given x-value.

28. $f(x) = x^2 - 4;\ x = -1$

29. $f(x) = x^2 - x;\ x = -4$

Find the derivative of each function.

30. the constant function: $f(x) = c$ $f'(x) = 0$

31. the linear function: $f(x) = mx + b$ $f'(x) = m$

32. If the height (in meters) of an object thrown into the air from a height of 2 m with an initial vertical velocity of 20 m/sec is modeled by $d(t) = -4.9t^2 + 20t + 2$, find the average vertical velocity for the first 2 sec.   10.2 m/sec

33. If the height (in feet) of an object dropped from a height of 200 ft is modeled by $d(t) = -16t^2 + 200$, find the average vertical velocity from its drop until it reaches 56 ft.   −48 ft/sec

34. If the height (in meters) of an object dropped from a height of 320 m is modeled by $d(t) = -4.9t^2 + 320$, find the object's instantaneous vertical velocity at 1 sec and at 4 sec.   −9.8 m/sec; −39.2 m/sec

CUMULATIVE REVIEW

Given $z_1 = 2$ cis $30°$ and $z_2 = 3$ cis $60°$, express each complex number in standard form. [6.7]

41. $z_1 z_2$ $6i$

42. $\dfrac{z_1}{z_2}$ $\dfrac{\sqrt{3}}{3} - \dfrac{1}{3}i$

Evaluate each determinant. [7.4]

43. $\begin{vmatrix} -4 & 5 \\ 6 & 1 \end{vmatrix}$ -34

44. $\begin{vmatrix} -7 & 1 & -2 \\ 5 & -8 & 0 \\ 3 & -3 & 3 \end{vmatrix}$ 135

Evaluate each infinite geometric series or state that the series is divergent. [9.3]

45. $\displaystyle\sum_{n=1}^{\infty} 3n$ divergent

46. $\displaystyle\sum_{n=1}^{\infty} 4\left(\dfrac{-1}{2}\right)^n$ $-\dfrac{4}{3}$

47. Find the general solution of the equation $\sin 2\theta + \sin \theta = 0$ (assuming $n \in \mathbb{Z}$). [5.5]   A
  A. $\theta = n\pi,\ n2\pi \pm \dfrac{2\pi}{3}$
  B. $\theta = \dfrac{n2\pi}{3}$
  C. $\theta = n\pi$
  D. $\theta = n2\pi \pm \dfrac{2\pi}{3}$
  E. none of these

35. If the height (in feet) of an object shot into the air from ground level with an initial vertical velocity of 120 ft/sec is modeled by $d(t) = -16t^2 + 120t$, find the object's instantaneous vertical velocity at 3 sec and at 6 sec.   24 ft/sec; −72 ft/sec

⟩ **C. Exercises**

Find the derivative of each function.

36. the quadratic function $f(x) = ax^2 + bx + c$ $f'(x) = 2ax + b$

37. the inverse variation $f(x) = \dfrac{k}{x}$ $f'(x) = -\dfrac{k}{x^2}$

38. the absolute value function $f(x) = \begin{cases} x & \text{if } x \ge 0 \\ -x & \text{if } x < 0 \end{cases}$

39. Examine a graph of the greatest integer function and state its derivative function. $f'(x) = 0$ for all $x \notin \mathbb{Z}$

40. The derivative of $f(x) = x^2$ is $f'(x) = 2x$. The derivative of $g(x) = x^3$ is $g'(x) = 3x^2$.
  a. Make a conjecture predicting the derivative of $h(x) = x^4$. $h'(x) = 4x^3$
  b. Use the definition of a derivative to confirm your conjecture.
  c. Predict the derivative of $j(x) = x^5$. $j'(x) = 5x^4$

48. Eliminate the parameter $\theta \in [0, 2\pi]$ to find the standard form equation for the ellipse $x = 2 \cos \theta$ and $y = 4 \sin \theta$. [8.5]   D
  A. $\dfrac{x^2}{2} + \dfrac{y^2}{4} = 1$
  B. $\dfrac{x}{2} + \dfrac{y}{4} = 1$
  C. $\dfrac{x^2}{4} - \dfrac{y^2}{16} = 1$
  D. $\dfrac{x^2}{4} + \dfrac{y^2}{16} = 1$
  E. none of these

49. Which expression is used to evaluate $\displaystyle\sum_{k=1}^{n} k$? [9.5]   A
  A. $\dfrac{n(n+1)}{2}$
  B. $\dfrac{n(n-1)}{2}$
  C. $\dfrac{n(n+1)(2n+1)(3n^2+3n-1)}{30}$
  D. $\dfrac{n^2(n+1)^2}{4}$
  E. $\dfrac{n(n+1)(2n+1)}{6}$

50. Find the equation of the asymptote of $f(x)$ implied by $\displaystyle\lim_{x \to -\infty} f(x) = 4$. [12.1]   C
  A. $x = 4$
  B. $x = 2$
  C. $y = 4$
  D. $y = 0$
  E. none of these

$$\dfrac{\Delta d}{\Delta t} = \dfrac{d(3) - d(0)}{3 - 0} = \dfrac{56 - 200}{3}$$
$$= -48 \text{ ft/sec}$$

34. $v_{inst}(t) = d'(t)$
$$= \lim_{h \to 0} \dfrac{-4.9(t+h)^2 + 320 - (-4.9t^2 + 320)}{h}$$
$$= \lim_{h \to 0} \dfrac{-4.9(t^2 + 2ht + h^2) + 320 + 4.9t^2 - 320}{h}$$
$$= \lim_{h \to 0} \dfrac{-4.9t^2 - 9.8ht - 4.9h^2 + 4.9t^2}{h}$$
$$= \lim_{h \to 0} \dfrac{h(-9.8t - 4.9h)}{h}$$
$$= \lim_{h \to 0} (-9.8t - 4.9h) = -9.8t$$
$$v_{inst}(t) = -9.8t$$
$$v_{inst}(1) = -9.8(1) = -9.8 \text{ m/sec}$$
$$v_{inst}(4) = -9.8(4) = -39.2 \text{ m/sec}$$

35. $v_{inst}(t) = d'(t)$
$$= \lim_{h \to 0} \dfrac{-16(t+h)^2 + 120(t+h) - (-16t^2 + 120t)}{h}$$
$$= \lim_{h \to 0} \left[\dfrac{-16t^2 - 32th - 16h^2 + 120t + 120h}{h} \right.$$
$$\left. + \dfrac{16t^2 - 120t}{h} \right]$$
$$= \lim_{h \to 0} \dfrac{-16t^2 - 32th - 16h^2 + 120t + 120h + 16t^2 - 120t}{h}$$
$$= \lim_{h \to 0} \dfrac{-32th - 16h^2 + 120h}{h}$$
$$= \lim_{h \to 0} \dfrac{h(-32t - 16h + 120)}{h}$$
$$= \lim_{h \to 0} (-32t - 16h + 120)$$
$$= -32t + 120$$
$$v_{inst}(t) = -32t + 120$$
$$v_{inst}(3) = -32(3) + 120 = 24 \text{ ft/sec}$$
$$v_{inst}(6) = -32(6) + 120 = -72 \text{ ft/sec}$$

36. $f'(x)$
$$= \lim_{h \to 0} \dfrac{a(x+h)^2 + b(x+h) + c - (ax^2 + bx + c)}{h}$$
$$= \lim_{h \to 0} \dfrac{a(x+h)^2 + bx + bh + c - ax^2 - bx - c}{h}$$
$$= \lim_{h \to 0} \dfrac{ax^2 + 2axh + ah^2 + bh - ax^2}{h}$$
$$= \lim_{h \to 0} \dfrac{2axh + ah^2 + bh}{h}$$
$$= \lim_{h \to 0} (2ax + b + ah) = 2ax + b$$

37. $f'(x) = \lim_{h \to 0} \dfrac{\frac{k}{x+h} - \frac{k}{x}}{h}$
$$= \lim_{h \to 0} \dfrac{\frac{kx}{x(x+h)} - \frac{k(x+h)}{x(x+h)}}{h} = \lim_{h \to 0} \dfrac{\frac{kx - kx - kh}{x^2 + xh}}{h}$$
$$= \lim_{h \to 0} \left(\dfrac{-kh}{x^2 + xh} \cdot \dfrac{1}{h} \right) = \lim_{h \to 0} \dfrac{-k}{x^2 + xh}$$
$$= -\dfrac{k}{x^2}$$

---

29. $f'(x) = \lim_{h \to 0} \dfrac{(x+h)^2 - (x+h) - (x^2 - x)}{h}$
$$= \lim_{h \to 0} \dfrac{x^2 + 2xh + h^2 - x - h - x^2 + x}{h}$$
$$= \lim_{h \to 0} \dfrac{2xh - h + h^2}{h} = \lim_{h \to 0} \dfrac{h(2x - 1 + h)}{h}$$
$$= \lim_{h \to 0} (2x - 1 + h) = 2x - 1$$
$$m = f'(-4) = 2(-4) - 1 = -9$$
$$f(-4) = (-4)^2 - (-4) = 19;\ (-4, 20)$$
$$y - 20 = -9(x - (-4))$$
$$y - 20 = -9x - 36$$
$$y = -9x - 16$$

30. $f'(x) = \lim_{h \to 0} \dfrac{f(x+h) - f(x)}{h} = \lim_{h \to 0} \dfrac{c - c}{h}$
$$= \lim_{h \to 0} \dfrac{0}{h} = 0$$

31. $f'(x) = \lim_{h \to 0} \dfrac{m(x+h) + b - (mx + b)}{h}$
$$= \lim_{h \to 0} \dfrac{mx + mh + b - mx - b}{h} = \lim_{h \to 0} \dfrac{mh}{h}$$
$$= m, \text{ the line's slope}$$

32. $d(0) = -4.9(0)^2 + 20(0) + 2 = 2$
$$d(2) = -4.9(2)^2 + 20(2) + 2 = 22.4$$
$$\dfrac{\Delta d}{\Delta t} = \dfrac{d(2) - d(0)}{2 - 0} = \dfrac{22.4 - 2}{2}$$
$$= 10.2 \text{ m/sec}$$

33.
$$56 = -16t^2 + 200$$
$$-144 = -16t^2$$
$$9 = t^2;\ t = 3 \text{ sec}$$

continued in Answers and Solutions Overflow

12.4 Basic Derivative Theorems

Objectives

1. To find derivatives by applying basic derivative theorems

2. To derive the power rule and apply it to find derivatives of polynomial functions

3. To apply derivative theorems in finding slopes and tangent lines

4. To find extreme values using $f'(x) = 0$ to solve real-world problems

Flash

Principia introduced Newton's three laws of motion and his proofs relating to gravity and its effect in our solar system. The text also contains his inverse square law, which agreed with Kepler's elliptical orbits. This British stamp was issued in 1987 to commemorate the 300th anniversary of the work.

Vocabulary

derivative of a constant function
derivative of a linear function
derivative of a power function
derivative of a scalar multiple
 of a function
derivative of a sum or difference
power rule

Reading and Writing Mathematics

The first derivative of a position-time function is velocity. Its second derivative is acceleration. Perform an Internet search to find what the third and fourth derivatives of a position-time graph are called and briefly describe each. The third derivative is jerk and the fourth derivative is snap. Just as velocity describes how quickly an object's position changes and acceleration describes how quickly velocity changes, jerk measures how quickly acceleration changes, and snap measures how quickly jerk changes.

12.4 Basic Derivative Theorems

Issac Newton's Principia is one of the most profound achievements in the field of science.

Isaac Newton and Gottfried Leibniz are credited with many of the theorems related to derivatives. These theorems provide a faster and easier way of finding derivatives than repeatedly applying the definition. This section develops theorems that can be used to find the derivatives of power and polynomial functions.

After completing this section, you will be able to

• find derivatives by applying basic derivative theorems.

• derive and apply the power rule to find derivatives of power and polynomial functions.

• apply derivative theorems in finding slopes of tangent lines.

• find extreme values using $f'(x) = 0$ to solve real-world problems.

Isaac Newton discovered many famous scientific laws, but did not publish his findings until he was persuaded to do so by Edmond Halley, the discoverer of Halley's Comet.

THE DERIVATIVE OF A CONSTANT OR LINEAR FUNCTION

The derivative of any constant function $f(x) = c$ is $f'(x) = 0$.
The derivative of any linear function $f(x) = mx + b$ is $f'(x) = m$.

The proofs of these theorems were completed in exercises 30–31 in Section 12.3. These theorems emphasize that constant and linear functions have a constant slope, either 0 or m.

The following theorem is extremely useful when finding derivatives of powers. The proof of the theorem when $n \in \mathbb{N}$ is demonstrated here, and the case of n being a negative integer is proved in exercise 43 of the next section. The theorem is also true for rational and irrational powers, but the proofs for these other cases will not be required in this text.

THE POWER RULE

If $f(x) = x^n$ and $n \in \mathbb{R}$, then $f'(x) = nx^{n-1}$.

Proof: (for $n \in \mathbb{N}$)

$$f'(x) = \lim_{h \to 0} \frac{(x+h)^n - x^n}{h}$$

$$= \lim_{h \to 0} \frac{\left(x^n + nx^{n-1}h + \frac{n(n-1)}{2}x^{n-2}h^2 + \cdots + h^n\right) - x^n}{h} \quad \text{(using the Binomial Theorem)}$$

$$= \lim_{h \to 0} \frac{h\left(nx^{n-1} + \frac{n(n-1)}{2}x^{n-2}h + \cdots + h^{n-1}\right)}{h}$$

$$= \lim_{h \to 0} \left(nx^{n-1} + \frac{n(n-1)}{2}x^{n-2}h + \cdots + h^{n-1}\right)$$

$$= nx^{n-1} + 0 + \cdots + 0 = nx^{n-1}$$

Example 1   Applying the Power Rule

Find the derivative function for each function.

a. $f(x) = x^{21}$ **b.** $g(x) = \dfrac{1}{x^6}$ **c.** $h(x) = \sqrt[3]{x^2}$

Answer

a. $f'(x) = 21x^{21-1} = 21x^{20}$ Apply the power rule for $n = 21$.

b. $g(x) = x^{-6}$ 1. Rewrite the function as a power.
$g'(x) = -6x^{-6-1} = -6x^{-7}$ 2. Apply the power rule for $n = -6$.

CONTINUED ➡

PRESENTATION

Lesson Opener

PRESENTATION

Lesson Opener

1. Use the limit of the difference quotient to find $f'(x)$ if $f(x) = 3x^2 - 6$. $f'(x) = 6x$

2. What does the answer to the above exercise represent? the equation of the line tangent to the curve at x

3. What does $f'(2)$ represent? the slope (instantaneous rate of change) of the line tangent to the curve at $x = 2$

Encourage the students by assuring them that the theorems in this section will simplify the process of finding a derivative.

Ask the students to describe the derivative of any constant function and any linear function of the form $y = mx + b$. *(0 and m as proved in Section 12.3 exercises 30–31)* Emphasize that both are constants representing the slope of the line, whether it is horizontal or not.

Introduce the *power rule* and its proof for powers that are natural numbers. Use Example 1 to illustrate this rule. In parts *b* and *c*, emphasize the need to rewrite the expression in terms of a power. The first step in finding the derivative of a function is often rewriting the function in an equivalent form.

c. $h(x) = \sqrt[3]{x^2} = x^{\frac{2}{3}}$

    1. Rewrite the function as a power.

$h'(x) = \frac{2}{3}x^{\frac{2}{3}-1} = \frac{2}{3}x^{-\frac{1}{3}}$ or $\frac{2}{3x^{\frac{1}{3}}} = \frac{2}{3\sqrt[3]{x}}$

    2. Apply the power rule for $n = \frac{2}{3}$.

SKILL ✔ EXERCISE 11

THE DERIVATIVE OF A SCALAR MULTIPLE OF A FUNCTION

If $f(x) = cg(x)$, where g is differentiable, then $f'(x) = cg'(x)$.

Proof: $f'(x) = \lim_{h \to 0} \frac{cg(x+h) - cg(x)}{h} = \lim_{h \to 0}\left(c \cdot \frac{g(x+h) - g(x)}{h} \right) = c \lim_{h \to 0} \frac{g(x+h) - g(x)}{h} = cg'(x)$

Combining the results of this theorem with the power rule produces a rule for the derivative of a power function.

THE DERIVATIVE OF A POWER FUNCTION

The derivative of any power function $f(x) = cx^n$ is $f'(x) = ncx^{n-1}$.

To find the derivative of a power function $f(x) = cx^n$, simply multiply the coefficient by the exponent and then decrease the exponent by 1.

Example 2   Finding the Equation of a Tangent Line

Find $f'(x)$ if $f(x) = 3x^4$. Then find the equation of the tangent to the curve at $x = 2$.

Answer

$f'(x) = (4)3x^{4-1} = 12x^3$ 1. Apply the power rule and simplify.

$m = f'(2) = 12(2)^3 = 96$ 2. Evaluate $f'(2)$ to find the slope of the tangent.

$y = f(2) = 3(2)^4 = 48$ 3. Evaluate $f(2)$ to find the point of tangency, P.

$\therefore P\,(2, 48)$

$y - y_1 = m(x - x_1)$ 4. Find the equation of the line through P with $m = 96$.

$y - 48 = 96(x - 2)$

$y - 48 = 96x - 192$

$y = 96x - 144$

Check

     5. Graph the function $f(x) = 3x^4$ and the line $y = 96x - 144$ and confirm that they intersect at $P\,(2, 48)$.

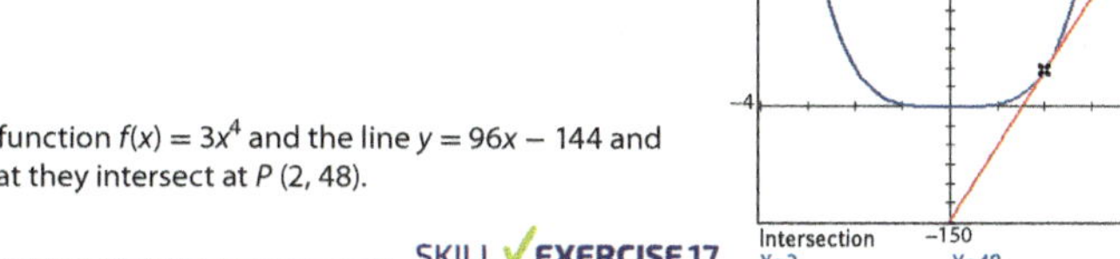

SKILL ✔ EXERCISE 17

Apply the following theorem describing the derivative of a sum or difference to quickly find the derivative of any polynomial function.

DERIVATIVE OF A SUM OR DIFFERENCE

If $f(x) = u(x) \pm v(x)$ when u and v are differentiable, then $f'(x) = u'(x) \pm v'(x)$.

Proof: $f'(x) = \lim_{h \to 0} \frac{[u(x+h) + v(x+h)] - [u(x) + v(x)]}{h} = \lim_{h \to 0} \frac{u(x+h) - u(x) + v(x+h) - v(x)}{h}$

$= \lim_{h \to 0} \left[\frac{u(x+h) - u(x)}{h} + \frac{v(x+h) - v(x)}{h} \right]$

$= \lim_{h \to 0} \frac{u(x+h) - u(x)}{h} + \lim_{h \to 0} \frac{v(x+h) - v(x)}{h}$

$= u'(x) + v'(x)$

Additional Exercises

Find the derivative of each function.

1. $f(x) = x^3$ $f'(x) = 3x^2$

2. $g(x) = \frac{2}{x^3}$ $g'(x) = -\frac{6}{x^4}$

3. $h(x) = \sqrt[4]{x^3}$ $h'(x) = \frac{3}{4\sqrt[4]{x}}$

4. $f(x) = 4x^3 - 3x^2 + 2x - 1$

     $f'(x) = 12x^2 - 6x + 2$

5. Find the equation of the tangent to the curve of $f(x) = 3x^2$ when $x = -1$.

     $y = -6x - 3$

Use derivatives to find the relative extrema of each polynomial function.

6. $f(x) = 16 - x^4$ max: $(0, 16)$

7. $g(x) = \frac{x^3}{3} - \frac{5x^2}{2} + 4x + 10$

     max: $(1, \approx 11.833)$; min: $(4, \approx 7.33)$

8. An open container can be made from a 16 in. square piece of metal by cutting congruent squares from each corner and bending up and welding the sides. Determine what size square should be cut from each corner to maximize the volume; then calculate the maximum volume.   square: $2\frac{2}{3}$ in.

     max. vol.: ≈ 303.4 in.^{3}

9. An object is shot from a height of 25 ft with an initial vertical velocity of 100 ft/sec.

     a. Determine its instantaneous vertical velocity after 3 sec.   4 ft/sec

     b. How long does it take to reach its maximum height? What is the maximum height of the object?

     3.125 sec; 181.25 ft

One-on-One A student may need reminding that if a radical expression in the denominator has a variable in the radicand, the expression does not have to be rationalized, because whether it is irrational or not depends on the value of the variable.

Introduce the *derivative of a scalar multiple of a function* and the commonly used rule for the *derivative of a power function*. This rule is applied in Example 2 to find the equation of a tangent to the curve. The power rule is applied to the function to find the derivative function $f'(x)$ that provides the slope of the tangent at $x = 2$. Evaluating $f(2)$ provides the point's y-coordinate.

Present the rule for the *derivative of a sum or difference* along with the proof of the derivative of a sum. Using Example 3, demonstrate how derivatives of polynomials can be found quickly without showing any work.

Example 4 provides a real-world maximization application. Point out that the tangent at any maximum or minimum is horizontal and has a slope of 0. The quadratic derivative function in step 3 represents the slope of the tangent to the curve. The zeros of the derivative function indicate possible relative minimum or maximum points. The margin tip illustrates that a point at which $f'(x) = 0$ could

be a point of inflection as well. Considering the sign of the slope (the derivative function) on both sides of any value of x where $f'(x)$ can help determine whether the point is a relative maximum, relative minimum, or point of inflection. In exercises 39–40 students will use the *second derivative* to classify the point. Consider having the students use the Internet keyword search *second derivative concavity* to further explore points of inflection.

Interactive Activity An interactive view of a function, its derivative, and its tangent can be found with the Internet keyword search *derivative applet*. Try using $f(x) = x^3$.

10. Use technology along with derivatives to find the relative extrema of $f(x) = x^5 - 3x^3 + 1$.

max: $\approx (-1.342, 3.898)$

min: $\approx (1.342, -1.898)$; $(0, 1)$ is a pt. of inflection since $f'(x) < 0$ as $x \to 0^{\pm}$.

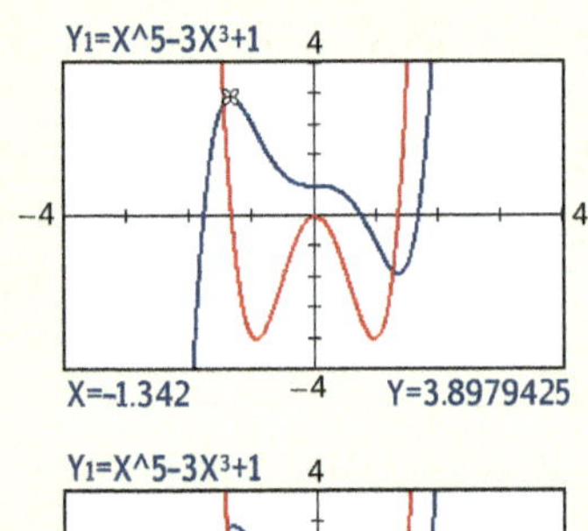

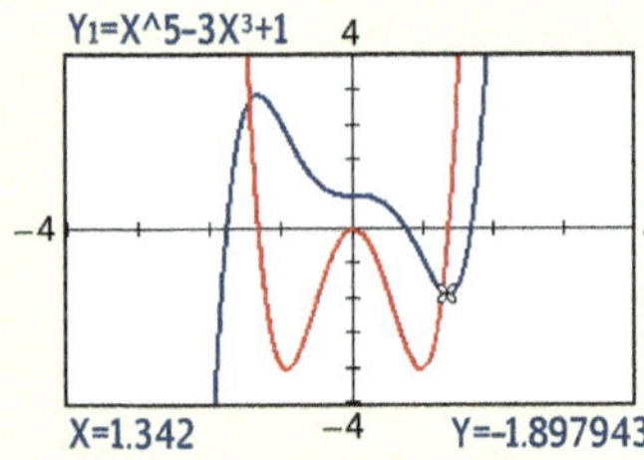

Assignments

- **Minimum:** 1–5; 7–11 odd; 13–14; 17–25 odd; 29(a–b), 30, 32, 43, 46, 48–50

- **Standard:** 1–2, 5, 7–8, 10–11, 14–15, 18–19, 23, 27, 28a, 29(a–c), 32–33, 38–39, 44–45, 49–50, 52

- **Extended:** 1, 6, 9–11, 13, 16, 20, 24, 27, 28b, 29(a–b, d); 32–36 even; 39–42, 44, 49, 51–52

Solutions

A. Exercises

2. $f'(x) = 12(2)x^{11} = 24x^{11}$

3. $f'(x) = 4(11)x^3 = 44x^3$

4. $h'(x) = 4(6)x^3 + 2(5)x = 24x^3 + 10x$

6. $h'(x) = 3(7)x^2 - 2(2)x + 5$
$= 21x^2 - 4x + 5$

7. $p'(x) = 4x^3 - 3(3)x^2 = 4x^3 - 9x^2$

8. Since e^5 is a constant, its derivative is 0.

9. $f'(x) = -15x^{-15-1} = -15x^{-16}$

10. $g(x) = x^{-4}$
$g'(x) = -4x^{-4-1} = -4x^{-5}$

11. $t(x) = x^{\frac{1}{3}}$
$t'(x) = \frac{1}{3}x^{-\frac{2}{3}} = \frac{1}{3\sqrt[3]{x^2}}$

12. $v(x) = x^{\frac{17}{5}}$
$v'(x) = \frac{17}{5}x^{\frac{12}{5}}$ or $\frac{17\sqrt[5]{x^{12}}}{5} = \frac{17x^2\sqrt[5]{x^2}}{5}$

The proof of the derivative of a difference is left as an exercise.

Example 3   Finding the Derivative of a Polynomial Function

Find the derivative of $f(x) = 2x^3 - 6x^2 + x - 5$.

Answer

$f'(x) = (3)2x^{3-1} - (2)6x^{2-1} + 1x^{1-1} - 0$
$= 6x^2 - 12x + 1$

Apply the derivative of a power function or the derivative of a constant function to each term and simplify.

SKILL ✔ **EXERCISE 7**

With practice you should be able to use these theorems to differentiate polynomial functions mentally without showing any work.

Calculus is frequently used in real-world applications to find relative extrema. For example, businesses seek to maximize profit and minimize expenses. Since the tangent to a curve is horizontal at any relative minimum or maximum point, the derivative of the function at these points is zero. The x-coordinates of any relative extrema can be found by setting the derivative of the function equal to zero and solving for x. The relative extrema can often be identified by evaluating the function at these values of x and by examining the overall shape of the function.

Since the slope of the graph changes from positive to negative at a relative maximum $(a, f(a))$, $P'(x) > 0$ as $x \to a^-$ and $P'(x) < 0$ as $x \to a^+$. The slope of the graph changes from negative to positive at a relative minimum $(b, f(b))$, so $P'(x) < 0$ as $x \to b^-$ and $P'(x) > 0$ as $x \to b^+$.

Example 4   Finding the Maximum for a Function

An open container can be made from a 20 in. square piece of metal by cutting congruent squares from each corner and bending up and welding the sides. Determine what size square should be cut from each corner to maximize the volume and then calculate maximum volume.

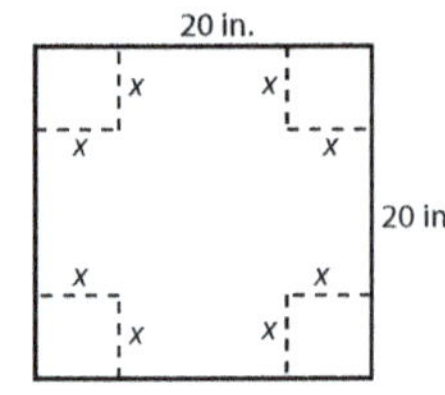

Answer

$V = lwh$ with $h = x =$ side of the square
$V(x) = (20 - 2x)(20 - 2x)x$

1. Write a function modeling the volume, the quantity that is to be maximized.

$V(x) = (400 - 80x + 4x^2)x$
$= 4x^3 - 80x^2 + 400x$

2. Express $V(x)$ as a polynomial function.

$V'(x) = 12x^2 - 160x + 400$

3. Apply the derivative theorems to find the derivative.

$12x^2 - 160x + 400 = 0$
$3x^2 - 40x + 100 = 0$
$(3x - 10)(x - 10) = 0$
$x = \frac{10}{3}, 10$

4. Solve $V'(x) = 0$.

CONTINUED ➡

> **TIP**
>
> The slope of the tangent to a graph can also be 0 at a point of inflection. Examine the graph of $f(x) = x^5 - x^3$.
>
> pt. of inflection
>
> rel. max.
>
> rel. min.
>
> Its derivative,
> $f(x) = 5x^4 - 3x^2$
> $= x^2(5x^2 - 3)$, is 0 if $x = \pm 0.77$ (the relative maximum and minimum) or if $x = 0$, where the slope of the function is zero but doesn't have opposite signs to the left and right of the point as it does at a relative maximum or minimum.

Before presenting Example 5, consider reminding the students that slope represents the rate of change, that is $\frac{\Delta y}{\Delta x}$. On a position-time graph, the rate of change, or velocity, is $\frac{\Delta d}{\Delta t}$. The rate of change in position at time t is $d'(t)$, therefore the instantaneous velocity at t is $d'(t)$. Consider explaining the process for Example 5 in general terms: define the function $f(x)$, find $f'(x)$, and then find $f'(2)$. To find the maximum value for $f(x)$, find where the derivative function, $f'(x)$, equals 0.

TIPS

Ex. 31 Consider informing students that they need to find an expression for v_0 and then substitute it into the height equation.

$V(10) = [20 - 2(10)][20 - 2(10)](10) = 0$

$V\left(\frac{10}{3}\right) = \left[20 - 2\left(\frac{10}{3}\right)\right]\left[20 - 2\left(\frac{10}{3}\right)\right]\left(\frac{10}{3}\right)$

$\phantom{V\left(\frac{10}{3}\right)} = \left(\frac{40}{3}\right)\left(\frac{40}{3}\right)\left(\frac{10}{3}\right) \approx 592.6$

The maximum volume of ≈ 592.6 in.^{3} is obtained by cutting a $3\frac{1}{3}$ in. square from each corner.

5. Evaluate $V(x) = (20 - 2x)(20 - 2x)x$ at each resulting value to identify the relative maximum.

SKILL ✔ EXERCISE 23

The height of a projectile was modeled (ignoring air resistance) in Section 1.6 with the quadratic function $h(t) = -\frac{1}{2}gt^2 + v_0 t + h_0$ where h_0 is the initial height, v_0 is the initial vertical velocity, and the magnitude of g, the acceleration due to gravity, is approximately 32 ft/sec^2 or 9.8 m/sec^2. The derivative of this function can be used to find the object's instantaneous vertical velocity. At the projectile's maximum height, its vertical velocity is 0.

In manufacturing, the instantaneous cost (the cost to produce the next item), is often modeled by an upward-opening parabola. Setting the derivative of this quadratic function equal to zero and solving yields the level of production where instantaneous cost is at a minimum, which is modeled by the vertex of the parabola.

Example 5 — Finding Instantaneous Vertical Velocity and Maximum Height

A ball is propelled from a height of 5 ft with an initial vertical velocity of 72 ft/sec.

a. Determine the ball's instantaneous vertical velocity after 2 sec.

b. Find the maximum height of the ball during its flight.

Answer

a. $h(t) = -\frac{1}{2}(32)t^2 + (72)t + (5)$

$ = -16t^2 + 72t + 5$

$h'(t) = 2(-16t^{2-1}) + 72 + 0$

$\therefore v(t) = h'(t) = -32t + 72$

$v(2) = h'(2) = -32(2) + 72 = 8$ ft/sec

1. Substitute $g = 32$ ft/sec^2, $v_0 = 72$ ft/sec and $h_0 = 5$ ft into $h(t) = -\frac{1}{2}gt^2 + v_0 t + h_0$.

2. Use the derivative theorems to find the instantaneous vertical velocity function, $h'(t)$.

3. Evaluate $h'(2)$ to find the instantaneous vertical velocity after 2 sec.

b. $0 = -32t + 72$

$32t = 72$

$t = 2.25$ sec

$h(2.25) = -16(2.25)^2 + 72(2.25) + 5$

$ = 86$ ft

1. Find the time it takes to reach the maximum height by setting $v(t) = 0$ and solving for t.

2. Evaluate $h(2.25)$ to find the maximum height.

SKILL ✔ EXERCISE 29

❯ A. Exercises

Find the derivative of each function by applying derivative theorems.

1. $f(x) = 10$ $f'(x) = 0$

2. $f(x) = 2x^{12}$ $f'(x) = 24x^{11}$

3. $f(x) = 11x^4$ $f'(x) = 44x^3$

4. $h(x) = 6x^4 + 5x^2 + 7$

5. $g(x) = x^5 + x^4 + x^3 + x^2 + x + 1$

6. $h(x) = 7x^3 - 2x^2 + 5x - 6$ $h'(x) = 21x^2 - 4x + 5$

7. $p(x) = x^4 - 3x^3 + 9$ $p'(x) = 4x^3 - 9x^2$

8. $n(x) = e^5$ $n'(x) = 0$

4. $h'(x) = 24x^3 + 10x$

5. $g'(x) = 5x^4 + 4x^3 + 3x^2 + 2x + 1$

9. $f(x) = x^{-15}$ $f'(x) = -15x^{-16}$

10. $g(x) = \frac{1}{x^4}$ $g'(x) = -4x^{-5}$

11. $t(x) = \sqrt[3]{x}$ $t'(x) = \frac{1}{3}x^{-\frac{2}{3}}$

12. $v(x) = \sqrt[5]{x^{17}}$ $v'(x) = \frac{17}{5}x^{\frac{12}{5}}$

Find the slope of the tangent to the graph of the given function at each value of x.

13. $f(x) = 3x^2$ at $x = 4$ and $x = -2$ $f'(4) = 24; f'(-2) = -12$

14. $m(x) = x^2 - 3x + 4$ at $x = 1$ and $x = 3$ $m'(1) = -1$; $m'(3) = 3$

15. $h(x) = \frac{1}{x}$ at $x = 1$ and $x = -3$ $h'(1) = -1; h'(-3) = -\frac{1}{9}$

16. $g(x) = \sqrt{x}$ at $x = 4$ and $x = 9$ $g'(4) = \frac{1}{4}; g'(9) = \frac{1}{6}$

13. $f'(x) = 6x$

$f'(4) = 6(4) = 24$

$f'(-2) = 6(-2) = -12$

14. $m'(x) = 2x - 3$

$m'(1) = 2(1) - 3 = -1$

$m'(3) = 2(3) - 3 = 3$

15. $h(x) = x^{-1}: h'(x) = -1x^{-2} = -\frac{1}{x^2}$

$h'(1) = -\frac{1}{(1)^2} = -1$

$h'(-3) = -\frac{1}{(-3)^2} = -\frac{1}{9}$

16. $g(x) = x^{\frac{1}{2}}; g'(x) = \frac{1}{2}x^{-\frac{1}{2}} = \frac{1}{2\sqrt{x}}$

$g'(4) = \frac{1}{2\sqrt{4}} = \frac{1}{4}$

$g'(9) = \frac{1}{2\sqrt{9}} = \frac{1}{6}$

❯ B. Exercises

17. $f'(x) = 6x$

$m = f'(1) = 6(1) = 6$

$f(1) = 3(1)^2 = 3$

through $(1, 3)$ with $m = 6$

$y - 3 = 6(x - 1)$

$y - 3 = 6x - 6$

$y = 6x - 3$

18. $g'(x) = 3x^2$

$m = g'(-1) = 3(-1)^2 = 3$

$g(-1) = (-1)^3 = -1$

through $(-1, -1)$ with $m = 3$

$y - (-1) = 3(x - (-1))$

$y + 1 = 3x + 3$

$y = 3x + 2$

19. $r(x) = x^{\frac{1}{2}}$

$r'(x) = \frac{1}{2}x^{-\frac{1}{2}}$ or $\frac{1}{2\sqrt{x}}$

$m = r'(4) = \frac{1}{2\sqrt{4}} = \frac{1}{4}$

$r(4) = \sqrt{4} = 2$

through $(4, 2)$ with $m = \frac{1}{4}$

$y - 2 = \frac{1}{4}(x - 4)$

$y - 2 = \frac{1}{4}x - 1$

$y = \frac{1}{4}x + 1$

20. $h(x) = 12x^{-2}$

$h'(x) = -24x^{-3} = -\frac{24}{x^3}$

$m = h'(-2) = -\frac{24}{(-2)^3} = 3$

$h(-2) = \frac{12}{(-2)^2} = 3$

through $(-2, 3)$ with $m = 3$

$y - 3 = 3(x - (-2))$

$y - 3 = 3x + 6$

$y = 3x + 9$

21. $P'(x) = x + 2 = 0$

$x = -2$

$P(-2) = \frac{1}{2}(-2)^2 + 2(-2) - 6 = -8$

min: $(-2, -8)$, the parabola's vertex

22. $P'(x) = 4x^3 + 4x = 0$

$4x(x^2 + 1) = 0$

$4x = 0$ or $x^2 + 1 = 0$

$x = 0$

$P(0) = (0)^4 + 2(0)^2 - 3 = -3$

$P'(x) < 0$ as $x \to 0^-$ and

$P'(x) > 0$ as $x \to 0^+$

$\therefore (0, -3)$ is a rel. min.

23. $P'(x) = 6x^2 + 6x - 12 = 0$

$x^2 + x - 2 = 0$

$(x + 2)(x - 1) = 0$

$x = -2, 1$

$P(-2) = 2(-2)^3 + 3(-2)^2 - 12(-2) - 5 = 15$

$P(1) = 2(1)^3 + 3(1)^2 - 12(1) - 5$

$ = -12$

max: $(-2, 15)$; min: $(1, -12)$

24. $P'(x) = 6x^2 + 18x - 108 = 0$

$x^2 + 3x - 18 = 0$

$(x - 3)(x + 6) = 0$

$x = 3, -6$

$P(3) = 2(3)^3 + 9(3)^2 - 108(3) - 240$

$ = -429$

$P(-6) = 2(-6)^3 + 9(-6)^2 - 108(-6) - 240 = 300$

max: $(-6, 300)$; min: $(3, -429)$

25. $P(x) = x^2(120 - x) = 120x^2 - x^3$
$P'(x) = -3x^2 + 240x = 0$
$-3x(x - 80) = 0$
$x = 0, 80$
$P(0) = 0^2(120 - 0) = 0$ (min)
$P(80) = 80^2(120 - 80)$
$\quad = 256{,}000$ (max)

26. $A(x) = x(40 - 2x) = 40x - 2x^2$
$A'(x) = 40 - 4x = 0$
$4x = 40; x = 10$ ft
$A(10) = 10(40 - 2(10)) = 200$ ft^2

27. $A(x) = \frac{1}{2}x(30 - x) = 15x - \frac{1}{2}x^2$
$A'(x) = 15 - x = 0$
$x = 15$
$A(15) = \frac{1}{2}(15)(30 - 15) = 112.5$ m^2

28a. $f(t) = -\frac{1}{2}(32)t^2 + 40 = -16t^2 + 40$
$f(t) = -16t^2 + 40 = 0$
$t^2 = 2.5; t \approx 1.58$ sec

28b. $f'(t) = 2(-16)t = -32t$
$v(t) = -32(1.5) \approx -48$ ft/sec

29a. $h(t) = -\frac{1}{2}(32)t^2 + 63t + 4$
$\quad = -16t^2 + 63t + 4$
$h(1) = -16(1)^2 + 63(1) + 4 = 51$ ft
$h(2) = -16(2)^2 + 63(2) + 4 = 66$ ft
$h(3) = -16(3)^2 + 63(3) + 4 = 49$ ft

29b. $v(t) = h'(t) = 2(-16)t + 63$
$\quad = -32t + 63$
$v(1) = -32(1) + 63 = 31$ ft/sec
$v(2) = -32(2) + 63 = -1$ ft/sec
$v(3) = -32(3) + 63 = -33$ ft/sec

29c. $h(t) = -16t^2 + 63t + 4 = 0$
$-(16t^2 - 63t - 4) = 0$
$-(16t + 1)(t - 4) = 0$
$t = -\frac{1}{16}, 4$

using $t = 4$ sec:
$v(4) = -32(4) + 63 = -65$ ft/sec

29d. $v(t) = -32t + 63 = 0$
$63 = 32t$
$t = \frac{63}{32} \approx 1.97$ sec

$h(1.97) = -16(1.97)^2$
$\quad\quad + 63(1.97) + 4 \approx 66$ ft

30. $h(t) = -4.9t^2 + 50t + 1.5$
$h'(t) = v(t) = -9.8t + 50 = 0$
$t = \frac{50}{9.8} \approx 5.1$ sec
$h(5.1) = -4.9(5.1)^2 + 50(5.1) + 1.5$
$\quad \approx 129.05$ m

31. $h(t) = -4.9t^2 + v_0t + 1.5$
$h'(t) = v(t) = -9.8t + v_0 = 0$
$v_0 = 9.8t$
$h(t) = -4.9t^2 + (9.8t)t + 1.5 = 30$
$4.9t^2 = 28.5$
$t^2 \approx 5.81$
$t \approx 2.41$ sec

Derive an equation of the tangent to the function at the given value of x.

17. $f(x) = 3x^2$ at $x = 1$ **18.** $g(x) = x^3$ at $x = -1$

19. $r(x) = \sqrt{x}$ at $x = 4$ **20.** $h(x) = \frac{12}{x^2}$ at $x = -2$

Use derivatives to find the relative extrema of each polynomial function.

21. $P(x) = \frac{1}{2}x^2 + 2x - 6$ min: $(-2, -8)$

22. $P(x) = x^4 + 2x^2 - 3$ min: $(0, -3)$

23. $P(x) = 2x^3 + 3x^2 - 12x - 5$ max: $(-2, 15)$; min: $(1, -12)$

24. $P(x) = 2x^3 + 9x^2 - 108x - 240$ max: $(-6, 300)$ min: $(3, -429)$

Use derivatives to solve exercises 25–32.

25. If the sum of two positive integers is 120, find the maximum value of the product of one integer and the square of the other. 256,000

26. What is the largest possible area for Jill's rectangular garden if she plants the garden next to her garage and uses 40 ft of fencing for the other three sides? 200 ft^2

27. Find the maximum area possible for a triangle if the sum of the base and the height is 30 m. 112.5 m^2

28. Jenna drops a ball off the roof of a building from a height of 40 ft.
  a. State the function modeling the ball's height and determine how long it takes to hit the ground.
  b. Derive a function for its instantaneous vertical velocity and determine its velocity after 1.5 sec.

29. John throws a ball upward with an initial vertical velocity of 63 ft/sec, releasing the ball when it is 4 ft high.
  a. State the function modeling the ball's height and determine its height after 1, 2, and 3 sec.
  b. Derive a function for its instantaneous vertical velocity and determine the vertical velocity of the ball after 1, 2, and 3 sec.
  c. Determine how long it takes for the ball to hit the ground and its vertical velocity as it hits.
  d. How long does it take for the ball to reach its maximum height and what is that maximum height (to the nearest foot)? $\approx$ 1.97 sec; 66 ft

30. Blake shot a ball into the air from a height of 1.5 m with an initial upward velocity of 50 m/sec. Find the maximum height. $\approx$ 129.05 m

31. Frank throws a ball straight up from a height of 1.5 m. Find the upward velocity required for the ball to reach the roof of a building that is 30 m high. at least 23.6 m/sec

32. If the vertical velocity of an object thrown upward from a height of 1.5 m is -14.7 m/sec after 4 sec, what was its initial vertical velocity? 24.5 m/sec

33. Find the derivative of $h(t) = -\frac{1}{2}at^2 + v_0t + h_0$. What does this derivative function represent? Compare your result to the formula for vertical velocity of a projectile presented in Section 1.2.

34. Prove: If $f(x) = u(x) - v(x)$ when u and v are differentiable, prove $f'(x) = u'(x) - v'(x)$.

Express each product or quotient as a sum or difference, then find the derivative of the function.

35. $p(x) = (x^2 - 5)(2x^6)$ **36.** $h(x) = 12x(x^3 + 5x)$

37. $q(x) = \frac{2x + 5}{x}$ **38.** $f(x) = \frac{-11x^3 - 7x + 2}{x^5}$

The *second derivative* of $f(x)$, denoted $f''(x)$, is the derivative of $f'(x)$. It can be used to classify a point where $f'(x) = 0$ as a relative minimum if $f''(x) > 0$, a relative maximum if $f''(x) < 0$, or a point of inflection if $f''(x) = 0$.

Identify any points where $f'(x) = 0$ and use $f''(x)$ to classify each point.

39. $f(x) = x^3 - 6x^2 + 9x + 1$ **40.** $f(x) = x^4 - 2x^3 + 1$

Instantaneous acceleration is defined as the derivative of the instantaneous velocity function (the second derivative of the displacement-time function).

41. The distance (in feet) that a ball rolls down a ramp after t seconds is modeled by $d(t) = 5t^2 + 4t$. Derive the functions that model its velocity and its acceleration down the ramp. $v(t) = (10t + 4)$ ft/sec; $a(t) = 10$ ft/sec^2

42. If a rocket's acceleration for several seconds after launch is modeled by $a(t) = 250t$ and its initial height and initial velocity are both zero, find the velocity function in m/sec. $v(t) = 125t^2$ m/sec

17. $y = 6x - 3$
18. $y = 3x + 2$
19. $y = \frac{1}{4}x + 1$
20. $y = 3x + 9$

28a. $f(t) = -16t^2 + 40; \approx 1.58$ sec
28b. $v(t) = f'(t) = -32t; \approx -48$ ft/sec
29a. $h(t) = -16t^2 + 63t + 4$; 51 ft, 66 ft, 49 ft
29b. $h'(t) = -32t + 63$; 31 ft/sec, -1 ft/sec, -33 ft/sec
29c. 4 sec; -65 ft/sec

35. $p'(x) = 16x^7 - 60x^5$
36. $h'(x) = 48x^3 + 120x$
37. $q'(x) = -5x^{-2}$
38. $f'(x) = 22x^{-3} + 28x^{-5} - 10x^{-6}$
39. rel. max.: $(1, 5)$; rel. min.: $(3, 1)$
40. pt. of inflection: $(0, 1)$ rel. min.: $(1.5, -0.6875)$

$v(t) = -9.8(2.41) + v_0 = 0$
$v_0 \approx 23.6$ m/sec

32. $h(t) = -4.9t^2 + v_0t + 1.5$
$v(t) = h'(t) = -9.8t + v_0$
$-14.7 = -9.8(4) + v_0$
$v_0 = 24.5$ m/sec

33. $h'(t) = -gt + v_0$; instantaneous velocity; They are identical.

34. $f'(x)$
$= \lim_{h \to 0} \frac{[u(x + h) - v(x + h)] - [u(x) - v(x)]}{h}$
$= \lim_{h \to 0} \frac{u(x + h) - u(x) - v(x + h) + v(x)}{h}$
$= \lim_{h \to 0} \left[\frac{u(x + h) - u(x)}{h} - \frac{v(x + h) - v(x)}{h} \right]$
$= \lim_{h \to 0} \frac{u(x + h) - u(x)}{h} - \lim_{h \to 0} \frac{v(x + h) - v(x)}{h}$
$= u'(x) - v'(x)$

35. $p(x) = 2x^8 - 10x^6$
$p'(x) = 16x^7 - 60x^5$

36. $h(x) = 12x^4 + 60x^2$
$h'(x) = 48x^3 + 120x$

37. $q(x) = \frac{2x}{x} + \frac{5}{x} = 2 + 5x^{-1}$
$q'(x) = -5x^{-2}$ or $-\frac{5}{x^2}$

38. $f(x) = \frac{-11x^3}{x^5} + \frac{-7x}{x^5} + \frac{2}{x^5}$
$\quad = \frac{-11}{x^2} - \frac{7}{x^4} + \frac{2}{x^5}$
$\quad = -11x^{-2} - 7x^{-4} + 2x^{-5}$
$f'(x) = 22x^{-3} + 28x^{-5} - 10x^{-6}$

43. State functions $u(x)$ and $v(x)$ such that
$f(x) = 3(x^4 + 2x^2 + 5)^{10} = (u \circ v)(x)$. [1.7]
$u(x) = 3x^{10};\ v(x) = x^4 + 2x^2 + 5$

44. Describe the graph of $g(x) = 3^{x+3} - 2$ as a transformation of $f(x) = 3^x$. Then sketch both $f(x)$ and $g(x)$. [3.1] translated 3 units left and 2 units down

45. Find the angle (to the nearest degree) between $\mathbf{u} = \langle 7, 2 \rangle$ and $\mathbf{v} = \langle 1, 9 \rangle$. [6.2] $68°$

46. Write and solve a system of equations to find the dimensions of a rectangle with a perimeter of 58 m and an area of 154 m^2. [7.1] $7\ \text{m} \times 22\ \text{m}$

Use the properties of limits to find the limit of each rational function. [12.2]

47. $\displaystyle\lim_{n \to \infty} \frac{6n^3 - 5n^2 + 7n - 3}{2n^3 + n^2 - 4n + 3}$ $\quad$ 3

48. $\displaystyle\lim_{x \to \infty} \frac{-4x^4 + 2x^2 - 3x}{3x^2 + 4x - 17}$ $\quad$ $-\infty$ (divergent)

49. Identify the function rule for $(f + g)(x)$ if $f(x) = -2x + 7$ and $g(x) = 5x^2$. [1.7] C

A. $(f + g)(x) = -10x^2 + 35x^2$
B. $(f + g)(x) = 20x^2 - 140x + 245$
C. $(f + g)(x) = -10x^2 + 7$
D. $(f + g)(x) = 5x^2 - 2x + 7$
E. none of these

50. Find c (to the nearest tenth) in $\triangle ABC$ in which $A = 65°$, $B = 68°$, and $a = 20$. [5.6] D

A. 47 $\qquad$ **C.** 20.5 $\qquad$ **E.** none of these
B. 22.1 $\qquad$ **D.** 16.1

51. Which point is the focus of the parabola $y = \frac{1}{8}(x - 2)^2 + 4$? [8.1] C

A. $(2, 4)$ $\qquad$ **C.** $(2, 6)$ $\qquad$ **E.** none of these
B. $(2, 2)$ $\qquad$ **D.** $(4, 2)$

52. Which formula is used to evaluate $\displaystyle\sum_{i=1}^{n} i^3$, the sum of the first n cubes? [9.5] C

A. $\dfrac{n(n+1)}{2}$ $\qquad$ **D.** $\dfrac{n^3(n+1)^3}{8}$

B. $\dfrac{n(n+1)(2n+1)}{6}$ $\qquad$ **E.** none of these

C. $\dfrac{n^2(n+1)^2}{4}$

---

44. $g(x)$ is translated 3 units left and 2 units down.

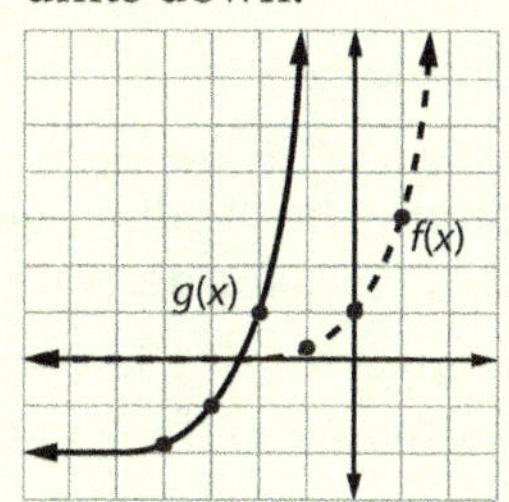

45. $\mathbf{u} \cdot \mathbf{v} = 7(1) + 2(9) = 7 + 18 = 25$

$|\mathbf{u}| = \sqrt{7^2 + 2^2} = \sqrt{49 + 4} = \sqrt{53}$

$|\mathbf{v}| = \sqrt{1^2 + 9^2} = \sqrt{1 + 81} = \sqrt{82}$

$\cos\theta = \left(\dfrac{25}{\sqrt{53} \cdot \sqrt{82}}\right)$

$\theta = \text{Cos}^{-1}\left(\dfrac{25}{\sqrt{53} \cdot \sqrt{82}}\right) \approx 68°$

46. $xy = 154$
$2x + 2y = 58$
$\qquad x + y = 29$
$\qquad\qquad x = 29 - y$
$\qquad (29 - y)y = 154$
$y^2 - 29y + 154 = 0$
$(y - 7)(y - 22) = 0$
$y = 7 \quad \text{or} \quad y = 22$
$x = 29 - 7 \quad x = 29 - 22$
$x = 22 \qquad\quad x = 7$
$7\ \text{m} \times 22\ \text{m}$

47. $\displaystyle\lim_{n \to \infty} \frac{6 - \frac{5}{n} + \frac{7}{n^2} - \frac{3}{n^3}}{2 + \frac{1}{n} - \frac{4}{n^2} + \frac{3}{n^3}} = \frac{6}{2} = 3$

48. $\displaystyle\lim_{x \to \infty} \frac{\frac{-4x^4}{x^2} + \frac{2x^2}{x^2} - \frac{3x}{x^2}}{\frac{3x^2}{x^2} + \frac{4x}{x^2} - \frac{17}{x^2}} = \lim_{x \to \infty} \frac{-4x^2 + 2 - \frac{3}{x}}{3 + \frac{4}{x} - \frac{17}{x^2}}$

$= \displaystyle\lim_{x \to \infty} \frac{-4x^2 + 2}{3} = -\infty$

49. $f(g(x)) = f(5x^2) = -2(5x^2) + 7$
$\qquad\qquad = -10x^2 + 7$

50. $C = 180 - 65 - 68 = 47°$

$\dfrac{c}{\sin 47°} = \dfrac{20}{\sin 65°}$

$c = \dfrac{20 \sin 47°}{\sin 65°} = 16.1$

51. $\dfrac{1}{4p} = \dfrac{1}{8}$
$4p = 8;\ p = 2$
$V(2, 4);\ F = (2, 4 + 2) = (2, 6)$

> **C. Exercises**

39. $f'(x) = 3x^2 - 12x + 9$
$f''(x) = 6x - 12$
$f'(x) = 3x^2 - 12x + 9 = 0$
$\qquad x^2 - 4x + 3 = 0$
$\qquad (x - 1)(x - 3) = 0$
$\qquad\qquad\qquad x = 1, 3$
$f(1) = 5$ and $f''(1) = -6\ (< 0)$
$\therefore (1, 5)$ is a rel. max.
$f(3) = 1$ and $f''(3) = 6\ (> 0)$
$\therefore (3, 1)$ is a rel. min.

40. $f'(x) = 4x^3 - 6x^2$
$f''(x) = 12x^2 - 12x$
$f'(x) = 4x^3 - 6x^2 = 0$
$\qquad 2x^2(2x - 3) = 0$
$\qquad\qquad x = 0, \frac{3}{2}$

$f(0) = 1$ and $f''(0) = 0$
$\therefore (0, 1)$ is a pt. of inflection.
$f\left(\frac{3}{2}\right) = -\frac{11}{16}$ and $f''(1.5) = 18\ (> 0)$
$\therefore \left(\frac{3}{2}, -\frac{11}{16}\right)$ is a rel. min.

42. Given $a(t) = v'(t) = 250t$,
$v(t)$ is a quadratic function (since its derivative is a linear function).
letting $v(t) = kt^2$:
$v'(t) = 2kt = 250t$ and $k = 125$
$\therefore v(t) = 125t^2$ m/sec

12.5 Product, Quotient, and Chain Rules

Objectives

1. To state and apply the product, quotient, and chain rules for derivatives
2. To derive formulas for the derivatives of trigonometric and exponential functions
3. To solve real-world problems using derivatives

Flash

Salmonella reproduces by cell division approximately every 40 min. These are one-celled, rod-shaped bacilli, discovered in the 1880s by a medical research scientist under the administration of American veterinary surgeon, Dr. Daniel E. Salmon, and later named in his honor. Salmonella infection is the most frequently reported foodborne illness; the bacteria cannot be killed by freezing food.

Vocabulary

chain rule
derivative of the natural exponential function
derivative of the sine function
product rule
quotient rule

12.5 Product, Quotient, and Chain Rules

How quickly can salmonella bacteria grow?

After completing this section, you will be able to

- apply the product, quotient, and chain rules to find derivatives.
- derive and apply derivatives of trigonometric and exponential functions.
- solve real-world problems using derivatives.

By 1680 Leibniz had discovered and published rules for differentiating products and quotients.

The formula for the derivative of the sum (and difference) of two functions was proved in Section 12.4. This section illustrates the application of the more complicated formulas for the product, quotient, and composition of functions. The proofs of many of these theorems are studied in calculus.

PRODUCT AND QUOTIENT RULES

Given the differentiable functions $u(x)$ and $v(x)$:

if $f(x) = u(x) \cdot v(x)$, then $f'(x) = u'(x) \cdot v(x) + u(x) \cdot v'(x)$. (product rule)

if $f(x) = \frac{u(x)}{v(x)}$ and $v(x) \neq 0$, then $f'(x) = \frac{u'(x) \cdot v(x) - u(x) \cdot v'(x)}{[v(x)]^2}$. (quotient rule)

The product rule states that the derivative of a product is the derivative of the first function times the second function plus the first function times the derivative of the second function.

Example 1 Finding the Derivative of a Product

Find the derivative of $f(x) = x^5(2x^3 - 1)$.

Answer

$u(x) = x^5$ and $v(x) = 2x^3 - 1$
$u'(x) = 5x^4$ and $v'(x) = 6x^2$

1. Identify the functions forming the product and determine their derivatives.

$f'(x) = 5x^4(2x^3 - 1) + x^5(6x^2)$

2. Substitute into $f'(x) = u'(x) \cdot v(x) + u(x) \cdot v'(x)$.

$= 10x^7 - 5x^4 + 6x^7$
$= 16x^7 - 5x^4$

3. Simplify.

Check

$f(x) = 2x^8 - x^5$
$f'(x) = 16x^7 - 5x^4$

4. Find the derivative of the equivalent polynomial function.

SKILL ✓ EXERCISE 7

In Example 1, finding the derivative of the polynomial function obtained by distributing was simpler than applying the product rule, but this rule is essential when finding the derivative of functions such as $f(x) = x^3 \cos x$ or $h(x) = e^x\sqrt{x - 5}$.

The derivatives of rational functions are found using the quotient rule.

PRESENTATION

Lesson Opener

State functions $u(x)$ and $v(x)$ such that $f(x) = (u \circ v)(x)$.

1. $f(x) = (x^3 - 4)^{12}$
   $u(x) = x^{12}; v(x) = x^3 - 4$
2. $f(x) = \cos^2 x$ $u(x) = x^2; v(x) = \cos x$
3. $f(x) = \sqrt{1 + e^x}$ $u(x) = \sqrt{x}; v(x) = 1 + e^x$

Finding the derivative of a polynomial is relatively easy. In this section students will learn to find the derivative of products, quotients, and the composition of functions, and then they will learn to apply these processes.

Present the *product* and *quotient rules* for finding derivatives and apply the product rule using Example 1. In this example, applying the distributive property and finding the derivative of the resulting polynomial would be easier, but this is not the usual case. Applying the product rule will often provide the easiest way to find the derivative.

Common Student Error A student may incorrectly think that if $f(x) = u(x) \cdot v(x)$, then $f'(x) = u'(x) \cdot v'(x)$. The product rule is more complicated than the sum rule.

To find the derivative of the quotient in Example 2, substitute the functions and their derivatives into the formula. Initially writing $u(x)$, $v(x)$, and their derivatives, $u'(x)$ and $v'(x)$, can be helpful, but most students eventually are able to substitute and find derivatives of the two functions in the same step.

Introduce the *chain rule* and apply it using Example 3. The derivative in part *a* could also be found by rewriting the function as a polynomial and then finding the derivative. This alternative method is often more difficult or even not possible, as in part *b*. Communicate any preferences regarding the form of the final answer.

Explain that the derivative of sin x is cos x and demonstrate how this fact is used to find the derivation of cos x in

Find the derivative of $f(x) = \dfrac{5x^3 - 2x^2 - 1}{3x}$.

Answer

$u(x) = 5x^3 - 2x^2 - 1$ and $v(x) = 3x$
$u'(x) = 15x^2 - 4x$ and $v'(x) = 3$

1. Identify the functions forming the quotient and determine their derivatives.

$f'(x) = \dfrac{(15x^2 - 4x)(3x) - (5x^3 - 2x^2 - 1)(3)}{(3x)^2}$

2. Substitute into $f'(x) = \dfrac{u'(x) \cdot v(x) - u(x) \cdot v'(x)}{[v(x)]^2}$.

$\quad = \dfrac{45x^3 - 12x^2 - (15x^3 - 6x^2 - 3)}{9x^2}$

3. Simplify.

$\quad = \dfrac{30x^3 - 6x^2 + 3}{9x^2} = \dfrac{10x^3 - 2x^2 + 1}{3x}$

SKILL ✔ **EXERCISE 11**

Recall that function composition involves substituting one function into another function. For example, if $u(x) = x^{25}$ and $v(x) = x^2 - 9$, then $u \circ v = u(v(x)) = (x^2 - 9)^{25}$. In this example $u(x)$ is often called the outside function and $v(x)$ is often called the inside function. The chain rule is used to find derivatives of functions expressed as a function composition.

▶ CHAIN RULE

If $f(x) = (u \circ v)(x) = u(v(x))$, where u and v are differentiable, then $f'(x) = u'(v(x)) \cdot v'(x)$.

Applying the chain rule requires finding the product of the derivative of the outside function (leaving the inside alone) and the derivative of the inside function. Therefore the derivative of

$f(x) = u \circ v = u(v(x)) = (x^2 - 9)^{25}$ is

$f'(x) = 25(x^2 - 9)^{24}(2x^1 - 0) = 50x(x^2 - 9)^{24}$.

Example 3 Applying the Chain Rule

Find $\dfrac{dy}{dx}$ for each function.

a. $y = 2(x^2 - 1)^3$ **b.** $y = \sqrt[3]{x^2 + 7x}$

Answer

a. $y = 2(x^2 - 1)^3$
$\quad = u \circ v$ where $u(x) = 2x^3$ and $v(x) = x^2 - 1$
$\therefore u'(x) = 6x^2$ and $v'(x) = 2x$

1. Identify the outer cubic function and the inner polynomial function and determine their derivatives.

$y' = \dfrac{dy}{dx} = 6(x^2 - 1)^2(2x)$

2. Apply the chain rule, substituting $v(x)$ into $u'(x)$ and multiplying by $v'(x)$.

$y' = 12x(x^4 - 2x^2 + 1)$
$\quad = 12x^5 - 24x^3 + 12x$

3. Simplify.

b. $y = (x^2 + 7x)^{\frac{1}{3}}$

1. Change the radical to exponential form.

$\quad = u \circ v$ where $u(x) = x^{\frac{1}{3}}$ and $v(x) = x^2 + 7x$

2. Identify the outer cube root function and the inner polynomial function and determine their derivatives.

$\therefore u'(x) = \dfrac{1}{3}(x)^{-\frac{2}{3}}$ and $v'(x) = 2x + 7$

CONTINUED ➡

Reading and Writing Mathematics

Give an example of two quadratic functions that have the same derivative. Then explain what having the same derivative implies geometrically and how these functions are related graphically.

Since the derivative of a constant is 0, as long as the x^2- and x-terms are the same, the derivatives will be the same. For example, $f(x) = 2x^2 + 5x - 2$ and $g(x) = 2x^2 + 5x + 9$ have the derivative of $4x + 5$. Geometrically, this implies that the tangent to the function at any given value of x is the same for each function. The family of quadratic functions with the same derivative is represented by vertical translations of the parent function.

Additional Exercises

1. Find the derivative of
$f(x) = 3x^4(2x^5 + 3)$.
$f'(x) = 54x^8 + 36x^3$

2. Find the derivative of
$g(x) = \dfrac{2x^2 + 3x + 1}{5x^2}$. $g'(x) = \dfrac{-3x - 2}{5x^3}$

Find $\dfrac{dy}{dx}$ for each function.

3. $y = 3(2x^2 - 3x)^2$
$y' = 48x^3 - 108x^2 + 54x$

4. $y = \sqrt[4]{(2x^2 + 4x)^3}$ $y' = \dfrac{3x + 3}{\sqrt[4]{2x^2 + 4x}}$

5. $y = \sec x$ $y' = \dfrac{\sin x}{\cos^2 x}$

6. $y = \tan x$ $y' = \sec^2 x$

Example 4a. Students should know the derivatives of these two functions since they are used to determining the derivatives of other trig functions, as illustrated in Example 4b. Note that the derivative operator, $\dfrac{d}{dx}$, is being used in Example 4.

Remind the students that y', $f'(x)$, $\dfrac{d}{dx} f(x)$, and $\dfrac{dy}{dx}$ are all commonly used notations for the derivative.

Interactive Activity Although the proof for the *derivative of the sine function* is beyond the scope of this course, it can be demonstrated graphically as illustrated in the margin tip. Note that by taking the derivative at $x = x$, you are taking the de-rivative at all points at which the function is defined. Check table values for the two functions for further confirmation. Use the Internet keyword search *sine cosine derivative desmos* or *sine derivative applet* to find an interactive presentation.

Introduce the *derivative of the natural exponential function e^x* and apply it with Example 5. Consider having the students confirm $\dfrac{d}{dx}[e^x] = e^x$ graphically as well.

The chain rule is used in a real-world application in Example 6. Consider having the students identify the function, $f(t)$ or $f'(t)$, that is needed for each part. (*a uses $f(t)$; b uses $f'(t)$; and c uses $f'(t)$.*)

TIPS

Ex. 5–6 Product and quotient rules could be used, but using the scalar multiple rule is easier.

7. $f(x) = 5x^2 \sin x$

$f'(x) = 10x \sin x + 5x^2 \cos x$

8. $y = 12e^x + e^3 \sin x$

$y' = 12e^x + e^3 \cos x$

9. The value of an investment over t years is modeled by $f(t) = 1250e^{0.05t}$.

  a. What was the initial investment?

    $1250

  b. What is the rate of growth after 12 yr? $113.88/yr

  c. When will the rate of growth reach $150/yr? ≈ 17.5 yr

Assignments

- **Minimum:** 1–5, 7, 9, 11–12, 15, 18–19, 22, 25, 27–28, 30; 31–35 odd; 45, 47–49, 51, 53
- **Standard:** 2–8 even; 9–13 odd; 16, 18–19, 22–23, 26–27, 29, 32–33, 35–37, 41–42, 44; 47–51 odd; 52
- **Extended:** 4–10 even; 11, 14, 17, 20–22, 24, 27, 29–30, 32–34, 36, 38–41, 43, 46, 50–51, 53

Assessment

- Quiz 12B covers Sections 12.3–12.5.

Solutions

❭ A. Exercises

1. The derivative of a constant is 0.

2. $f'(x) = 8 \dfrac{d}{dx}[\sin x] = 8 \cos x$

3. $f'(x) = e \dfrac{d}{dx}[\cos x]$

    $= e(-\sin x) = -e \sin x$

5. $f'(x) = 7 \dfrac{d}{dx}[e^x] = 7e^x$

6. $f'(x) = \dfrac{1}{5} \dfrac{d}{dx}[e^x] = \dfrac{e^x}{5}$

7. $u(x) = 2x^6; \; v(x) = x^2 - 5$

$u'(x) = 12x^5; \; v'(x) = 2x$

$h'(x) = 12x^5(x^2 - 5) + 2x^6(2x)$

$ = 12x^7 - 60x^5 + 4x^7$

$ = 16x^7 - 60x^5$

8. $u(x) = (x^2 - 9); \; v(x) = 3x - 5$

$u'(x) = 2x; \; v'(x) = 3$

$h'(x) = 2x(3x - 5) + (x^2 - 9)(3)$

$ = 6x^2 - 10x + 3x^2 - 27$

$ = 9x^2 - 10x - 27$

9. $u(x) = e^x; \; v(x) = \cos x$

$u'(x) = e^x; \; v'(x) = -\sin x$

$f'(x) = e^x \cos x + e^x(-\sin x)$

$ = e^x(\cos x - \sin x)$

10. $u(x) = 7x; \; v(x) = \sin x$

$u'(x) = 7; \; v'(x) = \cos x$

$g'(x) = 7 \sin x + 7x \cos x$

11. $u(x) = x^2 - 3; \; v(x) = x^2 + 5$

$u'(x) = 2x; \; v'(x) = 2x$

$f'(x) = \dfrac{2x(x^2 + 5) - (x^2 - 3)2x}{(x^2 + 5)^2}$

$ = \dfrac{2x^3 + 10x - 2x^3 + 6x}{(x^2 + 5)^2} = \dfrac{16x}{(x^2 + 5)^2}$

12. $u(x) = 3x; \; v(x) = 5x - 6$

$u'(x) = 3; \; v'(x) = 5$

$f'(x) = \dfrac{3(5x - 6) - 3x(5)}{(5x - 6)^2}$

$ = \dfrac{15x - 18 - 15x}{(5x - 6)^2} = -\dfrac{18}{(5x - 6)^2}$

$$y' = \frac{dy}{dx} = \frac{1}{3}(x^2 + 7x)^{-\frac{2}{3}}(2x + 7)$$

$$= \frac{2x + 7}{3\sqrt[3]{(x^2 + 7x)^2}}$$

  3. Apply the chain rule, substituting $v(x)$ into $u'(x)$ and multiplying by $v'(x)$.

  4. The result can also be written in radical form. (Rationalizing the denominator's radical is not necessary when it contains variables.)

SKILL ✔ EXERCISE 15

The chain and quotient rules enable the derivatives of the trigonometric functions to be derived from the derivative of the sine function, whose proof is beyond the scope of this text.

▌ DERIVATIVE OF THE SINE FUNCTION

If $f(x) = \sin x$, then $f'(x) = \cos x$.

Example 4 Finding the Derivatives of Trigonometric Functions

Find the derivative of each trigonometric function.

  a. $\dfrac{d}{dx}[\cos x]$ **b.** $\dfrac{d}{dx}[\cot x]$

Answer

a. $\dfrac{d}{dx}[\cos x] = \dfrac{d}{dx}\left[\sin\left(\dfrac{\pi}{2} - x\right)\right]$

  1. Use a cofunction identity to express $\cos x$ in terms of $\sin x$.

$\phantom{\dfrac{d}{dx}[\cos x]} = \cos\left(\dfrac{\pi}{2} - x\right)(-1)$

  2. Apply the chain rule, $f'(x) = u'(v(x)) \cdot v'(x)$, using $u(x) = \sin x$ and $v(x) = \dfrac{\pi}{2} - x$.

$\phantom{\dfrac{d}{dx}[\cos x]} = -\sin x$

  3. Apply a cofunction identity and simplify.

b. $\dfrac{d}{dx}[\cot x] = \dfrac{d}{dx}\left[\dfrac{\cos x}{\sin x}\right]$

  1. Express $\cot x$ in terms of $\cos x$ and $\sin x$.

$\phantom{\dfrac{d}{dx}[\cot x]} = \dfrac{\dfrac{d}{dx}[\cos x](\sin x) - (\cos x) \cdot \dfrac{d}{dx}[\sin x]}{(\sin x)^2}$

  2. Apply the quotient rule, $\dfrac{d}{dx}\left[\dfrac{u(x)}{v(x)}\right] = \dfrac{u'(x) \cdot v(x) - u(x) \cdot v'(x)}{[v(x)]^2}$.

$\phantom{\dfrac{d}{dx}[\cot x]} = \dfrac{(-\sin x)(\sin x) - (\cos x)(\cos x)}{\sin^2 x}$

$\phantom{\dfrac{d}{dx}[\cot x]} = \dfrac{-(\sin^2 x + \cos^2 x)}{\sin^2 x}$

  3. Simplify the expression.

$\phantom{\dfrac{d}{dx}[\cot x]} = -\dfrac{1}{\sin^2 x} = -\csc^2 x$

SKILL ✔ EXERCISE 13

Transformations of the exponential function $f(x) = e^x$ are frequently used to model growth in real-world applications. This function is the only function whose value at a point is always equal to the value of its derivative at that point.

▌ DERIVATIVE OF THE NATURAL EXPONENTIAL FUNCTION

If $f(x) = e^x$, then $f'(x) = e^x$.

This theorem's proof is studied in calculus.

TIP

To graphically verify that $\frac{d}{dx}[\sin x] = \cos x$, confirm that the graph of the numerical derivative of the original function (using MATH, 8:nDeriv) and the graph of the derivative function coincide.

13. $y = \tan x = \dfrac{\sin x}{\cos x}$

$\dfrac{dy}{dx} = \dfrac{\dfrac{d}{dx}[\sin x] \cos x - \sin x \dfrac{d}{dx}[\cos x]}{\cos^2 x}$

$y' = \dfrac{\cos x (\cos x) - \sin x (-\sin x)}{\cos^2 x}$

$ = \dfrac{\cos^2 x + \sin^2 x}{\cos^2 x} = \dfrac{1}{\cos^2 x} = \sec^2 x$

14. $y = \sec x = \dfrac{1}{\cos x}$

$\dfrac{dy}{dx} = \dfrac{\dfrac{d}{dx}[1](\cos x) - 1\left(\dfrac{d}{dx}[\cos x]\right)}{\cos^2 x}$

$y' = \dfrac{0(\cos x) - 1(-\sin x)}{\cos^2 x} = \dfrac{\sin x}{\cos^2 x}$

$ = \left(\dfrac{1}{\cos x}\right)\left(\dfrac{\sin x}{\cos x}\right) = \sec x \tan x$

15. $u(x) = x^5; \; v(x) = 2x + 1$

$u'(x) = 5x^4; \; v'(x) = 2$

$y' = 5(2x + 1)^4(2) = 10(2x + 1)^4$

<table>
<tr><td>

Example 5 Applying Derivatives of Special Functions

Find the derivative function for $f(x) = 15e^x + e^4 \cos x$.

Answer

$f'(x) = \frac{d}{dx}[15e^x] + \frac{d}{dx}[e^4\cos x]$ 1. Apply the derivative of a sum and the derivative of a scalar multiple.

$\quad = 15\frac{d}{dx}[e^x] + e^4\frac{d}{dx}[\cos x]$

$\quad = 15e^x + e^4(-\sin x)$ 2. Find the derivative of e^x and $\cos x$.

$\quad = 15e^x - e^4 \sin x$ 3. Simplify.

SKILL ✓ EXERCISE 9

The derivative of $f(x) = e^x$ is used to find the derivative of many other exponential functions.

Example 6 Applying the Chain Rule to Exponential Growth

A biologist models the number of bacteria in a culture after t hours with $f(t) = 250e^{0.6t}$.

a. How many bacteria were initially in the culture?

b. What is the rate of growth after 10 hr?

c. When will the rate of growth reach 750,000 bacteria/hr?

Answer

a. $f(0) = 250e^{0.6(0)} = 250(1) = 250$ bacteria   Evaluate $f(x)$ when $t = 0$.

b. $f'(t) = \frac{d}{dt}[250e^{0.6t}]$ 1. The derivative function describes the rate of change at any time t.

$\quad = 250\frac{d}{dt}[e^{0.6t}]$ 2. Apply the derivative of a scalar multiple.

Let $u(t) = e^t$ and $v(t) = 0.6t$.
$u'(t) = e^t$ and $v'(t) = 0.6$ 3. Identify $e^{0.6t}$ as a composition of u and v and find their derivatives.

$f'(t) = 250[e^{(0.6t)}(0.6)]$
$\quad = 150e^{0.6t}$ 4. Apply the chain rule and simplify.

$f'(10) = 150e^{0.6(10)}$
$\quad = 60{,}514$ bacteria/hr   5. Evaluate $f'(10)$ to find the rate of growth after 10 hr.

c. $750{,}000 = 150e^{0.6t}$ 1. Substitute 750,000 for $f'(t)$.

$5000 = e^{0.6t}$ 2. Solve for t.
$\ln 5000 = \ln e^{0.6t}$
$\ln 5000 = 0.6t$
$t = \frac{\ln 5000}{0.6} \approx 14.2$ hr

SKILL ✓ EXERCISE 33

❯ A. Exercises

Find the derivative of each function with respect to x.

1. $y = e^\pi$ $y' = 0$ **2.** $f(x) = 8 \sin x$

3. $f(x) = e \cos x$ **4.** $k(x) = 3e^x - 7x^3 + 10^9$

5. $f(x) = 7e^x$ $f'(x) = 7e^x$ **6.** $f(x) = \frac{e^x}{5}$ $f'(x) = \frac{e^x}{5}$

Use the product rule or quotient rule to find the derivative of each function.

7. $h(x) = 2x^6(x^2 - 5)$ **8.** $h(x) = (x^2 - 9)(3x - 5)$

9. $f(x) = e^x \cos x$ **10.** $g(x) = 7x \sin x$

11. $f(x) = \frac{x^2 - 3}{x^2 + 5}$ **12.** $f(x) = \frac{3x}{5x - 6}$

13. $y = \tan x$ $y' = \sec^2 x$ **14.** $y = \sec x$ $y' = \sec x \tan x$

</td><td>

22. $u(x) = x^2 + 4x + 2;\ v(x) = x^2 - x - 1$
$u'(x) = 2x + 4;\ v'(x) = 2x - 1$
$k'(x) = (x^2 + 4x + 2)(2x - 1)$
$\qquad\qquad + (2x + 4)(x^2 - x - 1)$
$\quad = (2x^3 + 7x^2 - 2)$
$\qquad\qquad + (2x^3 + 2x^2 - 6x - 4)$
$\quad = 4x^3 + 9x^2 - 6x - 6$

23. $y = \csc x = \dfrac{1}{\sin x}$

$\dfrac{dy}{dx} = \dfrac{\frac{d}{dx}[1](\sin x) - 1\left(\frac{d}{dx}[\sin x]\right)}{\sin^2 x}$

$y' = \dfrac{0(\sin x) - 1(\cos x)}{\sin^2 x} = -\dfrac{\cos x}{\sin^2 x}$

$\quad = \left(\dfrac{-1}{\sin x}\right)\left(\dfrac{\cos x}{\sin x}\right) = -\csc x \cot x$

24. $u(x) = \cos x;\ v(x) = x^2 + 3$
$u'(x) = -\sin x;\ v'(x) = 2x$
$k'(x) = \dfrac{-\sin x\,(x^2 + 3) - 2x \cos x}{(x^2 + 3)^2}$

25. $u(x) = x^2 - 5x + 1;\ v(x) = e^x$
$u'(x) = 2x - 5;\ v'(x) = e^x$
$f'(x) = \dfrac{(2x - 5)e^x - (x^2 - 5x + 1)e^x}{(e^x)^2}$

$\quad = \dfrac{e^x(2x - 5 - x^2 + 5x - 1)}{(e^x)^2} = \dfrac{-x^2 + 7x - 6}{e^x}$

26. $u(x) = \cos x;\ v(x) = 3e^x$
$u'(x) = -\sin x;\ v'(x) = 3e^x$
$f'(x) = \dfrac{(-\sin x)3e^x - (\cos x)3e^x}{(3e^x)^2}$

$\quad = \dfrac{-(\sin x + \cos x)}{3e^x} = -\dfrac{\sin x + \cos x}{3e^x}$

27. $u(x) = 5x^{50};\ v(x) = x - 8$
$u'(x) = 250x^{49};\ v'(x) = 1$
$g'(x) = 250(x - 8)^{49}(1)$
$\quad = 250(x - 8)^{49}$

28. $h(x) = (3x + 5)^{\frac{1}{2}}$
$u(x) = x^{\frac{1}{2}};\ v(x) = 3x + 5$
$u'(x) = \frac{1}{2}x^{-\frac{1}{2}};\ v'(x) = 3$
$h'(x) = \frac{1}{2}(3x + 5)^{-\frac{1}{2}}(3)$
$\quad = \frac{3}{2}(3x + 5)^{-\frac{1}{2}}$ or $\dfrac{3}{2\sqrt{3x + 5}}$

29. $h(x) = (2x + 1)^{\frac{7}{4}}$
$u(x) = x^{\frac{7}{4}};\ v(x) = 2x + 1$
$u'(x) = \frac{7}{4}x^{\frac{3}{4}};\ v'(x) = 2$
$h'(x) = \frac{7}{4}(2x + 1)^{\frac{3}{4}}(2)$
$\quad = \frac{7}{2}(2x + 1)^{\frac{3}{4}}$ or $\frac{7}{2}\sqrt[4]{(2x + 1)^3}$

30. $u(x) = x^{17};\ v(x) = 5x^7 - x^4 + 3x$
$u'(x) = 17x^{16};\ v'(x) = 35x^6 - 4x^3 + 3$
$f'(x) = 17(5x^7 - x^4 + 3x)^{16}$
$\qquad\qquad \cdot (35x^6 - 4x^3 + 3)$

31. $u(x) = x^2;\ v(x) = \sin x$
$u'(x) = 2x;\ v'(x) = \cos x$
$y' = 2(\sin x)(\cos x)$
$\quad = 2 \sin x \cos x = \sin 2x$

</td></tr>
</table>

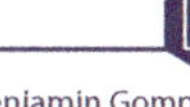

Benjamin Gompertz published a paper in 1825 introducing a growth function that could be applied to human mortality, a rate that increases exponentially with age. Gompertz's growth curve is often used today to model cancerous tumor growth.

2. $f'(x) = 8 \cos x$

3. $f'(x) = -e \sin x$

4. $k'(x) = 3e^x - 21x^2$

7. $h'(x) = 16x^7 - 60x^5$

8. $h'(x) = 9x^2 - 10x - 27$

9. $f'(x) = e^x(\cos x - \sin x)$

10. $g'(x) = 7 \sin x + 7x \cos x$

11. $f'(x) = \dfrac{16x}{(x^2 + 5)^2}$

12. $f'(x) = \dfrac{-18}{(5x - 6)^2}$

---

16. $u(x) = x^{100};\ v(x) = 3x + 5$
$u'(x) = 100x^{99};\ v'(x) = 3$
$y' = 100(3x + 5)^{99}(3)$
$\quad = 300(3x + 5)^{99}$

17. $u(x) = \sin x;\ v(x) = x^2 - 8$
$u'(x) = \cos x;\ v'(x) = 2x$
$f'(x) = \cos(x^2 - 8)(2x)$
$\quad = 2x \cos(x^2 - 8)$

18. $u(x) = x^5;\ v(x) = 3 + e^x$
$u'(x) = 5x^4;\ v'(x) = e^x$
$f'(x) = 5(3 + e^x)^4(e^x) = 5e^x(3 + e^x)^4$

❯ B. Exercises

19. $u(x) = x^3;\ v(x) = \cos x$
$u'(x) = 3x^2;\ v'(x) = -\sin x$

$f'(x) = 3x^2(\cos x) + x^3(-\sin x)$
$\quad = 3x^2 \cos x - x^3 \sin x$

20. $u(x) = \cos x;\ v(x) = \sin x$
$u'(x) = -\sin x;\ v'(x) = \cos x$
$y' = \cos x\,(\cos x) + (-\sin x)(\sin x)$
$\quad = \cos^2 x - \sin^2 x = \cos 2x$

21. $u(x) = 3x^2 + x;\ v(x) = x^{\frac{1}{2}}$
$u'(x) = 6x + 1;\ v'(x) = \frac{1}{2}x^{-\frac{1}{2}}$
$h'(x) = (6x + 1)x^{\frac{1}{2}}$
$\qquad\qquad + (3x^2 + x)\left(\frac{1}{2}x^{-\frac{1}{2}}\right)$
$\quad = 6x^{\frac{3}{2}} + x^{\frac{1}{2}} + \frac{3}{2}x^{\frac{3}{2}} + \frac{1}{2}x^{\frac{1}{2}}$
$\quad = \frac{15}{2}x^{\frac{3}{2}} + \frac{3}{2}x^{\frac{1}{2}}$ or $\left(\frac{15}{2}x + \frac{3}{2}\right)\sqrt{x}$

32. $u(x) = e^x$; $v(x) = x^2$
$u'(x) = e^x$; $v'(x) = 2x$
$k'(x) = e^{(x^2)}(2x) = 2xe^{x^2}$

33a. $f'(t) = 50\frac{d}{dt}[e^{0.8t}]$
$u(t) = e^t$; $v(t) = 0.8t$
$u'(t) = e^t$; $v'(t) = 0.8$
$f'(t) = 50e^{0.8t}(0.8) = 40e^{0.8t}$
$f'(5) = 40e^{0.8(5)} \approx 2184$ cells/day

33b. $1{,}000{,}000 = 40e^{0.8t}$
$25{,}000 = e^{0.8t}$
$\ln 25{,}000 = \ln e^{0.8t}$
$\ln 25{,}000 = 0.8t$
$t = \frac{\ln 25{,}000}{0.8} \approx 12.66$ days

34a. $f'(t) = 100\frac{d}{dt}[e^{0.4t}]$
$u(t) = e^t$; $v(t) = 0.4t$
$u'(t) = e^t$; $v'(t) = 0.4$
$f'(t) = 100e^{0.4t}(0.4) = 40e^{0.4t}$
$f'(20) = 40e^{0.4(20)}$
$\approx 119{,}238$ bacteria/day

34b. $500{,}000 = 40e^{0.4t}$
$12{,}500 = e^{0.4t}$
$\ln 12{,}500 = 0.4t$
$t = \frac{\ln 12{,}500}{0.4} \approx 23.6$ days

35. $A(t) = Pe^{rt} = 20{,}000e^{0.045t}$
$A'(t) = 20{,}000\frac{d}{dt}[e^{0.045t}]$
$u(t) = e^t$; $v(t) = 0.045t$
$u'(t) = e^t$; $v'(t) = 0.045$
$A'(t) = 20{,}000e^{0.045t}(0.045)$
$= 900e^{0.045t}$
$A'(7) = 900e^{0.045(7)} \approx \$1233.23$/yr

36. $A(t) = Pe^{rt}$
$A'(t) = 80{,}000\frac{d}{dx}[e^{0.025t}]$
$u(t) = e^t$; $v(t) = 0.025t$
$u'(t) = e^t$; $v'(t) = 0.025$
$A'(t) = 80{,}000e^{0.025t}(0.025)$
$= 2000e^{0.025t}$
$A'(10) = 2000e^{0.025(10)} \approx \$2568.05$/yr

C. Exercises

37. $u(x) = e^x$; $v(x) = x^3 - 4$
$u'(x) = e^x$; $v'(x) = 3x^2$
$f'(x) = e^{(x^3 - 4)}(3x^2) = 3x^2 e^{x^3 - 4}$

38. $u(x) = e^x$; $v(x) = (x + 5)^{\frac{1}{2}}$
$u'(x) = e^x$; $v'(x) = \frac{1}{2}(x + 5)^{-\frac{1}{2}}(1)$
$h'(x) = e^x(x - 5)^{\frac{1}{2}} + e^x\left[\frac{1}{2}(x - 5)^{-\frac{1}{2}}\right]$
factoring:
$= (x - 5)^{-\frac{1}{2}}\left[e^x(x - 5) + \frac{e^x}{2}\right]$
$= \frac{x - \frac{9}{2}}{\sqrt{x - 5}}e^x$

Use the chain rule to find the derivative of each function.

15. $y = (2x + 1)^5$

16. $y = (3x + 5)^{100}$

17. $f(x) = \sin(x^2 - 8)$

18. $f(x) = (3 + e^x)^5$

❭ B. Exercises

Find the derivative of each function.

19. $f(x) = x^3 \cos x$

20. $y = \cos x \sin x$

21. $h(x) = (3x^2 + x)\sqrt{x}$

22. $k(x) = (x^2 + 4x + 2)(x^2 - x - 1)$

23. $y = \csc x$

24. $k(x) = \frac{\cos x}{x^2 + 3}$

25. $f(x) = \frac{x^2 - 5x + 1}{e^x}$

26. $y = \frac{\cos x}{3e^x}$

27. $g(x) = 5(x - 8)^{50}$

28. $h(x) = \sqrt{3x + 5}$

29. $h(x) = \sqrt[4]{(2x + 1)^7}$

30. $f(x) = (5x^7 - x^4 + 3x)^{17}$

31. $y = \sin^2 x$

32. $k(x) = e^{x^3}$

Solve each exponential growth problem using the chain rule.

33. A researcher models the number of cells in a culture after t days with $f(t) = 50e^{0.8t}$.
≈ 2184 cells/day
  a. What is the rate of growth after 5 days?
  b. When will the rate of growth reach 1,000,000 cells per day? ≈ 12.66 days

34. A biologist models the number of bacteria in a culture after t days with $f(t) = 100e^{0.4t}$.
  a. What is the rate of growth after 20 days?
  b. When will the rate of growth reach 500,000 bacteria per day? ≈ 23.6 days

35. If \$20,000 is invested in an account at 4.5% APR compounded continuously, find the investment's instantaneous rate of growth (in \$/yr) 7 yr later.

36. If \$80,000 is invested in an account at 2.5% APR compounded continuously, find the investment's instantaneous rate of growth (in \$/yr) 10 yr later.

❭ C. Exercises

Find a simplified expression for the derivative of each function.

37. $f(x) = e^{x^3 - 4}$

38. $h(x) = e^x\sqrt{x - 5}$

39. $k(x) = 5x(x - 8)^{50}$

40. $f(x) = \left(\frac{x}{x^2 - 1}\right)^2$

41. $f(x) = \frac{x}{\sqrt{x^2 - 1}}$

42. Prove: Use the product and chain rules to prove the quotient rule.

43. Prove: Show that the power rule $\frac{d}{dx}[x^n] = nx^{n-1}$ is true if n is a negative integer by letting k be a natural number such that $n = -k$.

35. $\approx \$1233.23$/yr

36. $\approx \$2568.05$/yr

CUMULATIVE REVIEW

44. Find the y-intercept and zeros of $f(x) = 6x^2 - 2x - 4$.
[1.6] y-int.: $(0, -4)$; $x = -\frac{2}{3}, 1$

45. Find the dot product of $\mathbf{u} = \langle 3, 6\rangle$ and $\mathbf{v} = \langle 12, -6\rangle$; then state whether the angle between the vectors is $> 90°$, $< 90°$, or $= 90°$. [6.2] 0; $= 90°$

46. Use the distributive property to state a summation equivalent to $\sum_{i=1}^{n} cx_i$. [9.5] $c\sum_{i=1}^{n} x_i$

47. true or false: $\sum_{i=1}^{n}(x_i + y_i) = \sum_{i=1}^{n} x_i + \sum_{i=1}^{n} y_i$ [9.5]   true

Find and interpret the meaning of the z-score for each test score if the test scores have a mean of 118 and a standard deviation of 12. [10.5]

48. 123 $z \approx 0.42$

49. 97 $z = -1.75$

48. The score is 0.42 standard deviations above the mean.

49. The score is 1.75 standard deviations below the mean.

50. Identify the solution to $\frac{x^2 + x - 6}{2x^2 + 11x + 15} \geq 0$. [2.7]   A
  A. $(-\infty, -3) \cup (-3, -2.5) \cup [2, \infty)$
  B. $(-\infty, -2.5] \cup [2, \infty)$
  C. $(-\infty, -3) \cup (-3, -2.5) \cup (2, \infty)$
  D. $(-\infty, -3) \cup (-2.5, 2]$
  E. none of these

51. Evaluate $\sum_{n=1}^{20} 7n - 4$. [9.2]   B
  A. 136   **C.** 1420   **E.** none of these
  B. 1390   **D.** 2780

52. Evaluate $\lim_{x \to 4} 7x^2$. [12.2]   B
  A. 28   **C.** 56   **E.** none of these
  B. 112   **D.** 784

53. Which equation represents the tangent to the graph of $f(x) = 2x^2$ at $x = -1$? [12.4]   D
  A. $y = 2x + 4$ **C.** $y = 4x + 6$ **E.** none of these
  B. $y = -4x - 9$ **D.** $y = -4x - 2$

---

39. applying the product rule:
$k'(x) = \frac{d}{dx}[5x](x - 8)^{50}$
$\qquad\qquad + (5x)\frac{d}{dx}\left[(x - 8)^{50}\right]$
applying the chain rule:
$u(x) = x^{50}$; $v(x) = x - 8$
$\frac{d}{dx}\left[(x - 8)^{50}\right] = 50(x - 8)^{49}(1)$
$k'(x) = 5(x - 8)^{50} + 5x[50(x - 8)^{49}]$
$\qquad = 5(x - 8)^{49}[(x - 8) + 50x]$
$\qquad = 5(x - 8)^{49}(51x - 8)$

40. $u(x) = x^2$; $u'(x) = 2x$
$v(x) = \frac{x}{x^2 - 1}$
applying the quotient rule:
$v'(x) = \frac{1(x^2 - 1) - x(2x)}{(x^2 - 1)^2}$
$\qquad = \frac{x^2 - 1 - 2x^2}{(x^2 - 1)^2} = -\frac{x^2 + 1}{(x^2 - 1)^2}$
$f(x) = u(v(x))$ (chain rule)
$f'(x) = 2\left(\frac{x}{x^2 - 1}\right)\left(-\frac{x^2 + 1}{(x^2 - 1)^2}\right)$
$\qquad = -\frac{2x(x^2 + 1)}{(x^2 - 1)^3}$ or $-\frac{2x^3 + 2x}{(x^2 - 1)^3}$

continued in Answers and Solutions Overflow

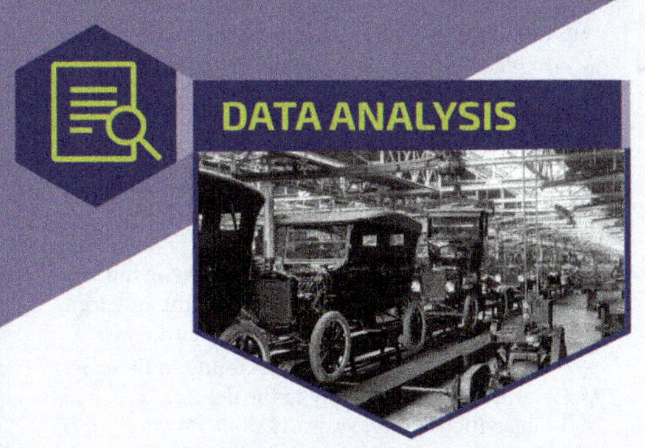

Marginal Analysis in Economics

The benefits (and liabilities) of assembly line mass production burst on the scene in the early twentieth century with Ford's Model T production line. Today it is hard to imagine a business not impacted by these developments. But how do businesses decide how many items to produce and the price at which they should be marketed? Once a process has been modeled with a mathematical equation, it can be investigated with calculus—the mathematics of change. Marginal analysis is the process of examining unit changes (instantaneous rates of change) of functions. Note these commonly used terms.

cost function—the cost of producing and marketing x units of an item; denoted $C(x)$

marginal cost—an estimate of the cost of producing one more item; denoted $C'(x)$

revenue function—the income from the sale of x units at a price of p each; denoted $R(x) = px$

marginal revenue—an estimate of the revenue obtained by selling one more item; denoted $R'(x)$

The graph illustrates a typical cost function. The total cost of manufacturing x items typically increases rapidly when only a few items are produced, increases at a lower rate as higher production rates benefit from economies of scale, and then increases rapidly again as production approaches the plant's capacity.

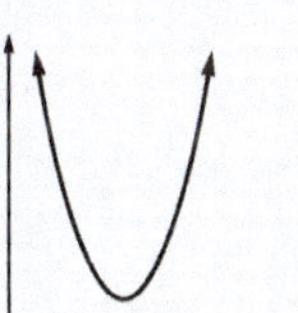

The marginal cost describes the rate of change in cost at a particular time, which is the slope (or derivative) of the cost function. The graph of the marginal cost is generally parabolic. The marginal cost is high when production is minimal, decreases as more items are produced, but then increases as production approaches plant capacity. Similarly, marginal revenue is the derivative of the revenue function.

Marginal functions allow businesses to understand the rate at which cost or revenue is changing at various production levels. Suppose a company produces an item where $C(x) = 0.04x^3 - 3x^2 + 320x + 150$. The marginal cost $C'(x)$ is a parabola whose vertex is a minimum. The minimum marginal cost is found by setting the derivative of $C'(x)$, the second derivative of $C(x)$, which is denoted $C''(x)$, equal to zero and solving for x. This provides the x-value of the point of inflection in the cost function, which represents the production level at which the cost to produce the next item is the lowest. $C'(x)$ is also used to find the cost of producing the xth item by simply substituting the desired value for x and evaluating.

1. Given $C(x) = 0.04x^3 - 3x^2 + 320x + 150$, find $C'(x)$.

2. Find the marginal cost of the twentieth and the thirtieth items. $C'(20) = \$248; C'(30) = \$248$

3. At what production level is the cost to produce the next item the lowest? What is that cost? 25 items; $245

The price at which a company can sell an item varies with the availability of the item. The price is often a decreasing function $p = f(x)$ that models the price people are willing to pay as the supply increases. Recall that the revenue function is $R(x) = px$. If the price function is linear, then the revenue function is parabolic with a maximum point that occurs where the marginal revenue, $R'(x)$, is zero.

4. Find $R(x)$ for a product whose demand equation is $p = 20 - 0.1x$. $R(x) = 20x - 0.1x^2$

5. Find $R'(x)$ and use it to determine the production level that produces the maximum revenue. Then state the maximum revenue. $R'(x) = -0.2x + 20; x = 100; R(100) = \$1000$

Objectives

1. To define marginal analysis
2. To find the marginal cost or revenue function, given the cost or revenue function
3. To evaluate marginal functions to determine minimum cost, maximum revenue, and maximum profit
4. To evaluate profitability from a biblical worldview

Assignments

- **Minimum:** optional
- **Standard:** 1–9; 10 (in class)
- **Extended:** 1–10

Solutions

1. $C'(x) = 0.12x^2 - 6x + 320$

2. $C'(20) = 0.12(20)^2 - 6(20) + 320$
$= 48 - 120 + 320 = \$248$
$C'(30) = 0.12(30)^2 - 6(30) + 320$
$= 108 - 180 + 320 = \$248$

3. $C''(x) = 0.24x - 6 = 0$
$x = 25$
$C'(25) = 0.12(25)^2 - 6(25) + 320$
$= 75 - 150 + 320 = \$245$

4. $R(x) = px = (20 - 0.1x)x$
$= 20x - 0.1x^2$

5. $R'(x) = -0.2x + 20 = 0$
$x = 100$
$R(100) = 20(100) - 0.1(100)^2$
$= 2000 - 1000 = \$1000$

PRESENTATION

Marginal analysis provides a wealth of opportunities to explore instantaneous rates of change with concrete, real-world examples. Whether starting with a cost function and using it to think through and sketch the marginal cost curve or vice versa, students are sure to gain a better grasp of derivatives through these exercises.

Cost functions are normally presented as the sum of variable costs and fixed costs (rent, insurance, permits, etc.). For simplicity's sake, and since fixed costs translate but do not change the shape of the curve, this feature does not specifically address these different parts of the cost function. For example, the constant term of 150 in exercise 1 represents a fixed cost that is independent of the number of items produced.

The following example illustrates the relationship between a derivative and the actual change between two points on a curve. Suppose total revenue is given by $R(x) = x^2 - 10x + 5$. To find the marginal revenue of the 101st item produced, find $R'(101)$. $R'(x) = 2x - 10$, so $R'(101) = 2(101) - 10 = 192$. Since this revenue function is quadratic, its derivative is linear. $R'(101) = 192$ is a point on the line tangent to the revenue curve. To find the exact revenue from the 101st item, we can find $R(101) - R(100) = [(101)^2 - 10(101) + 5] - [(100)^2 - 10(100) + 5] = 9196 - 9005 = 191$. The marginal revenue function is useful in evaluating general characteristics of the revenue function and in estimating marginal revenue for complicated functions.

6. $P(x) = (200 - 0.01x)x - (25x + 500)$
$$= -0.01x^2 + 175x - 500$$

7. $P'(x) = -0.02x + 175 = 0$
$$x = 8750$$
$$P(8750) = -0.01(8750)^2$$
$$+ 175(8750) - 500$$
$$= \$765{,}125$$

8. $p = 200 - 0.01(8750) = \$112.50$
$$C(x) = 25(8750) + 500 = 219{,}250$$
$Note$: $R(x) = 112.50(8750)$
$$= \$984{,}375$$

9. $\dfrac{219{,}250}{8750} \approx \$25.06/\text{item}$

$\dfrac{112.50}{25.06} \approx 4.49$ or 449%, implying

$\approx 349\%$ markup

sample answers:

10a. for higher profit margins: Research and development of new drugs is very expensive, time consuming, and involves considerable risks for the company. Less than 12% of drugs that go to clinical trial are approved, and it takes a minimum of 10 yr for a drug to reach the market. In our free market society, the higher profit margins drive research and innovation. FDA regulations make drugs safer but more expensive to develop in the United States.

for lower profit margins: Unregulated profit margins are fine for nonessential goods, but life-saving drugs must be affordable for everyone. Compensation for pharmaceutical company CEOs in the S&P 500 is over 70% higher than CEO compensation in other sectors. Drug prices are often significantly higher in the United States than in other developed countries, partially due to US laws passed as a result of the lobbying power of wealthy pharmaceutical companies.

Companies seek to find the production level that maximizes their profit, which is a function of revenue and costs: $P(x) = R(x) - C(x)$. Consider a product with $p = 200 - 0.01x$ and $C(x) = 25x + 500$.

6. Find $P(x)$ for a product if $C(x) = 25x + 500$ and $p = 200 - 0.01x$. $P(x) = -0.01x^2 + 175x - 500$

7. Use the derivative of $P(x)$ to find the production level that produces the maximum profit and state the maximum profit. $x = 8750$ items; $\$765{,}125$

8. Find the price, p, that the company should charge per item to maximize profit. Then find the cost of the production level required to maximize profit.

9. Determine the average cost of producing each item and the percent markup on each item. $\approx \$25.06$; $\approx 349\%$

8. $\$112.50$; $C(x) = \$219{,}250$

10. Discuss: Profit margins are the topic of heated debate in pharmaceuticals. Opinions range from those who maintain that companies should seek to maximize profits in order to fulfill their shareholder obligations to those who contend that life-saving drugs should be provided at manufactured cost or even at no cost to anyone who needs them.

a. Research arguments concerning pharmaceutical company profits. Summarize convincing arguments for increasing profit margins and arguments for decreasing profit margins.

b. Discuss how the principles found in these Scripture verses can apply to the debate.
Leviticus 25:14; Psalm 112:1–3; Proverbs 11:1; Ecclesiastes 5:19; Isaiah 10:1–2; Matthew 25:14–23; 1 Timothy 6:10

10b. Riches are a blessing from God and are good, not evil, as long as they are obtained legally and without oppression. Investing resources wisely to make a profit is expected of stewards. The love of riches is the root of all kinds of evil. We must live for God and not for wealth. We must treat others as we would want to be treated. Unjust gain is condemned.

Area Under a Curve and Integration

Archimedes was very close to discovering parts of calculus.

Previous sections demonstrated how limits are used to perform one of the fundamental tasks in calculus: finding the equation of a tangent to a curve. The final two sections address the second fundamental task: finding the area between a function's graph and the x-axis.

Since geometric area formulas frequently cannot be applied to find these areas, the area under the curve will be approximated using the sum of several rectangles. To estimate the area under the parabola $y = x^2$ over the interval $[0, 1]$ on the x-axis, we can divide (or *partition*) the interval into equal parts and draw several rectangles of equal width whose heights are determined by the function's value at the largest x-value (the right endpoint) in each partition. The area under the curve is then approximated by finding the sum of these rectangular areas. Partitioning $[0, 1]$ into 5 parts causes each rectangle to have a width of 0.2.

$R_1 = 0.2\,f(0.2) = 0.2(0.2)^2$
$R_2 = 0.2\,f(0.4) = 0.2(0.4)^2$
$R_3 = 0.2\,f(0.6) = 0.2(0.6)^2$
$R_4 = 0.2\,f(0.8) = 0.2(0.8)^2$
$R_5 = 0.2\,f(1.0) = 0.2(1.0)^2$
Sum $= 0.2\,(0.04 + 0.16 + 0.36 + 0.64 + 1.00) = 0.2(2.2) = 0.44\ \text{u}^2$

This approximation overestimates the area. Rectangles whose height is the function value at the smallest x-value (the left endpoint) of each partition could be used, but this approximation would underestimate the area. Rectangles whose height is the function value at the midpoint of each partition frequently provide a better estimate.

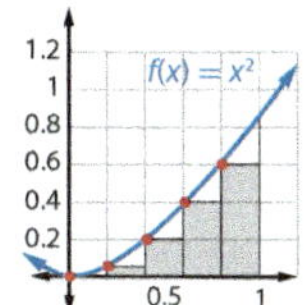
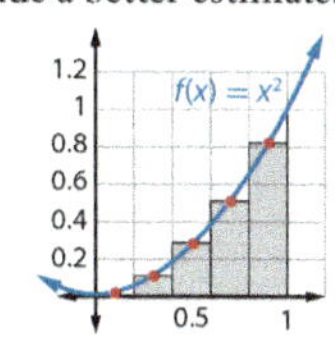

$Area \approx 0.2(0^2 + 0.2^2 + 0.4^2 + 0.6^2 + 0.8^2)$
$\quad\ = 0.2(0 + 0.04 + 0.16 + 0.36 + 0.64)$
$\quad\ = 0.2(1.2) = 0.24$

$Area \approx 0.2(0.1^2 + 0.3^2 + 0.5^2 + 0.7^2 + 0.9^2)$
$\quad\ = 0.2(0.01 + 0.09 + 0.25 + 0.49 + 0.81)$
$\quad\ = 0.2(1.65) = 0.33$

This process can be generalized to find the area beneath the function $f(x)$ over the interval $x \in [a, b]$. Each rectangle has a width of $\Delta x = \frac{b - a}{n}$, and the right endpoints of the partitions used to determine the rectangle's height can be expressed as $x_i = a + i\Delta x$ for $i = 1, 2, 3, \ldots, n$. The area of each rectangle is then found using $A = \Delta x \cdot f(x_i)$, and the sum of these rectangular areas approximates the area under the curve.

The Greek mathematician Archimedes accurately determined the area under a parabola around 200 BC with a method similar to those used today. Georg Riemann pioneered today's methods of integration in the early nineteenth century.

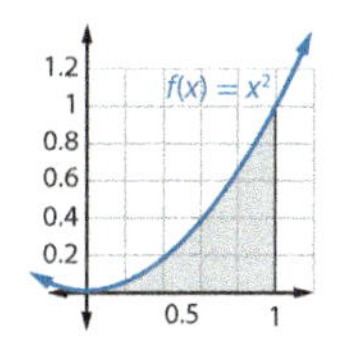

12.6 Area Under a Curve and Integration

Objectives

1. To approximate the area under a curve using rectangles

2. To use definite integrals to find the area under a curve

3. To calculate the area under a velocity-time graph to find the distance traveled

Flash

Archimedes (287–12 BC) was one of the greatest mathematicians and scientists of the classical age. He used *infinitesimals,* values so small that they cannot be measured, in his calculations of areas and volumes. His ideas foreshadowed the reasoning that became the framework for integral calculus about 1800 yr later.

Vocabulary

bounds
definite integral
lower limit
partition
upper limit

PRESENTATION

Lesson Opener

Find the derivative of each function.

1. $f(x) = 7x^3 - 3x^2 - 9$ $f'(x) = 21x^2 - 6x$

2. $f(x) = 3(x^2 - 1)^2$ $f'(x) = 12x^3 - 12x$

3. Simplify $\dfrac{12}{n^3} \cdot \dfrac{n(n + 1)(2n + 1)}{8}$.

$3 + \dfrac{9}{2n} + \dfrac{3}{2n^2}$

Consider discussing different approaches to simplifying the expression in the last Lesson Opener exercise. Students will encounter similar expressions in the examples and exercises throughout this section.

Review that differential calculus enables us to find the slope of a tangent at a given point on a curve, which leads to finding relative maximum and minimum points. Then explain that integral calculus enables us to find the area under a curve above an interval on the x-axis. This area can represent many different quantities, such as the total distance traveled.

This lesson presents areas only above the x-axis (considered to be positive). Areas below the x-axis and above the curve (considered to be negative) are discussed in Section 12.7.

Illustrate estimating the area under a curve using the figures in the text and

Example 1. Point out that 30 u^2 is only an estimate. The exact area over the interval $[0, 4]$ is found in Example 3. Emphasize that the estimated area approaches the actual area as the width of the rectangles decreases and the number of rectangles increases.

Present the summation properties, assuring the students they will use these throughout this section to simplify summations, as demonstrated in Example 2. Point out that the use of 1000 rectangles provides a fairly accurate value for the area, which is actually 2500 u^2.

Introduce the definition of and notation for a *definite integral*. The definite

Additional Exercises

1. Approximate the area between $f(x) = -x^2 + 6x$ and $[0, 4]$ on the x-axis using 8 rectangles. Use the right endpoint of each partition to determine each rectangle's height. **28.5 u^2**

2. Estimate the area under the curve $f(x) = 12x^2$ and over the interval $[0, 10]$ using 100 rectangles. **4060.2 u^2**

3. Use the definition of a definite integral to evaluate the following.

   a. $\int_0^{10} 12x^2 - 4x$ **3800**

   b. $\int_2^6 x^2 - 13x + 42$ **$\frac{88}{3}$**

4. The velocity of a dropped object (in ft/sec) is modeled by $v(t) = 32t$. How far did the object fall from 6 sec to 7 sec? **208 ft**

Assignments

- **Minimum:** 1, 3–7, 9–11, 13–15, 17, 21, 23, 25–26, 29, 31, 33(c–d), 39, 41–44, 47–48
- **Standard:** 1–3, 6–8; 10–14 even; 15, 18; 19–23 odd; 26–27, 30–31, 33–34, 40, 43–44, 46–47
- **Extended:** 4, 7, 10; 14–20 even; 21–22, 26–29, 32–33, 35–38, 40, 44–46

Solutions

❯ A. Exercises

1. $\Delta x = \frac{b - a}{n} = \frac{3 - 0}{6} = 0.5$

   $x_i = 0.5, 1, 1.5, 2, 2.5, 3$

Approximate the area between $f(x) = -x^2 + 6x$ and the x-axis over the interval $[0, 4]$ using four rectangles. Use the right endpoint of each partition to determine each rectangle's height.

Answer

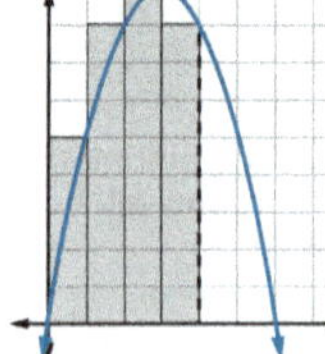

$\Delta x = \frac{b - a}{n} = \frac{4 - 0}{4} = 1$

$x_i = 1, 2, 3, 4$

1. Find the width of each rectangle and list the right endpoint of each partition.

$R_1 = 1 \times f(1) = 1[-(1)^2 + 6(1)] = 5 \text{ u}^2$
$R_2 = 1 \times f(2) = 1[-(2)^2 + 6(2)] = 8 \text{ u}^2$
$R_3 = 1 \times f(3) = 1[-(3)^2 + 6(3)] = 9 \text{ u}^2$
$R_4 = 1 \times f(4) = 1[-(4)^2 + 6(4)] = 8 \text{ u}^2$

2. Sketch the parabola and the rectangles used to approximate the area. Then determine the area of each rectangle using $A = \Delta x \cdot f(x_i)$.

$S = 5 + 8 + 9 + 8 = 30 \text{ u}^2$
Area $\approx 30 \text{ u}^2$

3. Find the sum of the rectangular areas to estimate the area under the curve.

SKILL ✔ EXERCISE 13

TIP

The partition's right endpoints can be described recursively using $x_1 = a + \Delta x$ and $x_j = x_{j-1} + \Delta x$. The left endpoints are described recursively using $x_1 = a$ and $x_j = x_{j-1} + \Delta x$. The midpoints are described recursively using $x_1 = a + \frac{1}{2}\Delta x$ and $x_j = x_{j-1} + \Delta x$.

Using a greater number of thinner rectangles can provide a better estimate of the area under a curve. The summation properties and formulas presented in Section 9.5 are used to simplify the calculation of the sum of the rectangular areas.

$$\sum_{i=1}^{n} c = nc \qquad \sum_{i=1}^{n} cx_i = c\sum_{i=1}^{n} x_i \qquad \sum_{i=1}^{n}(a_i + b_i) = \sum_{i=1}^{n} a_i + \sum_{i=1}^{n} b_i$$

$$\sum_{i=1}^{n} i = \frac{n(n+1)}{2} \qquad \sum_{i=1}^{n} i^2 = \frac{n(n+1)(2n+1)}{6} \qquad \sum_{i=1}^{n} i^3 = \frac{n^2(n+1)^2}{4}$$

Estimate the area under the polynomial curve $f(x) = x^3$ over the interval $[0, 10]$ using 1000 rectangles.

Answer

$\Delta x = \frac{b - a}{n} = \frac{10 - 0}{1000} = \frac{1}{100}$

$x_i = \frac{1}{100}, \frac{2}{100}, \frac{3}{100}, \cdots, \frac{1000}{100}$

1. Find the width of each rectangle and use $x_i = a + i\Delta x$ to find the right endpoint for each partition.

$S = \frac{1}{100}f\left(\frac{1}{100}\right) + \frac{1}{100}f\left(\frac{2}{100}\right)$
$\quad + \frac{1}{100}f\left(\frac{3}{100}\right) + \cdots + \frac{1}{100}f\left(\frac{1000}{100}\right)$

2. Using $f(x_i)$ for the height of the rectangles, write an expression for the sum of the rectangular areas.

$S = \frac{1}{100}\left(\frac{1}{100}\right)^3 + \frac{1}{100}\left(\frac{2}{100}\right)^3$
$\quad + \frac{1}{100}\left(\frac{3}{100}\right)^3 + \cdots + \frac{1}{100}\left(\frac{1000}{100}\right)^3$

3. Substitute for the function values.

$S = \left(\frac{1}{100}\right)^4(1^3 + 2^3 + 3^3 + \cdots + 1000^3)$

4. Factor out the common factor of $\left(\frac{1}{100}\right)^4$.

$S = \left(\frac{1}{100}\right)^4 \sum_{i=1}^{1000} i^3$

5. Write the sum in sigma notation.

$S = \left(\frac{1}{100}\right)^4\left(\frac{1000^2(1001)^2}{4}\right) = \frac{1001^2}{4(100)}$

6. Substitute with the formula for the sum of the first 1000 cubes and simplify.

Area $\approx 2505.0025 \text{ u}^2$

SKILL ✔ EXERCISE 17

integral represents a process in which the area is determined by finding limits as the number of rectangles approaches infinity. This formula is also known as the right Riemann sum. More details can be found using the Internet keyword search *Riemann sum applet* for an interactive presentation.

One-on-One A student may want to know why the word *definite* is used with integrals. Explain that definite integrals have defined *upper* and *lower* limits resulting in a real number that can represent the area under the curve. This can be contrasted with an indefinite integral (Section 12.7), for which there is no given interval and the result is a function.

A definite integral is evaluated in Example 3. No label is given in the answer because, in the context, we are finding the value of the definite integral and not the area under the curve.

Motivational Idea Consider evaluating $\int_0^{10} x^3 \, dx$ to determine the exact area that was estimated in Example 2. The next section illustrates how definite integrals can be calculated on the TI-84 using the 9:fnInt(function from the MATH menu. Finding the exact area that was estimated in Example 1 is demonstrated below.

$$\int_0^4 (-x^2 + 6x)\,dx$$
$$26.66666667$$

Common Student Error A student may think that the value found using a definite integral is an estimation of the area under the curve. Explain that, unlike using a real number of rectangles, a definite integral produces the exact area under the curve if $f(x) > 0$ over $[a, b]$, eliminating all errors of estimation.

The interval and the parabola in Example 4 are translated so a equals 0. Warn the students to be cautious when simplifying expressions like the one in step 7, where the denominator affects all three terms in the numerator. Alternative ways to view expressions such as

The sum in Example 2 can be written as $S = \sum_{i=1}^{n} f(x_i)\,\Delta x = \sum_{i=1}^{1000}\left(i\frac{1}{100}\right)^3 \frac{1}{100}$. More accurate estimates can be found by using even more rectangles. Note that as $n \to \infty$, $\Delta x \to 0$. Finding the limit of this sum yields the exact value for the area under the curve over the stated interval.

DEFINITION

The area of a region between the graph of a function $f(x)$ and the interval $[a, b]$ on the x-axis is given by the **definite integral** $\int_a^b f(x)\,dx = \lim\limits_{n\to\infty} \sum_{i=1}^{n} f(x_i)\,\Delta x$, where a and b are the *lower limit* and *upper limit* (or bounds), respectively, $\Delta x = \frac{b-a}{n}$, and $x_i = a + i\Delta x$ for $i = 1, 2, 3, \ldots, n$.

Notice how a definite integral is used to find the exact area that was approximated in Example 1.

Example 3 Evaluating a Definite Integral

Use the definition of a definite integral to evaluate $\int_0^4 (-x^2 + 6x)\,dx$.

Answer

$\int_a^b f(x)\,dx = \lim\limits_{n\to\infty} \sum_{i=1}^{n} f(x_i)\,\Delta x$

1. Apply the definition of a definite integral, noting that $\Delta x = \frac{b-a}{n} = \frac{4-0}{n} = \frac{4}{n}$.

$\int_0^4 (-x^2 + 6x)\,dx = \lim\limits_{n\to\infty} \sum_{i=1}^{n} (-x_i^2 + 6x_i)\left(\frac{4}{n}\right)$

$= \lim\limits_{n\to\infty} \sum_{i=1}^{n} \left(-\left(\frac{4i}{n}\right)^2 + 6\left(\frac{4i}{n}\right)\right)\left(\frac{4}{n}\right)$

2. Substitute for x_i using $x_i = a + i\Delta x = 0 + i\left(\frac{4}{n}\right) = \frac{4i}{n}$.

$= \lim\limits_{n\to\infty} \sum_{i=1}^{n} \left(-\frac{64i^2}{n^3} + \frac{96i}{n^2}\right)$

3. Distribute and simplify.

$= \lim\limits_{n\to\infty} \left(-\frac{64}{n^3} \sum_{i=1}^{n} i^2\right) + \lim\limits_{n\to\infty} \left(\frac{96}{n^2} \sum_{i=1}^{n} i\right)$

4. Apply the scalar and sum properties to rewrite the limit of the sum.

$= \lim\limits_{n\to\infty} \left(-\frac{64}{n^3} \cdot \frac{n(n+1)(2n+1)}{6}\right) + \lim\limits_{n\to\infty} \left(\frac{96}{n^2} \cdot \frac{n(n+1)}{2}\right)$

5. Substitute using summation formulas.

$= \lim\limits_{n\to\infty} \left(-\frac{32}{n^2} \cdot \frac{2n^2 + 3n + 1}{3}\right) + \lim\limits_{n\to\infty} \left(\frac{48(n+1)}{n}\right)$

6. Simplify, distributing the division.

$= \lim\limits_{n\to\infty} \left(-\frac{64}{3} - \frac{32}{n} - \frac{32}{3n^2}\right) + \lim\limits_{n\to\infty} \left(48 + \frac{48}{n}\right)$

$= \left(-\frac{64}{3} - 0 - 0\right) + (48 + 0) = \frac{80}{3}$ or $26.\overline{6}$

7. Evaluate each limit and simplify.

SKILL ✓ EXERCISE 21

The evaluation of definite integrals in which the lower limit is not zero can be simplified by horizontally translating the function and interval to make the lower limit zero.

---

2. $\Delta x = \dfrac{b-a}{n} = \dfrac{3-(-2)}{5} = 1$
$x_i = -1, 0, 1, 2, 3$

3. $\Delta x = \dfrac{b-a}{n} = \dfrac{2-(-1)}{6} = 0.5$
$x_i = -0.5, 0, 0.5, 1, 1.5, 2$

4. $\Delta x = \dfrac{b-a}{n} = \dfrac{4-2}{8} = 0.25$
$x_i = 2.25, 2.5, 2.75, 3, 3.25, 3.5, 3.75, 4$

5. $A = bh = (7)(6) = 42$

6. $A = \frac{1}{2}bh = \frac{1}{2}(6)(2) = 6$

7. $b_1 = f(3) = \frac{1}{2}(3) + 2 = 3.5$
$b_1 = f(7) = \frac{1}{2}(7) + 2 = 5.5$
$A = \frac{1}{2}h(b_1 + b_2)$
$\quad = \frac{1}{2}(4)(3.5 + 5.5) = 18$

8. $b_1 = f(-2) = -\frac{1}{4}(-2) + 2 = 2.5$
$b_2 = f(4) = -\frac{1}{4}(4) + 2 = 1$
$A = \frac{1}{2}h(b_1 + b_2)$
$\quad = \frac{1}{2}(6)(2.5 + 1) = 10.5$

9. $\Delta x = \dfrac{b-a}{n} = \dfrac{3-(-2)}{5} = 1$
$x_i = -1, 0, 1, 2, 3$
$R_i = \Delta x \cdot f(x_i)$
$R_1 = 1(-1)^2 = 1$
$R_2 = 1(0)^2 = 0$
$R_3 = 1(1)^2 = 1$
$R_4 = 1(2)^2 = 4$
$R_5 = 1(3)^2 = 9$
$S = 1 + 0 + 1 + 4 + 9 = 15 \text{ u}^2$

10. $\Delta x = \dfrac{b-a}{n} = \dfrac{6-2}{4} = 1$
$x_i = 3, 4, 5, 6$
$R_i = \Delta x \cdot f(x_i)$
$R_1 = 1[-0.5(3)^2 + 4(3)] = 7.5$
$R_2 = 1[-0.5(4)^2 + 4(4)] = 8$
$R_3 = 1[-0.5(5)^2 + 4(5)] = 7.5$

$\lim\limits_{n\to\infty} \frac{9}{n^2}\left(\frac{2n^2 + 3n + 1}{2}\right)$ include

$\lim\limits_{n\to\infty} \left[\frac{9}{2n^2}(2n^2 + 3n + 1)\right]$ or

$\lim\limits_{n\to\infty} \left[\frac{9}{n^2}\left(\frac{2n^2}{2} + \frac{3n}{2} + \frac{1}{2}\right)\right]$.

Example 5 illustrates a key concept in secondary-level physics: the area under the velocity-time graph represents the object's displacement. When the velocity function is constant, the area is rectangular and the area formula $A = bh$ is equivalent to $d = rt$.

Direct attention to the graph of the velocity-time function following Example 5. This problem could be solved by graphing the function and finding the area of the trapezoid. Emphasize that without calculus, most of our applications would be limited to solving problems in which acceleration (the change in velocity) is constant, as illustrated here, but definite integrals provide a method of finding values in intervals that involve changing velocity.

Note that the translation in Example 4 simplifies the process, but since the function in Example 5 is linear, there is no real advantage in translating the function and its interval.

TIPS

Ex. 9–20 You may want to share the exact value of each area (found by using a calculator or an Internet app) so students can compare their estimates.

$$R_4 = 1[-0.5(6)^2 + 4(6)] = 6$$
$$S = 7.5 + 8 + 7.5 + 6 = 29 \text{ u}^2$$

11. $\Delta x = \dfrac{b-a}{n} = \dfrac{7-1}{6} = 1$

$x_i = 2, 3, 4, 5, 6, 7$

$R_i = \Delta x \cdot f(x_i)$

$R_1 = (1)\sqrt{2}; \ R_2 = (1)\sqrt{3}$

$R_3 = (1)\sqrt{4} = 2; \ R_4 = (1)\sqrt{5}$

$R_5 = (1)\sqrt{6}; \ R_6 = (1)\sqrt{7}$

$S = \sqrt{2} + \sqrt{3} + 2 + \sqrt{5} + \sqrt{6}$
$$+ \sqrt{7} \approx 12.48 \text{ u}^2$$

12. $\Delta x = \dfrac{b-a}{n} = \dfrac{3-0}{6} = 0.5$

$x_i = 0.5, 1, 1.5, 2, 2.5, 3$

$R_i = \Delta x \cdot f(x_i)$

$R_1 = 0.5[(0.5)^2 + 2] = 1.125$

$R_2 = 0.5[(1)^2 + 2] = 1.5$

$R_3 = 0.5[(1.5)^2 + 2] = 2.125$

$R_4 = 0.5[(2)^2 + 2] = 3$

$R_5 = 0.5[(2.5)^2 + 2] = 4.125$

$R_6 = 0.5[(3)^2 + 2] = 5.5$

$S = 1.125 + 1.5 + 2.125 + 3$
$$+ \ 4.125 + 5.5 = 17.375 \text{ u}^2$$

13. $\Delta x = \dfrac{b-a}{n} = \dfrac{10-0}{5} = 2$

$x_i = 2, 4, 6, 8, 10$

$R_i = \Delta x \cdot f(x_i) = 2 \cdot f(x_i)$

$R_1 = 2 \cdot \dfrac{1}{10}(2)^2 = 0.8$

$R_2 = 2 \cdot \dfrac{1}{10}(4)^2 = 3.2$

$R_3 = 2 \cdot \dfrac{1}{10}(6)^2 = 7.2$

$R_4 = 2 \cdot \dfrac{1}{10}(8)^2 = 12.8$

$R_5 = 2 \cdot \dfrac{1}{10}(10)^2 = 20$

$S = 0.8 + 3.2 + 7.2 + 12.8 + 20$
$$= 44 \text{ u}^2$$

14. $\Delta x = \dfrac{b-a}{n} = \dfrac{10-0}{10} = 1$

$x_i = 1, 2, 3, 4, 5, 6, 7, 8, 9, 10$

$R_i = \Delta x \cdot f(x_i) = f(x_i)$

$R_1 = \dfrac{1}{10}(1)^2 = 0.1; \ R_2 = \dfrac{1}{10}(2)^2 = 0.4$

$R_3 = \dfrac{1}{10}(3)^2 = 0.9; \ R_4 = \dfrac{1}{10}(4)^2 = 1.6$

$R_5 = \dfrac{1}{10}(5)^2 = 2.5; \ R_6 = \dfrac{1}{10}(6)^2 = 3.6$

$R_7 = \dfrac{1}{10}(7)^2 = 4.9; \ R_8 = \dfrac{1}{10}(8)^2 = 6.4$

$R_9 = \dfrac{1}{10}(9)^2 = 8.1; \ R_{10} = \dfrac{1}{10}(10)^2$
$$= 10.0$$

$S = 0.1 + 0.4 + 0.9 + 1.6 + 2.5 + 3.6$
$$+ \ 4.9 + 6.4 + 8.1 + 10 = 38.5 \text{ u}^2$$

15. $\Delta x = \dfrac{b-a}{n} = \dfrac{4-0}{4} = 1$

$x_i = 1, 2, 3, 4$

$R_i = \Delta x \cdot f(x_i) = f(x_i)$

$R_1 = -\dfrac{1}{2}(1)^3 + 3(1)^2 = 2.5$

$R_2 = -\dfrac{1}{2}(2)^3 + 3(2)^2 = 8$

$R_3 = -\dfrac{1}{2}(3)^3 + 3(3)^2 = 13.5$

Use the definition of a definite integral to evaluate $\int_{-5}^{-2} (x^2 + 8x + 17)\,dx$.

Answer

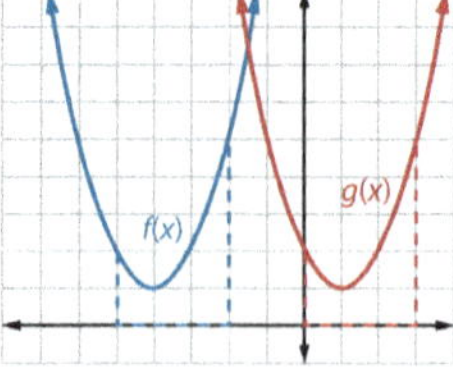

$f(x) = x^2 + 8x + 17$

$g(x) = f(x-5) = (x-5)^2 + 8(x-5) + 17$
$$= x^2 - 10x + 25 + 8x - 40 + 17$$
$$= x^2 - 2x + 2$$

1. Translate the interval and parabola 5 units right so the new interval is [0, 3].

$\displaystyle\int_0^3 (x^2 - 2x + 2)\,dx = \lim_{n\to\infty} \sum_{i=1}^{n} (x_i^2 - 2x_i + 2)\left(\dfrac{3}{n}\right)$

2. State the equivalent definite integral and apply the limit definition, noting that
$\Delta x = \dfrac{b-a}{n} = \dfrac{3-0}{n} = \dfrac{3}{n}$.

$\displaystyle = \lim_{n\to\infty} \sum_{i=1}^{n} \left[\left(\dfrac{3i}{n}\right)^2 - 2\left(\dfrac{3i}{n}\right) + 2 \right]\left(\dfrac{3}{n}\right)$

3. Substitute for x_i using $x_i = a + i\Delta x = 0 + i\left(\dfrac{3}{n}\right) = \dfrac{3i}{n}$.

$\displaystyle = \lim_{n\to\infty} \sum_{i=1}^{n} \left(\dfrac{27i^2}{n^3} - \dfrac{18i}{n^2} + \dfrac{6}{n} \right)$

4. Distribute and simplify.

$\displaystyle = \lim_{n\to\infty} \dfrac{27}{n^3} \sum_{i=1}^{n} i^2 - \lim_{n\to\infty} \dfrac{18}{n^2} \sum_{i=1}^{n} i + \lim_{n\to\infty} \dfrac{1}{n} \sum_{i=1}^{n} 6$

5. Apply the scalar and sum properties to rewrite the limit.

$\displaystyle = \lim_{n\to\infty} \dfrac{27}{n^3}\left(\dfrac{n(n+1)(2n+1)}{6} \right) - \lim_{n\to\infty} \dfrac{18}{n^2}\left(\dfrac{n(n+1)}{2} \right) + \lim_{n\to\infty} \dfrac{6n}{n}$

6. Substitute using summation formulas.

$\displaystyle = \lim_{n\to\infty} \dfrac{9}{n^2}\left(\dfrac{2n^2 + 3n + 1}{2} \right) - \lim_{n\to\infty} \left(\dfrac{9(n+1)}{n} \right) + 6$

7. Simplify and apply limit theorems.

$\displaystyle = \lim_{n\to\infty} \left(9 + \dfrac{27}{2n} + \dfrac{9}{2n^2} \right) - \lim_{n\to\infty} \left(9 + \dfrac{9}{n} \right) + 6$

$= (9 + 0 + 0) - (9 + 0) + 6 = 6$

8. Evaluate each limit and simplify.

SKILL ✔ EXERCISE 25

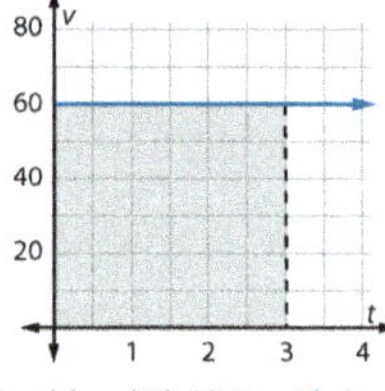

$A = bh = (3 \text{ hr})(60 \text{ mi/hr})$
$= 180 \text{ mi} = \text{distance traveled}$

In Section 12.3 you learned that instantaneous velocity is represented by the slope of the tangent to the distance-time graph and is described using the derivative of the function that expresses position in terms of time. In this velocity-time graph of a car traveling 60 mi/hr for 3 hr, the area under the curve represents the distance traveled. Definite integrals allow us to determine the distance represented by similar areas when the velocity-time graph is not constant.

If the velocity of a dropped object (in m/sec) is modeled by $v(t) = 10t$, how far did the object fall from 3 sec to 5 sec?

Answer

$\displaystyle\int_3^5 10t\,dt$

1. State the definite integral of the velocity-time function with the desired time interval (using t instead of x).

$\displaystyle = \lim_{n\to\infty} \sum_{i=1}^{n} (10t_i)\left(\dfrac{2}{n}\right)$

2. State the equivalent definite integral and apply the limit definition, noting that $\Delta t = \dfrac{5-3}{n} = \dfrac{2}{n}$.

CONTINUED ➡

$R_4 = -\dfrac{1}{2}(4)^3 + 3(4)^2 = 16$

$S = 2.5 + 8 + 13.5 + 16 = 40 \text{ u}^2$

16. $\Delta x = \dfrac{b-a}{n} = \dfrac{4-0}{8} = 0.5$

$x_i = 0.5, 1, 1.5, 2, 2.5, 3, 3.5, 4$

$R_i = \Delta x \cdot f(x_i) = \dfrac{1}{2} \cdot f(x_i)$

$R_1 = 0.5[-(0.5)^3 + 3(0.5)^2]$
$$\approx 0.34$$

$R_2 = 0.5[-(1)^3 + 3(1)^2] = 1.25$

$R_3 = 0.5[-(1.5)^3 + 3(1.5)^2]$
$$\approx 2.53$$

$R_4 = 0.5[-(2)^3 + 3(2)^2] = 4$

$R_5 = 0.5[-(2.5)^3 + 3(2.5)^2]$
$$\approx 5.47$$

$R_6 = 0.5[-(3)^3 + 3(3)^2] = 6.75$

$R_7 = 0.5[-(3.5)^3 + 3(3.5)^2]$
$$\approx 7.66$$

$R_8 = 0.5[-(4)^3 + 3(4)^2] = 8$

$S \approx 0.34 + 1.25 + 2.53 + 4 + 5.47$
$$+ \ 6.75 + 7.66 + 8 = 36 \text{ u}^2$$

> **B. Exercises**

17. $\Delta x = \dfrac{1-0}{1000} = \dfrac{1}{1000}$

$x_i = \dfrac{1}{1000}, \dfrac{2}{1000}, \dfrac{3}{1000}, \ldots, \dfrac{1000}{1000}$

$S = \dfrac{1}{1000} f\left(\dfrac{1}{1000}\right) + \dfrac{1}{1000} f\left(\dfrac{2}{1000}\right)$
$$+ \dfrac{1}{1000} f\left(\dfrac{3}{1000}\right) + \cdots + \dfrac{1}{1000} f\left(\dfrac{1000}{1000}\right)$$

$= \dfrac{1}{1000} \cdot 3\left(\dfrac{1}{1000}\right)^2$
$$+ \dfrac{1}{1000} \cdot 3\left(\dfrac{2}{1000}\right)^2 + \dfrac{1}{1000} \cdot 3\left(\dfrac{3}{1000}\right)^2$$
$$+ \cdots + \dfrac{1}{1000} \cdot 3\left(\dfrac{1000}{1000}\right)^2$$

$= 3\left(\dfrac{1}{1000}\right)^3 (1^2 + 2^2$
$$+ \ 3^2 + \cdots + 1000^2)$$

$$= \lim_{n \to \infty} \sum_{i=1}^{n} \left[10\left(3 + \tfrac{2i}{n}\right)\right]\left(\tfrac{2}{n}\right)$$

3. Substitute for t_i using $t_i = a + i\Delta t = 3 + i\left(\tfrac{2}{n}\right) = 3 + \tfrac{2i}{n}$.

$$= \lim_{n \to \infty} \sum_{i=1}^{n} \left(\tfrac{60}{n} + \tfrac{40i}{n^2}\right)$$

4. Distribute and simplify.

$$= \lim_{n \to \infty} \tfrac{60}{n} \sum_{i=1}^{n} 1 + \lim_{n \to \infty} \tfrac{40}{n^2} \sum_{i=1}^{n} i$$

5. Apply the scalar and sum properties to rewrite the limit of the sum.

$$= \lim_{n \to \infty} \tfrac{60}{n}(n) + \lim_{n \to \infty} \tfrac{40}{n^2}\left(\tfrac{n(n+1)}{2}\right)$$

6. Substitute using summation formulas.

$$= \lim_{n \to \infty} 60 + \lim_{n \to \infty} \tfrac{20n + 20}{n}$$

7. Simplify and apply limit theorems.

$$= \lim_{n \to \infty} 60 + \lim_{n \to \infty} \left(20 + \tfrac{20}{n}\right)$$

$$= 60 + (20 + 0) = 80 \text{ m}$$

8. Evaluate each limit and simplify.

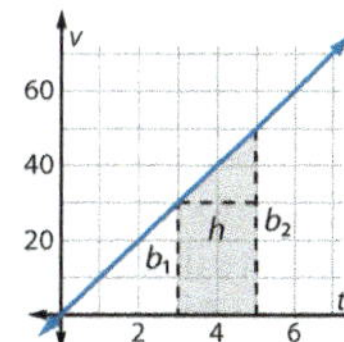

If the velocity-time function in Example 5 is graphed, it becomes apparent that the area under the curve in this case is a trapezoid and therefore could have been found using a formula from geometry.

$$A = \tfrac{1}{2}(b_1 + b_2)h = \tfrac{1}{2}\left(30\,\tfrac{\text{m}}{\text{sec}} + 50\,\tfrac{\text{m}}{\text{sec}}\right)(2\text{ sec}) = 80\text{ m}$$

Both of these velocity-time graphs illustrate constant accelerations (changes in velocity): 0 m/sec² in the first graph and 10 m/sec² in the second. In other cases the definite integral will likely need to be used instead of a geometric formula.

The Fundamental Theorem of Calculus, which is discussed in the next section, simplifies the solution to many integration problems just as theorems describing derivatives of basic functions simplified the differentiation of many functions.

7. $\int_{3}^{7}\left(\tfrac{1}{2}x + 2\right)dx$; 18

8. $\int_{-2}^{4}\left(-\tfrac{1}{4}x + 2\right)dx$; 10.5

❯ A. Exercises

State the value of Δx used to estimate the area under the graph of $f(x)$ over the interval $[a, b]$ on the x-axis with a sum of n rectangles. Then list the right endpoint of each partition, x_i.

1. $[0, 3]$; $n = 6$ **2.** $[-2, 3]$; $n = 5$

3. $[-1, 2]$; $n = 6$ **4.** $[2, 4]$; $n = 8$

Write a definite integral representing the area of each shaded region. Then evaluate the integral using a geometric formula.

5. 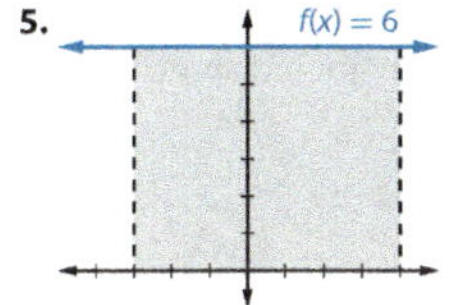

6. 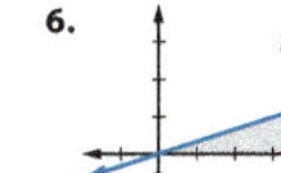

7. 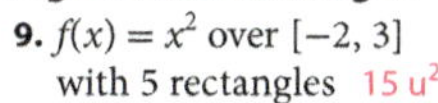

8. 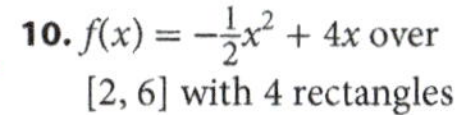

5. $\int_{-3}^{4} 6\,dx$; 42

6. $\int_{0}^{6} \tfrac{1}{3}x\,dx$; 6

Estimate the area of the shaded region between $f(x)$ and the x-axis using the stated number of rectangles. Use the right endpoint of each partition to determine the height of each rectangle.

9. $f(x) = x^2$ over $[-2, 3]$ with 5 rectangles 15 u²

10. $f(x) = -\tfrac{1}{2}x^2 + 4x$ over $[2, 6]$ with 4 rectangles 29 u²

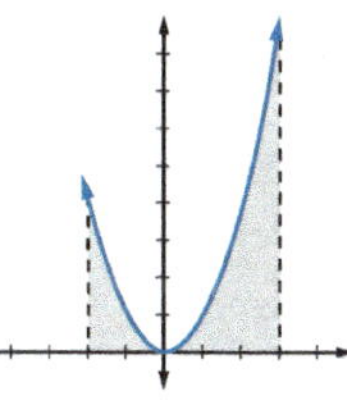

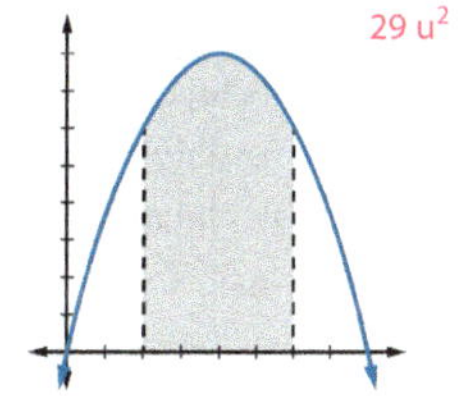

1. 0.5; $x_i = 0.5, 1, 1.5, 2, 2.5, 3$

2. 1; $x_i = -1, 0, 1, 2, 3$

3. 0.5; $x_i = -0.5, 0, 0.5, 1, 1.5, 2$

4. 0.25; $x_i = 2.25, 2.5, 2.75, 3, 3.25, 3.5, 3.75, 4$

$$= 2\left(\tfrac{1}{250}\right)^4 (1^3 + 2^3 + 3^3 + \cdots + 1000^3)$$

$$= 2\left(\tfrac{1}{250}\right)^4 \sum_{i=1}^{1000} i^3$$

$$= 2\left(\tfrac{1}{250}\right)^4 \cdot \tfrac{1000^2(1001)^2}{4} \approx 128.26\text{ u}^2$$

20. $\Delta x = \dfrac{5 - 0}{1000} = \dfrac{1}{200}$

$$x_i = \tfrac{1}{200}, \tfrac{2}{200}, \tfrac{3}{200}, \ldots, \tfrac{1000}{200}$$

$$S = \tfrac{1}{200}f\left(\tfrac{1}{200}\right) + \tfrac{1}{200}f\left(\tfrac{2}{200}\right) + \tfrac{1}{200}f\left(\tfrac{3}{200}\right) + \cdots + \tfrac{1}{200}f\left(\tfrac{1000}{200}\right)$$

$$= \tfrac{1}{200}\cdot 7\left(\tfrac{1}{200}\right)^2 + \tfrac{1}{200}\cdot 7\left(\tfrac{2}{200}\right)^2 + \tfrac{1}{200}\cdot 7\left(\tfrac{3}{200}\right)^2 + \cdots + \tfrac{1}{200}\cdot 7\left(\tfrac{1000}{200}\right)^2$$

$$= 7\left(\tfrac{1}{200}\right)^3 (1^2 + 2^2 + 3^2 + \cdots + 1000^2)$$

$$= 7\left(\tfrac{1}{200}\right)^3 \sum_{i=1}^{1000} i^2$$

$$= 7\left(\tfrac{1}{200}\right)^3 \cdot \tfrac{1000(1001)(2001)}{6}$$

$$= \tfrac{35(1001)(2001)}{6(200)^2} \approx 292.10\text{ u}^2$$

21. $\Delta x = \dfrac{3 - 0}{n} = \dfrac{3}{n}$

$$\lim_{n \to \infty} \sum_{i=1}^{n} (-x_i + 4)\left(\tfrac{3}{n}\right)$$

$$= \lim_{n \to \infty} \sum_{i=1}^{n} \left(-\left(\tfrac{3i}{n}\right) + 4\right)\left(\tfrac{3}{n}\right)$$

$$= \lim_{n \to \infty} \sum_{i=1}^{n} \left(-\tfrac{9i}{n^2} + \tfrac{12}{n}\right)$$

$$= \lim_{n \to \infty} \left(-\tfrac{9}{n^2} \sum_{i=1}^{n} i\right) + \lim_{n \to \infty} \tfrac{1}{n} \sum_{i=1}^{n} 12$$

$$= \lim_{n \to \infty} \left(-\tfrac{9}{n^2} \cdot \tfrac{n(n+1)}{2}\right) + \lim_{n \to \infty} \tfrac{12n}{n}$$

$$= \lim_{n \to \infty} \left(\tfrac{-9(n+1)}{2n}\right) + 12$$

$$= \lim_{n \to \infty} \left(\tfrac{-9n}{2n} + \tfrac{-9}{2n}\right) + 12$$

$$= \tfrac{15}{2} \text{ or } 7\tfrac{1}{2}$$

22. $\Delta x = \dfrac{5 - 0}{n} = \dfrac{5}{n}$

$$\lim_{n \to \infty} \sum_{i=1}^{n} 4x_i^2\left(\tfrac{5}{n}\right)$$

$$= \lim_{n \to \infty} \sum_{i=1}^{n} 4\left(\tfrac{5i}{n}\right)^2\left(\tfrac{5}{n}\right)$$

$$= \lim_{n \to \infty} \sum_{i=1}^{n} \tfrac{500i^2}{n^3} = \lim_{n \to \infty} \tfrac{500}{n^3} \sum_{i=1}^{n} i^2$$

$$= \lim_{n \to \infty} \tfrac{500}{n^3}\left(\tfrac{n(n+1)(2n+1)}{6}\right)$$

$$= \lim_{n \to \infty} \tfrac{250}{n^2}\left(\tfrac{2n^2 + 3n + 1}{3}\right)$$

$$= \lim_{n \to \infty} \left(\tfrac{500n^2}{3n^2} + \tfrac{750n}{3n^2} + \tfrac{250}{3n^2}\right)$$

$$= \tfrac{500}{3} \text{ or } 166\tfrac{2}{3}$$

$$= 3\left(\tfrac{1}{1000}\right)^3 \sum_{i=1}^{1000} i^2$$

$$= 3\left(\tfrac{1}{1000}\right)^3 \cdot \tfrac{1000(1001)(2001)}{6}$$

$$= \tfrac{(1001)(2001)}{2(1000^2)} \approx 1.00\text{ u}^2$$

18. $\Delta x = \dfrac{2 - 0}{1000} = \dfrac{1}{500}$

$$x_i = \tfrac{1}{500}, \tfrac{2}{500}, \tfrac{3}{500}, \ldots, \tfrac{1000}{500}$$

$$S = \tfrac{1}{500}f\left(\tfrac{1}{500}\right) + \tfrac{1}{500}f\left(\tfrac{2}{500}\right) + \tfrac{1}{500}f\left(\tfrac{3}{500}\right) + \cdots + \tfrac{1}{500}f\left(\tfrac{1000}{500}\right)$$

$$= \tfrac{1}{500}\left(\tfrac{1}{500}\right)^3 + \tfrac{1}{500}\left(\tfrac{2}{500}\right)^3 + \tfrac{1}{500}\left(\tfrac{3}{500}\right)^3 + \cdots + \tfrac{1}{500}\left(\tfrac{1000}{500}\right)^3$$

$$= \left(\tfrac{1}{500}\right)^4 (1^3 + 2^3 + 3^3 + \cdots + 1000^3)$$

$$= \left(\tfrac{1}{500}\right)^4 \sum_{i=1}^{1000} i^3$$

$$= \left(\tfrac{1}{500}\right)^4 \cdot \tfrac{1000^2(1001)^2}{4} \approx 4.01\text{ u}^2$$

19. $\Delta x = \dfrac{4 - 0}{1000} = \dfrac{1}{250}$

$$x_i = \tfrac{1}{250}, \tfrac{2}{250}, \tfrac{3}{250}, \ldots, \tfrac{1000}{250}$$

$$S = \tfrac{1}{250}f\left(\tfrac{1}{250}\right) + \tfrac{1}{250}f\left(\tfrac{2}{250}\right) + \tfrac{1}{250}f\left(\tfrac{3}{250}\right) + \cdots + \tfrac{1}{250}f\left(\tfrac{1000}{250}\right)$$

$$= \tfrac{1}{250}\cdot 2\left(\tfrac{1}{250}\right)^3 + \tfrac{1}{250}\cdot 2\left(\tfrac{2}{250}\right)^3 + \tfrac{1}{250}\cdot 2\left(\tfrac{3}{250}\right)^3 + \cdots + \tfrac{1}{250}\cdot 2\left(\tfrac{1000}{250}\right)^3$$

23. $\Delta x = \dfrac{3-0}{n} = \dfrac{3}{n}$

$\displaystyle \lim_{n\to\infty} \sum_{i=1}^{n} \left(x_i^2 + 1\right)\left(\dfrac{3}{n}\right)$

$\displaystyle = \lim_{n\to\infty} \sum_{i=1}^{n} \left(\left(\dfrac{3i}{n}\right)^2 + 1\right)\left(\dfrac{3}{n}\right)$

$\displaystyle = \lim_{n\to\infty} \sum_{i=1}^{n} \left(\dfrac{27i^2}{n^3} + \dfrac{3}{n}\right)$

$\displaystyle = \lim_{n\to\infty} \dfrac{27}{n^3} \sum_{i=1}^{n} i^2 + \lim_{n\to\infty} \dfrac{1}{n} \sum_{i=1}^{n} 3$

$\displaystyle = \lim_{n\to\infty} \dfrac{27}{n^3}\left(\dfrac{n(n+1)(2n+1)}{6}\right) + \lim_{n\to\infty} \dfrac{3n}{n}$

$\displaystyle = \lim_{n\to\infty} \dfrac{9}{n^2}\left(\dfrac{2n^2 + 3n + 1}{2}\right) + 3$

$\displaystyle = \lim_{n\to\infty} \left(\dfrac{9n^2}{n^2} + \dfrac{27n}{2n^2} + \dfrac{9}{2n^2}\right) + 3$

$= 9 + 3 = 12$

24. $\Delta x = \dfrac{3-0}{n} = \dfrac{3}{n}$

$\displaystyle \lim_{n\to\infty} \sum_{i=1}^{n} 6x_i^3\left(\dfrac{3}{n}\right)$

$\displaystyle = \lim_{n\to\infty} \sum_{i=1}^{n} 6\left(\dfrac{3i}{n}\right)^3\left(\dfrac{3}{n}\right)$

$\displaystyle = \lim_{n\to\infty} \sum_{i=1}^{n} \dfrac{486i^3}{n^4} = \lim_{n\to\infty} \dfrac{486}{n^4} \sum_{i=1}^{n} i^3$

$\displaystyle = \lim_{n\to\infty} \dfrac{486}{n^4}\left(\dfrac{n^2(n+1)^2}{4}\right)$

$\displaystyle = \lim_{n\to\infty} \dfrac{243}{n^2}\left(\dfrac{n^2 + 2n + 1}{2}\right)$

$\displaystyle = \lim_{n\to\infty} \left(\dfrac{243n^2}{2n^2} + \dfrac{243n}{n^2} + \dfrac{243}{2n^2}\right)$

$= \dfrac{243}{2}$ or $121\dfrac{1}{2}$

25. translating 2 units left: $[0, 2]$

$f(x) = (x - 2)^3$

$f(x + 2) = (x - 2 + 2)^3 = x^3$

$\Delta x = \dfrac{2-0}{n} = \dfrac{2}{n}$

$\displaystyle \int_0^2 (x^3)\,dx = \lim_{n\to\infty} \sum_{i=1}^{n} \left(\dfrac{2i}{n}\right)^3\left(\dfrac{2}{n}\right)$

$\displaystyle = \lim_{n\to\infty} \sum_{i=1}^{n} \dfrac{16i^3}{n^4}$

$\displaystyle = \lim_{n\to\infty} \dfrac{16}{n^4} \sum_{i=1}^{n} i^3 = \lim_{n\to\infty} \dfrac{16}{n^4}\left(\dfrac{n^2(n+1)^2}{4}\right)$

$\displaystyle = \lim_{n\to\infty} \dfrac{4}{n^2}(n^2 + 2n + 1)$

$\displaystyle = \lim_{n\to\infty} \left(\dfrac{4n^2}{n^2} + \dfrac{8n}{n^2} + \dfrac{4}{n^2}\right) = 4$

26. translating 3 units right: $[0, 5]$

$f(x) = x^2 + 6x + 4$

$f(x - 3) = (x - 3)^2 + 6(x - 3) + 4$

$\quad = x^2 - 6x + 9 + 6x - 18 + 4$

$\quad = x^2 - 5$

$\Delta x = \dfrac{5-0}{n} = \dfrac{5}{n}$

$\displaystyle \int_0^5 (x^2 - 5)\,dx$

$\displaystyle = \lim_{n\to\infty} \sum_{i=1}^{n} \left(x_i^2 - 5\right)\left(\dfrac{5}{n}\right)$

$\displaystyle = \lim_{n\to\infty} \sum_{i=1}^{n} \left(\left(\dfrac{5i}{n}\right)^2 - 5\right)\left(\dfrac{5}{n}\right)$

$\displaystyle = \lim_{n\to\infty} \sum_{i=1}^{n} \left(\dfrac{125i^2}{n^3} - \dfrac{25}{n}\right)$

---

11. $f(x) = \sqrt{x}$ over $[1, 7]$ with 6 rectangles $\approx 12.48\ \text{u}^2$

12. $f(x) = x^2 + 2$ over $[0, 3]$ with 6 rectangles $17.375\ \text{u}^2$

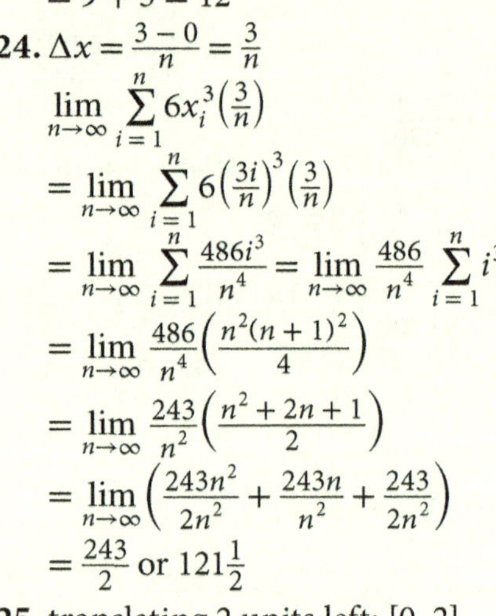

Estimate the area under $f(x)$ over the given interval on the x-axis using the stated number of rectangles. Use the right endpoint of each partition to determine the height of each rectangle.

13. $f(x) = \frac{1}{10}x^2$ over $[0, 10]$ with 5 rectangles $44\ \text{u}^2$

14. $f(x) = \frac{1}{10}x^2$ over $[0, 10]$ with 10 rectangles $38.5\ \text{u}^2$

15. $f(x) = -\frac{1}{2}x^3 + 3x^2$ over $[0, 4]$ with 4 rectangles $40\ \text{u}^2$

16. $f(x) = -\frac{1}{2}x^3 + 3x^2$ over $[0, 4]$ with 8 rectangles $36\ \text{u}^2$

> **B. Exercises**

Approximate the area (to the nearest hundredth) under the curve over $[0, b]$ on the x-axis using 1000 rectangles.

17. $f(x) = 3x^2$; $[0, 1]$ $1.00\ \text{u}^2$ **18.** $f(x) = x^3$; $[0, 2]$ $4.01\ \text{u}^2$

19. $f(x) = 2x^3$; $[0, 4]$ $128.26\ \text{u}^2$ **20.** $f(x) = 7x^2$; $[0, 5]$ $292.10\ \text{u}^2$

Use the definition of a definite integral to evaluate each definite integral.

21. $\int_0^3 (-x + 4)\,dx$ $\frac{15}{2}$ or $7\frac{1}{2}$ **22.** $\int_0^5 4x^2\,dx$ $\frac{500}{3}$ or $166\frac{2}{3}$

23. $\int_0^3 (x^2 + 1)\,dx$ 12 **24.** $\int_0^3 6x^3\,dx$ $\frac{243}{2}$ or $121\frac{1}{2}$

Evaluate each definite integral after applying a translation.

25. $\int_2^4 (x - 2)^3\,dx$ 4 **26.** $\int_{-3}^2 (x^2 + 6x + 4)\,dx$ $\frac{50}{3}$

27. $\int_{-3}^{-2} (-x^2 - 6x + 11)\,dx$ **28.** $\int_{-1}^3 (x^2 - 2x + 4)\,dx$

The function $v(t) = 10t$ models the velocity (in m/sec) of a dropped object.

29. How far did the object fall in the first 3 sec? $45\ \text{m}$

30. How far did the object fall from 3 to 5 sec? $80\ \text{m}$

27. $\frac{59}{3}$ or $19\frac{2}{3}$ **28.** $\frac{52}{3}$ or $17\frac{1}{3}$

The function $v(t) = 32t$ models the velocity (in ft/sec) of a dropped object.

31. How far did the object fall from 2 to 3 sec? $80\ \text{ft}$

32. How far did the object fall in the first 6 sec? $576\ \text{ft}$

33. Explore: Complete the following steps to compare the methods of using a geometric formula and evaluating a definite integral to find the area under $f(x) = -2x + 12$ over $[2, 5]$ on the x-axis.

  a. Graph the function and shade the desired region.

  b. Calculate the area using a geometric formula. $15\ \text{u}^2$

  c. State the definite integral representing this area.

  d. Use the definition of the definite integral to evaluate the integral. $15\ \text{u}^2$

  e. How do the answers for b and d compare? They are equal.

> **C. Exercises**

34. Explore: Estimate the area of the shaded region under $f(x) = -(x - 1)^2 + 9$ over $[0, 4]$ on the x-axis using 4 rectangles by using the following points in each partition to determine the height of the rectangles.

  a. the right endpoint $22\ \text{u}^2$

  b. the left endpoint $30\ \text{u}^2$

  c. the midpoint $27\ \text{u}^2$

  d. Compare each result and the average of the right-end and left-end sums to the exact area of $26\frac{1}{3}\ \text{u}^2$, found using $\int_0^4 (-(x - 1)^2 + 9)\,dx$.

35. Analyze: Use sketches of an increasing function $f(x)$ and a decreasing function $g(x)$ to explain when each sum overestimates the area under the graph and when it underestimates the area.

  a. right-end sum   **b.** left-end sum

36. Explain: Why doesn't it matter whether the left endpoint, right endpoint, or midpoint of each partition is used in the definition of the definite integral, $\int_a^b f(x)\,dx = \lim\limits_{n\to\infty} \sum\limits_{i=1}^{n} f(x_i)\,\Delta x$?

37. Evaluate: $\int_0^b (x^2 + 3)\,dx$ $\frac{1}{3}b^3 + 3b$

38. Use the result from exercise 37 to find an expression for $\int_a^b (x^2 + 3)\,dx$. $\frac{1}{3}(b^3 - a^3) + 3b - 3a$

33c. $\int_2^5 (-2x + 12)\,dx$

36. As $n \to \infty$, the rectangles get thinner and the left endpoint, the right endpoint, and the midpoint converge to the same point.

$\displaystyle = \lim_{n\to\infty} \dfrac{125}{n^3} \sum_{i=1}^{n} i^2 - \lim_{n\to\infty} \dfrac{1}{n} \sum_{i=1}^{n} 25$

$\displaystyle = \lim_{n\to\infty} \dfrac{125}{n^3}\left(\dfrac{n(n+1)(2n+1)}{6}\right)$

$\displaystyle \qquad - \lim_{n\to\infty} \dfrac{25n}{n}$

$\displaystyle = \lim_{n\to\infty} \dfrac{125}{n^2}\left(\dfrac{2n^2 + 3n + 1}{6}\right) - 25$

$\displaystyle = \lim_{n\to\infty} \left(\dfrac{125n^2}{3n^2} + \dfrac{125n}{2n^2} + \dfrac{125}{6n^2}\right) - 25$

$\displaystyle = \dfrac{125}{3} - 25 = \dfrac{50}{3}$ or $16\dfrac{2}{3}$

27. translating 3 units right: $[0, 1]$

$f(x) = -x^2 - 6x + 11$

$f(x - 3) = -(x - 3)^2 - 6(x - 3) + 11$

$\quad = -x^2 + 6x - 9 - 6x + 18 + 11$

$\quad = -x^2 + 20$

$\Delta x = \dfrac{1-0}{n} = \dfrac{1}{n}$

$\displaystyle \int_0^1 (-x^2 + 20)\,dx$

$\displaystyle = \lim_{n\to\infty} \sum_{i=1}^{n} \left(-x_i^2 + 20\right)\left(\dfrac{1}{n}\right)$

$\displaystyle = \lim_{n\to\infty} \sum_{i=1}^{n} \left(-\left(\dfrac{i}{n}\right)^2 + 20\right)\left(\dfrac{1}{n}\right)$

$\displaystyle = \lim_{n\to\infty} \sum_{i=1}^{n} \left(-\dfrac{i^2}{n^3} + \dfrac{20}{n}\right)$

$\displaystyle = \lim_{n\to\infty} \left(-\dfrac{1}{n^3} \sum_{i=1}^{n} i^2\right) + \lim_{n\to\infty} \dfrac{1}{n} \sum_{i=1}^{n} 20$

$\displaystyle = \lim_{n\to\infty} \left[-\dfrac{1}{n^3}\left(\dfrac{n(n+1)(2n+1)}{6}\right)\right]$

$\displaystyle \qquad + \lim_{n\to\infty} \dfrac{20n}{n}$

$\displaystyle = \lim_{n\to\infty} \left[-\dfrac{1}{n^2}\left(\dfrac{2n^2 + 3n + 1}{6}\right)\right] + 20$

$\displaystyle = \lim_{n\to\infty} \left(-\dfrac{n^2}{3n^2} - \dfrac{n}{2n^2} - \dfrac{1}{6n^2}\right) + 20$

$\displaystyle = -\dfrac{1}{3} + 20 = \dfrac{59}{3}$ or $19\dfrac{2}{3}$

39. Find B^{-1} if $B = \begin{bmatrix} 5 & 6 \\ -3 & 8 \end{bmatrix}$. [7.5] $\begin{bmatrix} \frac{4}{29} & -\frac{3}{29} \\ \frac{3}{58} & \frac{5}{58} \end{bmatrix}$

40. Solve $\begin{bmatrix} 3 & 7 \\ 2 & 4 \end{bmatrix} X = \begin{bmatrix} 4 \\ -8 \end{bmatrix}$. [7.5] $X = \begin{bmatrix} -36 \\ 16 \end{bmatrix}$

41. Determine the mean, median, and mode for home runs hit in a season by Henry Aaron. [10.3–10.4]

mean: ≈ 32.8
median: 34
mode: 44

Home Runs in a Season	
1	0 2 3
2	0 4 6 7 9
3	0 2 4 4 8 9 9
4	0 0 4 4 4 4 5 7

42. What is the range of home runs hit by Henry Aaron in a season? [10.4] 37

43. Given $f(x) = 2x$, evaluate $\lim\limits_{h \to 0} \dfrac{f(x+h) - f(x)}{h}$. [12.3] 2

44. Given $f(x) = 4x^2 + 3x - 6$, find $f'(-2)$. [12.4] $f'(2) = -13$

45. Find the fourth term in the expansion of $(5m + 2n)^4$. [9.4] D

  A. $16n^4$ **C.** $1000mn^3$ **E.** none of
  B. $40mn^3$ **D.** $160mn^3$ these

46. Find the probability that a randomly selected value in a normally distributed population with $\mu = 87$ and $\sigma = 8.2$ falls within the interval [76, 94]. [10.6] B

  A. 0.140   **C.** 0.051   **E.** none of
  B. 0.713   **D.** 1.000   these

47. Find $\dfrac{d}{dx} \sqrt{x}$. [12.4] A

  A. $\dfrac{1}{2\sqrt{x}}$ **C.** $-\dfrac{3}{2\sqrt{x}}$ **E.** none of
                                            these
  B. $-\dfrac{1}{2\sqrt{x}}$ **D.** $\dfrac{1}{\sqrt{x}}$

48. Find the derivative of $f(x) = \dfrac{2x^2 + 7}{x^2 - 3}$. [12.5] A

  A. $f'(x) = -\dfrac{26x}{(x^2 - 3)^2}$ **D.** $f'(x) = -\dfrac{2}{x^2 - 3}$

  B. $f'(x) = \dfrac{2x}{(x^2 - 3)^2}$ **E.** none of these

  C. $f'(x) = \dfrac{8x^4}{(x^2 - 3)^2}$

---

29. $\displaystyle\int_0^3 10t\, dt = \lim_{n \to \infty} \sum_{i=1}^{n} (10t_i)\left(\frac{3}{n}\right)$

$\displaystyle = \lim_{n \to \infty} \sum_{i=1}^{n} 10\left(\frac{3i}{n}\right)\left(\frac{3}{n}\right) = \lim_{n \to \infty} \sum_{i=1}^{n} \frac{90i}{n^2}$

$\displaystyle = \lim_{n \to \infty} \frac{90}{n^2} \sum_{i=1}^{n} i = \lim_{n \to \infty} \frac{90}{n^2}\left(\frac{n(n+1)}{2}\right)$

$\displaystyle = \lim_{n \to \infty} \frac{90n + 90}{2n} = 45 \text{ m}$

30. $\displaystyle\int_3^5 10t\, dt = \lim_{n \to \infty} \sum_{i=1}^{n} (10t_i)\left(\frac{2}{n}\right)$

$\displaystyle = \lim_{n \to \infty} \sum_{i=1}^{n} 10\left(3 + \frac{2i}{n}\right)\left(\frac{2}{n}\right)$

$\displaystyle = \lim_{n \to \infty} \sum_{i=1}^{n} \left(\frac{60}{n} + \frac{40i}{n^2}\right)$

$\displaystyle = \lim_{n \to \infty} \frac{60}{n} \sum_{i=1}^{n} 1 + \lim_{n \to \infty} \frac{40}{n^2} \sum_{i=1}^{n} i$

$\displaystyle = \lim_{n \to \infty} \frac{60}{n}(1)n + \lim_{n \to \infty} \frac{40}{n^2}\left(\frac{n(n+1)}{2}\right)$

$\displaystyle = \lim_{n \to \infty} 60 + \lim_{n \to \infty} \frac{20n + 20}{n}$

$\displaystyle = 60 + 20 = 80 \text{ m}$

31. $\displaystyle\int_2^3 32t\, dt = \lim_{n \to \infty} \sum_{i=1}^{n} (32t_i)\left(\frac{1}{n}\right)$

$\displaystyle = \lim_{n \to \infty} \sum_{i=1}^{n} 32\left(2 + \frac{i}{n}\right)\left(\frac{1}{n}\right)$

$\displaystyle = \lim_{n \to \infty} \sum_{i=1}^{n} \left(\frac{64}{n} + \frac{32i}{n^2}\right)$

$\displaystyle = \lim_{n \to \infty} \frac{64}{n} \sum_{i=1}^{n} 1 + \lim_{n \to \infty} \frac{32}{n^2} \sum_{i=1}^{n} i$

$\displaystyle = \lim_{n \to \infty} \frac{64}{n}(1n) + \lim_{n \to \infty} \frac{32}{n^2}\left(\frac{n(n+1)}{2}\right)$

$\displaystyle = \lim_{n \to \infty} 64 + \lim_{n \to \infty} \frac{16n + 16}{n}$

$\displaystyle = 64 + 16 = 80 \text{ ft}$

32. $\displaystyle\int_0^6 32t\, dt = \lim_{n \to \infty} \sum_{i=1}^{n} (32t_i)\left(\frac{6}{n}\right)$

$\displaystyle = \lim_{n \to \infty} \sum_{i=1}^{n} 32\left(\frac{6i}{n}\right)\left(\frac{6}{n}\right) = \lim_{n \to \infty} \sum_{i=1}^{n} \frac{1152i}{n^2}$

$\displaystyle = \lim_{n \to \infty} \frac{1152}{n^2} \sum_{i=1}^{n} i = \lim_{n \to \infty} \frac{1152}{n^2}\left(\frac{n(n+1)}{2}\right)$

$\displaystyle = \lim_{n \to \infty} \frac{576n + 576}{n} = 576 \text{ ft}$

---

28. translating 1 unit right: [0, 4]

$f(x) = x^2 - 2x + 4$
$f(x - 1) = (x - 1)^2 - 2(x - 1) + 4$
$= x^2 - 2x + 1 - 2x + 2 + 4$
$= x^2 - 4x + 7$
$\Delta x = \dfrac{4 - 0}{n} = \dfrac{4}{n}$

$\displaystyle\int_0^4 (x^2 - 4x + 7)\, dx$

$\displaystyle = \lim_{n \to \infty} \sum_{i=1}^{n} \left(x_i^2 - 4x_i + 7\right)\left(\frac{4}{n}\right)$

$\displaystyle = \lim_{n \to \infty} \sum_{i=1}^{n} \left(\left(\frac{4i}{n}\right)^2 - 4\left(\frac{4i}{n}\right) + 7\right)\left(\frac{4}{n}\right)$

$\displaystyle = \lim_{n \to \infty} \sum_{i=1}^{n} \left(\frac{64i^2}{n^3} - \frac{64i}{n^2} + \frac{28}{n}\right)$

$\displaystyle = \lim_{n \to \infty} \frac{64}{n^3} \sum_{i=1}^{n} i^2 - \lim_{n \to \infty} \frac{64}{n^2} \sum_{i=1}^{n} i$

$\displaystyle \qquad\qquad + \lim_{n \to \infty} \frac{1}{n} \sum_{i=1}^{n} 28$

$\displaystyle = \lim_{n \to \infty} \frac{64}{n^3}\left(\frac{n(n+1)(2n+1)}{6}\right)$

$\displaystyle \quad - \lim_{n \to \infty} \frac{64}{n^2}\left(\frac{n(n+1)}{2}\right) + \lim_{n \to \infty} \frac{28n}{n}$

$\displaystyle = \lim_{n \to \infty} \frac{32}{n^2}\left(\frac{2n^2 + 3n + 1}{3}\right)$

$\displaystyle \qquad\qquad - \lim_{n \to \infty} \frac{32(n+1)}{n} + 28$

$\displaystyle = \lim_{n \to \infty} \left(\frac{64n^2}{3n^2} + \frac{32n}{n^2} + \frac{32}{3n^2}\right)$

$\displaystyle \qquad\qquad - \lim_{n \to \infty} \left(\frac{32n}{n} + \frac{32}{n}\right) + 28$

$\displaystyle = \frac{64}{3} - 32 + 28 = \frac{52}{3} \text{ or } 17\frac{1}{3}$

33a.

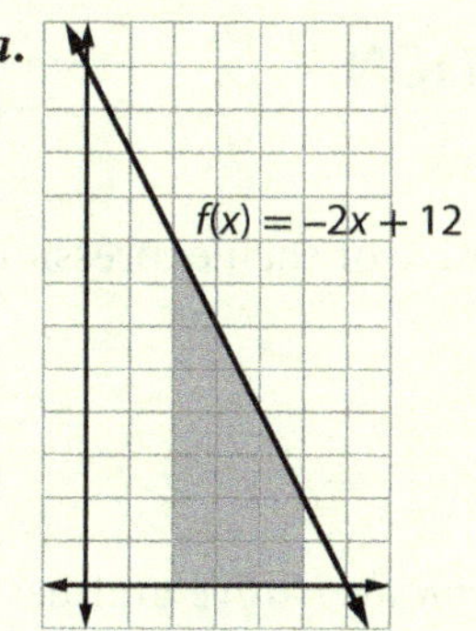

33b. $A = \frac{1}{2}(b_1 + b_2)h = \frac{1}{2}(8 + 2)(3)$
$\qquad = 15 \text{ u}^2$

33c. $\displaystyle\int_2^5 (-2x + 12)\, dx$

continued in Answers and Solutions Overflow

12.7 The Fundamental Theorem of Calculus

Objectives

1. To find antiderivatives of a function
2. To represent an indefinite integral as an antiderivative in general form
3. To apply the Fundamental Theorem of Calculus to find areas
4. To solve real-world problems involving integrals

Flash

Skydivers can vary their speed of descent by adjusting body position. The fastest skydiver freefall speed was achieved by Felix Baumgartner in 2012. He hit a maximum speed of 833.9 mi/hr, breaking the sound barrier.

Vocabulary

antiderivative
constant of integration
Fundamental Theorem of Calculus
general form of an antiderivative
indefinite integral

The distance fallen during a given time interval can be calculated with integration.

After completing this section, you will be able to

- find antiderivatives of a function using antiderivative rules.
- represent an indefinite integral as an antiderivative in general form.
- state the Fundamental Theorem of Calculus and apply it to find areas using substitution.
- solve real-world problems involving integrals.

When the height of a thrown ball is modeled by $h(t)$, its instantaneous vertical velocity (the slope of the tangent to the height-time graph) is represented by the derivative $h'(t)$ (see Section 12.4, Example 5). In this section we seek a way to reverse the process. If the function $v(t) = -32t + 72$ represents the ball's instantaneous vertical velocity, how can a function representing the ball's height be found? In general terms, given the function $f(x)$, how can a function $F(x)$ be found such that its derivative, $F'(x)$, is $f(x)$?

> **DEFINITION**
>
> If $F'(x) = f(x)$, the function $F(x)$ is an **antiderivative** of $f(x)$.

Example 1 Finding an Antiderivative

Find an antiderivative for each function.

a. $f(x) = 5$ **b.** $g(x) = 6x^2$

Answer

a. $F(x) = mx + b = 5x + b$
Note that b can be any constant.
$F(x) = 5x$ is the simplest antiderivative.

Recall that $\frac{d}{dx}(mx + b) = m$.
Therefore, the antiderivative of a constant is a linear function with that slope.

b. $G(x) = \frac{6}{2+1}x^{2+1}$
$G(x) = 2x^3$ is the simplest antiderivative.

Recall that $\frac{d}{dx}(cx^n) = ncx^{n-1}$. To reverse the process of taking the derivative, increase the exponent by one and divide the constant by the new exponent.

SKILL ✓ EXERCISE 3

Another antiderivative of $g(x)$ would be $G(x) = 2x^3 - 4$ since the derivative of a constant is zero. The variable C (the *constant of integration*) is commonly used for this real number constant that can vary depending on the application. Antiderivatives that include this constant are said to be in *general form* and actually represent an entire family of functions.

Example 2 Finding a Specific Antiderivative

If $f(x) = 2x$, find the antiderivative function passing through the point $(-2, 3)$.

Answer

$F(x) = \frac{2}{1+1}x^{1+1}$
$F(x) = x^2 + C$

1. Find the general form of the antiderivative. Then check your work with differentiation.
$\frac{d}{dx}(x^2 + C) = \frac{d}{dx}x^2 + \frac{d}{dx}C = 2x + 0 = 2x$

CONTINUED ➡

PRESENTATION

Lesson Opener

Find the derivative of each expression.

1. $2x^3 + 7$ $6x^2$

2. $2 \sin x - 5$ $2 \cos x$

3. Evaluate $\int_0^2 3x\, dx$. 6

Introduce the *antiderivative* and use Example 1 to demonstrate the process of finding the antiderivative of a function. In part *b*, point out that the exponent is increased by 1 and the coefficient is then divided by this new exponent. Have the students find the derivative of the antiderivative to check their work.

Emphasize that the antiderivative produces an entire family of functions. In order to know the specific function from which the given derivative originated, we would have to know the constant or find the constant using a specific point, as shown in Example 2. Simply substitute the x- and y-values into the antiderivative, $F(x)$.

Define the *indefinite integral*, emphasizing that an indefinite integral produces a family of functions while a definite integral produces a unique real number (a definite value). Discuss the indefinite integral rules. You may want to have students find the derivative of each resulting function $F(x)$. Consider asking the students to describe how the constant rule is actually a special case of the power function rule. (*Viewing $\int k\, dx$ as $\int kx^0\, dx$ and applying the power rule produces* $F(x) = \frac{k}{0+1}x^{0+1} + C = kx + C$.) Introduce the scalar multiple and sum rules and apply these rules using Example 3.

Present the *Fundamental Theorem of Calculus* and explain that it provides a shorter method of finding an antiderivative. Note that some texts will use the equivalent notation $F(x)\big|_a^b$. Use Example 4 to demonstrate evaluating a definite integral using the Fundamental Theorem of Calculus. Consider having students

$3 = (-2)^2 + C$
$3 = 4 + C$
$-1 = C$

2. Substitute $(-2, 3)$ into $F(x)$ and solve to find C.

$F(x) = x^2 - 1$

3. State the antiderivative passing through $(-2, 3)$.

SKILL ✔ EXERCISE 17

The family of antiderivatives for a function $f(x)$ is usually stated with integral notation.

> **DEFINITION**
>
> The **indefinite integral** of $f(x)$ is defined as $\int f(x)\,dx = F(x) + C$, where $F(x) + C$ is a family of antiderivative functions for $f(x)$ and C is the *constant of integration*.

You should know the following basic indefinite integral rules, which can be verified by differentiation.

Indefinite Integral Rules	
Constant	**Power Function**
$\int k\,dx = kx + C$	$\int kx^n\,dx = \dfrac{k}{n+1}x^{n+1} + C$
Sine Function	**Cosine Function**
$\int (\sin x)\,dx = -\cos x + C$	$\int (\cos x)\,dx = \sin x + C$
Natural Exponential Function	
$\int e^x\,dx = e^x + C$	

The scalar multiple and sum rules for integrals are also essential.

$$\int cf(x)\,dx = c\int f(x)\,dx \quad \text{and} \quad \int [f(x) \pm g(x)]\,dx = \int f(x)\,dx \pm \int g(x)\,dx$$

Example 3   Finding Indefinite Integrals

Find each indefinite integral.

a. $\int 12x^3\,dx$ **b.** $\int (2\cos x)\,dx$ **c.** $\int (3x^2 - e^x)\,dx$

Answer

a. $\int 12x^3\,dx = \dfrac{12}{3+1}x^{3+1} + C = 3x^4 + C$ Apply the power rule.

b. $\int (2\cos x)\,dx = 2\int (\cos x)\,dx$

1. Apply the scalar multiple rule to factor out the numerical coefficient.

$\qquad = 2\sin x + C$

2. Then use the cosine rule.

c. $\int (3x^2 - e^x)\,dx = \int 3x^2\,dx - \int e^x\,dx$

1. Apply the sum rule.

$\qquad = (x^3 + C_1) - (e^x + C_2)$

$\qquad = x^3 - e^x + C$

2. Apply the power and exponential rules and simplify by combining the constants.

SKILL ✔ EXERCISE 21

The notation indicating an indefinite integral is similar to that of the definite integral introduced in Section 12.6. This similarity reflects the fact that the antiderivative provides

Reading and Writing Mathematics

Explain how the use of 10 trapezoids would compare to the use of 10 rectangles for estimating the area between a curve and the x-axis. Since the top side of the trapezoids can be angled to follow the curve more closely, trapezoids would likely produce a more accurate estimate.

Additional Exercises

1. Find an antiderivative for each function.
   a. $f(x) = -3x + 2$
   $\quad F(x) = -\frac{3}{2}x^2 + 2x + C$
   b. $g(x) = 2x^2 + 3x + 1$
   $\quad G(x) = \frac{2}{3}x^3 + \frac{3}{2}x^2 + x + C$

2. If $f(x) = -3x$, find the antiderivative function passing through the point $(2, 4)$. $\; F(x) = -\frac{3}{2}x^2 + 10$

3. Find each indefinite integral.
   a. $\int 8x^3 - 4x\,dx$ $2x^4 - 2x^2 + C$
   b. $\int -2\sin x\,dx$ $2\cos x + C$
   c. $\int 2x^3 + 3e^x\,dx$ $\frac{1}{2}x^4 + 3e^x + C$

4. Evaluate $\int_{-5}^{-2} x^2 + 8x + 17$.   6

5. Evaluate each definite integral to find the area between the graph of each function and the interval on the x-axis.
   a. $\int_{-\frac{\pi}{2}}^{\frac{\pi}{2}} \cos x\,dx$ $2; 2\,\mathrm{u}^2$
   b. $\int_{\frac{\pi}{2}}^{\pi} \cos x\,dx$ $-1; 1\,\mathrm{u}^2$

---

perform the check on their calculators. While the calculator illustrates the area represented by the interval, it provides the answer without requiring students to find the antiderivative and apply the Fundamental Theorem of Calculus. Emphasize that these steps need to be shown in the students' work.

Caution students regarding the proper interpretation of the value of a definite integral as it relates to the area between the function and the x-axis. As seen in Example 5, if $f(x)$ is above the x-axis for the given interval, the positive value represents the area between the function and the x-axis. If $f(x)$ is below the x-axis for the given interval, the negative value is the opposite of the area between the function and the x-axis. If there are equal areas above and below the x-axis, the definite integral will have a value of 0. The area between the curve and the x-axis should be calculated using separate integrals over intervals in which $f(x) > 0$ or $f(x) < 0$, as shown in part c.

Remind students that the vertical velocity function for a projectile is the derivative of its height (position) function. In Example 6, the height function is found from the velocity function using integration. Encourage the students to check their work for $h(t) = \int v(t)\,dt$ by taking the derivative of $h(t)$. Graphing the height function confirms the integration and substitution steps for the constant term.

Motivational Idea Unlike previously-learned math processes, calculus gives us the ability to find values that are constantly changing. Consider sharing real-world uses of calculus or having the students use the Internet keyword search *real-world uses of calculus*. Some uses include AI, search engines, credit card statements, electrical engineering, and statistics.

Motivational Idea It is common for students to learn the processes for integra-

c. $\int_{-\frac{\pi}{2}}^{\pi} \cos x \, dx$ 1; 3 u²

6. Graph the height function $h(t) = -4.9t^2 - 8t + 377.5$. Then find the derivative at $t = 5$ and compare the result to the velocity predicted by $v(t) = -9.8t - 8$.

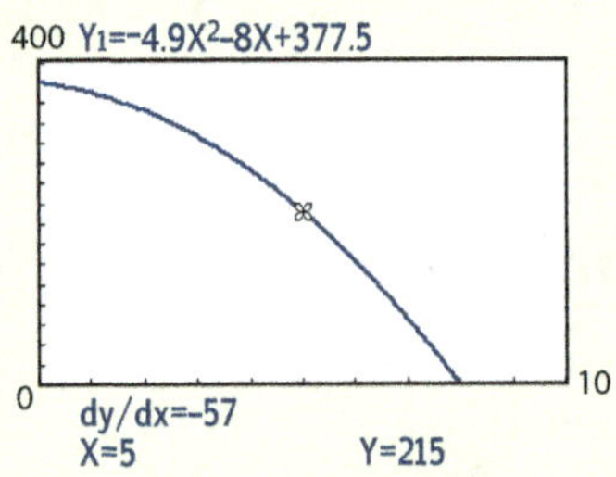

Both are −57.

7. Find and interpret the definite integral of $v(t) = -9.8t - 8$ from 0 to 5 sec. −162.5; the distance (in meters) an object has fallen during the first 5 sec of freefall

8. Use technology to graph $f(x) = \frac{1}{3}x^3$ and $g(x) = -x^2 + 2x + 12$. Then use the calculator's definite integral function, found on the CALC ([2nd], [TRACE]) menu, to find the area between the graphs of the functions over [0, 3].

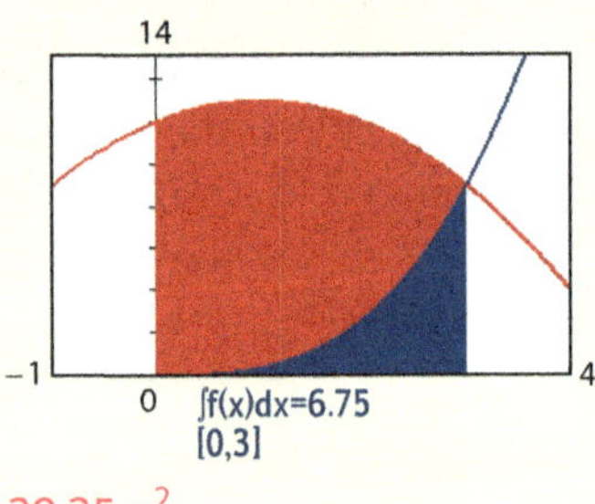

29.25 u²

---

a shorter alternative to using the limit of a sum to evaluate definite integrals. The Fundamental Theorem of Calculus describes this shorter method.

THE FUNDAMENTAL THEOREM OF CALCULUS

If $f(x)$ is continuous over $[a, b]$ and $F(x)$ is an antiderivative of $f(x)$, then $\int_a^b f(x)\, dx = F(b) - F(a)$.

The difference $F(b) - F(a)$ is often notated as $[F(x)]_a^b$ so that $\int_a^b f(x)\, dx = [F(x)]_a^b$. If possible, choose the simplest antiderivative function $F(x)$ by letting $C = 0$.

Example 4   Applying the Fundamental Theorem of Calculus

Evaluate $\int_2^5 (-x^2 + 6x + 1)\, dx$.

Answer

$$\int_2^5 (-x^2 + 6x + 1)\, dx = \left[-\frac{x^3}{3} + 3x^2 + x\right]_2^5$$

1. Use the sum rule to find the antiderivative. $\int -x^2\, dx = -\frac{x^3}{3}, \int 6x\, dx = 3x^2, \int 1\, dx = x$

$$= \left[-\frac{(5)^3}{3} + 3(5)^2 + 5\right] - \left[-\frac{(2)^3}{3} + 3(2)^2 + 2\right]$$

$$= -\frac{125}{3} + 75 + 5 + \frac{8}{3} - 12 - 2 = 27$$

2. Apply the Fundamental Theorem of Calculus, finding $F(b) - F(a)$ by substituting into the antiderivative and simplifying.

Check

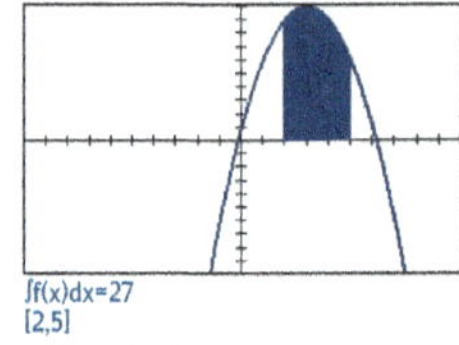

3. Graph $f(x) = -x^2 + 6x + 1$, select 7: ∫f(x)dx from the [CALC] menu ([2nd], [TRACE]), and enter the lower limit of 2 and upper limit of 5 to calculate the area below the parabola over [2, 5] on the x-axis.

SKILL ✔ EXERCISE 27

It is important to note that definite integrals can be positive, negative, or equal to zero. When the function $f(x)$ is nonnegative over $[a, b]$, each nonzero term in $\sum_{i=1}^{n} f(x_i)\, \Delta x$ represents the area of a rectangle above the x-axis and $\int_a^b f(x)\, dx$ represents the area under the graph of $f(x)$ and above $[a, b]$ on the x-axis. If $f(x)$ is nonpositive over $[a, b]$, the nonzero terms in $\sum_{i=1}^{n} f(x_i)\, \Delta x$ are negative and $\int_a^b f(x)\, dx$ is the opposite of the area above $f(x)$ and under $[a, b]$ on the x-axis. Therefore, special care should be taken when using integrals to find areas.

DEFINITE INTEGRALS AND THE AREA BETWEEN A CURVE AND THE X-AXIS

The area, A, between the graph of a continuous function $f(x)$ and the x-axis over the interval $[a, b]$ can be found using the definite integral.

$$A = \begin{cases} \int_b^a f(x)\, dx & \text{if } f(x) \geq 0 \text{ over } [a, b] \\ -\int_b^a f(x)\, dx & \text{if } f(x) \leq 0 \text{ over } [a, b] \end{cases}$$

---

tion while not obtaining a clear understanding of or an appreciation for their application. Consider using the following curve to illustrate another practical application.

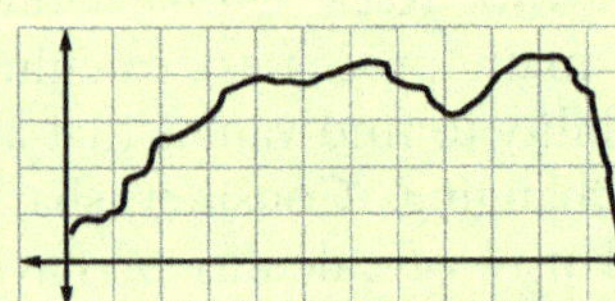

If the curve represents the download speed of a file in respect to time, the area of the rectangle determined by a value of the function during each infinitesimally small time interval, $i\Delta t$, represents the amount of the file downloaded during that time. The total accumulation of these values (equal to the total area under the curve) would represent the size of the downloaded file and could be obtained by finding the integral of the function.

TIPS

Ex. 3–4, 7, 11, 14 Although x is the commonly used independent variable for integration, there are many applications with other independent variables.

Ex. 12, 25, 28 Recall that the value of a definite integral can be positive, negative, or equal to 0.

Ex. 34–35 Students may need to be prompted to calculate the chain's weight per meter (the force required to lift the end of the chain varies as one end is lifted).

Ex. 38–39 Graphing the functions can help students find the solution.

Example 5 — Relating the Definite Integral Value to Area

Evaluate each definite integral.

a. $\int_0^{\pi}(\sin x)\,dx$ **b.** $\int_{\pi}^{2\pi}(\sin x)\,dx$ **c.** $\int_0^{2\pi}(\sin x)\,dx$

Then state the area between the graph of $y = \sin x$ and the x-axis over each interval.

d. $[0, \pi]$ **e.** $[\pi, 2\pi]$ **f.** $[0, 2\pi]$

Answer

1. Use the antiderivative to evaluate each definite interval.

a. $\int_0^{\pi}(\sin x)\,dx$

$= [-\cos x]_0^{\pi}$
$= -\cos \pi - (-\cos 0)$
$= -(-1) - (-1) = 2$

b. $\int_{\pi}^{2\pi}(\sin x)\,dx$

$= [-\cos x]_{\pi}^{2\pi}$
$= -\cos 2\pi - (-\cos \pi)$
$= -1 - 1 = -2$

c. $\int_0^{2\pi}(\sin x)\,dx$

$= [-\cos x]_0^{2\pi}$
$= -\cos 2\pi - (-\cos 0)$
$= -1 - (-1) = 0$

2. Use the values of the definite integrals to state each area.

d. $\sin x \geq 0$ over $[0, \pi]$

$A_{[0, \pi]} = 2\ u^2$

e. $\sin x \leq 0$ over $[\pi, 2\pi]$

$A_{[\pi, 2\pi]} = -(-2) = 2\ u^2$

f. $A_{[0, 2\pi]} = A_{[0, \pi]} + A_{[\pi, 2\pi]}$
$= 2 + 2 = 4\ u^2$

SKILL ✔ EXERCISE 25

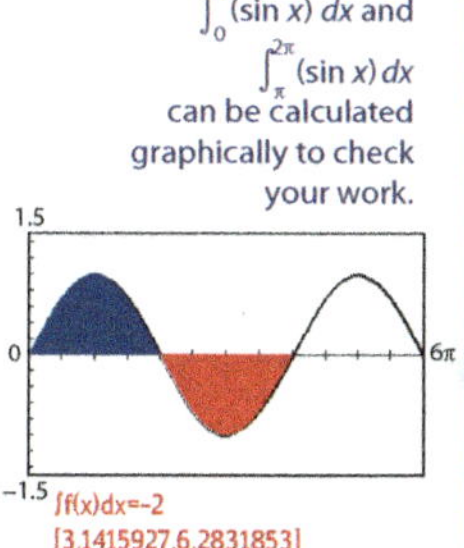

Example 5 illustrates an important property of definite integrals:

if $a < b < c$, then $\int_a^b f(x)\,dx + \int_b^c f(x)\,dx = \int_a^c f(x)\,dx$.

Example 6 examines the real-world application introduced at the beginning of this section.

Example 6 — Modeling Height from a Vertical Velocity Function

The vertical velocity (in ft/sec) of a ball propelled upward with an initial vertical velocity of 72 ft/sec is modeled by $v(t) = -32t + 72$. Find a function modeling its height above the ground (in feet) if it reaches its maximum height of 86 ft in 2.25 sec.

Answer

$h(t) = \int(-32t + 72)\,dt = -16t^2 + 72t + C$

1. since $v(t) = h'(t)$, $\int v(t)\,dt = h(t)$
   Note: C represents h_0, the height at $t = 0$.

$h(2.25) = -16(2.25)^2 + 72(2.25) + C = 86$
$81 + C = 86$
$C = 5$

2. Substitute into $h(t)$ using $h(2.25) = 86$ and solve for C.

$h(t) = -16t^2 + 72t + 5$

3. Write the function modeling the ball's height where $C = 5$ ft represents the ball's initial height.

SKILL ✔ EXERCISE 31

A graph of the height function $h(t)$ confirms that the height of the ball after 2.25 sec is 86 ft. At this time, the derivative of the height function $\frac{d}{dt}h(t)$ (the slope of the tangent to the curve) is 0, as predicted by $v(2.25) = -32(2.25) + 72 = 0$ ft/sec. The area below the graph of the velocity function $v(t) = h'(t) = -32t + 72$ and over $[0, 2.25]$ on the t-axis represents the 81 ft that the ball has risen during the first 2.25 sec of its flight.

Assignments

- **Minimum:** 1, 3, 6–7, 9, 11–13, 15–17, 21, 23–24, 26–27, 31, 35, 37
- **Standard:** 2–12 even; 15; 18–22 even; 25–29 odd; 30, 33(a–c); 35–39 odd; 40, 43, 47, 49–50, 53–54
- **Extended:** 6–7, 9, 12–13, 15, 19, 21, 24, 28–30, 32–34, 36, 38, 40–42, 44–45, 47–48, 50, 52

Assessment

- Quiz 12C covers Sections 12.6–12.7.

Solutions

> **A. Exercises**

10. $\int_1^{10} 2\,dx = [2x]_1^{10}$
$= 2(10) - 2(1) = 18$

11. $\int_0^3 y^2\,dy = \left[\frac{1}{3}y^3\right]_0^3$
$= \frac{1}{3}(3)^3 - \frac{1}{3}(0)^3 = 9$

12. $\int_{-2}^2 x^3\,dx = \left[\frac{1}{4}x^4\right]_{-2}^2$
$= \frac{1}{4}(2)^4 - \frac{1}{4}(-2)^4 = 0$

13. $\int_{-1}^1 e^x\,dx = [e^x]_{-1}^1 = e^1 - e^{-1} \approx 2.350$

14. $\int_0^{\pi} 2\sin\theta\,d\theta = [-2\cos\theta]_0^{\pi}$
$= -2\cos\pi - (-2\cos 0)$
$= -2(-1) + 2(1) = 4$

> **B. Exercises**

16. $F(x) = 4x^2 + C$
$1 = 4(-2)^2 + C$
$1 = 16 + C$
$C = -15$
$F(x) = 4x^2 - 15$

17. $G(x) = x^2 - 3x + C$
$5 = (-4)^2 - 3(-4) + C$
$5 = 16 + 12 + C$
$C = -23$
$G(x) = x^2 - 3x - 23$

18. $H(x) = -\frac{x^3}{3} + 3x^2 + C$
$0 = -\frac{3^3}{3} + 3(3)^2 + C$
$0 = -9 + 27 + C$
$C = -18$
$H(x) = -\frac{x^3}{3} + 3x^2 - 18$

19. $K(x) = -3x^3 - 4x + C$
$-2 = -3(7)^3 - 4(7) + C$
$-2 = -1057 + C$
$C = 1055$
$K(x) = -3x^3 - 4x + 1055$

24. $\int \frac{x^4 - 6x^3 - x^2}{x^2}\,dx = \int(x^2 - 6x - 1)\,dx$
$= \frac{x^3}{3} - 3x^2 - x + C$

25. $\int_{-5}^0 (3x + 4)\,dx = \left[\frac{3}{2}x^2 + 4x\right]_{-5}^0$
$= \frac{3}{2}(0)^2 + 4(0) - \left[\frac{3}{2}(-5)^2 + 4(-5)\right]$
$= -\left(\frac{75}{2} - 20\right) = -17.5$

26. $\int_1^3 (-3x^2 + 30)\,dx = [-x^3 + 30x]_1^3$
$= -3^3 + 30(3) - [-(1)^3 + 30(1)]$
$= 63 - 29 = 34$

27. $\int_0^{10} (4x^3 + 3x^2 + 2x)\,dx$
$= [x^4 + x^3 + x^2]_0^{10}$
$= 10^4 + 10^3 + 10^2 - (0^4 + 0^3 + 0^2)$
$= 10{,}000 + 1000 + 100 = 11{,}100$

28. $\int_1^2 (e^x - 9)\,dx = [e^x - 9x]_1^2$
$= e^2 - 9(2) - (e^1 - 9(1))$
$= e^2 - e - 9 \approx -4.329$

29. $\int_4^9 x^{\frac{1}{2}}\,dx = \left[\frac{1}{\left(\frac{3}{2}\right)}x^{\frac{3}{2}}\right]_4^9 = \left[\frac{2}{3}x^{\frac{3}{2}}\right]_4^9$
$= \frac{2}{3}(9)^{\frac{3}{2}} - \frac{2}{3}(4)^{\frac{3}{2}}$
$= \frac{2}{3}(27 - 8) = \frac{38}{3}$

30. $\int(-32t)\,dt = -16t^2 + C = h(t)$
$h(1) = -16(1)^2 + C$
$4 = -16 + C$
$C = 20 \text{ ft} = h_0$

31. $\int(-32t + 20)\,dt$
$-16t^2 + 20t + C = h(t)$
$h(2) = -16(2)^2 + 20(2) + C$
$14 = -64 + 40 + C$
$C = 38 \text{ ft} = h_0$

32. $\int_1^4 (-9.8t)\,dt = \left[-4.9t^2\right]_1^4$
$= -4.9(4)^2 - (-4.9(1)^2) = -73.5 \text{ m}$

33. $\int_a^b (-32t + 64)\,dt = \left[-16t^2 + 64t\right]_a^b$

33a. $-16(1)^2 + 64(1) - [-16(0)^2 + 64(0)]$
$= 48 - 0 = 48 \text{ ft}$

33b. $-16(2)^2 + 64(2) - [-16(1)^2 + 64(1)]$
$= 64 - 48 = 16 \text{ ft}$

33c. $-16(4)^2 + 64(4) - [-16(2)^2 + 64(2)]$
$= 0 - 64 = -64 \text{ ft}$

33d.

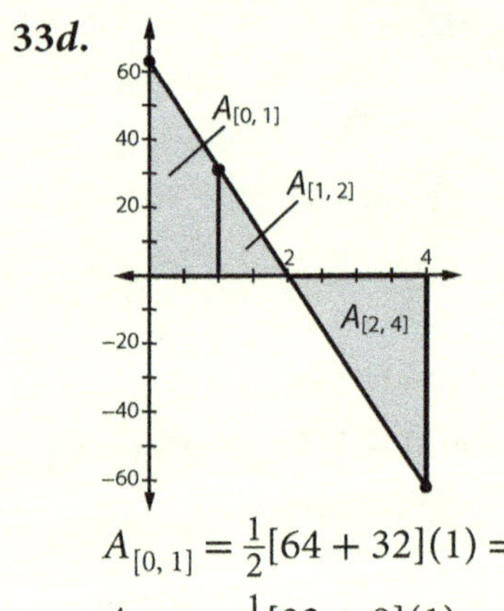

$A_{[0,\,1]} = \frac{1}{2}[64 + 32](1) = 48 \text{ ft}$

$A_{[1,\,2]} = \frac{1}{2}[32 + 0](1) = 16 \text{ ft}$

$A_{[2,\,4]} = \frac{1}{2}[64](2) = 64 \text{ ft}$

33e. Areas above the x-axis (positive) indicate gains in height. The area below (negative) indicates a loss of height. Combined, they indicate a return to the original height 4 sec after the kick.

> ## C. Exercises

34. weight/meter $= \frac{400 \text{ N}}{10 \text{ m}} = 40 \text{ N/m}$
Let x = height lifted.
force: $f(x) = 40x$
$\int_0^5 f(x)\,dx = \int_0^5 40x\,dx$
$= \left[20x^2\right]_0^5 = 20(5)^2 - 20(0)^2$
$= 500 \text{ N·m or } 500 \text{ J}$

Given a function $f(x)$, the derivative function $f'(x)$ represents the slope of the tangent to $f(x)$ at any x-value. Definite integrals can be used to find the area between the graph of a function and the x-axis. These definite integrals are evaluated using indefinite integrals (or antiderivatives). Who would have thought that the area between a curve and the x-axis could be found by "undoing" the slopes of tangent lines!

> ## A. Exercises

Find the simplest antiderivative for each function.

1. $f(x) = 10$ $F(x) = 10x$
2. $r(x) = 8x$ $R(x) = 4x^2$
3. $h(t) = t^4$ $H(t) = \frac{1}{5}t^5$
4. $g(\theta) = 6\sin\theta$ $G(\theta) = -6\cos\theta$

Find each indefinite integral.

5. $\int 3\,dx$ $F(x) = 3x + C$
6. $\int x^2\,dx$ $F(x) = \frac{1}{3}x^3 + C$
7. $\int 7\cos\theta\,d\theta$ $F(\theta) = 7\sin\theta + C$
8. $\int e^x\,dx$ $F(x) = e^x + C$
9. $\int 2x^{19}\,dx$ $F(x) = \frac{1}{10}x^{20} + C$

Evaluate each definite integral.

10. $\int_1^{10} 2\,dx$ 18
11. $\int_0^3 y^2\,dy$ 9
12. $\int_{-2}^2 x^3\,dx$ 0
13. $\int_{-1}^1 e^x\,dx$ ≈ 2.350
14. $\int_0^\pi 2\sin\theta\,d\theta$ 4

15. Classify each statement as *always*, *sometimes*, or *never* true.
  a. The constant of integration is zero. sometimes
  b. The antiderivative of a quadratic function is a linear function. never
  c. The value of a definite integral is positive. sometimes
  d. An indefinite integral is a family of functions. always

> ## B. Exercises

For each function, find the specific antiderivative that contains the given point.

16. $f(x) = 8x;\ (-2, 1)$ $F(x) = 4x^2 - 15$
17. $g(x) = 2x - 3;\ (-4, 5)$ $G(x) = x^2 - 3x - 23$
18. $h(x) = -x^2 + 6x;\ (3, 0)$ $H(x) = -\frac{1}{3}x^3 + 3x^2 - 18$
19. $k(x) = -9x^2 - 4;\ (7, -2)$ $K(x) = -3x^3 - 4x + 1055$

Find each indefinite integral.

20. $\int \frac{1}{2}x\,dx$ $\frac{1}{4}x^2 + C$
21. $\int (2x^2 - x + 1)\,dx$ $\frac{2}{3}x^3 - \frac{1}{2}x^2 + x + C$
22. $\int (e^x + \sin x - 2)\,dx$
23. $\int 7(\cos x + e^x)\,dx$
24. $\int \frac{x^4 - 6x^3 - x^2}{x^2}\,dx$ $\frac{1}{3}x^3 - 3x^2 - x + C$

22. $e^x - \cos x - 2x + C$
23. $7(\sin x + e^x) + C$

Evaluate each definite integral.

25. $\int_{-5}^0 (3x + 4)\,dx$ -17.5
26. $\int_1^3 (-3x^2 + 30)\,dx$ 34
27. $\int_0^{10} (4x^3 + 3x^2 + 2x)\,dx$ 11,100
28. $\int_1^2 (e^x - 9)\,dx$ ≈ -4.329
29. $\int_4^9 \sqrt{x}\,dx$ $\frac{38}{3}$

30. If the velocity (in ft/sec) of a wrench dropped from the roof of a building is modeled with $v(t) = -32t$ and its height after 1 sec is 4 ft, write a function modeling the wrench's height. $h(t) = -16t^2 + 20$

31. If the vertical velocity (in ft/sec) of a ball thrown from the top of a building is modeled by $v(t) = -32t + 20$ and its height after 2 sec is 14 ft, write a function modeling the ball's height.

32. If the velocity (in m/sec) of ball dropped from a drone is modeled by $v(t) = -9.8t$, what was its change in height from 1 to 4 sec? -73.5 m

33. If the vertical velocity (in ft/sec) of a kicked football is modeled by $h(t) = -32t + 64$, use definite integrals to determine the change in height for each time interval.
  a. from 0 to 1 sec 48 ft
  b. from 1 to 2 sec 16 ft
  c. from 2 to 4 sec -64 ft
  d. Draw a velocity time graph and find the area between the curve and the x-axis for each time interval above.
  e. Explain how the areas in step d relate to the changes in height for each time interval and the height of the ball after 4 sec.

> ## C. Exercises

Work is defined as the product of the applied force and the distance traveled in the direction of that force. It is frequently measured in Newton meters (N·m), which are also called joules (J). The force required to lift a portion of a chain is directly proportional to the length of the chain lifted. The work required to lift the chain is represented by the area under the force-distance graph and can be calculated using a definite integral.

34. Determine the work required to raise one end of a 10 m chain weighing 400 N to a height of 5 m. 500 J

31. $h(t) = -16t^2 + 20t + 38$

---

35. weight/meter $= \frac{250 \text{ N}}{5 \text{ m}} = 50 \text{ N/m}$
Let x = height lifted.
force: $f(x) = 50x$
$\int_0^3 f(x)\,dx = \int_0^3 50x\,dx$
$= \left[25x^2\right]_0^3 = 25(3)^2 - 25(0)^2$
$= 225 \text{ N·m or } 225 \text{ J}$

36a. $\int_0^{\frac{\pi}{2}} (\cos x)\,dx = [\sin x]_0^{\frac{\pi}{2}}$
$= \sin\frac{\pi}{2} - \sin 0 = 1 - 0 = 1$
$A_{[0,\,\frac{\pi}{2}]} = 1 \text{ u}^2$

36b. $\int_{\frac{\pi}{2}}^{\pi} (\cos x)\,dx = [\sin x]_{\frac{\pi}{2}}^{\pi}$
$= \sin\pi - \sin\frac{\pi}{2} = 0 - 1 = -1$
$A_{[\frac{\pi}{2},\,\pi]} = -(-1) = 1 \text{ u}^2$

36c. $\int_0^{\pi} (\cos x)\,dx = [\sin x]_0^{\pi}$
$= \sin\pi - \sin 0 = 0 - 0 = 0$
$A_{[0,\,\pi]} = A_{[0,\,\frac{\pi}{2}]} + A_{[\frac{\pi}{2},\,\pi]}$
$= 1 + 1 = 2 \text{ u}^2$

37a. $\int_{-2}^0 x^3\,dx = \left[\frac{x^4}{4}\right]_{-2}^0$
$= \frac{0^4}{4} - \frac{(-2)^4}{4} = -4$
$A_{[-2,\,0]} = -(-4) = 4 \text{ u}^2$

35. Determine the work required to raise one end of a 5 m chain weighing 250 N to a height of 3 m. 225 J

36. Evaluate each definite integral and find its associated area.
   a. $\int_0^{\frac{\pi}{2}}(\cos x)\,dx$ 1; 1 u²
   b. $\int_{\frac{\pi}{2}}^{\pi}(\cos x)\,dx$ −1; 1 u²
   c. $\int_0^{\pi}(\cos x)\,dx$ 0; 2 u²

37. Evaluate each definite integral and find its associated area.
   a. $\int_{-2}^{0}x^3\,dx$ −4; 4 u²
   b. $\int_0^{2}x^3\,dx$ 4; 4 u²
   c. $\int_{-2}^{2}x^3\,dx$ 0; 8 u²

38. Use symmetry to evaluate $\int_{-7}^{7}|x|\,dx$. 49

39. Use two definite integrals to evaluate $\int_3^5[x]\,dx$. 7

Use the Venn diagram illustrating the relationship of integrable (I), continuous (C), and differentiable (D) functions.

40. Classify each statement as true or false.
   a. All differentiable functions are continuous. true
   b. All integrable functions are continuous. false
   c. All continuous functions are integrable. true
   d. All continuous functions are differentiable. false

41. Name a type of function contained in each set but not its subset(s).
   a. C (continuous)
   b. I (integrable)
   c. F (functions)

42. Use technology to graph $f(x) = -\frac{1}{2}x^2 - x + \frac{15}{2}$. Then use the calculator's definite integral function, found on the [CALC] ([2nd], [TRACE]) menu, to find the area between the function's graph and the x-axis over the interval [1, 5]. 16 u²

43. Use technology to graph $f(x) = \frac{1}{2}x^2$ and $g(x) = x + 4$. Then use the calculator's definite integral function, found on the [CALC] ([2nd], [TRACE]) menu, to find the area between the graphs of the functions over the interval [−2, 4]. 18 u²

44. Find the simplest antiderivative for each formula and identify the resulting formula.
   a. circumference of a circle: $c(r) = 2\pi r$
   b. surface area of a sphere: $S(r) = 4\pi r^2$

45. Use technology to graph $f(x) = x^{-2}$.
   a. Find $\int x^{-2}\,dx$. $-\frac{1}{x} + C$
   b. Use the Fundamental Theorem of Calculus to evaluate $\int_{-1}^{1} x^{-2}\,dx$. −2
   c. Explain why the definite integral found in step b does not represent the area under the curve over [−1, 1] on the x-axis.

CUMULATIVE REVIEW

Write the slope-intercept form equation of each line. [1.2]

46. a line passing through (−1, 2) and (−9, −4)

47. a line passing through (0, 7) and perpendicular to the line $y = \frac{1}{3}x + 2$ $y = -3x + 7$

Find $a_0, a_1, a_2, a_3,$ and a_4 for each sequence. [9.1]

48. $a_n = \frac{4n!}{3n+1}$

49. $a_n = \frac{(-2)^n}{3(n+1)!}$

Find the slope of the tangent to the graph of the function at the given value of x. [12.4]

50. $f(x) = 5x^2$ at $x = 3$ $f'(3) = 30$

51. $f(x) = 2x^2 + 4x + 12$ at $x = -2$ $f'(-2) = -4$

52. Identify the conic $49x^2 + 16y^2 - 294x + 128y - 87 = 0$. [8.4] B
   A. parabola
   B. ellipse
   C. circle
   D. hyperbola
   E. cannot be determined

53. How many different committees of 7 people can be made from a group of 22 people? [10.1] A
   A. 170,544
   B. 859,541,760
   C. 2,494,357,888
   D. 2.23×10^{17}
   E. none of these

54. Find the derivative of $h(x) = (2x^2 - 4x + 5)(3x^2 - 2x + 11)$. [12.5] B
   A. $6x^4 - 16x^3 + 45x^2 - 54x + 55$
   B. $24x^3 - 48x^2 + 90x - 54$
   C. $24x^2 - 32x + 8$
   D. $6x^4 + 8x^2 + 55$
   E. none of these

55. Evaluate the definite integral $\int_2^5 2x\,dx$ using a geometric formula. [12.6] B
   A. 15
   B. 21
   C. 25
   D. 42
   E. none of these

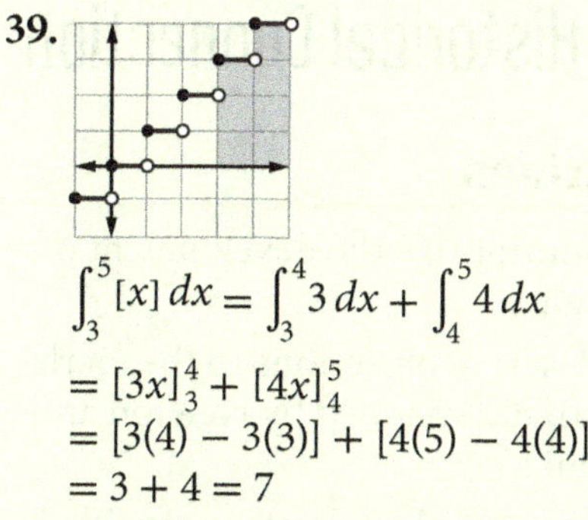

$$\int_3^5 [x]\,dx = \int_3^4 3\,dx + \int_4^5 4\,dx$$
$$= [3x]_3^4 + [4x]_4^5$$
$$= [3(4) - 3(3)] + [4(5) - 4(4)]$$
$$= 3 + 4 = 7$$

41a. absolute value functions that have a sharp point (not differentiable)

41b. piecewise functions, such as the greatest integer function, that have a vertical jump or hole (not continuous)

41c. rational functions that have a vertical asymptote (not integrable)

42. 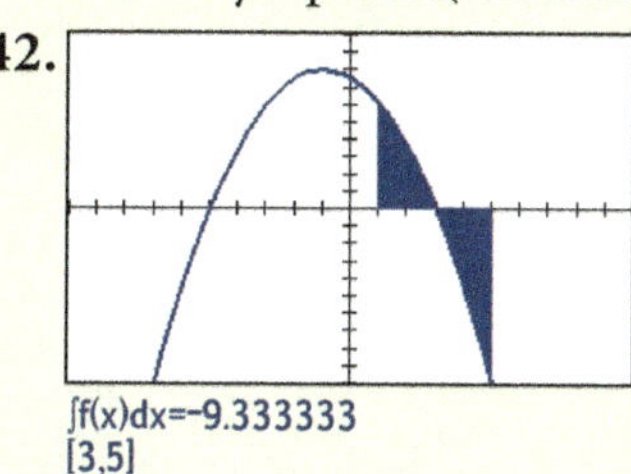
$\int f(x)\,dx = -9.333333$
[3,5]

$$A_{[1,\,5]} = A_{[1,\,3]} + A_{[3,\,5]}$$
$$= 6\frac{2}{3} + \left[-\left(-9\frac{1}{3}\right)\right] = 16 \text{ u}^2$$

43. 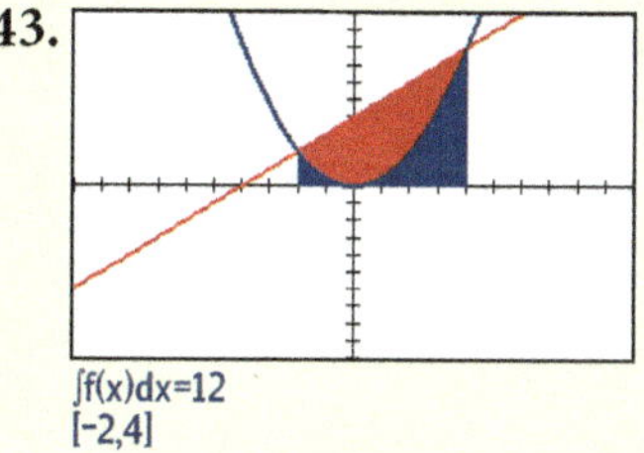
$\int f(x)\,dx = 12$
[-2,4]

$$A = \int_{-2}^{4}(x+4)\,dx - \int_{-2}^{4}\frac{1}{2}x^2\,dx$$
$$= 30 - 12 = 18 \text{ u}^2$$

44a. $\int c(r)\,dr = \int 2\pi r\,dr = \pi r^2$
area of a circle

44b. $\int S(r)\,dr = \int 4\pi r^2\,dr = \frac{4}{3}\pi r^3$
volume of a sphere

37b. 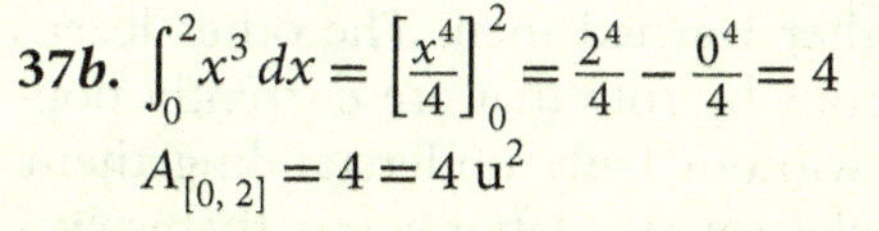 $\int_0^2 x^3\,dx = \left[\frac{x^4}{4}\right]_0^2 = \frac{2^4}{4} - \frac{0^4}{4} = 4$
$$A_{[0,\,2]} = 4 = 4 \text{ u}^2$$

37c.  $\int_{-2}^{2} x^3\,dx = \left[\frac{x^4}{4}\right]_{-2}^{2} = \frac{2^4}{4} - \frac{(-2)^4}{4} = 0$
$$A_{[-2,\,2]} = A_{[-2,\,0]} + A_{[0,\,2]}$$
$$= 4 + 4 = 8 \text{ u}^2$$

38. 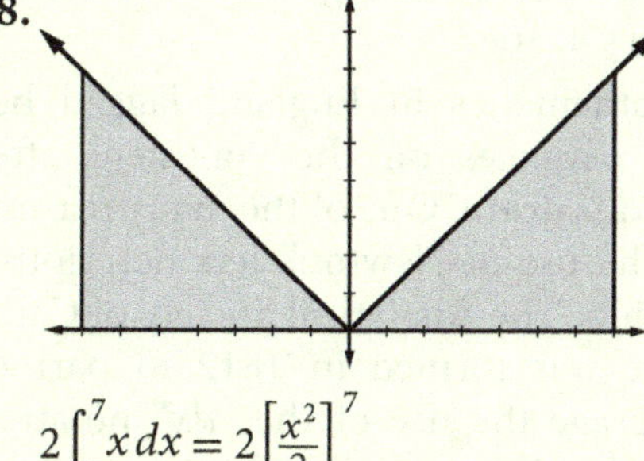
$$2\int_0^7 x\,dx = 2\left[\frac{x^2}{2}\right]_0^7$$
$$= 2\left(\frac{7^2}{2} - \frac{0^2}{2}\right) = 2\left(\frac{49}{2}\right) = 49$$

continued in Answers and Solutions Overflow

Historical Connection

Objectives

1. To summarize the development of calculus
2. To describe objections to the methods first developed by Newton and Leibniz
3. To evaluate the test of reasonableness from a biblical worldview as it relates to the Christian faith

Assignments

- **Minimum:** 1–3
- **Standard:** 1–3; 4 (in class)
- **Extended:** 1–3; 4 (in class)

THE ANALYST

Finding the area under a curve and finding a tangent to a point on a curve are both problems that have been studied and solved in various ways even before Euclid's contributions in these areas during the fourth and third centuries BC. But precise algebraic solutions and an understanding of how the problems relate to each other (what is now called the Fundamental Theorem of Calculus) are due primarily to the efforts of Isaac Newton (1643–1727) and Gottfried Leibniz (1646–1716). Most historians agree that the development of calculus by Newton in England and Leibniz in Germany is an example of near simultaneous independent discovery. Newton is usually credited with developing his method of fluxions before Leibniz developed the same ideas with infinitesimals, but Leibniz published first and his notation using $\frac{dy}{dx}$ was more widely used than Newton's $\dot{y}$. The bitter priority dispute that raged between Newton and Leibniz became an issue of national pride and effectively isolated the mathematicians in England from advancements being made on the European continent.

Despite the obvious power of the new method of "analysis" to solve complex problems, not everyone was convinced that the methods rested on a firm foundation. One of the most ardent critics was George Berkeley, an Irish bishop in the Anglican church and a prominent philosopher.

In his work *The Analyst; Or, A Discourse Addressed to an Infidel Mathematician* (1734), Berkeley argues that men who disdain religion because it calls for faith in things not fully understood were nevertheless willing to embrace mathematical procedures on faith alone. Using diagrams and displaying in-depth

1. Isaac Newton and Gottfreid Leibniz; finding the tangent to a curve at a point, finding the area beneath a curve

understanding of mathematical methods, he points out the absurdity of both Newton's fluxions and Leibniz's infinitesimals. He takes mathematicians to task for wholeheartedly embracing concepts they cannot explain to reasonable and intelligent men.

Cambridge scientist James Jurin, who wrote under the pen name Philalethes Cantabrigiensis ("Lover of truth at Cambridge University"), responded by publishing *Geometry No Friend to Infidelity: Or, A Defence of Sir Isaac Newton and the British Mathematicians, in a Letter to the Author of the Analyst* (1734). Berkeley countered Jurin's emotional and often invective critique with *A Defence of Free-Thinking in Mathematics* (1735), wherein he maintains that Jurin answered none of his objections and that any thinking man could recognize the fallacies inherent in fluxions and infinitesimals. The tone of Berkeley's rebuttal is evident from the first sentence: "WHEN I read your Defence of the British Mathematicians, I could not, Sir, but admire your Courage in asserting with such undoubting Assurance things so easily disproved."

Another 100 yr passed before the development of the concept of limits allowed mathematicians, principally Augustin-Louis Cauchy, Karl Weierstrass, and Riemann, to place calculus on a firm foundation. Others, particularly Jean le Rond d'Alembert, had espoused limits as a better approach, but they were unable to develop a precise definition that was both rigorous and useful. A modern calculus course typically reverses the order of development by beginning with limits, using limits to define derivatives and integrals, and then using derivatives and integrals to solve problems similar to the ones that led Newton and Leibniz to develop analysis in the first place.

2. George Berkeley; Mathematicians who claim to accept only what can be proved were accepting by faith methods that could not be proved while showing disdain toward religion because it requires faith.

COMPREHENSION CHECK

1. Which two persons are generally credited with developing calculus? What two basic problems does calculus address?

2. Name the author of *The Analyst* and state in your own words his primary objection to analysis.

3. Which mathematical concept was used to finally place calculus on a firm foundation? limits

4. The Fundamental Theorem of Calculus displays harmony, unity, and beauty. Since these are attributes of God, it is not surprising to find them in math that models His creation. Since we are made in the image of God, we can recognize and apply these principles as we seek to fulfill the Genesis mandate to exercise dominion over God's creation.

4. Discuss: The Fundamental Theorem of Calculus unites the mathematical ideas of tangents to a curve and the area under a curve in an unexpected way. How do the harmony of these inverse operations, their effectiveness in modeling real-world problems, and our ability to conceptualize them all reflect the glory of God and His attributes?

PRESENTATION

The three main manuscripts referenced in this feature, George Berkeley's *The Analyst* and *A Defence of Free-Thinking in Mathematics* and James Jurin's *Geometry No Friend to Infidelity*, are easily accessible on the Internet. The most quoted portion of *The Analyst* is probably Berkeley's description of fluxions as "the Ghosts of departed Quantities" at the end of paragraph XXXV.

According to the Julian calendar used by the British Empire at the time of his birth, Isaac Newton was born on December 25, 1642. When the British adopted the Gregorian calendar in 1752, eleven days were lost, making Newton's birthday January 4, 1643.

Mathematics in England lagged behind advances on the continent after Newton's death. One of the many reasons was the use of Newton's less descriptive notation. The Analytical Society of Cambridge was formed in 1812 in part to encourage the use of the "*dx*" notation instead of Newton's dots. In the words of its founder, Charles Babbage, the Society sought to encourage "*d*-ism" and oppose the "*dot*-age" of the university.

In Berkeley's response to Jurin he contrasts two types of learned men. One seeks truth and welcomes examination by other learned men. The other learns methods by rote that are currently popular without truly understanding them. Berkeley says the latter betray themselves by anger and surprise when their methods are questioned, clearly a reference to Jurin's emotional reply to *The Analyst*.

Chapter 12 Review

1. Use technology to graph $f(x) = \frac{1}{x+6}$ and find each limit. [12.1]

 a. $\lim_{x \to \infty} f(x)$ 0 **b.** $\lim_{x \to -5^-} f(x)$ 1 **c.** $\lim_{x \to -6} f(x)$ *does not exist*

2. Find each limit, if it exists. If it does not exist, explain why. [12.1]

 a. $\lim_{x \to 1} \frac{2}{x^2 - 9}$ $-\frac{1}{4}$ **b.** $\lim_{x \to -2} |x - 2|$ 4 **c.** $\lim_{x \to 4} \sqrt{x - 5}$ *undefined*

3. Use limits to describe the end behavior of each polynomial function. [12.1]

 a. $h(x) = x^3 + 2x^2 - 8$ **b.** $g(x) = 5x^2 + 35$

State the domain of each function. Then evaluate the limit of the function at the given value of x. [12.1]

4. $k(x) = [x + 9]; x = -\frac{1}{7}$ $D = \mathbb{R}; 8$

5. $g(x) = 2x^3 + 2x^2 + 8; x = -3$ $D = \mathbb{R}; -28$

6. Evaluate $\lim_{h \to 0} \frac{(7 + h)^2 - 7^2}{h}$. [12.1] 14

7. Find the limit of each sequence. [12.1]

 a. $a_n = \frac{4n + 3}{n - 2}$ 4 **b.** $b_n = \frac{3n}{5n^2 + 1}$ 0

8. A new diesel bus is purchased for \$495,000 with an estimated depreciation of 12.5% per year. [12.1]

 a. Write a function $f(x)$ modeling the value of the bus after x years. $f(x) = 495{,}000(0.875)^x$

 b. Find $\lim_{x \to \infty} f(x)$. 0

 c. What would be the value of the bus at the end of 7 yr? \$194,384

 d. If the owners want to sell the bus for at least \$100,000, what is the last year they could sell it? *twelfth year*

Evaluate each limit. [12.2]

9. $\lim_{x \to 4} 5x^3$ 320

10. $\lim_{x \to -2} \frac{x^3 - 5x}{x + 3}$ 2

11. $\lim_{\theta \to \frac{\pi}{2}} \csc \theta$ 1

12. $\lim_{h \to 3} \frac{x - 3}{x^2 - 2x - 3}$ $\frac{1}{4}$

13. $\lim_{x \to 0} \frac{\sqrt{x + 3} - \sqrt{3}}{x}$ $\frac{\sqrt{3}}{6}$

14. $\lim_{x \to -\infty} \frac{8x^4 - 3x + 12}{2x^4 + 2x^2 + 3}$ 4

15. $\lim_{x \to \infty} \frac{-9x^2 + 2x + 5}{2x^4 - 5x^3 - 9x}$ 0

16. Using $f(x) = 3x^2 + 14$, find the following. [12.3]

 a. $\lim_{h \to 0} \frac{f(x + h) - f(x)}{h}$ 6x **b.** $f'(2)$ $f'(2) = 12$

17. Using the limit definition of a derivative, show that the derivative of the direct variation $f(x) = kx$ is k, the constant of variation. [12.3]

18. Use the limit of the difference quotient to find the derivative function of $f(x) = \sqrt{5x}$. [12.3]

19. The height of an object thrown into the air from a height of h_0 meters with an initial vertical velocity of v_0 is modeled by the function $d(t) = -4.9t^2 + v_0 t + h_0$. Find the average velocity during the first 2 sec for an object thrown from a height of 12 m with an initial vertical velocity of 30 m/sec. [12.3] 20.2 m/sec

20. Find the instantaneous velocity (in m/sec) at 1.5 sec if the height of an object (in meters) is modeled by $d(t) = -4.9t^2 + 425$. [12.3] −14.7 m/sec

21. Find $\frac{dg}{dx}$ if $g(x) = 10x^{23} - 4x^{17} - 3x^{11} + 2x^5 - x$. [12.4]

22. Find the derivative of $f(x) = \sqrt[3]{x}$. [12.4]

23. Find the equation of the tangent to the graph of $f(x) = x^2 + 3$ at $x = -3$. [12.4] $y = -6x - 6$

24. Use the derivative of $P(x) = 2x^3 - 27x^2 + 120x - 150$ to identify its relative extrema. [12.4]

25. An object is launched from ground level with an initial upward velocity of 176.4 m/sec. [12.4]

 a. State the function modeling the height of the object and determine how long it takes to hit the ground. $h(t) = -4.9t^2 + 176.4t$; 36 sec

 b. State the function modeling the vertical velocity of the object and find its instantaneous velocity at 4 sec and at 20 sec.

 c. Give the maximum height of the object. 1587.6 m

Use the product rule or quotient rule to find each derivative. [12.5]

26. $f(x) = (3x - 4)(2x + 4)$

27. $h(x) = \frac{3x - 4}{2x + 4}$

28. $g(x) = \frac{2x^2 - 4}{3x^2 + 7}$

29. $y = 2x \cos x$

30. $f(x) = \frac{\sin x}{3x^2 - 4}$

3a. $\lim_{x \to \infty} h(x) = \infty; \lim_{x \to -\infty} h(x) = -\infty$

3b. $\lim_{x \to \infty} g(x) = \infty; \lim_{x \to -\infty} g(x) = \infty$

18. $f'(x) = \frac{5}{2\sqrt{5x}}$

21. $g'(x) = 230x^{22} - 68x^{16} - 33x^{10} + 10x^4 - 1$

22. $f'(x) = \frac{1}{3}x^{-\frac{2}{3}}$ or $\frac{1}{3\sqrt[3]{x^2}}$

24. max: (4, 26); min: (5, 25)

25b. $h'(t) = v(t) = -9.8t + 176.4$
$h'(4) = 137.2$ m/sec
$h'(20) = -19.6$ m/sec

26. $f'(x) = 12x + 4$

27. $h'(x) = \frac{20}{(2x + 4)^2}$

PRESENTATION

Be sure to communicate which theorems, properties, and rules the students will be able to access and which ones they will need to know for the chapter test. Consider providing a quick reference sheet.

TIPS

Ex. 46 Direct students to the explanatory paragraph that precedes Section 12.7 exercises 33–34 if they have not completed those exercises.

Chapter 12 Review

Objective

To prepare for evaluation

Vocabulary

See Appendix A.

Assignments

- **Minimum:** 1, 3a, 5–6, 7a, 8(a, c), 10, 13–14, 16, 20–23, 26–27, 29, 34, 37–38, 41, 42a, 44a, 45–46

- **Standard:** 1–47 odd; 2–46 even (in class)

- **Extended:** 1c, 3a; 4–8 even; 11–13, 17–19, 24–25, 29–30, 32, 34–36, 40, 42b, 45–47

Assessments

- Chapter 12 Test
- Fourth Quarter Exam covers Chapters 10–12.

Solutions

❯ A. Exercises

1.

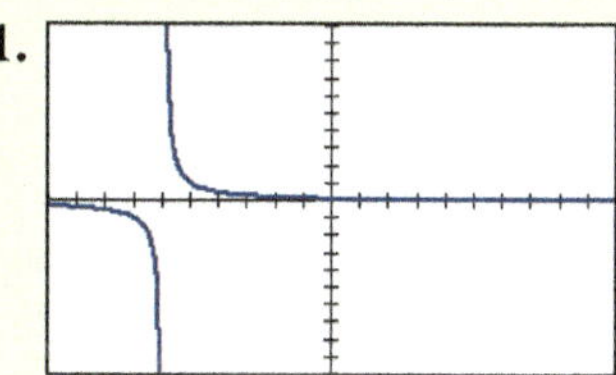

2a. continuous rational function except at $x = \pm 3$; $\frac{2}{1^2 - 9} = \frac{2}{-8} = -\frac{1}{4}$

2b. continuous absolute value function $|-2 - 2| = |-4| = 4$

2c. 4 is not in the domain of the function.

3a. The end behavior matches that of an odd power function.

3b. The end behavior matches that of an even power function.

4. continuous at all $x \notin \mathbb{Z}$ using direct substitution:
$$\left[-\frac{1}{7} + 9\right] = \left[8\frac{6}{7}\right] = 8$$

5. using direct substitution:
$$2(-3)^3 + 2(-3)^2 + 8 = -28$$

6. $\lim_{h \to 0} \frac{7^2 + 14h + h^2 - 7^2}{h} = \lim_{h \to 0} \frac{h(14 + h)}{h}$
$= \lim_{h \to 0} (14 + h) = 14$

7a. $a_{1000} = \dfrac{4(1000) + 3}{(1000) - 2} = \dfrac{4003}{998} \approx 4$

$a_{1,000,000} = \dfrac{4(1,000,000) + 3}{(1,000,000) - 2}$

$= \dfrac{4,000,003}{999,998} \approx 4$

$\displaystyle\lim_{n \to \infty} \dfrac{4n + 3}{n - 2} = 4$

7b. $b_{1000} = \dfrac{3(1000)}{5(1000)^2 + 1} = \dfrac{3000}{5,000,001} \approx 0$

$b_{1,000,000} = \dfrac{3(1,000,000)}{5(1,000,000)^2 + 1}$

$= \dfrac{3,000,000}{5,000,000,000,001} \approx 0$

8a. $f(x) = 495{,}000(0.875)^x$

8c. 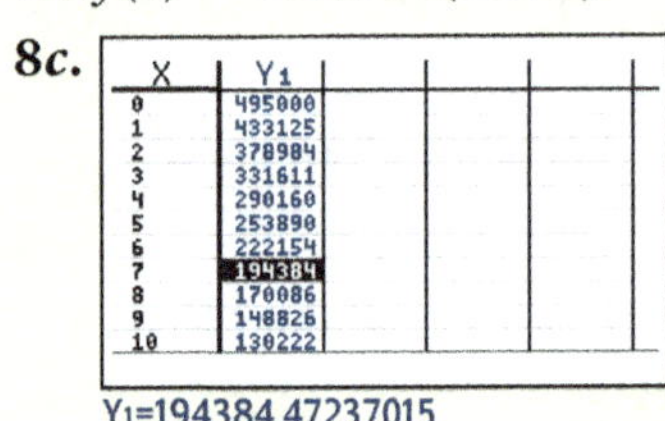

$Y_1 = 194384.47237015$

8d. 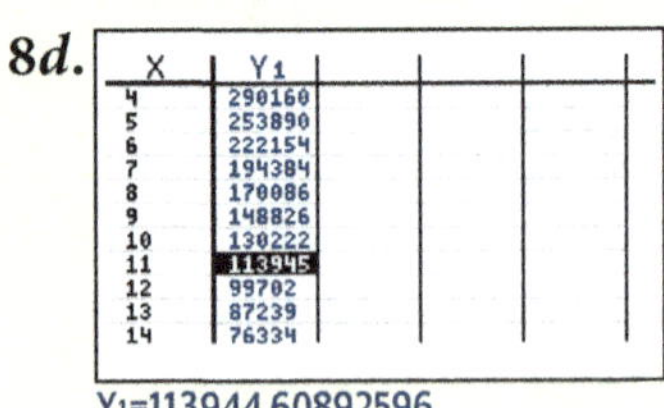

$Y_1 = 113944.60892596$

9. using direct substitution:

$5(4)^3 = 5(64) = 320$

10. using direct substitution:

$\dfrac{(-2)^3 - 5(-2)}{-2 + 3} = \dfrac{(-8 + 10)}{1} = 2$

11. $\csc \dfrac{\pi}{2} = \dfrac{1}{\sin \frac{\pi}{2}} = \dfrac{1}{1} = 1$

12. $\displaystyle\lim_{h \to 3} \dfrac{x - 3}{(x - 3)(x + 1)} = \lim_{h \to 3} \dfrac{1}{x + 1}$

$= \dfrac{1}{3 + 1} = \dfrac{1}{4}$

13. $\displaystyle\lim_{x \to 0} \dfrac{\sqrt{x + 3} - \sqrt{3}}{x}$

$= \displaystyle\lim_{x \to 0} \dfrac{(\sqrt{x + 3} - \sqrt{3})(\sqrt{x + 3} + 3)}{x(\sqrt{x + 3} + \sqrt{3})}$

$= \displaystyle\lim_{x \to 0} \dfrac{x + 3 - 3}{x(\sqrt{x + 3} + \sqrt{3})}$

$= \displaystyle\lim_{x \to 0} \dfrac{x}{x(\sqrt{x + 3} + \sqrt{3})}$

$= \displaystyle\lim_{x \to 0} \dfrac{1}{(\sqrt{x + 3} + \sqrt{3})} = \dfrac{1}{2\sqrt{3}} = \dfrac{\sqrt{3}}{6}$

14. $\displaystyle\lim_{x \to -\infty} \dfrac{\frac{8x^4}{x^4} - \frac{3x}{x^4} + \frac{12}{x^4}}{\frac{2x^4}{x^4} + \frac{2x^2}{x^4} + \frac{3}{x^4}} = \dfrac{\lim\limits_{x \to -\infty} \left(8 - \frac{3}{x^3} + \frac{12}{x^4}\right)}{\lim\limits_{x \to -\infty} \left(2 + \frac{2}{x^2} + \frac{3}{x^4}\right)}$

$= \dfrac{\lim\limits_{x \to -\infty} 8 - 3 \lim\limits_{x \to -\infty} \frac{1}{x^3} + 12 \lim\limits_{x \to -\infty} \frac{1}{x^4}}{\lim\limits_{x \to -\infty} 2 + 2 \lim\limits_{x \to -\infty} \frac{1}{x^2} + 3 \lim\limits_{x \to -\infty} \frac{1}{x^4}}$

$= \dfrac{8 - 3(0) + 12(0)}{2 + 2(0) + 3(0)} = \dfrac{8}{2} = 4$

Use the chain rule to find each derivative. [12.5]

31. $f(x) = (3x - 2x^2)^3$ **32.** $h(x) = \sqrt{x^2 + 4}$

Find each derivative. [12.5]

33. $\dfrac{d}{dx}(10e^x + e^2x^2)$ **34.** $\dfrac{d}{dx} \tan x$

35. A pathologist models the number of diseased cells in a culture after t days with $f(t) = 45e^{0.62t}$. Determine the instantaneous rate of growth after 4 days. When will the instantaneous rate of growth be 50,000 cells per day? [12.5] ≈ 333.2 cells/day; ≈ 12.08 days

36. State the value of Δx used to estimate the area under the graph of $f(x)$ over the interval $[-2, 4]$ on the x-axis using 12 rectangles. Then list the right endpoint of each partition, x_i. [12.6]

37. Write a definite integral representing the area of the shaded region. Then evaluate the integral using geometric formulas. [12.6]

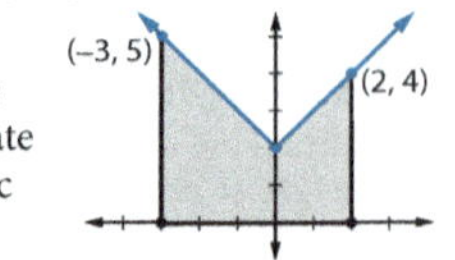

38. Approximate the area under the graph of $f(x) = 4x^2$ over $[1, 3]$ on the axis using 4 rectangles. Use the right endpoint of each partition to determine the height of each rectangle. [12.6] 43 u^2

39. Estimate the area (to the nearest hundredth) under $f(x) = 5x^2$ over $[0, 2]$ on the x-axis using 1000 rectangles. [12.6] 13.35 u^2

40. Use the limit definition to evaluate $\int_0^4 2x^3 \, dx$. [12.6] 128

---

28. $g'(x) = \dfrac{52x}{(3x^2 + 7)^2}$

29. $y' = 2 \cos x - 2x \sin x$

30. $f'(x) = \dfrac{(3x^2 - 4) \cos x - 6x \sin x}{(3x^2 - 4)^2}$

31. $f'(x) = (9 - 12x)(3x - 2x^2)^2$

32. $h'(x) = \dfrac{x}{\sqrt{x^2 + 4}}$

33. $10e^x + 2e^2x$

34. $\sec^2 x$

36. $0.5; x_i = -1.5, -1, -0.5, 0, 0.5, 1, 1.5, 2, 2.5, 3, 3.5, 4$

37. $\int_{-3}^{2} |x| + 2 \, dx; 16.5 \text{ u}^2$

---

41. Find the simplest antiderivative for each function. [12.7]

  a. $f(x) = 8$ $F(x) = 8x$ **b.** $h(t) = t^3$ $H(t) = \frac{t^4}{4}$

  c. $g(x) = 3e^x$ $G(x) = 3e^x$ **d.** $f(x) = 2x + 3 \cos x$ $F(x) = x^2 + 3 \sin x$

42. Find the specific antiderivative through the given point. [12.7]

  a. $g(x) = 4x + 7; (-2, 5)$ $G(x) = 2x^2 + 7x + 11$

  b. $p(x) = -6x^2 - 7; (3, -1)$ $P(x) = -2x^3 - 7x + 74$

43. Find each indefinite integral. [12.7]

  a. $\int 3y \, dy$ $\frac{3}{2}y^2 + C$ **b.** $\int 6 \sin \theta \, d\theta$ $-6 \cos \theta + C$

44. Evaluate each definite integral. [12.7]

  a. $\int_1^{12} 4 \, dx$ 44 **b.** $\int_{-3}^{3} y^2 \, dy$ 18

45. If the velocity (in ft/sec) of a wrench dropped from the roof of a building is modeled by $v(t) = -32t$ and its height after 2 sec is 18 ft, write a function modeling the wrench's height. [12.7] $h(t) = -16t^2 + 82$

46. Determine the work required to raise an end of a 10 m chain weighing 550 N to a height of 4 m. [12.7] 440 J

47. Use technology to graph $f(x) = x^3 - 4x^2 + 3x$. Then use the calculator's definite integral function, found on the [CALC] ([2nd], [TRACE]) menu, to find the area (to the nearest thousandth) between the function's graph and the x-axis over the interval $[0, 3]$. [12.7] 3.083 u^2

---

15. $\displaystyle\lim_{x \to \infty} \dfrac{\frac{-9x^2}{x^4} + \frac{2x}{x^4} + \frac{5}{x^4}}{\frac{2x^4}{x^4} - \frac{5x^3}{x^4} - \frac{9x}{x^4}}$

$= \dfrac{\lim\limits_{x \to \infty} \left(-\frac{9}{x^2} + \frac{2}{x^3} + \frac{5}{x^4}\right)}{\lim\limits_{x \to \infty} \left(2 - \frac{5}{x} - \frac{9}{x^3}\right)}$

$= \dfrac{-9 \lim\limits_{x \to \infty} \frac{1}{x^2} + 2 \lim\limits_{x \to \infty} \frac{1}{x^3} + 5 \lim\limits_{x \to \infty} \frac{1}{x^4}}{\lim\limits_{x \to \infty} 2 - 5 \lim\limits_{x \to \infty} \frac{1}{x} - 9 \lim\limits_{x \to \infty} \frac{1}{x^3}}$

$= \dfrac{-9(0) + 2(0) + 5(0)}{2 - 5(0) - 9(0)} = \dfrac{0}{2} = 0$

16a. $\displaystyle\lim_{h \to 0} \dfrac{3(x + h)^2 + 14 - (3x^2 + 14)}{h}$

$= \displaystyle\lim_{h \to 0} \dfrac{3(x^2 + 2xh + h^2) + 14 - 3x^2 - 14}{h}$

$= \displaystyle\lim_{h \to 0} \dfrac{3x^2 + 6xh + 3h^2 + 14 - 3x^2 - 14}{h}$

$= \displaystyle\lim_{h \to 0} \dfrac{6xh + 3h^2}{h} = \lim_{h \to 0} \dfrac{h(6x + 3h)}{h}$

$= \displaystyle\lim_{h \to 0} (6x + 3h) = 6x$

16b. $f'(2) = 6(2) = 12$

17. $f'(x) = \displaystyle\lim_{h \to 0} \dfrac{k(x + h) - kx}{h}$

$= \displaystyle\lim_{h \to 0} \dfrac{kx + kh - kx}{h} = \lim_{h \to 0} \dfrac{kh}{h} = k$

18. $f'(x) = \lim\limits_{h \to 0} \dfrac{\sqrt{5(x+h)} - \sqrt{5x}}{h}$

$= \lim\limits_{h \to 0} \dfrac{\sqrt{5(x+h)} - \sqrt{5x}}{h} \cdot \dfrac{\sqrt{5(x+h)} + \sqrt{5x}}{\sqrt{5(x+h)} + \sqrt{5x}}$

$= \lim\limits_{h \to 0} \dfrac{5x + 5h - 5x}{h\left(\sqrt{5(x+h)} + \sqrt{5x}\right)}$

$= \lim\limits_{h \to 0} \dfrac{5h}{h\left(\sqrt{5(x+h)} + \sqrt{5x}\right)}$

$= \lim\limits_{h \to 0} \dfrac{5}{\sqrt{5(x+h)} + \sqrt{5x}}$

$= \dfrac{5}{\sqrt{5x} + \sqrt{5x}} = \dfrac{5}{2\sqrt{5x}}$

19. $d(t) = -4.9t^2 + 30t + 12$

$d(0) = -4.9(0)^2 + 30(0) + 12 = 12$

$d(2) = -4.9(2)^2 + 30(2) + 12 = 52.4$

$\dfrac{\Delta d}{\Delta t} = \dfrac{d(2) - d(0)}{2 - 0} = \dfrac{52.4 - 12}{2}$

$= 20.2 \text{ m/sec}$

20. $v(t) = d'(t)$

$= \lim\limits_{h \to 0} \dfrac{-4.9(t+h)^2 + 425 - (-4.9t^2 + 425)}{h}$

$= \lim\limits_{h \to 0} \dfrac{-4.9(t^2 + 2ht + h^2) + 425 + 4.9t^2 - 425}{h}$

$= \lim\limits_{h \to 0} \dfrac{-4.9t^2 - 9.8ht - 4.9h^2 + 4.9t^2}{h}$

$= \lim\limits_{h \to 0} \dfrac{h(-9.8t - 4.9h)}{h}$

$= \lim\limits_{h \to 0} (-9.8t - 4.9h) = -9.8t$

$v(t) = -9.8t$

$v(1.5) = -9.8(1.5) = -14.7 \text{ m/sec}$

21. $g'(x) = 23(10)x^{22} - 17(4)x^{16}$
$\qquad\qquad - 11(3)x^{10} + 5(2)x^4 - 1$

$= 230x^{22} - 68x^{16} - 33x^{10}$
$\qquad\qquad\qquad\qquad + 10x^4 - 1$

22. $f(x) = x^{\frac{1}{3}}; f'(x) = \frac{1}{3}x^{-\frac{2}{3}}$ or $\dfrac{1}{3\sqrt[3]{x^2}}$

23. $f'(x) = 2x$

$m = f'(-3) = 2(-3) = -6$

$f(-3) = (-3)^2 + 3 = 12$

through $(-3, 12)$ with $m = -6$

$12 = -6(-3) + b$

$b = -6$

$y = -6x - 6$

24. $P'(x) = 6x^2 - 54x + 120 = 0$

$x^2 - 9x + 20 = 0$

$(x - 4)(x - 5) = 0$

$x = 4, 5$

$P(4) = 2(4)^3 - 27(4)^2 + 120(4) - 150$

$= 26$

$P(5) = 2(5)^3 - 27(5)^2 + 120(5) - 150$

$= 25$

max: $(4, 26)$; min: $(5, 25)$

25a. $h(t) = \frac{1}{2}(-9.8)t^2 + 176.4t + 0$

$h(t) = -4.9t^2 + 176.4t = 0$

$t(-4.9t + 176.4) = 0$

$t = 0, 36 \text{ sec}$

25b. $v(t) = h'(t) = 2(-4.9)t + 176.4$

$= -9.8t + 176.4$

$v'(4) = 137.2 \text{ m/sec}$

$v'(20) = -19.6 \text{ m/sec}$

25c. $h'(t) = -9.8t + 176.4 = 0$

$176.4 = 9.8t$

$t = 18 \text{ sec}$

$h(18) = -4.9(18)^2 + 176.4(18)$

$= 1587.6 \text{ m}$

26. $u(x) = 3x - 4; v(x) = 2x + 4$

$u'(x) = 3; v'(x) = 2$

$f'(x) = 3(2x + 4) + (3x - 4)2$

$= 12x + 4$

27. $u(x) = 3x - 4; v(x) = 2x + 4$

$u'(x) = 3; v'(x) = 2$

$h'(x) = \dfrac{3(2x+4) - (3x-4)2}{(2x+4)^2} = \dfrac{20}{(2x+4)^2}$

28. $u(x) = 2x^2 - 4; v(x) = 3x^2 + 7$

$u'(x) = 4x; v'(x) = 6x$

$g'(x) = \dfrac{4x(3x^2 + 7) - 6x(2x^2 - 4)}{(3x^2 + 7)^2}$

$= \dfrac{52x}{(3x^2 + 7)^2}$

29. $u(x) = 2x; v(x) = \cos x$

$u'(x) = 2; v'(x) = -\sin x$

$y' = 2\cos x + 2x(-\sin x)$

$= 2\cos x - 2x\sin x$

30. $u(x) = \sin x; v(x) = 3x^2 - 4$

$u'(x) = \cos x; v'(x) = 6x$

$f'(x) = \dfrac{(3x^2 - 4)\cos x - 6x\sin x}{(3x^2 - 4)^2}$

31. $u(x) = x^3; v(x) = 3x - 2x^2$

$u'(x) = 3x^2; v'(x) = 3 - 4x$

$f'(x) = 3(3x - 2x^2)^2(3 - 4x)$

$= (9 - 12x)(3x - 2x^2)^2$

32. $u(x) = x^{\frac{1}{2}}; v(x) = x^2 + 4$

$u'(x) = \frac{1}{2}x^{-\frac{1}{2}}; v'(x) = 2x$

$h'(x) = \frac{1}{2}(x^2 + 4)^{-\frac{1}{2}}(2x) = \dfrac{x}{\sqrt{x^2 + 4}}$

34. $\dfrac{d}{dx}[\tan x] = \dfrac{d}{dx}\left[\dfrac{\sin x}{\cos x}\right]$

$u(x) = \sin x; v(x) = \cos x$

$u'(x) = \cos x; v'(x) = -\sin x$

$\dfrac{d}{dx}\left[\dfrac{\sin x}{\cos x}\right]$

$= \dfrac{(\cos x)(\cos x) - (\sin x)(-\sin x)}{(\cos x)^2}$

$= \dfrac{\cos^2 x + \sin^2 x}{\cos^2 x} = \dfrac{1}{\cos^2 x} = \sec^2 x$

35. $f'(t) = 45\dfrac{d}{dt}[e^{0.62t}]$

$u(t) = e^t; v(t) = 0.62t$

$u'(t) = e^t; v'(t) = 0.62$

$f'(t) = 45e^{0.62t}(0.62) = 27.9e^{0.62t}$

$f'(4) = 27.9e^{0.62(4)} \approx 333.2 \text{ cells/day}$

$50{,}000 = 27.9e^{0.62t}$

$1792.1 = e^{0.62t}$

$\ln 1792.1 = \ln e^{0.62t}$

$= 0.62t$

$t = \dfrac{\ln 1792.1}{0.62} \approx 12.08 \text{ days}$

36. $\Delta x = \dfrac{b - a}{n} = \dfrac{4 - (-2)}{12} = 0.5$

37. $A_{[-3, 5]} = A_{[-3, 0]} + A_{[0, 2]}$

$= \frac{1}{2}(5 + 2)(3) + \frac{1}{2}(2 + 4)(2)$

$= 10.5 + 6 = 16.5 \text{ u}^2$

38. $\Delta x = \dfrac{3 - 1}{4} = 0.5$

$x_i = 1.5, 2, 2.5, 3$

$R_i = \Delta x \cdot f(x_i)$

$R_1 = 0.5[4(1.5)^2] = 4.5$

$R_2 = 0.5[4(2)^2] = 8$

$R_3 = 0.5[4(2.5)^2] = 12.5$

$R_4 = 0.5[4(3)^2] = 18$

$S = 4.5 + 8 + 12.5 + 18 = 43 \text{ u}^2$

39. $\Delta x = \dfrac{2 - 0}{1000} = \dfrac{1}{500}$

$x_i = \dfrac{1}{500}, \dfrac{2}{500}, \dfrac{3}{500}, \ldots, \dfrac{1000}{500}$

$S = \dfrac{1}{500}f\left(\dfrac{1}{500}\right) + \dfrac{1}{500}f\left(\dfrac{2}{500}\right)$

$\quad + \dfrac{1}{500}f\left(\dfrac{3}{500}\right) + \cdots + \dfrac{1}{500}f\left(\dfrac{1000}{500}\right)$

$= \dfrac{1}{500}(5)\left(\dfrac{1}{500}\right)^2 + \dfrac{1}{500}(5)\left(\dfrac{2}{500}\right)^2$

$\qquad\quad + \dfrac{1}{500}(5)\left(\dfrac{3}{500}\right)^2$

$\qquad\quad + \cdots + \dfrac{1}{500}(5)\left(\dfrac{1000}{500}\right)^2$

$= 5\left(\dfrac{1}{500}\right)^3$

$\qquad \cdot (1^2 + 2^2 + 3^2 + \cdots + 1000^2)$

$= 5\left(\dfrac{1}{500}\right)^3 \sum\limits_{i=1}^{1000} i^2$

$= 5\left(\dfrac{1}{500}\right)^3\left(\dfrac{1000(1001)(2001)}{6}\right)$

$= \dfrac{5(1001)(2001)}{3(500)^2} \approx 13.35 \text{ u}^2$

40. $\Delta x = \dfrac{4 - 0}{n} = \dfrac{4}{n}$

$= \lim\limits_{n \to \infty} \sum\limits_{i=1}^{n} 2x_i^3\left(\dfrac{4}{n}\right)$

$= \lim\limits_{n \to \infty} \sum\limits_{i=1}^{n} 2\left(\dfrac{4i}{n}\right)^3\left(\dfrac{4}{n}\right)$

$= \lim\limits_{n \to \infty} \sum\limits_{i=1}^{n} \dfrac{512i^3}{n^4} = \lim\limits_{n \to \infty} \dfrac{512}{n^4} \sum\limits_{i=1}^{n} i^3$

$= \lim\limits_{n \to \infty} \dfrac{512}{n^4}\left(\dfrac{n^2(n+1)^2}{4}\right)$

$= \lim\limits_{n \to \infty} \dfrac{128(n^2 + 2n + 1)}{n^2}$

$= \lim\limits_{n \to \infty} \left(128 + \dfrac{256}{n} + \dfrac{128}{n^2}\right) = 128$

continued on next page

42a. $G(x) = 2x^2 + 7x + C$
$$G(-2) = 2(-2)^2 + 7(-2) + C = 5$$
$$-6 + C = 5$$
$$C = 11$$
$$G(x) = 2x^2 + 7x + 11$$

42b. $P(x) = -2x^3 - 7x + C$
$$P(3) = -2(3)^3 - 7(3) + C = -1$$
$$-75 + C = -1$$
$$C = 74$$
$$P(x) = -2x^3 - 7x + 74$$

44a. $\displaystyle\int_1^{12} 4\,dx = [4x]_1^{12}$
$$= 4(12) - 4(1) = 44$$

44b. $\displaystyle\int_{-3}^3 y^2\,dy = \left[\frac{y^3}{3}\right]_{-3}^3 = \frac{3^3}{3} - \frac{(-3)^3}{3} = 18$

45. $\displaystyle\int(-32t)\,dt = -16t^2 + C = h(t)$
$$h(2) = -16(2)^2 + C$$
$$18 = -64 + C$$
$$C = 82 \text{ ft} = h_0$$

46. weight $= \dfrac{550\text{ N}}{10\text{ m}} = 55$ N/m
Let $x =$ height lifted.
force: $f(x) = 55x$
$$\int_0^4 f(x)\,dx = \int_0^4 55x\,dx = [27.5x^2]_0^4$$
$$= 27.5(4)^2 - 27.5(0)^2$$
$$= 440 \text{ N·m or } 440 \text{ J}$$

47.

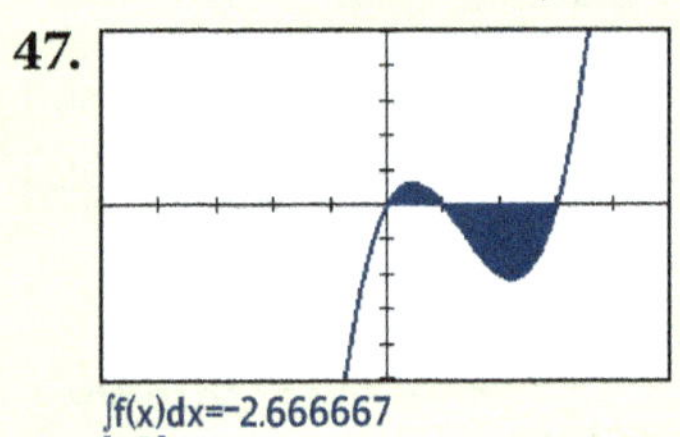

$$A_{[0,\,3]} = A_{[0,\,1]} + A_{[1,\,3]}$$
$$\approx 0.4167 + [-(-2.6667)]$$
$$\approx 3.083 \text{ u}^2$$

A.1 Sets and Real Numbers

A *set* is a collection of objects. Each object in the set is called an *element* of the set. A set with no elements is called the *empty set*, denoted $\varnothing$. Consider the following sets $A = \{1, 3, 7, 8, 10\}$ and $B = \{1, 4, 6, 10\}$. The fact that 7 is an element of set A but not of set B can be stated symbolically as $7 \in A$ and $7 \notin B$.

Union and *intersection* are the two main set operations. The union of two sets contains all elements that appear in both sets, while the intersection of the two sets contains only the elements common to both sets. Relationships between two or more sets are often represented with a *Venn diagram*.

Example 1 Finding the Union and Intersection of Sets

Given $A = \{1, 3, 7, 8, 10\}$ and $B = \{1, 4, 6, 10\}$, complete each set operation and illustrate the operation with a Venn diagram.

a. $A \cup B$ **b.** $A \cap B$

Answer

a. $A \cup B = \{1, 3, 4, 6, 7, 8, 10\}$

Combine all the elements into one set.

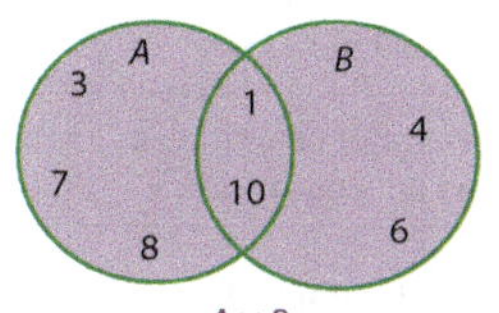

$A \cup B$

Shade all regions of both circles.

b. $A \cap B = \{1, 10\}$

List the elements in common to both sets.

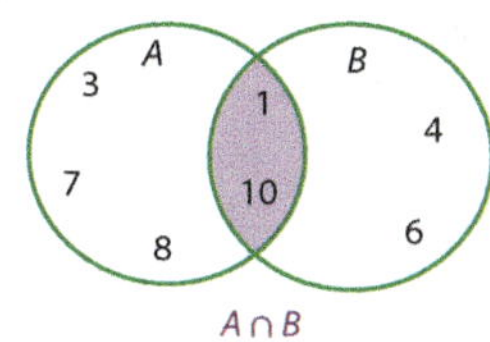

$A \cap B$

Shade the overlapping area.

——————————————— SKILL ✔ **EXERCISES 3, 11**

Set C is a *subset* of set D (denoted $C \subseteq D$) if every element of C is also in D. Sets D and E are *disjoint* since they have no common elements. The *universal set* U contains all possible elements for a problem. The *complement* of D, symbolized by D', is the set that contains everything in the universal set except the elements in set D.

Example 2 Determining Subsets

Determine whether each statement is true or false for the illustrated sets C, D, and E. Explain your reasoning.

a. $\{5, 9\} \subseteq D$ **b.** $D \cup E \subseteq D$ **c.** $2 \subseteq E$

d. $C \cap D = C$ **e.** $D \cap E = \varnothing$ **f.** $D \cup E = U$

Answer

a. true; Both 5 and 9 are in D.

b. false; $D \cup E = \{1, 2, 3, 5, 7, 8, 9\}$ includes 2 and 8, neither of which are in D.

c. false; $2 \in E$; it is not a subset of E. Instead, $\{2\} \subseteq E$.

d. true; $C \cap D = \{1, 3\}$, and these are all the elements of C.

CONTINUED ➡

After completing this section, you will be able to

- use set notation to describe relationships between groups of objects.
- perform unions, intersections, and complements of sets.
- use Venn diagrams to illustrate set operations and subsets.

TIP

Sets are named using capital letters and denoted using braces, { }, which are read "the set whose elements are."

A.1 Sets and Real Numbers

Objectives

1. To use sets to describe relationships between groups of objects
2. To perform unions, intersections, and complements of sets
3. To use Venn diagrams to illustrate set operations and subsets

Vocabulary

complement
disjoint sets
element
empty set
integer
intersection of sets
irrational number
natural number
rational number
real number
set
set-builder notation
subset
union
universal set
Venn diagram
whole number

PRESENTATION

This lesson reviews set terminology. Exercises 41–50 of Section 1.1 on page 8 can be used as an assessment to determine whether students need further instruction.

① Disjoint sets are sets whose intersection is the empty set.

② $A = \{51, 52, 53, \ldots\}$

③ In Example 3e, some students will benefit from reviewing the process for converting a repeating decimal to a fraction as follows.

Let $x = 0.1\overline{6}$ and $10x = 1.\overline{6}$.

$$10x = 1.6\overline{6}$$
$$-\ (x = 0.1\overline{6})$$
$$\overline{9x = 1.5}$$
$$x = \frac{15}{90} = \frac{1}{6}$$

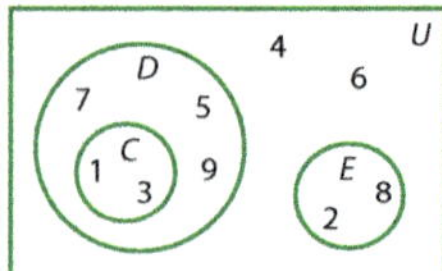

e. true; Sets D and E are disjoint.

f. false; $D \cup E = \{1, 2, 3, 5, 7, 8, 9\}$, but U also includes 4 and 6.

SKILL ✓ EXERCISE 19

Sets are often described using *set-builder notation*. The vertical bar, |, is read "such that" and is followed by the conditions for the variable expression that precedes the bar. For example, $A = \{n \mid n \in \mathbb{W} \text{ and } n > 50\}$ is read "Set A is the set of all n such that n is a whole number greater than 50." Can you list the numbers that belong in set A?

Classifying numbers with a shared characteristic enables mathematicians to make generalized statements about their relationships. Note the following subsets of the real numbers, $\mathbb{R}$, which are all the numbers found on a number line. Complex numbers were introduced in *Algebra 2* and will be reviewed in Appendix Lesson 5.

Number Set	Symbol	Description
natural	$\mathbb{N}$	counting numbers $\mathbb{N} = \{1, 2, 3, \ldots\}$
whole	$\mathbb{W}$	$\mathbb{W} = \{0, 1, 2, 3, \ldots\}$
integers	$\mathbb{Z}$	whole numbers and their opposites $\mathbb{Z} = \{\ldots, -3, -2, -1, 0, 1, 2, 3, \ldots\}$
rational	$\mathbb{Q}$	numbers that can be written as a ratio of integers $\mathbb{Q} = \left\{ \frac{a}{b} \mid a \in \mathbb{Z}, b \in \mathbb{Z}, \text{ and } b \neq 0 \right\}$
irrational	$\mathbb{Q}'$	real numbers that cannot be written as a ratio of integers on a number line
real	$\mathbb{R}$	all the numbers that can be graphed on a number line

Every counting number is a whole number. The integers are composed of the natural numbers, their opposites, and 0. Therefore, $\mathbb{N} \subseteq \mathbb{W} \subseteq \mathbb{Z}$. Since any integer can be written over 1, the integers are a subset of an even larger set of numbers, the rational numbers.

Decimals that either terminate or repeat are also rational numbers because they can be written as a ratio of integers. The repeating portion is signified by a line over the sequence of repeating digits. Non-terminating, non-repeating decimals such as π, $\sqrt{2}$, and e are irrational numbers. Every real number is either a rational or an irrational number.

15. false; Set A contains only odd natural numbers.

16. false; $0 \notin A$ or B.

19. false; Set D, having 3 elements, cannot be a subset of the empty set $B \cap C$.

27. $\mathbb{N}, \mathbb{W}, \mathbb{Z}, \mathbb{Q}, \mathbb{R}$

28. $\mathbb{Q}, \mathbb{R}$

29. $\mathbb{Z}, \mathbb{Q}, \mathbb{R}$

33. $\{n \mid n \geq -5\}$

34. $\{2n - 1 \mid n \in \mathbb{N} \text{ and } n \leq 50\}$

35. $\{2z \mid z \in \mathbb{Z} \text{ and } -5 < z < 24\}$

State the most specific named subset of real numbers to which each number belongs.

a. $\sqrt{6}$ **b.** $\sqrt{9}$ **c.** 3.14159

d. $\frac{\pi}{8}$ **e.** $0.1\overline{6}$ **f.** $0.121221222\ldots$

Answer

a. irrational $\sqrt{6} \approx 2.4494\ldots$; The decimal does not terminate and does not repeat.

b. natural $\sqrt{9} = 3$, a natural (or counting) number

c. rational   The decimal terminates.

d. irrational   This fraction cannot be expressed as a ratio of integers. The quotient of an irrational number and a rational number is irrational.

e. rational   The decimal repeats. $0.1\overline{6} = \frac{1}{6}$ ③

f. irrational   The decimal does not terminate or repeat.

SKILL ✓ **EXERCISE 27**

❯ Exercises

Use the Venn diagram to complete exercises 1–8.

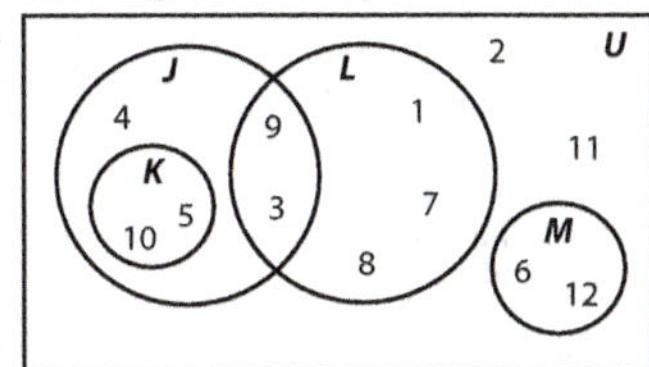

1. List the universal set. $U = \{1, 2, 3, \ldots, 12\}$

2. Which set is a subset of set J? K

3. Name two sets that are disjoint from set L. M and K

Determine the result of the set operations.

4. $J \cap L$ $\{3, 9\}$ 5. $L \cap M$ $\varnothing$ 6. $(J \cap L) \cup M$ $\{3, 6, 9, 12\}$

7. J' $\{1, 2, 6, 7, 8, 11, 12\}$ 8. $L' \cap M$ $\{6, 12\}$

Use $A = \{1, 3, 5, 7, \ldots\}$, $B = \{2, 4, 6, 8, \ldots\}$, $C = \{1, 5\}$, $D = \{5, 6, 7\}$, and the named subsets of the real numbers to complete exercises 9–14.

9. $D \cap B$ $\{6\}$ 10. $C \cup D$ $\{1, 5, 6, 7\}$

11. $B \cap \mathbb{Z}$ B 12. $\mathbb{N} \cup \mathbb{W}$ $\mathbb{W}$

13. State the smallest universal set for A, B, C, and D. $\mathbb{N}$

14. Given $U = \mathbb{N}$, find B'. A

21–26.

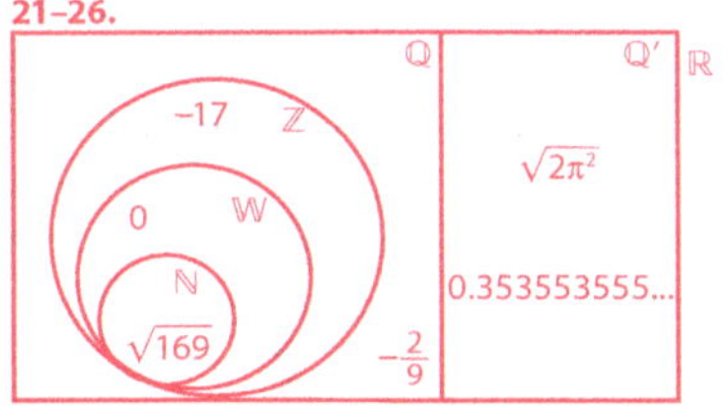

Use the sets given for exercises 9–14 to determine whether each of the following is true or false. If the statement is false, explain why.

15. $8 \in A$ 16. $A \cup B = \mathbb{W}$

17. $A \cap B = \varnothing$ true   18. $A \subseteq \mathbb{Z}$ true

19. $D \subseteq (B \cap C)$ 20. $\mathbb{Q}$ and $\mathbb{Q}'$ are disjoint.   true

Draw a Venn diagram representing the relationship between the named subsets of the real numbers. Then place each number within the smallest named subset to which it belongs.

21. $\sqrt{169}$ 22. -17

23. $\sqrt{2\pi^2}$ 24. $-\frac{2}{9}$

25. $0.353553555\ldots$ 26. 0

List all the number sets, $\mathbb{N}$, $\mathbb{W}$, $\mathbb{Z}$, $\mathbb{Q}$, $\mathbb{Q}'$, and $\mathbb{R}$, to which each number belongs.

27. $\sqrt{25}$ 28. $\frac{2}{3}$ 29. $-\frac{\sqrt{16}}{2}$

Write each set as a list of elements.

30. $\{2n \mid n \in \mathbb{N}\}$ $\{2, 4, 6, \ldots\}$

31. $\{2z + 1 \mid z \in \mathbb{Z}\}$ $\{\ldots, -5, -3, -1, 1, 3, 5, \ldots\}$

32. $\left\{ \frac{2a - 1}{3} \mid a \in \mathbb{W} \right\}$ $\left\{ -\frac{1}{3}, \frac{1}{3}, 1, \frac{5}{3}, \frac{7}{3}, \ldots \right\}$

Write each set using set builder notation.

33. all real numbers greater than or equal to -5

34. all odd natural numbers less than 100

35. all even integers between -10 and 48

36. $\{2, 5, 8, 11, \ldots, 44\}$ $\{3n - 1 \mid n \in \mathbb{N} \text{ and } n \leq 15\}$

Solutions

❯ Exercises

4. $\{3, 4, 5, 9, 10\} \cap \{1, 3, 7, 8, 9\}$
   $= \{3, 9\}$

5. $\{1, 3, 7, 8, 9\} \cap \{6, 12\} = \varnothing$

6. $J \cap L = \{3, 9\}$
   $\{3, 9\} \cup \{6, 12\} = \{3, 6, 9, 12\}$

8. $\{2, 4, 5, 6, 10, 11, 12\} \cap \{6, 12\}$
   $= \{6, 12\}$

9. $\{5, 6, 7\} \cap \{2, 4, 6, 8, \ldots\} = \{6\}$

10. $\{1, 5\} \cup \{5, 6, 7\} = \{1, 5, 6, 7\}$

11. $\{2, 4, 6, 8, \ldots\} \cap \mathbb{Z} = \{2, 4, 6, 8, \ldots\}$
    $= B$

12. $\{1, 2, 3, \ldots\} \cup \{0, 1, 2, 3, \ldots\}$
    $= B' = \mathbb{W}$

14. $B' = \{1, 3, 5, \ldots\} = A$

20. $\mathbb{Q} \cap \mathbb{Q}' = \varnothing$

21. $\sqrt{169} = 13 \in \mathbb{N}$

23. $\sqrt{2\pi^2} \approx 4.442882937\ldots \in \mathbb{Q}'$

A.2 Solving Equations and Inequalities Algebraically

Objectives

1. To solve linear equations by applying the properties of equality
2. To solve linear inequalities
3. To solve compound inequalities
4. To solve inequalities involving absolute values
5. To solve quadratic equations by factoring and by using the quadratic formula

Vocabulary

Absolute Value Inequality Theorem
compound inequality
conjunction
disjunction
equivalent equations
least common denominator
linear equation
linear inequality
quadratic equation
quadratic formula
Zero Product Property

After completing this section, you will be able to
- solve linear and quadratic equations.
- solve linear inequalities and inequalities involving absolute value.

A.2 Solving Equations and Inequalities Algebraically

A challenge in every mathematics course involves remembering and applying concepts from previous courses. Every rule becomes a tool in your toolbox to be used in solving various situations. For example, the properties of equality and real numbers are used to solve first-degree equations in one variable.

Properties of Equality $(a, b, c \in \mathbb{R})$		Properties of Real Numbers $(a, b, c \in \mathbb{R})$	
Reflexive	$a = a$	Commutative	$a + b = b + a$ $a \cdot b = b \cdot a$
Symmetric	If $a = b$, then $b = a$.	Associative	$a + (b + c) = (a + b) + c$ $a \cdot (b \cdot c) = (a \cdot b) \cdot c$
Transitive	If $a = b$ and $b = c$, then $a = c$.	Identity	$a + 0 = a$ and $0 + a = a$ $a \cdot 1 = a$ and $1 \cdot a = a$
Addition	If $a = b$, then $a + c = b + c$.	Inverse	$a + (-a) = 0$ $a \cdot \frac{1}{a} = 1; a \neq 0$
Multiplication	If $a = b$, then $ac = bc$.	Distributive (of Multiplication over Addition)	$a(b + c) = ab + ac$

The first step to solve a linear equation should be to simplify by combining like terms or removing parentheses. Applying any of these properties produces an *equivalent equation*, one that has the same solution as the original equation. When an equation contains more than one operation, these operations are "undone" by doing the inverse operations in reverse order on both sides of the equation.

Example 1   Solving a Linear Equation

Solve $-3(x + 2) = 5(x + 9) - 3$.

Answer

$-3x - 6 = 5x + 42$	1. Use the Distributive Property to simplify both sides.
$-6 = 8x + 42$ $-48 = 8x$	2. Add $3x$ to both sides and subtract 42 from both sides to collect the terms containing the variable on one side of the equation and constant terms on the other side.
$-6 = x$	3. Divide both sides by 8.
$x = -6$	4. Apply the Symmetric Property of Equality.
$-3[(-6) + 2] = 5[(-6) + 9] - 3$ $12 = 12$	5. Check mentally.

SKILL ✔ EXERCISE 3 ①

PRESENTATION

This lesson reviews solving equations and inequalities. Some students may only need to reference this material while others may need a more thorough review.

① This exercise requires students to distribute a negative factor. Remind students to change the sign of the second term. For example, $-(2a + 6) = -2a - 6$. See step 1.

② Remind students not to cancel terms. In step 1 some students might cancel the 12 with the 4 and then use the 3 to cancel the denominator of 3.

$$12^{\overset{3}{}}\left(\frac{x}{4} + \frac{x - 1}{3}\right) = x + (x - 1)$$

Emphasize that a factor of 12 is distributed to each term as demonstrated.

③ The common multiple used to clear the decimals is not necessarily the LCM. Students should multiply both sides of the equation by 10^n, in which n is the maximum number of place values behind the decimal.

④ After completing Skill Check exercises 17 and 19 for Example 4, ask the students what the answers would be if exercise 17 was a disjunction and exercise 19 was a conjunction. (*all real numbers; empty set*)

Equations often contain fractions or decimals. Even though calculators are useful in dealing with these types of equations, it is often helpful to eliminate the fractions or decimals before solving. An equation can be cleared of fractions by multiplying both sides by a common denominator, preferably the *least common denominator* (LCD).

Example 2   Solving by Clearing Fractions

Solve $\frac{x}{4} + \frac{x-1}{3} = \frac{x+3}{6}$.

Answer

$$12\left(\frac{x}{4} + \frac{x-1}{3}\right) = 12\left(\frac{x+3}{6}\right)$$
$$3x + 4(x-1) = 2(x+3)$$
1. Multiply both sides by the LCD. 

$$7x - 4 = 2x + 6$$
2. Simplify both sides.

$$5x = 10$$
$$x = 2$$
3. Solve.

$$\frac{2}{4} + \frac{2-1}{3} = \frac{2+3}{6}$$
4. Check the solution.

SKILL ✔ **EXERCISE 9** 

Mathematical inequalities compare unequal quantities. The *Trichotomy Property* states that every comparison of two real numbers a and b has one of three results: $a < b$, $a = b$, or $a > b$. While there are no reflexive or symmetric properties of inequality, there is a *Transitive Property of Inequality*: if $a > b$ and $b > c$, then $a > c$. This also is true for the other inequalities $<$, $\leq$, and $\geq$.

Solving *linear inequalities* is identical to solving linear equations, with one exception. When multiplying or dividing both sides of an inequality by a negative number, the inequality sign is reversed.

Example 3   Solving a Linear Inequality

Solve $0.05n + 0.1(20 - n) \geq 1.65$ and graph the solution set.

Answer

$$100[0.05n + 0.1(20 - n)] \geq 100(1.65)$$
$$5n + 10(20 - n) \geq 165$$
$$200 - 5n \geq 165$$
1. Multiply each term on both sides by 10^2 since there are at most 2 decimal places.

$$\frac{-5n}{-5} \leq \frac{-35n}{-5}$$
$$n \leq 7$$
2. Reverse the inequality sign whenever both sides are multiplied or divided by a negative number.

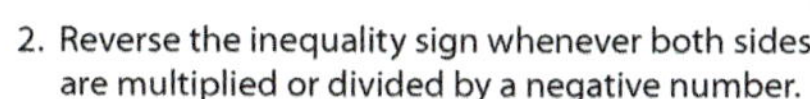

3. Use a solid dot to indicate that 7 is part of the solution.

SKILL ✔ **EXERCISE 11**

A *compound inequality* combines two inequalities into one statement that can be used to describe a range of numbers. A *conjunction* joins inequalities with the word *and* (or the symbol $\wedge$). Since the solution to a conjunction must satisfy both inequalities, it is the intersection of the inequalities' solutions. A *disjunction* joins two inequalities with the word *or* (or the symbol $\vee$). Since the solution to a disjunction can satisfy either inequality, the solution is the union of the inequalities' solutions.

TIPS

Ex. 25 When variables appear in all 3 portions of a combined inequality, consider solving the two inequalities separately.

Conjunction	Sketch the inequalities.	State the intersection.
$x > -2 \wedge x \leq 1$ $x > -2$ and $x \leq 1$		$-2 < x \leq 1$

Disjunction	Sketch the inequalities.	State the union.
$x \geq 2 \vee x > 4$ $x \geq 2$ or $x > 4$		$x \geq 2$

Example 4   Solving Compound Inequalities

Solve each inequality.

a. $5x + 9 \leq 24 \wedge -3x - 7 > 14$

b. $2x + 3 < -7 \vee 9x - 4 > 5$

Answer

a. $5x + 9 \leq 24$ and $-3x - 7 > 14$

$5x \leq 15 \qquad -3x > 21$

$x \leq 3 \qquad\quad x < -7$

1. Solve each inequality separately.

2. Find the intersection of the graphs of the two inequalities.

$x < -7$

3. State the solution to the conjunction.

b. $2x + 3 < -7$ or $9x - 4 > 5$

$2x < -10 \qquad 9x > 9$

$x < -5 \qquad\quad x > 1$

1. Solve each inequality separately.

2. Find the union of the graphs of the two inequalities.

$x < -5$ or $x > 1$

3. State the solution to the disjunction.

SKILL ✓ **EXERCISES 17, 19** ④

The conjunction $5x - 7 > 3$ and $5x - 7 < 23$ can be rewritten as $3 < 5x - 7 < 23$, which more clearly indicates that $5x - 7$ lies between 3 and 23. Disjunctions cannot be written this way.

When the same steps are used to solve each inequality, the properties can be used on each of the three parts of a combined conjunction as illustrated below.

Example 5   Solving a Compound Inequality

Solve $3 < 5x - 7 < 23$.

Answer

$10 < 5x < 30$

1. Add 7 to each expression.

$2 < x < 6$

2. Divide each expression by 5.

SKILL ✓ **EXERCISE 25**

The equation $|x| = 3$ can be interpreted as "x is 3 units from 0 on the number line," and has the solution of $x = -3$ or $x = 3$. The inequality $|x| > 3$ describes all the numbers more than 3 units from the origin, and the inequality $|x| < 3$ describes all the numbers less than 3 units from the origin. The solution to an absolute value inequality can be written as a compound inequality.

ABSOLUTE VALUE INEQUALITY THEOREM

If $c \geq 0$, then
$|x| < c$ is equivalent to the conjunction $-c < x < c$ and
$|x| > c$ is equivalent to the disjunction $x < -c$ or $x > c$.

Example 6   Solving Absolute Value Inequalities

Solve each inequality and graph its solution set.

a. $|2x - 9| < 4$ **b.** $|-4x + 9| - 11 \geq 4$

Answer

a. $-4 < 2x - 9 < 4$

1. Write a conjunction stating that $2x - 9$ is less than 4 units from the origin.

$5 < 2x < 13$

$\frac{5}{2} < x < \frac{13}{2}$

2. Solve the combined conjunction.

3. Graph the solution set.

-1 0 1 2 3 4 5 6 7 8

b. $|-4x + 9| \geq 15$

1. Isolate the absolute value.

$-4x + 9 \leq -15$ or $-4x + 9 \geq 15$

2. Write a disjunction stating that $-4x + 9$ is at least 15 units from the origin.

$-4x \leq -24$ or $-4x \geq 6$

$x \geq 6$ $x \leq -\frac{3}{2}$

3. Solve. Remember to reverse the inequality signs when dividing by a negative value.

$x \leq -\frac{3}{2}$ or $x \geq 6$

4. State the solution as a disjunction.

5. Graph the solution set.

-4 -2 0 2 4 6 8

SKILL ✓ **EXERCISES 27, 29**

A *quadratic equation* can be written in the standard form $ax^2 + bx + c = 0$, where a, b, and $c \in \mathbb{R}$ and $a \neq 0$. The first step in solving most quadratic equations is to write the equation in standard form. If the trinomial can be factored, the equation can be solved using the *Zero Product Property*.

ZERO PRODUCT PROPERTY

If $pq = 0$, then $p = 0$ or $q = 0$.

Solutions

Exercises

1. $-2x + 18 = 3x - 12 - 5$
$$-5x = -35$$
$$x = 7$$

2. $2n + 8 - 15n + 35 = -113$
$$-13n + 43 = -113$$
$$-13n = -156$$
$$n = 12$$

3. $3a + 12a - 18 - 2a - 6 = 41$
$$13a - 24 = 41$$
$$13a = 65$$
$$a = 5$$

4. $23x + 11x - 66 = 21x - 35 + 8$
$$13x = 39$$
$$x = 3$$

5. $21\left[\frac{b}{3} - \frac{8b}{7} = \frac{68}{21}\right]$
$$7b - 24b = 68$$
$$-17b = 68$$
$$b = -4$$

6. $9\left[\frac{x + 2}{9} - \frac{x}{3} = 10\right]$
$$x + 2 - 3x = 90$$
$$-2x = 88$$
$$x = -44$$

7. $24\left[\frac{3x}{8} + 5 = \frac{x + 17}{3}\right]$
$$9x + 120 = 8(x + 17)$$
$$9x + 120 = 8x + 136$$
$$x = 16$$

8. $3b + \frac{3}{2} - 8b - 3b + 27 = \frac{13}{2}$
$$2\left[3b + \frac{3}{2} - 8b - 3b + 27 = \frac{13}{2}\right]$$
$$6b + 3 - 16b - 6b + 54 = 13$$
$$-16b = -8$$
$$b = \frac{1}{2}$$

9. $1000(0.3x + 0.89 = 5.321)$
$$300x + 890 = 5321$$
$$300x = 4431$$
$$x = 14.77$$

10. $10[0.5(4x + 12) - 0.9(x - 3) = 18.6]$
$$5(4x + 12) - 9(x - 3) = 186$$
$$20x + 60 - 9x + 27 = 186$$
$$11x = 99$$
$$x = 9$$

11. $2x - x - 4 \neq 7$
$$x - 4 \neq 7$$
$$x - 4 + 4 \neq 7 + 4$$
$$x \neq 11$$

12. $-2x - 1 \leq 3x + 3$
$$-5x \leq 4$$
$$x \geq -\frac{4}{5}$$

13. $10a - 5a - 45 \leq 4a + 12$
$$5a - 45 \leq 4a + 12$$
$$a \leq 57$$

14. $-5x + 6x - 12 < 5x - 36$
$$-4x < -24$$
$$x > 6$$

15. $\quad 0.3y - 0.45 > 0.9y - 0.18 + 0.6y$
$$100(0.3y - 0.45 > 1.5y - 0.18)$$
$$30y - 45 > 150y - 18$$
$$-120y > 27$$
$$y < -\frac{9}{40}$$

16. $5(0.04y - 0.08)$
$$> 3(0.2y - 0.05) - 0.4y$$
$$0.2y - 0.4 > 0.6y - 0.15 - 0.4y$$
$$0.2y - 0.4 > 0.2y - 0.15$$
$$-0.4 > -0.15$$

17. $4x \leq -7$ and $x \geq \frac{25}{4} - 8$
$$x \leq -\frac{7}{4} \qquad x \geq -\frac{7}{4}$$

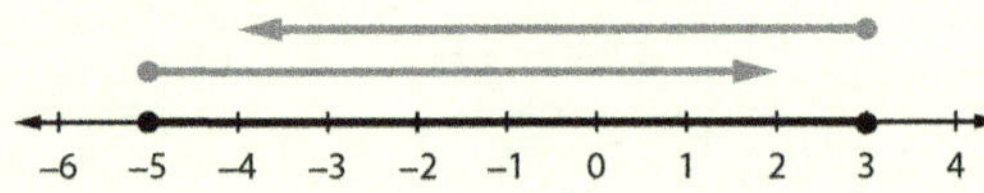

18. $4x \leq 12$ and $-3x \leq 15$
$$x \leq 3 \qquad\qquad x \geq -5$$

19. $-2x \geq 6$ or $2x + 7 \geq 17 - 3x$
$$x \leq -3 \qquad\qquad x \geq 10$$
$$x \geq 2$$

20. $8z - 16 + 4z + 4 \geq 4z$
$$12z - 12 \geq 4z$$
$$8z \geq 12$$
$$z \geq 1.5$$

$\quad$ or $5z < 7$
$$z < \frac{7}{5}$$

21. $6\left[\frac{1}{2}(d + 9) > \frac{1}{3}(6 - d)\right]$
$$3(d + 9) > 2(6 - d)$$
$$3d + 27 > 12 - 2d$$
$$5d > -15$$
$$d > -3 \text{ and}$$
$$6\left[\frac{1}{3}(d - 4) > \frac{1}{2}(d + 5)\right]$$
$$2(d - 4) > 3(d + 5)$$
$$2d - 8 > 3d + 15$$
$$-d > 23$$
$$d < -23$$

22. $9(z - 2) + 3(z + 1) \leq 4z$
$$9z - 18 + 3z + 3 \leq 4z$$
$$12z - 15 \leq 4z$$
$$8z \leq 15; \; z \leq 1.875$$

$\quad$ or $5z - 7 > 2$
$$5z > 9; \; z > \frac{9}{5}$$

23. $6 < 7x - 8 \leq 41$
$$14 < 7x \leq 49$$
$$2 < x \leq 7$$

24. $7 \leq 2x + 14 + 3x - 9 < 35$
$$7 \leq 5x + 5 < 35$$
$$2 \leq 5x < 30$$
$$\frac{2}{5} \leq x < 6$$

25. $a - 6 < 5a + 10$
$$\wedge 5a + 10 < 4a + 10$$
$$-16 < 4a \qquad\qquad a < 0$$
$$-4 < a$$
$$-4 < a < 0$$

26. $8y - 8 < -12$ or $8y - 8 > 12$
$$8y < -4 \qquad\qquad 8y > 20$$
$$y < -\frac{1}{2} \quad \text{or} \quad y > \frac{5}{2}$$

27. $4x - 8 < -(8 - x)$ or $4x - 8 > 8 - x$
$$4x - 8 < x - 8 \qquad\qquad 5x > 16$$
$$3x < 0 \qquad\qquad\qquad x > \frac{16}{5}$$
$$x < 0$$
$$x < 0 \text{ or } x > \frac{16}{5}$$

Solve each quadratic equation.

a. $6x^2 + 13x = -5$ $\qquad$ **b.** $2x^2 - 7x - 15 = 0$

Answer

a. $\quad 6x^2 + 13x + 5 = 0$ $\qquad$ 1. Write the equation in standard form.

$\quad (3x + 5)(2x + 1) = 0$ $\qquad$ 2. Factor the trinomial.

$3x + 5 = 0 \;$ or $\; 2x + 1 = 0$ $\qquad$ 3. Apply the Zero Product Property
$$3x = -5 \qquad 2x = -1$$ $\qquad\qquad$ and solve.
$$x = -\frac{5}{3} \qquad x = -\frac{1}{2}$$
$$\therefore x = -\frac{5}{3}, -\frac{1}{2}$$

b. $\quad 2x^2 - 7x - 15 = 0$ $\qquad$ 1. The equation is written in standard form.

$\quad (2x + 3)(x - 5) = 0$ $\qquad$ 2. Factor the trinomial.

$2x + 3 = 0 \;$ or $\; x - 5 = 0$ $\qquad$ 3. Apply the Zero Product Property and solve.
$$2x = -3 \qquad x = 5$$
$$x = -\frac{3}{2}$$
$$\therefore x = -\frac{3}{2}, 5$$

SKILL ✔ **EXERCISE 35**

When a trinomial cannot be factored, use the *quadratic formula*.

QUADRATIC FORMULA

If $ax^2 + bx + c = 0$, where a, b, and $c \in \mathbb{R}$ and $a \neq 0$, then $x = \dfrac{-b \pm \sqrt{b^2 - 4ac}}{2a}$.

Solve $7x^2 = 8 - 2x$.

Answer

$7x^2 + 2x - 8 = 0$ $\qquad$ 1. Write the equation in standard form and
$a = 7, b = 2, c = -8$ $\qquad\qquad$ identify a, b, and c.

$x = \dfrac{-b \pm \sqrt{b^2 - 4ac}}{2a} = \dfrac{-(2) \pm \sqrt{(2)^2 - 4(7)(-8)}}{2(7)}$ $\qquad$ 2. Substitute into the quadratic formula.

$\quad = \dfrac{-2 \pm \sqrt{4 - (-224)}}{14} = \dfrac{-2 \pm \sqrt{228}}{14}$ $\qquad$ 3. Simplify.

$\quad = \dfrac{-2 \pm 2\sqrt{57}}{14} = \dfrac{-1 \pm \sqrt{57}}{7}$

SKILL ✔ **EXERCISE 39**

Solve.

1. $-2(x-9) = 3(x-4) - 5$ $x = 7$

2. $2(n+4) - 5(3n-7) = -113$ $n = 12$

3. $3a + 6(2a-3) - (2a+6) = 41$ $a = 5$

4. $23x + 11(x-6) = 7(3x-5) + 8$ $x = 3$

5. $\frac{b}{3} - \frac{8b}{7} = \frac{68}{21}$ $b = -4$

6. $\frac{x+2}{9} - \frac{x}{3} = 10$ $x = -44$

7. $\frac{3x}{8} + 5 = \frac{x+17}{3}$ $x = 16$

8. $\frac{1}{2}(6b+3) - 8b - (3b-9) = \frac{13}{2}$ $b = \frac{1}{2}$

9. $0.3x + 0.89 = 5.321$ $x = 14.77$

10. $0.5(4x+12) - 0.9(x-3) = 18.6$ $x = 9$

11. $2x - (x+4) \neq 7$ $x \neq 11$

12. $-2x - 1 \leq 3(x+1)$ $x \geq -\frac{4}{5}$

13. $10a - 5(a+9) \leq 4a + 12$ $a \leq 57$

14. $-5x + 3(2x-4) < 5x - 36$ $x > 6$

15. $5(0.06y - 0.09) > 3(0.3y - 0.06) + 0.6y$ $y < -\frac{9}{40}$

16. $5(0.04y - 0.08) > 3(0.2y - 0.05) - 0.4y$ $\varnothing$

Solve each inequality and graph its solution set.

17. $4x + 9 \leq 2 \wedge x + 8 \geq \frac{25}{4}$ $x = -\frac{7}{4}$

18. $4x - 5 \leq 7 \wedge -3x + 4 \leq 19$ $-5 \leq x \leq 3$

19. $-2x + 9 \geq 15 \vee 2x + 7 \geq 17 - 3x$ $x \leq -3$ or $x \geq 2$

20. $8(z-2) + 4(z+1) \geq 4z \vee 5z - 7 < 0$ $z < 1.4$ or $z \geq 1.5$

21. $\frac{1}{2}(d+9) > \frac{1}{3}(6-d) \wedge \frac{1}{3}(d-4) > \frac{1}{2}(d+5)$ $\varnothing$

22. $9(z-2) + 3(z+1) \leq 4z \vee 5z - 7 > 2$ $\mathbb{R}$

Solve.

23. $6 < 7x - 8 \leq 41$ $2 < x \leq 7$

24. $7 \leq 2(x+7) + 3x - 9 < 35$ $\frac{2}{5} \leq x < 6$

25. $a - 6 < 5(a+2) < 4a + 10$ $-4 < a < 0$

26. $|8y - 8| > 12$ $y < -\frac{1}{2}$ or $y > \frac{5}{2}$

27. $|4x - 8| > 8 - x$ $x < 0$ or $x > \frac{16}{5}$

28. $|24z - 80| < 6z + 70$ $\frac{1}{3} < z < \frac{25}{3}$

29. $\left|\frac{x}{3} + \frac{5}{6}\right| + 4 \leq 9$ $-\frac{35}{2} \leq x \leq \frac{25}{2}$

30. $|10 - 7x| \leq -8$ no solution

Solve by factoring.

31. $x^2 - 6x + 5 = 0$

32. $y(y + 16) = -28$

33. $x^2 - 3x - 54 = 0$

34. $8x^2 + 8x = 3 - 15x$

35. $12z^2 + z - 6 = 0$

36. $2y^2 + 9y = -10$

Use the quadratic formula to solve.

37. $6x^2 + 13x = -5$

38. $2x^2 + 2x = 2x + 4$

39. $4x^2 + 5x - 3 = 0$

40. $9x + 6x^2 = 12$

31. $x = 1, 5$

32. $y = -14, -2$

33. $x = -6, 9$

34. $x = -3, \frac{1}{8}$

35. $z = -\frac{3}{4}, \frac{2}{3}$

36. $y = -2, -\frac{5}{2}$

37. $x = -\frac{1}{2}, -\frac{5}{3}$

38. $x = \pm\sqrt{2}$

39. $x = \frac{-5 \pm \sqrt{73}}{8}$

40. $x = \frac{-3 \pm \sqrt{41}}{4}$

35. $(4z+3)(3z-2) = 0$
$$4z + 3 = 0 \text{ or } 3z - 2 = 0$$
$$z = -\frac{3}{4}, \frac{2}{3}$$

36. $2y^2 + 9y + 10 = 0$
$$(y+2)(2y+5) = 0$$
$$y + 2 = 0 \text{ or } 2y + 5 = 0$$
$$y = -2, -\frac{5}{2}$$

37. $6x^2 + 13x + 5 = 0$
$$x = \frac{-13 \pm \sqrt{13^2 - 4(6)(5)}}{2(6)}$$
$$= \frac{-13 \pm \sqrt{49}}{12} = \frac{-13 \pm 7}{12}$$
$$= -\frac{1}{2}, -\frac{5}{3}$$

38. $2x^2 - 4 = 0$
$$x = \frac{0 \pm \sqrt{0^2 - 4(2)(-4)}}{2(2)} = \frac{\pm\sqrt{32}}{4}$$
$$= \frac{\pm 4\sqrt{2}}{4} = \pm\sqrt{2}$$

39. $x = \frac{-5 \pm \sqrt{5^2 - 4(4)(-3)}}{2(4)}$
$$= \frac{-5 \pm \sqrt{73}}{8}$$

40. $2x^2 + 3x - 4 = 0$
$$x = \frac{-3 \pm \sqrt{3^2 - 4(2)(-4)}}{2(2)}$$
$$= \frac{-3 \pm \sqrt{41}}{4}$$

---

28. $24z - 80 < 6z - 70$
$$30z < 10$$
$$z < \frac{1}{3} \text{ and}$$
$$24z - 80 < 6z + 70$$
$$18z < 150$$
$$z < \frac{25}{3}$$
$$\frac{1}{3} < z < \frac{25}{3}$$

29. $\left|\frac{x}{3} + \frac{5}{6}\right| \leq 5$
$$-5 \leq \frac{x}{3} + \frac{5}{6} \leq 5$$
$$-30 \leq 2x + 5 \leq 30$$
$$-35 \leq 2x \leq 25$$
$$-\frac{35}{2} \leq x \leq \frac{25}{2}$$

30. no solution (Absolute values cannot be negative.)

31. $(x-5)(x-1) = 0$
$$x - 5 = 0 \text{ or } x - 1 = 0$$
$$x = 5, 1$$

32. $y^2 + 16y + 28 = 0$
$$(y+2)(y+14) = 0$$
$$y + 2 = 0 \text{ or } y + 14 = 0$$
$$y = -2, -14$$

33. $(x+6)(x-9) = 0$
$$x + 6 = 0 \text{ or } x - 9 = 0$$
$$x = -6, 9$$

34. $8x^2 + 23x - 3 = 0$
$$(x+3)(8x-1) = 0$$
$$x + 3 = 0 \text{ or } 8x - 1 = 0$$
$$x = -3, \frac{1}{8}$$

Objectives

1. To solve equations of the form $f(x) = 0$ by finding zeros

2. To solve equations of the form $f(x) = g(x)$ by finding the x-coordinate of the point(s) of intersection of $f(x)$ and $g(x)$

3. To solve inequalities by comparing graphs of two functions

Vocabulary

zeros of a function

After completing this section, you will be able to

- solve equations of the form $f(x) = 0$ by finding zeros.
- solve equations of the form $f(x) = g(x)$ by finding the x-coordinate of the points of intersection.
- solve inequalities by comparing graphs of two functions.

A.3 Solving Equations and Inequalities Graphically

You can use your graphing calculator to solve or check the solution(s) to an equation. If the equation is written in the form of $f(x) = 0$, the x-coordinate of any x-intercept (a point where the graph intersects the x-axis) is a solution of the equation. These values are also called the *zeros* of $f(x)$. If the graph does not intersect the x-axis, then the equation has no real solution.

Example 1   Solving a Quadratic by Graphing

Solve the equation $x^2 - 2x - 8 = 0$ by finding the zeros.

Answer

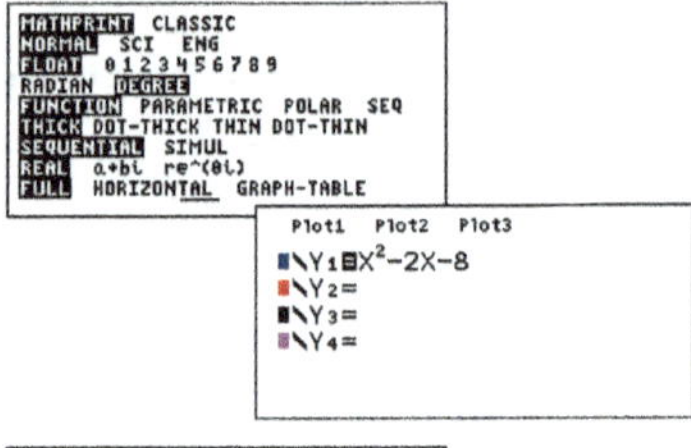

1. Press MODE and check that the calculator is in function mode. Then press Y= and enter the equation as Y_1. If Plot1, Plot2, or Plot3 is highlighted, move the cursor to that stat plot and press ENTER to turn it off.

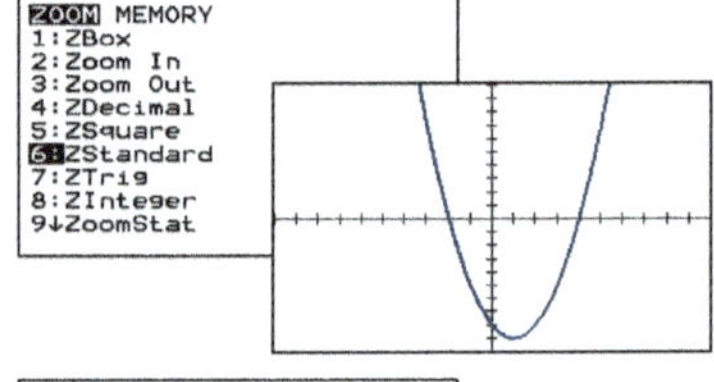

2. GRAPH the function and then select 6:ZStandard from the ZOOM menu. If necessary, adjust the WINDOW until all the x-intercepts are visible.

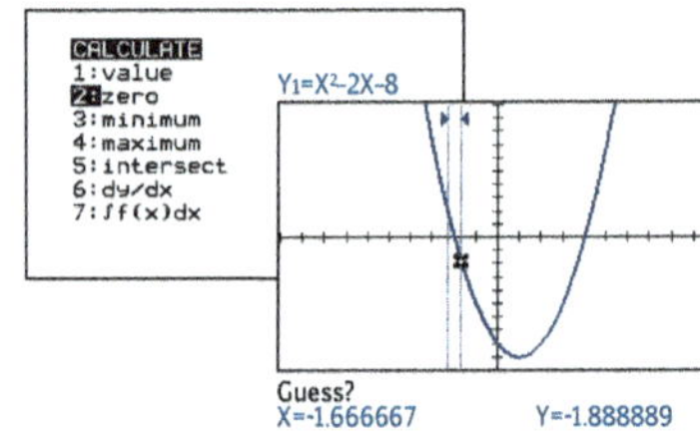

3. Press [CALC] (2nd, TRACE), 2:zero, ENTER to find the zeros. Set the left boundary by moving the cursor to the left of the zero and press ENTER. Similarly, set the right boundary and a guess. Pressing ENTER reveals the zero between the boundaries closest to the guess.

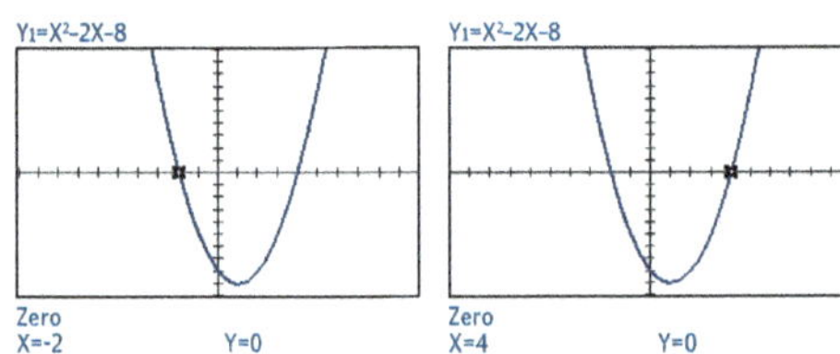

4. Repeat the process to find the second x-intercept.

   x-intercepts: $(-2, 0)$, $(4, 0)$
   solutions: $x = -2, 4$

SKILL ✓ EXERCISE 3

PRESENTATION

This lesson reviews solving equations and inequalities graphically. Be sure to communicate the type of solution expected for each topic and any restrictions on calculator use.

Equations containing nonzero expressions on both sides of the equality can be expressed as $f(x) = g(x)$ and solved by finding the x-coordinates of any points of intersection for the graphs of $f(x)$ and $g(x)$.

Example 2 Solving a Polynomial by Finding Points of Intersection

Solve the equation $x^3 - x^2 - 21x - 24 = x^2 + 9$ by finding the points of intersection.

Answer

1. Enter $Y_1 = x^3 - x^2 - 21x - 24$ and $Y_2 = x^2 + 9$, and then GRAPH the functions.

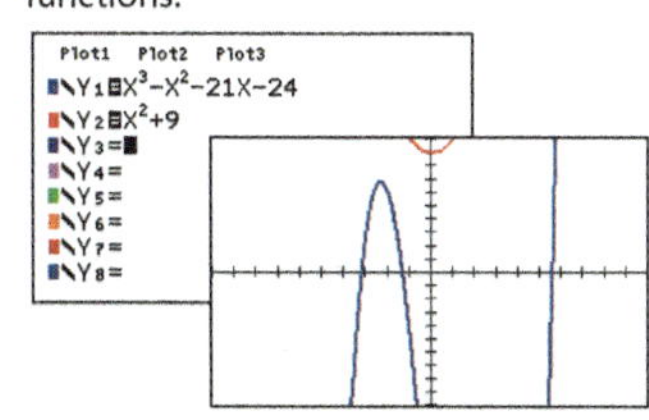

2. Adjust the WINDOW to view the points of intersection and press GRAPH. Notice that the Xscl and Yscl can be set to convenient values.

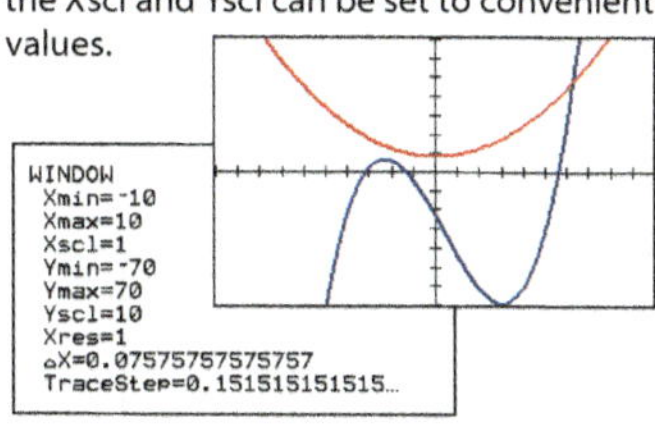

3. Select 5:intersect from the [CALC] menu ([2nd], [TRACE]). Use ▲ or ▼ and [ENTER] to select each function. Then move the cursor near the desired intersection and [ENTER] this guess. The x-value of the point of intersection is a solution to the equation.

solution: $x \approx 6.225$

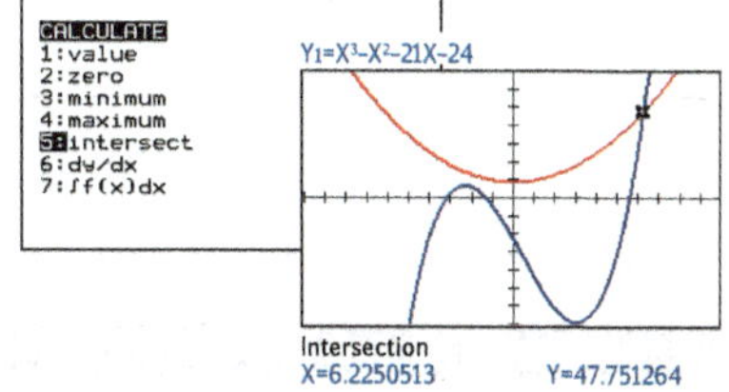

SKILL ✔ EXERCISE 7

It may be necessary to zoom out to verify that all points of intersection have been identified. The end behaviors of the functions in Example 2 indicate there is only one solution. An alternative method of solving Example 2 involves subtracting all the terms on one side of the equation to obtain an equivalent equation in the form of $h(x) = 0$ and finding its zeros, as illustrated in Example 1.

The method used in Example 2 can also be applied to solve equations of the form of $f(x) = k$, where k is a constant.

Solutions

Exercises

3. window: $[-3, 3]$ by $[-3, 3]$

4. window: $[-3, 3]$ by $[-3, 3]$
   $x = -\frac{3}{7} \approx -0.43$ or $x = \frac{1}{3} \approx 0.33$

5. 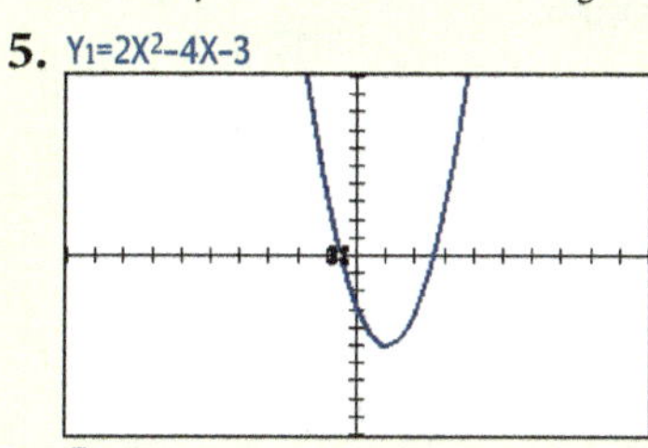
   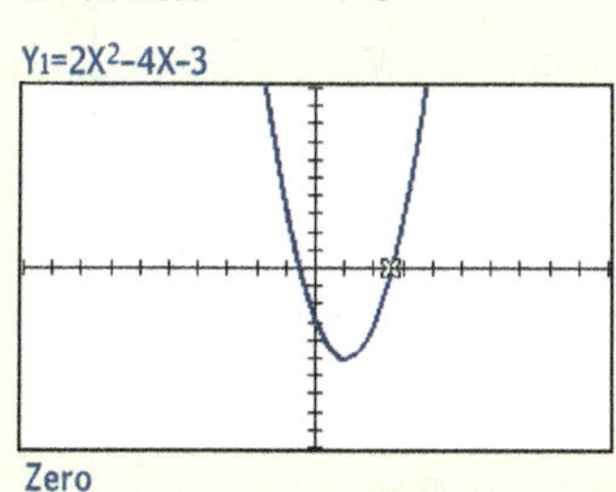

6. 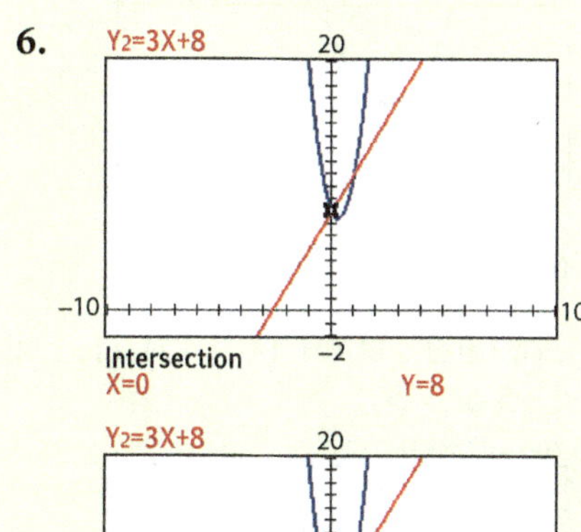
   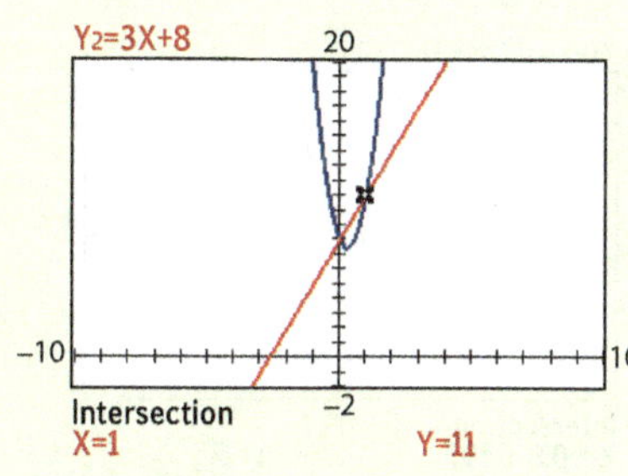

7. pt. of intersection: $\approx (1.383, -6.087)$

8. pt. of intersection: $\approx (3.332, 2.768)$

10. pts. of intersection:
    $\approx (-1.224, 0.050)$, $\approx (3.977, -3.714)$

11. Let $Y_1 = x^2 - 2x - 6$ and $Y_2 = 2$.

12. Let $Y_1 = 2x^2 + 5x - 4$ and $Y_2 = 3$.

13. Let $Y_1 = 5x^3 - 6x^2 - 9x + 5$ and $Y_2 = 3$.

14. Let $Y_1 = 25x^3 + 45x^2 - 16x - 15$ and $Y_2 = -3$.

15. Let $Y_1 = 4x^4 + 4x^3 - 7x^2 - 4x + 8$ and $Y_2 = 5$.

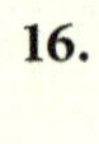

16. 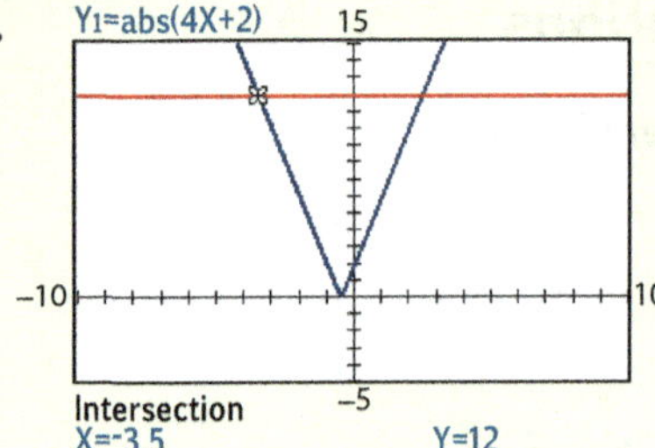

pts. of intersection: $(-3.5, 12)$, $(2.5, 12)$; $-3.5 < x < 2.5$

17. 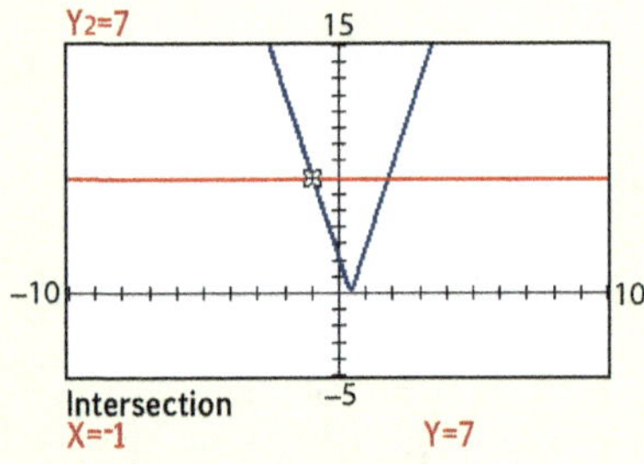

pts. of intersection: $(-1, 7)$, $(1.8, 7)$
$x \leq -1$ or $x \geq 1.8$

18. 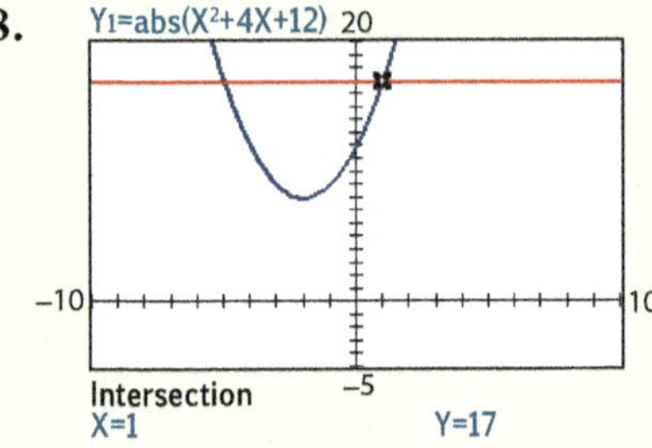

pts. of intersection: $(-5, 17)$, $(1, 17)$
$x < -5$ or $x > 1$

19. 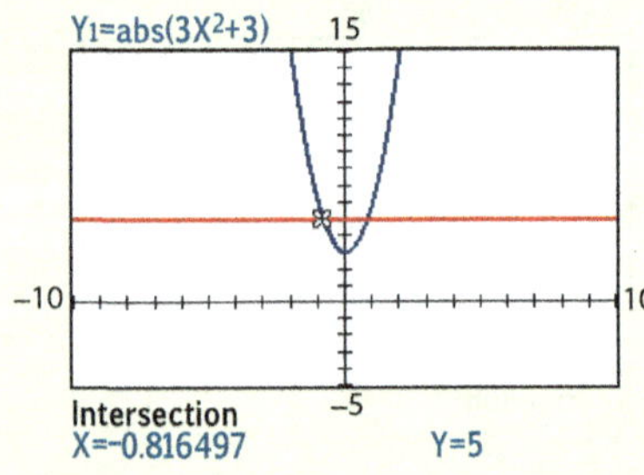

pts. of intersection: $\approx (\pm 0.816, 5)$
$-0.81 \leq x \leq 0.81$

20. 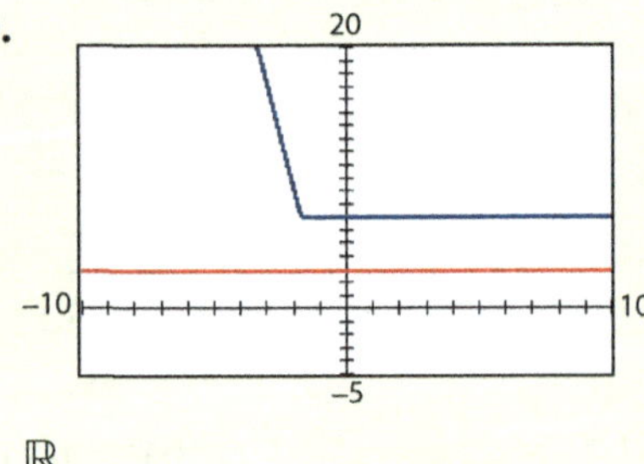

Solve $2x^2 - x - 2 = 4$.

Answer

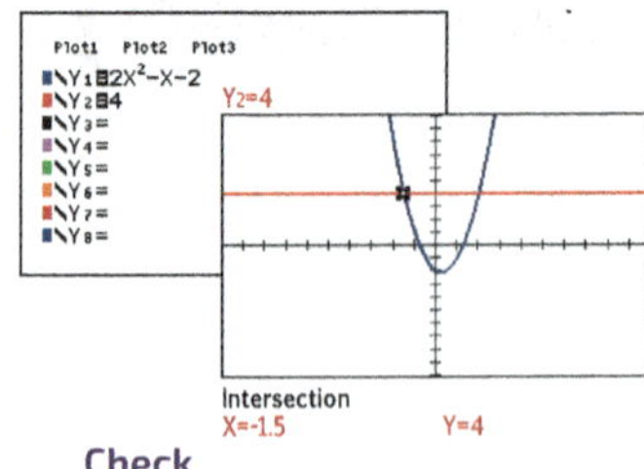

1. Define $Y_1 = 2x^2 - x - 2$ and $Y_2 = 4$ and graph the functions in the standard window as shown. Find the x-value of any intersection of the graphs using the 5:intersect function from the [CALC] menu.

The solution includes the illustrated solution of $x = -1.5$ and the x-coordinate of the second point of intersection, $x = 2$.

Check

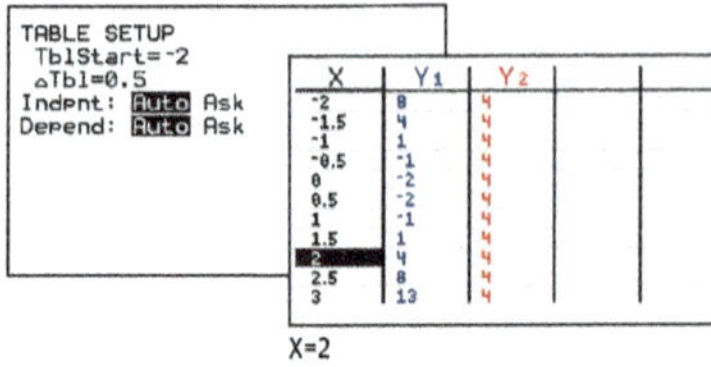

2. The TABLE SETUP menu is accessed using [TBLSET] ([2nd], [WINDOW]) and is used to define the TblStart and ΔTbl values. Select [TABLE] ([2nd], [GRAPH]) to generate a table that verifies that Y_1, the value of the left side of the equation, is 4 when $x = -1.5$ and when $x = 2$.

SKILL ✓ EXERCISE 13

These graphing techniques can also be used to solve inequalities. To solve inequalities of the form $f(x) > g(x)$, graph $Y_1 = f(x)$ and $Y_2 = g(x)$ and identify the interval(s) of x-values for which the graph of Y_1 is above the graph of Y_2.

Solve the inequality $|2x + 1| \geq 5$.

Answer

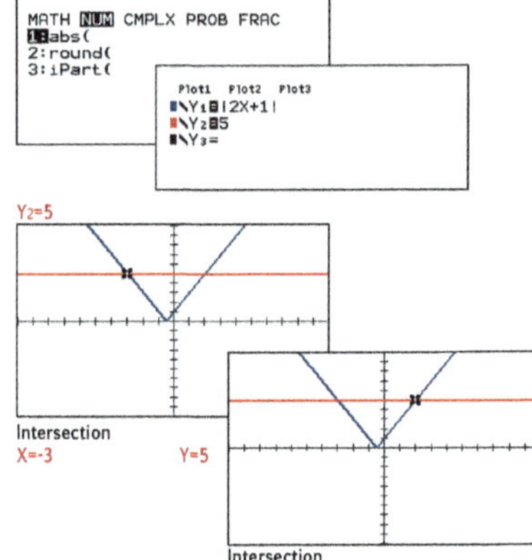

1. Use the absolute value function 1: abs on the [MATH], NUM submenu to enter $Y_1 = |2x + 1|$, and then enter $Y_2 = 5$. Then graph the functions in the standard window.

2. Use 5:intersect to identify the x-coordinates where $Y_1 = Y_2$. Since $Y_1 > Y_2$ to the left of $x = -3$ and to the right of $x = 2$, the solution to the inequality can be stated with the disjunction $x \leq -3$ or $x \geq 2$ or in interval notation as $(-\infty, -3] \cup [2, \infty)$.

SKILL ✓ EXERCISE 19

---

21. 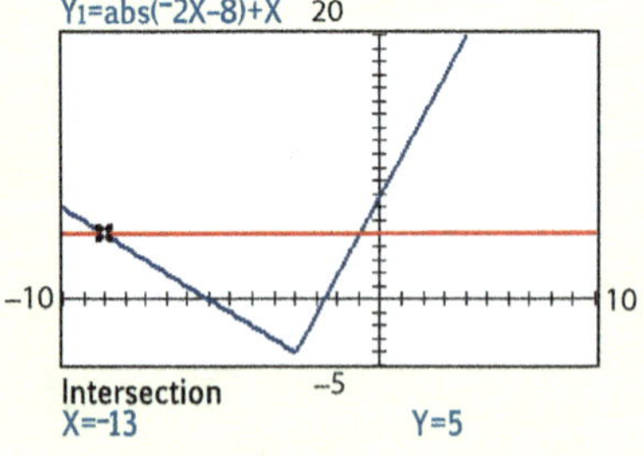

pts. of intersection: $(-13, 5)$, $(-1, 5)$
$-13 < x < -1$

22. 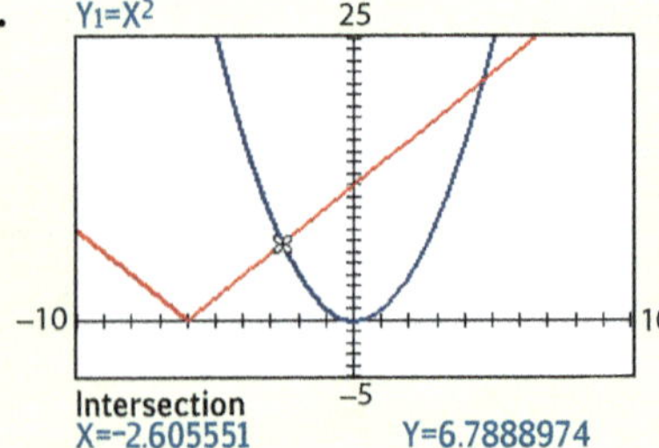

pts. of intersection:
$\approx (-2.606, 6.789)$, $\approx (4.606, 21.211)$
$x < -2.61$ or $x > 4.61$

Inequalities of the form $f(x) < g(x)$ can also be solved by applying the same graphing steps, and then identifying the interval(s) of x-values for which the graph of Y_1 is below the graph of Y_2.

❯ Exercises

Solve each equation graphically by finding zeros. Round to the nearest hundredth if necessary.

1. $x^2 - 2x - 15 = 0$ **2.** $-x^2 - x + 6 = 0$

3. $10x^2 - 3x - 4 = 0$ **4.** $21x^2 + 2x - 3 = 0$

5. $2x^2 - 4x - 3 = 0$

Solve each equation graphically by finding points of intersection. Round to the nearest hundredth if necessary.

6. $7x^2 - 4x + 8 = 3x + 8$ $x = 0, 1$

7. $3x^3 + 3x + 6 = x^2 - 8$ $x = -1.38$

8. $x^2 + 2x - 15 = \sqrt{5x - 9}$ $x = 3.33$

9. $-4x^2 + 8 = 2x^2 - 3$ $x = \pm 1.35$

10. $0.5x^2 - 2.1x - 3.27 = -2.37x^2 + 5.8x + 10.7$
$$x = -1.22, 3.98$$

Solve each equation graphically. Then use a table to verify the solution numerically.

11. $x^2 - 2x - 6 = 2$ $x = -2, 4$

12. $2x^2 + 5x - 4 = 3$ $x = -3.5, 1$

13. $5x^3 - 6x^2 - 9x + 5 = 3$ $x = -1, 0.2, 2$

14. $25x^3 + 45x^2 - 16x - 15 = -3$ $x = -2, -0.4, 0.6$

15. $4x^4 + 4x^3 - 7x^2 - 4x + 8 = 5$ $x = -1.5, -1, 0.5, 1$

Use graphing technology to solve the following inequalities.

16. $|4x + 2| < 12$ **17.** $|-5x + 2| \geq 7$

18. $|x^2 + 4x - 12| > 17$ **19.** $|3x^2 + 3| \leq 5$

20. $|4x + 7| - 4x > 3$ **21.** $|-2x - 8| + x < 5$

22. $x^2 > |2x + 12|$ **23.** $x^2 \leq |4x - 8|$

24. $|3x^2 - 2| > 1$ **25.** $8 < |5x - 7| < 20$

1. $x = -3, 5$

2. $x = -3, 2$

3. $x = -0.5, 0.8$

4. $x = -0.43, 0.33$

5. $x = -0.58, 2.58$

16. $(-3.5, 2.5)$

17. $(-\infty, -1] \cup [1.8, \infty)$

18. $(-\infty, -5) \cup (1, \infty)$

19. $[-0.81, 0.81]$

20. $\mathbb{R}$

21. $(-13, -1)$

22. $(-\infty, -2.61), (4.61, \infty)$

23. $[-5.46, 1.46]$

24. $(-\infty, -1) \cup (-0.58, 0.58) \cup (1, \infty)$

25. $(-2.6, -0.2) \cup (3, 5.4)$

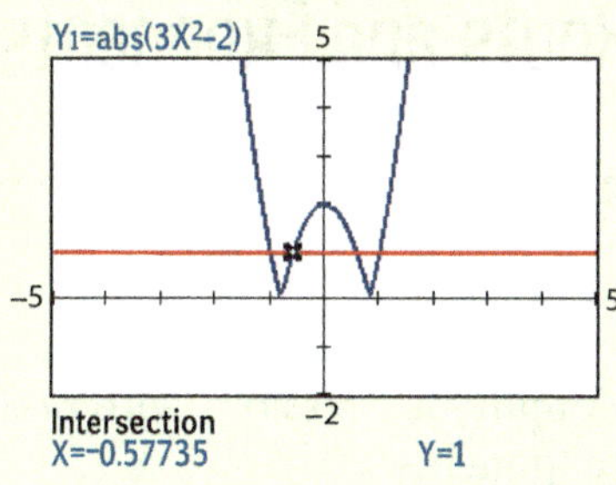

Use tests for symmetry to find the other x-values of ≈ 0.58 and 1.

$$x < -1 \text{ or } -0.58 < x < 0.58 \text{ or } x > 1$$

25.

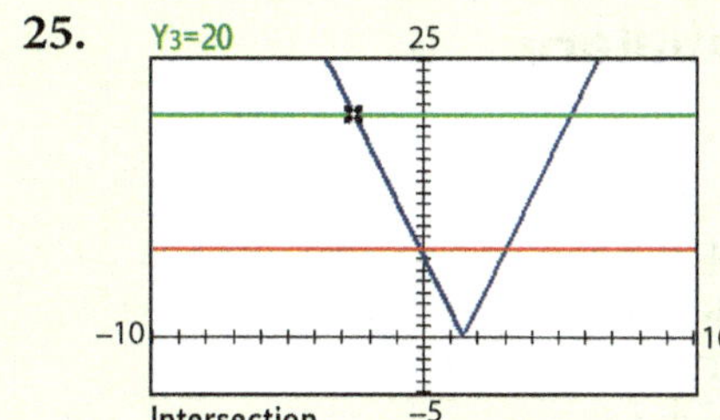

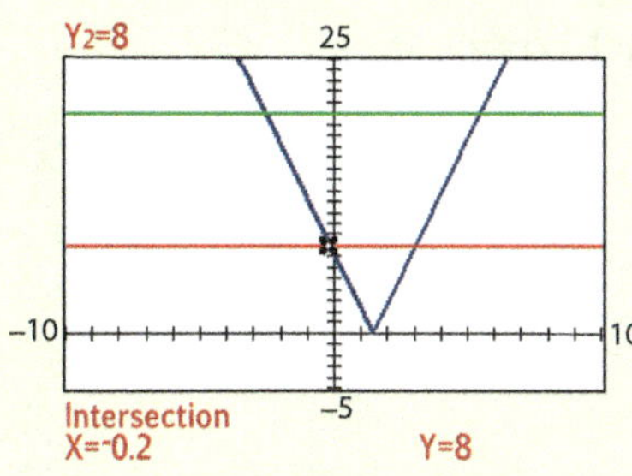

$$-2.6 < x < -0.2$$
or

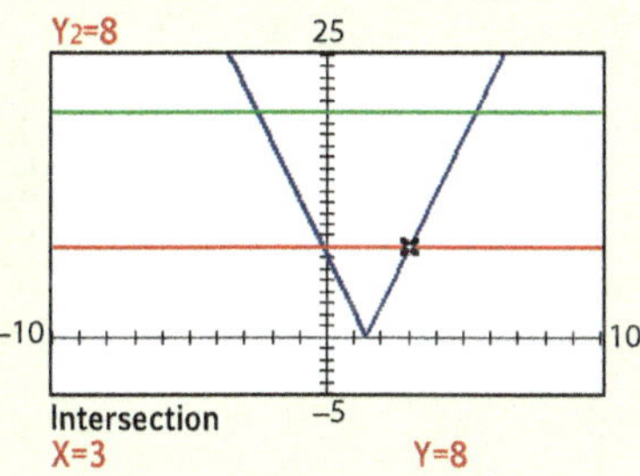

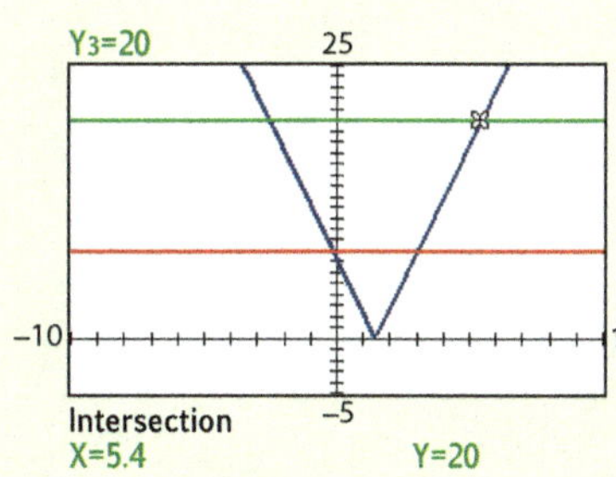

$$3 < x < 5.4$$

23.

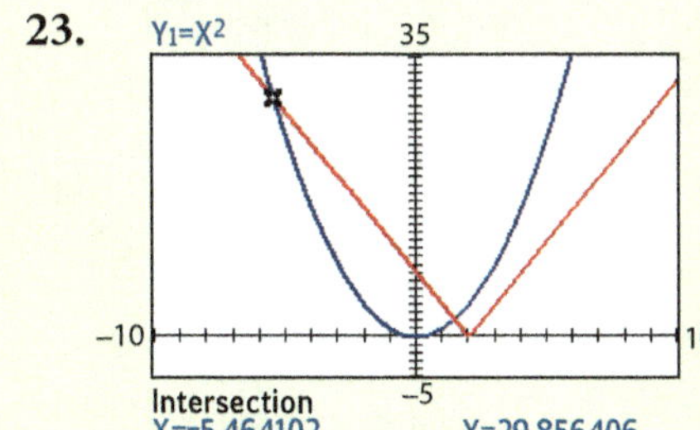

pts. of intersection:
$\approx (-5.464, 29.856), \approx (1.464, 2.144)$
$$-5.46 \leq x \leq 1.46$$

24.

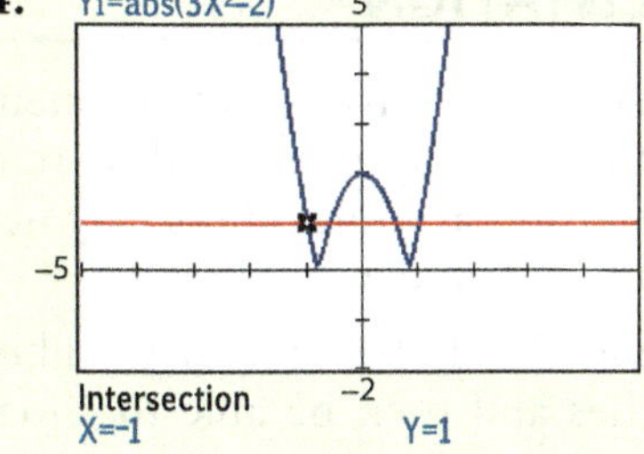

A.4 Roots and Exponents

Objectives

1. To simplify expressions with radicals
2. To simplify expressions with rational exponents
3. To convert between radical and exponential expressions

Vocabulary

nth root of b
nth root of x^n
principal root
Product Property of Radicals
properties of exponents
radical sign
radicand
rational exponent

A.4 Roots and Exponents

The ability to simplify radicals and expressions with rational exponents is crucial to success in advanced math courses. You will also need to know how to convert between radicals and expressions with rational exponents. We will review real roots in this lesson and complex roots in Appendix Lesson 5.

DEFINITIONS

For any real numbers a and b and any positive integer n, if $a^n = b$, then a is an **nth root of b**. The *radical sign* indicates a root. The expression under a radical sign is called the *radicand*.

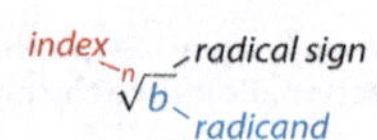

Real nth Roots of b		
Radicand	n is even	n is odd
$b > 0$	one positive and one negative	one positive
$b = 0$	one (0)	one (0)
$b < 0$	none	one negative

When n is even, the radical sign indicates the nonnegative or *principal root*.
$$\sqrt{9} = 3 \text{ and } -\sqrt{9} = -3$$

Example 1 Simplifying Numerical Radicals

Find each real number root. If no real root exists, state so.

a. $\sqrt[3]{-512}$ **b.** $\sqrt[4]{-81}$ **c.** $\sqrt{0.36}$ **d.** $\sqrt{10{,}000}$ **e.** $\sqrt[3]{\frac{64}{27}}$

Answer

a. $\sqrt[3]{-512} = \sqrt[3]{(-8)^3} = -8$ **b.** no real root exists

c. $\sqrt{0.36} = \sqrt{(0.6)^2} = 0.6$ **d.** $\sqrt{10{,}000} = \sqrt{10^4} = 10^2 = 100$

e. $\sqrt[3]{\frac{64}{27}} = \sqrt[3]{\frac{4^3}{3^3}} = \frac{4}{3}$

SKILL ✓ EXERCISES 1, 5

Other radicals are also simplified using the Product Property of Radicals.

PRESENTATION

This lesson reviews roots and exponents. Some students may only need to reference this material while others may need a more thorough review.

① Students should be familiar with both approaches and then be able to choose the more desirable or simpler approach. For example, it may be desirable to express $5^{\frac{3}{2}}$ as $\sqrt[2]{5^3} = 5\sqrt{5}$ or as $\left(\sqrt{5}\right)^3$.

② The coefficient is usually factored into primes and then the Power Property of Exponents is applied.

Product Property of Radicals

If $\sqrt[n]{x}$ and $\sqrt[n]{y}$ are real numbers, $\sqrt[n]{xy} = \sqrt[n]{x} \cdot \sqrt[n]{y}$.

Example 2 Simplifying Variable Radicals

Simplify each radical.

a. $\sqrt[3]{54x^7y^{12}}$ **b.** $\sqrt[4]{405a^{15}b^9}$ **c.** $\sqrt{-25c^{12}}$

Answer

a. $\sqrt[3]{54x^7y^{12}} = \sqrt[3]{(3)^3(2)(x^2)^3(x)(y^4)^3} = 3x^2y^4\sqrt[3]{2x}$

b. $\sqrt[4]{405a^{15}b^9} = \sqrt[4]{3^4(5)(a^3)^4 a^3(b^2)^4 b} = 3a^4b^2\sqrt[4]{5a^3b}$

c. $\sqrt{-25c^{12}}$ has no real roots since the radicand is negative and the index is even.
(The radical indicates the imaginary root $5c^6i$.)

SKILL ✔ **EXERCISE 9**

Be careful when finding an even root of an even power. When $x = -3$,
$\sqrt{x^2} = \sqrt{(-3)^2} = \sqrt{9} = 3$, which is $|x|$, not x.

▌**DEFINITION**

The **nth root of x^n** or $\sqrt[n]{x^n} = \begin{cases} |x| \text{ if } n \text{ is even.} \\ x \text{ if } n \text{ is odd.} \end{cases}$

The absolute value is used when n is even because the principal root is desired. The absolute value can be dropped when the result is an even power, which cannot be negative.

Example 3 Simplifying Variable Radicals by Using Absolute Values

Simplify each radical.

a. $\sqrt{25x^{14}}$ **b.** $-\sqrt{49y^2z^{12}}$ **c.** $\sqrt{320x^2yz^4}$

Answer

a. $\sqrt{5^2(x^7)^2} = 5|x|^7$ **b.** $-\sqrt{7^2y^2(z^6)^2} = -7|y|z^6$ **c.** $\sqrt{5(8)^2x^2yz^4}$
$= 8|x|z^2\sqrt{5y}$

SKILL ✔ **EXERCISE 11**

The properties of exponents apply to all integral and fractional exponents. In general, $\left(x^{\frac{1}{n}}\right)^n = x^{\frac{1}{n}(n)} = x^1 = x$; thus $x^{\frac{1}{n}} = \sqrt[n]{x}$. Note that the exponential form also indicates the principal root.

Properties of Exponents	
Property or Definition	**Rational Example**
Product $x^a \cdot x^b = x^{a+b}$	$x^{\frac{1}{2}} \cdot x^{\frac{1}{2}} = x^{\frac{1}{2}+\frac{1}{2}} = x^1 = x$
Quotient $\dfrac{x^a}{x^b} = x^{a-b}; x \neq 0$	$x^{\frac{7}{2}} \div x^{\frac{3}{2}} = x^{\frac{7}{2}-\frac{3}{2}} = x^{\frac{4}{2}} = x^2$
Power $(x^a)^b = x^{ab}$	$\left(x^{\frac{3}{4}}\right)^8 = x^{\frac{3}{4} \cdot \frac{8}{1}} = x^{\frac{24}{4}} = x^6$
Power of a Product $(xy)^a = x^a y^a$	$\left(x^2 y^3\right)^{\frac{1}{6}} = x^{\frac{2}{6}} y^{\frac{3}{6}} = x^{\frac{1}{3}} y^{\frac{1}{2}}$
Power of a Quotient $\left(\dfrac{x}{y}\right)^a = \dfrac{x^a}{y^a}; y \neq 0$	$\left(\dfrac{z^3}{25}\right)^{\frac{1}{2}} = \dfrac{z^{3\left(\frac{1}{2}\right)}}{5^{2\left(\frac{1}{2}\right)}} = \dfrac{z^{\frac{3}{2}}}{5^1} = \dfrac{z^{\frac{3}{2}}}{5}$
Zero Exponent $x^0 = 1; x \neq 0$	$\left(\dfrac{1}{4}\right)^0 = 1$
Negative Exponent $x^{-n} = \dfrac{1}{x^n}$ and $\dfrac{1}{x^{-n}} = x^n; x \neq 0$	$25^{-\frac{1}{2}} = \dfrac{1}{(5^2)^{\frac{1}{2}}} = \dfrac{1}{5^1} = \dfrac{1}{5}$ and $\left(\dfrac{1}{8}\right)^{-\frac{1}{3}} = 8^{\frac{1}{3}} = (2^3)^{\frac{1}{3}} = 2^1 = 2$

Evaluate each exponential expression.

a. $16^{\frac{1}{4}}$ **b.** $-125^{\frac{1}{3}}$ **c.** $9^{-\frac{1}{2}}$

Answer

a. $\sqrt[4]{16} = \sqrt[4]{(2)^4} = 2$ **b.** $-\sqrt[3]{125} = -\sqrt[3]{(5)^3} = -5$ **c.** $\dfrac{1}{9^{\frac{1}{2}}} = \dfrac{1}{\sqrt{9}} = \dfrac{1}{3}$

SKILL ✔ **EXERCISE 15**

Recognizing that $x^{\frac{m}{n}} = (x^m)^{\frac{1}{n}} = \left(x^{\frac{1}{n}}\right)^m$ leads to the following definition.

▶ RATIONAL EXPONENT

If the nth root of x is real and m is an integer, $x^{\frac{m}{n}} = \sqrt[n]{x^m} = (\sqrt[n]{x})^m$.

This definition indicates that you can evaluate either the power or the root first.

Evaluate $4^{\frac{3}{2}}$.

Answer

evaluating the power first: or   evaluating the root first:

$\sqrt[2]{4^3} = \sqrt[2]{64} = 8$ $(\sqrt[2]{4})^3 = 2^3 = 8$

SKILL ✔ **EXERCISE 21**

Combining the properties of exponents and the definition of a rational exponent enables the conversion between radical and exponential forms of an expression.

Example 6 Converting Between Radical and Exponential Forms

Convert each expression.

a. $\sqrt[4]{75a^3b}$ to exponential form

b. $2^{\frac{3}{5}}x^{\frac{4}{5}}y^{\frac{1}{2}}$ to radical form

Answer

a. The simplest exponential form should include only prime factors.

$$75^{\frac{1}{4}}a^{\frac{3}{4}}b^{\frac{1}{4}} = (3\cdot 5^2)^{\frac{1}{4}}a^{\frac{3}{4}}b^{\frac{1}{4}}$$
$$= 3^{\frac{1}{4}}\cdot 5^{\frac{2}{4}}a^{\frac{3}{4}}b^{\frac{1}{4}}$$
$$= 3^{\frac{1}{4}}\cdot 5^{\frac{1}{2}}a^{\frac{3}{4}}b^{\frac{1}{4}}$$

b. Rename each exponent with the lowest common denominator to rewrite the expression as a single radical.

$$2^{\frac{6}{10}}x^{\frac{8}{10}}y^{\frac{5}{10}} = \sqrt[10]{64x^8y^5}$$

SKILL ✔ EXERCISES 23, 27

The properties of exponents are used to simplify expressions with rational exponents and radical expressions.

Example 7 Simplifying Expressions with Rational Exponents

Simplify each expression, stating answers in the same form as the original expression.

a. $(36a^6b^{12})^{1.5}$ **b.** $\left(\dfrac{216x^9}{125}\right)^{\frac{1}{3}}$ **c.** $\sqrt[4]{1024}$ **d.** $\dfrac{\sqrt{8}}{\sqrt[3]{4}}$

Answer

a. $(6^2a^6b^{12})^{\frac{3}{2}}$
$= (6^2)^{\frac{3}{2}}(a^6)^{\frac{3}{2}}(b^{12})^{\frac{3}{2}}$
$= 6^3a^9b^{18} = 216a^9b^{18}$

b. $\left(\dfrac{6^3x^9}{5^3}\right)^{\frac{1}{3}}$
$= \dfrac{(6^3)^{\frac{1}{3}}(x^9)^{\frac{1}{3}}}{(5^3)^{\frac{1}{3}}} = \dfrac{6}{5}x^3$

c. $\sqrt[4]{2^{10}}$
$= \sqrt[4]{2^8 2^2}$
$= 2^2\cdot \sqrt[4]{2^2}$
$= 4\cdot 2^{\frac{1}{2}}$
$= 4\sqrt{2}$

d. $\dfrac{2^{\frac{3}{2}}}{2^{\frac{2}{3}}}$
$= 2^{\frac{3}{2}}\cdot 2^{-\frac{2}{3}}$
$= 2^{\frac{5}{6}} = \sqrt[6]{2^5} = \sqrt[6]{32}$

SKILL ✔ EXERCISE 31

❯ **Exercises**

Evaluate each expression without using a calculator. State "no real solution" if none exists.

1. $\sqrt{256}$ 16
2. $\sqrt[3]{-27}$ -3
3. $\sqrt{-324}$ no real solution
4. $-\sqrt{810,000}$ -900
5. $\sqrt[3]{-0.027}$ -0.3
6. $\sqrt[4]{\dfrac{625}{256}}$ $\dfrac{5}{4}$

Simplify each radical expression. Assume variables represent positive real numbers.

7. $\sqrt{128x^8y^2z^5}$
8. $\sqrt[4]{-64a^{10}b^4c}$
9. $-\sqrt{192x^{10}y^{13}}$
10. $\sqrt[3]{405a^4bc^6}$

Simplify each expression, using absolute values when needed.

11. $\sqrt{4x^2}$ $2|x|$
12. $-\sqrt{25x^8y^6}$ $-5x^4|y|^3$
13. $\sqrt{(3n-4)^2}$ $|3n-4|$
14. $\sqrt{81x^3y}$ $9|x|\sqrt{xy}$

Evaluate each expression.

15. $81^{\frac{1}{4}}$ 3
16. $(-27)^{\frac{1}{3}}$ -3
17. $125^{-\frac{1}{3}}$ $\dfrac{1}{5}$
18. $32^{\frac{1}{5}}$ 2
19. $8^{\frac{2}{3}}$ 4
20. $(-27)^{\frac{4}{3}}$ 81

21. $81^{-\frac{3}{2}}$ $\dfrac{1}{729}$
22. $-256^{\frac{3}{4}}$ -64

Convert each radical to exponential form. Assume variables represent positive real numbers.

23. $\sqrt{5xy^5}$ $5^{\frac{1}{2}}x^{\frac{1}{2}}y^{\frac{5}{2}}$
24. $\sqrt{25a^3b^2}$ $5a^{\frac{3}{2}}b$
25. $\sqrt[3]{12a^2b}$ $2^{\frac{2}{3}}3^{\frac{1}{3}}a^{\frac{2}{3}}b^{\frac{1}{3}}$
26. $\sqrt[5]{128x^7y^3z}$ $2^{\frac{7}{5}}x^{\frac{7}{5}}y^{\frac{3}{5}}z^{\frac{1}{5}}$

Convert each exponential expression to radical form. Assume all variables represent positive real numbers.

27. $3^{\frac{2}{3}}x^{\frac{1}{3}}$ $\sqrt[3]{9x}$
28. $2^{\frac{3}{4}}c^{\frac{3}{2}}$ $\sqrt[4]{8c^6}$
29. $7^{\frac{2}{5}}x^{\frac{3}{10}}$ $\sqrt[10]{2401x^3}$
30. $5^{\frac{1}{4}}x^{\frac{2}{3}}y^{\frac{5}{4}}$ $\sqrt[12]{125x^8y^{15}}$

Simplify each expression without using a calculator.

31. $32^{1.4}$ 128
32. $\left(\dfrac{64x^3}{27y^6}\right)^{\frac{1}{3}}$ $\dfrac{4x}{3y^2}$
33. $\sqrt[4]{3^6}$ $3\sqrt{3}$
34. $\dfrac{\sqrt{27}}{\sqrt[3]{9}}$ $\sqrt[6]{243}$

7. $8x^4yz^2\sqrt{2z}$
8. no real solution
9. $-8x^5y^6\sqrt{3y}$
10. $3ac^2\sqrt[3]{15ab}$

---

22. $-(2^8)^{\frac{3}{4}} = -(2^2)^3 = -4^3 = -64$

23. $(5xy^5)^{\frac{1}{2}} = 5^{\frac{1}{2}}x^{\frac{1}{2}}y^{\frac{5}{2}}$

24. $(5^2a^3b^2)^{\frac{1}{2}} = 5a^{\frac{3}{2}}b$

25. $(3\cdot 2^2a^2b)^{\frac{1}{3}} = 2^{\frac{2}{3}}\cdot 3^{\frac{1}{3}}a^{\frac{2}{3}}b^{\frac{1}{3}}$

26. $(2^7x^7y^3z)^{\frac{1}{5}} = 2^{\frac{7}{5}}x^{\frac{7}{5}}y^{\frac{3}{5}}z^{\frac{1}{5}}$

27. $(3^2x)^{\frac{1}{3}} = \sqrt[3]{9x}$

28. $2^{\frac{3}{4}}c^{\frac{3}{2}} = (2^3c^6)^{\frac{1}{4}} = \sqrt[4]{8c^6}$

29. $7^{\frac{2}{5}}x^{\frac{3}{10}} = (7^4x^3)^{\frac{1}{10}} = \sqrt[10]{2401x^3}$

30. $(5^3x^8y^{15})^{\frac{1}{12}} = \sqrt[12]{125x^8y^{15}}$

31. $32^{\frac{7}{5}} = (2^5)^{\frac{7}{5}} = 2^7 = 128$

32. $\dfrac{(4^3)^{\frac{1}{3}}(x^3)^{\frac{1}{3}}}{(3^3)^{\frac{1}{3}}(y^6)^{\frac{1}{3}}} = \dfrac{4x}{3y^2}$

33. $3^{\frac{6}{4}} = 3^{\frac{3}{2}} = 3\cdot 3^{\frac{1}{2}} = 3\sqrt{3}$

34. $\dfrac{27^{\frac{1}{2}}}{9^{\frac{1}{3}}} = \dfrac{(3^3)^{\frac{1}{2}}}{(3^2)^{\frac{1}{3}}} = 3^{\frac{3}{2}}\cdot 3^{-\frac{2}{3}} = 3^{\frac{3}{2}-\frac{2}{3}}$

$$= 3^{\frac{5}{6}} = \sqrt[6]{3^5} = \sqrt[6]{243}$$

Solutions

❯ **Exercises**

1. $\sqrt{256} = \sqrt{16^2} = 16$

2. $\sqrt[3]{-27} = \sqrt[3]{(-3)^3} = -3$

4. $-\sqrt{810,000} = -\sqrt{900^2} = -900$

5. $\sqrt[3]{-0.027} = \sqrt[3]{(-0.3)^3} = -0.3$

6. $\sqrt[4]{\dfrac{625}{256}} = \sqrt[4]{\dfrac{5^4}{4^4}} = \dfrac{5}{4}$

7. $\sqrt{2^7x^8y^2z^5} = 8x^4yz^2\sqrt{2z}$

9. $-\sqrt{3(2)^6x^{10}y^{13}} = -8x^5y^6\sqrt{3y}$

10. $\sqrt[3]{3^4 5a^4bc^6} = 3ac^2\sqrt[3]{3.5ab}$
$= 3ac^2\sqrt[3]{15ab}$

12. $-\sqrt{5^2x^8y^6} = -5x^4|y^3|$

14. $\sqrt{3^4x^3y} = 3^2|x|\sqrt{xy} = 9|x|\sqrt{xy}$

15. $(3^4)^{\frac{1}{4}} = 3$

16. $((-3)^3)^{\frac{1}{3}} = -3$

17. $(5^3)^{-\frac{1}{3}} = 5^{-1} = \dfrac{1}{5}$

18. $(2^5)^{\frac{1}{5}} = 2$

19. $\sqrt[3]{8^2} = \sqrt[3]{(2^3)^2} = 2^{\frac{6}{3}} = 2^2 = 4$

20. $((-3)^3)^{\frac{4}{3}} = (-3)^4 = 81$

21. $(9^2)^{-\frac{3}{2}} = 9^{-3} = \dfrac{1}{9^3} = \dfrac{1}{729}$

A.5 Complex Numbers

Objectives

1. To use set notation to describe relationships between groups of objects
2. To identify the subsets of the real number system, their relationship to each other, and examples of each
3. To perform the set operations of union, intersection, and complement
4. To use Venn diagrams to illustrate set operations and subsets

Vocabulary

conjugates
complex conjugates
complex number
imaginary number
imaginary unit
pure imaginary number
quadratic formula

After completing this section, you will be able to
- define the imaginary unit i, imaginary numbers, and complex numbers.
- simplify powers of i.
- add, subtract, and multiply complex numbers.
- write the quotient of two complex numbers in standard form.
- find complex roots of quadratic equations.

A.5 Complex Numbers

Solving $x^2 + 1 = 0$ by taking roots produces $x^2 = -1$ or $x = \pm\sqrt{-1}$. Since the result "$\sqrt{-1}$" is not a real number, it is designated as "imaginary." Leonard Euler (1707–83) first used i to represent $\sqrt{-1}$.

> **DEFINITION**
>
> The **imaginary unit**, i, is the square root of -1, so that $i = \sqrt{-1}$ and $i^2 = -1$.

The square roots of other negative numbers are expressed in terms of i.
$$\sqrt{-4} = \sqrt{4} \cdot \sqrt{-1} = 2i \text{ and } \sqrt{-24} = \sqrt{4} \cdot \sqrt{6} \cdot \sqrt{-1} = 2i\sqrt{6}$$

Numbers such as i, $2i$, and $2i\sqrt{6}$ are called *pure imaginary numbers*. Any *complex number*, z, can be expressed in the standard form $z = a + bi$, where $a, b \in \mathbb{R}$ and $i = \sqrt{-1}$. The set of complex numbers contains the real numbers, $a + 0i$, and the pure imaginary numbers, $0 + bi$. The set of imaginary numbers $a + bi$, $b \neq 0$ is the complement of the real numbers within the set of complex numbers. Mathematicians studied imaginary numbers for many years without any known real-life applications. Today, imaginary numbers are essential elements in electronics and engineering.

Notice the pattern in the powers of i that can be used to quickly simplify larger powers of i.

$$i = i \qquad\qquad i^5 = i^4 \cdot i = 1 \cdot i = i$$
$$i^2 = -1 \qquad\qquad i^6 = i^4 \cdot i^2 = 1 \cdot -1 = -1$$
$$i^3 = i^2 \cdot i = -1 \cdot i = -i \qquad\qquad i^7 = i^4 \cdot i^3 = 1 \cdot -i = -i$$
$$i^4 = i^2 \cdot i^2 = -1 \cdot -1 = 1 \qquad\qquad i^8 = i^4 \cdot i^4 = 1 \cdot 1 = 1$$

To simplify large powers of i, divide the exponent by 4. The remainder (which will be 3 or less) will give a power of i equivalent to the original.

Example 1   Simplifying a Power of i

Simplify i^{27}.

Answer

$27 \div 4 = 6\,R.\,3$

1. Divide the exponent by 4 to determine the number of groupings of i^4 and the number of factors of i leftover.

$i^{27} = (i^4)^6 \cdot i^3 = 1^6 \cdot i^3$
$= i^3 = -1$

2. Use the fact that each group of $i^4 = 1$ and simplify the remaining power of i.

SKILL ✔ EXERCISE 1

The set of complex numbers is governed by most of the properties that have been established for the set of real numbers. The Commutative, Associative and Distributive Properties hold true for the complex numbers. The Product Property of Radicals cannot be applied to negative radicands of even roots: $\sqrt{-1} \cdot \sqrt{-1} \neq \sqrt{1}$. Always write complex numbers in standard form before simplifying.
For example, $\sqrt{-4} \cdot \sqrt{-9} = 2i \cdot 3i = 6i^2 = -6$.

To add or subtract complex numbers, simply add or subtract their real and imaginary components. The same result is obtained by treating complex numbers as binomials and combining like terms. When the FOIL method is used to multiply complex numbers, the product of the last terms has a factor of i^2 that is simplified to -1 before that term is combined with the real product of the first terms. To divide complex numbers, write the division as a fraction and multiply both the numerator and the denominator $c + di$ by its *conjugate*, $c - di$. The product in the denominator, $(c + di)(c - di) = c^2 + d^2$, no longer contains the radical $i = \sqrt{-1}$.

Operation	Rule
addition	$(a + bi) + (c + di) = (a + c) + (b + d)i$
subtraction	$(a + bi) - (c + di) = (a - c) + (b - d)i$
multiplication	$(a + bi)(c + di) = (ac - bd) + (ad + bc)i$
division	$\dfrac{(a + bi)}{(c + di)} \cdot \dfrac{(c - di)}{(c - di)} = \dfrac{(ac + bd) + (bc - ad)i}{c^2 + d^2}$

Example 2 Performing Basic Operations with Complex Numbers

Complete each operation, given $z_1 = 3 + 4i$ and $z_2 = 5 - 6i$.

a. $z_1 + z_2$ **b.** $z_1 - z_2$ **c.** $z_1 \cdot z_2$ **d.** $z_1 \div z_2$

Answer

a. $(3 + 4i) + (5 - 6i)$
$= 8 - 2i$

Combine the real parts and imaginary parts as you would combine like terms.

b. $(3 + 4i) - (5 - 6i)$
$= -2 + 10i$

Combine the real parts and imaginary parts as you would combine like terms.

c. $(3 + 4i)(5 - 6i)$
$= 15 - 18i + 20i - 24i^2$
$= 15 + 2i + 24$
$= 39 + 2i$

Multiply using the FOIL method, substitute for i^2, and then simplify.   ①

d. $\dfrac{3 + 4i}{5 - 6i} \cdot \dfrac{5 + 6i}{5 + 6i}$
$= \dfrac{15 + 18i + 20i + 24i^2}{25 - 36i^2} = \dfrac{-9 + 38i}{25 + 36}$
$= \dfrac{-9 + 38i}{61} = -\dfrac{9}{61} + \dfrac{38}{61}i$

Multiply the numerator and the denominator by the denominator's conjugate and simplify, expressing the result in standard form.

SKILL ✔ **EXERCISES 9, 13, 15**

Recall that the product of conjugates is a difference of squares: $(a + b)(a - b) = a^2 - b^2$. This pattern is used to factor the difference of squares. It is also helpful to recognize that the product of *complex conjugates* is a sum of squares: $(c + di)(c - di) = c^2 + d^2$. This pattern can be used to factor a sum of squares into the product of two complex numbers.

$$9x^2 + 25 = (3x + 5i)(3x - 5i)$$

An alternate method of factoring a sum of squares is to rewrite the expression as an equivalent sum using i^2 as a factor in the second term before factoring as a difference of squares.

$$9x^2 + 25 = 9x^2 - 25i^2 = (3x + 5i)(3x - 5i)$$

This method provides insight into factoring expressions such as $x^2 + ix + 2$.

$$x^2 + ix + 2 = x^2 + ix - 2i^2 = (x + 2i)(x - i)$$

Solutions

❯ Exercises

1. $i^{15} = (i^4)^3 \cdot i^3 = -i$

2. $i^{46} = (i^4)^{11} \cdot i^2 = -1$

3. $i^{73} = (i^4)^{18} \cdot i = i$

4. $i^{156} = (i^4)^{39} = 1$

5. $(5 + 4) + (9 - 6)i = 9 + 3i$

6. $(8 + 3) + (1 - 7)i = 11 - 6i$

7. $(-5 + 11) + (11 - 23)i = 6 - 12i$

8. $11 + (5 - 5)i = 11$

9. $(-7 - 5) + (-5 + 2)i = -12 - 3i$

10. $(8 - 17) + (-15 - 12)i = -9 - 27i$

12. $35i + 21i^2 = -21 + 35i$

13. $25 - 9i^2 = 34$

14. $-32 - 24i - 20i - 15i^2 = -17 - 44i$

15. $\dfrac{6 + 5i}{2 - 3i} \cdot \dfrac{2 + 3i}{2 + 3i}$

$= \dfrac{12 + 18i + 10i + 15i^2}{3 - 9i^2}$

$= \dfrac{12 + 15(-1) + (18 + 10)i}{4 - 9(-1)}$

$= \dfrac{-3 + 28i}{13} = -\dfrac{3}{13} + \dfrac{28}{13}i$

16. $\dfrac{7 + 6i}{4 + 3i} \cdot \dfrac{4 - 3i}{4 - 3i}$

$= \dfrac{28 - 21i + 24i - 18i^2}{16 - 9i^2}$

$= \dfrac{46 + 3i}{25} = \dfrac{46}{25} + \dfrac{3}{25}i$

17. $x^2 - 49i^2 = (x + 7i)(x - 7i)$

18. $9x^2 - 16i^2 = (3x + 4i)(3x - 4i)$

19. $x^2 - 9xi + 14i^2 = (x - 7i)(x - 2i)$

20. $x^2 - 3xi - 28i^2 = (x - 7i)(x + 4i)$

21. $x = \dfrac{-(-6) \pm \sqrt{(-6)^2 - 4(1)(13)}}{2(1)}$

$= \dfrac{6 \pm \sqrt{-16}}{2}$

$= \dfrac{6}{2} \pm \dfrac{\sqrt{-16}}{2} = 3 \pm 2i$

22. $x = \dfrac{-5 \pm \sqrt{5^2 - 4(3)(13)}}{2(3)}$

$= \dfrac{-5 \pm \sqrt{-131}}{6}$

$= -\dfrac{5}{6} \pm \dfrac{\sqrt{131}}{6}i$

23. $x = \dfrac{-8 \pm \sqrt{8^2 - 4(-1)(-18)}}{2(-1)} = \dfrac{-8 \pm 2i\sqrt{2}}{-2}$

$= 4 \pm \sqrt{2}\,i$

24. $x = \dfrac{-(-10) \pm \sqrt{(-10)^2 - 4(4)(9)}}{2(4)}$

$= \dfrac{10 \pm 2i\sqrt{11}}{8}$

$= \dfrac{5}{4} \pm \dfrac{\sqrt{11}}{4}i$

Example 3 Factoring Over the Set of Complex Numbers

Factor each expression over the set of complex numbers.

a. $x^2 + 25$ **b.** $x^2 - 7xi - 10$

Answer

a. $x^2 + 25 = (x + 5i)(x - 5i)$ Apply the sum of squares factoring pattern.

b. $x^2 - 7xi - 10 = x^2 - 7xi + 10i^2$ 1. Replace the last term with an equivalent expression containing i^2.

    $= (x - 5i)(x - 2i)$ 2. Factor the resulting trinomial.

SKILL ✔ **EXERCISES 17, 19**

Many quadratic equations have complex solutions. Since these equations cannot be factored over the real numbers, they are usually solved using the *quadratic formula*.

▌QUADRATIC FORMULA

If $ax^2 + bx + c = 0$, where $a, b, c \in \mathbb{R}$ and $a \neq 0$, then $x = \dfrac{-b \pm \sqrt{b^2 - 4ac}}{2a}$.

Example 4 Using the Quadratic Formula

Solve $2x^2 + 4x + 7 = 0$. Express the roots in standard form.

Answer

$a = 2, b = 4, c = 7$ 1. Identify the values of a, b, and c.

$x = \dfrac{-4 \pm \sqrt{4^2 - 4 \cdot 2 \cdot 7}}{2 \cdot 2}$ 2. Substitute into the quadratic formula.

$x = \dfrac{-4 \pm \sqrt{-40}}{4} = \dfrac{-4 \pm 2\sqrt{10}\,i}{4}$ 3. Simplify.

$x = -1 \pm \dfrac{\sqrt{10}}{2}i$ 4. Express the complex roots in standard form.

SKILL ✔ **EXERCISE 23**

❯ Exercises

Simplify each power.

1. i^{15} $-i$ **2.** i^{46} -1 **3.** i^{73} i **4.** i^{156} 1

Perform each indicated operation. Express answers in standard form.

5. $(5 + 9i) + (4 - 6i)$ **6.** $(8 + i) + (3 - 7i)$

7. $(-5 + 11i) + (11 - 23i)$ **8.** $(11 + 5i) - 5i$ 11

9. $(-7 - 5i) - (5 - 2i)$ **10.** $(8 - 15i) - (17 + 12i)$

11. $13(4 + 2i)$ $52 + 26i$ **12.** $7i(5 + 3i)$ $-21 + 35i$

13. $(5 + 3i)(5 - 3i)$ 34 **14.** $(-8 - 5i)(4 + 3i)$

15. $\dfrac{6 + 5i}{2 - 3i}$ $-\dfrac{3}{13} + \dfrac{28}{13}i$ **16.** $\dfrac{7 + 6i}{4 + 3i}$ $\dfrac{46}{25} + \dfrac{3}{25}i$

5. $9 + 3i$ **6.** $11 - 6i$

7. $6 - 12i$ **10.** $-9 - 27i$

9. $-12 - 3i$ **14.** $-17 - 44i$

Factor each expression over the set of complex numbers.

17. $x^2 + 49$ **18.** $9x^2 + 16$

19. $x^2 - 9xi - 14$ **20.** $x^2 - 3xi + 28$

Solve each equation using the quadratic formula. Express answers in standard form.

21. $x^2 - 6x + 13 = 0$ **22.** $3x^2 + 5x + 13 = 0$

23. $-x^2 + 8x - 18 = 0$ **24.** $4x^2 - 10x + 9 = 0$

17. $(x + 7i)(x - 7i)$ **18.** $(3x + 4i)(3x - 4i)$

19. $(x - 7i)(x - 2i)$ **20.** $(x - 7i)(x + 4i)$

21. $x = 3 \pm 2i$ **22.** $x = -\dfrac{5}{6} \pm \dfrac{\sqrt{131}}{6}i$

23. $x = 4 \pm \sqrt{2}\,i$ **24.** $x = \dfrac{5}{4} \pm \dfrac{\sqrt{11}}{4}i$

Selected Odd Answers

Pages 667–82 in the *Student Edition* contain selected odd answers.

A link to a PDF of these selected odd answers can be found on AfterSchoolHelp.com. Click on "Math" from the "Subjects" menu and then choose "View Overview" for *Precalculus*.

Glossary

absolute value function (p. 18) A function of the form $f(x) = |x|$ where $f(x) = x$ if $x \geq 0$ and $f(x) = -x$ if $x < 0$.

acceleration due to gravity (p. 14) The constant g, where g is approximately 32 ft/sec^2 or 9.8 m/sec^2.

algebraic function (p. 128) A function that involves only the algebraic operations of addition, subtraction, multiplication, division of constants and an independent variable, and raising an independent variable to a rational power.

alternative (research) hypothesis (p. 573) A statement that is the negation of the null hypothesis and contains an inequality such as $>$, $<$, or $\neq$. Denoted H_a.

ambiguous case (p. 270) The SSA case, which does not always uniquely determine a triangle.

amplitude (p. 198) Half the difference of the maximum and minimum values of a periodic function.

anchor step (p. 477) The initial step in mathematical induction that shows $P(n)$ is true for $P(1)$.

angle of depression (p. 183) The angle formed by a horizontal line and the line of sight toward an object that is below the horizontal.

angle of elevation (p. 183) The angle formed by a horizontal line and the line of sight toward an object that is above the horizontal.

angular speed (p. 175) The ratio of the angle of rotation (in radians) per unit of time: $\omega = \frac{\theta}{t}$.

annual percentage rate (APR) (p. 130) The yearly interest rate expressed as a percentage.

annual percentage yield (APY) (p. 134) The equivalent annual rate for an investment at a given annual percentage rate (APR) compounded n times after 1 yr: $P(1 + APY) = P\left(1 + \frac{APR}{n}\right)^n$.

antiderivative (p. 638) If $F'(x) = f(x)$, the function $F(x)$ is an antiderivative of $f(x)$.

aphelion (pp. 407, 438) The point at which an orbiting object is at its farthest distance from the sun.

Arccosine (p. 217) The inverse function of cosine: $y = \text{Cos}^{-1} x$.

arc length formula (p. 174) A central angle of radian measure θ in a circle with radius r intercepts an arc with length $s = r\theta$.

Arcsine (p. 216) The inverse function of sine: $y = \text{Sin}^{-1} x$.

Arctangent (p. 218) The inverse function of tangent: $y = \text{Tan}^{-1} x$.

argument (p. 324) The angle θ of a complex number in polar form $z = r(\cos \theta + i \sin \theta)$.

arithmetic means (p. 458) The terms between any two given terms of an arithmetic sequence.

arithmetic sequence (p. 456) A sequence in which the difference of any two consecutive terms is a constant.

arithmetic series (p. 459) The sum of the terms of an arithmetic sequence.

asymptote (p. 104) A line that a graph continually approaches.

augmented matrix (p. 353) A combination of the coefficient and constant matrices of a linear system.

average velocity (p. 615) $\frac{\text{change in position}}{\text{change in time}} = \frac{\Delta d}{\Delta t}$.

axis of symmetry (pp. 39, 396) The line that divides a parabola into mirror-image halves.

bias (p. 585) A factor that systematically causes a sample to differ from the population.

binomial (p. 472) The sum of two unlike terms.

binomial coefficients (p. 472) The coefficients of the terms generated by expanding $(a + b)^n$.

binomial experiment (p. 474) A statistical experiment of n identical independent trials where each trial results in either a success or a failure.

binomial probability (p. 474) The probability of x successes in n trials of a binomial experiment is $P(x) = {}_nC_x p^x q^{n-x}$, where p is the probability of success and q, the probability of failure, is $1 - p$.

binomial probability distribution (p. 549) A table describing all the possible outcomes of a binomial experiment: $P(x) = {}_nC_x p^x q^{n-x}$.

bivariate data (p. 59) Data containing two related variables.

carrying capacity (p. 158) The limit of growth supported by the environment.

center (of a hyperbola) (p. 412) The midpoint of the transverse axis of a hyperbola.

center (of an ellipse) (p. 403) The midpoint of the major axis of an ellipse.

change of base formula (p. 145) $\log_b x = \frac{\log_a x}{\log_a b}$ for $a, b \in \mathbb{N}$ when $a \neq 1$ and $b \neq 1$.

circular function (pp. 191–92) A trigonometric function, such as $\tan \theta = \frac{y}{x}$, defined in terms of a point $P(x, y)$ on the unit circle where x, y, and θ are real numbers.

closed interval (p. 4) An interval that includes both endpoints. The interval $\{x \mid a \leq x \leq b\}$ is denoted $[a, b]$.

coefficient of determination (p. 60) (r^2 value) Describes the extent to which the variation in the dependent variable can be explained by variation in the independent variable.

cofactor (p. 362) The cofactor C_{ij} of entry a_{ij} in a square matrix with minor M_{ij} is $C_{ij} = (-1)^{i+j} M_{ij}$.

cofunction (pp. 180, 234) A trigonometric function of an angle's complement.

combination (pp. 473, 492) The number of different ways to choose r out of n items where order is not important, denoted ${}_nC_r$, $C(n, r)$, or $\binom{n}{r}$.

common difference (p. 456) The difference, d, between any two consecutive terms in an arithmetic sequence.

common logarithmic function (p. 137) The function $y = \log x$ where $10^y = x$.

common ratio (p. 463) The quotient, r, between any two consecutive terms in a geometric sequence.

completing the square (p. 40) Adding the constant $\left(\frac{b}{2}\right)^2$ to an expression in the form $x^2 + bx$ to obtain a perfect square trinomial.

complex number (p. 644) Any complex number, z, can be expressed in the standard form $z = a + bi$, where $a, b \in \mathbb{R}$ and $i = \sqrt{-1}$.

complex plane (p. 324) A coordinate plane with real and imaginary axes.

components (p. 287) The horizontal and vertical parts of a vector. Denoted $\mathbf{v} = \langle a, b \rangle$, where a represents the horizontal change $(x_2 - x_1)$ and b represents the vertical change $(y_2 - y_1)$ from the tail to the head of the vector.

component vectors (p. 296) The result of decomposing a vector into two orthogonal vectors, one of which is parallel to another given vector.

composition (p. 47) The composition of two functions f and g, denoted $f \circ g$, is defined as $(f \circ g)(x) = f(g(x))$.

compound inequality (p. 651) The combination of two inequalities into one statement using *and* or *or*.

compound interest (p. 130) The interest on both the principal and any interest accumulated.

conditional equation (pp. 233, 246) An equation that is true only for certain value(s) of the variable.

conditional probability (p. 499) The probability of an event, given the occurrence of a previous event.

confidence interval (p. 567) An interval around the point estimate in which it is reasonably certain that the actual parameter lies.

confidence level (p. 567) The probability that the confidence interval actually contains the population parameter.

conic section (conic) (p. 395) The intersection of a plane with a right circular conical surface.

conjugate axis (p. 412) The segment that passes through a hyperbola's center and is perpendicular to the hyperbola's transverse axis.

conjunction (p. 651) A compound mathematical sentence of the form $a \wedge b$ (or "a and b") that is true when both statements are true (the intersection of their solutions).

constant function (pp. 12, 25–26) A function of the form $f(x) = k$ whose value does not vary.

constant of variation (proportion) (p. 27) The nonzero constant relating the variables in a variation. In the functions $y = kx$ and $y = \frac{k}{x}$, k is the constant of variation.

continuity correction factor (p. 560) A correction applied when a continuous distribution is used to approximate a discrete distribution.

continuous function (pp. 19, 21) A function with no gaps or holes.

continuous on an interval (p. 21) A function is continuous on an interval (a, b) iff it is continuous at every point in (a, b).

continuous random variable (p. 546) A variable representing an uncountable number of outcomes within an interval in a probability experiment.

convergent sequence (p. 449) A sequence whose values approach a finite real number as the position number approaches infinity.

correlation coefficient (p. 60) Denoted r, where $-1 \leq r \leq 1$. Describes the degree to which two variables are related.

cosecant (pp. 179, 188) The trigonometric ratio that is the reciprocal of sine.

cosine (pp. 179, 188) The ratio of the leg adjacent to an acute angle and the hypotenuse of a right triangle. The ratio $\frac{x}{r}$, where $P(x, y)$ is a point on the terminal side of an angle in standard position and $r = \sqrt{x^2 + y^2}$.

cotangent (pp. 179, 188) The trigonometric ratio that is the reciprocal of tangent.

coterminal angles (p. 174) Two angles in standard position with the same terminal side.

co-vertices (pp. 403, 412) The endpoints of an ellipse's minor axis or of a hyperbola's conjugate axis.

critical (rejection) region (p. 575) The area where the difference between the test statistic and the proposed population parameter is enough to reject the null hypothesis.

critical value (p. 567) The z-value that corresponds to a certain confidence level.

damped harmonic motion (p. 225) Motion that occurs when friction reduces the displacement of an oscillating object. Modeled by $f(x) = ae^{-ct} \cos bt$ or $f(x) = ae^{-ct} \sin bt$ $(c > 0)$, where a is the maximum displacement.

damped sinusoid (p. 225) A function of the form $f(x) = g(x) \sin bx$ or $f(x) = g(x) \cos bx$.

damping constant (p. 225) The value of $c > 0$ in the equation $f(x) = ae^{-ct} \cos bt$ or $f(x) = ae^{-ct} \sin bt$ that affects the rate of decrease in the amplitude of oscillations.

damping factor (p. 225) A factor $g(x)$ that reduces the amplitude of a sinusoid model $f(x) = g(x) \sin bx$ or $f(x) = g(x) \cos bx$.

decay factor (p. 129) The base, b, in the exponential function $y = ab^x$, given $0 < b < 1$.

decomposition (p. 48) The process of finding simpler functions $g(x)$ and $f(x)$ such that the given function $h(x) = f(g(x))$.

decreasing function (pp. 25–26) A function in which the value of y decreases as x increases.

definite integral (p. 633) The area of a region between the graph of a function $f(x)$ and the interval $[a, b]$ on the x-axis is given by the definite integral $\int_a^b f(x)\, dx = \lim_{n \to \infty} \sum_{i=1}^{n} f(x_i)\, \Delta x$, provided the limit exists.

degenerate conic (p. 395) A point, a line, or two intersecting lines that result when a plane intersecting a conical surface contains the vertex.

dependent events (p. 490) Events in which the probability of one event changes because another event has occurred.

dependent variable (p. 3) The variable representing the values of the range in a relation.

depressed polynomial (p. 88) The quotient of a polynomial divided by one of its binomial factors.

derivative function (p. 613) The derivative of $f(x)$, denoted $f'(x)$, where $f'(x) = \lim_{h \to 0} \dfrac{f(x+h) - f(x)}{h}$.

descriptive statistics (p. 507) The organization of data and the determination and reporting of its numerical characteristics.

determinant (pp. 361, 363) A characteristic value associated with a square matrix that indicates whether the inverse of the matrix exists.

difference quotient (p. 612) The slope of a secant line through the given points $(x, f(x))$ and $(x + h, f(x + h))$ is $\dfrac{f(x+h) - f(x)}{h}$.

differentiable function (p. 613) A function that has a derivative at every point of its domain.

differential operator (p. 614) The notation $\frac{d}{dx}$ that indicates to find the derivative of a function.

directrix (p. 435) A fixed line such that all points on a conic are equidistant to a conic's focus and the line.

direct variation (proportion) (p. 27) A function of the form $y = kx$ or $k = \frac{y}{x}$, where k is a nonzero constant of variation.

discontinuity (p. 21) A gap or hole in a relation caused by excluded values.

discrete random variable (p. 546) A variable representing a countable number of outcomes of a probability experiment.

disjunction (p. 651) A compound mathematical sentence of the form $a \vee b$ (or "a or b") that is true when either statement is true (the union of their solutions).

divergent sequence (p. 449) A sequence whose values do not approach a finite real number as the position number approaches infinity.

domain (p. 2) The set of first elements of the ordered pairs in a relation.

dot product (pp. 294–95, 305) For $\mathbf{u}$ and $\mathbf{v}$, $\mathbf{u} \cdot \mathbf{v} = |\mathbf{u}|\, |\mathbf{v}| \cos \theta$, where θ is the angle between $\mathbf{u}$ and $\mathbf{v}$ when the vectors are placed in standard position $(0 \leq \theta \leq 180)$. Using vector components: if $\mathbf{u} = \langle u_1, u_2 \rangle$ and $\mathbf{v} = \langle v_1, v_2 \rangle$, then $\mathbf{u} \cdot \mathbf{v} = u_1 v_1 + u_2 v_2$.

eccentricity (pp. 407, 414, 435) The measure of a conic's deviation from a circle.

elementary row operation (p. 354) An operation that transforms an augmented matrix into an equivalent system by interchanging any two rows, multiplying or dividing each entry in a row by a nonzero number, or adding a multiple of one row to another.

eliminating the parameter (p. 52) The process of expressing y in terms of x when given a set of parametric equations.

ellipse (p. 403) The set of points in a plane such that the sum of the distances from two fixed points (the foci) is constant.

empty set (p. 647) A set containing no elements. Denoted $\varnothing$. Also called a *null set*.

end behavior (p. 25) A description of the values of a function as x approaches positive or negative infinity.

equivalent equations (p. 650) Equations that have the same solution set.

even function (p. 26) A function where $f(-x) = f(x)$ for all $x \in D$.

existence theorems (p. 94) Theorems that guarantee the existence of a quantity, but do not provide a means of finding it.

expected value (p. 547) The mean of a probability distribution. Denoted $E(x)$.

experiment (p. 497) A repeatable activity resulting in exactly one of a finite set of outcomes.

experimental (empirical) probability (p. 497) The ratio of the number of times an event occurs to the number of trials in an experiment.

explicit formula (pp. 447, 456) A formula in terms of a_n that allows direct computation of any term in the sequence.

exponential decay (p. 129) The decrease of a quantity by the same percentage per unit of time.

exponential equation (p. 149) An equation in which the variable occurs in the exponent.

exponential function (p. 128) A function of the form $f(x) = ab^x$ where $a \neq 0$, $b > 0$, and $b \neq 1$.

exponential growth (p. 129) The increase of a quantity by the same percentage per unit of time.

extraneous root (p. 74) A derived solution that is invalid because it does not make the original equation true.

extrapolation (p. 60) The process of making predictions outside the range of a data set.

factorial (p. 448) The product of a number, n, and every natural number less than that number. Denoted $n!$.

family of functions (p. 32) A group of functions whose graphs share similar characteristics.

five-number summary (p. 519) The minimum, lower quartile, median, upper quartile, and maximum values of a data set.

focal axis (p. 435) The line that passes through a conic's focus and is perpendicular to the conic's directrix.

focal distance (pp. 403, 412) The distance from the center of an ellipse or a hyperbola to each focus.

focal length (p. 396) The distance from the vertex of a parabola to its focus.

focus (p. 435) A fixed point such that the distances to that point from any point on the conic and from the conic's directrix are in a constant ratio, which is the conic's eccentricity, e.

frequency (p. 199) The reciprocal of the period of a function:
$$f = \frac{1}{p} = \frac{|b|}{2\pi}.$$

function (p. 2) A relation in which every element of the domain is paired with one and only one element of the range.

future value of an annuity (p. 467) The future value, FV, of an annuity consisting of n periodic payments of P with a periodic interest rate i is given by $FV = P\left[\dfrac{1 - (1 + i)^n}{1 - (1 + i)}\right]$ or $FV = P\left[\dfrac{(1 + i)^n - 1}{i}\right]$.

Gaussian elimination (p. 353) The process of eliminating variables from a system to find an equivalent system in row echelon form.

Gauss-Jordan elimination (p. 355) The process of applying elementary row operations to an augmented matrix until the unique reduced row echelon form is found.

geometric means (p. 465) The terms between any two given terms of a geometric sequence.

geometric sequence (p. 463) A sequence in which the ratio of any two consecutive terms is a constant.

geometric series (p. 465) The sum of the terms of a geometric sequence.

greatest integer function (p. 18) A function that returns the greatest integer less than or equal to x.

growth factor (p. 129) The base, b, in the exponential function $y = ab^x$, given $b > 1$.

growth rate (p. 131) The term r in the exponential growth function, $P(t) = P_0(1 + r)^t$, where $t =$ time.

half-life (p. 132) The time it takes for half a radioactive isotope to decay into a stable isotope of another element.

harmonic sequence (p. 458) The reciprocals of an arithmetic sequence.

histogram (p. 510) A graphic representation of a frequency distribution.

horizontal asymptote (p. 104) The line $y = c$ is a horizontal asymptote of $f(x)$ if $\lim\limits_{x \to -\infty} f(x) = c$ or $\lim\limits_{x \to \infty} f(x) = c$.

hyperbola (p. 412) The set of points in a plane such that the difference of the distances from two fixed points (its foci) is constant.

hypothesis testing (p. 573) The process of using statistics from a sample to examine a claim about the value of a population parameter.

identity (p. 233) An equation that is true for all values in the domain of the variable.

imaginary axis (p. 324) The vertical axis in the complex plane.

imaginary unit (p. 664) The imaginary unit, i, is the square root of -1, so that $i = \sqrt{-1}$ and $i^2 = -1$.

increasing function (pp. 25–26) A function in which the value of y increases as x increases.

indefinite integral (p. 639) The indefinite integral of $f(x)$ is $\int f(x)\, dx = F(x) + C$, where $F(x)$ is the antiderivative of $f(x)$ and C is a constant.

independent events (p. 490) Events in which the occurrence of one has no influence on the probability of the other events occurring.

independent variable (p. 3) The variable representing the values of the domain in a relation.

indeterminate form (p. 606) An expression obtained, such as $\frac{0}{0}$, that implies the function does not contain enough information to determine the limit at that point.

inferential statistics (p. 507) The process of interpreting the statistics of a sample to make predictions about a population's parameters.

infinite discontinuity (pp. 21, 105) An infinite discontinuity exists for $f(x)$ at $x = a$ if a one-sided $\lim\limits_{x \to a^{\pm}} f(x) = \pm\infty$.

infinite limit (p. 600) Describes a function's behavior, but the true limit does not exist.

instantaneous velocity (pp. 615–16) If the position of an object is represented by a function of time $f(t)$, then its instantaneous velocity at time t is $v_{inst}(t) = f'(t) = \lim\limits_{h \to 0} \dfrac{f(t+h) - f(t)}{h}$.

interpolation (p. 60) The process of making predictions within the range of a data set.

interquartile range (*IQR*) (p. 519) The difference between the third quartile and the first quartile of a data set.

intersection of sets (p. 647) The set of elements common to the given sets.

interval (p. 4) A continuous set of real numbers.

interval estimate (p. 566) A range of values used to estimate a population's parameter.

inverse functions (p. 56) A relation and its inverse that are both functions: $(f^{-1} \circ f)(x) = (f \circ f^{-1})(x) = x$.

inverse matrix (p. 370) The multiplicative inverse of square matrix A, denoted A^{-1}, is the matrix such that $AA^{-1} = A^{-1}A = I$, the identity matrix.

inverse relation (p. 53) If a relation R contains the ordered pair (a, b), its inverse relation, R^{-1}, contains the ordered pair (b, a).

inverse variation (proportion) (p. 28) A relation in which the product of the two variables is a nonzero constant, k. That is, $xy = k$, where $k \neq 0$.

invertible (nonsingular) matrix (p. 370) A matrix for which an inverse exists.

irrational number (p. 648) A real number that cannot be expressed as a ratio of two integers.

irreducible quadratic factor (p. 99) A second-degree polynomial that cannot be factored over the integers.

leading coefficient (p. 78) The constant factor of the term with the highest degree in a polynomial expression.

least common denominator (LCD) (p. 651) The smallest positive integer that is a multiple of two or more given denominators.

left-hand limit (p. 21) The left-hand limit of $f(x)$ as x approaches a, denoted $\lim\limits_{x \to a^-} f(x)$, is the number to which $f(x)$ gets closer and closer as x approaches a from the left.

level of significance (p. 575) The maximum allowable probability of making a type I error.

limit (pp. 21, 599) The limit of $f(x)$ as x approaches a, denoted $\lim\limits_{x \to a} f(x)$, is L iff $\lim\limits_{x \to a^-} f(x) = \lim\limits_{x \to a^+} f(x) = L$.

linear combination (p. 289) A sum of vectors in the form $a\mathbf{v} + b\mathbf{w}$, where a and b are scalars.

linear correlation (p. 60) Describes how closely data points are clustered along a trend line.

linear programming (p. 377) A technique used to maximize or minimize a particular quantity under a given set of constraints.

linear regression (p. 59) A statistical method of determining the line of best fit (trend line).

linear speed (p. 175) The ratio of arc length, or distance traveled, per unit of time: $v = \frac{s}{t}$.

line of best fit (p. 59) See **trend line**.

logarithmic equation (p. 149) An equation of the form $\log_b x = k$ in which the variable occurs in the argument (input) of a logarithmic expression.

logarithmic function (p. 136) The inverse of the exponential function $b^y = x$, denoted $y = \log_b x$, where x and b are positive and $b \neq 1$: $y = \log_b x$ iff $b^y = x$.

logistic growth function (p. 158) A function that can be written in the form $f(x) = \dfrac{c}{1 + ae^{-bx}}$ where a, b, and c are positive constants.

magnitude (pp. 287–88) The length of vector $\mathbf{v} = \langle a, b \rangle$ is $|\mathbf{v}| = \sqrt{a^2 + b^2}$.

major axis (p. 403) The chord containing the foci of an ellipse whose endpoints are the vertices of the ellipse.

margin of error (maximum error of the estimate) (p. 567) The greatest acceptable difference between the point estimate and the actual value of the parameter for a given confidence level, c.

mathematical induction (p. 477) A method of proving a proposition $P(n)$ is true for all natural numbers by showing that $P(1)$ is true and that $P(k + 1)$ is true when $P(k)$ is true.

matrix (p. 343) A rectangular array of numbers.

mean (p. 516) The arithmetic average obtained by dividing the sum of the values by the number of values.

median (p. 516) The middle value or the mean of the two middle values in a set of data arranged in numerical order.

midline (p. 201) The horizontal line halfway between the maximum and minimum values of a function: $y = \frac{max - min}{2}$.

minor (p. 362) The minor of entry a_{ij} in square matrix A, notated M_{ij}, is the determinant of the matrix formed by removing the ith row and jth column of A.

minor axis (p. 403) The chord perpendicular to the major axis that passes through the center of an ellipse and whose endpoints are the co-vertices of the ellipse.

mode (p. 516) The most frequently appearing value(s) in a data set.

modulus (p. 324) The absolute value of a complex number (the distance from the complex origin). If $z = a + bi$, then the modulus of $z = \sqrt{a^2 + b^2}$.

monomial functions (p. 78) Power functions of the form $f(x) = kx^n$, where n is a natural number, and constant functions of the form $f(x) = c$, where c is a real number.

multiplicative identity matrix (p. 346) A square matrix in which entries on the main diagonal are 1 and all other entries are 0.

multiplicity (p. 81) A root of a polynomial has a multiplicity of n if the polynomial contains n identical binomial factors.

natural base (p. 130) The number $e = \lim\limits_{x \to \infty} \left(1 + \frac{1}{x}\right)^x \approx 2.71828$.

natural exponential function (p. 130) The function $f(x) = e^x$ where $e \approx 2.71828$.

natural logarithmic function (p. 138) The inverse of the exponential function $e^y = x$, denoted $y = \ln x$, where x is positive: $y = \ln x$ iff $e^y = x$.

nonlinear asymptote (p. 107) Given a rational function $f(x) = \dfrac{a(x)}{b(x)}$ where degree of $a(x)$ is at least 2 greater than the degree of $b(x)$, the nonlinear asymptote is the quotient (without the remainder) when $a(x)$ is divided by $b(x)$.

nonrigid transformation (p. 33) A transformation that distorts the shape of the graph.

nth partial sum (p. 450) The sum of a sequence's first n terms, denoted S_n.

nth root of a complex number (p. 327) If $w = a + bi$ and z is a complex number such that $w^n = z$, then w is an nth root of z.

nth root of unity (p. 327) If $w = a + bi$ and $w^n = 1$, then w is the nth root of unity.

null hypothesis (p. 573) A statement about a population parameter that contains an equality such as $\geq$, $\leq$, or $=$. Denoted H_0.

objective function (p. 377) A linear function representing the quantity to be optimized in a linear programming problem.

octant (p. 303) One of eight regions created in space by a three-dimensional coordinate system.

odd function (p. 26) $f(-x) = -f(x)$ for all $x \in D$.

one-sided limit (p. 21) See **right-hand limit** or **left-hand limit**.

one-to-one function (p. 55) A function in which each element of the domain corresponds to exactly one element of the range.

open interval (p. 4) An interval that includes all the points between two endpoints but does not include the endpoints. The interval $\{x \mid a < x < b\}$ is denoted (a, b).

orthogonal vectors (p. 296) Two vectors $\mathbf{u}$ and $\mathbf{v}$ are orthogonal iff $\mathbf{u} \cdot \mathbf{v} = 0$. Nonzero orthogonal vectors are also perpendicular.

outlier (pp. 60, 519) A value in a data set that is significantly higher or lower than the other values in the set.

parabola (pp. 39, 396) The set of points in a plane that are equidistant from a fixed line (the directrix) and a fixed point (the focus) not on that line.

parameter (statistical) (p. 507) The numerical value for a characteristic of a population.

parametric curve (p. 429) A set of ordered pairs (x, y) defined by the parametric equations $x = f(t)$ and $y = g(t)$, where $f(t)$ and $g(t)$ are continuous functions of t on the parameter interval, I.

parametric equations (p. 52) A set of functions representing each coordinate in terms of one or more other independent variables (the parameters).

parent function (p. 32) The basic function used as a reference for transformed functions within a family of functions.

partial fraction decomposition (p. 383) The process of expressing a proper rational expression as a sum of simpler rational functions.

percentile (p. 520) When a distribution is divided into 100 equal parts, each percentile represents the amount of the distribution that lies below that percentile's value.

perihelion (pp. 407, 438) The point at which an orbiting object is at its closest distance to the sun.

period (pp. 193, 198) The horizontal length of one cycle in a periodic function.

periodic function (p. 193) A function f with a repeating pattern of y-values so that there exists a constant c such that $f(x + c) = f(x)$ for all x in the domain.

permutation (p. 491) The number of different ways to arrange (or order) r out of n items. Denoted $_nP_r$ or $P(n, r)$.

phase shift (p. 200) A horizontal translation in the graph of a trigonometric function.

piecewise function (p. 18) A function whose value is defined by multiple function rules for various intervals of its domain.

point estimate (p. 566) A single-value estimation of a population's parameter.

point (removable) discontinuity (pp. 21, 106) A single point on the graph of $f(x)$ that is not defined.

point-slope form (p. 12) A linear equation of the form $y - y_1 = m(x - x_1)$, where m is the slope of the line and (x_1, y_1) is a point on the line.

polar axis (p. 309) An initial ray drawn from the pole in the direction of the positive x-axis.

polar coordinates (p. 309) Coordinates of the form $P(r, \theta)$, where r is a directed distance from the pole to the point P and θ is a directed angle from the polar axis.

polar (trigonometric) form (p. 324) If $z = a + bi$, then $z = r(\cos \theta + i \sin \theta)$ or $z = r \operatorname{cis} \theta$ where r is the modulus and θ is the argument of z.

pole (p. 309) A fixed point, O, that serves as the origin for the polar coordinate system.

polynomial function (p. 78) A function of the form $p(x) = a_n x^n + a_{n-1} x^{n-1} + a_{n-2} x^{n-2} + \cdots + a_1 x + a_0$, where n is a nonnegative integer and the coefficients are real numbers.

population (p. 507) The larger overall set being considered in a statistical study.

power function (p. 25) A function of the form $f(x) = kx^n$ where k and the power n are nonzero constants.

primary solution (p. 246) A solution to a trigonometric equation that falls within the interval $[0, 2\pi)$.

principal root (p. 328) When $k = 0$ in the complex root of $\sqrt[n]{r} \operatorname{cis} \frac{\theta + k2\pi}{n}$.

principal value (p. 216) A value within a restricted domain of a trigonometric function.

probability distribution (p. 546) A table, graph, or equation that associates each possible value of a random variable, x, with its probability, $P(x)$.

projection (p. 296) If $\mathbf{w}_1$ and $\mathbf{w}_2$ are orthogonal component vectors of $\mathbf{u}$ and $\mathbf{w}_1$ is parallel to vector $\mathbf{v}$, then $\mathbf{w}_1$ is the projection of $\mathbf{u}$ onto $\mathbf{v}$, $\operatorname{proj}_\mathbf{v} \mathbf{u}$.

p-value (p. 577) The probability of obtaining a test statistic as extreme as or more extreme than the one found in the sample, assuming the null hypothesis is true.

quadrantal angle (p. 174) An angle in standard position whose terminal side lies along either the x- or y-axis.

quadratic function (p. 40) A function that can be written in the standard form $f(x) = ax^2 + bx + c$, where a, b, and $c \in \mathbb{R}$ and $a \neq 0$.

quartiles (p. 518) Values that divide ordered data into four equal sets.

radian (p. 173) The measure of a central angle that intercepts an arc whose length is the same as the circle's radius: $1 \text{ radian} = \frac{180°}{\pi} \approx 57.3°$.

radian measure (p. 173) The ratio of a central angle's intercepted arc's length, s, to the circle's radius, r: $\theta = \frac{s}{r}$.

radical equation (p. 74) An equation of the form $\sqrt[n]{x} = k$ that contains a radical expression with a variable in the radicand, or an equation of the form $x^{\frac{p}{r}} = k$ that contains a variable with a rational exponent.

radical function (p. 72) A function that contains a radical expression with the independent variable in the radicand.

random sample (pp. 507, 586) A sample for which every member of the population has an equal chance of being selected.

random variable (p. 546) A variable representing all possible numerical outcomes in a probability experiment.

range (of a data set) (pp. 510, 517) The difference between the maximum and minimum values.

range (of a function) (p. 2) The set of second elements of the ordered pairs in a relation.

rational equation (p. 112) An equation containing at least one ratio of polynomials, $\frac{a(x)}{b(x)}$, where $b(x) \neq 0$.

rational function (p. 104) A function that can be expressed as $f(x) = \frac{a(x)}{b(x)}$, where $a(x)$ and $b(x)$ are polynomials and $b(x) \neq 0$.

rational number (p. 648) A number that can be expressed as a ratio of two integers $\frac{a}{b}$, where $b \neq 0$.

real numbers (p. 648) All the numbers that can be graphed on a number line.

recursive formula (pp. 447, 456) A formula that states the first term(s) of a sequence and defines the remaining terms using the previous term(s).

reduced row echelon form (RREF) (p. 355) The form of a coefficient matrix when the first nonzero element in each row is 1 and any other elements in that column are 0.

reference angle (p. 189) The acute angle formed by the x-axis and the terminal side of an angle in standard position.

reference triangle (p. 189) The triangle formed by a perpendicular drawn to the x-axis from the terminal side of an angle in standard position.

reflection (p. 33) A transformation across a reference line that creates a mirror image.

relation (p. 2) Any set of ordered pairs.

relative extrema (p. 80) The maximum or minimum values relative to a local region of points.

residual (p. 60) The difference between the actual y-value of a data point and the value predicted by the modeling function.

resultant vector (vector sum) (p. 288) The vector that represents the sum of two vectors. If $\mathbf{u} = \langle u_1, u_2 \rangle$ and $\mathbf{v} = \langle v_1, v_2 \rangle$, $\mathbf{r} = \mathbf{u} + \mathbf{v} = \langle u_1 + v_1, u_2 + v_2 \rangle$.

right-hand limit (p. 21) The right-hand limit of $f(x)$ as x approaches a, denoted $\lim_{x \to a^+} f(x)$, is the number to which $f(x)$ gets closer and closer as x approaches a from the right.

rigid transformation (p. 33) A transformation that maintains the shape of the graph, such as a translation or reflection.

row echelon form (REF) (p. 353) A system of linear equations written in inverted triangular form in which all leading coefficients are 1.

sample (p. 507) A group selected from a population.

sample space (p. 497) All possible outcomes in an experiment.

sampling error (p. 586) An error that occurs when the sample is not representative of the population.

scalar (pp. 287–88) A real number.

scatterplot (p. 59) A graph of ordered pairs that illustrates the relationship between two sets of data.

secant (pp. 179, 188) The trigonometric ratio that is the reciprocal of cosine.

sector (p. 175) The region of a circle bound by two radii and the intercepted arc.

semi-perimeter (p. 278) One half the sum of a triangle's side lengths.

sequence (p. 446) A function whose domain is the set of natural numbers (infinite sequence) or the first n natural numbers (finite sequence) that is usually written as an ordered list of numbers.

series (p. 449) The sum of the terms in a sequence.

set (p. 647) A group or collection of numbers or objects.

sigma notation (p. 449) See **summation notation**.

simple harmonic motion (p. 225) Oscillations that continue with the same amplitude and period.

sine (pp. 179, 188) The ratio of the leg opposite an acute angle and the hypotenuse of a right triangle. The ratio $\frac{y}{r}$, where $P(x, y)$ is a point on the terminal side of an angle in standard position and $r = \sqrt{x^2 + y^2}$.

singular matrix (p. 370) A matrix for which no inverse exists.

sinusoidal function (p. 198) A function that can be written as $y = a \sin b(x - h) + k$ or $y = a \cos b(x - h) + k$.

slant asymptote (p. 107) Given a rational function $f(x) = \frac{a(x)}{b(x)}$ where the degree of $a(x)$ is one more than the degree of $b(x)$, the slant asymptote is the quotient (without the remainder) when $a(x)$ is divided by $b(x)$.

slope (pp. 10, 14) The ratio of a line's rise (vertical change) to its run (horizontal change). Denoted m.

slope-intercept form (p. 11) A linear equation in the form $y = mx + b$, where m is the slope of a line and $(0, b)$ is the y-intercept.

solution (p. 233) Any value of the variable for which a conditional equation or inequality is true.

square matrix (p. 343) A matrix that has an equal number of rows and columns.

standard deviation (pp. 517, 548) A measure of variance from the mean found by finding the square root of the variance.

standard position (of an angle) (p. 172) An angle with its vertex at the origin of a coordinate plane and its initial side coinciding with the positive x-axis.

standard position (of a vector) (p. 287) A vector with its initial point at the origin.

standard unit vectors (p. 289) $\mathbf{i} = \langle 1, 0 \rangle$ and $\mathbf{j} = \langle 0, 1 \rangle$.

statistic (p. 507) An estimate of a population's parameter based on a sample.

subset (p. 647) A set whose elements are all contained in another set.

summation notation (p. 449) A series represented as $\sum_{i=b}^{c} a_i = a_b + a_{b+1} + a_{b+2} + \cdots + a_c$ where the index (or counter) i has a lower bound of b and an upper bound of c.

sum of the squares (p. 517) The sum of the squares of the difference of each value from the mean.

synthetic division (p. 86) A shortened version of the long-division algorithm used when dividing a polynomial by a linear polynomial of the form $x - c$.

synthetic substitution (p. 89) Using synthetic division to evaluate a function.

tangent (pp. 179, 188) The ratio of the leg opposite an acute angle and the leg adjacent to that angle in a right triangle. The ratio $\frac{y}{x}$, where $P(x, y)$ is a point on the terminal side of an angle in standard position.

terminal side (p. 172) The side of an angle in standard position that rotates around the origin.

theoretical (classical) probability (p. 497) The ratio of the number of favorable outcomes, $n(E)$, to the number of possible outcomes, $n(S)$, where every outcome is equally likely.

transcendental function (p. 128) A function, such as an exponential, logarithmic, or trigonometric function, that goes beyond, or transcends, the basic algebraic operations.

translation (p. 32) A change in the position of a figure without rotation or resizing.

transverse axis (p. 412) The segment containing a hyperbola's vertices whose endpoints are the hyperbola's foci.

trend line (p. 59) A line drawn to model the relationship between two related variables and to make predictions related to other values of either variable.

type I error (p. 574) An error that occurs when the null hypothesis is rejected, but it is true.

type II error (p. 574) An error that occurs when the null hypothesis is not rejected, but it is false.

union of sets (p. 647) The set of all elements that appear in the given sets.

unit circle (p. 191) The circle centered at the origin with a radius of 1 unit: $x^2 + y^2 = 1$.

unit vector (p. 289) A vector with a magnitude of 1.

variance (pp. 517, 548, 550) The average of the sum of the squared deviations from the mean.

vector (p. 287) A quantity $\langle a, b \rangle$ modeled with a directed line segment having both magnitude and direction.

verifying identities (p. 240) Showing that both sides of an equation are equal for all values of the variable for which the expressions are defined.

vertex (of a conic section) (p. 435) Any intersection of a conic and its focal axis.

vertex (of a parabola) (pp. 39, 396) The minimum or maximum point (h, k) of the quadratic function $f(x) = a(x - h)^2 + k$ at the intersection of the parabola and its axis of symmetry.

vertical asymptote (p. 104) The line $x = c$ is a vertical asymptote of $f(x)$ if $\lim\limits_{x \to c^-} f(x) = \pm\infty$ or $\lim\limits_{x \to c^+} f(x) = \pm\infty$.

zero of a function (pp. 41, 656) A value of x such that $f(x) = 0$.

zero vector (pp. 288, 305) A vector having a length of 0 and no direction.

z-score (pp. 527–28) The number of standard deviations a value is from the mean.

Index

Cosecant 179, 188, 209
Cosine 179, 188
Cotangent 179, 188, 207
Coterminal angles 174
Co-vertices
  of a hyperbola 412, 414
  of an ellipse 403, 405
Cramer's rule 364
Critical region 575
Critical value 567

D

Damped harmonic motion 225
Damped oscillation 225
Damped sinusoid 225
Damping constant 225
Damping factor 225
Decay factor 129
Decomposition 48
Decreasing function 26
Definite integral 633, 640
Degenerate conic 395–96
Degree 172
De Moivre's Theorem 326
Dependent consistent system 338
Dependent events 490, 501
Dependent variable 3
Depressed polynomial 88
Derivative 612
  chain rule 625
  function 613
  of a constant function 618
  of a linear function 618
  of a power function 619
  of a scalar multiple of a function 619
  of a sine function 626
  of a sum or difference 619
  of the natural exponential function 626
  power rule 618
  product rule 624
  quotient rule 624
Descartes's Rule of Signs 96
Descriptive statistics 507
Determinant 361, 363
Deviation 517
Difference quotient 612
Differentiable function 613
Differential operator 614
Directrix
  of a conic section 435
  of a parabola 396
Direct variation (proportion) 27–28
Discontinuity
  infinite 21, 105
  jump 21
  point 21, 106
Discrete random variable 546
Discriminant 423–24
Disjoint sets 647

Disjunction 651
Distance formula 13
Distinguishable permutation 492–93
Divergent sequence 449
Divergent series 450, 467
Deviation 517
Domain 2
Dot product 294–95, 305
Double-angle identities 261

E

Eccentricity 407, 415, 435
Element
  of a conical surface 395
  of a matrix 343
  of a set 647
Elementary row operation 354
Eliminating the parameter 52
Ellipse 395, 403
  aphelion/perihelion 407, 438
  center of 403
  co-vertices of 403, 405
  discriminant of 424
  eccentricity of 407, 415, 435
  focal distance of 403
  focus of 403, 405
  major/minor axis of 403, 405
  semimajor/semiminor axis of 403
  standard form of 405
  vertex of 403
Empirical Rule 530
Empty set 647
End behavior 25
Equality Property of Logarithms 149
Equal matrices 343
Equation(s)
  conditional 233, 246
  equivalent 650
  exponential 149
  false 233
  linear 11, 14, 650
  logarithmic 149
  parametric 52, 429
  polar, for conics 436
  quadratic 39, 653
  radical 74
  rational 112
  system of 336
  trigonometric 246
Equilibrium 291
Equivalent equations 650
Equivalent system 353
Even function 26
Even/odd identities 234
Event(s) 497–98
  certain 498
  complement of 498
  compound/simple 499
  dependent/independent 490, 501

impossible 498
  mutually exclusive/inclusive 500–501
Existence theorems 94
Expected value 547
Experiment 497
Experimental probability 497–98
Experimental research 584
Explicit formula 447, 456
  of a geometric sequence 464
  of an arithmetic sequence 457
Exponential decay 129, 131
Exponential equation 149
Exponential function 128
Exponential growth 129, 131
Exponents, properties of 143, 662
Extraneous root 74
Extrapolation 60

F

Factor
  irreducible quadratic 99
  linear 94
  Theorem 89
Factorial 448
False equation 233
Family of functions 32
Feasible region 377
Finite sequence 446, 464
Finite series 449, 466
Five-number summary 519
Focal axis 435
Focal distance (length)
  of a hyperbola 412, 414
  of an ellipse 403
  of a parabola 396
Focal width 397
Focus
  of a conic 435
  of a hyperbola 412, 414
  of an ellipse 403, 405
  of a parabola 396
Frequency 199, 201
Frequency distribution table 510
Function(s) 2
  absolute value 18
  algebraic 128
  circular 191–92
  common logarithmic 137
  composition of 47
  constant 12, 25–26
  continuous 19, 21
  decomposition of 48
  decreasing 26
  derivative 613
  differentiable 613
  even/odd 26
  exponential 128
  family of 32
  greatest integer 18

Photo Credits

Key: (**t**) top; (**c**) center; (**b**) bottom; (**l**) left; (**r**) right

Cover

tl "Shoreditch, London, United Kingdom (Unsplash gBCHiVSfrYY)" by Tom Eversley/Wikimedia Commons/Public Domain; **tr** Photo by Kevin Crosby on Unsplash; **bl** Photo by Mike Enerio on Unsplash; **br** Photo by Mike Kononov on Unsplash

Front Matter

vt TCD/Prod.DB/Alamy Stock Photo; **vb** naddi/Shutterstock.com; **vit** Vallia/Shutterstock.com; **vicl** xtock/Shutterstock.com; **vicr** Olaf Schulz/Shutterstock.com; **vib** James Cavallini/Science Source; **vii** Guy Corbishley/Alamy Stock Photo

Chapter 1

1 Aspen Photo/Shutterstock.com; **2** Lyu Hu/Shutterstock.com; **10** Stas Vulkanov/Shutterstock.com; **14** alchemist no.9/Shutterstock.com; **25** MikhailSh/Shutterstock.com; **29** HomeArt/Shutterstock.com; **32** © iStock.com/simonbradfield; **37** casadaphoto/Shutterstock.com; **39** photofriday/Shutterstock.com; **46** © iStock.com/Poike; **52** MarcelClemens/Shutterstock.com; **59** Heiko Kiera/Shutterstock.com; **64** WitthayaP/Shutterstock.com; **67** AP Photo/Alex Gallardo

Chapter 2

71 KUPRYNENKO ANDRII/Shutterstock.com; **72** Pavel L Photo and Video/Shutterstock.com; **78** Syda Productions/Shutterstock.com; **80** brovkin/Shutterstock.com; **86** Vicki Vale/Shutterstock.com; **94** Concept Photo/Shutterstock.com; **102** agsandrew/Shutterstock.com; **104** bociek666/Shutterstock.com; **112** chombosan/Shutterstock.com; **114** Andrew Linscott/Shutterstock.com; **116** © iStock.com/Sladic; **118** etnika/Shutterstock.com; **122** "USA-17-flying-cropped" by Pedro de Arechavaleta/Wikimedia Commons/CC BY-SA 3.0; **124** © iStock.com/Daft_Lion_Studio; **126** Odua Images/Shutterstock.com

Chapter 3

127 © iStock.com/FotografiaBasica; **128** © iStock.com/BirdImages; **132** Fotos593/Shutterstock.com; **134** © iStock.com/naturediver; **136** Haje Jan Kamps/EyeEm/Getty Images; **143** AGCuesta/Shutterstock.com; **149** praphab louilarpprasert/Shutterstock.com; **156** f11photo/Shutterstock.com; **166** Michaelstockfoto/Shutterstock.com; **168** © iStock.com/fotoguy22

Chapter 4

171 kdshutterman/Bigstock.com; **172** Design Pics/Bilderbuch/Media Bakery; **177** umnola/Shutterstock.com; **179** © iStock.com/ChuckSchugPhotography; **185** (silhouette) © iStock.com/4x6; **185** (blimp) © iStock.com/luismmolina; **185** (hot air balloon) Miki Studio/Shutterstock.com; **185** (Statue of Liberty) © iStock.com/Samuil_Levich; **188** Alisa24/Shutterstock.com; **197** Debbie Steinhausser/Shutterstock.com; **204** Media Bakery; **206** Nano Calvo/age fotostock/Getty Images; **214** Phovoir/Shutterstock.com; **216** Don Pablo/Shutterstock.com; **223** rommma/Shutterstock.com; **228** Igor Zh./Shutterstock.com; **230l** starfotograf/Bigstock.com; **230r** robert cicchetti/Shutterstock.com

Chapter 5

232, 389 © iStock.com/Henrik5000; **233** © iStock.com/Bosca78; **238** arhendrix/Shutterstock.com; **240** Anthony Lynn/Alamy Stock Photo; **246** Copyright by Educational Innovations, INC. Reprinted by permission.; **253** Visharo/Shutterstock.com; **254** CobraCZ/Shutterstock.com; **261** Charles D. Winters/Science Source; **265** Nor Gal/Shutterstock.com; **269, 394** Johan Swanepoel/Shutterstock.com; **276** © iStock.com/zmeel; **282** BakerJarvis/Shutterstock.com

Chapter 6

286 Vallia/Shutterstock.com; **287** Guy Corbishley/Alamy Stock Photo; **294** mariakraynova/Shutterstock.com; **299** Africa Studio/Shutterstock.com; **301** xtock/Shutterstock.com; **303** sybanto/Shutterstock.com; **309** © iStock.com/FrozenShutter; **313** Photo Melon/Shutterstock.com; **315** Chursina Viktoriia/Shutterstock.com; **324** Desiree Walstra/Shutterstock.com; **331l** Olaf Schulz/Shutterstock.com; **331r** James Cavallini/Science Source

Chapter 7

335, 420 Ezume Images/Shutterstock.com; **336** asharkyu/Shutterstock.com; **343** wavebreakmedia/Shutterstock.com; **349** Andrey Suslov/Shutterstock.com; **351** sokolenok/Shutterstock.com; **353** Yasonya/Shutterstock.com; **361** Bibliothèque de Genève; **370** Wojciech Strozyk/Alamy Stock Photo; **374** Nerthuz/Shutterstock.com; **376** EtiAmmos/Shutterstock.com; **383** Tami Knisely

Chapter 8

395 James Schwabel/Alamy Stock Photo; **400** Xinhua/Alamy Stock Photo; **403** Artem Avetisyan/Shutterstock.com; **409** Jerrie McKenney; **412** John D Sirlin/Shutterstock.com; **417** "Cassegrain.en" by Szőcs Tamás Tamasflex/Wikimedia Commons/CC By-SA 3.0; **418l** kmls/Shutterstock.com; **418r** Michael Rosskothen/Shutterstock.com; **419** Graphical calculator images are powered by Desmos; **425** baimaple/Shutterstock.com; **428** Mick Harper/Shutterstock.com; **429** Sergey Nivens/Shutterstock.com; **433** omnimoney/Shutterstock.com; **434** Air Images/Shutterstock.com; **435** 3Dsculptor/Shutterstock.com; **441tl** Dotted Yeti/Shutterstock.com; **441bl** creativestockexchange/Shutterstock.com; **441br** Golden Sikorka/Shutterstock.com; **443l** LowePhoto/Alamy Stock Photo; **443r** Nomad_Soul/Shutterstock.com

Chapter 9

445 Adam Hart-Davis/Science Source; **446** Organic Matter/Shutterstock.com; **456** Theeraphong/Shutterstock.com; **463** William Potter/Shutterstock.com; **468** canbedone/Shutterstock.com; **472** Science Photo Library/Robert Brook/Media Bakery; **477** © iStock.com/Martin Barraud; **485** Oleksiy Mark/Shutterstock.com

Chapter 10

489 Pirotehnik/iStock/Getty Images Plus/Getty Images; **490** Thomas M Perkins/Shutterstock.com; **492** Cathy Keifer/Shutterstock.com; **494l** © iStock.com/Yevhenii Dubinko; **494r** abxyz/Shutterstock.com; **497t** Believe_In_Me/iStock/Getty Images Plus/Getty Images; **497b** TeddyandMia/Shutterstock.com; **498** Stock Up/Shutterstock.com; **501** Susan Schmitz/Shutterstock.com; **502l** julie deshaies/Shutterstock.com; **502tr** Lantern Works/Shutterstock.com; **502br** Sergiy Kuzmin/Shutterstock.com; **503** Westend61/GmbH/Media Bakery; **505** ALPA PROD/Shutterstock.com; **507** Oksana Kuzmina/Shutterstock.com; **510** cynoclub/Shutterstock.com; **516** © Mbr Images | Dreamstime.com; **523** Preston James Garbe/Shutterstock.com; **527** Basar/Shutterstock.com; **529** Pictorial Press Ltd/Alamy Stock Photo; **534** Mark Conlin/Alamy Stock Photo; **540** Nice_Media_PRODUCTION/Shutterstock.com

Text Acknowledgements

Chapter 1

9 Eves, Howard. *An Introduction to the History of Mathematics*, 6th ed. (Pacific Grove: Brooks/Cole, 1990), 264; **17** Wigner, Eugene. *Symmetries and Reflections: Scientific Essays* (Cambridge, MA: The MIT Press, 1970), 237; **17** Calvin, John. *Institutes on the Christian Religion*, vol. 1, trans. Henry Beveridge (Edinburgh: Calvin Translation Society, 1845), 318–19; **17** Hertz, Heinrich, quoted in E. T. Bell. *Men of Mathematics* (New York: Touchstone Books, 1986), 16; **17** Nickel, James. *Mathematics: Is God Silent?* (Vallecito, CA: Ross House Books, 2001), 223; **49** Fourier, Joseph. *The Analytical Theory of Heat*, trans. Alexander Freeman (Cambridge: The University Press, 1878), 1; **59** Alligator data from Example 2.20, Scheaffer, Richard L. *Probability and Statistics for Engineers*, 5th Edition (Boston: Brooks/Cole, 2011), 97; **64** Average US gas prices data from US Department of Energy; **65** Life expectancy data from the CDC/National Center for Health Statistics; **70** US Census Bureau.

Chapter 2

122 Formula from "International America's Cup Class," Wikipedia.

Chapter 3

134 National Marine Fisheries Service. 2013. Status Review of The Eastern Distinct Population Segment of Steller Sea Lion (Eumetopias jubatus). 144pp + Appendices. Protected Resources Division, Alaska Region, National Marine Fisheries Service, 709 West 9th St, Juneau, Alaska 99802; **134** "Fukushima Radiation: 800 Tera Becquerel of Cesium-137 to Reach West Coast of North America by 2016," Centre for Research on Globalization; **142** Cajori, Florian. *A History of Mathematics*, 2nd ed., rev. (New York: Macmillan, 1919), 149; **155** Kline, Morris. *Mathematics for the Nonmathematician* (New York: Dover Publications, 1967), 206–7; **155** Kepler, Johannes, quoted in James Nickel. *Mathematics: Is God Silent?* (Vallecito, CA: Ross House Books, 2001), 145; **155** Zimmerman, Larry L. "Mathematics: Is God Silent?," Part III, *The Biblical Educator*, 2:3, 2 (1980), n.p.; **155** Leibniz, Gottfried, quoted in Kline, Morris. *Mathematics: The Loss of Certainty* (New York: Oxford University Press, 1980), 60; **155** Fehr, Howard F. "Reorientation in Mathematics Education," *Teachers College Record*, May 1953, vol. 54, no. 8, p. 435; **163** Nesting pairs data from "Chart and Table of Bald Eagle Breeding Pairs in Lower 48 States," US Fish and Wildlife Service.

Chapter 4

214 Einstein, Albert, quoted in Morris Kline. *Mathematics and the Physical World* (New York: Dover, 1981), 464; **214** Einstein, Albert. "Physics and Reality," *Out of My Later Years* (New York: Citadel Press, 1950), 61; **214** Nickel, James. *Mathematics: Is God Silent?* (Vallecito, CA: Ross House Books, 2001), 225; **214** Lobachevsky, Nikolai, quoted in George E. Martin. *The Foundations of Geometry and the Non-Euclidean Plane* (New York: Springer, 1998), 225; **215** Gardner, Martin, quoted in Donald J. Albers and Gerald L. Alexanderson, eds. *Fascinating Mathematical People: Interviews and Memoirs* (Princeton: Princeton University Press, 2011), ix.

Chapter 5

235 Feynman, Richard. *The Character of Physical Law* (Cambridge, MA: The MIT Press, 1985), 34; **252** Fourier, Joseph. *The Analytical Theory of Heat*, trans. Alexander Freeman (Cambridge: The University Press, 1878), 8; **252** Russell, Bertrand. *Mysticism and Logic and Other Essays* (London: George Allen & Unwin Ltd, 1959), 61; **252** Gioia, Dana. "Beauty's Place in the Christian Vision" (chapel sermon), Biola University, February 8, 2012; **252** Nickel, James. *Mathematics: Is God Silent?* (Vallecito, CA: Ross House Books, 2001), 94; **252–53** Kline, Morris. *Mathematics in Western Culture* (New York: Oxford University Press, 1953), 512; **268** Bhatia, Rajendra. *Fourier Series* (Providence, RI: American Mathematical Society, 2004), 116; **268** Bracewell, Ronald, quoted in Prestini, Elena. *The Evolution of Applied Harmonic Analysis: Models of the Real World* (Boston: Birkhäuser, 2004), xii; **282** Data table from Smith, A. Mark. *Ptolemy's Theory of Visual Perception: An English Translation of the "Optics" with Introduction and Commentary* (Philadelphia: American Philosophical Society, 1996), 233.

Chapter 6

301 Hermite, Charles, quoted in Morris Kline. *Mathematics: The Loss of Certainty* (New York: Oxford Univ. Press, 1980), 345; **301** Galilei, Galileo, quoted in Morris Kline. *Mathematics and the Physical World* (New York: Dover, 1981), 206; **301** Kasner, Edward and James Newman. *Mathematics and the Imagination* (London: G. Bell and Sons, 1949), 359; **301** Jourdain, Philip E. B. *The Nature of Mathematics* (London: Jack and Nelson, 1919), 46; **301** Gauss, Carl, quoted in Ioan James. *Remarkable Mathematicians* (New York: Cambridge University Press, 2002), 67; **302** Zimmerman, Larry L. *Truth & the Transcendent* (Florence, KY: Answers in Genesis, 2000), n.p.

Chapter 7

351 Hardy, G. H. *A Mathematician's Apology* (Cambridge: Cambridge University Press, 2004), 94; **352** Bahnsen, Greg L. "The Great Debate: Does God Exist?," University of California, Irvine, CA 1985; **369** Huxley, Thomas, quoted in Cajori, Florian. "A Review of Three Famous Attacks upon the Study of Mathematics as a Training of the Mind," *Popular Science Monthly*, vol. 80, April 1912; **369** Sylvester, James Joseph, ibid.

Chapter 8

427 Wiles, Andrew. "Andrew Wiles on Solving Fermat," NOVA, 1 November 2007, PBS online; **442** Hawking, Stephen. Interview by Kenneth Campbell, *Reality on the Rocks*, 1995.

Chapter 9

454 Hales, Thomas, quoted in George Szpiro. *Kepler's Conjecture* (Hoboken, NJ: Wiley, 2003), 201; **454** Hawking, Stephen, quoted in Rory Cellan-Jones. "Stephen Hawking Warns Artificial Intelligence Could End Mankind.", 2 December 2014, *BBC News* online; **454** Volkers, Cody. "A Christian View of Artificial Intelligence," 1 March 2017, *Ecclesiam Journal* online; **455** ibid; **471** Pascal, Blaise. "Prayer to Ask of God the Proper Use of Sickness," *Minor Works*, trans. O. W. Wight (New York: P. F. Collier & Son, 1910), 371; **481** Van Til, Cornelius. A *Survey of Christian Epistemology* (Phillipsburg, NJ: P&R Publishing, 1969), 19.

Chapter 10

497 First paragraph ideas based on Tagliapietra, Ron. *Math for God's Glory* (Bloomington, IN: Xlibris, 2004); **526** Boyer, Carl B. *A History of Mathematics*, 2nd ed. (New York: John Wiley & Sons, 1992), 492; **526** Poythress, Vern S. *Chance and the Sovereignty of God: A God-Centered Approach to Probability and Random Events* (Wheaton, IL: Crossway, 2014), 101; **540** Hardy, G. H. *A Mathematician's Apology* (Cambridge: Cambridge University Press, 1969), 123–24; **540** Linnebo,

Øystein, "Platonism in the Philosophy of Mathematics," *The Stanford Encyclopedia of Philosophy* (Spring 2018 Edition), Edward N. Zalta (ed.), https://plato.stanford.edu/archives/spr2018/entries/platonism-mathematics/; **540** Kepler, Johannes, quoted in Morris Kline. *Mathematics: The Loss of Certainty* (New York: Oxford University Press), 31; **540** Du Sautoy, Marcus. *The Music of the Primes: Searching to Solve the Greatest Mystery in Mathematics* (New York: HarperCollins, 2003), 33–34.

Chapter 11

559 Graphs generated from the Controlling for Variables tool by SticiGui©. Philip B. Stark, University of California, Berkeley; **560** Teen data from Stage of Life LLC, 24 South Franklin Street, Dallastown, PA 17313. Copyright © 2011 All rights reserved; **562** Bureau of Labor Statistics, American Time Use Study; **562** Molla, Rani. "Why People Are Buying More Expensive Smartphones Than They Have in Years," January 23, 2018, *Recode* online. Vox Media, Inc.; **578** Ioannidis, John P. A. "Why Most Published Research Findings Are False," August 30, 2005, *PLOS Medicine* online; **582** Nickel, James. *Mathematics: Is God Silent?* (Vallecito, CA: Ross House Books, 2001), 59; **582** Eves, Howard. *An Introduction to the History of Mathematics*, 4th ed. (New York: Holt, Rinehart and Winston, 1976), 481; **588** Elbel, Brian, et al. "A Water Availability Intervention in New York City Public Schools: Influence on Youths' Water and Milk Behaviors," *American Journal of Public Health*, vol. 105, no. 2. Washington, DC: American Public Health Association, Feb. 2015; **597** Morning commute data from the US Census Bureau.

Chapter 12

601 Leibniz, Gottfried. *The Philosophical Works of Leibniz*, trans. George Martin Duncan (New Haven, CT: The Tuttle, Morehouse & Taylor Co., 1908), 71; **610** Hilbert, David, quoted in Eli Maor. *To Infinity and Beyond: A Cultural History of the Infinite* (Boston: Birkhäuser Boston, 1987), vii; **610** Weyl, Hermann, quoted in Peter Pesic. *Levels of Infinity* (Mineola, NY: Dover, 2012), 17; **644** Fraser, Alexander Campbell, ed. *The Works of George Berkeley, D.D.*, vol. 3 (Oxford: Clarendon Press, 1871), 301.

TEACHER EDITION

Chapter 3

155 Newton, Isaac, quoted in Florian Cajori. *Sir Isaac Newton's Mathematical Principles of Natural Philosophy and His System of the World*, vol. 2 (Berkeley: University of California Press, 1934), 669; **155** Kepler, Johannes, quoted in Morris Kline. *Mathematics: The Loss of Certainty* (New York: Oxford University Press, 1980), 31.

Chapter 4

214 Nickel, James. *Mathematics: Is God Silent?* (Vallecito, CA: Ross House Books, 2001), 51; **214** Ruffini, Remo, quoted in James Nickel. ibid., 69; **214** Kline, Morris. *Mathematics and the Physical World* (New York: Dover, 1980), 463.

Chapter 5

253 Lisle, Jason. "Fractals: Hidden Beauty Revealed in Mathematics," *Answers* magazine online, January 1, 2007; **268** Fourier, Joseph. *The Analytical Theory of Heat* (Cambridge: The University Press, 1878), 7–8; **268** Campbell, Lewis and William Garnett. *The Life of James Clerk Maxwell* (London: Macmillan, 1882), 416.

Chapter 7

351–52 Wiles, Andrew. "The Proof," NOVA, 28 October 1997, PBS Online.

Chapter 8

428 Wiles, Andrew. "Solving Fermat: Andrew Wiles," NOVA, n.d., PBS Online.

Chapter 9

454 Szpiro, George. *Kepler's Conjecture* (Hoboken, NJ: Wiley, 2003), 212; **454** Stewart, Ian. ibid; **454** Hardy, G. H. *A Mathematician's Apology* (Cambridge: Cambridge University Press, 1969), 113; **455** Volkers, Cody. "A Christian View of Artificial Intelligence," 1 March 2017, *Ecclesiam Journal* online.

Chapter 10

540–41 Kline, Morris. *Mathematics: The Loss of Certainty* (New York: Oxford University Press, 1980), 18; **541** Zimmerman, Larry L. *Truth & the Transcendent* (Florence, KY: Answers in Genesis, 2000), n.p.

Quick Reference

Functions

even	$f(-x) = f(x)$ for all $x \in D$	symmetrical with y-axis
odd	$f(-x) = -f(x)$ for all $x \in D$	symmetrical with origin

Transformations of Functions

	Translation	**Stretch or Shrink**	**Reflection**
vertical	$g(x) = f(x) + k$ up if $k > 0$ down if $k < 0$	$g(x) = af(x)$ stretch by a if $\lvert a \rvert > 1$ shrink by a if $0 \le \lvert a \rvert < 1$	$g(x) = -f(x)$ reflection across x-axis
horizontal	$g(x) = f(x - h)$ right if $h > 0$ left if $h < 0$	$g(x) = f(bx)$ shrink by a factor of $\frac{1}{\lvert b \rvert}$ if $\lvert b \rvert > 1$ stretch by a factor of $\frac{1}{\lvert b \rvert}$ if $0 \le \lvert b \rvert < 1$	$g(x) = f(-x)$ reflection across y-axis

Exponents and Logarithms

	of Exponents	**of Logarithms**	
Equality Property	$b^m = b^n$ iff $m = n$	$\log_b m = \log_b n$ iff $m = n$	
Product Property	$b^m \cdot b^n = b^{m+n}$	$\log_b mn = \log_b m + \log_b n$	**change of base**
Quotient Property	$\frac{b^m}{b^n} = b^{m-n}$	$\log_b \frac{m}{n} = \log_b m - \log_b n$	$\log_b x = \frac{\log_a x}{\log_a b}$
Power Property	$(b^m)^p = b^{mp}$	$\log_b m^p = p \log_b m$	

Vectors

magnitude	**reference angle**	**Direction (θ degrees from the positive x-axis)**	
$\lvert \mathbf{v} \rvert = \sqrt{a^2 + b^2}$	$\alpha = \text{Tan}^{-1} \left\lvert \frac{b}{a} \right\rvert$	Q I: $\theta = \alpha$ Q II: $\theta = 180° - \alpha$	Q III: $\theta = 180° + \alpha$ Q IV: $\theta = 360° - \alpha$

dot product
$\mathbf{u} \cdot \mathbf{v} = \lvert \mathbf{u} \rvert \, \lvert \mathbf{v} \rvert \cos \theta$
If $\mathbf{u} = \langle u_1, u_2 \rangle$ and $\mathbf{v} = \langle v_1, v_2 \rangle$, then $\mathbf{u} \cdot \mathbf{v} = u_1 v_1 + u_2 v_2$.

unit vector $\mathbf{u} = \frac{\mathbf{v}}{\lvert \mathbf{v} \rvert}$

projection of u onto v $\text{proj}_{\mathbf{v}} \, \mathbf{u} = \left(\frac{\mathbf{u} \cdot \mathbf{v}}{\lvert \mathbf{v} \rvert^2} \right) \mathbf{v} = \mathbf{w}_1; \ \mathbf{w}_2 = \mathbf{u} - \text{proj}_{\mathbf{v}} \, \mathbf{u}$

Complex Numbers

standard form	$z = a + bi$	**De Moivre's Theorem**
polar (trig) form	$z = r(\cos \theta + i \sin \theta)$ or $z = r \text{ cis } \theta$	$z^n = r^n \text{ cis } n\theta$
product	$z_1 z_2 = r_1 r_2 \text{ cis } (\theta_1 + \theta_2)$	**Corollary to De Moivre's Theorem**
quotient	$\frac{z_1}{z_2} = \frac{r_1 \text{ cis } \theta_1}{r_2 \text{ cis } \theta_2} = \frac{r_1}{r_2} \text{ cis } (\theta_1 - \theta_2)$	The n complex roots of z are $\sqrt[n]{r} \text{ cis } \frac{\theta + k2\pi}{n}$ where $k = 0, 1, 2, \ldots, n - 1$.

Sequences and Series

	Arithmetic	**Geometric**
recursive formula	$a_n = a_{n-1} + d$	$a_n = a_{n-1} \cdot r$
explicit formula	$a_n = a_1 + (n - 1)d$	$a_n = a_1 r^{n-1}$
finite series	$S_n = \frac{n}{2}(a_1 + a_n)$ or $S_n = \frac{n}{2}[2a_1 + (n - 1)d]$	$S_n = \frac{a_1 - a_n r}{1 - r}$ or $S_n = \frac{a_1(1 - r^n)}{1 - r}$
infinite series	The series diverges.	If $\lvert r \rvert < 1$, then $S = \frac{a_1}{1 - r}$. If $\lvert r \rvert \ge 1$, the series diverges.

Standard Form Equations of a Parabola

with vertex $V(h, k)$ and focal length $|p|$ where $a = \frac{1}{4p}$

$$y = \frac{1}{4p}(x - h)^2 + k \qquad\qquad x = \frac{1}{4p}(y - k)^2 + h$$

focus: $F(h, k + p)$; axis: $x = h$; directrix: $y = k - p$ $\qquad$ focus: $F(h + p, k)$; axis: $y = k$; directrix: $x = h - p$

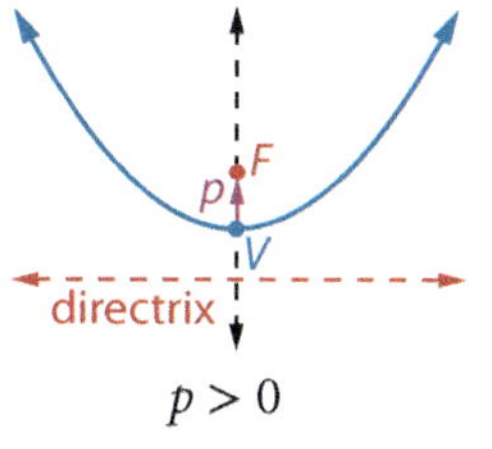

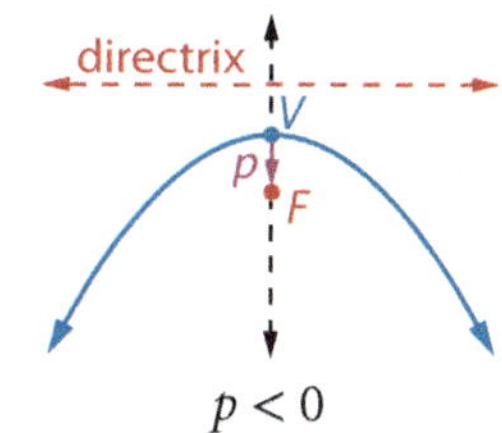

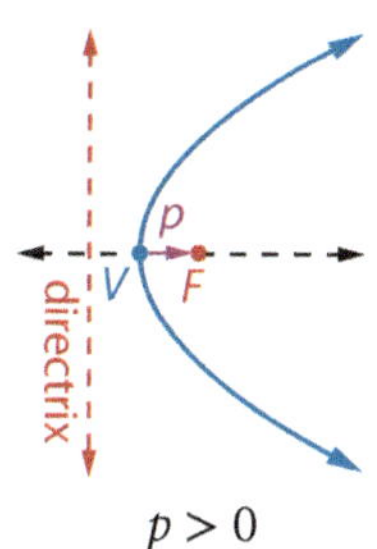

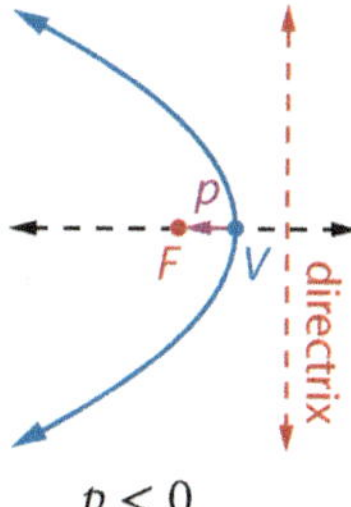

Standard Form Equations of an Ellipse

centered at $C(h, k)$ with $a > b$ and $c^2 = a^2 - b^2$

$$\frac{(x - h)^2}{a^2} + \frac{(y - k)^2}{b^2} = 1 \qquad\qquad \frac{(x - h)^2}{b^2} + \frac{(y - k)^2}{a^2} = 1$$

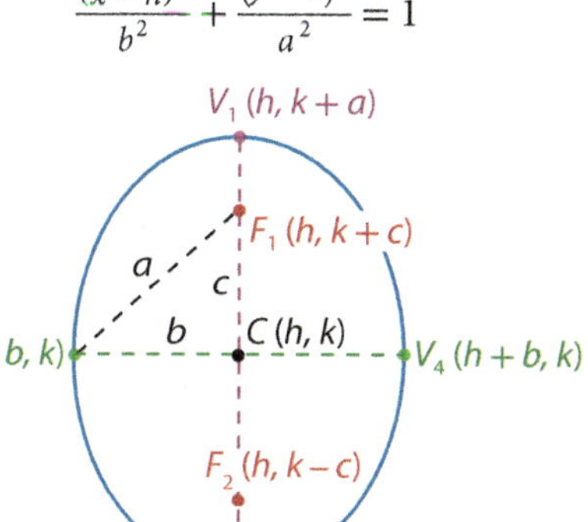

Standard Form Equations of a Hyperbola

centered at $C(h, k)$ with $c^2 = a^2 + b^2$

$$\frac{(x - h)^2}{a^2} - \frac{(y - k)^2}{b^2} = 1 \qquad\qquad \frac{(y - k)^2}{a^2} - \frac{(x - h)^2}{b^2} = 1$$

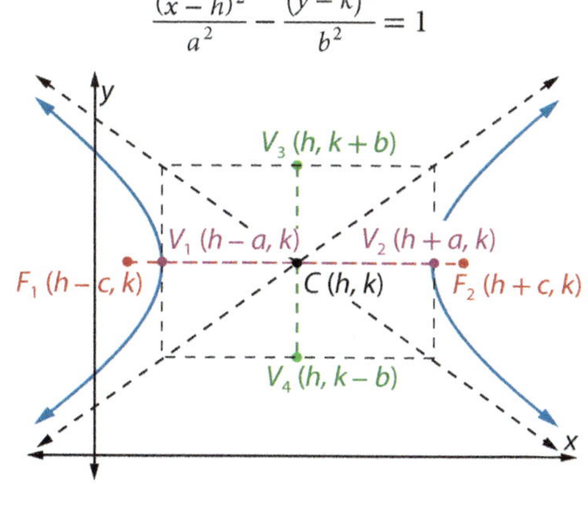

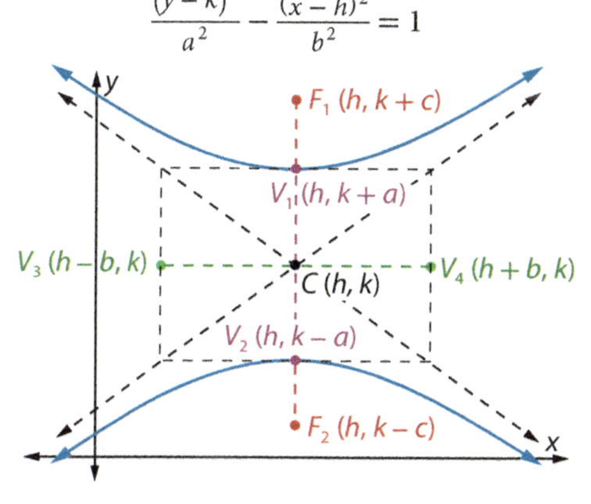

$$y = \pm\frac{b}{a}(x - h) + k \qquad \textbf{asymptotes} \qquad y = \pm\frac{a}{b}(x - h) + k$$

General Form Equation of Conics

$Ax^2 + Bxy + Cy^2 + Dx + Ey + F = 0$ with discriminant $B^2 - 4AC$ and eccentricity $e = \frac{c}{a}$

Circle	**Ellipse**	**Parabola**	**Hyperbola**
$B^2 - 4AC < 0$;	$B^2 - 4AC < 0$;	$B^2 - 4AC = 0$	$B^2 - 4AC > 0$
$B = 0$ and $A = C$	$B \neq 0$ or $A \neq C$	$e = 1$	$e > 1$
$e = 0$	$e < 1$		

Summations

Properties

$$\sum_{i=1}^{n} c = nc \qquad \sum_{i=1}^{n} ca_i = c \sum_{i=1}^{n} a_i \qquad \sum_{i=1}^{n} (a_i + b_i) = \sum_{i=1}^{n} a_i + \sum_{i=1}^{n} b_i \qquad \sum_{i=1}^{n} a_i = \sum_{i=1}^{p} a_i + \sum_{i=p+1}^{n} a_i \text{ where } 1 < p < n$$

Partial Sums of Powers of Integers

$$\sum_{k=1}^{n} k = \frac{n^2 + n}{2} = \frac{n(n+1)}{2} \qquad \sum_{k=1}^{n} k^2 = \frac{n(n+1)(2n+1)}{6} \qquad \sum_{k=1}^{n} k^3 = \frac{n^2(n+1)^2}{4} \qquad \sum_{k=1}^{n} k^4 = \frac{n(n+1)(2n+1)(3n^2+3n-1)}{30}$$

Probability

$$\text{permutation} \quad {}_nP_r = \frac{n!}{(n-r)!} \qquad\qquad \text{combination} \quad {}_nC_r = \frac{n!}{r!(n-r)!}$$

Binomial Theorem

$$(x+y)^n = {}_nC_0\, x^n + {}_nC_1\, x^{n-1}y + {}_nC_2\, x^{n-2}y^2 + {}_nC_3\, x^{n-3}y^3 + \cdots + {}_nC_n\, y^n = \sum_{r=0}^{n} {}_nC_r\, x^{n-r}y^r$$

theoretical $\quad P(E) = \frac{n(E)}{n(S)} \qquad$ **experimental** $\quad P(E) = \frac{\text{number of times the event occurs}}{\text{number of trials}} \qquad$ **complement of an event** $\quad P(E') = 1 - P(E)$

independent events $\quad P(A \text{ and } B) = P(A) \cdot P(B)$
dependent events $\quad P(A \text{ and } B) = P(A) \cdot P(B|A)$

mutually exclusive events $\quad P(A \text{ or } B) = P(A) + P(B)$
inclusive events $\quad P(A \text{ or } B) = P(A) + P(B) - P(A \text{ and } B)$

Statistics

	Population	Sample		Binomial Probability Distribution
mean	$\mu = \dfrac{\sum_{i=1}^{N} x_i}{N}$	$\bar{x} = \dfrac{\sum_{i=1}^{n} x_i}{n}$	**formula**	$P(x) = {}_nC_x\, p^x q^{n-x}$
			mean	$\mu = np$
			variance	$\sigma^2 = npq$
variance	$\sigma^2 = \dfrac{\sum_{i=1}^{N}(x_i - \mu)^2}{N} \text{ or } \dfrac{SS}{N}$	$s^2 = \dfrac{\sum_{i=1}^{n}(x_i - \bar{x})^2}{n-1} \text{ or } \dfrac{SS}{n-1}$	**standard deviation**	$\sigma = \sqrt{\sigma^2} = \sqrt{npq}$
standard deviation	$\sigma = \sqrt{\dfrac{\sum_{i=1}^{N}(x_i - \mu)^2}{N}}$	$s = \sqrt{\dfrac{\sum_{i=1}^{n}(x_i - \bar{x})^2}{n-1}}$		
z-score	$z = \dfrac{x - \mu}{\sigma}$	$z = \dfrac{x - \bar{x}}{s}$		

Limits

sum $\quad \lim_{x \to a}[f(x) + g(x)] = \lim_{x \to a} f(x) + \lim_{x \to a} g(x) \qquad$ **product** $\quad \lim_{x \to a}[f(x) \cdot g(x)] = \lim_{x \to a} f(x) \cdot \lim_{x \to a} g(x)$

scalar multiple $\quad \lim_{x \to a} cf(x) = c \lim_{x \to a} f(x) \text{ where } c \in \mathbb{R} \qquad$ **quotient** $\quad \lim_{x \to a} \dfrac{f(x)}{g(x)} = \dfrac{\lim_{x \to a} f(x)}{\lim_{x \to a} g(x)}$

power $\quad \lim_{x \to a}[f(x)]^n = \left[\lim_{x \to a} f(x)\right]^n \qquad$ **reciprocal power function at infinity** $\quad \lim_{x \to \pm\infty} \dfrac{1}{x^n} = 0 \text{ where } n \in \mathbb{N}$

Derivatives

constant	$\frac{d}{dx}[c] = 0$	**product**	$\frac{d}{dx}[u(x) \cdot v(x)] = u'(x) \cdot v(x) + u(x) \cdot v'(x)$
linear	$\frac{d}{dx}[mx + b] = m$	**quotient**	$\frac{d}{dx}\left[\frac{u(x)}{v(x)}\right] = \frac{u'(x) \cdot v(x) - u(x) \cdot v'(x)}{[v(x)]^2}$
power	$\frac{d}{dx}[cx^n] = ncx^{n-1}$	**chain rule**	$\frac{d}{dx}[(u \circ v)(x)] = u'(v(x)) \cdot v'(x)$
sum or difference	$\frac{d}{dx}[u(x) \pm v(x)] = u'(x) \pm v'(x)$	**basic trigonometric**	$\frac{d}{dx}[\sin x] = \cos x$
natural exponential	$\frac{d}{dx}[e^x] = e^x$		$\frac{d}{dx}[\cos x] = -\sin x$

Integrals

constant	$\int k \, dx = kx + C$	**scalar**	$\int cf(x) \, dx = c \int f(x) \, dx$
power	$\int kx^n \, dx = \frac{k}{n+1}x^{n+1} + C$	**sum**	$\int [f(x) \pm g(x)] \, dx = \int f(x) \, dx \pm \int g(x) \, dx$
sine	$\int (\sin x) \, dx = -\cos x + C$	**cosine**	$\int (\cos x) \, dx = \sin x + C$
natural exponential	$\int e^x \, dx = e^x + C$	**Fundamental Theorem of Calculus**	$\int_a^b f(x) \, dx = F(b) - F(a)$
definite integral	$\int_a^b (x) \, dx = \lim_{n \to \infty} \sum_{i=1}^{n} f(x_i) \, \Delta x$		

Other Formulas

distance formula	$d = \sqrt{(x_2 - x_1)^2 + (y_2 - y_1)^2 + (z_2 - z_1)^2}$
midpoint formula	$M = \left(\frac{x_1 + x_2}{2}, \frac{y_1 + y_2}{2}, \frac{z_1 + z_2}{2}\right)$
absolute value of a complex number	If $z = a + bi$, then $|z| = |a + bi| = \sqrt{a^2 + b^2}$.
Heron's formula	$Area = \sqrt{s(s - a)(s - b)(s - c)}$, where $s = \frac{a + b + c}{2}$
compound interest	$A(t) = P\left(1 + \frac{r}{n}\right)^{nt}$
continuously compounded interest	$A(t) = Pe^{rt}$
height of a projectile	$h(t) = -\frac{1}{2}gt^2 + v_0 t + h_0$, where $g = 32$ ft/sec^2 or 9.8 m/sec^2

Trigonometry

Basic Ratios

$$\sin A = \frac{opp.}{hyp.}$$

$$\cos A = \frac{adj.}{hyp.}$$

$$\tan A = \frac{opp.}{adj.}$$

Reciprocal Ratios

$$\csc A = \frac{1}{\sin A} = \frac{hyp.}{opp.}$$

$$\sec A = \frac{1}{\cos A} = \frac{hyp.}{adj.}$$

$$\cot A = \frac{1}{\tan A} = \frac{adj.}{opp.}$$

Reciprocal Identities

$$\csc \theta = \frac{1}{\sin \theta} \qquad \sec \theta = \frac{1}{\cos \theta} \qquad \cot \theta = \frac{1}{\tan \theta}$$

Quotient Identities

$$\tan \theta = \frac{\sin \theta}{\cos \theta} \qquad \cot \theta = \frac{\cos \theta}{\sin \theta}$$

Pythagorean Identities

$$\sin^2 \theta + \cos^2 \theta = 1 \qquad \cot^2 \theta + 1 = \csc^2 \theta \qquad \tan^2 \theta + 1 = \sec^2 \theta$$

Even/Odd Identities

$$\sin(-\theta) = -\sin \theta \qquad \cos(-\theta) = \cos \theta \qquad \tan(-\theta) = -\tan \theta$$

$$\csc(-\theta) = -\csc \theta \qquad \sec(-\theta) = \sec \theta \qquad \cot(-\theta) = -\cot \theta$$

Cofunction Identities

$$\sin \theta = \cos\left(\frac{\pi}{2} - \theta\right) \qquad \tan \theta = \cot\left(\frac{\pi}{2} - \theta\right) \qquad \sec \theta = \csc\left(\frac{\pi}{2} - \theta\right)$$

$$\cos \theta = \sin\left(\frac{\pi}{2} - \theta\right) \qquad \cot \theta = \tan\left(\frac{\pi}{2} - \theta\right) \qquad \csc \theta = \sec\left(\frac{\pi}{2} - \theta\right)$$

Sum/Difference Identities

$$\cos(\alpha \pm \beta) = \cos \alpha \cos \beta \mp \sin \alpha \sin \beta$$

$$\sin(\alpha \pm \beta) = \sin \alpha \cos \beta \pm \cos \alpha \sin \beta$$

$$\tan(\alpha \pm \beta) = \frac{\tan \alpha \pm \tan \beta}{1 \mp \tan \alpha \tan \beta}$$

Double-Angle Identities

$$\cos 2\theta = \cos^2 \theta - \sin^2 \theta$$

$$\sin 2\theta = 2 \sin \theta \cos \theta$$

$$\tan 2\theta = \frac{2 \tan \theta}{1 - \tan^2 \theta}$$

Half-Angle Identities

$$\sin \frac{\theta}{2} = \pm\sqrt{\frac{1 - \cos \theta}{2}}$$

$$\cos \frac{\theta}{2} = \pm\sqrt{\frac{1 + \cos \theta}{2}}$$

$$\tan \frac{\theta}{2} = \pm\sqrt{\frac{1 - \cos \theta}{1 + \cos \theta}} = \frac{1 - \cos \theta}{\sin \theta} = \frac{\sin \theta}{1 + \cos \theta}$$

Power-Reducing Identities

$$\sin^2 \theta = \frac{1 - \cos 2\theta}{2}$$

$$\cos^2 \theta = \frac{1 + \cos 2\theta}{2}$$

$$\tan^2 \theta = \frac{1 - \cos 2\theta}{1 + \cos 2\theta}$$

Product-to-Sum Identities

$$\sin \alpha \sin \beta = \tfrac{1}{2}[\cos(\alpha - \beta) - \cos(\alpha + \beta)]$$

$$\cos \alpha \cos \beta = \tfrac{1}{2}[\cos(\alpha - \beta) + \cos(\alpha + \beta)]$$

$$\sin \alpha \cos \beta = \tfrac{1}{2}[\sin(\alpha + \beta) + \sin(\alpha - \beta)]$$

$$\cos \alpha \sin \beta = \tfrac{1}{2}[\sin(\alpha + \beta) - \sin(\alpha - \beta)]$$

Sum-to-Product Identities

$$\sin \alpha + \sin \beta = 2 \sin \frac{\alpha + \beta}{2} \cos \frac{\alpha - \beta}{2}$$

$$\sin \alpha - \sin \beta = 2 \cos \frac{\alpha + \beta}{2} \sin \frac{\alpha - \beta}{2}$$

$$\cos \alpha + \cos \beta = 2 \cos \frac{\alpha + \beta}{2} \cos \frac{\alpha - \beta}{2}$$

$$\cos \alpha - \cos \beta = -2 \sin \frac{\alpha + \beta}{2} \sin \frac{\alpha - \beta}{2}$$

angle measure

$$\pi \text{ radians} = 180°$$

arc length

$$s = r\theta \text{ (in radians)}$$

angular speed

$$\omega = \frac{\theta}{t}$$

linear speed

$$v = \frac{s}{t} = r\omega$$

area of a sector
(θ in radians)

$$A_{sector} = \tfrac{1}{2}r^2\theta$$

Polar-to-Rectangular Equations

$$r^2 = x^2 + y^2 \qquad x = r \cos \theta \qquad y = r \sin \theta$$

Law of Sines

$$\frac{\sin A}{a} = \frac{\sin B}{b} = \frac{\sin C}{c} \quad \text{or} \quad \frac{a}{\sin A} = \frac{b}{\sin B} = \frac{c}{\sin C}$$

Law of Cosines

$$a^2 = b^2 + c^2 - 2bc \cos A, \quad b^2 = a^2 + c^2 - 2ac \cos B,$$
$$\text{and } c^2 = a^2 + b^2 - 2ab \cos C$$

$\in$	element
$\cup$	union
$\cap$	intersection
r	correlation coefficient
r^2	coefficient of determination
e	the number ≈ 2.71828
i	imaginary unit, $\sqrt{-1}$
$\theta, \alpha, \beta, \gamma$	theta, alpha, beta, gamma (angles)
Δ	delta, the change in
$\{x \mid \ldots\}$	the set of x such that $\ldots$
$\sqrt{x}$	principal square root of x
$\sqrt[n]{x}$	principal nth root of x
(a, b)	ordered pair
(a, b, c)	ordered triple
(a, b)	open interval
$[a, b]$	closed interval
$f(x) = \lvert x \rvert$	absolute value function
$f(x) = [x]$	greatest integer function
$(f \circ g)(x)$	function composition, $f(g(x))$
$f^{-1}(x)$	inverse function of $f(x)$
$f(x, y)$	function of two variables
$f(x) = \dfrac{a(x)}{b(x)}$	rational function
$\log_b x$	base b logarithm of x
$\log x$	common logarithm (base 10) of x
$\ln x$	natural logarithm (base e) of x
ω	omega, angular speed
$\mathbf{v}, \vec{v}$	vector $\mathbf{v}$
$\langle v_1, v_2 \rangle$	component form of vector $\mathbf{v}$
$\lvert \mathbf{v} \rvert$	magnitude of $\mathbf{v}$
$\mathbf{u} \cdot \mathbf{v}$	dot product
$\mathrm{proj}_{\mathbf{v}} \mathbf{u}$	projection of $\mathbf{u}$ onto $\mathbf{v}$
(r, θ)	polar coordinates
$r \operatorname{cis} \theta$	polar form of a complex number
$\lvert a + bi \rvert$	modulus of a complex number
$A_{m \times n}$	matrix dimensions
a_{mn}	entry in row m and column n
I	identity matrix
$\lvert A \rvert$ or $\det A$	determinant of matrix A
A^{-1}	inverse matrix
$\sum$	sigma, sum (summation notation)
$\displaystyle\sum_{i=1}^{\infty} a_i$	infinite series
$\displaystyle\lim_{n \to \infty} a_n$	limit of a sequence
$\displaystyle\sum_{i=1}^{n} a_i$	nth partial sum of a sequence (S_n)
$n!$	n factorial
$_nP_r$	permutation of r out of n objects
$_nC_r$	combination of r out of n objects
$P(E)$	probability of an event
$P(A\mid B)$	probability of A given B
μ	mu, mean of a population
$\bar{x}$	x-bar, mean of a sample
σ	sigma, standard deviation of a population
σ^2	sigma squared, variance of a population
s	standard deviation of a sample
s^2	variance of a sample
x_i	value of random variable x at ith position
n	sample size
N	population size
Q_n	quartile
$\mu_{\bar{x}}$	mean of the sample means
$\sigma_{\bar{x}}$	standard deviation of the sample means
z_c	critical value
CI	confidence interval
α	level of significance
p	p-value
E	margin of error for the mean
SS	sum of the squares
H_0	null hypothesis
H_a	alternative (research) hypothesis
$\displaystyle\lim_{x \to a} f(x)$	limit of a function
$\displaystyle\lim_{x \to a^-} f(x)$	left-hand limit
$\displaystyle\lim_{x \to a^+} f(x)$	right-hand limit
$f'(x), y', \dfrac{dy}{dx}, \dfrac{d}{dx} f(x)$	derivative of a function
$\displaystyle\int_a^b f(x)\, dx$	definite integral
$\displaystyle\int f(x)\, dx$	indefinite integral

Appendix A
Vocabulary

CHAPTER 7

additive identity matrix
additive inverse matrix
augmented matrix
boundary of a feasible region
cofactors
column matrix
constraints
Cramer's rule
dependent consistent system
determinant of a matrix
dimensions
entry (element)
equal matrices
equivalent system

feasible region
Gaussian elimination
Gauss-Jordan elimination
inconsistent system
independent consistent system
inverse matrix
invertible (nonsingular) matrix
linear programming
main diagonal
matrix
matrix addition
matrix product
minor
multiplicative identity matrix
multiplicative inverse matrix
objective function

order of a square matrix
partial fraction decomposition
partial fractions
reduced row echelon form (RREF)
row echelon form (REF)
row matrix
scalar multiplication
scalar product
singular matrix
solution to a system of equations
square matrix
system of equations
system of inequalities
Vertex Principle of Linear Programming
zero matrix

CHAPTER 8

aphelion
branches
center
chord
conical surface
conic section (conic)
conjugate axis
co-vertices
directrix
discriminant
eccentricity
elements
ellipse
focal axis

focal distance
focal length
focal width
focus
general form of a conic
generating curve
hyperbola
invariant under rotation
locus
major axis
minor axis
nappe
orientation
parabola
parameter interval

parametric curve
parametric equations
perihelion
right circular conical surface
rotation formulas
semiconjugate axis
semimajor axis
semiminor axis
semitransverse axis
standard form equation of a parabola
tangent
transverse axis
vertex

CHAPTER 9

anchor step
arithmetic mean
arithmetic sequence
arithmetic series
binomial
binomial coefficient
binomial experiment
Binomial Theorem
combination
common difference
common ratio

convergent
divergent
explicit formula
factorial
finite sequence
future value of an annuity
general term
geometric mean
geometric sequence
geometric series
harmonic sequence
inductive conclusion
inductive hypothesis

inductive step
infinite sequence
infinite series
limit of the sequence
mathematical induction
nth partial sum
recursive formula
sequence
series
summation (sigma) notation
term

Answers and Solutions Overflow

Chapter 7

7.1 C. Exercises

38. $m_{BC} = \dfrac{1+1}{-6-4} = -\dfrac{1}{5}$

$\overleftrightarrow{BC}: y - 1 = -\dfrac{1}{5}(x + 6)$

$\qquad y = -\dfrac{1}{5}x - \dfrac{1}{5}$

Letting D be the point of intersection, $\overleftrightarrow{AD}$ passes through $(-2, 5)$ with $m_{AD} = 5$.

$y - 5 = 5(x + 2)$

$\quad y = 5x + 15$

$\overleftrightarrow{AD} \cap \overleftrightarrow{BC}:$

$5x + 15 = -\dfrac{1}{5}x - \dfrac{1}{5}$

$25x + 75 = -x - 1$

$\quad 26x = -76$

$\qquad x = -\dfrac{38}{13}$

$y = 5\left(-\dfrac{38}{13}\right) + 15 = \dfrac{5}{13};\ \left(-\dfrac{38}{13}, \dfrac{5}{13}\right)$

39. $\dfrac{1}{2}$(1st) $\Rightarrow \sin x + 2\cos y = 2$

2nd $\Rightarrow \sin x - 2\cos y = 0$

$\overline{\hphantom{2\sin x \qquad = 2}}$

$\quad 2\sin x \qquad\ = 2$

$\qquad\quad \sin x = 1$

$\qquad\qquad x = \dfrac{\pi}{2}$

2nd: $\sin\dfrac{\pi}{2} - 2\cos y = 0$

$\qquad 1 = 2\cos y$

$\qquad \cos y = \dfrac{1}{2}$

$\qquad\quad y = \dfrac{\pi}{3}, \dfrac{5\pi}{3}$

$\therefore \left(\dfrac{\pi}{2}, \dfrac{\pi}{3}\right), \left(\dfrac{\pi}{2}, \dfrac{5\pi}{3}\right)$

40. Substitute each point into the general equation to produce a system.

$(3, -6) \Rightarrow 3A - 6B + C = -45$

$(-1, 2) \Rightarrow -A + 2B + C = -5$

$(6, 3) \ \Rightarrow 6A + 3B + C = -45$

-1(1st) $\Rightarrow -3A + 6B - C = \ \ 45$

3rd $\qquad \Rightarrow \ \ 6A + 3B + C = -45$

$\overline{\hphantom{(a)\qquad 3A + 9B \qquad = 0}}$

(a) $\qquad\ 3A + 9B \qquad\ = \ \ 0$

2nd $\qquad \Rightarrow -A + 2B + C = -5$

-1(3rd) $\Rightarrow -6A - 3B - C = \ \ 45$

$\overline{\hphantom{(b)\qquad -7A - B \qquad = 40}}$

(b) $\qquad -7A - \ B \qquad\ = 40$

(a) $\Rightarrow \qquad 3A + 9B = \ \ 0$

9(b) $\Rightarrow -63A - 9B = 360$

$\overline{\hphantom{-60A \qquad = 360}}$

$\qquad -60A \qquad\ = 360$

$\qquad\qquad A = -6$

(a) $3(-6) + 9B = 0$

$\qquad\quad B = 2$

2nd: $-(-6) + 2(2) + C = -5$

$\qquad\qquad\qquad\qquad C = -15$

41.

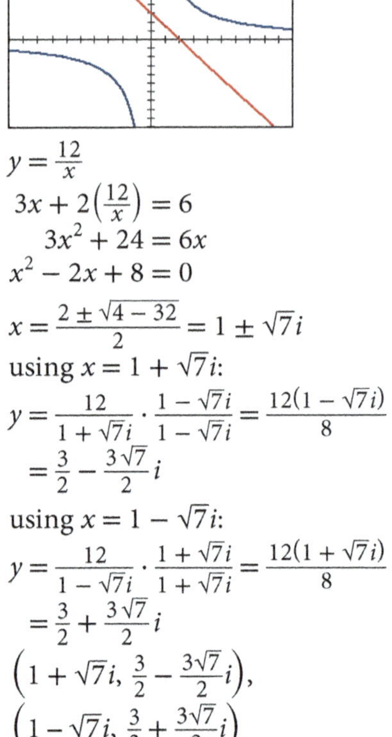

$y = \dfrac{12}{x}$

$3x + 2\left(\dfrac{12}{x}\right) = 6$

$3x^2 + 24 = 6x$

$x^2 - 2x + 8 = 0$

$x = \dfrac{2 \pm \sqrt{4 - 32}}{2} = 1 \pm \sqrt{7}\,i$

using $x = 1 + \sqrt{7}\,i$:

$y = \dfrac{12}{1 + \sqrt{7}\,i} \cdot \dfrac{1 - \sqrt{7}\,i}{1 - \sqrt{7}\,i} = \dfrac{12(1 - \sqrt{7}\,i)}{8}$

$\quad = \dfrac{3}{2} - \dfrac{3\sqrt{7}}{2}i$

using $x = 1 - \sqrt{7}\,i$:

$y = \dfrac{12}{1 - \sqrt{7}\,i} \cdot \dfrac{1 + \sqrt{7}\,i}{1 + \sqrt{7}\,i} = \dfrac{12(1 + \sqrt{7}\,i)}{8}$

$\quad = \dfrac{3}{2} + \dfrac{3\sqrt{7}}{2}i$

$\left(1 + \sqrt{7}\,i, \dfrac{3}{2} - \dfrac{3\sqrt{7}}{2}i\right),$

$\left(1 - \sqrt{7}\,i, \dfrac{3}{2} + \dfrac{3\sqrt{7}}{2}i\right)$

7.1 Cumulative Review

42. x-int.: $5x + (0) = -10; x = -2$

$(-2, 0)$

y-int.: $5(0) + y = -10; y = -10$

$(0, -10)$

$m = -\dfrac{A}{B} = -\dfrac{5}{1} = -5$

43. $m = \dfrac{-5 - 7}{4 - 2} = -6$

44. $f(x) = x^2$ is translated 4 units left and 3 units down.

45. $f(x) = |x|$ is reflected in the x-axis and translated 3 units right and 1 unit up.

46. $p(x) = (x - 5)(x + 5)(x - 3)(x + 3)$

zeros: $\pm 5, \pm 3$

47. $x = 7$ (multiplicity 4)

48. $\qquad 5y = 2x - 15$

$\qquad 2x - 5y = 15$

49. $Area = \dfrac{1}{2}(5)(8)\sin 40° \approx 12.9\ \text{ft}^2$

50. $M\left(\dfrac{-3 + 7}{2}, \dfrac{1 + (-4)}{2}, \dfrac{2 + (-2)}{2}\right)$

$\quad = \left(2, -\dfrac{3}{2}, 0\right)$

---

51. substituting $(r, -\theta)$ or $(-r, \pi - \theta)$:

A. passes using $(r, -\theta)$: $r = -\sin 3$

B. fails using $(r, -\theta)$:

$\quad r = -5\sin(-\theta)$

$\quad r = 5\sin\theta \ne -5\sin\theta$

fails using $(-r, \pi - \theta)$:

$\quad (-r) = -5\sin(\pi - \theta)$

$\quad r = 5\sin\theta \ne -5\sin\theta$

C. passes using $(-r, \pi - \theta)$:

$\quad (-r) = 2\sin 2(\pi - \theta)$

$\quad r = -2\sin(2\pi - 2\theta)$

$\quad r = -2\sin(-2\theta)$

$\quad r = 2\sin 2\theta$

D. passes using $(r, -\theta)$:

$\quad r^2 = 9\cos 2(-\theta) = 9\cos 2\theta$

7.3 C. Exercises

43.

$$\begin{array}{l} Eq_2 \Rightarrow \\ Eq_3 \Rightarrow \\ Eq_1 \Rightarrow \end{array} \begin{bmatrix} 1 & 1 & 1 & 4 \\ 4 & 2 & 1 & -3 \\ 9 & -3 & 1 & -8 \end{bmatrix}$$

$$\begin{array}{l} 4r_1 - r_2 \Rightarrow r_2 \\ 9r_1 - r_3 \Rightarrow r_3 \end{array} \begin{bmatrix} 1 & 1 & 1 & 4 \\ 0 & 2 & 3 & 19 \\ 0 & 12 & 8 & 44 \end{bmatrix}$$

$$\begin{array}{l} \tfrac{1}{2}r_2 \Rightarrow r_2 \\ 6r_2 - r_3 \Rightarrow r_3 \end{array} \begin{bmatrix} 1 & 1 & 1 & 4 \\ 0 & 1 & \tfrac{3}{2} & \tfrac{19}{2} \\ 0 & 0 & 10 & 70 \end{bmatrix}$$

$$\begin{array}{l} -r_2 + r_1 \Rightarrow r_1 \\ \\ \tfrac{1}{10}r_3 \Rightarrow r_3 \end{array} \begin{bmatrix} 1 & 0 & -\tfrac{1}{2} & -\tfrac{11}{2} \\ 0 & 1 & \tfrac{3}{2} & \tfrac{19}{2} \\ 0 & 0 & 1 & 7 \end{bmatrix}$$

$$\begin{array}{l} \tfrac{1}{2}r_3 + r_1 \Rightarrow r_1 \\ -\tfrac{3}{2}r_3 + r_2 \Rightarrow r_2 \end{array} \begin{bmatrix} 1 & 0 & 0 & -2 \\ 0 & 1 & 0 & -1 \\ 0 & 0 & 1 & 7 \end{bmatrix}$$

44. $f(1) = a + b + c = 150$

$f(2) = 4a + 2b + c = 400$

$f(3) = 9a + 3b + c = 750$

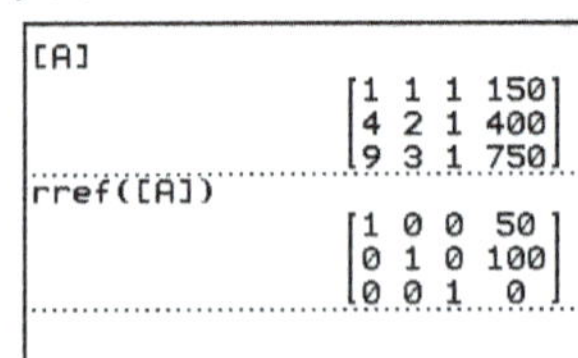

$f(x) = 50x^2 + 100x$

$f(4) = 50(4)^2 + 100(4) = \$1200$

7.3 Cumulative Review

45. $\tan 52° = \dfrac{x}{500};\ x \approx 640\ \text{ft}$

$h = x + 6 \approx 646\ \text{ft}$

46. $y = 3\left(x^2 - \frac{7}{3}x\right) + 1$

$y = 3\left(x^2 - \frac{7}{3}x + \left(\frac{7}{6}\right)^2\right) + 1 - 3\left(\frac{49}{36}\right)$

$y = 3\left(x - \frac{7}{6}\right)^2 - \frac{37}{12}$

$V\left(\frac{7}{6}, -\frac{37}{12}\right)$; min

47. $p = \frac{2\pi}{|b|} = \frac{\pi}{2}$; $|b| = \frac{4\pi}{\pi} = 4$

$f(x) = 3 \cos 4\left(x - \frac{\pi}{3}\right)$

48. $r = \sqrt{2^2 + (-2)^2} = 2\sqrt{2} \approx 2.83$

$\alpha = \text{Tan}^{-1}\left|\frac{-2}{2}\right| = \text{Tan}^{-1} 1 = \frac{\pi}{4}$

Q IV: $\theta = 2\pi - \frac{\pi}{4} = \frac{7\pi}{4}$

$\therefore z = 2\sqrt{2} \text{ cis } \frac{7\pi}{4}$

49. $r \cos^2 \theta = \sin \theta$

$r^2 \cos^2 \theta = r \sin \theta$

$(r \cos \theta)^2 = r \sin \theta$

$x^2 = y$

50. $A = \cos^{-1}\left(\frac{a \sin B}{b}\right) \approx 39.8°$

$B = \cos^{-1}\left(\frac{9^2 + 14^2 - 10^2}{2(9)(14)}\right) \approx 45.4°$

$C \approx 180 - 39.8 - 45.4 = 94.8°$

51. $-x^2 + 2x - 15 = 3x - 21$

$x^2 + x - 6 = 0$

$(x + 3)(x - 2) = 0$

$x = -3, 2$

If $x = -3$, $y = 3(-3) - 21 = -30$.

If $x = 2$, $y = 3(2) - 21 = -15$.

52. 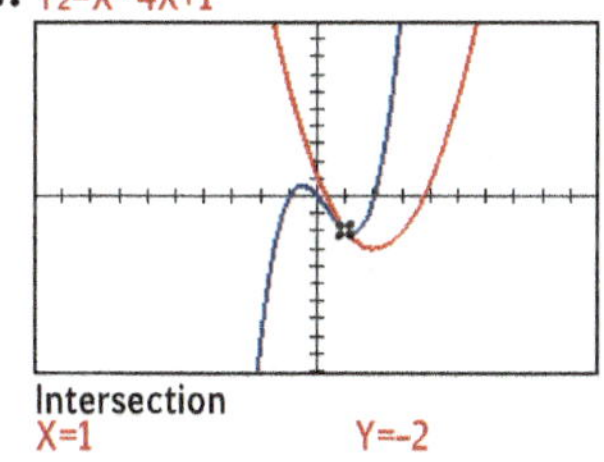

53. A. AB is a 2×2 matrix.

C. Matrix multiplication is not commutative.

54. $e_{23} = 2(6) + (-3)(-2) = 18$

7.5 C. Exercises

39. $U = CE^{-1} = \begin{bmatrix} 18 & 4 & 13 \\ 3 & 12 & 14 \\ 13 & 4 & 24 \end{bmatrix}$

$\Rightarrow$ SEND MONEY

40. $U = CE^{-1} = \begin{bmatrix} 1 & 4 & 7 \\ 14 & 13 & 4 \\ 18 & 19 & 1 \\ 4 & 10 & 8 \\ 13 & 3 & 23 \end{bmatrix}$

$\Rightarrow$ BE HONEST BE KIND

41. $[A|I] = \begin{bmatrix} a & b & | & 1 & 0 \\ c & d & | & 0 & 1 \end{bmatrix}$

$\frac{1}{a}r_1 \Rightarrow r_1 \begin{bmatrix} 1 & \frac{b}{a} & | & \frac{1}{a} & 0 \\ -cr_1 + ar_2 \Rightarrow r_2 & 0 & ad - bc & | & -c & a \end{bmatrix}$

$\frac{1}{ad - bc}r_2 \Rightarrow r_2 \begin{bmatrix} 1 & \frac{b}{a} & | & \frac{1}{a} & 0 \\ 0 & 1 & | & -\frac{c}{ad - bc} & \frac{a}{ad - bc} \end{bmatrix}$

$-\frac{b}{a}r_2 + r_1 \Rightarrow r_1 \begin{bmatrix} 1 & 0 & | & \frac{d}{ad - bc} & -\frac{b}{ad - bc} \\ 0 & 1 & | & -\frac{c}{ad - bc} & \frac{a}{ad - bc} \end{bmatrix}$

$= [I | A^{-1}]$

Note: $e_{13} = -\frac{b}{a}\left(-\frac{c}{ad - bc}\right) + \frac{1}{a} = \frac{bc}{a(ad - bc)} + \frac{ad - bc}{a(ad - bc)} = \frac{ad}{a(ad - bc)} = \frac{d}{ad - bc}$

$A^{-1} = \frac{1}{ad - bc}\begin{bmatrix} d & -b \\ -c & a \end{bmatrix} = \frac{1}{|A|}\begin{bmatrix} d & -b \\ -c & a \end{bmatrix}$

7.5 Cumulative Review

42. $p = k\frac{qr}{t}$ $8 = \left(\frac{3}{4}\right)\frac{q(12)}{20}$

$3q = k\frac{q(4t)}{t}$ $160 = 9q$

$3q = 4kq$ $q = \frac{160}{9}$

$k = \frac{3}{4}$

43. $4 \sin \theta \cos \theta = 2(2 \sin \theta \cos \theta)$

$= 2 \sin 2\theta$

45. $p(x) = x(x^2 - 1)$

$= x(x - 1)(x + 1)$

$x = 0, \pm 1$

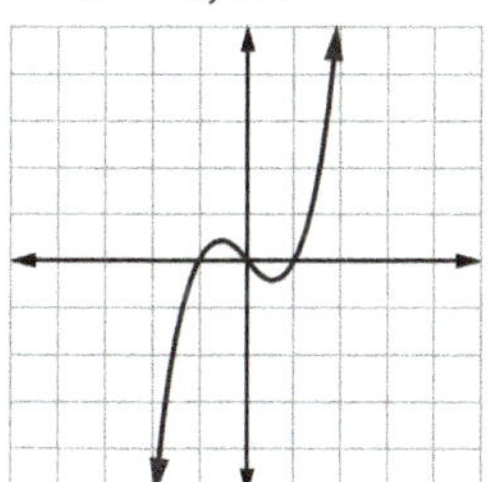

46. $4y + (3) = 19$; $y = 4$

$5x + 2(4) - 3(3) = 9$

$5x = 10$; $x = 2$

47. $|A| = 0$; The product of each entry and its cofactor is 0.

48. $\cos\left(\frac{3\pi}{12} + \frac{2\pi}{12}\right)$

$= \cos\frac{\pi}{4} \cos\frac{\pi}{6} - \sin\frac{\pi}{4} \sin\frac{\pi}{6}$

$= \left(\frac{\sqrt{2}}{2}\right)\left(\frac{\sqrt{3}}{2}\right) - \left(\frac{\sqrt{2}}{2}\right)\left(\frac{1}{2}\right)$

$= \frac{\sqrt{6} - \sqrt{2}}{4}$

49. $Area = \frac{1}{2}(3.2)(4.8) \sin 75° \approx 7.4 \text{ mi}^2$

50. $\sin A = \frac{7}{19}$

$A = \sin^{-1}\frac{7}{19} \approx 22°$

51. $\frac{2\pi}{12} = \frac{\pi}{6}$

7.6 C. Exercises

40. $P(x, y) = 175x + 275y$, where
$x = $ Zphone 7 and $y = $ Zphone 8
$x \geq 0; y \geq 0; x + y \leq 200$
$300x + 500y \leq \$75{,}000$

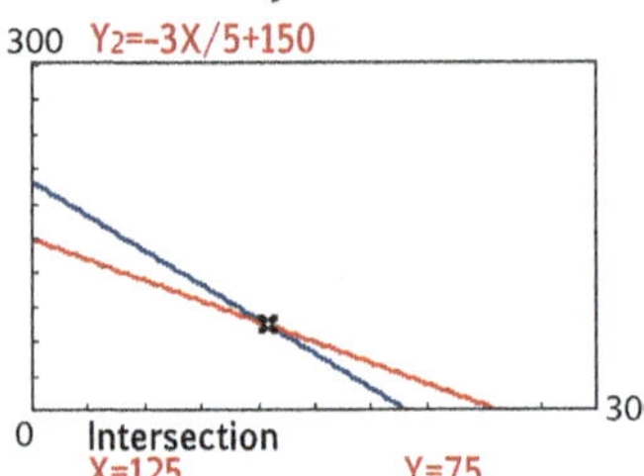

$\begin{bmatrix} X \\ Y \end{bmatrix} = \begin{bmatrix} 1 & 1 \\ 3 & 5 \end{bmatrix}^{-1} \begin{bmatrix} 200 \\ 750 \end{bmatrix} = \begin{bmatrix} 125 \\ 75 \end{bmatrix}$

$P(0, 0) = 175(0) + 275(0) = \$0$

$P(0, 150) = 175(0) + 275(150)$
$\qquad\qquad = \$41{,}250$

$P(125, 75) = 175(125) + 275(75)$
$\qquad\qquad = \$42{,}500;$ max

$P(200, 0) = 175(200) + 275(0)$
$\qquad\qquad = \$35{,}000$

41. $C(x, y) = 0.75x + 1.00y$
$x \geq 0; y \geq 0$
$10x + 30y \geq 60$ g protein
$45x + 15y \geq 90$ g carbohydrates

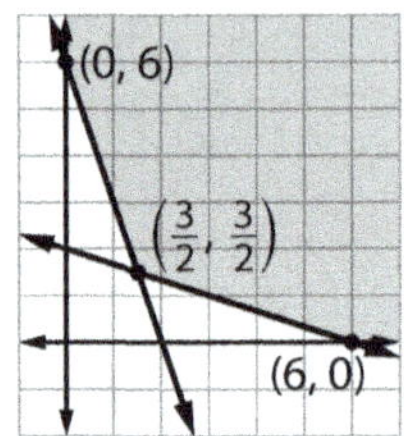

$45x + 15y \geq 90; 3x + y = 6$
$10x + 30y \geq 60; x + 3y = 6$

$\begin{bmatrix} X \\ Y \end{bmatrix} = \begin{bmatrix} 3 & 1 \\ 1 & 3 \end{bmatrix}^{-1} \begin{bmatrix} 6 \\ 6 \end{bmatrix} = \begin{bmatrix} 1.5 \\ 1.5 \end{bmatrix}$

$C(0, 6) = 0.75(0) + 1.00(6) = \$6.00$

$C(1.5, 1.5) = 0.75(1.5) + 1.00(1.5)$
$\qquad\qquad \approx \$2.63;$ min

$C(6, 0) = 0.75(6) + 1.00(0) = \$4.50$

7.6 Cumulative Review

42. $\log \dfrac{a^2 b^3}{c} = \log a^2 + \log b^3 - \log c$
$= 2 \log a + 3 \log b - \log c$
$= 2(2.3) + 3(1.4) - 0.9 = 7.9$

43. $10 \sin^2 \theta - 5 = -5(1 - 2\sin^2 \theta)$
$\qquad\qquad = -5 \cos 2\theta$

44. A polar equation of the form
$r = b \sin n\theta$ where n is odd will have
n petals.

45. $r = \sqrt{(-4)^2 + (-6)^2} = 2\sqrt{13} \approx 7.21$
Q III: $\theta = \pi + \text{Tan}^{-1} \dfrac{-6}{-4}$
$\qquad\qquad \approx \pi + 0.98 \approx 4.12$
$z \approx 7.21 \text{ cis } 4.12$

---

46. $4x - 6y = 24$
$\underline{5x + 6y = 30}$
$9x \qquad\;\; = 54$
$\qquad x = 6$
$5(6) + 6y = 30; y = 0$

47. $|A| = 0 - (-6) = 6$

$A^{-1} = \dfrac{1}{6} \begin{bmatrix} 7 & -3 \\ 2 & 0 \end{bmatrix} = \begin{bmatrix} \frac{7}{6} & -\frac{1}{2} \\ \frac{1}{3} & 0 \end{bmatrix}$

48. $\text{proj}_{\mathbf{h}} \mathbf{F} = |\mathbf{F}| \cos \theta$
$= |234| \cos 50° \approx 150.4$ lb

49. $r = \sqrt{10^2 + (-8)^2} = 2\sqrt{41} \approx 12.81$
Q IV: $\theta = 2\pi + \text{Tan}^{-1} \dfrac{-8}{10}$
$\qquad\qquad \approx 2\pi + (-0.67) \approx 5.61$
$z \approx 12.81 \text{ cis } 5.61$

50. $|A| = 7(-8) - (-2)(4)$
$\qquad = -56 + 8 = -48$

51. $\log_4 \dfrac{(x-2)(x+1)}{(x+1)} = 2$
$\qquad \log_4 (x - 2) = 2$
$\qquad\qquad x - 2 = 4^2; x = 18$

7.7 B. Exercises

30. $\left[\dfrac{3x^2 - 2x + 2}{(x-3)(x^2 + 4x + 2)} = \dfrac{A}{x-3} \right.$

$\qquad \left. + \dfrac{Bx + C}{x^2 + 4x + 2} \right] (x-3)(x^2 + 4x + 2)$

$3x^2 - 2x + 2$
$= A(x^2 + 4x + 2) + (Bx + C)(x - 3)$
$= (A + B)x^2 + (4A - 3B + C)x$
$\qquad\qquad\qquad\qquad + (2A - 3C)$

$A + B = 3$
$4A - 3B + C = -2$
$2A - 3C = 2$

```
[A]
        [ 1   1   0   3 ]
        [ 4  -3   1  -2 ]
        [ 2   0  -3   2 ]
rref([A])
        [ 1   0   0   1 ]
        [ 0   1   0   2 ]
        [ 0   0   1   0 ]
```

$\therefore \dfrac{1}{x-3} + \dfrac{2x}{x^2 + 4x + 2}$

31. $\left[\dfrac{3x^2 - 24x + 32}{x(x-4)^2} = \dfrac{A}{x} + \dfrac{B}{x-4} \right.$

$\qquad \left. + \dfrac{C}{(x-4)^2} \right] x(x-4)^2$

$3x^2 - 24x + 32$
$= A(x-4)^2 + Bx(x-4) + Cx$
$= Ax^2 - 8Ax + 16A + Bx^2 - 4Bx$
$\qquad\qquad\qquad\qquad\qquad + Cx$
$= (A + B)x^2 + (-8A - 4B + C)x$
$\qquad\qquad\qquad\qquad\qquad + 16A$

$A + B = 3$
$-8A - 4B + C = -24$
$16A = 32; A = 2$

$B = 3 - (2) = 1$
$C = -24 + 8(2) + 4(1) = -4$
$\therefore \dfrac{2}{x} + \dfrac{1}{x-4} - \dfrac{4}{(x-4)^2}$

32. $\left[\dfrac{10x^4 - 32x^2 + 2x + 27}{x(2x^2 - 3)^2} = \dfrac{A}{x} \right.$

$\qquad \left. + \dfrac{Bx + C}{2x^2 - 3} + \dfrac{Dx + E}{(2x^2 - 3)^2} \right] x(2x^2 - 3)^2$

$10x^4 - 32x^2 + 2x + 27$
$= A(2x^2 - 3)^2 + (Bx + C)(2x^3 - 3x)$
$\qquad\qquad\qquad\qquad\qquad + (Dx + E)x$
$= 4Ax^4 - 12Ax^2 + 9A + 2Bx^4 - 3Bx^2$
$\qquad\qquad\qquad + 2Cx^3 - 3Cx + Dx^2 + Ex$
$= (4A + 2B)x^4 + 2Cx^3$
$\qquad\qquad\qquad\qquad + (-12A - 3B + D)x^2$
$\qquad\qquad\qquad\qquad + (-3C + E)x + 9A$

$4A + 2B = 10; 2C = 0; C = 0$
$-12A - 3B + D = -32$
$-3C + E = 2; 9A = 27; A = 3$
$2B = 10 - 4(3); B = -1$
$E = 2 + 3(0) = 2$
$D = -32 + 12(3) + 3(-1) = 1$
$\therefore \dfrac{3}{x} - \dfrac{x}{2x^2 - 3} + \dfrac{x + 2}{(2x^2 - 3)^2}$

7.7 C. Exercises

33. $x^2 - x \overline{\smash{\big)}\,4x^2 + 0x + 3} \quad {\scriptstyle 4}$
$\qquad\quad \underline{-(4x^2 - 4x)}$
$\qquad\qquad\qquad\quad 4x + 3$

$\dfrac{4x^2 + 3}{x^2 - x} = 4 + \dfrac{4x + 3}{x(x - 1)}$

$\left[\dfrac{4x + 3}{x(x-1)} = \dfrac{A}{x} + \dfrac{B}{(x-1)} \right] x(x - 1)$

$4x + 3 = A(x - 1) + Bx$
$\qquad\quad = (A + B)x - A$

$A + B = 4$
$-A = 3; A = -3$
$B = 4 - (-3) = 7$
$\therefore 4 - \dfrac{3}{x} + \dfrac{7}{x - 1}$

34. $x^2 + 3x \overline{\smash{\big)}\,2x^2 - x + 3} \quad {\scriptstyle 2}$
$\qquad\quad \underline{-(2x^2 + 6x)}$
$\qquad\qquad\qquad -7x + 3$

$\dfrac{2x^2 - x + 3}{x^2 + 3x} = 2 + \dfrac{-7x + 3}{x(x + 3)}$

$\left[\dfrac{-7x + 3}{x(x+3)} = \dfrac{A}{x} + \dfrac{B}{(x+3)} \right] x(x + 3)$

$-7x + 3 = A(x + 3) + Bx$
$\qquad\qquad = (A + B)x + 3A$

$A + B = -7$
$3A = 3; A = 1$
$B = -7 - (1) = -8$
$\therefore 2 + \dfrac{1}{x} - \dfrac{8}{x + 3}$

35.
$$x^2 + 2x - 3\,\overline{\big)\,x^3 + 0x^2 - 8x + 11}$$
$$\underline{-\;(x^3 + 2x^2 - 3x)}$$
$$-2x^2 - 5x + 11$$
$$\underline{-\;(-2x^2 - 4x + 6)}$$
$$-x + 5$$

$$\frac{x^3 - 8x + 11}{x^2 + 2x - 3} = x - 2 + \frac{-x + 5}{(x + 3)(x - 1)}$$

$$\left[\frac{-x + 5}{(x + 3)(x - 1)} = \frac{A}{(x + 3)} + \frac{B}{(x - 1)}\right]$$
$$\cdot\,(x + 3)(x - 1)$$

$$-x + 5 = A(x - 1) + B(x + 3)$$
$$= (A + B)x + (-A + 3B)$$
$$A + B = -1;\ -A + 3B = 5$$
$$-A + 3(-1 - A) = 5$$
$$-4A = 8;\ A = -2$$
$$B = -1 - (-2) = 1$$
$$\therefore\ x - 2 - \frac{2}{x + 3} + \frac{1}{x - 1}$$

36.
$$x^2 - x - 12\,\overline{\big)\,3x^3 - 3x^2 - 33x - 19}$$
$$\underline{-\;(3x^3 - 3x^2 - 36x)}$$
$$3x - 19$$

$$\frac{3x^3 - 3x^2 - 33x - 19}{x^2 - x - 12} = 3x + \frac{3x - 19}{x^2 - x - 12}$$

$$\left[\frac{3x - 19}{(x - 4)(x + 3)} = \frac{A}{x - 4} + \frac{B}{x + 3}\right]$$
$$\cdot\,(x - 4)(x + 3)$$

$$3x - 19 = A(x + 3) + B(x - 4)$$
$$= (A + B)x + (3A - 4B)$$
$$A + B = 3;\ 3A - 4B = -19$$
$$3A - 4(3 - A) = -19$$
$$7A = -7;\ A = -1$$
$$B = 3 - (-1) = 4$$
$$\therefore\ 3x - \frac{1}{x - 4} + \frac{4}{x + 3}$$

37a.
$$\left[\frac{4x + 5}{x(x + 5)} = \frac{A}{x} + \frac{B}{x + 5}\right]x(x + 5)$$
$$4x + 5 = A(x + 5) + Bx$$
$$= (A + B)x + 5A$$
$$A + B = 4;\ 5A = 5;\ A = 1$$
$$B = 4 - (1) = 3$$
$$\therefore\ \frac{1}{x} + \frac{3}{x + 5}$$

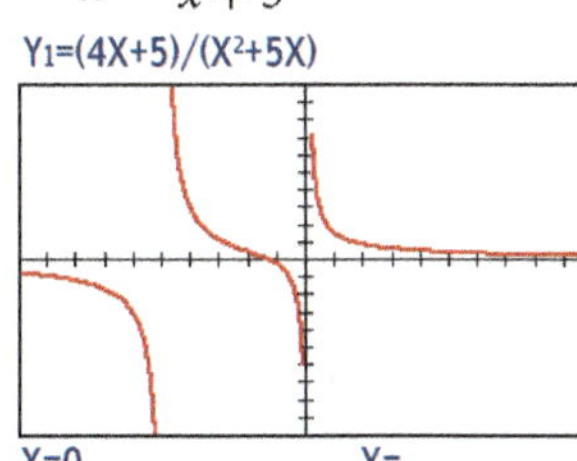

37b. 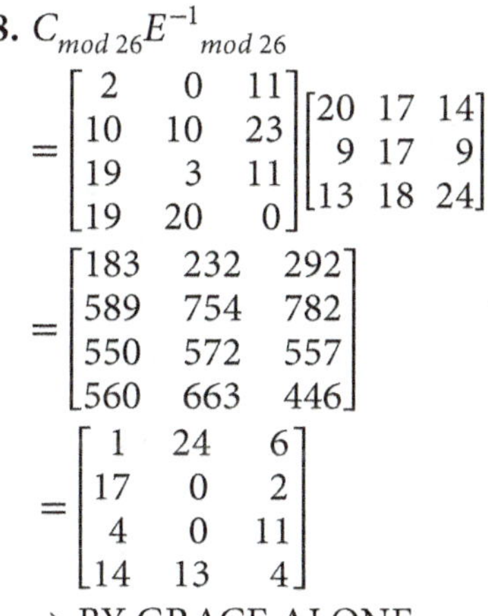
; no

38a.
$$\left[\frac{12}{16n^2 + 8n - 3} = \frac{A}{4n - 1} + \frac{B}{4n + 3}\right]$$
$$\cdot\,(4n - 1)(4n + 3)$$
$$12 = (4n + 3)A + (4n - 1)B$$
$$= (4A + 4B)n + (3A - B)$$
$$4A + 4B = 0;\ B = -A$$
$$3A - B = 12;\ 3A - (-A) = 12$$
$$4A = 12;\ A = 3$$
$$B = -A = -(3) = -3$$
$$\therefore\ \frac{3}{4n - 1} - \frac{3}{4n + 3}$$

38b.
$$\left(\frac{3}{4(1) - 1} - \frac{3}{4(1) + 3}\right)$$
$$+ \left(\frac{3}{4(2) - 1} - \frac{3}{4(2) + 3}\right)$$
$$+ \left(\frac{3}{4(3) - 1} - \frac{3}{4(3) + 3}\right)$$
$$+ \left(\frac{3}{4(4) - 1} - \frac{3}{4(4) + 3}\right)$$
$$= \left(1 - \frac{3}{7}\right) + \left(\frac{3}{7} - \frac{3}{11}\right)$$
$$+ \left(\frac{3}{11} - \frac{3}{15}\right) + \left(\frac{3}{15} - \frac{3}{19}\right)$$

7.7 Cumulative Review

39.
$$\mathbf{p} \cdot \mathbf{q} = -1(3) + 2(-3) = -9$$
$$|\mathbf{p}| = \sqrt{(-1)^2 + 2^2} = \sqrt{5}$$
$$|\mathbf{q}| = \sqrt{3^2 + (-3)^2} = \sqrt{18}$$
$$\cos\theta = \frac{-9}{\sqrt{5} \cdot \sqrt{18}} \approx -0.9487$$
$$\theta \approx \cos^{-1}(-0.9487) \approx 162°$$

40.
$$\left(3, -\frac{\pi}{3} + 2\pi\right) = \left(3, \frac{5\pi}{3}\right)$$
$$\left(-3, -\frac{\pi}{3} + \pi\right) = \left(-3, \frac{2\pi}{3}\right)$$
$$\left(-3, -\frac{\pi}{3} - \pi\right) = \left(-3, -\frac{4\pi}{3}\right)$$

41.
$$x = 3e^y;\ \frac{x}{3} = e^y$$
$$\ln\frac{x}{3} = \ln e^y$$
$$g(x) = \ln x - \ln 3$$

42.
$$AB = \begin{bmatrix} 1 & -3 \\ 2 & 4 \end{bmatrix}\begin{bmatrix} \frac{2}{5} & \frac{3}{10} \\ -\frac{1}{5} & \frac{1}{10} \end{bmatrix}$$
$$= \begin{bmatrix} \frac{2}{5} + \frac{3}{5} & \frac{3}{10} - \frac{3}{10} \\ \frac{4}{5} - \frac{4}{5} & \frac{6}{10} + \frac{4}{10} \end{bmatrix} = \begin{bmatrix} 1 & 0 \\ 0 & 1 \end{bmatrix}$$

$$BA = \begin{bmatrix} \frac{2}{5} & \frac{3}{10} \\ -\frac{1}{5} & \frac{1}{10} \end{bmatrix}\begin{bmatrix} 1 & -3 \\ 2 & 4 \end{bmatrix}$$
$$= \begin{bmatrix} \frac{2}{5} + \frac{6}{10} & -\frac{6}{5} + \frac{12}{10} \\ -\frac{1}{5} + \frac{2}{10} & \frac{3}{5} + \frac{4}{10} \end{bmatrix} = \begin{bmatrix} 1 & 0 \\ 0 & 1 \end{bmatrix}$$

43a. VA: $x = 0$; HA: $y = 0$
$D = \{x \mid x \neq 0\};\ R = \{y \mid y \neq 0\}$

43b. VA: $x = 0$; HA: $y = 0$
$D = \{x \mid x \neq 0\};\ R = \{y \mid y > 0\}$

43c. VA: $x = -3$ (PD at $x = -1$)
HA: $y = 0$
$D = \{x \mid x \neq -1 \text{ and } x \neq -3\}$
$R = \left\{y \mid y \neq 0 \text{ and } y \neq \frac{1}{2}\right\}$

44. $\det\begin{bmatrix} 1 & 0 \\ 0 & 1 \end{bmatrix} = (1 \cdot 1) - (0 \cdot 0) = 1$

$$\det\begin{bmatrix} 1 & 0 & 0 \\ 0 & 1 & 0 \\ 0 & 0 & 1 \end{bmatrix} = (1 + 0 + 0)$$
$$- (0 + 0 + 0) = 1$$

45. $\left(-4, \frac{5\pi}{6} - 2\pi\right) = \left(-4, -\frac{7\pi}{6}\right)$
$\left(4, \frac{5\pi}{6} + \pi\right) = \left(4, \frac{11\pi}{6}\right)$
$\left(4, \frac{5\pi}{6} - \pi\right) = \left(4, -\frac{\pi}{6}\right)$

46. $(-14)(-5.6) - (3.5)(-42) = 225.4$

47. $a^2 = 86^2 + 79^2 - 2(86)(79)\cos 47°$
$\approx 4370;\ a \approx 66$

48. 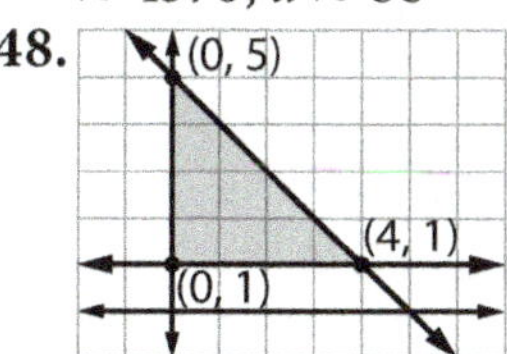

Data Analysis

8. $C_{\bmod 26}E^{-1}_{\bmod 26}$

$$= \begin{bmatrix} 2 & 0 & 11 \\ 10 & 10 & 23 \\ 19 & 3 & 11 \\ 19 & 20 & 0 \end{bmatrix}\begin{bmatrix} 20 & 17 & 14 \\ 9 & 17 & 9 \\ 13 & 18 & 24 \end{bmatrix}$$

$$= \begin{bmatrix} 183 & 232 & 292 \\ 589 & 754 & 782 \\ 550 & 572 & 557 \\ 560 & 663 & 446 \end{bmatrix}$$

$$= \begin{bmatrix} 1 & 24 & 6 \\ 17 & 0 & 2 \\ 4 & 0 & 11 \\ 14 & 13 & 4 \end{bmatrix}$$

$\Rightarrow$ BY GRACE ALONE

8.2 C. Exercises

8.2 Cumulative Review

43.

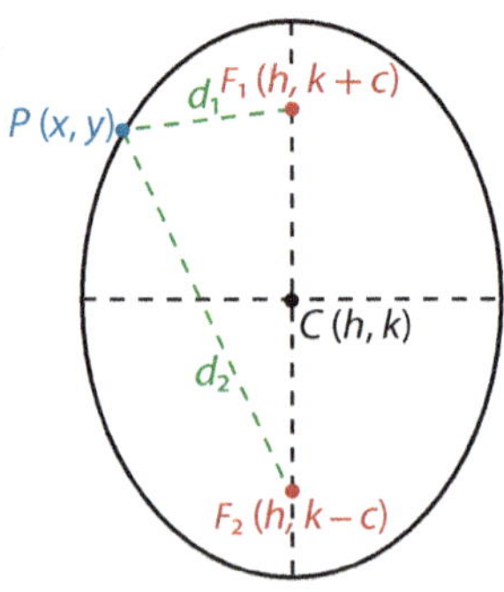

$PF_1 + PF_2 = 2a$

$\sqrt{(x - h)^2 + [y - (k + c)]^2} + \sqrt{(x - h)^2 + [y - (k - c)]^2} = 2a$

Regroup within the brackets.

$\sqrt{(x - h)^2 + [(y - k) - c]^2} + \sqrt{(x - h)^2 + [(y - k) + c]^2} = 2a$

Isolate a radical.

$\sqrt{(x - h)^2 + [(y - k) - c]^2} = 2a - \sqrt{(x - h)^2 + [(y - k) + c]^2}$

Square both sides.

$(x - h)^2 + [(y - k) - c]^2$
$$= 4a^2 - 4a\sqrt{(x - h)^2 + [(y - k) + c]^2} + (x - h)^2 + [(y - k) + c]^2$$

Subtract $(x - h)^2$ from both sides and expand the binomials outside the radical.

$(y - k)^2 - 2c(y - k) + c^2$
$$= 4a^2 - 4a\sqrt{(x - h)^2 + [(y - k) + c]^2} + (y - k)^2 + 2c(y - k) + c^2$$

Isolate the radical term and simplify.

$4a\sqrt{(x - h)^2 + [(y - k) + c]^2} = 4a^2 + 4c(y - k)$

$a\sqrt{(x - h)^2 + [(y - k) + c]^2} = a^2 + c(y - k)$

Square both sides and expand the left side, keeping $(x - h)$ and $(y - k)$ intact.

$a^2\{(x - h)^2 + [(y - k) + c]^2\} = a^4 + 2a^2c(y - k) + c^2(y - k)^2$

$a^2[(x - h)^2 + (y - k)^2 + 2c(y - k) + c^2] = a^4 + 2a^2c(y - k) + c^2(y - k)^2$

$a^2(x - h)^2 + a^2(y - k)^2 + 2a^2c(y - k) + a^2c^2 = a^4 + 2a^2c(y - k) + c^2(y - k)^2$

Collect all terms containing x or y on the left and other terms on the right.

$a^2(x - h)^2 + a^2(y - k)^2 - c^2(y - k)^2 = a^4 - a^2c^2$

Factor $(y - k)^2$ on the left and a^2 on the right.

$a^2(x - h)^2 + (a^2 - c^2)(y - k)^2 = a^2(a^2 - c^2)$

Substitute using $b^2 = a^2 - c^2$.

$a^2(x - h)^2 + b^2(y - k)^2 = a^2b^2$

$\dfrac{a^2(x - h)^2}{a^2b^2} + \dfrac{b^2(y - k)^2}{a^2b^2} = \dfrac{a^2b^2}{a^2b^2}$

$\dfrac{(x - h)^2}{b^2} + \dfrac{(y - k)^2}{a^2} = 1$

44.

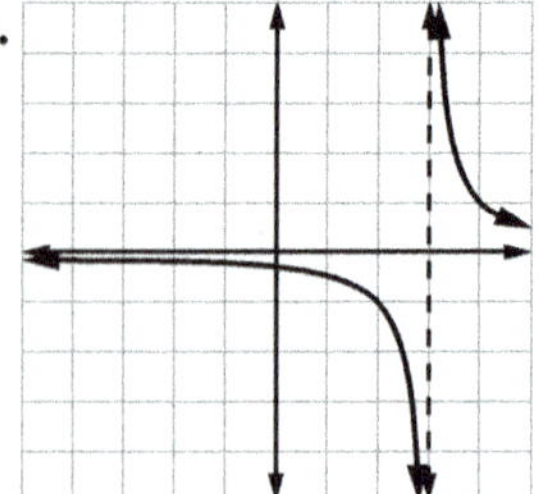

45.

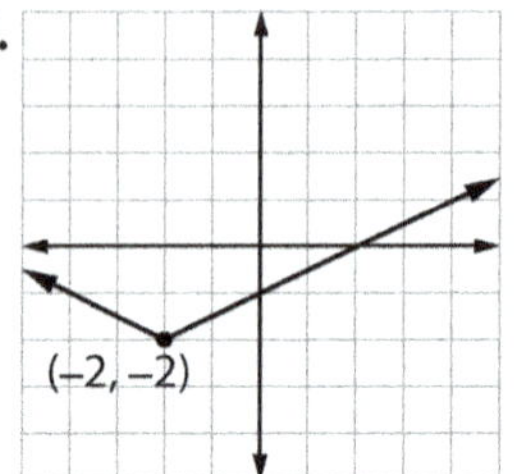

46. $V (2, -3); \ p = \dfrac{1}{4\left(\frac{1}{2}\right)} = \dfrac{1}{2}$

$F\left(2, -3 + \tfrac{1}{2}\right) = \left(2, -\tfrac{5}{2}\right)$

directrix: $y = -3 - \dfrac{1}{2} = -\dfrac{7}{2}$

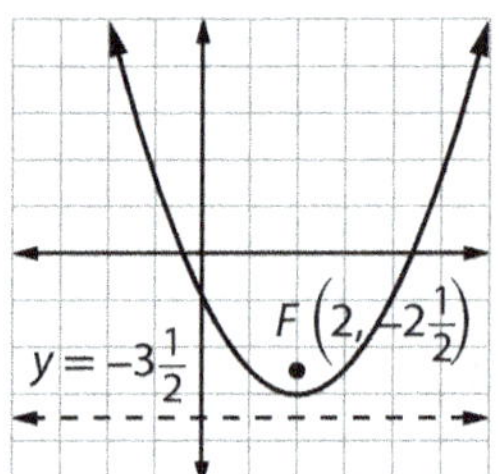

47. $V (-2, -1); \ p = \dfrac{1}{4\left(-\frac{1}{4}\right)} = -1$

$F (-2 - 1, -1) = (-3, -1)$

directrix: $x = -2 + 1 = -1$

48. $4; \ 3$

$y = 4x^2(x^2 - 9x^2 + 8)$
$\quad = 4x^2(x - 1)(x - 8)$
$x = 0, 1, 8$

49. $6; \ 5$

$y = 5x^4(x^2 - 5x - 14)$
$\quad = 5x^4(x + 2)(x - 7)$
$x = 0, -2, 7$

50.
$$r^2 = 2r\cos\theta$$
$$x^2 + y^2 = 2x$$
$$(x^2 - 2x + 1) + y^2 = 1$$
$$(x - 1)^2 + y^2 = 1$$

51. $A(5000) = 5000e^{(0.045)(10)}$
$$= \$7841.56$$

52. $t = 528\left(\dfrac{\pi}{180°}\right) \approx 0.9076$ radians
$$358 \approx r(0.9076)$$
$$r \approx 394 \text{ ft}$$

53. $xy = \left(\dfrac{\sqrt{3}\,u - v}{2}\right)\left(\dfrac{u + \sqrt{3}\,v}{2}\right)$
$$= \dfrac{\sqrt{3}\,u^2 + 2uv - \sqrt{3}\,v^2}{4}$$

8.3 C. Exercises

38b. $y = \pm\dfrac{4}{3}(x - 3)$
$$y = \dfrac{4}{3}x - 4 \text{ and } y = -\dfrac{4}{3}x + 4$$

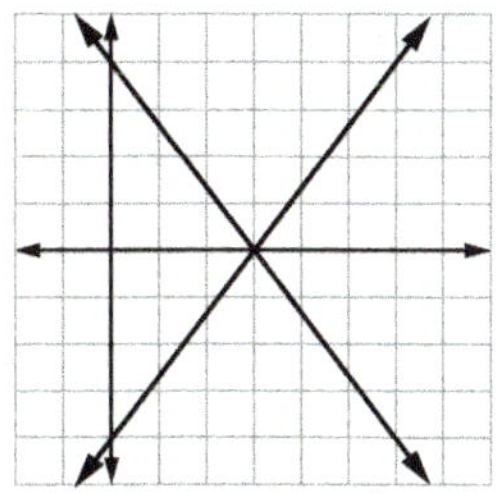

39. $C(0, 1)$; horizontal TA
$a = 3,\ b = 2$
$V\colon (\pm 3, 1);\ y = \pm\dfrac{2}{3}x + 1$

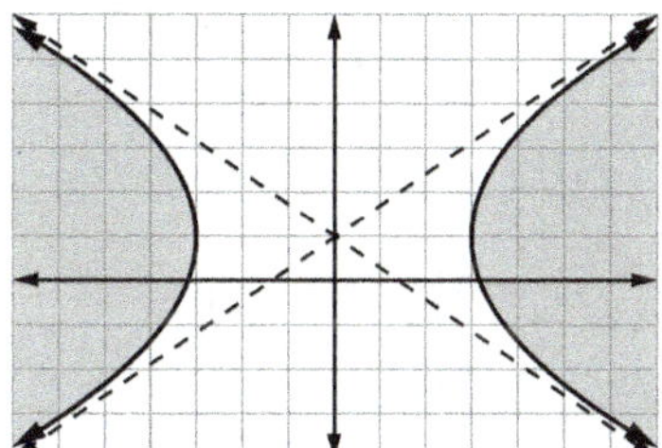

40. $\dfrac{y^2}{9} - \dfrac{x^2}{9} < 1$
$C(0, 0)$; vertical TA; $a = b = 3$
$V\colon (\pm 3, 0);\ y = \pm x$

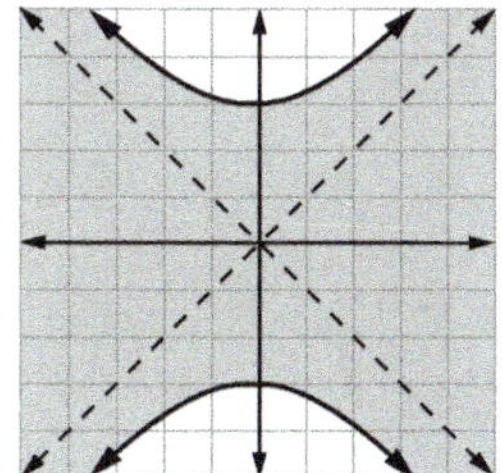

41a. $C(100, 0);\ c = 100$; horizontal TA
$$\Delta d = 300\,\tfrac{\text{km}}{\text{sec}}(400 \times 10^{-3}\text{ sec})$$
$$= 120 \text{ km} = 2a$$
$$a = 60;\ a^2 = 3600$$
$$b^2 = 100^2 - 60^2 = 6400$$
$$\dfrac{(x - 100)^2}{3600} - \dfrac{y^2}{6400} = 1$$

41b. $C(0, 180);\ c = 180$; vertical TA
$$\Delta d = 300\,\tfrac{\text{km}}{\text{sec}}(500 \times 10^{-3}\text{ sec})$$
$$= 150 \text{ km} = 2a$$
$$a = 75;\ a^2 = 5625$$
$$b^2 = 180^2 - 75^2 = 26{,}775$$
$$\dfrac{(y - 180)^2}{5625} - \dfrac{x^2}{26{,}775} = 1$$

41c. Since signals from S_1 and S_2 arrived earlier, the ship is on the branch closer to each substation.

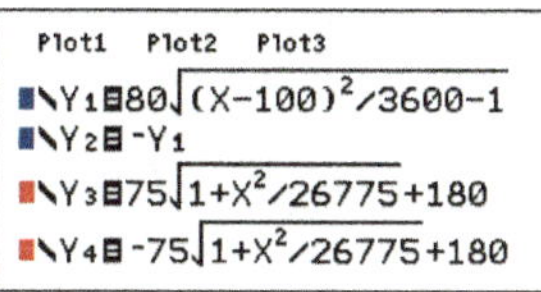

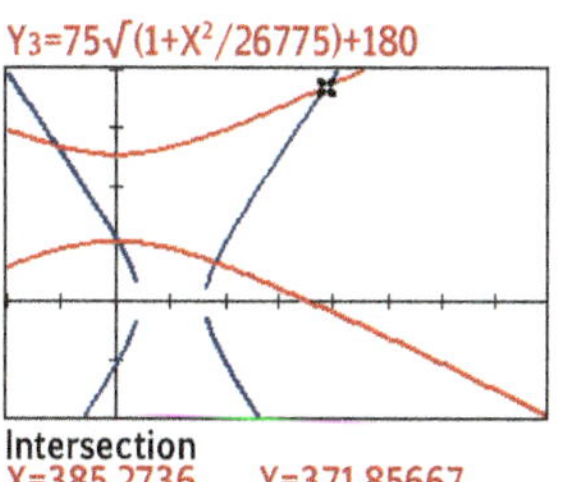

42. hyperbola: $\dfrac{x^2}{36} - \dfrac{y^2}{13} = 1;\ a = 6$
hyperbolic mirror at $(6, 0)$
$c^2 = 36 + 13 = 49;\ c = 7;\ F_h\colon (\pm 7, 0)$
Let $F_1 = F_p$. The parabolic mirror is located at $(-6, 0)$ and F_2 is 1 in. behind it.

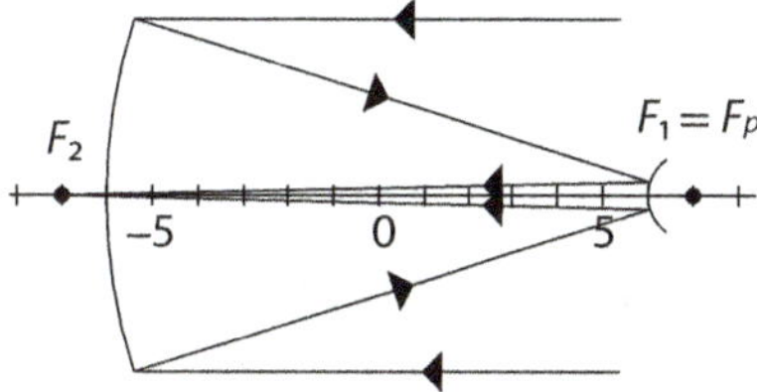

43.

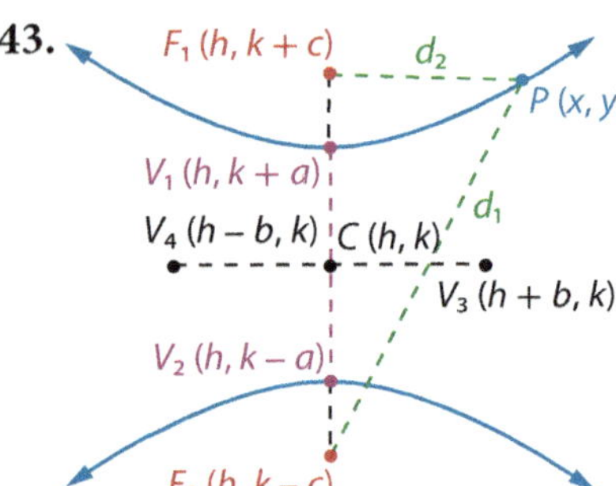

$PF_2 - PF_1 = 2a$

$\sqrt{(x-h)^2 + [y-(k-c)]^2} - \sqrt{(x-h)^2 + [y-(k+c)]^2} = 2a$

Regroup within the brackets.

$\sqrt{(x-h)^2 + [(y-k)+c]^2} - \sqrt{(x-h)^2 + [(y-k)-c]^2} = 2a$

Isolate a radical.

$\sqrt{(x-h)^2 + [(y-k)+c]^2} = 2a + \sqrt{(x-h)^2 + [(y-k)-c]^2}$

Square both sides.

$(x-h)^2 + [(y-k)+c]^2 = 4a^2 + 4a\sqrt{(x-h)^2 + [(y-k)-c]^2}$
$$+ (x-h)^2 + [(y-k)-c]^2$$

Subtract $(x-h)^2$ from both sides and expand the binomials outside the radical.

$(y-k)^2 + 2c(y-k) + c^2 = 4a^2 + 4a\sqrt{(x-h)^2 + [(y-k)-c]^2}$
$$+ (y-k)^2 - 2c(y-k) + c^2$$

Isolate the radical term and simplify.

$4c(y-k) - 4a^2 = 4a\sqrt{(x-h)^2 + [(y-k)-c]^2}$

$c(y-k) - a^2 = a\sqrt{(x-h)^2 + [(y-k)-c]^2}$

Square both sides and expand the right side, keeping $(x-h)$ and $(y-k)$ intact.

$c^2(y-k)^2 - 2a^2c(y-k) + a^4 = a^2\{(x-h)^2 + [(y-k)-c]^2\}$
$c^2(y-k)^2 - 2a^2c(y-k) + a^4 = a^2[(x-h)^2 + (y-k)^2 - 2c(y-k) + c^2]$
$c^2(y-k)^2 - 2a^2c(y-k) + a^4 = a^2(x-h)^2 + a^2(y-k)^2 - 2a^2c(y-k) + a^2c^2$

Collect all terms containing x or y on the left and all other terms on the right.

$c^2(y-k)^2 - a^2(y-k)^2 - a^2(x-h)^2 = a^2c^2 - a^4$

Factor $(y-k)^2$ on the left and a^2 on the right.

$(c^2 - a^2)(y-k)^2 - a^2(x-h)^2 = a^2(c^2 - a^2)$

Substitute using $b^2 = c^2 - a^2$.

$b^2(y-k)^2 - a^2(x-h)^2 = a^2b^2$

$\dfrac{b^2(y-k)^2}{a^2 b^2} - \dfrac{a^2(x-y)^2}{a^2 b^2} = \dfrac{a^2 b^2}{a^2 b^2}$

$\dfrac{(y-k)^2}{a^2} - \dfrac{(x-h)^2}{b^2} = 1$

8.3 Cumulative Review

44. $9x^2 + 16(y^2 + 8y + 16) = -112 + 16(16)$
$\qquad 9x^2 + 16(y+4)^2 = 144$

$\qquad \dfrac{x^2}{16} + \dfrac{(y+4)^2}{9} = 1$

$F: (\pm\sqrt{7}, -4) \approx (\pm 2.65, -4)$

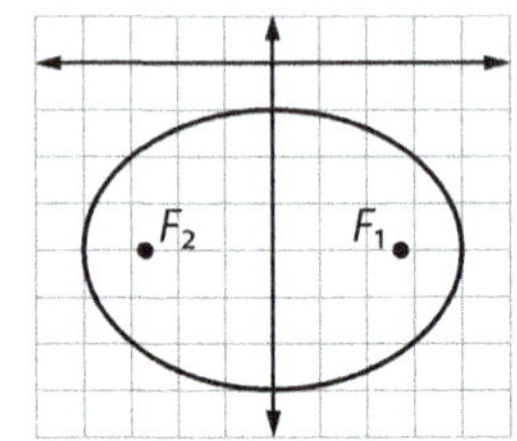

45. $2x = y^2 - 8y + 16$
$\qquad 2x = (y-4)^2$
$\qquad x = \tfrac{1}{2}(y-4)^2$
$\qquad V(0, 4); \ p = \dfrac{1}{4\left(\frac{1}{2}\right)} = \dfrac{1}{2}$
$\qquad F\left(0 + \tfrac{1}{2}, 4\right) = \left(\tfrac{1}{2}, 4\right)$

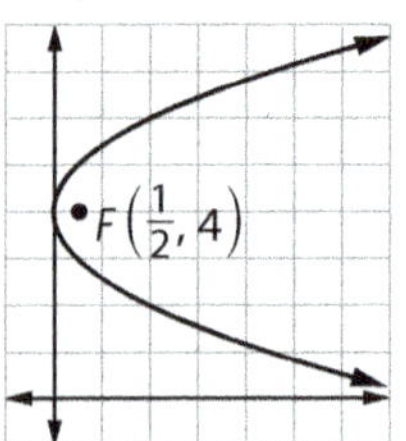

46. $\cot 2\theta = \dfrac{1}{\sqrt{3}}; \ \tan 2\theta = \sqrt{3}$
$\qquad 2\theta = \dfrac{\pi}{3}; \ \theta = \dfrac{\pi}{6}$

47. $\cot 2\theta = 0; \ \tan 2\theta$ is undefined.
$\qquad \tan\theta$ is undefined at $\dfrac{\pi}{2}$,
$\qquad$ so $\tan 2\theta$ is undefined at $\dfrac{\pi}{4}$.
$\qquad \therefore \theta = \dfrac{\pi}{4}$

48. $\cot 2\theta = \dfrac{5}{12}; \ \cos 2\theta = \dfrac{5}{13}$

$\qquad \cos\theta = \sqrt{\dfrac{1 + \cos 2\theta}{2}} = \sqrt{\dfrac{1 + \left(\frac{5}{13}\right)}{2}}$

$\qquad = \sqrt{\dfrac{18}{26}} = \dfrac{3}{\sqrt{13}} = \dfrac{3\sqrt{13}}{13}$

49. $\cot 2\theta = \dfrac{7}{24}; \ \cos 2\theta = \dfrac{7}{25}$

$\qquad \cos\theta = \sqrt{\dfrac{1 + \cos 2\theta}{2}} = \sqrt{\dfrac{1 + \left(\frac{7}{25}\right)}{2}}$

$\qquad = \sqrt{\dfrac{16}{25}} = \dfrac{4}{5}$

50. $x^2 = \left(\dfrac{\sqrt{3}u - v}{2}\right)\left(\dfrac{\sqrt{3}u - v}{2}\right)$

$\qquad = \dfrac{3u^2 - 2\sqrt{3}uv + v^2}{4}$

51. $xy = \left(\dfrac{\sqrt{2}u - \sqrt{2}v}{2}\right)\left(\dfrac{\sqrt{2}u + \sqrt{2}v}{2}\right)$

$\qquad = \dfrac{2u^2 - 2v^2}{4} = \dfrac{u^2 - v^2}{2}$

52. $3^{2(x+1)} = 3^{3(1-2x)}$
$\qquad 2x + 2 = 3 - 6x$
$\qquad 8x = 1; \ x = \dfrac{1}{8}$

53. $\log_4 2x = 2$
$\qquad 4^2 = 2x$
$\qquad 2x = 16; \ x = 8$

8.4 B. Exercises

26. $\cot 2\theta = \dfrac{621 - (-75)}{290} = \dfrac{12}{5}$; $\cos 2\theta = \dfrac{12}{13}$; $\theta = 11.3°$

$$\cos\theta = \sqrt{\dfrac{1 + \left(\frac{12}{13}\right)}{2}} = \dfrac{5}{\sqrt{26}}; \quad \sin\theta = \sqrt{\dfrac{1 - \left(\frac{12}{13}\right)}{2}} = \dfrac{1}{\sqrt{26}}$$

$$x = \dfrac{5u - v}{\sqrt{26}}; \quad y = \dfrac{u + 5v}{\sqrt{26}}$$

$$621\left(\dfrac{25u^2 - 10uv + v^2}{26}\right) + 290\left(\dfrac{5u^2 + 24uv - 5v^2}{26}\right)$$
$$- 75\left(\dfrac{u^2 + 10uv + 25v^2}{26}\right) = 2600$$
$$(15{,}525 + 1450 - 75)u^2 + (-6210 + 6960 - 750)uv$$
$$+ (621 - 1450 - 1875)v^2 = 2600(26)$$
$$16{,}900u^2 - 2704v^2 = 67{,}600$$
$$\dfrac{u^2}{4} - \dfrac{v^2}{25} = 1$$

V': $(\pm 2, 0)$

F': $(\pm\sqrt{29}, 0)$

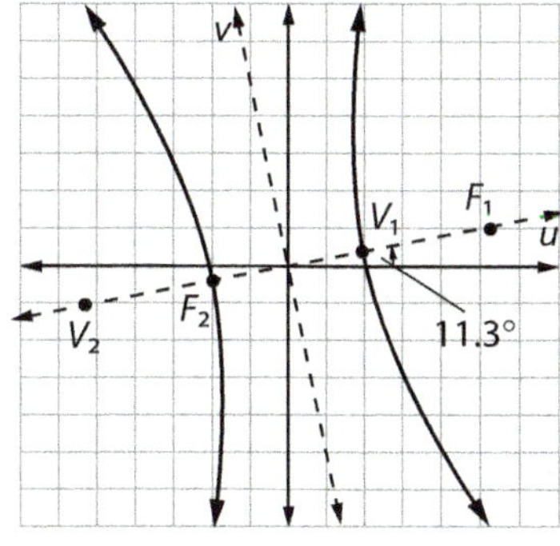

27. $u = \dfrac{\sqrt{2}x + \sqrt{2}y}{2}$; $v = \dfrac{\sqrt{2}y - \sqrt{2}x}{2}$

$$3\left(\dfrac{\sqrt{2}(x+y)}{2}\right)^2 + \left(\dfrac{\sqrt{2}(y-x)}{2}\right)^2 = 4$$
$$3(x^2 + 2xy + y^2) + (y^2 - 2xy + x^2) = 4(2)$$
$$4x^2 + 4xy + 4y^2 - 8 = 0$$
$$x^2 + xy + y^2 - 2 = 0$$
$$y^2 + xy + (x^2 - 2) = 0$$
$$y = \dfrac{-(x) \pm \sqrt{(x)^2 - 4(1)(x^2 - 2)}}{2(1)}$$
$$y = \dfrac{-x \pm \sqrt{x^2 - 4(x^2 - 2)}}{2}$$

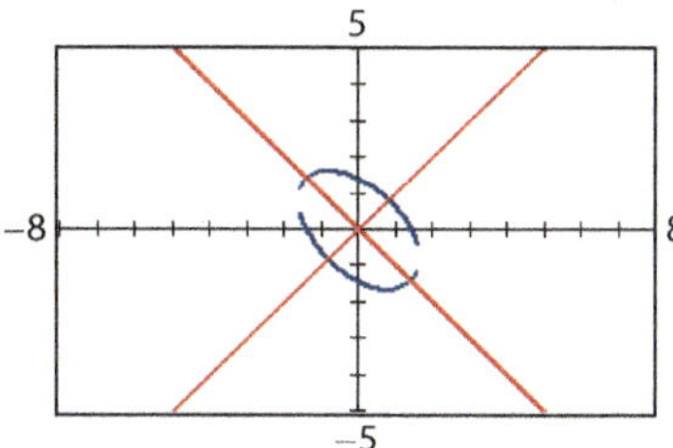

28. $u = \dfrac{\sqrt{3}x + y}{2}$; $v = \dfrac{\sqrt{3}y - x}{2}$

$$\left[\left(\dfrac{\sqrt{3}y - x}{2}\right)^2 = 6\left(\dfrac{\sqrt{3}x + y}{2}\right)\right]4$$
$$3y^2 - 2\sqrt{3}xy + x^2 = 12\sqrt{3}x + 12y$$
$$x^2 - 2\sqrt{3}xy + 3y^2 - 12\sqrt{3}x - 12y = 0$$
$$3y^2 - (2\sqrt{3}x + 12)y + (x^2 - 12\sqrt{3}x) = 0$$
$$y = \dfrac{-(-(2\sqrt{3}x + 12)) \pm \sqrt{\left(-(2\sqrt{3}x + 12)\right)^2 - 4(3)(x^2 - 12\sqrt{3}x)}}{2(3)}$$

$$y = \dfrac{2\sqrt{3}x + 12 \pm \sqrt{(2\sqrt{3}x + 12)^2 - (12x^2 - 144\sqrt{3}x)}}{6}$$

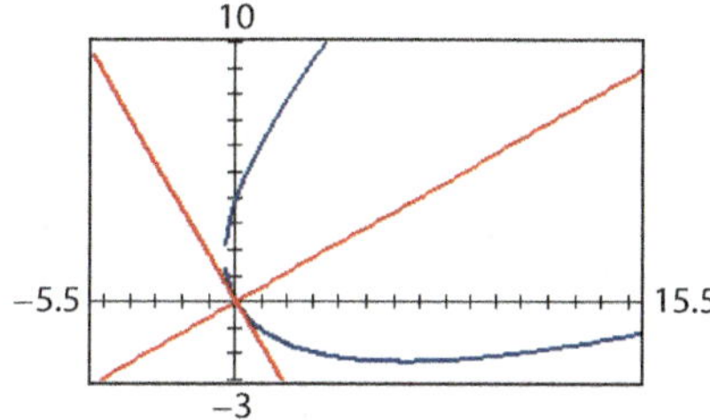

29. $u = \dfrac{\sqrt{2}x + \sqrt{2}y}{2}$; $v = \dfrac{\sqrt{2}y - \sqrt{2}x}{2}$

$$\left(\dfrac{\sqrt{2}(x+y)}{2}\right)^2 - 25\left(\dfrac{\sqrt{2}(y-x)}{2}\right)^2 = 25$$
$$(x^2 + 2xy + y^2) - 25(y^2 - 2xy + x^2) = 25(2)$$
$$-24x^2 + 52xy - 24y^2 - 50 = 0$$
$$12x^2 - 26xy + 12y^2 + 25 = 0$$
$$12y^2 + (-26x)y + (12x^2 + 25) = 0$$
$$y = \dfrac{-(-26x) \pm \sqrt{(-26x)^2 - 4(12)(12x^2 + 25)}}{2(12)}$$
$$y = \dfrac{26x \pm \sqrt{(-26x)^2 - 48(12x^2 + 25)}}{24}$$

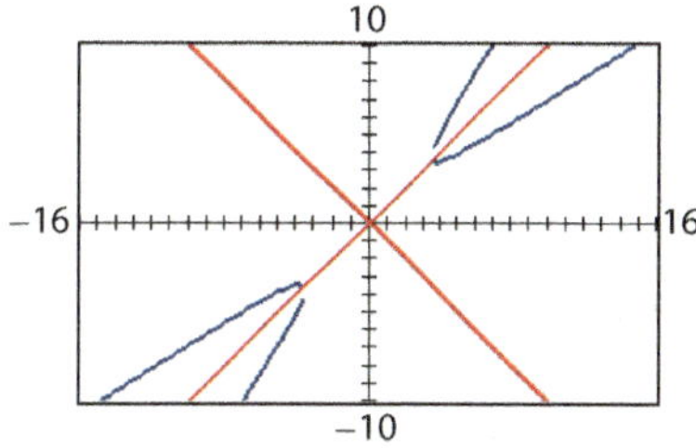

30. $u = \dfrac{x + \sqrt{3}y}{2}$; $v = \dfrac{y - \sqrt{3}x}{2}$

$$5\left(\dfrac{x + \sqrt{3}y}{2}\right)^2 + 4\left(\dfrac{y - \sqrt{3}x}{2}\right)^2 = 20$$
$$5(x^2 + 2\sqrt{3}xy + 3y^2) + 4(y^2 - 2\sqrt{3}xy + 3x^2) = 20(4)$$
$$17x^2 + 2\sqrt{3}xy + 19y^2 - 80 = 0$$
$$19y^2 + (2\sqrt{3}x)y + (17x^2 - 80) = 0$$
$$y = \dfrac{-(2\sqrt{3}x) \pm \sqrt{(2\sqrt{3}x)^2 - 4(19)(17x^2 - 80)}}{2(19)}$$
$$y = \dfrac{-2\sqrt{3}x \pm \sqrt{12x^2 - 76(17x^2 - 80)}}{38}$$

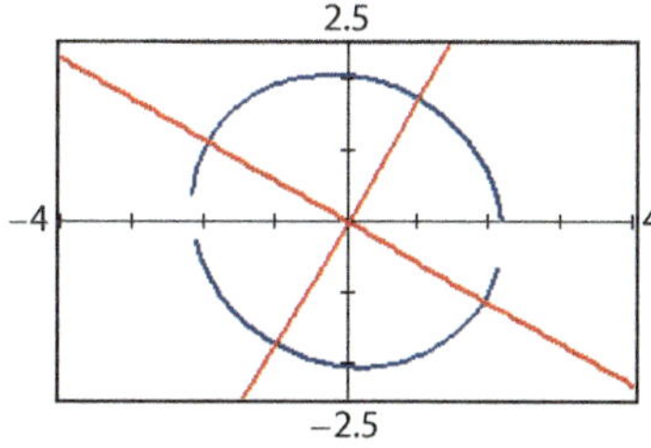

31. $(12)^2 - 4(4)(9) = 0$; parabola
$(2x - 3y)^2 = 0$; $2x = 3y$
$y = \dfrac{2}{3}x$; line

32. $(0)^2 - 4(16)(-25) > 0$; hyperbola
$16x^2 = 25y^2$; $y^2 = \dfrac{16}{25}x^2$
$y = \pm\dfrac{4}{5}x$; 2 intersecting lines

33a. $\cot 2\theta = \frac{1-1}{-2} = 0;\ 2\theta = \frac{\pi}{2};\ \theta = \frac{\pi}{4}$

$x = \frac{\sqrt{2}u - \sqrt{2}v}{2};\ y = \frac{\sqrt{2}u + \sqrt{2}v}{2}$

$\left(\frac{\sqrt{2}(u-v)}{2}\right)^2 - 2\left(\frac{\sqrt{2}(u-v)}{2}\right)\left(\frac{\sqrt{2}(u+v)}{2}\right) + \left(\frac{\sqrt{2}(u+v)}{2}\right)^2$
$\qquad - 10\sqrt{2}\left(\frac{\sqrt{2}(u-v)}{2}\right) - 10\sqrt{2}\left(\frac{\sqrt{2}(u+v)}{2}\right) = 0$

$\frac{u^2 + 2uv + v^2}{2} - (u^2 - v^2) + \frac{u^2 - 2uv + v^2}{2}$
$\qquad\qquad - 10(u - v) - 10(u + v) = 0$

$\left(\frac{1}{2} - 1 + \frac{1}{2}\right)u^2 + (1 - 1)uv + \left(\frac{1}{2} + 1 + \frac{1}{2}\right)v^2 - 20u = 0$

$2v^2 - 20u = 0$

$u = \frac{v^2}{10}$

33b. $a = \frac{1}{10};\ p = \frac{1}{4\left(\frac{1}{10}\right)} = \frac{10}{4}$ or 2.5 ft

34. $x = u\cos\theta - v\sin\theta;\ y = u\sin\theta + v\cos\theta$

$(u\cos\theta - v\sin\theta)^2 + (u\sin\theta + v\cos\theta)^2 = r^2$

$u^2\cos^2\theta - 2uv\cos\theta\sin\theta + v^2\sin^2\theta + u^2\sin^2\theta$
$\qquad\qquad + 2uv\cos\theta\sin\theta + v^2\cos^2\theta = r^2$

$u^2\cos^2\theta + u^2\sin^2\theta + v^2\sin^2\theta + v^2\cos^2\theta = r^2$

$u^2(\sin^2\theta + \cos^2\theta) + v^2(\sin^2\theta + \cos^2\theta) = r^2$

$u^2 + v^2 = r^2$

35. $x = u\cos\theta - v\sin\theta$

$y = u\sin\theta + v\cos\theta$

$A(u\cos\theta - v\sin\theta)^2$
$\qquad\qquad + B(u\cos\theta - v\sin\theta)(u\sin\theta + v\cos\theta)$
$\qquad + C(u\sin\theta + v\cos\theta)^2 + D(u\cos\theta - v\sin\theta)$
$\qquad\qquad + E(u\sin\theta + v\cos\theta) + F = 0$

Only the first three terms generate terms containing uv.

$A(u^2\cos^2\theta - 2uv\cos\theta\sin\theta + v^2\sin^2\theta)$
$\quad + B(u^2\sin\theta\cos\theta - uv\cos^2\theta - uv\sin^2\theta - v^2\sin\theta\cos\theta)$
$\quad + C(u^2\sin^2\theta + 2uv\sin\theta\cos\theta + v^2\cos^2\theta) + \cdots = 0$

collecting terms containing uv:

$[-2A\cos\theta\sin\theta + B(\cos^2\theta - \sin^2\theta) + 2C\sin\theta\cos\theta]uv$

and rearranging the coefficient:

$2(C - A)\sin\theta\cos\theta + B(\cos^2\theta - \sin^2\theta)$

36. Express as a matrix equation, $\begin{bmatrix}\cos\theta & -\sin\theta \\ \sin\theta & \cos\theta\end{bmatrix}\begin{bmatrix}u \\ v\end{bmatrix} = \begin{bmatrix}x \\ y\end{bmatrix}$,

and multiply both sides by

$A^{-1} = \frac{1}{\cos^2 x + \sin^2 x}\begin{bmatrix}\cos\theta & \sin\theta \\ -\sin\theta & \cos\theta\end{bmatrix}$.

$\begin{bmatrix}u \\ v\end{bmatrix} = \begin{bmatrix}\cos\theta & \sin\theta \\ -\sin\theta & \cos\theta\end{bmatrix}\begin{bmatrix}x \\ y\end{bmatrix}$

$u = x\cos\theta + y\sin\theta$

$v = -x\sin\theta + y\cos\theta$

8.4 C. Exercises

37. $\cot 2\theta = \frac{3}{4};\ \cos 2\theta = \frac{3}{5};\ \theta \approx 26.6°$

$\cos\theta = \sqrt{\frac{1 + \left(\frac{3}{5}\right)}{2}} = \frac{2}{\sqrt{5}};\ \sin\theta = \sqrt{\frac{1 - \left(\frac{3}{5}\right)}{2}} = \frac{1}{\sqrt{5}}$

$x = \frac{2u - v}{\sqrt{5}};\ y = \frac{u + 2v}{\sqrt{5}}$

$4\left(\frac{2u - v}{\sqrt{5}}\right)^2 - 16\left(\frac{2u - v}{\sqrt{5}}\right)\left(\frac{u + 2v}{\sqrt{5}}\right)$
$\qquad\qquad + 16\left(\frac{u + 2v}{\sqrt{5}}\right)^2 + 20\sqrt{5}\left(\frac{u + 2v}{\sqrt{5}}\right) = 0$

$4(4u^2 - 4uv + v^2) - 16(2u^2 + 3uv - 2v^2)$
$\qquad\qquad + 16(u^2 + 4uv + 4v^2) + 100(u + 2v) = 0(5)$

$100v^2 + 100u + 200v = 0$

$v^2 + 2v + u = 0$

$u = -(v + 1)^2 + 1$

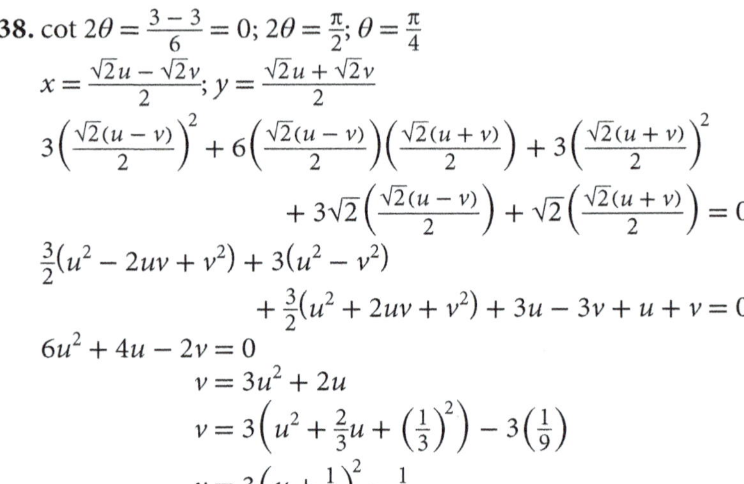

38. $\cot 2\theta = \frac{3 - 3}{6} = 0;\ 2\theta = \frac{\pi}{2};\ \theta = \frac{\pi}{4}$

$x = \frac{\sqrt{2}u - \sqrt{2}v}{2};\ y = \frac{\sqrt{2}u + \sqrt{2}v}{2}$

$3\left(\frac{\sqrt{2}(u-v)}{2}\right)^2 + 6\left(\frac{\sqrt{2}(u-v)}{2}\right)\left(\frac{\sqrt{2}(u+v)}{2}\right) + 3\left(\frac{\sqrt{2}(u+v)}{2}\right)^2$
$\qquad\qquad + 3\sqrt{2}\left(\frac{\sqrt{2}(u-v)}{2}\right) + \sqrt{2}\left(\frac{\sqrt{2}(u+v)}{2}\right) = 0$

$\frac{3}{2}(u^2 - 2uv + v^2) + 3(u^2 - v^2)$
$\qquad\qquad + \frac{3}{2}(u^2 + 2uv + v^2) + 3u - 3v + u + v = 0$

$6u^2 + 4u - 2v = 0$

$v = 3u^2 + 2u$

$v = 3\left(u^2 + \frac{2}{3}u + \left(\frac{1}{3}\right)^2\right) - 3\left(\frac{1}{9}\right)$

$v = 3\left(u + \frac{1}{3}\right)^2 - \frac{1}{3}$

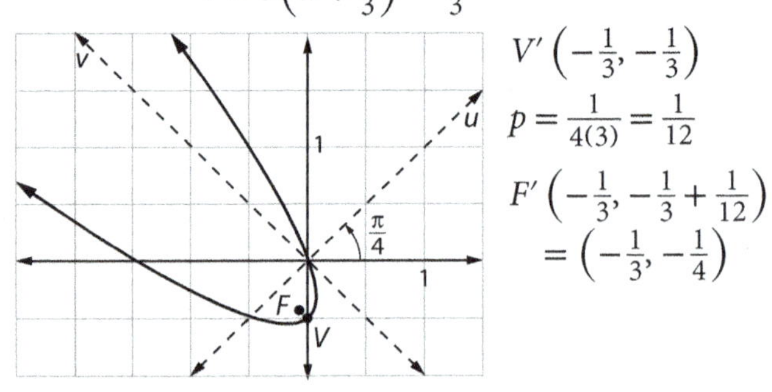

$V'\left(-\frac{1}{3}, -\frac{1}{3}\right)$

$p = \frac{1}{4(3)} = \frac{1}{12}$

$F'\left(-\frac{1}{3}, -\frac{1}{3} + \frac{1}{12}\right)$
$\quad = \left(-\frac{1}{3}, -\frac{1}{4}\right)$

39. $u = \frac{\sqrt{3}x + y}{2};\ v = \frac{\sqrt{3}y - x}{2}$

$4\left(\frac{\sqrt{3}x + y}{2}\right)^2 + 2\left(\frac{\sqrt{3}x + y}{2}\right) - \frac{\sqrt{3}y - x}{2} = 0$

$4(3x^2 + 2\sqrt{3}xy + y^2) + 4(\sqrt{3}x + y) - 2(\sqrt{3}y - x) = 0(4)$

$12x^2 + 8\sqrt{3}xy + 4y^2 + (4\sqrt{3} + 2)x + (4 - 2\sqrt{3})y = 0$

$6x^2 + 4\sqrt{3}xy + 2y^2 + (2\sqrt{3} + 1)x + (2 - \sqrt{3})y = 0$

$2y^2 + (4\sqrt{3}x + 2 - \sqrt{3})y + (6x^2 + 2\sqrt{3}x + x) = 0$

$y = \frac{-4\sqrt{3}x + \sqrt{3} - 2 \pm \sqrt{(4\sqrt{3}x - \sqrt{3} + 2)^2 - 8(6x^2 + 2\sqrt{3}x + x)}}{4}$

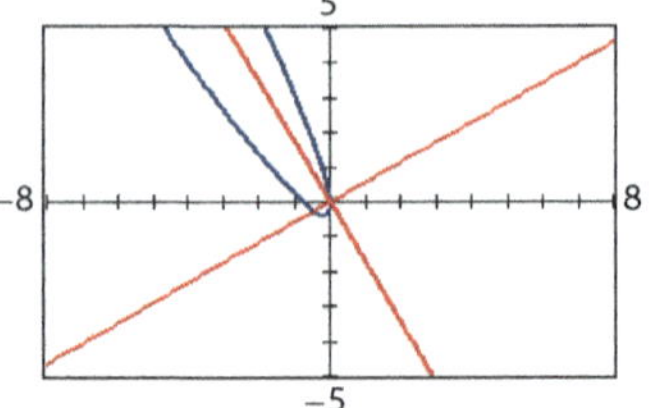

40. $u = \dfrac{\sqrt{2}x + \sqrt{2}y}{2}; \; v = \dfrac{\sqrt{2}y - \sqrt{2}x}{2}$

$u^2 - v^2 - 4\sqrt{2}u + 2\sqrt{2}v + 5 = 0$

$\left(\dfrac{\sqrt{2}(x+y)}{2}\right)^2 - \left(\dfrac{\sqrt{2}(y-x)}{2}\right)^2 - 4\sqrt{2}\left(\dfrac{\sqrt{2}(x+y)}{2}\right)$

$\qquad\qquad + 2\sqrt{2}\left(\dfrac{\sqrt{2}(y-x)}{2}\right) + 5 = 0$

$\dfrac{x^2 + 2xy + y^2}{2} - \dfrac{x^2 - 2xy + y^2}{2} - 4x - 4y + 2y - 2x + 5 = 0$

$2xy - 6x - 2y + 5 = 0$

$2xy - 2y = 6x - 5$

$y = \dfrac{6x - 5}{2x - 2}$

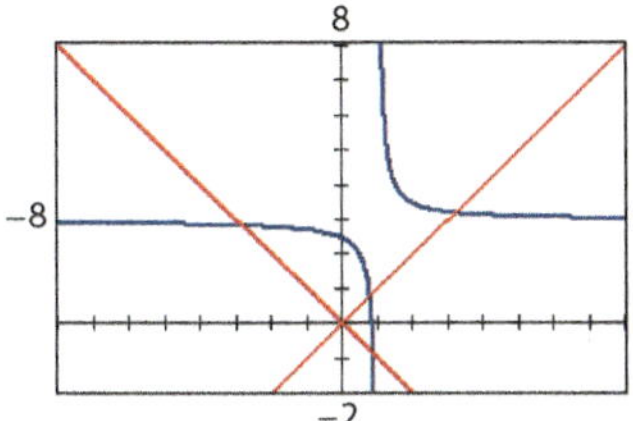

41. $A(u\cos\theta - v\sin\theta)^2$

$\qquad + B(u\cos\theta - v\sin\theta)(u\sin\theta + v\cos\theta)$
$\qquad + C(u\sin\theta + v\cos\theta)^2 + D(u\cos\theta - v\sin\theta)$
$\qquad\qquad + E(u\sin\theta + v\cos\theta) + F = 0$

$Au^2\cos^2\theta - 2Auv\cos\theta\sin\theta + Av^2\sin^2\theta$

$\qquad + Bu^2\cos\theta\sin\theta + Buv\cos^2\theta$
$\qquad - Buv\sin^2\theta - Bv^2\cos\theta\sin\theta$
$\qquad + Cu^2\sin^2\theta + 2Cuv\sin\theta\cos\theta + Cv^2\cos^2\theta$
$\qquad + Du\cos\theta - Dv\sin\theta + Eu\sin\theta + Ev\cos\theta + F = 0$

$\qquad (A\cos^2\theta + B\cos\theta\sin\theta + C\sin^2\theta)u^2$
$\qquad + (-2A\cos\theta\sin\theta + B\cos^2\theta$
$\qquad - B\sin^2\theta + 2C\sin\theta\cos\theta)uv$
$\qquad + (A\sin^2\theta - B\cos\theta\sin\theta + (D\cos\theta + E\sin\theta)u$
$\qquad + (-D\sin\theta + E\cos\theta)v + F = 0$

$A' = A\cos^2\theta + B\cos\theta\sin\theta + C\sin^2\theta$

$C' = A\sin^2\theta - B\cos\theta\sin\theta + C\cos^2\theta$

$D' = D\cos\theta + E\sin\theta$

$E' = E\cos\theta - D\sin\theta$

$F' = F$

42a. $F' = F$, so the constant term is unchanged by a rotation.

42b. $A' + C' = A\cos^2\theta + B\cos\theta\sin\theta + C\sin^2\theta$

$\qquad\qquad + A\sin^2\theta - B\cos\theta\sin\theta + A\cos^2\theta$

$\qquad = A(\cos^2\theta + \sin^2\theta) + C(\sin^2\theta + \cos^2\theta) = A + C$

8.4 Cumulative Review

43. $x = 2(2) - 1 = 3; \; y = (2) - 4 = -2$

44. $t = x - 5$

$y = 3(x - 5) + 1$

$y = 3x - 14$

45. $2(\cos^2\theta + \sin^2\theta) = 2(1) = 2$

46. $4(\tan^2\theta - \sec^2\theta) = 4(-1) = -4$

48. $\qquad 3x^2 + 18x + 6y^2 - 24y = -33$

$3(x^2 + 6x + 9) + 6(y^2 - 4y + 4) = -33 + 27 + 24$

$\qquad\qquad 3(x + 3)^2 + 6(y - 2)^2 = 18$

$\qquad\qquad \dfrac{(x+3)^2}{6} + \dfrac{(y-2)^2}{3} = 1$

49. $\tan 45° = \cot 45° = 1; \; 45 + 45 = 90°$

50. $\sin x \csc x + \sec x \tan x = \sin x \left(\dfrac{1}{\sin x}\right) + \dfrac{1}{\cos x}(\tan x)$

$\qquad\qquad = 1 + \dfrac{1}{\cos 30°}(\tan 30°)$

$\qquad\qquad = 1 + \dfrac{2}{\sqrt{3}}\left(\dfrac{1}{\sqrt{3}}\right)$

$\qquad\qquad = 1 + \dfrac{2}{3} = \dfrac{5}{3}$

51. $\dfrac{(\sin^2\theta + \cos^2\theta) + \tan\theta}{\sec^2\theta - \tan^2\theta} = \dfrac{(1) + \tan\theta}{(\tan^2\theta + 1) - \tan^2\theta}$

$\qquad = \dfrac{1 + \tan\theta}{1}; \; 1 + \tan\dfrac{\pi}{4} = 2$

52. $r = \sqrt{6^2 + 8^2} = 10$

$x = r\cos\theta; \; 6 = 10\cos\theta$

$\cos\theta = \dfrac{6}{10}; \; \theta \approx 53.1°$

$(r, \theta) = (10, 53.1°)$

8.5 B. Exercises

30. horizontal TA on $y = 3$

$C\left(\dfrac{-2 + 4}{2}, 3\right) = (1, 3)$

$a = 3; \; c = 5$

$b^2 = 5^2 - 3^2 = 16; \; b = 4$

$\dfrac{(x-1)^2}{3^2} - \dfrac{(y-3)^2}{4^2} = 1$

Using $\sec^2\theta - \tan^2\theta = 1$,

let $\sec\theta = \dfrac{x-1}{3}$ and $\tan\theta = \dfrac{y-3}{4}$.

$x = 1 + 3\sec\theta; \; y = 3 + 4\tan\theta$

31. vertical TA on $x = 1; \; C\left(1, \dfrac{-8 + 4}{2}\right) = (1, -2)$

$a = 10; \; c = 6$

$b^2 = 10^2 - 6^2 = 64$

$\dfrac{(x-1)^2}{8^2} + \dfrac{(y+2)^2}{10^2} = 1$

Using $\cos^2\theta + \sin^2\theta = 1$,

let $\cos\theta = \dfrac{x-1}{8}$ and $\sin\theta = \dfrac{y+2}{10}$.

$x = 1 + 8\cos\theta; \; y = -2 + 10\sin\theta$

32. horizontal TA on $y = 3; \; C\left(\dfrac{-6 + 4}{2}, 3\right) = (-1, 3)$

$a = 5; \; b = 3$

$\dfrac{(x+1)^2}{5^2} + \dfrac{(y-3)^2}{3^2} = 1$

Using $\cos^2\theta + \sin^2\theta = 1$,

let $\cos\theta = \dfrac{x+1}{5}$ and $\sin\theta = \dfrac{y-3}{3}$.

$x = -1 + 5\cos\theta; \; y = 3 + 3\sin\theta$

33a. $x = (70\cos 15°)t$

$\qquad y = -16t^2 + (70\sin 15°)t + 6.5$

33b. $y = -16(1)^2 + (70\sin 15°)(1) + 6.5 \approx 8.6$ ft

33c. $x = (70\cos 15°)(1.2) \approx 81.1$ ft

34. $x = (64\cos 12°)t$

$\qquad y = -16t^2 + (64\sin 12°)t + 6.3$

34a. $y = -16(0.5)^2 + (64\sin 12°)(0.5) + 6.3 \approx 8.95 > 7.5$

no; The ball will be above his maximum vertical reach.

or

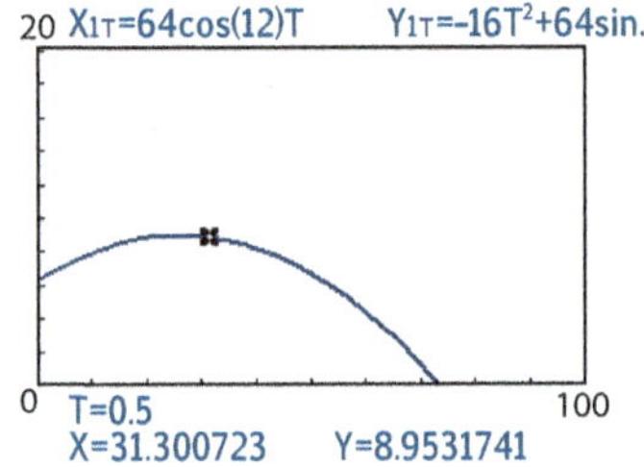

34b. $30(3) = (64 \cos 12°)t$

$$t = \frac{30(3)}{64 \cos 12°} \approx 1.44 \text{ sec}$$

$y \approx -16(1.44)^2 + (64 \sin 12°)(1.44) + 6.3 \approx -7.72$
no; The ball has already hit the ground (at ≈ 73 ft or 24 yd).

or

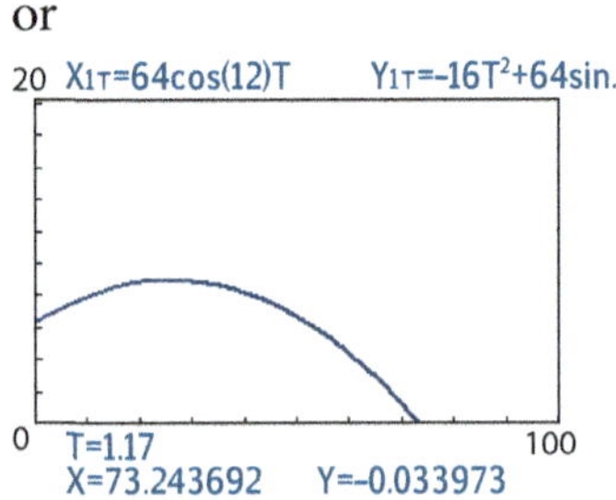

35. $x = (120 \cos 15°)t$
$y = -16t^2 + (120 \sin 15°)t$

35a.
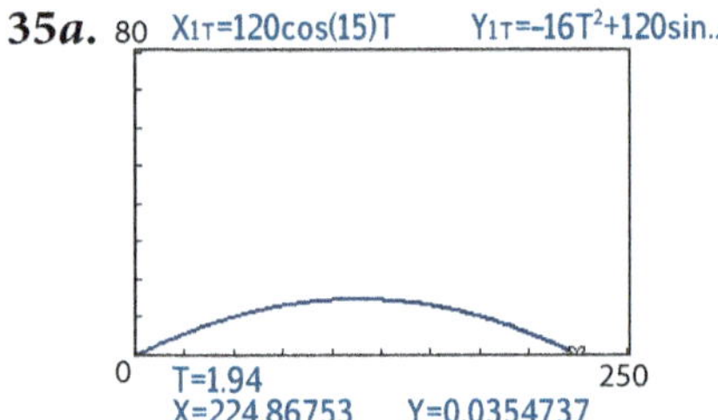

35b.
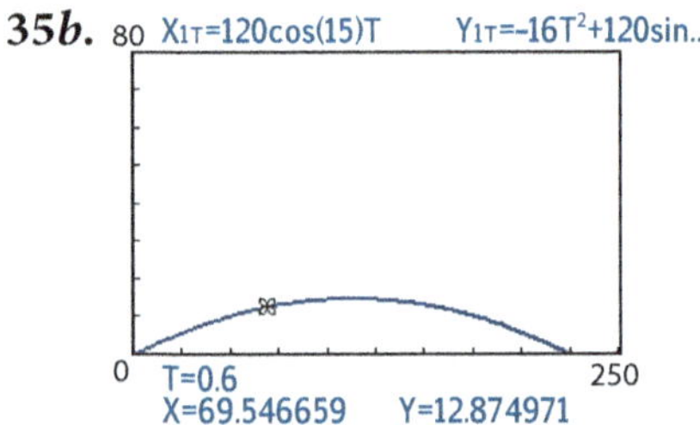

when $t \approx 0.6$, $x \approx 70$ and $y \approx 13$
$13 - 10 = 3$ ft

36. $x = (85 \cos 30°)t$
$y = -16t^2 + (85 \sin 30°)t$
Let $y = 0$ and solve for t.
$(-16t + (85 \sin 30°))t = 0$

$t = 0$ or $t = \dfrac{85\left(\frac{1}{2}\right)}{16} = \dfrac{85}{32} \approx 2.66$ sec

$x = (85 \cos 30°)\left(\dfrac{85}{32}\right) \approx 195.5$ ft

$\dfrac{195.5}{300} \approx 65\%$

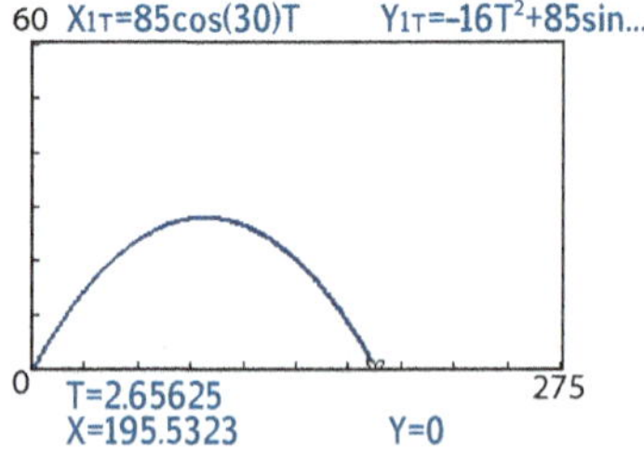

8.5 C. Exercises

37a. $\cos \theta = \frac{x}{r}$ and $\sin \theta = \frac{y}{r}$
using $\cos^2 \theta + \sin^2 \theta = 1$:
$$\left(\frac{x}{r}\right)^2 + \left(\frac{y}{r}\right)^2 = 1; \; x^2 + y^2 = r^2$$

37b. $\cos \theta = \frac{x-h}{r}$ and $\sin \theta = \frac{y-k}{r}$
using $\cos^2 \theta + \sin^2 \theta = 1$:
$$\left(\frac{x-h}{r}\right)^2 + \left(\frac{y-k}{r}\right)^2 = 1$$
$$(x - h)^2 + (y - k)^2 = r^2$$

38a. $x = 7 \cos \theta$; $y = 7 \sin \theta$

38b. $x = 6 \cos \theta + 3$; $y = 6 \sin \theta - 2$

39. (a) If $a = b$, it is a circle.
(b) If $a > b$, then the cosine term (x-component) is greater;
$\therefore$ it is an ellipse with its major axis on the x-axis.
(c) If $a < b$, then the sine term (y-component) is greater;
$\therefore$ it is an ellipse with its major axis on the y-axis.

40. Using $\cos^2 \theta + \sin^2 \theta = 1$,
let $\cos \theta = \frac{x-h}{a}$ and $\sin \theta = \frac{y-k}{b}$.
$x = h + a \cos \theta$; $y = k + b \sin \theta$

41. Using $\sec^2 \theta - \tan^2 \theta = 1$,
let $\sec \theta = \frac{x-h}{a}$ and $\tan \theta = \frac{y-k}{b}$.
$x = h + a \sec \theta$; $y = k + b \tan \theta$

42. $x = 40(3) = (70 \cos \theta)t$; $t = \dfrac{12}{7 \cos \theta}$
$y = -16t^2 + (70 \sin \theta)t$
at 15°: $t \approx 1.77$, $y \approx -18.2$ ft
at 30°: $t \approx 1.98$, $y \approx 6.6$ ft
at 45°: $t \approx 2.42$, $y \approx 26.0$ ft
at 60°: $t \approx 3.43$, $y \approx 19.8$ ft

8.5 Cumulative Review

43. $r^2 = 64$
$x^2 + y^2 = 64$; circle

45. $y = (r \cos \theta)^2$
$y = x^2$; parabola

46. ellipse; $C(h, k) = (2, 0)$
$a = 6$, $b = 4$
$V: (2 \pm 6, 0) = (-4, 0), (8, 0)$
$c^2 = 6^2 - 4^2 = 20$; $c = 2\sqrt{5}$
$F: (2 \pm 2\sqrt{5}, 0) \approx (-2.47, 0), (6.47, 0)$
$e = \dfrac{c}{a} = \dfrac{2\sqrt{5}}{6} \approx 0.75$

47. $\frac{x^2}{4} - \frac{y^2}{9} = 1$; hyperbola
$C(h, k) = (0, 0)$; $a = 2, b = 3$
V: $(\pm 2, 0)$
$c^2 = 9 + 4 = 13$; $c = \sqrt{13}$
F: $(\pm\sqrt{13}, 0) \approx (\pm 3.61, 0)$
$e = \frac{c}{a} = \frac{\sqrt{13}}{2} \approx 1.80$

48. $x = -(y - 1)^2 + 3$; parabola
$V(h, k) = (1, 3)$
$p = \frac{1}{4(-1)} = -\frac{1}{4}$
$F\left(1 - \frac{1}{4}, 3\right) = \left(\frac{3}{4}, 3\right)$; $e = 1$

49. $\cos\theta = \frac{\sqrt{3}}{2}$ at $30°$
Cosine is negative in Q II and Q III.
$\frac{5\pi}{6}$ and $210°$ have $30°$ reference angles in these quadrants.

50. Since $f(x) = \sin 2x$ has an amplitude of 1 and a period of π (a horizontal shrink by a factor of $\frac{1}{2}$), its relative minimum occurs at $\frac{3\pi}{4}$.

51. $p = -3$; $a = \frac{1}{4(-3)} = -\frac{1}{12}$
$y = -\frac{1}{12}(x - 2)^2 + 6$

52. $4x^2 + 12xy + 9y^2 - 16 = 0$
$12^2 - 4(4)(9) = 0$
Since the conic contains both an x^2- and a y^2-term and the determinant is 0, the conic is degenerate.

8.6 B. Exercises

33.
$$y^2 = 8x + 16$$
$$x^2 + y^2 = x^2 + 8x + 16$$
$$r^2 = (x + 4)^2$$
$$r = x + 4$$
$$r = r\cos\theta + 4$$
$$r - r\cos\theta = 4$$
$$r = \frac{4}{1 - \cos\theta}$$

34.
$$3(x^2 + 4x + 4) + 4y^2 = 48$$
$$4x^2 - x^2 + 12x + 12 + 4y^2 = 48$$
$$4x^2 + 4y^2 = x^2 - 12x + 36$$
$$x^2 + y^2 = \frac{1}{4}(x - 6)^2$$
$$r^2 = \frac{1}{4}(r\cos\theta - 6)^2$$
$$r = \frac{1}{2}r\cos\theta - 3$$
$$r\left(1 - \frac{1}{2}\cos\theta\right) = -3$$
$$r = -\frac{3}{1 - \frac{1}{2}\cos\theta}$$

35. $C(0, 5)$; $a = 4, b = 3$
V: $(0, 1), (0, 9)$
$c^2 = 4^2 + 3^2 = 25$; $c = 5$
$e = \frac{c}{a} = \frac{5}{4}$

---

using $r = \frac{ed}{1 + e\sin\theta}$ and $r\left(\frac{\pi}{2}\right) = 1$:

$$1 = \frac{\frac{5}{4}d}{1 + \frac{5}{4}\sin\frac{\pi}{2}}; \frac{9}{4} = \frac{5}{4}d; d = \frac{9}{5}$$

$$r = \frac{\frac{5}{4}\left(\frac{9}{5}\right)}{1 + \frac{5}{4}\sin\theta} = \frac{9}{4 + 5\sin\theta}$$

8.6 C. Exercises

36.
$$\left[\frac{(r\cos\theta)^2}{64} - \frac{(r\sin\theta)^2}{36} = 1\right]576$$
$$9r^2\cos^2\theta - 16r^2\sin^2\theta = 576$$
$$9r^2\cos^2\theta + 9r^2\sin^2\theta - 25r^2\sin^2\theta = 576$$
$$9r^2(\cos^2\theta + \sin^2\theta) - 25r^2\sin^2\theta = 576$$
$$r^2(9 - 25\sin^2\theta) = 576$$
$$r^2 = \frac{576}{9 - 25\sin^2\theta} \text{ or } r = \pm\frac{24}{\sqrt{9 - 25\sin^2\theta}}$$

37.

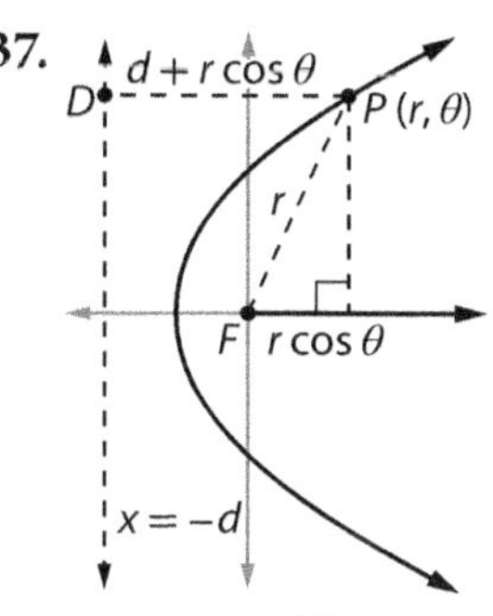

since $e = \frac{PF}{PD}$, $PF = e \cdot PD$
$$r = e(d + r\cos\theta)$$
$$r = ed + er\cos\theta$$
$$r - er\cos\theta = ed$$
$$r(1 - e\cos\theta) = ed$$
$$r = \frac{ed}{1 - e\cos\theta}$$

38.

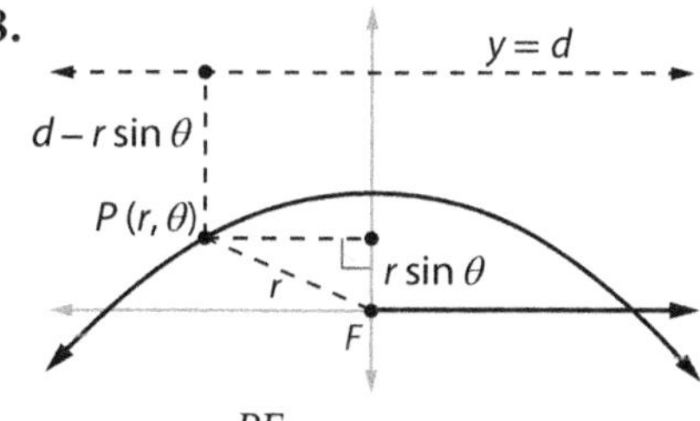

since $e = \frac{PF}{PD}$, $PF = e \cdot PD$
$$r = e(d - r\sin\theta)$$
$$r = ed - er\sin\theta$$
$$r + er\sin\theta = ed$$
$$r(1 + e\sin\theta) = ed$$
$$r = \frac{ed}{1 + e\sin\theta}$$

39.

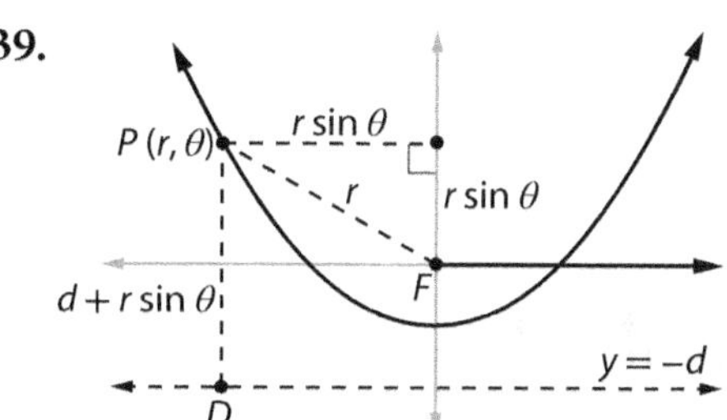

since $e = \frac{PF}{PD}$, $PF = e \cdot PD$
$$r = e(d + r\sin\theta)$$
$$r = ed + er\sin\theta$$
$$r - er\sin\theta = ed$$
$$r(1 - e\sin\theta) = ed$$
$$r = \frac{ed}{1 - e\sin\theta}$$

8.6 Cumulative Review

40. $\cot 2\theta = \frac{-1 - (-1)}{2} = 0$; $2\theta = \frac{\pi}{2}$; $\theta = \frac{\pi}{4}$
$x = u\cos\frac{\pi}{4} - v\sin\frac{\pi}{4} = \frac{\sqrt{2}}{2}u - \frac{\sqrt{2}}{2}v$
$y = u\sin\frac{\pi}{4} + v\cos\frac{\pi}{4} = \frac{\sqrt{2}}{2}u + \frac{\sqrt{2}}{2}v$

41. $\cot 2\theta = \frac{3 - 7}{-4\sqrt{3}} = \frac{1}{\sqrt{3}}$; $2\theta = \frac{\pi}{3}$; $\theta = \frac{\pi}{6}$
$x = u\cos\frac{\pi}{6} - v\sin\frac{\pi}{6} = \frac{\sqrt{3}}{2}u - \frac{1}{2}v$
$y = u\sin\frac{\pi}{6} + v\cos\frac{\pi}{6} = \frac{1}{2}u + \frac{\sqrt{3}}{2}v$

42. ellipse; vertical MA; $C(4, 1)$
$c^2 = a^2 - b^2 = 25 - 1 = 24$; $c = 2\sqrt{6}$
F: $(4, 1 \pm 2\sqrt{6}) \approx (4, 5.90), (4, -3.90)$
$e = \frac{c}{a} = \frac{\sqrt{24}}{5} \approx 0.98$

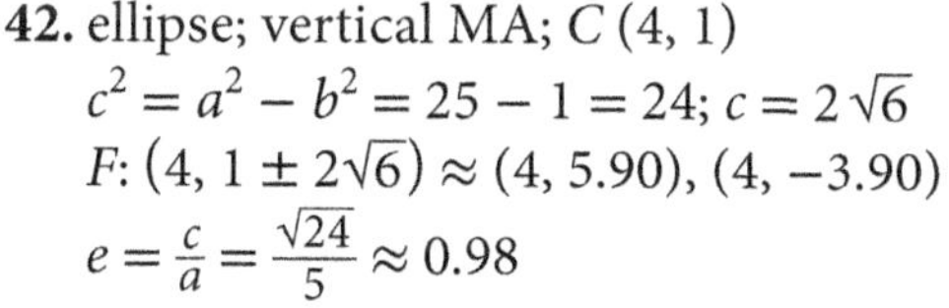

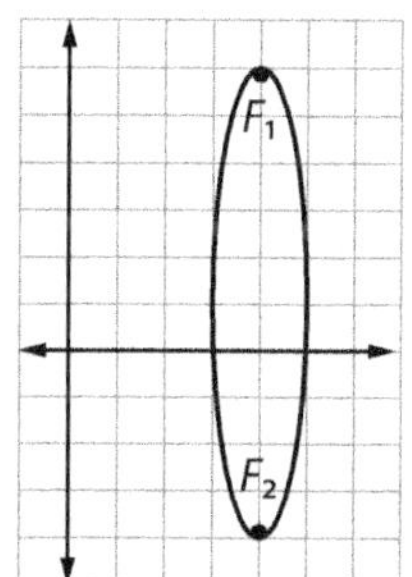

43. $a_n = a_{n-1} + 5$
44. $a_n = 4a_{n-1}$ or 4^n
45. $a_1 = 4(1) = 4$
$a_2 = 4(2) = 8$
$a_{10} = 4(10) = 40$
46. $\alpha = \text{Tan}^{-1}\sqrt{3} = 60°$
$r^2 = 1^2 + (\sqrt{3})^2 = 4$; $r = 2$
47. $AB = \begin{bmatrix} 2(0) + 1(3) & 2(-1) + 1(1) \\ 4(0) + (-2)3 & 4(-1) + (-2)1 \end{bmatrix}$
$= \begin{bmatrix} 3 & -1 \\ -6 & -6 \end{bmatrix}$

48. $t = x - 1$ where $-2 \leq x - 1 \leq 2$; $-1 \leq x \leq 3$

49. $t^2 = x + 3$
$$0 \leq x + 3 \leq 4$$
$$-3 \leq x \leq 1$$

Chapter 9

9.5 C. Exercises

37. $P(1)$: $\displaystyle\sum_{k=1}^{1} k^4 = 1^4 = \frac{1(1+1)[2(1)+1][3(1)^2 + 3(1) - 1]}{30} = \frac{1(2)(3)(5)}{30} = 1$

Assume $P(k)$: $1^4 + 2^4 + \cdots + k^4 = \dfrac{k(k+1)(2k+1)(3k^2 + 3k - 1)}{30}$.

Add $(k+1)^4$ to both sides.

$1^4 + 2^4 + \cdots + k^4 + (k+1)^4$

$$= \frac{k(k+1)(2k+1)(3k^2 + 3k - 1)}{30} + (k+1)^4$$

$$= \frac{k(k+1)(2k+1)(3k^2 + 3k - 1) + 30(k+1)^4}{30}$$

$$= \frac{(k+1)\left[k(2k+1)(3k^2 + 3k - 1) + 30(k+1)^3\right]}{30}$$

$$= \frac{(k+1)(6k^4 + 9k^3 + k^2 - k + 30k^3 + 90k^2 + 90k + 30)}{30}$$

$$= \frac{(k+1)(6k^4 + 39k^3 + 91k^2 + 89k + 30)}{30}$$

having worked backwards from the right side of $P(k+1)$:

$$= \frac{(k+1)(2k^2 + 7k + 6)(3k^2 + 9k + 5)}{30}$$

$$= \frac{(k+1)(k+2)(2k+3)\left[3(k^2 + 2k + 1) + 3k + 3 - 1\right]}{30}$$

$$= \frac{(k+1)[(k+1)+1][2(k+1)+1]\left[3(k+1)^2 + 3(k+1) - 1\right]}{30}$$

$$= P(k+1)$$

$\therefore P(n)$ is true for all $n \in \mathbb{N}$.

38. $S_1 = \frac{1}{4}$; $S_2 = \frac{1}{3}$; $S_3 = \frac{3}{8}$; $S_4 = \frac{2}{5}$; $S_5 = \frac{5}{12}$

S_n: $\dfrac{1}{4}, \dfrac{2}{6}, \dfrac{3}{8}, \dfrac{4}{10}, \dfrac{5}{12}, \ldots, \dfrac{n}{2(n+1)}$

$P(n)$: $\dfrac{1}{4} + \dfrac{1}{12} + \dfrac{1}{24} + \cdots + \dfrac{1}{2n(n+1)} = \dfrac{n}{2(n+1)}$

$P(1)$: $\dfrac{1}{4} = \dfrac{1}{2(1+1)} = \dfrac{1}{4}$

Assume $P(k)$: $\dfrac{1}{4} + \dfrac{1}{12} + \dfrac{1}{24} + \cdots + \dfrac{1}{2k(k+1)} = \dfrac{k}{2(k+1)}$.

$\dfrac{1}{4} + \dfrac{1}{12} + \cdots + \dfrac{1}{2k(k+1)} + \dfrac{1}{2(k+1)(k+2)}$

$$= \frac{k}{2(k+1)} + \frac{1}{2(k+1)(k+2)}$$

$$= \frac{k(k+2)}{2(k+1)(k+2)} + \frac{1}{2(k+1)(k+2)}$$

$$= \frac{k^2 + 2k + 1}{2(k+1)(k+2)} = \frac{(k+1)^2}{2(k+1)(k+2)}$$

$$= \frac{k+1}{2(k+2)}$$

$P(k+1)$: $\dfrac{1}{4} + \dfrac{1}{12} + \cdots + \dfrac{1}{2(k+1)(k+2)} = \dfrac{k+1}{2((k+1)+1)}$

$\therefore P(n)$ is true for all $n \in \mathbb{N}$.

39. $P(1)$: $\cos 1\pi = -1 = (-1)^1$

Assume $P(k)$: $\cos k\pi = (-1)^k$.

$$\cos k\pi (\cos \pi) = (-1)^k(\cos \pi)$$

$\cos k\pi \cos \pi - \sin k\pi \sin \pi = (-1)^k \cos \pi - \sin k\pi \sin \pi$

Apply the cosine of a sum identity on the left side and simplify on the right side.

$$\cos (k\pi + \pi) = (-1)^k(-1) - (\sin k\pi)(0)$$
$$\cos (k+1)\pi = (-1)^{k+1} - 0$$

$P(k+1)$: $\cos (k+1)\pi = (-1)^{k+1}$
$\therefore P(n)$ is true for all $n \in \mathbb{N}$

40a. 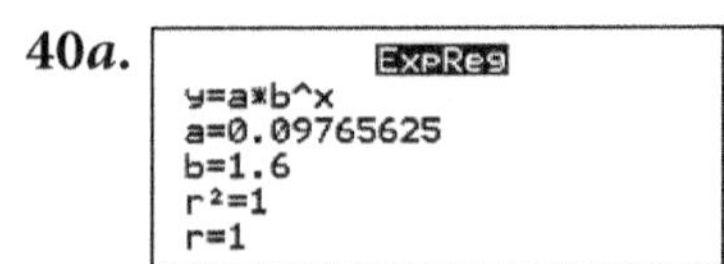

40b. 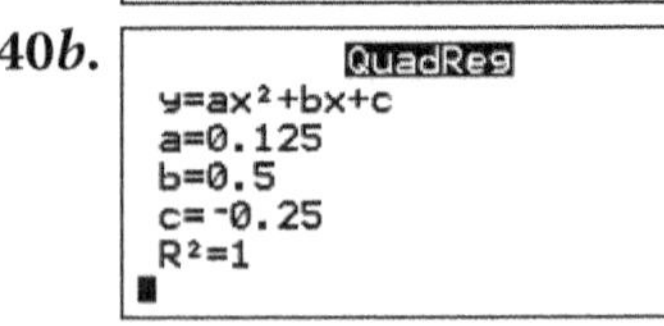

40c. 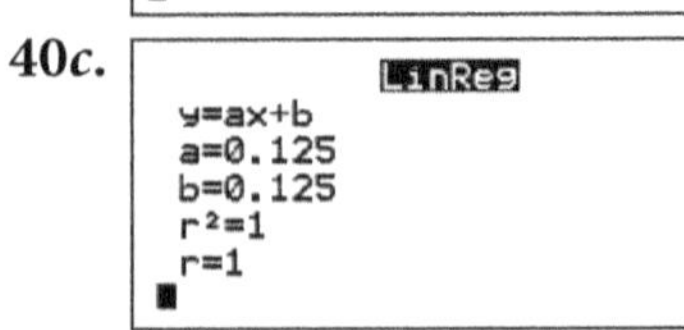

40d. arithmetic (linear): $r^2 \approx 0.8856$
quadratic: $r^2 \approx 0.9981$
geometric (exponential):
$r^2 \approx 0.9578$

41a. 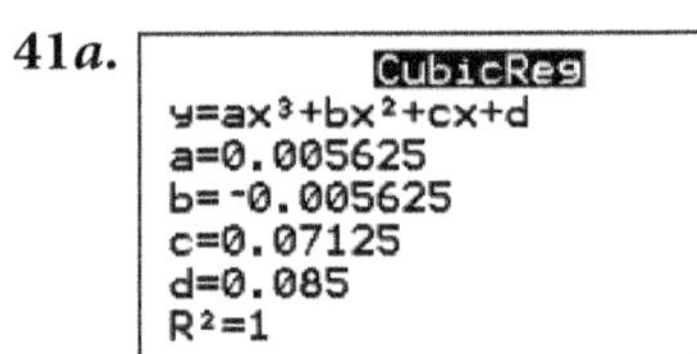

41b. 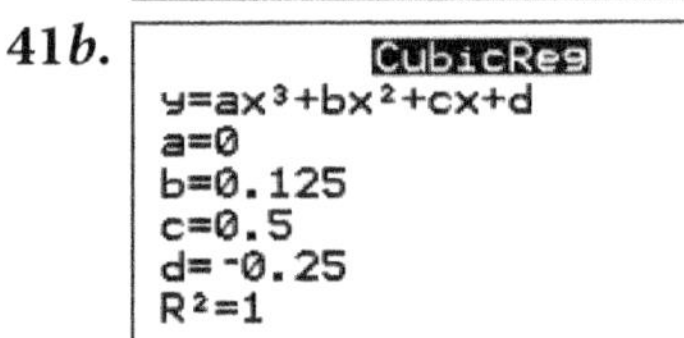

41c. 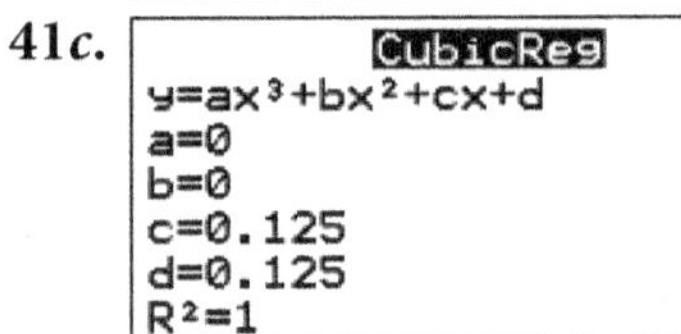

41d.

42. $P(1)$: $(x + y)^1 = {}_1C_0x^1y^0 + {}_1C_1x^0y^1 = x + y$

Assume $P(k)$: $(x + y)^k$

$= {}_kC_0x^k + {}_kC_1x^{k-1}y + {}_kC_2x^{k-2}y^2 + \cdots + {}_kC_{k-1}xy^{k-1} + {}_kC_ky^k.$

Evaluate the coefficients and multiply both sides by $(x + y)$.

$(x + y)(x + y)^k$

$= (x + y)[1x^k + kx^{k-1}y + \left(\dfrac{k(k-1)(k-2)!}{2!(k-2)!}\right)x^{k-2}y^2 + \cdots + kxy^{k-1} + 1y^k]$

$(x + y)^{k+1}$

$= x[1x^k + kx^{k-1}y + \left(\dfrac{k(k-1)}{2!}\right)x^{k-2}y^2 + \cdots + kxy^{k-1} + 1y^k]$

$\quad + y[1x^k + kx^{k-1}y + \left(\dfrac{k(k-1)}{2!}\right)x^{k-2}y^2 + \cdots + kxy^{k-1} + 1ky^k]$

$= 1x^{k+1} + kx^ky + \left(\dfrac{k(k-1)}{2}\right)x^{k-1}y^2 + \cdots + kx^2y^{k-1} + 1xy^k$

$\quad + 1x^ky + kx^{k-1}y^2 + \left(\dfrac{k(k-1)}{2}\right)x^{k-2}y^3 + \cdots + kxy^k + 1y^{k+1}$

$= 1x^{k+1} + (kx^ky + 1x^ky) + \left[\left(\dfrac{k(k-1)}{2}\right)x^{k-1}y^2 + kx^{k-1}y^2\right]$

$\hspace{6cm} + \cdots + (kxy^k + 1xy^k) + 1y^{k+1}$

$= {}_{k+1}C_0x^{k+1} + (k+1)x^ky + \left(\dfrac{k(k-1)}{2} + k\right)x^{k-1}y^2$

$\hspace{5cm} + \cdots + (k+1)xy^k + {}_{k+1}C_{k+1}y^{k+1}$

Since $k + 1 = {}_{k+1}C_1 = {}_{k+1}C_k$

and $\left(\dfrac{k(k-1)}{2} + k\right) = \left(\dfrac{k^2 - k + 2k}{2}\right) = \left(\dfrac{k^2 + k}{2}\right) = \left(\dfrac{(k+1)k}{2}\right)$

$= \left(\dfrac{(k+1)k(k-1)!}{2!(k-1)!}\right) = \left(\dfrac{(k+1)!}{2!((k+1)-2)!}\right) = {}_{k+1}C_2,$

$(x + y)^{k+1}$

$= {}_{k+1}C_0x^{k+1} + {}_{k+1}C_1x^ky + {}_{k+1}C_2x^{k-1}y^2 + \cdots + {}_{k+1}C_kxy^k + {}_{k+1}C_{k+1}y^{k+1},$

which is $P(k + 1)$.

$\therefore P(n)$ is true for all $n \in \mathbb{N}$.

9.5 Cumulative Review

44. $f(\sqrt{2}) = (\sqrt{2})^4 - 3(\sqrt{2})^2 + 2 = 4 - 3(2) + 2 = 0$

$\therefore \sqrt{2}$ is a zero of $f(x)$.

45. $AB = \begin{bmatrix} -3 & -6 \\ 1 & 3 \end{bmatrix}\begin{bmatrix} -1 & -2 \\ \frac{1}{3} & 1 \end{bmatrix}$

$= \begin{bmatrix} 3 - 2 & 6 - 6 \\ -1 + 1 & -2 + 3 \end{bmatrix} = \begin{bmatrix} 1 & 0 \\ 0 & 1 \end{bmatrix} = I_2$

$BA = \begin{bmatrix} -1 & -2 \\ \frac{1}{3} & 1 \end{bmatrix}\begin{bmatrix} -3 & -6 \\ 1 & 3 \end{bmatrix}$

$= \begin{bmatrix} 3 - 2 & 6 - 6 \\ -1 + 1 & -2 + 3 \end{bmatrix} = \begin{bmatrix} 1 & 0 \\ 0 & 1 \end{bmatrix} = I_2$

46. $AB = BA = \begin{bmatrix} -1 & 0 \\ -3 & 1 \end{bmatrix}\begin{bmatrix} -1 & 0 \\ -3 & 1 \end{bmatrix}$

$= \begin{bmatrix} 1 + 0 & 0 + 0 \\ 3 - 3 & 0 + 1 \end{bmatrix} = \begin{bmatrix} 1 & 0 \\ 0 & 1 \end{bmatrix} = I_2$

47. $a_{10} = a_8r^{10-8}$

$507 = 3r^2$

$169 = r^2; r = \pm 13$

$a_9 = a_8(\pm 13) = 3(\pm 13) = \pm 39$

48. $a_{41} = a_{38}r^{41-38}$

$3645 = -5r^3$

$-729 = r^3; r = -9$

$a_{39} = a_{38}(-9) = -5(-9) = 45$

$a_{40} = a_{39}(-9) = 45(-9) = -405$

49. Let $t =$ years since 2000.

$P(t) = ab^t; 416{,}474b^{10} = 418{,}981$

$b^{10} = \dfrac{418{,}981}{416{,}471}; b = \sqrt[10]{\dfrac{418{,}981}{416{,}471}}$

$\hspace{3cm} \approx 1.00060$

$r \approx 0.00060 \approx 0.06\%$

51. $\left[\dfrac{5x - 12}{(x - 3)^2} = \dfrac{A}{x - 3} + \dfrac{B}{(x - 3)^2}\right](x - 3)^2$

$5x - 12 = A(x - 3) + B$

$\hspace{1.5cm} = Ax - 3A + B$

$\hspace{1.5cm} = Ax - (3A - B)$

$A = 5; 3A - B = 12$

$\hspace{1cm} 3(5) - B = 12$

$\hspace{2cm} B = 3$

$\therefore \dfrac{5x - 12}{(x - 3)^2} = \dfrac{5}{x - 3} + \dfrac{3}{(x - 3)^2}$

52. arithmetic with $d = -3$

$a_{98} = -12 + (98 - 1)(-3) = -303$

Chapter 9 Review

24. $r = -\dfrac{3}{2}; a_1 = 4; a_n = -\dfrac{3}{2}a_{n-1}$

$a_n = 4\left(-\dfrac{3}{2}\right)^{n-1}$

$a_{10} = 4\left(-\dfrac{3}{2}\right)^{10-1}$

$\hspace{0.8cm} = -\dfrac{19{,}683}{128} \approx -153.8$

25. $a_7 = a_2r^{7-2}$

$80 = 2.5r^5$

$r^5 = 32; r = \sqrt[5]{32} = 2$

$a_n = 2.5(2)^{n-2} = \dfrac{5}{2}(2)^n(2)^{-2} = \dfrac{5}{8}(2)^n$

$a_{10} = \dfrac{5}{8}(2)^{10} = 640$

26. $a_{11} = a_8r^{11-8}$

$\dfrac{1}{4} = 16r^3$

$\dfrac{1}{64} = r^3; r = \dfrac{1}{4}$

$a_9 = 16\left(\dfrac{1}{4}\right) = 4; a_{10} = 4\left(\dfrac{1}{4}\right) = 1$

27. $a_1 = -2; r = \dfrac{3}{2}$

$S_{10} = \dfrac{-2\left(1 - \left(\frac{3}{2}\right)^{10}\right)}{1 - \left(\frac{3}{2}\right)} \approx -226.66$

28. $a_1 = -\dfrac{1}{3}; r = -\dfrac{1}{2}; S = \dfrac{-\frac{1}{3}}{1 - \left(-\frac{1}{2}\right)} = -\dfrac{2}{9}$

29. $i = \dfrac{0.03}{4} = 0.0075; n = 15(4) = 60$

$FV = 3000\left(\dfrac{1.0075^{60} - 1}{0.0075}\right)$

$\hspace{1cm} = \$226{,}272.41$

$226{,}272.41 - 60(3000) = \$46{,}272.41$

30a. $r = 0.75; n = 7$
$b_1 = 50(0.75) = 37.5$
$b_7 = 37.5(0.75)^{7-1} \approx 6.7$ ft

30b. $S_7 = \frac{37.5(1 - 0.75^7)}{1 - 0.75} \approx 130$
$2(b_0 + S_7) \approx 2(50 + 130) = 360$ ft

30c. $S = \frac{37.5}{1 - 0.75} \approx 150$
$2(b_0 + S) \approx 2(50 + 150) = 400$ ft

31. $a^5 + 5a^4(-3) + 10a^3(-3)^2$
$\quad + 10a^2(-3)^3 + 5a(-3)^4 + (-3)^5$
$= a^5 - 15a^4 + 90a^3 - 270a^2$
$\qquad\qquad + 405a - 243$

32. $(3m)^4 + 4(3m)^3(4n) + 6(3m)^2(4n)^2$
$\qquad\qquad + 4(3m)(4n)^3 + (4n)^4$
$= 81m^4 + 432m^3n + 864m^2n^2$
$\qquad\qquad + 768mn^3 + 256n^4$

33. $(2)^6 + 6(2)^5(-2z) + 15(2)^4(-2z)^2$
$\quad + 20(2)^3(-2z)^3 + 15(2)^2(-2z)^4$
$\qquad + 6(2)(-2z)^5 + (-2z)^6$
$= 64 - 384z + 960z^2 - 1280z^3$
$\qquad + 960z^4 - 384z^5 + 64z^6$

34. $\frac{7!}{3!4!} = \frac{7 \cdot 6 \cdot 5 \cdot 4!}{3! \cdot 4!} = 35$
$\frac{7!}{5!2!} = \frac{7 \cdot 6 \cdot 5!}{5! \cdot 2!} = 21$

35. $\frac{4!}{2!2!} = \frac{4 \cdot 3 \cdot 2!}{2! \cdot 2!} = 6;\ \frac{4!}{4!0!} = 1$

36. $_3C_0(3c)^3 + {_3}C_1(3c)^2(1)^1$
$\qquad + {_3}C_2(3c)^1(1)^2 + {_3}C_3(1)^3$
$= 1(27c^3) + 3(9c^2) + 3(3c) + 1(1^3)$
$= 27c^3 + 27c^2 + 9c + 1$

37. $_4C_0(\sqrt{2})^4 + {_4}C_1(\sqrt{2})^3(-3t)$
$\quad + {_4}C_2(\sqrt{2})^2(-3t)^2 + {_4}C_3\sqrt{2}(-3t)^3$
$\qquad\qquad + {_4}C_4(-3t)^4$
$= 1(4) + 4(2\sqrt{2})(-3t) + 6(2)(-3t)^2$
$\qquad + 4\sqrt{2}(-3t)^3 + 1(-3t)^4$
$= 4 - 24\sqrt{2}t + 108t^2$
$\qquad\qquad - 108\sqrt{2}t^3 + 81t^4$

38. $_9C_3(3a)^6(-2b)^3 = 84(729a^6)(-8b^3)$
$= -489{,}888a^6b^3$

39. The 8th term uses $r = 7$.
$_{11}C_7(w)^4(5)^7 = 330w^4(78{,}125)$
$= 25{,}781{,}250w^4$

40. $_{12}C_3\left(\frac{1}{6}\right)^3\left(\frac{5}{6}\right)^9 = 220\left(\frac{5^9}{6^{12}}\right) \approx 19.7\%$

41a. $P(1)$: 4 is a factor of $9^1 - 5^1 = 4$.

41b. $P(k)$: 4 is a factor of $9^k - 5^k$.

41c. $P(k + 1)$:
$\quad$ 4 is a factor of $9^{k+1} - 5^{k+1}$.

42a. $P(1)$: $(2(1))^2 = \frac{2(1)(1 + 1)(2(1) + 1)}{3}$
$\qquad\qquad = \frac{2(2)(3)}{3} = 4$

42b. $P(k)$: $2^2 + 4^2 + 6^2 + \cdots + (2k)^2$
$\qquad\qquad = \frac{2k(k + 1)(2k + 1)}{3}$

42c. $P(k + 1)$: $2^2 + 4^2 + \cdots + (2(k + 1))^2$
$\qquad = \frac{2(k + 1)((k + 1) + 1)(2(k + 1) + 1)}{3}$

43a. $5\sum_{k=1}^{100} a_k = 5(18) = 90$

43b. $2\sum_{k=1}^{100} b_k + 3\sum_{k=1}^{100} c_k$
$= 2(54) + 3(243) = 837$

44a. $\sum_{k=1}^{15} 2k^3 - \sum_{k=1}^{15} 2$
$= 2\left(\frac{15^2(15 + 1)^2}{4}\right) - 15(2)$
$= 28{,}800 - 30 = 28{,}770$

44b. $\sum_{k=6}^{20} k^4 = \sum_{k=1}^{20} k^4 - \sum_{k=1}^{5} k^4$
$= \frac{20(20 + 1)(2(20) + 1)(3(20)^2 + 3(20) - 1)}{30}$
$\quad - \frac{5(5 + 1)(2(5) + 1)(3(5)^2 + 3(5) - 1)}{30}$
$= 722{,}666 - 979 = 721{,}687$

45. $P(1)$: $1^2 + 1 + 4 = 6 = 2(3)$
2 is a factor.
Assume $P(k)$: $k^2 + k + 4 = 2c, c \in \mathbb{Z}$.
$P(k + 1)$: $(k + 1)^2 + (k + 1) + 4$
$= k^2 + 2k + 1 + k + 1 + 4$
$= (k^2 + k + 4) + 2k + 2$
$= 2c + 2(k + 1) = 2c + 2d = 2e, e \in \mathbb{Z}$
$\therefore P(n)$ is true for all $n \in \mathbb{N}$.

46. $P(1)$: $6^1 - 1 = 5 = 5(1)$
5 is a factor.
Assume $P(k)$: $6^k - 1 = 5c, c \in \mathbb{Z}$.
$P(k + 1)$: $6^{k+1} - 1 = 6 \cdot 6^k - 1$
$\quad = 6 \cdot 6^k - 6 - 1 + 6$
$\quad = 6(6^k - 1) - 1 + 6$
$\quad = 6(5c) + 5 = 5(6c + 1) = 5d$
$6^{k+1} - 1 = 5d, d \in \mathbb{Z}$
$\therefore P(n)$ is true for all $n \in \mathbb{N}$.

47. $P(2)$ is true.; $3^2 > 3(2)$
Assume $P(k)$: $3^k > 3k$ for $k > 1$.
$\therefore 3(3^k) > 3(3k)$ (Mult. Prop. of Ineq.)
For any $c > 1$, $3c > c + 3$
substituting $3k$ for c, $3(3k) > 3k + 3$
and $3(3^k) > 3k + 3$ (Transitive Prop.):
$P(k + 1)$: $3^{k+1} > 3(k + 1)$ for $k > 1$
$\therefore P(n)$ is true for all $n > 1, n \in \mathbb{N}$.

48. $P(1)$:
$1^2 = \frac{(1)(2(1) - 1)(2(1) + 1)}{3} = \frac{3}{3} = 1$
Assume $P(k)$: $1^2 + 3^2 + \cdots$
$\quad + (2k - 1)^2 = \frac{k(2k - 1)(2k + 1)}{3}$.
$1^2 + 3^2 + \cdots + (2k - 1)^2$
$\qquad\qquad + (2(k + 1) - 1)^2$
$= \frac{k(4k^2 - 1)}{3} + (2(k + 1) - 1)^2$
$= \frac{k(4k^2 - 1) + 3(2k + 1)^2}{3}$
$= \frac{4k^3 + 12k^2 + 11k + 3}{3}$
$= \frac{(k + 1)(2k + 1)(2k + 3)}{3}$
$= \frac{(k + 1)(2k + 2 - 1)(2k + 2 + 1)}{3}$
$= \frac{(k + 1)(2(k + 1) - 1)(2(k + 1) + 1)}{3}$
$= P(k + 1)$
$\therefore P(n)$ is true for all $n \in \mathbb{N}$.

49. $P(4)$ is true.; $\frac{4(4 - 3)}{2} = 2$
Assume $P(k)$: A k-gon has $\frac{k(k - 3)}{2}$
diagonals.
For every additional vertex, the polygon has an additional $(k + 1) - 2$ or
$k - 1$ diagonals.
$P(k + 1)$: $\frac{k(k - 3)}{2} + (k - 1)$
$= \frac{k(k - 3) + 2(k - 1)}{2} = \frac{k^2 - k - 2}{2}$
$= \frac{(k + 1)(k - 2)}{2} = \frac{(k + 1)((k + 1) - 3)}{2}$
$\therefore P(n)$ is true for all $n \geq 4, n \in \mathbb{N}$.

50. $S_1 = 2; S_2 = 6; S_3 = 12$
$S_4 = 20; S_5 = 30; S_n = n(n + 1)$
$P(n)$: $2 + 4 + \cdots + 2n = n(n + 1)$
$P(1)$: $1 = 1(1 + 1) = 2$ is true.
Assume $P(k)$:
$2 + 4 + \cdots + 2k = k(k + 1)$.
$2 + 4 + \cdots + 2k + 2(k + 1)$
$= k(k + 1) + 2k + 2$
$= k^2 + k + 2k + 2$
$= (k^2 + 2k + 1) + k + 1$
$= (k + 1)^2 + (k + 1)$
$= (k + 1)((k + 1) + 1) = P(k + 1)$
$\therefore P(n)$ is true for all $n \in \mathbb{N}$.

Chapter 10

Chapter 10 Review

31a. $\mu_{L2} = 5\mu_{L1} = 5(6) = 30$
$s_{L2} = 5s_{L1} = 5(2) = 10$

31b. $\mu_{L3} = \mu_{L1} + 4 = 6 + 4 = 10$
$s_{L3} = s_{L1} = 2$

32a. $z = \dfrac{100 - 85}{5} = 3$

32b. $z = \dfrac{78 - 85}{5} = -1.4$

33. $z_{SAT} = \dfrac{1206 - 1060}{195} \approx 0.75$
$z_{ACT} = \dfrac{26 - 21}{5.4} = 0.93$
$z_{ACT} > z_{SAT}$; His ACT score was relatively better.

34. $1.32 = \dfrac{x - 20.8}{4.7}$
$x - 20.8 = 1.32(4.7)$
$x \approx 27.0$

35. $1 - \dfrac{1}{2^2} = \dfrac{3}{4} = 0.75$ or 75%

```
         1-Var Stats
x̄=6.9
Σx=69
Σx²=531
Sx=2.469817807
σx=2.343074903
```

$\bar{x} \pm 2s \approx 6.9 \pm 2(2.47)$
$\approx [1.96, 11.84]$
Only one score out of 10 (10%) lies outside [1.96, 11.84], so at least 75% of the data lies within the interval.

36a. $[\mu - \sigma, \mu + \sigma]$; $0.34 + 0.34 = 0.68$

36b. $[\mu - 3\sigma, \mu + 2\sigma]$
$2(0.34 + 0.135) + 0.0235 = 0.9735$

36c. $[\mu + \sigma, \infty)$; $0.50 - 0.34 = 0.16$

The normalcdf function can be used in exercises 37–39 instead of ShadeNorm.

37. ShadeNorm$(-1.2, 0) \approx 0.3849$

38. ShadeNorm$(-0.75, 0.75) \approx 0.5467$

39. ShadeNorm$(-0.9, 1.85) \approx 0.7837$

40. normalcdf$(-4, -2.1) \approx 0.0178$

41. normalcdf$(-0.2, 1.35) \approx 0.4908$

42. $z_L = \dfrac{90 - 82}{7.8} \approx 1.026$
$z_U = \dfrac{95 - 82}{7.8} \approx 1.667$
normalcdf$(1.03, 1.67) \approx 0.1047$

43. $z_L = \dfrac{70 - 78}{8.6} \approx -0.930$
$z_U = \dfrac{89 - 78}{8.6} \approx 1.279$
normalcdf$(-0.93, 1.28) \approx 0.7234$

44a. $z_L = \dfrac{53 - 58}{4} = -1.25$
$z_U = \dfrac{62 - 58}{4} = 1$
normalcdf$(1.25, 1) \approx 0.7357$

44b. $z_L = \dfrac{66 - 58}{4} = 2$
normalcdf$(2, 4) \approx 0.0227$

45. $z_L = \dfrac{63 - 58}{4} = 1.25$
normalcdf$(-4, 1.25) \approx 0.8943$

46. $z = $ invNorm$(0.6) \approx 0.2533$
$\dfrac{x - 58}{4} \approx 0.2533$; $x \approx 59.0$

47. $z = $ invNorm$(0.75) \approx 0.6744$
$58 \pm 0.6744(4) \approx [55.3, 60.7]$

Chapter 11

11.1 Cumulative Review

46.

E, $70°$, $60°$, $50°$, D, F, 19 m

$\dfrac{ED}{\sin 50°} = \dfrac{19 \text{ m}}{\sin 70°}$
$ED = \dfrac{(19 \text{ m}) \sin 50°}{\sin 70°}$

48. $z_L = \dfrac{3.0 - 2.1}{1.6} \approx 0.56$
$z_U = \dfrac{4.0 - 2.1}{1.6} \approx 1.19$
$P(0.56 < z < 1.19)$
normalcdf$(0.56, 1.19) \approx 17\%$

Chapter 11 Review

21. $E = 2.58\left(\dfrac{3.87}{\sqrt{50}}\right) \approx 1.41$
CI: $19.6 \pm 1.41 = (18.19, 21.01)$
We are 99% confident that μ is in the interval (18.19, 21.01).

22. $\bar{x} \approx 42.67$; $s \approx 25.91$
$E = 1.645\left(\dfrac{25.91}{\sqrt{30}}\right) \approx 7.78$
CI: $42.67 \pm 7.78 = (34.89, 50.45)$

23. $\bar{x} \approx 65.78$; $s \approx 10.31$
$E = 1.96\left(\dfrac{10.31}{\sqrt{32}}\right) \approx 3.57$
CI: $65.78 \pm 3.52 = (62.21, 69.35)$

24. $E_{90} = 1.645\left(\dfrac{3}{\sqrt{18}}\right) \approx 1.16$
CI_{90}: $15.5 \pm 1.16 = (14.34, 16.66)$
$E_{95} = 1.96\left(\dfrac{3}{\sqrt{18}}\right) \approx 1.39$
CI_{95}: $15.5 \pm 1.39 = (14.11, 16.89)$

25. $E_{90} = 1.645\left(\dfrac{3}{\sqrt{180}}\right) \approx 0.37$
CI_{90}: $15.5 \pm 0.37 = (15.13, 15.87)$
$E_{95} = 1.96\left(\dfrac{3}{\sqrt{180}}\right) \approx 0.44$
CI_{95}: $15.5 \pm 0.44 = (15.06, 15.94)$

26. $E = 1.645\left(\dfrac{185.67}{\sqrt{70}}\right) \approx 36.51$
CI: $267.40 \pm 36.51 = (230.89, 303.91)$
We are 90% confident that the population mean is in the interval (230.89, 303.91).

27. $z_c = $ invNorm$(0.94) = 1.55$
$n = \left(\dfrac{z_c s}{E}\right)^2 = \left(\dfrac{1.55(7.64)}{2.3}\right)^2 \approx 26.7$
$n \geq 27$

28. $n = \left(\dfrac{1.96(1.39)}{0.27}\right)^2 \approx 101.82$; $n = 102$

34. using critical values:
H_0: $\mu \geq 420$ (claim); H_a: $\mu < 420$
$\alpha = 0.01$
left-tailed test with $z_c = -2.33$
critical region: $z < -2.33$
$z = \dfrac{\bar{x} - \mu_{\bar{x}}}{\frac{s}{\sqrt{n}}} = \dfrac{415 - 420}{\frac{24.18}{\sqrt{52}}} \approx \dfrac{-5}{3.35} \approx -1.49$
Since z is not within the critical region, fail to reject H_0. Based on this sample, the average earnings are not significantly lower than \$420/week at the 99% level of confidence.

35. using p-values:
H_0: $\mu = 26.1$; H_a: $\mu \neq 26.1$ (claim)
$\alpha = 0.05$
$z = \dfrac{\bar{x} - \mu_{\bar{x}}}{\frac{s}{\sqrt{n}}} = \dfrac{30.4 - 26.1}{\frac{8.2}{\sqrt{40}}} \approx \dfrac{4.3}{1.30} \approx 3.32$
two-tailed test:
$p \approx 2 \times$ normalcdf$(3.32, 4) \approx 0.0008$
Since $p < \alpha$, reject H_0 and accept H_a. Based on this sample, commute times in their city are significantly different from the national average at a 95% confidence level.

36. 1. Clarify the research question.
2. Plan the study.
3. Collect the data.
4. Analyze the data.
5. Interpret the data.

37. Experimental studies seek to establish a cause-and-effect relationship by comparing data from a control group to data collected from an experimental group to which a treatment was applied. Observational studies (including those that use a survey) attempt to determine correlation by comparing data collected from pre-existing groups without defining and applying a treatment.

41. self-selected; Sampling error may occur if individuals complete more than one survey. Researcher bias is possible since the questioner is not anonymous and participants may adjust their answers.

44a. married men in the United States; the age of men when they marry

44b. Is the average age of American men when they marry significantly older than that of women?

44c. H_0: $\mu \leq 27.4$; H_a: $\mu > 27.4$ (claim)

45a. observational

45b. Roll a die to determine the first marriage license selected and then choose every sixth marriage license thereafter.

45c. Unless the county is representative of the rest of the United States, there is a sampling error and the results from the county cannot be generalized to the population of the United States.

46a. $z = \dfrac{29.1 - 27.4}{\frac{1.59}{\sqrt{60}}} \approx \dfrac{1.7}{0.205} \approx 8.3$

$P(\overline{x} > 27.4) = P(z > 8.3) \approx 0$
Since $p < \alpha$, reject H_0 and accept H_a.

46b. The z-score of 8.3 and related p-value of almost 0 imply that the men's ages are much higher than the women's ages. However, the researcher should not attempt to apply the results to the US population.

Chapter 12

12.3 C. Exercises

38. At $x = 0$, there is a sharp point (no derivative).
If $x > 0$, then
$$f'(x) = \lim_{h \to 0} \frac{(x+h) - x}{h}$$
$$= \lim_{h \to 0} \frac{h}{h} = 1.$$
If $x < 0$, then
$$f'(x) = \lim_{h \to 0} \frac{-(x+h) - (-x)}{h}$$
$$= \lim_{h \to 0} \frac{-x - h + x}{h} = \lim_{h \to 0} \frac{-h}{h} = -1.$$
$$f'(x) = \begin{cases} 1 & \text{if } x > 0 \\ -1 & \text{if } x < 0 \end{cases}$$

40a, c. It appears that the exponent becomes the coefficient before the exponent is decreased by 1.

40b. $f'(x) = \lim\limits_{h \to 0} \dfrac{(x+h)^4 - x^4}{h}$
$$= \lim_{h \to 0} \frac{x^4 + 4x^3h + 6x^2h^2 + 4xh^3 + h^4 - x^4}{h}$$

$$= \lim_{h \to 0} \frac{h(4x^3 + 6x^2h + 4xh^2 + h^3)}{h}$$
$$= \lim_{h \to 0} (4x^3 + 6x^2h + 4xh^2 + h^3) = 4x^3$$

12.3 Cumulative Review

41. $(2 \text{ cis } 30°)(3 \text{ cis } 60°) = 6 \text{ cis } 90°$
$= 6(\cos 90 + i \sin 90) = 6(0 + i) = 6i$

42. $\dfrac{2 \text{ cis } 30°}{3 \text{ cis } 60°} = \dfrac{2}{3} \text{ cis } (-30°)$
$= \frac{2}{3}(\cos (-30°) + i \sin (-30°))$
$= \frac{2}{3}\left(\frac{\sqrt{3}}{2} - \frac{1}{2}i\right) = \frac{\sqrt{3}}{3} - \frac{1}{3}i$

43. $-4(1) - 6(5) = -34$

44. using column 3:
$(-1)^{1+3}(-2)\begin{vmatrix} 5 & -8 \\ 3 & -3 \end{vmatrix}$
$\qquad + (-1)^{3+3}(3)\begin{vmatrix} -7 & 1 \\ 5 & -8 \end{vmatrix}$
$= -2(9) + 3(51) = 135$

45. $3 + 6 + 9 + \cdots$;
Since $a_n \to \infty$, the sum $\to \infty$.

46. geometric; $a_1 = -2$; $r = -\frac{1}{2}$
$S = \dfrac{-2}{1 + \frac{1}{2}} = \dfrac{-2}{\frac{3}{2}} = -\dfrac{4}{3}$

47. $2 \sin \theta \cos \theta + \sin \theta = 0$
$\sin \theta (2 \cos \theta + 1) = 0$
$\sin \theta = 0$ or $2 \cos \theta = -1$
$\theta = 0 \qquad \cos \theta = -\frac{1}{2}$
$\qquad\qquad\qquad \theta = \pm\frac{2\pi}{3}$
$\theta = n\pi, \pm\frac{2\pi}{3} + n2\pi$

48. $\cos \theta = \frac{x}{2}$; $\sin \theta = \frac{y}{4}$
$\left(\frac{x}{2}\right)^2 + \left(\frac{y}{4}\right)^2 = 1$
$\dfrac{x^2}{4} + \dfrac{y^2}{16} = 1$

50. Since $y = f(x) \to 4$ as $x \to -\infty$, the horizontal asymptote is $y = 4$.

12.5 C. Exercises

41. applying the quotient rule:
$$f'(x) = \frac{\frac{d}{dx}[x](x^2-1)^{\frac{1}{2}} - (x)\frac{d}{dx}\left[(x^2-1)^{\frac{1}{2}}\right]}{\left[(x^2-1)^{\frac{1}{2}}\right]^2}$$

using the chain rule for $\frac{d}{dx}(x^2-1)^{\frac{1}{2}}$:
$u(x) = x^{\frac{1}{2}}$; $v(x) = x^2 - 1$
$u'(x) = \frac{1}{2}x^{-\frac{1}{2}}$; $v'(x) = 2x$

$$f'(x) = \frac{1(x^2-1)^{\frac{1}{2}} - x\left[\frac{1}{2}(x^2-1)^{-\frac{1}{2}}(2x)\right]}{\left[(x^2-1)^{\frac{1}{2}}\right]^2}$$

factoring $(x^2-1)^{-\frac{1}{2}}$ from the numerator:

$f'(x) = \dfrac{(x^2-1)^{-\frac{1}{2}}[(x^2-1) - x^2]}{(x^2-1)}$
$= \dfrac{-1}{(x^2-1)^{\frac{3}{2}}}$ or $-(x^2-1)^{-\frac{3}{2}}$

42. $f'(x) = u(x)[v(x)]^{-1}$
$= u'(x)[v(x)]^{-1} + u(x)\frac{d}{dx}[v(x)]^{-1}$
$= u'(x)[v(x)]^{-1}$
$\qquad\qquad + u(x)[(-1)[v(x)]^{-2}v'(x)]$
$= \dfrac{u'(x)}{v(x)} - \dfrac{u(x)v'(x)}{[v(x)]^2}$
$= \dfrac{u'(x)v(x) - u(x)v'(x)}{[v(x)]^2}$

43. Let k be a natural number such that $n = -k$.
$f(x) = x^n = x^{-k} = \dfrac{1}{x^k}$
$f'(x) = \dfrac{0(x^k) - 1(kx^{k-1})}{(x^k)^2} = \dfrac{-kx^{k-1}}{x^{2k}}$
$= -kx^{k-1-2k} = -kx^{-k-1} = nx^{n-1}$

12.5 Cumulative Review

44. y-int.: $f(0) = -6(0)^2 - 2(0) - 4 = -4$
zeros: $6x^2 - 2x - 4 = 0$
$3x^2 - x - 2 = 0$
$(3x + 2)(x - 1) = 0$
$x = -\frac{2}{3}, 1$

45. $3(12) + 6(-6) = 0$; $\theta = 90°$

48. $z = \dfrac{123 - 118}{12} = \dfrac{5}{12} \approx 0.42$

49. $z = \dfrac{97 - 118}{12} = -1.75$

50. $\dfrac{(x+3)(x-2)}{(x+3)(2x+5)} = \dfrac{(x-2)}{(2x+5)} \geq 0$ where $x \neq -3$
PD: $x = -3$; zero: $x = 2$; VA: $x = -\frac{5}{2}$

	PD		VA		zero	
	$p(x) > 0$	$p(x) > 0$		$p(x) < 0$		$p(x) > 0$
$(+)$	-3	$(-)$	-2.5	$(-)$	2	$(+)$
$(+)$		$(-)$		$(+)$		$(+)$

51. $a_1 = 3$; $a_{20} = 136$
$S_{20} = \dfrac{20(3 + 136)}{2} = 1390$

52. $\lim\limits_{x \to 4} 7x^2 = 7 \lim\limits_{x \to 4} x^2 = 7\left(\lim\limits_{x \to 4} x\right)^2$
$= 7(4)^2 = 7(16) = 112$

53. $f'(x) = 4x$
$m = f'(-1) = 4(-1) = -4$
$f(-1) = 2(-1)^2 = 2$
through $(-1, 2)$ with $m = -4$:
$y - 2 = -4(x - (-1))$
$y - 2 = -4x - 4$
$\qquad y = -4x - 2$

12.6 B. Exercises

33d. $\int_2^5 (-2x + 12)\, dx$

$\Delta x = \frac{5-2}{n} = \frac{3}{n}$

$= \lim_{n\to\infty} \sum_{i=1}^{n} \left(-2x_i + 12\right)\left(\frac{3}{n}\right)$

$= \lim_{n\to\infty} \sum_{i=1}^{n} \left[-2\left(2 + \frac{3i}{n}\right) + 12\right]\left(\frac{3}{n}\right)$

$= \lim_{n\to\infty} \sum_{i=1}^{n} \left(-\frac{18i}{n^2} - \frac{12}{n} + \frac{36}{n}\right)$

$= \lim_{n\to\infty} \left(-\frac{18}{n^2} \sum_{i=1}^{n} i\right) + \lim_{n\to\infty} \frac{1}{n}\sum_{i=1}^{n} 24$

$= \lim_{n\to\infty} \left(-\frac{18}{n^2} \cdot \frac{n(n+1)}{2}\right) + \lim_{n\to\infty} \frac{24n}{n}$

$= \lim_{n\to\infty} \frac{-18(n+1)}{2n} + 24$

$= \lim_{n\to\infty} \left(\frac{-9n}{n} + \frac{-9}{n}\right) + 24$

$= (-9 - 0) + 24 = 15\ \text{u}^2$

12.6 C. Exercises

34. $\Delta x = \frac{4-0}{4} = 1$

34a. $x_i = 1, 2, 3, 4$

$R_1 = 1[-(1-1)^2 + 9] = 9$
$R_2 = 1[-(2-1)^2 + 9] = 8$
$R_3 = 1[-(3-1)^2 + 9] = 5$
$R_4 = 1[-(4-1)^2 + 9] = 0$
$S = 9 + 8 + 5 + 0 = 22\ \text{u}^2$

34b. $x_i = 0, 1, 2, 3$

$R_1 = 1[-(0-1)^2 + 9] = 8$
$R_2 = 1[-(1-1)^2 + 9] = 9$
$R_3 = 1[-(2-1)^2 + 9] = 8$
$R_4 = 1[-(3-1)^2 + 9] = 5$
$S = 8 + 9 + 8 + 5 = 30\ \text{u}^2$

34c. $x_i = 0.5, 1.5, 2.5, 3.5$

$R_1 = 1[-(0.5-1)^2 + 9] = 8.75$
$R_2 = 1[-(1.5-1)^2 + 9] = 8.75$
$R_3 = 1[-(2.5-1)^2 + 9] = 6.75$
$R_4 = 1[-(3.5-1)^2 + 9] = 2.75$
$S = 2(8.75) + 6.75 + 2.75 = 27\ \text{u}^2$

34d. The right-end sum underestimates the area, and the left-end sum overestimates the area. The midpoint sum is much closer but still overestimates the area. The average of the right-end and left-end sums is the best estimate but still underestimates the area.

35a.

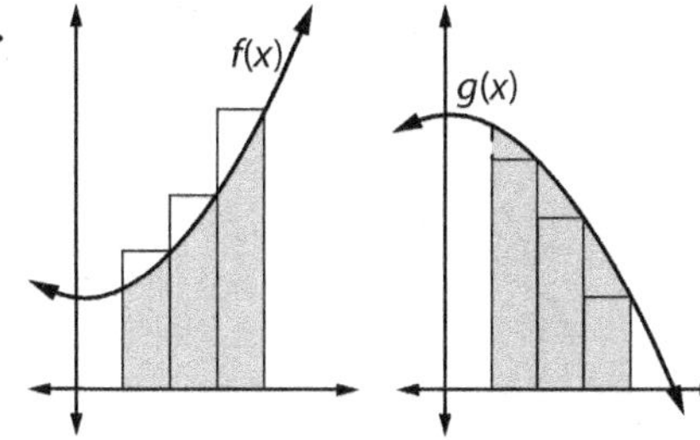

A right-end sum overestimates areas under increasing functions since each rectangle contains extra area to the upper left of the curve. It underestimates areas under decreasing functions since each rectangle misses part of the area under the curve.

35b.

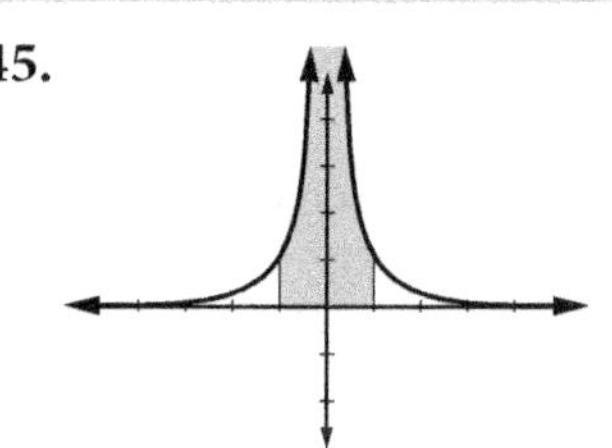

A left-end sum underestimates areas under increasing functions since each rectangle misses part of the area under the curve. It overestimates areas under decreasing functions since each rectangle contains extra area to the upper right of the curve.

37. $\Delta x = \frac{b-0}{n} = \frac{b}{n}$

$= \lim_{n\to\infty} \sum_{i=1}^{n} \left(x_i^2 + 3\right)\left(\frac{b}{n}\right)$

$= \lim_{n\to\infty} \sum_{i=1}^{n} \left(\left(\frac{bi}{n}\right)^2 + 3\right)\left(\frac{b}{n}\right)$

$= \lim_{n\to\infty} \sum_{i=1}^{n} \left(\frac{b^3 i^2}{n^3} + \frac{3b}{n}\right)$

$= \lim_{n\to\infty} \frac{b^3}{n^3} \sum_{i=1}^{n} i^2 + \lim_{n\to\infty} \frac{b}{n} \sum_{i=1}^{n} 3$

$= \lim_{n\to\infty} \frac{b^3}{n^3}\left(\frac{n(n+1)(2n+1)}{6}\right) + \lim_{n\to\infty} \frac{3bn}{n}$

$= \lim_{n\to\infty} \frac{b^3}{n^2}\left(\frac{2n^2 + 3n + 1}{6}\right) + 3b$

$= \lim_{n\to\infty} \left(\frac{b^3 n^2}{3n^2} + \frac{b^3 n}{2n^2} + \frac{b^3}{6n^2}\right) + 3b$

$= \frac{1}{3}b^3 + 3b$

38. $\int_0^a (x^2 + 3)\, dx = \frac{1}{3}a^3 + 3a$

$\int_a^b (x^2 + 3)\, dx$

$= \int_0^b (x^2 + 3)\, dx - \int_0^a (x^2 + 3)\, dx$

$= \frac{1}{3}b^3 + 3b - \left(\frac{1}{3}a^3 + 3a\right)$

$= \frac{1}{3}(b^3 - a^3) + 3(b - a)$

$= \frac{1}{3}(b^3 - a^3) + 3b - 3a$

12.6 Cumulative Review

39. $|B| = 40 - (-18) = 58$

$B^{-1} = \frac{1}{58}\begin{bmatrix} 8 & -6 \\ 3 & 5 \end{bmatrix} = \begin{bmatrix} \frac{4}{29} & -\frac{3}{29} \\ \frac{3}{58} & \frac{5}{58} \end{bmatrix}$

40. $|A| = 12 - 14 = -2$

$A^{-1} = -\frac{1}{2}\begin{bmatrix} 4 & -7 \\ -2 & 3 \end{bmatrix} = \begin{bmatrix} -2 & \frac{7}{2} \\ 1 & -\frac{3}{2} \end{bmatrix}$

$X = A^{-1}C = \begin{bmatrix} -2 & \frac{7}{2} \\ 1 & -\frac{3}{2} \end{bmatrix}\begin{bmatrix} 4 \\ -8 \end{bmatrix}$

$= \begin{bmatrix} -36 \\ 16 \end{bmatrix}$

41. $\bar{x} = \frac{\sum x_i}{n} = \frac{755}{23} \approx 32.8$

$x_{12} = 34$ is the middle value of 23 values. 44 occurs 4 times.

42. $47 - 10 = 37$

43. $f'(x) = \lim_{h\to 0} \frac{2(x+h) - 2x}{h} = \lim_{h\to 0} \frac{2h}{h} = 2$

44. $f'(x) = 8x + 3$

$f'(-2) = 8(-2) + 3 = -13$

45. $_4C_3(5m)^1(2n)^3 = 160mn^3$

46. $z_L = \frac{76 - 87}{8.2} \approx -1.34$

$z_U = \frac{94 - 87}{8.2} \approx 0.85$

normalcdf$(-1.34, 0.85) \approx 0.713$

47. $f(x) = x^{\frac{1}{2}}$

$f'(x) = \frac{1}{2}x^{-\frac{1}{2}} = \frac{1}{2\sqrt{x}}$

48. $u(x) = 2x^2 + 7;\ v(x) = x^2 - 3$

$u'(x) = 4x;\ v'(x) = 2x$

$f'(x) = \frac{4x(x^2 - 3) - (2x^2 + 7)2x}{(x^2 - 3)^2}$

$= \frac{4x^3 - 12x - 4x^3 - 14x}{(x^2 - 3)^2} = \frac{-26x}{(x^2 - 3)^2}$

12.7 C. Exercises

45.

45a. $\int x^{-2}\, dx = \frac{1}{-2+1}x^{-2+1} + C$

$= -\frac{1}{x} + C$

45b. $\int_{-1}^{1} x^{-2}\,dx = \left[-\frac{1}{x}\right]_{-1}^{1}$

$$= -\frac{1}{1} - \left(-\frac{1}{-1}\right) = -2$$

45c. The Fundamental Theorem of Calculus applies only to continuous intervals and $f(x) = \dfrac{1}{x^2}$ has a vertical asymptote at $x = 0$.

12.7 Cumulative Review

46. $m = \dfrac{-4 - 2}{-9 - (-1)} = \dfrac{-6}{-8} = \dfrac{3}{4}$

$$y - 2 = \frac{3}{4}(x + 1)$$
$$y - 2 = \frac{3}{4}x + \frac{3}{4}$$
$$y = \frac{3}{4}x + \frac{11}{4}$$

47. $m_G = \dfrac{1}{3}$; $m_\perp = -\dfrac{3}{1}$; y-int.: $(0, 7)$
$$y = -3x + 7$$

48. $a_0 = \dfrac{4(0!)}{3(0) + 1} = \dfrac{4(1)}{1} = 4$

$a_1 = \dfrac{4(1!)}{3(1) + 1} = \dfrac{4(1)}{4} = 1$

$a_2 = \dfrac{4(2!)}{3(2) + 1} = \dfrac{4(2 \cdot 1)}{7} = \dfrac{8}{7}$

$a_3 = \dfrac{4(3!)}{3(3) + 1} = \dfrac{4(3 \cdot 2 \cdot 1)}{10} = \dfrac{12}{5}$

$a_4 = \dfrac{4(4!)}{3(4) + 1} = \dfrac{4(4 \cdot 3 \cdot 2 \cdot 1)}{13} = \dfrac{96}{13}$

49. $a_0 = \dfrac{(-2)^0}{3(1!)} = \dfrac{1}{3(1)} = \dfrac{1}{3}$

$a_1 = \dfrac{(-2)^1}{3(2)!} = \dfrac{-2}{3(2)} = -\dfrac{1}{3}$

$a_2 = \dfrac{(-2)^2}{3(3!)} = \dfrac{4}{3(6)} = \dfrac{2}{9}$

$a_3 = \dfrac{(-2)^3}{3(4!)} = \dfrac{-8}{3(24)} = -\dfrac{1}{9}$

$a_4 = \dfrac{(-2)^4}{3(5!)} = \dfrac{16}{3(120)} = \dfrac{2}{45}$

50. $f'(x) = 10x$
$f'(3) = 10(3) = 30$

51. $f'(x) = 4x + 4$
$f'(-2) = 4(-2) + 4 = -4$

52. $A \neq C$
$B^2 - 4AC = 0^2 - 4(49)(16) < 0$
$\therefore$ an ellipse

53. $_{22}C_7 = \dfrac{n!}{r!(n - r)!} = \dfrac{22!}{7!(22 - 7)!}$
$$= 170{,}544$$

54. $u(x) = 2x^2 - 4x + 5$
$v(x) = 3x^2 - 2x + 11$
$u'(x) = 4x - 4$; $v'(x) = 6x - 2$
$k'(x) = (2x^2 - 4x + 5)(6x - 2)$
$$+ (4x - 4)(3x^2 - 2x + 11)$$
$$= (12x^3 - 28x^2 + 38x - 10)$$
$$+ (12x^3 - 20x^2 + 52x - 44)$$
$$= 24x^3 - 48x^2 + 90x - 54$$

55. 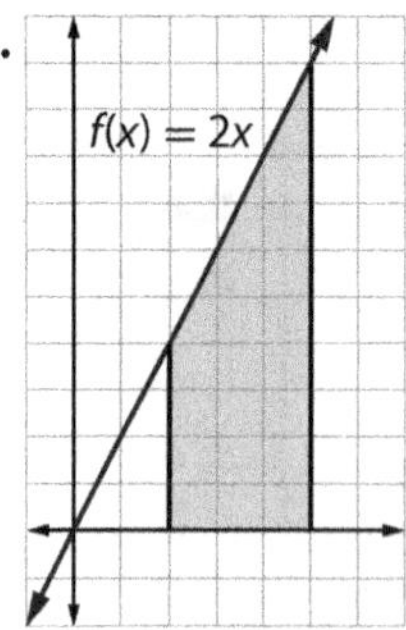

$$A = \frac{1}{2}h(b_1 + b_2) = \frac{1}{2}(3)(4 + 10) = 21$$